MARINE & FRESHWATER PRODUCTS HANDBOOK

HOW TO ORDER THIS BOOK

BY PHONE: 800-233-9936 or 717-291-5609, 8AM–5PM Eastern Time

BY FAX: 717-295-4538

BY MAIL: Order Department
Technomic Publishing Company, Inc.
851 New Holland Avenue, Box 3535
Lancaster, PA 17604, U.S.A.

BY CREDIT CARD: American Express, VISA, MasterCard

BY WWW SITE: http://www.techpub.com

Marine & Freshwater Products Handbook

EDITED BY

Roy E. Martin
National Fisheries Institute, Inc.

Emily Paine Carter
Virginia Polytechnic Institute & State University

George J. Flick, Jr.
Virginia Polytechnic Institute & State University

Lynn M. Davis
Virginia Polytechnic Institute & State University

TECHNOMIC PUBLISHING CO., INC.
LANCASTER • BASEL

Marine and Freshwater Products Handbook
a**TECHNOMIC**® publication

Technomic Publishing Company, Inc.
851 New Holland Avenue, Box 3535
Lancaster, Pennsylvania 17604 U.S.A.

Printed in the United States of America
10 9 8 7 6 5 4 3 2 1

Main entry under title:
Marine and Freshwater Products Handbook

A Technomic Publishing Company book
Bibliography: p.
Includes index p. 913

Library of Congress Catalog Card No. 00-101386
ISBN No. 1-56676-889-6

Dedicated

to the Future

of the

World's Water Resources

Marine and Freshwater Products Handbook

CONTENTS

Marine and Freshwater Products Handbook

Preface

The International Year of the Ocean was celebrated in 1998. It reminded people that the waters of the world are the beginning of life. It focused the attention of the public, governments, and decision-makers on the importance of marine and freshwater environments as resources that need to be sustained.

Oceans, without including freshwaters, cover more than 70 percent of the globe. The sea's immense biological diversity is critical to our planet's life support systems. Marine ecosystems are home to many phyla that live nowhere else; these systems protect shorelines from flooding, break down wastes, moderate climate, and maintain a breathable atmosphere. Moreover, marine species provide food, medicines, raw materials, a livelihood for millions of people, and recreation for billions.

Because the world's water resources are so intrinsically important to all living systems, this guide should prove to be a useful and valuable tool not just for those who work in the industry or the field of education, but also for the lay public. The book provides summary information on the foods and other products that come from our life-giving waters.

The last time such a book as this was published was in 1951. Called *Marine Products of Commerce,* it was a second edition of the first book and covered the acquisition, handling, biological aspects, and the science and technology of the preparation and preservation of marine products. Essentially, this book is the "third" edition.

In the time since the second edition was published, great advances have been made in the technology of fishery and other marine products, as the world has moved into a highly technical society. Entirely new industries have arisen. The manufacture of chemicals from the sea is one industry that continues to become more and more important to the future.

Biotechnology and molecular biology open up vast stores of new knowledge and applications. Sponges and soft corals, for instance, have been shown to produce compounds that inhibit the release of an enzyme that initiates a cascade of reactions leading to pain and inflammation. The discovery means new potent anti-inflammatory agents. Products from the sea are also being studied as potential medicines for the treatment of cancer and AIDS.

Some dinoflagellates and other microalgae have shown significant anti-tumor capabilities. Rhode Island Sea Grant researchers are developing large-scale aquaculture systems to grow these marine organisms so they can be studied without disrupting natural supplies and affected ecosystems. Successful mass culturing will enable scientists to conduct tests of promising potential drugs, as well as eliminate industry concern about the supply of the compounds in the event they become therapeutic drugs.

Virginia Sea Grant in its *Waterfront News* (1997) reported how the dinoflagellate known as *Pfiesteria piscicida* had been implicated in fish kills in coastal waters of Virginia and North Carolina. Many different species of dinoflagellates cause red tides, and red tides are responsible for many different effects. Research, enhanced by the advances made in molecular biology, is ongoing to determine exactly what happens.

In the wake of food poisoning incidents in recent years, the subject of food irradiation has been appearing in the news. This often misunderstood method of food preservation is dealt with in one of the chapters. Research that has been ongoing for more than 40 years currently focuses on what happens to the unique radiolytic products, the fragments left when radiation breaks down some of the chemical bonds. Most of the lytic products appear no different from those produced by other forms of food preparation and preservation.

New York Sea Grant is making waves in seafood safety. Researchers are developing an onboard toxin analyzer that can be used aboard fishing vessels to detect the presence of PSP (paralytic shellfish poisoning) toxin in shellfish. This analyzer will allow fishing vessels to detect potential contamination of shellfish beds before any harvesting is done and will preclude the need for lengthy lab-culturing techniques now required for toxin detection. New York Sea Grant researchers also are using DNA fingerprinting techniques to detect various strains of pathogenic *Listeria monocytogenes* in smoked salmon and in processing plants. Detecting various strains can be used to verify the efficacy of Hazard Analysis Critical Control Point (HACCP) programs of seafood safety, as well as to characterize certain strains that are avirulent and non-pathogenic for humans.

Mississippi-Alabama Sea Grant has found that multiple microbial pathogens can be detected simultaneously in a single-test polymerase chain reaction (PCR). PCR involves making millions of copies of a selected DNA segment. When the segment is unique to a particular pathogen, PCR is a fast, accurate method for detecting the presence of even minimal numbers of pathogenic cells. These findings have applications for detecting not only total coliform/*Escherichia coli* in shellfish-growing waters, but also total and pathogenic *Salmonella typhimurium*, *Vibrio cholerae*, and total and clinically important strains of *Vibrio vulnificus* and *Vibrio parahaemolyticus* in oysters collected from various shellfish-harvesting areas along the Gulf of Mexico and from local seafood restaurants and suppliers. This research will lead to a more reliable and rapid method of detecting shellfish-born pathogens.

The Sea Grant network helped pull together the National Seafood HACCP Alliance, with representation from the United States Food and Drug Administration, United States Department of Agriculture's Cooperative Extension services, National Marine Fisheries Service, National Fisheries Institute, National Food Processors Association, Interstate Shellfish Sanitation Conference, Association of Food and Drug Officials, and various state agencies that deal with health, food safety, and commerce.

HACCP lays out seven principles for food processors, intended to ensure food safety from the point of harvest to the consumer's table. Among other things, the system identifies critical points in the processing chain at which potential hazards can be controlled or eliminated. The buzzword in seafood processing for the last decade of the 20th century has certainly been HACCP, which is discussed in this book.

Research continues to advance the seafood and fish industry. In the 1980s, Oregon fishers and seafood processors began to set their sights on Pacific whiting, an abundant but not especially popular species. One problem became a hurdle: the fish softened under processing, making it unsuitable for many seafood products. Producers were especially interested in adapting whiting to the production of surimi, the extruded fish paste used to create increasingly consumer-popular imitation crab/lobster products. Oregon Sea Grant-funded research provided the breakthrough to solve "the texture problem" by developing food-based chemicals (protease inhibitors) that deactivate the enzymes that turn the fish to mush. Extension Sea Grant specialists, meanwhile, helped processors tackle the technical barriers, from waste disposal to processing technology and long-term cold storage of the final product. With Sea Grant's help, Oregon's whiting fishery and on-shore surimi processing had become a $30 million-a-year industry by the early 1990s.

Milestones in the University of Wisconsin's Sea Grant seafood technology research have included:

- Identification of the specific compounds that give different species of fish their unique flavor profiles, a breakthrough in flavor chemistry and a substantial contribution to food science in general.
- Identification of the bacterial processes that make fish smell and taste "fishy," and the development of new processing and packaging techniques for keeping fish fresher and safer over longer periods of time.
- Development of new processing techniques for utilizing so-called "trash fish" for both human and pet food products.
- Development of liquid fertilizer and composting techniques for the environmentally friendly means of disposing of fish wastes.

One of Alaska Sea Grant's most noteworthy processing successes is a patented new method of processing arrowtooth flounder, which contains an enzyme in its flesh that when exposed to heat turns the flesh to mush — a drawback making it difficult to sell as fillets or in processed foods. The development of this new method clears the way for processors to use a fish once considered trash but that is now one of the most abundant in the North Pacific, with an estimated harvestable biomass of 386,000 tons each year.

As for harvesting and the health of species, ongoing research runs the gamut. The Maine/New Hampshire Sea Grant Program has an innovative study using video cameras to examine the behavior of lobsters in and around traps. This is part of a larger effort to determine the relative health of the region's major fishery.

South Carolina Sea Grant scientists are studying the importance of oxygen and carbon dioxide on oyster mortality from *Perkinsus marinus* (Dermo) infections. Study results could lead to the development of alternatives for oyster recovery and enhance efforts to culture oysters in the southeastern United States. Researchers are also assessing methods of measuring high-energy phosphates, which represent fundamental units of biochemical energy for all living organisms. They are examining whether these compounds reflect nutritional and toxicological stress in estuarine plankton and fish. Development of these techniques could help resource managers determine the degree of stress in marine ecosystems. Marine scientists are developing aquaculture techniques for the southern flounder and evaluating broodstock for hatchery use; cultured flounder is a species that shows commercial promise for the aquaculture industry.

In a joint study with the Skidaway Institute of Oceanography in Georgia, South Carolina Sea Grant is studying how juvenile white shrimp are recruited to the southeastern estuaries. Results should allow for better predictions of survival rates of larval white shrimp, which would improve management of this fishery. Computer simulations suggest that jetties and other man-made structures may block larvae from entering inlets and thus may lower successful recruitment of shrimp into the adult fishery.

Such shrimp research is just one example of how the ubiquitous computer has also propelled innumerable changes in how resources from the sea and freshwaters are understood, harvested, used, and marketed. The future for innovative applications of the computer is wide open.

Research, of course, is limitless, as we continue to learn more about our gifts from the sea. This book merely marks a point in time.

Each author, an expert in his or her field, presents an overview of the topic and usually begins the chapter with an introduction, followed by text, charts, tables, and a bibliography. Many chapters are accompanied by photographs and illustrations. The authors and their collaborators have attempted to present the most up-to-date material available. Sometimes data are not collected and compiled on a yearly basis but on a longer term, so in those cases the most recent available statistics were used.

The editors hope that this new major reference source, long overdue, will serve as a helpful guide to those interested in the field of marine and freshwater products. The editors are deeply indebted to the authors, including Sea Grant colleagues, who contributed to this project.

Because we are all involved in research, we expect new discoveries constantly on the horizon. So the story told here does not stop.

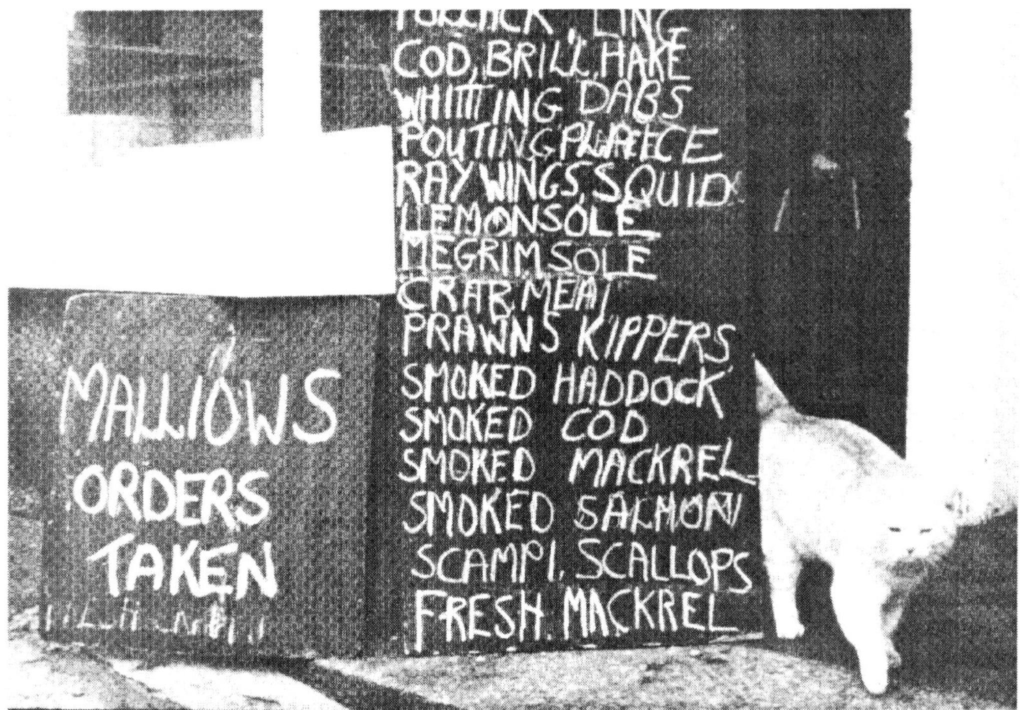

Acknowledgments

The support of Virginia Polytechnic Institute and State University (Virginia Tech) — particularly from the Department of Food Science and Technology and the College of Natural Resources — and of the National Fisheries Institute helped make this book possible, in partnership with the National and Virginia Sea Grant College Programs.

Authors and Contributors

Shigeru Akamatsu, Director, Pearl Research Lab., K. Mikimoto & Co. Ltd., Mie-ken, Japan.

Robert C. Bayer, Ph.D., University of Maine, Lobster Institute, Orono, Maine.

Anthony P. Bimbo, A.P. Bimbo & Associates, Kilmarnock, Virginia.

Roberto S. Chamul, M.S., Mississippi State University, Food Science and Technology.

John D. Cirino, Gulf Environmental Associates, Ocean Springs, Mississippi.

Charles W. Coale Jr., Ph.D., Professor, Virginia Tech, Agriculture and Applied Economics, Blacksburg, Virginia.

Robert L. Collette, M.S., National Fisheries Institute, Division of Science and Technology, Arlington, Virginia.

Paul G. Comar, M.S., NOAA, National Marine Fisheries Service, Charleston, South Carolina.

Angela I. Correa, B.A., Sea Grant Communicator, Virginia Tech, Blacksburg, Virginia.

Robert E. Croonenberghs, Ph.D., Virginia Department of Health, Division of Shellfish Sanitation, Richmond, Virginia.

Laurie M. Dean, City of Portsmouth Public Schools, Norfolk, Virginia.

Patricia A. Fair, Ph.D., NOAA, National Marine Fisheries Service, Charleston Laboratory, Charleston, South Carolina.

Custy F. Fernandes, Ph.D., Experimental Seafood Processing Laboratory, Mississippi State University, Pascagoula, Mississippi.

Roland Finch, M.S., retired, National Marine Fisheries Service, Portland, Oregon.

Robert A. Fisher, College of William and Mary, Virginia Institute of Marine Science, Gloucester Point, Virginia.

George J. Flick Jr., Ph.D., University Distinguished Professor, Virginia Tech, Food Science and Technology, Blacksburg, Virginia.

Ken Gall, M.S., Seafood Specialist, Cornell University and New York Sea Grant, Ithaca, New York.

William G. Gordon, former Director, National Marine Fisheries Service, Fairplay, Colorado.

Richard E. Gutting Jr., National Fisheries Institute, Arlington, Virginia.

Michael G. Haby, Texas A&M University, Corpus Christi, Texas.

Cameron R. Hackney, Ph.D., Virginia Tech, Department Head, Food Science & Technology, Blacksburg, Virginia.

Eric M. Hallerman, Ph.D., Associate Professor, Virginia Tech, Fisheries and Wildlife Sciences, Blacksburg, Virginia.

Roy C. Heidinger, Ph.D., Southern Illinois University, Director, Fisheries Research Lab, Carbondale, Illinois.

Chris M. Ireland, Ph.D., Brent R. Copp, Mark P. Foster, Leonard A. McDonald, Derek C. Radisky and J. Christopher Swersey, University of Utah, Department of Medicinal Chemistry, Salt Lake City, Utah.

John A. Keinath, Ph.D., Thomas Nelson Community College, Gloucester Point, Virginia.

Tyre C. Lanier, Ph.D., North Carolina State University, Food Science Department, Raleigh, North Carolina.

Robert J. Learson, R.J. Learson Associates, South Hamilton, Massachusetts.

Joseph J. Licciardello, Ph.D., formerly of National Marine Fisheries Service, Gloucester, Massachusetts.

Martha Llaneras, Kelco Company/Latin America, San Diego, California.

Richard T. Lovell, Ph.D., Distinguished University Professor, Auburn University, Fisheries Department, Auburn, Alabama.

Roy E. Martin, Emeritus, National Fisheries Institute, Arlington, Virginia.

John V. Merriner, Ph.D., National Marine Fisheries Service, Beaufort, North Carolina.

Russell J. Miget, Ph.D., Texas A&M University, Corpus Christi, Texas.

Peter D. R. Moeller, Ph.D., National Marine Fisheries Service, Charleston, South Carolina.

Michael W. Moody, Ph.D., Louisiana State University, Baton Rouge, Louisiana.

Carter Newell, Greater Eastern Mussel Farms Inc., Tenants Harbor, Maine.

Michael J. Oesterling, College of William & Mary, Virginia Institute of Marine Science, Gloucester Point, Virginia.

Jae W. Park, Ph.D., Oregon State University, Seafood Research Laboratory, Astoria, Oregon.

W. Michael Peirson, Ph.D., Cherrystone Aqua Farms, Eastville, Virginia.

Harriett M. Perry, Institute of Marine Science, Gulf Coast Research Laboratory, Ocean Springs, Mississippi.

Robert J. Price, Ph.D., University of California/Davis, Food Science and Technology Department.

Kolli P. Reddy, Cornell University, Institute of Food Science, Ithaca, New York.

Carrie E. Regenstein, Cornell University, Institute of Food Science, Ithaca, New York.

Joe M. Regenstein, Ph.D., Cornell University, Institute of Food Science, Ithaca, New York.

William L. Rickards III, Ph.D., University of Virginia, Virginia Graduate/Marine Science Consortium, Charlottesville, Virginia.

Thomas E. Rippen, Sea Grant Extension Program, Princess Anne, Maryland.

Brian Rudolph, Ph.D., Marine Biologist, Copenhagen Pectin A/S, Denmark.

Ronald J. Sasiela, Coldwater Seafood Corporation, Easton, Maryland.

Frank J. Schwartz, Ph.D., University of North Carolina, Professor of Marine Sciences, Institute of Marine Sciences, Morehead City, North Carolina.

Juan L. Silva, Ph.D., Mississippi State University, Food Science and Technology.

Bohdan M. Slabyj, Ph.D., Professor Emeritus, University of Maine, Food Science and Human Nutrition, Brewer, Maine.

Stephen A. Smith, D.V.M., Ph.D., Virginia Tech, Virginia-Maryland Regional College of Veterinary Medicine, Blacksburg, Virginia.

John Spinelli, Ph.D., formerly with Food Chemical and Research Foods, National Marine Fisheries Service, Seattle, Washington; deceased.

Donn R. Ward, Ph.D., North Carolina State University, Department of Food Science, Raleigh, North Carolina.

Jeffrey C. Wolf, D.V.M., Virginia Tech, Virginia-Maryland Regional College of Veterinary Medicine, Blacksburg, Virginia.

Cheryl M. Woodley, Ph.D., NOAA Fisheries, SEFSC Charleston Laboratory, Charleston, South Carolina.

Section I. General Introduction and Background

An oyster-dredging boat in the Chesapeake Bay.

Marine Fisheries and Management

John V. Merriner

Fisheries have been important in the history of the United States since early colonial times, as settlers were closely linked to fishes in the coastal streams, lakes, and rivers for food and fertilizer uses. Today, United States fisheries — both recreational and commercial — use living marine resources virtually worldwide with products including edible flesh, chemical extracts, and recreation. Worldwide, the United States ranked fifth in commercial fishery landings [5.5 million metric tons (mmT)], representing 4.5 percent of the global total (122.1 mmT) based upon 1997 data [National Marine Fisheries Service (NMFS), 1999]. China ranked first with 28.7 percent, followed by Peru (6.4 percent), Chile (5.5 percent), and Japan (5 percent). Landings of commercial products (edible and industrial) in the United States for 1998 were 9.2 billion pounds; finfish comprised 86 percent of that total. Recreational fisheries on the Atlantic, Gulf, and Pacific coasts were estimated to have landed (retained) 194.7 million pounds.

The evolution of these fisheries and concerns for the availability of these common property resources to all participants have led to an intricate system of resource management, spanning state, federal and international jurisdictions. Coincidentally, the public frequently cites confusion as to the appropriate managerial agency for data acquisition, fishery information, and/or rules promulgation. This chapter provides a synopsis of modern federal and interstate fishery management. Status of the United States marine resources are detailed periodically in government reports entitled "Our Living Oceans" [United States Department of Commerce (USDOC), 1996, 1993, 1999].

Extended marine fisheries jurisdiction by the United States, in the modern context, began with passage of the Magnuson Fisheries Conservation and Management Act (MFCMA) in 1976. This act provided the means to control harvest of valuable fisheries resources from continental shelf waters 3 to 200 miles off the United States coasts (i.e., beyond the internationally recognized 12-mile limit of jurisdiction at the time). Previously, fisheries on the "high seas" were either unregulated or controlled cooperatively (managed) by negotiated agreements (bi- or multilateral international treaties) through the United States State Department with technical advisory input from the NMFS. When the Magnuson Act was passed, there was strong public sentiment that the foreign fleets (trawlers and factory ship fleets off the East Coast) were taking excessive amounts of fish from the United States continental shelf. Additionally, United States fishermen were denied expansion of their own fisheries by that foreign harvest (allocation). A major purpose of the MFCMA was to eliminate foreign fishing activity in the waters of extended United States fisheries jurisdiction and to develop sustainable domestic fisheries on those resources exploited by the foreign fleet. Subsequently, other nations have declared their exclusive rights over the economic resources within 200 miles of their shores. Thus, United States fisheries management today has its foundation in the Magnuson Act of 1976 and subsequent amendments — most recently amended through October 11, 1996, in the Magnuson-Stevens Fisheries Conservation and Management Act (M-SFCMA) [also known as the Sustainable Fisheries Act (SFA)]. The SFA provides a framework for federal management of the United States domestic fishery in the Exclusive Economic Zone (EEZ) (Figure 1). It makes detailed provisions for foreign fishing, international fisheries treaties, foreign fishing permits, a national fishing vessel registration system, a fishery observer program, and fisheries and bycatch research. The Secretary of Commerce, with support by the NMFS, is charged with carrying out the domestic fishery management provisions of the SFA through Regional Fishery Management Councils. Through the Councils' activities, the public has direct input to decisions on management measures undertaken to conserve

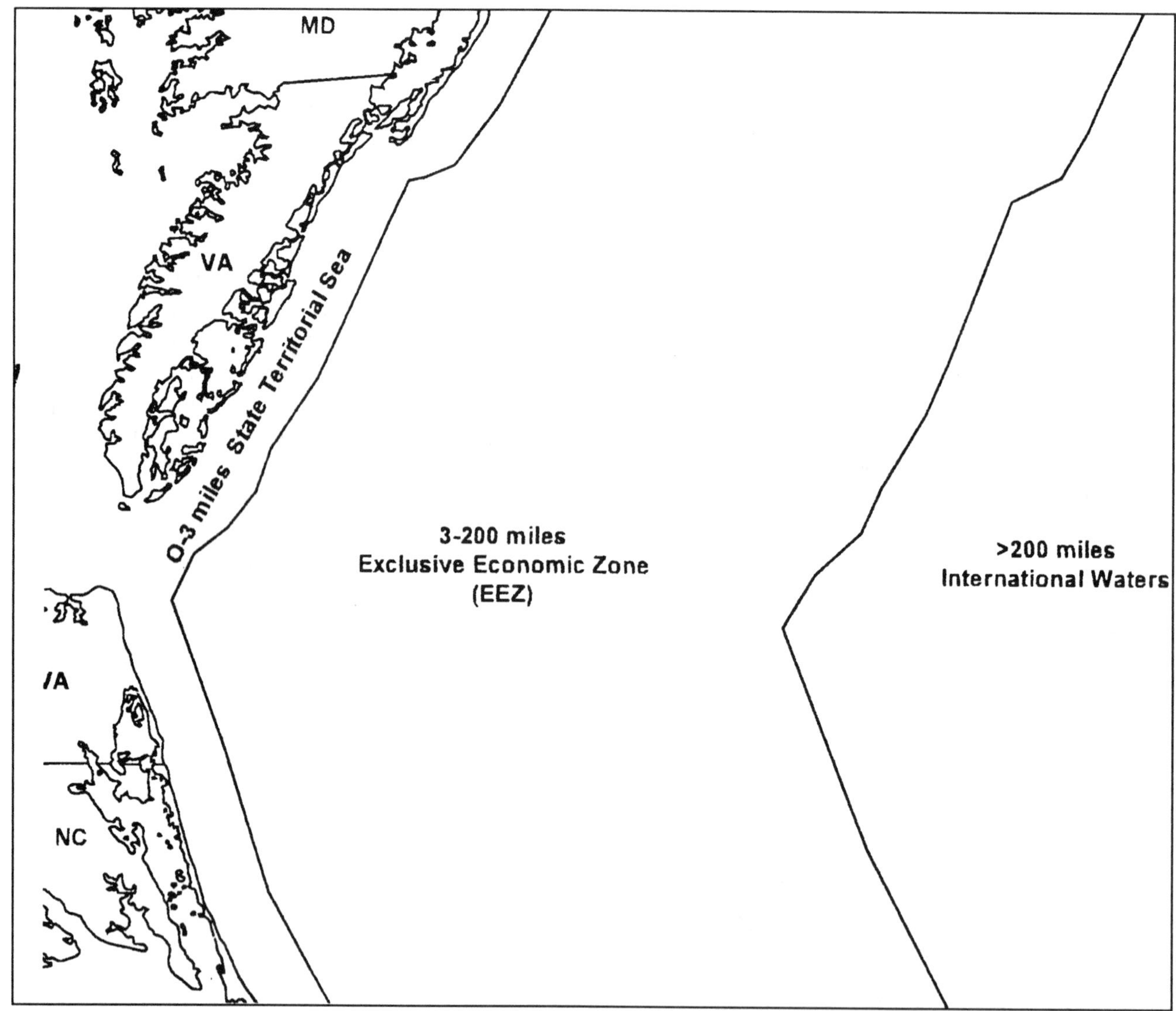

Figure 1. Fishery management jurisdictions in effect for U.S. fishery interests. Controlling factor is the distance from shore: 0-3 miles (=Territorial Sea), 3-200 miles (=Exclusive Economic Zone), and 200+ miles (=International Waters).

and sustain United States marine resources. The Council management process and membership criteria are described in federal regulations. Detailed accounts of the Council system are found in Wallace et al. (1994) and in the National Oceanic and Atmospheric Administration (NOAA) (1996). They may be referenced on the Internet, at <http://www.nmfs.gov>.

PRINCIPAL FISHERIES MANAGEMENT REGIMES

Principal fisheries management regimes are: 1) the Federal Council management system (from 3 to 200 miles, EEZ) under the Sustainable Fisheries Act (SFA,=M-SFCMA), 2) selected international fishery management organizations, and 3) the coordinated multi-state (Commission) system (from 0 to 3 miles, the territorial sea).

Regional Fisheries Management Councils

United States federal fisheries jurisdictions are apportioned under the SFA to eight regional fishery management councils (see Table 1 and Figure 2): three for the East Coast, one for the Caribbean, one for the Gulf of Mexico, and three for the West Coast and Pacific areas. Councils are quasi-governmental organizations whose function is the development of Fishery Management Plans

Table 1. Regional Fishery Management Councils

Council	Jurisdiction/States	Membership
New England 781-231-0422 (NEFMC)	ME, NH, MA, RI, CT	17 voting, 11 appointed
Mid-Atlantic 302-674-2331 (MAFMC)	NY, NJ, DE, PA, MD, VA, NC	21 voting, 13 appointed
South Atlantic 843-571-4366 (SAFMC)	NC, SC, GA, FL	13 voting, 8 appointed
Caribbean 787-766-5926 (CFMC)	USVI, PR	7 voting, 4 appointed
Gulf of Mexico 813-228-2815 (GMFMC)	FL, AL, MS, LA, TX	17 voting, 11 appointed
Pacific 503-326-6352 (PFMC)	CA, OR, WA, ID	14 voting, 8 appointed
North Pacific 907-271-2809 (NPFMC)	AK, WA, OR	11 voting, 7 appointed
Western Pacific 808-522-8220 (WPFMC)	HI, Am. Samoa, Guam No. Mariana Islands	13 voting, 8 appointed

(FMPs), as needed, for living marine resources and the fisheries occurring within the United States EEZ of their geographic juridictions. Some species are managed by two or more Councils under a joint FMP process (for example, the spiny lobster). Alternately, FMPs for some species, such as sharks, billfishes, and tunas, are specifically assigned to the Secretary of Commerce. Each Council gathers public input and advice from the NMFS on the status of the fishery resources in its jurisdiction and develops an FMP and management actions that will govern the fishery (species) to prevent overfishing while achieving optimum yield. The Councils gather public input, write the plans, and select the preferred actions for management. The NMFS provides scientific advice and data analyses on stock status, federal rules development and publication, and legal review to ensure proposed actions are consistent with federal laws. Federal authority for rules governing the fisheries in the EEZ resides with the Secretary of Commerce, while the United States Coast Guard and NMFS agents enforce the rules.

Council membership includes voting and nonvoting positions as specified by the SFA (Table 1). The Council members are from the states within a given Council's jurisdiction, and possess expertise in a particular fishery or fisheries. Voting members include the regional administrator of the NMFS, the marine fishery administrator for each state within the Council's jurisdiction, and citizens nominated by their respective state governors and

Figure 2. Fishery management councils and the FMPs administered by each.

appointed by the Secretary of Commerce. Voting Council members appointed by the Secretary "must be individuals who, by reason of their occupational or other experience, scientific expertise, or training, are knowledgeable regarding the conservation and management, or the commercial or recreational harvest, of the fishery resources of the geographical area concerned" (NOAA, 1996, section 302). This may include commercial or recreational fishermen, university scientists, conservationists, or business persons. Governors submit a list of candidates for open Council seats after determining that each individual is qualified and has, to the extent practicable, consulted with representatives of the commercial and recreational fishing interests of that state. Further, the governors are to nominate not less than three individuals for each vacancy and describe how each individual meets the qualifying requirements. Non-voting members are named by their agencies and represent the regional interstate fisheries commissions, the United States Department of State, the Fish and Wildlife Service, and other regional Councils. Voting members serve three-year terms, or less if desired by the Secretary. Maximum service is three consecutive terms. Information about fisheries of particular interest to the reader can be obtained from council members or state marine fisheries offices. More detailed information, special documents, and FMPs are available directly from the Council offices (Table 1) or via the Internet on web pages of the Council or NMFS.

Fishery Management Plans

Fishery management plans (FMPs) have their genesis in the expression of concern (usually about a particular fish stock) by recreational or commercial fishing interests to the Council. Often this is

Table 2. Summary of Fishery Status as Reported in the 1997 NMFS Report to Congress

Number of Stocks under Council Jurisdiction	Council	Overfished	Not Overfished	Approaching Overfished Condition	Unknown
26	NEFMC	12	7	2	5
11	MAFMC	5	6	0	0
9	NEFMC/MAFMC	1	1	0	7
84	SAFMC	14	5	0	65
69	GMFMC	4	6	1	58
10	SAFMC/GMFMC	1	5	0	4
179	CFMC	3	1	0	175
109	PFMC	13	16	6	74
64	WPFMC	3	47	1	13
105	NPFMC	0	63	0	42
1	PFMC/NPFMC	0	1	0	0
50	HMS	26	21	0	3
8	ASMFC	4	3	0	1
2	GSMFC	0	1	0	1
TOTAL		86	183	10	448

through letter(s) or comment(s) to the Council staff, or a council member citing individual experiences or evidence suggesting a need for management action. After informal discussion at the Council, a decision is made to determine the extent of the problem and possible solutions by conducting a series of "scoping" meetings throughout the Council's area of jurisdiction. These are typically held in concert with regularly scheduled Council meetings or public hearings on other FMP measures. The scoping sessions provide the background information to the public and collect guidance/input from the public for a better understanding of the basic problem, such as how extensive are the geographic impacts, how severe is the problem, and what management objectives or approaches might be effective in solving the issue. This information is also provided to the public for comment in the Federal Register (available in most public libraries, or the Internet site <http://www.access.gpo.gov/su_docs>). Additional input is solicited from NMFS, often as a status report/review on the species or fishery issue under consideration and a request for a future stock assessment report to the Council. After review, the Council may decide to develop a work plan and schedule for the FMP or amendment. Given Council approval, the staff prepares a Public Hearing Draft FMP document which lists various options and alternatives considered for the FMP, and where possible indicates Council's preferred alternatives. A Federal Register notice announces the availability of the draft document and the schedule/locations of public hearings throughout the area. After these hearings, the oral and written comments are reviewed at a Council meeting and the Council then selects final actions for the Draft FMP. This revised document, Final Draft FMP, is available to the public for comment and review (announced in the Federal Register). Hearing testimony and written comments on this document are received for approximately six weeks. At its next meeting, these comments are reviewed by the Council and a final public hearing is held. Final changes or minor clarifications to the document are approved by the full Council. The Council must approve the final document for submission to the Secretary of Commerce for formal review by a roll-call vote. The biological, social, or economic complexity and number of issues proposed in the documents often interact to prolong the timeline for amendment or FMP development. Up to this point in the document preparation and Council approval process, informal technical and editorial interchanges occur among Council staff, the NMFS, and the NOAA General Counsel, striving to perfect the document prior to official submission for agency review.

With Council's submission of the final document to the Secretary of Commerce (NMFS), a formal review schedule with specific time lines is followed within the NMFS, NOAA, and Department of Commerce (DOC) offices (total of 95 days' review after receipt by NMFS). Proposed rules are published in the Federal Register and public comments are received over a 45-day period. The NMFS must review these comments and address all issues raised in writing. On or before day 95, the Secretary of Commerce is to announce the federal government's decision on all or parts of the proposed rules. If approved by the Secretary, the draft final rules are forwarded to the Office of Management and Budget (OMB) for review/approval. On day 110, final rules are published with an effective date that is typically 30 days after publication. If the Secretary disapproves the Council's submission in whole or part, the rejected part is referred back to the Council with comments on the basis of the rejection. The Council reviews any rejected material and the Secretary's reasons for rejection, and decides to resubmit or withdraw the material from consideration. Resubmission by the Council may require a restarting of the 95-day time frame if substantial modification is required. Alternately, under new procedures adopted in 1997 the Council may request a fast-track determination. Major consideration by the NMFS and the NOAA General Counsel is given to the rationale and supporting justification for proposed measures, National Standards of the SFA, and the consistency of the proposed rules with other federal rules and law. In abbreviated form, the National Standards (NOAA, 1996):

- prevent overfishing while achieving optimum yield, on a continuing basis;
- take measures based on best scientific information available;
- manage an individual stock of fish as a unit throughout its range;
- do not discriminate among residents of different states: i.e., are fair and equitable;
- promote conservation, and allow no excessive share or privilege;
- are efficient in utilization, not having economic allocation as sole purpose;
- take into account variations and contingencies in fisheries, fishery resources, and catches;
- minimize costs and avoid unnecessary duplication;
- take into account the importance of fishery resources to fishing communities;
- minimize bycatch and mortality of any such bycatch; and
- promote safety of human life at sea.

In addition to the above Council FMP development and review process, the SFA now requires the Secretary of Commerce to report to Congress annually on the status of the United States fisheries within each Council's geographic area and to identify those fisheries that are overfished or approaching a condition of overfishing. In September 1997, the NMFS submitted such a report to Congress: 86 species are cited as overfished, 183 species are listed as not overfished, and 10 species are approaching an overfished condition. A larger group, 448 species, was cited as having an "unknown status" relative to an overfished condition. There were 39 approved and implemented FMPs at the time of the NMFS report to Congress, while five others were under development. Some plans address single or few species; others include more than 100 species. Management entities and stock status are shown in Table 2. The SFA also required the Councils to submit FMP amendments with new management measures by October 1998, including a 10-year rebuilding schedule for those species cited as overfished. This represents a major additional undertaking by the Councils and NMFS. Amendments, for example, have been made to most FMPs since their adoption; some plan amendments now number into the teens (e.g., Gulf of Mexico reef fish).

The FMP process outlined above is intense and involves many months of effort. Annual listings of stock status by the NMFS give additional breadth and detail to the Council process, and require an even greater investment of government, Council, and public effort in the management of marine fisheries resources under the SFA. For the next several years, the federal fisheries management process under the SFA will be very dynamic,

Table 3. Selected International Fishery-Related Organizations or Agreements in Which the United States Participates; See Text for Others (Contact the NMFS, Sustainable Fisheries Office, Silver Spring, MD, for Additional Information.)

International and Regional Management Organizations
- Convention on Biological Diversity: Offices in Montreal, Quebec, Canada
- Convention on International Trade in Endangered Species of Wild Fauna and Flora (CITES):Offices in Geneva, Switzerland
- International Convention for the Regulation of Whaling (IWC): Offices in Cambridge, UK
- Convention of Great Lakes Fisheries: Offices in Ann Arbor, Michigan
- Commission for the Conservation of Antarctic Marine Living Resources: Offices in Hobart, Tasmania, Australia
- Pacific Salmon Commission: Offices in Vancouver, British Columbia, Canada
- North Pacific Anadromous Fish Commission: Offices in Vancouver, British Columbia, Canada
- North Atlantic Salmon Conservation Organization: Offices in Edinburgh, Scotland

Scientific Organizations
- North Pacific Marine Sciences Organization: Offices in Sidney, British Columbia, Canada
- International Council for the Exploration of the Sea: Offices in Copenhagen, Denmark

Bi-Lateral Consultative Agreements
- United States -Mexico Fisheries Cooperative Program
- United States -Japan Consultative Committee on Fisheries
- United States -Russia Intergovernmental Consultative Committee
- United States -Canada Agreement on Fisheries Enforcement

Other Agreements (Research and Science-Related)
- International Oceanographic Commission (IOC)
- Global Ecosystem Dynamics (GLOBEC)
- Global Ocean Observing System (GOOS)

and the workload for Councils and NMFS will require extraordinary coordination. Under the SFA, opportunities for active public participation abound. Interested readers are urged to contact the nearest Council office for information, or seek information via their Internet site.

International Fisheries Agreements

Management of international fisheries that are exploited within the United States EEZ is coordinated through the Departments of State and Commerce, with the particular fisheries being subject to SFA regulations, Title II. Any such international arrangement is termed a Governing International Fishery Agreement (GIFA) and within that agreement the signatory nations acknowledge the exclusive management authority of the United States as set forth in the SFA. These agreements are a binding commitment of each nation and its fishing vessels for specific year(s) to abide by all regulations promulgated under the SFA (i.e., an FMP). Additional requirements include: vessel permits, access of United States enforcement on the fishing vessel for search, inspection, and seizure of illegal catches, placing transponders or observers to monitor performance (paid by the foreign nation), and repayment to United States fishermen for lost or damaged gear caused by foreign vessel activity. Each GIFA allows a specified amount of harvest of selected species for a given year of operation. In an important link to the SFA and United States fisheries, the total allowable level of foreign fishing (TALFF) available for GIFAs is only that portion of the "optimum yield" for such fishery that will not be harvested by permitted vessels of the United States (NOAA, 1996, Section 201d). All existing international fishing agreements are to be renegotiated to ensure that they meet the purposes, policies, and provisions of the SFA. Further, if renegotiation does not take place within a reasonable period of time, the United States shall withdraw from any such treaty. Particular mention was made of international fishery agreements which pertain to highly migratory species (HMS) fisheries. The Secretaries of State and Commerce are to evaluate all agreements pertaining to HMS and to consider whether each provides for:

1) adequate collection and analysis of information for management;
2) measures applicable for the fishery which are necessary and appropriate for conservation of the resource;
3) equitable access by fishing vessels of the United States to a portion of the allowable catch reflecting the traditional participation, enforcement and access throughout the range of the fishery and area of jurisdiction; and
4) adequate funding to implement the provisions of the agreement.

Congress exercises oversight of international fishery agreements. The President is to provide Congress with a draft of each new United States international fishery agreement at least 120 calendar days prior to its proposed effective date. Then, the House of Representatives and the Senate undertake a formal review and approval process for each agreement. Action by either chamber of Congress may prohibit, modify, or set other conditions for part or all of such international fishery agreements through passage of a "fishery agreement resolution."

International agreements concerning living marine resources of interest to the United States include those dealing with international or regional resource management of particular fisheries, bilateral consultation agreements, scientific organizations or councils, and other special purpose agreements (Table 3). At NMFS' Office of Sustainable Fisheries, staff provide liaison and technical support for most of these agreements and organizations. Examples of international organizations pertaining to United States fisheries interests are the International Pacific Halibut Commission (IPHC), Inter-American Tropical Tuna Commission (IATTC), International Convention for the Conservation of Atlantic Tunas (ICCAT), and the Northwest Atlantic Fisheries Organization (NAFO). Numerous others exist, but these provide an example of the agreements in place and the United States' role in the management of specific international fishery resources.

International Pacific Halibut Commission

The International Pacific Halibut Commission (IPHC) was formed by the Northern Pacific Halibut Act of 1982; members are the United States and Canada. The purpose of this organization is to conserve, manage, and rebuild halibut stocks to achieve and maintain maximum sustainable yield. Offices are in Seattle, Washington (web address <http://www.iphc.washington.edu>). This convention, and prior agreements between the United States and Canada, have provided joint management of Pacific halibut since 1923. Waters off the West Coast of the United States and Canada are under IPHC jurisdiction, including the southern and western coasts of Alaska. The United States has three commissioners on the IPHC, who are appointed by the President for a set term of service. Of these one must be a NOAA official, one a resident of Alaska, and the third a non-resident of Alaska. Further, one of the commissioners must be a voting member of the North Pacific Fishery Management Council. Official alternate commissioners to IPHC may be designated by the Secretary of State in consultation with the Secretary of Commerce. There are no formal advisory committees or groups serving the IPHC. Informal advisory groups of United States and Canadian industry representatives often attend the annual commission meetings and make recommendations. The IPHC staff coordinates a research program and develops procedures of the commission in consultation with the commissioners. The commission staff numbers about 25, mostly fishery biologists, inclusive of administrative and support staff. Quotas are set for the entire convention area. United States and Canadian vessels are excluded from fishing in waters of the other country. IPHC regulations include limits on recreational harvest, individual vessel quotas off Canada, individual fishing quotas off Alaska, and special catch-sharing rules promulgated by the NMFS for selected fishing areas off Washington, Oregon, and California.

Inter-American Tropical Tuna Commission

The Inter-American Tropical Tuna Commission (IATTC) was formed by a convention between the United States and Costa Rica in 1949, with implementing legislation, the Tunas Convention Act, passed in 1950. Member nations include Costa Rica, France, Japan, Vanuatu, Venezuela, and the United States. The IATTC offices are located at Scripps Institute of Oceanography in LaJolla, California. The IATTC was originally established to study the biology of tunas and related species in the Eastern Pacific Ocean (EPO) in an effort to determine the effects that fishing and natural factors had upon the abundance of the resource and to make recommendations for conservation measures that would result in maximum sustainable catches. In the mid-1970s, its studies were expanded to address the problems with dolphins and tuna schooling in the EPO. According to the Tunas Convention Act, United States representation to IATTC may not exceed four commissioners, appointed by the President. Other member nations have the same maximum membership constraint. For the United States, at least one commissioner must be a NOAA official, one must be from a non-governmental conservation organization, and all but one must reside in a state whose fishing vessels maintain a substantial fishery in the area of IATTC jurisdiction. The advisory committee from the United States (plus a like number for each other nation), specified in the enabling legislation, may be composed of not less than five, nor more than 15 persons representing groups in the EPO tuna fisheries and non-governmental conservation organizations. The advisors are appointed by the United States commissioners for set terms; they attend all non-executive sessions of the Commission, and provide input on all proposed programs, reports, recommendations and regulations. The IATTC organization is composed of national sections and a director of investigations. A Commission chairperson and secretary (from different national sections) are elected to one-year terms, with representatives from different national sections succeeding them. The director of investigations is appointed by the Commission and is responsible for budget, accounting and administrative support, development of a program of research, and direction of technical staff; and represents the IATTC when interacting with other organizations. Passage of resolutions, recommendations, reports or publications and other official IATTC actions require the unanimous vote of all member nations, with one vote per nation. The Commission conducts an extensive research program which includes an international staff of fishery experts. The primary research and management focus of IATTC to date is the yellowfin tuna fishery in the EPO and the incidental takes of dolphins in that fishery. Yellowfin tuna harvests are controlled by a catch quota system among the member nations. However, since 1979 member nations have not uniformly abided by the quotas, citing as reasons that non-member nations can harvest unrestricted amounts of yellowfin tuna from the IATTC area. IATTC efforts have brought major advances in our knowledge of yellowfin tuna, the dolphin association with yellowfin tuna, and techniques to lessen the mortality of dolphins in the tuna fishery.

International Commission for the Conservation of Atlantic Tunas

The International Commission for the Conservation of Atlantic Tunas (ICCAT) developed from the United States adoption of the Convention for Conservation of Atlantic Tunas, and was implemented through the Atlantic Tunas Conservation Act of 1975. Membership in the ICCAT includes over 25 nations with a historical record of harvest of Atlantic tunas, even though some member nations do not border on the Atlantic Ocean. The headquarters for the ICCAT is in Madrid. This organization was formed to maintain the populations of tuna and tuna-like fishes in the Atlantic Ocean at levels that would permit the maximum sustainable catch for food and other uses. Further, the Commission is to provide an effective international cooperative research program and recommend catch levels for each member nation fishing on the stocks. Up to three commissioners appointed by the President for three-year terms represent the United States on the ICCAT: one can be a salaried employee of any state or federal government for any number of terms; the second must be knowledgeable and experienced regarding commerical fisheries of the Atlantic Ocean, Caribbean Sea, or the Gulf of Mexico; and the third represents the recreational fishery. The commercial and recreational commissioners are limited to two consecutive terms of service. The United States commissioners appoint an advisory committee of no less than five, nor more than 20, persons, representing the various interest groups concerned with tuna fisheries in the Atlantic Ocean, plus the chairperson or designee of the five Fishery Management Councils having jurisdiction on the Atlantic and Gulf of Mexico coasts. Advisors serve two-year terms and may be reappointed without limit. Advisors attend all non-executive meetings of United States ICCAT Commissioners. They may examine and comment on all programs of study, reports, recommendations and regulations of the commission. The ICCAT's structure of commission, council, an executive director, and several special subject panels is more complex than either of the earlier examples. The commission is the full multi-national membership (three per nation); the council is an elected subset of the commission (chair, vice-chair, and 4 to 8 representatives from member nations). The executive secretary is elected by the commission and functions as the fiscal and administrative leader of ICCAT, and maintains oversight of its scientific programs and reports. Special panels of the commission focus on reviews of species or groups of species or geographical areas and may recommend joint actions for the ICCAT or scientific studies for the member nations. There are standing committees under the commission, such as the Standing Committee on Research and Statistics (SCRS). The ICCAT undertakes: 1) joint planning of research, evaluation of results, and coordination of research efforts among member nations, 2) collection of statistical data from the fishery and analysis in determination of stock status, and 3) joint development of recommendations which are submitted to the member nations for acceptance/implementation. To date, the ICCAT has taken conservation and management actions on bluefin, yellowfin, bigeye, and southern albacore tuna. Other actions have been recommended on Atlantic billfishes and sharks. For example, the United States has developed an intricate set of rules pertaining to the harvest of Atlantic bluefin tuna in the recreational and commercial fisheries to spread the United States quota equitably among the historical and traditional fisheries for the species.

Northwest Atlantic Fisheries Organization

The Northwest Atlantic Fisheries Organization (NAFO) was developed from the convention on Future Cooperation in the Northwest Atlantic Fisheries, which became effective 1 January 1979. The legislative basis for NAFO is the Northwest Atlantic Fisheries Convention Act of 1995. This organization is the successor of the International Commission of the Northwest Atlantic Fisheries (ICNAF). The NAFO is headquartered in Dartmouth, Nova Scotia, Canada, and has 16 member nations. The NAFO provides a forum for multinational consultation and cooperation relating to virtually all aspects of research, data collection, and assessment of fisheries in the convention area. Its mission is to conserve and manage fishery resources within its regulatory area. The area for NAFO includes waters of the Northwest Atlantic Ocean, approximately bounded by 35 degrees N latitude and 42 degrees W longitude. The United States may have not more than three commissioners appointed to NAFO by the Secretary of Commerce, each serving at his pleasure for a term not to exceed four years, and they may be reappointed. One commissioner must be an official of the United States government, at least one representative of the commercial fishing industry, and one voting (non-government) member of the New England Fishery Management Council. There are also three United States representatives appointed by the Secretary of Commerce to the NAFO Science Council with terms and reappointments similar to the commissioners. Science Council representatives must be knowledgeable and have experience in the scientific issues related to NAFO and one must be a United States government official. The Secretaries of State and Commerce also establish a Consultative Committee made up of representatives from the New England and Mid-Atlantic Fishery Management Councils, the states within those council jurisdictions, the Atlantic States Marine Fisheries Commission, the fishing industry, seafood processors, and others knowledgeable and experienced in management and conservation of fisheries of the Northwest

Atlantic. Administratively, NAFO has more committees, councils, and commission entities than the earlier-mentioned organizations. A General Council provides executive guidance to the Secretariat and serves as the forum for approval of member nations' management programs and regulations. The Scientific Council is the research and assessment body of the NAFO, providing guidance to member nations and scientific advice to the NAFO fisheries commission. The Fisheries Commission is responsible for management and conservation of the fishery resources. Standing committees address administrative aspects, research coordination, fisheries science, the fisheries environment, and publications (e.g., The Journal of Northwest Atlantic Fishery Science: Volume 20 contains a thorough discussion of fishery management systems in the North Atlantic, pre- and post-MFCMA). NAFO has maintained a set of management measures for fish stocks in the conservation area since 1979, built around Total Allowable Catches (TACs) for species and member nation quota allocations. The principal species managed under the NAFO are flounders, cod, redfish (*Sebastes*), American plaice, turbot, capelin, and northern shrimp. Recommended actions under the NAFO are non-binding on member nations, if that nation formally objects. Thus, over recent decades the fish stocks which were the mainstays of the United States and other northwest Atlantic groundfisheries underwent a progressive decline in abundance as they were overfished. Recognizing the dire status of major groundfish stocks and in an effort to recover the depressed stocks of groundfish, Canada declared a fishing moratorium, and the United States, under the SFA, has placed stringent regulations on harvests, with season and area closures throughout its historical fishing area.

UNITED STATES TERRITORIAL SEA FISHERIES: "THE COMMISSIONS"

United States territorial sea fisheries (0 to 3 miles) are managed by the process and procedures of the individual states as provided in their marine resource agency legislation. When applied to fish stocks that traverse state boundaries, this can lead to inconsistencies in the management measures of adjoining jurisdictions. Efforts to overcome this potential for confusion or public perception of cross-purposed regulations have included state conferences on particular fishery issues, bi- or multistate conferences (of legislators or managers or biologists or some combination thereof), formation of bi-state or river basin fishery management authorities, and other forums. Coordination for these types of endeavors was formalized in the founding of three interstate fisheries commissions, passage of enabling federal legislation, and approval of these compacts by the member states in the 1940s. Atlantic, Pacific, and Gulf States Marine Fisheries Commissions thus became the organizations for the conservation of the common property marine resources of the coastal states. These Commissions share a common purpose: coordination and promotion of policies and actions that foster the conservation and management of coastal marine fishery resources of importance to their member states. The authority to manage the resources is retained by the member states. As resource issues arose, the commissions held meetings, discussions, and made fishery-specific recommendations for coordinated management of selected resources. Implementation of the Commission's recommendations generally followed as individual states obtained management legislation requiring action by state fishery managers. On all commissions the membership consists of three commissioners per member state: the state fishery agency director, an appointee from the state's legislature, and an appointee of the governor. Funding for each commission includes a mixture of member states' contributions, federal grants, and special contracts.

Pacific States Marine Fisheries Commission

The Pacific States Marine Fisheries Commission (PSMFC) was authorized in 1947 and represents Alaska, Washington, Oregon, Idaho, and California. This commission is headquartered in Gladstone, Oregon. Pacific salmon issues are prominent in commission business. The PSMFC works to increase the public awareness of the importance of commercial and recreational fishing to local economies and to ensure that state and federal legislation reflects the needs of the marine industries. Important habitat issues are featured in the commission publication "Habitat Hotline," thereby helping fishermen and others participate in issues related to fish habitat protection and restoration or land-use planning programs.

The PSFMC has important projects in the field of coordinated fishery information and data management. Among them are StreamNet, covering all facets of anadromous fish information for the region; a passive integrated transponder (PIT) tag operations center (which provides users access to subsets of PIT tag data collection in the Columbia and Snake rivers); a regional mark-processing center (which provides coastwide coordination and management of coded wire-tag data for salmon on the West Coast); the Pacific Fisheries Information Network (PacFIN) (which maintains a central database for fisheries data off the West Coast states, Hawaii, and United States-flagged Pacific islands, giving estimates of catch groundfish species by gear, area, and month); and the Recreational Fisheries Information Network (RecFIN) (which provides coordination of state and federal field sampling of Pacific Coast recreational fisheries catch and effort data). Additional PSMFC activities focus on specific fishery needs such as the northern squawfish predator control program, the Northwest emergency assistance plan, habitat education program, and a gillnet recycling program.

Gulf States Marine Fisheries Commission

The Gulf States Marine Fisheries Commission (GSMFC) was approved by governors of Texas, Louisiana, Mississippi, Alabama, and Florida in 1949. The GSMFC headquarters are in Ocean Springs, Mississippi. Its concerns are the marine resources of the Gulf of Mexico and the coastal habitats that support them. The GSMFC serves as a forum for discussion of problems and issues of marine management, the fishing industry, and research, and seeks to develop regionwide policies for the improvement of public fishery resources. The GSMFC staff provide coordination and administrative support for cooperative state/federal programs dealing with marine fisheries resources: the interjurisdictional fishery program for development of fishery management plans for transboundary fish stocks that move between state and federal waters; the SouthEast Area Monitoring and Assessment Program (SEAMAP), a state/federal/university program for the collection, management, and dissemination of fishery independent data in the Southeast United States; and the Fisheries Information Networks for commercial (ComFIN) and recreational (RecFIN) programs that plan and conduct the collection, management and distribution of statistical data on the region's fisheries. The commission is responsive to the needs of the marine resources of the region and the state agencies charged with management of those resources.

Atlantic States Marine Fisheries Commission

The Atlantic States Marine Fisheries Commission (ASMFC) was created by the 15 Atlantic coastal states (Maine to Florida, plus Pennsylvania) and approved by an Act of Congress in 1942. Its headquarters are in Washington, D.C. The ASMFC focus is on five principal policy areas: research and statistics, interstate fisheries management, habitat conservation, sport-fish restoration, and law enforcement. The research and statistics program of the commission seeks to provide accurate and timely information to managers in support of federal, state, and interjurisdictional fisheries management. A management and science committee provides scientific advice to the ASMFC, as requested, covering such topics as aquaculture, fish stocking, fish health, protected species, and stock assessments. Relative to catch information, the Atlantic Coastal Cooperative Statistics Program (ACCSP) is developing a coastwide coordinated fisheries statistics system with a focus on fishery-dependent data. The goal is increased coastwide efficiency and cost-effectiveness of existing data collection efforts. The ASMFC also participates in SEAMAP-South Atlantic, maintains striped bass tagging files, and develops stock assessment evaluation and training protocols for fishery management plans of the commission.

The ASMFC's habitat conservation program includes policy development by inclusion of habitat information and research needs in the coastal fishery management plans and making state agencies responsible for habitat management aware of the fishery habitat needs. Public education takes the form of a newsletter "Habitat Hotline Atlantic," similar to that of the PSFMC.

Law enforcement activities of the ASMFC include data exchange, problem identification, and the means to resolve common problems among the member states, the Fish and Wildlife Service, and the NMFS. Enforcement and compliance issues are also incorporated in the commission fishery management plans.

The sport-fish restoration program of the ASMFC acts as liaison among state and federal agencies and non-governmental groups to promote cooperation on marine recreational fisheries programs, including artificial reef programs.

The InterState Fisheries Management Program (ISFMP) is the most complex and intensive program of the ASMFC and it is supported by other ASMFC activities. Components of the ISFMP are to identify priorities for fishery management; develop, implement, and monitor FMPs for priority species; recommend management measures for state and federal waters; and conduct short-term research related to the FMPs. Prior to 1993, the ASMFC actions recommended for management of coastal fisheries were not binding upon the member states — except for striped bass management, which was a special case in which a federal law was passed enforcing the ASMFC recommendations when states did not comply with the recommended ASMFC management-plan actions. Building on the success of the Striped Bass Act, Congress passed the Atlantic Coastal Fisheries Cooperative Management Act in 1993. This provides a mechanism to ensure that all Atlantic coastal states declaring an interest in the management of a particular ASMFC fishery (FMP) would be in compliance with all plan-mandated actions. Alternately, if a state is found out of compliance, the ASMFC informs the Secretary of Commerce, who in turn may, by federal rule, impose a harvest moratorium in the offending state for the species in question.

Under the rules and procedures adopted by the ASMFC for the interstate cooperative fishery management program, the various policy, technical, and advisory groups formed for each FMP meet periodically to conduct the business of FMP development, monitoring, and revision. Conceptually, the ISFMP fishery management plan process parallels that of the fishery Councils. Presently, 19 species FMPs are in place and 3 to 5 others are in development. Striped bass, northern lobster, summer flounder, weakfish, bluefish, and northern shrimp are among the more widely-known plans. Throughout the ASMFC's interjurisdictional fishery management process, numerous opportunities for public participation and input are provided.

SUMMARY

Many constituents feel the "system" of fisheries management at all levels and jurisdictions is not user-friendly, or is so fraught with acronyms to confuse all but those inside the system. This brief summary of fishery management systems, plan development and review processes, and selected international organizations pertaining to United States fisheries is intended to de-mystify fishery management programs and participants. Admittedly, it is complex and differs from one legal jurisdiction to another, and from FMP to FMP depending upon the objectives and problems addressed in the documents. However, one now has electronic access to a virtual library of fishery management information not readily available before. Participation of an informed public is crucial to the successful management of marine resources and fisheries.

REFERENCES

National Marine Fisheries Service. 1999. Fisheries of the United States, 1998. NMFS Current Fisheries Statistics. No. 9800. NMFS, Silver Spring, MD.

National Oceanic Atmospheric Administration. 1996. Magnuson-Stevens fishery conservation and management act. NOAA TECH MEMO. NMFS-F/SPO-23. NMFS, Silver Spring, MD.

United States Department of Commerce. 1993. Our Living Oceans: Report on the status of U. S. Living Marine Resources, 1993. NOAA TECH MEMO. NMFS-F/SPO-15. NMFS, Silver Spring, MD.

United States Department of Commerce. 1996. Our Living Oceans: The economic status of U. S. Fisheries, 1996. NOAA TECH MEMO. NMFS-F/SPO-22. NMFS, Silver Spring, MD.

Wallace, R.K., W. Hosking, and S.T. Szedlmayer. 1994. Fishery management for fishermen: a manual for helping fishermen understand the federal management process. Mississippi-Alabama Sea Grant Publ. MASGP-94-012. Auburn University Marine Extension and Research Center, Mobile, AL.

Harvesting

Robert L. Collette

INTRODUCTION

Fishing, born from the need to gather whatever food Nature provided, was one of the earliest methods of obtaining food. Little is known about how early fishing tools developed. It is known that ancient man fashioned hand spears for the land animals he hunted, and it is likely that these weapons also served as the first devices for capturing fish. Indeed, these simple spears served as a prototype for more advanced wounding gear.

A tool called the gorge is known to have been used in the Paleolithic era, half a million years ago. It was a short piece of wood, bone, or other material, straight or curved, and sharpened at both ends that we speculate was baited and attached to the end of a fiber line. When struck by a fish, the gorge would wedge in its mouth. Although not documented, it is supposed that during this same period our ancestors also learned to entrap fish in small rivers, bays, and inlets.

Historically, much of the basic gear used by modern fishermen was developed in neolithic times, roughly 10,000 BC. Barbed hooks, nets, gaffs, sinkers, and fiber lines were used by the Egyptians, Greeks, and Romans during this period. The Mayan and Aztec tribes of Central and South America reportedly used the hook-and-line, net, harpoon, and trap — as early as 1500 B.C. Chinese fishermen of the period were known to use spun-silk fishing lines.

The pace at which we moved from ancient fishing methods and primitive gear to today's modern techniques has varied considerably. Two factors influencing this evolution were the need to catch fish in bulk rather than singly, and the need to expand fisheries from shallow waters to greater depths. Today, many variations of the net, trap, and hook-and-line have evolved for specialized fisheries. The greatest changes have come in the materials and design of these nets and traps, in the methods of detecting fish, and in the boats rigged for fishing.

CLASSIFICATION OF HARVESTING TECHNIQUES

Techniques for harvesting fish involve both (1) fishing methods, the ways in which fish may be captured, and (2) fishing gear, the implements or tools used for that capture.

With hundreds of fish and shellfish species of commercial importance, each with its own characteristic habits and environment, modern fishermen have to use a variety of fishing gear and methods. One common way to classify harvesting techniques is to group them by the gear employed. Fishing gear used today may be grouped into the following categories:

1. Encircling or encompassing gear (seines)
2. Entrapment gear (pound nets, traps, and pots)
3. Lines (troll and longline)
4. Scooping gear (reef net and fish wheel)
5. Impaling gear (harpoon, spear)
6. Shellfish gear (dredges, rakes, tongs, etc.)
7. Entanglement gear (gillnet or trammel net)
8. Miscellaneous and experimental gear

Certain gear such as the harpoon, fish wheel, and some experimental devices are of little commercial significance in world fisheries so we do not discuss them in detail. Nets of all types (some of which were previously listed under separate gear categories) are grouped together and discussed first, since fish nets collectively take most of the world's commercial fisheries catch.

NETS

There are two main types of nets: nets that are used in motion that are drawn, hauled, or towed, and nets that remain stationary or static. Nets are designed for the habits and environment of the fish they are intended to catch. Therefore, some are designed to work at the bottom, others at midwater or at the surface.

Nets in Motion

The purse seine is a motion net designed for and especially effective in the capture of schooling fish such as mackerel, tuna, sardines, salmon, herring, and menhaden. All purse seines work on the same principle: a wall of net is used to encircle a school of fish. The top of the net is fitted with numerous corks or floats for support, and the bottom of the net is weighted to keep the wall of webbing in an upright position. A pursing cable (purse line) is threaded through rings sewn on the bottom of the net to allow the fisherman to close off or purse the bottom of the net, thus trapping the fish in an inverted umbrella-shaped enclosure. The basic operation of the purse seine is similar to closing off the drawstrings of an old-fashioned purse.

There are essentially two techniques used to set and haul purse seines. The two-boat system is commonly used on the East and Gulf coasts of the United States in the menhaden fishery (Figure 1). This system utilizes two small seine boats that are lashed side by side and towed behind a larger carrier or mother ship when on the fishing grounds.

The seine boats are shallow-draft, open boats usually constructed of aluminum and varying in length from 32 to 36 feet [9.7 to 10.9 meters (m)]. They are hung from the carrier vessel, one on each side. Half of the seine net is carried in each seine boat. The menhaden purse seines used in the two-boat system average 200 fathoms (1,200 feet/365.8 m) in length and 10 fathoms (60 feet/18.3 m) in depth.

Spotter aircraft accompany the seine boats in the search for schools of menhaden. When a school is spotted, the two seine boats begin setting their respective ends of the net. This operation is directed from the spotter aircraft. The seine boats deploy in almost opposite directions, eventually forming a circle around the school. When the two boats meet to form the circle, the ends of the purse line are run through pulleys on either side of the seine, which are attached to a heavy lead weight called a tom. The tom is sent overboard and the ends of the purse line are then hauled together with the aid of a hydraulic power block.

After pushing the seine, the tom is retrieved and the wings of the seine are hauled into the seine boats using power blocks in each boat. The fish are gradually concentrated in the bunt, a section of the net made of heavy twine, positioned between the two seine boats. Once the carrier vessel is alongside, the cork line is secured to its side to form a tight pocket from which the fish cannot escape. A flexible suction hose, attached to a centrifugal pump, is then lowered from the carrier into the net, and the fish are pumped into the hold.

One-boat seining is generally practiced on the Pacific Coast of the United States in the salmon, anchovy, mackerel, and tuna fisheries. The size and design of the West Coast seines differ depending on the species being harvested and local regulations. Seines carried on tuna boats are typically the largest, up to 3,600 feet (10973 m) long and 300 feet (91 m) in depth.

In one-boat seining, the net is carried aboard the carrier vessel (seiner), with an auxiliary boat assisting in setting and hauling it. When a school is sighted, the seiner is maneuvered into position to head off the school. The skiff is put down, and one end of the net, including the purse line, is put

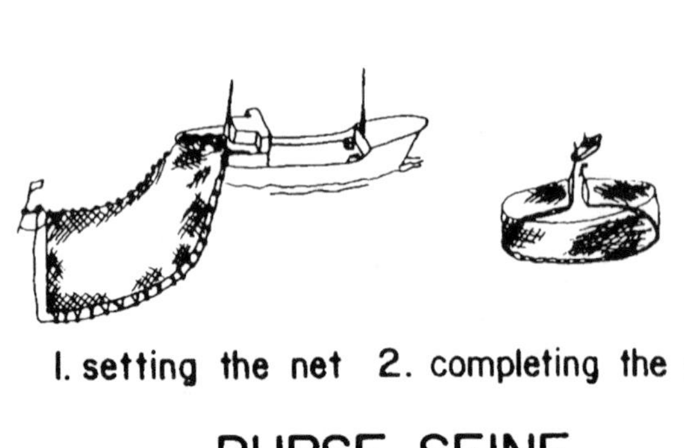

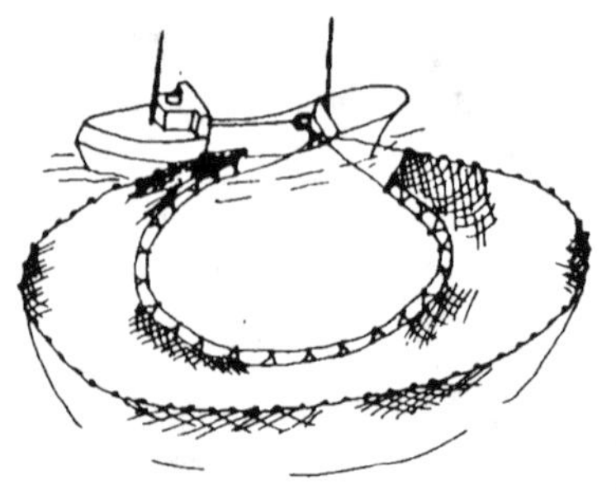

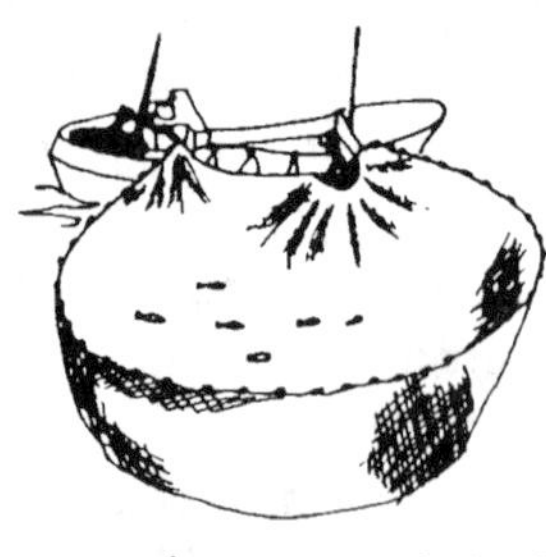

Figure 1. Setting and hauling purse seines.

into the water and held in position by the skiff. To surround the school of fish, the skiff begins towing away from the seiner as the larger vessel encircles the fish. At the close of the set, the skiff and the seiner turn toward each other. After the skiff comes alongside, the seine is pursed from the seiner and then hauled aboard using a power block. Fish are then brailed (captured with dip nets) or pumped into the hold.

Drum seining is a version of purse seining used exclusively in the Pacific Northwest. The operation of drum seining is similar to the one-boat operation. It differs only in that a large drum, usually hydraulically powered, is mounted at the stern of the seiner, and the entire net is spooled during hauling.

The lampara net is an encircling-type net which is considered the forerunner of the purse seine. The lampara was introduced by Italian fishermen in California in the late 1800s (Browning, 1974). It is shorter and shallower than the purse seine and can be set and hauled in less time and with less power. It also lacks rings but has a relatively large, simple bunt area and comparatively short wings. The mesh in the wings is generally large; in the bunt it is very small. The gear is set in a circular fashion, similar to the purse seine, and hauled by pulling both wings simultaneously. Lampara nets were used in the defunct California sardine fishery, along the West Coast to take bait for the tuna fishery, and throughout the West Coast mackerel and squid fishery.

Haul or beach seines are the simplest type of seining net. The haul seine, a long strip of netting with the head line (cork line) buoyed and the ground line weighted, is operated from shore. One end of the net is retained on land and the other is drawn through the water encircling the school of fish, and brought back to land. Smelts, shad, striped bass, croaker, bluefish, and weakfish can be caught with haul seines.

Trawling is the most important fishing method used to harvest demersal (bottomfish) species — such as cod, haddock, rockfish, flat fish, shrimp, etc. — that normally inhabit waters near the seabed. The trawl is a conical-shaped net with a wide mouth, and tapers to a socklike or cod end in which the fish collect. Trawl nets consist of the upper and lower nets, joined together at the sides with the upper net extending over the lower net like a roof.

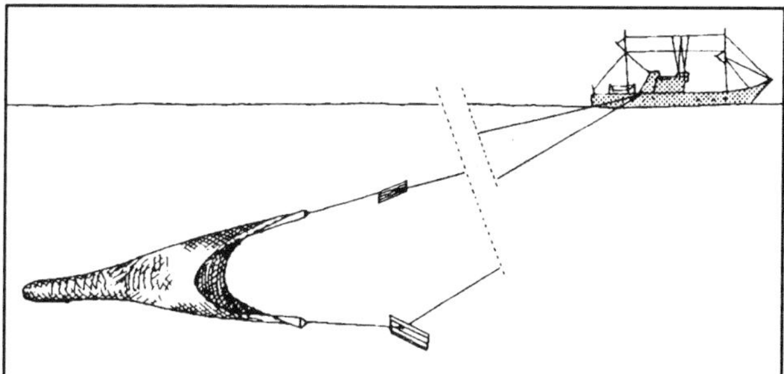

Figure 2. Otter trawl net.

Trawls are subdivided into several categories depending on the method used to spread the net mouth open. In the United States two trawling techniques are used: the beam trawl and the otter trawl, which is by far the more important (Figure 2). Trawl nets range up to 100 feet (30.5 m) across the mouth with a depth of 20 feet (6.1 m) and length of 150 feet (45.7 m). The otter trawl uses two otter boards (or doors) attached by bridles to the wings of the trawl. The boards spread the trawl mouth horizontally. Towing warps (cables) are attached to the opposite side of the boards.

Floats give the headrope buoyancy, which together with a weighted line at the footrope or groundrope keep the mouth of the trawl open. On rough fishing grounds the groundrope may be fitted with wooden or steel rollers to assist the net in getting over rocks and debris.

On traditional East Coast or Atlantic otter trawlers, large trawl winches are placed just forward of the house. The trawl nets are shot (set) and hauled from blocks secured to two heavy A-frame gallows mounted on one side of the vessel, usually starboard. In operation the cod end goes overboard first, followed by the midsection, wings, and doors. When the doors are properly set and spaced, the warps are paid out rapidly and the vessel gains speed slightly. When the net sets on the bottom, the warps are drawn together near the stern. Trawling speed is usually between 2 and 5 knots (3.7 to 9.3 kilometers/hour), depending on which fish species is being targeted. Trawling time ranges from 30 minutes to three hours. The process is reversed in hauling back. When the net is alongside the vessel, the doors are hooked to the gallows frames and the bulk of the net is hauled in with the help of quarter ropes. When enough of the trawl is aboard, the cod end is hoisted over the rail by a haul line. Once the cod end is aboard, the fish are released on deck by loosening its puckering strap.

On the West Coast a somewhat different technique is used in operating otter trawl gear. With the possible exception of some halibut schooner vessels, which by design may still use the side trawl method, a large majority of the Pacific trawls are shot over the stern. Typically trawling is carried out by Pacific combination vessels that operate both in the seining and trolling fisheries. When trawling, a pair of gallows frames are secured on each side of the stern with the winch mounted aft of the pilothouse. The combination boats may also use hydraulic drums to assist in setting and hauling the trawl, which is spooled onto the drum from the stern with only the cod end strapped over the side. On some modern vessels stern ramps have been constructed so that the entire trawl can be hauled from the stern.

In bottomfishing, otter trawls are sometimes fished using a two-boat system. Each vessel has one of the trawling warps attached to one wing of the trawl. This method is used only for very large nets and is rarely seen in the United States fisheries today.

In the early 1990s the shrimping industry switched from haul seines to trawls. Shrimp trawls are much like the East and West Coast otter trawl used for bottomfish, except they generally are smaller and lighter. Shrimp trawlers use double towing booms or outriggers secured to both the main mat and to a larger lifting boom. The towing booms can be rigged to pull one or two nets. In operation, the net(s) are towed from warps, which pass from a winch just behind the pilothouse through a block at the tip of the towing boom. In single-rig towing (one net) the two warps pass through blocks on the same boom, and in double-rig towing (two nets) a single warp for each net passes to blocks at the end of each boom.

Over the last few decades the beam trawl, from which the otter trawl evolved, has been largely replaced by the otter trawl, which is more efficient in handling and fishing. In beam trawling, a tapered wooden beam is used instead of otter boards to spread the mouth of net. U- or D-shaped runners are attached to the ends of the beam. The trawl, about 25 feet (7.6 m) long and tapering to a narrow pocket, is secured directly to the beam and runners. Beam trawls are operated similarly to other trawls, except they are always set from the ship's stern.

Stationary Nets

The gillnet is one of the oldest types of stationary net. Browning (1974) observed that gillnets apparently are a logical evolutionary development of the simple haul seine. As the name suggests, gillnets are designed to catch fish by the gills as they struggle to escape. The gillnet hangs vertically in the water, although some slack is built into the net to allow it to bulge. It is designed in this manner so fish swimming into a taut section of webbing do not bounce away from it but will entangle themselves.

Like seines, gillnets are vertical walls of webbing secured to a cork line and weighted lead line that keep the net upright in the water. The length, depth, and mesh size of gillnets vary with federal and state regulations as well as with the fish species being targeted. Two basic forms of the gillnet are commercially important in the United States: the drift net in its several forms and the set or anchor net. Drift nets are used typically on the high seas; set nets are used most often in inshore fisheries. The drift gillnet, designed for pelagic (surface or mid-water) fish such as mackerel, salmon, and saithe, is rectangular and is usually fished in a straight line. Drift nets are commonly set at the surface but can also be fished at intermediate depths or near the bottom. Different sets are made by varying the float and weight rigging.

Since gillnets typically are hung and float from the surface, they are visible to the fish, and therefore are usually fished at night. Once set, drift nets are allowed to drift anchored. Fishermen use marker buoys, or sometimes lanterns attached to floats to follow the drift of the net.

Set gillnets are put out along the sea floor to catch bottom-living fish, such as mullet cod and flatfish. These nets can be held in place by anchors or stakes when the sea nets are fished from shore or shallow water. Some gillnets are set so that they fish the middle-water column. These nets are anchored to the bottom and supported by buoys (Figure 3).

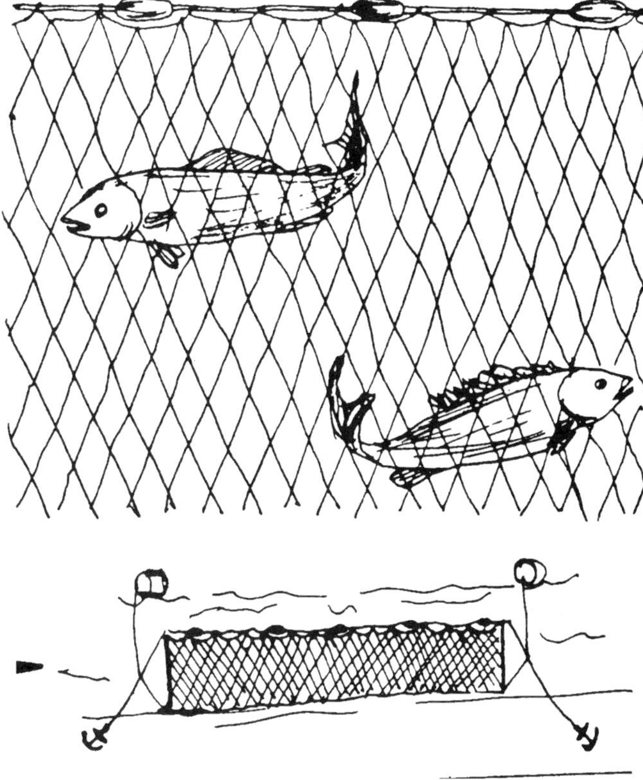

Figure 3. Gill nets.

Trammel nets, sometimes called tangle nets, are derived from the gillnet (Figure 4). They have three sheets of netting suspended from a common cork line and attached to a lead line at the bottom. The middle net is of fine mesh, loosely hung; the mesh of the two outer sheets is usually three times larger than the center-net mesh. Fish swimming against the trammel net draw the small inner mesh through one of the outer meshes and become trapped in the pocket which forms in the net. These nets are typically set in strings by one boat.

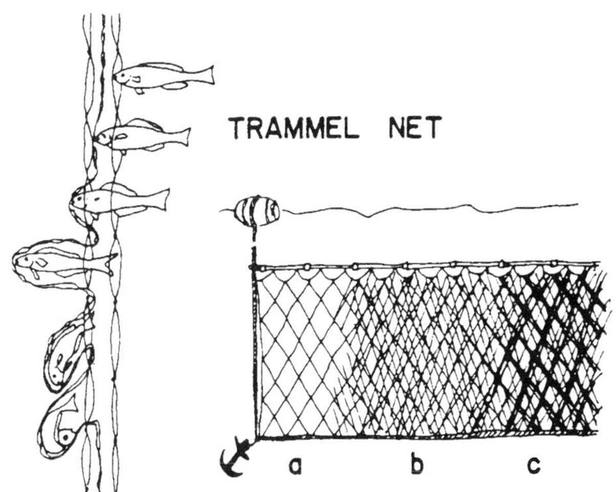

Figure 4. Trammel nets.

Gill and trammel nets usually remain set from 12 to 24 hours. They are hauled with the aid of a power roller, although the fisherman must manually assist to get a smooth winding of the net onto the reel. Most gillnet boats are rigged to work their nets from the stern. A noticeable exception exists in the Columbia River salmon gillnet fishery where small boats called bow-pickers are rigged to haul their nets from the bow.

TRAP AND POT GEAR

A wide variety of traps and pots are used to capture fish and shellfish. Capture by trap gear generally depends on attracting fish or shellfish to pots by means of bait or by leading fish into an enclosure, which is the case with pound nets, trap nets, and weirs. Numerous variations in the form and construction of the trap nets are used in American fisheries, but only a few of the more common types are described here.

Weirs are the primitive prototype of modern trap gears. They are usually set at a point of land that extends into the water for some distance or in channels, where the tide is strongest, to take advantage of the tendency of fish to stay in a strong current. The weir's main body is a large circular or heart-shaped enclosure, constructed by driving long, heavy posts into the bottom, with smaller posts set closely between. In the traditional weir, fine brush is then interwoven between these smaller posts, horizontally on the lower portion and vertically in the upper part, which remains visible at high tide. A leader of brush extends from the shore to the mouth of the trap. These leaders may be as long as 500 feet (152.4 m) and extend inside the mouth 5 feet (1.5 m) or more. The openings at the mouth are made wide enough for a dory to enter and work the trap.

In most applications, netting has replaced the brush formerly used in the heart of the weirs. During harvesting, a dory enters the weir and the mouth is closed by dropping a net. In some weirs, seines are used to capture the fish; in others, the trap bottom is raised by means of pulleys, and fish are herded into one section where they can be brailed or pumped on board.

Pound nets, which still see limited use in the harvesting of sardine, salmon, and other fish, are slightly more complex than the fish weir. In its simplest form a pound net consists of three parts: (1) the leader, extending from the shore or shallow water; (2) the heart or wings, a heart-shaped enclosure that deflects the fish; and (3) the pot (Figure 5). The fish are captured and removed from the pot, sometimes called the crib, pound, or pocket. Some pound nets are designed with two or more heart pockets.

Leaders used to guide fish through the series of progressively smaller compartments are up to 800 feet (244 m) long. The lead extends into the entrance of the heart, and fish move from the heart into the pot through an opening called a gate, situated directly in the center. The pound net's pot, which varies in size, is composed of small mesh netting supported by large, anchored poles. This section, like its equivalent in the weir, often has a net bottom secured to the sides. When harvested the bottom is raised, and the fish are brailed or pumped from the pot. In larger pound nets the fish are seined from the inner pocket.

Trap nets are similar in construction to pound nets except the former are supported by floats instead of by poles or stakes. The lead, heart, and pot of a trap net may extend 40 feet (12.2 m) up from the bottom but are completely submerged with only marker buoys visible at the surface.

Fish and shellfish pots or basket traps are one of the primary pieces of harvesting gear used for several commercially important species of crab and lobster. Other species commonly captured by this type of gear are shrimp, eel, crawfish, sea bass, and octopus. Dozens of variations in pot design exist, but only the more common pots used in the United States fisheries are discussed here.

Crab Pots

On the West Coast, an important type of pot is used in the dungeness crab fishery (Figure 6). The dungeness crab is harvested in estuaries, bays, and along coastal shorelines, usually where smooth, sandy bottoms are found.

The Dungeness pot is a circular, stainless steel frame, covered with soft stainless steel wire. It ranges in size from 36 to 48 inches (0.91 to 1.22 m) across and weighs 75 to160 pounds (34.1 to 72.6 kg). Usually two cone-shaped tunnels are placed at opposite sides of the pot's rounded surface. The tunnels are ramplike structures leading crabs to the opening (eye) and into the pot. The opening is constructed of small-diameter stainless steel rods equipped with single or double triggers, which are free-swinging, gatelike devices extending from the top of the opening downward across the bottom.

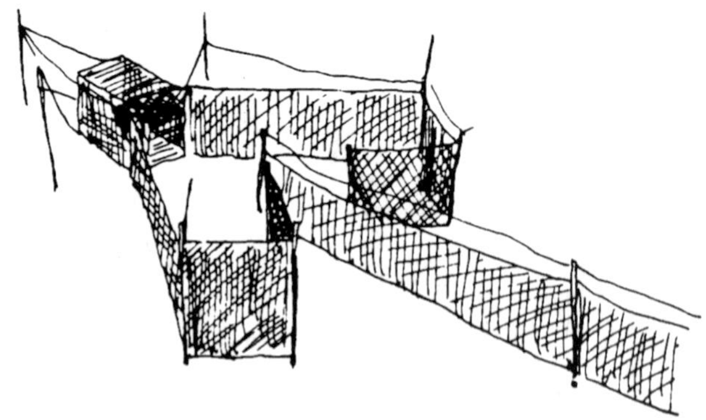

Figure 5. Pound trap net.

As the crab enters the pot, the trigger closes, preventing escape. The lower flat portion of the pot is weighted so that the pot will sink to the bottom. The lid, one-half of the top portion of the pot, is hinged and is held in place with steel hooks attached to rubber bands when closed. A small ring opening on the top or side gives undersized crabs an escape route.

Methods of fishing the pots are generally the same, with only the baits, time of the season, and vessel size differing. Pots are typically baited with herring, squid, or shad and are set in rows, with varying lengths and numbers of pots. A single line and cylindrical plastic buoy attached to each pot marks the position.

When hauling, the crab vessel usually travels against the current, allowing time to gather in the buoys and start hauling the pot by the time the vessel is over it. Using a crab power block, the pots are taken aboard, emptied, and rebaited before the next pot is hauled. An average boat can haul and reset about 300 pots a day.

Dungeness crab pots served as the basis for the development of the Alaskan king crab pot. The king crab pot is similar in construction but much larger and often rectangular rather than circular. The pots are 7 to 9 feet (2.1 to 2.7 m) square, 34 to 36 inches (76.2 to 91.4 cm) deep, and weigh 300 to

400 pounds (136.4 to 181.8 kg) each. Like the dungeness crab pots, king crab pots are fished singly from a buoy line.

Another important type of pot gear is used in the blue crab fishery, based on the East and Gulf Coasts of the United States. Blue crabs are smaller than dungeness or king crabs so the pot size and design are substantially different. Blue crab pots, introduced to the Chesapeake Bay blue crab fishery in the 1930s, usually are cubical in shape, 2 feet (61 cm) square, and constructed of wire mesh or a rigid metal frame. The pots are divided into a lower chamber, which contains a bait holder and funnels or passageways from the outside, and an upper or trap chamber. Crabs enter the bait chamber through funnels located at the pot's lower edges and then after taking the bait, swim upward through an opening into the trap chamber. Crabs are removed by spreading an opening in one seam at the top and shaking the crabs from the pot. Set singularly on buoy lines along the flat, sandy, or muddy edges or bays and channels in depths from 1 to 10 fathoms (1.8 to 18.3 m), the pots are usually lifted every day and are hauled by hand.

Lobster Pots

Three basic types of lobster pots used throughout the United States fishery are recognized by their shapes: the half-round pot, the rectangular pot, and the square pot. These pots are usually constructed of wooden laths or wire. Nearly all modern lobster pots consist of one or two funnels (heads) of coarse netting that slope upward toward the center of the net. The lobster enters through the funnels into the chamber compartment, or kitchen, in search of bait. After grabbing the bait, the lobster moves through a second funnel into another compartment, sometimes called a parlor, where it becomes trapped. It is removed through a door on top of the pot.

In the northern lobster fishery, pots are baited with fish such as salted herring, menhaden, skate, and scup. The pots are either set on a single buoy or several may be attached to a longline called a trawl line and weighted so that they rest flat on the bottom. Offshore lobster pots are set in trawl lines and are generally larger, heavier, and sturdier than those used in inshore waters.

Pots are hauled after soaking (fishing) one, two, or sometimes three nights, depending on the rate of deterioration of the bait and on the number of pots being fished. As with crabbers, lobster boats are rigged to store the catch live in seawater barrels or tanks.

HOOK-AND-LINE FISHING

Hook-and-line fishing has been used throughout the world for centuries. The objective of modern hook-and-line fishing is to orient fishing lines to obtain maximum geographic coverage by the hooks while minimizing the effort needed to handle the gear (Alverson, 1963).

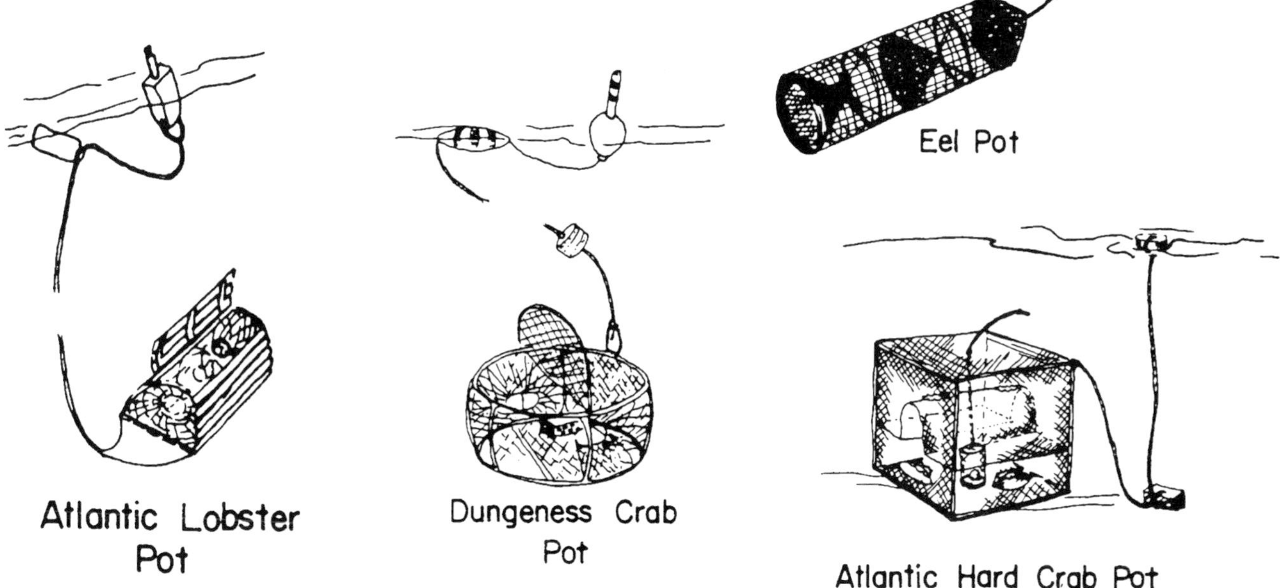

Figure 6. Various pots and traps used in harvesting seafood.

Hook-and-line fishing can be divided into four categories: (1) hand lines, (2) pole and lines, (3) troll lines, and (4) longlines or set lines.

Hand lines are important in the Gulf of Mexico snapper fishery and in some small inshore fisheries. In red-snapper fishing, lines about 60 fathoms (360 feet/110 m) long, with two hooks at the end and a lead sinker placed about 1 fathom (6 feet/2 m) above, are set from the deck of the vessel when fish are located. The lines are forked at the end, providing room for two hooks, and are held apart by wire spreaders. Artificial spoons, or sometimes herring, are used as bait and the lines are retrieved using hand reels.

In major commercial fisheries, the pole-and-line method is most prominent in catching tuna and mackerel-like fishes. The poles function to hold lines above deck level. A common rig has two or more poles set in sockets on each side of the boat, and two more lines are set from the stern. In the California tuna fishery, the ends of the poles are laced with a linen or nylon loop to which a 30- to 48-inch (76 to 112-cm) length of heavy cotton line is attached. From the cotton line a wire leader is added, with a barbless hook baited with herring or a feather jib (two or more hooks embedded in a small metal fishlike lure) attached to the edge of the leader. Vessels involved in pole and line fishing vary in size and construction from small motorized boats to larger diesel-powered vessels. In the United States, yellowfin, skip-jack, and albacore tuna are the principal species harvested by this method.

Trolling adds motion to the bait or lure being used. Simple trolling may be conducted by one line, although the modern troll fisheries, such as the fishery for Pacific salmon, use as many lines as possible. Large outriggers or spreader poles are used to space the lines, which are rigged with numerous lures or baited hooks. At the end of the trolling line "cannonballs," 10- to 50-pound (4.5 to 22.7-kg) weights, carry the lines to the desired depth and help prevent fouling. Several lines can be fished from each outrigger, and as many as 15 lures or baited hooks are attached to each line. The number of lines and lures trolled varies with the species sought and is sometimes governed by conservation laws.

Depending on the target species, lures or baited hooks can be fished from the surface down to 80 fathoms (240 feet/146 m). During fishing, the troller moves forward at the desired speed, giving action to the lures. In the salmon fishery, lines are hauled by reels or spools known as gurdies. These gurdy assemblies can be worked by hand but are more often powered by motor or hydraulics.

Longline or set line fishing uses a main or groundline that has a number of short branch lines (droppers, gangens, or offshoots) where baited hooks are attached. Longlines, which may be fished on the bottom, at intermediate depths, or near the surface, have the advantage of needing fewer people to handle the large number of hooks that are fished over a wide area (Figure 7). The halibut fishery of the Pacific Northwest is one of the world's principal longline fisheries. Here the groundline consists of a single string of ten skates (the primary unit of the longline) of 300 fathoms (1800 feet/549 m) each. Hooks are attached to gangens spaced along the groundline. Each skate, with 80 to 120 evenly spaced hooks, is coiled and baited prior to fishing.

When setting the gear, a flag marker — a bamboo pole with a light attached at the top — a buoy keg, and an anchor are put over as the vessel runs ahead. As the vessel continues on course, the longline is played out through a chute on the stern.

A line vessel may set any number of skates to form a string of gear. When the complete string has been set, another anchor line and float marker are dropped. After the gear has been adequately fished, one end of the longline is picked up and the gear retrieved using a power gurdy. As fish are brought to the surface they are gafted and lifted

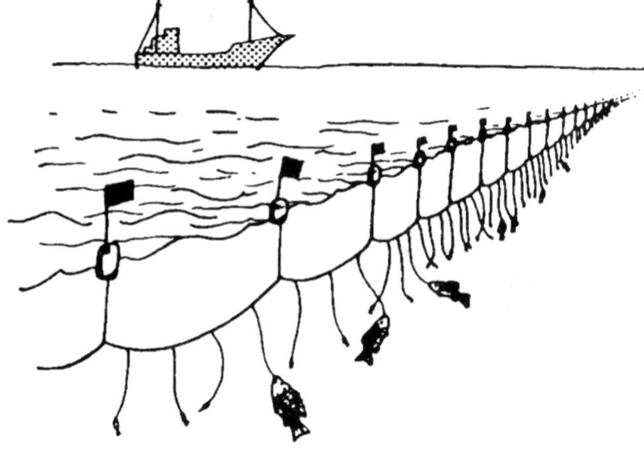

Figure 7. Longline sets.

aboard. During the process the lines are recoiled; baited with herring, octopus, or other baits; and readied for the next set. Pelagic species can be fished by longline by rigging the groundline with floats.

Schooners and West Coast combination vessels are the most often used vessels in the halibut longline fishery. However, longline gear can be fished from almost any properly equipped vessel.

SHELLFISH DREDGING AND SCOOPING GEAR

Shellfish such as oysters, clams, mussels, and scallops are sessile (not free-moving) marine animals and hence are harvested by different means than other marine species. Some simple devices popularly used for taking these shellfish include shovels, tongs, and rakes. A somewhat more sophisticated technique, which uses gear known as a dredge, is found in oyster, clam, and scallop fisheries.

Tongs consist of two rakes fixed to the ends of long wooden poles and hinged together with rakelike teeth facing each other. A basketlike frame is attached to each rake to collect the oysters. Oyster tongers typically fish oyster beds or reefs from small wooden, shallow-draft boats usually powered by outboard motors.

The fisherman stands on the deck of the boat, lowers the tongs to the bottom, and, by opening and closing the handles, scoops up a quantity of oysters between the heads (Figure 8). The tongs are lifted and the oysters piled on deck. Although slow and laborious, this method is still practiced in many areas because of the limited investment in gear. In some areas tonging is the only method available, as dredges are often considered destructive to natural reefs and are therefore prohibited on public reefs (Figure 9).

Oyster-dredging boats tow a dredge or drag consisting of a metal frame that has a toothed "raking" bar across the front, which in turn is attached to a bag-shaped net made of metal rings. The frame is connected to a towing cable by a triangular A-frame. As the dredge is towed across the reef, the rake bar dislodges the oysters, and they roll back across the drag bag into the metal mesh bag. When the bag is full the dredge is lifted and the contents dumped onto the deck.

Specialized suction and scoop dredges are also used in this fishery. A suction dredge creates suction at the head of the dredge. Oysters and water are suctioned up to a continuous conveyor located on the deck. The scoop-type harvester consists of a rakelike dredge with steel teeth. The dredge, which rests on runners to prevent excessive digging, is attached to a chain conveyor. In operation, oysters raked by the dredge are scooped onto the conveyor's loops and carried to the deck.

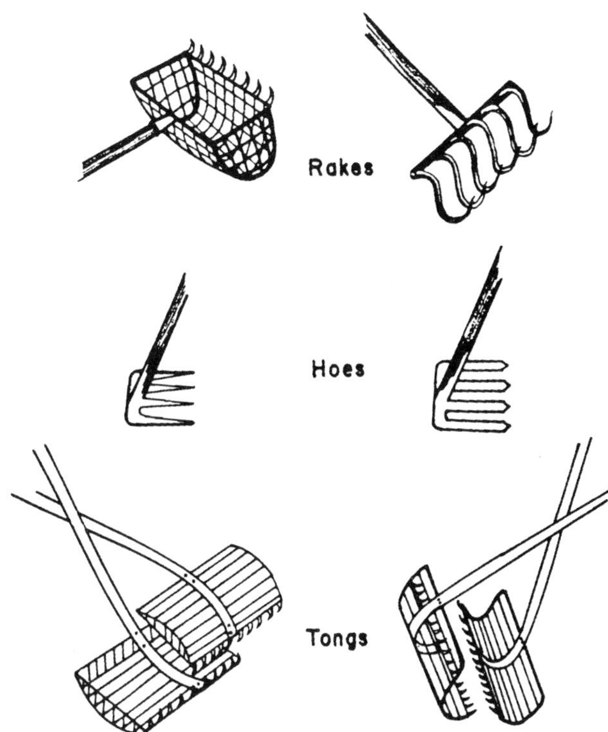

Figure 8. Hand-operated tongs and rakes.

Clams and mussels are taken by various types of rakes or by dredge. A regular clam rake is similar to a common steel garden rake except it is heavier and has longer, sharper teeth that are spaced about an inch apart and are curved distinctly upward. Clam rakes varying in widths and handle lengths are designed for shallow-water digging. A basket rake is one adapted for digging clams in deep water with a handle that may be 35 feet (11 m) long, much longer than the regular clam rake. The end of the basket-rake handle is fitted with a crosspiece to aid in dragging it across the clam bed, and a basket or wire or netting is attached to the back of the rake to hold the clams. A third rake used in harvesting clams, the bull rake, has a long handle like the basket rake but does not have a mesh basket attached. The bull rake is

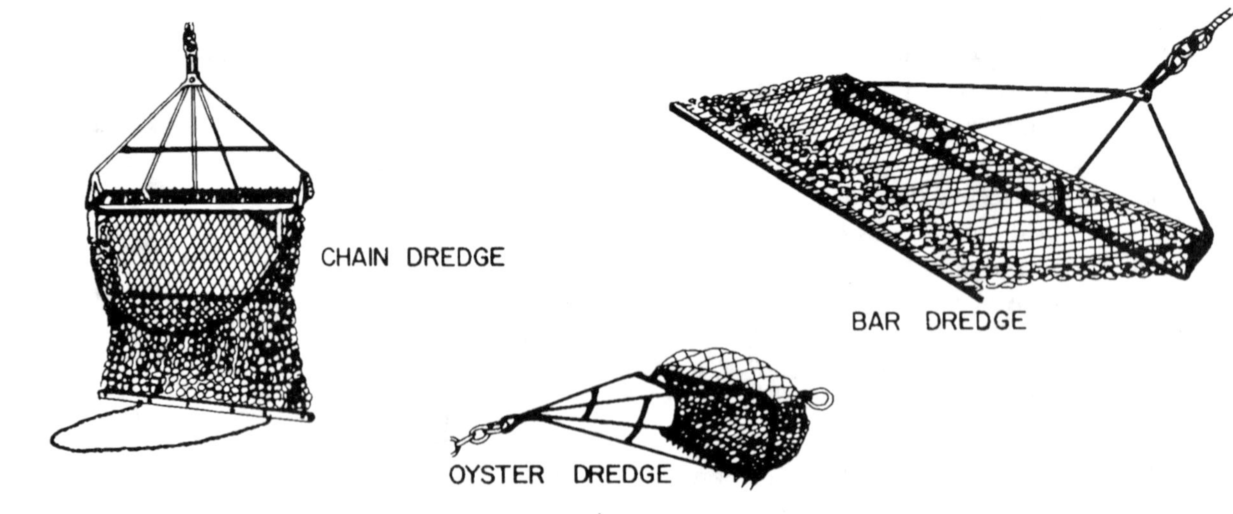

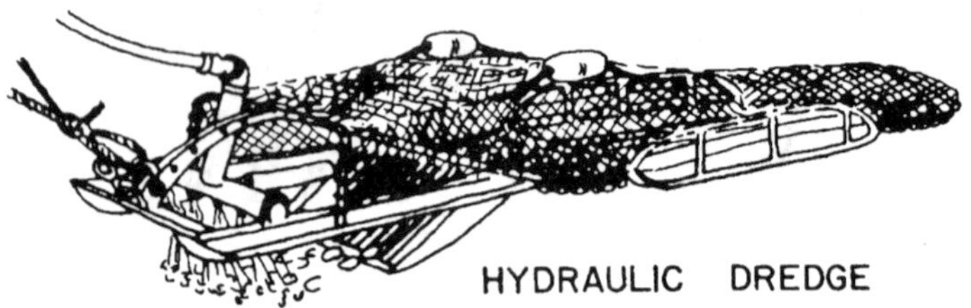

Figure 9. Dredges.

designed much like the regular clam rake, except it is wider and has more teeth. All clam rakes can be operated from small boats, but the regular clam rake can also be handled from shore. In either case, the teeth are worked into the sand or mud of the bottom, and the rake is then pulled in and lifted out of the water.

As in the oyster fishery, the majority of clams are taken by dredge. Clam dredges are operated with or without hydraulic equipment. The hydraulic or jet dredge is most often used for surf clams and ocean quahogs because of its effectiveness in extracting these mud-burrowing clams. During operation, pressurized water supplied by a hydraulic pump on board the vessel is pumped through jets located in front of the toothed bar. The jets of water loosen the bottom, allowing the clam to be scooped more efficiently. The hydraulic dredge may collect the clams in a metal ring bag or deposit them directly on deck via a conveyor.

Essentially all commercially harvested scallops are taken by dredge. The scallop dredge consists of an iron framework about 3 by 12 feet (91 by 46 cm), with an attached netting bag, which will hold one to two bushels of scallops. A single scallop boat often pulls several dredges across the scallop grounds.

Among the many scallop dredge styles, the scraper is one of the most popular. It has a rigid, triangular iron frame; a raised crossbar connecting the two arms; and a lower strip of iron, about 2 inches (5 cm) wide and set at an angle, for digging in the sand.

Abalone, sponges, and sometimes oysters are harvested by hand by fishermen who use various hand tools to gather the shellfish. The catch is stored in a net bag until it is full, then it is raised to the surface and emptied by a second fisherman tending the vessel. Although tedious, this method is extremely selective.

MISCELLANEOUS AND EXPERIMENTAL GEAR

Here we briefly describe some additional fishing methods, which either have a limited impact on commercial fisheries or are still in the experimental stage of development.

Jigging is a hook-and-line technique used most notably to harvest squid, although it is not used to any significant extent in United States fisheries. Jigging involves setting a line with baited hooks or lures, then a jigging machine provides a constant jerking motion on the line to induce fish to take the hook.

Harpooning is of great historical but declining economic importance. It is still the major method for taking whales, but whaling has been banned by most nations including the United States. Swordfish, shark, and tuna are still taken by harpoon, although longline methods have all but replaced it in those fisheries. In practice, harpoons can be thrown by hand or shot from a gun.

Cast nets (Figure 10), circular-shaped nets popular along the Gulf of Mexico and on the Pacific Coast, were once important gear in the shrimp fishery but today are more often used by sport fishermen. A cast net is draped over one arm and thrown so it spreads out in a flat circle; the weight of the lead line carries it to the bottom.

Gigging is another method used primarily by sport fishermen. The fisherman, using a spearlike instrument called a gig, wades through shallow waters and spears or gigs the fish when it has been spotted. Flounder are often caught by this method.

A number of fishing techniques rely on physical or chemical stimuli such as light, electricity, and odors. Most of these methods are experimental, although lights already have practical application and are used in a variety of fisheries to attract fish into traps or to aggregate them so they may be netted. Lights are used to capture bait fish and are used extensively in the shrimp and eel fisheries. Although electricity for shocking or guiding fish has been used experimentally since World War II, it has not yet been employed extensively in marine fisheries.

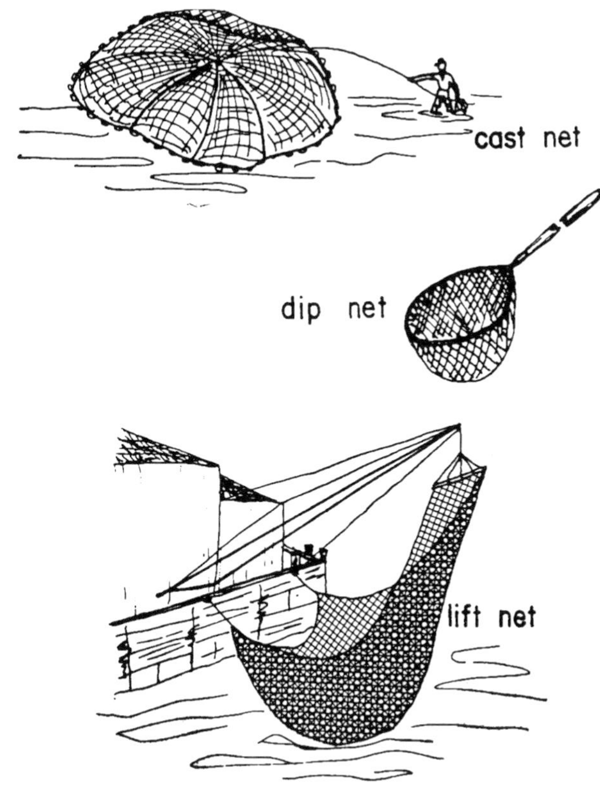

Figure 10. Hand nets.

REFERENCES

Alverson, Dayton L. 1963. Fishing gear and methods. In *Industrial Fishery Technology*. Maurice E. Stansby (ed.). Krieger, Huntington, NY.

Browning, Robert J. 1974. *Fisheries of the North Pacific*. Alaska Northwest, Anchorage, AK.

Schmidt, Peter G. 1960. Purse seining: deck design and equipment. In *Fishing Boats of the World*. Jan-Olof' Traung (ed.). Fishing News (Books) Ltd., London.

Tressler, Donald K. and James McW. Lemon. 1951. *Marine Products of Commerce*. Reinhold, New York.

Section II. Aquatic Life — Chemistry and Biology

Composition of Marine and Freshwater Finfish and Shellfish Species and Their Products

Juan L. Silva and Roberto S. Chamul

INTRODUCTION

The nutrient profiles of most animals, including fish and shellfish, vary both with and within species, location of harvest/source, whether wild or cultured, the season, environmental conditions, age or sexual maturity, sex, feed (especially in aquacultured species), genetic makeup, mode of preparation/analysis, and other factors. This chapter presents a survey of published data on the nutritional composition of the edible portion of the most common fish and shellfish. Included are variations in composition associated with various environmental and preparation/processing factors. Most of the data is from Exler (1987), Gooch et al. (1987), Watt and Merrill (1975), and Sidwell et al. (1974); binomial nomenclature is from Randolph and Snyder (1993).

Nutritional composition is but one of the hidden quality factors of importance to health-conscious consumers (Nettleton, 1992). Fish and shellfish are known for their high nutritional quality (Martin and Flick, 1990): they are relatively low in fat, saturated fat, and usually cholesterol; and high in polyunsaturated fatty acids, protein, and minerals such as calcium, phosphorus, sodium, potassium, and magnesium (Dean, 1990). Table 1 gives the ranges in individual nutrients of selected fish and shellfish, including daily recommended values (DRV).

Fish are usually classified as fatty (>10 percent fat), moderately fat (5 to 10 percent) and lean (<5 percent). Pelagic fish are usually fatty, whereas bottom-dwellers are lean.

PROXIMATE COMPOSITION OF FISH AND SHELLFISH

Table 2 shows the proximate composition of 84 species of fish. Moisture contents range from 68 to 83 percent, with the exception of caviar; this range is very similar to that of mammalian species. Total lipids vary among species, within species, within lots of the same species, and from one part of the fish to another — the tail is leaner, and dark muscles are fatter (Stansby, 1976; Stansby and Hall, 1967). Depth and latitude of fish catch also influence caloric density, fat, and the moisture content of fish (Bailey and Robinson, 1986). Fish from deep water and from northern latitudes have higher moisture and lower fat content. Moisture is usually replaced by fat deposition in fish, with older/larger fish and summer/fall-caught fish being higher in fat (Silva and Ammerman, 1993; Clark and Almy, 1920; 1918). Stansby (1962) reported an inverse relationship between fat and moisture of raw fish flesh, so that the sum averages approximately 80 percent. As fish grow, they eat more and thus their fat content increases. Wild-caught and aquacultured fish tend to have higher fat contents in the summer/fall months, and lower in the spring/winter months, when their metabolic processes are at a minimum (Nettleton et al., 1990; Clark and Almy, 1918). An exception to this are fish from tropical waters and "greenhouse" operations. Transplantation of fish species from the wild to a cultured environment usually leads to a change in nutritional composition (Amano et al., 1988). Product sampling (whole vs. fillet) also influences nutritional composition. Whole fish contain bones, skin, and viscera, decreasing the moisture content and increasing fat and ash (Silva et al., 1991).

Fish contain 8 to 20 percent protein and 1 to 2 percent ash. Among the species with high protein contents are shark, yellowtail, yellowfin tuna, and most salmon species. Greenland halibut, Pacific herring, orange roughy, sablefish, sea trout, shad,

Table 1. Ranges in Nutrients (raw, edible portion) of Selected Finfish and Shellfish

Nutrient, units per 100 g (DRV)*	Low Range	Medium Range	High Range
Ash,%	<1% Roughy, orange Sucker, white	1-2% Mullet Pollock Tuna Whiting	>2% Herring, Pacific Dolphinfish Salmon, Atlantic Turbot, European
Calcium, mg	<15 Mackerel	15-50 Bluefish Catfish Haddock Oysters	>50 Pollock Salmon Trout
Cholesterol, mg (300 mg)	<60 Clams Cod Catfish Flounder Trout	60-100 Lobster Mackerel Pollock	>100 Oysters Prawns Shrimp
Food Energy, calories (2,000 cal)	<100 Cod Roughy,orange Pollock Shrimp	100-200 Flounder Perch, Ocean Oysters Scallops	>200 Mackerel Salmon, Atlantic and sockeye
Iron, mg	<0.9 Cod Flounder Pollock	0.9-2.0 Carp Catfish Salmon Trout	>2.0 Clams Oysters Shrimp
Lipids,% (65 g)	<5% Cod Halibut Flounder Oysters Pollock Snapper, red Scallops Whiting	5-10% Bluefish Catfish Mullet	>10% Herring Trout, Lake Mackerel
Phosphorus, mg (1000 mg)	<200 Clams Flounder Oysters	200-300 Catfish Cod Pollock Trout	>300 Salmon Sardines

Table 1, continued			
Protein,% (50 g)	<17% Clam Halibut, Greenland Herring, Pacific Oysters Roughy, orange	17-20% Bass Catfish Cod Perch	>20% Anchovy Mackerel Roe Salmon, coho and sockeye Shrimp Tuna, Skip Jack Yellowtail
Sodium, mg (2,400 mg)	<60 Catfish Trout, rainbow Salmon	60-100 Cod Haddock Halibut Roughy, orange	>100 Crab, blue Oysters Pollock Scallops Shrimp
Water,%	<70% Eel Mackerel, Atlantic Salmon, Atlantic Shad, American	70-79% Bluefish Catfish Lobster Shrimp	>79% Cod Haddock Monkfish Oysters Hake, Pacific
* DRV: Daily Recommended Values			

and sturgeon average less than 16 percent protein — similar to chicken or beef. Ash content is higher in herring, salmon, turbot, yellowtail, and dolphinfish and less than 1 percent in orange roughy and white sucker. Some of the variability in ash content analysis may be attributed to the presence of bones and pinbones in the fillets.

Most shellfish, including mollusks, average moisture and protein contents similar to fish (Table 3). Oysters, octopus, clams, snails, and crayfish average over 80 percent moisture. Oysters and mussels contain less than 10 percent protein. Crustaceans and mollusks are lower in fat than fish and other muscle foods; these range from 0.4 percent (whelk and freshwater prawns) to about 2.4 percent for oysters. Some of these species contain a considerable amount of carbohydrate; whelk and abalone average 7.8 and 6 percent, respectively. Ash content varies between 1.2 and 2.2 percent.

Nutritional data on the 20 most commonly consumed seafoods, based on a three-ounce cooked serving, are listed in Table 4 [Food and Drug Administration (FDA), 1994]. These data are based on the assumption that most fish and shellfish will lose approximately 25 percent of moisture upon cooking, with little or no loss of fat; this assumption is one of the reasons why these data differ from most published data, including some of the data presented in this chapter. The assumption that only moisture is lost upon cooking is not correct, since it will depend upon the method of cooking and the product (Kinsella et al., 1975). Of 40 species studied, Gooch et al. (1987) reported five species (channel catfish, gag grouper, red porgy, American shad, and tilefish) to be lower in fat after cooking in steam to 70°C internal temperature. Many other species did not change significantly in fat content upon cooking. Silva and Allen (unpublished data, 1996) determined that the fat content of a four-ounce raw channel catfish fillet was about 5.4 percent and that, upon baking, there was not only moisture but also fat loss, even though the fat content of the baked fillets increased by weight, because fish fat is highly unsaturated

Table 2. Proximate Composition (g/100g raw basis) of the Edible Portion of the Most Common Finfish

Common Name	Scientific name	Water	Crude Protein	Total Lipids	Ash
Alewife	*Alosa pseudoharengus*	74.4	19.4	4.90	1.50
Anchovy	*Engraulis encrasicholus*	73.37	20.35	4.84	1.44
Bass, Black Sea	*Centropristis striata*	81.20	18.01	0.70	1.40
Bass, striped	*Morone saxatilis*[2]	78.86	20.70	2.94	1.19
Bass, hybrid striped[1]	*M. saxatilis* x *M. chrysops*[3]	78.14	18.47	2.75	0.95
Barracuda, great	*Sphyraena barracuda*	77.00	22.20	1.00	1.40
Bluefish	*Pomatomus saltatrix*	70.86	20.04	4.24	1.04
Bonito	*Sarda* spp.	71.30	24.70	4.50	1.50
Burbot	*Lota lota*	79.26	19.31	0.81	1.16
Butterfish	*Peprilus triacanthus*	74.13	17.28	8.02	1.46
Carp	*Cyprinus carpio*	76.31	17.83	5.60	1.46
Catfish, channel[1]	*Ictalurus punctatus*	76.39	18.18	4.26	1.26
Caviar, sturgeon (granular)		47.50	24.60	17.90	6.5
Cisco	*Coregonus artedii*	78.93	18.99	1.91	1.20
Cod, Atlantic	*Gadus morhua*	81.22	17.81	0.67	1.16
Cod, Pacific	*Gadus macrocephalus*	81.28	17.90	0.63	1.20
Croaker, Atlantic	*Micropogonias undulatus*	78.03	17.18	3.17	1.11
Cusk	*Brosme brosme*	76.35	18.99	0.69	1.25
Dolphinfish	*Coryphaena hippurus*	77.25	18.50	0.70	2.10
Drum, freshwater[1]	*Aplodinotus grunniens*	77.33	17.54	4.93	1.08
Drum, red	*Sciaenops ocellatus*	78.8	19.4	0.08	1.10
Eel	*Anguilla* spp.	68.26	18.44	11.66	1.41
Flatfish	*Bothidae* and *Pleuronectidae*	79.06	18.84	1.19	1.20
Flounder, Southern	*Paralichthys lethostigma*	82.4	17.4	0.60	0.90
Goosefish (monkfish)	*Lophius americanus*	83.24	14.18	1.52	1.21
Grouper	*Epinephelus* spp.	79.22	19.38	1.02	1.17
Haddock	*Melanogrammus aeglefinus*	79.92	18.91	0.72	1.21
Halibut, Atlantic and Pacific	*Hippoglossus hippoglossus* and *H. stenolepis*	77.92	20.81	2.29	1.36
Halibut, Greenland	*Reinhardtius hippoglossoides*	70.27	14.37	13.84	1.00
Herring, Atlantic	*Clupea harengus harengus*	72.05	17.96	9.04	1.46
Herring, Pacific	*Clupea harangus pallasi*	71.52	16.39	13.88	2.37
Ling	*Molva molva*	79.63	18.99	0.64	1.40
Lingcod	*Ophiodon elongatus*	81.03	17.66	1.06	1.21
Mackerel, Atlantic	*Scomber scombrus*	63.55	18.60	13.89	1.35
Mackerel, king	*Scombermorus cavalla*	75.85	20.28	2.00	1.28
Mackerel, Pacific and jack	*Scomber* and *Trachurus* spp.	70.15	20.07	7.89	1.62
Mackerel, Spanish	*Scombermorus maculatus*	71.67	19.29	6.30	1.27
Milkfish	*Chanos chanos*	70.85	20.53	6.73	1.14
Monkfish	*Lophius piscatorius*	83.24	14.48	1.52	1.21
Mullet, striped	*Mugil cephalus*	77.01	19.35	3.79	1.20
Ocean perch, Atlantic	*Sebastes marinus*	78.70	18.62	1.63	1.20
Perch, ocean	*Morone americana* and *Perca flavescens*	79.13	19.39	0.92	1.24

Table 2, Continued

Pike, northern	*Esox lucius*	78.92	19.26	0.69	1.20
Pike, walleye	*Stizostedion vitreum vitreum*	79.31	19.14	1.22	1.20
Pollock, Atlantic	*Pollachius virens*	78.18	19.44	0.98	1.41
Pollock, walleye	*Theragra chalcogramma*	81.56	17.18	0.80	1.21
Pompano, Florida	*Trachinotus carolinus*	71.12	18.48	9.47	1.10
Pout, ocean	*Macrozoarces americanus*	81.36	16.64	0.91	1.13
Rockfish, Pacific	*Sebastes* spp.	79.26	18.75	1.57	1.20
Roe		67.73	22.32	6.42	1.37
Roughy, orange	*Hoplostethus atlanticus*	75.90	14.70	7.00	0.90
Sablefish	*Anoplopoma fimbria*	71.02	13.41	15.30	1.05
Salmon, Atlantic	*Salmo salar*	68.50	19.84	6.34	2.54
Salmon, chinook	*Oncorhynchus tshawytscha*	73.17	20.06	10.44	1.37
Salmon, chum	*Oncorhynchus keta*	75.38	20.14	3.77	1.18
Salmon, coho	*Oncorhynchus kisutch*	72.63	21.62	5.95	1.21
Salmon, pink	*Oncorhynchus gorbuscha*	76.35	19.94	3.45	1.22
Salmon, sockeye	*Oncorhynchus nerka*	70.24	21.30	8.56	1.18
Scup	*Stenotomus chrysops*	75.37	18.88	2.73	1.21
Sea bass	*Centropristes striata* and *Lateolabrax japonicus*	78.27	18.43	2.00	1.09
Sea trout	*Cynoscion* spp.	78.09	16.74	3.61	1.26
Shad, American	*Alosa sapidissima*	68.19	16.93	13.77	1.32
Shark	Squaliformes	73.58	20.98	4.51	1.39
Sheepshead	*Archosargus probatocephalus*	77.97	20.21	2.41	1.09
Smelt, rainbow	*Osmerus mordax*	78.77	17.63	2.42	1.40
Snapper	Lutianidae	76.87	20.51	1.34	1.31
Spot	*Leisotomus xanthuras*	75.95	18.51	4.90	1.06
Sturgeon	*Acipenser* spp.	76.55	16.14	4.04	1.10
Sucker, white	*Catostomus commersoni*	79.71	16.76	2.32	0.46
Sunfish, pumpkinseed	*Lepomis gibbosus*	79.50	19.40	0.70	1.10
Swordfish	*Xiphias gladius*	75.62	19.80	4.01	1.48
Tilefish	*Lopholatilus chamaeleonticeps*	78.90	17.50	2.31	1.14
Trout	Salmonidae	71.42	20.77	6.61	1.17
Trout, rainbow[1]	*Salmo gairdneri*	71.48	20.55	3.36	1.30
Tuna, bluefin	*Thunnus thynnus*	68.09	23.33	4.90	1.18
Tuna, skipjack	*Euthynnus pelamis*	70.58	22.00	1.01	1.30
Tuna, yellowfin	*Thunnus albacares*	70.99	23.38	0.95	1.34
Turbot, European	*Scophthalmus maximus*	76.95	16.05	2.95	2.1
Whitefish	*Coregonus* spp.	72.77	19.09	5.86	1.12
Whiting	Gadidae, *Merluccius bilinearis*	80.27	18.31	1.31	1.30
Wolf fish, Atlantic	*Anarhichas lupus*	79.90	17.50	2.39	1.16
Yellowtail	*Seriola* spp.	74.52	23.14	5.24	1.09

[1] Aquacultured (freshwater) species
[2] Rawles et al., 1997.
[3] Silva et al., 1996.

Table 3. Proximate Composition (g/100 gram raw basis) of the Edible Portion of the Most Common Crustaceans and Mollusks

Common Name	Scientific Name	Water	Crude Protein	Total Lipids	Carbo-hydrates	Ash
Abalone	*Haliotis assimilis*	74.56	17.10	0.76	6.01	1.57
Clam, miscellaneous		81.82	12.77	0.97	2.57	1.87
Clam	*Venerupis semidecusata*	84.90	12.80	0.80	[2]	[2]
Crab, Alaskan king	*Paralithodes camtschatia*	79.57	18.29	0.60	0.0	1.8
Crab, blue	*Callinectes sapidus*	79.02	18.06	1.08	0.04	1.81
Crab, Dungeness	*Cancer magister*	79.18	17.41	0.97	0.74	1.70
Crayfish or crawfish[1]	*Astacur* and *Orconectes* spp.	80.79	18.66	1.06	0.00	2.00
Lobster, northern	*Homarus americanus*	76.76	18.80	0.90	0.50	2.20
Lobster, spiny	*Panulirus* spp.	74.07	20.60	1.51	2.43	1.39
Mussels	*Mytilus edulis*	80.58	11.90	2.24	3.69	1.59
Octopus	*Octopus vulgaris*	80.52	14.91	1.04	2.20	1.60
Oysters	Ostridae spp.	84.80	7.80	1.50	4.20	1.80
Oysters, Eastern	*Crassostrea virginica*	85.14	7.06	2.47	3.91	1.42
Oysters, Pacific	*Crassostrea gigas*	82.06	9.45	2.30	4.95	1.23
Prawns[1]	*Macrobrachium rosenbergii*	77.80	19.60	0.36	[2]	1.23
Scallop	Pectinidae spp.	78.57	16.78	0.76	2.36	1.53
Scallop, Atlantic Bay	*Pecten irradians*	80.70	15.40	0.50	1.70	1.40
Scallop, calico	*Aequipecten gibbus*	79.8	15.9	0.60	[2]	1.50
Shrimp	Penaeidae and Pandalidae	75.86	20.31	1.73	0.91	1.20
Snail	*Helix pomatia*	79.20	16.10	1.40	4.40	2.10
Squid	Loliginidae and Ommastrephidae	78.55	15.58	1.38	3.08	1.41
Whelk	Buccinidae	66.00	23.84	0.40	7.76	2.00

[1] Aquacultured species
[2] Negligible amount

Table 4. Nutritional Composition (3 oz. cooked or 4 oz raw weight) of the 20 Most Common Seafoods[1]

	Calories	Calories from Fat	Total Fat	Saturated Fat	Cholesterol	Sodium	Potassium	Total Carbohydrate	Dietary Fiber	Sugars	Protein	Vitamin A	Vitamin C	Calcium	Iron
		(g)		(mg)				(g)			(%DV)[2]				
Shrimp	80	10	1	0	165	190	140	0	0	0	18	0	0	2	15
Cod	90	0	0.5	0	45	60	450	0	0	0	20	0	0	2	2
Pollock	90	10	1	0	80	110	360	0	0	0	20	0	0	0	2
Catfish	140	60	9	2	50	40	230	0	0	0	17	0	0	0	0
Scallops	120	10	1	0	55	260	280	2	0	0	22	0	0	2	2
Salmon, Atlantic/coho	160	60	7	1	50	50	490	0	0	0	22	0	0	0	4
Salmon, chum/pink	130	35	4	1	70	65	410	0	0	0	22	2	0	0	2
Salmon, sockeye	180	80	9	1.5	75	55	320	0	0	0	23	4	0	0	2
Flounder/ Sole	100	14	1.5	0.5	60	90	290	0	0	0	21	0	0	2	2
Oysters, about 12 medium	100	35	3.5	1	115	190	390	4	0	0	10	0	0	6	45
Roughy, orange	80	10	1	0	20	70	330	0	0	0	16	0	0	0	0
Mackerel, Atlantic/Pacific	210	120	13	1.5	60	100	400	0	0	0	21	0	0	0	5
Perch, ocean	110	20	2	0	50	95	290	0	0	0	21	0	0	10	6
Rockfish	100	20	2	0	40	70	430	0	0	0	21	4	0	0	2
Whiting	110	25	3	0.5	70	95	320	0	0	0	19	2	0	6	0
Clams, about 12 small	100	15	1.5	0	55	95	530	0	0	0	22	10	0	6	60
Haddock	100	10	1	0	80	85	340	0	0	0	21	0	0	2	6
Blue crab	100	10	1	0	90	320	360	0	0	0	20	0	0	8	4
Trout, rainbow	140	50	6	2	60	35	370	0	0	0	214	4	6	2	
Halibut	110	20	2	0	35	60	490	0	0	0	23	2	0	4	4
Lobster	80	0	0.5	0	60	320	300	1	0	0	17	0	0	4	2
Swordfish	130	35	4.5	1	40	100	310	0	0	0	22	2	2	0	4

[1] Food and Drug Administration, 1994.
[2] % Daily Reference Value

(~ 75 percent) and has a low melting point. This fat is located just underneath the skin, and some of it will melt upon cooking.

The FDA data shows that only one serving (of mackerel) contributes more than 10 percent of the DRV of total calories, and that most fat in seafoods and freshwater fish is unsaturated. Oysters, clams, and shrimp are seen as significant sources of iron.

Amino Acids in Fish and Shellfish

Not only are fish and shellfish high in protein, but their protein is of a very high quality, as it contains significant amounts of all the essential amino acids (Table 5). Essential amino acids are found in slightly higher amounts in fish than shellfish, with freshwater species having a slightly higher content. Biological protein quality is high and exceeds the requirements for young children [Food and Agriculture Organization/World Health Organization (FAO/WHO), 1990]. Freshwater fish are high in sulfur, amino acids, and valine.

Fat, Fatty Acid Composition, and Cholesterol

Fish lipids occur primarily as triglycerides and phospholipids, the latter being highly unsaturated. Triglycerides are the main component of deposited fat, whereas phospholipids make up the bulk of fat in the muscle (Gruger, 1967). Most fish and shellfish contain about 0.7 percent phospholipids (Ackman, 1973). Approximately 25 percent of the fatty acids in fish are saturated, with the balance being mono- and polyunsaturated (Exler and Weihranch, 1976). Cold-water fish tend to deposit more unsaturated fat with lower melting points (Stansby, 1976). The relatively large amounts of polyunsaturated fatty acids in fish and shellfish offer unique nutritional and health benefits such as providing essential fatty acids, being a carrier of fat-soluble vitamins, and decreasing the risk of cardiovascular disease (Mitchell, 1986). However, one disadvantage of these oils is their susceptibility to oxidative rancidity (Erickson, 1993).

Table 6 shows that lipids in aquacultured shellfish contain 40 to 60 percent polyunsaturated fatty acids (PUFA). However, if one compares shrimp to prawns, the latter are lower in PUFA, a reflection of the profile of the feed (Silva et al., 1989) as well as other factors (Angel et al., 1981; Sidwell et al., 1974). The fatty acid profile and fat in fish tissue are related to the dietary-energy ratio and reflect the type and amount of lipids used in the feed. Seafoods are not only a good source of polyunsaturated fatty acids, but also provide ω3 fatty acids and have a good ω3/ω6 ratio. Aquacultured species are usually lower in ω3 PUFA, due to their feed source (Piggot, 1989; Lim and Lovell, 1986).

Minerals and Vitamins

Fish and shellfish contain considerable amounts of calcium, magnesium, phosphorus, potassium, and sodium (El-Faer et al., 1992). Nilson and Coulson (1939) stated that the mineral content of fish was similar to that of a beef round, except lower in iron and higher in iodine. Inclusion of bones in fillets or in some prepared products (such as canned salmon or sardines) will increase ash and thus mineral content. Shellfish are usually higher in minerals, including iodine, iron, and copper, than are fish (Table 7). Freshwater fish are the lowest in iodine content. On average, oysters, shrimp, and crab contain approximately half the calcium, five times the magnesium, and more phosphorus than an equal quantity of whole milk (Tressler and Lemon, 1951).

Most of the oil-soluble vitamins, especially vitamins A and D, and riboflavin, are concentrated in the fish liver and other parts (roe, offal) of the fish (Sautier, 1946). Shark species are important sources of vitamin A, as are cod, halibut, tuna, and whale. Good amounts of vitamin A are contained in most fish, such as halibut, herring, mackerel, some salmon, eels, and clams (Table 8). Fish and shellfish also contain fair amounts of thiamin, riboflavin, and niacin.

ENVIRONMENTAL AND SOURCE EFFECT

Aquacultured fish and shellfish composition is more predictable than that of wild-caught fish. Their composition is derived from the feed, and the influences of season, stocking rate, sexual maturity, size, and other genetic and environmental factors (Love, 1988). Amerio et al. (1996) reported that farm-raised eels, rainbow trout, sea bass, and sea bream had higher fat contents than their wild counterparts; they also had high biological protein value and a high content of essential amino acids. Many aquacultured species, such as shrimp and tilapia, are low in fat (Lovell, 1991). Lowering protein or increasing fat in the feed will usually

Table 5. Typical Amino Acid (g/100g) Profile of Selected Fish and Shellfish (edible portion, raw)

Amino Acid	Shrimp	Clams	Oysters	Scallops	Lobster	Cod	Pollock	Catfish[1]	Salmon	Trout/ Rainbow	Flounder	Mackerel
Tryptophan	0.283	0.143	0.709	0.188	0.262	0.199	0.218	0.204	0.222	0.230	0.211	0.208
Threonine	0.822	0.550	0.304	0.722	0.761	0.781	0.852	0.797	0.870	0.901	0.826	0.815
Isoleucine	0.985	0.556	0.307	0.730	0.911	0.821	0.896	0.838	0.914	0.947	0.868	0.857
Leucine	1.612	0.899	0.497	1.181	1.492	1.447	1.580	1.477	1.613	1.670	1.532	1.512
Lysine	1.768	0.954	0.528	1.254	1.636	1.635	1.786	1.669	1.822	1.887	1.731	1.708
Methione	0.572	0.288	0.159	0.379	0.529	0.527	0.576	0.538	0.587	0.608	0.558	0.551
Cysteine	0.228	0.168	0.093	0.220	0.211	0.191	0.208	0.195	0.213	0.220	0.202	0.199
Phenylalanine	0.858	0.458	0.253	0.601	0.794	0.695	0.759	0.710	0.775	0.802	0.736	0.726
Tyrosine	0.676	0.409	0.226	0.537	0.626	0.601	0.656	0.614	0.670	0.694	0.636	0.628
Valine	0.956	0.558	0.308	0.733	0.884	0.917	1.002	0.936	1.022	1.058	0.971	0.958
Arginine	1.765	0.932	0.515	1.224	1.642	1.066	1.164	1.088	1.187	1.229	1.128	1.113
Histidine	0.413	0.245	0.136	0.322	0.382	0.524	0.572	0.535	0.584	0.605	0.555	0.548
Alanine	1.151	0.772	0.427	1.015	1.065	1.077	1.176	1.099	1.200	1.242	1.140	1.125
Aspartic Acid	2.100	1.232	0.681	1.619	1.943	1.823	1.991	1.861	2.032	2.104	1.930	1.905
Glutamic Acid	3.465	1.737	0.961	2.282	3.207	2.658	2.903	2.713	2.962	3.067	2.813	2.777
Glycine	1.225	0.799	0.442	1.050	1.134	0.855	0.933	0.872	0.953	0.986	0.905	0.893
Proline	0.670	0.521	0.288	0.685	0.620	0.630	0.688	0.643	0.702	0.726	0.666	0.658
Serine	0.800	0.572	0.316	0.752	0.740	0.726	0.793	0.742	0.810	0.838	0.769	0.759

[1] Aquacultured

Table 6. Typical Fat Content (g/100 g), Fatty Acid Profile (g/100g), and Cholesterol (mg/100 g) of Selected Fish and Shellfish (edible portion, raw)

	Shrimp	Prawns	Oysters	Clams	Cod	Pollock	Catfish[1]	Flounder	Mackerel	Trout
Fat	1.73	0.36	2.47	0.97	0.67	0.98	4.26	1.19	13.89	3.36
Fatty acids										
Saturated	0.328	46.08	0.631	0.094	0.131	0.135	0.983	0.283	3.257	0.649
Monounsaturated	0.253	37.25	0.250	0.080	0.094	0.112	1.612	0.233	4.058	1.035
Polyunsaturated	0.669	36.67	0.739	0.282	0.231	0.483	1.002	0.329	4.761	1.199
14:0	0.021	0.39	0.109	0.013	0.009	0.005	0.054	0.048	0.674	0.068
16:0	0.184	29.64	0.435	0.060	0.091	0.084	0.691	0.158	2.125	0.384
16:1	0.083	02.61	0.074	0.022	0.016	0.012	0.165	0.080	0.727	0.158
18:0	0.103	15.61	0.060	0.018	0.030	0.043	0.214	0.071	0.423	0.132
18:1	0.147	34.64	0.097	0.034	0.061	0.067	1.377	0.120	2.283	0.618
18:2	0.028	14.56	0.049	0.016	0.005	0.009	0.393	0.008	0.219	0.249
18:3	0.014	0.80	0.037	0.004	0.001	—	0.039	0.008	0.159	0.119
18:4	0.006	—[2]	0.096	0.016	0.001	0.005	0.033	0.016	0.278	0.048
20:1	0.017	—	0.060	0.018	0.015	0.022	0.053	0.028	1.039	0.128
20:4	0.087	0.51	0.068	0.041	0.022	0.026	0.102	0.038	0.183	0.114
20:5	0.258	—	0.211	0.069	0.064	0.071	0.130	0.093	0.898	0.138
22:1	0.005	—	0.009	0.006	0.003	0.010	0.002	0.008	1.407	0.131
22:4	—	—	—	—	—	—	—	—	—	—
22:5	0.046	—	0.050	0.052	0.010	0.022	0.062	0.046	0.212	0.106
22:6	0.22	—	0.228	0.073	0.120	0.350	0.243	0.106	1.401	0.430
Cholesterol	152	109	34	43	71	58	48	70	57	

[1]Aquacultured
[2]Data not available or contains negligible amount

Table 7. Mineral Content of Selected Fish and Shellfish (edible portion, raw)

Mineral (mg/100g)	Shrimp	Oysters	Clams	Cod	Pollock	Salmon	Catfish[1]	Flounder	Mackerel	Trout[1]
Calcium	52	45	46	16	60	[2]	40	18	12	67
Iron	2.41	6.70	14.0	0.38	0.46	0.70	0.97	0.36	1.63	1.90
Magnesium	37	54	9	32	67	—	25	31	76	31
Phosphorus	205	139	169	203	221	—	213	184	217	250
Potassium	185	229	314	413	356	422	349	361	314	495
Sodium	148	112	56	54	86	46	63	81	90	27
Zinc	1.11	90.9	1.37	0.45	0.47	0.41	0.72	0.45	0.63	1.09
Copper	0.26	4.46	0.34	0.03	0.05	0.05	0.09	0.03	0.07	0.11
Manganese	0.05	0.45	0.50	0.01	0.01	0.01	0.01	0.02	0.01	0.70
Iodine (mg/kg)	0.45	1.16	0.37	0.24	0.12	—	0.01	—	0.14	0.01

[1] Aquacultured

Table 8. Vitamin Content of Selected Fish and Shellfish (edible portion, raw)

Vitamins (units)	Shrimp	Oysters	Clams	Lobster	Cod	Pollock	Catfish[1]	Flounder	Mackerel	Trout
Vitamin A (IU)	[2]	—	300	—	40	35	—	33	165	65
Vitamin C (mg/100 g)	—	—	—	—	1.0	—	—	—	0.4	3.6
Thiamin	0.028	—	—	—	0.076	0.047	0.045	0.089	0.176	0.074
Riboflavin	0.034	.166	0.213	0.002	0.065	0.185	0.106	0.076	0.312	0.185
Niacin	2.552	1.310	1.765	0.215	2.063	3.270	2.143	2.899	9.080	—
Pantothenic acid	0.276	0.184	0.362	0.360	0.153	0.358	0.57	0.503	0.856	—
Vitamin B_6	0.104	0.050	—	—	0.245	0.287	0.19	0.208	0.399	—
Folacin	3.00	9.9	—	—	—	—	—	—	—	—
Vitamin B12	1.161	19.13	49.44	0.031	0.908	3.191	—	1.520	8.711	—

[1] Aquacultured
[2] Data not available

result in an increase in fish muscle fat (Lovell and Mohammed, 1989). More importantly, adding fish oil supplement to the fish diet will increase the ω3/ω6 ratio of the fish muscle. The ω3/ω6 ratio for cultured catfish is still lower than for wild catfish and salmon, but can be increased through feed manipulation (Piggot, 1989; Robinson, 1989). Increasing protein in the feed does not influence product composition (Silva et al., 1994).

Sample type or product form has a tremendous influence on the composition of fish. Botta (1973) reported that the white/lean muscle of catfish contained 25.5 percent fat while the dark muscle (lateral line) contained 1.3 percent. Nuggets (belly flaps) contain up to 10 times more fat than fillets (Silva et al., 1994; Wang et al., 1977). Most of the fat in fish is in the dark/lateral tissue and the skin (Freeman, 1990). Whole fish is more fatty than dressed (deheaded, eviscerated, skinned) fish and this in turn is higher in lipids than fillets (Table 9).

Storebakken and Austreng (1987) reported that feed and ration level affected the composition of farmed Atlantic salmon. Larger-sized fish tend to deposit more fat in the muscle. Fish caught in the summer and fall tend to be higher in fat than those caught in the winter and spring. However, this is variable and depends on feeding rates, environmental conditions, degree of satiation, and species.

Not only is proximate composition affected but, more important, fatty acid contents and profiles are affected by genetic and environmental factors (Erickson, 1993; Lovell, 1991). The amino acid profile and essential amino acid index are also affected by sex and season (Dabrowski, 1982).

Table 9. Effect of Size, Diet, Season of Harvest, Form/Part of Fish on Composition of Farm-raised Channel Catfish

Attribute	Water	Protein	Fat	Ash
Diet[1]:				
Control	[6]	—	8.6	—
+ 6% fish oil	—	—	12.1	—
Size[2]:				
0.63 kg	68.1	17.0	10.3	1.0
0.18 kg	70.8	17.1	9.2	1.8
Season[3]:				
Fall	74.4	16.1	7.7	1.1
Winter	77.4	15.4	6.2	0.8
Spring	77.8	15.4	6.4	0.9
Summer	76.0	15.7	7.4	0.9
Tissue Type[4]:				
Lateral	70.7	14.9	13.7	1.2
Skin-side	70.5	16.1	12.1	1.0
Visceral-side	82.5	15.6	1.1	1.0
Internal	79.8	18.9	0.8	1.2
Form (sample type):				
Whole[5]	58.5	—	14.6	—
Dressed[2]	68.1	17.8	10.3	1.0
Fillet[3]	76.4	15.6	6.9	1.0

[1] Lovell and Mohammed, 1989
[2] Silva and Ammerman, 1993 (dressed fish)
[3] Nettleton et al., 1990 (fillets)
[4] Freeman, 1990 (fillets)
[5] Silva et al., 1991
[6] Data not available

STORAGE, PROCESSING, AND COOKING EFFECTS

Refrigeration and frozen storage have little effect on the proximate composition of fish (Ammerman, 1985). However, fatty acid composition can be somewhat affected by frozen storage due to oxidation of PUFA (Silva and Ammerman, 1993; Haard, 1992). Addition of vitamin E or other antioxidants to the feed or direct application on the harvested fish have shown little effect on fatty acid profile stability (Silva et al., 1994). Moist cooking has little effect on the fatty acid composition of fish (Gooch et al., 1987). However, dry cooking, frying, thermal processing, and drying/salting have a marked effect on the composition of fish and shellfish. Alkaline processing decreases fat content of fish (Silva et al., 1984).

Baking results in the loss/melting of fat and moisture. Baked catfish loses about 20 to 25 percent of it by weight, mostly due to loss of moisture. About 6 to 8 percent moisture is lost during baking, resulting in as little as 8 percent (Silva, unpublished data) to 25 percent increase in fat content (Table 10). Boiling has little effect on the composition of shellfish but this is a function of a relatively short cooking time. Cold smoking has little effect on product composition, while hot smoking reduces moisture and increases fat, protein, and ash. Canning of shellfish and fish has little effect on proximate composition but decreases vi-

Table 10. Effect of Processing and/or Cooking Methods on Some Nutrients of Selected Fish and Shellfish

Species/Process	Water	Protein	Fat	Saturated fat	Carbohydrates	Ash	Calcium	Iron	Vit. A
	(g/100g)						(mg/100g)		(IU)
SHRIMP									
Raw	75.86	20.31	1.73	0.33	0.91	1.20	52	2.41	—
Imitation (from surimi)[1]	74.91	12.39	1.47	[2]	2.11	9.13	19	0.60	—
Boiled	77.28	20.91	1.08	0.29	0.0	1.57	39	3.09	—
Canned in water (drain)	72.56	23.08	1.96	0.37	1.03	1.36	59	2.74	—
Breaded and fried	52.86	21.39	12.28	2.09	11.47	1.99	67	1.26	—
CRAB, BLUE									
Raw	79.02	18.06	1.08	0.22	0.04	1.81	89	0.74	—
Boiled/steamed	77.43	20.20	1.77	0.29	0.00	2.00	104	0.91	—
Canned (drained)	76.16	20.52	1.23	0.25	0.00	2.05	101	0.84	—
TUNA									
Raw (bluefin)	68.09	23.33	4.90	1.26	0.00	1.18	—	1.02	2184
Oven-cooked (bluefin)	59.09	29.91	6.28	1.61	0.00	1.51	—	1.31	2520
Canned, water (white)	69.48	26.66	2.46	0.65	0.00	1.58	—	0.60	—
Canned, oil (white)	64.02	26.53	8.08	0.00	2.18	4	0.65	—	
SALMON, SOCKEYE									
Raw	70.24	21.30	8.56	1.49	0.00	1.18	6	0.47	192
Oven-cooked	61.84	27.31	10.97	1.92	0.00	1.37	7	0.55	209
Canned (drain with bone)	68.72	20.47	7.31	1.64	0.00	2.70	239	1.06	176
Smoked (cold)	72.00	18.28	4.32	0.93	0.00	2.62	11	0.85	88
Smoked, Atlantic (hot)	58.90	21.60	9.3	—	—	9.4	—	—	—
CATFISH[1]									
Raw (aquacultured)	77.8	15.4	4.5	—	—	1.2	—	—	—
Oven-cooked/baked	71.1	20.4	6.3	—	—	1.4	—	—	—
Breaded and fried	62.7	19.0	7.9	—	—	1.6	—	—	—
Canned, tuna-style	75.0	21.0	3.0	—	—	1.2	—	—	—

[1] Aquacultured
[2] Data not available

tamin content and amino acid availability. Canned tuna products are usually low in fat, since they undergo a baking and subsequent skinning process prior to canning. Frying increases fat content and changes the composition of the product because of oil absorption, moisture loss, and the kind of ingredients added (Koo and Silva, 1996; Mustafa and Mederios, 1985).

REFERENCES

Ackman, R. G. 1973. Marine lipids and fatty acids in human nutrition. In *Fishery Products*, Kreuzer. R. (ed)., pp. 112-131. Surrey, England: Fishing News Books.

Amano, H., T. Fujiyoshi, and H. Noda. 1988. Çhanges in body components of whitefish *Coregonus muksun* after transportation in reservoir. *Bull. Jap. Soc. Fish* **54**(3):529-536.

Amerio, M, C. Ruggi, and C. Badini. 1996. Meat quality of reared fish: nutritional aspects. *Italian J. Food Sci.* **8**:221-230.

Ammerman, G. R. 1985. Processing. In *Channel Catfish Culture*. Tucker, C.S. (ed.), pp. 569-620. Elsevier, New York.

Angel, S., D. Baker, J. Kannan, and B. J. Juven. 1981. Assessment of shelf-life of freshwater prawns stored at 0°C. *J. Food Technol.* **16**:357-359.

Bailey, T. G. and B. H. Robinson. 1986. Food availability as a selective factor on the chemical composition of midwater fishes in the eastern North Pacific. *Mar. Biol.* **9**(1):131-141.

Botta, J. R. 1973. Chemical and organoleptic quality changes of frozen stored and canned channel catfish. Ph. D. Dissertation, Louisiana State University. Baton Rouge.

Clark, E. D. and L. H. Almy. 1918. A chemical study of food fishes. *J. Biol. Chem.* **33**:483-498.

Clark, E. D. and L. H. Almy. 1920. A chemical study of frozen fish in storage for long and short periods. *Ind. Eng. Chem.* **12**:656-663.

Dabrowski, K. R. 1982. Seasonal changes in the chemical composition of fish body and nutritional value of the muscle of the pollan *Coregonus-pollan* from Lough Neagh, Northern Ireland. *Hydrobiologia*. **87**(2):121-141.

Dean, L. M. 1990. Nutrition and preparation. In *The Seafood Industry*. R. E. Martin and G. J. Flick (eds.), pp. 255-267. Van Nostrand Reinhold, New York.

El-Faer, M. Z., T. N. Rawdah, K. M. Attar, and M. Arab. 1992. Mineral and proximate composition of some commercially important fish of the Arabian gulf. *Food Chem.* **45**(2):95-98.

Erickson, M. C. 1993. Compositional parameters and their relationships to oxidative stability of catfish. *J. Agric. Food Chem.* **41**:1213-1218.

Exler, J. 1987. Composition of foods: finfish and shellfish products. Agric. Handbook 8-15. United States Department of Agriculture, Washington, DC.

Exler, J. and J. L. Weihranch. 1976. Comprehensive evaluation of fatty acids in foods. VIII. Finfish *J. Am. Diet Assoc.* **69**:243-248.

FAO/WHO. 1990. Protein quality evaluation. FAO Food and Nutrition Papers, No. 51. Rome, Italy.

FDA. 1994. Food labeling; Nutrition labeling of raw fruit, vegetables and fish. 21CFR 101. Food and Drug Administration, Washington, DC.

Freeman, D. W. 1990. Chemical and dynamic headspace analysis of oxidative compounds in selected catfish (*Ictalurus punctatus*) fillet tissues as affected by phosphate and antioxidant injection and frozen storage. Ph. D. Dissertation, Mississippi State University, MS.

Gooch, J. A., M. B. Hale, T. Brown Jr., J. C. Bonnet, C. G. Brand, and L. W. Rogier. 1987. Proximate and fatty acid composition of 40 southeastern United States finfish species. NOAA Techn. Report NMFS 54. Springfield, VA.

Gruger, E. H. Jr. 1967. Fatty acid composition. In *Fish Oils: Their Chemistry,Technology, Stability, Nutritional Properties, and Uses*. M. E. Stansby (ed.), pp. 3-30. AVI Publ. Co., Westport, CT.

Haard, N. 1992. Biochemical reactions in fish muscle during frozen storage. In *Seafood Science and Technology*, E. G. Bligh (ed.), pp. 176-209. London: Fishing News Books.

Kinsella, J. E., L. Posati, J. Weihranch, and B. Anderson. 1975. Lipids in foods: problems and procedures in collating data. *CRC Crit. Rev. Food Technol.* **5**:299-324.

Koo, J. and J. L. Silva. 1996. Optimization of a process for preparing breaded heat-set and precooked catfish fillets. Presented at the Catfish Processors Workshop, October 9, Greenwood, MS.

Lim, C. and R. T. Lovell. 1986. Are omega-3 fatty acids in farm-raised catfish important? *Proceedings Catfish Processors Workshop*, Cooperative Ext. Service, Mississippi State University.

Love, R. M. 1988. *The Food Fishes: Their Intrinsic Variation and Practical Implications*. Van Nostrand Reinhold, New York.

Lovell, R. T. 1991. Foods from aquaculture. *Food Technol.* **45**(9):87-92.

Lovell, R. T. and T. Mohammed. 1989. Content of omega-3 fatty acids can be increased in farm-raised catfish. Highlight of Agric. Res., Alabama Agric. Exp. Sta. **35**:16.

Martin, R. E., and G. J. Flick (eds.). 1990. *The Seafood Industry*. Van Nostrand Reinhold, New York.

Mitchell, C. S. 1986. Lipid profile of selected tissue in channel catfish as affected by harvesting season and diet. *Proceedings Catfish Processors Workshop*, Cooperative Ext. Serv., Mississippi State University.

Mustafa, F. A. and D. M. Medeiros. 1985. Proximate composition, mineral content and fatty acids of catfish (*Ictalurus punctatus*, Rafinesque) for different seasons and cooking methods. *J. Food Sci.* **50**:585-588.

Nettleton, J. A. 1992. Seafood nutrition in the 1990s: Issues for the consumer. In *Seafood Science and Technology*, E. G. Bligh (ed.), pp. 32-39. London: Fishing News Books.

Nettleton, J. H., W. H. Allen Jr., L. V. Klatt, W.M.N. Ratnayake, and R. G. Ackman. 1990. Nutrients and chemical residues in one- to two-pound Mississippi farm-raised channel catfish (*Ictalurus punctatus*). *J. Food Sci.* **55**:954-958.

Nilson, H. W. and E. J. Coulson. 1939. The mineral content of the edible portions of some American fishery products. United States Bureau of Fisheries, Investigative Report 41, Washington, DC.

Piggot, G. M. 1989. The need to improve omega-3 content of cultured fish. *World Aquac.* **20**:63-65.

Randolph, S. and M. Snyder. 1993. *The Seafood List*. Food and Drug Administration, Washington, DC.

Rawles, D. D., C. F. Fernandes, G. J. Flick Jr., and G. R. Ammerman. 1997. Processing and food safety. In *Striped Bass and Other Morone Culture*, A. M. Harrell, (ed.), pp. 329-356. Elsevier, London.

Robinson, E. H. 1989. Channel catfish nutrition. *Rev. Aquatic Sci.* **1**:365-391.

Sautier, P. M. 1946. Thiamine assays of fishery products. *Commercial Fish. Rev.* **8**(2):17-19.

Sidwell, V. D., P. R. Foncannon, N. S. Moore, and J. C. Bonnet. 1974. Composition of the edible portion of raw (fresh or frozen) crustaceans, finfish, and mollusks. I. Protein, fat, moisture, ash, carbohydrate, energy value, and cholesterol. *Marine Fish. Rev.* **36**:21-35.

Silva, P. L., E. Robinson, J. O. Hearnsberger, and J. L. Silva. 1994. Effects of dietary vitamin E enrichment on frozen channel catfish (*Ictalurus punctatus*) fillets. *J. Applied Aquac.* **4**(3):45-55.

Silva, J. L., J. O. Hearnsberger, G. R. Ammerman, W. E. Poe, R. P. Wilson, G. Leigeber, and R. Wand. 1989. Composition of freshwater prawns. In *A Summary of Processing Research on Freshwater Prawns at Mississippi State University, 1984-1988*. J. L. Silva, G. R. Ammerman, J. O. Heransberger, and F. Hoya (eds.). Bull. 961, pp. 3-5. Mississippi Agric. For. Exp. Station, Mississippi State University.

Silva, J. L., J. O. Hearnsberger, G. Sudasna-Na-Ayudthya, and D. W. Freeman. 1991. Rapid fat analysis in whole channel catfish, pp. 14-16. *Proceedings Catfish Proc. Workshop*, Info. Bull. 203, Mississippi Agric. For. Exp. Station, Mississippi State University.

Silva J. L. and G. R. Ammerman. 1993. Composition, lipid changes and sensory evaluation of two sizes of channel catfish during frozen storage. *J. Applied Aquac.* **2**:39-49.

Silva, J. L., G. R. Ammerman, and C. W. Shannon. 1984. Chemical removal of the peritoneal membrane from channel catfish belly flaps. *Proceedings Catfish Proc. Workshop*, pp. 46-49. Cooperative Ext. Serv., Mississippi State University, MS.

Silva, J. L., S. W. Harrel, and J. O. Hearnsberger. 1996. Composition and quality changes of hybrid striped bass fillets during frozen storage. (Abstract) Presented at the Seafood Science and Technology Conference of the Americas, Clearwater, FL.

Stansby, M. E. 1976. Chemical characteristics of fish caught in the northeast Pacific Ocean. *Mar. Fish. Rev.* **46**:60-63.

Stansby, M. E. 1962. Proximate composition in fish. In *Fish in Nutrition*, E. Heen and R. Kreuzer (eds.), pp. 55-60. London: Fishing News Books Ltd.

Stansby, M. E. and A. S. Hall. 1967. Chemical composition of commercially important fish of the United States. *Fish. Ind. Res.* **3**(4):29-46.

Storebakken, T. and E. Austreng. 1987. Ration level for salmonids. I. Growth, survival, body composition and feed conversion in Atlantic salmon and salmonids. *Aquac.* **60**(3-4):189-206.

Tressler, D. K. and J. M. Lemon. 1951. Fish and shellfish as food. In *Marine Products of Commerce*, pp. 282-306. Reinhold, New York.

Wang, J. C. C., E. R. Amiro, and R. Selfridge. 1977. Chemical composition, storage life on ice and some physical characteristics of the Atlantic argentine (*Argentinasilus*). Tech. Rep. No. 725 Fish. Mar. Serv., Halifax, Nova Scotia, Canada.

Watt, B. K. and A. L. Merrill. 1975. Composition of foods. Agric. Handbook 8, United States Department of Agriculture, Washington, DC.

Osmoregulation in Freshwater and Marine Fishes

Eric M. Hallerman

Why do some fish normally live in fresh water and others in saltwater? How can some fish adapt to both?

Fishes must maintain in their tissues physiologically appropriate concentrations of salts. This need is metabolically challenging, especially when the salt concentration is very different in the external environment, or when a fish moves between salt- and freshwater.

Life in salt- or freshwater poses different challenges to fish. Saltwater fishes, with some exceptions noted below, maintain internal salt concentrations lower than their environment. They have to replenish water lost passively to the environment by diffusion and then excrete salt to the environment against the concentration gradient.

Freshwater fishes face the opposite problem. Salt concentration in their external environment is more dilute than in their tissues, so they can lose salt passively to the environment. Freshwater fishes must actively concentrate salt in their tissues. Because adaptation to one environment or the other is challenging, it is not surprising that rather few fish species can cope with moving between salt- and freshwater.

The regulation of internal water and salt concentrations is termed osmoregulation. Fishes take one of four osmoregulatory strategies. Two strategies take the approach of keeping internal and external salt concentrations equal:

Hagfishes — highly primitive, jawless fishes — practice no osmoregulation at all. They are tolerant of only a narrow range of salinities. They are the only vertebrates with internal inorganic salt concentrations roughly equal to seawater.

Marine sharks and the coelacanth, a "living fossil" species, maintain internal inorganic salt concentration at roughly one-third that of seawater. However, their internal osmotic concentration is brought to that of seawater with the organic salts, urea and trimethylamine oxide (TMAO). Urea usually denatures proteins, which would prove disastrous to cellular processes, but the 2:1 concentration of TMAO to urea protects cellular proteins.

Either of these strategies for balancing salt concentration also makes it easy to maintain water balance. Water diffuses easily across the skin of the fish, especially at the gills, as water will diffuse to a region of high salinity. Total salt concentration internally is equal to that externally, so passive water flow inward or outward is minimized.

Two other osmoregulatory strategies involve maintaining internal salt concentration different from external concentration:

Marine teleosts, the bony fishes (most species of fish), are hypo-osmotic; their internal salt concentration is about one-third that of seawater. These fishes tend to lose internal water to the environment. They must continually replace lost water by ingesting seawater. Hence the salts they take in must constantly be excreted at higher concentration than ingested. "Chloride cells" in the gills and skin of the gill chamber actively transport salts out of the body.

Freshwater teleosts and sharks are hyperosmotic; their internal salt concentration is greater than that in the environment. They tend to gain water through diffusion. Excess water, excreted as dilute urine, can be quite a lot, up to one-third of body weight per day. Hence, the kidneys of freshwater fishes are well developed. Some salts are lost through urine and diffusion through gills. Some replacement salts are taken in with food, but most replacement salts are actively transported through gills. The chloride cells of freshwater fishes pump salts in — the opposite of marine fishes. Fishes in soft, ion-deficient waters, such as rainbow trout, have especially well-developed chloride cells.

Special osmoregulatory challenges are posed by diadromy, or migration between fresh and salt water. Such migration involves both hormonal and physiological processes that are not entirely understood.

Adaptation to exposure to saltier water is affected by the steroid hormone cortisol, which is secreted by the kidney. Cortisol secretion peaks with exposure to hypertonic (saltier) water. Cortisol promotes: an increase in the number of chloride cells, increased activity of chloride cells, increased permeability of the urinary bladder to promote retention of water, and increased uptake of water in the gut due to drinking of seawater. In fishes that migrate to saltwater as part of their life cycle, ion regulatory ability is closely associated with life stage. For example, salmon undergo profound physiological changes to prepare for migration to sea.

Adaptation to exposure to freshwater is a process mediated by the hormone prolactin. Prolactin affects the morphology and physiology of the gills, gut, kidney, and urinary bladder, working counter to the action of cortisol. It minimizes passive loss of salt from the tissues of the fish upon migration from sea to freshwater, and prevents loss of salt by diffusion in freshwater-adapted fish.

The process of adaptation to water of different salinity is not instantaneous. For example, an eel moved from fresh- to saltwater can suffer an osmotic water loss of 4 percent of its body weight in 10 hours. Its water and salt equilibrium returns only after one or two days. If an eel is moved from salt- to freshwater, it will gain weight from passive water diffusion. Again, equilibrium returns in one or two days as dilute urine production increases.

Adaptation of fishes to life in salt- or freshwater, or to movement between waters of different salinities, is a complex process. Given the complexities, it is not surprising that most fishes are adapted to life in either salt- or freshwater, and that relatively few are adapted to both.

Biology of Certain Commercial Mollusk Species: Clams

W. Michael Peirson

INTRODUCTION

Clams belong to a large group of mollusks classified as Bivalvia. This class consists of at least 7000 species (Quayle and Newkirk, 1989), but only a few are commercially important. This chapter will review seven North American species for which a significant commercial or recreational fishery exists. The taxonomy for these species is presented in Table 1 and landings data in Table 2. The commercially important Manila clam (*Tapes philippinarum*), an introduced species intensively cultured on the West Coast of the United States, is not included in this discussion of North American species.

Bivalvia, as well as the older classification names Lamellibranchia and Pelecypoda, are all descriptive of this class. Respectively, they identify clams as having two valves or shells, plate-like gills, and a hatchet-shaped foot — all important adaptations in the lifestyle of clams. These and others are discussed in the following sections on anatomy and physiology of bivalves in general (see Barnes, 1980, for more detail). The final sections are highlights of what is known about the seven important clam species.

ANATOMY

Bivalves are more or less enclosed in a shell consisting of two valves which are hinged dorsally and attached by an elastic ligament. The hinge ligament tends to spring the shell open in the absence of muscular effort to keep the shell closed. The part of the shell rising above the hinge ligament is the umbo and represents the oldest part of the shell. Most bivalves have hinge teeth which serve to interlock the valves to prevent lateral slipping. Dentition along the shell margin serves the same purpose. The shell is composed

Table 1. Taxonomy

Phylum	Mollusca		
Class	Lamellibranchiata or Bivalvia		
Subclass	Heterodonta		
Order	Myoidea		
	Family	Myidae	
		Mya arenaria	Soft clam
	Family	Hiatellidae	
		Panope abrupta	Geoduck
		syn. *Panope generosa*	
Order	Veneroidea		
	Family	Mactridae	
		Spisula solidissima	Surf clam
	Family	Solenidae	
		Siliqua patula	Pacific razor clam
	Family	Veneridae	
		Arctica islandica	Ocean quahog
		Mercenaria mercenaria	Northern hard clam
		Mercenaria campechiensis	Southern hard clam
		Saxidomus giganteus	Butter clam

Table 2. 1995 Landings of the Major United States Bivalve Species

Species	Pounds	Dollars
Hard clams	7,200,000	41,800,000
Ocean quahogs	39,900,000	18,400,000
Soft clams	2,800,000	12,200,000
Surf clams	54,000,000	29,200,000

Source: *Fisheries of the United States, 1999,* U.S. Dept. of Commerce, NOAA, NMFS, No. 9800.

of an organic matrix and crystalline calcium carbonate (aragonite or a mixture of aragonite and calcite). The outer layer of the shell, the periostracum, is organic material secreted by the extreme margin of the mantle and serves as the substrate for the deposit of calcium carbonate and the organic matrix of the shell. The shell materials are secreted by epithelial cells of the mantle. These cells are joined to the shell only at the outer fold of the mantle and at the points of attachment of the pallial, adductor, and pedal muscles (Figure 1). At all other points on the shell, the organic matrix and calcium carbonate are secreted by the mantle into the minute space between the mantle and the shell. The shell constituents are thus deposited from the extrapallial fluid to form new shell material. The same process is responsible for the production of pearls when foreign matter is trapped in the mantle. The clams considered in this chapter are members of the subclass Heterodonta which do not produce an inner nacreous layer in the shell and so the pearls from these species have no commercial value.

To protect the soft tissues of the clam, the valves of the shell are held together by two more or less equal-sized adductor muscles. These muscles are composed of smooth and striated fibers so that they are capable of prolonged contraction as well as quick movement. The mantle is attached to the shell by muscle fibers along a semicircular line a short distance from the margin of the shell. The scars of this muscular attachment are seen in the shell as the pallial line (Figure 1). At the posterior end of the clam, the mantle fuses to form the inhalant and exhalant siphons. Their position is indicated in the shell by the scar of the siphon retractor muscles (the pallial sinus). Enclosed within the mantle cavity are the paired gills or ctenidia and the visceral mass (Figure 2). Within the visceral mass are the organs of digestion and excretion (stomach, crystalline style, digestive diverticula, nephridia). The outer layer of the visceral mass contains the gonads. The muscular foot of clams extends anteriorly and ventrally from the visceral mass and protrudes through the shell gape for burrowing into the substrate. The foot, which is capable of great extension in some species, is moved by a combination of muscular action and engorgement with blood. The scars of the muscles responsible for retracting the foot can be seen in the shell above the adductor scars (Figure 1).

Nervous System

The nervous system is arranged bilaterally. It consists primarily of three pairs of ganglia and two pairs of long nerve cords. The cerebropleural ganglia are situated on either side of the esophagus with short interconnections between them. From each cerebropleural ganglion extend the two nerve cords. The upper cord travels through the viscera to the visceral ganglion located on the surface of the posterior adductor muscle. The lower nerve cord connects to the pedal ganglion in the foot. The movement of the foot and anterior adductor muscle is under the control of the pedal and cerebropleural ganglia. The posterior adductor muscle and the siphons are controlled by the visceral ganglia. The coordination of foot and valve movements is provided by the cerebropleural ganglia. Most of the bivalve sense organs are located on the margin of the mantle. Tactile and chemosensory organs may be located here. Tentacles may also be present around the apertures of the siphons to sense and prevent entry of sand or other debris. Statocysts are usually found in the foot or the pedal ganglia. Pigment spot photoreceptors are found on the siphons of some species and are responsible for the shadow withdrawal response. Beneath the posterior adductor muscle in the exhalant chamber of the mantle cavity is a patch of sensory epithelium which may be involved in testing the water for particulates or in chemoreception. The gill filaments are connected by nerve fibers to the visceral ganglia and serotonin has been shown to have an excitatory effect on cilia (Owen, 1974).

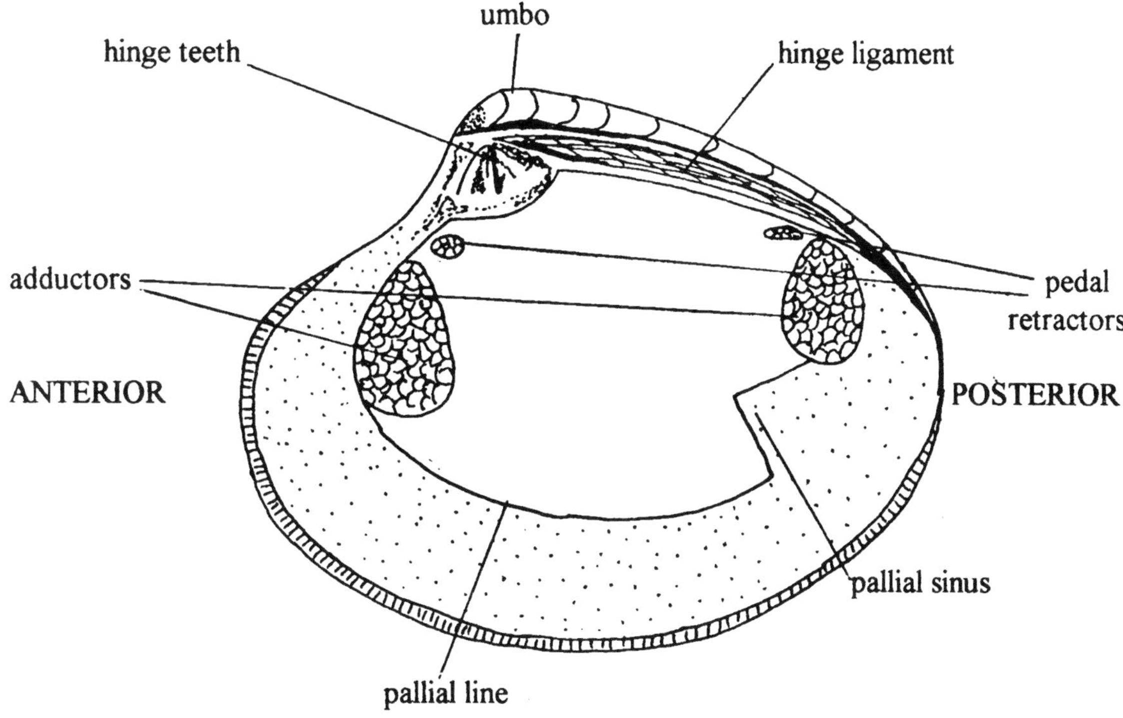

Figure 1. Interior of the right valve of the hard clam, Mercenaria mercenaria.

Excretion

The excretory system in bivalves consists of two nephridia located beneath the pericardial cavity (Figure 2). The nephridia are folded in the shape of a long U. The lower arm is glandular. It opens through the nephrostome to the pericardial cavity anterior to the heart. The upper arm forms a bladder. It empties through a nephridiopore into the suprabranchial cavity where wastes are removed by the water currents through the clam.

Circulation and Respiration

In clams the heart consists of a ventricle and left and right auricles with a single anterior aorta and a single posterior aorta. The heart develops around part of the digestive tract so that the pericardial cavity encloses the heart and part of the rectum (Figure 2). The circulatory system is classified as an open system since the blood is not always carried within blood vessels. Part of the circulation is through tissue sinuses. The typical circulation is from the heart, through tissue sinuses, through the nephridia, to the gills and back to the heart, although variations exist according to species.

The gills in clams are considerably oversized for their function as a respiratory organ. The relatively low metabolic requirements of clams and their ability to exchange gases through other body tissues (especially the mantle) seem to make large gills unnecessary. The explanation for the large gill size is that clams have adapted the gills for food gathering as well as respiration. Ciliary tracts on the gill filaments create water currents through the mantle cavity allowing the exchange of gases between the blood in the gills and the incoming water. Clams have evolved a complex system for utilizing suspended organic material (principally phytoplankton) in the incoming ventilating water as a food source.

Feeding and Digestion

The literature on bivalve feeding and digestion is vast. Much of the material in this section is taken from reviews by Winter (1978), Jorgensen (1975), Owen (1974), and Purchon (1968). The reader is advised to refer to these papers for more detail.

Clams are suspension feeders — that is, they filter suspended particulate matter from the water for their food. The gills of bivalves are highly

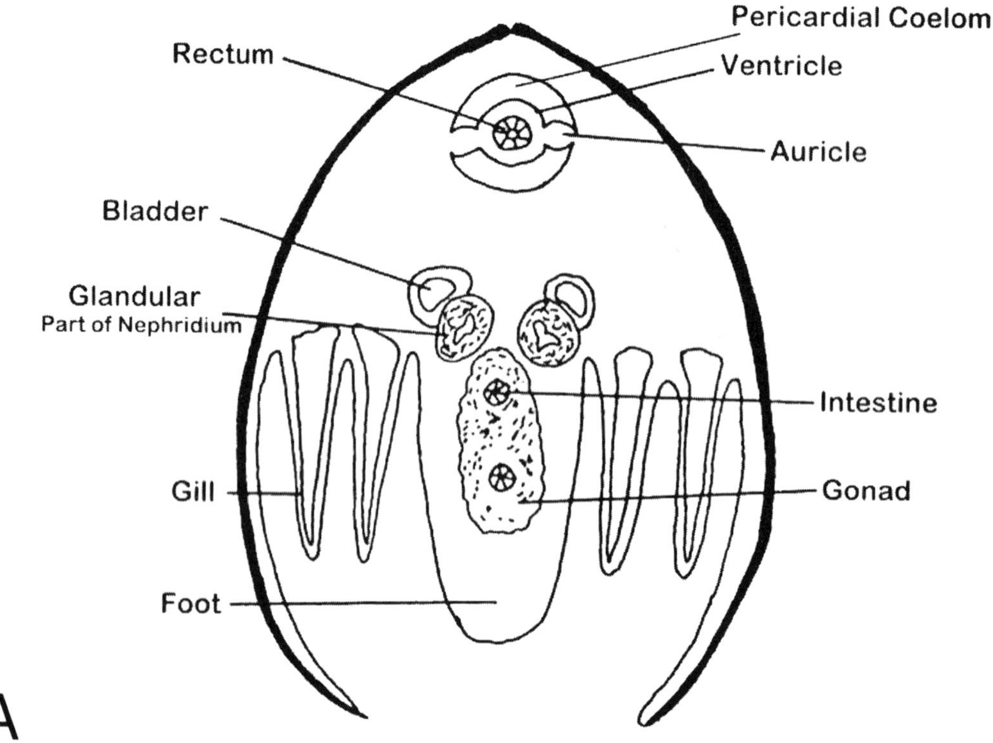

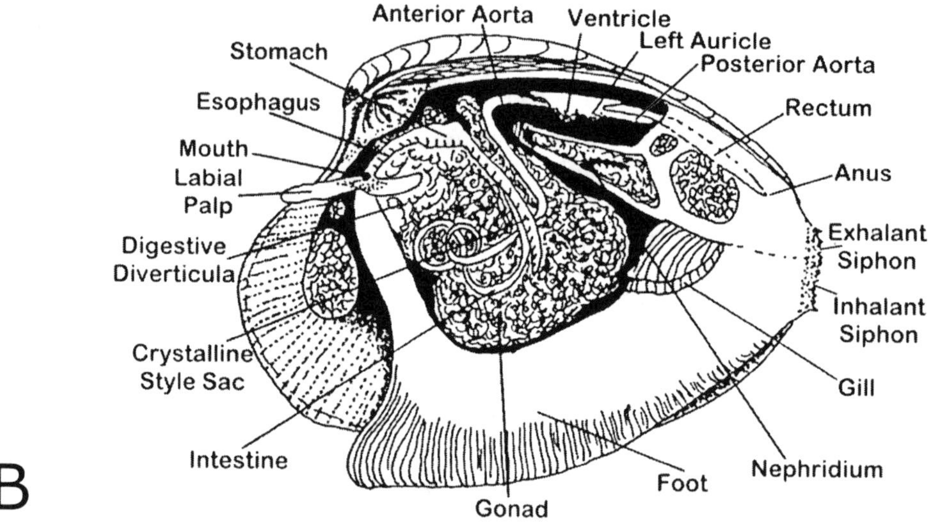

Figure 2. Anatomy of a hard clam.
A. Transverse section of a clam showing internal organs.
B. View of the internal anatomy of the hard clam (partial dissection with left valve, mantle, and most of gill removed). Redrawn from Barnes (1980).

specialized for the capture and movement of suspended particles in addition to their function as respiratory organs. The gill filaments possess three groups of cilia (Figure 3). The lateral cilia produce water currents through the gill filaments which cause water to be drawn from the outside of the clam through the inhalant siphon. The laterofrontal cirri (compound cilia), beating at right angles to the long axis of the gill filaments, form a straining grid which traps particulate matter (Moore, 1971). The frontal cilia pass the particulate matter in mucus strings along the gill filament to marginal grooves where additional ciliary tracts carry the material to the labial palps. There is much discussion in the literature as to the degree of particle sorting that can be accomplished by the bivalve feeding mechanisms. Heavy concentrations of food or large amounts of non-food particulates result in the production of pseudofeces (waste material that does not come from the gut). Production of pseudofeces can be initiated at the frontal cilia where the particles are passed to the edge of the gill and dropped into the mantle cavity for expulsion rather than to the labial palps for ingestion. Ciliary tracts on the labial palps can also direct particulate matter to the mantle cavity for rejection. Newell and Jordan (1983) suggest that the labial palps in American oysters can selectively reject non-food particles while passing food particles on to the mouth. They suggest that the palps can alter the viscosity of the mucus string so that individual particles can be actively selected, possibly on the basis of their nitrogen content (i.e., protein). This ability has not been identified for most bivalves.

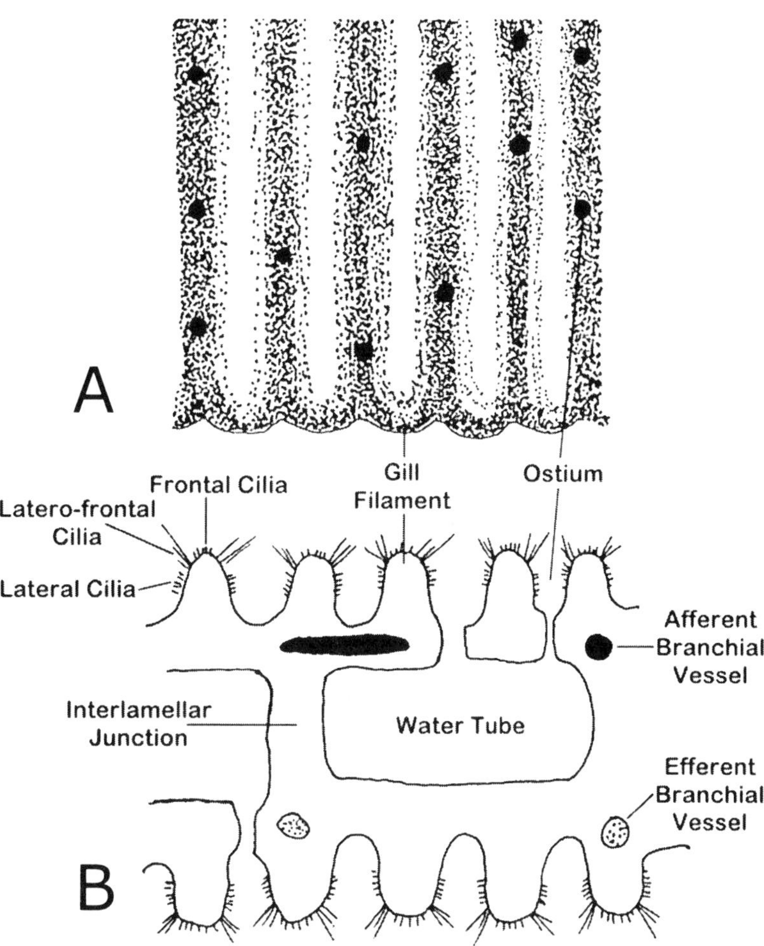

Figure 3. Surface view (A) and frontal section (B) of a portion of lamellibranch gill. Redrawn from Barnes (1980).

Feeding is assumed to be nearly continuous in subtidal bivalves under appropriate environmental conditions (Jorgensen, 1975). The measurement of filtration rates for various bivalves is highly variable. Filtration rates observed are influenced by body weight of the animal, temperature, suspension concentration, the nature of the suspended particles, and, unfortunately, by the various experimental methods used for measurement. Care must be taken in using values that were obtained where the test bivalves were under unusual stress from very unnatural conditions. Figure 4 shows the general relationship of food concentration to filtration rate and ingestion rate. It can be seen that within a certain range of food concentration the ingestion rate remains constant. At higher concentrations, pseudofeces are produced, which reduces the rate of ingestion. Determining energy balances for bivalves is further complicated by the multiplicity of destinations of ingested material.

The quantity of food particles that is removed from suspension by the processes described above and that ultimately reaches the mouth constitutes the ingested volume. Not all of this material is necessarily available to the clam as food. The particles

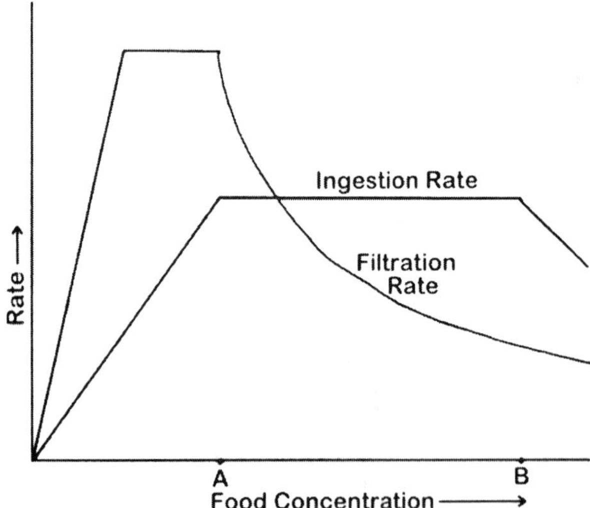

Figure 4. Generalized graph of the relationship of food concentration to filtration and ingestion rates. A. Food concentration at which maximum ingestion rate is reached. B. Food concentration at which ingestion rate begins to decrease due to the production of pseudofeces. *After Winter (1978).*

move down the esophagus, embedded in mucus strings, by means of ciliary action along the esophagus (Figure 5). The crystalline style rotating in its style sac sometimes acts as a capstan in winding up mucus strings and helping to pull them down the esophagus into the stomach. The crystalline style also may serve to grind food particles against the gastric shield to reduce their size for subsequent digestion The crystalline style may also have a role in releasing enzymes and changing the surface tension and the pH of the stomach contents. This latter function may assist in the sorting of particles in the stomach by altering the viscosity of the mucus strands so that individual particles may be freed for further sorting by ciliary tracts in the stomach and subsequent rejection to the midgut with no further digestion. This mechanism may protect the digestive diverticula against the entry of large particles. Small particles to be digested are carried into the digestive diverticula by countercurrents. These currents are caused by the loss of material in the tubules of the digestive diverticula by absorption and by excretion of particulates along ciliary tracts in the ducts (Purchon, 1968). The loss of this material causes a countercurrent to fill the void. The brush border epithelial cells lining the ducts absorb soluble and fine particulate matter (Owen, 1974). The bulk of digestion occurs in the tubule cells lining the sacs of the digestive diverticula (Figure 5). These cells ingest small particulates and absorb soluble matter from the lumen of the tubules. The digested nutrients are concentrated in vacuoles, migrate to the proximal end of the cell and discharge the nutrients to the hemolymph. Waste material is concentrated in spherules at the distal end of the cell, nipped off, carried by ciliary tracts to the stomach, and discharged to the midgut via the intestinal groove for elimination. While traveling from the midgut to the hindgut, the waste materials from both the stomach and the digestive diverticula are consolidated into feces. This is accomplished in part by a change in pH which increases the viscosity of the mucus. The feces are thus pelletized, preventing the material from being resuspended upon elimination, and preventing reingestion of waste material.

From the above discussion, it should be apparent that not all the material filtered from suspension is available as an energy supply for the clam. Depending on the type and concentration of the suspended material, varying amounts of the material may be eliminated without assimilation of the nutrients. Even phytoplankton known to be a food source can be ingested and subsequently eliminated in an intact (even live) state (Foster-Smith, 1975). When attempting to compute energy budgets for clams, one must remember to superimpose assimilation efficiencies over the graph in Figure 4. Thus, even though the ingestion rate may remain constant, if the digestive system is overwhelmed with food, part or all of the subsequent ingested material may be eliminated without digestion.

The tubule cells of the digestive diverticula disintegrate after performing their digestive function and must be regenerated. Clams feeding in high concentrations of phytoplankton can reach a point where all digestive tubules are either occupied in active digestion or are in the process of regenerating. At this point, subsequent ingested material cannot be digested until new tubule cells become available. This may be more common in aquaculture systems than in the natural environment but, in either case, must be taken into account when computing energy budgets.

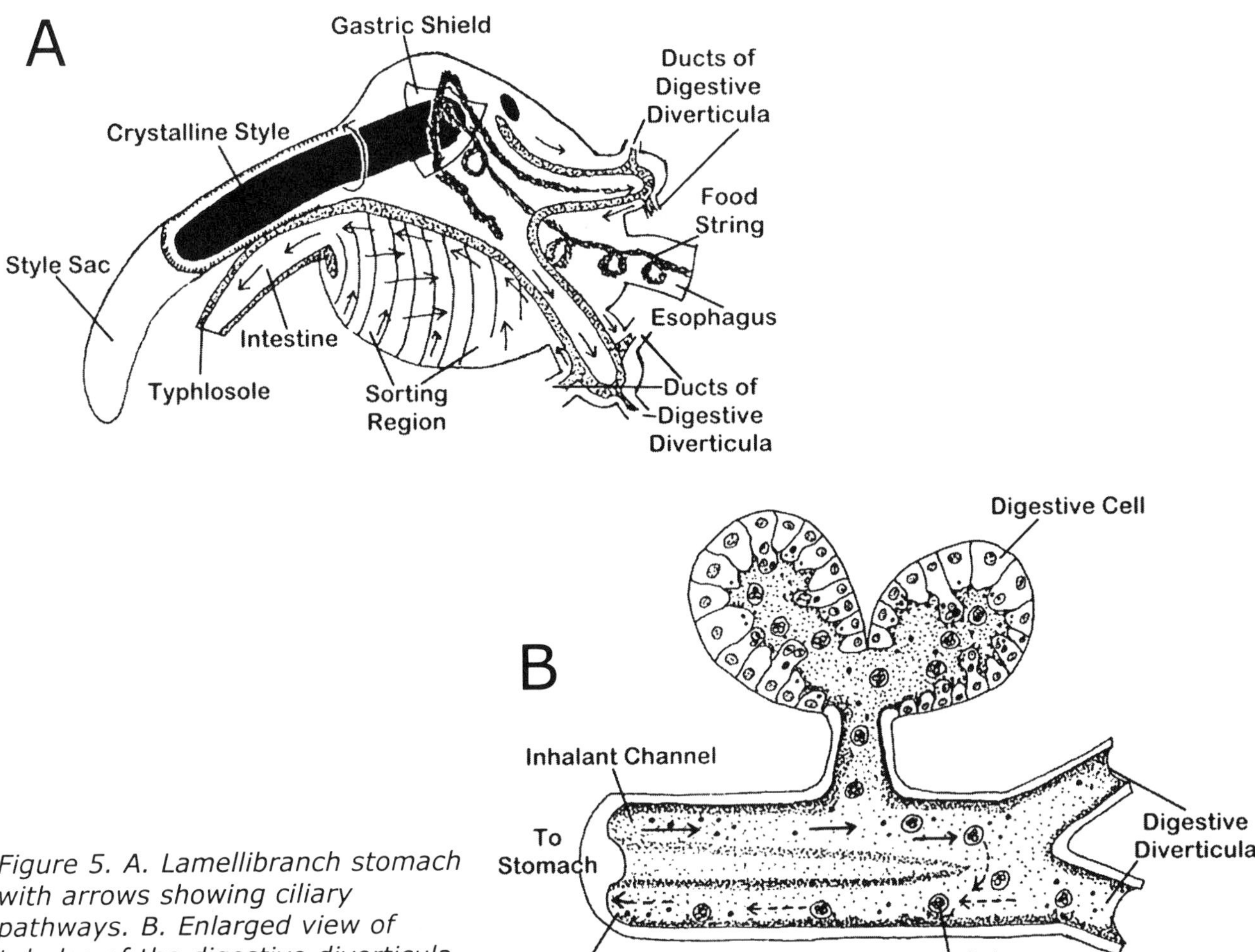

Figure 5. A. Lamellibranch stomach with arrows showing ciliary pathways. B. Enlarged view of tubules of the digestive diverticula. *Redrawn from Barnes (1980).*

Reproduction and Life Cycle

The clams considered here are all dioecious (having separate sexes), although quahogs appear to have both male and female sex cells during their first-year juvenile sexual phase (Eversole et al., 1980). By the second to the sixth year, depending on the age of maturity for the various species, clam populations tend to have an equal number of males and females. There are no gross morphological differences between sexes of clams, so that sex determination can only be made by sampling gonadal tissue or by observing spawning. Spawning occurs in species-specific and geographic-specific seasons (see later sections for each clam species). The eggs and sperm are ejected into the water through the exhalant siphon, and fertilization occurs in the open water. The fertilized egg begins to divide within an hour or two and within about 12 hours a short-lived trochophore larval stage is reached. At 24 to 48 hours the larva develops a bivalve shell and a ciliated velum which is used for both locomotion and food gathering. The hinge side of the shell is flat, so that the larval shell is shaped like the letter D. Clams at this stage are referred to as straight-hinged veliger larvae. The clam begins to feed at this point in its life on small phytoplankton (5-10 microns); the feeding can be verified by the onset of a dark color developing in the digestive diverticula, which can be seen through the transparent shell. Species of clams are hard to distinguish at this stage. However, scanning electron micrographs of the larval shells showing detail of the hinge teeth shows promise for identifying larvae in field samples (Lutz et al., 1982). As larval development progresses, the hinge side of the shell gradually curves to take on the shape of the umbo of the adult clam. Near the end of the larval period, a foot develops and the clam spends alternate periods swimming and crawling on the bottom. This pediveliger stage lasts for several days before the velum is lost and the clam metamorphoses into the plantigrade or juvenile form. The clam is then confined to crawling on the sea bottom with its ciliated foot. It is able to attach to and detach from the sea floor by means of a byssal thread until the clam has grown to several millimeters in length. The siphons develop over the next several weeks, and the clam is then able to bury itself into the sea bottom with its foot and extend its siphons from its protected burrow up to the water column.

SURF CLAMS

The surf clam fishery is the largest bivalve fishery in the United States, based on total weight of landings (Table 2). This resource has been heavily fished since the late 1940s and peaked at 96 million pounds in 1976 (Jones, 1981). Surf clams are found from the Gulf of St. Lawrence, Canada, to Cape Hatteras, North Carolina. They occur from shallow subtidal areas to depths of 128 meters. Commercial quantities are found primarily off New Jersey from the nearshore to 55 meters depth (Ropes, 1980).

The shell of the surf clam is oval, with the umbo midway between the anterior and posterior ends. The color varies from cream to tan. It is covered with a glossy periostracum and the valves gape slightly. The right valve has two partially-joined cardinal teeth. The siphons, which are united all the way to their tips, can be fully retracted into the shell.

The upper temperature limit for adult surf clams is 26°C to 28°C. The minimum salinity tolerance has been measured as 12.5 parts per thousand ppt for adults and 16 ppt for larvae (Castagna and Chanley, 1973). Low dissolved oxygen levels have been shown to cause massive mortalities. In 1976, low oxygen levels caused by weather, hydrologic conditions, and the die-off of a massive algal bloom contributed to the loss of about 62 percent of the New Jersey surf clam resource (Ropes, 1980). Low levels which may not be directly lethal can cause clams to leave their burrows, making them vulnerable to predation.

Surf clams grow rapidly for the first four to five years with offshore clams (> 5 kilometers from shore) growing faster than inshore clams (Ambrose et al., 1980). By the age of 10 years, offshore clams reach an average shell length of 152 millimeters (mm), while inshore clams average 116 mm. It is rare to find clams older than 11 years inshore and none has been found older than 16 years. Clams 25 years old and older are commonly found offshore. The possible explanations for these differences have been attributed to a more stable bottom offshore with more stable temperatures. Also, lower densities offshore may explain the higher growth rate for this population. The largest clam reported measured 226 mm in shell length. The estimated maximum age is 26 years (Ropes, 1980).

Surf clams are sexually mature as small as 45 mm and an age of one year. Most clams are fully mature in their second year. The ratio of male to female clams is 1:1. The major spawning period is mid-July to early August, with a minor spawn from mid-October to early November (off New Jersey). In some years there is only one spawn from mid-September to mid-October, believed to be caused by lower water temperatures. The larval period is 21 days under controlled 20°C laboratory conditions, but is probably longer in colder natural conditions. Surf clams can reach a shell length of 33 mm within six to eight months after setting. Densities as high as 8000 juveniles per square meter have been measured within 2.5 kilometers of the New Jersey shore (MacKenzie et al., 1985). These high levels of recruitment usually do not translate to high numbers of adult clams, since there is nearly 100 percent mortality by the next spring.

When surf clams are out of the substrate (which many are when at high densities), the force counteracting the tendency for the hinge to open the shell is reduced, thus demanding more work of the adductor muscles to keep the shell closed. Eventually, the muscles fatigue, causing gaping which exposes the soft body parts to predators. Clams that are buried normally may avoid predation by sensory cues from the siphons. The siphons have sensory papillae which allow closure or withdrawal when stimulated, as well as photoreceptors which allow withdrawal as a response to shadows. Clams out of the substrate have been observed performing an evasive jump by rapidly extending the foot. A large clam was observed jumping about 300 mm high and landing nearly a meter away. Observed or suspected predators include the moon snails, *Lunatia heros* and *Polynices duplicatus*; haddock, *Melanogrammus aeglefinus*; cod, *Gadus callarias*; lady crabs, *Ovalipes ocellatus*; and rock crabs, *Cancer irroratus*.

Other organisms which can affect surf clams include nematodes (*Sulcascari* and *Paranisakiopsis* infestations); coliform bacteria from nearshore populations exposed to sewage effluent; the bacterium *Vibrio parahaemolyticus*; and the agent for paralytic shellfish poisoning (PSP), *Gonyaulax tamarensis*. PSP toxin may be retained for prolonged periods after exposure, possibly from one year to the next (Prakash et al., 1971).

Controlled culture of the surf clam shows some promise. An extensive study of the aquaculture of surf clams was accomplished by Goldberg (1980) and some data on development can be found in Ropes (1980). Clams were conditioned (ripened for spawning) by holding them in flowing seawater at 15°C to 20°C for two weeks. The spawning temperature was dependent on the temperature at which the ripened clams were held. Clams held at 20°C spawned at 28°C, while those held at 10°C spawned at 22°C. The eggs are 56 microns in diameter. When held at 20° to 22°C, the fertilized egg will release polar bodies within 70 minutes with a swimming ciliated gastrula formed at five hours. The trochophore larval stage is reached at nine hours and straight- hinge larval stage at 20 hours. The straight-hinge larva measures 89 microns in shell length and 71 microns in shell height. Goldberg (1980) held larvae at 15 larvae per ml and fed them 100,000 cells of a mixture of *Monocrysis* and *Isochrysis* per milliliter of larval culture each day. At about 18 days the pediveliger stage develops and measures 245 microns in shell length and 230 microns in shell height. By metamorphosis at about 21 days, the clam has reached a shell length of 280 microns. *Spisula* is unusual among clams in that it does not form a byssal attachment and the newly-set clams do not adhere to culture vessels.

The rapid growth of post-set surf clams makes aquaculture of this species for commercial purposes a possibility. In 250-mm diameter downwelling cylinders with a flow rate of 750 ml per minute, clams grew from set to 3 mm in just three weeks and from 3 mm to 15 mm in an additional three weeks. Clams larger than this do not grow well out of substrate, so further growth was accomplished in flowing water raceways filled to a depth of 100 mm with sand. In this system clams reached a shell length of 55 mm in 20 weeks. Goldberg considered this size as a potentially marketable product that would be competitive with quahogs and soft clams of a similar size.

OCEAN QUAHOGS

The ocean quahog (*Arctica islandica*) fishery developed in 1975 to 1976 as a result of the overfishing of surf clams (Jones, 1981). Ocean quahogs are found from the southern coast of Newfoundland, Canada, to Cape Hatteras, North Carolina. They are also found around Iceland and off the coast of Europe. The fishery is concentrated in the Mid-Atlantic Bight. Off the coast of the United States, ocean quahogs are usually found further offshore than surf clams, but there is overlap of habitat in the range of 18 to 55 meters. This clam has been reported in depths ranging from 4 to 256 meters. It is an important commercial species in the United States (Table 2).

Ocean quahogs resemble quahogs (*Mercenaria mercenaria*) in the shape of the shell, but *Arctica* has a black periostracum in large specimens and a pale yellow to deep brown periostracum in smaller individuals. The shell is more robust than that of surf clams and also differs in that it completely encloses the body parts with no gapes.

Ocean quahogs are strictly marine species, generally found at considerable depth. It would be expected that their salinity and temperature tolerances would not vary appreciably from the limits of their deep water marine habitat, but little information is available in this area.

As might be expected from a clam that lives in cold, deep waters, growth of this species is very slow. It requires an average of 10 years to reach a 60-mm shell length. The annual increase in shell length is estimated at 6.3 percent at age 10 years, 0.5 percent at age 50 years, and 0.2 percent at age 100 years (Murawski et al., 1982). *Arctica* lives to a maximum age of at least 150 years. The age at sexual maturity is also fairly advanced. In sampling off Long Island, New York, Ropes et al. (1984) found undifferentiated gonads in clams with shell lengths ranging from 19 to 46 mm and ages of two to eight years. Fully mature males were found in clams 36 to 58 mm at ages of five to 18 years, and fully mature females in the 41 to 60 mm range at six to 16 years. Sexual maturity seems to be reached at an average size of 60 mm and an age of about six years. The ratio of males to females is highly variable and related to the size of the clam. In smaller clams that are less than fully mature, the sex ratio is 4:1 favoring males. In clams larger than 90 mm, the ratio is greater than 2:1 favoring females. Fritz (1991) found only 19 males in a sample of 98 clams larger than 100 mm. Possible explanations for this are that males have a shorter life span; that sex may be environmentally determined; or that a change of sex occurs at a certain stage (Ropes et al., 1984). The reason for varying sex ratios is unknown, but since hermaphroditic individuals are not found, it is unlikely that a sex change occurs.

Spawning in the population off New Jersey occurs from autumn to early winter. It is believed to coincide with the highest annual water temperature for this habitat of 13.5°C (Jones, 1981). In more northern populations, clams may not mature until they are eight to 14 years old with spawning occurring over a long period extending from July to winter (Rowell et al., 1990). The resulting larvae from fall spawns may spend the entire winter as larvae for a larval life of as much as six months (Mann, 1985).

There is not much commercial prospect for culturing a clam that grows as slowly as *Arctica*. Larvae have been cultured, usually by stripping the eggs from the clams and treating them with ammonium hydroxide to dissolve the egg capsule. Completely natural spawning has not been duplicated, but Gibbon et al. (1983) induced spawning by injecting serotonin into the anterior adductor muscle. Spawning occurred within 15 minutes at a water temperature of 15°C to 16°C. Larvae were raised at 13.5°C to the post-set stage. The eggs are slightly ovoid and 75 to 85 microns long. Metamorphosis occurred at an average shell length of 220 microns between 37 and 62 days.

QUAHOGS

Quahogs or hard clams are the highest-valued bivalve crop in the United States (Table 2). They are found from intertidal areas to depths of 15 meters from the Gulf of St. Lawrence down the entire Atlantic Coast and into the Gulf of Mexico to Texas. Hard clams have also been introduced in California and Europe. This species is generally referred to as quahogs in the North and hard clams in the South. *Mercenaria campechiensis*, a species that readily hybridizes with *Mercenaria mercenaria*, is found from North Carolina to Florida and into Texas. The species are very similar but have some important physiological differences.

Quahogs have a thick shell with valves that seal tightly together through the action of strong adductor muscles. The umbo is nearly at the anterior end of the valves and projects forward. The shell is sculpted by numerous concentric lines over the whole exterior. The interior of the shell is white except near the margins where it takes on a purple color, especially in larger specimens, though rarely seen in *M. campechiensis*. The purple part of the shell was cut into wampum by the American Indians, which prompted the species name *Mercenaria*. The ventral margin of the shell interior is crenulate (small, closely-spaced grooves perpendicular to the shell margin), and there are three hinge teeth on the valves. The siphons are short and fully retractable into the shell. The adductor muscles are capable of great force and can hold the valves tightly together for two to three weeks after removal from the water at proper temperatures (< 10°C).

As with the other estuarine species (*Mya*) discussed in this chapter, *Mercenaria* has fairly wide-ranging temperature and salinity tolerances. It is not as tolerant as *Mya* of low salinities, but it is more tolerant of high temperatures, giving the species a very broad geographic range. Quahogs are not generally found where bottom salinities are below 12 parts per thousand (ppt). They have been found off Texas in waters as saline as 48 ppt (Roegner and Mann, 1990). The lower salinity limit for eggs to develop normally is 20 ppt and for larvae about 17.5 ppt. The optimum salinity for larvae is 27.5 ppt. In the Chesapeake Bay, hard clams are not abundant in areas below 18 ppt (Roegner and Mann, 1990). Juvenile and adult clams may live in very low salinities for 14 to 28 days but no burrowing or water pumping occurs (Stanley and DeWitt, 1983). Tolerance of very low salinities is probably more a function of the clam's ability to seal its valves together rather than a physiological ability of its tissues to deal with these salinities. Adult clams can survive temperatures between -6°C and 45°C, while larvae cease growing below 12.5°C and tolerate temperatures into the mid-thirties (Roegner and Mann, 1990). Temperature can also affect the larval period. It can be as short as seven days at 30°C and as long as 24 days at 18°C (Loosanoff, 1959). There is little growth in adults at temperatures below 10°C.

Clams are fairly tolerant of low levels of dissolved oxygen. Embryos require only 0.5 milligram (mg) of oxygen per liter to survive. Growth occurs at 2.4 milligrams/liter but is best at 4.2 mg/l. Adult quahogs can tolerate one mg/l for three weeks and still rebury afterwards (Stanley and DeWitt, 1983).

The maximum shell length of hard clams is 135 mm. Clams can reach harvest size (38 to 51 mm shell length, depending on state law) in less than two years for fast-growing individuals in the Mid-Atlantic states to 13 years at the clam's northern geographic limit (Landry et al., 1993). The maximum age is estimated at up to 50 years.

Although the sex ratio of adult *Mercenaria* is 1:1 as it is for most other clams, this species differs from the others in possessing a juvenile sexual phase. Clams as small as 6 to 7 mm and only a few months old have functional gonads and are capable of spawning. In juvenile clams where the sex can definitely be established, males outnumber females by 9.5:1 (Eversole et al., 1980). Even in juveniles functioning as males, both male and female sex cells are present. This classifies *Mercenaria* as a consecutive hermaphrodite. At about 30 mm shell length, clams establish an equal number of males and females.

The spawning period for *Mercenaria* is highly variable, given its range from Canada to Florida. Even in the Chesapeake Bay, ripe clams can be found from April to October, depending on the area and the water depth. Spawning takes place when temperatures rise above the 23°C range. Spawning in South Carolina extends over a six-month period, with peaks in May and June and in September and October (Eversole et al., 1980), and in Long Island Sound spawning occurs in June and July (Bricelj and Malouf, 1980). Larger clams produce more eggs (up to 17 million over the spawning season); and there is no indication of reduced viability of spawn from older clams (Peterson, 1983; Bricelj and Malouf, 1980).

It has been estimated that clams in the planktonic phase of their lives may be dispersed from 2 to 25 kilometers from their spawning site. Larvae can move vertically 7 to 8 centimeters per minute by swimming and may control their vertical position in the water column to take advantage of tidal currents to control their dispersal (Stanley and DeWitt, 1983). Clams tend to prefer a sandy bottom over a muddy bottom. This may be a sediment preference or a function of physical dispersal by currents. Quahogs generally grow faster in a sandy substrate compared to mud, though Peterson et al. (1984) found higher densities and faster growth of clams in eelgrass (*Zostera marina*) beds compared to nearby sand flats. He hypothesized that the current baffling of the eelgrass beds caused suspended particulates to drop out of suspension, increasing the concentration of food available to the clams in the grass bed. Aside from this type of effect, clams generally grow faster in areas with greater water currents.

Because of the hard clam's commercial importance and widespread hatchery production, there is a large volume of information on predation. A fairly complete review of this information is presented by Gibbons and Blogoslawski (1989). From Cape Cod, Massachusetts, south, the primary predator is probably the blue crab, *Callinectes sapidus*. This crab can eat hundreds of juvenile *Mercenaria* per day and can prey upon clams as large as 40 mm shell length. In the South the stone crab, *Menippe mercenaria*, joins the blue crab as a major clam predator, while in the North the green crab, *Carcinus maenas*, is the dominant crustacean predator of hard clams. The green crab is capable of preying upon clams of up to 30 mm shell length. Additional predators of juvenile clams are the mud crabs (*Eurypanopeus depressus*, *Neopanope sayi*, *Panopeus herbstii*, *Rhithropanopeus harrissi*). In higher salinity waters the sea star, *Asterias forbesi*, can excavate and consume clams buried over 40 mm deep, though the clam is able to respond to the presence of the sea star by burrowing more deeply and reducing the ability of *Asterias* to prey upon it (Doering, 1982). A major predator of adult clams is the cownose ray, *Rhinoptera bonasus*. These rays traveling in large schools can destroy hundreds of thousands of market-size clams in commercial plantings in a day or two if the clams are not adequately protected with fences or nets. In intertidal areas gulls (*Larus* spp.) and oystercatchers (*Haematopus ostralegus* and *H. palliatus*) prey on adult clams. Gulls excavate the clams and carry them over a hard surface and drop them from heights sufficient to break the shell. Oystercatchers stab at the clams when their positions are revealed by squirting water from the siphons or they make multiple stabs until they locate the clam.

The hatchery culture of quahogs is a well-established practice. Commercial clam hatcheries exist from Maine to Florida and are becoming an important source for the commercial clam market. Because clams do not have a prolonged recuperative phase after spawning (as, for instance, oysters do), they can be conditioned to spawn any time of the year. Ripened clams can be spawned by a rise in water temperature to 24°C to 30°C. The actual temperature depends on the temperature acclimation of the clam and the geographic area from which the clam was derived. Eggs average about 75 microns in diameter but can range from 50 to 97 microns or 270 microns with the gelatinous membrane surrounding the egg (Bricelj and Malouf, 1980). Larger eggs tend to survive better than smaller eggs (Kraeuter et al., 1982). The embryo develops within the gelatinous membrane and at about 10 hours a spinning, ciliated gastrula escapes from the egg capsule. A trochophore larva is formed a few hours later, followed by the straight-hinge veliger larva at about 24 hours. The standard diet for quahog larvae is the alga *Isochrysis galbana*, with increasing rations as the larvae grow. At 22°C setting occurs at 12 to 14 days at a shell length of 200 to 210 microns. At elevated temperatures the larval period can be as short as seven days. Clams are generally kept in a flowing seawater nursery system until several millimeters long and then moved to a sea bottom nursery system. Final grow-out on the natural bottom usually commences when the juveniles have reached at least 12 mm shell length. They are planted at about 750 per square meter and the grow-out beds are covered with 6-mm mesh plastic netting to exclude predators. Quahogs generally reach a marketable size in 1.5 to 2.5 years.

Attempts to improve the rate of grow-out have included induced triploidy (Beaumont and Fairbrother, 1991) and hybridization of the northern and southern hard clams (Menzel, 1977). Triploid clams have been produced but are actually smaller at three years old than diploids. The southern hard clam, *M. campechiensis*, generally grows faster than the northern hard clam, *M. mercenaria*. The southern clam has not been widely used in culture because it is not tolerant of cold temperatures experienced in the winter from North Carolina northward, nor is it tolerant of low salinities (Menzel, 1989). In addition the southern clam will gape and die within several days of being removed from the water. Hybrids of the clams show an intermediate growth rate and an intermediate shelf life out of the water. The hybrids may have promise in the more southern regions of the United States. The primary breeding that is done in commercial hatcheries is to select fast-growing offspring and grow them out for future spawners. The other widely practiced breeding program is to incorporate *M. mercenaria* variant *notata* into the spawning stock. This genetic variant manifests its presence as reddish-brown bands (homozygous) or zigzag patterns (heterozygous) on the shell. This occurs in 0.76 percent to 2.25 percent of the population in Georgia and South Carolina (Humphrey and Walker, 1982). By incorporating this marking in commercially produced clams, growers can distinguish their product from naturally harvested clams. Thefts of cultured clams have been successfully prosecuted using the *notata* marking as evidence of hatchery ownership of the clams.

BUTTER CLAMS

The butter clam (*Saxidomus giganteus*) is found from Sitka, Alaska, to San Francisco Bay, California. The fishery began in 1930 in Alaska producing clams for canning. It was an important fishery for the area but collapsed due to the presence of PSP (paralytic shellfish poisoning) toxin in the canned product (Paul et al., 1976). The clams occur in Washington from the middle to lower intertidal zone and subtidally to depths of 18 meters. They are generally found in substrates composed of a mixture of sand, broken shell, and pea gravel and are often found mixed with the littleneck clam, *Protothaca staminea*. Over 0.5 million pounds were harvested in 1970 from Washington waters, but the landings declined substantially over the following 10 years [Oceanographic Institute of Washington (OIW), 1981].

The butter clam has a thick shell that varies in color from yellow in juveniles to gray-white in adults with a thick, black hinge ligament. The shell is marked with pronounced concentric rings and there is a slight gape between the valves at the posterior end. The umbo is located about midway between the anterior and posterior ends of the shell.

In British Columbia clams reach a shell length of 65 mm in five to six years. In Alaska the same size is reached in eight to nine years. The oldest clam examined in Alaska was 15 years old with a shell length of 100 mm (Paul et al., 1976). In Washington clams reach a commercial size of 50 to 75 mm after a few years' growth and a maximum size of 100 to 150 mm (OIW, 1981).

Saxidomus matures at three to five years old in British Columbia at an average 38 mm shell length. Spawning occurs in Oregon from March through June (Eversole, 1989). In Alaska spawning may not occur in some years due to low water temperatures (Paul et al., 1976).

Butter clams can be cultured in controlled conditions. Breese and Phibbs (1970) conditioned clams at a salinity of 25 ppt at 16°C to 18°C for one to two weeks. Upon transfer to 21°C water, the clams would usually spawn within eight hours. If natural spawning did not occur, the eggs could be stripped or potassium chloride could be added at a rate of one to two grams per liter of seawater to induce spawning. Spawning can also be induced by adding 2 to 2.5 million cells of *Pseudoisochrysis paradoxa* or *Thalassiosira pseudonana* per milliliter of seawater (Breese and Robinson, 1981). The conditioned clams could be spawned at a later date if placed in seawater at a temperature of less than 15°C. The eggs are round and 80 to 90 microns in diameter (230 microns with the membrane). At 18°C polar bodies form within 60 minutes of fertilization and the first cleavage occurs at 90 minutes. Trochophore larvae are formed at 24 hours and straight-hinge veliger larvae at 48 hours. Larvae were held at one larva per milliliter of culture and fed a mixture of *Monochrysis lutheri* and *Isochrysis galbana* three times per week at a density of 20,000 cells per milliliter of culture water. Larvae grew an average of seven microns per day and reached metamorphosis at 22 to 30 days. Set clams attached themselves to containers with byssal threads. During the 146 days of the study, clams grew an average of 18 microns per day.

RAZOR CLAMS

Razor clams (*Siliqua patula*) are found from the Aleutian Islands to Pismo Beach, California. They occur on broad, open, flat, sandy beaches and offshore to depths of 6 meters. The beach populations support a recreational fishery and a limited commercial fishery in Oregon, Washington, British Columbia, and Alaska. This clam has apparently been exploited for many years as evidenced by shells found in middens of American Indians of the Pacific Northwest. In 1940 the fishery was 80 percent commercial. By 1968 the last major razor clam-producing beach in Washington was closed to commercial clamming (Lassuy and Simons, 1989).

The shell of the razor clam is elongate, thin, flat, and smooth with a heavy, glossy, yellowish periostracum. The color may range from olive green in young clams to brown in older clams. There is a prominent rib on the inside of the valves which runs diagonally from the umbo to the margin. The umbo is closer to the anterior end of the shell. The body has a large, muscular foot which is unpigmented. The siphons are short and joined except at the tips. *Siliqua patula* is differentiated from other razor clams by the interior shell rib and the lack of pigment on the foot.

Razor clams require cold water and probably fairly high salinities. The lethal level for 50 percent of the sample clams experimentally exposed to high temperatures was 22.5°C for a four-hour exposure and 27.5°C for a one-hour exposure (Sayce and Tufts, 1972). At ambient temperatures of 5°C to 11.6°C, elevations in temperature of 11.2°C for four hours or 14.8°C for one hour proved fatal to 50 percent of the clams. Salinity data could not be found, but it may be a factor in mortalities of clams high on the beach after heavy rains and on the slower growth of clams on beaches near rivers. Razor clams probably have a high dissolved oxygen requirement as evidenced by the type of beaches they inhabit, with the surf providing continual water change over the clams. They are found mostly on beaches with sand grains averaging 0.2 mm in size and with a very low silt or organic matter content.

Razor clams grow to a maximum size of about 180 mm shell length, but this varies within the geographic range. Clams may reach a maximum of 120 mm in California and 160 mm in Alaska. The rate of growth also varies with location. A shell length of 120 mm is reached in four years in California and the Pacific Northwest, but requires six years in Alaska. Within the same area, faster growth occurs at points lower in the intertidal zone.

Sexual maturity seems to be reached when razor clams achieve a size of about 100 mm rather than at a particular age. Clams reach this size and mature at about two years in the Pacific Northwest and three to four years in Alaska. Males and females are found in equal numbers. Maximum age follows a geographic trend ranging from five years at Pismo Beach, California, to nine to 11 years in the Pacific Northwest, to 18 to 19 years in Alaska. The spawning season is mid-May through July in Washington and Oregon, and July to September in British Columbia. Spawning seems to be triggered by a rapid rise in ambient temperature. A small, not-fully-mature clam of 40 mm shell length may produce 300,000 eggs, while a 180-mm clam may produce over 100 million eggs. Depending on temperature and other factors, the resulting larvae will spend the next five to 16 weeks in the water column. Heavy sets of clams several months after metamorphosis may cover beaches in patches several inches deep and several acres in extent. Average density after a heavy set was nearly 40,000 per square meter, but after eight months the density was down to about 30 per square meter. The high densities slowed the growth of the clams and made little contribution to the population on a long-term basis (Tegelberg and Magoon, 1969). In normal sets, the highest densities occur in the lower third of the intertidal zone. Small clams can crawl along laterally and be redistributed by currents and the surf. Juveniles larger than 25 mm generally remain in place buried in the sand. Adult clams burrow to about 300 mm below the surface and can move vertically through the sand at a rate of at least 300 mm per minute. Clams have been found nearly 1.5 meters below the surface of the sand. There is also evidence that considerable populations of razor clams exist offshore and may provide recruitment for beach populations.

Predation is usually observed on the beach populations after dense sets of juveniles. These concentrated food sources attract gulls, crows, ducks, sandpipers, Dungeness crabs, and sturgeon which decimate the juvenile population. In Washington a prokaryotic pathogen referred to as nuclear inclusion X (NIX) caused losses of 95 percent of all clams on some beaches. This pathogen causes an inflammatory overgrowth of epithelial cells and gill damage and was responsible for the complete closure of the Washington razor clam fishery in 1984 and 1985. The organism was present in the Oregon populations but at sublethal levels. A nemertean worm (*Malacobdella grossa*) 25 to 50 mm long may live in the siphons of clams, but is considered harmless.

In the fall of 1991, the recreational fisheries in Oregon and Washington were closed when the neurotoxin domoic acid was found at levels above those considered safe for human consumption. Domoic acid causes amnesic shellfish poisoning (ASP). The toxin was found to be concentrated in the muscular tissues of the clam (adductor muscles, foot, siphon, and mantle) and to remain at elevated levels for at least three months after exposure (Drum et al., 1993). The toxin for paralytic shellfish poisoning (PSP) has also been found in razor clams from Washington and Alaska.

Hatchery culture of razor clams has been carried out on a fairly large scale. The state of Washington began hatchery operations in 1980 to produce juveniles to seed public beaches. Juveniles from subtidal populations have also been harvested and replanted in intertidal areas. Eggs can be stripped from naturally-ripened clams, but the survival rates to the straight-hinge larvae stage is only six to 35 percent. Natural spawning can be induced by exposing the clams to high concentrations of algal cells (2 to 2.5 million cells of *Pseudoisochrysis paradoxa* per milliliter). The survival to the straight-hinge larvae stage increases to 38 to 60 percent using this method. Breese and Robinson (1981) maintained razor clam cultures at 16.5°C and a salinity of 30 to 33 ppt. The 90-micron-diameter eggs developed to 110 micron shell length straight-hinge larvae within 48 hours. They were held at five larvae per milliliter of seawater with the water changed twice per week. Larvae were fed *Pseudoisochrysis paradoxa* at 20,000 cells/ml for the first week; 40,000 cells/ml for the second week; and 50,000 cells/ml for the third week. Larvae reached metamorphosis at 20 to 25 days at a shell length of 300 microns and were fed at 80,000 cells/ml post-set. This routine allowed clams to be grown to an average of 5 mm in three months, but experienced high mortalities at the larval metamorphosis stage.

SOFT CLAMS

Soft clams (*Mya arenaria*) are found from Labrador (Canada) to Cape Hatteras (North Carolina) and are occasionally found in the waters off South Carolina. They have been introduced to the West Coast of the United States from Alaska to San Francisco, California. In the northern areas of its distribution, *Mya* tends to be found in marine environments, while in the South it is more often found in estuarine environments. Soft clams occur intertidally in bays and sounds and to depths of 9 meters. It is an important commercial species in the New England states and in Maryland (Abraham and Dillon, 1986).

Mya has an egg-shaped shell that is thin and brittle. The shell gapes at both ends and is gray or chalky white with a brown periostracum. The left valve has an erect spoon-shaped tooth at the hinge and the right valve has a corresponding heart-shaped pit. The siphons are fused into a rigid siphonal process which cannot be completely withdrawn into the shell. The siphon is capable of considerable elongation and allows the clam to remain deeply buried.

Soft clams can survive wide and rapid salinity fluctuations with a tolerance down to 5 ppt. Maximum temperature for soft clam habitat is about 28°C. Juveniles less than 15 mm shell length are more heat tolerant than adults. The 24-hour temperature exposure that would kill 50 percent of the test specimens was between 32.5°C and 34.4°C in summer-acclimated clams in Virginia. Clams apparently were at the limit of their tolerance, as a one degree change could make the difference between no mortality and complete mortality. Temperatures at 150 mm below the surface where the clams would be burrowed never rose above 30.6°C (Lucy, 1976). The temperature limits may explain why soft clams are rarely intertidal in the south as they are in the north.

The oxygen consumption of *Mya* has been measured at 14°C as 30 to 40 microliters of oxygen per gram of body weight per hour. Lowe and Trueman (1972) found that reduced oxygen concentrations caused a reduction in water flow through the soft clam and a reduced heart rate. Unlike many other bivalves, increasing temperatures did not cause much change in water flow rate, but did increase the heart rate. In water with a reduced oxygen concentration, the increase in heart rate with increasing temperature was not as pronounced.

In the Chesapeake Bay clams reach a commercial size of 50 mm in 18 to 24 months, but in northern areas they may require as much as seven years to reach the same size. Soft clams reach a maximum size of 150 mm. A maximum life span of 10 to 12 years is estimated for the Chesapeake Bay and Long Island Sound populations. Mature clams are found generally after attaining a shell length of 40 mm. With such a wide-ranging animal it is difficult to define the spawning period. At Solomons, Maryland, clams ripen in late October or early November; spawning occurs while water temperatures drop to the range of 15°C to 12°C, which usually occurs over a period of 10 days. Clams ripen and spawn again from mid-April to May while water temperatures rise from 10°C to 14°C (Pfitzenmeyer, 1965). The reported spawning periods throughout the soft clam's range are presented in chart form (Brousseau, 1987). In some areas there is one spawning period in some years and two in others. This may be due to the rate of water temperature change in various years. The degree of maturation of eggs may be dependent on the number of days the water temperature remains in the correct range for gametogenesis. Late, cold winters followed by early summer temperatures may not allow eggs to develop in some years (Pfitzenmeyer, 1965). Older clams may be more likely to undergo multiple spawnings than younger clams (Brousseau, 1987). The sex ratio in *Mya* is 1:1.

The larval period in the Chesapeake Bay is four weeks in the spring and six weeks in the fall. The 79 microns-long egg develops through the larval stage to a pediveliger at about 200 microns. A byssal attachment is present at setting and persists until the juvenile is about 7 mm long. It requires about 35 days from setting to 2 mm and an additional 95 days to 12 mm at Gloucester, Massachusetts. Clams up to 12 mm only burrow 10 to 20 mm deep and are sometimes washed into dense aggregations where the beach slopes break abruptly into deeper water (Lucy, 1976). Because of their thin, brittle shells which do not completely seal the body, soft clams are prone to predation until they reach a size large enough to burrow deeply. Adult clams can burrow as deeply as 300 mm.

Juvenile clams have been placed in different substrates to test the effect on growth rate. *Mya* grew fastest in mud, slightly slower in sand, slower in gravel, and slowest in nets. In addition to affecting growth rate, the different treatments affected the shape of the clams. Those grown in nets and gravel took on a more globose shape, while those in sand grew longer and narrower than clams grown in mud. One possible explanation for the variations relates to the hinge of the valves and the muscular energy to maintain the normal 10° angle of shell gape in the soft clam. The hinge ligament in *Mya* does not exert sufficient force to open the shell against the force of coarser substrates, so that it requires muscular energy to maintain the proper shell gape. In net culture the clams expend muscular energy in keeping the shell closed to the proper angle with no help from the pressure of surrounding substrate (Newell and Hidu, 1982). The various substrates could also alter the positioning of the mantle margin, which would in turn affect the pattern of shell deposition.

Predators that have been identified include the nemertean worm, *Cerebratulus lacteus*, which is believed to be responsible for the complete destruction of a *Mya* flat in Nova Scotia (Rowell and Woo, 1990). In the Chesapeake Bay predators include: jellyfish, *Chrysaora quinquecirrha*, and comb jelly, *Mnemiopsis leidyi*, on clam larvae; oyster drill, *Urosalpinx cinerea*; thick-lipped oyster drill, *Eupleura caudata*; flatworm, *Stylochus ellipticus*; and crabs, including probably the most important predator, the blue crab, *Callinectes sapidus*. Soft clams capable of burrowing more than 100 mm below the surface are less prone to blue crab predation. Predators of lesser importance include: starfish, *Asterias* sp.; horseshoe crab, *Limulus polyphemus*; channeled whelk, *Busycon canaliculatum*; and moon snail, *Polynices duplicatus*. North of Cape Cod the green crab, *Carcinus maenas*, is a major predator. Also in Massachusetts, the mummichog, *Fundulus heteroclitus*, is a major predator of juveniles less than 12 mm. Larger predators include: cownose ray, *Rhinoptera bonasus*; American eel, *Anguilla rostrata*; winter flounder, *Pseudopleuronectes americanus*; herring gull, *Larus argentatus*; canvasback duck, *Aythya valisineria*; oldsquaw duck, *Clanqula hyemalis*; tundra swan, *Cygnus columbianus*; and raccoon, *Procyon lotor*.

Internal organisms associated with soft clams are pea crabs and the nemertean worm, *Malacobdella grossa*. Paralytic shellfish poisoning from *Gonyaulax tamarensis* toxin can also be found in *Mya*, but rarely occurs south of Cape Cod.

Hatchery culture of *Mya* has been investigated by Stickney (1964), and more recently by Hidu and Newell (1989). The primary problem with hatchery production is the inability to condition soft clams for spawning out of their natural spawning season. It may be related to the usual hatchery practice of holding bivalves for conditioning in tanks without substrate. Soft clams are stressed when out of substrate, probably due to muscle fatigue in holding the valves together. Hatchery production must, therefore, rely on a population of spawners taken from natural areas in the proper spawning season. Spawning may be induced in ripe clams by wide temperature fluctuations (Stickney, 1964) or by the addition of cultured algae and a suspension of stripped sex cells in heated (24°C) seawater (Hidu and Newell, 1989). Larvae are maintained at 20°C at 2.5 larvae/ml and fed appropriate amounts of *Thalassiosira pseudonana* with three water changes per week. After setting, juvenile clams grow rapidly and reach a size of 30 to 35 mm in the first year. If the clams are not placed in a substrate at this stage, they grow more globose with little increase in shell length (Hidu and Newell, 1989). Triploidy has been induced in hatchery-reared clams in the hopes of producing a faster-growing clam. The reasoning is that the sterile triploids will not use energy in gametogenesis and may, therefore, grow faster. However, triploid soft clams actually grow slower than diploids (Beaumont and Fairbrother, 1991).

GEODUCK

Geoduck clams (*Panope abrupta* or *P. generosa*) are found from Alaska to Baja California. They occur from the lower intertidal regions to 110 meters depths. They are abundant in Puget Sound, Washington, and in British Columbia, Canada. Clams are found in sand and mud bottoms at depths of 6 to 18 meters at an average density of

1.7 clams per square meter in Washington and an average of 4.9 per square meter in British Columbia with a maximum of 36 per square meter reported. The fishery in Puget Sound started in 1970 and produced a range of 1100 to 3900 metric tons annually between 1975 and 1987. The production in British Columbia was 5000 metric tons in 1987 (Goodwin and Pease, 1989).

The shell of *Panope* is more or less rectangular and highly variable in shape. It has a light brown periostracum. The hinge area has one large cardinal tooth in each valve, with the tooth on the left valve always larger. The pallial sinus scar on the inner surface of the valves is very broad. In adults the shell is widely gaped and does not fully cover much of the soft body. The siphon, which can extend as much as one meter in adults, has no tentacles except in early post-set juveniles. The foot is well developed in juveniles but becomes proportionately smaller as the clam grows. In adults the foot is too small to right a clam lying on its side or to allow the clam to burrow into the substrate from the surface. The mantle is large and fleshy and fused, except for a small slit for the pedal gape.

Salinity and temperature requirements are not fully known, but embryos of geoducks must be maintained at 27.5 to 32.5 ppt salinity in a temperature range of 6°C to 16°C for 70 percent of them to develop to straight-hinge larvae (Goodwin, 1973). Larvae and adults can withstand prolonged temperatures of 18°C, although those living at depths of 60 meters probably never experience temperatures greater than 10°C (Goodwin, 1976).

Geoducks grow to an acceptable commercial size of 0.7 kg in about six years. They can reach a size of over 200 mm shell length and 4.5 kg (Sloan and Robinson, 1984). Clams grow about 30 mm per year for the first three years. Their greatest annual weight gain occurs between the third and seventh year and they reach an average size of 158 mm at 10 years (Goodwin, 1976). Growth slows considerably after 10 years. The shell weight and hinge plate thickness correlate better with age than shell length or body weight (Sloan and Robinson, 1984). The oldest clam found is believed to be 146 years old. Clams begin to mature at five years old and are fully mature sometime after 10 years. In sampling in British Columbia, 90 percent of geoducks less than 10 years old were males, while 47 percent of clams greater than 51 years old were males. It is possible that males mature earlier or that there is a sex reversal at some point (Sloan and Robinson, 1984). In fully mature clams the sex ratio is 1:1. Ripe gonads have been observed in a 107-year- old clam, demonstrating a reproductive life of at least 100 years.

The spawning period is June and July in Washington and British Columbia and is apparently a response to increasing water temperatures. Spawning occurs when water temperatures reach 10.5°C to 14°C. Laboratory spawning can occur between 8.5°C and 16°C, but mostly between 12°C and 14°C. Females contain millions of eggs but normally release only one or two million per spawn, although 20 million eggs have been observed from one female in a single laboratory spawn. Individual clams in the laboratory may spawn three or four times in a season. In culture the larval period is 16 to 47 days, but is probably longer in the colder natural environment. Settlement of juveniles seems to cluster around existing adults. Post-set clams possess a byssal attachment and will dig down one shell-length deep and anchor to several sand grains. At a shell length of 1.5 to 2 mm, clams have developed siphons and can burrow into the substrate. At 8 mm geoducks still attach with a byssus and still almost completely withdraw into the shell. Since the soft body parts of larger clams are not fully enclosed by the shell, the ability to burrow deeply is important in escaping predation. Clams with shell lengths averaging 36.6 mm can dig to an average depth of 336 mm. The depth at which clams are believed to be safe from predation is estimated at about 600 mm. Fast-growing clams can achieve a size capable of burrowing to this depth when they are about two years old. Adult geoducks can burrow as deeply as one meter.

The state of Washington is actively involved in reseeding areas with hatchery-produced stock, giving researchers a first-hand look at predators feeding on the newly planted geoducks. These predators include: sunflower star, *Pycnopodia helianthoides*; lean basket whelk, *Nassarius mendicus*; coonstripe shrimp, *Pandalus danae*; red rock crab, *Cancer productus*; graceful crab, *Cancer gracilis*; starry flounder, *Platichthys stellatus*; English sole, *Parophyrs vetulus*; rock sole, *Lepidopsetta bilineata*;

sand sole, *Psettichthys melanostictus*; pile perch, *Rhacochilus vacca*; moon snails, *Polynices lewisii* and *Natica* sp.; starfish, *Pisaster brevispinus*; and sea otters, *Enhydra lutris*, which can excavate and eat adult geoducks. Spiny dogfish, *Squalus acanthia*; cabezon, *Scorpaenichthys marmoratus*; and halibut, *Hippoglossus stenolepis*, have been observed grazing upon the extended siphons of clams.

As mentioned above, hatchery production of geoducks is carried out by the state of Washington with the goal of planting 30 million clams per year to supplement natural recruitment. Hatchery production was described by Goodwin et al. (1979). Clams were conditioned at 9°C to 10°C for two weeks and spawned at 14°C to 15°C. An inflow of the alga *Monochrysis lutheri* or the addition of sperm suspension was used as a spawning stimulant. The eggs were held at 14°C at 4000 to 10,000 per liter. Initial mortalities of 80 to 100 percent of early larvae were subsequently controlled by the addition of 12 ppm tetracycline hydrochloride, which reduced mortalities to 30 to 50 percent. High (50 to 80 percent) mortalities were also seen at metamorphosis, with constant mortalities occurring for the first month. Subsequent studies reduced larval mortalities to 5 percent by improving the algal culture and reducing larval densities to 3000 straight-hinge per liter and 250 late larvae per liter. The pediveliger stage was reached at a shell length of 300 microns and metamorphosis at 395 microns. The larval period was as long as 47 days but was reduced to 30 days in subsequent studies. At setting the clam develops spines on the growing edge of the shell and crawls with its foot along the substrate. Juveniles were planted into natural areas when 8 to 20 mm long. Survival varied from 0 percent to 40 percent after transplanting.

REFERENCES

Abraham, B. J. and P.L. Dillon. 1986. Species profiles: Life histories and environmental requirements of coastal fishes and invertebrates (Mid-Atlantic): Soft clam. U. S. Fish and Wildlife Service Biological Report 82 (11.68).

Ambrose, W. G., D.S. Jones, and I. Thompson. 1980. Distance from shore and growth rate of the suspension feeding bivalve, *Spisula solidissima*. *Proceedings of the National Shellfisheries Association.* **70**(2): 207-215.

Barnes, R. D. 1980. *Invertebrate Zoology* (4th ed.). Saunders College, Philadelphia, PA.

Beaumont, A. R. and J.E. Fairbrother. 1991. Ploidy manipulation in molluscan shellfish: A review. *Journal of Shellfish Research.* **10**(1):1-18.

Breese, W. P. and F.D. Phibbs. 1970. Some observations on the spawning and early development of the butter clam, *Saxidomus giganteus* (Deshayes). *Proceedings of the National Shellfisheries Association.* **60:** 95-98.

Breese, W. P. and A. Robinson. 1981. Razor clams, *Siliqua patula* (Dixon): Gonadal development, induced spawning, and larval rearing. *Aquaculture.* **22**: 27-33.

Bricelj, V. M. and R.E. Malouf. 1980. Aspects of reproduction of hard clams (*Mercenaria mercenaria*) in Great South Bay, New York. *Proceedings of the National Shellfisheries Association.* **70**(2): 216-219.

Brousseau, D. J. 1987. A comparative study of the reproductive cycle of the soft-shell clam *Mya arenaria* in Long Island Sound. *Journal of Shellfish Research.* **6**(1): 7-15.

Castagna, M. and P. Chanley. 1973. Salinity tolerance of some marine bivalves from inshore and estuarine environments in Virginia waters on the western Mid-Atlantic coast. *Malacologia.* **12**: 47-96.

Doering, P. H. 1982. Reduction of sea star predation by the burrowing response of the hard clam, *Mercenaria mercenaria* (Mollusca: Bivalvia). *Estuaries.* **5** (4): 310-315.

Drum, A. S., T.L. Siebens, E.A. Crecelius, and R.A. Elston. 1993. Domoic acid in the Pacific razor clam *Siliqua patula* (Dixon, 1789). *Journal of Shellfish Research.* **12**(2): 443-450.

Eversole, A. G. 1989. Gametogenesis and spawning in North American clam populations, in *Clam Mariculture in North America.* J. J. Manzi and M. Castagna (eds.), pp. 75-109. Elsevier Science Publishers. New York.

Eversole, A. G., W.K. Michener, and P.J. Eldridge. 1980. Reproductive cycle of *Mercenaria mercenaria* in a South Carolina estuary. *Proceedings of the National Shellfisheries Association.* **70**(1): 22-30.

Fritz, L. W. 1991. Seasonal condition change, morphometrics, growth and sex ratio of the ocean quahog, *Arctica islandica* (Linneaus, 1767) off New Jersey, U.S.A. *Journal of Shellfish Research.* **10**(1): 79-88.

Foster-Smith, R. L. 1975. The effect of concentration of suspension and inert material on the assimilation of algae by three bivalves. *Journal of the Marine Biological Association U.K.* **55**: 411-418.

Gibbons, M. C. and W.J. Blogoslawski. 1989. Predators, pests, parasites, and diseases. *Clam Mariculture in North America.* J. J. Manzi and M. Castagna (eds.), pp. 167-200. Elsevier Science Publishers. New York.

Gibbons, M. C., J.G. Goodsell, M. Castagna, and R.A. Lutz. 1983. Chemical induction of spawning by serotonin in the ocean quahog, *Arctica islandica* (Linneaus). *Journal of Shellfish Research.* **3**(2): 203-205.

Goldberg, R. 1980. Biological and technical studies on the aquaculture of yearling surf clams. Part I: Aquaculture production. *Proceedings of the National Shellfisheries Association.* **70**(1): 55–60.

Goodwin, C. L. 1976. Observations on spawning and growth of subtidal geoducks (*Panope generosa*, Gould).

Proceedings of the National Shellfisheries Association. **65**: 49-58.

Goodwin, C. L. and B. Pease. 1989. *Species profiles: Life histories and environmental requirements of coastal fishes and invertebrates (Pacific Northwest). Pacific geoduck clam.* U. S. Fish and Wildlife Service Biological Report **82** (11.120).

Goodwin, L. 1973. Effects of salinity and temperature on embryos of the geoduck clam, *Panope generosa* (Gould). *Proceedings of the National Shellfisheries Association.* **63**: 93-95.

Goodwin, L., W. Shaul, and C. Budd. 1979. Larval development of the geoduck clam, *Panope generosa* (Gould). *Proceedings of the National Shellfisheries Association.* **69**: 73-76.

Hidu, H. and C.R. Newell. 1989. Culture and ecology of the soft-shelled clam, *Mya arenaria*, in *Clam Mariculture in North America.* J. J. Manzi and M. Castagna (eds.), pp. 277-292. Elsevier Science Publishers. New York.

Humphrey, C. M. and R.L. Walker. 1982. The occurrence of *Mercenaria mercenaria* form *notata* in Georgia and South Carolina: Calculation of phenotypic and genotypic frequencies. *Malacologia.* **23**(1): 75-79.

Jones, D. S. 1981. Reproductive cycles of the Atlantic surf clam, *Spisula solidissima*, and the ocean quahog, *Arctica islandica*, off New Jersey. *Journal of Shellfish Research.* **1**(1): 23-32.

Jorgensen, C. B. 1975. Comparative physiology of suspension feeding. *Annual Review of Physiology.* **37**: 57-79.

Kraeuter, J. N., M. Castagna, and R. Van Dessel. 1982. Egg size and larval survival of *Mercenaria mercenaria* (L.) and *Argopecten irradians* (Lamarck). *Journal of Experimental Marine Biology and Ecology.* **56**: 3-8.

Landry, T., T.W. Sephton, and D.A. Jones. 1993. Growth and mortality of northern quahog, *Mercenaria mercenaria* (Linnaeus, 1758), in Prince Edward Island. *Journal of Shellfish Research.* **12**(2): 321-327.

Lassuy, D. R. and D. Simons. 1989. *Species profiles: Life histories and environmental requirements of coastal fishes and invertebrates (Pacific Northwest): Pacific razor clam.* U. S. Fish and Wildlife Service Biological Report. **82** (11.89).

Loosanoff, V. L. 1959. The size and shape of metamorphosing larvae of *Venus* (*Mercenaria*) *mercenaria* grown at different temperatures. *Biological Bulletin.* **117**: 308-318.

Lowe, G. A. and E.R. Trueman. 1972. The heart and water flow rates of *Mya arenaria* (Bivalvia: Mollusca) at different metabolic levels. *Comparative Biochemistry and Physiology.* **41A**:487-494.

Lucy, J. A. 1976. *The reproductive cycle of Mya arenaria L. and distribution of juvenile clams in the upper portion of the nearshore zone of the York River, Virginia.* M. S. Thesis. The College of William and Mary, Williamsburg, Virginia.

Lutz, R., J. Goodsell, M. Castagna, S. Chapman, C. Newell, H. Hidu, R. Mann, D. Jablonski, V. Kennedy, S. Siddall, R. Goldberg, H. Beattie, C. Falmagne, A. Chestnut, and A. Partridge. 1982. *Journal of Shellfish Research.* **2**(1): 65-70.

MacKenzie, C. L., D.J. Radosh, and R.N. Reid. 1985. Densities, growth, and mortalities of juveniles of the surf clam (*Spisula solidissima*) (Dillwyn) in the New York Bight. *Journal of Shellfish Research.* **5**(2): 81-84.

Mann, R. 1985. Seasonal changes in the depth-distribution of bivalve larvae on the southern New England shelf. *Journal of Shellfish Research.* **5**(2): 57-64.

Menzel, W. 1977. Selection and hybridization in quahog clams (*Mercenaria* spp.). In *Proceedings of the World Mariculture Society.* **8**: 507-521.

Menzel, W. 1989. The biology, fishery and culture of quahog clams, *Mercenari.* In *Clam Culture in North America,* J. J. Manzi and M. Castagna (eds.), pp. 201-242. Elsevier Science Publishers. New York.

Moore, H. J. 1971. The structure of the latero-frontal cirri on the gills of certain lamellibranch molluscs and their role in suspension feeding. *Marine Biology.* **11**: 23-27.

Murawski, S. A., J.W. Ropes, and F.M. Serchuk. 1982. Growth of the ocean quahog, *Arctica Islandica*, in the middle Atlantic bight. *Fisheries Bulletin.* **80**(1) 21-34.

Newell, C. R. and H. Hidu. 1982. The effects of sediment type on growth rate and shell allometry in the soft-shelled clam, *Mya arenaria* L. *Journal of Experimental Marine Biology and Ecology.* **65**: 285-295.

Newell, R. I. E. and S.J. Jordan. 1983. Preferential ingestion of organic material by the American oyster *Crassostrea virginica. Marine Ecology — Progress Series.* **13**: 47-53.

Oceanographic Institute of Washington. 1981. *Clam and Mussel Harvesting Industries in Washington State,* Oceanographic Commission of Washington, Seattle.

Owen, G. 1974. Feeding and digestion in the Bivalvia. In *Advances in Comparative Physiology and Biochemistry,* O. Lowenstein (ed.), Vol. 5, pp. 1-35. Academic Press, New York.

Paul, A. J., J.M. Paul, and H.M. Feder. 1976. Age, growth, and recruitment of the butter clam, *Saxidomus gigantea*, on Porpoise Island, southeast Alaska. *Proceedings of the National Shellfisheries Association.* 66: 27-28.

Peterson, C. H. 1983. A concept of quantitative reproductive senility: application to the hard clam, *Mercenaria mercenaria* (Linnaeus). *Oecologia.* **58**: 164-168.

Peterson, C. H., H.C. Summerson, and P.B. Duncan. 1984. The influence of seagrass cover on population structure and individual growth rate of a suspension-feeding bivalve, *Mercenaria mercenaria. Journal of Marine Research.* **42**: 123-138.

Pfitzenmeyer, H. T. 1965. Annual cycle of gametogenesis of the soft-shelled clam, *Mya arenaria*, at Solomons, Maryland. *Chesapeake Science.* **6**, No. 1: 52-59.

Prakash, A., J.C. Mecof, and A.D. Tennant. 1971. Paralytic shellfish poisoning in eastern Canada. *Fisheries Research Board of Canada Bulletin 177.*

Purchon, R. D. 1968. *The Biology of the Mollusc.* Pergamon Press, Oxford.

Quayle, D. B. and G.F. Newkirk. 1989. *Farming Bivalve Molluscs: Methods for Study and Development.* The World Aquaculture Society, Baton Rouge, LA.

Roegner, G. C. and R. Mann. 1990. *Habitat Requirements for the Hard Clam, Mercenaria mercenaria, in Chesapeake*

Bay. Virginia Institute of Marine Science Special Report No. 126.

Ropes, J. W. 1980. *Biological and Fisheries Data on the Atlantic Surf Clam, Spisula solidissima (Dillwyn)*. U. S. Department of Commerce Technical Series Report No. 24.

Ropes, J. W., S.A. Murawski, and F.M. Serchuk. 1984. Size, age, sexual maturity, and sex ratio in ocean quahogs, *Arctica islandica* (Linnaeus), off Long Island, New York. *Fisheries Bulletin* **82**(2): 253-267.

Rowell, T. W., D.R. Chaisson, and J.T. McLane. 1990. Size and age of sexual maturity and annual gametogenic cycle in the ocean quahog, *Arctica islandica* (Linnaeus, 1767), from coastal waters in Nova Scotia, Canada. *Journal of Shellfish Research*. **9**(1): 195-203.

Rowell, T. W. and P. Woo. 1990. Predation by the nemertean worm, *Cerebratulus lacteus* (Verrill), on the soft-shell clam, *Mya arenaria* (Linnaeus, 1758), and its apparent role in the destruction of a clam flat. *Journal of Shellfish Research*. **9**(1): 291-297.

Sayce, C. S. and D.F. Tufts. 1972. The effect of high water temperature on the razor clam, *Siliqua patula* (Dixon). *Proceedings of the National Shellfisheries Association*. **62**: 31-34.

Sloan, N. A. and S.M.C. Robinson. 1984. Age and gonad development in the geoduck clam *Panope abrupta* (Conrad), from southern British Columbia, Canada. *Journal of Shellfish Research*. **4**(2): 131-137.

Stanley, J. G. and R. DeWitt. 1983. *Species profiles: Life histories and environmental requirements of coastal fishes and invertebrates (North America: Hard clam)*. U. S. Fish and Wildlife Service Biological Report **82** (11.18).

Stickney, A. P. 1964. Salinity, temperature and food requirements of soft-shell clam larvae in laboratory culture. *Ecology*. **45**: 283-291.

Tegelberg, H. C. and C.D. Magoon. 1969. Growth, survival and some effects of a dense razor clam set in Washington. *Proceedings of the National Shellfisheries Association*. **59**:126-135.

Winter, J. E. 1978. A review of the knowledge of suspension-feeding in lamellibranchiate bivalves, with special reference to artificial aquaculture systems. *Aquaculture*. **13**: 1-33.

Biology of Certain Commercial Mollusk Species: Oysters

Harriet M. Perry and John D. Cirino

INTRODUCTION

Oysters are members of the phylum Mollusca, class Bivalvia, and order Ostreoida (Turgeon et al., 1988). All commercial oyster species in the United States belong to the family Ostreidae. This family includes a large number of edible and nonedible oysters, with most species restricted to shallow waters between 44°S and 64°N latitude (Galtsoff, 1964). Members of this family possess pseudolamellibranch-type gills with inter-filamentous tissue junctions, lack a foot as an adult, lack a siphon, and have retained only the posterior adductor muscle (Barnes, 1980). Figure 1 shows the general anatomy of an adult oyster.

Although there are many species of oysters, only a few have commercial importance. In the United States two species support major commercial fisheries: the eastern oyster, *Crassostrea virginica* (Gmelin, 1791), and the Pacific oyster, *Crassostrea gigas* (Thunberg, 1793). The Olympia oyster,

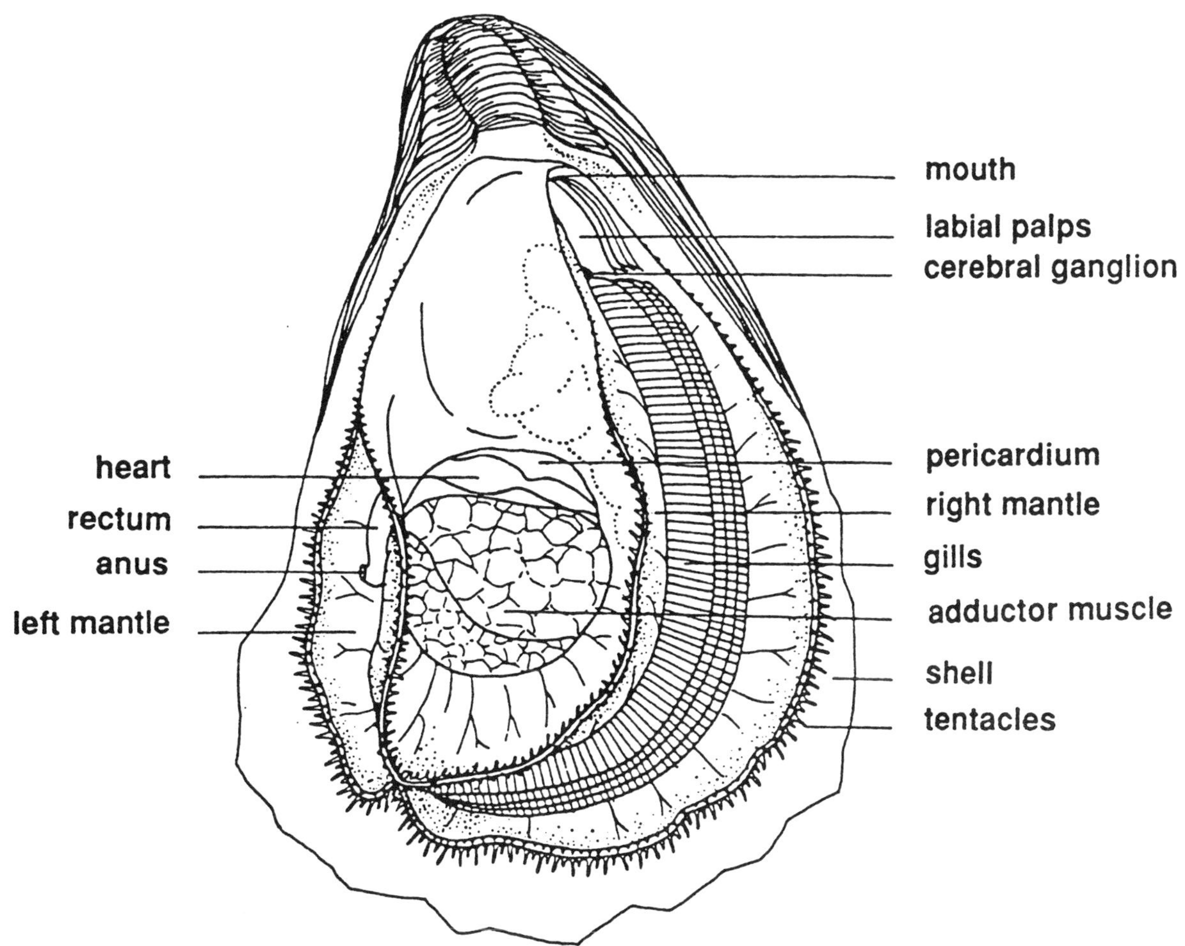

Figure 1. General anatomy of the eastern oyster (from Galtsoff, 1964).

Ostreola conchaphila (Carpenter, 1857) (formerly *Ostrea lurida*), and the European flat oyster, *Ostrea edulis* (Linnaeus, 1750), support limited, specialized fisheries.

Bahr and Lanier (1981) suggest that the most important characteristic of the genus *Crassostrea* is its ability to tolerate wide ranges of salinity, temperature, turbidity, and oxygen tension; this tolerance is responsible for its wide geographic distribution. Eastern oysters range from the Gulf of St. Lawrence in Canada through the Gulf of Mexico to the Bay of Campeche, Mexico, and into the West Indies (Andrews, 1981; Stenzel, 1971). Within that geographic range, King and Gray (1989) noted that biochemical evidence suggested that four racially distinct populations occur: Canadian, United States Atlantic, United States Gulf, and Bay of Campeche. In addition, they identified an "unusual" population of *C. virginica* from the lower Laguna Madre, Texas. Buroker (1983), however, reported a genetically homogeneous population between Cape Cod, Massachusetts, and Corpus Christi, Texas. The existence of a genetically distinct population of oysters in Laguna Madre was also noted by Buroker (1983). Data from biochemical studies suggest that morphological or physiological variability in eastern oyster stocks appears to be more related to environmental parameters than to genetic composition (Groue and Lester, 1982). The eastern oyster supports the largest of the oyster fisheries in the United States, with most of the production coming from the Gulf of Mexico (Berrigan et al., 1991).

The Pacific or Japanese oyster was first introduced to Pacific North America in the early 1900s; production now extends from British Columbia to central California (Chew, 1990). In Japan, this oyster exists in several different races or forms, depending upon the area of the country (Quayle, 1988). Contribution of Pacific oyster landings to total United States production generally averages less than 20 percent (Berrigan et al., 1991).

Current production of Olympia oysters is centered in Puget Sound in Washington State. While this oyster has a high market value and "gourmet" appeal, slow growth, high mortality, specialized culture requirements, and small size preclude mass culture and fishery expansion under present circumstances (Quayle, 1988).

The decline of soft-shell clam stocks in northern New England prompted the introduction of European flat oysters into that area in the mid-1900s, and a population was established in Maine (Chew, 1990). The European flat oyster has also been produced in California hatcheries and supports a small commercial harvest for the half-shell trade (Chew, 1984).

Because oyster production in the United States is primarily dependent upon the harvest of the eastern oyster, this paper will focus on the biology of *C. virginica*. Parasites and diseases will be discussed for the individual *Crassostrea* species. Morphological differences among *C. virginica*, *C. gigas*, and *O. conchaphila* are summarized in Table 1.

Table 1. Comparative characteristics of oyster species (modified from Quayle, 1988)

Scientific name	*Ostreola conchaphila*	*Crassostrea virginica*	*Crassostrea gigas*
Anatomy	w/o promyal chamber	with promyal chamber	with promyal chamber
Adductor	w/o pigment	dark purple or brown	mauve or white
Muscle scar	clearly outlined	sharply outlined	not clearly defined
Size	5 cm	7-10 cm	10-15 cm
Concentric growth	indistinct	flat but clear	projecting with flutings
Color inside shell	yellow-green	iridescent white	white
Color outside shell	gray with dark purple	yellow-brown	gray with purple
Hinge area denticles (chomata)	present	absent	absent
Radial grooves	not apparent	barely apparent	generally deep

HABITAT

Oysters grow and survive over a wide range of environmental conditions. They are most successful when attached to a firm, clean substrate in an area where water currents provide sufficient food and oxygen, dissipate wastes, disperse larvae, and prevent burial by siltation. For maximum feeding, current velocity must be high enough to exchange the water above a reef three times every hour (Galtsoff, 1964). Oysters have been known to set on virtually any hard surface, including shells, glass, concrete, rock, metal, wood, rubber, cans, crabs, and turtles (Martin, 1987; Burrell, 1986). Usually the most suitable cultch material is a clean unencrusted shell on the surface of an established, adequately elevated oyster reef (Stanley and Sellers, 1986; St. Amant, 1959; Gunter, 1938). Existing reefs generally provide the best and most attractive habitat and receive the greatest spat sets. The reef-building proclivity of oysters stems from the fact that oyster larvae attach to the shells of other oysters, and all are able to grow and survive for a period of time (Gunter, 1972). Live oysters may be found in several layers on a well-elevated reef, with the youngest oysters forming the top layer (Gunter, 1972). This type of growth leads to thick deposits of aggregated shells.

Temperature affects the life of an oyster by controlling feeding, respiration, gonadal development, and spawning (Galtsoff, 1964). Thermal requirements vary geographically; however, optimum water temperatures for growth, reproduction, and survival generally range from 20° to 30°C (Stanley and Sellers, 1986). Adult oysters exist over a range of water temperatures from -2°C (New England) to 36°C in the Gulf of Mexico and can survive internal temperatures from 46° to 49.5°C for brief periods (Ingle et al., 1971; Galstoff, 1964). Eastern oysters generally cease to feed as temperatures drop below 7°C (Galtsoff, 1964). Atlantic Coast stocks tolerate below-freezing water temperatures and freezing of their tissue (Loosanoff, 1965); however, Gulf stocks do not survive freezing (Cake, 1983). As temperatures approach 42°C, metabolic functions cease or are reduced to a minimum in eastern oysters (Galtsoff, 1964); the critical thermal maximum determined from laboratory studies is 48.5°C (Henderson, 1929).

Oyster populations flourish within a narrow range of salinity. Salinities less than 10 parts per thousand (ppt) through the spring and summer inhibit spawning and reduce larval survival, thereby resulting in insufficient numbers of mature oyster larvae. When salinities greater than 15 ppt predominate, mature larvae are abundant, but survival of recently-set oysters is poor because of increased numbers of fouling organisms, competitors, and predators. Oysters generally occur in salinities from 10 to 30 ppt but are capable of surviving salinities ranging from 3 to in excess of 40 ppt (Copeland and Hoese, 1966; Breuer, 1962; Gunter and Geyer, 1955). Galtsoff (1964) indicated that Atlantic Coast oyster populations found at the upper and lower limits of this range exist under marginal conditions, with growth and reproduction being inhibited; however, Breuer (1962) found that oysters were capable of spawning, setting, and rapid growth in salinities from 32 to 42 ppt, in the lower Laguna Madre, Texas. Copeland and Hoese (1966) noted that elevated salinities (in excess of 40 ppt) in combination with temperatures above 37°C caused mass mortality of oysters in shallow bays of the Laguna Madre. Salinities below 5 ppt affect oyster feeding and growth (Loosanoff, 1953). Butler (1949) found that low salinities affected gonadal development, and May (1972) noted that high freshwater inflow inhibited maturation and spawning. Prolonged exposure to low and high salinity regimes can be detrimental to oyster populations; however, periodic exposure may prove beneficial. Reduced salinities (0 to 15 ppt) for brief periods of time can benefit oyster stocks by reducing the abundance of some predators. Oyster drills *(Thais*, now *Stramonita)* and stone crabs (*Menippe* spp.) can pose serious threats to oyster populations, but they cannot tolerate average salinities below 15 ppt (Menzel et al., 1966). A short-term decrease in salinity can therefore help control these predators and lead to increased productivity.

Oysters are facultative anaerobes and are able to tolerate hypoxic conditions and survive brief exposures to anoxic conditions. Oysters can survive dissolved oxygen levels below one part per million (ppm) for up to five days (Sparks et al., 1958). Laboratory experiments indicate that the oxygen consumption rate for oysters is 303 milliliters/kilogram/hour for wet tissue (Hammen, 1969); however, oxygen requirements vary with salinity and temperature. Between water temperatures of 10° and 30°C and salinities of 7 ppt and 28 ppt, oxygen consumption increases with increasing temperature and decreasing salinity (Shumway, 1982).

REPRODUCTION

Eastern oysters are dioecious but exhibit protandrous hermaphroditism (Bahr and Lanier, 1981; Galtsoff, 1964). Development of maleness before the female phase often occurs, and young oysters are predominantly male (Menzel, 1951). Older oyster populations are primarily female (Burrell, 1986). Oysters that settle in unfavorable environments or are injured tend to develop as males, because functioning as a female requires more energy for gonadal development, and coping with environmental or physiological stress may limit the amount of energy available for reproductive development (Galtsoff, 1964). The sex of an individual oyster may change at least once during its life and can change annually (Bahr and Lanier, 1981; Galtsoff, 1964). A small percentage of oysters function as true hermaphrodites (Galtsoff, 1964).

Maturation periods fluctuate with changing environmental conditions. Temperature, salinity, and food availability may affect the time required for oysters to mature (Soniat and Ray, 1985). Under optimal environmental conditions, oysters in warmer waters are sexually mature and reproductively active four weeks after setting (Menzel, 1951). However, oysters that spawn shortly after setting generally do not contribute significantly to the year class because of low gamete production (Hayes and Menzel, 1981). The number of gametes released during each spawn is directly correlated with oyster size and gonadal development (Galtsoff, 1964). The number of eggs produced is highly variable, ranging from 23 to 86 million (Davis and Chanley, 1956). Among oysters of the same size, variability of fecundity is due primarily to differences in the physiological condition of the oysters (Galtsoff, 1964).

Spawning peaks are usually clearly defined and typically occur several times throughout the season. The release of sex cells from sexually mature oysters requires a stimulus. Gonadal maturation and spawning are usually associated with rising water temperature. Gametogenesis usually takes place in the early spring (depending on local climatic conditions). Spawning temperatures differ among populations, but most spawning is initiated and maintained when water temperature exceeds 15°C (Burrell, 1986).

Salinity also influences spawning, and most spawning occurs when salinities remain above 10 ppt. Davis and Calabrese (1964) found that larval growth and survival was highest at salinities above 12 ppt. Fertilization is external, and reproductive success is dependent on the close proximity of the sexes and their simultaneous response to spawning stimuli. Females are less responsive than males to rising temperature (Dupuy et al., 1977) and require chemical stimulation from sperm to ensure that eggs are not discharged without the presence of sperm. During heavy spawns, the water over shallow reefs may become "milky" with gametes. The duration and intensity of any spawning event depends on the physiological state of the oysters and ambient water conditions. The number of spawns per individual is variable; however, male oysters may spawn more often than females.

LARVAL DEVELOPMENT

Development after fertilization is rapid, with the embryo remaining demersal until the prototroch, or swimming organ, forms (Figure 2). The trochophore larval stage is attained four to six hours after fertilization. During this stage a powerful ciliated girdle is formed, and the larvae begin to swim. The trochophore stage lasts from one to two days and is followed by the veliger stage. The veliger stage is characterized by the development of the ciliated velum. The large cilia around the margin of the velum are used for swimming; the smaller cilia covering the base carry food particles to the mouth. While in the veliger stage, the shell is formed (straight-hinge stage), the larval "beak" develops (umbo stage), and the foot and byssus gland develop. Following the development of the foot, the larva is known as a pediveliger. During the last part of the pediveliger stage, a pair of pigmented eyes develop (eyed-pediveliger) and the oyster is ready to settle. The larval stage of *C. virginica* usually lasts from seven to 10 days but is dependent on water temperature (Bahr and Lanier, 1981). Larvae may remain planktonic for longer periods of time in cooler temperatures or in the absence of sufficient food.

The pelagic larvae of oysters aid in distribution of the species. Larval movement appears to

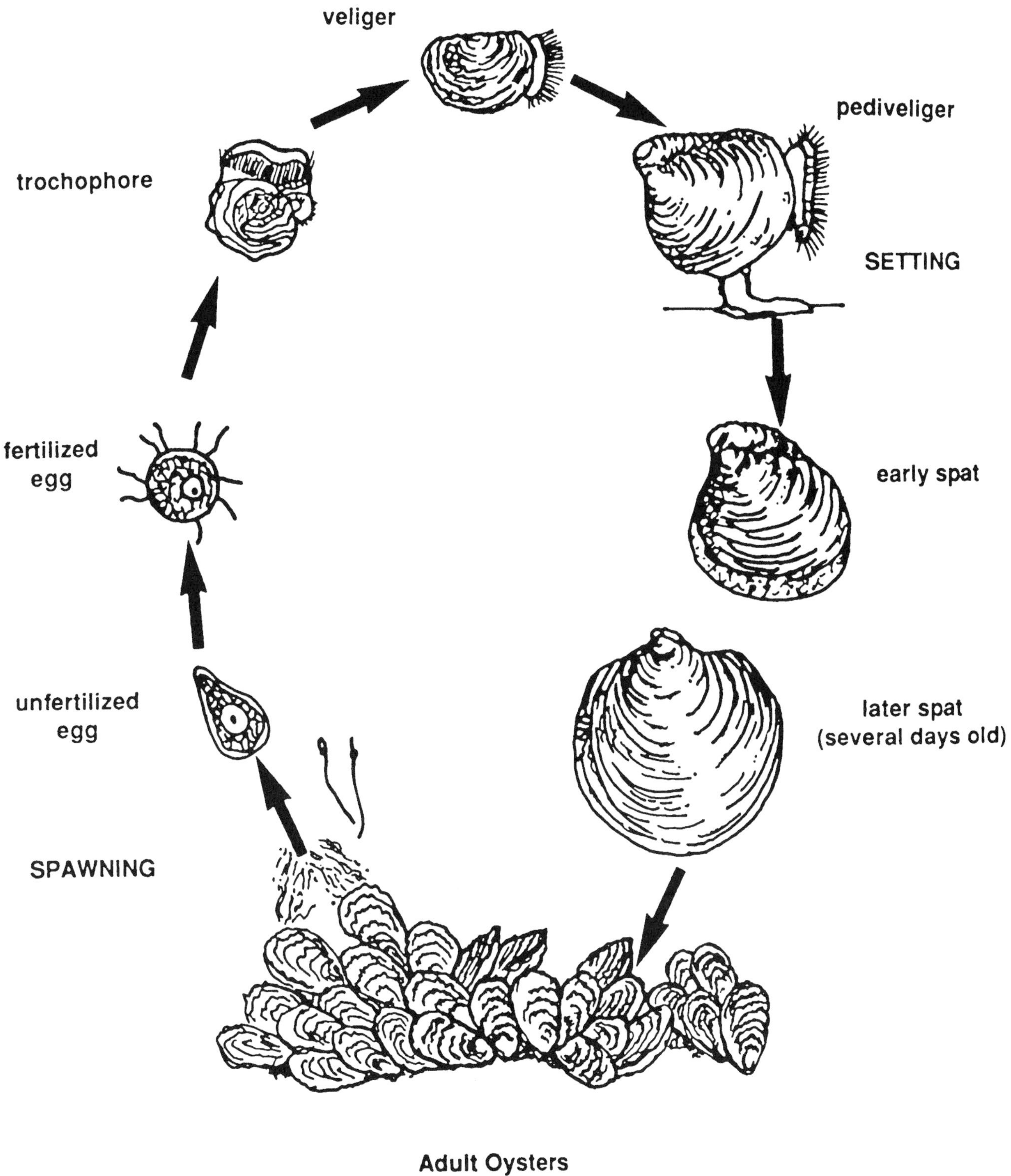

Figure 2. Life cycle of the eastern oyster (modified from Berrigan et al.*, 1991).*

be related to salinity and water currents. Based on plankton samples collected during all tidal stages, Andrews (1983) concluded that larvae swim continuously and that their dispersal and ultimate fates are strongly dependent on current regimes and flushing rates of estuaries. Veliger larvae are passively transported via water currents, and their dispersal is controlled by the hydrodynamic regime of the estuarine system. Vertical migration can help mitigate the dispersive effects of currents; however, larvae exert little control over the direction of their movement.

Prior to metamorphosis (settling and attachment), the veliger develops two eye spots (eyed-pediveliger) that aid in selecting an acceptable location for attachment. The eyed-pediveliger larva passively uses tidal currents, the salt wedge, and vertical migration to "select" the optimal environment for metamorphosis. It ceases to swim, drops to the bottom, and creeps over the substrate with its foot to locate a suitable attachment point. Larval set is higher in an area of reduced light (inside paired empty shells, on the underside of a piece of cultch). Shortly after metamorphosis, the newly-attached oyster (the spat) loses its velum, foot, and eye spots and begins a sessile life. Once attached to the cultch, the tiny oyster is referred to as a spat.

Oyster larvae are filter feeders. In the wild, their diet probably includes algae, bacteria, protozoa, and detritus (Bahr and Lanier, 1981). Hatchery-reared larvae are fed mixed cultures of "naked" flagellates (chrysophytes). As larvae develop, they are able to utilize chlorophytes as a food source (Davis, 1953).

SETTLEMENT

Successful setting is primarily dependent upon the availability of clean substrate. Larvae are selective and seek substrates free of detritus, soft mud, oil or grease, and biofouling organisms. At metamorphosis, the mature larva attaches its left valve to the cultch with a small amount of cementing fluid from the pedal byssus gland. Metamorphosing oyster larvae are gregarious and tend to attach in large groups on common cultch where other larvae have already attached or in the presence of mature oysters or remains of oysters. Hidu and Haskin (1971) noted that settlement was a function of two major factors. The physiographic and hydrographic characteristics of estuarine habitats that influence larval transport are the initial determinants of "prospective" spat set. Second, behavioral factors (vertical migration) and response to physico-chemical variables (salinity, light, temperature, pheromones, and metabolic products from live oysters) act to initiate settlement. Keck et al. (1971) found that biochemical constituents of oyster shell liquor and tissue were high set stimulants and suggested that the pheromones identified by Hidu (1969) were in fact constituents of shell liquor.

During the spat stage, shell growth is rapid and early shell growth generally follows the contour of the surface to which the spat is attached. Following the initial rapid growth phase, the shell starts to thicken and the shape of the young oyster begins to resemble that of an adult. A true juvenile stage does not exist because gonadal development and gametogenesis begin within a few weeks of metamorphosis and setting. Oysters may become adults within four to 12 weeks of settlement, thereby permitting spawning by young-of-the-year oysters and production of two or more generations of oysters per year.

GROWTH

Spat and mature oysters are filter feeders that feed primarily on planktonic organisms and organic detritus. The specific diet of adult oysters is not clearly understood. While oysters are known to ingest a wide variety of plant and animal material (bacteria, algae, dinoflagellates, diatoms, eggs and larvae of marine invertebrates, pollen grains from terrestrial plants, particulate organic matter, and detritus), little is known of their specific contributions to oyster nutriment (Quayle, 1988). Loosanoff (1949) demonstrated that oysters exhibited a degree of food selectivity and in some instances were able to select their food quantitatively and qualitatively.

Oyster growth is influenced by environmental conditions (temperature, salinity, food, turbidity, intertidal exposure), position on the reef, and population density (Stanley and Sellers, 1986). Maximum growth usually occurs from autumn through spring. Depending on geographic location, growth slows and may stop during summer as energy reserves are used for gamete production. Oysters expend as much as 48 percent of their annual energy budget on reproduction (Dame,

1976). Oysters exposed to fluctuating salinities within normal ranges were found to grow faster than those held at a relatively constant salinity (Pierce and Conover, 1954). Oysters that are exposed to air for short periods may grow as well as do continually submerged individuals (Gillmore, 1982); however, long air-exposure periods inhibit growth. Stanley and Sellers (1986) reported that oysters in reefs exposed approximately 20 percent of the time grew almost twice as fast as those oysters exposed 60 percent of the time. Oysters also grow faster in areas with higher phytoplankton densities (Manzi et al., 1977).

The growth rate of *C. virginica* is rapid until the onset of gametogenesis. In the Gulf of Mexico, oysters typically reach harvest size (76 millimeters or 3 inches) in 18 to 24 months from setting (Berrigan, 1990 and 1988; Martin, 1987). The time required to reach commercial harvest in colder waters may take from three to five years (Galtsoff, 1964). The life span of an oyster has not been determined; however, oysters are estimated to live for 25 to 30 years (Martin, 1987) and may exceed 350 millimeters or 14 inches in length (Galtsoff, 1964). Large shells taken from Indian middens may be more than 50 years old (McGraw, personal communication).

FACTORS AFFECTING SURVIVAL

Oyster populations are affected by a host of biotic and abiotic environmental parameters. Parasites, predators, and diseases have been implicated in oyster mortalities (Burreson and Calvo, 1996; Soniat, 1996; Ewart and Ford, 1993; Quayle, 1988; Overstreet, 1978; Butler, 1954). Heavy metals, petroleum hydrocarbons, pesticides, chlorine derivatives, sewage, freshwater runoff, and other pollutants also negatively affect oyster stocks. In the discussion that follows, references are specific to the eastern oyster unless otherwise stated.

Parasites

Parasitic infestation can reduce growth and inhibit development. Increased physiological stress resulting from parasitic invasions may make oysters more vulnerable to unfavorable environmental conditions and disease. Oysters can be infected with a variety of parasites; however, the majority are considered only mildly pathogenic to their hosts (Gauthier et al., 1990). Sporozoans, ciliates, flagellates, gregarines, copepods, trematodes, cestodes, and nematodes are commonly reported from oysters, but few have been associated with massive oyster mortalities (Quayle, 1988; Overstreet, 1978; Galtsoff, 1964).

The ectoparasitic gastropod *Boonea impressa* has been found to reduce juvenile growth rates in eastern oysters. White et al. (1984) noted that the reduction in growth of parasitized oysters, the aggregated distribution of the snail, and the preference of the snail for small oysters could have a significant impact on the population structure and health of oyster populations. White et al. (1987) demonstrated that this snail directly transfers the pathogen *Perkinsus marinus* from infected to uninfected oysters.

Diseases

Oysters can suffer from both noncontagious and infectious diseases. Symptoms of disease include slow growth, failure to fatten and develop gonads, recession of the mantle, and gaping valves (Galtsoff, 1964). The most serious diseases of eastern oysters are caused by the protozoan *Perkinsus marinus* and the haplosporidian *Haplosporidium nelsoni*.

The endoparasitic protozoan *Perkinsus marinus* is a virulent pathogen that has been implicated in mass mortalities of eastern oysters. Commonly called "dermo," a previous derivation from *Dermocystidium marinum* (*Labyrinthomyxa marina*), it infects oysters throughout the Gulf of Mexico and the Atlantic. The pathogen is known from Cape Cod, Massachusetts (Ford, 1996), through the Gulf of Mexico to Tabasco, Mexico (Soniat, 1996; Burreson et al., 1994). The distribution of *P. marinus*, its pathogenicity, and its relationship to environmental factors have received critical attention because of its association with extensive, warm-water mortality of oysters. Water temperature and salinity are important factors controlling the occurrence and effect of *P. marinus* infections. Increased mortality often occurs among larger oysters during late summer and early autumn months when elevated water temperatures and salinities and weakened physiological conditions exacerbate disease problems. Salinities in excess of 12 ppt and temperatures above 20°C increase prevalence and intensity of the pathogen (Burreson and Calvo, 1996). Transmission of the disease is direct from oyster to oyster. Andrews (1996) reviewed the history of this pathogen in the Chesapeake Bay, and Burreson and Calvo (1996) summarized the epizootiology of *P. marinus* for that area. *Perkinsus marinus* has been characterized as the most important pathogen of the eastern oyster along the East Coast of the United States.

This pathogen also infects eastern oysters in the Gulf of Mexico, and Soniat (1996) noted that distribution of the parasite in Gulf estuaries was coincident with its oyster host. Soniat (1996), in his review of the epizootiology of this disease in the Gulf of Mexico, suggested that temperature and salinity did not explain all of the variation of levels of parasitism and that other factors such as the interaction of temperature and salinity, the timing of the spring plankton bloom in relation to the increase in temperature, nutritional factors, and the health and condition of the host may be important in the disease process.

Mass mortalities of eastern oysters along the Atlantic Coast have also been attributed to another infectious disease, MSX. The causative agent, *Haplosporidium nelsoni,* was originally given the acronym "MSX" because it was designated as a multinucleated sphere (ms) with unknown affinity (X). The infective stage has not been found, and the means by which the disease is transmitted is unknown. Infections of *H. nelsoni* are more prevalent and intensify at salinities above 15 ppt (Ewart and Ford, 1993). Temperature also influences distribution and intensity; low temperatures may act to suppress disease activity. The Pacific oyster, *C. gigas*, is apparently immune to this disease. While MSX is lethal to oysters, it is not harmful to humans.

Bacterial species of *Aeromonas*, *Vibrio*, *Moraxella*, *Acinetobacter*, and *Pseudomonas* have been identified from eastern oysters (Vanderzant et al., 1973). Mortality of oyster larvae and spat has been attributed to infection with *Aeromonas* sp. (Tubiash et al., 1965). Infections of *Vibrio* sp. and *Pseudomonas* sp. have also been implicated in mortalities of eastern oysters (Stanley and Sellers, 1986).

Foot disease, identified in both eastern and Pacific oysters, is attributed to a fungal infection (Quayle, 1988; Overstreet, 1978; Galtsoff, 1964). The disease affects the area of attachment of the adductor muscle, and in its advanced stage the oyster has difficulty in closing its valves and becomes more vulnerable to predation (Quayle, 1988; Overstreet, 1978; Galtsoff, 1964).

Quayle (1988) listed amoebae, bacteria, and fungi as causative agents of disease in the Pacific oyster but noted that this oyster is relatively free of disease organisms. Mortalities of Pacific oysters in British Columbia have been linked to protozoans. Several amoeboid protozoans and a protozoan designated generally as a "microcell" were isolated from disease-infected oysters (Denmam oyster disease); however, these have not been positively identified as the causative agent of the disease.

Pollutants

Heavy metals in the environment affect oysters during all life cycle stages. These substances can stress or kill oysters by reducing their ability to withstand diseases and parasites (MacInnes, 1981; Calabrese et al., 1973). The presence of heavy metals in coastal waters can lead to the mortality of embryos and larvae, reduce growth of larvae and spat, reduce spat setting, and cause shell thinning in the eastern oyster (Cunningham, 1976; Boyden et al., 1975; Calabrese et al., 1973). Results of tests designed to determine contaminant concentrations at which 50 percent of oyster embryos die (LC_{50}) indicate that, of the heavy metals tested, mercury, silver, copper, and zinc were the most toxic. Nickel, lead, and cadmium have been classified as relatively toxic, while arsenic, chromium, manganese, and aluminum have been labeled as nontoxic to oyster embryos (Calabrese et al., 1973). Quayle (1988) noted that tributyltin, a constituent of antifouling paints, affected shell deposition, growth and conditioning, and survival of Pacific oyster larvae.

Oil pollution can affect oyster abundance by increasing mortality and reducing reproductive success. Determining the effect of petrochemical pollution on oysters is difficult, because oil is composed of a complex mix of hydrocarbons that exhibit different levels of toxicity. Crude oil is generally less toxic than partially refined oils (Anderson and Anderson, 1975). Petroleum hydrocarbons cause mortalities or negatively affect oyster physiology by reducing food intake or utilization, interfering with reproduction, and lowering resistance to parasites. Chronic exposure to oil-contaminated sediment at low concentrations results in a reduction in food intake or utilization, while exposure to higher concentrations of oil in the sediment can cause extensive mortalities (Mahoney and Noyes, 1982). Fertilization and developmental success is reduced in proportion to concentrations of water-soluble hydrocarbon fractions between one and 1,000 ppm (Renzoni, 1975). Incidence of parasites is higher in oysters chronically exposed to oil pollution than in unexposed oysters (Barszcz et al., 1978).

Pesticides reduce oyster growth, cause pathological tissue damage, interfere with egg development, and cause mortalities (Schimmel et al., 1975; Lowe et al., 1972; Rowe et al., 1971; Davis and Hidu, 1969). The extent to which pesticides affect oysters depends on the chemical, its concentration, and the oyster life stage.

Chlorine and chlorinated compounds affect oyster survival, growth, feeding, reproduction, and development. Chlorine may be used to purify municipal water supplies and disinfect sewage wastewater. It is also used as a biocidal antifouling agent in industrial cooling water. Chlorine and chlorine derivatives (chlorine-produced oxidants) have been demonstrated to affect the swimming behavior of oyster larvae (Haven et al., 1978). Roberts et al. (1975) found chlorine to be extremely toxic to oyster larvae at concentrations as low as 0.005 ppm. Exposure to chlorine also adversely affects adult oyster growth, food intake, and reproduction, and can be toxic at elevated levels (Scott and Vernberg, 1979).

Quayle (1988) identified wastewater from pulp mills and the effects of log-booming operations as primary industrial pollutants of concern to the Pacific oyster industry. Pulp mill operations affect oyster production through release of toxic effluents, increased biological oxygen demand (BOD) associated with decomposition of organic materials, and release of particulate materials (wood fibers, chips, and bark). Toxins in the effluent may directly or indirectly affect oysters. Direct effects include physiological changes in the oyster; indirect effects influence the biotic and abiotic environment of the oyster (food organisms and BOD). Particulate materials suspended in the water may clog the gills of oysters, resulting in reduced growth rate, reduced body condition or "fatness," and mortality. Log-booming operations near oyster leases may contribute to anaerobic conditions by decomposition of bark broken from logs.

Commensals, Competitors, and Fouling Organisms

Many different organisms are associated with oyster populations and compete with them for food and space. Algae, sponges, hydroids, anemones, flatworms, other mollusks, polychaetes, barnacles, bryozoans, and tunicates occur either on or in association with oysters. Invertebrate assemblages associated with oyster populations vary geographically and reflect the physiographic and hydrographic characteristics of the estuarine area. Common oyster associates include species of the boring sponge, *Cliona*. Boring sponges colonize the oyster's shell matrix, and extensive concentrations of these organisms may debilitate living populations and limit future generations by destroying valuable cultch. Species of *Cliona* are prevalent in the shells of both eastern and Pacific oysters. Quayle (1988) noted that while the boring sponge was a serious pest on the Atlantic Coast, it was rarely encountered on living Pacific oysters. The boring clam, *Diplothyra smithii*, is also common in the matrix of eastern oyster shells. Heavy infestations weaken the shell and make the oyster more vulnerable to predation.

Various species of worms are associated with oyster beds on the Atlantic, Gulf, and Pacific coasts. Species of polyclad flatworms (oyster leeches) occur on both *Crassostrea* species. The Japanese species *Pseudostylochus ostreophagus* sometimes reaches pest proportions on northwestern oyster beds (Quayle, 1988). Extensive losses of eastern oysters have been associated with the oyster leech, *Stylochus* (Overstreet, 1978; Menzel et al., 1966; Ingle and Dawson, 1953; Pearse and Wharton, 1938). Overstreet (1978) reported that the polyclad, *Stylochus ellipticus*, feeds on small oysters and barnacles and has been implicated in mortalities of eastern oysters. Although polyclad flatworms may cause serious damage, evidence suggests that they are secondary predators and generally cause harm in areas where oysters are already in a weakened condition. Large populations of polydorid polychaetes or mud worms, *Polydora websteri*, compete for space and food. These polychaetes accumulate mud, and heavy infestations have been implicated in oyster mortalities. Galtsoff (1964) noted that when large numbers of these worms settle on shells, they can smother an entire oyster population. Polydorids also cause ulcerations of muscle tissue and viscera, blistering of the internal surface of the shell, and shell damage resulting in poor closure of the valves.

Several species of crabs of the family Pinnotheridae (pea crabs or oyster crabs) are commensal with the eastern oyster and may occur in large numbers (Galtsoff, 1964). *Pinnotheres ostreum* is abundant in many estuaries along the Atlantic Coast. Female crabs settle on the oyster gills and may impair gill function. Although infestations are usually not fatal, damage to the gills may result in oysters of poor condition. Crustaceans associated with Pacific oysters include grapsid and majid crabs, mud shrimp (*Callianasa* and *Upogebia*), barnacles, isopods, and amphipods (Quayle, 1988).

While fouling species are primarily nuisance organisms, dense settlement of biofoulers can prevent spat settlement, can reduce oyster growth, and may cause mortalities. Fouling organisms compete with oysters for food and space, and smothering of oysters can occur in heavy accumulations. Species of slipper shells, *Crepidula*, are conspicuous biofoulers of eastern oysters (Galtsoff, 1964). Other important fouling organisms of eastern oysters include hooked mussels, polychaete tube worms, acorn barnacles, bryozoans, and tunicates (Galtsoff, 1964). Galtsoff (1964) collected 29 species of invertebrates from oysters suspended from a raft in Oyster River, Massachusetts, and noted that these species comprised 44 percent of the total weight of a string of oysters. Quayle (1988) listed algae, sponges, hydroids, anemones, mollusks, tube-dwelling polychaetes, acorn barnacles, bryozoa, and tunicates as fouling organisms associated with populations of Pacific oysters. He identified acorn barnacles, *Balanus*, as the most ubiquitous of the fouling organisms, with mussels second in importance.

Predators

Predation represents a serious threat to oyster populations, with severe consequences to commercial harvests. The planktonic larvae of many benthic invertebrates, larval fish, and pelagic ctenophores and cnidarians prey on oyster larvae. Benthic organisms with mucous or ciliary feeding mechanisms (e.g., hydroids, sea anemones, mollusks, barnacles) also ingest bivalve larvae, and the most efficient predators might be adult oysters themselves (Andrews, 1983).

Numerous species of gastropods, crustaceans, and fish prey on spat, juveniles, and adult oysters. The principal predators of eastern oysters are more abundant in high salinity waters (Gunter, 1955). In many areas where oyster populations flourish, critical fluctuations in daily and seasonal salinity patterns act to deter the establishment of predators with marine affinities.

Oyster drills *(Eupleura caudata, Urosalpinx cinerea,* and *Thais)* prey on eastern oysters along the Atlantic and Gulf coasts. The eastern oyster drill, *U. cinerea,* was introduced to the northwest from the Atlantic Coast with the importation of eastern oysters to the Pacific Coast. *Thais* is the most serious natural predator of oysters in the Gulf of Mexico, and heavy infestations may severely restrict oyster distribution (May, 1971; Butler, 1954). Butler (1985) estimated that this drill prevents profitable oyster culture on about half of the suitable oyster bottoms in the northern Gulf of Mexico. Southern oyster drills (*Thais haemastoma,* now *Stramonita haemastoma*) are euryhaline, but they are most abundant in salinities above 15 ppt (Pollard, 1973). Annual oyster mortalities from oyster drills range from 50 to 85 percent in Louisiana (Schlesselman, 1955) and from 50 to 100 percent in Mississippi (Chapman, 1959). Butler (1954) reported that losses due to *Thais* were incalculable and concluded that their voracious feeding habits, high fecundity, and widely distributed larval stages combined to make this snail the most destructive oyster predator in the Gulf environment. Quayle (1988) noted that oyster drills were among the most destructive predators of Pacific oysters, with the Japanese drill, *Ceratostoma inornatum,* the most widespread and damaging of the species.

Other gastropods that feed on eastern oysters in the Gulf of Mexico include whelks (*Busycon* spp.), the crown conch (*Melongena corona*), the moon snail (*Polynices duplicatus,* now *Neverita duplicata*), and the ectoparasitic snail (*Boonea impressa*) (White et al., 1984; Menzel et al., 1966; Menzel and Nichy, 1958; Butler, 1954; Ingle and Dawson, 1953). The levels of predation due to these snails are poorly understood but are generally considered to be less devastating than is *Thais*

(Menzel et al., 1966; Butler, 1954). Atlantic Coast populations of C. *virginica* are also subject to predation by whelks of the genus *Busycon* (Burrell, 1986). The European veined whelk (*Rapana venosa*) has recently been collected in Chesapeake Bay. Establishment of reproducing populations of this non-indigenous species poses a serious threat to shellfish resources, and the potential for ballast-carried transfer of *R. venosa* into other North American estuaries must be considered (Harding and Mann, 1999).

The most important predators of Pacific oysters are sea stars. Quayle (1988) noted that about 12 sea star species were predatory on Pacific oysters, but only four species were considered serious predators: *Pisaster ochraceus*, *Pisaster brevispinus*, *Evasterias troschelii*, and *Pycnopodia helianthoides*. Burrell (1986) included the common sea star, *Asterias forbesi*, among the predators of eastern oysters on the Atlantic coast.

Many crustaceans prey upon oysters. The stone crabs (*Menippe adina* and *M. mercenaria*) have been identified as major oyster predators, particularly in the Gulf of Mexico (Powell and Gunter, 1968; Menzel et al., 1966; Menzel and Hopkins, 1956). The blue crab (*Callinectes sapidus*) and smaller xanthid "mud" crabs also prey on oysters. Menzel and Nichy (1958), Menzel and Hopkins (1956), and Krantz and Chamberlin (1978) noted that blue crabs preyed on small oysters, and Menzel et al. (1966) attributed mortality of oysters to predation by *C. sapidus*. Crustacean predators of Pacific oysters include three species of *Cancer*: the Dungeness crab, *C. magister;* the rock crab, *C. productus;* and the graceful crab, *C. gracilis* (Quayle, 1988).

Known fish predators of eastern oysters and/or spat include black drum, *Pogonias cromis*; toad fishes, *Opsanus* spp.; Atlantic croaker, *Micropogonias undulatus;* spot, *Leiostomus xanthurus;* cownosed ray, *Rhinoptera bonasus;* sheepshead, *Archosargus probatocephalus;* and the striped burrfish, *Chilomycterus schoepfi* (St. John and Cake, 1980; Cave and Cake, 1980; Haven et al., 1978; Pearson, 1929). The extent to which these fish impact oyster stocks has not been determined; however, Louisiana oyster farmers lose significant numbers of transplanted seed oysters to schools of large black drum that can consume at least one commercial oyster per pound of body weight per day (Cave and Cake, 1980).

REFERENCES

Anderson, R.D. and J.W. Anderson. 1975. Oil bioassays with the American oyster, *Crassostrea virginica* (Gmelin). *Proc. Natl. Shellfish. Assoc.* **65**:38-42.

Andrews, J. 1981. *Texas Shells*. University of Texas Press, Austin.

Andrews, J. D. 1983. Transport of bivalve larvae in James River, Virginia. *J. Shellfish Res.* **3**(1):29-40.

Andrews, J. D. 1996. History of *Perkinsus marinus*, a pathogen of oysters in Chesapeake Bay 1950-1984. *J. Shellfish Res.* **15**(1):13-16.

Bahr, L. M. and W.P. Lanier. 1981. The ecology of intertidal oyster reefs on the South Atlantic coast: a community profile. *U.S. Fish and Wildlife Service, Biology Service Program.* **81**(15):1-105.

Barnes, R .D. 1980. *Invertebrate Zoology*. Saunders College. Holt, Rinehart and Winston, Philadelphia, PA.

Barszcz, C., P. P. Yevich, L. R. Brown, J. D. Yarbrough, and C. D. Minchew. 1978. Chronic effects of three crude oils on oysters suspended in estuarine ponds. *J. Environ. Path. Tox.* **1**:879-896.

Berrigan, M. E. 1988. Management of oyster resources in Apalachicola Bay following Hurricane Elena. *J. Shellfish Res.* **7**(2):281-288.

Berrigan, M. E. 1990. Biological and economical assessment of an oyster resource development project in Apalachicola Bay, Florida. *J. Shellfish Res.* **9**(1):149-158.

Berrigan, M. E., T. Candies, J. Cirino, R. Dugas, C. Dyer, J. Gray, T. Herrington, W. Keithly, R. Leard, J. R. Nelson, and M. Van Hoose. 1991. The oyster fishery of the Gulf of Mexico, United States: a regional management plan. *Gulf States Marine Fisheries Commission Publication.* **24**:1-220.

Boyden, C. R., H. Watling, and I. Thornton. 1975. Effect of zinc on the settlement of the oyster, *Crassostrea gigas. Mar. Biol.* **31**:227-234.

Breuer, J. P. 1962. An ecological survey of the lower Laguna Madre of Texas, 1953-1959. *Publ. Inst. Mar. Sci. Univ. Tex.* **8**:153-183.

Buroker, N. E. 1983. Population genetics of the American oyster, *Crassostrea virginica,* along the Atlantic Coast and the Gulf of Mexico. *Mar. Biol.* **75**:99-112.

Burrell, Victor G. 1986. Species profiles: life histories and environmental requirements of coastal fishes and invertebrates (South Atlantic) — American oyster. *U.S. Fish and Wildlife Service, Biological Report 82(11.57).* U.S. Army Corps of Engineers, *TR-EL-82-4*:1-17.

Burreson, E.M., R.S. Alvarez, V.V. Martinez, and L.A. Macedo. 1994. *Perkinsus marinus* (Apicomplexa) as a potential source of oyster *Crassostrea virginica* mortality in coastal lagoons of Tabasco, Mexico. *Dis. Aquat. Org.* **20**:77-82.

Burreson, E. M. and L. M. Ragone Calvo. 1996. Epizootiology of *Perkinsus marinus* disease of oysters in Chesapeake Bay, with emphasis on data since 1985. *J. Shellfish Res.* **15**(1):17-34.

Butler, P. A. 1949. Gametogenesis in the oyster under conditions of depressed salinity. *Biol. Bull.* **96**(3):263-269.

Butler, P. A. 1954. The southern oyster drill. *Proc. Natl. Shellfish. Assoc.* **44**:67-75.

Butler, P. A. 1985. Synoptic review of the literature on the southern oyster drill, *Thais haemastoma floridana*. *USDC, NOAA, NMFS Tech. Rep.* **35**:1-9.

Cake, E. W. Jr. 1983. Habitat suitability models: Gulf of Mexico American oysters. *U.S. Fish and Wildlife Service, Biological Report.* **82**(10.57).

Calabrese, A., R. S. Collier, D. A. Nelson, and J. R. MacInnes. 1973. The toxicity of heavy metals to embryos of the American oyster, *Crassostrea virginica*. *Mar. Biol.* **18**:162-166.

Cave, R.N. and E.W. Cake Jr. 1980. Observations on the predation of oysters by the black drum *Pogonias cromis* (Linnaeus) (Sciaenidae). *Proc. Natl. Shellfish. Assoc.* **70**(1):121 (Abstract).

Chapman, C. R. 1959. Oyster drill *Thais haemastoma* predation in Mississippi Sound. *Proc. Natl. Shellfish. Assoc.* **49**:87-97.

Chew, K. K. 1984. Recent advances in the cultivation of molluscs in the Pacific United States and Canada. *Aquaculture.* **39**:69-81.

Chew, Kenneth K. 1990. Global bivalve shellfish introductions. *World Aquacult.* **21**(3):9-22.

Copeland, B. J. and H. D. Hoese. 1966. Growth and mortality of the American oyster, *Crassostrea virginica*, in high salinity shallow bays in central Texas. *Publ. Inst. Mar. Sci. Univ. Texas.* **11**:149-158.

Cunningham, P. A. 1976. Inhibition of shell growth in the presence of mercury and subsequent recovery of juvenile oysters. *Proc. Natl. Shellfish. Assoc.* **66**:1-5.

Dame, R. F. 1976. Energy flow in an intertidal oyster population. *Est. Coast. Mar. Sci.* **4**(3):243-253.

Davis, H.C. 1953. On food and feeding of larvae of the American oyster, *C. virginica*. *Biol. Bull.* **104**(3):334-350.

Davis, H. C. and A. Calabrese. 1964. Combined effects of temperature and salinity on development of eggs and growth of larvae of *M. mercenaria* and *C. virginica*. *U.S. Fish and Wildlife Service Fish. Bull.* **63**(3):643-655.

Davis, H. C. and P. E. Chanley. 1956. Spawning and egg production of oysters and clams. *Proc. Natl. Shellfish Assoc.* **46**: 40-58.

Davis, H. C. and H. Hidu. 1969. Effects of pesticides on embryonic development of clams and oysters and on survival and growth of the larvae. *Fish. Bull.* **67**(2):393-403.

Dupuy, J. L., N. T. Windsor, and C. E. Sutton. 1977. Manual for design and operation of an oyster seed hatchery for the eastern oyster *Crassostrea virginica*. Virginia Institute of Marine Science, Special Report Number 142.

Ewart, J. W. and S. E. Ford. 1993. History and impact of MSX and Dermo diseases on oyster stocks in the northeast region. *Northeastern Regional Aquaculture Center, NRAC Fact Sheet* **200**:1-8.

Ford, S. E. 1996. Range extension by the oyster parasite *Perkinsus marinus* into the northeastern United States: response to climate change? *J. Shellfish Res.* **15**(1):45-56.

Galtsoff, P. S. 1964. The American oyster, *Crassostrea virginica* (Gmelin). *U.S. Fish and Wildlife Service Fish. Bull.* **64**:1-480.

Gauthier, J. D., T. M. Soniat, and J. S. Rogers. 1990. A parasitological survey of oysters along salinity gradients in coastal Louisiana. *J. World Aquacult. Soc.* **21**(2):105-115.

Gillmore, R. B. 1982. Assessment of intertidal growth and capacity adaptations in suspension-feeding bivalves. *Mar. Biol.* **68**(3):277-286.

Groue, K. J. and L. T. Lester. 1982. A morphological and genetic analysis of geographic variation among oysters in the Gulf of Mexico. *Veliger* **24**:331-335.

Gunter, G. 1938. A new oyster cultch for the Texas coast. *Proc. Texas Acad. Sci.* (**21**):14.

Gunter, G. 1955. Mortality of oysters and abundance of certain associates as related to salinity. *Ecology* 36(4): 601-605.

Gunter, G. 1972. Use of dead reef shell and its relation to estuarine conservation. *Transactions of 37th North American Wildlife Natural Resource Conference.* Pp. 110-121.

Gunter, G. and R. A. Geyer. 1955. Studies of fouling organisms in the northwestern Gulf of Mexico. *Publ. Inst. Mar. Sci. Univ. Texas* **4**(1):39-67.

Hammen, C. S. 1969. Metabolism of the oyster, *Crassostrea virginica*. *Amer. Zool.* **9**(2):309-318.

Harding, J.M. and R. Mann. 1999. Observations on the biology of the veined Rapa whelk, *Rapana venosa* (Valenciennes, 1846) in the Chesapeake Bay. *J. Shellfish Res.* 18(1): 9-17.

Haven, D. S., W. J. Hargis Jr., and P. C. Kendall. 1978. The oyster industry of Virginia: its status, problems and promise. Virginia Institute of Marine Science, Gloucester Point, VA.

Hayes, P. F. and R. W. Menzel. 1981. The reproductive cycle of early setting *Crassostrea virginica* (Gmelin) in the northern Gulf of Mexico, and its implications for population recruitment. *Biol. Bull.* **160**:80-88.

Henderson, J. T. 1929. Lethal temperatures of Lamellibranchiata. *Can. Biol.* **4**(25):397-411.

Hidu, H. 1969. Gregarious setting in the American oyster, *Crassostrea virginica* (Gmelin). *Ches. Sci.***10**:85-92.

Hidu, H. and H. H. Haskin. 1971. Setting of the American oyster related to environmental factors and larval behavior. *Proc. Natl. Shellfish. Assoc.* **61**:35-50.

Ingle, R. M. and C. E. Dawson. 1953. A survey of Apalachicola Bay. Florida Board of Conservation, Marine Research Laboratory, Education Series 10.

Ingle, R. M., E. A. Joyce Jr., J. A. Quick Jr., and S. W. Morey. 1971. Basic considerations in the evaluation of thermal effluents in Florida (Preface). In *A preliminary investigation: the effect of elevated temperature on the American oyster, Crassostrea virginica (Gmelin).* J. A. Quick Jr. (ed.), pp. vii-viii. Florida Department of Natural Resources, Professional Papers Series No. 15.

Keck, R., D. Maurer, J. C. Kauer, and W. A. Sheppard. 1971. Chemical stimulants affecting larval settlement in the American oyster. *Proc. Natl. Shellfish. Assoc.* **61**:24-28.

King, T. L. and J. D. Gray. 1989. Allozyme survey of the population structure of *Crassostrea virginica* inhabiting Laguna Madre, Texas, and adjacent bay systems. *J. Shellfish Res.* **8**(2):448.

Krantz, G. E. and J. F. Chamberlin. 1978. Blue crab predation on cultchless oyster spat. *Proc. Natl. Shellfish. Assoc.* **68**:38-41.

Loosanoff, V. L. 1949. On the food selectivity of oysters. *Science.* **110**:122.

Loosanoff, V. L. 1953. Behavior of oysters in water of low salinities. *Proc. Natl. Shellfish. Assoc.* **43**:135-151.

Loosanoff, V. L. 1965. The American or eastern oyster. *U.S. Fish and Wildlife Service Circ.* **205**:1-36.

Lowe, J. I., P. R. Parris, J. M. Patrick Jr. and J. Forester. 1972. Effects of the polychlorinated biphenyl aroclor 1254 on the American oyster, *Crassostrea virginica. Mar. Biol.* **17**:209-215.

MacInnes, J. P. 1981. Response of embryos of the American oyster, *Crassostrea virginica*, to heavy metal mixtures. *Mar. Environ. Res.* **4**:217-227.

Mahoney, B. M. S. and G. Noyes. 1982. Effects of petroleum on feeding and mortality of the American oyster. *Archives Environ. Contam. Tox.* **11**(5):527-531.

Manzi, J. J., V. G. Burrell, and W. Z. Carlson. 1977. A comparison of growth and survival of subtidal *Crassostrea virginica* (Gmelin) in South Carolina saltmarsh impoundments. *Aquaculture* **12**:293-310.

Martin, N. 1987. Raw deals. *Texas Shores* **20**(3):4-8.

May, E. B. 1971. A survey of the oyster and oyster shell resources of Alabama. *Alabama Mar. Res. Bull.* **4**:1-53.

May, E. B. 1972. The effect of floodwater on oysters in Mobile Bay. *Proc. Natl. Shellfish. Assoc.* **62**:67-71.

Menzel, R. W. 1951. Early sexual development and growth of the American oyster in Louisiana waters. *Science* **113**(2947):719-721.

Menzel, R. W. and F. E. Nichy. 1958. Studies of the distribution and feeding habits of some oyster predators in Alligator Harbor, Florida. *Bull. Mar. Sci. Gulf Caribb.* **8**(2):125-145.

Menzel, R. W., N. C. Hulings, and R. R. Hathaway. 1966. Oyster abundance in Apalachicola Bay, Florida, in relation to biotic associations influenced by salinity and other factors. *Gulf Res. Repts.* **2**(2):73-96.

Menzel, R. W. and S. H. Hopkins. 1956. Crabs as predators of oysters in Louisiana. *Proc. Natl. Shellfish. Assoc.* **46**:177-184.

Overstreet, R. M. 1978. Marine maladies? Worms, germs, and other symbionts from the northern Gulf of Mexico. *Mississippi-Alabama Sea Grant Consortium Report No. MASGP-78-021*:1-144.

Pearse, A. S. and C. W. Wharton. 1938. The oyster "leech" *Stylochus inimicus* Palombi associated with oysters on the coasts of Florida. *Ecol. Mono.* **8**(4):605-655.

Pearson, J. C. 1929. Natural history and conservation of redfish and other commercial sciaenids of the Texas coast. *Bull. Bur. Fish.* **44**:129-214.

Pierce, M. E. and J. T. Conover. 1954. A study of the growth of oysters under different ecological conditions in Great Pond. *Biol. Bull.* **107**:318.

Pollard, J. F. 1973. Experiments to reestablish historical oyster seed grounds and control the southern oyster drill. *Louisiana Depart. Wildl. Fish., Tech. Bull.* **6**:1-82.

Powell, E. H. and G. Gunter. 1968. Observations on the stone crab *Menippe mercenaria* Say, in the vicinity of Port Aransas, Texas. *Gulf Res. Repts.* **2**:285-299.

Quayle, D. B. 1988. Pacific oyster culture in British Columbia. *Can. Bull. Fish. Aquat. Sci.* **218**:1-241.

Renzoni, A. 1975. Toxicity of three oils to bivalve gametes and larvae. *Mar. Poll. Bull.* **6**:125-128.

Roberts, M. H. Jr., R. J. Diaz, M. E. Bender and R. J. Huggett. 1975. Acute toxicity of chlorine to selected estuarine species. *J. Fish. Res. Bd. Can.* **32**(12):2525-2528.

Rowe, D. R., L. W. Canter, P. J. Snyder, and J. W. Mason. 1971. Dieldrin and endrin concentrations in a Louisiana estuary. *Pesticides Monitoring J.* **4**(4):177-183.

St. John, L. and E. W. Cake Jr. 1980. Observations on the predation of hatchery-reared spat and seed oysters by the striped burrfish, *Chilomycterus schoepfi* (Walbaum) (Diodontidae). *Proc. Natl. Shellfish. Assoc.* **70**:130.

Schlesselman, G. W. 1955. The Gulf Coast oyster industry of the United States. *Geo. Rev.* **45**(4):531-541.

Schimmel, S. C., P. R. Parrish, D. J. Hansen, J. M. Patrick Jr., and J. Forester. 1975. Endrin: effects on several estuarine organisms. *Proc. 28th Ann. Conf. Southeast. Assoc. Game Fish Comm.* Pp. 187-194.

Scott, G. I. and W. B. Vernberg. 1979. Seasonal effects of chlorine produced oxidants on the growth, survival and physiology of the American oyster *Crassostrea virginica* Gmelin. *Marine Pollution: Environmental Responses.* Pp. 415-435.

Shumway, S. E. 1982. Oxygen consumption in oysters: an overview. *Mar. Biol. Let.* **3**:1-23.

Soniat, T. M. 1996. Epizootiology of *Perkinsus marinus* disease of eastern oysters in the Gulf of Mexico. *J. Shellfish Res.* **15**(1):35-43.

Soniat, T. M. and S. M. Ray. 1985. Relationships between possible available food and the composition, condition, and reproductive state of oysters from Galveston Bay, Texas. *Cont. Mar. Sci.* **28**:109-121.

Sparks, A. K., J. L. Boswell, and J. G. Mackin. 1958. Studies on the comparative utilization of oxygen by living and dead oysters. *Proc. Natl. Shellfish. Assoc.* **48**:92-102.

St. Amant, L.S. 1959. Successful use of reef oyster shells (mud shells) as oyster cultch in Louisiana. *Proc. Natl. Shellfish. Assoc.* 49:71-76.

Stanley, J. G. and M. A. Sellers. 1986. Species profiles: life histories and environmental requirements of coastal fishes and invertebrates (Gulf of Mexico) — American oyster. *United States Fish and Wildlife Service Biological Report 82 (11.64).* U.S. Army Corps of Engineers, *TR-EL-82-4*:1-25.

Stenzel, H. B. 1971. Oysters. In *Treatise on Invertebrate Paleontology. Part N, Volume 3, Mollusca 6.* K. C. Moore (ed.), pp. N953-N1224. Geological Society of America, Boulder, Colorado, and University of Kansas, Lawrence.

Tubiash, H. S., P. E. Chanley, and E. Leifson. 1965. Bacillary necrosis, a disease of larval and juvenile bivalve mollusks. I. Etiology and Epizootiology. *J. Bacteriology.* **90**:1036-1044.

Turgeon, D. D., A. E. Bogan, E. V. Coan, W. K. Emerson, W. G. Lyons, W. L. Pratt, C. F. E. Roper, A. Scheltema, F.G. Thompson, and J.D. Williams. 1988. *Common and Scientific Names of Aquatic Invertebrates from the United States and Canada: Mollusks.* American Fisheries Society Special Publication 16.

Vanderzant, C., C. A. Thompson Jr., and S. M. Ray. 1973. Microbial flora and level of *Vibrio parahaemolyticus* in oysters (*Crassostrea virginica*), water and sediment from Galveston Bay. *J. Milk Food Tech.* **36**:447-452.

White, M. E., E. N. Powell, and C. L. Kitting. 1984. The ectoparasitic gastropod, *Boonea (= Odostomia) impressa:* population ecology and the influence of parasitism on oyster *Crassostrea virginica* growth rates. *Mar. Ecol.* **5**(3):283-299.

White, M. E., E. N. Powell, S. M. Ray, and E. A. Wilson. 1987. Host-to-host transmission of *Perkinsus marinus* in oyster (*Crassostrea virginica*) populations by the ectoparasitic snail *Boonea impressa* (Pyramidellidae). *J. Shellfish Res.* **6**:1-5.

Biology of Certain Commercial Mollusk Species: Scallops

Robert A. Fisher

INTRODUCTION

Scallops are highly valued benthic marine mollusks, belonging to the class Bivalvia and the family Pectinidae. In the United States, there are seven species of scallops commercially harvested which include the Atlantic sea scallop (*Placopecten magellanicus*); the weathervane, or Pacific sea scallop (*Patinopecten caurinus*); the bay scallop, (*Argopecten irradians* spp.); the calico scallop (*Argopecten gibbus*); the pink scallop (*Chlamys rubida*); the spiny scallop (*Chlamys hastata*); and the rock scallop (*Crassadoma gigantea*, formally *Hinnites multirugosus*). Of these species, the sea, bay, and calico scallops are currently of greatest economic importance, and will be the focus of this chapter. Pink, spiny, and rock scallops are currently harvested and marketed only in localized areas along the Pacific Coast of the United States, and are not widely commercialized.

The scallop fishery in the United States may be considered a wasteful one because only the adductor muscle is commonly marketed. The adductor muscle makes up approximately one-third (by weight) of the total fleshy part of the scallop. Development of markets using more of the scallop body should be encouraged for added economic gain and wise resource utilization. International markets currently exist for mature scallop roe, either attached to the adductor muscle (roe-on), or loose. The thickened folds of the mantle of the sea scallop also warrant consideration for marketing in the form of "clam" strips. The smaller scallop species available in the United States have been marketed whole. The bay scallop industry has recently seen an increase in the utilization of the whole scallop, with efforts targeting the "steamer" market. Spiny and pink scallops have supported small, localized steamer markets in Washington and British Columbia for over a decade. When utilizing the roe or the whole body of any mollusk, the water quality of the harvesting area should be of primary concern. Pollutants and PSP (paralytic shellfish poisoning) organisms should be routinely monitored. With expanding scallop market forms, and the advancement of commercial scallop-culturing programs, an understanding of scallop anatomy and functional biology is necessary. Knowledge of scallop reproductive biology is important for the proper management of this natural resource.

GENERAL BIOLOGY

The biology and functional anatomy of scallops has a solid foundation in the literature (Sastry, 1979, 1975; Naidu, 1970; Bullock, 1965; Bourne, 1964; Pierce, 1950; Coe, 1943; Gutsell, 1931; Kellogg, 1915, 1892; Dakin, 1910a, 1910b; Drew, 1906) and has been thoroughly summarized in *Scallops: Biology, Ecology and Aquaculture*, edited by Shumway (1991). A brief description of general scallop biology and anatomical features is presented here, followed by a discussion of selected species.

SHELL

As representatives of the class Bivalvia (also known as Pelecypoda or Lamellibranchiata), scallops possess two valves (shells or discs) which are hinged along a margin defined by triangular auricles, or ears (Figure 1). The right valve is the bottom valve, or the side on which the scallop lies; the left valve is on the top. In most Pectinids, there is a notch (byssal notch) in the right, or in both, anterior auricle(s) that permits passage of the foot and byssus. Anatomically, "dorsal" refers to the hinge-line margin of both valves, and "ventral" to the margins of both valves farthest from the hinge-line. The valves are attached and articulate dorsally by means of a weak ligament running along the hinge-line and by a stronger, highly elas-

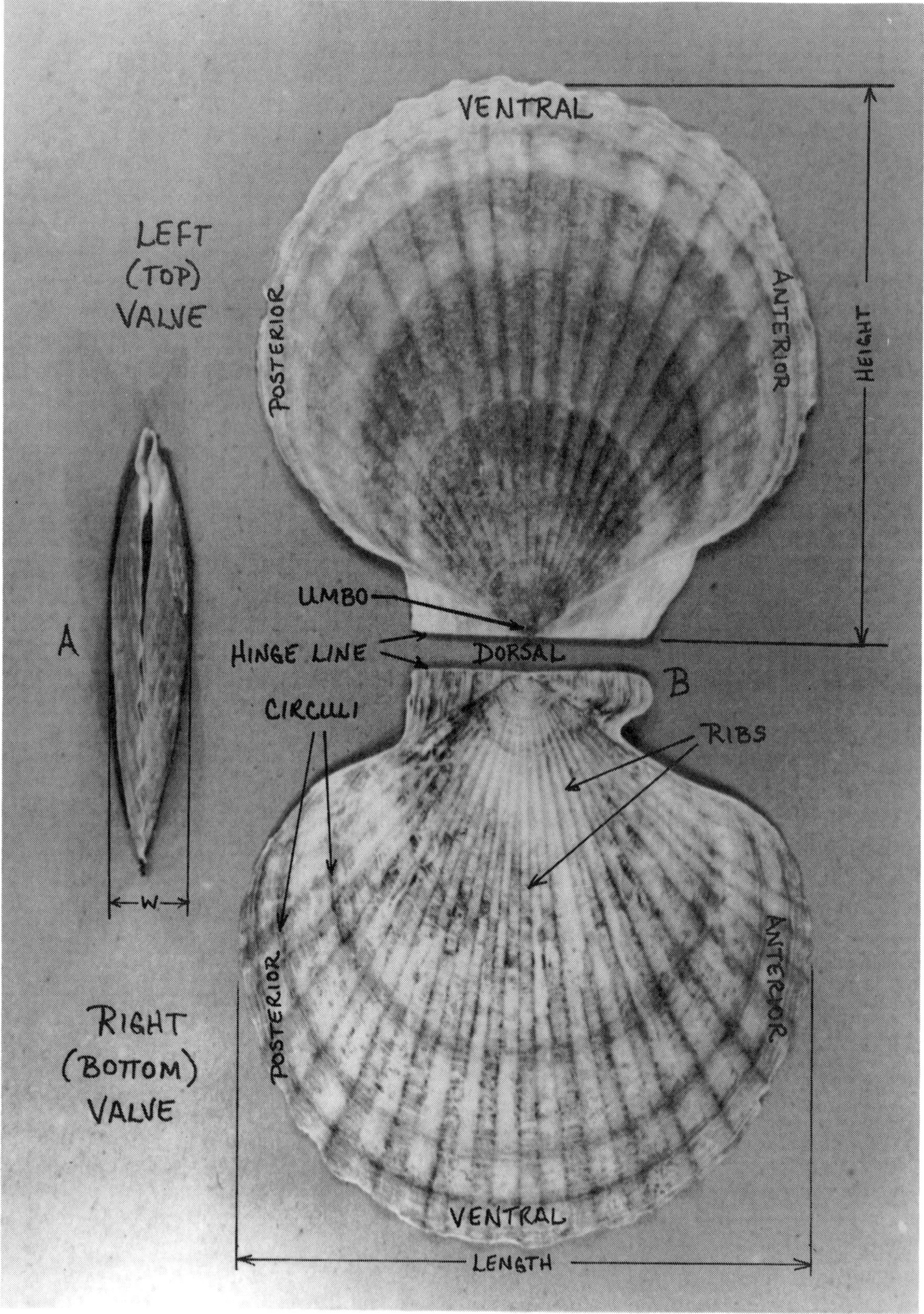

*Figure 1. Scallop measurement terms used in literature. A. Anterior view of closed weathervane scallop (*P. caurinus*); W=width, or inflation. B. Exterior of opened weathervane scallop.*

tic ligament (resilium) located just interior of the umbo. Additionally, the valves are connected internally by a large adductor muscle positioned near the center of each valve. The valves are closed by the action of this muscle, and are opposed by the action of the ligament (resilium), which tends to pull the valves apart.

The exterior surfaces of the valves possess both radial ribs, which extend outward from the umbo (the oldest part of the shell) to the valve margins, and concentric lines (circuli) resulting from the addition of new shell growth along the valve margins. The spacing between and thickness of circuli demarcate seasonal growth periods; therefore, the circuli are used to age scallops (Tan et al., 1988; Stevenson and Dickie, 1954). Oxygen isotope studies on *P. magellanicus* shells have also confirmed that growth lines are annual events, consistent with biological evidence (Tan et al., 1988). Annular rings are often difficult to discern, either being very weak or masked by false rings, or "shock rings." Scallops are sensitive to abrupt physical and environmental disturbances which, when encountered frequently, cause them to lay down concentric rings, or shock rings, on the shell (Posgay, 1950). The elastic hinge ligament (resilium) is also used for aging scallops (Merrill et al., 1965).

The left valves (top shell) of mid- to older-age scallops are often fouled by boring sponges (Medcof, 1949) and/or annelid worms (Merrill, 1967; Medcof and Bourne, 1964; Wells and Wells, 1962) which settle on the shell surface. The boring organisms penetrate the shell, weakening the shell structure, and frequently cause blisters to develop on the interior surface of the shell (Figure 2). Infested scallops expend energy to repair these physical damages at the expense of soft body components. Heavily-infested scallops display reduced meat yields. Meat quality is significantly compromised. Off-color, flaccid meats are commonly associated with infested scallops. Young *P. magellanicus* are not generally affected, probably because they are more active swimmers, which inhibits the fouling organisms from settling on their shells. Older scallops are less active, preferring to settle into depressions or furrows.

The interior surfaces of most scallop valves are smooth and white, unless altered by fouling organisms. Near the center of both valves is a circular depression, or scar, where the adductor muscle attaches. About one-third of the way from the valve margins there is a shallow narrow groove, the pallial line, which marks the attachment site of the mantle.

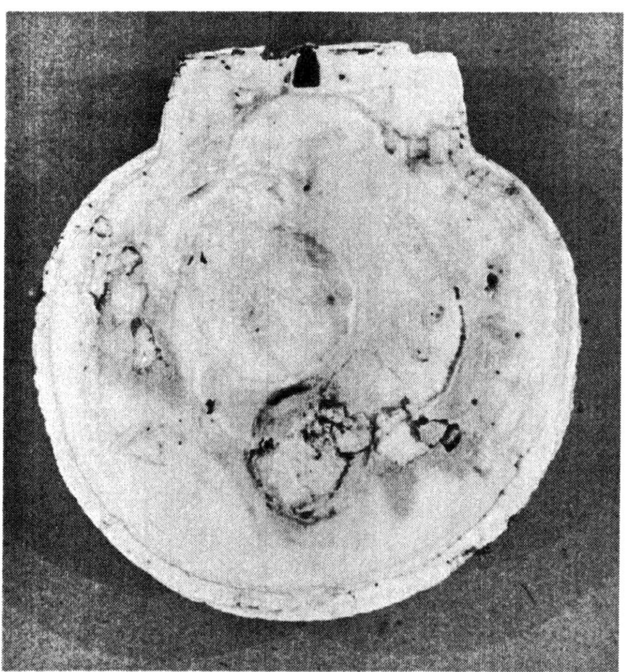

Figure 2. Internal surface of top (left) valve from a heavily fouled sea scallop, P. magellanicus. *Internal blisters resulting from boring organisms that penetrate the shell.*

Size reference to scallops is typically a measurement of shell height — a straight-line measurement of the greatest distance between the umbo — which is centrally located on the hinge-line and the ventral margin (Figure 1). Other common scallop measurements are shell length (the farthest distance measured between the anterior and posterior margins parallel to the hinge-line) and width (or inflation) [the greatest distance between the exterior surfaces of the top and bottom valves taken perpendicular to the line of junction of the two valves (commissure)]. Within the Pectinidae family, there is a great deal of interspecific variation of valve size, shape, convexity, sculpturing, and color. Of the species discussed here, the valves are discoidal to fan-shaped with the hinge-line defined by prominent anterior and posterior ears (auricles); the right valve is typically light-colored while the left valve maintains the species-specific color variations, and the valves range from nearly equiconvex (calico scallops) to largely left-convex (sea scallop, where the left valve is more convexed than the right valve).

Adductor Muscle

Unlike members of Bivalvia that possess anterior and posterior adductors such as clams (dimyarian bivalves), scallops are among the monomyarian bivalves which possess a single, more centrally-positioned posterior adductor muscle. The anterior adductor muscle in scallops is lost, and the posterior adductor migrates more centrally during the late larval stage of development and shortly after settlement (Cox, 1969). Due to the central location of the scallop adductor, it is commonly used to describe orientation of other scallop soft body parts (i.e., anteriorly or posteriorly located to the adductor muscle).

The scallop adductor muscle is divided by connective tissue into two portions which differ in size, structure, and function. The larger, rounded portion, referred to as the quick muscle, is cross-striated muscle, and is capable of multiple rapid contractions for relatively short periods of time. This phasic muscle functions, in part, during the rapid opening and closing of the valves for locomotion. In addition, this muscle powers the valve "clapping" action used, in part, to help create depressions in the substrate in which the scallop settles, and to facilitate the removal of feces and pseudofeces from the mantle cavity. The smaller, crescent-shaped portion of the adductor, usually referred to as the catch muscle by biologists and "sweet-meat" by industry, is a smooth muscle capable of providing a sustainable force with little expenditure of energy. The catch muscle, in part, is responsible for bringing and holding the valves together for long periods of time, thus closing the scallop in an attempt to escape predation. In the adult scallop, the adductor muscle comprises about a third of the weight of the soft body parts, and 10 to 18 percent of the total weight (Mottet, 1979). Usually only the larger portion of the adductor is represented on the scallop market — the catch muscle either being lost during processing and handling due to its weak connection to the larger muscle, or removed by retailers/restaurateurs due to its tough, chewy texture. In some localities, the adductor muscle is referred to as the "scallop eye," probably because of its anatomical shape and position within the scallop.

The adductor muscle is usually white to off-white in color; however, muscle discoloration does occur. Older sea scallops, and scallops with heavy shell infestations of fouling organisms, frequently display muscle coloration from light brown to gray. Orange-red adductor muscles are observed in some sexually mature female scallops of dieocious species (Bourne and Bligh, 1965), and yellow-orange muscles are observed periodically in sexually mature hermaphroditic species, both of which are associated with the color of the female gonadal tissue. Coloration is due to carotenoid pigments from marine plants, which accumulate in the scallop during feeding (Kantha, 1989). Carotenoid pigments associated with the roe become dispersed to surrounding somatic tissue, including the adductor muscle. In the Atlantic sea scallop, *P. magellanicus*, orange-tinged meats occur throughout its range, with the intensity of the color increasing with gonad maturation. The deepest-colored meats are found at periods of peak spawning, when the gonads are ripe. Orange-colored sea scallop meats are commonly referred to as "salmon scallops" in their northern range, and "pumpkins" in their southern range. Though this coloration in scallop meats is a natural phenomenon, and has no apparent difference in nutritional quality from white meats, domestic and international buyers and regulatory inspectors frequently reject these meats, citing visual quality defects. The adductor muscle has been reported to be an important energy storage site (glycogen and protein) that is utilized during the reproduction cycle (Barber and Blake, 1981).

Mantle

Internally, scallop anatomy is the same among pectinid species (Figure 3). The inner surface of each valve is lined with thin transparent tissue, the mantle lobe. The two lobes, one corresponding to each valve, are attached dorsally to the ligament along the hinge-line, and to the adductor muscle and digestive gland (Drew, 1906), thereby enclosing the visceral mass. From their dorsal attachment, the lobes extend ventrally to the valve margins, surrounding a relatively large space, the mantle (pallial) cavity. The margins of the mantle lobes thicken, and form three folds, each with a definite function. The outermost fold (shell fold) is responsible for shell formation; the middle fold

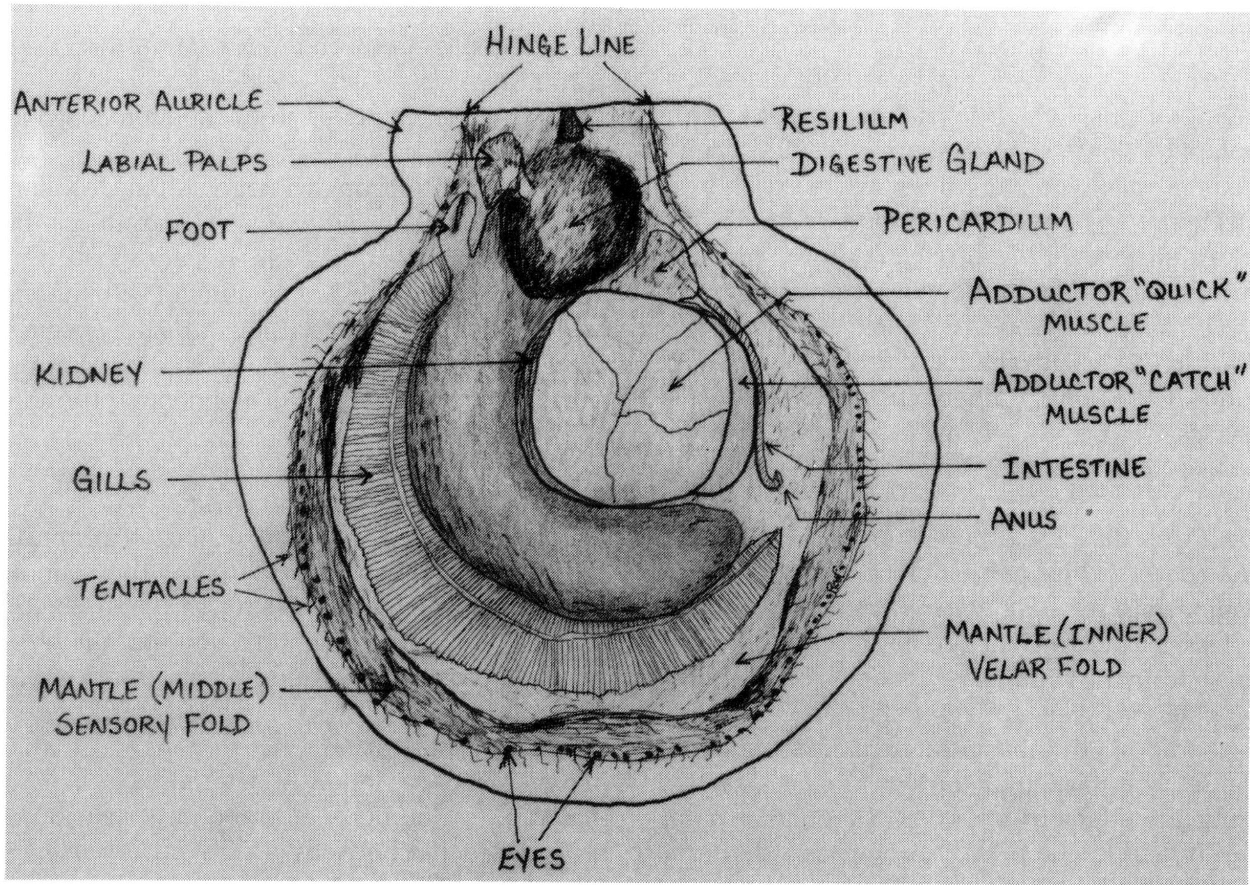

Figure 3. Internal anatomy of the sea scallop, P. magellanicus, *with left valve, mantle, and gills removed.*

(sensory fold) bears sensory organs such as tactile tentacles and numerous eyes; and the innermost fold, or mantle curtain, contains the pallial muscles which attach the mantle to the valves along the pallial line (a line or narrow band on the interior of valve positioned between the adductor muscle scar and the valve margin). The pallial muscles function by withdrawing the mantle from the margins when the valves are closed, and controlling the water flow into and out of the mantle cavity during periods of feeding and locomotion.

Gills

There are two plicate (folded) gills in the scallop, composed of long, slender, ciliated filaments which are folded over to form a W-shape in cross-section. The gills, or ctenidia, function in respiration, food collection, and the subsequent transport of food to the mouth along the ciliated gill filaments. The gills are situated just interior of the mantle lobes of the right and left valves, lying ventral to the adductor muscle in an anterior-to-posterior position. The gills are permanently attached by a thin membrane dorsally to the adductor muscle near the muscle-valve insertion point. The gills' ventral margin is not permanently attached. During periods of feeding, or at rest, the ciliated inner and outer folds of the gills ventral margin adhere to the gonad (visceral mass) medially and to the mantle laterally by interlocking cilia (Beninger and Le Pennec, 1991). By such connections, the gills effectively divide the mantle cavity into inhalant (ventral) and exhalant (dorsal) chambers, which maximize water transport for feeding and respiration. These connections are easily broken during periods of swimming, so as not to damage the gills during the rapid movement of water through the mantle cavity during the swimming action, but are readily re-connected when at rest. Water movement within these chambers during feeding is primarily the result of the beating action of cilia that cover the filaments.

Feeding

Scallops are active ciliary suspension feeders, relying on suspended detrital material and phytoplankton as their food source. Feeding begins with the opening of the valves, which exposes the gills to moving water. The amount of food available to suspension-feeding bivalves is a function of the volume of water transported across the gills, as well as the efficiency with which particles are retained by the gills (Bricelj and Shumway, 1991). Caddy (1968) and Mathers (1976) have reported that scallops position themselves within a current to maximize feeding. Scallops may also create their own water movement to facilitate feeding by the beating of cilia located along the gill filaments. The cilia covering the ctenidia vibrate in a concerted fashion, producing a water current which is directed into the inhalent opening, passing over the gills within the mantle cavity, and exiting through the exhalant opening. Suspended particulate matter is trapped by cilia and mucus on the gill filaments. The particulate-laden mucus is carried along the edges of the gills to the labial palps (fleshy, leaf-like appendages on each side of the mouth). Along with the gills, the labial palps sort the particles based on their suitability for ingestion. Rejected particles are deposited into the mantle cavity and removed by the exhalant water flow as pseudofeces. Food particles accepted by the palps reach the mouth by way of the oral groove. The oral groove is bordered by enlarged lips, which may function in retaining food particles (Gilmour, 1964).

The scallop digestive system is similar to other bivalves. The food-laden mucus is carried from the mouth to the stomach by cilia lining the esophagus. Extra-cellular digestion takes place in the stomach primarily by the physical and enzymatic action of the crystalline style, a long, narrow, transparent, gelatinous rod which rotates within the stomach pulling in food from the esophagus and grinding it against a chitinized part of the stomach, the gastric shield. The head of the style which contacts the gastric shield is slowly worn away, releasing enzymes that aid in extracellular digestion. When food particles are broken down to sufficient size in the stomach, the food is passed by ciliary action into the dark-colored digestive gland which surrounds the stomach and is the site of intra-cellular digestion. The digestive gland has also been reported as a storage site of food reserves (Sastry, 1968). The intestine (alimentary canal) extends from the stomach midventrally, passes through the digestive gland, loops ventrally through the gonad, and extends along the posterior side of the catch muscle before it terminates at the anus. The anus lies on the posterior side of the adductor muscle with the anal opening directing the feces towards the posterior valve margin, resulting in the feces being removed in the excurrent stream of water coming from the gills (Drew, 1906).

Cardiovascular System

The scallop circulatory system is generally considered to be an open system, where venous blood collects in numerous sinuses. Bivalve blood, or hemolymph, is a clear, transparent fluid which functions, in part, in wound repair, the transport and assimilation of food, excretion, and the conveyance of calcium carbonate for shell formation (Cox, 1969). Circulation is maintained by regular pulsation of the heart, and blood is distributed throughout the body by a network of arteries, capillaries, veins, and sinuses.

The heart is symmetrical with two auricles and one ventricle, and lies in the triangle-shaped pericardium (pericardial cavity) located dorsal to the adductor muscle (Drew, 1909). Two aortas leave the ventricle — a posterior aorta which supplies blood to the intestine and adductor muscle, and the anterior aorta which supplies blood to the rest of the body, including the mantle, digestive gland, stomach, foot, and reproduction organs. The veins collect the blood from the various organs and carry it to sinuses located alongside the kidneys where the blood is filtered and waste products removed. The blood exiting the kidneys supplies the gills and then is returned to the heart. Re-oxygenation of the blood occurs within the blood sinuses and during passage through the gills (Cox, 1969).

Excretion

The excretory system of bivalves functions by the process of filtration, and involves collaboration of the circulatory system, and the secretion and reabsorption functions of the paired kidneys, or nephridia (Martin, 1983). The kidneys, right and left, are elongated, brownish sacs attached to the anterior side of the adductor muscle (Figure 3).

Filtration occurs within the pericardial cavity with which the kidneys communicate. Waste is excreted through the renal opening, exiting into the mantle cavity near the ventral side of the adductor. In scallops, the renal opening also serves as the site where gametes are released from the gonad during spawning. Hence, the renal opening is commonly referred to as the reno-genital opening.

Foot

The pectinid foot in most species is more degenerate than in other bivalves, but may be more or less developed depending on adult life habit. The foot is a worm-like muscle extending from the anterior-dorsal side of the gonad. A single retractor muscle attaches the foot to the left valve at the valve-adductor muscle insertion point (Drew, 1906). The distal end of the foot is cleft, enabling the end to flatten out (sole) and function in the non-swimming locomotion of newly-settled scallops and, in some adult species, in holding the scallop onto a surface while securing byssal attachment. A byssal gland is located at the base of the foot and secretes a proteinaceous material that hardens when exposed to seawater. The hardened material is hair-like (byssal threads), and when bundled together forms the byssus. Some species, such as the spiny and pink scallop, attach themselves to a substrate by byssus throughout their lives; in free-living adult species (bay, calico, and sea scallops), the foot-byssal complex regresses in the juvenile stage (Beninger and Le Pennec, 1991).

Locomotion in free-living scallops is primarily achieved by the rapid closure of the valves, resulting in the propulsion of water from the mantle cavity. In the active swimming response, successive rapid closures of the valves brings water into and out of the mantle cavity, with the muscular inner mantle fold directing the water from the mantle cavity through the restricted openings (gapes) located at each auricle, thus providing a jet-propulsion that moves the scallop through the water with the ventral margin leading. Caddy (1968) reported ground speeds of 67 centimeters/second for distances up to 4 meters in *P. magellanicus*. Wilkens (1991) describes other locomotive responses that enable scallops to move in variable directions as a result of the strength of valve adduction and the direction of the jet current. Once into the water column, currents may serve as a means of migration for small scallops (Baird, 1954).

Nervous System and Sensory Structures

The nervous system in scallops is typical of bivalves, consisting of three pairs of ganglia — the cerebral, pedal, and visceral — and their nerves (Drew, 1906). Due to anatomical modifications in scallops, ganglia size and position differ from other bivalves. With the reductions in the anterior region of the scallop (including the foot, byssal complex, and pedal retractor muscle), the cerebral ganglia are reduced and closely related to the pedal ganglia (Beninger and Le Pennec, 1991). Likewise, with the centralizing of posterior organs (such as the adductor muscle) and the development of numerous sensory structures (such as eyes and tentacles), the visceral ganglia are larger and more complex in scallops; they are referred to as the visceroparietal ganglia by Bullock and Horridge (1965) and the parietovisceral ganglia by Beninger and Le Pennec (1991). The pedal and cerebral ganglia are linked to the visceroparietal ganglia by a pair of cerebro-visceral connections.

The cerebral and pedal ganglia are located ventral to the mouth and are linked by cerebropedal connectives. The cerebral ganglia innervate the anterior part of the mantle, oral palps, and statocysts (otocysts). Statocysts are ciliated organs located within a mass of connective tissue surrounding the pedal ganglia and function in the control of the swimming reflex (Wilkens, 1991). The pedal ganglia innervate the foot and the byssus retractor muscle.

The visceroparietal ganglia are located just beneath the skin on the antero-ventral surface of the adductor muscle. They innervate the visceral and parietal components of scallop internal anatomy, including the posterior part of the mantle and mantle margins (and associated sensory structures), adductor muscle, gills, kidneys, heart, gonads, and intestine. Collectively, the mantle and all its sensory structures are innervated by the circum-pallial nerve, which originates from both the cerebral and visceroparietal ganglia.

Eyes

The major sensory structures of the scallop are the eyes, epithelial sensory cells, the statocyst (statoreceptors), and tentacles. Early work by Dakin (1910a, 1910b), Drew (1906), Buddenbrock

(1915), and Gutsell (1930) provides detailed descriptions and functionality of scallop sensory structures.

The blue-pigmented eyes of scallops are located along the margin of each mantle lobe within the middle fold of the mantle lining. They are positioned at the end of short stalks and are innervated by optic nerves arising from specialized lobes of the visceroparietal ganglia (Dakin, 1910b). Scallop eyes may vary in size and number within and among species (Dakin, 1910a), and also within a single individual, with the upper mantle having more eyes than the lower mantle (Wilkens, 1981; Gutsell, 1930). Within an individual, additional eyes grow as mantle space is provided during scallop growth. Each eye contains a cornea, lens, retina, and optic nerve and is capable of image formation (Wilkens, 1991), allowing for the detection of movement and decreases in light intensity (Gutsell, 1930).

The ciliated epithelial sensory cells serve primarily as receptors for tactile and chemical stimuli (Wilkens, 1991). They are found throughout the scallop epidermis, but are concentrated on the long tentacles of the mantle's middle fold.

Tentacles

Scallop tentacles are located within the folds of the free edge of the mantle lobes. Each tentacle is covered with epithelial sensory cells and supplied with a large nerve originating from the visceroparietal ganglia. Tentacles possess longitudinal retractor muscles and numerous blood spaces (cavities), which serve to lengthen the tentacles by forcing blood into the blood spaces, and to shorten the tentacles by contracting the muscle (Drew, 1906). Short tentacles extend from the shell fold (Moir, 1977) and the inner fold. The function of these short tentacles has not yet been adequately addressed. Within the middle (sensory) fold are long sensory tentacles. The long tentacles function primarily in detecting predators and in exploring the surrounding substrate.

Beninger and Le Pennec (1991) summarize the neurosecretory role of ganglia in several bivalve species. They report that studies have shown the existence of neurosecretory cells in bivalve ganglia which influence reproduction and regulate somatic growth. The relationship of neurosecretion and gametogenesis in *Argopecten irradians* was examined by Blake (1972) and Blake and Sastry (1979).

SCALLOP SEXUALITY AND REPRODUCTION

Sexuality among bivalves is generally characterized by dioecism (separate sexes). Coe (1943) reported that 96 percent of the species in the class Bivalvia have separate sexes. Within the Pectinid family, however, the majority of the species are hermaphroditic, where an individual produces gametes of both the male and the female. Coe (1945) classified various forms of hermaphroditism: functional hermaphroditism (both types of gametes become mature at the same time); consecutive sexuality (an individual is of one sex when young and later changes to the opposite sex); rhythmical consecutive sexuality (an individual repeatedly alternates from one sex to another); and alternative sexuality (an individual unpredictably alternates sexuality between or during a reproductive season).

In functional hermaphrodites, the gonad is divided into a ventral ovary and a dorsal testis (Barnes, 1974). There is a general tendency in hermaphroditic species for protandry, the production of male gametes first, followed by the female gametes. With protandric behavior, the chance of self-fertilization is reduced (Coe, 1945). Hermaphroditism in exclusively dioecious species is rare, but has been identified in *P. caurinus* and *P. magellanicus* by Hennick (1971) and Naidu (1970), respectively.

Sexual dimorphism in dioecous scallops is rarely obvious apart from the examination of the mature or developing gonads. Maturing ovarian and spermatic gonadal tissue is most often differentiated by color (Coe, 1945). Ovarian tissue becomes pink, yellowish, or red, and spermatic tissue becomes white or cream-colored during gonadal maturation, thus facilitating identification of gonad tissue type. Shell size, color, or shape has not been used to sex scallops. Females tend to be more numerous than males in dioecous species and become more numerous as the age of the population increases (Fretter and Graham, 1964).

Gonad and Gametogenesis

The gonad is a crescent-shaped, sac-like structure that varies considerably in size and color throughout its reproduction period. It lies in a dorso-ventral position curving around the anterior margin of the adductor muscle and partially occluding the kidney (Figure 3). The gonad is attached dorsally to the adductor muscle with the ventral margin, or distal end, free. As noted by Beninger and Le Pennec (1991), the gonadal tissue in some species (for example, *H. multirugosus* and *P. magellanicus)* may extend dorsally into the labial palps or digestive gland. In the processing of *P. magellanicus* for roe-on product, the digestive gland has to be physically severed from the gonad, with care taken to completely remove digestive gland tissue to prevent contamination of the product.

The gonad is filled with numerous follicles (small sacs), each lined with gamete-producing cells, or gametogenic cells, which receive their nourishment directly from the surrounding vesicular connective tissue (Coe, 1943). Gametogenesis — the production of sex cells, of sperm (spermatogenesis) and eggs (oogenesis) — is similar for all bivalve mollusks (Naidu, 1970). As the gametes are produced and mature within the follicles, the gonad enlarges and takes on sex specific colorations. In some hermaphrodites both eggs and sperm may be formed in the same follicle, while in others gametes are produced in separate follicles (Coe, 1945). Gametogenesis is reported to be influenced by such factors as water temperature and food availability (MacDonald and Thompson, 1986, 1985; Sastry, 1968, 1963). The scallop gametogenic cycle may occur on an annual, semiannual, or more continual basis (Barber and Blake, 1991).

During gametogenesis a great deal of energy is expended. Energy reserves, in the form of lipids, glycogen, and proteins, are accumulated and stored in various soft-body parts prior to gametogenesis, and then expended during gamete production (Barber and Blake, 1981; Bayne, 1976). Reproductive development has been correlated with a decline in adductor muscle weight (Kirkley et al., 1991; Barber and Blake, 1983; Robinson et al., 1981). Faveris and Lubet (1991) report that, along with the adductor muscle, the mantle and digestive gland probably also contribute energy reserves for gonad development. Somatic growth is limited during periods of gonadal maturation, but increases after spawning. The overall quality of the adductor muscle as a marketed product may also become compromised during reproductive development and spawning. Meats shucked from *P. magellanicus* during spawning periods generally exhibit reduced meat yields per given shell height, higher moisture, and lower protein contents, and a loss of muscle integrity (Fisher, 1993, unpublished data). As scallops grow older, there is a shift in priority of energy partitioning from somatic growth towards gonadal development, resulting in decreased shell-heights and increased fecundity (MacDonald and Bourne, 1987).

As a marketed product, the mature roe of dioecious scallops (primarily females) are currently targeted for international markets. The gonads of scallops undergo significant seasonal changes in size and color, thus limiting the roe market to specific time periods. For this reason, and to provide a means for determining sex and state of reproduction, a brief description of the stages of gonad maturation seems warranted. Naidu (1970) describes ten stages of gonad development for *P. magellanicus*:

Undifferentiated (immature): Gonad inconspicuous and often barely visible, transparent and colorless, and only discernible by the intestinal loop within.

Differentiated (immature): Gonad still small and inconspicuous; flat and angular; transparent and colorless; intestinal loop distinct, follicles not visible. Beginnings of differentiation of follicles.

Developing (virgin): Gonad slightly larger, translucent but retains flat angular configuration. Loop of alimentary canal (intestine) visible through outer wall of gonad. Reproductive elements just visible to the naked eye; no visible differentiation into testis or ovary.

Maturing I (differentiated): Egg or early sperm development first evident. Gonad slightly larger but remaining flat and angular. Clear differentiation into testis (whitish) and ovary (light orange). Loop of alimentary canal still visible. Follicles small and sparse, substance of gonad loose.

Maturing IA (recovering): Gonad larger, depending on size of scallop. First recognizable stage of recovery after spawning. Flabby, containing much free water; assuming characteristic brighter colors (testis, pale white; ovary, light orange). Follicles larger and denser; loop of alimentary canal visible, sometimes protruding from the gonad surface.

Maturing II (filling): Gonad larger and thicker but somewhat flabby because of water within the gonad. Testis white, ovary light red; follicles clearly visible. Alimentary canal not usually visible; distinct visceral arteries.

Maturing III (half-full): Gonad still larger and thicker, firmer, becoming rounded in outline and containing little free water. Distal end pointed. Follicles packed and gonad appearing granular; testis creamy white and ovary brick-red. Loop of alimentary canal not usually apparent. Visceral arteries becoming prominent within gonad.

Mature (full): Gonads attain maximum size; thick, rounded and plump, containing no free water and rounded at tip. Follicles packed, assuming intense coloration; testis cream, and ovary brick-red to bright coral-red. Gonad surface smooth and glossy. Cut surfaces usually gaping, exposing reproductive elements which now have a thick gelatinous consistency. Alimentary canal indiscernible except when near gonad wall. Well-defined visceral arteries and veins.

Spawning (initial and partially spent): Gonads retain differentiation. Tissue becoming dull, angular, and flabby depending on the number of follicles emptied. During spawning peak the kidneys are often gorged with sex products. Gonad outline collapses if pierced, gonad shriveling up and losing its free water. Visceral mass and gills of females sometimes tinged orange in spawning females. Gonads of spawning males frequently show patches of translucent tissue, and regions of comparatively opaque tissue, indicating empty and full follicles respectively. Tip of gonad becoming progressively more pointed as spawning proceeds.

Spent: Gonad considerably shrunken in volume; generally dull, flaccid, and fawn-colored; containing much free water which is discharged when the gonad wall is pierced. Visible differentiation into testis and ovary sometimes lost. Un-emptied or partially spent follicles stand out against the dull background tissue as specks, especially near proximal end of gonad. Visceral arteries and veins on gonad no longer visible. Gonad tip becomes pointed. Loop of index alimentary canal usually evident.

Gonad maturation in hermaphroditic species follows the same gamete developmental stages as dioecious species, but the coloration of mature gonads may differ. Sastry (1966, 1963) examined gonadal condition and coloration during gametogenesis in *A. irradians*. At the initiation of gametogenesis in the bay scallop, the gonad is covered by a dark membrane. As the gametes mature into cream-colored testis and orange-colored ovaries, the dark membrane is lost. After spawning, the spent gonads become shrunken and pale brown in color (Sastry, 1966). Scallops' gonad color, as well as their size and shape, have been widely used to assess their stage of reproduction.

Quantitative description of scallop gonadal development can be performed using several techniques. Microscopically, using histologically prepared gonadal tissue, developing egg diameter can be monitored and gamete volume fractions (proportion of gonad occupied by gametes) can be derived (MacDonald and Thompson, 1986). Macroscopically, scallop gonad development can be quantified by generating a gonadal index, or gonosomatic index (Dibacco et al., 1995; Giese and Pearse, 1974). This index examines the weight relationship of the gonad to the total soft body, weight, expressed as a percent:

Gonadal Index =
gonad weight X 100/total soft body weight

Somatic body component indices are frequently used to indicate periods of gametogenesis versus periods of animal growth. Seasonal changes in the digestive gland have been correlated to gonadal maturation. Sastry (1966) indicated that a decline in the digestive gland index of *A. irradians* occurs during the period of gonad growth, and becomes higher than the gonadal index during resting periods in the reproduction cycle. Further, an inverse relationship between adductor muscle weight and gonad weight has been demonstrated for numerous species (Lauren, 1982; Robinson et al.,1981; Barber and Blake, 1981; Taylor and Venn, 1979).

Scallop fecundity, or the innate potential reproductive capacity of an individual scallop, has been related to food availability, or the amount of energy available after maintenance requirements have been met (Barber and Blake, 1991). Fecundity was observed to be greater for in-shore populations of *P. caurinus* and *P. magellanicus* (Barber et al., 1988; MacDonald and Thompson, 1986) than off-shore populations. Barber et al. (1988) and MacDonald and Bourne (1987) suggest that gametogenesis and fecundity are influenced by factors which vary with depth, such as food availability and temperature. Other factors related to the total number of gametes produced and released at spawning would include the species involved, the age and size of the individual, and species-specific gamete size. In long-lived scallop species, such as *P. magellanicus* and *P. caurinus*, reproductive effort increases and somatic growth decreases with age (Thompson and MacDonald, 1991). Further, relative fecundity in scallops is expected to be proportional to gonad size, thus the use of gonadal weight indices.

Sexual Maturity and Spawning

Sexual maturity in scallops is generally reached within one to two years in short-lived species, and two to three years in longer-lived species of temperate waters. Bay and calico scallops reach sexual maturity within their first year and successful spawning occurs, while *P. magellanicus* may reach sexual maturity at age one but initial spawning does not occur until the second year (Posgay, 1979; Naidu, 1970). In many bivalves sexual maturity is a matter of size rather than age (Quayle and Bourne, 1972). However, Hennick (1970) reported that age, not growth or size, is more directly related to time of maturity in *P. caurinus*. Of the species of commercial importance in the United States, age and shell height (size) at sexual maturity has been reported as follows: *A. gibbus*, 71 days, 20 mm (Miller et al., 1981); *A. irradians*, 12 months, 55 to 60 mm (Gutsell, 1930); *P. magellanicus*, one year, 23 to 75 mm (Naidu, 1970); and *P. caurinus*, three to four years, 70 mm (Hennick, 1970).

Various exogenous (temperature, salinity, and food supply) and endogenous (age, nutrient reserves, and neural secretions) factors determine the timing and duration of reproduction for a particular species at a particular location and time (Barber and Blake, 1991). Spawning may occur at any time of year, with some species spawning twice annually. Generally, populations from temperate waters spawn in late spring and early fall. Protracted spawns within populations frequently occur in response to environmental factors (MacDonald and Thompson, 1988; Roe et al., 1971; Hennick, 1970; Naidu, 1970). Naidu (1970) reports that the follicles within an individual *P. magellanicus* may not mature at the same time, thus allowing for protracted spawns. Langton et al. (1987) suggest that the timing of spawn can determine year class recruitment.

Spawning has been reported to be triggered by increasing or decreasing water temperatures (Taylor and Capuzzo, 1983; Miller et al., 1981; Castagna and Duggan, 1971; Sastry, 1963; Gutsell, 1930); lunar periodicity (Dickie, 1955; Amirthalingam, 1928), and strong winds and wave action (Naidu, 1970). In laboratory experiments on *P. magellanicus*, it was observed that spawning could be induced by first slowly increasing water temperature, then quickly dropping it (Posgay, 1953, 1950). In natural populations of *P. magellanicus*, Naidu (1970) reports spawning occurring with a rise of water temperature in the spring, and also with a decrease in water temperature in the fall. Similarly, the northern bay scallop population (*A. i. irradians*) spawns in conjunction with increasing water temperatures (Bricelj et al., 1987; Belding, 1910). The southern bay scallop population (*A. i. concentricus*) spawns with decreasing temperatures (Barber and Blake, 1983; Gutsell, 1930). If scallops are not stimulated to spawn, the reproductive elements degenerate and become reabsorbed within the gonad.

As in most bivalves, scallops are broadcast spawners which release their gametes into the surrounding water where fertilization takes place. The eggs and sperm are evacuated from the gonads through the collecting tubules via ciliated action. Evacuation of gametes may also be assisted by contractions of muscular connective tissue located in the outer layer of the gonad (Naidu, 1970; Mason, 1958). The gametes move into the kidney and are expelled through the reno-genital opening into the mantle cavity, from which they are removed by the exhalant water flow to the surrounding water. Scallops may facilitate the expulsion of the gametes by "clapping" their valves (Bourne, 1964). In hermaphroditic species, the eggs and sperm are typically released at different times to help prevent self-fertilization. In the bay scallop, spawning is generally protandrous: sperm is released before the eggs (Loosanoff and Davis, 1963). However, there is some evidence of self-fertilization (Castagna and Duggan 1971; Turner 1957).

All pectinids investigated undergo similar embryonic and larval developmental stages which include the formation of a blastula and gastrula first, followed sequentially by the trochophore, veliger, and pediveliger stages. General scallop early life history is illustrated in Figure 4 for *A. irradians*. Egg dimensions for commercial species are summarized by Cragg and Crisp (1991), with egg diameters of domestic scallops ranging from 60 to 105 micrometers. Upon fertilization, the zygote progresses through successive cleavages which give rise to a spherical, non-motile blastula. Gastrulation, by epiboly (Gutsell 1930; Drew 1909), or epiboly and invagination (Hodgson and Burke, 1988), produces a ciliated gastrula. The gastrula develops into a top-shaped trochophore larva.

During trochophore development, a band of cilia (the prototroch) forms, dividing the trochophore into two regions. At the apical region, tufts of long flagella are formed which provide motility. The prototroch of the trochophore larva later develops into the organ of locomotion, the velum. The main components of the larval musculature — velum, mantle, digestive tract, and foot — are formed at this stage out of spindle-shaped mesenchymal cells (Sastry, 1979). The shell gland is also formed, and shell secretion is initiated by the gland at the end of the trochophore stage (Malakhov and Medvedeva, 1986). Two discs (shell) are secreted and subsequently bridged at one end, forming a hinge. The discs become shell valves (prodissoconch I shell) and take a D-shape outline characterizing the beginning of the straight-hinged, or D-veliger larval stage of development. The time to reach this stage of development under laboratory conditions for commercial species in the United States ranges from 18 to 28 hours for *A. irradians* (Castagna and Duggan, 1971) to 96 hours for *P. magellanicus* (Culliney, 1974).

During the veliger stage, various organs attain further development (Sastry, 1979). Most characteristic of this stage is the emergence of the velum. The prototroch of the trochophore larva develops into a ciliated, bi-lobed velum which serves in feeding and active swimming locomotion. Energy requirements are provided by the egg up to the early veliger stage, where the digestive system becomes functional and the larva becomes planktotrophic. The digestive system becomes functional with the opening of the anus, and the development of the digestive diverticular and the crystalline style sac (Sastry, 1979). The cerebral, pedal, visceral, and parietal ganglia are also formed and statocysts emerge at the base of the foot. The shell continues to grow at the margins. The new shell being laid down (prodissoconch II shell) is secreted by the rim of the mantle (Waller, 1981). With additional shell growth, hinge teeth appear and the umbos begin to develop. As the veliger stage progresses, the enlarging umbo covers the straight-hinge line, the velum becomes reduced, the foot begins to develop, and the anterior adductor muscle is now present. The veliger displays photopositive and geonegative responses which cause them to swim upwards toward an illuminated surface (Sastry, 1979).

Towards the end of the veliger stage, and prior to metamorphosis, the larva becomes photonegative and geopositive and settles to the bottom in search of suitable substrate to "set" on. The term "setting" refers to the first byssal attachment of the larval scallop, which, upon setting, is commonly termed "spat." At this stage, the larva enters the pediveliger, or footed larva, stage of development. The foot is now functional and provides additional mobility and byssal attachment to substrate. Byssal attachment, which can be cast off at will by the scallop, prevents the scallop from being carried away by currents or tides to less favorable environments. The presence or absence of a suitable substrate can influence the duration of the pediveliger stage by postponing metamorphosis until preferred conditions are found (Sastry, 1979). Prominent characteristics of the pediveliger stage include the rounded shell, reduced velum, the appearance of primary gill filaments, and the enlarged foot. Yearly scallop distribution and abundance largely depend on both successful spawning and larval (spat) set.

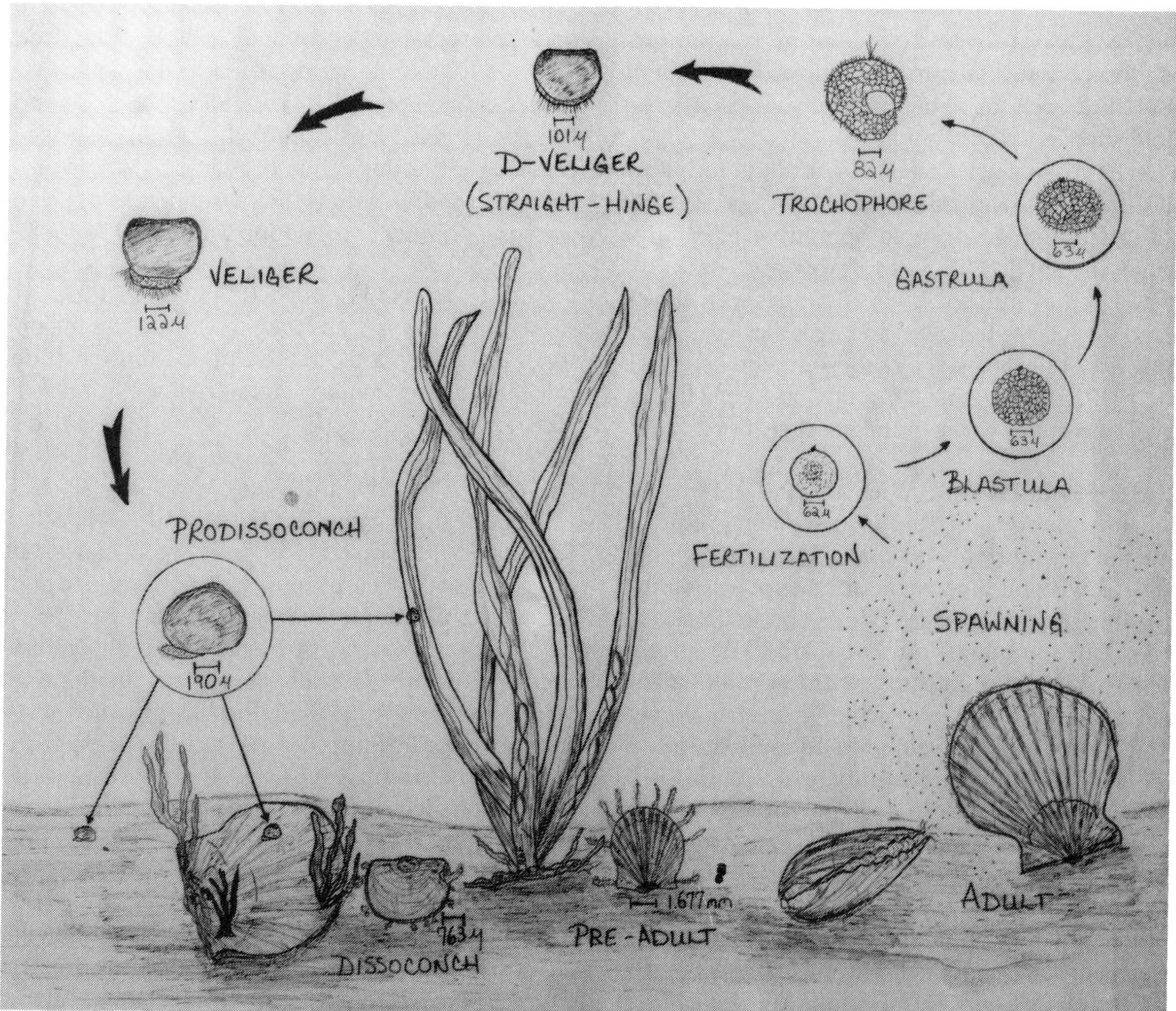

Figure 4. Life cycle of the bay scallop, A. irradians. *Developmental stages and measurements after Sastry (1965); illustrated by R. A. Fisher (1998). Approximate size at larval and post-larval stages is indicated below each drawing (based on Sastry, 1965).*

Metamorphosis begins within a few hours to several days after the pediveliger stage (Sastry, 1979). Most scallop larvae metamorphose on reaching a width within a range of 200 to 300 micrometers (Cragg and Crisp, 1991). Larval metamorphosis involves physiological and anatomical changes that initiate post-larval development. Some organs are lost, while others are rearranged and begin further development. The main organs lost during metamorphosis are the velum and the anterior adductor muscle. With the loss of the anterior adductor muscle, the posterior adductor assumes a more centralized position.

Following metamorphosis and byssal attachment, the post-larval (dissoconch) shell is secreted by the mantle edge. The hinge-line straightens again, the auricles appear, the byssal notch forms, the larval teeth gradually disappear, and the resilium forms under the umbo. Further, the mantle becomes bi-folded with the outer fold later dividing into the outer and middle marginal folds, the gill filaments increase in size and number and become capable of filter-feeding (Gruffydd and Beaumont, 1972), and the foot enlarges, providing the sole method of active movement on the bottom. The young scallop may detach and re-attach to substrate in search of preferred conditions. Scallops often attach to firm substrates, such as shell and rock, or to submerged aquatic vegetation, such as eelgrass and kelp.

The dissoconch stage lasts until the radiating plications (ribs) appear, marking the beginning of the plicate stage of development. During this stage, the shell continues to grow and attain the shape of the adult shell. The mantle is well-developed with large tentacles and eyes along the margin. The plicated-stage scallop retains its crawling mode of movement, but now assumes the active swimming mode of the adult scallop.

SCALLOP PREDATION/MORTALITY

Information pertaining to scallop predation and natural mortality is lacking. It can be assumed that, as with all bivalve broadcast spawners during the planktonic stage of development, massive predation occurs on scallop larvae by planktotrophic organisms, with mortalities also occurring as a result of adverse environmental conditions. Severe weather, sudden changes in water temperature and salinity, and pollution may cause mortalities during planktonic life. Loss of adequate setting substrate and protective nursery/grow-out vegetation (loss of eelgrass beds which provide protection for the bay scallop) may have a direct impact on larval and juvenile mortality. Predation on juvenile and adult scallops is more limited due, in part, to the mobile lifestyle of most species. Shell-boring worms, *Polydora*, may cause some mortalities by weakening the shell (MacDonald and Bourne, 1987), thereby increasing vulnerability to crushing by predators or fishing gear. Besides the obvious impact of fishermen, bivalve-consuming predators as starfish, various gastropods, rays, crabs, lobsters, sea fowl, and various finfish contribute to natural scallop mortalities. Jamieson et al. (1982) and Tettlebach and Feng (1986) indicate scallop mortality due to predation may be a function of size, reporting that the rate of crustacean predation on scallops was significantly higher on the smaller scallops. Annual survival of adult, long-lived species tends to be high.

Scallop mortality within commercial harvesting areas can also be attributed directly and indirectly to harvesting gear type and fishing practices. Both the crushing of the scallop by dredges, and the excessive handling of culled, undersized scallops that are to be returned to the sea, cause direct mortalities (Naidu, 1988; Medcoff and Bourne, 1964; Wells and Wells, 1962). Damage caused by the impact of dredges or trawls also results in indirect mortalities by increasing scallop susceptibility to predation (Caddy, 1973, 1968).

ECONOMICALLY IMPORTANT SPECIES

The scallop species that constitute the majority of product landed in the United States include the Atlantic sea scallop *(Placopecten magellanicus)*; the Pacific sea, or weathervane, scallop *(Patinopecten caurinus)*; the calico scallop *(Argopecten gibbus)*; and two bay scallop species *(Argopecten irradians irradians* and *Argopecten irradians concentricus)* (Figure 5). As a marketed resource, scallops are commonly sold by size and identified by common vernacular names as "bay" or "sea" scallops. Large meats (under 10- to 60-count) are typically sea scallops, and smaller (60- to 200+count) meats are either bay or calico scallops. As the name implies, bay scallops principally inhabit shallow inclosed bays, harbors, estuaries, and sounds where large variations in salinity and temperature may occur. Conversely, sea scallops inhabit the open seas with fewer salinity and temperature variations. The use of the terms "bay" and "sea" to distinguish scallop meat size can be misleading. For instance, the calico scallop inhabits open sea waters but constitutes some of the smallest scallop meats marketed, and are often referred to as bay scallops. Further, scallops entering the United States market from various international sources are also classified in this manner. For example, the 60- to 200-count scallop *Chlamys (Zygochlamys) patagonica* inhabits ocean waters off the coast of Argentina, but is typically marketed as a bay scallop here in the United States. To further complicate the issue of scallop size and market terminology, recent industry attempts at value-added product development have produced large "sea" scallops from the agglomeration of small "bay" scallops. In the wake of recent (1997) declines in domestically harvested 20- to 40-count sea scallops (*Placopecten magellanicus*), the industry is utilizing product-binding technologies to bind multiple whole 80- to 150-count Argentine (*C. patagonica*), Chinese (*A. irradians* and *C. farreri*),

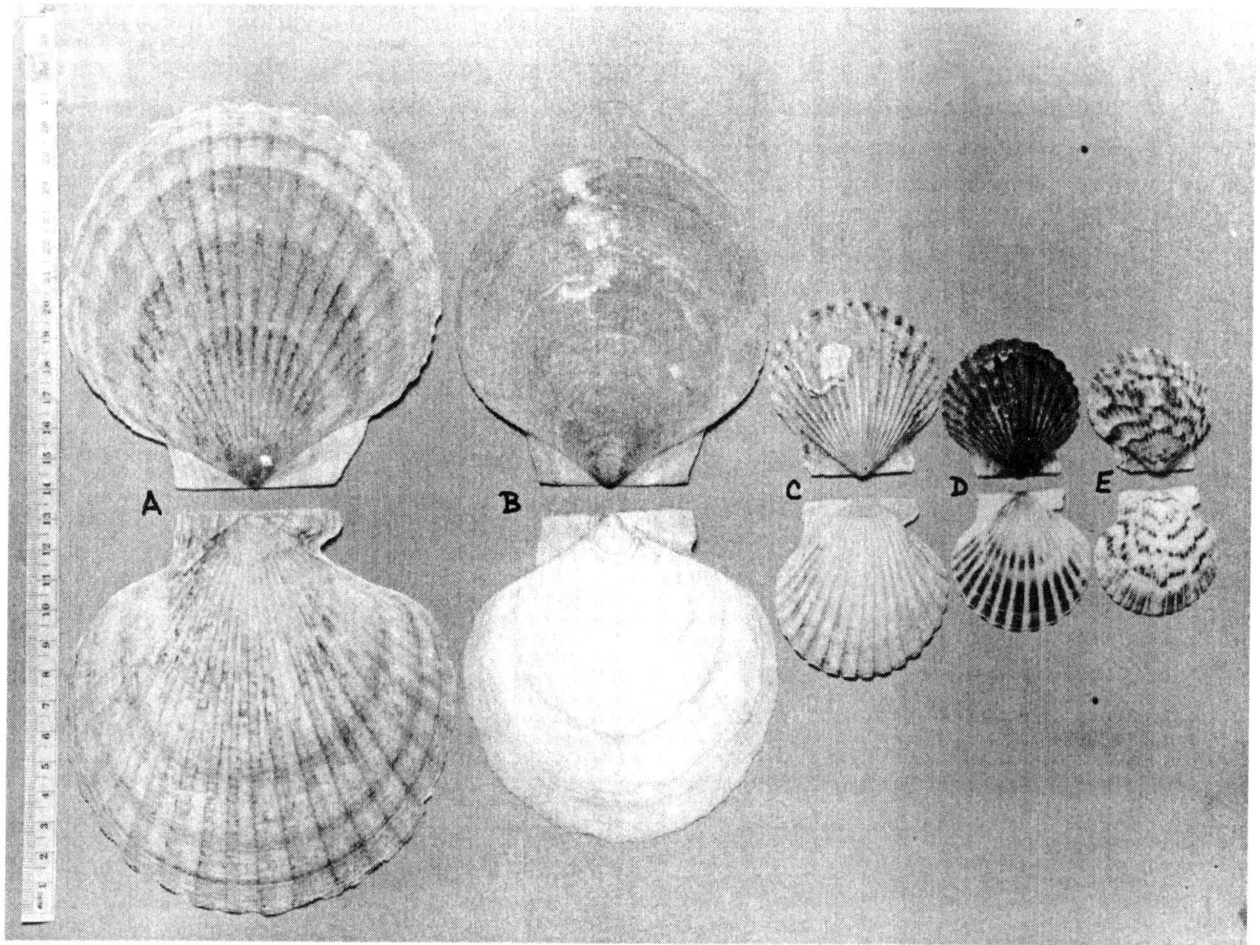

Figure 5. Major commercial scallop species in the United States. A. Weathervane scallop, Patinopecten caurinus. *B. Atlantic sea scallop,* Placopecten magellanicus. *C. Northern bay scallop,* Argopecten irradians irradians. *D. Southern bay scallop,* Argopecten irradians concentricus. *E. Calico scallop,* Argopecten gibbus.

or calico (*Argopecten gibbus*) scallops together to make a 30-count scallop.

All scallops entering the United States market are currently required by the Food and Drug Administration (FDA), Office of Seafood, to conform to either of two product identification statements for labeling which are based on the moisture content (by weight) of the scallop meat. Product with an average meat moisture content below 80 percent may be labeled "Scallops." Product containing an average moisture content between 80 to 83.9 percent must be labeled "X percent Water-Added Scallop Product," where X can be any value up to 25 percent which accurately represents added moisture levels above 80 percent. Product with moisture content of 84 percent and above is currently deemed adulterated by the Food and Drug Administration (FDA). This federal policy is currently an interim agreement between the FDA and the scallop industry. Reassessment will occur when adequate scientific research data have been collected on natural scallop moisture levels (including seasonal fluctuations) and on industry harvesting, processing, and distributing practices. The natural moisture content of scallop meat varies seasonally in relation to its reproductive cycle. Natural moisture contents in *P. magellanicus* meats over a two-year period off Virginia were observed to range from 76.5 to 80.6 percent (Fisher, 1993, unpublished). DuPaul et al. (1997) reported natural moisture in *P. magellanicus* at point of harvest ranging from 73.7 to 78.9 percent, and moisture levels of 74.2 to 82.5 percent at point of off-loading. Moisture addition to scallop meats harvested and shucked offshore and landed fresh is an artifact of domestic fisheries. Ice is routinely used for chilling onboard to preserve quality, and upon melting, adds water to the meats. Further, water is used shore-side in the processing of meats for distribution to various markets (Fisher et al.,1996, 1990; DuPaul et al., 1993, 1990).

THE ATLANTIC BAY SCALLOP
Argopecten irradians (Lamarck, 1819)

Subspecies:

Argopecten irradians irradians (Lamarck, 1819)
Argopecten irradians concentricus (Say, 1822)
Argopecten irradians amplicostatus (Dall, 1898)

Other names in literature:

Aequipecten irradians
Pecten irradians

Vernacular names:

English, Atlantic bay scallop
French, *Peigne baie de l'Atlantique*
Spanish, *Peine caletero atlantico*

The bay scallop, *A. irradians*, is distributed along the East coast of the United States from New England to Texas, inhabiting shallow, protected coastal bays and estuaries commonly associated with seagrass beds. Clark (1965) reported the southern extension of bay scallop populations reaches from Mexico to Colombia, South America (Atlantic side). Bay scallops are extremely temperature-tolerant, withstanding fluctuations from 0°C to 30°C (Kirbey-Smith, 1970; Gutsell, 1931). However, Rhodes (1991) reports that bay scallops are more sensitive to low salinity than most other bivalves, and are seldom found in areas of salinity below 20 parts per thousand (ppt).

A. irradians is divided into three recognized subspecies which have been distinguished by their physical differences (Abbot, 1954), genetic variations (Blake, 1994; Nei, 1972), and ecological and general geographical separation (Abbot, 1954; Dodge, 1952). The subspecies include the northern bay scallop, *A. i. irradians*, which is distributed from Cape Cod to New Jersey/Maryland (Clarke, 1965); the Atlantic, or southern bay scallop, *A. i. concentricus*, which occurs from New Jersey to Georgia and in the Gulf of Mexico from Florida Bay to Louisiana (Clarke, 1965); and the southernmost bay scallop, *A. i. amplicostatus*, found from Texas to Columbia. Subspecies of most commercial importance are *A. i. irradians* and *A .i. concentricus*. Little information is available on the contribution of *A. i. amplicostatus* to the United States fishery.

Such morphological characteristics as shell (valve) inflation and rib counts were used by Clarke (1965) to distinguish the subspecies *A. i. irradians*, which possesses the most compressed valves and has 17 to 18 ribs radiating from the ventral margins to the hinge-line on the upper valve. *A. i. amplicostatus* has the most inflated shell (most similar to the calico scallop shell inflation) and possess the fewest (12 to 18) ribs. Shell inflation of *A. i. concentricus* lies somewhere between the other two with a moderately inflated shell, and possesses 18 or more ribs. Fischer (1978) noted that the bottom valve within these subspecies becomes proportionally deeper going from the northern (*A. i. irradians*) through the central-located (*A. i. concentricus*) to the southernmost form (*A. i. amplicostatus*).

The distribution of *A. i. concentricus* is not continuous, with a large gap ranging from North Carolina to the west coast of southern Florida. Through mitochondrial DNA tests, Blake (1994) reports that no two geographically-separated populations of *A. irradians* were found to share a common gene pool. Populations of *A. i. irradians* are found integrated with those of *A. i. concentricus* along the mid-Atlantic. Morphological characteristics have traditionally been used to distinguish the two. More recent morphological distinctions have been supported through genetic testing by Blake (1994), who suggests that *A. i. concentricus* from the mid-Atlantic are genetically more closely related to *A. i. concentricus* from the Florida Gulf coast then *A. i. irradians* from New England.

Bay scallop shells are symmetric, or nearly so, with a distinct notch on the anterior edge of the hinge (Fay et al., 1993). The left (top) valve varies in color, but is generally dark grey to brown, with occasional reddish to yellow hues (Figure 6). The right valve (bottom) also varies in color, from all white to color approaching that of the top valve (Waller, 1969).

Bay scallops are a short-lived species with life expectancies of 20 to 30 months for populations off Massachusetts (Belding, 1910); approximately 14 to 18 months off North Carolina (Gutsell, 1930); and 12 to 18 months off western Florida (Barber and Blake, 1983). During their short life span, bay scallops have only one complete gametogenic cycle, with time of spawning varying latitudinally. Spawning of bay scallops has been reported to occur off New England from May to September (Taylor and Capuzzo, 1983; Sastry, 1970); off New York from June to August (Bricelj et al., 1987); off North Carolina from July to November (Sastry, 1966, 1970; Gutsell, 1930); and off the Florida west coast from September to November (Barber, 1984; Barber and Blake, 1981).

Bay scallops are functional hermaphrodites, where both types of gametes become mature at the same time, and generally protandrous (releasing male gametes first, followed by the female gametes), which minimizes self-fertilization (Loosanoff and Davis, 1963). Gametogenesis and spawning is correlated to water temperature and food supply. Mean gonadal indices (estimate of fecundity) of bay scallop populations were reported to decrease with decreasing latitude, from 20 percent in Massachusetts, to 17 percent in North Carolina, to 14 percent in Florida (Barber and Blake, 1983). *A. i. irradians* population in Massachusetts spawned with increasing water temperatures (14° to 16°C to 23°C), compared to those from North Carolina and Florida releasing gametes with decreasing water temperatures (25° to 17°C to 18°C) after an initial temperature increase (Sastry, 1963, 1966, 1970; Beldon, 1910).

As with most bivalves, bay scallops are broadcast spawners, releasing eggs and sperm into the surrounding water where fertilization takes place. The life cycle of the bay scallop is depicted in Figure 4. Average size of bay scallop eggs range from 60 to 62 micrometers (Sastry, 1965; Loosenoff and Davis, 1963). After fertilization, the trochophore stage of development is reached in ~24 hours., D-veliger stage by ~48 hours, with time of settlement ~14 days (Sastry, 1965). Time of settlement may range from 10 to 19 days depending upon environmental conditions (Castagna and Duggan, 1971). Ribbed adult forms are reached in 30 to 40 days post-fertilization (Sastry, 1965).

The bay scallop maintains a strong, high value domestic market in the United States, and is recognized as a prime candidate for commercial aquaculture. The bay scallop has been widely studied for its aquaculture potential, with hatchery and larval culture techniques well established (Barber and Blake, 1983; Castagna, 1975; Castagna and Duggan, 1972; Sastry, 1963; Turner and Hanks, 1960). However, the grow-out of scallop seed to a marketable size has had limited success in the United States, not biologically but economically. Grow-out techniques using pens (Castagna and Duggan, 1971), raceways (Rhodes et al., 1981), pearl and lantern nets (Rhodes and Widman, 1984), and trays (Oesterling, 1997) have successfully demonstrated the production of marketable size scallops (50 mm) within seven months. However, at the present time, costs typically exceed income under commercial conditions when targeting the adductor muscle (meat) market only. Labor and material costs and coastal usage rites are examples of impediments facing the potential commercialization of bay scallop production in the United States. The production of whole bay scallops for chowder, steamer, and/or half-shell markets (where the entire animal is consumed) may prove more lucrative within the short run, if marketing efforts are successful.

Blake (1996) outlined biological advantages and disadvantages for culturing the southern bay scallop *A. i. concentricus*. These include: (1) scallops are functional hermaphrodites, providing control over the amount of eggs and sperm in the water; (2) they are synchronistic spawners (when one starts they all spawn); (3) rapid growth, obtaining harvestable product in approximately seven months; and (4) feasible three-dimensional nursery application (scallops set on various substrate, including vertically-oriented grasses, diminishing crowding). Disadvantages cited include: (1) short life cycle/essentially one spawn, readiness required when the time comes); (2) extremely sensitive to environmental fluctuations; and (3) susceptible to high degree of fouling and predation.

Bay scallop production for the meat-only market has been quite successful for the Chinese. In 1982, 26 bay scallops grown in Virginia (Virginia Institute of Marine Science) were introduced into China and successfully spawned. Five years from the introduction of the bay scallop, a 50,000 metric ton live weight fishery was reported in China (Chew, 1990). By the early 1990s China became one of the leading suppliers of bay scallop meats to the United States market.

THE CALICO SCALLOP
Argopecten gibbus (Linnaeus, 1758)

Other names in literature: *Aequipecten gibbus*
Pecten gibbus

Vernacular names: English, calico scallop
French, *Peigne calicot*
Spanish, *Peine percal*

The calico scallop, *Argopecten gibbus*, occurs in the sub-temperate and tropical waters of the western Atlantic from Cape Hatteras, North Carolina, to Brazil, including Bermuda, the Greater Antilles, and the Gulf of Mexico (Waller, 1969). *A. gibbus* inhabits open ocean water along the continental shelf to depths of 370 m (Waller, 1969), with the major commercial stocks usually located in depths of 20 to 50 m on hard sand bottom (Blake and Moyer, 1991). However, Neumann (1965) reports finding calico scallops in a sound in Bermuda. *A. gibbus* was formerly known as *Pecten gibbus* and *Aquipecten gibbus*, but was placed into the genus *Argopecten* by Waller (1969). Related species within the calico scallop range include the nucleus scallop, *Argopecten nucleus*, and the bay scallop, *A. irradians*, however both are generally limited to shallow, protected waters. Dodge (1952) and Sastry (1962) provided morphological characteristics, and Blake (1994) provided genetic differences which distinguished *A. gibbus* from *A. irradians*.

The calico scallop derives its name from the mottled coloration of the valves (Figure 6). Colors may vary, but generally bright hues of red, purple-to-light brown on a white or yellow background dominates. Both valves typically display the same coloration, but the right (bottom) valve is whitish, or more lightly pigmented. The valves are well inflated and slightly left convex or equiconvex and ribbed. Waller (1969) reports 17 to 23 ribs on the right valve. Adults attain shell heights between 40 to 60 mm.

Calico scallops are functional hermaphrodites. Both sexual organs undergo maturation within a common gonad, and spawning is generally protandrous to prevent self-fertilization. Calico scallops spawn two or three times during their 18- to 24-month life span, with the first spawn generally occurring between six to nine months of age. Blake and Moyer (1991) report that age at first spawn appears to be less important than environmental conditions, since scallops as young as three

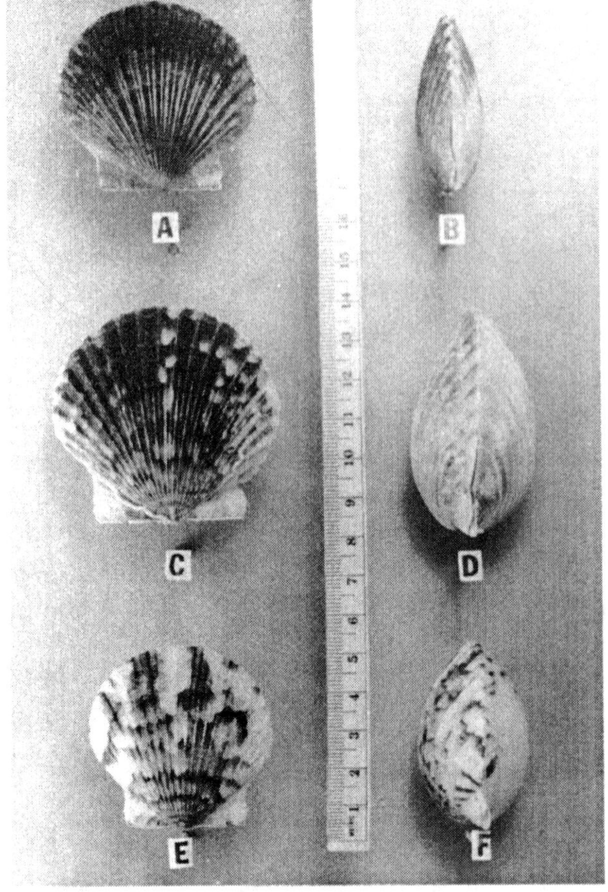

Figure 6. Commercially harvested bay scallops and the calico scallop.
A, B: The northern bay scallop, Argopecten irradians irradians.
C, D: The southern bay scallop, Argopecten irradians concentricus.
E, F: The calico scallop, Argopecten gibbus.

months occasionally spawn. Miller et al. (1981) reported age and size at sexual maturity in calico scallops of 71 days and 20 mm shell height, respectively. Ovarian color changes occur with gonad maturation. On the basis of ovarian color, Roe et al. (1971) concluded calico scallops from Florida's east coast begin gonadal maturation in August and spawn the following spring. However, these authors and Allen and Costello (1972) found small scallops throughout most of the year, suggesting protracted or multiple spawning. Moyer and Blake (1986) report two annual spawning periods for calico scallops under optimum environmental conditions, a predominant late spring spawn from April to June, and a minor fall spawn.

Water temperature and food availability are reported to control gametogenesis (Miller et al., 1981; Sastry, 1963), while spawning is induced by rapid fluctuations in environmental parameters, primarily temperature. Allen and Costello (1972) report spawning to be initiated by raising temperatures from 25° to 30°C. However, Blake and Moyer (1991) report that maturation stops and spawning does not occur at temperatures above 22°C. Miller et al. (1981) report major spawning occurred when bottom water temperatures ranged from 15° to 22.5°C at depths from 18 to 55 m.

Calico scallops are broadcast spawners, releasing both male and female gametes into the surrounding water where fertilization takes place. Successful fertilization, degree of mortality, and subsequent growth rates become dependant on environmental conditions. Under laboratory conditions, fertilized eggs remain planktonic for approximately 16 days and set when they attain a shell height of about 0.25 mm (Costello et al., 1973). During the pelagic stage, currents may distribute larvae great distances. Kirby-Smith (1970) suggested that scallop larvae spawned off Florida may be carried by the Gulf Stream northward and recruited into the population off North Carolina. Allen (1979) and Blake and Moyer (1991) estimate 10 mm shell heights are reached at 51 days; 28 mm at three months; and 38 to 40 mm (commercially desired size) at eight months.

Commercial stocks of *A. gibbus* are fished off North Carolina and the east and west coasts of Florida (Allen and Costello, 1972), but are often highly variable. Historically, the fishery has centered around commercial stocks off Cape Canaveral, Florida, where less stock variability occurs. Calico scallops are harvested and shellstocked at-sea, then mechanically shucked and processed in shore-based processing plants. Marketable meat counts (meats per pound) range from 100 to 400. The quality of the larger meats (100 count) tend to be lower, as these scallops are approaching the end of their lives (Blake and Moyer, 1991; Bullis and Cummins, 1961). Calico scallops serve as an intermediate host to a parasitic, hair-like nematode worm, *Sulcascaris sulcata*, which usually appears along the periphery of the adductor muscle (meat) curled within a yellowish cyst. The worms present a marketing problem, but are not harmful to humans (Blake et al., 1985).

THE ATLANTIC SEA SCALLOP

Placopecten magellanicus (Gmelin, 1791)

Other names in literature: *Pecten magellanicus*

Vernacular names:

English, Atlantic sea scallop
giant scallop
smooth scallop
French, *Pecten d'Amerique*
Spanish, *Vieira americana*

The Atlantic sea scallop, *Placopecten magellanicus*, is distributed in the Northwest Atlantic Ocean from the Strait of Belle Island, Newfoundland, Canada, to Cape Hatteras, North Carolina (Porter, 1974; Posgay, 1957), and commercially exploited throughout most of its range. Sea scallop resources are most abundant on Georges Bank and the mid-Atlantic Shelf, with major commercial landings occurring in Massachusetts, New Jersey, and Virginia. Georges Bank constitutes the largest single natural scallop stock and fishery in the world (Caddy, 1989). The sea scallop resource is primarily distributed on the continental shelf in depths ranging from 18 to 110 meters, but is also found in shallower waters in the Gulf of Maine and deeper waters off Virginia. Preferred bottom type is highly variable, but includes mud, sand, and various size rocks. The highest densities of sea scallops tend to be associated with substrate composed of glacial till (Caddy, 1989). Water temperature plays a significant role in sea scallop depth and longitudinal distribution, with the resource distributed within water temperatures less than 20°C and salinities characteristic of ocean waters (Mackenzie, 1979).

The sea scallop is one of the larger scallops in both shell height and adductor muscle (meat) weight. Naidu (1991) reported sea scallop measurements for shell height and meat weight of 211 mm and 231 g, respectively. The sea scallop shells (valves) are subequal, and singularly subcircular and compressed (Figure 7). The left (top) valve is more convexed than the right, delicately ribbed, and typically brownish in color; however, brighter yellow, pink, and white valves are also commonly observed. Valve color in juvenile and young adult scallops seem to be more variable, with the brighter colors more frequently observed. The right (bottom) valve is flatter than the left, smooth, and prevalently pale cream to white in color. On

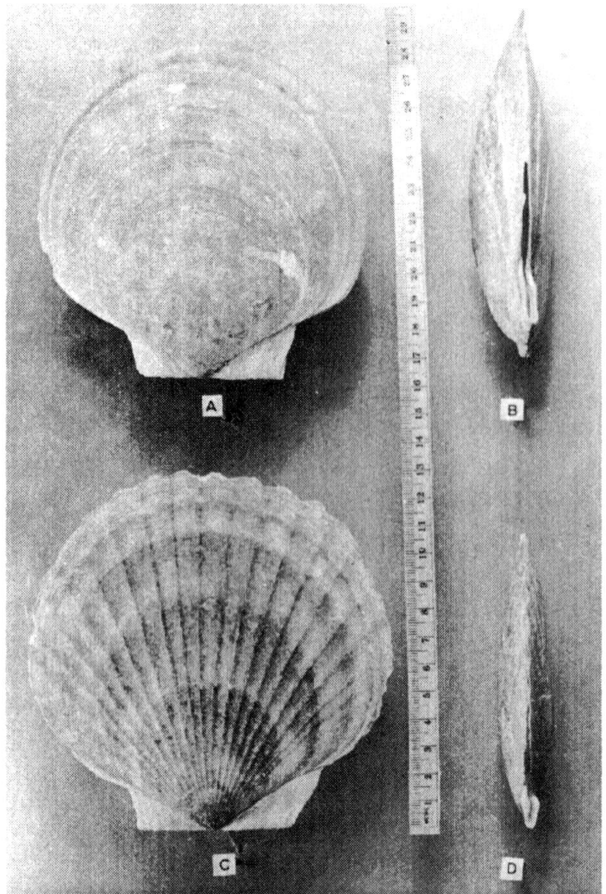

Figure 7. Commercially harvested sea scallops in the United States.
A, B: Atlantic sea scallop, Placopecten magellanicus.
C, D: Pacific sea scallop (or weathervane scallop), Patinopecten caurinus.

rare occasions, the right valve may be observed to be partially, or totally, colored as that of the left valve. The left valve is commonly used for age determination by counting the concentric rings (circuli).

The left valve in older, less mobile scallops is commonly fouled by various benthic organisms, including boring sponges, sea anemones, encrusting bryozoans, hydroids, barnacles, tube worms, and colonial ascidians. Invasion of these organisms results in a range of scallop deformities including irregular-shaped shell margins and blisters developing on the inner surface of the valve (Figure 2). Reduced adductor muscle (meat) size and weight for a given scallop shell height is frequently observed in scallops with internal blisters. Greyish or brownish, flaccid meats are also common in scallops with these blisters.

Sexual maturity in *P. magellanicus* is reached at age one, but initial spawning may not occur until the second year (Posgay, 1979; Naidu, 1970). Shell height at first spawn has been reported by Naidu (1970) to range from 23 to 75 mm. Spawning times vary between the northern and southern populations, as well as by depth (environmental conditions). Peak spawning periods have been reported to occur from August to September off Maine (Barber et al., 1988); from September to October off Georges Bank and Massachusetts (MacKenzie et al., 1978); from October to November off New Jersey (MacDonald and Thompson, 1988); and from April to May and again in October in the mid-Atlantic region (DuPaul et al., 1989). Within the biannual spawning event documented in the mid-Atlantic population, the spring spawn is the more predictable and dominant event, with the fall spawn more sporadic, if occurring at all, from year to year (Kirkley and DuPaul, 1991). Through histological analysis of the gametogenic cycle of *P. magellanicus* from the mid-Atlantic region, Schmitzer et al. (1991) observed greater fecundity per individual scallop during the spring spawning event than in the fall spawn. MacDonald (1984) reports a healthy, 10-year-old female may release up to 90 to 95 million eggs during spawning. Water temperatures associated with spawning varies among location. Water temperatures reported at spawning ranged from 8° to 11°C on Georges Bank (Posgay and Norman, 1958), and 6.5° to 11°C off the mid-Atlantic (MacKenzie et al., 1978).

Sea scallops release their gametes into the surrounding water where fertilization takes place. *P. magellanicus* eggs are 64 microns in diameter and upon fertilization hatch within 30 to 40 hours (Culliney, 1974). The scallop larvae remain planktonic for over a month after hatching (Posgay, 1979). The pediveliger stage is reached by the twenty-eighth day (292 microns) with metamorphosis initiated once preferred conditions or a suitable substrate are found (Culliney, 1974). Young adults over 10 mm long detach from the substrate and assume a free-living existence (Dow, 1969).

Growth is rapid in the first few years of life, with the first discernable growth ring appearing at age 1.5 years and about 20 mm long (Posgay, 1979). Mullen and Moring (1986) reported shell height increases of 50 to 80 percent between years three and five on Georges Bank, with meat weight increasing by 400 percent. The relationship between meat weight and shell height has provided a valuable means to analyze yield-per-recruitment for assessing management practices in *P. magellanicus* populations. Meat weight-to-shell height relationships have been reported to vary with respect to geographical area, seasonality, reproductive cycle, water depth, and shell size (Kirkley et al., 1991; Shumway and Schick, 1987; Serchuk, 1983, 1984). Values for shell height (mm) and meat weight (grams) at given ages (years) are reported by MacKenzie (1979) for scallops off the mid-Atlantic. Derived meats per pound (MPP) values were added to the table for marketing interest (Table 1).

The sea scallop is an excellent candidate for culture due to its low-temperature tolerance, being a dioecious species (roe-on market), and possessing a large, high-valued meat. MacDonald (1986) and Naidu and Cahill (1986) have reported success with culturing *P. magellanicus* using suspended culture techniques. Dadswell and Parsons (1991), using suspension culture and naturally produced spat, reported the production of commercial size scallops (90 mm) in 33 to 36 months, with overall growth rate of 0.08 mm/day. Current attempts are underway for open ocean bottom culturing of *P. magellanicus* off Massachusetts. The production of small sea scallops for the growing whole-bay scallop specialty markets (chowders, half-shell, and steamer) may help commercialize sea scallop culturing.

Table 1. Meat weight-to-shell height relationship in *Placopecten magellanicus* (after MacKenzie, 1979)

Age	Shell height	Meat weight	MPP
3	65	5.1	88.9
4	88	12.5	36.3
5	104	21.3	21.3
6	117	29.9	15.2
7	126	37.6	12.1
8	132	44.0	10.3
9	138	49.3	9.2

THE WEATHERVANE SCALLOP
Patinopecten caurinus (Gould, 1850)

Other names in literature: *Pecten caurinus*

Vernacular names:

English,	Weathervane scallop
	Pacific sea scallop
	Alaskan scallop
French,	*Pecten gant pacifique*
Spanish,	*Vieira gigante del Pacifico*

The weathervane scallop (*Patinopecten caurinus*) is distributed along the West Coast of North America, from central California (Point Sur), northward to the Gulf of Alaska, in depths of 10 to 200 m associated with sandy or mud bottoms (Kaiser, 1986; Mottet,1979; Grau, 1959). Major commercial stocks are fished in the Gulf of Alaska and the Bering Sea, with limited smaller quantities historically harvested off Washington and Oregon (Kaiser, 1986; Starr and McCrae, 1983; Capps, 1981). Sixty-eight to 100 percent of the calculated scallop biomass in the Gulf of Alaska was reported to be within 0 to 100 m of water, depending on the area (Ronholdt et al., 1977).

Within the Alaskan weathervane scallop fishery, incidental harvesting of *Chlamys* spp. (*C. pseudoislandica* and *C. rubida*) has been reported (Shirley and Kruse, 1995). The most closely related living species to *P. caurinus* is the Japanese scallop, *Patinopecten (Mizuhopecten) yessoensis,* which is similar in size and overall appearance, and also supports a large commercial fishery.

The weathervane scallop is a large, long-lived species with maximum age and size records of 28 years and 250 mm shell height, respectively (Hennick, 1973). Mottet (1979) states that *P. caurinus* is the largest scallop in the world. Growth studies indicate that inshore populations of *P. caurinus* grow faster than offshore populations, with mean shell height for each age class consistently greater for inshore scallops (Bourne, 1991; MacDonald and Bourne, 1987). Offshore scallops are more frequently infested with boring sponges and annelid worms which divert energy from shell growth to shell repair; however, a conclusive relationship between infestation and growth is lacking (Haynes and Hitz, 1971). MacDonald and Bourne (1987) concluded that poorer environmental conditions offshore are responsible for growth rate differences.

The valves of *P. caurinus* both display coloration, with the top (left) valve being pale brown to reddish, and the bottom (right) valve, yellowish white to light brown. The right valve is more convex, while the left valve is flat to concave (Figure 7). There are 18 to 22 flattened ribs on both valves which radiate from the umbo to the ventral margins (Bourne, 1991).

Sexes are separate in *P. caurinus* with only a single occurrence of hermaphroditism reported (Hennick, 1971). There are no external signs of sexual dimorphism, though sexing of mature individuals is achieved by examination of gonad color: males, cream-white; females, bright orange (Hennick, 1970). Gonadal maturation occurs during the autumn and winter periods, increasing steadily until spawning. Gamete production was reported by MacDonald and Bourne (1987) to be greater in inshore scallops than for offshore scallops of equivalent age.

Sexual maturity in *P. caurinus* is reported to be reached between the ages of three and four years (Hennick 1970). However, Bourne (1991) suggests sexual maturity in *P. caurinus* is attained at shell height of approximately 70 mm, regardless of age. Spawning occurs once a year, with spawning time and length of spawn varying throughout its distribution in accordance with environmental and geographical parameters. MacDonald and Bourne (1987) report a single but protracted spawn for inshore scallops off British Columbia beginning in mid-April and ending in mid-June, whereas offshore scallops spawned during July and August. Robinson and Breese (1984) documented a single protracted spawn off Oregon ranging from mid-January to June. In Alaska spawning occurs from early June through mid-July (Hennick, 1970). As with other species, Hennick (1970) suggests abrupt temperature changes trigger spawning in *P. caurinus*.

Weathervane scallops broadcast gametes into the surrounding water where fertilization takes place. Information on the weathervane scallop larval development and early growth is lacking. In a commercial hatchery operation, Bunting (1997) provides general observations of *P. caurinus* early development as follows: at 9° to 13°C all larvae reach D-stage by day five at 110 to 116 microns (at 16° to 17°C, D-stage was observed at day three); eye-spots appear by day 24 (245 to 270 microns) at 11° to 14°C; and setting occurs between day 29 to 30 (260 to 280 microns) at 11° to 14°C. Considerable attention has been directed to the culturing parameters of the closely related Japanese scallop, *P. yessoensis*. In this species, egg diameter ranges from 60 to 70 micrometers, with larval development reported to last for 18 to 35 days after fertilization (Kasyanov, 1991). Setting is reported to occur at ~40 days (Ito, 1991). Juvenile *P. yessoensis* are reported to remain attached for 1.5 to 4 months (Kasyanov, 1991) before assuming the free mode existence. At age one, young Japanese scallops are 20 to 50 mm in length; 50 to 90 mm at age two; 80 to 120 at age three; and 100 to 150 mm at age four (Ito, 1991). These growth rates parallel those reported for Alaskan populations of *P. caurinus* by Hammarstrom and Merritt (1985) and Hennick (1970).

The relationship between shucked meat weight-to-age and shell height varies within and among populations of *P. caurinus* (Hennick, 1973). Average shell heights (mm) and shucked meat weights (grams) per given age (years) from Kamishak Bay, Alaska, are summarized from Hammarstrom and Merritt (1985). Meats per pound (MPP) values were added for marketing interest (Table 2).

Average meat recovery (average meat weight divided by average live weight in percent) was greatest in the seven- and eight-year-old classes at 11.3 and 11.6 percent, respectively.

Table 2. Meat weight-to-shell height relationship in *Patinopecten caurinus* (after Hammarstrom and Merrit, 1985)

Age	Shell height	Meat weight	MPP
1	29.0	—	—
2	62.0	—	—
3	101.2	—	—
4	122.6	22.2	20.4
5	133.7	33.0	13.7
6	142.9	34.9	13.0
7	150.8	43.7	10.4
8	154.3	46.4	9.8
9	156.7	46.3	9.8
10	163.0	51.9	8.7
11	165.2	50.5	9.0
12	170.4	53.9	8.4

REFERENCES

Abbot, R. T. 1954. American seashells, First Edition, Van Nostrand Co., Princeton, NJ.

Allen, D. M. 1979. Biological aspects of the calico scallop, *Aropecten gibbus*, determined by spat monitoring. *The Nautilus.* **3**:107-119).

Allen, D.M. and T. J. Costello 1972. The calico scallop, *Argopecten gibbus.* NOAA Tech. Rep. NMFS Spec. Sci. Rep. Fish. No. 656.

Amirthalingam, C. 1928. On lunar periodicity in the reproduction of *Pecten opercularis* near Plymouth in 1927-1928. *J. Mar. Biol. Assoc. U.K.* **15**:605-641.

Baird, F. T. 1954. Migration of the deep sea scallop (*Pecten magellanicus*). Bull. Dept. Sea and Shore. Fish. Fisheries circular No. 14.

Barber, B. J. and N. J. Blake. 1981. Energy storage and utilization in relation to gametogenesis in *Argopecten irradians concentricus* (Say). *J. Exp. Mar. Biol. Ecol.* **52**:121-134.

Barber, B. J. and N. J. Blake. 1983. Growth and reproduction of the bay scallop, *Argopecten irradians* (Lamarck) at its southern distributional limit. *J. Exp. Mar. Biol. Ecol.* **66**:247-256.

Barber, B. J. 1984. Reproductive energy metabolism in the bay scallop, *Argopecten irradians concentricus* (Say). Ph.D. Thesis, University of South Florida, Tampa.

Barber, B. J., R. Getchell, S. Shumway, and D. Shick. 1988. Reduced fecundity in a deep-water population of the giant scallop, *Placopecten magellanicus*, in the Gulf of Maine, U.S.A. *Mar. Ecol. Prog. Ser.* **42**:207-212.

Barber, B. J. and N. J. Blake. 1991. Reproductive physiology. In *Scallops: Biology, Ecology and Aquaculture.* S.E. Shumway, ed., pp. 377-428. Elsevier, New York.

Barnes, R. D. (ed.) 1974. *Invertebrate Zoology*, 3rd ed. Saunders Company, Philadelphia, PA.

Bayne, B. L. 1976. Aspects of reproduction in bivalve mollusks. In *Estuarine Processes.* M. L. Wiley, ed., pp. 432-448. Academic Press, New York.

Belding, D. L. 1910. The scallop fishery of Massachusetts, including an account of the natural history of the common scallop. Mass. Div. Fish and Game. *Mar. Fish. Ser.* **3**:1-51.

Beninger, P. G. and M. LePennec. 1991. Functional anatomy of scallops. In *Scallops: Biology, Ecology and Aquaculture.* S.E. Shumway. ed., pp. 13-223. Elsevier, New York.

Blake, N. J. 1972. Environmental regulation of neurosecretion and reproductive activity in the bay scallop, *Argopecten irradians* (Lamarck). Thesis, University of Rhode Island, Kingston.

Blake, N. J. and A. N. Sastry. 1979. Neurosecretory regulation of oogenesis in the bay scallop *Argopecten irradians* (Lamarck). In *Cyclic Phenomena in Marine Plants and Animals.* E. Naylor and G. G. Hartnoll, eds., pp. 181-190. Pergamon Press, New York.

Blake, N. J. and M. A. Moyer. 1991. The calico scallop, *Argopecten gibbus*, fishery of Cape Canaveral, Florida. In *Scallops: Biology, Ecology and Aquaculture.* S.E. Shumway, ed., pp. 899-911. Elsevier, New York.

Blake, S. G. 1994. Mitochondrial DNA variation in natural and cultured populations of the bay scallop, *Argopecten irradians* (Lamarck), and the calico scallop, *Argopecten gibbus* (Dall). Masters thesis, Virginia Inst. of Marine Sci., College of William and Mary, Gloucester Point, VA.

Blake, N. J. 1996. Personal communication. Univ. of South Florida, St. Petersburg.

Bourne, N. 1964. Scallops and the offshore fishery of the Maritimes. *Fish. Res. Bd. Can. Bull.* **145**:1-61.

Bourne, N. and E. G. Bligh. 1965. Orange-red meats in sea scallops. *J. Fish. Res. Bd. Can.* **22**(3):861-864.

Bourne, N. 1991. West Coast of North America. In *Scallops: Biology, Ecology and Aquaculture.* S.E. Shumway, ed., pp. 925-942. Elsevier, New York.

Bricelj, V. M., J. Epp, and R. E. Malouf. 1987. Intraspecific variation in reproductive and somatic growth cycles of bay scallops *Argopecten irradians. Mar. Ecol. Prog. Ser.* **36:**123-137.

Bricelj, V. M. and S. Shumway. 1991. Physiology: Energy acquisition and utilization. In *Scallops: Biology, Ecology and Aquaculture.* S.E. Shumway, ed., pp. 305-376. Elsevier, New York.

Buddenbrock, W. V. 1915. Die statocyste von *Pecten*, ihre histologie und physiologie. *Zool. Jahrb. Abt. Allg. Zool. Physiol. Tiere* **35**:301-356.

Bullock, T. H. and Horridge, G. A. 1965. Mollusca: Pelecypoda. In *Structure and Function in the Nervous Systems of Invertebrates, Vol. II.* T.H. Bullock and G.A. Horridge, eds., pp. 1390-1431. W.H. Freeman and Company, San Francisco, CA.

Bunting, B. 1997. Personal communication. Island Scallop Ltd., Qualicum Beach, B.C. Canada.

Caddy, J. F. 1968. Underwater observations on scallop (*Placopecten magellanicus* (Gemlin). *J. Fish. Res. Bd. Can.* **25**(10):2123-2141.

Caddy, J. F. (1973). Underwater observations on the track of dredges and trawls and some effects of dredging on a scallop ground. *J. Fish. Res. Bd. Can.* **30**(2):173-180.

Caddy, J. F. (1989), A perspective on the population dynamics and assessment of scallop fisheries, with special reference to the sea scallop, *Placopecten magellanicus* (Gmelin). In *Marine Invertebrate Fisheries: Their Assessment and Management.* J. F. Caddy, ed., pp. 559-589. Wiley, New York.

Capps, J. 1981. Easterners initiate Oregon scallop fishery. *Nat. Fish.* **64**(4):15.

Castagna, M. 1975. Culture of the bay scallop, *Aropecten irradians*, in Virginia. *Mar. Fish. Rev.* **37**(1):19-24.

Castagna, M. and W. S. Duggan. 1971. Rearing of the bay scallop, *Aequipecten irradians. Proc. Natl. Shellfish. Assoc.* **61**:86-92.

Castagna, M. and W. S. Duggan. 1972. Mariculture experiments with the bay scallop, *Argopecten irradians*, in waters of the seaside of Virginia. *Bull. Amer. Malacological Union* **37**:21.

Chew, K. K. 1990. Global bivalves introductions. *World Aquaculture* **21**(3):9-22.

Clarke, A. H. 1965. The scallop superspecies *Aequipecten irradians* (Lamarck). *Malacologia,* **2**:161-188.

Coe, W. R. 1943. Development of the reproductive system and variations in sexuality in *Teredo navalis* and other pelecypod mollusks. *Biol. Bull.* **84**:178-187.

Coe, W. R. 1945. Development of the reproductive system and variations in sexuality in *Pecten* and other pelecypod mollusks. *Trans. Conn. Acad. Arts Sci.* **36**:673-700.

Costello, T. J., M. J. Hudson, J. L. Dupuy, and S. Rivkin. 1973. Larval culture of the calico scallop *Argopecten gibbus. Proc. Natl. Shellfish Assoc.* **63**:72-76.

Cox, L. R. 1969. General features of Bivalvia. In *Treatise on Invertebrate Paleontology.* R. C. Moore, ed., pp. 1-129. Univ. of Kansas.

Cragg, S. M. and D. J. Crisp. 1991. The biology of scallop larvae. In *Scallops: Biology, Ecology and Aquaculture.* S. E. Shumway, ed., pp. 75-132. Elsevier, New York.

Culliney, J. L. 1974. Larval development of the giant scallop *Placopecten magellanicus* (Gmelin). *Biol. Bull.* **147**:321-332.

Dadswell, M. J. and G. J. Parson. 1991. Potential for aquaculture of sea scallop, *Placopecten magellanicus* in Canadian Maritimes using naturally produced spat. In *Scallop Biology and Culture.* S.E. Shumway and P.A. Sandifer, eds., pp. 300-307. The World Aquaculture Society, Baton Rouge, LA.

Dakin, W. J. 1909. Pecten. *Liverpool Marine Biology Committee Memoirs* **17**:1-146.

Dakin, W. J. 1910a. The eye of *Pecten. Q. J. Microsc. Sci.* **55**:49-112.

Dakin, W. J. 1910b. The visceral ganglion of *Pecten*, with some notes on the physiology of the nervous system, and an inquiry into the innervation of the osphradium in Lamellibranchia. *Mitt. Zool. Statz. Neapel.* **20**:1-40.

Diabacco, C., G. Robert, and J. Grant. 1995. Reproductive cycle of the sea scallop *Placopecten magellanicus* on the northeastern Georges Bank. *J. Shell. Res.* **14**(1):59-69.

Dickie, L. M. 1955. Fluctuations in abundance of the giant scallop, *Placopecten magellanicus* (Gmelin), in the Digby area of the Bay of Fundy. *J. Fish. Res. Bd. Can.* **12**:797-857.

Dodge, H. 1952. The classes Loricata and Pelecypoda: Part 1 of a historical review of the mollusks of Linneaus. *Am. Mus. Nat. Hist. Tull.* **100**:182-183.

Dow, R. L. 1969. Sea scallop fishery. In *The Encyclopedia of Marine Resources.* F.E. Firth, ed., pp., 616-623. Van Nostrand Reinhold Co., New York.

Drew, G. A. 1906. The habits, anatomy, and embryology of the giant scallop (*Pecten tenuicostatus*, Mighels). *Univ. Maine Studies Series* **6**:1-89).

DuPaul, W. D., J. E. Kirkley, and A. C. Schmitzer. 1989, Evidence of a semiannual reproductive cycle for the sea scallop, *Placopecten magellanicus* (Gmelin), in the Mid-Atlantic region. *J. Shelf. Res.* **8**(1):173-178.

DuPaul, W. D., R. A. Fisher, and L. E. Kirkley. (1990). An evaluation of at-sea handling practices: Effects on sea scallop meat quality, volume and integrity. Gulf and South Atlantic Fish. Dev. Found. contract report, Virginia Inst. of Marine Sci. NA90AA-H-SK008.

DuPaul, W. D., R. A. Fisher, W. S. Otwell, and T. E. Rippen. 1993. An evaluation of processed Atlantic sea scallops (*Placopecten magellanicus*). Virginia Marine Res. Rep., Virginia Inst. of Marine Sci. **93**-1:127.

DuPaul, W.D., R. A. Fisher, and J. E. Kirkley. 1997. Natural and ex-vessel moisture content of sea scallops (*Placopecten magellanicus*). Virginia Marine Res. Rep. No. 96-5:18. Virginia Inst. of Marine Sci., College of William and Mary, Gloucester Point, VA.

Faveris, R. and P. Lubet. 1991. Energetic requirements of the reproductive cycle in the scallop, Pecten maximus (Linnaeus, 1758) in Baie De Seine (channel). In *Scallop Biology and Culture.* S.E. Shumway and P.A. Sandifer, eds., pp. 67-73. The World Aquaculture Society, Baton Rouge, LA.

Fay, C. W., R. J. Neves, and G. B. Pardue. 1983. Species profiles: life histories and environmental requirements of coastal fishes and invertebrates (mid-Atlantic) — the bay scallop. U.S. Fish and Wildlife Service FWS/OBS 82/11.2.

Fischer, W. 1978. FAO species identification sheet for fishery purposes: Western Central Atlantic (fishing area 31). *Bivalves* **6**.

Fisher, R. A., J. E. Kirkley, and W. D. DuPaul. 1990. Phosphate use in processing sea scallops, *Placopecten magellanicus,* in the mid-Atlantic region. *Proc. Fifteenth Annual Trop. and Subtrop. Fish. Tech. Conf. of the Americas.* SGR-105: 154-162.

Fisher, R. A., W. D. DuPaul, and T. E. Rippen. 1996. Nutritional, proximate, and microbial characteristics of phosphate processed sea scallops (*Placopecten magellanicus*). *J. Muscle Foods* **7**:73-92.

Fretter, V. and A. Graham. 1964. Reproduction. In *Physiology of Mollusca.* K.M. Wilbur and C.M. Yong, eds., pp. 127-164. Academic Press, New York.

Giese, A. D. and J. S. Pearse. 1974. Introduction: general principles. In *Reproduction of Marine Invertebrates.* A.C. Giese and J.S. Pearse, eds., pp. 1-49. Academic Press, New York.

Gilmour, T. H. J. 1964. The structure, ciliation and function of the lip-apparatus of *Lima* and *Pecten* (Lamellibranchia). *J. Mar. Biol. Assoc. U.K.* **44**:485-498.

Grau, G. 1959. Pectinadae of the eastern Pacific. Univ. Calif. Publ. Allan Hancock Found. Pac. Exp. 23.

Gruffydd, L.D. and A. R. Beaumont. 1972. A method of rearing *Pecten maximus* larvae in the laboratory. *Mar. Biol.* **15**:350-355).

Gutsell, J.S. 1931. Natural history of the bay scallop. *Bull. Bur. Fish. (U.S.).* **46**(193):569-632.

Hammarstrom, L. F. and M. F. Merritt. 1985. A survey of Pacific weathervane scallops (*Pecten caurinus*) in Kamishak Bay, Alaska. Alaska Dept. Fish and Game Inf. Leaf. No. 252.

Haynes, E. B. and G. C. Powell. 1968. A preliminary report on the Alaska sea scallop; fishery exploration, biology and commercial processing. Alaska Dept. Fish. Game. Info. Leaf. No. 125.

Haynes, E. B. and C. R. Hitz. 1971. Age and growth of the giant Pacific sea scallop, *Patinopecten caurinus*, from the Strait of Georgia and outer Washington coast. *J. Fish. Res. Bd. Can..* **28**:1335-1341.

Hennick, D. P. 1970. Reproductive cycle, size at maturity, and sexual composition of commercially harvested weathervane scallops (*Patinopecten caurinus*) in Alaska. *J. Fish. Res. Bd. Can.* **27**:2112-2119.

Hennick, D. P. 1971. A hermaphroditic specimen of weathervane scallop, *Patinopecten caurinus*, in Alaska. *J. Fish. Res. Bd. Can.* **28**:608-609.

Hennick, D. P. 1973. Sea scallop, *Patinopecten caurinus*, investigations in Alaska. Alaska Dept. Fish and Game, Div. of Comm. Fish. 5-23-R.

Hodgson, C. A. and R. D. Burke. 1988. Development of larval morphology of the spiny scallop, *Chlamys hastata*. *Biol. Bull.* **174**:303-318.

Ito, H. 1991. Japan. In *Scallops: Biology, Ecology and Aquaculture*. S.A. Shumway, ed., pp. 1017-1055. Elsevier, New York.

Jamieson, G. S., H. Stone, and M. Etter. 1982. Predation of sea scallops (*Placopecten magellanicus*) by lobsters (*Homarus americanus*) and rock crabs (*Cancer irroratus*) in underwater cage enclosures. *Can. J. Fish. Aquat. Sci.* **39**:499-505.

Kaiser, R. J. 1986. Characteristics of the Pacific weathervane scallop *Pecten (Patinopecten) caurinus* (Gould 1850) fishery in Alaska. Alaska Dept. Fish Game, Div. Comm. Fish., Kodiak, Alaska.

Kantha, S. S. 1989. Carotenoids of edible mollusks: A review. *J. Food Biochem.* **13**:429-442.

Kasynov, V. L. 1991. Development of the Japanese scallop *Mizuhopecten yessoensis* (Jay. 1985). In *Scallop Biology and Culture*. S.E. Shumway and P.A. Sandifer, eds., pp. 1-9. The World Aquaculture Society, Baton Rouge, LA.

Kellogg, J. L. 1892. A contribution to our knowledge of the morphology of the lamellibranchiate molluscs. *Bull. U.S. Fish Commis.* **10**:389-434.

Kellog, J. L. 1915. Ciliary mechanisms of lamellibranchs with descriptions of anatomy. *J. Morph.* **26**:625-701.

Kirbey-Smith, W.W. 1970. Growth of the scallop *Argopecten irradians concentricus* (SAY), as influenced by food and temperature, Ph.D. Thesis, Duke University. Durham, NC.

Kirkley, J. E. and W. D. DuPaul. 1991. Temporal variations in spawning behavior of sea scallops, *Placopecten magellanicus* (Gmelin, 1791), in the mid-Atlantic resource area. *J. Shell. Res.* **10**(2):389-394.

Kirkley, J. E., W. D. DuPaul, and A. Schmitzer. 1991. Factors affecting the relationship between meat weight and shell height of *Placopecten magellanicus* in the mid-Atlantic region. In *Scallop Biology and Culture*. S.E. Shumway and P.A. Sandifer, eds., pp. 134-139. The World Aquaculture Society, Baton Rouge, LA.

Langton, R. W., W. E. Robinson, and D. Schick. 1987. Fecundity and reproductive effort of sea scallops *Placopecten magellanicus* from the Gulf of Maine. *Mar. Ecol. Prog. Ser.* **37**:19-25.

Lauren, D. J. 1982. Oogenesis and protandry in the purple-hinge rock scallop, *Hinnites giganteus*, in upper Puget Sound, WA. *Can. J. Zool.* **60**:233-2336.

Loosanoff, V. L. and H. C. Davis. 1963. Rearing of bivalve mollusks. *Adv. Mar. Biol.* **1**:2-136.

MacDonald, B. A. 1984. The partitioning of energy between growth and reproduction in the giant scallop, *Placopecten magellanicus* (Gmelin). Ph.D. Thesis, Memorial University of Newfoundland. Canada.

MacDonald, B. A. 1986. Production and resource partitioning in the giant scallop *Placopecten magellanicus* grown on the bottom and in suspended culture. *Mar. Ecol. Prog. Ser.* **34**:79-86.

MacDonald, B. A. and R. J. Thompson. 1985. Influence of temperature and food availability on the ecological energetics of the giant scallop *Placopecten magellanicus*. II. Reproductive output and total production. *Mar. Ecol. Prog. Ser.* **25**:295-303.

MacDonald, B. A. and R. J. Thompson. 1986. Influence of temperature of food availability on the ecological energetics of the giant scallop *Placopecten magellanicus*. III. Physiological ecology, the gametogenic cycle and scope for growth. *Mar. Biol.* **93**:37-48.

MacDonald, B. A. and N. F. Bourne. 1987. Growth, reproductive output, and energy partitioning in weathervane scallops, *Patinopecten caurinus*, from British Columbia., *Can. J. Fish. Aquat. Sci.* **44:**152-160.

MacDonald, B. A. and R. J. Thompson. 1988. Intraspecific variation in growth and reproduction in latitudinally differentiated populations of the giant scallop *Placopecten magellanicus* (Gmelin). *Biol. Bull.* (Woods Hole) **175**:361-371.

MacKenzie, C. L. Jr., A. S. Merrill, and F. M. Serchuk. 1978. Sea scallop resources off the northeastern U.S. coast, 1975. *Mar. Fish. Rev.* **40**(2):19-23.

MacKenzie, C. L. Jr. 1979. Biological and fisheries data on sea scallop *Placopecten magellanicus* (Gmelin). NOAA Fish. Cent. (Sandy Hook Lab.) *Tech. Ser. Rep.* **19**.

Malakhov, V. V. and L. A. Medvedeva. 1986. Embryonic development in the bivalves *Patinopecten yessoensis* (Pectinida, Pectinidae) and *Spisula sachalinensis* (Cardiida, Mactridae). *Zool. Zh.* **65**:72-740.

Martin, A. W. 1983. Excretion. In *The Mollusca: Physiology*. A.S.M. Saleuddin and K.M. Wilbur, eds., pp. 353-405. Academic Press, San Diego, CA.

Mason, J. 1958. The breeding of the scallop (*Pecten maximus* L.) in Manx waters. *J. Mar. Biol. Assoc. U.K.* **37**:653-671.

Mathers, N. F. 1976. The effects of tidal currents on the rhythm of feeding and digestion in *Pecten maximus*. *J. Exp. Mar. Biol. Ecol.* **24**:271-283.

Medcof, J. C. 1949. Dark-meat and the shell disease of scallops, Prog. Rep. Atlant. Cst. Stns. **45**:3-6.

Merrill, A. S., J. A. Posgay, and F. E. Nichy. 1966. Annual marks on shell and ligament of sea scallop (*Placopecten magellanicus*). U.S. Fish. and Wildlife Ser., Fish. Bull. No. 65(2):299-311.

Merrill, A. S. 1967. Shell deformity of mollusks attributable to the hydroid, *Hydractinia echinata*. Fish. Bull. Fish Wildl. Serv. U.S. Vol. 66:273-279.

Miller, G. C., D. M. Allen, and T. J. Costello. 1981. Spawning of the calico scallop *Argopecten gibbus* in relation to season and temperature. *J. of Shell. Res.* **1**:17-21.

Moir, A. J. G. 1977. Ultrastructural studies on the ciliated receptors of the long tentacles of the giant scallop, *Placopecten magellanicus* (Gmelin). *Cell Tissue Res.* **184**:367-380.

Mottet, M. G. 1979. A review of the fishery biology and culture of scallops. State of Washington, Dept. Fisheries, Tech. Rep. No. 39.

Mullen, D. M. and J. R. Moring. 1986. Species profiles: Life histories and environmental requirements of coastal fishes and invertebrates (North America) — sea scallops. *U.S. Fish. Wildl. Ser. Biol. Rep.* **82**(11.67).

Naidu, K. S. 1970. Reproduction and breeding cycle of the giant sea scallop *Placopecten magellanicus* (Gmelin) in Port au Port Bay, Newfoundland. *Can. J. Zoo.* **48**:1003-1012.

Naidu, K. S. and F. M. Cahill. 1986. Culturing giant scallops in Newfoundland waters. *Can. MS Rep. Fish. Aquat. Sci.* **1876**.

Naidu, K. S. 1991. Sea scallop, *Placopecten magellanicus*. In *Scallops: Biology, Ecology and Aquaculture.* S.E. Shumway, ed., pp. 861-897. Elsevier, New York.

Nei, M.(1972. Genetic distance between populations. *American Naturalist* **106**:283-292.

Neumann, A. C. 1965. Processes of recent carbonate sedimentation in Harrington Sound, Bermuda. *Bull, Mar. Sci.* **15**:987-1035.

Oesterling, M.J. and W. D. DuPaul. 1993. Shallow water bay scallop, *Argopecten irradians*, Culture in Virginia. *Proceedings of the 9th International Pectinid Workshop, Nanaimo, B.C. Can.* **2**:58-65.

Oesterling, M. J. 1997. Personal communication. Virginia Institute of Marine Science, Gloucester Point, VA.

Pierce, M. E. 1950. *Pecten irradians*. In *Selected Invertebrate Types.* F.A. Brown, ed., pp. 321-324. John Wiley and Sons Inc., New York.

Posgay, J. A. 1950. Investigations into the sea scallop, *Pecten grandis*. Third report on investigations of methods of improving the shellfish resources of Massachusetts, pp. 24-30. Commonwealth of Massachusetts, Dept. Cons. Div. Mar. Fish.

Posgay, J. A. 1953. Sea scallop investigations. Sixth report on investigations of the shell fisheries of Massachusetts, pp. 9-24. Commonwealth of Massachusetts, Dept. Cons. Div. Mar. Fish.

Posgay, J. A. 1957. The range of the sea scallop. *The Nautilus* **71**:55-57.

Posgay, J. A. and K. D. Norman. 1958. An observation on the spawning of the sea scallop (*Placopecten magellanicus*) on Georges Bank. *Limnol. Oceanogr.* **3**(4):478.

Posgay, J. A. 1979. Population assessment of the Georges Bank Sea Scallop stocks in *Rapp. P.-v. Reun. Cons. Int. Explor. Mer.* **175**:109-113.

Quayle, D. B. and N. Bourne 1972. The clam fisheries of British Columbia. *Fish. Res. Bd. Can. Bull.* **179**.

Rhodes, E. W. 1991. Fisheries and Aquaculture of the bay scallop *Argopecten irradians*, in the eastern United States. In *Scallops: Biology, Ecology and Aquaculture.* S.E. Shumway, ed., pp. 913-924. Elsevier, New York.

Robinson, A.M. and W. P. Breese. 1984. Spawning cycle of the weathervane scallop *Pecten (Patinopecten) caurinus* (Gould) along the Oregon coast. *J. Shellf. Res.* **4**(2):165-166.

Robinson, W. E., W. E. Wehling, M. P. Morse, and G. C. McLeod. 1981. Seasonal changes in soft-body component indices and energy reserves in the Atlantic deep-sea scallop, *Placopecten magellanicus. Fish. Bull.* **79**:499-458.

Roe, R. B., R. Cummins, and H. R. Bullis. 1971. Calico scallop distribution, abundance, and yield off eastern Florida, 1967-1968. *Fish. Bull.* **69**:399-409.

Ronholt, L. L. and C. R. Hitz. 1968. Scallop explorations off Oregon. *Comm. Fish. Rev.* **30**(7):42-49.

Ronholt, L. L., H. H. Shippen, and E. S. Brown. 1977. Demersal fish and shellfish resources of the Gulf of Alaska from Cape Spencer to Unimak Pass, 1948-1976. NAAA/OCSSEAP **2**:1-955.

Sastry, A. N. 1962. Some morphological and ecological differences in two closely related species of scallop, *Aeqipecten irradians* (Lamarck) and *Aequipecten gibbus* (Dall) from the Gulf of Mexico. *Quart. J. Fla. Acad. Sci.* **25**:89-95.

Sastry, A. N. 1963. Reproduction of the bay scallop, *Aequipecten irradians concentricus* Lamarck. Influence of temperature on maturation and spawning. *Biol. Bull.* (Woods Hole) **125**:146-153.

Sastry, A.N. 1966. Temperature effects in reproduction of the bay scallop, *Aequipecten irradians* (Lamarck), in *Biol. Bull.* (Woods Hole) **130**:118-134.

Sastry, A. N. 1968. The relationships among food, temperature, and gonad development of the bay scallop, *Aequipecten irradians* (Lamarck). *Physiol. Zool.* **41**:44-53.

Sastry, A. N. 1970. Reproductive physiological variation in latitudinally separated populations of the bay scallop, *Aequipecten irradians* (Lamarck). *Biol. Bull.* (Woods Hole) **138**:56-65.

Sastry, A. N. 1979. Pelecypoda (excluding Ostreidae). In *Reproduction of Marine Invertebrates,* A.C. Geise and J.S. Pearce, eds., pp. 113-292. Academic Press, New York.

Schmitzer, A. C., W. D. DuPaul, and J. E. Kirkley. 1991. Gametogenic cycle of sea scallops *Placopecten magellanicus* (Gmelin, 1791) in the Mid-Atlantic Bight. *J. Shellfish Res.* **10**:221-228.

Serchuk, F. M. 1983. Seasonality in sea scallop shell height-meat weight relationships, review and analysis of temporal and spatial variability and implications for management measures based on meat count. NMFS, Woods Hole Lab. Ref. Doc. No. 83-05.

Serchuk, F. M. (1984. Fishing patterns and management measures regulating size at capture in the Georges Bank sea scallop fishery: A brief historical review. NMFS, Woods Hole Lab. Ref. Doc. 84-11.

Shirley, S.M. and G. H. Kruse. 1995. Development of the fishery for weathervane scallops, *Patinopecten caurinus* (Gould, 1850) in Alaska. *J. Shell. Res.* **14**(1):71-78.

Shumway, S. E. (ed.) 1991. *Scallops: Biology, Ecology and Aquaculture.* Elsevier, New York.

Shumway, S. E. and D. F. Schick. 1987. Variability of growth, meat count and reproductive capacity in *Placopecten magellanicus*: Are current management policies sufficiently flexible? Inter. Comm. Explor. Sea (ICES) C.M. No./K.2.

Starr, R. M. and J. E. McCrae, J.E. 1983. Weathervane scallop (*Patinopecten caurinus*) investigations in Oregon, 1981-1983. Oregon Dept. of Fish and Wldlf. Inf. Rep.. 83-10.

Stevenson, J. A. and L.M. Dickie. 1954. Annual growth rings and rate of growth of the giant scallop, *Placopecten magellanicus* (Gmelin), in the Digby area of the Bay of Fundy. *Fish. Res. Bd. of Can.* **11**(5):660-671.

Tan, F. C., D. Cai, and D. L. Roddick. 1988. Oxygen isotope studies on sea scallops, *Placopecten magellanicus*, from Browns Bank, Nova Scotia. *Can. J. Fish. Aquat. Sci.* **45**:1378-1386.

Taylor, A. C. and T. J. Venn. 1979. Seasonal variation in weight and biochemical composition of the tissues of the queen scallop, *Chlamys opercularis*, from the Clyde Sea area. *J. Mar. Biol. Assoc.* U.K. **59**:605-621.

Taylor, R. E. and J. M. Capuzzo. 1983. The reproductive cycle of the bay scallop, *Argopecten irradians irradians* (Lamarck), in small coastal embayment on Cape Cod, Massachusetts. *Estuaries* **6**(4):431-435.

Thompson, R. J. B. A. MacDonald. 1991. Physiological integrations and energy partitioning. In *Scallops: Biology, Ecology and Aquaculture.* S.E. Shumway, ed., pp. 347-376. Elsevier, New York.

Turner, H. J. Jr. 1957. Spawning and fertilization of the eggs of the bay scallop. Commonwealth of Massachusetts Dept. of Natural Resources, Division of Marine Fisheries Report on Investigation of the Shellfisheries of Massachusetts for 1957, pp. 15-16.

Turner, H. J. Jr. and Hanks, J.E. 1960. Experimental stimulation of gametogensis in *Hydroides dianthus* and *Pecten irradians* during the winter, in *Biol. Bull. Mar. Biol. Lab.* (Woods Hole) (No. 119-145-152).

Waller, T. R. 1969. The evolution of the *Argopecten gibbus* stock (Mollusca: Bivalvia) with emphasis on the tertiary and quarternary species of eastern North America, in *Jr. Paleont.* (Vol. 43 (suppl. to No. 5):1-125).

Waller, T. R. 1981. Functional morphology and development of veliger larvae of the European oyster, *ostrea edulis* (Linne) in *Smithson. Contrib. Zool.* **28**:1-70).

Wells, H. W. and Wells, M.J. (1962. The polychaete *Ceratonereis tridentata* as a pest of the scallop *Aequipecten* gibbus. *Biol. Bull. Mar. Biol. Lab.* (Woods Hole) **122**:149-159.

Wilkens, L. A. 1981. Neurobiology and the scallop. I. Starfish-mediated escape behaviors, in *Proc. R. Soc. Lond. B.* **211**:241-372.

Wilkens, L. A. (1991. Neurobiology of behavior of the scallop, in *Scallops: Biology, Ecology and Aquaculture.* S. E. Shumway, ed., pp. 429-469. Elsevier, New York.

Biology of Certain Commercial Mollusk Species:

Abalone

Robert J. Price

INTRODUCTION

Abalones are members of a large class (Gastropoda) of mollusks having one-piece shells. They belong to the family Haliotidae and the genus *Haliotis*, which means sea ear, referring to the flattened shape of the shell (Haaker et al., 1986).

Abalone shells are rounded or oval with a large dome towards one end. The shell has a row of respiratory pores. The muscular foot has strong suction power, permitting the abalone to clamp tightly to rocky surfaces. An epipodium, a sensory structure and extension of the foot that bears tentacles, circles the foot and projects beyond the shell edge in the living abalone (Haaker et al., 1986). Nine species of abalone occur in North America: black (*H. cracherodii*), flat (*H. walallensis*), green (*H. fulgens*), pink (*H. corrugata*), pinto (*H. kamtschatkana*), red (*H. rufescens*), threaded (*H. assimilis*), western Atlantic (*H. pourtalesii*), and white (*H. sorenseni)* abalone.

SPECIES OF ABALONE

Black Abalone

H. cracherodii have a black and smooth epipodium and tentacles. The shell surface is black or dark blue, and smooth. There are five to nine open pores, and the pores are flush with the shell surface.

Black abalone range from Mendocino County, California, to southern Baja California. They are found in intertidal and shallow subtidal zones down to a depth of about 20 feet. Black abalone reach 7.75 inches in length, but are commonly 5 to 6 inches long (Haaker et al., 1986).

Flat Abalone

H. walallensis have a mottled yellowish and brown epipodium, with a pebbly-appearing surface and lacy edge. The tentacles are greenish and slender. The shell is flattened, narrow, and marked with low ribs. There are five to six open pores, and the pore edges are moderately elevated above the shell surface.

Flat abalone range from British Columbia, Canada, to San Diego, California. They are found in the subtidal zone from 20 feet down to at least 70 feet. Flat abalone reach 7 inches in length, but are commonly under 5 inches (Haaker et al., 1986).

Green Abalone

H. fulgens have a mottled cream-and-brown epipodium, with tubercles scattered on the surface and a frilly edge. The tentacles are olive green. The shell is usually brown, and its surface marked

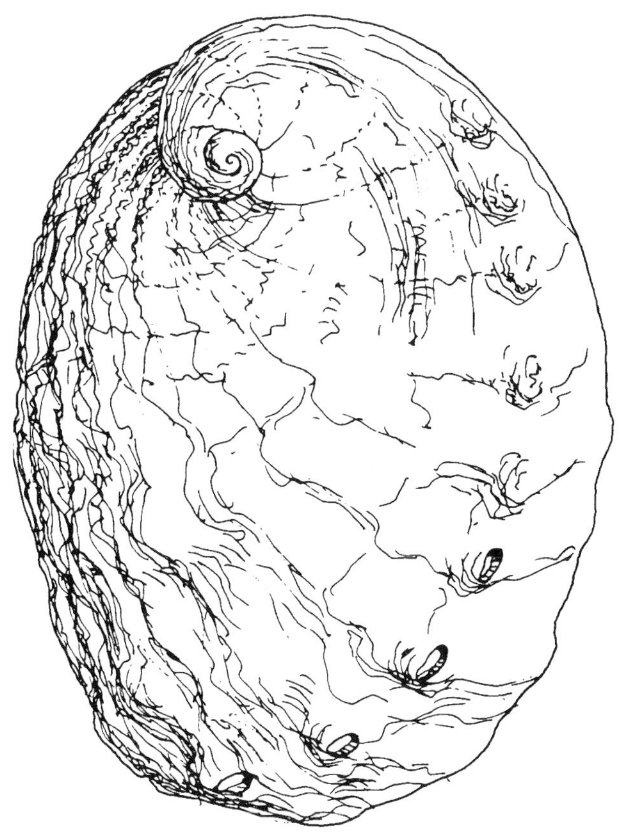

Figure 1. Abalone.

with many low, flat-topped ribs that run parallel to the pores. There are five to seven open pores, and the pore edges are elevated above the shell surface. A groove often parallels the outer edge of the line of pores.

Green abalone range from Point Conception, California, to Bahia Magdalena, Baja California. They are found in the intertidal and subtidal zones down to at least 30 feet. Green abalone are often found in crevices where surfgrass and algal cover is dense. They reach 10 inches in length, but are generally smaller (Haaker et al., 1986).

Pink Abalone

H. corrugata have a mottled black-and-white epipodium with many tubercles on the surface and a lacy edge. The foot is yellow to light orange. The tentacles are black. The shell is thick and its surface is marked with wavy corrugations. There are two to four open pores, and pore edges are strongly elevated above the surface.

Pink abalone range from Point Conception, California, to Santa Maria Bay, Baja California. They are found in the subtidal zone from 20 feet down to at least 120 feet, commonly in beds of giant kelp. Pink abalone reach 10 inches in length, but individuals over 7 inches long are now rare (Haaker et al., 1986).

Pinto Abalone

H. kamtschatkana have a mottled pale yellow to dark brown epipodium, with a pebbly appearing surface and a lacy edge. Tentacles are yellowish brown, or occasionally green, and thin. The shell is irregularly mottled and narrow. There are three to six open pores, and the pore edges are elevated above the shell surface. A groove often parallels the line of pores.

Pinto abalone range from Sitka, Alaska, to Monterey, California. They are found in the intertidal and subtidal zones down to at least 70 feet. Pinto abalone reach 6.49 inches in length, but are commonly 4 inches long. Pinto abalone are also known regionally as northern abalone (Haaker et al., 1986).

Red Abalone

H. rufescens usually have a black epipodium, but some specimens have a barred black-and-cream pattern on their epipodium. The surface of the epipodium is smooth and broadly scalloped along the edge. The area around the foot is black and the sole is tan to grey. The tentacles are black. The shell surface is generally brick red and the inside edge is often red. There are three to four open pores, and the pores are moderately elevated above the shell surface.

Red abalone range from Sunset Bay, Oregon, to Tortugas, Baja California. North of Point Conception, they are found in the intertidal and subtidal zones down to at least 60 feet. South of Point Conception, they are found in the subtidal zone down to over 100 feet. Red abalone reach 12.3 inches in length, but are commonly 7 to 9 inches long (Haaker et al., 1986).

Threaded Abalone

H. assimilis have a mottled pale yellow to dark brown epipodium with a pebbly appearing surface and a frilly edge. The tentacles are yellowish brown, short, and thin. The shell is oval and the surface is marked with prominent ribs interspersed with narrow ones. There are four to six open pores, and the pores are moderately elevated above the shell surface.

Threaded abalone range from San Luis Obispo County, California, to Bahia Tortugas, Baja California. They are found in the subtidal zone from 20 feet down to at least 80 feet, commonly on rock surfaces. Threaded abalone reach six inches in length, but are commonly smaller. Threaded abalone are considered a subspecies of the pinto abalone by some scientists (Haaker et al., 1986).

Western Atlantic Abalone

H. pourtalesii have a yellowish epipodium with large and small sensory tentacles. The sole of the foot is tan. The shell is reddish-orange. Western Atlantic abalone range from North Carolina through the Gulf of Mexico to Brazil. They are found from 187 feet down to at least 1,200 feet on hard substrates. The largest recorded shell had a length of about 1.2 inches (Titgen and Bright, 1985).

White Abalone

H. sorenseni have a tan and pebbly epipodium. The sole of the foot is orange. The shell is deep, thin, and oval. There are three to five open pores, and the edges of the pores are elevated above the shell surface.

White abalone range from Point Conception to Bahia Tortugas, Baja California. Most white abalone are found in the Channel Islands in California. White abalone are found in the subtidal zone down to at least 200 feet. They are commonly found in open, exposed areas. White abalone reach 10 inches in length, but are commonly 5 to 8 inches long (Haaker et al., 1986).

NATURAL HISTORY

Abalones reach sexual maturity at a small size, and fertility is high and increases exponentially with size. Sexes are separate and fertilization is external. The eggs and sperm broadcast into the water through the pores with the respiratory current. A 1.5-inch abalone may spawn 10,000 eggs or more at a time, while an 8-inch abalone may spawn 11 million or more eggs. The spawning season varies among species with black, green, and pink abalone spawning between spring and fall, and pinto abalone spawning during the summer. Red abalone in some locations spawn throughout the year. The fertilized eggs hatch into floating larvae that feed on plankton until their shells begin to form. Once the shell forms, the juvenile abalone sinks to the bottom where it clings to rocks and crevices with its single powerful foot. Settling rates appear to be variable. After settling, abalones change their diet and feed on macroalgae (Karpov and Tegner, 1992; Haaker et al., 1986).

Except for black abalone, hybridization between abalone species is not uncommon in areas where several species occur together. There are 12 recognized hybrids in southern California and northern Baja California (Owen et al., 1971).

Growth information is limited. Commercial sizes of 6.25 inches for pinks, 7 inches for greens and 7.75 inches for reds are reached after a minimum of 10 to 15 years in southern California (Karpov and Tegner, 1992). Pinto abalone reach about 2.5 inches in a minimum of six years (Farlinger and Campbell, 1992).

Juvenile abalones feed on rock-encrusting coralline algae and on diatom and bacterial films. Adult abalones feed primarily on loose pieces of marine algae drifting with the surge or current. Large brown algae such as giant kelp, bull kelp, feather boa kelp, and elk kelp are preferred, although other species of algae may be eaten at various times (Karpov and Tegner, 1992; Haaker et al., 1986).

Abalone eggs and larvae are consumed by filter-feeding fish and shellfish. Predators of juvenile abalones include crabs, lobsters, gastropods, octopuses, seastars, and fishes. The bat ray in southern California and the sea otter in central California prey selectively on larger abalones (Karpov and Tegner, 1992; Haaker et al., 1986).

PRODUCTION

In decreasing order of total catch between 1950 and 1995, red (46.6 percent), pink (41.2 percent), black (8.7 percent), green (3.5 percent), and white (>1 percent) abalones have all been harvested in California.

Aquaculture of red, pink, and green abalones occurs in California (Ebert, 1992). There is limited aquaculture of green and *H. diversicolor supertexta* abalones in Hawaii (Olin, 1994).

CALIFORNIA

The commercial fishery for abalones in California began in the 1850s. Chinese-Americans initially harvested intertidal green and black abalones with skiffs using long, hooked poles. This fishery was eliminated in California in 1900 by closure of shallow waters to commercial harvest. Japanese-American divers followed the Chinese-Americans as the fishery moved to the subtidal zone. Initially, free divers working from barrel floats harvested abalones. Later, hard-hat divers harvested abalones from deeper waters. In the late 1950s "hooka" gear, which supplied air from the surface to divers using light masks, fins, and wet suits, began replacing hard-hat gear. Since the 1970s, multi-hose hooka gear and specialized, high-speed, seaworthy boats have become common in the fishery (Karpov and Tegner, 1992).

In California, abalone divers used underwater diving gear consisting of an above-surface air pump operated from a boat and at least 100 feet of air hose, and had to be fully submerged while taking abalone. Abalones were taken only by hand or with abalone irons. An abalone iron is a flat device not more than 36 inches long and not less than 1/16 inch thick, with rounded smooth edges and a curve with a radius of less than 18 inches. The commercial abalone fishery in California was managed through size limits, limits on the number of permits for commercial abalone divers, and restrictions on harvesting areas. Minimum commercial size limits in California were: 7-3/4 inches for red abalone, 7 inches for green abalone, 6-1/4 inches for pink or white abalone, 5-3/4 inches for black abalone, and 4 inches for pinto, threaded, and flat abalone. Commercial harvesting was prohibited during January, February, and August. A moratorium on commercial harvesting of black abalone began in July, 1993. The California Department of Fish and Game proposed, and the Fish and Game Commission adopted effective January 1, 1995, a two-year closure on sport and commercial harvesting of pink, green, and white abalone. Prices to fishermen for red abalone were around $500 to $600 per dozen in 1993-94 [Duffy, 1997; California Department of Fish and Game (CDFG), 1994; Haaker, 1994; Wagner, 1994].

The California commercial abalone harvest reached a record 5.4 million pounds in 1957. Since then, commercial harvests have declined dramatically to about 224,792 pounds in 1996 (Table 1) (Haaker, 1994). Current stocks of most abalone species in central and southern California are overutilized. This is the combined result of commercial harvest efficiency, increased market demand, sport fishery expansion, an expanding population of sea otters, pollution of mainland habitat, unexplained mortalities of black abalone due to a condition known as "withering syndrome," and loss of kelp populations associated with El Niño events. Management efforts through size limits and limits on commercial harvesting permits have been ineffective. Reseeding experiments have not been successful. Commercial abalone harvesting in California may be eliminated if the sea otter range is not contained. Studies in a California fishery reserve have shown that even protected populations cannot support a fishery within the sea otter range in central California (Karpov and Tegner, 1992). In March 1997, commercial harvesting of abalone was banned (Duffy, 1997).

Alaska

The southeast Alaska commercial abalone fishery was sporadic and local prior to 1971. Shore picking was the primary harvesting method, but after 1960 some scuba gear was used. The fishery increased dramatically during the 1970s due to improved scuba gear, increased product demand, and the use of larger vessels. The Alaska abalone harvest reached a record 315,000 pounds in 1978-79, and then fell to about 14,352 pounds in 1995-96 when a minimum size limit was instituted

Table 1. United States abalone landings, 1977-98 (pounds)

Year	California	Alaska	British Columbia	Total
1977	1,436,154	6,981	1,046,754	2,489,889
1978	1,295,034	164,719	890,666	2,350,419
1979	992,499	315,187	434,751	1,742,437
1980	1,238,989	272,375	233,689	1,745,053
1981	1,109,651	263,394	206,352	1,579,397
1982	1,240,579	202,463	180,999	1,624,041
1983	840,112	81,654	117,506	1,039,272
1984	826,672	109,216	126,766	1,062,654
1985	762,070	67,616	96,562	926,248
1986	615,037	40,537	100,531	756,105
1987	763,056	61,224	102,294	926,574
1988	568,826	67,615	100,310	736,751
1989	730,890	76,100	105,822	912,812
1990	520,854	52,071	110,231	683,156
1991	376,980	68,386	closed	445,366
1992	519,103	44,034	closed	563,137
1993	461,376	35,988	closed	497,364
1994	327,019	34,852	closed	361,871
1995	264,334	22,879	closed	287,213
1996	224,792	14,352	closed	239,144
1997	112,751	closed	closed	112,751
1998	closed	closed	closed	0

(Table 1). The Alaska pinto abalone fishery is managed through guideline harvest ranges, a minimum legal size of 3.75 inches, a restrictive season, and local area closures for conservation and food fisheries. The fishery was closed in 1997 (Koeneman, 1994; Farlinger and Campbell, 1992).

British Columbia

Prior to 1971, the British Columbia commercial pinto abalone fishery was sporadic and local. Shore picking was the main harvest method, but after 1960 some scuba gear was used. The fishery accelerated rapidly during the 1970s due to improved scuba gear, reduced access to herring and salmon fisheries, acceptance of the pinto abalone in the Japanese market, increased product demand, and the introduction of larger vessels with freezer capacity. Abalone landings peaked in 1977 at 474.8 metric tons (1,047,000 pounds) and then declined rapidly as management of the fishery began. Landings in 1990 totaled 110,000 pounds. The British Columbia abalone fishery was managed through a minimum size limit of 100 mm (3.9 inches), vessel license limitations, vessel and fishery quotas, seasonal restrictions, and local permanent area closures. In 1991, the commercial abalone fishery was closed (Kostner, 1994; Farlinger and Campbell, 1992).

PRODUCTS

During the early years of the abalone fishery, abalones were dried and smoked, or canned for export, and sold fresh for local markets.Most abalones were exported to Japan, either fresh or frozen whole. The United States market is primarily in California for live abalone for the sashimi market, and for some fresh and frozen steaks for restaurants (Wagner, 1994; Dore, 1991).

A major change occurred in marketing United States abalones in 1993. Prior to 1993, black abalones were the primary export product. After the 1993 moratorium on black abalone harvesting, due to the "withering syndrome" that reduced black abalone stocks, red abalones took over the export market. Prices to the fishermen of $500 to $600 per dozen for red abalone made production of abalone steaks uneconomical for most markets. High prices for abalone may have also intensified illegal abalone fishing operations in closed areas (Wagner, 1994).

Abalone steaks are prepared by removing the abalone from the shell, cutting off the head and viscera, and hand-trimming the foot. Red and some green abalone are allowed to relax for 24 hours before the final trimming of the foot. This resting period weakens muscle contractions that can damage the flesh during tenderizing. The foot is then sliced horizontally across the grain of the meat. The steaks are tenderized by pounding, usually with wooden mallets, to break the tough fibers in the meat. The yield of steaks from a live abalone is about 15 percent.

The entire flesh of the abalone is edible. Traditional United States consumption has been primarily the muscle portion. The gonad, however, is considered a delicacy by the Japanese when it can be removed and eaten immediately from a live abalone. The trimmed muscles remaining after trimming for steak production were historically used for abalone burger production. As the price of abalone meat increased, these trimmings were canned. Today, they are used fresh or frozen in Asian restaurants for soups and other dishes. The primary use for abalone shells is in making mother-of-pearl inlays on furniture, produced principally in Korea. Abalone shells are also sold to shell collectors, sold as souvenirs, and used in making jewelry (Wagner, 1994; Dore, 1991; Talley, 1982).

REFERENCES

Botelho, C. 1997. Personal communication. Alaska Department of Fish and Game, Commercial Fisheries Management and Development, Southern Region, Douglas, Alaska.

CDFG. 1994. Commercial Fishing Provisions 94-01: Abalone Diving. State of California. The Resources Agency, Department of Fish and Game, Marine Resources Division. Sacramento, CA.

Dore, I. 1991. *Shellfish: A Guide to Oysters, Mussels, Scallops, Clams and Similar Products for the Commercial User*. Van Nostrand Reinhold, New York.

Duffy, J. M. 1997. Personal communication. State of California. The Resources Agency. California Department of Fish and Game, Marine Resources Division, Sacramento, CA.

Ebert, E.E. 1992. Abalone aquaculture: A North American regional review. In *Abalone of the world -- Biology, fishery and culture*. S.A. Shepherd, M.J. Tegner, and S.A. Guzmán del Próo (eds.), pp. 570-582. Fishing News Books, Cambridge, MA.

Farlinger, S. and A. Campbell. 1992. Fisheries management and biology of northern abalone, *Haliotis kamtschatkana*, in the Northeast Pacific. In *Abalone of the world -- Biology, fishery and culture.* S.A. Shepherd, M.J. Tegner, and S.A. Guzmán del Próo (eds.), pp.395-406. Fishing News Books, Cambridge, MA.

Haaker, P.L., K.C. Henderson, and D.O. Parker. 1986. California Abalone. Marine Resources Leaflet No. 11, State of California. The Resources Agency, Department of Fish and Game, Marine Resources Division. Long Beach, CA.

Haaker, P.L. 1994. Personal communication. State of California. The Resources Agency. Department of Fish and Game. Marine Resources Division. Long Beach, CA.

Haaker, P. O. 1997. Personal communication. State of California. The Resources Agency, Department of Fish and Game, Marine Resources Division, Long Beach, CA.

Koeneman, T. 1994. Personal communication. Department of Fish and Game. State of Alaska. Petersburg, AK.

Kostner, M. 1994. Personal communication. Department of Fisheries and Oceans. Vancouver, BC.

Karpov, K. A. and M. J. Tegner. 1992. Abalone. In *California's Living Marine Resources and Their Utilization.* W. S. Leet, C. M. Dewees, and C. W. Haugen (eds.), pp. 33-36. Sea Grant Extension Publication UCSGEP-92-12. Sea Grant Extension Program, Wildlife Conservation Department, University of California. Davis.

Olin, P. G. 1994. Personal communication. Hawaii Sea Grant College Program, University of Hawaii. Honolulu, HI.

Owen, B., J. H. McLean, and R.J. Meyer. 1971. Hybridization in the eastern Pacific abalones (*Haliotis*). Bulletin of the Los Angeles County Museum of Natural History Science. **9**:1-37.

Talley, K. 1982. Abalone: Question of survival. *Pacific Fishing.* March 1982: 47-51, 72-73, 75.

Titgen, R.H. and T.J. Bright. 1985. Notes on the distribution and ecology of the western Atlantic abalone, *Haliotis pourtalesii. Northeast Gulf Science.* **7**:147-152.

Wagner, M. 1994. Personal communication. Andrias Seafood Specialties. Ventura, CA.

Wilson, D. E. 1997. Personal communication. Department of Fisheries and Oceans, Burnaby, British Columbia, Canada.

Biology of Certain Commercial Mollusk Species: Pacific Snails

Robert J. Price

INTRODUCTION

Snails are members of the largest class (Gastropoda) of mollusks and have one-piece shells. This section excludes abalones, limpets, periwinkles, whelks, and conchs — all of which are gastropods (some are discussed in other sections; see Table 1).

Wavy turban *(Lithopoma undosum)* snails are in the Turbinidae or turban snail family. The shell is tan, large, and sculptured with wavy ridges and nodules. The base of the shell has strong spiral ribs. The upper parts of the shell are covered with a thick fibrous periostracum or protective layer. The wavy turban has an operculum or plate attached to the soft body parts that allows the snail to close the opening in the shell when the animal withdraws into its shell. Small wavy turbans up to 2 inches in height may be common at low water in rocky areas. Large wavy turbans reaching 4-1/3 inches are found on rocky bottoms, particularly in kelp beds. Wavy turbans range from Point Conception, California, to central Baja California (Turgeon et al., 1988; Hinton, 1987; McLean, 1969).

Moonsnails are members of the Naticidae family. The tall moonsnail *(Polinices altus)*, Lewis' moonsnail *(P. lewesii)*, and *P. draconis* are found on the Pacific Coast. The brown moonsnail *(P. hepaticus)*, milk moonsnail (*P. lacteus)*, and the white moonsnail (*P. uberinus*) are found on the Atlantic Coast. The pale moonsnail *(P. pallidus)* is found on both coasts (Turgeon et al., 1988). Several species of *Natica*, particularly the Arctic moonsnail *(N. clausa)*, occur in Alaskan waters (Table 2).

Moonsnails have a relatively low spire, a rounded body whorl, and an oval aperture. They live on sand or mud bottoms, and plow through the substrate with a greatly expanded foot that forms a shield over the head. The shells are somewhat glossy because the body envelops the shell. Moonsnails are carnivores and feed upon clams by drilling round holes with the radula (McLean, 1969). Because they are predators, moonsnails are capable of acquiring paralytic shellfish poisoning (Otto, 1994).

Moonsnail eggs are deposited in a collar-shaped structure made by cementing sand grains with mucus. Moonsnails range from 1 inch to about 5 inches in height (McLean, 1969).

Oregon triton snails (*Fusitriton oregonensis*) are members of the Ranellidae family (Turgeon et al., 1988). The shell has numerous varices and a hairy, brown epidermis. The inside of the shell is com-

Table 1. Other Mollusks of the Gastropoda Class, Some of Which Are Commercially Harvested

Abalones:
Haliotis

Whelks:
Beringius
Buccinum
Busycon
Colus
Engina
Exilioidea
Kelletia
Liomesus
Mohnia
Neptunea
Plicifusus
Ptychosalpinx
Searlesia
Volutopsius

Conchs:
Melongena
Pleuroploca
Strombus

Limpets:
Acmaea
Collisella
Diodora
Fissurella
Fissurellidea
Iothia
Lepeta
Lottia
Lucapina
Lucapinella
Notoacmea
Patelloida
Phodopetala
Problacmaea
Tectura

Periwinkles:
Algamorda
Littorina
Nodilittorina
Tectarius
Tectininus

Table 2. Other snails found in trawl surveys in the eastern Bering Sea (Otto, 1994; MacIntosh and Somerton, 1981)

Family		
Trochidae	*Margarites giganteus*	giant margarite
	M. costalis	boreal rosy margarite
	Solariella obscura	obscure solarelle
	S. micraulax	five-groove solarelle
	S. varicosa	varicose solarelle
Turritellidae	*Tachyrhynchus erosus*	eroded turretsnail
Epitoniidae	*Epitonium groenlandicum*	...
Calyptraeidae	*Crepidula grandis*	great slippersnail
Trichotropidae	*Trichotropis insignis*	gray hairysnail
	T. kroyeri	...
Naticidae	*Natica clausa*	Arctic moonsnail
Lamellariidae	*Velutina velutina*	smooth lamellaria
	V. lanigera	woolly lamellaria
	V. plicatilus	oblique lamellaria
Muricidae	*Boreotrophon clathratus*	clathrate trophon
	B. pacificus	elegant trophon
	B. muriciformis	...
Volutidae	*Arctomelon stearnsii*	Alaskan volute
Volumitridae	*Volumitria alaskana*	...
Cancellariidae	*Admete couthouyi*	northern admete
Turridae	*Aforia circinata*	keeled aforia
	Antiplanes perversa	
	Oenopota simplex	...
Pyramidellidae	*Odostomia* spp.	

monly pure white. The Oregon triton ranges from Alaska to San Diego, California (Johnson and Snook, 1927). Oregon tritons are scavengers and have been implicated in at least one case of paralytic shellfish poisoning. There are also undocumented reports of unpleasant hallucinations following consumption of the viscera of this species (Otto, 1994).

Threetooth snails (*Triodopsis picea* and *T. neglecta*) are members of the Polygyridae snail family (Turgeon et al., 1988). The shells are globe- or ball-shaped, with a characteristic semicircular toothed aperture, and a tight-fitting operculum. Threetooth snails are herbivores and browse on algae covering rocky substrate. Both species are called "pipipi" by Hawaiians who use the animals for food and the shells for leis.

Triodopsis picea, the Spruce Knob threetooth snail, has a black shell (occasionally vaguely marked with white), with incised spiral grooves.

They are about 0.6 inches in length and 0.4 inches in diameter. *Triodopsis picea* are abundant along Hawaiian shorelines on all rocky substrates from the splash zone to the high water mark. *Triodopsis picea* is rare in the Pacific, except in Hawaii and the Johnson Islands where it is abundant.

Triodopsis neglecta, the Ozark threetooth snail, has a smooth, black, flecked-with-white, shell. They are about 0.75 inches in length and 0.6 inches in diameter. *Triodopsis neglectas* are endemic to the Hawaiian Islands. They are able to live in waters with wide variations in salinity, and are found at the seaward edges of basalt and solution benches, and in tide pools and brackish water. They are always found immersed, both on the surface of the substratum and under rocks and rubble (Kay, 1979).

HARVESTING

Moonsnails, Oregon tritons, and other snails are commonly taken as by-catch in trawl fisheries or in crab pots. Several species of moonsnails, particularly the Arctic moonsnail, Oregon tritons, keeled aforia, and other snails are taken in small quantities in an eastern Bering Sea pot fishery for whelks (Otto, 1994). Hawaiian threetooth snails are taken by hand (Kay, 1979). Wavy turbans are considered a potential West Coast snail fishery.

PROCESSING

At present, snails are not processed before sale.

MARKET FORMS AND GRADES

Snails are sold live for food, and there is a small market for live snails for aquariums. Primary world markets for snails are in Japan, Chile, Peru, and several areas of Europe (Otto, 1994).

REFERENCES

Hinton, S. 1987. *Seashore Life of Southern California*, revised edition. University of California Press. Berkeley, CA.

Johnson, E. J. and H.J. Snook. 1927. *Seashore Animals of the Pacific Coast*. New York: Dover Publications Inc.

Kay, E. A. 1979. *Hawaiian Marine Shells*. Bernice P. Bishop Museum Special Publication 64(4), Bishop Museum Press. Honolulu, HI.

MacIntosh, R.A. and D.A. Somerton. 1981. Large marine gastropods of the eastern Bering Sea. In *Eastern Bering Sea Shelf: Oceanography and Resources*. D.W. Hood and J.A. Calder (eds.), pp. 1215-1227. Office of Marine Pollution Assessment, University of Washington Press. Seattle.

McLean, J. H. 1969. *Marine Shells of Southern California*. Science Series 24, Zoology No. 11. Los Angeles County Museum of Natural History. Los Angeles, CA.

Otto, R.S. 1994. Personal communication. National Marine Fisheries Service. Alaska Fisheries Science Center, Kodiak.

Turgeon, D.D., A.E. Bogan, E. V. Coan, W.K. Emerson, W.G. Lyons, W.L. Pratt, C.F.E. Roper, A. Scheltema, F.G. Thompson, and J.D. Williams. 1988. *Common and Scientific Names of Aquatic Invertebrates from the United States and Canada*. Special Publication 16, American Fisheries Society. Bethesda, MD.

Biology of Certain Commercial Mollusk Species: Octopus

Robert J. Price

INTRODUCTION

Octopuses are members of the Cephalopoda class of mollusks, which includes squids and cuttlefishes. They are in the Octopodidae family and the majority of the species are placed in the genus *Octopus* (Turgeon et al., 1988). Octopuses have a soft saclike body, a large head with a mouth on the undersurface, and eight arms bearing suckers. The mouth consists of a set of chitinous jaws or a beak equipped with a radula, or tongue-like oral structure, with rows of small teeth. The beak is used to tear up food and the radula helps take it into the mouth (Hochberg, 1994; Duffy, 1992; Paust, 1988).

North American West Coast octopuses include the California two-spotted octopuses (*Octopus bimaculatus* and *O. bimaculoides*), California bigeye octopus (*O. californicus*), Pacific pygmy octopus (*Paroctopus digueti*), North Pacific giant octopus (*Enteroctopus dofleini*), lilliput octopus (*O. fitchi*), red octopus (*O. rubescens*), and smoothskin octopus (*Benthoctopus leioderma*). North American East Coast octopuses include the Caribbean reef octopus (*O. briareus*), the white-spotted octopus (*O. macropus*), Atlantic common octopus (*O. vulgaris*), brownstriped octopus (*O. burryi*), Caribbean two-spotted octopus (*O. filosus*), longarm octopus (*Macroctopus defilippi*), and Atlantic pygmy octopus (*Paroctopus gabon*). The Cane's or day octopus (*O. cane*) and the white-striped or night octopus (*O. ornatus*) are found in Hawaiian waters. Locally, the Hawaiian octopuses are referred to as the "day squid" (he'e mauli) and the "night squid" (he'e pü loa) (Hochberg, 1994; Turgeon et al., 1988; Kay, 1979).

Sexes are separate in octopuses. During copulation, the male places packets of sperm (spermatophores) in the body cavity of the female using a modified third arm with a suckerless, groove-shaped tip. The male usually dies a few months after mating with the female. Females lay 100 to 100,000 or more eggs attached singly or in grape-like clusters to a hard substrate. Females brood their eggs for several weeks to several months. The females usually do not feed while guarding the nest and typically die shortly after the eggs hatch. The young of some species hatch out with well-developed ink sacs and arms capable of feeding and crawling, while the young of other species go through a planktonic larval phase after hatching (Hochberg, 1994; Duffy, 1992).

The life span of octopuses varies with species from about six months to three-to-five years. Adults range in weight from one ounce to over 400 pounds, and in size from two inches to about 30 feet in diameter (from arm tip to arm tip) (Hochberg, 1994; Duffy, 1992).

Octopuses are found in many diverse habitats, including rocks, kelp holdfasts, and soft bottom substrates such as sand and mud. Most octopuses live in rocky areas where small caves, crevices, and rocks serve as dens or homes. Some species live in sandy areas and apparently do not have permanent homes. Octopuses are thought to be territorial because some individuals are known to defend their territory against other octopuses entering their area (Hochberg, 1994; Kato, 1994; Duffy, 1992; Talley, 1982).

They feed on crustaceans, other mollusks, and small fishes. Some species feed exclusively at night, while others are only active during the day or at dawn or dusk (Hochberg, 1994; Duffy, 1992). Often they prey on crustaceans caught in traps on the North American West Coast (lobster, Dungeness crab, tanner crab, king crab, and spot shrimp) and on the southeast and Gulf coasts (stone crab) (Lang and Hochberg, 1997; Paust, 1988; Voss, 1985). Many species drill holes in the shells of crabs, clams, and snails, and inject a poison to paralyze or kill their prey and liquefy the muscle tissue.

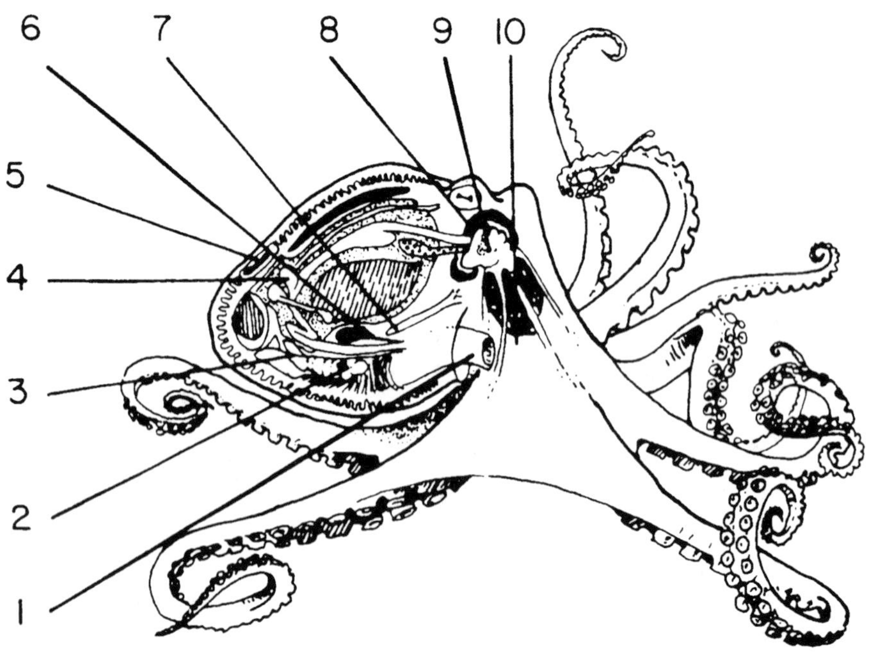

Figure 1. Octopus. 1-Funnel, 2-Gill, 3-Heart, 4-Stomach, 5-Shell rudiment, 6-Ink Sac, 7-Poison gland, 8-Skull, 9-Brain, 10-Beak.

Predators include marine birds, marine mammals, sharks, and fishes (Hochberg, 1994; Duffy, 1992).

HARVESTING

On the North American Pacific Coast, the California bigeye octopus and the red octopus contribute most of the catch south of Point Conception, California (Duffy, 1992). The North Pacific giant octopus makes up most of the catch from northern California through Alaska and the Aleutian Islands (Paust, 1988; High, 1976). The common octopus contributes most of the catch on the Atlantic and Gulf coasts (Voss, 1985). In Hawaii, the day octopus is the most common catch (Kay, 1979).

Historically, Hawaiians harvested both the day and night octopuses during the summer and fall. The day octopus was preferred for food. Octopuses are caught incidentally in fish and shrimp traps in the northwestern Hawaiian Islands, or with spears, traps, and lures in the main islands (Uchida and Uchiyama, 1986; Kay, 1979).

In northern California during the 1920s, octopuses were harvested with "devilfish" pots, cone-shaped wicker baskets with a funnel-shaped mouth. The traps were four to five-and-a-half feet high and about two to three feet in diameter at the wide end. They were constructed of rattan, which was deemed better than wire because of its flexibility when in contact with rocks. The catches were made in rocky areas in 20 to 30 fathoms of water. Most of the catch consisted of North Pacific giant octopus, with an average weight of 20 to 30 pounds (Duffy, 1992).

Several pot fisheries for octopus were attempted on the North American West Coast in the 1950s, but most were not successful. By the mid-1960s, increasing harvesting costs, decreasing market prices, and low-cost imported octopus destroyed the small commercial effort (Talley, 1982).

In the late 1970s and early 1980s, experimental pot, trap, and longline fisheries for octopus were attempted for giant octopus on the North American West Coast, and for the common octopus on the United States southeast coast and the Gulf of Mexico (Voss, 1985; Talley, 1982). Again, high fishing costs, erratic catches, low market prices, and competition from imported octopus limited any commercial fishery development.

On the North American West Coast, octopuses now are taken commonly as by-catch in trawl fisheries for bottomfishes, in longline (hook) fisheries, and in trap fisheries for crabs and shrimps. On the Atlantic and Gulf coasts, octopuses are incidental catches in trawl and trap fisheries (Lang and Hochbert, 1997; Duffy, 1992; Paust, 1988).

PROCESSING

Octopus is available live, fresh, frozen, and/or cooked. Processed octopus normally is cleaned by removing the viscera and eyes. The beaks may or may not be removed. The ink sac may be left with the animal (Voss, 1985). Spiced and boiled octopus (one to seven pounds dressed weight) and pickled octopus (seven to 15 pounds dressed weight) are processed in Asia (Paust, 1988).

MARKET FORMS AND GRADES

Frozen octopus is sold head-on in 10- to 50-pound frozen blocks. Grading is usually one-half to one pound, one to two, two to four, four to six, and six pounds and up. Frozen octopus blocks often are wrapped in plastic. Some frozen octopuses are individually plastic bagged. Imported cooked octopus often is packed with the arms turned back around the head and mantle, exposing the beak and suckers (Voss, 1985).

The major United States market for octopus, especially the North Pacific giant octopus, is for use as bait in the Pacific halibut fishery. A sizable portion of Japan's exports of octopus also is shipped to the United States and Canada for use as halibut bait. Domestic use of octopus as a food is limited, and it is purchased primarily by people of Chinese, Japanese, Spanish, Portuguese, Italian, or Greek descent (Paust, 1988). There is a small market for live octopus in biomedical research.

REFERENCES

Duffy, J. M. 1992. Octopuses. In *California's Living Marine Resources and Their Utilization*. W. L. Leet, C. M. Dewees, and C. W. Haugen (eds.), pp. 194-195. Sea Grant Extension Publication UCSGEP-91-12. Sea Grant Extension Program, Wildlife and Fisheries Biology Department, University of California, Davis.

High, W. L. 1976. The Giant Pacific Octopus. In *Marine Fisheries Review*. **88**(9):17-22.

Hochberg, F. G. 1994. Personal communication. Santa Barbara Museum of Natural History, Santa Barbara, CA.

Kato, S. 1994. Personal communication. Larkspur, CA.

Kay, E. A. 1979. Hawaiian Marine Shells. Reef and Shore Fauna of Hawaii, Section 4: Mollusca. Bernice P. Bishop Museum Special Publication 64(4). Bishop Museum Press. Honolulu, HI.

Paust, B. C. 1988. Fishing for Octopus: A Guide for Commercial Fishermen. Alaska Sea Grant Report No. 88-3. University of Alaska, Fairbanks.

Lang, M. A., and F. G. Hochberg. 1997. *Proceedings of the Workshop on the Fishery and Market Potential of Octopus in California*. Smithsonian Institution. Washington, DC.

Talley, K. 1982. The Fish of the Month: Octopus. In *Pacific Fishing Magazine*. 3(1):39-43.

Turgeon, D. D., A. E. Bogan, E. V. Coan, W. K Emerson, W. G. Lyons, W. L. Pratt, C. F. E. Roper, A. Scheltema, F. G. Thompson, and J. D. Williams. 1988. Common and Scientific Names of Aquatic Invertebrates from the United States and Canada: Mollusks. *American Fisheries Society Special Publication* 16.

Uchida, R. N. and J. H. Uchiyama. 1986. Fishery Atlas of the Northwestern Hawaiian Islands. NOAA Technical Report NMFS 38. National Marine Fisheries Service. Department of Commerce, Washington, DC.

Voss, G. L. 1985. Octopus Fishery Information Leaflet. Rosenstiel School of Marine and Atmospheric Science. University of Miami, FL.

Biology of Certain Commercial Mollusk Species:

Blue Mussels: A Case Study

Carter Newell

INTRODUCTION

The blue mussel, *Mytilus edulis*, belongs to the genus *Mytilus* (*mitilos*, Greek for sea mussel), and is part of the family Mytilidae which originated over 400 million years ago (Gosling, 1992). While mussels are most conspicuous in the intertidal zone along the New England coast of the United States, they occur from North Carolina to Canada on the East Coast and from California to Alaska on the West Coast, in depths up to 200 meters (Tressler and Lemon, 1951). Currently, more than 17 species are harvested and cultured worldwide (Lutz et al., 1991), but the annual worldwide production of over half a million metric tons (whole live weight) is dominated by the species *Mytilus edulis*. The global distribution of *Mytilus* based on genetic and morphological data (MacDonald et al., 1991) illustrates that *Mytilus edulis* is the species primarily harvested along the eastern United States, along eastern Canada, and in Europe; *Mytilus galloprovincialis* is common in southern Europe and along the Mediterranean, and is cultured in the Pacific Northwest; and *Mytilus trossulus* is common in the Baltic, northeast China, northeast Canada, and to a limited extent on the Northeast and Northwest coasts of the United States. The green-lipped mussel (*Perna viridis* and *Perna canaliculus*) also makes a significant contribution to world landings in China, India, eastern Asia, and New Zealand. Mussels are cultured successfully through bottom culture, suspended culture (ropes, rafts, longlines), and pole culture; they may even be scraped from offshore platforms.

Due to their ubiquitous appearance in temperate coastal waters, a wealth of information is available about the biology, ecology, and culture of the blue mussel (Gosling, 1992; Lutz et al., 1991, 1980; Bayne et al., 1976; Korringa, 1976). Mussels may grow to over 12 centimeters in shell length, and commonly live to be 18 to 24 years old (*Mytilus edulis*) (Thiesen, 1974). Some species (*M. californianus*) may live to be more than 50 years old (Suchanek, 1981). Beds of *M. edulis* attached to the ocean bottom may dominate the marine community in favorable locations, typically reaching a standing crop of over two kilograms dry tissue weight per square meter (Newell, 1990; Prins and Smaal, 1990), and up to more than 100,000 individuals per square meter after initial settlement from the plankton stage (Chalfant et al., 1980). Mussel beds may reach considerable thickness, over 10 centimeters deep in subtidal beds, serving both to increase vertical mixing over the mussel bed (Newell, 1993) and the supply of particulate food available to the population. As mussel communities age, they support a high diversity of other marine species [over 69 species in six marine invertebrate phyla (Tsuchiya and Nishibira, 1985; Suchanek, 1979)], and may accumulate sediment deposits up to 14 kilograms dry weight per square meter (Nixon et al., 1971).

FUNCTIONAL MORPHOLOGY

Because of their habit as epifauna, mussels may be found attached to rocks, piers, and floats, as well as to themselves and other shells in bottom cultures; attachment is made possible with strong byssal threads at the base of the foot. The shell is composed mostly of calcium carbonate (aragonite and calcite). The mussel shell is covered with a proteinaceous layer called the periostracum which may wear off in older specimens, revealing the blue and silver shell layers below. Market-sized mussels usually range between five to eight centimeters, but in good areas shell growth may exceed 10 centimeters. Like other bivalves, the mussels have an inhalant and exhalant siphon along the shell margin, or mantle (which secretes the shell). The body or visceral mass has pairs of mussel gills on each side for feeding and respiration.

Conspicuous at the front end of each gill are the labial palps, which help to sort particles of high organic content from the filtered suspension. Under normal conditions, the feeding mussel would be pointed with its mouth down and siphons up at the wide end of the shell, holding on to the substrate with a strong byssal thread and rapidly closing the valves on occasion in order to reject sediment particles or to protect the animal from wave action, predators, or desiccation during low tide. Since the gonad is in the mantle tissue, the composition and color of mussel meats may change in relation to natural cycles of nutrient storage and reproduction. During buildup of glycogen after spring algal blooms, the mantle appears plump and white and tastes sweet. After development and during spawning, the females may turn yellowish or orange due to egg development, while the males remain more or less white. After spawning, the meats are smaller, less sweet, and slightly saltier.

FEEDING

Mussels are filter-feeders, and may process large amounts of suspended particulate material (seston), which is dominated by silt and clay sediment particles, marine phytoplankton, and organic detritus. Gill retention efficiencies are low for bacteria but increase to nearly 100 percent for particles three to five microns in diameter and larger. Mussels can ingest rather large particles. Benthic diatoms as large as ¼ millimeter long were found in the gut analyses of subtidal mussels from the Gulf of Maine (Newell et al., 1989). Pelagic and benthic marine diatoms, ciliates, dinoflagellates, and nanoflagellates are the primary food source of the sea mussel. It is the trophic dynamics of the marine phytoplankton which limit the production of these filter feeders. In temperate climates such as Maine, total particle concentrations along shallow, subtidal coastal zones average about 10 to 20 million particles per liter of seawater, with the living phytoplankton cells comprising about 10 to 35 percent of the particles in early summer (Newell et al., 1989). With filtration rates ranging from one to five liters per hour per gram dry weight, a one-square-meter mussel bed might clear as much as 10,000 liters of the overlying water of all the particles between three and 250 microns diameter in one hour. On an ebb tide, a Maine mussel bed of 1.4 kilograms dry weight per square meter was observed to deplete the particulate organic carbon in the bottom 10 centimeters of the water column by 15 times (Muschenheim and Newell, 1992). Experiments with field benthic flumes over mussel beds have demonstrated not only high filtration rates on living plankton [over 50 milligrams of chlorophyll per hour per square meter in the Netherlands (Prins and Smaal, 1990; Dame and Dankers, 1988)], but also the significant release of inorganic nutrients (ammonia, phosphate, and silicate) by direct release from the bivalves and associated fauna — as well as the mineralization of feces and mucus-bound excess feed (pseudofeces). These nutrients are then taken up rapidly by the phytoplankton, resulting in a feedback loop. Thus, mussel populations serve to stabilize phytoplankton concentrations in shallow coastal waters. For example, a 5 percent increase in suspension feeder biomass would control a doubling of nutrient loadings (Small, 1994). In these natural "feed lot systems," the "herbivorous bivalves play . . . the same role as large mammals in cultivated terrestrial systems" (Dame, 1993), grazing on natural plant matter under both wild and cultured conditions. The relatively high levels of Omega-3 fatty acids in mussels attest to the high nutritional quality of their single-celled phytoplankton and detritus food source.

The mussel gill has been described as a "pump" (Jorgensen, 1991), with water currents developed by the beating of cilia on the ciliary tracts of the mussel gill. Maximum pumping rates are achieved when the mussel valves are fully open. Particles are cleared from suspension on the gills, where they are swept toward the mouth along ciliary tracts, and they may be sorted for organic rich particles prior to ingestion along specialized mechanisms to adapt both their feeding physiology and their feeding behavior to cyclic changes in seston quantity and quality (Bayne, 1993). When conditions of food and temperature are good, mussels may reach a market size of over 50 millimeters shell length in one year.

LIFE HISTORY

Mussel fertilization takes place in open water, and within 24 hours a ciliated trochophore stage has developed. A larval shell is then secreted. Mussel larvae swim during their planktotrophic stages using a ciliated swimming organ called the velum. After two to four weeks, depending on food and water temperature, at about 1/4 millimeter shell length, the larvae settle and metamorphose into a "juvenile" or plantigrade mussel seed or spat. Often the larvae will initially settle on filamentous algae or eelgrass, away from epibenthic predators. After growth to about 1 millimeter a month or so later, they develop, through special glands in the foot, the capability of secreting a single-fiber, long byssal thread (up to 20 centimeters long for a 1-millimeter mussel), which allows them to redistribute themselves to a final benthic habitat. On the flood tide, the "wandering juveniles" may detach from the location of primary settlement, secreting the drifting thread and casting it off to settle passively to the bottom to attach to rocks, shells, or an adult mussel bed. During certain periods of the year, larval mussel concentrations may exceed 5,000 per cubic meter in good seed collection areas (Newell et al., 1991; Newell, unpublished data). Recruitment to experimental collectors may exceed 1.5 million per square meter of collector at some sites. Individual mussels (of at least 7 centimeters in length) may produce eight million eggs a year, and larger individuals up to 40 million a year (Thompson, 1979). Therefore, the natural productivity of mussel populations is potentially enormous. Due to their ubiquitous appearance in the world's oceans, mussels are a preferred food source for many marine predators, both in the planktonic larval stage and on the bottom by epibenthic predators that include crabs, predatory snails, starfish, eider ducks and other shorebirds, sea otters, and the two-legged predator, man. Manipulations that result in higher survival rates of predator-prone small seed mussels or high individual growth rates of properly-thinned cultures have been successful in making mussel culture one of the leaders in marine aquaculture worldwide.

MUSSEL MODELING

In order to determine the optimal bottom densities for mussel farms in Maine, Great Eastern Mussel Farms has developed a computer simulation, or carrying capacity model, called MUSMOD©, which can be run on a personal computer. The model has inputs such as current speed, food concentration, and temperature, and determines through a series of equations the growth of mussel meat and shell in relation to their seeding densities. Thus, the mussel model MUSMOD (Campbell and Newell, 1998) can allow the mussel farmer to increase productivity by seeding low current areas at lower densities and increase growth rate and meat yields by reducing the "seston depletion effect" when too many mussels strip out the food for their upstream neighbors. Utilization of the model in eastern Maine in 1994 allowed for a doubling of growth rates on the farms, and has provided a new tool for evaluation of potential culture sites.

MUSSEL AQUACULTURE

It has been estimated that by the year 2000 over 25 percent of the food coming from the world's oceans will be from aquaculture. Mussels are one of the leaders in terms of world production in metric tons, with a variety of culture strategies. A review of world production in 1988 was summarized by Hickman (1992). The top eight producers of mussels, with their production methods, are summarized in Table 1.

In longline or raft culture, mussel seed (which is often obtained from collectors or attached to seaweed along the shore) is tubed up into a mesh sock and suspended off floating rafts, buoys, or lines held to the surface by buoys (longlines) at typical densities of about 400 to 600 per meter. The mussel seeds work their way to the outside of the sock, attaching with their powerful byssus threads, and grow at a rapid rate suspended off the bottom due to the relatively unlimited access to sus-

Table 1. Major Producers of Mussels (Hickman, 1992)

Country	Production Method	Annual Production (MT)
China	Longline, raft	429,000
Spain	Raft	210,000
Italy	Longline, raft	85,000
Netherlands	Bottom	78,000
Denmark	Bottom	73,000
France	Pole, longline	55,000
Thailand	Pole	35,000
Germany	Bottom	31,000

pended particles (seston). During harvesting, the ropes (usually 5 to 15 meters long) are pulled to the surface and the mussels are stripped off, washed, graded, and debyssed (i.e., excess byssal thread is removed). In pole culture, mussel seeds are attached by wrapping socks around poles driven in the mud. In areas with a high tidal range, the poles may be serviced by special amphibious vehicles which may drive out on the mudflat (France) or float at high tide. In bottom cultures, pioneered by the Dutch in the Osterschelde in large sailing scows, large mechanized boats use four 2-meter bottom drags, catching 100 metric tons a load, and pumping out seed at appropriate densities on the farm sites which are chosen for high current speeds and algal densities. Appropriately thinned cultures reach market size in about two years, with over 7 kilograms of meat per 25 kilograms during peak meat-yield periods. As the world realizes the potential of mussel culture from a variety of methods, this relatively efficient converter of phytoplankton biomass to tasty seafood will continue to lead world production in seafood aquaculture. Recent developments in mussel raft culture on both the east and west coasts of the United States will add significantly to the U.S. landings of farmed mussels in the upcoming years.

PLANTING SEED FOR TOMORROW'S OCEAN HARVEST: MAINE'S MUSSEL FARMERS

If you were to travel down the Saint George peninsula on the mid-coast of Maine, you would notice two things: the lobster traps and fishing boats sitting at the end of the many driveways, and the old granite quarries hugging the coastline. The granite industry died more than 50 years ago, and fishing is becoming more difficult. Much of the stocks of cod and haddock are on the verge of collapse due to overfishing, and only the state's 4,000-strong lobster boat fleet has had strong seasons recently.

On the site of an old granite quarry midway down the peninsula sits Great Eastern Mussel Farms Inc., a successful aquaculture operation

Figure 1. Harvesting half-grown seed mussels along the Maine coast. (Cindy McIntyre photo, courtesy of Great Eastern Mussel Farms.)

which is literally planting seed for the future. More than 200 million seed mussels were dropped on the ocean's bottom in 1998 by this innovative company. The seed will mature to market size using the bottom-culture techniques originally developed in Europe.

According to the United States Department of Agriculture (USDA), during the 1980s aquaculture production rose steadily from 200 million pounds to 800 million pounds. Unlike the wild fisheries, aquaculture has the advantage of consistent quality, price, and availability. This means that if a seafood item is placed in a supermarket feature or on the menu at a restaurant, the buyer is assured of obtaining good quality at the expected price.

This all begins with finding seed beds of mussels which are located in channels, along bars, and inside bays throughout the Gulf of Maine. These are one-year-old "baby" mussels which grow clumped together in shallow saltwater generally six to 20 feet deep at low tide. While one can spot these black streaks of mussel beds in the blue waters from aircraft, most of the time the boat captains locate them by a second sense developed from years on the water. They know where to look.

Captain "Hub" Bradford has the distinction of operating the F/V Saint George. There are numerous mussel or scallop draggers working the Maine Coast which can harvest shellfish, but the Saint George is the only vessel seeding mussels for the future.

The 60-foot, steel-hulled vessel is designed after the boats which seed and harvest bottom-cultured mussel beds in Holland. The Saint George has a large, deep hold which runs from her bow to midship. The hold can carry up to 30 tons of seed mussel. The seed is dredged up from the ocean's bottom with a metal meshed net (Figure 1) and dropped into the hold. The Saint George then plots a course to one of the many lease sites to plant the seed (Figure 2) in an area which provides optimum growth.

Since the State of Maine owns the ocean's bottom out to the three-mile federal limit, Great Eastern Mussel Farms must apply to the state's Department of Marine Resources in order to obtain lease sites. Once a public hearing is held and a lease is granted, only Great Eastern Mussels can use the area for aquaculture, although other activities such as recreational boating or fishing are still allowed.

Figure 2. Spreading seed mussels at a bottom-culture site in Maine. (Cindy McIntyre photo, courtesy of Great Eastern Mussel Farms.)

The most important factors for choosing a grow-out site are good water quality and fast currents which bring the maximum amount of food (plankton) to the shellfish. Sites must be deep enough not to be exposed at low tide so that the mussels filter-feed 24 hours per day for optimum growth rates. Also considered in picking a bottom site is a lack of natural predators. Starfish, for example, will wrap around a mussel shell and eat the meat; eider ducks will dive 30 feet to eat seed mussels, which they digest shell and all.

The seed mussel is broadcast over the site at just the right density; then the mussel farmers let nature do all the work. No food or chemicals are ever introduced to the site.

After 18 to 24 months, the seed mussel has grown out to market-size (about 2 to 3 inches in length). The Saint George or one of several independent mussel harvesting boats will return to the site to harvest. The mussels are hauled up with a meshed net, then placed in a cylindrical cage called a "stern washer" which returns any mud or small seed back into the ocean. The boats harvest only as much product as the sales force projects they will sell the following day at prices which are stable. This is in direct contrast to boats which catch as much as they can of wild stocks of fish to sell at fluctuating market prices.

When the mussels arrive at the processing plant in Tenants Harbor, they are first placed in 20,000 gallon re-watering tanks for 24 hours. They will filter saltwater through their systems, purging out any sand or grit. This process results in mussels that need no additional cleaning other than the removal of their beards (byssal threads). After being graded for size and inspected for breakage, the mussels are boxed and shipped via distributors to restaurants and supermarkets throughout the United States and Canada.

Seafood lovers who like clams will find mussels richer tasting, less expensive, and more versatile than their molluscan cousins. Mussels appear on menus steamed in wine or beer; breaded and deep-fried; served in sauces over pasta; and made into chowders as well as the classic Mediterranean shellfish stews: *bouillabaisse*, *cioppino*, and *paella*. Mussels are low in calories, fat, and cholesterol, while high in protein and heart-healthy Omega-3 fatty acids.

As the world's population increases, aquaculture will play an increasingly important role in providing the food we eat. Mussel aquaculture is particularly productive, yielding 50,000 pounds of mussels per acre of ocean bottom. This is 10,000 pounds of edible meat. Even the richest grasslands in the United States can produce only about 150 to 200 pounds of beef per acre, making mussel bottom culture 50 times more productive.

NUTRITION FROM MUSSELS

The heart-healthy qualities of shellfish have been widelyy publicized over the past few years. Study after study reveals the beneficial effects of Omega-3 fish oils, which are found in especially high quantities in edible blue mussels and certain other seafoods.

While mussels contain very little cholesterol (only 50 milligrams per 3.5-ounce serving), they feed on phytoplankton, which is rich in Omega-3 fatty acids. Recently published studies, including one from *The New England Journal of Medicine*, have shown that Omega-3 acids help prevent heart disease.

Mussels contain 0.5 percent Omega-3 fatty acids by weight, making them a good source of this beneficial oil. In fact, a 3.5-ounce serving of mussels provides 480 milligrams of Omega-3 fatty acids, more than is currently provided in fish oil concentrates.

The American Journal of Clinical Nutrition, citing a 1991 University of Washington study, indicated that mussels can dramatically lower fats in the blood. The ratio of "bad" (low-density lipoprotein) to "good" (high-density lipoprotein) cholesterol (LDL:HDL) decreased in the test subjects who ate a three-week diet of mussels instead of meat, eggs, and cheese. Of particular interest was a two-fold increase in the highly protective HDL-2 "good" cholesterol. A diet of clams, oysters, and crabs showed similar results.

Shellfish were once thought to contain too much cholesterol; previous measurements may have been misleading since plant and seafood cholesterol had been analyzed together.

Mussels have less than half the calories of lean roast beef or roast chicken, and one-third as many as hamburger. According to the USDA, mussels contain as much protein per serving as a T-bone steak, with only 172 calories and 60 percent less

Table 2. Nutritional Composition of Mussels

Cooked Meat 3.5 ounces (100 grams)	Common Blue Mussel	T-bone Steak (choice)
Calories	172	214
Protein	23.80 g	26.13 g
Fat	4.48 g	10.37 g
Carbohydrates	7.39 g	0.00 g
Cholesterol	56.00 mg	80.00 mg
Calcium	33.00 mg	7.00 mg
Magnesium	36.00 mg	26.00 mg
Phosphorus	285.00 mg	208 mg
Potassium	268.00 mg	407.00 mg
Iron	6.721 mg	3.00 mg
Omega 3s	782.00 mg	0.00 mg
Omega 6s	36.00 mg	290.00 mg

fat. Mussels are also extraordinarily high in vitamins.

When compared to other seafood items, mussels also rank high on the nutrition scale (Table 2). Mussels have fewer calories than shrimp or sole, less cholesterol, more vitamins, and equivalent amounts of protein, fat, and iron. The positive results from eating mussels can be obtained only with low-fat recipes. Deep-fat frying or heavy cream sauces will, of course, counteract the heart-healthy benefits.

COOKING METHODS

First-time consumers may hesitate when trying cultivated mussels, recalling barnacle-covered, gritty harvests of their own along ocean shorelines. But with high-quality standards, bottom- and surface-cultivation techniques, and 24-hour soaking in clean seawater after harvest, supermarket mussels are easier than ever to prepare. Here are some methods. [For a comprehensive list of mussel recipes, see the *Great Eastern Mussel Cookbook* (McIntyre and Callery, 1995).]

Steamed

Mussels are steamed in a small amount of water, wine, or beer with their own liquid adding to the broth, and can be steamed with onions, carrots, and garlic or chives. Common spices are saffron, parsley, thyme, Old Bay seasoning, and curry. Steam over a high heat in a covered pot about five minutes, shaking the pot two or three times if it boils over. The mussels are done when the shells pop open and the meat pulls away. DO NOT OVERCOOK, or the meats will get tough and dry. Pour the broth over mussels in individual bowls, garnish with lemon or parsley, and serve with French bread.

Microwaved

Use a shallow baking dish covered with plastic. Cook on high five to six minutes until the shells pop open. Dip in garlic butter or salsa.

Baked

After steaming and shucking mussels from their shells, the meats can be baked in seafood casseroles, au gratin dishes, Newburghs, or quiches.

Roasted

Mussels, clams, and oysters can be cooked in a 450° oven for about 15 minutes. Place mussels in an iron frying pan or shallow roasting dish and roast until the shells pop open.

Barbecued

Lay mussels on the grill in a single layer about four to five inches over the hot coals. They will steam in their own liquid and pop open when done (about five minutes). Remove them from the grill immediately, so as not to overcook.

Stove-top Lobster Bake

Take a large lobster and layer the following ingredients in order: Seaweed (rockweed), whole red potatoes, lobster, corn-on-the-cob (with only the outer layer of husk removed), and mussels. Add seaweed to the top and fill with two inches of salted water. Cover and steam for about 30 minutes until the potatoes are tender and the mussels pop open.

Stewed

Some of the greatest seafood recipes are the famous shellfish stews that originate from the Mediterranean. Mussels, clams, shrimp, scallops, or crab are mixed with Portuguese sausage or chunks of white fish in a spicy saffron tomato base. The Italian version is *zuppa di pesce* or *cioppino*; the French, *bouillabaisse*; the Portuguese, *alentejana* (with pork), or *mariscada*.

Marinated Salads

Cooked, shucked mussel meats can be mixed with oil and vinegar (or lemon juice), diced peppers, onions, celery, parsley, and cilantro or other herbs and spices to make a mussel salad vinaigrette. These seafood salads are often served inside a scooped-out tomato or avocado. The shortcut recipe is simply to marinate mussel meats in Italian or Caesar salad dressing and let them sit overnight. The marinade will preserve freshly-cooked mussel meats for about one week under refrigeration.

REFERENCES

Bayne, B.L. 1979. Marine Mussels: Their Ecology and Physiology. Cambridge University Press. Cambridge, England.

Campbell, D. E. and C. R. Newell. 1998. MUSMOD©: A mussel production model for use on bottom culture lease sites. *J. Exp. Mar. Biol. Ecol.* **219**:171-203.

Chalfant, J., T. Archamabault, and A.E. West. 1980. Natural stocks of mussels: growth, recruitment and harvest potential. In *Mussel Culture and Harvest: A North American Perspective. Developments in Aquaculture and*

Fisheries Science. 7. R.A. Lutz (ed.), pp. 38-68. Elsevier, Amsterdam.

Dame, R.F. and N. Dankers. 1988. Uptake and release of material by a Wadden Sea mussel bed. *J. Exp. Mar. Biol. Ecol.* **118**:207-216.

Dame, R.F. 1993. The role of bivalve filter feeder material fluxes in estuarine ecosystems. In *Bivalve Filter Feeders in Estuarine and Coastal Ecosystem Processes*. R.F. Dame (ed.), pp. 245-270. NATO ASI Series G. Vol. 33. Springer-Verlag, Berlin.

Gosling, E. 1992. The mussel *Mytilus*: ecology, physiology, genetics and culture. *Developments in Aquaculture and Fisheries Science*. **25**. Elsevier, Amsterdam.

Hickman, R. W. 1992. Mussel cultivation. In *The Mussel Mytilus: Ecology, Physiology, Genetics and Culture*. E. Gosling (ed.), pp. 465-510. Elsevier, Amsterdam.

Jorgensen, C. B. 1991. Bivalve filter feeding: hydrodynamics, bioenergetics, physiology and ecology. Olsen and Olsen, Fredensborg.

Korringa, P. 1976. Farming marine organisms low in the food chain; a multidisciplinary approach to edible seaweed, mussel and clam production. Elsevier Science Publishers, B. V., Amsterdam.

Lutz, R. A., K. Chalermwat, A. Figueras, R. G. Gustafson, and C. R. Newell. 1991. Mussel aquaculture in marine and estuarine environments through the world. In *Culture of Estuarine and Marine Bivalve Mollusks in Temperate and Tropical Regions*. W. Menzel (ed.), pp. 57-97. CRC Press Inc., Boca Raton, FL.

Lutz, R. A. (ed.) 1980. *Mussel Culture and Harvest: A North American Perspective*. Elsevier Science Publishers, B. V., Amsterdam.

MacDonald, J. H., R. Seed, and R. K. Koehn. 1991. Allozyme and morphometric characters of three species of *Mytilus* in Northern and Southern hemispheres. *Mar. Biol.* **111**:323-335.

Muschenheim, D. K. and C. R. Newell. 1992. Utilization of seston flux over a mussel bed. *Mar. Ecol. Prog. Ser.* **85**:131-136.

Newell, C.R., S. E. Shumway, T.L. Cucci and R. Selvin. 1989. The effects of natural particle size and type on feeding rates, feeding selectivity and food resource availability for the mussel *Mytilus edulis* (Linnaeus, 1758) at bottom culture sites in Maine. *J. Shellfish Res.* 8:187-196.

Newell, C.R. 1990. The effects of mussel [*Mytilus edulis* (Linnaeus, 1758)] position in seeded bottom patches on growth at subtidal lease sites in Maine. *J. Shellfish Res.* **9**:113-118.

Newell, C.R., H. Hidu, B.J. McAlice, G. Podnieskinski, F. Short and L. Kindblom. 1991. Recruitment and commercial seed procurement of the blue mussel, *Mytilus edulis*. *J. World Aquaculture Soc.* **22**:134-152.

Newell, C.R. and S.E. Shumway. 1993. Grazing of natural particulates by bivalve molluscans: a spatial and temporal perspective. In *Bivalve Filter Feeders in Estuarine and Coastal Ecosystem Processes*. R.F. Dane (ed.), pp. 85-1148. NATO ASI Series Vol G 33. Springer-Verlag, Berlin.

Newell, C. R. Unpublished data. Tenants Harbor, ME.

Nixon, S.W., C.A. Oviatt, C. Rogers, and K. Taylor. 1971. Mass and metabolism of a mussel bed. *Oceologia*. **69**: 341-347.

Prins, T.C. and A.C. Smaal. 1990. Benthic-pelagic coupling: the release of inorganic nutrients by an intertidal bed of *Mytilus edulis*. In *Trophic Relationships in the Marine Environment*.

Barnes and R.N. Gibson (eds.), pp. 89-103. Aberdeen Univ. Press, Aberdeen.

Smaal, A.C. 1991. The ecology and cultivation of mussels: new advances. *Aquaculture*. **94**:245-261.

Suchanek, T.H. 1979. The *Mytilus californianus* community: studies on the composition, structure, organization, and dynamics of a mussel bed. Ph.D. Thesis. University of Washington.

Suchanek, T.H. 1981. The role of disturbance in the evolution of life history strategies in the intertidal mussels *Mytilus edulis* and *Mytilus californianus*. *Oceologia*. **50**:143-152.

Theisen, B.F. 1973. The growth of *Mytilus edulis* (Bivalvia) from Disko and Thule district, Greenland. *Ophelia*. **12**:152.

Thompson, R.J. 1979. Fecundity and reproductive effort of the blue mussel *Mytilus edulis*, the sea urchin *Strongylocentrotus droebachiensis* and the snow crab *Chioneccles Opili*, from populations in Nova Scotia and Newfoundland. *J. Fish. Res. Board Can.* **36**:955-964.

Tsuchiya, M. and M. Nishibira. 1985. Islands of *Mytilus* as habitat for small intertidal animals: effect of island size on community structure. *Mar. Ecol. Prog. Ser.* **25**:71-81.

Biology of Certain Commercial Crustaceans:

Warm-Water Shrimp Fisheries of the United States

Russell J. Miget and Michael G. Haby

INTRODUCTION

Figures 1 through 4 illustrate the major contribution of the Gulf of Mexico and southeastern United States to the United States shrimp fishery, both in terms of product landed (81.3 percent) as well as value (93.4 percent). The significant difference between pounds and value is a reflection of the pricing structure of shrimp — the larger, warm-water shrimp landed in the Gulf and South Atlantic being worth considerably more per pound than the smaller cold-water species landed off New England and the North Pacific. In the past 20 years shrimp have ranked either first or second annually in landed value of all United States fisheries, even though production often ranked seventh or eighth, again attesting to the relatively high value of the product. Even with domestic production hovering around 300 million pounds

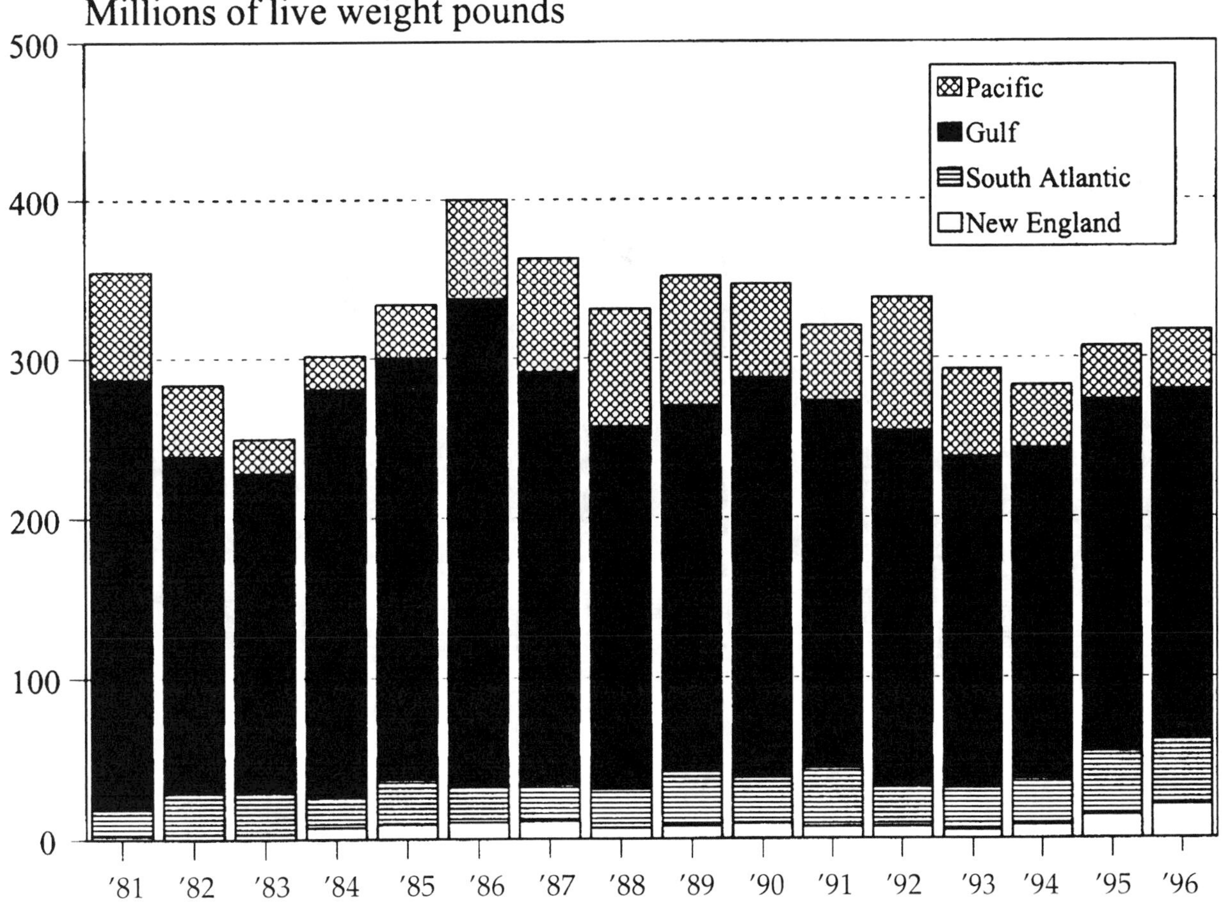

Figure 1. Annual production of shrimp by geographic region, 1981-1996.

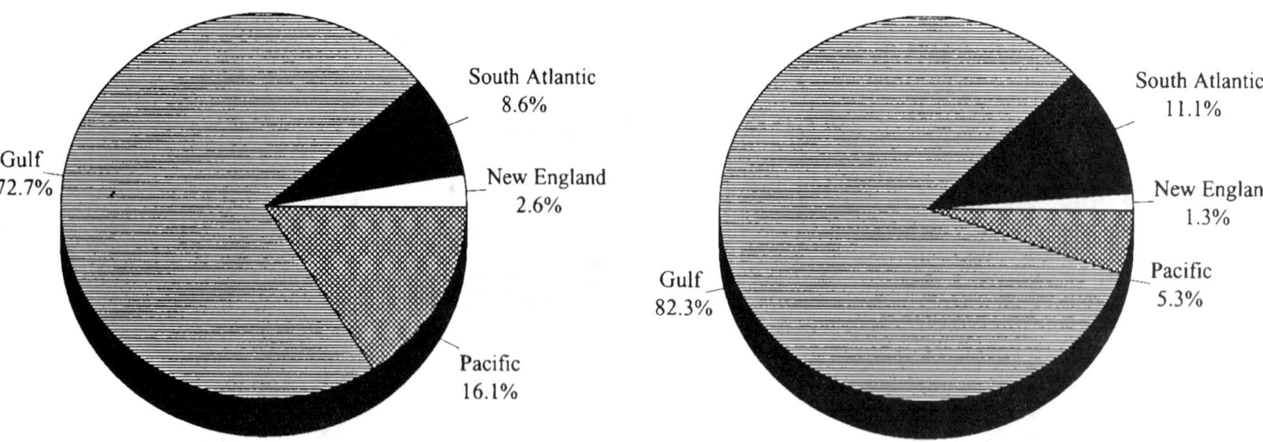

Figure 2. Average annual percentage contribution to total U.S. shrimp production by region.

Figure 4. Average annual percentage contribution to total U.S. shrimp ex-vessel value by region.

Millions of dollars

Pacific
Gulf
South Atlantic
New England

700
600
500
400
300
200
100
0

'81 '82 '83 '84 '85 '86 '87 '88 '89 '90 '91 '92 '93 '94 '95 '96

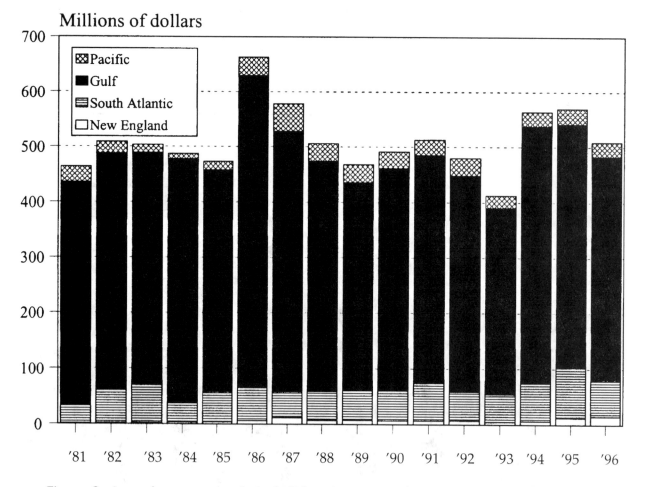

Figure 3. Annual ex-vessel value of shrimp by geographic region, 1981-1996.

live weight annually, the United States still must import 80 percent of the shrimp consumed.

Shrimp commercially harvested in the Gulf of Mexico and United States South Atlantic are comprised primarily of three members of the family Penaeidae (genus *Penaeus*) and include the white shrimp (*P. setiferus*), the brown shrimp (*P. aztecus*), and the pink shrimp (*P. duorarum*), along with lesser production of rock shrimp (*Sicyonia brevirostris*), seabobs (*Xiphopenaeus kroyeri*), and royal red shrimp (*Hymenopenaeus robustus*). In order to understand the development of the shrimp fishery in the southeastern United States and Gulf of Mexico, one must first understand the biology of the major species.

BIOLOGICAL LIFE CYCLE

The life cycles of the three most abundant commercial species are similar, although the various stages of development occur during different seasons of the year and in different locations (Figure 5). Adult white shrimp spawn in offshore waters from the beach out to 120 feet, primarily in March through June. Brown shrimp spawn principally in deeper waters 90 to 240 feet in the fall, whereas pink shrimp prefer intermediate depths of 90 to 150 feet and are found primarily in warmer waters off southwestern Florida and the Mexican Yucatan peninsula. They spawn year-round. Spawning areas generally coincide with offshore fishing areas, which in most fisheries would create concerns regarding recruitment overfishing. However, due to their fecundity, large numbers, and relatively short life span (12 to 18 months), harvesting the spawning stock below sustainable levels has been less of a management concern than environmental and habitat issues. Recently, however, regulatory agencies have begun to reconsider the effects of fishing pressure on the spawning stock of white shrimp.

Adult females release up to one million eggs per spawn. Eggs are fertilized via a sperm packet (spermatophore) attached to the female as they are released into the water. Within 24 hours eggs hatch and the animals begin a series of metamorphoses through 10 to 11 larval stages culminating after two to three weeks in miniature adults about ¼-inch long called postlarvae. Through a series of

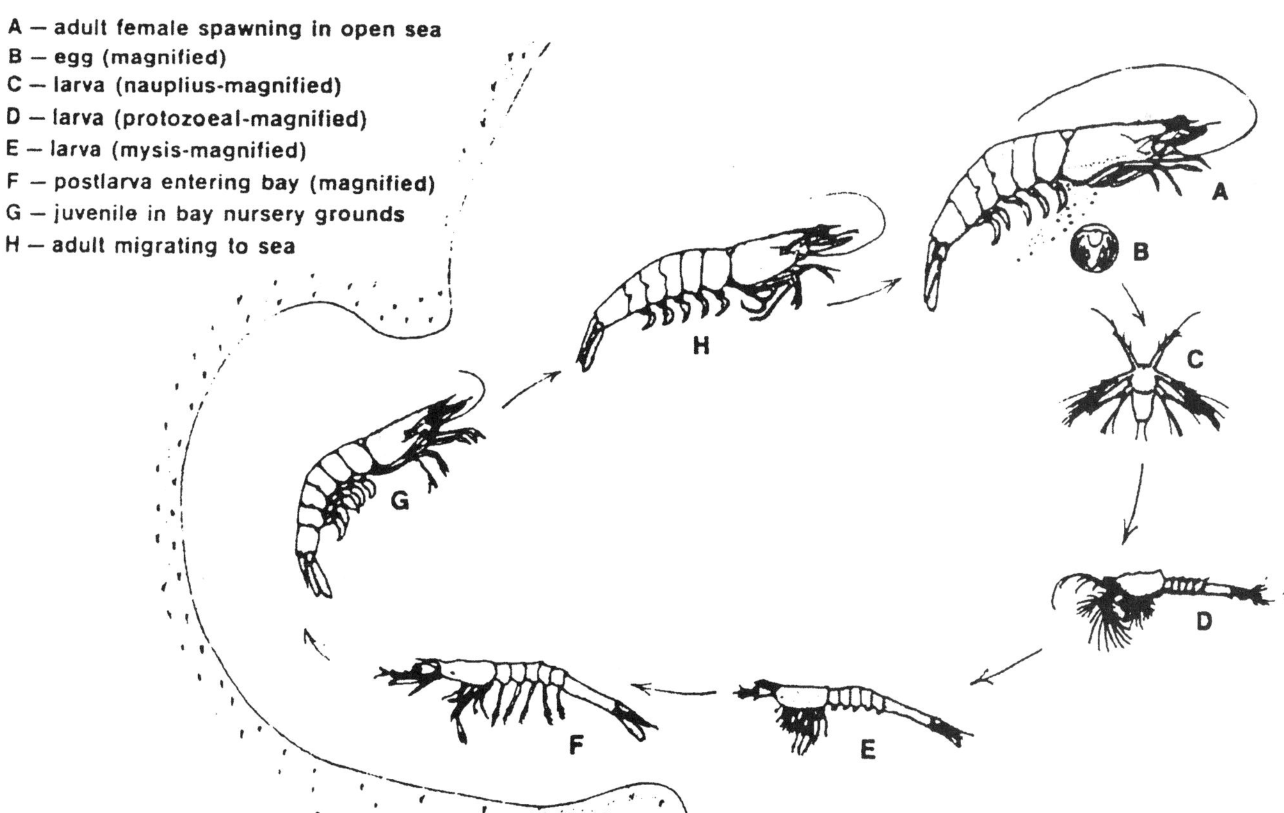

Figure 5. The life cycle of shrimp.

as-yet-unverified physical and chemical cues, the now-actively-swimming postlarvae make their way to passes leading into coastal embayments, ride the tides into the bays, and take up residence in the protective shallow bay and estuarine systems fringing the Gulf of Mexico and southeastern United States. Growth is rapid in these shallow, warm, faunal rich waters, and the juvenile shrimp reach 4 to 6 inches in less than six months, at which time they begin their emigration to offshore waters where they become sexually mature and repeat the cycle. As the shrimp grow they continue to move to deeper waters farther offshore. Although physiologically influenced by salinity, growth rate is determined largely by water temperature, the most rapid growth taking place in the warm summer months. For this reason all three species time their offshore spawning peaks so that offspring arrive in the estuaries during the warmer months of the year (Knoff, 1970).

This life cycle has resulted in the development of two somewhat distinct fisheries throughout the region, a bay and a Gulf fishery, each with its own trawler size and fishing methods as well as regulations governing fishing seasons and gear. These will be discussed in the context of the species descriptions which follow.

SHRIMP SPECIES

White Shrimp (*Penaeus setiferus*)

White shrimp were the principal species commercially harvested throughout the region until about 1948, at which time the offshore stocks of brown and pink shrimp were discovered and began to be exploited. Today white shrimp account for only about 30 percent of the total Gulf and southeastern United States harvest of warm-water shrimp. As indicated above, the entire life cycle of this species occurs relatively close to shore and in the bays; thus the white shrimp fishery was historically, and continues to be primarily, a somewhat local and inshore operation utilizing relatively small, often single-trawl vessels. White shrimp survive well in brackish waters, and appear to tolerate flood periods of fresh water better than brown or pink postlarvae and juveniles (Hildebrand and Gunter, 1953). The soft mud and sand bottoms off the Louisiana coast are especially productive white shrimp grounds. Likewise, white shrimp are often the dominant species caught off the southeastern United States. Fishing is primarily a daytime operation with abundance influenced strongly by season and geographic location. White shrimp production generally peaks in the fall and early winter from nearshore spawns which occurred in late spring/early summer.

Brown Shrimp (*Penaeus aztecus*)

Since the first commercial catches of brown shrimp were landed in Texas in 1947, this species has grown in importance in the Gulf shrimp fishery and now comprises over 60 percent of the total Gulf and southeastern United States catch. Brown shrimp are especially important off the Texas coast from Freeport down into Mexican waters, comprising over 80 percent of reported landings from these areas. The majority of these shrimp are somewhat older and therefore larger than white shrimp when harvested; thus, they bring a significantly higher price per pound. As brown shrimp grow, they move to deeper water and become less concentrated, thereby increasing harvesting expenses. Brown shrimp are almost exclusively harvested offshore at night as they come out of their shallow burrows and actively forage for food. Peak production occurs between July and October; however, due to their increase in size (i.e., higher price per pound), the offshore fishery for brown shrimp continues into the winter months even though production drops off dramatically.

Pink Shrimp (*Penaeus duorarum*)

Like brown shrimp, pink shrimp are generally bottom- and nocturnal-feeders. Although distributed throughout the range of the other commercial penaeids (North Carolina south through Florida and the entire Gulf of Mexico), pink shrimp are concentrated in areas where there is a wide shelf, protected waters, and a rather firm bottom of mud, silt, or coral sand. Thus pink shrimp are relatively scarce on the soft muddy bottoms from Mississippi to mid-Texas. In fact, Texas combines pink with brown shrimp landings data, having determined over several years that pinks constitute only about 5 percent of the brown/pink offshore harvest. Largest concentrations are found in 40 to 100 feet of water off the Florida Gulf coast near the Dry Tortugas and Sanibel Island as well as the Bay of Campeche off the Mexican Yucatan Peninsula. With the closure of Mexican waters to foreign fishing in 1978, United States catches have come primarily from the west coast of Florida. Interestingly, pink shrimp have been shown to be the only Penaeid to overwinter in the estuaries of North Carolina, having been collected in 6° C waters. However, even though they are more tolerant of low temperatures than are brown or white shrimp, the northern limit of their range is some 2 to 3 degrees latitude south of the others.

Rock Shrimp (*Sicyonia brevirostris*)

Rock shrimp are quite distinct from the *Penaeus* species as they have a thick shell that gives a stony appearance and a rostrum that does not extend past the eye. Although this species ranges from Norfolk, Virginia, throughout the Gulf of Mexico to the Yucatan and Campeche Banks, it is most frequently captured on sand and shell-sand bottoms between 50 and 200 feet. Adult rock shrimp are generally smaller than the *Penaeus* species, averaging between 4 to 5 inches long. Peak spawning off northeast Florida was observed November through January, while most abundant larvae were observed October through February off west Florida. Since this species is only occasionally caught in bays and estuaries, a life span of 20 to 22 months is assumed to occur entirely offshore. Centers of abundance that have been identified include: offshore Cape Lookout, North Carolina; Cape Canaveral and Cape San Blas, Florida; the Louisiana coast; and the Campeche Banks.

Seabobs (*Xiphopenaeus kroyeri*)

It is assumed that seabobs, like rock shrimp, spend their entire life cycle in offshore waters. This species does not attain the size of the major Penaeid species, usually growing to only 4 to 5 inches at harvest. While not a major contributor to the overall shrimp fishery, seabobs can contribute significantly to a shrimper's income during periods of slack production.

Royal Red Shrimp (*Hymenopenaeus robustus*)

In the 10-year period 1955 through 1965, commercial quantities of royal red shrimp were discovered in deep waters off the continental shelf in the Gulf of Mexico by the research vessel "Oregon." Initial assessment indicated not only commercial quantities, but a shrimp that was somewhat larger than the Penaeid species, thus worth more per pound. On the downside, however, was the investment in gear needed to capture these animals. Concentrated at depths between 1,000 and 2,000 feet, and more than 50 miles from shore, harvesting these shrimp presented a challenge for vessels accustomed to fishing in waters up to 200 feet deep. Special hydraulic winch systems, modified trawl doors capable of withstanding water pressure at 2,000 feet without warping, mud ropes designed to prevent the net from bogging down in extremely soft bottoms, and state-of-the-art navigational systems and depth recorders made, and continue to make, royal red shrimp-fishing an economically risky venture. Catch rates vary widely; therefore, boats have gotten into and out of the business quickly. As a result, no clear trend as to the economic potential of this fishery has emerged over the years. Maximum production of royal reds in United States waters has been from a relatively small area (200 square miles) offshore between St. Augustine and Ft. Pierce, Florida; a 100-square mile area southwest of Dry Tortugas (off the southwest Florida coast in the Gulf); and a major 700-square mile area off Mobile, Alabama. Royal red shrimp are thought to complete their entire life cycle offshore.

REGULATIONS

As with all marine products, shrimp are a public resource to be equitably divided among competing user groups. However, unlike marine finfish resources, the recreational demand for shrimp is minimal; therefore, regulations are principally designed to provide allocations within the shrimping industry itself. Such regulations generally fall into one or more of the following categories: fishing area closure, seasonal closure, restrictions on time of day or hours fished, and restrictions on gear designed to optimize the size of shrimp harvested and thereby prevent growth overfishing. Each state has the authority to set its own rules, which vary considerably based on the principal species harvested, as well as the historical development of the fishery and resulting makeup of the commercial fleet.

Recalling the life cycle of the major commercial species, virtually the entire crop of each major species of shrimp is concentrated in estuaries or in protected nearshore areas as juveniles and subadults, prior to moving offshore. For this reason individual states have totally restricted shrimping in specific "nursery" areas. They further regulate the "bay" shrimp fishery as to gear and season in order to allow a portion of the annual crop to migrate out of the bays and nearshore areas where they are eventually harvested at a larger, more valuable size by the offshore fleets. Setting the opening dates for both the inshore/bay fishery as well as the offshore season is based largely on estimated population size, growth rates, predicted market value for various-size shrimp, and even the lunar cycle as it influences migration.

With the passage of the Magnuson-Stevens Fishery Conservation and Management Act of 1976, the United States extended its jurisdiction over fishery management to 200 miles offshore, as opposed to the previous 3-mile limit. Fishing in these offshore waters is now regulated through plans promulgated by both the South Atlantic as well as the Gulf of Mexico Fishery Management Councils. These regional councils coordinate with the various states to develop plans for species which routinely migrate between federal and state waters, such as shrimp.

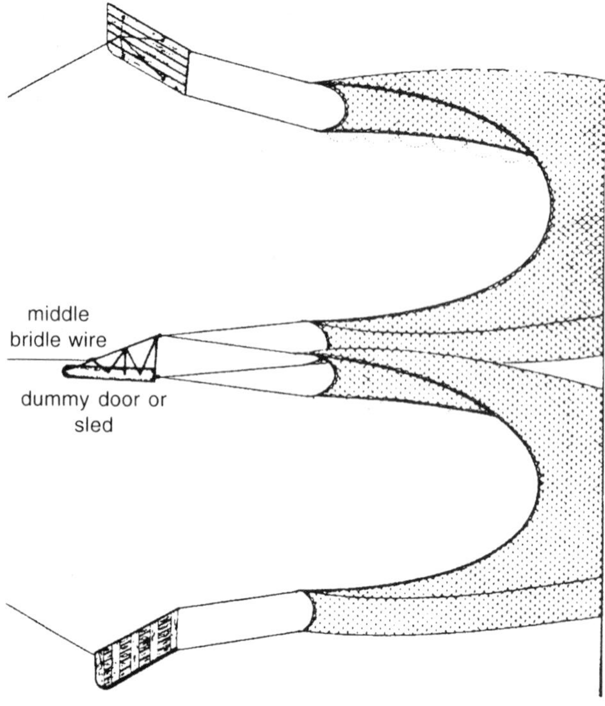

Figure 6. Illustration of a four-net rig.

THE SHRIMP-FISHING FLEETS

At the turn of the century, shrimp fishing was restricted to cast nets and haul seines. Seines were set from shore around schools of shrimp and hauled by hand. For this type of harvest, sufficiently large quantities were near shore only during very limited times of the year (Anderson, 1958). However, circa 1912 fishermen in North Carolina observed biologists catching shrimp in a small otter trawl. These fishermen constructed a larger version, and soon the first "shrimp" trawl was put into use near Fernandina Beach, Florida. It was not long before this gear technology was transferred to the Gulf of Mexico and the days of the cast net and haul seine were numbered.

The next big advance in shrimp-fishing technology came in the 1940s when the use of war surplus diesel engines allowed the construction of larger vessels. Fishermen began replacing their 25 foot "luggers" with 50- to 65-foot wooden boats capable of fishing 75- to 120-foot wide nets, and safely fishing further offshore for extended periods of time. The modern day offshore fleets are comprised of vessels which have basically "refined" this equipment. Vessel lengths have increased only slightly. Most boats are now constructed of fiberglass or steel rather than wood. Engine size has been increased to around 300 horsepower, allowing vessels that originally pulled one large net to switch to double rig trawling (two nets, one pulled from an outrigger on each side of the boat). In the 1980s fishermen began pulling rigs referred to as twin trawls (four nets, two pulled from each outrigger) (Figure 6). Another major innovation in shrimp-fishing efficiency occurred in the 1960s when shrimp nets, previously made of cotton, were constructed of nylon. A typical offshore trawler carries a crew of three: a captain, a rigman, and a header. A relatively few larger boats recently entering the fishery (80 to 90 feet) may carry a crew of four or five.

As previously mentioned, shrimp fishing in the early days was confined largely to the bays and nearshore areas of the Gulf and southeastern United States because of technological limitations (i.e., small vessels and haul seines). However, even as the larger offshore fleets began to develop, a number of fishermen were content to continue fishing the bays. The capital investment was not as great (a 25- to 40-foot vessel), trips were gener-

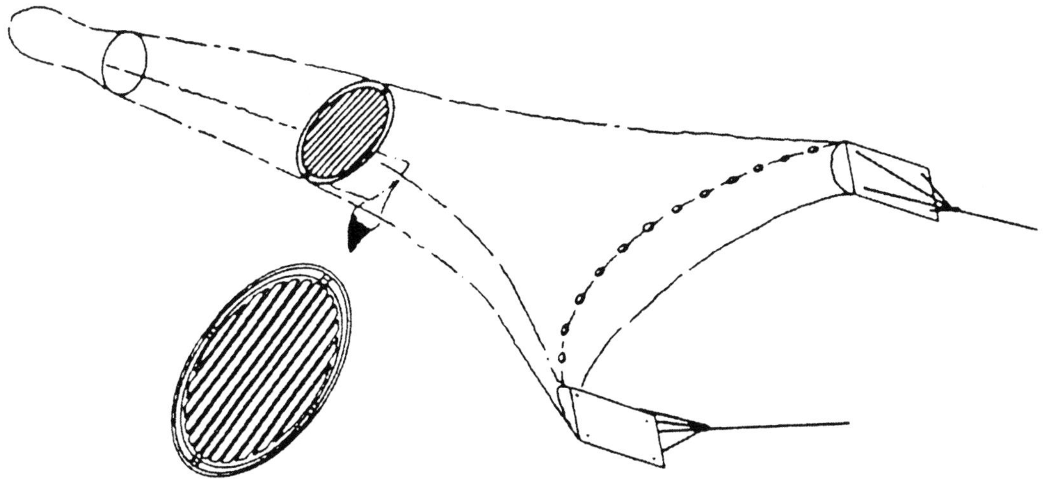

Figure 7. Illustration of a Turtle Excluder Device (TED) — a grid-like device that is placed in front of the bag or cod end of a shrimp trawl. When large objects — such as turtles, large fish, or jellyfish — enter the TED, they encounter bars slanting upward at a 45° angle. These bars force them up through a trap door and out of the shrimp net.

ally limited to a single day, and in most cases the operator was the owner. Bay boats are typically operated by the captain with a single deckhand. This bay fishery continues today. Its contribution to total production depends on the fishery in each particular state. For example, in Texas the bay fishery contributes, on average, 20 percent of annual landings with offshore Gulf production accounting for the remaining 80 percent, both in pounds and value. In Louisiana, however, the numbers are almost reversed, with the majority of landings coming from the bay/inshore fleet.

Over the years fishermen have continued to be the driving force in developing more efficient gear, as one might expect. However, recent concerns regarding the impact that shrimp trawling may be having on non-target species has prompted some gear development/modifications which are not primarily designed to improve the shrimp-harvesting efficiency of the trawl. The first of these came about in the early 1990s to reduce the impact of shrimp trawling on marine turtles. Turtle Excluder Devices (TEDs) (Figure 7) are now mandated gear for shrimp trawling. More recently regulatory agencies have been assessing various By-catch Reduction Devices (BRDs) (Figure 8) to determine which are most effective in reducing non-target species (bycatch) catch rates while simultaneously minimizing shrimp loss.

Figure 8. Illustration of a By-catch Reduction Device (BRD).

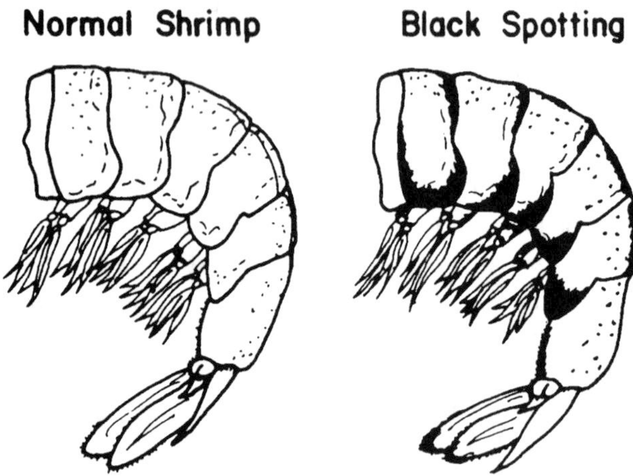

Figure 9. Normal and black spot in shrimp.

PROCESSING AND MARKETING SHRIMP

Historical Review

Investment in shore-side processing facilities was a natural reaction to increased, year-round shrimp harvests. Yet, shrimp processors initially faced a monumental product-preservation and materials-handling dilemma (Figure 9). But just as pioneering shrimp fishermen invented, adapted, and explored in boosting harvests and productivity, processors also found solutions to their problems.

Specially-designed freezing equipment enabled processors to inventory shrimp for sale throughout the year. Today, practically all shrimp are distributed and sold in the frozen state. Likewise, as processors began marketing shrimp according to count sizes (i.e., the number of tails per pound), some objective, efficient means of sorting shrimp into customary-size intervals was required. This equipment, developed in Harlingen, Texas, is still the standard means worldwide for classifying shrimp by size.

Early success in processing and marketing shrimp was measured by being able to meet year-round commitments. Typically the market for shrimp was best during the first half of the year, but production occurred during the second half. Thus, processors purchased shrimp when it was available, held it in frozen storage, and drew down inventories between January and June. With ex-vessel price inversely proportional to local abundance and knowledge that wholesale prices typically increased during the first half of the year, earnings from inventory speculation became a significant source of revenue stream for many processors.

Current Operating Conditions

Inventory speculation still occurs as a normal course of business operations, given the inherent lag between purchases of raw materials, processing, and sales. For example, the months of July and August typically account for about 40 percent of annual Texas landings. Therefore, processors which rely on domestic shrimp must accumulate it during the third quarter. However, holding inventory for the sole purpose of anticipated price increases is now the exception among most processors since several related conditions have diminished the advantages but increased the costs and risks of doing so. These conditions are: 1) greater availability of shrimp throughout the year; 2) less definitive information about (a) total quantities, (b) relative proportion of count sizes comprising overall supplies, and (c) price; and 3) less change in seasonal prices among certain size categories.

Greater year-round availability of shrimp facilitates purchase of various sizes from any tropical or subtropical coastal country worldwide. And with shrimp products such as shell-on, headless varieties (the raw material for most full-line processors) traded along commodity lines, large quantities can be procured with a Telex, fax, or phone call.

Shortly after World War II, United States shrimp prices were routinely set by the conditions manifested in the Gulf of Mexico shrimp industry. Today, however, local conditions are less important in the determination of prices. This makes the computation of appropriate offering prices (i.e., boat or ex-vessel price) more difficult because buyers must set a price high enough to attract producers but low enough to prevent amassing overpriced inventory. Limited information about quantity and count-size composition also subjects inventory to a greater risk of unexpected price changes.

Because of aquaculture, certain sizes of shrimp have become more prevalent in the market, thereby changing the traditional contribution each count size makes to the total supply. With larger, more stable supplies within some size categories, the amplitude of price changes over time has become less pronounced. This reduces the risk of adverse price changes and the potential margins from speculation.

Today, shrimp processing and marketing firms are judged successful not by the amount of raw materials they can stockpile, but rather by how efficiently products can be manufactured that meet market needs. These firms are mainstream players within the food processing sector. The extent of shrimp processing varies by facility. The least processed product sold through customary marketing channels (i.e., producer–processor–distributor–retail interests) is the shell-on, headless form that has been sorted by size, packed in 5-pound boxes, block frozen, and master-cartoned into 50-pound cases. These steps have not changed appreciably over the years, and many firms still specialize in this level of processing activity.

But with processing, marketing, and distribution functions becoming more technical, more complex, more regulated, more expensive, and more risky — but potentially more profitable — shrimp processing establishments have responded by becoming larger. Many processors maintain a full line of shrimp products as determined by the combination of: 1) the extent of convenience or value (e.g., peeled, breaded, ready-to-eat, etc.); 2) a freezing method which produces blocks of product or individually frozen shrimp; and 3) the mode of packaging (e.g., bulk or full-view consumer containers). Most of these firms have even diversified into other marine foods. This is a logical strategy for a successfully performing brand which increases sales opportunities and improves competitive advantage. Such diversification also improves the firm's weighted average gross margin and ensures more complete utilization of processing and distribution assets.

Historically, processors focused on serving the shrimp needs of the food-service sector, and most shrimp were consumed in away-from-home settings. This dependence on food service created a cyclical variation in demand, with periods of economic recovery and expansion accounting for significant sales gains for food-service establishments and thus the shrimp industry. However, the converse has also been true. Today, shrimp processors are better at balancing their account base between food service and the retail food sector. This is a prudent strategy for managing cyclical risk and partially insulates sales and earnings from uncontrollable macroeconomic conditions. The specific type of customer served in each sector is based, in part, on geography. Processors frequently deliver directly both food service and retail food firms within a certain geographic radius. Outside of their primary service area they distribute to wholesale firms which subsequently drop-ship to retail accounts. The decision to serve end-users or use intermediaries is primarily based on the processor's need to balance service against distribution cost.

Increased processing capacity has necessitated expansion of both geographic trading areas and product lines. Current assets are required to fuel this expansion, and most processors' capital needs are generally skewed towards short-term borrowing. All processing and marketing firms need credit to fund accounts receivable. Additionally, shrimp processors require significant short-term credit needs to fund inventories due to: 1) the high unit cost of purchases relative to other meats, and 2) the necessity of accumulating shrimp during peak production periods for subsequent processing and sale. The nature of such short-term borrowing suggests that lenders examine both an individual firm's credit policy and the extent to which price variation may impact inventory value.

The Components of Shrimp Value

The essence of shrimp value is size, with larger shrimp always commanding higher unit prices. Shrimp are classified according to count size, which refers to the number of shrimp comprising one pound. For example, with 16- to 20-count tails the buyer expects the pack to average 18 tails per pound and be comprised of shrimp which range in size from 16 per pound to 20 per pound. As shrimp size decreases, the range of individual sizes comprising each interval increases. This size range can be as small as 2 (for example, from 10 to 12 per pound or 13 to 15 per pound), and as wide as 100 (for example, 201 to 300 per pound). The shrimp market is segmented along a series of sizes, with each count size having its own niche in the market. Large shrimp (e.g., shrimp no smaller than 21 to 25 to the pound) are typically marketed in more exclusive restaurants while mid-sized shrimp (e.g., 31 to 35 through 41 to 50) are a mainstay in moderately-priced food service establishments and the retail food sector (Table 1).

The unit price is also influenced by market form, with more convenient forms such as shrimp peeled from 36- to 40-count tails commanding higher prices than the unpeeled 36- to 40-tail. While the two main determinants of value are size and market form within a count size, other factors such as country of origin, pack style, color, and species affect final price. Regardless of species, most users consider all penaeid shrimp of the same or contiguous count size to be equal.

Table 1. The Percentage Contribution Made by Trend, Seasonal, and Residual Variations in Explaining Monthly, Ex-Warehouse Prices between 1987 and 1991

Count Size	Percent Contribution to Total Variation by Type of Effect		
	Trend	Seasonal	Residual
16-20	48.28	15.94	35.78
21-25	49.56	18.63	31.82
26-30	24.96	36.97	38.07
31-35	16.32	58.89	24.79
36-40	11.45	66.71	21.84
41-50	16.52	44.08	39.40

Pricing

Ex-vessel and wholesale prices typically move in concert since the reported wholesale value is used as the benchmark from which boat prices are computed. However, fishermen typically land a distribution of sizes. Therefore, some means is required to categorize boat-run shrimp into customary size-intervals since each count size has a unique price. In Texas, the classification method used depends upon where the unloading occurs.

The ports of Brownsville and Port Isabel use the "pack out" method. In this approach, all shrimp pass through mechanical sorting equipment which objectively measures the diameter of the tail. Once sorted, the producer knows the exact quantity of each count size produced. Moving up the coast from the Rio Grande valley, another approach is used to compute ex-vessel value. This is known as the "box weight" method, and instead of using mechanical grading technology relies on a "grab" sample chosen from every 100-pound box the producer lands to determine count size. This classification method is a strong impetus for producers to sort shrimp aboard the vessel so that each box is as homogeneous as possible.

Wholesale prices are the result of a world-traded commodity entering the United States duty free. The price for each particular size is dependent upon current supply, supplies of similar-sized shrimp, and individual needs of buyers. Prices of most count sizes tend to move in the same direction over time. However, price variation within some count sizes can be dramatic. This typically occurs when the supply of a particular size abruptly changes (Figure 10). In 1989, the wholesale price of 16- to 20- and 21- to 25-count Gulf brown shrimp declined about 35 percent between January and December. This price drop was in response to significant quantities of black tiger shrimp (*Penaeus monodon*), a large shrimp cultured in Southeast Asia, being diverted to the United States instead of Japan, the traditional marketplace for "black tigers."

A situation similar to that in 1989 occurred in 1991. In the first half of 1991, shrimp prices continued the recovery which began in 1990. But beginning in May, prices for most sizes declined. Two factors account for this decrease: a significant proportion of shrimp imports in 1991 were 21 to 25 counts or larger, and 31 to 40 sizes. These large

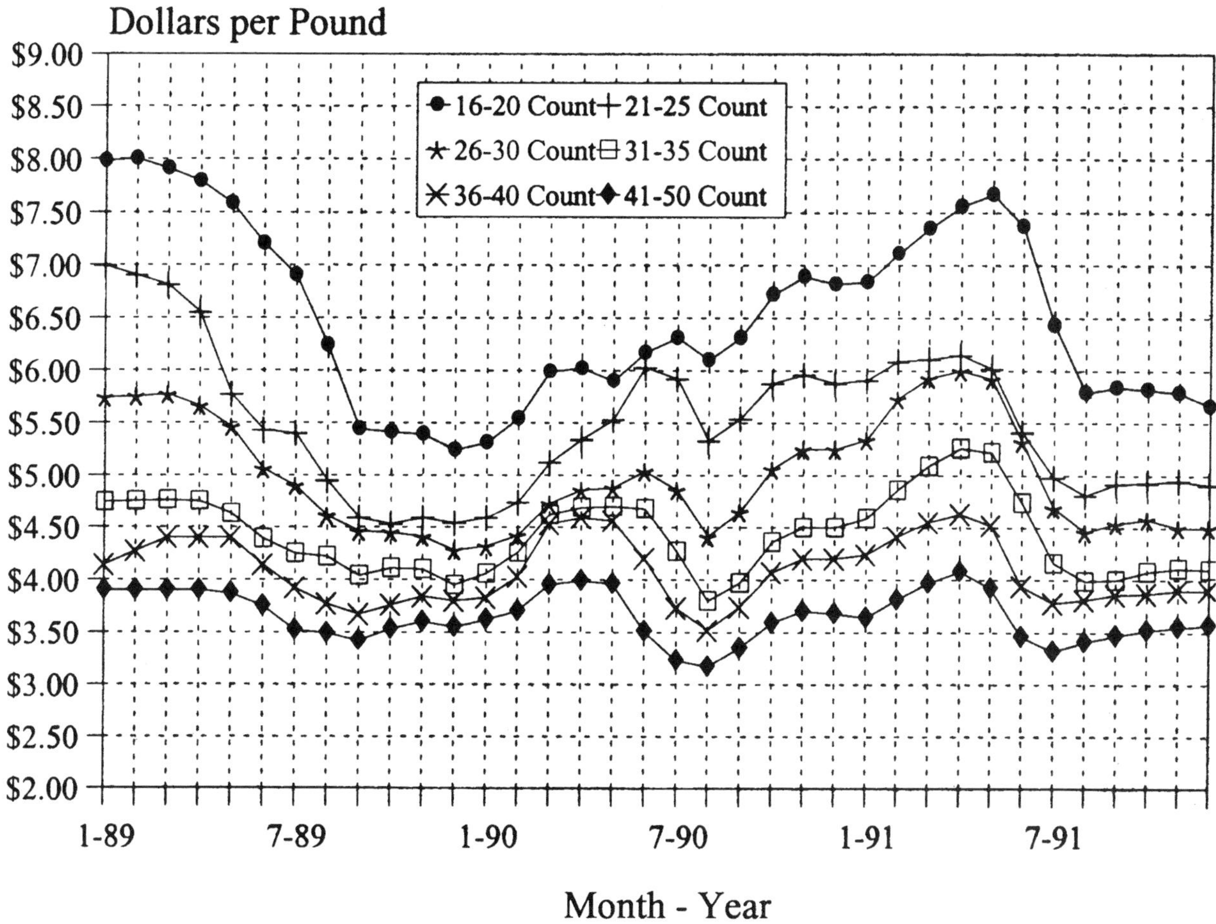

Figure 10. Computed average monthly wholesale prices for shell-on headless shrimp: 1989 to 1991.

shrimp accounted for 40 percent (121 million pounds) of all shell-on headless shrimp entering the United States. Besides imports of large-sized shrimp, local processors reported larger than normal shrimp being produced once the "Texas closure" was lifted. This further depressed the price for large shrimp. Mid-sized shrimp fared similarly since roughly 18 percent of all imported shell-on, headless product was in the 31 to 40 count range (54 million pounds). Such changes in supply with concomitant price responses provide graphic evidence that the only consistency in shrimp prices is their constant state of flux.

REFERENCES

Anderson, William W. 1958. The shrimp and shrimp industry of the southern United States. United States Department of Interior Fish and Wildlife Service Fishery Leaflet Number 472.

Hildebrand, H. H. and J. Gunter. 1953. Correlation of rainfall with catch of white shrimp *Penaeus setiferus*. *Trans. Am. Fish. Soc.* **82**:151-155.

Knoff, G. Miller. 1970. Opportunities in the shrimp fishing industry of the southeastern United States. Information Bulletin Number 3. University of Miami Sea Grant Institutional Program.

Biology of Certain Commercial Crustaceans: LOBSTERS

Bohdan M. Slabyj and Robert C. Bayer

AMERICAN LOBSTER

The American lobster *(Homarus americanus)* is a crustacean with two major body parts: a cephalothorax and a tail. It has two prominent antennae and four smaller ones, and its eyes are on movable stalks. The thorax has five pairs of legs. The first two have developed into two strong claws, one adapted for crushing and the other for ripping, while the rest are long walking legs. The abdomen carries four pairs of swimmerets, and the last tail segment is developed into a fan-like structure. This powerful tail is used to escape threatening situations, propelling the lobster backwards. The lobster can lose claws and legs to get away in a life-threatening situation; the lost appendage is replaced in time. A lost claw is usually replaced with a much smaller appendage.

The color of the exoskeleton varies from light to very dark brown. Blue- or red-colored lobsters occur, but are rare. American lobsters do not have any special markings on their bodies. During cooking all turn bright red. The largest lobster ever found was a female weighing about 40 pounds. Lobsters live singly in burrows and crevices close to shore most of the year, where the water is warmer and food more plentiful. As winter approaches, they move into deeper, warmer waters.

Lobsters are caught in two-chambered traps. In Maine, it is unlawful to catch lobsters other than by the conventional trap. The lobster must have at least 8.26 centimeters (3.25 inches) carapace length or a maximum length of 12.7 cm (5 inches). Most of the traps are now made from heavy, plastic-coated wire, although some regions in Canada continue using heavier wooden traps (see Figure 12, page 162). The wire-coated traps must have a "ghost panel" to release trapped lobsters in case the trap is lost. Traps made out of hardwood are required to have two slats of softwood. It may be interesting to note that, while the wire traps become disabled in six months in the ocean, the wooden traps may last up to two years. The wire traps must have an escape panel with 44.5 by 14.6 cm (1.75 by 5.75 inches) openings or two round escape vents 5.72 cm (2.25 inches) in diameter, located next to the bottom edge. The wooden traps can be adjusted by moving the bottom laths to leave the required escape space (Maine Department of Marine Resources, 1995). The traps are baited with fish scraps or whole fish. Herring scraps from sardine canneries are common, along with racks that remain after red fish (ocean perch) are filleted.

A buoy is attached to the trap as a marker of the location. The buoys and the boat must bear the color design that is on file with the license permit. Furthermore, the lobster traps and the boat must bear the license number of the permit. A fisherman may not have more then 1,200 traps in the water at a time, but there are additional, voluntary, regional restrictions. A lobsterman — as well as the processor, retailer, distributor, and wholesaler dealing with lobsters — must have an appropriate license, which ranges in fees from $46.50 to $417.00 (Maine, 1995).

The American lobster can be found in offshore waters along the Atlantic Coast from Labrador (Canada) to North Carolina. The early Puritan settlers utilized this crustacean locally, but could not exploit the resource because of lack of transportation. Starting in 1840, transportation began to improve rapidly, as well as the knowledge in packaging lobsters for live shipment. Thus by 1892 the lobster fishery had reached a peak landing of 10,800 metric tons valued at $1,062,392 (Tressler, 1951).

Annual lobster landings for the United States and Canada from 1928 to 1993 are shown in Figure 1. These data have been obtained from records of landings kept by the Maine Department of

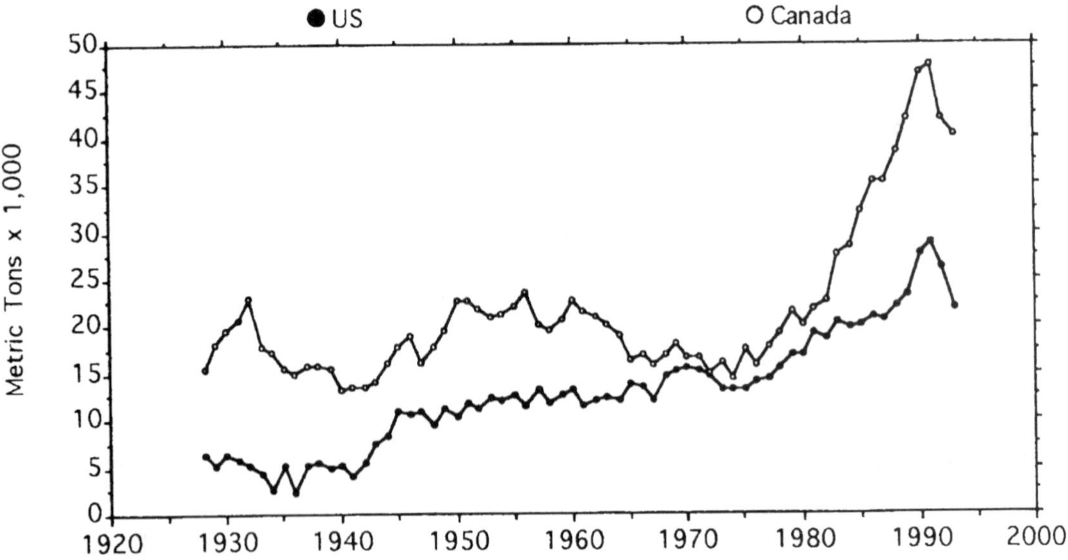

Figure 1. American lobster landings in the United States and Canada, 1928–1993 (Krouse, 1994a; Barrow 1994).

Marine Resources for the various states and provinces (Krouse, 1994a). It is interesting to note that between 1928 and 1944 lobster landings in the United States were below those of 1892. However, from 1945 on, landings increased steadily, reaching a peak of 28,771 metric tons in 1991. Table 1 shows landings by state, with Maine being number one, followed by Massachusetts and Rhode Island. The highest yield per trap was realized in Connecticut and the lowest in Maine (Krouse, 1994b). While the observed decrease in landings since 1991, both in the United States and Canada, appears to indicate overfishing, the estimate for Maine in 1994 was 18,614 metric tons, as compared to 14,004 metric tons that were landed by Maine fishermen in 1991 (Krouse, 1994b).

Essentially all lobsters in the United States are consumed fresh; very little is processed. While a large portion of this crustacean is consumed locally, a significant volume is shipped live to metropolitan areas. Only very recently were soft-shell lobsters sufficiently low-priced to allow commercial freezing (cooked, or raw/in shell, or shucked/whole, or tail-and-claws). It is estimated that 10 percent of total landings in Maine in 1994 were processed (Krouse, 1994b) (Table 1). Canada's lobster landings fluctuated between 13,000 and 24,000 metric tons from 1928 until 1982. Since then the landings have increased steadily, reaching a peak in 1991 of 47,600 metric tons (Figure 1). While most lobsters are caught in Nova Scotia (Table 2), Prince Edward Island and New Brunswick together contribute a similar volume of catch. In 1992, 32,500 metric tons of lobsters were processed, of which 72.7 percent were fresh/in shell, 12.1 percent were frozen/in shell (cooked), 8.3 percent were frozen/shucked (cooked), 0.5 percent were canned, 2.5 percent were frozen cooked tails, and 3.9 percent were miscellaneous cooked products (Barrow, 1994).

Six million pounds of lobster are held in either tidal pounds or floating wooden storage cars (81 x 53 x 33 centimeters) in Maine; approximately four million pounds are stored in Canadian pounds. Recently-molted lobsters have a low muscle content and do not withstand shipping very well. These lobsters are held in high density

Table 1. Lobster Landings in New England in 1992 (Krouse, 1994b)

State	Landing (m.t.)	Traps (kg/trap)	Effort
Maine	12,207	2,002,998	6.1
Massachusetts	6,315	477,836	13.2
Rhode Island	3,060	348,047	8.8
New York	1,565	131,630	11.9
Connecticut	1,066	65,123	16.4
New Hampshire	619	–	–
New Jersey	550	–	–

Figure 2. Lobster pound in Jonesport, Maine.

Table 2. Lobster Landings in Canada by Province in 1993 (Barrrow, 1994)	
Province	Landings (m.t.)
Nova Scotia	17,877
Prince Edward Island	8,794
New Brunswick	7,633
Quebec	3,572
Newfoundland	2,222

confinement for four to six months and are fed to increase their muscle mass. The lobster storage facility consists of a dam across a small cove to hold back five to eight feet of water during low tide. There is a slatted fence on top of the dam to allow water exchange with the tides (Figure 2). Mechanical aeration is provided so that the lobsters will not suffer from anoxia during low tide. Lobsters use oxygen at a high rate during the early autumn. Oxygen levels of one to three parts per million have been observed in the bottom few inches of a heavily loaded pound. When a pound is full, a density of one pound of lobster per square foot is common.

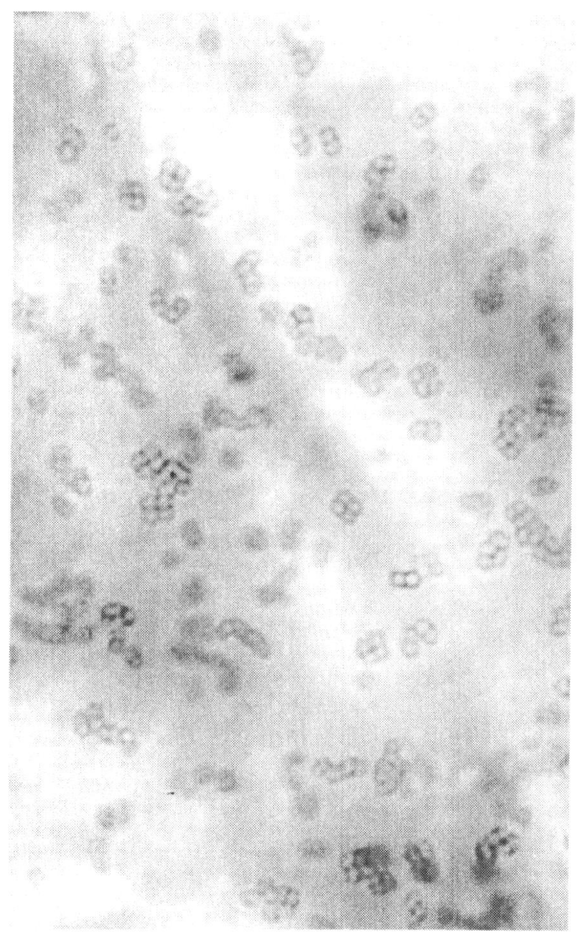

Figure 3. Aerococcus viridans *var.* homari *(1,000x).*

LOBSTER DISEASES

Three major diseases are of economic significance in lobsters held in high density confinement. Gaffkemia was the most pernicious of these diseases, but can now be controlled by medication. Ciliated protozoan disease and shell disease remain as problems for the lobster pound industry.

Gaffkemia is caused by a systemic infection of the bacterium *Aerococcus viridans* var. *homari* (Figure 3), an organism endemic to lobster stocks. Studies in Maine and Canada indicate an average infection in wild stocks of 5 to 7 percent (Vachon et al., 1981; Stewart et al., 1966). Outbreaks of gaffkemia are rare in the United Kingdom; they have been associated only with lobsters imported from Europe or North America (Edwards et al., 1981). The lobster contracts the disease when bacteria enter unhealed wounds. In pounds, lobsters have many opportunities for aggressive interaction, causing breaks in their integument. When the lobster has become infected, the bacteria multiply readily in the nutrient-rich hemolymph. Water temperature is a critical factor in determining the duration of a lobster's life, once infected with *Aerococcus viridans*. At 15°C, an infected lobster will have an average time to death of 12 days, while at 10°C average survival is 28 days. As temperature approaches zero degrees, lobsters will live with the infection for many months (Stewart et al., 1972). Oxygen level in the pound is critical to the survival time of lobsters with gaffkemia. The organism interferes with the ability of hemocyanin to bind oxygen, thus causing internal suffocation. Although it is possible to immunize lobsters against *Aerococcus viridans* by means of injection of a treated bacteria, this procedure is far too labor-intensive to be practical (Rittenburg and Bayer, 1980).

A control program has been developed that allows monitoring for incidence of the disease, treatment with an antibiotic, and monitoring for antibiotic residue. Gaffkemia will have to be controlled in egg-to-market lobster aquaculture as well. Lobsters in storage are sampled for the relative incidence of disease by a field blood culture technique. Samples that show growth are examined microscopically for the characteristic tetrad morphology of *Aerococcus viridans*.

Aerococcus viridans is sensitive to many different antibiotics *in vitro* (Stewart and Cornick, 1967). However, when given orally to lobsters, only oxytetracycline, novobiocin, and amoxycillin show activity in the hemolymph (Huang and Bayer, 1989). Based on efficacy, target species, safety, and human and environmental safety, oxytetracycline has been chosen as the treatment for gaffkemia in lobster.

Depending on the number of positive samples, oxytetracycline may be given in the form of a feed pellet for five to seven days. When lobsters are going to be fed medicated feed, they are fasted for a day, then fed three to six pounds of medicated feed containing 2.2 milligrams of oxytetracycline per gram of diet, per 454 kilograms of lobster. This cycle may be repeated, based on blood culture indication of the disease. Lobsters may accumulate in the pound in 10,000-20,000 increments (Bayer et al., 1996).

Administration of medicated feed requires withholding lobster from the market for at least 30 days. Tissue residue studies have shown that the oxytetracycline leaves the muscle almost immediately in summer and in six days in the autumn, followed by mid-gut at 13 and 27 days; the last tissue to be residue-free is the blood at 17 and 82 days, respectively (Bayer and Daniel, 1987).

The presence of antibiotic can be monitored by using the Delvo test, a test kit designed to detect antibiotic residues in milk. Lobster hemolymph is used in place of milk. The test requires the use of a temperature-controlled bath or heater and takes about 2.5 hours. The use of oxytetracycline is approved for use in lobster by the United States Food and Drug Administration (FDA) and Health and Welfare Canada (Anderson and Bayer, 1991).

Ciliated protozoan disease (Figure 4) outbreaks have occurred in Gulf of Maine lobsters since 1971 (Loughlin and Bayer, 1991; Sherburne and Bean, 1991; Aiken et al., 1973). Epizootics and associated mortalities have been noted in two to three lobster pounds annually between 1990 and 1992. Between January and April of 1993, outbreaks were noted in five Maine pounds and at least one Grand Manan, Canada, location. The source of the infection is not known, nor is there a treatment.

Shell disease syndrome is characterized by lesions of the exoskeleton, resulting in an unappeal-

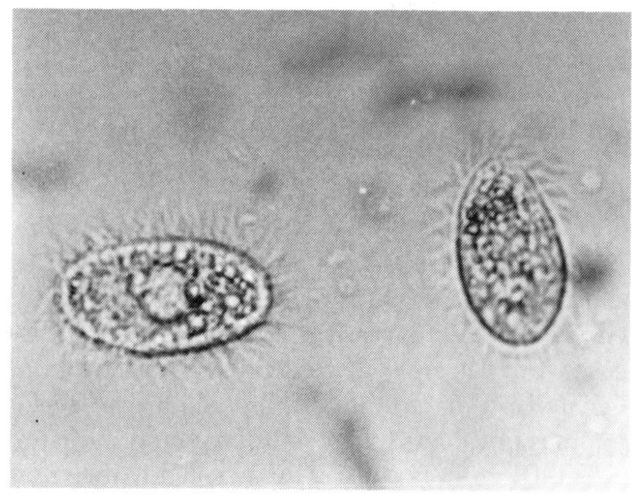

Figure 4. Ciliated protozoan.

ing product with reduced viability and value (Figure 5). This syndrome occurs primarily in southwest Nova Scotia. Shell erosions are most predominant on the dorsal carapace, abdomen, and claws, and are frequently bilaterally distributed. Erosions are usually confined to the calcified layers of the exoskeleton, but may be severe enough to affect underlying tissues. Melanization is often apparent within lesions, but is seldom extensive. Obvious signs of the disease appear in as few as three weeks after introduction of lobsters to the pound. As much as 25 percent of the impounded stock may be affected by shell disease.

Although known to occur in a variety of decapod crustaceans, the etiology of shell disease is poorly understood. The disease is presumably caused by ubiquitous, opportunistic bacteria and fungi that degrade the chitin component of the crustacean cuticle. Bacteria of the genera *Vibrio*, *Pseudomonas*, and *Aeromonas* are the microbes most commonly implicated as causative agents. Invasion of the exoskeleton by these microbes is thought to be initiated by mechanical damage to the shell. Infection also appears to be facilitated by environmental stresses such as pollution, overcrowding, and poor water quality. Physiological conditions including molt stage, molt frequency, and reproductive and nutritional status may contribute to development of the disease as well (Getchell, 1989).

A number of different types of bacteria appear to be responsible for shell lesions in winter-impounded lobsters. Most require salt and temperatures below 23°C. Some grow at temperatures less than 4°C. The majority have been identified as members of the genera *Vibrio*, *Pseudomonas*, and *Aeromonas*, and display chitinase activity. Two isolates were tentatively identified as *Vibrio alginolyticus* and *Pseudomonas maltophilia*. No isolates appear to be exclusive to lesions, because normal shells yield similar flora. Species compositions within lesions are also somewhat variable, even among lesions located on the same lobster.

Lobster shell is generally resistant to microbial attack. Resistance is due to the outermost shell layer, the epicuticle. This layer contains polyphenolic compounds that are impervious to most microbes (Dennell, 1947). Some bacteria, however, appear capable of initiating lesions by enzymatic degradation of the epicuticle (Lightner, 1988). Diverse assemblages of colonizing microorganisms have been observed on the surface epicuticle in histological section and scanning electron microscopy (SEM). Histology also provided evidence of erosion beneath colonized areas of the epicuticle. Epicuticular colonization and degradation were often most pronounced on sides of setal pits, as observed by Smolowitz et al. (1992). Invasion at the site of setae and other sensory structures could explain the bilateral occurrence of lesions.

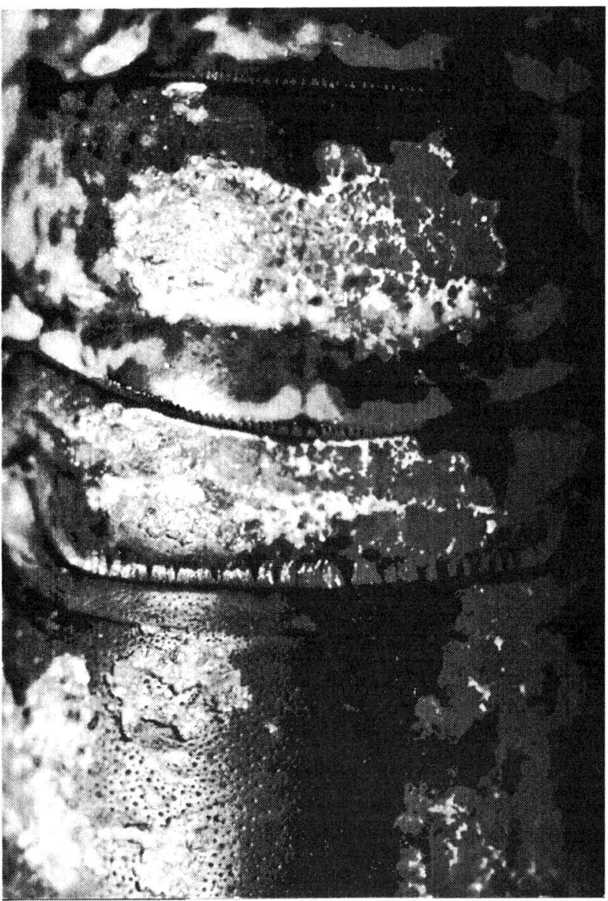

Figure 5. Characteristic lesions due to lobster shell disease.

Invading bacteria may also gain entry to chitinous shell layers through any of the numerous cuticular pores found in the lobster exoskeleton. Bacterial cells were observed inside such structures using a transmission electron microscope (TEM). The actual involvement of these bacteria in lesion development, however, was not determined.

Damage to the epicuticle from handling and storage provides another route of invasion for opportunistic microbes. Exoskeletons with obvious superficial abrasions often displayed signs of erosion when viewed in section. Malloy (1978) found lesions developed on experimental lobsters only after the shell had been abraded.

Once they had invaded the chitinous shell layers, bacteria were observed to proliferate and form pockets in the exo- and endocuticle. Within these pockets, degradative enzymes may accumulate and become uncoupled from the bacteria that produce them. Eventually, enzymatic activity in the pockets results in the loss of overlying shell layers, and the formation of a characteristic lesion. That lesions initially develop internally may help explain the unanticipated appearance of the disease in impoundments.

In addition to epicuticular damage, lesion development appears dependent on impaired host defenses. Preliminary experiments demonstrate that shell disease is difficult to create under controlled conditions. Usually, deliberate abrasion of the exoskeleton does not produce the disease in unaffected lobsters. Direct application of bacterial cultures to wounds is also ineffective. Lesions occur only when the shells of previously affected lobsters are similarly treated.

A number of conditions associated with pounds may cause immunodeficiencies in impounded lobsters. Winter pounding exposes the lobster to temperatures below its optimal range. Lower temperatures will slow down repair, presumably, because hemocyte migration to wounded sites will be retarded. Defenses of impounded lobsters may also be affected by improper nutrition and starvation. While it is common practice to feed impounded lobsters, there is considerable variation in the amount and frequency of feeding. Starvation results in the depletion of hemocytes and serum protein over time, events which undoubtedly make the impounded lobster more vulnerable (Stewart et al., 1967). Deficient diets have been directly linked to shell disease in hatchery-reared juvenile lobsters (Fisher et al., 1976). Bacterial endotoxins and poor water quality conditions present within pounds may interfere with defense mechanisms as well.

Although bacteria have been defined as the etiologic agents of winter-impoundment shell disease, the causes of host susceptibility remain incomplete. It is not known exactly what physiological changes result in impaired immunity in lobsters, but it is reasonable to assume that those changes involve some aspect of the prophenoloxidase system (Johansson and Soderhall, 1989). These changes could include decreased total hemocyte counts, shifts in populations of granular hemocytes, and inhibition of the system's enzymatic pathways.

MANAGEMENT OF THE MAINE LOBSTER FISHERY

Maine lobster management emphasizes protection of female lobsters with prohibition on harvesting berried or V-notch marked females (Figure 6). Fishermen are also prohibited from taking lobsters with a carapace length under 8.26 centimeters (3.25 inches) or over 12.7 centimeters (5 inches). The 12.7-centimeters law is unique to Maine, and the minimum carapace length is currently being debated among New England fisheries managers.

In Canada there are limited seasons for lobster fishing, defined by region. New England does not generally use seasons, with some limited exceptions. There is a 1,200-trap limit per fisherman with

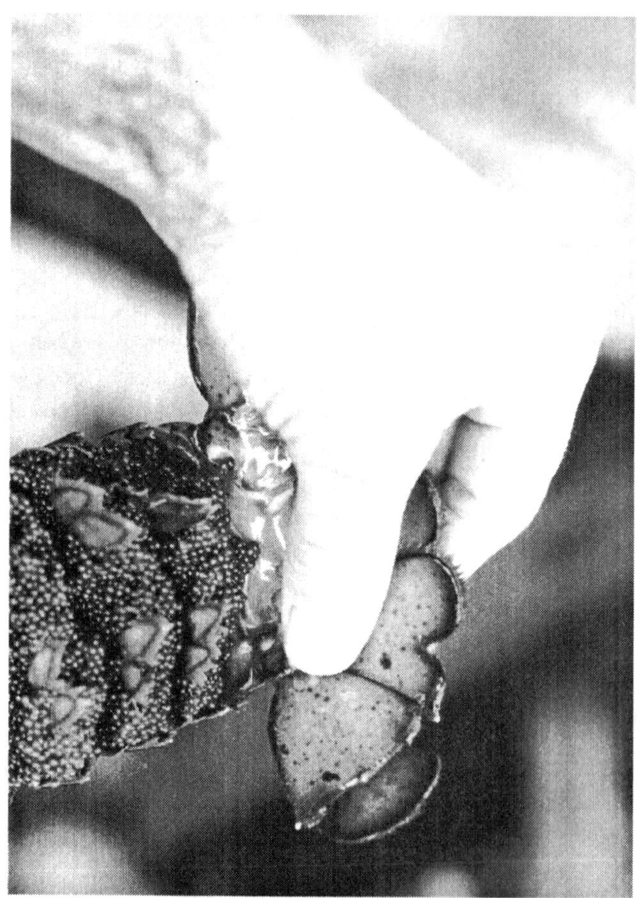

Figure 6. V-notched female lobster.

a few lower voluntary regional limits. Canada makes use of trap limits.

Current Maine law allows fishermen to place a V-shaped notching on the right uropod of ovigerous lobsters before returning them to the sea. The landing of a lobster in Maine with a mutilated or notched right uropod with or without eggs is illegal (Maine, 1964). The marking of ovigerous lobsters and subsequent protection of marked lobsters has been in practice in Maine since 1917. Originally, only the Commissioner of Marine Resources could hole-punch the telson of ovigerous lobsters purchased for that purpose (Maine, 1917). The law gradually evolved to its present form in the 1970s. The Maine Department of Marine Resources still buys some ovigerous and non-ovigerous female lobsters to V-notch and release in addition to the voluntary V-notching by lobstermen. The rationale for this management tool is to protect lobsters with proven reproductive capability, thus enhancing egg production and ultimately recruitment.

Despite the apparent attempt to strengthen the effectiveness of the law as a management device by allowing voluntary participation of lobstermen in the V-notching process, no attempts have been made to evaluate its impact on the fishery until very recently. Indeed, current fishery assessment procedures as outlined by Thomas (1973) sampled only landed lobsters and therefore ignored the possible significance of V-notched lobsters in augmenting the recruitment process. Because of increased fishing pressure in the Gulf of Maine, a proposal to increase minimum legal size of landings, and fishermen's concern that many V-notched lobsters leave Maine waters and move to unprotected areas, the Maine Lobstermen's Association (MLA) proposed that the V-notching law be extended to include the entire Gulf fishery. Concern about the complete lack of data that could be used to evaluate the success or failure of such a program led the MLA to initiate a survey of its members to determine the incidence of V-notched lobsters and of ovigerous lobsters with or without V-notches in Maine lobster traps. Based on the data collected to date, it is apparent that 29 percent of trapped nonovigerous females and 69 percent of trapped ovigerous lobsters were found to be V-notched, during a three-year survey (1982-1984) of trap hauls conducted in October by the Maine Lobstermen's Association. The size-frequency distribution of V-notched lobsters is biased towards larger females relative to that of unmarked females. Calculated numbers of annual eggs per V-notched lobsters were about nine times greater than for unmarked lobsters as a result of their larger size, hence a greater proportion of sexually mature lobsters with higher fecundity. These preliminary results indicate the need to include V-notched lobsters in Maine fishery surveys and assessments and to investigate the possibility of V-notching as a management tool for the Gulf of Maine fishery region (Daniel et al., 1989).

There is no sport fishery for lobster in the State of Maine or Canada. Lobster can be taken by diving in several of the states south of Maine.

COMPOUNDED DIET FOR LOBSTERS

A pelleted diet to meet the nutritional needs of the American lobster (*Homarus americanus*) is now available commercially. This first and only commercially available lobster feed is meant to supplement, not replace, the traditionally used diet of fish cuttings. Ten pounds of this feed is equivalent to about one bushel of fish cuttings in dry matter food value.

In the mid-1970s, a group of faculty and students in the Department of Animal and Veterinary Sciences at the University of Maine began to develop a balanced artificial diet suitable for pound feeding of the American lobster, which would be comparable in price to herring but not present the problems of quality and availability often associated with herring.

With support from the University of Maine/University of New Hampshire Sea Grant College Program, studies were conducted, beginning with a nutritional analysis of the contents of lobster stomachs. From a base study (Leavitt et al., 1979) of the natural dietary intake of adult lobsters, sample diets were formulated using commercially available feed ingredients. With additional funding from the Maine Agricultural Experiment Station, field testing of the synthetic diet has continued over a number of years.

The diet can be fed to American lobsters held in storage facilities such as lobster pounds and tank systems. Feeding of the pelleted diet to American lobsters stored in live cars is not recommended due to the current design of the live cars. Historically, the feeding of lobsters held in high-density confinement has been considered a sound investment because it reduces cannibalism, maintains weight, and hardens the shell of new (soft-shell) lobsters.

The results of a recent study comparing the weight gain and meat flavor of lobster fed the pelleted diet versus a diet of herring are discussed below. The effect of diet on meat flavor is always a concern — e.g., too much fish meal in a broiler diet yields fishy-tasting chicken meat. The objective of this study was to compare weight gain and meat flavor of lobsters fed synthetic diet versus a diet of herring. Salted herring scrap is the conventional ration fed to lobsters in high-density confinement in lobster pounds. The formulated synthetic diet was composed of fish meal (30 percent), brewer's yeast (10 percent), alfalfa (10 percent), kelp meal (3 percent), and wheat flour (47 percent).

Before the feeding trial began and after 48 days on the diet, the lobsters were weighed both suspended in water (SW) and in air (AW). The mean changes in wet weight were 16.0 grams (SW), 22.4 (AW), and 6.8 (SW), 11.7 grams (AW), for lobsters fed the formulated diet and the herring diet, respectively. The mean change in weight for lobsters fed the formulated diet was significantly ($P \leq 0.01$) greater for both weight values.

Sensory data revealed that the flavor of lobsters fed the synthetic diet was rated equal to that of the coded control sample. Meat from lobsters that had received a diet of herring scraps, however, was judged slightly poorer in flavor than both the freshly harvested control and the synthetic diet-fed lobsters. These differences were significant at the one percent level of detection (Bayer et al., 1981).

These results show that the flavor of lobsters fed the synthetic diet is significantly ($P \leq 0.01$) better as opposed to those fed the conventional diet. From this study it is safe to infer that the flavor of the lobster is not adversely affected when fed the pelleted diet.

Nutrition data for edible lobster meat are given in Table 3 and discussed in the section on Florida lobsters. Traditional cooking requires placing the lobster in boiling water and boiling for 12 minutes or longer (depending on the size of the lobster) after the water comes to a second boil.

HATCHERY TECHNOLOGY

Female lobsters molt and are inseminated by the male during a period that usually begins in August and ends in November. The following June to July the female extrudes eggs; at this time, they are fertilized. The eggs are attached to the ventral side of the tail of the lobster until hatching, or about six to nine months. There are four larval stages, the fourth being the benthic stage (Hughes and Matthiesson, 1962).

Lobsters can be hatched and reared to maturity, although not economically, or reared to stage IV for release into the fishery. The State of Massachusetts has operated a hatchery on Martha's Vineyard for more than 20 years. There are similar community hatcheries in Maine. The apparatus used for this purpose was developed by John Hughes and is the standard equipment for rearing lobsters through their pelagic larval stages. The rearing device allows sufficient turbulence to disperse the cannibalistic larval lobsters in the water column without harming them. Survival to stage IV of greater than 80 percent is usually obtained. These results are a considerable improvement over earlier attempts (1939-1945) at rearing lobsters in Maine, in which stage IV survival was only 19 percent (Taylor, 1950). If lobsters are to be reared further, individual cages are needed to prevent cannibalism (Figure 7).

Temperature is critical to a viable molt in larval lobsters. Successful development to stage IV benthic juvenile lobsters requires temperatures within a range of 6°–24°C (Templeman, 1936). A duration as short as 13 days to stage IV has been observed (Hughes and Matthiesson, 1962). In am-

Figure 7. Cages for rearing lobsters.

bient sea water in the Gulf of Maine, it should take approximately 60 days from hatch to stage IV. It would be advantageous to heat sea water serving the larval lobsters so as to increase the number of hatches in a culture-and-release operation.

SHIPPING LOBSTERS

Lobsters are relatively delicate crustaceans but can be kept in tanks for long periods of time if proper care is given. Lobsters can also be successfully shipped some distance by truck and around the world by air. Newly-molted lobsters, sometimes referred to as shedders or new-shell lobsters, are at a stage of life when their shells are soft and prone to cracks and punctures. These are the most vulnerable lobsters and the least desirable for shipping. They usually do not survive well in long distance transport. It is best to sell new-shell lobsters locally, or they can be hardened by storing them in pounds and feeding them for a week or two until the shells toughen up sufficiently for shipping. A lobster car could also be used for hardening, but some mortality (1 to 2 percent) from cannibalism should be expected.

Temperature is probably the most critical factor in shipping lobsters. Low temperatures are required to reduce the metabolic rate and oxygen use in order to prolong lobster survival in the absence of sea water. A good temperature range is 1.1°-4.4°C. During most of the year, some refrigeration and/or ice will be required to keep the lobsters cool . If ice is used, do not allow the lobsters at the bottom of the pile to rest in the fresh water from the melted ice, as they are likely to suffer osmotic shock and death. If lobsters are transported in subfreezing weather, heat should be provided to prevent freezing. Well-packed lobsters can be expected to survive at least 12 to 24 hours.

The same principles apply to transporting lobsters by air as by truck. They must be kept at a proper temperature by: (1) providing a well-insulated container and chilling the lobsters in cold sea water before packing; (2) providing sufficient sealed ice packs. Loose ice cannot be used since it will melt and remain with the lobsters; (3) seaweed or seawater-moistened paper is often packed with lobsters to provide moisture and cushioning during transportation.

TANK STORAGE

There are many commercially-available lobster tanks, or one can build a refrigerated tank. There are several requirements for a lobster tank system:

Adequate Aeration

Lobsters need oxygen, which they remove from the water. Aeration can be accomplished by splashing the water as it enters the tank from the pump, or by using air diffusion stones, or both (Figure 8).

Refrigeration System

It is desirable to keep the lobsters as cold as economics will allow, as cold makes the lobsters docile and will reduce cannibalism as well as the lobster's need to feed. Cold also reduces the lobster's oxygen requirement. From time to time, one may purchase lobsters that have gaffkemia, a disease fatal to lobsters but harmless to humans. The colder the water, the longer the infected lobsters will survive. A reasonable temperature to aim for is 4.4°-7.4°C. When selecting a compressor, someone with knowledge of refrigeration should be consulted to determine the correct capacity of the compressor and refrigeration coils.

Filtration

A filter should remove ammonia and organic and particulate material, serve as a biofilter, and also maintain pH. The simplest filter system is one in which water is pumped into a filter box on top of the tank, allowing the water to percolate through the filter. The filter should contain a form of activated carbon to remove organics, and a form of calcium carbonate (such as washed oyster shell)

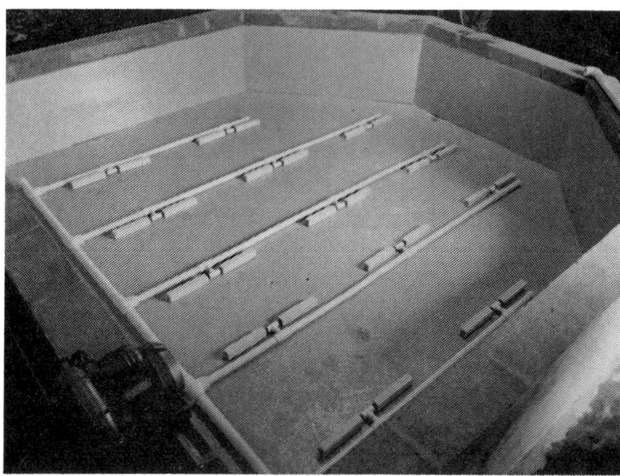

Figure 8. Diffusion stones used to aerate this tank.

Table 3. Lobster Nutrition Data

Components	*Homarus americanus*		*Panulirus argus*	*Panulirus marginatus*	*Panulirus interruptus*
	claw	tail	tail	tail	tail
Proximate composition (g/100g)					
Water	75.1(0.9)a	72.5(0.9)b	74.7(1.1)a	71.8(0.3)b	72.3(0.4)b
Protein (N x 6.25)	19.0(0.9)c	22.1(0.4)b	21.5(1.1)b	24.4(0.3)a	23.4(0.1)a
Total fat	0.63(0.11)d	0.74(0.08)cd	1.04(0.14)b	1.33(0.12)a	0.94(0.17)cd
Ash	1.79(0.07)a	1.55(0.05)b	1.79(0.04)a	1.75(0.07)a	1.71(0.04)a
Energy (Kcal/100g)	121(5)b	141(2)a	122(3)b	138(4)a	143(3)a
Minerals (mg/100g)					
Ca	63(9)a	37(4)bc	54(25)abc	58(11)ab	29(1)a
Cu	3.65(0.24)a	1.32(0.12)c	0.96(0.24)c	1.27(0.30)c	1.94(0.17)b
Fe	0.48(0.17)b	0.38(0.09)b	0.32(0.06)b	1.21(0.56)a	0.43(0.12)b
K	273(19)c	334(20)b	392(29)a	353(34)ab	399(31)a
Mg	39(1)c	40(2)c	59(3)a	63(4)a	49(3)b
Mn	0.065(0.012)a	0.048(0.008)b	0.016(0.008)c	0.023(0.006)c	0.020(0.012)c
Na	355(22)a	226(5)c	242(7)c	274(6)b	270(9)b
P	180(10)c	249(13)b	283(40)ab	301(11)a	184(2)c
Zn	5.54(0.28)a	2.91(0.12)bc	2.58(0.09)d	2.77(0.14)cd	3.14(0.02)b
Lipids (mg/100g) Cholesterol	55(9)c	127(29)a	89(19)b	96(16)ab	80(3)bc

Table 3. continued

Components		*Homarus americanus*		*Panulirus argus*	*Panulirus marginatus*	*Panulirus interruptus*
		claw	tail	tail	tail	tail
Fatty acids	14:0	0.006(0.001)b	0.008(0.001)b	0.021(0.012)a	0.012(0.004)ab	0.016(0.005)ab
	15:0	0.003(0.001)c	0.005(0.001)bc	0.007(0.003)b	0.012(0.002)a	0.008(0.001)b
	16:0	0.081((0.012)c	0.104(0.008)bc	0.140(0.033)ab	0.157(0.014)a	0.139(0.022)ab
	17:0	0.003(0.001)d	0.004(0.001)d	0.015(0.002)b	0.024(0.002)a	0.011(0.003)c
	18:0	0.032(0.008)c	0.035(0.005)c	0.127(0.008)b	0.182(0.007)a	0.110(0.024)b
	19:0	0.001(<0.001)d	0.001(<0.001)d	0.006(0.001)b	0.011(0.001)a	0.003(0.001)c
	20:0	0.003(0.001)d	0.003(0.001)d	0.010(0.002)b	0.013(0.002)a	0.007(0.001)c
	22:0	0.005(0.002)b	0.007(0.005)ab	0.009(0.002)ab	0.013(0.007)a	0.010(0.001)ab
	Total saturated	0.133 (0.022)c	0.166(0.011)c	0.335 (0.052)b	0.424(0.027)a	0.303(0.055)a
	15:1	0.006(0.004)b	0.003(0.001)b	0.012(0.004)a	0.006(0.003)b	0.005(0.001)b
	16:1	0.035(0.006)a	0.040(0.004)a	0.059(0.022)a	0.053(0.010)a	0.050(0.013)a
	17:1	0.008(0.003)bc	0.006(0.001)c	0.015(0.004)a	0.016(0.005)a	0.013(0.002)ab
	18:1	0.128(0.027)a	0.137(0.019)a	0.119(0.023)a	0.148(0.027)a	0.133(0.009)a
	19:1	0.002(0.001)c	0.003(0.001)c	0.006(0.002)ab	0.008(0.001)a	0.004(0.001)bc
	Total monounsaturated	0.180(0.038)a	0.188(0.022)a	0.211(0.055)a	0.231(0042)a	0.206(0.022)a
	18:2	0.005(0.002)c	0.007(0.001)c	0.025(0.003)a	0.026(0.002)a	0.011(0.003)b
	18:3	0.003(0.002)ab	0.004(0.001)ab	0.001(0.001)c	0.005(0.001)a	0.002(0.001)bc
	20:2	0.005(0.002)c	0.004(0.001)c	0.004(0.001)c	0.015(0.002)a	0.009(0.002)b
	20:4	0.020(0.007)c	0.020(0.003)c	0.128(0.027)b	0.303(0.051)a	0.117(0.012)b
	20:5	0.135(0.037)ab	0.155(0.022)ab	0.114(0.019)b	0.096(0.031)b	0.195(0.047)a
	22:4	0.001(<0.001)b	0.001(<0.001)b	0.003(0.001)b	0.012(0.001)a	0.012(0.006)a
	22:5	0.006(0.002)c	0.009(0.001)bc	0.012(0.002)b	0.019(0.003)a	0.010(0.003)bc
	22:6	0.057(0.001)b	0.090(0.011)a	0.098(0.036)a	0.050(0.006)b	0.047(0.006)b
	Total polyunsaturated	0.231(0.0361)c	0.289(0.039)c	0.384(0.040)b	0.526(0.071)b	0.404(0.055)b
	Omega-3	0.200(0.028)ab	0.256(0.036)a	0.217(0.058)ab	0.163(0.024)b	0.254(0.048)a

Values shown are averages of three determinations, with standard deviation in parentheses. Means followed by different letters are significantly different at the 5 percent level.

to act as a buffer. The system should be topped with a porous disposable filter material, such as unfaced fiberglass insulation, to remove particulate matter and allow microflora (which will remove ammonia) to become established.

FLORIDA SPINY LOBSTER

Commercial harvesting of Florida's spiny lobsters (*Panulirus argus*) started in the early 1800s in the lower Florida Keys. The bait lobster fishery was part of the fin fishery and fluctuated with the local economy (Labinsky et al., 1980). By the latter part of 1910, an estimated annual catch of 160 metric tons was recorded. It was estimated that 40 percent of the catch was consumed locally, 40 percent was shipped to market (as far North as New York and Boston), and 20 percent was used as bait (Schroeder, 1924).

Annual Florida landings (1925 to 1993) for *Panulirus argus* can be seen in Figure 9. Peak landings in the early 1970s are attributed to catches in international waters. Expansion of the Bahamas' offshore sovereign fishing grounds from three to 12 miles in 1969, and the complete closure of the Bahamas' continental shelf from foreign fishermen in 1975, is considered to have contributed to the reduction of spiny lobster landings in Florida (Prochaska, 1976).

Since 1978 monthly landing statistics indicate that about 77.9 percent of the catch is landed during the first four months of the season. This observation has been interpreted to imply that only the first half of the fishing season is productive (Harper, 1993). While several other spiny lobsters are found in Florida and surrounding waters, these account for less than 10 percent of the landings (McAvoy, 1994).

At present the commercial fishing season is set between August 6 and March 31. It is illegal to possess egg-bearing lobsters. Legal-sized lobster must have 7.6 centimeters (3 inches) carapace length. Undersized lobsters ("shorts") are illegal to keep, except if used as lobster bait. The limit on shorts is 50 per boat or one per trap on a boat. The regulations also require that the undersized lobsters be kept alive in a shaded well, with continuously circulated seawater, and returned to water alive each day. However, such handling is traumatic to the lobsters, resulting in more than 25 percent mortality (Stewart, 1989). Other attractants are pigs' feet, fish heads, cowhide, and punctured sardine cans. Some lobstermen prefer to use cowhide because it does not attract stone crab that may mutilate trapped lobsters (Labinsky et al., 1980). The traps must be 91 x 61 x 61 centimeters or volume equivalent. They are made out of slats

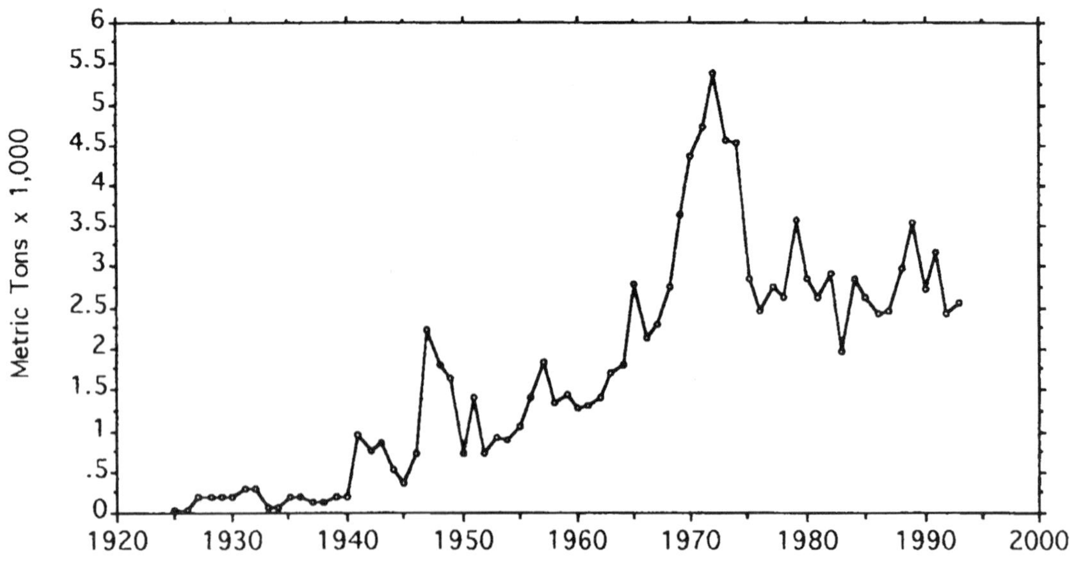

Figure 9. Spiny lobster landings in Florida, 1925–1993 (McAvoy, 1994; Stewart, 1989; Labinsky et al., 1980).

with 3.8 centimeters (1.5 inches) spacing. Since Florida law no longer allows the use of oils and paints to preserve the wood, plastic traps are increasing in popularity (McAvoy, 1994).

In 1992 a three-year restricted lobster fishery was initiated, in an attempt to reduce the number of people entering this fishery. For a person to obtain an annual license, the applicant must provide proof of having sold at least $5,000 worth of seafood product in at least one of the last three years, or of having earned at least 25 percent of annual income as a fisherman in one of the last three years. During this period each fisherman must also reduce the number of traps by 10 percent each year. Should a person wish to increase the number of traps in use, then he or she must purchase traps from another fisherman, who in turn must reduce his or her traps by the number of traps sold. The fishermen are required to put their license numbers on their traps. The state of Florida issued about 3,600 spiny lobster fishing licenses in 1992 (McAvoy, 1994). Some lobstermen have been known to work as many as 2,000 traps (Beardsley et al., 1975).

Commercial divers must use a "diver down" flag and have the boat marked with a permanent identification on a shield no smaller than 49 x 61 centimeters (16 x 20 inches). Divers fishing at night may not exceed the daily recreational limit. All divers must have a carapace-measuring device while in the water, and each measurement must take place in the water (Stewart, 1989).

Two additional days are allotted to the recreational fishery, which occur on the full weekend before August 1 each year. The person with the license can possess six lobsters per person or 24 lobsters per boat, whichever is larger. Method of catch is by hand or with a hand-held net. There is no off-water possession limit (McAvoy, 1994). It has been estimated that recreational catch accounts for 29 percent of total lobster landings (Bertelsen and Hunt, 1991).

Spiny lobsters can be purchased on the market as frozen, cooked tails, or whole lobsters. During the closed fishing season, a licensed Florida wholesaler must have a "Closed Season Permit" in order to be able to sell frozen, whole lobsters or tails from inventory to customers inside or outside the state. The above license is also required if a dealer imports spiny lobsters and sells these inside or outside the state. The permit can be obtained from the Florida Department of Law for $100 (McAvoy, 1994).

The recommended cooking of spiny lobsters is by dropping a live lobster or its thawed tail (if previously frozen) in boiling water, salted to taste. When the water comes to a second boil, maintain boiling for nine minutes. Other methods of cooking include barbecuing the tail or baking it in an oven at 260°C (500°F) for about eight minutes, after having split open the shell.

Nutritional composition is shown in Table 3. Fat was extracted according to Bligh and Dyer (1959), and fatty acids were methylated and identified by retention time using a Supelcowax-10 capillary column (30 meters x 0.53 millimeters ID). Cholesterol was analyzed according to Kovacs et al. (1979); food energy measured using the Oxygen Bomb Calorimeter, Model 1341 (Parr Instrument Co., Moline, Illinois); proximate composition according to Helrich (1990); and minerals using ICP, Model 950 (Jarrell-Ash, Franklin, Massachusetts).

There are no large differences in the proximate composition among the four species of lobsters, although significant differences at the 5 percent level can be seen. The tails of Hawaii's and California's lobsters are closer in composition as compared to tails of the American and Floridian lobsters. Total fat in the edible portions of boiled lobsters is low at 0.6–1.33 percent; nevertheless, Hawaiian lobster has the highest fat content, followed by Floridian, then Californian, and finally the American lobster. Energy values for the American, Hawaiian, and Californian lobster tails is similar at 140 kilocalories, while that of Floridian lobster tail and claws of the American lobsters are similar at 120 kilocalories.

Difference in mineral composition among the species examined has no special pattern, but individual elements are frequently significantly different. Perhaps one interesting observation may be that copper and zinc show less variability among the tail tissue of the various species as compared to the claw of the American lobster. In general, the mineral composition of the claw is frequently different from that of the tail of the American lobster, which probably reflects tissue functionality.

Analysis of lobster tails in the authors' laboratory over the years reveals that cholesterol content of tails varied between 70 and 170 milligrams per 100 grams. It also is appropriate to point out that the USDA Agriculture Handbook 8–15 (Exter, 1987) lists northern cooked lobster as having 72 milligrams/100 grams of cholesterol. Recent analysis shows the variability in findings: the tail meat measured 127 milligrams of cholesterol, and the claw meat 55 milligrams /100 grams. Hawaiian, Floridian, and Californian lobsters measured 96, 89, and 80 milligrams/100 grams, respectively.

Total saturated fatty acids were highest in Hawaiian lobsters (0.424 milligrams/100 grams) and lowest in American lobsters (0.166 milligrams/100 grams). Monounsaturated fatty acids varied between 0.180 and 0.231 milligrams/100 grams, but because of the large variability among triplicate samples, no significant difference was detected among the different species. Polyunsaturated fatty acids had a similar pattern to that of the saturated fatty acids (0.526 and 0.289 milligrams/100 grams, respectively). This observation reflects the total lipid content of each species. Despite the significant difference in the tail's fat content, no significant difference in omega-3 fatty acids among the different species was detected, except in Hawaiian lobster, which had significantly lower omega-3 fatty acids. The reduced omega-3 concentration in the Hawaiian lobster is due to the lower concentration of 20:5n3 and 22:6n3. This species instead has a higher concentration of 20:4n6, resulting in the highest level of polyunsaturated fatty acids among the four species. Lipid differences, although significant in many cases, are not large, because all the lobster species examined have a low total fat content.

Compositional differences observed apparently reflect species and dietary differences among the four species. Differences observed between the claw and the tail of the American lobster are probably due to muscle functionality.

In some instances people enjoy eating the digestive gland of a boiled lobster. The digestive gland is also used at times in the preparation of lobster pastes and lobster-based sauces. It should be pointed out that this tissue is frequently high in cadmium and consuming such a meal once a week would exceed a person's total allowable cadmium intake for that period (Uthe et al., 1982).

Spiny lobsters are called "spiny" because the spines on their exoskeleton make it difficult to handle the lobster by hand. They possess five pairs of legs and move freely on the ocean floor. They can swim backward rapidly, because of a powerful tail and the fan-like shape of the last tail segment. The abdomen consists of six segments; on the first four there are swimmerets. Males have a sperm duct opening at the base of the last pair of walking legs, and the swimmerets are single and paddle-like. Females have an opening at the base of the third pair of walking legs and the swimmerets are bi-lobed (Stewart, 1989). Spiny lobsters do not have claws as do American lobsters.

The color of spiny lobsters varies from very light to very dark (red-orange or almost black), depending on the surrounding. They also have two pairs of very pronounced large yellow spots with very dark-colored borders on two abdominal segments (second and sixth) and small spots on the remaining segments (Stewart, 1989).

Adult lobsters are gregarious, preferentially found around sponges, reefs, and rock outcroppings. They hide in the day and come out at night to hunt for food. They have been observed to participate in mass migration. Lobsters have been known to move 50 to 60 miles at a rate of two miles per day. In such migrations, the lead lobster sets the course and the pace. The remaining lobsters follow in a single file, keeping in contact with the lobster in front with their antennae. The purpose of these migrations is not understood. Nevertheless, these species are found from North Carolina to Brazil (Stewart, 1989).

An algal bloom was observed developing in 1991, feeding on decaying seagrass and resulting in the destruction of sponges. By 1993, a 300-square-mile area of sponges had been destroyed. It is not clear how algal bloom causes death of the sponge population (Keating, 1993). Nevertheless, absence of sponges creates an ecological problem for the young lobsters since they hide and feed among sponges. Labinsky et al. (1980) referred to a similar die-out of sponges in 1879.

The life cycle of the spiny lobster is complex. Adult lobsters mate from March through July. Fertilized eggs are carried about three weeks on the swimmerets of the female until hatching. Phyllosome larvae emerging from the eggs are very small, transparent, flat, and spider-like, except for eye pigmentation. They drift for about nine months in the ocean water, going through 11 stages, before entering a transparent, lobster-like stage, called puerulus. During the next three-to-four-week period, they do not feed but can swim. The puerulus swims toward the shore, where it will settle in a suitable habitat and molt into a pigmented juvenile spiny lobster, resembling the adult. The juvenile lobsters live in shallow water among algae, seagrass, and rocks. As they increase in size through molting and mature, they move to deeper waters. Reproductive maturity is reached in three years (Stewart, 1989).

CALIFORNIA SPINY LOBSTER

Spiny lobster *(Panulirus interruptus)* is found along the California coast from the Monterey area to the Mexican border. While the lobster fishery is relatively small, it plays an important role in the commercial fishery supplying area restaurants. Of equal economic importance is the recreational sports fishery. California citizens consider it their right to fish in state waters. This privilege was guaranteed in a constitutional amendment in 1910 (State of California, 1993).

Annual catches have been recorded since 1916, fluctuating around 150 metric tons up until 1945 (Figure 10). After World War II, landings increased to about 400 metric tons, but steadily declined thereafter, reaching a low point in 1975. The decline in catches has been attributed to illegal harvesting of undersized lobsters, increases in recreational catches, and loss of habitat due to urban development along the coast (Shaw, 1986). The seriousness of illegal catches of undersized lobsters can be appreciated from the frequency of citations involving commercial as well as sports fishermen and especially professional poachers who may or may not have a license (State of California, 1993). Commercial fishermen encounter problems with boaters, surfers, and other fishermen who will rob their traps (Shaw, 1986).

Commercial fishermen use rectangular traps made of 11-gauge wire mesh, but plastic traps are allowed (Shaw, 1986). The number of commercial licenses issued varies from year to year and ranges between 200 and 400 (State of California, 1993). Most fishermen use 100 to 300 traps, but some have as many as 500. The traps are marked with a buoy identified with the owner's permit number. Traps are baited with small fish or fish heads and frames. Most traps are placed in waters about 30 meters deep or less, but some traps have been set as deep as 100 meters (Barsky and Swartzell, 1992). Traps are serviced every one to two days along the coast and every two to four days around the islands (Shaw, 1986).

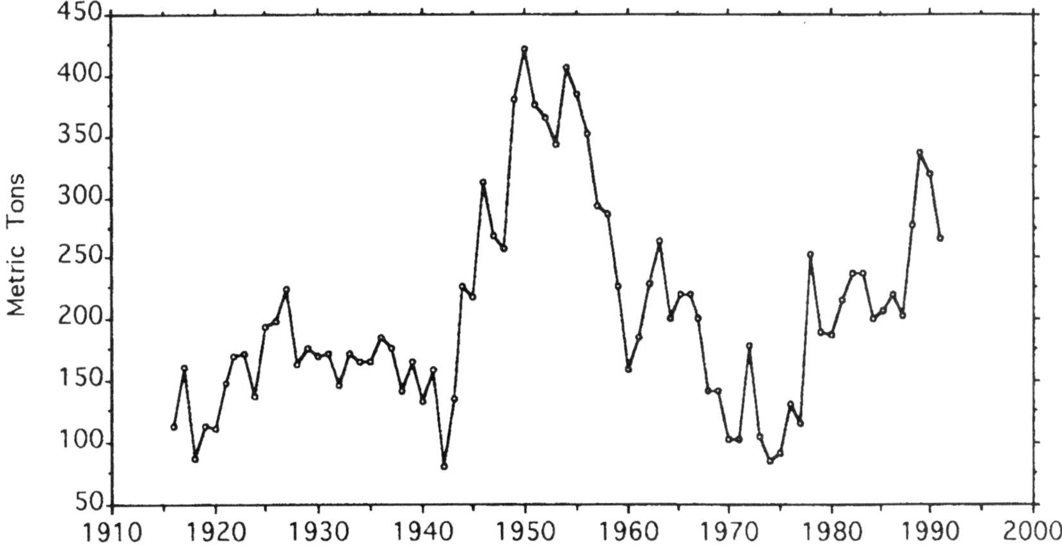

Figure 10. Spiny lobster landings in California, 1916–1993 (State of California, 1993; Barsky and Swartzell, 1992).

Odemar et al. (1975) reported that the ratio of trapped undersized lobsters to legal size was 5.4:1 before the escape window (5.1 centimeter diameter) was made mandatory and 0.8:1 after the law went into effect (Shaw, 1986). The current regulations require a rectangular escape window of 6.0 x 29.2 centimeter (Barsky and Swartzell, 1992). No evaluation is as yet available as to the effect that this regulation may have on lobster landings.

Harvested lobsters are kept in "live" boxes anchored in kelp beds until a full load is obtained to be taken to the market. Eighty-one percent of the catch is sold to the wholesaler, nine percent is sold to restaurants, seven percent is sold to private individuals, and three percent is kept by the fishermen (Odemar et al., 1975). At the fish market, lobsters can be purchased live or frozen.

The recreational season for spiny lobsters starts on the Saturday preceding the first Wednesday in October and ends on the first Wednesday after March 15. The daily bag limit is seven lobsters. During a trip of several days that must be previously declared, a three-day bag limit is allowed. Most lobsters are caught by skin divers and SCUBA divers using a gloved hand, but some sport fishermen use baited hoop nets. No spears, hooks, or other instruments are allowed. A hoop net consists of a ring made of steel rod with a three-foot diameter to which a shallow bag is attached. The bag is baited with dead fish and lowered from a pier or a boat to the ocean floor by a rope. After several minutes the bait is quickly taken to the surface, to avoid the escape of a lobster that may have entered the bag (State of California, 1993).

Recommended cooking for lobster tails is the same as for Florida spiny lobster. Nutrition data are listed in Table 3.

The closed season is aimed at protecting the berried female as well as all the molted lobsters during a period when they are very susceptible to damage by handling. The restriction of harvesting lobsters larger than 8.4 centimeters carapace length assures recruitment of adults, since spiny lobsters are sexually mature when they are about 6.4 centimeters. In fact, data reveal that almost all lobsters landed during the harvest season represent those that were below the legal size the previous season (State of California, 1993).

The Department of Fish and Game and other agencies provide recommendations to the State Commission regarding regulations to assure continued recruitment of adult lobsters. The general public also has the ability to provide input. The Commission meets in August every odd year at which time regulations are reviewed and modified.

Panulirus interruptus has no special body markings and its color varies with its surrounding from light red, through dark mahogany, to black. Species differentiation is on a morphological basis. Spiny lobsters are primarily found in rocky coastal waters, where giant kelp, ribbon kelp, coralline algae, and surf grass are present (Lindberg, 1955). They are also found in eel grass beds where rocks are not found. Divers have spotted lobsters in rock piles, crevices, sunken automobiles, and aircraft (State of California, 1993). Some investigators have reported as many as several hundred lobsters in a single den (Engle, 1979). Lobsters stay hidden during the day, but come out at night searching for food. They are omnivorous – consuming algae, dead fish, and various invertebrates. Predators that prey on lobsters include sheephead, cabezon, kelp bass, octopuses, California moray eels, horn sharks, leopard sharks, rockfishes, and giant sea bass (Barsky and Swartzell, 1992). Sea otters also prey on spiny lobsters (Shaw, 1986).

During winter months lobsters are found in deeper, warmer water (15 meters or more), avoiding the greater wave action of inshore waters. During the summer they move to warmer water closer to shore, where food is more plentiful and the time for egg development is shortened (State of California, 1993). Mating season is reported to be primarily December through March (Shaw, 1986). The male deposits a sticky spermatophore on the ventral side of the female's thorax. When the female spawns, the eggs are fertilized as they are extruded and become attached to hair-like structures on swimmerets. The number of eggs extruded depends on the size of the female and can range between 50,000 and 800,000. The eggs hatch to phyllosoma larvae in nine to ten weeks. The larvae drift in the ocean currents as they go through 11 developmental stages that last six to nine months (Shaw, 1986). These larvae have been found 480 kilometers out to sea and from the surface to a depth of 120 meters (Barsky and Swartzell, 1992). At the end of this period they metamorphose to the puerulus stage, which lasts about 2.5 months. The puerulus lobster exhibits directional swimming and is found to live in the near shore surf grass. Toward the end of the puerulus stage they settle on the ocean floor and molt, becoming juveniles. Sexual maturity is reached five to six years after hatching, at which time their carapace length is about 6.3 centimeters, but to reach the minimum legal size requires seven to ten years for males and about 12 years for females (Shaw, 1986).

Raising *Panulirus interruptus* in aquaculture is difficult, because of the complex phyllosome stages and lack of knowledge of their feeding habits (Shaw, 1986). The puerulus stage, however, was successfully reared in the laboratory to legal size in two years at 28°C and in three years at 22°C (Serfing and Ford, 1975). However, it is not known what may happen to the natural population if the surf grass were disturbed in an attempt to harvest the puerulus lobster in large numbers (Shaw, 1986).

HAWAIIAN SPINY LOBSTER

Commercial harvesting of spiny lobster *(Panulirus marginatus)* around the Northwestern Hawaiian Islands began in 1977, after research cruises in 1976 documented the potential size of such a fishery (Uchida and Tagami, 1984). The lobsters are caught in commercially manufactured traps made of polyethylene. They are circular, single chamber, flat cages with 4.5 x 4.5 centimeters mesh. The cages are 29.5 centimeters high and have a 77 centimeters-diameter roof with a 98 centimeters-diameter floor. There are two entry ports and two escape panels (opposite each other) with four exit windows (6.7 centimeters diameter). The traps are baited with chopped mackerel and are set at a depth of 20 to 70 meters in a string of several hundred traps. Midsize vessels (20-30 meters) carry a crew of five-nine members and can set 600 to 800 traps each day. Larger vessels are not economical, and smaller boats have operational difficulties. The two more productive areas are Maro Reef and Necker Island (Polovina, 1993).

Annual lobster landings are shown in Figure 11. These landings include spiny lobster *(P. marginatus)* and the common slipper lobster *(Scyllarides squammosus),* because both are caught in the same trap and so recorded in the logbook. Since 1988 it has been documented that 80 percent of the landings were spiny lobsters, although spiny lobster only slightly predominated the landings in 1985-1987. The fishermen are able to favor the catch for one species or the other by selecting the area and depth of setting the traps. In fact, between 1985 and 1987 the fishermen targeted the common slipper lobster and largely depleted that population. The size and shape of the escape windows has been chosen to allow exit of the two species with a minimum legal tail width of 5.0 centime-

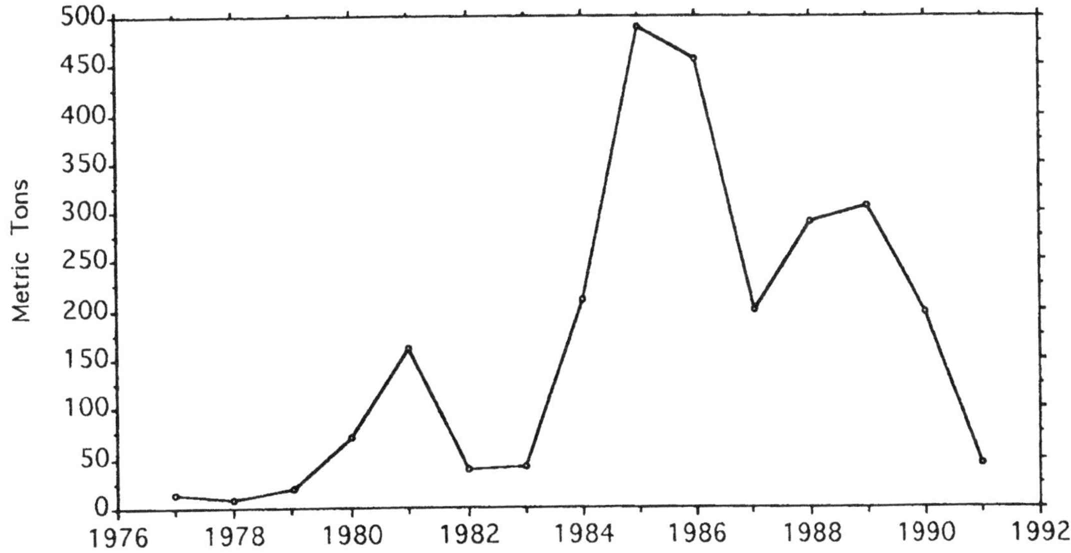

Figure 11. Lobster landings in Hawaii, 1977–1991 (Polovina, 1993).

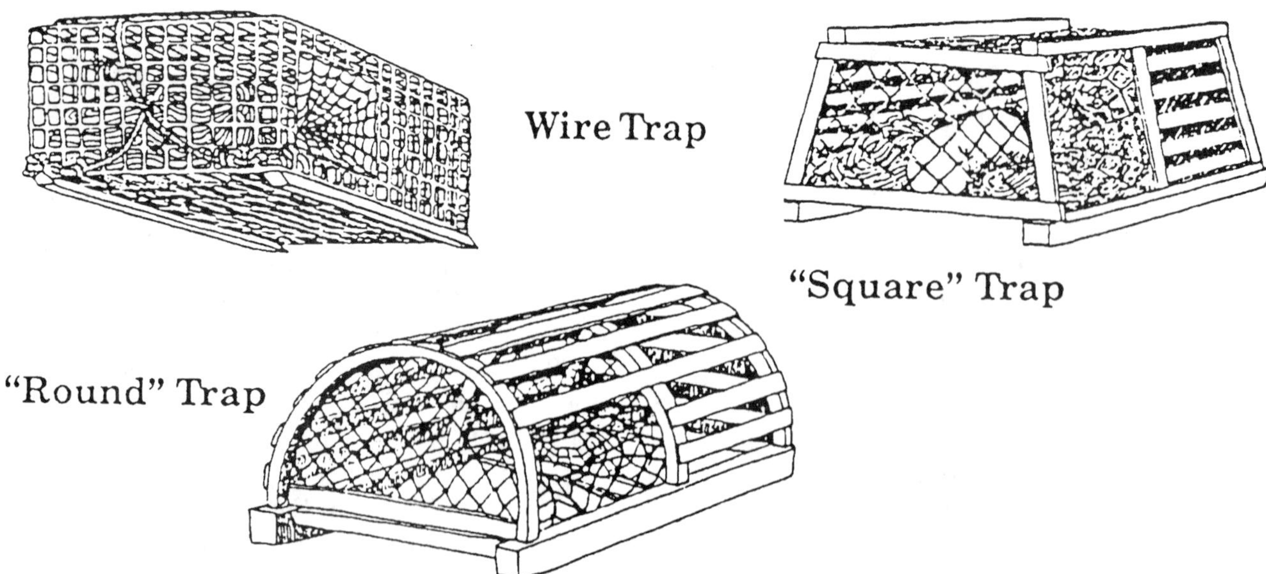

Figure 12. Lobster traps come in various sizes and shapes. Many lobstermen build their own rounded or rectangular traps from wood. Some traps are made of heavy plastic-coated wire, but the design is almost the same. Netting in the entrance of the trap leads to a metal circle through which the lobster enters when he tries to get the bait. Another funnel-shaped piece of netting at the back of the trap leads to a second compartment.

ters for the spiny lobster and 5.6 centimeters for the common slipper lobster (Polovina, 1993; Figure 12).

It is estimated that annually about 2,000 traps are lost. A study evaluating efficiency of the eight 6.7 centimeter-diameter escape windows revealed that they will allow all the lobsters that may have entered the trap to escape with time (Parrish and Kazama, 1992).

The initial increase in catches, peaking in 1985-1986, are considered to be the result of a normally growing fishery. Subsequent decline is thought to be due to low recruitment at some locations because of poor oceanographic conditions. Furthermore, to compensate for the low catches in some areas, fishing was intensified at the more productive banks. These factors have resulted in a serious decline in recruitment (Polovina, 1993).

Recruitment data at Maro Reef showed very good correlation to the relative sea level between French Frigate Shoals and Midway Island, four years earlier. The sea level variability is in turn related to the Subtropical Counter Current (El Niño). The connection between these phenomena is not clear (Polovina and Mitchum, 1992). MacDonald (1986) studied the puerulus lobster and reported highest settlement in the summer at Kure Atoll, in the winter at Oahu, and continuous through the year at the French Frigate Shoals. Polovina (1993) suggested that lobster density is not so much due to larval densities at different locations as due to "the amount of relief provided by the benthic habitat."

To protect spawning lobsters, emergency closure was implemented from May to October, 1991. Due to continued decline in catches, limited entry was imposed in March 1992, and a closed season declared from January through June. These measures did not improve recruitment and in the summer of 1994 a complete closure for one year was enforced.

The life cycle and the habitat of the Hawaiian spiny lobster is very similar to that of spiny lobster in California and Florida. The adult is dark red to black, depending on the environment, covered with different-sized, light-colored eye spots on the dorsal part of the thorax. A large pair of eye spots can be found on the first segment, and smaller eyes on all succeeding segments. In addition, there is a transverse light-colored line on each segment, which forms an almost-continuous line along the edges of the fan-shaped tail.

Recommended cooking is the same as for Florida spiny lobster. Nutrition data are listed in Table 3.

WORLD LOBSTER FISHERY

FAO data (1991) for the clawed lobsters account for 135.3 thousand metric tons and spiny lobsters 79.1 thousand metric tons (FAO, 1993) (see Table 4, p. 164). Of the clawed lobsters, North America provided 76.3 thousand metric tons and Europe 59.0 thousand metric tons. Although there is an overlap in catches of species among areas, it appears that *Panulirus argus* contributed essentially all of the New World landings of 38.8 thousand metric tons. *P. cygnus, Jasus verreauxi, J. edwardsii,* and *Metenephrops challengeri* accounted for about 17.3 thousand metric tons from Australia and New Zealand. Tropical species of *Panulirus* around the world along with Palinurid spiny lobster from Europe, Africa, and China accounted for 15.3 thousand metric tons. *Jasus lalandii, J. tristani, Palinurus delagoae, P. gilchristi,* and *Meteanephrops andamanicus* account for about 3.6 thousand metric tons from around Africa. Scyllaridae are caught in the Far East and account for about 2.8 thousand metric tons, while *Panulirus longipes* is primarily landed in Japan and contributed about 1.3 thousand metric tons to the 1991 world catch. France landed about 34 metric tons of *Palinurus elephas* and *P. mauritanicus.*

World spiny lobster fishery management as well as aquaculture and marketing are presented in a publication by Phillips et al. (1994), while identification of lobsters can be aided by the publication of Williams (1988).

ACKNOWLEDGMENTS

The authors thank the following for their assistance and for information provided: Philip T. Briggs, New York Dept. of Environmental Conservation, Stony Brook, NY; Glenda Duncan, Dept. of Agriculture, Fisheries and Forestry, Charlottetown, Prince Edward Island, Canada; Joseph R. O'Hop, Dept. of Environmental Protection, Florida Marine Research Institute, St. Petersburg, FL; David Parker, Dept. of Fish and Game, Long Beach, CA; Robert Ross, United States Dept. of Commerce, NMFS, Gloucester, MA; and Marylin Whetzel, The Florida Marine Research Institute, St. Petersburg, FL. Special thanks for providing data and helpful suggestions with data collection go to Jay Krouse, Maine Dept. of Marine Resources, Boothbay Harbor, ME. A note of special gratitude goes to William Gibbs, Keys Marine Lab., Long Key, FL, and Robert B. Moffitt, NMFS, Honolulu Laboratory, Honolulu, HI, for providing lobsters for analyses. At the University of Maine, Orono, ME, great appreciation is due Richard Work, for assistance with the calorimeter; L. Brian Perkins, for assistance with the gas chromatograph; William P. Cook, for performing protein, ash, and mineral analyses; Wayne B. Persons, for assistance with data analysis; and Alfred Bushway for reviewing the manuscript.

Maine Agricultural Experiment Station Publication No. 2370. This work was also supported by The Lobster Institute, University of Maine.

REFERENCES

Aiken, D.E., J.B. Sochasky, and P.G. Wells. 1973. Ciliate infestation of the blood of the lobster *Homarus americanus.* ICS CM1973/K:46.

Anderson, K. and R. C. Bayer. 1991. A rapid method for determination of oxytetracycline levels in the American lobster. *Prog. Fish Cult.* **53**:25-28.

Barrow, J. 1994. Dept. Fisheries and Oceans, Ottawa, Canada (personal communication).

Bayer, R. C., L. K. Good, R. H. True, T. M. Work, and M. L. Gallagher. 1981. Effect of feeding a synthetic diet on weight gain and meat flavor in lobsters. *Proc. Sec. Int. Conf. Aquacult. Nutr.*

Bayer, R. C. and P. C. Daniel. 1987. Safety and efficacy of oxytetracycline for control of gaffkemia in the American lobster. *Fish. Res.* **5**:71-81.

Bayer, R. C., H. Hodgkins, M. Loughlin, and D. Prince. Lobster health manual. Maine/New Hampshire Sea Grant Publication, MSG-E-97-7.

Barsky, K. C. and P. G. Swartzell. 1992. California spiny lobster. In *California's Living Marine Resources and their Utilization.* W.L. Leet, C.M. Dewees, and C.W. Haugen (eds.), pp. 22-24. Sea Grant Ext. Public. UCSGEP-92-12.

Bertelsen, R. D. and J. H. Hunt. 1991. Results of the 1991 mail surveys of recreational lobster fishermen. Florida Marine Fisheries Comm. Dec. 1991.

Beardsley G. L., T. J. Costello, G. E. Davis, A. C. Jones, and D.C. Simmons. 1975. The Florida spiny lobster. *Florida Sci.* **38**(3):144-149.

Bligh, E. G. and Dyer, W.J. 1959. A rapid method of total lipid extraction and purification. *Can. J. Biochem. Physiol.* **37**:911-917.

Daniel, P. C., R. C. Bayer, and C. Waltz. 1989. Egg production of V-notched American lobsters (*Homarus americanus*) along coastal Maine. *J. Crust. Biol.* **9**:77-82.

Dennell, R. 1947. The occurrence and significance of phenolic hardening in newly formed cuticle in

Table 4. World Lobster Landings (metric tons) (FAO, 1993)

	1982	1983	1984	1985	1986	1987	1988	1989	1990	1991
Panulirus longipes long-legged spiny lobster, primarily Japan	1274	1400	1331	1315	1510	1301	1074	1323	1341	1286
Panulirus argus spiny lobster, U.S. to Brazil	29603	28633	34738	36908	34480	33125	32423	34248	32741	38590
Panulirus cygnus Australian spiny lobster, Australia	10483	12456	10689	11264	11000	11025	11569	14150	11943	10676
Panulirus spp. tropical spiny lobster, around the world	7864	8352	8784	10054	12207	10967	10468	10567	11047	11159
Jasus lalandii cape rock lobster, Spain to South Africa	6558	6226	7095	7235	6123	6568	7145	4765	4306	2210
Jasus frontalis Juan Fernandez rock lobster, Chile	50	35	42	35	43	36	29	23	19	27
Jasus verreauxi green rock lobster, Australia and New Zealand	5232	5261	4996	5246	4658	5200	5457	2854	3114	3128
Jasus tristani Tristan da Cunha rock lobster, St. Helena	461	395	384	370	342	405	441	427	451	426
Jasus edwardsii red rock lobster, New Zealand	4750	4963	5442	5474	5259	4937	3594	3754	3120	2976
Palinurus mauritanicus pink spiny lobster, France	2	2	3	9	4	5	–	7	8	2
Palinurus elephas common spiny lobster, France	89	97	97	104	89	69	46	36	33	32
Palinurus delagoae natal spiny lobster, South Africa	115	66	43	32	43	89	25	32	24	18
Palinurus gilchristi south coast spiny lobster, South Africa	372	797	1049	450	1031	1820	880	572	1042	885
Palinurus spp. Palinurid spiny lobsters, Europe, Africa, and China	2374	2257	3066	3740	4561	4780	7738	5956	5776	4145
Scyllaridae slipper lobsters, Far East	1637	1134	1599	1866	1941	2326	1865	1873	2018	2813
Metanephrops andamanicus Andaman lobster, South Africa	219	95	107	352	430	270	298	306	283	51
Metanephrops challengeri New Zealand lobster, New Zealand	–	–	–	–	–	–	–	–	627	501
Nephrops norvegicus Norway lobster, Europe and North Africa	50145	54008	53531	61724	58832	60826	61566	55253	55460	56709
Homarus americanus American lobster, Canada and United States	40698	47708	48637	53574	58861	60095	62456	67092	74230	76329
Homarus gamarus European lobster, Europe and North Africa	2041	2287	2442	2229	1971	2286	2576	2776	2634	2277
Total	163967	176172	184075	201981	203385	206130	209650	206014	210217	214240

crustacean decapoda. *Proc. R. Soc.* (Lond.) B (Biol. Sci.) **134**:485-503.

Edwards, E., P.A. Ayres, and M.L. Cullum. 1981. Incidence of the disease gaffkemia in native (*Homarus gammarus*) and imported lobsters (*Homarus americanus*) in England and Wales. ICES CM1981/K:12.

Engle, J.M. 1979. Ecology and growth of juvenile California spiny lobster, *Panulirus interruptus* (Randall). Ph.D. Dissertation. University Southern California, Los Angeles.

Exleer, J. 1987. Composition of Foods. United States Dept. of Agric., Agric. Handbook No. 8-15.

FAO. 1993. Fishery statistics, 1991. Rome.

Fisher, W.S., T.R. Rosemark, and E.H. Nilson. 1976. The susceptibility of cultured American lobsters to a chitinolytic bacterium. *Proc. World Maricult. Soc.* **7**:511-520.

Getchell, R.G. 1989. Bacterial shell disease in crustaceans: A review. *J. Shellf. Res.* **8**:1-6.

Harper, DE. 1993. The 1993 spiny lobster monitoring report on trends in landings, CPUE, and size of harvested lobster. NOAA/NMFS, SE Fisheries Center, Miami, FL. MIA-92/93-92.

Helrich, K. (ed.). 1990. *Official Methods of Analysis* (15 Ed.), AOAC, Arlington, VA.

Huang, C. H. and R. C. Bayer. 1989. Gastrointestinal absorption of various antibacterial agents in the American lobster. *Prog. Fish Cult.* **51**:95-97.

Hughs, J. T. and G. C. Matthiesson. 1962. Observations on the biology of the American lobster, *Homarus americanus*. *Limnol. Oceanogr.* **7**:414-421.

Johansson, M. W. and J. Soderhall. 1989. Cellular immunity in crustaceans and the proPo system. *Parasit. Today* **5**:171-176.

Keating, D. 1993. Killer algae threatening Florida bay lobsters. *The Miami Herald*. February 17, 1993.

Kovacs, M.I.P., W.E. Anderson, and R.G. Ackman. 1979. A simple method for the determination of cholesterol and some plant sterols in fishery-based food products. *J. Food Sci.* **44**:1299-1301 and 1305.

Krouse, J. S. 1994a. Maine Dept. of Marine Resources, West Boothbay Harbor, ME (personal communication).

Krouse, J. S. 1994b. United States commercial and recreational American lobster, *Homarus americanus*, landings and fishing effort by state, 1970-1992. Maine Dept. Marine Resources. West Boothbay Harbor.

Labinsky, R. F., D. R. Gregory Jr., and J.A. Conti. 1980. Florida's spiny lobster fishery: An historic perspective. *Spiny Lobster Series* **5**(4):28-37.

Leavitt, D. F., R. C. Bayer, M. L. Gallagher, and J. H. Rittenburg. 1979. Dietary intake and nutritional characteristics in wild lobsters. *J. Fish. Res. Bd. Canada.* **36**:965-969.

Lightner, D.V. 1988. Milk or cotton disease of shrimps. In *Disease Diagnosis and Control in North American Marine Aquaculture*, C. Sinderman (ed.), pp. 48-51. Elsevier. New York.

Lindberg, R. G. 1955. Growth, population dynamics, and field behavior in the spiny lobster, *Panulirus interruptus* (Randall). *Zoology* **59**(6):157-248.

Loughlin, M. B. and R. C. Bayer. 1991. Scanning electron microscopy (SEM) of *Mugardia*, formerly *Anophrys*, a pathogenic protozoan of the American lobster. Abstracts. *J. Shellf. Res.* **10**:298.

MacDonald, C. D. 1986. Recruitment of the puerulus of the spiny lobster, *Panulirus marginatus*, in Hawaii. *Can. J. Aquatic Sci.* **43**:2118-2125.

Malloy, S.C. 1978. Bacterial induced shell disease of lobsters (*Homarus americanus*). *J. Wildl. Dis.* **14**:2-10.

Maine. 1917. Acts and Resolves. Seventy-Eighth Legislature. Chapter 255.

Maine. 1964. Revised Statutes Annotated. Vol. 6. Title 12. Statute 6451.

Maine. 1995. Compilation of lobster laws and regulations. Revised Oct. 1995. Dept. Marine Resources, Augusta.

McAvoy, H. 1994. Florida Dept. Agric. & Consumer Serv. Tallahassee, FL (personal communication).

Odemar, M. W., R. R. Bell, C. W. Haugen, and R. A. Hardy. 1975. Report on California spiny lobster, *Panulirus interruptus* (Randall) research with recommendations for management. *California Fish and Game Operations Research Branch* (Special Publication).

Parrish, F. A. and T. K. Kazama. 1992. Evaluation of ghost fishing in the Hawaiian lobster fishery. *Fishery Bull.* **90**:720-725.

Phillips, B. F., J. S. Cobb, and J. Kittaka. 1994. Spiny lobster management. Fishing News Books. Oxford, England.

Polovina, F.A and G.T. Mitchum. 1992. Variability in spiny lobster *Panulirus marginatus* recruitment and sea level in the Northwestern Hawaiian Islands. *Fishery Bull.* **90**:483-493.

Polovina, F.A. 1993. The lobster and shrimp fisheries in Hawaii. *Mar. Fisher. Rev.* **55**(2):28-33.

Prochaska, F.J. 1976. An economic analysis of effort and yield in the Florida spiny lobster industry with management considerations. In *Proceedings of the First Annual Tropical and Subtropical Fisheries Technological Conference*. B.F. Phillips, B.F. Cobb III, and A.B. Stockton (compilers), pp. 661-674. Texas A & M University.

Rittenburg, J. H. and R. C. Bayer. 1980. Lobster gaffkemia vaccine. U.S. patent 4,215,108.

Schroeder, W. C. 1924. Fisheries of Key West and the clam industry of southern Florida. *Appendix XII to the Report of the United States Commissioner of Fisheries for 1923*. Bureau of Fisheries Document 962. United States Printing Office, Washington, DC.

Serfing, S. A. and R. F. Ford. 1975. Laboratory culture of juvenile stages of the California spiny lobster *Panulirus interruptus* at elevated temperatures. *Aquaculture* **6**:377-387.

Shaw, W. N. 1986. Species profiles: Life histories and environmental requirements of coastal fishes and invertebrates (Pacific Southwest): Spiny Lobster. *United States Dept. Interior Biol. Report* **82**(11.47).

Sherburne, S. W. and L. L. Bean. 1991. Mortalities of impounded and feral Maine lobsters, *Homarus americanus* H. Milne-Edwards, 1837, caused by the protozoan ciliate *Mugardia* (Formerly *Anophrys = Paranophrys*), with initial prevalence data from ten locations along the Maine coast and one offshore area. *J. Shellf. Res.* **10**(2):315-326.

Smolowitz, R. M., R. A. Bullis, and D. A. Abt. 1992. Pathologic cuticular changes of winter impoundment shell disease preceding and during intermolt in the American lobster. *Biol. Bull.* **193**:99-112.

State of California, 1993. Supplemental environmental document. *Ocean Sport Fishing Regulations.* Section 27.00-30.10 Title 14, California Code of Regulations. Dept. Fish and Game.

Stewart, J. E. and J. W. Cornick. 1967. In vitro susceptibility of the lobster pathogen *Gaffkya homari* to various disinfectants and antibiotics. *J. Fish. Res. Bd. Canada.* **24**:2623-2626.

Stewart, J. E., J. W. Cornick, and D. I. Spears, 1966. Incidence of *Gaffkya homari* in natural lobster (*Homarus americanus*) populations of the Atlantic region of Canada. *J. Fish. Res. Bd. Canada.* **23**:1325-1330.

Stewart, J. E., B. M. Zwicker, B. Arie, and G. W. Horner. 1972. Food and starvation a factor affecting the time to death of the lobster *Homarus americanus* infected with *Gaffkya homari. J. Fish. Res. Bd. Canada.* **29**:461-464.

Stewart, J. E., J. W. Cornick, and J. R. Dingle. 1967. An electronic method for counting lobster (*Homarus americanus*) hemocytes and the influence of diet on hemocyte numbers and hemolymph protein. *Can. J. Zool.* **45**:291-304.

Stewart, V.N. 1989. Spiny lobster. *Sea-Stats. No. 11.* April 1988. Revised July 1989.

Taylor, C. C., 1950. A review of lobster rearing in Maine. Research Bull. #5. Dept. of Sea Shore Fisheries, Augusta, ME.

Templeman, W. 1936. Influence of the temperature, salinity, light, and food conditions on survival and growth of the larvae of the lobster. *J. Fish. Res. Bd. Can.* **2**:485-497.

Thomas, J. C. 1973. An analysis of the commercial lobster fishery along the coast of Maine, Aug. 1966 - Dec. 1970. NOAA Tech. Report. NMFS FFRF667 P1-57.

Tressler, D. K. 1951. *Marine Products of Commerce.* Reinhold Publishing Corp. New York.

Uchida, R. N. and D. T. Tagami. 1984. Biology, distribution, population structure, and pre-exploitation abundance of spiny lobster, *Panulirus marginatus* (Quoy and Gaimard, 1825), in the Northwestern Hawaiian Islands. In *Proceedings of the Second Symposium on Resource Investigations in the Northwestern Hawaiian Islands,* Vol. 1, pp. 157-197. University of Hawaii, HI.UNIHI-SEAGRANT-MR-84-01.

Uthe, F. J., H. C. Freeman, G. R. Sirota, and C. L. Chou. 1982. Studies on the chemical nature of and bioavailability of arsenic, cadmium, and lead in selected marine fishery products. In *Chemistry & Biochemistry of Marine Food Products,* R.E. Martin, G. J. Flick, C. E. Hebard, and D.R. Ward (eds.), pp. 105-113. AVI Publishing Co. Westport, CN.

Vachon, N.S., R.C. Bayer and J.H. Rittenburg. 1981. Incidence of *Aerococcus viridans* (var.) *homari* in American lobster populations from the Gulf of Maine. *Prog. Fish Cult.* **43**:49.

Williams, A.B. 1988. Lobsters of the world: An illustrated guide. Osprey Books. Huntington, NY.

Biology of Certain Commercial Crustaceans: CRABS

Michael J. Oesterling

INTRODUCTION

Crabs and crabmeat have become a favorite with the American consumer. This is evidenced by the many different species of crabs that are harvested throughout the United States. Each region of the country has a "local" crab that is the favorite of the area and professed to be the best-tasting of all species.

Crustaceans are members of the phylum Arthropoda, which also includes insects, spiders, centipedes, and millipedes. Among the 26,000 known species of crustaceans are some of the most popular and valuable seafood products: crabs, shrimp, and lobster.

The name crustacean is derived from the Latin word for shell. Indeed, the hard exoskeleton is such a prominent anatomical feature that crustaceans are often referred to as shellfish. Crustaceans have several other characteristic features: they have mandibles as mouth parts, possess two pairs of antennae, and breathe through gills derived from leg appendages. Crustaceans are found in great variety in both freshwater and saltwater habitats.

Among the crustaceans, the animals of the order Decapoda are of primary economic interest. The 8,500-plus species of decapods represent about one-third of all crustaceans. The name decapod means ten feet; all decapods have five pairs of thoracic appendages. In many decapods the first pair of legs is modified into chela (claws or pinchers) used for prey capture and defense. Many decapods qualify as important food resources; they are abundant, wholesome and accessible. Of particular interest in this section are the crabs.

There are two types (families) of crabs that are important as food resources, the brachyuran and anomuran. Many times the brachyuran are referred to as "true" crabs, referring to the fact that all 10 legs are well-developed. Prime examples of the brachyuran are the blue crab (*Callinectes sapidus*), Dungeness crab (*Cancer magister*) and stone crab (*Menippe mercenaria*). In anomuran crabs the fifth pair of legs is greatly reduced in size, making the crab appear to have only four pairs of legs. The best-known anomuran crab is the king crab, *Paralithodes camtschatica*.

Both brachyuran and anomuran crabs exhibit discontinuous growth. Because of their hard outer exoskeleton, in order for crabs to increase in size they must periodically shed (molt) this shell, to grow into a larger size. This important process, termed ecdysis, could be linked to critical life cycle events, such as mating, and forms the basis for entire industries as in the production of soft-shell blue crabs. The actual process of molting and growth is physiologically complex. Prior to the actual molt, a new shell will begin forming underneath the old one. In some species there are visible indications that molting is about to occur. As the time for molting approaches, the crab will resorb some carbohydrates, proteins, and calcium from the old shell to be stored within the body and used to help form the new shell. Muscle attachments to the old shell are loosened and reattached to the new shell. In most instances a portion of the stomach lining will also be lost during the molt; hence feeding will cease just prior to the molt. With the assistance of absorbed water, the old shell will split open along predetermined fracture lines and the crab will back out of the old shell to emerge as a soft, pliable animal. At this time the crab is very soft and defenseless. For this reason, molting takes place in hiding. Once out of the old shell, the soft crab will again absorb water to expand into a new, larger size and begin to reharden its shell. The amount of size increase is probably controlled by both genetics and by environmental conditions. Complete hardening will take hours or days depending upon the size of the crab and species. In general, the interval between molts increases with increasing size of the crab, with younger individuals molting more frequently than older ones.

GENERAL EXTERNAL ANATOMY

Crabs are covered by a hard outer shell known as the exoskeleton. The top, or dorsal, side of a crab is covered by a single heavy piece of exoskeleton called the carapace that covers the crab's head and thorax region. Stalked eyes are located at the front of the body and the rostrum, an extension of the exoskeleton, is located between the eyes.

The ventral (bottom) side of a crab looks nothing like the carapace. The most striking feature is the turned-under abdomen. The abdomen of the crab corresponds to the "tail" of the shrimp, crayfish, and lobster. Markings or the size of the abdomen reveal the crab's sex; the male generally has a narrower abdomen than the female of the species. Under the abdomen are fine legs called pleopods, commonly known as swimmerets. Female crabs use these swimmerets as a place for the attachment of their eggs. Besides the abdomen, the other notable features of the crab's ventral side are the thoracic divisions that appear as sections of the shell.

Crabs have a number of paired appendages, including five pairs of legs, or pereiopods. The first pair is modified to be the cheliped, ending in the chela, or pinchers. They function in defense and feeding. The remaining pairs of pereiopods are for locomotion and may also be used in food gathering.

Pereiopods are the most obvious of the crab's paired appendages, but there are others just as important. Most of these are centered around the crab's head region. Between the eyes are two pairs of filamentous, hairlike structures: the antennae and the antennules. The antennules are generally smaller than the antennae and are located directly on either side of the rostrum. Between the antennules and eyes are the antennae. Together the antennae and antennules are part of a crab's sense of smell, receiving chemical "odors" from the water. They are also sensitive to touch.

The first mouth appendages encountered are three pairs of maxillipeds, which are associated with feeding and food manipulation. Closer to the mouth, under the maxillipeds, are two pairs of maxillae. The maxillae are much thinner and more flexible than the maxillipeds. They assist somewhat in feeding, but their main function is in the crab's "breathing." The distal end of the maxilla is greatly expanded into a scaphognathite, or "gill bailer." By fluttering this gill bailer very rapidly, the crab circulates water through its gill chamber and over its gills where oxygen exchange takes place. The final pair of mouth appendages is the mandibles. These are the crab's "teeth" and are very hard and strong for holding and crushing food.

The internal systems of crabs are complex and are beyond the scope of this discussion. However, a brief comment on the most visible of these systems, the respiratory, is appropriate. Along the sides of the body cavity are the gill structures. These appear as eight pairs of frilly, fingerlike projections. The gills, often referred to as "dead man's fingers," are the sites where oxygen is obtained and waste materials such as carbon dioxide are removed. Water enters the gill chamber near the base of the claws, flows upward over the gills, and passes out at the sides of the mouth.

CRAB SPECIES OF ECONOMIC IMPORTANCE

Because of their abundance, wholesomeness, and accessibility many crab species are economically valuable. Within the United States there are relatively few species that are of prime importance. These are the blue crab (*Callinectes sapidus*), the jonah crab (*Cancer borealis*), the Dungeness crab (*Cancer magister*), the king crab (*Paralithodes camtschatica*), the snow or tanner crabs (*Chionoecetes bairdi*, *C. opilio*, or *C. tanneri*), the red or golden crabs (*Geryon quinquedens* or *G. fenneri*), and the stone crabs (*Menippe mercenaria* or *M. adina*).

As a group, the landings of all species of crabs in 1998 exceeded 552.7 million pounds, with an ex-vessel value of approximately $473.4 million [National Marine Fisheries Service (NMFS), 1999]. Ranking in terms of poundage landed in 1998 were: snow/tanner crabs (251.8 million pounds) blue crabs, all forms (224.2 million pounds); Dungeness crab (34.2 million pounds); king crab (24.1 million pounds); stone crabs (6.9 million pounds); jonah crab (2.7 million pounds); and red crab (2.1 million pounds) (Table 1). Value ranking is very similar, with the notable exception that the number one and two positions are reversed: blue crabs, all forms ($184.3 million); snow/tanner crabs ($145.0 million); Dungeness crab ($61.8 mil-

Table 1. Poundage and Value for Major Crab Species Landed in the United States during 1998 (National Marine Fisheries Service, 1999)

SPECIES	POUNDS (Ranking)	VALUE (Ranking)
Snow/tanner crabs	251,800,000 (1)	$145,000,000 (2)
Blue crab, all forms	224,200,000 (2)	$184,300,000 (1)
Dungeness crab	34,200,000 (3)	$61,800,000 (3)
King crab	24,100,000 (4)	$57,400,000 (4)
Stone crabs	6,900,000 (5)	$22,800,000 (5)
Jonah crab	2,700,000 (6)	$1,300,000 (7)
Red crab	2,100,000 (8)	$1,200,000 (8)

lion); king crab ($57.4 million); stone crabs ($22.8 million); jonah crab ($1.3 million); and red crab ($1.2 million) (Table 1). Based upon the relative importance of the commercial species, the jonah and red crabs will not be discussed further. Additionally, because of the life history similarities of the king and snow/tanner crabs (both anomuran crabs with similar habitats) and availability of information, only the king crab will be fully characterized.

Blue Crab (*Callinectes sapidus*)

The blue crab (*Callinectes sapidus*) supports a large commercial fishery along the eastern seaboard of the United States and the Gulf of Mexico (Figure 1). Actually there are two blue crab fisheries, one for hard-shelled crabs and one for soft-shelled crabs. Soft-shelled crabs are blue crabs that have recently shed (molted) their hard outer shell. These crabs command a premium price in the marketplace. Because of the blue crab's economic value a great deal of information is available on all aspects of the fishery and biology of the species.

It is found along the Atlantic coast from Nova Scotia to northern Argentina and throughout the Gulf of Mexico (Williams 1984, 1974). The blue crab is most abundant in the Chesapeake Bay, which annually produces close to 30 percent of the hard crab harvest and 50 percent of the soft crab harvest. Although not native to Europe, blue crabs have been found along the European coast and in the Mediterranean Sea (Williams, 1984). It is believed that these crabs "hitchhike" in ballast tanks, or cling to ocean-going vessels.

The blue crab typically inhabits coastal areas, from the shoreline to a depth of approximately 300 feet (91 meters) (Williams, 1974). It has been taken from freshwater environments such as Florida's Salt Springs and St. Johns River, and from hypersaline lagoons, such as Laguna Madre de Tamaulipas in Mexico (Williams, 1974). The blue crab's normal diet includes fishes, bottom invertebrates (clams, snails, worms, other crabs, etc.), and plant matter (Williams, 1974; Tagatz, 1968; Darnell, 1959; Van Engel, 1958). Although it is considered a scavenger, it is more properly classified as an opportunistic omnivore that prefers fresh to decaying flesh.

The blue crab generally lives two to three years, with the adult stage being reached after twelve to eighteen months (Fischler, 1965; Darnell, 1959; Van Engel, 1958; Churchill, 1919). Its life history begins with the mating of a sexually mature male and female. The female blue crab mates only once in her lifetime, just after the molt that marks her transition from juvenile to adult (Williams,

Figure 1. Dorsal view of the blue crab, Callinectes sapidus. *Photo courtesy of the Virginia Sea Grant College Program, Virginia Institute of Marine Science.*

1984; Van Engel, 1958; Churchill, 1919). Unlike the female, the male blue crab reaches sexual maturity before he is fully grown. A male blue crab may mate with more than one female and at any time during his last three growth stages (Williams, 1984; Van Engel, 1958).

Prior to the female's terminal molt, she moves to lower salinity waters and pairs with a male who will carry or cradle her underneath him (Futch, 1965; Williams, 1965). At this time both crabs are called doublers or buck-and-rider. The female completes her final molt in this cradled position and becomes an adult. While she is in the soft intermolt stage, copulation takes place. The male transfers his sperm to the female, which she stores in seminal receptacles within her body. The sperm are able to live for about one year (Williams, 1984). Following copulation, the male again cradles the female beneath him until her new shell hardens. This cradling serves two purposes. It assures that there will be a male present at the one stage in the female's life when she is able to copulate. It also serves to protect her while she is in the soft stage and extremely vulnerable to predators. Once the female's new shell has hardened, the male releases her.

Spawning (laying of eggs) usually takes place one to nine months after mating, usually in the spring and summer months (Tagatz, 1968). It is generally thought to occur in higher salinity waters at the mouth of estuaries and offshore areas (Williams, 1984; Van Engel, 1958). Egg-laying is quite rapid and may be completed within two hours. Eggs are passed from the ovaries through the seminal receptacles to be fertilized on their way to the outside of the female. As the eggs pass out of the body, they are attached to the small swimmeret appendages on the female's abdomen. When first laid, the eggs are orange, but as they mature, they turn yellow, then brown, and finally dark brown.

At the time of spawning, the female blue crab lays 700,000 to 2,000,000 eggs, but only about one ten-thousandth of 1 percent of the eggs will survive to become mature crabs (Van Engel, 1958; Churchill, 1919). The eggs are carried seven to 14 days, at which time they hatch into planktonic zoea, which are about 1/24 inch (1 millimeter) long (Fischler, 1965; Van Engel, 1958; Sandoz and Rogers, 1944) (Figure 2). The zoeal phase has seven stages and lasts 31 to 49 days, depending upon water temperature, salinity, and food availability (Costlow and Bookhout, 1959). The optimum ranges for development are 19° to 29°C (66° to 84°F) and 23 to 28 parts per thousand (ppt) salinity (Sandoz and Rogers, 1944). The zoea then metamorphose into the single megalops stage which has both planktonic and benthic (bottom) affinities (Sulkin, 1974; Williams, 1971) (Figure 3). After six to 20 days the megalops changes into the first crab stage, at which time the crab form is first seen (Costlow and Bookhout, 1959).

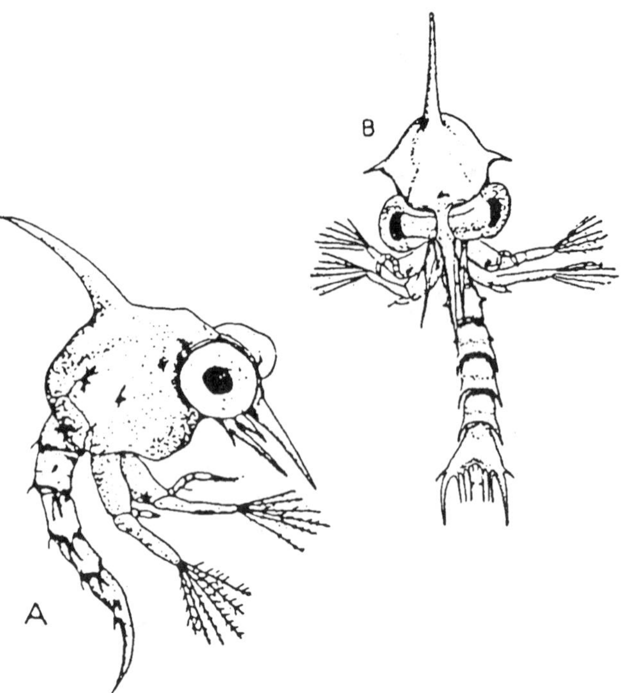

Figure 2. Zoea larvae of the blue crab. A. Side view. B. Front view. After Costlow and Bookhout, 1959, from Oesterling, 1985.

Larval (zoea) development takes place "offshore" in more saline waters than the confines of the estuary (McConaugha, 1988). The young crabs, however, spend the majority of their growing life within the nursery grounds of the estuary. During the megalops and first few crab stages, there is a movement shoreward toward the nursery grounds (Olmi, 1994; McConaugha, 1988). Figure 4 illustrates the movements of the blue crab during its life cycle.

Following the first crab stage, growth is rapid. Adulthood is reached 12 to 18 months after egg hatching (Fischler, 1965; Van Engel, 1959; Darnell, 1959; Churchill, 1919). After reaching the adult stage, blue crabs live about one year longer.

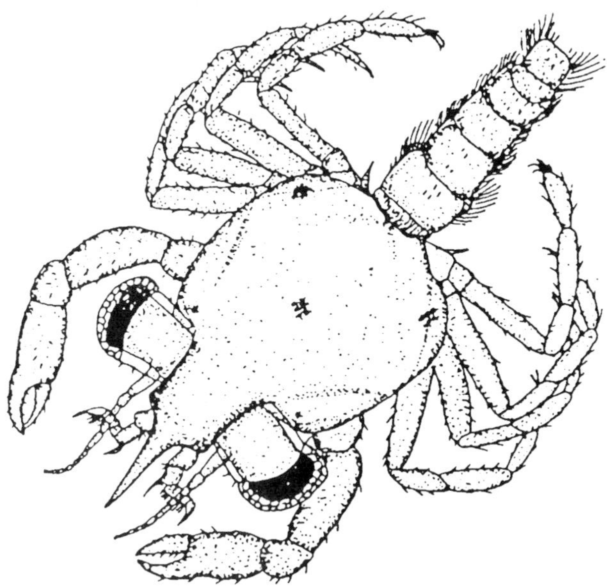

Figure 3. Megalops larval stage of the blue crab. After Costlow and Bookhout, 1959, from Oesterling, 1985.

Stone Crabs (*Menippe mercenaria* and *M. adina*)

The stone crabs (*Menippe mercenaria* and *M. adina*) form the basis for a commercial fishery centered in the Florida Keys, southeastern states, and the Gulf of Mexico (Figure 5). Indeed, in 1998 over 99 percent of the poundage and of the landed value of stone crabs were in Florida. Since most of these animals were *M. mercenaria*, its life history will be described.

Over the past 25 years a great deal of research has been focused on the Florida stone crab and in particular its expanding fishery (Zuboy and Snell, 1980; Costello et al., 1979; Sullivan, 1979; Bert et al., 1978; Savage et al., 1975; Bender, 1971). The fishery for the stone crab is unique among crab fisheries in that only the large claw of the stone crab is harvested, the animal being returned alive to regenerate the lost appendage. Savage et al. (1975) showed that nearly 10 percent of the stone crab claws landed along the west coast of Florida were regenerated claws.

The stone crab is found in warm temperate, subtropical, and tropical waters from Cape Lookout, North Carolina, and southward through the Gulf of Mexico to the Yucatan peninsula in Mexico, with reports from other Caribbean locations (Williams, 1984; Bert et al., 1978). It can be found in a variety of habitats, from grass beds to shell rubble or reefs, from the intertidal zone to depths approaching 200 feet (Williams, 1984; Bert et al., 1978). Adult stone crabs dig burrows (Williams, 1984); however, animals with less than a one-inch carapace width do not dig burrows, but live in deep channels, on grass flats, and under shell fragments or other structures offering protection (Costello et al., 1979).

The life cycle of the stone crab is very similar to other brachyuran crabs, possessing a planktonic larval zoea stage, a post-larval megalopa, and a benthic juvenile phase, all leading to the adult.

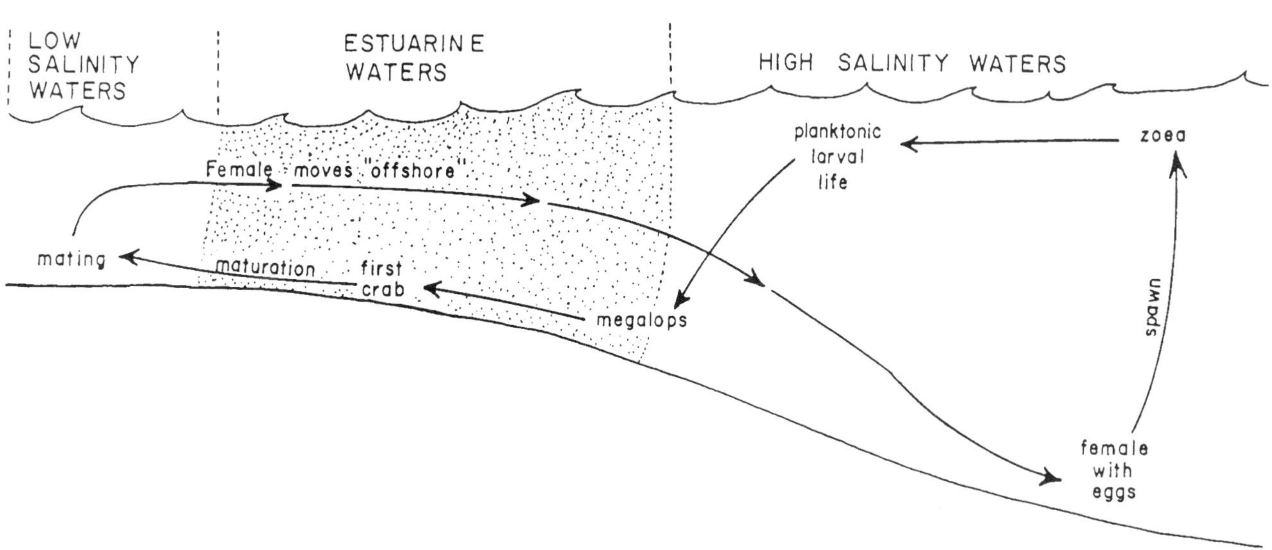

Figure 4. Generalized life cycle of the blue crab. From Oesterling, 1985.

Figure 5. Dorsal view of the stone crab, Menippe mercenaria *(adapted from Williams, 1965).*

Mating in the stone crab mimics that already described for the blue crab, in that male and female stone crabs "double." Copulation occurs immediately following a molt of the female, with the male in the superior position cradling an inverted female (Williams, 1984). Savage (1971) described mating under natural conditions, explaining how stone crabs will utilize burrow habitats during copulation. There is still some question as to whether or not there is a specific mating season (Bert et al., 1978). Bender (1971) identified November through March as the peak mating period for the Cedar Keys, Florida, region, while in the Florida Keys area Bert et al. (1978) state that November or December are peak mating periods.

There is no doubt that spawning (egg-laying) season for the stone crabs increases in length in the southern portion of its range, peaking in the spring-summer, but extending year-round in the Florida Bay area (Williams, 1984; Costello et al., 1979; Bert et al., 1978; Bender, 1971). From a single copulation, a female may produce many egg masses (sponge); over 10 broods of viable eggs from a single mating have been reported by Cheung (1968). At each spawning there can be between 100,000 and 1,000,000 eggs within the sponge, the number being closely related to the size of the female (Costello et al., 1979; Bert et al., 1978; McRae, 1950).

Porter (1960) described the zoeal stages of the stone crab. He identified a pre-zoeal and six zoeal stages; however, he considered the pre-zoea and last zoeal stage to be atypical. Later, Ong and Costlow (1970) refined the description of the larval stages and characterized optimum conditions for growth and survival. They found that in a salinity range of 30 to 35 ppt and a water temperature of 30°C (86°F), the megalopal stage could be reached in 14 days, and the first crab in 21 days with a survival between 60 percent to 72 percent. In the Cedar Key area, Bender (1971) reported that female stone crabs attempted to move to higher salinities for spawning. Below salinities of 10 ppt, Ong and Costlow (1970) reported no larval survival, regardless of water temperature.

Following the zoeal stages, megalopa and juvenile crabs migrate inshore to coastal estuaries (Costello et al., 1979). Juvenile stone crabs are opportunistic carnivores that will also consume plant material. Bender (1971) included polychaetes, small bivalves, oyster drills, other small snails and grass blades (*Thalassia*) as major diet items of juvenile stone crabs. Growth to sexual maturity can be rapid depending upon water temperature and salinity. Bender (1971) reported that sexual maturity could be attained in the fall after the first year of growth.

Bert et al. (1978) considered animals with a carapace width in excess of 1.25 inches (3.2 centimeters) to be adults. Growth to the adult stage will accelerate with increasing water temperatures. As with all crustaceans, in order for growth to occur, stone crabs must periodically molt. Male stone crabs may molt throughout the year, but female molting is generally restricted to the period between November and Spring (Bender, 1971). The female molting period apparently coincides with a time of increased mating activity (Bert et al., 1978; Bender, 1971).

Stone crabs live to be at least five years old (Costello et al., 1979). They enter the fishery sometime around their third year of life.

Dungeness Crab (*Cancer magister*)

Crabs of the genus *Cancer* occur worldwide in temperate regions (Bigford, 1979; Sastry, 1977; Williams, 1965; Waldron, 1958; MacKay, 1942). In the United States there are East and West Coast species that are harvested either in directed fisheries or as incidental catches. On the Pacific coast, the Dungeness crab (*Cancer magister*), named after a small fishing village, supports a large commercial fishery (Figure 6). Harvested to a lesser degree from the Atlantic coast are the jonah crab (*Cancer borealis*) and the rock crab (*Cancer irroratus*).

The ranges of these three species occur within the same latitudes on both coasts. The Dungeness crab is found from the Aleutian Islands southward to Magdalena Bay in Baja, Mexico. It is found from shore to approximately 300 feet (91.5 m) (Pauley et al., 1989). Jonah and rock crabs are found from Nova Scotia to the South Atlantic states (Williams, 1984; Marchant and Holmsen, 1975). The rock crab is found in shallower waters in the north and deeper (up to 1,887 feet/575 m) in the southern portion of its range. The jonah crab is generally found in deeper waters than the rock crab (to 2,625 feet/800 m), although their ranges overlap in places and at certain times of the year (Haefner, 1977). All three species are more abundant in the northern portions of their ranges.

The life cycles of the *Cancer* crabs are very similar. Because of its economic importance, the Dungeness crab will be used as a representative for this group. The following description is based on Dungeness crabs living in California waters. The life cycle events of those to the north follow the same general sequence but occur in different months, as a result of lower water temperatures.

Mature Dungeness crabs mate annually, completing an entire reproductive cycle each year. Mating between hard-shelled males and soft-shelled females occurs in oceanic waters from March through May (Pauley et al., 1989). As the time for mating approaches, the male Dungeness carries the female in a belly-to-belly embrace. This lasts for approximately seven days; on the eighth day the female struggles to escape (Snow and Neilsen, 1966). The male releases the female, permitting her to right herself. As the female sheds her shell, the male inserts his copulatory pleopods into her seminal receptacles. Copulation lasts from 30 to 120 minutes, after which the male again carries the female for several days (Pauley et al., 1989). Sperm are stored internally until October or November, at which time eggs are extruded and fertilization takes place. From 1,000,000 to 2,000,000 eggs are carried on the female's abdomen until late

Figure 6. Dorsal view of the Dungeness crab, Cancer magister. *Adapted from original photo by P.W. Wild, from Wild and Tasto (eds.), 1983.*

December or mid-January, when they hatch as zoea larvae (Wild, 1983).

There are five zoeal stages and one megalopal stage, which last for a total of 105 to 125 days (Reilly, 1985). The optimum environmental ranges for zoeal development are 10.0° to 13.9°C (50° to 57°F) and 25 to 30 ppt salinity (Reed, 1969). Due to the seaward movement of surface waters, the planktonic zoeae are transported offshore. Megalopae are found offshore during March but move shoreward and are found concentrated near shore in April (Reilly, 1985). Following the single megalops stage, the first crab stage occurs.

Young crabs abound in areas where currents are likely to concentrate megalopae until they are ready to settle out. Hence the youngest crabs are patchily distributed on the nursery grounds both in the ocean and in coastal bays. For the next several years growth and frequency of molting varies with crab size, sex, and location (bay or ocean). Crabs growing within bay systems molt more frequently than their counterparts in the ocean (Armstrong and Gunderson, 1985; Stevens and Armstrong, 1984; Tasto, 1983). During the first two years of life both sexes increase in size at the same rate. After this time, however, females shed less frequently and do not grow as large as males (Tasto, 1983). Dungeness crabs become sexually mature after approximately one year and may live to be six years old (Butler, 1961). Figure 7 presents a generalized life cycle for the Dungeness crab population in California.

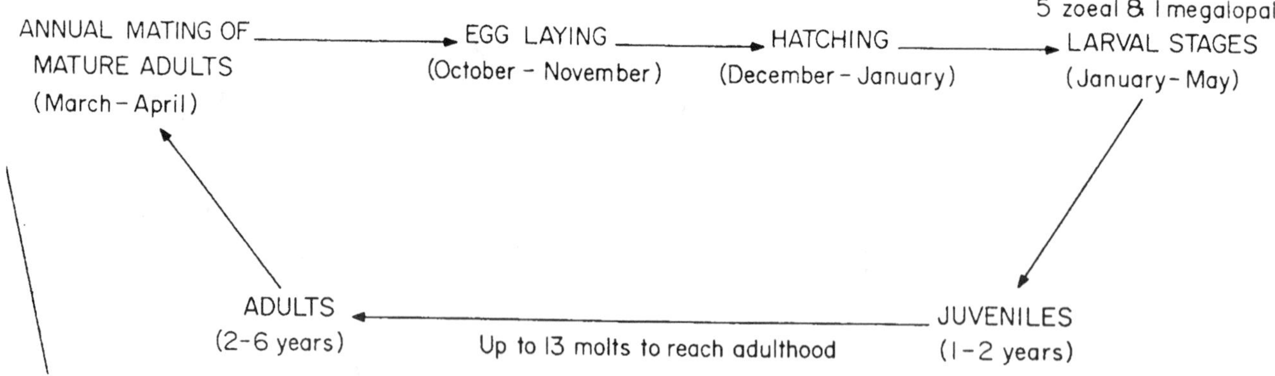

Figure 7. Life cycle of the Dungeness crab in California. *From Oesterling, 1985.*

King Crab (*Paralithoides camtschatica*)

The king crab (*Paralithoides camtschatica*) is one of the most recognizable of the crab species in the United States, primarily because of its size (Figure 8). It is the largest crab harvested, with an average weight of 10 pounds.

The range of the king crab extends from Korea and the Sea of Japan, northeastward to Kamchatka (Russia) and into the Bering Sea, eastward along the Aleutian Islands to Bristol Bay, into the Gulf of Alaska (Butler and Hart, 1962; MacGinitie, 1955). It is most abundant in the eastern Bering Sea and the northwestern Gulf of Alaska. Adult king crabs inhabit the deep waters of the continental shelf, occurring to depths greater than 600 feet (182 m) (Powell and Nickerson, 1965a). Male crabs tend to be found at greater depths (900 feet/274 m) than females. The king crab's normal diet consists of brittle stars, sea urchins, and starfish, and to a lesser degree, other crustaceans, polychaetes, and seaweed (Jewett and Feder, 1982). King crabs may live 20 to 25 years and grow to weigh more than 25 pounds (11.5 kg), with a leg span of over 6 feet (1.8 m) (McCaughran and Powell, 1977; Wallace et al., 1949).

The life cycle of the king crab begins in late winter or early spring when sexually mature adults migrate shoreward for mating. Mating takes place in waters shallower than 18.2 m (60 feet) from late March through early May (Powell and Nickerson, 1965a). Prior to actual copulation there is a courtship period during which the male grasps the chela of the female in a face-to-face position. Although grasping may last one to 16 days, the act of mating lasts only a few minutes (Powell et al., 1974). Mating occurs immediately following a molt by the female. The male king crab may even assist the female in getting out of her old shell. After the female has completed her molt, the male will flip her over and begin mating. The male will deposit spermatophore bands around the genital openings of the female (Powell and Nickerson, 1965a). Eggs will be fertilized externally as they exit the genital openings and as they become attached to the female's pleopods (McMullen, 1970; Powell and Nickerson, 1965a).

Following mating and the hardening of her shell, the female begins to move offshore for egg hatching, which occurs in approximately 46 m (150 feet) of water. The female carries the eggs for about eleven months, after which they hatch as zoea lar-

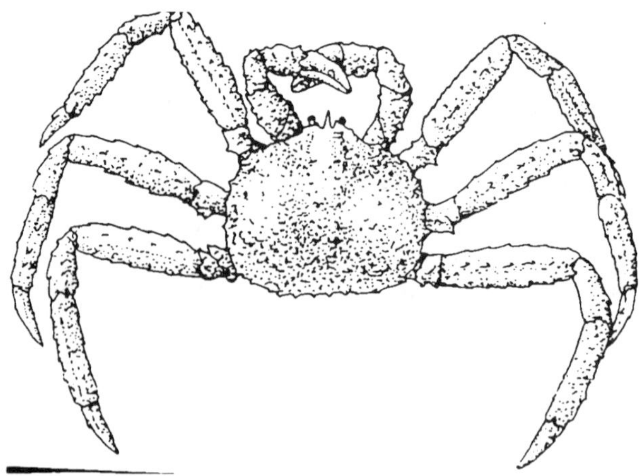

Figure 8. Dorsal view of the king crab, Paralithodes camtschatica. *Note the long walking legs and reduced chela. The last pair of legs is small and hidden from view.* *From Oesterling, 1985.*

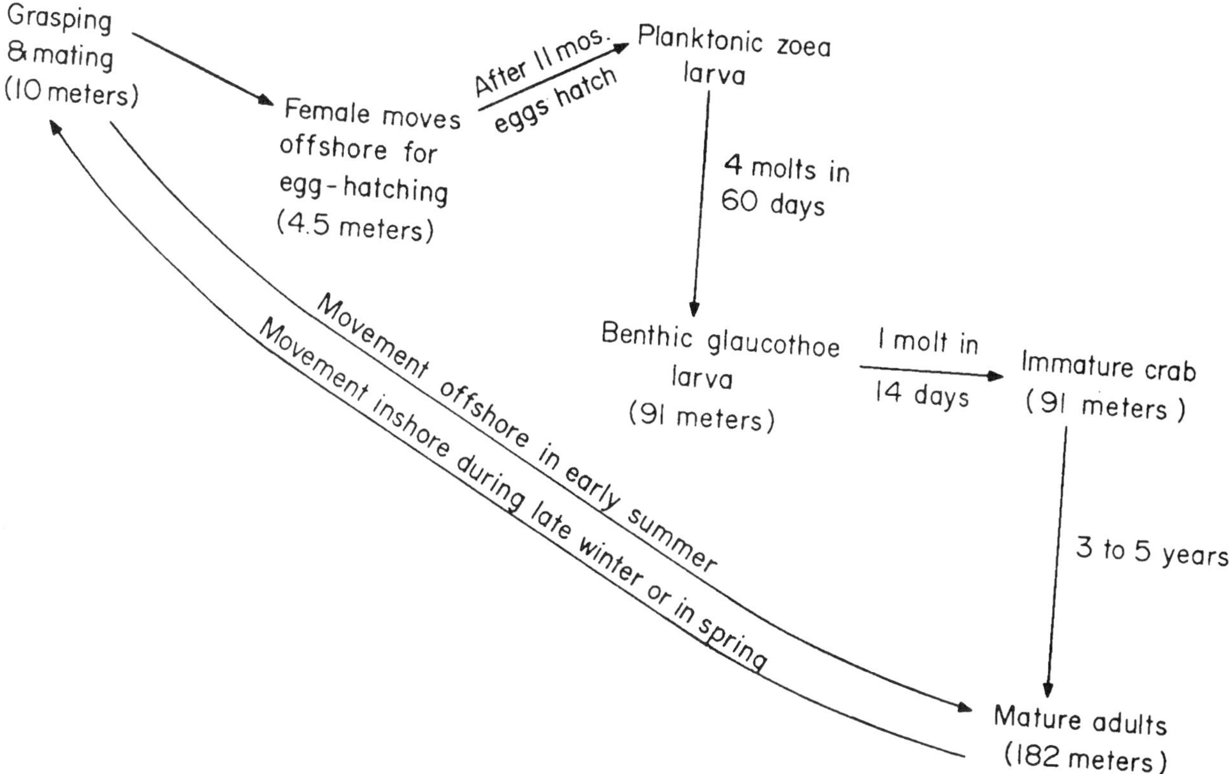

Figure 9. General life cycle of the king crab. Numbers in parentheses represent approximate water depth at which the events occur. From Oesterling, 1985.

vae only 0.8 mm (1/32 inch) long (Haynes, 1968). There are four planktonic zoeal stages that persist for approximately 60 days (Armstrong et al., 1981). Following the zoeal stages, is a benthic glaucothoe stage that lasts for about 14 days (Weber, 1967). The metamorphosis of the glaucothoe larva to the first-crab form occurs in water depths of about 91 m (300 feet). The immature king crab will molt (increase in size) several times a year until it reaches maturity; the molting period will then lengthen (Powell, 1967; Weber, 1967). A king crab becomes sexually mature at the age of five or six years, at a carapace width of 8.9 to 10.2 cm (3.5 inches) and a weight of about 1 kg (2.2 pounds) (Jewett and Onuf, 1988). The general life cycle for the king crab is represented in Figure 9.

One- and two-year-old king crabs exhibit a unique gregarious behavior of grouping together into "pods" (Powell and Nickerson, 1965b). The young crabs climb onto each other forming assemblages (pods) that may be 3.7 m (12 feet) long and comprise thousands of individuals. These aggregations move slowly along the bottom and may serve as protection against predators (Alaska Sea Grant Program, undated). Occasionally the crabs disband either to feed or to change location. During this time there is a gradual movement toward the deeper waters inhabited by the adults. As they get older, king crabs abandon the habit of podding; three- and four-year-old crabs spend more time grazing and tend not to form pods. Older adult crabs, however, may also form aggregates. In contrast to the juvenile pods that lack any orientation, older adults, in groups of 2,000 to 6,000, pile on top of another, each facing outward from the center of the group. The reasons for this action are unknown.

REFERENCES

Alaska Sea Grant Program. Undated. King crab (*Paralithodes camtschatica*). University of Alaska, Marine Science Curriculum Aid No. 4.

Armstrong, D.A. and D.R. Gunderson. 1985. The role of estuaries in Dungeness crab early life history: a case study in Grays Harbor, Washington. In *Proceedings of the Symposium on Dungeness Crab Biology and Management*. Univ. Alaska Sea Grant Rep. 85-3, pp. 145-170. Fairbanks.

Armstrong, D.A., L.S. Incze, D.L. Wencker, and J.L. Armstrong. 1981. Distribution and abundance of decapod crustacean larvae in the southeastern Bering Sea with emphasis on commercial species. U.S. Dept. Commerce, NOAA, OCSEAP Final Rep. **53** (1986):479-878.

Bender, E.S. 1971. Studies of the life history of the stone crab, *Menippe mercenaria* (Say), in the Cedar Key area. M.S. Thesis, Univ. of Florida, Gainesville.

Bert, T.M., R.E. Warner, and L.D. Kessler. 1978. The biology and Florida fishery of the stone crab, *Menippe mercenaria* (Say), with emphasis on southwest Florida. Florida Sea Grant Tech. Pap. No. 9, Univ. of Florida, Gainesville.

Bigford, T.E. 1979. Synopsis of biological data on the rock crab, *Cancer irroratus* (Say). NOAA Tech. Rep. NMFS Circ. 426. National Oceanic and Atmospheric Administration. Washington, DC.

Bliss, D.E. 1982. *Shrimps, Lobsters and Crabs.* New Century Publishers Inc., Piscataway, NJ.

Butler, T.H. 1961. Growth and age determination of the Pacific edible crab, *Cancer magister* (Dana). *J. Fish. Res. Board Can.* **18** (5):873-889.

Butler, T.H. and J.F.L. Hart. 1962. The occurrence of the king crab, *Paralithodes camtschatica* (Tilesius), and of *Lithodes aequispina* (Benedict) in British Columbia. *J. Fish. Res. Board Can.* **19**(3):401-408.

Cheung, T.S. 1968. Trans-molt retention of sperm in the female stone crab, *Menippe mercenaria* (Say). *Crustaceana* **15** (1):117-120.

Churchill, E.P. 1919. Life history of the blue crab. *U.S. Bur. of Fish. Bull.* **36**:95-128.

Costello, T., T.M. Bert, D.G. Cartano, G. Davis, G. Lyon, C. Rockwood, J. Stevely, J. Tashiro, W.L. Trent, D. Turgeon, and J. Zuboy. 1979. Fishery management plan for the stone crab fishery of the Gulf of Mexico. In *Stone Crab Fishery; Plan Approval and Proposed Regulations.* Federal Register, **44** (65), book 1:19444-19496.

Costlow, J.D. and C.G. Bookhout. 1959. The larval development of *Callinectes sapidus* (Rathbun) reared in the laboratory. Biol. Bull. **116**:373-396.

Darnell, R.M. 1959. Studies of the life history of the blue crab [*Callinectes sapidus* (Rathbun)] in Louisiana waters. *Trans. Am. Fish. Soc.* **88**(4):294-304.

Fischler, F.J. 1965. The use of catch-effort, catch-sampling, and tagging data to estimate a population of blue crabs. *Trans. Am. Fish. Soc.* **94**(4):287-310.

Futch, C.R. 1965. The blue crab in Florida. Salt Water Fish. Leaflet No. 1. Florida Board of Conservation Marine Lab. St. Petersburg.

Haefner, P.A., Jr. 1977. Aspects of the biology of the jonah crab, *Cancer borealis* (Stimpson), 1859 in the mid-Atlantic Bight. *J. Nat. Hist.* **11**:303-320.

Haynes, E. 1968. Relation of fecundity and egg length to carapace length in the king crab, *Paralithodes camtschatica. Proc. Natl. Shellfish. Assoc.* **58**:60-62.

Jewett, S.C. and H. M. Feder. 1982. Food and feeding habits of the king crab, *Paralithodes camtschatica*, near Kodiak Island. *Alaska. Mar. Biol.* **66**:243-250.

Jewett, S.C. and C.P. Onuf. 1988. Habitat suitability index models: red king crab. U.S. Fish. and Wildlife Serv. Biol. Rep. **82**(10.153).

Kaestner, A. 1970. *Invertebrate Zoology, Volume III. Crustaceans.* Interscience Publishers, New York.

MacGinitie, G.E. 1955. Distribution and ecology of the marine invertebrates of Point Barrow, Alaska. Smithsonian Misc. Collect. **128**:1-201.

MacKay, D.C.G. 1942. The Pacific edible crab, *Cancer magister.* Fish. Res. Board Can. Bull. No. 62.

Marchant, A. and A. Holmsen. 1975. Harvesting rock and jonah crabs in Rhode Island: some technical and economic aspects. Resource Economics, University of Rhode Island, Marine Memorandum No. 35.

McCaughran, D.A. and G.C. Powell. 1977. Growth model for Alaska king crab, *Paralithodes camtschatica. J. Fish. Res. Board Can.* **34**(7):989-995.

McConaugha, J.R. 1988. Export and reinvasion of larvae as regulators of estuarine decapod populations. In *Larval fish and shellfish transport through inlets.* M.P. Weinstein (ed.), pp. 90-103. *Am. Fish. Soc. Symp.* 3.

McMullen, J.C. 1970. Aspects of early development and attachment of fertilized king crab eggs. Alaska Dep. Fish Game Inf. Leafl. 140.

McRae, E.D., Jr. 1950. An ecological study of the Xanthid crabs in the Cedar Key area. M.S. Thesis, Univ. of Florida, Gainesville.

National Marine Fisheries Service. 1999. Fisheries of the United States, 1998. Current Fishery Statistics No. 9800, U.S. Dept. Commerce, Washington, DC.

Oesterling, M.J. 1985. The seafood industry — a self-study guide: Shellfish: Crustaceans. Virginia Sea Grant College Program, VPI-SG-85-03.

Olmi, E.J. III. 1994. Vertical migration of blue crab, *Callinectes sapidus* megalopae: implications for transport in estuaries. *Mar. Ecol. Prog. Ser.* **113**:39-54.

Ong, K. And J.D. Costlow. 1970. The effect of salinity and temperature on the larval development of the stone crab, *Menippe mercenaria* (Say), reared in the laboratory. *Ches. Sci.* **11**(1):16-29

Pauley, G.B., D.A. Armstrong, R. Van Citter, and G.L. Thomas. 1989. Species profiles: life histories and environmental requirements of coastal fishes and invertebrates (Pacific Southwest) — Dungeness crab. *U.S. Fish. and Wildlife Serv. Biol. Rep.* **82**(11.121). U.S. Army Corps of Engineers, TR EL-82-4.

Porter, H.J. 1960. Zoeal stages of the stone crab, *Menippe mercenaria* (Say). Ches. Sci. **1**(3-4):168-177.

Powell, G.C. 1967. Growth of king crab in the vicinity of Kodiak Island, Alaska. Alaska Dep. Fish Game Inf. Leafl. 92.

Powell, G.C. and R.B. Nickerson. 1965a. Reproduction of king crabs, *Paralithodes camtschatica* (Tilesius). *J. Fish. Res. Board Can.* **22**(1):101-111.

Powell, G.C. and R.B. Nickerson. 1965b. Aggregations among juvenile king crab [*Paralithodes camtschatica* (Tilesius)], Kodiak, Alaska. Anim. Behav. **13**(2-3):374-380.

Powell, G.C., K.E. James, and C.L. Hurd. 1974. Ability of male king crab, *Paralithodes camtschatica*, to mate repeatedly, Kodiak, Alaska, 1973. *U.S. Fish. and Wildlife Serv. Fish. Bull.* **72**:171-179.

Reed, P.N. 1969. Culture methods and effects of temperature and salinity on survival and growth of Dungeness crab (*Cancer magister*) larvae in the laboratory. *J. Fish. Res. Board Can.* **26**(2):389-397.

Reilly, P.N. 1985. Dynamics of Dungeness crab, *Cancer magister*, larvae off central and northern California. In *Proceedings of the Symposium on Dungeness Crab Biology and Management*. Univ. of Alaska Sea Grant Rep. 85-3, Fairbanks.

Sandoz, M. and R. Rogers. 1944. The effect of environmental factors on hatching, moulting, and survival of zoea larvae of the blue crab, *Callinectes sapidus* (Rathbun). *Ecology* **25**(2):216-228.

Sastry, A.N. 1977. The larval development of the rock crab, *Cancer irroratus* (Say), 1817, under laboratory conditions (Decapoda; Brachyura). *Crustaceana* **32**(2):155-168.

Savage, T. 1971. Mating of the stone crab, *Menippe mercenaria* (Say) (Decapoda; Brachyura). *Crustaceana* **20**(3):315-317.

Savage, T., J.R. Sullivan, and C.E. Kalman. 1975. An analysis of stone crab (*Menippe mercenaria*) landings on Florida's west coast, with a brief synopsis of the fishery. Florida Mar. Res. Pub. No. 13, St. Petersburg.

Schmitt, W.L. 1965. *Crustaceans*. University of Michigan Press, Ann Arbor.

Snow, C.D. and J.R. Neilsen. 1966. Premating and mating behavior of the Dungeness crab [*Cancer magister* (Dana)]. *J. Fish. Res. Board Can.* **23**(9):1319-1323.

Stevens, B.G. and D.A. Armstrong. 1984. Distribution, abundance and growth of juvenile Dungeness crabs, *Cancer magister*, in Grays Harbor Estuary, Washington. *U.S. Natl. Mar. Fish. Serv. Fish. Bull.* **82**(3):469-483.

Sulkin, S.D. 1974. Factors influencing blue crab population size: nutrition of larvae and migration of juveniles. Chesapeake Biol. Lab. Annu. Rep. Ref. No. 74-125. Center for Environmental and Estuarine Studies, Solomons, MD.

Sullivan, J.R. 1979. The stone crab, *Menippe mercenaria*, in the southwest Florida fishery. Florida Mar. Res. Pub. No. 36, St. Petersburg.

Tagatz, M.E. 1968. Biology of the blue crab, *Callinectes sapidus* (Rathbun), in the St. Johns River, Florida. *U.S. Fish. and Wildlife Serv. Fish. Bull.* **67**(1):17-33.

Tasto, R.N. 1983. Juvenile Dungeness crab, *Cancer magister*, studies in the San Francisco Bay area. Pp. 135-154, In *Life history, environment, and mariculture studies of the Dungeness crab,* Cancer magister, *with emphasis on the central California fishery resource.* P.W. Wild and R.N Tasto (eds.). Calif. Dept. Fish and Game, Fish. Bull. 172.

Van Engel, W.A. 1958. The blue crab and its fishery in Chesapeake Bay. Part 1. Reproduction, early development, growth, and migration. *Commer. Fish. Rev.* **20**(6):6-17.

Waldron, K.W. 1958. The fishery and biology of the Dungeness crab [*Cancer magister* (Dana)] in Oregon waters. Fish Commission of Oregon Contrib. No. 24, Portland.

Wallace, M.M., C.J. Pertuit and A.H. Hvatum. 1949. Contributions to the biology of the king crab, *Paralithodes camtschatica* (Tilesius). U.S. Fish. and Wildlife Serv., Fish Leaflet **340**:1-49.

Warner, G.F. 1977. *The Biology of Crabs*. Van Nostrand Reinhold Company, New York.

Warner, W.W. 1976. *Beautiful Swimmers*. Little, Brown and Company, Boston, MA.

Waterman, T.H. (ed.). 1960. *The Physiology of Crustacea, Volume 1: Metabolism and Growth*. Academic Press, New York.

Waterman, T.H. (ed.). 1961. *The Physiology of Crustacea, Volume 2: Sense Organs, Integration, and Behavior.* Academic Press, New York.

Weber, D.D. 1967. Growth of immature king crab, *Paralithodes camtschatica* (Tilesius). *Int. N. Pac. Fish. Comm. Bull.* **21**:21-53.

Wild, P.W. 1983. The influence of seawater temperature on spawning, egg development, and hatching success of the Dungeness crab, *Cancer magister*. Pp. 197-214, In *Life history, environment, and mariculture studies of the Dungeness crab,* Cancer magister, *with emphasis on the central California fishery resource.* P.W. Wild and R.N. Tasto (eds.). Calif. Dept. Fish and Game, Fish. Bull. 172.

Wild, P.W. and R.N. Tasto (eds.). 1983. *Life history, environment, and mariculture studies of the Dungeness crab,* Cancer magister, *with emphasis on the central California fishery resource.* Calif. Dept. Fish and Game, Fish. Bull. 172.

Williams, A.B. 1965. Marine decapod crustaceans of the Carolinas. *U.S. Fish. and Wildlife Serv. Fish. Bull.* **65**(1):1-298.

Williams, A.B. 1971. A ten-year study of meroplankton in North Carolina estuaries: annual occurrence of some brachyuran development stages. *Chesapeake Sci.* **12**(2):53-61.

Williams, A.B. 1974. The swimming crabs of the genus *Callinectes* (Decapoda: Portunidae). *U.S. Fish Wildl. Serv. Fish. Bull.* **72**(3):685-798.

Williams, A.B. 1984. *Shrimps, Lobsters, and Crabs of the Atlantic Coast of the Eastern United States, Maine to Florida.* Smithsonian Institution Press, Washington, DC.

Zuboy, J.R. and J.E. Snell. 1980. Assessment of the Florida stone crab fishery. U.S. Dept. of Commerce, NOAA Tech. Memo. NMFS-SEFC-21.

Major Marine Finfish Species

Frank J. Schwartz

INTRODUCTION

Oceans cover about 70 percent of the earth's surface and account for 97 percent of its waters. Exactly how many species of fishes exist is difficult to determine, for new species are or will be described as our knowledge of fishes expands and is refined. Nearly 25,000 species of fishes are currently known. About 14,500 species live all their lives in marine waters; 9,966 live in freshwaters of the world. Perhaps 200 marine species are diadromous, living part of their lives in freshwaters and part in marine waters (e.g., American eels, *Anguilla rostrata*). Many marine fishes (10,465) may venture into fresh waters to spawn, feed, etc. (e.g., salmon, sturgeon) or are tolerant of brackish waters (a mixture of fresh and saline waters) into which they are carried by various currents or may follow for food.

Marine fishes occupy all niches of the world's waters from the shallowest to the greatest ocean depths (nearly 11,800 meters). Many more species occur in tropical, subtropical, or temperate waters than in polar or subarctic waters. About 11,300 species occupy coastal or littoral habitats and 2,900 species occupy waters deeper than 200 meters. The greatest variety of marine fishes occurs in Asian waters, while 130 species are circumtropical in distribution. The smallest fish is an eight- to 10-millimeter (standard length) marine goby, *Trimmatom nasus*, found in Indian waters, while the largest is the 12-meter whale shark, *Rhincodon typus*, that roams the world's oceans.

Marine fishes have adapted and can exist in most marine waters. Adaptations are: pelagic eggs, external and internal fertilization and development, enlarged body features (such as fins, keels, head and fin spines to increase or facilitate flotation or swimming speed and buoyancy), large or small eyes, symbiotic or parasitic relationships, special foods, and a host of other features (Figure 1). Marine fishes that tolerate wide ranges of water temperatures are considered eurythermal or stenothermal. Fishes that tolerate wide water salinities are called euryhaline or stenohaline fishes. A few endothermal fishes can control their body temperatures, while most are influenced by the external water temperatures in which they swim. Marine fishes can be found in waters of 0 to 36 parts per thousand (ppt) salinity. They also can tolerate water temperatures from -2°C to those near deep ocean hot thermal vents. Only recently has man begun to consider the effects of natural events, El Niño, hurricanes, and long-term water cooling and heating influences to the well-being, abundances, or occurrences of marine fishes. This chapter treats the biology, ecology, and status of selected marine species from North American marine waters.*

WHAT IS A MARINE FISH?

Marine fishes can be defined by the type of lifestyle they exhibit. That may be diadromous, catadromus, complementary, sporadic, primary, or secondary. Examples are the bull shark (*Carcharhinus leucas*) or salmon (*Oncorhynchus* spp.), species that travel hundreds of kilometers into freshwater to spawn or live out portions of their lives. The American eel is an example of a fish that grows up in freshwater, following birth in the ocean, and migrates into saline waters to spawn prior to death.

While all fishes possess gills through which they obtain oxygen to sustain them, what distinguishes a freshwater from a marine fish? One way to distinguish between the two groups is in how they utilize the waters which they do or do not drink. Freshwater fishes, living in a high water/low saline environment, must eliminate most of the water they drink via excretion across their gills and kidneys in order to conserve salts. Conversely, marine fishes, living in a low water/high saline ion environment, must eliminate or reduce large amounts of salts from the water they drink, yet

* See "Table of Terms" on page 199 for definitions of some terms not explained in the text.

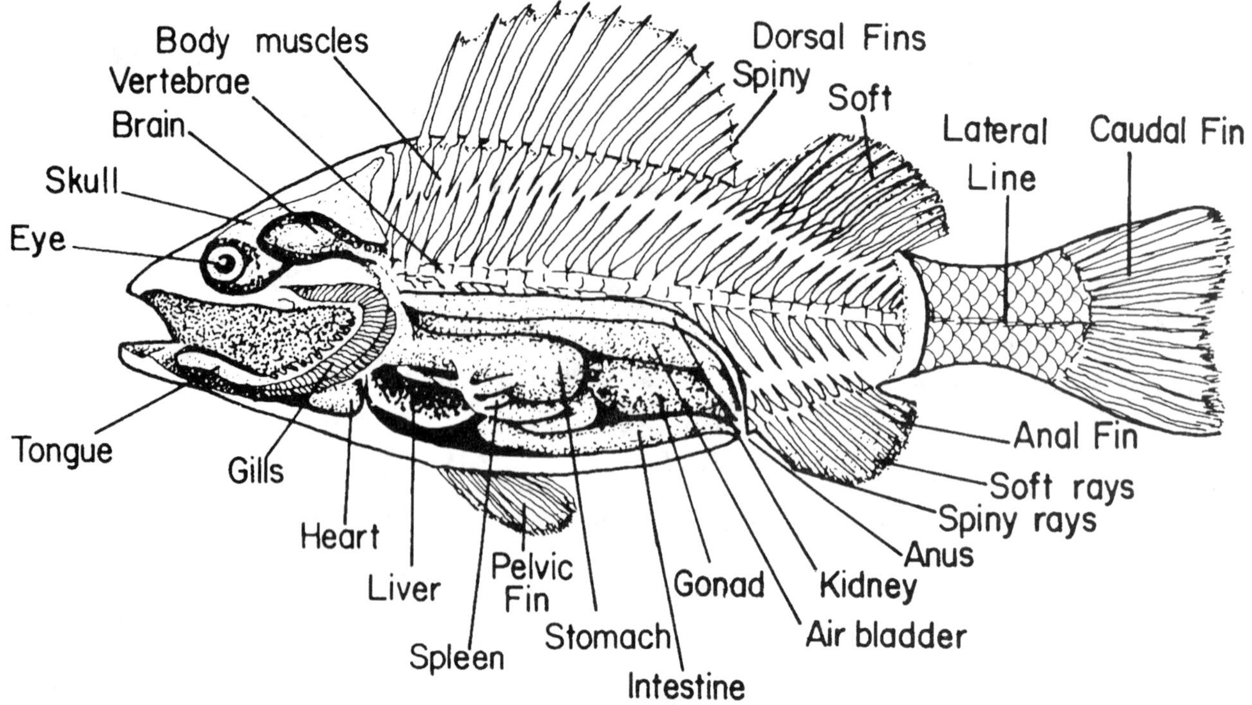

Figure 1. Anatomy of a fish.

retain some water for good physiological balance. Most marine fishes possess glomerular kidneys to help with water-salt excretions; aglomerular fish such as the oyster toadfish (*Opsanus tau*) and pipefish (*Syngnathus* spp.) lack such features and use other means to excrete unwanted substances.

Suggested Reading

Ali, 1975 (vision); Bone et al., 1994 (biology); Briggs, 1974 (zoogeography); Cohen, 1970 (numbers); Denton, 1990 (deep sea environment); Eschmeyer et al., 1983 (Pacific fishes); Fiedler, 1986 (El Niño); Finch, 1990 (age); Gabull, 1992 (climate effects); Greenwood, 1992 (species numbers); Haedrick and Merritt, 1988 (deep sea); Hanel, 1987 (sizes); Hardy, 1978b (larvae); Hoar and Randall, 1969-1993 (physiology); Johnson, 1986 (larval development); Jones et al., 1978 (young development); Kelley, 1994 (climate effects); Marshall, 1984, 1972, 1954 (deep sea fishes); Martin and Drewry, 1978 (larval development); Merritt and Marshall, 1981 (deep sea ecology); Moyle and Cech, 1982 (ichthyology, all aspects); Murawski, 1993 (climate effects); Myers, 1951, 1949a,b, 1938 (definitions); Nelson, 1994 (taxonomy, zoogeography); Olson et al., 1994 (environment); Rass, 1974 (greatest depths occupied); Robins and Rae, 1986 (Atlantic Coast fishes); Schwartz, 1995, 1989, 1981 (low salinity occurrences); Spotte, 1992 (physiology); Springer, 1982 (plate tectonics); Stoskoff, 1993 (physiology, ecology); Swain and Kryner, 1995 (climate effects); Tritos, 1984 (climate effects); Winterbottom and Emery, 1981 (smallest fish); and Wootton, 1990 (ecology).

WHAT IS MARINE WATER?

Just as it is difficult to define what is a marine fish, it remains unsettled as to what constitutes marine water. Some researchers consider freshwaters those to be possessing 0 to 0.5 parts per thousand (ppt) salinities and ocean waters 34 to 36 ppt salinities. Others define marine waters as those with at least 0 to 14 ppt salinity. Intrusion of saline water into estuaries or rivers confounds attempts at definition stability, since many freshwater or marine fishes can be found, for short or extended periods, in waters originally considered fresh or marine — depending on freshwater downstream inflow or upstream intrusion during summer drought periods or as influenced by winds and currents. In general, estuarine waters are usually considered to possess 20 ppt salinities, yet designations are as varied as the localities considered.

Suggested Reading

Bulger et al., 1980; Schwartz, 1990 (classification); Gunter, 1942; Gunter et al., 1974 (marine species intrusions into freshwater); Hoar and Randall, 1969-1989 (physiology); Kinne, 1967, 1964 (definitions); Schwartz, 1995, 1989 (intrusions into low salinities); Symposium, 1958 (classification).

CHONDRICHTHYES

Sharks, skates, and rays (826 species), of which 370 are sharks, comprise the chondrichthyes — or cartilaginous-skeletoned fishes (Figure 2). One hundred ten species of sharks are known from North American marine waters. They range in size from the small 20- to 25-centimeter pygmy shark (*Euprotomicrus bispinatus*) to the large 12-meter whale shark. They may be sluggish (dogfish) or fast swimmers (mako sharks), terete (white shark, *Carcharodon carcharias*) to flattened (angel shark, *Squatina dumeril*). Most sharks feed on fishes and other foods, while basking sharks (*Cetorhinus maximus*) and whale sharks are filter feeders. Some, such as the cookie-cutter shark (*Isistius brasiliensis*), may even feed on the flesh of other fishes, whales, or sharks. Sharks occupy every habitat and exhibit every type of behavior imaginable — from resting in freshwater caves to regulating body temperatures. Sharks are apex feeders, have sensitive olfactory receptors, may see in dim light, or can detect foods at a distance by pressure and sensory cues. Most sharks, skates, and rays frequent marine waters, yet a few enter freshwaters for parts of their lives (e.g., bull sharks).

Figure 2. Two chondrichthyes species. Top: little skate; bottom: spiny dogfish shark.

Reproductively, sharks, skates, and rays have internal fertilization via insertion of modified male copulatory organs (claspers) that develop as part of the pelvic fins. Skates lay external eggs. Whale sharks, which were once believed to lay external egg cases, are now known to produce living young following early eggcase development. Most develop their young internally. The number of young produced per female varies from two to near 100. There are three basic types of reproduction: oviparity, ovoviparity, and viviparity. Oviparity is a primitive condition where large-yolked eggs are laid in leathery cases (e.g., whale sharks, skates). No parental care takes place. Oviparity is known in four families of sharks: bullhead, nurse, cat, and whale sharks.

Ovoviparity, also known as a placental viviparity, is a common form of reproduction. Eggs hatch in the uterus before the eggs develop fully. Embryos, nourished by the yolk, grow in the uterus but do not form a placental connection to the uterus. Such embryos then feed, usually one per uterus, on other eggs prior to hatching. Examples are found in these sharks: cow (*Hexanchus griseus* and *Notorhynchus cepedianus*); sand tiger (*Carcharias taurus*); frill (*C. anguineus*); mackerel (*Isurus* spp.); basking, thresher (*Alopias* spp.); false cat (*Pseudotriakis microdon*); saw (*Pristiophorus schroederi*); angel, squaloid, ribbontail catshark (*Eridacnis barbouri*); some nurse sharks; some dogfishes; and some other cat sharks.

Viviparity is the common form of development in most sharks and rays. An internal placental attachment occurs to the uterus. Broods are small (2 to 6), with some pelagic species producing 300 young. Gestation may be six months to two years in length. Maturity can take as long as 18 years (dogfishes). Much remains to be learned, as sexes usually travel in segregated schools to specific unknown breeding grounds. Viviparity is found in smooth dogfish, requiem sharks (most), hammerheads, and rays.

For many years sharks and some skates were neglected resources, as man preferred to catch and eat other fishes, did not like the ammonia smell often associated with dead sharks, or was fearful of the species because of dramatic reports of shark attacks (the movie *Jaws*). Fishermen were encouraged to fish for sharks, etc., once other bony fishes and fisheries declined (see also other species accounts). Soon school lunch programs and the public were clamoring for this healthy, tasty, and low-cost food. Commercial and sport catches of sharks and some skates soared. Once declines in populations of some sharks and skates became evident, the federal government devised shark management plans in order to prevent overfishing, exploitation, or extinction of a species, in hopes of restoring population levels. Quotas, trip catch and poundage limits, prohibition of fishing, and seasons were set (often based on little basic data) for the commercial and sport fisheries by the Federal Management Planning Committee (1991) in hopes of restoring population levels. Most management plan concerns revolved around species with slow growth, which take years to mature, have long reproduction cycles, usually produce few young, and are highly migratory, often transoceanic. These concerns were reinforced by heavy fishing and rapid declines in soupfin (*Galeorhinus zygopterus*), spiny dogfish (*Squalus acanthias*), thresher sharks (*Alopias* spp.), and some skates (*Raja* spp.) in California. The sharks were caught for their livers (in order to produce Vitamin A) or, in the case of skates, for use as false scallops; catch prices rose to $1653/ton in 1942 from $11 in 1938. Similarly on the Atlantic Coast, the porbeagle (*Lamna nasus*) fishery landings, existing from the 1960s, soared to 93 metric tons by 1964 before collapse. As a result, fisheries on both coasts targeted other sharks (mako, *Isurus* spp.; blacknose sharks, *Carcharhinus acronotus*; or sandbar sharks, *C. plumbeus*). It is uncertain if recent catches were the result of natural declines or the result of regulating the number of commercial fishing boats fishing for sharks, for catches declined from thousands to just a hundred or so metric tons landed. Other compounding aspects have been the surge in fishing pressure for sharks by sport fishermen, now numbering far more than commercial fishermen.

Today, sharks, skates, and some rays are used as food. Shark fins of the mako, thresher sharks, etc., are relished by the Asian peoples as ingredients in costly soups. Medical researchers are experimenting with shark cartilage as a possible cure for cancer and as anticlotting agents. Tooth replacement aspects are of interest to the dental industry, while shark teeth, jaws, and ray tail spines are highly prized by the trinket trade. Vertebrae may be used as a fine powder by Japanese geishas.

Continued research has shown sharks can travel long distances and return to specific habitats. Shark attacks are associated with time of day, water temperature, color of swimsuit, or resemblance of surf boards or people to otters, seals, etc.

Although the world demand for shark meat, fins, etc., has increased, sharks have survived for 400 million years; this resourceful group will be with us for many more years. Some species may be overfished, but others will endure as the "silent shadows in the sea."

Suggested Reading

NMFS-N0AA management plan, 1991; Anderson, 1990 (West Coast fishes); Bigelow and Schroeder, 1953a, 1948 (classic studies on biology, taxonomy, sharks, skates, and rays); Branstetter, 1993 (biology); Caillet et al., 1992 (angel sharks); Casey and Holey, 1985 (biology); Castro, 1983 (North American sharks); Compagno, 1990, 1988, 1984, 1973 (sharks of the world); Dingerkus, 1993 (phylogeny); Ellis, 1991 (great white shark); Fisher and Ditton, 1933 (Gulf fishery); Griffith et al., 1984 (skates and rays as food); Holts, 1988 (commercial fishery); Klimley, 1993, 1987, 1981 (sexual reproduction, movements); Joung et al., 1996 (whale shark embryology); Klimley, 1994 (white shark); Klimley and Ainley, 1996 (white shark); Martin and Zorzi, 1993 (skates); McCormick et al., 1963 (general shark information); Mysah, 1986 (senses); Paust and Smith, 1986 (North Pacific fishery); Pratt et al., 1990 (living resources); Pratt and Casey, 1983 (biology); Prince and Pulos, 1983 (aging, biology); Richards, 1987 (biology); Saunders and McFarlane, 1993 (dogfish); Schwartz, 1995, 1990, 1984 (biology, development, zoogeography); Springer and Gold, 1989 (shark information); Squire, 1987b (environmental effects); Wourms and Demski, 1993 (embryology); and Zeiner and Wolf, 1993 (skate fishery).

OSTEICHTHYES

Sturgeons

Sturgeons are some of the largest and long-lived cartilaginous fishes in the world. Twenty-six species are found in Northern Hemisphere waters, 14 of which live their entire lives in freshwater. Eight species of sturgeon occur in North America: shortnose *Acipenser brevirostris*, lake *A.*

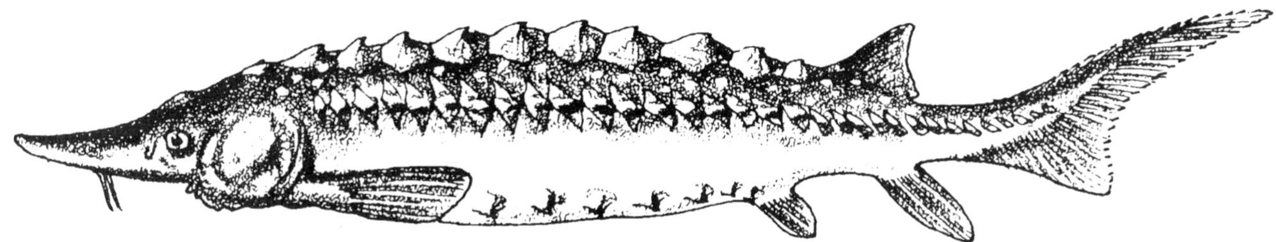

Figure 3. Atlantic sturgeon.

fulvescens, green *A. mediorostris*, Atlantic *A. oxyrinchus* (Figure 3), and Gulf of Mexico subspecies *A.o. desotoi*, white *A. transmontanus*, pallid *Scaphirhynchus albus*, shovelnose *S. platorhynchus*, and Alabama *S. suttkusi.* Five species are listed as endangered or threatened as a result of overfishing and habitat destruction: shovelnose, green, pallid, Alabama, and Gulf sturgeons. North American sturgeons exhibit three types of life history patterns: four species prefer fresh water during their entire lives (lake, pallid, Alabama, and shovelnose), two have adults that enter brackish waters during some portion of their lives (white, shortnose), and two species live the greater portion of their lives in the sea (green, Atlantic/Gulf). Confusion still persists regarding the taxonomic status of the Gulf sturgeon: some consider *A. o. desotoi* a full species, others a subspecies of the Atlantic sturgeon. Regardless, all sturgeons spawn in freshwater.

Sturgeons have existed in North America, virtually unchanged, since the Upper Cretaceous. Atlantic sturgeons once attained sizes of 3.1 meters (largest, 4.3 meters) and 363 kilograms (kg), while white sturgeons of the Pacific Coast grew to 6.1 meters and 550 kg. Some species live more than 100 years.

Sturgeons are triangular- to flat-sided fishes that possess five rows of body scutes or plates (one row dorsally, one along the middle of each side, and one row low on each side), and have a pointed-to-blunt snout, small eyes, and four barbels located on the ventrally-positioned protrusible toothless mouth. Some species (Atlantic, shortnose) can be distinguished by the number of rows of shield-like scales located between the pelvic and anal fins (one or two). The tail is heterocercal (upper lobe longer and extended, as in sharks). The swim bladder is large; the skeleton, cartilaginous. No external sexual dimorphism is evident. Sturgeons are a valuable resource for their meat (smoked), eggs (caviar), and oil. For many years at the beginning of the twentieth century, their swim bladders and notochords (within the backbone) were used to produce isinglass.

Sturgeon reproduction is similar throughout all species. Feeding seems to stop during spawning. Adults ascend freshwater rivers, where spawning occurs over hard substrates containing little sediment when water temperatures are about 12° to 21°C, depth one meter. Females rub against the rough substrate to help extrude sticky encapsulated eggs that become attached to nearby objects. Eggs may account for one-third of the female's body weight and, by species, may consist of 3+ million eggs. Several males attend a female during the spawning act. Adults leave the spawning site and return, with no parental care being exhibited, to their feeding grounds elsewhere. Age at first spawning varies by species, but usually males are nine to 10 years and females 13 to 14 years old. Larvae hatch in three to seven days, depending on water temperature. Growth is rapid, first by feeding on their body yolk and then, after attaining lengths of 22 millimeters, on plankton. Sturgeons, whether in fresh or marine waters, feed on bottom foods.

Gill nets have been the major capture method of sturgeons, although hook-and-line with cut bait, pound nets, trawls, traps, hoops, and seines are other methods. Tagged specimens have been recaptured, often years after tagging, up to 1400 kilometers away. Atlantic sturgeon occur from Labrador, Canada, to the St. John's River, Florida. Major fisheries once existed in Delaware and along the Hudson River; today they are primarily on the Hudson. Commercial landings decreased from 3350 metric tons in 1890 to 82 metric tons in

1992. Gulf sturgeons occur in the Gulf of Mexico from Tampa, Florida, to the Mississippi River area of the Gulf of Mexico. White sturgeon, the largest freshwater fish in North America, occurs from northern California to northwestern Alaska as does the green sturgeon, the latter spending much of its life near river mouths and less in freshwater. Meat and roe of the white sturgeon is highly regarded, while that of the green sturgeon is of little food value.

With declining sturgeon catches on both North American coasts, regulations to control fishing for sturgeons were implemented as early as 1891 in New Jersey. Regulations vary state to state. Most recently, many states have instituted moratoria prohibiting fishing for sturgeon in hopes of restoring this proud group of species to once-thriving populations. Even hatcheries have been instituted by some states to improve matters, but with the slow growth of sturgeons many years of fishery regulation may be necessary before positive results are achieved. Meanwhile, with man's increased clamor to use coastal rivers, waterways, and the oceans, habitat and benthic foods therein continue to be destroyed. Today an occasional 2.5-meter specimen is encountered; however, captured specimens continue to be smaller and smaller. The days of the sturgeon may soon be one only of memory.

Suggested Reading

Beamesderfer and Rein, 1993 (white); Birnstein, 1993 (conservation); Clugston and Foster, 1993 (Gulf); Dees, 1961 (biology); Dovel and Berggren, 1983 (Atlantic); Gilbert, 1992 (Florida); Jury et al., 1994 (distribution); Mason and Clugston, 1993 (Gulf, growth, food); Mason et al., 1993 (Gulf); Mason et al., 1992 (*A. o. desotoi*); Miller, 1972 (migration); Murawski and Pacheco, 1977 (Atlantic); North et al., 1993 (white distribution); Parsly et al., 1993 (white); Pacific States Marine Fisheries Commission, 1992 (white); Reynolds, 1993 (Gulf); Rochard et al., 1990 (threats); Smith, 1990, 1985 (culture, enhancement, management); Smith and Dingley, 1984 (Atlantic culture); Smith et al., 1984, 1982 (biology); Species profile, 1989; Taub, 1990 (management plan); Timoshkin, 1969 (Atlantic, at sea); Tsepkin and Sokolov, 1972 (size); Wooley, 1985 (taxonomy), and International Conference, 1977.

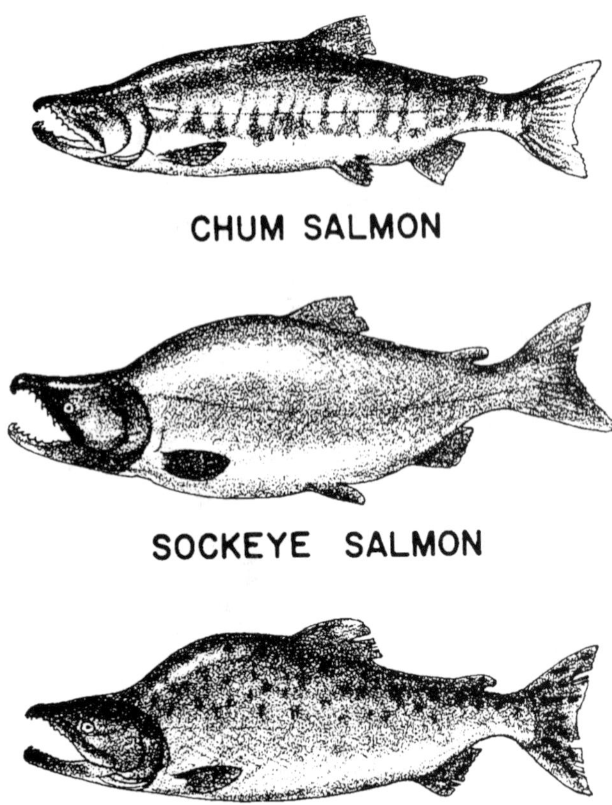

Figure 4. Three salmon species.

Salmons

Sixty-six species comprise the family Salmonidae (Figures 4 and 5). Continued confusion persists regarding the taxonomy of many members. Many researchers base their conclusions on strict morphological observations, while new studies of DNA, etc., are being used to distinguish among species, races, and/or populations.

Salmons are freshwater and marine species that inhabit the world's Northern Hemisphere waters. Several species live most of their lives in marine waters, migrating to freshwaters to spawn, as do all salmonids. All salmons have an adipose fin (extra fleshy/fatty fin located far back on the back), gill membranes that extend far forward and are free of the isthmus to permit wide flaring of the gill covers (opercula), and a fleshy process located in the axil of each pelvic fin. There are seven to 20 branchiostegal rays (rib-like structures) that help flare and support the gill cover; the mouth possesses teeth; and some fish attain lengths of 1.5 meters. Internally 11 to 210 pyloric coeca

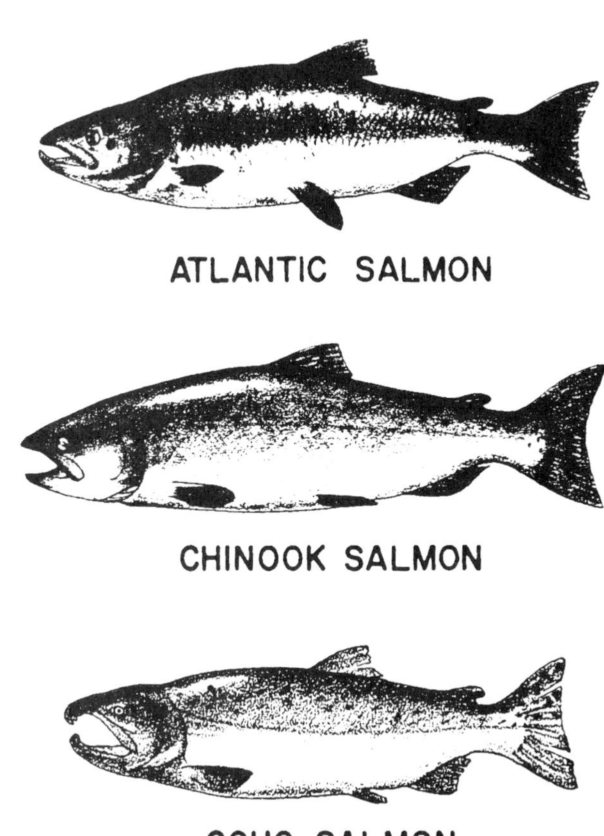

Figure 5. Three salmon species..

(pouches) exist along the digestive tract to increase digestion.

All salmons are highly valued as food fishes and support valuable sport and commercial fisheries. Many species have been introduced, to the detriment of local species, throughout the world in order to increase local sport fisheries. Many species have been hybridized in hopes, by the sportsmen, of producing faster- and heavier-growing specimens and species.

Salmonids that live the greater portion of their lives in marine waters are: Atlantic salmon, *Salmo salar*, and the five Pacific salmon: pink (*Oncorhynchus gorbuscha*); coho (*O. kisutch*); rainbow, with several marine and freshwater forms [(*O. mykiss*), sockeye (*O. nerka*), and chinook (*O. tshawytscha*)]. Brown trout (*Salmo trutta*) and Arctic charr (*Salvelinus alpinus*) have some freshwater populations, but may spend considerable time in marine waters.

Much has been written of how salmon, especially the Pacific salmon, navigate hundreds of miles across ocean waters and up inland rivers and streams to spawn in rocky, pebbly stream beds. Such navigation feats have been shown to be linked to strong homing instincts. Most salmons are fall spawners, and most die after spawning. Pink salmon live for two years and have even- and odd-year spawning stocks; some stocks even spawn at the same time in the same stream. Sockeye salmon live for eight years. Sockeye salmon have even given rise to a freshwater-inhabiting form, kokanee. Pink and chum salmon exhibit parental care and often spawn in the lower reaches of rivers.

Today, inland migrating salmon find the task of reaching their spawning grounds hampered by many, often consecutive, high level dams spanning major rivers once used by vast hordes of spawning salmon (Columbia, Fraser, etc.). The same is true for Atlantic salmon. Where hundreds of streams were once ascended by Atlantic salmon, today only a few stream systems remain suitable for spawning. Excessive logging, especially in the West, has clogged streams and rivers, increasing siltation of once pristine streams and river beds. Pollution on land and sea has drastically affected marine stocks. Conservation efforts have tried to alleviate the damage by building elaborate and complicated fishways (many stories high) or by trucking fish around dams in order to aid the salmon during their upstream migrations. Offshore, disputes exist over how far salmon migrate in the Pacific, which country has jurisdiction of the fishing area, or who owns them or from which country did they migrate. High-sea fisheries seem to have overfished all species and stocks.

Recent efforts to restore salmon stocks have resulted in a proliferation of hatcheries. While seemingly beneficial, disputes have arisen concerning their waste pollution and the value of such efforts for whoever owns the returning fish, whether they can be distinguished from wild native fish, and whether inland life history patterns are compatible. Does restocking have an impact on native fish and their fisheries? Compounding all these efforts is the impact of historic Native American fishing and their rights to salmon, regardless of regulation.

Meanwhile, Pacific and Atlantic salmon continue to struggle to survive and migrate upstream to spawn. Their spectacular leaps to breach falls, dams, etc., diminish with each passing year. Each year fewer spawning grounds exist where severely jaw-deformed males can tend a female as she digs a nest (redd) in a clear cool stream, and in which the large eggs are deposited to develop and hatch, depending on water temperature, days later. Salmon that survive nature's and man's impediments must also contend with birds (eagles) and bears before dying. In the past, high recruitments resulted from the large number of successfully spawning fishes. Today recruitment has diminished drastically so that stocks no longer exist in many western and eastern rivers.

Young (parr) that hatch, usually species-dependent, remain in freshwater for varying periods before moving downstream to the sea. Parr and young must negotiate all the hazards adults encountered (logs, silt, dams, predators). Once in the sea, migrations of several hundred miles are common, as is fast growth. Salmon return to repeat the life cycle again after several years at sea, often in their natal stream.

Suggested Reading

Balon, 1980 (*Salvelinus*); Behnke, 1972, 1992 (taxonomy); Blaxter, 1993 (rarity); Buck, 1989 (taxonomy); Ewing and Ewing, 1995 (survival); Fisher and Pearcy, 1988 (ocean growth); Foerster, 1968 (sockeye); Foster and Sehorn, 1989 (Atlantic salmon); Groot and Margolis, 1991 (Pacific life histories); Hanson et al., 1993 (survival); Burgner, 1991 (sockeye); Dizon et al., 1973 (homing); Dorofeeva (homing); Hasler, 1966 (homing); Hasler and Scholz, 1983 (olfaction); Heggeberget et al., 1993 (homing); Kendall and Behnke, 1984 (ontogeny); Mills and Piggins, 1988 (Atlantic salmon); Naiman et al., 1987 (anadromy); Pearcy, 1992 (ecology); Pearcy and Fisher, 1990 (sea distribution); Quin, 1989 (stocks); Quin et al., 1989, 1987 (movements); Rounsefell, 1962 (meristics); Schwartz, 1993 (movements); Simon and Larkin, 1972 (rare mating); Stearsly and Smith, 1993 (phylogeny); and Wedemeyer et al., 1980 (environmental effects).

CODS AND HAKES

Fishes of the cod and hake family (Figure 6) are some of the most important sizable and edible commercial food fishes in the world. Important cods and hakes found in temperate-to-arctic waters are: Atlantic cod (*Gadus morhua*), haddock (*Melanogrammus aeglefinus*), Atlantic tomcod (*Microgadus tomcod*), pollock (*Pollachius virens*), walleye pollock (*Theragra chalcogramma*), silver hake (*Merluccius bilinearis*), and Pacific hake (*M. productus*).

Cods and close relatives have three dorsal fins, two anal fins, no spines in the fins, chin barbel, pelvic fins located forward of the pectoral fins, scales cycloid, and a slightly-forked caudal fin; they can attain sizes of 1.8 meters.

Hakes (about 12 species) possess two dorsal fins, one anal fin, no chin barbels, a first dorsal fin with a spine, a terminal mouth, no internal pyloric caeca, an extremely long body, and forked tail. Sizes are up to 0.75 meter.

Atlantic Cod

Because of their size and weight, cod are one of the world's most important food fishes, especially in America and Europe. Atlantic cod occur in Northern Hemisphere waters to 400 meters deep from Greenland to Cape Hatteras, North Carolina (Figure 6). Capture was first with hand lines, subsequently by long lines, trawls, and/or gill nets. Atlantic cod usually frequent western Atlantic Ocean rough substrates and waters of 10° to 12°C. Spawning occurs during winter months on Georges Bank (east of Cape Cod, Massachusetts) and other nearby banks. Between 3 to 9 million buoyant eggs possessing oil globules are spawned. The young hatch out in about six days. Feeding begins in 10 to 11 days on planktonic copepods and, with growth, on bottom foods, crabs, fishes, etc. Spiny dogfish and pollock are known enemies of the young.

Turn-of-the-century commercial fishery catches consisted of thousands of kilograms of cod by large foreign and American fishing fleets. The cod was so important to the livelihoods of many that as early as 1784 a likeness — which still hangs — was hung in the Massachusetts Capitol Building in tribute. Today only a few thousand kilograms are landed. Some blame the decline on overfishing. Others charge that global warming (increased) or water temperature (decreased) changes since the turn of the century have caused the declines, since eggs, larvae, the young, and perhaps adults may have been impacted by such temperature changes. Others say the declines are nothing more than another recurring fifty-year

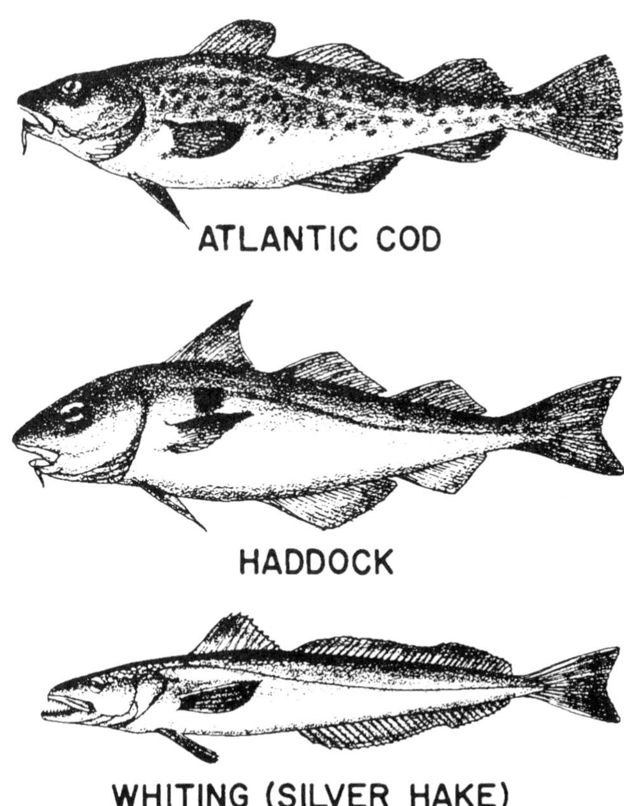

Figure 6. Three finfish species of major importance worldwide.

cycle, and temperatures and overfishing have not been the culprits causing declines in population and survival, and instability. Arguments will rage on until the populations re-emerge to former levels naturally or as a result of man's attempts to regulate fishing, quotas, number of boats, area closures, etc. Whether continued long-term pollution of the seas by oil, debris, etc., is detrimental to cod survival remains unresolved.

Atlantic Tomcod

The Atlantic tomcod is a smaller species that can be found in estuaries and streams as well as the ocean. It occurs from southern Labrador to North Carolina and attains a size of 0.4 meter. Feeding is mostly at night. Spawning of eggs containing an oil globule occurs from November to February. Hatching takes 24 days and young and adults prefer to remain nearer the substrate than do cod.

Haddock

Haddock, a species that occurs from Newfoundland to Cape Hatteras, is another bulky-bodied 1.1-meter, 17-kg fish that can be easily distinguished by the black lateral line and a conspicuous dark spot at the level of the pectoral fins on the middle of each side (Figure 6). Haddock never enter estuaries or brackish waters, prefer 275- to 830-meter water depths, and occur over smooth substrates of sand, gravel, or clay. Generally, haddock prefer deeper waters than do cod. In spring to early summer occurs the spawning of up to one million buoyant eggs, without oil globules, over broken or rough substrates or sand. The young hatch in almost 10 days and feed similarly to cod. Larger fish feed on fishes, squid, etc. A definite movement into deeper waters occurs with size. Once plentiful, today less than 100 metric tons are landed by the commercial fishery.

Pollock

Pollock are active fish found anywhere in the water column from Labrador to North Carolina; they can attain sizes of 1.1 meter and 32 kg. Pollock are usually found in shallower waters and down to 180 meters depths, and prefer cooler waters (11°C) than other Atlantic-inhabiting cods. It is a spring spawner of about 200,000 eggs which contain an oil globule. Hatching occurs in six days at 9°C. Pollock is another species experiencing recent declining commercial catches of only about 4,000 metric tons.

Walleye Pollock

The walleye pollock is a Pacific Ocean open-water inhabitant that occurs from Japan to British Columbia and attains sizes of 0.6 meter, 1.4 kg. Trawling is the preferred method of capture, usually of three-year-old mature fish, within waters not over 180 meters. A number of separate stocks seem to exist throughout its range, three in North America. Considered one of the world's largest fisheries, as much as six million metric tons of walleye pollock were landed in 1984, and used in Japan as surimi. Populations are influenced by shifts in El Niño and its effects on sea water temperatures and other food species abundances.

Silver Hake

The silver hake is a long, slender fish of 0.7 meter and 2.3 kg that occurs from the Gulf of St. Lawrence to South Carolina in waters to 535 meters deep (Figure 6). The silver hake has a black area at the axil of the pectoral fin, the lower jaw projects, its teeth are large, it lacks a chin barbel, the rear portions of the second dorsal and anal fins are wider than the rest of the fins, its eyes are large, and the two dorsal fins are separate. It feeds at night. Silver hake is a spring spawner of buoyant eggs with one oil globule, over pebbly substrates. Silver hake is abundant, now occupying areas and contributing to catches once dominated by cod. Catches, once 14 million metric tons in 1919, now amount to about 2,000 metric tons (1992).

Pacific Hake

The Pacific hake is another long, slender fish found from Alaska to Mexico, often in feeding aggregates. Most are captured by trawls, usually in waters greater than 90 meters. Spawning occurs off northern Mexico to southern California between November and August. Pacific hakes contribute to a large groundfish resource off Washington to California. Since the 1960s Pacific hake have supported a sizable commercial foreign fishery. Today catches are much less. Populations are stratified latitudinally by age and size, the youngest and smallest fish inhabiting waters of the southern portion of its range. Larger fish are found over the continental shelf. Wide population and year-class strength fluctuations occur every three to four years by area.

Whether cods and hakes will continue to contribute to world commercial catches remains unresolved as they are subject to changing fishery practices, regulation, pollution, changing world environments and conditions, and world food demands. Attempts at management may prove futile since nature, with a few good spawning periods, may restore or affect their abundance more easily than those attempted by man.

Suggested Reading

Anderson et al., 1980 (silver hake); Beacham, 1983 (pollock); Bigelow and Schroeder, 1953 (cod and hake); Blaxter, 1993 (cod ranching); Brander, 1993 (cod spawning); Brodeur et al., 1996 (walleye pollock); Claireaux et al., 1995 (climate); Clay et al., 1989 (pollock); Cohen, 1980 (names); Cohen, 1989a,b (Gadiformes); Cohen et al. 1990 (Gadiformes, world); Dark and Wilkins, 1994 (Pacific groundfish); Dark et al., 1980 (Pacific pollock, whiting); Dickson and Brandeis, 1993 (environmental effects); Earl, 1980 (silver hake); Gabull, 1992 (climate); Grant and Stahl, 1988 (evolution); Herrfurth, 1987 (walleye pollock); Hughes and Hirschhorn, 1979 (walleye pollock); Hutchings and Myers, 1994 (fishery); Kelley, 1994 (climate); Koslow et al., 1987 (cod recruitment); Langton and Bowman, 1980 (food); Mackie and Ritchie, 1980 (differentiation); Mayo et al., 1989 (pollock); Mork et al., 1985 (cod variation); Murawski, 1993 (climate); Myers et al., 1995 (climate, cod); Olson et al., 1994 (Pacific species); Renaud et al., 1986 (cod speciation); Rose and Leggett, 1988 (environmental effects); Shaw and McFarlane, 1983 (walleye pollock, Canada); Stauffer, 1985 (Pacific whiting); Swan and Kremer, 1995 (climate); Thomas et al., 1980 (Pacific whiting); Tritos, 1984 (climate); and Whittaker, 1980 (hake utilization).

REDFISHES OR ROCKFISHES

Redfish or rockfish are a large group of marine fishes that inhabit temperate and tropical seas. Capture is with trawls, usually in deep waters of the Atlantic and Pacific oceans. Three species of the reddish-colored, spiny fishes occur in the Atlantic and at least 62 in the eastern Pacific. Their life histories are poorly known, other than the *Sebastes* redfish, or rockfish as they are often called. Redfish are ovoviporous and give birth to living young about five months after fertilization. Actual spawning behavior, at what age, and where it occurs are unknown. Most seem to spawn in May/June.

Redfishes are compressed-bodied fishes possessing ctenoid (spiny) scales and a single continuous notched dorsal fin; the dorsal fin has 11 to 17 spines and eight to 18 rays; the anal fin possesses one to three spines and three to nine soft rays. Both head and body are covered with many fleshy flaps and spines, and an occipital pit (behind the eyes) may or may not be present. Growth varies by species and sex. Longevity seems to vary, by species, between seven to 20 years.

Much confusion persists regarding the identity of the redfishes of the Atlantic since several species were marketed as rosefish or ocean perch. Only recently have morphometric and electrophoretic analyses revealed that three species were involved in Atlantic commercial catches: golden redfish (*Sebastes norwegicus*, earlier as *S. marina*), deepwater redfish (*S. mentella*), and Acadian redfish (*S. fasicatus*). Golden redfish occur from western Greenland to New Jersey and attain sizes of 0.5 meters, 4 kg. Deepwater redfish occur from Baffin Bay to Nova Scotia in waters of 200 meters and attain sizes of 0.4 meter, 3 kg. Acadian redfish, 0.3 meter, occur from the Gulf of St. Lawrence to Nova Scotia in shallow to deep waters. Two other related redfish, also of commercial interest, and also sold as rosefish or ocean perch, occur in the Atlantic: blackbelly rosefish (*Helicolenus dactylopterus*), a 0.3-meter species that occurs from Maine to Argentina in 110 to 735 meter waters; and the spinycheek scorpionfish (*Neomerinthe hemingwayi*), a 0.4-meter fish that inhabits 45 to 230 meter waters from New Jersey to the northern Gulf of Mexico.

The Pacific Ocean perch (*Sebastes alutus*) is one of the most abundant (of many) Pacific scorpionfishes supporting a northeastern Pacific commercial and sport fishery. Spawning varies by season, locality, and depth. What is known of their biology is similar to that of Atlantic species.

Suggested Reading

Dark and Wilkins, 1994 (Pacific groundfish); Fahrig and Atkinson, 1990 (stocks); Gulder et al., 1980 (Pacific ocean perch); Gunderson and Lenarz, 1980 (Pacific distribution); Gunderson and Sample, 1980 (Pacific distribution); Mayo et al., 1981 (age); Ni and McKone, 1983 (Atlantic distribution); Ni and Sandman, 1984 (morphology); Ni and Templeman, 1985 (reproduction); O'Brien et al., 1993 (age, landings); Sevigny and de Lafontaine, 1992 (parasites); and Washington et al., 1984 (systematics).

SABLEFISH

The sablefish (*Anoplopoma fimbria*), a nearly cylindrical blackish-gray (above) to white (below) fish, is a 1-meter, 51-kg inhabitant of the Pacific Ocean from the Bering Sea to Isle Cedros, Baja California. It is found over soft substrates to depths of 1800 meters. Sablefish have two widely-separated dorsal fins (the first with 17 to 30 spines), a sharp-pointed snout, no spines in the anal fin, and a forked tail. Spawning occurs in winter at the edge of the continental shelf at depths of 300 meters.

Eggs are translucent, pelagic, and lack an oil globule. Hatching occurs in two to three weeks. Larvae immediately swim to the surface and feed on copepods, etc., growing rapidly in size. Large individuals are voracious feeders of fishes. Much remains to be learned about sablefish life history, growth, and population size in order to manage a large trawl fishery.

Once neglected, interest developed in their capture as other West Coast fisheries declined. Hatcheries have attempted to increase their abundance. Hatchery production problems pertain to lack of early development and larval and young feeding needs. How long sablefish will sustain their fishery depends on many factors of fishing pressure and recruitment. The large turn-of-the-century fishery declined, once fishermen learned they could be easily captured by trawls and longlines.

Suggested Reading

Wilkins and Saunders, 1997 (biology and management); Kendall and Matarese, 1987 (early life history); McFarlane and Beamish, 1983a,b (biology).

STRIPED BASS

Striped bass, *Morone saxatilis*, occur in rivers, bays, estuaries, and the Atlantic Ocean from the St. Lawrence River, Canada, to St. John's River, Florida (Figure 7). Fish from New Jersey were stocked in San Francisco Bay near Martinez, California, in the 1870s and since then into many freshwater lakes and dams throughout the country. In recent years, striped bass have been hybridized with the white perch (*M. americana*), white bass (*M. chrysops*), and yellow bass (*M. mississippiensis*) in order to produce faster-growing fish for the sport fisheries in inland lakes and dams. Hybrids are believed to be hardier and better fighters than the original parents.

The striped bass is an anadromous fish that migrates from marine waters inland, often up to 160 kilometers, to spawn. Spring spawning occurs in 15° to 18°C waters over rocky or sandy areas of

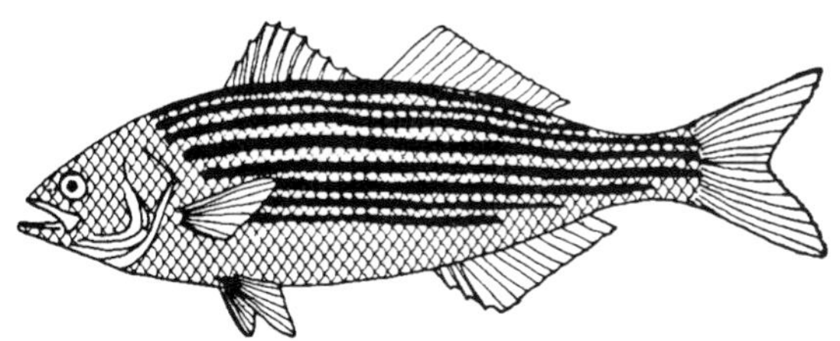

Figure 7. Striped sea bass.

some coastal streams and rivers. A large female "cow" is usually accompanied by several smaller males during the highly agitated spawnings. Males are about two years old and females four to five years old when first spawning. Large adult females may spawn four million light yellowish eggs in April to June, now believed every other year, whereas small 0.5-meter young females may spawn 1.5 million eggs yearly. Hatching occurs in 74 hours at 15°C or 30 hours at 21°C. A current is necessary to carry the semi-buoyant eggs downstream as they incubate and hatch in lower river and estuarine areas where food is plentiful. Growth is rapid in estuarine waters with eventual migration into the ocean or other rivers and bays.

The striped bass is a large fish attaining sizes of 1.8 meters and 56 kg. They possess two separated dorsal fins, the first with eight to 10 spines, the second 10 to 12 soft rays. The anal fin has three spines and nine to 12 rays. Two spines are on the opercle (gill cover). The olive-green-colored (dorsally) to silvery-white (ventrally) body is covered by large scales (65 to 70), and seven to eight longitudinal stripes along the upper lateral body scales. Stripes in hybrids are usually broken instead of continuous and vary in number of rows along each side.

Historically, major areas that contained the largest populations of striped bass were the Hudson River (New York), Delaware River and Bay (Delaware), Chesapeake Bay (Maryland-Virginia), and Roanoke River and Albemarle Sound (North Carolina). Migration of Atlantic Coast striped bass among areas is common. Striped bass in the San Joaquin system of California or Coos Bay, Oregon, exhibit little movement among areas.

Striped bass populations declined drastically in the 1960s to 1980s as a result of increased pollution. Damming of many rivers drastically reduced downstream flows. This seriously affected spawning sites, eggs, and resultant larvae and food supplies. Pollution abatement programs, as well as fishing regulations and moratoria, were instituted to restore various populations. Today, with flow regulations in effect along with pollution abatement practices, striped bass populations have exhibited remarkable resilience. Several good year classes and spawnings have occurred, indicating some populations may again attain prior levels of abundance. Hatcheries have sprung up and are adding thousands of young, especially in the Chesapeake Bay, to help the restoration. Sport and commercial fishing interests acknowledge that with cooperation, the species may be helped by nature to restore a once-plentiful fish.

Suggested Reading

Chadwick, 1964 (California stocks); Kirby, 1993, 1986 (hybrids); Kirby et al., 1983 (culture); Secor et al., 1995 (Chesapeake Bay); Smith and Wells, 1977 (biology); Stevens et al., 1985 (California decline); Waldram and Fubrizio, 1994 (stock identification).

SEA BASSES AND GROUPERS

Sea basses and groupers are members of a large group of perch-like marine tropical and temperate fishes. Sixty-one species occur in North American oceans, with four of the large species occurring in the Atlantic. Four of the 17 occurring in the Pacific are members of the genus *Epinphelus.* Nine additional species, two in the Pacific and seven in the Atlantic, are members of the genus *Mycteroperca.* All have a single dorsal notched spinous and soft-rayed dorsal fin, the mouth is large, teeth are present in the jaws and mouth, and they lack a fleshy axillary flap or process at the base of the pelvic fin. The opercle is free and has three flat spines. The lateral line is continuous and extends to the base of the caudal fin. Many species are hermaphroditic; transformation from female to male occurs at early ages. Most are coastal species occurring over reefs and hard substrates, but some can be found at great depths. Few enter sounds and estuaries as young. Some species attain sizes of three meters and 400 kg.

Sea Basses

Centropristes ocyurus, C. philadelphia, and *C. striata* — all called blackfish — are Atlantic ocean inhabitants from Maine to the eastern Gulf of Mexico. Sea bass attain sizes of 0.6 meter and 2.7 kg, are black-colored and heavy-bodied, and have wide, bulky heads. Spawning occurs in spring and summer in waters 8° to 37°C and 18 to 45 meters deep. Eggs are colorless and buoyant, and have one oil globule. Hatching occurs in 75 hours at 16°C. Life span may be 20 years. Being protogynous hermaphrodites, females begin to transform into males at about 0.3 to 0.6 meter. Males older than six years develop a head hump bulge, and have several elongate caudal and dorsal fin rays. Important commercial fisheries exist using longlines, traps, and crab pots to catch sea basses. Catches have declined drastically since the 1920s. Sea bass are a large component of the hook-and-line sport fishery.

GROUPERS

Groupers vary in size from small to large bulky specimens up to 400 kg (Figure 8). Habitat preference is reefs and shelf breaks around hard structures and substrates. Like sea basses, groupers are protogynous hermaphrodites, changing sex with size and age. Some species live up to 20 years and are found from Massachusetts to Rio de Janeiro in 27- to 101-meter water. Most Atlantic species spawn in March to April, with Pacific forms spawning March to May. Life histories are like those of sea basses. All groupers are also overfished by sport and commercial fishermen. Federal management plans have been/are being devised that will open/close seasons, determine sizes by species and geography, etc.

Suggested Reading

Baldwin and Johnson, 1993 (phylogenetics); Brule et al., 1994 (red grouper); Bullock and Smith, 1991 (sea bass); Heemstra and Randall, 1993 (identification); Hood and Schlieder, 1992 (gag); Hood et al., 1994 (sea bass); Kendall, 1977 (sea bass); Kendall, 1984 (systematics); Manooch, 1987 (grouper); Manooch and Haimovici, 1978 (age); Matheson et al., 1986 (biology, gag, scamp); NMFS, 1983 (management plan); O'Brien et al., 1993 (life history); Pawson and Pickett, 1993 (sea bass); Shapiro, 1987 (reproduction); Shephard and Terceiro, 1994 (sea bass); Smith, 1971 (systematics); Smith, 1981 (identification sheets); and Wenner et al., 1986 (biology).

BLUEFISH

The bluefish, *Pomatomus saltatrix*, is a voracious 1.1-meter, 12-kg, elongate, greenish-hued, slabsided, silvery, forked-tail fish that occurs in the western Atlantic from Nova Scotia to Bermuda to Argentina, but is absent from southern Florida to northern South America. Bluefish also occur in the Mediterranean, parts of Africa, Australia, and Malaysia. Several features distinguish this highly migratory commercial and sport fish. It possesses separate dorsal fins; the first is spiny (seven to eight), the second has one spine and 13 to 28 rays, and an anal fin has two to three spines and 12 to 27 soft rays. Ctenoid (spiny-edged) scales cover the body and dorsal and anal fins. A lateral line is present. The opercle ends in a flap and a black spot occurs at the base of each pectoral fin. The lower jaw projects and sharp teeth occur on the jaws, tongue, and various jaw bones.

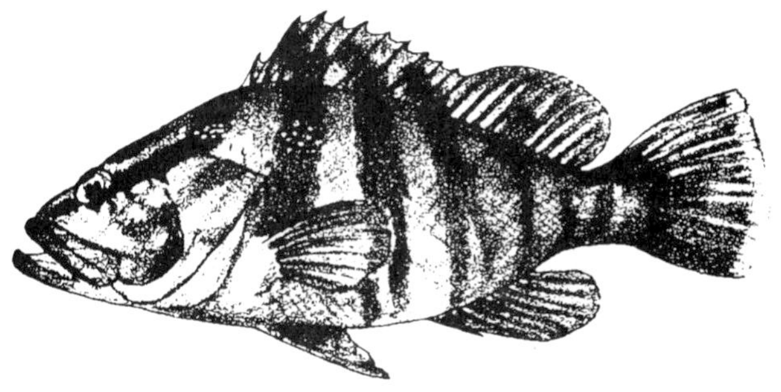

Figure 8. Nassau grouper.

Controversy persists as to whether two or three stocks occur and whether one or two spawning areas exist in the western Atlantic. Some believe one group spawns from North Carolina to southern Florida in April to May; another off the continental shelf from June to August. Females that mature in two years may spawn up to one million buoyant eggs that possess an amber-colored oil globule. Hatching occurs in 48 hours at 21°C. Adults are heterosexual and can be 14 years old when 0.8 meter in length. Young and adults are voracious feeders, feeding singly or in schools, on menhaden, croakers, or anything that gets in their way. Such feeding frenzies frequently result in complete decimation; much may be left uneaten, bitten in two, or may be stranded on the beach as the prey attempt to elude the passing marauding bluefish. Bluefish are often prey of sharks, tunas, swordfish, and wahoo, and are food competitors like striped bass, Spanish and king mackerel, and large weakfish. Bluefish are highly susceptible to parasites and fin rot and other effects of predation or the environment, such as pollution.

Bluefish make long spring migrations northward and late fall to early winter southward coastal movements. Commercial and sport fishermen have long sought this valuable food fish. The flesh is dark and slightly oily, but broiled or fried rates as high as flounder, striped bass, and other food fishes in palatability and acceptability.

Commercial catches amounted to 41 million kilograms in 1965 and 540 million kilograms in 1970. Since then, bluefish catches have declined or are absent. Big "Hatteras blues" and other sizes seemingly have disappeared. Presently needless alarm prevails. Overfishing is blamed as the cause of the present sharp decline in abundance. However, natural cycles, perhaps influenced by long-term environmental features, may be the real causal agents causing the depletion, rather than overfishing. Few realize bluefish exhibit five- and 50-year abundance cycles. Perhaps as with certain other fishes (such as striped bass and weakfish) bluefish levels will return to previous abundances that can be fished by gill net, pound net, haul seines, hook-and-line, otter trawl, and a host of other fishing gear.

Suggested Reading

Davy, 1994 (migration); Graves et al., 1992 (stocks); Hare and Cowen, 1993 (ecology); Kendall and Walford, 1979 (egg distribution); Lund, 1961 (stocks); Lund and Maltezos, 1970 (stocks); McBride et al., 1993 (recruitment); Olla and Studholine, 1971 (environmental effect); Smith et al., 1994 (spawning, pollution); Stone et al., 1994 (profile); and Wilk, 1977 (biology).

DRUMS

Drums, members of the Sciaenid family, include 33 North American species that occur in fresh, estuarine, and marine waters. Many are highly valued and esteemed sport and commercial fishes. Important Atlantic species are: black drum (*Pogonias chromis*), red drum (*Sciaenops ocellatus*), sea mullets (*Menticirrhus* spp.), croaker (*Micropogonias undulatus*), weakfish (*Cynoscion regalis*), and spotted sea trout (*C. nebulosus*). Pacific species are white seabass (*Atractoscion nobilis*) and shortfin corvina (*C. parvipinnis*). All have long notched dorsal fins with six to 13 spines and 26 to 35 soft rays, and an anal fin with one to two spines and six to 13 rays. Some species possess conspicuous chin barbels and/or pores on the lower jaw. The most distinctive feature is that the lateral line extends out to the end of the caudal fin. Scales are usually ctenoid. Most can produce a drumming sound by activating special muscles along the large and/or various-shaped swim bladder (some females cannot drum).

Black Drum

Black drum are heavy-bodied ctenoid-scaled fish possessing four to five dark vertical body bands and many short chin barbels. Large molar teeth occur at the rear of the jaws. Black drum are found from the Bay of Fundy (Canada) to Argentina and are highly adaptable to life in inland brackish waters of 10°C or ocean sea waters of 32°C. With two to six oil globules that, during development, become one, 6 million eggs are externally fertilized in spring (May to June) in 17°C waters. Hatching occurs in 24 hours at 20°C. Individuals mature at an early age and adults can live at least 20 years.

Black drum are important sport and commercial fishes that are caught from spring to autumn as large schools migrate along the coast feeding on crabs, fishes, mollusks, etc. They once supported a fishery of 18.9 million kg; landings in 1990 were far fewer.

Red Drum

The red drum is an elongate, reddish golden-colored fish of 1.3 meters, 42 kg, that is highly esteemed as a sport and commercial food fish (Figure 9). It is found in dense coastal schools that frequent estuaries, low saline waters, and coastal waters from New York to Mexico over sandy/muddy substrates. Red drum lack chin barbels and are conspicuously characterized by a large ocellus (spot) at the base of each side of the tail. These ocelli may number up to 500. Spawning of up to 8 million eggs occurs in July to October in inshore to coastal areas over sandy/muddy substrates. Hatching occurs at 25°C in 30 ppt waters. Young and adults feed on crustaceans, fishes, and other foods, changing preference with age and size.

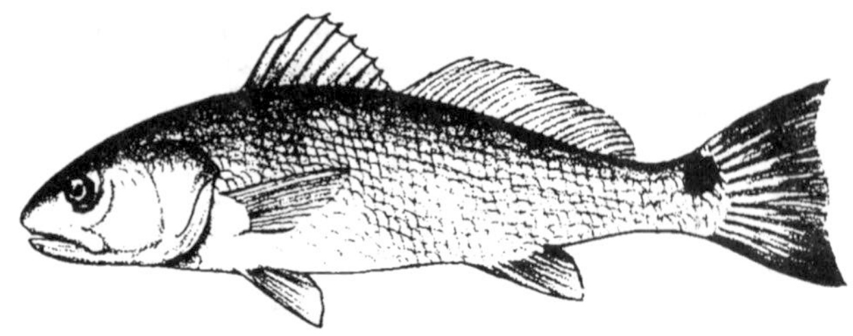

Figure 9. Red drum.

Young are found in estuaries and young and adults are easily responsive to water temperatures, salinity, and photoperiod changes. Red drum can be induced to spawn in hatcheries by manipulating water temperatures, photoperiod, and food. Large schools migrate North and South along the coast in Spring and Fall and are important elements of sport and commercial fishermen catches. Capture can be by haul seines, gill nets, pound nets, hook and line, etc. Catches have declined from 1945 when 788 metric tons were landed in the Atlantic and 1594 metric tons in the Gulf of Mexico. Many state and federal agencies have imposed size, season, and catch limits in hopes of reviving the species and fisheries. Recently hatcheries have been influential in helping restock and revive the Gulf of Mexico stocks and commercial fisheries.

White Sea Bass

The white sea bass is a Pacific Coast, 1.5-meter, 38-kg fish that supports a good sport and commercial fishery. It is found inshore and to 20 meter depths from Juneau, Alaska, to the Gulf of California. An elongate, slightly compressed fish that is blue-gray above, silvery below, the white sea bass possesses a black spot at the base of the pectoral fin, has pores on chin and upper jaw, rather than a chin barbel, and has no large canine teeth on the middle of the upper jaw. Two dorsal fins are joined by a membrane, the anal fin possesses two spines, and the lateral line extends to the tip of caudal fin. It spawns from April to November during inshore congregation excursions; migrates north and south along the coast; and feeds on squid, sardines, and anchovies.

Spotted Sea Trout

The spotted sea trout is a 0.7-meter, 7.3-kg fish with black spots on its sides and dorsal fins. It lacks a barbel and pores on the chin, but possesses two large canines in the upper jaw. Spotted sea trout are found in estuarine and coastal waters of the Atlantic over sandy/muddy substrates. Both young and adults enter estuaries and can often be caught in or near grass beds. Sea trout (highly esteemed as a sport food fish) are captured by hook-and-line, trawls, pound nets, and gill nets. This enigmatic species does not seem to follow consistent patterns of migration and abundance.

Weakfish

The weakfish is a 0.5-meter, 8-kg fish that possesses oblique streaks on each side and is found from Nova Scotia to Florida and Mexico. Two large canine teeth occur in the upper jaw. Ctenoid scales cover the body, cycloid scales the head, and scales one-half of the base of the soft dorsal fin. Weakfish are found over mud and sandy substrates of estuaries, sounds, and oceans. Habits are similar to spotted sea trout. Weakfish have supported large sport and commercial fisheries, but present populations have declined drastically or are depleted possibly by overfishing. Conservation efforts are in place to restore populations that may be cyclic in abundance.

Shortfin Corvina

This Pacific, 0.9-meter, sport and commercial fish is found from Santa Barbara Island to Mazatlan, Mexico. Habits are similar to other drums.

Kingfishes

These are fish with an elongate conical snout and tapering body with dark oblique body bands. Three species are found in the Atlantic [southern kingfish (*Menticirrhus americanus*); gulf kingfish (*M. littoralis*); northern kingfish (*M. saxatilis*)] and one in the Pacific [California corvina (*M. undulatus*)]. All are about 0.5 meter, 1 kg. Each has a small horizontal mouth; a chin with a single short, stout barbel; and no canine teeth. Scales are small and the lateral line extends to an odd-shaped caudal fin, not forked or square. Most species are found in the surf zones along coastal beaches. Atlantic species occur from Massachusetts to Florida, in the Gulf of Mexico, and to Yucatan, Mexico. Spawning occurs in bays and sounds from June to August. Foods are mostly mollusks and crustaceans. Kingfish are caught in gill nets, by trawls, and by hook-and-line. Kingfish are one of the most tasty food fishes.

Croaker

Croakers are found from Massachusetts to south Florida and the northern Gulf of Mexico (Figure 10). The silvery, multi-colored, slightly compressed fish has somewhat narrow dark lines or rows of spots on the upper body. Dorsal fins are spotted. A strong spine occurs at the angle of

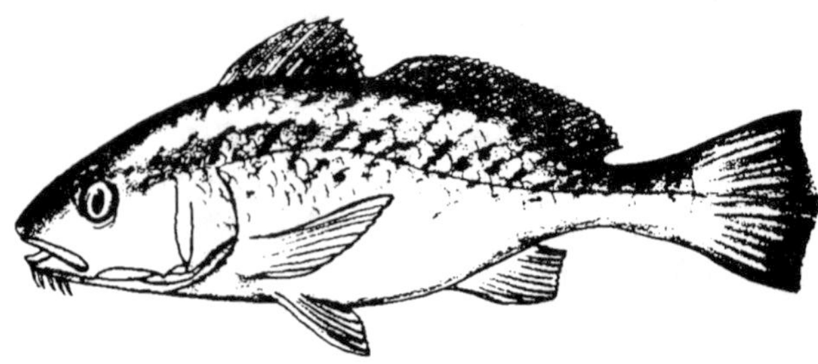

Figure 10. Croaker.

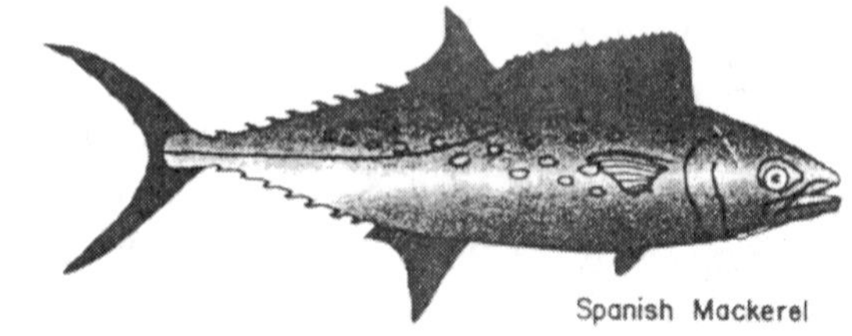

Figure 11. Two species of mackerel.

the preopercle. Three to five tiny barbels and five pores occur in a row along the inner edge of the lower jaws. The size is 0.5 meter, 1.8 kg; the lateral line extends to the caudal fin. Croakers are found over sand and mud of coastal and estuarine waters. Estuaries are vital nursery habitats for larvae and young. Croakers spawn in late fall or early winter in inshore sounds and offshore waters. Both sexes can produce drumming sounds. They feed on bottom organisms such as crustaceans, fishes, etc. Catches at the turn of century often reached 18.5 million kg; recently catches have fluctuated widely. Croakers are one of the most common sport and commercial fishes of the Atlantic.

Suggested Reading

Arnold et al., 1988 (aquaculture); Chao, 1978 (systematics); Chao, 1986 (zoogeography); Chamberlain et al., 1990 (aquaculture); Connaryhton and Taylor, 1995 (sound production); Daniel and Graves, 1994 (eggs, identification); Linton et al., 1990 (bibliography); Matlock, 1990 (black drum); McEachran and Daniels, 1995 (red drum); Mercer, 1983, 1984a,b (species profiles); Murphy and Taylor, 1989 (black drum); Nieland and Wilson, 1993 (reproduction); Murphy and Taylor, 1981 (black drum); Ross et al., 1983 (taxonomy); Ross et al., 1995 (age, growth); Saucier and Blatz, 1993 (black drum); Silverman, 1979 (profiles); Sutter et al., 1986 (species profiles); Trewavas, 1977 (systematics); Wenner, 1992 (red drum); and Wilk, 1979 (profile).

MACKERELS

Spanish (Figure 11) and king mackerels and tunas are members of the Scombrid family, of which 23 species occur in North America in one or more oceans: 17 in the Atlantic, and 14 in the Pacific Ocean (Figure 11). All are fast swimmers, apex carnivores, and valued sport and commercial food fishes that inhabit the upper waters (epipelagic) of the world's warm oceans. Mackerels are oval to round in body shape and smooth-skinned, although some cycloid scales are found in some young. Unique features that permit mackerels to achieve sustained fast swimming speeds are: the dorsal fin can depress into a groove in the back; a series of dorsal and anal small finlets occur on the body behind the dorsal and anal fins; and a keel along each side of the narrow caudal peduncle helps to stabilize the rear portion of the body during continuous swimming. Speeds of 75 kilometer/hour are possible for wahoo (*Acanthocybium solandri*).

Spanish Mackerel

This compressed, bluish-silver, 0.9-meter, 5-kg fish occurs from Cape Cod to the Gulf of Mexico. Sides possess large brownish-yellow spots. An important distinguishing feature is the lateral line that slopes gradually, at the level of the second dorsal fin, down along the side to the tail. A blackish area prevails at the front of the first dorsal fin. Scales are present on the entire body, but not the pectoral fin. The swim bladder is absent.

Spanish mackerel have a prolonged spawning period from May (earlier in the South) to October throughout its range. Sexes are separate. During spawning two million eggs with one oil globule can be spawned. Hatching occurs in 20 hours at 29°C. Adults live up to eight years and dense schools exhibit long seasonal migrations when water temperatures are > 20°C. Sport and commercial fishermen catch large quantities of Spanish mackerel using hook-and-line (trolling), gill nets, and a variety of other gears. As catches of other species declined, fishing pressures shifted to catch mackerels. For many years catches by both fishing groups were high. Declining catches alarmed various states and fishermen, causing a proliferation of management plans or bans on using gill nets and/or seasons, etc. Many states may follow with similar action or laws. The problem could be resolved by limiting catches. No sportsman should be free to catch hundreds of fish, nor should unlimited numbers of commercial fishermen be permitted to overfish for one species because of a decline in another resource.

King Mackerel

Similar in appearance to Spanish mackerel, the silvery-sided compressed king mackerel, 1.7 meters, 45 kg, occurs from Maine to Brazil. Spots are absent on the sides except in the young up to 0.5 meter, called snake mackerel. Scales occur on the body, the second dorsal, and the anal fins; the pectoral fin is scaleless. The mouth is large and well-toothed with sharp triangular-shaped teeth. The most important distinguishing feature is the abrupt downward bend of the lateral line, at the level of the second dorsal fin, to resemble a "Z," as it continues to the tail. A keel occurs on the caudal peduncle. The air bladder is absent. Spawning occurs, region-dependent, from July to November. King mackerel migrate long distances in large, dense schools, and provide catches to sport and commercial interests. Similar conservation problems and remedies to preserve stocks and populations are being applied to king as for Spanish mackerel.

Tunas

A number of tunas occur in Atlantic and Pacific warm waters of North America (Figure 12). All are pelagic, ocean-swimming, apex feeders; however, a few can be caught near shore (for example, little tunny, *Euthynnus alleteratus*). Dense schools of these torpedo-shaped, fast-swimming fishes migrate long distances locally and internationally. Most possess a distinguishing corselet (patch of scales) at the front of the sides just behind the head. This corselet serves to increase turbulence and/or to reduce drag on the scaleless body. Smallest tunas are the bullet (*Auxis rochei*) and frigate mackerel (*Auxis thazard*), 3 kg. Bluefin tuna (*Thunnus thynnus*) are the largest, with some fishes often weighing 700 kg. Canadian fishermen (Atlantic Coast) often catch large bluefin tuna in pound nets and then keep them alive, "fattening" them for sale to Japan, where one fish may sell for $25,000. Yellowfin tuna (*Thunnus albacares*) often attain weights >182 kg.

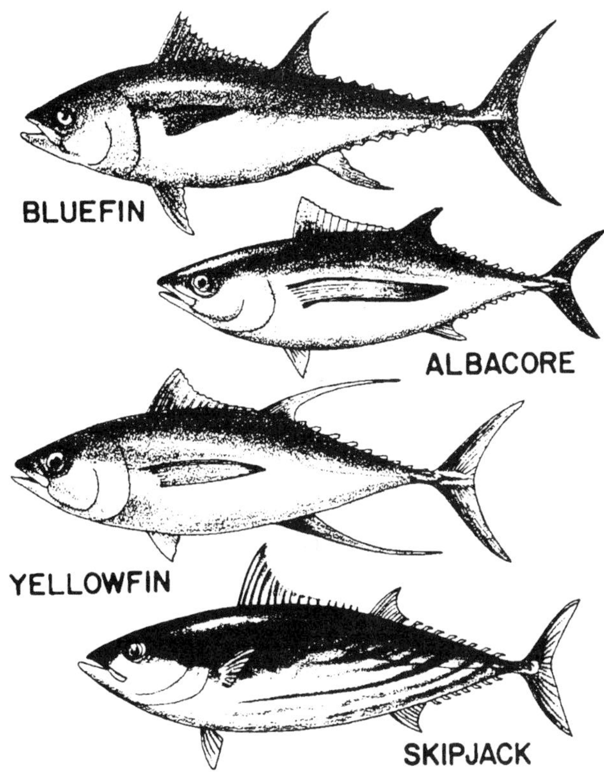

Figure 12. Tunas.

Important features of tunas are their possession of heat exchanger systems that help them reduce or maintain body heat loss, and a reduced or lost swim bladder. They always swim with open mouths to permit a constant flow of oxygenated waters over their gills, ensuring fulfillment of their high body-function needs. Tunas are voracious fish eaters. Additional distinguishing features are grooves into which the dorsal fins can depress during high-speed swimming, and a series of dorsal and anal finlets. Tunas have protracted spawnings during autumn to spring. Females may spawn from 100,000 to 6.5 million one oil-globule eggs/kg body weight. Hatching of the pelagic eggs occurs in 30 hours. Growth is rapid. Capture methods include longlines, hook-and-pole, and purse nets (baglike encircling nets). Until recently commercial catches of tuna accounted for 25 percent of the world's food fish catches. Problems arose when thousands of porpoises (also called dolphins), which associate with tunas, were also caught and died before release from the purse nets. Boycotts and bans on fishing developed. Presently, flaps or slits are installed or required in purse nets, and changes in hauling procedures (backing down to lower the top of the net so porpoises can escape) are in effect. This has greatly reduced the tension between fishermen and conservationists. A perfect solution is yet to be achieved.

SWORDFISH

Swordfish occur in the world's warmer open oceans as well as the Atlantic and Pacific temperate, semitropical, and tropical waters of North America (Figure 13). Distinguishing features are the widely-separated dorsal fins; long, flattened sword (bill one-third of body length); lack of pelvic fins; one caudal peduncle keel; a large, ridged first dorsal fin in adults (movable in young); no bars or spots on a silvery-gray body; no teeth in the jaws; a large eye; and a lack of scales in adults. Swordfish are important sport and commercial food fish that attain 4.5 meters, 59 kg, occur from Nova Scotia to Brazil, and can be caught in surface- to 600 meters-deep waters. Swordfish in northern cooler waters like to bask near the surface — hence, their easy capture by harpoon. Capture in southern waters is deeper, as swordfish seem to be found nearer the thermocline (cold/warm water interface area). Sportsmen troll while commercial fishermen fish with longlines to capture these fast-swimming, squid- and fish-eating fishes. Little is known of larval or other spawning aspects, as young are difficult to distinguish from other scombrids.

Decreasing catches, as with other pelagic species, have caused alarm and implementation of management and conservation plans. Fortunately swordfish rarely frequent inshore coastal waters (except near New England). Seasons, with limits, and shifts to other targeted species may resolve the declines in this highly esteemed food fish.

BILLFISH

Seven billfishes occur in North American temperate, subtropical, and tropical waters (four in the Atlantic and four in the Pacific). Most inhabit open ocean habitats of the world. Billfishes are some of the longest and fastest bony fishes in the world. Black marlin, 4.8 meters, 1300 kilograms (*Makaira indica*), have been caught in the Pacific off Kenya; blue marlin (*Makaira nigricans*) attain sizes of 4.3 meters, 901 kg, off the United States. Sailfish (*Istiophorus platypterus*) are known to attain swimming speeds of 110 kilometers/hour.

Billfishes are large, flat-sided fishes with a rounded, in cross-section bill (one-fourth of its body length), lacking teeth; the color is dark blue dorsally, silvery below; the body may possess bars; two keels occur on sides of the caudal peduncle; dorsal, pectoral and pelvic fins fold into grooves in the body; pelvic fins are long, thin, and located below the pectoral fins. Scales occur on the second dorsal and anal fins. Spawning occurs from May to August. Young blue marlin, up to 0.5 meters, lack a bill, and have a high sail-like first dorsal fin; the bill elongates, with growth of the

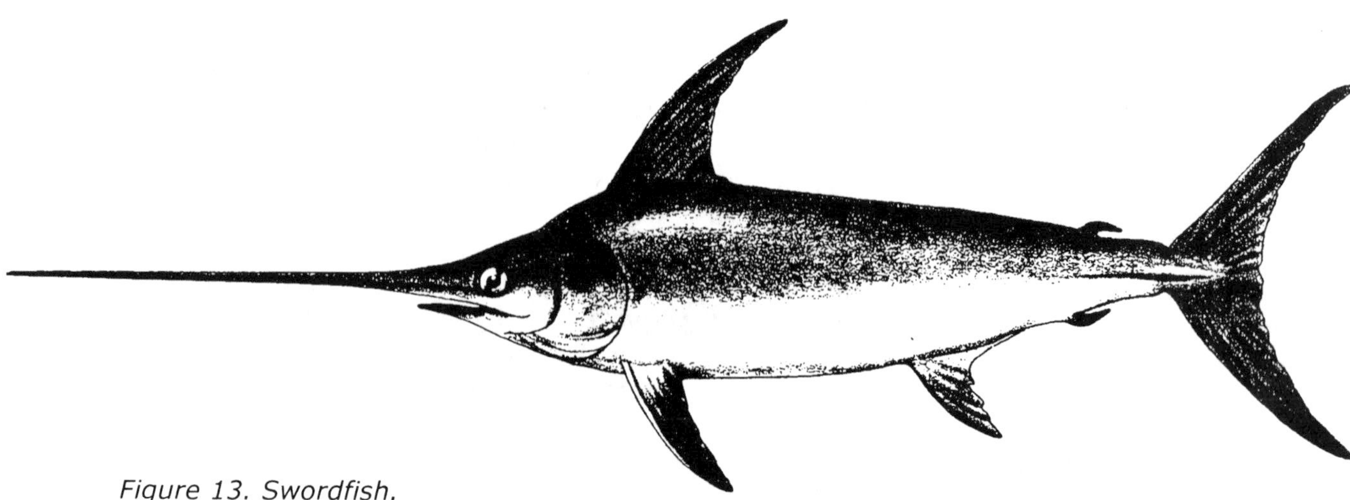

Figure 13. Swordfish.

upper jaw. Young stages are difficult to identify. Unique melanophore (spot colorations) patterns are becoming useful ways in distinguishing among young of each species.

Most billfishes, spectacular jumpers when caught, are considered inedible by United States fishermen, while they are highly prized (smoked or raw) in other parts of the world. No commercial fishery exists for billfishes. Billfish support highly lucrative sport and taxidermy industries. Like other species, concern prevails that too many billfishes, once landed and discarded, now released during tagging programs, still die and add to the woes of the species. Capture by longline is incidental when capturing other species. Management plans have been devised to help restore populations. Answers seem to rest with control and capture limits, whether during sport or tournament fishing.

Suggested Reading

Berrien and Finan, 1977a,b (Spanish mackerel); Carpenter et al., 1995 (classification); Collette and Nauen, 1983 (Scombrids); Collette and Russo, 1984, 1985 (Spanish mackerel); Collette et al., 1984 (relationships, phylogeny); Graves and McDowell, 1994 (striped marlin); Johnson, 1986 (phylogeny); Johnson et al., 1994 (stocks); Joseph et al., 1988 (tuna, billfish); Klawe, 1988, 1980, 1977 (tunas); Manooch et al., 1978 (Scombrid bibliography); Manooch et al., 1987 (king mackerel); Nakamura, 1985 (billfish); Palko et al., 1987, 1981 (Spanish mackerel); Potthoff and Kelley, 1982 (development); Richards, 1974 (billfish); Richard and Klawe, 1972 (billfish); Richardson and McEachran, 1989 (mackerel larvae); Schmidt et al., 1993 (Spanish mackerel); Shomura and Williams, 1974-1975 (identification); Squire, 1987 a,b (billfish tagging, environment); and Yoshida, 1979 (tuna).

FLOUNDERS

Two major families of flounders, lefteye (Bothidae, 33 species) and righteye (Pleuronectidae, 31 species), occur in North American waters (Figure 14). Most are benthic inhabitants of temperate, tropical, or Arctic coastal and ocean waters; a few enter estuaries. Most flounders are too small to be of interest, but some, like the halibuts, are large and support vast sport and commercial fisheries. Sizes range from 5 centimeters (cm) to 3 meters and >320 kg. Flounders prefer to lie in wait, often covered with substrate, for their prey. Newly-born flounders swim upright and have one eye on each side of the flattened body. The eye migrates across the skull, when young are about 25 millimeters long, to occupy the colored side of the fish. Failure of the eye to move across, because of a skull bone obstruction, often causes piebald, anomalous-colored portions to the head, body, and/or tail. Adult fish swim horizontally, are brown dorsally and white-yellow ventrally, and possess the ability to rapidly change body color patterns to match the substrate, in order to enhance their camouflage. Most species may have ctenoid, cycloid, or tuberculate scales, even ctenoid dorsally, cycloid ventrally, in one species. Sexual dimorphism is evident in many species as males develop elongations to some dorsal and pectoral fin rays. All species possess long dorsal or anal fins that either begin over the eye or behind the pelvic fins. The lateral line is usually arched at the level of the pectoral fin, straight in a few species.

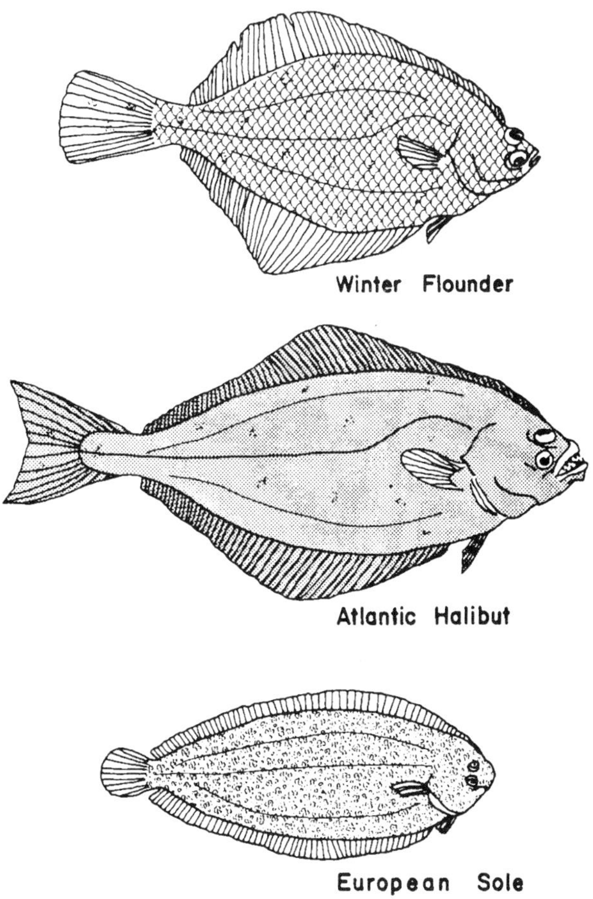

Figure 14. Three species of flounders.

Lefteye flounder eggs possess one oil globule in the egg, while righteye flounder eggs possess two oil globules. Problems of identity exist, for right- and left-sided individuals may be found within the same species.

Lefteye Flounders

California halibut (*Paralichthys californicus*), a 1.5-meter, 33-kg sport fish, occur from Washington to Baja California in waters to 180 meters. Sport catches amount to 300,000 kg; commercial to 900,000 kg.

The summer flounder (*Paralichthys dentatus*) is a 0.9-meter, 12-kg, Atlantic species that occurs from Maine to southern Florida. It possesses five conspicuous, ocellated spots on the color side, three in a triangle pattern. Young are found in estuaries where they grow up. With age they move into the ocean and begin a five-year growth and northward movement pattern towards New England, returning to North Carolina after five years to spawn offshore from November to February. Summer flounder are widely sought by sport, public, and commercial interests. Conflicts exist between the winter fishery and offshore incidental captures of sea turtles. Use of excluder escapement devices for sea turtles and fishes are in effect as ways to resolve the fish vs. sea turtle conflicts. Perhaps abundances of this flounder reflect, as do others, a natural 20-year cycle of abundance rather than overfishing.

The southern flounder (*Paralichthys lethostigma*) is a 0.9-meter, 9-kg spotted but not ocellated flounder that occurs from North Carolina to Texas. Young, like *P. dentatus*, enter sounds to grow, but prefer mud substrates, rather than the sand preferred by *P. dentatus*. Movement with growth is to the ocean, where adults spawn in nearshore sandy areas from October to February. Southern flounders may make long southerly migrations from North Carolina into the Gulf of Mexico. This highly-valued sport and commercial fish is subject to the same pressures and problems confronting *P. dentatus*.

Righteye Flounders

The Pacific halibut (*Hippoglossus stenolepis*), a 2.7-meter, 360-kg fish, occurs in water up to 1050 meters deep from the Bering Sea to the Islands off Santa Rosa, California. Pacific halibut support a valuable sport fishery, especially off Alaska to Washington.

Greenland halibut (*Reinhardius hippoglossoides*) are 0.8-meter, 11-kg fish that support a large sport and commercial fishery from the Bering Sea to northern Baja California in the Pacific, and Arctic to New Jersey in the Atlantic. Most Greenland halibut are caught by trawls to 960-meter depths. Greenland halibut possess a straight lateral line. Its common name has provoked controversy: some prefer it be called simply halibut, especially in the Pacific.

The English sole (*Pleuronectes vetulus*) is a diamond-shaped 0.6-meter flounder that lacks scales on its fins. English sole are highly migratory and support a vast commercial trawl fishery from the Bering Sea to Bahia de San Cristobal, Baja California, in depths to 90 meters.

The winter flounder (*Pleuronectes americanus*) is a small-headed, 0.6-meter flounder that possesses ctenoid scales on the eyed side, smooth cycloid scales on the white side. Winter flounders were once highly prized as a food fish, being captured from Labrador to Georgia. Catches have diminished drastically in recent years. The yellow tail flounder (*Pleuronectes ferrugineus*), a once-abundant, easily recognized, 0.6-meter species possessing many spots and yellow edges to the dorsal, caudal, anal fin, and caudal/peduncle, was the mainstay of the commercial trawl fishery off New England. Highly prized as a food fish, this southern Labrador to Chesapeake Bay species prefers mud or sand substrates.

Suggested Reading

Able and Kaiser, 1994 (summer flounder); Ahlstrom et al., 1984 (development); Tech. Rep. International Pacific Halibut Comm., 1987 (Pacific halibut); Best and St. Pierre, 1986 (food); Chapleau, 1993 (relationships); Chopin and Aramoto, 1995 (gear); Gilbert, 1986 (summer flounder); Gutherz, 1967 (field guide); Haugen, 1990 (California halibut); Hensley and Ahlstrom, 1984 (relationships); Klein-MacPherson, 1978 (winter flounder); Liu et al., 1993 (Pacific halibut); Nelson et al., 1991 (southeastern estuaries); Nickerson, 1978 (Atlantic halibut); Nielson et al., 1993 (Atlantic halibut); Parker, 1989 (Pacific halibut); Powell and Schwartz, 1972 (development in estuary); Sakamoto, 1984 (interrelationships); Safrit and Schwartz, 1998; Shephard and Terceiro, 1994 (summer flounder); Smith et al., 1975 (distribution); Stone et al., 1994 (profiles); Van Housen, 1989 (stocks); and White and Stickney, 1973 (bibliography).

CONCLUSIONS

The biology, ecology, abundances, and social problems that beset marine fishes are just as varied as the marine fishes that inhabit the world's oceans. Declines in their abundances have been all too common as a result of pollution, drastic environmental influences (El Niño, hurricanes, cold snaps, changing air and water temperatures, etc.), habitat destruction (oil, debris, dams, etc.), and effects of man (overfishing, faulty management plans and manipulation, feeble conservation efforts, etc.). Man once assumed that the oceans and marine waters harbored vast resources that could be used indefinitely for food, by all public, sport, and commercial interests. We now are rapidly learning that that infinite situation does not exist, and that it is rapidly being decimated and destroyed. Concerted action by all will be necessary if many species are to survive, be enhanced, be helped by the environment, or be restored to once-proud levels for to all to capture and enjoy.

ACKNOWLEDGMENTS

Thanks are extended to Glen Safrit (UNC Institute of Marine Sciences); to Pattie Marraro (librarian, NMFS Southeast Lab, Beaufort, NC) for assistance in electronically retrieving some literature references; and to Laura White (UNC-IMS) for undertaking the laborious task of transcribing my unique hieroglyphics into a final readable manuscript.

REFERENCES

Able, K.W. and S.C. Kaiser. 1994. Synthesis of summer flounder habitat parameters. NOAA Coastal Ocean Prog. *Decision Analyses*. Ser. 1.

Ahlstrom, E.H., K. Amoaka, D.A. Hensley, H.G. Moser, and B.Y. Sumida. 1984. Pleuronectiforms: development. In H.G. Moser et al. (eds.), pp. 640-669. *Ontogeny and Systematics of Fishes*. Spec. Publ. 1, Am. Soc. Ichs. Herps.

Ali, M.A. 1975. Vision in Fishes. *New applications in vision*. Academia Press. New York.

Anderson, E.D. 1990. Estimates of large shark catches in the western Atlantic and Gulf of Mexico, 1960-1986. In H.L. Pratt Jr., S.H. Gruber and T. Tanuichi (eds.), pp. 443-454. Elasmobranchs as living resources: advances in their biology, ecology, systematics, and the status of the fishery. *NOAA Tech. Rep. NMFS* 90.

Anderson, E.D., F.E. Lux, and F.P. Almeida. 1980. The silver hake stocks and fishing off the northeastern United States. *Mar. Fish. Rev.* **42**(1):12-20.

Arnold, C.K., G.J. Holt, P. Thomas, and D.E. Wohlschlag (eds.). 1988. Red drum aquaculture. Proc. Symp. on Culture of Red Drum and other warm water fishes. *Contr. Mar. Sci. Suppl. to Vol. 30*. Mar. Sci. Inst. Univ. Tex. Pt. Arkansas, TX.

Baldwin, C.C. and G.D. Johnson. 1993. Phylogeny of the Epinepheline Teleostei Serranidae. *Bull. Mar. Sci.* **52**(1): 240-283.

Balon, E.K. (ed.). 1980. Charrs, Salmonid fishes of the genus *Salvelinus*. W. Junk, The Hague, Netherlands.

TABLE OF TERMS

Adipose fin: a fleshy fin posterior to dorsal fin
Anadromous: migrates from marine waters upstream to spawn in freshwater
Axil: point where pectoral fin joins body
Axillary process: an elongate scale inserted between pelvic fin base and body
Barbel: fleshy projection found near mouth or chin
Ctenoid scale: scale with spines along rear exposed edge of scale
Cycloid scale: scale with exposed rear edge smooth
Keel: ridge located on caudal peduncle in front of or onto tail
Lateral line: a line along side of body that may pass through scales as tubes or pores, located from behind opercle to base or onto caudal fin
Molar: broad flat teeth located in rear of jaws for crushing food
Opercle: flap covering gill chamber
Oviparity: a primitive condition where large-yolked eggs are laid in leathery cases
Ovoviviporous: young are born alive from eggs hatched internally, no maternal nourishment involved
Pectoral fins: paired fins attached to shoulder girdle located posterior to gill cover
Pelvic fins: paired fins on lower part of body located behind or before pectoral fins
Preopercle: part of opercle (gill flap) located below and behind the eye, often spiny

Beacham, T.D. 1983. Variability in size or age at sexual maturity of white hake, pollock, longfin hake, and silver hake in the Canadian maritime area of Northwest Atlantic Ocean. *Can. Tech. Rep. Fish. Aq. Sci.* No. 1157.

Beamesderfer, R.C. and T.A. Rien. 1993. Dynamics and potential production of white sturgeon populations in three Columbia River reservoirs. In R.C. Beamesderfer and A.A. Nigro (eds.), pp. 175-204. Status and habitat requirements of the white sturgeon populations in the Columbia River downstream from McNary Dam. Vol. 1. Bonneville Power Adm. Portland, OR.

Behnke, R.J. 1972. The systematics of salmonid fishes of recently glaciated lakes. *J. Fish. Res. Bd. Can* . **29**:639-671.

Behnke, R.J. 1992. Native trouts of western North America. *Am. Fish. Soc. Monogr.* 6. Bethesda, MD.

Berrien, P. and D. Finan. 1977a. Biological and fishery data on Spanish mackerel, *Scomberomorus maculatus* (Mitchill). *NMFS Tech. Ser. Rep.* 9. Highlands, NJ.

Berrien, P. and D. Finan. 1977b. Biological and fishery data on king mackerel, *Scomberomorus cavalla* (Cuvier). *NMFS Tech. Ser. Rep.* **8**:40. Highlands, NJ.

Best, E.A. and G. St. Pierre. 1986. Pacific halibut as predator and prey. *Tech. Rep. Intern. Pacific Halibut Comm.* No. 21. Seattle, WA.

Bigelow, H.B. and W.C. Schroeder. 1948. Fishes of Western North Atlantic: Sharks. Pp. 59-576 In *Fishes of the Western North Atlantic.* Pt. 1. Mem. Sears Found. Mar. Res. New Haven, CT.

Bigelow, H.B. and W.C. Schroeder. 1953a. Fishes of the Western North Atlantic: Sawfishes, guitarfishes, skates, and rays. Pp. 1-514. In *Fishes of the Western North Atlantic*: Pt. 2. Mem. Sears Found. Mar. Res. New Haven, CT.

Bigelow, H.B. and W.C. Schroeder. 1953b. Fishes of the Gulf of Maine. Fish. Bull. 53.

Birnstein, V.J. 1993. Sturgeon and paddlefish: threatened fishes in need of conservation. *Cons. Biol.* **7**:773-787.

Blaxter, J.H.S. 1993. Summary of symposium on sea ranching of cod and other marine fish species, Arendal, Norway 15-18, 1993. In D.S. Danielsen, B.P. Howell, and Moksness (eds.), pp. 259-264. Internat. Symp. sea ranching of cod and other marine fish species. *Suppl. to Vol. 25. Aquaculture.*

Bone, O., N.B. Marshall, and J.S. Blaxter. 1994. *Biology of fishes* (2nd ed.). Chapman and Hall, New York.

Brander, K. 1993. Comparisons of spawning characteristics of cod, *Gadus morhua*, stocks in the North Atlantic. *Northwest Atl. Fish. Oceanogr. Sci. Council Study* 18, pp. 13-20.

Branstetter, S. 1993. Conservation biology of elasmobranchs. *NOAA Tech. Rep.* NMFS 115.

Briggs, J.C. 1974. *Marine zoogeography.* McGraw-Hill, New York.

Brodeur, R. D., P. A. Livingston, T. R. Loughlin, and A. B. Hollowed. 1996. Ecology of juvenile walleye pollock, *Theragra chaleogramma.* NOAA Tech. Rep. NMFS 126, 227.

Brule, T., D.O. Avila, M.S. Crespo, and C. Deniel. 1994. Seasonal and diet changes in distinguishing of juvenile red grouper, *Epinephelus morio*, from Campeche Bank. *Bull. Mar. Sci.* **55**(1):255-262.

Buck, R. 1993. Silver swimmers: the struggle for survival of the wild Atlantic salmon. *Lyons & Burford Publ.*, New York.

Bulger, A.J., B.P. Hayden, M. McCormick, M.E. Monaco, and D.M. Nelson. 1990. A proposed estuaries classification analysis of specific salinity ranges. ELMR Rep. 5. *Strategy assessment Br. NOS/NOAA.* Rockhill, MI.

Bullock, L.H. and G.B. Smith. 1991. Sea basses (Pisces: Serranidae). Mem. Hourglass Cruises. Mar. Res. Lab. Fl. Dept., St. Petersburg, FL. *Nat. Res.* **90**:243-249.

Burgner, R.L. 1991. Life history of the sockeye salmon, *Oncorhynchus nerka.* In Groot C. and L. Margolis (eds.), pp. 1-117. *Pacific Salmon Life Histories.* Univ. Bt. Col. Press. Vancouver, Canada.

Cailliet, G.M., H.F. Mollet, G.G. Petranger, P. Bedford, and L.J. Natansa. 1992. Growth and demography of the Pacific angel shark, *Squatina californica*, based upon tag returns off California. *Austr. J. Mar. Freshwater Res.* **83**:1313-1330.

Carpenter, K.E., B.B. Collette, and J.L. Russo. 1995. Unstable and stable classification of Scombroid fishes. *Bull. Mar. Sci.* **56**(2):379-405.

Casey, J.G. and J.J. Hoey. 1985. Estimated catches of large sharks by United States recreational fishermen in the Atlantic and Gulf of Mexico. NOAA Tech. Rep. NMFS 31, pp. 15-19.

Castro, J.I. 1983. *The Sharks of North American Waters.* Texas A&M Univ. Press, College Station, TX.

Chadwick, H.K. 1964. Annual abundance of young striped bass, *Roccus saxatilis*, in the Sacramento San Joaquin Delta, California. *Calif. Fish Game.* **50**(2):69-99.

Chamberlain, G.W., R. J. Miget, and M.G. Haby. 1990. Red drum aquaculture. Texas A&M Univ. Sea Grant Coll. Prog.

Chao, L.N. 1978. A basis for classifying western Atlantic Sciaenidae (*Teleostei perciformes*). NOAA Tech. Rep. NMFS Circ. 415.

Chao, L.H. 1986. A synopsis on zoogeography of the Sciaenidae. Pp. 570-589. In T. Uyeno, R. Arai, T. Taniuchi, and K. Matsuura (eds.). Indo-Pacific Fish Biol. Tokyo.

Chapleau, F. 1993. Pleuronectiform relationships: a cladistic reassessment. Bull. Mar. Sci. **52**(1):516-540.

Clareaux, G., P.M. Webber, S.R. Kerr, and R.G. Boutelier. 1995. Physiology and behavior of free-swimming Atlantic Cod (*Gadus morhua*) facing fluctuating temperature conditions. *J. Exp. Zool.* **198**(1):61-70.

Clay, D., W. T. Stobo, B. Beck, and P.L.F. Hurley. 1989. Growth of juvenile pollock, *Pollachius virens* L., along the Atlantic coast of Canada, with influences of inshore-offshore movements. *J. NW Atlantic Fish. Sci.* **9**(1):37-43.

Clugston, J.P. and A.M. Foster. 1993. Gulf sturgeon, *Acipenser oxyrinchus desotoi*, in the Suwannee River, Florida, USA. Proc. 2nd. Intern. Sympo. Sturgeon. Moscow, Russia 6-11 Sept. 1993.

Cohen, D.M. 1970. How many recent fishes are there? *Proc. Calif. Acad. Sci. Ser.* 4. **38**:341-345.

Cohen, D.M. 1980. Names of fishes. Mar. Fish. Rev. **42**(1):2-3.

Cohen, D.M. 1989a. Gadiformes: overview. Pp. 259-265 In Ontogeny and Systematics of Fishes. H.G. Moser et al. (eds.). Spec. Pub. 1 Am. Soc. Ichs. Herps.

Cohen, D.M. 1989b. Paper on the systematics of gadiform fishes. Nat. Hist. Mus. Los Angeles Co. Sci. Ser. 32.

Cohen, D.M., T. Inada, T. Iwamoto, and N. Scialabba. 1990. FAO Species Catalog. Vol. 10. Gadiform fishes of the world, order Gadiformes. FAO Fish Synop. 125 (Vol. 10).

Collette, B.B. and C.E. Nauen. 1983. FAO Species Catalogue. Vol. 2. Scombrids of the World. Annotated and illustrated catalogue of tunas, mackerels, bonitos, and related species known to date. FAO Fish Synop. 125 (Vol. 2).

Collette, B.B. and J.L. Russo. 1984. Morphology, systematics, and biology of the Spanish mackerel *Scomberomorus, Scombridae*. Fish. Bull. **82**(4):545-692.

Collette, B.B., T. Potthoff, W.J. Richards, S. Ueyanagi, J.L. Russo, and Y. Nishikawa. 1984. Scombroidei: Development and relationships. Pp. 591-620. In H.G. Moser (eds.), Ontogeny and Systematics of Fishes. Spec. Pub. 1, Am. Soc. Ichs. Herps.

Compagno, L.J.V. 1973. Interrelationships of living elasmobranchs. Pp. 15-61. In Interrelationships of fishes. P.H. Greenwood, R.S. Miles, and C. Patterson (eds.). *J. Linn. Soc.* (Zool.) **53** Suppl. 1, Academic Press, New York.

Compagno, L.J.V. 1984. Sharks of the world. FAO Species Catalog. Fish Synop. 125 (Vol. 4, 2 pts.). Rome.

Compagno, L.J.V. 1988. Sharks of the order Carcharhiniformes. Princeton Univ. Press, Princeton, NJ.

Compagno, L.J.V. 1990. Shark exploitation and conservation. Pp. 391-414. In Elasmobranchs as Living Resources: Advances in the Biology, Ecology, Systematics, and the Status of the Fisheries. H.L. Pratt Jr., S.H. Gruber, and T. Taniuchi (eds.). NOAA Tech. Rep. NMFS 90.

Chopin, F.S. and T. Arimoto. 1995. The condition of fish escaping from fishing gear — a review. Fish. Res. **21**:315-327.

Connaryhton, M.A. and M.H. Taylor. 1995. Seasonal and daily cycles in sound production associated with spawning in the weakfish by *Cynoscion regalis*. Env. Biol. Fish. **42**(3):233-240.

Daniel, L.B. and J.E. Graves. 1994. Morphometric and genetic identification of eggs of spring-spawning Serranids in lower Chesapeake Bay. Fish. Bull. **92**(2):254-261.

Dark, T.A. and M.E. Wilkins. 1994. Distribution, abundance, and biological characteristics of groundfish off the west coast of Oregon and California. 1977-1986. NOAA Tech. Rep. NMFS 117.

Dark, T.A., M.O. Nelson, J.J. Traynor, and E.P. Nunnallee. 1980. The distribution, abundance, and biological characteristics of Pacific whiting, *Merluccius productus*, in the California-British Columbia region during July-September 1977. Mar. Fish. Rev. **44**(3-4):17-33.

Davy, K.B. 1994. South Carolina marine game fish tagging program. 1974-1992. S.C. Dept. Nat. Res. Tech. Rep. 83.

Dees, L. 1961. Sturgeons. Fish. Leaflet 526. United States Fish Wildl. Serv. Wash., DC.

Denton, E.J. 1990. Light and vision at depths greater than 200 meters. Pp. 127-145. P.J. Herring, A.K. Campbell, A.K. Whitfield, and L. Maddock (eds.). Light and Life in the Sea. Cambridge Univ. Press, New York.

Dickson, K.E. 1995. Unique adaptation of the metabolic biochemistry of tunas and billfishes for life in the pelagic environment. Env. Biol. Fish. **42**(1):65-97.

Dickson, R.R. and K.M. Brander. 1993. Effects of changing wind field on cod stocks of the North Atlantic Fish Oceanogr. **2**(3-4):124-153.

Dingerkus, G. 1993. Phylogenetic relationships and classification of the living Chondrichthyian fishes, class Chondrichthyes. Mem. Mus. Nat. Hist. Nat. Paris. In press.

Dizon, A. E., R. M. Horrall, and A.D. Hasler. 1973. Long-term olfactory "memory" in coho salmon. *Onchorynchus kisutch*. Fish. Bull. **71**(1):315-317.

Dorofeyeva, E.A. 1989. Basic principles of classification and phylogeny of the salmonid fishes (Salmoniformes: Salmonoidei: Salmonidae). Pp. 5-16. In V.M. Korovinoi (ed.), Biology and Phylogeny of Fishes. *Proc. Zool. Int. USSR Akad. Nauk. SSSR*. St. Petersburg, Russia.

Dovel, W.L. and T.J. Berggren. 1983. Atlantic sturgeons of the Hudson River estuary, New York. *New York Fish Game J*. **30**(2):140-172.

Earl, P.M. 1980. Silver hake: a prospectus. Mar. Fish Rev. **42**(1):26-31.

Ellis, R. and J.L. McCosker. 1991. *Great White Shark*. Harper- Collins Publ., New York, N.Y.

Eschmeyer, W.N., E.S. Herald, and H. Hammann. 1983. *A Field Guide to Pacific Coast Fishes of North America*. Houghton Mifflin, Boston, MA.

Ewing, R.D. and S.K. Ewing. 1995. Review of the effects of rearing density on survival to adulthood for Pacific salmon. *Prog. Fish. Cult.* **57**(1):1-25.

Fahrig, L. and B. Atkinson. 1991. Uncertainty in a mixed stock fishery: a redfish similation study. Management under circumstances related to biology and assessment with case study on some north Atlantic fisheries. No. 16. Pp. 25-37.

Fiedler, B.C. 1986. Effects of California El Niño, 1982-1984 on the northern anchovy. *Mar. Fish. Rev.* **44**(2):317-338.

Finch, C. 1990. *Longevity, Senescence, and Genome*. Univ. Chicago Press. Chicago, IL.

Fisher, J.P. and W. G. Pearcy. 1988. Growth of juvenile Coho salmon *Oncorhynchus kisutch*, in the ocean off Oregon and Washington, U.S.A., in years of differing coastal upwelling. *Can. J. Fish. Sci.* **105**:1036-1044.

Fisher, M.R. and R.B. Ditton. 1933. A social and economic characterization of the United States Gulf of Mexico recreational shark fishing. *Mar. Fish. Rev.* **55**(1):21-27.

Foerster, R.E. 1968. The sockeye salmon, *Oncorhynchus nerka*. *Fish. Res. Bd. Can. Bull.* **162**.

Foster, R. and C. B. Schom. 1989. Imprinting and homing of Atlantic salmon *Salmo salar* adults. *Can. J. Fish. Aq. Sci.* **46**(4):714-719.

Gabull, W.L. 1992. Persistence of demersal fish assemblages between Cape Hatteras and Nova Scotia, Northwest Atlantic. *J. Northwest Atl. Fish. Sci.* **14**:28-46.

Gilbert, C.R. 1986. Southern Gulf and summer flounders (south Florida). Spec. Profile Biol. Rep. **82**(11-54). TR EL 82-4 USFWS.

Gilbert, C.R. 1992. Atlantic sturgeon. Pp. 31-39. R.A. Ashton (ed.). Rare and endangered biota of Florida. Vol. 2. Univ. of Florida, Gainesville, FL.

Grant, W.S. and G. Stahl. 1988. Evolution in Atlantic and Pacific cod; loss of genetic variation and gene expression in Pacific cod. *Evolution.* **42**(1):138-140.

Graves, J.E. and J.R. McDowell. 1994. Genetic analysis of striped marlin. *Tetrapterus audax* populations in the Pacific Ocean. *Can. J. Fish. Aq. Sci.* **51**(8):1762-1768.

Graves, J.E., R.J. McDowell, A.M. Beardsley, and D.R. Scoles. 1992. Stock structure of the bluefish *Pomatomas saltatrix_* along the mid-Atlantic Coast. *Fish. Bull.* **90**:703-710.

Greenwood, P.H. 1992. Are the major fish faunas well-known? *Neth. J. Zool.* **42**(43):131-138.

Griffith, D., J. Johnson, and J. Murray. 1984. Skates and rays, the scallop impersonators. Univ. Calif. Sea Grant Progr. Sea Grant Publ.

Groot, C. and L. Margolis (eds.). 1991. Pacific salmon life histories. Univ. Bt. Col. Press, Vancouver, Canada.

Gunderson, D.R. and W.H. Lenarz. 1980. Cooperative survey of rockfish and whiting resources off California, Washington, and Oregon. 1977. Introduction. *Mar. Fish. Rev.* **42**(3-4):1.

Gunderson, D.R. and T.M. Sample. 1980. Distribution and abundance of rockfish off Washington, Oregon, and California during 1977. *Mar. Fish. Rev.* **42**(3-4):2-16.

Gunter, G. 1942. A list of fishes of the mainland recorded from both freshwater and sea water. *Am. Midl. Nat.* **28**(2):305-326.

Gunter, G., B.S. Ballard, A. Venkataramiah. 1974. A review of salinity problems of organisms in central states coastal areas subject to the effects of engineering works. *Gulf. Res. Rep.* **4**(3):374-380.

Gutherz, E.J. 1967. Field guide to the flatfishes of the family Bothidae in the western north Atlantic. USFWS Circ. 263.

Haedrick, R.L. and N.R. Merritt. 1988. Summary atlas of deep-diving demersal fishes in the North Atlantic basin. *J. Nat. Hist.* **22**(8):1325-1362.

Hanel, L. 1987. The smallest fish in the world. *Akvarium Terrarium.* **30**(4):5.

Hanson, L.P., N. Johnson, and B. Jonsson. 1993. Oceanic migrations in homing Atlantic salmon. *Animal Behav.* **45**(5):927-941.

Hardy, J.D., Jr. 1978a. Development of fishes in the mid-Atlantic bight. Vol. 2, Anguillidae through Syngathidae. USFWS Biol. Surv. Progr. FWS/OBSW/ 78-12.

Hardy, J.D., Jr. 1978b. Development of fishes of the mid-Atlantic bight. Vol. 3. Aphredoderidae through Rachycentridae. USFWS Biol. Surv. Prog. FWS/OBS/ 78-12.

Hare, J.A. and R.K. Cowen. 1993. Ecology and evolutionary implication of the larvae transfer and reproductive strategy of bluefish, *Pomatomus saltatrix. Mar. Ecol. Prog. Ser.* **98**:1-6.

Hasler, A.D. 1966. Underwater guideposts — homing in salmon. Univ. Wisc. Press, Madison, WI.

Hasler, A.D. and A.T. Scholz. 1983. Olfactory imprinting and homing in salmon: Investigations into the mechanics of the imprinting process. Springer-Verlag, New York.

Haugen, C.W. (ed.). 1990. California halibut. *Paralichthys californicus.* Calif. Fish Game Fish. Bull. 174.

Heemstra, P.C. and J.E. Randall. 1993. Groupers of the world (family Serranidae, subfamily Epinephelinae) an annotated and illustrated catalogue of the grouper, rockcod, hind, coral grouper, a lyretail species known to date. *FAO Synop.* **125**(16):1-382.

Heggeberget, T.G., F. Okland, and O. Ugedel. 1993. Distribution and migratory behavior of adult wild and farmed Atlantic salmon, *Salmo salar*, during return migration. *Aquaculture.* **118**(1-2):73-83.

Hensley, D.A. and E.H. Ahlstrom. 1984. Pleuronectiformes: relationships. Pp. 670-687. In Ontogeny and systematics of fishes. H.G. Moser et al. (eds). Spec. Publ. 1, Am. Soc. Ichs. Herps.

Herrfurth, G. 1987. Walleye pollock and its utilization and trade. *Mar. Fish. Rev.* **49**(11):61-68.

Hoar, W.S. and W.J. Randall (eds.), 1967-1987. *Fish physiology.* Vols. 1-9. Academic Press, New York.

Holts, D. 1988. Review of the West Coast commercial shark fisheries. *Mar. Fish. Rev.* **80**(1):1-8.

Hood, P.B., M.E. Godcharles, and R.S. Barco. 1994. Age, growth, reproduction and the feeding ecology of black sea bass, *Centropristis striata* (Pisces: Serranidae), in the Gulf of Mexico. *Bull. Mar. Sci.* **54**(1):27-37.

Hood, P.B. and R.A. Schlieder. 1992. Age, growth, and reproduction of gag, *Mycteroperea microlepis* (Pisces: Serranidae) in the eastern Gulf of Mexico. *Bull. Mar. Sci.* **51**(3):337-352.

Howard. W. 1997. One fish, two fish, redfish, new fish? *Palm Beach Post* (Florida). April 3, 1997.

Hughes, S.E. and G. Hirschhorn. 1979. Biology of walleye pollock, *Theragra chalcogramma*, in the western Gulf of Alaska, 1973-1975. *Fish Bull.* **77**(1):263-274.

Hutchings, J.M. and R.A. Myers. 1994. What can be learned from the collapse of a renewable resource? Atlantic cod, *Gadus morhua*, of Newfoundland and Labrador. *Can. J. Fish. Aq. Sci.* **51**(9):2126-2146.

International conference on sturgeons, biodiversity, and conservation. Am. Mus. Nat. Hist. 28-30. July 1994. (Birstein et al., 1977).

Johnson, A.G., W.A. Fable Jr., C.B. Grimes, C. Trent, and J.V. Perez. 1994. Evidence for district stocks of king mackerel, *Scomberomorus cavalla*, in the Gulf of Mexico. *Fish. Bull.* **92**(1):91-101.

Johnson, G.D. 1986. Development of fishes in the mid-Atlantic bight. Carangidae through Epippidae. USFWS Biol. Serv. Progr. FWS/OBS/78-12, Vol. 4.

Johnson, G.D. 1986. Scombroid phylogeny: an alternative hypothesis. *Bull. Mar. Sci.* **39**(1):1-41.

Jones, P.W., F.D. Martin, and J.D. Hardy Jr. 1978. Development of fishes of the mid-Atlantic bight. An atlas of eggs, larval, and juvenile stages. Acipenseridae through Ictaluridae USFWS Biol. Serv. Progr. FWS/OBS/78-12, Vol 1.

Joseph, J., W. Klawe, and A. Murphy. 1988. Fish without a country. Intern. Am. Trop. Tuna Comm., La Jolla, CA.

Joung, S. J., C. T. Chen, E. Clark, S. Uchida, W. Y. P. Huang. 1996. The whale shark, *Rhincodon typus*, is a live bearer: 300 embryos found in one "megamama" supreme. *Env. Biol. Fish.* **46**(3):219-223

Jury, S.H., J.D. Field, S.L. Stone, D.M. Nelson, and M.E. Monaco. 1994. Distribution and abundance of fishes and invertebrates in North Atlantic estuaries. ELMR. Rep. 13, NOAA/NOS Strategic Env. Assess. Div., Silver Springs, MD.

Kelley, K. 1994. Colder oceans put a chill on northern fishes. *Nat. Fisherman.* **74**(1):22-23.

Kendall, A.W. Jr. 1977. Biological and fisheries data on black sea bass, *Centropristis striata* (Linnaeus). NOAA Tech. Ser. Rep. 7. Highlands, NJ.

Kendall, A.W. Jr. 1984. Serranidae: development and relationships. Pp. 499-510. H.G. Moser (ed.). Ontogeny and Systematics of Fishes. Spec. Publ. 1. Am. Soc. Ichs. Herps.

Kendall, A.W. Jr. and A.J. Behnke. 1984. Salmonidae: development and relationships. Pp. 142-149 H.E. Moser et al. (eds.). Ontogeny and Systematics of Fishes. Spec. Publ. 1 Am. Soc. Ichs. Herps.

Kendall, A.W. Jr. and A.C. Matarese. 1987. Biology of eggs, larvae, and epipelagic juveniles of sablefish, *Anoplopoma fimbria*, in relation to their potential use in management. *Mar. Fish. Rev.* **49**(1)1-13.

Kendall, A.W. Jr. and L.A. Walford. 1979. Sources and distributions of bluefish, *Pomatomus saltatrix*, and juveniles off the East Coast of the United States. *Fish. Bull.* **77**(1): 213-228.

Kerby, J.H. 1986. Striped bass and striped bass hybrids. Pp. 127-147. R.R. Stickney (ed.), Culture of nonsalmonid freshwater fishes. CRC Press, Boca Raton, FL.

Kerby, J.H. 1993. The striped bass and its hybrids. Pp. 251-306. R.R. Stickney (ed.). Culture of nonsalmonid freshwater fishes. CRC Press, Boca Raton, FL.

Kerby, J.H., C.C. Woods III, and M.T. Huish. 1983. Culture of the striped bass and its hybrids: a review of methods, advances, and problems. Pp. 23-54. In R.R. Stickney and S.P. Meyers (eds.). *Proc. Warm Water Fish Culture Workshop.* March 1982.

Kinne, O. 1964. The effects of temperature and salinity on marine and brackish water animals. Vol. II. Salinity and temperature salinity combinations. *Oceanogr. Mar. Biol. Ann. Rev.* **2**:281-399.

Kinne, O. 1967. Physiology of estuarine organisms with special references on salinity and temperature. General aspects. *Am. Assoc. Adv. Sci. Publ.* **83**:525-540.

Klawe, W.L. 1977. What is a tuna? *Mar. Fish. Rev.* **39**(1):1-5.

Klawe, W.L. 1980. Classification of the tunas, marlins, billfishes and their geographical distinctions. Intern. Am. Trop. Tuna Comm. Spec. Rep. **2**:5-16.

Klawe, D.L. 1988. Tunas and billfishes — fish without a country. Inter. Am. Trop. Tuna Comm. La Jolla, CA.

Klein-Mac Phee G. 1978. Synopsis of biological data for the winter flounder, *Pseudopleuronectes americana* (Walbaum). NOAA Tech. Rep. NMFS Circ. 414.

Klimley, A.P. 1981. Grouping behavior in the scalloped hammerhead. *Oceanus.* **24**(4):65-71.

Klimley, A.P. 1987. The determination of sexual segregation in the scalloped hammerhead, *Sphyrna lewini. Env. Biol. Fish.* **18**(1):29-40.

Klimley, A.P. 1993. Highly directional swimming by the scalloped hammerhead shark, *Sphyrna lewini*, and subsurface irradiance, temperature, bathymetry, and geomagnetic field. *Mar. Biol.* **117**:1-22.

Klimley, A.P. 1994. The predatory behavior of the white shark. *Am. Sci.* **82**:122-133.

Klimley, A. P. and D. G. Ainley. 1996. *Great White Sharks: The Biology of Carchardon carcharius.* Academic Press, New York.

Koslow, J.A., K.R. Thompson, and W. Silvert. 1987. Recruitment to northwest Atlantic cod *Gadus morhua* and haddock *Melanogrammus aeglefinus* stocks: influence of stock size and climate. *Can. J. Fish. Aq. Sci.* **44**(1):26-39.

Langton, R.W., and R.E. Bowman. 1980. Food of fifteen northwest Atlantic gadiformes fishes. NOAA Tech. Rep. NMFS. SSSRF 740.

Linton, T.L., J.H. Clark, and J.M. Boslet. 1990. Appdx. 6. Annotated bibliography of the red drum, *Sciaenops ocellata*. Pp. 214-225. G.W. Chamberlain, R.C. Miget, and H.G. Haby (eds.). Red Drum Aquaculture. Texas A&M Univ. Univ. Sea Grant College Progr.

Liu, H.W., R.R. Stickney, and W.W. Dickhoff. 1993. Neutral buoyancy salinity of Pacific halibut, *Hippoglossus stenolepus*, eggs and larvae. *J. World Aquac. Soc.* **24**(14):486-492.

Lund, W.A. Jr. 1961. A racial investigation of the bluefish *Pomatomus saltatrix* (Linnaeus) of the Atlantic coast of North America. Boletin del Inst. Oceanogr. Univ. Oriente, Cumana, Venezuela. **1**(1):73-129.

Lund, W.A. Jr. and G.C. Maltezos. 1970. Movements and migration of the bluefish, *Pomatomus saltatrix*, tagged in waters of New York and southern New England. *Tr. Am. Fish. Soc.* **99**(4):719-725.

Mackie, I.M. and A.H. Ritchie. 1981. Differentiation of Atlantic cod, *Gadus morhua morhua*, and Pacific cod, *Gadus macrocephalus*, by electrophoresis and isoelectric focusing of water-soluble proteins of muscle tissue. *Comp. Biochem. Physiol.* **68B**(2):173-175.

Manooch, C.S., III. 1987. Age and growth of snappers and groupers. Pp. 329-373. J.J. Polivina and S. Ralston (eds.). Biology and management of snappers and groupers. Westview Press, Boulder, CO.

Manooch, C.S. III, and M. Haiimovici. 1978. Age and growth of gag, *Mycteroperca microlepsis*, and size-age composition of the recreational catch off the southeast United States. *Tr. Am. Fish. Soc.* **107**(2): 234-240.

Manooch, C.S., E.L. Nakamura, and A.B. Hall. 1978. Annotated bibliography of four Atlantic scombrids, *Scomberomorus brasiliensis, S. cavalla, S. maculatus*, and *S. regalis.* NOAA Tech. Rep. NMFS Circ. 418.

Manooch, C.S., III, S.P. Naughton, C.B. Grimes, and L. Trent. 1987. Age and growth of king mackerel. *Mar. Fish Rev.* **99**(2):102-108.

Marshall, N.B. 1954. *Aspects of Deep-Sea Biology*. London.

Marshall, N.B. 1972. Swim bladder organization and depth responses of deep-sea teleosts. *Symp. Soc. Exp. Biol.* **26**:261-272.

Marshall, N.B. 1984. Progenetic tendencies in deep-sea fishes. Pp. 91-111. In G.W. Potts and R.J. Wootten (eds.). *Fish Reproduction: Strategies and Tactics*. Academic Press. Orlando, FL.

Martin, F.D. and G.E. Drewry. 1978. Development of fishes of the mid-Atlantic bight. Vol. 6. USFWS Serv. Progr. FWS/OBS/78-2.

Martin, L. and G.D. Zorzi. 1993. Status and review of California skate fishery. Pp. 39-52. S. Branstetter (ed.), conservation biology of elasmobranchs. NOAA Tech. Rep. 115.

Mason, W.T. Jr. and J.P. Clugston. 1993. Foods of the Gulf sturgeon in the Suwannee River, Florida. Tr. *Am. Fish. Soc.* **122**:378-385.

Mason, W.T. Jr., J.P. Clugston, and A.M. Foster. 1993. Growth of laboratory-held Gulf of Mexico sturgeon, *Acipenser oxyrinchus desotoi. Prog. Fish. Cult.* **54**(1):59-61.

Matheson, R.H. III, G.R. Huntsman, and C.S. Manooch, III. 1986. Age, growth, mortality, food, and reproduction of the scamp, *Mycteroperca phenax*, collected off North Carolina and South Carolina. *Bull. Mar. Sci.* **38**(2):300-312.

Matlock, G.C. 1990. Maximum length and age of black drum, *Pogonias cromis* (Osteichthyes, Serranidae) off Texas. *NE Gulf Sci.* **11**(2):171-173.

Mayo, R.K., V.R. Gifford, and A. Jearld Jr. 1981. Age validation of redfish, *Sebastes marinus* (L.), from the Gulf of Maine, Georges Bank region. *J. NW Atl. Fish. Sci.* **13-19**.

Mayo, R.K., J.M. McGlade, and S.H. Clark. 1989. Patterns of exploitation and biological status of pollock, *Pollachius virens* L., in the Scotian Shelf, Georges Bank and Gulf of Maine area. *J. NW Atl. Fish. Sci.* 9(1):13-36.

McBride, R.S., J.L. Ross, and D.O. Conover. 1993. Recruitment of bluefish, *Pomatomus saltatrix*, to estuaries of the United States, South Atlantic bight. *Fish. Bull.* **91**(2):398-395.

McCormick, H.W., T. Allen, and W. Young. 1963. Shadows in the sea, the sharks, skates, and rays. Chilton Co., Phila., PA.

McEachran, L.W. and K. Daniels. 1995. Red drum in Texas: a success story in partnership and commitment. *Fisheries* **20** (3):6-8.

McFarland, G.A. and R.J. Beamish. 1983a. A preliminary observation in the juvenile biology of sablefish, *Anoplopoma fimbria*, in waters off the west coast of Canada. Pp. 119-106 *Proc. Intern. Sablefish Symp.* 29-31 March 1983. Anchorage, AK. No. 83. Alaska Sea Grant Coll. Progr.

McFarland, G.A. and R.J. Beamish. 1983b. Biology of adult sablefish, *Anoplopomus fimbria*, in waters off western Canada. Pp. 59-80. *Proc. Intern. Sablefish Symp. 29-31 March 1983*. Anchorage, AK. Alaska Univ. Sea Symp. No. 83-8.

Mercer, L.P. 1983. A biological and fisheries profile of weakfish, *Cynoscion regalis*. N.C. Dept. Nat. Res. Div. Spec. Sci. Rep. 40.

Mercer, L.P. 1984a. A biology and fisheries profile of spotted seatrout, *Cynoscion nebulosus*. N.C. Dept. Nat. Res. Div. Spec. Sci. Rep. 40.

Mercer, L.P. 1984b. A biological and fisheries profile of red drum, *Sciaenops ocellatus*. NC Dept. Nat. Res. Spec. Serv. Rep. 41.

Merritt, N.R. and N.B. Marshall. 1981. Observations on the ecology of deep-sea bottom living fishes, collected off northwest Africa (08°-27°N). *Progr. Oceanogr.* **9**:185-244.

Miller, L.W. 1972. White sturgeon population characteristics in the Sacramento-San Joaquin estuary as measured by tagging. Calif. Fish Game. **58**(2):94-101.

Mills, D. and D. Piggins (eds.). 1988. Atlantic salmons: Planning for the future. Timber Press, Portland, OR.

Mork, J., N. Reyman, G. Stahl, F. Utter, and G. Sundner. 1985. Genetic variation on Atlantic cod, *Gadus morhus*, throughout its range. *Can. J. Fish. Aq. Sci.* **42**(10):1580-1587.

Moyle, P.B. and J.J. Cech. 1982. *Fishes: an introduction to ichthyology*. Prentice Hall, Englewood, NJ.

Murawski, S.A. 1993. Climate/change and marine fish distribution forecasting from historical analogy. *Tr. Am. Fish. Soc.* **122**(5): 647-658.

Murawski, S.A. and A.L. Pacheco. 1977. Biological and fisheries data on Atlantic sturgeon, *Acipenser oxyrhynchus* (Mitchill). NMFS Tech. Ser. Rep. 10. Sandy Hook Lab. Highlands, NJ.

Murphy, M.D., and R.G. Taylor. 1989. Reproduction and growth of black drum, *Pogonias chromis*, in northeast Florida. *NE Gulf Sci.* **10**(2):127-137.

Myers, G.S. 1938. Freshwater fishes of the West Indian zoogeography. Ann. Rep. Smiths. Inst. Pub. 3451:339-364 for 1937.

Myers, G.S. 1949a. Use of anadromous, catadromous, and allied terms for designating fishes. *Copeia*. 1949(2):89-96.

Myers, G.S. 1949b. Salt-tolerance of freshwater fish groups in relation to zoogeographical problems. *Biolgr. Dierkunde* (Leiden). **27**:315-322.

Myers, G.S. 1951. Freshwater fishes and east Indian zoogeography. *Stanford Ich. Bull.* **4**(1):11-21.

Myers, R.A., N.J. Barrowman, and K.R. Thompson. 1995. Synchrony of recruitment across the North Atlantic: an update (or, "Now you see it, now you don't"). *ICES J. Mar. Sci.* **52**(1):103-110.

Mysah, L.A. 1986. El Niño interannual variability and fisheries in the northeast Pacific Ocean. *Can. J. Fish. Aq. Sci.* **43**(2):464-497.

Naiman, R.J., S.D. McCormick, W.L. Montgomery, and R. Morin. 1987. Anadromous brook charr, *Salvelinus fontinalis*: opportunities and constraints for population enhancement. *Mar. Fish. Rev.* **49**(4):1-13.

Nakamura, I. 1985. FAO Species Catalogue. Vol 5. Billfishes of the world. An annotated and illustrated catalogue of marlins, sailfish, spearfishes, and swordfish known to date. FAO Fish. Synop. 125.

National Marine Fisheries Service South Atlantic Fisheries Management Council. 1983. Fishery management plan, regulation impact reviews and final environmental impact status for the snapper-grouper fishes of the south Atlantic bight area. Survey DOE. Feb. 8.

Neilson, J.D., J.F. Kearney, P. Perley, and H. Sampson. 1993. Reproductive biology of Atlantic halibut, *Hippoglossus hippoglossus*, in Canadian waters. *Can. J. Fish. Aq. Sci.* **50**(3):551-563.

Nelson, D.M., M.E. Monaco, C.D. Williams, T.E. Czepley, M.E. Pattillo, L. Coston-Clements, L.C. Settle, and E.A. Irlandi. 1991. Distributional abundances of fishes and in vertebrates in southeastern estuaries. ELMR Rep. 9 NOAA/NOS Strategic Env. Assess. Div. Rockville, MD.

Nelson, J. 1994. *Fishes of the World*. John Wiley and Sons, New York. 3rd ed.

Nelson, J.D., K.C. Waiwood, and S.J. Smith. 1989. Survival of Atlantic halibut, *Hippoglossus hippoglossus*, caught by long line and other trawl gear. *Can. J. Fish. Aq. Sci.* **46**:887-897.

Ni, I.H. and W.D. McKone. 1983. Distribution and concentration of redfishes in Newfoundland and Labrador coasts. NATO-SCI Counc. Stud. **6**:7-14.

Ni, I.H. and E.J. Sandman. 1984. Size at Maturity for Northwest Atlantic Redfishes, *Sebastes*. *Can. J. Fish Aq. Sci.*

Ni, I.H. and W. Templeman. 1985. Reproductive cycles of redfishes, *Sebastes*, in southern Newfoundland waters. *J. NW Atl. Fish. Sci.* **6**(1):57-63.

Nickerson, J.T.R. 1978. The Atlantic halibut and its utilization. *Mar. Fish. Rev.* **40**(7):21-28.

Nieland, D.L. and C.A. Wilson. 1993. Reproductive biology and annual variation of reproductive variables of black drum in the northern Gulf of Mexico. *Tr. Am. Fish. Soc.* **122**(3):318-327.

NMFS-NOAA. Shark fishery management plan for the Atlantic Ocean. April, 1991.

North, J.A., R.C. Beamesderfer, and T.A. Rien. 1993. Distribution and movements of white sturgeon in three lower Columbia River reservoirs. *Northwest Sci.* **67**:105-111.

O'Brien, L., J. Burnett, and R.K. Mayo. 1993. Maturation of 19 species of finfish off the northeast coast of the United States. 1985-1990. NOAA Fish. Rep. NMFS 113.

Olla, B.L. and A.L. Studholine. 1971. The effect of temperature on the activity of bluefish, *Pomatomus saltatrix* (Linnaeus). *Biol. Bull.* **141**(2):337-349.

Olson, D.B., C.L. Hitchurch, A.J. Moriss, C.J. Ashigawa, G. Penz, R.W. Nero, and G.P.F. Panosta. 1994. Life on the edge: Marine life and fronts. *Oceanography* **7**(2):52-60.

Palko, B.J., L. Trent, and H.A. Brusher. 1987. Abundance of Spanish mackerel, *Scombromorus maculatus*. *Mar. Fish. Rev.* **49**(2):27-77.

Palko, B.J., G.L. Beardsley, and W.J. Richards. 1981. Synopsis of the biology of the swordfish, *Xiphias gladius* (Linnaeus). NOAA Tech. Pap. NMFS Circ. 441.

Parker, K.S. 1989. Influence of oceanographic and meteorological processes on the recruitment of Pacific halibut, *Hippoglossus stenolepus* in the Gulf of Alaska. Intern. Recruitment Investig. in Subarctic. *Intern. No. Pac. Fish. Comm. Symp.* Vancouver, Bt. Col. 26 Oct., 1987, No. 108. Pp. 221-237.

Parsly, N.M., L.G. Beckman, and G.T. McCabe Jr. 1993. Spawning and rearing habitat use by white sturgeon in the Columbia River downstream from McNary dam. *Tr. Am. Fish. Soc.* **122**:217-228.

Paust, B.C. and R. Smith. 1986. Salmon shark, *Lamma ditropis*. The development of a commercial salmon shark, *Lamma ditropis*, fishery in the north Pacific. Univ. Alaska Sea Grant Rep.86, 01.

Pawson, M.G. and G.D. Pickett. 1993. Sea bass, biology, exploitations, and conservation. Chapman-Hall, New York.

Pearcy, W.G. 1992. Ocean ecology of North Pacific salmonids. Wash. Sea Grant Progr. Seattle, WA.

Pearcy, W.G. and J.P. Fisher. 1990. Distribution and abundance of juvenile salmonids off Oregon and Washington, 1981-1985. NOAA Tech. Rep. NMFS 93.

PSMFC (Pacific States Marine Fisheries Commission). 1992. White sturgeon management fishery plan. Portland, OR.

Potthoff, T. and S. Kelley. 1982. Development of the vertebral column, fins and fin support, branchiostegal rays, and squamation in the swordfish, *Xiphias gladius*. *Fish. Bull.* **80**:161-186.

Powell, A.B. and F.J. Schwartz. 1972. Anomalies of the genus *Paralichthyes* (Pisces: Bothidae) including an unusual double-tailed southern flounder, *Paralichthyes lethostigma*. *J. Elisha Mitchill Sci. Soc.* **88**(3):155-161.

Pratt, H.L. Jr. and J. Casey. 1983. Age and growth of shortfin mako, *Isurus oxyrinchus*, using four methods. *Can. J. Fish. Aq. Sci.* **40**:1947-1957.

Pratt, H.L. Jr., S.H. Gruber, and T. Taniuchi (eds.). 1990. Elasmobranchs as living resources: advances in the biology, ecology, systematics, and the status of the fishery. NOAA Tech. Rep. NMFS 90.

Prince, E.P. and L.M. Pulos (eds.). 1983. Proceedings of the international workshop on age determination of oceanic pelagic fishes: tunas, billfishes, and sharks. NOAA Tech. Rep. NMFS 8.

Quin, T.P. 1989. Estimating swimming speeds of migrating adult sockeye salmon. *Can. J. Zool.* **66**:2160-2163.

Quin, T.P., E.L. Brannon, and A.H. Dittman. 1989. Special aspects of imprinting and homing in coho salmon, *Oncorhynchus kisutch*. *Fish Bull.* **87**(4):769-774.

Quin, T.P. and B.A. Teshart. 1987. Movements of adult sockeye salmon, *Oncorhynchus nerka*. In British Columbia coastal waters in relation to temperature and salinity stratification: Ultrasonic telemetry results. Pp. 61-77. H.D. Smith, L. Margolis, and C.C. Wood (eds.).

Rass, T.S. 1974. Fishes of the greatest depths of the ocean. *Dokl. Akad. Nauk. SSR.* **217**(1):207-212.

Renaud, C.B., L.I. Spears, S.U. Qadrix, and D.E. McAllister. 1986. Biochemical evidence of speciation in the cod genus, *Gadus*. *Can. J. Zool.* **64**(7):1563-1566.

Reynolds, C.R. 1993. Gulf sturgeon sightings; a summary of recent responses. Publ. No. PG. 70-FR 93-01.

Richards, N.J. 1974. Evaluation of identification methods for young billfishes.Pp. 62-72. In R.S. Shomura and F. Williams (eds.). *Proc. Internat. Billfish Symp.* Kailua-Kona, Hawaii 9-12 Sept. 1972. Pt. 2 Rev. and Contr. Pap.

Richards, J.B. 1987. Developing a localized fishery: the Pacific angel shark. Pp. 147-160. S. Cook (ed.). Sharks: an inquiry into biology, behavior, fishery, and use. Oregon St. Univ. Corvallis, OR.

Richards, W.J. and W.L. Klawe. 1972. Index bibliography of the eggs and young of tuna and other scombrids (Pisces: Scombridae) 1880-1972. NOAA Tech. Rep. NMFS SSRF 652.

Richardson, S.L., and J.D. McEachran. 1986. Identification of small (< 3 mm) larvae of king and Spanish mackerel, *Scomberomorus cavalla* and *Scomberomorus maculatus*. *NE Gulf Sci.* **5**(1):75-79.

Robins, C.R. and G.C. Ray. 1986. A field guide to Atlantic Coast fishes of North America. Houghton-Mifflin, Boston, MA.

Rochard, E., G. Castelnaud, and M. Lepage. 1990. Sturgeons (Pisces: Acipenseridae): threats and prospects. *J. Fish. Biol.* 37A. (Suppl. **9**):123-132.

Rose, G.A. and W.L. Leggett. 1981. Atmosphere-ocean coupling and Atlantic cod migration and effects of wind force variation in sea temperature and currents on nauplae distribution and catch rates of *Gadus morhua*. *Can. J. Fish. Aq. Sci.* **45**(7):1244-1248.

Ross, J.L., J.S. Pavelea, and M.E. Chittenden Jr. 1983. Seasonal occurrence of black drum, *Pogonias chromis*, and red drum, *Sciaenops ocellatus*. *NE Gulf Sci.* **6**(1):67-70.

Ross, J.L., T.M. Stevens, and D.S. Vaughn. 1995. Age, growth, mortality, and reproductive biology of red drum in North Carolina waters. *Tr. Am. Fish. Soc.* **124**(1):37-54.

Rounsfell, G.A. 1962. Relationships among North American salmonidae. *Fish. Bull.* **62**(209):235-270.

Safrit, G. and F.J. Schwartz. 1998. Age and growth, weight, and gonadosomatic indices for female southern flounder, *Paralichthys lethostigma*, from Onslow Bay, North Carolina. *J. Elisha Mitchell Sci. Soc.* 114(3): 137-148.

Sakamoto, K. 1984. Interrelationships of the family Pleuronectidae (Pisces: Pleuronctiformes). *Mem. Fac. Fish.* Hokkaido Univ. **31**(1/2):95-215.

Saucier, M.H. and D.M. Baltz. 1993. Spawning site selection by spotted sea trout, *Cynoscion nebulosus*, and black drum, *Pogonias cromis*, in Louisiana. *Env. Biol. Fish.* **36**(3):257-272.

Saunders, M.W. and G.A. McFarlane. 1993. Age and length at maturity of the female spiny dogfish, *Squalus acanthias*, in the Strait of Georgia, British Columbia, Canada. *Env. Biol. Fish.* **38**(1-3):49-57.

Schmidt, D.J., M.R. Collins, and D.M. Wyanski. 1993. Age, growth, maturity and spawning of Spanish mackerel, *Scomberomorus maculatus* (Mitchill), from the Atlantic coast of the southeastern United States. *Fish. Bull.* **41**(3):526-533.

Schwartz, F.J. 1981. Effects of freshwater runoff on fishes occupying the freshwater and estuarine coastal watersheds of North Carolina. Pp. 282-294. In R.D. Cross and D.L. Williams (eds.), *Proc. Nat. Symp. Freshwater Inflow to Estuaries*. Vol. 1. USFWS Biol. Serv. Progr. FWS/OBS/81-04.

Schwartz, F.J. 1984. Occurrences, abundance, and biology of the blacknose shark, *Carcharhinus sacronotus*, in North Carolina. *NE Gulf Sci.* **7**(1):29-43.

Schwartz, F.J. 1989. Zoogeography and ecology of fishes utilizing North Carolina marine waters to depths of 600 meters. Pp. 333-372. R.Y. George and A.W. Hurlbert (eds.). *Carolina Coastal Oceanography*. NOAA-NURP Res. Rep. 89-2.

Schwartz, F.J. 1990. Mass migratory congregations and movements of several species of cownose rays, genus *Rhinoptera*: a worldwide review. *J. Elisha Mitchill Sci. Soc.* **106**(1):10-13.

Schwartz, F.J. 1993. A steelhead rainbow trout, *Oncorhynchus mykiss*, from coastal estuarine Peddeford Creek, Carteret County, North Carolina. *J. Elisha Mitchill Sci. Soc.* **109**(2):115-118.

Schwartz, F.J. 1995. Elasmobranchs frequenting fresh and low saline waters of North Carolina during 1971-1991. *J. Aquaricult. Aq. Sci.* **7**:45-51.

Secor, D.H., E.D. Houde and D.M. Monteleone. 1995. A mark-release experiment on larval striped bass, *Morone saxatilis*, in a Chesapeake bay tributary. ICES *J. Mar. Sci.* **52**(1):87-101.

Sevigny, J.M. and Y. de Lafontaine. 1992. Identification of redfish parasites in the Gulf of St. Lawrence using genotype specific variation. *Can. Tech. Rep. Fish. Aq. Sci.* 1890:69-73.

Shapiro, D.Y. 1987. Reproduction in groupers. Pp. 295-327. J. Polovename and S. Ralston (eds.). Tropical snappers and groupers: biology and fishery management. Westview Press, Boulder, CO.

Shaw, W. and G.A. McFarlane. 1983. Biology, distribution, and abundance of walleye pollack, *Theregra chalcogramma*, off the west coast of Canada. Pp. 262-283. *Intern. Groundfish Symp.* Anchorage, AK. 26-28 Oct., 1983.

Shephard, G.R. and M. Terceiro. 1994. The summer flounder, scup, and black sea bass fishery of the middle Atlantic bight and southern New England waters. NOAA Tech. Rep. NMFS 122.

Shomura, R.S. and F. Williams (eds.). 1974-75. Proceedings of international billfish Symposium. Kailua-Kona, Hawaii. 9-12 August, 1972. U.S. NASA Tech. Rep. NMFS-SSFR 675.

Silverman, M.J. 1979. Biological and fisheries data on black drum, *Pogonias cromis* (Linnaeus). NMFS Tech. Ser. Rep. 22.

Simon, R.G., and P.A. Larkin (eds.). 1972. The stock concept for Pacific salmon. H.R. Miller lectures in fisheries. *Intern. Animal Res. Ecol.* Univ. Bt. Col., Vancouver, Canada.

Smith, C.L. 1971. A revision of the American groupers, *Epinephalus* and allied genera. *Bull. Am. Mus. Nat. Hist.* **146**:71-241.

Smith, C.L. 1981. Serranidae: 34, 47. Vol. 4, W. Fischer (ed.). Synopsis identification sheets. Western Coastal Atlantic Fishery Area 31.

Smith, T.I.J. 1985. The fishery biology and management of the Atlantic sturgeon, *Acipenser oxyrhynchus*, in North America. *Env. Biol. Fish.* **14**(1):61-72.

Smith, T.I.J. 1990. Culture of North American sturgeons for fishery management. Pp. 19-27. A.K. Sparkes (ed.). Marine farming and enhancement. Proc. 15th U.S.-Japan Meeting Aquacult. Kyoto, Japan. NOAA Tech. Rep. NMFS 85.

Smith, T.I.J. and E.K. Dingley. 1984. Review of biology and culture of Atlantic, *Acipenser oxyrhynchus*, and shortnose sturgeon, *A. brevirostrum*. *J. World Maricult. Soc.* **15**:210-218.

Smith, T.I.J., D.E. Marchette, and R.A. Smiley. 1982. Life history, ecology, culture, and management of Atlantic sturgeon, *Acipenser oxyrhynchus oxyrhynchus* (Mitchill), in South Carolina. S.C. *Wildlf. Mar. Resource. Res. Dept. Final Rep.* USFWS Ser. Progr. AFS-9.

Smith, T.I.J., D.E. Marchette, and G.F. Ulrich. 1984. The Atlantic sturgeon fishery in South Carolina. *N. Am. J. Fish. Mgt.* **4**:164-176.

Smith, W., P. Berrien, and T. Potthoff. 1994. Spawning pattern of bluefish, *Pomatomus saltatrix*, in the northwest continental shelf ecosystem. *Bull. Mar. Sci.* **54**(1):8-15.

Smith, W.G., J.D. Sibunka, and A. Wells. 1975. Seasonal distribution of larval flatfishes, Pleuronectiformes, on the continental shelf between Cape Cod, Massachusetts, and Cape Lookout, North Carolina 1965-66. NOAA Tech. Rep. NMFS SSRF 691.

Squire, J.L. 1987a. Striped marlin, *Tetrapterus audax*, migrations, patterns, and rates in the northeastern Pacific Ocean as determined by a cooperative tagging program and its relation to resource management. *Mar. Fish. Rev.* **49**(2):26-43.

Squire, J.L. 1987b. Relation of sea surface temperature changes during the 1983 El Niño to the geographical distribution of some important recreational pelagic fishes and their catch temperature parameters. *Mar. Fish. Rev.* **49**(2):44-57.

Stauffer, G.D. 1985. Biology and life history of the coastal stocks of Pacific whiting, *Merluccius productus*. *Mar. Fish. Rev.* **47**(2):2-7.

Stearley, R.F. and G.R. Smith. 1993. Phylogeny of the Pacific trouts and salmon, *Oncorhynchus*, and genera of the family Salmonidae. *Tr. Am. Fish. Soc.* **122**(1):1-33.

Stevens, D.E., D.W. Kohlhorst, L.W. Miller, and D.W. Kelley. 1985. The decline of striped bass in the Sacramento-San Joaquin estuary, California. *Tr. Am. Fish. Soc.* **114**:12-30.

Stone, S.L., T.A. Lowery, J.D. Field, S.H. Jury, D.M. Nelson, M.E. Monaco, C.D. Williams, and L.A. Andressen. 1994. Distribution and abundance of fishes and invertebrates in the mid-Atlantic estuaries. ELMR Rep. 12 NOAA/NOS Strategic Env. Assess. Div. Silver Springs, MD.

Stoskoff, M.K. 1993. *Fish Medicine*. W.B. Saunders Co., Phila, PA.

Sutter, F.G., R.S. Waller, and T.D. McLlwain. 1986. Species profiles: life histories and environment requirements of coastal fishes and vertebrates (Gulf of Mexico) black drum. Biol. Rep. USFWS.

Swain, D.P. and W.L. Kramer. 1995. Animal variation in temperature selection by Atlantic cod, *Gadus morhua*, in the southern Gulf of St. Lawrence, Canada, and its relation to population size. *Mar. Ecol. Prog. Ser.* **116**:11-24.

Symposium on the classification of brackish waters. 1958. Venis 8-14 April Centro Naz. Studia Talassograf. Conseglio Naz. Richerche XI, Suppl. Pp. 5-248.

Taub, S.H. 1990. Fishery management plan for Atlantic sturgeon, *Acipenser oxyrhynchus*. Final Mgt. Rep. 17 *Atl. St. Mar. Fish Comm.*

Tech. Rep. Internat. Pacific Halibut Comm. No. 22. 1987. Pacific halibut: biology, fishery, and management. Seattle, WA.

Thomas, A.D., M.O. Nelson, J.J. Trayner, and E.P. Numallee. 1980. The distribution, abundance, and biological characteristics of Pacific whiting, *Merluccius productus*, in California - British Columbia region during July-September 1977. *Mar. Fish. Rev.* **42**(3-4):17-33.

Timoshkin, V.P. 1969. Atlantic sturgeon, *Acipenser sturio* L., caught at sea. *Vopr. Ikthiol.* **8**(4) 57:749.

Tritos, R.W. 1984. Sea surface temperature trends and patterns in the southwest Atlantic. Biol. Rev. 84. Pp. 18-40. *Bedford Inst. Oceanogr.* Canada.

Trewavas, E. 1977. The sciaenid fishes (croaker and drums) of the Indo-West Pacific. *Tr. Zool. Soc.* London. **32**:253-541.

Tsepkin, Y.E.A. and K.I. Sokolov. 1972. The maximum size and age of some sturgeons. *Vopr. Ikthiol.* **11**(3)68:541-42.

Van Housen, G. 1989. Electrophoretic stock identification of summer flounder, *Paralichthys dentatus*. M.A. Thesis. Coll. William & Mary, Williamsburg, VA.

Waldrum, J.R. and F.C. Fabrizio. 1994. Problems of stock identification in estimating selective controllability of Atlantic striped bass in the coastal fishery. *Tr. Am. Fish. Soc.* **123**(5):766-778.

Wedemeyer, G.A., D.J. McLeay, and C.P. Goodyear. 1990. Assessing the tolerance of fish and fish populations to environmental stress: the problems and methods of monitoring. *Am. Fish. Soc. Symp.* **8**:164-195.

Wedemeyer, G.A., R.L. Saunders, and W.C. Clarke. 1980. Environmental factors affecting smoltification and early marine survival of anadromous salmonids. *Mar. Fish. Rev.* **42**(6):1-14.

Wenner, C.A. 1992. Red drum, natural history and fishing techniques in South Carolina. *S.C. Mar. Res. Ser. Educ. Rep.* 17.

Wenner, C.A., W.A. Roumillat, and C.W. Waltz. 1986. Contribution to the life history of black sea bass, *Centropristis striata*, off the southeastern United States. *Fish. Bull.* **84**(7):727-741.

White, D.B. Jr. and R. R. Stickney. 1973. A bibliography of flatfish, Pleuronectiformes, research and partial annotation. *Tech. Rep. Ser.* **73**-6. Ga. Mar. Sci. Center. Skidaway Island, GA.

Whittaker, D.R. 1980. World utilization of the hake. *Mar. Fish. Rev.* **42**(1):4-7.

Wilk, S.J. 1977. Biological and fisheries data on bluefish, *Pomatomus saltatrix* (Linnaeus). NMFS *Tech. Ser. Rep.* **11**. Highlands, NJ.

Wilk, S.J. 1979. Biological and fisheries data on weakfish, *Cynoscion regalis* (Bloch and Schneider). NOAA *Tech. Ser. Rep.* **21.** Highlands, NJ.

Wilkins, M. E. and M. W. Saunders (eds.) 1997. Biology and management of sablefish, *Anoplopoma fimbria*. NOAA Tech. Rep. NMFS 130.

Winterbottom, R., and A.R. Emery. 1981. A new genus and two species of gobiid fishes, Perciformes, from the Chagos Archipelago, Central Indian Ocean. *Env. Biol. Fish.* **6**(2):139-149.

Wooley, C.M. 1985. Evaluation of morphometric characters used in taxonomic separation of Gulf of Mexico sturgeon, *Acipenser oxyrhynchus desotoi*. Pp. 93-107. F.P. Binkowski and S.I. Doroshov (eds.). North American sturgeon: biology and aquaculture potential. Development and environmental biology of fish. Vol. 6. W. *Junk Publ.* Netherlands.

Wootten, R.J. 1990. Ecology of teleost fishes. Chapman & Hall, London.

Wourmes, J.P. and L.S. Demski (eds.). 1993. The reproduction and development of sharks, skates, and ratfishes. *Env. Biol. Fish.* **38**(163):7-24.

Yoshida, H.O. 1979. Synopsis of the biological data on tunas of the genus *Euthynnus*. NMFS *Tech. Rep. Circ.* 429.

Zeiner, S.J. and P. Wolf. 1993. Growth characteristics and estimates of age at maturity of two species of skates, *Raja binoculata* and *Raja rhina*, from Monterey Bay, California. Pp. 87-99. S. Branstetter (ed.). Conservation biology of elasmobranchs. NOAA Tech. Rep. NMFS 115.

Commercial Harvest and Biology of Major Freshwater Finfish Species

Roy C. Heidinger

INTRODUCTION

More than 50 species of fish make up the freshwater commercial harvest in North America; this does not include the bait minnow industry. In addition to a brief discussion of trends in the harvest, the basic biology of 24 species is discussed in this chapter. Since year-class strength is often determined at an early age, their reproductive biology is emphasized. Most factors that are responsible for determining year-class strength are the same for many species. Some of these factors are discussed in more detail for channel catfish, yellow perch, and walleye than for the other species. Many anadromous species such as salmon and alewife are also harvested in freshwater; however, since many more pounds of these species are usually harvested in saltwater, most of them are not discussed in this chapter. Unless specifically noted, all weights and measurements are in the English system and all fish lengths are total lengths (TL). In North America, less than 1 percent of the total commercial harvest of finfish comes from freshwater. Nevertheless, this natural resource is extremely important for the individual whose livelihood depends on these fish (Figure 1a).

With the exception of the Laurentian Great Lakes (Canada), it is difficult to obtain statistical information on freshwater commercial harvest from natural populations in North America. The National Marine Fisheries Service (NMFS), United States Department of Commerce (USDC), publishes a yearly summary called "Fisheries of the United States," but freshwater landings were not listed separately until 1995. The Food and Agriculture Organization of the United Nations (FAO) yearbook of fishery statistics tends to combine aquaculture landings with the commercial wild-caught fish. The author was able to determine some trends in the commercial harvest by surveying the United States state agencies and comparing these data to a survey made in 1975, and to use selected FAO data and data on the Great Lakes landings for comparison purposes.

Average yearly harvest of selected freshwater fishes from 1982 to 1984 compared to the average yearly harvest from 1995 to 1997 indicates a 61 percent reduction in harvest in the United States and a 20 percent reduction in Canada. Mexico, on the other hand, almost tripled its production (Table 1). The large increase in Mexico's landings between 1982 and 1984 may be due to differences in how the data were reported. From 1984 to 1997 Mexico reported a 28 percent increase in harvest. Commercial production by Canada and the United States from the Great Lakes continues to be around 90 million pounds (Table 2). The species composition of the catch in the various Great Lakes has changed dramatically over the years (Jude and Leach, 1993).

Flickinger (1976) stated that commercial fishing in inland waters of the United States reached over 80 million pounds in the 1950s, declined to around 55 million pounds in the 1960s, and then began to increase. His 1975 survey indicated a

Figure 1a. Commercial fishing in Illinois.

Table 1. Approximate commercial harvest of Selected Freshwater Finfish Species in Tons (FAO, 1991; 1997)

Species	1982	1983	1984	1985	1986	1987	1988	1989	1990	1991	1992	1993	1994	1995	1996	1997
United States																
Common carp	12406	13220	13188	13435	8889	13837	1358	1163	856	1046	859	1219	908	939	1139	1186
Buffalo	17642	15553	11222	11560	11995	12712	2565	1856	776	1394	1156	1082	987	529	874	1092
Catfish	17928	16966	13184	14508	14195	15428	7596	6918	5426	5591	5408	6020	5799	5801	4529	7573
Gizzard shad	443	508	1287	1423	1204	1001	631	496	249	314	586	853	1239	998	1402	2192
Lake herring	11	156	142	132	86	136	205	177	281	390	308	428	380	348	249	237
Lake trout	332	388	504	472	583	558	615	582	534	456	345	357	328	447	315	542
Lake whitefish	2620	4094	5242	4342	4738	4628	5374	4916	5304	4974	5073	5688	5631	5852	5810	6438
paddlefish	990	676	659	705	730	762	--	--	--	--	296*	--	--	--	--	--
Sturgeon	354	354	529	256	602	572	346	256	305	222	242	239	162	162	346	314
White perch	68	155	192	239	241	368	375	372	346	804	901	1030	823	884	1096	1283
Freshwater drum	--	--	--	--	--	--	1015	1038	484	687	566	549	673	518	565	605
Whitefish	1216	1286	1304	2028	1768	1916	2149	1771	1792	1998	2104	2026	2684	2475	1559	1660
Yellow perch	2186	1372	1270	1644	1674	1952	1334	1861	1674	1818	1726	1556	1528	1373	783	685
Total	56196	54728	49123	50744	46705	53870	23563	21406	18027	19694	19570	21047	21142	20326	18667	23807
Canada																
Common carp	774	708	816	481	485	661	580	373	830	883	544	245	715	799	714	603
Catfish	565	584	467	62	219	77	68	78	77	59	59	43	47	41	41	32
Lake trout	772	660	748	954	970	1014	1017	1204	683	626	530	634	760	722	767	683
Lake herring	1858	1462	1987	1189	912	1108	1668	1232	1077	350	770	670	1434	1206	996	1113
Lake whitefish	8652	9119	9212	10228	9752	9370	10655	10342	9452	9943	9960	8518	10447	10105	10342	9228
Northern Pike	5143	3694	3586	3780	3696	4101	4072	4246	2902	3496	3599	3773	1969	2713	2720	2610
Pond smelt	21834	14802	8256	12256	9208	12764	10253	8137	8944	10121	6357	8788	5318	6080	4387	6572
Sturgeon	122	114	160	121	33	28	304	380	292	280	298	497	422	331	440	464
Yellow perch	5757	4174	5966	5540	6471	6172	897	9357	7936	8034	8494	8992	8312	8779	8613	8566
Total	45477	35317	31198	34611	31746	35295	37604	35349	32193	33742	30611	32160	29424	30776	29020	29871
Mexico																
Common carp	2756	7975	11120	12326	23061	28847	23594	19037	21615	22282	22473	20024	17838	24990	23557	18499
Catfish	371	1452	1976	1644	3071	2608	1892	2024	2936	3115	3823	4204	3488	4874	5905	5169
Chichlidae	1162	63429	70051	59226	72256	82752	65990	65036	86854	77883	79745	83570	82676	82260	81938	82445
Total	4289	32113	83147	73196	98388	114207	914767	86097	111405	103280	106041	107798	104002	112124	111400	106113

* Oven and Fiss, 1996.

Table 2. Commercial Canadian and United States Harvest from the Great Lakes (Eshenroder, 1994, Great Lakes Fisheries Commission, personal comm.; Baldwin et al., 1979)

Year	Pounds[a]	Year	Pounds
1879	79,057	1940	99,314
1890	143,406	1950	96,072
1899	146,617	1960	104,651
1908	139,266	1970	110,643
1920	106,375	1980	90,440
1930	116,246	1990	92,902

[a] times 1,000.

harvest of 132.0 million pounds when all of the harvest from the Great Lakes was added to his figure. In my 1994 survey, 42 states provided sufficient information on harvest (88.2 million pounds) for comparison to Flickinger's survey. These same 40 states harvested 111.9 million pounds of fish in the 1975 survey versus 88.2 million pounds in 1994, a 21.1 percent decrease. If this 21.1 percent were subtracted from Flickinger's 132.0 million pound estimate, the estimated total harvest in 1994 would be 104.2 million pounds. In the 1994 survey, 33.1 percent of the harvest was from the Great Lakes (29.2 million pounds) and 33 percent was from the state of Arkansas (29 million pounds). Harvest data was not available for each state in 1993. Depending upon the state, harvest data used in these calculations were from 1987 to 1993. The price paid for fish in the round varies both by species and by location. Pricings paid for selected species range from $0.05 per pound to $2.06 per pound (Table 3).

To place this freshwater harvest in perspective, one only needs to realize that the 1998 commercial harvest of salmon from Alaska was 713 million pounds. The United States aquaculture trout industry produced 54 million pounds in 1996 (USDA, 1996), and the channel catfish (*I. punctatus*) aquaculture industry produced 507 million pounds in 1996 (USDA, 1997).

BIOLOGY OF SELECTED SPECIES

STURGEON

Depending upon the authority, either seven or eight species of sturgeon in the family Acipenseridae are recognized in North America. Four are marine but spawn in freshwater. These anadromous species are the white sturgeon (*Acipenser transmontanus*) and green sturgeon (*A. medirostris*), both on the Pacific Coast, and the Atlantic sturgeon (*A. oxyrhynchus*) and shortnose sturgeon (*A. brevirostrum*), both on the Atlantic Coast. A reproducing population of shortnose sturgeon is apparently landlocked in the Holyobe Pool of the Connecticut River (Taubert, 1980). The rare and federally-endangered pallid sturgeon (*Scaphirhynchus albus*) and the more common shovelnose (*S. platorhynchus*) and lake sturgeon (*A. fulvescens*) inhabit freshwater throughout their lives. Williams and Clemmer (1991) have described what they called the Alabama sturgeon (*S. suttkusi*) from Mobil Bay Basin. The status of this species which is very morphologically similar to the shovelnose sturgeon is still in question. The pallid sturgeon is found in the Missouri River and its impoundments and in the lower Mississippi River (Pflieger, 1975). The biology of the lake sturgeon and shovelnose sturgeon will primarily be discussed in this chapter. All sturgeon are protected from harvest in at least some states (Johnson, 1987).

A 68-pound pallid sturgeon was caught in Lewis and Clark Lake, North Dakota (Walburg, 1964). Pallid sturgeon from Oahe Lake, South Dakota, at ages one, five, and 10 grew to 11, 25, and 35 inches in total length, respectively.

Evidently pallid sturgeon, lake sturgeon, and possibly, to a lesser extent, the shovelnose sturgeon, are highly K-adapted species. It appears to be very difficult to recover their populations once they have been overharvested. Sturgeon are harvested in the United States by gill and trammel nets. Characteristics of shovelnose and lake sturgeon are listed in Table 4. Freshwater sturgeon are taken for their meat, which is frequently smoked, and for their roe, which may constitute 18 percent of their body weight, as well as for sport. They sell for $0.25 to $0.60 per pound (Table 3). Their roe recently sold for $50 per pound.

PADDLEFISH

The paddlefish (*Polyodon spathula*) family Polyodontidae is one of the largest freshwater fish in North America. Paddlefish were once considered a new genus of sharks (Becker, 1983). This zooplankton filter feeder was historically found in the large streams and rivers of the Mississippi

Table 3. Approximate Price per Pound (round) of Selected Species (in cents)

Taxon	Illinois 1993 (Dufford, 1994)	Michigan 1993 (Kinunen, 1993)	Minnesota 1992-93 (Johanness, pers. comm., MDNR)	Missouri 1992 (Robinson, 1994)
Alewife	—	25	—	—
American eel	12-32	—	—	18
Blue catfish	36-75	—	—	54
Bowfin	7-15	—	—	7
Buffalofishes	19-35	—	19	24
Bullhead	23-50	21	—	24
Burbot	—	35	5	—
Carp	7-35	9	7	12
Channel catfish	44-75	—	35	55
Chub	—	63	64	—
Crappie	—	206	—	—
Fat lake trout	—	24	—	—
Flathead catfish	35-75	—	44	54
Freshwater drum	9-40	14	12	15
Gar	15-50	25	—	10
Gizzard shad	—	10	—	—
Goldfish	—	25	—	—
Grass carp	7-25	—	8	21
Herring	—	37	34	—
Lean lake trout	—	44	150	—
Other Asian carp	7-25	—	—	—
Paddlefish	20-31	—	—	30
Quilback carpsucker	7-50	33	7	19
Round smelt	—	34	—	—
Shovelnose sturgeon	25-60	—	—	25
Smelt	—	14	20	—
Suckers	7-20	6	5-10	9
Walleye	—	129	—	—
White bass	—	81	—	—
White fish	—	65	150	—
White perch	—	60	—	—
Yellow perch	—	199	—	—

Valley and adjacent gulf slope drainage and in Lake Erie (Gengerke, 1986). The only other species in this family, *Psephurus gladius*, is in the Yangtze-Kiang River in China; it is a fish-eater (Becker, 1983). Protein electrophoretic and mitochondris DNA analysis indicates the possibility of one stock throughout its range in the United States (Epifano et al., 1989). Their historical range included 26 states; they have been extirpated from four, are completely protected in 11, and are considered a sport species in 14 states and a commercial species in six additional states (Oven and Fiss, 1996).

Growth of paddlefish is variable and depends upon food supply. Females grow faster and larger than males (Russell, 1986). In Mississippi River Pool 13, they reach 24.2 inches in total length at age one; 39.9 inches at age five; 51.0 inches at age 10; and 59.5 inches in total length at age 15 (Meyer, 1960). Because their paddle is often damaged, in recent studies body length (BL), which is measured from the anterior edge of the eye to the fork

Table 4. Distribution and Selected Biological Characteristics of Two Freshwater North American Sturgeon

Characteristic	Lake sturgeon	Shovelnose sturgeon
Distribution	North to Hudson Bay in large rivers and lakes, south in the Mississippi River drainage to Alabama and (Vladykov and Greeley, 1963)	Mississippi River drainage basin throughout USA (Becker, 1983)
Spawning	Broadcast spawners in stream over rocks in shallow water (1-15 feet), adhesive eggs (Priegel and Wirth, 1971)	Same as lake sturgeon (Christenson, 1975)
Parental care	None	None
Spawning temperature (°F)	53-59° (Priegel and Wirth, 1971)	67-70 (Christenson, 1975)
Incubation period (days)	4-8 at 55-63°F (Ceskleba et al., 1985)	7-13 at 55-62°F (Snyder, 1994)
Fecundity	4,333-5,960 eggs per pound of female (Harkness and Dymond, 1961)	13,908-51,217 for 24-34 inch fork length (Helms, 1974)
Egg diameter (mm)	3.5-4.0 (Ceskleba et al., 1985)	2.0-2.4 (Christenson, 1975)
Food of young	Small crustaceans (Eddy and Underhill, 1974)	NA
Food of adult	Insect larvae, leeches, snails (Schenberger and Woodburg, 1944)	Insects, fish eggs, crustaceans (Helms, 1974; Held, 1969)
Sexually mature	Some females at 41 inches (Ceskleba et al., 1985); 55 inches, 24-26 years old (Priegel, 1964)	Some males and females at 14 in. (Shields, 1956)
Frequency of spawnings	Females every 4-6 years (Priegel, 1964)	Females do not spawn every year (Helms, 1974)
Total length (inches) at age	Age I-II, age V-31, age X-45, age XX-54, age XXX-62, age XL-68 (Priegel, 1973)	Age I-8.4, age V-15.7, age X-19.8 (Fogler, 1963); faster in Mississippi River (Helms, 1974)
Total length-weight relationship	Log W=-4.037+3.253 Log L. (Priegel and Wirth, 1975)	NA
Maximum weight (pounds)	310, age 100-plus years old (Becker, 1983)	11.2 (IDOC, 1989)

of the tail, is normally reported (Rulle and Hudson, 1977). The length (centimeters)-weight (kilograms) relationship for paddlefish in the Missouri River is Ln W= 2.77 Ln BL –10.14 (Rosen et al., 1982).

Paddlefish are long-lived; 15- to 20-year-old fish are common and some live 30 years or more. Individuals in excess of 50 pounds are common, and 90-to-100-pound fish are frequently reported (Russell, 1986). A 163-pound individual was taken from Lake Manitau, Indiana (Jordan and Evermann, 1923). Spawning is associated with increased water levels and water temperatures of 50°F to 63°F (Purkett, 1961). Their adhesive eggs attach to the gravel that they spawn over (Russell, 1986). There is no parental care. Mature males spawn every year, but females require two or more years to develop ova (Elser, 1976; Meyer, 1960; Sprague, 1959). However, under culture conditions, females have been spawned in consecutive years (Semmans, 1986). Relative fecundity is 6,800 to 15,900 ova per pound of female and the ovaries may be 15 to 25 percent of the body weight (Purkett, 1961).

Paddlefish eggs hatch in seven to 12 days at 51°F to 61°F; they switch to exogenous feeding on zooplankton three to five days after hatching (Russell, 1986). Larval and juvenile fish are particulate feeders (Ruelle and Hudson, 1977), and adults use their long gillrakers to filter feed indiscriminately (Rosen and Hales, 1981). Adults may feed at night on upward migrating aquatic insects (Ruelle and Hudson, 1977).

Hubert et al. (1986) constructed habitat suitability models for paddlefish. Optimum models suggest an annual minimum of at least a 21-day period of rising water temperature between 50°F and 61°F with yearly upstream access to 123 acres or more of spawning area consisting of more than 80 percent substrate with 0.6 to 4-inch diameter gravel within 125 miles of winter habitat; an average spring water rise of 10 to 20 feet with a current velocity one foot above the spawning substrate between 1.3 and 1.6 feet per second; dissolved oxygen in the spawning area of six to eight parts per million; a minimum of 250,000 acres of possible summer and winter habitat with an average river channel width of at least 3,300 feet; 100 percent of water area continuous with summer and winter habitat with a current velocity of less than 0.2 feet per second; and 30 or more eddies per 0.63 miles of summer or winter channel habitat.

Very few estimates have been made of paddlefish density. In the Barkley Lake rotenone project the standing stock of paddlefish was 8.0 pounds per acre (Table 5).

A combination of loss of spawning areas, overharvesting, and the loss of backwater areas and lakes in the river system has led to a significant reduction in paddlefish populations. Dams prevent paddlefish from reaching their spawning areas and contribute to silting in of traditional spawning areas. Large rivers in the central United

Table 5. Standing Stock (pounds/acre) of Selected Commercial Freshwater Finfish Determined from a 115-acre Rotenone Sample of Douglas Reservoir, Tennessee (Hayne et al., 1967); a 210-acre Rotenone Sample of Barkley Lake, Kentucky (Aggus et al., 1979); and a Review of the Literature by Carlander (1955)

Species	Douglas Lake rotenone	Barkley Lake rotenone	Carlander range *
Bigmouth buffalo	<1	73	1-1,016
Black buffalo	<1	8	2-89
Blue catfish	8	6	--
Bullheads	—	1	<1-292
Carp	31	136	1-2,222
Carpsuckers	<1	8	—
Channel catfish	4	46	1-57
Flathead catfish	1	2	1-92
Freshwater drum	8	161	1-82
Gizzard shad	27	286	1-468
Northern pike	—	—	<1- 42
Paddlefish	—	8	—
Smallmouth buffalo	22	62	1-290
Walleye	—	—	<1- 54
White perch	—	—	6-18
Yellow perch	—	—	<1-184

* Does not include data from small ponds.

States tend to be very turbid and do not produce large standing stocks of zooplankton. Considerable amounts of zooplankton are produced in the fertile, clear, backwater areas of rivers, but these areas are rapidly being lost through siltation and dikes. Most male paddlefish mature in seven to nine years while females require 10 to 12 years to mature (Russell, 1986). Many immature paddlefish are removed by the 3-inch bar mesh gillnets that are frequently used by commercial fishermen.

A few paddlefish are being cultured in the United States, mostly for the aquarium trade. Ten states have a stocking program (Oven and Fiss, 1996). Anglers snag this species with treble hooks below dams where the fish concentrate in large numbers at certain times of the year. Commercial fishing is primarily done with floating or set gill or trammel nets. Paddlefish are harvested for their flesh, which is often smoked, and their roe for caviar. In 1993 wholesale prices in Illinois paid for fish in the round was $0.21 to $0.31 per pound (Table 3). In 1989 the much more valuable roe sold for $50 to $90 per pound (Pitman, 1991).

Gizzard Shad

Gizzard shad (*Dorosoma cepedianum*), family Clupeidae, occur in lakes and streams in the Mississippi River, Lake Michigan drainage basins, and northeastern Mexico (Becker, 1983). Gizzard shad have been introduced into some of the southeastern states. This species is not a sport species, but it is very important as a forage fish for predators such as the largemouth bass (*Micropterus salmoides*) (Heidinger, 1984).

Gizzard shad are shoreline broadcast spawners. Their adhesive eggs adhere to any firm substrate. They spawn in the spring at water temperatures between 61°F and 85° F (Storch et al., 1978; Dendy, 1945). Sexual maturity is reached at seven inches in total length (Miller, 1963), which is equivalent to age two in many waters (Shelton, 1972; Jester and Jenson, 1972; Bolton, 1966). Actual spawning activity is primarily at night or in the early morning (Shelton, 1972; Bodola, 1966). Shelton (1972) found the sex ratio of spawning schools to be three males to one female. Gizzard shad lay 39,000 to 544,800 eggs per pound of female (Kilambi and Baglin, 1969; Boldola, 1966). The eggs hatch in three days.

As shad grow, they become more pelagic (Van Den Avyle and Fox, 1980; Edward et al., 1977). They tend to school during the day and become more uniformly distributed at night (Netsch et al., 1971). In general, their food habits are as follows: larval fish <10 millimeters (mm) eat unicellular protozoans, and some unicellular algae; larval fish 10 to 20 mm eat zooplankton; larvae 20 to 30 mm eat zooplankton and phytoplankton; adults eat detritus plus additional organisms associated with detritus (Jude, 1973; Baker et al., 1971; Bodola, 1966).

Gizzard shad frequently produce standing crops of 50 to 250 pounds per acre (Table 5). Normally they do not live longer than seven years and most adult fish in a population will be two or three years old (Heidinger, 1984). In Lake Erie gizzard shad reached 7 inches in total length by age one and 19 inches by age five (Bodola, 1966). The standard length-weight relationship is log W = – 3.580 + 3.170 log total length (Anderson and Gutreuter, 1983).

Habitat suitability index models have been constructed for gizzard shad. According to Williamson and Nelson (1985), optimum conditions include the following: total dissolved solids of 1000 parts per million (ppm); 280 to 320 frost free-days per year; mean weekly summer water temperature between 72°F and 85°F; dissolved oxygen between 6 to 8 ppm; water level rising and inundating vegetation during spawning season; and 15 to 30 percent of vegetated area less than 6.6 feet deep during the spawning season. At northern latitudes large numbers of gizzard shad die during the winter (Heidinger, 1984).

In 1994, 1239 tons of gizzard shad were harvested from United States waters (Table 1). These fish are not consumed by humans; they are sold primarily for catfish bait and as bait for crab and crayfish traps. They are usually harvested with dip nets below dams where they congregate in huge numbers, especially in the spring. They sold for $0.10 a pound (Table 3).

American Shad

American shad (*Alosa sapidissima*) is an anadromous species. It is the largest member of the family Clupeidae. On the Atlantic Coast, its range extends from the Sand Hill River, Labrador (off

Canada), to the St. Johns River in Florida. In 1871, it was introduced into the Sacramento and Columbia rivers on the Pacific Coast. It has become established from Alaska south to Todos Santos Bay, Mexico. A landlocked population exists in Millerton Lake, California (Page and Burr, 1991; Scott and Crossman, 1973). Historically, the American shad probably once spawned in almost all rivers on the North American East Coast.

On average, male and female shad return from the ocean to spawn between their fourth and fifth (range two to six years) year of age (Leggett, 1969). Males tend to mature earlier than females. In the St. Johns River, Florida, males mature at 12.0 to 17.6 inches fork length (FL) and females at 15.1 to 19.1 inches FL (Walburg, 1960). Shad spawn from November until March in the St. Johns River and from May through July in Canadian waters (Scott and Crossman, 1973; Walburg, 1960). Spawning temperatures range from 57°F to 76°F (Ross et al., 1993). American shad exhibit a latitudinal cline in post-spawning survival. Individuals from southern populations tend to die after spawning, whereas individuals from northern populations do not die after spawning. An inverse relationship appears to exist between the number of repeat spawners and fecundity (Leggett and Carscadden, 1978). Fecundity ranges between 58,000 and 659,000 eggs per female (Facey and Van Den Avyle, 1986; MacKenzie et al., 1985; Roy, 1969; Walburg, 1960). For the St. Johns River, the relationship is $Y = 8.084\ X + 65.404$, where Y = ova production in thousands and X = weight in ounces ($r = 0.900$). Shad often ascend to the headwater areas of streams to spawn, which may involve distances of hundreds of miles (Glebe and Leggett, 1981). Males tend to arrive on the spawning grounds before the females. Males and females swim near or at the surface and the approximately 1.7-mm diameter eggs are released into the water column. The 2.5- to 3.8-mm diameter, semidermersal and initially adhesive, fertilized eggs are carried downstream by the current and gradually sink (March, 1976; Scott and Crossman, 1973; Marcy, 1972; Walburg and Nichols, 1967; Ryder, 1887). Eggs hatch in 12 days, seven days, and three days at 52°F, 63°F, and 75°F, respectively (Scott and Crossman, 1973; Bigelow and Schroeder, 1953). At hatching, the larvae are 5.7 to 10.0 mm long (March, 1976; Leim, 1924). They absorb their yolk sacs in four to seven days at approximately 12.2 mm in total length (Walburg and Nichols, 1967). Larval shad are plankton feeders. They reach the juvenile stage in about 30 days at approximately 1 inch in total length (Jones et al., 1978). Juveniles spend their first summer in rivers before migrating down to the estuaries (Chittenden, 1972).

According to MacKenzie et al. (1985), "the American shad in the mid-Atlantic region form large schools and undertake extensive ocean migration" (Leggett and Whitney, 1972). Shad from all Atlantic Coast rivers spend the summer and fall in the Gulf of Maine (Talbot and Sykes, 1958). This congregation includes immature shad and spawned-out adults from rivers north of Chesapeake Bay. Shad are apparently scattered along the mid-Atlantic Coast during the winter (Walburg and Nichols, 1967; Talbot and Sykes, 1958.) Migrating American shad seek bottom water temperatures between 37.4°F and 59°F, but probably prefer temperatures between 44.6°F and 55.4°F (Neves and Depres, 1979).

In early spring, the schools of shad migrate toward the coast. Those returning to rivers south of Cape Hatteras follow the Gulf Stream to remain within the 37.4°F and 59°F bottom isotherm. Those migrating to rivers north of Cape Hatteras later in the spring follow a route farther seaward into the Middle Atlantic Bight where water temperatures have risen sufficiently. Tag returns indicate that some schools migrate north along the coast (Neves and Depres, 1979).

American shad migrate as much as 13 miles/day in the Chesapeake Bay and the Bay of Fundy (Leggett, 1977). Some migrate up to 1,860 miles during the spring or fall migration.

A large majority of American shad return to their natal rivers to spawn. Homing behavior involves both oflaction and rheotaxis (Dodson and Leggett, 1974). The homing mechanism is sufficiently robust to perpetuate migrations even after major changes in water flow, such as below a hydroelectric dam.

Selected habitat suitability index models developed by Stier and Crance (1985) indicate that ideal spawning season temperatures should range between 57°F and 68°F, water velocity during spawning is one to three feet/second, mean surface temperature during egg and larval development is 59°F to 77°F, and percentage of estuary supporting the area is 5 to 20 feet. In the Delaware River, the highest spawning activity was in the runs and the lowest in pools. Some fish spawned in no current (Ross et al., 1933a). Ross et al.(1993b) developed habitat suitability index models for premigratory juveniles. Optimum values include temperatures between 67°F and 76°F and turbidity less than 2.2 Jackson Turbity Units (JTU).

Females appear to grow only slightly faster in length than males. Mean fork length at annulus from six rivers by three authors cited in Carlander (1960) is age one: 6 to 8 inches; age two: 10.9 inches; age three: 13.8 inches; age four: 15.9 inches; age five: 17.4 inches; age six: 18.9 inches; and age seven: 19.4 inches. Shad live at least to age 11 (Cating, 1953). Shad have been reported to weigh up to 14 pounds on the Pacific Coast and historically up to 15 pounds on the Atlantic Coast. The average size is considerably smaller now.

Shad have been primarily harvested commercially in gill nets, pound nets, haul seinees, weirs, fyke nets, throw nets, and dip nets (MacKenzie et al., 1985). In 1896, 25,000 tons of shad were harvested from the United States Atlantic Coast. By the 1930s, the harvest had dropped to approximately 5,500 tons (Walburg and Nichols, 1967). In 1982, Canada and the United States harvested 2,696 tons; by 1994, the commercial catch for both countries was reduced to 1,264 tons (FAO, 1994; 1991). The decline in the shad populations has been attributed to dams which block their spawning runs, overfishing, and pollution (Chittenden, 1969). Shad have a delicate light meat. The name *sapidissima* means "most delicious." Females are highly sought after for their roe. Shad liver oil contains from 500 to 800 USP of Vitamin A, and 50 to 100 I.U. of Vitamin D per gram (Scott and Crossman, 1973). According to Moring (1993), "the American shad was probably the first species of anadromous fish to be aided by North American fish cultural programs because, along with Atlantic salmon, it was the first anadromous species to be severely affected by human impact." Efforts are underway to restore shad to river basins such as the Susquehanna by stocking larvae (Johnson and Dropkin, 1992), fingerlings and adults (Moring, 1993).

Alewife

The Alewife (*Alosa pseudoharengus*) is an anadromous species in the family Clupeidae. It is distributed on the Atlantic from Red Bay, Labrador, to South Carolina (Page and Burr, 1991). Numerous landlocked populations have been established either through purposeful stockings or by unintentional introduction. Alewives were found in Lake Ontario in 1873; they migrated into Lake Erie and Lake Huron by the 1930s. In the 1950s they became established in Lake Michigan and Lake Superior (Jude and Leach, 1993).

On the Atlantic Coast, adult alewife migrate into their spawning streams from March in the southern portion of their range through July in the northern portion of their range (Mullen et al., 1986). They exhibit a homing behavior cued by olfaction (Thumbery, 1971), but considerable mixing appears to occur (Messiah, 1977). Repeat spawners increase from 10 percent in southern population to approximately 60 percent in more northern populations (Mullen et al., 1986, citing various authors).

Cianci (1969) reports a minimum spawning temperature of 51°F . Typical Atlantic Coast population spawn at temperatures of 54°F to 60°F (Bigelow and Schroeder, 1953) while Lake Michigan fish spawn at 63°F to 70° F (Norden, 1967). They spawn in a wide variety of habitats, including standing water, backwater, and midstream (Kissil, 1974). Sex ratio of males to females on the spawning grounds is 1:1 (Kissil, 1974). Adults return to the sea after spawning.

The 0.8- to 1.3-mm-diameter, water-hardened eggs are somewhat adhesive until they water harden, after which they become non-adhesive (cited in Mullen et al., 1986; Mansueti, 1962; 1956). Alewives are broadcast spawners and do not guard their offspring. Females on the Atlantic Coast lay from 48,000 to 360,000 eggs while landlocked females are smaller and lay from 11,000 to 49,300 eggs (Nigro and Ney, 1982; Cole et al., 1980; Norden, 1967). The eggs hatch in 3.9 days at 69°F. According to Edsall (1970) the number of days to hatch (D) can be determined by the equation: $D = 6.335 \times 10^6 \times t^{-3.122}$ where t is in Fahrenheit. The upper lethal temperature limit for eggs is 85°F (Kellogg, 1982). At hatching, the larvae average 5.1 millimeters in total length. They require two to five days to adsorb the yolk (Mansueti, 1962,

Table 6. Distribution and Selected Biological Characteristics of Three Species of *Coregonus* (Family Salmonidae) Frequently Taken in Commercial Catches

Characteristic	Cisco or Lake Herring	Lake Whitefish	Bloater
Distribution	Most of Canada, north central and eastern United States (Scott and Crossman, 1973)	Northern United States, Canada, Yukon Territory, Alaska, British Columbia (Scott and Crossman, 1973)	Great Lakes basin except Lake Erie (Scott and Crossman, 1973)
Spawning time	Fall 39-43°F (John, 1956)	October-January, 40-43°F (Wells and McLain, 1973; McNight et al., 1970)	January-March (Emery and Brown, 1978)
Spawning location	Broadcast spawners, shallow water to pelagic, various bottom types (Becker, 1983 citing other authors)	Broadcast spawners over gravel and rock 6-60 feet deep (Wells and McLain, 1973)	Broadcast eggs at depths of 120-300 feet over various bottom types (Dryer and Beil, 1968)
Fecundity	13,472 per pound (Dryer and Beil, 1964)	4,000–16,134 per pound (Mesa, from Carlander, 1969)	19,856 per pound, 30 percent of body weight (Dryer and Beil, 1968); log F=1.649+1.034 log W (Emery and Brown, 1978)
Egg diameter (mm)	1.9 (Dryer and Beil, 1964)	2.3-3.2 (Brooke, 1970; Hart, 1930)	2.0 (Brooke, 1970)
Incubation (days)	90 at 50°F (Colby and Brooke, 1970)	80-140 at 39-33°F (Price, 1940)	120 (Wells, 1966)
Sexual maturity	I-IV years (Smith, 1957) between 8.5 and 12 inches (Bryer and Beil, 1964)	Non-dwarf races II-VII between 8.2-17.4 inches (Carlander, 1969; Mraz, 1964; Dryer, 1963)	II-III years, smallest male 6 inches, smallest female 7 inches (Dryer and Beil, 1968)
Growth (inches)	Age I-4.6, age V-11.6, age X-15.7 (mean from Carlander, 1969)	Lake Huron and Lake Michigan average from six studies age I-5.3, age V-18.4, age X-25.1, age XV-29.3 (Carlander, 1969)	Age I-3.9, age V-8.5, age X-11.7 from Lake Superior (mean from Dryer and Beil, 1968)
Maximum size (pounds)	8 from Lake Erie in 1949 (Scott and Crossman, 1973)	42, 26 (Van Oosten, 1946)	Less than 1
Diet of young	Zooplankton (Cahn, 1927)	Zooplankton (Reckahn, 1970)	Zooplankton for fish less than 7 inches (Wells and Beeton, 1963)
Diet of adults	Crustaceans, insects, fish eggs (Dryer and Beil, 1964)	Crustaceans, insects, fish, mollusca, fish eggs (Rechkahn, 1970)	Zooplankton, larger crustaceans, clams, fish eggs (Dryer and Beil, 1968; Wells and Beeton, 1963)

cited in Mullen et al., 1986). Transformation to the juvenile stage occurs at approximately 20 millimeters total length (Hildebrand, 1963).

In freshwater, young alewives feed primarily on diatoms (Jenkins and Burkhead, 1993). Young-of-the-year (YOY) migrate seaward in the fall. After they enter salt water, larger fish shift to a diet of shrimp and fish (Hildebrand, 1963; Hildebrand and Schroeder, 1928). In Lake Michigan, adult alewives feed heavily on *Mysis relicta* (Janssen and Brandt, 1980). Feeding behavior is reduced below 44°F (Colby, 1971). Alewife exhibit diel vertical migrations. They move up in the water column at night and down during the day (Neves, 1981). Landlocked populations inhabit open water during the day and move shoreward to feed at night (Forester and Goodbred, 1978). Their size-selective predation can alter the size and species composition of zooplankton communities (Kohler and Ney, 1981).

In the Atlantic Ocean, alewives are typically found on the New England Continental Shelf in 180 to 360 feet of water (Neves, 1981). The ultimate minimum low temperature for adult alewife is 37.4°F (Otto et al., 1976).

In a given population, such as those in the Atlantic Ocean, males tend to mature one year ahead of females. Atlantic Ocean populations mature in three to five years (Jenkins and Burkhead, 1993, citing seven authors). Landlocked populations mature in one to three years (Jenkins and Burkhead, 1993, citing five authors). Ten-year-old fish have been reported (O'Neil, 1980). The mean fork length of males and females from five bodies of water cited in Mullen et al. (1986) was 8.7, 9.0, 10.9, 11.3, 11.6, and 12.2 inches at age three through eight, respectively. The length (fork length in mm) - weight (grams) equation for male alewives from the St. John River, New Brunswick, is log W = 3.235 x log FL –5.420. For females, the relationship is log W = 3, 192 x log FL –5.294 (Messieh, 1977). A variety of commercial gear including weirs, trap nets, pound nets, and trawls are used to harvest alewife. Approximately 90 percent of the United States commercial catch was made within three miles of the coastline (NMFS, 1983). Mullen et al. (1986) reported that the Maine harvest was used primarily to bait lobster traps. Roe has been canned for caviar (Merriner, 1978). A large portion of the harvest is used as a source of fish meal, although some direct human consumption also occurs. The mean United States and Canadian harvest from 1982 to 1991 was 20,049 tons; 29,209 tons in 1985; 114,394 tons in 1988. By 1994, the combined harvest dropped to 4,821 tons (FAO, 1994; 1991).

Salmon, Whitefish, Cisco, and Lake Trout

In North America, a number of species in the family Salmonidae are taken in freshwater commercial catch. Many species of salmon have been stocked and are now landlocked in the Great Lakes, but their biology is basically the same as the coastal forms. The biology of the anadromous salmon species (*Onchorhynchus*) is covered in another chapter. Numerous species of freshwater coregonids occur in North America. They all are primarily harvested in gill nets. Biological characteristics of the cisco or lake herring (*Coregonus artedi*), lake whitefish (*C. clupeaformes*), and bloater (*C. hoyi*) are outlined in Table 6. Recent sale prices are listed in Table 3. Characteristics of the round whitefish (*Prosopium cylindraceum*), sometimes listed as the menominee in catch reports, is given in Table 7. Rainbow trout (*Oncorhynchus mykiss*) is frequently cultured in North America, but it is not often taken commercially from the wild.

The biology of the lake trout (*Salvelinus namaycush*) is discussed below. The largest of the char, the lake trout, is native to North America with its natural range closely associated with the limits of Pleistocene glaciation (Lindsey, 1964). This species is found in cool, deep lakes. Two subspecies are recognized: the lake trout (*S. n. siscowet*) and the fat lake trout (*S. n. siscoiwet*). The siscoiwet is found only in Lake Superior; however, integrades are known to exist (Becker, 1983). The fat content of siscowet varies from 32.5 to 88.8 percent, while fat content from lake trout ranges from 6.6 to 52.3 percent (Eschmeyer and Phillips, 1965). Siscowet are found at depths below 300 feet, normally between 495 to 590 feet, but have been found at depths of 1800 feet (Eddy and Underhill, 1974). Lake trout spawn in the fall at night over boulder or rubble bottoms. In inland lakes, they spawn in water from one to 40 feet deep, but in the Great Lakes they tend to spawn deeper. They spawn at temperatures of 46°F to 57°F (Becker, 1983; Scott and Crossman, 1973; Martin, 1957). There is evidence of homing to the spawning area (Martin,

Table 7. Distribution and Selected Biological Characteristics of Round Whitefish (Family Salmonidae)

Distribution	New England states to Labrador, northern New Brunswick, Ungava west through parts of Quebec, Ontario, Great Lakes except Lake Erie, Alaska, but discontinuous in Ontario-Manitoba border area (Scott and Crossman, 1973)
Spawning time	October-December (Scott and Crossman, 1973) 37-40°F (Normandeau, 1969)
Spawning location	Broadcast spawners over gravel and rock in rivers and lakes at depths of inches to 90 feet (Becker, 1983; Normandeau, 1969)
Fecundity	2,461-10, 459 eggs in 12-17 inch fish (Bailey, 1963)
Egg diameter (mm)	2.4-4.6 unfertilized-fertilized (Bailey, 1963)
Incubation (days)	140 at 36°F (Normandeau, 1969)
Sexual maturity	In Lake Michigan between 12-15 inches, age III (Mraz, 1964)
Growth (inches)	Lake Superior age I-4.7, age V-12.0, age IX-15.5 (Bailey, 1963)
Maximum size (pounds)	Approximately 5 (Keleher, 1961; Koelz, 1929)
Diet of young	NA
Diet of adults	Opportunistic bottom-feeder on invertebrates, snails, crayfish, fish, fish eggs, zooplankton (Armstrong et al., 1977; Normandeau, 1969)

1960). Males clean the spawning area, but they do not build a redd. Lake trout are polygamous. The eggs sink into the crevices among the rocks (Eschmeyer, 1957). The five-to six-millimeter-diameter eggs hatch in 105 to 147 days at temperatures of 32.5°F to 33.8°F. Relative fecundity is in the range of 400 to 1200 eggs per pound of female (Martin, 1957).

Lake trout are fairly slow-growing, but they have reached over 100 pounds in weight. They reach sexual maturity in six to 13 years (Scott and Crossman, 1973). An average growth rate of six lake trout populations given in Scott and Crossman (1973) indicates that age one fish reach 4.5 inches; age five: 14.8 inches; age 10: 21.0 inches; and age 15 fish obtain 32.2 inches in total length.

In the Great Lakes, crustaceans such as *Mysis* are important food items in the diet of young lake trout. Larger lake trout feed on a wide variety of fishes — for example, ciscoes, smelt, and cottids (Dryer et al., 1965).

Factors such as slow growth, coupled with their susceptibility to sea lamprey predation and overfishing have probably all lead to the reduction of lake trout stocks throughout the Great Lakes. Many of the deep water spawning stocks have been lost. Lake trout were eliminated from Lake Michigan by the mid-1950s. Since 1965, a lake trout rehabilitation program has stocked several million fingerlings per year. They were re-established, but natural recruitment has not been documented.

Lake trout are primarily harvested with gill nets. In 1992-93 lean lake trout sold for $0.44 to $1.50 per pound. Fat lake trout command a lower price of $0.24 per pound (Table 3). In 1994, the FAO reported 328 tons of lake trout taken from the United States and 828 tons harvested from Canadian waters (Table 1). Most of the poundage taken commercially in the United States was from tribal waters or from Lake Superior.

Pond Smelt

In North America, pond smelt (*Hypomesus olidus*), family Osmeridae, occur from the Cooper River north to the Kobuk River in Alaska and disjunctly in the Yukon Territory's Peel and Mackenzie Rivers (McAllister, 1963). Very little data on the biology of this species were found in the published North American literature. Spawning occurs in littoral areas of streams in the late spring (Scott and Crossman, 1973). Only a few individuals live longer than three years. At age one, they average 2.4 inches and at age two, 3.2 inches in total length (McPhail and Lindsey, 1970). McAllister (1963) reported a four-inch North American record, but the Russian record is a 7.3-inch individual. Young pond smelt feed on rotifers. Older fish feed on rotifers, algae, insects, and crustaceans (Sato, 1952). Pond smelt are thought to feed pelagically (McPhail and Lindsey, 1970). From 1982 to 1994 Canada has harvested 21,834 to 6,171 tons of pond smelt per year (Table 1).

Rainbow Smelt

Rainbow smelt (*Osmerus mordax*), family Osmeridae, is primarily an anadromous marine species. According to Page and Burr (1991), its North American range includes the Atlantic drainages from Melville, Newfoundland, to the Delaware River in Pennsylvania, and west through the Great Lakes. In the Arctic and Pacific, it occurs from Bathurst Inlet, Northwest Territories, to Vancouver Island, British Columbia. Smelt have been introduced into many North American lakes, including the Great Lakes. They seasonally occur in the Missouri, Ohio, Illinois, and Mississippi Rivers. From the 1920s to 1935, smelt became established in all of the Great Lakes (Jude and Leach, 1993). Cahill (1994) did not find any discreet stocks of rainbow smelt in Lake Michigan. He used protein electrophoresis and mitochrondrial DNA techniques in his analysis.

Smelt are broadcast spawners and neither sex guards the eggs. They move into their spawning area, above the tide head in coastal streams, before ice breakup (McKenzie, 1964), from late March through late May (Clayton, 1976). On the East Coast, spawning temperatures range from 39°F to 48°F (Clayton, 1976), except in the Miramichi Estuary, New Brunswick, where they spawn at temperatures of 50°F to 59°F (McKenzie, 1964). In Lake Michigan, spawning occurs at 50°F to 64°F (Jilek et al., 1979). Most smelt spawn in streams, but they are known to spawn in several feet of water over gravel bars in lakes (Rupp, 1965). They also spawn over gravel in a few inches to six feet of water in coastal streams (Murawaski et al., 1980). The eggs are adhesive and immediately attach to the substrate by a short pedicel formed from the outer shell membrane (Rupp, 1965).

In Lake Superior smelt [7.3 to 8.8 inches total length (TL)] had a mean fecundity of 31,338 (21,534 to 40,894) eggs per female (Bailey, 1964). Fertilized eggs range in diameter from 0.9 to 1.3 millimeters (Cooper, 1978; Crestin, 1973; Van Oosten, 1940). Spawning usually takes place in less than two feet of water at night. Early in the season males may predominate over females four to one; the sex ratio becomes more equal as the spawning season progresses (Becker, 1983). Salinities of 12 to 14 ppt will kill the eggs [Buckley, 1985, citing Unanian and Soin (1963)]. After spawning, smelt return to saltwater or the estuary, usually within 1.25 miles of shore and in waters less than 20 feet deep (Bigelow and Schroeder, 1953). In the Great Lakes they are found at much greater depths.

Eggs hatch in 29 days at 43°F to 45°F; 19 days at 48°F to 50° F; 11 days at 54° F; and eight days at 61° F (Cooper, 1978; McKenzie, 1964). At hatching, the photonegative larvae are five to six mm long (Cooper, 1978; Clayton, 1976; Rup, 1965; McKenzie, 1964). The larvae drift downstream and are usually near the bottom during the day and at the surface at night. Thus, it is hypothesized that they are maintained in the estuary by the two-way transport system (Rogers, 1939). Larval and juvenile smelt feed on planktonic crustaceans. Larger fish feed on euphausiids, amphipods, polychaetes, *Mysis*, fly larvae and a variety of fish (Foltz and Norden, 1977; Burbridge, 1969; Gordon, 1961).

Male and females mature at one to four years of age. The majority mature at two to three years of age (Murawski, 1976; Bailey, 1964; McKenzie, 1964). Fish at more southerly latitudes tend to grow faster and mature earlier.

After reaching sexual maturity, females grow faster than males. Smelt in marine populations usually grow faster than those in freshwater populations (Buckley, 1989). The mean total length of smelt from the Parker River, Massachusetts (Murawski, 1976), Great Bay, New Jersey (Warfel et al., 1943), Miramichi River, Nebraska, and western Lake Superior (Bailey, 1964) is 3.8, 6.1, 7.2, 8.7, 8.9, 10.1 and 12.2 inches at ages one to seven, respectively. Optimum temperatures for the growth of smelt are 43°F to 56°F (Becker, 1983).

Smelt are harvested as a "sport fish" by hook-and-line and by small seines and dip nets. Smelt have been introduced into a number of lakes to serve as forage for various sport fishes. Since they eat other larval fish, this management technique has been the subject of considerable controversy.

Their introduction into a Colorado lake to supply forage for walleye seems to be working well (Jones et al., 1994). Many people consider smelt to be a renowned food fish. Fresh smelt have an odor resembling freshly cut cucumbers which may be responsible for their name (Scott and Crossman, 1973). Smelting reached a carnival-like atmosphere in the 1930s when thousands of people came to Wisconsin during the spawning runs. There were dances, parades, and evening "smelting" matches where wrestlers fought in rings covered with two tons of smelt to see who could stuff the greater number of smelt into his opponents' pants (Becker, 1983).

Commercially, large numbers of smelt have been taken by trawling. From 1982 to 1994, the United States and Canada have harvested 2,491 to 5,220 tons of smelt annually (FAO, 1994).

Northern Pike

Northern pike (*Esox lucius*), family Esocidae, are native circumpolar north of 40°N latitude (Toner and Lawler, 1969). Their range in North America has been expanded primarily to the Great Plains and Rocky Mountain states. In North America, lakes account for the majority of their habitat (Carlander et al., 1978). Machniak (1975) summarized much of the biological data on northern pike.

In the early spring around ice-out, adult northern pike migrate into shallow, often newly-flooded, marsh areas and broadcast their adhesive 2.3- to 3.2-mm-diameter eggs over shallow vegetation, (Kleinert and Mraz, 1966; Franklin and Smith, 1963). Adults frequently move more than 18 miles to their spawning areas (Moen and Heneger, 1971). Spawning occurs during the day (Scott and Crossman, 1973) and may be retarded by cloudy weather (McNamara, 1937). In North America, spawning temperatures have been reported to range between 34°F and 63°F and spawning depths between 2 to 24 inches (Machniak, 1975). One to three males accompany each female. Preigel and Krohn (1975) report fecundity to be 10,300 eggs per pound of female.

Northern pike usually hatch in 12 to 14 days, which is equivalent to 210 to 270 degree-days above 32°F (Becker, 1983). Newly hatched larvae have a sucker-type membrane on top of their head by which they attach to vegetation for four to 15 days until they absorb the yolk-sac (Franklin and Smith, 1963). Initially (11 to 13 mm TL) they feed on zooplankton; after 16 to 18 days (20 mm TL) they start to feed on insects; by the time they reach 45 to 50 mm in total length, they are feeding on fish (Machniak, 1975; Threinen et al., 1966; Franklin and Smith, 1963). "Water is required in the spawning marsh for at least three months after egg deposition for the best survival" (Becker, 1983). Adult northern pike feed primarily on fish, but they will prey on almost any organism found in the water, including young ducks.

An average growth rate obtained from the numerous studies listed in Carlander (1969) indicates that northern pike reach 8.4 inches in total length at age one; 23.8 inches at age five; 32.5 inches at age 10; 33.7 inches at age 15; and 42.2 inches at age 20. The standard length-weight relationship for northern pike is log weight = –3.727 + 3.059 log total length (Willis, 1989). Northern pike weighing over 50 pounds have been reported, especially from Europe (Scott and Crossman, 1973). In aquaria, they have lived for 75 years (Lagler et al., 1977).

Inskip (1982) prepared a set of habitat index models for northern pike. According to his models, optimum conditions include a spawning area with loose vegetation greater than 80 percent; dense vegetation throughout the six inches of water column above the substrate; less than a two-inch drop in water level during embryo and fry stages; 25 to 75 percent of midsummer area with emergent or submerged aquatic vegetation; pH from six to nine; 120 to 170 frost-free days; maximal weekly average temperatures between 67°F and 79°F with oxygen greater than 1.5 ppm; 100 percent backwaters and pools; and in streams less than four feet per mile of gradient.

Northern pike are no longer harvested commercially in the United States except from certain tribal waters. Canadians harvested 2,513 tons of northern pike in 1994 (Table 1). A few northern pike are harvested to stock as a sportfish in other bodies of water. The capture gear most frequently used are gill nets, fyke nets, and trammel nets.

Common Carp

Carp (*Cyprinus carpio*), family Cyprinidae, was originally from Asia, and it reached Europe prior to 1758 (Burns, 1966). According to McGeachin (1993):

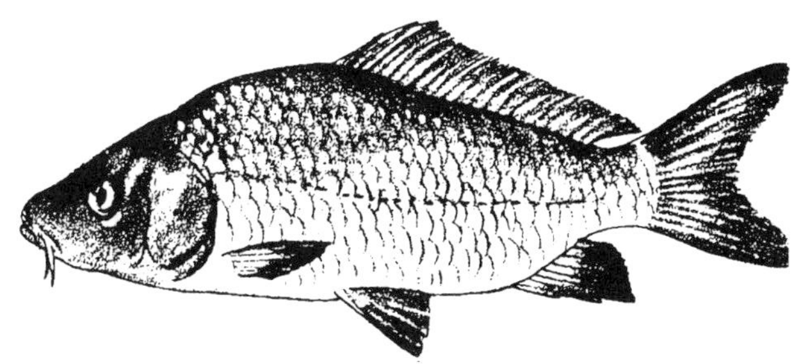

Figure 1b. Carp.

The common carp was first brought to North America in 1831 (Balon, 1974), and the first domestic carp were successfully introduced by Rudolph Hessel in 1877 for the newly-created United States Fish Commission (Doughty, 1980; Bowen, 1970). The 227 leather-and-mirror carp and 118 scaled carp in the shipment were temporarily kept in Baltimore and then moved to Washington, D.C., where they were bred in the reflecting ponds of the Washington Monument.

The United States Fish Commission started a program to rejuvenate the rapidly depleting inland fisheries of the United States by stocking carp in public waters and fish ponds on the request of private individuals. The intent was to provide inexpensive protein in the face of depleted wildlife and fish stocks and rapidly rising livestock prices. Carp was particularly popular with, and requested by, European immigrants and their descendants who were familiar with the fish and its culture. The first carp were distributed in 1879, and by 1882, some 79,000 had been stocked in 298 of the 301 Congressional districts (Doughty, 1980; Bowen, 1970). By the end of the program in 1896, 2.2 million fingerlings had been distributed to about 50,000 individuals throughout the United States, and about 250,000 fish had been stocked directly into public waters (Doughty, 1980).

Carp spread rapidly and are now found in North America below the 50th parallel down into the northern fourth of Mexico, except in central and southern Florida (Figure 1b).

Habitat

Carp abound in large rivers, streams, and reservoirs. They thrive in eutrophic waters where they can tolerate low levels of oxygen and perturbances in temperature and pollution. In fact, they, along with goldfish (*Carassius auratus*), often are the only fish species found in many highly polluted waters. Young carp can tolerate up to 6,000 milligrams/liter chloride and their eggs can tolerate 4,500 milligrams/liter (Nakamura, 1948).

Reproduction and Early Development

Carp move into the shallows in the spring and broadcast their eggs most actively between 65°F and 68°F (Sigler, 1958). They move into shallow water where one to six males per female compete to fertilize the adhesive eggs which fall on vegetation, rocks, limbs, etc. Fecundity tends to range from 50,000 to 125,000 eggs per pound (Carlander, 1969). The 0.9- to 2.0-mm-diameter eggs hatch in three to five days at 68°F (Mansueti and Hardy, 1967). Young carp feed on small crustaceans; larger carp feed on animal material, primarily insect larvae, crustaceans, mollusca, and annelids (Moen, 1953). Small carp are very vulnerable to largemouth bass predation, and for this reason they do not recruit well in small ponds containing bass; they grow very rapidly to a nonvulnerable size. Averaged data from Carlander (1969) indicate they reach 6.6 inches in total length at age one; 14.7 inches at age three; 21.6 inches at age six; and 24.9 inches at age nine. They have been known to live for 47 years (Emig, 1935). The largest carp caught in the United States was 57.8 pounds, but the world record was an 83.5-pound fish caught near Pretoria,

Figure 1c. Trammel netting for carp and buffalo. *Courtesy of Illinois Department of Natural Resources.*

dard length-weight relationship is log weight = –3.059 + 2.859 log total length (Stephen, 1978). Standing stocks of carp are often several hundred pounds per acre (Table 5).

Initially the introduction of carp into the United States was considered very successful. Later the negative ecological consequences became more clear. The carp, with the help of deteriorating habitat, has displaced many native species. Often it is the most abundant species in terms of both numbers and biomass (Heidinger, 1989; Becker, 1983). Carp feed on the bottom and they tend to uproot vegetation and maintain silt in suspension. They feed on the same organisms as other species such as the bluegill *(Lepomis macrochirus)* and the redear (*L. microlophus*), but carp have the ability to feed on the infauna portion of the population and not only those organisms that are exposed in the water column. Thus they eat the principal as well as the interest.

Harvest

Since carp compete heavily with both sport species and non-game species, many states allow it to be commercially harvested in unlimited quantities and with no size restriction. In the United States, state agencies often allow the commercial harvest of carp and buffalo in large reservoirs. In theory, harvest is permitted in order to utilize resources that are not normally used and to reduce intraspecific competition. Normally this fishery has limited entry and is only permitted during winter months, thus reducing conflict between commercial and sport fishermen. A variety of fishing gear is used including gill nets, trammel nets, hoop nets, and large haul seines. Carp tend to school during the winter months. A single seine haul from Lake Koshkonong, Wisconsin, in 1973 produced 750,000 pounds of carp (Becker, 1983). In addition to marketing carp for human consumption at $0.07 to $0.35 per pound in the round (Table 3), large carp are sold for use in payfishing ponds. In 1991, world-wide production of both wild-caught and cultured carp was 1,793,171 tons (FAO, 1994). Currently, carp are not cultured in North America (McGeachin, 1993) (Figure 1c).

Buffalo

Several species in the family Catostomidae (suckers) are taken commercially from freshwaters of North America: bigmouth buffalo (*Ictiolus cyprinellus*), smallmouth buffalo (*I. bubalus*), black buffalo (*I. niger*), quillback (*Carpiodes cyprinus*), river carpsucker (*C. carpio*), highfin carpsucker (*C. velifer*), redhorses (*Moxostoma* spp.), and others. Due to space requirements, only the bigmouth and smallmouth buffalo, the most frequently-harvested species in the family Catostomidae, are covered in this chapter. All of these broadcast spawners primarily inhabit the Mississippi, Missouri, Illinois, and Ohio River drainages.

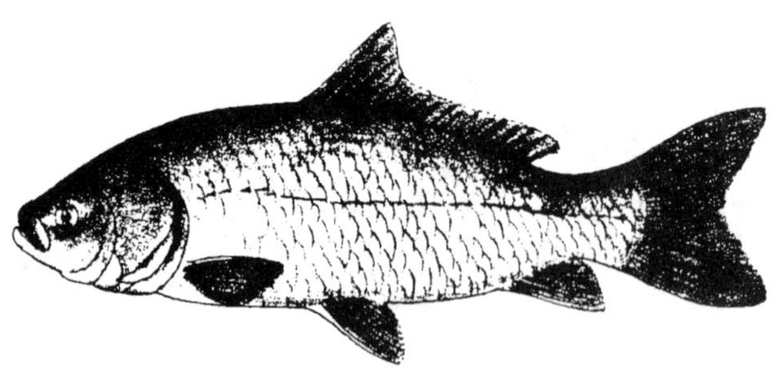

Figure 2. Buffalofish.

Only the bigmouth buffalo is cultured in North America and mostly in Arkansas (McGeachin, 1993). According to the habitat suitability index models of Edwards (1983), ideal conditions for bigmouth buffalo are 50 to 75 percent pools/backwater/marsh; monthly turbidity averaging less than 60 JTU; pH between 6.5 and 8.5; average current less than 11.8 inches/second; maximum salinity less than 4.5 parts per thousand (ppt); abundant inundated terrestrial, submergent, or emergent vegetation in spawning area; 25 to 75 percent littoral area; substantial rise in water level before spawning and stable level after spawning; 200 ppm or more total dissolved solids; and 25 to 75 percent vegetative cover.

The smallmouth buffalo is not cultured in the United States. Habitat suitability models indicate optimum conditions are 25 to 75 percent pool; average turbidity less than 40 JTU; pH 6.5 to 8.0; average summer water temperature for adult 71°F to 79°F, for embryo 62°F to 24°F, for fry and juvenile 75°F to 86°F; average minimum dissolved oxygen 6.0 ppm; average current velocity 60 to 80 centimeters/second for adults; average pool velocity less than 20 centimeters/second; and vegetation cover 30 to 40 percent (Edwards and Twomey, 1982).

The largest bigmouth buffalo known is an 80-pound individual taken from Spirit Lake, Iowa (Harland and Speaker, 1956). A 68.5-pound, smallmouth buffalo was taken from Lake Hamilton, Arkansas. Most bigmouth buffalo males reach sexual maturity at 14 inches and females at 19 inches in total length (Johnson, 1963). In Arkansas, these measurements correspond to age one males and age two females (Swingle, 1957). In South Dakota where growth is slower, some 90 percent of age three males and 95 percent of age eight females were sexually mature (Walburg and Nelson, 1966). Fifty percent of both male and female smallmouth buffalo are mature when they reach 14 inches in total length (Shields, 1958).

Bigmouth spawn in the spring at 54°F to 77°F (Becker, 1983) and smallmouth at 66°F to 82°F (Jester, 1973). Usually, upon a rise in water level they broadcast their eggs, which adhere to any substrate. Two or more males accompany each female. They prefer to move into newly-flooded vegetation. Smallmouth buffalo may spawn in slightly deeper water (four to 20 feet) than do bigmouth (Becker, 1983; Jester, 1973). There is no parental care. Based on only a few fish, both species have a relative fecundity of approximately 46,000 eggs per pound (Becker, 1983; Jester, 1973; Harland and Speaker, 1956). Ova average 1.5 mm in diameter and may make up 15 to 20 percent of body weight (Minckley, 1969, cited in Jester, 1973; Harland and Speaker, 1956). The eggs hatch in eight to 14 days (Nord, 1967; Harland and Speaker, 1956). Carlander (1955) reported a mean standing stock of bigmouth buffalo of 174 pounds per acre and smallmouth buffalo of 51 pounds per acre (Table 5).

Both the bigmouth and smallmouth buffalo are somewhat opportunistic feeders. Copepods and cladocerns are very important in the diet of bigmouth buffalo and zooplankton and attached algae are often eaten by smallmouth buffalo. A large variety of other organisms such as mollusca, insects, protozoa, etc., have been found in both species (Minckley et al., 1970; Cross, 1967; McCamish, 1967; Walburg and Nelson, 1966).

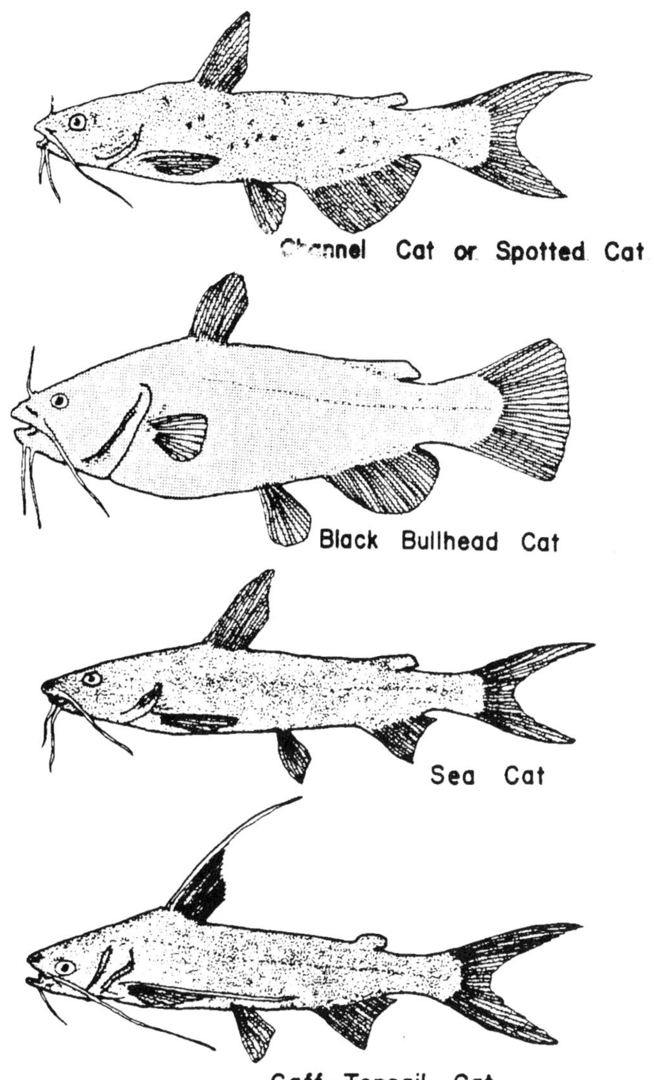

Figure 3. Catfish.

A summarization of age and growth data presented by Carlander (1969) indicates that the bigmouth buffalo grows faster than the smallmouth buffalo. Age one to 10 bigmouth buffalo grew to a mean of 6.3, 11.2, 14.4, 16.4, 18.0, 20.4, 23.3, 24.8, 27.0, and 27.7 inches in total length, respectively; whereas age one to nine smallmouth buffalo grew to a mean of 4.6, 8.2, 11.2, 13.1, 14.6, 16.6, 17.9, 19.7, and 21.2 inches in total length, respectively. Large individuals must be very old. The standard length-weight relationship for bigmouth buffalo is log weight = –3.263 + 3.092 log total length. For smallmouth buffalo it is log weight = – 3.383 + 3.092 log total length (Stephen, 1978). In 1982, more than 17,600 tons of buffalo were harvested from the United States; by 1994, the catch was reduced to 987 tons (Table 1). Buffalo are primarily harvested by gill nets, trammel nets, and hoop nets. Only a few anglers fish for buffalo.

Current prices range from \$0.19 to \$0.35 per pound in the round (Table 3).

Channel Catfish

The channel catfish *Ictalurus punctatus* (Rafinesque), family Ictaluridae, is an important species both for recreational angling and aquaculture as well as in the commercial wild-capture fishery (Becker, 1983). Americans ate 0.98 pounds of catfish in 1993, ranking it sixth behind salmon (0.99), cod (1.0), Alaskan pollock (1.2), shrimp (2.5), and tuna (3.5) (Talley, 1994). Most of the catfish eaten was channel catfish.

Distribution

Its native range includes most of the central United States east of the Rocky Mountains to the Appalachians, south to the Gulf Coast and into northern Mexico, with the northern limit including small portions of central and eastern Canada (Scott and Cross, 1973). The channel catfish is one of the most widely stocked and managed species in the United States (Smith and Reeves, 1986), and since it was first artificially propagated at Durant Hatchery in 1925 (Finnell and Jenkins, 1954), its range has been gradually expanded to all states except Alaska, Hawaii, and Maine. Smith and Reeves (1986) reported that as of 1984 to 1985, 35 states were stocking channel catfish (Figure 3).

Habitat Requirement

While channel catfish are found in a wide variety of habitats, they prefer moderate- to swift-moving streams, but also commonly inhabit lakes and reservoirs (Miller, 1966). McMahon and Terrell (1982), after an extensive literature review, formulated both a riverine and a lacustrine habitat suitability index model for channel catfish. A total of 18 variables were used in the models. The optimum variables are in summer 40 to 60 percent riverine pools and 15 to 30 percent littoral zone in lakes; in summer 40 percent cover (logs, boulders, brush, timber); rubble, gravel and boulders in riffle-run area and greater or equal to 30 percent vegetation cover in pools; average midsummer water temperature within pools, backwater, or littoral zone for adults 82.4°F to 85.1°F, juveniles 84.2°F to 86°F, fry 82.4°F to 86°F and embryo 78.8°F to 84.5°F; maximum salinity during summer for adults less than one ppt, fry less than 5 ppt, and embryo less than 3 ppt; 200 or more frost-free days per year; monthly average turbidity 100 to 110 JTU; average minimum dissolved oxygen 7 ppm within pools and backwaters; impoundments with a storage ratio of 0.8 to 0.95; impoundments with monthly average level of total dissolved solids during summer from 100 to 350 ppm; maximum impoundment retention time of six days while fry are present; and average current velocity in cover areas during average summer flow less than 0.5 feet per second.

Spawning

Age at maturity is highly variable and appears to be a factor of size rather than age, with most channel catfish maturing between 10 and 12 inches TL (Miller, 1966). Appelget and Smith (1951) reported that none of the channel catfish from Pool Nine of the Mississippi River was mature before age five, but all the males and 90 percent of the females were mature by age eight. In the southern portions of their range where growth is faster, channel catfish mature earlier. In Texas ponds, they have been reported to mature in 18 months (McClellan, 1954), and by age two in Kansas (Davis, 1959).

Spawning generally occurs during late spring and early summer as the water temperature reaches 75°F to 85°F (Scott and Crossman, 1973). Optimum temperature appears to be 80°F (Becker, 1983). In a power cooling pond that maintains a near-constant 80°F, they spawned two months before a nearby native riverine population (Scott and Crossman, 1973).

Channel catfish are cavity spawners and the male generally constructs the nest in a dark secluded area, such as an undercut bank, a hole in a rock pile, a log jam, or a hollow log. The male builds the nest by fanning the area with his tail (Davis, 1959). Yellow eggs (0.13 to 0.15 inches in diameter) are laid in an adhesive mass in the nest. They are fertilized and guarded by the male (Carlander, 1969). Fecundity is between 3,000 and 4,000 eggs per pound of female depending on size. Females spawn once per season, whereas a male may spawn multiple times (Scott and Crossman, 1973). During the five- to 10-day incubation period (at 70°F to 85°F), the male guards the nest from predators and fans the eggs with his body (Becker, 1983; Miller, 1966). A good description of the complete spawning activity is given by Clemens and Sneed (1957).

Young fry remain in the nest for two to five days, utilizing the nutrients in their large yolk-sacs, before swim-up (Scott and Crossman, 1973). After the school disperses, young juveniles (1-inch) feed in shallow areas over sand bars, around drift piles, and among rocks (Becker, 1983). Young channel catfish are very vulnerable to predation by such species as the largemouth bass.

The diet of channel catfish, according to Carlander (1969), varies with age. Fry feed primarily on aquatic insects and benthic arthropods until they reach about four inches TL, at which time the young become omnivorous or piscivorous, depending on available forage. The diet of older channel catfish may include mayflies, caddisflies, chironomids, mollusca, crayfish, crabs, green algae, large water plants, tree seeds, and other fishes (Scott and Crossman, 1973). Adult catfish feed typically on the bottom in a random pattern, using both touch and smell to detect food (Becker, 1983). They are most active during the period from sunset to midnight (Miller, 1966), but they do feed during daylight hours as well (Scott and Crossman, 1973). Finnell and Jenkins (1954) suggested that intraspecific rather than interspecific competition was more responsible for stunting. This idea was supported by Regier (1963) when he reported that the growth of stocked fingerlings was inversely correlated to the density of the population. Large channel catfish feed heavily on fish. The culture methods for channel catfish are fairly well-developed (Stickney, 1993).

Age, Growth, and Standing Stock

Pectoral spines are commonly used to age channel catfish (Sneed, 1961). Age and growth data on channel catfish from numerous areas across the United States were reviewed and tabulated by Carlander (1969). Growth rates from selected studies are given in Table 8. The standard relative length-weight equation for channel catfish is log weight = –3.749 + 3.243 log total length (Anderson, 1980). The world record channel catfish was caught in 1964 from the Santee Cooper Reservoir, South Carolina, and weighed 58 pounds [Illinois Department of Conservation (IDOC), 1989].

Most standing crop data has been obtained by rotenoning small lakes or small coves. Average standing crop from 26 bodies of water (ponds, reservoirs, rivers) was 28 (0.2 to 300) pounds per acre (Miller, 1966). Two large rotenone studies have been conducted (Table 5). One was a 115-acre arm of the 30,000 acre Douglas Reservoir in Tennessee in which the standing crop of channel catfish was four pounds per acre (Hayne et al., 1967). In a similar rotenone study on a 210-acre arm of Barkley Lake (57,200 acres) on Kentucky's portion of the Cumberland River, biologists collected 46 pounds per acre of channel catfish (Aggus et al., 1979).

Movement and Escapement

Channel catfish are often stocked for recreational fishing in ponds and lakes, from which they move into rivers. Stewart and Murawski (1973) reviewed numerous lacustrine stockings in New Jersey waters which received either hatchery-reared and/or wild-river stock channel catfish. Their findings indicated that the wild river stock had a much greater tendency to escape from the stocked lakes than did hatchery-reared fish. Other authors (Powell, 1976; Tiemeier, 1966; Brown, 1965) have also reported escapement of channel catfish, especially during periods of high water.

In an effort to quantify the escapement of stocked channel catfish, Storck and Newman (1986) caught fish escaping from a 15-acre Illinois impoundment. During one year, 22 percent of the four-inch, 12 percent of the six-inch, and 9 percent of the eight-inch TL stocked channel catfish escaped either through a tower system drawing bottom water or over the spillway. No studies have specifically measured the loss of stocked channel catfish from larger reservoirs, but examinations of populations from the stilling basins below some reservoirs have given some indications. Canton Reservoir in Oklahoma has a stilling basin below the dam which has rock piles downstream of the basin that restricts most upstream migration. It was, therefore, assumed that most, if not all, of the fish harvested from the stilling basin emigrated from the reservoir. Channel catfish made up 20 percent of the total standing crop in the basin, both in terms of number and weight. Of the channel catfish harvested from the basin, by far the greatest number were in the 5- to 9-inch TL range (76 percent), and these accounted for 50.5 percent of the total channel catfish biomass. Channel catfish larger than 10 inches TL made up 23.3 percent by number and 49.2 percent by weight, while those 4 inches and less accounted for less than 1 percent

Table 8. Growth Rate of Channel Catfish from Selected Waters of North America

Body of Water	Back calculated total length in inches at annulus										Reference
	I	II	III	IV	V	VI	VII	VIII	IX	X	
Salt River, Missouri	2.4	4.4	6.7	8.9	10.7	12.5	14.0	15.8	19.9	—	Purkett, 1958
Des Moines River, Iowa	1.8	4.9	7.7	10.1	12.3	15.0	17.4	19.3	21.5	24.3	Muncy, 1959
Mississippi River, Iowa	2.9	6.3	9.1	11.7	14.1	16.6	19.1	21.1	24.0	26.6	Appleget and Smith, 1991
Lake Havasu, California	3.0	5.4	6.6	7.7	9.3	11.2	13.8	15.3	17.9	19.8	Kinsey et al., 1957
Lake of the Ozarks, Missouri	2.1	4.3	6.1	7.1	9.2	10.4	11.5	13.0	—	—	Marzolf, 1955
Lake Moultrie, South Carolina	3.4	7.3	11.2	14.5	17.4	20.9	23.7	26.2	28.6	30.4	Stevens, 1959

either by weight or number (Moser and Hicks, 1970). Walburg (1971) examined the loss of age zero from Lewis and Clark Reservoir, a main stream reservoir on the Missouri River. He reported a peak 24-hour loss of 170,000 channel catfish. Most were less than one inch in length and the numbers increased when the flushing rate was less than seven days. Water was normally drawn from 10 feet below the surface, and the typical flush rate was seven to 10 days, with a maximum of 5.5 days. In the event of high water, spillway gates (at a depth of 26 feet) could be opened and the flush rate could be increased to as little as four days. The author reported that age zero channel catfish escapement was associated with the flush rate as well as the time of year; channel catfish escapement peaked in late July to mid-August. Tagged channel catfish are known to move considerable distances in rivers. Many fish have been shown to move 100 to 200 miles (Ranthum, 1971; Hubley, 1963). One fast-moving individual traveled 111 miles in 36 days.

Stocking

Under controlled experimental conditions, the size of stocked channel catfish as it relates to the amount of predation by largemouth bass has been evaluated. Mestl (1983) reported that a largemouth bass can consume a channel catfish whose total length does not exceed 60 percent of the largemouth bass. Channel catfish, 2 to 8 inches TL, were stocked in both aquaria and ponds with 2-pound largemouth bass to determine size vulnerability (Krummrich and Heidinger, 1973). The authors found that 7- to 8-inch channel catfish were relatively immune to predation. Spinelli et al. (1985) tested not only size vulnerability of channel catfish to largemouth bass, but also the effect of an additional vulnerable forage on the predation rate. Three sizes of channel catfish (2.5 to 3.5, 5.5 to 6.5, and 8 to 9 inch TL) were stocked at 10 each with three sizes of largemouth bass (11 to 12.8, 14.8 to 16.3, and 17.6 to 20 inch TL) with one bass per trial. The presence of minnows as a vulnerable forage greatly reduced predation of channel catfish by largemouth bass. By Lawrence's (1958) mouth-gap calculations, all sizes of channel catfish used in the study were vulnerable when their spines were retracted; however, during most of the experiment, the author noted that the catfish kept their spines extended while in the presence of a largemouth bass. None of the large channel catfish (8- to 9-inch TL) was killed or eaten.

In 1977 and 1978, 7-inch TL channel catfish were stocked at 10 per acre in Iowa's Lake Anita. In 1979 and 1980, 4-inch TL channel catfish were stocked at 40 per acre. Ninety percent of the 7-inch fish were creeled in the first two years following stocking. The 4-inch fish were totally absent from the creel for the two years following stocking (Adair, 1981).

Storck and Newman (1987) tested the relative vulnerability of different size channel catfish to largemouth bass in a 15-acre Illinois lake where harvest and escapement were known. In addition, the lake was drained at the study's completion to recover remaining fish. When two sizes of channel catfish were stocked (greater than and less than 8 inches TL) the natural mortality of the smaller channel catfish was 0.85, whereas those stocked at a size greater than 8 inches had a mortality rate of 0.31. Dudash (1986) stocked equal numbers of 3- to 4-, 4- to 5-, 5- to 6-, 6- to 7-, and 7- to 8-inch TL marked channel catfish into a 60-acre lake in Jackson County, Illinois. Relative survival rates were 0.33, 0.67, 7.3, 14.0 and 18.0 percent, respectively. Thus the 7- to 8-inch size group was recovered 50 times more frequently than the 3- to 4-inch size groups, 25 times more frequently than the 4- to 5-inch size group, 2.5 times more frequently than the 5- to 6-inch size group and 1.3 times more frequently than the 6- to 7-inch size group.

Harvest

Most channel catfish are commercially harvested year-around with trot lines, hoop nets, gill nets, and trammel nets. Slat traps are used primarily during the spawning season. In the Mississippi River drainage system, it is often the second or third most frequently captured species after buffalos and carp. In 1994, the FAO reported that excluding production from aquaculture, a total of 5,799 tons of all species of catfish were caught in the United States (Table 1). The price of wild-caught channel catfish ranged from $0.35 to $0.75 a pound (Table 2). To place this in perspective, during 1994 sales of cultured channel catfish in the United States were expected to be between 435 and 445 million pounds, a 3 to 5 percent decrease from 1993. Farm level prices averaged $0.785 a pound, for a gross farm revenue of $340 million (USDA, 1994).

Blue Catfish

The blue catfish (*Ictalurus furcatus*), family Ictaluridae, is very similar in biology to the channel catfish. It is a large catfish native to the major rivers such as the Mississippi, Missouri, and Ohio of the central and southern United States. Its range extends into Mexico and northern Guatemala and it has been introduced on the Atlantic slope and into California (Jenkins and Burkhead, 1993).

Blue catfish are a migratory, big-river species (Jordan and Evermann, 1916). They inhabit deep swift chutes and flowing pools (Pflieger, 1975). Blue catfish can tolerate salinities of at least 11.4 percent (Perry, 1968).

The blue catfish is a cavity spawner much like the channel catfish (Harlan and Speaker, 1956). Young and small juveniles feed on invertebrates, and individuals above 9 to 12 inches feed on insects, mollusca, and fish (Harlan and Speaker, 1956). According to Jenkins and Burkhead (1993, citing other authors), blue catfish reach sexual maturity at approximately 5 pounds, which is 16 to 24 inches in total length and four to eight years in age. A 97-pound individual was taken from the Missouri River, South Dakota, in 1959, and fish over 200 pounds have been reported (Pflieger, 1975). Growth rates are variable; the mean from four studies cited in Carlander (1969) is age one: 6.0; age five: 18.5; and age 10: 38.7 inches in total length. Few standing crop values for blue catfish are available in the literature. Two large rotenone projects yielded 6 and 8 pounds per acre (Table 5). Harvest methods are similar to those used for channel catfish and flathead catfish. Blue catfish also wholesale for a similar price of $0.36 to $0.75 per pound (Table 3).

White Catfish; Black, Brown, and Yellow Bullhead

There are seven freshwater species in the genus *Ameiurus*, family Ictaluridae. The most frequently captured commercially are the white catfish (*A. catus*), black bullhead (*A. melas*), brown bullhead (*A. nebulosus*), and yellow bullhead (*A. natalis*). Their distribution and selected biological characteristics are given in Table 9. All of these species tend to be found in lakes and rivers. Slow flows, logs, vegetation, etc., appear to increase their abundance.

Stuber (1982) proposed habitat suitability models for the black bullhead: optimum variables include 50 to 75 percent pools or backwaters; greater than 25 percent cover; less than four cm/second current velocities; maximum midsummer water temperatures 64°F to 85°F; pH 6.5 to 8.5, salinity less than two ppt; 25 to 100 ppm turbidity; 100 to 600 ppm total dissolved solid; and greater than 25 percent littoral zone. Standing crops of bullheads range from less than one to 300 pounds per acre (Table 5). Bullheads are primarily harvested by set lines, slat traps, and basket traps (Dufford, 1994). The wholesale price of bullheads in the round range between $0.21 and $0.50 per pound (Table 3). Some bullheads are sold to payfishing operators.

Flathead Catfish

According to Lee and Terrell (1987), the flathead catfish, *Pylodictis olivaris*, family Ictaluridae, is native to large rivers in the Mississippi River and Rio Grande drainages, including northeastern portions of Mexico. It has been introduced into Florida, South Carolina, Idaho, Oregon, Washington, Arizona, and California (Gholson, 1970). Flathead catfish are primarily a stream or river species, but have been introduced into a number of man-made lakes. The species has also been successful in reservoirs created by damming large rivers. Adult flathead catfish are often found around submerged logs, rocks or other cover (Smith, 1979; Pflieger, 1975).

Habitat suitability index models of Lee and Terrell (1987) suggest that optimum conditions include stream grades of less than two m/km; highest turbidity of 70 to 140 JTU more than 50 percent of the year; mean summer velocities less than three m/second; 15 to 30 percent riffles; less than 25 percent runs; and 60 to 90 percent pools. Spawning habitats also need to be available. Flathead catfish are cavity spawners. They spawn in holes in the bank, under rock outcropping and in hollow logs, etc., where the male guards the nest.

Males mature between 13 and 18 inches in total length, which is reached between three to five years of age, and females mature at 18 to 23 inches between four to seven years of age (Turner, 1977; Turner and Summerfelt, 1971, 1970; Barnickol and Starrett, 1951). Flatheads spawn in the late spring at temperatures of 68°F to 82°F (Turner and

Table 9. Distribution and selected biological characteristics of four catfish species in the genus *Ameiurus*

Characteristic	White catfish	Black bullhead	Yellow bullhead	Brown bullhead
Distribution	Eastern coastal U.S. streams from New Jersey to Florida. Stocked in California and midwest (Hubbs and Lagler, 1958; Troutman, 1957)	Temperate North America (Scott and Crossman, 1973)	Eastern and central U.S., southern Ontario, northeastern Mexico. Introduced into California and some southeastern states (Becker, 1983; Scott and Crossman, 1973)	Eastern and central North America. Introduced into California (Becker, 1983)
Nest type	Circular in gravel, sand or silt in 1-5 foot of water (Fowler, 1917)	Circular beneath obstructions in 1-5 feet of water (Wallace, 1967)	Saucer-shaped beneath obstructions in 1.5-5 feet of water (Adams and Hankinson, 1926)	Circular nest near obstructions and in cavities (Breder, 1939)
Parental care	Male (Prather and Swingle, 1960)	Female constructs nest; both guard (Wallace, 1967; Breder and Rosen, 1966)	Male (Eddy and Surber, 1947)	Both guard (Breder, 1939)
Spawning temperature (°F)	70 (Murphy, 1951)	68 (Breder and Rosen, 1966)	70	70-77 (Mansueti and Hardy, 1967)
Fecundity	3,500/12 inch fish (Menzel, 1945)	3,845 9-10-inch fish (Forney, 1955)	3,950-4,270 10-11-inch fish (Vessel and Eddy, 1941)	2,000-13,000 8-13-inch fish (Scott and Crossman, 1983)
Egg diameter (mm)	4.0-4.5 (Menzel, 1945)	1.2-1.6 (Becker, 1983)	2.2-3.0 (Becker, 1983)	3 (Breder, 1935)
Days to hatch (°F)	6-7 at 75-85 (Prather and Swingle, 1960)	5-10 (Wallace, 1967)	5-10 (Becker, 1983)	6-9 at 69-74 (Breder, 1935)
Food of young	Insects (Goodson, 1965)	Zooplankton (Applegate and Mullan, 1966; Alson, 1971)	NA	Zooplankton, insect larvae (Rayney and Webster, 1940)
Food of adults	Omnivorous (Miller, 1966)	Omnivorous (Stuker, 1982)	Omnivorous	Omnivorous
Sexual maturity (inches)	7-8, 3-4 years (McCammon, 1957; Menzel, 1945)	7-10, 3-5 years (Dennison and Buckley, 1972; Carlander, 1969)	Less than 7 inches, 2-3 years (Dill, 1944)	Less than 8 inches, 2-3 years
Total length at age (inches)	Age I-3.2, age V-12.8, age X-18.6 (Stevens, 1959)	Age I-3.7, age V-12.3 (Houser and Collins, 1962)	Age I-2.7, age V-14.1 (mean from Carlander, 1969)	Age I-3.9, age V-12.5 (mean from Carlander, 1969)
Maximum weight (pounds)		8.0 (IDOC, 1989)	5.2 (IDOC, 1989)	5.5 (IDOC, 1989)

NA: Not Available

Summerfelt, 1970; Henderson, 1965; Snow, 1959). Fecundity of 2.5- to 25.6-pound females ranged from 4,076 to 31,579 eggs per female (Summerfelt and Turner, 1971). After the eggs are laid, the male drives the female away (Fontaine, 1944). The adhesive 2.8- to 3.7-mm-diameter eggs are laid in a mass that the male fans and physically moves around with its fins and mouth. The eggs require six to nine days to hatch at 75°F to 82°F (Guidice, 1965; Snow, 1959). At hatching the fry are four to 11 mm long. The male guards the fry for several days (Breder and Rosen, 1966; Fontaine, 1944). The young feed mainly on aquatic insects (Cross, 1967). Juveniles between eight to 20 inches in total length feed on benthic macroinvertebrates, and fish longer than 20 inches are piscivorous (Lee and Terrell, 1987; Becker, 1983).

Twenty- to 40-pound fish are commonly harvested. The record fish is a 98.6-pound individual caught from Lake Lewisville, Texas, in 1989 (IDOC, 1989). Gammon (1973) proposed an optimum temperature range of 89°F to 92°F for adults. During winter months, they may go into a "pseudo-hibernation state, embedding themselves in muddy holes in 40 to 60 feet of water" (Bachay, 1944). From a number of populations and studies cited in Carlander (1969), the mean total length of flathead catfish at one, five, 10, and 11 years of age is 4.3, 20.0, 35.3 and 43 inches, respectively. Murphy et al. (1991), citing Laue, state that the standard length-weight relationship is

$$\log \text{weight} = -3.485 + 3.082 \log \text{total length}.$$

For a big-river fish, flathead catfish seem to move very little. Seventy-one percent of tagged adult catfish moved less than five miles in a year (Funk, 1957).

Flathead catfish are harvested with set lines using live fish as bait and in hoop nets and entanglement nets. Their meat is considered to be of high quality and they are also used to some extent as a trophy fish in payfishing operations. Recent wholesale prices paid for flathead catfish in the round ranged from $0.35 to $0.75 per pound (Table 3).

White Perch

White perch (*Morone americana*), family Percichthyidae, have an estuarine native range from Nova Scotia to South Carolina (Scott and Crossman, 1973). They have been stocked or landlocked by dams in many freshwater systems east of the Appalachians, and have spread through the Great Lakes into the Mississippi River drainage. In 1994 personnel from Southern Illinois University collected this species from the Mississippi River near the mouth of the Ohio River.

White perch broadcast their adhesive eggs (Todd, 1986) in the spring at water temperatures of 50°F to 61°F (Mansueti, 1964; 1961). They typically reach sexual maturity at total lengths greater than approximately five inches and at age two to four (Mansueti, 1961). Their 0.92-mm- diameter eggs hatch in four days at 59°F and 30 hours at 68°F (Scott and Crossman, 1993). Fecundity of 6-inch to 10-inch perch ranged from 15,740 to 247,681 ova (Sheri and Power, 1968). They typically live for four to seven years, but a 17-year-old individual was reported from Maine (Jenkins and Burkhead, 1993). White perch feed on a variety of organisms including microcrustaceans, grass shrimp, crayfish, insect larvae, and fish (Jenkins and Burkhead, 1993, citing five articles).

A 4.75-pound individual was reported by Carlander (1953). In Lake Ontario they reached 3.9 inches at age one and 10.1 inches in total length at age five (Scott and Crossman, 1973). The white perch is both a commercial and sport species. In 1994, the United States reported a harvest of 823 tons (Table 1). In 1993, they were valued at $0.60 per pound (Table 3).

Walleye

The ease with which walleye (*Sizostedion vitreum*), family Percidae, can be artificially propagated and their desirability as a sport fish have resulted in massive annual stocking programs by many state, federal, and private hatcheries. Over 900 million walleye were stocked in North America during 1984 (Conover, 1986). According to Conover, the five leading states in terms of numbers stocked during 1983 and 1984 were Minnesota, New York, Iowa, Wisconsin, and Illinois. The success of these introductory, maintenance, and supplemental stockings has varied considerably. Literature on the life history of the walleye has been reviewed by a number of authors (McMahon et al., 1984; Colby et al., 1979; Machniak, 1975; Scott and Crossman, 1973) and is summarized in the following section. Extensive bibliographies have been compiled by Addison and Ryder (1977) and Ebbers et al. (1988).

Distribution

The original distribution of the walleye is "limited to the fresh waters of Canada and the United States with rare occurrences in brackish water" (Scott and Crossman, 1973). Its northward penetration into Canada approximates the 55.4°F-mean July isotherm. It ranges from near the Arctic Coast in the MacKenzie River southeastward through Quebec and the St. Lawrence River and southward to the Gulf Coast in Alabama. Much of the distribution is restricted to east of the foothills of the western Cordillers and west of the Appalachian Mountains, although there is a residual, apparently native, stock along the Atlantic Coast from Pennsylvania to North Carolina (Colby et al., 1979). According to Carlander and Payne (1977), Clear Lake, Iowa, was the southernmost lotic environment of the walleye. Due to stocking, walleye populations now exist as far south as Texas.

Habitat Requirements

McMahon et al. (1984) formulated habitat suitability index models for walleye. Optimum suitable habitat for the variables examined are as follows: Secchi reading depth during the summer of 5 to 12 feet; relative abundance of small (<4.7 inches) forage during spring and summer of greater than 400 mg-prey/m; percent of area with cover and dissolved oxygen levels greater than three ppm during spring and summer between 25 and 45 percent; pH levels between 6.5 and 8.4; minimum dissolved oxygen in the pools or above the thermocline during the summer, for adults and juveniles greater than four ppm, for fry greater than five ppm, and for embryos greater than 6.5 ppm; mean weekly water temperature during the summer for adults and juveniles between 70°F to 77°F; for fry, 64.4°F to 75.2°F; and for embryos, 51.8°F to 64.4°F; number of degree days between 39°F and 50°F from October 3 through April 15 between 750 to 1100; water level in pool or impoundment during spawning season either rising or stable at normal level; mesotrophic conditions; and spawning area consisting of gravel and rubble incorporating at least 20 percent of the area greater than one foot but less than five feet.

Spawning

Walleye are broadcast spawners; their adhesive eggs are laid on clean gravel in the ripples of streams and on shorelines or offshore reefs in lakes. A few populations in Wisconsin spawn in marshes (Priegel, 1970). "As a consequence of their tendency to return to the chosen spawning sites, stocks of walleye are probably composed of discrete sub-populations identified with particular spawning areas" (Crowe, 1962). Spawning begins shortly after ice-out over a temperature range of 42°F to 52°F (Scott and Crossman, 1973). Walleye normally spawn in water 3 to 10 feet deep (Breder and Rosen, 1966). Relative fecundity is variable and has been reported to range from 11,959 to 55,832 eggs per pound of body weight (Wolfert, 1969; Libbey, 1969). The relationship between number of eggs and weight tends to be linear (Wolfert, 1969).

A number of factors appear to affect relative fecundity. Baccante (1988) found that three years of exploitation (12 to 65 percent) of two Ontario lakes resulted in a statistically significant increase in the walleye's relative fecundity (from 18,709 per pound to 25,906 per pound in Henderson Lake; from 17,984 per pound to 19,569 per pound in Savanne Lake). He also noted that fecundity of Ontario walleye related to yearly lake conditions and overall food availability two years prior to spawn. Spangler et al. (1977) also reported that fecundity is dependent on severity of exploitation. As exploitation increases, walleye growth rate increases and age at first maturity decreases. Chevalier (1977) explained how over-exploitation by commercial fishermen on Rainy Lake walleye resulted in a decrease in the population. The severe exploitation caused an increase in the growth rate that resulted in the susceptibility of younger fish to the commercial fishermen's gill nets. Although the walleye were maturing at a younger age, their fecundity was less than that of older walleye. The commercial fishermen caught larger numbers of the smaller fish to compensate for their size, and the result was a drastic reduction in the quantity and quality of the walleye population. Anthony and Jorgensen (1977) reported the same problem in Lake Nipissing, Ontario. The young walleye were growing at a rate that caused them to be susceptible to gill nets before they could mature for spawning.

Low water discharge can cause egg retention which, in turn, probably lowers the hatch rates (Bush et al., 1975). Low temperatures at incubation cause slower embryo development, resulting in greater susceptibility of the eggs to predation and siltation. Nelson and Walburg (1977) reported that mean flow rates, changes in flow rates, and air temperature accounted for 55 percent of the variation of the adult year-class strength for a 10-year period (1965 to 1974) in Lake Oake, North Dakota. Mean flow rates accounted for 70 percent of the 55 percent variation for the river spawning walleyes. Chevalier (1977) found that spring water levels and broodstock numbers accounted for 65 percent of the variation of catch per unit effort by commercial fishermen in Rainy Lake. In Rainy Lake, most of the walleye spawn on beaches with gravel, rubble, or shingle rock in water less than one foot deep. In low water years, there is less suitable spawning substrate and poor egg survival. If there is a rapid lowering of water levels in the spring, a large percentage of the eggs are exposed and destroyed.

In Lake Winnibigoshish, Minnesota, walleye utilize substrates of soft muck, sand, gravel, rubble, and boulders (Johnson, 1961). Egg survival varied from 0.6 percent in the soft muck to 35.7 percent in improved gravel-rubble substrate. The latter were sand beaches where rocks and rubble had been added to create more optimum spawning habitat. In those areas egg abundance increased 10 times over the number of eggs found on the beach substrates. Walleye generally selected gravel bottom substrates in water 12 to 30 inches deep when they were available. They avoided sandy beaches and often crowded into gravel areas only a few feet in diameter. Low water levels decreased the availability of spawning habitat and caused many walleye to use lower quality substrates for egg deposition.

Wind was also reported to be a factor in egg survival (Johnson, 1961). Waves dislodged eggs, and they were either caught in filamentous algae, swept on shore, or deposited into deeper water where they settled on the muck substrate. Johnson noted that the effect of the wind action was dependent on lake size, amount and duration of the wave action, and the substrate on which the eggs were deposited. Forney (1976) reported that walleye utilize gravel and rubble substrates in Oneida Lake and its tributaries. Wind action on the lake caused waves to scour the shoreline and dislodge the eggs. The tributaries were protected from the wind and waves. Forney postulated that even though there were fewer eggs deposited in those areas, their contribution to the walleye recruitment in Oneida Lake was higher than that of the lentic portion of the lake.

Bush et al. (1975) believed wind and temperature affected spawning success and subsequent year-class strength in Western Lake Erie from 1960 to 1970. Walleye started spawning at temperatures of 35.6°F and peaked at 44.6°F. The rate of temperature increase affected the length of the spawning period. They noted that temperature reversals also caused fully ripe females to hold their eggs, resulting in physiological damage to the eggs. If the eggs were held too long, the females began to absorb them. Strong winds also accounted for a substantial loss of eggs. The winds caused currents that moved the eggs and deposited them in mud bottom substrates where the mud-water interface had low levels of oxygen. Koenst and Smith (1976) reported that in Red Lake, Minnesota, optimum fertilization temperatures are 42.8°F to 53.6°F and 48.2°F to 59.0°F was best for incubation. The upper lethal limit for those walleye eggs was 69.8°F. Incubation lasted 50 days at temperatures of 41.0°F to 42.8°F and only six days at 68°F (Ney, 1978).

Recruitment, Food, and Growth

The factors responsible for natural recruitment of walleye fry and subsequent strong year-classes are much the same as those involved in optimal spawning success. Temperature is a major factor for determining walleye fry survival as it affects metabolism, forage abundance, and stress. Hokanson [1977, citing Smith and Koenst (1975)] reported walleye larvae required temperatures greater than 59°F to initiate feeding at swim-up. Subsequent growth was positively correlated until temperatures approach 88.9°F, which is the upper lethal limit for 4.5 inch walleye that have been acclimated at 78.4°F and subjected to a temperature increase of 5.4°F to 7.2°F per hour (Ferguson,1958). Koenst and Smith (1976) reported 78 to 98 percent survival from hatch to egg sac absorption at temperatures from 42.8°F to 69.8°F. Survival of fry to juveniles (3.2 to 3.9 inches) was one percent or less at temperatures from 47.2°F to 64.4°F. Eight percent of the fry survived to juveniles at 69.8°F. They reported upper lethal temperatures for juvenile walleye were 80.6°F to 88.9°F and that the wide range was due to acclimation temperatures. Forney (1977) stated that strong year-classes develop when growth is rapid because high-prey densities that enhance growth also reduce predation and cannibalism of young of the year (YOY) walleye. Abundant forage also acts as a buffer for YOY walleye predation as older walleye and other potential predators may concentrate on the most available prey (Ritchie and Colby, 1988; Ney, 1978; Forney, 1976). Koonce et al. (1977) stated that the temperature-dependent timing of the specific prey availability was critical for walleye fry development. A late-arriving prey population would result in more predation and cannibalism on walleye fry. They also noted that a strong year-class may prevent subsequent strong year-classes if prey availability is low and older fish must resort to cannibalism.

Forney (1965) noted that not all walleye of the same age class consumed the same diet, and that there was an increase in growth rates for walleye that preyed on fish as opposed to invertebrates. Even different invertebrates have different caloric content. For example, mayflies have a higher caloric value than amphipods (Kelso, 1973).

Walleye fry will generally convert to a fish diet during their first year of life. The conversion is dependent on available forage types, size, and numbers. Smith and Pycha (1960) reported that in Red Lake, Minnesota, larval walleye less than 2 inches ate copepods, cladocera, and insect larvae. Forage fish and YOY yellow perch comprised the bulk of the diet for the remainder of the first year. The spottail shiner (*Notropis hudsonius*) was the most common diet for walleye between 2.0 and 2.8 inches, and yellow perch fry was second. Rawson (1956) reported that in Lac La Ronge, Saskatchewan, YOY walleye consumed mostly amphipods but also preyed on crayfish, caddisflies, mayflies, and stonefly larvae until they reached 2 to 5 inches. They then switched to fish diets. Coscoes (*Coregonus* spp.) (56 percent), ninespine sticklebacks (*Pungitius pungitius*) (31 percent), white suckers (7 percent), and yellow perch (3 percent) were most utilized as prey. In Clinton Reservoir, Kansas, YOY walleyes' food preferences were largely size dependent. Walleye less than 0.5 inches (TL) preyed on cyclopoid copepods and to a lesser extent *Chaoborus* spp. and chironomids (Schademann, 1987). As the YOY walleye grew they switched to the larger calanoids, and *Daphnia* spp. gradually replaced *Bosminia* spp. Walleye longer than 0.9 inches had completely switched to a fish diet.

In Caesar Creek Lake, Ohio, YOY walleye fed primarily on copepods and cladocerans until they reached 1.8 inches in the summer, when they became almost entirely piscivorous on gizzard shad (Hurley and Austin, 1987). However, in the spring yearling walleye preyed on invertebrates due to a lack of forage.

Female walleye tend to grow larger, and live longer, than male walleye. Walleye, like many other fish species, tend to grow faster in the southern portion of their range than in the northern portion. They also initially grow more rapidly in newly impounded waters. Colby et al. (1979) summarized the growth rates of walleye from various authors from more than 70 bodies of water. One-year-old fish ranged from 64 to 460 mm; five-year-olds, 290 to 271 mm; and 10-year-olds, from 457 to 752 mm in total length. The standard length-weight relationship for walleye is log W = –3.642 + 3.180 log total length (Murphy et al., 1990). The largest walleye was caught in 1960 from Old Hickory Lake, Tennessee, and weighed 25 pounds (IDOC, 1989).

Movement and Escapement

Walleye movement can be separated into a diel cycle and a yearly migratory cycle. The diel cycle is dependent on factors such as temperature, light intensity, wind, turbidity, oxygen levels, and time of year. The yearly movements are a function of their age, the time of year, temperature regimes, and habitat availability. Walleye older than age zero show an inverse relationship between overhead light intensity and their vertical positioning in the water column (Scherer, 1976).

The walleye daily cycle of movement is primarily a result of the transformation of their retinal structure (Ryder, 1977). After hatch, and for much of their first year, walleye have a retinal response that is positively phototaxic. Before their second year, the walleye's subretinal "tapetum lucidum" causes them to become sensitive to bright light and improves their visual acuity under dim light situations. Their movement decreases during the lighter hours of the day, and they often move inshore during the evening hours to feed. They generally spawn at night, but have been known to stay on the spawning grounds during the day. The time they spend in the shallows other than at spawn is determined by water turbidity and light intensity during the day. Walleye will increase daytime activity in shallow water on days of low light intensity, windy days, turbid waters, or on cloudy days (Hokanson, 1977; Carlander and Cleary, 1949).

Yearly migrational movement for YOY walleye starts at hatch when they are photopositive and pelagic (Ney, 1978). They evade dispersal on windy days by moving into deeper water and concentrating in bays. They become demersal at about 1.0 to 1.6 inches (TL). Grinstead (1971) reported that in Canton Reservoir, Oklahoma, newly hatched walleye migrated from their spawn site in a nonstratified distribution throughout the reservoir during April and May, and were found along the shoreline in June and July. They were found offshore at greater depths in open water from August through October and concentrated near the bottom in the deeper part of the reservoir from November through February.

Forney (1976) reported that fry are pelagic until late June in Oneida Lake, New York. They usually make the transition to a more demersal mode in June when they attain a length of 1.4 inches, and it is complete by mid-July when they have grown to approximately 2.4 inches. Fingerling and adult walleye are widely distributed over both shoal and deeper mud-bottom areas during late summer and fall.

Evidently, walleye movements are partly regulated by homing. The phenomenon is learned or imprinted as adults, and its intensity varies as to the heterogeneity of stimuli between spawning locations and the walleye's ability to perceive and react to that stimuli. Walleye "home" to their spawning areas (Forney, 1963; Crowe, 1962).

Escapement of walleye from the target area, be it a river pool or an impoundment, can undermine a stocking program. LaJeone (1988) reported that the stocking of some upstream pools on the Rock River, Illinois, contributed more to the downstream pool than direct stocking of the downstream pool.

Jernejcic (1986) found that of the tagged age zero and one walleye released in Tygart Lake, 9 percent of those were captured below the spillway while only 6 percent were harvested in the lake. Loss of age zero fish from Lewis and Clark Lake, a mainstream reservoir on the Missouri River, was significant and one of the most frequently captured species (Walburg, 1971). During summer collections, it was estimated that up to 700,000 age zero walleye and sauger were lost from Lewis and Clark Lake on the Missouri River within a 24-hour period (Walburg, 1971).

Harvest

Carlander (1977) summarized standing stock data on 23 walleye lakes. Mean standing stock was 18 pounds per acre (range 1.2 to 41). Mean annual sportfishing yield from 48 bodies of water was 4.1 pounds per acre (0.2 to 14.4), and mean commercial harvest from 22 bodies of water was 1.3 pounds per acre (0.05 to 3.4). Walleye are normally commercially harvested in gill nets, poundnets, fyke nets, and trap nets. They are also speared at night on the spawning grounds in waters subject to tribal treaties. In 1996 the average wholesale price of walleye in the north central region of the United States was $2.21 per pound (Riepe, 1998). Their mild-tasting, white, flaky flesh is considered excellent table fare in North America.

Yellow Perch

Introduction

In North America, the yellow perch (*Perca flavescens*), family Percidae, is valuable as a sport, forage, and commercial food fish. By the early 1900s, various agencies cultured these fish to the fingerling stage for stocking. Eggs were obtained from the wild and incubated in jars, screened-bottom floating boxes, or wire baskets suspended in streams (Muncy, 1962; Leach, 1928). Leach (1928) reported that in 1927 15 states stocked 12 million eggs, 194 million fry and 1.25 million fingerlings from United States federal hatcheries. Due to these early stockings, the perch became naturalized over a very large area, and as a result, the demand for stocking decreased so that by 1983 only one-half million eggs, fry, and fingerling perch were distributed by federal hatcheries.

Distribution

Originally, yellow perch occurred in eastern North America from Labrador to Georgia and west to the Mississippi River (Collete and Banarescu,

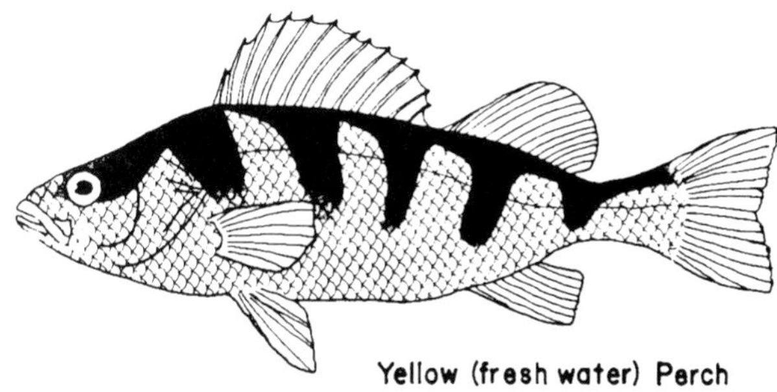

Figure 4. Yellow perch.

1977; Scott and Crossman, 1973). The present established range includes much of the United States and most of Canada. On the East Coast, it extends from Nova Scotia south to Florida, then west to upper Missouri and eastern Kansas, northwest to Montana and north to Great Slave Lake (63°N Latitude), then southeast to James Bay, Quebec, and New Brunswick. In the United States, it has been introduced into nearly all states west and south of its former range. Its extensive range and diversity of habitat is a reflection of the great ecological adaptability of this species.

Spawning

Yellow perch are annual spawners with synchronous oocyte growth during fall through winter, culminating in a spring spawning season of approximately two weeks. Just prior to spawning, the gonadal somatic index (GSI) of mature females exceed 20 percent of body weight (Clugston et al., 1978; Brazo et al., 1975; Lagler et al., 1962). The GSI of males may reach 8 percent (Lagler et al., 1962).

West and Leonard (1978) found that males mature at 98 to 165 mm (mean 108) and females at 140 to 191 mm (mean 158). Clugston et al. (1978) reported a 92-mm male and a 129-mm female to be sexually mature. Relative fecundity ranges from 79 to 223 eggs per gram of female (West and Leonard, 1978; Sheri and Power, 1969). Based on the Clugston et al. (1978) regression equation (log fecundity = 4.21565 + 3.58816 log total length), a 130-mm female would have approximately 3,000 eggs and a 250-mm female 109,000 eggs. A few males mature after one year of age and some females at age two (Muncy, 1962; Herman et al., 1959), while most three- and all four-year-old fish are mature (Clady, 1976).

In the spring, males move to the shoreline first, followed by females. Harrington (1947) described the spawning behavior. A female is accompanied by two to 25 males as she drags the unique transparent, gelatinous, accordion-folded hollow egg tube through the milt. The egg tube unfolds from the female like a long concertina. It may be several meters long and 10 cm wide (Mansueti, 1964). Access to the eggs by the sperm and aeration of the eggs is partially accomplished by water circulating through holes in the gelatinous matrix of the egg tube to the central canal (Worth, 1892). It takes several minutes at 57°F to 59°F, and up to several days at temperatures below 41°F, to extrude the entire egg mass (Kayes, 1977). No protection is given the egg mass or young by either parent.

Spawning has been reported at temperatures ranging from 37°F to 68°F. Optimum incubation temperatures from fertilization to hatch are considered to be from 50°F to 68°F. Within the range of 55°F to 63°F, 85 to 90 percent of the eggs hatch and 70 to 75 percent reach the swim-up stage. Embryos can be incubated up to 72°F after the neural keel is formed (Hokanson, 1977a; Hokanson and Kleiner, 1974). Yellow perch appear to require a cooling period (chill period) for maturation of the eggs. Hokanson (1977b) states that the minimum chill period is 160 days at approximately 50°F or less. He did not obtain viable eggs when the perch were held at 54°F. The optimum chill period was 43°F or lower for 185 days.

Kayes and Calbert (1979) and Hokanson (1977b) postulate that within a certain temperature range the onset of spawning depends more on the intrinsic maturation state of the gonads than on photoperiod or temperature cues.

Before water hardening, the diameter of fertilized eggs ranges from 1.6 to 2.1 mm; after hardening, the eggs increase in diameter to 1.7 to 4.5 mm (Mansueti, 1964). Hatching time ranges widely, depending on temperature, requiring 51, 27, 13, and 6 days to hatch at 42°F, 46°F, 61°F, and 67°F, respectively (Wiggins et al., 1983; Hokanson and Kleiner, 1974; Mansueti, 1964).

Prior to hatching, the egg envelope, which initially has a specific gravity slightly greater than water, loses rigidity and becomes flaccid; a gold iris pigmentation surrounds the melanin in the eye; larval movement decreases; and bubbles accumulate in the envelope, giving the mass a tendency to rise.

Larvae

Newly hatched prolarvae are 4.7 to 6.6 mm in total length (Hokanson and Kleiner, 1974; Scott and Crossman, 1973; Mansueti, 1964). The literature conflicts on when swimbladder inflation occurs. Mansueti (1964) indicates it occurs when the prolarvae are approximately seven mm long, while Hokanson (1977b) believes swimbladder inflation takes place during the swim-up stage, which at water temperatures above 55°F occurs on the day of hatching and below 55°F within two days of hatching. He further postulates that the prolarvae must fill their swimbladders with air at the surface. However, Ross et al. (1977) found that at temperatures above 68°F the swimbladder inflates in seven to 10 days after hatching. Prolarvae reach the larvae stage at 13 to 14 mm. Although all of the fins are present at this size, they are not complete until the larvae are 25 to 30 mm in total length (Mansueti, 1964). Perch from eight to 50 mm in total length are attracted to light (Manci et al., 1983).

Houde (1969) determined that perch larvae are better swimmers than walleye larvae for length classes less than nine mm; swimming ability of the two species is equal between nine and 15 mm. Velocities that larvae under 9.5 mm can sustain are less than three cm/second. Larger larvae can sustain current velocities of three to four body lengths/second for at least one hour.

In tanks, optimum temperatures for rearing and feeding of larval perch are between 68°F and 75°F. This range corresponds very closely to their thermal preference (Ross et al., 1977; McCauley and Read, 1973). Perch larvae have been raised experimentally on mixed zooplankton obtained from lakes (Hale and Carlson, 1972). According to Hale and Carlson (1972), 250 zooplankton organisms per larvae per day are required to obtain 50 percent survival during the first three weeks. Egestion time for larval perch held between 68°F and 73°F is less than one hour (Hokanson, 1977a).

The thermal tolerance of successive embryonic and larval stages of yellow perch increases with morphological differentiation (Hokanson and Kleiner, 1974). The optimum temperature for feeding and rearing juvenile perch is 75°F to 84°F (McCormick, 1976; 1974). This is slightly higher than the 68°F to 75°F range found by Huh (1975). The upper incipient lethal temperature for juveniles is 84°F and for adults 91°F (Hokanson, 1977b). Deformities occurred at 90°F, and all fish died within seven days at 93°F. Little growth occurred below 46°F (McCormick, 1976).

In the wild, yellow perch initially (six mm) feed on zooplankton such as copepod nauplii; as they get larger they feed on insects and fish (Becker, 1983, citing various authors).

Growth

A mean growth rate of yellow perch from natural populations in North America, calculated from data given in Carlander (1950), indicates that they reach 2.9, 5.2, 7.1, 8.0, and 9.0 inches, respectively, after one to five years. The standard length-weight relationship is log W = –3.506 + 3.230 log total length (Murphy et al., 1991).

In lakes containing only yellow perch, standing stocks ranged from 44 to 241 pounds per acre. In 18 lakes containing other species, standing stocks ranged from one to four pounds per acre (Carlander, 1977). Annual production of perch from Red Deer Lake, Ontario, was 25 pounds per acre (Chadwick, 1976). Yellow perch over four pounds have been caught (Scott and Crossman, 1973), but fish weighing more than 2 pounds are now rare.

Female yellow perch grow considerably faster and reach a greater ultimate size than male perch (Scott and Crossman, 1973; Carlander, 1950). Under laboratory conditions, the females commenced to grow faster than the males at 4.3 inches (0.03 pounds) (Schott et al., 1978).

Harvest

By hand-filleting wild fish and leaving the belly tissue intact (butterfly fillet), dress-out yields of 45 percent are obtained; with machine processing the yield is 42 percent (Lesser, 1978). The dress-out weight of cultured fish is approximately 5 percent higher than that for wild fish (Calbert and Huh, 1976). The current market is for at least an 8-inch (0.33 pound) fish (Heidinger and Kayes, 1993).

Demand for cultured perch would depend upon season and price in relation to competing marine species such as cod and ocean perch. Traditionally, people in Wisconsin consumed 75 percent (17.6 to 24.2 million pounds) of yellow perch commercially caught from the Great Lakes. The remaining 25 percent was consumed in other states that border the Great Lakes (Follett, 1975). Following the record 37 million-pound commercial harvest in 1969, the ex-vessel price was $0.11/lb; in 1974, it increased to $0.32/lb; in 1976, to $0.74/lb (Lesser and Vilatrup, 1979). In 1984 the price was approximately $1.00/lb (Heidinger and Kayes, 1993). In 1998 the wholesale price ranged from $2.30 to $3.00 per pound (Malison, 1999).

As the market demand for yellow perch exceeded the supply, interest increased in developing an economical culture method for this species (Downs, 1975). Interest continues to be high today, but less than 200,000 pounds are cultured.

Freshwater Drum

Freshwater drum (*Aplodinotus grunniens*) is the only completely freshwater member of the family Sciaenidae in North America. It probably has the greatest latitudinal range distribution of any freshwater fish in North America, occurring in lakes and rivers from the Hudson Bay and central United States south to the Rio Usumacinta system of Guatemala (Becker, 1983). The freshwater drum is sometimes caught by sport fishermen, but no state or Canadian province has placed a size limit or daily quota limit on them. Freshwater drum are not currently cultured in North America. Freshwater drum have a pair of very large otoliths in their inner ear. Historically, these otoliths were worn around the neck as a cure for colic (Priegel, 1967). North American Indians apparently used otoliths as jewelry and as money.

Habitat suitability index models have not been developed for this species. However, it is clear that the slightly warmer water in the backwaters as opposed to the channel borders of the middle and upper Mississippi River are necessary for the winter survival of freshwater drum (Bodensteiner and Lewis, 1992).

Males mature as small as 7.5 inches and females at 8.9 inches in total length (Bur, 1984). Mature males make a drumming sound with muscles and a tendon associated with their gas bladder. Many non-biologists believed this sound was made with the otolith. Freshwater drum broadcast their eggs pelagically away from shore (Wirth, 1958). Its wide distribution may in part be accounted for by the fact that its egg floats, an unusual characteristic for a native North American freshwater fish. The one-mm diameter egg contains a large oil droplet. At spawning temperatures of 66°F to 72°F (Nord, 1967; Butler, 1965) the eggs hatch in 24 to 48 hours (Warburg, 1976). The gonadal somatic index may reach 12 percent (Daiber, 1953). In Pool 14 of the Mississippi River, an 11-year mean for relative fecundity was 84,580 eggs per pound (Lawler, Matushy, and Shelly, 1993).

Drum less than one inch in total length feed primarily on small crustaceans such as copepods and cladocerns. One- to 2-inch fish feed on insects, leeches and crayfish (Wirth, 1958). Adult drum have large pharyngeal plates that can be used to crush mollusca; large drum feed on fish.

The largest drum on record is a 54.5-pound individual caught in 1972 from Nickajack Lake, Tennessee. Based on the size of otolith found in Indian middens, much larger individuals were once caught (Witt, 1960). The average total length-at-age from Oklahoma lakes, reservoirs, and streams is: age one: 4.6 inches; two: 9.0 inches; three: 11.9 inches; four: 15.3 inches; five: 17.5 inches; six: 20.3 inches; seven: 22.5 inches; eight: 260 inches; nine: 27.5 inches; and 10: 28.9 inches (Houser, 1960). Growth rate is considerably slower in Lake Erie where they reach only 13.2 inches in total length at age five and 18.1 inches at age 10 (Edsall, 1967). Part of this difference in growth rate may be due to the cooler temperature in Lake Erie. Gammon (1973) suggests that the optimum growing temperature is 84°F to 88°F. Frequently, large numbers of a population will be very slow-growing, but a few individuals grow large. Murphy (1991, citing Brown) states that the standard length-weight relationship is log weight = –3.583 + 3.208 log total length.

Carlander (1955) gave a mean-standing stock of 20 pounds-per-acre for freshwater drum in lakes (Table 5). Aggus et al. (1979) found 315 pounds per acre in Cumberland River's Lake Barkley, Tennessee. The 10-year mean for Pool 14 of the Mississippi River was 122 pounds-per-acre (Lawley, Matusky and Shelly, 1993).

Freshwater drum are harvested with seines, trap nets, trammel nets, gill nets, and trot lines (set lines). In 1997, 605 tons of freshwater drum were harvested in the United States (Table 1). In many states within the Mississippi River watershed the freshwater drum ranks in the top five species harvested by weight. They have recently sold for $0.09 to $0.40 per pound (Table 3).

FUTURE

Even though the United States continues to import (40 to 60 percent) of its edible fish products, and total demand is increasing, at least in the United States it is not likely that the commercial freshwater wild-caught finfish harvest will increase significantly. Most of the populations are probably being overharvested or are being harvested at their maximum rate. For example, in 1992 commercial fishing for catfish was banned on most of the Missouri River. Large rivers have been channelized, leveed, and dammed. Much fish production in these systems has been lost because of silting and the loss of the adjacent floodplains. Contaminants such as PCBs, mercury, and chlordane are also a problem. In many bodies of water, agencies have closed or restricted the use of fishes for food.

Other forces negatively influence the potential for commercial harvest of finfish. One of the strongest is sportfishing. One out of four people sportfish in the United States; thus, anglers far outnumber commercial fishermen. Most of the species that are commercially harvested are also sought after by anglers, and the rest probably serve as forage for the sport species. For example, of the 278 species covered in this chapter, only the gizzard shad and pond smelt do not support any sportfishing. There is a growing attitude that from an economic standpoint these species are much more valuable as a component of the sportfishing industry than of the commercial fishing industry. This attitude is also gaining considerable support in the political arena.

REFERENCES

Adair, B. 1981. Open-water creel survey at Lake Anita, 1979-1981, with emphasis on effectiveness of channel catfish maintenance stockings. Iowa Conservation Commission. Des Moines.

Adams, C.C. and T.L. Hankinson. 1926. Annotated list of Oneida Lake fish. Bulletin, New York State College of Forestry. *Roosevelt Wildlife Annual*. **1** (1-2):283-542.

Addison, W.D. and R.A. Ryder. 1977. An independent bibliography of North America *Stizostedion* (Pisces, Percidae) species. Ontario Department of Land Forests. Research Information. *Paper Fisheries*. **38**:1-318. Ontario, Canada.

Aggus, L.R., D.C. Carver, L.L. Olmsted, L.L. Rider, and G.L. Summers. 1979. Evaluation of standing crops of fishes in Crooked Creek Bay, Barkley Lake, Kentucky. *Proceedings of the Annual Conference, Southeastern Association of Fish and Wildlife Agencies*. **33**:710-722.

Anderson, R.O. 1980. Proportional stock density and relative weight. (Wr): interpretive indices for fish populations and communities. In *Practical Fisheries Management: More with Less in the 1980s*, S. Gloss and B. Shupp (eds.), pp. 27-33. New York Chapter, American Fisheries Society. Bethesda, MD.

Anderson, R.O. and S.J. Gutreuter. 1983. Length, weight, and associated structural indices. In *Fisheries Techniques*, L.A. Nielsen and D.L. Johnson (eds.), pp. 283-300. American Fisheries Society. Bethesda, MD.

Anthony, D.D. and C.R. Jorgensen. 1977. Factors in the declining contributions of walleye (*Stizostedion vitreum vitreum*) to the fishery of Lake Nipissing, Ontario, 1960-1976. *Journal of the Fisheries Research Board, Canada*. **34**:1703-1709.

Applegate, R.L. and J.W. Mullan. 1966. Food of the black bullhead, *Ictalurus melas*, in a new reservoir. *Proceedings of the Annual Southeastern Association of Game and Fish Commissioners*. **20**:288-292.

Appelget, J. and L.L. Smith. 1951. The determination of age and rate of growth from vertebrae of the channel catfish, *Ictalurus lacustris punctatus*. *Transactions of the American Fisheries Society*. **80**:119-139.

Armstrong, J.W., C.R. Liston, P.I. Tack, and R.C. Anderson. 1977. Age, growth, maturity, and seasonal food habits of round whitefish, *Prosopium cylindraceum*, in Lake Michigan near Ludington, MI. *Transactions of the American Fisheries Society*. **106**(2):151-155.

Baccante, D.A. 1988. Fecundity changes in two exploited walleye populations. *North American Journal of Fisheries Management*. **8**(2):199-209.

Bachay, G.S. 1944. Mississippi catfish. *Wisconsin Conservation Bulletin*. **9**:(12):9-10.

Bailey, M.M. 1963. Age, growth, and maturity of round whitefish of the Apostle Islands and Isle Royale regions, Lake Superior. *U.S. Fish and Wildlife Service Fishery Bulletin*. **63**(1):63-75.

Baker, C.D., D.W. Martin, and E.H. Schmitz. 1971. Separation of taxonomically identifiable organisms and detritus taken from shad foregut contents using density-gradient centrifugation. *Transactions of the American Fisheries Society*. **100**(1):138-139.

Baldwin, N.S., R.W. Sallfeld, R.A. Roas, and H.J. Buettner. 1979. Commercial fish production in the Great Lakes 1867-1977. Technical Report Number 3:1-187. Great Lakes Fishery Commission. Ann Arbor, MI.

Balon, E.K. 1974. Domestication of the carp, *Cyprinus carpio L.* Royal Ontario Museum Life Sciences Miscellaneous Publication. Toronto, Canada.

Barnickol, P.G. and W.C. Starrett. 1951. Commercial and sport fishes of the Mississippi River between Caruthersville, Missouri, and Dubuque, Iowa. *Bulletin of Illinois Natural Historical Survey*. **25**(5):267-350.

Becker, G.C. 1983. Fishes of Wisconsin. University of Wisconsin Press. Madison.

Bigelow, H.B. and W.C. Schroeder. 1953. Fishes of the Gulf of Maine. *U.S. Fish and Wildlife Services Fisheries Bulletin.* **53**(51):395-398.

Bigelow, H.B. and W.C. Schroeder. 1963. Family Osmeridae. *Fishes of the Western North Atlantic. Memorial Sears Foundation Marine Research (1).* Part 3:553-597.

Bodensteiner, L.R. and W.M. Lewis. 1992. Role of temperature, dissolved oxygen, and backwaters in the winter survival of freshwater drum (*Aplodinotus grunniens*) in the Mississippi River. *Canadian Journal of Fisheries and Aquatic Sciences.* **49**(1):173-184.

Bodola, A. 1966. Life history of the gizzard shad, *Dorosoma cepedianum* (LeSueur), in western Lake Erie. *United States Fish and Wildlife Service Fisheries Bulletin.* **65**(2):391-425.

Booke, H.E. 1970. Speciation parameters in coregonine fishes. Part I, Egg-size. Part II, Karyotype. In *Biology of Coregonid Fishes*, C.C. Lindsey and C.S. Woods (eds.), pp. 61-66. University Manitoba Press. Winnepeg, Canada.

Bowen, J.T. 1970. A history of fish culture as related to the development of fisheries programs. In *A Century of Fisheries in North America.* N.G. Benson (ed.), pp. 71-93. American Fisheries Society. Washington, DC.

Brazo, D.C., D.I. Tack, and C.R. Liston, 1975. Age, growth, and fecundity of yellow perch, *Perca flavescens*, in Lake Michigan. *Transactions of the American Fisheries Society.* **104**:726-730.

Breder, C.M. Jr. 1935. The reproductive habits of the common catfish, *Ameiurus mebulosus* (LeSueur), with a discussion of their significance in ontogeny and phylogeny. *Zoologica.* **19**(4):143-185.

Breder, C.M. Jr. 1939. Variations in the nesting habits of *Ameiurus mebulosus* (Le Sueur) *Zoologica.* **24**(3):367-378.

Breder, C.M. Jr. and D.E. Rosen. 1966. *Modes of Reproduction in Fishes.* Natural History Press. Garden City, NY.

Buckley, J. 1989. Species profiles: life histories and environmental requirements of coastal fishes and invertebrates (North Atlantic) — rainbow smelt. *U.S. Fish and Wildlife Service Biological Report.* **82** (11.106). U.S. Army Corps of Engineers. TREL-82-4.

Burbridge, R.G. 1969. Age, growth, length-weight relationship, sex ratio, and food habits of American smelt, *Osmerus mordax* (Mitchill), from Gull Lake, Michigan. *Transactions of the American Fisheries Society.* **98**(4):631-640.

Burns, J.W. 1966. Carp. In *Inland Fisheries Management,* A. Calhoun (ed.), pp. 510-515. California Department of Fish and Game. Sacramento.

Bur, M.T. 1984. Growth, reproduction, mortality, distribution, and biomass of freshwater drum in Lake Erie. *Journal Great Lakes Research.* **10**(1):48-58.

Bush, W.D.N., R.R. Scholl, and W.L. Harman. 1975. Environmental factors affecting the strength of walleye (*Stizostedion vitreum vitreum*) year classes in western Lake Erie, 1960-1970. *Journal of the Fisheries Research Board Canada.* **32**:1733-1743.

Butler, R.L. 1965. Freshwater drum in the navigational impoundments of the upper Mississippi River. *Transactions of the American Fisheries Society.* **94**(4):339-349.

Cahill, J.R. 1994. Genetic variation assessment of rainbow smelt in Green Bay and Western Lake Michigan. Master's Thesis. University of Wisconsin, Stevens Point.

Cahn, A.R. 1927. An ecological study of the southern Wisconsin fishes. The brook silverside, *Labidesthes sicculus*, and the cisco, *Leucichthys artedi*, in their relations to the region. *Illinois Biological Monographs.* **11**(1):1-51.

Calbert, H.E. and H.T. Huh. 1976. Culturing yellow perch, *Perca flavescens*, under controlled environmental conditions for the upper midwest market. *Proceedings of the World Maricultural Society.* **7**:137-144.

Carlander, K.D. and R.E. Cleary. 1949. The daily activity patterns of some freshwater fishes. *American Midland Naturalist.* **41**:447-452.

Carlander, K.D. 1950. *Handbook of Freshwater Fisheries Biology.* William C. Brown Company. Dubuque, IA.

Carlander, K.D. 1953. *Handbook of Freshwater Fishery Biology with the First Supplement.* William C. Brown Company. Dubuque, IA.

Carlander, K.D. 1955. The standing crop of fish in lakes. *Journal of the Fisheries Research Board, Canada.* **12**(4):543-570.

Carlander, K.D. 1969. *Handbook of Freshwater Fishery Biology.* Vol. I. Iowa State University Press. Ames.

Carlander, K.D. 1977. Biomass, production, and yields of walleye, *Stizostedium vitreum vitreum*, and yellow perch, *Perca flavescens*, in North American Lakes. *Journal of the Fisheries Research Board, Canada.* **34**(10):1602-1612.

Carlander, K.D. and P.M. Payne. 1977. Year-class abundance, population, and production of walleye, *Stizostedion vitreum vitreum*, in Clear Lake, Iowa, 1948-1974, with varied fry stocking rates. *Journal of the Fisheries Research Board, Canada.* **34**:1792-1799.

Carlander, K.D., J.S. Campbell, and R.J. Muncy. 1978. Inventory of percid and esocid habitat in North America. *American Fisheries Society Special Publication.* **11**:27-38.

Casting, J.P. 1953. Determining the age of the Atlantic shad from their scales. *U.S. Fish and Wildlife Service Fishery Bulletin.* **54**(85):187-189.

Ceskleba, D.G., S. AveLallemant, and T.F. Thuemler. 1985. Artificial spawning and rearing of lake sturgeon, *Acipenser fulvescens*, in Wild Rose State Fish Hatchery, Wisconsin, 1982-1983. *Environmental Biology of Fishes.* **14**(1):79-85.

Chadwich, E.M.P. 1976. Ecological fish production in a small Precambrian shield lake. *Environmental Biology Fish.* **1**:13-60.

Chevalier, J.R. 1977. Changes in walleye, *Stizostedion vitreum vitreum*, populations in Rainy Lake and factors in abundance, 1924-1975. *Journal of the Fisheries Research Board, Canada.* **34**:1696-1703.

Chittendum, M.E. Jr. 1969. Life history and ecology of the American shad, *Alosa sapidissima*, in the Delaware

River. Ph.D. Dissertation. Rutgers University. New Brunswick, NJ.

Christenson, L.M. 1975. The shovelnose sturgeon, *Scaphirhynchus platorhynchus* (Rafinesque), in the Red Cedar-Chippewa River system, Wisconsin. *Wisconsin Department of Natural Resources Research Report.* **82**:1-23.

Cianci, J.M. 1969. Larval development of the alewife and the glut herring. Master's Thesis. University of Connecticut, Storrs.

Clady, M.D. 1976. Influence of temperature and wind on the survival of early stages of perch, *Perca flavescens*. *Journal of the Fisheries Research Board, Canada.* **33**:1187-1893.

Clayton, G.R. 1976. Reproduction, first year growth, and distribution of anadromous rainbow smelt, *Osmerus mordax*, in the Parker River and Plum Island Sound Estuary, Massachusetts. Master's Thesis. University of Massachusetts, Amherst.

Clemens, H.P. and K.E. Sneed. 1957. The spawning behavior of the channel catfish *Ictalurus punctatus*. *United States Fish and Wildlife Service Special Scientific Report.* **219**:1-15. Washington, DC.

Clugston, J.P., J.L. Oliver, and R. Ruell. 1978. Reproduction, growth, and standing crops of yellow perch in southern reservoirs. In *Selected Coolwater Fishes of North America*, R.L. Kendall (ed.). American Fisheries Society. Washington, DC. Special Publication. **11**:89-99.

Colby, P.J. and T.L. Brooke. 1970. Survival and development of lake herring, *Coregonus artedic*, eggs at various incubation temperatures. In *Biology of Coregonid Fishes*, C.C. Lindsey and C.S. Woods (eds.), pp. 417-428. University of Manitoba Press. Winnipeg, Canada.

Colby, P.J. 1971. Alewife dieoffs: Why do they occur? *Limnos.* **4**(2):18-27.

Colby, P.J., R.E. McNicol, and R.A. Ryder. 1979. Synopsis of biological data on the walleye, *Stizostedion vitreum vitreum*, Mitchill 1818. Contribution 77-13:1-123. Ontario Ministry of Natural Resources. Fisheries Research Section of the Food and Agriculture Organization of the United Nations.

Cole, C.F., R.J. Essig, and O.R. Sainelle. 1980. Biological investigation of the alewife population, Parker River, Massachusetts. *U.S. National Marine Fish Wildlife Service Fishery Bulletin.* **73**(2):375-396.

Collette, B.B. and P. Banarescu. 1977. Systematics and zoogeography of the family Percidae. *Journal of the Fisheries Research Board, Canada.* **34**:1450-1463.

Conover, M.C. 1986. Stocking cool-water species to meet management needs. In *Fish Culture in Fisheries Management.* R.H. Stroud (ed.), pp. 31-39. American Fisheries Society. Bethesda, MD.

Cooper, J.E. 1978. Identification of eggs, larvae, and juveniles of rainbow smelt, *Osmerus mordax*, with comparisons to larval alewife, *Alosoa pseudohargenus*, and gizzard shad, *Dorosma cepedianum*. *Transactions of the American Fisheries Society.* **107**(1):56-62.

Crance, J.H. and L.G. McBay. 1966. Results of tests with channel catfish in Alabama ponds. *Progressive Fish-Culturist.* **28**:193-200.

Crestin, D.S. 1973. Some aspects of the biology of adults and early life stages of the rainbow smelt, *Osmerus mordax*, from the Wewantic River Estuary, Wareham-Marion, MA. Master's Thesis. University of Massachusetts, Amherst.

Cross, F.B. 1967. *Handbook of Fishes of Kansas.* University of Kansas Museum of Natural History. Lawrence. **45**:1-357.

Crowe, W.R. 1962. Homing behavior in walleyes. *Transactions of the American Fisheries Society.* **91**:350-354.

Daiber, E.C. 1956. A comparative analysis of winter feeding habits of two lenthic stream fishes. *Copeia.* 1956 (**3**):141-151.

Davis, O. 1959. Management of channel catfish in Kansas. University of Kansas Museum of Natural History. Miscellaneous Publication. **21**:1-56. Lawrence.

Dendy, J.S. 1945. Fish distribution, Norris Reservoir, Tennessee. II. Depth distribution of fish in relation to environmental factors, Norris Reservoir. *Journal of the Tennessee Academy of Science.* **20**(1):114-135.

Dennison, S.G. and R.V. Bulkley. 1972. Reproductive potential of the black bullhead, *Ictalurus melas*, in Clear Lake, Iowa. *Transactions of the American Fisheries Society.* **101**(3):483-487.

Dill, W.A. 1944. The fishery of the lower Colorado River. *California Fish and Game.* **30**(3):109-211.

Dodson, J.J. and W.C. Leggett. 1974. Role of olfaction and vision in the behavior of American shad, *Alosa Sapidissima*, homing to the Connecticut River from Long Island Sound. *Journal of the Fisheries Research Board, Canada.* **31**(10):1607-1619.

Doughty, R.W. 1980. Wildlife conservation in late nineteenth-century Texas: The carp experiment. *Southwestern History Quarterly.* **84**:169.

Downs, W. 1975. Wisconsin: the dairy state takes a look at fish farming. Raising perch for the Midwest market. *Commercial Fish Farming.* **1**(5):27.

Dryer, W.R. 1963. Age and growth of the whitefish in Lake Superior. *U.S. Fish and Wildlife Service Fishery Bulletin.* **63**(1):77-95.

Dryer, W.R. and J. Beil. 1964. Life history of the lake herring in Lake Superior. *U.S. Fish and Wildlife Service Fishery Bulletin.* **63**(3):493-530.

Dryer, W.R., L.F. Erkkila, and C.L. Tetzloff. 1965. Food of lake trout in Lake Superior. *Transactions of the American Fisheries Society.* **94**(2):169-176.

Dryer, W.R. and J. Beil. 1968. Growth changes of the bloater, *Coregonus hoyi*, of the Apostle Islands region of Lake Superior. *Transactions of the American Fisheries Society.* **97**(2):146-148.

Dudash, M.J. 1986. Comparative survival and growth of various size channel catfish stocked into a lake containing largemouth bass. Master's Thesis. Southern Illinois University, Carbondale.

Dufford, D.W. 1994. Illinois 1993 commercial catch report exclusive of Lake Michigan. Division of Fisheries. Illinois Department of Conservation. Springfield.

Ebbers, M.A., P.J. Colby, and C. A. Lewis. 1988. Walleye-sauger bibliography. *Minnesota Department of Natural Resources Investigative Report.* **396**:1-201. St. Paul.

Eddy, S. and T. Surber. 1947. *Northern Fishes.* 2nd edition. Charles T. Brandford Company. Newton Centre, MA.

Eddy, S. and J.C. Underhill. 1974. *Northern Fishes.* University of Minnesota Press, Minneapolis.

Edsall, T.A. 1967. Biology of the freshwater drum in western Lake Erie. *Ohio Journal of Science.* **67**(6):321-340.

Edsall. T.A. 1970. The effect of temperature on the rate of development and survival of alewife eggs and larvae. *Transactions of the American Fisheries Society.* **99**(2):376-380.

Edwards, E.A. 1983. Habitat suitability index models: bigmouth buffalo. United States Fish and Wildlife Service. FWS/OBS-82/10.34.

Edwards, E.A. and K. Twomey. 1982. Habitat suitability index models: smallmouth buffalo. United States Fish and Wildlife Service. FWS/DBS-82/10.13.

Edwards, T.J., W.H. Hunt, and L.L. Olmsted. 1977. Density and distribution of larval shad, *Dorosoma* spp., in Lake Norman, North Carolina, entrainment at McGuire Nuclear Station. In *Proceedings, First Symposium on Freshwater Larval Fish.* L.L. Olmsted (ed.), pp. 143-158. Duke Power Company. Huntersville, NC.

Elser, A.A. 1976. Paddlefish investigations. Montana Department of Fish and Game. *Federal Aid Report in Fish Restoration.* Project F-30-R, Job 2-A:1-13. Helena.

Emery, L. and E.H. Brown Jr. 1978. Fecundity of the bloater, *Coregonus hoyi,* in Lake Michigan. *Transactions of the American Fisheries Society.* **107**(6):785-789.

Emig, J.W. 1935. Further notes on duration of life in animals. (I) Fishes determined by otolith and scale readings and direct observations on live individuals. *Proceedings of the Zoological Society, London.* **2**:265-304. London, England.

Epifano, J., M. Nedbal, and D.P. Philipp. 1989. A population genetic analysis of paddlefish, *Polydon spathula.* Illinois Natural History Survey, Aquatic Biology Section. Technical Report 89/4. Champaign.

Eschmeyer, P.H. 1957. The lake trout, *Salvelinus namaycush. U.S. Fish and Wildlife Service Fishery Leaflet.* **441**:1-11.

Eschmeyer, P.H. and A.M. Phillips Jr. 1965. Fat content of the flesh of siscowets and lake trout from Lake Superior. *Transactions of the American Fisheries Society.* **94**(1):62-74.

FAO (Food & Agriculture Organization of the United Nations). 1991. Yearbook of fishery statistics catches and landings. Vol. 72. Food and Agriculture Organization. Rome, Italy.

FAO. 1997. Yearbook of fishery statistics capture production. Vol. 84. Food and Agriculture Organization, Rome, Italy.

Facey, D.E. and M.J. Van Den Avyle. 1986. Species profiles: life histories and environmental requirements of coastal fishes and invertebrates (South Atlantic) — American shad. *U.S. Fish and Wildlife Service Biological Report.* **82** (11.45). U.S. Army Corps of Engineers. TREL-82-4.

Ferguson, R.G. 1958. The preferred temperature of fish and their midsummer distribution in temperate lakes and streams. *Journal of the Fisheries Research Board, Canada.* **15**:607-624.

Finnell, J.C. and R.M. Jenkins. 1954. Growth of channel catfish in Oklahoma waters. 1954 revision. *Oklahoma Fisheries Research Laboratory Report.* Number 41. Oklahoma City.

Flickinger, S. 1976. Survey of inland commercial fisheries of the United States. *Proceedings of the Fifth Inland Commercial Fisheries Workshop.* Nebraska Game and Parks Commission.

Foerster, J.W. and S.L. Goodbred. 1978. Evidence for a resident alewife population in the northern Chesapeake Bay. *Estuarine and Coastal Marine Science.* **7**:431-444.

Fogler, N.W. 1963. Report of fisheries investigations during the fourth year of impoundment of Oake Reservoir, South Dakota. 1962. South Dakota Department of Game Fish and Parks. Dingell-Johnson Project F-1-R12:1-43. Cited in Carlander 1969.

Foltz, J.W. and C.R. Norden. 1977. Food habits and feeding chronology of rainbow smelt, *Osmerus mordax,* in Lake Michigan. *National Marine Fishery Service Bulletin.* **75**(3):637-640.

Follett, R. 1975. Raising perch for the Midwest market. Advisory Report 13:1-87. University of Wisconsin Sea Grant Program. Madison.

Fontaine, P.A. 1944. Notes on the spawning of the shovelheaded catfish, *Pilodictus olivaris* (Rafinesque). *Copeia.* 1944 (**1**):50-51.

Forney, J.L. 1955. Life history of the black bullhead, *Ameirus melas* (Rafinesque), of Clear Lake, Iowa. *Iowa State College Journal of Science.* **30**(1):145-162.

Forney, J.L. 1963. Distribution and movement of marked walleyes in Oneida Lake, New York. *Transactions of the American Fisheries Society.* **92**:47-52.

Forney, J.L. 1965. Factors affecting growth and maturity in a walleye population. *New York Fish Game Journal.* **12**:217-232.

Forney, J.L. 1976. Year-class formation in the walleye, *Stizostedion vitreum vitreum,* population of Oneida Lake, New York, 1966-1973. *Journal of the Fisheries Research Board, Canada.* **33**:783-792.

Forney, J.L. 1977. Evidence of inter and intraspecific competition as factors regulating walleye, *Stizostedion vitreum vitreum,* biomass in Oneida Lake, New York. *Journal of the Fisheries Research Board, Canada.* **34**:1812-1820.

Fowler, H. 1917. Some notes on the breeding habits of local catfishes. *Copeia.* **42**:32-36.

Franklin, D.R. and L.L. Smith. 1963. Early life history of the northern pike, *Esox lucius* L., with special reference to the factors influencing the numerical strength of year classes. *Transactions of the American Fisheries Society.* **92**:91-110.

Funk, J.K. 1957. Movement of stream fishes in Missouri. *Transactions of the American Fisheries Society.* **85**:39-57.

Gammon, J.R. 1973. The effect of thermal input on the populations of fish and macroinvertebrates in the

Wabash River. Purdue University Water Research Center. Technical Report. **32**:1-106.

Gengerke, T.W. 1986 Distribution and abundance of paddlefish in the United States. In *The Paddlefish: Status, Management and Propagation*. J.G. Dillard, L.K. Graham and T.R. Russell (eds.), pp. 22-35. Special Publication Number 7. American Fisheries Society, North Central Division. Bethesda, MD.

Gholson, K.W. 1970. Life history study of the flathead catfish, *Polyodictus olivaris*. Fisheries Investigations. Texas Parks Wildlife Department. Federal Aid Project. Region 5-A. F-9-R17:1-29. Austin, TX.

Glebe, B.D. and W.C. Leggett. 1981. Temporal, intra-population differences in energy allocation and use by American Shad, *Alosa Sapidissima*, during the spanning migration. *Canadian Journal of Fisheries and Aquatic Science*. **38**(7):795-805.

Goodson, F.L. Jr. 1965. Diets of four warmwater game fishes in a fluctuating, steep-sided California reservoir. *California Fish and Game*. **51** (4):259-269.

Grinstead, B.G. 1971. Reproduction and some aspects of the early life history of walleye in Canton Reservoir, Oklahoma. In *Reservoir Fisheries and Limnology*, F.F. Fish et al. (eds.), pp. 41-51. Special Publication Number 8. American Fisheries Society. Washington, DC.

Guidice, J.J. 1965. Investigations on the propagation and survival of flathead catfish in troughs. *Proceedings of the Annual Conference Southeast Association of Game Fish Commissioners*. **17**:178-180.

Hale, J.G. and A.R. Carlson. Culture of the yellow perch in the laboratory. *Progressive Fish-Culturist*. **34**:195-198.

Harkness, W.J.K. and J.R. Rymond. 1961. The lake sturgeon. The history of its fishery and problems of conservation. Ontario Department of Lands, Forests, and Fish Wildlife, pp. 1-121.

Harlan, J.R., and E.B. Speaker. 1956. Iowa Fish and Fishing. State of Iowa. Iowa Conservation Commission.

Harrington, R.W. 1947. Observations on the breeding habits of the yellow perch, *Perca flavescens* (Mitchill). *Copeia*. **1947**:199-200.

Hart, J.L. 1930. The spawning and early life history of the whitefish, *Coregonus clupeaformis* (Mitchill), in the Bay of Quinte, Ontario. *Contribution Canadian Biology Fishery*. **6**(7):165-214.

Hayne, D.W., G.H. Hall, and H.M. Nichols. 1967. An evolution of cove sampling of fish populations in Douglas Reservoir, Tennessee. In *Reservoir Fishery Resources Symposium*, pp. 244-297. Southern Division American Fisheries Society.

Heidinger, R.C. 1984. Life history of gizzard shad and threadfin shad as it relates to the ecology of small lake fisheries. In *Proceedings Small Lake Management Workshop, "Pros and Cons of Shad,"* pp. 1-8. Iowa Conservation Commission and Sport Fishing Institute. Des Moines.

Heidinger, R.C. 1989. Fishes in the Illinois portion of the upper Des Plaines River. *Transactions of the Illinois State Academy of Science*. **82**(1 and 2):85-96.

Heidinger, R.C. and T.B. Kayes. 1993. Chapter 7, Yellow perch. In *Culture of Nonsalmonid Freshwater Fishes*, Second Edition, R.R. Stickney (ed.), pp. 215-229. CRC Press, Boca Raton, FL.

Held, J.W. 1969. Some early summer foods of the shovelnose sturgeon in the Missouri River. *Transactions of the American Fisheries Society*. **98**(3):514-517.

Helms, D.R. 1974. Age and growth of shovelnose sturgeon, *Scaphirhynchus platyorynchus* (Rafinesque), in the Mississippi River. *Proceedings of the Iowa Academy of Science*. **81**(2):73-75.

Henderson, H. 1965. Observation on the propagation of flathead catfish in the San Marcos State fish hatchery, Texas. *Proceedings Southeastern Association of Game and Fish Commissioners*. **17**:173-177.

Herman, E., W. Wisly, L. Wiegert, and M. Burdick. 1959. The Yellow Perch: Its Life History, Ecology, and Management. Wisconsin Conservation Department Publication. **228**:1-14.

Hildebrand, S.F. and W.C. Schroeder. 1928. Fishes of Chesapeake Bay. *U.S. Bureau of Fisheries Bulletin 43*. Reprinted 1978. Tropical Fish Hobbycat Publications. Neptune City, NJ.

Hildebrand, S.F. 1963. Family Clupeidai. In *Fishes of the Western North Atlantic*. Part III, pp. 257, 385, 397-442, 452-454. Sears Foundation for Marine Research. Yale University, New Haven, CT.

Hokanson, K.E.F. 1977a. Temperature requirements of some percids and adaptations to the seasonal temperature cycle. *Journal of Fisheries Research Board, Canada*. **34**:1524-1550.

Hokanson, K.E.F. 1977b. Optimum culture requirements of early life phases of yellow perch. In *Perch Fingerling Production for Aquaculture*. R.W. Soderberg (ed.). **421**:24-38. University of Wisconsin Sea Grant Advisory Report.

Hokanson, K.E.F. and C.F. Kleiner. 1974. Effects of constant and rising temperatures on survival and developmental rates of embryonic and larval yellow perch, *Perca flavescens* (Mitchill). In *The Early Life History of Fish*. J.S. Blaxter (ed.), pp. 437-448. Springer-Verlag, New York.

Houde, E.D. 1969. Sustained swimming ability of larvae of walleye, *Stizostedion vitreum*, and yellow perch, *Perca flavescens*. *Journal of the Fisheries Research Board, Canada*. **26**:1647-1659.

Houser, A. 1960. Growth of freshwater drum in Oklahoma. Oklahoma Fishery Research Laboratory Report. Number 78:1-15.

Houser, A. and C. Collins. 1962. Growth of black bullhead catfish in Oklahoma. Oklahoma Fisheries Research Laboratory Report. Number 79:1-18.

Hubbs, Carl L. and Karl F. Lagler. 1958. Fishes of the Great Lakes Region. Cranbrook Institute of Science. Bulletin **26**:1-213.

Hubert, W.A., S.H. Anderson, P.D. Southall, and J. Crance. 1984. Habitat suitability index models and instream flow suitability curves: paddlefish. United States Department of the Interior Fish and Wildlife Service. FWS/DBS-82/10.80.

Hubley, R.C. Jr. 1963. Movement of tagged channel catfish in the upper Mississippi River. *Transatlantic American Fisheries Society*. **92**(2):165-168.

Huh, H.T. 1975. Bioenergetics of food conversion and growth of yellow perch, *Perca flavescens*, and walleye, *Stizostedion vitreum vitreum*, using formulated diets. Ph.D. Dissertation. University of Wisconsin, Madison.

Hurley, S.T. and M.R. Austin. 1987. Evaluation of walleye stocking in Caesar Creek Lake. Ohio Department of Natural Resources. Division of Wildlife. Federal Aid and Restoration project F-29-R-23. Columbus.

IDOC. 1989. Illinois Department of Conservation. 1989. *Guide to Illinois Fishing Regulations*. Springfield, IL.

Inskip, P.D. 1982. Habitat suitability index models: Northern Pike. U.S. Department of the Interior Fish and Wildlife Service. FWS/OBS-82/10.17.

Islands region of Lake Superior. *Transactions of the American Fisheries Society*. **97**(2):146-158.

Janssen, J. and S.B. Brandt. 1980. Feeding ecology and vertical migration of adult alewife in Lake Michigan. *Canadian Journal of Fisheries and Aquatic Sciences*. **37**:177-184.

Jenkins, R.E. and N.M. Burkhead. 1993. *Freshwater Fishes of Virginia*. American Fisheries Society. Bethesda, MD.

Jernejcic, F. 1986. Walleye migration through Tygart Dam and angler utilization of the resulting tailwater fisheries. In *Reservoir Fisheries Management:Strategies for the 80s*, G.E. Hall and M.J. Van Den Avyle (eds.), pp. 294-300. American Fisheries Society. Bethesda, MD.

Jester, D.B. 1973. Life history, ecology, and management of the smallmouth buffalo, *Ichtiobus bubalus* (Rafinesque), with reference to Elephant Butte Lake. New Mexico State University. *Agricultural Experimental Station Research Report*. **261**:1-111.

Jester, D.B. and B.L. Jensen. 1972. Life history and ecology of the gizzard shad, *Dorosoma cepedianum* (LeSueur), with reference to Elephant Butte Lake, New Mexico. *University Agriculture Experiment Station Research Report*. **218**:1-56.

Jilek, R., B. Cassell, D. Peace, Y. Garza, L. Riley, and T. Siewart. 1979. Spawning population dynamics of smelt, *Osmerus mordax*. *Journal of Fish Biology*. **115**(1):31-35.

John, K.R. 1956. Onset in spawning activities of the shallow water cisco, *Leucichthys artedi* (LeSueur), in Lake Mendota, Wisconsin, relative to water temperature. *Copeia*. 1956.(**2**):116-118.

Johnson, F.H. 1961. Walleye egg survival during incubation on several types of bottom in Lake Winnibigoshish, Minnesota, and connecting waters. *Transactions of the American Fisheries Society*. **90**:312-322.

Johnson, J.E. 1987. Protected fishes of the United States and Canada. American Fisheries Society. Bethesda, MD.

Johnson, J.H. and D.S. Dropkin. 1992. Predation on recently released larval American shad in the Susquehanna River Basin. *North American Journal of Fisheries Management*. **12**(3):5-4-508.

Johnson, R.R. 1963. Studies on the life history and ecology of the bigmouth buffalo, *Ictiobus cyprinellus* (Valenciennes). *Journal of the Fisheries Research Board, Canada*. **20**(6):1397-1429.

Jones, P.W., F.D. Martin, and J.D. Hardy Jr. 1978. Development of fishes of the Mid-Atlantic Bight. In *An atlas of egg, larval and juvenile stages*, pp. 98-104. Vol. I. U.S. Fish and Wildlife Service Biological Service Program. GSW/OBS-78/12.

Jones, M.S., J.P. Goettl Jr., and S.A. Flickinger. 1994. Changes in walleye food habits and growth following a rainbow smelt introduction. *North American Journal of Fisheries Management*. **14**(2):409-414.

Jordan, D.S. and B.W. Evermann. 1916. *American Food and Game Fishes*. Doubleday, Page and Company. New York.

Jordan, D.S. and B.W. Evermann. 1923. *American Food and Game Fishes*. Doubleday, Page and Company, New York.

Jude, D.J. 1973. Food and feeding habits of gizzard shad in Pool 19, Mississippi River. *Transactions of the American Fisheries Society*. **102**(2):378-383.

Jude, D.J. and J. Leach. 1993. The Great Lakes. In *Inland Fisheries Management in North America*. C.C. Kohler and W.A. Hubert (eds.), pp. 517-551. American Fisheries Society. Bethesda, MD.

Kayes, T. 1977. Reproductive biology and artificial propagation methods for adult perch. In *Perch Fingerling Production for Aquaculture, Advisory Report*. R.W. Soderberg (ed.). 421:6-18. University of Wisconsin Sea Grant Program, Madison.

Kayes, T.B. and H.E. Calbert. 1979. Effects of photoperiod and temperature on the spawning of yellow perch, *Perca flavescens*. *Proceedings of the World Maricultural Society*. **10**:306-316.

Keleher, J.J. 1961. Comparison of largest Great Slave Lake fish with North American records. *Journal of the Fisheries Research Board, Canada*. **18**(3):417-421.

Kellogg, R. L. 1982. Temperature requirements for the survival and early development of the anadromone alewife. *The Progressive Fish Culturist*. **44**(2):63-73.

Kelso, J.R.M. 1973. Seasonal energy changes in walleye and their diet in Blue Lake, Manitoba. *Transactions of the American Fisheries Society*. **102**:363-368.

Kelso, J.R.M. 1976. Diel movement of walleye (*Stizostedion vitreum vitreum*) in West Blue Lake, Manitoba, as determined by ultrasonic tracking. *Journal of the Fisheries Research Board, Canada*. **33**:2070-2072.

Kilambi, R.V. and R.E. Baglin Jr. 1969. Fecundity of the gizzard shad, *Dorosoma cepedianum* (LeSueur), in Beaver and Bull Shoals reservoirs. *American Midland Naturalist*. **82**(2):444-449.

Kimsey, J.B., R.H. Hagy, and G.W. McCammon. 1957. Progress report on the Mississippi threadfin shad, *Dorosoma petenensis atchafaylae*, in the Colorado River for 1956. California Department of Fish and Game. *Inland Fisheries Administration Report*. **57-23**:1-48.

Kinnunen, R.E. 1994. Commercial fisheries newsline. XIII(2):1-21. Michigan Sea Grant Extension Service.

Kissil, G.W. 1974. Spawning of the anadromous alewife in Bride Lake, Connecticut. *Transactions of the American Fisheries Society*. **103**(2):312-317.

Kleinert, S.J. and D. Mraz. 1966. Life history of the grass pickerel, *Esox Americianus vermiculatus*, in Southeastern Wisconsin. *Wisconsin Conservation Department Technical Bulletin*. **37**:1-40.

Koelz, W. 1929. Coregonid fishes of the Great Lakes. *Bulletin U.S. Bureau Fisheries*. **43**(1927) Part II:297-643.

Koenst, W.M. and L.L. Smith Jr. 1976. Thermal requirements of the early life history stages of walleye, *Stizostedion vitreum vitreum*, and sauger, *S. canadense*. *Journal of the Fisheries Research Board, Canada*. **33**:1130-1138.

Kohler, C.C. and J.J. Ney. 1981. Consequence of an alewife die-off to fish and zooplankton in a reservoir. *Transactions of the American Fisheries Society*. **110**(3):360-369.

Koonce, J.F., T.B. Bagenal, R.F. Carline, K.E.F. Hokanson, and M. Nagiec. 1977. Factors influencing year-class strength of percids: a summary and model of temperature effects. *Journal of the Fisheries Research Board, Canada*. **34**:1900-1909.

Krummrich, J.T., and R.C. Heidinger. 1973. Vulnerability of channel catfish to largemouth bass predation. *Progressive Fish-Culturist*. **35**:173-175.

Lagler, K.F., J. E. Bardach, and R. R.Miller. 1962. *Ichthyology*. John Wiley and Sons, New York.

Lagler, K.F., J.E. Bardach, B.R. Miller, and D.R.M. Passino. 1977. *Ichthyology*. 2nd edition. John Wiley and Sons, New York.

LaJeone, J.J. 1988. An evaluation of experimental stocking in the Rock River, Illinois, 1985-1987. Commonwealth Edison Company. Environmental Affairs Department. Chicago, IL.

Lawler, Matusky and Shelly Engineers. 1983. Quad Cities Aquatic Program 1992 Annual Report. Volume I:1-268. Prepared for Commonwealth Edison Company. Chicago, IL.

Lawrence, J.M. 1958. Estimated sizes of various forage fishes largemouth bass can swallow. *Proceedings of the Annual Conference of the Southeastern Association of Game Fish Commissioners*. **11**:220-225.

Layher, W.G. 1976. Food habits and related studies of flathead catfish. Master's Thesis. Emporia State College, Emporia, KS.

Leach, G.C. 1928. Propagation and Distribution of Food Fishes, Fiscal Year 1927. *Report to the U.S. Commissioner of Fisheries*. U.S. Government Printing Office, Washington, DC, pp. 683-736.

Lee, L.A. and J.W. Terrell. 1987. Habitat suitability index models: flathead catfish. *United States Fish and Wildlife Service Biological Report*. **82**(10.152).

Leggett, W.C. 1969. Studies on the reproductive biology of the American shad, *Alosa sapidissima* (Wilson). A comparison of populations from four rivers of the Atlantic seaboard. Ph.D. Dissertation. McGill University, Montreal, Canada.

Leggett, W.C. 1976. The American Shad, *Alosa Sapidissima*, with special reference to its migration and population dynamics in the Connecticut River. American Fisheries Society. Monograph. **1**:169-225.

Leggett, W.C. 1977. Ocean migration rates of American shad, *Alosa sapidissima*. *Journal of the Fisheries Research Board, Canada*. **34**(0):1422-1426.

Leggett, W.C. and J.E. Carscadden. 1978. Latitudinal variation in reproductive characteristic of American shad *Alosa sapidissima*: evidence for population specific life history strategies in fish. *Journal of the Fisheries Research Board, Canada*. **35**(11):1469-1478.

Leggett, W.C. and R.R. Whitney, 1972. Water temperature and the migration of American shad. *U.S. National Marine Fisheries Service Fishery Bulletin*. **70**(3)659-670.

Lesser, W.H. 1978. Marketing Systems for Warm Water Aquaculture Species in the Upper Midwest. Ph.D. Dissertation. University of Wisconsin, Madison.

Lesser, W.H. and R. Vilstrup. 1979. The Supply and Demand for Yellow Perch 1915-1990. University of Wisconsin. *College of Agriculture Life Science Research Bulletin* R3006.

Libbey, J.E. 1969. Certain aspects of the life history of the walleye, *Stizostedion vitreum vitreum* (Mitchill), in Dale Hollow Reservoir, Tennessee, Kentucky, with special emphasis on spawning. Master's Thesis. Tennessee Technical University.

Lindsey, C.C. 1964. Problems in zoogeography of the lake trout, *Salvelinus namaycush*. *Journal of the Fisheries Research Board, Canada*. **21**(5):977-994.

Lovell, R.T. 1973. Put catfish offal to work for you. *Fish Farm Industries*. October-November: 22-24.

Machniak, K. 1975. The effects of hydroelectric development on the biology of northern fishes. (Reproduction and population dynamics) III. Yellow walleye, *Stizostedion vitreum vitreum*, (Mitchill). A literature review and bibliography. *Department of Environmental Fisheries Development Directorate Technical Report*. **529**:1-82.

MacKenzie, C., L.S. Weiss-Glanz, and J.R. Moring. 1985. Species profiles: life histories and environmental requirements of coastal fishes and invertebrates (mid-Atlantic) — American shad. *U.S. Fish and Wildlife Service Biological Report*. **82**(11.37). U.S. Army Corps of Engineers. TREL-82-4.

Malison, J.A. 1999. A white paper on the status and needs of yellow perch aquaculture in the north central region. USDA North Central Regional Aquaculture Center. Michigan State University.

Manci, W.E., J.A. Malison, T.B. Kayes, and T.E. Kuczynaki. 1983. Harvesting photopositive juvenile fish from a pond using a lift net and light. *Aquaculture*. **34**:157-164.

Mansueti, R.J. 1956. Alewife herring eggs and larvae reared successfully in lab. *Maryland Tidewater News*. **13**(1):2-3.

Mansueti, R.J. 1961. Movements, reproduction, and mortality of the white perch, *Roccus americanus*, with comments on its ecology in the estuary. *Chesapeake Science*. **5**:3-45.

Mansueti, R.J. 1962. Eggs, larvae, and young of the hickory shad, with comments on its ecology in the estuary. *Chesapeake Science*. **3**:173-205.

Mansueti, R.J. 1964a. Eggs, larvae, and young of the white perch, *Roccus americanus*, with comments on its ecology in the estuary. *Chesapeake Science*. **5**:3-45.

Mansueti, R.J. 1964b. Early development of the yellow perch, *Perca flavescens*. *Chesapeake Science*. **5**:46-66.

Mansueti, R.J. and J.D. Hardy. 1967. Development of fishes of the Chesapeake Bay Region. University of Maryland, Baltimore. Natural Resources Institute. Part I:1-202.

Marcy, B.C. Jr. 1972. Spawning of the American shad, *Alosa sapidissima*, in the lower Connecticut River. *Chesapeake Science*. **13**(2):116-119.

Marcy, B.C. Jr. 1976. Early life history studies of American shad in the lower Connecticut River and the effects of the Connecticut Yankee Plant. *American Fisheries Society Monograph*. **1**:141-168.

Martin, N.V. 1957. Reproduction of lake trout in Algonquin Park, Ontario. *Transactions of the American Fisheries Society*. 1956. **86**:231-244.

Martin, N.V. 1960. Homing behavior in spawning lake trout. *Canadian Fish Culturist*. **26**:3-6.

Marzolf, R.C. 1955. Use of pectoral spines and vertebrae for determining age and rate of growth of the channel catfish. *Journal of Wildlife Management*. **19**(2):243-249.

McAllister, D.E. 1963. A revision of the smelt family, *Osmeridae*. Bulletin National Museum. Canada 191. *Biological Series*. **71**:1-53.

McCammon, G.W. 1954. A tagging experiment with channel catfish in the lower Colorado River. *California Fish and Game*. **42**(4):323-335.

McCammon, G.W. 1957. Anglers "purr" as Delta catfish study pays off. *Outdoor California*. **18**(3):12-14.

McCauley, R.W. and Read, L.A.A. 1973. Temperature selection by juvenile and adult yellow perch, *Perca flavescens*, acclimated to 24°C. *Journal of the Fisheries Research Board, Canada*. **30**:1253-1255.

McClellan, W.G. 1954. A study of the southern spotted channel catfish, *Ictalurus punctatus* (Rafinesque). Master's Thesis. North Texas State College, TX.

McComish, T.S. 1967. Food habits of bigmouth and smallmouth buffalo in Lewis and Clark Lake and the Missouri River. *Transactions of the American Fisheries Society*. **96**(1):70-74.

McCormick, J.H. 1974. Temperatures suitable for the well-being of juvenile yellow perch during their first summer-growing season. *Annual Report*. ROAP-16AB1. National Water Quality Laboratory. Duluth, MN.

McCormick, J.H. 1976. Temperature effects on young yellow perch, *Perca flavescens* (Mitchill), pp. 1-18. EPA-600/3-76-057. U.S. Environmental Protection Agency Ecology Research Service. Washington, DC.

McGeachin, R.B. 1993. Carp and buffalo. In *Culture of Nonsalmonid Freshwater Fishes*, 2nd edition, R.R. Stickney (ed.), pp. 118-143. CRC Press. Boca Raton FL.

McGill, E. 1966. A chronological history of walleye stocking in Lake Geeson (Narrows Reservoir), Pike County, Arkansas. Arkansas Game and Fish Commission. D-J Federal Aid Project F-1-R. Lonoke.

McKenzie, R.A. 1964. Smelt life history and fishery in the Miramichi River, New Brunswick. *Bulletin of the Fisheries Research Board, Canada*.

McKnight, T.C., R.W. Wendt, R.L. Theis, L.E. Morehouse, and M.E. Burdich. 1970. Cisco and whitefish sport fishing in Northeastern Wisconsin. Wisconsin Department of Natural Resources. *Bureau of Fish Management Report*. **41**:1-15.

McMahon, T.E. and J.W. Terrell. 1982. Habitat suitability index models: Channel catfish. United States Department of the Interior Fish and Wildlife Service. FWS/OBS 82/10.2.

McMahon, T.E., J.W. Terrell, and P.C. Nelson. 1984. Habitat suitability information: walleye. United States Department of the Interior Fish and Wildlife Service. FW/OBS-82/10.56.

McNamara, F. 1937. Breeding and food habits of the pikes, *Esox lucius and Esox vermiculatus*. *Transactions of the American Fisheries Society*. **66**:372-373.

McPhail, J.D. and C.C. Lindsey. 1970. Freshwater fishes of northwestern Canada and Alaska. *Fisheries Research Board Canada Bulletin*. **173**:1-381.

Menzel, R.W. 1945. The catfish fishery of Virginia. *Transactions of the American Fisheries Society*. **73**:364-372.

Merriner, J.V. 1978. Anadromous fishes of the Potomac Estuary. *Virginia Institute of Marine Science Contributors 696*. Gloucester Point.

Messieh, S.N. 1977. Population structure and biology of alewives and blueback herring in the Saint John River, New Brunswick. *Environmental Biology of Fisheries*. **2**(3):195-210.

Mestl, G.E. 1983. Survival of stocked channel catfish in small ponds. Master's Thesis. Oklahoma State University. Stillwater.

Meyer, E.P. 1960. Life history of *Marsipometra hastata* and the biology of its host, *Polyodon spathula*. Ph.D. Dissertation. Iowa State University, Ames.

Miller, E.E. 1966a. Channel catfish. In *Inland Fisheries Management*, A. Calhoun (ed.), pp. 440-463. California Department of Fish and Game, Sacramento.

Miller, E.E. 1966b. White catfish. In *Inland Fisheries Management*, A. Calhoun (ed.). California Department of Fish and Game, Sacramento.

Miller, R.R. 1963. Genus *Dorosoma* (Rafinesque, 1820). Gizzard shads, threadfin shads. Fishes of the western North Atlantic. *Memorial Sears Foundation Marine Resource*. **1** (Part 3):443-451.

Minckley, W.L., J.E. Johnson, J.N. Rinne, and S.E. Willoughby. 1970. Foods of buffalofishes, genus *Ictiobus*, in central Arizona reservoirs. *Transactions of the American Fisheries Society*. **99**(2):333-342.

Moen, T.E. 1953. Food habits of the carp in northwest Iowa lakes. *Proceedings of the Iowa Academy of Science*. **60**:655-686.

Moen, T.E. and D. Henegar. 1971. Movement and recovery of tagged northern pike in Lake Oake, South and North Dakota, 1964-68. In *Reservoir fisheries and limnology*. G. Hall (ed.), pp. 85-93. Special Publication Number 8. American Fisheries Society. Bethesda, MD.

Moring, J.R. 1993. Anadromous stocks. In *Inland Fisheries Management in North America*. C.C. Kohler and W.A. Hubert (eds.), pp. 553-580. American Fisheries Society. Bethesda, MD.

Moser, B.B. and D. Hicks. 1970. Fish populations of the stilling basin below Canton Reservoir. *Proceedings of the Oklahoma Academy of Science.* **50**:69-74.

Mraz, D. 1964. Age, growth, sex ratio, and maturity of the whitefish in central Green Bay and adjacent waters of Lake Michigan. *U.S. Fish Wildlife Service Fishery Bulletin.* **63**(3):619-634.

Mullen, D.M., C.W. Fay, and J.R. Moring. 1986. Species profiles: life histories and environmental requirements of coastal fisheries and invertebrates (North Atlantic) — Alewife/blueback herring. *U.S. Fish and Wildlife Service Biological Report.* **82** (11.56).

Muncy, J.R. 1959. Age and growth of channel catfish from the Des Moines River, Boone County, Iowa, 1955 and 1956. *Iowa State College Journal of Science.* **34**(2):127-137.

Muncy, R.J. 1962. Life history of the yellow perch, *Perca flavescens*, in estuarine waters of Seven River, a tributary of Chesapeake Bay, Maryland. *Chesapeake Science.* **6**:545-555.

Murawski, S.A. 1976. Population dynamics and movement patterns of anadromous rainbow smelt, *Osmerus mordax*, in the Parker River estuary. Master's Thesis. University of Massachusetts, Amherst.

Murawski, S.A., G. Clayton, R.J. Reed, and C.F. Cole. 1980. Movement of spawning rainbow smelt, *Osmerus mordax*, in a Massachusetts estuary. *Estuaries.* **3**(4):308-3114.

Murphy, G.I. 1951. The fishery of Clear Lake, Lake County, California. *California Fish and Game.* **37**(4):439-484.

Murphy, B.R., M.L. Brown, and T.A. Springer. 1990. Evaluation of the relative weight (Wr) index, with new applications to walleye. *North American Journal of Fisheries Management.* **10**:85-97.

Murphy, B.R., D.W. Willis, and T.A. Springer. 1991. The relative weight index in fisheries management: status and needs. *Fisheries.* **16**(2):30-38.

National Marine Fisheries Service. 1983. Fisheries of the United States. 1982. U.S. Department of Commerce. *Current Fisheries Statistics.* Number 8300.

Nelson, W.R. and C.H. Walburg. 1977. Population dynamics of yellow perch, *Perca flavescencs*; sauger, *Stizostedion canadense;* and walleye, *Stizostedion vitreum vitreum*, in four main stream Missouri River reservoirs. *Journal of the Fisheries Research Board, Canada.* **34**:1748-1763.

Netsch, N.F., G.M. Kersh Jr., A. Houser, and V. Kilambi. 1971. Distribution of young gizzard and threadfin shad in Beaver Reservoir. In *Reservoir fisheries and limnology*, G. Hall, (ed.), pp. 95-105. Special Publication. Number 8. American Fisheries Society. Bethesda, MD.

Neves, R.J. and L. Depres. 1979. The oceanic migration of American shad, *Alosa sapidissima*, along the Atlantic Coast. *U.S. National Marine Fishery Service Bulletin.* 77(1):199-212.

Neves, R.J. 1981. Offshore distribution of alewife, *Alosa pseudoharengus*, and blueback herring, *Alosa alstivalis*, along the Atlantic coast. *U.S. Fish and Wildlife Services Fishery Bulletin.* **79**(3):473-485.

Ney, J.T. 1978. A synoptic review of yellow perch and walleye biology. In *Selected Coolwater Fishes of North America*, R.L. Kendall (ed.), pp. 1-12. American Fisheries Society Special Publication. Number 11. Washington, D.C.

Nigro, A.A. and J.J. Ney. 1982. Reproduction and early life accommodations of landlocked alewives to a southern range extension. *Transactions of the American Fisheries Society.* **111**(5):559-569.

Nord, R.R. 1967. A compendium of fishery information on the upper Mississippi River. Upper Mississippi River Conservation Committee.

Norden, C.R. 1967. Age, growth and fecundity of the alewife, *Alosa pseudohaarengus* (Wilson), in Lake Michigan. *Transactions of the American Fisheries Society.* **96**(4):387-393.

Normandeau, D.A. 1969. Life history and ecology of the round whitefish, *Prosopium cylindraceum* (Pallas), of Newfound Lake, Bristol, New Hampshire. *Transactions of the American Fisheries Society.* **98**(1):7-13.

Olson, D.E. and W.J. Scidmore. 1962. Homing behavior of spawning walleyes. *Transactions of the American Fisheries Society.* **91**:355-361.

Olson, D.E. 1971. Life history and relative abundance of three species of bullhead in Lake Sallie. Minnesota Department of Conservation, Job Progress Report. F-26-R-1:1-9.

O'Neil, J.T. 1980. Aspects of the life histories of anadromous alewife and the blueback herring, Margaree River and Lake Ainsle, Nova Scotia, 1978-1979. Master's Thesis. Acadia University, Wolfville. Nova Scotia, Canada.

Otto, R.G., M.A. Kitchel, and J.O. Rice. 1976. Lethal and preferred temperatures of the alewife, *Alosa pseudoharengus*, in Lake Michigan. *Transactions of the American Fisheries Society.* **105**(1):96-106.

Oven, J. H. and F. C. Fiss. 1996. MICRA national paddlefish research. 1995 Interim Report. Bettendorf, IA.

Page, L.M. and B.M. Burr. 1991. *A Field Guide to Freshwater Fishes.* Houghton-Mifflin Company. Boston, MA.

Perry, W.G. 1968. Distribution and relative abundance of blue catfish, *Ictalurus furcatus*, and channel catfish, *Ictalurus punctatus*, with relation to salinity. *Proceedings of the Annual Conference of the Southeastern Association of Game and Fish Commissioners, 1967.* **21**:436-444.

Perry, W.G. Jr. and J.W. Avault Jr. 1970. Culture of blue, channel and white catfish in brackish water ponds. *Proceedings of the Annual Conference of the Southeastern Association of Game and Fish Commissioners.* **23**:592-604.

Pflieger, W.L. 1975. The fishes of Missouri. Missouri Department of Conservation.

Pitman, V.M. 1991. Synopsis of paddlefish biology and their utilization and management in Texas. *Special Report.* Fisheries and Wildlife Division. Texas Parks and Wildlife Department. Austin.

Powell, D.H. 1976. Channel catfish as an additional sport fish in Alabama's state-owned and managed public fishing lakes. *Proceedings of the Annual Conference of Southeastern Association of Game and Fish Commissioners.* **29**:265-272.

Prather, E.E. and H.S. Swingle. 1960. Preliminary results on the production and spawning of white catfish in ponds. *Proceedings of the Annual Conference of the Southeastern Association of Game and Fish Commissioners.* **14**:143-145.

Price, J.W. 1940. Time-temperature relations in the incubation of the whitefish, *Coregonus clupeaformis* (Mitchill). *Journal of General Physiology.* **23**(4):449-468.

Priegel, G.R. 1964. Review in Lake Poygan sturgeon management problems. *Wisconsin Conservation Department.* Oshkosh.

Priegel, G.R. 1967. The freshwater drum — its life history, ecology and management. *Wisconsin Department of Natural Resources Publication.* **236**:11-115.

Priegel, G.R. 1970. Reproduction and early life history of the walleye in the Lake Winnebago region. *Wisconsin Conservation Department Technical Bulletin.* **45**:1-105.

Priegel, G.R., and T.L. Wirth. 1971. The lake sturgeon, its life history, ecology and management. *Wisconsin Department of Natural Resources Publication,* pp. 240-270.

Priegel, G.R. 1973. Lake sturgeon management on the Menominee River. *Wisconsin Department of Natural Resources Technical Bulletin.* **67**:1-20.

Priegel, G.R. and D.C. Krohn. 1975. Characteristics of a northern pike spawning population. *Wisconsin Department of Natural Resources Technical Bulletin.* **86**:1-18.

Priegel, G.R. and T.L. Wirth. 1975. Lake sturgeon harvest, growth, and recruitment in Lake Winnebago, Wisconsin. *Wisconsin Department of Natural Resources Technical Bulletin.* **83**:1-25.

Purkett, C.A. Jr. 1958. Growth rates of Missouri stream fishes. Missouri Conservation Committee. Game and Fish Division. D-J Service Number 1:1-46.

Purkett, C.A. Jr. 1961. Reproduction and early development of the paddlefish. *Transactions of the American Fisheries Society.* **90**:125-129.

Ranthum, R.G. 1971. A study of the movement and harvest of catfish tagged in the lower Trempealeau River and Trempealeau Bay. Wisconsin Department of Natural Resources Bureau of Fish Management. Report Number 59:1-21.

Raney, E.C. and D.A. Webster. 1940. The food and growth of the young of the common bullhead, *Ameiurus nebulosus nebulosus* (LeSueur), in Cayuga Lake, New York. *Transactions of the American Fisheries Society.* **69**:205-209.

Ranthum, R.G. 1971. A study of the movement and harvest of catfish tagged in the lower Trempealeau River and Trempealeau Bay. Wisconsin Department of Natural Resources Bureau of Fish Management. Report Number 59:1-21.

Rawson, D.S. 1956. The life history and ecology of the yellow walleye, *Stizostedion vitreum,* in Lac La Ronge, Saskatchewan. *Transactions of the American Fisheries Society.* **86**:15-37.

Reckahn, J.A. 1970. Ecology of young lake whitefish, *Coregonus clupeaformis,* in South Bay, Manitoulin Island, Lake Huron. In *Biology of Coregonid Fishes,* C.C. Lindsey and C.S. Woods (eds.), pp. 437-460. University of Manitoba Press, Winnepeg, Canada.

Regier, H.A. 1963. Ecology and management of channel catfish in farm ponds in New York. *New York Fish and Game Journal.* **10**:170-185.

Riepe, J.R. 1998. Walleye markets in the north central region: results of a 1996/97 survey. Technical Bulletin Series #113. Illinois Indiana Sea Grant Program. Purdue University.

Ritchie, B.J. and P.J. Colby. 1988. Even-odd year differences in walleye year-class strength related to mayfly production. *North American Journal of Fisheries Management.* **8**:210-215.

Robinson, J.W. 1994. Missouri's commercial fishery harvest, 1992. Missouri Department of Conservation. Final Report.

Rogers, H.M. 1939. The estuary as a biological habitat, with special reference to the smelt, *Osmerus mordax.* Ph.D. Dissertation. University of Toronto, Canada.

Rosen, R.A. and D.C. Hales. 1981. Feeding of paddlefish, *Polyodon spathula. Copeia.* **1981**:441-455.

Rosen, R.A., D.C. Hales, and D.G. Unkenholz. 1982. Biology and exploitation of paddlefish in the Missouri River below Gavins Point Dam. *Transactions of the American Fisheries Society.* **III**:216-222.

Ross, J., P.M. Powles, and M. Berrill. 1977. Thermal selection and related behavior in larval yellow perch, *Perca flavescens. Canada Field National.* **91**:406-410.

Ross, R.M., R.M. Bennett, and T.W.H. Backman. 1993a. Habitat used by spawning adult, egg, and larval American shad in the Delaware River. *Rivers.* **4**(3):227-238.

Ross, R.M., T.W.H. Backman, and R.M. Bennett. 1993b. Evaluation of habitat suitability index models for riverine life stages of American shad, with proposed models for premigratory juveniles. Fish and Wildlife Service. *U.S. Department of Interior Biological Report.* **14**.

Roy, J. 1969. The American shad and the alewife. *Fishes of Quebec.* Album 8.

Ruelle, R. and P.L. Hudson. 1977. Paddlefish, *Polydon spathula*: growth and food of young-of-the-year and a suggested technique for measuring length. *Transactions of the American Fisheries Society.* **106**:609-613.

Rupp, R.S. 1965. Shore-spawning and survival of eggs of the American smelt. *Transactions of the American Fisheries Society.* **94**(21):160-168.

Russell, T.R. 1986. Biology and life history of the paddlefish — a review. In *The Paddlefish: Status, Management and Propagation,* J.G. Dillard, L.K. Graham, and T.R. Russell (eds.), pp. 2-21. *Special Publication Number 7.* American Fisheries Society, North Central Division. Bethesda, MD.

Ryder, J.A. 1887. On the development of osseus fishes, marine and freshwater forms. *U.S. Commission Fisheries Report* (1885). **13**:489-604.

Ryder, R.A. 1977. Effects of ambient light variations on behavior of yearlings, subadults, and adult walleyes, *Stizostedion vitreum vitreum. Journal of the Fisheries Research Board, Canada.* **34**:1481-1491.

Sato, R. 1952. Biological observation on the pond smelt, *Hypomesus olidus* (Pallas), in Lake Kogawara, Aomori Prefecture, Japan II. Early life history of the fish. *Tohoku Journal of Agricultural Research.* **3**(1):175-184.

Schademann, R. 1987. Food habits, growth, distribution of walleye in Clinton Reservoir, Kansas. Kansas Department of Wildlife Parks. D.J. Project FW-0-P-5. ID 87-2.

Scherer, E. 1976. Overhead light intensities and vertical positioning of the walleye, *Stizostedion vitreum vitreum. Journal of the Fisheries Research Board, Canada.* **33**:289-292.

Schneberger, E. and L.A. Woodbury. 1944. The lake sturgeon, *Acipenser fulvescens* (Rafinesque), in Lake Winnebago, Wisconsin. *Transactions of the Wisconsin Academy of Science, Arts, Letters.* **36**:131-140.

Schott, E.F., T.B. Kayes, and H.E. Calbert. 1978. Comparative growth of male versus female yellow perch fingerlings under controlled environmental conditions. *Selected Coolwater Fishes of North America.* R.L. Kendall (ed.). **11**:181-186.

Scott, W.B. and E.J. Crossman. 1973. Freshwater Fishes of Canada. Fisheries Research Board of Canada Bulletin. Number 184. Ottawa, Canada.

Semmens, K.J. 1986. Evaluation of paddlefish hypophysis, carp hypophysis and LHRH analogue to induce ovulation in paddlefish, *Polyodon spathula.* Ph.D. Dissertation. Auburn University, Auburn, AL.

Shelton, W.L. 1972. Comparative reproductive biology of the gizzard shad, *Dorosoma cepedianum,* and threadfin shad, *D. petenense* (Gunther), in Lake Texoma, Oklahoma. Ph.D. Dissertation. University of Oklahoma, Norman.

Sheri, A.N. and G. Power. 1968. Reproduction of white perch, *Roccus americanus,* in the Bay of Quinte, Lake Ontario. *Journal Fisheries Research Board,Canada.* **25**(10):2225-2231.

Sheri, A.N. and G. Power. 1969. Fecundity of the yellow perch, *Perca flavescens* (Mitchill), in the Bay of Quinte, Lake Ontario. *Canadian Journal of Zoology.* **47**:55-58.

Shetter, J.T. 1956. Report of fisheries investigations during the third year of impoundment of Fort Randall Reservoir, South Dakota, 1955. South Dakota Department of Game, Fish, and Parks. D.J. Project F-1-R-5.

Shields, J.T. 1958. Report of fisheries investigations during the third year of impoundment of Gavins Point Reservoir, South Dakota, 1957. D.J. F-I-R-7. South Dakota Department of Game, Fish, and Parks. Pierre.

Sigler, W.F. and R.R. Miller. 1963. Fishes of Utah. Utah State Department of Fish and Game. Salt Lake City.

Smith, B.W. and W.C. Reeves. 1986. Stocking warmwater species to restore or enhance fisheries. In *Fish Culture in Fisheries Management.* R.H. Stroud (ed.), pp. 17-29. Bethesda, MD.

Smith, P.W. 1979. The Fishes of Illinois. *Illinois State Natural Historical Survey.* University of Illinois Press. Urbana.

Smith, L.L. Jr. and R.L. Pycha. 1960. First-year growth of the walleye, *Stizostedion vitreum vitreum* (Mitchill), and associated factors in the Red Lakes, Minnesota. *Limnology and Oceanography.* **5**:281-290.

Smith, L.L. Jr. and W.M. Koenst. 1975. Temperature effects on eggs and fry of percoid fishes. United States Environmental Protection Agency Project. EPA-660/3-75-017.

Smith, S.H. 1957. Evolution and distribution of the coregonids. *Journal of the Fisheries Research Board of Canada.* **14**(4):599-604.

Sneed, K.E. 1951. A method for calculating the growth of channel catfish, *Ictalurus lacustris punctatus. Transactions of the American Fisheries Society.* **80**:174-183.

Snow, J.R. 1959. Notes on the propagation of the flathead catfish, *Pilodictus olivaris* (Rafinesque). *Progressive Fish-Culturist.* **21**(2):75-80.

Snyder, D.E. 1994. Morphological development and identification of pallid, shovelnose, and hybrid sturgeon larvae. Contribution number 71:1-48. Larval Fish Laboratory, Department of Fishery and Wildlife Biology. Colorado State University, Fort Collins.

Spangler, G.R., N.R. Payne, J.E. Thorpe, J.M. Bryne, H.A. Regier, and W. J. Christie. 1977. Responses of percids to exploitation. *Journal of the Fisheries Research Board Canada.* **34**:1983-1988.

Spinelli, A.J., B.G. Whiteside, and D.G. Huffman. 1985. Aquarium studies on the evaluation of stocking various sizes of channel catfish with established largemouth bass. *North American Journal of Fisheries Management.* **5**:138-145.

Sprague, J.W. 1959. Report of fisheries investigations during the sixth year of impoundment of Fort Randall Reservoir, South Dakota, 1958. South Dakota Department of Game, Fish, and Parks. Federal Aid Report in Fish Restoration. Project F-1-R. Jobs 105. Progress Report. Pierre.

Stephen, J.L. 1978. Marketable fisheries investigations. Kansas Fish and Game Commission. National Marine Fisheries Service. Project 2-272-R. Final Report. Pratt.

Stevens, R.E. 1959. The white and channel catfishes of the Santee-Cooper Reservoir and tailrace sanctuary. *Proceedings of the Annual Conference of the Southeastern Association of Game and Fish Commissioners.* **13**:203-219.

Stewart, R.W. and W.S. Murawski. 1973. Evaluation of the channel catfish, *Ictalurus punctatus* (Rafinesque), in New Jersey. New Jersey Department of Environmental Protection Division Fish Game Shellfish. D-J Federal Aid Project F-3-R, F-9-R, F-23-R. Miscellaneous Report Number 37. Trenton.

Stickney, R.R. 1993. Channel catfish. In *Culture of Nonsalmonid Freshwater Fishes, Second Edition,* R.R. Stickney (ed.), pp. 33-79. CRC Press. Boca Raton, FL.

Stier, D.J. and J.H. Crance. 1985. Habitat suitability index models and instream flow suitability curves: American shad. *U.S. Fish and Wildlife Service Biological Report.* **82**(10.88).

Storck, T.W., D.W. Dufford, and K.T. Clement. 1978. The distribution of limnetic fish larvae in a flood control reservoir in central Illinois. *Transactions of the American Fisheries Society.* **107**(3):419-424.

Storck, T.W. and D. Newman. 1987. Survival and harvest of channel catfish stocked at various sizes in a 6.1 hectare impoundment. *Illinois Natural History Survey.* Sullivan.

Stuber, R.J. 1982. Habitat suitability index models: black bullhead. United States Fish and Wildlife Service. FWS/OBS-82/10.14. Washington, DC.

Swingle, H.S. 1957. Revised procedures for commercial production of the bigmouth buffalo fish in ponds in the southeast. Proceedings for commercial production of the bigmouth buffalo fish in ponds in the Southeast. *Proceedings of the Annual Conference of the Southwestern Association of Game and Fish Commissioners.* **11**:162-165.

Talbot, G.B. 1961. The American shad. Fishery Leaflet. Washington, DC.

Talbot, G.B. and J.E. Sykes. 1958. Atlantic coastal migrations of American shad. *U.S. Fish and Wildlife Service Fishery Bulletin.* **58**(142):473-490.

Talley, K. 1994. American consumers eating more seafood. *Northern Aquaculture.* **10**(5):7.

Taubert, B.D. 1980. Reproduction of the shortnose sturgeon, *Acipenser brevirostrum,* in Holyoke Pool. Connecticut River, MS. *Copeia.* **1980**:114-117.

Threinen, C.W., C. Wistrom, B. Apelgren, and H. Snow. 1966. The northern pike, life history, ecology, and management. *Wisconsin Conservation Department Publications.* **235**:1-16.

Tiemeier, O.W. 1966. Kansas farm ponds. Kansas State University Agriculture Experiment Station. Bulletin Number 488. Manhattan.

Thunberg, B.E. 1971. Olfaction in parent stream selection by the alewife. *Animal Behavior.* **19**:217-225.

Todd, T.N. 1986. Occurrence of white bass-white perch hybrids in Lake Erie. *Copeia.* 1986:196-199.

Toner, E.D. and G.H. Lawler. 1969. Synopsis of biological data on the pike, *Esox lucius* (Linnaeus, 1758). FAO Fisheries Synopsis. Number 30, Revision 1.

Trautman, M.B. 1957. *The Fishes of Ohio.* Ohio State University Press.

Turner, P.R. 1977. Age determination and growth of flathead catfish. Ph.D. Dissertation. Oklahoma State University, Stillwater.

Turner, P.R and R.C. Summerfelt. 1970. Food habits of adult flathead catfish, *Pylodictus olivaris* (Rafinesque), in Oklahoma Reservoirs. *Proceedings of the Annual Conference of the Southeastern Association of Game and Fish Commissioners.* **24**:387-401.

Turner, P.R and R.C. Summerfelt. 1971. Reproductive biology of the flathead catfish, *Pylodictus olivaris* (Rafinesque), in a turbid Oklahoma Reservoir. In *Reservoir Fisheries and limnology,* G.E. Hall (ed.), 107-119. A Special Publication Number 8. American Fisheries Society. Washington, DC.

United States Department of Agriculture. 1996. Aquaculture Outlook. LDP-AQS-4. Economic Research Service. United States Department of Agriculture.

United States Department of Agriculture. 1997. Aquaculture Outlook. LDP-AQS-5. Economic Research Service. United States Department of Agriculture.

Van Den Avyle, M.J. and D.D. Fox. 1980. Diel, vertical and horizontal variations in abundance of larval *Dorosoma* spp. in Center Hill Reservoir, Tennessee. In *Proceedings of the 4th Annual Larval Fish Conference,* pp. 116-122. U.S. Fish and Wildlife Service. FWS/OBS-80/43.

Van Oosten, J. 1946. Maximum size and age of whitefish. *The Fisherman.* **14**(8):17-18.

Van Oosten, J. 1960. The smelt, *Osmerus mordax* (Mitchill). Michigan Department of Conservation.

Vessel, M.F. and S. Eddy. 1941. A preliminary study of the egg production of certain Minnesota fishes. *Minnesota Department of Conservation Fisheries Research Investigation Report.* **26**:1-26.

Vladykov, V.D. and J.R. Greeley. 1963. Order Acipenseroidei. In *Fishes of the Western North Atlantic. Memorial Sears Foundation for Marine Research.* **1**(3):24-60.

Walburg, C.H. 1960. Abundance and life history of the shad, St. Johns River, Florida. *U.S. Fish and Wildlife Service Fishery Bulletin.* **60**(177):487-501.

Walburg, C.H. 1964. Fish population studies, Lewis and Clark Lake, Missouri River, 1956-1962. *U.S. Fish and Wildlife Service Special Scientific Report: Fish.* 482.

Walburg, C.H. and W.R. Nelson. 1966. Carp, river carpsucker, smallmouth buffalo, and bigmouth buffalo in Lewis and Clark Lake, Missouri River. *United States Department of the Interior Fish and Wildlife Service Research Report.* **69**:1-30.

Walburg, C.H. and P.R. Nichols. 1967. Biology and management of the American shad and status of the fisheries. Atlantic coast of the United States, 1960. *U.S. Fish and Wildlife Service Special Scientific Report: Fish. 550.*

Walburg, C.H. 1971. Loss of young fish in reservoir discharge and year-class survival, Lewis and Clark Lake, Missouri River. *American Fisheries Society Special Publication.* **8**:441-448.

Walburg, C.H. 1976. Changes in the fish populations of Lewis and Clark Lake, 1956-1974, and their relation to water management and the environment. *United States Wildlife Service Research Report.* **79**:1-34.

Walden, H.T. 1964. *Familiar Freshwater Fishes of America.* Harper and Row, New York.

Wallace, C.R. 1967. Observations on the reproductive behavior of the black bullhead, *Ictalurus melas. Copeia.* 1967(4):852-853.

Warfel, H.E., T.P. Frost, and W.H. Jones. 1943. The smelt, *Osmerus mordax*, in Great Bay, New Hampshire. *Transactions of the American Fisheries Society.* **72**:257-262.

Wells, L. 1966. Seasonal and depth distribution of larval bloaters, *Coregonus hoyi,* in southeastern Lake Michigan. *Transactions of the American Fisheries Society.* **95**(4):388-396.

Wells, L. and A.M. Bekton. 1963. Food of the bloater, *Coregonus hoyi*, in Lake Michigan. *Transactions of the American Fisheries Society.* **92**(3):245-255.

Wells, L. and A.L. McLain. 1973. Lake Michigan: man's effects on native fish stocks and other biota. *Great Lakes Fishery Commission Technical Report.* **20**:1- 55.

West, G. and J. Leonard. 1978. Culture of yellow perch with emphasis on development of eggs and fry. In *Selected Coolwater Fishes of North America,* R.L. Kendall (ed.). **11**:172-176.

Wiggins, T.A., T.R. Bender, V.A. Mudrak, and M.A. Takacs. 1983. Hybridization of yellow perch and walleye. *Progressive Fish-Culturist.* **45**:131-132.

Williams, J.D. and G.H. Clemmer. 1991. *Schaphirhynchus suttbusi*, a new sturgeon (Pisces: Acipenseridae) from the Mobile Basin of Alabama and Mississippi. *Bulletin of the Alabama Museum of Natural History.* **10**:17-31.

Williamson, K.L. and P.C. Nelson. 1985. Habitat suitability index models and instream flow suitability curves: Gizzard shad. *United States Fish and Wildlife Service Biological Report.* **82**(10.11.2).

Willis, D.W. 1989. Proposed standard length-weight equation for northern pike. *North American Journal Fisheries Management.* **9**:203-208.

Wirth, T.L. 1958. Lake Winnebago freshwater drum. *Wisconsin Conservation Bulletin.* **23**(5):30-32.

Witt, A., Jr. 1960. Length and weight of ancient freshwater drum, *Aplodinotus grunniens*, calculated from otoliths found in Indian middens. *Copeia.* 1960 (3):181-185.

Wolfert, D.R. 1969. Maturity and fecundity of walleyes from the eastern and western basins of Lake Erie. *Journal of the Fisheries Research Board, Canada.* **26**:1877-1888.

Worth, S.G. 1892. Observations on hatching of yellow perch. *Bulletin of the U.S. Fish Commissioners for 1890.* **10**:331-334.

Other Aquatic Life of Economic Significance: Eels

William L. Rickards III

INTRODUCTION

Foods and other products manufactured from freshwater eels (Family Anguillidae) are commonplace in many countries, but particularly in Japan and the nations of northern Europe. Most eel products in Japan are derived from cultured or farmed eels. European countries depend largely upon wild-caught eels, although eel farming is emerging as a significant contributor in that part of the world. While the predominant product from eels is food derived from the body muscle as a smoked, jellied, or steamed/grilled fillet, nearly every part of the fish finds a use in one country or another.

SPECIES AND DISTRIBUTION

Species of freshwater eels of the genus *Anguilla* inhabit rivers, streams, lakes, and estuaries throughout most subtropical and temperate regions of the world. Notable exceptions include the West Coast of Africa, the West Coast of North America, and nearly all of Central and South America (Usui, 1991; Forrest, 1976).

Taxonomists have classified the freshwater eels into at least 17 separate species, but there are cases where true separation is in question. Insofar as human food and other products are concerned, the most heavily utilized species are:

* Japanese eel, *Anguilla japonica*, found in Japan and China;
* European eel, *Anguilla anguilla*, found in western Europe, northern Africa and Iceland;
* American eel, *Anguilla rostrata*, found along the East Coast of North America and in Greenland; and
* Australian eels, *Anguilla bicolor*, found in East Africa, India, Indonesia, and northwestern Australia, and *A. australis* found in Eastern Australia and New Zealand (Usui, 1991).

LIFE HISTORY SUMMARY

All of the Anguillid eels have the same general life history pattern: catadromous. In this type of life cycle, spawning occurs in the ocean, and the young migrate into estuarine and inland waters where they grow and begin to mature; in a final phase they migrate back to the ocean spawning area as maturation is completed (Tesch, 1977).

In the case of the American eel, spawning occurs late in the calendar year at considerable depth in the ocean area known as the Sargasso Sea. Females release the eggs into the water where they are fertilized by the males. The eggs develop and hatch in one to two days, and the resulting larvae drift with the currents for several months (Moriarty, 1978; Tesch, 1977; Sinha and Jones, 1975). Depending on which part of the East Coast is to become their nursery and juvenile habitat, this phase may last just a few months (for young that colonize southeastern estuaries and streams in a January-to-March immigration) to several months (for young moving into New England or Canadian waters in April through June). As the larvae drift closer to coastal waters, they metamorphose from the early larval form (called a leptocephalus) to the glass eel stage in which they become "eel"-shaped but lack pigment.

As the glass eels move into estuaries and streams, they undergo internal changes that will allow them to live in low salinity and fresh water; they also become pigmented (Moriarty, 1978). At this point, they are called elvers. The elvers continue moving upstream and take up residence where they will feed and grow into yellow eels. Female eels grow to a larger average body size than the males. Over a period of several years as juvenile or yellow eels, they reach the appropriate size for the adult stage to begin. As this oc-

Figure 1. Male and female American eel, Anguilla rostrata.

curs, the eels begin moving downstream into the estuaries and out to the ocean in late fall (Moriarty, 1978; Tesch, 1977).

During the migration back to the ocean, the body form alters to adapt to conditions at the spawning area: the eyes enlarge, the body coloration becomes more bronzish (rather than the greenish-yellow seen in the inland habitat), and the gonads mature. As females mature, the body cavity swells due to its greatly enlarged ovaries. Since spawning of eels in the ocean has never been observed, it is presumed that the males and females pair off and spawning occurs several thousand feet below the surface (Tesch, 1977).

The same life history pattern has been observed in all species of Anguillid eels, but the spawning areas are not precisely known.

Researchers in several countries have attempted to complete our understanding of this life cycle with captive eels through manipulation of water conditions and hormone injections. In some cases, mature eels have been developed and spawning has resulted in a limited number of larvae. However, none of the larvae has survived beyond a few days, because the conditions and foods necessary for their survival are not known (Chiu et al., 1991; Wang et al., 1980; Yamauchi et al., 1976).

Because of our inability to fully control the life cycle of eels in captivity, all eel culture is dependent upon the capture of glass eels or elvers as the source of seed stock for the farms. Thus, wild eel stocks are harvested at their youngest stages as well as during their entire life in brackish or freshwater.

PRODUCT FORMS AND SOURCES

Nearly all stages of the freshwater eels are sought by fishermen; the resulting products reflect ethnic preferences and ways of using the animal.

Wild Harvested Eels

The youngest stages which are harvested — glass eels and elvers — are destined mainly for eel-farming operations. However, glass eels are eaten in a variety of dishes in Spain and Portugal.

Glass eels and elvers are captured by a variety of small-meshed, fixed nets (i.e., fyke nets), traps, weirs, and dip netting. All of the techniques take advantage of the behavioral patterns of the young eels as they migrate into brackish and fresh waters. At this time, the eels frequently group together and move mostly at night, utilizing tidal currents to assist their upstream movements even though they are relatively good swimmers at this stage (Usui, 1991; Lane, 1978; Moriarty, 1978; Forrest, 1976).

As the young eels are moving upstream, the harvesting gear is set so as to intercept them. No active fishing gear (i.e., trawls, etc.) is used for the young eels since they can be injured easily and will quickly become infected with bacterial diseases. Dip nets and push nets are used because they can be emptied frequently, avoiding injury to the the elvers (Usui, 1991).

Captured glass eels and elvers are generally held by the harvester in large flow-through tanks until shipped to an eel farm. Since this period lasts only a few days, feeding is not conducted and therapeutic chemicals are not applied. These activities are initiated once the young eels reach the farm (Gousset, 1992).

Juvenile eels are harvested in fyke nets and various traps for use as fishing bait. The eels may be sold live to recreational fishermen in pursuit of carnivorous fishes, especially striped bass and flatfish. The eels may also be frozen or salted for use as bait in crab pots. This latter product form is common in the mid-Atlantic states, especially in the Chesapeake Bay where crab potting is very common (Berg, Jones, and Crow, 1975).

Larger eels (e.g., over one-quarter pound in weight) are captured by fixed gear, generally traps or eel "pots" in the eastern United States. These pots may be approximately three feet in length, with one-half by one inch wire mesh forming a cylinder and a wire mesh funnel-shaped entrance at one end and a cloth tail bag at the other end, through which the eels are removed from the pot. Traps may also be square with the same mesh dimensions as the pot, and an entrance funnel low on one side leading to a capture chamber containing the bait. Complete descriptions of eel pots and traps are given in Lane (1978); Forrest (1976); Berg, Jones, and Crow (1975); and Sinha and Jones (1975).

Most larger eels captured in the United States are destined for sale in European countries. Eels greater than one-third to one-half pound are readily sought in nearly all northern European markets (Lane, 1978). In other countries, the larger eels may be captured in traps, weirs, and stationary nets. Most harvesting occurs at night when the eels are actively moving about in search of food, or when they are moving downstream in late fall (Moriarty, 1978; Forrest, 1976; Sinha and Jones, 1975).

The most sought-after larger eel is the silver eel, which is captured as it is beginning its spawning migration. This eel is prized because it is ideal for smoking due to its higher fat content at the time of capture. These eels can be caught in unbaited traps and fixed nets (Lane, 1978).

Larger eels are generally held in flow-through tanks for a few days before being shipped to market. This allows for clearing of the digestive tract and intestines of any residual food so that shipping containers are not fouled, or, if shipped frozen, the eels do not begin deteriorating due to enzymatic activity (Lane, 1978; Berg, Jones, and Crow, 1975).

Eels may be shipped live in tanks of aerated water if the distance and cost are not too great, or via air in polyethylene bags with a small amount of ice. As the ice melts, it not only cools the eels, but also provides moisture for their skin and gills, enabling them to obtain enough oxygen to survive shipping.

Since the 1970s, large quantities of frozen eels have been shipped by air from the eastern United States to European processors who then prepare the eels according to local market preferences, using proprietary smoking recipes (Lane, 1978).

Cultured Eels

Since they are readily accessible, farmed eels may be harvested at the time of optimal size and market demand. In Japan, the preferred size is about one-fourth pound (115 grams); in Europe, the optimal size is between one-third and one-half pound (150-230 grams) (Usui, 1991; Forrest, 1976).

Harvesting of cultured eels depends on the type of enclosure being used to raise them: ponds are generally seined while tanks or concrete enclosures may be pumped or drained. In any case, preparation for shipping is generally the same as for wild-harvested eels, with product being transported alive (in water or moist air) or frozen. Handling and preparation of the final product for smoking or other further processing occurs at the processor level, or with the consumer if the eel is purchased directly for home use (Forrest, 1976).

PROCESSING AND MARKETING

United States

Marketing of eels in the United States has not developed any uniquely American styles. Rather, the limited sale of eels (either wild-caught or imported from eel-farming nations) is targeted to the ethnic preferences of the consumers making the purchases. Eel consumption in this country derives mainly from ethnic cuisine brought by immigrants into the nation's cities. People of European heritage may purchase smoked eel, most of which is imported, or fresh eel if it is available in the market, and prepare it for home use. Likewise, those of Japanese heritage may purchase product which has been prepared abroad and imported for marketing in cities where demand is sufficient.

There are no eel processing and cooking plants in the United States, but some eel smoking does occur, as processors will custom-smoke product upon demand from a customer.

Japan

Nearly 91 percent of all the eel consumed in Japan is a prepared product derived from very highly-managed culture and marketing systems (Gousset, 1992).

Cultivation of eels includes feeding them a prepared diet which has a considerable influence on body composition, especially of fat and fatty acid components (Otwell and Rickards, 1982; 1981). In the final days before harvesting cultured eels, feeding is halted so that the gut is cleared and the body fat content decreases. This conditioning period results in some loss of body weight and a change in the texture of the meat, but this is still preferred. Following 10 to 20 days of conditioning, the eels are harvested and sent to the processing plant (Matsui, 1979).

At the processing plant, the eels are kept alive and cool by slowly trickling water through the containers in which they have been shipped.

Processing of eels is involves a single method within Japan, but there are some geographic variations that give rise to preparations with different names. The following description applies to the product known as "kabayaki" (Gousset, 1992; Lane, 1978).

The eel is removed from the holding barrel and laid on the cutting board, where the belly is cut and viscera removed. The eel is pinned to the board with a slender spike held in the left hand. The right hand then passes along the eel to straighten it on the board, and the left hand initiates the first filleting cut just behind the head and passes on to the tail. The meat is set aside.

A second pass of the knife is used to remove the backbone, leaving the fillet from the other side of the eel. Both fillets of meat are cut to produce four nearly equal-sized pieces through which bamboo skewers are inserted. The meat is now ready for cooking. Variations in cooking result in different product forms. The process in general involves repeated basting with a soy sauce mixture, and roasting. This product is sold fresh or frozen in vacuum packaging (Gousset, 1992).

Nearly every part of the eel is used for some form of product:

* Skins are cured and tanned for use in eel leather products such as wallets and briefcases.
* In some areas of Japan, the eel heads are placed on bamboo sticks and sold as snack food in bars.
* The viscera may be roasted with soy sauce or canned for addition to the typical Japanese broths or soups.
* the backbones are dried by roasting and salted for sale as snack food.

The Japanese prepare a great variety of dishes based upon steamed, roasted, or broiled eel including combinations with bean curd (tofu), rice, salad greens, eggs, and yams (Tomiyama and Hibiya, 1977).

European Countries

Eel products in European countries are based largely on use of the body musculature, rather than the varied uses seen in Japan. The body meat may be smoked, steamed, fried, broiled, jellied, baked, or grilled. In Germany and the Netherlands, in particular, smoked eel is very popular, and there are a great number of regional preferences for certain flavors or textures imparted to the meat by recipes that are closely-guarded business secrets. The meat may be either hot- or cold-smoked, but the hot-smoked product is much more popular (Lane, 1978).

Because of their greater fat content, silver eels are preferred for smoking, but a great deal of smoked yellow eel is also produced. The final product may be sold chilled, frozen, canned, or vacuum-packed. Chilled smoked eel has a relatively short shelf life and must be refrigerated at all times (Lane, 1978).

In general, food preparations using the methods described above are most successful with an eel larger than one-half pound in weight.

This size preference establishes the primary distinction between eel markets in Japan and Europe. Several representative recipes for preparing eel dishes may be found in Usui (1991), Lane (1978), Forrest (1976), and Berg, Jones, and Crow (1975).

REFERENCES

Berg, D.R., W.R. Jones, and G.L. Crow. 1975. The case of the slippery eel or: how to harvest, handle and market wild eels. University of North Carolina (Raleigh) Sea Grant Program. Publication No. UNC-SG-75-20.

Chiu Liao, I., Cheng-Sheng Lee, and Mao Sen Su, ed. 1991. Finfish hatcheries in Taiwan. In: *Finfish Hatchery in Asia. Proceedings of Finfish Hatcher in Asia '91*. Tungkang Marine Laboratory, Keelung (Taiwan), 1993, no. 3, pp. 1-25. *TML Conference Proceedings*, Tungkang.

Forrest, D.M. 1976. *Eel Capture, Culture, Processing and Marketing*. Fishing News (Books). Oxford, England.

Gousset, B. 1992. Eel Culture in Japan. *Bulletin de l'Institut oceanographique, Monaco.* Special number **10**:1-128.

Lane, J.P. 1978. Eels and their utilization. *Marine Fisheries Review*. **40**(4): 1-20.

Matsui, I. 1979. *Theory and Practice of Eel Culture*. Oxonian Press Pvt. Ltd. Faridabad, India.

Moriarty, C. 1978. *Eels: A Natural and Unnatural History*. Universe Books, New York.

Otwell, W.S. and W.L. Rickards. 1981. Cultured and wild American eels, *Anguilla rostrata*: fat content and fatty acid composition. *Aquaculture*. **26**(1981/1982 issue): 67-76.

Sinha, V.R.P. and J.W. Jones. 1975. *The European Freshwater Eel*. Liverpool University Press. England.

Tesch, F. W. 1977. *The Eel: Biology and Management of Anguillid Eels*. Chapman and Hall. London.

Tomiyama, T. and T. Hibiya, eds. 1977. *Fisheries in Japan: Eel*. Japan Marine Products Photo Material Association, Tokyo.

Usui, A. 1991. *Eel Culture*. Fishing News (Books), 2nd edition. Oxford, England.

Wang, Y., C. Zhao, Z. Shi, Y. Tan, Y. Li, Y. Yang, and Y. Hong. 1980. Studies on the artificial inducement of reproduction in common eel. *J. Fish. China* **4**(2): 147-156.

Yamauchi, K., M. Nakamura, H. Takahashi, and K. Takano. 1976. Cultivation of larvae of Japanese eel. *Nature*. **263**: 412. London.

Other Aquatic Life of Economic Significance: Turtles

John A. Keinath

INTRODUCTION

Turtles are characterized by a dorsal (carapace) and ventral (plastron) shell made up of ribs fused in bone with a covering of keratin (scutes). Turtles are the only extant animals that have the pelvic and pectoral girdles situated inside the ribcage. Land turtles (tortoises) have elephantine feet; freshwater turtles have webbed feet with more than three nails, and use the rear feet for swimming; sea turtles have oar-shaped feet (flippers) with fewer than three nails, and use the front flippers for swimming.

Turtles have been utilized by humans since the earliest times, and probably "every species . . . has become reduced in numbers to some extent" (Pritchard, 1979). The hard parts have been used for worship items, ornaments, and jewelry. The skin was used for leather, and the flesh and eggs as food. Body fluids, especially the oils, have been utilized for cosmetics and lotions, and for waterproofing of vessels. In addition, turtles are kept as pets (Humane Society of the United States, 1994; Cervignon et al., 1993; Halliday and Adler, 1986; Carr, 1984, 1954; Pritchard, 1979; Liner, 1978; Bustard, 1972; Tressler and Lemon, 1951).

All turtles nest on land. Their slow movements, stereotypical nesting behavior, and predictable appearance at nesting areas make turtles easy targets for human and animal exploitation (Halliday and Adler, 1986). Because few offspring survive to sexual maturity [which occurs at an "old age" relative to most other vertebrates (Burke et al., 1994)], exploitation has driven many populations to, and sometimes over, the brink of extinction.

SEA TURTLES

Sea turtles are found throughout the tropics, with some species migrating into temperate, and even boreal, waters. Sea turtles have been among the most economically important reptiles, due to their food value (both eggs and meat); leather (among the most durable); oil (for lubrication, preservation of boats, cosmetics, and body lotions); and products for ornamentation, particularly "tortoise shell," which is derived primarily from hawksbill turtles (Cervigon et al., 1993; Sternberg, 1982; Mack et al., 1981; Pritchard, 1979; Rebel, 1974; Bustard, 1972).

Carr (1984, 1954) described the importance of sea turtles used for food during the exploration of the New World: they were kept on their backs in the holds of ships, where they would stay alive for months, thus providing fresh meat that needed no refrigeration. "Only the turtle could take the place of spoiled kegs of beef and send a ship on for a second year of wandering or marauding. All early activity in the new world tropics (exploration, colonization, buccaneering, and the maneuverings of naval squadrons) was in some way dependent on the turtle" (Carr, 1984). Sea turtles were an important resource even into the twentieth century (Rebel, 1974); however, due to over-exploitation and interaction with human activities, most sea turtle populations have been severely depleted.

Sea turtles can be captured by hand (diving), hook and line, tangle nets, gillnets, trawls, seines, and harpoons. A more common practice is to collect females when they come ashore to nest. In addition, eggs are collected as they are deposited, or nests are discovered by probing with a sharp stick until a soft area is found (Cervigon et al., 1993; Dodd, 1988; Carr, 1984; Rebel, 1974; Irvine, 1947). Irvine (1947) and Rebel (1974) describe an unusual method used to capture sea turtles: A remora is attached to a string and the fish released in the vicinity of turtles. The fish attaches to a turtle and the turtle is retrieved.

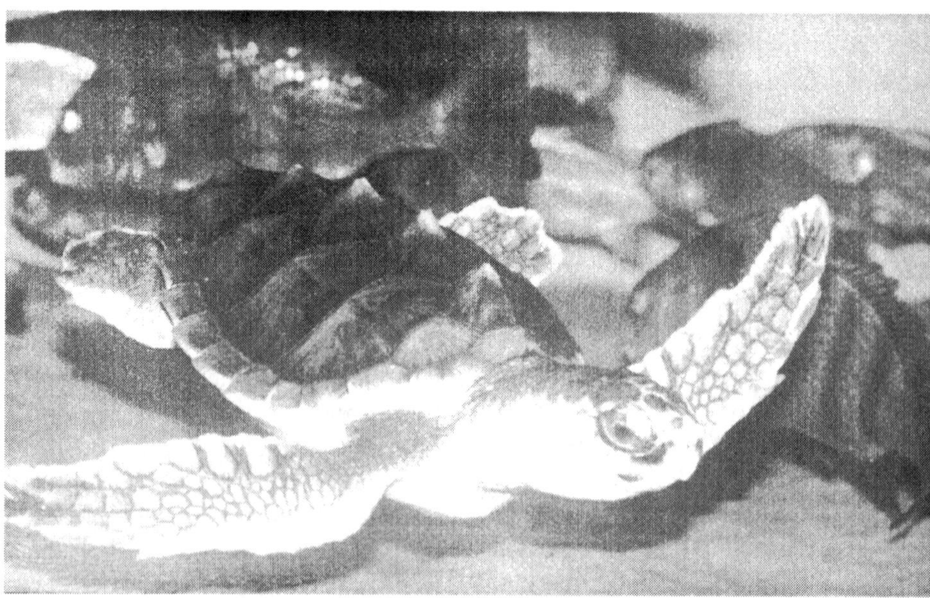

Figure 1. A loggerhead turtle.

Bustard (1972) recounts the harvest of turtles in Australia during the 1920s. Nesting turtles were turned on their backs ("turning turtle") and left, sometimes for days, until gathered. The turtles were then taken to the slaughterhouse where they were decapitated and butchered. The meat was removed from the bones and boiled in vats. The resulting soup was poured into tins, sterilized for 40 minutes, and sealed. Bustard writes: "Twenty-two to twenty-five turtles, a good day's catch, produced about nine hundred tins of soup. Last season (1924–1925), thirty-six thousand tins were prepared." He concluded that this amount of soup represented about 1000 turtles, "which approximated the size of the entire breeding population at the present time. Clearly the removal of virtually the whole breeding population each year was the surest way to wipe out the resources!" Cato et al. (1978) described the harvest and use of "sea turtles" (but included diamondback terrapins in the statistics) from 1948 through 1976 (mostly as canned meat, stew, or soup) in the United States and worldwide, and Tressler and Lemon (1951) described butchering of turtles in Key West, Florida, and the taking of turtles in Costa Rica.

Halstead (1980) stated that although "poisoning from (ingestion) of marine turtles is one of the lesser-known types of intoxications produced by marine organisms, the cases that have been reported are sufficiently severe to be impressive." This may be especially true of the liver.

Sea turtles have been the most heavily exploited of the turtles worldwide, and many populations have gone extinct or are declining rapidly due to human activities (Pritchard, 1980). Because of the precarious status of these species, many countries have legally protected sea turtle species. Six sea turtle species are protected in the United States under the Endangered Species Act (the flatback is not) (CFR, 1994). Violation — which includes killing, harassment, and possession of any part of these animals — can incur large fines and prison terms. Six of the seven species are listed in the International Union for the Conservation of Nature and Natural Resources (IUCN) Red Data Book, and all seven species are listed under Appendix I of the Convention on International Trade in Endangered Species of Wild Fauna and Flora (CITES) (Marine Turtle Specialist Group, 1994).

Seven species of sea turtles are currently recognized. These include the leatherback (*Dermochelys coriacea*), Kemp's ridley (*Lepidochelys kempii*), olive ridley (*Lepidochelys olivacea*), loggerhead (*Caretta caretta*) (Figure 1), hawksbill (*Eretmochelys imbricata*), Australian flatback (*Natator depressus*), and green sea turtle (*Chelonia mydas*). The east Pacific black sea turtle (formerly *Chelonia agassizi*) is now considered a subspecies of the green turtle by most sea turtle biologists. The Australian flatback has a very restricted range off northern Australia.

Leatherback Turtle

Leatherbacks are the most wide-ranging reptile, and cannot be mistaken for any other sea creature. Although leatherbacks range well into boreal waters, they nest primarily in the tropics. Leatherbacks reach carapace lengths of 170 centimeters (cm) and weigh up to 900 kilograms (kg) (Committee on Sea Turtle Conservation, 1990). The plastron and ventral surfaces are white, and the carapace and dorsal surfaces are black, often with white and pink spots. There are raised longitudinal ridges along the carapace and plastron, and there are no claws or scutes (except in hatchlings). Leatherbacks eat soft-bodied invertebrates (salps, siphonophores, and jellyfish) exclusively. Leatherbacks have a large fat reserve, and their oil was previously used as varnish or to waterproof canoes (Cervigon et al., 1993; Frazier, 1981; Carr, 1952). Although leatherbacks are protected in most areas, turtles are still poached for their flesh and eggs (Keinath, 1990; Frazier, 1981; Ross, 1981; Ernst and Barbour, 1972). Carr (1952) and Halstead (1980) stated that leatherback meat may be poisonous, perhaps because of the toxins contained in its prey.

Kemp's Ridley Turtle

Kemp's ridley turtles are found in the Gulf of Mexico and the North Atlantic Ocean. Kemp's ridleys reach carapace lengths up to 70 cm and weigh up to 45 kg (Committee on Sea Turtle Conservation, 1990). The plastron and ventral surfaces are creamy white, and the carapace and dorsal surfaces are dark grey. Except for presumed waifs, Kemp's ridleys nest exclusively at Rancho Nuevo, Tamaulipas, Mexico (Hildebrand, 1963). Like the olive ridley, females aggregate off the beach, and when the correct weather conditions occur (presumably windy weather), the females emerge from the sea *en masse* — this behavior is termed an "arribada," or arrival. Kemp's ridleys historically have been used as food. Tressler and Lemon (1951) and Carr (1952) stated, "I have eaten several and would rate the young 'chicken' ridley along with the young of any of the others." Rebel (1974) reported that ridley meat was marketed as loggerhead in Florida, and that the eggs were eaten by local residents. Hildebrand (1981) stated that egg collecting and slaughter for meat and leather were once commonplace. Kemp's ridley is the most severely endangered sea turtle (Ross et al., 1989; Carr, 1977), with population estimates of adults at less than 3000 individuals. The precarious outlook for survival of this species has produced an unprecedented amount of legal protection.

Olive Ridley Turtle

Olive ridleys occur in the Pacific and Indian oceans, and in the south Atlantic. Olive ridleys nest entirely in tropical areas. Olive and Kemp's ridleys reach similar sizes (Pritchard, 1979). The plastron and ventral surfaces are creamy white, and the carapace and dorsal surfaces are olive green. Like the Kemp's ridley, olive ridleys also nest in large aggregations (arribadas), where hundreds or thousands of turtles emerge from the surf *en masse* (Hildebrand, 1963). Since many turtles can be taken in a short period of time, local populations may be decimated quickly. Although olive ridleys are protected in most areas, the flesh and eggs are still eaten (Carr, 1952); moreover, the valuable leather obtained from the flippers also contributes to this species' decline (Cervigon et al., 1993; Mack et al., 1981; Pritchard, 1979).

Loggerhead Turtle

Loggerheads occur throughout the tropics into temperate areas. Loggerheads are the only sea turtles that nest on temperate beaches (anti-tropical) (Pritchard, 1979). Loggerheads reach carapace lengths over 100 cm and weigh up to 270 kg (Pritchard, 1979). The plastron and ventral surfaces are creamy yellow, and the carapace and dorsal surfaces are mahogany brown, often with many encrusting organisms such as barnacles (Dodd, 1988).

Although loggerheads are protected in most areas, eggs are eaten, but the meat is not considered very good by many (Cervigon et al., 1993; Dodd, 1988; Tressler and Lemon, 1951). The eggs were used in the southeastern United States, especially in cakes (Carr, 1952), which remained moister than those made with chicken eggs because the whites do not harden when cooked. In the Caribbean the meat was ground and made into "turtle balls" which are "delicious if properly prepared; moreover, good soup is made of the shell, flippers, and meat" (Carr, 1952). Carr stated that loggerhead eggs were a staple for many Caribbean communities. Directed fisheries have occurred until recently in Cuba, Mexico, and Madagascar (Dodd, 1988). Cuban loggerheads are still eaten and utilized for leather (Dodd, 1988), where the eggs are dried and smoked while in the oviduct, and sold like sausage (Carr, 1952). In Colombia the eggs were made into "dulce," a "curious form of candy" (Carr, 1952). In some cultures an "oxidizing oil" is made from loggerheads for use as varnish, and the scutes have been used as a "poor substitute" for hawksbill tortoise shell (Carr, 1952). See Dodd (1988) and Rebel (1974) for a complete description of the utilization of loggerheads.

Hawksbill Turtle

Hawksbills occur throughout the tropics, usually associated with coral reefs. They nest entirely in tropical areas, usually on islands. Hawksbills reach carapace lengths up to 95 cm and weigh up to 125 kg (Committee on Sea Turtle Conservation, 1990; Pritchard, 1979). The plastron and ventral surfaces are creamy yellow, and the dorsal surfaces are brown and dark green. Scutes of the carapace and plastron overlap like shingles, except in large individuals. The carapace scutes usually have beautiful brown, tan, and yellow markings, and are the source of "tortoise shell," also called "bekko." Carr (1952) described the removal of the scutes from live turtles: heat is applied in various fashions, and the scute is separated from the underlying bone. Many believe turtles will survive this operation; however, many die (Obst, 1986).

Demand for items made from bekko is the most significant threat to the species survival, along with collection of eggs (Obst, 1986; Mack et al., 1981; Carr, 1952). Asia has been the major consumer of bekko, which can be heated and "welded" into blocks and made into trinkets (Carr, 1952; Tressler and Lemon, 1951). However, the 1994 announcement that Japan (the greatest consumer of tortoise shell) will no longer import bekko (*Turtle Newsletter*, 1994b) will undoubtedly assist the recovery of hawksbills. Although hawksbills are protected in most areas, the flesh and eggs are sometimes eaten (Cervigon et al., 1993; Rebel, 1974; Carr, 1952). However, the flesh is sometimes poisonous (Halstead, 1980; Rebel, 1974; Carr, 1952), probably because of the ingestion of toxins contained in its preferred diet of sponges (Meylan, 1988).

Green Turtle

Green turtles occur throughout the tropics, with some small individuals venturing into temperate areas, probably as lost waifs. Greens nest almost entirely in tropical areas, with a few nests on warm/temperate beaches. Green turtles vary considerably in size throughout their range, but in the United States they reach carapace lengths of 100 cm and weigh up to 130 kg (Committee on Sea Turtle Conservation, 1990). The plastron and ventral surfaces are creamy white, and the carapace and dorsal surfaces are mahogany brown and dark green, often with yellow on intermediate areas. Green turtles are the only predominantly herbivorous sea turtle, and some populations make extensive reproductive migrations.

Although greens are protected in most areas, the flesh and eggs are held in high esteem (Obst, 1986; Carr, 1952). The fat and cartilage from under the plastron (calipee) is the main ingredient in clear turtle soup (Carr, 1952; Tressler and Lemon, 1951); other parts are also included in the soup. The green turtle's edibility was a major factor in local extinctions (King, 1981; Mack et al., 1981; Carr, 1952). Oliver (1955) reported that a plant in New York City processed 3,000 turtles per year, mostly for soup (Cervigon et al., 1993; Rebel, 1974). Green turtles have been reported as poisonous in some locations (Halstead,1980). The use of green turtle skins for leather (Mack et al., 1981) and the scutes for bekko (Obst, 1986) in Asia is also a threat.

REFERENCES

Burke, V.J., N.B. Frazer, and J.W. Gibbons. 1994. Conservation of turtles: The chelonian dilemma. Pp. 35-38 in B.A. Schroeder and B.E. Witherington (compilers), *Proc. 13th Ann. Symp. on Sea Turtle Biology and Conservation*. NOAA Tech. Mem. NMFS-SEFSC-341.

Bustard, R. 1972. *Sea Turtles: Natural History and Conservation*. William Collins Sons and Co. Ltd. London.

Carr, A. 1952. *Handbook of Turtles: The Turtles of the United States, Canada, and Baja California*. Comstock Pub. Assoc. Cornell Univ. Press. Ithaca, NY.

Carr, A. 1954. The passing of the fleet. *A.I.B.S. Bull.* Oct. 1954:17-19.

Carr, A. 1977. Crisis for the Atlantic ridley. *Mar. Turtle Newsletter.* **4**:2-3.

Carr, A. 1984. *The Sea Turtle: So Excellent a Fish.* Univ. of Texas Press. Austin.

Cato, J.C., F.J. Prochaska, and P.C.H. Pritchard. 1978. An analysis of the capture, marketing and utilization of marine turtles. Rept. to Environ. Asses. Division. Nat. Mar. Fish. Serv. St. Petersburg, FL.

Cervigon, F., R. Cipriani, W. Fischer, L. Garibaldi, M. Hendricks, A.J. Lemus, R. Marquez, J.M. Poutiers, G. Robaina, and B. Rodriguez. 1993. Field guide to the commercial marine and brackish-water resources of the northern coast of South America. FAO. Rome.

CFR. 1994. Endangered and threatened wildlife and plants. Title 50. 17.11.

Committee on Sea Turtle Conservation. 1990. *Decline of the Sea Turtles: Causes and Prevention.* Nat. Acad. Press. Washington, DC.

Dodd, C.K. Jr. 1988. Synopsis of the biological data on the loggerhead sea turtle *Caretta caretta* (Linnaeus, 1758). *USFWS Biol. Rept.* **88**(14). Washington, DC.

Ernst, C.H. and R.W. Barbour. 1972. *Turtles of the United States.* Univ. Press Kentucky. Lexington.

Ernst, C.H. and R.W. Barbour. 1989. *Turtles of the World.* Smithsonian Inst. Press. Washington, DC.

Frazier, J. 1981. Subsistence hunting in the Indian Ocean. Pp. 391-396 in K. Bjorndal (ed.), *Biology and Conservation of Sea Turtles.* Smithsonian Inst. Press. Washington, DC.

Halliday, T. and K. Adler. 1986. *The Encyclopedia of Reptiles and Amphibians. Facts on File.* New York.

Halstead, B.W. 1980. Dangerous marine animals that bite, sting, shock, and are non-edible. Cornell Maritime Press. Centerville, MD.

Hildebrand, H.H. 1963. Hallazgo del area de anidacion de la tortuga marina "lora," *Lepidochelys kempi* (Garman), en la costa occidental del Golfo de Mexico. *Ciencia* (Mexico City) **22**:105-122.

Hildebrand, H.H. 1981. A historical review of the status of sea turtle populations in the western Gulf of Mexico. Pp. 447-453 in K.A. Bjorndal (ed.), *Biology and Conservation of Sea Turtles.* Smithsonian Inst. Press. Washington, DC.

Humane Society International of the United States. 1994. Live freshwater turtle and tortoise trade in the United States. Preliminary report by The Humane Society of the United States. Humane Society International.

Irvine, F.R. 1947. *The Fishes and Fisheries of the Gold Coast.* Univ. Press. London.

Keinath, J.A. and J.A. Musick. 1990. Life history: *Dermochelys coriacea* (leatherback sea turtle). Migration. *Herp. Rev.* **21**:92.

King, F.W. 1981. Historical review of the decline of the green turtle and hawksbill. Pp. 183-188 in K.A. Bjorndal (ed.), *Biology and Conservation of Sea Turtles.* Smithsonian Inst. Press. Washington, DC.

Liner, E.A. 1978. A herpetological cookbook: How to cook amphibians and reptiles. Privately printed by the author. Houma, LA.

Mack, D., N. Duplaix, and S. Wells. 1981. Sea turtles, animals of divisible parts: International trade in sea turtle products. Pp. 545-563 in K. Bjorndal (ed.), *Biology and Conservation of Sea Turtles.* Smithsonian Inst. Press. Washington, DC.

Mar. Turtle Newsletter. 1994. Tortoiseshell trade: End of an era? **66**:16-17.11.

Marine Turtle Specialist Group. 1994. A global strategy for the conservation of marine turtles (draft).

Meylan, A.B. 1988. Spongivory in hawksbill turtles: A diet of glass. *Science.* **239**:393-395.

Obst, F.J. 1986. *Turtles, Tortoises and Terrapins.* St. Martin's Press. New York.

Oliver, J.A. 1955. *The Natural History of North American Amphibians and Reptiles.* D. Van Nostrand Co. Inc. Princeton, NJ.

Pritchard, P.C.H. 1979. *Encyclopedia of Turtles.* T.F.H. Pub. Neptune, NJ.

Pritchard, P.C.H. 1980. The conservation of sea turtles: Practices and problems. *Am. Zool.* **20**:609-617.

Rebel, T.P. 1974. Sea turtles and the turtle industry of the West Indies, Florida, and the Gulf of Mexico. Univ. Miami Press. Coral Gables, FL.

Ross, J.P. 1981. Historical decline of loggerhead, ridley, and leatherback sea turtles. Pp. 189-195 in K. Bjorndal (ed.), *Biology and Conservation of Sea Turtles.* Smithsonian Inst. Press. Washington, DC.

Ross, J.P., S. Beavers, D. Mundell, and M. Airth-Kindree. 1989. The status of Kemp's ridley. Center for Marine Conserv. Washington, DC.

Sternberg, J. 1982. Sea turtle hunts throughout the world. Sea Turtle Rescue Fund. Center for Environmental Education. Washington, DC.

Tressler, D.K. and J.M. Lemon. 1951. *Marine Products of Commerce.* Reinhold Pub. Corp. New York.

Other Aquatic Life of Economic Significance: Sea Urchins

Robert J. Price

INTRODUCTION

Sea urchins are members of a large group of marine invertebrates in the phylum Echinodermata (spiny-skinned animals), which also includes starfish, sea cucumbers, sea lilies, and brittle stars (Turgeon et al., 1988; Kato and Schroeter, 1985). All sea urchins have a hard calcareous shell called a test, which is covered with a thin epithelium and is usually armed with spines. The spines are used for locomotion, for protection, and for trapping drifting algae for food. Between the spines are tube feet that are used in food capture, for locomotion, and for holding on to the substrate. Sea urchins also have small pinchers, called pedicellariae, that are used for defense and for clutching food (Parker and Kalvass, 1992; Kato and Schroeter, 1985).

The mouth is located on the underside. It consists of a complex array of skeletal elements, plates, and teeth arranged in five-symmetry called "Aristotle's lantern." The mouth leads to the digestive tract, which empties through the anus located on the top of the test (Kato and Schroeter, 1985).

Five skeins of roe are the most prominent structures in the internal cavity of sea urchins. Among the skeins of roe are gill-like structures that are part of the water vascular system — important in movement, respiration, and food gathering. The gut is dark and often filled with partially digested plant material (Kato and Schroeter, 1985).

The sexes are generally separate in sea urchins. Females release up to several million eggs into the sea, where fertilization takes place. After a multistage planktonic existence, the larvae settle and metamorphose to the characteristic form (Parker and Kalvass, 1992).

Sea urchins feed mainly on sloughed or broken kelp. If the preferred food is absent, they will eat other algae, dead fish, lobster bait, and small animals (Chenoweth, 1994; Parker and Kalvass, 1992). Predators include sea otters, starfishes, crabs, wolf eels, lobsters, and fishes (Kato, 1994; Parker and Kalvass, 1992).

The major commercially valuable sea urchin species in the United States are the red (*Strongylocentrotus franciscanus*), the purple (*S. purpuratus*), and the green (*S. droebachiensis*) sea urchins. Several other species also occur, but are either too small or too rare to be of economic significance (Kato and Schroeter, 1985).

RED SEA URCHIN

The red sea urchin is one of the largest species of sea urchins in the world, growing to a test diameter of about seven inches. Its test and spine color is usually dark purple. Not infrequently, however, either the test or the spines, or both, are reddish or light purple (Kato and Schroeter, 1985).

It occurs on the West Coast of North America from the tip of Baja California to Sitka and Kodiak, Alaska (Kato and Schroeter, 1985). Occurrences have also been reported along the Asiatic Coast as far south as the southern tip of Hokkaido Island, Japan (McCauley and Carey, 1967). Recent studies, however, indicate that the red sea urchin does not occur along the Asiatic Coast (Bazhin, 1994). Bazhin (1994) also suggests that the single report of a red sea urchin off Hokkaido Island in 1943 may have been a misidentification.

The red sea urchin usually occupies shallow waters, from the mid-to-low intertidal zones to depths in excess of 164 feet, but it has been found as deep as 410 feet (McCauley and Carey, 1967). Individuals prefer rocky substrates, particularly ledges and crevices, and avoid sand and mud (Kato and Schroeter, 1985).

Red sea urchins are sexually mature at 1.5 to 2.0 inches in test diameter. Spawning times can vary from year to year and at different locations, but the episodes appear to be cyclic (Parker and

Kalvass, 1992; Kato and Schroeter, 1985). In southern California most spawning occurs in winter. In northern California spawning occurs in spring and summer (Parker and Kalvass, 1992). In Puget Sound, Washington, spawning occurs in the spring (Mottet, 1976).

Some red sea urchins take four to five years to reach a size of 3.5 inches (Parker and Kalvass, 1992), while others take eight to 16 years (Schroeter et al., 1994). Red sea urchins are comparatively long-lived, and some live for at least 30 years. In southern California the giant kelp (*Macrocystis pyrifera*) is preferred for food (Leighton, 1966). In northern California sea urchins feed on bull and brown kelp (Parker and Kalvass, 1992).

PURPLE SEA URCHIN

The purple sea urchin, which has relatively short spines, is usually light purple or lavender in color (Kato and Schroeter, 1985). The purple urchin is the common intertidal sea urchin of exposed and semi-protected rocky areas on the West Coast of North America from Baja California to Sitka, Alaska (Mottet, 1976). It is found on almost every rocky outcrop along the entire coast, and has been dredged at depths up to 210 feet (McCauley and Carey, 1967). The purple sea urchin is well adapted to pounding surf, where it usually lives in crevices or holes. Purple sea urchins spawn in the winter (Mottet, 1976).

GREEN SEA URCHIN

The green sea urchin has a circumpolar range. In the eastern Pacific, it is found as far north as Point Barrow, Alaska, and southward to Washington. It occurs in the Aleutian Islands and westward to Kamchatka, Korea, and Hokkaido, Japan. In the North Atlantic, it is found on the East Coast of the United States and Canada, and in Greenland, Iceland, and northern Europe (Mottet, 1976).

In the Gulf of Maine, the green sea urchin is common from intertidal pools to depths of 80 to 90 feet. It is most abundant in the shallow subtidal zone on rocky, gravelly, or shelly bottoms. Green sea urchins from the Gulf of Maine sexually mature in about three years at a diameter of one to 1½ inches. They spawn during the winter or early spring, usually in April, and the larval stage lasts four to six weeks. Kelps of the genus *Laminaria* are preferred for food (Chenoweth, 1994).

HARVESTING

On the United States Pacific Coast, red, purple, and green sea urchins are commercially harvested. On the Atlantic Coast, only green sea urchins are commercially harvested.

West Coast sea urchins are commercially harvested by divers using "hooka" diving gear, consisting of a low-pressure air compressor that feeds air through a hose from the vessel to the divers. Most vessels are 25 to 40 feet long and are capable of holding one to three tons of sea urchins, an amount usually harvested in a day's fishing by one to three divers. Harvesting takes place at depths of five to 100 feet, with most dives taking place in 20 to 60 feet. Sea urchins are harvested from the ocean bottom with a hand-held rake or hook and put into a hoop-net bag or wire basket. The basket is winched onto the boat and emptied into a larger net bag. In areas far from port, a larger "pick-up" vessel may take the catch from several harvesting vessels back to port (Parker and Kalvass, 1992).

East Coast sea urchins are commonly harvested with small bottom drags or by scuba diving. The divers pick the urchins off the bottom and place them in a mesh bag. The boat operator hauls the bag up and empties it into a bin. Urchins are harvested with a drag in areas where diving is not feasible (Chenoweth, 1994).

ALASKA

Sea urchin harvesting began in Alaska in 1980, and landings peaked at 755,000 pounds in 1987. Ketchikan in the southeast, Kodiak Island in the western gulf, and Homer in the Cook Inlet area are the three principal sea urchin harvesting areas. Primarily red sea urchins have been harvested near Ketchikan, and mostly green sea urchins have been harvested in Kodiak and Homer (Savikko, 1997).

Thus far, sea urchin fisheries in Alaska are in development stages, and the potential for a sea urchin industry in Alaska has not been determined. Management has utilized harvesting permits that have specified harvest areas, times, and size limits. Future management of red sea urchins in the southeast region will be based on area-specific harvest quotas following stock assessment surveys. Expanding populations of sea otters, a major sea urchin predator, may ultimately limit fisheries in parts of Alaska (Davidson, 1994) (Table 1).

Washington

Populations of red and green sea urchins in Washington are moderately abundant along the Strait of Juan de Fuca. Most sea urchins have been taken from the San Juan Islands and Port Angeles (Phu, 1990). The sea urchin fishery began in Washington in 1971. In 1976, the harvest increased to about 1.5 million pounds, prompting more active management by the state. Following surveys in 1976, Washington implemented the following regulatory measures: 1) fishing districts were rotated every three years to allow recovery; 2) upper and lower size limits were established; and 3) seasons were restricted (Bradbury, 1992).

The sea urchin fleet grew rapidly from 1986 to 1988. Landings in 1988-89 reached 9.4 million pounds, prompting the first emergency closure. Sea urchin divers and the state jointly authored a

Table 1. United States Sea Urchin Landings by State (thousand pounds)
(Blevins, 1999; Bradbury, 1999; Hensleigh, 1999; Robertson, 1999; Savikko, 1999; Lamb, 1997; Wood, 1997; CDFG, 1994; Wood, 1994)

Year	Alaska	California	Maine	Massachusetts	New Hampshire	Oregon	Washington
1971	–	<1	52	–	–	–	–
1972	–	76	50	–	–	–	2
1973	–	3,595	128	–	–	–	15
1974	–	7,108	47	–	–	–	58
1975	–	7,567	42	–	–	–	31
1976	–	11,106	36	–	–	–	1,548
1977	–	16,536	57	–	–	–	904
1978	–	14,428	8	–	–	–	1,030
1979	–	20,559	3	–	–	–	1,004
1980	–	22,167	33	–	–	–	43
1981	–	26,434	4	<1	–	–	269
1982	–	19,441	–	–	–	–	203
1983	–	17,757	–	–	–	–	482
1984	86	14,979	51	–	–	–	605
1985	125	19,998	–	–	–	–	884
1986	282	34,134	–	–	–	56	3,501
1987	755	46,062	1440	–	–	203	4,908
1988	191	51,998	6222	–	–	1,971	9,358
1989	169	51,189	9,426	–	111	7,843	5,740
1990	101	45,270	13,459	–	60	9,321	6,839
1991	204	41,938	20,535	<1	48	4,737	5,686
1992	468	32,528	26,438	3	102	2,857	3,298
1993	283	26,055	41,625	734	46	2,183	1,868
1994	110	23,683	38,297	563	12	1,790	2,038
1995	2,133	22,260	34,269	172	4	1,504	1,036
1996	937	19,959	25,743	103	10	820	1,224
1997	6,458	18,109	19,489	334	18	490	1,048
1998	3,125	10,550	15,054	408	–	345	691

limited-entry law that took effect during the 1989-90 season and set a goal of 45 vessels in the fishery (Bradbury, 1997; Kraxberger, 1992) (Table 1).

Oregon

The Oregon sea urchin fishery developed when southern California catches declined in 1986 (Phu, 1990). Annual landings increased from 56,000 pounds in 1986 to 9.3 million pounds in 1990, and then declined. Port Orford, Coos Bay, and Gold Beach are the main landing ports (Wood, 1994; McCrae, 1992; Phu, 1990).

The commercial sea urchin fishery has changed from essentially no regulations to a detailed limited entry system. In 1988, the first management regulations were enacted. Now the management system includes: a limited number of permits (46); a renewal requirement of 20,000 pounds of urchins landed in the previous year; temporary permit transfers for medical reasons; a minimum harvest depth of 10 feet from the mean-lower-low water; a minimum size limit of 3.5 inches; and seasonal buffer zones to protect sea lion rookeries (McCrae, 1992) (Table 1).

California

Sea urchins have been used by native Americans in California for thousands of years, but the modern commercial fishery did not begin until 1972 (Parker and Kalvass, 1992). Prior to this time, sea urchins were considered a pest. Quicklime (calcium oxide) was used to control sea urchins in commercial kelp beds and groups of recreational divers smashed sea urchins with hammers (Dewees, 1991).

The commercial fishery for red sea urchins began in southern California as part of a National Marine Fisheries Service (NMFS) program to develop fisheries for underutilized species. From 1971 to 1981, landings increased rapidly, reaching more than 26 million pounds in 1981. Prior to 1985, almost all landings were made in southern California, with most sea urchins being harvested from the Channel Islands. Starting in 1985, the northern California fishery expanded rapidly and the total California landings peaked at about 52 million pounds in 1988 (Parker and Kalvass, 1992) (Table 1).

In 1987, the California Department of Fish and Game and the Sea Urchin Advisory Committee negotiated a management system in an attempt to maintain a sustainable fishery. The management program, which began in 1988, includes a reduction of harvesting permits from 915 to 400; minimum size limits and season restrictions; requiring divers to maintain and turn in log books; and closure of some study areas to harvesting (Dewees, 1991). In 1994, minimum size limits were 3¼ inches test diameter in southern California and 3½ inches in northern California [California Code of Regulations (CCR), 1994].

A fishery for purple sea urchins began in 1990 as landings for red sea urchins continued to decline and harvest restrictions for red sea urchins expanded (Dewees, 1991). The harvest of purple sea urchins is currently not regulated, except that a red sea urchin permit is required.

Maine

Commercial landings of sea urchins were first recorded on the East Coast in 1929 when several thousand pounds were taken in Connecticut. Maine, whose commercial green sea urchin fishery began in 1933, was the only East Coast state with a commercial sea urchin fishery until the 1990s (Palmer, 1994). In the early days, sea urchins were harvested with dip nets with handles 10 feet or longer. The nets were used to scrape submerged ledges or rocks to dislodge the sea urchins (Scattergood, 1961).

The Maine sea urchin fishery was of minor importance until the mid-1980s when a Japanese market developed. The fishery then expanded gradually until landings reached more than 41 million pounds in 1993. Today, most sea urchins are harvested by divers. Sea urchin management regulations, which began in 1994, included a closed season from May 15 to August 15, a minimum size limit of two inches exclusive of spines, and the prohibition of night harvesting (Chenoweth, 1994) (Table 1).

Initially, almost all green sea urchins taken in Maine were shipped whole to Japan, but in 1991 the industry began moving toward stateside extraction of roe to counteract rising air freight costs (Chenoweth, 1994).

Massachusetts / New Hampshire

Commercial fisheries for green sea urchins began in New Hampshire in 1989 and in Massachusetts in 1991. Landings in Massachusetts reached 734,000 pounds in 1993. Landings in New Hampshire varied between 4,000 and 111,000 pounds from 1989 to 1996 (Blevins, 1997; Palmer, 1994) (Table 1).

PROCESSING

Sea urchin roe is primarily sold fresh rather than frozen. Packing methods are different, and a chemical treatment is used to improve the appearance of the roe. Only a small amount of roe is salted, steamed, baked, or frozen (Kato and Schroeter, 1985).

Sea urchin processing is labor-intensive. The test or shell is cracked into halves by hand using duckbill-like pliers that open when the handles are squeezed. Cracking is accomplished by driving the duckbill into the peristomal membrane and mouth area in the bottom of the test, expanding the plier by squeezing the handle, and causing the test to separate along the vertical axis. The cracked urchins are distributed to a group of "spooners" who use long-handled spoons or spatulas to gently remove the five pieces of gonad from the test and place them in a plastic or metal mesh tray. The gonads of both sexes are equally valuable and are referred to as roe ("uni" in Japanese). The roe is rinsed in cold saltwater to remove viscera and extraneous matter. Final cleaning of attached membranes is done with tweezers or small forks (Singh, 1990; Kato and Schroeter, 1985).

From this point, processing methods depend on the type of product produced. The Japanese name is given for each of the products described here (Kato and Schroeter, 1985).

1) **Fresh roe** ("uni" or "nama uni"):
   a) The roe is placed in stackable plastic strainers.
   b) The strainers are placed in tanks containing a solution of anhydrous potassium alum, $KAl(SO_4)_2$, in cold saltwater until the roe becomes firm. Concentrations used vary from 0.4 to 0.7 percent, and soak times vary from 15 minutes to one hour.
   c) The roe is then drained and packed in small wooden trays. At least 250 grams (g), and up to 350 g, of roe are packed in a standard tray (Juntz, 1994). A medium-size tray containing about 170 g is gaining in popularity, however. Occasionally, smaller trays holding 100, 50, or 30 g are also used. The standard trays measure about 9 cm x 16 cm x 1.3 cm deep (inside dimensions).
   d) Alternately, the drained roe is bulk-packed in larger, perforated foam trays, 32 cm x 40 cm x 2 cm deep, sometimes lined with absorbent cloth to prevent sliding and damaging of roe in transit. Only one layer of roe is packed in a tray. These trays normally hold about one kilogram of roe.
   e) The wooden trays are tied in bundles of eight to 13 trays, with a wooden cover over the top. These are placed in a plastic bag, and the roe is allowed to drain further in a refrigerator. It is important that the roe not be exposed to drafts while draining.
   f) Just prior to shipment, the trays are placed in insulated master cartons, each holding 50 to 54 trays. The bulk-pack foam trays are also stacked and placed in insulated master cartons with about eight to nine trays in each carton. Artificial coolant (commonly called "jelly ice") is added prior to shipment (about 1.4 kg per carton in winter and twice as much in summer).
2) **Salted roe** ("shio uni"): Methods of salting vary, depending on the requirements of buyers. Generally, the steps are as follows:
   a) Layers of cheesecloth are placed on a wire rack.
   b) A layer of roe is placed on the cheesecloth and covered thoroughly with salt; about 25 percent salt to weight of roe is used.
   c) More layers of roe and salt are placed on the rack, sometimes with cheesecloth between layers, until the thickness reaches about five centimeters.
   d) The roe is allowed to drain for several hours or overnight; about 40 to 50 percent of the moisture is removed; and salt uptake is 10 to 15 percent.
   e) The salted roe is packed in plastic-lined wooden kegs or plastic containers, sometimes with the addition of 10 percent by weight of ethyl alcohol (95 percent). An alternate type of shio uni has been produced using less salt but requiring freezing for preservation. This product apparently brings a high price if good quality roe is used.

3) **Steamed roe** ("mushi uni"):
   a) Fresh roe is placed in wood or screen containers of various sizes.
   b) The containers are stacked and placed in a large steamer.
   c) The roe is steamed for about 30 minutes; about 20 to 30 percent of the moisture is removed during the process. Some processors steam roe under pressure, reducing cooking time to 15 minutes or less.
   d) The roe is bulk-packed or packed in small wooden or plastic trays, and frozen.

4) **Baked roe** ("yaki uni"):
   a) Fresh roe is placed in shallow oven-proof dishes.
   b) The roe is baked in an oven at 190°C for 30 minutes. About 30 to 40 percent of the moisture is removed in the process.
   c) The cooked roe is then packed in small wooden trays (around 30 g) or in plastic imitation "scallop shells" and frozen.

5) **Frozen roe** ("reito uni"):
   a) Fresh roe of good quality is packed in standard wooden trays or bulk plastic trays.
   b) The trays are stacked and inserted in a plastic bag, then frozen at -17°C.
   c) The frozen roe, still in the plastic bag, is stored in insulated master cartons in the freezer.

This method is used when the product is to be sold later as raw-thawed sea urchin roe, and only good roe is acceptable. If the roe is destined to be salted or processed further, second-grade roe may be used, and it is often simply placed in plastic bags and frozen in bulk.

California firms have experimented with other processing methods in attempts to service specialty markets. Two such items produced were canned roe and freeze-dried roe. In addition, sea urchin roe has been used as feed for aquarium animals, particularly for sea anemones and other invertebrates.

Trays used for the Japanese market were formerly imported from Japan because only certain types of wood are acceptable, as they impart some odor or flavor to the roe. Now, however, a firm in Los Angeles manufactures nearly all wooden trays used by the United States industry (Kato and Schroeter, 1985).

Sanitation is important because sea urchin roe, which is usually sold fresh out of the shell, can easily pick up bacterial and fungal contamination. To counteract bacteria, some processors use ultraviolet-treated water in all phases of processing, and disinfectants are used by all to maintain a sanitary environment. Good refrigeration is also critical, especially during summer (Kato and Schroeter, 1985).

Labor costs are rather high in an operation that calls for meticulous cleaning, sorting, and packing. Thus, several processors turned to bulk-packing methods. In addition to savings in labor costs, bulk-packing also offered lower costs for trays (plastic instead of wood), as well as lower shipping costs, because the plastic trays are considerably lighter, and only a few are needed compared to wooden trays. Once in Japan, the roe is repacked in the traditional wooden trays and marketed through normal channels (Kato and Schroeter, 1985).

The proximity of the markets and the use of experienced labor ensured that the appearance of the repacked product was good. The only drawback was the extra time and cost needed to transport and repack the roe. Presently, fresh red sea urchin roe is being shipped to Japan primarily in wooden trays, but some foam bulk trays are also used. The product form is largely determined by the current prices of both, and recently, roe packed in wooden trays has generally yielded higher profits (Kato and Schroeter, 1985).

Several attempts have been made to process sea urchins aboard vessels at sea. The principal advantages are proximity to the sea urchin supply, access to clean saltwater, and easy disposal of waste products. Apparently the cost of maintaining crews and workers aboard the vessels for extended periods proved uneconomical (Kato and Schroeter, 1985).

Most red and purple sea urchins harvested in California, Washington, and Oregon are shipped to Japan as roe. Some Washington and Oregon sea urchins are shipped whole to Canada where they are processed into roe and shipped to Japan. Green sea urchins from Maine, Washington, and Alaska are either sent whole to Japan where the roe is extracted and packed, or are processed into roe stateside (Chenoweth, 1994; Parker and Kalvass, 1992; Phu, 1990).

Although most sea urchin roe is exported, an increasing amount of fresh roe is being packed for the growing domestic sushi trade. For the domestic market, wooden trays and plastic "oyster" trays — which are partitioned, have five or 10 depressions, and hold 100 or 200 grams of roe — are used for sale through retail markets. Small amounts packed in plastic oyster trays are also sold in United States retail outlets. Sea urchin roe for the domestic market is shipped in insulated cartons containing coolants similar to those used in overseas shipment. Both truck and air transport are used, depending on distance (Kato and Schroeter, 1985).

Since fresh roe is highly perishable, transporting time is critical for maintaining quality. Air freight arrangements are made soon after sea urchins are received by processing plants. Occasionally, fresh roe is shipped on passenger flights. Upon arrival in Japan, cargo is unloaded within 30 minutes to an hour. One to two hours are needed for clearing customs. Usually, it takes six to seven hours after arrival before the cargo is released to truckers. Thus, the products will not be available for auction sale on the day of arrival, but rather on the following trading day (Phu, 1990).

MARKET FORMS AND GRADES

The best quality roe is reserved for the fresh product, which brings the best prices. Secondary products are made from broken roe, or roe that is off-color, too large, or leaking fluids excessively. Salted roe is usually produced in the summer, when the price of fresh roe is low in Japan (Kato and Schroeter, 1985).

Because appearance is important to the Japanese (who are said to "eat with their eyes"), the packing process is critical. In Japan as well as in the United States, most of the roe is bought by "sushi" shops, which are Japanese seafood restaurants specializing in fresh seafood. Customers in sushi shops usually sit at a counter in front of refrigerated showcases which contain many seafoods, mostly raw, in plain view. Sea urchin roe is displayed in the same wooden trays used by the processor. To maintain good appearance, broken pieces of roe are placed on the bottom, and only whole, firm roe is placed on the top layers of wooden trays. The best size is 40 to 50 millimeters, but California sea urchin roe is usually larger. Skeins of roe are separated according to their color. Very large skeins are used in other than fresh products, or broken up into smaller pieces and packed on the bottom of trays. Bright yellow roe was historically considered the highest quality in Tokyo, although consumers in different areas of Japan often prefer bright orange roe. Since the late 1970s, orange roe has equalled yellow roe in price in Tokyo.

The quantity and quality of roe contained in sea urchins is vital to processors. Quantity is by and large a seasonal phenomenon, as the amount of roe depends in part on the reproductive state of the sea urchins. However, nutritional state is important, and areas devoid of preferred algae produce sea urchins with poor yield or poor color (Kato and Schroeter, 1985).

Roe color is exceedingly important in marketing. Clear, bright yellow, or orange roe is best for the fresh market. As many as eight color grades are used by California processors, and all dark or discolored roe is discarded. For salted roe, the preferred color is orange, which is the color of high quality bottled salted roe made with Japanese sea urchins (Kato and Schroeter, 1985).

In the California fishery, Kato and Schroeter (1985) noted that orange roe was exclusively from male red sea urchins, while yellow roe was usually found in females. Dark brown roe represents gonads degenerated because of starvation.

Table 2. Proximate Analyses of Red and Purple Sea Urchin Roe (percent of total roe weight) (Kato and Schroeter, 1985)[1]

	Red Sea Urchin		Purple Sea Urchin	
Item:	1	2	3	2
Moisture	70.0	70.8	68.6	71.8
Protein	7.7	9.6	9.5	12.3
Lipid	7.6	8.3	5.4	5.2
Ash	1.6	1.5	1.3	1.7
Glycogen	1.3	–	1.7	–
Nonprotein nitrogen	0.1	0.5	0.1	0.5

[1] Sources:
1 — Modified after Greenfield et al., 1958
2 — From Kramer and Nordin, 1979
3 — Modified after Giese et al., 1958

In addition to having good color and appearance, the best quality roe is firm, small (less than five cm), and free of leaking fluids. Some processors feel that poor quality roe tends to occur in greater frequency in large (old) urchins, and for this reason they discourage harvest of the very large red sea urchins.

Sea urchin roe contains an assortment of nutrients. Major components of the roe of red and purple urchins are given in Table 2. The percentages of various constituents are likely to change, depending on the nutritional and reproductive states of the urchins (Kato and Schroeter, 1985).

REFERENCES

Bazhin, A. 1994. Personal communication. Kamchatka Research Institute of Fisheries and Oceanography, Laboratory of Fishery Invertebrates and Seaweeds. Naberezhnaya, Russia.

Blevins, J. 1999. Personal communication. United States Department of Commerce, National Oceanic and Atmospheric Administration. National Marine Fisheries Service. Northeast Fisheries Center, Woods Hole, MA.

Bradbury, A. 1992. A History of Red Sea Urchin Management in Washington: The Manager's Perspective. Pp. 33-34 in *Sea Urchins, Abalone, and Kelp: Their Biology, Enhancement and Management*. C.M. Dewees and L.T. Davies (ed.), Report No. T-CSGCP-029. California Sea Grant College, University of California. La Jolla.

Bradbury, A. 1999. Personal communication. Washington Department of Fish and Wildlife, Point Whitney Shellfish Laboratory. Brinnon.

California Department of Fish and Game. 1994. California Commercial Fish Landings by Region — December. 1993. State of California, The Resources Agency, CDFG, Marine Resources Division. Long Beach.

Chenoweth, S. 1994. The green sea urchin in Maine — fishery and biology. Maine Department of Marine Resources. West Boothbay Harbor.

CCR. 1994. Taking of Sea Urchins for Commercial Purposes. Title 14, California Code of Regulations. Section 120.7.

Davidson, W. 1994. Personal communication. Alaska Department of Fish and Game, Commercial Fisheries Management and Development Division. Sitka.

Dewees, C. M. 1991. California Urchin Fishery: Lessons for Alaska. Pp. 2-4 in *Alaska Marine Resource Quarterly*. 6:2. University of Alaska Marine Advisory Program. Anchorage.

Giese, A.C., L. Greenfield, H. Huang, A. Farmanfarmaian, R. Boolootian, and R. Lasker. 1958. Organic productivity in the reproductive cycle of the purple sea urchin. *Biol. Bull. Woods Hole* **116**(1):49-58.

Greenfield, L., A.C. Giese, A. Farmanfarmaian, and R.A. Boolootian. 1958. Cyclic biochemical changes in several echinoderms. *J. Exp. Zool.* **139**(3):507-524.

Hensleigh, J. 1999. Personal communication. Oregon Department of Fish and Wildlife. Portland.

Juntz, R.S. 1994. Personal communication. Marusan Enterprises Inc., a division of Ocean Fresh Seafood Products. Fort Bragg, CA.

Kato, S. 1994. Personal communication. Larkspur, CA.

Kato, S. and S.C. Schroeter. 1985. Biology of the red sea urchin, *Strongylocentrotus franciscanus*, and its fishery in California. *Marine Fisheries Review*. **47**(3):1-20.

Kramer, D.E. and D.M.A. Nordin. 1979. Physical data from a study of size, weight and gonad quality for the red sea urchin [*Strongylocentrotus franciscanus* (Agassiz)] over a one-year period. *Can. Fish. Mar. Serv.* Vancouver Lab., Manuscr. Rep. Ser. 1372.

Kraxberger, R. 1992. A harvester's perspective of Washington State's urchin regulations. Pp. 34-35 in *Sea Urchins, Abalone, and Kelp: Their Biology, Enhancement and Management*. C. M. Dewees and L.T. Davies (eds.). Report No. T-CSGCP-029. California Sea Grant College, University of California. La Jolla.

Lamb, J. 1997. Personal communication. State of California, Department of Fish and Game, Marine Resources Division, Southern Operations. Long Beach.

Leighton, D.L. 1966. Studies of food preferences in algivorous invertebrates in southern California kelp beds. *Pacific Science*. **20**(1):104-113.

McCauley, J. E. and A. G. Carey Jr. 1967. Echinoidea of Oregon. *J. Fish. Res. Bd. Can.* **24**(6):1365-1401.

McCrae, J. 1992. Oregon sea urchin fishery, 1986-1991. p. 32 in *Sea Urchins, Abalone, and Kelp: Their Biology, Enhancement and Management*. C.M. Dewees and L.T. Davies (eds.). Report No. T-CSGCP-029. California Sea Grant College. University of California. La Jolla.

Mottet, M.G. 1976. The fishery biology of sea urchins in the family Strongylocentrotidae. Technical Report No. 20. Washington Department of Fisheries. Olympia.

Palmer, J. 1994. Personal communication. United States Department of Commerce, National Oceanic and Atmospheric Administration. National Marine Fisheries Service. Northeast Fisheries Center. Woods Hole, MA.

Parker, D. and P. Kalvass. 1992. Sea urchins. Pp. 41-43 in *California's Living Marine Resources and Their Utilization*. W. L. Leet, C. M. Dewees, and C. W. Haugen (eds.). Sea Grant Extension Publication. UCSGEP-91-12, Sea Grant Extension Program. Wildlife and Fisheries Biology Department. University of California. Davis.

Phu, C.H. 1990. The United States Sea Urchin Industry and Its Market in Tokyo. NOAA-TM- NMFS-SWR- 025. National Marine Fisheries Service. Southwest Region. Terminal Island, CA.

Robertson, J. 1999. Personal communication. State of California, Department of Fish and Game, Marine Region, Southern Operations. Long Beach.

Savikko, H. 1999. Personal communication. Alaska Department of Fish and Game, Commercial Fisheries Management and Development Division, Juneau.

Scattergood, L. W. 1961. The sea urchin fishery. Fishery Leaflet 511. United States Department of the Interior. Bureau of Commercial Fisheries. Washington, DC.

Schroeter, S. C., J. D. Dixon, and T. A. Ebert. 1994. Growth of red sea urchins (*Strongylocentrotus franciscanus*) at two sites in southern California during 1993-1994. Final report to the Pacific States Marine Fisheries Commission. August 20, 1994. Contract 93-75.

Singh, P. 1990. Design and Development of a Sea Urchin Processing System. California Sea Grant Progress Report. Project R/F-118. California Sea Grant College Program. University of California. La Jolla.

Turgeon, D. D., A. E. Bogan, E. V. Coan, W. K. Emerson, W. G. Lyons, W. L. Pratt, C. F. E. Roper, A. Scheltema, F.G. Thompson, and J.D. Williams. 1988. *Common and Scientific Names of Aquatic Invertebrates from the United States and Canada.* Special Publication 16, American Fisheries Society. Bethesda, MD.

Wood, C. 1997; 1994. Personal communication. Oregon Department of Fish and Wildlife, Newport.

Other Aquatic Life of Economic Significance: Alligators

Michael W. Moody

INTRODUCTION

The American alligator (*Alligator mississippiensis*) is highly valued not only for its skin as a source of prized leather, but also for its meat, which is considered to be a delicious seafood. The American alligator is native to the southeastern portion of the United States and today is extremely abundant in Louisiana and Florida. Historically, alligators have been relentlessly hunted for their hides, and as trophies and pets. In 1967, the alligator was classified as "Endangered" throughout its range (Department of Interior, 1981). In 1981, the Fish and Wildlife Service reclassified alligators in 52 Louisiana parishes from Endangered to Threatened under the Similarity of Appearance provisions of the Endangered Species Act of 1973. Alligators were brought back from the brink of extinction through good resource management practices and through strict harvesting and processing regulations. Because of the rise in their numbers and of the health of the stocks, the Louisiana Department of Wildlife and Fisheries began a controlled harvest program of wild alligators in 1972 (Palmisano, 1973). This successful program continues today (Moody et al., 1981).

Alligators may be harvested from wild populations or from aquaculture facilities. In Louisiana, approximately 26,000 alligators were taken from wild stocks in 1997, and approximately 122,000 came from aquaculture. Wild alligators may be harvested from privately- or publicly-owned lands. All wild-harvested alligators must be properly tagged. State wildlife officials assess both privately- and publicly-owned lands for the number of harvestable animals and issue tags based on that assessment. The tags are issued directly to the private land-owner or auctioned off to the public in the case of state-owned lands. The wild harvest season lasts approximately 30 days during September. Alligators can be harvested throughout the year from aquaculture facilities.

Lane and King (1989) provide a brief description of alligator production in Florida. They state that the first commercial alligator farm was established in 1891. These authors discuss two methods of captive alligator production: farming and ranching. According to their description, alligator farming requires adult breeding stocks to produce eggs. The eggs are collected and incubated, and the resulting hatchlings are raised to market size. Ranching does not require adult breeding stocks. The eggs are collected from wild alligator nests and incubated, and the hatchlings are raised to market size.

Some states also have provisions for the harvest of nuisance alligators. Generally, special tags are issued or qualified individuals are designated to take the animals. Joanen and McNease (1971) reviewed the propagation of alligators in captivity. They evaluated the various parameters related to reproductive success of pen-held alligators. In a separate study, Joanen and McNease (1977) evaluated the parameters related to egg collection and transportation, incubation, hatching, and posthatching culture.

PROCESSING

Alligator processing involves using nearly the whole carcass. When commercial processing facilities were established in Louisiana in the early 1970s, the primary product was skins. There was little market for the meat, and most was discarded as waste. Today, the meat is a highly desirable seafood product and is considered as economically important as the skins. In addition, the feet, head, and teeth are sold to specialty markets for trinkets, jewelry, and souvenirs. Only the viscera are considered waste.

Proper processing begins with the slaughter operation. Typically, wild alligators are captured using a baited hook and dispatched with a firearm. Farm-raised alligators may be dispatched

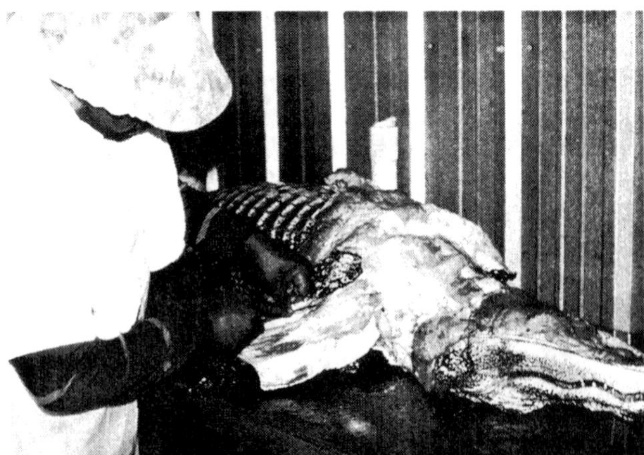

Figure 1. Alligator skinning prior to processing meat.

with a firearm or, more commonly, by cutting the spinal cord behind the head. Warwick (1990) evaluated the effectiveness of both methods and concluded that the most effective and humane method of alligator slaughter was the use of free-projectile firearms or a captive-bolt system rather than spinal cord severance at the atlanto-occipital area, termed "nape-stab." One of the major drawbacks in shooting alligators with a free-projectile (bullet), however, is the possibility of the bullet lodging in an edible portion of the meat.

Alligators delivered to processing facilities are generally dead if caught from wild stocks and alive if farm-raised. There can be great variation in animal size. Generally most animals are at least 4 feet long and may weigh less than 50 pounds (whole animal carcass weight), but some can be as large as 14 feet long and weigh in excess of 350 pounds.

The first processing step involves removing the skin. Since the skin is an extremely important product, great care is taken to ensure that damage is minimal. This is a hand-operation and is usually conducted on a large table. Usually one individual will work on a single animal, but two skinners may work on a large animal. A skinning knife is used to separate the skin from the carcass. Other devices such as air pressure and red-meat rotary skinning knives may also be employed.

During the skinning operation, great care is taken to ensure that the viscera are not cut or spilled on the meat (Figure 1). Evisceration is not conducted until after the skin has been removed intact. As a note, skinning may involve removing the side and belly skin intact or splitting the belly

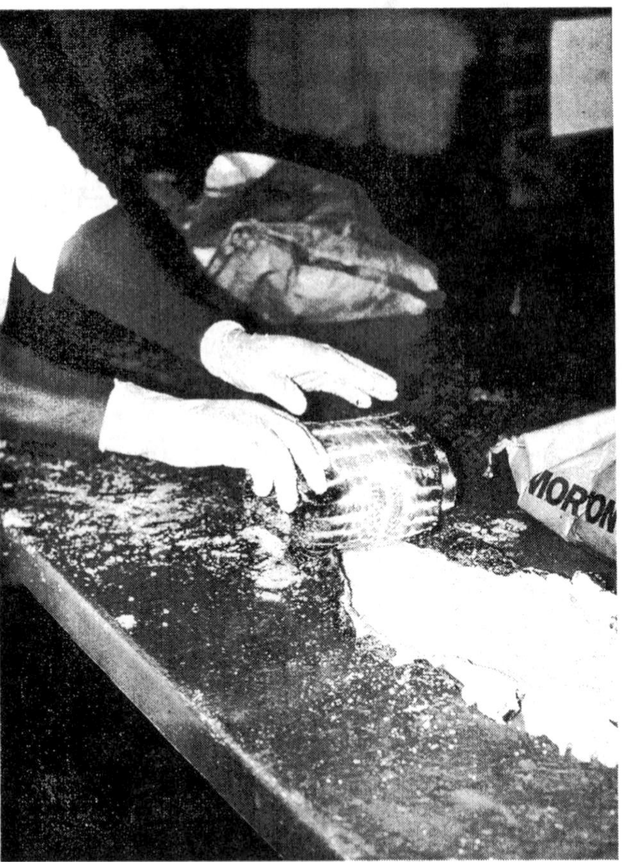

Figure 2. Preparing hides.

and producing a "hornback" hide (Figure 2). Hides must be carefully salted, rolled, and stored to minimize harmful deterioration (Figure 3). (The Louisiana Fur and Alligator Advisory Council provided detail steps in the skinning and care of alligator hides.)

After skinning and removal of the skin from the table, the belly is split and the viscera, feet, and head are removed. There are basically four general categories of meat: tail, torso, jaw, and leg. Moody et al. (1981) conducted a pioneering study on alligator meat yields based on the size of the alligator and the cut of meat. They described the tail meat as being white to light pink in color with internal bands of hard white fat. This fat appeared circular in cross section and ran lengthwise near the tail bone. They also described the torso meat as similar to the tail meat, except that it did not have the bands of fat. The jaw meat was described as lean with no fat deposits, and the leg meat as darker with scattered fat deposits and substantial amounts of connective tissue and tendons. In gen-

Figure 3. Alligator skin properly salted and stored.

eral, the meat yield was approximately 60 percent of total live weight. Deboning the dressed carcass resulted in an 80 percent meat yield.

The general practice in the industry is to remove as much fat as possible from alligator meat prior to packaging, since it is suspected of imparting an undesirable off-flavor. Personal observations have shown that wild-harvested alligators tend to have a stronger-tasting fat than farm-raised alligators. Leak et al. (1988) showed significant differences in fat from farm-raised animals harvested from different farms and summarized that this difference could be attributed to differences in diet.

PRODUCT FORMS

Alligator meat is available as fresh or frozen. Most alligator meat is deboned and cut into cubes prior to packaging. Packaging techniques include such traditional materials and techniques as plastic bags and vacuum packaging. Some meat processors may take steps to tenderize the meat prior to packaging.

Leak et al. (1988) determined that alligator meat was of an acceptable quality after four months of frozen storage. In addition, this same study showed that vacuum-packaged fresh alligator meat has an approximate 14-day shelf life.

Processing of alligators yields raw meat intended to be cooked prior to consumption. The microbiological profile of the meat is an important indicator of sanitation during the processing operation. Oblinger et al. (1981) evaluated alligator meat for microbial levels and types of microorganisms. Their findings showed that the typical microorganisms recovered included *Corynebacterium, Staphylococcus, Micrococcus, Flavobacterium, Pseudomonas, Acinetobacter, Arthrobacter,* and yeast. It is interesting to note that the authors did not detect any *Salmonella*. The study further showed that the Aerobic Plate Count (APC) of the fresh meat was low (2.88 to 3.02 logs/g).

NUTRITIONAL INFORMATION

One of earliest nutritional composition studies was conducted by Moody et al. in 1981. The data show that alligator meat is low in fat. Results from this study are shown in Table 1. Leak et al. (1988) also conducted a comprehensive nutritional study on alligator meat, the results of which agreed with the finding of Moody et al. (1981). In addition, they showed that alligator tail meat is low in saturated fatty acids, high in monounsaturated fatty acids, and relatively high in polyunsaturated fatty acids. Cholesterol was 64.8 mg/100g. This study also showed that alligator meat is a good source of omega-3 fatty acids, phosphorus, potassium, niacin, vitamin B12, and a high-quality protein.

HACCP CONSIDERATIONS

Because alligator meat is a seafood product under the definition of 21CFR Part 123, Fish and Fishery Products, it is subject to the provisions of that regulation. Alligator processors are required to have approved facilities capable of providing good sanitary operations. Under the Hazard Analysis Critical Control Point (HACCP) concept, seafood processors must identify safety hazards and preventive methods for specific seafood products. Alligator meat is a raw product that will be fully cooked by the consumer prior to consump-

Table 1. Composition of Alligator Meat

Cut of Meat	Crude Protein	Crude Fat	Moisture	Ash
Tail	21.3	1.5	76.5	1.3
Torso	21.1	1.2	73.0	1.3
Jaw	22.3	1.2	75.9	1.3
Leg	21.1	1.0	76.8	1.3

tion. Consequently, there are few potential safety hazards associated with it. The following discussion, however, is useful for establishing a HACCP plan for alligator processing.

Alligators are available to processing plants from two sources: wild stocks and aquaculture facilities. Alligators harvested from wild stocks are commonly dispatched with firearms that could potentially create a physical hazard in the meat in the form of metal fragments from bullets or buckshot. Alligators obtained from farms may have been treated with medications or drugs. This chemical hazard should be considered when obtaining farm-raised animals for processing. *Salmonella* sp. presents the most obvious biological hazard for alligators since it is closely associated with the processing of reptiles and birds. This is not a cooked-ready-to-eat product, and *Salmonella* is better controlled through the use of an effective sanitation program as provided for in the Sanitation Standard Operating Procedures (SSOP).

REFERENCES

Department of the Interior. 1981. Reclassification of the American alligator in Louisiana. *Federal Register.* **46:**153:40664-40669.

Joanen, T. and L. McNease. 1971. Propagation of the American alligator in captivity. *Proceedings of the 25th Annual Conference of the Southeastern Association of Game and Fish Commissioners.*

Joanen, T. and L. McNease. 1977. Artificial incubation of alligator eggs and post-hatch culture in controlled environment chambers. *Proceedings of the Eighth Annual Meeting of the World Mariculture Society, 1977.*

Lane, T.J. and F.W. King. 1989. Veterinary Medicine Fact Sheet, VM-52, Alligator Production in Florida. Florida Cooperative Extension Service, University of Florida, College of Veterinary Medicine, Institute of Food and Agricultural Sciences.

Leak, F.W., J.W. Lamkey, D.D. Johnson, and M.O. Balaban. 1988. Aquaculture Report Series. A further analysis of Florida alligator meat as a wholesome food product. Florida Department of Agriculture and Consumer Services, Division of Marketing, Tallahassee.

Louisiana Fur and Alligator Advisory Council. Alligator Hide Care.

Moody, M.W., P.D. Coreil, and T. Joanen. 1981. Alligators: Harvesting and Processing. Louisiana Cooperative Extension Service, LSU, Publication No. LSU-TL-81-002.

Oblinger, J.L., J.E. Kennedy Jr., E.D. McDonald, and R.L. West. 1981. Microbiology analysis of alligator (*Alligator mississippiensis*) meat. *Jour. of Food Protection.* **44**: 2:98-99.

Palmisano, A.W., T. Joanen, and L.L. McNease. 1973. An analysis of Louisiana's 1972 experimental alligator harvest program. *Proceedings of the 27th Annual Conference of the Southeastern Association of Game and Fish Commissioners.*

Warwick, C. 1990. Crocodilian slaughter methods, with special reference to spinal cord severance. *Texas Jour. of Sci.* **42:2**:191-198.

Other Aquatic Life of Economic Significance:

Frogs and Frog Legs

Roy E. Martin

REGULATION

Frog legs imported into the United States are automatically detained by the Food and Drug Administration (FDA) until they have been tested and found free of *Salmonella* (FDA, 1985). The Convention on International Trade in Endangered Species (CITES) specifies that products from threatened species of fish or shellfish (listed in its Appendix II) may be imported into the United States but only under a permit issued by the United States Fish and Wildlife Service (FWS). Appendix II includes the Asian (*Rana hexadacyta*) and Indian bullfrog (*Rana tigerina*).

FWS regulations require that each import of a threatened species product must have a valid foreign "export permit" issued by "the managing authority of the country of re-export." These documents must be obtained prior to importation. Permits or certificates must be displayed, but are not collected at United States ports of entry. Imports of protected species from a nation that is not a party to the CITES must be accompanied by documents from that nation that contain all of the information normally required.

Imports of threatened species also must enter the United States through a designated Customs' port unless an "Exception to Designation Port" permit is obtained. Importers may apply for a permit to allow for importation at a non-designated port or ports during a specified period of time not to exceed two years. Permits may be granted either to "minimize deterioration or loss" or "to alleviate undue economic hardship." Finally a "Declaration for Importation or Exportation of Fish and Wildlife" (Form 3-177) must be filed at the Customs' port of entry. This form is available from the Customs Service or the FWS [United States Department of the Interior (USIA)].

INTRODUCTION

Consumption of edible frogs can be traced back to cavemen. These excellent-tasting animals have commanded an intense interest by man down through the centuries. The American bullfrog (*Rana catesbeiana*) is so highly esteemed that it has been introduced into Mexico, South America, several Asian countries, the Pacific islands, and Europe. There exist over 20 other large species of frogs equal to the American bullfrog in quality (Culley, 1984).

The international human-food market is principally supplied with wild-caught frogs. Just how many species contribute to this pool is unknown, but the dominant species include *Rana catesbeiana, R. tigrina, R. hexadactyla, R. esulenta,* and possibly *R. ridibunda*. Demand for the legs has never lagged; supply is the problem.

Even though a considerable body of knowledge on bullfrog culture has been generated (Culley, 1986, 1984), commercial scale systems have been few.

Development is slow because of technological complexity, not because of the widely-held myth of a slow growth-rate. With adequate food and optimum environmental conditions, bullfrogs grow rapidly. For example, research-quality bullfrogs can be produced in three to five months, and edible-size frogs in five to eight months post-metamorphosis.

Components of an intensive culture system include the following: breeding area, tadpole hatchery, tadpole growout facilities, frog culture area, tadpole-feed preparation area, frog-feed culture facility (often involving two or three types of living feed), feed preparation area for the living feeds produced, disease diagnostic laboratory, isolation quarters, abundant water supply, frog processing area, storm drains (if culture pens are outside), refrigerated feed storage area, supply room, shop, and headquarters. In addition to the prepared feeds used for tadpole culture, specific natural feeds are produced in the tadpole culture area to ensure that complete nutrition is achieved. Each of these areas is critical if the system is to be well managed.

If proper attention is given to each of the major components (breeding center; tadpole and frog culture areas; and feed production centers), the system can work at 60 to 70 percent efficiency.

BREEDING

Frog breeding can be divided into extensive, semi-intensive, or intensive systems. As in other types of culture, each of the levels has a peculiarity which depends both on the natural characteristics of the selected site and on economic factors — *e.g.*, the cost of raw materials and manpower. The entrepreneur who considers going into frog breeding should ensure that all the biological, economic, and human factors are conducive. If the economic projections are correct, the venture should have technical and economic success.

Extensive Breeding

This type of frog breeding involves the restocking of tadpoles in vast open areas where the overcollection of frogs has resulted in depletion of natural populations (for example, as in India and China). It also involves the capture of frogs and tadpoles from fish ponds or ricefields. In many fish farms, frogs are considered pests because they feed on natural fish food and also directly on the fish. Many fish farmers do not realize that they can obtain an extra source of income from exploiting the frogs and tadpoles that live in their ponds.

All extensive culture systems are characterized by the absence of enclosures such as fencing; consequently the survival rate is very low, presumably under 5 percent. The extensive system of production also depends on weather and climatic conditions and does not provide a homogeneous good-quality frog product.

In developing countries, the introduction of adult frogs into ricefields can produce a remarkable quantity of tadpoles and a better ecological balance. It must also be remembered that ricefields and marshes which are subject to pollution or chemical treatment are not suitable for this type of breeding because of the obvious danger to the health of the frogs and humans.

A female frog will mature in two years, and will spawn about 30,000 eggs. These will hatch in about eight days. The eggs are laid in shallow water, enclosed among weeds. Frogs do not take care of their young. After they hatch, the tiny tadpoles adhere to plant life until they are large enough to swim. In sexing frogs, the female is mottled gray and white under the throat. She is a bit smaller than the male.

Semi-Intensive Breeding

Semi-intensive systems represent the first step towards a more certain and dependable production system in which one can better control environmental and biological factors. This type of breeding system began approximately 20 years ago and the technique has remained practically unchanged since. Production remains relatively low but is higher than that of the extensive breeding system. To protect frogs from external predators and to prevent their escape, barriers are installed. Frogs are, by nature, animals which prefer wide open areas and whenever an opportunity arises they escape from a breeding site following their natural instincts (Culley et al., 1981).

Semi-intensive culture systems produce a considerable increase in the amount of tadpoles in a central pond where, after metamorphosis, they will remain until they reach slaughter size. As can be imagined, due to the different farm size configurations and feeding systems, cannibalism is a common occurrence. The frogs feed mainly on insects attracted by the water and lights strategically placed inside the enclosure. Farmers add insect larvae, fish, other insects, and byproducts as supplementary feeds. The water is heavily fertilized so that phyto- and zooplankton can develop to provide food for young tadpoles. When they have sufficiently grown, the tadpoles like tender, appetizing foliage such as large-leafed waterweed, which they tear with their horny jaws and digest through their long guts.

The tadpoles are collected with nets or by hand in a little depression in the lower part of the pond.

All the various developmental stages of frogs are concentrated in the same facility, which is well protected with fences from outside predators.

Intensive Closed-Cycle Breeding

In the last few years, intensive frog breeding has become more popular and capture from the wild has declined. In the extensive and semi-intensive types of breeding systems, a high-percentage survival and, in particular, profitable production is not usually attained. With new and better breeding systems, it is possible to get a higher survival rate and better quality frog meat for the domestic and export markets. The largest breeding farms are now in the developing countries because of favorable environmental conditions and low costs of production. The current situation also permits the development of profitable frog farms in developed countries.

The intensive system features a clear separation of the various growth stages of the amphibians, so as to increase production and survival. The frog's growth cycle can be divided into different stages: mating, spawning, larval development, intermediate growth, metamorphosis, and growth to different sizes of a mature adult frog. Each of the stages needs a separate compartment to minimize stress and to increase performance. Breeders sometimes develop original technical solutions to these life stages, which they safeguard as "secrets." It must be remembered that spawning and metamorphosis are particularly critical stages of the intensive breeding system.

All the stages need a complete integrated feed. Part of the feed could also consist of insect larvae or any live-moving food. Live food is often employed in the final fattening stage. The frogs bred currently are generally collected from the wild.

Intensive frog breeding can be divided into an aquatic and a terrestrial phase. Each of these phases presents its own problems which have to be solved to obtain good production performance. The intensive frog farm needs divisions between one growing stage and another. As often as possible, frogs of the same size and age should be kept together. This leads to better production performance, fewer disease problems, and a lower incidence of cannibalism (Negroni, 1996).

FEEDING

Food and feeding are critical concerns in frog breeding. Frogs which breed naturally feed on live food after metamorphosis. There are two things to consider: first, the live food must have a high protein content; second, it must move. These two conditions must be reproduced artificially for farmed frogs to ensure success. Since high animal protein feed is expensive, vegetable protein is used to lower the cost and yet retain a high protein content.

Different techniques have been used to tackle the requirement for moving feed. Somehow the frogs must be fooled into believing that the feed is alive. Pepping up artificial feeds with wriggling earthworms seemed promising at first, but the frogs were not so easily tricked. They soon learned to pick out the worms, leaving the lifeless food untouched. Investigators then found that feed placed on a moving tray enticed the frogs into giving it a try; however, once in the frog's mouth, the lifeless pellets again betrayed their identity. Another approach was a food pellet covered with ciliary hairs. When mouthed by the frogs, these produce the required titillating sensation of live food.

Middle Eastern experts train the frogs to feed themselves with non-living food; others employ a physical or mechanical strategy to give "life" to the food directly; others use live food such as fish, insect larvae, etc. This later method yields good results, but has a high cost as production because live food is expensive. The cannibalistic nature of frogs must also be taken into account. Available local resources of food permit breeders to choose food with the best performance:price ratio.

DISEASES

Another problem in frog breeding is that the animal can easily be affected by disease, especially when they are confined in high concentrations and hygienic conditions are not maintained. The delicate skin of the frog presents a convenient route for the passage of disease organisms. Many frog farms, during the fattening stage, experience some problems with bacterial disease (e.g., *Flavobacterium* sp.) and high loss of animals. It is very important to keep frogs in excellent health during this stage.

FEEDING THE LARVAE (TADPOLES)

A diet is made by sifting the following ingredients through a 0.5 millimeter (mm) mesh screen.

3.6 pounds fish meal
2.3 pounds rice bran
1.8 pounds crawfish meal
16.0 ounces yeast
8.2 ounces soy protein
8.2 ounces bone flour
8.2 ounces whey
4.8 ounces alginate
3.0 ounces vitamin premix
0.7 ounces linolenic acid

The crawfish meal is the dried and ground carapace and fat from the crawfish.

The newly hatched tadpoles are fed this mixture by sprinkling it on the water. After the tadpoles begin to swim freely and cling to the walls, grass and other debris is removed and the incubators cleaned. The tadpoles are then switched over to semi-solid food.

The semi-solid food is prepared by mixing 20 ounces of the above ingredients with 30 ounces of water in which 6 milligrams (mg) of sodium hexametaphosphate has been dissolved to promote solidification. This makes a mixture with the consistency of dough. The mixture is spread onto screens (1.5 to 3 mm mesh) to a thickness of approximately 2 to 4 mm. The covered screens are placed in water containing calcium chloride ($CaCl_2$) at a 5:1 ratio for approximately one minute; this allows the alginate-calcium reaction to occur, firming the food on the screens.

When the tadpoles begin feeding on the screens they are moved to growing tanks. These tanks contain nets in which the tadpoles are placed. This allows the feces to pass through the net and to be removed from the bottom of the tank.

POST-METAMORPHIC FROG

This is the most critical stage for the frogs in a growing operation. The frogs are just learning to use their newly-developed tongues for feeding. Instinctively they attempt to grab moving food with their tongues but often miss. When they do capture live food, the food's movement stimulates the frogs to swallow and ingest the food. This instinct to eat only living food is the fundamental problem to be overcome when raising frogs in captivity.

A mechanical device has been used to train the juvenile frogs to eat non-living food. This device simulates living food by pulling a piece of food on a thread past the frogs. The juvenile frogs attempt to capture this moving food. When they are successful, the food stays attached to the thread until the frog can pull it off. The resistance from the tread simulates a live creature struggling to escape and increases the frog's feeding instinct.

That food is a mixture of the following ingredients:

4.0 pounds crawfish meal
3.6 pounds fish meal
4.3 ounces bone meal
2.5 ounces alginate
2.5 ounces vitamin premix

Two and one-half pounds of the above mixture is combined with 1 ounce of horse hair (from the mane or tail) and 2.75 pounds of water. Fifteen mg of sodium metaphosphate is dissolved in the water to promote solidification. This dough is extruded through 3-mm circular holes (an ordinary meat grinder works well) and sprayed with the 5:1 mixture of H_2O: $CaCl_2$. After drying, the long strings of food are broken into 3 to 5 mm long pieces. The hair protrudes from the food and helps stimulate the frog's reflex to swallow. The food is similar in appearance to an insect.

After the small frogs have been trained to eat the food dragged from threads, they will take the food when it is dropped near them while feeding. An improved dropping feeder has been designed to dispense the food in the feeding area.

The area where the food is dropped has a removable 1/8 inch-mesh galvanized wire cloth which provides a means for recovering uneaten food. During feeding, the frogs battle for the food, urinating and defecating. The urine passes through the mesh as does the fecal matter when it is washed from the food. The food can then be dried and reused. The screen serves a second purpose: when the frogs jump on the mesh it moves, causing the food to move so that the frogs can respond to instinct and grab it.

CANNIBALISM

Cannibalism definitely occurs among starving frogs but does not occur among well-fed, similar-sized frogs. In the battle for food, it is not uncommon for one frog to catch another by the foot or leg and then be unable to release it voluntarily. However, an active healthy frog can extricate himself from the clutches of a similar-sized frog. Therefore, as the frogs grow they are moved to larger-sized areas containing similar-sized juvenile frogs.

GROWING JUVENILES

Naturally, the growth of the trained juvenile is affected by the amount of food it eats. Therefore when the juvenile frogs begin eating the dropped food rather than the dragged food, they are moved to a separate area because dragging the food is more difficult than dropping the food. These juvenile frogs are fed the same food more easily using a dropping feeder. Feeding time is normally in the early evening.

Another frog-feeding technique has been used occasionally in the past with limited success. Floating catfish food (Delta Western, Indianola, MS 38751) is thrown into the water. The mosquito fish attempt to eat it, causing the food to bobble and dance. This attracts the frog's attention, and the food is eaten. This food is 32 percent protein and approximately 7 to 9 mm in diameter. If the frogs are feeding well, it is readily eaten. When the frogs weigh about 5 to 6 ounces, they are approaching sexual maturity and are moved again to a larger pond.

Why frogs eat has little to do with hunger. Many factors contribute to the likelihood of the frog's desiring to eat, including temperature, barometric pressure and/or the direction of barometric pressure movement, amount of sunlight, and, while not as well-defined, the phase of the moon and the solar period. Regular feeding times and having multiple frogs vying for the food appear to help induce feeding. Besides the necessity for the frog to have the desire to eat, it is necessary for the food to move in a way that attracts the frog's attention and allows it to capture, swallow, and digest the food (Ingle, 1975).

FOOD CONVERSION RATIO

A simple measurement of the food conversion ratio using crawfish heads indicated that approximately 4.9 pounds of crawfish was converted into 2.2 pounds of frogs. In a study, a 7-pound group of frogs was fed only crawfish heads throughout the winter and the following spring. They were weighed at the beginning and at the end of the study, and the conversion ration of 2.2:1 was determined.

CULTURE SYSTEM CONCERNS

The frog in culture is faced with many problems: (1) predation by mammals, birds, reptiles, and small aquatic insects, (2) cannibalism, (3) diseases, (4) an inadequate supply of living food, (5) poor water quality, (6) sporadic egg production, (7) seasonal temperatures, (8) poor sanitation, and (9) season availability of frogs. Items 3 and 4 are the major deterrents to the development of commercial systems.

In semi-natural systems all of the above problems are magnified. Predation by aquatic insects is particularly devastating to young tadpoles. Effective techniques for control are well-known and must be practiced. Since the frogs cannot be easily observed to detect conditions of health, diseased frogs cannot be located or treated, the culture area cannot be readily changed, and loss by cannibalism is serious.

In controlled laboratory-culture systems or in commercial facilities using a building or partially protected concrete ponds, the same problems exist but are not as acute. In totally enclosed systems, tadpole nutrition must be complete, and unless great care is taken in planning the feed supply, a large percentage of the tadpoles will be deformed and die (Culley and Sotiaridis, 1984). A dominant problem in commercial systems is the difficulty of maintaining continuous production. Unless a facility is totally enclosed and temperature is controlled, culture will be confined to the tropics. Breeding and live food production pose serious problems in temperate climates.

In order to obtain a continuous supply of eggs in temperate climates, spawning must be induced by hormone injections. This is no simple task and requires professionally trained personnel. In tropical climates (even on the Equator) breeding is, surprisingly, cyclic as in temperate climates. To overcome this cycle, a large number of breeders must be used, and continuously replaced with young, new breeders. Young frogs approaching the breeding condition for the first time will breed at six to nine months of age at any time of the year. The same condition holds true under laboratory conditions or in an enclosed commercial facility, but breeding must be induced with hormones.

Production of live food is the major concern of the frog culturist. Use of such foods as fish or tadpoles can require a great deal of space for ponds (often several acres). During the warm season, several thousand ponds of feed/acre can be produced. However, extra ponds must be well-stocked with fish to provide the frogs with food during the winter months and well into the spring. Repopulating of depleted ponds begins in the spring. However, the ponds may not accumulate sufficient fish for use until three or four months after reproduction begins. Considerable planning must go into feed production if aquatic animals are the major food supply.

Terrestrial invertebrates can be cultured as a source of feed for frogs. Worms, crickets, and fly pupae have been used with varying degrees of success. Worm and cricket culture is very labor-intensive and requires much time. Temperature control is mandatory. Fly pupae are the simplest feed to culture and a clean, odorless product can be produced quickly in great abundance. The technique is not labor-intensive and is economical. Interestingly enough, fly pupae may be a good feed for other aquacultured animals that require living feed, or high-quality protein, at some stage in their growth. Fly pupae are one of the easiest animals to culture quickly in large quantities (several hundred pounds/day) in a limited space.

ECONOMICS AND SUPPLY

In developed countries, predominantly in the United States and Europe with their high land and labor costs, production of bullfrogs for leg consumption is uneconomical due to the low cost of processed wild frogs exported from Third World countries. However, mass production of frogs for the research market in developed countries appears economical. In Third World countries, production of frogs for human consumption will be economical, but production will be low. Seminatural-type systems will dominate in these countries (Culley, 1986).

INSTALLATION SYSTEMS

Another important aspect of the culturing process is providing suitable tanks for housing the growing frogs. This is complicated by the fact that young frogs are not really frogs at all, but rather tadpoles. Unlike their semi-aquatic elders, they require a completely aquatic environment. Agricultural engineers working on housing designs feel that the best approach is to use separate tanks for the two different stages. Furthermore, they have found that tadpoles excrete a growth-inhibiting substance, making it necessary to maintain continuous water exchange in the tadpole tanks.

There are different types of installation systems for frog breeding, depending on the type of breeding system employed (intensive, semi-intensive, or extensive). One can use galvanized iron or plastic enclosures, containers of wood or plastic, or enclosures made of cement with several compartments. For the larval stages, tanks of different shape are utilized. Based on the tadpoles' physiological needs and sizes, about 20 tadpoles can be stocked in each liter of water.

For metamorphosis, tanks with appropriate facilities are used; metamorphosis and spawning stages are vulnerable to high mortality and, consequently, to economic losses. In intensive systems, between two and 50 adult stage frogs can be kept in each square meter, according to the size of the frog and type of installation. It is preferable that the water is of good quality. Tadpoles and adult frogs are more resistant than fish to low-quality water.

PROCESSING AND MARKETING

Many frog farms have processing plants annexed to the farm to add value to the product through frog processing. Generally there is a specialized frog slaughterhouse and a packaging plant with a low-temperature freezer to store the final products. The bull frog (reaching 8 inches in body length) is the most preferred species for food production.

Frogs can be processed into frog legs, meat, and even sausages for human consumption. The skins of frogs can be processed into leather which can be transformed into attractive products such as handbags, belts, etc. Live frogs can also be sold for use in education and biomedical research.

Usually the back or jumping legs are the only part of the frog eaten. They are dressed for market by being skinned; the two large legs remain attached by a small portion of the body meat. The legs are packed well-iced or frozen.

OTHER PROCESSING CONSIDERATIONS

Because of the problem of *Salmonella* mentioned earlier, the Food and Agriculture Organization (FAO) of the United Nations developed a circular in 1976 concerning the processing of frog legs for human consumption. It outlines good commercial practices that should be followed in the processing of frog legs for human consumption. It deals with the subject in a general manner, without referring to any particular species of frogs or geographical area. However, it must be acknowledged that most of the practical information has been gained from major frogleg-producing countries of the Indian Ocean and South China Sea regions.

THE FROGLEG INDUSTRY

In many countries frog legs have become an important food product, either as an export item or as a food delicacy which is very much in demand. Countries such as India, Pakistan, Bangladesh, Mexico, and Cuba have been the main exporters, while the United States, Japan, and Europe are presently the major markets. The world consumption of frog legs has been constantly increasing as more and more people acquire a taste for this delicious and highly nutritious food.

For some developing countries the frogleg industry has played an important role in the national economy by providing employment and earning foreign exchange. It utilizes readily available resources, which in most cases would not have been used by the local population because of popular aversion, customs, prejudices, or religious beliefs.

EXISTING DIFFICULTIES

The most serious difficulty, which has faced the frogleg industry for years, is contamination with *Salmonella* — the microorganism dangerous to human health.

Millions of dollars have been lost every year by the industry of the exporting countries because of salmonella. Buyers and importers insist on products free of this contamination. *Salmonella* is a genus of small microorganisms, invisible to the naked eye, found frequently in frogs' intestines and on their skin. It is not present in healthy frogs' muscles (flesh) or in their veins, unless these were contaminated during processing. *Salmonella* does not make frogs sick, just as *Salmonella* does not make chickens sick.

WHAT CAN BE DONE ABOUT *SALMONELLA*?

The *Salmonella* problem should be resolved by the use of good manufacturing procedures and the strict application of sanitary practices. These must be rigidly enforced throughout the processing line and require the full understanding and cooperation of plant management and every employee.

ORGANIZATION OF THE OPERATION

This undoubtedly is the most important task which should be undertaken by management. From the time the frogs are caught to the time the frog legs are frozen and packaged there should be a constant reduction in the number of *Salmonella* organisms. Killing of frogs should preferably be done under the strict control of the processing plant.

If butchering is carried out at a collection point or cutting center away from the processing plant, it should be done under strict supervision and under sanitary conditions similar to those of the processing plants; such a center should be officially registered. The processing line should be designed and the operations so arranged as to secure the orderly flow of material without any overcrowding of equipment or personnel.

It is the responsibility of management to assign each employee his definite place and duty on the processing line in order to prevent intermingling or movement of employees from more-contaminated areas to less-contaminated areas. All employees should be made aware that every stop in the processing line must result in progressive reduction in the number of microorganisms and that no equipment, utensils, or material should be moved from the more-contaminated areas to the less-contaminated areas until it has been thoroughly cleaned and disinfected. Special attention should be given to frequent cleaning of utensils, trays, tanks, table surfaces, and washing of hands.

Ice should be supplied along the processing line only by the employees assigned for this purpose, using clean containers and being fully aware of the danger of cross-contamination. Any leftover ice should be discarded.

Any containers and utensils used for ice, water, chlorine, or salt solutions or containing raw material should be kept off the floor. Small elevated platforms or stands which can be cleaned easily should be provided.

All equipment and utensils used in the processing of frog legs should be exclusively assigned for this purpose. It is extremely important that the processing of frog legs is carried out as a separate operation divorced entirely from shrimp or any other fish and shellfish operation. Areas where live frogs are received, stored, and butchered should be separated from areas in which final product preparation (skinning, bleeding, washing, trimming, etc.) or packaging is conducted.

Adequate and conveniently located handwashing facilities for employees should be provided. Chlorine water dips [clean water with 100 parts per million (ppm) available chlorine] for dipping gloves or hands should be made available at several points in the processing line. An ample supply of clean water with a residual chlorine that can be detected easily by smell should be available at several points in the processing line, as should clean ice, preferably flaked or well-crushed.

Premises should be well-lit. The illumination in any part of a working room should not be less than 35 foot-candles.

RAW MATERIAL REQUIREMENTS

Frogs obtained from habitats which may be polluted should be subjected to washing in running clean water for at least 24 hours. For this purpose, a clean, cemented tank, with an outlet at the bottom or an overflow pipe, may be employed.

Any frogs that are dead, diseased, damaged, or inactive should be removed and discarded.

Live frogs, before being placed into a holding tank, should be washed (hosed down or immersed in rapidly changing water) to remove soil, feces, and slime. Only clean water should be used for this purpose. It is preferable that live frogs be held in running, clean water for at least 24 hours.

PREPARATION AND HANDLING OF FROGS AT THE POINT OF CUTTING (BUTCHERING)

In preparation for butchering, the live frogs should be put into a 10 percent solution of common salt containing 250 ppm of chlorine for 15 minutes. By treatment in brine solution the live frogs become anesthetized and thus are relieved from pain during the cutting.

The hind legs should be cut at the abdomen not more than 2.5 centimeters (cm) above the waist and in such a manner that the intestines are left intact.

Immediately after the cutting, the legs should be washed thoroughly under the running (chlorine) water to remove blood, remnants of viscera, slime, feces, and other extraneous materials.

Care should be taken not to squeeze the legs to facilitate bleeding as this practice, because of the elasticity of the blood vessels, will remove the blood but, at the same time, suck into the vessels the outside water containing *Salmonella.*

Immediately after washing, the legs are immersed for a period of two minutes in chilled water (chilled by addition of crushed ice) containing 500 ppm of chlorine. This treatment will reduce the number of *Salmonella* organisms found on the surface of the skin.

Removal of skin and clipping of feet should be carried out on clean surfaces and with a minimum of delay.

Once again, the legs should be carefully washed in a copious amount of running water and bled immediately by placing them into a container of ice and chlorinated water (20 ppm of chlorine) for a period of not less than 20 minutes.

After the bleeding, the legs should be trimmed, removing bits of membrane, hanging pieces of flesh, and a remaining portion of the cloaca. During this operation, the dressed material should be carefully examined for parasites, bruises, blood spots, and other defects.

This operation should be followed by washing the legs thoroughly in a copious amount of running water and then immersing them again in a container of ice and chlorinated water (500 ppm of chlorine) for 15 minutes. The legs should then be taken out and washed in four or five changes of chilled, chlorinated (20 ppm) water.

During packaging, extreme care should be taken not to contaminate the product. The legs should either be wrapped individually in polyethylene film or preferably inserted into small polyethylene bags. The wrapping material or the bags should be dipped into clean water containing 20 ppm of chlorine. It is not necessary to dip the rubber bands in chlorine solution. It is preferable that size-grading be done before freezing. After freezing, the material should be transferred into cold storage, the temperature of which should not be higher than -18°C.

WHAT ARE GOOD-QUALITY FROG LEGS?

Good quality frog legs should, first of all, be free from *Salmonella* contamination. They should be well-trimmed, reasonably free from blood clots, and should not have black discoloration due to spoilage. In appearance they should be of ivory-white or creamy white in color. (The "blue" variety sells at a lower price. Prominent blue patches on the muscles or at the joints downgrade the product.) They should be tender when cooked and should not have any off-odor or off-taste. On presentation in the package, they should be of the same size, well-wrapped, and frozen without distorting their natural shape.

The FAO suggests that when tested by appropriate methods of sampling and analysis, the frozen frog legs should comply with the following requirements (FAO, 1976):

1. Total bacterial count at 37°C, per gram, maximum 500,000
2. *E. coli*, count per gram, maximum 10
3. Coagulase-positive *Staphylococcus*, count per gram, maximum 100
4. *Salmonella* or *Arizona*, per 25 gram analytical units 0

MISCELLANEOUS

The history of frog culture has been a turbulent one. Periodic recurrences of what might be called the great frog fraud attest to the success of promoters in selling unworkable frog culture schemes to a gullible public at grossly inflated prices. The 1920s and 1930s were full of such stories.

In 1948 the United States depended on Cuba for 766,262 pounds of fresh and frozen frogs. In the 1970s, 15 to 20 million frogs were needed for education and research purposes, 80 percent of which came from Mexico.

Figures for 1996, from the National Marine Fisheries Service (NMFS), show imports of fresh and frozen frog legs at 3,200,000 pounds. Indonesia and Taiwan are responsible for 85 percent of the imports, with lesser amounts from China, Hong Kong, India, Mexico, and Vietnam.

The United States does not have a significant industry for the production of edible frog legs.

REFERENCES

Culley, D.D. 1986. Bullfrog Culture — Still a High Risk Venture, *Aquaculture Magazine* Sept./Oct.: 28-35.

Culley, D.D. 1986. Bullfrog culture. In *World Animal Science* (Nash and Gall, eds.). Production of Aquatic Animals. Elsevier, New York.

Culley, D.D. 1984. Edible frogs. In *Evaluation of Domestic Animals* (I.L. Mason, ed.), pp. 370-74. Longman, New York.

Culley, D.D., W. J. Baldwin, and K. J. Roberts. 1981. The feasibility of mass culture of the bull frog in Hawaii. In Louisiana State University Sea Grant (LSU-81-004), Baton Rouge.

Culley, D.D. and P. K. Sotiaridis. 1984. Progress and problems associated with bullfrog tadpole diets and nutrition. In *Nutrition of Captive Wild Animals, Third Annual Dr. Schall Conference* (Meehan and Thomas, eds.), Lincoln Park Zoological Society. Chicago, IL.

Food and Agriculture Organization of the United Nations (FAO). *Draft Code of Hygienic Practice for Processing of Frog Legs*, CX/FH 75/6. Rome, Italy.

Food and Drug Administration. 1993. *Automatic Detention of Frog Legs*. FDA Division of Import Operations and Policy. Import Alert, No.16-12.

Ingle, D. 1975. Focal attention in the frog: behavioral and physiological correlates. *Science*. **188**: 1033-34.

National Marine Fisheries Service. Private correspondence. Silver Spring, MD.

Negroni, G. 1996. The basics of breeding frogs. *Infofish International*. 4:34.

United States Department of Interior. 1985. Moratorium on the Enforcement of Appendix II — Species of Frogs. File No. ENF-4-02-004376, Washington, DC.

Section III.
Processing and Presentation

Processing and Preservation of North Atlantic Groundfish

Robert J. Learson and Joseph J. Licciardello

INTRODUCTION

North Atlantic groundfish represent a variety of bottom-dwelling species including cod, haddock, pollock, hakes, and flatfish.

Spoilage begins the moment fish are taken out of the water. The rate of subsequent deterioration is affected by certain intrinsic factors such as species, size, season, fishing grounds, etc., and by extrinsic factors, such as handling practices, which are subject to human control. Spoilage of fish during storage at temperatures above freezing is the composite result of three different activities: l) bacterial decomposition; 2) autolytic enzyme action, either from tissue or digestive enzymes or from certain feeds that may be in the gut, leading to torn bellies and softening of the flesh; and 3) oxidation of lipid material, resulting in rancidity. In lean fish which have been gutted, spoilage invariably results from bacterial action. In fatty fish, loss of quality due to oxidative rancidity may precede bacterial spoilage.

FACTORS AFFECTING QUALITY

Washing

It is extremely important that the fish be iced as soon as possible after harvesting. However, prior to icing, the fish should be washed to remove mud, blood, and intestinal contents that may have been expelled as a result of the intense pressure developed when the nets are hoisted out of the water. Unless the washing operation is executed thoroughly and efficiently, the catch effort can be a waste of time and resources. If the washing is to be conducted manually with a hose, a copious amount of water under pressure should be used. Mechanical cylindrical washing machines are efficient and offer the advantage of being geared to a conveying device for transporting the fish to the hold for stowage. This reduces handling and allows for proper icing (Waterman, 1965).

Bleeding

Although not feasible with large catches of small fish, bleeding prior to evisceration is a desirable practice. It results in a lighter-colored flesh and also removes heme compounds which promote oxidative rancidity. Bleeding is usually accomplished by cutting the throat. It is recommended that the fish be allowed to bleed in seawater for about 15 minutes before washing.

Gutting

Gutting of larger groundfish species such as cod, haddock, pollock, and hakes should be carried out immediately after harvest. Gutting flounders and other flatfish is generally too labor-intensive and the time required may have a deleterious effect on quality, especially during summer months. Rupture of the intestines could contaminate the gut cavity with intestinal contents and accelerate spoilage. Failure to remove the last few inches of intestine, which remain attached to the vent, can result in obnoxious odors in that part of the fillet. Extension of the knife cut beyond the vent into the muscle can cause a more rapid deterioration in that area of the fillet (Castell et al., 1956). It is often the practice to remove the gills from large fish, particularly in the summertime; this action retards the development of off-odors and spoilage when the fish are examined as whole gutted fish. The benefit to be gained from washing fish after gutting appears to be related to the efficiency with which the evisceration is performed. Nevertheless, it has been observed that washing eviscerated fish in chlorinated [50 to 60 parts per million (ppm)] seawater, under pressure, rinses the blood and slime off the fish more effectively than does plain seawater (Linda and Slavin, 1960). The odor and color of the flesh were not affected.

Handling

Some species of groundfish are particularly susceptible to bruising. Consequently, rough handling aboard the vessel should be avoided. Bruising of freshly-caught live fish usually results in discoloration of the flesh. Although this effect is not apparent with freshly-caught dead fish, it has been shown that bruised fish do not keep as well as undamaged fish (Castell et al., 1956).

Effect of Rigor Mortis

Rigor mortis in muscle is characterized by the formation of lactic acid from glycogen with a subsequent lowering of pH and a stiffening of the muscle due to contraction. Bacterial growth on fish does not start until rigor mortis has begun. To optimize quality, it is necessary that the time to resolution of rigor be maximized.

Lowering the body temperature close to freezing delays the onset of rigor. The duration of rigor is also a function of body temperature and glycogen content at death. A high glycogen content and a low temperature above freezing both serve to prolong rigor. Rigor mortis is of longer duration if the fish has exerted less muscular activity prior to death. This results in a higher glycogen content. Results of several studies demonstrated that rigor mortis developed earlier and disappeared sooner in trawl-caught fish compared to fish caught with hand lines. This was attributed to struggling, crushing, and anoxemia in the trawl-caught fish. It is also believed that the shorter the trawler haul is, the better the fish will keep. In any given catch of trawl-caught fish, some of the fish may still be alive when the nets are hoisted on deck; thus, there may be a variation among the fish in the amount or degree of struggle expended prior to death. This could account in part for the variability in keeping-quality within the same catch of fish. The onset and duration of rigor vary with different species. In general, flatfish exhibit a more extensive rigor than round fish.

Effect of Temperature

The most important single factor controlling spoilage of fresh fish is storage temperature. Temperature regulates the onset of rigor mortis and also the lag period and growth rate of spoilage microorganisms.

The flesh of freshly-caught fish is sterile (Procter and Nickerson, 1935). The microorganisms that are present are located on the skin in the slime layer, gills, and gut. Initially, the numbers on the skin are essentially low, averaging about 103 to 105 per square centimeter (Spencer, 1961; Georgala, 1958), but through mishandling and contamination from the deck and the holding pens, the bacterial load can increase quite rapidly. The types of bacteria that eventually induce spoilage in iced North Atlantic (temperate waters) fish are termed psychrophilic or psychrotrophic, which signifies a tolerance for low temperatures. They constitute the natural microflora of newly caught fish and may also be picked up during subsequent handling (Shaw and Shewan, 1968; Shewan, 1961). These bacteria are capable of growing at temperatures slightly below freezing; however, they grow most rapidly in the temperature range of 20° to 25°C (68° to 77°F) (Hess, 1950). Although the relationship between storage temperature and spoilage rates of fish has been shown to be approximately linear within certain temperature limits (Ronsivalli and Licciardello, 1975; Spencer and Baines, 1964) as the temperature is lowered and approaches 0°C (32°F), the growth rate of fish-spoilage bacteria is drastically retarded. Consequently, it is necessary for the fisherman to rapidly lower and maintain the temperature of the fish as close to freezing as is possible in order to obtain maximum shelf life. This can be accomplished by the judicious use of various cooling media.

Chilling

Freshwater Ice

The amount of ice required for a trip will depend on the season, length of trip, size of the catch, and insulation of the boat. For a five-day summer trip in North Atlantic waters, it has been calculated that 0.113 kg (0.25 pound) of ice is adequate for cooling 0.454 kg (1 pound) of fish from 12.8° to 0°C (55° to 32°F), and an additional 0.182 kg (0.4 pound) is required to maintain the fish away from the holding pen surfaces, to cool the hold, and to remove heat leaking into the hold (MacCallum, 1955).

From these requirements, a ratio of 0.454 kg (1 pound) of ice to 0.681 kg (1.5 pounds) of fish was recommended to ensure landing high-quality fish. In practice, in northern waters in uninsulated holds of wooden vessels, a ratio of 1:2 (ice to fish) is recommended (Dassow, 1963).

Ice of small particle size such as that produced in a flake-ice machine or finely-crushed block ice is recommended since it permits more intimate contact with the fish for more efficient cooling, and is less damaging to the flesh as compared to large chunks of ice. A mechanical refrigerating system installed in the hold can lessen the requirement for the amount of ice to be carried. However, air temperature at the pens should not be allowed to go below freezing since one of the benefits of ice, in addition to cooling and providing aerobic conditions around the fish, is that the melt water gradually washes away blood and bacteria-laden slime. The holding pens should be designed to allow melt water to escape into the bilges.

Storage of ice in a refrigerated compartment is desirable, because crushed ice held at temperatures above freezing tends to fuse into a solid mass which has to be broken up manually or put through a crusher prior to use.

The precise method of icing varies with the construction of the hold and layout of the pen for a particular vessel. A satisfactory method for most vessels employing bulk storage is as follows (American Society for Refrigeration Engineering, 1959): Cover the floor of the pen with a layer of ice 20 to 30 centimeters (cm) (8 to 12 inches) deep. similar amount should be placed along the sides of the pen. A layer of fish not exceeding 15 cm (6 inches) deep should then be placed on the ice and covered with a 20 cm (8 inches) layer of ice. Fish and ice should be mixed together.

Successive layers of ice and fish should then be built up in the same manner until an overall depth of 1.2 m (4 feet) is reached, at which point shelf boards are inserted and the stowage process repeated. Gutted fish should be placed with the belly cavity down. In the case of large fish, the cavity should be filled with ice.

For proper cooling, it is important that intimate contact be made between fish and ice. When fish were piled in layers 38 to 46 cm (15 to 18 inches) deep and interspersed with thinner layers of ice, the fish at the center of these layers often required 24 to 36 hours to cool to approximately the temperature of melting ice (Castell et al., 1956).

Boxing at Sea

Some of the problems encountered in bulked pen storage can be eliminated by boxing in ice at sea (Waterman, 1964). The crushing effect of excessive pressure is reduced; in addition, there is a greater opportunity for the catch to be carefully and speedily handled during stowage and after discharge from the boat. There is a greater space requirement aboard the vessel for boxed fish. As with pen stowage, the fish should not be packed in direct contact with the surfaces of the box; otherwise the "bilgey" type of spoilage may result.

Refrigerated Seawater

The benefits of using refrigerated seawater (RSW) or refrigerated brine for storing fresh fish on board a fishing vessel are: 1) greater speed of cooling, 2) less textural damage due to reduced pressure upon the fish, 3) lower holding temperature, 4) greater economy in handling the fish due to time and labor saved, and 5) longer effective storage life of the fish. The real advantage of RSW compared with freshwater ice appears to be that the brine temperature can be maintained at about -1°C (30°F), which is just above the freezing point of fish.

Although it is now generally regarded that the storage life of whole fish is longer in RSW than in ice, shelf life is limited by the uptake of water and salt, particularly with lean fish species, by the development of oxidative rancidity.

A problem common with all fish species held in RSW is the eventual growth of spoilage bacteria in the brine, producing foul odors which can be imparted to the fish.

Chilled Seawater

For the small-boat fisherman, the benefits of RSW storage can be obtained without a mechanical refrigeration system through the use of chilled seawater (CSW) or slush ice; that is, stowage of the fish, ice, and seawater in tanks at the ratio of 3:1:1. The exact proportions for maintaining temperatures slightly below freezing (0°C) depend on the temperature of the fish and seawater and the duration of the trip. Successful results have been reported with herring and mackerel (Hume and

Baker, 1977). With this system, it is important that ice and seawater be mixed together just prior to loading with fish and that efficient circulation be maintained.

In general the use of RSW or CSW is not recommended for long fishing trips unless the eventual market is a filleted product. Both RSW and CSW systems produce a bleaching effect, and whole fish will exhibit cloudy eyes, bleached skin, and bleached gills which will reduce acceptability on the dressed-fish market. However, in terms of fillet quality, the product is highly acceptable and fillet shelf-life will often exceed that of traditionally iced fish. Some work carried out at the NMFS Gloucester (Massachusetts) Laboratory demonstrated that the addition of 1 percent potassium sorbate to the water phase of CSW systems greatly reduces bacterial growth.

FREEZING AT SEA

Over a period of 40 years much research and many commercial ventures have been undertaken relating to freezing North Atlantic groundfish at sea.

In the past few years, because of the relative collapse of groundfish stocks in the waters of the Northeast United States and eastern Canada, frozen at sea (FAS) product has become highly marketable.

The basic principles and technology for producing high quality FAS groundfish have not changed dramatically over the years. The fish must be of high quality and properly bled, gutted, and washed. Freezing dressed fish and/or fillets should be carried out in blast or plate freezers, and the packaging or glaze should be sufficient to prevent freezer burn and dehydration. For best quality, the product should be stored at -25°C, or below.

A major problem with FAS product is the effect of rigor mortis. If the product is frozen prerigor or during rigor, the fish must be allowed to pass through rigor upon thawing.

Freezing fillets cut from pre-rigor fish is generally not recommended, especially for fish blocks and shatter packs. If the fillets are not allowed to pass through rigor prior to cooking, the result will be a relatively flavorless and rubbery textured product.

PROCESSING SHORESIDE

The ice used for storage at sea contains blood and slime — an excellent medium for the growth of spoilage bacteria. Therefore, upon receipt at either dockside or the processing plant, all fish should be immediately re-iced.

Fish that are pre-rigor or in rigor should be iced down and allowed to pass through rigor prior to filleting.

Before filleting, the fish should be washed. Research reported by the Virginia Polytechnic Institute demonstrated that an agitated wash significantly reduced the level of spoilage bacteria. The use of 10 to 15 ppm sodium hypochlorite in the wash water was effective.

All filleting operations should be carried out quickly, preferably in a refrigerated environment. The fillets, steaks, etc., should be trimmed, candled, and washed in a continuous process. Ideally, the product should not be exposed to temperatures above 10°C for more than two hours. Temperature profiles of filleting operations in plant temperatures of about 13°C indicate that the final fillet temperature should not exceed 5°C.

USE OF BRINE

It is common practice to wash or rinse groundfish fillets with a solution of 2 to 5 percent saltwater (brine). This process eliminates blood, scales, and slime from the fillets or steaks prior to packing. This practice, with proper temperature controls and routine sanitation maintenance, is generally recommended. However, the improper use of brine systems can be detrimental to fish quality. Since brine solutions solubilize proteins, the brine quickly becomes an excellent medium for bacterial growth, especially without temperature control. Care should be taken to refrigerate brine systems, and the brine should be routinely discarded and replenished several times per day.

Another issue related to the misuse of brine systems is economic fraud. Brine tanks, especially in conjunction with the addition of sodium tripolyphosphate, can be used to "soak" fillets to add weight to the product. Soaking fillets in brine also acts to mask poor quality. Since the added water is not bound to protein, the result to the consumer is excessive shrinkage due to drip loss in storage and subsequent cooking.

PACKING AND SHIPPING

All products should be pre-chilled prior to packing and shipping. Based on research carried out at the NMFS Laboratory in Gloucester, Massachusetts, a general "rule of thumb" is that for every 24 hours fish are exposed to temperatures one or two degrees above 0°C, the product will lose one day of quality shelf-life. Fillets, steaks, etc., not chilled prior to packing or not properly refrigerated during distribution, could lose several days of quality shelf life during shipment. For example, fish fillets packed and transported at 5°C could lose as much as three days of quality shelf-life in a 24-hour period.

Modified Atmosphere (MAP), Vacuum Packing (VP), Irradiation

After death and the time immediately after rigor mortis, the primary changes in fish quality are related to biochemical reactions where autolysis (self digestion) is the major factor and bacterial activity is of little consequence.

Since MAP and VP only relate to the retardation of spoilage bacteria, there is little to be gained in preserving high quality products using MAP and VP.

These packaging and processing procedures will increase total edible shelf life but only in the intermediate quality range. The same is true for irradiation where the treatment will eliminate spoilage bacteria, but the resulting shelf-life extension will only be in the marginal quality range.

The use of MAP and VP for bulk shipment with good temperature control, however, is an acceptable practice. In bulk shipments modified atmosphere packing reduces spoilage bacteria during distribution. When the shipment is broken down for final packing and storage, etc., the product will undergo its normal spoilage pattern. Because bacterial degradation or spoilage is significantly retarded during the distribution period, several days of edible shelf-life can be gained.

MAP and VP of fresh product may represent problems, especially in retail packs. Since the product is not undergoing normal aerobic bacterial spoilage, most of the key indicators of degradation such as odor and discoloration are not readily evident. A vacuum-packed fillet or steak may have all the visual characteristics of a high-quality product. However, if the product is degraded to the point where severe anaerobic spoilage has occurred, the consumer may be assaulted by foul odors.

There also exists the potential development of *Clostridium botulinum* toxin in MAP or VP products if they are temperature-abused (above 42°F). In North Atlantic groundfish the incidence of this organism is extremely low and possible health hazards are minimal. However, with extreme temperature abuse, there could be a potential public health problem.

FREEZING AND COLD STORAGE

Only high-quality groundfish products should be frozen. The freezing process only preserves the initial quality, which will never improve beyond the quality of the original raw material. In reality, although rapid freezing and cold storage at temperatures below -25°C have shown superior quality retention, the United States consumer still continues to demonstrate his/her perception that frozen products are inferior to fresh products in terms of quality.

For frozen groundfish the packaging materials are important for quality retention. Moisture- and oxygen-impermeable materials are essential. During frozen storage, groundfish are susceptible to dehydration and "freezer burn" which will result in textural toughening and oxidative rancidity.

Products packaged simply with sealed bags ("air packs") are not recommended. These will very quickly become "frosted" and the quality will degrade rapidly.

For food service, "layer" or "shatter" packs are preferable. In this procedure, the fillets/steaks are packed in single layers on sheets of plastic film. This allows the end user to peel off layers of frozen fillets or steaks without thawing the entire container.

All fish products should be frozen as quickly as possible. Research has shown that rapid freezing reduces the size of ice crystals in the flesh which results in a better product texture. It is also recommended that the product should only be frozen to an internal temperature equivalent to the intended cold storage temperature. Fish products frozen to temperatures below the intended cold storage temperature will increase the relative size of ice crystals when the temperature equilibrates to the higher frozen storage temperature. This can result in textural changes in the finished product.

Cold-storage holding temperatures should be as low as possible. In general, -18°C storage is not recommended for seafoods. High quality shelf-life has been shown for non-fatty species such as cod and haddock stored at –18°C from three to five months. Fatty species — e.g., mackerel — will only retain high quality for two to three months. In comparison, storage at -30°C will retain high quality eight to 10 months for lean species and six months for fatty species. For export to Japan, storage temperatures of -40°C or below are recommended.

Fluctuating cold-storage holding temperatures should be avoided. Fluctuating temperatures result in a cycle of increasing and decreasing ice crystal size in the frozen product. This constant cycle eventually will produce increased drip loss and textural toughening of the product.

THAWING

Frozen fish can be thawed in a number of ways, including in water or air, and also by cooking directly from the frozen state. For preserving high quality, defrosting should be carried out under refrigeration. Thawing at warm temperatures can result in dehydration and potential bacterial spoilage. Thawing in cold water is recommended for packaged products. Unpackaged products should not be thawed in water, since the water will solubilize proteins and result in significant flavor loss.

Some products, especially fillets, steaks, or portions, can be defrosted by using microwaves. Microwave tunnels at 915 Megahertz (MHz) are commonly used to temper fish fillet blocks to about –10°C for automated slicing into fish sticks or fish portions. Most home or restaurant microwave ovens have an output of 2450 MHz. Since the heating chracteristics of this frequency are much faster than 915 MHz and the depth penetration is shallower, this method is not generally recommended. However, many microwave ovens are equipped with "defrost cycles" where the frozen product is subjected to successive pulses of microwave energy. Using the "defrost" cycle, microwave ovens can be very effective for defrosting seafoods. It is recommended that microwaves should only be used as a method to partially defrost seafoods. Attempting to fully defrost by microwaves usually produces "hot spots" where some parts of the product become cooked.

REFERENCES

American Society for Refrigeration Engineering. 1959. Fresh fishery products. In *Refrigeration Applications — Air Conditioning, Refrigeration Data Book*. New York.

Bramnsaes, F. 1979. Quality and stability of frozen seafood. In *Quality and Stability of Frozen Foods*. W. B. Van Arsdel, M. J. Copley, and R. L. Olson (eds.), pp. 217-236. Wiley Interscience, New York.

Castell, C. H., W. A. MacCallum, and E. H. Power. 1956. Spoilage of fish in vessels at sea. *Journal of the Fishery Research Board of Canada* **13**:21-39.

Dassow, J. A. 1963. Handling of fresh fish. Pp. 275-287 in *Industrial Fishery Technology*. M. E. Stansby and J. A. Dassow (eds.). Reinhold Publishing Co., New York.

Dyer, W. J. 1971. Speed of freezing and quality of frozen fish. In *Fish Inspection and Quality Control*. R. Kreuzer (ed.), pp. 5-81. Fishing News (Books) Ltd., London.

Dyer, W. J. and J. Peters. 1969. Factors influencing quality changes during frozen storage and distribution of frozen products, including glazing, coating and packaging. Pp. 317-322 in *Freezing and Irradiation of Fish*. R. Kreuzer (ed.). Fishing News (Books) Ltd., London.

Georgala, D. L. 1958. The bacterial flora of the skin of the North Sea cod. *Journal of General Microbiology* **18**:84-91.

Heen, E. and O. Karsti. 1965. Fish and shellfish freezing. Pp. 353-418 in *Fish as Food*, vol. 4. G. Borgstrom (ed.). Academic Press, New York.

Hess, E. 1950. Bacterial fish spoilage and its control. *Food Technology* **4**:477-480.

Hume, S. E. and D. W. Baker. 1977. Chilled seawater system for bulk holding sea herring. *Marine Fisheries Review* **39**(3):4-9.

Huss, H. H. 1995. Quality and quality changes in fresh fish. FAO Fisheries Technical Paper No. 348. Fisheries Department, FAO, Rome.

Jason, A. C. 1982. Thawing. In *Fish Handling and Processing*. Aitken, Mackie, Merritt, and Windsor (eds.). Torry Research Station, Aberdeen, Scotland.

Johnston, W. A., F. J. Nicholson, A. Roger, and G. D. Stroud. 1994. Freezing and refrigerated storage in fisheries. FAO Fisheries Technical Paper No. 340. Fisheries Department, FAO, Rome, Italy.

Lane, J. Perry. 1964. Time-temperature tolerance of frozen seafoods. *Food Technolology* **18**:1100-1106.

Learson, R. J. and J. J. Licciardello. 1986. Literature reporting of shelf-life data: What does it all mean? *Rev. Int. Froid.* **91**:179-181.

Licciardello, J. J. 1980. Handling whiting aboard fishing vessels. *Marine Fisheries Review*. Jan. 1980.

Licciardello, J. J., and D. O. Entremont. 1987. Bacterial growth rate in iced fresh or frozen-thawed Atlantic cod. *Journal of Food Protection* **49**(4):43-45.

Licciardello, J. J. 1990. Freezing. In *The Seafood Industry*. Van Nostrand Reinhold, New York.

Linda, A. H. and J. W. Slavin. 1960. Sanitation aboard fishing trawlers improved by using chlorinated seawater. *Commercial Fisheries Review* **22**(1):19-23.

MacCallum, W. A. 1955. Fish handling and hold construction in Canadian North Atlantic trawlers. *Fisheries Research Board of Canada Bulletin* 103-161.

Nicholson, F. J. 1973. The freezing time of fish. Torry Advisory Note No. 62. Torry Research Station. Aberdeen, Scotland.

Olavie, E. Nikkil, and Reino R. Linko. 1956. Freezing, packaging, and frozen storage of fish. *Food Res.* **21**(1)42-46.

Pottinger, S. R. 1951. Effect of fluctuating storage temperatures on quality of frozen fish fillets. *Comm. Fish. Rev.* **13**(2):19-27.

Procter, B. E. and J. T. R. Nickerson. 1935. An investigation of the sterility of fish tissues. *Journal of Bacteriology* **30**:377-382.

Ronsivalli, L. J. and D. W. Baker. 1981. Low temperature preservation of seafood: a review. *Marine Fisheries Review* **43**(4).

Ronsivalli, L. J. and J. J. Licciardello. 1975. Factors affecting the shelf life of fish. *U. S. Atomic Energy Commission Activities Report* **27**(2):34-42. Oak Ridge, TN.

Shaw, B. G. and J.M. Shewan. 1968. Psychrophilic spoilage bacteria of fish. *Journal of Applied Bacteriology* **31**:89-96.

Shewan. J. M. 1961. The microbiology of seawater fish. Pp. 487-560 in *Fish as Food*, vol. 1. G. Borgstrom (ed.). Academic Press Inc., New York.

Slavin, J. W. 1963. Freezing and cold storage. In *Industrial Fishery Technology*, M. E. Stansby (ed.). Krieger, Huntington, NY.

Slavin, J. W. and J. A. Dassow (eds.). 1971. Fishery products. In *ASHRAE Guide and Data Book*. American Heating, Refrigeration and Air Conditioning Engineers, New York.

Spencer, R. 1961. The bacteriology of distant water cod landed at hull. *Journal of Applied Bacteriology* **24**:4-11.

Spencer. R. and C. R. Baines. 1964. The effect of temperature on the spoilage of wet fish. *Food Technology* **18**:769-773.

Torry Research Station. 1965. Quick Freezing of Fish. Torry Advisory Note No. 27, Aberdeen, Scotland.

Waterman, J. J. 1964. Bulking, shelving, or boxing? Torry Advisory Note #15. Torry Research Station, Aberdeen, Scotland.

Waterman, J. J. 1965. Handling wet fish at sea and onshore. Pp. 133-148 in *Fish Handling and Preservation: Proceedings of Meeting on Fish Technology*. Scheveningen, Sept. 1964. Organization for Economic Cooperation and Development, Paris.

Processing Blue Crab, Shrimp, and King Crab

Donn R. Ward

INTRODUCTION

Crustaceans comprise a relatively small proportion of the marine food products marketed. However, given the diversity of species, the high prices they command, and the fact that species are marketed as cooked, ready-to-eat products, they are an extraordinarily important group.

BLUE CRAB

The scientific name of the blue crab, *Callinectes sapidus*, describes three notable attributes of the species. "Calli" is the Latin word for beautiful; "nectes" and "sapidus" the Latin words for swimmer and savory, respectively. The blue crab is thus a beautiful swimmer crab and is also tasty to eat.

Processing

Crabs arrive at processing plants either directly from boats or in trucks which have transported the crabs from other landing sites. The crabs are weighed, then dumped into large stainless steel baskets. During the winter dredging season, the crabs are run through a tumble spray washer prior to being dumped into the baskets. The washing step is essential for dredged crabs because they are covered with sand and grit from being buried in the sandy bottom. While some processors will wash only dredge crabs, others wash all incoming crabs irrespective of season (Figures 1, 2, 3).

Although subsequent handling and cooking methods vary depending on regional customs and state laws, the processing of blue crabs has not changed dramatically since fresh crabmeat was first marketed in the late 1800s. It is a very labor-intensive industry, with most of the picking still done by hand. (See the end of this chapter for steps involved in removing the meat from the crab.) This is a major problem for the industry. In recent years it has been very difficult for processors to find local workers with the essential skills, or interest in developing the skills, to remove the meat from crabs. As a consequence, many processors have resorted to bringing in foreign nationals as seasonal labor.

Currently, the industry processes live crabs either under steam pressure or in boiling water. Some states, such as Maryland, North Carolina, and Florida, have regulations which stipulate that "crabs shall be cooked only under steam pressure." Some regulations go so far as to stipulate cold point temperature minimums; for example, the rules governing crabmeat operations in North Carolina declare, "Crustacea shall be cooked under steam until the internal temperature of the

Figures 1. Cooking live crabs in a horizontal retort.

Figure 2. Performing thermal penetration studies with thermocouples to establish an adequate cooking process.

Figure 3. Air-cooling cooked crabs prior to refrigerated storage.

center-most crab reaches 235°F (112.8°C)." The regulations for processing crabs in Texas simply state, "Crabs shall be cooked so as to provide a sterile crab."

Cooking live crabs in a steam retort is the most common processing method. Traditionally, there has been a general lack of uniformity among processors with respect to the times and temperatures used in the steam cook process, state regulations notwithstanding. Regulatory authorities prefer long cook times at high temperatures due to the destruction of microorganisms, which are found as part of the natural microflora of the crab. From the processor's perspective the issue is not just killing microorganisms; it is also economics. The higher the temperature and/or the longer the cook time, the lower the yield of picked meat. The reduction in yield is the result of moisture loss from the edible tissues. Short cook times cause less drying of the meat and, therefore, produce greater yields. Since the average yield of picked meat from a blue crab is approximately 10 percent, this is a problem to which crab processors are acutely sensitive. Furthermore, processors in states that allow crabs to be boiled are clearly at an advantage with respect to yield. Cooking under pressurized steam results in significant moisture loss; this is not an issue when cooking in boiling water.

After cooking, crabs are moved to a cooling room and air-cooled to ambient temperature. The cooling room is usually a screened area or well-ventilated room with exhaust fans to remove the steam rising from the hot crabs. Before crabs are moved to the cooked crab cooler (33° to 40°F/0.6° to 4.4°C), they must be cooled to the extent that steam is no longer rising from them. If cooked crabs were moved immediately to the cooler, without a precooling period, steam rising from the crabs would condense on the ceiling of the cooler and drip back on the crabs. This could potentially contaminate the crabs and result in what is termed a "sour crab." Furthermore, this could serve as a source for contamination with the bacterial pathogen *Listeria monocytogenes*.

In some states, subsequent to ambient cooling, the practice is to deback (remove the top shell), eviscerate (remove internal organs) and wash the crabs before placing the crab cores in refrigerated coolers. After overnight cooling, whole debacked crabs (or crab cores) are taken to the picking room where pickers hand-pick the meat from the crabs (Figure 4).

Blue crabmeat is available in several different forms. The Product Quality Code of the Southeastern Fisheries Association defines these forms as:

Jumbo: largest, white pieces or chunks of crabmeat, typically from the backfin.

Lump or Backfin: large, white pieces or chunks of crabmeat which can include backfin.

Special, Flake, Regular, or Deluxe: smaller, white pieces or chunks of crabmeat which usually exclude backfin.

Claw: only includes meat from the crab claw. This meat has a darker, brownish tint than the other forms of meat taken from the crab body.

Minced: crabmeat removed and/or separated from the shell by a physical process that actually minces the meat.

Mixed: any combination of meat as requested.

Cocktail Claws: clawmeat intact on the claw with the shell removed, except for the forward tip to be used as a handle.

As mentioned previously, the technology used in processing blue crabs has changed little in the past century. Most of the picking is still done by hand. In recent years, however, increasing efforts have been made to perfect various machines for picking meat from the blue crab. Cockey (1980) noted that the different design principles for extracting meat from the crab include vacuuming, squeezing the meat from the legs and bodies with rollers, throwing the meat from the cores by centrifugal force, shaking meat from the core by vibration, or crushing in a hammer mill and separating the meat from the shell by brine flotation. These machines and methods have met with varying degrees of success. Since the mid-1980s, several companies have used meat/bone separators to recover residual meat remaining on the waste from the hand-picking operation. Typically, the material recovered is sold to pet food manufacturers.

One major disadvantage of the machines mentioned is that none allows removal of the backfin portion as one large lump. Inasmuch as consumers desire to purchase crabmeat in lump form and are willing to pay a premium price, it is doubtful that hand-picking will give way completely to machine-picking in the near future.

There have been several attempts at developing "imitation lump," by binding the smaller flake meat pieces into larger pieces. The author is not aware of any firms currently involved in manufacturing such a product. This type of product is different from the imitation crabmeat (usually imitating meat from king crab legs) often seen on salad bars. This latter product is the result of surimi technology.

Pasteurization

Most muscle protein foods, particularly seafoods, are quite perishable. Under normal refrigeration, fresh crabmeat has a shelf-life of approximately seven to 10 days. This relatively short shelf-life places responsibility on all those involved in the processing and marketing channels to handle the product appropriately and quickly. One alternative to the shelf-life problem is freezing. Although blue crabmeat can be frozen successfully, few processors do it because it has traditionally suffered from the same stigma that has affected all frozen seafoods. Another alternative has been pasteurization.

Traditionally, pasteurization of blue crabmeat was understood to involve heating hermetically sealed containers to a cold point temperature (slowest heating point) of 185°F (85°C) and holding for at least one minute. This understanding worked well as long as the industry pasteurized meat in one-pound (454 g), 401x301 cans [4 1/16 inches (10.3 cm) in diameter by 3 1/16 inches (7.8 cm) in height]. In recent years the industry has begun to pasteurize in containers of various sizes, shapes, and composition. This change made it necessary to define crabmeat pasteurization in a more scientifically precise manner. Information on pasteurization processing standards can be obtained from the Blue Crab Industry Association, 1901 N. Fort Meyer Drive, Suite 700, Arlington, VA 22209 (Figures 5 , 6, 7).

Pasteurization extends the shelf life of blue crab meat by destroying the bacteria that would cause spoilage of the fresh product under normal refrigeration conditions. Pasteurized crabmeat,

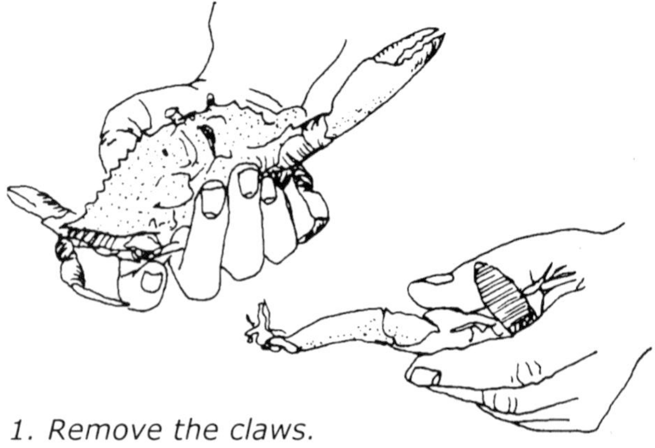

1. Remove the claws.

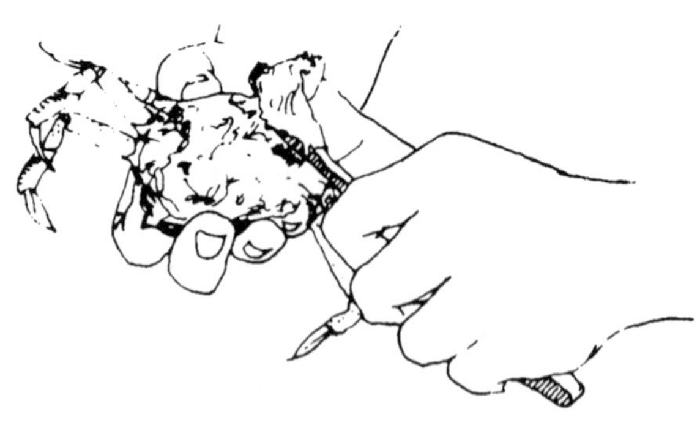

4. Make a straight cut directly over the crab's legs.

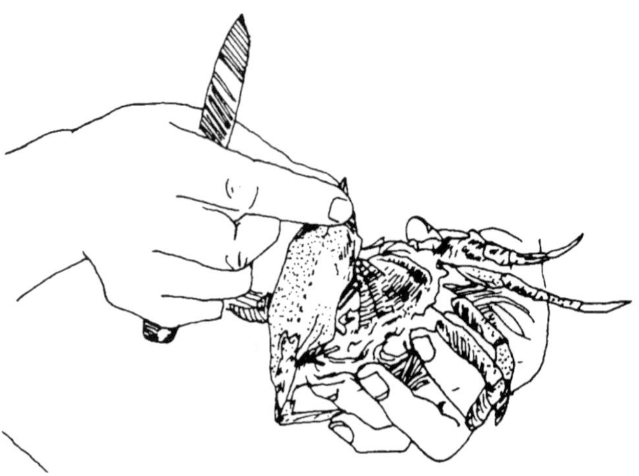

2. Remove the top shell.

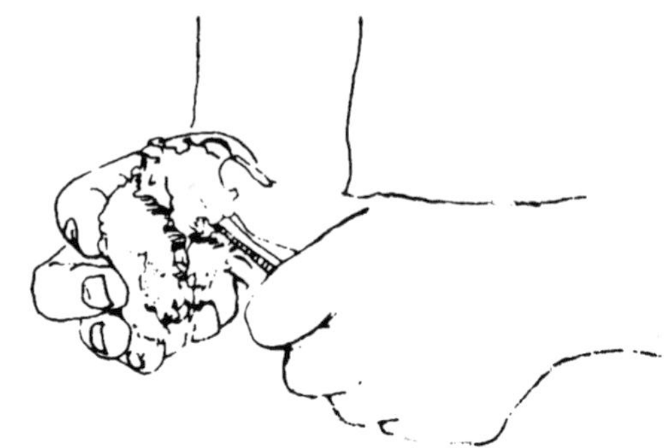

5. Remove the legs.

6. Remove the backfin and flake meats.

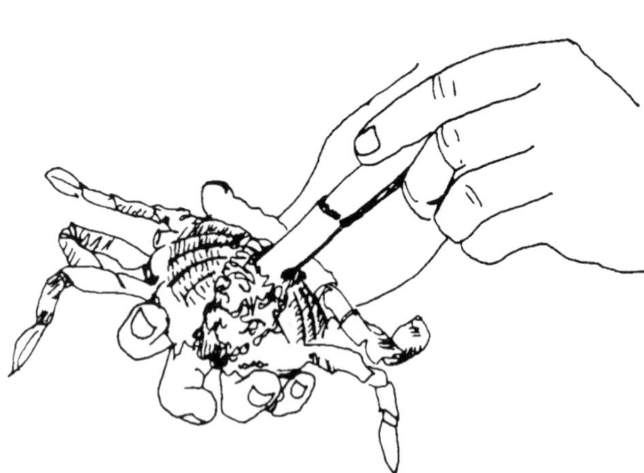

3. Remove the internal organs

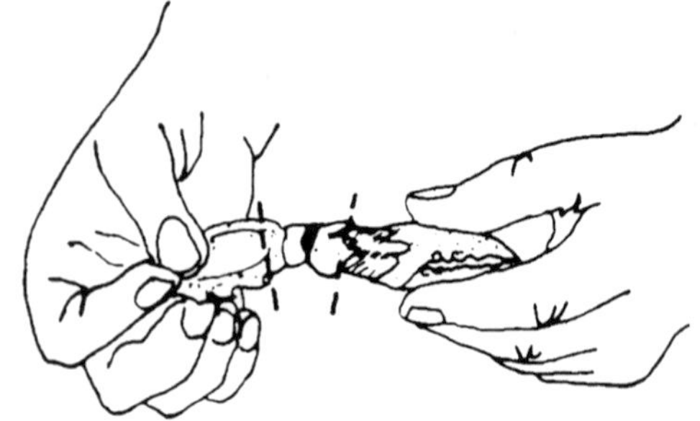

7. Crack the claws and remove the dark claw meat.

Figure 4. Picking a blue crab.

Figure 5. Removing one-pound cans from the pasteurization tank.

like pasteurized milk, must be kept refrigerated. However, since the normal spoilage microorganisms have been destroyed, pasteurized crabmeat has a shelf life of at least six months. Once the can has been opened, the crabmeat should be used within five to seven days.

Figure 6. Pasteurization of crabmeat in flexible films.

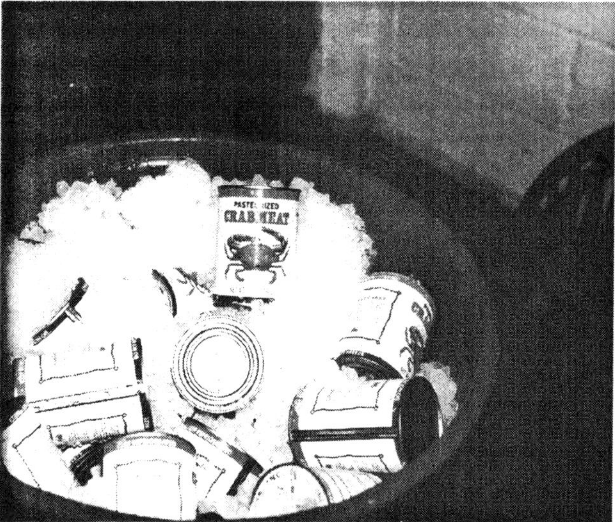

Figure 7. Cooling pasteurized crabmeat in wet ice.

Pasteurization allows processors to inventory product during periods of crab abundance. Furthermore, it allows consumers, particularly those in inland markets, to purchase crabmeat year-round. And just as importantly, the pasteurized product is almost indistinguishable in taste and texture from the fresh product.

Regulatory agencies are currently enforcing a zero tolerance for the microorganism, *Listeria monocytogenes*, on all cooked, ready-to-eat foods such as crabmeat. Since normal pasteurization procedures destroy the organism, some processors concerned with the regulatory repercussions associated with isolation of the bacterium from their product are trying to establish greater market interest in pasteurized crabmeat.

SHRIMP

The commercial shrimp industry is one of the largest seafood industries in the United States both in value and quantity of product caught. There are two commercially important shrimp fisheries: the North Pacific and Alaska area, and the southern fishery located in the South Atlantic and Gulf of Mexico.

In the southern fishery, there are three major commercial species: the white shrimp *(Penaeus setiferus)*, the brown shrimp (*Penaeus aztecus*), and the pink shrimp (*Penaeus durarum*). In some areas vessels return to port daily; boats equipped with freezers or adequate ice storage may remain on the fishing grounds for two to three weeks. In recent years, the domestic penaeid shrimp industry has reached a steady-state production condition in terms of available wild resources, and they must compete with an increasing amount of cultured penaeid species from international sources.

Handling of shrimp on board the vessel is critically important. On completion of a tow, the contents of the net are dumped on the deck of the vessel, and the shrimp are separated from the "trash" (i.e., anything that is not shrimp). The latter is usually discarded overboard. Crew members quickly remove the heads of the shrimp by hand, and the "headed" shrimp are then shoveled into baskets and washed with a stream of water.

It is important to head and wash the shrimp before storage. Although the procedure is referred to as heading, in fact crew members remove not just the head but the entire cephalothorax section, which contains the gills and many of the organs associated with the digestive tract. Studies have shown that removal of this section removes a significant source of bacteria as well as active enzymes that can hasten deterioration of the shrimp. Thorough washing is important since this further reduces bacteria and enzymes.

Once headed and washed, two different methods can be used to preserve the catch: icing or brine freezing. Boats using ice usually immerse each basket of shrimp in a "dip" solution prior to placing the shrimp on ice in the vessel's hold. The dip retards the formation of black spot (see later discussion).

Another method of preserving shrimp at sea is brine freezing. According to an article written by Bruce Cox of Texas A&M University, the use of freezers aboard shrimp vessels has been both cursed and applauded by boat owners. The process can produce very favorable results, but attention to detail is important. The following information on brine freezing was adapted from Cox's article.

The proper use of freezer brines is very important. An efficient brine should rapidly freeze shrimp or bring them close to freezing, so that they can be completely frozen in the boat's hold. Proper brine freezing helps prevent black spot and dehydration. Correct mixtures of salt, corn syrup, and dip powder (sodium bisulfite) effectively retard dehydration and black spot formation.

Salt in the proper concentration (23 percent) reduces the freezing point of a brine tank -6°F (-21°C), whereas only slightly less salt significantly affects the freezing capabilities of the brine tank. Corn syrup in the brine mixture coats the shrimp with an elastic coating and helps prevent black spot and dehydration. Regular table sugar becomes brittle and will flake off during the trip. Corn syrup acts as a chemical reducing agent, robbing the black spot enzymes of oxygen which is required to complete the black spot reaction. The syrup coating also holds moisture inside the shrimp, preventing dehydration. Sodium bisulfite (dip powder), like corn syrup, is a chemical reducing agent that effectively binds oxygen so that it is unavailable to the enzymes responsible for black spot formation.

Approximately 50 pounds (22.7 kg) of headed shrimp are placed in open mesh sacks and then submerged in the brine tanks. Shrimp should not be allowed to soak for more than 15 or 20 minutes; otherwise the shrimp will become too salty and eventually toughen. Once brine freezing is complete, the bags are placed in the freezer hold.

At the dock, boats that stored shrimp on ice will flood the hold to melt the ice. The shrimp are then removed by vacuum pumps to wash tanks in the processing plant. Bags of brine-frozen shrimp are off-loaded from the boats and emptied into thaw tanks. Shrimp remain in these tanks for five to 10 minutes to allow the frozen shrimp to separate.

From this point on, whether the shrimp were iced or frozen, the process is much the same. The shrimp are graded according to size. Size grades of shrimp are expressed as "count," meaning the average number of shrimp to the pound. Following are the common commercial size categories:

Less than 10	
10-15	41-45
16-20	46-50
21-25	51-55
26-30	56-60
31-35	61-70
36-40	More than 70

After grading, the shrimp are packed in 5-pound (2.3-kg) boxes and frozen in a blast freezer or plate freezer. After the product is thoroughly frozen, it is removed from the freezer, the top of the box is opened, and about 8 ounces (237 ml) of water is sprayed on the shrimp. The lid is closed and the box inverted. This method allows a solid block of ice shrimp to form, protecting the shrimp from freezer burn. Alternatively, shrimp are cryogenically frozen and placed in bags or boxes for retail and institutional sale (Figure 8).

Black Spot

Black spot, also called box ring, ice burn or ringer shrimp, is a dark discoloration that may form on stored shrimp. Black spot is caused by a biochemical reaction, called melanosis, that is produced from naturally occurring chemicals in the shrimp shell and is similar to the reaction that takes place when a person gets a suntan.

Black spot is *not* caused by excessive levels of spoilage bacteria. In fact, large numbers of actively growing bacteria may reduce the formation of black pigment, by depleting oxygen that the black spot reaction requires.

The black spot reaction is similar to suntanning; thus, exposure to sunlight speeds up the process. For example, after 11 days of storage in ice, shrimp that had been placed immediately in the hold had 14 percent black spot; shrimp exposed to the sun for two hours before storage had 55 percent. Six hours of exposure resulted in 98 percent black spot. Interestingly, holds that are over-insulated or over-iced can also increase the occurrence of black spot. In order to reduce black spot development, the ice must melt. Melting ice washes away some of the compounds associated with the reaction, as well as decreases the oxygen contact with shrimp.

Figure 8. Separating raw shrimp prior to cryogenic freezing.

Shrimp molting cycles also influence susceptibility to black spot. As shrimp prepare to molt, they build up materials needed for a new shell. One of these materials is an amino acid called tyrosine, an essential compound in black spot production. As a result, shrimp are more susceptible to black spot development just prior to molting and least susceptible just after molting.

Prevention of Black Spot

Efficient handling on deck; immediate, thorough washing; and storage on good quality melting ice are the most natural and effective means of controlling black spot. Rapid handling on deck reduces exposure to sunlight and elevated temperatures which speed up the chemical reaction leading to black spot.

Chemicals have also been used to control black spot development. The chemicals evaluated include sodium sulfite, sodium bisulfite, sodium metabisulfite, ascorbic acid, ethylenediaminetetraacetic acid (EDTA), baking soda and others. The most commonly used is sodium bisulfite, often referred to as "dip." Sodium bisulfite is a strong reducing agent which competes with tyrosine for molecular oxygen, thus preventing the oxygen from entering the reaction which leads to black spot.

Any shrimper or processor using sodium bisulfite must exercise care and caution. There is a small percentage of the population which exhibits an allergic response to sulfites, and in some instances these reactions can be quite severe. As a result, there is controversy regarding the use of sulfiting agents in foods. Although it is legal to add sulfiting agents to shrimp to retard the formation of black spot [up to 100 parts per million (ppm), measured as SO_2 on the edible meat], there is active research to find an acceptable alternative. Also, if a sulfiting agent is used in excess of 10 ppm, the current limit of detection, it must be listed on the product label. For these reasons, researchers are actively seeking suitable alternatives to the use of sulfites. Although there are several possibilities, one of the more promising is 4-hexyl resorcinol (McEvily et al., 1991).

KING CRAB

The king crab *(Paralithudes camschatica)* is an extremely large crab, weighing as much as 24 pounds (11 kg). Fishing areas extend in a large crescent from Southeastern Alaska to the Bering Sea side of the Aleutian Peninsula and island chain.

Only healthy male king crabs are kept, because conservation regulations prohibit the taking of females. The crabs are held live on board the boats in tanks containing either circulating seawater or refrigerated seawater.

At the processing plant, crabs from the fishing vessels are processed immediately or placed in holding tanks of circulating seawater similar to those found on the boats. Only live crabs are processed. During the butchering operation, the back shell is pulled off, the crab cut in half, and the viscera and gills removed. The butchered sections are then thoroughly washed to remove blood and viscera.

Two methods of precooking the crab sections are commonly used, depending on whether the crabmeat is to be canned or frozen. The first method, the two-stage cook, is most commonly used by canners. After the first wash, the crab sections are placed on a wire mesh conveyor that transports them through a tank of water heated to 155° to 159 ° F (68.3° to 70.6°C). The conveyor speed is controlled so that the sections remain in the heated water for 10 to 12 minutes. After emerging from the first cook tank, the crab sections are thoroughly washed with cold water to reduce product temperature and remove uncoagulated blood. The removal of the blood is an extremely important step, particularly for crabmeat that is to be canned. Should blood remain, the copper in the blood can produce a blue or black discoloration, often referred to as "bluing," in the canned product.

After the first cook and wash, but while the sections are still warm, meat is removed by passing the sections between rollers. The squeezing action breaks the shell just enough to force the meat out while the shell proceeds on through the rollers. The meat is then thoroughly washed again to remove traces of blood, shell particles, and traces of viscera.

After the second wash, the meat is cooked again to fully coagulate the protein and shrink the meat so that proper fills may be obtained in the cans. A continuous cooker is normally used, with the meat passing in approximately four minutes through a water tank or steam tunnel maintained at 210° to 212°F (98.9° to 100°C). The meat is then spray-washed, inspected and passed to the packing table, where cans are filled by hand. Packers cut the large leg sections to fit the can and fill most packs with one layer of leg meat, followed by smaller sections and shoulder meat. Parchment paper ends are commonly used to minimize possible tin-plate discoloration and to prevent meat from sticking to the can lid.

Cans of packed meat are thermally processed to render the product commercially sterile. The time and temperature used will vary, depending on the can size and shape. Canned king crabmeat should be stored at a relatively cool temperature to minimize color and flavor changes. It should be marketed within one to two years for best quality.

The second method is the single-stage cook, which is primarily used by processors who freeze the final product, although some canners also use this method. In the single-stage process the crabs are cooked only once. Following butchering and washing, the crabs are cooked at 210° to 212°F (98.9° to 100°C) for 20 to 22 minutes. The sections are spray-washed to cool the product, and the meat is removed from the shell by techniques similar to those described for the two-stage process.

Once the meat has been removed, it is thoroughly washed. However, because of the more rigorous initial heat treatment, the blood is coagulated and therefore very difficult to remove. Most canners do not use this method of processing, since the likelihood of bluing is greatly increased. Meat from this processing method is, however, quite suitable for freezing.

In general, king crab meat is much more suitable for freezing than other crab species. It can be stored up to a year at 0°F (-18°C) with good acceptability. King crab meat is usually frozen as a large block, incorporating 250 ounces (7.1 kg) of crabmeat and 24 ounces (0.68 kg) of water to fill the voids. The blocks are inspected and packaged in a suitable film and wax-board carton, then frozen and glazed. Institutional products are made from these large blocks by sawing the frozen blocks into 1- or 1.5-pound (454 or 681 g) units.

Frozen king crab sections (meat in the shell) have become very popular with retail and institutional buyers. To provide this product, the butchered sections are chilled, thoroughly washed, trimmed, and divided into uniform 10-pound (4.5-kg) lots. The sections are then frozen and glazed. Extra glazing is necessary at the shoulder end, where the meat is exposed, to prevent dehydration. Signs of dehydration are yellowing and then honeycombing (spongy appearance) of the meat. The shoulder of the crab is enclosed by a yellowish membrane so it is important to distinguish between the natural yellow of the membrane and yellowing of the meat due to dehydration and rancidity.

REFERENCES

Cockey, R.R. 1980. Bacteriological assessment of machine-picked meat of the blue crab. *Journal of Food Protection* **43**:172.

Cox, B. Undated. Freezing Shrimp at Sea. Texas A&M University, College Station.

McEvily, A.J., R. Iyengar, and W.S. Otwell. 1991. Sulfite alternative prevents shrimp melanosis. *Food Technology* **45**:80.

Phillips, S.A., and J.T. Peeler. 1972. Bacteriologic survey of the blue crab industry. *Applied Microbiology* **45**:80.

Ulmer, D.H.B. Jr. 1964. Preparation of chilled meat from Atlantic blue crab. *Fishery Industrial Research*. **2**:12.

Van Engel, W.A. 1962. The Blue Crab and Its Fishery in Chesapeake Bay. *Commercial Fisheries Review*. **24**:9.

Handling and Processing Crawfish

Michael W. Moody

Freshwater crawfish have been a seafood specialty in the United States for decades. Most crawfish harvested and consumed in this country are located in Louisiana and adjoining states. Internationally, crawfish are an important product of commerce. The total annual commercial harvest of freshwater crawfish is more than 110,000 metric tons. The United States produces 55 percent of that volume, and the People's Republic of China, 36 percent. Europe and Australia also contribute to the production of freshwater crawfish (Huner, 1989). Although there are more than 300 species of crawfish, only a few dozen have any commercial importance. The species *Procambarus clarkii* is the single most important species in North America, making up more than 70 percent of all harvested species (Moody, 1994) (Figure 1).

Most processing of North American crawfish for meat and other value-added products is done in Louisiana, although there are some minor production and processing efforts in other states such as Texas, Mississippi, Florida, California, Wisconsin, Oregon, and Washington (Moody, 1994). Louisiana is also the major site for crawfish processing development. Crawfish processing started as a "salvage" operation for unsold or unused live crawfish. Since many crawfish are purchased whole and alive for recreational crawfish boils, producers and fishermen have traditionally been able to sell their daily harvest near the end of the week and on weekends with little problem. Early in the week, however, live crawfish do not sell as well, even though production levels may be the same. To minimize this loss, crawfish were cooked and the meat picked and refrigerated for later consumption. Today, even though these same consumption patterns still exist, Louisiana has dozens of crawfish processing facilities that process crawfish for meat on a daily basis.

Figure 1. A crawfish.

The industry of crawfish meat processing began more than 50 years ago near Henderson, Louisiana (a small settlement about 10 miles outside of Lafayette). These early processing efforts involved nothing more than boiling the crawfish in water heated with gas jets and then hand-peeling the meat. Packaging consisted of placing several pounds of meat in boxes and icing it down. Today, although the basics of crawfish processing remain the same, they have been refined and improved with technological advancements and scientific evaluations. This chapter will focus on each phase of the processing of freshwater crawfish. Although foreign processing plants (especially in the People's Republic of China) are competing in what was once an industry dominated by Louisiana interests, most of this discussion will focus on methods and procedures used in Louisiana.

DESCRIPTION

Freshwater crawfish are crustaceans, characterized for processing purposes by a hard exoskeleton, a cephalothorax, two large claws, and a meat-filled tail. To processors, the two most important species are *Procambarus clarkii* and *Procambarus zonangulus*. *P. clarkii* is commonly known as the red swamp crawfish because of its characteristic mottled red coloration. *P. zonangulus* is commonly called the white river crawfish. Both are harvested in the southeastern United States in sufficient quantities to be of significant commercial value. Both are relatively large, averaging 8.0 to 9.0 centimeters long and weighing approximately 20 grams at maturity (Avery, 1994). Moody (1994) describes the physical differences between them. Each species is easily distinguished by markings on the carapace and on the ventricle side of the tail. The two species are generally harvested together. Although the red swamp crawfish makes up the majority of the catch, in some crawfish aquaculture ponds (especially in the northern ranges), the white river crawfish may dominate. There is no economical way to separate the species; therefore, both are processed together. The red swamp crawfish is generally the species of preference.

The hepatopancreas (called "fat" by consumers) is often consumed along with the meat and is an important ingredient in regional crawfish recipes and cooking. The color of the cooked hepatopancreas in red swamp crawfish ranges from yellow to dark orange and is very pleasing to the eye. In white crawfish, this material may be yellow to green and does not have the same eye appeal. In addition, the meat of the white crawfish appears to be lighter in color. Some consumers claim that there is also a taste difference, the white crawfish having a less desirable flavor. Marshall (1988) did a comprehensive study on many of the factors affecting the processing of freshwater crawfish, including a detailed evaluation of flavor and other characteristics. Marshall conducted the study using both instrumental analysis and an untrained taste panel and compared the results. Color, texture, and flavor of both the meat and hepatopancreas were examined. The study showed no significant difference in meat texture between the two species. Mean Instron shear force values for red swamp crawfish and white river crawfish were 1.56 kilograms/gram and 1.60 kg/g, respectively. These findings agreed with those of the panelists, in that they could not distinguish any significant differences in texture. Surprisingly, the panelists could not distinguish any significant differences between the color of the meat, appearance of the hepatopancreas, or the taste of the product. An instrumental analysis of color, using a Hunterlab Tristimulus Colorimeter, showed a highly significant difference in both the meat and hepatopancreas — supporting claims made by consumers that the red swamp crawfish has redder meat and the white crawfish has lighter-colored meat. The instrumental analysis of the hepatopancreas showed that the red crawfish typically has a red-orange hepatopancreas that changed little over a 20-hour storage period, but that the hepatopancreas of the white crawfish is greener at the time of cooking and that the green color intensified over this 20-hour storage period.

Crawfish are quite active and are kept alive until processed. Like other crustaceans, they have internal gills. Moisture can be regulated to keep the gills healthy and functioning, even when the crawfish are removed from water for an extended period. Although crawfish are opportunistic feeders and will feed on a variety of plant and animal matter, decaying plant material is generally considered to be their staple. Like all crustaceans, crawfish must periodically molt to grow and mature.

HARVESTING

Like most seafood, the freshwater crawfish fishery is highly seasonal. There are two primary sources of crawfish: traditional "wild" crawfish from the many swamps and river systems of the lower Mississippi Valley and cultured or pond-grown crawfish located in the same regions. Growing crawfish in ponds is an emerging industry that had its start in the early 1950s. Because of the unpredictability in volume of wild crawfish harvested, aquaculture is seen as a stabilizing factor in the quest to provide a consistent source of crawfish to consumers and processing plants. In 1995 there were approximately 110,000 acres of crawfish ponds in Louisiana. Louisiana once had more than 130,000 acres (Roberts, 1995). Crawfish grown in ponds are not subject to many of the factors that can affect the wild harvests — such as low rain or snowfall amounts in the upper reaches of the Mississippi Valley, predation, and poor water conditions. However, during high-production years for wild crawfish, the resulting drop in prices can make the expense of maintaining ideal pond conditions difficult to meet. One reason for crawfish aquaculture, other than increasing total production, is making an early season crawfish available. Wild crawfish are typically marketed in small quantities beginning in middle-to-late winter (January and February). Wild crawfish volumes dramatically increase as spring runoff fills swamps and temperatures rise. The peak season is during April and May. The crawfish life cycle of burrowing, laying eggs, and growth depends on water levels and other factors. By draining ponds and reflooding them at specified times, crawfish aquaculture farmers can manipulate these natural cycles within the confines of the pond. Draining ponds in late spring forces burrowing and brooding. Reflooding in early fall releases the young to begin feeding months ahead of wild crawfish. Often pond crawfish are available to customers as early as November, several months ahead of the competitive wild crop. Typically, most crawfish harvesting and processing ends by the middle of June. There have been years when wild crawfish production has lasted considerably longer. The season depends on water conditions and other factors. Huner (1994) discusses in great detail the life cycles and culture of commercial freshwater crawfish.

Both wild and cultured crawfish are harvested using wire-baited traps. This procedure is highly labor intensive and has been estimated to make up 60 to 80 percent of the annual farm operating costs (Dellenbarger et al., 1987). Among other responsibilities, harvesters are required to check traps every day. During this "running of the traps", the crawfish are removed and traps rebaited. Although small, motorized boats are used to assist in these activities, harvesting is quite labor intensive. Traps placed by a single harvester may number in the hundreds. Huner (1994) suggests as many as 50 to 75 traps per hectare (2.471 acres) are required to harvest crawfish effectively. Crawfish are attracted to several types of bait; low commercial value fish or pieces of fish such as shad or menhaden are commonly used. Recently, a manufactured pellet bait has been successfully used.

Unlike many commercial aquatic species, there are currently no restrictions on the quantity or size of crawfish that may be harvested from ponds or wild stock in Louisiana. Fishing pressure has not been shown to be a factor in native populations. In fact, there is evidence that whole populations may have stunted growth and may decimate large, vital, vegetated areas when fishing pressure is inadequate.

It is important that certain precautions be taken during harvesting to prevent contamination of the crawfish (Moody, 1989). For example, vessels should be cleaned regularly and crawfish, both sacked and unsacked, should not come in contact with bilge water, fuel, oil, or other foreign materials associated with the operation of a powered vessel.

Several studies have shown that pesticides or heavy metal contamination have not been identified as a problem with crawfish and crawfish products (Madden et al., 1991; Finerty et al., 1990). However, crawfish aquaculture activities should be conducted only in areas free of persistent chemical contamination.

HANDLING HARVESTED CRAWFISH

Upon being harvested, crawfish are alive and active. Traditionally, harvesters sack crawfish in large vegetable or onion sacks. These sacks are pliable, and the mesh material has many advantages. The sacks can be easily handled and stacked. Each sack holds nearly 50 pounds of crawfish. More important, the mesh sack allows for tight packing of the crawfish. This restricts the movement of this naturally aggressive animal, and allows buyers to observe the size and condition of the crawfish. In addition, these sacks are surprisingly good for preventing escape. They come in a variety of colors and are used by some producers to color-code grade sizes. The mesh sack also allows quick penetration of moisture and makes it easier to maintain conditions that promote keeping crawfish alive and in good condition.

Moody (1989) discusses factors conducive to maintaining live crawfish in good condition. When live crawfish are held out of water successfully, they must be kept cool and moist, and have plenty of fresh air. Many processors and harvesters hold sacked crawfish in coolers at temperatures between 40° to 50°F (4° to 10°C). Cooling crawfish quickly removes heat and slows metabolism, minimizing stress and extending shelf life. The sacks are stacked horizontally on pallets. Processors and harvesters take care to limit the number of sacks stacked on top of each other. Early in the season when shells are thin and tender, excessive stacking could crush or injure the crawfish.

The sacks are also stored to facilitate adequate air circulation around the sacks — an extremely important factor. Some processors have placed low velocity fans in their coolers to improve air circulation. There have been reports that crawfish mortality is higher in sacks that have a solid label attached to a portion of the sack, which restricts penetration of oxygen. Others have also found it useful to place a thin layer of crushed or flaked ice on top of sacks of crawfish placed in coolers. As the ice melts, the cool water drips through the sack onto the crawfish, keeping delicate gills moist. As far as practical, air humidity should be at or near 100 percent. Live crawfish coolers should be used only for the storage of live crawfish. They must be kept clean and free from odors by regular and scheduled sanitation and maintenance. For sanitation reasons, bait for traps should be stored in separate coolers.

Crawfish in good condition can be stored in coolers for several days prior to processing. McClain (1994) stored live crawfish at 4°C for six days and evaluated the results. The mortality during cold storage was 6.88 percent, 11.07 percent, and 19.94 percent for the control group, graded crawfish, and purged and graded crawfish, respectively.

Crawfish that have undergone stress prior to storage can have a remarkably short shelf life. Stress can come from several sources. Harvesting on hot days with little protection for the crawfish can increase mortality. Improper transport of sacked crawfish to storage coolers can cause losses. Transporting crawfish in open-air trailers or truck beds can cause dehydration and should be avoided. One of the most stressful situations for crawfish is one over which the harvester has little control: low oxygen in harvesting waters. Crawfish can die quickly when held in traps with poor dissolved oxygen conditions. There have been reports of crawfish dying *en masse* from this type of stress shortly after having been received at the processing facility. A little-understood type of stress is created when crawfish go through mass molting while in the sacks, but not much can be done to control this situation (Romaire, 1994).

It is interesting to note that as the quantity of crawfish increases in the spring, the quality also improves. The overall quality of crawfish is judged on several parameters. As was previously mentioned, the yearly life cycle of the crawfish has a profound effect on its growth patterns and activities. In the early season, crawfish are small but are growing rapidly. During this period the exoskeleton can be soft and is easily damaged by rough handling or stacking. As the season progresses and the crawfish become more mature, the shell becomes thicker and harder. Late season crawfish exoskeletons can be extremely hard, making peeling and handling difficult. Many times the ideal exoskeleton toughness is achieved during the peak harvesting times of April and May. When processors note crawfish shell hardening, it is a sign that the season is ending. Meat can also vary in quality and condition at this time. Generally, the meat fills out the tail segments of the exoskeleton during most of the harvest season, unlike some crustaceans such as blue crab where "skinny" crabs can be a problem, even during the peak harvest season. One exception, however, are the old "holdovers" that are sometimes caught early in the season. These crawfish may have so little meat in the tail that they are referred to as "hollow tails." There is no published data on the changes in the texture of the meat throughout the season. Marshall and Moody (1985) showed that the amount of adhering hepatopancreas also varied with the season. Adhering hepatopancreas is defined as the hepatopancreas that is removed during the hand motions required to extract the meat. It is desirable to have ample hepatopancreas to pack with the peeled meat. The findings showed that the amount of hepatopancreatic tissue increases with the progression of the wild crawfish season. The range of adhering hepatopancreas tissue ranged from 2.2 to 13.1 percent throughout the season with a seasonal average of 8.14 percent. There have been no studies on the flavor changes that occur in hepatopancreas over the season. Many consumers will confirm that near the end of the season, the hepatopancreas will develop a stronger and more undesirable flavor, as well as a darkened color.

PROCESSING FRESHWATER CRAWFISH

Moody et al. (1988) compiled information on seafood processing facilities in Louisiana. Of 670 processing permits issued by the state, 130 were for processing crawfish. Family-owned businesses are by far the most common. The size and capacity of crawfish processing businesses vary greatly.

Some may process only a few hundred pounds of live crawfish daily, while others may process upward of 50,000 pounds a day. A typical plant probably has the capacity to process between 10,000 to 15,000 pounds a day. As noted earlier, because the supply of crawfish is season-dependent, most plants process crawfish from January or February until late May or mid-June. Between seasons the plant must either close or process another product using similar processing techniques, such as blue crabs.

Most crawfish processing plants have five general physical areas: the live crawfish storage area, the cooking room, the peeling room, the packaging room, and the finished product coolers and freezers. Each area will be discussed in the following sections.

Live Storage Area

Most considerations for maintaining crawfish alive have been discussed previously in detail. The live storage areas may be within the confines of the plant or may be completely detached. In some cases, refrigerated trailer tractor rigs have been used for storage. Upon receiving, the live crawfish are weighed in the sack and a weight tag is attached for future reference in the event of sack sales to consumers. Crawfish are generally received by the processor in the afternoon of the same day of harvest; they may be placed in dry cooler storage for later processing or they may be processed immediately.

Purging

True crawfish purging is a technique that is used to remove the content of the entire gut. An intestine runs the length of the tail meat and, in freshly harvested crawfish, is often filled with an unappetizing black material. In the course of processing meat, the intestine, commonly called the vein, is removed by hand along with this dark material. Consumers who buy live crawfish for home consumption sometimes desire purged crawfish, since less gut material is present. McClain (1994) studied the effects of grading, depuration, and time on live crawfish during cold storage. He defines purging as a system where feed is withheld for 24 to 48 hours in captive crawfish. The purging system is an aquatic environment capable of supporting the live crawfish. In this research, a flow-through system was used. Results showed that losses from purging were 11.6 percent. This differed from some earlier studies which suggested that the rate should be lower than 10 percent. Some consumers erroneously believe that a brief soak in saltwater will purge crawfish. This is not an industry practice and has not been shown effective.

Grading

Grading is a recent practice that has become commonplace in most processing facilities. The demand for large crawfish for the whole, cooked, and seasoned Scandinavian market was instrumental in the acceptance of grading. Difficulty in peeling undersized crawfish was also a factor that ushered in grading on a large scale. Moody (1994, 1989) gives an overview of the different types of graders currently used by processors. The widest part of the crawfish hard exoskeleton is the cephalothorax area. Since the growth of this area is representative of the overall growth rate (however, the cephalothorax and claws grow disproportionately faster than the tail) of the crawfish, graders use it as a measuring stick to separate the different sizes. The grading concept most commonly used is a system of variably spaced bars or slots; when the crawfish pass over an opening that will permit the passage of the cephalothorax, the crawfish will fall through. These slots gradually increase in size as the crawfish are moved down the grader; this separates the smaller crawfish.

Overall, the system is fairly efficient for a few grades, but there are some limitations. For example, when crawfish are in a defensive position, with legs and claws extended, they will not fall through the proper slots. Sometimes crawfish clump together and do not grade properly. Gentle sprays of water have been used to calm crawfish and increase efficiency. Another drawback to some of the current graders is the mechanical damage caused by the movement of the machine during the grading process. Several types of graders have been successfully employed in commercial crawfish processing. A large rotating drum constructed of spaced bars is commonly used. Vegetable graders have also been useful for the grading of crawfish. Rollason and McClain (1994) developed a crawfish grader that takes advantage of the natural instincts of the animal; this water-based system allows the crawfish to grade themselves using their own movements and also washes the crawfish prior to sacking.

The crawfish industry has unofficially adopted five basic size grades of crawfish, as shown in Table 1. The evolution of grading has been of considerable benefit not only to the industry but to consumers as well. Sizing and pricing permit customers to tailor those requirements to specific needs.

Table 1. Grading of Crawfish

Jumbo	15 or fewer per pound
Large	16 to 20
Medium	21 to 25
Peeler	26 or more per pound
Field run	a random count or ungraded lot of crawfish

Washing and Inspecting

Crawfish are not washed or cleaned on the harvesting vessels, and consequently, they may be sacked with unwanted debris such as bait or pond vegetation. Washing removes this material as well as dirt and soil picked up during harvesting. A wash tank similar to those used to de-ice shrimp seems to produce good results. Sacks of crawfish are cut open and emptied into one end of the wash tank; the live crawfish immediately settle to the bottom where a conveyor belt gently moves them through the tank and lifts them out at the opposite end. As the crawfish emerge from the water, they can be easily inspected for cleanliness and for any dead or injured specimens. Not all processors use this system. Sometimes the crawfish are emptied directly into cooking baskets and washed with running water to remove dirt and debris.

A crusty layer of brown-to-black material is sometimes encountered in mid-to-late season on large, mature crawfish. These crawfish are often intended to supply the whole, cooked market. This defect has been a major problem for the industry. There is no published data on procedures to remove the stain. This is an area where further research is needed.

Cooking

For crawfish meat to be extracted, it is necessary to cook the whole, live crawfish (Figure 2). Although attempts have been made in the past to remove meat from raw or semi-cooked crawfish, an undesirable meat mushiness generally develops in the meat. This texture can be directly attributed to enzymatic action. Sufficient cooking will minimize the action of these naturally occurring proteolytic enzymes. The degree of cooking, however, is extremely important. As noted, undercooking will result in meat with poor texture. On the other hand, overcooking may result in crawfish that are difficult to handpeel. Overcooked crawfish meat may not slip from the shell as easily as properly cooked crawfish, and the intestine running the length of the meat has a tendency to break rather than to slip easily from the meat.

Naturally occurring proteolytic enzymes in the hepatopancreas of crawfish have been isolated, purified, and characterized, and their properties have been evaluated (Kim et al., 1994, 1992). The use of these enzymes as a commercial extract has been shown to improve flavor extractability from a crab processing by-product (Kim et al., 1994a).

Cooking procedures used by the industry can vary greatly. When crawfish processing began years ago in Louisiana, the typical plant used a natural gas flame to heat water in a cooking vessel. Live crawfish were added directly to the boiling water for a pre-determined period of time. Today, this same procedure may be used, although utilizing more sophisticated flame spreaders and a more efficient cooking vessel. In many plants today, steam-heated water is used instead to cook

Figure 2. Boiling whole crawfish for peeling.

crawfish. Steam may be injected directly into the water using a spreader. Steam-jacketed kettles have also become popular in some processing facilities. The water used to cook crawfish is not seasoned and is generally used more than once. Depending on the cleanliness of the crawfish and other factors, cooking water should be routinely changed. Crawfish to be cooked in boiling water are placed live into cooking baskets; these baskets are designed to fit snugly into the cooking vessel. A large cooking basket may hold several hundred pounds of crawfish. The cooking vessel may hold several baskets at a time. Prior to hoisting the basket into the boiling water, the crawfish may be given a final washdown. During the cooking process, the crawfish must be completely submerged under boiling water. Many processors will periodically stir the crawfish mass to promote even heating. Upon completion of cooking, the basket is hoisted from the boiling water and emptied for cooling. Many baskets are outfitted with a simple bottom-release mechanism so the hot crawfish can be safely discharged.

Not all crawfish are cooked in boiling water. Some facilities have opted to use steam chambers for heat processing. There can be some advantages to this procedure: boiling crawfish in water is a batch system, whereas crawfish cooked on a conveyor belt going through a steam chamber is a continuous system. There is little published yield or microbiological quality data comparing the various cooking procedures discussed. Baskin and Wells (1990) conducted a study comparing the effects of a modified commercial steamer versus conventional boiling methods on crawfish meat. Although a decrease in yield would be expected, the meat yield from the steamed meat actually increased. There was little difference in the quality of the meat from a physical or microbiological standpoint. One of the biggest advantages to the system was efficiency. With the steaming process, cooking time was shorter, and a computer simulation showed that a 350 percent increase in throughput was achievable.

Marshall (1988) conducted the most comprehensive study on processing factors that contribute to the quality of crawfish meat to date. This work examined many factors related to the processing of crawfish in Louisiana, including cooking times and freezing methods. Crawfish meat characteristics such as texture, color, flavor, microbiological quality, and drip loss were evaluated. Marshall et al. (1987) examined color, texture, and flavor of both crawfish species. An extremely important part of this study was an initial evaluation of the proper blanching or cooking time for whole, live crawfish. Over- or undercooking problems have been previously discussed. Many factors contribute to inconsistencies in cooking time, including initial temperatures of product, blanching water-to-product ratios, heating capacities, and size and density of the product. Time-temperature measurements should be part of any heat treatment involved in the processing of food. In addition, tests that verify the desired results should be used when available. Marshall et al. (1987a) developed a simple test using gelatin to confirm enzyme deactivation in blanched crawfish. This is a useful test for processors since it clearly and visually demonstrates at which point the enzymatic activity of the hepatopancreas has been minimized. The test is easily conducted, and most plant personnel can be trained in the procedures. The equipment required for the test is readily available.

The cooking step in crawfish processing should be monitored closely and recorded. Under the current guidelines and concepts of HACCP (Hazard Analysis Critical Control Point), the cooking step for most seafood is considered to be a Critical Control Point (CCP). There is a need to further evaluate the cooking or blanching step in crawfish processing. Parameters such as yield, microbiological quality, and shelf life should be studied as they relate to the various commercially-available cooking techniques.

Cooling

Cooked crawfish are generally cooled only long enough to be safely handled before the meat is extracted by hand. Usually this cooling is done in the open air. Because of this relatively short period, refrigeration is not required to accelerate the cooling process. Some processors, however, may use a water spray to speed cooling. The cooking operation is timed so that the crawfish are peeled at approximately the same rate that they are cooked. Cooking faster than the peelers can work will create a bottleneck in the cooling operation and an accumulation of cooked product waiting for peeling. The holding of warm, cooked product for an extended period is undesirable under most circumstances. Recent FDA Hazard Analysis and Critical Control Point (HACCP) policies concerning cooling times and temperatures require that crawfish processors carefully monitor post-cooking cooling procedures. Crawfish processors may have as little as two hours to peel, package, and refrigerate finished product. Processing times depend upon temperature of the post-cooked crawfish when touched by workers or food-contact surfaces. Obviously, the higher the temperature, the shorter the time period for post-cooking processing. Consequently, it is recommended that crawfish processors implement accelerated cooling assistance technology prior to touching the product. Procedures utilizing water spray and/or ice show promise in removing cooking heat to permit additional processing time. Post-cooking cross contamination is a major consideration at this step. Employees handling live crawfish or other contaminated materials should not be in contact with equipment or personnel associated with the cooked product. In fact, the discharge of cooked product from the cooking baskets should be physically separated from the live handling area and its employees. Only employees approved to handle cooked and finished product should be allowed to handle the post-cooked crawfish. When sufficiently cooled, the crawfish are delivered to the peeling tables for meat removal. Precautions should be taken to ensure that this transfer is accomplished in a sanitary and efficient way. All equipment involved in the process should be clean and sanitary, and employees should be trained in all aspects of food handling procedures (Figure 3).

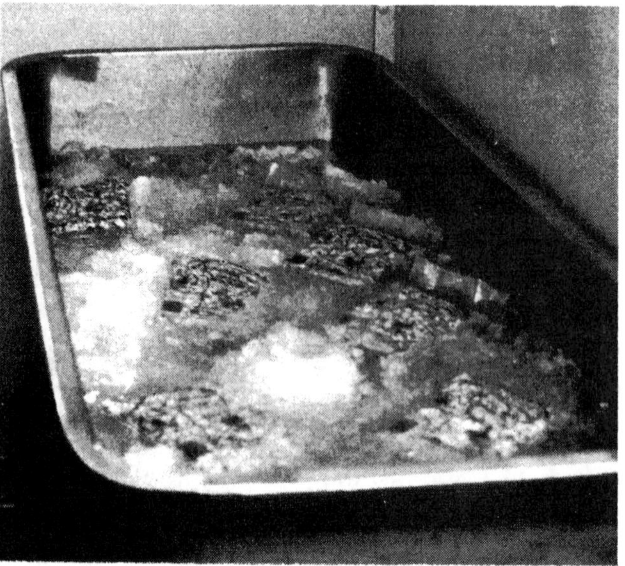

Figure 3. Ice slushing packaged meat.

Peeling

There has been little change in the methods of crawfish meat removal since the inception of the industry. It is basically a manual operation performed by skilled workers. Using photographs, Moody (1980) illustrates the technique that is generally employed by Cajun workers. The establishment and success of a crawfish processing facility depend on a large and stable work force familiar with meat removal techniques. There have been numerous attempts to develop and market mechanized peeling devices; but, to date, there has been no general acceptance by the industry. Moody (1994, 1989) gives a brief history of some of the techniques that have been employed by the industry. Basically, three types of peeling machines have been developed. Each is based on a different principle. The Rutledge peeler (Rutledge, 1978), patented in 1978, uses a roller device to remove the meat from previously frozen and thawed raw crawfish. Another device uses air pressure to force the meat from the shell. A pressure vessel has also been used to assist in the peeling of crawfish meat. This method cooks the crawfish under pressure and then quickly releases the pressure. The resulting steam loosens the shell from the meat. Development of a mechanical peeling technique is a major consideration for the continued viability of the crawfish industry. The manual labor required to remove the meat from crawfish is the largest cost factor in the final meat price. Developing a mechanized peeler that would produce meat quality equal to that of handpeeled and that is acceptable to the industry would be a major achievement.

There can be significant variations in the yield of meat and its quality throughout the crawfish season. Generally, meat yield should approach 15 percent. This figure includes the hepatopancreas (the edible, liver-like organ that is peeled and packaged with the crawfish meat). One of the factors that greatly influences the meat yield is crawfish maturity. As a general rule, small or immature crawfish have a higher meat yield than do large mature crawfish, probably because crawfish grow disproportionately. As a crawfish matures, the cephalothorax and claws grow faster than the meat-containing tail section. Huner (1988) showed that immature crawfish had meat yields that were consistency 4-5 percent higher than mature crawfish. Ironically, processors will not peel crawfish that are excessively small because of the increased labor required. The very large mature crawfish are most often graded out and sold live to consumers or are marketed as a whole, cooked product intended for the international market.

Huner (1988) also showed that the species being processed can affect meat yield. Of the two commercially important crawfish species in Louisiana, *Procambarus clarkii* had a meat yield 3 to 5 percent higher than *P. zonangulus*. Other factors such as seasonal variations, condition of the crawfish, and processing techniques used can also affect yield. For example, Marshall and Moody (1986) found that cooking times and meat removal methods used by workers can influence overall yield. These yield variations, however, may be attributed more to the amount of adhering hepatopancreas rather than to actual meat yields. Marshall and Moody (1985) showed that cooking procedures, peeling methods, and season affected hepatopancreas yield. When proper cooking times were employed, hepatopancreas yields were the highest. This study also showed that, as the season progressed, there was a slight trend for increases in hepatopancreas yield. By comparison, Huner (1988) suggests that the overall hepatopancreas yield will decrease as the season progresses because the percentage of low-yield, mature males in the population increases. Environmental conditions, cultural and management practices, and meteorological changes may all affect any one season's crop.

Most crawfish processing plants have a separate room dedicated to handpeeling of meat from whole, cooked crawfish. The typical crawfish peeling room is furnished with one or more peeling tables. These peeling tables are long and narrow and are generally made of stainless steel or aluminum and are constructed in such a way as to be easily cleaned. Workers line either side of the table. The number of peelers depends on the quantity of crawfish to be processed. Crawfish are delivered to the peeling table from the cooling step. Crawfish are placed on the table so that peelers can selectively peel the crawfish from the pile. Generally, only enough crawfish are piled onto a table to occupy the peelers for a short period of time. To maximize the quality of the finished product, it is also recommended that all crawfish from the pile be peeled before additional crawfish are placed on the table.

After removing the meat by hand, workers place the single piece of tail meat into a cleaned and sanitized colander. Meat is placed in the colander until it is nearly full or until a predetermined period has lapsed. Meat colanders are then collected from tables and weighed, and newly cleaned and sanitized colanders are given to the workers as replacements to restart the procedure. Most of the edible meat of crawfish is located in the strong, muscular tail, with a certain portion of that meat extending into the cephalothorax. Properly peeled crawfish meat will be in one solid piece and is removed along with the naturally adhering hepatopancreas from the cephalothorax. The unappetizing intestine runs the length of the meat near the dorsal side. Workers remove the intestine as part of the peeling process. An excessive number of intestine parts in the final product is considered a defect. There is additional meat in the claws. However, because of the difficulty of removal, it is not practical to remove it by hand. The use of a mechanical deboner may have some application in recovering claw meat. The texture and color of claw meat is quite different from tail meat. Claw meat may have some application as an ingredient in certain types of prepared dishes.

Waste Management

A major consideration for all crawfish processing facilities is the large amount of waste product generated. As mentioned earlier, the meat yield from whole crawfish is only about 15 percent. This means 85 percent of the crawfish must be discarded as waste material. Disposal of this material is becoming an increasingly important consideration. There have been several research efforts to assess the potential usefulness of crawfish waste by-products for commercial applications. Lovell et al. (1968) determined the compositional profile of crawfish meal. There may be some application for the use of these meals as a supplement in animal feeds. Rutledge (1971) provided a procedure for reducing the amount of ash in the dried meal. Crawfish shells have been shown to have significant concentrations of astaxanthin pigment (Chen and Meyers, 1983; Meyers and Bligh, 1981). The oil-extracted crawfish pigment has been used in the aquaculture of finfish for enhanced coloration. No et al. (1989) demonstrated that crawfish shells are an excellent source of chitin. On a dry-weight basis, the chitin content of the shell waste is 23.5 percent.

The development of commercial flavor compounds from crawfish waste products has potential for an economically viable food ingredient. Hseih et al. (1989) analyzed boiled crawfish and hepatopancreas for its volatile components, and more than 100 compounds were identified. Vejaphan et al. (1988) also characterized volatile flavor components from boiled crawfish; a wide variety of aromas were detected. Tanchotikul (1989) and Cha et al. (1992) examined the volatile flavor components from crawfish processing waste, and many compounds were isolated.

As discharge and waste treatment requirements become more restrictive, use of crawfish byproducts should increase.

Packaging

From the peeling room, the freshly-peeled meat is delivered to a separate packaging room where the colanders are emptied and the meat is inspected for defects such as bits of shell or pieces of intestine. It is important that the packaging step proceed at the same rate as peeling room delivery. If the packaging room cannot keep up with the meat being delivered, then a bottleneck may occur where meat may be held for an unacceptably long period before packaging (Figure 4). Other than inspecting, no other processing step is required prior to placing the meat into its final package. There are no food additives or preservatives traditionally added to the product. Generally, packaging is a hand operation. Individual bags (usually designed to hold one pound of crawfish meat) are hand-filled, weighed, and sealed. The bags are usually flexible pouches made of an array of materials, depending on the intended final use of the product. Under the provisions of HACCP, crawfish processors must consider *Clostridium botulinum* as a potential hazard in packaged crawfish meat. Fresh crawfish meat is generally packaged in an air-permeable material. Meat for freezing may be vacuum-packaged. Meat to be sold fresh may be packaged in polyethylene bags. Air is pressed out of the bags by hand before heat-sealing. Sealed bags are immediately submerged in slush ice to chill the final product quickly. For extended storage, the chilled, fresh product may be iced and boxed. The time required for processing fresh crawfish from the time the crawfish are cooked until the final product is chilled to near 32°F may take less than two hours. The final shelf life of fresh crawfish meat will obviously depend on the initial quality of the product and the conditions under which it is held. It is estimated that meat of good microbiological quality can be held from seven to 12 days (Moody et al., 1994; Moody, 1989).

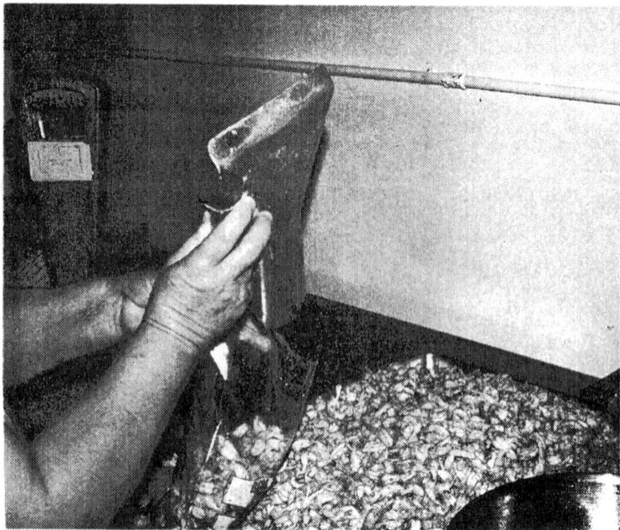

Figure 4. Packaging crawfish meat.

There have been several studies on the microbiological quality of peeled crawfish meat (Grodner and Novak, 1975; Lovell, 1968). Since crawfish are processed under similar conditions as other cooked crustaceans, such as blue crab, microbiological guidelines should be similar. The presence of pathogens in cooked shellfish is always a concern. Several studies examined processing conditions and the growth of pathogens in crawfish meat (Dorsa et al., 1993; Dorsa et al., 1993a; Ingham, 1990). Because crawfish are cooked and are handled in several processing steps after cooking, there is an opportunity for cross-contamination. Consequently, processors should closely monitor each processing step to minimize the opportunity for microbiological contamination.

Gerdes et al. (1989) evaluated the effects of carbon dioxide (CO_2) on the shelf life of fresh crawfish meat under storage. The data showed that crawfish packaged with 80 percent CO_2 and 20 percent air had a lower microbiological count and better scores on certain chemical parameters than those samples stored with 100 percent CO_2 or 100 percent air. The quality of the 80/20 product was comparable to fresh meat.

Crawfish meat can also be successfully frozen. There are, however, several considerations when producing frozen crawfish meat. The hepatopancreas has been associated with oxidative rancidity in frozen crawfish meat and its resultant undesirable off-flavor (Lovell, 1968). Consequently, steps should be taken to minimize those conditions that would contribute to the development of rancidity. Some processors remove the hepatopancreas prior to freezing; it is easily removed with agitation and clean water. Some processors feel that frozen crawfish meat can be safely packaged with the hepatopancreas, provided that the package is vacuum sealed to remove oxygen and that special laminated freezer bags are used to prevent the intrusion of air. Another common problem associated with frozen crawfish meat is a darkening or bluing of the meat after it has been thawed and reheated in the final product preparation. Moody and Moertle (1986) evaluated several chelating agents for their effectiveness in eliminating the problem. Frozen crawfish is becoming increasingly important to the success of this industry. In view of the seasonal nature of the animal and the year-round demand for the product, freezing crawfish is a consideration for many processing facilities. Crawfish meat from other nations is generally frozen with good results. Marshall (1988) conducted a comprehensive study on the microbiological quality, freezing methods, frozen storage temperatures, texture changes, and drip loss on frozen crawfish meat. This work is an essential reference for crawfish processing.

OTHER FRESHWATER CRAWFISH PRODUCTS

Although fresh-peeled crawfish meat has been the primary finished product of the industry since its inception, other important crawfish products have evolved in recent years. One of the most significant has been **cooked, whole, frozen crawfish.** One of the largest markets for this product has been Scandinavia, where the consumption of the product is tied to tradition. Other local and international producing areas are not able to meet the demand; consequently, the Louisiana industry has geared up with innovative technology and processing techniques to produce a high-quality whole, cooked, and frozen crawfish. Only the largest and most select crawfish can be used for this product. All legs and claws must be attached, and the animals must be thoroughly cleaned. Although techniques can vary in production practices, the process usually involves cooking the whole, live animal with steam or in boiling water and chilling. The boiling water may or may not be seasoned with salt, dill, or other specified ingredients, depending on the destination of the final product. The whole, cooked crawfish are hand-packed into plastic trays (1 kilogram is a common pack); additional chilled seasoned water is added to immerse the crawfish completely, and a lid or cover is sealed air-tight over the tray. The tray is quickly frozen after packaging. Currently, most packers of the product use cryogenic freezing to increase efficiency and to maximize quality. Cole and Kilgen (1987) and Godber et al. (1989) confirmed the higher quality of cryogenically frozen whole crawfish when they evaluated the various freezing techniques used to package this particular product. Their findings show that cryogenically frozen whole, cooked crawfish were superior to all other freezing methods tested. Because the cryogenically frozen samples retained the most moisture and had the least expressible moisture and high degree of extractable meat, it appeared that the least amount of freezer damage occurred in these samples. The samples that were "still"-frozen at -23°C showed the most damage. In a continuing study of whole, cooked crawfish, Cole and Kilgen (1988) determined that cryogenically and conventionally frozen products were acceptable to a taste panel throughout a 48-week test. Plants that pack whole, cooked crawfish must process relatively huge quantities of product in a short period because of the extremely limited season and availability of acceptable product. Consequently, the equipment and methods that are employed generally involve a continuous system rather than a batch system.

Figure 5. Molting tray for soft-shell crawfish.

Figure 6. Tray-packaged soft-shell crawfish.

Soft-shell crawfish has also been established as a product of commercial significance. Unlike other crawfish products, soft-shell crawfish are offered raw, alive or frozen, for cooking by consumers or restaurants. Moody and Culley (1991) completed an extensive study on the processing parameters of soft-shell crawfish. Like most crustaceans, crawfish periodically molt the hard exoskeleton in order to grow (Figure 5). When freshly molted, the crawfish exoskeleton is entirely soft and the whole animal can be made into a cooked product. The process begins when hard crawfish are placed into special molting trays filled with recirculating water. Over time the crawfish will molt and must be removed quickly before the exoskeleton begins to harden. Once removed from the molting trays, the soft-shell crawfish are placed in refrigerated water at 2°C until packaging. The most common method for preserving soft-shell crawfish is by freezing. The typical procedure is by packaging in water and using conventional freezing methods such as blast or still-air freezing. Because of the low volume of production, cryogenic freezing is not generally used. Properly cleaning the soft-shell crawfish prior to preparation for cooking is essential. This is generally done before packaging and freezing (Figure 6). Moody (1994) states that the stomach stones or gastrolith (there are two in each crawfish) are removed by making a cut just behind the eye stalks to expose the stones, which can then be easily pushed out. It is important that these stones be removed completely since they are hard and could cause injury to consumers. There are several options for cleaning and preparing soft-shell crawfish prior to cooking. Most involve the presentation of the cephalothorax or the removal of some organs such as the hepatopancreas.

Currently, there has not been a developmental effort for a **commercially canned** or **pasteurized crawfish** meat product. One reason is that frozen crawfish meat, when properly prepared and packaged, is an excellent product that most customers have readily accepted. This area of research should be evaluated thoroughly.

Another potential area is preservation of crawfish meat by **irradiation**. Although there are no commercial facilities currently treating crawfish with irradiation, research should be conducted to determine its feasibility.

HACCP CONCEPTS AND CRAWFISH PROCESSING

A major consideration for all seafood processors is product safety and quality; this is especially true of seafood products that are cooked ready-to-eat. The Hazard Analysis Critical Control Point (HACCP)-based system has found favor with regulatory agencies, industry associations, and other groups as being a concept that will establish a means for ensuring food safety and preventing problems from occurring. In December 1995, the United States Food and Drug Administration (FDA) promulgated sweeping new HACCP-based, mandatory seafood inspection regulations (21 CFR Part 123). HACCP requires that seafood processors, in this case crawfish processors, be knowledgeable in the areas of food safety, sanitation, and technology. When using a HACCP-based system, processors must determine the hazards involved in specific products and the likely locations to control the hazards, called Critical Control Points (CCPs). Monitoring of the CCPs requires the use of instruments that measure both time and temperature. These data must be recorded at the time of the measurement and maintained as a permanent record for FDA inspections. Since the enforcement of HACCP regulations beginning in December 1997, the crawfish industry has seen some changes in processing procedures. As previously mentioned, the industry currently uses oxygen-permeable bags for packaging fresh meat to prevent the growth of *Clostridium botulinum* in fresh meat. Vacuum packaging in oxygen-impermeable bags is still commonly used in frozen meat since *C. botulinum* cannot grow at freezing temperatures. More processors are also using cooling-assisted techniques for post-cooked crawfish to provide more processing time. Guidelines for post-cook processing times for cooked, ready-to-eat foods such as crawfish meat are provided by the FDA. Crawfish processors must also monitor sanitation procedures and conditions under the provisions of the Sanitation Standard Operating Procedures (SSOPs). These are eight key areas of sanitation that are extracted from the Good Manufacturing Practices (GMPs) (21 CFR 110).

REFERENCES

Avery, J. 1994. Personal communication.

Baskin, G.R. and J.H. Wells. 1990. Evaluation of alternative cooking schemes for crawfish processing. *Journal of Shellfish Research* **9**:389-393.

Cha, Y.J., H.H. Baek, and T.C. Hsieh. 1992. Volatile components in flavor concentrates from crayfish processing waste. *J. Sci. Food Agric.* **58**:239-248.

Chen, H.M. and S.P. Meyers. 1983. Ensilage treatment of crawfish waste for improvement of astaxanthin pigment extraction. *Journal of Food Science* **48**(5): 1516-1520 & 1555.

Cole, M.T. and M.B. Kilgen. 1988; 1987. Characterization of the quality and shelflife of whole frozen crawfish. The Gulf and South Atlantic Fisheries Development Foundation Inc., Tampa, FL.

Dellenbarger, L.E., L.R. Vandeveer, and T. M. Clarke. 1987. Estimated investment requirements, production costs, and break-even prices for crawfish in Louisiana, 1987. Department of Agricultural Economics and Agribusiness Research Report No. 670, Louisiana State University Agricultural Center, Baton Rouge.

Dorsa, W.J., D.L. Marshall, M.W. Moody, and C.R. Hackney. 1993. Low temperature growth and thermal inactivation of *Listeria monocytogenes* in precooked crawfish tail meat. *J. Food Prot.* **56**(2):106-109.

Dorsa, W. J., D.L. Marshall, and M. Semien. 1993a. Effect of potassium sorbate and citric acid sprays on growth of *Listeria monocytogenes* on cooked crawfish (*Procambarus clarkii*) tail meat at 4°C. *Lebensm.-Wiss. u.-Technol.* **26**:480-482.

Finerty, M.W., J.D. Madden, S.E. Feagley, and R.M. Grodner. 1990. Effect of environs and seasonality on metal residues in tissues of wild and pond-raised crayfish in Southern Louisiana. *Arch. Environ. Contam. Toxicol.* **19**:94-100.

Gerdes, D.L., J.J. Hoffstein, M.W. Finerty, and R.M. Grodner. 1989. The effects of elevated CO_2 atmospheres on the shelf-life of freshwater crawfish (*Procambarus clarkii*) tail meat. *Lebensm.-Wiss. u.-Technol.* **22**:315-318.

Godber, J.S., J. Wang, M.T. Cole, and G.A. Marshall. 1989. Textural attributes of mechanically and cryogenically frozen whole crayfish (*Procambarus clarkii*). *J. Food Sci.* **54**(3):564-566.

Grodner, R.M. and A F. Novak. 1975. Microbiological guidelines for freshwater crawfish (*Procambarus clarkii*, Girard). In *Freshwater Crawfish.* Papers from the Second International Symposium on Freshwater Crawfish. J.W. Avault (ed.), LSU Division of Continuing Education, LSU, Baton Rouge, LA.

Hsieh, T.C., W. Vejaphan, S.S. Williams, and J.E. Matiella. 1989. Volatile flavor components in thermally processed Louisiana red swamp crayfish and blue crab. In *Thermal Generation of Aromas.* T.H. Parliment, R.J. McGorrin and C.T. Ho (eds.), American Chemical Society, Washington, DC.

Huner, J.V. 1988. Comparison of the morphology and meat yield of red swamp crawfish and white river crawfish. *Crawfish Tales* **7**:2.

Huner, J. V. 1989. Overview of international and domestic freshwater crawfish production. *J. Shellfish Res.* **8**:259-266.

Huner, J.V. 1994. Cultivation of freshwater crayfishes in North America, Section I: freshwater crayfish culture. In *Freshwater Crayfish Aquaculture in North America, Europe, and Australia,* J.V. Huner (ed.), pp. 5-89. Food Products Press, New York.

Ingham, S.C. 1990. Growth of *Aeromonas hydrophila* and *Plesiomonas shigelloides* on cooked crayfish tails during cold storage under air, vacuum, and a modified atmosphere. *J. Food Prot.* **53**(8):665-667.

Kim, H.R., H.H. Baek, S.P. Meyers, K.R. Cadwallader, and J.S. Godber. 1994. Crayfish hepatopancreatic extract improves flavor extractability from a crab processing by-product. *J. Food Sci.* **59**(1):91-96.

Kim H.R., S.P. Meyers, and J.S. Godber. 1992. Purification and characterization of anionic typsins from the hepatopancreas of crayfish, *Procambarus clarkii.* Comp. Biochem. Physiol. **103B**(2):391-398.

Kim H.R., S.P. Meyers, J.H. Pyeun, and J.S. Godber. 1994a. Enzymatic properties of anionic typsins from the hepatopancreas of crayfish, *Procambarus clarkii. Comp. Biochem. Physiol.* **107B**(2): 197-203.

Lovell, R. 1968. Development of a crawfish processing industry in Louisiana. U.S. Dept. of Commerce,

Economic Development Administration, Technical Assistance Project.

Lovell, R.T., J.R. Lafleur, and F.H. Hoskins. 1968. Nutritional value of freshwater crayfish waste meal. *J. Agric. Food Chem.* **16**:204.

Madden, J.D., R.M. Grodner, S.E. Feagley, M.W. Finerty, and L.S. Andrews. 1991. Minerals and xenobiotic residues in the edible tissues of wild and pond-raised Louisiana crayfish. *J. Food Safety* **12**:1-15.

Marshall, G.A. 1988. Processing and freezing methods influencing the consistency and quality of fresh and frozen peeled crawfish (*Procambarus* sp.) meat. Ph.D. Dissertation. Louisiana State University, Baton Rouge.

Marshall, G.A. and M.W. Moody. 1986; 1985. Department of Food Science, Louisiana State University, Baton Rouge. Unpublished data.

Marshall, G.A., M.W. Moody, and C.R. Hackney. 1987. Differences in color, texture, and flavor of processed meat from red swamp crawfish (*Procambarus clarkii*) and white river crawfish (*P. acutus acutus*). *J. Food Sci.* **52**:1504-1505.

Marshall, G.A., M.W. Moody, C.R Hackney, and J.S. Godber. 1987a. Effect of blanch time on the development of mushiness in ice-stored crawfish meat packed with adhering hepatopancreas. *J. Food Sci.* **52**(6):1500-1503.

Meyers, S.P. and D. Bligh. 1981. Characterization of astaxanthin pigments from heat-processed crawfish waste. *J. Agric. Food Chem.* **29**:505.

McClain, W.R. 1994. Evaluation of grading, depuration, and storage time on crawfish mortality during cold storage. *J. Shellfish Res.* **13**:217-220.

Moody, M.W. 1980. Louisiana seafood delight -- the crawfish. Sea Grant Publication LSU-TL-80-002, Louisiana State University.

Moody, M.W. 1989. Processing of freshwater crawfish: a review. *J. Shellfish Res.* **8**:293.

Moody, M.W. and G.M. Moertle. 1986. An evaluation of the effectiveness of specific chemical compounds on the prevention of discoloration development during the cooling of frozen crawfish meat. Annual Meeting of the Institute of Food Technologists, Dallas, TX.

Moody, M.W. and D. Culley. 1991. Evaluation of soft-shell crawfish processing and packaging parameters. Gulf and South Atlantic Fisheries Development Foundation, Final Report.

Moody, M.W., L.C. Douglas, and W.S. Otwell. 1996. Total quality assurance and Hazard Analysis Critical Control Point (HACCP) manual for crawfish processing. Louisiana Agricultural Center, Louisiana State University.

Moody, M.W. 1994. Cultivation of freshwater crayfishes in North America, Section II, freshwater crayfish processing. In *Freshwater Crayfish Aquaculture in North America, Europe, and Australia.* J.V. Huner (ed.), pp. 91-115. Food Products Press, New York.

Moody, M.W., K. Roberts, and C.D. Harper. 1988. Louisiana seafood processing: an overview. Louisiana State University Agricultural Center, LSU, Baton Rouge.

No, H.K, S.P. Meyers, and K. S. Lee. 1989. Isolation and characterization of chitin from crawfish shell waste. *J. Agric. Food Chem.* **37**:575.

Roberts, K.J. 1995. Personal communication.

Rollason, S.H. and W.R. McClain. 1994. Development of a water-based grading apparatus for live crayfish. Louisiana Agricultural Experiment Station, LSU, Baton Rouge.

Romaire, R. 1994. Personal communication.

Rutledge, J.E. 1971. Decalcification of crustacean meals. *J. Agric. Food Chem.* **19**(2):236.

Rutledge, J.E. 1978. U.S. patent 4, 121, 322.

Tanchotikul, U.and T.C. Hsieh. 1989. Volatile flavor components in crayfish waste. *J. Food Sci.* **54**(6):1515-1520.

Vejaphan W., T.C. Hsieh, and S.S. Williams. 1988. Volatile flavor components from boiled crayfish (*Procambarus clarkii*) tail meat. *J. Food Sci.* **53**(6): 1666-1670.

The Molluscan Shellfish Industry

Cameron R. Hackney and Thomas E. Rippen

BIVALVE MOLLUSKS HARVESTED IN THE UNITED STATES

Clams

Clams are bivalve (two-shelled) mollusks found mostly in shallow waters worldwide. They are characterized by two adductor muscles that open or close the shells and siphons (necks) for pumping water. They are found in both fresh and salt water, usually buried in the mud or sand. Although approximately 20,000 clam species exist, most of which are edible, only about 50 are commercially harvested. The principal species harvested and processed in the United States are described in this chapter (Figure 1). [For a more detailed biological description of species, see other chapters in this book. European, as well as Central and North American, molluscan fisheries are covered in a recent National Marine Fisheries Service (NMFS) publication; see NMFS, 1997, in references.]

The Surf Clam (*Spisula solidissima)*

Other common names: skimmer, hen clam, sea clam, giant clam, bar clam.

While surf clams range from the Gulf of St. Lawrence to South Carolina, the center of the fishery is in the Mid-Atlantic region, most notably New Jersey. They are large and somewhat triangular in shape, attaining shell lengths up to 9 inches. Surf clams are found in near-shore areas out to 200 feet. This is a major fishery, producing 63 million pounds of meats in 1996, accounting for nearly half of the clam meat production from all species. They are harvested primarily with large, highly efficient hydraulic dredges (Figure 2). Surf clams are used extensively to produce clam strips and are retailed only in processed form.

Ocean Quahog

Arctica islandica. Other common names: mahogany clam, mahogany quahog, black quahog, black clam.

Figure 1. Clamming.

Figure 2. Surf clam dredging.

Ocean quahogs resemble the common quahog or hard clam with a few exceptions — namely, they have extremely brittle shells, which have a black or dark brown skin-like outer covering (periostracum), and they lack a purplish mark on the inside of their shells. Ocean quahogs are found in European waters and along the Eastern seaboard from Canada to North Carolina, usually in depths from 100 to 250 feet. Ocean quahog harvests have increased and, at 46 million pounds, now account for more than 38 percent of the United States clam meat production. Like surf clams, they enter retail markets as processed products. They are a long-lived species; thus, they may become vulnerable to overfishing in the future.

Figure 3. Hard clam.

The Hard Clam

Mercenaria mercenaria. Other common names: hardshell clam, bay quahog, northern quahog, and quahog.

The hard clam (Figure 3) is the most widely-utilized of the clam species in the United States, although commercial landings (10 million pounds of meats in 1996) are exceeded by the surf clam and ocean quahog harvests. Hard clams inhabit brackish bays as well as ocean environments. They have heavy, rounded, white or light-colored shells, which measure up to 5 inches across and can be completely closed. The inside of each shell is distinguished by a purplish mark and two adductor muscle scars. "Quahog," the name used in New England, is of Native American derivation. Hard clams live in relatively shallow waters and inhabit sheltered bays and coves from Canada to Texas; however, they are found only sporadically north of Cape Cod, Massachusetts.

Hard clams support a substantial recreational and commercial fishery. Unlike the automated surf clam and ocean quahog fishery, commercial harvesting methods are mostly limited to tonging. Recreational fishermen gather hard clams with rakes and hoes, or simply by feeling for clams by probing the bottom with their feet and toes. The minimum legal size varies from state to state, but is usually one inch. Since the early 1990s, the wild harvest has been supplemented with a limited number of aquaculture sources. These enterprises contribute a large percentage of the littleneck (small hard clam) production in some areas. Hard clams grow to market size up to three times faster in warm southern waters than in northern mid-Atlantic waters. Commercial landings in the southern United States often include a related species, the southern quahog, *Mercenaria campechiensis*.

The Soft-Shell Clam

Mya arenaria. Other common names: soft clam, manninose, longneck clam, long clam, steamer, squirt clam, sandgaper, old maid, Ipswich clam, belly clam.

Soft-shell clams are distinguished by their thin, elongated, off-white shells and their long siphons. The siphon, which cannot be fully retracted within the shell, is commonly referred to as the "neck." Soft-shell clams grow to a shell length of 5 inches but are usually harvested before they attain this size. A minimum legal harvest size exists, which varies by state. This clam inhabits fine sand or sandy mud substrates from the intertidal zone to about 30 feet. The soft-shell clam is found from Labrador (Canada) to North Carolina and in a number of scattered locations on the Pacific Coast. Their occurrence is sporadic south of Maryland on the Atlantic Coast.

The Geoduck Clam

Panopea generosa.

The largest of all burrowing clams, the geoduck is found in shallow waters of the Pacific from Southeastern Alaska to southern California. It occasionally attains a shell length from 7 to 9 inches, a weight up to 8 pounds, and a siphon measuring nearly 4 feet when extended. Even the mantle bulges out of the shell, which is always far too small to contain the entire clam.

Razor Clams

Ensis directus and *Siliqua patula*. Other common name: jackknife clam.

The razor clam common to the Atlantic Coast (*E. directus*) does not have a typical clam shape but is long and thin with squared ends. Its appearance is similar to an old-fashioned straight razor. This clam is approximately six times longer than it is wide, and it can grow to 10 inches in length. The shells are covered with a thin, yellowish-brown periostracum. Razor clams inhabit the lower intertidal zone and out to 120 feet. They are found from Labrador to Georgia burrowed deeply in very fine sand or sandy mud bottoms and often live in colonies.

The Pacific razor clam (*Siliqua patula*) is the more important species commercially. It occurs from California to the Aleutian Islands in the intertidal zone of sandy beaches and out to depths up to 180 feet. The Pacific clam possesses an elongated, yellow-brown to brown appearance similar to *E. directus*, except that it is more oval in shape. The periostracum is often peeled and discontinuous in older animals. Pacific razor clams are popular with recreational clam diggers, although clam populations fluctuate significantly.

The Manila Clam

Tapes philippinarium. Other common name: Japanese littleneck.

Manila clams are, at times, an important commercial species on the Pacific Coast although harvests are small compared with the major Atlantic clam fisheries. It is an exotic species in the United States, introduced from Japan. Manila clams inhabit sand and gravel beaches, commonly high in the intertidal zone.

Mussels

Mytilus edulis.

Blue mussels have smooth, dark-bluish shells that are elongated and somewhat pear-shaped. The inside of the shell is pearly violet or white, and quite attractive. Mussels average 2 to 4 inches in length and live in colonies or beds in very shallow waters, predominantly in the rocky intertidal zone. They are distributed worldwide in most polar and temperate seas. Off North America, they occur from Canada to North Carolina (commercially just to New Jersey) and have been transplanted to the Pacific Coast. Mussels are able to attach themselves to nearly any surface with a tough bundle of brown fibers extending from the shell, called byssal threads or byssus, more commonly known as the beard.

As demand increased in the late 1980s, wild populations were supplemented by a growing aquaculture industry. This industry is located in New England, which contributes most of the United States catch. Mussels are cultured on ropes or long lines secured between buoys, and by bottom methods. The industry relies on natural supplies of larval mussels, which set in large numbers on the ropes or other collectors that are placed in late spring and early summer. The mussels are regularly cleaned and thinned as they grow. Despite the growth in aquaculture, the wild fishery experienced a resurgence in the 1990s as methods improved for cleaning and sorting the clams after harvest. Currently, wild sources appear to be holding down prices and limiting the expansion of mussel farms.

Of all the shellfish, mussels are the most efficient and indiscriminate feeders, consuming virtually everything in the 10 to 15 gallons of water they filter each day. Hence, they are very susceptible to pollution and the accumulation of paralytic shellfish toxins. Extensive monitoring of shellfish and their growing waters assures their safety.

OYSTERS

Eastern Atlantic Oysters

Crassostrea virginica. Common names: Eastern, Atlantic, or American oyster.

The Eastern or Atlantic oyster (Figure 4) ranges from the Gulf of Saint Lawrence (Canada) to the Gulf of Mexico. It is more abundant south of Cape Cod, and most of the production occurs in the Gulf Coast states. *C. virginica* currently accounts for the majority of oyster production in the United States. The American oyster has declined in numbers for several reasons, including diseases fatal to oysters (Dermo and MSX), pollution (nutrient enrichment), storms, shoreline development, and natural predators. Oyster production peaked in 1908 with 152 million pounds, whereas only 38 million pounds were harvested in 1996. However, the Eastern oyster remains the predominant oyster species marketed in the United States.

Figure 4. Eastern oyster.

The flavor of oysters varies greatly, depending on the type of algae fed upon and the salinity of the water where they are harvested. Oysters frequently take on the name of their harvesting region. Names familiar on the East and Gulf coasts are Chesapeake, Blue Point, Long Island, Chincoteague, Gulf, and Louisiana. Oysters are often referenced by their origin on restaurant menus.

The color of meat varies with the color of the algae that the oyster feeds upon. The typical color of fresh-shucked oyster meats is cream, tan, or gray. However, depending upon the diet, other colors may be observed, including green and red. These pigments are safe and unrelated to toxic algae blooms, such as "red tides." The red colorant is destroyed (becomes clear) when the oyster is heated to 120°F for a few minutes. Although favored in some foreign markets, the bright pigments are generally considered defects in the United States.

The Pacific or Japanese Oyster

Crassostrea gigas

The Pacific oyster, a transplant from Japan, is found along the West Coast from British Columbia to northern California. It is the second most common oyster in the United States. Production is increasing (nearly 7 million pounds per year), while production of the Eastern oyster is generally decreasing. It grows quickly and possesses a very large shell. Having a greater tolerance to high salinity, it is often bedded far from the influence of rivers. Its natural habitat is the intertidal region. Pacific oysters are an intensively cultured species, with spawning, setting, and growout steps managed by private growers, primarily in the state of Washington.

C. gigas can be differentiated from *C. virginica* by their larger size, dark mantle (referred to as the flavor edge by producers) and white color. In the past, it was recommended that thorough cooking provides the best-tasting Pacific oyster. Pacific oysters can be precooked by dropping them into boiling water and simmering for five minutes. After pre-cooking they are ready for use in typical recipes. As the markets for *C. gigas* have extended eastward, they increasingly are marketed in a manner similar to *C. virginica*. They now are more frequently harvested at smaller sizes and are finding their way into the raw market. As with any muscle food, raw consumption increases safety concerns.

The Western or Olympia Oyster

Ostrea lurida

The oyster native to the Pacific Coast waters is the small species known as the Olympia. It rarely reaches 2 inches in length and is regarded as a true delicacy. Harvesting is restricted and volumes preclude marketing beyond local markets.

The European or Belon Oyster

Ostrea edulis

The European oyster has been successfully transplanted to the Casco Bay area of Maine and is also cultured elsewhere in Maine, in New Hampshire, and on the Pacific Coast. This oyster has a somewhat flattened, wavy-edged shell and a rounded shape. As with the Pacific oyster, the culture of *O. edulis* is highly managed, with the spawning and setting (attachment to a hard substrate) of larval oysters conducted in controlled aquaculture facilities. After setting, the tiny juvenile oysters are transferred to upweller tanks where they are fed a diet of cultured algae. As they grow, they are moved to ocean growout sites, first in trays, then to enclosed mesh nets or bags. The meat is darker than that of Eastern oysters and has a distinct black edge along the perimeter.

SCALLOPS

Scallops are bivalve mollusks with scallop-edged, fan-shaped shells. The shells are further characterized by radiating ribs or grooves (less apparent in sea scallops) and concentric growth rings. The latter are used to determine a scallop's age. Near the hinge, where the two valves meet, the shell is flared out on each side to form small "wings" or "ears." This flaring-out creates a square, straight hinge. Just inside each valve along the edge of the mantle is a row of short sensory tentacles and a row of small jewel-like eyes. The shells are opened and closed by a single, oversized adductor muscle that is sometimes called the "eye." Scallops usually rest on the bottom but are quite mobile, swimming by contracting the adductor muscle.

Unlike clams, oysters, and mussels, scallops are not commonly eaten whole in the United States, and only the marshmallow-shaped adductor muscle is retained in the shucking process. In Europe the muscle with the orange-pink roe attached is considered a delicacy, and this market form is now served in some gourmet restaurants in the United States. Occasionally the white gonad of male scallops is also marketed with the meat. Scallops are primarily harvested by dredging and are shucked soon after capture. Although scallops are harvested in the North Pacific, the traditional commercial resources occur on the East Coast.

The Sea Scallop

Placopecten magellanicus

Sea scallops have large, flattish, saucer-shaped shells that are finely ribbed and grow to 8 inches across. The upper shell is reddish brown or tan, sometimes rayed with white, and the lower valve is pinkish white. The inside of both shells is glossy white and marked by a prominent muscle scar. The marketed muscle meats average 1 to 3 inches across. This scallop inhabits waters from 12 to 900 feet on firm sand or gravel bottoms from Labrador to Virginia. The sea scallop industry is based in Massachusetts and Virginia. Large steel dredges are used in this fishery, usually in pairs. Landings declined in the late 1970s, rebounded in the late 1980s, then declined again in recent years. This species supports the predominant scallop fishery, accounting for more than 90 percent of United States landings (18 million pounds of meats in 1996).

The Bay Scallop

Argopecten irradians

Bay scallops have deeply ribbed shells and grow to 4 inches across. Their shells are mottled and vary in color — brown, white, tan, orange, reddish, or purplish. Marketed muscle meats average 1/2 to 3/4 inches across. Once common in protected coastal waters of the United States, much of the production now comes from aquaculture.

The Calico Scallop

Argopecten gibbus. Other common names: Bay or cape scallop.

The Calico scallop is a fast-maturing species that can grow to commercial size in about six months. It inhabits deep waters from North Carolina to Brazil. It is named for the mottled or calico appearance of its shells. The markings are usually red or maroon on a white or yellow background. The calico scallop is slightly smaller than the bay scallop and has an average shell height of 1-1/2 to 2-1/3 inches. A commercial fleet based at Cape Canaveral, Florida, produces most of the national landings (nearly 1 million pounds of meats in 1995). The fishery is highly cyclical; no meats were produced in 1996.

The Weathervane Scallop

Patinopecten cauimus

The Pacific weathervane scallop is a large, somewhat-flattened scallop, harvested in Alaskan waters of the Pacific. The shell is brown and heavily ribbed, and may exceed 8 inches in length. Weathervane scallops are found at depths from 150 to 600 feet. Harvest methods are similar to those used by the sea scallop industry. Commercial production is variable but significant in some years.

UNIVALVE MOLLUSKS HARVESTED IN THE UNITED STATES

Abalone

This large univalve (single shell) mollusk resembles a human ear in shape — thus, its generic name *Haliotis* from two Greek words meaning "sea ear." A univalve's shell protects the animal's body, yet permits the muscular foot to maneuver along the bottom or cling with great suction. Other univalves such as conch, whelk, and limpet are utilized as food but none has the culinary status of abalone. Many species of abalone occur in the warm seas of the world; wherever they are found, the resource has been thoroughly exploited. Approximately 100 species of abalone exist worldwide, of which eight occur along the United States Pacific Coast. Popular species include the black abalone (*H. cracherodii*) and red abalone (*H. rufescens*). The red abalone is the largest species, growing to one foot in length and 8 pounds. In California wild abalone is carefully managed, and efforts to farm abalone are underway.

Conch and Whelks

Strombidae and *Buccinidae*. Other common names: periwinkle, scungilli.

Although the terms "whelk" and "conch" are often used interchangeably, they actually are gastropod shellfish from different biological families and geographic locations. True conchs belong to the family Strombidae, and whelks to the family Buccinidae. Whelks typically inhabit temperate waters, while the edible pink or green conchs (*Strombus gigas*) occur in southern latitudes, and in this country are a significant resource only off the Florida Keys.

Queen conch is primarily imported from Belize, followed by the Caribbean countries of Honduras, Haiti, and Mexico. The Caribbean fishery exceeds 8 million pounds per year and provides a staple food in some fishing communities. The species possesses an attractive shell, although these are often destroyed during processing. Whelks are available from the East Coast, notably in the New England states. They are available year-round frozen, and fresh from spring through the fall. Production from the mid-Atlantic (Delaware to Virginia) has increased appreciably in recent years in response to demand for fresh-shucked meats in ethnic cuisines.

Whelks have a thick-walled, spiral-shaped shell that grows to 9 inches in height. The outside of the shell is dull white, tan, or yellowish gray, and the inside usually is yellow or orange. The edible part of a conch or whelk is the foot, which is removed from the shell by breaking the shell where the muscle is attached.

The Periwinkle

Littorina littorea. Other common names: Winkles, sea snails.

Periwinkles are small, marine, snail-like mollusks found in large numbers along the shoreline of the northern Atlantic. This same periwinkle was found originally in the European Atlantic but has spread around the eastern coast of North America from Canada during the past two centuries and now occurs as far south as Delaware Bay. Nearly 300 species are known throughout the world, but relatively few of these reach edible sizes. Periwinkles are a common food in Europe but are not harvested commercially in large volumes.

PROCESSING OF MOLLUSKS

Nearly all mollusks are processed prior to use. Processing may vary from boxing bivalves for the live market to further-processed frozen, canned, and pickled products (Figure 5).

Processors of molluscan shellfish vary from small to moderate in size. The industry depends largely on natural stocks, which fluctuate from year to year. This unpredictable supply often discourages processors from expanding their operations. Thus, the industry is slow to adopt new technology or to make major investments in new product development. With a couple of exceptions, the processing of mollusks is labor-intensive with relatively little mechanization. Mollusks are processed to convert the raw material to a more desirable form, to preserve the product, to maintain quality, to more fully utilize the raw product, and to assure food safety. Mollusks such as clams and oysters are shucked to provide a useable market form. Molluscan shellfish are smoked, frozen, or canned to prolong shelflife and stabilize quality.

Processing for the Live Market

Bivalve mollusks, including oysters, mussels, and clams, often are presented to the consumer in the live state; however, they are often processed for the live market. Bivalves can survive out of the water for extended periods, allowing them to go through the processing chain.

Temperature control is the most critical factor for providing a good product, yet few harvesting boats provide refrigerated storage for bivalves. Soft-shell clams are harvested primarily during the warm summer months, and although oysters and clams are primarily harvested during the winter and spring, some are also harvested in the summer. The on-deck temperature in the summer often exceeds 86°F (30°C), a temperature that can re-

Figure 5. Clam processing.

duce product quality, even though the animals are alive. For optimum quality and shelf life, the bivalves should be cooled and stored at temperatures lower than 50°F (10°C).

Processing for the live market is usually limited to washing, sorting, and packing. Oysters are often received as clusters or clumps covered with mud. Because single oysters are desired for the half-shell market, the clusters are broken into singles manually and the shell debris is discarded. Once the oysters are broken apart, they are often sorted by size; a medium-size oyster is preferred for the half-shell market. Some restaurants prefer the outside shell to be relatively free of mud, although the mud serves to keep the oyster moist and tends to keep the oyster alive longer. A variety of washing methods, from hand-scrubbing to mechanical pressurized washing, can be used to clean the shellstock. Simply hosing down a pile of oysters is rarely an effective cleaning method. Once the oysters are sorted and washed (if desired), they are put into boxes and shipped to market.

Hard clams (*Mercenaria mercenaria*) are often sold in the shell by size, with smaller clams priced higher per piece. The clams are sorted either manually or mechanically, washed, and boxed for shipment. Mechanical-size-grading is achieved with equipment that directs clams among a declining set of diverging rollers. Small clams fall between the rollers first, followed by progressively larger clams. Catchments below the rollers direct the clams through electronic counters. Counts are necessary because in-shell hard clams are commonly marketed by the piece rather than by weight or volume. Hard clams are marketed according to size, usually in three categories:

Littlenecks, or little necks (named after Little Neck Bay on Long Island, once the center of the half-shell trade) are also known as "necks." They range from the minimum legal size, which varies by state (usually one inch), to 2 inches across, and yield approximately 450 to 600 clams per bushel.

Cherrystones or "cherries" are medium-sized and medium-priced. They range from 2 to 3 inches across and yield about 300 to 400 shellfish per bushel.

Chowders, sometimes referred to as quahogs, are more than 3 inches across and yield an average of 125 to 180 clams per bushel. They have the toughest texture and are the least expensive of the size categories.

Buyers should be aware that no standards of identity apply to these size grades. To prevent confusion, buyers should specify size ranges when requesting quotes. Also, clams harvested from certain warm water areas are known within the trade to possess a shorter shelf life than that of clams harvested from other sites or in cooler months. Recent work indicates that slow (or stepwise) cooling reduces dead-loss compared to direct transfer from harvest boats to 40°F storage. Hard clams that are stored dry and directly exposed to refrigerated air, appear to survive longer than when covered with damp burlap. Cooling of shellfish is specified in the Interstate Shellfish Sanitation Conference Manual of Operations.

The Manila clam (*Tapes philippinarium*) is handled in a manner similar to the Mercenaria clam. The soft-shell clam (*Mya arenaria*) also is washed and boxed before sale. The clams are often placed in clean water for a short time to remove grit and sand from their intestinal tract.

Mussels destined to be sold live are washed, graded, and processed by removing the "beard." Live mussels sell very well along the Eastern Seaboard and to a lesser extent along the Pacific Coast. Mussels represent one of the best current opportunities for seafood market development, as the product is relatively inexpensive and plentiful, thanks to the growth of aquaculture production and improved wild harvest quality.

Wet storage of bivalves is limited but may be practiced more extensively in the future. Requirements for wet storage are similar to those for depuration and are covered under Part II of the National Shellfish Sanitation Program (NSSP) Manual of Operations. Wet storage can be used not only to prolong shelf life but also to improve flavor. For example, "salty" oysters are preferred for the half-shell market, but much of the available shellstock is harvested from low salinity waters. Oysters are osmoconformers (their internal salt content reflects that of the surrounding water), so when shellstock is placed into higher salinity water, their salt content quickly changes to match that of the environment.

Processing for the Fresh Market

Mollusks are processed for the fresh market in a variety of ways, from the shucking of bivalves to the cleaning of squid. Most processing is performed manually. Mechanical means of shucking bivalves can be used, but mechanically shucked products are most often further processed.

Bivalves

Shucking is the process of separating the meat of bivalves from their shells. Bivalves destined for the fresh market are usually shucked by hand, a labor-intensive job requiring considerable skill. Each bivalve requires a slightly different method of shucking, and shucking methods can vary from one region to another. For example, the most common way to shuck oysters is to insert the shucking knife through the lip or bill of the oyster, cut the adductor muscle, and remove the body from the shell. In some areas, however, oysters are opened by "popping the hinge," in which the knife is inserted into the hinge of the oyster and twisted to break the hinge apart. In general, the first method provides the better product, and a number of variations of this method are used commercially. Many shuckers use a hammer to chip a small piece of shell from the lip of the shell to facilitate the entering of the knife. In other plants, the bill of the oyster shell is chipped off using a device similar to a grinder wheel with an electric motor. Shell thickness and shape often dictate the method of choice.

As oysters are removed from the shell, they are sorted by size. Eastern oysters (*Crassostrea virginica*) are normally packed according to the number of meats per gallon (Table 1). Price normally increases with size. In practice, these size grades are not absolute and may vary by the size of shellstock (in-shell mollusks) available at the time of processing.

After shucking, the meats are washed and packed. Washing is often performed by blowing, a process in which air is pumped into the bottom of a chilled water tank to agitate the oyster meats. After 10 to 15 minutes the air is shut off; then water is added and allowed to overflow the tank. Grit and shell particles settle to the bottom. When the water is clear, the oysters are removed and packaged. Oysters have a standard of identity that requires that they not be exposed to water for more than 30 minutes or be blown for more than 15 minutes, and the oysters must be packed dry. After packing, they tend to lose liquid, which is called free liquor. The amount of free liquor depends on the season of harvest, the condition of the oyster, and the geographic location of harvest. Oysters

Table 1. Standard of Identity for Size (number in one gallon) (Office of Federal Register)

Eastern (*C. virginica*)	Pacific (*C. gigas*)
Extra Large (counts) Not more than 160 oysters	————————
Large (extra selects) More than 160 but not more than 210 oysters	Large Not more than 64 oysters
Medium (selects) More than 210 but not more than 300 oysters	Medium More than 64 but not more than 96 oysters
Small (standards) More than 300 but not more than 500 oysters	Small More than 96 but not more than 144 oysters
Very Small More than 500 oysters	Extra Small More than 144 oysters

Table 2. Characteristics of Eastern and Pacific Oysters		
	Eastern	Pacific
Color of liquor	clear	milky
Color	creamy to tan	white with black mantle
Cooking recommendation	light	thorough*

* Western oysters are substituted for Eastern oysters in many applications, and older recommendations for pre-cooking are being dropped.

may lose up to 30 percent of their weight as free liquor, although losses of 5 to 15 percent are normal.

Until the mid-1980s, the Eastern oyster dominated the fresh oyster market; however, in recent years, *C. gigas* has increased its market share significantly, even in the East and Gulf Coast areas. *C. gigas* tend to be large and are usually sold as extra selects or counts. Their appearance is somewhat different than that of the Eastern oyster. They tend to be larger and whiter, and often have a dark area around the gills. Also, the liquor in the containers is often cloudy, which is considered a defect in packs of *C. virginica* but is normal for packs of *C. gigas* (Table 2).

More than 22 species of clams are listed by the United States Food and Drug Administration (FDA) as being harvested commercially or recreationally in the United States. Only five of these species are of commercial importance: the hard clam (*Mercenaria mercenaria*), the surf clam (*Spisula solidissima*), the ocean quahog (*Arctica islandica*), the soft-shell clam (*Mya arenaria*) and the geoduck (*Panopea generosa*). These five species account for 99 percent of the commercial catch.

Most clams are shucked commercially by hand. The hard clam is held in the palm of the hand with the shell hinge against the palm. A strong, slender knife is inserted between the halves and the shell is pried open, the adductor muscles cut, and the meat removed from the shell. Clams are washed (sometimes by blowing as described previously for oysters), packaged, and marketed. Most hard clams that are commercially shucked are chowder clams, which are large and destined to be minced for clam chowder or clam sauces. Hand-shucking is often facilitated by placing the sharp edge of the hard clam against a grinding or chipper wheel.

Geoduck clams are often very large, averaging three pounds, but grow to 15 pounds. The geoduck is mostly neck (siphon), and the meat yield is about half the original weight.

Surf clams, ocean quahogs, and occasionally large hard clams are processed in automated facilities where product is conveyed and flumed between operations. The clams are either placed on a metal link, flat conveyor belt and passed through an open gas flame or are steamed at high temperature in pressure vessels. This heating step, which often partially cooks the meat, greatly weakens the muscle-shell bond and the meat is freed from the shell by tumbling or shaking. The meats and shell pieces fall into a brine flotation tank to be separated. The meats are washed and agitated to remove sand and inspected prior to packing. However, very little clam meat processed in this manner is marketed fresh. More commonly it is flumed to size reduction equipment to be sliced, chopped, or ground, and then frozen. The clam liquor released during heat-shucking is sometimes collected, concentrated, and marketed as clam juice or natural clam flavor.

Currently, in the United States there are four commercially important species of scallops: the sea scallop (*Placopecten magellanicus*), the bay scallop (*Aequipecten irradians*), the calico scallop (*Argopecten gibbus*), and the Pacific sea or weathervane scallop (*Patinopecten caurinus*). In the United States only the adductor muscle is consumed, but Europeans consume the entire scallop. There is also a demand for the adductor muscle with the roe attached.

Sea scallops usually are processed on-board ship. The hand-shucking step is similar to that for oysters except that scallops do not close their shells tightly, making for easier knife entry. The soft body parts are removed, leaving only the "eye" or adductor muscle meat. In some cases, sea scallops destined for export to European markets are processed so that the roe (gonads) remain attached to the eye. In Europe, the roe is esteemed as a delicacy and may be even more desired than the meat. Meat color may range from white to grey or bluish and even yellowish or pinkish colorations are quite common and acceptable.

The shucked meats are washed, packed into cloth bags, and stored surrounded with ice in the hold. In hot weather, the scallops are sometimes pre-cooled in an ice slush prior to stowage. Management plans for sea scallops usually include size requirements, such as a minimum 3-inch shell diameter and an average of no more than 30 meats per pound with a 10 percent variance (20 percent variance in the winter). When sea scallop vessels return to port, the bagged scallops are usually transferred to a processing plant. The meats are separated and washed in a tank of chilled water, then inspected and graded by size prior to packing in plastic or metal containers. A bright orange stain, usually on one area of the meat, occasionally occurs due to the growth of pigmented bacteria during storage at sea. These scallops are removed and are rarely seen in the marketplace. Commonly the meats are treated with sodium tripolyphosphate before they are packed to reduce drip losses.

Bay and calico scallops usually are shucked on land. Their small size makes then uneconomical to shuck by hand on-board ship.

Calico scallops are shucked mechanically. Machines that employ a shock-heat-shock method have been used both ashore and on ships, although currently most are used ashore. In this process, the scallops are passed through a sorter to remove trash and then are fed into a tank of water heated to 176° to 215°F (80° to 100°C) or through a steam tunnel. Rollers that revolve in opposite directions grip the shells and sling them with considerable force against a steel baffle slanted at a 45° angle. They are removed by conveyor and undergo a second shock-heat-shock treatment. They are then dropped onto a vibrating screen that separates the meat and viscera from the shells. The meat then goes to an eviscerator, basically a number of paired rollers, that grips and pulls the viscera from the meat. The meat is then washed or it may be placed in a brine tank to remove shell fragments. The meats are inspected and washed. There are many variations of these procedures. For example, slinger rollers may or may not be used.

Gastropods

Two species of abalone, *Haliotitis rufescens* and *H. corrusala*, are harvested and processed commercially. Most often they are sold fresh. The foot of the abalone is the only part consumed. The muscle is very tough and is usually tenderized by hitting the meat with a mallet.

The queen conch (*Strombus gigas*), which is principally harvested from Central American waters, is usually cut from the shell on-board and then taken to a shoreside facility for final processing. Conch meat is produced by opening an elongated hole in the spire. A very sharp blade is then inserted and the animal cut free of its attachment to the central axis of the shell. The viscera and other soft parts are removed from the foot, and the tough dark skin is removed. The marketable meat yield is about half the total in-shell weight. Ocean conch imported into the United States is frozen.

Most of the conch processed in the United States is actually whelk, a distant relative. Two species are important in the United States: the knobbed whelk (*Busycon carica*) and the channeled whelk (*B. canaliculatum*). They are harvested in the Atlantic with traps or by fishing at night with dredges. Until recently, nearly all were sold as precooked meats. Currently, a growing ethnic market exists for raw meats. The animal is released from its shell by cutting the attachment site, then either distributed whole for later processing or eviscerated and packed as fresh or frozen meats. The once-common practice of retorting (pressure steam-cooking) prior to meat extraction currently finds limited practice. Another species, the waved whelk (*Buccinum undatum*) is found from New England to Northern Europe, achieving most of its commercial importance in the United Kingdom.

FURTHER PROCESSING

Batter and Breading Operations

Many molluscan products are sold battered and breaded. Coating seafood with batter and/or breading before cooking is a common practice of homemakers, food processors, and commercial food service establishments. Commercial battering and breading of seafood, including mollusks, followed by freezing, offers a widely accepted convenient product to the consumer. The United States is the world's largest consumer of breaded seafood products. Breaded mollusks such as oysters and scallops are sold to luxury consumer markets or to the restaurant trade. Breaded and battered molluscan products include oysters, clam strips, and scallops.

In this chapter, a batter refers to a liquid mixture consisting of water, flour, starch, and seasoning into which seafood products are dipped. A breading is defined as a dry mixture of flour, starch, and coarse seasoning that is applied to moistened and battered products before cooking. There are a variety of batter and breading formulations. The flour and starch level is usually higher in batter mixes, which account for 80 to 90 percent of the total. In breading mixes, flour and starch usually account for 70 to 80 percent, with salt and other seasonings making up the remainder.

Batters and breadings enhance product appearance (for example, color), texture, and flavor. They also act as a moisture barrier, holding in natural juices, thus often making the product more tender.

Applying batter and breading to molluscan products is mostly automated. The seafood enters the batter machine on a conveyor, which has a large number of openings so that excess batter can fall through. The seafood first passes through a pool of batter that coats the bottom, then travels under a film of falling batter that coats the rest of the product. Excess batter drips from the product. Breading is applied in the same manner, except it is dry. The product is vibrated after application to remove loose breading.

In the 1960s, the concept of batter fry was developed and today is widely used. In this process the seafood is predusted with flour or dry batter mix and then conveyed through a special batter applicator. The product is pre-fried to set the batter and to impart the desired frying oil content for enhanced texture and eating quality. The selected batter can be leavened (tempura) or non-leavened. Production rates are slower than those that produce a battered/breaded product.

Examples of battered and breaded products, which are usually sold frozen, are scallops, oysters, clam strips, and clam cakes.

Freezing

Frozen molluscan products available commercially include most of the types described for the fresh market as well as battered and/or breaded products. In many cases, freezing is a normal part of processing; nearly all commercial battered and breaded products, for example, are frozen. However, some mollusks freeze more successfully than others, and frozen storage life varies with the species. Other important considerations include packaging, rate of freezing, and storage temperature. In general, oxygen-impermeable packaging is most effective, but packaging cost should be balanced against inventory turnover rate and product stability characteristics. The highest quality results from fast freezing rates and maintaining storage temperatures below -5°F. It is important that the product to be frozen be of good quality.

Oysters

Most shucked oysters are sold fresh, but some food service establishments prefer frozen oysters because they provide for ease of storage and a constant supply. Oysters change composition, especially after spawning in the summer, and these changes affect yields. Also, demand patterns are such that shortages in supply often occur during Thanksgiving and Christmas holidays. Although oyster appearance and quality are affected, freezing can provide an adequate supply of good quality oysters year-round.

The free-liquor content of previously frozen oysters may be as high as 20 to 30 percent — twice as high as fresh oysters, resulting in reduced yields. Freezing also may cause the oyster to darken. The amount of darkening depends on the freezing rate. Slow freezing greatly increases both the darkening and the free-liquor content. Oysters should not be frozen in one-gallon metal cans because the freezing rate is slow and metal ions may accelerate darkening. It is better to freeze them in oxygen-barrier bags or pouches, which are laid flat to increase surface area. Both Eastern oysters and Western oysters may be frozen for as long as 10 months and still maintain acceptable quality.

Clams

Only a limited segment of the total clam harvest is frozen, principally those destined for the chowder market and for further processing. Frozen shelf life is limited to four to six months at 0°F (-18°C) because of rancidity and toughening. As with oysters, the packaging type and integrity, freezing rate, and storage temperature greatly influence quality.

Surf clams, ocean quahog, and large hard clams are often frozen after mechanical shucking. Cryogenic freezing methods (carbon dioxide or liquid nitrogen) are now preferred. In addition to surf clams, management practices have caused many plants also to process ocean quahogs and, more rarely, the hard clam.

Scallops

Scallops are often marketed either frozen in blocks with plate freezers or individually quick frozen (IQF) with or without breading. In 1978, almost 62 million pounds of scallop meats were breaded and frozen, but by 1980 the amount of breaded and frozen scallops decreased to 50 million pounds, largely because of a decrease in the availability of scallops. Scallops have a frozen shelf life of about 12 months at 0°F (-18°C).

Conch and Whelk

Almost all conch meat and some whelk is frozen, especially the queen conch. Once the animal is processed, it is frozen in 5- to 10-pound bags.

Abalone

As mentioned previously, only the foot of the abalone is consumed. In preparation for freezing, the muscle is sliced across the grain into 2-inch steaks. The critical processing step is tenderizing. The steak slices are put on a table and allowed to relax, then hit with a mallet. Most abalone meat

Figures 6 and 7. Oyster-processing plant in Newport News, Virginia.

is consumed in California, where regulations prohibit it from being shipped beyond state boundaries. Abalone is found in other countries, including Korea and South Africa, and products from these nations occasionally find their way to American markets. Aquaculture may expand availability in the future.

Canning Molluscan Shellfish

As with other canned foods, shellfish canning involves heat processing in a hermetically sealed container to achieve commercial sterility. These products are shelf-stable. Many species of molluscan shellfish are canned throughout the world. In the United States, clams and oysters are the most important molluscan shellfish for commercial canning. Other canned species include mussels, cockles, scallops, snails, and abalone.

Oysters

Heat processing of oysters in the United States (Figures 6 and 7) began as early as 1820. Various standards for canned oysters are covered by the FDA. The standards of identity, which includes standard fill of container, require a drain weight of at least 59 percent of capacity. Currently, oysters are canned in three areas of the United States: the East Coast from Maryland to Florida; the state of Washington on the West Coast; and the Gulf Coast, which produces most of the domestic canned oysters. The volume of oysters that are canned is small compared to the quantity sold fresh, but canned oysters are still a popular item and can be found in most supermarkets. Imports of canned oysters, mostly *C. gigas*, exceed domestic production.

Oysters for canning may be shucked either mechanically or by hand, as previously described. For mechanical shucking, the oysters are first washed, often with high pressure, to remove mud and debris. They then enter a steam tunnel or retort. The best results are obtained by using a preheating step followed by briefly cooking at high temperatures and pressure in special retorts. This process degrades the hinge material, thus causing the shell to gap open. After the oysters are steamed, they are conveyed to shucking units where the meats are separated from the shells. Mechanical shucking has the advantage that all oysters are shucked, including clustered shellstock that is difficult to shuck by hand. Because the oysters are cooked slightly, they are unsuitable for the fresh market, although they are excellent for canned products.

The oysters are packed by hand into "C" enameled cans, usually Number 2 or Number 95, which are the most popular sizes for oysters. Fill-in weights depend on the conditions and composition of the oysters, which changes with season, water salinity, and other environmental factors. If the oysters are mechanically shucked, the amount of cooking incurred during steaming affects the fill-in weight since the oysters may lose considerable weight during heating.

After the oysters are placed in cans, a weak brine solution (1 percent salt) is added at near the boiling point. The cans are then closed and retorted.

Four different styles of canned oysters are on the market: whole, stew oysters, oyster stew base, and smoked oysters. The process for whole oysters is described above. Canned stew oysters are prepared by chilling the oyster meat to firm the flesh, then slicing the meats in a mechanical slicer. Fifty grams of meat are put in a 10-ounce can and, just prior to sealing, a mixture of milk, salt, monosodium glutamate, and disodium phosphate is added. Some packers also add oyster nectar, which is prepared by boiling whole oysters. Finally, a small amount of butter is added, and the cans are vacuum-closed and retorted. Canned oyster stew base consists of sliced oysters and nectar. The user then makes a stew as desired. This product is usually canned for the institutional market in large 404 x 700 cans [1 to 3 kilograms (kg)].

Smoked oysters are made from precooked, whole, small oysters or from sliced, large oysters. The meats are placed into a 20°-salometer brine for three to four minutes, then drained, spread onto racks, and smoked two to three hours. Sugar can be added to the brine for flavor. The meats develop a dark golden color and smoked flavor. The smoked meats are commonly packed into 301 x 106 (110g) cans or glass jars to which vegetable oil is added. The containers are vacuum-sealed and processed. Many smoked oyster products are imported from Korea where *C. gigas* is used. Smoked oysters canned on the East and Gulf coasts of the United States are usually *C. virginica*.

Clams

At least 14 species of clams are canned worldwide. In the United States, six species are canned in various states, with hard clams, soft-shell clams, and razor clams (*Siliqua patula*) accounting for most of the production. Most canned clams marketed in the United States are minced and canned as formulated chowders and soups. Surf clams and, to a lesser extent, ocean quahogs (mahogany clams) are the species of choice.

Maine, Maryland, Massachusetts, and Florida account for most of the processing of hard- and soft-shell clams on the East Coast. These are the principal species canned as unformulated meats. The clams are washed and retorted under pressure, removed from the steamer, and sorted by size (small or large) and by color (light or dark).

The clams are packed into "C" enameled cans, including sizes 211 x 400, 307 x 409, and 300 x 404, the first two being preferred. Clams shrink considerably during processing, so "fill weights" are greater than "drained weights." Dark discoloration is occasionally encountered with canned clams, although advances in processing and container coatings have greatly reduced this problem.

On the West Coast, particularly Oregon, Washington, and Alaska, the razor clam is the principal clam selected for canning, usually as minced clams. The clams are washed and scalded. The meats are shaken out of the shells and split along one side to remove sand and mud. They are washed a second time, and the siphon, body side walls, and stomach are removed. The remainder is chopped and packed into cans. Brine or clam juice is added; then the cans are closed and heat processed.

Other Mollusks

Although clams and oysters account for most of the canned mollusk production in the United States, other species, including mussels, scallops, abalone, cockles, donax, and snails are also canned. The canning of mussels (*Mytilus edulis*) is a small but growing industry on the East Coast. They are first steamed 5 to 10 minutes, then shucked and packed into cans. Juice from the steaming process and salt are added.

Scallops are canned almost exclusively in Japan. The scallop meat is packed with a 2 percent brine and processed. Abalone of the family Haliotidae are sometimes canned abroad. Formerly, abalone foot muscle was minced and canned in the United States, but the practice has been abandoned. In Japan and some other countries, the animal is removed from the shell and the visceral mass and mantle fringe are trimmed. The meat is then washed, dry-salted for 24 to 48 hours, rubbed to remove mucous substances, and canned.

Canning of cockles (*Cardium* spp.) is a minor industry in the northwestern United States but an important industry in Western Europe. In Spain, cockles are washed, steamed, shucked by agitation, and canned. In France, they are pickled for three days in a 3 percent vinegar solution containing 3 percent salt, prior to canning.

Land snails (escargots) are also canned with and without shells. Donax (coquina) clams are small and are used primarily in soups prepared by boiling the entire mollusk. The southern coquina (*Donax variabilis*) is found on sandy beaches from Virginia to the Gulf of Mexico, attaining a length up to 1 inch. They are common in Florida. The smaller northern coquina (*Donax fossor*) is found along the New York and New Jersey shoreline, and attains a length of less than 1 inch. *Donax laevigata* occurs on the southern California coast.

Pickled Mollusks

Pickling seafood with vinegar and spices is an ancient form of food preservation. (For a further discussion, see the chapter on Specialty Seafood Products.) It is more often practiced with fish, but some mollusks are pickled. In the 1800s, pickled molluscan products, especially oysters, were prepared commercially in most of the United States Atlantic coastal region. They are not nearly as popular now but are prepared in some areas, especially in Virginia and Louisiana, for local consumption.

Pickled mussels are becoming a popular seafood. Mussels are plentiful and inexpensive because of greatly increased aquaculture production, and the acceptance of pickled mussels has increased accordingly. Clams are also pickled but to a lesser extent than are mussels and oysters.

Pickled cockles are common in France, where the meat is dipped in a 3-percent salt brine, drained, then covered for three days with a solution containing 3 percent vinegar and 3 percent salt. The cockles are drained, packed into jars, covered with a spiced vinegar solution, and processed.

REFERENCES

Applewhite, L., S. Otwell, and L. Sturmer. 1996. Survival of Florida aquacultured clams in refrigerated storage. Tropical and Subtropical Seafood Science and Technology Society of the Americas Abstracts, p. 9.

Brownell, W. and J. Stevely. 1981. The biology, fisheries and management of the Queen Conch, *Strombus gigas*. *Marine Fish. Rev.* **43**:No. 7:1-12.

Borgstrom, G. 1965. *Fish as Food Vol. 4*. Academic Press, Orlando, FL.

Castagna, M. 1990. Shellfish — mollusks. In *The Seafood Industry*. R. Martin and G. Flick (eds.), pp. 77 - 87. Van Nostrand Reinhold, New York.

Desrosier, N. and D. Tressler. 1977. *Fundamentals of Food Freezing*. AVI Publishing Co., Westport, CT.

Fisher, R.A., W. D. DuPaul, and T. E. Rippen. 1996. Nutritional, proximate, and microbial characteristics of phosphate processed sea scallops (*Placopecten magellanicus*). *J. Muscle Foods* **7**(1):73-92.

Lopez, A. 1987. *A Complete Course in Canning — Book III — Processing Procedures for Canned Food Products*. The Canning Trade Inc. Baltimore, MD.

Martin, R. 1990. A history of the seafood industry. In *The Seafood Industry*, R. Martin and G. Flick (eds.), pp. 1-16. Van Nostrand Reinhold, New York.

NMFS. 1995. Fisheries of the United States. National Marine Fisheries Service. United States Government Printing Office, Washington, DC.

NMFS. 1997. *The History, Present Condition, and Future of the Molluscan Fisheries of North and Central America and Europe*. National Marine Fisheries Service. United States Government Printing Office, Washington, DC.

Office of the Federal Register. 1989. Code of Federal Regulation 21. United States Government Printing Office, Washington, DC.

Peters, J. 1978. Scallops and their utilization. *Marine Fish. Rev.* **40**: No. 11:1-9.

Rippen, T.E., H. C. Sutton, P. F. Lacey, R. M. Lane, R. A. Fisher, and W. D. DuPaul. 1996. Functional, microbiological and sensory changes in ice-stored sea scallops (*Placopecten magellanicus*) treated with sodium tripolyphosphate. *J. Muscle Foods* **7**(1):93-108.

Suderman, D. and F. Cunningham. 1983. *Batter and Breading*. AVI Publishing Co., Westport, CT.

Wheaton, F. and T. Lawson. 1985. *Processing Aquatic Food Products*. John Wiley and Sons, New York.

Wheelen, J. and C. Hebard. 1981. *Seafood Product Resource Guide*. Mid-Atlantic Fisheries Development Foundation, Annapolis, MD.

Preservation of Squid Quality

Robert J. Learson

INTRODUCTION

There are about 30 families representing 460 species of squid and cuttlefish known to exist in the world's oceans. In the United States, the primary commercial species in the Northwest Atlantic are *Loligo pealei* and *Illex illecebrosus,* and on the Pacific Coast *Loligo opalescens*.

Between 1975 and 1979, United States squid landings averaged only about 6.3 million pounds. In 1996 United States landings were 240 million pounds, with the West Coast accounting for over 70 percent of the catch. United States squid product exports in 1996 were 137 million pounds.

Because of their physiology, squid require special handling during harvesting, refrigeration, and processing to retain the highest quality. Being a "soft-bodied" species, squid are prone to physical and mechanical damage such as crushing, torn mantles, and loose skin — all of which result in poor quality and lower market value.

Squid also contain multi-colored chromatophores on the skin surface. These alter the color of the squid to match the ambient color when swimming in the ocean. However, after death, these chromatophores can expand or contract depending upon processing and storage temperatures.

Since many markets, especially the prime Japanese market, grade squid quality based largely on color, the handling and storage techniques must be designed to provide the required surface color and texture for the target market. In general, surface color is important for whole squid and significantly less important for a product which is destined for skinning and further processing.

HARVESTING AND PROCESSING AT SEA

The premier quality squid is harvested by jigging. The jigging operation essentially harvests a live product and since there is no bulk storage on deck, very little crushing or mechanical damage occurs. Trawl- or net-harvested squid often exhibit severe physical damage. Crushing, due to the weight of the squid in the net, produces ruptured ink sacs, which can result in staining of the mantle. Handling on deck during sorting, icing, etc., can result in torn mantles and loose skin. Physical damage also increases the spoilage rate of fresh squid.

Freezing at sea is the recommended processing method for the whole, frozen squid market. Immediately after capture the squid should be placed in refrigerated seawater (RSW) at 0° to 4°C. The cold seawater prevents the squid from drying out and eliminates discoloration of the skin surface. The squid should then be quickly sorted and packed, and either blast- or plate-frozen at -30° to -40°C.

After freezing, the squid are glazed and placed in cold storage at -25° to -35°C. The glazing process can be carried out on shore. However, during long fishing trips this is not recommended due to the potential of dehydration. Rapid freezing and constant storage at low temperatures are essential to avoid changes in color and adverse textural changes. Fluctuating storage temperatures can also result in reddening of the product surface which reduces market value.

Squid are very perishable, and extra care should be taken if the squid are to be landed fresh. Bulk icing in pens can result in physical damage resulting from shoveling the catch, crushing, and damage from the use of coarse crushed ice. Boxing at sea is recommended over bulk icing in pens, in order to reduce the amount of handling and physical damage. In either case, since squid have a tendency to "bridge" and become a single mass of flesh material, harvesters should be careful to intermix the squid with ice to ensure proper chilling of the product.

The ratio of squid to ice is important for maintaining quality. Studies carried out by the National Marine Fisheries Service (NMFS) resulted in a recommendation of squid-to-ice ratios of a minimum

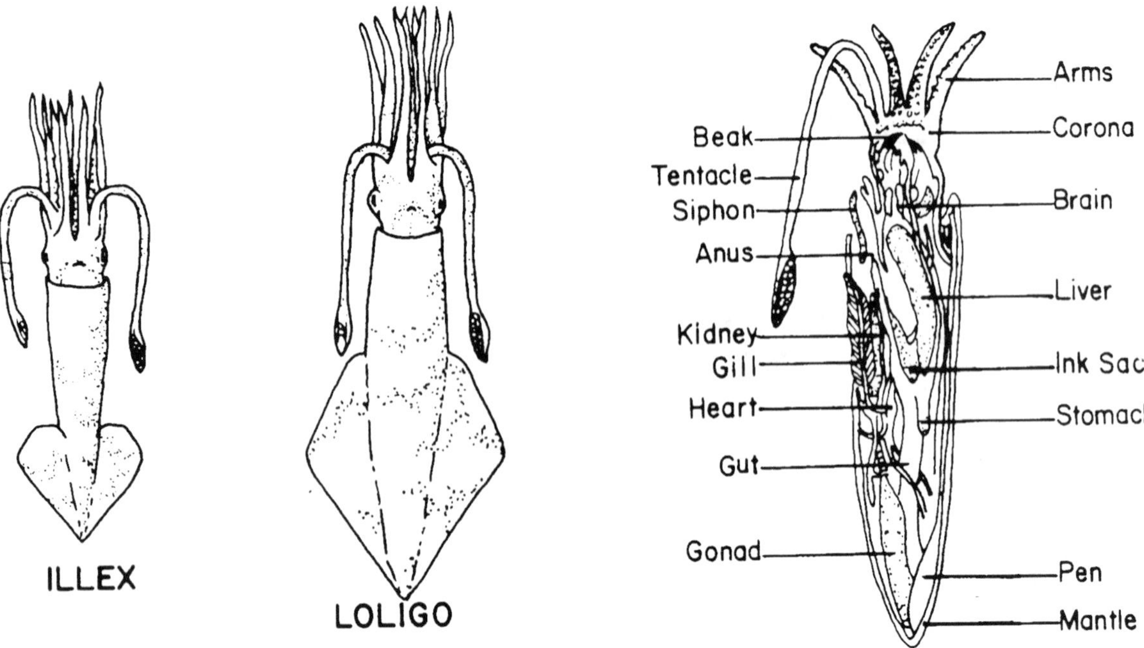

Figure 1. Squid.

of 2:1. Fresh squid shelf life averaged 6.3 days, compared to normal icing at a ratio of 5:1 where the shelf life averaged 4.7 days.

Depending upon season and trip length, the Japanese recommend a 1:3 ratio for a three-day summer trip and a 3:1 ratio for a three-day winter trip. In any case, harvesters and processors should be aware of the marketing requirements associated with iced squid. For a product which is to be processed into rings, strips, or skinless tubes, properly handled squid has been shown to be readily acceptable regardless of surface discoloration and slight physical damage. For whole, fresh product, damaged, stained, or discolored squid should be culled.

Preferred alternatives to iced squid are both refrigerated seawater (RSW) and chilled seawater (CSW). Both methods provide immediate rapid chilling, and since the product is "floating," the prevalence of crushing and other physical damage is greatly reduced. Research carried out by the NMFS Gloucester (Massachusetts) Laboratory and others have shown a significant increase in fresh shelf life using either RSW or CSW systems over standard icing practices. A shelf life of up to eight days can be expected. The Japanese and others have indicated a preference for RSW over CSW. The claim is that there is better color retention due to increased salt content and less abrasion of the skin. CSW systems need no mechanical refrigeration and represent an excellent alternative for small inshore vessels. Using a CSW mixture of 300 pounds of squid to 100 pounds of ice and 100 pounds of seawater, the NMFS report an average shelf life of seven to eight days with the squid retaining rigor after three to four days at sea. Although the quality exceeded that of iced squid, there was some bleaching of the product that would reduce the value of the product on the whole, frozen market. The Japanese have reported that squid held in one part crushed ice to two parts seawater are acceptable for the sliced raw sashimi market only after one or two days at sea.

PROCESSING

All processing of squid should be carried out under refrigeration. The recommended squid temperatures should be no greater than 4°C to reduce spoilage and discoloration. In Japan the use of RSW is recommended for shore facilities to maintain color. Thawing of frozen squid should be carried out under refrigeration in coolers or in cold water thaw tanks. Since squid can discolor during thawing and the mantle flesh can become stained, it is recommended that the squid be skinned immediately after thawing.

Processed squid products should be packaged and well iced for fresh markets. For frozen markets the products should be rapidly frozen in plate or blast freezers and stored at -25°C or below for inventory.

Squid generally have excellent frozen quality characteristics. High-quality frozen squid can maintain very good quality and texture in frozen storage for well over one year at -25°C or below. A study performed by the NMFS demonstrated that whole, frozen squid could be held frozen for periods up to one year, then thawed, cleaned, processed into breaded strips, and refrozen, with an expected shelf life of six to 12 months (Ampola and Learson, 1979).

QUALITY CRITERIA

For whole, frozen squid the premium market quality is based primarily on color and texture. Although the squid flesh may be of excellent quality, that market, particularly in Asia and Spain, demands a lack of any discoloration and a firm texture.

Fresh squid undergoes rapid spoilage, usually beginning with a red discoloration of the body surface and then the development of sour off-odors. The Japanese have detected the presence of several organic acids such as formic, acetic, and isobutyric acid. Phenyl acetate was also detected. Several researchers, including Licciardello from the United States and Tanigawa from Japan, have reported correlation of quality versus total volatile basic nitrogen (TVBN). Data reported indicated that TVBN gradually increased through time and correlated well with sensory testing up to 30 mg percent nitrogen.

For frozen squid there appear to be no chemical indices agreed upon for quality.

REFERENCES

Ampola, V. G. and R. J. Learson. 1979. Comparison of the quality of squid held in chilled seawater after capture versus conventional shipboard handling. Unpublished report. National Marine Fisheries Service. Gloucester, MA.

Ampola, V. G. 1974. Squid — its potential and status as a United States food resource. *Marine Fisheries Review* **36**:12.

Kreuzer, R. 1984. Cephalopods, handling, processing, and products. FAO Fisheries Technical Paper No. 254. FAO. Rome, Italy.

Proceedings of the International Squid Symposium. August 1981. New England Fisheries Development Foundation. Boston, MA.

Scott, D. N. 1986. Squid research, processing and products in Europe. DSIR Fish Processing Report. Auckland, New Zealand.

Sugiyama, M., S. Kovso, M. Hanabe, and Y. A. A. Balkema Okuda. 1986. *Utilization of Squid*. Rotterdam, The Netherlands.

Optional Processing Methods

Roy E. Martin, Custy F. Fernandes, and Robert J. Learson

INTRODUCTION

The great variety of fish and shellfish, whether freshwater, farm-raised or ocean-caught, while delicate in taste, are more perishable than most muscle protein foods. The attributes of appearance, smell, taste, and texture begin to change rapidly after harvesting. These changes may be due to chemical, microbial, or biochemical reactions. To produce acceptable quality seafood these post-harvest changes must be reduced or brought under some control.

Traditionally, seafood has been preserved by canning, glazing, freezing, smoking, salting, or drying. These processes are discussed in other sections of this book. To enhance and extend these processes, new technologies, treatments, and techniques are being developed to supplement the desire for extending shelf lives and reaching additional distant markets.

USE OF ENZYMES

Enzymes are biochemical molecules bio-synthesized primarily by living cells. Enzymes (e.g., proteases, lipases, and catalases) occur naturally in the muscle and intestinal tracts of fish and shellfish. Enzymes have been used as processing aids in the food industry for a very long time. While they find limited use in the seafood industry, they do find use if the following factors are considered: (1) reactant concentration (2) operational conditions (e.g., temperature, pH, and time) and (3) reaction modulators (e.g., inhibitor/activator).

In various ways protein-splitting enzymes can be useful in the manufacture of protein products from fish raw materials. One example is the production of fish meal (covered extensively in another chapter). In this process the fish is pressed in order to reduce its moisture content before further processing into fish meal. The press water or "stick water" which contains valuable protein is evaporated as much as technically feasible in multi-stage evaporators.

The limit to this very economical form of evaporation is reached when the product has become so viscous that heat transfer is hampered and the product tends to stick to the walls of the evaporator. Thereafter the rest of the water must be removed by spray-drying or drum-drying, which causes a relatively large consumption of energy (steam) before the dried product can be mixed into the fish meal.

If a protein-splitting enzyme is added, the product may be rendered less viscous, and it can, therefore, be evaporated to a higher dry-matter content than otherwise possible. Hereby an improvement of the process economy is achieved. Proteases are added when the solids content in stick water exceeds 20 percent.

By using enzymes it is further possible to convert fish protein into protein products which are so easily digestible that they can be used as milk replacers for the feeding of, for example, young calves and piglets (Haard, 1992; Novo Industries, 1984).

Enzymes are also helping to produce high-quality caviars with much higher yields than traditional mechanical processes. These newer caviars are available from a growing number of fish species, including the lumpsucker, bream, coalfish, cod, capelin, carp, herring, mullet, pike, tuna, rainbow trout, and salmon.

The roe particles that are released as a result of the enzyme action can be separated from connective tissues by sedimentation/flotation in an enzyme-processing tank.

The tissues can be skimmed off from the surface of this tank and the caviar from the bottom. Any dead or damaged roe particles can be removed in the flotation tank by controlling the salinity. The roe is then drained in specially-designed boxes and packed.

With the enzyme method, caviar can be extracted from roe sacs that are much firmer than those that are closer to spawning. Trying to produce caviar from firm roe sacs by mechanical means results in considerable damage to the eggs and therefore considerable loss in yield.

Caviar produced by the enzyme method is completely free from connective tissue and damaged roe particles. The product can be stored frozen for two to three months without discernible loss of quality. It can also be stored as a half-conserve with added preservatives and antioxidants.

The enzyme used in the processing tank is isolated from the fish and is added in minute quantities. The subsequent flotation, sedimentation and resuspension steps ensure efficient removal of any trace of the enzyme (*Seafood Processing and Packaging Magazine*, 1986).

Enzymes have been used to produce a skinless squid product. The enzymes used in the process selectively attack the skin of the squid without affecting the muscle. The muscle tissue of some large squid, especially the North Atlantic species, can be a little too rubbery and tough to be well accepted on some markets. This too can be adjusted by using enzymes which gently attack the muscle fibers and make the tissue more tender.

The squid is dipped in a weak sodium chloride solution containing the exogenous enzyme papain at a low temperature (Hempl, 1983).

These processes can be combined in the production of skinless squid tubes, which are often further processed into breaded or battered rings, and the de-skinning and tenderizing of tentacles and wings.

Fish protein concentrates (FPC), while not presently commercially produced, can be made using enzymes (Hale, 1974). (See also the chapter on FPC in this book.) This technique leads to the subjects of hydrolysis and hydrolysates, and their finished products.

Fish raw material is mechanically de-boned. The skin and bones can be reduced to meal or fertilizer. The fish flesh is homogenized with water and placed into an enzyme reactor. Pepsin is used for this process, since it allows for good yields of soluble protein and has a low pH optimum (pH2), resulting in minimal bacteriological problems during processing. The enzyme breaks up the natural long-chain, high-molecular-weight fish proteins into smaller poly-peptides, peptides, and amino acids with an average length of three or four amino acid residues. These shorter molecules are very soluble in water and can be mechanically separated by centrifugation. The soluble protein fraction, after centrifugation, is still at low pH and must be neutralized before being dried. This is done in an ion-exchange reactor, where the H+ ions are removed from the solution. The ion exchange resin is regenerated and recycled. The neutralized liquid is then concentrated and spray-dried.

This process produces a functional, soluble, high-quality protein product that could have immediate marketability as a milk solids substitute in animal feed formulations and a good potential as a human food additive. Other proteolytic enzymes (such as papain and bromelin) have also been used with some success (Hoyle and Merritt, 1994; Hyung et al., 1992).

If the final product remains in the liquid state it is called fish silage or fish hydrolysate (Bucove and Pigott, 1977). Fish silage has been produced for years in Scandinavia and is used as a protein supplement in animal feed (Gildberg and Almas, 1986).

Membrane removal from cod livers and cod liver swim bladders has been successful using hydrolytic enzymes (Harrd, 1992). This technique is used prior to canning to help remove sealworms that infest this membrane. The through-put of liver during this processing can be increased 20-30 percent with the aid of enzymes as compared to manual processing methods. An eight-fold increase in process efficiency can be gained using similar enzymes in removing the membrane surrounding cod-liver swim bladders.

Endogenous enzymes are responsible for the production of fermented fish foods. Endogenous enzymes gradually hydrolyze the fish muscle in the presence of 20 to 35 percent sodium chloride. Salt penetrates into the fish flesh and water diffuses from the tissue, forming a brine. Slowly an equilibrium is attained between the salt concentration in the tissue and the salt concentration in the surrounding brine. Endogenous enzymes hydrolyze proteins and increase soluble nitrogeneous compounds (e.g., peptides and amino acids), with concomitant changes in the tissue structure. The resulting broth is high in free amino acids as well as peptides which contribute to an excellent flavor. The traditional fermentation process is slow and ripening is completed over several months (Varlik et al., 1993). These techniques yield fish sauces, fermented flavors, and seafood flavor concentrates. Fish sauces have also been prepared using an accelerated fermentation process by actually adding proteolytic enzymes to the fish raw material (Chaveesuk et al., 1993) (Figure 1).

MELANOSIS CONTROL

Shrimp melanosis or blackspot is an objectional surface discoloration caused by enzymatic formation of precursors of insoluble polymeric pigments. The endogenous shrimp enzyme that catalyzes this reaction, polyphenol oxidase, remains active during refrigeration, ice storage, and post-freeze thawing. Melanosis is a problem in most commercial shrimp species, and has a negative impact upon the economic value and consumer acceptance of shrimp (Haard, 1992).

Sodium bisulfite is used to control melanosis in shrimp. Shrimp are dipped in a bisulfite solution before storage aboard the vessel. Federal regulations state that residues of bisulfite above 10 parts per million (ppm) must be labeled to indicate the presence of the substance. Sulfite-sensitive consumers should read all shrimp labels to avoid an adverse reaction. 4-Hexyl resorcinol has been used as a bisulfite substitute with some success (Otwell et al., 1991). (For further discussion, see the chapter on the shrimp industry.)

PROCESSING AIDS

Acidulants both mineral, and in particular, organic (acetic, lactic, citric, proprionic, etc.) have found use in some processing areas. Acids can function through the improvement of quality, microbiological control, and shelf-life extension. The extension of shelf life is due to the bacteriocidal activity of the weak organic acids, because they do not readily dissociate.

Processing with acidulants alters the flavor characteristics of fishery products, resulting in the development of marinated seafoods. Most marinades are a blend of fish, spices and seasonings, and the acidulants (e.g., acetic/vinegar).

The use of polyphosphates in muscle-food systems has been investigated since the 1950s. The polyphosphates, also called condenses or combines phosphates, are manufactured from orthophosphates. The polyphosphates are straight-

Figure 1. Research leads to improved processing methods. Here, Editor George J. Flick Jr. works in his laboratory at Virginia Tech.

chain (long or short) or ring structures. The combines or condenses polyphosphates have the broadest application in the seafood processing industry. Food-grade polyphosphates (e.g., sodium tripolyphosphate or sodium hexametaphosphate) have many functional uses in seafood products, including, but not limited to, the retention of moisture and flavor, the prevention of lipid oxidation by chelation of heavy metals, shelf-life extension, and cryoprotection. Polyphosphates have a profound effect on the functional properties of freshwater and marine fish products (Lampila, 1992). Mahon (1962) obtained the first United States patent on the use of polyphosphates to inhibit thaw drip and cooking losses in frozen fish.

Shrimp dipped in tripolyphosphate have lower drip loss during frozen storage and cooking. Atlantic cod fillets dipped in tripolyphosphate or metaphosphate have decreased thaw and cooked drip loss and result in higher weight content of cooked and raw product.

Struvite [$Mg(NH_4)PO_4$] results in some processed seafood products. In canned tuna and crab, struvite crystals are often mistaken for pieces of glass. Sodium acid pyrophosphate or hexametaphosphate can be used to prevent struvite formation. The phosphates are used to sequester magnesium ions.

Phosphates have also been used in the mechanical peeling of cold-water shrimp (Crawford, 1980). Phosphate action hardens the flesh and facilitates the loosening of the shell by water jets, as well as assists in shell removal by mechanical rollers. While sodium tripolyphosphate is the phosphate of choice, blends of condenses phosphate have also been used.

Canned salmon develops a surface curd (denatured protein) if either frozen prior to canning or held on ice for a prolonged period of time. The curd may constitute up to 4 percent weight of the pack and is viewed unfavorably by many consumers. Curd formation was decreased by dipping sockeye salmon steaks for two to 120 seconds in solutions containing 15 to 20 percent sodium hexametaphosphate.

MICROWAVE PROCESSING

The conventional heating process is an external thermal process; heat is chiefly transferred through conduction (i.e., solids) and convection (i.e., liquids). However, microwave processing is an internal thermal process and is of considerable interest to the food industry. The heat is transferred through dielectric and conductive modes. The rate and mechanism of heat transfer through microwaves are qualitatively and quantitatively different from those of conventional heating. Hence, the microwave heating process is not a simple extension of the optimal conventional heating process.

Microwaves are electromagnetic waves and are generated by a magnetron. The electromagnetic waves produce electromagnetic fields with an alternating polarity of a frequency of 2,450 MHz (in the United States and Europe) and 915 MHz (in the United Kingdom). The alternating electromagnetic field stimulates the oscillation of dipole molecules in foods. The penetration of microwaves into the foods instantaneously results in an increase in food-product temperature. Water is easy to activate, and since almost every food consists of 40 to 80 percent water, microwave technology has many advantages in treating food products. Some considerations that need to be considered in microwave heating processing are electrical, physical, and thermal properties, as well as any microbial hazard.

A microwave system has three components: (a) Magnetron module, (b) microwave-focusing system and (c) microwave heating-chamber. The magnetron is the heart of the microwave, and it converts electrical energy into microwave energy. Large industrial magnetrons [e.g., 1.2 kilowatts (KW)] are electronically controlled and fabricated to deliver microwave energy, depending on the processing requirements. Each module contains several magnetrons (5 to 50), the array of which depends on the number of processing lanes in the microwave heating-chamber. The power (10 to 100 percent) of each magnetron can be programmed, thus assuring reproducibility. Generally, electromagnetic fields are concentrated at edges, corners, and peaks; this causes overheating at the corners of a food-product tray. An oscillating or pulsing energy supply overcomes the aforementioned drawbacks, producing a higher temperature at the center of a tray than at the edges (Schlegel, 1992). A microwave-focusing system focuses the microwave energy on the food-product tray. Focusing the energy directly on a small food-product surface area increases the efficiency as well as the life of the magnetron. The food-product tray is exposed to the microwaves as it is conveyed through the microwave heating-chamber. The magnetrons are installed exactly above and below the conveyed food-product tray. This allows individual treatment of the separate compartments of a three-compartment food-product tray, for example, so that the fish compartment will get more energy than the noodles, and the vegetable compartment less than the noodles. The entrance and exit of the microwave heating-chamber are protected by microwave traps and absorber sections.

Microwave processing of foods offers product as well as processing advantages over traditional processing methods. Spencer (1950, 1949) developed the microwave heating process and obtained the first patents for the microwave oven and microwavable food.

THE PROCESS OF FOOD IRRADIATION

The normal source of ionizing radiation used for food irradiation is a gamma ray emission from the isotopes cobalt-60 and cesium-137. The cobalt-60 isotope emits two gamma rays of 1.17 and 1.33 million electron volts (MeV), whereas cesium-137 emits a 0.66 MeV gamma ray. The half-life of cobalt-60 is 5.3 years; the half-life of cesium-137 is 30.2 years. Energy levels of greater than 10 to 15 MeV are required to induce significant radioactivity in an irradiated food. Thus, cobalt-60 (1.17 and 1.33 MeV) and cesium-137 (0.66 MeV) are below the 10 to 15 MeV necessary to induce radioactivity, and their use in a food irradiation technique results in no induced radioactivity in the food product. Cobalt-60 is perhaps a better choice when selecting a radionuclide for use in a food irradiator, as cesium-137 is water soluble — with a greater potential for environmental contamination. X-rays and electron beams constitute other sources of ionizing radiation that could be used in a food-irradiation system (Urbain, 1986).

One important parameter to consider in a food-irradiation process is radiation dose. Radiation dose is usually expressed in terms of how much radiation has been delivered to a food product. In older literature, the unit used to express radiation dose was the "radiation absorbed dose," or rad. One rad is equivalent to 100 ergs of absorbed energy per gram. The International System of Units (SI) uses the unit Gray (Gy). One Gray is equal to 100 rads and 1 joule of energy per kilogram (k) of food mass (Urbain, 1986). At this time, the United States Food and Drug Administration (FDA) suggests that doses up to 1.0 kiloGray (kGy) be used in irradiation processes. This is one-tenth of the 10 kGy that has been suggested by the Codex Alimentarius Committee of the United Nations. Ten kGy is the thermal equivalent of 2.4 calories. Exposure doses of this or any level from cobalt-60 and cesium-137 do not induce radioactivity in foods in any way (Urbain, 1986). The exception of this 10 kGy maximum rule occurs with the irradiation of spices, in that they are approved to irradiation up to 30 kGy (Webb et al., 1987).

Depending on the dose that is chosen, irradiation can lead to a delay in the ripening of a food product, or it can lead to complete sterilization of a product. Below 1 kGy, food undergoes what is termed radurization. Radurization can prevent sprouting in vegetables, delay ripening in fruits, and sterilize insects in grains. If the radiation dose delivered falls in the range of 1 to 10 kGy, food undergoes what is termed as radicidation. At this level, insects are not only sterilized, but also killed. Furthermore, this level of radiation results in a significant reduction in the numbers of bacteria, yeasts, and molds in a food product. If food is irradiated above 10 kGy, it undergoes what is called radappertization. Radappertization results in the complete sterilization of a food, as all bacteria and viruses are eliminated (Urbain, 1986).

Gamma irradiation has been proposed as a method of food preservation in that it can lead to DNA and cellular damage that is lethal to food spoilage microorganisms. The main advantage of a food irradiation technique is that it can lead to a destruction in the numbers of pathogenic bacteria, while not elevating the product temperature. Therefore, if the product temperature remains relatively constant, there should be no significant sensory or quality changes (Kamplemacher, 1983).

AN EARLY HISTORY OF FOOD IRRADIATION

Food irradiation is a process in which food products are exposed to specified doses of ionizing radiation so that product shelf-life and product safety are enhanced. Food irradiation is not considered a new processing technique. Roentgen discovered X-rays in 1895, and Becquerel discovered radioactivity in 1896 (Josephson and

Peterson, 1982). In 1916, the Swedes used irradiation in the processing of strawberries (Webb et al., 1987). In 1921, the first United States patent was submitted by Schwartz, who used X-rays to eliminate *Trichinella spiralis* from pork, and subsequently in 1930, the first patent was submitted in France, by Wust, who wanted to irradiate all foods (Josephson and Peterson, 1982). However, progress in the field of food irradiation was very slow to develop, and it was not until 1953 that any significant research in food irradiation was performed by professor J.T. R. Nickerson and colleagues at the Food Technology department of the Massachusetts Institute of Technology (MIT).

The United States Program

The MIT effort was soon joined by the nearby Gloucester Laboratory of the then-Bureau of Commercial Fisheries (BCF) of the Department of the Interior, predecessor of the National Marine Fisheries Service (NMFS) of the Department of Commerce's National Oceanographic and Atmospheric Administration. Radiation research on the preservation of fishery products spread to other BCF-NMFS labs, notably Ann Arbor, Michigan, for the Great Lakes, and Seattle, Washington, for the Pacific Northwest. Other universities [e.g., Louisiana State University (LSU), Oregon State, California, and Washington] also joined in the effort with Atomic Energy Commission (AEC) support. Much of this work was directed through the United States Army Quartermaster food irradiation program studying radiation sterilized foods, including seafood, for new shelf-stable armed-forces field rations.

The United States was not alone in these efforts. In Canada, fishery product radiation preservation research and development was largely concentrated at the then-Fisheries Research Board's Halifax and Vancouver Laboratories, which generated important pioneering work in this field. Outside North America parallel efforts were undertaken at the then-Torry Research Station, United Kingdom, and in research centers in the Soviet Union, Japan, Australia, and continental Europe. The government fisheries research program actually began in 1960 at the National Marine Fisheries Service/Gloucester Laboratory. In 1965 the construction of the Marine Products Development Irradiator (MPDI) was completed. With the availability of a semi-commercial irradiator, the Gloucester laboratory became a center of seafood irradiation research in the United States, expanding cooperative research efforts, nationally and internationally.

The Marine Products Development Irradiator (MPDI) was one of a family of irradiators built or funded by the Atomic Energy Commission (AEC) in the early to mid-1960s. It differed from its predecessors in that it was designed to be a development facility rather than a research irradiator. The purpose of the MPDI was to determine the commercial feasibility of irradiating fresh seafoods on a large scale and shipping them by common carrier under prevailing conditions of transportation to distant markets while retaining a high degree of freshness. The source originally consisted of 250,000 curies of cobalt-60 and was made of six replaceable plaque units, each containing 16 Brookhaven national laboratory Mark 1 strips of cobalt-60. The normal dose for fillets was about 2,000 Grays at a production rate of one ton/hour, with a maximum-to-minimum dose ratio of 1:4.

Much of the early research on seafood irradiation was on shelf-life extension (Table 1), primarily involving sensory testing and microbiology. Table 1 shows shelf-life data from a number of optimum dose studies for several species of fish and shellfish. Optimum dose was defined as that which gives the longest shelf life without altering the normal characteristics of the product. Maximum dose was that which produced the first significant change in the sensory characteristics of the product. In general, most low-fat species can be irradiated at higher doses than the fattier species. Fatty species such as smelt, halibut, and mackerel require vacuum packing prior to irradiation. Highly pigmented species such as salmon and some high-fat species (e.g., dogfish) may not be acceptable for irradiation treatment because of bleaching and rancidity development. The MPDI was the first available source for large-scale studies on the commercial feasibility of irradiation of seafoods. Although the shelf-life extension studies indicated technical feasibility, it was recognized that these studies were carried out under closely controlled conditions with, in most cases, pristine quality product. For determining commercial feasibility, research plans were developed for producing data on how existing industry processing, han-

Table 1. Shelf Life and Optimum Dose Research Results Reported by NMFS Gloucester Laboratory

Species	Optimum Dose (kGy)	Maximum Dose (kGy)	Iced Shelf Life (days)
Monkfish, *Lophius americanus*	1.5	–	20
Butterfish, *Peprilus triancanthus*	2.3	4.6	49
Cod, *Gadus morhua*	1.5	4.6	30
Dogfish, *Squalus acanthias*	2.0	2.0	7[4]
Winter flounder, *Pleuronectes americanus*	4.5	9.3	22
English sole, *Parophus vetulus*	2–3	–	28–35[1]
Gray sole, *Glyptocephalus cynoglossus*	1–2	–	29
Atlantic halibut, *Hippoglossus hippoglossus*	2–3	5.0	30[2]
Petrale sole, *Eopsetta jordani*	2–3	3.0	28–49
Yellowtail flounder, *Limanda ferruginea*	1–2	–	21–25[1]
Haddock, *Melanogrammus aeglefinus*	1.5–2.5	6.7	30-35
Herring smelt, *Argentina silus*	0.5–1.0	–	15
Mackerel, *Scomber scombrus*	2.5	–	30-35[2]
Ocean perch, *Sebastes marinus*	1.5–2.5	–	30[2]
Pollock, *Pollachius virens*	1.5	–	28-30
Whiting, *Merluccius billincaris*	1.2	2.0–2.5	24–28
Soft-shell clam, *Mya arenaria*	4.5	–	30
Surf clam, *Spisula solidissima*	4.5	–	40
Oysters (not specified)	2.0	8.0	21
Shrimp	1.5–2.0	5.0	21–30[3]

[1] In cooperation with NMFS Seattle Laboratory
[2] Vacuum-packed
[3] In cooperation with Louisiana State University
[4] Became rancid after seven days on ice

dling, and distribution systems would perform with irradiated seafoods. Questions to be answered included:

- What minimum quality level would be acceptable for irradiation to extend shelf life?
- What percentage of New England groundfish was of suitable quality to benefit pasteurization?
- Does the existing distribution system provide adequate temperature control to ensure shelf-life extension?
- How would the fresh-fish industry deal with a product that had double or triple the normal iced shelf-life?

The first studies, which were conveniently named "Pre-and Post-Studies," were designed to determine the maximum period of iced storage of round/gutted fish which would result in an acceptable quality fillet for irradiation preservation. Results of these studies indicated that good commercial practices produced a generally acceptable product at dockside seven to nine days on ice after catch. It should be noted here that for the commercial feasibility studies research, effort was concentrated on the most abundant groundfish species such as cod, haddock, and flounder.

A second series of studies was designed to determine the volume of fish at dockside which met the quality criteria developed from the "Pre- and Post-Studies." A dockside grading program (the "Captlog Survey") was carried out to estimate the percentage of landed product which would be of sufficient quality for irradiation preservation. During the survey more than 4,500 individual fish at dockside, over three seasons of the year, were examined from both the top and the bottom of the catch. The results of this survey showed that 78.6 percent of cod and haddock landed at the Boston Fish Pier were of high enough quality to be suitable for irradiation.

Parallel studies were conducted on truck and rail shipments of fresh fillets, noting the average and maximum temperatures in normal commercial distribution channels. From these it was determined that the majority of fish transported by rail or refrigerated trucks could maintain temperatures of less than 40°F during shipment.

Other distribution/shipping studies on irradiated products included shipments of irradiated fish and shellfish from Iceland, English sole and petrale sole from Seattle, Gulf shrimp from Louisiana State University and some freshwater fish from the Great Lakes. In general, all of these studies indicated that the commercial and economic feasibility of seafood irradiation was positive.

During the same period basic research on the biochemistry of irradiated seafoods was being carried out to determine the biochemical effects of irradiation on seafoods. Determinations were made of vitamin destruction (Vitamin B) and volatile compounds produced by irradiation. Other laboratory studies were done on packaging, synergistic effects of additives, precooking, and re-irradiation.

A shipboard irradiator was also tested in the North Atlantic, the Gulf of Mexico, and the North Pacific to determine the feasibility of irradiating whole or dressed fish at sea. Although results indicated that fish could be irradiated at sea with significantly lower dose levels for increased shelf life, the feasibility of putting commercial-scale irradiators on typical commercial trawlers was not probable.

The research results on radiation sterilization of seafood done in collaboration with the United States Army Natick Laboratories were negative. Irradiation at the 40 to 50 kGy dose levels produced unacceptable products. Serious problems with bleaching, rancidity, browning, and autolysis were experienced. The only marginally acceptable products were sterilized halibut steaks treated by microwave energy to inactivate the enzymes, frozen and irradiated at dry ice temperatures (-80°C). Obviously, the processing costs would be prohibitive, and it was generally agreed that radiation sterilization of seafoods was not commercially feasible.

Table 2. Estimated Storage Times (days) at which Haddock Fillets Became Organoleptically Unacceptable to Consumers at the 100% Rejection Level Based on Irradiation Dose and Storage Temperature (six replicate tests)

Storage Temperature (BF)	Irradiation Dose (kGy)			
	0	0.5	1.0	2.0
33	35	35	44	57
38	18	20	30	31
40	15	20	30	31
42	11	16	20	26
45	9	13	15	20
47	9	0.5	15	19
50	6	9	13	16
55	5	8	10	12
60	3	5	6	8

Table 3. Some Toxin Outbreak Times in Irradiated Haddock and Petrale Sole Fillets Packed in Oxygen-Impermeable Films and Stored at 42°F

Irradiation Dose (kGy)	Toxic Outbreak (days)	Inoculum Spores per g	Substrate
0	55	10^4	Haddock
1.0	55	10^4	Haddock
2.0	55	10^4	Haddock
0	34	10^2	Petrale sole
0.5	34	10^2	Petrale sole
1.0	32	10^2	Petrale sole
2.0	21	10^2	Petrale sole

Safety Issues

One of the most controversial issues associated with low-dose irradiated seafoods was safety relative to a potential botulism hazard. Research on this problem was carried out by Licciardello and Nickerson at MIT and Graikowski and Ecklund at the NMFS Laboratories in Ann Arbor and Seattle. In Gloucester a study was conducted to determine the maximum shelf life of irradiated haddock. The premise of this research was to determine the time of universal rejection. It was proposed that by comparing the reported toxin outbreak times to the time of 100 percent rejection by consumer panels, one could determine a margin of safety. Two groups of panelists were used for the study — one at Michigan State University and one at Gloucester. Table 2 shows the results of six replicate storage studies. Table 3 shows the corresponding toxin outbreak values for irradiated haddock and petrale sole fillets stored at 42°F as reported by Ecklund et al. Comparing the two data sets, one can determine that irradiated haddock appears to have a good safety margin, whereas irradiated petrale sole does not if the maximum shelf life is similar to haddock. However, one can improve the margin of safety by simply reducing the irradiation dose and plan on a shorter shelf-life extension.

In 1968, the Atomic Energy Commission's (AEC) Division of Isotopes Development (DID) had $10 million for source development and facility construction and The Division of Biology and Medicine (DBM) had $3.3 million for contracts on microbiology, etc. In 1972, funding had fallen to $100,000 for DID and $300,000 for DBM. In 1973, DID had no funding, and DBM had $100,000. In the last days of the AEC and through the Department of Energy, funding was given to the Gloucester MPDI in its new role with the International Atomic Energy Administration, a division of the United Nations. From 1973 to 1976 the prime role of the MPDI was to supply irradiated cod and ocean perch for two animal-feeding studies. During this period, the MPDI produced over seven tons of irradiated fish and an equal amount of non-irradiated control samples.

During the 1970s and 1980s, irradiation research continued at Gloucester as a relatively low priority, and the focus of the research changed from shelf-life extension to seafood safety. Research was carried out on inactivating cod worms and other parasites and on the synergistic effects of food additives with low doses of irradiation. Results of these studies indicated that fillets treated with potassium sorbate and irradiated at 1.0 kGy had an extended shelf life equivalent to fillets subjected to 1.5-2.0 kGy.

During the 1980s the source, having gone through four half-lives, became usable for only small-scale irradiation work. The last research involving the MPDI was carried out by Dr. Joe Licciardello on the radiation resistance of pathogens in clam and mussel tissue. Table 4 indicates some of the D-values obtained. This project was done in collaboration with the University of Massachusetts/Lowell, where oysters and clams were irradiated alive to determine lethality and sensory changes in the product as a function of dose.

During the late 1980s, the MPDI facility was surveyed to determine the approximate costs of refueling and modernization of the MPDI. The possibility of developing agreements with other federal agencies was investigated relative to joint operation of the facility. Although both the FDA and the USDA expressed considerable interest in refurbishing the MPDI, the necessary funds

Table 4. Decimal Reduction Dose Determined from Recovery on Different Media for Various Bacteria in a Clam or Mussel Homogenate

Bacteria	Clam	Mussel	Recovery
	D_{10} (Krad)		Media
E. coli	39	41	TSYE agar
	39	47	MacConkey agar
	42	48	VRB agar
S. typhimurium	60	63	TSYE agar
		58	XLD agar
Strept. fascalis	100	105	TSYE agar
	97	105	KF agar
Sh. flexneri	35	34	TSYE agar
	30	24	XLD agar
	32	29	Hektoen agar

(~$250,000) could not be committed, due to budget restraints. A decision was then made to decommission the MPDI, and finally in April 1989 the source material was removed and transported to LSU, bringing to a close the 24-year history of the MPDI.

As with the later work done at Gloucester, emphasis worldwide has shifted from spoilage delay and prevention to control of parasites and microbial pathogens, thereby reducing fishery product consumption-related illness.

It should be noted that isotopes are not the only means of delivering irradiation doses. Electron beam accelerators for research have been in use since the 1960s, and, for industrial purposes other than food applications, in use since the 1980s. Newer versions offer accelerated electron energy conversion to highly penetrating x-ray (Bremmstrahlung) energy.

International Aspects

Canada and the United States have yet to permit any commercial fishery product irradiation applications, although both have contributed immensely to its research, development, and safety testing. However, several other countries, including Bangladesh, Belgium, Brazil, Chile, France, India, The Netherlands, Syria, Thailand, and Vietnam, have approved one or more fishery product irradiation applications, including dried fish insect disinfestation as of 1993 (Table 5). Virtually all came about as a result of the Codex Alimentarius Commission's 1983 adoption of the Codex General Standard for Irradiated Foods and Recommended International Code of Practice for the Operation of Radiation Facilities Used for the Treatment of Foods. The FAO recommends the unconditional use of radiation doses up to 1 kGy for insect disinfestation of dried "teleost fish and fish products," and up to 3.2 kGy for general and pathogenic microbial reduction of the same.

Parasites are not mentioned; however, many, if not all, can be inactivated at or near 2.2 kGy; some at much lower doses. In 1991 the Joint FAO/WHO/IAEA International Consultative Group on Food Irradiation, the Secretariat of which resides at the Food Preservation Section, Joint FAO/IAEA Division, IAEA, Vienna, published a "Code of

Table 5. Current Approval of Irradiated Fish and Seafood

Country	Product Name	Type of Clearance	Date of Clearance	Dose (kGy) Min.	Dose (kGy) Max.
Bangladesh	Fish	Unconditional	83.12.28	–	2.20
Bangladesh	Shrimp	Conditional	83.12.28	–	5.00
Belgium	Shrimp	Conditional	88.11.30	3.0	5.00
Brazil	Fish	Unconditional	85.03.07	–	2.20
Brazil	Fish products	Unconditional	85.03.07	–	2.20
Chile	Fish	Unconditional	82.12.29	–	2.20
Chile	Teleost fish	Unconditional	82.12.19	–	2.20
France	Shrimp	Unconditional	90.10.02	–	5.00
Korea, Rep. of	Fish powder	Unconditional	91.12.14	–	7.00
Korea, Rep. of	Seafood powder	Unconditional	91.12.14	–	7.00
Netherlands	Frozen shrimp	Unconditional	88.10.20	–	7.00
Netherlands	Fresh shrimp	Conditional	88.10.20	–	1.00
Netherlands	Fish fillets	Conditional	88.10.20	–	1.00
South Africa	Fresh fish	Conditional	87.03.09	0.500	2.00
Syria	Teleost fish	Unconditional	86.08.02	–	2.20
Thailand	Fish products	Unconditional	86.12.04	–	2.20
Thailand	Frozen shrimp	Unconditional	86.12.04	–	5.00
United Kingdom	Fish	Unconditional	91.01.01	–	3.00*
United Kingdom	Shellfish	Unconditional	91.01.01	–	3.00*
Vietnam	Dried fish	Conditional	89.11.03	–	1.00

* Overall average.

Good Irradiation Practice of Insect Disinfestation of Dried Fish and Salted-and-Dried Fish," and another ". . . for the Control of Microflora in Fish, Frog Legs and Shrimps" that covers both spoilage and pathogenic bacteria. Here, again, parasites are not covered except for a passing mention in the "Scope" statement of the latter that "when irradiating fresh fish for bacterial reduction, any parasites present are rendered non-infective," with the caveat that "some parasites such as the agnostics may be resistant at doses recommended in this Code" (i.e., 1 to 2 kGy for fresh, non-frozen items). Thus, again, the required dose depends on the parasite.

As mentioned earlier, several countries had tested fish irradiation onboard-ship at sea, and it has been rumored that the former Soviet Union and possibly other long-range factory ship operators have irradiated commercial quantities of seafood on-board factory ships during extended fishing cruises. It is known with certainty, however, that irradiator operators in Belgium, France, and The Netherlands have been regularly treating sizeable quantities of imported frozen fishery products in recent years for microbial pathogen control. A serious shigellosis outbreak in The Netherlands around a decade ago, traced to the consumption of Asian frozen shrimp, prompted Dutch public health authorities to require that such products must be irradiated as a condition for marketing since the irradiation capacity was available and its effectiveness was well understood and appreciated. Ironically, this rendering of marketable seafood more hygienic and safe gave rise to the anti-food irradiation activist-coined derogatory term "dutching" after it was learned several years ago that Asian frozen shrimp legally irradiated in Holland was illegally brought into the United Kingdom by the owners of the product.

Apart from the foregoing, there is as yet very little commercial fishery product irradiation to report on. Small quantities of radiation-disinfested (of insect pests) dried fish have recently been test-marketed in Bangladesh, as has radurized fresh fish in Chile in recent years. India permits the irradiation of frozen shrimp and frog legs for export, which no doubt has been and is being done using irradiators there. Irradiated prawns were test-marketed in Australia years before food irradiation became the political controversy there that it is today, resulting in a three-year moritorium. The very positive results of an Australian government-sponsored World Health Organization reevaluation of irradiated food safety were reported out at WHO Headquarters, Geneva, in late May, 1992. It has since become a "political football" in Australia, and it remains to be seen if its government and other remaining doubters will once and for all accept the inescapable fact of safety, proven beyond reasonable doubt and yet again reconfirmed, so that potential users may decide whether or not to use it unimpeded by this lingering non-issue. Worldwide fishery product irradiation growth is inevitable, the only question being at what rate.

CONCLUSION

Irradiation is a safe, scientifically-sound technology approved for some uses in more than 25 countries (Table 5). It will:

(1) Improve the quality of a seafood product by delaying deterioration.
(2) Increase industry's options for handling and distributing different species of seafood.
(3) Help reduce global post-harvest losses.
(4) Provide a treatment to reduce pathogenic organisms.
(5) Reduce refrigeration requirements.

Authorized sources include:

(1) Gamma rays – cobalt-60
(2) Gamma rays – cesium-137
(3) Machine sources (electrons) not to exceed 10 million electronvolts
(4) Machine sources (x-rays) not to exceed 5 million electronvolts

As a final note, it is ironic that United States regulations require irradiation applications to go through a food-additive petition process while our astronauts dine on a variety (ham, turkey, beef steak, corned beef, and bread) of irradiated foods in space.

REFERENCES

Bucove, G.O. and G.M. Pigott. 1977. Production of a Functional Protein from Fish Waste by Enzymatic Digestion. *Proceedings 7th National Symposium on Food Processing Wastes.*

Chaveesuk, R., J. P. Smith, and B.K. Simpson. 1993. Production of fish sauces and acceleration of sauce fermentation using protealytic enzymes. *J. Aquatic Food Prod. Technol.* **2**(3):59-77.

Crawford, D. L. 1980. Meat Yield and Shell Removal Functions of Shrimp Processing. Oregon State University, Cooperative Extension Special Report #597. Newport, OR.

Gildberg, A. and K.A. Almas. 1986. Utilization of fish waste. *Food Engineering and Process Applications.* M. Le Maguer and P. Jelen (eds.). Vol. 2, p. 383. Applied Science Publishers, London.

Haard, N.F. 1992. A review of protealytic enzymes from marine organisms and their application in the food industry. *J. Aquatic Food Prod. Technol.* **1** (1):17-35.

Hale, M.B. 1974. Using enzymes to make fish protein concentrate. *Marine Fish Review* **36**(2):15-18.

Hempl, E. 1983. Taking a short cut from the laboratory to industrial scale production. *Information Marketing Digest* **4**:18.

Hoyle, N.T. and J.H. Merritt. 1994. Quality of fish protein hydrolysate from herring. *J. Food Science.* **59**(1):76-79.

Hyung, J.S., H. Lee, Y. C.Hong, and C.Y. Han. 1992. Production of protein hydrolysates and plastein from Alaska pollock. *J. Korean Agri Chem Soc.* **35**(5):339-345.

Josephson, E.S. and M.S. Peterson. 1982. *Preservation of Food by Ionizing Radiation.* CRC Press, Boca Raton, FL.

Lampila, L.E. 1992. Functions and uses of phosphates in the seafood industry. *J. Aquatic Food Prod.Technol.* **1**(3/4):29-41.

Mahon, J.H. 1962. Preservation of Fish. U.S. Patent 3,036,923.

National Academy of Sciences. 1965. *Radiation Preservation of Foods, A Proceedings.* Washington, DC.

Novo Industries. 1984. Enzymes at Work. Corporate communication. Denmark.

Otwell, W.S., R. Iyengar, and A.J. McEvily. 1991. Inhibition of shrimp melanosis by 4-Hexyl Resorcinol. *J. Aquatic Food Prod. Technol.* **1**(1):53-65.

Seafood Processing and Packaging Magazine. **40**(6):18. 1986. Enzymes work wonders for caviar and squid.

Spencer, P. 1949. Method of Treating Food Stuffs. U.S. Patent 2,495,429.

Spencer, P. 1950. Means for Treating Food Stuffs. U.S. Patent 2,605,383.

Urbain, W.M. 1986. In *Food Irradiation.* Academic Press Inc. Orlando, FL.

Webb, T., T. Lang, and K. Tucker. 1987. In *Food Irradiation: Who Wants It?* Thorsons, London.

Varlik, C. N. Gokogluand, and H. Gun. 1993. The effect of temperature on the penetration of vinegar and salt. *GIDA.* **18**(4):223-228.

Further Processed Seafood

Ronald J. Sasiela

INTRODUCTION

The decision to process seafood further is often complex and requires a sound understanding of the interrelationship of the substrate, the equipment that will be performing the transformation, the added ingredient(s), and regulatory considerations. Further processing of seafood encompasses the transformation of a raw aquatic commodity into an item that is designed to meet a consumer need. These needs can be extremely diverse — driven by any number of reasons (Table 1). Factors such as portion control, taste, intended method of cooking, cost, and resource availability certainly play obvious and important roles in this endeavor. Less apparent factors such as competitive activity and legal considerations (regulatory, patent, etc.) can also be as much a driving force.

Seafoods that have historically been subjected to further processing by being coated with batter and breading include *finfish* (e.g., cod, haddock, pollock, perch, catfish); *crustacea* (e.g., shrimp, crab cakes, crawfish); *mollusks* (e.g., clams — both whole and stripped — oysters, scallops, squid); as well as an assortment of specialty items (such as fish cakes, surimi, and stuffed portions). Perhaps the most ubiquitous product is the fish stick ("fish finger" in Europe) which has been credited as being developed by Clarence Birdseye in New England in the 1940s. Since that time, significant advances have taken place in the refinement of

Table 1. Typical Factors to Consider for a Further Processed Product

1. Portion Weight
2. Species
3. Plate Coverage
4. Seafood Content
5. Cost
6. Organoleptic
   a. Crust color
   b. Aroma
   c. Sound
   d. Touch
   e. Taste
      aa. Flavor
      bb. Texture
7. Government Inspection Services
   a. Grade A
   b. P. U. F. I.
   c. Child Nutrition Statement
8. Nutritional Profile
9. Raw Material (Seafood) Availability
10. Plant Equipment
11. Legal Considerations
    a. Patent(s)
    b. Trademark
    c. Copyright
    d. Regulatory

11. d. —aa. Label approval
    bb. CFR
    cc. U. S. D. C. operating guidelines
    dd. Ingredient Statement
    ee. EPA
    ff. OSHA
12. Customer Cooking Equipment
13. Market Research Data
14. Competitive Activity
15. Packaging
    a. Net Weight
    b. Graphics
    c. Cooking Directions
    d. Nutritional Facts Panel (if retail)
16. Defects Level
17. Supplier Specifications
18. Shelf-Life Testing
19. Consumer "Abuse" Testing
20. Product Scheduling Efficiency
21. Coating Ingredients Characteristics
    a. Dustiness
    b. Foaminess
    c. Hygroscopic Tendency
    d. Resistance of Crumbs to Breakdown
22. Microwaveability

Table 2. Minimum Flesh Content Requirements for USDC-Inspected Products

In order to assure that all users of USDC-inspected fishery products are aware of the minimum flesh requirements required in USDC-inspected breaded and battered products, the following list of minimum flesh requirements for standardized and nonstandardized breaded and battered products is provided. These requirements apply to all species of battered and breaded fish and shellfish.

Products	USDC Grade Mark	PUFI Mark
Fish Fillets		
Raw Breaded Fillets	—*	50%
Precooked Breaded Fillets	—	50%
Precooked Crispy/Crunchy Fillets	—	50%
Precooked Battered Fish Fillets	—	40%
Fish Portions		
Raw Breaded Fish Portions	75%	50%
Precooked Breaded Fish Portions	65%	50%
Precooked Battered Fish Portions	—	40%
Fish Sticks		
Raw Breaded Fish Sticks	72%	50%
Precooked Breaded Fish Sticks	60%	50%
Precooked Battered Fish Sticks	—	40%
Scallops		
Raw Breaded Scallops	50%	50%
Precooked Breaded Scallops	50%	50%
Precooked Crispy/Crunchy Scallops	—	50%
Precooked Battered Scallops	—	40%
Shrimp		
Lightly Breaded Shrimp**	65%	65%
Raw Breaded Shrimp**	50%	50%
Precooked Crispy/Crunchy Shrimp	—	50%
Precooked Battered Shrimp	—	40%
Imitation Breaded Shrimp***	—	No Minimum Encouraged to put % on label
Oysters		
Raw Breaded Oysters****	—	50%
Precooked Breaded Oysters****	—	50%
Precooked Crispy/Crunchy Oysters****	—	50%
Precooked Battered Oysters****	—	40%
Miscellaneous		
Fish and Seafood Cakes	—	35%
Extruded and Breaded Products	—	35%

* No USDC grading standard currently exists.

** FDA standards of identity require that any products with a USDC minimum of 50% shrimp flesh by weight, if labeled "lightly breaded," must contain not less than 65% shrimp flesh.

*** Any product with a standard of identity which contains less flesh than the standard calls for must be labeled imitation.

**** Flesh content on oyster products can only be determined on an input weight basis during production.

quick-freezing techniques, seafood distribution, and the technology of coating.

The coatings of these different seafood items have several elements in common, and will therefore be treated collectively. Where coating differences do occur because of the particular substrate, they will be identified and highlighted in the text or by means of a separate flow chart. These altered processing techniques are ordinarily associated with the preparation of the seafood substrate prior to, rather than at, the coating application step(s).

U. S. DEPARTMENT OF COMMERCE INSPECTION

The United States Department of Commerce is the federally authorized agency responsible for the grading of seafood products. They offer either continuous or "lot inspection" services that are voluntarily contracted by a firm or individual for a fee. If all the necessary regulations are found to be in compliance, the company's packages will be eligible to display the "PUFI" (Packed/Processed Under Federal Inspection) symbol. A "Grade A" shield is also available, provided certain defect limits and minimum seafood content levels are maintained (Table 2).

The agency uses an assortment of inspection forms to record its data. The inspector's signature on the form serves as documentation with respect to size of the lot, species, sampling plan, identifying codes, net weight, frozen and cooked condition, etc. (Chart 1).

All child nutrition (CN)-labeled products are required to display the PUFI mark (discussed more fully later in this chapter) (Figure 1). The off-site "lot inspection" program is restricted to what an interested party wants independently evaluated. It is also used for import and export documentation.

FISH-BLOCK UTILIZATION

Since many of the further processed fish items are produced from commercial fish blocks, an understanding of how these blocks are handled will be necessary.

Fish blocks offer the processor unsurpassed flexibility in portion size and design at a lower cost than any other frozen form. Blocks can deliver a seemingly endless assortment of shapes.

Figure 1. Example of a seafood Child Nutrition Label for 1-ounce fish sticks.

Chart 1. Inspection Form for Recording Data (U.S. Department of Commerce)

NOAA FORM 89-837
(8-85)

U. S. DEPARTMENT OF COMMERCE
NATIONAL OCEANIC AND ATMOSPHERIC ADMINISTRATION

SCORE SHEET - FROZEN BREADED FISH

TYPE	SAMPLING PLAN	PAGE OF PAGES
☐ PORTIONS ☐ RAW ☐ STICKS ☐ FRIED	☐ 260 ☐ 105D	DATE

COMPANY CODE OR APPLICANT	INSPECTOR'S NUMBER	LABEL CODE	REF. LOT NO.	POUNDAGE	CONTAINER SIZE	SPECIES	TYPE OF INSPECTION	COUNTRY OF ORIGIN
			LOT SIZE		SAMPLE SIZE	STYLE	☐ LOT ☐ CONTRACT	

	ITEMS INSPECTED		SAMPLE NUMBER (INSERT AS NEEDED)													AVERAGE
GENERAL	CONTAINER IDENTIFICATION	CARTON CODE														
		CASE CODE														
	NET WEIGHT IN OUNCES															
	NUMBER OF PORTIONS PER CONTAINER															
FROZEN	CONDITION OF CONTAINER OIL ☐ BREADING ☐ FROST ☐	SMALL														
		LARGE														
	EASE OF SEPARATION	MINOR														
		MAJOR														
	BROKEN PORTION															
	DAMAGED PORTION	MINOR														
		MAJOR														
	UNIFORMITY OF SIZE															
	UNIFORMITY OF WEIGHT															
COOKED	DISTORTION	MINOR														
		MAJOR														
	COLOR	MINOR														
		MAJOR														
	COATING DEFECTS CURDS ☐ BREAKS ☐ BLISTERS ☐ RIDGES ☐	MINOR														
		MAJOR														
	BLEMISHES SKIN ☐ DISCOLORATIONS ☐ BLOODSPOTS ☐ BRUISES ☐	MINOR														
		MAJOR														
	BONES															
	TEXTURE OF COATING SOGGY ☐ DOUGHY ☐ TOUGH ☐ DRY ☐ OILY ☐	SMALL														
		LARGE														
	TEXTURE OF FISH FLESH RUBBERY ☐ TOUGH ☐ DRY ☐ SOFT ☐ MUSHY ☐	SMALL														
		LARGE														
TOTAL DEDUCTIONS																
TOTAL SCORE																
(CN) RAW INPUT																
PERCENT FISH FLESH (SCRAPE END PRODUCT)																
FLAVOR AND ODOR																
FINAL GRADE																

REMARKS (*Reasons for degrading product*)

SIGNATURE (*Official Inspector*)

Integrate this with the various two- or three-dimensional shaping technologies and it becomes evident that there is no lack of variety available to suit consumer needs (50 CFR 264.101 sets forth the standards for fish blocks). Figure 2 shows the theoretical dimensions of the most commonly traded 16 1/2-pound fish block produced for international commerce. The density of the fish block can be calculated from the dimensions in Figure 2:

- 19" x 10" x 2½" = 475 cubic inches
- 16½ pounds x 16 ounces/pound = 264 ounces

Density = 264/475 ounces/cubic inch
= **0.556 ounces/cubic inch**

This density value is useful in estimating the weight from the shape's volume of different block cuts made with saws, knives, or chopping blades.

Fish blocks are typically subdivided using specialized saws. Photo 1 shows the interior of such a high-quality band saw. Its design, made of tenzaloy aluminum casting for strength and rigidity, has rounded corners and surfaces easy to sanitize. The precision-balanced tensioned wheels, adjustable height blade guide, ball-bearing fish-block support table, safety-shielded blade, high-speed motor, and oversized dual ball-bearing shaft allows this machine to track its blade without wavering. With a saw kerf less than 0.03 inches, traveling at 6750 feet per minute, it provides long blade and guide life.

The fish block is initially subdivided by the saw into an assortment of first cuts, as illustrated in Figures 3 and 4. Using the slabs of Figure 4, for example, one may want to further subdivide it to create, with a second cut, a fish piece 3/8-inch thick (Figure 5). The volume of this final fish rectangle, measuring 2½" x 2 3/8" x 3/8", is 2.226 cubic inches. Multiplying that value by the density will show that the piece weighs 1.24 ounces (2.226 x .556). The fish block will deliver 208 (8 slabs x 26 pieces/slab) of these 1.24-ounce pieces (neglect-

Photo 1. Fish block saw with opened housing being swabbed by a microbiologist.

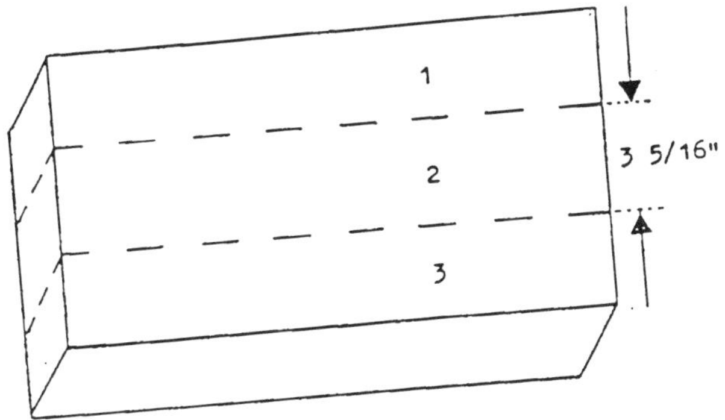

Figure 3. Fish block cut lengthwise into three equal bars.

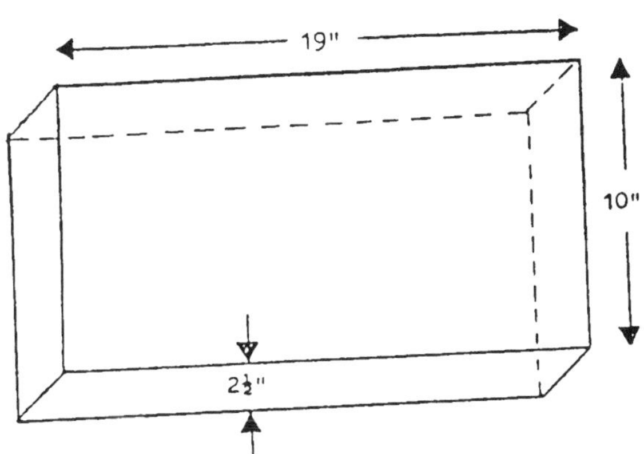

Figure 2. Dimensions of a 16½-pound fish block.

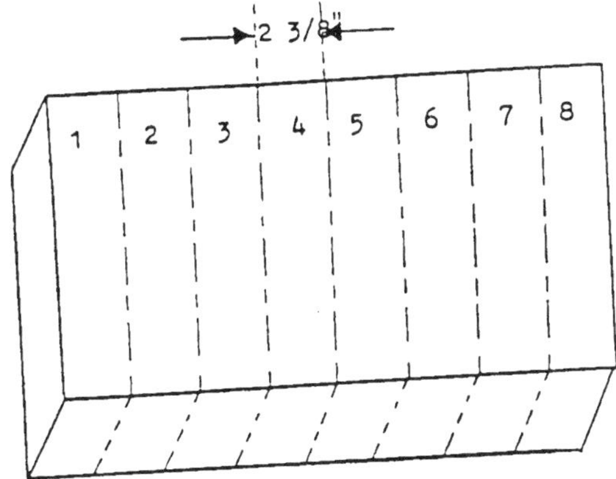

Figure 4. Fish block cut crosswise into eight equal bars.

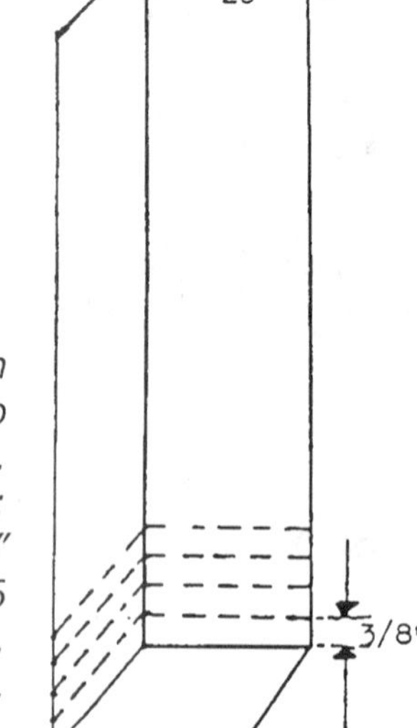

Figure 5. A 2 3/8" fish bar being cut into 26 final pieces. Portion size: 2½" x 2 3/8" x 3/8" Portion cube: 2.226 Portion weight: 1.24 oz. 62% input = 2 oz. coated portion.

ing any cutting waste). If we were to coat it to a final weight of 2.0 ounces, the finished product would contain 62 percent fish, or "input." Figures 6 through 10 show, in an abbreviated form, the block-cutting pattern for traditional fish sticks and portions (a 4¾ inch-long wedge suitable for batter coating); a square portion ideal for placement on a 4-inch sandwich bun; a 2½ inch-wide diamond-shaped portion; and a narrower 1 inch-wide diamond-shaped strip.

When fish blocks are designed for subdivision, consideration should be given to the manner in which the fillets were arranged in the block at the time of freezing. The two most common fillet arrangement methods are **lengthwise**, i.e., oriented with the 19-inch block dimension, or **crosswise,** with the 10-inch dimension. When using modern

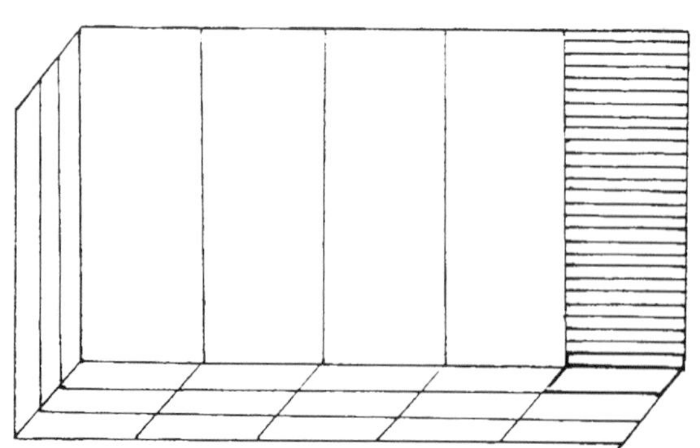

Figure 6. Typical block-cutting diagram to manufacture a fish stick that is 3¾"L x 13/16"W x 3/8"T, at 0.65 ounce (405 sticks).

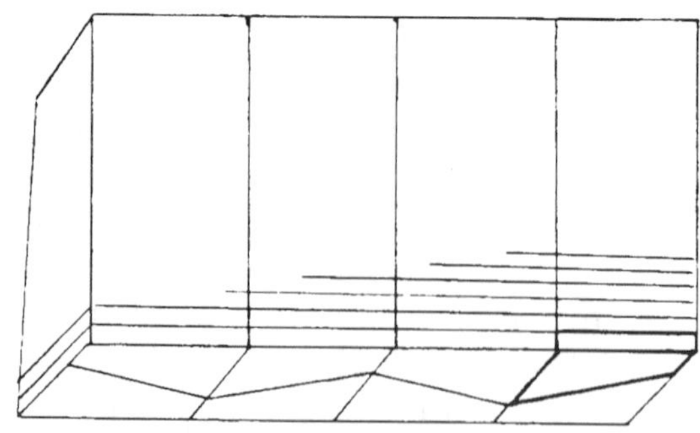

Figure 7. Typical block-cutting diagram to manufacture a wedge-shaped portion 4¾"L x 1¾"W x ¾"W x ½"T, at 1.65 ounces (160 pieces).

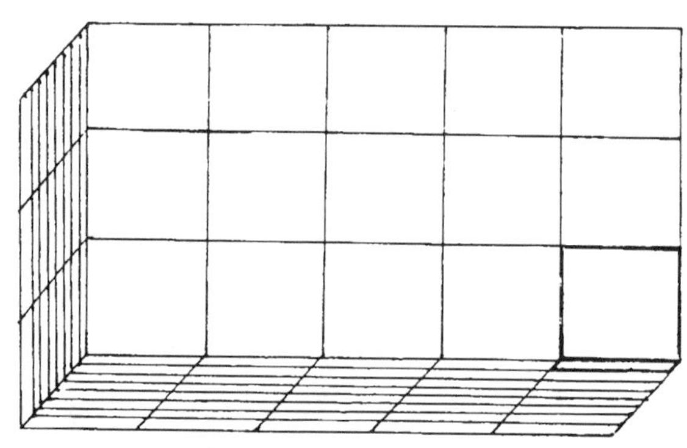

Figure 8. Typical block-cutting diagram to manufacture a nearly-square portion 3¾"L x 3 5/16"W x 5 1/6"T, at 2.2 ounces (120 pieces).

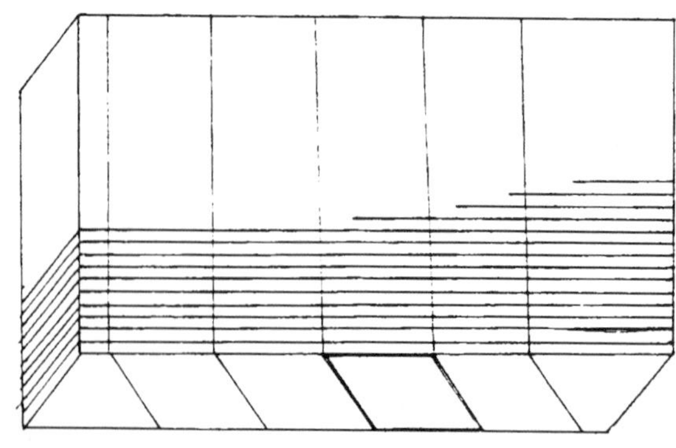

Figure 9. Typical block-cutting diagram to manufacture a wide diamond-shaped portion 1½"W x 3/8"T, at average 1.7 ounces each (156 pieces). Note ratio of **4** center *diamond shapes:* **2** end *rhomboid shapes per shim.*

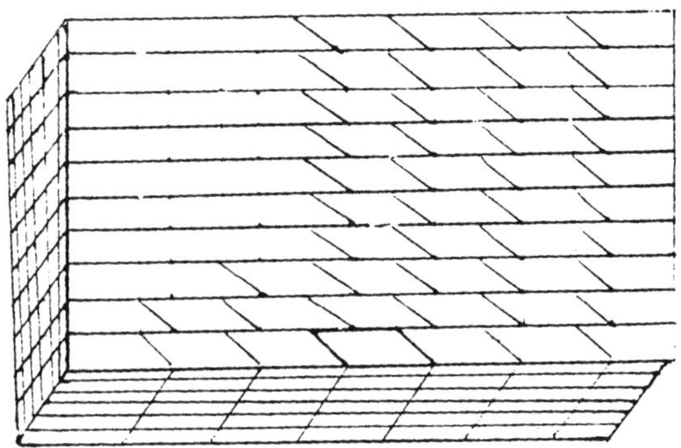

*Figure 10. Typical block-cutting diagram to manufacture a narrow diamond-shaped strip 1"W x 1 3/32"T, at average 0.63 ounce (420 pieces). Note ratio of **5** diamond shapes: **2** rhomboid shape per slab.*

fish-block blade-slicing techniques, this is a vital selection requirement in order to make full use of the block raw material and avoid loss of yield. Cutting with, rather than across, the grain of the fish reduces breakage and increases the fish piece cohesiveness. Photo 2 shows the solid continuous nature of the fish fillets in a frozen 19-inch long fish block. Also visible is the wax carton that serves as a freezing container; this design retards dehydration during storage and facilitates its quick and complete removal immediately before the block's use at the beginning of the production line.

BATTERED AND BREADED PRODUCTS

There are five general categories of coated seafood products:

1. Precooked (for oven finishing)
2. Raw Breaded (for fryer finishing)
3. Battered (for fryer finishing)
4. Battered (for oven finishing)
5. Sauced (for oven finishing)

Within these groupings, there is a wide assortment of variations. In attempts to combine categories in an effort to consolidate products, a compromise is often made in expecting one product to do two tasks well. The result is that the product may fail to perform as well in either task. For example, developing a single-coated portion for both frying and baking may result in a fried item that is excessively oily with a dark crust color, while the baked version will have a pale, "uncrisp" crust. However, market variability and equipment or supply limitations may require such a product to be made in order to fit a larger section of the target market.

Coating Ingredient Considerations

Flow Chart 1 depicts the typical process arrangement for a precooked seafood product. The type of coating generally used for this system is a corn or wheat flour-based aqueous cohesive batter, along with either a cracker-meal or breadcrumb breading. Since the time that the product is in the hot (375°F to 400°F) frying oil is limited to only about 30 seconds, the crumb coatings used commercially are generally colored with natural vegetable extracts to assist in creating an attractive, golden-brown crust appearance. These coloring agents include paprika, annatto, turmeric, or caramel. Crumbs have also been toasted to generate a desired color. Synthetic certified food colors can also be used, but are not preferred by con-

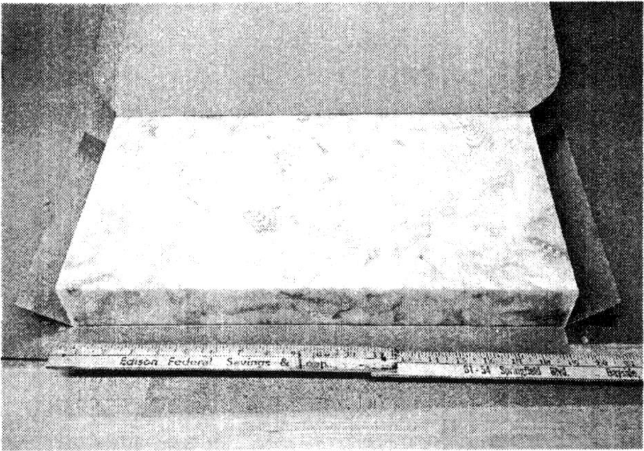

Photo 2. A 19" long fish block illustrating fillet cohesiveness.

Flow Chart 1 Typical Precooked Process
Seafood Substrate
Dry Predust (Optional)
Liquid Batter
Medium-Mesh Breading
Hot Frying Oil
In-Line Freezer
Packaging and Packing
Storage Freezer

sumers. By keeping the coated product's time in the fryer relatively short, actual heat transfer to the coated product is restricted to the coating's surface, while the fish core itself remains frozen. Little quality is lost, since the entire product is immediately brought back down to below freezing within minutes by an in-line freezer.

During the precooking step important changes take place within the coating. Among these are:

- starch gelatinization;
- protein coagulation;
- browning reactions;
- fat absorption;
- leavening release;
- moisture reduction; and
- flavor development.

The rates at which these individual events take place are a function of the types and amounts of ingredients used in the batter and breading, frying oil conditions, and process line parameters. Adjustment of these variables has traditionally been skillfully practiced by technical service food technologists employed by coating mix manufacturing firms, and partly accounts for the wide variety of tasty finished products available in the marketplace. Many processors use coating and seasoning mix companies to supply them with the large quantities needed for production. Therefore, it becomes very convenient to have the firm that was initially involved in the coating's development use its raw materials and special breading manufacturing equipment when preparing samples for screening. It is also rare for a seafood plant to undertake manufacture of its own coating needs. The low usage in a single plant does not justify the heavy costs of bulk flour handling. Concerns over quality control, labor, purchasing, and warehousing all combine to suggest "leaving it to the experts." Many coating and seasoning firms also maintain pilot plants to prepare sufficient quantities of prototype products for testing purposes. The pilot plant is also a more controlled and less costly environment in which to establish process controls and to explore whether any serious problems would develop during full-line scale-up. Issues of confidentiality should be clarified early in the relationship with ingredient and equipment vendors, so that misunderstandings are less likely to develop later. A basic understanding of some of the more important formulation and process parameters will better equip the seafood development technologist in communicating his needs to a coating firm. This primer would also be helpful, should technologists elect to conduct tests at their own facilities.

The formulation of a coating must satisfy two primary criteria that must be established before any further consideration can be given to secondary formulation issues. The issues are determined by answering these questions:

- How will the finished food product be cooked?
- What type of texture is desired?

The first answer, for example, can be a regular oven, microwave oven, flat griddle, deep-fat fryer, convection oven, steamer, conveyorized flame broiler, high-intensity quartz oven, jet-air impingement oven, etc. The second answer can be granular (coarse to fine mesh), liquid batter (tempura and corn-dog style), flour consistency duster, dough-like (fried pies), wrapped (egg rolls), etc.

Secondary product issues would include items such as the seafood-to-coating ratio, finished weight, shape, raw and cooked crust color, flavor and seasonings, regulatory concerns, and nutritional considerations.

The manner in which a further processed seafood item is actually cooked is often dictated by the cooking appliance used by the consumer. This is generally either a household still-air or microwave oven. The range of preferred heating appliances is, however, broader in the food service sector. There, deep-fat frying or convection baking are the primary choices, with rapid preparation time being an all-important criterion.

One of the controlling factors of coating texture is the granulation of the outermost coating. Figures 11 to 14 depict the combined influence that this single coating variable has on the four elementary attributes of crust color, crispness, coverage, and pick-up. Photo 3 shows some weighed, blended bread crumbs about to be placed on the top-stacked sieve, with increasingly finer mesh screens lower in the stack. As the stack is vibrated for a predetermined length of time (five to 15 minutes), the coarsest particles are retained on the upper screens and the finer mesh particles pass to the lower screens. By weighing the crumbs on each screen, one can establish the proportion of each in the starting sample and draw conclusions about

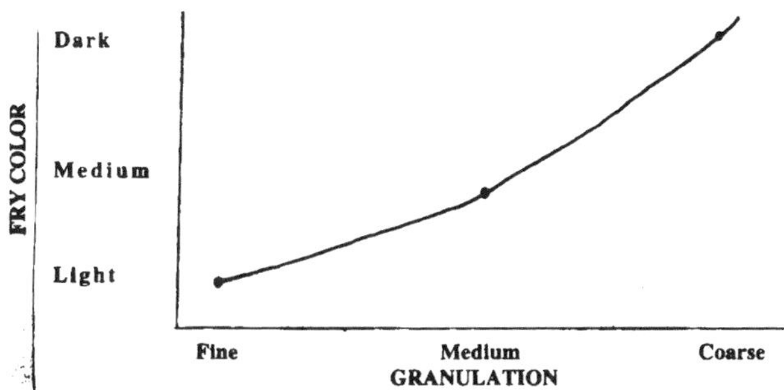

Figure 11. Breading granulation can influence crust color.

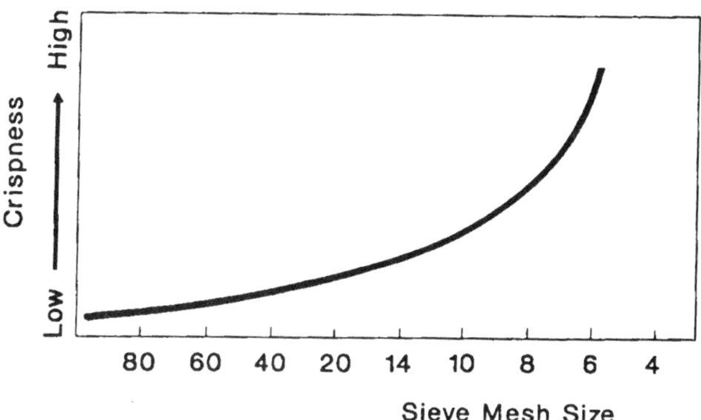

Figure 12. Breading granulation affects crispness.

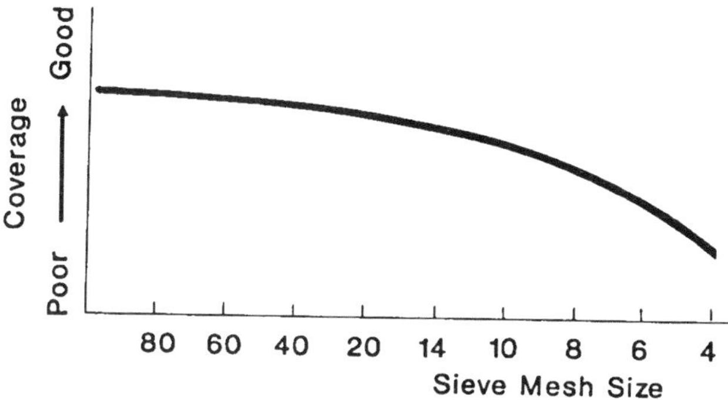

Figure 13. Breading granulation influences coverage.

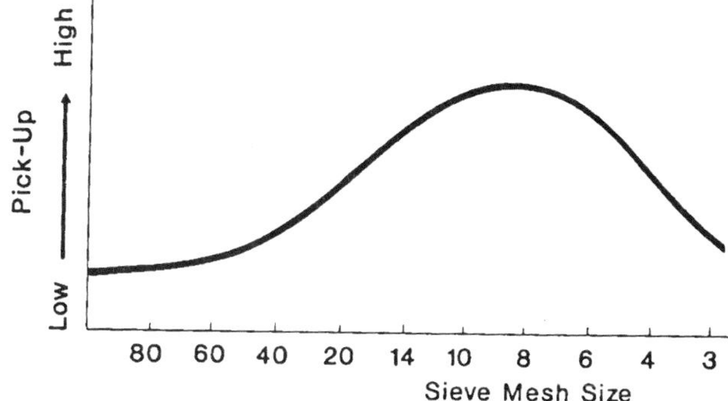

Figure 14. Breading granulation contributes to pick-up.

Photo 3. Breading particle size being fractionated onto six screens using vibratory shaker apparatus.

the coating's influence on crust color highlighting, coverage, pick-up, and crispness potential. The effect of different breading machines can be likewise assessed to determine which has the least effect on altering the particle size spectrum compared to the non-recirculated control. A decision can then be made as to which breading machine model will best serve the product's purposes.

Sensory Issues Dictate Formulation

When initially viewed by the consumer, the food begins to make a visual impression. For example, even without the ability to see the food beneath the coarse, Italian-seasoned, golden brown coating of a calamari ring, the consumer's senses of sight, and smell have been heightened in anticipation. As the ring is touched to the mouth, the brain begins to register some components of the food's texture.

Next, what sound does the food make? Coated foods are often described as either being crispy or crunchy. The former has been ascribed to high-pitched sound that travels outside the mouth to the ear, while crunchy food is characterized by a lower-pitched sound that vibrates through the jaw bone to the inner ear.

Taste is the most important and complex — a mixed sense of flavor and texture. Is the food crispy, soggy, peppery, brittle, sweet, soft, salty, mushy, sour, crusty, bitter, firm, gummy, oily, etc.? Are the coating and the substrate complementary or competing? Does the coating overwhelm the subtlety of the seafood? The trend has been to increase the flavor intensity of foods as a way of exciting the palate, yet often a "perfect" balance is found by avoiding the extremes.

Figure 15 is a helpful tool for understanding typical agents utilized in the formulation of batter and breading systems. It can be thought of as a see-saw where one lever arm accommodates the "strengtheners" and the other the "tenderizers." Moving out from the fulcrum, the influence of each ingredient is relatively greater. For example, by increasing the egg white by one percent, a coating would have several times more firming effect on the coating than a similar increase in the wheat flour or starch. Likewise, by reducing the leavening content in the batter, a marked toughening of the coating might be realized. In other words, a coating can be made firmer by reducing the tenderizers and/or increasing the strengtheners. Cost would play a role in the approach to be taken. How firm or tender the coating needs to be is a function of several factors, such as market research data, process considerations, taste panel results, and, of course, the characteristics of the seafood substrate itself.

Figure 16 can be used as an aid to control the crust color development in a coated food product. Its balance has been intentionally tilted so the lighter fry color elements "outweigh" the darker fry color factors. CO_2 represents leavening content. Low amylase wheat flour can help control excessive browning, particularly if there is anticipated freeze-thaw abuse to the product.

The condition of the frying oil, along with granulation effects (Figure 11), also plays a role in this important color attribute. Photo 4 illustrates the proper method of using a crust color comparison chart. The use of an accurate timing device (in the background) to measure the cook time in the oil, and a calibrated thermometer (in the foreground) to check the oil bath's temperature assure

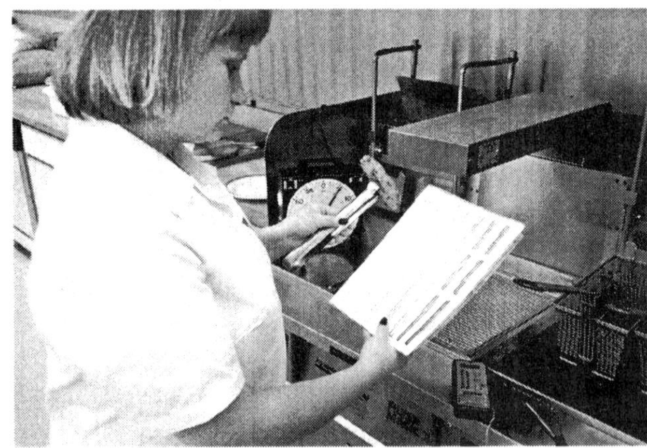

Photo 4. Food technologist comparing fried fish strip crust to reference color chart.

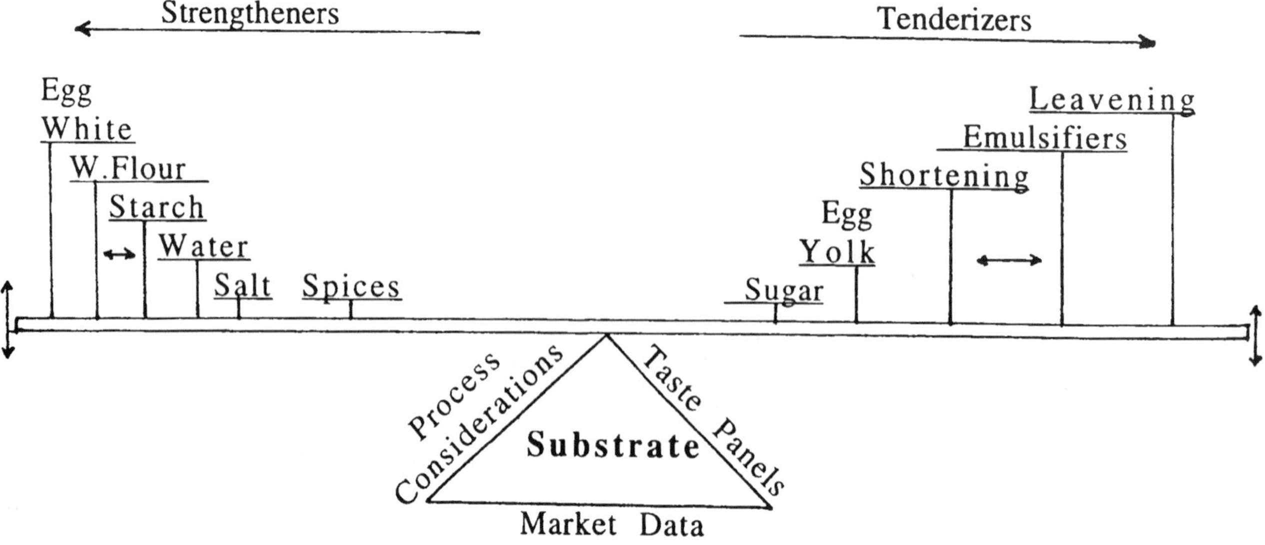

Figure 15. The coated product's formulation balance.

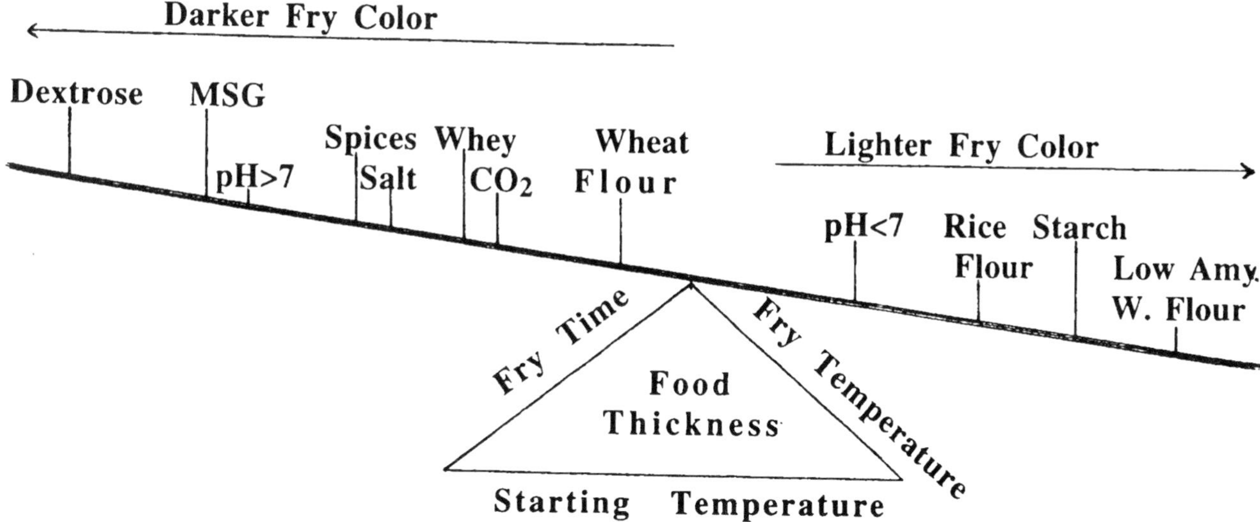

Figure 16. Fry color considerations.

that a true objective assessment of the fry color has been made. Formulating the coating to be slightly lighter than desired will result in a more satisfactory product for the consumer, since any temperature abuse will tend to accelerate browning.

Raw breaded products, in particular, tend to develop additional crust color with freezer-storage and freeze-thaw abuse. Coating systems that rely on raw flour based or very fine granulation (mesh<60) breading mixes require extra attention to avoid excessive dust in the production department. Issues of employee safety caused by the fine, air-borne particles, slipperiness of the floor adjacent to the breading machine, and sanitation concerns all dictate that dust be kept to an absolute minimum. Figure 17 documents several variables that need to be considered with this breading system.

PRECOOKED

The Code of Federal Regulations describes frozen fried fish portions and sticks as "uniformly shaped, unglazed masses of cohering pieces of fish flesh coated with breading and *partially* cooked" (emphasis added). The expression "precooked" should therefore not be confused with "fully cooked," the latter being a further modification of any of the earlier-mentioned five broad categories of coated seafood.

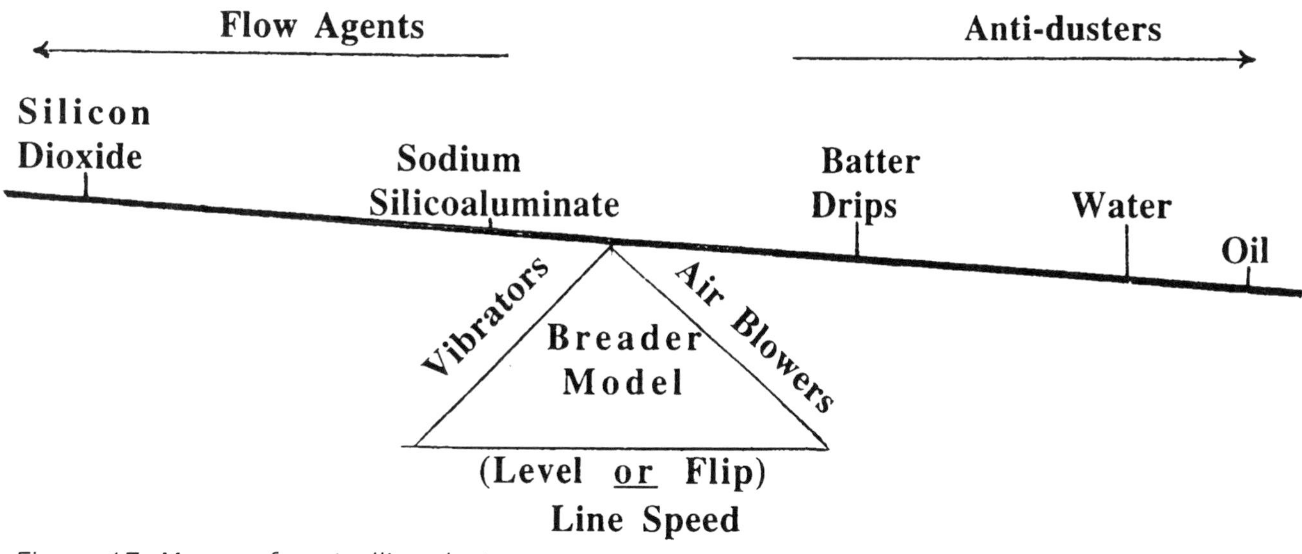

Figure 17. Means of controlling dust.

Clam Strip Manufacture

A unique specialty within the precooked category is the clam strip. It was introduced to the American market by the Howard Johnson's restaurant chain in the 1950s and quickly became a versatile menu item. It was popularly sold in a squared-off hotdog bun or as part of a combination fried seafood platter. The corporation also marketed the item in a retail carton that further solidified their place in history as the food's pioneer. The clam strip continues to find its way into many seafood restaurants, as well as being available in supermarket freezers under many brand names. The product consists of thin (approximately 5/32-inch-thick) ribbons cut from large surf clams' "tongues" that are then coated with special equipment expressly designed to separate the clumped strips from each other. The coating application equipment also assures complete coverage; a very short pre-fry step causes the strips to twist, curl, fold, and shrink, leading to their distinctive appearance and continued consumer appeal (see Flow Chart 2). Most of the processing plants for this item are along the Mid-Atlantic coast of the United States, where the raw material is harvested. Because the clam meat can become excessively tough and chewy, measures are often taken to tenderize it by enzyme treatment or phosphate application which helps to maintain moistness. The use of woven belting for transfers between equipment and drum-type dry coating applicators prevents loss of the thin strips through conventional flat flex belting. The tumbling action within the rotating drum assists in separating clumps. The flat and moist strips have a tendency to stick together. Additional extruded forms of these clam strips can be prepared from chopped clams and a binder.

Flow Chart 2
Clam Strip Breading Process

Harvested Surf Clam "Tongues"
Inspection for Shell Material
Orientation of Clam Meat
Slicing of Clam into Strips
Desanding of Strips in Agitated Tank
Rinsing and Dewatering
Weighing and Sodium Phosphate Treatment (optional)
Predusting with Dry Batter Mix (Drum Breading Applicator)
Liquid Batter Application (Woven belting conveyor)
Final Fine-Mesh Breading (Drum Breading Applicator)
Hot Frying Oil (10-15 seconds)
In-Line Freezer
Bulk or Portion-Controlled Bags
Final Packing of Bags
Storage Freezer

Fat-Reduced Precooked Products

With the increased emphasis on reduced fat content of food products, research has intensified for coated products that can deliver the same sensory appeal as their full-fat counterpart. Techniques that have met with commercial success have generally involved application of any one of a number of available protein, starch, or gum aqueous pre-fryer dips marketed by ingredient or coating mix manufacturers. These are intended to create a continuous film around the breaded seafood product as it contacts the hot frying oil. The low-solids dip will quickly evaporate its water component, leaving behind a nearly colorless, odorless, and tasteless invisible membrane that resists fat penetration. A fat reduction level using this technique can be as high as 40 percent according to product literature, with limited impact reported on the subsequently cooked product's sensory characteristics compared to an untreated control. Reformulation of the base coating can further minimize any changes such as flavor loss or lack of crust color development that may have occurred due to the fat reduction technique.

Another technique commonly used to create a low-fat breaded portion involves using a batter-and-breading combination that already contains a controlled amount of fat, and then simply freezing the product immediately, without any oil blanching step. In this situation the characteristics of the coating must be carefully orchestrated to capture the texture, flavor, and cohesiveness desired. Specialized application equipment may be required, particularly if an ingredient spray or dip, such as oil or protein, is being utilized. Proprietary systems are also marketed by coating manufacturers that claim to address all these issues.

Raw-Breaded Processing

As the name implies, this category of further processed seafood has its coating applied over a raw seafood item, either unfrozen or frozen, that is then immediately frozen after the coating's application (see Flow Chart 3). This category of processed seafood is designed to be cooked in a deep-fat fryer by the end-user (a restaurant, fast-food shop, etc.). The double-pass coating sequence allows a gradual buildup of the coating on the food's surface, thereby avoiding an excessively damp surface before the freezing process. The process can vary from a single pass of batter-breading to a triple pass depending upon the target ratio of coating to substrate. For example, a "Grade A" fish product or "lightly breaded" butterfly shrimp requires no less than 75 percent or 65 percent seafood content respectively. In these products the maximum 25 percent or 35 percent coating content can usually be achieved by a single coating pass. Conversely, the production of imitation shrimp, defined in 21 CFR as containing less than 50 percent shrimp content, would require a triple application of batter-breading, particularly if the shrimp was peeled and deveined. Being flatter and thinner, the round shape would have less available surface area for the coating to cling to, compared to a butterfly version.

When **frozen** products, such as shrimp, scallops, or fish block portions, are processed in the raw breaded version, the selection of the proper batter is extremely important. Because the seafood may typically be about 0°F, there could be a tendency for the liquid batter to freeze to the frozen substrate before the item is conveyed to the breading machine. If this occurs, then the breading will either not stick to the product's surface or be very loosely bound to it. The result is that much of it will eventually fall off during frozen distribution. The phenomenon just described can be avoided by slowing down the "drying rate" of the liquid batter. For example, adding higher levels of wheat flour or a vegetable gum (such as guar) to the batter formulation can extend the length of time that a batter is still fluid. Adhesion-type starch batters used as the first contact with the frozen seafood are generally formulated to contain a major component of modified food starch to vastly improve the adhesion of the coating to the seafood. Photo 5 shows the improvement in the coating to fish bond when using such a starch-type batter. Note the cavity between the fish and coating on the lower portion prepared without this type of specialty starch batter. Because starch has very little hydration potential, the batter can be prepared at moderate solids ratios: (30 to 40 percent) without creating any appreciable batter viscosity. The higher solids ratio also aids in less moisture remaining to be absorbed by the breading and improves the batter-breading bond.

The low viscosity of this batter has a tendency to cause the batter solids to settle out of suspension if the batter is not kept constantly agitated. To avoid having this occur on a production line, a **recirculation** batter applicator is the design of choice (see Figure 18). The batter is pumped to the top of the machine and deposited into a trough with a weir feature. When the batter flows evenly across the width of the weir tray, a curtain is formed which completely wraps around the seafood passing beneath it. Excess batter is removed

Flow Chart 3 Typical Double-Pass Raw Breaded Process
Seafood Substrate
1st Liquid Batter
Fine-Mesh Dry Predust
2nd Liquid Batter
Coarse-Mesh Bread Crumbs
In-Line Freezing
Packaging and Packing (Either Place or Bulk)
Storage Freezer

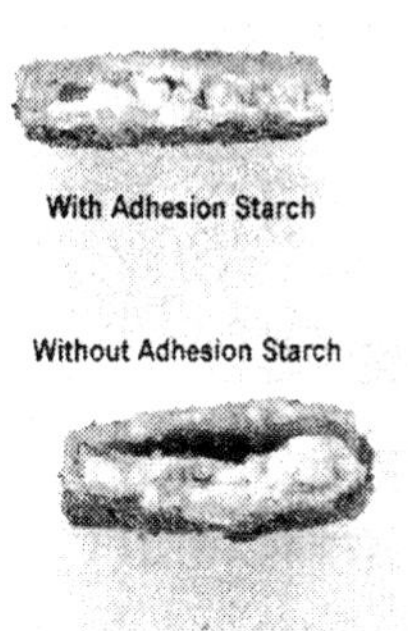

Photo 5. Adhesion to fish portion is superior with starch-type batter.

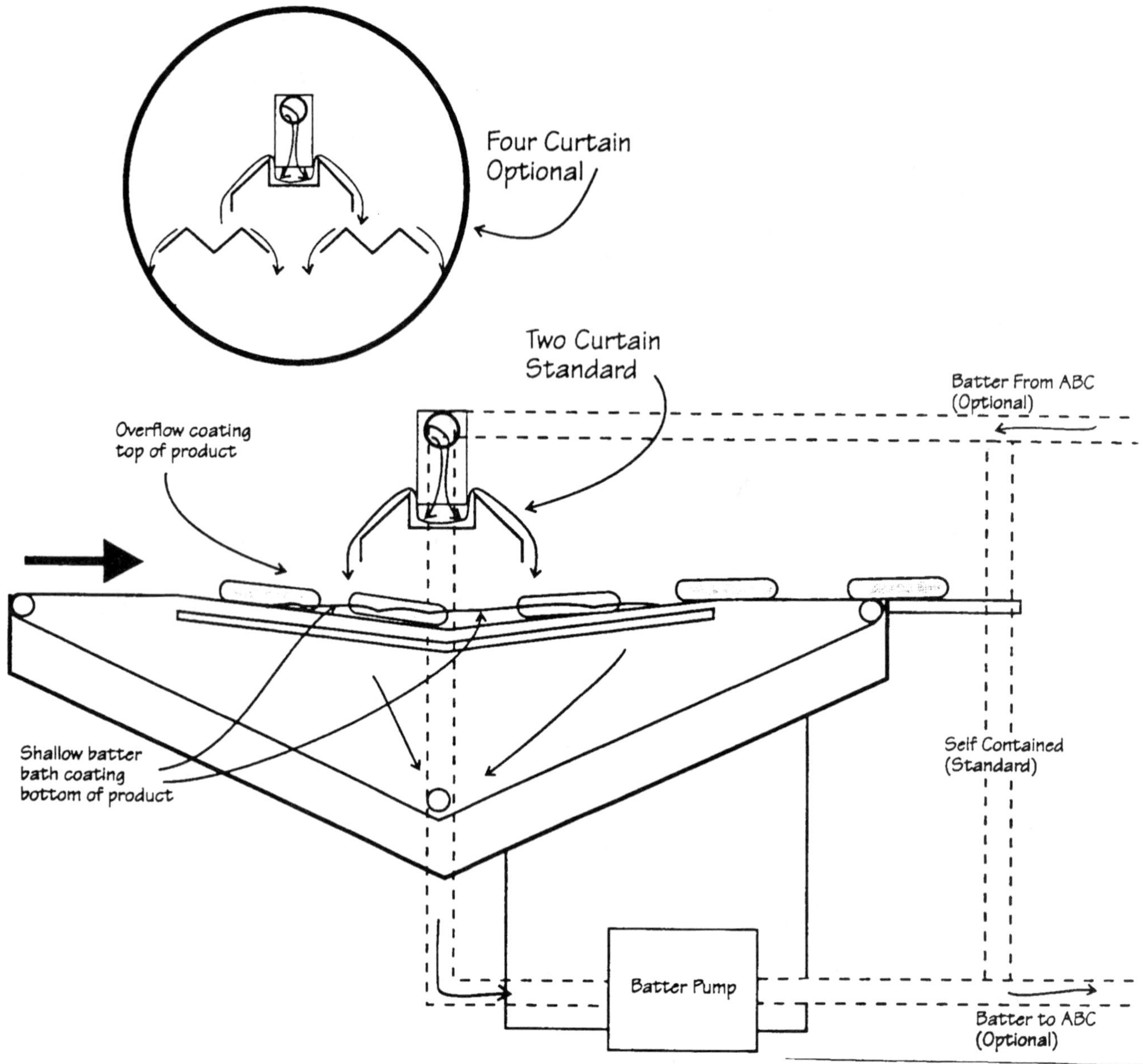

Figure 18. Typical recirculating batter applicator flow pattern.

by an air knife and directed to a reservoir for a pump that returns batter to the trough.

Provision can be made to automatically keep the batter at a predetermined temperature and viscosity. As a reference, the batter viscosity can be checked manually and an automatic batter machine set to that standard. Photo 6 illustrates a simple batter flow cup being timed with a stopwatch. As the batter drains through the hole at the bottom of the cup, the timer is stopped at the instant the stream breaks. These Zahn-type cups are available in a variety of fluid capacities and drain-hole diameters to accommodate the range of batter viscosities commonly encountered with starch-type batters, as well as those used for precooked processes. This drain cup measurement device cannot, however, be used for very thick batters because of excessive flow times and lack of a clearly identifiable endpoint to the stream flow.

Photo 7 depicts another suitable instrument for the measurement of batter viscosity. It is a rotating spindle design and can be useful for a very wide range of viscosity measurements. Note the assortment of spindle shapes to the right of the

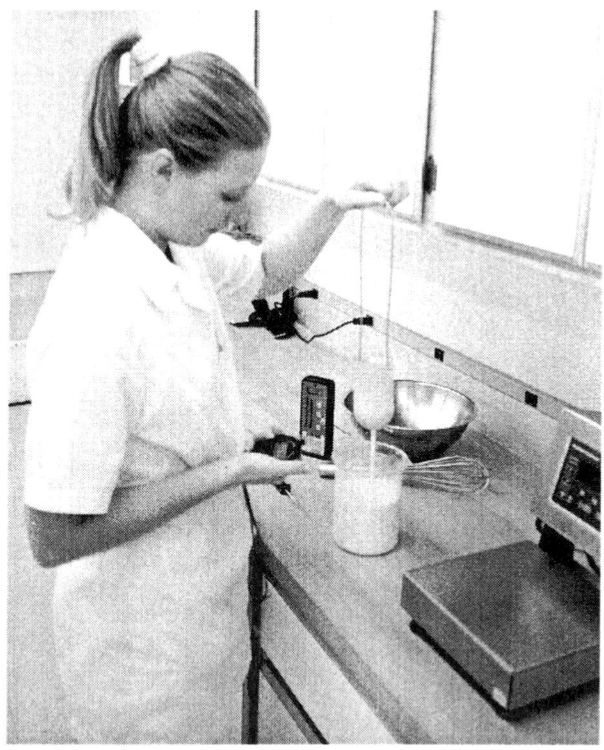

Photo 6. *Zahn-type viscosity cup often used for lower batter thickness measurements.*

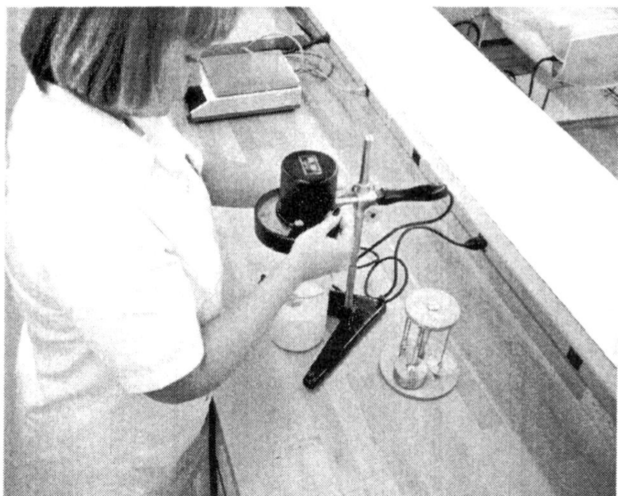

Photo 7. *Brookfield viscometer measurements span the entire range of batter viscosities.*

instrument. Their interchangeability, along with a speed-changing mechanism, provides for accurate measurement of any batter. (The unit is marketed by the Brookfield Engineering Laboratory located in Stoughton, Massachusetts.) If attempting to translate the Brookfield viscosity measurements to a coating supplier, it is necessary to specify the model, spindle number, speed, hydration ratio and exact mixing procedure in order to develop meaningful correlation between the two labs. Along with viscosity, batter temperature is critical to record since it has a direct bearing on the thickness of the batter. Colder batter temperatures will result in higher viscosity measurements, while warmer temperatures will cause batters to thin out or lose viscosity.

Photo 8 shows still another viscosity measurement tool, the Bostwick consistometer. It operates by having a spring-loaded gate at one end of a ¼ centimeter scale-etched trough. A stopwatch is started at the moment the spring gate is released, allowing the 75 milliliters of batter to flow down the trough. At a predetermined time (for instance, 10 to 30 seconds), the leading curved edge of the batter's flow is recorded directly from that centimeter scale. The viscosity is reported, for example, as 12.25 centimeters (15 seconds). Note the spirit level that was initially used to adjust the two directional inclines (along and across the trough) of the Bostwick. Relying on the permanently installed spirit level at the far end of the instrument can introduce error into a measurement since the Bostwick's frame is subject to distortion and may not, in fact, be plumb level when its circular bubble level so indicates. The Bostwick means of viscosity measurement is rugged, allowing it to be used in the production area, unlike the Brookfield unit which is a more delicate instrument, typically restricted to a laboratory. The rheology of batter

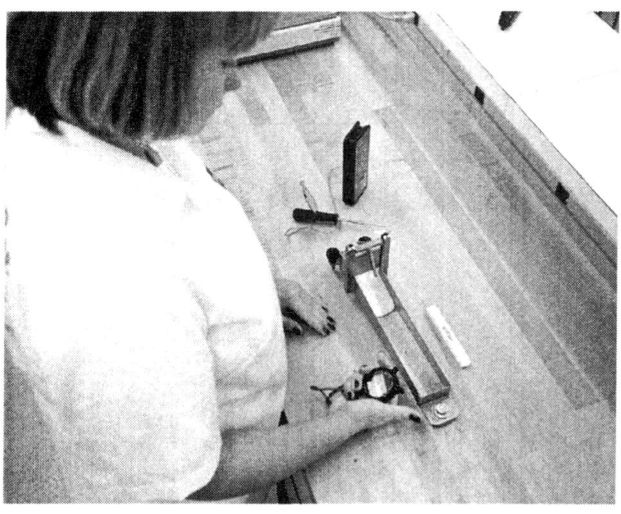

Photo 8. *Bostwick consistometer is used for thicker tempura-type batter viscosity measurements.*

draining from a coated food item closely resembles the batter's experience in the Bostwick trough.

The process step after the first batter application is referred to as a fine-mesh dry predust (Flow Chart 4). This breading material is generally white in color and made from economical, screened, baked, crackermeal stock. A sieve maximum size of approximately 20 (a screen with about 20 openings per inch) is selected for this coating component; this assures complete coverage of the first batter and absorption of the batter's unbound moisture by the predust layer. Excess breading is removed from the portion using air velocity or by a flipping of the portion (see Figure 19). This avoids breading material from being carried over into the next batter applicator where it could cause unwanted thickening from "wash off"; it could also create bare spots caused by a lack of hydration of the first breading that makes the second batter bead. The second batter can typically be prepared with a higher viscosity since it will be sandwiched between the underbreading and the top coat breading and will be simultaneously absorbed by both. The second batter composition is usually a non-starch type to avoid excessive toughness. The outer dry breading material is generally over 10 mesh. This coarseness allows for the final fried food to exhibit a high level of crispness. The coarser crumbs also have the potential to highlight, meaning they develop light- and darker-colored sites on the surface, adding to the product's visual appeal. Granulation of this outer breading can come from crackermeal breadings, so-called Japanese-style bread crumbs, or American-style bread crumbs.

Flow Chart 4 Common Shrimp-Breading Process
Remove 5-pound Block of Shrimp from Carton
Place Block into Thawing Tank of Water
Rinse and Dewater Shrimp
Size Grade
De-Shell, De-vein
Butterfly (Optional)
1st Liquid Batter
Fine-Mesh Dry Predust
2nd Liquid Batter
Coarse-Mesh Japanese-Style Bread Crumbs
IQF Freezing (Optional)
Packaging into Trays (Optional)
Tray Freezing (Optional)
Packing into Cartons
Carton Freezing (Optional)
Packing into Master Cases
Storage Freezer

The in-line freezing process can sometimes be avoided if the substrate is solidly frozen and of less than 25 percent coating content. A high-starch batter will set up swiftly and allow the product to be handled by packers, without creating any finger marks in the already firm coating. The cased product would rely on the storage freezer to drop the entire product's temperature to 0°F or below within 24 hours. However, the most common freezing techniques are cryogenic or mechanical, operating from -30°F to -150°F, after which the hard-frozen product is packaged. Using this approach also allows for the use of automatic bulk packaging equipment since the coating is less tender.

Batter-Coated Process

Commercial, batter-coated seafood products were pioneered in 1968-69 by Kris Gunnarsson of Coldwater Seafood Corporation's Maryland facility (see Flow Chart 5). The frozen substrate is handled in either of two ways today, depending on whether a low-viscosity liquid batter is first applied. A light predust followed by a single dip in a thick leavened batter before deep-frying is referred to as a true English-style "fish-and-chips" or "tempura" coating. By contrast, the process that includes first a dip in a low-viscosity batter (often a starch-type), followed by a coating of fine-mesh crackermeal, and then a moderately thick final batter before the fryer, is called a batter-fry system. Because of the layer of breading between the two batters used in the batter-fry system, the finished consumer product can often have a more doughy interface compared to the traditional English version.

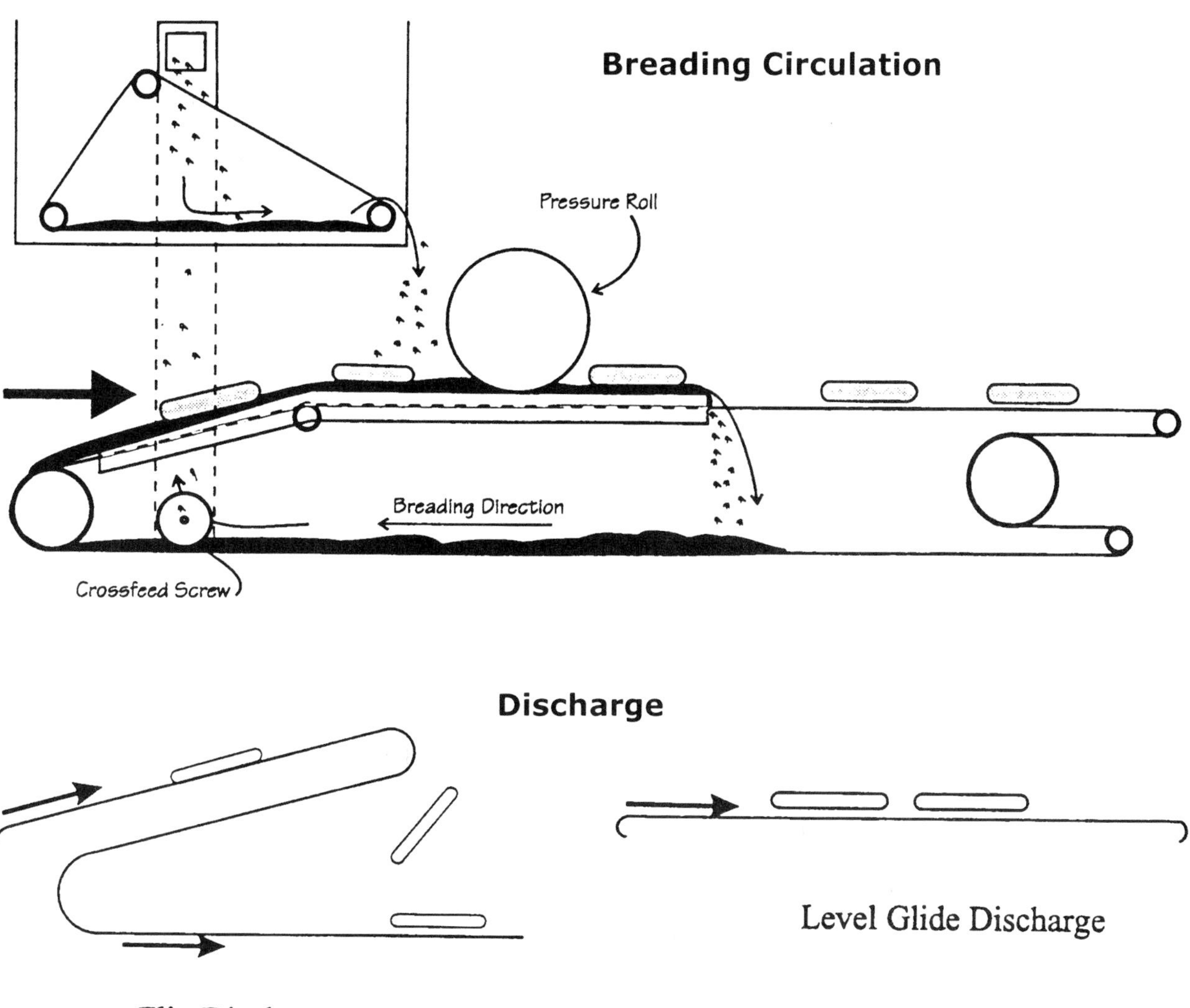

Figure 19. Flip-type versus level-type breading machines.

Flow Chart 5 Typical Batter-Coated Process
Seafood Substrate
1st Low-Viscosity Liquid Batter (Optional)
Dry Predust (Optional)
2nd High-Viscosity Puff Batter
Hot Frying Oil
In-Line Freezing
Packaging and Packing
Storage Freezer

The final batter applied before the fryer has a well-defined level of baking powder of 1 to 3 percent, a soft wheat flour content of about 45 percent, a corn-flour content of 45 to 50 percent, other functional seasoning (salt, monosodium glutamate, pepper, onion, etc.), and coloring agents, such as dextrose, whey, paprika, or turmeric. The baking powder, along with the thick viscosity of the batter, gives the finished product its smooth, puffed, crispy surface. When preparing this type of puff batter, it is important to use mixing equipment that will not overwork the batter (see Photo 9). The batter can be mixed with iced water as a means to inhibit leavening action from activating in the mixing bowl, until the

Photo 9. Thick tempura batter being gently stirred in automatic mixer.

Photo 10. Raw batter-coated product leaves the applicator and proceeds under an air tube knife towards blanch fryer.

coated product enters the hot frying oil (Photo 10). As further refinement, the tempura applicator is of a nonrecirculating design so that the thick batter is not overworked (see Figure 20). The seafood-to-coating ratio is primarily controlled by batter viscosity and the intensity of the air knife. How the portion enters the frying oil is critical to the finished product's acceptance. It usually includes a special adapter at the discharge end of the tempura applicator called a star roller. The small pointed wheels avoid damaging the liquid coating and allow the batter to flow into the wire belt strand marks to improve coverage. The fryer also includes a special Teflon drop plate that not only sets the batter, but also allows it to release easily without sticking, and to travel through the frying medium (see Photo 11). The batter-coated process remains something of an art form, because the process itself is a dynamic one, unlike the precooked and raw breaded processes. The wet batter-coated product generates a tremendous amount of boiling activity at the entrance to the fryer, where a large amount of free water is converted to steam upon hitting the hot frying oil. Leavening activation, coating expansion, crumb formation, and drop-plate impressions all take place within seconds of the product contacting the oil. A hold-down conveyor is used to keep the now-buoyant, batter-coated piece beneath the oil for a more uniform crust color. Continuous filtration systems for the oil remove crumbs as they are formed and thereby prevent unsightly black carbon specks from depositing on the food's surface. The fryer conditions of dwell time and temperature are similar to those discussed in the "precooked process" section.

Oven-ready tempura batter-coated products are processed in a similar fashion as shown in Flow Chart 5, except that the batter composition differs significantly from that described above for the fryer-finished item. Here the batter is "richer," meaning it is more highly leavened, often has shortening and egg incorporated, contains a higher browning agent level and uses a soft wheat flour exclusively, with about 15 to 30 percent starch as a diluent, in order to create a tender, delicate, baked-batter coating. Since this item is designed to be solely reheated in the oven rather than the fryer, its oil absorption tends to be slightly higher in the frozen product than that found in the fryer-reheated version, since the former will not have a second opportunity to contact frying oil again before being consumed.

Often, for food service applications in particular, the holding time of the batter-coated, fried portion is an important issue. This can be evaluated by checking the product's performance in the applicable holding environment (see Photo 12). Handling tolerance can be tested using tongs on the item and observing if any breakage takes place over time when the food is shaken in the tong's grip. This technique can provide an early warning of excessive tenderness that may point to a need to reformulate the coating.

Batter Circulation (no circulating pump — Still system)

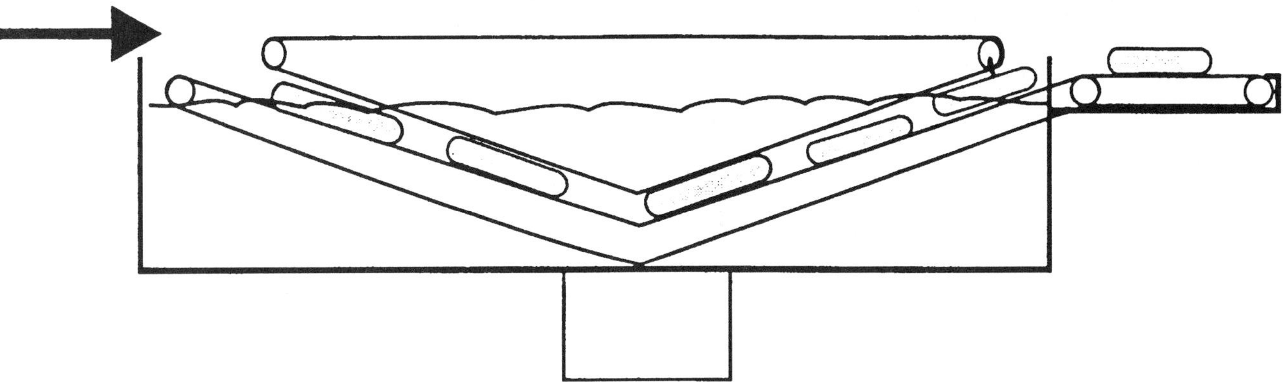

Transfer

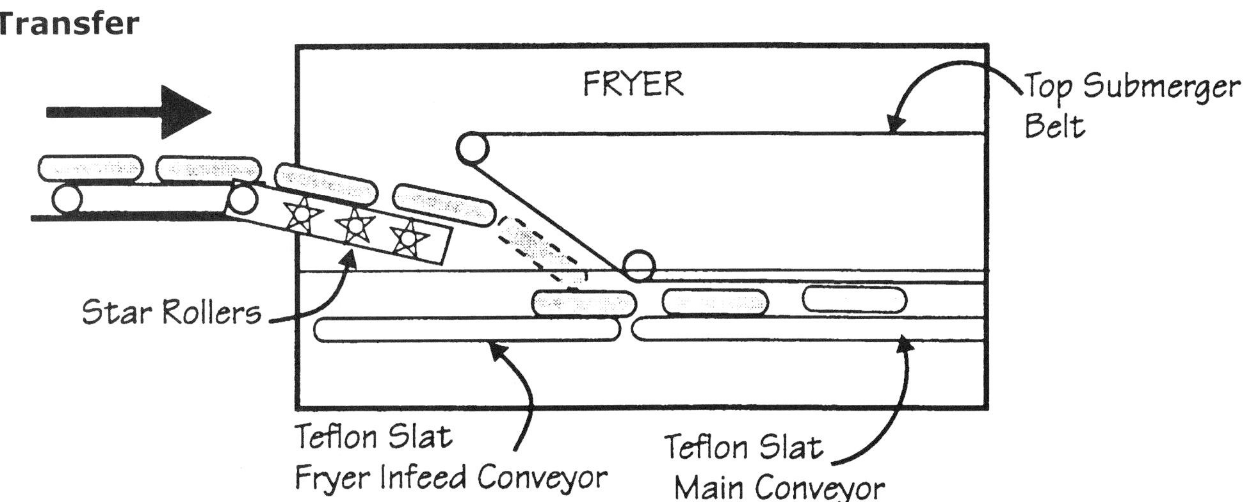

Figure 20. Tempura batter application design.

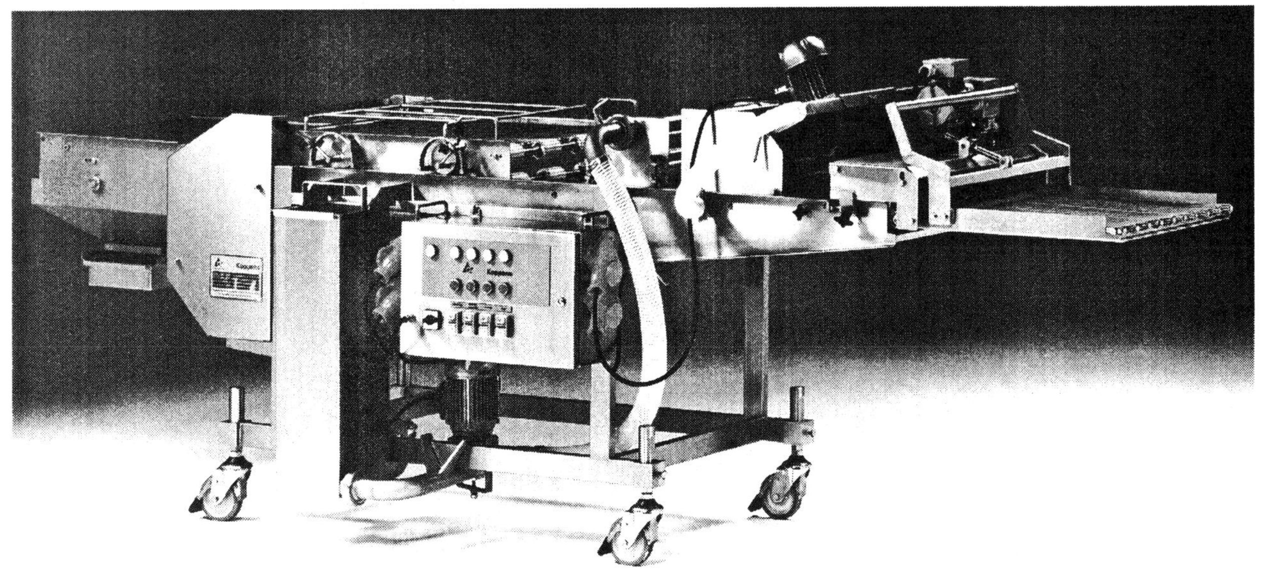

Photo 11. Stainless steel batter-fry applicator.

Photo 12. Tempura-coated products leave the par-fryer towards the in-line freezer.

Flow Chart 6 Typical Sauce-Coated Process
Grill Marking (Optional)
Sauce Application
In-Line Freezing
Packaging and Packing
Storage Freezer

Sauced Process

The introduction of grill-marked seafood at the March 1991, International Boston Seafood Show caught the eye of the food editor of *The New York Times* and heralded the beginning of a new category of further processed seafood (see Flow Chart 6). That initial product was a size-graded natural fillet portion of North Atlantic pollock, a species with a firm, steak-like texture. It was sear-marked and prebasted with pure olive oil (Figure 13) and provided a means for food service and retail consumers to enjoy the visual impact of seeing attractive grill marks on the surface of the fish without having to deal with the many disadvantages of using an open flame grill. Those problems included: (1) sticking to the char-broiler; (2) breakage from handling flake-prone fish; (3) coping with an uneven heat pattern across the flame-grill's surface; and (4) overcooking the fish due to the brief window of time within which the fish should be properly cooked. Morphologically, both the lack of natural fat (typical of seafood), and its low levels of connective tissue, compared to commonly char-grilled beef or chicken, are the primary reasons for the sensory and handling problems just mentioned. The new product eliminated those concerns entirely, and allowed the fish to be baked in a regular oven, on a flat grill, or in a microwave oven. The unique use of olive oil acted as a natural flavor enhancer to the mild seared notes, and prevented dehydration of the fish during bulk frozen distribution. While cooking, the fish was first basted with olive oil and then was shed into the cooking tray. This sequence significantly reduced the fat content of the cooked fish. A level below the United States Surgeon General's dietary guideline of no greater than 30 percent of calories from fat was achieved. It further allowed nutrient content claims to be made relative to conventionally processed, high-fat battered and breaded seafood.

Photo 13. Seared fish portions being conveyed under sauce curtain.

The use of a flavored sauce over the grill-marked fish is, in many ways, simply a form of flourless batter. The sauce is generally composed of water, seasonings, oil, emulsifier, gum stabilizer, and color, but can vary widely depending upon the taste and nutritional profile being sought. A typical sauce formulation would be:

Ingredient	Percentage
Water	40 - 60
Vegetable oil	10 - 50
Seasonings	5 - 25
Gum thickener	0.2 - 1

In order to enhance the visual impact of the sauce, particulates are added to the sauce as a component of the seasonings mix. The more popular are whole ground herbs, parsley flakes, coarsely cracked black pepper, dehydrated red bell peppers, and sesame seeds.

Popular sauce flavors include lemon-pepper, Cajun (hot and spicy), butter-garlic, teriyaki, Italian herb, Polynesian (pineapple sauce sprinkled with toasted coconut), smoked barbecue, and tomato. High salt levels that might make freezing the sauce difficult should be avoided. The sugar content, if kept low, will also help avoid unsightly charring and sticking of the sauce upon cooking. The gum stabilizer serves a dual function in the sauce. Initially, it helps form a stable emulsion that keeps the large particles from settling out of the sauce, and, second, it controls the shedding of the sauce from the top surface of the food during cooking. The sauce is applied to the seafood using conventional batter recirculating equipment, similar to that shown in Figure 18. The sauce is deposited on the grill-marked fish and its quantity is easily controlled by adjusting the intensity of the air-knife that removes its excess. If the fish has had grill marks seared into the surface or if the fish piece is convex, as might be the situation for a pressed shape, then the application equipment is generally operated without the bottom shallow batter bath pan illustrated in Figure 18. This results in only the top and sides of the fish being coated with the sauce, and would define the (uncoated) surface that was placed in contact with the cooking surface.

The packaging of these sauced items can represent a challenge because the coating will be soft if allowed to temper, unlike breaded or battered versions, which tend to be much more tolerant of a rise in temperature. If a softened, sauce-coated product is allowed to come in contact with another and is then refrozen, the pieces can stick together. The softened sauce may also transfer to the package film and not be recovered to flavor the baked product afterwards — leaving it less tasty. Solutions for controlling these concerns have led to individually pouching each portion before it is cartoned, using skin-sealing trays, or placing each portion in a compartmentalized tray that keeps the piece, with its sauce coating, physically separated from an adjacent piece. Adequate packaging provision must also be made to retard dehydration of an aqueous-based sauce during frozen storage.

Shrimp Processing Techniques

Flow Chart 4 depicts the typical process arrangement found in the commercial shrimp industry. Because of the large price spread of shrimp based on their count per pound, size grading is the rule by which breaded shrimp is marketed. Much of the shrimp processed in the United States arrives in frozen, five-pound bulk cartons that are first thawed in order to separate the individual shrimp. It is important to avoid copper or brass piping for the water system in the plant because it has been demonstrated to accelerate the development of "black spot" on shell-on shrimp. Semi-automatic machines are available to deshell, devein, and butterfly the shrimp. This splitting process exposes more surface area for the breading application and creates a more visually appealing item for the consumer.

Three categories of raw, breaded shrimp are recognized based on the percentage of shrimp meat present:

Lightly breaded:
 65 percent minimum shrimp meat
Breaded:
 50 percent minimum shrimp meat
Imitation:
 <50 percent shrimp meat

Because of the high price paid for breaded shrimp, the breading of choice has been the premium coarse-granulation, Japanese-style bread crumbs. The coated pieces are often placed in protective blister cavity trays that help to retain the shrimp's characteristic shape. The tray also prevents damage to the breading.

CHILD NUTRITION DEVELOPMENT CONSIDERATIONS

The technique of using **raw weight input** is the official method in preparing seafood products that will meet the requirements of the United States Department of Agriculture's (USDA) Child Nutrition program — commonly called "CN." Various cooking yields are applied to this input weight: for instance, 78 percent for regular fish blocks, 62 percent for peeled and cleaned frozen shrimp. (See Table 3 for a complete listing of "Of-

Table 3. Official Cooked Fish Yields for Seafood

In order to assure that all users of USDC-inspected fishery products are aware of the official cooked-yield factors that have been approved to date for seafood products, the following list has been developed for use with both standardized and non-standardized products. These factors are to be used when establishing the meat/meat alternate yields for USDA/FNS-approved Child Nutrition statements.

These factors have been developed using standard AOAC* cooking procedures and can be used to determine seafood yield by converting the raw edible seafood weight to cooked seafood weight by multiplying the weight of raw seafood in the product by its yield factor (i.e., 2.6 oz. fish x .78 = 2.0 oz. cooked meat alternate).

Cooked yield factors for seafoods based upon standard AOAC cooking procedures are:

1 – Products cut from regular frozen fish blocks	78%
2 – Fresh or frozen fish fillets, skin-on or skinless	70%
3 – Products made from frozen minced fish blocks	75%
4 – Products made from fresh or frozen minced clams	66%
5 – Products made from fresh or frozen minced shrimp	58%
6 – Rehydrated dry salted cod (before cooking)	75%

Cooked yield factors for seafoods as listed in the USDA Food Buying Guide for School Food Service:

Seafood, fresh or frozen	
1 – Clams, shucked	28%
2 – Crabmeat, cooked	97%
3 – Fish steaks, frozen	66%
4 – Oysters, shucked, undrained	38%
5 – Scallops, frozen	53%
6 – Shrimp, fresh, cooked, peeled and cleaned	100%
7 – Shrimp, frozen, cooked, peeled and cleaned	83%
8 – Shrimp, fresh or frozen, raw, peeled and cleaned	62%
9 – Shrimp, raw, in shell	54%
Seafood, canned (drained weight)	
1 – Clams, minced	43%
2 – Mackerel	58%
3 – Sardines	88%
4 – Salmon	81%
5 – Shrimp	100%
6 – Tuna, water-packed, chunk	77%
7 – Tuna, grated or flaked	92%

*AOAC = Association of Official Analytical Chemists, 13th edition, Section 18.003.

ficial Cooked Fish Yields for Seafood.") Additionally, if enriched or whole grain flour is used to prepare the coating, the finished product may also qualify for credit as a "bread alternate" serving. The USDA has published an informative 25-page booklet, describing in detail the calculations and methods available for submitting labels for approval with their agency. The label approval process is handled initially through the National Marine Fisheries Service (NMFS). (See Figure 1 as an example of a Grade A Child Nutrition approved label.) Additionally, the USDA has issued regulations for the use of a nutrient database system. It states that the school meal program must limit its weekly weighted nutritional profile to no more than 30 percent of calories from fat, as recommended by a number of health authorities.

EXTRUDED AND FORMED PROCESSED PRODUCTS

The term "extruded" generally refers to the class of smaller-sized, battered and breaded seafood items prepared from minced or chopped substrate that has been mixed with a binder, forced through an extrusion die, and cut into portion-controlled pieces that can retain their unique shape. Examples of these consist of: shrimp shapes made from small shrimp; clam strips made from minced clams; and calamari rings made from diced squid.

The process to manufacture this type of food is shown in Flow Chart 7. The process lends itself to a high degree of automation, since the extrusion apparatus can cycle at speeds over 200 strokes per minute and a large production line can accommodate a ten-nozzle extruder. In one technique the pieces are bound together using an assortment of vegetable gums, the most popular being sodium alginate, present at about ¾ percent in the mash. As the food piece is extruded from the die, a 4 to 6 percent recirculating solution of calcium chloride is sprayed at the extrusion nozzle that causes a thin, shape-retaining skin of calcium alginate to encapsulate the shape. The food piece is adequately sturdy, at this point, to withstand transfers between batter, breading, and prefrying operations. Once cooked and frozen the products can be automatically packaged. The FDA has mandated special package labeling for these reformed foods, requiring them to have the expression "Made from Minced (shrimp, fish, etc.)" in type size no smaller than one-half that of the product's name.

Flow Chart 7 Extruded Product Processing
Grinding Seafood
Mixing with Binder System
Transfer to Extruder Hopper
Adjust Portion Size
Apply Batter
Apply Breading
Pre-Fry in Hot Oil
In-Line Freezer
Automatic Packaging
Packing
Storage Freezer

FORMED SEAFOOD

The need for natural shaping of fish products has resulted in the development of new forming technologies that retain the natural flaky texture of the fish while utilizing the efficiencies and portion control advantages of fish blocks. The techniques either rely on reshaping large sections of the blocks which are then subdivided, or initially cutting the block into smaller individual pieces that are then pressed into a single or multiple shape in a single shaping cavity. Keeping the temperature of the process below 10°F helps to pre-

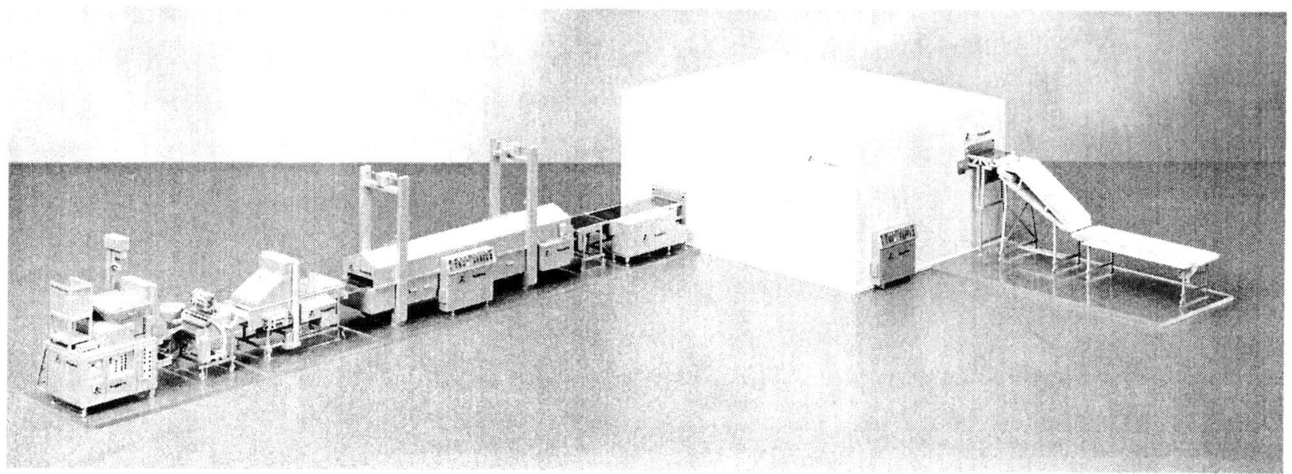

Photo 14. Illustration of forming machine, batter applicator, breading unit, fryer, and spiral freezer.

Photo 15. Close-up of forming machine capable of producing three-dimensional shaped pieces.

serve the desired flaky texture while delivering a contoured profile.

Photo 14 illustrates an arrangement of processing equipment that is often used to prepare shaped and coated fish products from unfrozen fillets. The fish, typically at 30°F to 40°F, is placed in the hopper, where augers force it into cavities pre-established in a forming plate (see Photo 15). These shapes can be as simple as a round, disc-like shape (Photo 16) to an assortment of nautical-theme shapes shown entering a batter applicator in Photo 17. The shapes made in this fashion can have flavoring and binding agents added as a means of

Photo 16. Formed fish cakes traveling to breader machine after being batter-coated.

Photo 17. Nautical-themed shapes shown entering batter applicator.

imparting some special characteristic to the product.

ACKNOWLEDGMENTS

The author thanks Austin Nute, a United States Department of Commerce inspector, for reviewing the section on regulatory compliance; Tetra Laval-Koppens for equipment photos; Robert Swackhamer of Stein Inc. for permission to use several equipment drawings; Harold Cohen and Neil Trager for their early teachings on the subject; the late John Urban for his inspirational guidance; and all the other professional peers with whom he has had an opportunity to be associated, whose input has become a part of the chapter in one manner or another. He dedicates this work to his two daughters, Lora Jane and Christy Ann.

REFERENCES

Advances in Nutritional and Fat Reduction Technologies. 1996. Food Technology Intelligence, Inc. Midland Park, NJ.

A Food Technologist's Guide to Methocel Premium Food Gums. (undated) Booklet 192-1037. The Dow Chemical Company. Midland, MI.

Akesson, Y. R., E.M. Andersen, and M. Olofssson. 1988. Process for preparing frozen rolled cod tail. U. S. Patent 4,780,328.

Ang, J. F., W. B. Miller, and I. M. Blais. 1991. Fiber additives for frying batters. U. S. Patent 5,019,406.

Attenburrow, G. E., R. M. Goodband, and L. W. A. Melles. 1992. High protein crumbs for coating foodstuffs. U. S. Patent 5,171,605.

Baker, R. C., J. M. Regenstein, and J. M. Darfler. 1976. Seafood Chowders. New York Sea Grant Institute. Albany, NY.

Baker, R. C., J. M. Regenstein, and J. M. Darfler. 1977. Seafood Crispies, New York Sea Grant Institute. Albany, NY.

Bernacchi, D. B. and R. J. Loewe. 1988. Process for preparing readily reconstituted frozen comestibles and frozen comestibles produced thereby. U. S. Patent 4,744,994.

Bernacchi, D. B., R. J. Loewe, and D. L. Immel. 1990. Process for preparing storage stable, readily reconstituted frozen comestibles and frozen comestibles produced thereby. U. S. Patent 4,948,603.

Bhardwaj, S. C. 1990. Method of cooking involving high protein frying batter that eliminates the need for breading and produces crispy and chewy crust. U. S. Patent 4,963,378.

Child Nutrition Labeling for Seafood Products. (undated) USDA. Food and Nutrition Service. Washington, DC.

Clairouin, N., A. Gueroult, and R. Lemoyne. 1990. Process for the manufacture of reformed fish products. U. S. Patent 4,948,620.

Crisp Coat SC, UC, & MC. 1992. Technical Service Bulletin. National Starch Corp. Bridgewater, NJ.

Fabricant, F. 1991. Food Notes —Just Heat and Serve. *The New York Times.* March 20.

Feeney, R. D., S. G. Haralampu, and A. Gross. 1993. Potato and other food products coated with edible oil barrier films. U. S. Patent 5,217,736.

Flick, G. J., Y. Gwo, R. L. Ory, W. L. Baran, R. J. Sasiela, J. Boling, C. H.Vinnett, R. E. Martin, G. C. Arganosa. 1989. Effects of Cooking Conditions and Post-Preparation Procedures on the Quality of Battered Fish Portions. *Journal of Food Quality.* **12**:227-242.

Food Code, Part 3-401-11. 1995. U. S. Public Health Service. Food and Drug Administration. Washington, D.C.

Fuller, D.B. and R. T. Parry (eds.). 1987. Savoury Coatings. Elsevier Applied Science. Essex, England.

Furesik, S. L. and E. D. DeBoer. 1992. AEDU batter starch for deep fat fried food. U. S. Patent 5,120,562.

Garrison, Y. J. 1972. A Study of Effects of Repeated Deep Fat Frying Shrimp Coated with Batters Made of Different Thickeners on the Stability of Cooking Oil. M. S. Thesis. University of Maryland.

Gerrish, T., C. Higgins., and K. Kresel. 1997. Method of making battered and/breaded food compositions using calcium pectins. U. S. Patent 5,601,861.

Grider, J. (ed.). 1993. Cereal Foods World — Batters & Breadings. September. AACC. St. Paul, MN.

Gustafsson, M. 1994. The U. S. Foodservice Market for Groundfish. Presentation at the Groundfish Forum. October 27. London, England.

Koop, C. E. 1988. The Surgeon General's Report on Nutrition and Health. U. S. Department of Health and Human Services. Public Health Service. Washington, DC.

Krochta, J. M., E. A. Baldwin, M. O. Nisperos-Carriedo. (eds.). 1994. Edible Coatings and Films to Improve Food Quality. Technomic. Lancaster, PA.

Kulp, K. and R. Loewe (eds.). 1990. Batter and Breadings in Food Processing. AACC. St. Paul, MN.

Jarrett, G. A., M. J. McBride, S. A. Symien, P. Walker, and A. Wooten. 1995. Process for producing a coated food product. U. S. Patent 5,478,583.

Langlands, I. H. 1973. Process and apparatus for shaping frozen flesh. U. S. Patent 3,728,136.

Larsson, I. C., A. M. Linse-Loefgren. 1988. Preparation of frozen food product. U. S. Patent 4,741,933.

Marinades/Glazes. (undated) Advertising Bulletin. NewlyWeds Foods. Chicago, IL.

Methocel Premium Food Gums in Structured Foods. (undated) Booklet 192-949. The Dow Chemical Company. Midland, MI.

Monagle, C., and J. Smith. 1985. Process for breading food. U. S. Patent 4,518,620.

Miller, M. E. 1990. Breader for coating edible food products with fresh bread crumbs. U. S. Patent 4,936,248.

Newly Crisp. (undated) Advertising Bulletin. NewlyWeds Foods. Chicago, IL.

Nishikawa, K., T. Kuwazuru, and H. Mizushima. 1984. Method of forming fillets. U. S. Patent 4,474,823.

NOAA Fishery Products Inspection Manual, Part 1-Inspection, Part II-Grading, Part III-Certification. 1995. Handbook No. 25. United States Department of Commerce, National Marine Fisheries Service. Washington, DC.

Olewnik, M., R. Rudd, and K. Kulp. 1988. Batter and Breading. Research Department Technical Bulletin X-8, August. American Institute of Baking. Manhattan, KS.

Olson, S. and R. Zoss. 1985. Fried foods of reduced oil absorption and methods of preparation employing spray of film-forming agent. U. S. Patent 4,511,783.

Pickford. 1992. Microwaveable Coatings. PCT Application GB92/01559.

Product Formulation-Batter & Breadings. 1985. National Starch Corp. Bridgewater, NJ.

Sasiela, R. J. 1978. Formed Clam Strip Expands Market, Increases Clam Meat Utilization. *Quick Frozen Foods.* February. New York.

Sasiela, R. J., and J. D. Smith. 1979. Apparatus for the preparation of extruded food products. U. S. Patent 4,152,102.

Sasiela, R. J., J. J. Urban, and B. Y. Chiang. 1979. The Effects of Conventional Breading Application Equipment on Oriental-Style Bread Crumbs. Presentation at the 24th Annual Atlantic Fisheries Technological Conference. October 17. Danvers, MA.

Schiffmann, R. 1996. Microwave Technology For Coated Food Products. Presentation at A.A.C.C. Short Course on Batter and Breading Technology. October. Chicago, IL.

Smadar, Y. 1972. Extruded food products and method of producing same. U. S. Patent 3,650,766.

Suderman, D. F. and E. Cunningham (eds.). 1983. *Batter and Breading Technology*, Avi Publishing Co. Westport, CT.

Stypula, R. J., L. Buckholz Jr. 1990. Process for preparing oil-impervious, water retaining silicon dioxide derivative-containing food products. U. S. Patent 4,948,608.

Tempura Machines. (undated) Technical Bulletin. Koppens Industries Inc. Stone Mountain, GA.

The Complete Guide to Coating Systems. 1995. Griffith Laboratories. Alsip, IL.

The Processor's Guide to Coating & Cooking. (undated) Stein Inc. Sandusky, OH.

Urban, J., R. J. Sasiela, and R.Klein. 1982. Improvements in batter-coated food products. UK Patent Application 8102046.

U. S. Food and Drug Administration. *Code of Federal Regulations*, 21 CFR, 1995.

U. S. Food and Drug Administration, *Code of Federal Regulations*, 50 CFR, 1995.

U. S. Food and Drug Administration. 1996. *Staphylococcus aureus* Toxin Formation in Batter Mixes. *Fish and Fishery Products Hazards and Controls Guide*, Chapter 15.

Varela, G, A. E. Bender, and I. D. Morton. (eds.). 1988. *Frying of Food Principles, Changes, New Approaches.* VCH. Ellis Horwood, England.

Wadell, L. G. A. 1993. Foodstuff coating process. U. S. Patent 5,248,511.

YELLOW PERCH:

An Example of How Processing Can Create a Value-Added Product

Yellow perch (*Perca flavescens)* is an esteemed food fish in Europe, the West Indies, and along the Great Lakes of the United States, where it is renowned for its sweetness. It is the fish of choice at the Friday night fish fries along the Great Lakes during Lent.

During the summer it is caught in great numbers. Because a yellow perch is small, it is relatively difficult to cut the fillets from the fish. Professional cutters along the lakes make a living cutting this small fish for the sport angler. These same cutters also fillet commercially-caught fish for fish fries.

The yellow perch, which thrives in a lake habitat, is found not only in the Great Lakes region but also in the mid-Atlantic estuaries as well from mid-January to mid-March.

The color pattern of the yellow perch — generally classified as a panfish — distinguishes this member of the grouper family from the others. It is olive-green on the back, blending into a golden-yellow on the sides, and white on the belly. Six to eight dark bands extend from the back to below the lateral line.

The average catch weighs ¼ to ¾ pound.

Various methods of raising the yellow perch commercially have been studied at Virginia Tech. The Virginia Sea Grant College Program in concert with the Virginia Tech Commercial Fish and Shellfish Technologies (CFAST) group have worked with fish processors to provide the desired fillet product to a wider market. The tasty fish has good export value and can be packaged according to needed specifications by processors.

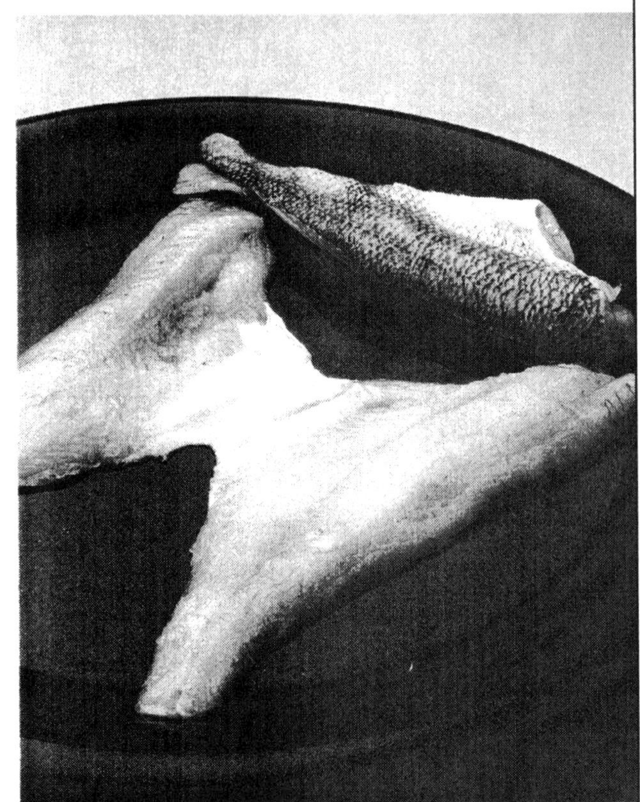

A yellow perch fillet.

George J. Flick, Jr.

Smoked, Cured, and Dried Fish

Michael W. Moody, George J. Flick Jr., Roy E. Martin, and Angela I. Correa

INTRODUCTION

Before refrigeration and canning, humans preserved food caught in times of plenty to use in times of scarcity by taking advantage of environmental conditions, both induced and natural. Commonly, they used naturally occurring preservatives such as salt and smoke. Undoubtedly one of the first foods cooked on an open wood fire was some form of fish.

Although the origin of fish smoking is obscured by antiquity, aboriginal men and women must have developed this method of preserving their catch shortly after they discovered how to make fire (Paparella, 1979; Crance, 1955). Experience soon told them that the barbecuing process made food keep longer and added a distinctive flavor. As time continued, these prehistoric cooks noticed that the flavor varied with the kind of wood burned, and other improvements gradually followed.

Proper timing and correct temperature, which to primitive peoples meant the correct position of the fish over the fire, were determined. Becoming more familiar with using salt, they found that a preliminary salting or brining further improved the flavor and the keeping qualities. Present methods of hot-smoking, or barbecuing, as it is sometimes called, surely evolved from these crude beginnings.

About the time humans were learning how to barbecue fish, they discovered the possibilities of drying fish in the open air. They also found that a smudge fire burning under their hanging fish not only preserved the fish but also imparted a smoky flavor. With certain types of fish the smoke flavor was preferred; thus, a smudge fire under the drying fish became an essential part of the process. Use of proper wood, regulation of the fire's heat, the density of the smoke, and a preliminary salting or brining completed the evolution of what is known as fish-smoking.

It was not until the development of controlled smokehouses that any significant advances were made over those early preservation methods. Controlled processes for both cold- and hot-smoking based on scientific principles have just begun to come into use in fish smoking; today, smoking is being changed from an art to a science.

The primary curing ingredient is still salt, but additional curing ingredients such as sugar, spices, and, in some products, sodium nitrite may be used. The cured products may or may not be subsequently smoked, smoke-flavored, and/or partially dried. In this discussion, our emphasis is on fresh- and saltwater species of fish that are both cured and smoked or smoke-flavored. Preparation of these food products has long been a tradesman's art. Applying scientific principles to this ancient art can produce a safer and more consistent, appetizing, and wholesome product.

Development of modern refrigeration has meant that there is no longer the need for the high salt content previously required. Although less salt is now used in curing smoked fish, it still contributes to the finished product's shelf life, safety, and flavor. The modern mild-cured products, which may be cold- or hot-smoked, are made possible by adherence to good sanitation practices and proper refrigeration during processing, distribution, and storage. *These products require proper refrigeration for preservation.*

The smoked-fish industry has achieved an excellent reputation for producing high-quality, wholesome products. However, the industry has experienced some problems that have resulted in serious economic loss, and to a degree, an erosion of consumer confidence. There have been periodic food poisoning outbreaks associated with cured and cured-and-smoked fish products, some of which have resulted in fatalities. Case studies of these relatively infrequent outbreaks, which often involved *Clostridium botulinum* toxin, are consistently related to improper processing procedures applied by inexperienced or unknowledgeable processors, inadequate sanitation, abusive product storage conditions (primarily by the consumer), and sometimes to many consumers' erroneous belief that smoking negates the need for refrigeration or that the product has an unlimited shelf life.

It is the processor's responsibility to adhere to manufacturing procedures that result in products that are not only appealing in appearance and flavor, but are wholesome, prepared under sanitary conditions, and safe to eat.

ECONOMIC IMPORTANCE

Today, people continue to smoke, cure, and dry seafood. This is not so much for preservation (modern preservation techniques yield an excellent product more closely associated to fresh) but for the delightful taste and texture and because of cultural preferences associated with smoked or cured products. The selection of smoked fish products is extensive, and regional preferences exist for both the type of fish and the style of preparation. In some countries salted and dried fish remain an important commercial product.

PRINCIPLES OF SMOKING, DRYING, AND CURING

The two most common reasons for food spoilage are the actions of bacteria and autolytic enzymes. The major emphasis of this chapter is on controlling the adverse effects of spoilage organisms in smoked, cured, and dried fish.

Water is the basic ingredient of all foods. Every food contains water in varying amounts. For the most part, the amount of available water in a food determines how rapidly that food will spoil. Foods with a high available water content (such as meats, seafood, milk, etc.) spoil quickly; foods with a low available water content (such as flour, honey, cereal grains, etc.) may last for years even at room temperature (Troller and Christian, 1978) (Table 1).

Bacteria and other spoilage microorganisms must have a minimum level of available water before they can carry out essential metabolic functions. If there is not sufficient available water, the bacteria die or become inactive. The amount of available water in a food is measured by the water activity (a_w) — not to be confused with the percentage of water in a food. Water activity is a measurement of the water available for microorganisms to use for metabolism. Salt and sugar can "tie up" water so that it is not available for microorganisms to use. Consequently, by adding enough salt to fish, the growth and destructive action of bacteria can be minimized. Salt draws moisture from the tissue of the fish by osmotic pressure to make less water available (Moschiar et al., 1984; Curran and Poulter, 1983).

Pure water has a water activity measure of 1.0. As the amount of available water decreases, so does the a_w value. Most bacteria must have a water activity of 0.95 or higher to grow. Table 1 lists some common foods and their a_w values.

Note that fresh fish has a water activity close to 1.0 but after heavy salting and drying, the water activity is between 0.80 and 0.70, a level far below the threshold at which normal bacteria can grow. However, there are some bacteria, called halophiles (i.e., salt-loving), that can grow at a water activity value as low as 0.75 (Table 2). Occasionally, these bacteria can cause considerable spoilage even in salted fish. Some contain a red pigment, and fish contaminated with them are referred to as "pink" spoiled (Troller and Christian, 1978). The preservation qualities of dried and salt-cured fish are obtained through the removal or displacement of available water to prevent bacterial growth.

SMOKED FISH PROCESSING

Although general operations in all smoked fish processing plants are similar, the specific processing procedures can vary considerably. This variability relates to differences in equipment, regional and ethnic consumer preferences, raw materials, and tradition (Faturoti, 1984).

Effective sanitation considerations, including plant design, construction, water, and personal hygiene must be carefully evaluated to minimize bacterial contamination (Dougherty and Seagran, 1967; Crance, 1955).

The unit operations involved in the processing of smoked fish can be categorized as follows:

1. Purchasing and receiving
2. Raw material storage (refrigerator and freezer)
3. Raw material preparation

Table 1. Approximate a_w Values of Some Foods and of Sodium Chloride and Sucrose Solutions

a_w	NaCl (%)	Sucrose (%)	Foods
1.00-0.95	0-8	0-44	Fresh meat, fruit, vegetables, canned fruit in syrup, canned vegetables in brine, frankfurters, liver, sausage, margarine, butter, low-salt bacon
0.95-0.90	8-14	44-59	Processed cheese, bakery goods, high-moisture prunes, raw ham, dry sausage, high-salt bacon, orange juice concentrate
0.90-0.80	14-19	59-saturation	Aged cheddar cheese, sweetened condensed milk, Hungarian salami, jams, candied peel, margarine
0.80-0.70	19-saturation	—	Molasses, soft dried figs, heavily salted fish
0.70-0.60	—	—	Parmesan cheese, dried fruit, corn syrup, licorice
0.60-0.50	—	—	Chocolate, confectionery, honey, noodles
0.40	—	—	Dried egg, cocoa
0.30	—	—	Dried potato flakes, potato crisps, crackers, cake mixes, pecan halves
0.20	—	—	Dried milk, dried vegetables, chopped walnuts

Source: Troller and Christian, 1978

4. Salting: dry and brine
5. Drying
6. Smoking
7. Cooling
8. Packaging
9. Finished product storage
10. Distribution and sale

Purchasing and Receiving

A processor should be aware that various state and federal statutes exist concerning the purchase and possession of various fish species and size. It is important that purchasers be cognizant of, and in compliance with, these regulations.

Table 2. Lowest a_w Values Permitting Spoilage-Organism Growth

Group of Microorganisms	Minimal a_w Value
Bacteria	0.91
Yeasts	0.88
Molds	0.80
Halophilic bacteria	0.75

Selection and Initial Preparation

It is imperative that only fresh, properly prepared fish be used for smoking. Smoking will not mask off-flavors or otherwise make a poor quality or spoiled fish acceptable. Smoking enhances the flavor and texture of fish. Fish with a high oil or fat content are generally more suitable for smoking than lean fish. Some examples of high-fat fish are salmon, eels, whitefish, catfish, sturgeon, chugs, mackerel, mullet, and bluefish. Examples of low-fat fish are flounder and snapper.

When selecting fish for smoking, choose only those that are of high quality and free from bruises, torn skin, or other physical damage. Fish may be smoked whole, headed-and-gutted, filleted, staked, or chunked, depending on the species and desired style. Large fish should be filleted or cut into smaller pieces so that proper cooking temperatures can be achieved without overcooking the outside of the fish.

Receiving

Production of a quality finished product starts at the receiving department. This operation not only involves physical control of incoming raw materials and supplies but is usually the first inspection point.

It is essential that all processing begin with a high-quality product. Containers and fish should be inspected on arrival at the plant. All fish must have been shipped at suitably low temperatures, and they should be free from adulteration and not have detectable off-odors or flavors. Special attention should also be given to firmness of flesh, eye condition, and gill color.

It is highly recommended that all incoming fish be divided into lots and given an identification tag which will accompany the product throughout the entire process and become part of the process records. Information on the tag should include point of product origin; date received; condition of fish (physical state as well as appropriate quality attributes); lot number (if desired); and size and type of fish.

Freshwater and saltwater finfish species represent the major tonnage of incoming materials. Depending on the species and ultimate use, fish may be received whole, gutted, or headed-and-gutted, and may be fresh and iced or frozen.

Inspections at the receiving department should include other edible raw materials used in the processing of smoked or smoke-flavored fish, including salt, sugar, spices, and, in some products, smoke flavoring, artificial color, sodium nitrite, vegetable oil, and other ingredients. Wood-smoking materials such as chips and excelsior, packaging materials, and cleaners and sanitizers also should be examined.

Raw Material Storage

All reasonable precautions should be taken to ensure that all products and raw materials are handled in a manner that will not contribute to their contamination or deterioration. Lots should be appropriately identified to assure their timely use on a first-in/first-out basis whenever possible.

Fish that are not smoked immediately can be iced and/or refrigerated for a short period. Fresh fish which are not to be processed immediately should be refrigerated at a temperature near 32°F (0°C). Frozen fish should be either thawed and processed promptly or stored at a temperature that will maintain them in a frozen state. Fish that are to be kept for an extended period prior to cooking and smoking should be properly frozen and stored to maintain desired quality. Usually a temperature no higher than 0°F (-18°C) should be used, with -20°F (-29°C) preferred. Proper chilling and/or freezing retards bacteria growth and enzyme activity, the major causes of spoilage in fishery products.

Dry and nonperishable food ingredients and package materials should be stored in a dry area in a manner that protects against contamination and deterioration. Cleaning and sanitation chemicals should be stored in a separate area.

Sodium nitrite requires special handling and should be stored in a locked area restricted to those who will use this ingredient and who have been properly instructed regarding its use and potential hazard. Only quantities needed in a given brine should be permitted out of the room.

Raw Material Preparation

Proper cleaning and preparation prior to salting or smoking improves product quality. Immediately before processing begins, both fresh and thawed fish should be thoroughly washed using a vigorous water spray or a continuous water flow system with a chlorination level of 25-50 parts per million (ppm). This process helps increase shelf life by removing blood and reducing bacterial populations. Washing fish after brining is not as effective, because once the product has been brined, a water-soluble protein layer covers the fish surface, making it more difficult to remove entrapped bacteria.

Whole fish should be thoroughly washed prior to gutting and dressing to remove external debris, blood, and the natural slime that encases most fish. Generally slime can be easily removed by washing the fish in cool water and rubbing the slime off. At times, slime can be difficult to remove, especially from heavily slimed fish such as eels. To make removal easier, one of the following four methods can be used:

1. Soak the fish in a heavy brine solution for a few minutes. Often, this will quickly separate the slime from the fish.
2. Wash the fish in a chlorine solution (Dudley et al., 1973) made up of one tablespoon of liquid hydrochlorite bleach for each four gallons of water. Make sure that fish washed in chlorine solution are thoroughly rinsed in fresh, clean water.

3. Quickly dip the fish in hot water (about 180°F/82°C) to coagulate the slime (Dudley et al., 1973).
4. Freeze the fish. When fish are thawed in preparation for smoking, the slime often is loosened and can be easily rinsed off.

The thawing or defrosting of frozen fish should be carried out in a sanitary manner and by methods that will not adversely affect the wholesomeness of the fish. Whole fish should not be mixed with gutted fish during thawing. Different species of fish should be thawed separately.

To maintain quality, fish should be thawed in air at a temperature of 45°F (7°C) or below so that no part of the fish exceeds 45°F (7°C). If the fish are thawed in water, then a continuous chilled water-overflow tank, spray system, or other process which provides frequent water exchange should be used. The fish should not remain in the tank longer than is needed to sufficiently thaw them for further processing, preferably no more than half an hour after the fish are completely defrosted. Care should be taken to ensure that fish entering the thaw tank are completely free of packaging liner material. Cleaning and sanitizing the tank is essential to maintain sanitary conditions and should be conducted as often as necessary.

Fish should be eviscerated before salt-curing and smoking. Whole fish should be eviscerated with a minimum disturbance of intestinal tract contents. Cut the fish and thoroughly wash the cavity in fresh, clean water. Be sure to remove all organs without puncturing or cutting them. The kidneys are usually lodged along the backbone and require extra effort to remove sufficiently. After evisceration, the fish (including the body cavity) are given a second thorough wash with a vigorous chlorinated water spray or a continuous waterflow system. All offal should be placed in suitable covered containers and removed at least once a day or more frequently if necessary. Depending on offal refuse pickup schedules, offal containers may require refrigeration.

Consider the cut of fish to be smoked. Style and form depends on the size of fish, desired product, and personal preference. Small fish such as chubs, whitefish, and eels are usually smoked whole, eviscerated, and gills removed. Larger fish such as salmon, sturgeon, sable, and bluefish are cut into steaks, fillets, split, or butterflied. The cuts may or may not be skinless. Always keep cuts of fish packed in ice or properly refrigerated at all times.

Salting

One of the most difficult, but most important, steps in preparing smoked fish is obtaining the desired concentration of salt or other preservatives in all parts of the product. Uniform salt concentrations are important. Depending on the concentration, salt can slow the growth of spoilage microorganisms and some food-poisoning bacteria. However, the main purpose of adding salt is to impart flavor, because the amount used in modern smoked products has little effect on keeping quality. The main caveat is that smoked fish is perishable and requires refrigeration.

Factors contributing to salt variation in smoked fish include fish size; species; fat content; condition (fresh or frozen, skin on or skin off, state of rigor); method of salt application; brine concentration; brine temperature; brining time; brine-to-fish ratio; circulation of brine; and section of fish used. Several of these factors are discussed in more detail later in this chapter.

In recent years there has been a trend toward lowering the salt content of processed foods as a means of reducing dietary sodium. Smoked fish processors have been very sensitive to this issue, because they are interested in adjusting to consumer tastes without sacrificing product safety or market share.

The salt used should be of food-grade quality, low in calcium and magnesium and essentially free of iron and copper. The application of salt to fresh or thawed fish is carried out prior to either hot- or cold-smoking by exposing the fish to dry salt, or more commonly, to salt brine. Some processors use a combination of the two procedures. Although there is no hard-and-fast rule that dictates the use of one procedure over the other, salt brines are most widely used because they are easier to handle and offer better control.

With dry salting, the amount of salt, time, and temperature should be carefully controlled to attain a desirable result. The ratio of salt-to-fish by weight may vary from 1:8 for light salting, 1:3 for split fish, or 1:1 for heavy salting. Dry salting should be carried out at a temperature not exceeding 38°F (3°C). (In the preparation of some Nova

Scotia salmon by the dry-salting procedure, brown sugar is also sprinkled in with salt.) Because of the variations possible, only through experimentation and experience can the proper curing be ascertained.

Brining or Curing

Brining serves three purposes: it firms the texture of the fish, provides seasoning or flavor, and acts as a preservative in some types and styles of smoked fish. But brining must be carried out under the most careful conditions to prevent making an unpalatable salty product.

Liquid salt solutions, or brines, are an important step in processing smoked fish and require some precision and vigilance on the part of the preparer. A saturated brine is made from good quality bulk or bagged salt in equipment available from the major salt companies. Salt storage and brine making can be confined to one location and the brine pumped to the points of use.

A brief explanation of the arithmetic of salt solutions is appropriate. Quantity of salt is usually stated in pounds. One gallon of fully saturated brine at 60°F (15°C) contains exactly 2.987 pounds of salt and tests at 100°S (100° salometer). In making brine, the concentration of the solution must be measured reasonably accurately (within 5°S) to predict the proper amount of time to soak the fish. The best and most common way to measure brine concentration is with a floating salometer, which measures the density of the brine according to its buoyancy when placed in the brine. The denser the solution, the more buoyant the salometer will be. The scale gives readings in salometer degrees from 0° to 100°S (corresponding to 0 to 100 percent saturation: 0°S for pure water and 100°S in fully saturated brine. For example, a 40°S brine is 40 percent saturated). When using a salometer, it is crucial that the brine temperature be considered. Ordinarily, salometers are scaled for reading at a temperature of 60°F (15°C). Table 3 gives an accurate conversion of salometer degrees to salt concentration when the brine is 60°F (15°C). However, if the temperature is above or below 60°F (15°C), adjustments must be made to give an accurate reading. Adjustments are made by adding or subtracting one salometer degree for each 10°F (5.6°C) that the brine temperature deviates from 60°F (15°C). For example, if the temperature of the brine is 80°F (27°C), you would *add* 2°S to the reading on the chart. If the temperature of the brine is 45°F (7°C), you would *subtract* 1.5°S (Bankston, 1973; Hilderbrand, 1973).

The typical brine used to soak fish for smoking varies from 30° to 50°S although higher concentrations can be used. Refer to Table 3 to determine the pounds of salt/gallons of water to be dissolved for a particular salometer degree brine. For example, an 80°S mixture would be a 21.116 percent sodium chloride solution, and would require 2.229 pounds (1.012 kg) of salt for each 1 gallon (3.785 l) of water. To make 100 gallons of solution, one would add 222.9 pounds (2.229 x 100) of salt to 100 gallons of water. The resulting solution would have a total volume greater than 100 gallons because water molecules would be displaced by the dissolved salt.

Diluted brine solutions are commonly made in the smoked fish industry by volumetric dilution of fully saturated brine (100°S). To make 100 gallons of a 40°S brine requires 40 gallons of fully saturated brine and 60 gallons of water, with proper adjustments for temperature. Diluted brines should be checked for salometer reading.

If a diluted brine is to be prepared from dry salt and water, a brine table should be used to establish the amounts of salt and water necessary to make the volume of diluted brine required. The salt should be completely dissolved before using the brine.

When making brine, several principles should be kept in mind:

1. To achieve a rapid and predictable salt penetration, use the purest salt (NaCl) possible. Calcium and magnesium, which are common salt impurities, hinder the proper penetration of salt into fish tissues. Improper or delayed salt penetration can cause spoilage. In addition, these impurities may cause a chalky, bleached-out, or unnatural color that detracts from the product's appearance. Consequently, use of sea salts or other salts with known impurities should be avoided.
2. Salt used for brining should be fine textured so it will dissolve quickly. Table salt has a suitable texture, but rock salt may take an inordinate amount of time to dissolve. For consistency, it is important that an accurate measurement of salt be added to the water and that it be completely dissolved prior to

Table 3. Brine Conversion Tables (60ºF)

Salometer Degrees	Percentage Sodium Chloride	Pounds Salt Per Gallon of Water by Weight
0	0.000	0.000
2	0.528	0.044
4	1.056	0.089
6	1.586	0.134
8	2.112	0.179
10	2.640	0.226
12	3.16	0.273
14	3.695	0.320
16	4.223	0.367
18	4.751	0.415
20	5.279	0.464
22	5.807	0.512
24	6.335	0.563
26	6.863	0.614
28	7.391	0.665
30	7.919	0.716
32	8.446	0.768
34	8.974	0.821
36	9.502	0.875
38	10.030	0.928
40	10.558	0.983
42	11.086	1.039
44	11.614	1.094
46	12.142	1.151
48	12.670	1.208
50	12.198	1.266
52	12.725	1.325
54	14.253	1.385
56	14.781	1.444
58	15.309	1.505
60	15.837	1.568

Table 3. continued

Salometer Degrees	Percentage Sodium Chloride	Pounds Salt Per Gallon of Water by Weight
62	16.365	1.629
64	16.893	1.692
66	17.421	1.756
68	17.949	1.822
70	18.477	1.888
72	19.004	1.954
74	19.532	2.022
76	20.060	2.091
78	20.588	2.159
80	21.116	2.229
82	21.644	2.300
84	22.172	2.372
86	22.700	2.446
88	23.338	2.520
90	23.755	2.594
92	24.283	2.670
94	24.811	2.745
96	25.339	2.827
98	25.86	2.906
100	26.395	2.98

Note: Special salometers are available, scaled for reading at 38ºF. Also, special brine tables at 38ºF are available to extrapolate salometer degree at 38ºF to the percent sodium chloride by weight, weight per gallon of brine freezing point, and other information.

use as a brine. When salt is initially added to water, it begins to dissolve quickly, but the rate may slow considerably as brine strength increases.

3. Stirring or agitating the brine will increase the rate at which the salt dissolves. Mechanical agitation by a pump, sparge, or propeller is highly recommended, but manual agitation with a paddle or stirrer may be sufficient. Agitation helps dissolve the salt and is also useful during the brining process to maintain consistent absorption. This process is discussed in more detail later.
4. As the water's temperature is increased, so is the rate at which the salt dissolves. Dissolving all the salt in non-chilled water and then chilling the solution to the proper brining temperature may save some effort. Prior to adding fish, it is important that the brine solution be chilled to at least 40°F (4.5°C), which maintains the quality of the fish during the procedure. Although the rate of salt penetration is retarded by chilling the brine, fish quality should never be sacrificed for the sake of speed.

During the brining procedure several phenomena occur:

1. Water migrates out of the fish tissues due to osmotic pressure. This water loss causes some weight loss but will favorably affect the texture of the fish.

2. The salt concentration in the tissue increases with soaking time. Consequently, if the fish are brined for longer than the recommended time, the final product may be unpalatable because of a high salt content.
3. Salt from the brine is absorbed by the fish, and water from the fish is expelled into the brine. The end result is diluted brine. Since brine is generally changed between batches of fish, this probably has little impact on brine quality.

As mentioned earlier, numerous factors influence the rate that fish absorbs salt (Bligh and Rendell, 1986; Burgess et al., 1967). Among the most significant are:

1. *Exposed flesh.* Salt penetrates skinless and/or filleted areas much more rapidly than areas protected by skin.
2. *Fat content.* As the percentage of fat increases, the rate at which the salt penetrates the fish flesh decreases. Consequently, fat content and the species of fish brined have a significant impact on required brine concentrations and brining time.
3. *Size and shape of fish or pieces of fish.* All other factors being equal, the thinner the fish portions, the faster the salt penetration. Fish of reasonably uniform size should be brined together. Fish of different species should not be mixed in the same tank.
4. *Agitation.* Stirring the brine results in a more even penetration of salt. Pockets of diluted or concentrated brine are continuously blended to give an overall, even concentration.
5. *Strength of brine.* Brine strength is important to ensure a uniform and standard product from batch to batch. For example, if one vat of brine has a 45°S value and another vat for the same product has an 80°S value because of a careless worker, then the final products will not be the same. As a general rule, the stronger the brine, the shorter the brining time. Typically, brine concentrations of between 30°S and 50°S are used (Bannerman, 1980).
6. *Submersion.* For a fish to brine properly and uniformly, it must be completely submerged in the brine. Floating fish or too many fish (which prevents complete coverage) cause uneven salt penetration and a substandard product.
7. *Weight ratio.* As the brine-to-fish ratios increase, the amount of salt per unit weight of fish increases. A longer refrigerated brining time (eighteen to thirty-six hours) with a more dilute brine (20° to 45°S) often results in a more uniform salt concentration than a short brining time (two to six hours) in a more concentrated brine (over 45°S). However, either brining procedure is acceptable.
8. *Temperature control.* The brining process should be performed so that the temperature of the fish and brine does not exceed 60°F (15°C) at the start of brining. If the fish are between 38°F (3°C) and 50°F (10°C) at the start of brining, the temperature should be continuously lowered to 38°F (3°C) or below within twelve hours. If the temperature is between 50° and 60°F (10° and 15°C) at the start of the brining process, the temperature should be continuously lowered to 50°F (10°C) or below within two hours and to 38°F (3°C) or below within the following ten hours. Once the brining process reaches 38°F, the temperature should be constantly maintained during the entire operation.

Water-Phase Salt Content

Salt content in the finished smoked fish product is typically determined on the loin muscle of the fish and is expressed as percentage of salt in the water phase. To determine the salt content in the water phase of the muscle, it is necessary to remove and analyze the loin muscle for both moisture content and salt content.

$$\% \text{ Salt in Water Phase} = \frac{\% \text{ Salt}}{\% \text{ Moisture} - \% \text{ Salt} \times 100}$$

A dry salting or brining process is established to attain no less than a minimum water-phase salt content appropriate for the product. There has been considerable debate over what constitutes an appropriate minimum water-phase salt content. The minimum water-phase salt content considered appropriate for smoked fish products can vary based on use of sodium nitrite, heat processing, type of packaging, intended shelf life, and expected storage conditions.

Typically the rate or amount of salt uptake increases with increased brine concentration, higher brine temperature, greater brine-to-fish ratio, smaller fish, longer brining period, a soft (post-rigor) condition, aged fish, and prefrozen fish. Although some factors increase the salt uptake, this should not be construed as optimum brining conditions. Brining conditions are selected to produce a uniform and high-quality product within a reasonable production schedule.

Brine tanks should be cleaned and made sanitary before each use. Brines should not be re-used unless a suitable procedure, such as ultrafiltration, is used to return the brine to an acceptable microbiological level.

Before the fish are removed from the brining process they are generally evaluated by a chemical and sensory analysis. If the fish do not have an acceptable salt level, brining time is extended. If too salty, the product is rinsed in water to remove some of the salt. Both of these processes must be carefully monitored to ensure that the fish are removed at the correct time. The actual required brining time can vary, and is sometimes quite rapid (Childs et al., 1976). It is common practice to rinse fish after brining to remove surface salt. This operation must be done quickly and cautiously to avoid leaching excessive salt from the product.

Ingredients

Other ingredients, such as sugar, other flavor ingredients, coloring agents, and sodium nitrite can be added during the brining process. Special brining formulas are proprietary to many smoked-fish processors. All ingredients must be generally recognized as safe and approved. The use of sweeteners, liquid smoke, and coloring agents is at the discretion of the processor and should be in accordance with good commercial practices appropriate for the product that is being prepared. Finished products should be labeled in accordance with regulatory requirements to reflect all ingredients used.

Some fish are placed into a tank or tub of coloring agent to achieve a desired color or the dye may be added to the brine, combining the two operations in one. The concentration of dye used in dipping is greater than when it is incorporated into the brining solution. In some cases, dye may be injected into the fish. For coloring consistency, some types of smoked fish are dipped after preslicing. Dye solutions, approved by the Food and Drug Administration, are made up in strengths dictated by the experience of the individual curer, and it is not possible to establish exact rules which will apply universally. Processors must determine their requirements through experimentation and attention to the desires of their customers. The use of dyes in smoked fish and the labeling required for such processes are subject to specific state and federal regulations.

Sodium nitrite is a curing agent currently permitted only in smoked chub; smoked, cured salmon; smoked, cured sablefish; smoked, cured shad; and smoked, cured hena as a preservative and color fixative. Sodium nitrite enhances the inhibitory effectiveness of salt against the outgrowth of *Clostridium botulinum* Type E spores. The use level in brines, either in the premix or pure form, is adjusted to attain in the finished smoked chub not less than 100 ppm and not greater than 200 ppm in the loin muscle. In finished smoked sablefish, salmon, or shad, the level should not exceed 200 ppm. A rule of thumb is to add twice as much sodium nitrite to the brine as desired in the final product. For example, to obtain 200 ppm in the final product, add 400 ppm to the brine. In a 35°S brine at 60°F (15°C), the amount of sodium nitrite (400 ppm) per 50 gallons (189.3 liters) of brine would be approximately 2¾ ounces (78 grams). The actual amount required should be established by controlled experimentation. In practice, it should be thoroughly dissolved in the brine before the fish are added.

Sodium nitrite can be broken down by bacteria present on smoked fish; thus, its inhibitory effect can be diminished. To retain its maximum effectiveness, good sanitation procedure must prevail, and the product should be stored at 32° to 34°F (0° to 2°C). It is strongly recommended that a processor obtain the services of a reputable supplier or qualified consultant for the development of safe and effective brining mixtures and their applications.

DRIED, SALTED FISH

Dried, salted fish is not commonly consumed in the United States. The process for its preparation can be described as follows (Burgess et al., 1967): lean fish, such as cod, are normally used for dry salting. The first step involves splitting the headed fish, usually from head to tail along the backbone, and removing the backbone. By splitting the fish, salt penetration is more easily controlled. Salting is done by stacking the fish in layers, alternating each layer with a layer of salt. The pile should be restocked and resalted periodically to assure a consistent cure. After the fish have cured this way, the fish are in the "green-cure" state, at which point the water content has been reduced to two-thirds of its original amount and the salt has penetrated throughout the fish and has saturated the remaining fluids. These green-cured fish must now be dried to get the final water content down to 25 to 38 percent. The fish are hung to dry, either by using the sun and breeze, or more likely, by artificial means such as warm air circulating within an indoor drying chamber. The final salt content may be as much as one-third of the weight of the finished product.

Burgess et al. (1967) describe cured fish as follows. Fatty fish such as herring, mackerel, or anchovies are best suited for brining or curing. First the guts and gius are removed, then the remaining whole fish is packed into barrels or casks, alternating a layer of fish with a layer of salt. The salt removes water from the fish by osmotic pressure and forms a brine. Removing water from the fish causes the fish to shrink considerably. Consequently, after about ten days of curing, the barrel must be repacked by first draining off excess brine and then adding additional cured fish to make up for the shrinkage. Additional brine may have to be added to displace air introduced as a result of disturbing the barrel. The barrel is tightly capped and sealed. At this point, the fish have been stabilized and are suitable for storage.

PICKLED FISH

Pickled fish rely on salt and the action of acetic acid in vinegar as a preservative. Prior to shipping to a packing plant, herring, the most popular pickled fish, is normally cured several days in an 80° to 90°S brine with 2.5 percent 120-grain distilled vinegar. They are shipped to the packing plant in barrels with a 70°S brine. The herring may be cut into the desired shape and freshened overnight in water. Prior to packing, the pieces are generally cured three days in a solution containing 3 percent white distilled vinegar and 6 percent salt. The last step is the final cutting and packing into jars with the desired curing solution.

DRYING FISH

After brining, fish are hung or laid on racks for drying, smoking, and heat processing. If these processes cannot be conducted within two hours after removal from the brine, fish should be stored in a refrigerator at 38°F (3°C) or below. Drying allows for good color formation, forms a "skin" that holds in juices, and gives the strength needed to keep fish from falling from hooks, rods, or other holding devices.

POSITIONING FISH FOR DRYING AND SMOKING

Fish are usually dried in the same position in which they will be smoked, and obviously fish should be smoked with as much flesh exposed as possible. For example, a split fish should be hung or presented in such a way that split halves are exposed. Depending on the size and cut of fish, there are three common methods for hanging or holding fish for drying and smoking.

1. Rods may be used to thread fish through the head, gills, or mouth. This method is good for smaller fish and for some fish that have been headed and split. Fish normally hang tails down when placed on a rod.
2. Hooks or nails allow the fish to be hung in any position, although usually they are hooked through the head. If they are hooked through the fleshy part of the body, the hooks leave an undesirable mark. In hanging fish, there should be sufficient space between the fish to prevent faulty processing. Overcrowding or overloading the smokehouse could result in an inferior product and should be avoided.
3. Racks are useful for pieces of large fish such as chunks, steaks, blocks, sides, or fillets that do not have skin or skeletal structure to support hanging. Racks should be made of large mesh screens or other materials that allow even exposure to smoke and air circulation (Dudley et al., 1973).

Mesh bottom trays, made of half-inch mesh, may be used for some products, such as fillets and steaks. The trays are sometimes coated with edible oil to prevent pieces of fish from sticking. Pieces in the tray should be close to the same size and should not touch each other. Any free liquid should be removed, because pools of brine lying in depressions in the fish will be slow to dry, and these damp areas may spoil rapidly during subsequent refrigerated storage. The trays may be placed directly on fixed racks in the smokehouse, or they may be conveyed into a cage which is placed into the smokehouse after it has been filled.

During hanging, dissolved proteins in the brine solution dry on the fish surface and produce the familiar glossy skin which is one of the commercial criteria for quality. When properly dried, this *pellicle* will form and the outer surface will be smooth, dry, and glossy (Crance, 1955). Without proper brining and drying, a glossy pellicle will not form. A properly formed pellicle helps give the finished product an attractive appearance and flavor, because smoke readily adheres to it. A poorly formed pellicle may allow the outer surface of the fish to burst or erupt, emitting coagulated body fluids, and resulting in an unattractive appearance.

Drying is an important step that must be controlled to produce a high-quality product. In almost all instances, fish must first be dried to remove a certain degree of surface moisture prior to heating or smoking. When properly done, drying makes the flesh firmer and prepares it for further cooking and smoking.

Forced drying (raising the temperature and drawing a current of air through the smokehouse) reduces the overall time required for the smoking process. Drying time depends on such factors as air circulation, temperature, and the relative humidity of the air. It usually takes several hours. It is generally recommended that fish be dried in a cool place with circulating air from a blower or fan, especially when home-smoking fish. Some commercial processors dry fish in the smoking chamber at an elevated temperature. The processor must be sure, however, that all air vents and doors are open to provide good circulation of air and temperatures low enough so that the fish are not cooked.

If too much humidity builds up during this step, or if processing takes too long and the meat and bone are exposed to heat before drying, the product will fall apart. To prevent this, the protein must be set or denatured with low-temperature drying *before* applying higher temperatures. If the air is too hot and moving too quickly, the surface of the fish will be damaged and will not dry properly. Fish flesh, like that of other animals, is primarily composed of protein. When proteins are dried too fast, they harden, or "denature," and the skin forms a hard case. When this happens, water cannot escape from the core of the fish and the outside forms a crust. To prevent this problem, known as *case hardening*, the oven must dry the fish slowly enough to prevent this process, but fast enough to avoid deterioration caused by bacterial and enzymatic activity. If the surface of the fish is over-dried, it will crumble later and smoke color formation will be poor because of inadequate smoke absorption. If the fish are not properly dried and the fish is smoked while too moist, the smoke will not be evenly absorbed, resulting in a "streaky" product. The longer the drying process, the greater the protein degradation. Control of time, temperature, and air flow are of critical importance in drying.

Smokehouses vary in design and many variables must be taken into consideration. A discussion of some of the key factors follows:

1. *Air circulation*. The air inside a processing chamber must have a certain velocity and volume to achieve efficient, even processing. The volume of air determines the distribution of heat and moisture. If the air velocity too high or too low, the resultant product may be defective. In addition, the air must flow evenly, reaching all areas of the smokehouse. Driving the air for too great a distance will result in uneven processing.
2. *Heating and cooling*. An efficient way to heat the air must be designed into the system. Optimum design places the heating element directly into the chamber, eliminating heat loss during transfer and using all the heat for processing. The shorter the distance from the heating element to the processing area, the more efficient the system. Air used in processing must be clean, with controls for heat and moisture content.
3. *Humidity*. Humidity control is an important but often misunderstood factor in producing smoked fish. Relative humidity is expressed as a percentage and measures the water content of air at a particular temperature compared with the maximum amount of water the air could hold at that temperature.

The higher the temperature, the more water the air can hold. Relative humidity must be expressed as a percentage at a given temperature. To determine wet-bulb temperature, take the air temperature with a wet sock placed over the bulb end of the thermometer. The water evaporates off the sock, cools the thermometer, and results in a temperature reading lower than actual air temperature. The drier the air, the faster the evaporation, the quicker the thermometer cools, and the lower the reading. Comparing the difference between the air temperature (dry bulb) and the evaporation temperature (wet bulb) on a prepared scale gives relative humidity. Air must blow at and around the wet bulb at a specific rate [approximately 800 ft. (244 m) per minute] to give a correct reading. Also, the wet wicking material should be kept free of contamination (cleaning chemicals, smoke particles, tar, creosote, or organic vapors) which may retard water evaporation. If the velocity is too low, there will not be enough evaporation and the wet-bulb temperature will be too high. If the air velocity is too great, the evaporation will be too fast and the temperature reading too low. In either case, the moisture reading will be incorrect.

This discussion brings us back to the importance of air velocity. The product is saturated with water and acts much like the wet-bulb thermometer at a specific temperature and air velocity. The location of the wet-bulb thermometer during the reading is crucial. If the reading is taken near the supply ducts, it will measure the humidity of the air before it picks up or gives off moisture to the product. If a reading is taken at the exhaust duct, it will measure the humidity of the air after processing has taken place. Every point in the chamber will have a different reading because of evaporation or condensation. Depending on circumstances, however, readings eventually equalize.

Because fish products contain a high percentage of water, the effect of the product on the air is as important as the effect of the air on the product. The product gives off or absorbs moisture at a rate dependent on its current temperature and pressure differentia. As the product absorbs or gives off heat or dries, that rate changes, even if processing air values stay the same. Thus the relative humidity is only a guide to the type of air to use. There may be times when it is desirable to reach or even exceed the saturation point of air, called *steaming*. Moisture condenses on the product, passing its energy content to the product.

Sometimes dehumidification at a specific point is more important than humidification. It is also possible that a slowly rising or falling humidity over a certain time interval is more important than maintaining a constant humidity. To get this effect, humidity must be removed or added on a time interval rather than at a constant rate. When excess humidity is present, a dehumidification cycle is automatically initiated to reduce the moisture content of the air in the processing chamber. This dehumidification can be accomplished in a number of ways. The simplest, fastest, and most economical method is to introduce fresh, ambient air into the oven or smokehouse while removing some of the moist air, until the humidity drops to the desired level. It is important that the system be properly engineered because smoke air pollution may occur. If the community has stringent air pollution requirements, a small afterburner, electric precipitator, or water scrubber can be used to further reduce the pollution in the exhaust air.

Another way to dehumidify is with a semi-closed system with water spray dehumidification. A fine spray of water is introduced into a chamber through which air is recirculated during the dehumidification cycle. The water cools the air, forcing out moisture. However, there are numerous disadvantages in this system. First, the water must be colder than the air at all times, resulting in high water use even if recirculating pumps are used. Second, the smoke is removed by the water spray, causing a reduction in smoke concentration and an increase in water pollution. Third, maintenance requirements are greatly increased because of its costs and limitations.

When temperatures exceed 212°F (100°C), relative humidity readings are not possible. However, air moisture content can still be controlled by adding moisture to the air, and the easiest way to add moisture is steam. Using water droplets or fog can cause water droplets to form on the product. When using steam, care must be taken. If the steam is too hot (at pressures over 7 per square inch) the air will be overheated and unwanted excess condensation will occur, causing spotting, burning, and separation in the flesh. The air must have time to absorb the moisture introduced. Thus, injections of moisture that are small and fast are preferred over one continuous injection.

It should be noted that the shrinkage of fish muscle along the dimension parallel to muscle fibers is significantly lower than that perpendicular to the fibers during air drying. In other words, length shrinks less than width or depth (Bulaban and Pigott, 1986).

Highly significant differences in proteins between protein efficiency ratio (PER) net protein utilization (NPU) and biological value were observed at different drying temperatures. The PER and NPU of fish dried at 60°C were significantly higher than those dried at 50° or 70°C (Raghunath et al., 1995).

SMOKING

Smoking fish, like salting, dates back to early times. Smoking probably was an ancillary benefit in the drying of salt-cured fish over open fires. In addition to facilitating preservation, the smoke contributed a pleasing flavor. Although many refinements have been made in smoking fish, the overall process retains many of the original attributes. Methods for smoking fish vary greatly, depending on such factors as the type of fish, desired flavor, desired texture, cooking method employed, and cultural preferences.

Today distinctions are made between smoked and smoke-flavored fish, between hot- and cold-smoked fish, and among the various methods of applying the smoke constituents.

Traditionally, smoke applied to fish came from burning hardwoods such as maple, oak, alder, hickory, birch, and fruitwoods. Liquid smoke was prepared as early as the late 1800s, but only within the last ten to fifteen years has it been widely used in the smoked fish industry. Even today, smoke-*flavored* fish represents only a small percentage of the total industry output.

The three kinds of smoke-flavor additives are natural smoke extracts, synthetic smoke flavors, and substances unconnected with smoke, such as yeast derivatives, that have a smoky flavor and smell. Smoke-flavored fish, made using liquid smoke, can be prepared by including liquid smoke in the brine, applying it as a dip after brining, or as an atomized spray within a modern automatic smokehouse. There is no single accepted method of application; use depends on the processor's objectives, ingenuity, and facilities. However, recent research suggests that spraying is preferable to dipping as a means of accurately controlling flavor and acceptability. It may be necessary to dry some products after the flavor has been added in order to obtain a texture comparable with traditional products. During spray application, there are several positive results if air is circulated at a low velocity, if air temperature is raised by 10° to 15°F (5.6° to 7.2°C), and if the liquid is injected in short bursts as a fine fog. This method provides improved flavor, a better, more even color throughout the load, and faster processing. In addition, less liquid smoke is used, lowering costs.

In general, smoking of fish is carried out as a cold-smoking process or as a hot-smoking process, and the equipment used in either process can be a traditional gravity oven or a modern electric oven.

Cold-smoking is exactly that. The fish are *not* cooked, and temperatures during smoking generally do not exceed 110°F (43°C). If the temperature is allowed to rise, muscle texture could be adversely affected. To maintain proper temperature during smoking and to ensure uniform drying and desired color, it is necessary to use an indirect source of heat and smoke. Proper cold-smoking often takes less than 24 hours. The cold-smoking process is primarily used for salmon. Other traditional cold-smoked items include black cod (sablefish) and herring.

The curing process may differ slightly since both liquid and dry cures are used. Sugar and salt may be added to the cure mixture. Achieving quality results in cold-smoking is somewhat more difficult than in hot-smoking. Drying, an important part of the process, must be slow and the humidity carefully controlled to attain the desired surface hardening. The drying and cold-smoking process typically requires eighteen to twenty-four hours.

With traditional gravity ovens it is particularly difficult, if not impossible, to carry out cold-smoking on warm and humid days unless the system is specially modified to cool and dehumidify. Often heat and smoke are produced separately and then carried to the fish by fan or pipe. Careful temperature controls with thermometers, ventilators, dampers, and fire controls are necessary to achieve a good finished product. Because of the flexibility of control in modern automatic ovens, they are adaptable to cold-smoking. Again, equipment installation is a factor. In most processing plants a number of ovens are required to simultaneously process different items, and equipment cost must be considered.

The temperatures and times used in processing cold-smoked fish are very favorable for the proliferation of food spoilage and food-poisoning types of microorganisms. Therefore, particular attention to sanitation, proper brining, limitation of the process to specific product types, careful handling, process control, and prompt refrigeration after smoking are essential. Although the finished product has not been cooked, it has excellent keeping properties because it has been dehydrated sufficiently to retard most bacterial growth.

Hot-smoking is the process used in the majority of smoked fish products. A hot-smoked product has been fully cooked and may reach temperatures as high as 180°F (82°C). Because of the higher temperature, hot-smoking takes only a short time, depending on the internal temperature of the product. Different species of fish tolerate heat differently; consequently, the hot-smoke process is not the same for all products. The process must be tailored to the species, the processing equipment used, market demand, distribution considerations, and regulatory requirements. Hot-smoked fish are moist and juicy when properly finished. Because of this, they have a relatively short shelf life and must be refrigerated.

In processing, the intention is to cook the fish as well as smoke it. Processing temperatures used in the industry vary significantly, but as with cold-smoked fish, it is necessary to maintain careful temperature controls and to always use a thermometer to monitor the coldest part of the fish (Hilderbrand, 1980; Dudley et al., 1973). Because elevated temperatures are used in hot-smoking, fish are generally close to the heat source.

The minimum internal temperature for adequate processing of hot-smoked fish has been a major issue of concern for years. Generally, the main considerations when determining process times and temperature are water-phase salt content, use of sodium nitrite, and the use of vacuum packaging. Proper heating helps to eliminate food-poisoning bacteria and to extend the fresh shelf-life. Both state and federal regulations that are currently in effect will dictate actual hot-smoked process minimum temperatures. Processors should always consult appropriate regulatory and health sources before the establishment of process times and temperatures.

In traditional gravity ovens, heat may be produced by charcoal briquets supplemented with gas burners. The rate of heat application, the type of fuel, and the sequence of its use are variables adjusted by the skilled smoker. Experienced smokers can usually produce acceptable results with these ovens.

Modern smoke ovens, available in many designs and sizes, can be equipped with as much automation as a processor is willing to pay for. Heat is generated by electricity, gas, or oil. The choice is dictated largely by convenience and cost. Heat transferred to the oven is usually in the form of steam and controlled air flow. The ovens can be automatically programmed to conduct a series of sequenced operations to accomplish the necessary drying, smoking, heating, and cooling. Humidity control and air flow rates during the sequenced operations also can be programmed.

The basic processing cycle includes drying, smoking, and heating. If the fish is dried a little more in the beginning and the humidity is raised a bit in the smoking step, the need for smoke venting to reduce the humidity within the smoke can be eliminated. During cooking, the moisture in the air can be increased to raise the temperature inside the fish. At the end of the cycle, smoke can be vented up the chimney slowly to minimize air pollution problems. The processor should have the ability to program these functions into the unit.

Electronic advances in modern smokehouses permit exact control of the operation. A skilled smoker can, with assistance from the equipment manufacturer, carry out the controlled testing necessary to attain the desired programs for processing various products. Processing still requires judgment from the smoker because other factors such as brining and fish size must be considered on a case-by-case basis. Because of the control advantages of this equipment, the industry trend is certainly toward greater use of more automated methods. Wood is no longer used as a heat source in the cooking stage because it is too expensive and makes the moisture content of the air too difficult to control. Most modern machines use steam generated by electricity, oil, or gas, but selecting a heat source is largely a matter of convenience.

Smoke Generation

Today's smoke generators are often separate from the processing unit. The primary function of smoke components is to provide a desirable color, aroma, and flavor to smoked products and to contribute to product preservation by acting as an effective bactericide and antioxidant agent. The kinds and quantities of chemicals present in smoke depend on the type of wood, its water content, the temperature to which it is heated, and the precise manner in which it is heated. For example, there is a substantial difference between smoke produced from slowly smoldering sawdust and from the same sawdust heated to a high temperature by blowing a strong current of air over it.

The different volatile chemical compounds in wood smoke are known to have varying levels of bacteriostatic and bactericidal effects. Smoke's effect on microorganisms is heightened by increasing its concentration and temperature. The effect also varies with the kind of wood used. It has been reported that the residual effect of smoke is greater against bacteria than against molds. However, the preservative properties are not nearly as important as strict hygienic requirements, modern packaging, and continuous refrigeration.

Aroma and flavor are a blend of smoke components. The influence of wood variety on smoke flavor is caused by the basic pattern of smoke compounds formed during thermal degradation of the wood. Each type of wood gives a different quality or taste and some even make the product inedible. Softwoods often cause flavor problems and are not generally used. Hardwoods are usually used and are legally required in many cases. Hardwoods impart good color and flavor, but the process takes longer than when softwoods are used, all other processing factors being equal. The type of wood chosen ultimately depends on the quality of the product desired and market preference.

The amount and chemical composition of curing smoke is strongly influenced by the temperature of smoke generation and by smoking technology. The composition of liquid smoke shows an extremely wide variation. During the normal smoking process, the smoke compounds penetrate into the product only a fraction of an inch (a few millimeters), but if liquid smoke preparations or other smoked ingredients are added to the curing mixture, these compounds can be found in the center of the product.

Keeping the product at the proper temperature and moisture and carefully controlling the smoke addition are critical points and usually are best controlled by an automatic system which consistently monitors and records these factors. Smoke is acidic and can over-dry or even denature the product if added at an improper temperature. A high smoke temperature can impart a burned odor and flavor to the product. The concentration of smoke per cubic foot of air has a direct effect on the process: the higher the concentration, the quicker the deposition.

Composition of Smoke

The deposit of smoke on fish is responsible for the golden color and delightful flavor of the finished product; however, there has been much discussion and much written about the composition of smoke and the possible side-effects on consumers of smoked food products. The composition of smoke is complex, with hundreds of chemical compounds. The physical state of smoke, however, is composed of gases and droplets. Two chemical compounds of special concern are polycyclic aromatic hydrocarbons and nitrosamines. Both are considered carcinogenic by scientists. By using certain techniques, including lower temperatures during smoking, the use of an electrostatic filter, or the use of liquid smoke in place of actual smoking, it has been shown that the polycyclic aromatic hydrocarbons can be reduced.

Low-Temperature Smoking (LTS)

Variations in smoked fish products and processing have made it possible to offer smoked-flavored fish that have many of the characteristics of hot-smoked fish with less smoking time and lower processing temperatures (Otwell et al., 1980). LTS products require additional cooking before consumption.

Cooling

After the smoking operation, whether cold-smoking or hot-smoking, the product must be promptly cooled. Proper cooling is essential in hot-smoked fish (Eklund, 1982). The finished product should be cooled to a temperature of 50°F (10°C) or below within three hours after cooking and further cooled to a temperature of 38°F (3°C) or below within twelve hours. This temperature should be maintained during all subsequent storage and distribution. Smoked fish should never be packaged hot because excessive condensation may form inside the package.

INTRODUCTION TO OTHER DRIED, SALTED PRODUCTS:

Dried-Salted Fish

In many developing countries where chilling and freezing facilities are often lacking, traditional curing methods are still the main method of preserving fish which cannot be sold while fresh (Mukherjee et al., 1990). In these countries, fish is an important source of low-cost dietary protein, and dried-salted fish is part of the staple diet. In Indonesia, for example, 34 percent of inland fish and 48 percent of marine fish are cured, with 71 percent processed by salting and drying (Buckle et al., 1988). The Indonesian government has included dried-salted fish as one of a group of nine essential food items (Souness, 1988). The main problems attributed to dried, salted fish vary, but most frequently are low overall quality in the finished product, its high salt content and the rapid rate of deterioration during transport, distribution, and storage (Smith and Hole, 1991; James, 1983; FAO, 1981; Waterman, 1976).

The demand for dried, salted fish has become widely established in many countries within Southeast Asia because of the simplicity of the salting and sun-drying process, which requires a minimal capital investment and therefore provides a cheap source of high-quality protein. A major factor preventing a wider acceptance of these product is their strong salty taste, due to fish being stored frequently in saturated salt brines for lengthy periods to ensure their preservation when weather conditions for sun-drying are not satisfactory. Some processors rinse or dip the heavily-salted fish in fresh water to prevent the formation of salt crystals on the surface upon drying and to give the appearance of a higher-value product. The strong salty taste of the dried product is acceptable to fewer consumers as their incomes increase and a broader range of foods become available. Still, use of heavily salted fish is possible as an ingredient in soups and blended dishes.

Yellowtail (*Trachurus mucculloch*i) was wet-salted using three brine solutions (15 percent, 21 percent, and saturated salt) and dried at 35°C and 50 percent relative humidity (RH), 45°C and 30 percent RH and 55°C and 18 percent RH. Brine concentration during salting and the drying conditions had a significant effect on the drying rate. Fish brined in saturated salt and dried at 55°C was of a lower sensory quality (Berhimdon et al., 1990).

Skipjack tuna (*Katsuwonus pelamis*) was used to produce Indonesian dried, salted fish (DSF), by soaking the flesh in 150 G liter $^{-1}$ or 250 G liter $^{-1}$ brine solution for 24 hours and then drying in an artificial drier at 45°C for 40 hours. The final products were stored for six months at 28°C. Male rats were used to evaluate the nutritional quality of the DSF protein. Feed conversion efficiency, protein efficiency ratio and net protein ratio rats fed DSF were significantly lower than the control (casein) group. True protein digestibility (TPD), biological value (BV) and net protein utilization (NPU) of rats fed DSF were also significantly lower than the control group. The values of TPD, BV, and NPU were all > 90 percent, suggesting that the DSF still had good protein nutritional quality even after six months' storage (Astawan et al., 1994).

Salt-dried mackerel (*Scomber scombrus*) was prepared using fresh fish or fish which had been stored in ice for 75 hours and by four methods of salting, namely 15 percent brine, saturated brine, pickle, and dry-salting. Samples were analyzed for available lysine, thiamine, and riboflavin before salting, after salting, after drying and after one month of storage at 20°C. Final losses of each of these nutrients were lowest when the fish were pickle-cured. Losses were also lower when fresher raw material was used. The use of good quality fish and a pickle-salting method is therefore recommended (Surono et al., 1994). The chemical and nutritional quality of dried, salted mackerel has also been studied with subsequent decreases found in net protein utilization (NPU) after 20 weeks storage at 30°C (Maruf et al., 1990).

Dried Squid

Dried squid in Southeast Asia has traditionally been produced by sun-drying. Water activity and temperature affect non-enzymatic browning of squid amino acids. A set of quality indices was established regarding desirable and undesirable qualities of dried squid that include sweetness, meaty aroma, color of mantle, texture and ammonia off-odor. The Milliard nesition directly causes undesirable darkening, toughening, and reduced rehydration (Kugino et al., 1993; Tsai et al., 1991).

Minced Fish Cakes

Dried, salted fish cakes were prepared using minced mackerel by adding salt (10 to 40 percent) to the unwashed mince. The dried cakes were analyzed for peroxide and TBA values, total counts and sensory attributes after preparation and a three-month storage period. Salted cakes with 15, 20, 30, and 40 percent salt were found to be stable with little sensory deterioration over the three-month storage period (Akande et al., 1988; Dhatemwa et al., 1985). Gulf by-catch from shrimping operations has also been minced and converted into dried, salted fish cake products (Poulter and Poulter, 1984). The Central Institute of Fisheries Technology in India prepared and determined the shelf life of semi-dried fish cake from several traditional species (Sankar et al., 1992).

PACKAGING

Shipping containers, retail packages, and shipping records should indicate by appropriate labeling the perishable nature of the product and should specify that the product be shipped, stored, and held for sale at 38°F (3°C) or below until consumed. Permanently legible code marks should be placed on both the outer layer of every finished product package and on the master carton. Such marks should identify, at least, the plant where packed, the date of packing, and the oven load. Records must be maintained as to positive identification of the process, procedures used, and the finished product's distribution.

It cannot be overemphasized that most smoked fish products, unless canned and sterilized by retorting, have about the same or just slightly longer shelf life than fresh fish. Consequently, smoked fish should be handled, packaged, and stored much like fresh fish. It should be kept frozen or under refrigeration just above freezing temperatures. Vacuum-packed smoked fish makes a beautiful package, but potentially it can be hazardous because the organisms that normally provide visual and odor indications of spoilage are retarded in growth, and certain food-poisoning organisms, if present, are favored in outgrowth.

Currently scientists recommend that vacuum packaging be restricted to cured salmon (lox), cold-smoked salmon (nova), hot-smoked salmon (kippered salmon), and sablefish. These products should be processed with nitrate, have the required water-phase salt content, and meet all prevailing regulations pertaining to heating and cooling. Products should be cooled to 38°F (3°C) before packaging, and maintained at 38°F (3°C) or below, or frozen during storage. Individual packages should be prominently marked with a use-by date. Noncommercial smoked fish products are popular and are commonly made at home (Richards and Price, 1979; Bradley, 1977; Berg, 1965; Waters and Bond, 1960; Crance, 1955), however, the same requirements and safety precautions taken by commercial producers should be observed by home producers.

STORAGE, DISTRIBUTION, AND SALE

The need for proper refrigeration cannot be overemphasized. The finished product should not be distributed until it has been properly cooled to 38°F (3°C) or below. Furthermore, because of the perishable nature of smoked fish, it is imperative that the finished product be maintained in a refrigerated condition at 38°F (3°C) or below until consumed. Most food-poisoning outbreaks related to smoked fish have been tied to abusive storage temperature conditions (Srikar et al., 1993; Plahar et al., 1991).

SPOILAGE AND CONTAMINATION OF SMOKED FISH

Smoked fish is a perishable food, so to maintain its good quality and to prevent food-borne illness, it must be preserved after smoking by processing techniques. *Clostridium botulinum* is a spore-forming bacterium. Because it forms heat-resistant spores, it is not destroyed by heat as easily as other nonspore-forming bacteria. In addition, *C. botulinum*, Type E, has been shown to grow and produce toxin at temperatures as low as 38°F (3°C). Consequently, it is important that proper processing techniques are strictly followed to prevent the rare but potentially dangerous growth of *C. botulinum* in smoked fish.

Equipment

The smoked fish industry has realized the need for modernization of its physical facilities. Although this industry has been, and continues to be, quite labor intensive, improvements have been made in materials handling, processing, and packaging. Among the most evident changes are the expansion and improvements in refrigeration and the transition to sophisticated, programmed, and environmentally controlled smoking ovens. However, gravity ovens are still the traditional means of processing smoked fish in the industry today.

With refrigeration, each freezer and cold storage compartment used to store food should be fitted with an indicating thermometer and should have a temperature-recording device installed to show the temperature accurately within the compartment. It also should be fitted with automatic controls for regulating temperature or with an automatic alarm system to indicate a significant temperature change in a manual operation. Thermometers and other temperature-measuring devices must have an accuracy of ± 2°F (1.1 °C).

Generally, smoked fish have been processed in a traditional smokehouse or in the more consistent mechanical smoking chamber. Many disadvantages are associated with the traditional smokehouse, which is basically a room with a large chimney to vent the smoke and heat, making uniform temperature and smoke control a problem. Cold spots or hot spots occur, making it difficult to produce a consistent batch. In fact, fish sometimes must be shifted to get an even cook and smoke. In addition, traditional smokehouses depend heavily on climatic conditions, so that a product smoked on a cool-dry day will not be the same as a product smoked on a warm, humid day. Significant adjustments in cooking time and temperature must be made to compensate for weather changes. These gravity smoke ovens, a tradition for more than a century, are being replaced by progressively more sophisticated systems.

Modern smoke ovens vary in size, design, sophistication, and, accordingly, cost. They can be used for both cold- and hot-smoking and can be programmed to control temperature, time, humidity, air circulation, cooking rates, cooling rates, and smoke density in an infinite number of combinations and sequences and can clean themselves when they are emptied. In addition, mechanical smoking usually has a separate smoke generator and, consequently, can introduce the correct amount of smoke evenly over the fish. Many times mechanical smoking also includes electrostatic filters that can remove unwanted components from the smoke.

All equipment and utensils used in smoked fish processing plants should be constructed of suitable materials and be designed for adequate cleaning and proper maintenance.

Government Regulations

Obviously, commercially smoking fish requires that a processor meet all state and federal regulations applicable to the general food processing industry. The manufacturing of smoked fish products in the United States is subject to the same general governmental regulations as other foods. If the processor's operation is solely intrastate, then that state's laws apply; if, on the other hand, the enterprise involves interstate commerce, compliance with both state and federal law is required.

A brief introduction to the regulatory aspects of manufacturing food, and smoked fish in particular, is considered a prerequisite. It is the manufacturer's responsibility to be familiar with the regulations. The laws and regulations pertaining to most of the seafood industry are discussed in detail in Title 21, Code of Federal Regulations.

In 1969 the Food and Drug Administration promulgated regulations to establish criteria for current good manufacturing practice (sanitation) in the manufacture, processing, packing, or holding of human foods (21 CFR 128). This regulation soon became known as the Food GMP. Later that same year, the FDA promulgated a food additive regulation entitled "Sodium nitrite used in processing smoked chub." The regulation not only specifies the limits of use of sodium nitrite in smoked chub (100 to 200 ppm in the loin muscle) but specifies other processing requirements for brining, cooking, and cooling that must be followed if sodium nitrite is used in the product.

In June 1986 the Food GMP (21 CFR 110), or "Umbrella GMP," as it was known, was revised. This regulation should be familiar to all those currently in the industry or contemplating entering the smoked-fish industry. As a result of this revision, as well as a United States Court of Appeals decision, the FDA has revoked the smoked-fish GMP. Manufacturing practices for smoked fish are now covered under the Seafood Hazard Analysis and Critical Control Points (HACCP) regulations (see chapter on HACCP), although the guidelines for use of sodium nitrite are still in effect.

Personnel

Compliance with good manufacturing procedures in any food establishment is only as good as that demanded by plant management and supervisory personnel. Food handlers and supervisors should receive appropriate training in proper handling techniques and food protection principles and should be informed of the dangers created by poor personal hygiene and unsanitary practices. Any person who is ill or has open sores, boils, or infected wounds, or who might reasonably contaminate food, food-contact surfaces, or food packaging should be excluded from operations until the condition is corrected.

All persons working in direct contact with food, food-contact surfaces, or food packaging are expected to conform with hygienic practices necessary to protect the food from contamination. Some of these guidelines are as follows:

1. Wear outer garments suitable to the operation.
2. Maintain personal cleanliness.
3. Wash and sanitize hands before starting work and after each absence from the work station.
4. Remove all insecure jewelry and other objects that might fall into the food during preparation.
5. Wear impermeable gloves as appropriate and maintain them in an intact, clean, and sanitary condition.
6. Do not store clothing or other personal belongings in areas where food is exposed or where equipment or utensils are washed.
7. Refrain from eating food, chewing gum, drinking beverages, or using tobacco where food may be exposed or where equipment and utensils are washed.
8. Take any other precautions to avoid contamination of foods, food-contact surfaces, and food packaging.

The National Fisheries Institute has produced a videotape on personal hygiene that should be used for training plant workers.

Buildings and Facilities

Although no two smoked-fish processing plants are the same, they share a number of characteristics. The buildings and facilities of smoked-fish processing plants, like all food-processing plants, should meet GMP regulations and HACCP guidelines. In the GMP, consideration is given to outside grounds, plant construction and design, general maintenance, substances used in cleaning and sanitizing, storage of toxic materials, pest control, sanitation of food-contact surfaces, and the storage and handling of cleaned equipment and utensils. These regulations also extend to sanitary facilities and controls that include the water supply, plumbing, sewage disposal, toilet facilities, hand-washing facilities, and rubbish and offal disposal.

Plants and Grounds

Unloading platforms should be made of readily cleanable materials and equipped with drainage facilities to accommodate all seepage and wash water. In order to prevent cross-contamination between raw and finished products, the following processes should be carried out in separate rooms or facilities:

1. Receiving or shipping
2. Storage of raw fish
3. Pre-smoking operations, including such processes as thawing, dressing, and brining
4. Drying and smoking

Cooling and packaging processes must be carried out in a room or separate facility from the storage of the final product. The product should be processed to prevent contamination by exposure to areas, equipment, or utensils involved in earlier processing, or to refuse or other objectionable substances.

Sanitary Facilities

Adequate hand-washing and sanitizing facilities should be located in the processing room(s) or in one area easily accessible from the processing room(s). Readily understandable signs directing employees to wash and sanitize their hands must be posted conspicuously in the processing room(s) and other appropriate areas. Debris or refuse should not be allowed to accumulate and should be placed in suitable covered containers for removal at least once a day or more frequently if necessary.

Sanitary Operations

Before beginning the day's operation, all utensils and product-contact surfaces must be rinsed and sanitized. Containers used to convey or store fish should not be nested while they contain fish or otherwise handled during processing or storage whereby their contents may become contaminated. Cleaning and sanitizing of utensils and equipment should be conducted in an area designated for this purpose and performed in a manner to prevent contamination of the fish or fish products.

QUALITY CONTROL

A meaningful quality control program should be in effect in every smoked fish processing operation. The program should be under the supervision of knowledgeable management personnel and should encompass all phases of operation from receiving inspection to finished product quality. Appropriate records should be maintained to document incoming raw material, brining procedures, cold-smoking and hot-smoking procedures used with each fish lot, storage temperature monitoring, and quality control or quality assurance testing. Testing should include both microbiological and chemical examination.

Microbial examination of in-line and finished product samples should be conducted with sufficient frequency to assure that processing steps and sanitary procedures are adequate. Microbial evaluation should include total aerobic plate count, *Salmonella* count, and coliform count. The finished product should be chemically analyzed often enough to assure that the fish has the proper salinity and that sodium nitrite, if used, is present at authorized levels.

The seafood-processing industry as a whole is regulated by HACCP, and this concept is now the primary method of achieving processing, sanitation, and quality standards. It is evident that HACCP has come to play an integral role in the smoked-fish industry. HACCP offers a means of assuring consistently produced products that will minimize the microbiological hazards associated with smoked fish.

ACKNOWLEDGMENT

The authors acknowledge and express appreciation for the contributions and suggestions of Dr. George W. Bierman, and Herbert V. Shuster Inc.

REFERENCES

Akande, G. R., M. J. Knowles, and K. D. Taylor. 1988. Improved utilization of flesh from mackerel as salted dried fish cakes. *International Journal of Food Science and Technology* **23**(5):495.

Astawa, M., M. Wahyuni, K. Yamada, T. Tadokoro, and A. Maekawa. 1994. Changes in protein-nutritional quality of Indonesian dried-salted fish after storage. *Journal of Science Food Agriculture* **66**(2): 155-161.

Balaban, M., and G. M. Pigott. 1986. Shrinkage in fish muscle during drying. *Journal of Food Science*. **51**(2): 510.

Bankston, D. 1973. Brine freezers. Marine Advisory Leaflet, Louisiana Cooperative Extension Service, Louisiana State University.

Bannerman, A. 1980. Hot smoking of fish. Torry Advisory Note No. 82. Ministry of Agriculture, Fisheries and Food. Torry Research Station, Aberdeen, Scotland.

Berg, I. I. 1965. *Smokehouses and the Smoke Curing of Fish*. Washington State Department of Fisheries, Olympia.

Berhimpon, S., R. A. Souness, K. A. Buckle, and R. A. Edwards. 1990. Salting and drying of yellowtail. *International Journal of Food Science and Technology* **25**(4):409-419.

Bligh, E. G. and R. D. Rendell. 1986. Chemical and Physical Characteristics of Lightly Salted Minced Cod. *Journal of Food Science* **51**(1):76-78.

Bradley, R. L., C. M. Dunn, M. E. Mennes, and D. A. Stuiber. 1977. Home Smoking and Pickling of Fish. Pub. No. 2000-318A009-77. The University of Wisconsin Sea Grant Program.

Buckle, K. A., R. A. Souness, S. Putro, and P. Wuttijumnong. 1988. Studies on the stability of dried salted fish. In *Food Preservation by Moisture Control*. Seow, Teng, and Quah (eds.), pp. 103-115. London: Elsevier Applied Science.

Burgess, G. H. O., C. L. Cutting, J. A. Lovern, and J. J. Waterman. 1977. *Fish Handling & Processing*. New York: Chemical Publishing Company.

Childs, E. A., F. Al-Dabbagh, O. G. Sanders, and T. L Sheddan. 1976. Techniques for Smoking Rough Fish. *Tennessee Farm and Home Science*.

Crance, J. H. 1955. Smoked fish. Texas A&M University. Fact Sheet #L-1043.

Curran, C. A. and R. G. Poulter. 1983. Isohoic sorption isotherms: Application to a dried salted tropical fish. *Journal of Food Technology* **18**(6):739-746.

Dhatemwa, C. M., S. W. Hanson, and M. J. Knowles. 1985. Approaches to the effective utilization of *Haplochromis* spp. from Lake Victoria: Production and utilization of dried, salted minced fish cakes. *Journal of Food Science and Technology* **20**(1):1-8.

Dougherty, J. B., and H. L. Seagran. 1967. Steps to Effective Sanitation in Smoked-Fish Plants. Washington, DC: Bureau of Commercial Fisheries.

Dudley, S., J. T. Graikoski, H. L. Seagran, and P. M. Ear. 1973. Sportsman's Guide to Handling, Smoking and Preserving Coho Salmon. Fishery Facts #5. National Marine Fisheries Service Extension Publications.

Eklund, M. W. 1982. Significance of *Clostridium botulinum* in fishery products preserved short of sterilization. *Journal of Food Technology and Technology* **12**:107.

FAO. (1981). The prevention of losses in cured fish. FAO Fisheries technical paper #219:1-87. Rome, Italy: Food and Agriculture Organization.

Faturoti, E. O. 1984. Biological utilization of sundried and smoked African catfish. *Nut. Reports Intl.* **30**(6):1395.

Hilderbrand, K. S. 1980. Smoking fish at home safely. SG-66. Oregon State University, Extension Marine Advisory Program.

James, D. 1983. The production and storage of dried fish. FAO Fisheries Report 279. Rome, Italy: Food and Agriculture Organization.

Kannan, D. and S. Bandyopadhyay. 1995. Drying characteristics of a tropical marine fish slab. *Journal of Food Science and Technology* **32**(1):13-16.

Kugino, M., K. Kugino, and Z. H. Wu. 1993. Rhenlogical properties of dried squid mantle change on softening. *Journal of Food Science and Technology* **58**(2):21.

Mukherjee, S., S. Bandyopadhyay, and A. N. Bose. 1990. An improved solar dryer for fish drying in the coastal belt. *Journal of Food Science and Technology* **27**(3):175-177.

Maruf, T. W., D. A Leoward, R. J. Neale, and R. G. Poulter. 1990. Chemical and nutritional quality of Indonesian dried-salted mackerel. *International Journal of Food Science and Technology* **25**(1):66-77.

Moschiar, S. M., J. P. Fardin, and R. L. Boeri. 1984. Sorption isotherms of dried-salted hake. *LebensmBWiss. U-Technol.* **17**(2):86-89.

Otwell, W. S., J. A. Koburger, and R. L. Degner. 1980. Low temperature smoked fish filets: A potential new product form for Florida fish. Technical Paper #19. Florida Sea Grant.

Paparella, M. 1979. Information Tips. University of Maryland Sea Grant.

Plahar, W. A., R. D. Pace, and J. Y. Lu. 1991. Effect of storage conditions on the quality of smoke-dried herring (Sardinella EBA). *Journal of Science Food Agriculture* **57**(4):597-610.

Poulter, N. H., and J. M. Poulter. 1984. Composition of shrimp by-catch fish from the Gulf of California and effects on the qualities of the dried salt fish cake product. *Journal of Food Science and Technology* **19**(2):151-161.

Price, R. J., and G. K. York. 1980. Smoking Fish at Home. Leaflet 2669. Division of Agricultural Sciences, University of California.

Raghunath, M. R., T. U. Sankar, K. Ammu, and K. Devadasan. 1995. Biochemical and nutritional changes in fish proteins during drying. *Journal of Science Food Agriculture* **67**(2):197-204.

Rauesi, E. M. and J. Krzynowek. 1991. Variability of salt absorption by brine dipped fillets of cod, blackback flounder, and ocean pencli. *Journal of Food Science and Technology* **56**(3): 648-652.

Richards, J. B., and R. J. Price. 1979. Smoked Shark and Shark Jerky. Leaflet 21121. Division of Agricultural Sciences, University of California.

Sankar, T. U., A. Ramachandran, and K. K. Solanki. 1992. Preparation and shelf life of semi-dried fish cake from DHOMA. *Journal of Food Science and Technology* **29**(2):123-124.

Souness, R. A. 1988. Reducing post-harvest losses associated with dried fish production in Indonesia. *Infofish Intl.* **6**:38-40.

Smith, G. and M. Hole. 1991. Browning of salted sun-dried fish. *Journal of Science Food Agriculture* **55**:291-301.

Srikar, L. N., B. K. Khunita, G. U. S. Reddy, and B. R. Srinivasa. 1993. Influence of storage temperature on the quality of salted mackerel and pink pencli. *Journal of Science Food Agriculture* **63**:319-322.

Surono, K. D., A. Taylor, and G. Smith. 1994. The effect of different salting procedures and qualities of raw material on some nutrients during processing and storage of salted-dried mackerel. *International Journal of Food Science and Technology* **29**(2):179-183.

Tsai, C. H., M. S. Kong, and B. Sun Pan. 1991. Water activity and temperature effects on nonenzymatic browning of amino acids in dried squid and simulated model system. *Journal of Food Science and Technology* **56**(3):665.

Tsai, C. H, B. Sun Pan, and M. S. Kone. 1991. Browning behavior of taurene and prolene in model and dried squid systems. *Journal of Food Biochemistry* **15**(1):67-77.

Troller, J. A., and J. H. B. Christian. 1978. *Water Activity and Food*. New York: Academic Press.

Waterman, J. J. 1976. The production of dried fish. FAO Fisheries Technical paper 160, p. 41. Rome, Italy: Food and Agriculture Organization.

Waters, M. E., and D. J. Bond. 1960. Construction and Operation of an Inexpensive Fish Smokehouse. *Commercial Fisheries Review* **22**(8):8.

Specialty Seafood Products

Ken Gall, Kolli P. Reddy, and Joe M. Regenstein

INTRODUCTION

Because foods in their natural state remain sound and edible for only a comparatively short time, food preservation has engaged the attention of mankind from earliest times. Salt curing, drying, pickling, and smoking are among the oldest methods of fish preservation. A variety of specialty seafood products whose production involves one or more of these preservation methods are popular around the world. Seafood products preserved by pickling or marinating with or without salt and an assortment of different spices are widely consumed in Europe, Asia and, to a lesser extent, in North and South America. Fish and shellfish products that are transformed by fermentation processes are common in Asia and Africa. Large amounts of various small fatty fish species abundant throughout the world are preserved in many different ways, including salting and canning, to produce the large number of sardine and anchovy products consumed in almost every part of the world. Fish roe and caviar are also important specialty seafood products in great demand in many areas of the world. All of these products represent a significant portion of global consumption of preserved seafood products.

PICKLED, SPICED, AND MARINATED FISH

Pickling with vinegar and spices is a very ancient form of food preservation. Spicing and pickling is an important method of fish preservation in Europe. A great many fishery products are prepared and preserved using vinegar, salt, and spices. Spiced and pickled herring preparations and other pickled fish are widely consumed in northern and eastern Europe. In Spain and other parts of southern Europe, dishes such as escabeche are prepared by frying fish in oil with bay leaves and spices, then marinating in vinegar and oil. In the Latin American countries of Central and South America, ceviche is a popular product prepared with fish such as sea bass or sea trout fillets that are diced and mixed with thin slices of onions, garlic, and hot chilies marinated in lime or other acidic citrus juices. Spiced and pickled fish are also widely consumed in Asia. In Japan, the Philippine Islands, and China, pickled fish is an important source of animal protein. In this part of the world, fish are pickled in hot or spiced vinegar without salting, in soy sauce, in sake, or in a fermenting rice paste. Spicing and pickling is a minor method of preservation in North America. Spiced and pickled fish have been prepared in the United States and Canada on a small scale for many years. The demand for pickled fish specialty products has been increasing, especially in major cities with large foreign populations.

According to the Food and Agricultural Organization of the United Nations (FAO) Fisheries Commodity statistics, world production of "spiced and marinated fish" was about 90 thousand metric tons in 1994. Germany produced about 85 percent of these products, and other producers were the Czech Republic and the Russian Federation. In addition, more than 60,000 metric tons of "herring preparations" were produced in 1994 (Annual Yearbook of Fisheries Statistics). In 1996 almost 160,000 pounds of prepared or preserved herring were produced, according to FAO statistics. Poland was the leading producer, followed by Denmark, Sweden, Germany, and the Netherlands. World production of spiced and marinated fish steadily decreased during the ten-year period from 1985 to 1994, while production of herring preparations increased slightly from 1990 to 1994.

Pickling or Preservation with Vinegar

Pickling is the term generally used to describe a process in which vinegar or vinegar and spices are used to preserve fish. Brine-salted fish is often called "pickled," but it is a misnomer because salt rather than organic acids is the primary mode of preservation. Organic acids such as acetic — the active agent in vinegar — lower the pH of the fish and other ingredients to be pickled, thus retarding the growth of spoilage molds and bacteria. The preservative effect of the vinegar is directly related to its acid content. An acid content of 15 percent or higher is required to stop bacterial growth completely. Since most commonly available vinegars contain 6 percent acetic acid or less, fishery products pickled in vinegar are only temporarily preserved. Most pickled products need to be stored refrigerated. Pickled fish products containing 3 percent acetic acid can generally be held in chilled storage for several months.

The shelf life of pickled products depends on the fish and other ingredients, and the storage conditions. Only fish of the highest quality should be used for pickling. All fish, especially those caught at times when they are feeding heavily, should be eviscerated, washed, and packed in ice as quickly as possible after they are caught. Other ingredients commonly used in pickled fish products include water, salt, sugar, spices, and herbs. High-quality ingredients with few impurities are preferred in order to minimize the development of off-flavors and odors in pickled fish products.

Uncooked Pickled and Spiced Products

In Europe and, in particular, Germany, herring is the most important pickled or marinated fishery product. In most of these products whole eviscerated, sliced, filleted, or boned herring are cured with salt or salt and vinegar solutions, rinsed or soaked, drained, and then packed in a solution of vinegar, salt, sugar, and spices. These products are prepared from salted fish that has not been cooked, as compared to other spiced and pickled fish preparations, such as escabeche, in which the fish is cooked prior to pickling.

Vinegar/Salt-Cured Herring Products

Spiced herring products such as cut-spiced herring, Bismarck herring, Kaiser-Friederich herring, and rollmops are produced from herring that are specially cured in salt and vinegar. Herring cured with traditional cures that use only salt, such as the Scotch cure, are generally not used for these products. Fresh herring are washed and scaled in a special washing machine with a large revolving drum and a water spray. The fish are eviscerated and beheaded, and the belly cavity is thoroughly washed. A variety of curing methods can be used. The following is the vinegar/salt-curing process described by Jarvis in *Curing of Fishery Products* (1987). The fish are placed in a curing tank where they are covered with brine testing 80° to 90°F with a salinometer and distilled vinegar to provide an acidity of 2.5 percent. The fish are left in this brine until the salt has completely penetrated the flesh. The length of the cure depends on the temperature, freshness, and size of the fish and can vary from three to seven days. Cured fish are stored in barrels filled with a salt/vinegar brine testing 70°F with a salinometer and shipped to the place where final manufacturing will occur. The herring are then either cut into fillets or the backbone of the dressed herring is removed; they are repacked in kegs with a solution of distilled vinegar and water at 3 percent acidity and sufficient salt to produce a 35°F salinometer reading. The final process involves soaking the herring in cold water for eight to 10 hours, after which they are drained and placed in a vinegar/salt solution consisting of one gallon of 6 percent white distilled vinegar and one pound of salt per gallon of water. Cut-spiced herring, Bismarck herring, Kaiser-Friedreich herring, and rollmops are then made from this vinegar/salt-cured herring.

Cut-spiced herring is produced from vinegar/salt-cured herring that has been cut into pieces one to two inches long. The herring pieces are packed with spice mixtures that may include bay leaves, mustard seeds, allspice, peppercorns, and cloves. Onions, lemons, pimiento, or chili peppers may also be added. The fish pieces, spices, and other ingredients are packed in a solution of vinegar, sugar, and salt. The finished product should be refrigerated and may have a shelf life of up to six months.

Bismarck herring is produced from dressed herring with the backbone removed so that the two sides of the fish remain joined together. The herring are cured under refrigeration for 10 days in a mixture of vinegar, sugar, salt, and spices similar to the mixture used for cut-spiced herring. The cured fish are packed vertically in jars with lemons, bay leaf, or other spices. The jars are then filled with distilled vinegar diluted to 3 percent acidity. Kaiser-Friedreich herring are prepared in the same way as are Bismarck herring, except that the cured herring is packed in mustard sauce instead of vinegar.

Rollmops are vinegar/salt-cured herring fillets rolled around a piece of dill pickle or a pickled onion and fastened with a wooden toothpick. The rollmops are cured for about 10 days in a solution of vinegar, sugar, and spices similar to that used for other spiced herring. The cured rollmops are packed in glass containers and covered with a solution of vinegar, salt, and sugar. The final product should be stored under refrigeration.

Herring in wine sauce is made from vinegar/salt-cured herring or ordinary salt-cured herring. Vinegar/salt-cured herring are rinsed, drained, and packed into jars with the wine sauce. Ordinary salt herring is freshened in water, drained, and cured in vinegar with 3 percent acidity for at least 48 hours. Dry wines are preferred rather than sweet wines. A dry white or burgundy-type red wine is generally used. In general, wine sauce formulas are like the standard spice sauce formulas, but about half or more of the vinegar is replaced by wine. The wine sauce is made by boiling a mixture of vinegar, wine, sugar, onions, and spices such as nutmeg, ginger, black and/or white pepper, allspice, cloves, bay leaves, and cinnamon. The cooked solution is then allowed to sit for a period of time to blend the flavors and spices. The sauce is strained and poured over the cured fish, which has been packed in jars or other containers.

Herring in sour cream sauce is a popular product that is prepared with herring fillets that are pickled in a spiced sauce with vinegar, wine, and cream. Salt-cured herring are filleted, soaked in cold water, and then drained. A pickling solution of distilled vinegar, white wine, and spices is boiled for several minutes. The mixture is cooled and strained to remove the spices. The solution is then mixed with sour cream, sweet cream, and the fish milt, which has been rubbed through a fine sieve. (The sour cream must be properly prepared so that it does not curdle when the other ingredients are added.) The fillets are then packed in a large container; sliced onions may be added; and the fillets are covered with the sour cream pickling sauce. The herring is pickled for approximately one week and then repacked into smaller jars or other containers with the sour cream sauce. The final product is stored at refrigeration temperatures.

Matjes herring, Gewürz herring, and **Russian sardines** are spiced and pickled products made from whole herring that are gutted but still contain their milt or roe. The gills are removed from the fish, and the intestines are pulled out through the gill opening so that neither the throat nor the belly is cut open. To prepare Matjes herring, cleaned fish are soaked in a seven percent solution of white wine vinegar for about 12 to 18 hours. The fish are dried and rolled in a mixture of salt, brown sugar, and saltpeter. The herring are packed in layers in a small keg or other container, and the curing mixture is sprinkled over each layer. The curing lasts for one to two days, after which the fish are repacked in the brine that was produced. The repacked product is stored under refrigeration for at least one month before it is used. **Gewürz herring** is prepared in the same way as Matjes herring, but ground spices such as black pepper, chili pepper, coriander, ginger, cloves, nutmeg, and cinnamon are added to the curing mixture. **Russian sardines** are prepared from small herring gutted and cured in a manner similar to Matjes herring or from small herring that have had the head and viscera removed in one stroke, leaving the belly whole, after an initial brine cure sufficient to produce a thoroughly salted product. After the initial pickling, the fish are rinsed, drained, and dredged in a spice/salt mixture that includes spices such as allspice, black pepper, bay leaves, cloves, and ginger. The spiced herring are packed in layers with onion, horseradish, and capers sprinkled among the layers. Vinegar is added and the product is refrigerated for several weeks before use.

Scandinavian anchovies, unlike the traditional highly salted anchovies from southern Europe, are prepared from fresh brisling or sprat that are salted and spiced. The fresh fish are salt-cured in a strong brine, then drained. They are then cured in a brine with sugar and spices or are rolled in a dry brine, sugar, and spice mixture. Spices such as cloves, nutmeg, allspice, black pepper, mace, cinnamon, ginger, and bay leaves may be used. The spiced fish are packed in barrels or other large containers for from two to eight weeks. The curing barrels are rolled or inverted periodically during the curing process. The cured product is then repacked in smaller containers with the filtered brine used in the curing process, and are stored at refrigeration temperatures.

Ceviche (sometimes spelled "seviche") is a variation of the Spanish dish "escabeche" that is widely used in Central and South America. With traditional escabeche, the fish is fried in oil before being pickled with vinegar and spices. To prepare ceviche, the fish is not cooked, but simply marinated in citrus juice and spices. Citric acid from citrus fruits, instead of the acetic acid in vinegar, provides the preservative effect. A variety of fish species such as corvina (drum), bass, sea trout, red snapper, and other species, as well as such shellfish as scallops and shrimp, are used. To prepare the dish, pieces of fish generally are marinated in lime or lemon juice with a variety of spices that could include hot red or jalapeño chili peppers, red or black pepper, bay leaf, onions, bell peppers, pimiento, or other ingredients. There are many variations, depending on regional tastes and preferences. The fish is generally marinated for at least eight to 24 hours, and should be kept refrigerated at all times.

Cooked Pickled and Spiced Products

A variety of fish and shellfish are cooked and then pickled or preserved with vinegar and spices. Fish species such as mackerel, salmon, sturgeon, eel, trout, pike, and pickerel are prepared in northern Europe. These larger fish are generally filleted and cut into smaller pieces before pickling. In most preparations the fish is cooked before pickling or cooked in the spiced pickling solution. Some fish are broiled or fried after an initial salt cure, before they are pickled in a solution of vinegar, spices, onions, and lemon. Shellfish can also be pickled with spices, and pickled oysters and shrimp are two common products. Pickled oysters are blanched before pickling. After blanching, the oyster liquor is heated with vinegar and such spices as cloves, allspice, black pepper, and mace. The oysters are packed in glass jars with a bay leaf and a thin slice of lemon, and covered with the vinegar and spice solution. Pickled mussels can be prepared in a similar manner. Pickled and spiced shrimp are another common product. The shrimp are boiled in a pickling solution of vinegar, salt, and spices, and then cooled and packed in jars.

Escabeche is one of the oldest spiced and pickled fish preparations. Its origins have been traced to the Greeks and Romans who brought it to Spain. It remains one of the most widely used methods of fish preparation in Spain and other Spanish-speaking parts of the world, such as Latin America. A variety of different fish species can be used to prepare escabeche; mackerel, tuna, and corvina are most often used. The fish are filleted and cut into small pieces, washed thoroughly, drained, soaked in a salt solution containing 20 percent brine for half an hour, and then dried. The salted fish is fried with garlic, bay leaves, and chili peppers, and removed to cool. Onions are then fried in the same pan and cooked until they are yellow. Additional spices such as black pepper, cumin seed, and marjoram are added along with vinegar, and the pickling mixture is cooked slowly for 15 to 30 minutes and then cooled. The fried fish is packed into jars with bay leaves and chili peppers, the pickling sauce is added, and the jars are closed. The pickled and spiced product can be used after one day but improves with age when stored at refrigeration temperatures. The shelf life is dependent on the strength of the vinegar that is used and the storage temperature.

Fried Marinated Herring

Fried, **spiced**, and **pickled** fish are also prepared from a variety of small fish such as herring, sardines, and anchovies in Europe. Meyer (1965) described the following method for preparing fried marinated herring: Salted herring are washed in fresh water and allowed to drain. The fish are then soaked in a breading mixture composed of rye and wheat flour in a 1:1 ratio. The fish can be fried in an open pan, or cooked with commercial frying devices with conveyor belts, or with immersion pans in which the breaded fish are placed on metal sieves immersed in hot oil. The fried fish are cooled and packed into cans, glass jars, barrels, kegs, or other containers with a brine containing 2 to 3.5 percent acetic acid and 3 to 5 percent salt. Spices such as mustard seeds, pimiento, and cloves are also added to the brine. Vinegar strength and storage temperature determine the product's shelf life.

FERMENTED FISH PRODUCTS

A variety of fermented seafood products are produced in Asia and Africa. Fermented fish products are most common in Southeast Asia, especially the Philippine Islands. Van Veen (1953) reported that the most common ways of preserving fish in the underdeveloped countries are drying, salting, and making fish sauces and pastes (fermenting). Fully fermented products are reduced to a liquid or paste over a period of several months to form fish sauces and pastes. These products are produced primarily in Southeast Asia, and can be compared with other fermented foods such as cheese, sauerkraut, and yogurt. In Africa a number of different, partially fermented products are consumed. These products are preserved by salting and drying accompanied by fermentation. The fermentation period lasts a few days with only partial breakdown of the muscle. These products have a strong odor and are called "stink" fish.

According to FAO Fisheries Commodity Statistics, more than 205,000 metric tons of fermented fish were produced in 1994. Approximately 80 percent of the fermented fish was produced in Vietnam. Other major producers of fermented fish are the Philippines and Thailand. According to FAO statistics, world production of fish paste was 164,000 metric tons in 1994. Over 75 percent of the total reported fish paste production was produced in Myanmar (formerly Burma). Other fermented fish products reported in FAO statistics include more than 5,000 metric tons of shrimp paste produced mainly in Malaysia, and fermented sea urchin in brine produced in Japan and Korea. Updated statistics are not available. Since 1994, the FAO has not reported these figures; all products are grouped together in a category called "Other Fish Preparations."

Fermentation is the transformation of organic materials into simpler compounds by the action of enzymes and microorganisms. In the fermentation of fish, the degradation of proteinaceous material is controlled primarily by using salt at sufficient concentrations to inhibit the growth of putrefactive and potentially pathogenic microorganisms. Owens and Mendoza (1985) have described fermented products as those involving deliberate growth of fermentative microorganisms, and products whose manufacture primarily involves the activity of indigenous or added enzymes. The products of fermentation in the traditional sense are extremely varied, and they can take the form of a liquid (*nuoc-mam*), a paste (*prahoc*), or whole fish (anchovy).

Fish Sauces

In many Southeast Asian countries, fermented fish products are manufactured in large quantities for human consumption. One of the most common fermented fish products is fish sauce, which is known as *nuoc-mam* in Cambodia and Vietnam. Fish sauce is a clear liquid, straw yellow to amber in color, that has a mild cheesy flavor and fishy odor. The fermentation is due to the proteolytic enzymes from the viscera.

Nuoc-mam is prepared from a variety of small marine fish, mainly species of *Clupeoids* and *Carangids* such as herrings, sardines, and herring-like fish. The fresh fish are mixed with salt without washing and placed in vats at a fish-to-salt ratio of 3:2 for three days. Then the liquid *(nuoc-hoi)* is drained off and the fish are covered with a small portion of liquid *nuoc-mam*; the mixture is covered with large bamboo wickerwork and heavily weighted. The fish are left in the vat for about 12 to 18 months to mature. After maturation, the liquid or pickle is collected and is considered as first quality *nuoc-mam* or *nuoc-nhut*. The residual fish is called *xat-mam* or *nuoc-xat* and is sold as manure. Guillerm (1928) reported that, if the pH of *nuoc-mam* is high (6.8 to 7.2), it cannot be kept for long.

In Thailand, fish sauces are called *nam-pla* and are made from anchovies. Nam-pla is prepared similarly to *nuoc-mam*, but the maturation period is only about six months. In Malaysia, fish sauces are prepared by the fermentation of small anchovies. The fish sauce is called *budu* in Malaysia: *ngam-pya-ye* in Burma; *kecap ikon* or *ketjap ikon* in Indonesia; and *shottsuru, ikanago, konago, ika shoyu*, and *ishiru* in Japan.

Fish Pastes

Fish pastes are also common in Southeast Asia, and are generally used as a condiment for rice dishes. Fish paste is produced in large quantities in the Philippines, where it is called *bagoong*. It is generally made from anchovies and ambassids that are cleaned and mixed with coarse-grained salt at a ratio of one part salt to three parts fish. The fish are left to mature in vats. Fish enzymes are the primary cause of the protein breakdown that occurs when making fish paste.

In Cambodia, fish paste is called *prahoc*. To make this product, fish are put into wicker baskets that are trampled by foot to remove the scales and press out the entrails. The fish are then washed in river water, and stirred by hand until the scales are completely removed. The upper part of the basket is covered with banana leaves and weighted with stones for a day. The following day, the fish are mixed with coarse salt for one hour, and then dried in the sun for a day. The salted and dried fish are pounded in a wooden mortar. The resulting paste is then placed in open earthen jars in the sun. The jars are covered in the evenings to keep out insects; during this period, fermentation or ripening occurs and the pickle or liquid gradually appears on the top of the paste. The liquid is removed every day. After a month, when no more pickle appears, the *prahoc* is finished and is ready for consumption — for example, in the preparation of soups. In Indonesia, fish paste is called *trasi ikan;* in Myanmar, *nga-pi* or *nappi*; and in Thailand, *plaa-raa*.

Partially Fermented Fish Products in Africa

In Africa, smoking, sun-drying, salting, fermentation, grilling, and frying are the predominant and most important methods of fish preservation. These processes may be used either alone or combined in order to achieve the desired product. Fermentation alone does not preserve fish. Most African fermented products are also salted and/or dried to reduce water activity and retard or eliminate the growth of proteolytic and putrefying bacteria. In Southeast Asia, the fermentation process lasts for months, and the final product is usually a paste, sauce, or liquid. In Africa, fish fermentation lasts from a few hours to about two weeks. Under such conditions, fermentation is usually partial, and the fish retains its original form.

Fermented fishery products in Africa, particularly Ghana, are called *momone*, an Akon word that literally means stinking. Watts (1965) described "stink" fish of Sierra Leone, which developed a strong odor within 24 hours of capture; they were salted for about four days and then dried. Watanabe (1982) described the fermented fishery products of Senegal as highly salted and semi-dried with an obnoxious odor and a cheesy flavor. The salts that are widely used in Africa for the salting and fermentation of fish are solar salt, rock salt, and vacuum salt. Solar salt is the most widely used salt in fish curing.

Essuman described and characterized a variety of fermented fish products produced and consumed in Africa in a 1992 FAO Fisheries Technical paper. The commonly-practiced fish preservation techniques identified in this report are produced by: fermentation with salting and drying (*momone, kako, tambadiang*); fermentation and drying without salting (*ndagala, dagaa, kejeick, salanga, yeet*); or fermentation with salting but without drying (*terkeen, fessiekh*). The following descriptions of the products and processes used to produce these products, and the products themselves are summarized from this 1992 FAO report.

Momone is prepared from fresh fish such as catfish, barracuda, sea bream, threadfin, croaker, mackerel, herring, squid, and ribbon fish. The dressed fish is thoroughly washed with freshwater or seawater. The raw fish is either left overnight before salting or dry salted immediately after washing. Salting and fermentation lasts for one to six days, after which the fish is dried on the ground, grass, nets, stones, or raised platforms for one to three days. *Momone* is a soft product with a very strong, pungent, and sometimes offensive, smell. This product is produced in Ghana.

Kako is prepared from species such as sharks, skates, and rays in Ghana. The raw fish are dressed, thoroughly washed, and dry-salted by rubbing salt into the gills, into the belly cavity, and on the surface. The fish are then arranged in alternate layers with salt and are allowed to ferment for two to three days before being dried for two to four days. More salt may be sprinkled on the fish during drying. *Kako* is a dried product with a mild to strong odor.

Tambadiang is commonly prepared from fish such as bonito in Senegal. The fish is washed, placed in concrete vats with alternate layers of salt, left to ferment for one to three days before being de-scaled, washed again, and then dried for three to five days on raised platforms. The final product is grey in appearance and mild-smelling. *Tambadiang* requires an occasional re-drying to ensure a longer shelf-life.

Ndagala is produced in Burundi from fresh fish that are sun-dried after harvest without salting. Slight fermentation takes place during sun drying. The final product is hard, dry, and brittle with a silvery color.

Dagaa is produced in Uganda from fresh fish dried by passing a stick through the eyes of 10 to 12 individual fish. Ten such sticks of fish are joined to form a mat that is hung in the open air for the fish to dry. The drying process takes three to five days. Slight fermentation occurs during drying.

Kejeick is produced from large fish that are gutted, split dorsally, scaled, headed, and washed. Smaller species of fish are dried whole or split at the belly to remove the entrails. In northern Sudan, the fish is dried immediately after it has been washed, either by hanging it on sticks or by laying it on stones, grass, or mats. In the South, the fish may be sprinkled with salt or dipped into a strong salt solution and fermented for two to three days before drying for up to seven days. The final product is hard-dried and can be stored for several months.

Salanga is a fermented product from Chad for which Nile perch and tilapia are generally used. The fish are scaled, gutted, and washed. Larger species of fish are split dorsally. In one process, the dressed fish is dried immediately after washing, and fermentation takes place during drying. In another preparation, the fish is left to ferment for 12 to 24 hours before drying. Fish are dried on mats or grass spread on the ground or may be dried by being hung vertically or horizontally on drying lines, either by passing a stick through the head or by tying the fish with a thread along the stick. Drying takes about three to six days, and the final product is light brown in color with a dry, firm texture.

Yeet is a fermented product made from sea snails. The flesh is removed from the shell, separated from the viscera, and split into two to four parts. It is then placed in fermentation tanks, jute bags, or sacks and allowed to ferment for two to four days before being washed and dried on raised platforms for two to four days. It is a semi-dry, light-brown product with a strong smell.

Terkeen is another fermented fish product made from Nile perch and tilapia. Fish are washed, salted, and arranged in alternate layers with salt in an earthenware pot or barrel. Pepper or other spices may be added. The salt-to-fish ratio is 1:5. Pickling and fermentation takes place for nine to 15 days, after which the fermentation vat or pot is placed in the sun or near a fireplace to speed up fermentation. This process continues for about four days; the fish is stirred daily until it is broken into a pasty mixture of muscle and bones that has a dark color and strong odor. The final product is a wet, pasty mixture of fish muscle and bones that loses moisture during storage and becomes more viscous and dark.

Fessiekh is a fermented fish product from the Sudan that is produced in temporary sheds to provide shade or a cool environment. Whole, fresh fish is washed, covered with salt, and arranged in alternate layers with salt either on a mat, in a basket, or in perforated drums, to ferment for three to seven days. Liquid exudate from the fish is allowed to drain off. After this period the fish is transferred into larger fermentation tanks where more salt and new batches of fish are added. The fermentation tanks are covered with jute sacks or polyethylene sheets, and weights are placed on top to press the fish. The fish is allowed to ferment for about 10 to 15 days, after which they are transferred into vegetable oil cans, covered with polyethylene sheets and sealed. It is a wet, salted, fermented product, soft in texture with a strong pungent smell and a shiny, silvery appearance.

ANCHOVY AND ANCHOVY PRODUCTS

Anchovies are small, oily, sardine-like fish found in both the Atlantic and Pacific Oceans. Many different species of small pelagic fish are called anchovies. The United States Food and Drug Administration (FDA), in their 1993 list of acceptable market names for seafood, lists 21 species in five different genera whose accepted market name is anchovy: *Anchoa*, *Anchoviella*, *Engraulis*, *Stolephorus*, and *Cetengraulis*. The Mediterranean-cured anchovy products that are most familiar in the marketplace are prepared from the European anchovy, *Engraulis encrasicolus*. Other *Engraulis* species include the California and Japanese anchovy. Anchovies travel in dense schools and are generally caught commercially with purse seines.

A wide variety of different anchovy products are made from whole or gutted fish and fillets. These products are salt-cured, pickled, or even fermented to produce an anchovy paste. A fermented anchovy sauce called *garum* was first popularized by the Romans. These fermented anchovy preparations continue to be used today throughout the Mediterranean. A variety of different anchovy product forms are produced around the world, including frozen and dried anchovies, European salted anchovies, canned anchovies, canned anchovy fillets, and anchovies boiled in saltwater. World production of these products from 1990 to 1994 is summarized in Table 1; data are unavailable at this level of detail after 1994.

Japan was the leading producer of frozen anchovies in 1994, followed by Italy and Greece. European anchovy production occurs in several countries that border the Mediterranean Sea; the largest producers in 1994 were Italy, Spain, and Turkey. Argentina, Malaysia, and Korea were the leading producers of dried anchovies. Anchovies boiled in saltwater were also produced in Korea. The countries surrounding the Mediterranean were also major producers of canned anchovy products. Spain was the leading producer of canned anchovies in 1994, followed by Argentina, South Africa, France, and Greece. Italy was the leading producer of canned anchovy fillets, followed by Portugal and France.

Mediterranean-Style Anchovies

Mediterranean-style anchovies are the most common and well-known product in the European and North American marketplace. These products are salt-cured. Cured fillets or whole dressed fish are often packed or canned in oil. Cured anchovies are generally used as condiments or garnish to add a distinctive flavor to a variety of dishes.

The following description of the process used to produce Mediterranean style anchovies is adapted from the one given by Kaylor and Learson in the book *The Seafood Industry* (1990). Fresh anchovies are beheaded using a technique that removes the entrails at the same time. The fish are packed in special barrels in layers with a layer of salt between each layer of fish. When the barrel is nearly full, a final layer of salt is added and a weighted cover is placed on the packed fish to keep them pressed down. After a few days the mass of fish will have shrunk, and the cover and top layer of salt is removed. More layers of fish and salt are added until the barrel is full. A final layer of salt is added, and the weighted cover is again placed on top of the fish. The fish are weighted to minimize contact of the fish with air or air bubbles in the liquid produced from the salted fish. The curing process, which involves both salting and fermentation, takes at least six months at a temperature between 60° to 68°F (15° to 20°C). When the curing process is determined to be complete, it is stopped by chilling the fish. The final product is judged by color, flavor, and odor. The desired color is a deep red. After curing, they are washed in brine, the skin and tail are removed, each fish is filleted by hand, any remaining bones are removed, and the fillets are blotted dry. The finished fillets are packed individually in cans or other containers with olive oil, with or without other ingredients such as vinegar and mustard, capers, and olives or other small veg-

Table 1. World Production of Anchovy and Anchovy Products, 1990 to 1994

Product	Annual World Production in Metric Tons				
	1990	1991	1992	1993	1994
Canned anchovies	13,775	14,200	13,045	13,590	12,006
Canned anchovy fillets	17,062	18,893	17,427	13,067	14,851
Salted European anchovies	16,334	17,342	13,060	17,590	17,980
Anchovies dried, salted, or in brine	16,719	16,571	20,217	17,338	16,639
Anchovies boiled in salt water	15,617	18,436	18,832	24,990	24,990
Frozen anchovies	90,826	147,447	123,806	103,920	111,553

Source: FAO Annual Fishery Statistics Commodities Yearbook, Volume 79, 1994. Information at this level of detail is not available in FAO Reports after 1994.

etables, and are then sealed. Anchovy paste may be produced using a similar curing process in which the fish are weighted down, the liquid that is generated is removed, and the process continued until a paste is formed.

Scandinavian Anchovies

Scandinavian anchovies are not true anchovies, but are produced from sprats, *Clupea sprattus*, or herring, *Clupea harengus*. The following process for producing this product is summarized from the description provided by Alm in *Fish as Food* (1965): fresh sprats are rinsed, drained, mixed thoroughly with a salt-sugar-spice mixture, and placed in barrels containing a layer of the salt-sugar-spice mixture on the bottom. After one to two days, a brine is formed, and the barrel is filled with additional brine mixture, sealed, and allowed to mature. A matured product is soft and has a smooth consistency, and the backbone is easily removed from the flesh. After maturation, the fish are cut into fillets and packed into containers, which are then filled with a solution of salt and sugar in which spices have been boiled. Some of the salt-sugar-spice mixture is poured into the bottom of the container and also sprinkled between each layer. The containers are allowed to stand for two days, during which time the sugar and salt are dissolved. Then the containers are filled with extra brine and sealed. Further maturation occurs during storage at 54° to 59°F (12° to 15° C) and the finished product is stored at refrigeration temperatures.

SARDINES

A variety of different species of small fatty fish are referred to as sardines in various parts of the world. These include pilchards, sardines, sprats, small sea herring, and other related species. The more important species used to produce sardines around the world are the European pilchard or sardine (*Sardina pilchardus*), the Japanese pilchard (*Sardinops melanostictus*), the South American pilchard (*Sardinops sagax*), the European sprat (*Sprattus sprattus*), and the small Atlantic herring (*Clupea harengus*). In the Seafood List that provides acceptable market names for seafood sold in the United States, the FDA includes nine species from two genera, *Harengula* and *Sardinella*, whose acceptable market name is sardine, and seven species from two genera, *Sardina* and *Sardinops*, whose acceptable market name is pilchard or sardine.

Sardines are consumed in a variety of forms around the world. Fresh sardines as well as lightly salted, dried, and frozen sardines can be found in world markets, but the canned sardine is the most important and widely known product. According to 1994 FAO fisheries commodity statistics, world production of canned sardine products included: 160,033 metric tons of canned European sardines primarily produced in the Mediterranean countries of Morocco, Portugal, Spain, France, Tunisia, Algeria, Italy, and Libya; 188,037 metric tons of canned pilchards primarily produced in Namibia, Mexico, Ecuador, Peru, Brazil, Japan, Chile, and South Africa; 11,800 metric tons of canned sprat primarily produced by Poland, Norway, and Sweden; 75,229 metric tons of canned Atlantic herring, which includes small herring canned as sardines as well as other canned herring products produced in Sweden, Denmark, Canada, and Germany; and 54,267 metric tons of canned *Clupeoid* fish species primarily produced in Venezuela, with lesser amounts produced in Latvia, Norway, Argentina, Spain, Iran, and Israel. United States production of canned sardines occurs primarily in Maine. In 1998, 11.8 million pounds valued at $19.5 million were canned in Maine (Fisheries of the United States, 1998). World production of various forms of sardines and pilchards for the years 1990 to 1994 is summarized in Table 2.

Canned Maine Sardines

A general description of the canning processes used for Maine sardines can be found in Wheaton and Lawson's *Processing Aquatic Food Products* (1985), and in the chapter on pelagic fish by Kaylor and Learson in *The Seafood Industry* (1990). The following description of sardine canning is based on these references. Immature herring for canning are generally caught with purse seines, stop seines, weirs, or even trawls. The fish are kept alive in nets when possible for one to two days to allow them to purge themselves. Preferably, the fish are transferred to a carrier boat from the nets with a fish pump, where they are held in a refrigerated seawater system. Upon reaching shore, the fish are transferred from refrigerated seawater-holding tanks using a fish pump to holding tanks in the processing plant. They are held in brine, then

Table 2. World Production of Sardine and Pilchard Products, 1990 to 1994

Product	Annual World Production in Metric Tons				
	1990	1991	1992	1993	1994
Canned pilchards	357,405	291,824	211,013	191,990	188,037
Canned European sardines	183,596	185,607	163,333	156,913	160,033
Canned Atlantic herring	73,121	72,356	70,170	72,429	75,229
Canned sprats	99,165	78,272	11,036	12,508	11,800
Canned clupeoids	51,732	63,902	63,519	81,310	54,267
Frozen pilchards	1,762,739	1,640,002	1,352,103	1,203,963	915,465
Frozen pilchard fillets	12,660	9,279	3,626	1,413	56
Frozen European sardines	26,879	28,275	22,183	15,976	11,773
Frozen sprat	48,301	43,898	22,364	15,576	11,867
Frozen clupeoids	68,193	59,519	2,066	210	1,283
Pilchards, dried, salted, or in brine	105,859	95,192	51,269	42,482	36,430
Dried, salted clupeoids	43,395	33,228	4,240	6,271	6,415
Dried, unsalted pilchards	52,071	57,984	46,553	46,387	44,874
Dried, unsalted clupeoids	32,115	38,602	30,392	27,683	28,407
Smoked sardines & sprats	20,221	22,168	15,749	14,525	14,348
Smoked clupeoids	79,753	81,006	37,982	37,542	37,000

Source: FAO Annual Fishery Statistics Commodities Yearbook, Volume 79, 1994. Information at this level of detail is not available in FAO Reports after 1994.

scaled, and the head and fins are removed either by hand or by machine. The viscera may or may not be removed, depending on the size of the fish and whether they were properly purged. Wheaton and Lawson describe a drying step that occurs for one to one-and-one-half hours after brining to reduce moisture content; the drying toughens the skin so that the fish does not fall apart during the cooking process. At this point the fish are either packed in open cans for preliminary cooking or are cooked and then packed. Six different cooking processes involve steaming the fish in cans with or without brine or on trays, frying in hot oil, baking in a steam-heated oven, or a combination of steaming and oven baking on a continuous conveyor. In each case the fish are drained and allowed to cool before the final packing step. After cooking, the cans packed with fish are filled with oil. Kaylor and Learson reported that most Maine packs use soybean oil, but olive oil or other oils may be used. Some products are also packed in mustard, tomato, pepper, or other sauces. The oil or sauce is added hot, the cans are immediately sealed, and then they are retorted (cooked under pressure) for the proper time and temperature necessary for sterilization to produce a shelf-stable product.

Variations of this basic process are used for canning sardines and sprat in northern and southern Europe, and in other parts of the world. Species differences and harvesting logistics dictate how the fish are handled and what product form is canned. Final product forms for canned sardines could include very small whole fish, fish with heads and fins and/or viscera removed, steaks, or fillets. There are also variations in terms of the oil or sauce that the sardines are packed in, and some products may receive other treatments such as smoking before they are canned.

FISH ROE AND CAVIAR

Fish Roe

The roe or the ovaries of a variety of marine fish, shellfish, and some anadromous and freshwater fish are used for food and are considered to be a delicacy in many parts of the world. The ovary is the female reproductive organ, and the roe consists of a pair of ovaries or membranous sacs that contain individual eggs held together by connective tissues that weaken as the ovary matures.

Figure 1. The eggs from sturgeon are considered the premium fish roe and are used to produce the most highly prized caviar.

Fresh whole or sliced roe is found in world markets as well as a variety of roe products that are salted, dried, marinated, smoked, cured, fried, or pressed. Other culinary roe products such as paté are prepared using roe, salt, flavoring agents, thickeners, and fillers. These products may be frozen, pasteurized, or canned. Individual fish eggs separated from the ovaries are commonly referred to as caviar. Caviar and caviar products or substitutes are generally salted and may be further processed by adding flavors, color, or other ingredients. Some caviar or caviar products may also be pasteurized.

The roe of some fish species is more highly valued than others. Sturgeon roe is considered to be the premium fish roe, and is used to produce the most highly prized and valuable caviar and caviar products (Figure 1). Products made from the roe of salmon, herring, cod, Alaskan pollock, lumpfish, capelin, and other species are also in demand. Although precise estimates of world roe production and consumption are not readily available, information on the 1994 production of various fish roe products, contained in the United Nations Food and Agriculture Organization (FAO) Annual Fishery Commodity Statistics Yearbook, provides an overview of some of the more important commercial products (see Table 3).

According to FAO commodity statistics for 1994, the production of fish roe or caviar products was concentrated in Asia, northern Europe, the former Soviet Union, the United States, and Canada. The Russian Federation was the leading producer of caviar and caviar substitutes. Iceland was the major producer of salted lumpfish roe. Japan was the major producer of cured salmon roe, with the United States the other primary source of this product. Japan was also the leading producer of pickled Alaskan pollock roe and cured herring roe; Korea and the United States were the other major producers of these products. China, Norway, Indonesia, other Asian countries, and the United States were the leading producers of unspecified roes that were dried, salted, or packed in brine. Production of canned roes and livers from unspecified fish species were primarily produced in the Faeroe Islands.

Caviar

The term "caviar" is used to refer to fish eggs prepared in a specific manner. The word is most likely derived from the Turkish word "khavyar," which means fish eggs. Caviar can be made from many kinds of fish roe, and specific products are generally identified by including the name of the fish species the eggs were taken from and in some cases the body of water where the fish was caught. This is particularly true for the most prized and expensive products — for example, Caspian beluga caviar. In the United States, the Food and Drug Administration (FDA) regards the term "caviar" to be synonymous with "sturgeon caviar." For a caviar product to be properly labelled, the common name of the fish species that the eggs are derived from must precede the word caviar for products made from all species except sturgeon. For example, caviar from salmon, lumpfish, or whitefish must be labelled "salmon caviar," "lumpfish caviar," or "whitefish caviar," while sturgeon caviar can simply be labeled "caviar." A more comprehensive definition of caviar is suggested in the reference work *Caviar: The Resource Book* (Sternin and Dore, 1993): "Caviar-type products constitute ready-to-consume, separated and salted fish eggs obtained by passing ovaries through screens or otherwise. The eggs may further undergo different processes like dyeing, pasteurization, flavoring, or pressing."

Table 3. World Production of Caviar and Fish Roe Products, 1990 to 1994

Product	Annual World Production in Metric Tons				
	1990	1991	1992	1993	1994
Caviar	1,331	1,258	953	544	512
Caviar substitutes	13,008	13,063	30,583	41,955	42,792
Cured salmon roe	17,518	17,883	14,472	17,257	14,948
Cured herring roe	15,006	16,892	17,395	18,278	18,328
Lumpfish roe, salted	965	745	617	387	527
Cod roe, salted or sugar-salted	3,311	2,473	2,370	1,340	1,338
Alaskan pollock roe, pickled	40,806	36,717	37,063	30,303	31,598
Fish roes, dried, salted, or in brine	1,627	1,790	1,841	1,813	1,492
Prepared fish roes	25,640	24,081	2,746	3,557	3,927
Canned fish livers & roe	3,442	3,291	478	360	405
Frozen cod roe	1,638	1,316	1,620	1,585	971
Frozen Alaskan pollock roe	852	1,996	1,971	1,206	1,206
Frozen capelin roe	2,250	3,954	2,062	724	5,641
Frozen fish roes	3,112	5,183	4,062	3,175	3,201

Source: FAO Annual Fishery Statistics Commodities Yearbook, Volume 79, 1994. Information at this level of detail is not available in FAO Reports after 1994.

The traditional and most highly-valued caviar is the dark- (blackish-) colored eggs of the sturgeon and paddlefish families. There are four different genera and more than 20 different sturgeon species. These species inhabit both freshwater and saltwater areas throughout the Northern Hemisphere. The smallest species are less than three feet at maturity, while the largest may be over 20 feet long and weigh over 1500 pounds. There are only two species of paddlefish, one of which lives in the Mississippi River basin in the United States and the other in the Yangtze River in China.

The Black and Caspian Seas and the rivers that drain into them have been the central area for sturgeon caviar production for centuries. Caviar "booms" with large sturgeon harvests and caviar production occurred in many parts of Europe and North America in the nineteenth and early twentieth centuries. However, Russia, Iran, and the former Soviet republics — all of which border the Black and Caspian Seas — remain the leading producers of sturgeon caviar today. Only the Russian Federation, Iran, and Kazakstan were identified as producers of caviar in airtight containers in the (1994) FAO Fishery Commodity Statistics.

Granular Caviar Processing

Caviar is produced from fresh mature ovaries, preferably extracted from living, stunned, and bled fish. The roe is removed carefully with attention to sanitation and handling to minimize contamination. Ovaries are graded according to species and quality. In grading, the stage of ovary maturity (egg size, shape, opacity or translucency, firmness of attachment, etc.) and freshness (color, firmness, odor, etc.) are important factors. The best quality ovaries are used to produce caviar, and inferior ovaries or parts of ovaries are salted whole. The eggs are separated from the ovary by a process called screening, which is generally done manually but may be automated for some species. Manual screening involves pressing the ovary through a stainless steel wire or other mesh material using a circular motion that allows the eggs to pass through the mesh and the connective tissue to remain on top. Automated systems that utilize mechanical force or enzymatic systems for separating the eggs from the roe have been developed for such species as cod, pollock, and salmon. The eggs are salted after separation. Salt is used to preserve the eggs and to obtain the desired shape and elasticity. Sturgeon eggs are dry-salted to achieve 3 to 3.5 percent salt in the final product. Salmon and other roes may be salted in a saturated brine solution. The salting process is critical and occurs rapidly. Salting may take from two to 20 minutes depending on the species, egg size and quality, and desired final salt content. After salting, the eggs are placed on a screen in a thin layer to drain. The eggs are then packed into metal, glass, or other containers.

Variations to this basic process depend on the fish species and the desired end product. If preservatives or other additives such as coloring agents are used, they are added after draining. Packaged products may also be pasteurized. To maintain quality, air should be removed from the final container, and the product must be refrigerated at all times between 32°-35°F (0° to 1°C).

Pressed Sturgeon Caviar

Pressed sturgeon caviar is a common form of caviar consumed in Russia and other former Soviet republics. It is generally made from ovaries that do not meet top quality standards for granular caviar, and there are generally two grades, first and second grade. Pressed caviar can be made from eggs of several species and sizes. Unlike granular caviar, which is dry-salted, this product is salted in a warm saturated brine between 100° to 115°F (37° to 46°C) for one to two minutes. The salted eggs are drained, packed in linen bags, and pressed. The pressed mass is cooled and gently mixed to even out the salt content and then packed in tins or kegs. The final product is spreadable or sliceable and can be stored refrigerated or frozen.

Salmon Caviar

Current production of salmon caviar is difficult to assess, but it is likely to play an increasingly important role in the caviar marketplace in the future. Chum salmon and pink salmon are the species most favored for caviar because of the size, taste, and shelf life of their eggs and the nature of the fisheries for these species. Salmon roe is processed in a manner similar to that used to produce caviar from other species, except that salting occurs in a saturated brine solution rather than with dry salt. Both pasteurized and unpasteurized roe that has been stored under refrigeration or frozen are sold in both retail and bulk packages.

Lumpfish Caviar

Lumpfish are a cold-water marine species found throughout the northernmost parts of the Atlantic Ocean. Iceland and Canada are the largest producers of salted lumpfish roe, which can be further processed into caviar products. Lumpfish roe is traditionally heavily salted (10 to 12 percent salt), packed in barrels, and refrigerated. This salt-preserved roe can be stored for one year or more. To produce lumpfish caviar, the eggs are desalted, flavored, and dyed. The final product may also be pasteurized. Lumpfish caviar is popular in many northern European countries as a less expensive alternative to sturgeon caviar.

Flavored Caviar

A variety of different types of flavored caviar is produced in the United States and other countries. These products have been made from the roe of lake sturgeon, hackleback sturgeon, paddlefish, bowfin, chinook and coho salmon, trout, and whitefish. The fish eggs are infused with flavors by hand. Caviar pepper and caviar citron are flavored with citron and pepper vodka. Smoked caviar is also produced from species such as bowfin and whitefish, and distinct flavors are imparted using a variety of different woods such as fruit- and hardwoods and mesquite.

Salted and Smoked Cod Roe

Salted and smoked cod roe is popular in Norway and other northern European countries. The roe is carefully removed, washed in seawater, and salted until the color turns red. The salt-to-roe ratio is about 1:10, and it is important that the roe be handled carefully to keep them intact. After the roe acquire the desired color, they are washed in seawater and hung out to air dry on wire racks for a day. After drying, the roe are hung on sticks or laid on wire mesh trays for smoking. The roe is cold-smoked for two to three days until they acquire a dark brown color. After smoking, the membrane of each roe sac is split, and the separated eggs are packed in barrels, closed tightly, and stored in a cool place (55° to 60° F or 13° to 15°C) for about a month. During this time the product will begin to ferment. This process is stopped by adding additional salt in a ratio of one part salt to seven parts roe. After thorough mixing, the product is packed into containers and sealed (Jarvis, 1987).

Dried Mullet Roe

Roe from a variety of species around the world is preserved by drying and salting. Along the Southeastern coast of the United States, mullet roe is preserved by dry-salting and smoking. Production of this product was characterized in the following way in *Marine Products of Commerce* (Tressler and Lemon, 1951): Roe is washed, drained, and then either rolled in fine salt or soaked in brine. About five quarts of salt are added for each 100 pounds of roe, and the roe are salted for 10 to 12 hours. The roe is drained and dried in the sun for about one week. Each day the roe is left in the sun, and brought indoors during the evening. The drying process is complete when the roe is reddish-brown and feels hard. Dried roe is dipped in melted beeswax, cooled, wrapped in waxed paper, packed in wooden or metal boxes, and refrigerated. The mullet roe may also be cold-smoked after the initial brining step.

REFERENCES

Alm, F. 1965. Scandinavian anchovies and herring tidbits. In *Fish as Food.* Vol. 3, G. Borgstrom (ed.), pp. 195-218. Academic Press. New York.

Annual Yearbook of Fishery Statistics 1994: Commodities. 1995. FAO Statistics Series No. 113 (Vol. 79). Food and Agricultural Organization of the United Nations. Rome.

Collins, C. 1994. Caviar is for everyone. *The Fisherman.* **46**(1):10-11, 18.

Dore, I. 1993. *The Smoked and Cured Seafood Guide.* Urner Barry Publications. Toms River, NJ.

Essuman, K. 1992. Fermented fish in Africa: A study on processing, marketing and consumption. FAO *Fisheries Technical Paper* No. 329. Rome.

Fisheries of the United States, 1995. 1996. National Marine Fisheries Service Current Fishery Statistics No. 9500. Washington, DC.

Food and Drug Administration. 1993. The seafood list: FDA's guide to acceptable market names for seafood sold in interstate commerce. U.S. Food and Drug Administration. Washington, DC.

Guillerm, J. 1928. Le "nuoc-mam" et l'industrie saumuriere en Indochine. *Arch. Inst. Pasteur Indochine.* 7: 21-60.

Jarvis, N D. 1987. *Curing of Fishery Products.* Teaparty Books. Kingston, MA.

Kaylor, J.D. and R.J. Learson. 1990. Pelagic Fish. *The Seafood Industry.* R. Martin and G. Flick (eds.), pp. 67-76. Van Nostrand Reinhold. New York.

McMohan, Jack. 1994. Stalking the golden egg: Pink salmon caviar gets a boost. *The Lodestar.* **12**(1): 3.

Meyer, V. 1965. Marinades. In *Fish as Food.* Vol. 3, G. Borgstrom (ed.), pp. 165-194. Academic Press. New York.

Owens, J.D. and L.S. Mendoza. 1985. Enzymatically hydrolyzed and bacterially fermented fishery products. *Journal of Food Technology.* **20**(3): 273-293.

Lassen, S. 1976. The herring, sardine, mackerel, and anchovy fisheries. In *Industrial Fishery Technology.* M.E. Stansby (ed.), pp. 124-129. Krieger Publishing. Huntington, NY.

Steinkraus, K.H. 1979. Nutritionally significant indigenous food involving an alcoholic fermentation. *Fermented Food Beverages in Nutrition.* C.F. Gastineau, W. J. Darby, and T.B. Turner (eds.), pp. 35-59. Academic Press. New York.

Steinkraus, K.H. 1983b. Lactic acid fermentation in the production of foods from vegetables, cereals and legumes. *Antonie van Leeuwenhoek.* **49**(3): 337-348.

Steinkraus, K.H. 1993. Comparison of fermented food of the East and West. In *Fish Fermentation Technology.* C.H. Lee, K.H. Steinkraus and P.J. Alan Reilly (eds.), pp. 1-12. United Nations University Press. Korea.

Stenstrom, M.D. 1965. Scandinavian sardines. In *Fish as Food.* Vol. 4. G. Borgstrom (ed.), pp. 265-290. Academic Press. New York.

Sternin, V. and I. Dore. 1993. *Caviar: The Resource Book.* Cultura Enterprises. Stanwood, WA.

Tressler, K.D. and J. McW Lemon. 1951. Miscellaneous Processes of Preserving Fish. In *Marine Products of Commerce.* Reinhold Publishing Corporation. New York.

van Veen, A.G. 1953. Fish preservation in Southeast Asia. In *Advances in Food Research.* Vol. 4. E.M. Mrak and G.F. Stewart (eds.), pp. 209-232. Academic Press. New York.

van Veen, A G. 1965. Fermented and dried seafood products in Southeast Asia, In *Fish as Food.* Vol. 3. G. Borgstrom (ed.), pp. 227-250. Academic Press. New York.

Watanabe, K. 1982. Fish handling and processing in tropical Africa. In *Proceedings of the FAO Expert Consultation on Fish Technology in Africa, Casablanca, Morocco.* June 7-11. FAO Fisheries Report No. 268. Rome.

Watts, J. C. D. 1965. Some observations on the preservation of fish in Sierra Leone. In *Bulletin de MPANT XXVII Sci.* No. 1.

Wheaton, F. and T. Lawson. 1985. Heat processing and canning. In *Processing Aquatic Food Products.* John Wiley and Sons. New York.

Processing of Surimi and Surimi Seafoods

Jae W. Park and Tyre C. Lanier

INTRODUCTION

Surimi is stabilized myofibrillar protein from fish muscle. More simply, it is mechanically deboned fish flesh that has been washed with water and blended with cryoprotectants to ensure a good frozen shelf life. This intermediate product is used to manufacture a variety of surimi seafoods, from the traditional kamaboko products of Japan to the shellfish substitutes now popular in Western countries. Prior to the development of stabilized surimi as we know it today, washed fish mince was made and used within a few days as a refrigerated raw material because freezing induced protein denaturation and poor functionality of the muscle proteins.

Nishiya et al. (1960) at Hokkaido Fisheries Research Station of Japan discovered a technique by which freeze denaturation of proteins in Alaskan pollock (*Theragra chalcogramma*) muscle could be prevented. This technique required the addition of low molecular weight carbohydrates such as sucrose and sorbitol in the dewatered, washed mince prior to freezing. Actomyosin of Alaskan pollock, which is remarkably unstable during frozen storage (Scott et al., 1988), was protected from the loss of functional (gelling) properties by the addition of these carbohydrates. This discovery revolutionized the industry, since it no longer had to depend on the variability of supplies of fresh fish. Kamaboko processing plants in Japan were able to extend their base beyond local supplies and to draw on fish resources which had been unavailable until then because of distance.

Alaskan pollock of the North Pacific Ocean and the Bering Sea had been a largely unexploited resource, but was now available in the quantity required to meet the large expansion that occurred in the Japanese kamaboko industry in the 1960s and 1970s [Okada, 1992; Alaska Fisheries Development Foundation (AFDF), 1987]. Production and sales of surimi-based foods rose in Japan, and eventually extended to Western countries such as the United States. The surimi industry became Americanized in the late 1980s, with annual United States production of 150,000 to 190,000 metric tons (MT) (Figure 1). United States production of Pacific whiting surimi began at sea in 1991 and in shore-side plants in 1992.

World surimi production grew to over 500,000 MT in 1992 (Figure 2). Japan's direct involvement decreased with the increased surimi production in the United States. Domestic production of surimi seafoods, primarily crab substitutes, began in the early 1980s. It became popular as an ingredient for cold salads and experienced an annual growth rate of 10 to 100 percent, reaching 72,500 MT annual production in 1994 (Figure 3). Consumption levels decreased in 1991 due to a temporary rise in the price of surimi from less than \$1.00 to \$2.50 per pound (Figure 4). Present market growth is about 3 to 5 percent every year (Figure 3) (Park, 1994b).

Even though the surimi industry is still largely based on Alaskan pollock, since 1991 efforts to develop surimi from other species have been successful. The most suitable species for surimi pro-

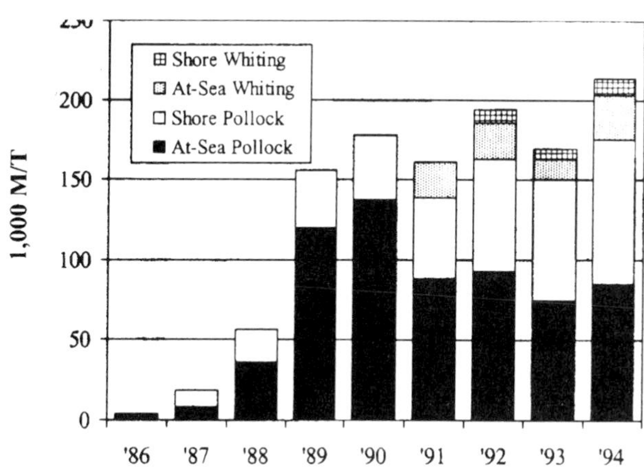

Figure 1. Surimi production in the United States. "At-sea" denotes surimi processed by on-board vessels, while "shore" indicates land-based plants.

800
700
600
500
400
300
200
100
0
1,000 M/T
Japan
Thailand
Others
USA
New Zealand
Korea
Russia
1988
1990
1992

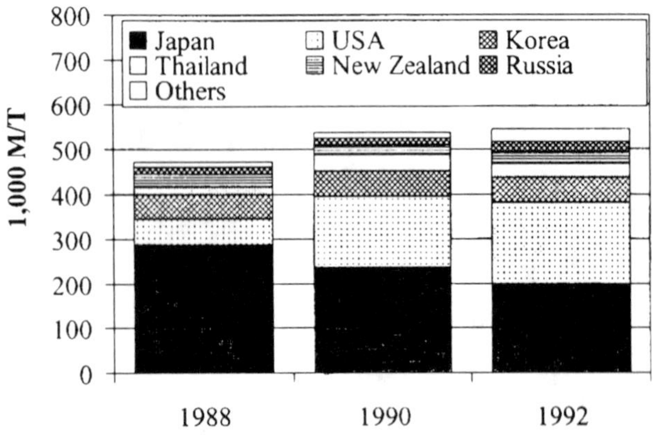

Figure 2. World surimi production.

cessing are those with white flesh and low fat content. These presently include Pacific whiting (*Merluccius productus*) from the Pacific coast; hoki (*Macruronus navaezelandiae*) from New Zealand; Southern blue whiting (*Micromesistius australis*) from Chile, Argentina, and New Zealand; Northern blue whiting (*Micromesistius poutassou*) from European Economic Community (EEC) waters; threadfin bream (*Nemipterus japonicus*) from Thailand; yellow croaker (*Pseudosciaena manchurica*) from the south of Japan; and yellow sole (*Buglossidium luteum*) from Alaska. Even species which have a higher proportion of red or dark muscle and/or a higher content of fat are used for the production of low-grade surimi. These include pink salmon (*Oncorhynchus gorbuscha*), atka mackerel (*Pleurogrammus azonus*), Japanese sardine (*Sardinops melanostrictus*), and Chilean jack mackerel (*Trachurus murphyii*).

In the United States and other Western countries surimi-based seafood products have expanded their markets in food service applications, in the retail deli, and in salad bars. The success of these surimi-based crab, shrimp, lobster, and scallop products is mainly due to their low price in relation to that of their natural counterparts. They

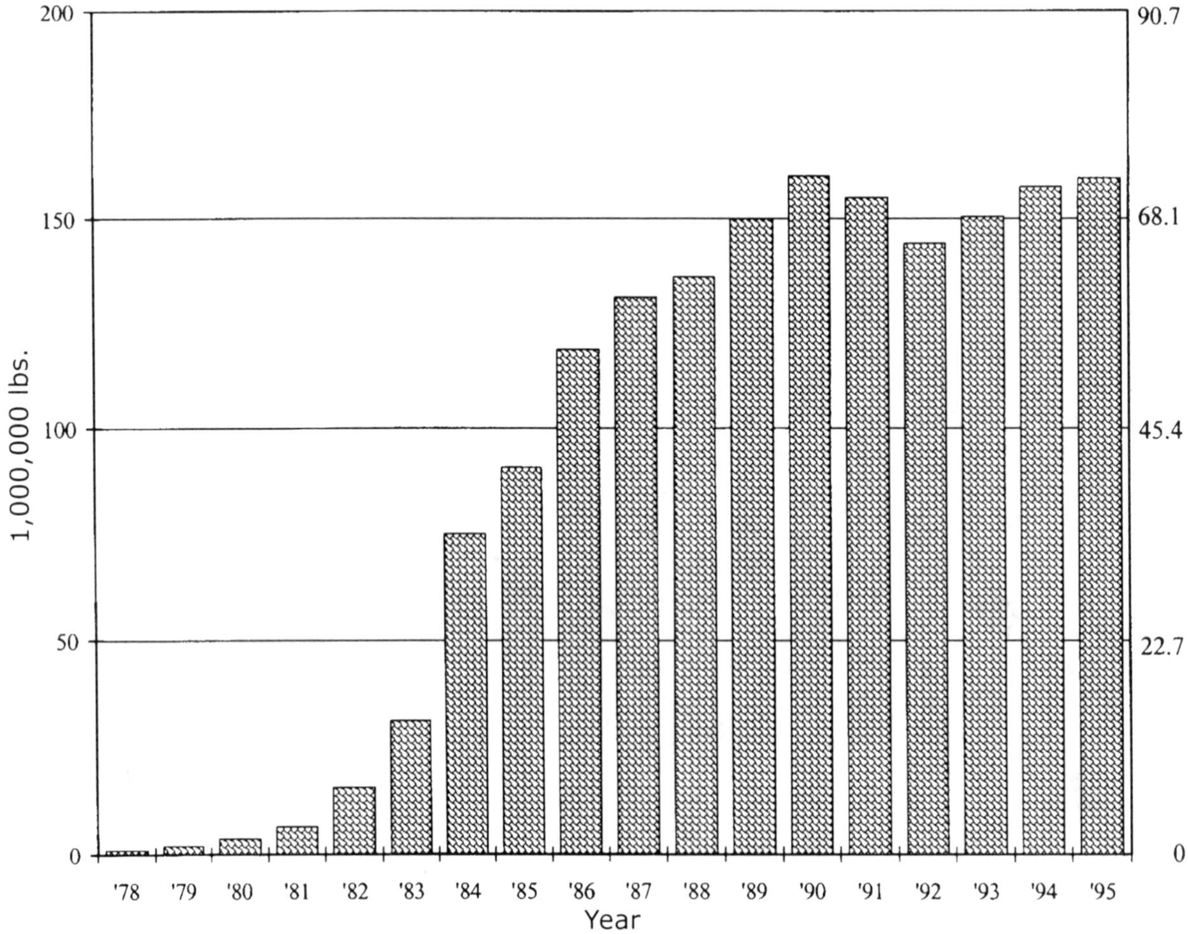

200
150
100
50
0
1,000,000 lbs.
90.7
68.1
45.4
22.7
0
1,000 MT
'78
'79
'80
'81
'82
'83
'84
'85
'86
'87
'88
'89
'90
'91
'92
'93
'94
'95
Year

Figure 3. Surimi seafood consumption in the United States.

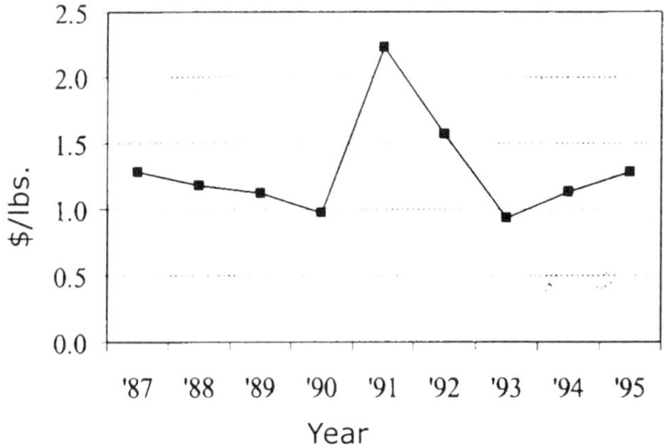

Figure 4. Market price of high grade (FA) Alaskan pollock surimi, FOB Seattle.

are also convenient for preparation, and shell-free. Nutritionally, they can be prepared to be fat-free. Surimi-based seafoods also do not fluctuate in availability of supply or quality as can the higher-priced natural shellfish.

In the United States surimi-based shellfish products must be legally labeled as "imitations." The size of the word "imitation" on the package may not be smaller than one-half that of the brand name. The FDA has offered guidance for labeling surimi seafoods: If the surimi-based product purports to be, or is represented as, any specific type of natural seafood, including shape or form representations, but is nutritionally inferior to that seafood, it must be labeled as imitation in accordance with 21 CFR 101.3 (FDA, 1992).

Surprisingly, the United States is the only country in the world with such a strict regulation requiring that the word "imitation" be used in labeling surimi-based shellfish products. For the last five years, the United States surimi seafood industry has tried to obtain approval from the FDA for the use of the term "surimi seafood" as a common name. According to a recent survey regarding consumer attitudes toward crab substitutes (Park, 1994b), almost 25 percent of those surveyed were extremely concerned by the presence of the word "imitation" on the label.

The United States surimi seafood industry is also losing a competitive edge in global markets due to a strict labeling regulation which tends to restrict the use of surimi produced from two or more species of fish. For example, when a certain formulation contains 50 percent pollock surimi, 30 percent water, and the remaining 20 percent other ingredients, the ingredient statement in descending order is "Alaskan pollock, water, etc." But if half the pollock were replaced by Pacific whiting surimi, a current rule requires that the descending order of ingredients be "water, Alaska pollock, Pacific whiting, etc." Should water be the first ingredient in this case? A more acceptable ingredient statement would allow the label to read "surimi (Alaskan pollock, Pacific whiting), water, etc." Despite the growth of the surimi-based shellfish business in the United States, there are obviously changes which could be made to further increase its consumption.

The remainder of this chapter will review the current practices of surimi manufacture and surimi seafood processing, with discussion of problems and solutions which may include new processing technologies.

THE MANUFACTURE OF SURIMI

Processing Technology and Sequence

Harvesting, Transportation, and Holding

When fishermen tow mid-water trawl nets, it causes a compacting of the fish against the net which impacts on the yield and quality of recovered flesh. Because factory trawlers or mother ships process fish at sea, holding time on the vessel is not as long in duration as the holding time required for shore-side delivery and processing. En route to port, fish are kept cool in the vessel by either refrigerated sea water (RSW), slush ice, or champagne ice (slush ice with aeration to mix the ice and water thoroughly). Time/temperature recorders are installed on some vessels to determine the rate at which the fish are cooled. Figure 5 is an example of a representative trip with RSW and champagne ice (Peters and Morrissey, 1994). The champagne ice system cools fish more rapidly than the RSW system. It takes approximately one hour to cool fish to 1°C to 2°C using champagne ice, but more than 20 hours using an RSW system. The champagne system is also highly recommended for holding due to delayed processing at shore-side plants.

Heading, Gutting, and Deboning

Mechanical meat separators for fish — developed primarily by companies in Japan, Germany, and the United States — are modern sanitary ma-

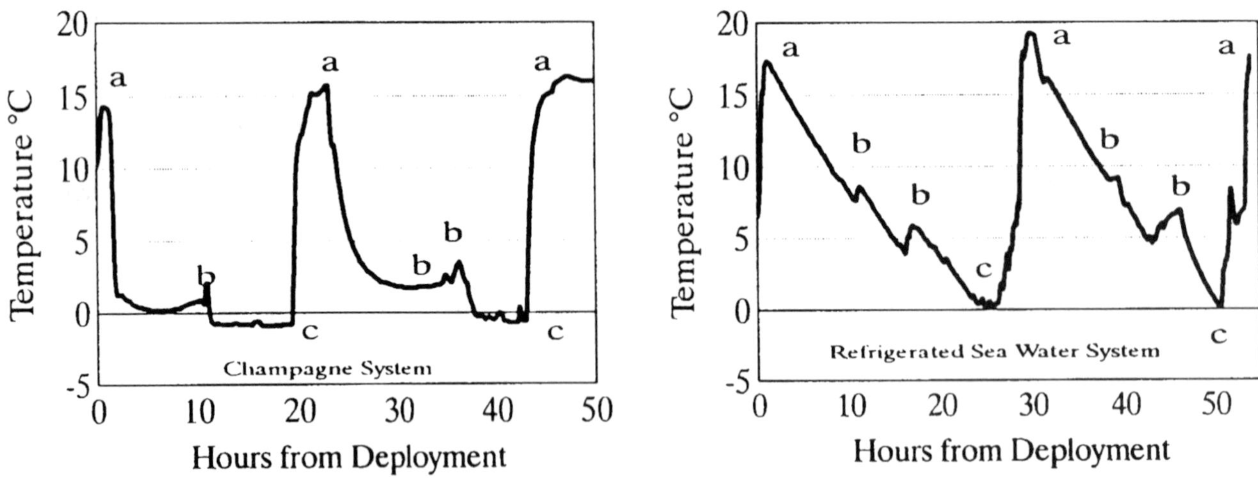

Figure 5. Time-temperature relationship of fish-holding tanks.

chines that remove virtually all of the flesh from the frame of a properly prepared fish. According to Pigott (1986), there are several methods of preparing fish for deboning. One is to remove the head, degut, and thoroughly clean the gut cavity prior to deboning the carcass. The other is to fillet the fish and then debone the fillet which still contains skin and bones. Quantities of water are used in either process. On factory trawlers, where costly desalinization must be used for water generation, refrigerated sea water is commonly used for these initial steps, whereas all shore-side operations use refrigerated fresh water. In the manufacture of surimi from Pacific whiting, fillets that are black-fleshed due to infection by myxosporidean parasites must be removed to prevent downgrading of the product due to higher visual contaminant content. Morrissey et al. (1993) reported that 4 to 5 percent of the Pacific whiting harvest is infected with these black hair-like striations. Although they present no health hazard, they are easily seen and are unacceptable for aesthetic reasons.

Deboning (Mincing)

All deboning machines, regardless of design, depend on the principle of forcing soft portions through a perforated plate or screen into the interior of a drum while leaving bones, hard cartilage, or skin on the exterior to be scraped away. The diameter of the drum perforations should not be larger than three to four millimeters to prevent the skin from passing through (Takeda, 1971). Deboning of fillets gives a cleaner minced meat since there are no membranes, blood, or other contaminants present. However, deboning of cleaned carcasses allows a higher final yield of minced flesh (Pigott, 1986).

Washing (Leaching) and Dewatering

Efficient washing, or leaching, is the most important step in surimi processing. This will ensure maximum gelling potential; likewise, problems of color, taste, and odor that develop in minced meat are minimized or eliminated when the minced meat is properly washed. The protein content of minced fish meat is approximately two-thirds derived from myofibrillar proteins, the main structural proteins of muscle, which are sparingly soluble in water. These proteins are primarily responsible for gel formation when manufacturing surimi-based foods. The remaining one-third of the protein in the fish mince consists of blood, myoglobin, and sarcoplasmic proteins — components that are more soluble in water and that would lower the gelling ability of the surimi if not removed. Washing/leaching thus effectively concentrates the functional myofibrillar proteins, and also extends their functional frozen storage life by removing many reactive components.

The washing process (Figure 6) involves mixing the minced meat with cold water (~5°C), followed by removing the added water via screening or centrifuging to about 5 to 10 percent solids. This process is repeated two or three times. The

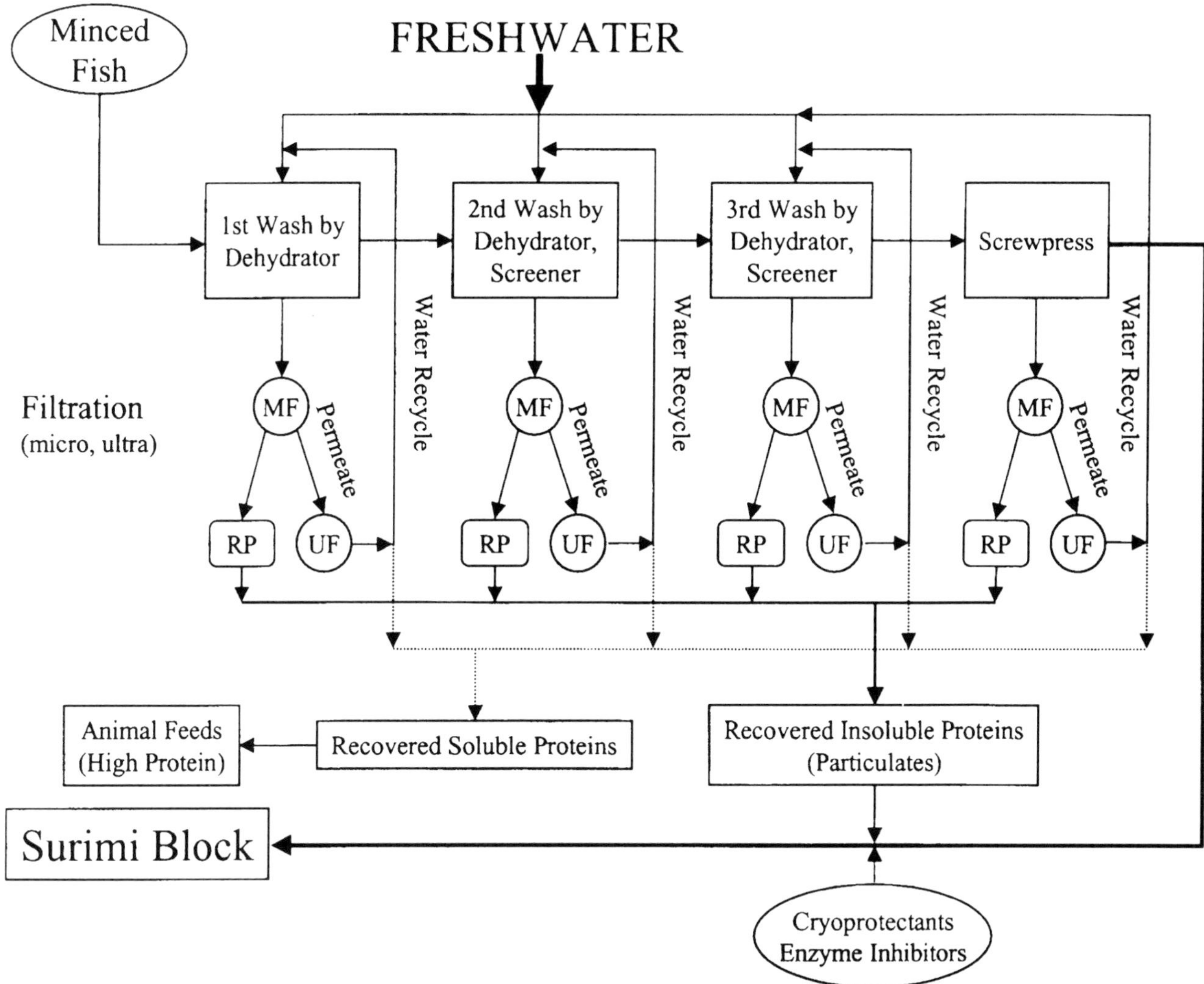

Figure 6. Protein recovery and water recycling from surimi processing waste water. Adapted from *Lin* et al. *(1995).*

removal of solubles can be greatly increased by proper dewatering between washing steps, particularly after the first wash. Undesirable particles such as fine bones, scales, and connective tissues are removed by refining through a fine-screened, centrifugal-fed machine just prior to final dewatering by a screw press. The screw press employs compression force against screens of 0.5 to 0.8 mm perforations to lower the moisture content 80 to 82 percent, which is similar to that of a fish fillet. A mixture of NaCl and $CaCl_2$, 0.1 to 0.3 percent total concentration, is commonly added in the final washwater to facilitate water release from the mince particles.

Water/mince ratios vary from five to 10 in a batch process; however, in most continuous commercial processes it is impossible to measure the exact ratio of water to mince. Lin et al. (1995) reported that 29.1 kilograms of waste water were generated to produce one kilogram of surimi in a shore-side operation, but this figure includes other operations besides leaching. Less freshwater is used in factory trawler operations. The number of washing cycles and the volume of water will vary with fish species, the freshness of fish, the type of washing unit, and the desired quality of surimi.

Swafford et al. (1990) introduced in-line washing, which employs the placement of static mixing devices at intervals within a continuous pipeline. This technique minimizes contact time between wash water and flesh while significantly reducing the amount of wash water required. However, few manufacturers have commercially installed this system and there is still need to investigate its possible benefits and drawbacks.

Stabilizing Surimi with Cryoprotectants

The addition of cryoprotectant compounds is important to ensure maximum gelling potential of frozen surimi. Freezing induces protein denaturation and aggregation, resulting in poor gelling ability. Sucrose and sorbitol, alone or mixed and added at approximately 9 percent ω–ω to dewatered fish meat, are most commonly used. A mixture (1:1) of sodium tripolyphosphate and tetrasodium pyrophosphate at 0.2 to 0.3 percent is also commonly added as a synergist to the cryoprotective effects of the carbohydrate additives. Minor quantities of other gelling and/or color enhancing ingredients may also be added. For example, calcium carriers (calcium lactate, calcium sulfate, calcium citrate, calcium caseinate) are often added. Apparently these can help cross-link the proteins through ionic bonding or their effect on muscle cross-linking enzymes; presumably, however, too much calcium limits long-term storage by promoting these reactions. Mono- and diglycerides (0.2 to 0.3 percent) as well as hydrogenated canola oil (0.1 to 0.15 percent) are also often added, presumably as cryoprotectants; however, these additives also improve surimi color through their effect on light scattering and thereby whiteness (Park and Morrissey, 1994). Proteolytic enzyme inhibitors such as beef plasma protein are commonly added along with cryoprotective ingredients in the manufacture of Pacific whiting surimi. These could equally improve the quality of surimi from other species, including pollock. Their addition at the point of surimi manufacture is strictly a convenience, as they would be equally effective when added during the manufacture of final food products from surimi.

Cryoprotectant incorporation is accomplished with a silent cutter or ribbon blender. The temperature of the mix must not exceed 10°C at this step, or any previous step, to avoid damage to the proteins. The leached fish meat mixed with cryoprotectants and other additives, now properly termed surimi, is formed into a 10-kilogram block within a plastic bag and placed on a stainless steel tray for freezing.

Freezing

Surimi blocks, each wrapped in a plastic bag, in stainless steel trays are placed in a contact plate freezer and held for approximately two and one-half hours or until the core temperature reaches -25°C. Two 10-kilogram frozen blocks are packed in a corrugated cardboard shipping box. Drum freezing of surimi, by application of unfrozen paste onto a refrigerated rotating drum, has offered the prospect of more rapid freezing (Lanier et al., 1992). This could enhance quality and offer surimi in a more convenient flaked form. At-sea production would require compression of frozen flakes into easily shattered blocks to minimize storage space required.

PARAMETERS AFFECTING SURIMI QUALITY

The quality of surimi is significantly affected by the intrinsic properties of fish and the processing factors such as equipment, processing sequences, and chemical/physical treatments.

Effects of Species

Surimi usually varies in compositional and functional properties according to fish species. Functional properties of surimi are only partially dependent on gross composition (fat, moisture, protein content), and thus cannot generally be predicted from such analysis.

The recent development of Pacific whiting as a surimi source has emphasized the importance of the enzymatic activities present. Flesh and surimi from this species possess a heat-stable, protein-degrading enzyme system that significantly softens texture during cooking. An et al. (1994) identified the responsible enzymes as cathepsins B, H, and L in this species. Cathepsins B and H are easily removed during surimi processing, while cathepsin L remains in the muscle tissue in significant quantities. The latter displays optimum protein degrading activity at 55°C. Therefore, surimi from this species requires the addition of enzyme inhibitors unless it is cooked very rapidly to high temperature (>70°C), such as by microwaving or ohmic heating (Yongsawatdigul et al., 1995). Typical food-grade inhibitors used include beef plasma protein, egg white, white potato extracts, and milk whey proteins. These vary in both cost and effectiveness, as well as taste and color contribution.

Arrowtooth flounder (*Atheresthes stomias*) is another species that requires addition of enzyme inhibitors to minimize the textural deterioration due to its heat-stable enzyme system (Greene and Babbitt, 1990). Gel weakening occurring during processing at 55° to 60°C has also been reported in threadfin bream (*Nemipterus bathybius*) (Toyohara and Shimizu, 1988), Atlantic menhaden (*Brevoorti tyrannus*) (Boye and Lanier, 1988), white croaker (*Micropogon opercularis*), and oval filefish (*Navodon modestus*) (Toyohara et al., 1990). Even pollock surimi can benefit in its gelling properties by the addition of inhibitors (Hamann et al., 1990).

Special technology must be employed to make bland- and light-colored surimi from oily/dark or red-flesh fish, such as mackerel, sardine, menhaden, and salmon. The red color of dark muscle is due to the high content of the heme proteins myoglobin and hemoglobin. Fat oxidation in the dark muscle is also promoted by the presence of these heme proteins; consequently an offensive rancid odor may develop in storage (Tokunaga and Nishioka, 1988). It is suggested that 0.1 to 0.5 percent sodium bicarbonate be added to the first wash to maintain a higher pH and thereby protect proteins from denaturation, and that a centrifugal decanter be employed to remove excess oil released during washing. Addition of 0.05 to 0.1 percent sodium hexametaphosphate (or other antioxidant) and the use of a vacuum during washing are also recommended for better fat stability and fat removal, respectively. For lightest color, fish muscle should be reduced in particle size for leaching, which may also necessitate microfiltration or centrifugation to ensure adequate protein recovery.

Effects of Seasonality/Sexual Maturity

Compositional properties of fish may change as the fishing season progresses. Alaskan pollock (AFDF, 1992) exhibit the highest protein content (19.0 percent) in November and the lowest (16.5 percent) in May, while moisture contents are highest (82.3 percent) in July and lowest (80.2 percent) in November. Morrissey (1993) conducted a seasonal analysis of the protein, moisture, fat, and ash contents of Pacific whiting (Figure 7). The highest moisture content (84.5 percent) was in April, while the lowest (80 percent) occurred at the end of October. Protein content was at its lowest (14 percent) in April and then increased and held relatively steady (15.5 to 16.5 percent) for the remainder of the season. Fat levels were relatively constant (0.5 to 1.5 percent) until September and then increased to over 2.5 percent by October.

Generally, fish harvested during the feeding period produces the highest quality surimi. During this period fish muscle has the lowest moisture content and pH, as well as the highest total protein (Overseas Fishery Cooperation Foundation and Japan Deep Sea Trawlers Association, 1984). Thus fish harvested during and just after the spawning season produces the lowest quality surimi. The pH of muscle obtained from spawning fish is relatively higher and the leached muscle, therefore, tends to retain more water, making water removal more difficult. Lowering the pH or increasing the salinity of the final wash water can alleviate this somewhat (Lee, 1986), yet yields of surimi protein recovery are decreased nonetheless.

Fat content of fattier fish species varies considerably throughout the year. Tokunaga and Nishioka (1988) found that in August the fat content of sardines harvested in the Pacific Ocean was as high as 33 percent, compared to a low value of 3 percent in April.

Effects of Freshness or Rigor

Freshness of fish is time- and temperature-dependent, and can affect surimi quality through the action of enzymes (inherent or microbial), and due to the stage of rigor mortis. Alaskan pollock is processed within 12 hours on factory vessels, or within 24 to 48 hours in shore-side operations. Due to endogenous proteolytic enzymes which are activated by rising temperatures, Pacific whiting must be processed more rapidly. The biochemical and biophysical changes occurring during the development of rigor mortis induce significant changes in the functional properties of the muscle proteins. Fish should be processed as soon as possible after it has gone through rigor. Prior to passing into rigor, up to about five hours post-harvest in the case of Alaskan pollock, it is reportedly difficult to remove the "fishy" odor and various membranes and other contaminants that affect the product quality (Pigott, 1986). However, Park et al. (1990) reported that significantly higher protein content and yield, reduced cook losses, and

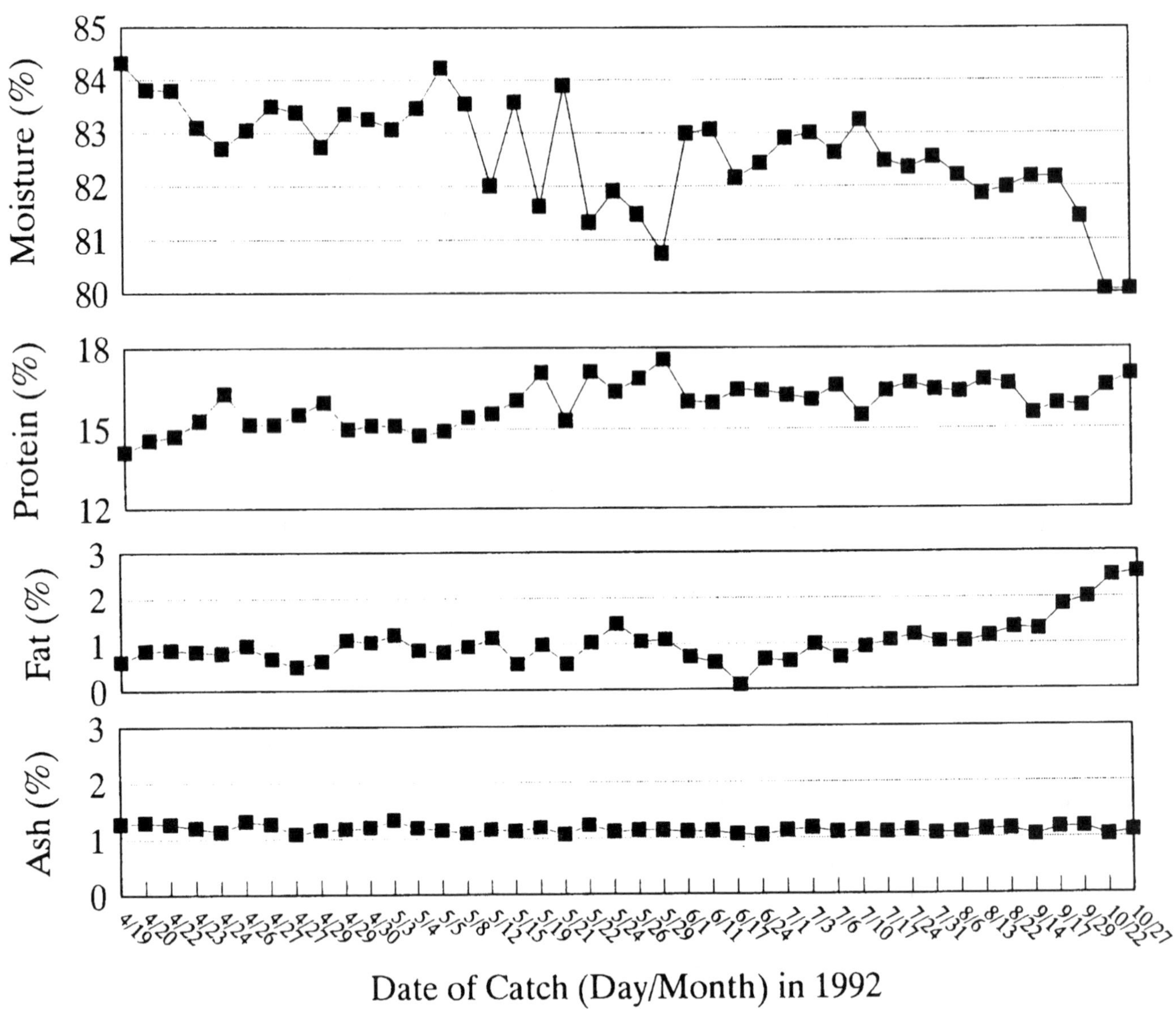

Figure 7. Seasonal changes of compositional properties of Pacific whiting.

enhanced gel-forming ability were associated with surimi processed from pre-rigor tilapia fish.

Other Factors Affecting Surimi Quality

The water-leaching process is critical to surimi quality. The important parameters are temperature, hardness or mineral content, pH, and salinity. The level of chlorination may also exert a bleaching or deodorizing effect (Lee, 1986). The water must be refrigerated to a temperature low enough to ensure that the fish muscle proteins can retain maximum functional properties. This will vary somewhat based on the thermostability of the fish proteins. Warm-water fish can tolerate a higher water/process temperature than cold-water fish without a reduction of protein functionality (Arai et al., 1973). Generally the recommended process temperature for maximum quality is 5°C or less.

Soft water, with a minimum level of minerals such as calcium, magnesium, iron, and manganese, is recommended for washing. On factory trawlers, where fresh water is generated using steam distillation, the water is always relatively low in mineral content. Most shore-side operations using either municipal water or purified well water employ water softening by ion exchange or reverse osmosis membrane systems. The presence of minerals in the water can cause deterioration of quality in texture and color during frozen storage. Calcium and magnesium are responsible for changes in product texture, while iron and manganese can induce color changes (Lee, 1990).

Ease of moisture removal gradually increases with an increase in salt concentration of the wash water (Lin and Park, 1995; Lee, 1990). However, too much salt in the wash water could cause solubilization of myofibrillar proteins, which would result in a loss in yield and/or poor gelling properties.

PROCESSING AUTOMATION: ON-LINE SENSORS

On-line sensors for the measurement of various physical and chemical characteristics of food can be powerful tools for process and quality control. Physical attributes such as pressure, weight, temperature, and flow rate; and chemical attributes such as moisture, fat, protein, and pH can be monitored to control a process and ensure product quality (Giese, 1993). Maintaining target moisture levels during surimi production is of the utmost importance to any processor. Variations in moisture content during processing can unknowingly alter finished surimi properties such as gel strength and whiteness if the surimi is tested without adjustment of moisture content, as in traditional Japanese methods (Lanier, 1992).

A new technology using near-infrared (NIR) energy in the transmission mode (i.e., passing NIR energy completely through the surimi) can monitor moisture content of surimi within the process pipeline. A percent moisture content is instantly calculated (Brown, 1993) and sent to a display so that manual adjustments to the process can be made. The system could be equipped with a series of calibrated analog output signals that could be sent to control key process steps affecting moisture content, such as the screwpress. Some processors have successfully monitored power input to the screwpress as an indicator of moisture content, as greater power input is needed to press a drier feed material.

There are other areas where sensors could be successfully used for surimi processing. Sensors for sugar content could control automatic injection of the cryoprotective mixture into a silent cutter or ribbon blendor. Automatic rejection of parasite-contaminated whiting fillets before mincing, using video sensors could eliminate problems of visual impurities in surimi and thus increase its overall grade.

WASTEWATER PROBLEMS

Surimi processing requires a large amount of chilled fresh water during the leaching/washing process. Minced fish solids (about 40 to 50 percent) are lost during washing and dewatering (Pacheco-Aguilar et al., 1989; Adu et al., 1983). According to Lin et al. (1995), as illustrated in the wastewater streams of a typical surimi processing line (Figure 6), the majority (about 75 percent) of the waste water is discharged from screening following each leach cycle. Protein contents of wastewater streams were between 0.46 to 2.34 percent. The wastewater at successive discharge points in the process contained decreasing amounts of protein, nonprotein nitrogen, fat, and ash. Myofibrillar proteins of white-fleshed fish, previously considered as insoluble in water, are nonetheless lost to some degree during leaching (Lin et al., 1995; Wu et al., 1991). Loss of myosin is small in the first wash cycle but increases substantially in the second cycle, remaining nearly constant throughout the rest of the process. Extensive washing, which may cause myofibrillar proteins to be dissolved in water to some extent, and mechanical force during dewatering, which increases the loss of myofibrillar proteins in the form of particulates, are primarily responsible for losses in functional myofibrillar proteins during surimi processing.

Wash water could be recycled by the principle of countercurrent washing (Figure 8) as proposed by Green et al. (1985). Theoretically, recycling of the wash water, except for water from the first wash cycle, could reduce water usage by two-thirds. Water from the first wash cycle (initial contact of meat with water) must be discarded due to its higher level of undesirable impurities. Recycling would reduce the cost of waste management and also save in energy costs related to fresh water production and its refrigeration. Thus far, however, this approach has been neglected by surimi manufacturers. With increasingly stringent requirements by environmental agencies, this process needs to be thoroughly investigated on a commercial scale for economic feasibility.

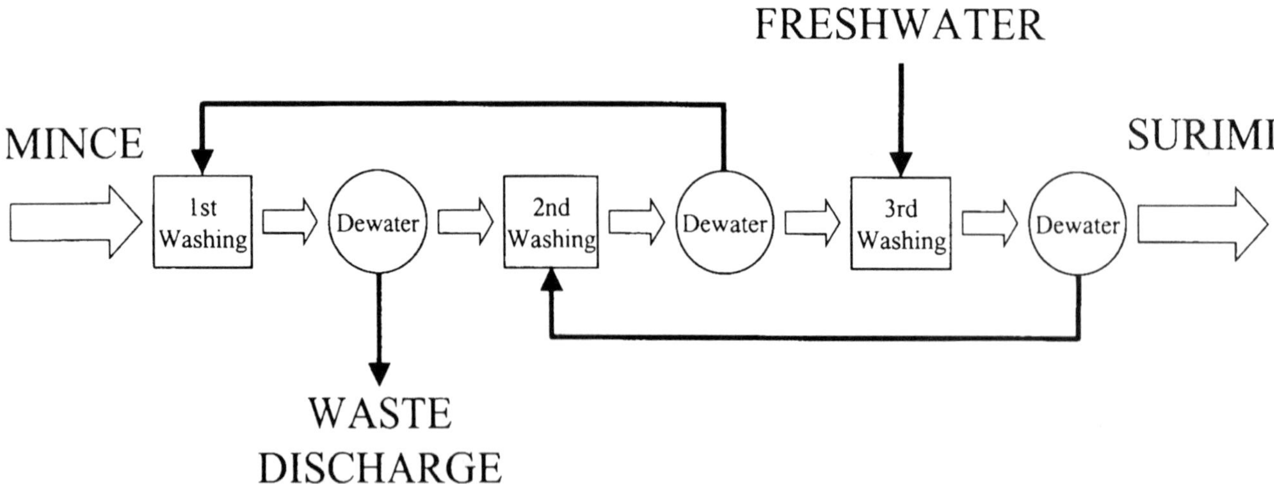

Figure 8. Water recycling with countercurrent flow. Adapted from Lee (1986).

In the past decade, innovative processing technologies have been applied to surimi processing, enhancing the efficiency of the process and improving product quality and consistency. Decanter centrifuges are now commonly used to recover the fine particles formerly lost through the dewatering screens and screw press (Lanier et al., 1992; Babbitt, 1990; Swafford et al., 1990). The recovered meat is either returned to a main processing line or marketed alone as a lower-grade surimi. More recently, Lin et al. (1995) successfully employed microfiltration using a perforated stainless steel screening device as a low-cost alternative to centrifugation for recovery of particulates (Figure 6). Surimi containing 10 percent of meat recovered using microfiltration had the same gel quality as regular surimi with respect to gel hardness, elasticity, water retention, and color. Such recovery increased the yield of the process by 1.7 percent (based on finished product). As a means of recovering soluble protein from leachwater, and to purify this water for possible reuse in the process, Lin et al. (1995) successfully employed ultrafiltration to reduce the aerobic plate count, chemical oxygen demand value, turbidity, and protease levels (Figure 6). However, soluble proteins recovered and concentrated by ultrafiltration had a dark color and strong odor, rendering them usable only as an animal feed additive.

A NEW PRODUCT FROM AN OLD CONCEPT: FRESH (UNFROZEN) SURIMI

All washed mince used in Japanese kamaboko production was fresh (unfrozen) prior to the discovery of cryoprotectants in 1960 (Okada, 1992). In Japan and Korea, where shore-side processing plants are located near kamaboko manufacturers, a certain amount of fresh surimi is still commercially produced on a small scale using locally available fish species. The geographic distances separating Alaskan surimi production sites and the manufacturers of surimi seafoods precluded use of fresh surimi until the establishment of shore-side surimi production on the Oregon coast. Pipatsattayanuwong et al. (1995) hypothesized several benefits of using fresh surimi over the frozen form: fresh surimi could be produced at lower cost without additives or freezing; it could be used to develop no- or low-sugar products; it might add market appeal to specialty "never-frozen" surimi seafood; and it may exhibit better gelling functionality than any frozen form, enabling it to be used at a reduced level. The maximum shelf life of fresh Pacific whiting surimi was subsequently shown to be five days with respect to microbiological and functional properties. When a 9 percent cryoprotectant (a 1:1 mixture of sucrose and sorbitol) was incorporated in frozen and fresh surimi, shear stress values of fresh surimi were almost three times higher than that of the frozen form, although shear strain values of fresh surimi were not as high as those of frozen surimi.

SURIMI PROCESSING FROM STABILIZED MINCE

MacDonald et al. (1990), followed by Simpson et al. (1994), demonstrated that high quality surimi could be produced from frozen unwashed mince of New Zealand hoki and Pacific whiting, respectively, to which cryoprotectants had been added prior to freezing. Such an approach could extend shore-side surimi manufacturing operations beyond the normal harvest season, and/or enable smaller vessels to process fresh fish at sea without the requirement for water generation and a surimi processing line on-board. Surimi made from unwashed mince containing 12 percent sucrose and 0.2 percent sodium polyphosphate after six months storage at -20°C was equivalent in quality to surimi processed initially from fresh mince and held for the same period in frozen storage (Simpson et al., 1994). An economic analysis needs to be conducted to determine the overall feasibility of this approach, which certainly has several technical advantages.

SURIMI SEAFOOD PROCESSING

Processing Technology and Sequence

Receiving and Storage of Raw Materials

Surimi is the major ingredient in manufacturing surimi-based seafoods, which include the many varieties of kamaboko products in Japan (steamed, baked, puffed, fried, etc.) and the popular shellfish substitutes (crab, scallop, shrimp, etc.) in Western countries such as the United States. The functional and compositional properties of the surimi used thus largely influence the quality of the finished products.

Upon receiving, the temperature of randomly selected surimi blocks must be measured to reveal any temperature abuse that may have occurred during transportation. Prior to use, the functional and compositional properties must be determined to permit efficient blending of surimi lots in order to control the quality and consistency of finished products.

In addition to surimi, other raw materials should also be inspected carefully upon receiving and properly stored. These include dry starches, protein additives, chemical additives (salt, phosphates, etc.), flavorings, and colorings. Most of these can be stored at room temperature, with the exception of frozen or refrigerated egg whites and volatile flavorings.

Thawing or Size Reduction

Before frozen blocks of surimi are to be comminuted, they must be either partially thawed or broken into smaller pieces using a frozen meat breaker or a hydrauflaker. This step reduces the load on the equipment during the subsequent fine comminution step. Thawing is typically accelerated by placing frozen blocks between warmed plates. Special attention must be given to avoid over-thawing, which may induce protein denaturation. If a frozen meat breaker is used, care must be taken to remove imbedded plastic film from the bag which encased the block.

Comminution and Ingredient Blending

The sequence of incorporation of ingredients during comminution and mixing affects the textural quality of the final product and varies with the type of equipment used. Two major functions of chopping are, first, to solubilize the myofibrillar (salt soluble) proteins as much as possible to yield a smooth-textured paste and, second, to incorporate filler and other ingredients into the surimi paste uniformly. The maximum solubilization of available myofibrillar proteins is dependent on time, temperature, and other mechanical functions of silent cutters such as blade configuration, sharpness, vacuum, etc. Open (non-vacuum) bowl ("silent") cutters have been commonly used in the surimi seafood industry. A 300-liter bowl can accommodate up to 650 pounds of finished paste. Approximately 25 minutes chopping time is normally required with open bowl cutters, as compared to only 15 minutes using a vacuum silent cutter. Thus there has been a trend within the industry to adopt the more efficient vacuum-cutting equipment. A 500-liter bowl unit is commonly used, which can hold 900 pounds of paste.

Comminution first reduces the particle size of the meat almost to a powder. In this first stage of comminution, salt is added and chopping proceeds to facilitate solubilization of myofibrillar proteins. In the second stage of comminution the remaining ingredients are added and blended.

With non-vacuum equipment, it is very important that dry ingredients be carefully sprinkled onto the surimi paste, and that starch be premixed with water before addition. This is not required for vacuum equipment, into which all ingredients may be simultaneously added and blended in a short period of time.

The final paste temperature at the completion of chopping is generally near 10°C in industrial practice. Lee (1984) suggested that the temperature of paste must be kept at or below the temperature above which fish actomyosin becomes unstable — i.e., about 10°C for Alaska pollock and 16°C for Spanish mackerel. Park (1994c) compared the effects of temperature during chopping on shear stress and strain values of Alaskan pollock surimi gels (Figure 9). During this experiment, the temperature of the paste was controlled by circulating a mixture of ethylene glycol and water through the double-walled bowl of the chopping equipment. The gelling functionality of surimi paste chopped at 20°C was extremely low. Maintaining chop temperatures between 0° and 5°C appeared to yield optimum gelling functionality.

Extrusion or Molding

During the forming step of a surimi seafood, the blended surimi-based paste is extruded, using a meat transfer unit commonly equipped with a diaphragm pump or a stuffing machine, onto a conveyor belt of the cooking machine. The size and shape of the extrusion nozzle depends on the type of finished product. Crab and scallop products which exhibit aligned fibers in the finished product begin as a very thin (1.5 to 2.5 millimeters) sheet of paste extruded on a stainless steel or teflon belt either 80 or 100 centimeters in width.

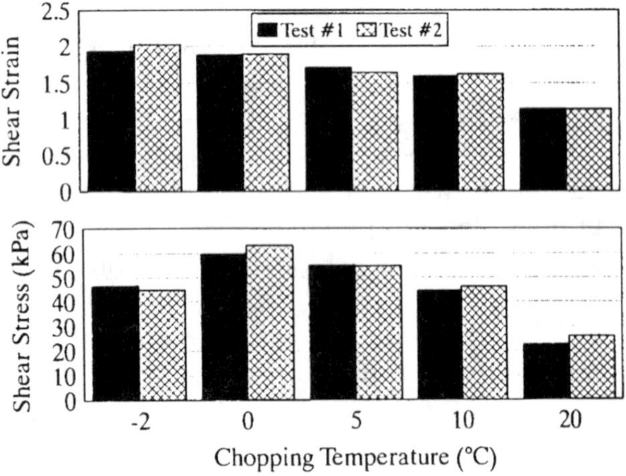

Figure 9. Effects of chopping temperatures on gel functionality of surimi.

Alternatively, paste can be formed in a molding machine or cold-extruded in three dimensional shapes, such as in the manufacture of shrimp- or lobster-shaped surimi seafoods. Shredded precooked product, prepared from cooked blocks of paste, is often blended with fresh paste prior to extrusion or molding to enhance the textural appeal of the finished product (random-fiber formed product).

Cooking and Cooling

Most western-style (shellfish substitutes) surimi-based seafoods are manufactured on machinery produced by one of four major companies. These are IKM, Yanagiya, Bibun, and Tono, all from Japan. There are slight differences among these machines, but the basic process steps are similar.

The type and sequencing of heating elements can vary. Most commonly the sequence is radiant heat, followed by steam heat, and then radiant heat again. Another common option consists of steam heat followed by radiant heat. Some machines employ steam heat only. The ambient temperature in the tunnel where the paste is exposed to radiant or steam heat is near 90°C. Total cooking time is dependent on the product specification, generally varying from 50 to 100 seconds for the thin sheets required in making the aligned-fiber type crab products. This short cooking process induces gelation of the surimi proteins, but is not sufficient to gelatinize (swell) the starch filling the paste. This occurs during a later pasteurization step, adding rigidity and water-holding properties to the gel.

Immediately upon completion of this first cooking step the product is cooled by air at room temperature or below. The cooling system employed is very different among manufacturers.

Fiberization

For shellfish products designed with aligned fibers, fiberization is accomplished by passing through two rollers with teeth, which make elongated cuts running lengthwise on the gelled sheet. The space between the teeth controls the number and width of individual fibers. A water mist or vegetable oil drip is commonly used to facilitate smooth passage of the sheet through the cutting rollers.

For random fiber products, a cooked (gelled) block of paste is shredded into short fibers prior to blending with fresh paste and is then extruded or molded.

Bundling, Color Application, and Wrapping

Bundling is a process for rolling the cooked product sheets tightly (aligned fiber products) into a rope shape. This rope is then passed through a color application machine. Polyethylene wrapping plastic, onto which colored fresh paste is applied, wraps around the product rope, which is then cut to a specified length. This colored paste is set to a gel by cooking under steam for 15 to 30 minutes, followed by rapid chilling.

Alternatively, the colored paste may be co-extruded at the edge of the sheet of uncolored surimi paste prior to the initial cooking step. This obviates the need for an intermediate wrapping step and subsequent unwrapping. Color application via a plastic wrapping step often causes a problem of color flaking from the finished product, whereas co-extruded color may sometimes bleed or transfer into the white portion of the product.

Extruded or molded products are colored by spraying the mold or formed product with a colorant solution.

Cutting and Packaging

Aligned-fiber products may be cut to different dimensions while in the rope shape. The most popular form of crab product in the United States is termed the flake cut. The rope is cut diagonally at a 25° to 30° angle in 5 to 8 centimeters long pieces, tip to tip. Another popular shape is the leg or stick shape. This is made by cutting straight across the rope into pieces 7.5 or 12.5 cm long.

Most surimi-based seafoods are packed in plastic films under full or partial vacuum, to enhance appearance of the product and extend shelf life.

Pasteurization and Rapid Chilling

The pasteurization step assures the microbiological quality of products, assuming sanitary ingredients and packaging materials have been used. Under the United States Department of Commerce (USDC) guidelines for pasteurization of PUFI (Packed Under Federal Inspection) sealed surimi seafoods, it is stated that vacuum-packed products must be heated at 85°C (internal temperature) for 20 minutes, followed by rapid chilling. The chilling step should bring the product temperature to 4°C within 30 minutes. This strict guideline centers on the potential for outgrowth and toxin production by the microorganism *Clostridium botulinum* which can occur under anaerobic conditions of vacuum packaging at temperatures as low as 3.3°C (Comar, 1987). In addition, a water-phase salt level of 2.4 percent in the formulation is recommended by Eklund (1993) to further prevent occurrence of this toxin.

Park (1993a) surveyed the pasteurization practices within the United States surimi seafood industry. Surprisingly, there was quite a variation in terms of the time and temperature of pasteurization within the industry, presumably with widely varying effects on the eating quality, and possibly the safety, of individual products. The United States industry would be wise to standardize the pasteurization process, based on good studies to determine the minimum process needed to assure safety and thereby maximize product quality.

Rapid Freezing and Storage

Refrigerated products are packed into cartons after chilling individually packaged units to 4°C. Most refrigerated (vacuum packaged) products for normal supermarket distribution have a shelf life of 60 to 90 days. For longer distribution times, the frozen form is recommended. Chilled product is typically frozen in a blast freezer or in a liquid nitrogen tunnel. To reach an internal temperature of -18°C for a 2½ pound package, usually 60 minutes in a blast freezer or 20 minutes by liquid nitrogen is required. Freezing rate affects product quality and shelf life. At customary freezer temperatures (-10° to -20°C), about 90 percent of the moisture freezes out, resulting in approximately a tenfold increase in the concentration of solutes in the remaining free liquid (Powrie, 1973). Fast freezing, which generates numerous small ice crystals rather than fewer large ice crystals, is very important to maintain the texture of products during long-term storage. Low fluctuation of storage temperature is also helpful in this regard. Often there must be formulation adjustments made, which can vary depending upon the freezing systems; these lessen the damaging effects of freezing on product quality. For example, the use of modified starches, especially those prepared by hydroxypropylati1on and cross-linking, instead of native starches can reduce the weepage (moisture release) that often accompanies the thawing of frozen products.

FUNCTIONS OF INGREDIENTS

Three major functional attributes of surimi seafoods are color, flavor, and texture. The former two can be controlled relatively easily by simple addition of ingredients, while the latter can be more difficult to control. The main ingredients affecting the development or modification of texture in surimi seafoods are the surimi, water, starch, protein additives, and hydrocolloids. The range of textural properties which are potentially obtainable by the addition of different ingredients are represented in the "texture map" of Figure 10, developed originally by Lanier (1986) to illustrate the relation between measureable physical parameters and human sensory characteristics. In the mouth one is able to perceive the two textural parameters of hardness (stress) and deformability (strain) of a product, and the relationship between the two ultimately determines the textural description. For example, a gel of high hardness but low deformability produces a "brittle" sensation in the mouth. Conversely, highly deformable gels of low hardness would be perceived as being "rubbery." For surimi-based crabmeat in which a relative balance exists between the textural hardness and deformability, the overall magnitude of the two textural parameters places the textural description on a continuum moving from a perception of "mushy" at low values upwards to one of "toughness" (Lanier et al., 1985). A desirable surimi seafood would of course have a texture that would be described in terms intermediate to these extremes: neither too "mushy," "rubbery," "brittle" or "tough."

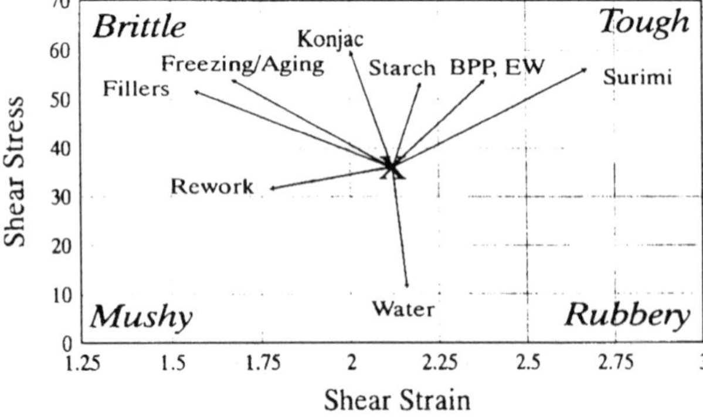

Figure 10. Texture map, illustrating effects of ingredient additions and storage on texture of a typical surimi gel. BPP: beef plasma protein; EW: egg white. Adapted from Park (1994a).

An understanding of why each of the common ingredients of surimi seafoods has its particular effect on surimi gel texture (illustrated in Figure 10) can be helpful in the proper choice of ingredients to modify texture.

SURIMI

Surimi, despite its relatively higher price per pound of dry protein, offers significantly greater functionality in terms of its gelling properties than competing proteins. It is recognized that the ability of the surimi seafood to bind water, fat, and particles in a matrix having a desirable texture are linked to the formation of a stable gel network (Lanier, 1986; Acton et al., 1983). High-quality surimi is able to form very deformable gels over a wide range of shear stresses (hardness). Preincubation (low temperature holding prior to cooking) greatly increases the strength (shear stress) and, to a somewhat lesser degree, the cohesive or deformable nature (shear strain) of gels. Effects of preincubation on gel properties are very much dependent on species. Kim et al. (1986) found that surimi prepared from Atlantic croaker (*Micropogan undulatus*) and Alaskan pollock, in which the gel texture of both responded similarly to a 40°C preincubation, responded differently to a 4°C overnight preincubation. The strongest pollock surimi gels in this study were obtained with a 4°C presetting for 24 hours, while the strongest croaker surimi gels were formed by 40°C pre-setting for 30 minutes before final heating at 90°C for 15 minutes; croaker did not show any response to setting at 4°C. Kamath et al. (1992) reported that maximum production of cross-linked proteins occurred at the optimum setting temperatures — i.e., at 25°C for Alaskan pollock surimi and 40°C for Atlantic croaker. The optimum gelling abilities of Alaskan pollock, Southern blue whiting, and New Zealand hoki surimi gels were all obtained by presetting gels at 25°C for about four hours (MacDonald et al., 1994). Park et al. (1994) similarly found that the optimum temperature of presetting for Pacific whiting surimi is 25°C. The optimum temperature for pre-setting effects seems mainly affected by the habitat temperature of the fish (Park et al.,1994; Misima et al., 1993; Kim, 1987; Arai, 1973). This is because the proteins adapt in their thermal stability to the body temperature of the animal: cold-water fishes have proteins that denature and react with an inherent crosslinking enzyme (transglutaminase) more readily at cold temperature (Joseph et al., 1994).

Since surimi forms the base protein matrix in which all other ingredients are embedded as a filler or interact with as a binder, the surimi blend used largely establishes the color, flavor, and texture of surimi-based crabmeat. There have been efforts to use computer-assisted (linear programming-based) least cost formulation as an effective means of selecting a blend of surimi (from different grades or lots) which will simultaneously maintain predetermined quality parameters in the finished food while minimizing overall ingredient costs (Park, 1993b; Lanier and Park, 1990). Figure 11 demonstrates a practical application of blended surimi for the production of consistent quality products. In formula optimization, ingredients are the independent variables. The dependent variables are always the entity to be optimized (i.e., maximized or minimized): these include functional properties, such as shear stress, shear strain, and whiteness; compositional properties, such as moisture and protein content, pH, visual impurities; and, especially, cost.

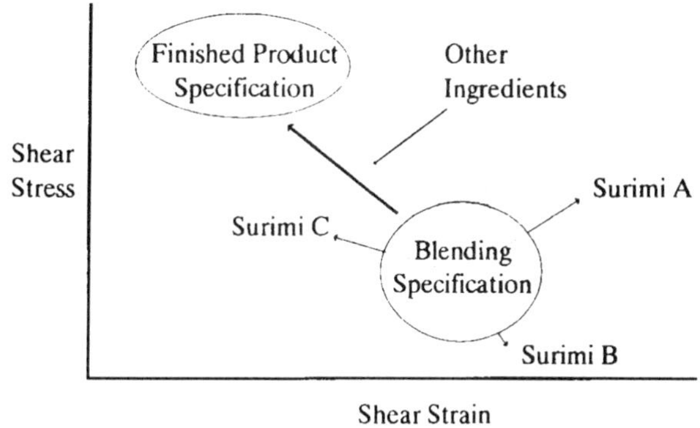

Figure 11. Control of finished product gel property specification by blending lots of varying gel properties.

WATER

Water is generally the second most prevalent ingredient in formulations for surimi seafoods. The quality of water used affects the color, texture, and flavor characteristics of finished products. Lee (1990) emphasized the importance of the mineral content of water. Calcium (Ca^{++}) and magnesium (Mg^{++}) affect texture development (gelling properties), while iron (Fe^{++}) and manganese (Mn^{++}) may cause color changes. Okada (1981) hypothesized that the superior quality of kamaboko produced in Odawara, Japan may be linked to the water quality (calcium content of 20 to 40 milligrams/100 grams).

As the water content of a formulation is increased, the general effect is a reduction in shear stress (gel hardness), primarily due to the reduction in protein concentration (Figure 10). Changes of a few percent in water content have little effect on shear strain (gel cohesiveness) (Lanier et al., 1985). Addition of water to surimi also affects the color of gels (Figure 12). Lightness (L* value) increases and yellow hue (b* value) decreases as the content of water increases (Park, 1995). A difference in moisture content of 7.5 percent induced an increase in L* value of about three units and a decrease in b* value of about two units in both pollock and whiting surimi gels. Water content

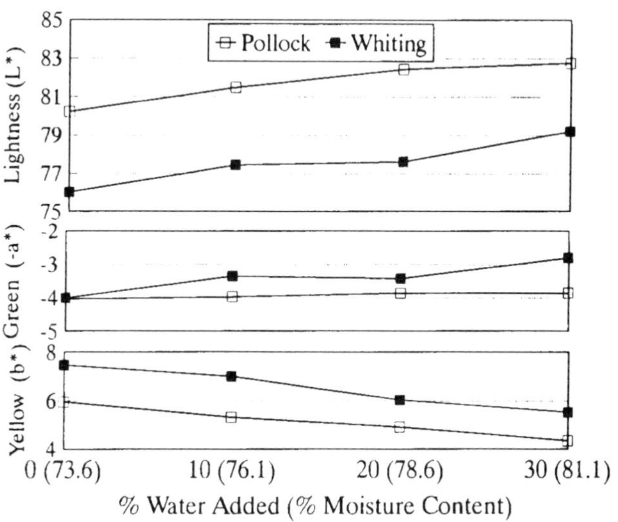

Figure 12. Effects of moisture content on surimi gel color parameters. Adapted from Park (1995).

effects on redness (the a* value) were not pronounced.

Starch

Starch is an indispensable ingredient in most surimi seafood formulations for both cost reduction and texture modification. The texture, water-holding properties, and color of surimi gels is profoundly affected by the type of starch(es) used. The most commonly used native (not chemically or physically modified) starches are wheat, corn, potato, waxy maize, and tapioca. For modified starches, the nature of the modification, such as crosslinking, hydroxypropylation, and acetylation, also determines the functional properties of the starch. Starch is commonly used either to maintain gel strength with the reduction of surimi content (cost reduction), to enhance the freeze-thaw stability of frozen products, or to modify the texture of the surimi gel. The functional properties of starches are quite different, as shown in Table 1. High amylose (linear polymer) starches such as corn, wheat, and potato form somewhat brittle gels, while high amylopectin (branced polymer) starches such as tapioca and waxy maize form adhesive and cohesive gels. High amylose starches modified by hydroxypropylation and crosslinking also yield adhesive and cohesive gels. While the latter have better freeze-thaw stability in frozen surimi seafood applications, they can impart an undesirable stickiness to the product at higher levels of usage. Because of its large granule size, potato starch has the highest swelling (and water-holding) capability, followed by tapioca, waxy maize, corn, and wheat. Potato starch also initiates swelling at lower temperature than other native starches, so that its swelling is more complete when cooking time is brief. Colors of high amylose starch gels are opaque, while high amylopectin starch gels are translucent. Crosslinking a starch makes its gels more translucent or transparent.

Most commonly, a 4 to 12 percent level of combined starches is used for surimi-based seafoods. However, this level of starch imbibes a great deal of water by swelling during cooking, such that hydrated starch constitutes a large fraction of the gel volume. Understanding the chemical and physical properties of starches is, therefore, very important to proper formulation of surimi-based seafoods. It is not uncommon to see very sticky and soft-textured products as a result of the replacement of surimi by cheap starches, without due regard to optimizing their interaction with other components of the formulation. Oversubstitution of surimi by starch can also result in a starchy taste and a sticky mouth feel in the product.

During refrigerated storage the expressible moisture and the hardness of surimi gels increases when starches of higher amylose content are added. This has been attributed to retrogradation

Table 1. Compositional and Functional Properties of Native Starch

	Wheat	Corn	Potato	Waxy Maize	Tapioca
Protein, %	0.4	0.6	0.06	0.15	0.1
Fat, %	0.8	0.44	0.05	0.15	0.1
Ash, %	0.2	0.1	0.4	1.1	0.2
Phosphorus, %	0.06	0.02	0.095	0.01	0.02
Amylose, %	25	26	25	1	17
Average Size, υm	30	15	40	15	20
Gelatinizing Temp, °C	77	67	61	72	65
Peak Viscosity, BU	80(6%)	250(6%)	800(3%)	600(6%)	600(6%)
Solubility (95°C), %	41	25	82	23	48
Swelling, g water/g	21	24	>1000	64	71
Paste Clarity	Opaque	Opaque	Opaque	Translucent	Translucent
Paste Texture	Short	Short	Long	Short	Long
Paste Flavor	Cereal	Cereal	Mild	Cereal	Mild

of the gelatinized starch (Kim and Lee, 1987). Retrogradation is the association of the starch molecules into a rigid structure that excludes water. During frozen storage, native high amylose starches undergo severe retrogradation, which results in gels that more freely express water. This loss of water-holding ability is also accompanied by an increased brittleness of texture (Yoon and Park, 1995). Hydroxypropylated and crosslinked waxy maize starch (1.0 to 2.0 percent) is popularly used to overcome severe expressible moisture (dripping) problems upon thawing of cooked surimi seafoods. Acetylated potato starch is also used to extend the shelf life of frozen products. However, this starch is not as effective in controlling water loss from thawed product.

Starch also plays an important role in the textural behavior of surimi gels when they are reheated (Lee, 1986). The rubberiness of a heated surimi-based product can be somewhat reduced by increasing the level of starches, especially by the addition of hydroxypropylated and crosslinked waxy maize starch.

Wu et al. (1985) demonstrated that higher salt and, to a lesser extent, higher sucrose concentration will delay starch gelatinization (swelling) and affect the rigidity modulus (firmness). Starch gelatinization is also affected by the availability of water. Yoon and Park (1995) reported that shear strain values of surimi gels with 4 percent starch [a mixture (1:1) of wheat and modified waxy maize] continued to increase as water increased from 30 to 45 percent. However, when 8 or 12 percent starch was incorporated, shear strain values increased and maximized at 35 percent and then decreased with more than 40 percent water added. Other factors affecting gelatinization of starch are inadequate gelling time and/or temperature. This certainly results in some granules being unswollen (not gelatinized) and can affect the flavor, color, and texture of the final product.

Thus the optimum level of starch for a formulation must be decided based upon its effects on texture, mouth feel, color, flavor, stability, and cost of the product.

Protein Additives

Protein additives serve to contribute either a functional, nutritional, and/or economic benefit. With respect to functional properties, protein-water, protein-protein, and protein-lipid-water identities are very important for the formulation of the stable gel network structure (Regenstein, 1984). Park (1994a) evaluated seven commercially available protein additives to investigate their interaction with surimi gels in the presence of 2 percent salt. Frozen egg white, dried egg white, and beef plasma protein acted as very functional binders, while wheat gluten, soy protein isolate, whey protein concentrate, and whey protein isolate were functional fillers. The functional binders increased both shear stress and shear strain values of surimi gels, while those designated as functional fillers increased shear stress but decreased shear strain values of gels (Figure 10). However, more reactive whey proteins have now been developed which enhance shear strain (Lanier and Thomsen, 1993).

All of the protein additives tested had an impact on the color of surimi gels, most causing slight reduction of L* value (lightness) and a greater increase of b* value (yellowish hue). The overall whiteness index, L*-3b*, is the best indicator of the overall effect on surimi color by protein additives (Park, 1994a).

Lanier (1991) reviewed the interactions of muscle and non-muscle protein additives (non-muscle proteins) which affect heat-set gel rheology. Gel matrix development by myofibrillar proteins can be directly influenced by chemical interactions between the protein additives and myofibrillar proteins, as well as indirectly by changes in the molecular environment (contribution to total protein concentration, water state and availability, ionic strength and types, pH) brought on by the presence of protein additives. Added proteins can act as active or passive fillers. Active fillers not only fill the interstitial spaces of the surimi gel and absorb water (such as starch and other passive fillers), but also chemically interact with the surimi proteins.

Unlike starch, protein additives have the potential to provide nutritional fortification of surimi-based seafoods. One of the main reasons that surimi seafoods must be labeled "imitation" is their lower protein content as compared to the natural shellfish meats they simulate. The protein contents of common protein additives (on a dry basis) vary from about 60 to 95 percent.

Regardless of what benefits a protein additive might impart, the overriding factor is economic. If a benefit can be obtained with no adverse effects in using the protein for no additional cost, then that addition becomes attractive. Typically there is great interest in using protein additives when surimi prices are high.

Hydrocolloids

Carrageenan and konjac are typical hydrocolloids used as ingredients in surimi seafoods. Carrageenan is a gelling agent extracted from certain species of red algae. It is widely used in the food industry for its unique stabilizing and texturizing effects, and also as a fat replacement. Three common types of carrageenan are commercially available: kappa carrageenan extracted from *Eucheuma cottonii*, iota carrageenan from *Eucheuma spinosum*, and lambda carrageenan from *Gigartina acicularis*. The extracts are primarily sulfated polysaccharides of varying ester contents. Their molecular weights generally range between 100,000 to 500,000. To compensate for the poor solubility of carrageenans, they should be dispersed into water with agitation. Heating (82° to 85°C) assures full solubility of the polysaccharide. (For a further discussion of carrageenans, see the chapter on Red Algae.)

Kappa and iota carrageenan form gels because of their lower sulfate content (lower surface charge, which inhibits polymer interaction), while lambda carrageenan is used for thickening and suspending applications. Kappa carrageenan forms hard, brittle gels while iota carrageenan gels are softer yet more deformable. The strongest gels of kappa carrageenan can be formed by addition of potassium ions, while the most elastic gels of iota carrageenan are formed with calcium ions. Kappa carrageenan gels have poor freeze/thaw stability, while iota carrageenan gels are more stable. Iota carrageenan provides the best functionality in surimi seafood applications, considerably improving the gelling potential and freeze/thaw stability of Atlantic pollock (*Pollachius virens*) surimi gels (Llanto et al., 1990), as well as of gels prepared from Alaskan pollock and red hake (*Urophycis chuss*) surimi (Bullens et al., 1990). Da Ponte et al. (1985) also reported that iota and lambda (but not kappa) carrageenans improved the water retention capacity of raw minced cod, and decreased toughening during frozen storage. Iota carrageenan also exhibits good synergy with starch in muscle protein gels (Tye, 1988).

Konjac flour is primarily a glucomannan polysaccharide, and is very water soluble. It can not form a gel alone, even if heated and cooled to room temperature. The polysaccharide of konjac flour contains acetyl groups which inhibit the molecules from forming a gel network, much as the sulfate groups of lambda carrageenan inhibit its gelation. However, introducing a mild alkali to a konjac solution cleaves many of the acetyl groups from the polysaccharide chain, resulting in a strong, elastic gel which retains its structure under various heating conditions, such as boiling water, retorting, and microwaving (Tye, 1991; Ohta and Maekaji, 1980). The traditional Asian method of preparing thermally stable konjac noodles requires that the konjac paste be soaked in mild lime water. Wu et al. (1991) patented the use of konjac to simulate the strong, resilient texture of shellfish meats using rapid freezing as a gelling method. Park (1996) investigated changes of gel shear stress and strain values upon cooling, reheating, and freezing/thawing of surimi-konjac gels at various test temperatures. Konjac at 5 percent addition reinforced gel hardness by eight to 10 times in both whiting and pollock surimi gels (Figure 13). Surimi gels with a minimum of 2 to 3 percent konjac were preferred when reheated at 55° and 75°C. These gels also exhibited an ability to maintain consistent shear strain values in gels despite freeze/thaw treatment. Up to 2 percent konjac increased lightness of surimi gels, gradually increasing yellow hue at higher levels.

Vegetable Oils

Soybean and canola oil are commonly used in surimi seafoods as a texture modifier, color enhancer, or processing aid. Lee and Abdollahi (1981) suggested that the addition of fat modifies the texture of fish protein gels in the following manner: it prevents a sponge-like texture development during extended frozen storage; it reduces rubberiness; and it minimizes textural variations resulting from cooking. Vegetable oil addition also makes products appear whiter through a light scattering effect by the emulsion created when it is comminuted with surimi proteins and water. Vegetable oil is often used as a processing aid on

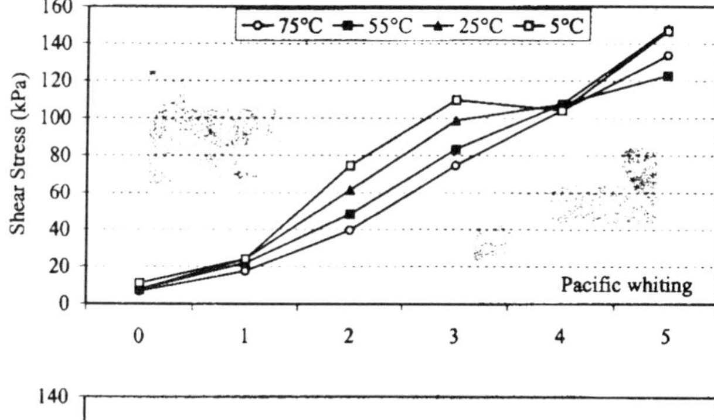

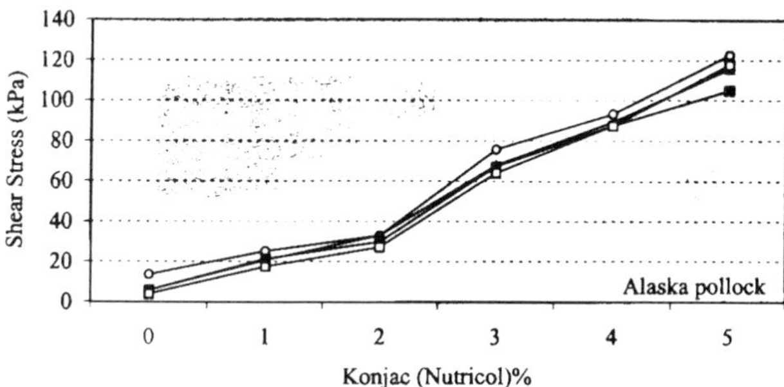

Figure 13. Effects of konjac addition on shear stress values of surimi gels. Temperatures indicated are those of the sample at testing. Nutricol is a commercial brand of konjac from FMC Corp. (Princeton, NJ). Adapted from Park (1996).

the cooking machine, especially when the paste becomes sticky due to increased starch contents.

Gel-Enhancing Agents

Oxidizing agents are commonly used in bread dough to improve textural properties by the formation of disulfide (S-S) bonds through oxidation of sulfhydryl (-SH) groups. A similar effect was proposed in the gelation of surimi (Lee et al., 1992; Nishimura et al. 1990; Yoshinaka et al., 1972). The use of L-ascorbic acid, sodium ascorbic acid, and erythorbic acid have been suggested for gel-strengthening by oxidation in thermally-induced surimi gels. Lee et al. (1992) found that sodium-L-ascorbate (SA) significantly improved the compressive force and sensory firmness of molded and fiberized products with the maximum effect at a 0.2 percent level. However, SA promoted freeze-induced syneresis (weepage) and brittle texture development. The gel-enhancing effect of ascorbate in surimi produced from the spawning season was not as high as in the other surimi. This would suggest that the effects of ascorbate are related to the quality of the surimi to which it is added. Pacheco-Aguilar and Crawford (1994) measured an increased oxidation of sulfhydryl groups in myofibrillar proteins induced by potassium bromate. Though used for a similar purpose in bread flours, potassium bromate has not been approved as an additive in surimi products.

Calcium compounds are also commonly used as gelation enhancers, especially for Pacific whiting surimi. Spurred by the introduction of a patent for the use of a mixture of sodium bicarbonate, calcium citrate, and calcium lactate as quality improving agents for surimi (Yamamoto et al., 1991), 0.1 to 0.2 percent of calcium-containing compounds such as calcium sulfate, calcium caseinate, calcium citrate, and calcium lactate are now used alone or in combination to improve gel strength. Lee and Park (1997) studied the action of various calcium compounds in surimi gelation, finding that the primary effect is to stimulate the calcium-dependent activity of the transglutaminase enzyme in surimi that is responsible for the setting (*suwari*) of salted surimi pastes at low temperatures (0° to 40°C).

Transglutaminase is a natural constituent of surimi from most species, although its activity and optimum temperature evidently varies by species. The low temperature setting (suwari) reaction of surimi that builds shear stress in the gel is probably the result of the action of this enzyme, which permanently crosslinks the proteins (Joseph et al., 1994). Because the natural transglutaminase is calcium dependent for activity, additives that affect the level of free calcium in surimi will consequently affect the setting rate. These include not only calcium compounds that stimulate transglutaminase activity, but also phosphates and EDTA or EGTA, compounds that chelate calcium and thus reduce transglutaminase activity.

More recently a microbial source of transglutaminase has been commercially developed in Japan, soon to be marketed in the United States This enzyme is similar in activity to the natural enzyme except that it is calcium insensitive. While it is somewhat reactive even during fast cooking (but likely not at the speed of cooking used in the manufacture of string-type crab substitute) it has its greatest effect when a lower temperature setting step is used (Figure 14). This is because the rate of the reaction is more dependent upon the species of the fish muscle which

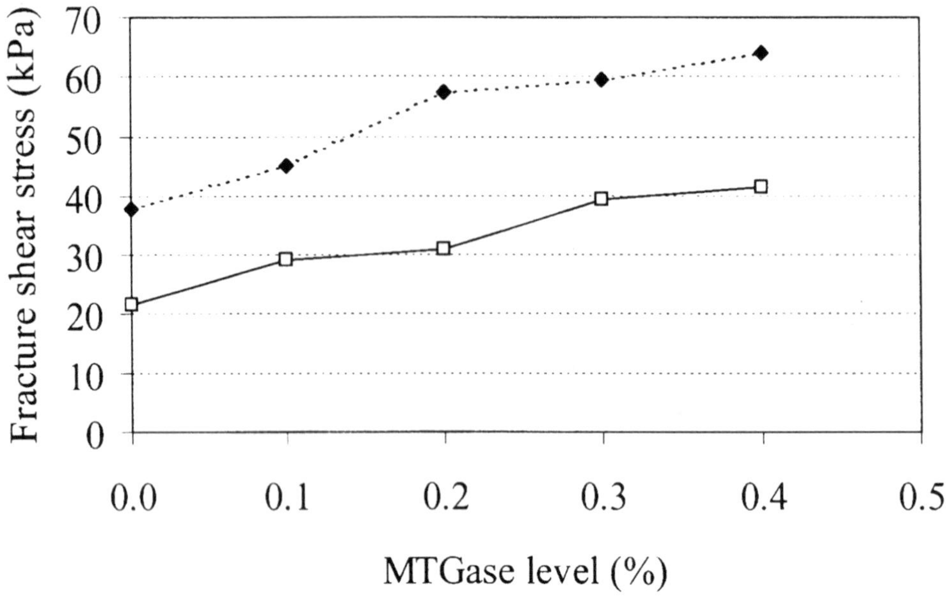

Figure 14. Shear stress at failure of gels prepared with varying levels of microbial transglutaminase (MTGase) in Alaskan pollock surimi.

the enzyme acts upon rather than the source of the enzyme (Joseph et al., 1994). A European company reportedly will market a calcium-sensitive microbial transglutaminase in the future.

Flavorings

Flavor creation requires a combination of highly sophisticated analytical chemistry and the creative artistry of highly trained flavorists. Seafood flavors, like other savory flavors such as beef, chicken, and cheese, are very complex chemically. While a good fruit flavor can be made with only 15 to 20 ingredients, savory flavors are much more complex in nature. When seafood is harvested, oxygenation of the highly polyunsaturated fatty acids is begun by enzymatic reaction of lipoxygenase yielding fatty acid hydroperoxides, which are further broken down to alcohols by lipase enzymes. Oxidase enzymes then act to convert the alcohol to volatile flavor compounds. While the lipids play a major role in the formation of a seafood flavor, other precursors and types of flavor compounds such as some sulfur compounds, isoprenoid-related compounds, amines and hydrocarbons also play significant roles. This complex mixture may include over 140 different chemical compounds (Gordon, 1994).

There are two key elements in seafood flavors: 1) base notes or body, and 2) volatile notes or aroma. One element without the other will not result in a complete flavor profile. In the manufacture of surimi-based crab analog product it is common to combine a natural crab extract, natural/artificial seasonings, and artificial aroma compound. Other flavor systems that simulate shrimp, lobster, and scallop are also available. Because of the multi-stage heat process employed for many surimi seafoods, the volatile notes are easily lost during cooking and particularly during pasteurization. Thus it is important to use seafood flavors developed with consideration of this heating process in mind.

Colorings

The current Nutrition Labeling and Education Act requires a food manufacturer to identify each certified colorant individually in the ingredient statement on the package. However, natural colorants are exempted and can be grouped generically on the label under the term "colorants." To be considered "natural," the colorants must: 1) be from agricultural/biological sources; 2) be extracted without chemical reaction; and 3) have a long history of use. The colorants meeting this definition must be chosen from those permitted for use in foods, as listed in 21 Code of Federal Register (CFR) 73, Color Exempt from Certification, Part A (FDA, 1992; Lauro, 1991).

There are two categories of color application in manufacturing surimi-based crab analog products: 1) applied surface colorants, normally of red to orange hue; and 2) whiteners admixed with the bulk of the paste during manufacture. Pigments commonly used for surface application include carmine, paprika oleoresin, caramel, and canthaxanthin. All of these are natural pigments, with the exception that canthaxanthin is actually a synthetic but "nature-identical" colorant and therefore does not meet the CFR definition for a natural colorant. There is also some doubt that caramel is a natural color, because it is manufactured in its simplest sense by the heating of sugar solution and thus is a reaction end product. Some pigments are approved as being "natural" by certain countries but not by other countries. Monascus is not listed as an approved natural color in the United States even though it meets the CFR definition, while carmine is not allowed in Japan.

Calcium carbonate and titanium dioxide are the most common whitening agents added to surimi seafoods. These are not soluble, and must be uniformly distributed during comminution. As a result of their insolubility, they impart a somewhat chalky or opaque whiteness which is less like the appearance of real shellfish meat. Vegetable oil can also contribute improved whiteness through a light-scattering effect.

ADVANCED PROCESSING TECHNOLOGY

Ohmic Heating

Ohmic heating employs alternating electrical current passed through an electrically conducting food product (Figure 15). Heat is internally generated due to the electrical resistance of the food sample, resulting in a rapid heating rate. This differs from microwave heating in which energy is transferred to polar molecules (mainly water) in food, resulting in a rise in temperature (Buffler, 1993). This may cause uneven heating in foods which contain solid chunks. In particulate food products, ohmic heating provides uniform temperature distribution because both liquid and solid phases are simultaneously heated (Parrot, 1992).

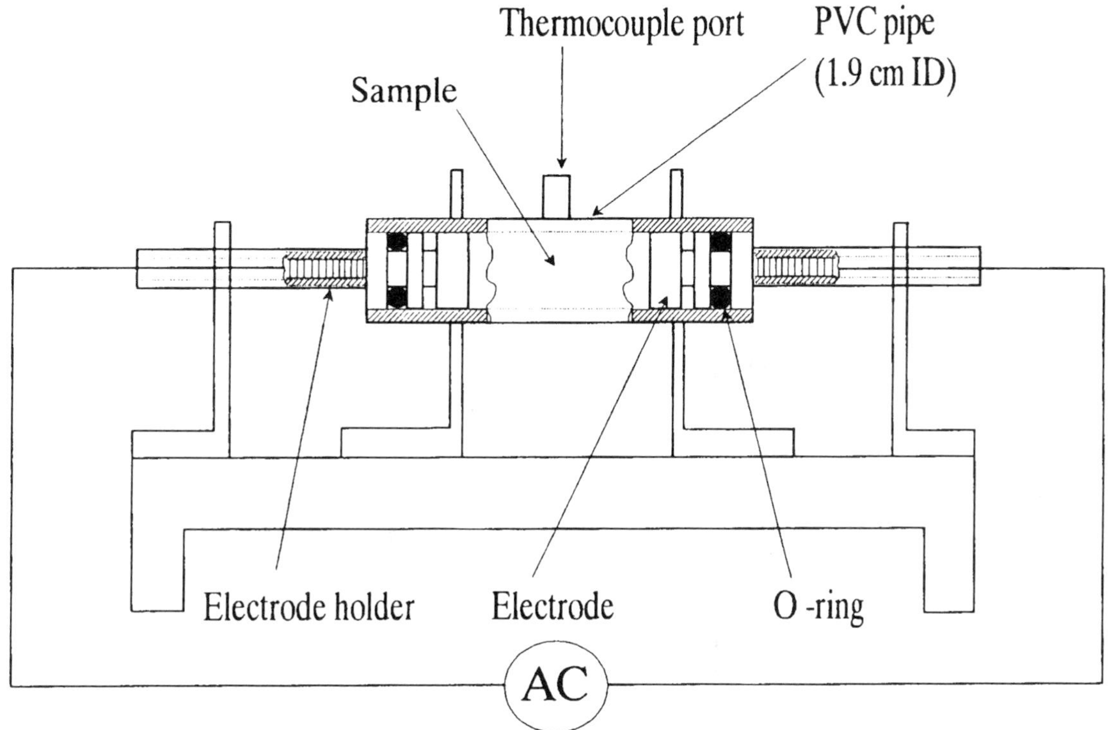

Figure 15. Diagram of ohmic cooking device for preparing surimi gels for testing. AC = alternating current. Adapted from *Yongswatdigul* et al. *(1995)*

These developments have led to the development of commercial ohmic sterilization for particulate foods such as chunky soups and stews (Biss et al., 1989).

It was found that the gel strength of Alaskan pollock, threadfin bream, and sardine surimi was improved when pastes were heated ohmically as compared with traditional heating in a 90°C water bath (Shiba, 1992; Shiba and Numakura, 1992). Yongsawatdigul et al. (1995) investigated the feasibility of ohmic heating to maximize the gelling functionality of Pacific whiting surimi without addition of enzyme inhibitors. The ohmically-heated gels showed more than a two-fold increase in shear stress and shear strain over gels heated in a water bath (Figure 16). Shear stress values of ohmically heated pollock surimi gels were also better than those of gels cooked in a water bath (90°C for 15 minutes), but shear strain values were not affected. Addition of beef plasma protein negated any effects of the heating method, indicating that rapid heating improved gel properties mainly due to rapid inactivation of muscle protease enzymes. This was verified by the finding that degradation of myosin and actin was minimized by ohmic heating (Yongsawatdigul, 1995). The use of ohmic heating for surimi which has high levels of heat-stable endogenous proteases can overcome the negative aspects of adding food-grade enzyme inhibitors, such as their added cost, and possible contribution of off-odors, off-colors, and labeling concerns (beef plasma is not kosher and is not esthetically acceptable to many consumers).

High Hydrostatic Pressure

High hydrostatic pressure (HHP) is a simple processing concept. The food product is sealed in a plastic bag, inserted into a chamber flooded with water, and hydraulically subjected to pressures up to 7,000 atmospheres (atm). For comparison purposes, a typical autoclave cook is at 2.5 atm (120°C), and an extruder cooker operates at 50 atm (180°C). A French press goes as high as 1,500 atm and the deepest part of the ocean has approximately 1,200 atm of pressure (Morrissey et al., 1994). A HHP unit applies extremely high pressures causing physical and chemical changes without the addition of heat. These pressures affect cell membranes, microorganisms, and enzymes, all of which are important constituents of foods (Farr, 1990). A number of food materials such as egg yolks, meats, and soy proteins have been shown to gel under HHP. Okazaki and Nakamura (1992) have shown that HHP-induced gelation of sarcoplasmic protein from different fishes varies depen-

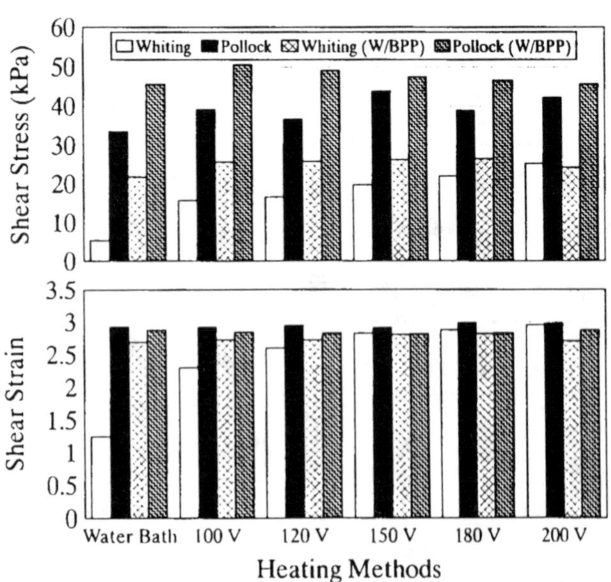

Figure 16. Shear stress and strain of ohmically-heated surimi gels. W/BPP: indicates addition of 1 percent beef plasma protein to these gels.

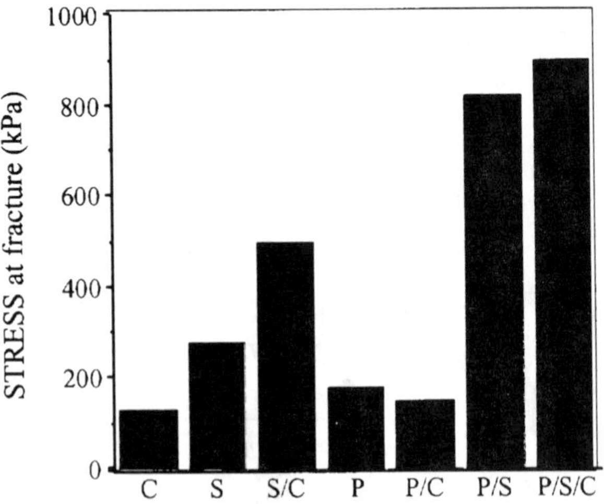

*Figure 17 - Shear stress at fracture of surimi gels prepared by cooking (**C**; 90°C 15 minutes), setting (**S**; 25°C 2 hours), high pressure treatment (**P**; 3000 atm 30 minutes 5°C), or combinations of these treatments in the order given.*

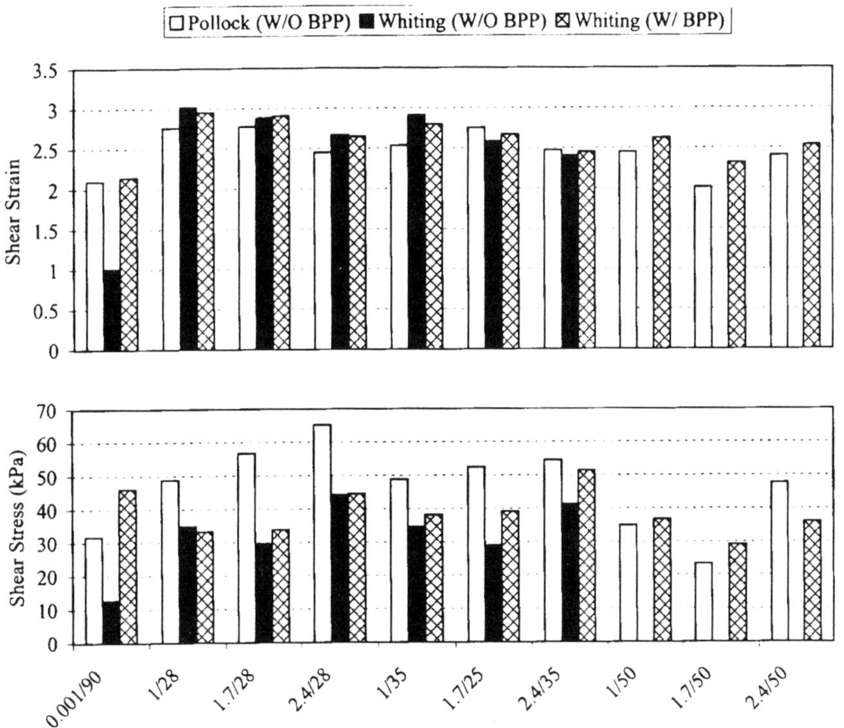

Figure 18. Shear stress and strain of high pressure-treated surimi gels, prepared at varying temperatures. W/O BPP: indicates no addition of beef plasma protein; W/ BPP: indicates addition of 1 percent beef plasma protein to these gels.

dent on species, pH, protein concentration, and pressure treatment. Surimi readily gels at pressures as low as two Kbar (2,000 atmospheres). Gelation of pollock surimi by HHP has been attributed to increased crosslinking of the myosin heavy chain (Shoji et al., 1990). Recently, Gilleland (1995) found that HHP-induced gelation of surimi results in increased disulfide bonding of the proteins, as well as an assumed extensive hydrophobic intermolecular bonding. They also discovered that, once surimi was gelled by HHP, a subsequent low temperature preincubation led to a gelled product which was up to five times stronger than gels prepared by heating alone (Figure 17). It may be that addition of microbially-derived transglutaminase (Nonaka et al., 1989), which can accelerate and accentuate the gelation that occurs during preincubation, could be combined with HHP to yield a super-gelation of surimi. This could be useful in reducing the surimi content of surimi seafoods without loss in textural properties, thus making them more economical than ever.

Chung et al. (1994) investigated the effects of HHP on gel strength of Pacific whiting and Alaska pollock. A three-fold increase in strain and stress was found for HHP-induced whiting gels made without an enzyme inhibitor (beef plasma protein) as compared with gels heated in a 90°C water bath (Figure 18). However, when pressure treatment was undertaken at 50°C, Pacific whiting surimi gels without beef plasma protein were too weak to test. The protease enzyme in Pacific whiting surimi has been shown to have a temperature optimum at 55°C (An et al., 1993). These results would indicate that the enzyme was still very active during pressure treatments near 55°C. Pollock surimi also showed significant increases in both strain and stress values for all pressure treatments except for those run at 50°C. A direct relationship was demonstrated between pressure level of the process and stress values of the resultant gels, while strain measurements were, for the most part, inversely related to the amount of pressure used (Chung et al., 1994).

The freezing point of water at two Kbar is -20°C. Thus a pressurized product might be kept at temperatures below 0°C without the formation of damaging ice crystals. The economics of pressurized holding seem prohibitive at present, however. A more feasible use of HHP in freezing would be its use to supercool products without freezing, followed by rapid release of pressure which rapidly induces formation of small ice crystals at atmospheric pressure (pressure-assisted freezing).

HHP represents a potential processing technology for surimi-based seafood. At present, the only HHP treatments used on a commercial scale are for the preservation of highly-valued jams and jellies in Japan.

SUMMARY

The development of surimi technology has revealed many unique opportunities heretofore unknown in the food industry: the ability to cryoprotect labile proteins in frozen storage; transglutaminase-mediated gelation of proteins at low temperature, and the associated ability of this low temperature gelation to enhance the gel strength of cooked and high pressure-induced gels; and the ability to modify gel texture by the application of various proteins, enzyme inhibitors, and hydrocolloids. Many of these discoveries may have spinoffs in improving the quality of intact fish and meat products. The surimi manufacturing process is now being applied to other meat materials, such as mechanically-deboned beef, pork, poultry, and mutton, as a means of improving functionality and color, lowering fat content, and improving flavor.

The surimi seafoods industry has grown tremendously in the past few decades, spurred at first by exploitation of the Alaskan pollock resource by Japan, and later by the introduction of surimi seafoods to Western markets. At present, world surimi production is near a plateau due to limits on the natural resource. The challenge for the industry in the future will be to add more value to the limited raw material available. One approach is to make more innovative and attractive food products from surimi. Another is to consider fish as a source for producing a variety of specialized food ingredients, surimi (as we now know it) being but one of these (Lanier, 1994). Other challenges include waste reduction in the industry and general increases in production yield. Better management systems for fisheries will, one hopes, stabilize supplies of fish in the future, and cross-utilization of fish for both fillet-type products and surimi (from deboned frames and trimmings) will also reduce waste and increase supplies of raw materials for surimi production.

REFERENCES

Acton, J.C., G.R. Ziegler, and D.L. Burge. 1983. Functionality of muscle constituents in the processing of comminuted meat products. *CRC Crit. Rev. Food Sci. Nutri.* **18**(2):99-127.

Adu, G.A., J.K. Babbitt, and D.L. Crawford. 1983. Effect of washing on the nutritional and quality characteristics of dried minced rockfish flesh. *J. Food Sci.* **48**:1053-1060.

Alaska Fisheries Development Foundation. 1987. *Surimi—It's American Now.* Anchorage, AK.

Alaska Fisheries Development Foundation. 1992. Ground fish quality chart. Anchorage, AK.

An, H., T. Seymour, P.S. Hartley, J.W. Wu, and M.T. Morrissey. 1993. Tissue softening and proteolysis of Pacific whiting. Paper #596, presented at the 53rd Annual Meeting of IFT. July 10-14. Chicago, IL.

Arai, K., K. Kawamura, and C. Hayashi. 1973. The relative thermostabilities of the actomyosin-ATPase from the dorsal muscles of various fish species. *Bull. Jap. Soc. Fish.* **39**:1077-1082.

Babbitt, J. 1990. The use of a decanter centrifuge to prepare Alaska pollock surimi. In *Evaluation of Factors Affecting the Consistency, Functionality, Quality, and Utilization of Surimi.* J.S. French and J. Babbitt (eds.). Alaska Fisheries Development Foundation. Anchorage, AK.

Biss, C.H., S.A. Coombes, and P.J. Skudder. 1989. The development and application of ohmic heating for the continuous processing of particulate food stuffs. In *Processing Engineering in the Food Industry.* R.W. Field and J.A. Howell (eds.), pp. 17-27. Elsevier Applied Science Publ., Essex, England.

Brown, D. 1993. On-line moisture measurements during production. Presented at Surimi Conference, University of Washington. October 25-27. Seattle, WA.

Boye, S.M. and T.C. Lanier. 1988. Effect of heat stable alkaline protease activity of Atlantic menhaden (*Brevoorti tyrannus*) on surimi gels. *J. Food Sci.* **53**:1340-1342,1398.

Buffler, C.R. 1993. *Microwave Processing and Cooking.* Van Nostrand Reinhold, New York.

Bullens, C.W., M.G. Llanto, C.M. Lee, and J.J. Modliszewski. 1990. The function of a carrageenan-based stabilizer to improve quality of fabricated seafood products. In *Advances in Fisheries Technology and Biotechnology for Increased Profitability.* M.N. Voigt and J.R. Botta (eds.), pp. 313-324. Technomic Publishing Co., Lancaster, PA.

Chung, Y.C., A. Gebrehiwot, D.F. Farkas, and M.T. Morrissey. 1994. Gelation of surimi by high hydrostatic pressure. *J. of Food Sci.* **59**:523-524, 543.

Comar, P. 1987. Personal communication. National Seafood Inspection Laboratory. Pascagoula, MS.

Da Ponte, D.J.B., J.M. Herft, J.P. Roozen, and W. Pilnik. 1985. Effects of different types of carrageenans and carboxymethyl cellulose on the stability of frozen stored minced fillets of cod. *J. of Food Technol.* **20**:587-590.

Eklund, M. 1993. Personal communication. National Marine Fisheries Service, Seattle, WA.

Farr, D. 1990. High pressure technology in the food industry. *Trends in Food Sci. and Tech.* **1**(1):14-16.

FDA. 1992. Code of Federal Regulations 21. Food and Drug Administration, Washington, DC.

Giese, J. 1993. On-line sensors for food processing. *Food Technology* **47**(5):88-95.

Gilleland, G. 1995. Investigation into the Mechanism of Gelation of Surimi Pastes Treated by High Isostatic Pressure. M.S. Thesis, North Carolina State University, Raleigh.

Gordon, R.J. 1994. Seafood flavor creation—the science and art. Presented at the 2nd Annual OSU Surimi Technology School, March 28-30, Astoria, OR.

Greene, D.H. and J. Babbitt. 1990. Control of muscle softening and protease parasite interactions in arrowtooth flounder (*Atheresthes stomias*). *J. Food Sci.* **55**:579-580.

Green, D.P. and T.C. Lanier, A.C. Chao, and T. Akahane. 1985. Batch washing of minced fish under lateral and reverse flow processing techniques. 25th Abst. Atlantic Fisheries Technol Conf. Boston, MA.

Hamann, D.D., P.M. Amato, M.C. Wu, and E.A. Foegeding. 1990. Inhibition of modori (gel weakening) in surimi by plasma hydrolysate and egg white. *J. Food Sci.* **55**:665-670.

Joseph, D., T.C. Lanier, and D.D. Hamann. 1994. Effect of pH and temperature on the transglutaminase catalyzed setting of crude fish actomyosin. *J. Food Sci.* **59**:1018-1023.

Kamath, G.G., T. C. Lanier, E. A. Foegeding, and D. D. Hamann. 1992. Nondisulfide covalent cross-linking of myosin heavy chain in "setting" of Alaska pollock and Atlantic croaker surimi. *J. of Food Biochem.* **16**:151-172.

Kim, B.Y. 1987. Rheological Investigation of Gel Structure Formation by Fish Proteins During Setting and Heat Processing. Ph.D. Dissertation, North Carolina State University, Raleigh, NC.

Kim, B.Y., D.D. Hamann, T.C. Lanier, and M.C. Wu. 1986. Effects of freeze-thaw abuse on the viscosity and gel-forming properties of surimi from two species. *J. Food Sci.* **51**:951-956,1004.

Kim, J.M. and C.M. Lee. 1987. Effect of starch of textural properties of surimi gel. *J. Food Sci.* **52**:722-725.

Lanier, T.C. 1986. Functional properties of surimi. *Food Technology* **40** (March):107-114,124.

Lanier, T.C. 1991. Interactions of muscle and nonmuscle proteins affecting heat-set gel rheology. In *Interactions of Food Proteins*, N. Parris and R. Barford (eds.), pp. 268-284. American Chemical Society, Washington, DC.

Lanier, T.C. 1992. Measurement of surimi composition and functional properties. In *Surimi Technology*. T.C. Lanier and C.M. Lee (eds.), pp. 123-166. Marcel Dekker, New York.

Lanier, T.C. 1994. Functional food protein ingredients from fish. In *Seafood Proteins*, B.S. Pan, F. Shahidi, and Z.E. Sikorski (eds.), pp. 127-159. Van Nostrand Reinhold, New York.

Lanier, T.C. and B. Thomsen. 1993. A whey protein texturizer for surimi replacement. Pac. Fish. Technol. Conf., Seattle, WA.

Lanier, T.C., D.D. Hamann, and M.C. Wu. 1985. *Development of Methods for Quality and Functionality Assessment of Surimi and Minced Fish.* Alaska Fisheries Development Foundation, Anchorage, AK.

Lanier, T.C., P.K. Manning, T. Zettering, and G.A. MacDonald. 1992. Process innovations in surimi manufacture. In *Surimi Technology*. T.C. Lanier and C.M. Lee (eds.), pp. 167-179. Marcel Dekker, New York.

Lanier, T.C. and J.W. Park. 1990. Application of surimi quality measurements to least cost linear programming of surimi product formulations. *Alaska Fisheries Development Foundation.* Anchorage, AK.

Lauro, G.J. 1991. A primer on natural colors. *Cereal Foods World.* **36**(11):949-953.

Lee, C.M. 1984. Surimi process technology. *Food Technol.* **38** (November):69-80.

Lee, C.M. 1986. Surimi manufacturing and fabrication of surimi-based products. *Food Technol.* **40** (March):115-124.

Lee, C. M. 1990. A pilot plant study of surimi-making properties of red hake. In *Engineered Seafood Including Surimi.* R.E. Martin and R.L. Collette (eds.), pp. 225-243. Noyes Data Corp., Park Ridge, NJ.

Lee, C. M. and A. Abdollahi. 1991. Effect of hardness of plastic fat on structure and material properties of fish protein gels. *J. Food Sci.* **56**:1755-1759.

Lee, H.G., C.M. Lee, K.H. Chung, and S.A. Lavery. 1992. Sodium ascorbate affects surimi gel forming properties. *J. Food Sci.* **57**:1343-1347.

Lee, H.G., T.C. Lanier, D.D. Hamann, and J.A. Knopp. 1997. Transglutaminase effects on low temperature gelation of fish sols. *J. Food Sci.* **62**(1):20-24.

Lee, N. and J.W. Park. 1997. Calcium compounds to improve gel functionality of Pacific whiting and Alaska pollock surimi. Unpublished results.

Lee, N.G. and J.W. Park. 1997. Calcium compounds to improve gel functionality of pollock and whiting surimi. Presented at the annual IFT. June 15-18. Orlando, FL.

Lin, T.M. and J.W. Park. 1995. Study of myofibrillar protein solubility during surimi processing: Effects of washing cycles and ionic strength. Submitted for the presentation at the IFT annual meeting. Anaheim, CA.

Lin, T.M., J.W. Park, and M.T. Morrissey. 1995. Recovered protein and reconditioned water from surimi processing waste. *J. Food Sci.* **60**(1): 4-9.

Llanto, M.G., C.W. Bullens, J.J. Modliszewski, and A.A. Bushway. 1990. Effects of carrageenan on gelling potential of surimi prepared from Atlantic pollock. In *Advances in Fisheries Technology and Biotechnology for Increased Profitability.* M.N. Voigt and J.R. Botta (eds.), pp. 305-312. Technomic Publishing Co. Lancaster, PA.

MacDonald, G.A., N.D. Wilson, and T.C. Lanier. 1990. Stabilised mince: an alterative to the traditional surimi process. In *Chilling and Freezing of New Fish*

Products. Intl. Inst. of Refrigeration (United Kingdom), pp. 57-63.

MacDonald, G. A., J. Stevens, and T.C. Lanier. 1994. Characterization of New Zealand hoki and Southern blue whiting surimi compared to Alaska pollock surimi. *J. of Aquat. Food Product Technol.* **3**(1):19-38.

Misima, T., H. Mukai, Z. Wu, K. Tachibana, and M. Tsuchimoto. 1993. Resting metabolism and myofibrillar Mg^{2+} -ATPase activity of carp acclimated to different temperatures. *Nippon Suisan Gakkaishi*. **59**:1213-1218.

Morrissey, M.T. 1993. Unpublished data. Oregon State University Seafood Laboratory, Astoria, OR.

Morrissey, M.T., J.W. Park, and J. Yongsawatdigul. 1994. Innovative processing in the seafood industry: the potential for ohmic heating and high hydrostatic pressure. Presented at the First International Symposium of Biochemical Engineering and Food Technology at ITESM - Campus Queretaro. October 13-15. Quertaro, Mexico.

Morrissey, M.T., G. Peter, and G. Sylvia. 1993. Quality issues in the Pacific whiting fisheries. In *Pacific Whiting: Harvesting, Processing, Marketing, and Quality Assurance*. G. Sylvia and M.T. Morrissey (eds.), pp. 9-16. Oregon Sea Grant OERSU-W-92-001. Corvallis, OR.

Morrissey, M.T., J.W. Wu, D.D. Lin, and H. An. 1993. Effect of food grade protease inhibitor on autolysis and gel strength of surimi. *J. Food Sci.* **58**:1050-1054.

Nishimura, K., M. Ohtsuru, and K. Nigota. 1990. Mechanism of improvement effect of ascorbic acid on the thermal gelation of fish meat. *Nippon Suisan Gakkaishi*. **56**:959-964.

Nishiya, K., F. Takeda, K. Tamoto, O. Tanaka, and T. Kubo. 1960. Hokkaido Fisheries Research Laboratory, Fisheries Agency, Japan. Report 21, p.44.

Nonaka, M., H. Tanaka, A. Okiyama, M. Motoki, H. Ando, K. Umeda, and A. Matsuura. 1989. Polymerization of several proteins by Ca^{2+} - independent transglutaminase derived from microorganisms. *Agric. Biol. Chem.* **53**:2619-2623.

Ohta, Y. and K. Maekaji. 1980. Preparation of konjac mannan gel. *Nippon Nogeikagaku Kaishi*. **54**:741-746.

Okada, M. 1981. Varieties of fish and paste products. In *Kneaded Seafood Products*. M. Okada, T. Imaki, and M. Yokozeki (eds.), pp. 169-177. Koesisha-Kosikaku Publishing, Tokyo.

Okada, M. 1992. History of surimi technology in Japan. In *Surimi Technology*. T.C. Lanier and C.M. Lee (eds.), pp. 3-21. Marcel Dekker Inc., New York.

Okazaki, E. and K. Nakamura. 1992. Factors influencing texturization of sarcoplasmic protein of fish by high pressure treatment. *Nippon Suisan Gakkaishi*. **58**:2197-2206.

Overseas Fishery Cooperation Foundation and Japan Deep Sea Trawlers Assoc.1984. Primary processing of surimi, secondary processing from surimi and its marketing. Presented at surimi workshop in Seattle, WA. October 4-5.

Pacheco-Aguilar, R. and D.L. Crawford. 1994. Potassium bromate effects on gel forming ability of Pacific whiting surimi. *J. Food Sci.* **59**:786-791.

Pacheco-Aguilar, R., D.L. Crawford, and L.E. Lampila. 1989. Procedures for the efficient washing of minced whiting (*Merluccius productus*) flesh for surimi production. *J. Food Sci.* **54**:248-252.

Park, J.W. 1993a. Unpublished data (Industry survey). OSU Seafood Laboratory, Astoria, OR.

Park, J.W. 1993b. Use of various grade of surimi with an application of least cost formulation. In *Pacific Whiting: Harvesting, Processing, Marketing, and Quality Assurance*. G. Sylvia and Morrissey, M.T. (ed.), pp. 17-19. Oregon Sea Grant, Corvallis.

Park, J.W. 1994a. Unpublished data (industry survey). OSU Seafood Laboratory, Astoria, OR.

Park, J.W. 1994b. Functional protein additives in surimi gels. *J. Food Sci.* **59**:525-527.

Park, J.W. 1994c. Unpublished data. OSU Seafood Laboratory, Astoria, OR.

Park, J.W. 1995. Surimi gel colors as affected by moisture content and physical conditions. *J. Food Sci.* **60**(1): 15-18.

Park, J.W. 1996. Temperature-tolerant fish protein gels using konjac flour. *J. Muscle Foods*. **7**:167-174.

Park, J.W., R.W. Korhonen, and T.C. Lanier. 1990. Effects of rigor mortis on gel-forming properties of surimi and unwashed mince prepared from tilapia. *J. Food Sci.* **55**:353-355,360.

Park, J.W. and M.T. Morrissey. 1994. The need for developing surimi standard. *In Quality Assurance and Quality Control for Seafood*. G. Sylvia and M.T. Morrissey (eds.), pp. 64-71. Oregon Sea Grant, Corvallis, OR.

Park, J.W., J. Yongsawatdigul, and T.M. Lin. 1994. Rheological behavior and potential cross-linking of Pacific whiting (*Merluccius productus*) surimi gel. *J. Food Sci.* **59**:773-776.

Parrot, D.L. 1992. Use of ohmic heating for aseptic processing of food particulates. *Food Technol.* **46**(12): 68-72.

Peters, G. and M.T. Morrissey. 1994. Processing parameters affecting the quality of Pacific whiting. Presented at the meeting of Oregon Trawl Commission, Astoria.

Pigott, G.M. 1986. Surimi: the "high tech" raw materials from minced fish flesh. *Food Reviews International*. **2**:213-246.

Pipatsattayanuwong, S., J.W. Park, and M.T. Morrissey. 1995. Functional properties and shelf life of fresh surimi from Pacific whiting. *J. Food Sci.* **60**:1241-1244.

Powrie, W. D. 1973. Characterization of food myosystem and their behavior during freeze-preservation. In *Low Temperature Preservation of Foods and Living Matter*. O.R. Fennema, W.D., Powrie, and E.H. Marth (eds.), pp. 282-336. Marcel Dekker, New York.

Regenstein, J.M. 1984. Protein-water interactions in muscle foods. *Reciprocal Meat Conference Proceedings*. **37**:44-51.

Scott, D.N., R.W. Porter, G. Kudo, R. Miller, and B. Koury. 1988. Effect of freezing and frozen storage of Alaska pollock on the chemical and gel-forming properties of surimi. *J. Food Sci.* **58**:353.

Shiba, M. 1992. Properties of kamaboko gels prepared by using a new heating apparatus. *Nippon Suisan Gakkaishi.* **58**:895-901.

Shiba, M. and T. Numakura. 1992. Quality of heated gel from walleye pollock surimi by applying joule heat. *Nippon Suisan Gakkaishi.* **58**:903-907.

Shoji, S., H. Saeki, A. Wakemeda, M. Nakamura, and M. Nonaka. 1990. Gelation of salted paste of Alaska pollock by high pressure and change in myofibrillar protein in it. *Nippon Suisan Gakkaishi.* **56**:2069-2076.

Simpson, R., E. Kolbe, G. MacDonald, T. Lanier, and M.T. Morrissey. 1994. Surimi production from partially processed and frozen Pacific whiting (*Merluccius productus*). *J. Food Sci.* **59**:272-276.

Swafford, T.C., J. Babbitt, K. Reppond, A. Hardy, C.C. Riley, T.K.A. Zetterling. 1990. Surimi process yield improvement and quality contribution by centrifuging. In *Engineered Seafood Including Surimi.* R.E. Martin and R.L. Collette (eds.), pp. 483-496. Noyes Data Corp. Park Ridge, NJ.

Takeda, F. 1971. Technological history of frozen surimi industry. *New Food Ind.* **13**:27-31.

Tokunaga, T. and F. Nishioka. 1988. The improvement of technology for surimi production from fatty Japanese sardine. In *Proceedings of a National Technical Conference, Fatty Fish Utilization: Upgrading from Feed to Food.* UNC Sea Grant Publication 88-04. Raleigh, NC.

Toyohara, H. and Y. Shimizu. 1988. Relation between the modori phenomenon and myosin heavy chain breakdown in the threadfin-bream gel. *Agric. Biol. Chem.* **52**:255-257.

Toyohara, H., T. Sakata, K. Yamashita, M. Kinishita, and Y. Shimizu. 1990. Degradation of oval-filefish meat gel caused by myofibrillar proteinase(s). *J. Food Sci.* **55**:364-368.

Tye, R. J. 1988. The rheology of starch/carrageenan systems. *Food Hydrocolloids.* **2**:259-264.

Tye, R. J. 1991. Konjac flour: properties and applications. *Food Technology.* **45**(3):88-92.

Wu, M.C., D.D. Hamann, and T.C. Lanier. 1985. Rheological and calorimetric investigations of starch-fish protein systems during thermal processing. *J. Texture Stud.* **16**:53-59.

Wu, M.C. and T. Suzuki. Process of forming simulated crustacean meat. United States Patent 5,028,445.

Wu, Y.J., M.T. Atallah, and H.O. Hultin. 1991. The proteins of washed, minced fish muscle have significant solubility in water. *J. Food Biochem.* **15**:209-218.

Yamamoto, Y. T. Okubo, S. Hatayama, M. Naito, and T. Ebisu. 1991. Frozen surimi product and process for preparing. United States Patent No. 5,028,444.

Yongsawatdigul, J., J.W. Park, E. Kolbe, Y. Abu Dagga, and M.T. Morrissey. 1995. Ohmic heating maximizes gel functionality of Pacific whiting surimi. *J. Food Sci.* **60**(1): 10-14.

Yoon, W.B. and J.W. Park. 1995. Least cost linear programming of surimi-based seafoods. Abstract. Presented at the annual meeting of IFT, June 3-7, Anaheim, CA.

Yoshinaka, R., M. Shiraishi, and S. Ikeda. 1972. Effect of ascorbic acid on the gel formation of fish meat. *Bull. Jap. Soc. Sci.* Fish. **38**:511-556.

Transportation, Distribution, and Warehousing

Roy E. Martin

Through the efforts of everyone associated with it, the American food supply has become the best, safest, and cleanest in the world. The public has come to expect high standards. It thus becomes the everyday responsibility of people in many diverse industries to see that our food is produced, processed, and packed under clean conditions and that it is kept that way throughout the distribution chain. The public health ramifications of handling shipments of food and related products demand such attention.

TRANSPORTATION

For shipping and receiving the industry uses truck, rail, and air to move shipments of seafood. The Food and Drug Administration (FDA) states under its Good Transportation Practices Model Code that establishments engaged in the processing, packing, and storage of human food are subject to comprehensive government regulations which require that the food be prepared, packed, and held under sanitary conditions. With few exceptions, persons engaged in transporting food are only indirectly subject to these controls. Compliance with the regulations described here should assist those in the transportation industry in assuring that foods for human consumption are handled and shipped under conditions that (1) prevent contamination; (2) protect against product deterioration and container damage; and (3) assure that conveyances intended, offered, or used for transporting food are suitable for that purpose.

The criteria in these regulations apply in determining whether conveyances, storage facilities, methods, practices, and controls used in transporting food are in conformance with, or are operated or administered in conformity with, good transportation practices. These regulations apply to all persons engaged in the transportation of human foods, including manufacturers, processors, distributors, common carriers, contract haulers, and private individuals. When used solely for the purpose of transporting raw agricultural commodities from the field to a point of initial storage and/or processing, conveyances are excluded from the requirements of these regulations, provided that special regulations covering these exclusions may be adopted whenever necessary to protect the public health.

The FDA defines the following transportable classes of food: (1) "Perishable food" is food which includes, but is not limited to, fresh fruits, fresh fish, fresh vegetables, and other products which need protection from extremes of temperatures in order to avoid decomposition by microbial growth or otherwise; (2) "Readily perishable food" is food or a food ingredient consisting in whole or in part of milk, milk products, eggs, meat, fish, poultry, or other food or food ingredient which is capable of supporting rapid and progressive growth of infectious or toxigenic microorganisms; and (3) "Frozen food" is food which is intended to be sold in the frozen state.

Other definitions used in the code include the following:

"Delivery equipment" means any truck, railcar, ship, barge, aircraft, or other conveyance together with its appurtenances, used or offered for the transportation of food.

"Special purpose delivery equipment" means those conveyances that are not designed for general purpose transportation, but are built specifically for the handling of foodstuffs and which in themselves may be immediate containers.

"Storage facility" means any warehouse, freight terminal, or other storage facility, including all loading docks and other appurtenances associated with and used in the storage of food during transportation.

"Carrier" is any person who owns, operates, or controls delivery equipment or storage facilities.

"Shipper" is any person of record who initiates the transportation of food from one place to another.

"Carrier controlled equipment" means any delivery equipment, the movement of which is controlled exclusively by a common carrier to a contract hauler.

"Shipper controlled equipment" means any delivery equipment which the carrier has assigned to a shipper for his exclusive use.

"Sanitize" means adequate treatment of surfaces by a process that effectively destroys cells of pathogenic microorganisms and substantially reduces other microorganisms. Such treatment shall not adversely affect the product and shall be safe for the consumer.

"Adequate" means that which is needed to accomplish the intended purpose in keeping with good public health practice.

DELIVERY EQUIPMENT DESIGN AND CONSTRUCTION

All delivery equipment shall be constructed of material that will withstand repeated cleaning and shall be designed to be easily cleaned and to protect the food being handled from dust, dirt and other contaminating materials. In addition, delivery equipment used or intended for handling perishable food shall be constructed to protect such food from temperatures which may cause or permit damage.

Special purpose delivery equipment used for transportation of processed or partially bulk food shall be constructed of smooth, corrosion-resistant, nontoxic materials, and shall be so designed and constructed as to be easily cleanable.

Delivery equipment used to handle readily perishable food requiring refrigeration, in addition to the requirements just outlined, shall be provided with mechanical aeration equipment or other methods or facilities capable of maintaining a product temperature of 45°F (7°C) or below. Delivery equipment used for handling frozen food shall be capable of maintaining the product temperature at 0°F (-18°C) or lower.

Delivery equipment used for delivery of readily perishable or frozen foods shall be equipped with a thermometer or other appropriate means for measuring and indicating the air temperature in the shipping compartment. The dial or reading element of the temperature measuring device must be located where it can be easily read from the outside of the conveyance.

This is a good point at which to interject the concept of HACCP (Hazard Analysis Critical Control Point). While not mandatory for the transportation, distribution, and warehousing sections of the food industry, it is strongly suggested that these parts of the industry voluntarily adopt the HACCP concept. This would be beneficial because the first critical control point in any other part of the food system will be *receipt* of raw material. The first receiver of product will require the transporter to produce a record of proper handling, so that the record can be made part of the processing company's own HACCP program. Proper temperature control will be the basis for most of the critical point record-keeping for both the transportation and warehousing industries.

Preloading Controls

All conveyances which are under the carrier's control and are offered to shippers for the purpose of transporting food shall, at time of delivery, be in a clean and sanitary condition, be in good repair, and be of adequate design and construction for the intended purpose. The carrier shall take all reasonable precautions, including the following, to assure that such conveyances will not contribute to contamination or deterioration of food products:

1. Effective measures shall be taken to remove and exclude all vermin (including, but not limited to, birds, rodents, and insects). The use of pesticides for this purpose shall include precautions to prevent contamination of food or packaging material with illegal chemical residues.
2. The interior of each conveyance shall be cleaned as needed to ensure removal of all debris, filth, mold, toxic chemicals, undesirable odors, or any other objectionable condition that may result in the contamination of food. When appropriate to control microbiological contamination, food-

contact surfaces shall be sanitized with a safe and effective sanitizing agent.

3. All doors and hatches shall be kept in good repair, be tightfitting, and when closed and sealed shall be capable of excluding rodents, birds, and other pests.
4. All refrigeration equipment shall be in proper working condition, and capable of holding transported food products at temperatures specified by the shipper.
5. Except in the case of assigned equipment, the carrier at time of delivery shall certify to the shipper that the equipment has been inspected in accordance with and conforms to these regulations. Such inspection records shall be retained by the carrier for one year from the date of issuance.

The shipper shall inspect all equipment offered or intended for food loading to determine if said equipment is in acceptable physical condition or whether it contains any potential food contaminant. The shipper shall refrain from loading food into any equipment deemed unacceptable until such time as all noted defects are corrected. When defects noted in carrier-controlled equipment cannot be corrected at the shipper's plant, the shipper shall reject such defective equipment to the carrier, stating the reasons for said rejection. The shipper shall maintain inspection records of all defects that cause equipment to be rejected and shall maintain such records for one year.

Loading Controls

Food products to be loaded shall be free from contaminants which may contribute to adulteration during transit of other products in the load or which may result in the contamination of the conveyance.

All packaged food products shall be loaded in such a manner as to minimize physical damage while in transit.

All containers used for transporting food shall be of such design and construction to protect the contents from damage and/or contamination under usual conditions of loading, shipment and transshipment.

In the loading of food products, adequate precautions shall be taken to minimize contamination of the vehicle through hatches, pipes, hoses, vents, conveyors, or other potential routes of contamination. Food products shall not be loaded into the same vehicle or shipped with fungicides, insecticides, rodenticides, or any other poisonous, toxic, or deleterious industrial chemicals.

Before and after closing doors or hatches of the loaded conveyance, persons responsible for the loading operation shall take all other precautions as may be appropriate to protect the integrity of both the vehicle and its contents.

The trans-shipment and en route storage of food products should be under such conditions as will prevent contamination and will protect against undesirable deterioration of the product or containers.

Unloading Controls

All incoming conveyances shall be carefully examined upon arrival at the delivery point to determine if doors or hatches are intact and untampered. Where appropriate, the seal numbers of the doors and hatches shall be recorded prior to their removal. Any broken or damaged seals shall be noted and reported to the carrier.

Upon opening and prior to unloading of the food, the interior of the conveyance shall be examined for evidence of any detectable signs of potential contaminants and adulterants, including, but not limited to, insects, rodents, mold, or undesirable odors. The examination shall continue during the entire unloading operation.

Before unloading refrigerated products, the internal temperature of the food products shall be taken and recorded (Critical Control Point).

A record shall be kept indicating the type and disposition of damaged, adulterated, and deteriorated products or conveyance. Such record shall indicate the disposition of the defective products and/or conveyance.

All food products, dunnage, debris, and other materials connected with the inbound shipment shall be completely removed from the conveyance before returning or releasing such conveyance to the carrier.

SPECIAL HANDLING AND PROTECTION OF PERISHABLE, READILY PERISHABLE, AND FROZEN FOODS

All perishable foods shall be protected at all times from extremes of temperature that may cause or permit damage or deterioration of the food.

All readily perishable food shall be transported and handled in transit at a product temperature of 45°F (7°C) or lower, except that during loading and unloading the product temperature shall not exceed 60°F (17°C).

All frozen food shall be transported and handled in transit at a product temperature of 0°F (18°C) or lower, except that during defrost cycles, loading and unloading such product temperature shall not exceed 10° F(-10°C).

Any variation from these temperature limits due to failure or faulty operation of temperature control equipment during transportation should be reported by the carrier to the nearest office charged with enforcing these regulations.

SPECIAL CONCERNS: RAILCARS

Because of the special nature of food, special categories of railcars are now evolving for use with food shipments Of particular interest, in relation to these guidelines, is the XF boxcar. This car, specially prepared with an easily-cleaned, FDA-approved interior white coating, can be effective in protecting food shipments if it is maintained in good condition. All users of these specialized railcars should realize that extra care in maintaining the condition and cleanliness of this equipment is an investment in protecting the food supply.

Three types of railcars are used to transport food: the free-running car; the car dedicated to food or related-product service; and the car assigned to the use of a particular shipper. Although there are some differences in the specific responsibilities of shippers, carriers, and receivers when different cars are considered, the basic principles remain the same. A car must be clean and in good repair in order to protect the food.

A clean car is free from evidence of vermin infestation (including, but not limited to, birds, rodents, and insects); and free from debris, filth, visible mold, undesirable odors, and evidence of residues of toxic chemicals. A car in good repair should have structurally sound interiors and exteriors, including doors and hatches that are tight-fitting and, when closed and sealed, are capable of excluding rodents, birds, and other pests.

For many new users of railcar service, and for some experienced users, questions arise over who takes responsibility for certain activities involved in transporting food by rail. The following information is intended to provide guidance as to the responsibility of each shipper, carrier, and receiver.

Ordering a car(s) is the shipper's responsibility, of course. The shipper should place car orders with the appropriate railroad personnel, specifying in each order:

1. The type and size car required (e.g., Class A-50 boxcar, airslide car, etc.);
2. The commodity to be loaded;
3. Whether commodity is bulk or packaged;
4. Date required;
5. Location (track and door number if applicable) where car is to be spotted;
6. Load destination and route (if known).

Furnishing Car

Where cars are required for transportation of food or related items, the carrier is responsible for furnishing cars suitable for the intended purpose and for providing cars that are in a clean condition, in good repair, and of adequate design and construction for the intended purpose.

In furnishing free-running cars, the carrier must take necessary precautions to ensure that the car is suitable for the intended purpose. Where cars are dedicated to food or related product category use, the carrier is to furnish cars with doors and hatches closed and sealed.

Although most responsibilities in furnishing cars fall on the carrier, the shipper does assume one responsibility: where a car is assigned to a shipper's exclusive use, the shipper must inspect and maintain the car in a clean and sanitary condition.

Car Loading

It is the shipper's responsibility to inspect all railcars offered or intended for loading to determine if they are clean and in good repair. Shippers should refrain from loading any railcar deemed unsuitable until such time as all noted defects that may contribute to contamination are corrected. Such defects may include:

1. Damage to floors, walls, ceilings, doors, and hatches;
2. Protruding nails or bolts;
3. Dunnage trash, or other debris;
4. Residue of contamination by prior toxic material loading;
5. Evidence of contamination by prior toxic material loading;
6. Vermin infestation or visible mold; and
7. Objectionable odors.

Whenever defects noted in a railcar are not corrected at the shipper's plant, the shipper should reject the defective railcar to the rail carrier, stating the reasons for rejection, and maintain inspection records of all defects causing railcars to be rejected.

The shipper should load only products which are themselves uncontaminated and are free from substances or components which are likely to contribute to contamination of other products in the load during transit or are likely to result in contamination of the railcar. All packaged food products are to be appropriately packaged and loaded in order to minimize physical damage or contamination under reasonable transportation conditions and procedures.

In loading products, shippers should take adequate precautions to minimize contamination of the railcar through hatches, pipes, hoses, vents, conveyors, or other potential routes of contamination. They also must see that persons responsible for the loading operation take all other precautions as may be appropriate to protect the integrity both of the transportation equipment and its contents. Once the car is loaded, the shipper is responsible for closing and sealing all doors and hatches and then tendering the billing instructions to the carrier.

Car Transporting and Delivery

The carrier should remove the car(s) from the shipper's siding and transport cars to their destination with all due care for the integrity of the loading. The carrier also must use reasonable diligence to prevent the unauthorized entry into cars and maintain all seals intact and all doors and/or hatches secured.

In the event of derailment or other type of major accident or damage, natural catastrophe (such as flood), or detection of unauthorized entry into car, it is the carrier's responsibility to notify the shipper and receiver promptly.

Once the car has reached its destination, the carrier will notify the receiver that the car has arrived and will spot the car according to the receiver's instructions. The carrier also will notify the receiver if a car is in the shipper's assigned service or is dedicated to food or related product category use.

Car Unloading

The receiver is responsible for examining all incoming railcars carefully to determine if doors, hatches, and seals are intact and untampered with. Whoever is inspecting the cars should record the seal numbers of the doors and hatches prior to their removal, and the receiver should note any broken or damaged seals and report such findings to the rail carrier and shipper.

Once these steps are completed, but before unloading begins, the receiver should examine the railcar's exposed interior for any evidence of potential contaminants and adulterants including, but not limited to, insects, rodents, mold, or undesirable odors. This examination should be continued during the entire unloading operation. In the event contaminants and/or adulterants are noted, the receiver should:

1. Notify the rail carrier to make an inspection and provide an inspection report.
2. Notify the shipper for disposition.
3. Do not permit the product to enter the building if the shipment contains contamination or damage which could lead to contamination of the receiver's establishment; in other cases, separate damaged or contaminated product from the remainder of the load.
4. Keep a record indicating the type and disposition of damaged, adulterated, and deteriorated product.

Before releasing a car to the carrier as an "empty," the receiver must completely remove all products, including a damaged or refused product, dunnage, debris, and other materials connected with the inbound shipment from the railcar. Two additional steps that should be completed before the car is released to the carrier are (1) reporting all contamination, physical damage, or other conditions incompatible with further use of the car for food-related products to the carrier, and (2) replacing and/or securing all bulkheads and other appurtenances that are a part of the railcar.

Once these pre-release conditions are met, the receiver should close all doors and hatches. If a car is in the shipper's assigned service or is dedicated to food or related product service, the receiver should seal the car after complete unloading.

Removing an Empty Car and Subsequent Handling

It is the carrier's responsibility to determine that the car has been completely unloaded and emptied as required. A car sealed by the receiver is considered to have been completely unloaded and emptied. The carrier should not remove a car if it is not completely unloaded and free from product, dunnage, or other debris.

If the carrier is notified by the receiver that a car contains contamination or physical damage, it must take necessary action for cleaning and/or upgrading the car before returning it to food or related product category use.

SPECIAL CONCERNS: AIR SHIPPING

The air cargo industry is responsible for transporting about 7 percent of the world's annual catch of seafood (Delta, 1996). Forecasts suggest that in the coming decades the air cargo industry can expect a growing demand for air shipment of fish and seafood.

Air transport provides the essential link between landlocked communities and the world's great fishing ports. However, successful air transport of fishery products requires special care in preparation and handling of the shipments and excellent communication among the shipper, carrier, and consignee. This section contains voluntary guidelines — developed through the cooperation of the Air Transport Association of America (ATA) and the National Fisheries Institute (NFI) — for the handling, packaging, and acceptance of fresh fish and seafood.

Seafood shippers need to keep a number of issues in mind. Narrow-body aircraft are common in United States fleets, as are hub airports that may necessitate cargo transfers under tight schedules. In addition, the reliance on combination passenger-cargo aircraft in many markets, as well as volatility in pricing, entry, and exit in all markets, can influence the transport of fishery products. Finally, and most importantly, the air shipment of improperly packaged fishery products is a safety hazard because of potential damage to the interiors and control mechanisms of aircraft. For example, in one recorded case, a major United States carrier removed an aircraft from service for an entire week, spending $750,000 to repair damage that resulted from leaking seafood packages.

For seafood shippers, these factors mean that air carriers must seek to eliminate leakage from seafood shipments. In addition, airlines are likely to require shipments packed in lighter-weight units that can be transferred easily and handled manually within the confines of the smaller aircraft in a carrier's fleet. Many airlines will prefer gross weight limits of 60 to 80 pounds, although individual carriers will accept units of 100 to 150 pounds or perhaps more.

The air shipping environment is the sum of all conditions affecting a shipment: scheduling, weather, shock variables, handling techniques and equipment, and vulnerability to theft and pilferage. A shipper must follow an approach to packaging, handling, and tendering seafood for air transport that accounts for all aspects of this environment. The ATA/NFI voluntary guidelines seek to assist shippers by guiding them in the selection of packaging materials and development of practices to preserve their perishable but highly valuable cargo in prime condition through air shipment to final destination, and to prevent leakage and damage to expensive aircraft interiors, other cargo, or passenger baggage.

ATA and NFI recognize that many suitable packaging systems have been developed for air transport of fishery products. These guidelines are not intended to exclude the use of proven shipping containers or those that may be developed in the future. Therefore, if a shipping container that lies outside the scope of these guidelines is being considered, the shipper must communicate with the carrier to establish whether the packaging is acceptable for its intended use.

General Considerations for Packaging

Whole or dressed fish should be cooled to 32°F (0°C) before packing. Several practices are used to reduce temperatures, including icing, brine chilling, and other chilling methods. Time is a major factor when reducing product temperature because cooling is a gradual process; therefore, random temperature checks are recommended regardless of the cooling method used. By cooling fish, a shipper can slow spoilage and reduce the melting of the refrigerant used in shipping containers.

Pre-chilling of shipping containers before packing will prevent fish from absorbing heat from the packaging. Take care to avoid overfilling the package, which increases the risk of damaging the product, the package itself, and the aircraft during shipment. When packed in shipping containers, all fish should be near 32°F (0°C). The temperature of packed fish can be effectively maintained but not easily reduced; therefore, packing procedures should be quick and efficient to minimize temperature rise. Coolants, such as gel refrigerants, dry ice, and wet ice sealed in polyethylene bags, should be placed along the bottom and at the top of the container to absorb heat from the outside. Poor placement of coolant will reduce its effectiveness. An absorbent pad should be placed in the package to absorb possible leakage, unless packaging design ensures that liquids cannot escape.

Fish packed for shipment should be placed on vehicles as soon as possible for transport to the airport. If a delay is anticipated, it is recommended that packages be placed in refrigerated storage. However, even when held in refrigeration, the time between packing and shipment should be minimized.

As with whole and dressed fish, handling procedures for fillets should be rapid and well organized. Fillets cut from small- and medium-sized fish are not very thick and although this means that they chill very quickly, it also means that they warm up rapidly as well. The need to precool fillets to 32°F (0°C) before packing is equally important as with whole and dressed fish. In some cases, fillets may be chilled by brief immersion in ice water. Chilling by short exposure to subfreezing temperature is another satisfactory cooling method, but care should be taken to avoid freezing the fillets.

A wide variety of packing materials and styles of packing fillets for distribution is utilized. These include special tubs, tins, and other containers, as well as polyethylene bags, polyethylene sleeves, tray packs, and so on. These styles are all widely accepted for air shipment when they are subsequently packed and handled by methods and in shipping containers similar to those described for whole and dressed fish.

The successful air shipment of live seafood depends on factors similar to those involved in whole/dressed fish and fillets. However, there are some additional considerations, because the product is live. For instance, adequate air is essential for live product. Do not seal bags containing live seafood.

One general set of conditions applies to the handling and shipping of a wide variety of live seafood, such as crabs, lobsters, crawfish, clams, mussels, and oysters. The method of storing the animals before shipment may vary by shipper or species but should in any case be capable of maintaining temperatures between 34° and 45°F (2° and 7°C). The cool temperature beneficially slows body metabolism. Only the healthiest of animals should be selected for shipment.

As with other fishery products, packing procedures should be quick and efficient to minimize handling time and temperature rise. Refrigerants should be placed at the bottom of containers. A layer of moist packing material (burlap, seaweed, and synthetic products are common) should be placed over the refrigerant to protect the animals from direct contact with refrigerant. This will also provide the high relative humidity needed to prevent mortality. The packaging material and container should be prechilled.

Live seafood should be carefully packed in successive layers. The final layer of animals should be topped with a layer of moist packing material. Generally, it is recommended that an additional layer of the refrigerant be added before closing the container. Once packed, the same considerations for shipment of whole/dressed fish and fillets apply.

A summary of important handling and packing considerations for all seafood follows:

1. Select appropriate packing materials according to durability, watertightness, and insulation.
2. Pre-chill product before packing to preserve low temperatures for as long as possible.
3. Prechill live seafood to reduce body metabolism. Adequate air for live product is essential. Do not seal bags containing live seafood.
4. Choose proper cooling media; for example, gel refrigerant, wet ice in sealed bags, or dry ice (check regulatory compliance for dry ice).
5. Place cooling media to absorb heat entering package from top and bottom.
6. Minimize time between packing and shipment.

FISH AND SEAFOOD ACCEPTANCE BY AIR CARRIERS

Acceptable Weights Per Box or Carton

Differences in the kinds of aircraft and ground support equipment used by the various air carriers require that each airline have its own limitations on weight dimensions of acceptable shipments. The most common maximum acceptable weight per box for carriage on passenger aircraft is 150 pounds; however, many airlines have the capability to accept heavier weights per box or container and some have lower acceptable weights. A shipper's ability to pack in 60-to-80 pound increments facilitates handling in aircraft. In designing containers for heavier weights, provision should be made for a pallet base to accommodate a forklift. Shippers should verify each carrier's limitations on size and weight of containers.

Acceptable Refrigerants and Insulation

Most air carriers prefer that shippers use chemical coolants or dry ice; however, many air carriers will also accept wet ice if it is contained in sealed polyethylene bags. Information on acceptable refrigerants may be obtained from each carrier. (Note: Under normal conditions, wet ice by itself will melt five times faster than chemical/gel-type refrigerants.)

The refrigerant used, in combination with insulation, should protect the product for the length of exposure to ambient temperatures, taking into consideration the time required for consignee pickup. Additional protection should be provided for shipments requiring transfer to connecting carriers because of longer transit times, possible exposure to higher temperatures, more frequent handling, weather delays, and so on. Every attempt is made to keep seafood shipments refrigerated at airport facilities; however, because of differences in available facilities, refrigeration cannot be guaranteed without specific arrangements, thus making proper insulation essential.

With regard to insulation, the following selected materials are listed in descending order of their ability to insulate:

1. Urethane foam
2. Polystyrene foam
3. Shredded paper
4. Double-wall corrugated cardboard
5. Excelsior

(Note: The insulating abilities of materials are additive, so that a packaging system assembled of various components — *e.g.*, a fiberboard box with expanded polystyrene inserts — would have insulation properties from the expanded polystyrene, and a small amount from the fiberboard.)

Dry Ice

Because it transforms from solid to gaseous carbon dioxide, dry ice has the ability to displace oxygen in enclosed spaces such as aircraft interiors and cargo holds. Dry ice is therefore considered dangerous goods for air transport, and is subject to governmental regulations controlling dangerous goods. Among these controls are restrictions on placing packages containing dry ice in compartments with live animals, such as pets accompanying passengers.

A shipper who uses dry ice must comply with specific governmental regulations, which specify that packages containing dry ice be designed to permit carbon dioxide gas to escape without rupturing the package. In addition, shippers using dry ice must supply specific information on the air waybill (these special requirements are discussed in detail later) and mark the net quantity of dry ice on each package (see the next section on Markings and Labeling). It is strongly suggested that a shipper make advance arrangements with the carrier when the net quantity of dry ice exceeds five pounds per package. This information is essential, as carriers must notify pilots of the presence of dangerous goods and must obey federal regulations restricting the total amount of dry ice to 440 pounds per inaccessible cargo compartment. By stating this information on the air waybill and marking it on the outside of each package, shippers enable carriers to determine the total amount of dry ice present in all shipments aboard an aircraft.

External Markings and Labeling

Seafood transported as air cargo should be identified on the outside of the carton by markings or labels which state:*"PERISHABLE SEAFOOD"* or *"PERISHABLE FISH."*

Arrows, such as the ISO standard arrows or *"THIS SIDE UP"* markings, should be used to indicate the upright position. Also, packages containing live product should carry special *"LIVE SEAFOOD"* warnings for extra care in handling.

It is essential that complete content information be displayed on the outside of the carton. Contact information should include a twenty-four-hour telephone number for the shipper, which should also be included on the air waybill. Labels should be designed to adhere to the external surface material. Indicate on the outside of the package whether the contents are *"LIVE," "FRESH,"* or *"FROZEN."*

In addition to the preceding guidelines, there are special dry ice markings requirements. According to governmental regulations, the net weight of dry ice (carbon dioxide, solid) must be marked on the outside of each package in which it is used as a refrigerant, so that carriers may monitor the quantity of dry ice loaded aboard their aircraft. For example, a package containing five pounds of dry ice should be marked *"DRY ICE, United Nations (UN) designation 1845, NET WEIGHT 5 LBS."* (Note: Some states require the name of the species and total weight of each species be stated on the shipping unit.)

Additional marking requirements may be imposed in the United States under the Federal Lacey Act Amendments of 1981 [16 USC 3376(a) (2)]. Rules have been proposed (51 FR 24559) which would require containers of fish to be marked with the term FISH or the common name of the species, and be accompanied with a "readily accessible" document containing the following information:

1. Name and address of the shipper and consignee;
2. Total number of packages in the shipment;
3. Common name of each species in the shipment;
4. Number of weight of each species in the shipment.

The term "fish" as defined in the Lacey Act encompasses shellfish as well as other types of fish. The marking/labeling guidelines described in these guidelines are subject to change in accordance with the issuance of final federal regulations.

Banding

Banding and other types of external sealing materials should be designed not to cut or damage the container or other packages with which it may come into contact. Cartons should have a minimum of two bands around the width of each box.

FACTORS INVOLVED IN PACKAGING DESIGN

Inside Packaging

Inside the box, the product should be completely enclosed in a sealed polyethylene bag of sufficient thickness to resist puncture and retain liquids. A polyethylene bag of at least 3 mil should be sufficient. In special cases, polyethylene bags may be omitted if the container design ensures against leakage (i.e., combinations of paper/fiberboard and molded polystyrene). In appropriate cases, protective padding, absorbent materials, or wrapping such as seaweed or polystyrene (EPS) inserts should be used to assure a puncture-proof inner package. Exposed fins, claws, or other sharp objects should never be in direct contact with an inner bag.

Adequate absorbent material or padding should be used between the sealed polyethylene product bag and the inner wall of the outer packaging, unless packaging design ensures that liquids cannot escape.

Polyethylene bags containing the product (excluding live fish or shellfish) should be large enough to overlap and fold closed. Polyethylene bags used for live seafood must not be sealed.

Outside Packaging

Outer boxes should be constructed of corrugated paper board or solid fiberboard. In some cases, the various plys of paperboard should be wax-saturated, impregnated, wax-coated, or treated by other water-resistant processes. Treatments or coatings to the paperboard are necessary to provide wet strength in case of exposure to moisture. A gussetted style is recommended whenever container design permits.

Containers of molded expanded polystyrene are virtually leakproof. However, the combination of a corrugated box and a molded foam box is recommended. Of course, other methods of combining insulation, impermeability, and external strength are also acceptable.

Box and container design should take into account the density of the product to be transported. The ability to withstand the strain of dense inside weight is as important as the ability to withstand external compression and strains. External puncture-resistance is critical to assuring that the shipping container will remain leakproof in transport. The complete package should be designed to withstand shock, handling, and stacking to at least five units high without damage to the package. [Note: Governmental regulations require that packages containing dry ice be designed to release carbon dioxide gas. In general, this is not a problem with seafood shipments, since leakproof container designs are not normally airtight. A possible exception is when "barrier" (gas-impermeable) bags are used inside containers of vacuum or modified atmosphere stored shipments; for these packages, dry ice is not recommended.]

Shipments in Unit Devices

Shippers should contact individual carriers for specific policy on accepting and loading shipments packed in ULDs (large shipping containers).

Transportation from Packing House to Airport

Complete package design should provide conditions suitable for maintaining the product temperature as near as possible to 32°F (0°C). It is essential that the packaged fish reach the airport quickly. Transporting shipments in refrigerated or insulated vehicles is useful when packages may be exposed to elevated temperatures or when long trips to the airport are expected. Packages should be loaded in transport vehicles so as to minimize movement and susceptibility to dropping. In addition, any stacks of seafood packages should be planned to avoid tilted or overhanging boxes that would place undue stress on any package structure. Methods and equipment used to load and unload shipments must protect package integrity.

Although advance arrangements are not essential for all seafood shipments, they are advisable for large loads. The carrier will be better able to accommodate such a shipment if advance arrangements have been made.

Air Waybill

Inclusion of a twenty-four-hour telephone number of the shipper is essential on the air waybill, as well as on the container.

Information about the contents of shipments, such as whether the seafood is live, fresh, or frozen should be noted in the "Handling Information" box of the air waybill. Other details should also be stated in the Handling Information box. Note the following examples:
"In case of delay, please refrigerate if available."
"Hold in cooler for pickup, if available."

Limits of liability are shown in the carrier's "Contract for Carriage" on the air waybill. Such limits vary among carriers, and shippers may wish to declare the value of a shipment for insurance above the carrier's limit. There is usually no additional charge for such declarations.

Air Waybill Requirements for Dry Ice

Dry ice, when shipped by air, is considered dangerous goods. (See previous comments on dry ice and special marking requirements for dry ice.)

No "Shipper's Declaration for Dangerous Goods" form (a special dangerous goods air waybill) is required when dry ice is used as a refrigerant. Instead, the normal air waybill must be filled out so that the entry for handling information shows the words "Dangerous Goods—Shipper's Declaration Not Required." The entry for the "Nature and Quantity of Goods" box should describe the seafood and present the following dangerous goods information (in this sequence):

1. Proper shipping name for the dangerous goods, which in this case is either "carbon dioxide, solid (dry ice)" or "dry ice";
2. Hazard class or division number for the dry ice, which is 9;

3. UN identification number for the dry ice, which is "UN1845";
4. The number of packages containing dry ice;
5. Net quantity of dry ice per package;
6. UN Packing Group for the dry ice, which is "III."

For example, the entry in the "Nature and Quantity of Goods" box for a shipment of four packages of fresh fish, each containing five pounds of dry ice, would read:

"FRESH FISH, DRY ICE 9 UN 1845, 4 X 5 POUNDS III."

Seafood Claims

Every effort is made by the air carrier to meet delivery needs and arrival notification within operational constraints. However, if an unforeseen delay or other problem results in a delayed shipment, a consignee should nevertheless be prepared to take delivery and remember that settlement procedures exist for resolution of any claims following final delivery of the cargo. Since there is a potential for a loss, all relevant records should be kept by all parties involved.

Final Delivery

As with delivery to the air carrier from the packing house, timely transport of the seafood shipment from the air carrier to the consignee is vital to assuring seafood freshness and customer satisfaction.

Conclusions

ATA and NFI believe that use of the guidelines described will promote customer satisfaction in the seafood shipping, packaging, manufacture, and air transport industries. With packaging that meets the needs of the air transport environment and shipments prepared to remain in prime condition throughout the journey, air carriers can provide a reliable link between quality products and distant markets.

DISTRIBUTORS THAT TAKE OWNERSHIP OF A PRODUCT

To ensure product wholesomeness and proper sanitation, the food distributor must have the commitment of top management. That commitment must be implemented by operating supervisors and supported by the entire food distribution staff. Preventive sanitation (the performance of inspection, sanitation, building maintenance, and pest control functions designed to prevent insanitation in preference to correcting it) should be an important goal of food distribution management and of food distribution operations.

ORGANIZATION AND PROGRAMS

A program to ensure continued success in safeguarding the wholesomeness of food and in providing good sanitation will ordinarily include:

1. An organization chart showing chain of authority and responsibility;
2. A flow diagram of receiving, storage, and shipping operations;
3. Regular maintenance schedules;
4. Regular sanitation programs;
5. Regular pest control programs;
6. A HACCP Plan;
7. An effective program of follow-up and control including reports to responsible executive officer(s).

CHECKPOINTS AND ADDITIONAL GUIDES

Grounds

1. Keep nearby grounds free of liquid or solid emissions that could be sources of contamination.
2. Prevent grounds from providing conditions for insect or rodent harborage.
3. Check paving, drainage, weed, and litter control regularly.
4. Materials that are stored in the open should be stacked neatly and away from buildings. Racks above ground level are recommended where feasible.
5. Use No-Vegetation strips around exterior building walls and at property lines adjacent to properties containing potential harborage. These are helpful for discovering and discouraging travel by rodents.

Buildings

1. Provide separate and sufficient space for placement of equipment and storage materials necessary for proper operations.
2. Separate activities that might cause contamina-

tion of stored foods by chemicals, filth, or other harmful material.

3. Check structural conditions, pest barriers, and repair of windows, screens, and doors continuously.
4. Seal and clean floor-and-wall junctions and fill holes and cracks; a painted inspection strip is also recommended.
5. Keep offices, including overhead offices, clean and do not permit them to become attractants or harborage for insects or vermin. Include them in the pest control program.
6. Check false ceilings for harborage of insects and possibly rodents.
7. Give basements, attics, elevators, and rail sidings special attention.

Sanitary Operations

1. Keep walls, ceilings, and rafters free of soil, insect webbing, mold, and similar materials.
2. Do not leave unscreened doors and windows open unnecessarily.
3. Do not permit dust to accumulate.
4. Keep floors free of product spillage, oil drippage, and buildup in all areas.
5. Provide proper trash and refuse storage and removal.
6. Store tools and equipment properly.
7. Clean and flush floor drains regularly.
8. Maintain railroad and truck courts free of debris, and properly patrol them for pest control.
9. Keep eating and break areas, locker room, and so on, clean and orderly. Vending machines are often overlooked; keep them and adjacent areas clean and sanitary. Maintain equipment in a properly functioning condition and do not permit it to serve as a source of sanitation or harborage problems.

Receiving and Inspection

1. Inspect the material that is being received for evidence of damage-temperature abuse at a (HACCP) Critical Control Point; insect, bird, rodent, or other vermin infestation; and moisture, odor, or chemical contamination.
2. Exclude contaminated materials, including products, pallets, and slip sheets from the building.
3. If damaged merchandise is accepted, segregate it for special handling.
4. Make sure that incoming and outgoing vehicles are free of conditions that could contaminate product. No birds, rodents, insects, spillage, or objectionable odor should be evident.
5. At the receiving point, code or mark food received to ensure proper stock rotation.
6. To facilitate handling of rejected and suspect product, it is a good idea to develop procedures with individual shippers, carriers, and/or manufactures for reinspection, returns, and so on.

Storage

1. Store products in an orderly manner and with date codes visible for proper rotation.
2. Generally it is desirable to stack food on pallets or racks (or on slip sheets, where a clamp truck operation is utilized) and away from walls to allow for inspection aisles between stacks and walls. Painting inspection aisles in a light color is often helpful in maintaining their effectiveness. Where full inspection aisles are not provided, take special care (such as more frequent inspection, rotation, and removal of product for cleaning) to ensure sanitary, pest-free conditions.
3. Separate bagged foods to provide visibility between stacks.
4. Dispose of contaminated or infested merchandise, or otherwise promptly remove it from the premises.
5. Promptly remove damaged merchandise and broken containers from general food storage areas. Handle and process salvageable merchandise separately in an area isolated from general food storage; this area probably will require extra sanitation and pest control attention.
6. If salvage operations include the repackaging or other manipulation of exposed foods, conduct such operations in compliance with good food sanitation practices, guidelines, or regulations.

7. Do not intermingle chemicals, including pesticides, with food or food products. Such products are best separated by an aisle.

Pest Control

1. Maintain written schedules, log activity, and monitor traps and bait stations regularly.
2. Use covered bait stations which are of such types and so located as to reduce the danger of spillage; and, where appropriate, use moisture-proof bait stations.
3. Keep pesticides used in the facility secure and separate from foods. Permit their use only by properly trained personnel. Use only types registered and approved by an appropriate government agency for the intended use.
4. Check especially for rodent burrows in nearby grounds, activity at floor-wall junctions and doorways, and insect crawl marks in dust accumulation, especially on overhead pipes, beams and windowsills and around flour, sugar, and pet food storage.
5. Where feasible, seal load levelers at the dock to prevent accumulations and rodent harborage and entry, and clean them frequently.
6. Look for insect activity in folds of bagged ingredients.
7. Use black light, supplemented with means for distinguishing other chemicals that fluoresce, to check for rodent urine stains; use flashlights to check for other evidence of contamination.

Shipping

1. Make sure that transportation equipment into which food is loaded is maintained in a sanitary condition comparable to that of a food warehouse.
2. Make sure that railcars, trailers, and trucks are free of birds, rodents, and insects or contamination from them; are free of odors, nails, splinters, oil, and grease; are free of accumulations of dirt or dunnage; and are in good repair and have no holes, cracks, or crevices that could provide entrances or harborage for pests.

Follow-up: Exercise follow-up and control programs to ensure that your employees, consultants, and outside services are doing their jobs effectively.

WAREHOUSING

As mentioned earlier, if proper care is not given to good handling and warehousing practices, all the effort put into harvesting and processing will have been for naught. This important aspect of the food handling system is guided by the following rules for food warehouse workers:

1. Promote personal cleanliness among employees.
2. Provide proper toilets and hand-washing facilities.
3. Adopt good housekeeping practices.
4. Keep food handling equipment clean.
5. Reject all incoming contaminated foods.
6. Maintain proper storage temperature according to Hazard Analysis Critical Control Point (HACCP) Plans.
7. Store foods away from walls.
8. Rotate stock and destroy spoiled foods.
9. Do not use or store poisonous chemicals near foods.
10. Maintain an effective pest control program:
    - Assign inspection and reporting duties to a dependable employee.
    - Keep buildings insect-, bird-, and rodent-proof.
    - Keep doors closed when not in use.
    - Follow label directions exactly when applying insecticides or rodenticides.
    - Use highly toxic rodenticides only in locked bait boxes.
    - Remove and prevent litter around buildings.
    - Be alert for signs of rodents and insects.

The goal is to protect the public health and avoid economic loss for both distributors and customers.

The Federal Food, Drug, and Cosmetic Act requires that foods be clean, free from insect, bird, rodent or other animal filth, and free from chemicals which may render the food harmful to health. Warehousemen have the burden for compliance if they receive, ship, or store foods in interstate commerce.

Failure to comply may result in seizure of adulterated foods and prosecution of responsible individuals. FDA inspectors leave a written report of objectionable conditions.

BUILDINGS AND GROUNDS

Maintain Grounds Around Food Warehouses in a Sanitary Manner

1. Maintain the grounds around food warehouse building under the control of the operator in a well-drained condition, and free from conditions that are likely to lead to contamination of foods in the food warehouse, leaving the warehouse, or being delivered to the warehouse.
2. Keep grounds, including wharf areas, within the immediate vicinity of the food warehouse — which may provide breeding places or harborage for rodents, insects, and other pests — clean and free of discarded equipment, lumber, litter, waste, refuse, and uncut weeds or grasses.
3. Locate outside waste disposal containers on properly drained areas, clean them as needed, and keep them covered between uses.
4. Maintain and surface driveways, truck aprons, parking areas, and rail sidings at receiving and shipping areas to facilitate good drainage and to minimize dust and dirt being blown or tracked into the food warehouse. Maintain them in a clean, well-drained condition.
5. If the food warehouse buildings are closely bordered by grounds not under the operator's control, exercise special care in the food warehouse. Use inspection, extermination, or other means to exclude and control pests, dirt, and other potential contaminants originating from such noncontrolled grounds.

Maintain and Operate Food Warehouse Buildings and Structures in a Sanitary Manner

1. Provide floors and interior walls that can be adequately cleaned and keep them clean and in good repair.
2. Suspend fixtures, ducts, and pipes that pass over working areas in a way that prevents drips or condensate from contaminating food or food packages.
3. Maintain adequate separation (by location or other effective means) for those operations which may cause contamination of foods with undesirable chemicals, filth, or other extraneous material.
4. Provide adequate lighting in areas where food is received, stored, held, or assembled for delivery, in order to facilitate handling, processing, and examination of merchandise and to permit adequate inspection, cleanup, and repair of the buildings and their structures.
5. Provide adequate lighting in hand-washing areas, dressing and locker rooms (if present), and toilet rooms.
6. Employ appropriate special efforts to maintain sanitation whenever necessitated by unique features of structure or design.
7. In a food warehouse utilizing light bulbs, light fixtures, skylights, or other glass over exposed food, use safety bulbs or shielded fixtures to prevent food contamination in case of breakage.

Fixtures and Equipment

Provide food warehouse equipment that is suitable for such use and is of design, material, and workmanship that permit it to be adequately cleaned and properly maintained by the methods used at the establishment. Use and maintain the equipment so as to prevent the adulteration of foods with lubricants, fuel, metal fragments, contaminated water, or any other contaminants. Install and maintain equipment in a manner which will facilitate its cleaning and the cleaning of adjacent spaces.

Sanitary Facilities

Provide the food warehouse with adequate sanitary facilities and accommodations.

Water Supply

From an adequate source, provide a water supply which is sufficient for the food warehouse operations.

Sewage

Dispose of sewage into an adequate sewerage system or through other appropriate means.

Plumbing

Install and maintain plumbing of adequate capacity and design and in accordance with applicable governmental sanitation requirements, if any, so as to provide sufficient quantities of water to required locations through the food warehouse, and to properly convey sewage and liquid disposable waste from the food warehouse.

Toilet Facilities

Provide toilet facilities that are adequate, kept in good repair, conveniently located, well ventilated, and in compliance with applicable governmental sanitation requirements, if any. They should have self-closing doors and walls, ceilings, and floors which are tightfitting and of a material which can be easily cleaned and kept in good repair. Maintain them in a clean condition, furnish with toilet tissue, and post signs instructing employees to wash their hands with soap or detergent before returning to work.

If toilet rooms are located near areas where exposed foods might be subjected to airborne contamination, provide the rooms with self-closing doors that do not open directly into such areas.

Handwashing Facilities

Provide adequate handwashing facilities in the toilet rooms or in places convenient to the toilet rooms for handwashing after use of the toilets. Furnish such facilities with hot and cold running water, hand-cleansing soaps or detergents, sanitary towels, or other suitable drying devices.

Provide adequate receptacles, with covers, for disposal of hand-drying articles or waste material. Maintain the washing facilities and the surrounding areas in a clean condition.

Dressing and Locker Areas

If dressing and locker areas are present, provide them with adequate ventilation and lighting, and maintain them in a clean and orderly condition.

Provide lockers with sufficient ventilation to keep them dry for the retardation of mold and odors, and maintain them in a clean condition, free from trash, litter, or food scraps which serve as insect or rodent attractants. Keep the tops of lockers clean and do not use them as surfaces for the storage of materials.

Eating Areas

If there are eating areas in the food warehouse, enclose them adequately or locate them in areas away from operations. Provide adequate space, light, and ventilation. Clean eating areas regularly, and provide a sufficient number of covered receptacles for disposal of meal trash. Clean such trash receptacles regularly and do not permit them to become insect or rodent attractants.

Clean and inspect vending machines and surrounding areas at regular and frequent intervals to detect and correct unsanitary conditions which may exist. If drinking fountains are provided, locate them conveniently and clean them regularly.

SANITARY OPERATIONS

Keep Buildings and Equipment Sanitary

1. Maintain buildings, fixtures, equipment, and other physical facilities of the food warehouse in good repair and in sanitary condition.
2. Conduct cleaning operations in such a manner as to minimize the danger of contamination. For cleaning and sanitizing procedures, utilize detergents, sanitizers, and other supplies which are safe and effective for their intended uses.
3. Exclusive of packaged products held for distribution, store and use only such toxic materials as are required for necessary activities, such as for maintaining sanitary and pest-free conditions; for use in laboratory testing procedures; or for food warehouse and equipment maintenance and operation. Identify and use such products only in such manners and under such conditions as will be safe.
4. Use pesticides only under such precautions and restrictions as will prevent the contamination of food and food packaging materials.
5. Convey, store, and dispose of rubbish in a manner which will minimize the development of odor, prevent waste from becoming an attractant and harborage or breeding place for vermin, and prevent contamination of warehoused food, food containers, ground surfaces, and water supplies.

Necessary Pest Control Measures

1. Establish and maintain positive control programs designed to exclude and eliminate pests

from the food warehouse and to deny them harborage, in order to protect against the contamination of food in or on the premises by animals, birds, and vermin (including, but not limited to, rodents and insects).

2. Keep trained security dogs out of actual storage areas to avoid excreta contamination of food stored at floor level. Keep cats out of the food warehouse.
3. Implement these programs as an integral part of the construction, maintenance, operational, and personnel programs.

PROCEDURE AND CONTROLS

Conduct operations in the receiving, inspection, transporting, handling, segregating, recouping, and storing of foods in accordance with appropriate sanitation principles. Implement overall sanitation under the supervision of an individual assigned responsibility for this function. Take reasonable precautions, including the following, to assure that warehouse procedures do not contribute to contamination of foods by harmful chemicals, objectionable odors, or other objectionable materials:

1. Incoming product shipments: The integrity of the sanitation program requires that the materials, including foods and their packaging, do not expose the food warehouse to contamination by reason of infestation by insects, birds, rodents, or other vermin, or by introduction of filth or other contaminants. It is often useful to work with suppliers and shippers in advance to establish guidelines for acceptance, rejection, and, where appropriate, reconditioning of a particular product, taking into consideration factors such as the nature, method of shipment, and ownership of the product, in order to effectively implement these programs.
   a. Within a reasonable time after arrival of a car or truck and before unloading, the product should be inspected to the extent permitted by the loading of the vehicle for evidence of damage or of insect or rodent infestation, objectionable odor, or other forms of contamination. Where an adequate inspection has not been possible prior to unloading, further inspect such product during and immediately after unloading.
   b. If a damaged product has been accepted, keep it separate and recondition or otherwise handle it as necessary in a manner which will not expose foods or the food warehouse to contamination or infestation.
   c. If the inspection reveals evidence of temperature abuse, infestation or contamination, determine whether the condition is only "suspect" or is superficial (e.g., surface infestation of flying insects may be on a package, but may not have penetrated, soiled, or compromised the integrity of the packaging) and might be fully correctable by fumigation or other means. In each such case, remove the product from the food warehouse area, utilizing the vehicle in which it arrived, if feasible, after closing and sealing it. In case of contamination, if rejection is appropriate (based on the origin and ownership of the product), promptly notify the carrier and shipper of the time, and circumstances of the rejection. After removal of the product from the food warehouse because of suspect and/or superficial conditions, concentrated efforts can be made to evaluate further the actual condition of the product, and to recondition it when possible.
   d. Give special attention to a product that has previously been rejected, or has otherwise been removed from the food warehouse because of suspect and/or superficial conditions, to assure that the product and packaging are fully acceptable on reinspection.
   e. In the event of a serious question, or a failure to agree with the shipper or carrier as to condition or reconditioning, consider requesting evaluation of the suspect or rejected product by appropriate federal, state, or local authorities.
2. Store the product properly: Place foods received into the food warehouse for handling or storage in a manner that will facilitate cleaning and the implementation of insect, rodent, and other sanitary controls and maintain product wholesomeness.
3. Proper stock rotation: Adopt and implement effective procedures to provide stock rotation appropriate to the particular food.

4. Contaminated or damaged foods: Unless promptly and adequately repaired or corrected at or near the point of detection, promptly separate foods that are identified as being damaged or otherwise suspect from other foods slated for further inspection, sorting, and disposition. Promptly destroy or remove from the food warehouse any product determined to present a hazard of contamination to foods already in the warehouse.

5. Hazardous nonfood products: Nonfood products which present hazards of contamination—undesirable odors, toxicity, or otherwise—to foods in the warehouse should be handled and stored in a manner which will keep them from contaminating the foods. Take special measures to safeguard from damage and infestation those foods which are particularly susceptible to such risks.

6. Avoid damage to packaging: Exercise care in moving, handling, and storing product to avoid damage to packaging which would affect the contents of food packages, cause spillage, or otherwise contribute to the creation of unsanitary conditions.

7. Shipping: Prior to loading with foods, inspect railcar and truck trailer interiors for general cleanliness and for freedom from moisture; from foreign materials that would cause product contamination (such as broken glass, oil, toxic chemicals, etc.) or damage to packaging and contents (such as boards, nails, harmful protrusions, etc.) and from wall, floor, or ceiling defects that could contribute to unsanitary conditions.

   Clean, repair, or reject transports as necessary to protect foods before loading. Exercise care in loading foods to avoid spillage or damage to packaging and contents. Maintain docks, rail sidings, truck bays, and driveways free from accumulations of debris and spillage.

8. Warehouse temperatures: Maintain warehouse temperatures (particularly for refrigerated and frozen food storage areas) in compliance with applicable governmental temperature requirements, if any, for maintaining the wholesomeness of the particular foods received and held in such areas.

9. Housekeeping, sanitation, and inspection: Establish a regularly scheduled program of general housekeeping, sanitation, and inspection to maintain floors, wall fixtures, equipment, and other physical facilities in a state of sanitation sufficient to protect foods from contamination or adulteration and to prevent waste from becoming an attractant and harborage or breeding place for vermin.

   In addition, develop and implement an effective program and procedure for timely cleanup of any debris and spillage resulting from accidents or other unscheduled occurrences.

10. Pest control measures: Implement pest control measures designed to prevent the entrance of pests, to deny them harborage, and to detect and eliminate them, with such scheduled instructions and procedures, and by such trained and qualified personnel or professional representatives as may be necessary, based on the nature of the foods and other products handled, the structure and condition of the building and equipment, and the surroundings and environment of the warehouse.
    Monitor traps and bait stations, whether inside or outside of buildings, on a regular basis. Use covered interior bait stations designed, located, or protected to prevent spillage. Where appropriate, use bait stations constructed of moisture-proof material.

11. Pesticides: Use only pesticides with labels showing USDA or EPA registration numbers, and only for the uses specified in the labeling. Have them applied only by responsible personnel in accordance with the manufacturer's labeling instructions and in a manner which prevents contamination of foods. While not in use, clearly mark and store pesticides in a secure place apart from foods.

12. Audit food warehouse sanitation programs: Establish programs internally and/or through outside consultants for effectively auditing the foods warehouse sanitation program.

PERSONNEL

Employee Practices

1. Prohibit employees affected by communicable disease while carriers of such disease or while afflicted with boils, sores, infected wounds, or

other abnormal sources of bacterial infection, from working in the food warehouse in capacities in which there is a likelihood of food becoming contaminated or of disease being transmitted to other persons.

2. Prohibit clothing or other personal belongings from being stored, food and beverages from being consumed, and tobacco from being used in areas where foods are handled or stored.
3. Instruct employees who are working in direct contact with exposed or partially exposed foods to maintain personal cleanliness and to conform to hygienic practices to avoid contamination of such foods with microorganisms or foreign substances such as human hair, perspiration, cosmetics, tobacco, chemicals, and medicants. If gloves are used in handling such foods, use only gloves which are of an impermeable material and maintain them in a clean and sanitary condition.

Management Responsibilities

1. Assign responsibility for the overall food warehouse sanitation program and authority commensurate with this responsibility to persons who, by education, training, and/or experience are able to identify sanitation risks and failures and food contamination hazards.
2. Instruct employees in sanitation and hygiene practices appropriate to their duties and the locations of their work assignments. Instruct employees to report observations of infestations (such as evidence of rodents, insects, or harborage) or construction defects permitting entry or harborage of pests, or other developments of unsanitary conditions.
3. Exercise programs of follow-up and control to ensure that employees, consultants, and outside services are doing their jobs effectively.

TEMPERATURE CONTROL AND HANDLING PRACTICES

Foods for Freezing

1. Quick freezing seldom changes original quality; hence, only sound and wholesome raw materials at an optimum level of freshness should be frozen.
2. Freezing should be performed with appropriate equipment in such a way as to minimize physical, biochemical, and microbiological changes. With most products this goal is best achieved by ensuring that the product passes through the temperature range of maximum crystallization (for most products +30° to +23F -1° to 5°C) in an appropriate time.
3. On leaving the freezing apparatus, the product should be minimally exposed to humidity and warm temperatures and moved into a cold warehouse as quickly as practical and then allowed an adequate dwell time for temperature equilibration.
4. Where a processor has his own freezer and warehouse, product should leave the warehouse at 0°F (-18°C) or lower.

Packaging and Identification of Frozen Foods

1. Packaging and outer cases for frozen foods should be of good quality in order to prevent contamination, ensure the integrity of the product during normal transit and storage, and minimize dehydration.
2. Package coding should be adequate for effective identification.
3. Outer case coding is useful to enable proper stock rotation of individual cases. It can be preprinted on shipping cases, leaving the number to be applied at the moment of packaging, if necessary. It may also be printed on an adhesive label or applied to the case at the moment of packing. Ideally, it should appear on two or three sides of the shipping case.
4. Lot, pallet, or unit load identity is useful in enabling loads to be properly rotated while the identity of the load is maintained. Electronic Inventory Control is an advantage.

Warehouse Equipment

1. Each warehouse should have adequate capacity and should be equipped with suitable mechanical refrigeration to maintain, under anticipated conditions of outside temperature and peak loading, a reasonably steady air temperature of 0°F (-18°C) or colder, in all cold storage areas when frozen foods are stored.
2. Each storage area should have an accurate temperature measuring device installed to reflect correctly the average air temperature. Every day the warehouse is open, temperatures of each

area should be recorded and dated, and a file of such temperatures should be maintained for a period of at least two years. Warehouse customers will require temperature records for their HACCP programs.

Warehouse Handling Practices

1. The warehouse operator should record the product temperature of each lot of frozen food received and should accept custody only in accordance with good general practice. Lot arrival temperature records should be retained for a period of at least one year.
2. Whenever frozen food is received with product temperatures of 15°F (-9°C) or warmer, the warehousemen should immediately notify the owner or consignee and request instructions for special handling. These procedures may consist of any available method for effectively lowering temperatures such as blast freezing, placement in low temperature areas with air circulation, and proper use of dunnage or separators in stacking.
3. Before a shipment of frozen food is placed in storage, it should be code-marked for effective identification.
4. Frozen food should be moved promptly over loading and unloading areas to minimize exposure to humidity, elevated temperatures, or other adverse conditions.
5. During defrosting, product should be effectively covered or removed from beneath areas of accumulated frost.
6. Frozen food going into a separate break-up room for order assembly should be moved out promptly unless the break-up room is maintained at a reasonably uniform temperature of 0°F (-18°C) or colder.
7. As many operations as practicable (casing, pelletizing, etc.) should be carried out in the cold storage area to reduce the heat gain and concomitant quality deterioration, energy, and dollar loss resulting from the exposure of frozen product to ambient temperatures.
8. If slip sheets are employed, the bottom unit load should be spaced from the floor of the cold warehouse by pallet or other means. To permit air circulation, sufficient space must be allowed between stacks and walls.

Transportation

1. All vehicles used to transport frozen foods (for example, trucks, trailers, or containers, railcars, ships and aircraft) should be:
   - so constructed, properly insulated, and equipped with appropriate refrigeration continuously to maintain product temperature of 0°F (-18°C) or colder;
   - equipped with an appropriate temperature recording device to measure accurately the air temperature inside the vehicle. The dial or reading element of the device should be mounted in a readily visible position;
   - equipped with tight-fitting doors and suitable closure for drain holes to prevent air leakage;
   - clear and free from dirt, debris, offensive odors, or any substances that could, with reasonable possibility, contaminate the food;
   - pre-cooled prior to loading. The object of pre-cooling is to establish a gradient across the insulation from 0°F (-18°C) on the inner surface to the prevailing temperature on the outer skin. If the interior of the truck is exposed to warm, humid air during loading, pre-cooling is not recommend, since it leads to condensation on internal surfaces.
2. Product temperatures should be measured and be at 0°F (-18°C) or colder when tendered to the carrier for loading. The carrier should not accept product tendered at a temperature warmer than 0°F (-18°C).
3. The shipper, consignor, or warehouseman should not tender to a carrier any container which has been damaged or defaced to the extent that it is in unsalable condition.
4. Free air circulation all around the load is essential during transport. Slip sheets should be supported on a pallet (not loaded directly onto the floor of the vehicles) to allow for adequate air circulation under the load.
5. The thermostat on the vehicle's refrigeration unit should be set to maintain an air temperature of 0°F (-18°C).

Storage on Retail Premises

1. Frozen food storage facilities should be capable of maintaining a reasonably steady product temperature of 0°F (-18°C) or colder. In addition, they should be of sufficient size to provide for proper stock control.
2. Frozen food storage facilities should have sufficient circulation of refrigerated air. Cases of frozen food should be on a pallet or other means of providing adequate air circulation between the bottom case and the floor. To permit air circulation, sufficient space should be allowed between stacks and walls.
3. Frozen food storage facilities should be equipped with a thermometer (accurate to +/- 2°F/1.1°C) which is easily read and sited to measure representative air temperatures.
4. Frozen food storage should be defrosted, as necessary, to maintain refrigeration efficiency.

Temperature Measurement

1. Measuring temperature without opening packages: Select seven cases of frozen foods. Stack any three of the seven on the floor area of the natural cold environment for the lot being sampled. Cut the sidewall of the top case (number three of stack) at either end with a sharp knife. Bend the cut tab outward. Insert the probe of the temperature measurement device at about the center of the first stack of packages and between the first and second layers of packages so that all of the sensing element is in firm contact with the package walls. Stack the other four cases on top of the case containing the probe. Read and record the temperature less for a dial thermometer. Close and tape the cut sidewall areas of the case. For solid pack products, cut the sidewall of the case at either end and insert the probe at the approximate center of the first stack and between the first and second layers of the walls. For poly bags, insert the probe in the same direction as the length of the bag and deep enough for firm contact between bags. For products in paperboard packages with metal ends, turn the case on its side to give an end view, cut the sidewall of the case, and follow the same procedure as above.
2. Measuring temperature by opening packages: Whenever there is doubt about product temperatures measured without opening packages, the following procedure should also be used. (It is also recommended for the product packed in cans, because the bead rim of cans does not allow for firm contact of the probe and sidewall surfaces.) Cut the cover of the case to expose a package or can that is surrounded by other packages. Using a sharp instrument such as an ice pick, punch a hole through the cut portion of the case wall and into the central area of the exposed package or can. Insert the probe so that all of the sensing element is in the central portion of package and record a steady temperature. Replace the punctured container with a good container from a case reserved for this purpose. Close and tape the case cover.
3. Another approach: Choose a reliable, accurate (+/– 1°F) temperature with a short response time (time required to reach a steady reading) which must be calibrated frequently. Calibration can most easily be carried out by immersing in melting ice 32°F/0°C). Mercury-in-glass or alcohol-in-glass thermometers are available with either flat-blade or needle probes. Bimetal dial thermometers, which can easily be calibrated, are also suitable. Highly satisfactory digital thermometers are available with either flat-blade or needle probes.
   Before reading a temperature, precool the probe by inserting it between two packets of frozen food and waiting until a steady reading is reached. If the product is in bulk and an internal temperature is required, precool the drill before boring the hole for the thermometer probe. To obtain a reading, insert the pre-cooled probe either into a hole bored in the product or between packets, ensuring that good contact is made with the packages. The temperature of an individually quick frozen product exiting from a freezing tunnel is best measured by filling a previously precooled vacuum flask with the product closely packed surrounding the probe and reading the thermometer when a steady reading is reached.

It is strongly suggested that the transportation, distribution, warehousing sections of the Seafood Industry play their parts in maintaining a safe seafood supply as part of a HACCP program.

REFERENCES

American Frozen Food Institute. 1978. A Manual of Recommended Practices for the Warehousing of Frozen Foods. Washington, DC.

Cooperative Food Distributors of America. 1976. Voluntary Transportation Guidelines. Washington, DC.

Delta Air Lines. 1996. Success in Shipping Seafood on Delta ir Lines. Bulletin #0432-04035, Atlanta, GA.

FDA Publication. No. 81-2138. Warehouse Sanitation Handbook. Washington, DC.

Grocery Manufacturers of America. 1974. Voluntary Industry Sanitation Guidelines for Food Distribution Centers and Warehouse. Washington, DC.

National Fisheries Institute. 1982. *Proceedings of First National Conference on Seafood Packaging and Shipping*. Washington, DC.

National Fisheries Institute & Air Transport Association of America. 1988. Guidelines for the Air Shipment of Fresh Fish & Seafood (2nd ed.). Washington, DC.

Section IV.
Regulation

Introduction to the HACCP
(Hazard Analysis Critical Control Point)
Food Safety Control System

Donn R. Ward

WHAT IS HACCP?

The National Advisory Committee on Microbiological Criteria for Foods (NACMCF, 1992) explained HACCP as follows:

> The Hazard Analysis Critical Control Point (HACCP) concept is a systematic approach to the identification, assessment of risk and severity, and control of the microbiological, chemical, and physical hazards associated with a food operation. In contrast to the traditional end product inspection approach to ensuring food product safety, HACCP is a proactive strategy that anticipates food safety hazards in a process or practice and identifies the critical control points (CCPs) at which the hazard can be managed. A HACCP system will emphasize the industry's role in continuous problem prevention and problem solving rather than relying on traditional facility inspections to detect loss of control. HACCP encompasses support systems based on physical, chemical, biological, and visual testing criteria to provide for "real time" monitoring procedures to assess the effectiveness of control. HACCP plans reflect the uniqueness of a food, its method of processing and the facility in which it is prepared. HACCP is becoming an integral part of the safety assurance plans of food companies throughout the world, focused cost-effectively on the CCPs that address issues of food safety, not quality.

The implications of HACCP as defined above are enormous. It represents a shift in philosophy of the entire food inspection and regulatory system. The basis of the previous "compliance" system was problem detection, whereas with HACCP it is problem prevention. Additionally, the roles and responsibilities of regulators and the regulated industries are not clearly delineated under the compliance system. Under a HACCP system, the roles are better defined and provide a framework for functions that are complementary as opposed to adversarial.

HACCP'S HISTORY

Research and development activities associated with the United States space program have produced innovations prevalent in today's consumer products — from pocket calculators to cookware. During the 1960s, the unique requirements associated with putting astronauts in space and feeding them safe foods created the need for an innovative approach to food safety control.

Traditional food safety control programs are designed to find problems and then react to them. While this approach has merit, it relies on discovery of the problem, which leads to the question of "how hard do you look?" With respect to the space program, NASA wanted an extremely high level of confidence in the safety of foods that were to be consumed in space. Meeting these requirements using traditional methods would have resulted in a tremendous expenditure of time and energy in destructively testing each batch of product produced. In fact, the testing requirements were so rigorous that most of the product produced in each batch would have been used for testing, therefore leaving very little available for actual consumption by astronauts. Clearly, a better system had to be developed.

The Pillsbury Company, working cooperatively with NASA and the United States Army's Natick Laboratories (Massachusetts), is credited with the actual development of HACCP. Realizing the impracticalities associated with the traditional system of problem detection, Pillsbury turned the focus from detection to prevention (Bauman, 1987). This seemingly small change in focus altered the philosophical direction of food safety control by 180 degrees. It also created the need for an imaginative, yet practical, approach to implementing a prevention-based food safety control program. As the following discussion will indicate, there are several key elements associated with implementation of a HACCP system.

While the NACMCF has applied HACCP only to safety concerns, this perspective has not been followed by other governmental organizations. For example, the National Marine Fisheries Service, U.S. Department of Commerce, has included safety, quality, and economic fraud in their HACCP-based fish and shellfish inspection programs.

IMPLEMENTATION OF HACCP

The original vision for the implementation of HACCP was that food production must be treated as a total system (Bauman, 1987), from growing and harvesting to distribution and preparation (Figure 1).

While the "total system" concept may be the ultimate objective, the primary focus for the evolution of HACCP principles as they are currently defined has been at the processing level. With this in mind, many of the following procedures and principles for the implementation of a HACCP system are taken directly from the document developed by the NACMCF (1992).

Initial Steps for Development of a HACCP Plan

1. Assemble the HACCP Team

The saying that "two heads are better than one" applies to the HACCP team concept. The team approach helps assure that nothing of relevance to the HACCP plan is overlooked and also aids in fostering a sense of ownership in those who will be involved in the plan's implementation. The team may require an outside expert who is knowledgeable in the potential microbiological or other public health risks associated with the product and process.

2. Describe the Food and its Method of Distribution

A separate HACCP plan must be developed for each product that is being processed in the establishment. To assist in determining the specific hazards associated with a product, it is useful to fully describe the product, including its recipe or formulation as well as information on whether the food is to be distributed frozen, refrigerated, or shelf-stable.

3. Identify the Intended Use and Consumers

Describe the intended normal use of the food by end-users. The manner in which the food is used by consumers has a significant impact on the potential hazards associated with the food; for example, a cooked ready-to-eat (RTE) product may be substantially more hazardous than a raw product that is intended to be cooked by the end-user. Furthermore, the intended consumers can have significant implications. Foods targeted for infants or the elderly may require extra precautions.

4. Develop a Flow Diagram

Chart all steps in the manufacture of the product that are directly under the control of the processing establishment. The diagram will be helpful to the HACCP team and others in subsequent work.

5. Verify Flow Diagram

The HACCP team should inspect the actual processing operation to verify the accuracy and completeness of the flow diagram. Experiences of people involved in developing HACCP plans have consistently demonstrated the importance of this verification step. It is remarkable how easy it is to overlook a small, but significant, step involved in the manufacturing process.

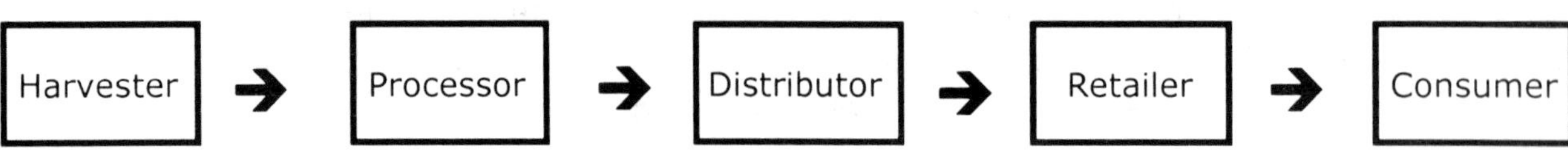

Figure 1. Total system for food production.

SEVEN PRINCIPLES OF HACCP

Principle No. 1: Conduct a hazard analysis. Prepare a list of steps in the process where significant hazards occur and describe the preventive measures.

The hazard analysis consists of asking a series of questions that are appropriate to the specific food or process. (Examples of the types of questions that may be considered are provided in the document developed by the NACMCF, 1992.) For inclusion in the list of potential hazards, the hazards must be of such a nature that their **prevention**, **elimination**, or **reduction** to acceptable levels are essential to the production of a safe food. Hazards which are of low risk and not likely to occur would not require further consideration. The team must then consider what preventive measures, if any, can be applied to each hazard. Preventive measures are physical, chemical, or other interventions that can be used to control an identified health hazard. More than one preventive measure may be required to control a specific hazard. More than one hazard may be controlled by a specified preventive measure.

During the hazard analysis, safety concerns must be differentiated from quality concerns. A hazard is defined as a biological, chemical, or physical property that may cause a food to be unsafe for consumption. As applied to HACCP, the NACMCF believes very strongly that the term hazard applies only to **safety**. Hence, the HACCP team must make a determination whether a potential problem is a safety concern or a quality concern. If the issue is quality, then it is not addressed in the HACCP plan.

Principle No. 2: Identify the CCPs in the process.

The definition of a critical control point is: a step or procedure at which control can be applied and a food safety hazard can be prevented, eliminated, or reduced to acceptable levels. To facilitate identification of CCPs, refer to the CCP decision tree (Figure 2). Examples of CCPs may include, but are not limited to: cooking, chilling, specific sanitation procedures, product formulation control, prevention of cross-contamination, and certain aspects of employee and environmental hygiene.

It is important to note that hazards and critical control points can be process-specific, not just commodity- or product-specific. Consequently, if the process is modified, then it may be that the hazard(s) have changed, and thus the CCP(s) must be modified accordingly.

Principle No. 3: Establish critical limits for preventive measures associated with each identified CCP.

A critical limit is defined as a criterion that must be met for each preventive measure associated with a CCP. Each CCP will have one or more preventive measures that must be properly controlled to assure prevention, elimination, or reduction of hazards to acceptable levels. Thus, each preventive measure has associated with it critical limits that serve as boundaries of safety for each CCP. Critical limits may be set for preventive measures such as temperature, time, physical dimensions, humidity, moisture level, water activity (a_w), pH, titratable acidity, salt concentration, available chlorine, viscosity, preservatives, or sensory information such as texture, aroma, and visual appearance. Critical limits may be derived from sources such as regulatory standards and guidelines, literature surveys, experimental studies, and experts.

Whenever a critical limit is exceeded, it is indicative that the process is out of control and that a potential hazard exists. Moreover it indicates that some action is needed to bring the process back under control. It does not mean, however, the automatic destruction of products produced while the situation was deemed "out of control." This will be further discussed under Principle 5, corrective actions.

Principle No. 4: Establish CCP monitoring requirements. Establish procedures for using results of monitoring to adjust the process and maintain control.

In order to determine that the process at a CCP is under control and that the critical limits have not been exceeded, some system of surveillance is essential. In HACCP this is achieved by monitoring, which is understood as a planned sequence of observations or measurements to assess whether a CCP is under control and to produce an accurate record for future use in verification. Monitoring serves three main functions. First, monitoring is essential to food safety management in that it tracks the system's operation. If monitoring indicates that there is a trend towards loss

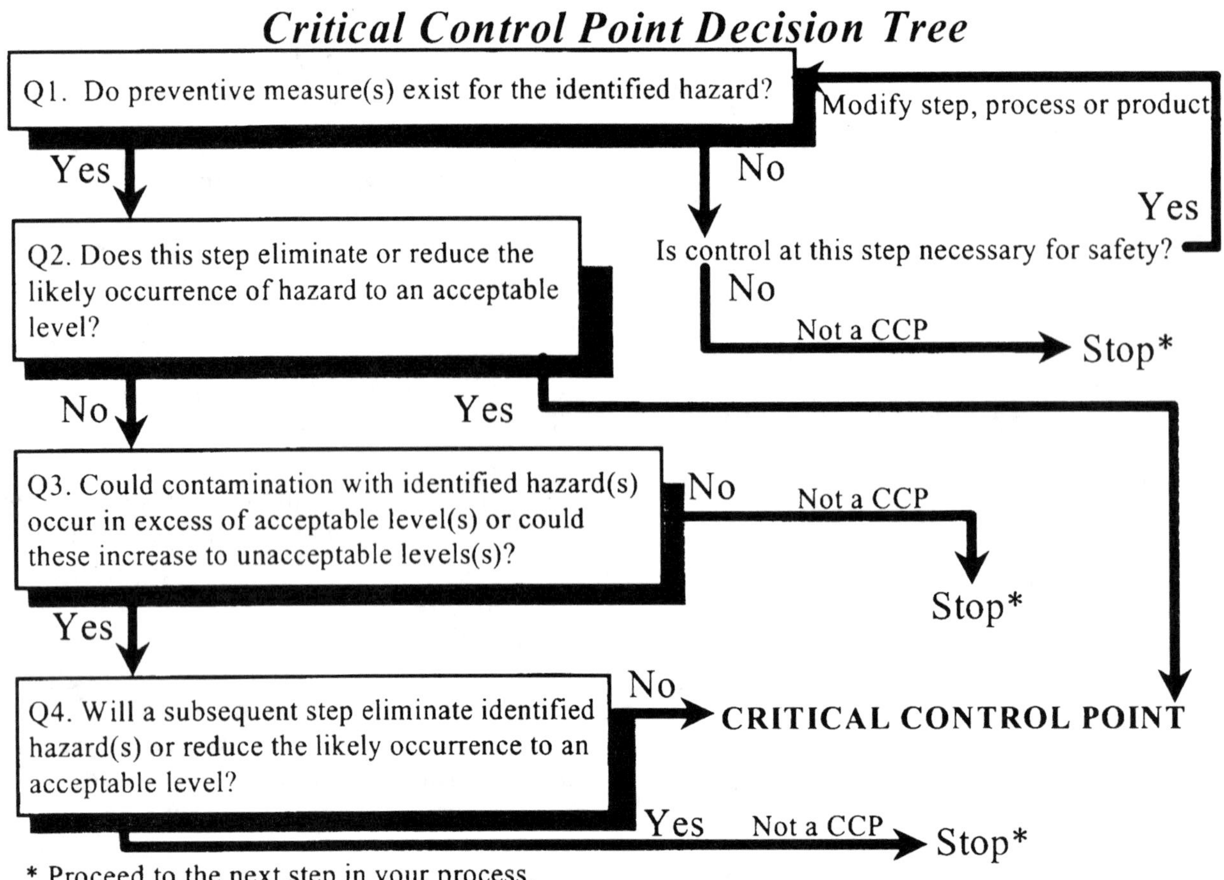

Figure 2. CCP Decision Tree. Apply at each step of the process with an identified hazard.

of control, such as exceeding a target level, then action can be taken to bring the process back into control before a deviation occurs. Second, monitoring is used to determine when there is loss of control and a deviation occurs at a CCP, i.e., exceeding a critical limit (corrective action must be taken). Third, as will be discussed, monitoring requires written documentation — thus, monitoring provides the documentation necessary for verification of the HACCP plan.

Ideally, monitoring should be at the 100 percent level, or, in other words, continuous. Some CCPs can be easily monitored on a continuous basis: e.g., the time and temperature of a crab cooking and/or pasteurization process. Freezer or cooler temperatures, pH measurements, and concentrations of brine solutions are other examples of processing steps that can be monitored on a continuous basis. There are situations for which continuous monitoring is not readily feasible. Under these circumstances, there are two important considerations: 1) specific assignment of monitoring responsibilities and 2) monitoring intervals. With respect to the personnel monitoring CCPs: they must be trained in the technique used to monitor each preventive measure; fully understand the purpose and importance of the monitoring; have ready access to the monitoring activity; be unbiased in monitoring and reporting; and accurately report the monitoring activity. The person responsible for monitoring must also report a process or product that does not meet critical limits so that immediate corrective action can be taken. As for monitoring intervals, it is crucial to establish monitoring intervals and procedures that will provide reliable proof that the process is under control, and therefore the hazard is being controlled. Statistically designed data collection or sampling systems lend themselves to this purpose.

Finally, because of the importance of "real time" information, monitoring must be done rapidly. Processing line situations can change quickly;

to be effective, monitoring procedures must detect these changes as they happen, or very soon thereafter. For this reason, microbiological testing is not recommended for **routine** monitoring. Physical and chemical measurements are preferred because they may be done rapidly and can indicate the conditions of microbiological control in the process.

Principle No. 5: Establish corrective action to be taken when monitoring indicates that there is a deviation from an established critical limit.

Food manufacturing processes are not ideal, and as an inevitable consequence deviations from desired operational parameters will occur. When these process deviations exceed critical limits at CCPs, procedures, called corrective actions, must be in place to: (a) determine the disposition of non-compliant product, (b) fix or correct the cause of non-compliance to assure that the CCP is under control, and (c) maintain records of the corrective actions that have been taken where there has been a deviation from the critical limits. Quite simply, corrective actions require that a firm plan in advance how it intends to handle problems for each CCP.

Principle No. 6: Establish procedures for verification that the HACCP system is working correctly.

There are four processes involved in verification. The first is the scientific or technical process to verify that the critical limits for control measures at CCPs are satisfactory. This process can be complex. The process consists of a review of the critical limits to verify that the limits are adequate to control the hazards that are likely to occur.

The second process of verification ensures that the facility's HACCP plan is functioning effectively. A functioning HACCP system requires little end-product sampling, since appropriate safeguards are built in early in the process. Therefore, rather than relying on end-product sampling, firms rely on frequent reviews of their HACCP plan, verification that the HACCP plan is being correctly followed, review of CCP records, and determinations that appropriate risk management decisions and product dispositions are made when process deviations occur.

The third process consists of documented periodic revalidations, independent of audits or other verification procedures, that must be performed to ensure the accuracy of the HACCP plan. Revalidations are performed by the HACCP team on a regular basis and/or whenever significant product, process, or packaging changes require modification of the HACCP plan. The revalidation includes a documented on-site review and verification of all flow diagrams and CCPs in the HACCP plan. The HACCP team modifies the HACCP plan as necessary.

Last, but certainly not least, is the government's role. The fourth process of verification deals with regulatory agency responsibility and actions to ensure that the establishment's HACCP system is functioning satisfactorily.

The NACMCF listed the following as examples of verification activities:

A. Verification procedures may include:
   1. Establishment of appropriate verification inspection schedules.
   2. Review of HACCP plan.
   3. Review of CCP record.
   4. Review of deviations and dispositions.
   5. Visual inspections of operations to observe if CCPs are under control.
   6. Random sample collection and analysis.
   7. Review of critical limits to verify that they are adequate to control hazards.
   8. Review of written record of verification inspections that certifies compliance with the HACCP plan or deviations from the plan and the corrective actions taken.
   9. Validation of HACCP plan, including on-site review and verification of flow diagrams and CCPs.
   10. Review of modifications to the HACCP plan.

B. Verification inspections should be conducted:
   1. Routinely, or on an unannounced basis, to assure selected CCPs are under control.
   2. When it is determined that intensive coverage of a specific commodity is needed because of new information concerning food safety.
   3. When foods produced have been implicated as a vehicle of foodborne disease.
   4. When requested on a consultative basis or when established criteria have not been met.

5. To verify that changes have been implemented correctly after a HACCP plan has been modified.

C. Verification reports should include information about:
 1. Existence of a HACCP plan and the person(s) responsible for administering and updating the HACCP plan.
 2. The status of records associated with CCP monitoring.
 3. Direct monitoring data of the CCP while in operation.
 4. Certification that monitoring equipment is properly calibrated and in working order.
 5. Deviations and corrective actions.
 6. Any samples analyzed to verify that CCPs are under control. Analysis may involve physical, chemical, microbiological, or organoleptic methods.
 7. Modifications to the HACCP plan.
 8. Training and knowledge of individuals responsible for monitoring CCPs.

Principle No. 7: Establish effective recordkeeping procedures that document the HACCP system.

A defining element of HACCP is recordkeeping; without it there is no documentation that the HACCP system is functioning. While recordkeeping is a fundamental part of HACCP, records access is one of the most sensitive issues in HACCP implementation. The NACMCF recommends that the records used in the total HACCP system include:

1. The HACCP Plan:
 - Listing of the HACCP team and their assigned responsibilities
 - Description of the product and its intended use
 - Flow diagram for the entire manufacturing process, indicating CCPs
 - Hazards associated with each CCP and preventive measures
 - Critical limits
 - Monitoring system
 - Corrective action plans for deviations from critical limits
 - Recordkeeping procedures
 - Procedures for the verification of the HACCP system
2. Records obtained during the operation of the plan:
 - Ingredients
   a. Supplier certification documenting compliance with processor's specifications
   b. Processor audit records verifying supplier compliance
   c. Storage temperature record for temperature-sensitive ingredients
   d. Storage time records of limited shelf life ingredients
 - Records relating to product safety
   a. Sufficient data and records to establish the efficacy of barriers in maintaining product safety
   b. Sufficient data and records establishing the safe shelf life of the product, if age of product can affect safety
   c. Documentation of the adequacy of the processing procedures from knowledgeable process authority
 - Processing
   a. Records from all monitored CCPs
   b. Records verifying the continued adequacy of the process
 - Packaging
   a. Records indicating compliance with specifications of packaging
   b. Records indicating compliance with sealing specifications
 - Storage and distribution
   a. Temperature Records
   b. Records showing no product shipped after shelf life date on temperature-sensitive products
 - Deviation and corrective action records
 - Validation records and modification to the HACCP plan indicating approved revisions and changes in ingredients, formulations, processing, packaging, and distribution control, as needed
 - Employee training records

CONCLUSION

The seven principles outlined above are the heart and soul of the HACCP system. There is nothing complicated or mysterious about any of the principles. Any difficulties associated with HACCP are usually not in understanding the concept, but in its implementation. Some have attributed these difficulties to the "cultural change" associated with the shift in emphasis from detection to prevention. No doubt, the cultural change will create challenges for many processors. However, perhaps the greatest challenge to successful integration of HACCP into the food industry will be commitment. In order for the value of HACCP to be realized, commitment to the precepts of HACCP is essential from both management of the food-processing industry and government regulatory agencies.

REFERENCES

Bauman, H.E. 1987. The Hazard Analysis Critical Control Point Concept. In *Food Protection Technology*. C.W. Felix (ed.), pp. 175-179. Lewis Publishers Inc., MI.

National Advisory Committee on Microbiological Criteria in Foods. 1992. Hazard Analysis and Critical Control Point System. United States Department of Agriculture Food Safety and Inspection Service. Washington, DC.

Religious Food Laws and the Seafood Industry

Joe M. Regenstein and Carrie E. Regenstein

Both Jewish (kosher) and Muslim (halal) laws control the dietary habits of their adherents, affecting the purchase and use of foods. The food industry must often adjust its processing, handling, sales, and marketing plans to accommodate these groups. This chapter will briefly review some of these laws as they might impact on the seafood industry. For more complete information about kosher, see Regenstein and Regenstein (1992; 1990a,b; 1988; 1979); and about halal, see Chaudry (1992) and Awan (1988).

Other religious and philosophical groups also have dietary laws that affect the marketplace. In many cases, these groups will use the kosher dietary law system in place in the United States to guide their purchases. Included in this category are the Seventh Day Adventists (Bosley and Hardinge, 1992); vegetarians; and people with allergies or aversions to dairy, shellfish, and some grains (Regenstein and Regenstein, 1994).

The kosher dietary laws, "kashruth," are observed to varying degrees by members of the Jewish faith. There are three broad branches of Judaism: the Orthodox are the most traditional, the Conservatives are in the middle, and the Reform are the least traditional. It is estimated that some 500,000 families of the almost six million Jews in the United States — whether Orthodox, Conservative, or Reform — observe some form of these rules that affect their purchase of food products. Worldwide, there are about 14 million Jews; the other major population centers are Europe and Israel.

Estimates of the number of Muslims in the United States vary widely: from one million to 14-15 million. Most knowledgeable observers think that four to six million is a reasonable estimate. However, worldwide the number reaches 800 million to over one billion, covering many countries throughout Asia, Europe, and Africa. Overall, the religious/philosophically-based food market in the United States is estimated to include about 25 million people.

Interpretation and guidance regarding procedures for producing kosher and halal products are available from rabbis and imams (mullals) respectively, and from other religious officials specializing in food processing (Figure 1).

THE ORIGIN OF KOSHER AND HALAL LAWS

The system of kosher laws is part of the larger religious and historical context of Judaism. Kosher laws are not "health laws" (Regenstein, 1994). Many of them may make sense from a health point

HALAL **Natural** Symbol for product/s containing neither meat, dairy, or fish ingredient or flavor.

HALAL **Dairy** Symbol for product/s containing dairy ingredient or flavor.

HALAL **Meat** Symbol for product/s containing meat ingredient or flavor.

HALAL **Fish** Symbol for product/s containing fish ingredient or flavor.

HALAL **Dairy & Meat** Symbol for product/s containing dairy and meat ingredients or flavors.

HALAL **Dairy & Fish** Symbol for product/s containing dairy and fish ingredients or flavors.

HALAL **Meat & Fish** Symbol for product/s containing meat and fish ingredients or flavors.

HALAL **Dairy, Meat & Fish** Symbol for product/s containing meat, dairy, and fish ingredients or flavors.

Figure 1. HALAL symbols on labels of certified products.

of view, especially historically, but that has not been their justification.

Grunfeld (1972) explains the dietary laws in their ethical context:

> And ye shall be men of a holy calling unto Me, and ye shall not eat any meat that is torn in the field' (Exodus XXII:30). Holiness or self-sanctification is a moral term; it is identical with . . . moral freedom or moral autonomy. Its aim is the complete self-mastery of man.
>
> To the superficial observer it seems that men who do not obey the law are freer than law-abiding men, because they can follow their own inclinations. In reality, however, such men are subject to the most cruel bondage; they are slaves of their own instincts, impulses, and desires. The first step towards emancipation from the tyranny of animal inclinations in man is, therefore, a voluntary submission to the moral law. The constraint of law is the beginning of human freedom....Thus the fundamental idea of Jewish ethics, holiness, is inseparably connected with the idea of Law; and the dietary laws occupy a central position in that system of moral discipline which is the basis of all Jewish laws.
>
> The three strongest natural instincts in man are the impulses of food, sex, and acquisition. Judaism does not aim at the destruction of these impulses, but at their control and their sanctification. The law spiritualizes these instincts and transfigures them into legitimate joys of the life.

The Jewish dietary laws control and sanctify man's food impulse.

Four Major Concepts in the Kosher Laws

Plants

Man was initially a vegetarian in the Garden of Eden. All plants grown in the United States are kosher. [There are, however, special rules governing grape juice and wine (which will be discussed later).]

Nevertheless, Grunfeld discusses numerous fine points of law relating to agricultural practices involving plants (such as prohibiting grafting certain types of plants) (1972). In some cases these subtleties restrict the growing of some plants by people of the Jewish faith, but do not preclude their using them once grown.

There are also a number of laws dealing with products of Israel, involving both the required biblical tithes and the observance of the sabbatical year for the land (where it is to remain fallow). Special attention must, therefore, be paid to products grown and/or produced in Israel.

The most important Muslim regulation concerning plants is that the Muslim community has a total ban on the use of alcohol. Thus, food products that use alcohol as a solvent are prohibited. Vinegar, however, is permitted as long as the alcohol is fully fermented. The Glorious Quran, Chapter V, Verse 90 states:

> O ye who believe! Strong drinks and games of chance, and idols and divining arrows are only an infamy of Satan's handiwork. Leave it aside in order that ye may succeed.

Flesh and Blood

Leviticus XVII:12-14 states:

> Therefore I said unto the children of Israel: No soul of you shall eat blood, neither shall any stranger that sojourneth among you eat blood. And whatsoever man there be of the children of Israel, or of the stranger that sojourn among them, that taketh in hunting any beast or fowl that may be eaten, he shall pour out the blood thereof and cover it with dust. For as to the life of all flesh, the blood thereof is all one with the life thereof; therefore I said unto the children of Israel: Ye shall eat the blood of no manner of flesh; for the life of all flesh is the blood therein; whatsoever eateth it shall be cut off.

These passages present the concept of blood as life, and thus prohibit it as a food. Laws concerning the slaughtering process ensure that this precept is properly fulfilled (Grandin and Regenstein, 1994; 1992). Notice that only the beast and fowl are cited; this restriction does not apply to fish, making it a more flexible item in the kosher market. However, Judaism stresses the importance of kindness to animals, and these more general humanitarian regulations also apply to fish.

The Muslim community is also concerned with the method of slaughter of animals and the proper removal of blood. In the Glorious Quran, Chapter II, Verse 173 we find:

> He hath forbidden you only carrion, and blood, and swine flesh, and that which hath been immolated to any other than ALLAH .
>
> . . .

Leviticus XI:39-40 states:

> And if any beast of which ye may eat, die, he that toucheth the carcass thereof shall be unclean until the even.
>
> And he that eateth of the carcass of it shall wash his clothes, and be unclean until the even; he also that beareth the carcass of it shall wash his clothes, and be unclean until the even.

Thus, an animal that dies naturally or that was hunted wild cannot be used for food; apparently this law and the one that follows do not include fish. As shown above, the Muslim community also restricts the use of animals that have not been properly slaughtered according to Muslim requirements.

In Deuteronomy XIV:21:

> Ye shall not eat of any thing that dieth of itself; thou mayest give it unto the stranger that is within thy gates, that he may eat it; or thou mayest sell it unto a foreigner; for thou art a holy people unto the Lord thy God.

In this passage, as in others, it is clear that kosher laws are restricted to Jews. Nonkosher meat may be offered to others. In practical terms, this provision is extremely important, because the rules concerning the kosher processing of the hindquarters of ruminants are so complex that it is generally easier to sell the hindquarter to non-Jews.

The Muslim community also requires that animals be ritually slaughtered. However, in the absence of Muslim (halal) meat, the community will sometimes accept meat ritually slaughtered by Christians or Jews (People of the Book). In practice, this means that Muslims may purchase kosher meat.

Milk and Meat

The third major concept of the kosher laws concerns the separation of milk and meat. There is no Muslim equivalent. It appears first in Exodus XXIII:19: "Thou shalt not seethe a kid in its mother's milk."

The *Five Books of Moses* repeats this concept three times. At a humanitarian level, this avoids cruelty: a mother and her offspring should not be killed at the same time. This rule has been read as a directive toward maintaining a separation of meat and milk, necessitating two sets of dishes, cookware, eating utensils, and processing equipment: one for "dairy," and one for "meat." The dairy and meat foods must be eaten separately, ruling out foods such as cheeseburgers and the use of dairy (coffee) creamers at meat meals. Though customs vary, most kosher Jews wait at least four hours (usually six) after a meat meal to eat dairy foods; from dairy meals to meat meals there may be no wait (but a ritual washing of the mouth is required) to a wait of one hour. [In the case of aged "sharp" cheeses (over six months aging) eaten cold, many observe the custom of waiting as long after eating such cheese as they would wait after a meat meal.]

Products that are neither meat nor dairy are referred to as "pareve" or "parve." Eggs, fish, and vegetable products are all pareve.

Acceptable Animal Species

The fourth major concept includes the designation of which species of animals are fit to eat, and which of these must be made kosher ("koshered"). The relevant laws appear twice, in Leviticus and Deuteronomy; the section in Leviticus includes much more detail. The Muslim community also defines acceptable and unacceptable animals. Of particular significance to both groups is the pig, which is prohibited as a food to the Jews and which is totally prohibited by the Muslims.

Meat: The mammals accepted as kosher both chew their cud and are cloven- or split-hoofed. These animals are all herbivores. Muslim law, for example, prohibits as food such animals as: carnivorous animals, elephants, monkeys, apes, donkeys, and mules.

Fowl: Domestic fowl are permitted. They are considered meat and must be slaughtered and treated accordingly (see below). Muslim law prohibits vultures, eagles, and the hoopoe.

Fish: Fish are not considered to be "meat." Thus, they can be used as the "center of the plate" protein source for a dairy meal. Most Orthodox Jews permit the mixing of fish and dairy, although a few prohibit it. Although fish can be used at a meat meal, it must be eaten separately. Eating utensils (silverware and plates) are traditionally collected and washed before further use. Products containing fish, e.g., Worcestershire sauce (anchovies), cannot be used directly on meat. The dietary laws do not prescribe particular slaughter practices for fish such as those defined for meat. Leviticus XI:9-12 states that some fish are kosher, others nonkosher:

> These may ye eat of all that are in the water: Whatsoever hath fins and scales in the waters, in the seas, and in the rivers, them may ye eat. And all that have not fins and scales in the seas, and in the rivers, of all that swarm in the waters, and of all the living creatures that are in the waters, they are a detestable thing unto you.
>
> And they shall be detestable things unto you; ye shall not eat of their flesh, and their carcasses ye shall have in detestation.
>
> Whatsoever hath no fins or scales in the waters, that is a detestable thing unto you.

Therefore, fish with fins and removable scales are acceptable. On the other hand, all shellfish (molluscan and crustacean) are prohibited. Rabbi Yakov Lipschutz of National Kosher in Monsey, New York (1988), offers the most complete list of kosher and nonkosher fish. The list has been prepared for many years by Dr. James W. Atz (Curator Emeritus, Department of Herpetology and Ichthyology, American Museum of Natural History, New York). Appendix 1 contains a list of nonkosher fish. [See also Regenstein and Regenstein (1981) for a list of kosher and nonkosher fish.]

Dr. Atz's list includes as kosher those fish with ctenoid scales (those with minute spiny projections at their exposed edge): for example, black bass; and those with cycloid scales (lacking the minute spines and having rounded edges): for example, carp and herring. The kosher status of fish with ganoid scales is more controversial (those with heavier and thicker scales: for example, sturgeon). Finally, fish with placoid scales (firmly attached to the skin and with tiny spinous projections: for example, shark) are nonkosher. Thus, within the category of finfish (Pisces), some fish are clearly acceptable, some are clearly unacceptable, and a few others are a matter of interpretation.

Within the same species, some fish may be covered with scales, while others may have as few as two scales. The carp, the standard kosher fish used to make the Jewish delicacy "gefilte fish" [a fish ball held together with eggs and matzo meal (bread crumbs made from unleavened flour) and boiled for about an hour] is usually covered with scales; however, on occasion it has been found without scales (Ginsburg, 1961).

The Orthodox and Conservative rabbis maintain separate definitions of acceptable fish. Juvenile swordfish have scales that are specialized and therefore unique. Conservative rabbis accept swordfish because the fish had juvenile scales while still in the water (Freedman, 1970). According to Freedman, the Orthodox rabbis contend that the juvenile scales of the swordfish are not real scales, but mere "bony tubercles or expanded, compressed, platelike bodies."

These scales are rough, having spinous projections at the surface, and they do not overlap one another as do scales in most other fishes. The scales disappear as the fish grows and the larger fish, including those sold in the market, have no scales (Ginsburg, 1961).

Furthermore, the Orthodox contend that even if the juvenile scales were real scales, no religious rulings allow for eating a species that has scales only as a juvenile. Similar arguments apply to sturgeon and some other fish. For example, the Union of Orthodox Jewish Congregations of America (the largest kosher-certifying agency in the United States) also considers North Atlantic turbot (*Pasetta maximus*) to be nonkosher.

According to Isaac Ginsburg (1961), a systematic zoologist with the former Bureau of Commercial Fisheries (Fish and Wildlife Service, United States Department of the Interior), the sturgeon has some large ganoid scales and some small tuberculoid scales. The freshwater eel also has scales, apparent only when its skin dries. Orthodox rabbis do not accept these species as kosher.

Surprisingly, tuna, now a staple of the kosher home, was once considered nonkosher because it generally does not have scales by the time it is sold. (Tuna and mackerel scales drop off when the fish thrashes in the catch process.) The yellowfin tuna has only a few rows of scales along the nape. Some rabbis are concerned that mutants might arise without scales. Tuna fish processing plants, like many plants producing kosher products, are not under continual rabbinical inspection.

Fish represents an important protein food in the kosher home. Unlike meat and poultry that require special slaughter procedures and subse-

Table 1. Examples of Nonkosher Fish

Before making a final decision, a competent authority should be consulted. Given the various names used for fish, care should be used in interpreting this list. (This list has been reorganized from that by Dr. James W. Atz in Lipschutz, 1988. For Latin names, please check the original reference.)

Alligator gar	Electric rays	Lumpsuckers	Sculpins
Alligatorfishes	Elephantfishes	Mako shark	Sea catfishes
American eel	European eel	Mantas	Sea raven
Angel sharks	European turbot	Maori trout	Sea-squab
Anglers	Filefishes	Marlin	Shannys
Armored gurnards	Five-bearded rockling	Midshipman	Sharks
Australian eel	Flying gurnards	Minnow	Sharksucker
Beluga	Fringeheads	Molas	Sharpnose puffers
Billfishes	Frogfish	Monkfish	Skate
Black durgon	Frostfishes	Moorish idols	Smoothhounds
Blenny	Fugu	Morays	Snailfishes
Blowfishes	Galaxias	Mountain trout	Soupfin shark
Boxfishes	Gars	Noodlefish	Spearfishes
Brotulas	Goby	Ocean catfish	Spoonbill cat
Bullhead	Goosefish	Ocean pout	Stargazers
Burrfishes	Grayfish	Ocean sunfish	Stingrays
Butterfly rays	Guitarfishes	Oilfish	Sturgeons
Butter jew	Gunnel	Paddlefish	Suckerfishes
Buttersnook	Hairtail	Pejesapo	Surgeonfishes
Cabezon	Icefish	Pikeblennies	Swellfishes
Catfishes	Inanga	Porcupine fishes	Swordfish
Channel catfish	Japanese eel	Pout	Tang
Chimaeras	Jollytail	Prickleback	Toadfishes
Clingfishes	Kalas	Puffers	Tobies
Cockscombs	Kelpfishes	Queenfishes	Torsk
Conger eel	Kihikihis	Ratfish	Triggerfishes
Cowfish	Kingklip	Ray	Trunkfishes
Cusk-eels	Klipfishes	Remora	Unicorn fishes
Cutlass fishes	Kokopus	Rockskippers	Viviparous blenny
Doctorfish	Lae	Rocksucker	Walking catfishes
Dogfish	Lampreys	Roughies	Warbonnets
Dragonets	Leatherback	Sailfish	Whitebait
Eaglerays	Leatherjacket	Sandfish	Wolf-eel
Eels	Lings	Sawfishes	Wolf fishes
Eelpout	Lumpfish	Scabbardfishes	

quent religious processing (thereby raising its price), fish has no such requirements. Thus, fish can be sold at its normal market price. Some Orthodox consumers may request that any gutting, dressing, or filleting be done using a knife that they may supply so that they can be assured that the same knife was not used for processing nonkosher fish.

Another concern for the kosher consumer is the fish parasite. Although some rabbis accept such parasites, the average kosher consumer is probably less tolerant of parasites than the general population.

The Muslim community is also concerned with fish, and some sects do not generally consume shellfish. Muslim consumers do, however, accept all teleost (bony-skeletoned) fish.

THE IMPLICATION FOR PROCESSOR AND MARKETS

With the advent of modern food technology, the problems of following a kosher or halal diet have multiplied, but the opportunities to obtain kosher and halal foods have also increased. Food additives, increased food processing, biotechnology (Chaudry and Regenstein, 1994), and the use of sophisticated machinery have all made keeping kosher or halal more complex than ever.

Listed in this section are a few of the foods and ingredients that cannot be taken at "face value." These are some of the items that force kosher and halal homemakers to think as food technologists do, and that must be considered by producers of kosher or halal products. Indeed, these products must be processed and handled carefully with knowledge of the origin and prior handling of all ingredients. Furthermore, the equipment used for processing must be properly koshered or prepared for halal production. Among other things, there must be no intermingling with prohibited items. To ensure proper handling of kosher fish, rabbinical supervision and a system of certification is required; the Muslim community is developing a similar system. In many cases, the fish industry may not need as much rabbinical supervision as the meat industry does.

On the other hand, because fish are often processed such that they lose their identity (e.g., processed into fillets or mince), kosher certifiers must assure themselves that only kosher fish are processed within each batch. Skinned fillets, minces, and similar products, therefore, require continuous rabbinical supervision during their production, and subsequent monitoring throughout any further processing.

At least one kosher supervision agency permits the production of raw fillet blocks (e.g., overseas product) without continuous supervision. Secular controls serve as a disincentive for cheating; such blocks still require supervision of further processing such as cutting and breading.

CONSIDERATION OF FOOD INGREDIENTS

The following ingredients are the greatest concerns in producing fish products:

Rennin is an extract from animal stomachs. While it is derived from an animal or "meat" source, it can be used in making kosher cheese — a dairy product — if it has been obtained from a kosher-slaughtered animal. Muslim law requires that animal rennin come from halal-slaughtered animals.

Some rabbis require that the stomach be fully dried before extraction. There is disagreement among rabbis concerning the acceptability of rennin from nonkosher-killed animals. This ruling affects the status of cheese and is thus important for those products in which cheese, cheese sauces, etc., are mixed with fish. With the advent of biotechnology, both the Jews and the Muslims have accepted the biotechnology-derived chymosin (the main enzyme in rennin), so that obtaining kosher cheese has become less difficult. Acid-set cheeses (even if some rennin is used) have no such requirements.

There is a requirement by many of the mainstream kosher certification groups that the coagulant be added to each batch by a Sabbath-observing Jew. (Some attempts to permit this process to be done remotely are currently being researched.)

Gelatin, i.e., cooked collagen, is usually made from bones and skins, which are considered "pareve" or neutral; therefore, gelatin can be used with meat or dairy products. Some rabbis do not feel it is possible to make pareve gelatin from mammals or fowl. Other authorities feel that the gelatin must be derived from kosher-killed animals or at least from very dry bones such as one may obtain from, for example, India. And still others argue that since gelatin processing includes a stage in which it is unfit to eat, it therefore becomes a chemical without any source identity. Clearly, gelatin and rennin are controversial and require special thought by both kosher consumers and manufacturers. Kosher fish gelatin similar to isinglass must be prepared exclusively from kosher fish. Muslims also require that mammalian gelatin be produced from religiously-slaughtered animals.

A number of companies both in the United States and abroad have brought kosher and halal fish gelatin to the marketplace. There is also one beef gelatin made exclusively from hides obtained from kosher-killed animals.

Lactose and **sodium caseinate** are dairy products (Regenstein and Regenstein, 1983). Confusingly, secular rulings demand that certain products with these ingredients (for example, caseinates) be marked "nondairy" while the kosher community also requires that they be marked "KM"

Table 2. Kosher Supervision Agencies	
The "OU"	The "Star K"
The "OK"	The "Kof-K"
The "cRc"	The "National K"
The "KVH"	

[a]A comprehensive list of agencies with names and addresses is available from the authors.

Table 3. Halal Supervision
The "Crescent M"

[a]A comprehensive list of agencies with names and addresses is available from the authors.

or "KD" ("kosher milchig"; that is, kosher dairy). Regardless of the nondairy designation, these foods can be eaten only as dairy foods in kosher homes.

Another product, whey (or whey solids), is obtained as a byproduct of cheese manufacture and therefore is a dairy product. If a kosher rennin was used — even if it was not added by a Sabbath-observing Jew — the whey can be acceptable as a kosher product. However, if it is obtained from a nonkosher cheese (note the rennin problem), then, clearly, the whey cannot be kosher. Lactic acid derived from whey has recently become available. According to most rabbis, it would be considered a dairy product and would also require appropriate kosher certification.

Emulsifiers and other functional ingredients — such as sodium or magnesium stearate, mono- and diglycerides, glycerine, polysorbates, and monostearates — can be derived from plant or animal sources, a significant distinction for the kosher homemaker. Furthermore, these compounds must be manufactured on kosher equipment. Unfortunately, they may be added to products such as "pure vegetable oil," regardless of their source, without a separate label. Many fish products are packed in oil. In addition, many foods can be baked in pans in which the "pan grease" used is unacceptable, although pan grease need not be listed on the product ingredient label. The food grease used for cooking or for machinery must be kosher. Thus, kosher supervision of the plant is often needed, even when all of the "ingredients" are kosher. The Muslim community is also concerned about the source and types of oils. Unlike the kosher laws that provide some limited ability to annul trace amounts of forbidden fat and other materials, Muslim laws do not.

A further note: a company selling ingredients to another manufacturer would be wise to assure themselves that they are using a "mainstream" kosher or halal certification organization that will be widely accepted by the certification agencies supervising their customers' plants. At retail, use of a less-than-mainstream certification may limit the market-base, but may be justified by the costs/problems associated with meeting mainstream requirements. Table 2 includes a list of the major certification organizations that (to the best of the authors' knowledge) are interchangeable with each other under normal circumstances. The major current Muslim certification organization is shown in Table 3.

Because many oils are transported in heated trucks or tanks onboard ships, the transportation of kosher and halal oils between manufacturing and end-user is of concern. In addition, since many production plants may also produce nonkosher and non-halal lards and tallows, supervision of oil is important.

Vitamins can be prepared from unapproved sources such as nonkosher fish oils (for example, shark) and nonkosher and non-halal bone meal. The United States Food and Drug Administration (FDA) does not require ingredient labels for these products, but they must be from kosher or halal sources if the product is to be sold as kosher or halal, respectively.

Eggs must be from kosher birds. According to United States Department of Agriculture (USDA) regulations for egg-breaking plants, blood spots must be removed but the rest of the egg may be used. Orthodox kosher law, however, requires that, in the case of blood spots, the entire egg must be discarded. USDA inspectors are allowed to certify for kosher marketing, upon request, that eggs with blood spots have been completely removed from a breaking plant's liquid egg supply. Kosher homemakers generally break eggs into individual containers to ensure that no blood spots are present. Small blood spots are often missed in commercial handling. There is no problem, however, in eating a hard-boiled egg that may have had a blood spot. (If the blood spot is observed, it should be removed.) Eggs are generally considered pareve. Note that the USDA also permits use of eggs and egg yolks from slaughtered animals; however, only eggs with fully-formed shells from kosher-killed animals are permitted by Jewish law; they are considered meat. Eggs from slaughtered animals must be labeled separately as ova.

Grape products, such as wine, grape juice, wine vinegar, and related products made from grape juice (which may be used for products such as herring in wine sauce) must be produced completely by Jews to be kosher. Once these are heated (for example, pasteurized), however, they may be handled by non-Jews. Thus, any blend of juices that includes these ingredients must be kosher. Cordials and liqueurs may also have a grape base and therefore require certification. Raisins are permitted, but any coating of animal-derivation used to aid in drying them would be prohibited.

Fruit and vegetable products generally do not need rabbinical supervision. However, since tomato products and bean products are often further processed into nonkosher products (for example, pork and beans) within the same plant and on the same equipment, an indication that tomato products and bean products are kosher is required. Muslim laws require assurance that produce is not processed on equipment that may have come in contact with prohibited materials. Recently, rabbis have been paying more attention to steam lines. According to Jewish law, steam on one side of a metal container interacts with the product on the other side. Thus, if the steam is used to heat a jacketed kettle containing nonkosher product, the steam becomes nonkosher and, if recirculated, could not be used to heat a kosher vessel.

Natural flavors derived from animal sources such as civet, castoreum, and ambergris are not permitted by either Muslim or Jewish authorities. Most rabbis do permit lac resin (shellac) on the same basis that honey is permitted.

Equipment kosherization requires that the manner in which a utensil became nonkosher or of a different kind (i.e., meat, dairy, or pareve) determines how it can be made kosher — assuming that the material it is made of is permitted to be koshered. Equipment for cold products can usually be made kosher by a good washing procedure. Cooked moist products generally require that the equipment be well-cleaned, kept idle for 24 hours and then covered with very hot water (ranging from 180°F to 212°F). Baking equipment generally requires heating to some degree of "glowing." The Muslim community also requires a washing of equipment prior to commencing halal production.

Passover foods: The week of Passover presents special problems, not only for the kosher homemaker, but also for the food processor.

Exodus XIII:7 states:

> Unleavened bread shall be eaten throughout the seven days; and there shall no leavened bread be seen with thee, neither shall there be leaven seen with thee, in all thy borders.

Regular wheat for leavened bread is prohibited, as are all grains (biblically), legumes, and a specific list of additional items (added over the years by the rabbis), such as peas, beans, rice, corn, and mustard. Special products, mainly unleavened bread ("matzo"), must be made for this week; and generally, matzo meal (a coarse, cooked, and baked "flour") is the only flour that can be used for processing baked goods for Passover.

A plant producing Passover products must be specially cleaned for the purpose. Although during the rest of the year the supervising agency may try to be accommodating, it is required by religious law to be quite strict at Passover. Therefore, even the gefilte fish made for Passover would require a special matzo meal that has been specially prepared in accordance with the laws of Passover.

KOSHER CERTIFICATION ORGANIZATIONS

Clearly, some system of labeling is needed to help the consumer and the industry when transferring products between manufacturer and user. Is a product "kosher?" Is it meat, dairy, or pareve? Is it acceptable for Passover? Such a label marking is called a "hechsher."

The different symbols, mostly trademarked, ensure that a food fulfills the kosher laws as interpreted by specified rabbis or organizations. Meat and poultry often are marked kosher by carvings and stamps in the carcass, along with an affixed metal tag called a "plumba." Other consumer groups that follow dietary laws, such as Seventh Day Adventists, often use these kosher markings as guides, as well. The conscientious consumer must be familiar with a variety of markings. Some, such as one rabbi's trademarked "S" (encircled), are esoteric hechshers that few kosher consumers recognize. In the United States and particularly in New York State, the marking of a package as "kosher" has a legal definition. Some other states (e.g., New Jersey) also have laws dealing with kosher foods. New Jersey has indicated that the letter "P" should be for Passover and not for "pareve." Unfortunately, three different supervision agencies use the symbol of a square box with a "K" in it. This situation leads to serious consumer confusion.

New York specifically recognizes the Orthodox dietary laws of kashruth for enforcement purposes, and foods improperly marked may be seized as "misbranded"; stores must mark whether they sell kosher foods exclusively or both kosher and nonkosher foods. If they sell both kosher and nonkosher meat and poultry, the food must be separated within the store. New York kosher food inspectors recently have been enforcing these laws more strictly.

A New York State law requires that the manufacturers of kosher products must file the name of the certifying rabbi with the Department of Agriculture and Markets. (The Department is preparing a list of such products that should be available soon, along with an 800 number —800-966-0009, New York State only — that interested consumers can call to obtain this information.) Orthodox laws have been upheld legally, specifically in a case involving seafood (Freedman, 1970). A Conservative fish seller, supervised by a Conservative rabbi, sold swordfish as kosher; after legal hearings, he was ordered to stop selling the fish as kosher.

Kosher laws in New Jersey were declared unconstitutional (a violation of the separation of church and state) by the New Jersey Supreme Court, and a law in the city of Baltimore was declared unconstitutional by a federal court. New Jersey has since established new regulations that focus completely on the consumers' right to know and truth-in-labeling. A series of disclosure posters must be posted in stores, giving consumers the information they require to make an informed purchase. Any standard is permitted as long as it is declared, and the kosher supervision agency and the seller remain consistent within the regulations. Such a system could easily be extended to other religious and philosophical issues (e.g., vegetarian, natural, organic, and "Heart-Healthy") and would permit the state to protect consumers without getting into the issues of defining the standards.

HOW TO OBTAIN CERTIFICATION

Obviously, processing plants must make special efforts to comply with all these regulations and inspections. "Mixed" plants must scrupulously keep kosher and nonkosher (or in some cases, meat and dairy) products completely separate. So, for example, all fish-handling equipment for nonkosher fish must be kept separate from kosher fish, or the equipment must be koshered before each kosher use. Generally, the procedure for koshering equipment (when used hot) is to let it sit for 24 hours and then run boiling water through it until it overflows. The procedures for koshering equipment in a plant, food service institution, or home are numerous. Some of the suggested readings offer more details on this subject, and a rabbi involved in kosher certification can be consulted for further information. If the processing is only done cold, then the clean-up procedures are less stringent in changing over.

Many rabbis and rabbinical organizations provide certification services, and many major cities have some form of kosher certification system. Rabbinical councils in those cities, as well as rabbis at individual synagogues, can assist the interested processor or consumer.

There are too many certifying rabbis and organizations to provide a complete list here. KASHRUS Magazine (Wikler, 1999) puts out an updated list annually, including more than 360 symbols, not counting the generic K used by many other rabbis.

Companies seeking kosher supervision should consider a number of points when selecting a kosher certifying agency:

1. *Acceptance of the kosher certifying agency by the consumer.*
2. *Acceptance of the kosher certifying agency by other kosher-certifying agencies:*

This is particularly important with respect to ingredients. In general, we recommend that ingredient companies seek a mainstream or more "traditional" certification.

3. *Compatibility and service*: Working with a kosher-certification agency is a long-term relationship. Like any vendor, their service record and ability to meet the needs (e.g., timeliness) of the company should be considered.
4. *Cost:* Prices vary, but to some extent you also "get what you pay for." You have every right to question and to shop around before selecting an agency.

Many companies are finding that the new markets opened by becoming kosher or halal are worth the time and effort. While only about one-third of the market is actually Jewish, adherence to kosher laws also opens markets to many other ethnic, religious, and philosophical groups.

Many companies make the commitment to produce kosher or halal products, despite extra handling and efforts, because: (1) kosher and halal raw materials often exist at no extra cost; (2) companies are reassured by the support the Jewish and Muslin communities offer in the form of rabbis, meshgiachs, imams, etc.; and (3) entrepreneurs are willing to invest in a new market that may help cut unit costs.

The Jewish and Muslim communities have demonstrated their interest in fulfilling the kosher and halah laws in an atmosphere of community harmony. It is the authors' hope that communicating this type of information to fish specialists will help encourage mutually beneficial undertakings by food processors and the American Jewish and Muslim communities.

REFERENCES

Atz, J. W. Curator Emeritus, Department of Herpetology and Ichthyology, American Museum of Natural History. New York.

Awan, J.A. 1988. Islamic Food Laws -- I. Philosophy of the prohibition of unlawful foods. *Science and Technology in the Islamic World.* **6**(3): 151-165.

Bosley, G.C. and M.G. Hardinge. 1992. Seventh-day Adventists: dietary standards and concerns. *Food Technol.* **46**(10):112-113.

Chaudry, M.M. 1992. Islamic food laws: philosophical basis and practical implications. *Food Technol.* **46**(10): 92-93,104.

Chaudry, M.M. and J.M. Regenstein. 1994. Implications of biotechnology and genetic engineering for kosher and halal foods. *Trends Food Sci. Technol.* **5**:165-168.

Freedman, S.E. 1970. *The Book of Kashruth, A Treasury of Kosher Facts.* Bloch Pub. Co., New York.

Ginsburg, I. 1961. *Food Fishes with Fins and Scales,* Fishery Leaflet 531. Bureau of Commercial Fisheries, U.S. Dept. of Interior, Washington, DC.

Grandin, T. and J.M. Regenstein. 1994. Slaughter; religious slaughter and animal welfare: a discussion for meat scientists. *Meat Focus International.* **3**:115-123.

Grunfeld, I. 1972. *The Jewish Dietary Laws.* Soncino Press, London.

Lipschutz, Y. 1988. *Kashruth: A Comprehensive Background and Reference Guide to the Principles of Kashruth.* Mesorah Publications, Brooklyn, NY.

Pickthall, M.M. 1984. *Text and Explanatory Translation of "The Glorious Quran."* Library of Islam. Des Plaines, IL.

Regenstein, J.M. 1994. Health aspects of kosher foods. *Activities Report and Minutes of Work Groups & Sub-Work Groups of the R & D Associates.* **46**(1): 77-83.

Regenstein, J.M. and T. Grandin. 1992. Religious slaughter and animal welfare -- an introduction for animal scientists. *Proceedings 45th Annual Reciprocal Meat Conference.* **45**:155-159.

Regenstein, J.M. and C.E. Regenstein. 1979. An introduction to the kosher dietary laws for food scientists and food processors. *Food Technol.* **33**(1): 89-99.

Regenstein, J.M. and C.E. Regenstein. 1981. *Old Laws in a New Market — The Kosher Dietary Laws for Seafood Processors.* New York Sea Grant Bulletin, Albany.

Regenstein, J.M. and C.E. Regenstein. 1983. Lactose intolerant? Try kosher . . . . *Food Technol.* **37**(2):24.

Regenstein, J.M. and C.E. Regenstein. 1988. The kosher dietary laws and their implementation in the food industry. *Food Technol.* **42**(6): 88-94.

Regenstein, J.M. and C.E. Regenstein. 1990a. Looking in on kosher supervision of the food industry. *Judaism.* **39**(4): 408-426.

Regenstein, J.M. and C.E. Regenstein. 1990b. Kosher certification of vinegar: a model for industry/ rabbinical cooperation. *Food Technol.* **44**(7): 90-93.

Regenstein, J. M. and C.E. Regenstein. 1992. The kosher food market in the 1990s -- a legal view. *Food Technol.* **46**(10): 122-124,126.

Regenstein, J.M. and C.E. Regenstein. 1994. Kosher foods and allergies. *Food Allergy News* **3**(3):1, 7.

Wikler, Y. 1999. 1999-2000 Kosher Supervision Guide. *Kashrus Magazine.* **20**(1): 40-72.

Design of Quality and Safety Management Systems for Full-Service Retail Seafood Departments

Michael G. Haby and Russell J. Miget

INTRODUCTION

Full-service seafood departments undertake many of the same activities found in full-service meat departments in processing various products into more convenient market forms. If improperly handled, these activities can have a dramatic impact upon spoilage and avoidable shrinkage since the seafood product line found in virtually all retail operations consists of a microbiologically dissimilar mix of products. One must consider that many products arrive with 75 to 80 percent of their shelf life already used up, while other items enter the retail establishment with 80 percent of their shelf life left. This disparity in remaining shelf life makes premature spoilage of fresher products a chronic concern.

Just as seafood departments have physically evolved into contemporary retail settings, the nature of department operations also has changed. Historically, the product line consisted of a wide mix of raw seafood products, but in response to a demand for more convenience, full-service departments are now offering an expanded line of ready-to-eat products. Many are prepared on-site. Both raw and ready-to-eat items are available and prepared within relatively tight physical confines. Most often the same individual routinely handles both types of products. The employee, therefore, becomes the potential for literally hundreds of different contamination opportunities in a single day; such handling may in turn compromise the safety of otherwise wholesome, ready-to-eat ingredients or foods. For example, that same employee who began steaming shrimp (custom cooking) for one customer must break away from that task to select, pack, and weigh an order of raw product for another customer, and then return to the seasoning, packaging, and labeling steps required to complete the custom-cooking order. Without proper procedures, this necessary crossover substantially increases the opportunity to cross-contaminate cooked, ready-to-eat shrimp with raw products or to recontaminate ready-to-eat shrimp with unclean food contact surfaces such as hands, gloves, countertops, utensils, trays, etc. The presence of ready-to-eat foods dramatically increases a company's exposure to a variety of negligence practices. Most departmental employees work without direct supervision; thus, the part-time employee who may have unwittingly recontaminated the cooked shrimp places the company in a virtually indefensible position unless a comprehensive quality and safety management program is in place.

To affect sharp reductions in avoidable shrinkage and compromised product safety, three objectives must be simultaneously and continuously maintained during retail stewardship. These include: (1) maintaining optimally cold product temperatures, (2) minimizing various contamination avenues, and (3) selling older merchandise first. These handling, holding, and rotational objectives are a preventive strategy. Generating this preventive strategy can be translated into a set of operational procedures that precisely define, in stepwise fashion, how to perform each of the various activities that comprise full-service departmental operations.

Designing and implementing such procedures can have many positive effects on the company. First, stores should be able to meet profitability expectations by sharply reducing premature spoilage, a significant source of avoidable shrinkage. Because shrinkage is a major cost (10 to 15 percent of departmental sales), reducing it proportionally boosts the contributions a department

Photo 1. Fish department in a grocery store, preparing seafood for case.

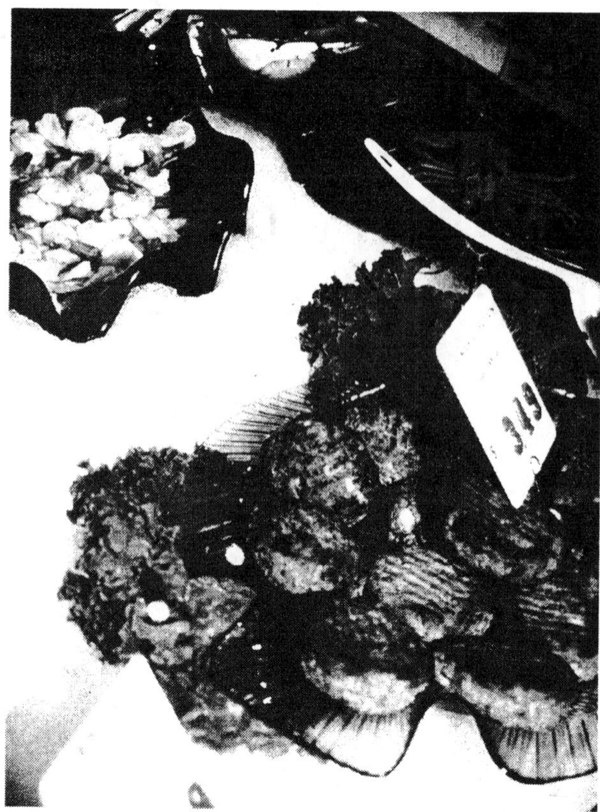

Photo 2. Crab cakes in seafood case at a food store.

makes to total store profit. Second, such procedures establish a defensible position for companies should challenges about the firm's quality and safety program arise from potential plaintiffs or opportunistic media. Third, such procedures establish a consistent way of managing complex service departments throughout multi-unit operations with scarce managerial resources.

This chapter presents a set of research-based Standard Operating Procedures (SOPs) for several activities found in full-service seafood departments. These SOPs were developed to correct a variety of errors that contributed to premature spoilage (i.e., avoidable shrinkage) and compromised product safety. These particular handling and holding errors were evaluated in a series of Standardized Quality and Safety Audits conducted with cooperating food chains between 1993 and 1995. The Standardized Quality and Safety Audit project was designed with two objectives: (1) to learn precisely how quality and safety were managed by departmental employees, and (2) to correct observed and potential quality and safety errors by creating more effective, efficient procedures for operating the department.

One way to demonstrate the steps, decisions, and linkages among different components of a procedure is to use a **process flow chart**. These charts take a general function and break it into its most basic steps. This adds clarity and structure to procedures, and teaches the correct way to complete a task. It also properly defines a comprehensive management structure. Such charting may also help with compliance checks by management. Finally, the program required to produce such a process chart indicates to regulatory authorities that management understands the steps and sequences necessary to ensure quality and safety on a day-to-day basis. Process flow charts are used throughout this chapter to document the SOPs.

Display is the first function addressed by an SOP. This section describes the procedures necessary to implement a preventive strategy during the variety of activities that comprise setting up a full-service case and closing down the department. The chapter then builds upon this SOP by present-

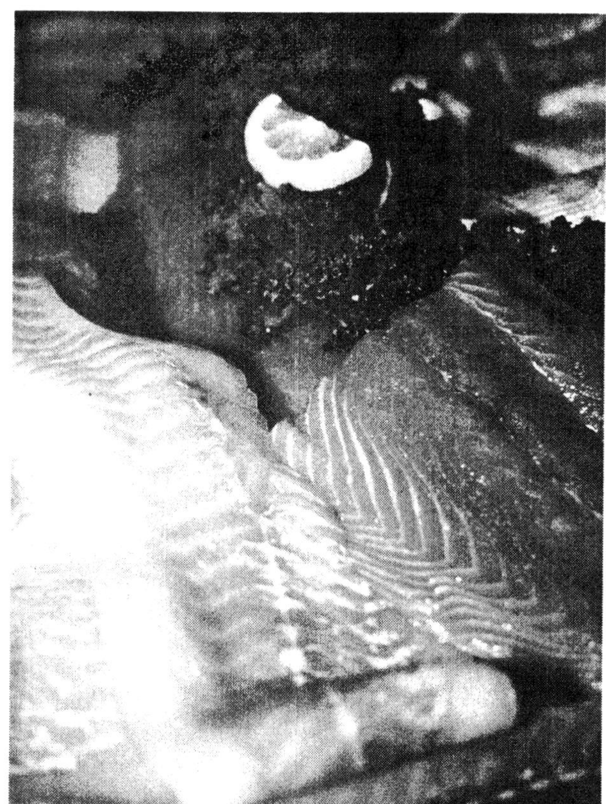

Photo 3. Fish fillets in seafood case.

Photo 4. Salmon steaks in seafood case.

ing two common situations within most display functions. The first is designing an SOP for fabricating a ready-to-eat product, which becomes another storekeeping unit (SKU) in the display case and, like all products, cycles between display and storage until it sells or is discarded. The second is development of an SOP for custom-cooked shrimp for an individual customer. These SOPs highlight the observed and potential quality and safety errors encountered within the display function and demonstrate how to achieve a preventive strategy with relatively simple, time-efficient procedures that unsupervised personnel can implement.

A STANDARD OPERATING PROCEDURE FOR BULK-PACKED INVENTORY SOLD THROUGH A REFRIGERATED SERVICE CASE

Historically, the display objective in seafood service departments has been to maximize eye appeal by constructing attractive, interesting product presentations. Eye appeal is certainly a key merchandising objective, but there are other equally important considerations within the display function. These are quality and safety management. Regulations are designed to ensure public health and must be incorporated into the department's display strategy. These regulations typically specify the separation of raw and ready-to-eat products so that cross-contamination between these product categories can be avoided. The display function should also incorporate three other objectives that reduce opportunities for spoilage. First, contamination among dissimilar raw products should be prevented. Second, low product temperatures must be maintained during case residence time so that the remaining shelf life of all products can be maximized. Third, products with the least amount of remaining shelf life need to be positioned to sell first.

Product display is traditionally viewed as a static event; however, meeting each of the aforementioned objectives requires various activities, such as loading the case in the morning, selecting products for customers throughout the day, and undertaking some type of close-down procedure at the close of the business day. The way that these activities are designed and implemented can have a significant impact on product quality and safety, since even the fastest turnover items are handled at least twice.

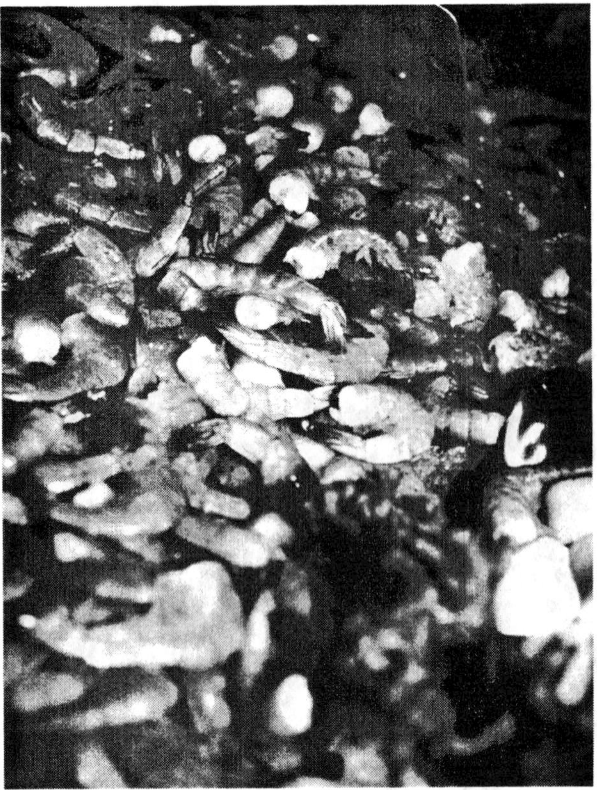

Photo 5. Shrimp in seafood case.

Although there are opportunities to make instantaneous, permanent quality and safety errors, the display function — when properly organized — can meet both merchandising and quality/ safety objectives. Table 1 highlights the quality and safety errors often made in various segments of the display step and offers a Standard Operating Procedure that can prevent those errors from occurring. Most of the SOP recommends slight changes or modifications to what is now routinely done. However, one component of the SOP recommends that the common practice of washing or rinsing seafoods prior to placing them on display be discontinued because the recommended SOP does not contain a wash step.

Eliminating a Needless Procedure

Most departments performed some type of washing, rinsing, or dipping prior to stocking inventory in the display case. Washing has been a common practice among seafood departments because of perceived benefits. These include (1) further reductions in product temperatures, and (2) an extension of shelf life by physically removing spoilage bacteria and bacterial metabolites from product surfaces.

Washing to chill is ineffective because product temperatures were quite cold just prior to removal from refrigerated storage. Across all cooperating stores, the average product temperature held overnight in refrigerated storage was 33.3°F, close to optimal holding conditions. Likewise, the prevalent thought that various spraying or dipping procedures effectively remove spoilage bacteria, thereby extending product shelf life, is not supported by controlled studies. Even when conducted under optimal handling conditions and the use of a shower spray at 30 pounds per square inch (psi), the study found only slight reductions in spoilage flora on skinless fillets that, within several hours, returned to pretreatment levels (Miget et al., 1987). This study also documented that no differences could be detected between washed and unwashed items and the sensory parameters of eye appeal or odor. Other studies have reported similar results. In a broadscale study to find ways of maximizing retail shelf life, researchers at Virginia Tech reported that dipping product, either in plain water or in solutions of various bactericides, produced no microbial or shelf life benefits (Samuels et al., 1984). The conclusion is that even under optimal handling procedures the "wash" step was ineffective in chilling products, extending shelf life, or improving the sensory quality of skinless fillets or steaks.

There are several unintended, negative consequences of washing. An improper washing technique can actually increase product temperatures. In one store the additive effects of a 1.5-hour exposure to ambient store temperatures and a rinse under 75° F tap water increased product temperatures by 8° F. In those departments where each species was passed through the same water bath, it only contaminated newer items with those having less shelf life. In standardized quality and safety audits conducted among cooperating retailers, auditors observed dipping procedures which did not consider remaining shelf life in sequencing products for washing. Furthermore, employees did not first clean and sanitize the sink compartment. Finally, product was subjected to additional physical damage from the repetitive handling required to move items into and out of the

Table 1. Solving Current Quality and Safety Errors within the Bulk-Packed Product Line

SOP Objective	Current Approach Revealed through the Audits	Proposed Solution
Ensure Proper Rotation of Inventory	Previously displayed merchandise was generally stocked first, with the balance added from new (never displayed) inventory. Long vertical ribbons did not facilitate a FIFO rotational plan because previously displayed merchandise was not distinguished from products having more shelf life.	Place each storekeeping unit (SKU) in a pan. When loading pans, use a cleaned, sanitized pan, and load it first with new product, then "top off" with previously displayed product. This facilitates the sale of previously displayed merchandise.
Minimize Elapsed Time between Roll-out and Stocking	Several hours elapsed between roll-out and stocking in the refrigerated case.	Where possible, load pans in the cooler using the sequence specified above, then roll out and stock filled pans in the case.
Minimize Contamination across Different SKUs	Departmental employees generally washed (i.e., dipped, sprayed, or rinsed) all inventory to be displayed prior to placing it in the service case. All inventory was washed *en masse*. Likewise, no cleaning and sanitation step was used between different SKUs.	Discontinue the wash step.
	Loading (unloading) the case was typically accomplished by completing all like tasks at the same time. Thus, all SKUs were loaded (unloaded) as one step, without a hand wash or glove change between different SKUs.	Except for customer service, only the most experienced staff member handles individual bulk-packed inventory items (i.e., individual fillets, steaks, etc.) during the morning loading procedure. At close-down, the evening staff member does not handle individual items; rather, he places lids on pans, removes them from the service case, and returns unsold items to the walk-in cooler. During the sales day, pans also provide a physical separation among different products.
Maintain Cold Product Temperatures	Generally, cold holding temperatures during display were maintained. However, a common approach is needed that achieves low product temperatures regardless of case types and ambient settings.	When stocking, embed pan in ice up to the lip.

sink and the physical agitation in the water baths. Washing products before stocking simply consumes employees' time as well as the shelf life of fresher products. This step is unnecessary and should be eliminated.

Making Modifications to Necessary Procedures

A common theme runs throughout the proposed solution: the use of pans. In fact, pans are the "heart and soul" of this SOP because their proper use facilitates each objective — by ensuring a first in/first out (FIFO) rotation and reducing the frequency and time required to disassemble and clean the display case. Steam table-type pans, available in a variety of dimensions, best meet these needs. Each pan must have a lid (required during close-down and overnight storage) and a perforated insert which holds the product off the pan bottom by one-fourth of an inch. This insert is an important element in the SOP because it separates the product from any drip. Instead of having to disassemble the case to clean and sanitize, employees need only to clean and sanitize the display pans daily, which minimizes off-odors.

This SOP requires two pans and two inserts for every SKU: one set in use, and another cleaned, sanitized, and available for use. Although several pan depths are available, at least six-inch-deep pans should be used because the lid and insert use about 1.5 inches of vertical capacity (Figure 1).

Historically, steam table pans were manufactured from stainless steel. However, with advances in plastics technology, steam table pans are now available in both stainless steel and high-density, transparent plastic. Both material types can be used for holding hot and cold items. The choice of container material is an individual one. Studies have shown that maintenance of low product temperatures is less dependent upon container material type and more on how deeply the container is embedded in ice (Figure 2). As the figure illustrates, only a slightly higher average temperature was recorded during an eight-hour trial for products held in a plastic pan (43.5° F) as compared to those products held in a stainless steel pan (42.3°F).

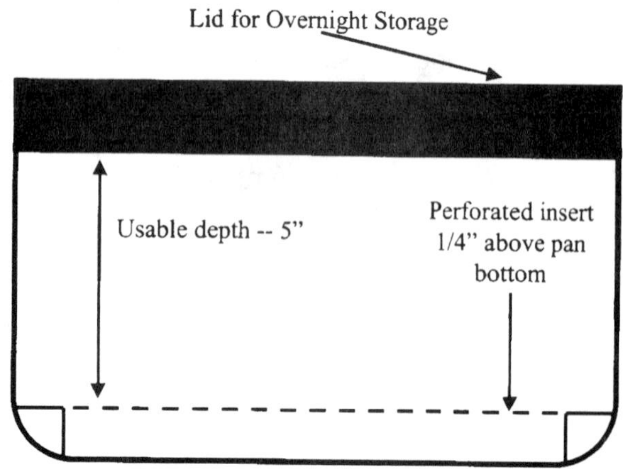

Figure 1. Cross sectional of pan.

Two important conclusions can be drawn from this comparison. First, the difference in average product temperature (1.2° F) was minimal, which suggests that over the eight-hour trial virtually the same amount of shelf life would be lost among identical products when placed in pans constructed of either material. Second, this small difference was obtained in a refrigerated case preset to maintain a 50°F average case airspace temperature. Given the colder case airspace temperatures observed in most retail seafood departments (an average of 38°F across 11 stores), the choice of container material would have even less effect on product temperature. The decision about whether to use metal or high-density plastic should be based on meeting cosmetic objectives (i.e., eye appeal).

Ensuring a First-In/First-Out (FIFO) Stock Rotation Plan

The morning employee charged with readying the department is initially faced with at least two classes of inventory: items that were displayed but unsold at the close of business the previous evening, and merchandise held in refrigerated or frozen storage since receipt. To prepare each storekeeping unit (SKU) for display while ensuring a FIFO rotation sequence, the procedure outlined in Figure 3 is required. Note that the sequence of activities begins anew with each SKU, requiring a glove change prior to handling the next SKU. The first decision to make is whether all or part of the previously displayed merchandise can be displayed again. If it cannot, the employee should set that pan aside, obtain a cleaned, sanitized pan and insert, and fill it with "new" merchandise. If more than one batch of an SKU exists in inventory, the employee should check the arrival dates

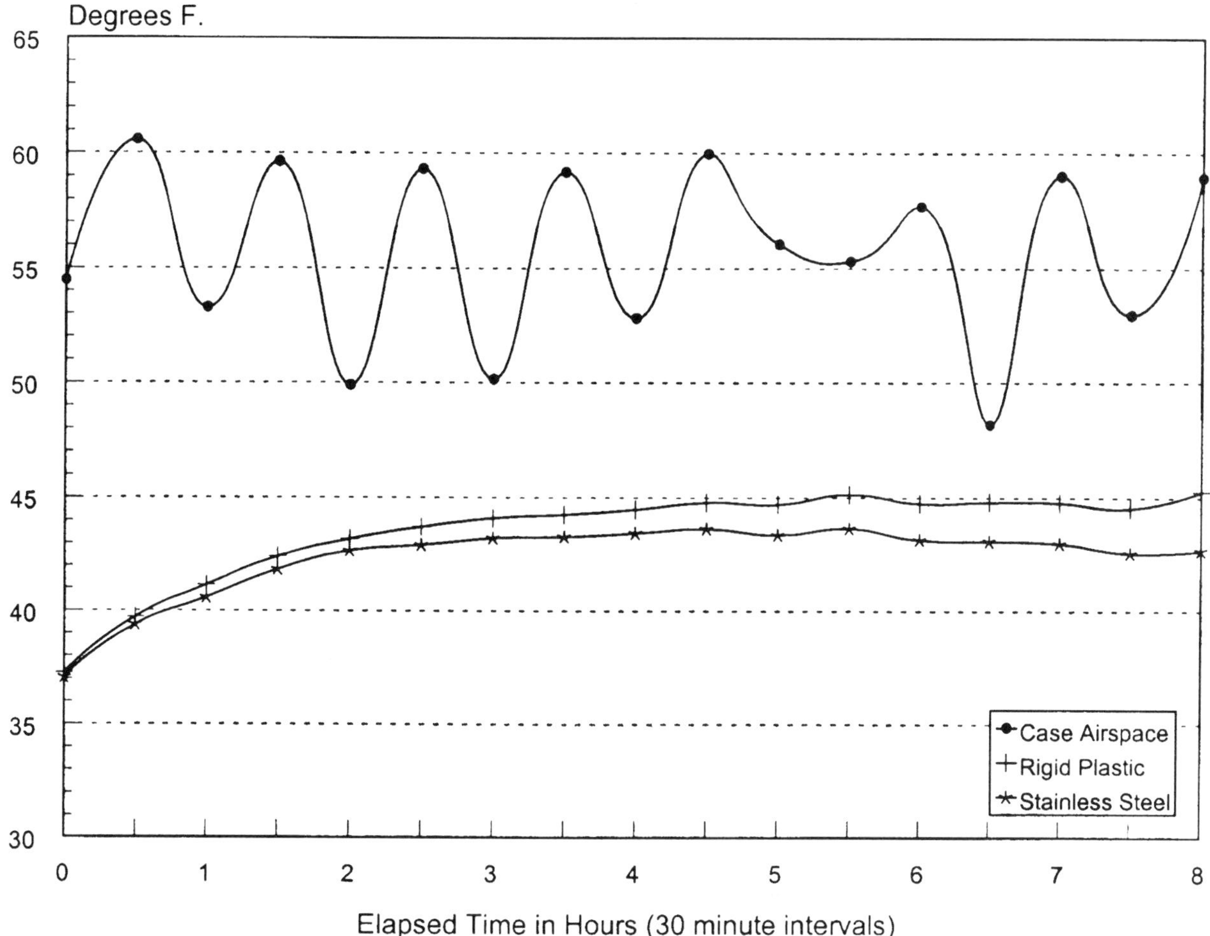

Figure 2. The influence of container material on product temperature when pans are embedded in ice in a refrigerated case set to average at 50°F.

so that the one delivered first is used first. Assuming that the employee judges that previously displayed merchandise can be redisplayed another day, the employee estimates the quantity of "new" product needed to fill the pan which currently contains the previously displayed but unsold merchandise. He then obtains a cleaned, sanitized pan and insert, adds the new merchandise, places cling film or deli wrap over the new merchandise, and finally adds the previously displayed items from the other pan. Using this procedure, the previously displayed merchandise is loaded last so it can exit first. Figure 4 details the mechanics of the pan loading procedure, assuming both previously displayed and new merchandise comprise an SKU.

The cling film or deli wrap serves two purposes. First, it provides a barrier between previously displayed but unsold merchandise and the newer items contained below. Second, the barrier acts as a visual cue to employees about how much of a previously displayed but unsold SKU remains after a subsequent day on display. If the store had a policy that items would only be displayed for two days and then discarded, the barrier would prevent the accidental discard of items that had been displayed for only one day.

Minimizing Contamination across Different Store-Keeping Units

Contamination across different SKUs during set-up or close-down generally occurs because the employee completes similar tasks at the same time. Standardized quality and safety audits conducted in cooperating retail establishments revealed that during morning set-up, each SKU was evaluated, "washed," and stocked without a hand wash or glove change between batches. With the shelf life dissimilarities found across the broad seafood product line, handling the entire line as one large

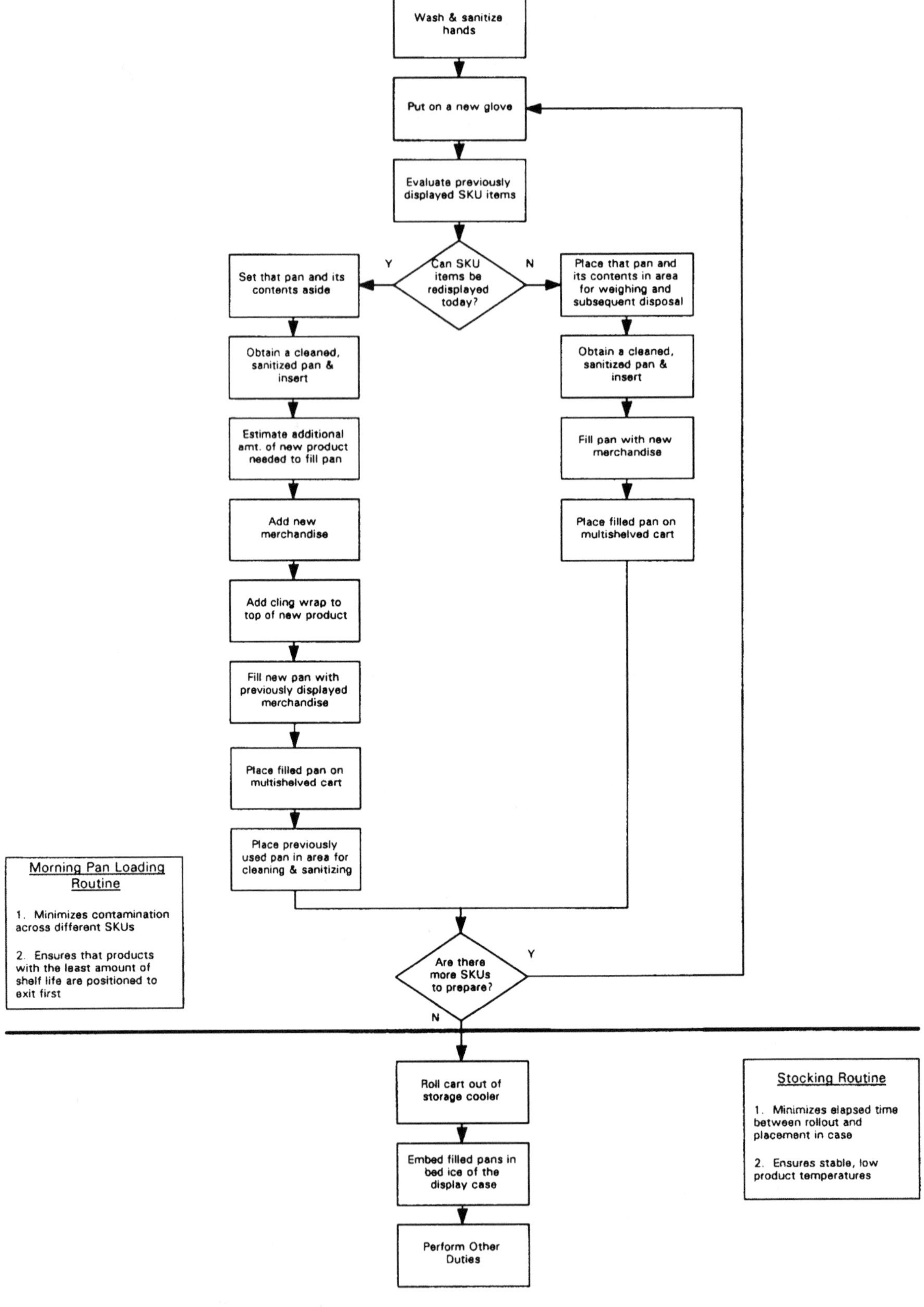

Figure 3. Stepwise procedure for loading pans from a walk-in cooler that minimizes inadvertent contamination among SKUs and ensures a FIFO rotation sequence.

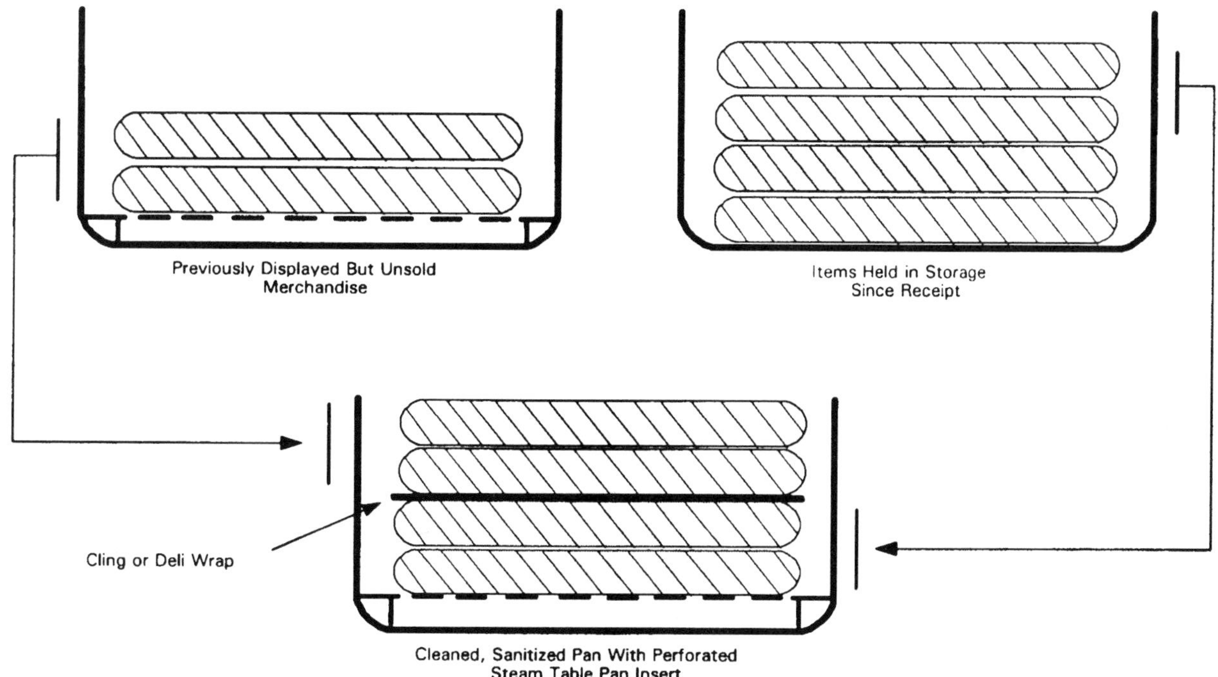

Figure 4. Ensuring a FIFO rotation plan when each SKU to be displayed is comprised of both previously displayed and "new" merchandise.

step may be expeditious for the employee, but unfortunately the time savings are often traded away by avoidable, premature spoilage through inadvertent contamination. Use of the pan loading routine outlined in Figures 3 and 4 utilizes an SKU approach that eliminates the inadvertent crossover and subsequent contamination among different products during set-up.

Virtually the same work pattern was observed during close-down, with each item removed from the display case as a single step without a hand wash or glove change between SKUs. Again, pans prevent contamination among SKUs during close-down (Figure 5). Rather than sequentially removing each saleable item one piece at a time, the employee can place a lid on each pan, remove it from the case, and place it on the multi-shelved rolling cart that can be wheeled into the walk-in cooler. Using this approach, the lack of food handler experience and the managerial expectation to be "off the clock" upon closing are not seen as diametrically opposing goals because the display case can be unloaded — one panload at a time as opposed to one saleable item at a time — in just a few minutes without compromising quality or safety.

Maintaining Optimally Low Product Temperatures

Many retail managers assume that refrigerated spaces effectively maintain products at ideal temperatures. Properly operating refrigeration systems certainly maintain product temperatures below the maximum required holding temperature for refrigerated foods. However, even at temperatures that meet public health requirements, multiple shelf life hours can still be lost with each elapsed hour. Because many seafoods arrive with relatively few shelf life hours remaining, and because of the 18-to-21-hour time lag generally required before previously displayed items sell, small increases in the rate that shelf life is consumed during display can add up to significant losses that are, for the most part, avoidable. Therefore, maintaining low product temperatures during case residence time is a key objective of the Standard Operating Procedure.

Ice is most effective at removing heat when it melts over a product. Melting ice chills products almost 10 times faster than cold air does (Bramsnaes, 1965). However, allowing ice to melt over ready-to-eat items and processed raw prod-

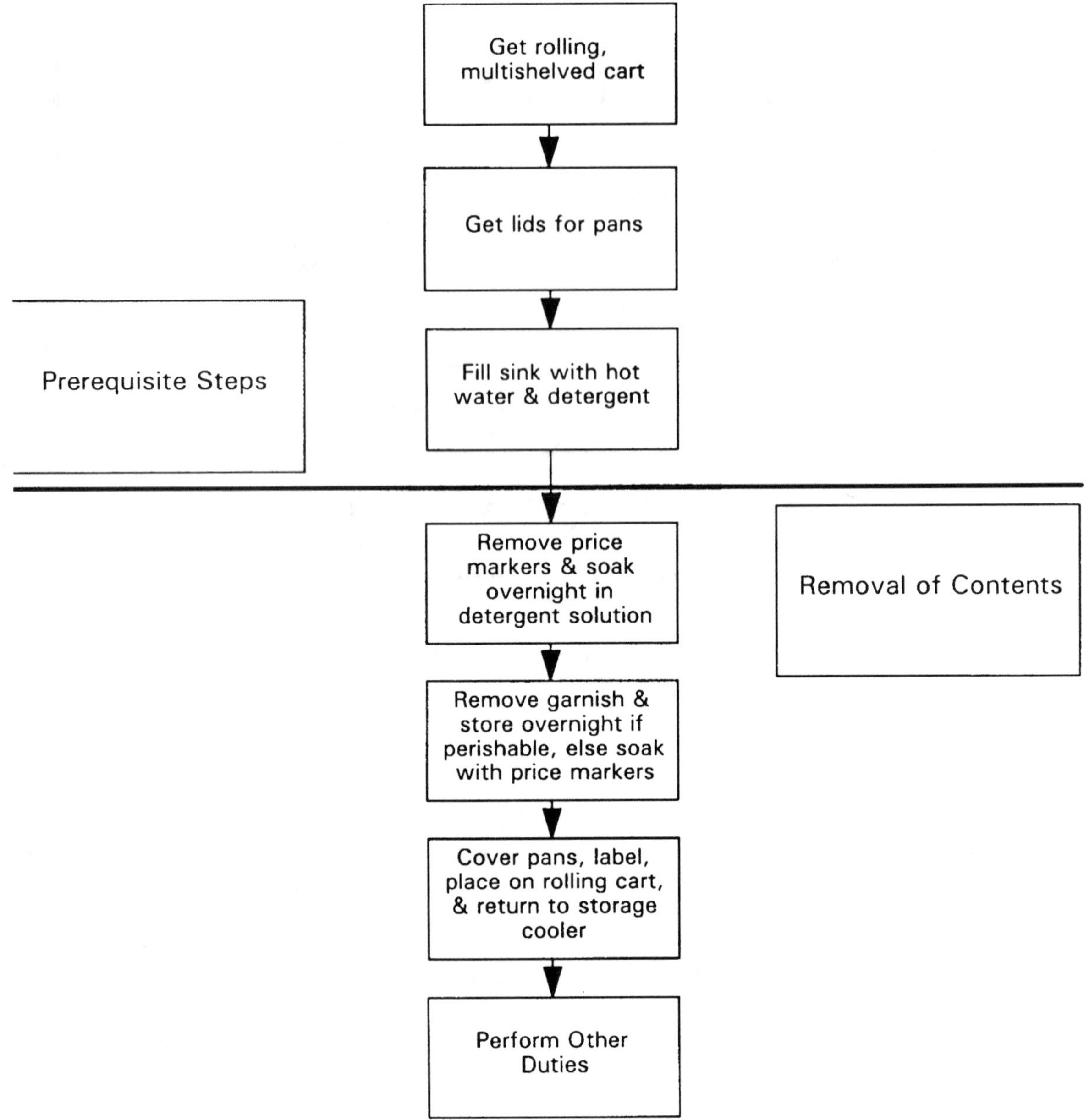

Figure 5. Close-down routine for the service case.

ucts such as skinless fillets, steaks, peeled shrimp, etc., is a mistake because these market forms can absorb some of this ice melt. Products and ice need to be physically separated with a pan. Once ice and product are separated, chilling from ice must occur through indirect means.

The first indirect effect is conduction of heat out of the product and into the bed of ice, which acts as a heat sink. Accessing this heat sink requires some type of heat exchanger. A heat exchanger can be any dense material such as stainless steel or high-density plastic steam table-type pans. Conduction occurs from the product, through the container material, and into the bed of ice. Because the effects of conduction are limited to that surface area which contacts ice, the efficiency of the pan as a heat exchange device is proportional to the surface area exposed to the heat sink. Thus, the more surface area exposed to a bed of ice — in this case the sides as well as the bottom of the pan — the better the potential heat removal.

Embedding pans in ice, as opposed to placing them on ice, is important for another, less obvious, reason. As Figure 6 illustrates, when air is chilled in a refrigeration system, it becomes dense and cascades downward toward the product. At the air/ice interface this air becomes further chilled and heavier. Being dense, this chilled air sinks into the display pans if they are embedded in ice. When most SKUs such as fillets, steaks, shrimp, etc., are loaded into pans, airspaces within the pan are unavoidably created. So long as these pans are embedded in ice, cold air will pool in these airspaces and buffer the product from warmer, cycling ambient case temperatures.

Pans placed **on** ice take limited advantage of the heat sink offered by a bed of ice. A pan with dimensions 11" x 9" x 6" has 339 square inches of surface area available as a heat exchanger. However, most of the pan surface is in the sides (71 percent) so these deep pans must be embedded in ice to take full advantage of this heat transfer potential. There is another reason for embedding the pan in ice. One to two inches above the bed, ambient case temperatures prevail. These ambient conditions can dramatically fluctuate over time as the compressor cycles (see Figure 2). Placing a six-inch deep pan on ice exposes more than 2/3 of the vertical capacity of the pan to this zone of the case.

The efficacy of this stocking procedure was evaluated over a range of case airspace conditions (e.g., controls were set to average 40°F, 50°F, and 60°F). Each of these ambient environments varied over time — particularly at the lower settings — but each setting returned an average temperature roughly equivalent to the planned, preset case airspace temperature. Despite an almost 20°F difference between the highest and lowest case airspace temperatures, embedding pans in ice resulted in product temperatures that, on average, varied by just 3.7°F (Figure 7) (Haby and Miget, 1991). Even at the 60°F setting, the average temperature of

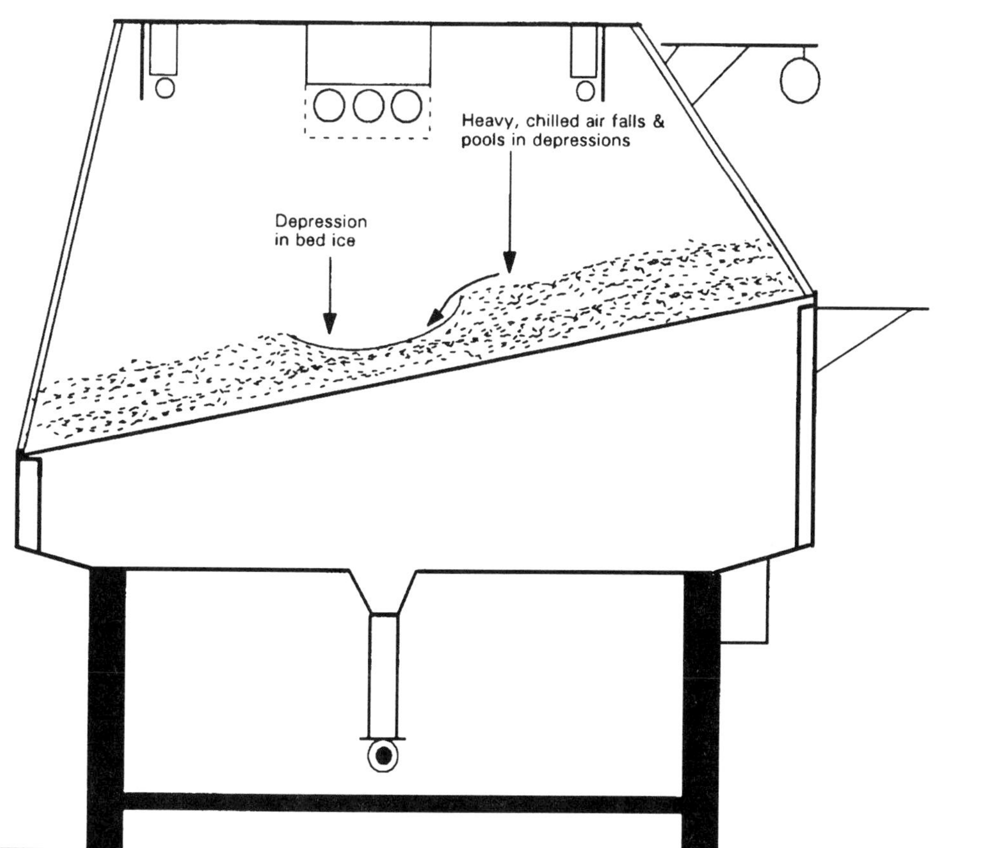

Figure 6. Movement of chilled air into depressions created in bed ice by embedding display pans in ice.

product over the 9.5-hour trial was below the maximum permissible holding temperature for refrigerated foods established by public health authorities.

This evaluation indicates that constant, low temperatures can be achieved when products are placed in pans that are then embedded in ice up to the lip of the pan. This finding is significant because multi-unit operations seldom have identical equipment that returns identical case airspace temperatures; no matter, the same stocking procedure can maintain constant, low temperatures regardless of ambient case airspace conditions.

Summary

Historically, product display has been considered a static function, but in this phase of the retail inventory cycle numerous repetitive steps are required to set the case in the morning, make selections for customers throughout the day, and close down the department at day's end. To minimize both avoidable shrinkage and the opportunity for otherwise wholesome product to become unsafe, three conditions must prevail from the time that the first employee arrives until the last one is "off the clock." First, handling procedures must be used that prevent inadvertent contamination between products and environmental surfaces and among the dissimilar mix of seafoods merchandised in the department. Second, items with the least amount of remaining shelf life must be positioned so they can be sold first. Third, stocking procedures must maintain optimally cold product temperatures once on display. Each of these criteria is met with the SOP presented above.

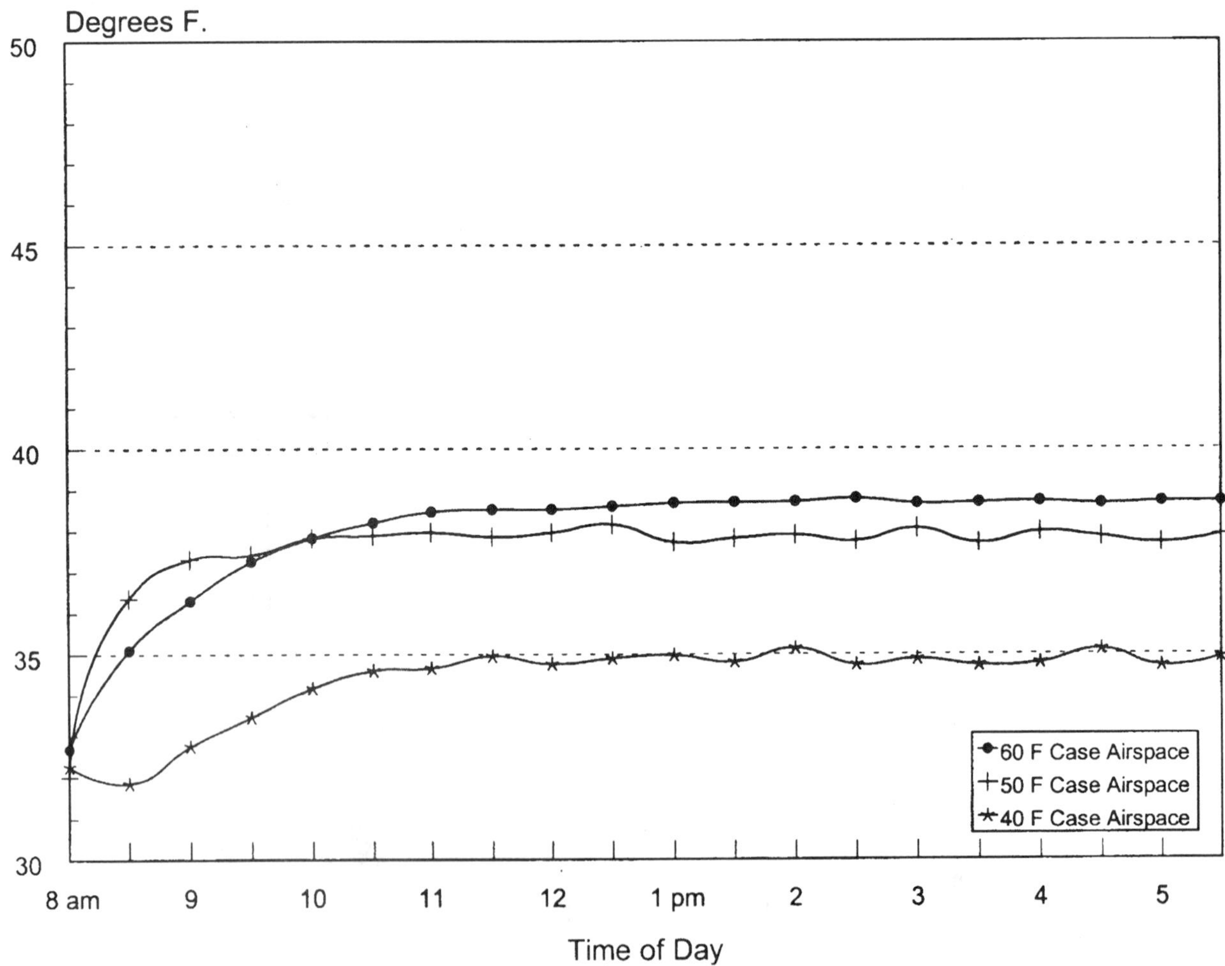

Figure 7. Temperature histories of products placed in steam table pans and embedded in ice over three different case airspace settings.

A STANDARD OPERATING PROCEDURE FOR MIXING, HOLDING, AND PROPERLY ROTATING READY-TO-EAT SEAFOOD SALADS

Among the ready-to-eat items prepared on-site in many full-service seafood departments are refrigerated seafood salads made from commercially available ingredients. Fabricating such items is a relatively simple process since ingredients are merely mixed together (e.g., a prepared base, vegetables, and imitation or natural crabmeat). Properly fabricating and holding a mixture made from various ingredients will provide the grocer with different shelf lives. Besides offering ultimate convenience for customers, some seafood salads can have a long shelf life and can generate a relatively high gross margin with a moderate unit price — all welcome additions to most full-service seafood departments. However, the safety and quality risks inherent in fabricating and holding refrigerated, ready-to-eat items must be managed. Spoilage is a constant concern with perishable inventory, but the overriding issue with ready-to-eat foods is maintenance of product safety.

Ready-to-eat foods have been a component of the retail product mix for years. Most seafood departments offer picked crabmeat, both live and shucked molluscan shellfish, smoked fish, and other cooked products. However, a ready-to-eat item fabricated on-site separates this particular type of product from other ready-to-eat SKUs in the seafood product line. In effect, stores become food processors, with many of the same process controls required. Once prepared, this SKU becomes another component of the product line that cycles between storage and display, and thus is governed by the SOP outlined in the previous section. For example, to maintain low product temperatures, containerized products should be embedded in ice up to the lip of the pan. However, there are some differences in handling and holding a ready-to-eat product. Table 2 enumerates the quality and safety errors observed in the fabrication and holding of ready-to-eat seafood salads and describes an SOP that prevents these errors.

Minimizing Contamination

Seafood salads are prepared from mixed ingredients and are sold and consumed chilled. Thus, there is no final cook step that can undo or negate the effects of contamination and/or time/temperature abuse. Preventing cross-contamination or recontamination is essential because some pathogens (such as *Listeria*) grow well at refrigerated temperatures. An imitation crab salad contaminated with such a bacterial pathogen poses a genuine safety risk since temperature control does *not* eliminate the possibility of food-borne illness. Extreme care must be used to prevent contamination throughout retail stewardship.

Fabrication

The best assurance against inadvertently cross-contaminating or recontaminating ready-to-eat products is to begin the procedure with clean, sanitized equipment and hands (Figure 8). Food contact surfaces such as pans, spoons, spatulas, counter tops, etc., should be detergent-cleaned, rinsed, and dipped in (or sprayed with) a sanitizing solution prior to use. Furthermore, the *Food Code* specifically states that unprotected hands cannot contact any ready-to-eat product. Therefore, gloves are essential in fabricating ready-to-eat seafood salads. These prerequisites are key to preventing inadvertent cross-contamination or recontamination. The primary reason is the inherent crossover employees and equipment make among various raw and ready-to-eat SKUs in the retail seafood department. Performing the actual fabrication steps without first completing the required prerequisites leaves the retail firm with few assurances that the fabricated product is as safe as the unopened ingredients.

Employees and equipment are the common links between raw and ready-to-eat items; separation between these two product categories is essential if product safety is to be maintained. Pans, tools, utensils, and gloves should be dedicated for either raw or ready-to-eat serving, *but not both*. The best way to reinforce the idea of separation is to color-code utensils, pans, and even disposable gloves. Color-coding is important for two reasons. First, it should reinforce the need to utilize different utensils and gloves when working with ready-to-eat or raw foods. This rule — "use white gloves and utensils for raw products and red gloves and utensils for ready-to-eat items" — will facilitate a glove change and the use of different equipment when moving between raw and ready-to-eat products. Second, color-coding presents an obvious visual cue to the employee and

Table 2. Solving Current Safety and Quality Errors Encountered in Mixing and Holding Refrigerated, Ready-to-Eat Seafood Salads

SOP Objective	Current Approach Revealed through the Audits	Proposed Solution
	Fabrication:	
Preventing Various Contamination Venues	Detergent handwashing not observed at initiation of mixing routine. Disposable gloves not used for the mixing operation. Just prior to mixing, utensils and bowls were not detergent-cleaned and sanitized. Pans and utensils were in constant use across raw and ready-to-eat lines.	The first prerequisite step is washing and sanitizing hands. In addition, the *Food Code* states that employees cannot handle ready-to-eat products with unprotected hands, so employees must don new disposable gloves before mixing salad. Utensils and pans need to be detergent-cleaned and sanitized prior to use. Consider using color-coded pans and utensils for raw and ready-to-eat lines.
	Close-down and Overnight Storage:	
	Several sheets of cling wrap were placed over the mixing/display bowl. Bowl was then removed from the service case and placed in storage cooler. Storage location was randomly chosen, and film did not provide a water-tight barrier for contents.	Use a display pan as described in Figure 1. Upon close-down, attach lid, and place in storage cooler in topmost location. Vertical location prevents cross-contamination originating from drip or splash from raw products, and the lid prevents recontamination from condensation.
Establishing a Correct Sales Interval and Ensuring Proper Rotation of Inventory	No fabrication date or sell-by date was marked on the bowl used to mix and display salad. The lack of a sell-by or fabrication date on the mixing/display container does not communicate elapsed time since the batch was mixed. Not knowing the age of the batch may enable an employee to mix a new batch on top of the remainder of an existing batch. This commingling would reduce the shelf life of the new batch to the remaining shelf life of the earlier batch.	The *Food Code* states that the maximum sales window is 10 days or the manufacturer's sell-by date, whichever occurs first. Once mixed, the batch should be transferred to a display pan appropriately sized for the anticipated daily sales level. The remainder should be transferred to a cleaned, sanitized container. This becomes the reserve stock of the batch. This container should be labeled with the fabrication date and the sell-by date, placed in a tote, surrounded with ice, then returned to the storage cooler. The display container can be refilled each morning with salad from the Reserve Stock container of the same batch.

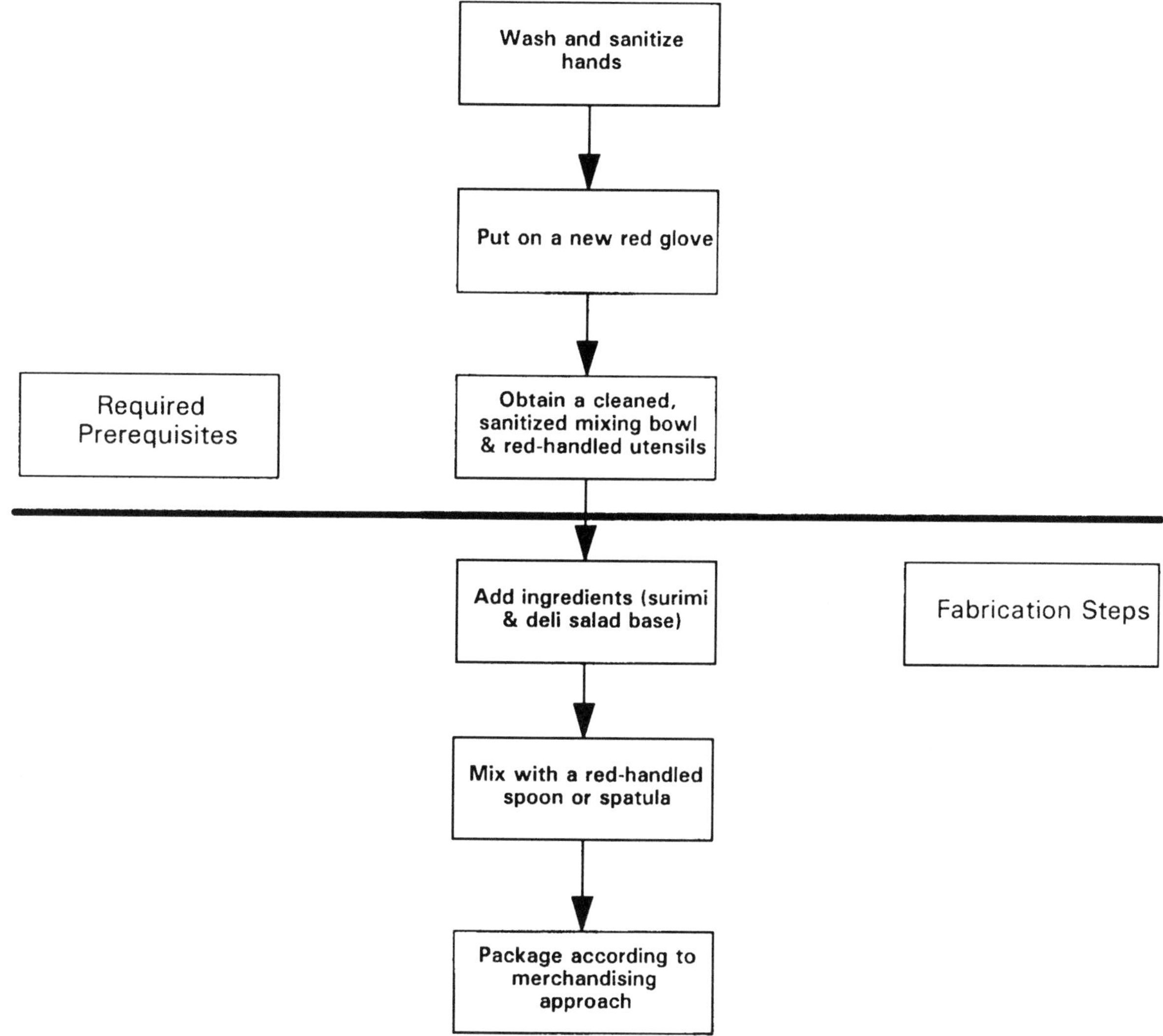

Figure 8. An SOP for preparing seafood salads that includes both prerequisites and actual fabrication steps.

management and facilitates simple, quick compliance checks and immediate corrective action if necessary.

Staging

Staging refers to the preparatory steps required before the case is loaded. Refilling a bulk display pan with salad held in storage, using the prerequisite steps outlined in Figure 8, prevents an uncleaned utensil from contacting the salad.

Display

Bulk displayed salad merchandised through the service case is vulnerable to a variety of contamination anywhere in the retail inventory cycle. To prevent accidental contamination in the service case, all ready-to-eat items should be segregated from raw inventory. Most firms have such a policy in place. In fact, separation between raw and ready-to-eat was always maintained during the standardized quality and safety audits with the cooperating firms. Perhaps the bigger concern is contamination during loading and unloading of the display case. Contamination can be sharply reduced if salads are displayed in the type of deep steam table type pans described in Figure 1. This type of pan is ideal for a homogeneous, fluid-like product such as a salad. The only modification would be to eliminate the perforated insert. When a lid is attached, the pan contents are protected from the drip or splash that can occur during loading/unloading.

Close-down and Overnight Storage

Since seafood salads should be packed in the display containers illustrated in Figure 1, the close-down procedure presented in Figure 5 should also be used. Seafood salads should be placed in the top-most location of the storage cooler to prevent drip or splash from raw products. The rigid top also prevents contamination of salad from condensate that may form in the cooler.

Estabishing a Correct Sales Interval, Ensuring Proper Rotation of Inventory, and Preventing Commingling of Different Batches

To market safe, fresh, long-lasting seafood salads, retail departments need procedures which (1) ensure compliance with regulatory guidance for the maximum length of time an on-site prepared item can remain saleable, (2) guarantee that proper rotational sequences are respected, and (3) prevent commingling of different batches. This procedure must address all these possible situations yet be simple to understand and implement since, at some point, all employees (even the least experienced team members) are responsible for ensuring proper rotational sequences.

The *Food Code* states that once an ingredient is opened, it must be discarded if not sold or served within 10 calendar days or the manufacture's use-by date, whichever comes first. To ensure that salads do not exceed this maximum 10-day shelf life, each batch needs to be labeled with the fabrication date, or the use-by date, or both.

To fabricate, display, and store a batch, three containers are necessary. These include a mixing bowl, a display pan, and a container which can hold that portion of the batch which remains in refrigerated storage until needed (i.e., the reserve stock). Once a batch is prepared in the mixing bowl, the quantity needed for the day should be transferred to the display pan, which is then embedded in the bed of ice in the service case. The remainder — the reserve stock — should be transferred to a cleaned, sanitized plastic container with a snap lid; be marked with a fabrication date, or a sell-by date, or both; then be placed in a tote, surrounded with ice, and held under refrigerated storage until needed.

The standardized quality and safety audits demonstrated that there was virtually no temperature difference between items embedded in ice of a refrigerated display case and items surrounded with ice and held in refrigerated storage. Since holding temperatures are about equal, each product class has the same amount of remaining shelf life at any point in time. Therefore, the fabrication date of the batch is the controlling feature in the stock rotation plan. Until the sell-by date is reached, employees can add reserve stock to previously displayed product from the same batch. As sales are made, the display container can be refilled from the reserve stock of that batch held in storage. If, at the beginning of day three, the employee realizes that there is not enough of Batch I (both previously displayed merchandise and reserve stock) to cover anticipated daily sales, the remainder should be packed in end-user containers so it can be sold first or used as samples. At that point another batch can be prepared (Figure 9). This approach prevents commingling of different batches, ensures maximum shelf life of each batch, and guarantees a first made/first out rotation system. Figure 9 illustrates a similar routine when the sell-by date of a batch expires. At the start of business on the day that the sell-by date for Batch II expires, all remaining components of the batch are treated alike. The remainder can be (1) packed in self-service containers, marked down for immediate sale, and placed in a self-service case; (2) used for samples; or (3) discarded as corporate policy dictates. The disposition of this merchandise depends upon store policy. Those items marked down that do not sell by the end of the sell-by date should be discarded.

A STANDARD OPERATING PROCEDURE FOR CUSTOM-COOKING PRODUCTS

Many seafood departments cook (steam) products to order. Cooking-to-order provides a significant convenience to customers since they are presented with a custom-cooked, possibly seasoned, ready-to-eat product. Typically, a quantity of raw or live product (shrimp, lobster, crawfish, crab, etc.) is selected, weighed, priced, and then steamed. Once an order is placed, customers normally continue with other shopping and return later to retrieve their purchases.

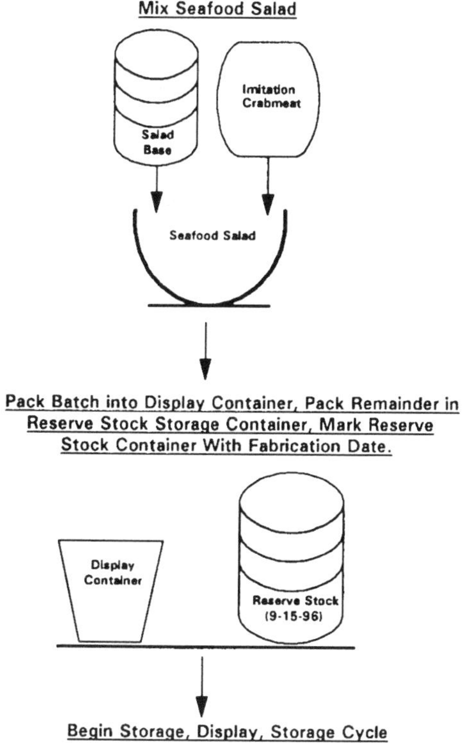

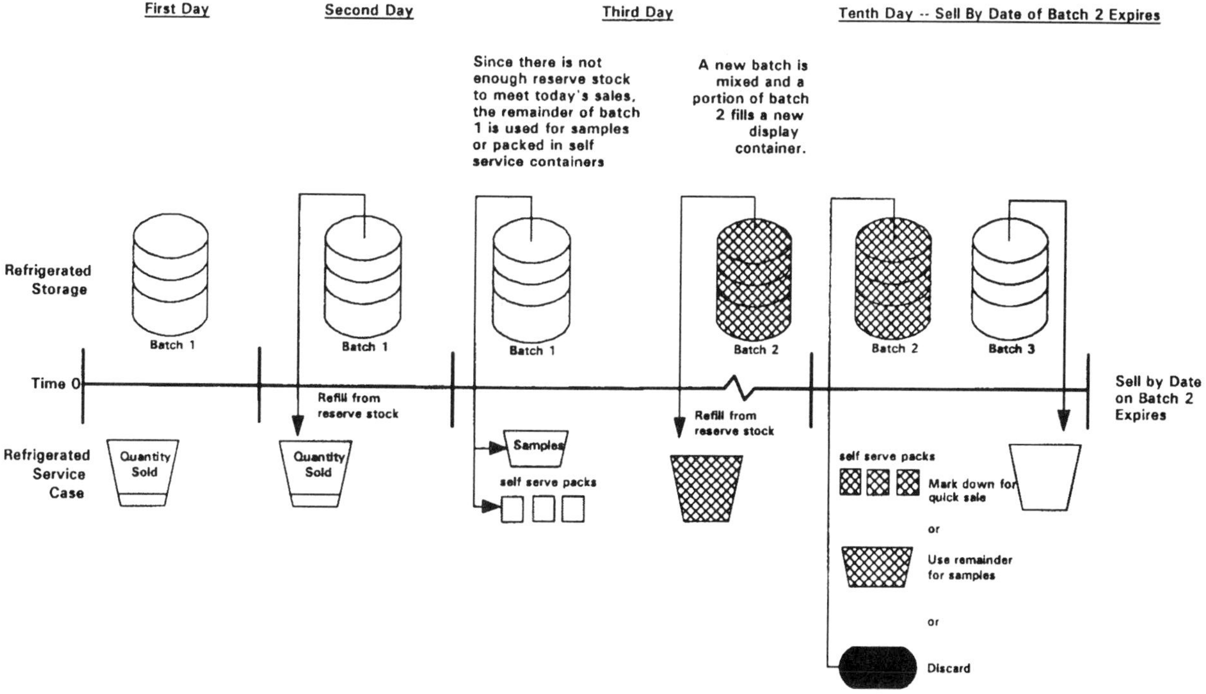

Figure 9. A routine that prevents commingling batches of bulk-packed seafood salads displayed in a refrigerated service case, ensured by tracking the sell-by date.

By design, other SOPs described in this chapter are structured so that experienced personnel are responsible for the more complex, production-oriented aspects of departmental operations (see Figure 3). Conversely, the primary role of the afternoon/evening employee is to wait on customers and fill their orders, and then to close down the department at day's end. The close-down procedure specified in Figure 5 is simple and straightforward. The SOP for custom cooking varies from the overall design because it specifies precisely how to accomplish a relatively complex task. This task must be conducted during a fairly busy time at the seafood counter by a less-experienced, often part-time, staff member.

This SOP is also unique because it addresses a sequence of steps required once a customer opts for a steamed product. Unlike other SOPs which address how best to meet the goals of a preventive strategy over several days of repetitive cycling between display and storage, this procedure focuses on a relatively short time frame — generally less than 30 minutes.

Because consumers purchase a cooked-to-order, ready-to-eat product, the custom-cooking function must be effectively designed and managed to minimize food safety risks. Custom cooking is an important service, and in all likelihood will expand in response to demand for more convenience. This ultimately means providing employees with necessary equipment and correct procedures. Table 3 presents the errors observed in the custom-cooking routine, and a proposed SOP that minimizes the potential of these errors.

The custom-cooking process is actually comprised of four distinct steps: (1) selecting and weighing raw or live products; (2) cooking; (3) post-cooking handling, which includes an optional seasoning step; (4) packaging; and (5) holding for subsequent pick-up by the customer.

Potentially, all steps beyond selecting the raw product can have an impact on product safety. Cooking should be sufficient to kill both spoilage and pathogenic bacteria yet not be so severe (i.e., too long) that the product is overcooked, since most customers would consider this to be a quality defect. Once the product is fully cooked, safety can be compromised by cross-contaminating cooked product with raw items or by recontaminating the cooked product with uncleaned food contact surfaces, hands, etc. Either of these contamination routes can be instantaneous, and most often results from an accidental or inadvertent action by an employee.

Nowhere is the opportunity to make these mistakes greater than when products are custom cooked in the seafood department, primarily because recontamination and cross-contamination potential is high in these settings. This is to be expected when a mix of raw and cooked products exists in the same space. While the custom-cooking process itself is straightforward, each order must be separately selected, priced, steamed, perhaps seasoned, and packaged. That is, the custom-cooking function is a batch process. It is not uncommon for the employee to handle multiple batches. For example, as a tray of cooked product is removed, another one is placed in the steamer, and then the employee returns to season and package the first batch. This type of crossover between raw and cooked product dramatically increases the potential for contamination.

Inadvertent contamination of any ready-to-eat food increases the probability of food-borne illness. For those pathogens with a high infective dose, maintenance of low temperatures after cooking can minimize the opportunity for food-borne illness because microbial growth is slowed at low temperatures. On the other hand, low temperatures are not an effective control mechanism for pathogens with low infective dosages. In most custom-cooking operations there is no temperature control after the batch exits the steamer. Thus, cooked products remain within a temperature range that supports rapid growth of microorganisms if they are not quickly refrigerated (Figure 11). As illustrated, the elapsed time products remain in this zone (40°F to 140°F) could be several hours. This time span occurs because cooked shrimp are allowed to cool on the counter (at ambient room temperature) and are then passed to customers in a warm (or hot) state. Once received, consumers may or may not refrigerate the product before using it. If contamination of cooked product occurs, there is no temperature control that minimizes outgrowth. In reality, the time and temperature range demonstrated in Figure 11 is optimal for bacterial outgrowth. Further complicating the picture is the assumption that the cooked product will not be reheated prior to con-

Table 3. Solving Current Quality and Safety Problems Encountered When Steaming Shrimp for Individual Customers

SOP Objective	Current Approach Revealed through the Audits	Proposed Solution
Adequate Cooking to Destroy Pathogens while not Overcooking	Shrimp was the only SKU custom-cooked during the audits. Employees generally used a *per se* rule that stated so many minutes of cooking per pound.	The *Food Code* states that raw animal foods such as eggs, fish, poultry, and meat be cooked to heat all parts of the food to 145°F or above for 15 seconds.
	Shrimp steamed for 5 minutes per pound resulted in a very safe product (average product temperature 200° F), but one that was judged to be over-cooked, thereby resulting in a tough, somewhat dry, texture.	Reaching this critical temperature depends upon several factors including: (1) initial temperature of the raw product, (2) the size of the item (i.e., 16 to 20 count shrimp vs. 31 to 40 count shrimp), and (3) the quantity to be cooked. Baseline cooking schedules need to be established for each of the various products expected to be steamed to order.
Preventing Various Contamination Venues	Cross-contaminating cooked shrimp by repacking them in the same bag used to select and weigh raw product.	Consider using color-coded gloves, pans, packaging, and utensils for raw and ready-to-eat products.
	Cross-contaminating the cooked shrimp with the same glove used to select and weigh raw product or recontaminating cooked product with uncleaned, unsanitized hands.	The first prerequisite step is washing and sanitizing hands. The *Food Code* states that employees cannot handle ready-to-eat products with unprotected hands, so employees must use new disposable gloves before handling cooked product.
	Cross contaminating cooked product from splash which resulted when thawing raw product alongside cooked items.	Establishing locations for thawing raw products or staging items to be cooked that are separate from cooked products.
	Recontaminating cooked product with unclean utensils used to mix shrimp and seasoning.	Holding frequently used utensils such as those used to mix shrimp and seasoning in sanitizing dips.
	Contaminating seasoning by using unclean utensils to scoop seasoning (from a bulk container) and subsequently mix with cooked shrimp.	Using a shaker to apply seasoning instead of dipping it from a bulk container.
Ensuring Proper Handling and Holding by Consumers	Shrimp are sold warm to customers who then assume control of the product. The holding temperature is an excellent incubation range for outgrowth of pathogens. Thus, product must be rapidly used or promptly refrigerated.	Use of adhesive, advisory labels for each package of cooked product to specify rapid use or prompt refrigeration.

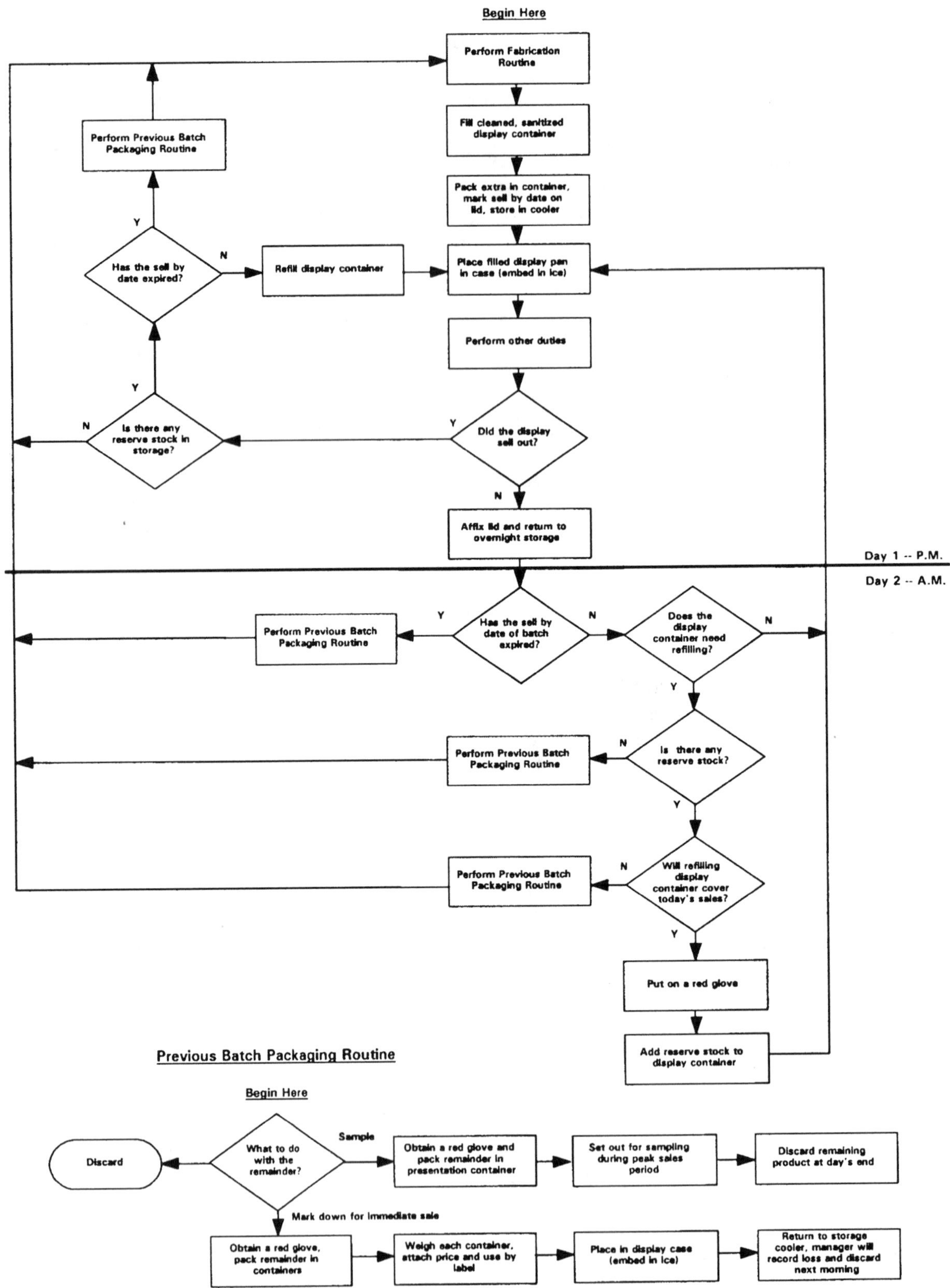

Figure 10. Fabrication, packaging, and stock rotation plan for chilled, ready-to-eat seafood salads sold through refrigerated, full-service cases and packed to order.

sumption. Thus, a busy day, multiple batches in progress, customer distractions, no set procedure to prevent contamination, and a high holding temperature after cooking all combine to compromise product safety. The impact upon the store and the consumer could be serious.

Developing Proper Cooking Schedules

To meet public health requirements while not compromising the quality of a steamed product, cooking trials need to be done. Because of the variation in steamer equipment and the products to be cooked, it is difficult to provide many specific recommendations. Trials should be carried out, with the product having the same initial temperature as an item selected from the service case. Cooking trials should be conducted using the approximate quantities generally ordered. Basing heat penetration on a relatively small quantity, when most of the time a larger volume is requested, could lead to inadequate cooking that may not kill all the pathogens. Additionally, the individual size of the item to be steamed — such as the count size of shrimp — should be considered, because heat penetration is dependent upon thickness. The *Food Code* states that raw animal foods must be cooked to a temperature of at least 145° F for 15 seconds. These figures should be considered as the minimum thermal processing needed.

Minimizing Various Contamination Venues

As Table 3 demonstrates, contamination of cooked product can occur through various means. To minimize the risk, several elements should be added to departmental operations management. The first of these is establishment of predefined, standard, separate locations for: (1) staging raw product prior to placement in the steamer, (2) cooling, seasoning, and packing cooked product, and (3) storing those tools that repeatedly contact cooked product throughout the day, such as perforated steam trays and mixing utensils. Second, use color-coded utensils, packaging materials, and gloves for handling raw and cooked product. Third, incorporate a comprehensive, simple-to-understand set of sequenced steps for completing the custom cooking routine where product safety might be compromised (e.g., placement of cooked product for cooling, seasoning, packaging cooked product, and holding the warm item for the customer).

Predefined, Standard Locations

While every retail store in a chain is different with respect to available square footage, wall space, etc., establishing predefined, standard, separate locations for raw and cooked product is an important first step in preventing contamination. A hypothetical work station highlighted in Figure 11 demonstrates the design elements necessary to establish standard, separate locations for both raw and cooked products, and the gloves, utensils, and packaging materials used with both raw and cooked products.

A spacious department with enough square footage for each required function is most often the exception. However, wall shelving, racks, pegs, hooks, etc., can accommodate many of the tools, pans, utensils, and packaging materials normally used in a retail operation. In this diagram the work station consists of a work table and a wall-mounted shelf or rack. The work table has space provided on each side of the steamer for holding cooked and raw material. All raw product to be steamed is staged to the right of the steamer. All functions involving cooked product are completed to the left of the steamer. These functions include air cooling, seasoning, and packaging.

A similar theme is repeated on the wall-mounted shelf or rack. Clear disposable gloves, clear bags, and white foam trays used to handle, select, and weigh the raw product are stored on the right side of the shelf directly above the work table location. Likewise, red disposable gloves and red overwrap trays (or bags) are stored at the left above the table for cooling, seasoning, and packaging cooked products. Logically grouping, handling, and packaging materials for raw and cooked product above standard locations reinforces proper use and facilitates time efficiency since the appropriate items are positioned for easy access.

There were two related contamination concerns associated with seasoning cooked product. Some stores used a large bulk pack of dry seasoning mix as the primary dispensing container. To dispense seasoning onto cooked product a mix-

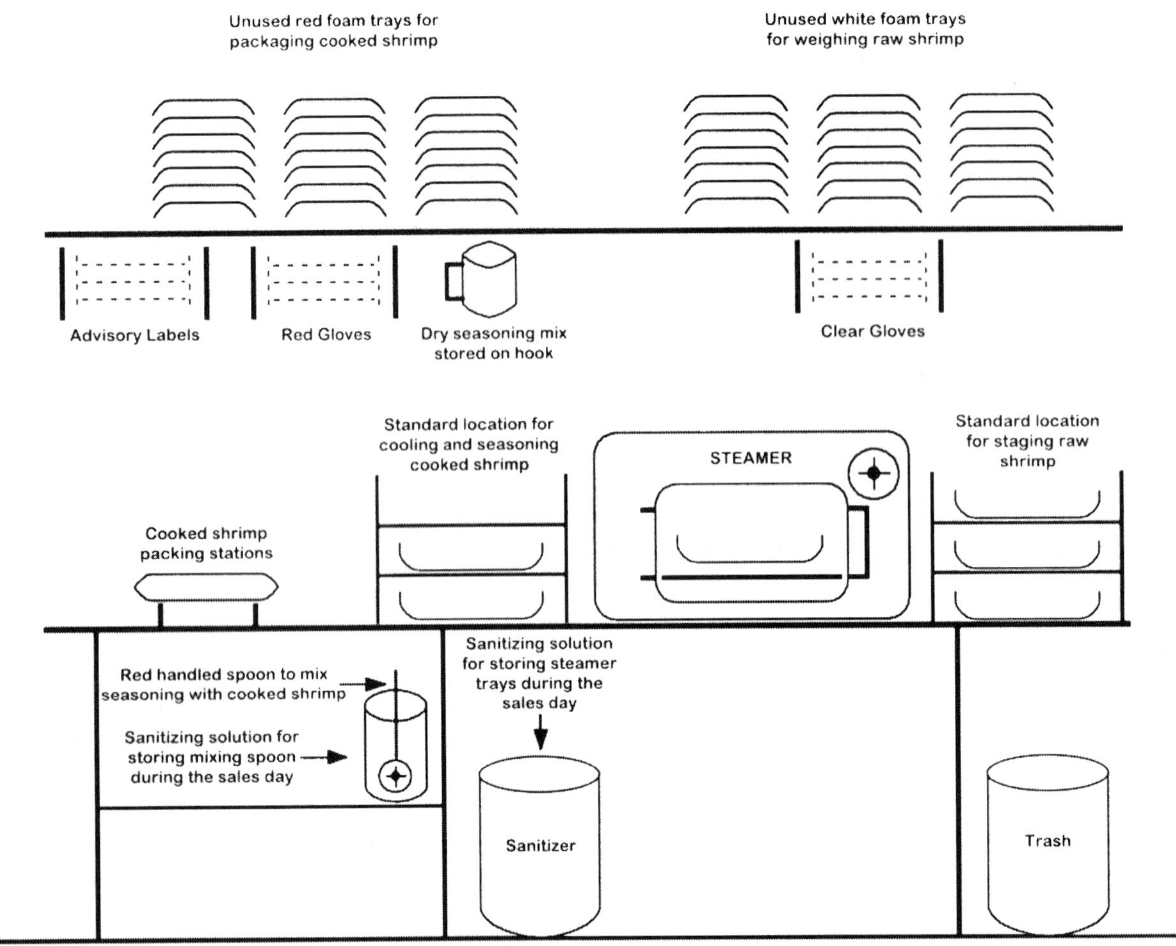

Figure 11. Hypothetical workstation for steaming shrimp or other seafoods within a full-service seafood department.

ing spoon was used. Typically, the employee picked up a mixing spoon and dipped powdered seasoning out of a large container. This mixing spoon was picked up from numerous locations including drawers, on top of the steamer, in the seasoning, etc. Auditors never saw the spoon being cleaned and sanitized prior to contact with cooked product or ready-to-eat seasoning. Failure to begin with a clean, sanitized spoon effectively contaminated both the seasoning and the cooked product. To prevent the contamination of dry seasoning, a red shaker can be used to apply the spice mix to cooked product. This red-colored shaker is stored on a hook alongside other red handling and packaging materials. Ensuring that repetitively used mixing utensils do not recontaminate the cooked product is addressed in the following paragraph.

As Figure 11 illustrates, two containers below the steamer table hold solutions of sanitizer. These containers are designated as standard locations for equipment and utensils repetitively used in the custom cooking operation. Holding mixing spoons in a sanitizing solution ensures a sanitary utensil surface each time the spoon is used to mix seasoning with cooked product. Obviously, steamer trays undergo the same cooking schedule as the shrimp, crawfish, etc. The process of repetitive steaming tends to create a build-up of proteinaceous soil on the pans. These perforated pans have a large surface area that makes effective cleaning and sanitizing difficult. The audit team found that repetitive use throughout the day resulted in a cooked-on film that was difficult to remove with commonly used detergents. By designating a container of sanitizing solution at a standard location for these pans, microflora are effectively controlled, and any build-up on pan surfaces remains hydrated throughout the day. This practice also significantly reduces the time to clean perforated steam trays at the end of the day.

Sanitizing solutions can be selected from among several classes of compounds: hypochlorites, iodophors, and quaternary ammonium compounds. Initially, the retail seafood quality and safety audit team believed that the best all-purpose sanitizer for retail seafood departments was the quaternary ammonium type of product. It controls odor, is effective against biofilms, and works well as a solution in which to store tools, utensils, and even clean-up articles such as mops, etc. However, the concentration of a sanitizer is key to its effectiveness. Because department personnel typically work unsupervised, the use of procedures and products that lend themselves to rapid, simple compliance checks is recommended. Therefore, iodophors seem to be the best compound for storing utensils and steam trays during the day since their concentrations can be visually assessed by the presence of a golden, yellow color. The concentration of an iodophor as either a hand dip or for storing tools and utensils is 25 to 50 parts per million (ppm). Assuming that most cooking occurs in the afternoon, the department manager should be responsible for preparing these sanitizing solutions just before leaving. Regardless of the sanitizing compound selected, the most effective way to ensure that the proper concentrations are mixed is with a system that automatically meters the compound with water.

Color Coding

Color coding supports the standard locations for raw and cooked product and the necessary gloves, trays, bags, and utensils. Requiring the use of different colored foam trays, bags, and gloves when working with raw or cooked product helps prevent accidental contamination by making such an error immediately obvious. This idea of color-coding is continued with the red-colored seasoning shaker and the red-handled spoon used to mix seasoning with cooked product. By virtue of their colors, employees are cued that these items can only be used for cooked product, and only when hands are protected with red gloves. Color coding also facilitates training in the proper use of handling and packaging materials, and provides for quick, accurate compliance checks by management.

A Stepwise Procedure Set for Custom Cooking

Standard locations and color-coded items are important tools, but alone they cannot ensure against inadvertent contamination. Contamination of the cooked product often occurs because the employee forgot a key task or did something out of sequence. An example would be handling a cooked product with the same gloved hand that originally selected and weighed the raw product. The process flow chart precisely defines the necessary tasks and their proper sequence. Knowing what to do and when to do it is essential in making sharp reductions in contamination opportunities (Figure 12).

Of the four elements that comprise the custom- cooking function, cooking is the most complex step in the diagram because it addresses handling multiple batches, performing other departmental duties while the steamer cycles, and the inherent crossover between raw and cooked inventory. This chart specifies which functions to complete with a particular color of glove, and precisely when a glove change is required. It also outlines how time/temperature abuse of the finished product can be minimized.

Consumer Advisory Labels

The fourth element in the SOP for custom cooked seafoods is the use of advisory labels on the packaged, cooked product that highlight proper consumer handling necessary to maintain product safety. This label highlights the customer's responsibility for rapid use or prompt refrigeration. Such labels are currently required by USDA for raw beef and poultry products.

Summary

Meeting *Food Code* requirements by establishing cooking schedules based on initial temperature, expected quantities, and size of item to be cooked (i.e., the count size of shrimp) is a prerequisite step that forms the basis for producing a safe, high quality product. After the cook step, it is imperative that the commercial sterility created via the heat treatment not be compromised by accidental contamination since there is generally no temperature control after the cook step. The next three components of a management plan — (1)

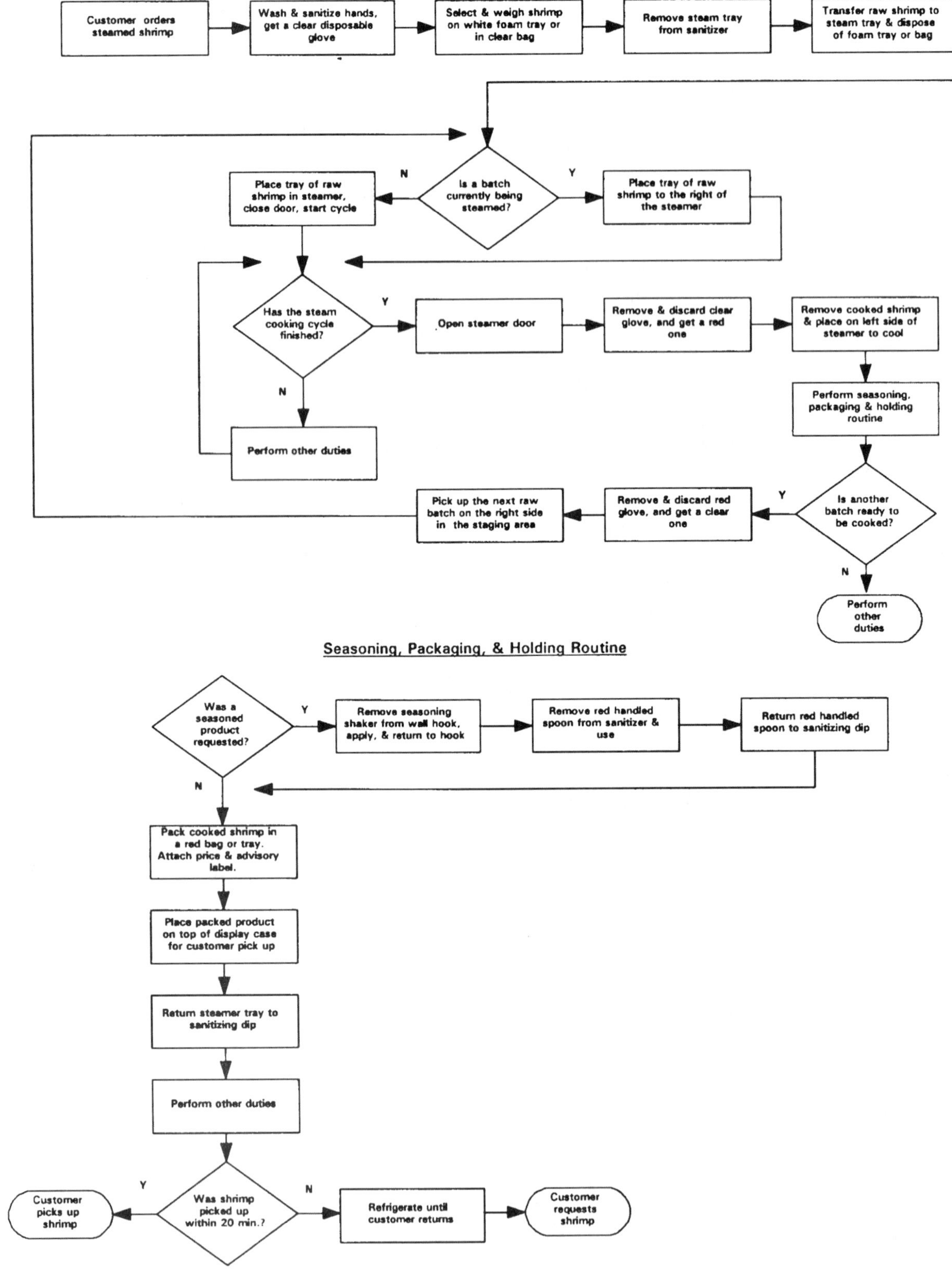

Figure 12. Custom-cooking routine in a full-service seafood department.

creating a standard, separate location for raw and cooked product; (2) color coding of gloves, packaging material and utensils; and (3) a stepwise process for completing the custom cooking function — are reflected in the behavior required by all employees who perform custom cooking. In particular, these steps should reinforce the following two points: (1) whenever a raw product is handled, the gloves, bags, and overwrap trays should be clear or white and (2) whenever cooked product is being handled, any clear glove(s) should be discarded, and a red glove obtained since only red-gloved hands should handle the product, the red shaker, the red-handled spoon, or red bags and foam trays. Color coding makes compliance checks simple, and provides a straightforward message to the employee charged with doing the work.

For quality and safety, the first three elements of the management plan "error proof" the custom cooking function. Collectively, they incorporate a key concept, *"What would you have me do differently from what I am doing now?"*

This procedure set dictates a positive skill, and breaks the cycle of improper practices being used. Equally important, these three components generate time efficiency among employees. When incorporated into retail operations, these components sharply reduce food safety risks that may result from inadvertent employee actions.

The last element of the plan — use of advisory labels on cooked product — provides a message to customers that they also share in the responsibility to ensure a safe product. While this step may appear as strictly a defensive tactic, its utility should not be discounted.

CONCLUSIONS

A wide mix of unique products are brought together in most full-service seafood departments. Those working in those departments handle a diverse product line and do more with it than their counterparts in the meat department. Whereas meat operations focus on preparing ready-to-cook retail cuts from either subprimals or converting one market form to another (e.g., steaks to cubed beef to ground beef), seafood department personnel not only merchandise a variety of seafoods, each with a different amount of shelf life, but they also prepare and handle various ready-to-eat items. Given the unique shelf life among different species and the heightened food safety risks implicit with on-site prepared, ready-to-eat items, there is even more reason for a standardized approach to retail operations which: (1) simplifies processes, (2) utilizes a science-based "recipe" approach in completing tasks, and (3) provides periodic training of those principles. Unfortunately, such an approach seems to be the exception among most retail seafood departments.

To market safe, fresh, long-lasting products through full-service departments, three objectives must be addressed. First, the interactions among products, workers, and food contact surfaces must be managed to prevent accidental contamination, cross-contamination, or recontamination. Second, the passive steps must employ stocking procedures that ensure maintaining low product temperatures regardless of equipment differences or preset ambient conditions. Third, procedures must be designed so that inventory with the least amount of remaining shelf life is positioned to exit the department first.

All of these considerations focus on the practices and procedures used by employees. Therefore, human skill and ability are necessary to reach quality and safety targets in full-service departments. Expressed another way, quality and safety are linked to employee management, not to the application of technologies. There are two primary reasons. First, full-service departments are predicated on intensive, repetitive handling of a product line between receipt and sale or discard. Second, sharp reductions in both premature spoilage and the probability of compromised food safety result when errors are prevented.

Sustained quality and safety improvement occur only when plans meet two objectives. All plans should be based on preventive strategies. Procedures which specify how to complete a given task should be based on performance proven through scientific validation. A plan that just satisfies the product-oriented criteria is but half complete, since real improvement happens only when employees implement a plan according to specification. Therefore, quality and safety management plans must also respect the limitations imposed by employees. For many full-service seafood departments, a growing proportion of the employment base is comprised of part-time, relatively unskilled individuals who typically work unsupervised. Thus, each procedure should be crafted so that opportunities for careless actions, inadvertent mistakes, or omission of key steps are sharply reduced (or eliminated).

A set of standard operating procedures (SOPs) designed for each function would give structure to the various departmental tasks, and reduce the opportunities for mistakes or omissions that compromise quality and safety. Furthermore, standard operating procedures would facilitate the development of those positive, human abilities that are so essential to maintain quality and safety improvements in service departments. Finally, the development and implementation of SOPs provide an excellent defensible program. Such activities are key to retail Hazard Analysis Critical Control Point (HACCP) programs. Specifically, the four broad categories that retail HACCP will have to consider are: (1) developing product specifications, (2) certifying or qualifying vendors, (3) adapting and implementing in-store processes, and (4) communicating with consumers through advisory labels, etc. Of these, the greatest challenge to management is perfecting the in-store processes.

The food retailing community is not alone in its need for effective, simple procedures. Across all industries, deficient operational processes are the single most common reason for poor quality (e.g., high levels of waste or shrinkage, various manufacturing defects, etc.) (Garvin, 1992). Building research-based, workable procedures to combat quality and safety errors and putting these in the hands of employees is a significant management challenge, one which offers the richest economic rewards for those companies committed to finding and implementing solutions.

ACKNOWLEDGEMENTS

This chapter is excerpted and summarized from a comprehensive reference for the grocery industry that addresses how best to design and implement a quality and safety management program for full-service seafood departments; the program sharply reduces avoidable shrinkage due to premature spoilage and simultaneously minimizes the opportunities to compromise food safety through retail negligence. The comprehensive reference was prepared with financial support from the Northeast Regional Office of the National Marine Fisheries Service via NOAA Award NA26FK0396, the Cooperative States Research, Education, and Extension Service via USDA Award 94-EFSQ-1-4143, the Texas Agricultural Extension Service, and the Sea Grant College Program at Texas A&M University.

REFERENCES

Bramsnaes, F. 1965. Handling of Fresh Fish. In Borgstrom, Georg (ed.). *Fish as Food*, Volume IV Processing: Part 2. Academic Press. New York.

Garvin, D. 1992. *Operations Strategy: Text and Cases*. Prentice-Hall. Englewood Cliffs, NJ.

Haby, M. and R. Miget. 1991. The effect of stocking procedure on consumption of shelf life in refrigerated seafood displayed in full-service departments. *J. Food Dist. Res.* **21**:69-79.

Miget, R., T. Wagner, and M. Haby. 1987. Determination of optimal display procedures for fresh and previously frozen seafoods using ice-only cases and seafood cases which require supplemental refrigeration. Final Report to the Gulf and South Atlantic Fisheries Development Foundation Inc.

Samuels, R., A. DeFeo, G. Flick, D. Ward, T. Rippen, J. Riggins, C. Coale, C. Smith, C. Armory, T. Morris, J. Bordinaro, and R. Martin. 1984. Demonstration of a quality maintenance program for fresh fish products. VPI-SG-84-04R.

U.S. Department of Health and Human Services. Public Health Service. Food and Drug Administration. 1993. *Food Code*. National Technical Information Service. Springfield, VA.

Section V. Other Useful Products from the Sea

Seaweed Products: Red Algae of Economic Significance

Brian Rudolph

INTRODUCTION

Seaweeds have been utilized by mankind for several hundreds of years — for food, for medicinal purposes, and as a fertilizer. Utilization of seaweed was first recognized in Asia, where it found use as food and also as medicine (Humm, 1951). In Asia edible seaweed is still consumed in large amounts.

In Europe, utilization of red seaweeds may have started with the use of *Palmaria palmata* and *Chondrus crispus*, for direct consumption when dried, and for the gelling of milk puddings when boiled, mainly in Ireland and around Brittany (Bretagne). The use of seaweeds for fertilizer/soil improvement was also well-known in Europe, and both large brown algae and calcified red algae have been collected for this purpose (Blunden, 1991). On small islands on the European side of the North Atlantic, it has been common practice to feed sheep and cattle on harvested seaweeds of the intertidal brown algae *Ascophyllum nodosum* (Indergaard and Jensen, 1991).

Of the three most important phycocolloids (hydrocolloids originating from algae) – agar, carrageenan and alginate – agar was the first to be developed. It was first used as a food and a gelling agent. In 1881 it was discovered that it could also be used as a gelling agent for microbial culture media (Booth, 1979). The first country to develop and use agar was Japan. Because of World War II, Japan lost its leadership in agar production, but is still a major producer, with numerous small agar factories and a very high per capita consumption. The world's largest agar factory today is in Chile, producing 10 percent of the world's production. There are also factories in China, Indonesia, New Zealand, Spain, Morocco, Ireland, and other countries.

Carrageenan was developed in Ireland as an extract from *Chondrus crispus*, made by boiling seaweed in water and drying the resulting liquid. A non-purified product containing carrageenan was also sold in the form of bleached Irish moss, the trade name for *Chondrus crispus*. The colonists brought the knowledge of using *Chondrus crispus* with them to America, and after many years of import from Ireland, it was discovered that *Chondrus crispus* also grows along the coasts north of Cape Cod (Massachusetts) and Canada. Today the major part of the *Chondrus crispus* used commercially is harvested in the maritime region of Canada: Prince Edward Island, New Brunswick, and Nova Scotia. Small quantities are harvested in France, which was the first nation to promulgate federal laws regulating seaweed harvest. Carrageenan is produced in Denmark, Ireland, France, the United States, Chile, Japan, South Korea, the Philippines, and a few other countries. There are approximately 25 carrageenan factories in the world.

Algin and alginate originating from brown algae were discovered in the 1880s in England. Alginate is produced in California, Norway, France, China, and a few other places. It is principally derived from *Macrocystis* and *Laminaria*, but today *Lessonia* and *Durvillea* from Chile and Tasmania are also used. China produces alginate from *Laminaria japonica*, which is cultivated in large open sea farms. Alginate is separated into industrial, food, and pharmaceutical grades, with increasing demands for purity.

The above-mentioned phycocolloids are all sold as highly developed and specialized white to yellow-brown powders standardized with sugar and food grade salts to meet specific applications.

CLASSIFICATION OF THE ALGAE

The term algae comprises several taxonomic divisions without any obvious ancestor, covering both macro- and microscopic algae (Bold and Wynne, 1985). Seaweeds are those algae that usually can be recognized macroscopically, and consist of multiple, different cells organized in a specific structural manner (Harlin and Darley, 1988). All algae reproduce by single-celled spores and lack any form of flower, fruit, or seed. The seaweeds include the following four divisions:

Cyanophyta, or the blue-green algae, have a wide distribution on land and in both fresh- and marine water. Blue-green algae can grow on many kinds of substrates and can be found on ice and on soil. Some blue-green algae are utilized for fish-feed (*Spirulina*); others are used in soups in China; there is also some limited use as a nutritional supplement in the United States. Generally, there is no commercial utilization of blue-green algae, perhaps because the algae are microscopic, although some form colonies.

Chlorophyta, or the green algae, have their widest distribution in freshwater, but are also widespread in marine and brackish waters. Green algae are frequently used as feed for animals and as a fertilizer (*Ulva*). Some is used for human consumption (*Caulerpa*), mainly in the western Pacific region and in Hawaii. Another relatively new use is as a nutrient-removing organism in wastewater treatment plants (Schramm, 1991), and as a source for biomass energy conversion (Morand et al., 1991).

Phaeophyta, or the brown algae, are principally marine, and are found most frequently in cold water, where they form dense beds in the sub-tidal zone. The best-known brown algae is probably *Macrocystis pyrifera* or the giant kelp, growing along the West Coast of the United States. They can reach a size of 43 meters (Neushul and Harger, 1987). The brown algae are used for the extraction of alginate, and were used to produce iodine and potash before other alternatives were developed. (The brown algae and their products are considered in another chapter.)

Rhodophyta, or the red algae, are a very diverse group of approximately 2,500 species. Distributed mainly in tropical marine water, they are also frequently found in cold marine water, and a few species can even be found in freshwater (Sheath, 1984). A characteristic of the red algae is that they all contain a red pigment, phycoerythrin. The actual color ranges from bright green, to yellow-red, to almost black. This is due to the content of other pigments. Red algae are the most valuable of all the algae, producing the phycocolloids agar and carrageenan. Belonging to this group is *Porphyra*, better known as nori, which is eaten in Japan and supports a large industry. Most red algae develop unique complex colloidal carbohydrates in their cell walls. Several of these properties are utilized industrially; for example, some give hard-gelling phycocolloids, and others, viscous phycocolloids that are unable to gel. Phycocolloids are generally soluble in hot water and can be extracted by boiling the algae.

The taxonomy of red algae undergoes continuous changes as new identification methods are developed and old methods for defining similarities among related species improve. One of the more recent methods developed is the use of genetic codes of particular gene units that can be found in all species of red algae (Goff et al., 1994). When these codes are compared, the more they are alike, the more closely related are the species. There are, however, difficulties with this method, because differences within a species sometimes exceed the difference between two closely-related species. These new techniques do prove to be very valuable in distinguishing different taxa, however (Bird et al., 1994).

EDIBLE ALGAE

From earliest times man has used algae as food, both for direct consumption (fresh or dried), or as an ingredient in other foods. Today, seaweed is still used as food in many parts of the world, especially in the Far East and Japan. Red, brown, and green algae are consumed. They are all used either fresh or dried in soups and salads.

In Hawaii a health industry cultivates seaweeds for restaurants, while in Ireland *Palmaria palmata* has traditionally been consumed (Abbott, 1988). With the increasing focus given to health and nutrition, edible algae have found a new use in the western world, with consumption steadily increasing — though still far below that of Japan on a per capita basis.

THE NORI INDUSTRY

The largest seaweed-based industry in the world is the nori industry in Japan. The value of nori sold in 1984 was approximately 500 million dollars (Mumford and Miura, 1988; McHugh, 1987), which today has increased to more than $2 billion.

The most widespread *Porphyra* species cultivated for nori production is *P. yezoensis*, originating from Japan.

In the 1950s it was discovered that *Porphyra* has a heteromorphic life cycle, with one phase growing as branched filaments in shells of marine mussels and snails (the "*Conchocelis*"-stage), and the other phase producing the upright thallus that is used for nori (Hansen et al., 1981). This discovery was used to create a very advanced cultivation system, in which shells inoculated with *Porphyra* are kept in basins under controlled conditions. By control of temperature and other parameters, it is possible to induce sporulation at a specific time. The spores are collected on lines entwined on frames, and when sufficient spores have settled on the lines, they are cultured under controlled conditions until they reach a size where they can be transferred to the sea. *Porphyra* is cultivated in areas near the coast, where there are tidal changes. *Porphyra* easily survives being out of the water for several hours, and this ability minimizes serious problems with epiphytes. New research has shown that it is possible to store the *Conchocelis* cells by freezing, facilitating strain preservation (Kuwano et al., 1994).

When the plants are ready for harvest, the lines are harvested by hand. *Porphyra* is traditionally washed several times in clean fresh water, then chopped or ground, and the resulting paste spread out in an even layer on sheets and dried in the sun. Today the same is done mechanically (Oohusa, 1993). When dried, the now-finished nori is sorted into different qualities and packed. There are more than 100 different qualities of nori in Japan, where important features are taste, color, and "mouth feel."

The main use of nori is as a wrapping for rice to make rolls or small packages, in the same manner as Europeans use cabbage and grape leaves.

Table 1. Contents of Red Algae

Content	% of dried algae (except for water)
Water	80-85% of live algae
Carbohydrates (agar, carrageenan)	15-50%
Cellulose	3-5%
Proteins	8-25%
Lipids/fat	2-4%
Minerals	15-30%

(Simplified after Indergaard and Minsaas, 1991; McHugh, 1987; Schmid, 1969; and own data.)

COMPOSITION OF RED ALGAE

Beyond the content of cell wall carbohydrates, which will be dealt with later, it is well known that red algae have a very high content of minerals as compared to land plants (Table 1).

Algae are very efficient in actively taking up ions from the surrounding water. This feature may be useful for cleanup of areas polluted with heavy metals, or the algae may be used as monitoring organisms of environmental conditions (Schramm, 1991; Sharp et al., 1988). Algae from polluted areas may cause health problems if used for human consumption. Very little is known about the toxic effects of red algae; only a very few incidents have been reported (United States Centers for Disease Control and Prevention, 1995).

COMMERCIAL SEAWEEDS

Historically, the first commercial use of red algae, excluding edible algae, was the production of agar from *Gracilaria* and *Gelidium* species. This was followed by the production of carrageenan from *Chondrus crispus*, and subsequently from several other species from the families Solieriaceae and Gigartinaceae. Today, the most important commercial seaweed for carrageenan production is the cultivated species *Kappaphycus alvarezii*, better known under the tradename "Cottonii." *Eucheuma denticulatum*, or Spinosum, is another important species cultivated in the same manner and places as Cottonii. *Gigartina radula* is harvested in Chile, and *Hypnea musciformis* is harvested in Brazil (Table 2).

Table 2. World Carrageenan and Agar Seaweed Production in 1993

Carrageenan	Metric tons	Agar	Metric tons
Cottonii	85,000	*Gracilaria*	21,000
Spinosum	5,000	*Gelidium*	18,500
Chondrus	5,500	*Pterocladia*	500
Gigartina	16,000	-	-
Hypnea	300	-	-
Total	111,800	-	40,000

Gracilaria is a genus containing more than 200 species, of which many are used commercially. The taxonomy is very difficult and can only be handled by experts (Abbott, 1994). Generally the commercially-used *Gracilaria* has 1 to 3 millimeters-thick branches and grows to a size of 50 to 100 centimeters from a basal disc. *Gracilaria* is harvested in many countries, especially Chile, Indonesia, Namibia, Japan, Thailand, Taiwan, and Vietnam.

Gelidium is another genus with many commercial uses. Some common features are that the species are fairly small [up to 30 centimeters (cm)], have many thin and rigid branches, and grow on wave-exposed coasts. *Gelidium* species can be found in most parts of the world, but commercial amounts are found in Morocco, Senegal, New Zealand, Japan, Chile, Venezuela, Spain, and Portugal (Santelices, 1988).

Pterocladia is similar to *Gelidium* in habitat and appearance, and can be found in commercial amounts in Portugal, Brazil, and Egypt.

Chondrus crispus has a flattened, dark violet thallus reaching a size of up to 25 centimeters, having numerous dichotomous branches. *Chondrus* grows from a basal disc adhered to rocks or stones from the upper subtidal and down to a depth of 5 to 6 meters. *Chondrus* grows from Morocco to Norway, and from Newfoundland, Canada, to Cape Cod, Massachusetts (Harvey and McLachlan, 1973).

Gigartina radula has flat, dark red leaves, growing from a basal disc adhered to rocks or stones. It can reach a size of up to 3 meters, but most often ranges between 25 to 100 centimeters. *Gigartina radula* can be found from the intertidal zone down to 15 meters. It is distributed from central Chile and south around Cape Horn up to Rio Negro in Argentina. It can also be found in the Falkland Islands and some other islands in the southern Atlantic (Lewis et al., 1988).

Cottonii (*Kappaphycus alvarezii*) has a bushy thallus consisting of numerous round branches. The surface can be rough or smooth, and the color varies from bright green (though it is a red algae) to dark brown. It can grow to a size of more than one meter in diameter, but is normally 20 to 30 centimeters in diameter. It has a natural distribution in the Philippines, Indonesia and East Africa, and grows on the inner side of coral reefs in the upper subtidal zone (Doty, 1988).

Spinosum (*Eucheuma denticulatum*) has the same general morphology as Cottonii, except the branches have spines of 3 to 4 millimeters all over the surface of the thallus. It can reach a size of 75 centimeters in diameter, and generally has more slender and smaller branches than Cottonii (Trono and Ganzon-Fortes, 1988). The distribution is the same as for Cottonii.

Hypnea musciformis has a bushy, but elongated, growth with many thin branches (2 to 3 millimeters) that have a characteristic curve at the tips. It can grow entangled in other algae without being attached to a substrate, and can reach a considerable size, though with little biomass. It is found in the Caribbean Sea, Indian Ocean, and the tropical Pacific Ocean, and may be considered a real globetrotter within the seaweed family (Mshigeni, 1978). The main focus on *Hypnea* is in Brazil (Berchez et al., 1993).

SEAWEEDS FOR INDUSTRY

Methods of Harvesting

There are many different methods of harvesting seaweeds. The simplest one is collecting the seaweeds thrown ashore after a storm. There is a natural limit to the amount that reaches the shore, and a limit to how fast it can be collected, as seaweeds decompose within a few days when exposed to sunlight and rain on the shore. Therefore, techniques have been developed allowing gathering of the seaweed before it is cast on the shore. One early method used a net, which was pulled by a horse in the water along the coastline, collecting seaweeds lying loose, but still submerged.

Another method is the raking technique developed in Canada for harvest of *Chondrus crispus*.

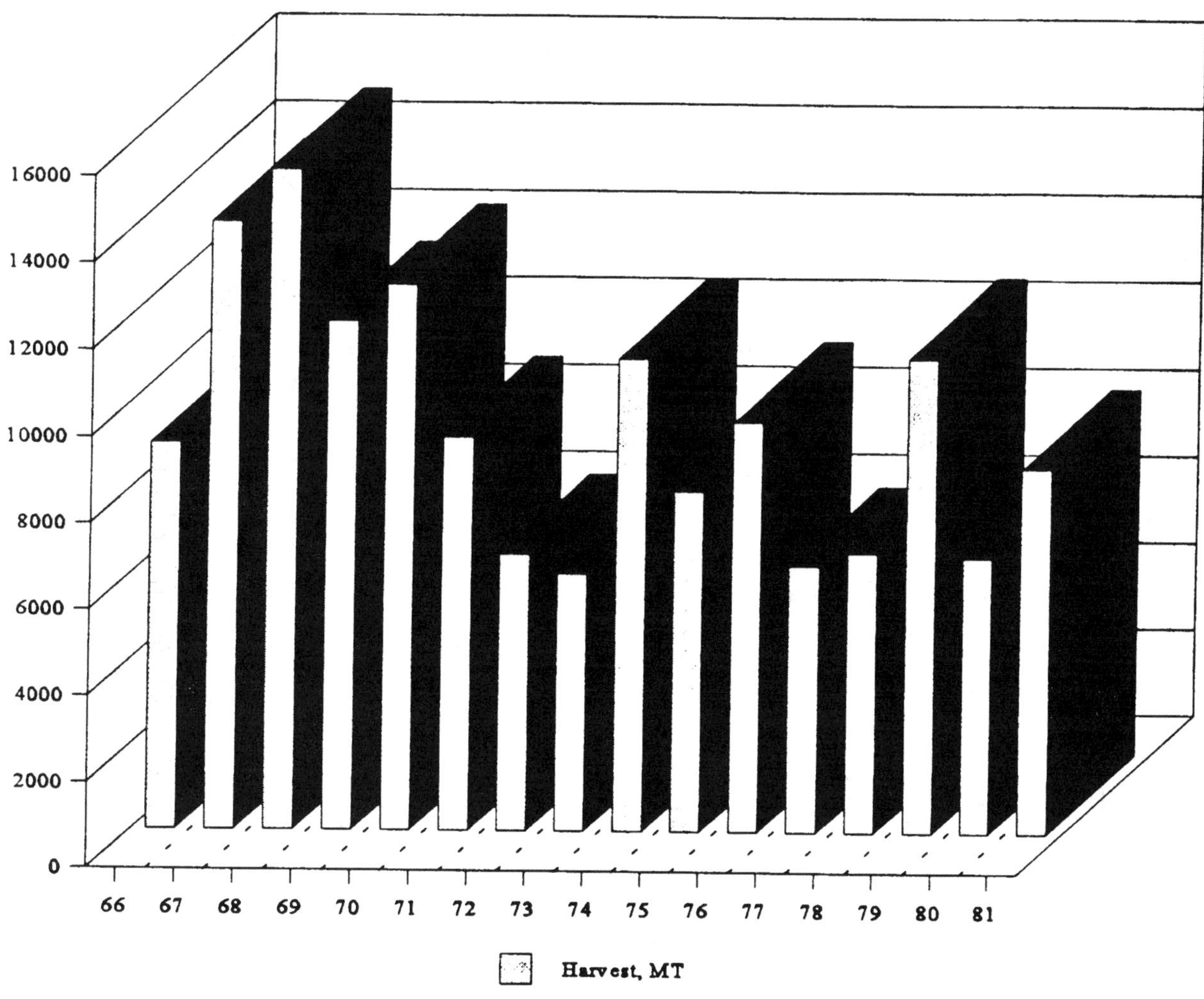

Figure 1. Harvest of Chondrus, *Prince Edward Island, Canada, 1966–1981.*

A harvester rakes from a small boat at high tide and collects fresh *Chondrus crispus* of good quality. The rake has a long shaft and a special head designed for retaining the *Chondrus* as it is pulled off rocks and stones (French, 1970).

When demand of high-quality *Chondrus* increased, some harvesters started to drag a large rake behind lobster-boats; this very efficient method, called dragraking (Pringle et al., 1987), increased production of *Chondrus* to its highest levels in 1968 of 15,000 tons (Pringle et al., 1990) (Figure 1).

In recent years, some of the large natural *Chondrus* beds have become problems for the harvesters, as they are being overtaken slowly by the less-wanted *Furcellaria lumbricalis* (Sharp et al., 1993; Bird et al., 1991). *Furcellaria* has for many years been used for production of what had been known as Danish Agar. This is now accepted as carrageenan, and not an agar variety. Today the use of *Furcellaria* is limited. It was harvested extensively in Denmark in the 1960s, until over-harvesting reduced production to a very low level. Harvesting was done partly from boats, and partly from gathering the shore, as for *Chondrus* (Bird et al., 1991).

In Chile another cold water species, known in the trade as *Gigartina radula*, is harvested for its high content of carrageenan. The harvest methods used range from simple collection at low tide of the seaweed growing on rocks, to advanced diving after large plants in deep water. The basic harvest method is as described for *Chondrus*, but due to the large size of the plants in Chile, diving is a very efficient way of collecting the seaweed growing submerged. The system used is called

Hookah, and consists of a low-pressure compressor pumping air through a thick-walled plastic hose to the diver. Two divers can be fed with air by one compressor, and many boats operate with two compressors. With this system the divers can work three to four hours a day in water depths down to 12 meters. The same setup is used by the fishermen when collecting clams, sea urchins, and abalone.

Another method that should be mentioned is the harvest of *Gelidium* species on exposed shores. Because several of the *Gelidium* species have their habitat in very exposed parts of rocky shores, harvesting is a very hazardous and dangerous job. The equipment used is the same used by mountain climbers, as there is often a risk of big waves making the harvester lose his foothold on the rocks. The *Gelidium* species growing submerged is harvested by diving or raking.

Depending on the local conditions under which the plants exist, the above-mentioned harvest methods are the most commonly used, and they can all be found in many varieties adapted to the specific seaweed in question.

Cultivation

In the Far East, different *Gracilaria* species are harvested for human consumption and agar production. Cultivation of these species in ponds for fish or shrimp farming has been developed through many years, as co-cultivation (cultivation with more than one species). *Gracilaria* is spread in a thin layer at the bottom of the pond and then uses the nutrients produced by the fish to grow to a harvestable size (Enander and Hasselstrom, 1994; Chaoyuan et al., 1993; Trono and Ganzon-Fortes,1988).

Gracilaria chilensis has traditionally been harvested from wild populations in Chile, but due to over-exploitation, decreases in harvested amounts have been seen in most areas. For this reason, and because of increased demand, several cultivation methods have been developed. The most well-known is the method whereby *Gracilaria* is held down against the soft bottom with plastic tubes filled with sand. This method has some environmental consequences, however, as the plastic bags often end up on the beaches after having been ruptured and destroyed by storms (Santelices and Ugarte, 1987; Pizarro and Barrales, 1986). Other similar methods stick small plants of *Gracilaria* into the substrate, or wrap a plant around a stone and plant it on the bottom.

In recent years, a new type of cultivation of *Gracilaria* has been developed. This new method takes advantage of the abundant spore production that can be induced in *Gracilaria* under specific controlled conditions. The spores are allowed to settle on a substrate, where they are cultivated until they reach a size that can be transplanted. This method makes it possible to cultivate *Gracilaria* in areas without good bottom facilities. It is also very valuable in boosting natural populations that have been over-harvested, because the plants on the substrate produce spores when they mature. This very interesting technique is also used for *Porphyra* and some brown algae.

The most important species of red algae being cultivated for phycocolloids is without doubt the tropical *Kappaphycus alvarezii* (Cottonii). Its cultivation method was developed in the 1970s by Dr. Maxwell S. Doty of the University of Hawaii, "the father of *Eucheuma* cultivation." The relatively thick and rigid branches of the *Kappaphycus* make this plant suitable for tying onto a line. During trials, a variety was found that grows rapidly and has a good resistance to diseases. The technique is to tie seedlings (plants of approximately 50 to 100 grams) to lines placed 30 centimeters above the bottom, between sticks 5 to 10 meters apart. The sticks are made of hardwood or bamboo, and the lines may be nylon or similar material. Several lines in a sub-sea field are called a module, and a farm consists of many modules with plants at different stages of growth. The cultivation site must have fine sand, good water exchange, and a depth of 30 to 50 centimeters at low tide — a condition found on the inside of coral reefs in many tropical waters. It is common practice that a farm is handled by a whole family, with women and children tying seedlings and cleaning off epiphytes and the like, whereas the men harvest and hammer the sticks into the sandy reef bottom. When the seedlings have grown to a size of 0.8 to 1.2 kilograms, the plants are harvested, and healthy tips are selected and used as new seedlings. It takes six to 10 weeks for the plants to reach a harvestable size, and the farm can be operated year-round, as there is no distinct seasonality, though a change in growth rates may be seen between the rainy and the dry season. Another species cultivated by the same method is *Eucheuma denticulatum*, or "Spinosum" as it is called in the trade (Lirasan and Twide, 1993; Edo-Sullano, 1988; Lim, 1982).

The most technically advanced, proven method of production is intensive land cultivation of mainly *Chondrus crispus*, as done in tanks or basins in Nova Scotia, Canada. Factors such as temperature, nutrients, salinity, pH, and light intensity are controlled by the design of the tanks/ basins and the continuous monitoring of these parameters. Temperature and salinity can be controlled via flow-through of seawater taken from different depths. Nutrients and pH can be controlled with the addition of nutrient salts and carbon dioxide (CO_2) to lower pH. Light intensity can be controlled with shades and by operating with a depth (approximately one meter) that allows the plants to circulate at a specific frequency, giving optimal light exposure. The plants are circulated by bubbling air through the water (Bidwell et al., 1985).

Handling the Raw Seaweed

After harvest it is very important that the seaweeds be dried quickly. In the tropics drying is done by spreading the seaweed out on mats, nets, or other material, to keep the seaweed away from sand and other possible contaminants. During drying, seaweeds that are not wanted, as well as stones, pieces of coral, and objects that may have gotten into the seaweed during harvest, are removed by hand. In temperate regions, it is normal to dry the seaweed for a short time in the sun, but because of less sun, lower temperatures, and more frequent rain, it is difficult to reach a sufficiently low moisture level in the seaweed. Therefore, it is necessary to use mechanical dryers, such as drum dryers, operating on wood, oil, or gas, to get a high quality seaweed. Depending on the species, the seaweed is dried until it has a crisp feel, assuring that it can withstand long storage without rotting or losing quality.

After drying, the seaweed is packed in bales or bags made of polypropylene or similar material. It is stored in a warehouse until it can be loaded into a container for shipment.

The quality parameters on which seaweed is traded are moisture content, amount of impurities (sand, stones, sticks, line-pieces, corals, etc.), and color (indicating whether the seaweed has been dried quickly or has been exposed to rain). Another factor is the time of harvest, especially from temperate regions, as the yield and quality often change through the season.

CARRAGEENAN

Carrageenan is obtained by the extraction of certain species of red algae with water or alkali (Therkelsen, 1993). Carrageenan is a hydrocolloid consisting of the potassium, sodium, magnesium, and calcium sulphate esters of galactose and 3,6-anhydro-galactose copolymers.

Manufacture

Carrageenan is extracted with water at high temperatures. First the liquid is coarsely filtered to remove impurities and undissolved seaweed residues, and then it is filtered in a fine filter to obtain a clear liquid, followed by concentration by evaporation. The resulting liquid may either be dried directly to a powder (drum drying), be mixed with salts to form a gel that can be pressed or freeze dried, or mixed with alcohol to precipitate the carrageenan. The important step is getting the carrageenan dissolved and releasing it from the seaweed matrix that consists of proteinaceous matter and cellulosic substances. Depending on the method of extraction, the carrageenan may contain various amounts of salts from the extract liquid, giving the final product different characteristics in solubility and clarity of the gels produced. Most of the carrageenan produced today is manufactured through alcohol precipitation or gel pressing (Pedersen, 1990). The alcohols used during precipitation are restricted to methanol, ethanol, and isopropanol. The commercial products classified as carrageenan are standardized with sugars and food grade salts blended to obtain desired gelling or thickening characteristics (Figure 2).

The carrageenan molecule is a linear polysaccharide consisting of about one thousand galactose residues linked with alternating alpha (1-4) and beta (1-3) linkages. The galactose units linking alpha (1-3) in the general structure often oc-

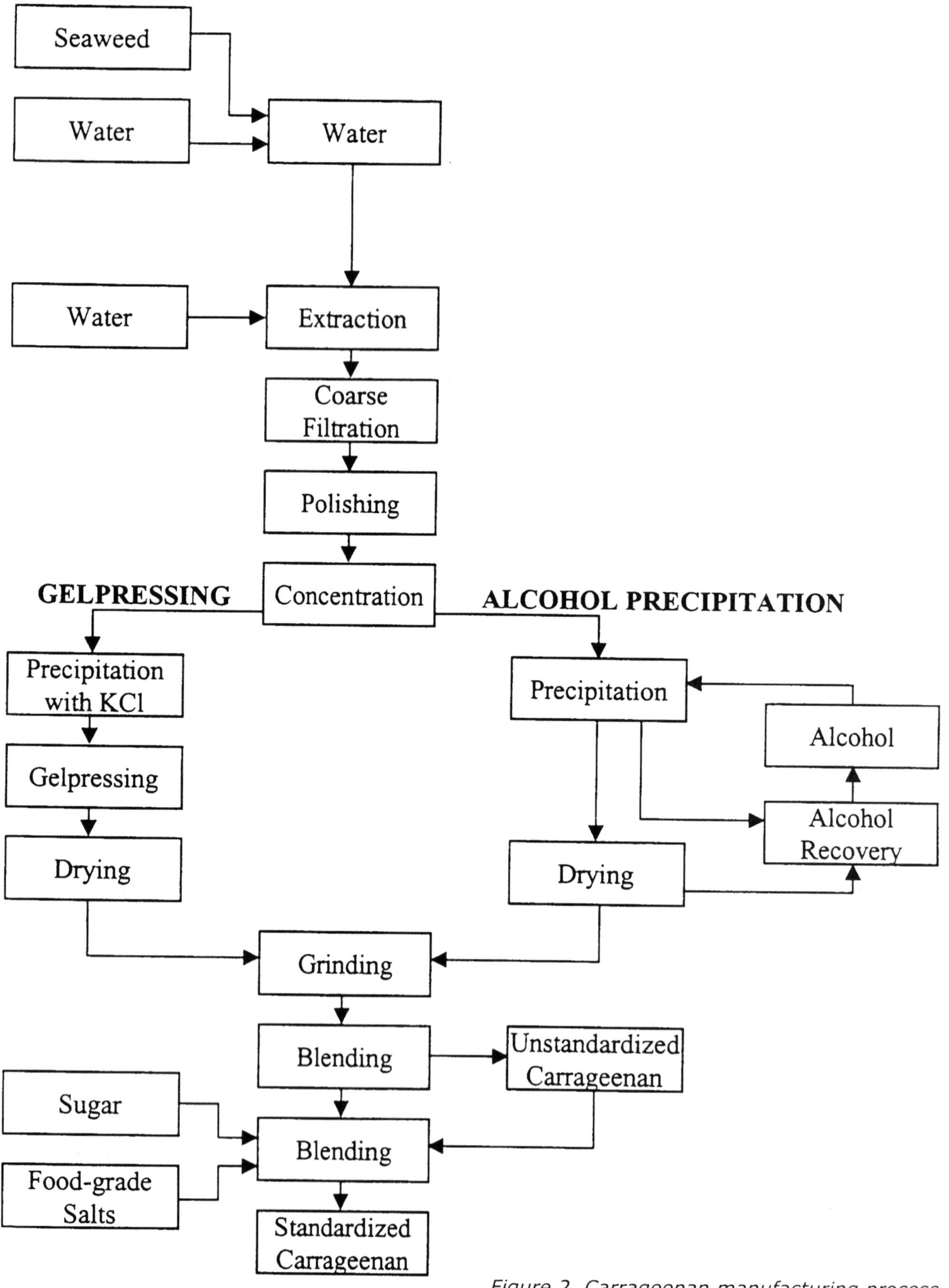

Figure 2. Carrageenan manufacturing process.

cur as 3,6-anhydro-D-galactose. Sulphate groups may be present on some or all galactose units.

Types

Pure commercial carrageenans are normally divided into three main types — kappa, iota, and lambda carrageenans — which are molecules assigned definitive structures, and which also represent specific properties (Painter, 1983). There is probably not one seaweed producing these ideal carrageenans, but more likely a range of intermediate structures. Several other carrageenan molecules exist; among these are beta, gamma, mu, and nu. Some of these are considered precursors that, when exposed to alkali conditions, are modified into kappa or iota carrageenan through formation of the 3,6-anhydro-galactose.

The main difference among kappa, iota, and lambda carrageenan at the molecular level is the amount and position of the sulphate ester groups. The position of the sulphate ester groups is determined with an infrared spectrophotometer (Figure 3).

Properties

Carrageenan is water soluble and insoluble in most organic solvents. The solubility characteristics of carrageenan are influenced by a number of factors:

1. the type of carrageenan
2. ions present
3. other solutes
4. temperature
5. pH

Lambda carrageenan has no 3,6-anhydro-galactose that is hydrophobic, but has three sulphate ester groups that are hydrophilic, making this carrageenan readily water-soluble under most conditions. Kappa carrageenan has a 3,6-anhydro-galactose and only one sulphate ester group, making this carrageenan less hydrophilic and less soluble in water. Iota carrageenan is intermediate with a 3,6-anhydro-galactose and two sulphate ester groups.

Kappa carrageenan in the potassium form is practically insoluble in cold water, whereas in the sodium form it readily dissolves (Oakenfull and Scott, 1990). The potassium form of iota carrageenan is also insoluble in cold water, although it swells markedly. Lambda carrageenan is water soluble in all its salt forms.

Kappa carrageenan is likewise the least soluble in the presence of other solutes. The effect of other solutes is that they compete for the available water and thus alter the hydration of the polysaccharide.

Being a water-soluble polysaccharide, carrageenan is difficult to disperse in water because a protective film layer forms around each carrageenan particle, and these particles then form large agglomerates or lumps, which are difficult for the water molecule to enter.

The less soluble the carrageenan, the easier the dispersion. For example, a potassium kappa carrageenan, being insoluble in cold water, is easier to disperse in cold water than is a sodium kappa carrageenan or a lambda carrageenan.

To avoid agglomeration, the carrageenan is often premixed with other ingredients such as sugar, with one part of carrageenan to 10 parts of sugar. If premixing is not possible, stirring with a high speed mixer together with slow addition of carrageenan can prevent agglomeration.

Acid and oxidizing agents may hydrolyze carrageenan in solution, leading to loss of properties through cleavage of glycosidic bonds. This cleavage is a function of pH, temperature, and time. Carrageenan in solution has maximum stability at pH 9 and should not be heat processed at pH-values below 3.5. At pH 6 or above, carrageenan solutions withstand normal processing used in food products — for example, meat and fish sterilization in cans.

Carrageenan is strongly negatively charged over the entire pH-range encountered in food. Carrageenan may therefore interact with other charged macromolecules — e.g., proteins — to give various effects such as increased viscosity, gel formation, or precipitation. When protein is interacting with carrageenan, the pH of the system is important, as it determines the isoelectric point of the protein.

The gelling mechanism of kappa and iota carrageenan is the formation of double helices that bind segments of the molecules in a three-dimensional network called a gel. This binding takes place only when gelling cations are present. By changing the concentration of gelling cations, it is possible to control the strength of the gel

Derivative of 3-linked galactose unit

Derivative of 4-linked galactose unit

Kappa Carrageenan

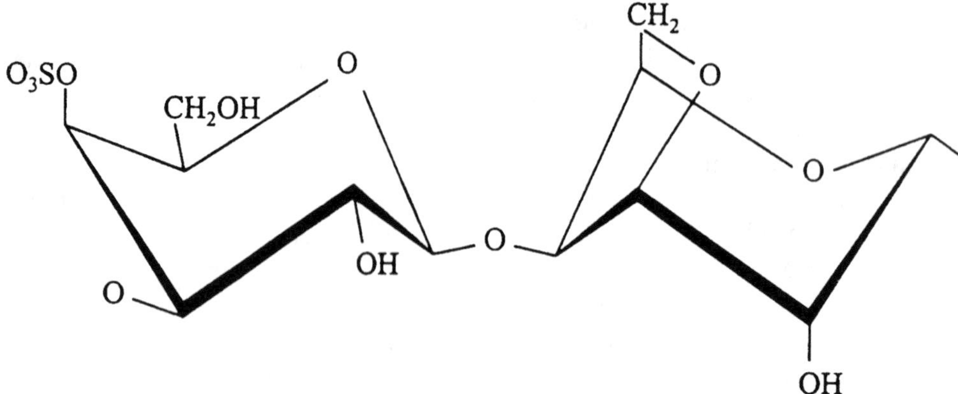

Iota Carrageenan

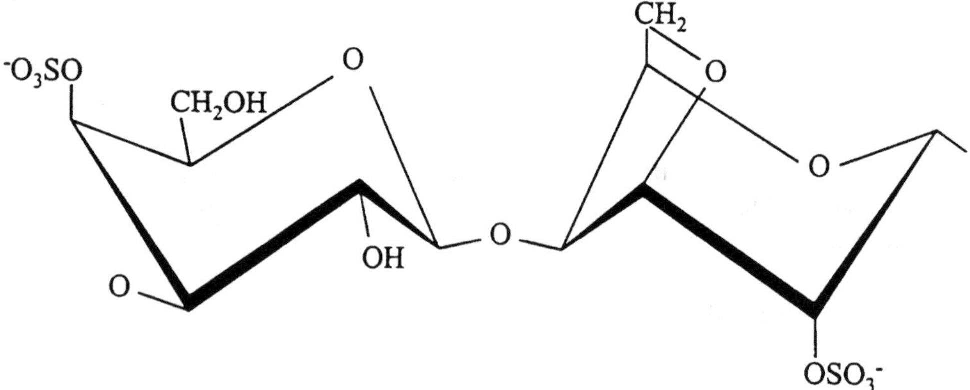

Lambda Carrageenan

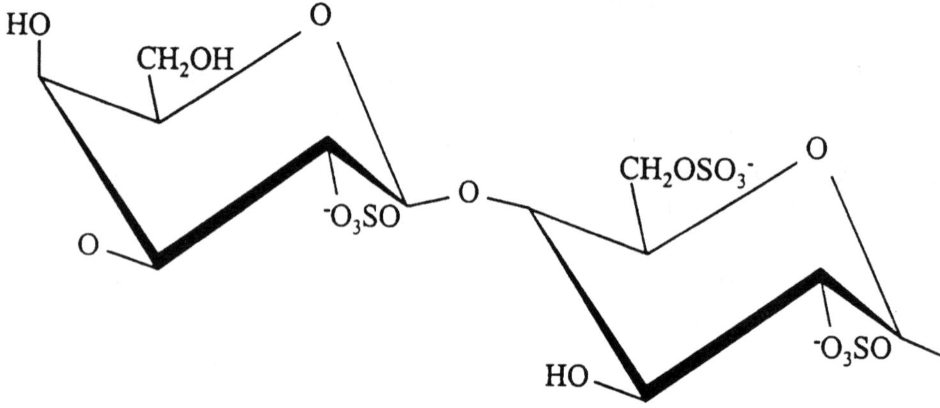

Figure 3. Ideal structures.

Table 3. Gel Characteristics of Carrageenan

Medium	Kappa	Iota	Lambda
Hot water	Soluble above 60°C	Soluble above 60°C	Soluble
Cold water	Sodium salt soluble. Potassium and calcium salt insoluble	Sodium salt soluble. Calcium salt gives thixotropic dispersions	Soluble
Hot milk	Soluble	Soluble	Soluble
Cold milk	Sodium, calcium, and potassium salt insoluble, but swells markedly	Insoluble	Soluble
Gelation	Gels, strongest with potassium salt	Gels, strongest with calcium salts	No gelation
Concentrated sugar solutions	Soluble hot	Soluble, with difficulty	Soluble hot
Concentrated salt solutions	Insoluble	Soluble hot	Soluble hot

(Oakenfull and Scott, 1990). Carrageenan gels are thermo- reversible; i.e., the carrageenan will redissolve when heated and form a gel again when cooled. The structure of the lambda carrageenan molecule does not allow helix formation, and therefore it does not gel (Moirano, 1977) (Table 3).

The iota carrageenan has a special property called thixotropic, meaning that a gel that has been broken will reform if left for some time without disturbance. Iota carrageenan is also freeze/thaw stable. A kappa carrageenan gel will not reform if broken, and is not freeze/thaw stable.

In practice, carrageenans are tailored to the properties needed in specific food applications through mixing of carrageenan types, sugars, and salts.

Carrageenan can be used in concentrations from 0.005 percent to 3 percent in a broad variety of food products.

Uses

The use of carrageenan may be separated into several areas utilizing different characteristics of the carrageenan. Some of the uses are in water-based, milk-based, and meat-based foods.

In order to make carrageenan gel in water-based foods, potassium or calcium salts must be added. Carrageenan will gel instantly in the presence of these ions when cooled below the gelling temperature. As there is no method for releasing potassium slowly, potassium must be added to the system before cooling below the gelling temperature.

Carrageenan may be used in instant preparations as a powder to be mixed with cold water, where a thickening effect is caused by the swelling of the carrageenan.

In solutions with a high content of soluble solids (above 50 percent), the gelling temperature is increased to a level limiting the use of carrageenan.

Though agar gels are stronger than carrageenan, carrageenan is often preferred in food products because a versatile texture range can be produced.

To achieve stronger gels, and thereby expand the texture range, a synergistic effect between kappa carrageenan and locust bean gum is often used.

Iota carrageenan is often used in cold-filled, ready-to-consume desserts. Carrageenan gelation can be delayed through stirring — even below the gelling temperature — while filling, and the thixotropic nature of the iota carrageenan will then form the gel after filling.

In milk-based products, where gelation or structural viscosity is required, carrageenan is preferred to other gums for functional and economic reasons. Carrageenan (especially the kappa type) is widely used in gelled milk products, both in ready-to-consume desserts and in powder preparations for making flans or puddings.

Stabilization of cocoa particles and fat suspension in chocolate milk is obtained through the addition of as little as 0.02 to 0.03 percent kappa carrageenan. An even smaller amount can be used to prevent separation of ice cream when it is thawing.

A functional property of carrageenan, besides creating a gel, is a synergistic reaction with milk proteins that makes use of very small amounts of carrageenan.

In meat-based products such as hams, carrageenan is used to improve texture by adding firmness, increasing sliceability, decreasing liquid separation, and producing a juicier product. Also, in meat products there is a synergistic effect between the meat protein and carrageenan. The relatively high salt content of most meat products is beneficial for carrageenan gelation. In the poultry industry carrageenan is used to make reformed products from the meat that is left when legs and breasts have been removed. These products, in the form of chicken or turkey rolls, sausages, or cutlets, all benefit from the addition of small amounts of carrageenan, and have proven very popular in the modern kitchen.

Carrageenan is also being incorporated into seafood products, especially to hold in juiciness (Jensen, 1993).

There are also many non-food applications, such as the thickening of toothpaste and shampoo. Carrageenan may be used as a tablet coating agent. Encapsulation of enzymes or living cells is another possibility that utilizes the rapid diffusion of small molecules (nutrients, sugars) that can take place through a carrageenan gel.

Carrageenan may also be used for bacteriological media, but because these require gelation at temperatures relatively far below the normal gelling temperature for carrageenan, gelation has to be made chemically by mixing special carrageenans with water before applying gelling ions. Agar is normally employed for this application because of the properties described in the next section.

A possible future use of carrageenan is as an agent that can prevent or inhibit development of viral infections transferred sexually. That is, carrageenan may help in the struggle against the HIV-virus (Ebinitim, 1995; Phillips et al., 1994).

AGAR

Agar is a gelling hydrocolloid obtained by extraction with water or alkali of certain species of red algae. It is a linear hydrocolloid (galactan), consisting primarily of alternating D-galactose and L-3,6-anhydrogalactose monomers, with a small amount of charged substituting groups.

Agar is recovered by freeze/thaw concentration, gel pressing, or drum drying.

Today, agar is available as a cream-colored powder, but older forms such as flakes or strips can still be found.

MANUFACTURE

The agar manufacturing process normally involves a washing procedure and an alkaline treatment of the seaweed. The alkaline treatment serves to enhance the gel strength of the agar extract. Upon treatment, the seaweed is extracted in boiling water, in which the agar is totally dissolved. The resulting liquid is filtered thoroughly and then cooled to make the agar form a gel. The gel is dewatered by pressing, perhaps in combination with a freeze/thaw procedure that accelerates dewatering. The dewatered gel is finally dried, ground, and standardized for commercial use (Selby and Whistler, 1993; Armisen and Galatas, 1987).

The agar molecule is a linear polysaccharide with a molecular weight around 100,000 Daltons. Structurally, the molecular chain is based on the repeating disaccharide agaro-biose, which may occur substituted or modified in various ways.

TYPES

The most important variations in practice are the following:

Methoxylation ($-O-CH_3$), mainly on C-6 on the D-galactose. Agar may contain up to 3 to 4 percent methoxyl. Substitution by methoxyl is positively correlated to gelling temperature and is found to a significant degree in agar based on *Gracilaria*.

Sulphation ($-O-SO_3^-$), mainly on C-2 of the L-galactose. Agar may contain up to 5 to 6 percent sulphate. Substitution by sulphate is negatively correlated to gelling strength.

Pyruvation ($-O-C\ (CH_3)(COOH)\ -O-$), as a ketal coupled onto C-4 and C-6 of the D-galactose. Agar may contain up to 3 percent pyruvate. Substitution by pyruvate is not believed to affect gelling properties very much.

Methoxylation occurs evenly along the molecular chain, whereas sulphation and pyruvation seem to occur in regions separate from each other.

It is possible to isolate an agar fraction that has a very low content of charged substituting groups and that therefore approaches the ideal structure of agaro-biose units. This product is known as **agarose**.

Properties

Agar is insoluble in cold water and soluble in boiling water. When it is cooled, a 1.5 percent agar solution will form a gel in the range of 32° to 45°C, depending on the seaweed species. After the gel is formed, it will not melt until heated to no less than 85° to 95°C, again depending on the seaweed species. This large difference in gelling and melting temperature is called hysteresis, and is one of the properties that makes agar commercially valuable.

Viscosity of an agar solution is influenced by seaweed raw material and processing conditions (Selby and Whistler, 1993). The viscosity at 45°C is relatively constant in the pH range from 4.5 to 9. The amount of salts in the solution does not affect viscosity seriously. At lower temperatures viscosity increases with the beginning of gelation.

Agar and agarose form strong gels, with agar gels being more elastic and less firm than agarose gels. Gel strength is normally measured as resistance against a plunger on a 1.5 percent solution at 20°C and is in the range of 200 to more than 1,000 grams per cm^2. As agar is nearly neutrally charged, it does not react strongly with proteins or other charged molecules.

A new agar type called quick-soluble agar has recently been developed, with the characteristic that it dissolves at lower temperatures than ordinary agar. This product is obtained by drying the agar directly from a solution instead of making the agar gel before drying (Lebbar, 1992).

Uses

1. Agar is separated into food-grade agar and bacteriological grade agar.
2. Food-grade agar is used as a stabilizer in many foods such as canned meat, confectionary, and glazing/icing for the baking industry.
3. Bacteriological-grade agar is used for stabilization of media for bacteriological growth. It can also be used for immobilization and encapsulation, as ions and nutrients easily diffuse through the gel (Jensen, 1990).
4. Agarose is mainly used in the pharmaceutical industry, where it can be used for separation and purification of proteins such as enzymes. These proteins can be made to travel through an agarose gel because they are charged differently and have different sizes, and by different speeds they become separated.

References

Abbott, I. A. 1988. Food and food products from seaweeds. In *Algae and Human Affairs*. C.A. Lembi, J. R. Waaland (eds.), pp. 136-146. Cambridge University Press, New York.

Abbott, I. A. (ed.). 1994. Taxonomy of Economic Seaweeds. With reference to some Pacific species. In *California Sea Grant College*. Vol. IV, pp. 81-83.

Armisen, R. and F. Galatas. (1987). Production, Properties and Uses of Agar. D.J. McHugh (ed.). University of New South Wales. Australian Defense Force Academy. Campbell, ACT 2600, Australia. *FAO Fisheries Technical Paper* 288.

Berchez, F. A. S., R. T. L. Pereira, and N. F. Kamiya. 1993. Culture of *Hypnea musciformis* (Rhodophyta, Gigartinales) on artificial substrates attached to linear ropes. *Hydrobiologia*. **260/261**: 415-420.

Bidwell, R. G. S., J. McLachlan, and N. D. H. Lloyd. 1985. Tank Cultivation of Irish Moss, *Chondrus crispus* Stackh. *Botanica Marina*. **XXVIII:** 87-97.

Bird, C. J., M. A. Ragan, A. T. Critchley, E. L. Rice, and R. R. Gutell. 1994. Molecular relationship among the *Gracilariaceae* (Rhodophyta): further observations on some undetermined species. *Eur. J. Phycol.* **29:** 195-202.

Bird, C. J., G. W. Saunders, and J. McLachlan. 1991. Biology of *Furcellaria lumbricalis* (Hudson) Lamouroux (Rhodophyta: Gigartinales), a commercial carrageenophyte. *Journal of Applied Phycology*. **3:** 61-82.

Blunden, G. 1991. Agricultural uses of seaweeds and seaweed extracts. In *Seaweed Resources in Europe: Uses and Potential*. M.D. Guiry and G. Blunden (eds.). John Wiley & Sons Ltd.

Bold, H. C. and M. J. Wynne. 1985. Introduction to the algae. Prentice-Hall, Englewood Cliffs, NJ.

Booth, E. 1979. The history of the seaweed industry. Part 4: a miscellany of industries. *Chemistry and Industry*. June, 1979.

Chaoyuan, W., R. Li, G. Lin, Z. Wen, L. Dong, J. Zhang, X. Huang, S. Wei, and G. Lan. 1993. Some aspects of the growth of *Gracilaria tenuistipitata* in pond culture. *Hydrobiologia*. **260/261:** 339-43.

Copenhagen Pectin. 1994. Carrageenan, General Description, B1.

Doty, M. S. 1988. *Prodromus ad Systematica Eucheumatoideorum: A Tribe of Commercial Seaweeds Related to* Eucheuma *(Solieriaceae, Gigartinales) in Taxonomy of Economic Seaweeds.* With reference to some Pacific and Caribbean Species, Volume II. Isabella A. Abbott (ed.). *A Publication of the California Sea Grant College Program.* Report No. T-CSGCP-018.

Edo-Sullano, M. 1988. Seaweed Farming. Technoguide. Central Visayas. Consortium for Integrated Regional Research and Development.

Enander, M. and M. Hasselstrom. 1994. An experimental wastewater treatment system for a shrimp farm. *Infofish International.* **4.**

Enomoto, Y. c/o Fuji Latex Co. Ltd. 1995. Condom coated with acidic polysaccharides. EP 0 661 028 A1.

French, R. A. 1970. A current appraisal of the Irish moss industry. Departments of Fisheries, Nova Scotia, New Brunswick, Prince Edward Island and the Industrial Development Branch, Fisheries Service. Department of Fisheries and Forestry of Canada.

Goff, L. J., D. A. Moon, and A. W. Coleman. 1994. Molecular delineation of species and species relationships in the red algal agarophytes *Gracilariopsis* and *Gracilaria* (Gracilariales). *J. Phycol.* **30:** 521-37.

Hansen, J. E., J. E. Packard, and W. T. Doyle. 1981. Mariculture of Red Seaweeds. *A California Sea Grant College Program Publication.* University of California. Pp. 1-42.

Harlin, M. M. and W. M. Darley. 1988. The Algae: an overview. In *Algae and Human Affairs.* C. A. Lembi, J. R. Waaland (eds.), pp. 3-29. Cambridge University Press. New York.

Harvey, M. J. and J. McLachlan (eds.). 1973. *Chondrus crispus,* Nova Scotian Institute of Science. Halifax, Nova Scotia.

Humm, H. J. 1951. The Red Algae of Economic Importance: Agar and Related Phycocolloids. In *Marine Products of Commerce.* D. K. Tressler and J. McW. Lemon (eds.), pp. 47-93. Reinhold Publishing Corporation. New York.

Indergaard, M. and A. Jensen. 1991. *Utnyttelse av Marin Biomasse.* Norges Tekniske Høgskole.

Indergaard, M. and J. Minsaas. 1991. Animal and Human Nutrition. In *Seaweed Resources in Europe: Uses and Potential.* M.D. Guiry and G. Blunden (eds.). John Wiley & Sons Ltd.

Jensen, J. 1993. Fancy Fish Products, A New Trend. *Food Marketing & Technology,* August.

Jensen, T.W. 1990. Fractionation of *Gracilaria verucosa* and *Gigartina skottsbergii* and characterization of extracted phycocolloids. (Part of studies at the Royal School of Pharmacy, Copenhagen.)

Kuwano, K., Y. Aruga, and N. Saga. 1994. Cryopreservation of the conchocelis of *Porphyra* (Rhodophyta) by applying a simple prefreezing system. *J. Phycol.* **30:** 566-70.

Lebbar, R. 1992. Quick Soluble Agar: An improvement in agar properties. *Food Ingredients Europe, Dusseldorf.* 25-27 November, Session 10: Confectionery.

Lewis, G. L., N. F. Stanley, and G. G. Guist. 1988. Commercial production and applications of algal hydrocolloids. In *Algae and Human Affairs.* C. A. Lembi and J.R. Waaland (eds.), pp. 205-237. Cambridge University Press.

Lim, J. R. 1982. Farming the Ocean (The Genu Story). Historical Conservation Society, Manila. XXXVI.

Lirasan, T. and P. Twide. 1993. Farming *Eucheuma* in Zanzibar, Tanzania. *Hydrobiologia.* **260/261:** 353-355.

McHugh, D. J. (ed.). 1987. Production and utilization of products from commercial seaweeds. *FAO Fisheries Technical Paper* 288. Pp. 1-189.

Moirano, A. A. 1977. Sulfated Seaweed Polysaccharides in *Food Colloids.* H.D. Graham (ed.).

Morand, P., B. Carpentier, R.H. Charlier, J. Mazé, M. Orlandini, B.A. Plunkett, and J. de Waart. 1991. Bioconversion of Seaweeds. In *Seaweed Resources in Europe: Uses and Potential.* M.D. Guiry and G. Blunden (eds.). John Wiley & Sons Ltd.

Mshigeni, K.E. 1978. The Biology and Ecology of Benthic Marine Algae with Special Reference to *Hypnea* (Rhodophyta, Gigartinales): a Review of the Literature. In *Bibliotheca Phycologica,* Band 37. J. Cramer, A.R. Gantner Verlag Kommanditgesellschaft (eds.). FL-9490 Vaduz.

Mumford, T.F., Jr., and A. Miura. 1988. *Porphyra* as food: cultivation and economics. In *Algae - Utilization.* C.A. Lembi and J.R. Waaland (eds.), pp. 87-119. Cambridge University Press. New York.

Neushul, M. and B. W. W. Harger. 1987. Nearshore Kelp Cultivation, Yield and Genetics. In *Seaweed Cultivation for Renewable Resources.* K.T. Bird and P.H. Benson (eds.). Developments in Aquaculture and Fisheries Science, 16.

Oakenfull, D. A. and Scott. 1990. The role of the cation in the gelation of kappa-carrageenan. *Gums and Stabilisers for the Food Industry 5.* G.O. Phillips, P.A. Williams, D.J. Wedlock (eds.). IRL University Press. Oxford.

Oohusa, T. 1993. Recent trends in nori products and markets in Asia. *Journal of Applied Phycology.* **5:** 155-159.

Painter, T. J. 1983. Algal Polysaccharides. In *The Polysaccharides* (vol. 2). G.O. Aspinall (ed). Academic Press. *Molecular Biology,* An International Series of Monographs and Textbooks. B. Horecker, N.O. Kaplan, J. Marmur, H.A. Scheraga (eds.).

Pedersen, J. K. 1990. Seaweed extracts: sources and production methods. *Gums and Stabilisers for the Food Industry 5.* G.O. Phillips, P.A. Williams, D.J. Wedlock (eds.). IRL University Press. Oxford.

Phillips, D., R. Pierce-Pratt, C. Elias, and S. Waldman. 1994. Carrageenan to Combat HIV Virus. In *Hydrocolloid Review,* Vol. 2. IMR International.

Pizarro, A. and H. Barrales. 1986. Field Assessment of Two Methods for Planting the Agar-containing Seaweed, *Gracilaria,* in Northern Chile. *Aquaculture.* **59:** 31-43.

Pringle, J. D., R. Ugarte, and R. E. Semple. 1990. Annual net primary production calculated from eastern Canadian Irish moss fishery data. *Hydrobiologia.* **204/205:** 317-23.

Pringle, J. D., D. J. Jones, and R. E. Semple. 1987. Fishing and catch characteristics of an eastern Canadian Irish moss (*Chondrus crispus* Stackh.) dragraker. *Hydrobiologia*. **151/152:**. 341-47.
Santelices, B. and R. Ugarte. 1987. Production of Chilean *Gracilaria*: problems and perspectives. *Hydrobiologia*. **151/152:** 295-99.
Santelices, B. 1988. Synopsis of Biological Data on Seaweed Genera *Gelidium* and *Pterocladia (Rhodophyta)*. FAO Fisheries Synopsis No. 145.
Schmid, O.J. 1969. Various Substances. In *Marine Algae, A Survey of Research and Utilization*. T. Levring, H.A. Hoppe and O.J. Schmid (eds.). Cram, De Gruyter & Co., Hamburg.
Schramm, W. 1991. Seaweeds for Waste Water Treatment and Recycling of Nutrients. In *Seaweed Resources in Europe: Uses and Potential*. M.D. Guiry and G. Blunden (eds.), pp. 149-69. John Wiley & Sons Ltd.
Selby, H. H. and R. L. Whistler. 1993. Agar. In *Industrial Gums: Polysaccharides and Their Derivatives*, 3rd ed. Roy L. Whistler, James N. BeMiller (eds.).
Sharp, G. J., H. S. Samant, and O. C. Vaidya. 1988. Selected Metal Levels of Commercially Valuable Seaweeds Adjacent to and Distant from Point Sources of Contamination in Nova Scotia and New Brunswick. *Bull. Environ. Contam. Toxicol.* **40:** 724-730.
Sharp, G. J., C. Têtu, R. Semple, and D. Jones. 1993. Recent changes in the seaweed community of Western Prince Edward Island: implications for the seaweed industry. *Hydrobiologia*. **260/261:** 291-296.
Sheath, R.G. 1984. The biology of freshwater red algae. *Prog. Phycol. Res.* **3:** 89-157.
Therkelsen, G. 1993. Carrageenan. In *Industrial Gums: Polysaccharides and Their Derivatives*, 3rd ed. Roy L. Whistler, James N. BeMiller (eds.).
Trono, G. C., E. T. and Ganzon-Fortes. 1988. *Philippine Seaweeds*. National Book Store Inc. Publishers. Metro Manila, Philippines.

Seaweed Products: Brown Algae of Economic Significance

Martha Llaneras

The *Phaeophyceae*, or brown algae, are primarily saltwater plants. They are commercially important because of their polysaccharide content and availability in quantities sufficient to support a sizable industry. Algin is the polysaccharide derived from the brown seaweeds. This polysaccharide is specific to seaweeds and is not found in land plants. Algin has many applications in food, pharmaceutical products, and industrial products.

Early historical utilization of seaweeds was as food for man and his domesticated animals and as fertilizers. With the onset of the Industrial Revolution and its demand for raw materials, seaweeds were exploited for their chemical content. In the seventeenth century, it was discovered that the ash from burned seaweed contained soda needed in pottery glazing and for the manufacturing of glass and soap.

In the first half of the nineteenth century, barilla (saltwort) and the Le Blanc process replaced seaweed as a source of soda ash. The discovery of iodine in kelp ash in 1811 and the need for potash kept the kelp industry alive, though at a much reduced level.

It was E.C.C. Stanford, the great pioneer in seaweed research, who first patented a process for extracting algin in 1881.

Krefting (1896) prepared a pure alginic acid. Later, investigators determined many of the properties of the salts of this acid. However, large-scale commercial production of alginates did not begin until the Kelco Company was founded in 1929 (Steiner and McNeely, 1954).

The brown seaweeds, all of which contain algin, are an amazingly varied family of plants. They grow on rocky shores or in ocean areas having a clean, rocky bottom. Some species are found at the high-tide line; others exist in a belt along the shore wherever the depth is less than 125 feet, the limit of sunlight penetration.

The adaptability of these seaweeds to their environment is quite remarkable. In areas of minimal wave action, plants may have a 15-year life span *(Laminaria hyperborea)*. Others which grow in areas of yearly storm cycles are annuals (*Nereocystis luetkeana*). Some varieties can survive although wet only at high tide, while others must remain submerged at all times. Some species are found in frigid waters north of the Arctic Circle and others in the very warm, almost sterile, waters of the Sargasso Sea.

Phaeophyceae usually have root-like holdfasts anchoring them to the rock base. On the deep-water algal plants, stipes (stems) grow to the surface out of the holdfast in a large bundle and may reach a length of 200 feet. From the stipes (or from the holdfast of tidal species) grow leaf-like blades, which usually contain a float bladder (Figure 1). Fronds originate at the base of the plant, near the holdfast, and eventually grow to the surface. The fronds consist of stipes, bladders, and blades.

The holdfast of a mature *Macrocystis pyrifera* plant will normally be one to three feet in diameter; the stipe bundle seems like a giant underwater tree trunk. The older stipes reaching the surface form a canopy sheltering the young growth. Beds not harvested become littered with dead and decaying fronds.

Since algae do not possess true roots, stems, or leaves, separate mechanisms for their nourishment and growth have evolved. No nourishment is obtained through the holdfast. Nourishment is acquired from sunlight and from mineral nutrients in the ocean. The brown skin that covers the entire surface of the blades absorbs sunlight and is the primary area of photosynthesis. The bulk of the plant (stipes and holdfast) is non-photosyn-

Figure 1. Macrocystis pyrifera. *(A: 1/64 natural size; b: 1/4 natural size). The giant kelp is shown in the left part of the plant in a natural pose with the long leafy stipes rising to the sea surface from the massive holdfast. On the right is one of the leaf-like fronds showing the gas-filled float bladder at its base and the distinctive teeth along the margin.*

thetic and is fed by translocation. Growth rates may be as high as two feet a day, but will vary directly with the availability of nutrients, sunlight, and water temperature.

A vegetative reproduction process results in the growth of new fronds from a holdfast. The new fronds, which are very slender and fragile, are entwined into the existing stipe bundles for protection against damaging wave action. Under the canopy of an established kelp bed, there is insufficient light intensity for photosynthesis, and the new fronds of *Macrocystis pyrifera* are fed by translocation from the older frond on or near the surface. Because growth in the vegetative reproduction process is continuous throughout the year, multiple harvests are possible.

Macrocystis pyrifera also has a sexual method of reproduction that is the source of all new plants. Through this reproductive cycle, growth is initiated in open or denuded areas of the ocean. The cycle begins with the release of microscopic zoopores (motile spores) from sporophylls (spore-producing leaves) growing at the base of a mature plant. The zoospores have approximately a 24-hour life span in which to attach themselves to a hard surface. Male and female gametophytes (reproductive cells) develop from the zoospores, and the sporophyte (a new asexual plant) is the result of a fertilized egg cell. At least a year is required before sexually-produced plants mature and provide a canopy. Embryonic kelp plants (sporophytes) that have been fertilized in the laboratory have been successfully transplanted to the ocean floor.

Since light under the canopy of an established bed is inadequate for photosynthesis, other species of algae are greatly reduced in beds of *Macrocystis pyrifera*. Other types of algae, such as elk kelp (*Pelagophycus porra)* are found around the edges.

Giant kelp, *Macrocystis pyrifera*, which grows in abundance along the coasts of North and South America, New Zealand, Australia, and Africa, is one of the principal sources of the world's algin supply. Also utilized are *Ascophyllum nodosum* and several varieties of *Laminaria* and *Ecklonia*.

The location of the algin component varies among the varieties of brown seaweeds. In *Macrocystis*, most of the algin is found in the blade. The algin in *Laminaria*, however, is mainly in the stipe.

Although the primary commercial interest in kelp is in its algin content, other ingredients make up a major portion of the plant. The trace mineral content is of value when the kelp is used as food, as a cattle food supplement, or as fertilizer. A typical chemical analysis of dried *Macrocystis pyrifera* is given in Table 1.

MANUFACTURE

Macrocystis pyrifera, the brown seaweed that is the most common source of algin, has growth characteristics making it an ideal raw material for modern technology.

Table 1. Analysis of *Macrocystis pyrifera* (partially dried)

Moisture	10–11%
Ash	33–35%
Protein	5–6%
Crude fiber (cellulose)	6–7%
Fat (ether extract)	1–1.2%
Algin, mannitol, and other carbohydrates	39.8–45%

Potassium	9.5%	Barium	None Detected
Sodium	5.5%	Cobalt	None Detected
Calcium	2.0%	Nickel	None Detected
Strontium	0.7%	Zinc	None Detected
Magnesium	0.7%	Selenium	None Detected
Iron	0.08%	Titanium	None Detected
Aluminum	0.025%	Chloride	11%
Rubidium	0.001%	Nitrogen	0.9%
Copper	0.003%	Phosphorus	0.29%
Chromium	0.0003%	Iodine	0.13%
Manganese	0.0001%	Boron	0.008%
Silver	0.0001%	Bromine	0.0002%
Vanadium	0.0001%	Fluorine	None Detected
Lead	0.0001%	Arsenic	None Detected

Macrocystis grows in relatively calm waters and in large, dense beds. Only mature beds are cut. At the time of harvesting, a dense mat of fronds floats on the ocean surface. If left uncut, these older fronds eventually slough off, breaking loose from the parent plant to rot in the water or on the nearest beach.

Cutting the dense mat on the surface allows light to penetrate the water and reach immature fronds, stimulating their growth. Harvesting is actually a massive pruning of the kelp bed: many old stems are removed in favor of newer, healthier growth.

Off the California coast, kelp beds are harvested under the supervision of the State of California Department of Fish and Game. The kelp is harvested by means of an underwater mowing machine carried by a motor-driven barge. These vessels are specifically designed for efficient harvesting. Underwater blades mow the kelp approximately three feet below the water surface. The cut kelp is automatically conveyed into the hold of the barge by a moving belt (North, 1972). The modern harvesters are able to move in reverse at slow speeds through the kelp beds while cutting.

Harvesting for other brown seaweeds is done quite differently. Some varieties such as the *Laminaria* are harvested manually. Mature plants are constantly dislodged and float away. Most are swept out to sea to decompose in deep water trenches, but a portion is washed ashore where it is collected manually. Other cast weeds, such as the bull kelp of Tasmania, *Durvillea potatorum*, are so big and heavy (40–50 feet long and weighing 150–250 pounds) that they have to be hooked and winched onto special collecting vehicles.

The fundamental steps of a typical production of sodium alginate are shown in Figure 2. In the seaweed, the algin is present as a mixed sodium and/or potassium, calcium, and magnesium salt. The exact composition varies considerably with the type of seaweed, but does not affect processing.

Alginic acid may also be neutralized with bases to give salts, or may be reacted with propylene oxide to make propylene glycol alginate.

STRUCTURE

The term algin is used to describe alginic acid and its various inorganic salt forms, derived from brown seaweeds (*Phaeophyceae*). The monovalent

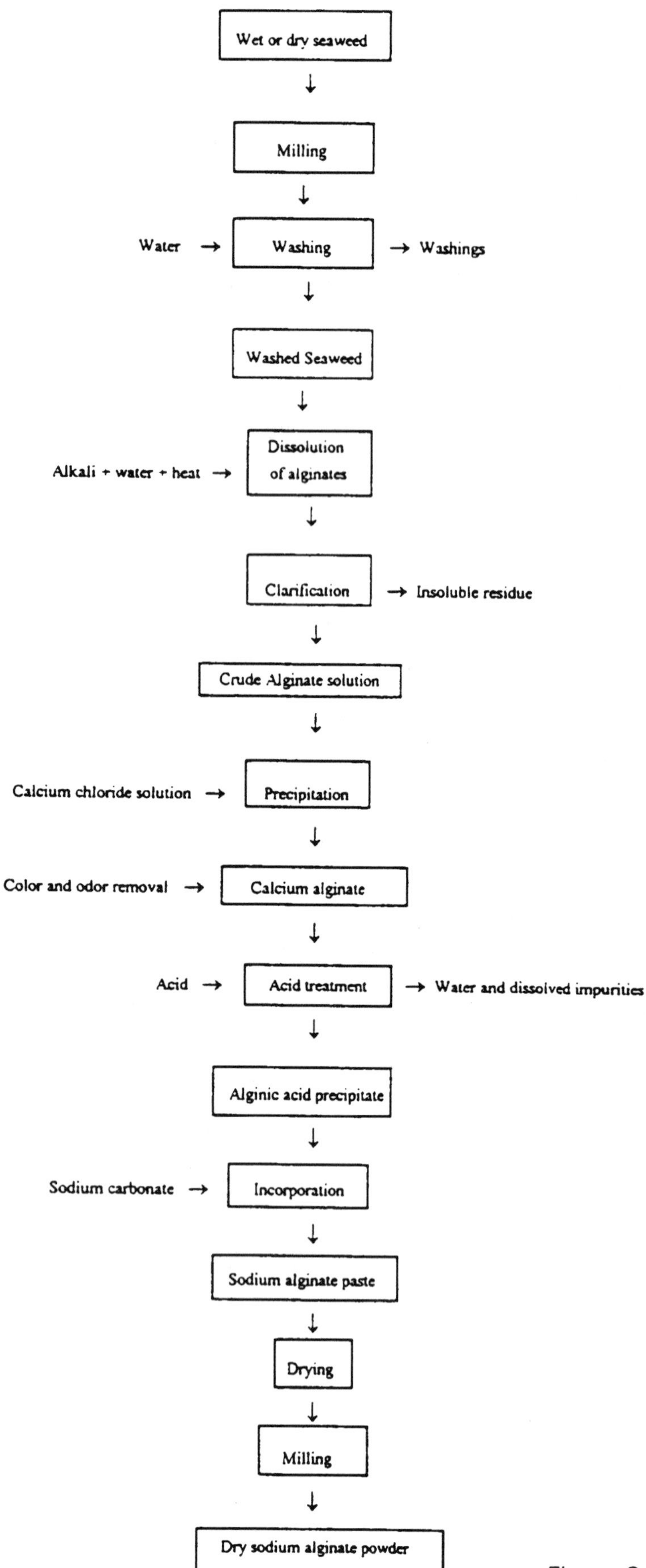

Figure 2. Sodium alginate manufacturing process.

salts, often referred to as alginates, are hydrophilic colloids and these, especially sodium alginate, are widely used in food, pharmaceutical, and industrial applications. One of the most important and useful properties of alginates is the ability to form gels by reaction with calcium salts. These gels, which resemble a solid in retaining their shape and resisting stress, consist of almost 100 percent water (normally 99.0 to 99.5 percent water and 0.5 to 1.0 percent alginate). In a great number of applications, the now well-known reactivity of alginate with calcium ions is utilized.

Alginate is a linear co-polymer composed of two monomeric units, D-mannuronic acid and L-guluronic acid. These monomers occur in the alginate molecule as regions made up exclusively of one unit or the other, referred to as M-blocks or G-blocks, or as regions in which the monomers approximate an alternating sequence. The calcium reactivity of alginates is a consequence of the particular molecular geometries of each of these regions. The shapes of the individual monomers are shown in Figure 3. The D-mannuronic acid exists in the C1 conformation and in the alginate polymer is connected in the ß-configuration through the 1- and 4-positions; the L-guluronic acid has the 1C conformation and is a-1, 4-linked in the polymer. Because of the particular shapes of the monomers and their modes of linkage in the polymer, the geometries of the G-block regions, M-block regions, and alternating regions are substantially different. Specifically, the G-blocks are buckled while the M-blocks have a shape referred to as an extended ribbon, as shown in Figure 4. If two G-block regions are aligned side by side, a diamond-shaped hole results. This hole has dimensions that are ideal for the cooperative binding of calcium ions. When

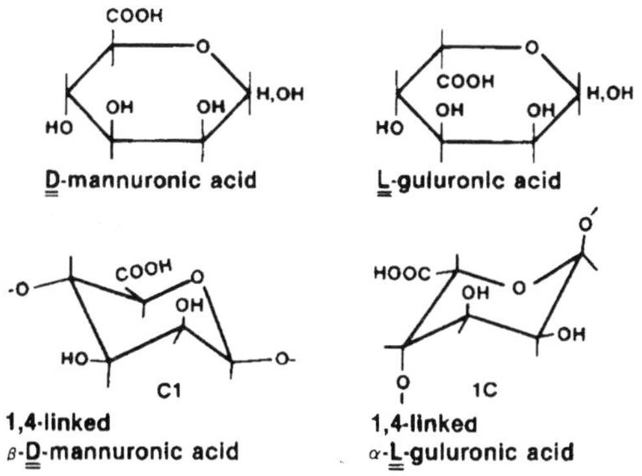

Figure 3. D-mannuronic acid and L-guluronic acid (the algin monomer).

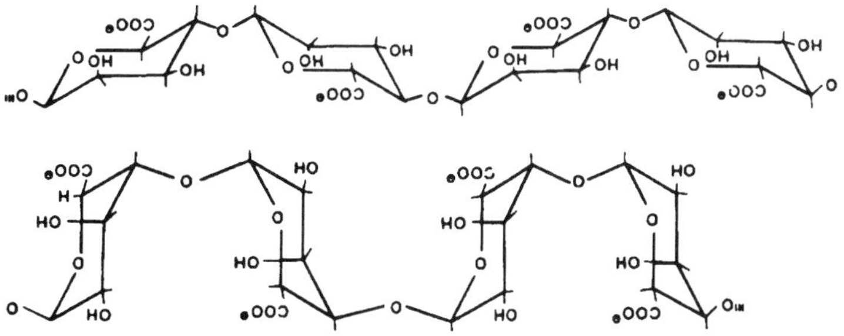

Figure 4. Block shapes in alginates.

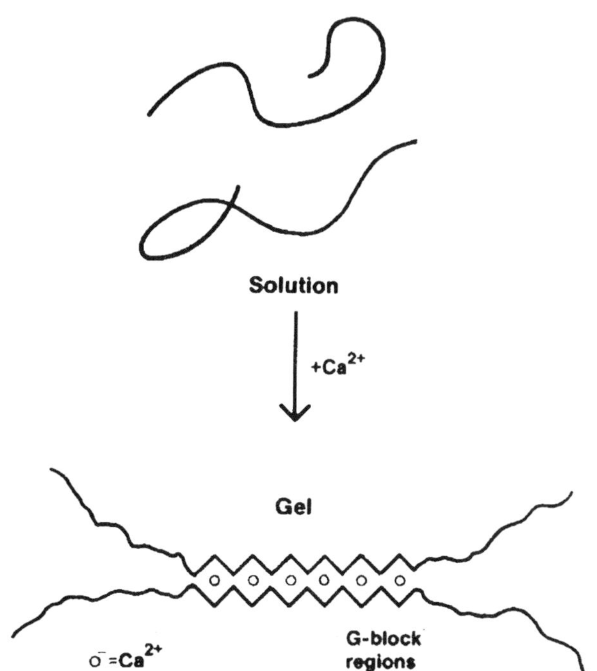

Figure 5. Egg-box model for alginate gel.

calcium ions are added to a sodium alginate solution, such an alignment of the G-blocks occurs, and the calcium ions are bound between the two chains like eggs in an egg box, as shown in Figure 5. Thus the calcium reactivity of algins is the result of calcium-induced dimeric association of the G-block regions. Depending on the amount of calcium present in the system, these interchain associations can be either temporary or permanent. With low levels of calcium, temporary associations are obtained, giving rise to highly viscous, thixotropic solutions. At higher calcium levels, precipitation or gelation results from permanent association of the chains.

Commercial alginates are derived from a variety of weed sources. Since different weeds yield alginates that differ in monomeric composition and block structure, a given alginate has its own characteristic calcium reactivity and gelation properties. The differences in composition and fine structure indicated in Tables 2 and 3 account for the differences in properties and functionality of alginates isolated from different species of brown algae. Although the ratio of mannuronic acid to guluronic acid and the ratio of M-blocks to G-blocks can be determined experimentally, the detailed molecular compositions of alginates in terms of block lengths and block distributions are still unresolved. As a result, alginates are usually referred to as "high M" or "high G," depending on the proportions of mannuronic acid and guluronic acid they contain. Most commercial products are of the high-M type, the best example being the alginate obtained from the giant kelp of *Macrocystis pyrifera*. *Laminaria hyperborea*, with a large percentage of polyguluronate segments, forms rigid, brittle gels which tend to undergo syneresis, or loss of bound water. In contrast, alginate from *Macrocyctis pyerifera* or *Ascophyllum nodosum* forms elastic gels that can be deformed and that have a markedly reduced tendency toward syneresis.

PROPERTIES

When dissolved in distilled water, pure monovalent alginate salts form smooth solutions with long flow properties. The solution properties are dependent on both physical and chemical variables.

The physical variables that affect the flow characteristics of alginate solutions include tempera-

Table 2. Mannuronic Acid (M) and Guluronic Acid (G) Composition of Alginic Acid Obtained from Commercial Brown Algae

Species	Mannuronic Acid Content (%)	Guluronic Acid Content (%)	M/G Ratio	M/G Ratio Range
Macrocystis pyrifera	61	39	1.56[a]	—
Ascophyllum nodosum	65	35	1.85 (1.1[a])	1.40–1.95[b]
Laminaria digitata	59	41	1.45[a]	1.40–1.60[b]
Laminaria hyperborea (*stipes*)	31	69	0.45[a]	0.40–1.00[b]
Ecklonia cava and *Eisenia bicyclis*	62	38	1.60[a]	—

[a] Data of Haug (1964) and Haug and Larsen (1962) for commercial algin samples. Of the two ratios shown for *Ascophyllum nodosum*, the algin sample manufactured in Canada has the higher M/G value; the lower ratio corresponds to a European sample.

[b] Data of Haug (1964), showing the range in composition for mature algae collected at different times at each of several locations.

ture, shear rate, polymer size, concentration in solution, and the presence of miscible solvents. Chemical variables affecting alginate solutions include acid, sequestrants, monovalent salts, polyvalent salts, and quaternary ammonium compounds.

Temperature

Alginate solutions, like those of most other polysaccharides, decrease in viscosity with an increase in temperature. Over a limited range, the viscosity of an alginate solution decreases approximately 12 percent for each 5.6°C (10°F) increase in temperature.

Temperature reduction causes a viscosity increase in an alginate solution, but usually does not result in gel formation. A sodium alginate solution that has been frozen and then thawed will not have its appearance or viscosity changed.

Water-Miscible Solvents

Alginates, as hydrophilic colloids, form aqueous solutions. The addition of increasing amounts of nonaqueous water-miscible solvents (alcohols, glycols, acetone, etc.) to an alginate solution results in viscosity increases and eventual precipitation. The source of the alginate, the degree of polymerization, the cations present, and the concentration in solution all affect the solvent tolerance of the solution.

Effect of pH

The curves in Figure 6 illustrate the effect of pH on the viscosity of solutions of several types of alginates. Sodium alginates with some residual calcium content (Curve A) gel at a pH of five. Sodium alginates with minimal calcium content do not gel until the pH reaches three to four (Curve B).

Esterification, by reducing the ionic character of the alginate molecule, increases its tolerance to low pH (Curve C). Propylene glycol alginates are compatible and stable at pH three to four, but alkaline pH causes de-esterification and, depending on pH, loss of viscosity.

Long-term stability is poor above about pH 10.

Table 3. Proportions of Polymannuronic Acid, Polyguluronic Acid, and Alternating Segments in Alginic Acid Isolated from Brown Algae*

Source	Polymannuronic Acid Segment (%)	Polyguluronic Acid Segment (%)	Alternating Segment (%)
Macrocystis pyrifera	40.6	17.7	41.7
Ascophyllum nodosum	38.4	20.7	41.0
Laminaria hyperborea	12.7	60.5	26.8

* Data of Penman and Sanderson, 1972.

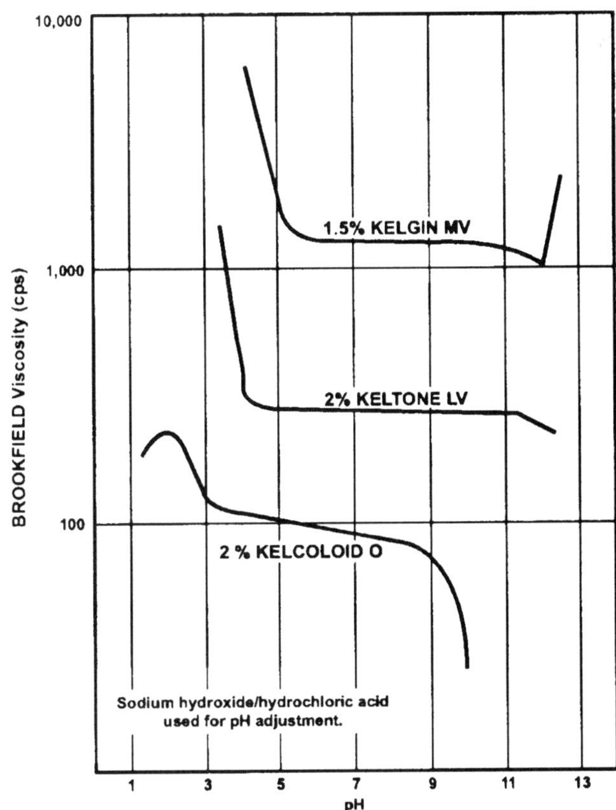

Figure 6. Effect of pH on alginate solutions.

EFFECT OF SEQUESTRANTS

Sequestrants are used in alginate solutions either to prevent the alginate from reacting with polyvalent ions in the solution or to sequester the calcium inherent in the alginate.

The use of sequestrants is common in most applications of alginates. Polyvalent ion contaminants can occur in water, chemicals, pigments, or almost any material of natural origin.

EFFECT OF MONOVALENT SALTS

The viscosity of dilute sodium alginate solutions is usually depressed by the addition of monovalent salts. As is typical with polyelectrolytes, the algin polymer contracts as the ionic strength of the solution is increased. However, the maximum viscosity effect is attained at a salt level of approximately 0.1N.

The salt effects vary with the source of the alginate, the degree of polymerization, the concentration in solution, and the type of salt used.

POLYVALENT CATIONS AND GELS

Polyvalent cations will react with and, in some cases, cross-link alginate polymers. As the polyvalent ion content in solution is increased, thickening, gelation, and finally precipitation will occur. As with other properties of alginate solutions, the mannuronic/guluronic acid ratio, the amount of monovalent salts in the solution, solution temperature, degree of polymerization, and the polyvalent ion itself will vary the properties of the reacted alginate.

The polyvalent cation most used to change the rheology, viscosity, and gel characteristics of algin solutions is calcium. Calcium also may be used in the formation of insoluble filaments and films.

The calcium reaction can be observed by injecting a stream of a sodium alginate solution into a calcium chloride solution. Almost instantly, calcium alginate will form on the surface of the sodium alginate, but over a period of time the soluble calcium will diffuse into the center and form a complete calcium alginate structure.

The reaction between calcium ions and alginate can be expressed in simple terms as:

$$2NaAlg + Ca^{++} \Leftarrow\Rightarrow CaAlg_2 + 2Na^+$$

As calcium ions are added to the system, the reaction proceeds to the right until all the alginate is precipitated as calcium alginate. In most applications, control of the calcium reaction is desired and is accomplished in several ways:

1. Variation in pH may be used to control calcium salt solubility.
2. Readily soluble sequestrants may be used to adjust setting and to vary the final gel texture.
3. Variations in the solubility of various acids can be utilized to control the reaction rate or setting time of the gel.
4. High levels of sugar solids inhibit reaction and result in soft gels.

The physical characteristics of the sodium-calcium alginate system vary with the amount of available calcium. Initially, viscosity increases will be evident along with shorter flow characteristics. Further addition of calcium ions results in gel formation and finally precipitation.

Stoichiometrically, 7.2 percent calcium is required based on the weight of sodium alginate for complete substitution. Gels are formed with about

30 percent of this amount, and thickened, flowable solutions with less than 15 percent.

Quaternary Ammonium Compounds

Normally, anionic polymers such as sodium alginates react with cationic organic ammonium compounds to form insoluble adducts. Two mechanisms can be used to prevent this reaction: (1) pH may be reduced to the level at which the cationic compound is not reactive; (2) a salt (such as NaCl) may be added at a concentration sufficient to interfere with the activity of the cationic compound.

Alginates in solution are compatible with a wide variety of materials, including other thickeners, synthetic resins, latices, sugar, oils, fats, waxes, pigments, and various surfactants.

APPLICATIONS

Industrial Applications

Alginates are widely used for surface treatment of paper and paperboard to improve the surface characteristics. In paper sizing, alginates are used for their excellent film-forming ability which results in uniformity of ink, wax, solvent, and/or coating acceptance. Other industrial applications, such as textiles, utilize sodium alginates as thickeners for fiber-reactive dyes. Fiber-reactive dyes produce brilliant shades and high color yields. However, because of the high reactivity of this dyestuff class, the thickeners used must be carefully selected. If the dye reacts with both thickener and fabric, the thickener will be chemically bonded to the fabric, washout will be incomplete, print quality will be reduced, and the finished product will have a stiff, brittle feel. However, the use of sodium alginates eliminates these problems. Sodium alginates provide excellent color yield and print quality.

Pharmaceutical Applications

Alginates are used in many pharmaceutical preparations, such as emulsions, tablets, ointments, jellies, and dental impression materials. Alginic acid has uses as a tablet disintegrant due to its swelling capabilities. Alginic acid is insoluble in water but will swell to many times its weight, thus breaking the tablet apart. Sodium and potassium alginates are often used in the manufacture of dental impression materials, which utilize the algin/calcium reaction to form a high-quality, rapid-setting gel. Alginates can also be used to modify drug -elease characteristics to a required profile in sustained release tablets. Other products which sometimes use alginates are shampoos and lotions.

Food Applications

Alginates have various uses in the food industry. Propylene glycol alginate is typically used as a stabilizer in pourable salad dressings. It provides viscosity and good flow characteristics, and acts as a secondary emulsifier (due to the ester groups).

Sodium alginate is used in the production of structured fruits, vegetables, meats, and fish. Since structuring is a means of raw materials extension, cost savings can be obtained. Sodium alginate is used in structured pimiento strips for olives. In this application, the pureed pimiento is added to an alginate solution; the mixture is then pumped onto a conveyor, which carries it as a sheet into a calcium chloride setting bath. The sheet is "set" or gelled by diffusion of the calcium ions. After setting, the sheet is cut into thin strips, which are allowed to equilibrate for several days in a solution of salt and calcium chloride. After aging, the strip is ready for use in the olive-stuffing machine.

In the late 1950s alginates were frequently used as a stabilizer in ice cream. Alginate-stabilized ice creams have small ice crystals. Sodium alginate also protects against heat shock, which can occur during distribution. Alginate also helps in the development of overrun by increasing the viscosity of the mix. Today, however, alginates have largely been replaced by less expensive stabilizers such as xanthan gum, carboxymethyl cellulose, guar gum, and locust bean gum.

Because of their unique properties, alginates have many applications in food and industrial products. Tables 4 and 5 illustrate how some of the principal properties apply to typical products.

GOVERNMENT REGULATIONS

Ammonium alginate, calcium alginate, potassium alginate, and sodium alginate are included in a list of stabilizers that are affirmed as Generally Recognized As Safe (GRAS) under 21 CFR 184.1133; 184.1187; 184.1610; and 184.1724, respectively. Propylene glycol alginate is approved as a food additive under 21 CFR 172.858 for use as an emulsifier, stabilizer, or thickener in foods.

The Food Chemicals Codex contains monographs on alginic acid, ammonium alginate, calcium alginate, potassium alginate, sodium alginate, and propylene glycol alginate (National Academy of Sciences, 1981). Monographs on sodium alginate, alginic acid, and propylene glycol alginate are also included in the National Formulary (United States Pharmacopeial Convention, 1984).

Alginic acid and its edible salts, as well as propylene glycol alginate, are included in the approved emulsifier/stabilizer list published by the European Economic Community. The FAO/WHO Joint Expert Committee on Food Additives has an unspecified Acceptable Daily Intake (ADI unspecified) for alginic acid and its edible salts, and 25 milligrams/kilograms for propylene glycol alginate based on the propylene glycol content.

Table 4. Food Applications of Brown Algae

Property	Product	Performance
Water-holding	Frozen foods	Maintains texture during freeze-thaw cycle.
	Pastry fillings	Produces smooth, soft texture and body.
	Syrups	Suspends solids, controls pouring consistency, provides body.
	Bakery icings	Counteracts stickiness and cracking.
	Dry mixes	Absorbs water or milk quickly upon reconstitution.
	Meringues	Stabilizes meringue texture and prevents syneresis.
	Frozen desserts	Provides heat-shock protection, improved flavor release, and superior meltdown.
	Relish	Stabilizes brine, allowing uniform filling.
Gelling	Instant puddings	Produces firm pudding with excellent body and texture; better flavor release.
	Cooked puddings	Stabilizes pudding system, firms body, and reduces weeping.
	Chiffons	Provides tender gel body that stabilizes instant (cold make-up) chiffons.
	Pie and pastry fillings	Provides cold-water base for dry-mix fillings. Develops soft gel body with broad temperature tolerance; improved flavor release.
	Dessert gels	Produces clear, firm, quick-setting gels with hot or cold water.
	Structured foods	Provides a unique system that gels under a wide range of conditions.
	Bakery jellies	Prevents boil-out, enhances flavor and color, improves depositing properties, and provides versatility with hot or cold make-up, range of solids, gel texture and fruit content.
Emulsifying	Salad dressings	Emulsifies, stabilizes, and modifies flow properties in pourable dressings.
	Meat and flavor sauces	Emulsifies oil and suspends solids.
Stabilizing	Beer	Maintains beer foam under adverse conditions.
	Fruit juice	Stabilizes pulp in concentrates and finished drinks.
	Fountain syrups, toppings	Suspends solids, produces uniform body.
	Whipped toppings	Aids in developing overrun, stabilizes fat dispersion, and prevents freeze-thaw breakdown.
	Sauces and gravies	Thickens and stabilizes in a broad range of applications.

Table 5. Industrial Applications of Brown Algae

Property	Product	Performance
Water-holding	Paper coating	Controls rheology of coatings; prevents dilatency at high shear.
	Paper sizing	Improves surface properties, ink acceptance, and smoothness.
	Adhesives	Controls penetration to improve adhesion and application.
	Textile printing	Produces very fine line prints with good definition and excellent washout.
	Textile dyeing	Prevents migration of dyestuffs in pad-dyeing operations. (Algin is also compatible with most fiber-reactive dyes.)
Gelling	Air freshener gel	Firm, stable gels are produced from cold-water systems.
	Explosives	Rubbery, elastic gels are formed by reaction with borates.
	Toys	Safe, nontoxic materials are made for impressions or putty-like compounds.
	Hydro-mulching	Holds mulch to inclined surfaces; promotes seed germination.
	Boiler compounds	Produces soft, voluminous flocs easily separated from boiler water.
Emulsifying	Polishes	Emulsifies oils and suspends solids.
	Antifoams	Emulsifies and stabilizes various types.
Binding	Ceramics	Imparts plasticity and improves green strength.
	Welding rods	Improves extrusion characteristics and green strength.
Filming	Warp sizing	Improves warp lubricity.
	Paper sizing	Improves surface properties, ink acceptance, smoothness, and hold-out.

REFERENCES

Alginate Products for Scientific Water Control. 1984. 3rd edn., Kelco, Div. Merck U.S.A.

Food Chemicals Codex. 1981. 2nd edn. with supplements. *National Academy of Sciences*, Washington, DC.

Haug, A. 1964. Composition and Properties of Alginates. Report No. 30. *Norwegian Institute Seaweed Research*. Trondheim, Norway.

Haug, A., and B. Larson. 1962. Quantitative determination of the uronic acid composition of alginates. *Acta. Chem. Scand.* **16**: 1908-18.

Krefting, A. 1896. An Improved Method of Treating Seaweed to Obtain Valuable Product Therefrom. Brit. Patent 11,538.

National Formulary. 1984. 16th edn. *United States Pharmacopeial Convention*. Washington, DC.

North, W. J. 1972. Giant Kelp: Sequoias of the Sea. *Nat. Geographic*. **142**: 251-69.

Penman, A., and G. R. Sanderson. 1972. A method for the determination of uronic acid sequence in alginates. *Carbohydrates Research*. **25**:273-82.

Stanford, E. C C. 1883. *Chemical News*. British Patent. 47,254.

Steiner, A. B., and W.H. McNeely. 1954. Algin in Review. In Natural Plant Hydrocolloids. Advan. Chem. Ser. No. 11. *American Chemical Society*. Washington, DC. Pp. 68-82.

Structured Foods With the Algin/Calcium Reaction. 1994. Technical Bulletin F-83, Kelco, Div. Merck U.S.A.

Fish Meal and Oil

Anthony P. Bimbo

INTRODUCTION

Fishing is one of America's first industries and can be traced back to the early Plymouth, Massachusetts, Colony (McCay, 1980). Lee (1952) mentions that the Indians used fish for fertilizer long before the colonists arrived. Frye (1978) describes a legend of how the early Plymouth settlers learned of the use of fish as fertilizer from an Indian called Squanto or Tisquantum — supposedly saving the colony from disaster. However, McCay (1980) suggests that perhaps the Indians learned this technique from the colonists through the observations and travels of Squanto. In any event, the practice was well-rooted in the colonies, and references to such use can be found in colonial writings of 1621.

The worldwide landings of fish and shellfish (including freshwater production) now exceed 100 million metric tons annually (James, 1994). Figure 1 shows the growth in total marine and freshwater aquaculture landings of fish and shellfish from 1953 to 1997. The 12-year average of catch by geographical area is shown in Figure 2. Worldwide, approximately 28 percent of the landings are pro-

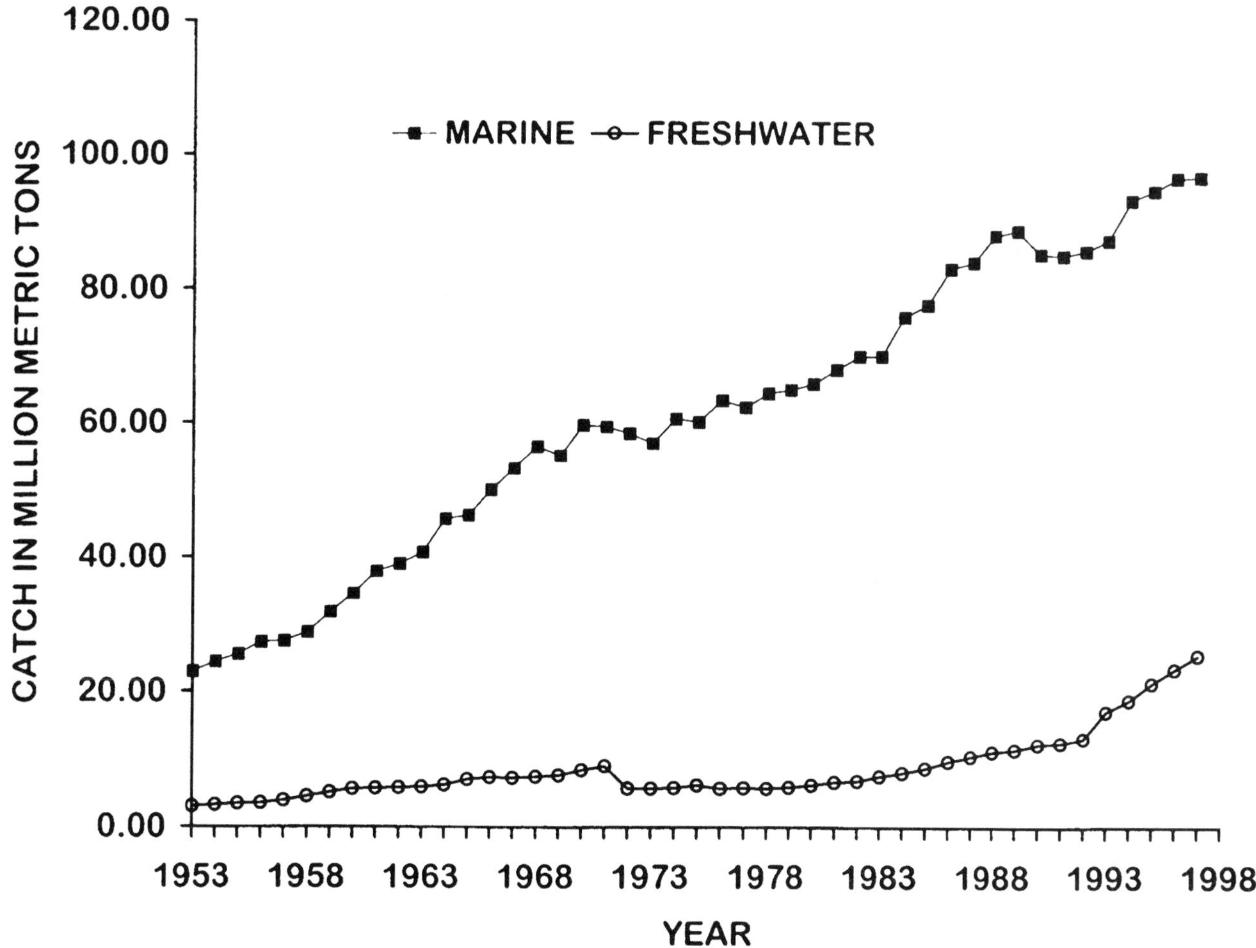

Figure 1. Marine and freshwater (aquaculture) landings of fish and shellfish, 1955 to 1997.

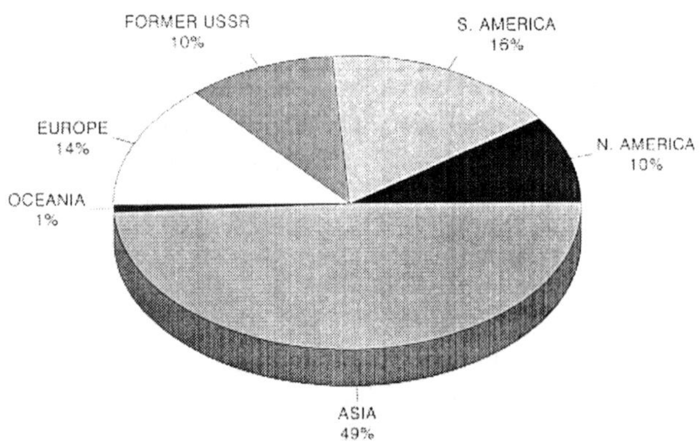

Figure 2. The 12-year average breakdown of catch by geographical area.

Figure 4. Photograph of a typical fishing vessel leaving port for the fishing grounds.

cessed into fish meal and oil, while in the United States the figure is closer to 20 percent. The 12-year average disposition of the world catch by end use for the period 1986 to 1997 is shown in Figure 3. It has been conservatively estimated that only about 40 percent of the edible landings are actually used as food. The remaining portion of the fish is waste (trimmings) and offers the potential for increased utilization in fish meal and oil production, silage, and other uses.

Fishing is the last major industry that still relies on the hunting-and-gathering technique. Figure 4 shows a typical fishing vessel leaving port for the fishing grounds.

Fish protein competes with oilseed protein; Figure 5 compares fish landings with harvests of the major oilseed crops around the world for the period 1992 to 1997 (forecast). From this perspective, fish as a source of protein are one of the main raw materials in the world. Fishing is the main industry in many countries, and the production of fish meal and oil is considered one of the principal sources of protein and fat commodities in world trade.

The fish meal and oil industry started in northern Europe and North America at the beginning of the nineteenth century and was principally based on the inshore herring fisheries. This was essentially an oil production activity with the oil used industrially for illumination and in the tanning of animal skins, the production of soap, the waterproofing of wooden ships, and other non-food uses. The cake residue was originally used as fertilizer, but since the early 1900s has been dried and ground into fish meal for animal feeding. Its main use is in the diets of poultry, pigs, and fish, which need a higher quality protein than do other farm animals such as sheep and cattle (FAO, 1986).

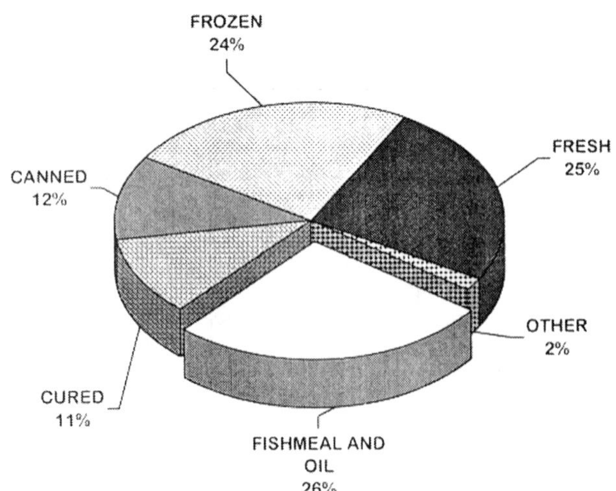

Figure 3. The 12-year average disposition of the world catch by end-use, 1986 to 1997.

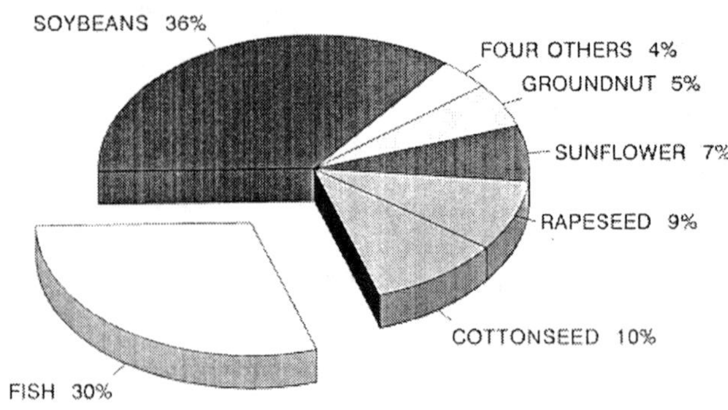

Figure 5. Fish landings compared with major oilseed crops harvested worldwide, 1992 to 1997F.

THE UNITED STATES MENHADEN INDUSTRY

The United States production of fish meal and oil accounts for approximately 10 percent of the world's production; the menhaden, *Brevoortia* spp., accounts for about 98 percent of the United States' production of fish meal and oil. Menhaden landings represent approximately 30 percent of the total pounds of fish and shellfish landed in the United States annually. In 1985 a record 2.7 billion pounds landed accounted for 44 percent of the commercial fishery landings in the United States [United States Department of Commerce (USDC) 1995]. Figure 6 shows the United States' menhaden landings for the Atlantic and Gulf of Mexico from 1946 to 1998. Table 1 gives the breakdown of fish meal, oil, and solubles produced in the United States from 1984 to 1998.

The menhaden, cousins of the herring, are characterized by big heads, a slight hump on their backs, and no teeth. Like the herring, they are dependent upon plankton for their food. Unlike the herring, they are seldom eaten as food, but are widely relied upon for other uses (Bimbo, 1970).

No one is sure of the geographical origin of the menhaden industry, but a wealthy Long Island farmer experimented with menhaden as a side dressing in his potato fields as early as 1801. In Rhode Island (1812) the first crude process for menhaden oil recovery was developed, followed by production facilities in Maine in 1850 and in Monmouth County, New Jersey, in the 1850s. By 1877 there were 14 plants in Maine, 13 in Rhode Island, five in Connecticut, four in the Chesapeake Bay, 23 on Long Island (New York), and five in New Jersey (McCay, 1980). The early plants were small, very crude, and usually located in out-of-the-way places. As the process developed, a tub drilled with holes was used to hold the fish under pressure from heavy rocks. Finally in the 1850s the mechanical screw press was developed and the production of menhaden fish oil reached the small factory operation (Lee, 1952).

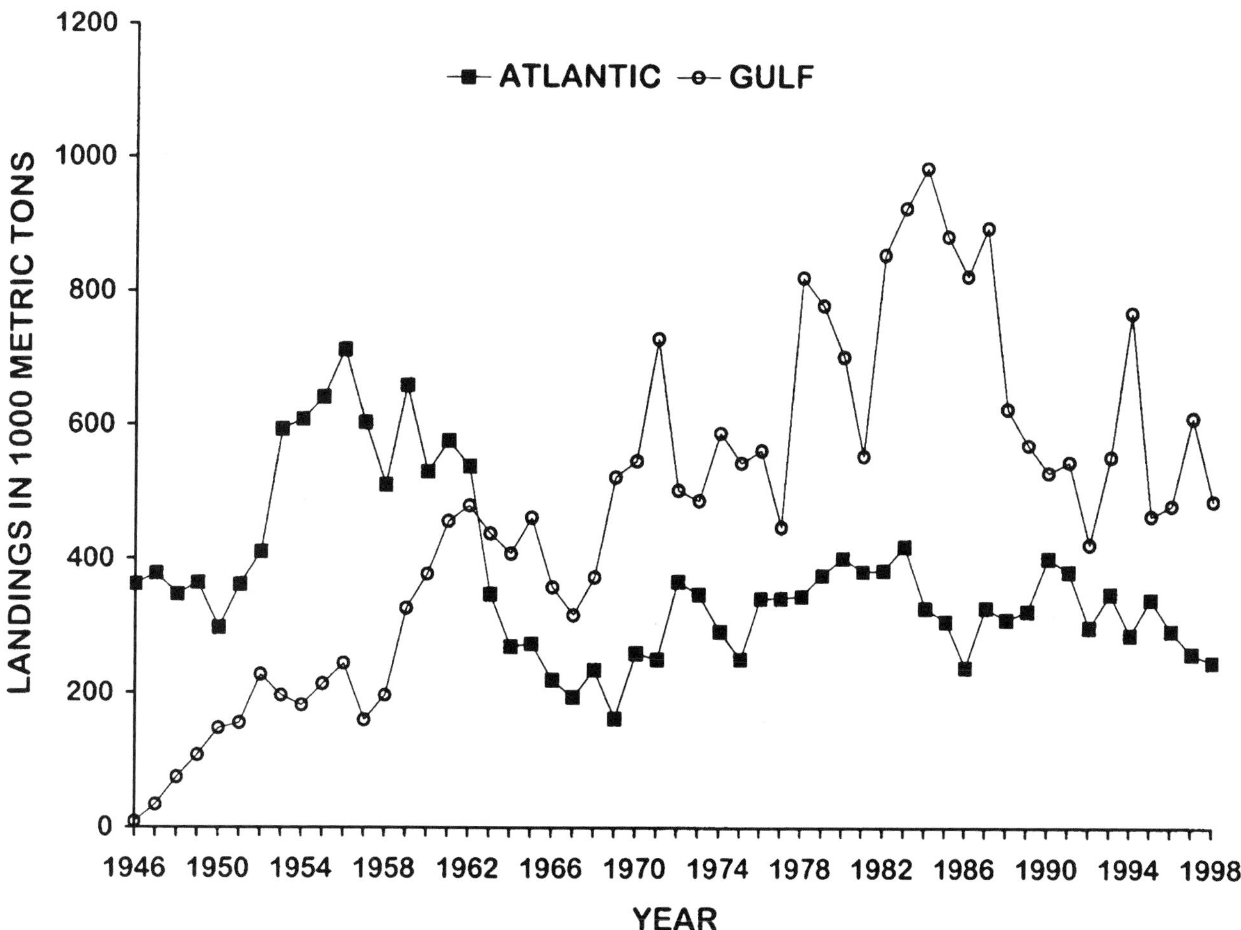

Figure 6. United States menhaden landings for the Atlantic and Gulf of Mexico, 1946 to 1998.

Table 1. Fish Meal, Oil, and Solubles Produced in the United States, 1984 to 1998

Year	Menhaden Landings	Menhaden Oil	Menhaden Fish Meal	Menhaden Solubles
1984	1173	126	286	114
1985	1242	151	280	147
1986	1100	134	269	89
1987	1230	99	303	113
1988	946	99	229	101
1989	902	124	210	106
1990	890	124	203	84
1991	897	120	211	77
1992	746	81	176	42
1993	900	132	214	58
1994	1054	131	259	66
1995	838	108	204	41
1996	796	113	292	39
1997	920	129	329	54
1998	774	101	269	28

Figures in 1000 metric tons. Source: U.S. Department of Commerce, 1999.

The fish were unloaded with pitchforks from the boats into tanks or directly into small wooden tram cars holding 20 barrels each. These cars were hauled to the upper floor of the plant where the fish were dumped into large reservoirs. From here, the fish flowed by gravity into the cooking tanks. These were constructed of wood staves, sometimes with a false bottom, and had perforated pipes in the bottom for the introduction of steam. The tanks held from 50 to 75 barrels of fish and were filled to a depth of six inches to a foot with sea water, which was sometimes preheated before addition of the fish. Cooking time was usually a half an hour, though in one plant the fish were simmered for five hours. After cooking, the hot water and oil were drawn off, and the mass of fish allowed to drain and cool. A man then climbed into the tank and pitchforked the fish into "curbs" that confined the fish during the pressing operation. The curbs were built of heavy wooden slats, iron-bound or made of iron with eighth-inch holes, and held from three to 10 barrels of fish each. They usually were mounted on small trucks running on tracks leading to the press. The first presses received their pressure by weighted rocks or by using a lever to squeeze the oil from the fish. Oil and water draining from the cooking tanks and presses ran to a series of settling tanks. The operators were aware of the fact that the oil contained finely divided fleshy material which settled out more slowly than the press water and that for the best grade oil this should be removed before putrefaction started. The top oil was skimmed off and held in open tanks for one to two weeks to sun bleach for the best grade of white oil. The lower levels of oil were run off into another tank and yielded progressively poorer grades of oil.

The wet press cake in the curbs was dumped through a trap door in the floor to a room or an open space beneath the plant. In some plants it was allowed to accumulate until fall or winter before disposal (Stansby, 1978). Frye (1978) says that by the middle of the nineteenth century disposal of the residue after the oil was removed had become a problem; subsequently, the residue was removed and given to farmers to fertilize fields, replacing the use of whole fish. By 1917, because of animal feed shortages, the fish scrap began to find use in animal feeds; by 1935, virtually all of it was used in this way. Some of the menhaden scrap continued to be sold as fertilizer until the late 1920s.

Smith (1940) described the rapid changes that were beginning to take place in the industry along the Atlantic Coast. Prior to 1940 the process consisted of cooking the fish in large vats, pressing the cooked fish, and catching the pressed liquor in a series of settling tanks. By a series of skimmings, the oil was separated from the water phase and an emulsion layer was recooked to recover oil. The remaining "gurry" was then allowed to decompose to form a dark low-grade oil which was removed by hand dippers. In 1940, the industry began to use centrifuges to separate oil from water and thus to eliminate this emulsion phase. The oil, the most valuable product recovered from the fish, was used in soap, paints, and linoleum. It was this introduction of the centrifuge that started the rapid increase in production of menhaden oil in the United States, the "Industrial Revolution" of the menhaden industry. With the discovery in 1949 of vitamin B12 as a source from animal protein, menhaden scrap's use in animal feeds was assured. About this same time, the water from the pressing operation was also found to be rich in Vitamin B12 and, after concentration to 50 percent solids, this water became a new byproduct: condensed fish solubles (Frye, 1978; Lee, 1952).

THE WET RENDERING PROCESS, PART 1: THE PRODUCTION OF FISH MEAL

RAW MATERIAL

Fish used in the production of meal and oil can be divided into three categories:

1. Fish caught specifically for fish meal production, such as menhaden in the United States; anchovy, jack mackerel, and sardines in Latin America; capelin, mackerel, and sandeel in Scandinavia; and sardines in Japan.
2. By-catches from another fishery, such as shrimp by-catch.
3. Fish offal and cuttings from edible operations, such as from filleting plants, surimi operations, and canneries. Tuna, groundfish, and catfish meals are examples of this.

In all cases the fish are usually small pelagic species, oily and bony, that swim in large schools; or the raw material is waste of no edible value; or the fish are classified as industrial and of no economic value when compared with the major fishery in which they are caught [Food and Agriculture Organization of the United Nations (FAO) 1986].

There continue to be many suggestions on how to upgrade these industrial species of fish to human foods. The literature abounds with papers on this subject (James, 1994; Barlow and Pike, 1994; Billy and Dreosti, 1983; Ackman, Eaton, and Ratnayake, 1981; Hansen, 1981; Jensen and Teutcher, 1980). Production and harvesting costs continue to make fish meal production a low margin/break-even industry. Research is in progress worldwide to upgrade markets for these fish. As this concept develops, we will see more of the traditional fish meal coming from cuttings as opposed to whole fish, but we will not see an end to fish meal production. In general, the industrial species are relatively small and oily. They are subject to rapid bacterial and enzymatic spoilage and to rapid oxidation and rancidity. The fish must be degutted or, at least, the gut contents must be removed, and the fish must be deboned [FAO/IDRC (Industrial Development Research Center), 1982; Luna, 1981; Kreuzer, 1974]. All these operations produce waste or offal for the fish meal plant; it is therefore envisioned that small packaged fish meal plants will be an integral part of any large edible operation. These packaged plants require little space, produce high quality products, are automatic, and require only one person on a part-time basis to operate the system. Plants with capacities of as little as 10 tons per day have been built and installed on factory trawlers. They can be fitted into a room 3.4 x 5.4 meters x 3 meters high (*World Fishing Magazine*, 1993).

HARVESTING

The method used to catch fish depends largely upon the feeding habits of the species. For fish that school near the surface, such as menhaden, tuna, mackerel, herring, sardines, and anchovies, the easiest and most efficient method of capture is the purse seine. For fish that live near the floor of the ocean or in mid-water, trawling is the most efficient method. The type of fishing gear used determines, at least in part, the kind of fishing vessel employed. Purse seiners and trawlers are large vessels with a carrying capacity that can exceed 500 tons; they can travel several days from the fishing grounds. Some factory ships are capable of staying on the fishing grounds for many months, using smaller fishing vessels to deliver the catch to them for processing. Finished product is then shipped home on cargo vessels (Lee and Sanford, 1960).

Unloading

The discharge of industrial fish from the vessel at the processing facility has presented many challenges to the fishing industry over many years. The methods employed must not only be economical and pollution-free, but the fish must arrive at the plant in good shape as quickly as possible, since the vessel must be able to get back on the fishing grounds while fish are available. Global environmental regulations in the late 1960s and early 1970s accelerated the development of new methods of discharging fish at the factory so that the industry could meet newly developing environmental standards. In 1977 the International Association of Fish Meal Manufacturers (IAFMM) organized a symposium to cover this issue. In his keynote paper Sola (1978) outlined the Norwegian requirements for an unloading system that would address strict anti-pollution requirements, restricted working hours, and high labor and energy costs. These requirements are still valid today:

1. No pollution of bays, rivers, or harbor water
2. Easy operation, mainly non-manual
3. Complete discharge process, without final manual stage
4. No risk to the operators
5. Easy clean-up
6. Low energy requirements
7. No addition of water
8. Unaffected by different fish sizes
9. Unaffected by tidal changes
10. Unaffected by condition of raw material
11. Large capacity
12. Installable on the fishing vessel
13. Reasonable initial cost
14. Low noise and odor level
15. Unaffected by climatic variations

There are only two ways to transport fish from the fishing vessel to the factory for processing. The fish can be moved either wet or dry. Within these two categories there are at least seven possibilities for unloading fish from the vessel to the plant. These can be categorized as follows:

WET UNLOADING	DRY UNLOADING
Wet pumps	Bucket elevator
Containers	Grab with crane
	Air suction
	Vacuum
	Dry pumps

The bucket elevator consists of a bucket that moves the fish to a conveyor, elevating the fish out of the hold of the vessel and finally to a receiving bin in the factory. While the bucket elevator is less likely to cause pollution of the harbor and surrounding waters, it is affected by tidal changes, cannot be installed on the fishing vessel, and is labor-intensive. Konge and Rasmussen (1978) described the chainpump-elevator system that was used in Denmark; it acts like a pump when the fish are soft and as a bucket elevator when the fish are fresh and firm. It was replaced by the grab because of problems with large tidal changes in the North Sea and North Atlantic.

The grab consists of a clamshell or grab at the end of a crane that is attached to a dock. Its disadvantages are the spillage and drippage of fish and liquid onto the dock and into the surrounding waters. It is labor-intensive and requires some other method to remove the last traces of fish from the vessel hold. Neither the grab nor the bucket elevator is used by the fish meal industry today. Both have been replaced by more environmental- and labor-friendly methods.

Konge and Rasmussen (1978) described a dry vacuum, or pneumatic off-loading system (IRAS system, after the name of the Danish company) that was designed to handle different species as well as different degrees of freshness and quality. The fish and air are sucked from the hold to a separation section where the fish slide down a tube that is closed by a flap valve. When the weight of the fish in the tube is heavy enough to overcome the vacuum, the flap opens and the fish slide out. The air escapes from the separating section to a cyclone where smaller particles of fish are collected. The air is then either discharged or recycled back to the fish. Capacities of one to two tons/minute on a 200 HP plant have been achieved. The unit is maneuverable, and the hose can be moved from hatch to hatch; the noise level is also reduced. The IRAS system is used in Denmark today.

Beugelink (1978) described a dry off-loading system that is being used in South Africa. Handlers have unloaded an average of 50 to 100 tons anchovies/hour, with breakage in the range of 2 to 3 percent. The fish enter the system by a suction nozzle and are conveyed through pipes to the separator. The fish are discharged from the separator through a slide box valve. According to the author, power requirements of 0.7 to 2.5 horsepower/ton of fish are needed.

The disadvantage of the vacuum unloading system is the limited distance that the fish can be transported. If the vessel cannot tie up adjacent to the factory (as is the case in many plants in South America) so that the discharge of fish can be conveyed into the raw holding bins, then it would be necessary to install the unloading systems in a harbor, unload the fish at the port, and deliver them to the factory by truck. This procedure has been tried with a factory in Mexico (no longer in operation) and at several locations in Chile. It is better suited to small plants, because with larger plants the volume of fish would require many trucks running from the unloading area back to the factory.

Porse (1978) described the Superfos hydraulic transport pump, which was developed in Denmark in 1973. In looking for an alternative to the wet system, they required:

1. Minimum destruction of raw material
2. Transportation without the addition of water
3. Closed system
4. Minimum maintenance.

The pump is a double-acting piston pump with a four-way valve as the central point. The action of the pistons and the rotary motion of the valve flap allow only one motion to happen at a time. No air or water is used and the actual capacity is 60 to 80 cubic meters/hour.

Gibbs and Green (1978) described the Hidrostal pump, which has a special centrifugal screw impeller that was developed in Peru specifically for pumping fish. It is the wet pump that is now used throughout the Peruvian fishery and in some plants in Chile. The only connection between the vessel and the shore or platform is a flexible suction pipe and water hoses, so there are no problems with tidal movements. The standard pump has a capacity of 50 to 100 tons of fish/hour, which can be varied by adjusting the fish-to-water ratio.

Nordstrom (1977) described the Nemo or Netzsch pump. This is a Mono pump that has been used in various factories for pumping liquids and semisolid materials. The pump consists of a metallic rotor and elastic stator. It is a positive displacement pump with the quantity of material pumped proportional to the speed of the pump. The pump has variable speeds and is reversible so that it can be cleaned out. The maximum quantity is about 250 cubic meters/hour. For unloading fish, the pump must be moved around the hold of the vessel; this can be done with a crane. This pump is being evaluated at several plants in Peru. One of the disadvantages of the pump is its weight, which makes it difficult to maneuver.

Sola and Tronstad (1978) described the Myrens dry pump, which offers two alternatives for mounting. The pump can be handled by a crane and lowered into the hold of the vessel, or it can be mounted on the vessel. The pump is a positive displacement pump with a rotating valve. This is a dry system, except for the water that is in the hold of the vessel with the fish; no extra transport water is used. At 45 revolutions per minute the pump will move 70 to 80 cubic meters per hour. Pump bearings and seals are of plastic. The Myrens pump is currently used in Iceland with excellent results. Only a small amount of water (10 percent) is needed to prime the pump. Pumps of this type are also used in the factory for moving fish around the plant. The Myrens pump is no longer manufactured, but an Icelandic company is negotiating for the manufacturing rights.

In the United States, fish meal plants use Humphreys piston pumps to unload fish. These are similar to the Hidrostal pump in that water is used to convey the fish from the vessel into the factory but, unlike other systems, the water is screened and recycled and finally evaporated to become part of the finished product.

Hansen et al. (1977) described work done in Chile. Fishing vessels that bring in fish for food use and also fish for fish meal use can be handled using containers and an icing and recirculating technique developed in the United Kingdom. Containers, stacked in the central part of the fish hold, reach from the floor to the deck; they are separated from the rest of the hold by walls of wood or aluminum. The containers are prefilled with ice before the vessel departs for the fishing grounds, an equal amount of sea water is added, and air circulation is started before filling up with fish. The containers are removed from the vessel by a crane.

Bozzo and Alister (1995) described experiments with a pressure/vacuum pump (TransVac) being evaluated in Chile. The project developed because fish were being discharged from the vessels at a congested port and then trucked from the port through the city to the factories. The pump is reported to be capable of moving fish approximately 1600 meters with less water and less degradation. The pipelines are of high-density polyethylene, thermofused together and floating on the surface. Fish are conveyed a distance of 1150 meters in the water and 450 meters on land to the plant. The capacity of the pump is 200 tons/hour. The pumps are electrically powered so power cables must be run to the unloading station. The pump has been successful, unloading 35,000 tons of fish with a ratio of 1:1 to 1:3 of fish to water. Degradation of the fish is comparable to other unloading methods. Today, the Chilean and Peruvian fish meal industry is moving towards the use of these types of pumps.

As environmental issues continue to grow, the discharge of the pump water has become more critical, and some companies have begun to use different methods to remove solids from the water stream before discharge. One such device is the rotating hydro-sieve which rotates opposite to the liquid flow that is introduced at a tangential angle, creating a higher shear velocity. The milliscreen separates small suspended particles from the water and has been successfully used in many applications, including the fish meal industry. In Peru, dissolved air flotation (DAF) systems are successfully removing oil from the pumpwater, thereby increasing oil yields as much as 20 percent.

Cooking

Cooking denatures the fish protein and makes it possible to separate the fat by mechanical means. A number of factors influence the quality of the cooked fish:

1. Heating temperature
2. Heating time
3. pH of the fish
4. Freshness of the fish
5. Particle size of the fish pieces
6. Type of fish

During the cooking process, the protein is coagulated into a firm mass capable of withstanding the pressure required to press out the stickwater and oil. During coagulation, a high proportion of the bound water is liberated and deposits of fat are released from the tissues and thus removed by water and oil separation.

Figure 7. Photograph of a typical cooker used in the wet reduction process.

The cooker is a cylinder with a steam-heated jacket and a hollow steam-heated auger. The cooker is equipped with covers for inspection and cleaning purposes and may have nozzles for the direct addition of steam to the fish mass. Figure 7 shows a typical cooker used in the wet reduction process.

Cooking is not an exacting operation in production and at times it is difficult to control. The production of cooked material which can be readily pressed depends on the quality of the raw material and the process conditions. Good cooking results in good pressability of the mass, which leads to proper removal of press liquor and efficient recovery of oil, giving a low-fat meal. Overcooking affects pressing also, and causes the formation of fines or suspended particles in the stickwater, which makes evaporation difficult (FAO, 1986; Ward et al., 1977).

Pressing

The liquid portion of the fish can account for 80 percent of the fish mass. Therefore, deoiling and dewatering are two of the major steps involved in the manufacture of fish meal and oil. The objective of the pressing and screening operation is to produce a meal with the lowest possible oil content. There are a number of methods by which this objective is achieved, but the major method uses presses (Kroken and Utvik, 1978).

Two types of continuous presses are used in the fish meal industry. The single-screw press works on the principle of a helical-screw conveyor rotating in a cylindrical cage, provided with perforations for the drainage of pressliquor. The screw, designed with a taper, exerts an increasing pressure on the mass by reducing the volume during its passage through the cage. Some problems are experienced when the press is used with poor raw materials. Slipping of the soft fish material may occur, and the screw is unable to convey and effectively press the material. This difficulty may be minimized by incorporating special devices in the press or by using twin screw presses.

In the twin screw press, pressing is carried out in a press chamber consisting of two hollow interlocked cylinders. The press consists of a stationary part (the stator), and a rotating part (the rotor or the screw), and the gearbox with motor. The free space between the rotor and the stator decreases from inlet to outlet in a taper, so that material introduced at the inlet will be compressed and pressure built up as the product travels to the outlet. The built-up pressure causes the liquid to be squeezed through the perforated stator while the solids remain inside (Onarheim and Utvik, 1979). A typical press used in the fish meal industry is shown in Figure 8.

A variation involves two-stage or double pressing when certain types of raw materials that are difficult to press are processed. This variation is also used when a lower fat content is required in the final meal for specialized end uses. The method involves pressing the cooked fish in the normal manner, cooling the presscake to 50°C by adding chilled stickwater, and then pressing again. Fat reductions of 2 to 4 percent in the final meal can be achieved (Onarheim, 1978). Another variation on the pressing technique involves the use of decanters to separate liquid from the cooked fish. This alternative finds merit when the fish are old and in poor condition. The decanter appears to be able to deliver a meal product of consistent quality independent of the raw material quality (Christensen, 1978). The meal is not as dry as that produced with a press and thus more energy is required to dry the meal, but the decanter lends itself readily to sanitary cleaning and thus would find applications in areas where fish protein might be used in human food products (Rask, 1979).

Figure 8. Photograph of a typical press used in the fish meal industry.

Drying

The prime reason for drying fish is to reduce the moisture content of the non-aqueous material to such a level that insufficient water remains to support the growth of microorganisms (Jason, 1980). Two types of dryers are used: direct and indirect. In direct dryers (also called convection dryers), heat transfer is accomplished by direct contact between the wet solid and hot gases. The vaporized liquid is carried away by the drying medium — i.e., the hot gases. In indirect dryers, heat for drying is transferred to the wet solid through a retaining wall. The vaporized liquid is removed independently of the heating medium. The rate of drying depends upon contact of the wet material with hot surfaces. Indirect dryers are also called conduction or contact dryers.

Several factors must be considered in the selection of a dryer system:

1. The dryer must handle all types of fish in all types of conditions.
2. The dryer must be able to handle stickwater concentrate (solubles).
3. The dryer must give a maximum meal yield.
4. The dryer must give a high-quality meal.
5. The dryer system must have an effective deodorizing system.
6. The dryer must have a reasonable cost and energy consumption (Hetland, 1980).

In direct dryers, the quality of the meal is influenced by the dryer inlet temperature, which should be below 600°C. The dryer consists of a large rotary tube in which presscake is tumbled rapidly in a stream of very hot air at inlet temperatures of 600°C. Tumbling is provided by the rotating action of the dryer and a number of flights within the dryer that provide a cascading action and good air-to-particle contact. The hot air is provided by a current of flue gases from oil combustion together with diluting secondary air. The particles of meal do not reach this high temperature because of the rapid evaporation of the water from the surface of each particle, causing cooling by the loss of heat of evaporation. The temperature of the meal is normally 80°C. The fish and air move through the dryer in the same direction, and the rapid flow of hot air tends to help carry the meal particles through the dryer. Thus air velocity becomes another important factor in direct drying.

The indirect dryer is also a rotary dryer consisting of a large cylindrical drum in which the presscake is dried, but the heat is supplied indirectly by contact with steam or hot-air heated discs, tubes, coils, or a jacket. A current of air is blown through the dryer to remove the water vapor produced, but the air itself is not normally heated and travels countercurrently to the meal flow. The rotary action of the discs, coils, or tubes together with a series of flights within the dryer causes agitation of the meal and enhanced drying. Blades or scrapers are often necessary to prevent sticking of the product to the drying surface and a subsequent deterioration of drying efficiency. The temperature of the drying surface is decided by the temperature of the heating medium within the discs, coils, or tubes. This medium is normally steam, and its temperature is related to its pressure (Windsor and Barlow, 1981).

There is very little nutritional difference between meals dried by direct or by indirect means. Properly controlled cooking and drying procedures will produce products that are nutritionally sound with no deleterious effect on quality (FAO, 1986). Figure 9 shows a typical indirect dryer, and Figure 10 shows a direct-fired flame dryer.

Compared with other methods of drying fish meal, low temperature drying is a relatively new concept within the industry and has had its greatest acceptance in the production of fish meal for the aquaculture industry, early-weaned pig market, mink feeds and, more recently, pet foods. Low temperature drying developed as the industry sought new ways to eliminate air pollution and recover wasted energy from the dryers. By taking the existing hot air direct-fired dryers and adding a heat exchanger between the furnace and the dryer drum, it was possible to lower the drying temperature of the fish meal, recover the wasted heat that was going up the stack, and reduce air pollutants. Temperatures of meal exiting in these dryers is low, normally 65°C or less, as compared with 90°C to 95°C in steam or direct-fired dryers. The nutritional quality of these meals is better since the protein has not been exposed to higher temperatures. This type of meal has found markets in areas where quality can demand a premium. However, capital costs to install these dryers is high compared to direct-fired dryers, primarily because of the low volume throughput (Gundersen and Wiedswang, 1993). Figure 11 shows the solids flow of a typical fish meal plant.

Figure 9. Photograph of a typical indirect dryer.

Figure 10. Photograph of a direct-fired flame dryer used in the fish meal industry.

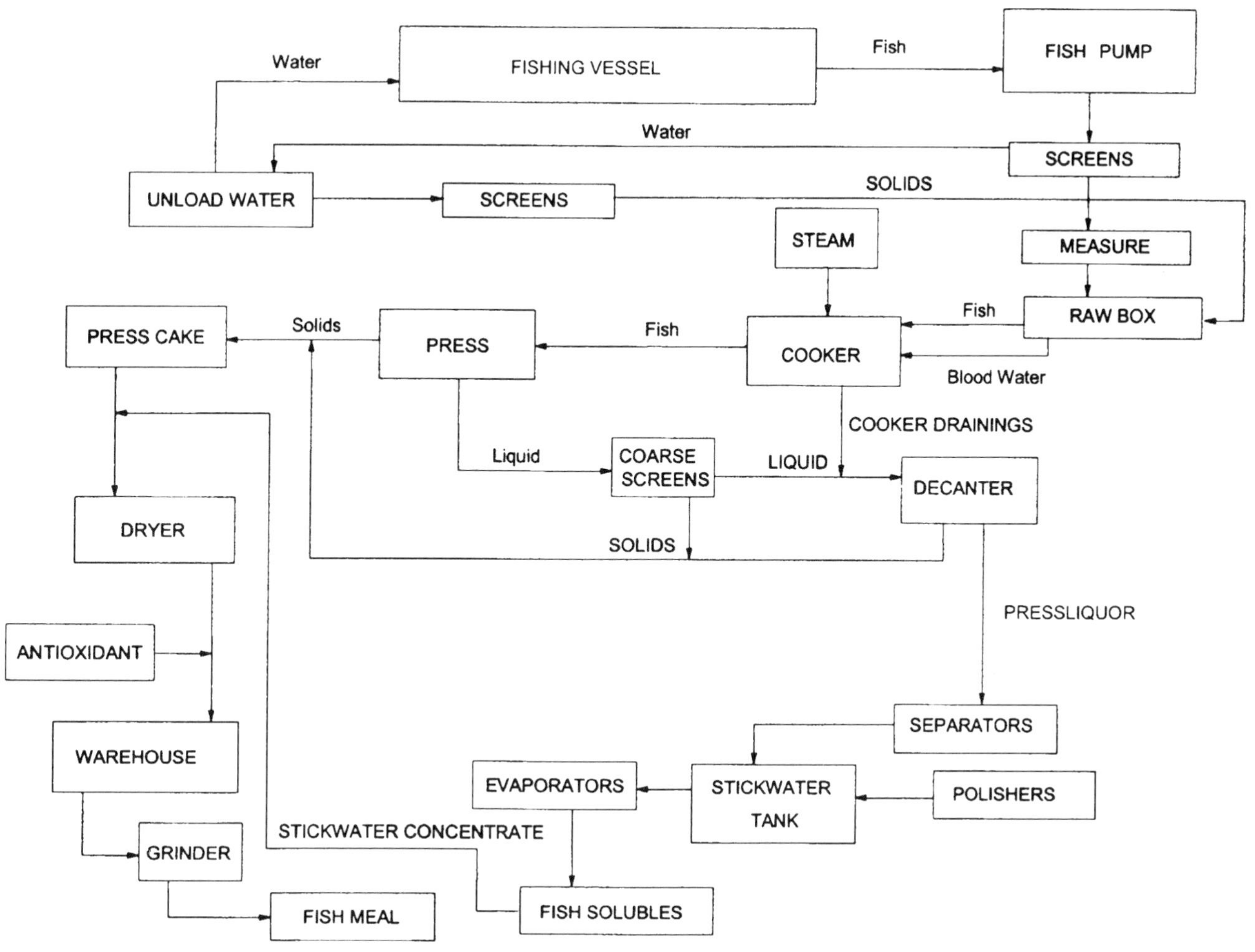

Figure 11. Solids flow in a typical fish-meal plant.

Antioxidant Addition

All fish meals will react with oxygen to some degree. The degree of reactivity of the meal is dependent upon a number of items, but usually the amount of fat and its level of unsaturation are the key issues. Reactive fish meals are stabilized by the addition of antioxidants immediately after the meals leave the dryers. The meal is first cooled to a temperature below the vaporization temperature of the antioxidant, and then stabilized with the antioxidant. The amount of antioxidant required to prevent this spontaneous heating depends upon the type of fish that has been processed and the degree of unsaturation of the lipid (fat) portion of the meal. Northern species of fish with relatively low unsaturation in the fat, such as herring or capelin, require low concentrations of antioxidant, while southern species such as anchovy, pilchard, and menhaden require higher concentrations of the antioxidant. Very careful control is necessary when the antioxidant is added because the amount is quite small (typically 0.75 to 2.0 pounds/ton of fish meal), and dispersion of the chemical in the meal becomes a critical factor. Normally, the antioxidant is added to the meal in a screw conveyor so that there is thorough mixing as the meal is conveyed. Automatic control devices and variable speed pumps are used to assure that changes in the rate of production receive the corresponding changes in antioxidant addition. Meal exiting from the dryer is usually too hot for the addition of antioxidant and it is first necessary to cool the meal below the vaporization point of the chemical. This cooling assures that the correct dosage remains in the fish meal. It has been reported that some factories add the antioxidant to the presscake prior to drying, but very little data is available.

Antioxidants are free radical acceptors that break the peroxide reaction chains, perhaps the best way to stabilize fish meal. Oxidation is checked and the lipids in the fish meal remain fully available in the finished feed. The effectiveness of the antioxidant is measured by how quickly the product can be stored in bulk or bags with little or no turning of the piles. Ethoxyquin, 1-2 dihydro-6 ethoxy-2,2,4-tri-methylquinoline, is the antioxidant of choice throughout the fish-meal industry.

Studies conducted throughout the fish-meal industry, not only in the United States but also in South Africa, Canada, and Peru during the mid-1960s, indicated that ethoxyquin was about eight to 10 times more effective in stabilizing fish-meal than was BHT (butylated hydroxy-toluene), which was then being used (Bimbo, 1990). Since that date, no other antioxidants have been used for the stabilization of fish meal, although many others have been and will continue to be evaluated (Chahine, 1978). In recent years certain users of fish meal have indicated a desire to have a meal with a natural antioxidant present. Research on this possibility continues in many areas, but so far no system has been found as effective as ethoxyquin.

Storage and Shipping

Storage methods for fish meal depend on many factors, including climatic conditions, production capacity, use of antioxidants, and transport and marketing arrangements. Factories usually have a storage capacity for a reasonable quantity of finished product. During times of difficult marketing conditions or glut catches of fish, it may be necessary to go to "outside" locations away from the actual processing plant. Fish meal must be stored in weatherproof, well- ventilated spaces, with a clear space between walls and the product piles or stacks.

Over half of the world's fish meal is stored in bulk, either in sheds or silos. Bulk storage is advantageous because:

1. All handling, from production to loading, becomes simpler and cheaper, and results in considerable savings in manpower and maintenance over the facilities used in bagging fish meal.
2. Most transport vessels and international receiving centers are geared to the handling of bulk, so factory bulk storage is compatible. Facilities for bulk storage are either of the open type (access of air through doors and windows) or of the sealed type such as silos. The open shed or warehouse predominates in the fish-meal industry today and is equipped with floor and overhead conveyors for turning the fish meal. The sheds may be of single or multi-unit construction with concrete walls and floors.

Silos offer good protection to meal in storage. Specially designed silos keep the meal in motion by continuously extracting the meal from the bottom of the pile and returning it to the top by means of automatic conveyor systems. This procedure keeps the meal from compacting and/or bridging.

In addition to bulk storage, fish meal may be bagged and stored on pallets. The bags usually hold 50 kilograms or 100 pounds, and may be open-ended and stitched, or have valves that are tucked in. Bag material ranges from Hessian to multilayer paper to woven plastic, with and without a plastic liner. Hessian and julep bags are being replaced by woven plastic bags. The paper bag keeps insects and rodents out, retards oxidation and absorption of moisture, and does not leak unless broken. While the barrier offers excellent protection of the product from external sources, the product could become moldy if the moisture content of the meal in the bag is 8 percent or higher. The moisture tends to migrate outward and condense, forming wet spots that can cause mold growth. Wooden pallets holding approximately one ton are often used to facilitate handling and storage of the bagged meal. Pallets may be stacked three high using forklift trucks after the meal has cooled (FAO, 1986; Dreosti, 1980). New packaging systems called bulk bags or ocean-going totes are designed to hold one ton of product in a protected package, suitable for export. Since the bag holds so much product, there is less human handling; the bags are dumped or ripped open to allow the product to fall into a conveyor for storage in bins or for direct feed to the blending operation.

THE WET RENDERING PROCESS, PART 2: THE PRODUCTION OF CRUDE FISH OIL

Liquid Flow

During the pressing operation, two intermediate products are produced: presscake and pressliquor. The pressliquor that is squeezed from the cooked fish contains coarse particles of fish and bone that must be removed before the liquor can be centrifuged. Removal of these solids is accomplished by passing the liquor over a vibrating screen with 5- to 6-millimeter perforations. The recovered solids go back on the presscake and are dried.

Separation of the screened presswater is then carried out in three steps. These steps involve:

1. Decanters: remove fine suspended solids, including sand, from the pressliquor in order to obtain a liquor suitable for the separation step.
2. Separators: remove as much oil as possible from the pressliquor and thus produce a stickwater with the least amount of fat.
3. Polishers or purifiers: remove the final traces of moisture and impurities from the oil prior to its pumping to storage.
4. In some plants there is also a fourth step: the subsequent separation of oil (deoiling) from the solubles after partial evaporation (Bimbo, 1990a).

Solids Removal

Decanters are cylindrical bowls with a cylindrical conveyor turning inside. The pressliquor is pumped into the bowl, and the solids are forced to the outer periphery and conveyed out of the system. The conveyor turns at a slower speed than the bowl and, through a combination of conveyor speed and liquid depth, the desired clarifications can be achieved (Gloppestad, 1979). Solids removed from the decanters are mixed with the presscake and dried.

Oil/Water Separation

Pressliquor discharged from the decanters is pumped into a holding tank, heated if necessary, and then fed either by gravity or by pump to the separators. Today, the separator is a modern, high-capacity machine capable of handling many times more gallons of feed per hour than the machines described by Smith in the 1940s. These machines

Figure 12. Photograph of a typical oil room with self-cleaning separators in a modern fish factory.

are self-cleaning; that is, either they have a sensor in the bowl capable of interrupting the separation cycle long enough to discharge the solids accumulated in the bowl or they discharge the contents of the bowl on a timed cycle. The machines make a three-phase separation of the pressliquor into an oil phase, a water phase, and a sludge phase. The sludge phase is pumped away either to the cooker or to the presscake where it is dried back on the fish meal (Bimbo, 1990a). Figure 12 shows a typical oil room with self-cleaning separators in a modern fish-meal factory.

Oil Polishing or Purification

The oil phase recovered from the separators is continuously washed and separated into two phases, water and crude fish oil. The water phase is mixed with the stickwater and evaporated. The fish oil is pumped into storage tanks where it is tested and sold for a variety of uses. The crude oil can also be further refined and processed into a number of other food and industrial raw materials (Bimbo, 1989, 1989b, 1988). A typical flow diagram of the liquid phase of fish reduction is shown in Figure 13.

THE WET RENDERING PROCESS, PART 3: THE PRODUCTION OF CONDENSED FISH SOLUBLES

The third part of the wet rendering process is the production of condensed fish solubles or stickwater concentrate. Production of fish solubles begins with the separation of oil and water in the production of fish oil as previously described.

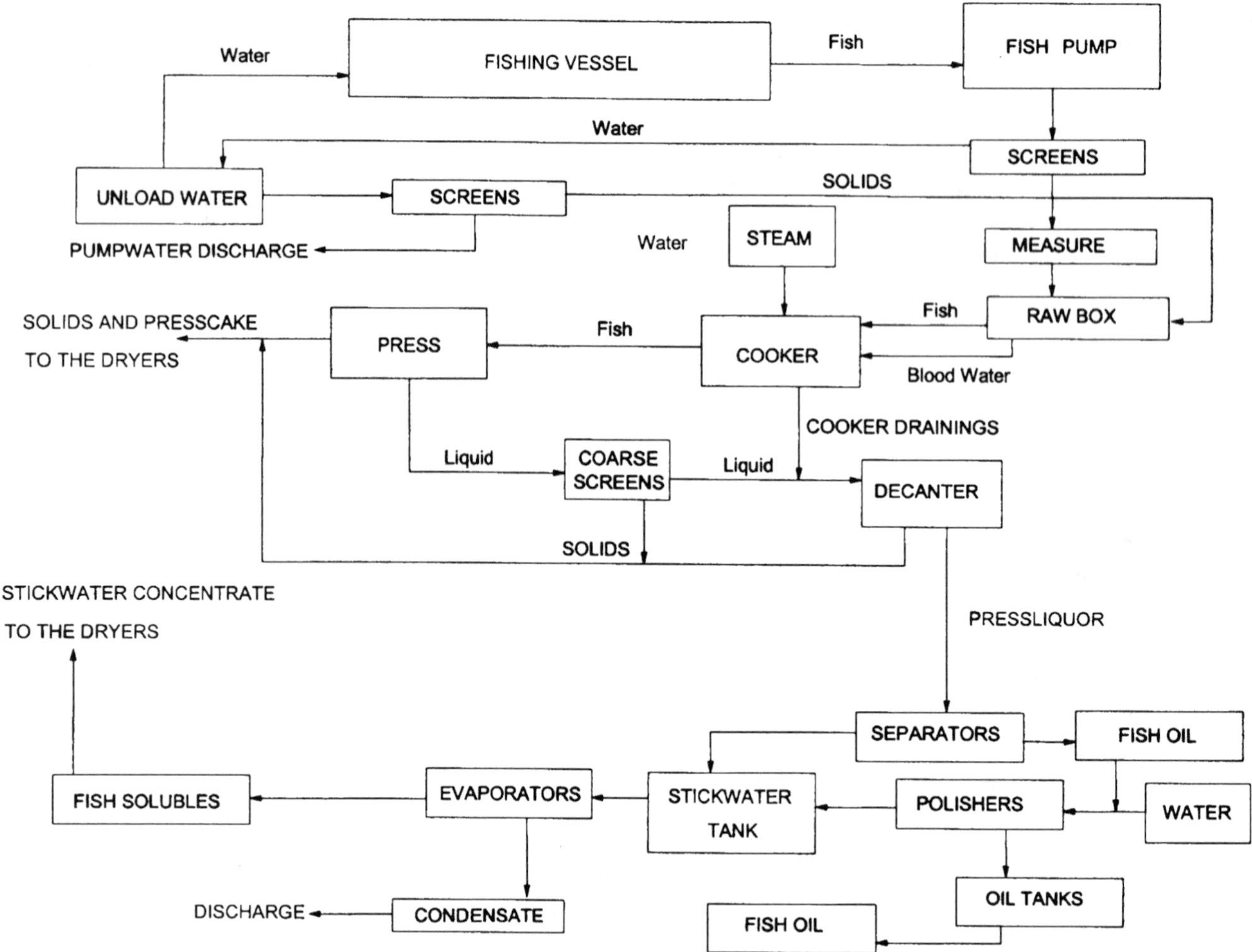

Figure 13. Flow diagram of the liquid phase in a fish reduction plant.

Evaporation

The water phase, which is rich in soluble proteins and vitamins, is stored in large holding tanks and then pumped to continuous vacuum evaporators that concentrate the solids content to 50 percent. This concentrate is called fish solubles. Today both high-speed multi-effect and falling film evaporators are capable of producing product as fast as the water is produced, thus ensuring that fresh product is added back onto the fish meal. A typical falling film evaporator can be seen in Figure 14. These evaporators are usually balanced with the steam dryers so that steam exhaust from the dryers heats the initial stage of the evaporator. The stickwater concentrate may then be added back on the presscake and dried to produce whole or full meal (Hovad and Lorentzen, 1978; Onarheim, 1978).

Oil separation from partly-concentrated stickwater is practiced by some manufacturers. The density of the stickwater is higher in the concentrated than in the dilute state. This greater difference between the densities of oil and stickwater produces an increase in the centrifugal potential and thus contributes to extra oil removal. Consequently, oil removal from the concentrate leads to a leaner wholemeal and increases the oil yield. The separated oil tends to be rather dark in color, high in free fatty acids, and of less value than oil separated from the pressliquor (FAO, 1986).

A typical flow diagram of a complete wet reduction fish processing plant is shown in Figure 15.

Figure 14. Photograph of a typical falling film evaporator.

OTHER PRODUCTION METHODS

Several alternative production methods are primarily designed to produce fish meal or its equivalent, with the oil phase as a minor byproduct. They are mentioned here because they have some general use for the recovery of protein and oil from fish and fish waste; occasionally these uses come up for review.

Dry Rendering

The dry rendering process that is commonly used to prepare meat and bone meal is not normally used in the manufacture of fish meal and oil. If used at all, it would find some limited use when the raw material is extremely low in oil. The process is usually batch type, involving a combined cooking and drying step. The cooker is usually a steam-jacketed, paddle-stirred rotary-operated device, sometimes operated under vacuum. The cooked and dried product is then conveyed

FISHING VESSEL
Fish
FISH PUMP
Water
Water
SCREENS
UNLOAD WATER
SCREENS
SOLIDS
DISCHARGE
Water
STEAM
MEASURE
PRESS CAKE
Solids
PRESS
Fish
COOKER
Fish
RAW BOX
COOKER DRAININGS
Liquid
COARSE SCREENS
LIQUID
DECANTER
DRYER
SOLIDS
PRESSLIQUOR
ANTIOXIDANT
CONDENSATE
SEPARATORS
FISH OIL
WAREHOUSE
EVAPORATORS
STICKWATER TANK
POLISHERS
WATER
STICKWATER CONCENTRATE
GRINDER
OIL TANKS
FISH MEAL
FISH SOLUBLES
FISH OIL

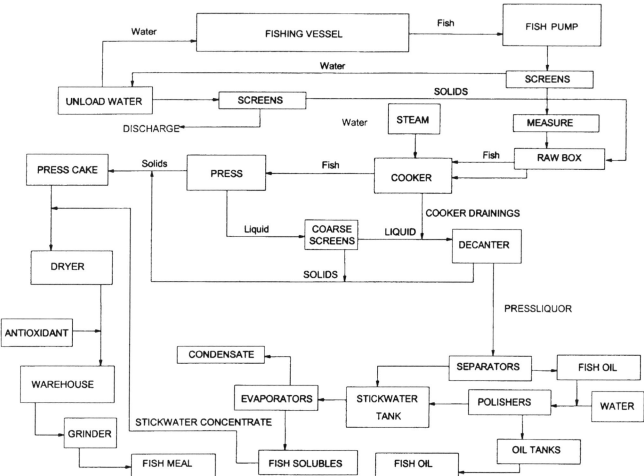

Figure 15. Flow diagram of a complete wet reduction fish-processing plant.

to a hydraulic press where the oil is removed. The recovered oil is generally dark in color and must be further refined before it can be used (Bimbo, 1990b; Pigott, 1967).

Silage Production

Fish silage is liquified fish stabilized against bacterial decomposition by an acid. The process involves mincing of the fish, followed by the addition of an acid for preservation. The enzymes in the fish break down the fish proteins into smaller soluble units, and acid helps to speed up their activity while preventing bacterial spoilage. Formic, propionic, sulfuric, and phosphoric acids have been used. Normally, about 3 to 4 percent of acid is added so that the pH remains near 4.0. Strong mineral acids require neutralization before feeding the final product. The silage process is outlined in Figure 16. Silage might be defined as a crude form of hydrolyzate.

Silage made from white fish offal does not contain much oil, but when made from fatty fish such as herring it is necessary to remove the oil. The composition of the silage will be very similar to the material from which it is made. Fish silage of the correct acidity is stable at room temperature for at least two years without decomposition. The protein becomes more soluble, and the amount of free fatty acids increases in any fish oil present during storage (Tatterson and Windsor, 1974). Silage production offers a solution to the handling of fish waste when the logistics of delivering to a fish reduction plant are not economical. Silage can be produced in large or small containers both on the vessel and on shore (Bimbo, 1990b).

POLLUTION CONTROL

Pollution control in the fish-meal industry can be divided into water and air pollution categories.

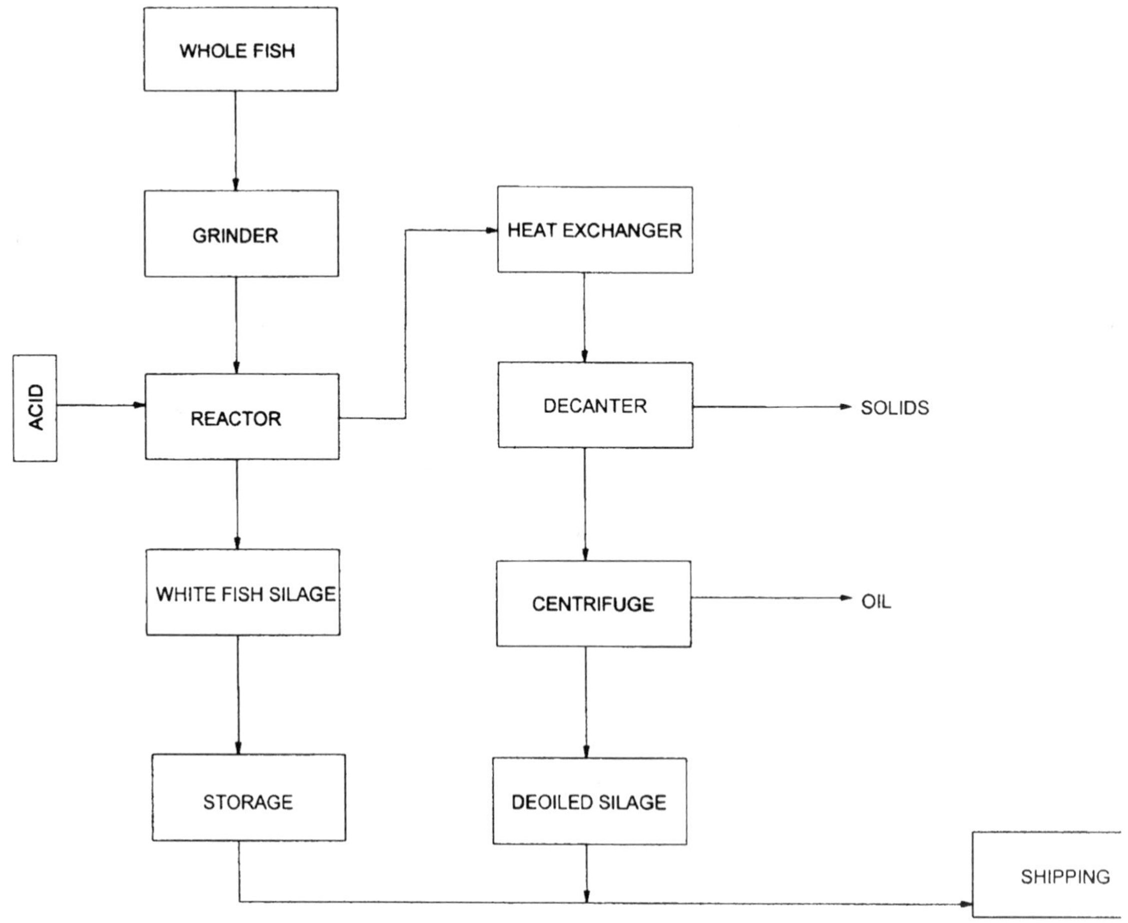

Figure 16. Flow diagram of the silage process.

Water Effluent Control

The sources of wastewater produced in the fish-meal processing industry can be summarized as follows:

1. Pump water (fish-unloading water)
2. Stickwater
3. Blood water
4. Equipment wash-up waters
5. Boiler blowdown water
6. Evaporator condensate water
7. Evaporator cooling water or condenser water
8. Air-scrubber water
9. Miscellaneous waters

Whether a plant will have all these types of discharges or not depends upon the type of process being used, and the degree of modernization of the plant. Normally, liquid effluent from fish-meal plants is contaminated with proteins and fats that exert a biochemical oxygen demand on the body of water into which they are discharged. Control of these effluents therefore usually involves recovery of these valuable commodities. Today, legislation is becoming more restrictive and new methods are being developed to further polish the final liquid effluents being discharged from fish-meal plants (FAO, 1986; Claggett, 1978; Cullinane, 1978; Nachenius, 1978).

In addition to the environmental issues, the discharge of effluent that contains valuable byproducts does not make good economic sense. Depending upon the condition of the raw material when the vessels arrive at the factory, the pumpwater could contain as much as 20 percent of the solids from the fish, and the blood water (the liquid material that exits the fish while it is in storage) could contain another 5 to 10 percent. Recovery of these solids usually returns a payback in less than one year.

Gaseous Effluent Control

Smells emitted from fish-meal plants and other processing factories formerly caused less concern than they do today. Fish-meal plants may be threatened by closure for this reason alone, though public reaction is sometimes tempered by the dependence of the local community on fishing activities. Production of bad smells affects the public image of the industry. Permits to construct new plants are sometimes difficult or even impossible to obtain. It is not that the emissions from the process are harmful to health; the odorous substances are present at such low concentrations that the question of toxicity does not arise, but the substances responsible are so intensely odorous that they can often be detected far from the factory. The sources of gaseous effluent in a fish-meal plant are as follows:

1. The raw material unloading, transfer of fish to the factories, and storage conditions at the factory.
2. Processing: cooking, pressing, de-oiling, and evaporation are carried out at elevated temperatures, and odorous compounds are produced.
3. Drying: probably accounts for 60 to 80 percent of the total emissions from a fish-meal plant. Direct flame dryers produce more gaseous effluent than do indirect steam-heated dryers.
4. Pneumatic conveying and grinding.

Odor control in fish-meal plants can include one or several of the following steps:

1. Condensation of the dryer exhaust to remove water and decrease the air volume.
2. Incineration in the boilers.
3. Chemical oxidation.

All of these methods have been reviewed [FAO, 1986; Cambell, 1978; Hansen, 1978; Onarheim and Utvik, 1978; Wignall, 1978; International Association of Fish Meal Manufacturers (IAFMM) News Summary, 1977].

MARKETS

Fish Meal

Fish meal is a major source of protein and as such competes with other protein sources worldwide. Figure 17 compares the average world fish-meal production with that of other major protein meals for the period 1992 to 1997 (forecast).

Fish-meal production is a major industry in many countries. The major fish meal-producing countries together with their five-year production average can be seen in Figure 18.

Fish meal is used in the feeds of poultry, pigs, ruminants, fish, crustaceans, pets, and fur-bearing animals because it increases productivity and improves feed efficiency. Figure 19 shows the estimated world consumption of fish meal by species being fed, estimated for 1995 and 2010 (Pike,

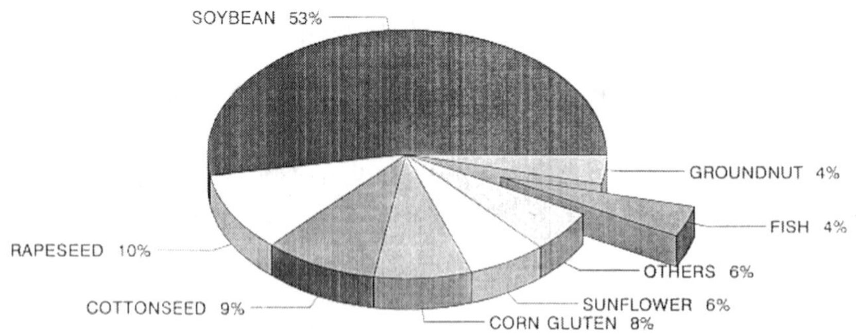

Figure 17. The average world fish-meal production vs. major oilseed protein meals, 1992 to 1997F.

1991). Fish meal has been used as a feed ingredient for farm animals in the United States for more than a century. It provides a unique balance of essential amino acids, energy, vitamins, minerals, and trace elements that complement the deficiencies of other feed ingredients. Fish meal is also a good source of the amino acid taurine and the essential fatty acid arachidonic acid, which are needed for cat nutrition. In addition to being a major source of energy, the residual fat in fish meal is a rich source of omega-3 fatty acids, which represent over 30 percent of the total fatty acids present

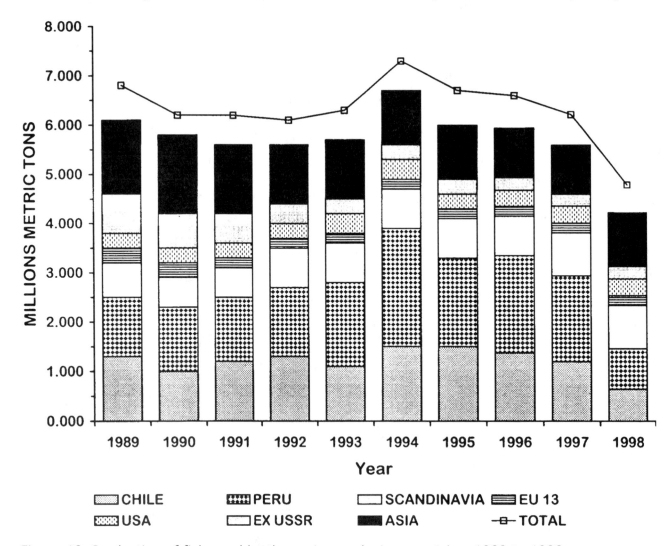

Figure 18. Production of fish meal by the major producing countries, 1989 to 1998

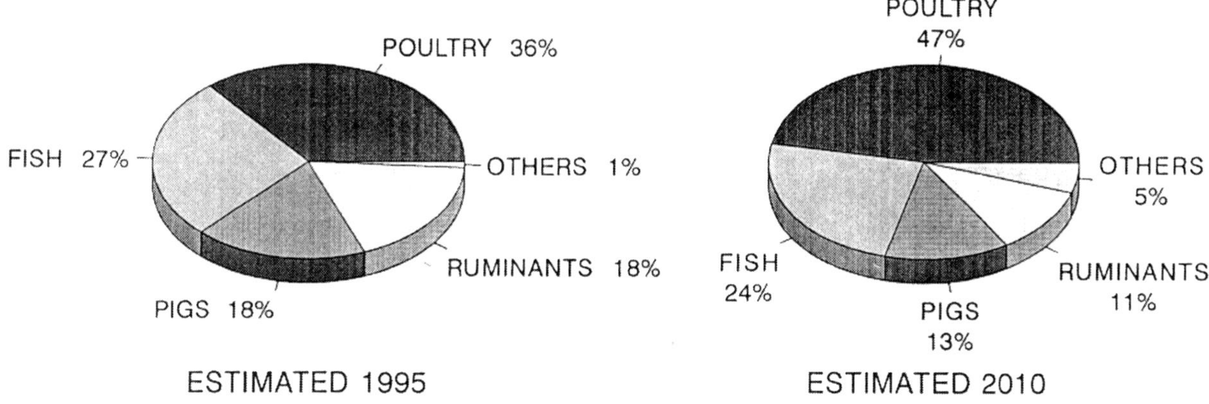

Figure 19. World consumption of fish meal by species for 1995 and 2010 (estimated).

(Bimbo, 1992a, 1989a). Table 2 gives typical composition data for a number of different classes of fish meal currently available on the market (National Fish Meal and Oil Association, 1995; Bimbo, 1992; Mantysaari, 1989; Pike, 1987; Opstvedt, 1985; Barlow and Windsor, 1984; Rumsey, 1982).

The introduction of new equipment and processing techniques in the production of fish meal continues to provide the industry the flexibility to produce proteins that are tailored for particular animals. Special quality fish meals are now available for ruminants, farmed fish, and early-weaned pigs. The freshness of raw material is an extremely important criterion for all these special meals since the profitability of feed use can vary with the freshness of the raw material used. In Denmark, a penalty payment system in which Danish fishing vessels are paid according to the quality of the raw material landed at their factory was initiated in 1986. Originally designed to reduce plant odors from poor quality fish, the system soon demonstrated a higher-quality finished product with higher yields for less cost (Madsen, 1986).

Apart from freshness, however, monogastrics and ruminants have quite different requirements for feed ingredients. Some producers now make a special quality fish meal using very fresh raw material processed through cookers and dryers, at temperatures 10°C to 20°C below normal processing temperatures. Trials in Norway have demonstrated that the processing temperature affects the digestibility of fish meal when young mink are used as the test animal. In other experiments in Norway using Atlantic salmon, drying temperature affected weight gain over 18 weeks of feeding (Pike, 1990).

Poultry-Broilers

The common broiler chicken has been the major consumer of fish meal, and has not been considered a rich source of omega-3 fatty acids in the diet. However, recent research in the United States and Canada indicates that chicken can be comparable to cod as a source of omega-3 fatty acids (Hulan et al., 1990, 1989, 1988; Ackman et al., 1988; Hulan et al., 1988, 1984). These researchers fed high levels of herring (white and red) fish meals to broilers over extended periods of time and then evaluated the carcass meat for flavor and fatty acid composition by gas chromatography. Their results indicate that significant levels of omega-3 fatty acids can be incorporated into poultry meat without affecting meat flavor.

United States Department of Agriculture (USDA) researchers evaluated United States food composition data and reported that the decline in the amount of the fish omega-3 fatty acids in the diet contributed by fatty fish has been offset by increased consumption of these fatty acids from poultry. They evaluated data for random periods from 1935 to 1985 and concluded that the increased poultry omega-3 content came from the use of fish meal in the poultry feeds. Thus, low concentrations of omega-3 fatty acids in foods consumed in large quantities can make an important contribution to overall omega-3 intake (Raper and Exler, 1990).

Table 2. Typical Composition Data for Different Classes of Fish Meal				
	Herring	White Fish	Anchovy	Menhaden
Proximate Analysis, %				
Crude Protein	71.9	64.5	66.4	61.25
Ether Fat	7.5	4.9	9.7	9.1
Moisture	8.4	10	8.6	8
Ash	10.1	20	15.4	18.8
Protein Characteristics, % of Crude Protein				
Rumen degradable	48.8	53.3	48.5	50.5
Water soluble	19.8	8.9	18.3	15.5
Energy content, MJ/KG				
Poultry, M.E.	13.7	11.6	13.5	12.8
Pigs, D.E.	18.1	15.6	16.9	16.5
Ruminants, M.E.	16.4	13.4	13.1	12.8
Fish, M.E.	17.0	16.5	16.5	16
Amino Acids, % of Protein				
Lysine	7.73	6.90	7.75	7.43
Methionine	2.86	2.6	2.95	2.63
Cystine	0.97	0.93	0.94	0.90
Tryptophan	1.15	0.94	1.20	0.78
Arginine	5.84	6.37	5.82	6.01
Phenylalanine	3.91	3.29	4.21	3.55
Threonine	4.26	3.85	4.31	3.98
Minerals				
Calcium, %	1.95	8	3.95	4.87
Phosphorus, %	1.5	4.8	2.6	2.93
Sodium, %	0.42	0.77	0.87	0.61
Magnesium, %	0.11	0.15	0.25	0.20
Potassium, %	1.20	0.90	0.65	0.81
Iron, ppm	150	300	246	1019
Copper, ppm	5.4	7.0	10.6	5.5
Zinc, ppm	120	100	111	84
Manganese, ppm	2.4	10	9.7	41
Selenium, ppm	2.78	1.5	1.39	2.21
Vitamins, ppm				
Panthothenic Acid	30.6	15	9.3	8.8
Riboflavin	7.3	6.5	2.5	4.8
Niacin	126	50	95	55
Choline	4396	4396	4396	4396
B12	0.25	0.07	0.18	0.06
Biotin	0.42	0.08	0.26	0.26
Essential Fatty Acids, % of Fatty Acids				
C18:2n-6	2	1	1	1
C18:3n-3	1	1	1	1
C18:4n-3	2	2	2	2
C20:4n-6	1	na	1	1
C20:5n-3	6	12	16	12
C22:5n-3	1	2	2	3
C22:6n-3	13	19	14	9
Total N-3 Fatty Acids	23	35	34	30

Ref: National Fish Meal and Oil Association, 1995; Bimbo, 1992; Mantysaari and Sniffen, 1989; Pike, 1987; Opstvedt, 1985; Barlow and Windsor, 1984; and Rumsey, 1982.

Ruminants

Ruminants have always been important in the utilization of land for food use because of the microflora that exist in their fore-stomach or rumen. Ruminants cannot efficiently utilize nutrients with a low fiber content such as grain and oilseed meals. The microorganisms allow the ruminant to utilize nutrients with a high crude fiber that cannot be used by monogastric animals. The microorganisms convert a substantial part of the fiber in the feed to volatile carboxylic acids. Proteins are also broken down into ammonia that is then used by the rumen microorganisms (Pike, 1981). These products of microbial metabolism are absorbed directly through the rumen wall, while the remaining microbial biomass and feed are passed to the abomassum where they are utilized in the same manner as in monogastric animals.

Feeding fish meal to ruminants has been done for some time in the United Kingdom and northern Europe, but is a relatively new concept in the United States. Cooked proteins tend to be relatively resistant to degradation by rumen microorganisms; thus that protein escapes or bypasses the rumen, and provides the animal with a source of high quality protein. Since fish meal is cooked, it has a fairly high "bypass" value. In cases where fish meal has been added to the diets of high-producing animals, some dramatic responses have been seen.

Cornell researchers Beermann et al. (1986) demonstrated significant increases in muscle size and the lean/fat ratio when 2 to 3 percent fish meal was added to the diets of lambs. Other researchers, Thonney and Hogue (1986), evaluating 3 percent menhaden fish meal in the diets of feedlot cattle, found that the fish meal-fed cattle took 30 days less time to reach market weight. Orskov (1981) reported that Scottish researchers who fed lambs and heifers straw or straw supplemented with fish meal found that the animals would utilize body fat for maintenance. Those on straw lost weight as fat and lean tissue; those on the fish-meal supplement lost fat but gained lean tissue, thus maintaining body weight. The lambs were given 75 grams/day of fish meal, while the heifers received 400 grams/day of fish meal.

The dairy cow presents a unique problem for fish-meal producers in the United States. Fish meal initially appeared to depress butterfat production under some feeding circumstances. However, studies (Broderick and Satter, 1990; Broderick, 1989; Dhiman and Satter, 1989) have demonstrated that fish meal can be used successfully with high-yielding dairy cows fed high alfalfa silage diets. Cows produced an additional three pounds or 1.4 quarts of milk daily when fed alfalfa silage mixed with one pound per day of fish-meal protein. The cows also produced milk with 4 percent more protein (*Food Business*, 1989).

Fish meal has an excellent amino acid profile, close to that required for growth and milk production, according to a review of animal byproducts as protein supplements for cattle (Stern and Marsfield, 1989).

British researchers Pike and Esslemont (1991) reported on trials in Israel and Northern Ireland in which fish meal improved the fertility of dairy cows. Higher milk production, improved income from more calves, and reduced veterinary charges were documented. In the Northern Ireland trial, the conception rate improved from 44 to 64 percent. The authors calculated that the combined cost benefit of improved fertility and improved production efficiency was worth £160-165 per cow per lactation.

Spain and Polan (1995) have reported that fish meal alters milk fat percentage when fed in diets based on corn silage. Increased concentrations of omega-3 fatty acids were measured with the inclusion of fish meal, suggesting a protection of these polyunsaturated fatty acids (PUFA) from microbial metabolism in the rumen.

Pigs

Intensive pig production methods require pigs to be early weaned at three to four weeks of age. At this age, young pigs are very sensitive to the form of protein in their diets. Many dietary proteins give an allergenic response in which diarrhea, reduced growth, and increased mortality can result. With proteins from fish, this response is low. Fish meal made from very fresh raw material is well tolerated by these young animals because it is highly digestible and palatable.

Studies conducted at Kansas State University evaluated the effect of a special quality fish meal as a protein source in starter diets for pigs. Results from the study indicated that the addition of 8 percent of this fish meal to the diets of three-week-old weaned pigs resulted in an 11.5 percent increase in average daily gain by the end of the fifth week (Stoner et al., 1986; IAFMM Fish Meal Flyer, 1987).

Aquaculture

The global use for fish meal in aquaculture in 1994 was 1,151,000 metric tons. Figure 20 compares fish-meal consumption by species (Tacon, 1996). It is estimated that aquaculture accounted for 15 percent of the consumption of the world's fish-meal production in 1994. By the year 2010 that figure is expected to rise to 24 percent. There are predictions that aquaculture could someday yield more fish than the oceans (Weddig, 1991). But there won't be a shortage of fish meal for the traditional market because the increased production of edible fish from aquaculture will naturally generate more cuttings and offal that must be converted back into fish meal.

Fish differ from domestic animals in several ways. Since their body temperature varies according to water temperature, they do not require energy to maintain a constant body temperature. They are efficient at eliminating waste through the gills in the form of soluble ammonium compounds; therefore, high-protein feeds are readily digested and have higher metabolizable energy values for fish than for warm-blooded animals (Pike, 1991, 1990).

Farmed fish, especially young fast-growing coldwater species such as salmon and trout, are very sensitive to dietary protein quality since they require very high protein diets (40 to 50 percent). Fish meal is used at high levels in these diets: 50 to 60 percent for salmon and around 30 percent for trout. The quality of the fish meal is critical: it should be produced from very fresh fish. Most fish, especially coldwater fish, have a requirement for the long chain omega-3 polyunsaturated fatty acids present in fish oil and in the residual fat present in fish meal; these products also supply omega-3s in the feeds of fish and crustaceans. Fish products have contributed importantly in animal feeding by supplying various vitamins and minerals not adequately supplied by cereal grains or plant protein supplements. Fish products have been used widely in starting rations for pigs and poultry, and in breeder rations for poultry because of their "unidentified factor" activity. Because of the rapid nutritional discoveries of recent years, followed by the equally vigorous role of the manufacturing chemist, nearly all of the vitamins and minerals critical for animal feeding are now readily and economically available. This is true for some of the amino acids as well.

Fish Oil

The fishermen of Iceland, Greenland, Scotland, and Norway have used fish oil for thousands of years. Cod liver oil was known to have some therapeutic value as early as 1657, when it was found that something in the oil helped alleviate the causes of night blindness. It was also reported during the Middle Ages that cod oil could be used to treat rickets. Between 1752 to 1784 Dr. Samuel Kay, a physician at Manchester Infirmary in England, conducted extensive clinical tests on the treatment of bone disease and rheumatism. In a paper given before the British Medical Society circa 1770, he reported that cod liver oil was effective in treating arthritis.

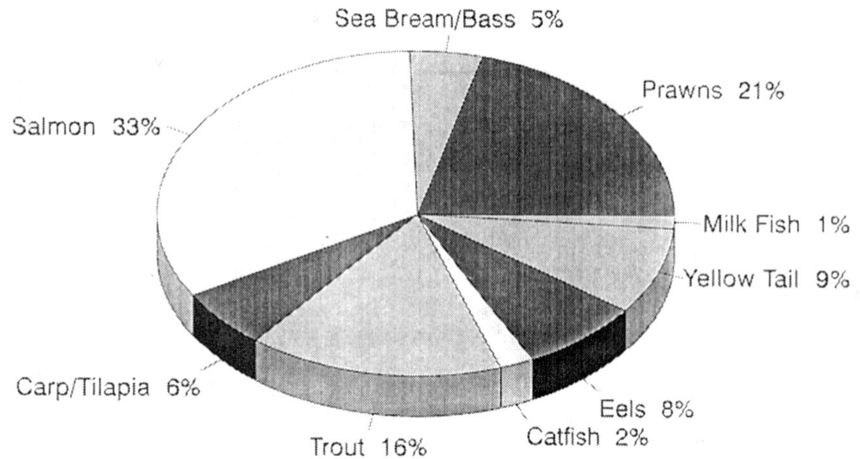

Figure 20. Estimated 1994 fish-meal consumption by aquaculture species.

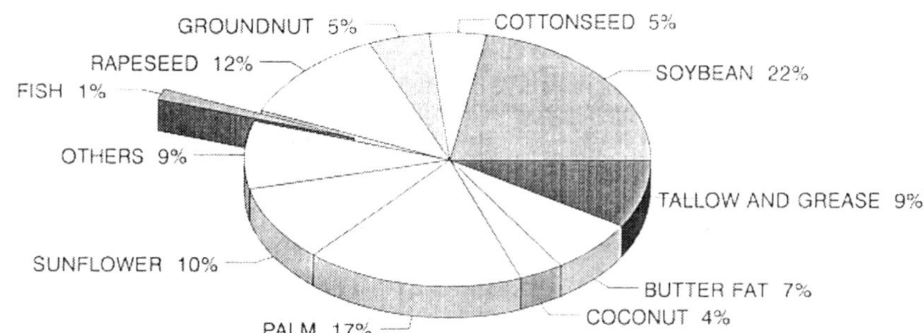

Figure 21. Average world production of fats and oils, 1992 to 1997.

The world produces more than 85 million metric tons of fats and oils in an average crop year. Figure 21 gives the average for 1992 to 1997 forecast. About 1.5 million metric tons of fish oil are produced in an average year. The predominant portion of the world's marine oil production is used in Europe, South America, and Japan. Figure 22 compares the production of fish oil by the major producing countries for 1989 to 1995. Fish oils compete with other fats and oils on the world commodity market. Figure 23 compares the price of fish oil with that of the four major vegetable oils and lard on the world market. It is evident that fish oil is the least expensive of the available edible fats and oils.

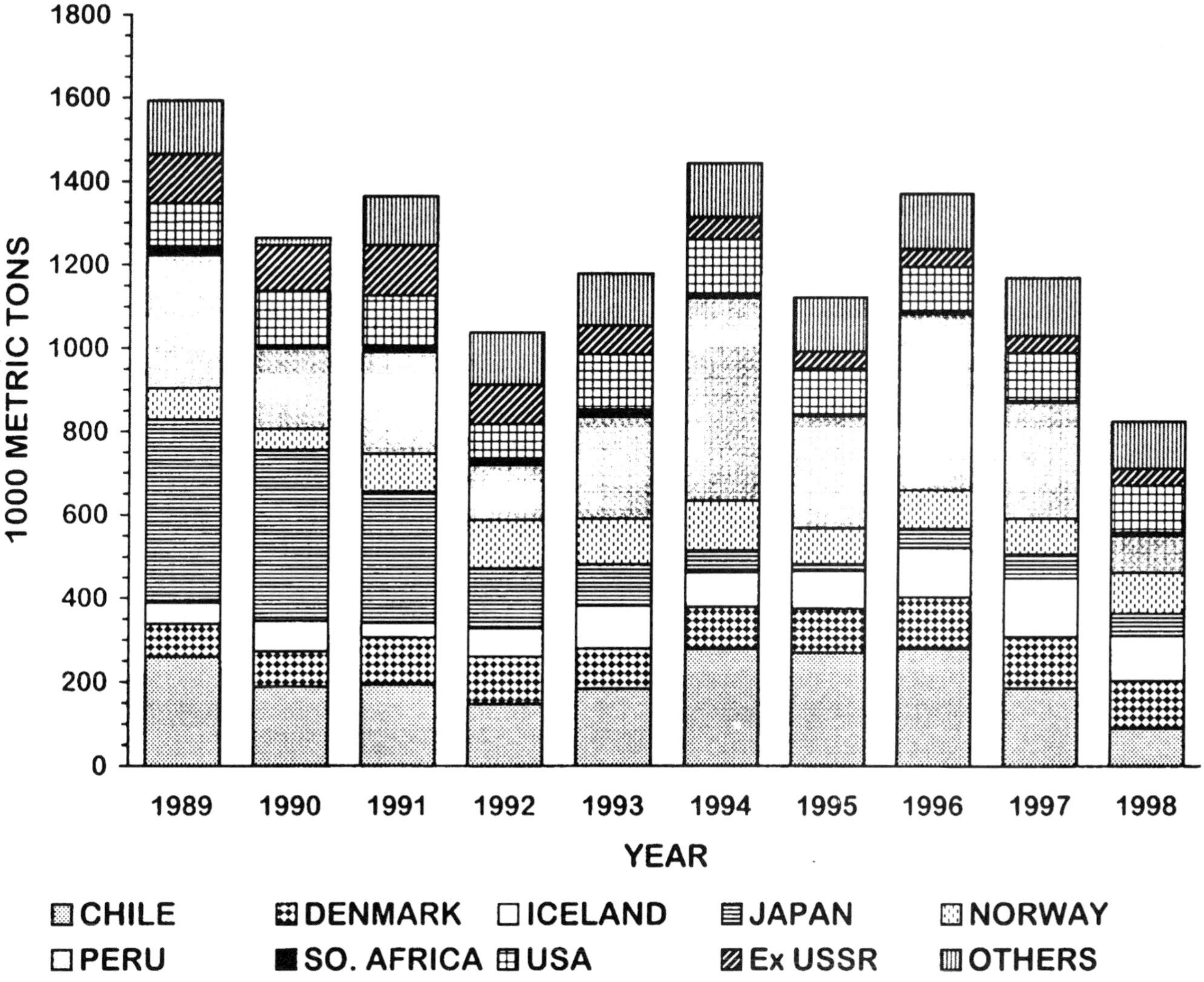

1000 METRIC TONS

YEAR

CHILE DENMARK ICELAND JAPAN NORWAY
PERU SO. AFRICA USA Ex USSR OTHERS

Figure 22. Production of fish oil by the major producing countries, 1989 to 1998.

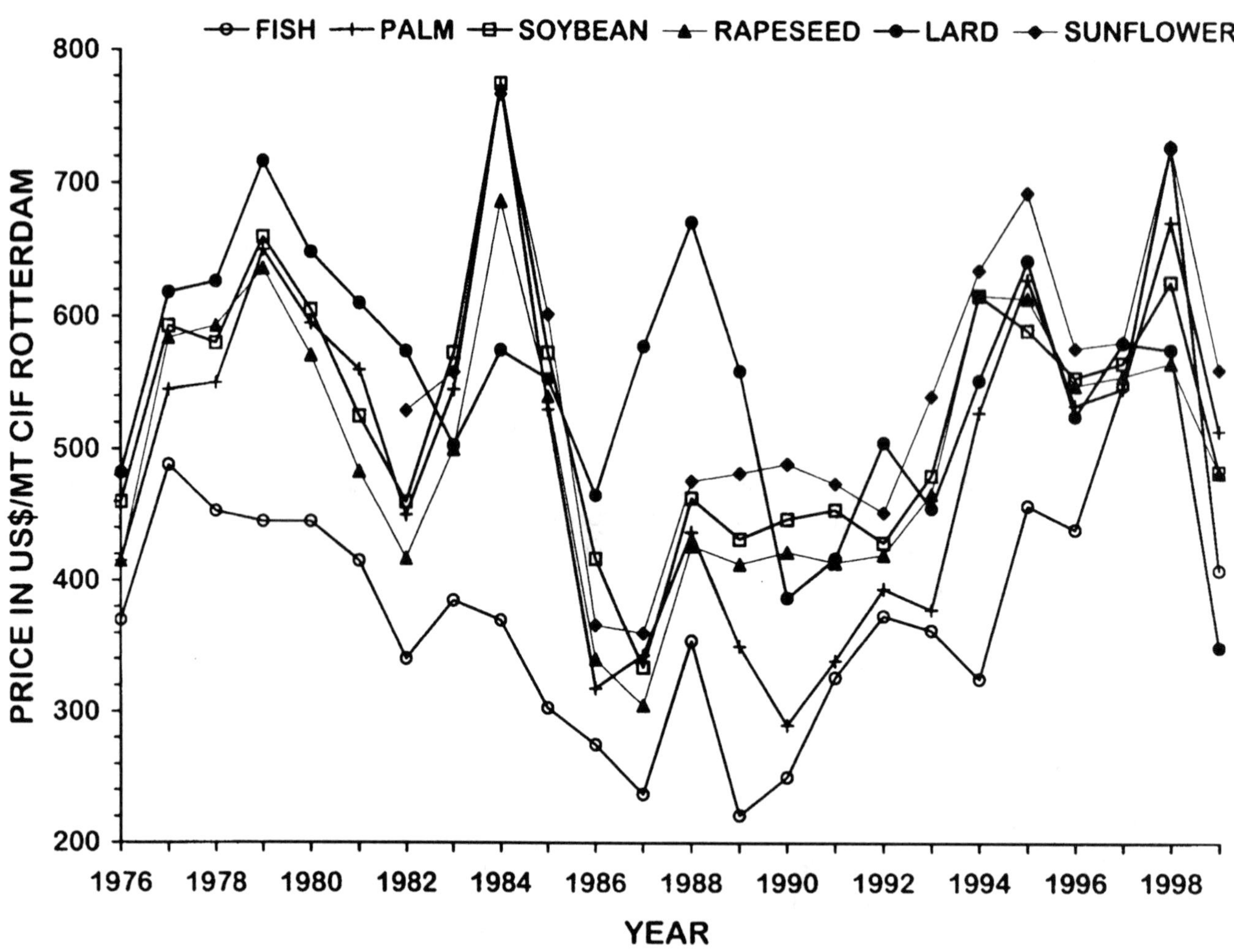

Figure 23. Price of fish oil compared with vegetable oil and lard on the world market, 1976 to 1999.

Fish oil is a versatile product that finds many applications in the food, feed, and technical industries of the world (Bimbo, 1987). Table 3 compares typical fatty acid profiles for some of the marine oils of commerce. The largest use for fish oil is in the partially hydrogenated form in Europe in the baking industry; over 80 percent is exported to Europe for edible use. The Generally Recognized as Safe (GRAS) affirmation of partially hydrogenated menhaden oil (PHMO) [United States Food and Drug Administration (FDA), 1989] has not yet resulted in a major market in the United States (even though studies have clearly shown that products produced from menhaden oil shortenings are functional) (Stauffer and Bimbo, 1994) and stable (Chapman et al., 1996). GRAS approval for refined (non-hydrogenated) menhaden oil was achieved in June 1997 (FDA, 1997), thus completing a task which took 20 years.

Edible Use

Fish oils have long been a constituent of the diet of man since they are a part of the edible portion of fish. Although food fish contain substantially less fat than land animals do, they represent a source of a different type of fat from that supplied by animals and plants. Marine fatty acids of the n-3 type have different impacts on our physiology than do the n-6 type of polyunsaturated fatty acids that are present in grains and cereals. These two families of fatty acids appear to serve different functions in human health and disease. In recent years much research has been conducted to evaluate the possible pharmaceutical uses of fish oils; this work has been reviewed several times (Endres et al., 1995; Barlow and Pike, 1991; Nettleton, 1991; Barlow et al., 1990; Simopoulos, 1988). As this research continues, it is expected that unique forms of these omega-3 oils and products

Table 3. Fatty Acid Profiles for Some of the Commercially Available Marine Oils

	Anchovy	Horse Mackerel	Menhaden	Sardine/ Pilchard	Capelin	Herring	Mackerel	Norway Pout	Sand Eel	Sprat	Tuna
C14:0	9	8	9	8	7	7	8	5	7	–	3
C15:0	1	1	1	1	–	–	–	–	1	–	1
C16:0	17	18	19	18	10	17	14	12	13	17	22
C16:1	13	8	12	10	10	6	7	4	5	7	3
C17:0	1	1	1	1	–	–	–	–	–	–	1
C18:0	3	3	3	3	1	2	2	3	2	2	6
C18:1	10	16	11	13	14	14	13	10	7	16	21
C18:2	1	1	1	1	1	1	1	1	2	2	1
C18:3	1	1	1	1	1	2	1	1	1	2	1
C18:4	2	2	3	3	3	3	4	3	5	–	1
C20:1	1	2	1	4	17	15	12	13	12	10	1
C22:1	1	1	–	3	15	19	15	17	18	14	3
C20:5	22	13	14	16	8	6	7	9	11	6	6
C22:5	2	2	2	2	–	1	1	1	1	1	2
C22:6	9	15	8	9	6	6	8	14	11	9	22
Others	7	8	14	7	7	1	7	7	4	14	6

Other fatty acids: C16:2, C16:4, and C20:4. Source: Bimbo, 1998.

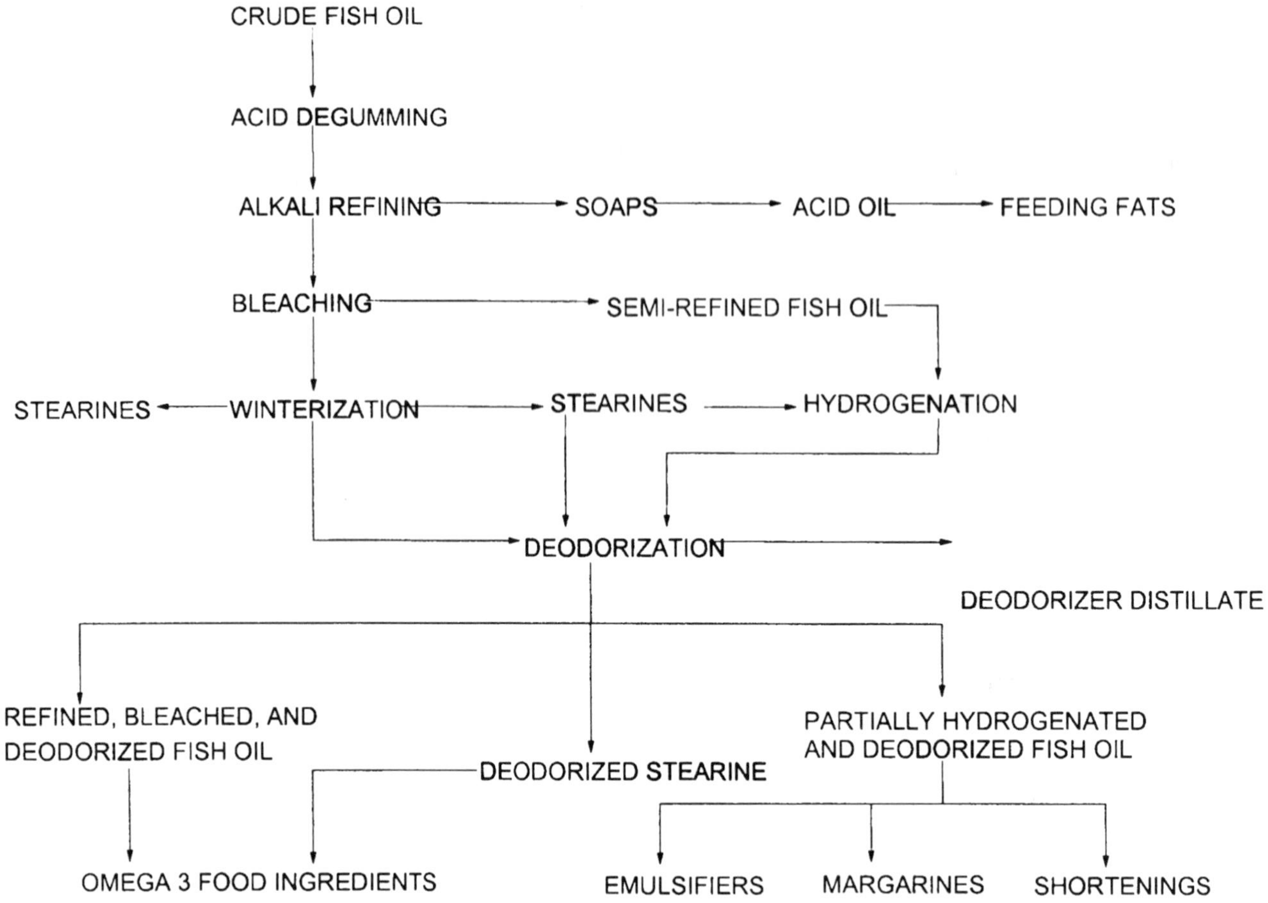

Figure 24. Processing steps needed for the conversion into food products.

will reach the marketplace (Hjaltason, 1989). This development will depend upon the capability of industry to produce crude oils that contain no oxidation or contamination (*World Fishing Magazine*, 1993a). If such challenges are met, then it is likely that the processing steps needed to prepare the crude fish oils for conversion into these unique products will follow the flow as outlined in Figure 24. Once the refined fish oils have been prepared, they would be converted into one or more of the fractions or further purified fatty acids or esters, according to the flow in Figure 25.

Hydrogenated Oils. —The majority of the fish oil used for edible purposes is first hydrogenated. During the hydrogenation process many complex and competing reactions take place, influenced by the process conditions, the extent and type of catalyst used, and the condition of the raw material. Generally speaking, as hydrogenation progresses, the melting point of the oil increases and its iodine value (measure of unsaturation) decreases. These reactions include saturation of the double bonds, geometric isomerization, and positional isomerization. Since all of these reactions can take place simultaneously, and since fish oils have a very wide range of fatty acids, it is possible to produce a family of partially hydrogenated products with unique functional properties.

The major food uses for fish oil are as ingredients in margarine, table spreads, cooking fats, salad oils, emulsifiers, and industrial baking fats used to make bread, pastries, cakes, cookies, biscuits, and imitation creams (Young 1986, 1986a, 1985). The crystal structure of a solid fat, designated by the Greek letters alpha, beta, and beta prime, is critical in some areas of fat utilization. Physically, fats in the alpha form are waxy, in the beta form coarsely crystalline, and in the beta prime form fine-grained. The beta prime form is the most desirable form because it produces a smooth texture in the end product and enhances the creaming performance (air entrainment) in industrial usage. The chain lengths of the fatty acids in marine oils extend from C14 to C24. Be-

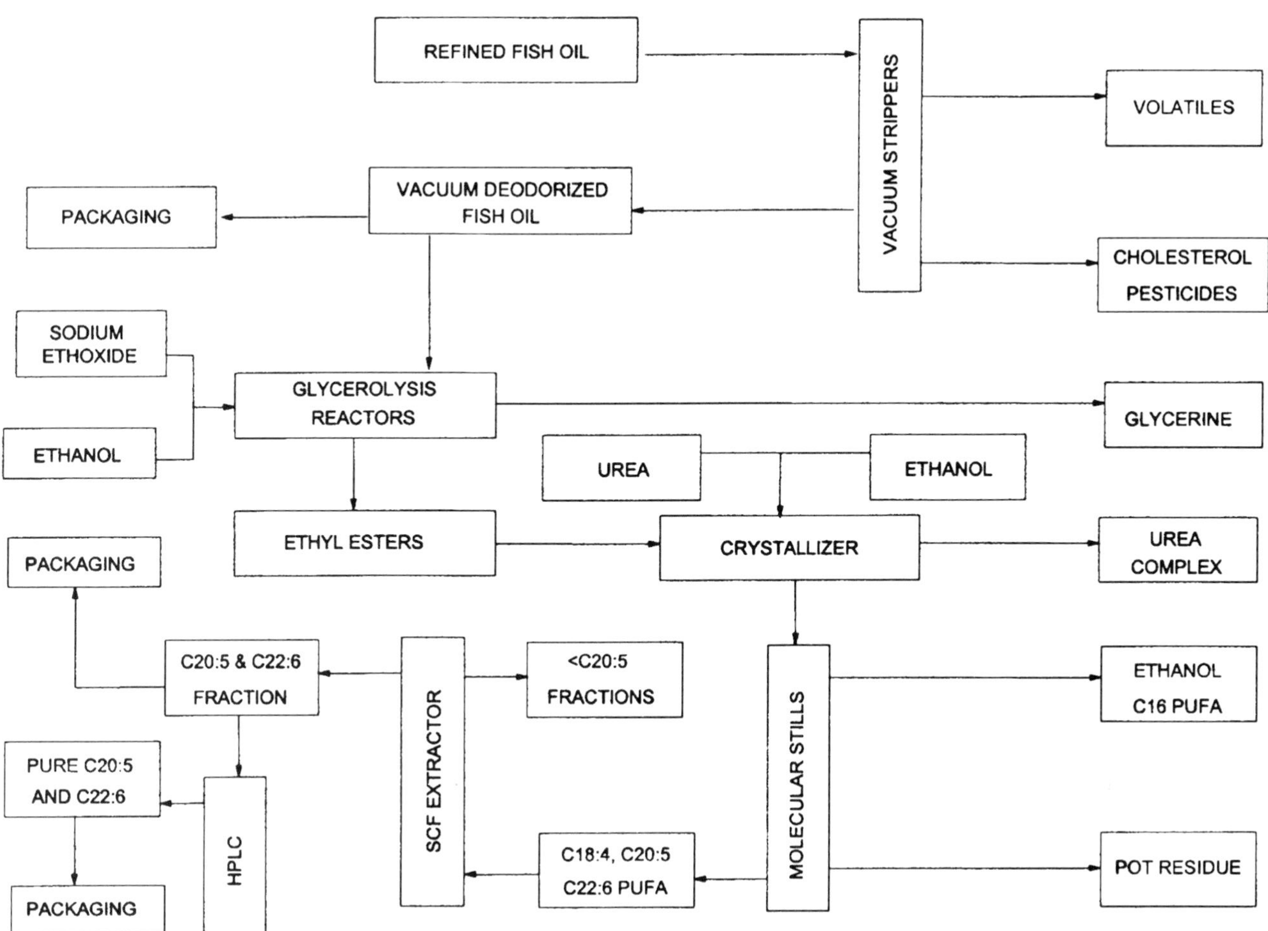

Figure 25. Process for converting refined marine oils into fractions, purified fatty acids, and ester.

cause they contain highly unsaturated fatty acids, the most unsaturated of which is docosahexaenoic acid C22:6 n-3, marine oils are usually hydrogenated for human food use. Because of the diverse nature of marine oils and the effect of different hydrogenation conditions, data depicting the principal melting points used with the respective iodine values and solid fat contents are only generally indicative of these products. As a result of the wide range of fatty acids and triglycerides formed during hydrogenation, the beta prime stabilizing effect of fish oil is very good. Addition of hydrogenated fish oils to bakery fats enhances performance of most products in which they are used. These enhancements include:

1. The incorporation of air during the creaming procedure of cake batter production, thus giving a baked product of better volume and crumb structure than is obtained from blends not containing fish oil.
2. The production of puff or flaky pastry that requires a fat or margarine that is plastic and workable but resistant to work softening (Young, 1986b).

For melting points below 38°C, the fats are very useful for retail margarines and as the middle-melting components of industrial cake and creaming products. The higher melting point-hardened marine oils increase the plastic range of shortenings. Marine oils with iodine values over 140 can be selectively hydrogenated to yield a product with an iodine value of 110 to 120. This product will have all of its fatty acid radicals with less than three double bonds. When this oil is winterized, a liquid fraction is obtained that can be blended with another soft oil such as soybean oil and used as a salad oil. This type of product is widely used in South America. The salad oil product can also be used in single-use shallow frying. Some examples of the varied uses of hydrogenated marine oils are shown in Table 4 (Bimbo, 1989b, 1989c, 1987a; Young, 1985a).

Omega-3 Refined Oils. — While the major portion of the food use of marine oils is in the partially hydrogenated form, there is continued interest in producing food systems where the n-3 fatty acids remain unchanged. Bimbo (1988), Newton (1996), and Young et al. (1993, 1991) have described the incorporation of long-chain PUFA from fish oils into various food products. There are numerous products on the market in Europe that contain fish oils. One of the most popular appears to be the omega breads that are being sold in Iceland, Denmark, Norway, and the United Kingdom. Clearly such uses offer the most viable approach to utilizing the unique nutritional properties of marine oils. There are still many challenges (Jafar et al., 1994), but progress has been made.

Animal Feeds. Fish oil is a normal constituent of many animal feeds, which often contain fish meal as a protein source. Fish meal usually contains up to 12 percent fish oil, which is metabolized and utilized as an energy source. As a nutritive component, fish oil possesses at least three potentially beneficial properties. It is a concentrated source of calories, it contains essential fatty acids, and it provides highly unsaturated fat (Gauglitz, 1973; Kifer and Miller, 1969).

Layer Hens. The consumption of shell eggs has decreased during the past decade because of recommendations by health professionals to reduce dietary cholesterol. Attempts to reduce the cholesterol level in eggs over the last 30 years has not been successful. A side effect of this work has been a change in the fatty acid composition of the yolk.

In 1987 an interesting area of feed-fat research emerged. This research involved the attempted production of low cholesterol, high omega-3 fatty acid eggs (Oh et al., 1994; Stadelman, 1989; Hulan, 1988; Yu and Sim, 1987). Some of these eggs are currently on the market. While the reduction in cholesterol never occurred, feeding 1.5 to 3 percent highly-refined menhaden oil in layer hen diets did increase the omega-3 levels in the egg yolk. Flavor is not affected until levels of 6 percent are used (Oh et al., 1994).

Lin et al. (1995) reported that dietary fats tend to produce a fatty acid pattern in eggs similar to the oil source in the diet. Most fatty acid profiles in egg yolks were altered within 15 days after hens began to consume the diets.

Table 4. Some Examples of the Various Edible Uses of Hydrogenated Marine Oils

Type of Fat	Uses
Iodine Value 100–120	
	Salad Oil, single-use shallow frying
Melting Point in degrees C	
30/32	Economic replacement for brush hydrogenated soya
32/34	Margarine, shortening, and baking fats
34/36	Deep-fat frying blends, margarine, shortening
	Biscuit cream filling, puff pastry fats
36/38	Bread fats and emulsions
40/42	Baking fats, bread fats and emulsions, industrial retail shortening, puff pastry compound fats
46/48	Baking fats, stick table margarines, industrial cake and creaming margarines, high-speed dough mixing
35/39	Danish pastry
36/40	Shortenings for biscuit dough
40/44	Industrial puff or flaky pastry margarines
46/52	Baked products eaten when hot
50/54	Anti-staling agent in bread dough
54/58	Emulsifiers

Source: Adapted from Young, 1986b.

In studies at Texas A&M University, 3 percent menhaden oil fed to layer hens over an 18-week period was compared to an isocaloric diet containing no added fat. The menhaden oil did not alter egg production, egg weight, total yolk fat, or yolk cholesterol, but the content of omega-6 and omega-3 fatty acids was altered. After one week the eicosapentaenoic (EPA) level increased, and the final ratio of omega-6 to omega-3 changed from 18 to 3 (Hargis et al., 1991). Cooking of these omega-3 eggs did not alter the fatty acid composition of the eggs, nor were functional properties of the eggs affected. Panelists were able to differentiate the n-3 enriched eggs from controls when scrambled, but not when hard-cooked (Van Elswyk et al., 1992).

Oh et al. (1991) reported that the effects of dietary eggs enriched with omega-3 fatty acids on the lipid concentrations in plasma and lipoproteins and blood pressure in 11 men and women indicated that, while cholesterol concentrations were increased in controls, there was no change in the test subjects. Both systolic and diastolic blood pressures were significantly lowered in the omega-3 groups, compared to the controls; plasma triglycerides were also decreased, compared to controls. Van Elswyk (1994) reported on the protocol of a human clinical study at Texas A & M University in which 45 subjects are consuming four eggs per week from hens fed either flaxseed or fish oil as part of the ration. The results are not yet available.

Researchers at the University of Rhode Island (Huang et al., 1990) fed diets containing up to 3 percent menhaden oil to layer hens for four weeks. Feed consumption, weight gain, and feed conversion of hens in four different groups were measured. They concluded that omega-3 PUFA in the egg yolk can be increased without causing a fishy flavor by feeding up to 3 percent menhaden oil. They attributed the lack of fishy flavor in the eggs to the fact that the oil was stabilized with an antioxidant.

Maurice (1994) reported on a series of experiments conducted at Clemson University to produce designer eggs enriched with EPA and DHA. Studies were conducted under laboratory and commercial conditions. Levels up to 4 percent of fish oil were fed to layer hens. A panel evaluated the organoleptic properties of samples of scrambled eggs from hens fed various concentrations of fish oil and could not detect changes in aroma, taste, aftertaste, or overall quality differences between controls and fish oil-modified eggs.

Broilers. Some researchers are taking a different approach to animal nutrition. Dietary lipids are one class of nutrients that offer tremendous opportunity for modifying immune responses in animals. Research indicates that the prostaglandin E2 is immunosuppressive and that the omega-3 fatty acids in menhaden oil reduce its biosynthesis. Commercial application of this modulation will require further research, especially with poultry, since this is a new area of work. Potential improvements in flock health and disease resistance may be realized by manipulating the amounts and types of fat in the diet (Watkins, 1991). Infectious diseases cause serious economic losses to the poultry industry every year. Evidence suggests that current nutrient guidelines for poultry do not optimize immune responses nor disease resistance. The immune response of chickens has been shown to be influenced by a number of nutrients. Dietary fats, in addition to supplying energy, are modulators of the immune response. A group at the University of Missouri fed linseed, menhaden, lard, and corn and canola oils and found that feeding chicks a diet containing 7 percent menhaden oil significantly enhances antibody production (Fritsche, 1991). Further work is underway, including plans for evaluations under commercial conditions at a number of other institutions (Hall, 1990).

Fish oil also affects the composition of broiler meat. Different dietary lipid sources (fish oil, safflower oil, and beef tallow) were fed to broilers for 18 weeks. The compositions of breast, thigh, and skin lipids were evaluated. Lipids in the meat reflected the lipid fed. Chickens fed the menhaden oil reflected the fish fatty acids not found in chickens fed the other diets (Nir, 1990; Phetteplace and Watkins, 1990, 1989; Adams et al., 1989).

In another study at Virginia Polytechnic Institute and State University, researchers fed linseed, menhaden, soybean, or chicken fat to chickens for 56 days. They found that linseed oil and menhaden oil resulted in similar levels of EPA in the tissue. Chicken fat gave the highest level of saturates, and soybean oil gave the highest levels of omega-6 fatty acids in the tissue (Phetteplace and Watkins, 1989).

Ruminants. Because of the bad publicity about red meats, several researchers have been attempting to decrease saturated fatty acids and incorporate omega-3 fatty acids into the flesh of red meat animals. At Southern Illinois University researchers infused 1 to 4 percent refined menhaden fish oil into the abomassum of beef cattle over a 60-day period. The results indicated that while there was no increase in carcass body fat, the omega-3 fatty acids were extensively incorporated into the muscle tissue at the expense of saturated fatty acids. There were no taste studies done and further research using protected fats is underway (Young, 1990; Barry, 1988).

A United States patent (Christensen and Storm, 1989) describes "split feeding" as a way to regulate the content and composition of fats in milk. The cows are trained to activate their esophageal groove reflex so that they can take liquid feed directly into the abomassum or concentrate into the rumen. In this way the unsaturated oils are not hydrogenated in the rumen and even with fish oils there is no off-flavor in the milk.

Pigs. Supplementing practical sow diets (corn/soy) with menhaden oil at 3.5 to 7 percent of the diet late in gestation can significantly increase the content of omega-3 fatty acids in the sow's serum, colostrum, and milk (Fritsche, 1991). The fatty acid profile of serum and tissues of the piglets was significantly affected by the fat source provided to the sow. Substituting menhaden oil for lard in a sow's late gestation and lactation diet greatly elevates the content of omega-3 fatty acids in the nursing pig and reduces PGE2 production of its immune cells, but has no effect on primary and secondary antibody responses of weanling piglets.

Aquaculture. — Animals with all the food provided under the management of a farmer have minimal choice of diet and will acquire a ratio of omega-6/omega-3 that is controlled by the farmer. Thus farmers can influence the omega-6/omega-3 balance in the people who eat their farm products, and an aquaculturist's decisions about feeding fat to fish are of more importance than just to meet the growth needs of the farmed animal. The supplies of longchain omega-3 fats that are currently obtained by hunting fish in the oceans may not be adequate in the next century. But the marine food chain offers some unique ways to increase omega-3s in the diet by the feeding of fish (Pike, 1990a; Lands, 1989).

Lipids from wild fish, particularly marine fish, contain comparatively high levels of omega-3 PUFA which they obtain in the diet by consuming plankton, algae, and other fish. Numerous investigators have demonstrated that fish are what they eat. The composition and flavor of the fat in the fish can be easily adjusted by the type of fat fed. Lousiana State University researchers Chanmugam et al. (1983) reported on the difference in fat composition of freshwater prawn and marine shrimp. Canadian researchers Ackman et al. (1990) have indicated that consumers are confused about whether to consume more high fat fish in order to increase their intake of omega-3 fatty acids or to avoid the extra fat and eat only lean white-fleshed fish. Ackman et al. state that farm-raised salmon should provide at least 5 grams of fat containing 400 milligrams of DHA and 200 milligrams of EPA per 100-gram serving in order to mimic wild salmon.

Tacon (1996) estimates that about 346,000 metric tons of fish oils were used in aquafeeds in 1994. (Figure 26 shows the estimated consumption of fish oil by species for 1994.) Many farm-raised fish are low in omega-3 fatty acids because their diets are formulated primarily from agriculture products. This deficiency can be eliminated by adding fish oil containing high levels of omega-3 to their diet. The compositions of shrimp, crayfish, catfish, eel, trout, and carp have been reviewed and it is apparent that omega-3s are going to have an ef-

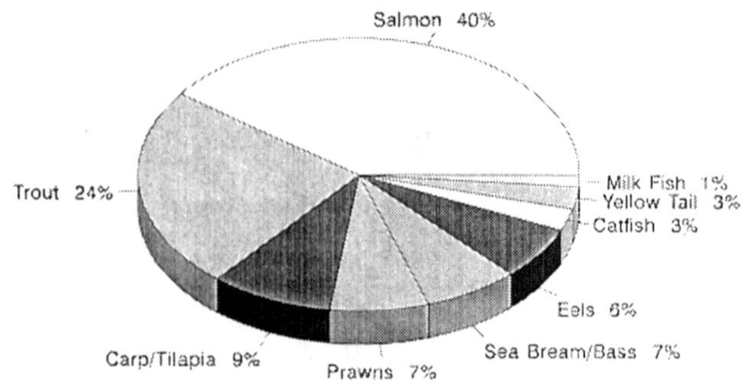

Figure 26. Estimated 1994 fish oil consumption by aquaculture species.

fect on the future of all commercial aquaculture (Pigott, 1989). The public image of aquaculture could suffer if these omega-3 fatty acids are not controlled in the fish diet. It is important for the aquaculture industry to investigate the possibility of economically altering fish-feeding programs to ensure that the omega-3s are available in their products.

An in-depth study of the nutrients and chemical residues in Mississippi farm-raised channel catfish has been reported. The total omega-3 content of these fish, 100 milligrams/100 grams, makes them an extremely poor source of these fatty acids, somewhat less than lean fish (Nettleton et al., 1990).

In a 12-week trial in Texas, catfish were fed menhaden oil at levels ranging from 1.5 to 6 percent. A practical diet supplemented with 3 percent menhaden oil is suitable for achieving maximum growth. At 6 percent menhaden oil and in the control treatments, fish did not grow as well as with the 3 percent treatment. A trained taste panel said that the catfish on 3 percent menhaden oil tasted more fishy but the flavor was not objectionable (Brown, 1990).

The consumption of fish and the fish oil it contains is believed to be beneficial to health. Some of the fatty acids in fish, in particular EPA and DHA, would appear to be of particular value. Using good quality fish oils that are adequately protected against oxidation, it should be possible to increase the content of these fatty acids in the fish lipids of farmed fish to similar or even higher concentrations than are found in the wild species (IAFMM Technical Bulletin 25, 1990).

Veterinary Use. — A new area of research with the n-3 fatty acids that crosses all species is the effect of these fatty acids on the immune system of the animals being fed the fish oil. Fritsche and Cassity (1992) and Fritsche et al. (1991) have reported on the effect of dietary fats on the immune systems of chickens. Rigau et al. (1994) and Fritsche et al. (1993) have reported on similar studies with pigs. Similar research is being done with a variety of pets, including horses.

Industrial Uses. — For many years, countries throughout the world have recognized the value of oils from marine sources for both edible and industrial purposes, and segments of the fishing industry have been developed to provide these products. The supply of marine oils has come from several sources. The tremendous quantities of whale oil, once accounting for as much as 75 percent of the total aquatic animal oil and fat production, today accounts for less than 2 percent of that production. Because of the depletion of species and the enactment of various endangered species conservation laws, whale oil has declined to the point of commercial insignificance (Gauglitz, 1973).

At the World Conference on Oleochemicals held in Switzerland in 1983, oleochemistry was described as a mature branch of chemistry with many applications for its products, but with few completely new fields. The challenge and the opportunities for oleochemistry lie in ever-changing economic and ecological conditions. The industrial applications of marine oils have not received much attention in recent years because of the overwhelming interest in n-3 fatty acids for nutritional supplements and food use. Because of the current demand for biodegradable and/or environmentally-friendly sources of raw materials, it would be good to mention some of the historical uses of fish oils that still exist or have existed in the past when different economic or ecological conditions were in place. In may cases, the marine oils were replaced by mineral oil-based products.

In the early stages of their preparation for industrial use, marine oils generally contain minor amounts of non-triglyceride substances that may prevent their use because of their effect on color, drying properties, or reactivity. These oils are therefore further processed to remove these substances while retaining their desirable features. The undesirable substances include: free fatty acids, trace metals, stearines, waxes, moisture, insoluble impurities, sulfur, halogen and nitrogen compounds, pigments, and sterols (Young 1985, 1978).

The processing steps used include: winterization to remove low melting point triglycerides for clarity at low temperatures and better drying properties; degumming to remove phosphatides and proteinaceous materials; neutralization to remove free fatty acids and some pigments; and bleaching to remove pigments, oxidation products, trace metals, and soaps (Bimbo, 1990a, 1990c). Figure

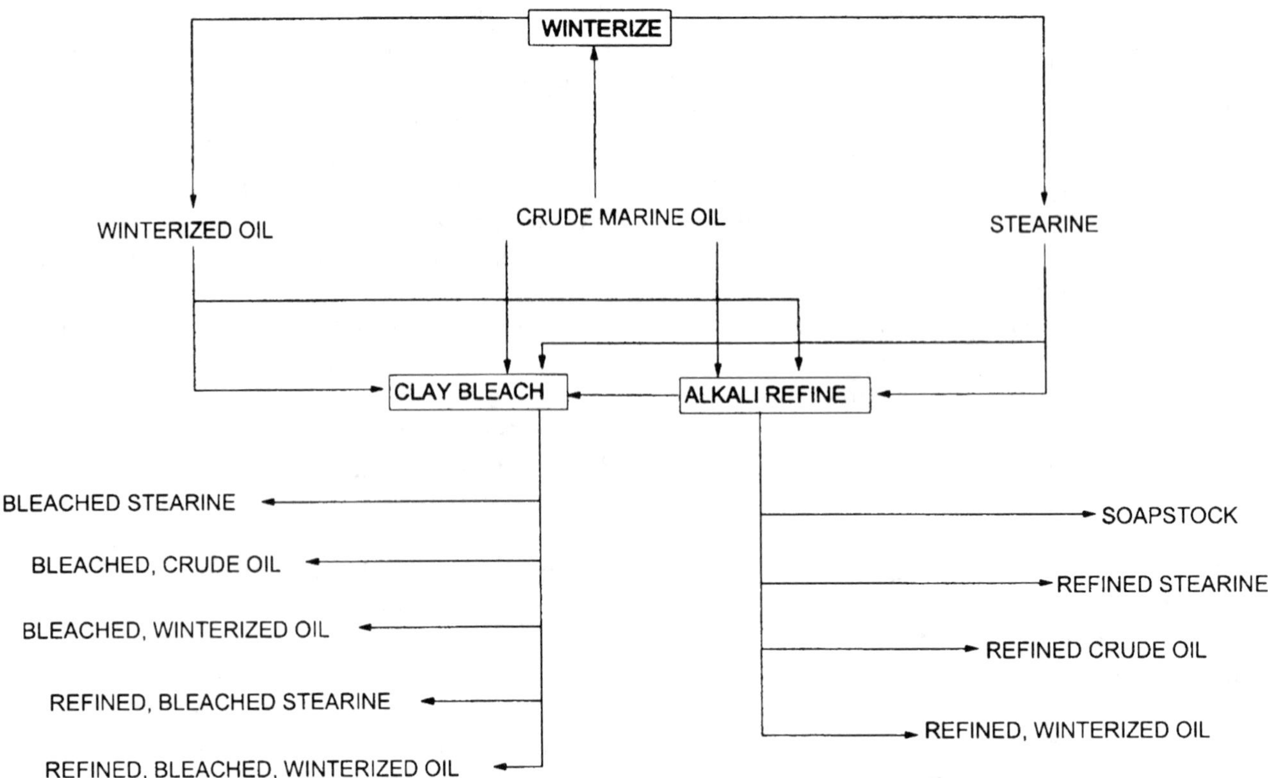

Figure 27. Processing steps needed to convert crude fish oils into various industrial products.

27 provides a flow diagram of these processing steps and the products produced (Bimbo, 1989).

Once prepared by one or several of these processes, fish oil can be chemically modified by hydrogenation, oxidation (blowing), polymerization (heat bodying), double processing (blowing and heat bodying), or copolymerization to give the oil unique industrial properties (Gruger, 1963).

Blown or oxidized oils are produced by blowing air through the oil at an elevated temperature, with or without mechanical agitation, for as long as 36 hours. The final color and acid number (2 x FFA) is dependent upon the processing temperature, the initial color, and the free fatty acids (FFA) of the starting oil. The usual temperature of blowing will range from 80°C to 93°C, and viscosities of 46 to 1066 stokes (Gardner Z3-Z10) can be obtained, providing highly viscous oils at low cost. This air treatment gives the oil unique and valuable properties not available with other varieties of blown oils. These properties include excellent wetting, non-penetration over porous surfaces, improved surfaces, improved drying, flexibility, and adhesion (all of value to the paint and coatings formulator). The addition of considerable amounts of water is possible to produce stable, easy-brushing, and economical paints with no water-sensitive emulsifiers or wetting agents present. This type of oil is suited for metal primers. Blown fish oils dry with a high gloss and can be blended with resins to produce excellent anti-fouling paints where the gradual erosion of

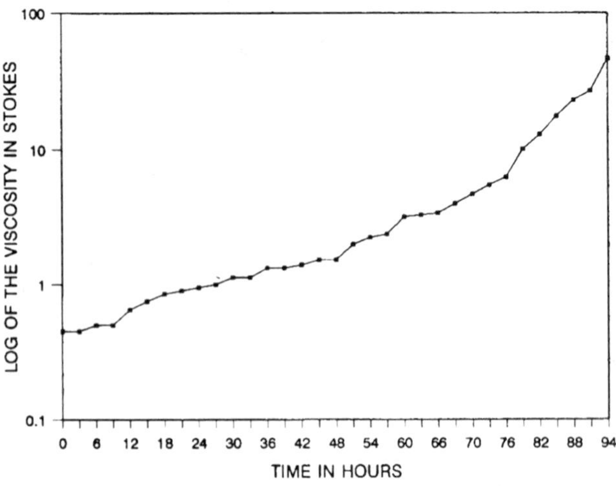

Figure 28. Logarithmic change in the viscosity of menhaden oil during the oxidation process.

the coating is of prime importance for continued fungicidal activity (Bimbo, 1992, 1989).

The oxidation process causes polymerization at the double bonds with cross-linking of the fatty acids. The viscosity increases, as seen in Figure 28. Color lightens during the early stages of the reaction but then darkens as the reaction time increases; free fatty acids increase very slowly (Figure 29). The iodine value decreases over the entire reaction period (Figure 30). The two primary measures of oxidation, peroxide value and anisidine number, combined and expressed as the totox number (2 x peroxide value + anisidine number), show an increase during the initial reaction period, leveling off during the latter stages of reaction as the polyunsaturates are bound up (Figure 31).

The process known as heat- or kettle-bodying is actually a polymerization process. The viscosity and the acid number of the oil become progressively higher as the iodine value decreases, due to the formation of dimers, trimers, and higher polymers. The processing temperature for fish oil is 290°C in closed kettles under vacuum or inert gas. For very high viscosity oils, a refined oil is used, since the heat necessary to increase the viscosity causes a rise in the free fatty acid content, and this in turn slows down the polymerization process. The bodied oil is water resistant and durable and generally used in exterior house paint formulations (Bimbo, 1992).

The process of hydrogenation consists of reacting hydrogen with the oil in the presence of a catalyst so that the hydrogen adds to the unsaturated bonds in the fatty acid chain. This addition of hydrogen saturates the double bonds of the oil and in essence hardens it. During the process, the iodine value decreases and the melting point increases. There is a relationship between the iodine value and the index of refraction, and the course of the hydrogenation reaction can be followed by monitoring the refractive index (Figure 32).

Oils and fats may be fully or partially hydrogenated, depending upon the final use of the end product. Industrial uses of hydrogenated fish oil include fatty acids of the C20 and C22 range that have been incorporated into adhesives, buffing compounds, crayons, pump packings, polishes, calcium and sodium greases, waxes, and textile chemicals. Hydrogenated fatty acids find similar

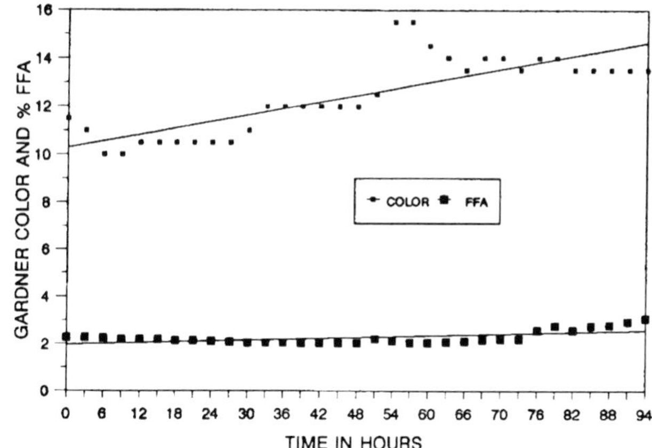

Figure 29. The change in free fatty acid content and color during the industrial oxidation of fish oils.

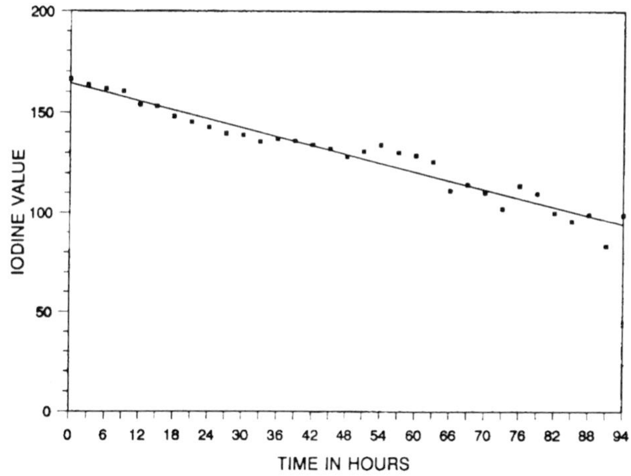

Figure 30. Change in the iodine value during the oxidation of fish oils.

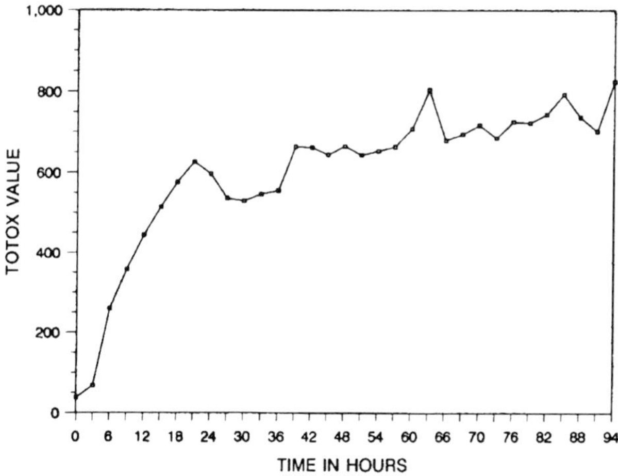

Figure 31. Change in totox number during the industrial oxidation process.

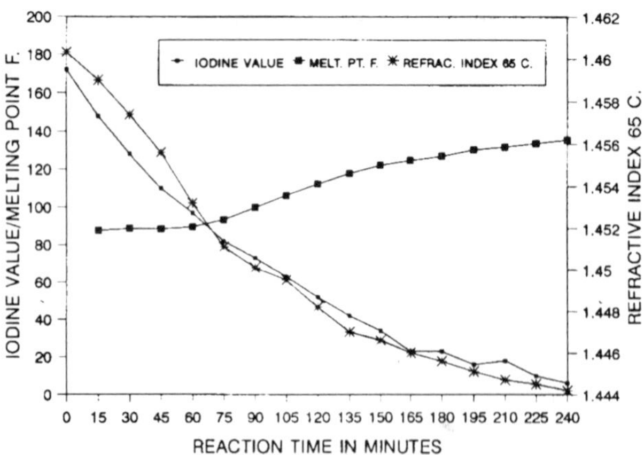

Figure 32. Relationship between iodine value and index of refraction and the change in the melting point during hydrogenation of refined fish oil.

uses; in addition, they are a source of raw material for the manufacture of fatty nitrogen products such as amides, amines, and quaternary ammonium compounds. Table 5 gives specification ranges for some typical chemically modified marine oils and industrial grade menhaden oils (Bimbo, 1992, 1989). With renewed pressure on petroleum-based products, marine oils offer an alternative family of products that are economical, renewable, and environmentally friendly. Table 6 summarizes many of the varied uses of marine oils that exist now or have existed in the past. A new and exciting use is as a hydraulic fluid replacement for equipment situated on or near the marine environment. In such cases, leakage of the fish oil into the marine environment should not have the same effect as that of petroleum-based products (Christensen and Bimbo, 1996).

OTHER MARINE OILS

At one time, marine mammal oils were the predominant marine oil of commerce, accounting for approximately 75 percent of the total production of the aquatic animal oils. The oils had a variety of uses but production gradually changed over the years, primarily due to conservation and protection of endangered species. Table 7 shows

Table 5. Specification Ranges for Some Typical Chemically Modified Marine Oils and Industrial-Grade Menhaden Oils

Oil	Gardner Color Range	Gardner/Holdt Viscosity Range	Free Fatty Acid Range	Iodine Value Range	Cold Test Hrs.@0°C
Crude	11-13	A	2-3	175-200	NA
Refined & bleached	7-9	A	0.1-0.5	175-200	2
Kettle-bodied	5-8	U-Z3	1-5	100-125	NA
Blown	9-13	Z3-Z7	4-6	95-115	NA
Double-processed	8-10	Z2-Z6	1.7-3	80-90	NA
Co-polymerized	8-11	M-Z6	1-2	-	NA

Source: Bimbo, 1989.

Table 6. Past and Present Industrial Uses of Marine Oils

Fatty Acids	Fatty Chemicals	Refractory Compounds
Soaps	Leather Tanning	Cutting Oils
Protective Coatings	Lubricants and Greases	Plasticizers
Rubber Compounds	Ore Floatation	Printing Inks
Caulking Compounds	Insecticidal Compounds	Linoleum
Glazing Compounds	Fermentation Substrates	Presswood Fiber Boards
Automotive Gaskets	Illuminating Oils	Oiled Fabrics
Core Oils	Fuel Oils	Ceramic Deflocculants
Tin Plating Oils	Mushroom Culture	Attractants and Lures
Rust Proofing	Fire Retardants	Polyurethanes
Specialty Chemicals	Oil Field Chemicals	Mold Release Agents

Source: Bimbo, 1992.

Table 7. Change in Distribution of Aquatic Animal Oil Production, 1938–1997

Year	Whale Oil	Sperm Whale Oil	Seal Oil	Fish Body Oil	Dolphin Oil	Sea Animal Oil	Liver Oils	Miscel-laneous*	Totals
1938	595	24	5	239	–	–	76	16	955
1948	356	58	5	175	–	1	65	16	676
1951	396	107	11	269	11	–	83	22	899
1952	418	76	7	248	9	–	78	16	852
1954	414	73	6	321	9	–	70	15	908
1955	378	91	7	310	9	–	73	15	883
1982	9	1	3	1259	–	–	22	1	1295
1983	6	0	1	1041	–	–	20	66	1134
1984	6	0	1	1391	–	–	26	93	1517
1985	5	0	0	1378	–	–	31	80	1494
1986	2	0	0	1549	–	–	25	90	1666
1987	1	0	1	1326	–	–	34	87	1449
1988	1	0	1	1413	–	2	40	101	1558
1989	0	0	0	1488	–	1	32	113	1634
1990	0	0	0	1264	–	1	24	122	1411
1991	0	0	0	1236	–	–	26	114	1376
1992	–	–	–	1038	–	–	20	3	1061
1993	0	0	0	1193	–	–	24	1	1218
1994	0	0	0	1482	–	–	21	1	1505
1995	0	0	0	1358	–	–	24	3	1387
1996	0	0	0	1384	–	–	26	3	1418
1997	0	0	0	1185	–	–	23	4	1213

*Beginning in 1983 FAO adjusted the reporting of the former USSR.
Source: FAO, 1999.

the change in distribution of aquatic animal oils from 1938 to 1997.

Fish Liver Oils

Fish liver oils were used in the treatment of rickets in the Middle Ages. As early as 1657 certain fish liver oils were believed to contain something that could cure night-blindness. With the development of vitamin chemistry early in the twentieth century, it was shown that night-blindness and rickets are largely caused by a dietary deficiency in vitamins A and D, respectively. Both vitamins A and D are found in certain fish liver oils at varying levels (Bimbo, 1988b).

The oil is contained in the protein of the livers and is sometimes easily removed by extraction; at other times the oil is removed from the livers after digestion of the liver protein, and a more complicated process of extraction is needed. Special processing is required for the extraction and separation of the vitamin-bearing material from low-fat proteinaceous liver tissue (Brody, 1965).

The most important raw material for the production of liver oils comes from the fisheries for cod, coalfish, and haddock. The livers of ling, tusk, several species of shark (such as dogfish, Greenland shark, and basking shark), and halibut, whale, and tuna have been used in the production of liver oils. In order to obtain high-quality, light-colored oils with good flavor and odor, containing a minimum of free fatty acids, it is important to eviscerate the fish and recover the livers so that they can be processed as quickly as possible (Windsor and Barlow, 1981).

In the early history of the industry, the livers were allowed to stand in wooden vats or barrels and the oil was removed as it floated to the surface after being released by autolysis. The first two batches removed in this fashion were used as medicinal oil, and the remainder as industrial oil. The residue was heated in underfired iron kettles, giving dark oil that contained large amounts of free fatty acids.

The method of heating the fish livers with steam, applied either directly in contact with the liver mass or indirectly, was introduced about 1950. This method gave lighter oils of good quality. Many methods developed later are based on this same principle. The most generally accepted way of extracting oil from cod livers is by means of a steam cooker. The method of applying the steam has been changed from time to time to meet the demand for a higher grade of oil. Low-pressure steam is piped into a tank containing the livers. The heat of the steam cooks the livers, it condenses, and a layer of hot water is produced upon which the oil floats. The oil is then either dipped off or the tank is fitted with an overflow into an oil storage tank. Some liver oils are extracted on board vessels since the ships must remain at sea for long periods of time. The Icelandic process was outlined in a private communication from Baldur Hjaltason of Lysi HF: Iceland developed a process that treats the liver residue with caustic soda after the initial steam treatment. The primary medicinal grade oil is floated off and the residue is then treated with caustic soda, which destroys the protein and releases the secondary veterinary grade cod liver oil. This grade is darker in color but contains a higher level of vitamins than that of the medicinal grade.

The livers are ground and pumped over magnets to remove tramp metal. The ground livers are heated and allowed to stand to allow the proteins to break down. The livers are then decanted to

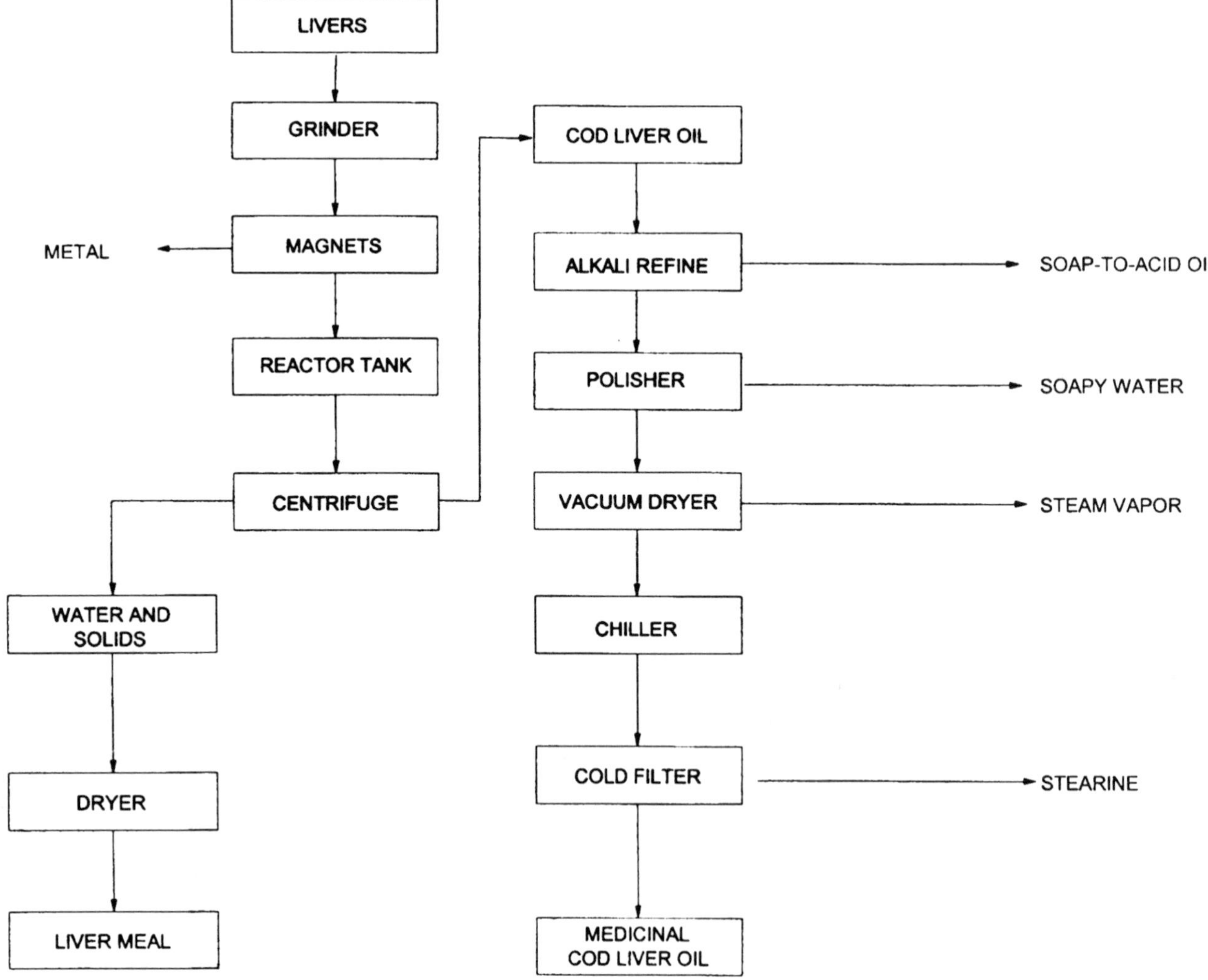

Figure 33. Flow diagram of the Icelandic cod liver oil process.

remove solids, and the liquor is collected in kettles, heated to 95°C, and then separated. Modern three-way separators are used, and the crude cod liver oil is collected and pumped to the refinery. In the refinery the oil is alkali-refined to remove free fatty acids, washed, vacuum-dried, and then winterized to remove stearines. The result is medicinal grade cod liver oil. (A diagram of the Icelandic process for producing cod liver oil is shown in Figure 33.)

MARINE MAMMAL OILS

Gilmore (1960) reported that the first great whaling nation of the world was England, followed closely by Holland. In the seventeenth and eighteenth centuries the English and Dutch caught right and bowhead whales worldwide. The early American industry initially caught right and bowhead whales but soon began searching for the sperm whale. In 1846 there was as many as 746 whaling ships registered in America fishing out of New England ports. The Norwegians began to use the harpoon-cannon and winch on steamships in the 1860s and this made whaling both profitable and efficient. As mentioned above, the reduction in production is primarily due to conservation and a worldwide moratorium on whaling. The United States production of whale oil was at its peak between 1936 and 1940. From the information in Table 7 it can be seen that marine mammal oils have virtually disappeared, although there has been some renewed interest in processing seals in Newfoundland and rendering the fat to produce a seal oil.

REFERENCES

Ackman, R.G., C.A. Eaton, and W.M.N. Ratnayake. 1981. Considerations of fatty acids in menhaden from the northern limits of the species. *Proc. N.S. Inst. Sci.* **31**:207-215.

Ackman, R., F. Lamothe, H. Hulan, and F. Proudfoot. 1988. *N-3 News Unsaturated Fatty Acids and Health* **3**:(1).

Ackman, R.G., S.M. Polvi, R.L. and S.P. Saunders. 1990. *Bull. Aquacul. Assoc. Canada.* **90**(4):45-49.

Adams, M.H., M. Colberg, M.A. Reiber and A.L. Izat. 1989. *Poultry Science.* **68**:166. Supplement 1.

Barlow, S. and M. Windsor. 1984. Fishery By-Products. *International Association of Fish Meal Manufacturers Technical Bulletin* **19**.

Barlow S.M., F.V.K.Young, and I.F. Duthie. 1990. Nutritional recommendations for n-3 polyunsaturated fatty acids and the challenge to the food industry. *Proceedings of the Nutrition Society* **49**:13-21.

Barlow, Stuart and Ian H. Pike. 1991. Humans, animals benefit from omega-3 polyunsaturated fatty acids. *Feedstuffs* **63**(19):18-26.

Barlow, Stuart M. and Ian H. Pike. 1994. Upgrading the uses of lower species to provide a source of omega-3 fatty acids in the human diet. *Omega-3 News* **IX**(4/94).

Barry, M. 1988. Thesis for the Master of Science degree. Southern Illinois University.

Beermann D., D. Hogue, V. Fishell, R. Dalrymple, and C. Ricks. 1986. *J. Anim. Sci.* **62**:370.

Beugelink, F. 1978. Dry off-loading of fish. *IAFMM News Summary* **42**:143.

Billy, T.J. and G.M. Dreosti. 1983. Potential food products from industrial fish. *IAFMM Technical Report* **1**.

Bimbo, A.P. 1970. The menhaden industry: yesterday, today, and tomorrow. Paper presented to the University of Rhode Island. Dept. Chemical Engineering. Oct. 8.

Bimbo, A.P. 1987. The emerging marine oil industry. *JAOCS* **64**:706-715.

Bimbo, A.P. 1987a. Marine oils — perspectives on the United States industry. Paper read at the American Institute of Baking Technical Seminar. Fish Oil (Omega-3 Fatty Acids) and Other Unconventional Oils. Manhattan, Kansas. May 11-12.

Bimbo, A. P. 1988. Fish oils: future challenges and opportunities. In *Seafood Technology — Preparing for Future Opportunities.* Marvin Kragt and Donn Ward, eds. Chicago: Institute of Food Technologists, pp. 167-203.

Bimbo, A.P. 1989. Technology of production and industrial utilization of marine oils. In *Marine Biogenic Lipids, Fats and Oils.* Volume II. R.G. Ackman (ed.), pp. 401-433. Boca Raton: CRC Press Inc.

Bimbo, A.P. 1989a. Recent advances in upgrading industrial fish to value added products. In *New Technologies for Value-Added Products from Protein and Co-Products.* Lawrence A. Johnson (ed.). Symposium III Changing Resources and Needs. Champaign: American Oil Chemists' Society.

Bimbo, A. P. 1989b. Fish oils as foods: challenges and opportunities. In *Fats and Oils in Bakery Products.* Okkyung Kim Chung (ed.), pp. 282-308. St. Paul: Am. Assoc. of Cereal Chemists.

Bimbo, A. P. 1989c. Fish oils: past and present food uses. *JAOCS* **66**:(12):1717-1726.

Bimbo, A. P. 1990. Historical Perspectives on the Stabilization of Fish Meal in the United States of America. *Symposium on the Antioxidant Stabilization of Fish Meal and Fish Oil. IFOMA News Summary* **64**.

Bimbo, A. P. 1990a. Fish meal and oil. In *The Seafood Industry.* Roy E. Martin and George J. Flick (eds.), pp. 325-350. Van Nostrand Reinhold, New York.

Bimbo, A. P. 1990b. Production of fish oil. In *Fish Oils in Nutrition.* Maurice E. Stansby (ed.), pp. 141-180. Van Nostrand Reinhold, New York.

Bimbo, A. P. 1990c. Processing of fish oils. In *Fish Oils in Nutrition.* Maurice E. Stansby (ed.), pp. 181-225. Van Nostrand Reinhold, New York.

Bimbo, A. P. 1992. Marine oils: fishing for industrial uses. *Inform.* **3**(9):968-1001.

Bimbo, A. P. 1992a. Fish meal and oil: current uses. *JAOCS* **69**(3):221-227 .

Bimbo, A. P. 1993. United States Food and Drug Administration petition on refined fish oil — update 1993. Paper presented at the 33rd Annual Conference of IFOMA. November 29-Dec. 3. New Orleans, LA.

Bimbo, A.P. 1998. Guidelines for characterizing food-grade fish oil. *Inform* 9(5): 473-483.

Bowman, S.J. 1995. Digest of selected statistics presented at the 1995 Annual IFOMA Conference. Santiago, Chile.

Bozzo B., Andres and Daniel A. Martin 1995. *Descarga de pesca con sistema de presion — vacio a larga distancia en Chile.* Paper presented at the 1995 Annual IFOMA Conference. Santiago, Chile.

Broderick, G. 1989. United States Dairy Forage Research Center Summaries. Pg. 59.

Broderick, Glen A. and Larry D. Satter. 1990. *Feedstuffs* **62**(38):20.

Brody, J. 1965. *Fishery By-Products Technology.* AVI Publishing Co., Westport, CN.

Brown, Robert H. 1990. *Feedstuffs* **62** (October):1.

Cambell, M. 1978. Boiler incineration of odourous gases. *IAFMM News Summary* **43**:122.

Chahine, M.H. 1978. Antioxidants to stabilize fish meal. *Feedstuffs* **28** (May).

Chanmugam, P., J. Donovan, C.J. Wheeler, and D.H. Hwang. 1983. *J. Food Science* **48**:1440-1462.

Chapman, Kathryn W., Ifat Sagi, Joe M. Regenstein, Tony Bimbo, and Clyce E. Stauffer. 1996. Oxidative stability of hydrogenated menhaden oil shortening blends in cookies, crackers, and snacks. *JAOCS.* **73**(2):167-172.

Christensen, B. and E. Storm. 1989. Split feeding. United States Patent No. 459-119-9. February 2.

Christensen, S. 1978. Control of decanters and separators. *IAFMM News Summary.* **44**:104.

Christensen, Thomas E. and A. P. Bimbo 1996. Fish oil for use as hydraulic oil. *Bioresource Technology* 56: 49-54.

Claggett, F.G. 1978. Practical measures of liquid effluent control. *IAFMM News Summary.* **44**:1.

Cullinane, M.J. Jr. 1978. Water pollution control in the fish meal processing industry -- a rational approach. *IAFMM News Summary.* **44**:22.

Dhiman T., and L. Satter. 1989. Supplementation of high forage diets with energy or protein. Abstract of paper presented at the American Dairy Science Association and American Society of Animal Science. Lexington, KY.

Dreosti, G.M. 1980. Spontaneous heating. *IAFMM News Summary.* **49**:154.

Effect of a selected menhaden fish meal in starter diets for early weaned pigs. 1987. Kansas State University. *International Association of Fish Meal Manufacturers Fish Meal Flyer.* **15**.

Endres, S., R. De Caterina, E. B. Schmidt, and S.D. Kristensen. 1995. n-3 polyunsaturated fatty acids update 1995. *European Journal of Clinical Investigation.* **25**:629-638.

FAO. 1955. Yearbook of Fishery Statistics, Production and Fishing Craft. 5(f4, f46). Food and Agriculture Organization of the United Nations. Geneva.

FAO. 1980. Yearbook of Fishery Statistics. **51**:148-152. Food and Agriculture Organization of the United Nations. Geneva.

FAO. 1991. Yearbook of Fishery Statistics: catches and landings, 1990. Food and Agriculture Organization of the United Nations. Geneva.

FAO. 1998. Yearbook of Fishery Statistics: catches and landings, 1997. Food and Agriculture Organization of the United Nations. Geneva.

FAO. 1986. Fisheries Industries Division. The production of fish meal and oil. FAO Fisheries Technical Paper (142) Rev. **1**:63.

FAO and International Development Research Center. 1982. *Fish By-Catch Bonus From the Sea.* IDRC, Ottawa, Ontario, Canada, IDRC-198e.

Food and Drug Administration. 1989. *Federal Register.* **54** (178):38219-38223.

Food and Drug Administration. 1997. *Federal Register.* **62** (108):30751-30.

Food Business. 1989. September 4.

Fritsche, Kevin L. 1991. Speech presented at the 32nd Annual National Fish Meal and Oil Association Fisheries Symposium. March 20. Baltimore, MD.

Fritsche, Kevin L., Nancy A. Cassity, and Shu-Cai Huang. 1991. Effect of dietary fats on the fatty acid compositions of serum and immune tissues in chickens. *Poultry Science* **70**:1213-1222.

Fritsche, Kevin L., Nancy A. Cassity, and Shu-Cai Huang. 1991. Effect of dietary fat source on antibody production and lymphocyte proliferation in chickens. *Poultry Science* **70**:611-617.

Fritsche, Kevin L. and Nancy A. Cassity. 1992. Dietary n-3 fatty acids reduce antibody-dependent cell cytotoxicity and alter eicosanoid release by chicken immune cells. *Poultry Science* **71**:1646-1657.

Fritsche, Kevin L., David W. Alexander, Nancy A. Cassity, and Shu-Cai Huang. 1993. Maternally-supplied fish oil alters piglet immune cell fatty acid profile and eicosanoid production. *Lipids* **28**:677-682.

Frye, John. 1978. *The Men All Singing.* Donning. Norfolk, VA.

Gauglitz et al. 1973. Application of fish oils in the food industry. Paper presented at the technical conference on fishery products held in Tokyo. Apr. 11-12. FAO DOC FII: FP/73/E-19.

Gibbs, C.R. and C.F.A. Green. 1978. Off-loading and transportation of fish -- the Hidrostal system. *IAFMM News Summary* **42**:137.

Gilmore, Raymond M. 1960. The whaling industry — whales, dolphins, and porpoises. In *Marine Products of Commerce.* D.K. Tressler and J. McW. Lemon (eds.), pp. 680-715. Van Nostrand Reinhold. New York.

Gloppestad, E. 1979. Techniques for the use of decanters in separating liquid from solids in the fish meal industry. *IAFMM News Summary* **44**:71.

Gruger, E.H. Jr. 1963. Uses of industrial fish oil. In *Industrial Fishery Technology.* M. Stansby (ed.), pp. 260-271. Reinhold Publishing Corp. New York.

Gundersen, R. and H. Wiedswang. 1993. Hot air dryers and the production of low temperature fish meal — a review. *World Fishing Magazine* **42**(3):40-41.

Hall, Kenneth N. 1990. Speech presented at the 31st Annual National Fish Meal and Oil Association Fisheries Symposium, March 21. Baltimore, MD.

Hansen, Knud, W. 1978. Ventilation systems for chemical treatment of odour. *IAFMM News Summary* **44**:66.

Hansen, P., B. Feldstedt, L. Quiroga, and T. Ettrup Petersen. 1977. Container and bulk stowage aboard purse seiners. *Fishing News International* **16**(2).

Hansen, P., J. Jensen, and Teutscher, F. 1980. Bulk handling of purse seine catches for the sardine canning industry. *Scandinavian Refrigeration* **3**.

Hansen, P. 1981. Alternative uses of industrial fish. Report of the Technological Laboratory. Lyngby, Denmark. May.

Hargis, P.S., M.E. Van Elswyk, and B.M. Hargis. 1991. Dietary modification of yolk lipid with menhaden oil. *Poultry Science* **70**:874-883.

Hetland, J.D.Y. 1980. Techniques and economics of direct (flame) and indirect drying. *IAFMM News Summary* **49**:35.

Hjaltason, Baldur. 1989. New frontiers in the processing and utilization of fish oil. In *Nutritional Impact of Food Processing: Bibl. Nutr. Dieta.* J. C. Somogyi and H.R. Müller (eds.), pp. 96-106. Karger, Basel.

Hovad, H. and J. Lorentzen. 1978. Automatic control of stickwater evaporators. *IAFMM News Summary* **44**:120.

Huang, Zhi-Bin, Henry Leibovitz, Chong M. Lee, and Richard Millar. 1990. *J. Agric. Food Chem.* **38**:743-747.

Hulan, H.W., F G. Proudfoot, and D.M. Nash. 1984. *Poultry Science.* **63**:324.

Hulan, H. 1988. *Poultry Science* **67**:99.

Hulan, H., R. Ackman, W. Ratnayake, and F. Proudfoot. 1988. *Can. J. Anim.Sci.* **68**:533-547.

Hulan, H., R. Ackman, W. Ratnayak, and F. Proudfoot. 1989. *Poultry Science* **68**:153-162.

Hulan, H.W. and R.G. Ackman. 1990. *Proceedings of the Eleventh Western Nutrition Conference*. September 13-14. University of Alberta. Calgary, Canada.

IAFMM. 1997. Reducing odor in fish meal production. Torry Advisory Note No. 72. News Summary No. 41, Page 52.

IAFFM Technical Bulletin **25**. 1990.

IAFMM. 1987. Fish Meal Flyer.

Jafar, Sajida S., Herbert O. Hultin, Anthony P. Bimbo, and Stuart M. Barlow. 1994. Stabilization by antioxidants of mayonnaise made from fish oil. *Journal of Food Lipids.* **1**:295-311.

James, David G. 1994. Fish as food: present utilization and future prospects. *Omega-3 News* **IX**(4/94):1-4.

Jason, A.C. 1980. General theory of drying fish. *IAFMM News Summary* **49**:5.

Joseph, Jeanne D. (ed.). 1989. *Biomedical Test Materials Program: Production Methods and Safety Manual.* NOAA Technical Memorandum. NMFS-SEFC -234, 120 p.

Kifer, R.R. and D. Miller, 1969. Fish oils -- fatty acid composition, energy values, metabolism and vitamin content. *Fisheries Indus. Res.* **1**(5).

Konge, J. and I. Rasmussen. 1978. Pneumatic off-loading of fish -- The IRAS system. *IAFMM News Summary* **42**:155.

Kreuzer, R., (ed.). 1974. *Fishery Products*, Fishing News Ltd. Surrey, England

Kroken, E. and A. O. Utvik. 1978. Two-stage pressing to reduce oil content in fish meal. *IAFMM News Summary* **43**:65.

Lands, W.E. and A. P. Bimbo. 1983. Possible beneficial effects of polyunsaturated fatty acids in maritime foods. Special report for IAFMM. Sept.

Lands, William E.M. 1989. *World Aquaculture* **20** (1):59-62.

Lee, C.F. 1952. Menhaden industry past and present. Reprinted from Fish Meal and Oil Industry. International Yearbook.

Lee, C.F. and F. B. Sanford.1960. United States Fish-Reduction Industry. United States Department of the Interior. *Commercial Fisheries.* TL14. Washington, DC. Pg. 8.

Lin, Jenn H., Dan E. Pratt, R L. Adams, and W.J. Stadelman. 1995. Influence of dietary menhaden fish oil on fatty acid composition of the egg. *Journal of Food Quality* **18**:149-165.

Luna, J. (ed.). 1981. *Non-Traditional Fish Products for Massive Human Consumption.* Inter-American Development Bank. Washington, DC. Vol. 1 and 2. Apr.

Madsen, Hans Berg. 1986. *Feed International* 7(7):16-20.

Mantysaari, P. and C. Sniffen. 1989. *Feedstuffs* **61**:18. January 23.

Marti, Cristina, Marlene Roeckel, Estrella Aspe, and Harumatsu Kanda. 1994. Recovery of proteins from fishmeal factory wastewaters. *Process Biochemistry* **29**:39-46.

Maurice, Denzil V. 1994. Feeding to produce designer eggs. *Feed Management* **45**(2):29-32.

McCay, Bonnie J. 1980. A footnote to the history of New Jersey fisheries: Menhaden as food and fertilizer. *New Jersey History-Fall-Winter. 1980.* NJSG Pub. No. NJSG-83-115.

Mielke, Thomas (ed.). 1999. *Oil World Annual 1999.* ISTA Mielke GmbH.

Nachenius, R.J. 1978. Water pollution in the fish meal and oil industry. *IAFMM News Summary* 44:1.

National Fish Meal and Oil Association. 1995. *Fish Meal Notes. Menhaden Meal Composition Data.* Washington, DC.

Nettleton, Joyce A., W.H. Allen Jr., L.V. Klatt, W.M.N. Ratnayake, and R.G. Ackman. 1990. *J. Food Sci.* **55**(4):954-958.

Nettleton, Joyce A. 1991. *Journal of the American Dietetic Association* **91**:331-337.

New packaged fishmeal plants from Stord. 1993. *World Fishing Magazine* **42**(3):37-38.

Newton, Ian S. 1996. Food enrichment with long-chain n-3 PUFA. *Inform.* **7**(2):169-176.

Nir, I. 1990. *Poultry Science* **69**:99, Supplement 1.

Nordstrom, R. 1978. Offloading and transportation of fish by mono pumps. *IAFMM News Summary* **42**:175.

Oh, Suk Y., Jehong Ryue, Chia-Hong Hsieh, and Dean E. Bell. 1991. Eggs enriched in *w*-3 fatty acids and alterations in lipid concentrations in plasma and lipoproteins and in blood pressure. *Am. J. Clin. Nutr.* **54**:689-695.

Oh, S., Chia-Hong Hsieh Lin, J. Ryue, and D. Bell. 1994. Eggs enriched with omega-3 fatty acids as a wholesome food. *J. Appl. Nutr.* **46**(12):14-25.

Onarheim, R. and A. O. Utvik. 1978. A system of complete effluent control. *IAFMM News Summary* **44**:75.

Onarheim, Roald. 1978. Stickwater evaporator designed for cleaning during operation. *IAFMM News Summary* **43**:140.

Onarheim, R. (ed.) 1978. Two-stage pressing. *Stord Bartz Review* **78**(4):30.

Onarheim, R., and A. O. Utvik. 1979. The design and operation of screens and presses. *IAFMM News Summary* **46**:105.

Opstvedt, J. 1985. Fish lipids in animal nutrition. *International Association of Fish Meal Manufacturers. Technical Bulletin* **22**.

Orskov, E.R. 1981. *Feedstuffs* **63**(4):67.

Phetteplace, H.W. and B.A. Watkins. 1989. Effects of various ω-3 lipid sources on fatty acid compositions in chicken tissues. *Journal of Food Composition and Analysis* **2**:104-117.

Phetteplace, Hope W. and Bruce A. Watkins. 1990. Lipid measurements in chickens fed different combinations of chicken fat and menhaden oil. *J. Agric. Food Chem.* **38**(9):1848-1853.

Pigott, G.M. 1967. Production of fish oil. In *Fish Oils*. M. E. Stansby (ed.). The AVI Publishing Co. Inc. Westport. Pp. 183-192.

Pigott, George M. 1989. *World Aquaculture* **20**(1):63-68.

Pike, I. 1981. Fish meal for ruminants — dietary protein and degradability in the rumen. *The Feed Compounder*. January. Pg. 1.

Pike, Ian H. 1987. *The Feed Compounder*. February. Pp. 13-14.

Pike, I.H. 1990. The role of fish meal in diets for salmonids. *International Association of Fish Meal Manufacturers Technical Bulletin* **24**.

Pike, I.H., Gudrid Andorsdottir, and H. Mundheim. 1990. The role of fish meal in diets for salmonids. *IAFMM Technical Bulletin* **24.**

Pike, I.H. 1990. The role of fish oil in feeds for farmed fish — estimated current and potential use. *IAFMM Technical Bulletin* **25.**

Pike, I.H. and R.J. Esslemont. 1991. *International Association of Fish Meal Manufacturers. Fish Meal Flyer* **19**.

Pike, Ian H. 1991. Increasing importance of aquaculture to the fish meal industry. *International Association of Fish Meal Manufacturers. Technical Market Report.* P. 2.

Porse, C.U. 1978. Pumping and transportation of fish — the Superfos system. *IAFMM News Summary* **42**: 182-184.

Production of virgin fish oils for dietary purposes. 1993a. *World Fishing Magazine* **42**(3):42-43.

Raper, N.R. and Jacob Exler. 1990. In Proceedings of the II International Conference on the Health Effects of Omega-3 Polyunsaturated Fatty Acids in Seafoods. June 20-23, 1990. Washington, DC.

Rask, F. 1979. Techniques for the use of decanters in separating liquid from solids in the fish meal industry. *IAFMM News Summary* **46**: 87.

Reducing odour in fish meal production. 1977. Torry Advisory Note No.72. *IAFMM News Summary* **41**:52.

Rigau, A. Perez, M.D. Lindermann, E.T. Kornegay, A.F. Harper, and B.A. Watkins. 1995. Role of dietary lipids on fetal tissue fatty acid composition and fetal survival in swine at 42 days of gestation. *J. Animal Science* **73**:1372-1380.

Roeckel, Marlene, M. Cristina Marti, and Estrella Aspe. 1994. Clean technology in fish processing industries. *J. Cleaner Prod.* **2**(1):31-35.

Rumsey, G. 1982. Nitrogen requirement and metabolism in fish: a comparative perspective. *Proceedings of the 43rd Minnesota Nutrition Conference.* University of Minn. Sept. 1982. Pg. 46.

Simopoulos, Artemis P. 1988. *w*-3 fatty acids in growth and development and in health and disease. *Nutrition Today.* May/June. 1988:**12**-18.

Smith, J. Howard 1940. Advances in menhaden reduction. *Chemical and Metallurgical Engineering* **47**:99-100.

Smith, Joseph W. and The Population Dynamics Team. 1999. Forecast for the 1999 Gulf and Atlantic menhaden purse-seine fisheries and review of the 1998 fishing season. March 1999.

Sobstad, Geir E. 1992. Marine oil separation, purification technology. *Inform.* **3**(7):827-830.

Sola, E. 1978. Discharge of fish for meal and oil production. *IAFMM News Summary* **42**:117.

Spain, J.N. and C.E. Polan. 1995. Evaluating effects of fish meal on milk fat yield of dairy cows. *J. Dairy Science* **78**:1142-1153.

Stadelman, W. 1989. Paper presented at the 30th Annual Fisheries Symposium sponsored by the National Fish Meal and Oil Association. March 15. Baltimore, MD.

Stansby, M. 1978. Development of fish oil industry in the United States. *JAOCS* **55**(2):238-243.

Stauffer, C.E. and A.P. Bimbo. 1994. Hydrogenated menhaden oil shortening in cookies, crackers and snacks. *Cereal Foods World.* **39**(9):688-690.

Stern, Marshall D. and Howard R. Mansfield. 1989. *Feedstuffs* **61**(42):16-26.

Stoner, G., G. Allee, J. Nelssen, and M. Johnston. 1986. Effect of a selected menhaden fish meal in starter diets for pigs. Kansas State University Swine Day. Progress Report No. 507.

Tacon, Albert G.J. 1996. *Global trends in aquaculture and aquafeed production.* FAO report presented in Dublin, Ireland. June.

Tatterson, I. N. and M. L.Windsor. 1974. Fish silage. Torry Advisory Note No. 64. Torry Research Station. Aberdeen, UK.

The role of fish oil in feeds for farmed fish — estimated current and potential use. 1990. International Association of Fish Meal Manufacturers. Tech. Bull. 25.

Thonney, M. and D. Hogue 1986. *J. Dairy Sci.* **69**:1648-1651.

Tronstad, I.M. 1978. Myrens raw material pump-BRP 1. *IAFMM News Summary* **42**:161.

United States Dept. of Commerce. 1999. National fishery statistics program: processed fishery products. *Fisheries of the United States, 1998.* CFS 9800.

Van Elswyk, M.E., A.R. Sims, and P.S. Hargis. 1992. Composition, functionality, and sensory evaluation of eggs from hens fed dietary menhaden oil. *Journal of Food Science.* **57**(2):342-349.

Van Elswyk, Mary. 1994. Looking ahead: will eggs become a dietary alternative to fish? *Egg Industry.* September/October. Pp. 20-24.

Ward, A., Wignal, J., and Windsor, M.L. 1977. The effect of cooking temperature and applied pressure on the release of liquor from fish. *IAFMM News Summary* **41**:24.

Watkins, Bruce A. 1991. Speech presented at the 32nd Annual National Fish Meal and Oil Association Fisheries Symposium. March 20. Baltimore, MD.

Weddig, Lee. 1991. *Meat and Poultry* March. Pg. 25.

Wignall, J. 1978. Odour pollution in the fish meal and oil industry. *IAFMM News Summary* **44**:44.

Windsor, M. and Barlow, S. 1981. *Fish Meal Production, Introduction to Fishery By-Products* 1st ed. Fishing News Books Ltd. Surrey, England.

Young, A.W. 1990. Speech presented at the 31st Annual National Fish Meal and Oil Association Fisheries Symposium. March 21. Baltimore, MD.

Young, F.V.K. 1978. Processing of oils and fats. *Chemistry and Industry* 16 September. 1978:692-703.

Young, F.V.K. 1985. The refining and hydrogenation of fish oil. International Association of Fish Meal Manufacturers Fish Oil Bulletin. **17**.

Young, F.V.K. 1985a. Interchangeability of fats and oils. *JAOCS* **62**:372-376.

Young, F.V.K. 1986. The chemical and physical properties of hydrogenated fish oils for margarine and shortening manufacturers. International Association of Fish Meal Manufacturers Fish Oil Bulletin. **19.**

Young, F.V.K. 1986a. Formulation in a multi-feedstock situation. Paper read at The AOCS Short Course No. 2 on Hydrogenation. May 1986. Honolulu, HI

Young, F.V.K. 1986b. The use of hydrogenated fish oils in margarines, shortenings and compound fats. International Association of Fish Meal Manufacturers Fish Oil Bulletin. **20**.

Young, F.V.K. 1986c. The chemical and physical properties of crude fish oils for refiners and hydrogenators. International Association of Fish Meal Manufacturers Fish Oil Bulletin. **18.**

Young, F.V.K., Vibeke From, S.M. Barlow, and J. Madsen. 1990. Using unhydrogenated fish oil in margarine. *Inform.* **1**(8):731-741.

Young, F.V.K., S.M. Barlow, and J. Madsen. 1993. Unhydrogenated fish oil in low-calorie spreads. *Inform.* **4**(10):1140-1146.

Yu, M. and J. Sim. 1987. *Poultry Science* **66**:195.

Fish Protein Concentrate

Roland Finch

INTRODUCTION

"Fish protein concentrate (FPC) is a stable product suitable for human consumption, prepared from whole fish or other aquatic animals or parts thereof. Protein concentration is increased by the removal of water and in certain cases of oil, bones and other materials. Traditionally-dried or other traditionally-preserved products do not fall within this guideline" [Food and Agriculture Organization (FAO)/World Health Organization (WHO)/United Nations International Children's Emergency Fund (UNICEF) Protein Advisory Group, 1971)].

THE ORIGINS OF FISH PROTEIN CONCENTRATE

Research to produce FPC started as early as the 1890s in Norway (Bakken, 1962), in 1937 in South Africa (Dreosti), and in 1938 in Iceland (Hannesson). After World War II, a wider interest was shown in the concept of fish protein concentrate, largely due to a growing awareness of the rapid and possibly exponential growth of the world's population and the resulting danger of insufficient food being available to provide adequate nutrition to many peoples. This eventuality was already apparent in some regions where already too little food was available to provide for rapidly multiplying local populations or where the food supply had an inadequate nutritional balance.

A special concern was that the food available to many people consisted principally of carbohydrates, such as rice, and contained an insufficient amount and quality of protein to maintain adequate nutrition. In 1963 the Food and Agriculture Organization of the United Nations published the results of a survey that gave estimates of animal protein available and needed in Third World countries (FAO, 1962). It indicated a shortfall at that time of 40, 30, 40, and 24 percent of the short-term targets in the Far East, Near East, Africa, and Latin America (excluding the River Plate countries), respectively. The deficiency was even greater for the long-term targets. Also, these figures were averages and did not show the variations within those surveyed, so that many peoples would have considerably lower protein intakes than even the low figures found in the survey. In 1967 the United States President's Science Advisory Committee, considering the information then available, projected that world protein needs that had been estimated for 1965 would increase to around 150 percent by 1985 (United States Government Printing Office, 1967). In 1968 the United Nations concluded that, although there were wide differences in the estimates of world protein deficiency, a gap existed between the need and the supply of protein and that the gap was increasing (United Nations, 1968).

An obvious solution to this problem would be to provide supplemental protein to the diets of those living on primarily carbohydrate diets. Such a supplement would increase the amount of protein in the diet and, if it had the right amino acid composition, would also enrich the quality of the protein already being consumed. Fish as a good source of high-quality animal protein was an obvious candidate. Its high content of the essential amino acids (in which vegetable proteins are deficient) was seen by many to make it an especially valuable supplement, especially if it could be economically converted to a stable and acceptable form that could be readily stored and distributed to those in need.

An important question was "why choose fish in the form of FPC?" The answer was economics. Where fish is available locally at low cost, it provides the best solution, but where it has to be stored and distributed over large areas — often in tropical climates — some form of preservation is required. Fish can be stabilized readily by such established processes as canning and freezing. However, it was projected then that FPC would offer a considerably lower delivered cost-per-unit of protein.

Additionally, the concept of upgrading the uses and value of fish meal, already manufactured on a large scale and widely employed in animal nutrition, was an attractive prospect to a large established industry in several countries.

Many of these concerns and the actions taken in several countries were brought to a focus in 1961 in an International Conference on Fish in Nutrition, held under the auspices of the United Nations' FAO in Washington, DC (Heen and Kruezer, 1962). The report of this meeting provides an excellent summary of knowledge and opinion at that time of the role of fish in human and animal nutrition, the nutritive components in fish, their contribution to national diets, the effects of processing on their dietary quality, the demand for fish as human food, and the possibilities for increased consumption.

DEFINITIONS OF FISH PROTEIN CONCENTRATE (FPC)

For many years, FPC was known widely as "fish flour." This term was discarded generally in response to the concern of cereal manufacturers that the term "flour" is generally understood to mean "a fine meal or flour made from grain," and that "fish flour" could be misleading or confusing. The FAO adopted the term "fish protein concentrate" in 1961 (Snyder, 1968). A proposed Standard of Identity for Fish Flour, to be made from fish fillets, was published by the United States Food and Drug Administration (FDA) in 1962 (Federal Register, 1962), but was subsequently withdrawn. In 1967, the FDA, in a revised standard, changed the name of the product to "Whole Fish Protein Concentrate" (Federal Register, 1967).

Various definitions have been proposed for FPC, all of which largely cover the same ground. One difference is that some forms are prepared from whole fish, some from the edible portions of fish, and others from fish scraps and trimmings. Snyder looked at a broad concept: ". . . a product variously known as fish flour, edible fish meal or FPC is generally considered to be a dehydrated, partially or almost completely defatted product which may or may not be flavorless or odorless" (Snyder, 1968). An FAO specification (Fishing News International, 1962) proposed three types of FPC. Type A was a deodorized, low-fat form, containing less than 0.75 percent lipids. Type B was an extracted fish meal containing not more than 3 percent lipids. Type C was fish meal prepared under sanitary conditions. All three types had to be free of *Enterococci, Salmonella/Shigella,* coagulase-positive *Staphylococci,* and pathogenic anaerobes. The FDA's 1967 regulation for whole fish protein concentrate described a product that was essentially an FAO Type A product, made by solvent extraction of "hake or hake-like fish," having a maximum of 0.5 percent residual lipids and free from excessive odors and flavors. The regulation was amended three years later to remove earlier packaging restrictions, permit the use of menhaden and herring, and impose a fluoride limit on the product (Federal Register, 1970). In 1970 the Canadian Government promulgated a somewhat similar standard (*Canada Gazette*, 1970).

After World War II, much of the research on FPC was directed to the removal of the fish lipids and therefore of all or part of the "fishy" odor and flavor by solvent extraction, giving Type A products. Some work was also done on preparing hydrolysates of fish using commercial proteolytic enzymes or microorganisms. Such products have been known to man for many years, particularly in Asia where preparations made by digestion of fish, such as *nuoc mam* of Vietnam and *bagoong* of the Philippines, are widely used mainly as flavoring rather than as nutritional supplements and probably have limited nutritive value as used. Another form of fish protein now widely in use is found in the minced fish products (*surimi*) of Japan. Fish protein isolates form a further and more refined series of products (reviewed by Spinelli et al., 1974). However, such forms are used as foods rather than as low-cost nutritional supplements and will not be considered here. The following review will therefore be limited mainly to fish protein concentrate prepared by the solvent extraction of fish or fish parts either before or after drying.

THE PREPARATION OF FISH PROTEIN CONCENTRATE

The term "preparation" used for this section is used advisedly, since relatively little FPC has been manufactured on a commercial scale and used in commerce. Most of the work reported has been performed on laboratory or pilot plant levels that hardly qualify as "manufacturing." This FPC is, of course, distinct from fish meal, which has been the basis for a large and well-established world-wide industry for many years. Both experimental and larger scale work is described in the following.

General Considerations for an FPC Process

From the definitions given above, it may be seen that there are four main technical objectives of any process for making FPC. The first is to remove water so as to concentrate the protein and make an end product that remains stable to flavor and odor alterations during transport and storage. The second objective usually sought is to remove all or most of the fish lipids which, being highly unsaturated, are particularly susceptible to oxidation; oxidation would lead to the development of undesirable odors and flavors when FPC is added as a supplement to foods. Some exceptions to this objective are noted under Types B and C. The third objective is to retain the nutritional value of the proteins during the process. A fourth, in common with all food processes, is to ensure the safety of the final product. Fifth, there is always the economic objective of manufacturing such products at a low enough price to make their intended purpose attainable.

Holding and Preparation of Raw Material

Technically, there is no limitation to the species of fish from which FPC can be made, and it has been produced experimentally from a wide variety. However, there are clearly some practical limitations in producing a protein supplement to be distributed on a large scale to people who have little means of support. Such a supplement must be inexpensive. A major factor in production cost is that of the raw material. One pound of end product will take at least five pounds of fish, and much more if it is deboned or filleted before processing. Therefore, only species that can be caught in large quantities and face no competition from the food market can be used. Moreover, for a plant to be operated on an economic basis, it must be supplied with large volumes of fish for a good many days of the year. It has been estimated that the production cost of FPC made from hake by isopropyl alcohol extraction in a plant operating 150 days a year would increase by 66 percent if the plant capacity were reduced from approximately 1,200 tons to 120 tons of raw fish daily. For a 200-tons-a-day capacity plant, the cost would increase by 70 percent if its operating time were reduced from about 230 to 60 days/year (Finch, 1974). While estimates of cost for FPC differ, this comparison illustrates the impact of plant size on production cost. However, few fisheries could support the delivery of cheap fish on such a scale and for so long a season.

FPC has been made from both lean and fatty fish. The complete extraction of lipids from fatty fish, which may amount to over 15 percent (Stansby and Olcott, 1963), may complicate the process. Many of the species that are used for making fish meal are largely fatty in nature, as the considerable volume of fish oil produced adds substantially to the value of the plant output. Such species are potentially valuable for FPC, as they can be harvested in large amounts at a reasonable cost. Among these caught off the United States are the Atlantic and Gulf menhaden (*Brevoortia* spp.), northern anchovy (*Engraulis mordax)*, the Atlantic herring (*Clupea harengus harengus*), and the Pacific sardine (*Sardinops sajax*). Atlantic and Pacific herrings have also been fished in other parts of the North Pacific, the Atlantic off Canada, and the North Sea for reduction to fish meal, condensed fish solubles, and oil. Large fisheries for reduction exist off South Africa, based mainly on herring and maasbanker. Canadian regulations for making FPC permit the use of fresh, whole edible fish of the order Clupeiformes, families Clupeidae and Osmeridae; and the order Gadiformes, family Gadidae; or from trimmings resulting from the filleting of such fish when eviscerated.

As FPC is to be added to human foods, the fish used in its preparation must be stored and handled in accordance with food processing standards. This may not always be an easy task, as landings of fish are usually irregular: "feast or famine." Yet for a plant to work efficiently and cost-effectively, it needs to operate at a reasonably constant rate. Sufficient buffer storage must, therefore, be available to be able to accept surges in supply and to even out the flow entering the process. Sometimes this will mean holding fish for quite long periods in an environment that protects it from spoilage.

FPC has been prepared from whole fish, deboned fish, fish fillets, and various forms of fish offal. Whole fish has perhaps been most generally investigated since it provides the lowest cost form of raw material, gives the highest yield for a given weight of raw material, and does not require the additional, costly steps of boning or filleting. However, made from whole fish, it is lower in protein than when all or part of the bone is removed. There is also a potential danger in using whole fish: that the fluorine found in the bone will become concentrated to an excessive level, sufficient to cause mottling to teeth. This possibility was recognized by the FDA, which established a limit of 100 parts per million (ppm) of fluoride for FPC, although Scrimshaw noted that the Committee of Aquatic Food Resources of the United States National Academy of Sciences/National Research Council (NAS/NRC) did not consider that the levels likely to be encountered in use would constitute a food hazard (Scrimshaw, 1974). The use of whole fish also gave FPC high levels of calcium and phosphorus, and possibly also some trace elements such as arsenic and lead present in the original bone. (See "Safety" section below.) The use of fish offal and fish trimmings would provide a "free" source of raw material and eliminate a disposal cost where no fish meal plant is available. The quality and organoleptic properties of FPC made from such waste would vary with its composition.

Other preparative methods will be discussed below in the context of the type of FPC under consideration.

Type C Fish Protein Concentrate

Type C Fish Protein Concentrate is fish meal that has been prepared under sanitary conditions and meeting the FAO requirement for freedom from pathogenic organisms. As prepared commercially, these products would be considerably lower in cost than those prepared by more sophisticated means, but they would normally have a strong flavor and odor that is likely to limit their acceptance, even when added to foods at a low level. However, not everyone objects to these fishy odors and flavors. For example, Roels (1957), working in the former Belgian Congo, found that fish meal made by simple means using fish from Lake Tanganyika was quite acceptable to local populations. Bakken (1962) describes early work in Norway in which FPC made from haddock was found acceptable. Later products made mainly from cod and haddock and from filleting waste without solvent extraction were shown to be satisfactory in quality.

Type B Fish Protein Concentrate

Extensive work has been carried out in several countries on FPC made by the extraction of fish meal. The appeal of this form of FPC is obvious. Fish-meal plants have been in use for a long time in many parts of the world and form the basis of a well-established industry. They operate both from supplies of fish caught almost exclusively for reduction, such as the menhaden from the East and Gulf coasts of the United States, and from the byproducts of fish being prepared for human consumption. The fish meal operation provides a relatively low margin of profit unless it can be conducted on a large scale. So it is natural that the institutions and companies concerned have sought ways to produce more profitable products, one of which was FPC. There is the added attraction that fisheries, plants, and a wide technical and marketing expertise were already in place and could readily be brought to bear on new developments.

The simplest process would be solvent extraction of fish meal prepared by standard processes modified, if needed, to ensure safe and sanitary products. Dreosti (1972) points out that an advantage of such a process is that fish meal, as an intermediate product, is relatively stable, so that buffer stocks can be established to enable extraction plants to operate on a continual basis to eliminate problems raised by the erratic nature of the fishing operation. Moreover, with a continual flow of raw material, smaller extraction plants may be used, reducing the capital cost. In South Africa extensive trials using a large number of solvents, both polar and non-polar, to extract lipids from fish meal found countercurrent extraction with ethanol to be the most effective. Somlai (1967, 1968) was granted two patents for the extraction of fish meal with various solvents, but as far as is known, these have never been put into commercial operation for making FPC.

FPC has also been prepared from fish press cake, an intermediate product in the manufacture of fish meal. Fish is cooked, then much of the water, oil, and fish solubles are pressed out prior to drying into fish meal. This procedure involves taking an additional step before solvent extraction, but it has the compensating advantage of removing much of the water and lipids before extraction.

Type A Fish Protein Concentrate

It is possible that the FAO specification of Type A FPC could include some products made from fish meal if their residual lipids were reduced to 0.75 percent or less, but this section will cover only FPC made by solvent extraction of raw or cooked fish without pre-drying. Reference will also be made to the azeotropic process, which falls within the scope of the FDA regulation.

A wide range of processes have been proposed for the pretreatment of fish prior to solvent extraction, including acid or alkali digestion as well as cooking and pressing. These processes appear to have offered no advantages to the process and were not widely pursued. Finch (1974) summarized a number of processes reported for extracting water and fat from raw, cooked, or otherwise pretreated fish, often using ethyl, isopropyl, or butyl alcohols, sometimes mixed with a nonpolar solvent, or sometimes followed by separate extraction with a nonpolar solvent. The lipids in fish are complex and contain free fatty acids, phospholipids, sterols and sterol esters, and sometimes waxes, in addition to triglycerides. The amounts of these lipid components present in fish vary with the species and maturity of the fish, the season, and other factors. Alcohols such as ethyl and isopropyl can remove water as well as most of these potentially unstable components of the lipid fraction. However, in the process of removing water, they become diluted, and their ability to dissolve lipids is reduced. This is one reason that a countercurrent or a countercurrent batch extraction is needed when alcohols are used. Nonpolar solvents may extract lipids other than triglycerides incompletely. Residual phospholipids from nonpolar solvent extraction may degenerate with storage, giving off undesirable odors and flavors. Sidwell et al. (1970) found that differences in flavor in five lots of cookies containing 10 percent FPC each made by isopropyl alcohol extraction of a different species of fish did not correlate with the levels of residual lipid (0.09 to 0.24 percent) in the FPCs used. Nevertheless, it is likely that the removal of most of the fish lipids is necessary to give stability and ensure a reasonable absence of flavor in FPC. Indeed, specifications for FPC usually give limits for residual lipids, such as the 0.5 percent maximum given in the FDA regulation for FPC.

Drying of the extracted fish has been carried out by both batch and continuous operations, but it is important to avoid high temperatures that may reduce protein digestibility and quality. Several reports seem to agree that drying at temperatures not exceeding 100°C has no adverse effects on the protein efficiency ratio (PER). The drying process must also provide for the removal and recovery of the solvent(s) used.

The "Halifax" process developed in Canada (Power, 1962) and applied to a large-scale plant built at Canso, Nova Scotia (although apparently never put into production) (Regier, 1974), was based on isopropyl alcohol extraction of fish. A 50/ton day demonstration plant built in Aberdeen, Washington, under the authorization and funding of the United States Congress (PL 89-701), employed a three-stage, batch, countercurrent extraction process with isopropyl alcohol.

The technique of extraction has varied considerably, much of the experimental work having been carried out on a batch extraction basis. However, countercurrent extraction is more efficient, especially since in any large-scale operation the volume of the solvent(s) used is such an important cost factor that it must be recovered and recycled.

An ingenious process used for preparing concentrates of agricultural proteins has been applied to fish. This "azeotropic process" extracts fats from fish with a boiling solvent that is immiscible with water. As applied in Levin's "Viobin" process (Levin, 1952, 1950; Worsham and Levin, 1950), ground fish is fed into a reactor containing boiling 1,2-dichloroethane, commonly called EDC. The azeotropic mixture of water and EDC distills off at 71°C to a separating chamber, from which the EDC is recirculated to the reactor and reused continuously. The dehydrated and largely defatted fish sinks to the bottom of the reactor, from which it is withdrawn and washed with another solvent such as isopropyl alcohol to remove residual EDC, then finally dried. This azeotropic process is the only one known to have been used commercially in the United States, although other such methods have been proposed.

BIOLOGICAL PROCESSES FOR MAKING FPC

As mentioned earlier, the digestion of fish by biological processes has been known and used for years, although the products seem largely to be used as flavoring rather than nutritional additions to foods. These traditional processes have been supplemented with a considerable amount of research, none of which appears to have moved them into commercial use for making protein supplements. Since they do not come within the scope of this chapter, no detailed review will be made here. [See Snyder (1968) for an overview of a number of these processes, and Finch (1970) for a summary of work published through 1970.]

THE SAFETY AND NUTRITION OF FPC

The Safety of FPC

The safety of protein concentrates made by the solvent extraction of fish has been widely investigated. Extensive data, including the results of animal feeding tests on FPC, were provided to the FDA by the Bureau of Commercial Fisheries (later the National Marine Fisheries Service) and by the Viobin Corporation. Considering these and other available results, the FDA concluded that FPC prepared by the extraction of whole "hake and hake-like fish" with isopropyl alcohol or with EDC followed by isopropyl alcohol was safe for use as a supplement in foods, provided it met stated specifications of composition, limits to solvent residues, nutritive value, and specified microbiological requirements. This regulation was subsequently amended to include FPC made from menhaden and herring of the genus *Clupea*. The safety of FPC made by solvent extraction of other species (herring, *Sardinella*, menhaden, as well as from fish meal) has also been established by several other investigators using animal growth and reproduction studies. Stillings (1974) cites toxicological studies on the safety of FPC made by solvent extraction of hake, menhaden, herring, sardine meal, and fish-meal extraction. No evidence of toxicity was found. FPC made by solvent extraction of fish is now generally recognized as a safe (GRAS) food supplement, provided it is free from chemical and bacterial toxins.

Toxins might arise in FPC from several sources. The first is from the fish itself. As described in another chapter, some species of fish such as lampreys, hagfish, ratfish, or moray eels may contain natural toxins, although such species are unlikely to be used for making FPC. Other species are known to convey ciguatera poisoning, especially at certain locations and times. Puffers, ocean sunfish, and porcupine fish can contain especially virulent toxins, although these too are unlikely to be used for FPC. Another source is the presence of toxins accumulating in fish from their environment. Organic compounds, such as PCBs or DDT, have been widely reported in fish from around the world, but these substances can be expected to be removed almost entirely in the solvent extraction process. More insidious could be the presence of lead, arsenic, or mercury in the raw fish used, and these must be watched for in the final products. The Canadian standard sets limits of 0.5 ppm for lead and 3.5 ppm for arsenic.

A third hazard could be the fluorine content of FPC resulting from that occurring in the raw fish, especially in the bones. Fluoride levels in FPC made experimentally from whole fish were shown to run from 46 to 70 parts per million in Atlantic menhaden (Finch, 1970), to 143 to 240 ppm in cod, and even higher in some cartilaginous species: 372 parts per million in skate, and 761 parts per million in dogfish (Ke et al., 1970). The FDA sets a limit of 100 ppm in FPC. However, the preamble to the regulation notes that a subcommittee of the Committee on Marine Protein Resource Development of the National Academy of Sciences had reviewed existing information on the fluoride content of FPC and concluded that the likelihood of health benefits from the consumption of FPC by man greatly exceeds the risk of any tooth disfigurement. The Canadian regulation permits up to 150 parts per million. The FDA level may be met by the deboning of fish prior to its conversion to FPC, but the additional step adds considerably to its cost.

Residual solvents could prove harmful also, although small amounts of alcohol are found naturally in some foods such as bread. Wills et al. (1969) found no adverse effects produced by the daily ingestion by men of 6.4 mg/kg body weight of isopropyl alcohol. They calculated that levels in excess of 6,000 ppm in 20 g/day of FPC could safely be consumed by children weighing 30 kg. Nevertheless, FDA limits residual levels of isopropyl alcohol to 250 ppm. Canadian regulations set no specific level, but limit the residual isopropyl alcohol to that resulting from good manufacturing practice.

At one time, much controversy surrounded the use of EDC to make FPC because of its possible carcinogenicity. Morrison and Munro made FPC experimentally using EDC (Morrison and Munro, 1967). They reported that the process resulted in reduced histidine and cystine and that the availability of some amino acids was lowered, but they noted that the significance of the findings to commercial practice was uncertain. Certainly these adverse results were produced under conditions considerably more severe than occur in the manufacturing process, and Levin noted that no adverse effects had been reported during the use of thousands of tons of FPC prepared using EDC (Levin, 1968). The FDA established a maximum residual level of 5 ppm. More recently, EDC has been listed as a carcinogen (State of California, 1985; United States Public Health Service, 1988).

In common with all foods, the microbiological quality of FPC is important, and good plant sanitation is essential. Fish as landed can be host to a wide range of microflora, which have been widely documented. Most often studies have related the role of microorganisms in the spoilage of fish, but the presence of pathogens has also been widely noted. The solvent extraction process can be expected to greatly reduce the level of organisms occurring in the raw material. Goldmintz and Hull (1970) found that when samples of menhaden were converted to FPC by isopropyl alcohol extraction, the average total plate count (TPC) of 70,000 fell to 300. The average TPC of 20,000 in samples of raw hake fell to 200. The FDA limits the TPC to 10,000. *Salmonellae, E. coli,* and other pathogenic organisms must be absent. The Canadian TPC limit is the same, with no *E. coli* permitted. As mentioned above, FAO microbiological recommendations for all types of FPC also require a specific absence of *Enterococci, Salmonella/Shigella,* and coagulase-positive *Staphylococci.*

The Composition of FPC

Although a high protein content is common to all FPC, the level can vary considerably, mainly dependent on how much bone is left in the product. A typical solvent-extracted whole fish FPC might contain 78 percent protein (total nitrogen calculated as protein), 14.6 percent ash, 7 percent moisture, and 0.4 percent residual lipids. Type A FPC made from completely deboned fish can have a protein content of 96 percent. Made from whole fish, it is more typically 75 to 80 percent, and a minimum of 75 percent is required by both Canadian and United States regulations. Residual lipids can be as low as 0.2 percent or even less, although the result for any sample depends on the method used for its determination. Limits for lipids in FPC have been proposed or set as follows: United States, 0.5 percent; South Africa, 0.7 percent; FAO, 0.75 percent (Type A). A maximum of 10 percent moisture is specified by the FDA.

The Nutritive Value of FPC

Stillings (1974) has made an excellent review of the nutritive value of FPC. A common index of protein quality is its Protein Efficiency Ratio (PER). This is a measure in which the weight increase in test animals produced by a diet incorporating the test protein is compared with that produced by a similar diet using casein as a reference protein. So measured, fish protein is shown to be of a high quality. Beveridge (1947) showed the protein efficiency ratio (PER) of six species of fish to run from 104 to 135 percent that of beef. When converted to FPC, fish protein retains its high PER. Stillings summarizes results that show that samples of FPC made by isopropyl alcohol extraction of eight species of fish ranged from 99 to 112 percent that of casein as a reference, and that samples of FPC made from various species of fish by several methods ranged from 103 to 117 percent that of casein, with the exception of that made by the EDC extraction of red hake, which was 89 percent.

The key to the value of FPC as a protein supplement lies in the high quality of its amino acid composition. It is especially high in lysine, methionine, and tryptophan, as compared with most vegetable proteins, which are generally deficient in these compounds compared, for example, with human milk and other animal proteins. A purely vegetable diet, therefore, tends to be limited regarding these amino acids, and its PER is lower. A vegetable diet with a lower PER does not promote as high a level of growth for the amount fed as one with a higher PER. FPC not only has a higher PER in itself, but also has sufficient additional lysine, methionine, and tryptophan to enrich the quality of vegetable protein to which it is added and raise its PER also. Thus Stillings (1974) found that when FPC was added to wheat flour at 5 percent, 10 percent, and 15 percent of the weight of flour, the PER of the flour was raised from the original 31 percent to 69, 89, and 103 percent, respectively, of that of the reference casein.

Scrimshaw (1974) evaluated the use of FPC in infant and child feeding. He summarized results of feeding tests made by a number of investigators under considerably different circumstances in several parts of the world. He concluded that "there is extensive evidence that FPC is an effective source of protein for human consumption and that it can be used interchangeably with milk and vegetable proteins of good quality to meet the protein needs of infants and young children." Young and Scrimshaw (1974) studied the protein quality of FPC for the maintenance of adults. They found that FPC compares favorably with other good quality protein sources.

THE USE, ECONOMICS, AND FUTURE OF FPC

The Use of FPC

In the 1960s and 1970s, there was an increasing awareness of the growing number of malnourished people in the world. Many showed evidence of protein deficiencies in their diets, demonstrated by the wide-spread incidence of kwashiorkor, a protein deficiency disease. Rao (1962), surveying data on food supplies to different areas of the world, concluded that more than half of the earth's population was undernourished or malnourished and that protein malnutrition was the most serious nutrition problem in the critical areas. Many believed that protein supplementation was the answer to this problem.

Success in applying a protein supplement to alleviate these conditions imposes certain practical requirements. There must be:

An identifiable community suffering from protein malnutrition.

A means of distributing the protein supplement to that community, probably already incorporated into a food such as flour or bread, and of educating consumers as to its use.

An economical, low-cost supply of the supplement that is safe and meets minimal quality standards.

Sufficient raw material resources to maintain the supply on a continuing basis.

Programs to pay for the supplements, their incorporation into an adjunct such as cereal flour, their distribution, and education in their use and benefits. It is improbable that those in need will have the means to do all this, or will choose to pay extra for protein supplementation of their food.

Failure to meet any one of these conditions is likely to make use of the supplement impractical. For example, Dreosti (1972) noted that a program to provide FPC incorporated in brown bread in South Africa to improve the nutritional status of those in need was discontinued after about 1,000 tons had been produced. Although the product was acceptable, "it did not reach those most in need of improvement of their diets, i.e., non-European infants and young children."

The Economics of FPC

Although FPC is a high-quality protein supplement, it would be in competition with other protein supplements that, although less nutritious, may be significantly less expensive than FPC and are, therefore, economically preferable. If it is to be successful, then, it must be very low in cost. In addition, since transportation is expensive, it would probably have to be made at or reasonably near the point of use. As stated earlier, these requirements mean that there must be an abundant supply of cheap fish available for much of the year reasonably close to the area of use.

The economics of FPC have been examined in some detail by Crutchfield and Deacon (1974). They reviewed various estimates and concluded that production costs in the range of \$0.16 -\$0.18 per pound were attainable in Peru and \$0.245 to \$0.265 per pound in Southern California. Pariser (1974) examined the economics of manufacturing

FPC by isopropyl alcohol extraction, taking as his model a plant making FPC that has less than 0.2 percent residual lipid and meets FDA requirements, a plant that is capable of processing 200 tons of fish in 24 hours and operating for at least 200 days/year. He concluded that the direct operating cost would be $0.336 per pound of FPC. A final cost would be higher by $0.10 to $0.15, for a total of about $0.35 to $0.50 per pound (all presumably in 1974 dollars).

On the marketing side, a computerized study by Devanney et al. (1970) estimated the amounts of different supplements that would have to be added to two typical Chilean diets to enable them to meet adequate calorie and protein requirements. Using their results, Crutchfield and Deacon calculated the break-even cost in cents per pound at which FPC would provide equivalent protein fortification to alternative protein supplements to meet 150 and 200 percent of FAO/WHO recommended protein requirements. At the 150 percent level, the FPC break-even costs would be: for dry skim milk ($0.36), peanut flour ($0.11), cottonseed flour ($0.22), soy flour ($0.17), and lysine and methionine ($0.21). However, they note that these figures may be complicated by several factors such as availability, acceptability, etc., and that the relative applicability of supplements in any given situation cannot be judged solely by these costs.

The Future of FPC

It would be tempting to say that the future of FPC has passed and, depending on how it is viewed, such a statement could be true. Finch (1977) reviewed the changing circumstances in which FPC arrived and departed from the fisheries and nutritional scenes. A crescendo of research, mainly technological and nutritional, was stirred up in several countries, notably Canada and the United States of America, in the 1960s and 1970s. It was fueled in part by the serious humanitarian concerns over widespread protein deficiencies expressed by the FAO/WHO, the United States NAS, and Congress. Considerable field activities took place, principally in Chile, Peru, India, Norway, and South Africa. In 1966, Congress passed an act (PL 89-701) to conduct research on FPC, principally though the construction of an "experiment and demonstration plant" that would develop practicable and economic means for the production by the commercial fishing industry of FPC. This act was amended in 1968 but was not immediately funded, so that by the time it expired in 1973, only one-and-a-half out of a projected three years' operation of the Demonstration Plant had been completed. Congress decided not to renew the authorizing Act.

Nevertheless, the project had demonstrated that FPC could be made on a large scale by isopropyl alcohol extraction. The principal product in this case was made from Pacific hake, which had to be deboned to meet the FDA fluoride requirement, an operation that considerably increased the cost and might not be needed. Smaller amounts of FPC were also made from menhaden and California anchovies. The experimental operation brought to light a number of practical problems, not all of which could be solved in the limited time available. These included handling and holding raw fish to match the plant flow, the loss of fine particles and solvent, and difficulties in oil recovery and in the very fine grinding of the FPC. The program was reviewed and summarized by North (North Services Inc., 1973). Several commercial ventures had been made into FPC. Astra Nutrition AB of Sweden developed a process using isopropyl alcohol extraction and made considerable quantities of FPC that were used in a number of large-scale studies, many in the FAO's World Food Program. Cardinal Proteins constructed a large plant in Canso, Nova Scotia, but due to fish supply and financing problems, production was never started. Alpine Geophysical Corporation built and operated a Viobin EDC extraction plant in New Bedford, Massachusetts, for two years, supplying FPC to the United States Agency for International Development. Smaller-scale operations appeared in other parts of the world. However, most of these ventures were short-lived, and it is believed that none of them is still active.

In the early 1970s, as more information became available, some of the original premises on which the concept of FPC was based appeared questionable. Several factors described below led to a growing loss of faith in the viability of the FPC concept and doubt as to its commercial feasibility. In 1973, the United States government program was terminated; the commercial ventures into FPC began to phase out at about the same time. Some of the initial assumptions in question were:

First, it was assumed that large resources of fish were readily available at a low cost. Some expectations of boundless or at least very large ocean resources that were widely held in the late 1960s have not been borne out in the light of advancing knowledge and experience. Fishing capacity and technology have grown enormously since then, so that world fish catches have gone from 48 million metric tons (mmT) in 1963 to 60 mmT in 1967, 70 mmT in 1977, and 100 mmT in 1989, although the diminution in the harvest of wild marine fishes has been augmented in recent years by the growth of commercial aquaculture. Nevertheless, competition for fish supplies has increased, driving fish prices up. The prospect in the early 1970s of large supplies of cheap fish needed to make FPC at acceptably low prices has virtually vanished. There could be one or two exceptions, such as the use of the vast Peruvian anchovy fishery or of United States menhaden, but, overall, the prospects have been discouraging for investment in commercial FPC ventures.

The second factor was a re-evaluation of the incidence of protein deficiency in the world. While the occurrence of malnutrition remains widespread, more recent examination seems to indicate that the proportion of protein-deficient diets is much smaller than was thought. It is now believed that many cases of apparent protein deficiency observed are due to protein-calorie malnutrition, which may often be overcome by an adequate total food intake, as sufficient protein of adequate quality is usually provided by the vegetable components of the diet. However, in limited areas where the principal energy source is provided by starchy foods such as manioc, true protein deficiencies may occur. Also the greater needs of pregnant women and children have to be considered. Still, the perception of need for protein supplementation of any kind, as opposed to the simple provision of a sufficient diet, has diminished considerably.

Third, it became apparent that the cost of FPC would be considerably greater than had been originally estimated. This miscalculation was due in part to the higher price of fish, which is the single largest cost factor in manufacturing FPC, and also to more realistic manufacturing cost estimates developed in the light of experience.

The introduction of FPC was also undoubtedly made more difficult by unreasonable requirements imposed by the FDA on use of FPC in the United States (Olcott, 1974), requirements that do not apply to other countries but sometimes strongly influence them. A comment sometimes made to proposals for nutritional supplementation programs in Latin America using FPC was "Don't expect us to use a product that your own FDA will not permit you to supply to your own people." Devanney and Mahnken (Pariser, 1974) commented that "... these restrictions were as economically prohibitive as direct prohibition." There was also considerable active opposition to FPC from sources of competitive protein resources, especially from dairy interests.

Absent some unexpected change of circumstances, it appears likely that the use of FPC as a nutritional supplement is history. An exception might be forms with functional proteins or others that might be developed for particular uses. As an example, with the demise of FPC as a nutritive supplement, fish protein research and development took off in different directions, concentrating more on flesh separation techniques, already highly developed in Japan.

The origins of FPC were driven by a mixture of motives, the dominant being, probably, a deep concern for suffering mankind. The work done considerably increased our knowledge of the chemistry and nutrition of fish, but its lack of success is a warning to both government and industry to examine prospects more carefully and dispassionately before expending substantial emotion, money, and effort in future research and development programs, however well-meaning and imaginative.

REFERENCES

Bakken, K. 1962. Technological Developments in Scandinavia. In *Fish in Nutrition.* Fishing News (Books) Ltd., pp. 419-424. London, England.

Beveridge, J. M. 1947. The nutritive value of marine products. XVI. The biological value of fish proteins. *J. Fish. Res. Bd. Can.* **7:** 35.

Canada Gazette. Oct. 14, 1970.

Crutchfield, J. A. and R. Deacon. 1974. The Economics of FPC. In *The Economics, Marketing and Technology of FPC.* Tannenbaum, S.R., Stillings, B.R., and Scrimshaw, N.S. (eds.), pp. 355-438. M.I.T. Press, Cambridge, MA.

Devanney, J. W. et al. 1970. *The Economics of Fish Protein Concentrate.* M.I.T. Sea Grant Program. Report MITSG 71-3. M.I.T. Cambridge, MA.

Dreosti, G.M. 1972. Technological Developments in South Africa. *Fish in Nutrition.* Fishing News (Books) Ltd., pp. 425-431. London, England.

FAO. 1962. *Third World Food Survey.* Basic Study No.11. Rome.

FAO. 1971. PAG Guideline for fish protein concentrates for human consumption. WHO/UNICEF Protein Advisory Group. United Nations, NY.

Federal Register. Fish Flour Identity. 1962. Washington, DC.

Federal Register. Whole Fish Protein Concentrate. 1970. Washington, DC.

Federal Register. Whole Fish Protein Concentrate. 1967. Washington, DC.

Finch, R. 1970. Fish proteins for human food. *Critical Reviews in Food Technology.* **1**:552-554.

Finch, R. 1974. FPC Processes. In *The Economics, Marketing and Technology of FPC.* Tannenbaum, S. R., B. R. Stillings, and N. S. Scrimshaw (eds.), pp. 45-101. MIT Press. Cambridge, MA.

Finch, R. A. 1977. Whatever happened to fish protein concentrate? *Food Technology.* Inst. Food Tech, pp. 44-53. Chicago, IL.

Fishing News International. 1962. Product Specifications for Concentrated Fish Protein. London, England.

Goldmintz, D. and J. C. Hull. 1970. Bacteriological aspects of fish protein concentrate production. *Dev. Ind. Microbiology.* **11**: 265.

Hannesson, G. Technological Developments in Iceland. *Dev. Ind. Microbiology.* **11**:416-418.

Heen, E. and R. Kreuzer (eds.). 1962. *Fish in Nutrition.* Fishing News (Books) Ltd. London, England.

Ke, P.J., H.E. Power, and L.W. Regier. 1970. Fluoride Content of FPC and Raw Fish. J. *Sci. Food and Agr.* **21**:108.

Levin, E. 1950. U.S. Patent 2,505,313.

Levin, E. 1952. U.S. Patent 2,619,425.

Levin, E. 1968. In *Proc. Conference on Fish Protein Concentrate,* Can. Fish Report No. 10. Dept. Fisheries, Canada, p. 107.

Morrison, A. B. and C. I. Munro. 1967. Factors affecting the nutritional value of fish flour. IV. The reaction between 1-2 dichloroethane and protein. *Can. Journal of Biochemistry.* **43:** 33.

National Center for FPC. 1970. Bureau of Commercial Fisheries (unpublished results). Quoted by Finch, R., in Fish Proteins for Human Foods. *Critical Reviews in Food Technology.* Chem. Rubber Co., Cleveland, OH.

North Services Inc. 1973. Fish protein concentrate information package. Parts I, II, and III. Publication 245-346. *Nat. Tech. Info. Svc.* Dept. Commerce, Washington, DC.

Olcott, H. S. 1974. Introductory remarks in *The Economics, Marketing and Technology of Fish Protein Concentrate* . Tannenbaum, S.R., B. R. Stillings, and N. S. Scrimshaw (eds.), pp. 1-6. M.I.T. Press. Cambridge, MA.

Pariser, E. R. 1974. An FPC manufacturing operation. In *The Economics, Marketing and Technology of Fish Protein Concentrate.* Tannenbaum, S. R., B. R. Stillings, and N. S. Scrimshaw (eds.), pp. 446-467.

Power, H. E. 1962. An improved method for the preparation of fish protein concentrates from cod. *J. Fish. Res. Bd.* **19:**1039-1045. Canada.

Rao, K.K.P.N. 1962. Food intake, nutritional requirements and incidence of malnutrition. In *Fish in Nutrition.* Fishing News (Books) Ltd. London, pp. 237-246.

Regier, L. W. 1974. Canadian FPC research and development. In *The Economics, Marketing and Technology of Fish Protein Concentrate.* Tannenbaum, S.R., B.R. Stillings, and N.S. Scrimshaw (eds.), pp. 103-110. M.I.T. Press. Cambridge, MA.

Roels, O. A. 1957. *Bull. Agr. Congo Belg.* **48**:423.

Scrimshaw, N. S. 1974. FMC for infant and child feeding. In *The Economics, Marketing and Technology of FPC.* Tannenbaum, S. R., B. R. Stillings, and N.S Scrimshaw (eds.), pp. 212-233. M.I.T. Press, Cambridge, MA.

Sidwell, V. D., B. R. Stillings, and G. M. Knobl. 1970. The fish protein concentrate story. *Food Technology.* (**10**) 24:876.

Snyder, D. G. 1970. *Food Technology.* (**10**)24:411.

Snyder, D. G. 1968. *Proc. Western Hemisphere Nutrition Conference II, San Juan, Puerto Rico,* p. 99.

Somlai, I. 1967. French Patent 492,228.

Somlai, I. 1968. British Patent 1,157,676.

Spinelli, J., H. J. Groninger Jr., and B. Koury. 1974. *Preparation and Properties of Chemically and Enzymically Modified Protein Isolates for Use as Food Ingredients in Fishery Products.* R. Kreuzer (ed.). Fishing News (Books) Ltd. London, England.

Stansby, M. E. and H. S. Olcott. 1963. Composition of Fish. *Industrial Fisheries Technology.* M.E. Stansby (ed.). Reinhold, New York.

State of California. 1988. Regulations under *Safe Drinking Water and Toxic Enforcement Act of 1986.* Health and Welfare Agency. Sacramento, CA.

Stillings, B. R. 1967. Nutritional evaluation of fish protein concentrate. In *Activities Report.* Research and Development Associates. Natick, MA. **19**: 109.

Stillings, B. R. 1974. Nutritional and Safety Characteristics of FPC. In *The Economics, Marketing and Technology of FPC.* Tannenbaum, S.R., B.R. Stillings, and N.S. Scrimshaw (eds.), pp. 165-211; Table 13. M.I.T. Press, Cambridge, MA.

United Nations. 1968. *International Action to Avert the Protein Crisis.* New York.

U.S. Government Printing Office. 1967. *Third World Food Problem.* A Report of the President's Science Advisory Committee. Washington, DC.

U.S. Public Health Service. 1985. *Carcinogens.*

Wills, J. H., E. M. Jameson, and F. Coulston. 1969. Effects on man of daily ingestion of small doses of isopropyl alcohol. *J. Toxicol. Applied Pharmacol.* **15:** 560.

Worsham, E. M. and E. Levin. 1950. U.S. Patent 2,503,312.

Young and Scrimshaw. 1974. FPC for adults. In *Technology of FPC.* Tannebaum, S. R., B. R. Stillings, and N. S. Scrimshaw (eds.), pp. 212-251. M.I.T. Press. Cambridge, MA.

Cultured Pearls

Shigeru Akamatsu

Few records on pearl culturing remain, with the Chinese invention the most significant among them. In the twelfth century (A.D.), the Chinese made tiny carved images of Buddha and cemented them inside freshly carved mussels; after several months, a deposit of nacre covered the whole image (Figure 1). Strictly speaking, this is not a true pearl but a part of the nacre of a shell. This technique has been superseded by the modern method of hemispherical "Mabe" pearl culturing.

In Europe the Swedish naturalist Linnaeus (Carl von Linné: 1707-1778) was the first person to suggest the possibility of pearl culturing. In 1748 he sent a letter to the Swiss anatomist Von Haller, writing: "At length I have ascertained the manner in which pearls originate and grow in shells; and in the course of five or six years I am able to produce, in any mother-of-pearl shell the size of one's hand, a pearl as large as the seed of the common vetch."

The Japanese were also interested in pearls. A collection of more than 4,000 natural pearls of the eighth century — kept in the Imperial Treasure House "Shosoin" in Nara — indicates the close relationship of the Japanese to natural pearls from these ancient times. Thanks to Kokichi Mikimoto, Japan may now be thought of as the land of cultured pearls (Figure 2). As a child he saw many natural pearls sold at a considerable price in the trading centers of Yokohama and Tokyo, and determined to culture pearls himself. In 1892, after painstaking efforts, he succeeded in culturing hemispherical pearls. It is not certain whether Mikimoto knew of the Chinese technique, but basically his culturing method was nearly the same as the old image-of-Buddha technique. Following the success of hemispherical pearl culturing, it took about twelve years for Mikimoto to culture whole, round pearls.

Stimulated by Mikimoto's success, many people joined the whole-round-pearl culturing race. Three Japanese (Kokichi Mikimoto, Tatsuhei Mise, and Tokichi Nishikawa) reached the same conclusion at nearly the same time: the most important substance in the production of pearl was not a nucleus as an irritant, but a part of the epithelial tissue of the mantle. When this tissue is transplanted into a living oyster together with a nucleus, it develops to form a "pearl sac" with the nucleus in it. Then inside the sac a secretion of calcium carbonate begins and numerous calcium

Figure 1. Chinese images of Buddha in the 1100s.

Figure 2. Akoya pearl farm in Japan.

crystals, usually in the form of aragonite, are deposited on the surface of the nucleus. Of the three men trying to make the most of this mechanism, Nishikawa, who was a young Tokyo Imperial University graduate scientist (and later Mikimoto's son-in-law), introduced an epoch-making "piece method." This method requires cutting out the part of the mantle tissue which has the ability to secrete nacre, and to make small (3 x 3 millimeter) pieces. When a nucleus, with a piece firmly attached to it, is inserted into an oyster, the piece develops into a pearl sac. The piece method proved to be the most efficient and became the essential technique of the pearl culturing industry; it has been inherited by modern pearl culturers. In the case of freshwater pearl culturing, only the piece is inserted into a mussel – proving that a nucleus is not necessarily an indispensable substance.

Since the development of the piece method, various improvements, though minor, have been made. Among them the following two are important:

1. **Nucleus-inserting location**: Pearl culturers found that the gonad is the best location for placing a big nucleus, and thus the culturing of bigger pearls, larger than 10 millimeters in diameter, has become possible (Figure 3).
2. **Suspending culture**: At the beginning of pearl culturing, culturers kept their oysters on the ocean floor, so they needed many (women) divers to gather them (Figure 4).

When red tide (an extraordinary growth of harmful plankton) occurred, culturers suffered great losses from the death of oysters. But after introducing the suspending culturing method — which kept oysters in wire nets, suspended in the water from floating rafts — pearl culturing improved remarkably. They could move rafts to a place where the sea was warm, calm, and rich with food — all while incurring fewer labor costs.

From its beginnings in Japan, over the last 100 years the industry has expanded worldwide. In Japan most pearl is produced from the Akoya oyster *(Pinctada fucata)*. The black-lip oyster *(Pinctada margaritifera)*, the Mabe or wing-pearl oyster *(Pteria penguin)*, and the freshwater mussel *(Hyriopsis schlegili)* are used for pearl culturing, but the amount of pearl produced is not as great. Although China has a long pearl-culturing history, the modern industry started rather recently. In 1970 full-scale research into freshwater pearl culture began, and by 1980 a mass production system was established. Using the vast areas along huge rivers such as the Chang, Yellow, and Zhu Jiang, China produces most of the world's freshwater pearls. Marine pearl culturing is also expanding. Using the same Japanese Akoya oyster, production is increasing yearly; the quality is not far from gem. Aus-

Figure 3. Nucleus-inserting operation.

Figure 4. Pearl culturers working on rafts: cleaning shells, protecting them from parasites, moving them to the warmer pearl farm in winter, etc.

tralia is noted for production of large fine-cultured South Sea pearls, ranging from 10 to 20 millimeters. Pearl culturers use silver-lip oysters *(Pinctada maxima)*, big bivalves which are distributed on the northwestern coast of Australia and the Arafura Sea. In addition to Australia, cultured South Sea pearls are produced by Indonesia, the Philippines, and Myanmar.

As for black-pearl culturing using black-lip oysters, Okinawa produced very fine black pearls in the 1970s. In the 1980s, Tahiti's production increased remarkably. Now the main black-pearl culturing areas are Tahiti, Cook Island, and New Caledonia. Black-lip pearl oysters inhabit the Indian and the southwest Pacific oceans. The pearls produce a beautiful black color, and grow to sizes of more than 10 millimeters.

There are three different kinds of cultured pearls: nucleated, non-nucleated, and hemispherical Mabe-type pearls. Nucleated pearls have nuclei in the cores. At an early stage of pearl culturing, various materials such as stone, silver, clay, and coral were tried as nuclei; finally, shell beads made of freshwater mussels from the Mississippi and Tennessee Rivers were most satisfactory. Every year Japanese nucleus manufacturers import freshwater mussel shells from the United States to make nuclei. The manufactured nuclei are then sold in Japan and exported to pearl culturers in foreign countries. In nucleated cultured pearls, the thickness of the nacre is a very important factor in creating pearl quality. If the nacre layer is thin, it can easily be melted away by sweat, exposing the surface of the nucleus. The thickness of the pearl layer is bound to the culturing period. Recently, because of the shortening of the culturing period, very thinly-coated pearls — especially cultured Akoya pearls — have been brought to the market.

The difference in body structure of freshwater mussels from marine bivalves makes it difficult for pearl culturers to produce nucleated freshwater pearls. Instead of inserting a nucleus and a piece, they cut the parts of the mantle of the mussel to make pockets. Then only pieces of mantle are transplanted into them. As this freshwater pearl is made without a round nucleus, the shape is not round; it is often referred to as "rice" or "potato." With an improvement in culturing techniques, nearly round Chinese freshwater pearls are sold on the market.

The hemispherical Mabe-type pearl has quite a different structure from both nucleated and non-nucleated pearls. As mentioned before, strictly speaking, it is not a pearl but a part of shell nacre. The culturing technique is nearly the same as the old Chinese image-of-Buddha. Semispherical nuclei made of shell, agalmatolite, or plastics are glued onto shell surfaces. After a certain culturing period (six to 10 months), the oyster coats the nucleus with nacre. Later this is cut out from the shell and the nucleus is removed. The cap-like nacre is filled with resin and then backed with mother-of-pearl. Mabe *(Pteria penguin)* is the most common oyster for this kind of pearl culturing. Hence, even though the pearl is cultured with white-lip oyster or abalone (genus *Haliotis)*, it is often called "Mabe pearl" (Figure 5).

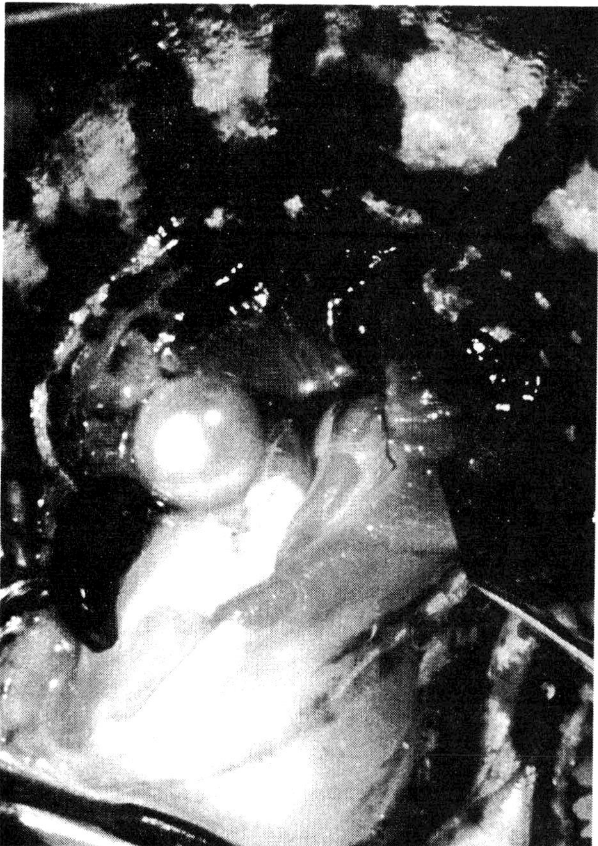

Figure 5. A Japanese pearl harvest in winter — the best season, since beautiful calcium carbonate crystals which improve pearl luster are then formed.

Industrial Products: Leather Production from Fish Skins

George J. Flick Jr. and Roy E. Martin

INTRODUCTION

"Leather" is the skin or hide of an animal chemically processed to prevent its deterioration by microbes, while "skin" refers to the outer coating of small land animals such as sheep, goats, calves, or of fishes such as salmon or cod. "Hide" refers to the external covering of large land animals — e.g., cattle and horses.

The skins of fishes can be converted into leather by methods that are similar and generally applicable to leather production from land animals. This aquatic source of skins would be particularly desirable for inhabitants of those areas that are not blessed with extensive green pastures and lush meadows necessary for grazing and maintaining large herds of cattle or wherever extremes of climate and topography are unsuited for raising goats and sheep.

The shortage of leather from land animals is further complicated in some countries by religious restrictions. For example, Hindus regard cows as sacred. Muslims and the Semitic people are forbidden to eat pork. These dietary restrictions reduce the supply of animal hides and skins for the production of leather.

For most of those countries which cannot produce a sufficient number of land animals to supply needed protein, it is fortunate that they are surrounded by large bodies of water which abound in fish. This supply of oceanic life, if it were systematically tapped, could serve as a source of skins that would supplement, and possibly even exceed, the supply of leather from land animals.

Before there was an endangered species set and marine mammal environmental protection concerns, the hides of walruses, porpoises, seals, and dolphins were used to make leather. Commercially the industry in the United States was never large. We do know that seal leather was produced on a commercial basis since before the turn of the century and that sharkskin leather for shoes and other prestige leather goods was produced into the 1950s. These two exotic leathers were classed in production and use at the time with alligator skin leather, which was the best known of the three (Moresi, 1964).

Other fish that have the potential to supply skins are salmon, cod, carp, pollock, shark, and hake. Provided the economics of the process would permit, substitution leather from aquatic sources could supplement some types of land-animal leather.

Since the late 1950s interest in fish skins has centered on salmon. One aspect of this accomplishment is cited in *Leather and Shoes* (1951) and by Joy (1953). This development was pioneered by John Metz in cooperation with the then-known Ocean Leather Corporation (Metz, 1953). Metz succeeded after years of experimentation in producing leather from salmon skins for shoe uppers.

More recently a new 90-day process developed by Hiland Processing, a tannery in Los Angeles, in cooperation with Key West Sandal Company, is producing salmon and salt-water skins. The key to the process is the ability to take out the oily material and eliminate odor. The company has also developed a method to sew salmon skins together for better dye-cutting (Martin, 1998).

HISTORICAL BACKGROUND OF THE LEATHER INDUSTRY

Leather-making is regarded as the world's oldest industry, dating back to prehistoric times. Archaeological records indicate that crude leather was "manufactured" more than 20,000 years BC (*Boot and Shoe Recorder*, 1963). It even preceded the housing and food industries; since early man lived in caves and had no facilities for fire-making and cooking, he was accustomed to eating his food raw but needed a protective covering for his body. Accordingly, he made leather in a crude way from animal hides by drying them first and pounding them against rocks to soften them. According to the Biblical account of approximately 5700 years ago, Adam and Eve used skin clothes (Genesis 3:21).

The ancient Arabs are credited with having devised a unique vegetable tanning-process. The Hebrews discovered oak-bark tanning which was used for many centuries and was still about the only tanning material used in medieval times. Throughout the ancient history of China, Persia, Egypt, Greece, and Rome there are references to leather-making and to its many uses, including those for military purposes.

The recently discovered Dead Sea scrolls were made from tanned skins of sheep and goats. The tanning of these skins was so skillfully carried out that the scrolls have withstood the elements for over 2000 years and are still intact.

The first major problem confronting the early leather producers was the preservation or curing of perishable skins and hides. Sun-drying constituted the earliest method of preservation. Later, skins and hides were treated with oils to prevent their deterioration. Subsequently, skins and hides were preserved by vegetable tanning, which was carried out with the aid of aqueous extracts of twigs, leaves, pods, fruits, shrubs, plants, barks, and roots.

Up to the turn of the twentieth century, progress in leather manufacturing was slight, being based primarily on trial and error. Development of the chrome-tanning process early in the twentieth century marked a great advance in this industry (Roberts, 1964). It has greatly accelerated the tanning operation, reducing it from four months required for vegetable tanning, to one month for chrome tanning. It has also increased the strength of the leather. About 20 percent of leather is still manufactured by the old vegetable-tanning technique, which is particularly applicable to heavy leather, such as sole leather or leather belting (Kaine, 1964).

FUNDAMENTAL PRINCIPLES AND PROCEDURES OF LEATHER MANUFACTURE

The converting of hides and skins into leather is a broad subject and it is impossible to specify in this summary the tanning materials and chemicals used, the time schedules of the operations, testing, temperatures, etc. There is substantial variance, depending upon the nature of the raw stock and upon the end uses of the leather (Moresi, 1964).

The fundamental principles involved in the production of leather from hides and skins of land animals are applicable, with modifications, to similar raw material from aquatic sources. Since each type of skin or hide requires special procedures best suited to its characteristics, manufacturing details differ in the production of the many different types of leather (Moresi, 1964). Nevertheless, regardless of the origin of the raw material, the production procedures are all governed by the same basic rules.

The similarity in properties, characteristics, and composition of the two kinds of raw material seems to justify the above indicated analogy in respect to their processing. The similarity in composition of skins can be seen from the fact that every skin or hide is divided into three distinct layers: (1) "epidermis" is the outer thin layer of epithelial tissue; (2) "derma" or "corium" is a much thicker layer of connective and other tissues referred to as the "true skin." Its principal constituent is collagen, which is present in the largest amount in the skins and is the chief component responsible for the formation of leather when combined with tanning agents after epidermis and flesh have been removed; and (3) "flesh" is fatty tissue, loose connective tissue, and some muscle tissue.

The major operations in the manufacture of leather (Kaine, 1964; Moresi, 1964) are discussed below.

Curing of Hides and Skins

As soon as the skins are removed from the animal or fish and adhering flesh is trimmed, preservatives are added; otherwise, incipient decomposition would set in, which would have deleterious effect on the properties of the finished leather. When salt is used as the preservative, it is usually spread over the flesh side and the skin is stored in such a way as to allow the brine to drain. In tropical countries, such as India and Java, skins are usually preserved by dehydration.

The tanning operations required to convert hides and skins into leather depend upon a number of variable factors such as the kind of raw stock being treated and the ultimate use of the leather. Fundamentally, according to Moresi (1964), the following operations are involved: (1) beam house, (2) tanning, (3) fat-liquoring and dyeing, (4) drying, (5) finishing, and (6) measuring and shipping.

Beam-house operations involve the preparation of the raw stock for tanning. They consist of (1) washing and soaking, (2) liming and sulfiding, and (3) splitting and bating.

Washing and Soaking: The hides and skins are washed and soaked with water by either drum-washing, paddle-washing, or soaking in vats. The washing and soaking operation plumps up the skins and hides, restoring them to their original texture and consistency.

Liming and Sulfiding: Before skins and hides are converted into leather, all hair and epidermis (composed chiefly of the protein keratin) must be removed. A saturated solution of calcium hydroxide, which has a maximum pH of 12.5, is most suitable for liming because it attacks the keratin almost exclusively. Thus, the hair and epidermis are softened and loosened. Collagen — which is the principal constituent of leather — is not appreciably damaged at pH 12.5 (Wilson, 1948).

Liming Procedure: The liming vats, either wooden or concrete, contain water, to which is added lime in amounts equal to 10 percent of the weight of the hides or skins. Two percent of the lime weight is sodium sulfide which is used to accelerate the reaction for the removal of hair and epidermis.

The washed and rehydrated skins or hides are then transferred to the liming vats. The hides are subjected to several fresh limings during the first few days. Fresh lime is used for each batch (Moresi, 1964). After the lime treatment, the skins are placed in a vat of warm water to facilitate hair removal.

Splitting and Bating: Adhering flesh is "split" or trimmed off. Next, bating hydrolyzes the elastin fibers in the skins, which, before hydrolysis, interferes with the proper swelling of the skins. Bating is carried out with the aid of proteolytic enzymes, particularly trypsin (Moresi, 1964; Wilson, 1948).

Since the enzyme acts best at pH 8.5, the bating solution is acidified to that pH with lactic acid. Elimination of the elastin fibers and lime softens the skin, prior to tanning (Kaine, 1964).

Tanning preserves hides and skins, rendering them resistant to deterioration. The two principal types of tannages are (1) vegetable, which is an extract from either bark or whole trees, such as oak, hemlock, chestnut, quebraco, or sumac, and (2) mineral, of which the principal example is chromium. There is also an alum tannage and a raw hide tannage.

Vegetable tanning applies principally to heavy leathers used for luggage, wallets, belts, fine leather goods, soles, and belting, while chrome tanning is used for shoe uppers (Moresi, 1964).

Tanning operations are carried out in drums or paddles, into which tan liquors and skins are placed. The strength or concentration of the tanning solutions is gradually increased in order to prevent plugging of surface pores which, in turn, would interfere with tanning efficiency.

Tanning may require from one to four weeks, depending upon the types of hides or skins being tanned, their thickness, and other factors.

The chrome-tanning process and the properties of the finished leather may be affected by a large number of variables such as temperature, time, basic chromium sulfate concentration, and degree of agitation.

Fat-liquoring and dyeing are carried out in drums. During the fat-liquoring operation, the leather is lubricated with oils and greases, all designed to replace the natural oil which has been removed during the beam-house operations. The strength and mellowness of the leather are improved by fat-liquoring. Dye-stuffs are used for coloring the skins.

During or after these operations, the hides and skins are shaved, or "split" to the desired thickness on automatic shaving machines and are put through wringers, or setting-out machines. These machines remove surplus water, smooth out wrinkles in the leather, and prepare the leather for drying.

Drying is done on tacking boards, toggle machines, or pasting units. The leather is pasted on large glass frames which automatically go through a tunnel drier.

Finishing operations are done either by hand or by machine. The finishing colors are applied by hand swabbing, by spraying in booths, or by spray machines attached to a tunnel drier. Dry splitting machines level the leather to the desired thickness. Machines are also used for ironing, glazing, and staking. Machine equipment includes large plating and embossing presses.

Measuring machines are designed to measure the square footage, irrespective of the shape and size of the leather.

For detailed information concerning tanning, coloring, and finishing processes, the reader is referred to O'Flaherty et al. (1956-1963).

EXAMPLES OF LEATHER PRODUCTION FROM THREE AQUATIC SOURCES

The application of the above principles, with modifications, to the following three types of skins from aquatic origin will be briefly described here: (1) salmon skins, (2) sharkskins, and (3) groundfish skins.

Principal Differences between Tanning Fish Skins and Land-Animal Skins

The tanning of fish skins varies from that of land-animal skins in the following respects. Scale is removed from the former, while hair is removed from the latter. The scale, in the case of shark hides, is removed by chemical processes — not by scraping. It is carried out in paddles or drums, and is a delicate operation. The chemicals used or the heat and time elements involved are closely-held trade secrets.

Vegetable tannages are generally used on shark and other fish skins, such as salmon and alligator.

Salmon Skins

One of the difficulties experienced with earlier aquatic leathers was that during the removal of scales and skins the surface would be scratched and torn, rendering it unsuited for leather production. A large canning company in the Pacific Northwest remedied this difficulty with salmon skins; with their patented process, skin and backbone can be removed from salmon undamaged.

Tanned salmon skins can be converted into a smooth, flexible leather of fine texture and durability, retaining its natural and variable pattern. This tends to give an "exclusive individuality" to each processed skin.

Notwithstanding the soft and pliable texture of salmon skins, they are, nevertheless, quite strong and durable — another asset in their application to footwear. Salmon skins are of medium to light weight and can be shaved to a required thickness. The skins are irregular in shape, and run, on the average, about 6 inches wide and 15 inches long (Mertz, 1953).

General Manufacturing Procedure

The skins are soaked and the scales are removed mechanically, leaving only the "scale pattern." The scales are cut off at the "roots," thereby eliminating the raspy surface condition which is prevalent in most aquatic type skins. The mechanical de-scaling process results in a smooth finish.

It is possible to dye salmon skins a wide variety of colors. The surface of the skin exhibits a two-tone effect, due to the fact that the skin surface and scale follicles take on different depths of color in the dyeing process. Salmon skins are aniline-dyed. Skins are first drum-dyed in suitable dye baths, then hung to dry to allow the dye to set completely. After the dye has set, the skins are rewetted, pasted out, seasoned, glazed, and finished in the usual way.

As to porosity and comfort against the foot, salmon skin compares favorably with the best leather. The leather can be cleaned with any good leather polish and the original colors hold well (Metz, 1953).

Sharkskin

The protective coating of sharkskin which corresponds to scales of other fishes consists of a calcareous deposit known as shagreen. The shagreen must be removed to render the sharkskin suitable for leather manufacture. Earlier scraping methods were not economical because they damaged most skins.

Progress in sharkskin leather production increased with the appearance of the patented processes of Kohler (1925) and Rogers (1922, 1921, 1920).

It was not until about 1925 that shark leather became suitable for use in shoes and fine leather goods. One of the first uses on a volume basis resulted from the introduction of sharkskin leather for the tips of the famous *shark-tip shoes*. These shoes were widely publicized by most of the best manufacturers of children's shoes and by leading retailers throughout the country. Sharkskin tips replaced metal tips and, because of their long-wearing and non-scuffing qualities, they proved ideal for this purpose. In the early thirties, shark tips were considered to be the most important development in shoes for boys, misses, and children. The production of the leather was limited but sufficient quantities were produced to interest a few high-grade shoe manufacturers. Today, although the quantity of leather available is still comparatively small, practically all of the high-grade men's shoe manufacturers have sharkskin shoes in their "line." Also, sharkskin leather is used by prestige manufacturers of belts, watch bands, wallets, and other fine leather goods (Moresi, 1964).

The identification and evaluation of the various types of sharks and the methods of catching them, removing the skins, and preserving them are discussed in a United States Government publication (United States Fish and Wildlife Service, 1945).

Tanning Operations

The preliminary preparation of sharkskins for tannage is similar to that of animal skins. First, the skins are freshened with water to wash out the salt and to soften them. Subsequently, the skins are limed, fleshed on a machine, and bated. The skins are then ready for tanning.

The procedure for removing the skin covering is the principal difference between shark leather and other leathers in the tanning operation. Because of the procedure practiced in the removal of the shagreen, chrome tannage cannot be used on sharkskin.

Groundfish Skins

The tanning of groundfish skins such as cod, haddock, cusk, and pollock is somewhat similar to that of sharkskin. The sequence of operations is as follows: scale removal, salt leaching, and treating skins with sodium carbonate to saponify small amounts of fat. Bating is carried out as described earlier. Tanning is accomplished by either the vegetable or chrome method. For these types of skins, chrome tanning is generally used, since it produces soft and pliable leather.

The follicles remaining after the removal of scales from the fish skins affect the thickness of the skins. The difference in color depth results in attractive color designs.

Up to the present time, no leather from groundfish skins is being produced in the United States because it cannot compete with leather produced from land-animal skins or synthetic leather. However, in seacoast countries in which the livestock industry is not highly developed, the production of leather from groundfish appears to be potentially feasible.

CHEMICAL REACTIONS INVOLVED IN TANNING

Liming

In removing the hair or plumping the stock a solution of milk of lime is generally used. Lime is obtained by "burning" limestone or calcium carbonate, whereby the gas (carbon dioxide) is driven off and calcium oxide remains. This calcium oxide, when treated with water, is converted into calcium hydroxide which, when mixed with water, forms a suspension known as milk of lime. Calcium hydroxide acting upon the hide substance forms soluble compounds with the albuminous materials, thus causing partial hydrolysis with a resulting swelling and opening of the fiber bundles. To a certain limited extent, also, the lime tends to saponify the fatty matter, thus making its removal less difficult in the subsequent operations.

Bating

The action of the bate through the ammonium chloride present is to remove the excess of lime, and a depleting effect upon the hide substance is brought about by the enzymes which it contains. The removal of the lime and the depleting action produce the desired softness and flaccidity required before actual tanning.

Vegetable-Tanning Materials

Tannic acid is the primary substance contained in tanning solutions. It combines with the gelatinous substance of the hide, forming an insoluble, nonputrescible product. Tannic acid alone, however, will not make a satisfactory leather as it requires the presence of other substances, called "non-tans," to secure the desired results. Sources of vegetable-tanning agents are: the barks of numerous trees, suck as oak, chestnut, and hemlock; the wood of some trees such as quebracho; and the leaves of certain shrubs such as sumac. The real tanning substances are obtained by extraction with water from all these materials.

Chrome Tannage

In the production of leather by the chrome process the active substance is basic sulfate of chromium. This compound may be produced in various ways, but in practically all cases it results from the reduction of sodium dichromate in the presence of sulfurous acid. The tannage is due to a deposition of chromium hydroxide and lower basic substances about the fiber bundles. Chrome tannage has the advantage over vegetable tannage in that the time required is much less and the resulting leather has characteristics which are very desirable for use in the manufacture of shoes.

Dyeing and Finishing

In dyeing leather, natural colors as well as coal-tar dye-stuffs are employed. For bright shades the basic dyes are most commonly used, but for certain purposes both acid and direct colors are applied. Pigment finishes have met with considerable favor and are now being used to a very large extent. To give leather the proper strength, fullness, and desired appearance it is necessary to introduce certain fats and oils at various stages in the process of manufacture, and to apply special finishes for producing a bright or dull appearance as the demand may indicate. The most common oils and fats are cod oil, neatsfoot oil, and degras, while the finishing materials consist of gum tragacanth, Irish moss, egg albumin, shellac, gelatin, and various alginates.

REFERENCES

Boot and Shoe Recorder. 1963. The wondrous world of American leathers. **164**(2):69-93.

Joy, A. F. 1953. Case of the tanned leather. *New England Homestead* **126**(9):31-32.

Kaine, E. J. 1964. Personal communication. Wabun, MA. Fishery By-Products Technology. AVI Publishing Co. 1965.

Kohler, T. H. 1925. Tanning and cleaning fish skins. United States Patent 1,524,039.

Leather and Shoes. 1951. Salium skin leather shoes. **122**(6):31-32.

Martin, Roy E. 1998. Personal communication. Key West Sandal Factory (Miami, FL). Alexandria, VA.

Metz, J. 1953. Personal communication. Boston, MA. Fishery By-Products Technology. AVI Publishing Co. 1965.

Moresi, L. R. 1964. Personal communication. Newark, NJ. Fishery By-Products Technology. AVI Publishing Co. 1965.

O'Flaherty, F., W. T. Roddy, and R. M. Lollar (eds.). 1956-1963. *The Chemistry and Technology of Leather.* Vols. 1-4. Reinhold Publishing Co., New York.

Roberts, I. B. 1964. Personal communication. Boston, MA. Fishery By-Products Technology. AVI Publishing Co. 1965.

Rogers, A. 1920. Sharkskins: preparation to tanning. United States Patent 1,338,531.

Rogers, A. 1921. Preserving sharkskins and the like. United States Patent 1,395,773.

Rogers, A. 1922. Treating sharkskins and the like. United States Patent 1,412,968.

United States Fish and Wildlife Service. 1945. Guide to commercial shark fishing in the Caribbean area. Leaflet No. 135. Washington, DC.

Wilson, J. W. 1948. *Modern Practice in Leather Manufacture.* Second printing. Reinhold Publishing Co., New York.

Industrial Products: Gelatin and Isinglass Production

George J. Flick Jr. and Roy E. Martin

GELATIN

INTRODUCTION

Gelatin prepared from a byproduct of the land-animal industry has many useful applications in foods, pharmaceuticals, photographic, and industrial products (Idson and Braswell, 1957). Depending on the details of the extraction process, gelatins with varying properties are obtained (Leuenberger, 1991). The principal use of gelatin is for increasing the viscosity of aqueous solutions and in the formation of solidified gels. It is used in the food industry in conjunction with other edible ingredients as a gelling, stabilizing, emulsifying, dispersing, or thickening agent (Selby, 1951). For example, in the manufacture of gelatin desserts, gelatin is combined with sugar, acidulants, flavor, and color. The entire mixture is dissolved in warm water and allowed to set at refrigeration temperatures; in this instance, gelatin serves as the gelling agent. In the production of ice cream, gelatin serves as the stabilizing agent.

From a nutritional standpoint, gelatin is an incomplete protein, since it lacks tryptophan, an essential amino acid (Dakin, 1920). An animal or human being could not survive if it had to subsist on gelatin as its sole source of protein. However, gelatin is relatively high in lysine and methionine.

Fish skins and fish bones could be used as a source of gelatin, although it is well known that gelatin from aquatic sources is not as good as gelatin from animal sources insofar as its gelling properties are concerned.

Norland (1990) has characterized the applications of gelatins into the following four uses:

1. Edible gelatin: free of heavy metals and aesthetically pleasing for eating.
2. Industrial gelatin: where the chemical and physical properties are uniquely suitable for an industrial application. A good example would be gelatin used in the microencapsulation of dye precursors for carbonless paper.
3. Photographic gelatin: the requirements are extremely critical. Photographic film requires a long shelf life. Gelatin has a major impact on silver halide chemistry and the ability to take a picture and develop it later under standard developing conditions.
4. Glue: especially for adhesive or gluing applications.

Fish gelatin, with the exception of photographic film, has been used in all these applications. Fish gelatin is also used as the base for a light-sensitive coating (or photoresist) for the electronics trade.

Distinction between Gelatin and Glue

The difference between gelatin and glue is explained by Bogue (1922a): "In commercial parlance a gelatin differs from a glue only in that the former is a very high grade product, is of high jelly strength, is light in color, gives solutions that are reasonably clear, is sweet, and does not contain excessive impurities."

PRELIMINARY TREATMENT OF RAW FISH STOCK FOR GELATIN MANUFACTURE

In the British Patent 235,635, Joseph C. Kernot (1924) describes a process for pretreating raw fish stock, such as skins, in order to obtain a gelatin which is both deodorized and improved in its gel strength. Essentially, the process consists of washing the stock with water, treating it first with dilute base solutions, then following it with dilute acid solutions and rinsing the stock with water after the base and acid treatments. A more detailed description follows.

The raw fish skins are put in suitable tanks and washed for three to four hours with running water. The water is then drained off and the skins are softened for six to eight hours with a dilute alkaline or alkali solution.

The solution should be very dilute; i.e., it should not exceed a 0.5 percent sodium hydroxide concentration. This low alkalinity is necessary in order to avoid excessive swelling of the stock. The solution is drained off at the end of the first softening period and a fresh basic solution is added. This treatment is repeated at least three times.

Subsequently, it is again washed thoroughly for three to four hours in running water. The water is drained off and the raw fish stock is softened with a dilute weak acid, such as sulfurous acid, the same number of times as it was subjected to in the alkaline or alkali treatment. Acidity and alkalinity concentrations should be equivalent. After the acid treatments, the fish skins are thoroughly washed with water for the third time. The skins are now ready for gelatin extraction. The extracted liquors are concentrated and dried.

The mild alkaline treatment, according to Kernot, liberates all of the volatile bases which are the causative agents of the characteristic fish-gelatin odors. The subsequent acid treatments remove the liberated bases. This process also imparts a stronger gel to the finished product since it removes albumins.

ALTERNATIVE PROCEDURES FOR THE PRELIMINARY TREATMENT OF FISH SKINS FOR GELATIN MANUFACTURE

Geiger et al. (1962) experimentally established the following conditions as being optimum for preparing dogfish gelatin: (1) wash skins in running tap water at 45°F for 24 hours, (2) treat skins at 68°F for 24 hours, changing the 0.2 percent sodium hydroxide solutions twice, followed by 15 minutes of washing in running tap water; subsequently, repeat the treatment twice with 0.2 percent of sulfurous acid solutions at 63°F for 24 hours, and (3) wash skins for four hours in running tap water.

The extraction of gelatin from treated fish skins was obtained by adding two parts of water for each part of pretreated material from 158° to 176°F for two consecutive 30-minute periods.

Before arriving at the conclusion that the above described procedure produces gels with the highest melting points, Geiger et al. evaluated extraction temperatures between 122° and 225°F, and extraction periods from 30 minutes to four hours. Bogue (1922) also describes the manufacture of gelatin, including isinglass.

In 1997, Gudmundsson and Hafsteinsson reported on the effects of varying chemical treatments on gelatin obtained from cod skins. Their process was a modification of the procedure developed by Grossman and Bergman in 1992. Their objective was to optimize the yield of gelatin from fish skins and retain a gelatin with the highest possible quality. The Grossman and Bergman process had been designed to eliminate the unpleasant fish odor from gelatin. Gudmundsson and Hafsteinsson found that various concentrations of sodium hydroxide and sulfuric and citric acids used in processing gelatin from cod skins affected both yield and quality. The highest yield of gelatin was obtained when low concentrations (0.1 to 0.2 percent) of sulfuric acid and sodium hydroxide were applied to the skins followed by treatment with 0.7 percent citric acid. The effects on bloom value, viscosity odor, clarity, color, and pH of the gelatin varied. However, the use of 0.7 percent citric acid in different combinations with sodium hydroxide and sulfuric acid usually provided the best results. Freezing gelatin had a considerable higher bloom value than air-dried gelatin.

The authors observed that the relative proportions of amino acids in the fish gelatins from various process treatments showed only minor variations. Different treatments had little effect on the amino acid composition of the gelatins. Thus, the different viscosities were concluded to be due to molecular weight differences. Cod and tilapia gelatin had the greatest compositional difference. The greatest difference was in the ratio of glycine to glutamic acid, which was higher for cod gelatin, and in proline and hydroxyproline, which was higher for tilapia.

Fresh but not cleaned cod skins could be kept up to three days in a refrigerator at 7°C without affecting the quality of the gelatin, especially odor. Seven-day-old skins made gelatins with unacceptable odor. There was no difference in the quality of gelatins whether fresh or frozen cod skins were used.

METHOD USED TO EXTRACT GELATIN FROM LUMPFISH

Osborne et al. (1990) described a process of extracting gelatin from lumpfish (*Cyclopeterus lumpus*) which included an alkaline wash (0.1 N NaOH) followed by an acid wash (0.1 N HCl), and a pH adjustment (pH 4 to 5) as required. The mixture was subsequently heated at 55°C for one hour and the temperature raised to 70°C and filtered. The filtrate was dried using a combination of a rotary evaporator and air dryer and then was ground.

PRODUCTION OF GELATIN FROM LUMPFISH

Three sources of lumpfish *(Cyclopterus lumpus)* — lumpfish skins (S), headed and gutted lumpfish (HG), and gutted lumpfish (G) — were used to produce gelatin (Osborne et al., 1990). The moisture content and pH for the gelatin were very similar to the reference, Knox gelatin. However, the content of ash was much higher and increased proportionally with the inclusion of raw material containing increasing quantities of bone. The amino acid content of the HG sample most closely resembled the reference. The best source, based on difficulty of preparing the raw material, amount of raw material per fish, content of amino acids, and the yield, was the HG fish. The powdered lumpfish gelatin was darker in color than the reference. Comparative sensory evaluations, conducted on strawberry-flavored jelly made from lumpfish gelatin and Knox gelatin, indicated the reference had a much higher gel strength and melting point than did the product prepared with lumpfish gelatin; however, the jelly prepared from the lumpfish gelatin tasted sweeter, since it melted at a faster rate in the mouth.

SIMILARITIES BETWEEN FISH AND ANIMAL GELATIN

The uses of fish gelatin in photoengraving led to its adoption in other interesting photographic applications. A considerable amount is used in nameplate manufacturing, where an image can be recessed by etching and then either filled in with lacquer or treated chemically to give a color that contrasts with the raised-metal surface. This nameplate is practically indestructible, since the recessed image can be read even though the lacquer has worn off.

Fish gelatin has been used more recently in the chemical etching of metal parts. Here it is possible to duplicate photographically a complicated metal part on thin metal, and to chemically etch unwanted metal to make a precision part. This process allows interesting metal fabrication that would be impossible to do mechanically. Additionally, the optical industry has found an interesting application of fish gelatin in what is called a glue-silver process.

Another major application is the formulation of coating for light-sensitive papers such as blueprint papers. Diazo-type materials are used as the light sensitizer with the addition of other chemicals that can be darkened with proper developing agents. The finished product is a coated paper that can be exposed photographically, washed out with water to remove the soluble areas, and then developed to darken the insoluble areas.

Thus, fish gelatin offers the unique properties of a liquid material that can be light sensitized and coated on almost any surface to give a photographically active coating.

PHYSICAL PROPERTIES

A correlation between Bloom number and Newtonian viscosity for mammalian gelatins has been identified. The Bloom number approximately increases with increasing viscosity. A correlation also exists between the activation energy *I* and the Newtonian viscosity. With increasing viscosity, the I-values roughly decrease for the mammalian gelatins. The I-values of the fish gelatins are compatible with the curve for mammalian gelatins. This indicates that the gelation process is similar for both types of gelatins.

A correlation has been shown between the gelation temperature T_c and the Newtonian viscosity. The T_{c} values increase with increasing viscosity for the mammalian gelatins. This increase is not constant but is largest at small viscosities. The fish gelatins do not duplicate mammalian gelatins. This is true for fish gelatins, which possess small T_{c} values but extremely large viscosities. However, both mammalian and fish gelatins have similar hydrodynamic radii (R_h) (Leuenberger, 1991).

Fish gelatins provide gelatins with different properties compared to mammalian gelatins: The standard fish gelatin has a solution viscosity that corresponds to a Bloom-30 gelatin, but the gelation temperature is even lower than that of the hydrolyzed gelatin. It is interesting that the activation energy of the standard fish gelatin corresponds to that of a Bloom-140 gelatin. This could indicate that the gelation properties of the two gelatins are similar. High-viscosity fish gelatins have viscosities higher than photographic gelatin. The gelation temperature, however, corresponds to a hydrolyzed gelatin.

The basic differences between fish and mammalian gelatins are probably due to their chemical composition. Fish gelatins have lower proline and hydroxyproline concentrations.

Fish gelatins may expand the technological application field of gelatins through their interesting physical properties. They provide gelatin qualities with rather higher-solution viscosities but low gelation temperatures when compared to mammalian gelatins. Fish gelatins may have direct application where high-solution viscosities are desired without concomitant gel formation.

Fraga and Williams (1985) studied the thermal properties of dehydrated "hot-cast" gelatin films obtained from hake skin (*Merluccius hubbsi*) using differential scanning calorimetry (dsc) and thermal mechanical analysis (tma). Two glass transition temperatures at 120°C and at 180° to 190°C were obtained. The low-temperature transition was assigned to the devitrification of blocks rich in the alpha-amino acids. The high-temperature transition was assigned to the devitrification of blocks rich in imino acids, proline, and hydroxyproline. For hydrated gelatins both transitions are shifted to lower temperatures. Mammalian gelatins have a high imino acid content that determines the extension of the protein secondary structure. On the other hand, fish gelatins have the lowest imino acid content, which produces a weak secondary structure.

A schematic diagram describing the mechanical modulus of fish gelatins as a function of temperature and moisture is contained in Figure 1. The first glass-transition transforms the material from a glass to a reinforced rubber. The second glass-transition leads to an ordinary rubber that eventually may begin to flow. In the case of mammalian gelatins, Region I (glass) and III (toughened glass) have almost the same mechanical modulus. Region IV represents a true overall glass-rubber transition.

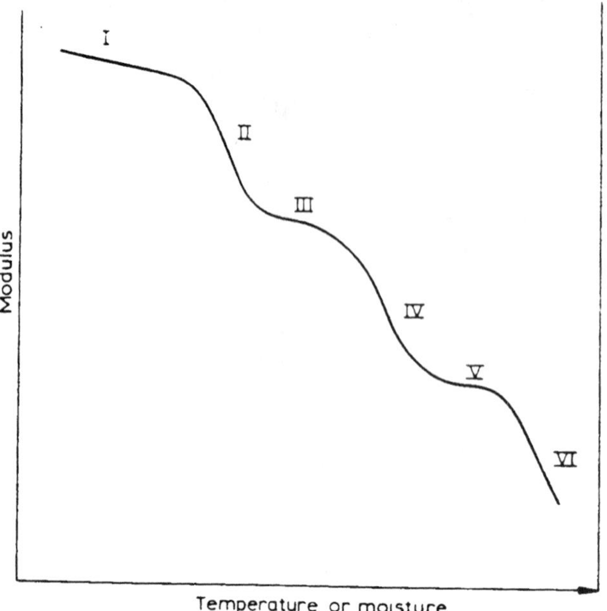

Figure 1. Schematic diagram of the mechanical modulus of fish gelatin as a function of temperature or moisture content. I. glass; II, glass transition of soft blocks; III, reinforced rubber; IV, glass transition of rigid blocks; V, rubber; VI, rubbery flow region.

The viscosity and gelation properties of three fish gelatins (Norland Products Ltd., Nova Scotia) were studied by Leuenberger (1991). The three fish gelatins showed a rather low gelation temperature, but relatively high solution viscosities. The fish gelatins did not correlate with any of the investigated mammalian gelatins. It was concluded that the fish gelatins provided a type of gelatin with different technological properties providing increased opportunities for the use of these gelatins. Of particular interest was use in applications where high solution viscosities without gel formation was desirable.

AMINO ACID COMPOSITION OF ELASMOBRANCH AND FISH GELATINS

Species of the elasmobranchs and fishes have been in existence for a long period, during the course of which they have been subjected to a range of environmental conditions, particularly with respect to temperature (Eastoe and Leach,

1977). From the perspective of gelatin composition, the groups can be divided into four categories: elasmobranchs (sharks, skates, and rays); cold-water fish, warm-water fish, and lungfishes which can be regarded as hot-water fish since they are able to survive relatively high temperatures.

The gelatins from these groups cover a wider range of composition than that exhibited by mammalian gelatins (Table 1). As previously stated, the hydroxyproline and proline contents are lower than those of the mammalian gelatins. The lowest hydroxyproline contents are exhibited by the cold-water species: 53 residuals per 1000 total residues for cod. The reduction of hydroxyl groups is offset by the increased serine content and to a lesser extent by increases in threonine and hydroxylysine, thereby resulting in an almost constant hydroxyl content for gelatins. The imino acid content is related to thermal stability of the collagen molecule. The amino acid content for crocodile skin is not substantially different from that of the other aquatic species described.

ISINGLASS

INTRODUCTION

Isinglass is an excellent raw material for producing fish gelatin. The word "isinglass" is derived from a similar-sounding word in Dutch and German meaning sturgeon's air bladder or swimming bladder. It consists of collagen which can readily be hydrolyzed to a high-grade fish gelatin or adhesive by heating the swimming bladder in water for a short period of time. The air bladders should be relatively large and easily detachable from the fish. Not all fish air bladders are suitable for isinglass production. In North America, air bladders obtained from large deep-water hake constitute the best raw material for isinglass production. One ton of hake yields about 45 pounds of dry isinglass containing approximately 85 percent gelatin, while one ton of cod yields about 18 pounds of a poorer-quality product, containing only 50 percent gelatin.

STRUCTURE AND FUNCTION OF THE FISH AIR BLADDER

The fish air bladder, also known as the fish sound or swimming bladder, is a compressible sac containing air. It is located in the abdominal cavity below the vertebral column, and consists of several external membranous layers or coatings, the innermost layer being the thinnest. The layer adjacent to the innermost layer is thick and fibrous and is high in collagen.

The function of the swimming bladder is to regulate the specific gravity of the fish, enabling it to maintain its position at any level in the water. In some fishes, such as hake, sturgeon, and carp, the swimming bladder is relatively large and well developed. Therefore, air bladders obtained from these fishes constitute a good source of isinglass.

PRODUCTION OF ISINGLASS FROM HAKE SWIMMING BLADDERS

Preliminary Treatment of Hake Swimming Bladders in Preparation for Isinglass Production

Hake swimming bladders are readily detached from the backbone during the dressing operation aboard ship. They are usually scraped and washed, although the latter two operations could be omitted without materially affecting the quality of the isinglass. During transit, the air bladders are temporarily preserved by salting. After they are delivered ashore, they are slit open and thoroughly washed, and the black outer membrane is removed by scraping. The swimming bladders are then air-dried.

Manufacturing Procedure for Isinglass

The cleaned, desalted, air-dried, and hardened swimming bladders are softened by immersing them in water for several hours. The swimming bladders, or sounds, are then mechanically cut into small pieces and rolled or compressed between hollow iron rollers that are cooled by water and are provided with a scraper for the removal of any adhering dried material. This rolling process converts the isinglass into thin strips or sheets of from 1/8 to 1/4 inch in thickness.

The sheets are further compressed by ribbon rollers into ribbons about 1/64 inch thick. These thin isinglass ribbons are easily dried while suspended in warm rooms for several hours. Finally the ribbons are rolled into coils.

FORMS OF ISINGLASS PRODUCED

Isinglass is produced either as leaf- or book-isinglass. To prepare it in the former form, the

Table 1. Composition of Selected Aquatic Animal Gelatins
(Values are expressed as numbers of amino acid residues per 1000 total residues)

	Crocodile Skin	Shark Skin[a]	Cod Skin[b]	Lungfish Skin[a]	Carp Skin[b]	Carp Scale Gelatin	Carp Swim Bladder Collagen	Carp Skin Gelatin	Cod Skin Gelatin	Pike Skin Gelatin	Calf Skin Gelatin
Alanine	114.0	119.0	107	126.0	120	119	126	120	107	114	112
Glycine	324	333	345	327	317	326	325	317	345	328	320
Valine	15.4	21.9	19	17.8	19	18	18	19	19	18	20
Leucine	20.1	23.9	23	19.8	25	22	21	25	23	20	25
Isoleucine	11.4	19.4	11	9.6	12	11	10	12	11	9.2	11
Proline	127.9	113.4	102	129.0	124	117	116	124	102	129	138
Phenylalanine	17.7	13.9	13	13.7	14	14	14	14	13	14	13
Tyrosine	3.3	1.4	3.5	0.8	3.2	3.3	2.0	3.2	3.5	1.8	2.6
Serine	42.1	44.5	69	42.1	43	43	37	43	69	41	36
Threonine	22	25.8	25	23.6	27	25	29	27	25	25	18
Cystine (half)	–	–	<1	–	<1	<1	<1	<1	<1	<1	<1
Methionine	6.5	10.0	13	3.4	12	14	13	12	13	12	4.3
Arginine	49.5	50.3	51	53.5	53	52	53	53	51	45	50
Histidine	4.7	7.4	7.5	4.5	4.5	5.2	3.8	4.5	7.5	7.4	5.0
Lysine	25.3	24.3	25	23.5	27	25	26	27	25	22	27
Aspartic acid	45.5	42.6	52	43.9	47	48	47	47	52	54	45
Glutamic acid	72.8	65.8	75	76.5	74	69	71	74	75	81	72
Hydroxyproline	92.8	78.5	53	78.2	73	82	81	73	53	70	94
Hydroxylysine	4.9	4.7	6.0	6.3	4.5	7.1	7.4	4.5	6.0	7.9	7.4
Amide N	–	–	–	–	–	(30)	(38)	(26)	(33)	(42)	(46)
Amide	25.5	29.4	33	42.3	26	–	–	–	–	–	–
Recovery by wt. (%)	99.8	96.2	–	96.1	–	–	–	–	–	–	–
Total N (%)	18.3	18.2	–	18.2	–	–	–	–	–	–	–
Recovery of N (%)	102.0	99.9	–	101.2	–	–	–	–	–	–	–
Shrinkage Temp., °C	–	–	–	–	–	–	54	57	40	55	65
Process	Acid	Acid	Auto-clave	Acid	Auto-clave						

References: [a]Eastoe (1957); [b]Piez and Gross (1960).

swimming bladders are immersed in warm water to remove dirt and mucous membranes, and are then opened and air-dried.

Book-isinglass is similarly processed but the swimming bladders are folded and covered with a damp cloth. Upon cooling, a 10 percent hot water solution of either type of isinglass sets to a good, firm gel.

PROPERTIES OF ISINGLASS

Isinglass dissolves readily in most dilute acids or alkalis, but is insoluble in alcohol. In hot water isinglass swells uniformly, producing an opalescent jelly with a characteristic fibrous structure which is absent in a gelatin sol. This characteristic behavior of isinglass in hot water serves as a method for differentiating it microscopically from a gelatin sol.

USES OF ISINGLASS

Isinglass is used as a clarifying agent for beverages such as cider, wine, beer, and vinegar by the entrapment of suspended impurities in the fibrous structure of the swollen isinglass (Tressler and Lemon, 1951).

Isinglass can also be used as an adhesive base. When dissolved in acetic acid, it forms a strong cement especially useful for glass or pottery. Modified isinglass can be used for repairing leather belts, or as a size for textiles. It serves as an essential ingredient in the manufacture of India ink, and was used in the production of some confectionery products (Bogue, 1922b; White, 1917).

As an adhesive, or as a thickening, emulsifying, dispersing, or jelling agent, isinglass cannot compete with gelatin or liquid fish glue in the United States or in other technologically advanced countries. It could, nevertheless, be advantageously used for local consumption by less-developed seacoast countries. This is particularly true in that the plant setup for isinglass production is inexpensive, and its manufacturing procedure is relatively simple as compared to the cost of large plants and up-to-date production methods for the manufacture of similar commodities in technologically advanced countries.

REFERENCES

Alexander, J. 1923. *Glues and Gelatin.* 1st Edition. Chemical Catalog Co., New York.

Ames, W. M. 1947. Heat degradation of gelatin. *J. Soc. Chem. Ind.* **66:** 279-284.

Ames, W. M. 1952. The conversion of collagen to gelatin and their molecular structures. *J. Sci. Food Agr.* **3**: 454-463.

Becher, C. Jr. 1950. Isinglass and fish glue. *Seifen-Oele-Fette-Wachse.* **76**: 391-392.

Bello, J. and J. Vinograd. 1955. Fundamental studies on gelatin. U. S. Army Contract No. DA-49-007-MD-298.

Bender, A. E., D. S. Miller, and E. J. Tunnah. 1953. The biological value of gelatin. *Chem. and Ind.* **799**.

Boedtker, H. and P. M. Doty. 1954. A study of gelatin molecules, aggregates and gels. *J. Phys. Chem.* **58**:968-983.

Bogue, R. H. 1920. Properties and constitution of glues and gelatin. *Chem. Met. Eng.* **23**:154-158.

Bogue, R. H. 1922a. *Chemistry and Technology of Gelatin and Glue.* 1st Edition. McGraw-Hill Book Co., New York.

Bogue, R. H. 1922b. Isinglass. *Chem. Age* **30**: 183-184.

Bowes, J. H. 1951. Composition of skin-collagen and the effect of alkalis on collagen. *Research* **4**:155-162.

Calwell, J. R. 1952. Acid solutions of proteins. U. S. Patent 2,592,120.

Cassel, J. M. and J. R. Kanagy. 1949. The purification of collagen. *J. Res. Natl. Bur. Std.* **42**:557-565.

Cohn, E. J. and J. T. Edsall. 1943. *Proteins, Amino Acids and Peptides.* 1st Edition. Reinhold Publishing Corp., New York.

Conrad, L. J. and H. S. Stiles. 1954. Improving the whipping properties of gelatin. U. S. Patent 2,692,201.

Dakin, H. D. 1920. Amino acids of gelatin. *J. Biol. Chem.* **44**:524-525.

Decoudon, C. and R. G. H. Decoudon. 1951. Process for manufacture of gelatin and glue. French Patent 992, 174.

Eastoe, J. E. and A. A. Leach. 1977. In *The Science and Technology of Gelatins*, A. G. Ward and A. Courts (eds.), pp. 79-80. Academic Press, London.

Fraga, A. N. and R. J. J. Williams. 1985. Thermal properties of gelatin films. *Polymer* **26**:113-118.

Ferguson, C. S. 1950. Composition for preserving articles of food and the like. U. S. Patent 2,532,489.

Gallop, P. M. 1955. Studies on a parent gelatin from ichthyocol. *Arch. Biochem. and Biophys.* **54**:501-512.

Geiger, S. E., E. Roberts, and N. Tomlinson. 1962. Dogfish gelatin. *J. Fisheries Res. Board Can.* **19**. No. 2, 321-326. Vancouver, BC.

Glass, J. V. S. 1939. Manufacture of gelatin. U. S. Patent 2,184,494.

Grettie, D. P. 1950. Gelatin desserts. U. S. Patent 2,519,961.

Gustavson, K. H. 1956. *The Chemistry and Reactivity of Collagen.* 1st Edition. Academic Press, New York.

Gudmundsson, M. and H. Hafsteinsson. 1997. Gelatin from cod skins as affected by chemical treatments. *J. Food Sci.* **62**(1):37-39, 47.

Highberger, J. H. 1953. Methods of forming fibers from collagen. U. S. Patent 2,631,942.

Idson, B. and E. Braswell. 1957. Gelatin. *Advances in Food Research* 7:235-338.

Kanagy, J. R. 1947. Chemistry of collagen. Natl. Bur. Stds. Circ. C458.

Kernot, J. C. 1924. Improvements in and relating to the manufacture of glue, gelatin and meal from fish and other offal of marine origin. Brit. Patent 235,635.

Lesparre, J. N. 1948. A coating composition for meat products. U. S. Patent 2,440,517.

Leuenberger, B. H. 1991. Investigation of viscosity and gelation properties of different mammalian and fish gelatins. *Food Hydrocolloids* 5(4):353-361.

Neuman, R. E. 1949. Amino acid composition of gelatins, collagens, and elastins from different sources. *Arch. Biochem.* **24**: 289-298.

Neuman, R. E. and M. A. Logan. 1950. The determination of hydroxyproline. *J. Biol. Chem.* **184**:299-306.

Norland, R. E. 1964. Personal communication. Cranbury, NJ.

Norland, R. E. 1990. Fish gelatin. In *Advances in Fish Fisheries Technology and Biotechnology for Increased Profitability,* M. N. Voight and J. K. Botta (eds.), p. 325. Technomic Publishing Co., Lancaster, PA.

Osborne, K., M. N. Voight, and D. E. Hall. 1990. Utilization of lumpfish carcasses for production of gelatin. In *Advances in Fisheries Technology and Biotechnology for Increased Profitability*, M. N. Voight and J. K. Botta (eds.), p. 143. Technomic Publishing Co., Lancaster, PA.

Randall, J. T. (ed.). 1953. *The Nature and Structure of Collagen*. Academic Press, New York.

Selby, J. W. 1951. Uses of gelatin in food. *Food*. **20**: 284.

Sheppard, S. E. and R. C. Houck. 1946. Methods of refining gelatin. U. S. Patent 2,400,375.

Sifferd, R. H. 1951. Preparation of gelatin. U. S. Patent 2,580,049.

Smith, P. I. 1943. *Glue and Gelatin*. 1st Edition. Chemical Publishing Co., New York. Revised edition of a work published in London, 1929.

Suyama, M., T. Hirano, and T. Suzuki. 1988. Nitrogenous constituents in hot water extracts of snapping turtle. *Nippon Suisan Gakkaishi* **54**(3):505-509.

Suyama, M., T. Hirano, and T. Suzuki. 1988a. Changes in gelatin in hot water extracts of snapping turtle during heating. *Nippon Suisan Gakkaishi* **54**(3):635-638.

Tressler, D. K. and J. M. Lemon. 1951. *Marine Products of Commerce.* 2nd Edition. Reinhold Publishing Corp., New York.

Veis, A. and J. Cohen. 1955. The degradation of collagen. *J. Am. Chem. Soc.* **77**:2364-2368.

Wallerstein, L. and J. Pfanmuller. 1942. Art of producing gelatin and glue. U. S. Patent 2,290,081.

White, G. F. 1917. Fish Isinglass and Glue. U. S. Bureau of Fisheries Doc. 852. Washington, D.C.

Industrial Products: Insulin

George J. Flick Jr. and Roy E. Martin

INTRODUCTION

Insulin is a hormone which maintains the blood-sugar level relatively constant in humans under normal metabolic conditions. It is produced by islets located in the epithelial tissue of the pancreas and is secreted internally by these islets and passed into general circulation to help store and oxidize glucose and convert it into energy for the use of various cells and tissues. These islets have been named the "islets of Langerhans" in honor of the man who first described them in 1869. Their exact function, however, was not known until 1922.

RELATION OF INSULIN TO DIABETES

When the islets of Langerhans do not produce enough insulin, or make available a sufficient supply to maintain the blood-sugar level within a normal range, the condition of diabetes mellitus sets in. While this disease, when unchecked, represents primarily a malfunction of carbohydrate metabolism, it also controls the proper metabolism of proteins and fats.

The old adage in nutrition chemistry that "fat burns in the flame of carbohydrates" indicates the interlocking mechanism of these two dietary essentials from which the body derives energy. Similarly, protein metabolism and formation also becomes affected. Unchecked diabetes — i.e., allowing excessive amounts of unused sugar in the blood stream and urine — may lead to a number of metabolic complications which, if medically neglected, become irreversible, causing permanent damage to the body through atherosclerosis, arteriosclerosis, or vascular disorders.

Some of the manifestations of this insidious disease are excessive thirst and urination, rapid loss in weight as a result of the increased catabolism of proteins and fats, loss of sugar in the urine, excessive hunger, fatigue, drowsiness, visual disturbances, skin infections, and the formation of an increased amount of ketonic substances such as acetone, acetoacetic acid, and B-hydroxybutyric acid. In fact, the physiological processes of every cell in the body are affected by diabetes.

Daily subcutaneous insulin injections correct the abnormal metabolic condition by lowering increased blood and urinary sugar levels. The administration of insulin under proper medical supervision is vital to life and health (Beaser, 1964).

HISTORY OF INSULIN DISCOVERY

The final discovery of the immediate source of insulin, its purification, and administration to humans is directly related to research on fish carried out in the Physiology Department of the University of Toronto, Canada. As early as 1889, J. von Hehring and O. Minkowsky showed that dogs develop symptoms of diabetes mellitus after removal of their pancreas. This work indicated that the pancreas has the additional function of controlling carbohydrate metabolism besides digestive enzymes.

McCormick (1924A) gave an account of the epoch-making discovery of insulin in his bulletin "Insulin from Fish." According to McCormick, Dr. Frederick G. Banting and Charles H. Best obtained a potent extract from a mammalian pancreas which, when injected subcutaneously in a diabetic animal or man, completely corrected and controlled their deranged carbohydrate metabolism.

This life-giving extract was named "insulin." They did not, however, pinpoint the exact focus in the pancreas from which insulin is extracted. The "missing link" for establishing the exact source of insulin was supplied by subsequent research on fish (McCormick, 1924B).

The pancreas of the cartilaginous fishes (Elasmobrachii) such as the skate, shark, and dogfish, have a regular structure with the typical islets of Langerhans embedded in their epithelial tissue and are very similar in this respect to mammalian islets. In the bony fishes (Teleosti) such as cod, haddock, halibut, and pollock, there are no sharply-defined islets.

Diamere, in 1896, observed that in the latter species the islet tissue was shifted in its location from the pancreas and appeared as a single agglomeration readily visible to the naked eye. For example, in the cod, the insulin-containing tissue appears as a cap at the tip of the gallbladder. In the pollock, the cap is often found at the side of the gallbladder. These caps, discernible to the naked eye, can be readily clipped off from the gallbladder.

Rennie, in 1901, described the large islets or "principal islets" of insulin in several dozen species of fishes. Frequently, some smaller islets, or secondary islets also occurred.

Injecting a diabetic patient with an extract prepared from the large islet of a monkfish reduced excessive blood sugar to the normal range.

Professor J.J.R. Macleod (1922), in whose laboratory insulin was discovered, further investigated with the object of proving conclusively the immediate source of insulin. He continued this phase of his work in the summer of 1922 at the Atlantic Biological Station, St. Andrews, New Brunswick, extracting large quantities of insulin from the principal and secondary islets of monkfish, sculpin, and flounder, along with a number of other fishes described by Rennie as having principal and some secondary islets filled with insulin.

As a result of this work, Macleod (1922) demonstrated the following: first, the amount of insulin extracted for the principal and secondary islets, excluding the pancreas, per pound of fish is directly proportional to the amount obtained from a mammalian pancreas per pound of body weight. Second, the remainder of the pancreas of those fishes contains only traces of insulin, since their pancreas had only a few scattered microscopic islets, which also resemble those found in the mammalian pancreas.

Consequently, Professor Macleod's experiments proved conclusively that in the mammalian pancreas, insulin is produced by the islets of Langerhans.

CHEMISTRY OF INSULIN

Insulin is a protein. Sanger (1949) and associates have shown that the insulin molecule consists of two polypeptide chains, one of 21 amino acids and the other of 30 joined by disulfide (-S-S-) linkages between their cysteine components. This structure accounts for the high sulfur content of the insulin molecule (3.3 percent).

Insulin contains all the commonly occurring amino acids except methionine, tryptophan, and hydroxyproline. Because of its polypeptide content, insulin cannot be taken by mouth since the strongly proteolytic enzymes of the stomach would readily hydrolyze it, splitting its peptide linkages and irreversibly inactivating the insulin.

The insulin molecule has no metal ions but readily combines with zinc. It is precipitated with a weak base from a strongly-buffered acetic acid solution at the isoelectric point of the insulin, where it is least soluble (Jensen, 1938).

The speed of absorption of crystalline insulin from the tissues is relatively fast, early use necessitating several daily injections for the average diabetic. The annoyance of insulin therapy has been overcome by the development of several long-acting and intermediate types of insulin: protamine insulin (Hagedorn et al., 1936), globin insulin, NPH insulin, and the Lente series of insulin. The rate of absorption of these is slower, and one injection is generally sufficient for a 24-hour period (Beaser, 1964). Cutfield et al. (1986) isolated and sequenced insulin from the teleost fish *Cottus scorpios* (daddy sculpin). Purification involved acid/alcohol extraction, gel filtration, and reverse-phase high-performance liquid chromatography to yield nearly 1 miligram pure insulin/gram wet-weight islet tissue. The biological potency was estimated as 40 percent compared to porcine insulin. The sculpin insulin crystallized in the absence of zinc ions, although zinc is known to be present in the islets in significant amounts. Two other hormones, glucagon and pancreatic polypeptide, were co-purified with the insulin. The primary structure of the insulin showed a number of sequence changes unique so far among teleost fish. The changes occurred at A14 (Arg), A15 (Val) and B2 (Asp). The B-chain contained 29 amino acids and there was no N-terminal extension as seen with other fish. Sculpin insulin did not readily form crystals containing zinc insulin hexamers.

PREPARATION OF INSULIN FROM FISH

In 1924, it at first seemed less expensive to produce insulin from fish than from mammalian sources such as beef or pig pancreas. This economy of insulin production from fish appeared particularly true when the gallbladders of cod, halibut, and pollock were used as the raw material (McCormick, 1924A).

Their large islets are readily noticeable and are very easily snipped off with scissors. One person can clip off 150 to 200 caps per hour from gallbladders of fresh offal.

In the United States, with a highly developed livestock industry, insulin has been and still is being produced as a byproduct of the meat-packing industry.

EARLY TECHNIQUES FOR COLLECTING INSULIN MATERIAL

Directions for collecting insulin material and preparing the insulin extract as practiced by McCormick (1924a) are essentially as follows. Deposit the offal of fresh fish in a receptacle. With a pair of scissors, immediately clip off the caps from the gallbladder and preserve the insulin-containing tissue in 95 percent alcohol, acidified with 0.3 percent mineral acid, preferably hydrochloric acid. Do not include any bile, gallbladder, or pancreatic tissue with the insulin.

Collect the tissue in a 4-ounce wide-mouth bottle half filled with the preserving fluid, and forward it daily to a central collecting station. Do not expose it to direct sunlight. This size container will hold insulin islets from approximately 4000 pounds of fish.

The alcohol concentration should not exceed 75 percent. Should delays in shipment arise, store bottles in cracked ice during transit for not more than 18 to 24 hours. The islets of cod, pollock, halibut, etc., may all be collected in the same container.

Preserving the insulin tissue in an acidified alcoholic solution was the method used during the early days of insulin production. A more desirable alternative is to freeze the tissue with solid carbon dioxide and ship it in frozen form (Fisher, 1964).

PREPARATION OF THE INSULIN EXTRACTS

The alcohol mixture is filtered through muslin and the tissue squeezed to dryness, weighed, and ground thoroughly in a mortar and re-extracted in supernatant alcohol, which has been adjusted to a 75-percent alcohol concentration. After the tissue is completely macerated, the mixture is allowed to remain at room temperature for about 1½ hours and the alcohol is again drained off through the muslin.

The residue is extracted three times in this fashion. The composite of the three alcoholic extracts is filtered through thin paper into porcelain trays and the alcohol is evaporated by vacuum distillation. The residual oil found in the aqueous insulin solution is extracted several times with ether, in a separator funnel.

While this insulin preparation has proven highly potent, in a few instances it showed undesirable local irritations. To prevent this, the alcoholic extract is further purified by conversion into insulin hydrochloride by the Dudley's process. Subsequently, the extract is concentrated and adjusted to a suitable potency determined by a standard procedure using rabbits or mice (Hill and Howitt, 1936).

Extensive chemical and physiological tests carried out on cod insulin showed it to be similar to insulin extracted from beef or pork. Further clinical tests carried out by Dr. W.R. Campbell on diabetic patients in the Toronto General Hospital proved very successful (McCormick, 1924A).

NEW CONSIDERATIONS: TRANSGENIC FISH FOR ISLET PRODUCTION

Fish guts may seem an unlikely source of organ transplants for humans, but Canadian scientists are trying to turn genetically-engineered fish into donors for pancreatic cells which could cure childhood diabetes. If the approach is a success, the scaly organ donors could mean an end to daily injections for millions of children.

When a healthy person eats sugar, cells in the islets of Langerhans secrete insulin, allowing tissue such as muscle to take up the sugar. But in the most serious form of diabetes, juvenile diabetes, the islets are destroyed and patients require daily injections of insulin for survival. Injections, however, cannot mimic the finely-tuned response of normal islets to blood sugar, and so fluctuating levels of sugar and insulin eventually damage diabetics' eyes, hearts, nerves, and kidneys.

For a decade scientists have been trying various ways to transplant healthy islets. A few people who had had their pancreases removed for other reasons have had their own islets successfully reimplanted. The cells are incorporated into liver tissue. Transplanting islets from other people has also worked for a few very ill diabetics, but this type of transplant can never be a routine treatment for children, says Jim Wright of the Izaac Walton Killam-Grace Health Center in Halifax, Nova Scotia. Transplant patients must take immunosuppressant drugs to prevent rejection of foreign tissue; these drugs increase the risk of developing other diseases, including infections and cancer. Putting a young diabetic on insulin is much safer, despite the long-term problems.

It also takes three human pancreases to supply enough islets for one transplant.

There are fewer than 10,000 pancreas donors in North America each year and 3 million diabetics.

The supply problem could theoretically be solved by using animal donors; the most widely-studied potential donor so far is the pig. But Wright says that even if tests show that pig islets are suitable, supplying enough cells will be a huge problem. "To get the 1 to 4 million islet cells needed for a human transplant, you need 10 pigs. To treat 10,000 diabetics a year, you need a million pigs," he says.

The animals would have to be reared in sterile conditions for two years. If each pig is given four square meters of space, 200 pig houses of 20,000 square meters, would be necessary to treat just 10,000 people. "The costs would be astronomical," says Wright.

Enter the tilapia, a freshwater fish widely farmed in the tropics. Unlike pigs, fish can be raised cheaply at high density in small spaces. Tilapia could also solve some of the other major problems with the cross-species transplantation. The first of these is finding a way to stop the patient's immune system from attacking the animal tissue without resorting to immunosuppressant drugs.

Scientists have previously transplanted rat and mouse islets between the two species without using drugs by encapsulating the islets in gel (MacKenzie, 1996). Wright's team uses gels derived from seaweed, called alginates. Their pores let sugar and oxygen in and insulin out, but exclude the cells and large molecules of the immune system.

In theory, encapsulated islets could be given to young diabetics with little or no immunosuppression, says Wright, but blood vessels cannot grow into the capsules, and many transplanted cells suffocate. Tilapia live in warm ponds with low levels of oxygen, so their cells need only one-fifth of the oxygen that human cells need. Thus, they should survive encapsulation.

The other big bonus with tilapia is that the fish have two pancreases, one for digestive enzymes and the other solely for insulin. Because one pancreas is basically an agglomeration of islets, it is much easier to isolate the cells from other tissue. For human or pig islets the extraction process costs $3000, which is 90 percent of the cost of human transplants.

Wright's team has transplanted fish islets into mice and rats, where they successfully produced insulin in response to changing blood-sugar levels. But there is still one big obstacle: Fish insulin works poorly because it differs from the human hormone by 17 amino acids. Pig insulin has only one different amino acid.

Wright's colleague Bill Pohajdak has cloned and modified the tilapia insulin gene to produce human insulin and the team is now injecting the gene into tilapia eggs. Wright says that previous experience with genetically-modified fish suggests that with enough injecting and screening, some of the animals will express the human gene in sperm or eggs. The team then hopes to breed a stable line of fish that produce only human insulin.

Fish availability in underdeveloped countries could potentially be a local source for insulin production. Local production could keep costs low enough for those who otherwise could not afford treatment.

REFERENCES

Beaser, S.B. Personal communication. Boston, MA.

Cutfield, J.F., S.M. Cutfield, A. Came, S.O. Emdin, and S. Falkmer. 1986. The isolation, purification and amino-acid sequence of insulin for the teleost fish daddy sculpin. *J. Biochem.* **158**:117-123.

Fisher, A.M. 1964. Personal communication. Toronto, Canada.

Hagedorn, H.C., B.N. Jensen, N.B. Krarup, and I. Woodstrup. 1936. Protamine insulinate. *J. Am. Med. Assoc.* **106**:177-180

Hill, D.W. and F.O. Howitt. 1936. Insulin: Its Production, Purification and Physiological Action. Hutchinson, London.

Jensen, H.F. 1938. Insulin: Its Chemistry and Physiology. Commonwealth Fund, New York.

Macleod, J.J.R. 1922. The source of insulin. *J. Metabolic Res.* **2** (2):149-172.

McCormick, N.A. 1924A. Insulin from fish. Bull Biol. Board. Canada.

McCormick, N.A. 1924B. The distribution and structure of the island of Langerhans in certain fresh-water and marine fishes. Trans. of the Roy. Can. Inst.

McCormick, N.A. and E.C. Noble. 1924. Insulin from fish. *J. Biol. Chem.* **59**. Proc. Soc. Biol., New Series, 2.

MacKenzie, D. 1996. Doctor's farm fish for insulin. *New Scientist.* **2**(2056):20.

Sanger, F. 1949. Terminal peptides of insulin. *Biochem. J.* **45**:563-574.

Tanikawa, E. 1965. Marine Products in Japan. Published by the author, Professor of Marine Food Technology, Faculty of Fisheries, Hokkaido University. Hakodate, Japan.

Industrial Products: Hydrolysates

George J. Flick Jr. and Roy E. Martin

INTRODUCTION

Hydrolysis is an enzyme, acid, or autolytic digestion process wherein protein materials are liquefied and fermented in a manner similar to that which occurs in the human system. Hydrolyzed fish protein or hydrolysate, oil, and coarse bone meal are produced. The hydrolysate can be dewatered to form a paste or dried to a shelf-stable powder. Oils can be centrifuged off and either added back to the hydrolysate or preserved for further refining. Bone meal, separated during the initial stage of hydrolysis, can reap its own benefits if dried and ground for agricultural markets.

The two common methods of hydrolysis produce two slightly different products. Acid hydrolysis changes the pH levels of the material, which alters the protein. Enzymatic hydrolysis involves separation of the protein, keeping its functional properties intact. Enzymatic hydrolysis is conducted at cooler temperatures, produces a slightly more nutritious product (higher in protein), and is considered to be gentler to the protein.

Hydrolysate differs from traditional fish meal in a couple of ways. Most important is that it is produced at low temperatures, so the valuable proteins remain intact and the ash content is lower than in regular fish meal. Hydrolyzed protein may also contain more fat — up to 35 percent in powder form, depending on the species and the process. Feed manufacturers could use high-fat hydrolysate in creating diets that are nutritionally balanced without introducing additional fats. The lipids in hydrolysate bind with the protein, so it can be compressed into a solid mass and easily divided into a fine-grained powder again, making hydrolyzed protein a good candidate for feeds.

The process of hydrolysis was developed a few thousand years ago as a simple way of reducing fish to its bare essentials: water and protein. Today it has been refined by a number of equipment companies around the world. Using fairly simple technology, fish processors now could install hydrolyzing equipment at a cost ranging from $150,000 to $1 million, depending on the system, and increase a plant's processing alternatives and profit potential (Lure, 1991).

AN AUTOMATED ENZYMATIC PROCESS

How Fish Are Handled

Whole fish are unloaded from trawlers by a pneumatic conveying system. This equipment consists of a pneumatic conveying line with a nominal diameter of twelve inches. Air, liquids, and whole fish are transported to a cyclone-type separator. Here, the fish are discharged through an elastomer-lined valve to the atmosphere at intervals as dictated by the static weight of fish and fluid above the valve. Conveying air, mists, and small particles leave the top of the cyclone separator to a secondary collector before reaching the positive displacement-type blower. Again, separated liquids are released to the atmosphere by the static weight acting on a sleeve valve. Dry air enters the blower, producing a vacuum in the order of one-half an atmosphere.

Process instruments are high-level indicators, switches, and alarms in both the unloading receiver and the secondary collectors. These level switches should alarm and shut down the blower in case of any malfunction of the separator discharge valves.

Raw fish is discharged from the unloading receiver in 250- to 500-pound increments onto a dewatering conveyor. The fish then move onto an inclined, belted, and cleated transport conveyor which rises from ground level to the second, or operating, level of the adjacent manufacturing building.

A magnetic conveyor sorts ferrous scrap metals from the raw fish in the building. This step is precautionary and is intended to protect "downstream" equipment. Raw fish is inspected and hand-sorted on a sorting conveyor. Non-magnetic trash and undesirable fish species such as sharks and rays are removed manually at this point. The total quantity of rejected material is not expected to exceed one weight percent of the incoming feed.

Following this hand-sort, the unloading rate is indicated as a continuous cumulative sum on a dual weigh scale. This device will consist of two parallel hoppers receiving product from the sorting conveyor. Each hopper will be sized to receive, weigh, and momentarily hold up to 300 pounds of whole fish. The weight sensors will alternately activate the diverting valve and discharge gates of each hopper.

Whole fish drop directly into an auger-fed prebreaker, where they are reduced in size to pieces approximating 2-inch cubes. Process dilution water is also added to the pre-breaker to assist in washing and carrying particles through this machine. The pre-breaker slurry is fed continuously into a hammer mill-type grinder, where the fish particles are further reduced to a maximum size of one-quarter inch and below, the bulk of which should be below one-eighth inch. This slurry will consist of fish tissues, dilution water, contained water, dissolved solids, and insoluble bones and scale.

Enzymatic Digestion Process

The finely-ground slurry flows into a feed-surge hopper with a capacity to contain 1,000 pounds of the mixture. The hopper is mounted directly on an auger feeder, which forces the slurry into the suction side of a sanitary rotary pump. The rotary speeds of both the auger feeder and the digester feed pumps are synchronized and regulated by adjustable speed controllers. The speed controllers are activated by a liquid-level control located in the feed hopper.

A high-level alarm signal will automatically shut down the pneumatic unloading system and the transport conveyor. A low-level alarm signal will stop the auger feeder and the digester feed pump. However, these alarms may be bypassed for start-up, testing, and shut-down operation. Additionally, electrical sequence circuits are to be established so that an overload on the grinders or other downstream equipment will alarm and shut down all upstream equipment.

Ground fish-slurry is continually pumped into the first of five jacketed, steam-heated, agitated, stainless-steel digestor vessels where the fish protein is enzymatically hydrolyzed. The digestion process is continuous. The fish slurry is further conditioned by the addition of enzymes and dilute caustic soda solutions.

Enzyme solution is metered into the first digestor with a proportioning pump. Low pressure steam in the heating jackets will raise the feed streams to the desired temperature and maintain this temperature in the digestion train. A magnetic or other slurry service sanitary flow meter will set a constant flow by controlling the variable speed digestor product pumps. The design flow rate will be such as to accommodate a 200-ton-per-day whole-fish slurry.

Digested slurry containing the undissolved bone and scale fraction is pumped at a constant rate from the last digestor to the first of two vibratory separating screens. Here, the bone and scale solids are removed from clear digested liquors. Two screens are employed: the first makes the primary separation between solids and process fluids; the second permits washing of the solids with process water.

A spiral-type preheater is used to raise the slurry to the desired temperature before screen separation. Low-pressure steam is used as the heating medium. This rise in temperature will stop further hydrolysis, inhibit bacterial decomposition, and preheat the liquors for subsequent evaporation.

Liquors are digested and washed, following separation, and flow by gravity into one of two evaporator feed tanks. The surge capacity in these vessels is sufficient to accumulate digest liquors from the digestion cycle and supply feed to the evaporators at a predetermined daily rate.

Bone Fraction Treatment

Separated, screened solids are transported into a dryer feed hopper with a nominal capacity of 350 cubic feet. The solids are then conveyed into a steam-heated dryer. Wet feed is admixed with recycled product before entry into the dryer itself. This is done to facilitate movement of the wet bone and scale fraction over the heating surfaces. Product temperatures will not exceed 200°F. The dryer operating cycle, at peak design rates, is 24 hours per day.

Dried product is discharged from the dryer recycle conveyor through an adjustable speed rotary airlock and moved to a pneumatic air conveying system. The product is simultaneously cooled and transported into one of two storage bins, each of which can hold a full day's production of dried material. Two bins are utilized to permit cleaning and/or inventory accumulation of in-process solids prior to final grinding, packaging, and shipping.

Dried solids are held in the storage bins and discharged through a rotary air lock into a bone/scale grinder. Live-bin bottoms are employed on the storage hoppers to ensure free flow of all materials. Bones and scale are reduced in the grinder to small mesh particles and are then pneumatically conveyed into a finished storage bin prior to packaging.

Evaporation System

The evaporator feed tanks are to be used alternately, thus permitting in-place cleaning as required. Clear digested liquors are pumped in a constant pressure loop around the two parallel evaporator banks. The feed pumps are sanitary centrifugals. The evaporator modules work in parallel. Digested liquors are concentrated to the desired percentage of dissolved-solids product under moderate vacuum, with steam as the heating medium.

The temperature of the material flowing out of each barrel is measured and indicated. Abnormally high operating temperatures in the vapor/liquid separator will alarm and automatically shut off the incoming steam supply to minimize thermal decomposition of the product on the heating surface.

The vacuum producers are water-sealed, electrically-driven mechanical pumps. Recovered process water is used as the sealing medium and is discharged through the pumps as a waste material to the process drain system. Air vented from the vacuum pumps will contain traces of odorous ammonia-related volatile elements stripped from the digestor liquors during evaporation.

Recovered water is collected in two process condensate collection vessels, each servicing one evaporator bank. This material is pumped hot, under level control, to the process water storage tank.

Concentrated product is pumped from the product collection tanks, under level control, to two parallel acidification systems for subsequent pH adjustment. The product collection tanks are jacketed and heated with warm water to maintain the proper temperature. These tanks are maintained at operating pressures and are vented to the vacuum system for this purpose. Process piping to and from the condensate receivers is electrically traced to facilitate flow and drainage of the viscous product.

Prior to storage and shipment, a pH adjustment is necessary. This is accomplished with dilute food-grade sulfuric acid in each of two agitated acidification tanks. Continuous pH indicating controllers adjust acid flow through the acid-metering pumps. Either system may be used to spare the other through parallel operation in the intended normal mode. High shear agitators minimize local organic decomposition when acid is admixed with the product.

Acidified liquor is pumped continuously under level control to one of two product check tanks. One eight-hour shift product accumulation is held for a quality and inventory check before final bulk storage and shipment. Each tank is used alternately, one being filled and checked, the second emptying and/or being cleaned.

Bone/Scale Fraction

One day's production of ground, cooled bone/scale solids is stored in a live bottom bin which feeds a bag-packaging unit on an eight-hour daily shift schedule. Ground product may be withdrawn or pneumatically fed into this bin simultaneously. The bag-packaging unit is designed to fill, under manual control, a minimum of 400 bags containing 50 pounds of dried product per shift.

Auxiliary Systems

The plant and manufacturing operations are self-sufficient with regard to process water. That is, aside from start-up requirements, all process water is obtained as a component of the raw material: whole fish. This water is made available for other uses after it is recovered in the evaporation step. Recovered process water will also be used for caustic soda solutions, enzyme solutions, vacuum pump seals (5 gpm), and CIP solutions; it is discharged as plant waste into the municipal sewer system.

Granular enzymes are supplied in 100-pound fiber drums, and the enzymes are added to the digestors as a metered solution. Enzyme solutions are made up by pneumatically conveying a drum's contents into a solution-tank system which will consist of a receiving hopper with dust collector, a solids-metering screw, and an agitated dissolving tank. The quantity of enzymes added on a continuous basis is set by manual adjustment of the solids-metering screw. Process water is added as a diluent to the agitated dissolving tank. The concentration of enzyme solution is established by the volumetric flow rate of the metering pumps. The dissolving tank capacity is such as to minimize solution hold-up. A small industrial vacuum system is employed to unload solids from the supply drums.

Caustic soda is supplied as a 50-weight per solution in a minimum 1,300 gallon lots by tank truck. Flow of dilute caustic soda into the digestors is manually set by flow rotometer indication for pH control. Dilute caustic soda solution may also be transferred to the CIP system as required.

Sulfuric acid is used to neutralize and acidify the concentrated digest product liquors following evaporation. Sulfuric acid will be supplied in tank truck quantities and kept in the concentrated sulfuric acid storage tank.

A three-tank CIP system is provided to clean 30 individual CIP circuits. Two supply pumps permit cleaning of two selected circuits simultaneously. The three-tank system permits acid, caustic, and rinse solution use. Normal operations include circuit isolation, air purging, and pre-rinse. The pre-rinse is discharged directly into a process drain. The second step is a heated, recirculating, alkaline wash. On selective circuits, a third step is a low-temperature acid rinse. Each circuit cleaning is completed with a cool water rinse.

Sulfuric acid, caustic soda, and recovered process water are used to manually make up the cleaning solution in three storage tanks located in the CIP room in the process building. Temperatures are controlled by direct injection of low-pressure steam. Manual flow distribution plates are used for contiguous circuits with the exception of single circuits with multiple supply points. These circuits will be sequence-controlled from a CIP local control panel (*Food Engineering*, 1997; Keyes and Meinke, 1996).

CHOOSING ENZYMES

Proteolytic enzymes are classified in several ways. They may be classified according to the conditions under which they work (for example, acid proteases or neutral proteases). They may be classified according to how they work, and divided into exo- and endo-proteases, depending upon the location of the peptide bonds they attack. They may also be classified according to the kind of organism that they come from; for example, proteases are commercially available from fungus, bacteria, animals, and vegetables. Thus papain, which is a popular enzyme for fish protein hydrolysis, might be described as a neutral, random endoprotease of vegetable origin. It is neutral because it works best in an environment which is neither very acidic nor very alkaline. It is an endoprotease because it attacks peptide bonds within the protein, rather than the ones at the end of the protein. It is random because it appears to attack a variety of peptide bonds rather than always attacking certain types (for example, those at one specific amino acid), and it comes from a tropical fruit: the papaya.

How important are these qualities? The fact that papain is derived from a vegetable is unimportant, but the fact that it is an endoprotease is very important.

Exoproteases work from the end of the protein molecule, snipping off one amino acid at a time. Ultimately, they would reduce the entire protein to single amino acids. But, if one is developing a product that will be fed to animals, free-amino acids are undesirable. Single-amino acids and very small peptides (a relatively small protein or a group of amino acids of any size which has been chopped off from a complete protein) can be extremely bitter; instead of enhancing palatability (which is a strong selling point for hydrolysates), they may actually reduce it.

In terms of the nutritional value of free-amino acids, many nutritionists believe that they are less easily absorbed and utilized by the body than are small peptides. So, for reasons of both palatability and nutrition, hydrolysed fish protein destined for feed markets should not contain free-amino acids, and should consist of the largest possible peptides consistent with liquefaction.

For these reasons, a neutral endoprotease for a fish protein hydrolysis should be chosen for the manufacture of a dry product.

An acidified product can be evaporated down to a relatively low moisture content (about 10 percent), but it cannot be completely dried. The acid holds on to a certain amount of water. One could do an acid hydrolysis and then neutralize the solution, but these steps represent considerable added effort and expense. In addition, neutralization forms salts that are undesirable in some products. For these reasons, an enzyme that works optimally at neutrality is needed if the product will be dried. If one intended to produce a wet or semimoist (condensed) acidic product, and wanted to use a commercial enzyme instead of depending upon the enzyme in the fish itself, there would be two options available: (1) Do a neutral digest, pasteurize, remove the bones and oil, and then acidify. Or, (2) acidify in the beginning and then digest with an acid protease.

How would one evaluate these two options? One way is to determine palatability. Since two enzymes are unlikely to be precisely the same, there might be a palatability difference between the products of the two methods. Establishing this difference would require fairly sophisticated testing, and the results would have to be fairly significant to tip the balance in one enzyme's favor. Another way is to compare the costs of the processes (Goldhor, 1998).

OTHER OPTIONS

Pigott et al. (1998) have shown that a high-quality fish protein hydrolysate can be commercially produced from whole fish or fish waste economically and with relatively simple engineering. This product, produced by pepsin hydrolysis, has excellent emulsification and foam formation qualities that make it useful in many formulations.

There has been a considerable amount of research work to develop or modify enzyme processes for specific applications or to overcome certain disadvantages of the final product. Protein hydrolysates characteristically have a bitter taste due to certain peptides that are produced. Lalasides and Sjoberg (1978) have reviewed this problem and have shown that azeotropic secondary butyl alcohol (SBA) extraction of enzymatic protein hydrolysates removed the bitter compounds. However, up to 10 percent of the hydrolysate is removed for total removal of bitterness. Adler-Nissen (1976) has prepared an excellent review paper that shows the requirement for concentrated proteins having high solubility. He points out that enzymatic hydrolysis is the best means of preparing products meeting this requirement but that the formation of bitter peptides is the main problem encountered.

Fujimaki et al. (1973, 1971) have studied the cause of bittering and other off-tastes, and how they can be removed. They have reported that a method of enzymatic hydrolysis followed by enzymatic resynthesis to produce a new substance called plastein results in a insoluble product that has a bland odor and taste. Yamashita et al. (1976) have used the plastein reaction to produce a low-phenylalanine, high-tyrosine dietetic food for phenylketonurics.

Heveia and Olcott (1977) have shown that the off-flavors in a fish protein hydrolysate prepared using the proteolytic enzymes bromelain and ficin were due to bitterness and the taste of glutamic acid. Techniques are becoming available for removing these flavors, particularly bitterness, when removal is essential for the intended use. Conversely, Pigott et al. (1978) have determined that bitterness can be reduced to a degree that is acceptable for many applications by controlled processing.

Iseki et al. (1969) used commercial enzymes to produce a high-quality functional product. Digestion temperatures of 50° to 60°C were maintained for four hours, with 0.3 percent enzyme (based on fish weight), giving yields of over 80 percent of the original protein. They found, however, that activated carbon was necessary to remove off-colors, and steam distillation was required for odor removal.

ACID HYDROLYSIS

About the time of World War II, it was discovered that waste fish and offal could be ground and preserved with acid while digestion occurred, resulting in a reasonably stable liquid product which could be a feed ingredient for livestock. This method is now known as "acid hydrolysis" or "fish silage." The fish is sufficiently pickled to inhibit microbial growth but not enough to stop enzyme activity.

The advantages of this method of acid hydrolysis are as follows:

(1) By varying the typem of acid used, the product can be made palatable to a wide range of animals, including cows, sheep, chicken, pigs, fish, and fur-bearing animals such as mink.
(2) Acid hydrolysis offers processors different options. It can be a very inexpensive and simple process, or it can be complex and costly. A number of steps can be added to the basic process that would increase production costs, but, in its simplest form, acid hydrolysis requires no fuel and almost no equipment. This procedure can be carried out with no more than a grinder, acid, a means of mixing acid into the minced or ground fish (a hand-held wooden paddle could suffice), and acid-proof containers in which to make and store the product.
(3) Acid hydrolysis products are inherently stable. Because these products are preserved by acid, they can be kept, literally for years, with only relatively small changes in nutritional content. However, the oil content must be preserved against oxidation (this is very easily accomplished with modern chemicals, and will be discussed at greater length in the section on stabilization). Once stabilized to prevent oxidation, acid hydrolysates will smell and taste fine to a wide range of animals for a remarkably long time.
(4) Another advantage is that the process produces essentially no odor. While sticking one's nose into a fresh batch can be quite unpleasant, the odors do not seem to carry; they diminish rapidly, and become more pleasant over time.
(5) Finally, this is one of those rare processes in which the yield is greater than 100 percent. When the product is sold in its completely wet state, it has all of the wastes plus the amount of acid added. In this instance the processor can expect a yield of about 120 percent.

An interesting variant of acid hydrolysis exists in which the acid is not added by the processor but is produced by a lactobacillus fermentation. Lactobacillus is the same bacterium that turns milk into yogurt. Unlike milk, fish do not contain sugar for the lactobacilli to eat, so sugar must be added, often in a form such as molasses, dried whey, or fruit pomace (pomaces are the wastes from juice making). The lactobacillus culture must also be added in proper quantities to produce sufficient lactic acid to preserve the digest. As in the case of yogurt, a distinctive flavor is produced, which seems to be advantageous in feeding some, although not all, animals (Goldhor, 1988).

Advantages offered by the enzyme hydrolysis over acid hydrolysis include the following:

One is no longer dependent upon the natural fish enzymes, which are variable and sometimes inadequate.

One can process batches rapidly, so very large amounts of waste can be handled with one equipment set-up.

A more consistent product can be offered.

One can offer a dried product (an acidified hydrolysate can be concentrated by not totally drying). Although drying is expensive, storage and transportation of dry materials is cheaper and simpler than that of wet ones. Most important, some markets will only accept dry materials.

Finally, the product has higher nutritional value because enzyme activity is stopped before it can break the protein down too far.

USES

A major investment and interest in North America commercial fish hydrolysing has been in producing feed ingredients (Lurie, 1991; Jaswai, 1990; Canadian Fishing Consultants, 1989; Shqueir et al., 1984). Interest has also been shown in the production of fish fertilizer (*Feed Management*, 1990; Athansas, 1989) and silage (Jangaard, 1987; Austreng et al., 1984; Raa et al., 1983, 1982).

It should not be forgotten that this technology can also be applied to supplemental human feeding (Castaneda et al., 1986; Galvez et al., 1986; Owens et al., 1985; Yanez et al., 1976; Amano, 1962).

REFERENCES

Adler-Nissen, J. 1976. Enzymatic hydrolysis of proteins for increased solubility. *Journal of Agriculture and Food Chemistry*. **24**(6):1090-1093.

Amano, K. 1962. The influence of fermentation on the nutritive value of fish with special reference to fermented fish products of Southeast Asia. In *Fish in Nutrition*. Fishing News (Books) Ltd., E. Heen, P. Kreuzer (eds.).pp. 180-187.

Athanas, R.P. 1989. Fish silage fertilizer potential for selected Northeast agricultural and horticultural crops Vol. II. New England Fishery Development Foundation. Boston, MA.

Austreng, E. and T. Asgard. 1984. Fish Silage and Its Use. Presented at the Verona Conference. Available from the authors at: Institute of Aquaculture Research, Agricultural University of Norway, 1432 AS-NLH, Norway.

Biocycle. 1990. Feeding fish wastes to cranberries.

Canadian Fishery Consultants Ltd. 1989.Aquaculture Feeds and Feed Ingredients. New England Fisheries Development Foundation. Boston, MA.

Castaneda, M.D., F. Pegeot, and J. Brisson. 1986. Nutrition value of leine-treated corn and hydrolyzed fish protein mixtures. *Nutrition Review International*. **33**(5):811-820.

Feed Management. 1991. New source of quality protein. **42**(11):22.

Food Engineering. 1997. Liquid hydrolyzed fish protein from enzymatic process. **49**(10):13-16.

Fujimaki, M., S. Arai, and M. Yamashita. 1971. Enzymatic protein hydrolysis and plastein synthesis: Their application to producing acceptable proteinaceous food materials. Paper presented at the Symposium of Microbial Foods. Kyoto, Japan.

Fujimaki, M., S. Arai, M. Yamashita, H. Kato, and M. Naguchi. 1973. Taste peptide fractionation from a fish protein hydrolysate. *Agriculture and Biological Chemistry* **37**(12):2891-2898.

Galvez, A., J.L. Morales, and H.B. Rodruguez. 1986. Development of an enzymatic fish hydrolysate and its use in soup bases. *Archiv. Latino Americanos De Nutricion*. **35**(4).

Goldhor, S.H. 1988. Fish Protein Hydrolysis: A User's Guide. New England Fisheries Development Foundation. Boston, MA.

Hevia, Patricio, and H. S. Olcott. 1977. Flavor of enzyme-solubilized fish protein concentrate fractions. *Journal of Agriculture and Food Chemistry*. **25**(4):772-774.

Iseki, S., T. Watanabe, and T. Kinumaki. 1969. Studies on "liquefied fish protein" IV — Examination of processing conditions for industrial production. *Bulletin of the Tokai Regional Fisheries Resources Lab*. **59**:81-99.

Jangaard, P.M. 1987. Fish Silage: A Review and Some Recent Developments. Paper No. 118. Fisheries Development Branch, Department of Fisheries and Oceans. Scotia-Fundy Region, PO Box 550, Halifax, Nova Scotia, B3J 2S7, Canada.

Jaswal, A.S. 1990. Amino acid hydrolysate from crab processing waste. *Journal of Food Science*. **55**(2):379-81.

Keyes, C.W., and W. Minke. 1966. Fish Protein — Enzymatic Process. US Patent 3,249,442.

Lalasides, G., and L.B. Sjoberg. 1978. Two new methods of debittering protein Hydrolysate and a fraction of Hydrolysates with exceptionally high content of amino acids. *Journal of Agriculture and Food Chemicals*. **26**(3):742-749.

Lure L. 1991. Fish Hydrolysis in Alaska. Special Bulletin. Alaska Fisheries Development Foundation. Anchorage, AK.

Owens, J.D., and L.S. Mendoza. 1985. Enzymatically hydrolysed and bacterially fermented fishery products. *Journal of Food Technology*. **20**:273-329.

Pigott, E., B. Bucove, and J. Ostrander. 1978. Engineering a plant for enzymatic production of supplemental fish protein. *Journal of Food Processing and Preservation*. **2**:33-54.

Raa, G. and A. Gildberg. 1983. Fish Silage: A Review. *CRC Critical Reviews in Food Science and Nutrition*. **14**:383-419.

Raa, G., A. Gildberg, and T. Stromg. 1983. Silage Production: Theory and Practice. Upgrading Waste for Feeds and Food, pp. 117-132. Ledward, D.A, A.J. Taylor, and R.A. Lawrie (eds.). University of Nottingham.

Shquier, A.A., R.O. Kellems, and D.C. Church. 1984. Effects of Liquefied Fish Cottonseed Meal and Feather Meal on *in vivo* and *in vitro*. Characteristics of Sheep. *Canadian Journal of Animal Science*. **64**:881-888.

Yamashita, Michiko, S. Arai, and M. Fujimaki. 1976. A low phenylaline, high-tyrosine plastein as an acceptable dietetic food. *Journal of Food Science*. **41**:1029-1032.

Yanez, E., D. Ballester, and F. Monckeberg. 1976. Enzymatic fish protein hydrolysate: chemical composition, nutrition value, and use as a supplement to cereal protein. *Journal of Food Science*. **41**(5):1289-1292.

Industrial Products: Chitin and Chitosan

George J. Flick Jr. and Roy E. Martin

INTRODUCTION

Chitin was first described in 1811 by Professor Henri Braconnot while he was Professor of Natural History and Director of Botanical Gardens at the Academy of Sciences in Nancy, France. He called it simply a "substance that is resistant to degradation (the process of being broken down)." In 1859 Professor C. Rouget discovered chitosan (Muzzarelli and Pariser, 1978; Muzzarelli, 1977). Over the next 150 years, a great deal of fundamental research took place on chitin. An intense interest in new applications grew in the 1930s and early 1940s, as evidenced by almost 50 patents; however, the lack of adequate supplies and competition from synthetic polymers hampered commercial development. Renewed interest in the 1970s was encouraged by the need to find uses for shellfish waste to avoid expensive disposal required by environmental regulations. Scientists worldwide began to chronicle the more distinct properties of chitin and its derivatives and to understand the potential of these nature-based polymers. Since then, numerous research studies have been undertaken to find ways to use this ubiquitous substance that for hundreds of years has gone untapped as a natural resource. Principal centers of United States research on chitin and chitosan include:

Massachusetts Institute of Technology

1. Basic chemistry research
2. Study of business opportunities. Marine Industries Business Strategy Program.

University of Delaware

Utilization of chitin.

University of Georgia

Recovery of food waste products.

University of Washington

New sources of chitin.

Rutgers: The State University

Medical applications.

University of Southwest Louisiana

Research on crawfish wastes.

University of California

Bioconversion of shellfish waste.

Two seafood firms in Japan are producing chitosan commercially. The Kyowa Oil and Fat Company in Tokyo pioneered commercial production in 1972. And in 1974, the Kyokuyo Company, also in Tokyo, entered the chitosan business by starting up a plant in the Chiba Prefecture. Total output of the two companies is approaching 100 tons per month of chitosan, all of which is used domestically. The demand for chitosan in Japan comes mainly from the food industry, which uses the material mainly as a flocculent in the treatment of wastewaters. This is mandated by law in Japan. Only those chemicals can be used in treatment that have been shown to have no deleterious effect on the marine environment. Also, in the treatment of this waste from the food industry (seafood, poultry, meat, cheese, etc.) substantial amounts of protein are recovered as a byproduct for an animal feed supplement. Because it is nontoxic and biodegradable, chitosan is allowed as an unintentional additive in animal feed. Chitosan is also being used to some extent for sludge dewatering of municipal and other industrial wastes (Paparella, 1978).

The world market for chitin and chitosan is estimated at about 2,000 metric tons. Japan, with an estimated 1.5 million pounds per year, is by far the largest user. Europe is said to use about 500,000 pounds, whereas the United States uses only about 150,000 pounds (*Chemical Marketing Reporter*, 1987).

CHITIN

Both cellulose and chitin are polysaccharides or polymers: long chains of smaller sugar molecules strung together. Sometimes, more than

5,000 of these smaller molecules can be found in one large chitin molecule. In nature, chitin serves as the "glue" for the chemical component that makes up the delicate wings of insects and the crunchy coats of crustaceans. Most prominent in crustacean shells, chitin occurs in a matrix with calcium carbonate and protein. These shells contain 14 to 35 percent chitin by dry weight, depending upon the species source. To obtain the pure chitin, the calcium carbonate and protein are removed by chemical processing of seafood shell waste. This chitin can then be used as is or broken down into numerous chitin derivatives (*Chemical Marketing Reporter*, 1995).

Chitin Structure and Applications

Chitin is a cellulose polymer present in fungal cell walls and exoskeletons of arthropods such as insects, crabs, shrimps, and lobsters. The polymer structure consists of N-acetyl-D-glucosamine (NAG) units linked by ~(1-4) glycosidic bonds that impart characteristics similar to cellulose (Figure 1). For example, chitin is insoluble in water and many commercial solvents. In pure form, it has limited food applications. Its uses are extended when chemically treated.

Microcrystalline chitin produced by controlled acid hydrolysis is suitable for use as a food thickener and stabilizer (Dunn and Lomita, 1974). Viscosity and emulsion stability of products containing commercial microcrystalline cellulose compared to those containing microcrystalline chitin are 10 to 20 times higher in the chitin product. This fact suggests applications in mayonnaise, peanut butter, and other emulsion-type foods.

Additional uses of chitin include enzyme purification and immobilization. Through the action of NAG sub-units, chitin purifies enzymes pertinent to food processors: wheat germ agglutinin (Bloch and Burger, 1974), lysozyme-like enzymes (Imoto and Yagishita, 1972), and glucose isomerase (Stanley, 1976). The polysaccharide also provides a solid support system for immobilization of glucose isomerase (Chang and Tsai, 1997; Bough, 1977; Averbach, 1975), lactase, a-chymotrypsin, and acid phosphatase (Stanley, 1975) by reaction with glutataldehyde.

Chitosan Properties

Unlike chitin, chitosan (deacetylated chitin, Figure 2) is readily soluble in various acidic solvents forming viscous non-Newtonian solutions. Solution viscosity depends on molecular weight, degree of deacetylation, concentration of polymer, concentration and types of solvents, and temperature. As solution temperature increases, viscosity decreases. Viscosity retention is compatible with that of carboxymethyl cellulose.

CHITOSAN PRODUCTION FROM SHELLFISH WASTE

Currently, the bulk of chitosan produced today is obtained by the chemical deacetylation of chitin from shellfish waste and requires a multistep process (Figure 3). After the butchering process, 70 to 80 percent of the crab and up to 80 per-

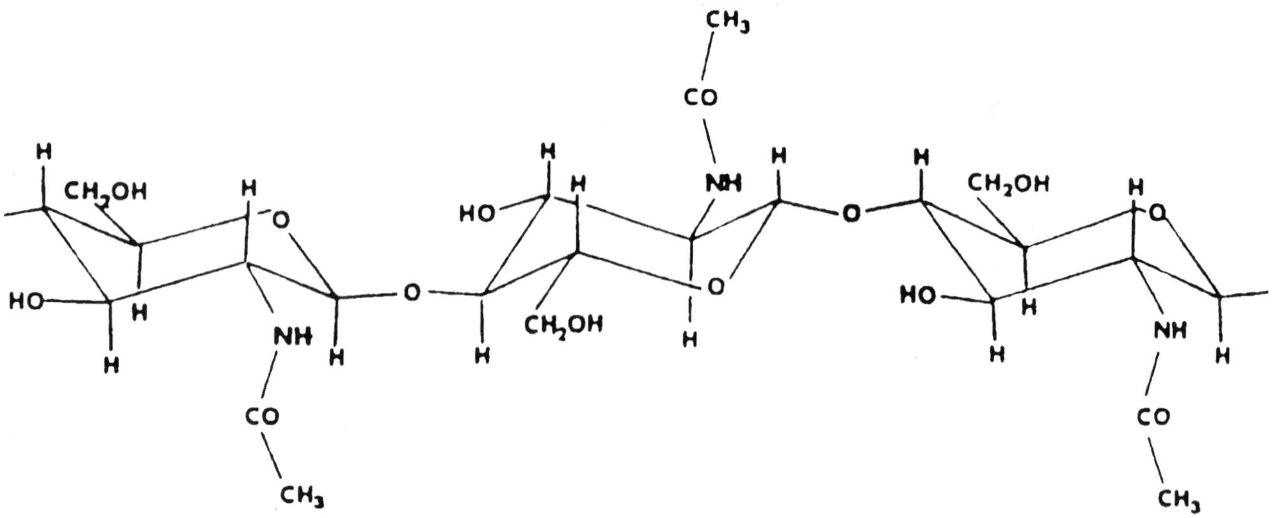

Figure 1. Chemical structure of chitin.

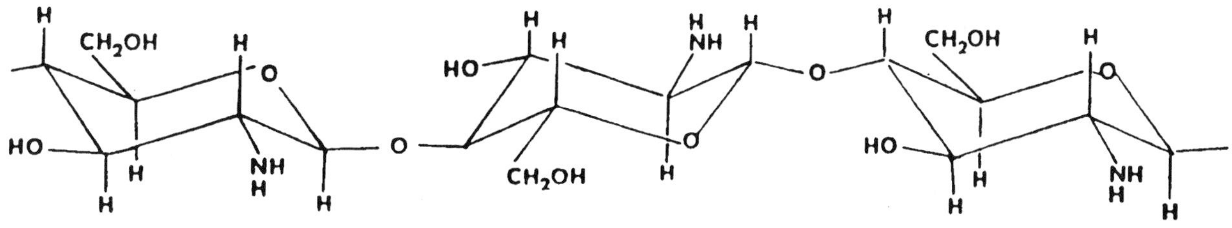

Figure 2. Chemical structure of chitosan.

cent of the shrimp biomass remains as waste (Gallanger, 1983). In a typical processing scheme the waste material is ground, extracted with sodium hydroxide, centrifuged, and washed to recover the associated proteinaceous material . This material has commercial value as a high-quality animal feed (Perceval, 1978). The material remaining after deproteinization is then demineralized with .6 N HCl. By filtering and washing the acid-treated material, crude chitin is recovered from the waste material (Anderson et al., 1978). The crude chitin is then deacetylated with 50 percent NaOH at 100°C for four hours (Figure 4). The chitosan formed by this method has a highly variable acetyl content ranging from almost 0 percent to 50 percent (Foster and Webber, 1960). This variability in acetyl content is most likely due to variable conditions used to hydrolyze chitin. It has been shown that many factors such as alkali concentration, reaction temperature and time, and presence or absence of oxygen will affect the extent of deacetylation (No et al., 1989; Hayes and Davies, 1978; Muzzarelli and Pariser, 1978; Muzzarelli, 1977).

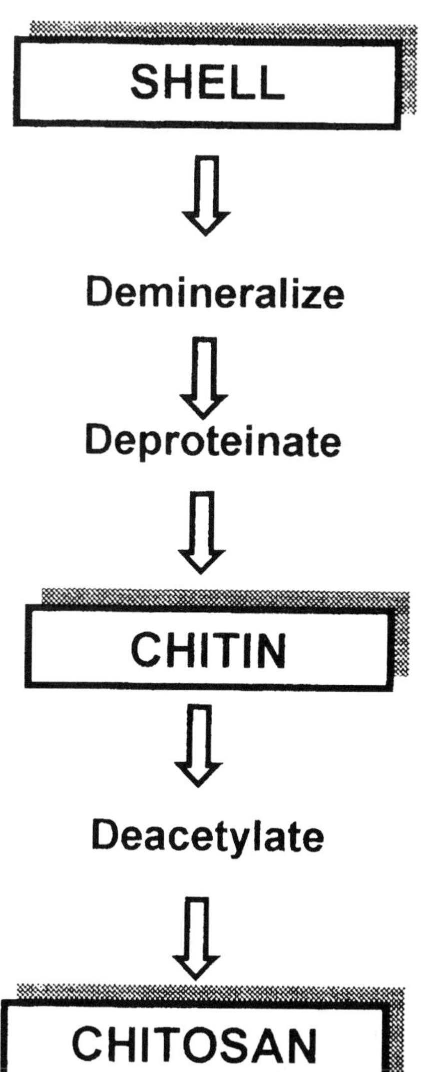

Figure 3. Steps in the processing of chitosan.

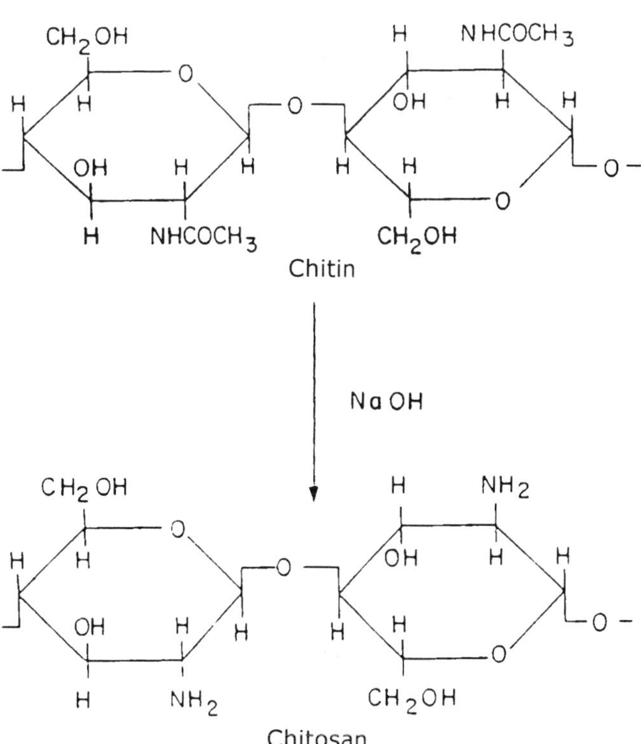

Figure 4. The deacetylation of chitin to chitosan.

COMMERCIAL APPLICATIONS OF CHITOSAN

(according to *Chemical Marketing Reporter*, 1995; Knorr, 1991; Ashford et al. 1977)

Organic Material and Mineral Recovery

Chitosan has the capability of coagulating organic waste materials from processing foods such as poultry, egg, meat, shrimp, cheese, and vegetables. In industrial and municipal wastewater treatment plants, a chitin/chitosan-derived polymer has been used to achieve sludge dewatering. Chitosan has also been used to remove toxic metallic ions from industrial waste water.

Wound-Healing Acceleration

Chitin, chitosan, and partially depolymerized chitin were found to accelerate wound healing when used as a coating for sutures or bandages. The active component in this process is thought to be the oligomers or dimers of N-acetyl-glucosamine derived from chitin or chitosan by the action of lysozyme (*Washington Post*, 1986b; Balassa, 1975, 1972).

Textile Additive

The application of chitosan as a sizing on wool helps to prevent shrinkage of the material. In addition, the treatment of glass and plastic fabrics with chitosan allows them to be dyed. The polymer is also used in the preparation of highly polymerized dyes for synthetic materials.

Film Formation

Chitosan membranes have high tensile strength, thus making them suitable for use as a food wrap. In addition, chitosan membranes are permeable to urea and creatine and impermeable to serum proteins, thus making them suitable for use as artificial dialysis membranes (Kifume, 1992; Austin and Brine, 1977).

Chromatographc Support

Chitosan can be used as a support for ion-exchange, chelation, affinity, and thin-layered chromatography.

Coagulation of Suspended Solids

Chitosan effectively coagulates suspended solids in wastewaters from processing poultry (Bough et al., 1974), eggs (Bough, 1975a), cheese (Bough and Landes, 1976), meat and seafood (Bough, 1976), and vegetables (Bough, 1975b). In addition, it functions as a conditioning agent for dewatering activated sludge suspensions resulting from biological treatment of vegetable, dairy, and brewery wastes (Olsenetal, 1996; Asano et al., 1977; Bough et al., 1976).

Functional Food Additive

Chitin Poly-N-acetyl-D-glucosamine, chitosan (deacetylated chitin) and microcrystalline chitin (redispersible chitin powder) were compared with microcrystalline cellulose to examine the use of those cellulose-like biopolymers as functional additives for potential application in food formulations. Water-binding, fat-binding, and emulsifying properties were studied. Baking tests were performed with 0.5 to 2.0 percent (flour basis) of microcrystalline chitin added to wheat flour bread or to potato protein fortified (8 percent potato protein concentrate) white bread. Water-binding capacity and fat-binding capacity of chitin, chitosan, and microcrystalline chitin ranged from 230 to 440 percent weight/weight (w/w) and from 170 to 315 percent (w/w). Chitosan and chitin did not produce emulsions but microcrystalline chitin showed good emulsifying properties and was superior to microcrystalline cellulose. Increasing concentration of microcrystalline chitin (0.12 to 0.8 g/100 ml water) has a positive effect on emulsion stability. Addition of microcrystalline chitin increased specific loaf volume of white bread and protein-fortified breads. Water addition of 65 percent (flour basis) was found to be optimum for "chitin breads" (Knorr, 1982).

Adsorption of Metals (Table I)

Wastewater containing heavy metals results from many sources. Industries, such as metal-plating operations or fertilizer manufacturers, lose heavy metal to the environment. Since such metals will not be degraded, accumulation and distribution of these metals to our environment occur — a concern to the public (Jha et al., 1988).

The binding of metals to naturally occurring polymers obtained from seafood processing wastes was investigated. Experimental results show that chitin, chitosan, and scales from three species of fish such as porgy, flounder, and cod are potentially useful materials to remove metals from contaminated water. Such metals include

Table 1. Chelation of Metal Ions by Chitosan

Metallic Ion	Percentage Absorbed
Antimony	100*
Arsenic	100
Cadmium	100
Cerium	100*
Cesium	100
Chromium (III)	54
Chromium (VI)	100
Cobalt	100
Copper	100
Europium	45
Gold	00
Hafnium	65
Indium	100
Iron (II)	100
Iron (III)	100
Lead	100
Iridium	43
Manganese	38
Mercury	100
Molybdenum	100*
Nickel	100
Niobium + zirconium	100*
Palladium	100
Phosphorus	100*
Ruthenium	95
Scandium	100
Silver	100
Strontium	100*
Thulium	43
Tin	96
Titanium	100*
Tungsten	73*
Uranium	100*
Vanadium	100
Zinc	100
Zirconium	98 (on chitin)

*Special chemical conditions must be met to achieve these percentages.

copper, zinc, chromium, cadmium, and lead. Batch methods were used to study equilibrium isotherms and adsorption kinetics. The experimental data of adsorption equilibrium from solutions of metals have been found to correlate well with the three-parameter isotherm equation. The study of adsorption kinetics shows that the rates of adsorption of the metals toward the natural polymers are best interpreted in terms of intraparticle diffusion as a rate-limiting step. Although the natural polymers mentioned are useful for treating waste water, it is advantageous to use the cross-linked form to counteract product solubilization in an acid environment (Plonski et al., 1990; Yang and Zall, 1984).

RECOVERY OF AMINO ACIDS AND ORGANIC COMPOUNDS FROM SEAFOOD-PROCESSING WASTEWATER

Crawfish chitosan, prepared from crawfish shell chitin, was demonstrated to be an effective ligand-exchange column material for recovery of amino acids. The amino acids recovered by this treatment have potential application as seafood flavors in terms of their sensory attributes (No and Meyers, 1989a; 1989b).

REMOVAL OF LIPIDS FROM CHEESE WHEN USING CHITOSAN

Addition of 0.01 to 0.02 percent chitosan to cheddar cheese whey at pH 4.5 resulted in formation of a chitosan-fat globule membrane complex. The complex flocculated and precipitated when incubated at ambient temperature for about 10 to 30 minutes. Centrifugation of the treated whey resulted in a clear supernatant that contained almost all the whey proteins but with a lipid content of less than 0.26 g/100 g of protein. No detectable amount of residual chitosan was present in the whey protein isolate. Preconcentration of whey to about four-fold volume/volume (v/v) only slightly increased the minimum concentration of chitosan needed to flocculate fat-globule membranes. On the basis of these results, a simple industrial method to remove fat from cheese whey is proposed. Preliminary economic analysis indicated that the additional cost involved in the manufacture of fat-free whey protein isolate was about $0.50/kg of whey protein (Hwang and Damodaran, 1995).

COLOR REMOVAL FROM AQUEOUS SOLUTIONS

Chitin and chitosan prepared from lobster shell wastes were used as adsorbents for the removal of various dyes from aqueous solutions. It was shown that high adsorption capacities were observed for reactive dyes. The adsorption equilibrium data could be well described by the Langmuir equation under the concentration range investigated (50~500 mg/L). On the basis of

Langmuir adsorption capacity, chitosan obtained in this work was found to be more effective compared to chitin (Juang et al., 1996).

FRUIT DECAY PREVENTION

The effect of chitosan coating (1.0 and 1.5 percent w/v) in controlling decay of strawberries at 13°C was compared to a fungicide, iprodione. Chitosan coating significantly reduced decay of berries compared to the control. There was no significant difference between chitosan and fungicide treatments up to 21 days in storage. Thereafter the fungicide-treated berries decayed at a higher rate than the chitosan-coated berries. Chitosan-coated berries stored at 4°C were firmer, higher in titratable acidity, and synthesized anthocyanin at a slower rate than the fungicide-treated or non-treated berries. Chitosan coating decreased respiration rate of the berries with a greater effect at higher concentration (Ghaouth et al., 1991).

OTHER APPLICATIONS

The commercial culture of marine organisms is an increasingly important industry. One of the bottlenecks still remaining for its complete development is the limited number of hatcheries and nurseries around the world. Success depends upon microalgae that are required for larval feeding. One of the most promising harvesting techniques of the algae produced appears to be some type of flocculation technique. Results obtained with chitosan flocculation appear superior to other chemical flocculants (Lubian, 1989; Morales et al., 1985).

The effect of chitin and chitosan as diet supplements on the growth of cultured fish was measured (Kono et al., 1987). Results suggested that in fishes which possess high activity of chitinase in their digestive gland, chitin is digested and utilized. This same effect was not rated with chitosan, however.

Research at the United States Department of Agriculture's (USDA) Southern Regional Research Center developed N-carboxymethylchitosan (NCMC) from shellfish waste as a meat-flavor preservative.

NCMC is effective at less than 0.1 percent concentration in meat (500 to 1000 parts per million).

The cost of manufacturing NCMC using modern technology is estimated at about $25.00 a pound as a food-grade material.

As a kitchen aid, preparations of NCMC in granular form with inert ingredients can be sold as are other food seasonings that can be sprinkled on gravies or meat, etc., for pantry convenience, at a very low cost to the consumer because it is used in such small amounts.

NCMC is very useful in preserving microwavable or quickly prepared foods, as well as in preventing flavor deterioration of leftovers or institutional foods, long termed "warmed-over flavor."

NCMC is itself tasteless and blends well into foods as a colorless ingredient while being non-toxic or non-allergenic.

In flash-freezing many vegetables, muscle foods, or prepared frozen entrees, an application of NCMC with the glaze formulation will inhibit surface oxidation for better frozen shelf-life storage.

Meat and poultry processors can use NCMC as a post-slaughter perfusion as a long-term flavor and storage preservative (St. Angelo and Vercellotti, 1991).

The Environmental Protection Agency (EPA) exempted from a tolerance poly-s-glucosamine (chitosan) in or on wheat as a growth regulator.

The EPA explained that chitosan is a naturally occurring substance and is produced from chitin extracts of crustacean shells. It is intended for use as a seed treatment of wheat to stimulate plant root growth and enhance the strength of wheat stems, thereby preventing "lodging," which is the falling over of the plants because of weak stems. The agency said wheat plants which lodge are difficult to harvest and may decrease crop yields.

The product is designed to augment the function of plant genes by enhancing the immunity system of plants. The EPA noted that chitosan "(1) is not toxic to humans and animals; (2) naturally occurs in the environment in large concentrations; (3) has been exempted from regulation by FDA when used as a food or feed additive; and (4) has been approved by the State of Oregon for use in unrestricted amounts as a soil amendment (fertilizer)."

Studies "consistently show that chitosan is not toxic," the EPA said, adding that a United States Fish and Wildlife Service study on fish "showed that chitosan is relatively non-toxic to small echo fingerlings."

The agency said residue information shows that use of the aqueous chitosan solution on wheat seed at 0.4 ounces/100 pounds of seed "is not expected to result in detectable levels of chitosan in food or feed." Noting that the substance is commonly found in insects, fungi, micro-fauna, and plankton, the EPA said, "Compared to the naturally-occurring concentrations of chitosan, it appears that the contribution from treated seed would be insignificant." It also added that most of the chitosan does not enter the plant but remains in the soil and decays with the expended seed.

The agency added that scientific literature shows that chitosan products are permitted to be used in food as hypocholesterolemic agents, as dietary fiber, in low-calorie diets, and as agents to increase the specific loaf volume of bread (*Chemical News*, 1986a). Use on soybeans was also approved (*Pesticide and Toxic Chemical News*, 1989).

In 1988 the EPA exempted Poly-N-Acetyl-D-Glucosamine (chitin) from a tolerance intended to act in the soil as a biological control agent to stimulate the growth of beneficial saprophytic nematodes and of other normal soil microorganisms which act to protect plants (*Pesticide and Toxic Chemical News*, 1988; Hadwiger et al., 1984).

Additional work was reported on using chitinous materials from the blue crab for control of root-knot nematodes (Kabana et al., 1989).

Chitosan has also been explored for its ability to inhibit mold (Fang et al., 1994).

REFERENCES

Anderson, C. G., N. de Pablo, and C. R. Romo. 1978. Antarctic krill (*Euphausia superba*) as a source of chitin and chitosan. R. A. A. Muzzarelli and E. R. Pariser, (eds.), pp. 54-63. *Proceedings of the First International Conference on Chitin/Chitosan*. MIT Sea Grant Report MITSG 78-7 Index No. 78-307-Dmb. Massachusetts Institute of Technology, Cambridge.

Asano, T. et al. 1977. Chitin-chitosan derived polymer applications in wastewater sludge dewatering. Presented at First International Conference on Chitin/Chitosan, April 11-13. Boston, MA.

Ashford, N. A., D. Hattis, and A. E. Murray. 1977. Industrial prospects for chitin and protein from shellfish wastes. Mass. Inst. of Tech. Report No. MITSG-77-3, Cambridge.

Austin, P. R. and C. J. Brine. 1977. Chitin Films and Fibers. United States Patent No. 4,029,727.

Averbach, B. L. 1975. The Structure of Chitin and Chitosan. MIT Sea Group Program Report No. MITSG-75-17, Index No. 75-317-11W, Mass. Inst. of Tech., Cambridge.

Balassa, L. L. 1972. Use of Chitin for Promotion of Wound Healing. United States Patent No. 3,632,754.

Balassa, L. L. 1975. Promoting Wound Healing with Chitin Derivatives. United States Patent No. 3,911,116.

Bloch, R. and Burger. 1974. Purification of wheat germ agglutinin using affinity chromatography on chitin. *Biochem. and Biophys. Research Communications* **58:**1:13.

Bough, W. A. 1975. Coagulation with chitosan — an aid to recovery of byproducts from egg-breaking wastes. *Poultry Science* **54**:1904.

Bough, W. A. 1975. Reduction of suspended solids in vegetable canning waste effluents by coagulation with chitosan. *Food Science.* **40**:297.

Bough, W. A., 1976. Chitosan — a polymer from seafood waste, for use in treatment of food processing wastes and activated sludge. *Process Biochem.* **11**:1:13.

Bough, W. A. 1977. Shellfish components could represent future food ingredients. *Food Prod. Dev.*, **II** (10):90.

Bough, W. A. and D. R. Landes. 1976. Recovery and nutritional evaluation of proteinaceous solids separated from whey by coagulation with chitosan. *J. Diary Sci.* **59**:1874.

Bough, W. A. et al. 1974. Use of chitosan for the reduction and recovery of solids in poultry processing waste effluents. *Poultry Science* **54**: 992.

Bough, W. A. et al. 1976. Utilization of chitosan for recovery of coagulated by-products from food processing wastes and treatment systems. *Proceedings Sixth National Symposium on Food Processing*. United States Environmental Protection Agency, EPA-600/2-76-224, p. 22.

Chang, K. L. and Tsai, G. 1997. Response surface optimization and kinetics of isolating chitin from pink shrimp shell waste. *J. Agri. Food Chem.* **45**(5):1900-1904.

Chemical Marketing Reporter. 1995. Chitin is still on launch pad. January 23.

Chemical Marketing Reporter. 1987. Chitin, derivative products ready to come out of shell. March 30. Schnell Pub. Philadelphia, PA.

Dunn, H. J. and P. M. F. Lomita. 1974. Microcrystalline chitin. United States Patent 3,847,897.

Fang, S. W., C. F.Li, and D.Y. C. Shih. 1994. Antifungal activity of chitosan and its preservative effect on low-sugar candied kumquat. *J. Food Prot.* **56**(2): 136-140.

Foster, A. B. and J. M. Webber. 1960. Chitin. M. L. Wolfrom and Tipson (eds.). In *Advances in Carbohydrate Chemistry*, pp. 371-393. Academic Press Inc., New York.

Gallanger, S. M. 1983. Wastewater treatment and by-product recovery in seafood processing using coagulant addition. Master of Science Thesis, University of Alaska, Fairbanks.

Ghaouth, A. E., J. Arul, R. Punnampalam, and M. Boulet. 1991. Chitosan coating effect on storability and quality of fresh strawberries. *J. Food Sci.* **56**(6):1618.

Hadwiger, L. A., B. Frislensky, and R. C. Riggleman. 1984. Chitosan, a natural regulator in plant-fungal pathogen interactions, increases crop yields. Joint U.S./Japan Seminar in Advances in Chitin, Chitosan and Related Enzymes. J. P. Zakis (ed.) Academic Press, New York.

Hayes, E. R. and D. H. Davies. 1978. Characterization of chitosan. II: The determination of the degree of acetylation of chitosan and chitin, pp. 406-420. In R. A. A. Muzzarelli and E. R. Pariser (eds). *Proceedings of the First International Conference on Chitin/Chitosan.* MIT Sea Grant Report MITSG 78-7 Index No. 78-307-Dmb. Massachusetts Institute of Technology, Cambridge.

Hwang, D. C. and S. Damodaran. 1995. Selective precipitation and removal of lipids from cheese whey using chitosan. *J. Agri Food Chem.* **43**(1): 33-37.

Imoto, T. and K. Yagishita, 1972. Chitin-coated cellulose as an adsorbent of lysozyme-like enzymes: some applications. *Agr. Biol. Chem.* **37**:5:1191.

International Commission on Natural Health Products. 1995b. Chitin and Chitosan. Personal communication. Atlanta, GA.

Jha, I. N., L. Iyengar, and V. S. P. Rao. 1988. Removal of cadmium using chitosan. *J. Env. Eng.* **114**(4): 962-974.

Junng, R. S., R. L. Tseng, R. S. Wu, and S. J. Lin. 1996. Use of chitin and chitosan in lobster shell wastes for color removal from aqueous solutions. *J. Environ. Sci. Health,* **31**(2): 325-338.

Kabana, R. R., D. Boure, and R. W. Young. 1989. *Nematropica* **19**(1): 53-74.

Kifune. 1992. *Clinical application of chitin with artificial skin, M: advances in chitin and chitosan.* Brine, Sandford, and Zikakis (eds.). Elsevier Applied Science, London.

Knorr, D. 1982. Functional properties of chitin and chitosan. *J. Food Sci.* **47**(2).

Knorr, D. 1991. Recovery and utilization of chitin and chitosan in food processing waste management. *Food Tech.* **45**(1): 116-121.

Lubian, L. M. 1989. Concentration of cultured marine microalgae with chitosan. *Aqua. Eng.* **8**: 257-265. Great Britain.

Morales, J., J. Noue, and G. Picard. 1985. Harvesting marine microalgae species by chitosan flocculation. *Agri. Eng.* **4** :257-270.

Muzzarelli, R. A. A. and E. R. Pariser (eds.). 1978. *Proceedings of the First International Conf. on Chitin/Chitosan.* MIT Sea Grant Report MITSG 78-7 Index No. 78-307 Dmb—Massachusetts Institute of Technology, Cambridge.

Muzzarelli, R. A. A., 1977. *Chitin.* Pergamon Press, New York.

No, H. K. and S. P. Meyers. 1989a. Recovery of amino acids from seafood processing wastewater with a dual chitosan-based ligand exchange system. *J. Food. Sci.* **54**(1): 60-62.

No, H. K. and S. P. Meyers. 1989b. Crawfish chitosan as a coagulation aid in recovery of organic compounds from seafood processing streams. *J. Agri Food Chem.* **37**(3):580-583.

No, H. K., S. P. Meyers, and K. S. Lee. 1989. Isolation and characterization of chitin from crawfish shell waste. *J. Agric. Food Chem.* **37**(3): 575-579.

Olsen, S. E., H. C. Ratnaweera, and R. Pehrson. 1996. A novel treatment process for dairy wastewater with chitosan produced from shrimp-shell waste. *Water Sci. Tech.* **34**(11): 33-40.

Paparella, M. 1978. Information Tips. Bull. 78-3, University of Maryland, Crisfield.

Perceval, P. M. 1978. The economics of chitin recovery and production. In Mazzarelli and Pariser (eds.). *Proceedings of the First International Conference on Chitin/ Chitosan.* MIT Sea Grant Report MITSE 78-7 Mass. Inst. Tech., Cambridge. Pp. 45-53.

Pesticide and Toxic Chem. News. 1986a. Poly-D-Glucosamine exempted from tolerance. October 8. Washington, DC.

Pesticide and Toxic Chem. News. 1988. Tolerance exemption granted biochemical nematocide. April 6. Washington, DC.

Pesticide and Toxic Chem. News. 1989. Chitosan exempted from tolerance requirements on soybeans. March 29. Washington, DC.

Plonski, B. A.,Y. H. V. Loung, and E. J. Brown. 1990. Arsenic sorption by chitosan and chitin deacetylase production by *Mucor ruoxii. BioRecovery* **1**:239-253.

St. Angelo, A. J. and J. R. Vercellotti. 1991. Preparation and use of food grade N-carboxymethylchitosan to prevent meat flavor deterioration. In *Food Science and Human Nutrition,* Charalambous (ed.) Elsevier Sci. Pub. B. V.

Stanley, W. L., et al. 1975. Lactase and other enzymes bound to chitin with glutaraldehyde. *Biotech. and Bioengineering* **17**: 315.

Stanley, W. L. et al. 1976. Immobilization of glucose isomerase on chitin with glutaraldehyde and by simple adsorption. *Biotech. and Bioengineering* **28**:439.

Washington Post. 1986b. Researchers turn crabshells into superior suture thread. November 10. Washington, DC.

Yang, T. C. and R. R. Zall. 1984. Absorption of metals by natural polymers generated from seafood processing wastes. *MD. Eng. Chem. Prod. Res. Dev.* **23**(1): 168.

Industrial Products: Waste Composting

Angela J. Correa and Roy E. Martin

INTRODUCTION

Recent years have brought changes in the preferable or accepted methods for the disposal of fish or shellfish offal. Traditional disposal methods, such as disposal at sea or landfilling, are falling under increased regulatory scrutiny in many areas (Line, 1992), and are seen as a danger to the environment, or simply too costly to support. Viable recycling uses also exist, such as the use of waste for the manufacture of fish meal, or fertilizer, or as a raw material for rendering. While these operations can utilize nearly all of the offal that comes to them, they have also come under criticism due to the obnoxious odors that characterize their operation. Similarly, use of fish wastes as a directly-applied manuring agent in land cultivation is rarely preferred, due to the particularly atrocious scent of putrefying fish (Mathur et al., 1986). As the trend towards wider diversification of previously single-industry commercial fishing communities continues, the tolerance within these communities for environmentally risky or potentially offensive strategies for management of fish wastes continues to decrease (Mathur et al., 1988).

As a result of these pressures, researchers and industry groups have sought to develop and refine a variety of methods for the composting of fish and shellfish wastes. These wastes have a high nutrition value and are readily degradable, and are thus a valuable raw material for composting. Composting is the "biochemical degradation of organic material to a sanitary, nuisance-free, humus-like material" (American Public Works Association, 1970). The end product of composting is blended into potting soil or used directly as a soil additive or as an ingredient in animal feeds (Gouin, 1978; Golueke, 1977, 1972). Successful composting produces a highly useful organic substance, along with a number of other benefits, including the control or elimination of unpleasant odors, destruction of pathogens, a sustainable cost/revenue ratio, and low labor requirements. Since composting is understood to be ecologically responsible, it also provides the company or municipality that undertakes it the less tangible advantage of being a "green" technology for the remediation of fishery wastes. In fishing industry areas where environmental lapses or unendurable odors have historically created a community relations problem, a responsibly-run composting operation could accomplish a great deal in addition to the production of a high-quality soil adjuvant.

Conservative estimates on the amount of waste generated by the seafood industry indicate that over 2 million metric tons of waste are generated annually in the United States alone (Mathur et al., 1988). The Canadian codfish industry alone produces 140 to 200 thousand tons of fish scrap every year (Fisheries and Oceans Canada, 1981). Commercial fishing operations commonly lose anywhere from 30 to 60 percent of the weight of their catches as waste. Scrapped exoskeletons of lobster or snow crab frequently weigh four times as much as the usable flesh recovered (Mathur et al., 1988). Waste includes the head, fins, tail, backbone, skin, cartilage, viscera, and shell, as well as whole unprocessed fish that have been deemed unfit for use.

COMPOSTING METHODS

Decisions regarding the selection of a composting method depend on a variety of factors: the quantity of waste to be handled, the type of bulking agent to be used, the availability of adequate space and shelter from wind and rain, the availability of labor and equipment, and time constraints brought on by market or economic concerns.

Static versus Turned Composting

Traditional composting of biosolids requires that the materials be turned several times during the maturation process to ensure that both aeration and temperature are uniform, and to break up any clumps of material that may be resistant to decomposition. Turning during composting reduces the total amount of time required for full maturation of the material (Line, 1992), and therefore makes it possible to handle larger volumes of waste in less space, and concomitantly produce more compost more quickly. However, the organized turning of large heaps of material poses significant challenges for a small operation (Brinton, 1994). Specialized equipment and additional labor can be prohibitive, and the gains that can be made by maturing compost weeks or months earlier are not always sufficient justification for the investment of this additional capital. In this case, static composting is a feasible alternative.

A static compost heap is constructed in a manner similar to one that is intended for turning, but includes a number of horizontal, perforated ventilation pipes that allow the passive flow of air into the center areas of the pile (Brinton, 1994). This design makes use of the natural convection currents created in a compost pile whereby the heated air released at the top of an active pile draws in ambient air from the sides (Brinton, 1994). Static compost heaps require the same protection from the elements as turned heaps, but very little specialized equipment or other effort is necessary once they are constructed. Maturation of composts usually takes somewhat longer when using the static method (Line, 1992), and the longer retention time means that additional space is required to handle large volumes of waste.

Piles, Windrows, and Container Composting

Large-scale composting is best accomplished through the construction of windrows, which are long heaps of compost at different stages of maturation. Windrows make good use of space, and facilitate turning by specialized machinery. They can be built all at once or sequentially. Compared to windrows, piles take more space and are prone to greater variability in achieved temperatures. However, when specialized equipment is not available and the compost is to be turned using a backhoe or front-loader, a pile may prove to be the better alternative, being accessible from all sides. Both windrows and piles require adequate drainage, and protection from wind and excessive wetness.

Container composting offers some protection from the weather as well as from vermin, and may be the only sustainable alternative in very windy or unprotected areas. However, due to the difficulty of using equipment to facilitate turning, container composting is problematic for large-scale operations, unless the static method is used. Container use is also an excellent alternative for small operations that produce waste sporadically or in small quantities, such as resorts or fish-cleaning centers that cater to individual fishermen.

Bulking Agent Alternatives / Additives

In order to control odors, adsorb complex cations, and inhibit anaerobic decomposition, a bulking agent is required (Mathur et al., 1988). Bulking agents are blended with the wastes to provide structural support, thus promoting aeration and providing a thermal barrier to retain the heat generated by the action of thermophilic microbes within the decomposing material (Liao et al., 1997).

A variety of suitable bulking agents are available. Most are cellulosic, and can be very cost effective if they also happen to be the waste product of another industry. Common examples include wood chips, waste straw, leaves, sawdust, bark, and peat moss. Different bulking agents have differing C:N ratios, pH, and decomposition rates, and consideration of these factors is a key factor in the success of a composting operation. The bulking agent can be chosen to supplement the quantity of carbon in the mixture (to optimize the C:N ratio), and control ammonia volatilization (Mathur et al., 1986). Other additives such as anhydrous acid (ferrous sulfate) and citrus peels, or urea and lime sludge, can also be added to stabilize pH (Cathcart, 1986) or to neutralize a previous addition of acid. Analysis of test samples is required to determine the tendency towards acidity or alkalinity in the end product.

Basic Construction Strategies

Widely accepted strategies for composting of fish wastes have the following characteristics in common: the use of an initial base of bulking agent of about 6 to 12 inches deep, and the necessity of covering the material to be composted with another 6- to 12-inch layer of bulking agent (Liao,

1993; Frederick, 1991; Mathur, 1988). Surrounding the waste material in this manner with the more innocuous bulking agent provides protection from vermin, holds in the heat of decomposition, and creates an active layer for the absorption of odors and ammonia. A small amount of uncured compost, manure, or commercially available inoculum is sometimes used to introduce thermophilic bacteria into the mixture (Lo, 1992; Mathur, 1986).

There are several techniques for the incorporation of the fish waste into the compost bed. One technique involves the alternation of thin layers of undiluted fish scrap with thin layers of the bulking agent, until the pile reaches 4 to 8 feet in height (Liao et al., 1993).

Another, more commonly employed, technique requires the blending of fish waste with a quantity of bulking agent (Frederick, 1991; Lo, 1991; Mathur, 1988), and depositing a single thick layer of this blended material on the base of bulking agent. The ratio of fish waste to bulking agent varies depending on the consistency of the waste. Depending on the type of fish and the processing techniques applied, wastes can be dry and bony, pastelike, liquefied, or a combination of these. Blending ratios should be modified to produce a material "fluffy" enough to allow adequate aeration, and to break up any large pockets of undiluted waste. Beyond the question of ease of handling, the bulking ratio will largely determine the ultimate nutritive value of the compost, with a higher fish content resulting in a higher nutrient content.

Once this "filling" of bulking agent and fish scrap is placed on the base layer, the material should be well covered by an adequate quantity of unblended bulking agent. In cases where the bulking agent used is prone to blowing away (for example, straw or sawdust), a net or other porous covering can be used until the material begins to settle (Mathur, 1986).

Determination of Complete Maturation

As decomposition accelerates within the windrow, the internal temperature rises to between 55° and 65°C (132° to 156°F). Achievement of this temperature signals the start of active decomposition, and must be maintained for three to four days, depending on the method used. These temperatures and exposure times are required for adequate destruction of parasites, viruses, and pathogenic bacteria (Hay, 1996). Additional time may be required to fully break down shells, cartilage, or bony portions of the waste. As the process of decomposition winds down, levels of phytotoxic organic acids and phenols in the compost decrease, and can be used as an index to measure the maturity of the compost (Liao, 1993). Typically, compost is allowed to mature for six to eight weeks, and is then moved to larger piles for final mixing, end-stage maturation, and further processing. At the end of the six-week period, when the compost is uncovered, there should be no problematic odor, and few recognizable bits of shell or bone.

QUALITY ANALYSIS AND PATHOGEN DESTRUCTION

Concerns

In order for the composting process to be beneficial over the long term, the end-product needs to provide a good concentration of nutrients, and be relatively free of pathogens, toxins, or phytoinhibitory substances. Any compost generated to be sold or publicly applied must meet the control and monitoring requirements for pathogens outlined in 40 CFR Part 503, the federal biosolids technical regulations (Hay, 1996).

Temperature / Time Monitoring

In the laboratory, conditions for composting tend to be tightly controlled, and time and temperature requirements derived in the laboratory may not reflect the variability of composting operations in the field. Poor mixing, clumping of materials, and inconsistent practices in turning the piles may lead to suboptimal conditions for the maturation of compost (Hay, 1996). For these reasons, it is important to track the temperatures achieved within the compost pile, as well as the length of time that these temperatures are maintained. For effective destruction of most pathogens, temperatures must be at or above 55°C (131°F) for a minimum of four days (Hay, 1996; Liao, 1995). After this time, temperatures should begin to decrease slowly, eventually returning to ambient. Temperatures should be taken with a thermometer inserted into various areas of the pile in order to detect any cold spots that may exist. Compost should be left to mature no longer than 60 days for windrows or static piles, or 20 days for in-vessel methods (Hay, 1996). When decomposition is complete, there remains the possibility of the survival of certain heat-resistant strains of bacteria, although these appear to be inactivated during periods of curing or storage (Hay, 1996).

Temperature/time monitoring gives a fairly good indication of the safety of compost, but should always be supplemented by testing for indicator bacteria or the pathogens themselves.

Testing for Indicators of Pathogen Destruction

A fecal coliform test is an effective indicator of the presence of *Salmonella* and other pathogens, and is simple enough to be used in daily monitoring. Studies have shown that concentrations of less than 1,000 MPN/g indicated a high probability of adequate control of pathogenic organisms. Testing for specific pathogens should be performed in cases where the fecal coliform indicates some cause for concern.

Composts must also comply with additional requirements for reducing the tendency of biosolids to attract disease vectors. These requirements stipulate an additional 14 days of aerobic processing at an average temperature of 45°. This additional processing time drastically affects the capacity of the compost to attract vectors, and also helps to prevent the regrowth of harmful bacteria (Hay, 1996).

Growth Tests

The use of different methods, waste materials, and bulking agents results in highly variable results in the quality of composts. Generally speaking, it is highly feasible to produce high-quality adjuvants from fishery wastes, with an excellent chemical profile and nutrient analysis.

MARKET CONCERNS

Successful production and marketing of fishery-waste composts depends on a variety of factors. The location of facilities at or near the site of waste generation is crucial to control both costs and the potential odor. Likewise, the availability of sufficient quantities of a cost-effective bulking agent is a prime concern. Ideally, the bulking agent would also be available as a waste material, thereby minimizing costs (Line, 1992).

Composting operations should be managed by individuals with a thorough understanding of the regulatory issues involved. Consistent procedures and accurate record-keeping is essential. Proper procedures for constructing heaps, monitoring decomposition, curing, bacterial testing, storage, and packing are all crucial to the production of a safe and beneficial product.

Preliminary market analysis shows that fish waste composts sell equally as well as standard composts in most markets (Frederick, 1991). Most consumers do not appear to have a strong preference, and cost is often the deciding factor. In areas with a large fishing industry, locally produced fish waste composts may actually be preferred over standard composts, since they have a positive impact on waste handling in the local area.

SUMMARY

Composting is essentially a process of controlled, aerobic decomposition of organic waste. Composting is an excellent alternative for the remediation of fishery wastes, neutralizing offensive odors and mitigating problems with bacterial growth. Fishery wastes are well suited to the rapid decomposition that occurs within the compost heap, and even bony fish or crab carapaces usually break up and disappear completely over the course of active decomposition. Methods and components of the composting process can be highly variable, but as long as C:N ratios are controlled and a good time/temperature profile is attained, the end-product is generally a beneficial additive for soils or animal feeds. The higher the nutrient content of the original wastes and bulking agents, the higher the quality of the resultant compost can be expected to be. Controlling transportation and materials costs is crucial to the development of a sustainable service. Marketing composts as an environmentally-friendly answer to the problem of waste management in fishery areas meets with higher rates of consumer and community approval and solves a variety of problems encountered with other waste management practices such as landfilling, direct application, and open dumping of wastes at sea.

REFERENCES

American Public Works Association, Institute for Solid Waste. 1970. Municipal Refuse Disposal. Public Administration Service, Chicago, IL.

Brinton, R. Low cost options for fish waste. *BioCycle*. March 1994:68-70.

Cathcart, T.P., F.W. Wheaton, and R.B. Brinsfield. 1986. Optimizing variables affecting composting of blue crab scrap. *Agricultural Wastes.* **15**:269-287.

Frederick, L. 1991. Turning fishery wastes into saleable compost. *BioCycle.* September 1991:70-71.

G.J.M. 1992. Small-scale fish composting trials. *BioCycle.* February 1992:62.

Golueke, C. 1972. Composting: A Study of the Process and its Principles. Rodale Press, Emmaus, PA.

Gouin, F. 1978. New organic matter source for potting mixes. *Compost Science/Land Utilization,* Jan./Feb. **19**(1), 26-28.

Hay, J. C. 1996. Pathogen destruction and biosolids composting. *BioCycle.* June 1996:67-76.

L. G. 1991. Composting fish wastes along the Great Lakes. *BioCycle.* May 1991:54.

Liao, P.H., A. Chen, A.T. Vizcarra, K.V. Lo. 1993. Evaluation of the Maturity of Compost Made from Salmon Farm Mortalities. Silsoe Research Institute.

Liao, P.H., L. Jones, A.K. Lau, S. Walkemeyer, B. Egan, and N. Holbek. 1996. Composting of fish wastes in a full-scale in-vessel system. *Bioresource Technology.* **59**:163-168.

Liao, P.H., A.T. Vizcarra, and K.V. Lo. 1994. Composting of salmon-farm mortalities. *Bioresource Technology.* **57**:67-71.

Line, M.A. 1992. Elimination of food industry wastes by commercial composting. *Food Australia.* **44**(3):124-125.

Line, M.A. 1994. Recycling of seastar (*Asterias amurensis*) waste by composting. *Bioresource Technology.* **49**:227-229.

Lo, K.V. and P.H. Liao. 1992. Composting of salmon farm mortalities. *Canadian Agricultural Engineering.* **34**:401-404.

Mathur, S.P., J. Y. Daigle, J.L. Brooks, M. Levesque, and J. Arsenault. 1988. Composting seafood wastes. *BioCycle.* September 1988:44-49.

Mathur, S.P., J.Y. Daigle, M. Levesque, and H. Dinel. 1986. The feasibility of preparing high quality composts from fish scrap and peat with seaweeds or crab scrap. *Biological Agriculture and Horticulture.* **4**:27-38.

Mathur, S.P. and W.M. Johnson. 1987. Tissue-culture and suckling mouse tests of toxigenicity in peat-based composts of fish and crab wastes. *Biological Agriculture and Horticulture.* **4**:235-242.

Bioactive Compounds from the Sea

Chris M. Ireland, Brent R. Copp, Mark P. Foster, Leonard A. McDonald, Derek C. Radisky, and J. Christopher Swersey

INTRODUCTION

Over the last 20 years, chemists and pharmacologists alike have given increasing attention to marine organisms and the secondary metabolites they produce. As with the chemical studies of terrestrial plants, the emphasis has shifted from phytochemical evaluations to possible applications — especially regarding control of agricultural pests and treatment of human disease (Albizati et al., 1990; Fautin, 1988). So, too, have pharmacological investigations evolved from studies of toxins: to those of antitumor, cytotoxin, and anti-inflammatory properties; to studies of regulatory processes in cells; to a multitude of experiments based on whole animal models and receptor-binding assays.

This chapter — while not a complete review of bioactive marine natural products (see Faulkner, 1984) — will pay particular attention to those metabolites which have been or are being developed as pharmaceuticals, and to those compounds which have significantly aided the understanding of cellular processes at the biochemical level. A distinction between these two categories is difficult to make, as what usually distinguishes an effective drug from a molecular probe is the degree of toxicity associated with said compound.

Certain trends in this developing field have emerged: from 1977 to 1987 approximately 2500 new metabolites were reported from a wide variety of marine organisms, ranging from prokaryotic microbes and soft-bodied invertebrates to vertebrate fish. These metabolites are equally diverse in their biosynthetic origins, spanning mevalonate, polyketide, and amino acid metabolic pathways (Ireland et al., 1989, 1988). An analysis of the phyletic distribution of these compounds (Figure 1) shows that the majority (93 percent) are confined to four groups (macro-algae, coelenterates, echinoderms, and sponges), largely due to the abundance and ease of collection of these organisms. By comparing these results with those of an earlier (1977 to 1985) survey, several trends are evident (Ireland et al., 1988) (Figure 2): while in each breakdown the same four groups dominate (accounting for greater than 92 percent of the compounds in each case), their relative contributions change. The contribution from macro-algae is decreasing significantly, whereas sponges are becoming the dominant source of compounds. The increase in studies of sponges explains, in large part, the increasing diversity of structural types reported from marine organisms; sponges have a wider range of biosynthetic capabilities than any other group of marine invertebrates. The number of compounds reported from echinoderms has also increased dramatically, due largely to recent comprehensive investigations of saponins produced by starfish (Bruno et al., 1990). Unfortunately, these compounds tend to be generally toxic to erythrocytes.

Although the contributions from ascidians and microbes remain modest, there is increasing interest in these groups. Interest in the former re-

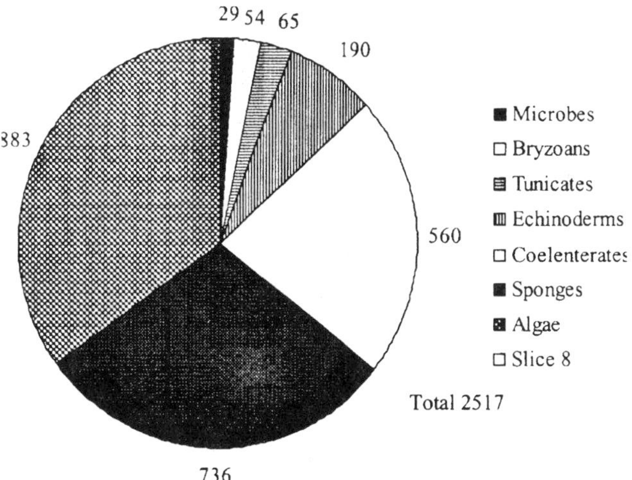

Figure 1. Phyletic distribution of secondary metabolites for 1977–1987.

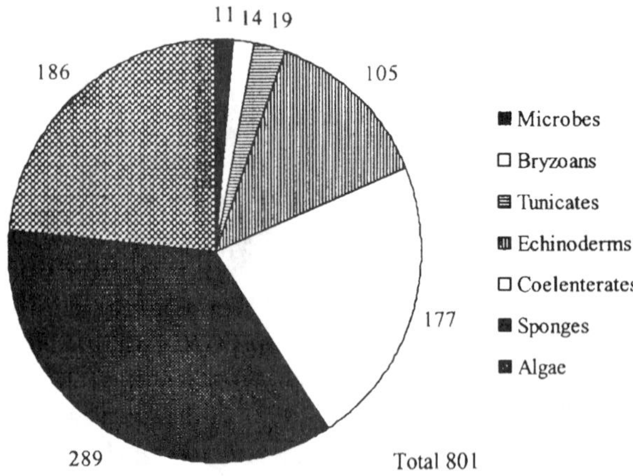

Figure 2. Phyletic distribution of metabolites for 1986–1987.

flects recognition that, in spite of their small size and general difficulty of collection, ascidians produce an array of very potent natural products. Didemnin B, the only marine natural product currently in human clinical trials, is from an ascidian (Dorr et al., 1988, 1986; Stewart et al., 1986). Microbes are being recognized for the nearly unlimited potential of culturable marine Actinomycetes, bacteria, and cyanobacteria as renewable sources of pharmaceuticals. It is likely that in the next decade these two groups will assume important roles in marine biomedical studies. It is also important to note that the number of compounds reported annually is increasing steadily, indicating that marine organisms will continue to be significant sources of natural products. While the above analysis can only be retrospective and is inherently flawed because it partially reflects the specific interests of investigators in the field, it also serves to indicate which groups of marine organisms are currently receiving the greatest attention from the perspective of drug development. It also provides some insights into which groups will play important roles in future biomedical studies of marine natural products.

Looking back over the past several decades, biomedical investigations of marine natural products have focused mainly on a few areas: CNS-membrane active toxins and ion channel effectors, anticancer and antiviral agents, tumor promoters, anti-inflammatory agents, and metabolites that control microfilament-mediated processes. The important developments within these areas are highlighted in the following sections.

MARINE NATURAL PRODUCTS THAT ACT AT MEMBRANE RECEPTORS

Perhaps the most important molecules (from a cellular physiology and pharmacology standpoint) derived from marine sources to date are the very potent and specific sodium (Na^+) channel blockers tetrodotoxin and saxitoxin. Pharmacological studies with tetrodotoxin and saxitoxin played the major role in developing the concept of Na^+ channels in particular and membrane channels in general. These compounds are small, structurally-unrelated organic cations that act as Na^+ channel occluders, with no effect on the gating mechanism of the blocked channel. They interact only on the external surface of Na^+ channels, interrupting passive inward flux of Na^+ ions. Numerous reviews on tetrodotoxin and saxitoxin have been published, covering their history, chemistry, physiology, pharmacology, and developments pertaining to the characterization of Na^+ channels (Hall et al., 1990; Narashashi, 1988, 1974; Kao, 1986, 1966; Catterall, 1980).

Although these compounds have been important for conceptualizing the existence of ion channels, the mechanism by which they act as occluders remains in question. Two models have been proposed to explain the action of these toxins. The simplest occluder model entails binding of the inhibitor to the pore, blocking ion transport; the second model involves binding of the toxin to a receptor on the exterior of the protein, causing a distinct change in channel permeability. In fact, stoichiometry and binding studies suggest that tetrodotoxin and saxitoxin bind at or near the external mouth of the channel, and their blocking action is inhibited by a variety of monovalent and divalent cations. Recent evidence has put into question the simple "plug" model. The structural dependence for activity of the toxins appears to be more complicated than necessary for simple pore blockage, and physiological behavior extends beyond simple occlusion (Strichartz and Castle, 1990; Kao, 1986; Strichartz et al., 1986).

These toxins have proven extremely useful and popular chemical tools for neurophysiology and neuropharmacology studies. Numerous experiments have been conducted using tetrodotoxin and saxitoxin to study the biophysics of Na^+ channel action (e.g., Cai and Jordan, 1990), Na^+ channel expression and assembly during cellular development (e.g.,Wollner et al., 1988), and the potential antiarrhythmic properties of tetrodotoxin against occlusion-induced arrhythmias (Abraham et al., 1989). Binding studies with ^3H-labeled toxins were used to determine the density of Na^+ channels in various tissue membranes (Narahashi, 1990). The toxins have also proven applicable in monoclonal antibody studies of lymphocyte activation (e.g., Pinchuk and Pinchuk, 1990).

The potency and selectivity of tetrodotoxin and saxitoxin for Na^+ channels has led to their common use in voltage clamp analyses in which it is desirable to separate the transient current (carried mostly by Na^+ ions) and the steady-state current (carried by K^+ ions) (Narahashi, 1974). Tetrodotoxin and saxitoxin will undoubtedly continue to be important fundamental tools for the study of Na^+ channels and excitatory phenomena.

The venoms of the predatory cone snails (*Conus* spp.) contain small, pharmacologically active peptides, which are targeted to various ion channels and receptors (Table 1). Due to the ability of the conotoxins to discriminate between closely-related receptor sub-types, they have become valuable tools for research in neuroscience (Olivera et al., 1990a). μ-Conotoxin GIIIA (also known as Geographutoxin II or GTX II), isolated from the cone snail *Conus geographus*, is a 22-residue polycyclic peptide bearing three hydroxyprolines and three disulfide bridges (Olivera et al., 1990a; 1990b). It was shown to bind and, hence, to occlude site one of the sodium channel in skeletal muscles. μ-Conotoxin GIIIA is an important tool for studying skeletal muscular neurotransmission because it has no discernable effect on sodium channels in cardiac, brain, and nervous tissue (Strichartz and Castle, 1990; Cruz et al., 1989; Hong and Chang, 1989).

ω-Conotoxin GVIA (ω-CgTx), a 27-amino acid peptide isolated from the piscivorous cone snail *Conus geographus* (Olivera et al., 1990a; 1990b; Yoshikami et al., 1989), is a neuron-specific antagonist with high affinity for voltage-sensitive calcium channels (VSCC) that control neurotransmitter release in response to depolarization of nerve termini. μ-Conotoxin GVIA interacted with N- and/or L-type channels, depending upon tissue type, and was also instrumental in characterizing these channel subtypes and their distribution in various tissue (Karschin and Lipton, 1989; Plummer et al., 1989; Yoshikami et al., 1989). In conjunction with specific L-type calcium channel-blocking ligands, the relative importance of N- and L-type VSCCs to the release of various neurotransmitters has been assessed in parallel experiments using μ-conotoxin GVIA and nifedipine (Wessler et al., 1990), nilvadipine (Takemura et al., 1989a), and PN 200-110 (Fredholm, 1990; Mangano et al., 1990; Dutar et al., 1989; Herdon and Nahorski, 1989; Keith et al., 1989). ^{125}I-labeled toxin was used to determine distribution of calcium channels in rat brain tissue (Takemura et al., 1989b) to identify N- and L-type VSCCs in bovine adrenomedullary plasma membranes (Jan et al., 1990; Ballesta et al., 1989), and to visualize toxin-binding sites in brain tissue of normal and cerebellar mutant mice (Maeda et al., 1989).

Maitotoxin, one of the principal toxins responsible for ingestive ciguatera fish poisoning, is a high-molecular weight polyether neurotoxin produced by the marine dinoflagellate *Gambierdiscus toxicus* (Santostasi et al., 1990) that causes death in mice [170 nanograms/kilogram (ng/kg)] when

Table 1. The Various Conotoxins and Their Linear Peptide Sequences. P: Hydroxyproline

Name of Peptide	Sequence
α-Conotoxin GI	ECCNPACGRHYSC-NH_2
α-Conotoxin GIA	ECCNPACGRHYSCGK-NH_2
α-Conotoxin MI	GRCCHPACKNYSC-NH_2
α-Conotoxin SI	ICCNPACGPKYSC-NH_2
μ-Conotoxin GIIIA	RDCCTPPKKCKDRQCKPQRCCA-NH_2
ω-Conotoxin GVIA	CKSPGSSCSPTSYNCCRSCNPYTKRCY-NH_2

injected intraperitoneally. Numerous pharmacological effects are elicited by maitotoxin in a dose-dependent manner, typically at concentrations in the pico- to nano-molar range. These effects include increased cellular calcium uptake, neurotransmitter/hormone release, phosphoinositide breakdown, contraction of smooth and skeletal muscle and stimulation effects on the heart (Gusovsky and Daly, 1990). Most effects of maitotoxin appear to be due to either its interaction with extra-cellular calcium or the enhanced influx of calcium (Gusovsky and Daly, 1990; Ohizumi and Kobayashi, 1990; Wu and Narahashi, 1988). These conclusions were supported by a recent study, which showed that all muscle-contracting effects of maitotoxin were profoundly suppressed or abolished by calcium ion entry blockers and polyvalent cations (Ohizumi, 1990). The potential of maitotoxin as a unique pharmacological tool for studying calcium transport is yet to be fully realized.

The nicotinic/acetylcholine receptor (nAChR) represents a focal point to understanding the process of signal transduction in biological systems. Characterization of this receptor/ion channel is critical to understanding voluntary neuromuscular transmission, as it is present at the postsynaptic endplates of skeletal muscle. The nAChR also plays an important role in signal transduction in the autonomic and central nervous systems. Signal transduction occurs by binding of agonists to their receptors on the α-subunits (one receptor per subunit) producing a transient opening of the ion channel, resulting in membrane depolarization (Stroud et al., 1990). Several phylogenetically and structurally diverse marine natural products have proven useful probes for studying the nAChR.

The α-conotoxins, isolated from cone shells (Olivera et al., 1985), contain from 13 to 15 amino acids and possess a consensus sequence that presumably preserves some structural, and thereby functional, homology (Pardi et al., 1989). The α-conotoxins (so named because of their pharmacological similarity to the snake α-toxins) were shown to act as antagonists of the nAChR at the endplate region of neuromuscular junctions blocking neuromuscular transmission. This binding was reversible and inhibited by preincubation with other nAChR antagonists, such as α-bungarotoxin (Olivera et al., 1990). Whole-animal studies showed that the α-conotoxins were selective for nACh receptors at neuromuscular junctions: α-conotoxins-GI and -MI had no effect on blood pressure, heart rate, or responses to vagal and preganglionic stimulation at concentrations that induced paralysis (Marshall and Harvey, 1990). More interestingly, α-conotoxin-SI, which has a proline at position nine instead of a positively-charged residue, displayed phylogenetic selectivity towards nACh receptors, indicating that structurally these receptors are a non-homogenous family of proteins (Zafaralla et al., 1988). α-Conotoxins have been used in conjunction with monoclonal antibodies, and other agonists and antagonists of the nAChR, to probe the structure of the agonist receptor sites on the α-subunits (Dowding and Hall, 1987). The results from this investigation supported a model with two binding sites for ACh on the receptor, and indicated that these sites are structurally distinct. The high selectivity of α-conotoxins to nACh receptors at neuromuscular junctions suggests they will continue to be useful probes of this system (Olivera et al., 1990; 1988; Kobayashi et al., 1989; Wu and Narahashi, 1988).

Lophotoxin and its analog-1 [diterpene lactones isolated from gorgonian corals, *Lophogorgia* spp. (Fenical et al., 1981)] are paralytic toxins that produce an irreversible postsynaptic blockade at neuromuscular junctions (Culver et al., 1985; Culver and Jacobs, 1981). Lophotoxin irreversibly inactivated the nAChR by preferential binding to tyrosine^{190} (Abramson et al., 1988) at one of the two primary agonist sites (Culver et al., 1984), via Michael addition or Schiff base formation (Culver et al., 1985). In addition, lophotoxin was shown to be a selective, high-affinity antagonist at the neuronal nAChR, blocking nicotinic transmission in autonomic ganglia. Lophotoxin is an important tool for studying this receptor because most α-neurotoxins demonstrate little or no physiological activity at this receptor (Sorenson et al., 1987). The importance and biomedical uses of lophotoxin have been reviewed (Taylor et al., 1988; Jacobs et al., 1985).

Neosurugatoxin, a reversible nAChR antagonist isolated from the Japanese ivory mollusk, *Babylonia japonica* (Kosuge et al., 1981), was useful for characterizing the ganglionic nAChR into two different subclasses. It is generally accepted that there are two subclasses of nACh receptors in nervous tissue, the so-called high- and low-affinity agonist receptors (Billiar et al., 1988). Neosurugatoxin binds with high selectivity at the high affinity agonist site, as indicated by the observation that it did not inhibit α-bungarotoxin binding, a known antagonist at the low-affinity site (Billiar et al., 1988; Bourke et al., 1988). The selectivity of neosurugatoxin for specific receptors in nervous tissue illustrated the pharmacological distinction between ganglionic nAChRs and those at the neuromuscular junction (Rapier et al., 1990; Wada et al., 1989; Hayaski et al., 1984; Hayashi and Yamada, 1975).

A study was conducted using α-conotoxins GIA and MI, lophotoxin and its analog-1, neosurugatoxin, and α-bungarotoxin to probe the structural differences among neuronal nAChR subunit combinations expressed in *Xenopus* oocytes (Luetje et al., 1990). The results indicated that neosurugatoxin and the conotoxins distinguished between muscle and neuronal subunit combinations, whereas lophotoxin and analog-1 distinguished among neuronal subunit combinations on the basis of differing α-subunits.

Onchidal, an acetate ester isolated from the mollusk *Onchidella binneyi* (Ireland and Faulkner, 1978), was shown to be an irreversible substrate inhibitor of acetylcholinesterase with a novel mechanism of action, with no effect on α-bungarotoxin binding to the nAChR (Abramson et al., 1989). Irreversible inhibition was prevented by coincubation with agents that block or modify the ACh-binding site. Thus, onchidal may prove useful for the design of new anticholinesterase insecticides and in identifying the active site residues that contribute to binding and hydrolysis of ACh.

The importance of adenosine as a second messenger in signal transduction has led to considerable interest in compounds that bind adenosine receptors. A pair of modified nucleosides that act as adenosine analogs have been used to study adenosine receptors in a variety of systems. (5'-deoxy-) 5-Iodotubercidin, isolated from the red alga *Hypnea valetiae* (Kaslauskas et al., 1983), inhibited adenosine uptake in brain tissue, and is one of the most potent and specific inhibitors of adenosine kinase reported to date (Davies et al., 1984). The specificity and potency of this molecule towards adenosine kinase led to its wide application in studies of purine/nucleotide metabolism and regulation (e.g., Kather, 1990; Bontemps and Van den Berghe, 1989; Das and Steinberg, 1988; Dawicki et al., 1988; Weinberg et al., 1988). 1-Methylisoguanosine, an orally active adenosine analog isolated from the marine sponge *Tedania digitata* (Baird-Lambert et al., 1980), exhibited properties as a muscle relaxant, inducing hypothermia and cardiovascular effects similar to adenosine, and was also used to aid in the characterization of these receptors (Williams et al., 1987).

Dysidenin, a hexachlorinated alkaloid isolated from the sponge *Dysidea herbacea* (Biskupiak and Ireland, 1984; Charles et al., 1980; 1978), inhibits iodide transfer in thyroid cells. This molecule may provide insight into the mechanism of the elusive "iodide pump" as it inhibits iodide transport by a different mechanism than that of ouabain (Van Sande et al., 1990).

ANTITUMOR COMPOUNDS

Spongothymidine (arabinosyl thymine; araT) and spongouridine (arabinosyl uracil; araU) were isolated in the early 1950s (Bergmann and Burke, 1955; Bergmann and Feeney, 1951) from the Caribbean sponge *Cryptotethya crypta*. This discovery ultimately led to synthesis of a new class of arabinosyl nucleoside analogs. One such analog, arabinosyl cytosine (araC), displayed *in vivo* antileukemic activity (Evans et al., 1964; 1961). AraC derived its activity by conversion to arabinosyl cytosine triphosphate, incorporation into cellular DNA, and subsequent inhibition of DNA polymerase (Cohen, 1977). It is currently in clinical use for treatment of acute myelocytic leukemia and non-Hodgkin's lymphoma (de Andrea et al., 1990; Geller et al., 1990; Minigo et al., 1990). The arabinosyl nucleosides served a catalytic role in promoting the search for antitumor compounds from marine sources. Over four decades, several research programs have focused on this objective. Although many compounds of marine origin have been isolated that possess *in vitro* activity, only didemnin B has been evaluated in clinical trials.

The didemnins, a series of new depsipeptides isolated from the ascidian *Trididemnum solidum*, showed promising cytotoxic, antitumor, antiviral (Rinehart et al., 1981a; 1981b ; 1981c), and immunosuppressive activities (Montgomery et al., 1987; Montgomery and Zukoski, 1985). Didemnin B, the most potent of this group, exhibited cytotoxicity against L1210 murine leukemia cells *in vitro* at a dose of 0.001 micrograms/milliliters (Rinehart et al., 1981b). Didemnin B was also reported active in stem cell assays against a variety of human tumors including mesothelioma, sarcoma, hairy cell leukemia, and carcinomas of the breast, ovary, and kidney (Rosof et al., 1983; Jiang et al., 1983) . Didemnin B also showed significant *in vivo* activity against intraperitoneally implanted B16 melanoma (T/C 160 @ 1 milligram/kilogram) and P388 leukemia cells (T/C 199 @ 1 milligram/kilogram) (Rinehart et al., 1983; 1981b). At the highest nontoxic dose, didemnin B effected a 45 percent survival rate with a 2.7-fold life extension of rats challenged with *Yoshida ascites* tumor cells (Famiani, 1987). These promising *in vivo* results led to preclinical toxicology studies and ultimately to evaluation of Didemnin B in phase I clinical trials in the United States as an anticancer agent (Dorr et al., 1988; 1986; Stewart et al., 1986). Preclinical toxicological studies suggested that the major target organs for toxicity were the lymphatics, gastrointestinal tract, liver, and kidney. Didemnin B caused hepatic-toxicity with a decrease in clotting factors (Chun et al., 1986; Stewart et al., 1986). Based on these results, it was anticipated that the maximum tolerated dose in humans would be defined by gastrointestinal- or hepato-toxicity. Didemnin B completed phase I clinical trials where maximum tolerable clinical doses were established. It is slated for phase II trials where it is hoped that objective tumor response will be observed.

Didemnin B was reported to inhibit protein synthesis more than DNA synthesis, and to a lesser extent RNA synthesis in B16 and L1210 cells (Crampton et al., 1984; Li et al., 1984). Also, didemnin B reportedly suppressed phosphorylation of a 100 kDa epidermal cytosolic protein along with immunological and inflammatory reactions induced by the tumor promoter 12-O-tetradecanoylphorbol-13-acetate (TPA) (Gschwendt et al., 1987). It was further noted that inhibition of protein synthesis was not due to decreased amino acid uptake since cytotoxicity was not reversed by high concentrations of several amino acids. Furthermore, flow cytometric results indicated that low doses of didemnin B (10 nanogram/milligram) inhibited progression of the cell cycle from G_1 to S phase (Crampton et al., 1984).

In vivo evaluation of several marine natural products yielded several promising leads for future development. Mycalamide A, isolated from a *Mycale* sp. sponge (Perry et al., 1988), exhibited marked *in vitro* cytotoxicity and moderate *in vivo* activity against P388 (T/C 140 @ 10 mg/kg), B16 melanoma (T/C 245 @ 30 mg/kg), and M5076 ovarian carcinoma (T/C 233 @ 60 mg/kg) (Burres and Clement, 1989).

Halichondrin B, one of several polyether macrolides isolated from the marine sponge *Halichondria okadai*, showed excellent activity *in vivo* against P388 leukemia (T/C 323 @ 10 mg/kg) and B16 melanoma (T/C 244 @ 5 mg/kg) (Hirata and Uemura, 1986). These promising results (T/C of 300 is considered a cure in this protocol) have generated much interest in the potential of the halichondrins as clinically-useful anticancer agents. Unfortunately, their development has been hampered by limited supplies of material.

Dolastatin 10, a cytotoxic peptide from the opisthobranch mollusk *Dolabella auricularia*, was reported to show exceptional antineoplastic activity *in vivo* against PS leukemia (T/C 169-202 @ 1-4 mg/kg) and B16 melanoma (T/C 142-238 @ 1.44-11.1 mg/kg); it effected 17 to 67 percent curative responses at 3.25 to 26 mg/kg against human melanoma xenographs in nude mice (Pettit et al., 1987). Complete evaluation of dolastatin 10 was hindered by limited supplies until the total synthesis of dolastatin 10 was reported (Pettit et al., 1989). The ability of dolastatin 10 to inhibit the growth of L1210 murine leukemia cells *in vitro*, in nanomolar concentrations, has been attributed to its ability to act as an antimitotic agent by causing cells to become arrested in metaphase. This arrest resulted from interaction with tubulin, inhibiting polymerization and preventing microtuble assembly (Bai et al., 1990a). Dolastatin 10 noncompetitively inhibited the binding of radiolabeled vincristine to tubulin and inhibited nucleotide exchange on tubulin without displacing nucleotide present in the exchangeable site. This result suggested close proximity of the dolastatin 10 binding site to those of the nucleotide and the vinca alkaloids on the same subunit of tubulin (Bai et al., 1990b).

The long-standing reports of *in vivo* antitumor activity in extracts of *Ecteinascidia turbinata* have recently been attributed to the ecteinascidins (Rinehart et al., 1990; Wright et al., 1990). Ecteinascidin 743 was reported to effect a half-maximal inhibition of L1210 leukemia cells at 0.5 ng/mL, while exhibiting notable activity in vivo against P388 leukemia (T/C 167 @ 15 mg/kg). The most active agent, ecteinascidin 729, showed potent activities against both P388 murine leukemia (T/C 214 @ 3.8 mg/kg) and B16 melanoma (T/C 246 @ 10 mg/kg) (Rinehart et al., 1990).

Dercitin, a new acridine alkaloid isolated from a *Dercitus* sp. sponge (Gunawardana et al., 1988), was found to prolong the life of P388-bearing mice (T/C 170 @ 5 mg/kg). Preliminary data suggested dercitin derives its activity by intercalating DNA (Burres et al., 1989).

Aeroplysinin-1, isolated from the sponges *Aplysina aerophoba* (Fattorusso et al., 1970), and *Ianthella ardis* (Fulmor et al., 1970) was recently reported to have significant antileukemic activity in the L51781y cell/NMRI mouse system (T/C 338 @ 50 mg/kg daily for five days) (Kreuter et al., 1989).

TUMOR PROMOTERS

Protein kinase C (PKC), an enzyme ubiquitous in eukaryotic cells, mediates transduction of extracellular signals into intracellular events, and is the major receptor for tumor-promoting phorbol esters. Structurally, PKC consists of a single polypeptide chain (M.W. ~80 kd) with two functionally-different domains: a hydrophobic domain that may function in membrane attachment, and a hydrophilic domain containing the catalytic site. Functionally, PKC is a large family of isozymes (reviewed in Nishizuka, 1988), differing in their distribution within the cell. They have different molecular weights, are non-equivalent to monoclonal antibodies, and have different auto-phosphorylation sites.

When an extracellular ligand binds a cell surface receptor, secondary messengers such as diacylglycerol (DAG) are produced. Effector proteins, such as PKC, respond to secondary messengers and activate a variety of biochemical systems within the cell. In the case of PKC, this action involves protein phosphorylation. Formation of a diacylglycerol-PKC complex initiates a protein phosphorylation cascade that, among other functions, controls cell regulation and proliferation. Proliferating and transformed cells display higher PKC activity. Unregulated protein phosphorylation has been linked with tumor promotion. Phorbol esters, a class of natural products that promote tumors in animals exposed to carcinogens, activate PKC. TPA mimics the action of DAG, resulting in uncontrolled proliferation and transformation of the initiated cell. Discovery that PKC contains the phorbol ester receptor has greatly aided the study of PKC signal transduction and understanding of the mechanism of phorbol ester tumor promotion. The structure and function of PKC, as well as its activation by phorbol esters, have been the topic of many recent reviews (Boivier, 1990; Farago and Nishizuka, 1990; Jaken, 1990; Hannun and Bell, 1989; Nakadate, 1989; Nishizuka, 1989).

Further understanding of the function of PKC in cellular biochemistry was made possible by the discovery that the marine natural products lyngbyatoxin (teleocidin A-1) and aplysiatoxin bind to the phorbol receptor (Fujiki and Sugimura, 1987). Although lyngbyatoxins, aplysiatoxins, and TPA are structurally quite dissimilar, they compete for the same receptor on PKC, evidence that the tumor-promoting activities ascribed to TPA are the direct consequence of activating PKC. Insight to the electrostatic structure of the phorbol ester receptor was gained by comparing superimposed computer models of the promoters TPA, teleocidin, and aplysiatoxin (Thomson and Wilkie, 1989; Itai et al., 1988; Irie et al., 1987; Jeffrey and Liskamp, 1986; Wender et al., 1986). Although this approach produced conflicting results, use of these data to design synthetic activators and repressors could provide powerful probes of the mechanism of action of PKC. It is hoped that this information will lead to future design of useful tumor-suppressing compounds.

TPA-type tumor promoters have proven useful tools in studying the many functions of PKC. Recently, teleocidin was used to study the regulation of tumor necrosis factor alpha, as influenced by expression of interleukin-1 (IL-1) and IL-2 receptors, and interactions among B-cell lymphokines (Benjamin et al., 1990). Teleocidins and TPA were also shown to directly simulate IL-2 function in cytotoxic T cells (Gavériaux and Loor, 1989), to potentiate the activity of decay-accelerating factor (Bryant et al., 1990), and to stimulate B-cell gamma-interferon production (Benjamin et al., 1986).

Bryostatin B, isolated from the bryozoan *Bugula neritina* (Pettit et al., 1982), is a cyclic macrolide with an oxygen-rich cavity reminiscent of cyclic ionophore antibiotics. In general, these antibiotics cause tumor destruction at the cellular level by transporting chelated cations such as K^+, Ca^{2+}, or Na^+ across the cell membrane. Based on preliminary evidence that bryostatin B chelated Ag^+, it was administered to quiescent Swiss 3T3 cells (Berkow and Kraft, 1985), and the intra- and extra-cellular ion gradient monitored. Although bryostatin B failed to induce any cation flux, it was observed to activate DNA synthesis, a response previously observed with phorbol esters. Furthermore, pre-incubation of 3T3 cells with phorbol 12-myristate 13-acetate (PMA) rendered the cells unresponsive to treatment with bryostatin B; likewise, bryostatin B desensitized the cells to TPA. Bryostatin B prevented receptor binding of [^{3}H]-phorbol 12-,13-dibutyrate to Swiss 3T3 cells. What makes bryostatin so valuable is that although administration of TPA-type promoters causes tumors in animals, administration of the bryostatins does not, and indeed they inhibit tumor promotion by phorbol esters (Hennings et al., 1987). The bryostatins show enormous potential for inhibiting PKC tumor promotion, and may have use as antitumor agents.

The cytotoxic macrolides bistratene A [also known as bistratamide A (Gouiffès et al., 1988a; 1988b)], and bistratene B (Degnan et al., 1989), isolated from the ascidian *Lissoclinum bistratum*, were shown to enhance the phospholipid-dependent activity of type II protein kinase C (Watters et al., 1990). These compounds may also prove useful tools for probing the mechanisms of cell regulation. The structure of bistramide A has recently been revised (Foster et al., 1992).

A new group of non-TPA-type tumor promoters (okadaic acid class) has been reported. Their mechanism of action is distinct from TPA-type promoters in that they do not bind to or activate protein kinase C, although they do modulate protein phosphorylation. While TPA- type promoters have illuminated many aspects of cellular regulation by enhancing protein phosphorylation, okadaic acid class promoters have made similar contributions by virtue of their specific inhibition of phosphatases 1 and 2A (PP-1 and PP2A) and resulting effects on protein dephosphorylation (Cohen et al., 1990; Suganuma et al., 1990; 1988; Ishihara et al., 1989; Bialojan and Takai, 1988).

Okadaic acid, a polyether derivative of a C_{38} fatty acid, initially isolated from the sponges *Halichondria okadai* and *H. melanodocia* (Tachibana et al., 1981), was subsequently shown to be produced by several species of dinoflagellates and to be concentrated in a variety of filter feeders (Fujiki et al., 1988). [^{3}H]-okadaic acid was reported to bind specifically to particulate and cytosolic fractions of mouse skin, suggesting interaction with both membrane-bound and cytosolic macromolecular receptor(s) (Suganuma et al., 1989). This binding was not inhibited by TPA-type tumor promoters. Okadaic acid inhibited phosphatase activity in a dose-dependent manner in experiments using cytosolic fractions of homogenized mouse brain that retain both kinase and phosphatase activity (Sassa et al., 1989; Suganuma et al., 1989). In experiments with smooth muscle, okadaic acid induced muscle fiber contraction due to enhanced myosin light chain phosphorylation (Obara et al., 1989; Bialojan et al., 1988; Takai et al., 1987; Kodama et al., 1986; Shibata et al., 1982). This hyper-phosphorylation was shown to be due to the inhibitory effects of okadaic acid on protein phosphatases (Takai et al., 1987). Okadaic acid also induced hyperphosphorylation of a 60 kDa protein in primary human fibroblasts (Issinger et al., 1988), which has been shown to be specifically dephosphorylated by PP-1 and PP-2A (Schneider et al., 1989). It was further shown that okadaic acid induced phosphorylation of the 100 kDa elongation factor 2 (EF-2) protein, a substrate of the Ca^{+2}/ Calmodulin-dependent protein kinase III (Redpath and Proud, 1989). Okadaic acid played a significant role in elucidating cellular processes involving phosphorylation/dephosphorylation (Haystead et al., 1989).

In a surprising development, okadaic acid reverted the phenotype of NIH 3T3 cells transformed by *raf* and *ret*-II oncogenes (Sakai et al., 1989). This result fueled speculation that phosphatases regulate interconversion of normal and malignant cells. Okadaic acid induced cdc2 kinase activation in *Xenopus* oocytes by inhibiting PP-2A during interphase (Félix et al., 1990; Kipreos and Wang, 1990; Goris et al., 1989). The cdc2 gene product, a 34 kDa protein kinase, is responsible for activating a major protein phosphorylation cascade that is important in regulating the cell cycle. While the level of cdc2 kinase is constant, there is a cyclic rise and fall in its activity during the cell cycle, thought to result from a balance between kinase phosphorylation and phosphatase dephosphorylation (Kipreos and Wang, 1990). These findings provided important clarification of the two-stage tumor promotion model (Fujiki and Sugimura, 1987). Tumor promotion apparently results from hyperphosporylation of protein kinase (e.g., PKC) effector proteins that initiate cell proliferation; dephosporylation of these proteins by PP1 or PP2A down-regulates cell proliferation. In the context of this model, phosphatases cells function as tumor suppressors in normal cells.

There have been many examples of the use of okadaic acid as a probe for identifying physiologically-relevant protein phosphatases and for identifying cellular processes that are regulated by phosphorylation/dephosphorylation, such as ion channel function (Kume et al., 1989; Hescheler et al., 1988) and the process of vision (Palczewski et al., 1989). Okadaic acid was also used in combination with inhibitor proteins to develop a method for quantification and identification of phosphatases (Cohen, 1989; Cohen et al., 1989a; 1989b; MacKintosh and Cohen, 1989; Takai et al., 1989). Dinophysistoxin 1 (35-methylokadaic acid) (Fujiki et al., 1988) and acanthifolicin (9,10-episulfide okadaic acid) (Schmitz et al., 1981) inhibited PP-1 and PP-2A with similar potencies to okadaic acid (Cohen et al., 1990; Fujiki et al., 1989a).

Calyculin A, isolated from the marine sponge *Discodermia calyx* (Kato et al., 1986) binds to okadaic acid receptors, and was a potent tumor promoter on mouse skin (Suganuma et al., 1990). Calyculin A was reported to be a more potent inhibitor of PP-1 and PP-2A than okadaic acid (Ishihara et al., 1989).

Palytoxin, a very potent toxin from the coelenterate *Palythoa toxica* (Moore and Bartolini, 1981; Uemura et al., 1981), was shown to stimulate arachidonic acid metabolism synergistically with TPA-type promoters, suggesting that palytoxin activates an alternative signal transduction pathway (Fujiki et al., 1986; Levine et al., 1986; Levine and Fujiki, 1985). Palytoxin down-regulated the epidermal growth factor (EGF) by reducing the number and affinity of EGF-binding sites. This down-modulation requires extracellular sodium, and a correlation exists between palytoxin-induced sodium uptake and inhibition of EGF binding. These results suggest that palytoxin activates a sodium pump, and that sodium may act as a second messenger in this signal transduction pathway. Palytoxin should therefore provide an additional useful tool for probing cellular regulation processes (Wattenberg et al., 1989a; 1989b; 1987).

Sarcophytol A, an oxygenated cembrane isolated from the soft coral *Sarcophyton glaucum* (Kobayashi et al., 1979), was recently shown to be antitumorigenic. Sarcophytol A inhibited development of N-methyl-N-nitrosourea-induced large bowel carcinomas in rats. It also suppressed sodium deoxycholate induction of ornithine decarboxylase, a marker for tumor promotion, in the large bowel mucosa (Narisawa et al., 1989). Sarcophytol A also inhibited tumor promotion by teleocidin in a two-stage carcinogenesis experiment on mouse skin (Fujiki et al., 1989b). Thus, sarcophytol may prove to be an important probe for studying the mechanism(s) of carcinogenesis.

ANTI-INFLAMMATORY/ANALGESIC COMPOUNDS

Several marine natural products with anti-inflammatory and analgesic properties have found use in studying the roles of arachidonic acid metabolism and calcium mobilization in inflammation. Pro-inflammatory stimuli induce their effect ultimately through the mobilization/release of calcium ions (Ca^{2+}) from intra- or extra-cellular stores. Mobilization of calcium is believed to be initiated by binding of an agonist to its receptor. The receptor transduces a signal via a guanine nucleotide-binding protein (G-protein), activating a phospholipase (e.g., PLA_2, PLC) that hydrolyzes membrane phospholipids to produce a series of second messengers [e.g., inositol triphosphate (IP_3) and arachidonic acid (AA)]. Inositol triphosphate binding to its receptor on the rough endoplasmic reticulum results in Ca^{2+} release from intracellular stores. Extracellular calcium release is believed to depend on release of arachidonic acid. This fatty acid is metabolized via the cyclooxygenase pathway to prostaglandins, prostacyclins, and thromboxanes, or the lipoxygenase pathway to tetraenoic acids, leukotrienes, and lipoxins. Binding of these metabolites to their receptors on hormone-activated Ca^{+2} channels leads to Ca^{+2} mobilization from extracellular sources. Many agents that mediate inflammation (accompanied by pain) and proliferation exert their physiological effects via modulation of phospholipid metabolism and Ca^{+2} mobilization. Consequently, compounds that inhibit a phopholipase and/or Ca^{+2} mobilization are anti-inflammatory, and thereby analgesic (Wheeler et al., 1988; Mayer and Jacobs, 1986).

Manoalide, a non-steroidal sesterterpene isolated from the sponge *Luffariella variabilis* (de Silva and Scheuer, 1980), emerged as a potent tool for studying inflammation. Manoalide irreversibly inhibited PLA_2 (Jacobson et al., 1990; Mayer et al., 1988; Glaser and Jacobs, 1986), inhibiting arachidonic acid release and thus its subsequent metabolism to prostaglandins and leukotrienes. It blocked phorbol ester (e.g., PMA)-induced inflammation but not arachidonic acid- induced response. This property made manoalide very important for elucidating the role of PLA_2 in arachidonic acid release for eicosanoid biosynthesis. Recent studies localized the manoalide-binding site on PLA_2 (Glaser et al., 1988), defined the pharmacophore responsible for PLA_2 activation (Glaser et al., 1989), and examined the range of phospholipases inhibited by manoalide (e.g., Jacobson et al., 1990; Lister et al., 1989; Ulevitch et al., 1988).

In addition to inhibiting PLA_2, manoalide inhibited 5-lipoxygenase (de Vries et al., 1988), leading to speculation that its anti-inflammatory activity was due in part to inhibition of leukotriene biosynthesis. However, the most important factor contributing to the anti-inflammatory activity of manoalide was attributed to its inhibitory effect on Ca^{+2} channels (Wheeler et al., 1988). Interestingly, at low concentrations manoalide inhibited calcium channels with no effect on phosphoinositide metabolism. The ability of manoalide to dissect these two components of the inflammation process may prove to be its most useful attribute in studying the role of Ca^{+2} signaling in inflammation and proliferation (e.g., Barzaghi et al., 1989).

Luffarielolide, an analog of manoalide isolated from the same organism, also exhibited anti-inflammatory activity, was slightly less potent than manoalide, and was a partially reversible PLA_2 inhibitor (Albizati et al., 1987).

Pseudopterosins A and E, members of a family of diterpene ribosides isolated from the gorgonians *Pseudopterogorgia bipinata* and *P. elizabethae*, exhibited potent anti-inflammatory and analgesic activities and act as reversible inhibitors of lipoxygenase and PLA_2 (Luedke, 1990; Look et al., 1986).

While the previous compounds all exhibited anti-inflammatory activity, 15-acetylthioxy-furodisinin lactone, isolated from a *Dysidea* sp. (Carté et al., 1989), elicited the opposite response. The compound caused intracellular Ca^{+2} mobilization that was blocked by LTB_4 receptor antagonists. 15-acetylthioxy-furodisinin lactone binds LTB_4 receptors with high affinity and activates the receptor-mediated signal transduction processes related to LTB_4. Thus, it will likely be of use in studying the role of leukotrienes in inflammation (Mong et al., 1990).

ANTIVIRAL AGENTS

The search for viral chemotherapeutic agents from marine sources has been disappointing to date. The only compound reported thus far to show significant therapeutic activity is ara-A, a semisynthetic based on the arabinosyl nucleosides isolated from the sponge *Cryptotethya crypta* (Bergmann and Burke, 1955; Bergmann and Feeney, 1951). A number of marine metabolites have shown very promising *in vitro* activity; however, only the didemnins demonstrated *in vivo* activity.

The didemnins, depsipeptides isolated from the Caribbean ascidian *Trididemnum solidum* (Rinehart et al., 1981a), were reviewed earlier in the antitumor section. In addition to their antitumor activity, they exhibited *in vitro* and *in vivo* antiviral properties. Didemnin B significantly reduced the yields of DNA and RNA viruses by 10^4-10^5 at 0.5 μg/ml, and in *in vivo* tests protected over 70 percent of a population of mice from lethal vaginal doses of *Herpes simplex* type 2 virus (intravaginal application three times per day) (Rinehart et al., 1983), and a lethal subcutaneous challenge of Rift Valley fever virus with a 90 percent survival rate at 0.25 mg/kg (Canonico et al., 1982).

Patellazole B, isolated from the ascidian *Lissoclinum patella* (Zabriskie et al., 1988), exhibited very potent antiviral activity against *Herpes simplex* viruses. In an *in vitro* assay, patellazole B inhibited viral replication of HSV-1 and HSV-2 at 0.5 and 60 ng/ml respectively. Patellazole B was cytotoxic towards the host Vero cells at concentrations 1000 times greater than it was active against HSV-1 (Ireland and Maiese, 1990).

The eudistomins, a class of ß-carbolines isolated from the Caribbean ascidian *Eudistoma olivacea,* were active in shipboard antiviral assays (Rinehart et al., 1981c). In subsequent experiments five of the eudistomins showed enhanced activity after activation with UVA light (Hudson et al., 1988).

Several derivatives of the sesquiterpene hydroquinone avarol isolated from sponges of the genus *Dysidea* were recently shown to inhibit HIV-1 reverse transcriptase. Avarone E and avarol F, the two most effective derivatives, showed two- to tenfold greater DNA polymerase activity than RNase H activity. Further results for one derivative, avarone E, suggested that HIV-1 RT binding occurred at sites different from those of DNA synthesis substrates, dGTP, and primer template (Loya and Hizi, 1990).

A class of sulfonic acid containing glycolipids that inhibit HIV-1 was isolated from two species of microcultured blue-green algae (Gustafson et al., 1989). The extract of *Lyngbya lagerheimii* contained glycolipids A and B, whereas the extract of *Phormidium tenue* contained C and D. The degree of HIV-1 protection varied substantially between different host cell lines, but seemed to be relatively independent of acyl chain length and degree of unsaturation. These glycolipids are structural components of chloroplast membranes, and although they occur widely in higher plants, algae, and photosynthetic microorganisms, they have not been previously associated with HIV-1 inhibitory activity (Gustafson et al., 1989).

Algal-sulfated polysaccharides, including the *Laminaria* sp. kelp (brown algae) metabolite fucoidin (fucoidan), were shown to inhibit a variety of DNA- and RNA-enveloped viruses, including *Herpes simplex* virus and HIV (Baba et al., 1988). Very important, fucoidin exhibited antiviral activity at concentrations several orders of magnitude lower than its anticoagulant threshold. *In vitro* studies with peripheral mononuclear cells from AIDS patients suggested fucoidin binds the *env* protein of HIV. This result suggested that HIV invaded target cell cytoplasm slowly and that fucoidin can still react with HIV in the cell membrane (Sugawara et al., 1989).

METABOLITES THAT AFFECT MICROFILAMENT-MEDIATED PROCESSES

Transformation of chemical bond energy into motion plays a key role in many cellular processes. Microfilaments are responsible for this transformation that occurs in fertilization and early development processes, phagocytosis, protein synthesis, and microorganismal propulsion (Stryer, 1981). Several marine-derived metabolites have contributed to our understanding of these cellular and molecular biochemical processes in unique ways.

The latrunculins, 2-thiazolidinone-containing macrolides isolated from the Red Sea sponge *Latrunculia magnifica* (Kashman et al., 1980), were reported to disrupt the organization of microfilaments but not microtubules in cultured cells (Spector and Shochet, 1983). Latruculin A was an order of magnitude more potent than cytochalasin, and affected different components of the actin-based cytoskeleton (Spector et al., 1989). The latrunculins currently represent the only alternative to the cytochalasins in pharmacological studies of both actin polymerization *in vitro* and actin organization and function in living cells.

Latrunculin A was also used to determine which fertilization processes are microfilament-mediated in different species (Schatten et al., 1986; Schatten and Schatten, 1986). The results suggested that the acrosome reaction of mouse sperm does not require microfilament activity, whereas that of sea urchins does. Latrunculin A was also found to inhibit macrophage phagocytosis without interfering with cell viability (de Oliveira and Mantovani, 1988), strengthening the case for participation of microfilaments in the mechanism of phagocytosis. Latrunculin A is expected to continue to make important contributions to the study of microfilament-mediated processes.

Purealin, isolated from the sponge *Psammaplysilla purea* (Nakamura et al., 1985), was shown to affect various myosin ATP-ases. Purealin enhanced the stability of thick filaments of dephosphorylated gizzard myosin against the disassembling action of ATP, suggesting it acted on the ATP-binding site of myosin (Takito et al., 1986), possibly by directly affecting the myosin heads (Nakamura et al., 1987). Furthermore, purealin increased the actin-activated ATPase activity of myosin (Takito et al., 1987a). These results indicated that purealin inhibited myosin phosphorylation by acting as a calmodulin antagonist, inhibiting formation of the calmodulin-myosin light chain kinase complex (Takito et al., 1987b). Purealin has provided the research tool required to investigate the structure and conformation of various forms of myosin (Takito et al., 1986).

Fucoidin, an L-fucose-rich sulfated heteropolysaccharide produced by brown algae of genus *Laminaria* (kelp), exhibited antithrombin (Church et al., 1989), anticoagulant, fibrinolytic, onco-inhibitory (Maruyama et al., 1987), and antitumor activities (Chida and Yamamoto, 1987), and was used as a probe of lymphocyte membranes (Brandley et al., 1987). Perhaps the most interesting and useful attribute of fucoidin was its inhibition of sperm-egg binding. Fucoidin was useful in elucidating the molecular basis of mammalian sperm-egg recognition, as it inhibits guinea pig sperm-egg binding by interacting with the inner acrosomal membrane and equatorial segment domains of guinea pig spermatozoa (Huang and Yanagimachi, 1984). These results implied that the mechanisms of sperm-egg adhesion and acrosomal reaction are distinct. Further experiments with fucoidin in mice suggested that an L-fucose component of the sperm surface is involved in sperm-egg recognition (Boldt et al., 1989). Thus, fucoidin provides important insight concerning the mechanism of sperm-egg recognition. Fucoidin was also found to be a potent inhibitor of tight binding of spermatozoa to the human zona pellucida in the human hemizona assay, an assay with demonstrated predictive value for human *in vitro* fertilization (Oehninger et al., 1990).

THE FUTURE

The huge resource of marine natural products has played a key role in the explosive growth of biomedical science over the last two decades — especially in the development and formulation of ion channel and tumor promotion models. Investigation of antitumor compounds has progressed to the extent that one compound, didemnin B, is now in clinical trials, and a second compound, bryostatin-1, will soon begin clinical trials. In spite of these contributions it is clear that this resource has yet to be fully utilized.

No longer restricted to chemical studies, the marine natural products field is encouraging an interdisciplinary approach between chemists and pharmacologists. Pharmacologists realize that these metabolites can provide key insights into complex cellular events. Marine natural products will continue to play an important role as molecular probes in studying these types of biochemical events and unraveling their roles in cell regulation. The establishment of interactive collaborations between chemists and pharmacologists will be essential in the future to ensure that this resource meets its full potential.

REFERENCES

Abraham, S., G.N. Beatch, B.A. MacLeod, and M.J. Walker. 1989. Antiarrhythmic properties of tetrodotoxin against occlusion-induced arrhythmias in the rat: a novel approach to the study of the antiarrhythmic effect of ventricular sodium channel blockade. *J. Pharmacol. Exp. Ther.* **251**:1166-1173.

Abramson, S.N., P. Culver, T. Cline, Y. Li, P. Guest, L. Gutman, and P. Taylor. 1988. Lophotoxin and related coral toxins covalently label the α-subunit of the nicotinic acetylcholine receptor. *J. Biol. Chem.* **263**:18568-18573.

Abramson, S.N., Y. Li, P. Culver, and P. Taylor. 1989. An analog of lophotoxin reacts covalently with Tyr^{190} in the a-subunit of the nicotinic acetylcholine receptor. *J. Biol. Chem.* **264**:12666-12672.

Abramson, S.N., Z. Radic, D. Manker, D.J. Faulkner, and P. Taylor. 1989. Onchidal: a naturally occurring irreversible inhibitor of acetylcholinesterase with a novel mechanism of action. *Mol. Pharmacol.* **36** 349-354.

Albizati, K F., T. Holman, D.J. Faulkner, K.B. Glaser, and R.S. Jacobs. 1987. Luffariellolide, an anti-inflammatory sesterterpene from the marine sponge *Luffariella* sp. *Experientia.* **43**:949-950.

Albizati, K.F., V.A. Martin, M.R. Agharahimi, and D.A. Stolze. 1990. Synthesis of Marine Natural Products. In *Bioorganic Marine Chemistry,* Volume 5, P.J. Scheuer (ed.). Springer-Verlag. Berlin, Heidelberg.

Baba, M., R. Snoeck, R. Pauwels, and E. de Clerq. 1988. Sulfated polysaccharides are potent and selective inhibitors of various enveloped viruses, including *Herpes simplex* virus, cytomegalovirus, vesicular stomatitis virus, and human immunodeficiency virus. *Antimicrobial Agents and Chemotherapy.* **32**(11):1742-1745.

Baden, D. G. 1989. Brevetoxins: Unique polyether dinoflagellate toxins. *FASEB Journal.* **3**:1807-1817.

Bai, R., G.R. Pettit, and E. Hamel. 1990a. Dolastatin 10, a powerful cytostatic peptide derived from a marine animal. Inhibition of tubulin polymerization mediated through the vinca alkaloid domain. *Biochem. Pharmacol.* **39**:1941-1949.

Bai, R., G. R. Pettit, and E. Hamel. 1990b. Binding of dolastatin 10 to tubulin at a distinct site for peptide antimitotic agents near the exchangeable nucleotide and vinca alkaloid sites. *J. Biol. Chem.* **265**:17141-17149.

Baird-Lambert, J., J. F. Marwood, L. P. Davies, and L. Taylor. 1980. 1-methylisoguanosine: an orally active marine natural product with skeletal muscle and cardiovascular effects. *Life Sci.* **26**:1069-1077.

Ballesta, J. J., M. Palmero, M. J. Hidalgo, L. M. Gutierrez, J. A. Reig, S. Viniegra, and A.G. Garcia. 1989. Separate binding and functional sites for ω-conotoxin and nitrendipine suggest two types of calcium channels in bovine chromaffin cells. *J. Neurochem.* **53**:1050-1056.

Barzaghi, G., H. M. Sarace, and S. Mong. 1989. Platelet-activating factor-induced phosphoinositide metabolism in differentiated U-937 cells in culture. *J. Pharm. Exp. Ther.* **248**:559-566.

Benjamin, D., D.P. Hartmann, L.S. Bazar, and R.J. Jacobsen. 1986. Burkitt's cells can be triggered by teleocidin to secrete interferon-gamma. *Am. J. Hematology.* **22**:169-177.

Benjamin, D., S. Hooker, and J. Miller. 1990. Differential effects of teleocidin on TNF-alpha receptor regulation in human B cell lines: relationship to coexpression of IL-2 and IL-1 receptors and to lymphokine secretion. *Cellular Immunology.* **125**:480-497.

Bergmann, W. and D. C. Burke.1955. Contributions to the study of marine products, XXXIX. The nucleosides of sponges, III: Spongothymidine and spongouridine. *J. Org. Chem.* **20**:1501-1507.

Bergmann, W. and R. J. Feeney. 1951. Contributions to the study of marine products. XXXII. The nucleosides of sponges, I. *J. Org. Chem.* **16**:981-987.

Berkow, R. L. and A. S. Kraft. 1985. Bryostatin, a non-phorbol macrocyclic lactone, activates intact human polymorphonuclear leukocytes and binds to the phorbol ester receptor. *Biochem. Biophys. Res. Comm.* **131**:1109-1116.

Bialojan, C. and A. Takai. 1988. Inhibitory effect of a marine-sponge toxin, okadaic acid, on protein phosphatases. *Biochem. J.* **256**:283-290.

Bialojan, C., J.C. Rüegg, and A. Takai. 1988. Effects of okadaic acid on isometric tension and myosin phosphorylation of chemically skinned guinea-pig tenia coli. *J. Physiology.* **398**:81-95.

Billiar, R.B., J. Kalash, V. Romita, K. Tsuji, and T. Kosuge. 1988. Neosurugatoxin: CNS acetylcholine receptors and leutinizing hormone secretion in ovariectomized rats. *Brain Res. Bull.* **20:** 315-322.

Biskupiak, J.E. and C.M. Ireland. 1984. Revised absolute configuration of dysidenin and isodysidenin. *Tetrahedron Lett.* **25**:2935-2936.

Boldt, J., A.M. Howe, J.B. Parkerson, L.E. Gunter, and E. Kuehn. 1989. Carbohydrate involvement in sperm-egg fusion in mice. *Biol. Reprod.* **40**(4): 887-896.

Bontemps, F. and G. van den Berghe. 1989. Mechanism of adenosine triphosphate catabolism induced by deoxyadenosine and by nucleoside analogs in adenosine deaminase-inhibited human erythrocytes. *Cancer Res.* **49**:4983-4989.

Bourke, J.E., S.J. Bunn, P.D. Marley, and B.G. Livett. 1988. The effects of neosurugatoxin on evoked catecholamine secretion from bovine adrenal chromaffin cells. *Br. J. Pharmacol.* **93**:275-280.

Bouvier, M. 1990. Cross-talk between secondary messengers. *Ann. N. Y. Acad. Sci.* **594**:120-129.

Brandley, B.K., T.S. Ross, and R.L. Schnaar. 1987. Multiple carbohydrate receptors on lymphocytes revealed by adhesion to immobilized polysaccharides. *J. Cell. Biol.* **105**(2):991-997.

Bruno, I., L. Minale, and R. Riccio. 1990. Starfish Saponins, Part 43. *J. Nat. Products.* **53**:366-374 and references therein.

Bryant, R.W., C.A. Granzow, M.I. Siegel, R.W. Egan, and M.M. Billah. 1990. Phorbol esters increase synthesis of decay-accelerating factor, a phosphatidylinositol-anchored surface protein, in human endothelial cells. *J. Immunol.* **144**:593-598.

Burres, N.S. and J.J. Clement. 1989. Antitumor activity and mechanism of action of the novel marine natural products mycalamide-A and -B and onnamide. *Cancer Res.* **49**:2935-2940.

Burres, N.S., S. Sazesh, G.P. Gunawardana, and J.J. Clement. 1989. Antitumor activity and nucleic acid binding properties of dercitin, a new acridine alkaloid isolated from a marine *Dercitus* species sponge. *Cancer Res.* **49**:5267-5274.

Cai, M. and P.C. Jordan. 1990. How does vestibule surface charge affect ion conduction and toxin binding in a sodium channel? *Biophys. J.* **57**:883-891.

Canonico, P.G., W.L. Pannier, J.W. Huggins, and K.L. Rinehart. 1982. Inhibition of RNA viruses *in vitro* and in rift valley fever-infected mice by didemnins A and B. *Antimicro. Ag. Chemother.* **22**:696-697.

Carmichael, W. W., N. A. Mahmood, and E. G. Hyde. 1990. Natural toxins from cyanobacteria (blue-green algae). In *Marine Toxins,* ACS Symposium Series 418, S. Hall and G. Strichartz (eds.), pp. 87-106. American Chemical Society, Washington, DC.

Carte, B., S. Mong, B. Poehland, H. Sarau, J.W. Westley, and D.J. Faulkner. 1989. 15-Acetylthioxy-furodysinin lactone, a potent LTB_4 receptor partial agonist from a marine sponge of the genus. *Dysidea, Tetrahedron Lett.* **30**:2725-2726.

Catterall, W.A. 1980. Neurotoxins that act on voltage-sensitive sodium channels in excitable membranes. *Ann. Rev. Pharmacol. Toxicol.* **20**:15-43.

Charles, C., J.C. Braekman, D. Dalose, and B. Tursch. 1980. The relative and absolute configuration of dysidenin. *Tetrahedron.* **36**:2133-2135.

Charles, C., J. C. Braekman, D. Daloze, B. Tursch, and R. Karlson. 1978. Isodysidenin, a further hexachlorinated metabolite from the sponge. *Dysidea herbacea, Tetrahedron Lett.* **17**:1519-1520.

Chida, K. and I. Yamamoto. 1987. Antitumor activity of a crude fucoidan fraction prepared from the roots of kelp (*Laminaria* species). *Kitasato Arch. Exp. Med.* **60**(1-2):33-39.

Chun, H.G., B. Davies, D. Hoth, M. Suffness, J. Plowman, K. Flora, C. Grieshaber, and B. Leyland-Jones. 1986. Didemnin B: the first marine compound entering clinical trials as an antineoplastic agent. *Invest. New Drugs.* **4**:279-284.

Church, F. C., J. B. Meade, R. E. Treanor, and H. C. Whinna. 1989. Antithrombin activity of fucoidin. The interaction of fucoidin with heparin cofactor II, antithrombin III, and thrombin. *J. Biol. Chem.* **264**(6):3618-3623.

Cohen, P. 1989. The structure and regulation of protein phosphatases. *Annu. Rev. Biochem.* **58**:453-508.

Cohen, P., C.F.B. Holmes, and Y. Tsukitani. 1990. Okadaic acid: a new probe for the study of cellular regulation. *Trends Biochem. Sci.* **15**:98-102.

Cohen, P., S. Klumpp, and D.L. Schelling. 1989a. An improved procedure for identifying and quantitating protein phosphatases in mammalian tissues. *FEBS Lett.* **250**:596-600.

Cohen, P., D.L. Schelling, and M.J.R. Stark. 1989b. Remarkable similarities between yeast and mammalian protein phosphatases. *FEBS Lett.* **250**:601-606.

Cohen, S.S. 1977. The mechanisms of lethal action of arabinosyl cytosine (araC) and arabinosyl adenine (araA). *Cancer.* **40**:509-518.

Crampton, S.L., E.G. Adams, S.L. Kuentzel, L.H. Li, G. Badiner, and B.K. Bhuyan. 1984. Biochemical and cellular effects of didemnins A and B. *Cancer Res.* **44**:1796-1801.

Cruz, L.J., G. Kupryszewski, G.W. LeCheminant, W.R. Gray, B.M. Olivera, and J. Rivier. 1989. μ-Conotoxin GIIIA, a peptide ligand for muscle sodium channels: chemical synthesis, radiolabelling and receptor characterization. *Biochemistry* **28**:3437-3442.

Culver, P. and R.S. Jacobs. 1981. Lophotoxin: a neuromuscular acting toxin from the sea whip (*Lophogorgia rigida*). *Toxicol.* **19**:825-830.

Culver, P., M. Bursch, C. Potenza, L. Wasserman, W. Fenical, and P. Taylor. 1985. Structure-activity relationships for the irreversible blockade of nicotinic receptor agonist sites by lophotoxin and congeneric diterpene lactones. *Mol. Pharm.* **28**:436-444.

Culver, P., W. Fenical, and P. Taylor. 1984. Lophotoxin irreversibly inactivates the nicotinic acetylcholine receptor by preferential association at one of the two primary agonist sites. *J. Biol. Chem.* **259**:3763-3770.

Das, D.K. and H. Steinberg. 1988. Adenosine transport in the lung. *J. Appl. Physiol.* **65**:297-305.

Davies, L.P., D.D. Jamieson, J.A. Baird-Lambert, and R. Kaslauskas. 1984. Halogenated pyrrolopyrimidine analogues of adenosine from marine organisms: pharmacological activities and potent inhibition of adenosine kinase. *Biochem. Pharmacol.* **33**:347-355.

Dawicki, D.D., K.C. Agarwal, and R.E. Parks Jr. 1988. Adenosine metabolism in human whole blood. *Biochem. Pharmacol.* **37**:621-626.

de Andrea, M.L., B. de Camargo, and R. Melaragno. 1990. A new treatment protocol for childhood non-Hodgkin's lymphoma: preliminary evaluation. *J. Clin. Oncol.* **8**:666-671.

de Oliveira, C.A. and B. Mantovani. 1988. Latrunculin A is a potent Inhibitor of Phagocytosis by Macrophages. *Tetrahedron Lett.* **43**:1825-1830.

de Silva, E.D. and P.J. Scheuer. 1980. Manoalide, an antibiotic sesterterpenoid from the marine sponge *Luffariella variabilis* (Polejaeff). *Tetrahedron Lett.* **21**:1611-1614.

de Vries, G.W., L. Amdahl, A. Mobasser, M. Wenzel, and L.A. Wheeler. 1988. Preferential inhibition of 5-lipoxygenase activity by manoalide. *Biochem. Pharmacol.* **37**: 2899-2905.

Degnan, B.M., C.J. Hawkins, M.F. Lavin, E.J. McCaffrey, D.L. Parry, and D.J. Watters. 1989. Novel cytotoxic compounds from the ascidian *Lissoclinum bistratum. J. Med. Chem.* **32**:1355-1359.

Dorr, A., R. Schwartz, J. Kuhn, J. Bayne, and D.A. Von Hoff. 1986. Phase I clinical trial of didemnin B (Abstract). *Proc. Am. Assoc. Cancer Res.* **5**:39.

Dorr, F.A., J.G. Kuhn, J. Phillips, and D.D. Von Hoff. 1988. Phase I clinical and pharmacokinetic investigation of didemnin B, a cyclic depsipeptide. *Eur. J. Cancer Clin. Oncol.* **24**:1699-1706.

Dowding, A.J. and Z.W. Hall. 1987. Monoclonal antibodies specific for each of the two toxin-binding sites of *Torpedo* acetylcholine receptor. *Biochemistry.* **26**:6372-6381.

Dutar, P., O. Rascol, and Y. Lamour. 1989. ω-Conotoxin GVIA blocks synaptic transmission in the CA1 field of the hippocampus. *Eur. J. Pharmacol.* **174**:261-6.

Edwards, R.A., P.L. Lutz, and D.G. Baden. 1989. Relationship between energy expenditure and ion channel density in the turtle and rat brain. *Am. J. Physiol.* **257**:R1354-358.

Evans, J.S., E.A. Musser, L. Bostwick, and G.D. Mengel. 1964. The effect of 1-b-D-arabinofuranosylcytosine hydrochloride on murine neoplasms. *Cancer Res.***24**:1285-1293.

Evans, J.S., E.A. Musser, G.D. Mengel, K.R. Forsblad, and J.H. Hunter. 1961. Antitumor activity of 1-b-D-arabinofuranosylcytosine hydrochloride. *Proc. Soc. Exp. Biol. Med.* **106**:350-353.

Famiani, V. 1987. *In vivo* effect of didemnin B on two tumors of the rat. *Oncology.* **44**:42-46.

Farago, A. and Y. Nishizuka. 1990. Protein kinase C in transmembrane signalling. *FEBS Lett* 2 68:350-354. S. Fattorusso, L. Minale, and G. Sodano. 1970. Aeroplysinin-I, a new bromo-compound from *Aplysina aerophoba. J. Chem. Soc. Perkin I*, **12**:751-752.

Faulkner, D.J. 1984. *Natural Products Reports.* **1**: 551-98 and subsequent volumes.

Fautin , D.G. (ed.). 1988. *Biomedical Importance of Marine Organisms.* Memoirs of the California Academy of Sciences Number 13. California Academy of Sciences, San Francisco.

Fenical, W., R.K. Okuda, M.M. Banduraga, P. Culver, and R.S. Jacobs. 1981. Lophotoxin: a novel neuromuscular toxin from pacific sea whips of the genus *Lophogorgia*. *Science.* **212**:1512-1514.

Félix, M., P. Cohen, and E. Karsenti. 1990. Cdc2 H1 kinase is negatively regulated by a type 2A phosphatase in the *Xenopus* early embryonic cell cycle: evidence from the effects of okadaic acid. *EMBO J.* **9**:675-683.

Foster, M. P., C. L. Mayne, R. Dunkel, R. J. Pugmire, D. M. Grant, J. M. Kornprobst, J. F. Verbist, J. F. Biard, and C. M. Ireland. 1992. The revised structure of Bistramide A (Bistratene A): Application of a new program for the automated analysis of 2D INADEQUATE Spectra. *J. Amer. Chem. Soc.* **114**:1110.

Fredholm, B.B. 1990. Differential sensitivity to blockade by 4-aminopyridine of presynaptic receptors regulating [^{3}H]acetylcholine release from rat hippocampus. *J. Neurochem.* **54**:1386-1390.

Fujiki, H. and T. Sugimura. 1987. New classes of tumor promoters: teleocidin, aplysiatoxin, and palytoxin. *Adv. Cancer Res.* **59**:223-264.

Fujiki, H., M. Suganuma, M. Nakayasu, H. Hakii, T. Horiuchi, S. Takayama, and T. Sugimura. 1986. Palytoxin is a non-12-O-tetradecanoylphorbol-13-acetate type tumor promoter in two-stage mouse skin carcinogenesis. *Carcinogenesis.* **7**:707-710.

Fujiki, H., M. Suganuma, H. Suguri, S. Yoshizawa, K. Takagi, and M. Kobayashi. 1989b. Sarcophytols A and B inhibit tumor promotion by teleocidin in two-stage carcinogenesis in mouse skin. *J. Cancer Res. Clin. Oncol.* **115**:25-28.

Fujiki, H., M. Suganuma, H. Suguri, S. Yoshizawa, K. Takagi, T. Sassa, N. Uda, K. Wakamatsu, K. Yamada, T. Yasumoto, Y. Kato, N. Fusetani, K. Hashimoto, and T. Sugimura. 1989a. New tumor promoters from marine sources: the okadaic acid class. In *Mycotoxins and Phycotoxins.* S. Natori, K. Hashimoto and Y. Ueno (eds.), pp. 453-460. Elsevier Science Publishers. B.V., Amsterdam.

Fujiki, H., M. Suganuma, H. Suguri, S. Yoshizawa, K. Takagi, N. Uda, K. Wakamatsu, K. Yamada, M. Murata, T. Yasumoto, and T. Sugimura. 1988. Diarrhetic shellfish toxin, dinophysistoxin-1, is a potent tumor promoter on mouse skin. *Japan. J. Cancer Res. (Gann)* **79**:1089-1093.

Fulmor, W., G. E. Van Lear, G. P . Morton, and R. D. Mills. 1970. Isolation and absolute configuration of the aeroplysinin 1 enantiomorphic pair from *Ianthella ardis. Tetrahedron Lett.* **52**:4551-4552.

Gavériaux, C. and F. Loor. Proliferation of interleukin 2 dependent cytotoxic T cell line cells. *Int. Arch. Allergy Appl. Immunol.* **88**:294-6.

Geller, R. B., R. Saral, J. E. Karp, G. W. Santos, and P. J. Burke. 1990. Cure of acute myelocytic leukemia in adults: a reality. *Leukemia.* **4**:313-315.

Glaser, K. B. and R. S. Jacobs. 1986. Molecular pharmacology of manoalide. *Biochem. Pharmacol.* **35**:449-453.

Glaser, K. B., M. S. de Carvalho, R. S. Jacobs, M. R. Kernan, and D. J. Faulkner. 1989. Manoalide: structure activity studies and definition of the pharmacophore for phospholipase A_2 inactivation. *Mol. Pharmacol.* **36**:782-788.

Glaser, K. B., T. S. Vedvivk, and R. S. Jacobs. 1988. Inactivation of phospholipase A_2 by manoalide. Localization of the manoalide binding site on bee venom phospholipase A_2. *Biochem. Pharmacol.* **37**:3639-3646.

Goris, J., J. Hermann, P. Hendrix, R. Ozon, and W. Merlevede. 1989. Okadaic acid, a specific protein phosphatase inhibitor, induces maturation and MPF formation in *Xenopus laevis* oocytes. *FEBS Lett.* **245**:91-94.

Gouiffès, D., M. Juge, N. Grimaud, L. Welin, M.P. Sauviat, Y. Barbin, D. Laurent, C. Roussakis, J.P. Henichart, and J.F. Verbist. 1988a. Bistratamide A, a new toxin from the urochordata *Lissoclinum bistratum* sluiter: isolation and preliminary characterization. *Toxicol.* **26**:1129-1136.

Gouiffès, D., S. Moreau, N. Helbecque, J.L. Bernier, J.P. Hénichart, Y. Barbin, D. Laurent, and J.F. Verbist. 1988b. Proton nuclear magnetic study of bistratamide A, a new cytotoxic drug isolated from *Lissoclinum bistratum* sluiter. *Tetrahedron.* **44**:451-459.

Gschwendt, M., W. Kittstein, and F. Marks. 1987. Didemnin B inhibits biological effects of tumor promoting phorbol esters on mouse skin, as well as phosphorylation of a 100 kD protein in mouse epidermis cytosol. *Cancer Lett.* **34**:187-191.

Gunawardana, G.P., S. Kohmoto, S.P. Gunasekera, O.J. McConnell, and F.E. Koehn. 1988. Dercitin, a new biologically active acridine alkaloid from a deep water marine sponge, *Dercitus* sp. *J. Am. Chem. Soc.* **110**:4856-4858.

Gusovsky, F. and J.W. Daly. 1990. Maitotoxin: A unique pharmacological tool for research on calcium-dependent mechanisms. *Biochem. Pharmacology.* **39**:1633-1639.

Gustafson, K.R., J.H. Cardellina, R.W. Fuller, O.S. Weislow, R.F. Kiser, K.M. Snader, G.M. Patterson, and M.R. Boyd. 1989. AIDS-antiviral sulfolipids from cyanobacteria (blue-green algae). *J. Natl. Canc. Inst.* **81**(16):1254-1258.

Hall, S., F. Strichartz, E. Moczydlowski, A. Ravindran, and P.B. Reichardt. 1990. The saxitoxins: sources, chemistry and pharmacology. In *Marine Toxins*, ACS Symposium Series 418. S. Hall and G. Strichartz (eds.), pp. 29-65. American Chemical Society. Washington, DC.

Hannun, Y. A. and R.A. Bell. 1989. Regulation of protein kinase C by sphingosine and lysophingolipids. *Clin. Chem. Acta.* **185**:333-345.

Hatanaka, Y., E. Yoshida, H. Nakayama, and Y. Kanaoka. 1990. Synthesis of m-conotoxin GIIIA: A chemical probe for sodium channels. *Chem. Pharm. Bull. Tokyo* **38**:236-8.

Hayashi, E. and S. Yamada. 1975. Pharmacological studies on surugatoxin, the toxin principle from Japanese ivory mollusk (*Babylonia japonica*). *Br. J. Pharmacol.* **53**:207-251.

Hayashi, E., M. Isogui, Y. Kagawa, N. Takayanagi, and S. Tamada, 1984. Neosurugatoxin, a specific antagonist of nicotinic acetylcholine receptors. *J. Neurochem.* **42**:1491-1494.

Haystead, T.A.J., A.T.R. Sim, D. Carling, R.C. Honnor, Y. Tsukitani, P. Cohen, and D.G. Hardie. 1989. Effects of the tumor promoter okadaic acid on intracellular protein phosphorylation and metabolism. *Nature.* **337**:78-81.

Hennings, H., P.M. Blumberg, G.R. Pettit, C..L. Herald, R. Shores, and S.H. Yuspa. 1987. Bryostatin 1, an activator of protein kinase C, inhibits tumor promotion by phorbol esters in SENCAR mouse skin. *Carcinogenesis.* **8**:1343-1346.

Herdon, H. and S.R. Nahorski. 1989. Investigations of the roles of dihydropyridine- and ω-conotoxin-sensitive calcium channels in mediating depolarization-evoked endogenous dopamine release from striatal slices. *Naunyn-Schmiedebergs Arch. Pharmacol.* **340**:36-40.

Hescheler, J., G. Mieskes, J.C. Rüegg, A. Takai, and W. Trautwein. 1988. Effects of a protein phosphatase inhibitor, okadaic acid, on membrane currents of isolated guinea-pig cardiac myocytes. *Pflügers Arch.* **412**:248-252.

Hirata, Y. and D. Uemura. 1986. Halichondrins: antitumor polyether macrolides from a marine sponge. *Pure Appl. Chem.* **58**:701-710.

Hong, S.J. and C.C. Chang. 1989. Use of geographutoxin II (μ-conotoxin) for the study of neuromuscular transmission in mice. *Br. J. Pharmacol.* **97**:934-940.

Huang, T.T.F. and R. Yanagimachi. 1984. Fucoidin inhibits attachment of guinea pig spermatazoa to the zona pellucida through binding to the inner acrosomal membrane and equatorial domains. *Exp. Cell. Res.* **153**:363-373.

Hudson, J.B., H. Saboune, Z. Abramowski, G.H. Towers, and K.L. Rinehart. 1988. The photoactive antimicrobial properties of eudistomins from the caribbean tunicate *Eudistoma olivaceum. Photochem. Photobiol.* **47**(3):377-381.

Ireland, C.M. and Maiese, W.M. 1992. Unpublished results.

Ireland, C. M., T.F. Molinski, D.M. Roll, T.M. Zabriskie, T.C. McKee, J.C. Swersey, and M.P. Foster. 1989. Natural Product Peptides from Marine Organisms. In *Bioorganic Marine Chemistry,* Volume 3. P. J. Scheuer (ed.), pp. 1-46. Springer-Verlag. Berlin, Heidelberg.

Ireland, C.M., D.M. Roll, T.F. Molinski, T.C. McKee, T.M. Zabriskie, and J.C. Swersey. 1988. Uniqueness of the Marine Chemical Environment: Categories of Marine Natural Products from Invertebrates, *Biomedical Importance of Marine Organisms,* Memoirs of the California Academy of Sciences Number 13. D.G. Fautin (ed.), pp. 41-57. California Academy of Sciences. San Francisco.

Ireland, C. and D. J. Faulkner. 1978. The defensive secretion of the opithobranch mollusk *Onchidella binneyi. Bioorg. Chem.* **7**:125-130.

Irie, K., N. Hagiwara, H. Tokuda, and K. Koshimizu. 1987. Structure-activity studies of the indole alkaloid tumor promoter teleocidins. *Carcinogenesis.* **8**:547-552.

Ishihara, H., B.L. Martin, D.L. Brautigan, H. Karaki, H. Ozaki, Y. Kato, N. Fusetani, S. Watabe, K. Hashimoto, D. Uemura, and D.J. Hartshorne. 1989. Calyculin A and okadaic acid: inhibitors of protein phosphatase activity. *Biochem. Biophys. Res. Commun.* **159**:871-877.

Issinger, O., T. Martin, W.W. Richter, M. Olson, and H. Fujiki. 1988. Hyperphosphorylation of N-60, a protein structurally and immunologically related to nucleolin after tumor-promoter treatment. *EMBO J.* **7**:1621-1626.

Itai, A., Y. Kato, N. Tomioka, Y. Iitaka, Y. Endo, M. Hasegawa, K. Shudo, H. Fujiki, and S. Sakai. 1988. A receptor model for tumor promoters: rational superposition of teleocidins and phorbol esters. *Proc. Natl. Acad. Sci. USA.* **85**:3688-3692.

Jacobs, R.S., P. Culver, R. Langdon, T. O'Brien, and S. White. 1985. Some pharmacological observations on marine natural products. *Tetrahedron.* **41**:981-984.

Jacobson, P.B., L.A. Marshall, A. Sunf, and R.S. Jacobs. 1990. Inactivation of human sinovial fluid phospholipase A_2 by the marine natural product manoalide. *Biochem. Pharmacol.* **39**:1557-1564.

Jaken, S. 1990. Protein kinase C and tumor promoters. *Curr. Opin. Cell. Biol.* **2**:192-197.

Jan, C. R., M. Titeler, and A. S. Schneider. 1990. Identification of ω-conotoxin binding sites on adrenal medullary membranes: possiblility of multiple calcium channels in chromaffin cells. *J. Neurochem.* **54**:355-358.

Jeffrey, A. M., R. M. L. Liskamp. 1986. Computer assisted molecular modeling of tumor promoters: rationale for the activity of phorbol esters, teleocidin B, and aplysiatoxin. *Proc. Natl. Acad. Sci. USA.* **83**:241-245.

Jiang, T. L., R. H. Liu, and S. E. Salmon. 1983. Antitumor activity of didemnin B in the human tumor stem cell assay. *Cancer Chemother. Pharmacol.* **11**:1-4.

Kao, C.Y. 1966. Tetrodotoxin, saxitoxin, and their significance in the study of excitation phenomena. *Pharmacol. Rev.* **18**:997-1049.

Kao, C. Y. 1986. Structure-activity relations of tetrodotoxin, saxitoxin and analogues. In *Tetrodotoxin, Saxitoxin and the Molecular Biology of the Sodium Channel.* C.Y. Kao and S.R. Levinson (eds.), pp. 52-67. Annals of the New York Academy of Science Volume 479. New York.

Karschin, A. and S.A. Lipton. 1989. Calcium channels in solitary retinal ganglion cells from post-natal rat. *J. Physiol.* **418**:379-396.

Kaslauskas, R., P.T. Murphy, R.J. Wells, J.A. Baird-Lambert, and D.D. Jamieson. 1983. Halogenated pyrrolo [2,3-d] pyrimidine nucleosides from marine organisms. *Aust. J. Chem.* **36**:165-170.

Kather, H. 1990. Pathways of purine metabolism in human adipocytes. *J. Biol. Chem.* **265**:96-102.

Kato, Y., N. Fusetani, S. Matsunaga, and K. Hashimoto. 1986. Calyculin A, a novel antitumor metabolite from the marine sponge *Discodermia calyx. J. Am. Chem. Soc.* **108**:2780-2781.

Keith, R.A., T.J. Mangano, and A.I. Salama. 1989. Inhibition of N-methyl-D-aspartate- and kainic acid-induced neurotransmitter release by ω-conotoxin GVIA. *Br. J. Pharmacol.* **98**:767-772.

Kipreos, E.T. and J.Y.T. Wang. 1990. Differential phosphorylation of c-Abl in cell cycle determined by *cdc2* kinase and phosphatase activity. *Science* **248**:217-220.

Kobayashi, M.J. Kobayashi, and Y. Ohizumi. 1989. Cone shell toxins and the mechanisms of their pharmacological action. In *Bioorganic Marine Chemistry,* Volume

3 P.J. Scheuer (ed.), pp. 71-84. *Springer-Verlag*. Berlin, Heidelberg.

Kobayashi, M., T. Nakagawa, and H. Mitsuhashi. 1979. Marine terpenes and terpenoids. I. Structures of four cembrane-type diterpenes: sarcophytol-A, sarcophytol-A acetate, sarcophytol-B, and sarcophytonin-A, from the soft coral, *Sarcophyton glaucum*. *Chem Pharm. Bull.* (Tokyo) **27**:2382-2387.

Kodama, I., N. Kondo, and S. Shibata. 1986. Electrochemical effects of okadaic acid isolated from black sponge in guinea-pig ventricular muscles. *J. Physiol.* **378**:359-373.

Kosuge, T., K. Tsuji, and K. Hirai. 1981. Isolation and structure determination of a new marine toxin, neosurugatoxin, from the Japanese ivory shell, *Babylonia japonica*. *Tetrahedron Lett.* **22**:3417-3420.

Kreuter, M. H., A. Bernd, H. Holzmann, W. Müller-Klieser, A. Maidhof, N. Weibmann, Z. Kljajic, R. Batel, H.C. Schröder and W.E.G. Müller. 1989. Cytostatic activity of aeroplysinin-1 against lymphoma and epithelioma cells. *Z. Naturforsch.* **44c**:680-688.

Kume, H., A. Takai, H. Tokuno, and T. Tomita. 1989. Regulation of Ca^{2+}-dependent K^+-channel activity in tracheal myocytes by phosphorylation. *Nature.* **341**:152-154.

Laporte, D.C., B. M. Wierman, et al. 1980. Calcium-induced exposure of a hydrophobic surface on Calmodulin. *Biochem.* **19**:3814-3819.

Levine, L. and H. Fujiki. 1985. Stimulation of arachidonic metabolism by different types of tumor promoters. *Carcinogenesis.* **6**:1631-1634.

Levine, L., D. Xiao, and H. Fujiki. 1986. A combination of palytoxin with 1-oleoyl-2-acetyl-glycerol (OAG) or insulin or interleukin-1 synergistically stimulates arachidonic acid metabolism, but combination of 12-O-tetradecanoylphorbol-13-acetate (TPA)-type tumor promoters with OAG does not. *Carcinogenesis.* **7**:99-103.

Li, L. H., L. G. Timmins, T. L. Wallace, W. C. Krueger, M. D. Prairie, and W. B. Im. 1984. Mechanism of action of didemnin B, a depsipeptide from the sea. *Cancer Lett.* **23**:279-288.

Lin, Y. Y., M. Risk, S. M. Ray, D. van Engen, J. C. Clardy, J. Golik, J. C. James, and K. Nakanishi. 1981. Isolation and structure of brevetoxin-B from the "red tide" dinoflagellate *Ptychodiscus brevis* (*Gymnodinium breve*). *J. Am. Chem. Soc.* **103**:6773-6775.

Lister, M. D., K. B. Glaser, R. J. Ulevitch, and E. A. Dennis. 1989. Inhibition studies on the membrane-associated phospholipase A_2 *in vitro* and prostaglandin E_2 production *in vivo* of the macrophage-like P388D_1 cell. *J. Biol. Chem.* **264**:8520-8528.

Look, S. A., W. Fenical, R. S. Jacobs, and J. Clardy. 1986 The pseudopterosins: anti-inflammatory and analgesic natural products from the sea whip *Pseudopterogorgia elisabethae*. *Proc. Nat. Acad. Sci. USA.* **83**:6238-6240.

Loya, S., and A. Hizi. 1990. The inhibition of human immunodeficiency virus type 1 reverse transcriptase by avarol and avarone derivatives. *FEBS.* **269**:131-134.

Luedke, E. S. 1990. The identification and characterization of the pseudopterosins: anti-inflammatory agents isolated from the gorgonian coral *Pseudopterogorgia elizabethae*, Ph.D. Thesis, University of California. Santa Barbara.

Luetje, C. W., K. Wada, S. Rogers, S. N. Abramson, K. Tsuji, S. Heinemann, and J. Patrick. 1990. Neurotoxins distiguish between different neuronal nicotinic acetylcholine receptor subunit combinations. *J. Neurochem.* **55**:632-639.

MacKintosh, C. and P. Cohen. 1989 Identification of high levels of type 1 and type 2A protein phosphatases in higher plants. *Biochem. J.* **262**:335-339.

Maeda, N.,K. Wada, M. Yuzaki, and K. Mikoshiba. 1989. Autoradiographic visualisation of a calcium channel antagonist, [^{125}I]ω-conotoxin GVIA, binding site in the brains of normal and cerebellar mutant mice (*pcd* and *weaver*). *Brain Res.* **489**:21-30.

Maldonado, E., J. A. Lavergne, and E. Kraiselburd. 1982. Didemnin A inhibits the *in vitro* replication of dengue virus types 1, 2 and 3. *Puerto Rico Health Science Journal.* **1**:22-25.

Mangano, T. J., J. Patel, A. I. Salama, and R. A. Keith. 1990. Glycine-evoked neurotransmitter release from rat hippocampal brain slices: evidence for the involvement of glutaminergic transmission. *J. Pharm. Experi. Ther.* **252**:574-80.

Marshall, I. G. and A. L. Harvey. 1990. Selective neuromuscular blocking properties of α-conotoxin *in vivo*. *Toxicon.* **28**:231-234.

Maruyama, H., J. Nakajima, and I. Yamamoto. 1987. A study on the anticoagulant and fibrinolytic activities of a crude fucoidan from the edible brown seaweed laminaria religiosa, with special reference to its inhibitory effect on the growth of sarcoma-180 ascites cells subcutaneously implanted in mice. *Kitasato Arch. Exp. Med.* **60**(3):105-121.

Mayer, A. M. S., K. B. Glaser, and R. S. Jacobs. 1988. Regulation of eicosanoid biosynthesis *in vitro* and *in vivo* by the marine natural product manoalide: a potent inactivator of venom phospholipases. *J. Pharm. Exp. Ther.* **244**:871-878.

Mayer, A. M. and Jacobs, R. S. 1988. Manoalide: an anti-inflammatory and analgesic marine natural product. In *Biomedical Importance of Marine Organisms*, Memoirs of the California Academy of Sciences Number 13. D.G. Fautin (ed.), pp. 133-1142. California Academy of Sciences. San Francisco.

Minigo, H., D. Nemet, A. Planinc-Peraica, V. Bogdanic, B. Labar, B. Jaksic, and E. Hauptmann. 1990. High dose ara-C as induction therapy of acute myeloid leukemia. *Folia Haematol.* **117**:135-140.

Mong, S., B. Votta, H.M. Sarau, J.J. Foley, D. Schmidt, B.K. Carte, B. Poehland, and J. Westley. 1990. 15-Acetylthioxy-furodysinin lactone, isolated from a marine sponge, *Dysidea* sp., is a potent agonist to human leukotriene B_4 receptor. *Prostaglandins.* **39**:89-97.

Montgomery, D.W. and C.F. Zukoski. 1985. Didemnin B: a new immunosuppressive cyclic peptide with potent activity *in vitro* and *in vivo*. *Transplantation Proc.* **40**:49-56.

Montgomery, D.W., A. Celniker, and C.F. Zukoski. 1987. Didemnin B: a new immunosuppressive cyclic peptide that stimulates murine antibody responses *in vitro* and *in vivo*. *Transplantation Proc.* **19**:1295-1296.

Moore, R.E. and G. Bartolini. 1981. Structure of palytoxin. *J. Am. Chem. Soc.* **103**:2491-2494.

Nakadate, T. 1989. The mechanism of skin tumor promotion caused by phorbol esters: possible involvement of arachidonic acid cascade/lipoxygenase, protein kinase C and calcium/calmodulin systems. *Japan J. Pharmacol.* **49**:1-9.

Nakamura, H., H. Wu, J. Kobayashi, Y. Nakamura, Y. Ohizumi, and Y. Hirata. 1985. Purealin, a novel enzyme activator from the okinawan marine sponge *Psammaplysilla purea*. *Tetrahedron Lett.* **26**(37):4517-4520.

Nakamura, H., K. Yoshito, M.A. Pajares, and R.R. Rando. 1989, Structural basis of protein kinase C activation by tumor promoters. *Proc. Natl. Acad. Sci. USA.* **86**:9672-9676.

Nakamura, Y., M. Kobayashi, H. Nakamura, H. Wu, J. Kobayashi, and Y. Ohizumi. 1987. Purealin, a novel activator of skeletal muscle actomyosin ATPase and myosin EDTA-ATPase that enhanced the superprecipitation of actomyosin. *Eur. J. Biochem.* **167**:1-6.

Narahashi, T. 1974. Chemical tools in the study of excitable membranes. *Physiol. Rev.* 54: 813-889.

Narahashi, T. 1988. Mechanism of tetrodotoxin and saxitoxin action. In *Handbook of Natural Toxins,* Volume 3: Marine Toxins and Venoms. A.T. Tu (ed.), pp. 185-210. Mercel Dekker Inc. New York.

Narisawa, T., M. Takahashi, M. Niwa, Y. Fukaura, and H. Fujiki. 1989. Inhibition of methylnitrosourea-induced large bowel cancer development in rats by sarcophytol A, a product from a marine soft coral, *Sarcophyton glaucum*. *Cancer Res.* **49**:3287-3289.

Nishizuka, Y. 1988. The molecular heterogeneity of protein kinase C and its implications for cellular regulation. *Nature.* **334**:661-5.

Nishizuka, Y. 1989. The Albert Lasker Medical Awards. The family of protein kinase C for signal transduction. *JAMA.* **262**:1826-32.

Obara, K., A. Takai, J.C. Rüegg, and P. de Lanerolle. 1989. Okadaic acid, a phosphatase inhibitor, produces a Ca^{2+} and calmodulin-independent contraction of smooth muscle. *Phlügers Arch.* **414**:134-138.

Oehninger, S., A. Acosta, and G.D. Hodgen. 1990. Antagonistic and agonistic properties of saccharide moieties in the hemizona assay. *Fertil. Steril.* **53**(1):143-149.

Ohizumi, Y. and M. Kobayashi. 1990. Ca-dependent excitatory effects of maitotoxin on smooth and cardiac muscle. In *Marine Toxins.* ACS Symposium Series 418. S. Hall and G. Strichartz (eds.), pp. 233-143. American Chemical Society, Washington, DC.

Olivera, B. M., W.R. Gray, and Cruz, L. J. 1988. Marine snail venoms. In *Handbook of Natural Toxins.* Volume 3: Marine Toxins and Venoms. A. T. Tu (ed.), pp. 327-354. Mercel Dekker Inc. New York.

Olivera, B.M., W.R. Gray, R. Zeikus, J.M. McIntosh, J. Varga, J. Rivier, V. deSantos, and L.J. Cruz. 1985. Peptide neurotoxins from fish-hunting cone snails. *Science.* **230**:1338-1343.

Olivera, B.M., D.R. Hillyard, J. Rivier, S. Woodward, W.R. Gray, G. Corpuz, and L.J. Cruz. 1990a. Conotoxins: targeted peptide ligands from snail venoms. In *Marine Toxins,* ACS Symposium Series 418. S. Hall and G. Strichartz (eds.), pp. 256-278. American Chemical Society. Washington, DC.

Olivera, B.M., J. Rivier, C. Clark, C.A. Ramilo, G.P. Corpuz, F.C. Abogadie, E.E. Mena, S.R. Woodward, D.R. Hillyard, and L.J. Cruz. 1990b. Diversity of *Conus* neurotoxins. *Science.* **249**:257-263.

Palczewski, K., J.H. McDowell, S. Jakes, T.S. Ingebritsen, and P.A. Hargrave. 1989. Regulation of rhodopsin dephosphorylation by arrestin. *J. Biol. Chem.* **264**:15770-15773.

Pardi, A., A. Galdes, J. Florance, and D. Maniconte. 1989. Solution structures of a-cono-toxin G1 determined by two-dimensional NMR spectroscopy. *Biochemistry.* **28**:5494 - 5501.

Perry, N.B., J.W. Blunt, M.H.G. Munro, and L.K. Pannell. 1988. Mycalamide A, an antiviral compound from a New Zealand sponge of the genus *Mycale*. *J. Am. Chem. Soc.* **110**:4850-4851.

Pettit, G. R., Y. Kamano, C. L. Herald, A. A. Tuinman, F. E. Boettner, H. Kizu, J. M. Schmidt, L. Baczynskyj, K. B. Tomer, and R. J. Bontems. 1987. The isolation and structure of a remarkable marine animal antineoplastic constituent: dolastatin 10. *J. Am. Chem. Soc.* **109**:6883-6885.

Pettit, G. R., S. B. Singh, F. Hogan, P. Lloyd-Williams, D. L. Herald, D. D. Burkett, and P. J. Clewlow. 1989. The absolute configuration and synthesis of natural(-)-olastatin 10. *J. Am. Chem. Soc.* **111**:5463-5465.

Pettit, G. R., C. L. Herald, D. L. Doubek, and D. L. Herald. 1982. Isolation and structure of Bryostatin 1. *J. Am. Chem. Soc.* **104**:6846-6848.

Pinchuk, L. N. and G. V. Pinchuk. 1990. Monoclonal antibody to brain cytoplasmic tetrodotoxin-sensitive protein detects an epitope associated with lymphocytes and involved in lymphocyte activation (MedLine Abstract). *J. Neuroimmunol.* **27**:71-78.

Plummer, M. R., D. E. Logothetis, and P. Hess. 1989. Elementary properties and pharmacological sensitivities of calcium channels in mammalian peripheral neurons. *Neuron.* **2**:1453-63.

Rapier, C., G. G. Lunt, and S. Wonnacott. 1990. Nicotinic modulation of [^{3}H]-dopamine release from striatal synaptosomes: pharmacological characterization. *J. Neurochem.* **54**:937-945.

Redpath, N.T. and C.G. Proud. 1989. The tumor promoter okadaic acid inhibits reticulocyte-lysate protein synthesis by increasing the net phosphorylation of elongation factor 2. *Biochem. J.* **262**:69-75.

Rinehart, K.L., T.G. Holt, N.L. Fregeau, J.G. Stroh, P.A. Keifer, F. Sun, L.H. Li, and D.G. Martin. 1990. Ecteinascidins 729, 743, 745, 759A, 759B, and 770: potent antitumor agents from the Caribbean tunicate *Ecteinascidia turbinata*. *J. Org. Chem.* **55**:4512-4515.

Rinehart, K.L., J.B. Gloer, J.C. Cook, S.A. Mizsak, and T.A. Scahill. 1981a, Structures of the didemnins, antiviral and cytotoxic depsipeptides from a Caribbean tunicate. *J. Am. Chem. Soc.* **103**:1857-1859.

Rinehart, K.L., J.B. Gloer, R.G. Hughes, H.E. Renis, J.P. McGovren, E.B. Swynenberg, D.A. Stringfellow, S.L. Kuentzel, and H. Li. 1981b. Didemnins: antiviral and antitumor depsipeptides from a Caribbean tunicate. *Science.* **212**:933-935.

Rinehart, K.L., J.B. Gloer, G.R. Wilson, R.G. Hughes, L.H. Li, H.E. Renis, and J.P. McGovren. 1983. Antiviral and antitumor compounds from tunicates. *Fed. Proc.* **42**:87-90.

Rinehart, K.L., P.D. Shaw, L.S. Shield, J.B. Gloer, G.C. Harbour, M.E.S. Koker, D. Samain, R.E. Schwartz, A.A. Tymiak, D.L. Weller, G.T. Carter, M.H.G. Munro, R.G. Hughes, H.E. Renis, E.B. Swyneberg, D.A. Stringfellow, J.J. Vavra, J.H. Coats, G.E. Zurenko, S.L. Kuentzel, L.H. Li, G.J. Bakus, R.C. Brusca, L.L. Craft, D.N. Young, and J.L. Connor. 1981c. Marine natural products as sources of antiviral, antimicrobial, and antineoplastic agents. *Pure Appl. Chem.* **53**:795-817.

Rossof, A.H., P.A. Johnson, B.D. Kimmell, J.E. Graham, and D.L. Roseman. 1983. *In vitro* phase II study of didemin B in human cancer (Abstract*). Proc. Am. Assoc. Cancer Res.* **24**:315.

Sakai, R., I. Ikeda, H. Kitani, H. Fujiki, F. Takaku, U. Rapp, T. Sugimura, and M. Nagao. 1989. Flat reversion by okadaic acid of *raf* and *ret*-II transformants. *Proc. Natl. Acad. Sci. USA.* **86**:9946-9950.

Santostasi, G., R.K. Kutty, A.L. Bartorelli, T. Yasumoto, and G. Krishna. 1990. Maitotoxin-induced myocardial cell injury: calcium accumulation followed by ATP depletion precedes cell death. *Toxicol. Appl. Pharmacol.* **102**:164-173.

Sassa, T., W.W. Richter, N. Uda, M. Suganuma, H. Suguri, S. Yoshizawa, M. Horita, and H. Fujiki. 1989. Apparent "activation" of protein kinases by okadaic acid class tumor promoters. *Biochem. Biophys. Res. Commun.* 159:939-944.

Sawyer, P.J., J.H. Gentile, and J.J. Sasuer Jr. 1968. Demonstration of a toxin from *Aphanizomenon flos-aque* (L.) Ralfs. *Can J. Microbiol.* **14**:1199-1204.

Schatten, G., H. Schatten, H. Spector, C. Cline, N. Paweletz, C. Simerly, and C. Petzelt. 1986. Latrunculin inhibits the microfilament-mediated processes during fertilization, cleavage and early development in sea urchins and mice. *Exp. Cell Res.* **166**:191-208.

Schatten, H. and G. Schatten. 1986. Motility and centrosomal organization during sea urchin and mouse fertilization. *Cell Motil. Cytoskel.* **6**:163-175.

Schmitz, F.J., R.S. Prasad, Y. Gopichand, M.G. Hossain, and D. van der Helm. 1981. Acanthifolicin, a new episulfide-containing polyether carboxylic acid from extracts of the marine sponge *Pandaros acanthifolium. J. Am. Chem. Soc.* **103**:2467-2469.

Schneider, H.R., M. Mieskes, and O. Issinger. 1989. Specific dephosphorylation by phosphatases 1 and 2A of a nuclear protein structurally and immunologically related to nucleolin. *Eur. J. Biochem.* **180**:449-455.

Shibata, S., Y. Ishida, H. Kitano, Y. Ohizumi, J. Habon, Y. Tsukitani, and H. Kikuchi. 1982. Contractile effects of okadaic acid, a novel ionophore-like substance from black sponge, on isolated smooth muscles under the condition of Ca deficiency. *J. Pharmacol. Exp. Ther.* **223**:135-143.

Shimizu, Y., H. Bando, H. N. Chou, G. van Duyne, and J.C. Clardy. 1986b. Absolute configuration of brevetoxins. *J. Chem. Soc., Chem. Commun*, pp. 1656-1658.

Shimizu, Y., H.-N. Chou, H. Bando, G. van Duyne, J.C. Clardy. 1986a.

Sorenson, E. M., P. Culver, and V.A. Chiappinelli. 1987. Lophotoxin: selective blockade of nicotinic transmission in autonomic ganglia by a coral neurotoxin. *Neuroscience.* **20**:875-884.

Spector, I. and N.R. Shochet. 1983. Latrunculins: novel marine toxins that disrupt microfilament organization in cultured cells. *Science.* **219**:493-495.

Spector, I., N.R. Shochet, D. Blasberger, and Y. Kashman. 1989. Latrunculins — novel marine macrolides that disrupt microfilament organization and affect cell growth: I. Comparison with cytochalasin D. *Cell Motil. Cytoskel.* **13**:127-144.

Stewart, J.A., W.P. Tong, J.N. Hartshorn, and J.J. McCormack. 1986. Phase I evaluation of didemnin B (NSC 325319) (Abstract). *Proc. Am. Soc. Clin. Oncol.* **5**:33.

Strichartz, G. and N. Castle. 1990. Pharmacology of marine toxins: effects on membrane channels. In *Marine Toxins,* ACS Symposium Series 418. S. Hall and G. Strichartz (eds.), pp. 2-20. American Chemical Society, Washington, DC.

Strichartz, G., T. Rando, S. Hall, J. Gitscher, L. Hall, B. Magnani, and B. Hansen Bay. 1986. On the mechanism by which saxitoxin binds to and blocks sodium channels. In *Tetrodotoxin, saxitoxin and the molecular biology of the sodium channel.* C. Y. Kao and S. R. Levinson (eds.), pp. 96-112. New York Academy of Science.

Stroud, R.M., M.R. McCarthy, and M. Shuster. 1990. Nicotinic acetylcholine receptor superfamily of ligand-gated ion channels. *Biochemistry.* **29**:11009-11023.

Structure of brevetoxin A (GB-1 Toxin), the most potent toxin in the florida Red Tide organism *Gymnodinium breve* (*Ptychodiscus brevis*). *J. Am. Chem. Soc.* **108**:514-515.

Stryer, L. 1981. "Biochemistry." *Biochemistry* 2 ed., W. H. Freeman and Co., New York.

Suganuma, M., H. Fujiki, H. Furuya-Suguri, S. Yoshizawa, S. Yasumoto, Y. Kato, N. Fusetani, and T. Sugimura. 1990. Calyculin A, an inhibitor of protein phosphatases, a potent tumor promoter on CD-1 mouse skin. *Cancer Res.* **50**:3521-3525.

Suganuma, M., H. Fujiki, H. Suguri, S. Yoshizawa, M. Hirota, M. Nakayasu, M. Ojika, K. Wakamatsu, K. Yamada, and T/ Sugimura. 1988. Okadaic acid: an additional non-phorbal-12-tetradecanoate-13-acetate-type tumor promoter. *Proc. Natl. Acad. Sci. USA.* **85**:1768-1771.

Suganuma, M., M. Suttajit, H. Suguri, M. Ojika, K. Yamada, and H. Fujiki. 1989. Specific binding of okadaic acid, a new tumor promoter in mouse skin. *FEBS Lett.* **250**:615-618.

Sugawara, I., W. Itoh, S. Kimura, S. Mori, and K. Shimada. 1989. Further characterization of sulfated homopolysaccharides as anti-HIV agents. *Experientia.* **45**(10):996-998.

Tachibana, K., P.J. Scheuer, Y. Tsukitani, H. Kikuchi, D. Van Engen, J. Clardy, Y. Gopichand, and F.J. Schmitz. 1981. Okadaic acid, a cytotoxic polyether from two marine sponges of the genus *Halichondria. J. Am. Chem. Soc.* **103**:2469-2471.

Takai, A., C. Bialojan, M. Troschka, and J.C. Rüegg. 1987. Smooth muscle myosin phosphatase inhibition and force enhancement by black sponge toxin. *FEBS Lett.* **217**:81-84.

Takai, A., M. Troschka, G. Mieskes, and A.V. Somlyo. 1989. Protein phosphatase composition in the smooth muscle of guinea-pig ileum studied with okadaic acid and inhibitor 2. *Biochem. J.* **262**:617-623.

Takemura, M., J. Kishino, A. Yamatodani, and H. Wada. 1989a. Inhibition of histamine release from rat hypothalamic slices by ω-conotoxin GVIA, but not by nilvadipine, a dihydropyridine derivative. *Brain Res.* **496:**351-6.

Takemura, M., H. Kiyama, H. Fukui, M. Tohyama, and H. Wada. 1989b. Distribution of the ω-conotoxin receptor in rat brain. An autoradiographic mapping. *Neurosci.* **32**:405-416.

Takito, J., H. Nakamura, J. Kobayasi, and Y. Ohizumi. 1987a. Enhancement of the actin-activated ATPase activity of myosin from canine cardiac ventricle by purealin. *Biochim. Biophys. Acta.* **912**:404-407.

Takito, J., H. Nakamura, J. Kobayasi, Y. Ohizumi, K. Ebisawa, and Y. Nonomura. 1986. Purealin, a novel stabilizer of smooth muscle myosin filaments that modulates ATPase activity of dephosphorylated myosin. *J. Biol. Chem.* **261**(29):13861-13865.

Takito, J., Y. Ohizumi, J. Kobayasi, K. Ebisawa, and Y. Nonomura. 1987b. The mechanism of inhibition of light-chain phosphorylation by purealin in chicken gizzard myosin. *Eur. J. Pharmacol.* **142**:189-195.

Tamplin, M.L. 1990. A bacterial source of tetrodotoxins and saxitoxins. In *Marine Toxins,* ACS Symposium Series 418. S. Hall and G. Strichartz (eds.), pp. 78-86. American Chemical Society, Washington, DC.

Taylor, P., P. Culver, S. Abramson, L. Wasserman, T. Kline, and W. Fenical. 1988. Use of selective toxins to examine acetylcholine receptor structure. In *Biomedical Importance of Marine Organisms.* Memoirs of the California Academy of Sciences Number 13. D.G. Fautin (ed.), pp. 109-114. California Academy of Sciences. San Francisco.

Thomson, C. and J. Wilkie. 1989. The conformations and electrostatic potential maps of phorbol esters, teleocidins and ingenols. *Carcinogenesis.* **10**:531-540.

Trainer, V. L., R.A. Edwards, A.M. Szmant, A.M. Stuart, T.J. Mende, and D.G. Baden. 1990. Brevetoxins: unique activators of voltage-sensitive sodium channels. In *Marine Toxins,* ACS Symposium Series 418. S. Hall and G. Strichartz (eds.), pp. 166-175. American Chemical Society. Washington, DC.

Uemura, D., K. Ueda, and Y. 1981. Further studies of palytoxin. II. Structure of palytoxin. *Tetrahedron Lett.* **22**:2781-2784.

Ulevitch, R.J., Y. Watanabe, M. Sano, M.D. Lister, R.A. Deems, and E.A. Dennis. 1988. Solubilization, purification, and characterization of membrane-bound phospholipase A_2 from the P388D_1 macrophage-like cell line. *J. Biol. Chem.* **263**:3079-3085.

van Duyne, G.D. 1990. X-ray crystallographic studies of marine toxins. In *Marine Toxins,* ACS Symposium Series 418. S. Hall and G. Strichartz (eds.), pp. 144-165. American Chemical Society. Washington, DC.

van Sande, J., F. Deneubourg, R. Beauivens, J.C. Breakman, D. Daloze, and J.E. Dumont. 1990. Inhibition of iodide transport in thyroid cells by dysidenin, a marine toxin, and some of its analogs. *Mol. Pharmacol.* **37**:583-589.

Wada, A., Y. Uezono, M. Arita, K. Tsuji, N. Yanagihara, H. Kobayashi, and F. Izumi, 1989. High affinity and selectivity of neosurugatoxin for the inhibition of ^{22}Na influx via nicotinic receptor-ion channel in cultured bovine adrenal medullary cells: comparative study with histrionicotoxin. *Neuroscience.* **33**:333-339.

Wattenberg, E.V., K.L. Byron, M.L. Villereal, H. Fujiki, and M.R. Rosner. 1989a. Sodium as a mediator of non-phorbol tumor promoter action. Down-modulation of the epidermal growth factor receptor by palytoxin. *J. Biol. Chem.* **264**:14668-14673.

Wattenberg, E.V., H. Fujiki, and M.R. Rosner. 1987. Heterologous regulation of the epidermal growth factor receptor by palytoxin, a non-12-O-tetradecanoylphorbol-13-acetate-type tumor promoter. *Cancer Res.* 47:4618-4622.

Wattenberg, E.V., P.L. McNeil, H. Fujiki, and M.R. Rosner. 1989b. Palytoxin down-modulates the epidermal growth factor receptor through a sodium-dependent pathway. *J. Biol. Chem.* **264**:213-219.

Watters, D., K. Marshall, S. Hamilton, J. Michael, M. McArthur, G. Seymour, C. Hawkins, R. Gardiner, and M. Lavin. 1990. The bistratenes: new cytotoxic marine macrolides which induce some properties indicative of differentiation in HL-60 cells. *Biochem. Pharmacol.* **39**:1609-1614.

Weinberg, J.M., J.A. Davis, A. Lawton, and M. Abarzua. 1988. Modulation of cell nucleotide levels in isolated kidney tubules. *Am. J. Physiol.* **254**:F311-F322.

Wender, P.A., K.F. Koehler, N.A. Sharkey, M.L. Dell'Aquila, and P.M. Blumberg. 1986. Analysis of the phorbol ester pharmacophore on protein kinase C as a guide to the rational design of new classes of analogs. *Proc. Natl. Acad. Sci. USA.* **83**:4214-8.

Wessler, I., D.J. Dooley, H. Osswald, and F. Schlemmer. 1990. Differential blockade by nifedipine and ω-conotoxin GVIA of a_1- and b_1-adrenoceptor-controlled calcium channel motor nerve terminals of the rat. *Neurosci. Lett.* **108**:173-178.

Wheeler, L.A., G. Sachs, D. Goodrum, L. Amdahl, N. Horowitz, and W. de Vries. 1988. Importance of marine natural products in the study of inflammation and calcium channels. In *Biomedical Importance of Marine Organisms,* Memoirs of the California Academy of Sciences Number 13. D.G. Fautin (ed.), pp. 11125-132. California Academy of Sciences. San Francisco.

Williams, M., M. Abreu, M.F. Jarvis, and L. Noronha-Blob. 1987. Characterization of adenosine receptors in the PC12 pheochromocytoma cell line using radioligand binding: evidence for A-2 selectivity. *J. Neurochem.* **48**:498-502.

Wollner, D.A., R. Scheinman, and W.A. Catterall. 1988. Sodium channel expression and assembly during development of retinal ganglion cells. *Neuron.* **1**:727-739.

Wright, A.E., D.A. Forleo, G.P. Gunawardana, S.P. Gunasekera, F.E. Koehn, and O.J. McConnell. 1990. Antitumor Tetrahydroisoquinoline Alkaloids from the Colonial Ascidian *Ectinascidia turbinata. J. Org. Chem.* **55**:4508-4512.

Wu, C.H. and T. Narahashi. 1988. Mechanism of action of novel marine neurotoxins on ion channels. *Ann. Rev. Pharmacol. Toxicol.* **28**:141-161.

Yoshikami, D., Z. Bagabaldo, and B.M. Olivera. 1989. The inhibitory effects of ω-conotoxins on Ca channel and synapses. *Ann. N. Y. Acad. Sci.* **560**:230-248.

Zabriskie, T.M., C.L. Mayne, and C.M. Ireland. 1988. Patellazole C: A Novel Cytotoxic Macrolide from *Lissoclinum patella. J. Amer. Chem. Soc.* **110**:7919-7920.

Zafaralla, G.C., C. Ramilo, W.R. Gray, R. Karlstrom, B.M. Olivera, and L.J. Cruz. 1988. Phylogenetic specificity of cholinergic ligands: α-conotoxin S1. *Biochemistry.* **27**:7102-7105.

This chapter adapted in part from *Marine Biotechnology,* Vol. 1. Permission granted by Kluwer Academic/Plenum Publishers.

For further information on bioactive compounds from the sea, see *Marine Biotechnology,* Vol. 1, ISBN 0-3-6-44174-8, edited by D. Attaway and R. Zaborsky and published by Plenum Publishers. Contact Kluwer Academic Publishers, Customer Service Department, Hingham MA 02018-0358; (781) 871-6600; E-mail: kluwer@wkap.com

Section VI. Seafood Safety

Contamination in Shellfish-Growing Areas

Robert E. Croonenberghs

INTRODUCTION

In this chapter the term shellfish will refer only to animals in the phylum Mollusca, class Bivalvia, which means that they are mollusks with two shells such as oysters, clams, mussels, and scallops. Although crabs, crayfish, shrimp, and lobsters have shells, they are members of the phylum Arthropoda, class Crustacea, and are not included as shellfish in this discussion.

Bivalve mollusks feed by filtering microscopically-sized particles out of the water. They pump water via the inhalant siphon through their gills, which sorts out particles of the correct size and "feel," and directs selected particles into their gut. The uneaten particles are collected in mucus, and ejected via the exhalent siphon as pseudofeces, along with the digested gut contents.

It is this filter-feeding mechanism, and the relatively large volume of water the bivalves pump through their siphons, that causes such public health concern about pollutants in shellfish-growing waters. Three primary mechanisms for uptake and concentration of pollutants in bivalves thus exist: particulate ingestion, absorption, and ionic attraction. Depending on the type of pollutant, one or more of these mechanisms may come into play.

Particulate ingestion is the manner in which shellfish concentrate viruses and bacteria from the water. Viruses and bacteria are too small for the shellfish's gills to detect or reject, but due to their small size in relation to such food items as algae and small particles suspended in the water, they tend to attach to these surfaces. The smaller the particle, the larger the surface area relative to its size and, because of the tremendous individual numbers of food items held in the shellfish's gut at any one time, the number of bacteria and viruses concentrated in that location is considerably larger than those present in an equal volume of the surrounding water. Other pollutants of public health concern that shellfish concentrate basically via this mechanism are many types of pesticides, toxic algae, and ions of heavy metals attached to particulate matter.

Shellfish absorb soluble lipophilic ("fat-loving") pollutants directly from the large volumes of water they pump through their bodies. Lipophilic substances have a low polarity; there is almost no electrical charge on the molecule because all ionic charges are balanced. However, water is quite polar: the oxygen end of the molecule has considerable negative charge, and the hydrogen end has considerable positive charge. Pollutants of low polarity (e.g., oils) do not dissolve well in water, so they tend to move out of water and into nonpolar fats when they get the opportunity. Due to the lipid content of shellfish, relatively nonpolar pollutants such as petroleum hydrocarbons, certain organic metallic complexes, and some pesticides are primarily concentrated in this manner.

Ionic substances (such as heavy metals) and highly polar substances are not only ingested when attached to particulate clays, but are also dissolved in the water. Again, due to the high pumping rate of the shellfish, large numbers of these dissolved substances come into contact with their body, and are taken up via an ion exchange reaction in the shellfish.

NATIONAL SHELLFISH SANITATION PROGRAM

The shellfish sanitation program began at the national level in the United States in 1925 after a series of typhoid epidemics spread by the consumption of raw shellfish threatened the collapse of the industry. The program has changed over the years and is now called the National Shellfish Sanitation Program (NSSP); it is run at the national level by the United States Food and Drug Administration (FDA). The principles behind this program protecting the consumer from bacterial and viral human pathogens have changed little over the years. Due to the ability of shellfish to concentrate pathogens and pollutants out of the surrounding waters, nearly pristine water is required. For example, the NSSP standard for the average concentration of fecal coliforms in shellfish waters is 14 fecal coliforms per 100 milliliters (ml). This is roughly 14 times more stringent than the 200 fecal coliforms per 100 ml used by many states as a safe swimming standard. The shellfish standard is even more strict because there is a second part to the standard, depending on the method of sample collection. It requires that either no more than 10 percent of the samples exceed 49 most probable number (MPN), fecal coliforms/100 ml for a three-tube test, or that the ninetieth percentile for the data set not exceed 49. As a result, states producing shellfish are required to conduct onshore surveys to check for sources of sanitary waste, and are required to collect numerous growing water samples for analysis of the indicator bacteria used (i.e., fecal coliforms).

Introduction of Pollutants into Waterways

When interpreting fecal coliform concentrations in shellfish-growing areas, one must consider numerous factors controlling the distribution of the bacteria in order to understand what the data mean. Fecal coliforms are used as an indicator of the possible presence of human pathogens, since they originate in the gut of humans and warm-blooded animals. However, pathogens do not necessarily occur in direct proportion to the number of fecal coliforms present. Fecal coliforms may die off at a rate different than that of accompanying pathogens. In addition, there is little documentation to indicate that pathogens from wild animals in the United States have caused human illness through the consumption of raw shellfish. In rural areas, wild animals are quite apt to be the primary source of fecal coliforms found in runoff. As such, an evaluation of the watershed for potential sources is quite important because the use of the fecal coliform indicator alone is not sufficient to assure safe growing waters.

The major human pathogens of concern in shellfish waters usually originate on land, except for sewage discharges from boats and sewage treatment facility (STF) outfalls. In most areas, except those heavily impacted by an STF or a marina, the primary vector transporting fecal coliforms to shellfish waters is runoff from the land. Runoff varies with rainfall, soil types, degree of saturation, etc., and, as such, is periodic. Therefore, one or a few samples in most areas may indicate very little about the pollution potential in that area. Numerous samples must be collected under all weather conditions, or one must sample under conditions that are most likely to transport fecal pollution to the area (e.g., after heavy rains, high tides, etc.).

Generally in estuarine areas most of the runoff enters a tributary through its headwaters. Since the runoff contains most of the bacteriological load, and the larger body of water holds fewer bacteria, the highest concentrations tend to occur in the headwaters. As the influent is carried downstream by the tidal action, the associated mixing steadily reduces the concentrations of fecal coliforms. Of course, where feeder streams enter, concentrations may increase. Also, higher counts tend to occur closer to shore due to runoff from the shore, and these higher counts tend to remain near shore for some distance downstream, since there is rarely a strong cross-channel mixing gradient. Illegal discharges from on-site sanitary facilities into shellfish waters can produce a dramatic local increase in fecal coliform concentrations.

PATHOGENS

Enteric Viruses Associated with Fecal Pollution

In the United States, it is widely believed that viral agents, as opposed to bacteria, have been the primary cause of shellfish-borne disease outbreaks as a result of pollution over approximately the last 20 years (Rippey, 1994). Part of the reason that viruses are considered to be the culprit may well be that sewage is more thoroughly treated now than in the past, and the more resilient viruses survive our sanitation procedures better than do bacteria. Thus, when shellfish become contaminated with fecal pollution, the water tends to have been treated, and primarily only viruses are present.

Shellfish programs in all coastal states have been conducting shoreline surveys to locate and correct sources of human fecal pollution in shellfish waters. In addition, local municipalities have been mandated by the federal government to upgrade sewage treatment facilities (STF). The result is that in coastal areas the majority of human feces-associated pathogens are either kept in the soil, or are subjected to an STF. While bacteria are quite sensitive to chlorination in a properly operating STF, viruses are significantly more resistant. States usually design permits for STF based on fecal coliform concentrations in the effluent. Thus, large numbers of viruses can be discharged into shellfish waters by a legally operating STF [Lewis et al., 1986; World Health Organization (WHO), 1979].

Other factors may also contribute to the predominance of viruses in recent United States shellfish-associated disease outbreaks. Once discharged into marine waters, human enteric viruses can survive for extended periods of time (La Belle and Gerba, 1982; Feachem et al., 1981). Some viruses are extremely infectious; for example, Norwalk virus has a theoretical infectious dose of from one to ten particles. Hard clams (*Mercenaria mercenaria*) have been shown to bioconcentrate hepatitis A virus 900 times over the concentration in water (Goswami et al., 1993); in addition, they show a marked seasonal increase in the concentration of bacteriophages (viruses that attack bacteria) in the spring, which may indicate a propensity to do so for other viruses (Burkhardt et al., 1992). Viruses can be retained within actively pumping shellfish much longer than bacteria (Enriquez et al., 1992; De Mesquita et al., 1991; Power and Collins, 1990a & b; Lewis et al., 1986). Viruses can be absorbed intracellularly into shellfish, as opposed to merely being held in the gut, and as such are much more difficult for the shellfish to eliminate (Hay and Scotti, 1987). Once harvested and refrigerated, Coxsackie B3 and polioviruses have been shown to survive basically unreduced for 28 days at 5°C (Tierney et al., 1982; Metcalf and Stiles, 1965). Other work has shown gradual inactivation up to approximately 90 percent for poliovirus over 30 days when refrigerated, and over 12 weeks when frozen (DiGirolamo et al., 1970). Many enteric human viruses are fairly heat-resistant (hepatitis A and Norwalk); even if partially cooked, the viruses may survive (Millard et al., 1987; Truman et al., 1987; Morse et al., 1986; Peterson et al., 1978).

Thus, viruses represent a class of human pathogens that are legally discharged in large quantities into coastal waters; they can be long-lived in the environment; and, when taken up by shellfish, they can remain in the animal for long periods of time after the pollution event and then again after harvesting. Many viruses tend to survive mild cooking and are able, in some cases, to cause infection with only a few particles. It is no wonder that viruses cause such a high percentage of the United States shellfish-related disease outbreaks.

Human pathogenic viruses cannot replicate in shellfish, so those present in shellfish at the time of harvest will not increase, despite temperature abuses that might occur. Therefore, when widespread viral-associated outbreaks occur due to shellstock (as opposed to shucked, which are more intensely handled), the growing area immediately becomes the prime suspect.

Hepatitis A

Hepatitis is the most serious viral disease associated with the consumption of raw shellfish, and the hepatitis A virus (HAV) is the most common agent. HAV is a 27-nanometer (nm) picornavirus, and the incubation period is 15 to 50 days, with an average of 28 to 30 days. The disease causes fever, malaise, anorexia, nausea, and abdominal discomfort, followed by jaundice. It is usually fairly mild and lasts one to two weeks, though it can be severely disabling and last for months. Complete recovery is often prolonged. The fatality rate is 0.6 percent.

HAV is spread via the fecal-oral route. The period of maximum shedding of the viral particles by infected persons appears to be during the latter half of the incubation period and into the onset of jaundice (Benenson, 1995; Hackney et al., 1992). In developing countries, children usually contract a mild form of the disease, which then provides them with a lifelong immunity. Incidence of hepatitis A due to the consumption of raw or undercooked shellfish has been steadily decreasing in the United States over the past 15 years, though in the period from 1961 through 1980, more than 1000 cases due to shellfish were reported along the Gulf and East Coasts (Glatzer, 1999; Rippey, 1994).

HAV can survive long periods of time in the estuarine environment. Though it survives longer in freshwater than estuarine water, it has been shown in the laboratory to survive three months in estuarine water. HAV is considerably longer-lived in water and other varied environmental conditions than is the polio virus, which has often been used as a model for HAV. Cooler temperatures (5°C) tend to enhance survival over warmer temperatures (25°C), as does particulate matter in water. HAV has been shown to survive in dried feces held at 5°C and 25°C for one month (Sobsey et al., 1988). HAV is quite heat stable and can survive 140°F (60°C) for four hours (Kilgen and Cole, 1991).

HAV appears to be one of the slower viruses to depurate from shellfish (Hackney et al., 1992). Greater than 10 percent of HAV taken up by the eastern oyster (*Crassostrea virginica*) remains after five days of depuration, though less than one percent of polio virus remains (Sobsey et al., 1988). Mussels (*Mytilus chilensis*) have been shown to concentrate HAV 100 times over that in the water, and to retain the virus for seven days under optimal feeding and filtration conditions, with longer retention under poorer feeding conditions (Enriquez, 1992).

Hepatitis E

Hepatitis E disease has also been linked to shellfish, and the active agent is called hepatitis E virus (HEV). The symptoms are very similar to hepatitis A. The incubation period is 15 to 64 days, though the average has varied from 26 to 42 in different epidemics. While the fatality rate is 0.1 to 1 percent, in pregnant women it may reach 20 percent in the third trimester (Hackney et al., 1992; Benenson et al., 1990).

Norwalk, Norwalk-like, and Other Viruses

Gastroenteritis is the now most common disease in the United States caused by the consumption of raw or undercooked shellfish from sewage-contaminated waters (Rippey, 1994). The disease is normally self-limited, with symptoms of nausea, vomiting, diarrhea, and abdominal pain that last normally 24 to 48 hours. Norwalk virus is one of the most common agents of this disease, and is a small, 27 to 32nm calicivirus. Norwalk viruses are extremely infectious, and may infect with as few as one to 10 particles; an outbreak in 1993 was traced to one harvester dumping his feces and vomitus overboard in a remote and wide-open harvest area (Kohn et al., 1995). Norwalk viruses are also quite heat resistant. Other viruses implicated in shellfish-associated illnesses that cause similar symptoms are Norwalk-like viruses, small round viruses (SRV) (Gray and Evans, 1993; Haruki et al., 1991), and Snow Mountain agent (Truman et al., 1987). The incubation period is normally 24 to 48 hours, although volunteer studies with Norwalk virus have shown 10 to 50 hours. Some immunity develops after exposure, but is variable in duration (Benenson, 1990), and some people may be naturally immune (Blacklow et al., 1979).

Enteric Bacteria Associated with Fecal Pollution

Salmonella

Typhoid fever was the dominant recognized form of United States shellfish-borne disease in the first half of this century. Indeed, an outbreak of 1,500 cases in 1924 and another outbreak the following year were the impetus for the development of the precursor to the NSSP. Typhoid fever is caused by *Salmonella typhi*, and though it has not been associated with shellfish since 1954, approximately 500 cases occur in the United States each year, mainly due to importation from endemic areas outside the United States; therefore, a slight potential for a shellfish-borne outbreak still exists. Symptoms include sustained headache, fever, malaise, and anorexia. Incubation typically lasts one to three weeks (Benenson, 1990).

Salmonellosis due to shellfish consumption is not very common, and has tended to occur sporadically, with only eight cases in the interval 1984 to 1993 (Rippey, 1994). Sufficient cells are needed to cause the syndrome, usually between 10^5 and 10^8, though the infective dose in high-fat foods can be much less. It is not uncommon to find salmonellae in shellfish, but the typical concentrations of approximately 2/100g of shellfish are believed to be too low to cause illness, and may represent part of the free-living microflora of shellfish (Hackney et al., 1992).

Symptoms of salmonellosis include headache, abdominal pain, diarrhea, nausea, and occasionally vomiting. Dehydration can be dangerous, especially in children, and can last several days. The incubation period can last from six to 72 hours, but is usually 12 to 36 hours (Benenson, 1995). All serotypes of *Salmonella* are considered to be pathogenic, but *S. typhimurium* and *S. enteritidis* are the most often reported types associated with the disease. *S. infantis* was associated with two cases of shellfish-borne disease in Maine (Rippey, 1991). Hackney et al. (1992) conclude from the literature that *Escherichia coli* (a type of fecal coliform) is a suitable indicator for the presence of *Salmonella*.

Campylobacter

Campylobacteriosis is a major form of food poisoning, and of late is being reported more commonly as a shellfish-borne disease, probably as a reflection of improved analytical capability. Symptoms include diarrhea, abdominal pain, malaise, fever, nausea, and vomiting. Bloody stools are common. The incubation period normally lasts three to five days, and the illness frequently ends within two to five days, though it can be prolonged in adults.

Although cattle, swine, sheep, goats, cats, and dogs all carry *C. jejuni*, poultry have some of the highest carriage rates. *C. jejuni* is a leading cause of gastroenteritis and is transmitted via food and water. Ingestion of only a few hundred cells can cause infection, and often illness (Benenson, 1995; Doyle, 1990a). Therefore, fecal pollution of shellfish waters by animal waste is a concern with this disease (Hackney et al., 1992). In the interval 1984 to 1993, there were 12 cases of shellfish-borne illness attributed to *Campylobacter* in the United States, though none from 1994 to 1999 (Glatzer, 1999; Rippey, 1994).

Shigella

Shigellosis associated with shellfish is not particularly common, though there have been four United States outbreaks and several individual cases in the past 25 years (Glatzer, 1999; Rippey, 1994). Symptoms of the disease include diarrhea, fever, nausea, and occasionally toxemia, vomiting, cramps, and tenesmus. The bacteria invade the intestinal tract, and can cause the discharge of blood, mucus, and pus (dysentery) with the stool. As few as 10 to 100 organisms can cause the disease, which has a usual incubation period of one to three days, and generally lasts four to seven days (Benenson, 1990). *Shigella* rarely occurs in animals; it is a disease of man and higher primates. Convalescent carriers are a major reservoir, and usually shed the bacteria for three to five weeks after symptoms subside (Doyle, 1990b). Outbreaks due to shellfish have been traced to sewage pollution, though the bacteria do not survive well in the environment (Hackney et al., 1992). One shellfish-borne outbreak was traced to the overboard disposal of feces from harvest boats (Reeve et al., 1989).

Escherichia

In their review of pathogenic *Escherichia coli* (*E. coli*), Hackney et al. (1992) indicate that there are four pathogenic strains: enterotoxigenic, enteropathogenic, enteroinvasive, and hemorrhagic. The first three strains are usually associated with human feces, but the hemorrhagic is usually found due to farm animals. The hemorrhagic strain is the most dangerous, sometimes causing death in children and the elderly, and is caused by the type *O157:H7*. The authors report that pathogenic *E. coli* have been isolated from oysters and mussels, and that when analyzed in certain environmental samples, the concentrations correlated well to the total number of *E. coli* present. No United States cases of shellfish-borne illness are attributable to these organisms (Glatzer, 1999; Rippey, 1994).

Vibrio cholerae O1

Vibrio cholerae exist in the estuarine environment both as a result of human fecal pollution and as a naturally-occurring organism. This species is normally classified as two serogroups, O1 and non-O1 *V. cholerae*, though there is a third serogroup, atypical O1, which does not produce cholera toxin. Epidemics of cholera and large-scale shellfish-borne outbreaks are due to fecal pollution of serogroup O1, though this group probably also exists naturally in estuarine waters.

V. cholerae O1 contains two biotypes, classical (*cholerae*) and *El Tor*, both of which contain toxigenic and non-toxigenic strains. These two biotypes can be further divided into serotypes Inaba, Ogawa, and Hikojima. The toxigenic strains produce cholera toxin or similar toxins, and it is these toxins that cause the illness. Toxigenic O1 does not cause localized infections outside of the intestines. The non-toxigenic strains can cause diarrhea and extraintestinal infections. *El Tor* is the current biotype of concern in the Americas, and individual organisms can live in water for long periods of time. *V. mimicus* is closely related and some strains can elaborate a toxin identical to *V. cholerae* (Roderick, 1991; Benenson, 1990; Doyle and Cliver, 1990; Blake, 1983). There have been only a few sporadic cases of shellfish-borne *V. cholerae* O1 in the United States (Glatzer, 1999; Rippey, 1994).

Symptoms of cholera include a sudden onset of profuse and painless, watery diarrhea, occasional vomiting, rapid dehydration, acidosis, and circulatory collapse. Asymptomatic infection is much more common than clinical illness, and mild cases with only diarrhea are common. The incubation period can be as little as a few hours to five days, though the average is two to three days (Benenson, 1990). The infective dose is estimated to be 10^8-10^9 cells, though antacids and medication to lower stomach acidity and lower the infective dose to 10^3 cells for some strains of *El Tor* (Doyle and Cliver, 1990).

Enteric Bacteria not Associated with Fecal Pollution

There are a number of species of bacteria that naturally occur in estuarine waters that can, on occasion, be pathogenic to humans. The largest number of these opportunistic pathogens occurs in the family Vibrionaceae, and most occur seasonally in largest numbers when the water is warm.

Vibrio cholerae Non-O1

Vibrio cholerae Non-O1 usually accounts for a few individual cases of shellfish-borne illness in the United States each year, occasionally causing an outbreak in a few people (Rippey, 1994). This sickness is characterized by diarrhea, often with abdominal pain and fever (Morris and Black, 1985), but to call this "cholera" is inaccurate (Benenson, 1990). Morris and Black (1985) indicate that less than 5 percent of the Non-O1 serogroup members produce cholera toxin. Members of the Non-O1 serogroup have not been associated with large epidemics or pandemics. Hackney et al. (1992) report that almost all the cases of Non-O1 illness in the United States have been due to the consumption of raw oysters. Non-O1 organisms occur most often in marine waters of a salinity reduced to between four to 17 parts per thousand (ppt) (Colwell and Kaper, 1978). While *V. cholerae* Non-O1 appear to survive and multiply in the estuarine environment, it is not clear whether the strains that cause human illness are spread by human feces or not (Blake et al., 1980).

Vibrio parahaemolyticus

Vibrio parahaemolyticus also regularly accounts for a few individual cases of shellfish-borne illness in the United States every year, and occasionally for outbreaks (Rippey, 1994). In Japan it accounts for a significant portion of the bacterial food-borne illness, and is due to consumption of raw seafood in the summer months. The symptoms typically include diarrhea, abdominal cramps, nausea, vomiting, headaches, fever, and chills. The infective dose is normally from 10^5 to 10^7 cells, though a reduction of gastric acid may reduce the threshold for infection. The onset time is nine to 25 hours, and normally lasts 2.5 to three days. The actual symptoms are due to heat-stable toxin production that can occur either in the food prior to consumption, or in the intestines (Hackney et al., 1992; Benenson, 1990; Blake et al., 1980).

Once contracted via oyster consumption, *V. parahaemolyticus* can cause death, accounting for six deaths in the interval 1984-1993 in the United States (Rippey, 1994). *V. parahaemolyticus* also causes extraintestinal infections (Armstrong et al., 1983; Blake, 1983). Levine et al. (1993) have pointed out the danger of *V. parahaemolyticus* to healthy persons, noting the high incidence of otherwise healthy persons contracting gastroenteritis from raw shellfish.

V. parahaemolyticus occurs in estuarine waters around the world (Blake et al., 1980), being present in highest numbers during the summer months. Though it is common in seafood, isolates pathogenic to man (Kanagawa positive) are rare. Estuarine and seafood isolates are predominantly Kanagawa negative (Hackney et al., 1992; Blake et al., 1980).

Recently, however, shellfish-borne disease outbreaks in the United States due to *V. parahaemolyticus* have become more prevalent and have raised concerns for this naturally occurring pathogen in oysters. In 1997 one outbreak occurred on the West Coast. In 1998 three outbreaks occurred, one on each coast of the United States, and in 1999 a small outbreak also occurred in British Columbia (Glatzer, 1999). One strain previously known only in Asian countries, serotype O3:K6, was identified for the first time in 1998 as a cause of two (Northeast and Gulf coasts) of the three U.S. outbreaks. The emergence of this strain as a major cause of outbreaks in Japan and the United States in recent years, along with other serotypes such as O4:K8, also prevalent in Japan, have raised concerns about the pathogenicities and abilities of such strains to cause outbreaks. A thermostable direct hemolysin (tdh) is produced by virtually all pathogenic strains of *V. parahaemolyticus*, and the presence of the causative gene currently is being used to screen isolates from shellfish to indicate the presence of pathogenic strains. Some data from the recent U.S. outbreaks suggest that the number of pathogenic cells required to cause illness may be as low as 100 to 1,000 cells; however, there remains much uncertainty about an infectious dose. In at least one instance, though, illness resulted from the consumption of a single contaminated oyster (Watkins, 1999).

Vibrio vulnificus

In the past decade, no other bacterium has raised more controversy with respect to shellfish than *Vibrio vulnificus*. From 1988 to 1999 approximately 263 people contracted *Vibrio vulnificus* infections due to the consumption of shellfish, of which 132 people died (Glatzer, 1999). Persons at highest risk are those with pre-existing liver disease, especially alcoholic cirrhosis, and also persons with iron-storage disorders (hemochromatosis, hemocirrhosis), immunosuppression, renal failure, and diabetes. Persons with decreased stomach acidity, steroid-dependent asthma, or rheumatic disorders are also at a significantly greater risk than the general population (Whitman, 1994; Levine et al., 1993; Klontz et al., 1988; Morris, 1988; Tacket et al., 1988; Blake et al., 1979).

Typically, *V. vulnificus* infects a person either through consumption with resulting primary septicemia, or through a preexisting wound. Primary septicemia is the most dangerous form of infection, and the disease can rapidly progress into intractable shock; more than 50 percent of these cases result in death. Bulbous skin lesions characteristically appear. *V. vulnificus* may occasionally cause gastroenteritis (Morris, 1988; Morris and Black, 1985). The infectious dose is not known, but in high-risk populations, fatal infections have occurred due to consumption of only one oyster (Whitman, 1994). Recent limited work indicates that infectious doses may be far less than earlier believed, on the order of 1,000 cells/gram (g) of oyster tissue (Tamplin, 1994a).

V. vulnificus occurs naturally throughout estuarine areas of the United States, preferring temperatures of 20° to 30°C and salinities of five to 25 ppt. Tamplin (1994b) examined water, sediment, and shellfish throughout the United States and found the lowest concentrations on the Pacific Coast, probably due to colder temperatures and higher salinities; the Northern Atlantic estuaries had the next highest concentrations, followed by the Southern and Middle Atlantic, with the Gulf Coast estuaries having the highest concentrations. As part of this work, Tamplin reported the concentrations of the more virulent opaque strains (encapsulated) to be 10 times or more prevalent than the weakly-virulent translucent strains (unencapsulated) in Gulf Coast waters. The encapsulated strains may be more virulent due to the protection provided to the bacteria by the capsule (Morris, 1994; 1988). Many of the wild strains of *V. vulnificus* have the capability to switch back and forth between encapsulated and unencapsulated forms (Morris, 1994).

Temperatures of approximately 13°C or greater are needed for *V. vulnificus* to grow, whether in shell oysters or in a growth medium. At summertime temperatures in the Gulf Coast, *V. vulnificus* can grow significantly in shell oysters in four hours, though icing shell oysters at the harvest site does prevent growth (Tamplin, 1994c). Icing of shellstock has been shown to reduce slowly, over the course of seven to 14 days, the levels of culturable cells (Cook and Ruple, 1992).

At temperatures below 13°C, *V. vulnificus* begins to enter a dormant state, where the cells are alive but cannot be grown on media. This state is called viable-but-non-culturable (VBNC). During the winter months, the number of *V. vulnificus* cells appears to decrease in the water, sediment, and oysters, but conclusive research now reveals that these cells, rather than dying, are simply in the VBNC state. These cells, when placed in warm conditions, revert back to normal culturable cells in one to two days, and apparently do so in shellfish as well. Encapsulated cells retain their capsules when they enter the VBNC state, a factor that aids in their resistance to the human immune system. Experiments with mice made susceptible to *V. vulnificus* show that these VBNC cells do kill the mice, though the longer the cells have been in the VBNC state, the more cells are required. Whereas seven to eight cells of recently transformed VBNC *V. vulnificus* can kill susceptible mice, 10^6 cells are needed once they have been in the VBNC state for seven days or more. This may help explain why human deaths are so reduced during the cold months (Oliver, 1994), and partially explains why iced, shucked oysters are less of a hazard.

Shucking, washing, blowing (agitation with compressed air bubbles), and then packing oysters (without additional fresh water) does not appear to decrease the number of *V. vulnificus* in oysters. However, keeping shucked oysters on ice usually reduces the number of culturable cells by a 1-log and 2-log unit over three and seven days respectively (Ruple and Cook, 1992). Commercially blowing oysters for extended periods of time and adding substantial amounts of fresh water to packed oysters may disrupt the osmotic pressure and kill significant quantities of *V. vulnificus* (Kilgen, 1994). Freezing, whether individually quick frozen (IQF), carbon dioxide (CO_2) at -30°C, blast freezing at -23°C, or conventional freezing at -18°C, reduces *V. vulnificus* but does not eliminate it. *V. vulnificus* is very heat-sensitive and normal cooking will eliminate it, but steaming oysters to the point of gaping will not totally eliminate it (Cook, 1994). Heating oysters to 50°C for 10 minutes will kill *V. vulnificus;* the process does not impart a noticeable cooked appearance or taste to the oysters, and may prove to be an acceptable method for eliminating the bacteria from oysters (Cook and Ruple, 1992). Depuration has not proven successful in eliminating *V. vulnificus* from oysters, because the cells are so tightly associated with the tissue. Rather than decrease in a depuration system, *V. vulnificus* has consistently grown (Tamplin, 1994d). Some interesting work in New Hampshire suggests that relay to high salinity waters of 25 ppt may help reduce *V. vulnificus* levels in oysters (Jones, 1994). Irradiation of shellstock oysters, though effective in killing *V. vulnificus*, shortens the shelf life of the product too much for commercial application (Tamplin, 1994d). It should be noted that recent work on the irradiation of oysters may have developed an acceptable approach to solve the shelf life problem (Martin, 1996).

The VBNC state and the ability for *V. vulnificus* to resist depuration from oysters have significant implications for the public health aspects of the transplantation of oysters. Since particular areas (Florida and Louisiana) account for most of the deaths due to *V. vulnificus*, it is not unreasonable to suspect more virulent strains exist in certain areas. Therefore, transplantation at any time is of concern. We now know that the individual *V. vulnificus* cells in an oyster do not die during the winter, but simply enter the VBNC state. Since the cells bind so tightly to the tissue of the oyster, it is quite possible that winter harvested and apparently "safe" oysters containing particularly virulent VBNC cells will grow out the virulent strains months later in warm weather. Therefore, to move these oysters from one area to another, even from one region to another, and to leave them for months may not decrease their potential for causing disease. If this practice continues, we may well see deaths attributable to oysters from previously believed-innocuous areas.

Other Vibrios

Numerous other species of the genus *Vibrio*, which occur naturally in the estuarine environment, cause some shellfish-borne gastroenteritis. These species include *V. alginolyticus, V. fluvialis, V. furnissi, V. hollisae,* and *V. mimicus*. As mentioned above, *V. mimicus* is very similar to *V. cholerae*, and can elaborate a toxin identical to *V. cholerae*. In addition, *V. mimicus*-like *V. cholerae* can live in freshwater. The other species of *Vibrio* mentioned here require some salt, probably prefer less than full-strength seawater, and probably exist in higher concentrations in warmer water. *V. furnissi* used to be classified as a biovar of *V. fluvialis*, and most of the shellfish-borne illnesses attributed to these two species may be due to *V. furnissi* (Hackney et al., 1992). One death in 1990 is attributed to *V. fluvialis* from the consumption of shellfish (Rippey, 1994). *V. metschnikovii* has not yet been associated with shellfish-borne disease, but since it is widely distributed in estuaries and has been shown to cause diarrhea (Roderick, 1991), it is a potential candidate. In their review of Rippey's (1994) data, Wittman and Flick (1995) indicate that 95 percent of the deaths attributable to shellfish-borne disease are attributable to the non-cholera vibrios (including *V. vulnificus*).

Pleisomonas shigelloides

Pleisomonas shigelloides is a member of the family Vibrionacae, and it occurs seasonally during the warmer months. Unlike most vibrios, it primarily occurs in freshwater, but does occur in marine waters. Symptoms of the disease resulting from consumption of food include diarrhea, abdominal pain, nausea, chills, fever, headache, and vomiting. The onset time is normally 24 to 50 hours, and usually lasts 24 to 48 hours. They survive well in shell oysters held at refrigeration temperatures (Hackney et al., 1992). There have been 23 shellfish-borne illnesses attributed to *P. shigelloides* over the years 1984 to 1993 (Rippey, 1994) and none from 1994 to 1999 (Glatzer, 1999). However, Roderick (1991) indicates that it is not a proven cause of gastroenteritis, since the disease has not been induced in human volunteers or provided a positive reaction in animal tests.

Aeromonas hydrophila

Aeromonas hydrophila is not accepted as an enteric pathogen, though it has been associated with many cases of diarrhea. It may well be an opportunistic pathogen, but the epidemiological data is conflicting (Roderick, 1991). In the past 15 years, two shellfish-borne outbreaks were attributed to *A. hydrophila*: one involved six persons and the other seven persons in 1984 (Glatzer, 1999; Rippey, 1994). In 1982 a shellfish-borne outbreak involving 472 persons was listed as being of an unknown cause (Rippey, 1994), though *A. hydrophila* was isolated from frozen samples and no other diarrhetic shellfish poison nor pathogens were found. *A. hydrophila* is widely distributed in freshwater and saltwater.

Clostridium perfringens

In their review of *Clostridium perfringens*, Hackney et al. (1992) indicate that although this bacteria is widely distributed in the environment and has often been isolated from shellfish, it is of little importance as a seafood-borne pathogen. Approximately 10^6 to 5×10^8 are needed to cause illness, and levels accumulated by shellfish from the environment are apparently less than that. *C. perfringens* are readily depurated from shellfish, though their spores survive well in the environment, and for these reasons have been proposed as an excellent indicator for depuration systems (Hackney et al., 1992). Vegetative cells do not survive long in the environment, and since they are much more likely to be present in elevated concentrations in fresh sewage, they have been suggested as an indicator (Kator and Rhodes, 1991). There are no recorded cases of shellfish- borne sickness attributed to *C. perfringens* in the United States (Glatzer, 1999; Rippey, 1994).

Bacillus cereus

Bacillus cereus is a common food-borne pathogen in Europe, but it is rarely reported in the United States. The bacteria produce two types of toxins: one that is heat stable and induces nausea and vomiting, while the other type (heat-labile) causes colic and diarrhea. Levels of $>10^5$ or $>10^6$ are needed in food to induce sickness. The onset time for vomiting is one to six hours, and is six to 24 hours for cases of diarrhea. Sickness usually lasts no more than 24 hours. It is a ubiquitous soil organism. Problems with cooked food generally occur after it has been held at ambient temperatures and thus allows proliferation of the bacteria (Benenson, 1995; Hackney et al., 1992). No cases of shellfish-borne illness attributed to *B. cereus* are listed by Rippey (1994) or by Glatzer (1999); however, Hackney et al. (1992) indicate a concern for the diarrheal-type sickness attributable to bacterial growth when shellstock are held out of water at warm temperatures.

Klebsiella pneumoniae

Klebsiella pneumoniae has been implicated as an enterotoxigenic agent causing diarrhea in people, though apparently not often; only some strains are potent histamine producers (Stratton and Taylor, 1991). In their evaluation of the entero-pathogenicity of environmental strains of *K. pneumoniae* from Louisiana oysters, Boutin et al. (1986) determined that even relatively high levels of the organism (3×10^3 CFU/100g) in oysters would be safe for noncompromised persons. No cases of shellfish-borne illness due to *K. pneumoniae* have been reported (Glatzer, 1999; Rippey, 1994).

Listeria monocytogenes

Listeria monocytogenes occurs in estuarine and marine waters primarily as runoff from the land, as a result of sewage discharges. However, it is so widespread in waters that one should not link it to pollution in a traditional sense, because most estuarine waters in the United States contain the bacteria (Motes, 1991). *L. monocytogenes* rarely occurs in raw shellfish; this may either be because the numbers of the bacteria in estuarine waters are fairly low (Embarek, 1994), or because of antimicrobial action by the shellfish, or because the shellfish are a poor substrate for growth.

Listeriosis is usually manifested as meningoencephalitis and/or septicemia, though milder flu-like symptoms and asymptomatic cases occur. Fetuses and newborn infants are highly susceptible; immunocompromised persons, the elderly, and pregnant women are at a higher risk than the general public. The infection is rarely diagnosed in the United States. The disease occurs extremely rarely worldwide as a result of seafood; Embarek (1994) reports three confirmed sporadic cases and two suspected outbreaks. At least one of the suspected outbreaks was due to smoked products (mussels) (Lennon et al., 1984) and is not conclusive (Fuchs and Surendran, 1989). The author could not determine whether any of the cases were linked to raw shellfish.

TOXIC SUBSTANCES

The concentration of lipids in a particular bivalve varies, due not only to species, but also to season, state of the spawning cycle, etc.; this concentration greatly affects uptake of lipophilic substances. The size of bivalves also affects uptake, because smaller individuals pump more water per body weight and may tend to concentrate to higher levels (Stronkhorst, 1992; Roberts et al., 1979). The concentration of the pollutant in the water, sediment, and suspended particulate matter all likewise affect the ultimate body burden of a bivalve.

Pesticides

As a general rule in the United States, pesticides in shellfish from growing waters do not pose a significant public health risk for several reasons. Many of the persistent pesticides such as chlordane and DDT have been banned. Also, although pesticides are widely used, a more educated public is keeping most of the pesticides on the land; and when they do run off into the waterways, they are quickly diluted by the typically large volume of water surrounding shellfish beds. When a pesticide is a problem, it is usually in a small local area and often of short duration.

As indicated in the 1993 Revision of Part II of the NSSP, the FDA will take action to exclude shellfish in interstate markets if they exceed the following concentrations: 0.1 parts per million (ppm), mirex; 0.3 ppm, aldrin + dieldrin, chlordane, chlordecone (Kepone[R]), heptachlor + heptachlor epoxide; 5.0 ppm, toxaphene, DDT+DDE+TDE. These are all long-lasting, chlorinated hydrocarbon pesticides that were of much concern in the 1960s and 1970s. Their use has been either banned or strictly curtailed in the United States, and now they are usually found only in low background concentrations well below regulation levels, though the potential for hot spots exists.

Organochlorine pesticides such as these are rather insoluble in water; as a whole they tend to adsorb onto clay particles and can be fairly rapidly buried in estuarine environments. In the James River of Virginia, Allied Chemical released an estimated 9,070-18,140 kg of Kepone[R] from the late 1960s until mid-1975. In the downstream estuarine portions of this four-mile-wide river, Kepone[R] concentrations reached a maximum of 0.5 ppm (wet weight) in oysters with an average value of approximately 0.2 ppm in 1976; by 1987 they averaged well below 0.02 ppm, though one outlier of 0.37 ppm did appear [Virginia Department of Health (VDH) data].

DDT and its breakdown products of DDD and DDE, along with chlordane, are usually the dominant pesticides present in shellfish. United States studies on the West Coast (Phillips and Spies, 1988), Gulf Coast (Sericano et al., 1993; 1990), South Carolina (Markus and Renfrow, 1990), Chesapeake Bay (VDH data, Murphy, 1990), and Long Island Sound (Robertson et al., 1991) all indicate concentrations of DDT and its metabolites, chlordane, and the other persistent organochlorine pesticides to be well below NSSP requirements. Chlorinated hydrocarbons in the Western Mediterranean also appear to be decreasing (Sole et al., 1994).

Chlordane is a resistant chlorinated hydrocarbon pesticide mixture of more than 140 compounds. The primary components are *a*-chlordane, *g*-chlordane, heptachlor, and *trans*-nonachlor. All sales and uses were banned in the United States in 1988; however, due to its widespread use in the past, it is ubiquitous in the marine environment, with higher concentrations near urban areas. Analysis of oysters in the Gulf of Mexico indicates that the dominant components bioconcentrated were *a*-chlordane and *trans*-nonachlor, along with lesser concentrations of heptachlor and its epoxide. The five-year trend (ending 1991) in the Gulf of Mexico oysters indicates little change in total chlor-dane concentrations, though heptachlor and heptachlor epoxide have been increasing since 1987 (Sericano et al., 1993).

While in the environment, chlorinated pesticides break down slowly into various metabolites. DDT has an estimated environmental half-life of 20 years (Woodwell et al., 1971); DDE, which is its major hydrolysis product, appears even more persistent than that (Wolfe et al., 1977). Aldrin may be degraded into dieldrin, and heptachlor is enzymatically changed to its epoxide by organisms.

Carbamate and organophosphorus pesticides are in wide use today. These pesticides are generally shorter-lived, and have not been found to accumulate to high concentrations in shellfish to date, though the data is limited (Markus and Renfrow, 1990; Hale, 1989). Many types of bacteria and higher-order animals, including the brackish water clam (*Rangia cuneata*) and the hard clam (*Mercenaria mercenaria*), can hydrolyze organophosphate pesticides to nontoxic forms (Landis, 1991).

Polychlorinated Biphenyls

Polychlorinated biphenyls (PCBs) represent a mixture of 209 possible congeners — that is, 209 ways that chlorine atoms can be attached to the biphenyl molecules. Even if world production were to stop, the marine environment would continue to receive inputs from numerous sources (Duursma, 1989). Concentrations of PCBs in shellfish worldwide seem to be either not decreasing (Stronkhorst, 1992; Phillips and Spies, 1988) or reaching a new slower rate of decrease (Sole et al., 1994; Picer and Picer, 1990). These data should be sobering because there is considerable concern about some of the specific congeners present in PCB mixtures.

PCB congeners of particular concern are called non-*ortho*-substituted congeners, of which there are 20 types, and, to a lesser extent, the mono-*ortho*-substituted congeners. With this particular type of chlorine substitution, the molecules can attain a planar structure, which makes them similar to the highly toxic dibenzo-*p*-dioxins and dibenzofurans (de Boer et al., 1993; Hansen, 1987; McKinney et al., 1985). The PCBs of particular concern are the ones with four, five, or six chlorines in the non-*ortho* positions because they mimic 2,3,7,8-tetrachlorodibenzo-*p*-dioxin (TCDD) and 2,3,7,8-tetrachlorodibenzofuran (TCDF) (Safe, 1990; Goldstein and Safe, 1989). These planar PCBs are slower to be taken up by shellfish; however, once taken up, they are slower to be depurated than the other PCBs of the same number of chlorine atom substitutions (Sericano et al., 1992; Kannan et al., 1989).

There is some indication in the literature that PCBs may not accumulate in marine animals in the classic food chain concept, but rather as a direct partitioning among the various compartments, such as water, sediment, particulate matter, and organisms (Mackay, 1991; Duursma, 1989). Mussels have been shown to comply with the concept that they take up PCBs mainly by equilibrium partitioning between PCBs in the surrounding water and the lipid pool of the mussel. As such, they are good indicators of the concentrations of PCBs in water. The uptake of these PCBs is thought to occur primarily through the gills and across the body surfaces, with ingestion of additional amounts in food playing a significantly lesser role. PCBs with fewer chlorine atoms are more hydrophilic, and, as such, reach equilibrium in uptake and depuration quicker in these mussels than the more highly chlorinated and thus more hydrophobic congeners (Bergen et al., 1993; Tanabe et al., 1987). Bivalves tend to accumulate the lesser chlorinated PCBs, especially the 4-, 5-, and 6-chlorine substitutions (Pruell et al., 1993; Sericano et al., 1990). Oysters tend to concentrate PCBs 10 to 20 times that of their surrounding sediment concentrations (Wade et al., 1988). Also, bivalves tend not to metabolize PCBs (Pruell et al., 1993; Sericano et al., 1990; Boon et al., 1989).

The NSSP requires that total PCBs in shellfish be less than 2.0 ppm on a wet-weight basis. In most of the United States, shellfish are substantially below the NSSP limit (Robertson et al., 1991; Murphy, 1990; Sericano et al., 1990; Phillips and Spies, 1988; Steimle et al., 1985; VDH data). However, there are some local areas of considerably higher concentrations.

Dioxins and Furans

Dioxins and furans represent classes of chemical compounds that can be chlorinated at numerous sites on different molecules producing a suite of congeners as in PCBs. These chemicals are regarded as among the most toxic chemicals in the marine environment (Tanabe, 1988b), and dioxins are believed to be about 10 times as toxic as furans (Waldichuk, 1990). Two of these congeners of particular concern due to their toxic effects to mammals are 2,3,7,8-TCDD and 2,3,7,8-TCDF. The FDA has not established advisory limits for dioxins and furans in seafood.

These compounds are released into the estuarine environment by bleached kraft mill effluents, from lumber treated with pentachlorophenol, from PCBs used as heat transfer fluids, and from the combustion of various types of waste. Though the number of isomers of polychlorinated dibenzo-*p*-dioxins and polychlorinated dibenzofurans are lower in pulp and paper mill effluents, the most toxic isomers 2,3,7,8-TCDD and 2,3,7,8-TCDF are a prominent fraction present (Tanabe, 1988b). There is little research concerning the uptake of these TCDDs or TCDFs by bivalves from the estuarine environment, but two studies (Pruell et al., 1993; Petreas et al., 1992) indicate that clams tend not to accumulate these toxins very much over that in bottom sediments, in contrast to PCBs. Clams apparently reach equilibrium concentrations with their environment within two months. Mollusks may be able to survive in TCDD- and TCDF- contaminated environments better than many other animals since they may not contain a biological receptor that causes the genetic expression of toxic effects (Hahn et al., 1993).

Polyaromatic Hydrocarbons

Hydrocarbons are organic (i.e., containing carbon atoms) molecules, such as oils. They are classified as to such structures as chain types (e.g., alkanes) and ring structures (e.g., naphthenes, aromatics). The simplest of these common ring structures is the six-carbon benzene ring. Benzene is one of the few documented human carcinogens; however, since it is so light, it evaporates from spilled gasoline, etc., and does not accumulate in the estuarine environment. Polyaromatic hydrocarbons (PAHs) basically are a series of these six-carbon benzene rings linked together in the molecule. Other names for PAH are polynuclear aromatics (PNAs), polycyclic aromatic compounds (PACs), etc. Though PAHs of two rings can accumulate in the benthos (Farrington et al., 1986), typically molecular structures of three to five rings are the types that are heavy enough to remain in the aquatic system (Farrington and Quinn, 1973) and are most toxic. PAHs are of particular concern because many of them have the potential to be carcinogenic or mutagenic. Sixteen types of PAHs are listed as priority pollutants by the World Health Organization (WHO) and the United States Environmental Protection Agency (EPA). The FDA has no consumption standards for seafood containing PAHs at this time.

PAHs are ubiquitous in the environment, and are due to the incomplete combustion of organic matter such as fossil fuels and wood and the use of petroleum products (Wade et al., 1988; Neff et al., 1979). PAHs are introduced into the estuary via river transport, nonpoint source runoff, industrial discharges, atmospheric deposition (which can be from remote sources), and spills. The usual low-level background concentrations (Steimle et al., 1985) appear to be due primarily to burning, whereas locally elevated concentrations tend to be due to spills, sewage treatment plant discharges, or urban runoff (Helmstetter and Alden, 1994; Jackson et al., 1992; Pereira et al., 1992; Wade et al., 1988; Hoffman et al., 1984). PAHs from urban highway and industrial areas are a significant source to the estuary; they are similar to PAHs in atmospheric fallout, and may primarily be due to fallout. PAHs from municipal STFs are not weathered like those from atmospheric deposition, and as such contain more, lighter-molecular-weight PAHs such as naphthalenes (Hoffman et al., 1984). The ratio of phenanthrene to anthracene is helpful in distinguishing between petroleum and nonpetroleum sources of PAHs in shellfish (Wade et al., 1988).

When released into the environment, PAHs partition into the various "compartments," both biotic and abiotic, due to their physicochemical properties such as vapor pressure, water solubility, and sediment/water partitioning coefficient (Porte and Albaiges, 1993). Microbial degradation and evaporation may remove most of the light-weight PAHs from the water column, whereas photooxidation and sedimentation may remove most of the heavier PAHs from the water column (Wade et al., 1988). Since PAHs are very hydrophobic, upon entering the estuary they tend to attach quickly to suspended particulate matter and sediments, which can then settle out on and be incorporated into the benthos. Release rates of PAHs from bottom sediments are inversely related to the organic carbon content, and directly correlated to the water solubility of the individual PAH (Helmstetter and Alden, 1994; Karickhoff et al., 1979).

Originally, mollusks were thought to be unable to metabolize PAHs. However, cytochrome P-450 and associated mixed-function oxidase enzyme systems capable of metabolizing PAHs have been found to operate in many mollusks, though at a slower rate than in fish and crustaceans (Neff et al., 1976), and much slower than in vertebrates. When metabolized, PAHs tend to be changed into more polar compounds which then mix better with water and are more easily eliminated by the organism (Hellou et al., 1993). There is some evidence that oysters may be able to metabolize benzo(a)pyrene into mutagenic or carcinogenic derivatives (Pittinger et al., 1987).

Oysters serve as excellent sentinel organisms for PAH pollution since they do not move, and they readily accumulate and release PAHs in close correlation with the amount in their environment. When transplanted into a more highly contaminated area, or when moved from a contaminated area to a cleaner area, oysters reach their new equilibrium within 14 days (Pittinger et al., 1987) to about a month (Jackson et al., 1994; Bender et al., 1987). Oysters have been shown to accumulate PAHs from around marinas, and areas of creosote pollution (Elder and Dresler, 1988; Bender et al., 1987; Marcus and Stokes, 1985). Oysters and the brackish water clam *Rangia cuneata* do not tend to highly concentrate PAHs over that found in the adjacent sediments, but concentrate from less than one to approximately four times that of the sediment (in contrast to organochlorine pesticides and PCBs) (Pendoley, 1992; Wade et al., 1988; Bender et al., 1987, 1986). However, particular PAHs can be significantly bioaccumulated (Bender et al., 1987). Blue mussels (*Mytilus edulis*) have been shown to rapidly take up PAHs from contaminated sediments, and to depurate them with half-lives ranging from 12 to 30 days depending on the type of aromatic present (Pruell et al., 1986). However, hard clams (*Mercenaria mercenaria*) do not appear to depurate certain PAHs even after 45 days or more (Tanacredi and Cardenas, 1991; Boehm and Quinn, 1976). Mollusks tend to have the highest concentrations of PAHs in the visceral mass, which represents the net balance between the uptake and elimination of the contaminants. The lowest levels of PAHs in mollusks occur in the muscular tissue, which reflects only uptake.

Variations in concentrations of PAHs among individuals of a molluscan species in nearby areas occur due to many factors. Smaller organisms tend to accumulate to higher concentrations, possibly due to the increased pumping of water and increased gill size in relation to their body. Organisms with large numbers of gametes may attain higher concentrations. Other factors include differences in pumping rate, filtration rate, lipid content, and the concentration of the PAHs in the sediment in the immediate vicinity of the organism (Hellou et al., 1993; Bender et al., 1986).

Heavy Metals

Generally, heavy metal pollution in United States coastal areas is restricted to local areas, and most United States shellfish are not above FDA advisory levels (Morse et al., 1993; Robertson et al., 1991; Turgeon et al., 1991; Lauenstein et al., 1990; Murphy, 1990; Farrington et al., 1983; Croonenberghs, 1974; VDH data). The primary source of heavy metals to marine waters is runoff. Due to the low solubility of metals in water, metals tend to attach to sediments and settle out locally around areas of high input. Sediments tend to concentrate heavy metals by three to five orders of magnitude above amounts in water.

Even though local areas may receive elevated inputs of heavy metals, shellfish in the area may not contain significantly increased amounts due to the complexity of factors that come into play. As a general rule, shellfish do reflect concentrations of metals in their environment, but these concentrations are the net result of many biological and chemical processes.

Most metals exhibit several pathways for uptake by bivalves. Metals will partition among dissolved phases (in several oxidation states) and phases associated with particles such as clays. They are often sequestered by either organic ligands (chelated) or precipitated out of the water by iron sulfides, again in several oxidation states. In addition, some metals are alkylated by bacteria into more organically soluble forms (e.g., methyl mercury) that can greatly increase the potential for uptake by organisms.

The biology of the bivalve also greatly affects its body burden of heavy metals. As is true for many other contaminants, smaller animals of a species tend to concentrate metals to higher levels. Different species concentrate metals to varying levels; as a general rule, oysters, with their high pumping rate, have higher concentrations of metals than other bivalves. Deposit-feeding bivalves (e.g., *Mya arenaria, Macoma balthica*) ingest sediment and therefore are exposed to different concentrations and forms of metals than suspension feeders (*Crassostrea virginica, Mytilus edulis*).

Following the lead of Bryan and Langston (1992) in their detailed review, the following brief summation of the different heavy metals in relation to shellfish will center on those generally perceived to be of concern in the estuarine environment. They are listed in order of their appearance in the periodic table for ease of comparison with their behavior (bioavailability, toxicity) and chemical similarity: copper (Cu), silver (Ag) (Group IB); zinc (Zn), cadmium (Cd), mercury (Hg) (Group IIB); tin (Sn), lead (Pb) (Group IVA); arsenic (As) (Group VA); selenium (Se) (Group VIA); chromium (Cr) (Group VIB).

Copper

Copper (Cu) is generally considered to be of little concern to humans in concentrations found in shellfish. However, copper is toxic to marine organisms and as such is widely used in antifouling paint on boat hulls. As such, one would expect to find locally elevated concentrations around marinas. Copper dissolved in water as the Cu^{2+} ion is the most available and toxic form, and oysters readily accumulate it in this form (Zamuda et al., 1985; Zamuda and Sunda, 1982). However, the dissolved ion is easily chelated by organic compounds and made less available for uptake. Also, the ionic form seems to compete with calcium (Ca) and magnesium (Mg) in seawater, such that less is taken up in higher salinity waters by the soft clam (*Mya arenaria*) (Wright and Zamuda, 1987). The eastern oyster (*Crassostrea virginica*) accumulates significantly higher Cu concentrations in lower salinity water than in higher salinity (Phelps et al., 1985; Huggett et al., 1973; VDH data), and tends to concentrate to higher levels in smaller organisms (Phelps et al., 1985). Oysters have been shown to be capable of taking up Cu incorporated into algae. In sediments, Cu is associated with iron oxides and humic acids (Bryan et al., 1992). Copper concentrations in bivalves may be increasing slightly in the United States (Stephenson and Leonard, 1994; Turgeon and O'Connor, 1991). The FDA has not established advisory limits for Cu in shellfish.

Silver

Silver (Ag) probably occurs primarily in seawater as $AgCl_2$, though the most available form to biota may be as the neutral monochloro-complex AgCl. In remote estuarine areas, Ag in oxidized sediments is associated with humic acids which seem to interfere with uptake of the metal by the biota. In sewage effluents, the organics present also bind the Ag, but extracellular polymers released by bacteria may increase the availability for uptake. Oysters seem to take up silver primarily from water since they do not ingest significant quantities of sediment, and since once incorporated into algae it is too tightly bound by organics for incorporation into the oyster. The deposit-feeding bivalve *Macoma balthica* is reported to accumulate Ag from sediments. In solution Ag is regarded as one of the most toxic of heavy metals (Bryan and Langston, 1992; Abbe and Sanders, 1990). The FDA has not set a specific advisory level for Ag in shellfish.

Zinc

Zinc (Zn) is most available to biota as the Zn^{2+} ion, and this form is often the most abundant of the dissolved species. Bryan and Langston (1992) conclude that loosely-bound Zn in sediments and dissolved Zn in interstitial water (among sediment grains) and overlying waters are important sources to biota. Oysters do not regulate the uptake of Zn, and depuration is minimal, although the mussel *Mytilus edulis*, with its more active kidney, strongly regulates Zn concentrations. The short-term uptake of Zn by the oyster is probably a function of its pumping rate, and its total body burden increases over its lifetime, although the rate of increase decreases as the oyster grows (Mo and Neilson, 1993). The levels of Zn in oysters is inversely related to salinity, such that the highest concentrations are found in oysters from the lowest salinity, unless local inputs exist (Phelps et al., 1985; Huggett et al., 1973; VDH data). Zinc is not regarded as particularly toxic, and the FDA has not set a specific advisory level for Zn in shellfish.

Cadmium

Cadmium (Cd) is most abundant in seawater as chloride complexes, but the most bioavailable form is as the Cd^{2+} ion, which increases with decreasing salinity (Bryan and Langston, 1992). Oysters in lower salinity waters concentrate Cd to higher levels (Phelps et al., 1985; Huggett et al., 1973; VDH data), which may be in response to the varying concentrations of the Cd^{2+} ion with salinity. The most important source of Cd to biota associated with sediment may be dissolved Cd that has dissociated from the sediment, rather than the sediment itself. The strength with which Cd combines to organics in phytoplankton may prevent its incorporation into the tissue of feeding bivalves as is true for Ag (Abbe and Sanders, 1990). Metallothionein-like Cd-binding proteins are thought to offer protection to biota from the toxic effects of the metal by sequestering the toxic ion and, when present in a species, the concentrations of Cd are often higher than for species without these proteins (Bryan and Langston, 1992).

Cadmium is not an essential element for humans, and high chronic intake can lead to kidney dysfunction. However, Cd is widely distributed in foods, and low levels are tolerated by the body. FDA levels of concern for total cadmium in molluscan shellfish on a mean wet-weight basis are 6 µg/g for males and females of all ages two years and older, and 5µg/g for males and females 18 to 44 years (Adams et al., 1993a).

Mercury

Mercury (Hg) is rapidly adsorbed onto suspended particulates once released into the water column, and tends to attach to particulate carbon compounds. Thus, in estuaries hot spots are created as a result of the flocculation and sedimentation patterns before dilution can occur. Due to the strong binding forces on particulate carbon, little Hg is released back into the water column once trapped in the sediment. As a result, once an area is contaminated with Hg, high concentrations are apt to remain for a considerable time. Decomposing rooted plant material such as *Spartina* can have Hg concentrations as high as 30 times that of phytoplankton, and so detritus-based food chains can be exposed to a higher flux of Hg than phytoplanktons (Bryan and Langston, 1992). Due to these high concentrations of Hg in detritus, on other organic particles, and on sediment, the main source of Hg to benthic invertebrates is from particulate matter rather than the relatively bioavailable dissolved forms (Bryan and Langston, 1992; King and Davies, 1987).

Inorganic Hg is slowly converted to methyl mercury (CH_3Hg^+), primarily in sediments; the degree to which this occurs is modified by several environmental factors such as temperature, salinity, pH, and redox/sulfide levels. Therefore, concentrations of methyl/mercury do not necessarily reflect total concentrations of Hg in sediment, though inorganic Hg is normally the dominant form. However, due to the increased lipid solubility and long biological half-life of the compound, as well as the increased longevity of top predators, methyl/mercury provides one of the few examples of biomagnification in food chains. This is unfortunate, since methyl/mercury is by far the most toxic form of the metal. While benthic invertebrates may or may not have dominant methyl/mercury concentrations, the Hg contained in the muscle tissue of fish and birds is predominantly methyl/mercury (Bryan and Langston, 1992). Field studies have shown the total Hg concentration of oysters to rapidly increase and decrease when moved into and out of Hg-contaminated areas, though equilibrium concentrations were not reached after 51 days (Palmer et al., 1993).

Methyl mercury is quite toxic to humans, causing severe nerve damage.The worst case of mercury poisoning due to shellfish occurred in Minimata Bay, Japan, the symptoms of which were named "Minimata disease." There is no evidence that mercury pollution is increasing in the United States, and current hot spots are well-known. The NSSP restricts the sale of shellfish with a methyl mercury concentration of more than 1.0 ppm wet weight.

Tin

Tin (Sn) is highly dependent on speciation for its bioavailability. Inorganic Sn is most commonly present as $SnO(OH_3)$ in seawater, and casserite

(SnO_2) is the most common form in sediment. Upon release to estuaries, inorganic Sn is rapidly scavenged by sediments and release to the water is minimal. Sediment-bound Sn and SnO_2 are not particularly available to the biota. As such, inorganic Sn is normally not a threat to the biota nor to man. However, organic Sn can be considerably more toxic. Tin can be methylated by microorganisms in sediment, and may be the major source of mono-, di-, and tri-methyltin (MMT, DMT, and TMT) in estuaries. Humic and fulvic acids may also produce MMT abiotically. Due to the strong adsorption of Sn to sediments, methyltin production may be limited, as it is usually found only at very low concentrations (Bryan and Langston, 1992). The FDA has not established advisory limits for tin in seafood.

Tributyltin (TBT) is the most toxic form of organic tin, and has been manufactured for use as an antifouling component in boat hull paints. It is particularly toxic to *Crassostrea gigas* and to the dogwhelk. Its breakdown products, dibutyl-(DBT), monobutyl-(MBT) and inorganic Sn are progressively less toxic. The presence of TBT in estuarine waters is related to boating activities; with the restrictions now placed upon its use as a boat paint, a decrease in the estuarine environment is expected. There is evidence that this is slowly occurring in the United States (Uhler et al., 1993; Hall et al., 1992; Wuertz et al., 1991) and in the United Kingdom (Evans et al., 1994; Dowson et al., 1993). Tributyltin has a short half-life in estuarine water of four to 14 days; microalgae may play a major role in its degradation (Lee et al., 1989). However, TBT's half-life in sediment is much longer, on the order of 20 weeks (Wade et al., 1990), and sediments may provide a long-term source of TBT to the biota (Unger et al., 1988; Huggett et al., 1986).

Though the FDA has no published safety limits for TBT in shellfish, Schweinfurth and Gunzel (1987) have suggested a 1.9 μg/g TBT limit. Wade et al. (1990, 1988) collected oysters and mussels from 36 sites on all three United States coasts and Hawaii, and all samples were below this concentration. All but one of the samples contained TBT, and the average concentration of total butyltin was 18 times that of the sediment. While mollusks seem to have a limited ability to metabolize TBT into less toxic forms, crustacea, fish, and mammals can, and food chain magnification does not appear to occur at these trophic levels (Bryan and Langston, 1992).

Lead

Lead (Pb) occurs dissolved in water primarily as $PbCO_3$ and $PbOH^+$, though the most bioavailable inorganic form is probably Pb^{2+}. However, most of the inorganic Pb in estuarine water is associated with particles, due to its strong affinity for them (Morse et al., 1993; Bryan and Langston, 1992). Tri- and dialkyl Pb compounds are not strongly associated with particles, and can form the highest concentrations of dissolved Pb species in estuarine waters. While tetra-, tri-, and diethyl Pb in the estuary are usually associated with leaded gasoline, there is evidence that methylated forms are produced either chemically or biologically in sediments. Apparently, environmental concentrations of inorganic lead are not particularly toxic to benthic invertebrates, but the organic forms are certainly more toxic. It is possible that high levels of inorganic Pb in sediments can give rise to environmentally significant concentrations of trimethyl Pb (Bryan and Langston, 1992). Fortunately, the concentrations of total Pb in mussels and oysters from around the United States is decreasing, as a result of the decreased use of tetraethyl Pb in gasoline (Stephenson and Leonard, 1994; Turgeon et al., 1991; Lauenstein et al., 1990).

Lead impairs the nervous system, and is generally of concern through elevated chronic exposure, since it is not easily incorporated into the body. Adults absorb five to fifteen percent of ingested lead, yet retain less than 5 percent; children absorb about 50 percent. FDA levels of concern for total lead in molluscan shellfish on a mean wet-weight basis are 1.5 μg/g for males and females two to five years; 2.1 μg/g for pregnant women; and 6.3μg/g for males and females 18 to 44 years (Adams et al., 1993b).

Arsenic

Arsenic (As) is dissolved in aerobic estuarine water primarily as arsenate (As^{5+}); however, under reducing conditions in interstitial water, it will predominate as arsenite (As^{3+}) which may be more toxic. Small amounts of monomethylarsonate and dimethylarsinate are also often present. The primary source of As to estuaries is usually via rivers, due to the natural weathering of rocks and ores. Higher contributions can come from mining and smelting, but its use as a wood preservative and in various manufactured goods apparently does not contribute significant percentages of As to estuaries (Bryan and Langston, 1992). Organisms growing directly on treated wood, however, can pick up elevated levels of leached As, and concentrations of treated wood may locally influence areas (Weis et al., 1993). Moderately elevated concentrations of As may occur in sediments around heavily industrialized areas due to discharges, but sewage discharges are not significant inputs of As (Bryan and Langston, 1992).

At the interface of fresh- and saltwater, arsenic can adsorb onto precipitating iron oxyhydroxides and settle out into the sediment. In settling zones the As in the sediment is continually buried by new sediment. Since it is easily reduced in anoxic sediments into the soluble As^{3+} form, it tends to move upward to the surface of the sediments. If the surface layers of the sediment are oxic, then the As tends to precipitate out in this area; thus the superficial oxidized layer of sediment tends to have a higher concentration of As than that of deeper layers. However, should the bottom sediments be anoxic to the surface, then As^{3+} will tend to escape into the overlying water. Tubes and holes in the bottom resulting from benthic invertebrates can greatly increase the escape of ionic As from sediments (Bryan and Langston, 1992; Riedel et al., 1987).

While most of the arsenic in sediments is present as inorganic As, most of the arsenic present in marine plants and animals exists as organic As, which is the least toxic form. It appears that most estuarine animals do not obtain their As primarily as dissolved inorganic As, but instead incorporate it into their bodies from food. Estuarine plants such as phytoplankton, macroalgae, and rooted higher plants take up dissolved ionic As and incorporate it into numerous organoarsenic compounds, possibly since it is chemically similar to phosphorous. Some of the compounds made by the plants are arsenosugars, and this may be the primary form incorporated into the bodies of animals. Most of the As in animals is present as complex organic compounds, which are relatively nontoxic to animals since they are primarily excreted when eaten. As a result, arsenic does not biomagnify in the food chain, and is *usually* relatively nontoxic and easily excreted when consumed in seafood (Bryan and Langston, 1992; Sanders et al., 1989; Phillips and Depledge, 1985). FDA levels of concern for total arsenic in molluscan shellfish on a mean wet weight basis are 130 µg/g for males and females of all ages two years and older, and 110 µg/g for males and females of two to five years and of 18 to 44 years (Adams et al., 1993c).

Selenium

Selenium (Se) concentrations in estuarine and coastal waters are generally fairly constant, even in watersheds with naturally high Se concentrations in the soil or ores. Urban areas do not usually have greatly increased concentrations unless they receive refinery discharges or Se- containing industrial effluents. Elevated concentrations usually only occur in the immediate vicinity of these discharges. Dissolved inorganic Se usually exists in coastal areas as selenite (Se^{4+}) and selenate (Se^{6+}), though the largest percentages may be organically bound. Se^{4+} tends to be scavenged by iron oxyhydroxides in the fresh- and saltwater mixing zone, and is also preferentially accumulated by algae and marine invertebrates over Se^{6+}. Most of the Se in estuaries occurs in the sediment as particulate elemental and selenide forms, and most of the Se in estuarine organisms is organically bound (Bryan and Langston, 1992).

Algae primarily take up selenium in the ionic form from the water, but most estuarine animals receive their major intake from food. Marine animals can take up some dissolved Se, and sediments also appear to be important for some species such as the deposit feeding bivalve *Macoma balthica* (Bryan and Langston, 1992; Johns et al., 1988). Though there is some potential for biomagnification in estuarine food chains, this is not a problem in most areas. Since Se is considered potentially dangerous to higher organisms at levels only slightly above those needed in the diet, care is needed in contaminated areas. The FDA has not established an advisory limit for Se in seafood.

Chromium

Chromium (Cr) is dissolved in coastal waters primarily as Cr^{6+}, Cr^{3+}, and sometimes as an organically-bound fraction. Cr^{3+} is far less available and toxic than Cr^{6+}. Cr^{3+} tends to be scavenged by iron oxides or by particle surfaces. Elevated concentrations in the estuarine environment are due to industrial discharges, effluents from waste dumps, etc. It is unclear how Cr in sediment is taken up by benthic organisms.

Cr^{3+} is a necessary micronutrient, and no specific adverse human effects have been associated with it. Cr^{6+} may cause adverse effects in the liver and kidney. FDA levels of concern for Cr in molluscan shellfish are 20 µg/g, wet-weight basis, for males and females of all ages two years and older, and 17 µg/g, wet weight basis, for males and females 18 to 44 years (Adams et al., 1993d).

HARMFUL ALGAE BLOOMS

Red Tides

Toxins of concern to shellfish consumers are those that are concentrated by mollusks to such a degree that their consumption poses a public health threat. The major phycotoxins of human health significance are paralytic shellfish poison (PSP, saxitoxins), neurotoxic shellfish poison (NSP, brevetoxins), amnesic shellfish poison (ASP, domoic acid), and diarrhetic shellfish poison (DSP, okadiac acid). Most of these toxins are produced by dinoflagellates, though the primary producers of ASP are diatoms.

The term "red tide" has traditionally been used to describe massive blooms of dinoflagellates, since they are often reddish brown in color, and often grow to such densities that they darken the water. Typical concentrations of phytoplankton cells during blooms are in the range of 1 to 3 x 10^5 cells/l (Brycelj et al., 1990). The term is now being expanded in the scientific community to generally include all nuisance algal blooms. Most dinoflagellates do not contain human or fish phycotoxins, so most red tides are not toxic; however, when the bloom dies, it can greatly reduce the dissolved oxygen in the water due to their large numbers. Thus, these dinoflagellate blooms are at least a nuisance, if not a harbinger of a potential health hazard. "Harmful algal bloom" is another term being used to describe the condition.

When toxigenic phytoplankton are in the water in sufficient numbers, shellfish can concentrate the toxins in their bodies, often with little or no obvious problems to the shellfish. One of the dangerous aspects of these toxic red tides is that shellfish exposed to them can become toxic before the phytoplankton reach a sufficient density to color the water and alert one to their presence. Often a discoloration never occurs or occurs out of sight below the surface. Another concern is that blooms of toxic phytoplankton may be occurring more often and in places where they had never been reported in the past. As a result, the NSSP requires that each certified state maintain a monitoring program either to test for the presence of toxic phytoplankton species, or to test for the toxins themselves.

It seems that more red tides are occurring than in the past, although this has not been proven. If so, coastal pollution may play the major role, especially sewage discharges and some forms of industrial waste (Shumway, 1990). It may well be that nitrogen and phosphorous inputs largely affect the amount of total biomass produced in the blooms, and that micronutrients affect which species dominates. Organometallics, such as iron and other trace elements chelated by humic acids, the presence and amount of B vitamins, and other dissolved organics appear to greatly affect their growth. The presence of suppressors or zooplanktonic grazers preventing the growth of the normal flora in some instances may also play a large role. In addition, it is quite possible that ballast water discharged from ships traveling around the world can seed an area with cysts (the resistant, resting phase of dinoflagellates) that sink to the bottom sediment and grow out when proper conditions occur (Anderson, 1989; Iwasaki, 1989; Smayada, 1989).

Paralytic Shellfish Poison

Paralytic shellfish poison (PSP) is distributed worldwide, and its distribution in North America is along the northeast Atlantic Coast from Cape Cod to at least the Gulf of Saint Lawrence, in the Gulf Coast, in the Gulf of California, and along the northwest United States and Canadian coasts up through Alaska. The major world species of algae producing PSP include *Alexandrium catenella*, *A. fraterculus*, *A. minutum*, *A. tamarense* (also known as *Protogonyaulax tamarensis*, *Gonyaulax tamarensis*, *G. excavata*, *G. tamarensis* var. *Gessnerium tamarensis*, or *A. fundyense* by several authors), *Gonyaulax acatenella*, *Gymnodinium catenatum*, *Prorocentrum micans*, *Prorocentrum minimum*, and *Pyrodinium bahamense* var. *compressum*. The cysts of these species are also highly toxic (Shumway, 1990).

The major toxigenic phytoplankton responsible for PSP in the northeast United States are *Alexandrium excavatum*, *A. tamarense*, and *A. fundyense* (or simply *A. tamarense* by some authors), causing high PSP concentrations in intertidal and subtidal shellfish including clams, mussels, oysters, and scallops. The toxicity of these species may decrease somewhat in a North-South direction. *A. tamarense* may be responsible for offshore toxicity in Georges Bank off the coast of Massachusetts (Cembella et al., 1993). These blooms in the Northeast are primarily of concern from April through October for mussels and clams, though intoxications can occur at any time of the year. Whole scallops, with their slow detoxification and unpredictable seasonal variations, are never really safe along the coast of Maine or Canada (Shumway et al., 1988; Hurst and Yentsch, 1981).

A. catenella is the dominant PSP-producing phytoplankon along California and the Pacific Northwest, though other toxigenic species also exist, including *A. tamarense* (Taylor et al., 1994; Rensel, 1993). Concentrations of 1,000 *A. catenella* cells l^{-1} are believed to be the threshold level needed to intoxicate shellfish (Taylor et al., 1994). *Gymnodinium catenatum* has caused an outbreak of PSP poisoning on the Mexican Coast on the Gulf of California (Mee et al., 1986). *Gymnodinium acatenella* caused several cases of PSP, resulting in one death in British Columbia (Shumway, 1990).

On the West Coast of the United States, dangerously high PSP concentrations are apt to occur in mussels at any time of the year. In the southern counties of California, unacceptable concentrations are equally likely to occur at any time, though in Northern California these concentrations tend to occur more from May through October. The most likely period for high levels of PSP in shellfish in Oregon and Washington is from April through October, though butter clams and whole pink scallops in certain areas are toxic all year. All Alaskan beaches are considered to be at risk year-around, partly because butter clams are usually toxic and partly because the shoreline is so extensive that adequate monitoring is not possible. Alaskan bay mussels, littlenecks, and cockles are seasonally highest in toxins in June and July (Nishitani and Chew, 1988).

PSP toxins are neurotoxins and are among the most potent phycotoxins of the marine environment (Cembella et al., 1993). One 31-year-old man in Canada became totally paralyzed after eating only a thumbnail-sized piece of a siphon tip of a butter clam (its most contaminated body part) (Taylor, 1992). The PSP toxins include at least 18 toxins, including sulfocarbamoyl and carbamate toxins (saxitoxin, neosaxitoxin, gonyautoxins) (Cembella et al., 1993; Bricelj et al., 1990; Shumway, 1990). Symptoms of intoxication begin with a numbness of the lips and mouth, progressing to vomiting, followed by partial and then total paralysis in extreme cases, which prevents breathing and is the cause of death.

PSP toxins from eastern Canada have been shown to be partially unstable to heat. Steaming or boiling seems to reduce the potency somewhat, though some of the toxin is only lost to the liquor. Frying seems to reduce the potency by about 70 percent (Medcof, 1947).

The rapidity and degree of PSP uptake and toxicity exhibited by shellfish is species-specific. Mussels have long been known as being excellent sentinel organisms for PSP since they accumulate it more rapidly than others such as oysters (*Crassostrea gigas*), hard clams (*Mercenaria mercenaria*), soft clams (*Mya arenaria*), littleneck clams (*Protothaca staminea*), and butter clams (*Saxidomus giganteus*). Mussels also detoxify more rapidly than such species as sea scallops (*Placopecten magellanicus*) and surf clams (*Spisula solidissima* and *S. giganteus*), which are slower than most species to detoxify (Cembella et al., 1993; Bricelj et al., 1990). Blue mussels (*Mytilus edulis*) have been shown to be capable of exceeding the NSSP limit of PSP toxin expressed as saxitoxin of 80 µg $100g^{-1}$ in one hour of controlled exposure to *A. fundyense* under optimal but realistic conditions as found in the environment (Brycelj et al., 1990).

West Coast razor clams (*Siliqua patula*) and geoducks (*Panope generosa*) tend to concentrate PSP toxins in the digestive gland and gut respectively, which are removed prior to eating or processing (Nishitani and Chew, 1988). Though hard clams have been reported to remain nontoxic during PSP red tides in Maine (Shumway, 1990), Lee and Brycelj (1990) have reported PSP toxin uptake under controlled laboratory conditions. In July, 1989, the roe-on scallop fishery, but not the adductor muscle-only fishery, on the Grand Banks was temporarily closed due to PSP; in August the surf clam fishery on the southern Grand Banks also was temporarily closed. The adductor muscle of the sea scallop has been known for years not to concentrate PSP during blooms (Medcof et al., 1947). The potency of the phycotoxins in shellfish can be changed over those originating in the phytoplankton by species-specific capabilities to convert the PSP toxins to more toxic decarbanoyl analogues or to less toxic forms (Cembella et al., 1993; Shimizu and Yoshioka, 1981).

Neurotoxic Shellfish Poison

NSP red tides have occurred only in the Caribbean, in the Gulf of Mexico, and along the southeastern United States coast. These blooms can occur at any time of year (Steidinger and Joyce, 1973). There is only one recorded extension of NSP sufficient to intoxicate shellfish north of Florida, and that occurred in North Carolina from Myrtle Beach to Cape Hatteras in the fall of 1987 (Tester et al., 1991).

The dinoflagellate responsible for NSP red tides is *Ptychodiscus brevis*. The species has a fairly high salinity requirement of at least 24 ppt, and as such is rarely found in estuaries unless they are seasonally low in freshwater runoff with high resulting salinities. *P. brevis* normally occurs at concentrations below 1,000 cells/liter (l), where they are harmless. The dinoflagellate responsible for the outbreak in New Zealand was a new species of *Gymnodinium* (*Ptychodiscus*) (Chang, 1994).

When cell counts reach 5,000 cells/l they are thought to pose a health hazard via the consumption of shellfish. Fortunately, NSP toxins are significantly less potent than PSP toxins, and no deaths have been reported as a consequence of consumption of even large quantities (e.g., 45 oysters) of intoxicated shellfish. Symptoms include tingling of the mouth and extremities, motor incoordination, hot-cold flashes, diarrhea, nausea, headache, etc. Swimming in the red tide sometimes causes rashes. Spray from the surf will cause burning eyes and upper respiratory tract irritation. The toxin is heat-stable to such forms of cooking as steaming, baking, boiling, and frying (Morris et al., 1991; Steidinger and Joyce, 1973). The toxins in the New Zealand species are similar to the toxins of *P. brevis*, though not identical (Chang, 1944). The eastern oyster (*Crassostrea virginica*) has been found to eliminate NSP toxin in 2-6 weeks (Shumway, 1990).

Diarrhetic Shellfish Poison

Diarrhetic shellfish poisoning (DSP) has primarily been a problem due to the consumption of mussels in Europe, though it has occurred in the Western Pacific basin, in Chile, and on the Atlantic Coast of Canada. Symptoms are diarrhea, nausea, vomiting, abdominal pain, and occasionally chills. Onset is usually between three and seven hours, with total recovery by the third day. No related deaths are known. Exposure does not provide immunity, but there is considerable variation in human tolerance. The toxins are known tumor promoters. While there are several fat-soluble, acidic DSP toxins, the primary toxins of concern are okadaic acid, along with its derivatives known as dinophysistoxin-1 and dinophysistoxin-3. These toxins are not destroyed by normal cooking. A minimum of 12 mouse units will cause human illness (one mouse unit corresponds to 3.2 μg of dinophysistoxin-1 or 4.0 μg of okadaic acid) (Shumway, 1990; Lee et al., 1989; Freudenthal and Jijina, 1988).

Many dinoflagellate species of *Dinophysis* can produce DSP toxins, including *D. acuminata*, *D. acuta*, *D. fortii*, *D. sacculus* and *D. norvegica*, all of which have been implicated in DSP poisoning. *D. rotundata* and *D. tripos* can also produce DSP toxins. The toxicity of specific *Dinophysis* spp. varies considerably with time and area. While the dinoflagellates *Prorocentrum lima*, *P. concavum* and *P. hoffmannianum* have been proven capable of producing okadaic acid, only *P. lima* has been implicated in cases of DSP sickness (Morton et al., 1994; Taylor et al., 1994; Subba Rao et al., 1993; Lee et al., 1989; Freudenthal and Jijina, 1988). Shumway (1990) indicates that *Ceratium furca* var. *berghii* may have caused the first outbreak of DSP in South Africa.

Many species of *Dinophysis* are prevalent in North American coastal waters. The most common species present on the northeast coast include *D. acuminata*, *D. acuta* and *D. norvegica*, though *D. fortii* is present. Along British Columbia and the northwest United States coast, *D. fortii* and *D. acuminata* are common.

DSP has not been much of a problem in the United States or Canada, though the potential exists on both coasts; a few cases have been reported on the Atlantic Coast. One incidence of DSP has been reported in cultured mussels from Nova Scotia, which may have been due to *P. lima* (Marr et al., 1992); others in the northeast United States may have occurred due to *Dinophysis* spp. (Freudenthal and Jijina, 1988; Maranda and Shimizu, 1987). No references were found for any cases of DSP poisoning on the Pacific Coast of the United States or Canada. Concentrations of *Dinophysis* spp. greater than 200 cells/l, which is the minimal level thought necessary to induce DSP intoxication in shellfish, have been observed on the coast of British Columbia (Taylor et al., 1994) and in the Bedford Basin of eastern Canada (Subba Rao et al., 1993). These authors expressed concern for potential DSP poisoning in the late summer in British Columbia and from June to November in the North Atlantic region.

DSP red tides are difficult to monitor for several reasons. *Dinophysis* spp. apparently need a stratified water column to bloom; when they do, they concentrate at depth in the lower layer of the water column (Taylor et al., 1994; Subba Rao et al., 1993). *P. lima* is primarily a bottom-dwelling dinoflagellate. Therefore, surface sampling is inadequate. The toxin production of both of these genera apparently changes due to environmental stress (Morton et al., 1994; Subba Rao et al., 1993). Furthermore, in a species such as scallops that may accumulate and depurate okadaic acid (and other toxins) slowly, the concentration of the toxin in the shellfish can increase at the same time that the concentration in the phytoplankton is decreasing (Subba Rao et al., 1993). Concentrations of as little as 100 *P. lima* cells/l may produce high levels of DSP in mussels (Shumway, 1990).

Okadaic acid has been found to be depurated slowly by some species of bivalves. The DSP toxins from *D. acuta* and *D. norvegica* have been found to persist in blue mussels for up to five months after accumulation in Swedish waters. The toxins have been known to persist in other areas for many months (Shumway, 1990).

Amnesiac Shellfish Poison

Prior to 1987, red tides capable of poisoning humans were thought to be limited to dinoflagellates. However, an outbreak of a new shellfish-borne intoxication led to the discovery of domoic acid (DA) in certain diatoms. Domoic acid is a water-soluble, heat-stable amino acid that destroys nerves by causing them to fire repetitively until they rupture. Symptoms of DA poisoning can begin within 30 minutes to 24 hours of consumption of contaminated shellfish, and include vomiting, diarrhea, abdominal cramps, and headache (Horner and Postel, 1993). More severe symptoms include neurological disorders such as dizziness, disorientation, and memory loss within 48 hours, and can result in death. Short-term memory loss is limited to events that happen after eating the contaminated shellfish; hence the term Amnesiac Shellfish Poisoning (ASP) has been applied (Horner and Postel, 1993). The FDA and Canadian limit for DA in shellfish is 20 μg/g wet weight.

ASP has been limited largely, perhaps entirely, to the North American continent. The major species involved are diatoms and include *Pseudo-nitzschia australis, P. pungens* forma *multiseries, P. delicatissima,* and *P. pseudodelicatissima*. Certain species of marine red algae can also produce domoic acid, including *Chondria armata, Chondria baileyana* and *Alsidium corallinum* (Horner et al., 1993; Horner and Postel, 1993; Villac et al., 1993). *Amphora conffeaeformis* is also a known DA producer (Villac et al., 1993).

Preliminary data indicates that certain environmental conditions may be more likely to trigger toxic ASP blooms. An extended dry summer with nitrogen limitation in the fall, followed by heavy rains resulting in significant nitrogen-laden runoff that then remains in the area, seems to be capable of triggering a bloom. Bloom conditions, at least on the West Coast, are considered to be 10^6 cells/l. Not all blooms are toxic, however. Toxin production seems to start and be maximally produced after the major growth phase begins to level off. Toxin production later in the stationary phase and in the die-off phase seems to be reduced (Smith, 1993; Villac et al., 1993; Wohlgeschaffen et al., 1992; Martin et al., 1990).

The first ASP outbreak occurred on Prince Edward Island (Canada) in 1987 and involved more than 100 illnesses and three deaths (Todd, 1990). The shellfish responsible was the blue mussel (*Mytilus edulis*) and the species of diatom was *Pseudonitzschia pungens* f. *multiseries* (Bates et al., 1989). In 1988 smaller amounts of DA were found in the soft clam (*Mya arenaria*) and in the blue mussel; it is believed that *P. pseudodelicatissima* was responsible (Martin et al., 1990).

In 1991 DA became a public health issue for the first time on the West Coast. In September, 1991, cormorants and brown pelicans began dying in Monterey Bay. The cause was traced to anchovies feeding on *P. australis;* it is considered the main source of DA on the West Coast. *P. pungens* f. *multiseries*, *P. delicatissima*, and *P. pseudodelicatissima* also occur on the West Coast (Villac et al., 1993). In late October, razor clams (*Siliqua patula*) from the surf zone of Oregon and Washington were found to have dangerously high concentrations of DA, and that fishery was closed to harvest. By early December, DA was found in the viscera of Dungeness crabs (*Cancer magister*) from Oregon and Washington beaches. This fishery was closed for a few weeks because cooking released the DA into the body meat (Horner and Postel, 1993). Once it was discovered that removal of the viscera prior to cooking eliminated the DA from the body meat, proper practices could be instituted. It is not known how the DA got in the Dungeness crabs, but it may have been due to their consumption of razor clams.

No poisonings or harvesting restrictions have occurred in the Gulf Coast due to DA, though *P. pungens* f. *multiseries* does occur there and has increased following cold fronts in the winter (Villac et al., 1993).

The uptake and release of DA by shellfish is highly species-dependent. Blue mussels rapidly take it up, concentrating it in the digestive gland (Wright et al., 1989) and rapidly eliminate it, with 90 percent eliminated within 24 to 72 hours. There is no evidence that mussels translocate significant amounts of DA for storage in other tissues (Novaczek et al., 1992; 1991). Mussels on the West Coast were never found to be contaminated with DA (Horner and Postel, 1993), perhaps because they quickly depurated DA before being tested. However, razor clams are very slow to eliminate DA and retain unacceptable concentrations in edible tissues (the foot having the highest concentration) for a minimum of three months (Drum et al., 1993; Horner et al., 1993). Razor clams on the West Coast retained dangerously high concentrations of DA for over six months. Sea scallops accumulate and depurate DA slower than mussels, but it is concentrated in the digestive tissues and gonads, and not in the adductor mussel. Thus the whole scallop and roe-on scallop industry could be hurt by a DA bloom, whereas the muscle-only industry should be unaffected (Wohlgeschaffen et al., 1992). Depuration of DA is rapid in soft-shell clams (Wohlgeschaffen et al., 1992). Oysters on the West Coast never became toxic (Horner and Postel, 1993).

Brown Tides

"Brown tides" is the name that has been given to blooms of a small alga that does not cause human health concerns, but which is of concern to bay scallop populations. The alga is a small chrysophyte, *Aureococcus anophagefferens*, which has developed densities of more than 10^9 cells/l that block light to eel grass beds and cause bay scallops to starve. Bay scallop fisheries in Long Island Sound, Narragansett Bay, and Barnegat Bay have been severely hurt. The lack of light to the eel grass kills it, which destroys the cover for bay scallops. In addition, bay scallops do not feed during these blooms and, therefore, die (Tettlebach and Wenczel, 1993; Cosper et al., 1987).

REFERENCES

Abbe, R.G. and J.G. Sanders. 1990. Pathways of silver uptake and accumulation by the American oyster (*Crassostrea virginica*) in Chesapeake Bay. *Estuarine, Coastal and Shelf Science.* **31**: 113-123.

Adams, M.A., M. Bolger, C.D. Carrington, C.E. Coker, G.M. Cramer, M.J. DiNovi, and S. Dolan. 1993a. *Guidance Document for Cadmium in Shellfish*, United States Food and Drug Administration. Washington, DC.

Adams, M.A., M. Bolger, C.D. Carrington, C.E. Coker, G.M. Cramer, M.J. DiNovi, and S. Dolan. 1993b. *Guidance Document for Lead in Shellfish*, United States Food and Drug Administration. Washington, DC.

Adams, M.A., M. Bolger, C.D. Carrington, C.E. Coker, G.M. Cramer, M.J. DiNovi, and S. Dolan. 1993c. *Guidance Document for Arsenic in Shellfish*, United States Food and Drug Administration. Washington, DC.

Adams, M.A., M. Bolger, C.D. Carrington, C.E. Coker, G.M. Cramer, M.J. DiNovi, and S. Dolan. 1993d. *Guidance Document for Chromium in Shellfish*, United States Food and Drug Administration. Washington, DC.

Anderson, D.M. 1989. Toxic algal blooms and red tides: a global perspective. In *Red Tides Biology, Environmental*

Science and Toxicology. T. Okaichi, D.M. Anderson, and T. Nemoto (eds.). Elsevier, NY.

Armstrong, C.W., J.L. Lake, and G.B. Miller Jr. 1983. Extraintestinal infections due to halophilic vibrios. *Southern Medical Journal*. **76**: 571-574.

Baker, J.E., S.J. Eisenreich, and B.J. Eadie. 1991. Sediment trap fluxes and benthic recycling of organic carbon, polycyclic aromatic hydrocarbons and polychlorobiphenyl congeners in Lake Superior. *Environmental Science and Technology*. **25**: 500-509.

Bates, S.S., C.J. Bird, A.S.W. de Freitas, R. Foxall, M. Gilgan, L.A. Hanic, G.A. Johnson, A.W. McCulloch, P. Odense, R. Pocklington, M.A. Quilliam, P.G. Sim, J.C. Smith, D.V. Subba Rao, E.C.D. Todd, J.A. Walter, and J.L.C. Wright. 1989. Pennate diatom *Nitzschia pungens* as the primary source of domoic acid, a toxin in shellfish from eastern Prince Edward Island, Canada. *Canadian Journal of Fisheries and Aquatic Science*. **46**: 1203-1215.

Bender, M.E., P.O. deFur, and R.J. Huggett. 1986. Polynuclear aromatic hydrocarbon monitoring in estuaries utilizing: oysters, brackish water clams and sediments. *IEEE Oceans '86 Conference Proceedings*. Pp. 791-796.

Bender, M.E., R.J. Huggett, and H.D. Slone. 1987. Uptake of polynuclear aromatic hydrocarbons by oysters (*Crassostrea virginica*) transplanted to an industrialized estuarine system. In *IEEE Oceans '87 Conference Proceedings*. Pp. 1561-1565.

Benenson, A.S. (ed.). 1995. *Control of Communicable Diseases in Man*, American Public Health Association. Washington, DC.

Bergen, B.J., W.G. Nelson, and R.J. Pruell. 1993. Bioaccumulation of PCB congeners by blue mussels, *Mytilus edulis*, deployed in New Bedford Harbor, Massachusetts. *Environmental Toxicology and Chemistry*. **12**: 1671-1681.

Blacklow, N.R., G. Cukor, M.K. Bedigian, P. Echeverria, H.B. Greenberg, D.S. Schreiber, and J.S. Trier. 1979. Immune response and prevalence of antibody to Norwark enteritis virus as determined by radioimmnoassay. *Journal of Clinical Microbiology*. **10**: 903-909.

Blake, P.A., P.A. Weaver, and D.G. Hollis. 1980. Diseases of humans, other than cholera, caused by vibrios. *Annual Review of Microbiology*. **34**: 341-352.

Blake, P.A. 1983. Vibrios on the half shell: what the walrus and the carpenter didn't know. *Annals of Internal Medicine*. **99**: 558-559.

Blake, P.A., M.A. Merson, R.E. Weaver, D.G. Hollis, and P.C. Heublein. 1979. Disease caused by a marine *Vibrio*: clinical characteristics and epidemiology. *New England Journal of Medicine*. **300**: 1-5.

Boon, J.P., F. Eijgenraam, J.M. Everaarts, and J.C. Duinker. 1989. A structure-activity relationship (SAR) approach towards metabolism of PCBs in marine animals from different trophic levels. *Marine Environmental Research*. **27**: 159-176.

Boutin, B.K., P.L. Spaulding, and R.M. Twedt. 1986. Evaluation of the enteropathogenicity of *Klebsiella pneumoniae* isolates from summer-harvested Louisiana oysters. *Journal of Food Protection*. **49**: 442-444.

Burkhardt, W., W.D. Watkins, and S.R. Rippey. 1992. Seasonal effects on accumulation of microbial indicator organisms by *Mercenaria mercenaria*. *Applied and Environmental Microbiology*. **58**: 826-831.

Bricelj, V.M., J.H. Lee, A.D. Cembella, and D.M. Anderson. 1990. Uptake kinetics of paralytic shellfish toxins from the dinoflagellate *Alexandrium fundyense* in the mussel *Mytilus edulis*. *Marine Ecology Progress Series*. **63**: 177-188.

Bryan, G.W. and W.J. Langston. 1992. Bioavailability, accumulation, and effects of heavy metals in sediments with special reference to United Kingdom estuaries: a review. *Environmental Pollution*. **76**: 89-131.

Cembella, A.D., S.E. Shumway, and N.I. Lewis. 1993. Anatomical distribution and spatio-temporal variation in paralytic shellfish toxin composition in two bivalve species from the Gulf of Maine. *Journal of Shellfish Research*. **12**: 389-403.

Colwell, R. and J. Kaper. 1978. *Vibrio cholerae, Vibrio parahaemolyticus* and other vibrios: occurrence and distribution in Chesapeake Bay. *Science*. **198**: 394-397.

Connor, D.J. and J.P. Connolly. 1980. The effect of concentration of adsorbing solids on the partition coefficient. *Water Research*. **14**: 1517-1523.

Cook, D. 1994. Time-temperature factors. In *Proceedings of the 1994 Vibrio vulnificus Workshop. June 15-16, 1994*. Washington, DC. Pp. 63-68.

Cook, D.W. and A.D. Ruple. 1992. Cold storage and mild heat treatment as processing aids to reduce the numbers of *Vibrio vulnificus* in raw oysters. *Journal of Food Protection*. **55**: 985-989.

Cosper, E.M., W.C. Dennison, E.J. Carpenter, V.M. Bricelj, J.G. Mitchell, S.H. Kuenstner, D. Colflesh, and M. Dewey. 1987. Recurrent and persistent brown tide blooms perturb coastal marine ecosystem. *Estuaries*. **10**: 284-190.

Croonenberghs, R.E. 1974. Heavy metal concentrations in the clam *Rangia cuneata* from the Rappahannock and James rivers. Master's Thesis. University of Virginia, Charlottesville.

de Boer, J., C.J.N. Stronk, W.A. Traag, and J. van der Meer. 1993. Non-ortho and mono-ortho substituted chlorobiphenyls and chlorinated dibenzo-p-dioxins and dibenzofurans in marine and freshwater fish and shellfish from the Netherlands. *Cemosphere*. **26**: 1823-1842.

de Mesquita, M.M.F., L.M. Evison, and P.A. West. 1991. Removal of fecal indicator bacteria and bacteriophages from the common mussel *Mytilus edulis* under artificial depuration conditions. *Journal of Applied Microbiology*. **70**: 495-501.

DiGirolamo, R., J. Liston, and J.R. Matches. 1970. Survival of virus in chilled, frozen, and processed oysters. *Applied Microbiology*. **20**: 58-63.

Dowson, P.H., J.M. Bubb, and J.N. Lester. 1993. Temporal distribution of organotins in the aquatic environment five years after the 1987 UK retail ban on TBT based antifouling paints. *Marine Pollution Bulletin.* **26**: 487-494.

Doyle, M.P. 1990a. *Campylobacter jejuni.* In *Foodborne Diseases.* D.O. Cliver (ed.), pp. 2117-2222. Academic Press Inc., San Diego, CA.

Doyle, M.P. 1990b. *Shigella.* In *Foodborne Diseases.* D.O. Cliver (ed.), pp. 205-208. Academic Press Inc., San Diego, CA.

Doyle, M.P. and D.O. Cliver. 1990. *Vibrio.* In *Foodborne Diseases.* D.O. Cliver (ed.), pp. 24102-24105. Academic Press Inc., San Diego, CA.

Drum, A.S., T.L. Siebens, E.A. Crecelius, and R.A. Elston. 1993. Domoic acid in the Pacific razor clam *Siliqua patula.* Dixon: 1789. *Journal of Shellfish Research.* **12**: 443-450.

Duursma, E.K., J. Nieuwenhuize, and J.M. Van Liere. 1989. Polychlorinated biphenyl equilibria in an estuarine system. *The Science of the Total Environment.* **79**: 141-155.

Elder, J.F. and P.V. Dresler. 1988. Accumulation and bioconcentration of polycyclic aromatic hydrocarbons in a nearshore estuarine environment near a Pensacola, Florida, creosote contamination site. *Environmental Pollution.* **49**: 117-132.

Embarek, P.K.B. 1994. Presence, detection and growth of *Listeria monocytogenes* in seafoods: a review. *International Journal of Food Microbiology.* **23**: 17-34.

Enriquez, R., G. G. Froesner, V. Hochstein-Mintzel, S. Riedmann, and G. Reinhardt. 1992. Accumulation and persistence of hepatitis A virus in mussels. *Journal of Medical Virology.* **37**: 174-179.

Evans, S.M., S.T. Hawkins, J. Porter, and A.M. Samosir. 1994. Recovery of dogwhelk populations on the Isle of Cumbrae, Scotland, following legislation limiting the use of TBT as an antifoulant. *Marine Pollution Bulletin.* **28**: 15-17.

Farrington, J.W., E.D. Goldberg, R.W. Risebrough, J.H. Martin, and V.T. Bowen. 1983. U.S. "Mussel Watch" 1976-1978: an overview of the trace-metal, DDE, PCB, hydrocarbon and artificial radionucleide data. *Environmental Science and Technology.* **17**: 490-496.

Farrington, J.W. and J.G. Quinn. 1973. Petroleum hydrocarbons in Narragansett Bay I. survey of hydrocarbons in sediments and clams. *Estuarine and Coastal Marine Science.* **1**: 71-79.

Feachem, R., H. Garelick, and J. Slade. 1981. Enteroviruses in the environment. *Tropical Disease Bulletin.* **78**:185-230.

Freudenthal, A.R. and J.L. Jijina. 1988. Potential hazards of *Dinophysis* to consumers and shellfisheries. *Journal of Shellfish Research.* **7**: 695-701.

Fuchs, R.S. and P.K. Surendran. 1989. Incidence of *Listeria* in tropical fish and fishery products. *Letters in Applied Microbiology.* **9**: 49-51.

Glatzer, M.B. 1999. Senior shellfish specialist, USFDA. Personal communication.

Goldstein, J.A. and S. Safe. 1989. Mechanism of action and structure-activity relationships for the chlorinated dibenzo-*p*-dioxins and related compounds. In *Halogenated Biphenyls, Terphenyls, Naphthalenes, Dibenzodioxins and Related Products.* Kimbrough and Jensen (eds.), pp. 239-293. Elsevier Science Publishers. New York.

Goswami, B.B., W.H. Koch, and T.A. Cebula. 1993. Detection of hepatitis A virus in *Mercenaria mercenaria* by coupled reverse transcription and polymerase chain reaction. *Applied and Environmental Microbiology.* **59**: 2765-2770.

Gray, S.F. and M.R. Evans. 1993. Dose-response in an outbreak of non-bacterial food poisoning traced to a mixed seafood cocktail. *Epidemiology and Infection.* **110**: 583-590.

Hackney, C.R., M.B. Kilgen, and H. Kator. 1992. Public health aspects of transferring mollusks. *Journal of Shellfish Research.* **11**: 521-533.

Hackney, C.R., B. Ray, and M.L. Speck. 1980. Incidence of *Vibrio parahaemolyticus* in and the microbiological quality of seafoods in North Carolina. *Journal of Food Protection.* **4**: 113-123.

Hale, R.C. 1989. Accumulation and biotransformation of an organophosphorous pesticide in fish and bivalves. *Marine Environmental Research.* **28**: 67-71.

Hall, L.W., Jr., M.A. Unger, M.C. Ziegenfuss, J.A. Sullivan, and S.J. Bushong. 1992. Butyltin and copper monitoring in a northern Chesapeake Bay marina and river system in 1989. An assessment of tributyltin legislation. *Environmental Monitoring Assessment.* **22**: 15-38.

Hahn, M.E., A.E. Poland, E. Glover, and J.J. Stegeman. 1993. The Ah receptor in marine animals: Phylogenetic distribution to cytochrome P450IA inducibility. *Marine Environmental Research.* **34**: 87-92.

Hansen, L. G. 1987. Environmental toxicology of polychlorinated biphenyls. In *Polychlorinated Biphenyls (PCBs): Mammalian and Environmental Toxicology.* S. Safe and O. Hutzinger (eds.), pp. 15-48. Springer-Verlag, Berlin.

Haruki, K., Y. Seto, T. Murakami, and T. Kimura. 1991. Pattern of shedding of small round-structured virus particles in stools of patients of outbreaks of food-poisoning from raw oysters. *Microbiology and Immunology.* **35**: 83-86.

Hay, B. and P. Scotti. 1987. Evidence for intracellular absorption of virus by the Pacific oyster *Crassostrea gigas. New Zealand Journal of Marine and Freshwater Research.* **20**: 655-660.

Hellou, J., C. Upshall, J.F. Payne, S. Naidu, and M. A. Paranjape. 1993. Total unsaturated compounds and polycyclic aromatic hydrocarbons in mollusks collected from waters around Newfoundland. In *Archives of Contamination and Toxicology.* **24**: 249-257.

Helmstetter, M.F. and R.W. Alden. 1994. Release rates of polynuclear aromatic hydrocarbons from natural sediments and their relationship to solubility and octanol-water partitioning. *Archives of Environmental Contamination and Toxicology.* **26**: 282-291.

Horner, R.A., M.B. Kusske, B.P. Moynihan, R.N. Skinner, and J.C. Wekell. 1993. Retention of domoic acid by Pacific razor clams, *Siliqua patula.* Dixon, 1789: preliminary study. *Journal of Shellfish Research.* **12**: 451-456.

Horner, R.A. and J.R. Postel. 1993. Toxic diatoms in western Washington waters (U.S. West Coast). *Hydrobiologia.* **269/270**: 197-205.

Horner, R.A., J.R. Postel, and J.E. Rensel. 1990. Noxious phytoplankton blooms in western Washington waters, a review. In *Toxic Marine Phytoplankton.* E. Graneli, B. Sundstrom, L. Edler, and D.M. Anderson (eds.), pp. 171-176. Elsevier, New York.

Huggett, R.J., M.E. Bender, and H.D. Slone. 1973. Utilizing metal concentration relationships in the Eastern oyster, *Crassostrea virginica*, to detect heavy metal pollution. *Water Research.* 7: 451-495.

Hurst, J.W. and C.M. Yentsch. 1981. Patterns of intoxication of shellfish in the Gulf of Maine coastal waters. *Canadian Journal of Fisheries and Aquatic Sciences.* **38**: 152-156.

Iwasaki, H. 1989. Recent progress of red tide studies in Japan: an overview. In *Red Tides Biology, Environmental Science, and Toxicology.* T. Okaichi, D.M. Anderson and T. Nemoto (eds.). Elsevier, New York.

Jackson, T.J., T.L. Wade, T.J. McDonald, D.L. Wilkinson, and J.M. Brooks. 1994. Polynuclear aromatic hydrocarbon contaminants in oysters from the Gulf of Mexico, 1986-1990. *Environmental Pollution.* **83**: 291-298.

Johns, C., S.N.V. Luoma, and S. P. Elrod. 1988. Selenium accumulation in benthic bivalves and fine sediments of San Francisco Bay, the Sacramento-San Joaquin Delta and selected tributaries. *Estuarine and Coastal Shelf Science.* **27**: 381-196.

Jones, S. 1994. Relaying. In *Proceedings of the 1994 Vibrio vulnificus Workshop. June 15-16, 1994.* Washington, DC. Pp. 73-74.

Karickhoff, S.W., D.S. Brown, and T.A. Scott. 1979. Sorption of hydrophobic pollutants on natural sediments. *Water Research.* **13**:241-248.

Kannan, N., S. Tanabe, R. Tatsukawa, and D.J.H. Phillips. 1989. Persistence of highly toxic coplanar PCBs in aquatic ecosystems: uptake and release kinetics of coplanar PCBs in green-lipped mussels, *Perna viridis Linnaeus. Environmental Pollution.* **55**: 65-76.

Kator, H. and M.W. Rhodes. 1991. Indicators and alternate indicators of growing water quality. In *Microbiology of Marine Food Products.* D.R. Ward and C.R. Hackney (eds.), pp. 135-196. Van Nostrand Reinhold, New York.

Kilgen, M.B. 1994. Discussion section. In *Proceedings of the 1994 Vibrio vulnificus Workshop. June 15-16, 1994.* Washington, DC.

Kilgen, M.B. and M.T. Cole. 1991. Viruses in seafoods. In *Microbiology of Marine Food Products.* D.R. Ward and C.R. Hackney (eds.), pp. 197-209. Van Nostrand Reinhold, New York.

King, D.G. and I. M. Davies. 1987. Laboratory and field studies of the accumulation of inorganic mercury by the mussel *Mytilus edulis* (L.). *Marine Pollution Bulletin.* **18**: 40-45.

Klontz, K.C., S. Lieb, M. Schreiber, H.T. Janowski, L.M. Baldy, and R.A. Gunn. 1988. *Vibrio vulnificus* infections in Florida, 1981-1987; clinical and epidemiologic features. *Annals of Internal Medicine.* **109**: 318-323.

Kohn, M.A., T.A. Farley, R. Ando, M. Curtis, S.A. Wilson, Q. Jin, S.S. Monroe, R.C. Baron, L.M. McFarland, R.I. Glass. 1995. An outbreak of Norwalk virus gastroenteritis associated with eating raw oysters. Implications for maintaining safe oyster beds. *Journal of the American Medical Association.* **273**: 466-471.

La Belle, R.I. and C.P. Gerba. 1982. Investigation into the protective effect of estuarine sediment on virus survival. *Water Research.* **16**: 469-478.

Landis, W.G. 1991. Distribution and nature of the aquatic organophosphorus acid anhydrases: enzymes for organophosphate detoxification. *Reviews in Aquatic Sciences.* **5**: 267-285.

Lauenstein, G.G., A. Robertson, and P. O'Connor. 1990. Comparison of trace metal data in mussels and oysters from a mussel watch program of the 1970s with those from a 1980s program. *Marine Pollution Bulletin.* **21**: 440-447.

Lee, J.S., T.S. Igarashi, Fraga, E. Dahl, P. Hovgaard, and T. Yasumoto. 1989. Determination of diarrhetic shellfish toxins in various dinoflagellate species. *Journal of Applied Psychology.* **1**: 147-152.

Lee, R.F., A.O. Valkirs, and P.F. Seligman. 1989. Importance of microalgae in the biodegradation of tributyltin in estuarine waters. *Environmental Science and Technology.* **23**: 1515-1518.

Lennon, D., B. Lewis, C. Mantell, D. Becroft, B. Dove, K. Farmer, S. Tonkin, N. Yeates, R. Stamp, and K. Mickleson. 1984. Epidemic perinatal listeriosis. *Pediatric Infectious Diseases.* **3**: 30-34.

Levine, W.C. and P.M. Griffin. 1993. *Vibrio* infections on the Gulf Coast: results of first year of regional surveillance. *Journal of Infectious Diseases.* **167**: 479-483.

Lewis, L, M.W. Loutit, and F.J. Austin. 1986. Enteroviruses in mussels and marine sediments and depuration of naturally accumulated viruses by green-lipped mussels, *Perna canaliculus. New Zealand Journal of Marine and Freshwater Research.* **20**: 431-437.

Maranda, L. and Y. Shimizu. 1987. Diarrhetic shellfish poisoning in Narragansett Bay. *Estuaries.* **10**: 298-302.

Markus, J.M. and R.T. Renfrow. 1990. Pesticides and PCBs in South Carolina Estuaries. *Marine Pollution Bulletin.* **21**: 96-99.

Markus, J.M. and T.P. Stokes. 1985. Polynuclear aromatic hydrocarbons in oyster tissue around three coastal marinas. *Bulletin of Environmental Contamination and Toxicology.* **35**: 835-844.

Marr, J.C., A.E. Jackson, and J.L. McLachlan. 1992. Occurrence of *Prorocentrum lima*, a DSP toxin-producing species from the Atlantic coast of Canada. *Journal of Applied Psychology.* **4**: 17-24.

Martin J.L., K. Haya, L.E. Burridge, and D.J. Wildish. 1990. *Nitzschia pseudodelicatissima* — a source of domoic acid in the Bay of Fundy, eastern Canada. *Marine Ecology Progress Series.* **67**: 177-182.

Martin, R. E. 1966. Personal communication.

McKinney, J.D., K. Chae, E.E. McConnell, and L.S. Birnbaum. 1985. Structure-induction versus structure-toxicity relationships for polychlorinated biphenyls and related aromatic hydrocarbons. *Environmental Health Perspectives.* **60**: 57-68.

Means, J.C., S.G. Wood, J.J. Hassett, and W.L. Banwart. 1982. Sorption of amino- and carboxy-substituted polynuclear aromatic hydrocarbons by sediments and soils. In *Environmental Science and Technology.* **16**: 93-98.

Medcof, J.C., A.H. Leim, A.B. Needler, and A.W.H.Needler. 1947. Paralytic shellfish poisoning on the Atlantic Coast. In *Bulletin of the Fisheries Research Board of Canada.* Vol. 75: 1-32.

Mee, L.D., M. Espinosa, and G. Diaz. 1986. Paralytic shellfish poisoning with a *Gymnodinium catenatum* red tide on the Pacific coast of Mexico. *Marine Environmental Research.* **19**: 77-92.

Metcalf, T.G. and W.C. Stiles. 1965. The accumulation of enteric viruses by the oyster *Crassostrea virginica. Journal of Infectious Disease.* **115**: 68-76.

Millard, J., H. Appleton, and J.V. Parry. 1987. Studies on heat inactivation of hepatitis A virus with special reference to shellfish, Part 1. Procedures for infection and recovery of virus from laboratory-maintained cockles. *Epidemiology and Infection.* **98**: 397-414.

Mo, C. and B. Neilson. 1993. Weight and salinity effects on zinc uptake and accumulation for the American oyster, *Crassostrea virginica* Gmelin. *Environmental Pollution.* **82**: 191-196.

Morris, J.G., Jr. 1988. *Vibrio vulnificus* — a new monster of the deep? *Annals of Internal Medicine.* **109**: 261-263.

Morris, J.G. 1994. Pathogenesis of *V. vulnificus.* In *Proceedings of the 1994 Vibrio vulnificus Workshop. June 15-16, 1994.* Washington, DC. Pp. 33-38 and 40-42.

Morris, J.G., Jr. and R.E. Black. 1985. Cholera and other vibrioses in the United States. *New England Journal of Medicine.* **312**: 343-350.

Morris, P.D., D.S. Campbell, T.J. Taylor, and J.I. Freeman. 1991. Clinical and epidemiological features of neurotoxic shellfish poisoning in North Carolina. *American Journal of Public Health.* **81**:471-474.

Morse, D.L., J.J. Guzewich, J.P. Hanrahan, R. Stricof, M. Shayegani, R. Deibel, J.C. Grabau, N.A. Nowak, J.E. Herrmann, G. Cukor, and N.R. Blacklow. 1986. Widespread outbreaks of clam- and oyster-associated gastroenteritis, role of Norwalk virus. *New England Journal of Medicine.* **314**: 678-681.

Morse, J.W., B.J. Presley, R.J. Taylor, G. Benoit, and P. Santschi. 1993. Trace metal chemistry of Galveston Bay: water, sediments and biota. *Marine Environmental Research.* **36**: 1-37.

Morton, S.L., J.W. Bomber, and P.M. Tindall. 1994. Environmental effects on the production of okadaic acid from *Prorocentrum hoffmannianum* Faust I. temperature, light, and salinity. *Journal of Marine Biology and Ecology.* **178**: 67-77.

Motes, M.L., Jr. 1991. Incidence of *Listeria* spp. in shrimp, oysters, and estuarine waters. *Journal of Food Protection.* **54**: 170-173.

Murphy, D.L. 1990. Contaminant levels in oysters from the Chesapeake Bay, USA 1981-1985. *Journal of Shellfish Research.* **8**: 487.

Neff, J.M. 1979. *Polycyclic Aromatic Hydrocarbons in the Aquatic Environment: Sources, Fates and Biological Effects.* Applied Science Publishers Ltd. London.

Neff, J.M., B.A. Cox, D. Dixit, and J.W. Anderson. 1976. Accumulation and release of petroleum derived aromatic hydrocarbons by four species of marine animals. *Marine Biology.* **38**: 279-289.

Novaczek, I., M.S. Madhyastha, R.F. Ablett, G. Johnson, M.S. Nijjar, and D.E. Sims. 1991. Uptake, disposition and depuration of domoic acid by blue mussels, *Mytilus edulis. Aquatic Toxicology.* **21**: 103-118.

Novaczek, I., M.S. Madhyastha, R.F. Ablett, G. Johnson, M.S. Nijjar, and D.E. Sims. 1992. Depuration of domoic acid from live blue mussels, *Mytilus edulis. Canadian Journal of Fisheries and Aquatic Sciences.* **49**: 312-318.

Oliver, J. 1994. Animal models, and viable-but-non-culturable states. *Proceedings of the 1994 Vibrio vulnificus Workshop. June 15-16, 1994.* Washington, DC. Pp. 43-52.

Palmer, S.J., B.J. Presley, R.J. Taylor, and E.N. Powell. 1993. Field studies using the oyster *Crassostrea virginica* to determine mercury accumulation and depuration rates. *Bulletin of Contamination and Toxicology.* **51**: 464-470.

Pendoley, K. 1992. Hydrocarbons in Rowley Shelf, Western Australia, oysters and sediments. *Marine Pollution Bulletin.* **24**: 210-215.

Pereira, W.G., F.D. Hostettler, and J.B. Rapp. 1992. Bioaccumulation of hydrocarbons derived from terrestrial and anthropogenic sources in the Asian clam *Potamocorbula amurensis* in San Francisco Bay estuary. *Marine Pollution Bulletin.* **24**: 103-109.

Peterson, D.A., G. Wolfe, E.P. Larkin, and F.W. Deinhardt. 1978. Thermal treatment and infectivity of hepatitis A virus in human feces. *Journal of Medical Virology.* **2**: 201-206.

Petreas, M.X., T. Wiesmuller, F.H. Palmer, J.J. Winkler, and R.D. Stephens. 1992. Aquatic life as biomonitors of dioxin/furan and coplanar polychlorinated biphenyl contamination in the Sacramento-San Joaquin River delta. *Chemosphere.* **25**: 621-631.

Phelps, H.L., E.A. Wright, and J.A. Milhursky. 1985. Factors affecting trace metal accumulation by estuarine oysters *Crassostrea virginica. Marine Ecology Progressive Series.* **22**: 187-197.

Phillips, D.J.H. and M.H. Depledge. 1985. Metabolic pathways involving arsenic in marine organisms: a unifying hypothesis. *Marine Environmental Research.* **17**: 1-12.

Phillips, D.J.H. and R.B. Spies. 1988. Chlorinated hydrocarbons in the San Francisco estuarine ecosystem. *Marine Pollution Bulletin.* **19**: 445-453.

Picer, N. and M. Picer. 1990. Long-term trends of DDT and PCB concentrations in mussels. *Chemosphere.* **21**: 153-158.

Pittinger, C.A., A.L. Buikema Jr., and J. O. Falkinham. 1987. *In-situ* variations in oyster mutagenicity and tissue concentrations of polycyclic aromatic hydrocarbons. *Environmental Toxicology and Chemistry.* **6**:51-60.

Porte, C. and J. Albaiges. 1993. Bioaccumulation patterns of hydrocarbons and polychlorinated biphenyls in bivalves, crustaceans, and fishes. *Archives of Contamination and Toxicology.* **26**: 273-281.

Power, U.F. and J.K. Collins. 1990a. Tissue distribution of a coliphage and *Escherichia coli* in mussels after contamination and depuration. *Applied and Environmental Microbiology.* **56**: 803-807.

Power, U.F. and J.K. Collins. 1990b. Elimination of coliphages and *Escherichia coli* from mussels during depuration under varying conditions of temperature, salinity and food availability. *Journal of Food Protection.* **53**: 208-212.

Pruell, R.J., J.L. Lake, W.R. Davis, and J.G. Quinn. 1986. Uptake and depuration of organic contaminants by blue mussels *Mytilus edulis* exposed to environmentally contaminated sediment. *Marine Biology.* **91**: 497-508.

Pruell, R.J., N.I. Rubinstein, B.K. Taplin, J.A. LiVolsi, and R.D. Bowen. 1993. Accumulation of polychlorinated organic contaminants from sediment by three benthic marine species. *Archives of Environmental Contamination and Toxicology.* **24**: 290-297.

Reeve, G., D.L. Martin, J. Pappas, R.E. Thompson, and K.D. Green. 1989. An outbreak of shigellosis associated with the consumption of raw oysters. *New England Journal of Medicine.* **321**: 224-227.

Rensel, J. 1993. Factors controlling paralytic shellfish poisoning, PSP, in Puget Sound, Washington. *Journal of Shellfish Research.* **12**: 371-376.

Riedel, G.F., J.G. Sanders, and R.W. Osman. 1987. The effect of biological and physical disturbances on the transport of arsenic from contaminated estuarine sediments. *Estuarine and Coastal Shelf Science.* **25**: 693-706.

Rippey, S.R. 1994. *Seafood Borne Disease Outbreaks.* Department of Health and Human Services, Public Health Service, Food and Drug Administration, Office of Seafood. North Kingstown, RI.

Roberts, J.R., A.S.W. de Freitas, and M.A.S. Gidney. 1979. Control factors on uptake and clearance of xenobiotic chemicals by fish. In *Animals as Monitors of Environmental Pollutants.* National Research Council of Canada. Ottawa. Pp. 3-13.

Robertson, A., B.W. Gottholm, D.D. Turgeon, and D.A. Wolfe. 1991. A comparative study of contaminant levels in Long Island Sound. *Estuaries.* **14**: 290-298.

Rodrick, G.E. 1991. Indigenous pathogens: Vibrionaceae. In *Microbiology of Marine Food Products.* D.R. Ward and C.R. Hackney (eds.), pp. 285-300. Van Nostrand Reinhold, New York.

Ruple, A.D. and D.W. Cook. 1992. *Vibrio vulnificus* and indicator bacteria in shellstock and commercially processed oysters from the Gulf Coast. *Journal of Food Protection.* **55**: 667-671.

Safe, S. 1990. Polychlorinated biphenyls, PCBs, dibenzo-*p*-dioxins, PCDDs, dibenzofurans, PCDFs, and related compounds, environmental and mechanistic considerations which support the development of toxic equivalency factors, TEFs. *Critical Reviews in Toxicology.* **21**: 51-88.

Sanders, J.G., R.W. Osman, and G.F. Riedel. 1989. Pathways of arsenic uptake and incorporation in estuarine phytoplankton and the filter-feeding invertebrates *Eurytemora affinis, Balanus improvisus* and *Crassostrea virginica. Marine Biology.* **103**: 319-325.

Schweinfurth, H.A.P. and M. W. Gunzel. 1987. The tributyltins: mammalian toxicity and risk evaluation for humans. *Proceedings of Oceans '87 Conference Record. Organotin Symposium*, Institute of Electrical and Electronics Engineers. New York. **4**: 1421-1432.

Sericano, J.L., E.L. Atlas, T.L. Wade, and J.M. Brooks. 1990. NOAA's status and trends mussel watch program: chlorinated pesticides and PCBs in oysters, *Crassostrea virginica*, and sediments from the Gulf of Mexico, 1986-1987. *Marine Environmental Research.* **29**: 161-203.

Sericano, J.L., T.L. Wade, J.M. Brooks, E.L. Atlas, R.R. Fay, and D.L. Wilkinson, D.L. 1993. National status and trends mussel watch program: chlordane-related compounds in Gulf of Mexico oysters. *Environmental Pollution.* **82**: 23-32.

Sericano, J.L., T.L. Wade, A.M. El-Husseini, and J.M. Brooks. 1992. Environmental significance of the uptake and depuration of planar PCB congeners by the American oyster, *Crassostrea virginica. Marine Pollution Bulletin.* **24**: 537-543.

Shimizu, Y. and M. Yoshioka. 1981. Transformation of paralytic shellfish toxins as demonstrated in scallop homogenates. *Science.* **212**: 547-549.

Shumway, S.E. 1990. A review of the effects of algal blooms on shellfish and aquaculture. *Journal of the World Aquaculture Society.* **21**: 65-104.

Shumway, S.E., S. Sherman-Caswell, and J.W. Hurst. 1988. Paralytic shellfish poisoning in Maine: monitoring a monster. *Journal of Shellfish Research.* **7**: 643-652.

Smayda, T.J. 1989. Homage to the International Symposium on Red Tides: the scientific coming of age of research on *akashiwo*; algal blooms; *flos-aquae; tsvetenie vody; wasserblute.* In *Red Tides Biology, Environmental Science, and Toxicology.* T. Okaichi, D.M. Anderson and T. Nemoto (eds.), pp. 23-30. Elsevier, New York.

Smith, J.C. 1993. Toxicity and *Pseudonitzschia pungens* in Prince Edward Island, 1987-1992. *Harmful Algae News.* **6**: 1.

Sobsey, M.D., P.A. Shields, F.S. Hauchman, A.L. Davis, V.A. Rullman, and A. Bosch. 1988. Survival and persistence of hepatitis A virus in environmental samples. In *Viral Hepatitis and Liver Disease.* A.J. Zuckerman (ed.), pp. 121-124. Alan R. Liss. New York.

Sole, M., C. Porte, D. Pastor, and J. Albaiges. 1994. Long-term trends of polychlorinated biphenyls and organochlorinated pesticides in mussels from the Western Mediterranean coast. *Chemosphere.* **28**: 897-903.

Steidinger, K.A. and E.A. Joyce Jr. 1973. Florida Red Tides. *State of Florida, Department of Natural Resources Educational Series* No. 17.

Steimle, F.W., P.D. Boehm, V.S. Zdanowicz, and R.A. Bruno. 1986. Organic and trace metal levels in ocean quahog, *Arctica islandica* Linne, from the northwestern Atlantic. *United States National Marine Fisheries Service Fishery Bulletin.* **84**: 133-140.

Stephenson, M.D. and G.H. Leonard. 1994. Evidence for the decline of silver and lead and the increase of copper from 1977 to 1990 in the coastal marine waters of California. *Marine Pollution Bulletin.* **28**: 148-153.

Stronkhorst, J. 1993. Trends in pollutants in blue mussel *Mytilus edulis* and flounder *Platichthys flesus* from two Dutch estuaries. 1985-1990. *Marine Pollution Bulletin.* **24**: 250-258.

Subba Rao, D.V., Y. Pan, V. Zitke, G. Bugden, and K. Mackeigan. 1993. Diarrhetic shellfish poisoning, DSP, associated with a subsurface bloom of *Dinophysis norvegica* in Bedford Basin, eastern Canada. *Marine Ecology Progress Series.* **97**: 117-126.

Tacket, C.O., F. Brenner, and P.A. Blake. 1984. Clinical features and an epidemiological study of *Vibrio vulnificus* infections. *Journal of Infectious Diseases.* **149**: 558-561.

Tamplin, M.L. 1994a. The ecology of *Vibrio vulnificus. Proceedings of the 1994* Vibrio vulnificus *Workshop. June 15-16, 1994.* Washington, DC. Pp. 53-61.

Tamplin, M.L. 1994b. *The Seasonal Occurrence of* Vibrio vulnificus *in Shellfish, Seawater, and Sediment of United States Coastal Waters and the Influence of Environmental Factors on Survival and Virulence.* Final Report to the Saltonstall-Kennedy Program. January 27, 1994.

Tamplin, M.L. 1994c. Icing at harvest. *Proceedings of the 1994 Vibrio vulnificus Workshop. June 15-16, 1994.* Washington, DC.

Tamplin, M.L. 1994d. Depuration and irradiation. *Proceedings of the 1994* Vibrio vulnificus *Workshop. June 15-16, 1994.* Washington, DC. Pp. 69-71.

Tanabe, S. 1988. Dioxin problems in the aquatic environment. *Marine Pollution Bulletin.* **19**: 347-348.

Tanabe, S., R. Tatsukawa, and D.J.H. Phillips. 1987. Mussels as bioindicators of PCB pollution: a case study on uptake and release of PCB isomers and congeners in green-lipped mussels (*Perna viridis*) in Hong Kong waters. *Environmental Pollution.* **47**: 41-62.

Tanacredi, J.T. and R.R. Cardenas. 1991. Biodepuration of polynuclear aromatic hydrocarbons from a bivalve mollusk, *Mercenaria mercenaria* L. *Environmental Science and Technology.* **25**: 1453-1461.

Taylor, F.J.R. 1992. Artificial respiration saves two from fatal PSP in Canada. *Harmful Algae News.* No. 3.

Taylor, F.J.R., R. Haigh, and T.F. Sutherland. 1994. Phytoplankton ecology of Sechelt Inlet, a fjord system on the British Columbia coast. II. Potentially harmful species. *Marine Ecology Progress Series.* **103**: 151-164.

Tettelbach, S.T. and P. Wenczel. 1993. Reseeding efforts and the status of bay scallop *Argopecten irradians.* Lamarck, 1819, populations in New York following the occurrence of "brown tide" algal blooms. *Journal of Shellfish Research.* **12**: 423-431.

Tester, P.A., R.P. Stumpf, F.M. Vukovich, P.K. Fowler, and J.T. Turner. 1991. An expatriate red tide bloom: transport, distribution, and persistence. *Limnology and Oceanography.* **36**: 1053-1061.

Tierney, J.T., R. Sullivan, J.T. Peeler, and E.P. Larkin. 1982. Persistence of polioviruses in shellstock and shucked oysters stored at refrigeration temperatures. *Journal of Food Protection.* **45**: 1135-1137.

Todd, E.C.D. 1990. Amnesiac shellfish poisoning — a new seafood toxin syndrome. In *Toxic Marine Phytoplankton.* E. Graneli, D.M. Anderson, L. Edler, and B. Sundstrom (eds.), pp. 504-508. Elsevier, New York.

Truman, B.I., H.P. Madore, M.A. Menegus, J.L. Nitzkin, and R. Dolin. 1987. Snow mountain agent gastroenteritis from clams. *American Journal of Epidemiology.* **126**: 516-525.

Turgeon, D.D. and T.P. O'Connor. 1991. Long Island Sound: distributions, trends, and effects of chemical contamination. *Estuaries.* **14**: 279-289.

Uhler, A.D., G.S. Durell, W.G. Steinhauer, and A.M. Spellacy. 1993. Tributyltin levels in bivalve mollusks from the east and west coasts of the United States, results from the 1988-1990 national status and trends mussel watch project. *Environmental Toxicology and Chemistry.* **12**: 139-153.

Unger, M.A., W.G. MacIntyre, and R.J. Huggett. 1988. Sorption behavior of tributyltin on estuarine and freshwater sediments. *Environmental Toxicology and Chemistry.* **7**: 907-915.

VDH. Data collected by the Virginia Department of Health, Division of Shellfish Sanitation, and analyzed by the Virginia Division of Consolidated Laboratories.

Villac, M.C., D.L. Roelke, F.P. Chavez, L.A. Cifuentes, and G.A. Fryxell. 1993. *Pseudonitzschia australis* Frenguelli and related species from the West Coast of the U.S.A.: occurrence and domoic acid production. *Journal of Shellfish Research.* **12**: 457-465.

Wade, T.L., E.L. Atlas, J.M. Brooks, M.C. Kennicutt II, R.G. Fox, J. Sericano, B. Garcia-Romero, and D. DeFreitas. 1988. NOAA Gulf of Mexico status and trends program: trace organic contaminant distribution in sediments and oysters. *Estuaries.* **11**: 171-179.

Wade, T.L., B. Garcia-Romero, and J.M. Brooks. 1988. Tributyltin contamination in bivalves from United States coastal waters. *Environmental Science and Technology.* **22**: 1488-1493.

Wade, T.L., B. Garcia-Romero, and J.M. Brooks. 1990. Butyltins in sediments and bivalves from U.S. coastal areas. *Chemosphere.* **20**: 647-662.

Waldichuck, M. 1990. Dioxin pollution near pulpmills. *Marine Pollution Bulletin.* **21**: 365-366.

Watkins, W.D. 1999. Director, Regulatory Operations Office, USFDA. Personal communication.

Weis, P., J.P. Weis, and E. Lores. 1993. Uptake of metals from chromated-copper-arsenate CCA-treated lumber by epibiota. *Marine Pollution Bulletin.* **26**: 428-430.

WHO Scientific Group on Human Viruses in Water. Waste Water and Soil. 1979. Human viruses in water, waste water and soil. *WHO Technical Report Series.* 639.

Whitman, C. 1994. Overview of the important clinical and epidemiologic aspects of *Vibrio vulnificus* infections. *Proceedings of the 1994* Vibrio vulnificus *Workshop. June 15-16, 1994.* Washington, DC. Pp. 11-21.

Wittman, R.J. and G.J. Flick Jr. 1995. Microbial contamination of shellfish: prevalence, risk to human health, and control strategies. *Annual Review of Public Health.* **16**:123-140.

Wohlgeschaffen, G.D., K.H. Mann, D.V. Subba Rao, and R. Pocklington. 1992. Dynamics of the phycotoxin domoic acid: accumulation and excretion in two commercially important bivalves. *Journal of Applied Psychology.* **4**: 297-310.

Wolfe, N.L., R.G. Zepp, D.F. Paris, G.L. Boughman, and R.C. Hallis. 1977. Methoxychlor and DDT degradation in waters: Rates and products. *Environmental Science and Technology.* **11**: 1077-81.

Woodwell, G.M., P.P. Craig, and A.J. Horton. 1971. DDT in the biosphere: Where does it go? *Science.* **174**: 1011-7.

Wright, D.A. and C.D. Zamuda. 1987. Copper accumulation by two bivalve mollusks: salinity effect is independent of cupric ion activity. *Marine Environmental Research.* **23**: 1-14.

Wright, J.L.C., R.K. Boyd, A.S.W. de Freitas, M. Falk, R.A. Foxall, W.D. Jamieson, M.V. Laycock, A.W. McCulloch, A.G. McInnes, P. Odense, V.P. Pathak, M.A. Quilliam, M.A. Ragan, P.G. Sim, P. Thibault, J.A. Walter, M. Gilgan, D.J.A. Richard, and D. Dewar. 1989. Identification of domoic acid, a neuroexcitatory amino acid, in toxic mussels from eastern Prince Edward Island. *Canadian Journal of Chemistry.* **67**: 481-490.

Wuertz, S., M.E. Miller, M.M. Doolittle, J.F. Brennan, and J.J. Cooney. 1991. Butyltins in estuarine sediments two years after tributyltin use was restricted. *Chemosphere.* **22**: 1113-1120.

Zamuda, C.D. and W.G. Sunda. 1982. Bioavailability of dissolved copper to the American oyster *Crassostrea virginica.* I. Importance of chemical speciation. *Marine Biology.* **66**: 77-82.

Zamuda, C.D., D.A. Wright, and R.A. Smucker. 1985. The importance of dissolved organic compounds in the accumulation of copper by the American oyster, *Crassostrea virginica. Marine Environmental Research.* **16**: 1-12.

Illnesses Associated with Consumption of Seafood: Introduction

Peter D. R. Moeller

Seafood is regarded as a healthy component of the American diet, providing an excellent source of protein, and low saturated fat, sodium, calories, and cholesterol (Nettleton, 1985). Seafood contains omega-3 fatty acids, a unique polyunsaturated fatty acid that has been shown to be beneficial for prevention and treatment of heart, autoimmune, and inflammatory diseases (Simopoulos et al., 1990). Seafood is also generally regarded as a safe food product. As with risks associated with consumption of raw or undercooked meat, there are also risks associated with eating raw or undercooked seafood.

A recent study by the National Academy of Sciences (NAS) (Ahmed, 1991) concluded that most seafoods are wholesome and unlikely to cause illness to consumers. However, some areas of risk are recognized, especially reef fishes containing ciguatoxin, certain scombroid fish that can contain hazardous levels of histamine at the time of consumption, and raw shellfish.The continued popularity of seafood, averaging 15 pounds per person in 1995 [National Marine Fisheries Service (NMFS), 1996], mirrors an increase in the concern over seafood safety. Much of the information on seafood illness is from the Centers for Disease Control and Prevention (CDC) Foodborne Disease Outbreak Surveillance Program and a Food and Drug Administration (FDA) database on shellfish-associated foodborne cases. The illness cases and outbreak data collected by the CDC come from state health departments, which report to the CDC when two or more illnesses are linked to the same food source. The FDA collects data on shellfish-borne illness from books, news accounts, CDC reports, and a number of other sources. The NAS study concluded that the two databases did not consistently correlate reports of outbreaks and cases of the same pathogens, due to the inherent differences in each system. Since these systems are the only available national databases on finfish- and shellfish-associated diseases, they do provide a basis to assess the relative importance of illnesses from seafood. Further, while the CDC underestimates the total foodborne illness in the United States because thousands of single illnesses are not reported, it is a consistent database to compare foodborne illnesses across food groups over time.

Among all food products, seafood has the most diverse and complex array of microbes, toxins, and parasites, due to their representation by hundreds of genera and species living in a wide range of habitats. This chapter is directed at toxins, pathogenic bacteria and viruses, and parasites found in marine or freshwater organisms, and the allergies and intolerances associated with them. A National Oceanic and Atmospheric Administration (NOAA)-funded report by Sue Kuenster and the New England Fisheries Development Association (1991) comprehensively covered the issues related to seafood and health, with the exception of allergies and intolerances. This report (with permission) served as a basis for developing and updating the information in this chapter on the sections for toxins, pathogenic bacteria, viruses, and parasites; Wolf and Smith's section on human pathogens also gives an overview of these topics.

REFERENCES

Ahmed, F.E. 1991. *Seafood Safety.* Committee on Evaluation of the Safety of Fishery Products. National Academy Press. Washington, DC.

Kuenstner, S. 1991. Seafood and Health: Risks and Prevention of Seafood-Borne Illness. New England Fisheries Development Association, S-K Cooperative Agreement NA-89-EAD-SK005.

NMFS. 1996. Fisheries of the United States, 1995. Current Fishery Statistics No. 9500. U.S. Dept. of Commerce, National Oceanic and Atmospheric Administration, National Marine Fisheries Service, U.S. Government Printing Office. Washington, DC.

Nettleton, J.A. 1985. *Seafood Nutrition. Facts, Issues and Marketing of Nutrition in Fish and Shellfish.* Osprey Books. Huntington, NY.

Simopoulos, A.P., R.R. Kifer, R.E. Martin, and S.M. Barlow. 1990. *Health Effects of ω3 Polyunsaturated Fatty Acids in Seafood.* Karger, NY.

Human Pathogens in Shellfish and Finfish

Jeffrey C. Wolf and Stephen A. Smith

INTRODUCTION

Inherent in the current trend of increased reliance upon shellfish and finfish to nourish the world's population is a correspondingly elevated potential for seafood-borne microbes and helminths to cause human disease. This statement should not be regarded as a general indictment of aquatic products, but as a challenge to the seafood industry as a whole to generate procedural standards that effectively minimize consumer exposure to infectious and toxigenic organisms. Clearly, the development of appropriate preventive measures must be based upon comprehensive biological and medical knowledge of aquatic pathogens, and the identification of natural and anthropogenic factors responsible for the presence of infectious agents within raw and processed seafoods. Accordingly, this chapter has been partitioned according to the putative origins of hazardous organisms found in aquatic products (Table 1). Such origins include: fish, shellfish, and their natural environments; fresh, brackish, and marine waters altered by human and animal pollution; and the post-harvest contamination of naturally occurring or cultivated aquatic food species.

The interactions that exist among infectious organisms, seafood products, and humans are quite complex. For example, some human pathogens have been microbiologically, serologically, or epidemiologically associated with fish or shellfish, but cause no apparent disease in their respective teleost or invertebrate hosts (Table 2). The ability of healthy shellfish to accumulate and transmit human enteric viruses illustrates this point. Alternatively, *Vibrio* spp. such as *V. vulnificus* are not only capable of causing disease in both seafood animals and humans, but also there is evidence of communicability between the two groups. Finally, for a few bacterial species that have been associated with disease in both humans and fish, transmission of infection is either not established or is not suspected (e.g., *Citrobacter freundii*).

Any discussion of zoonotic diseases would be incomplete if it did not address the relevant routes of human infection. Pertaining to seafood-borne zoonoses, pathogens may be grouped according

Table 1. Origins of Human Pathogens in Shellfish and Finfish

Pathogens Ubiquitous in Aquatic Animals or Environment	Pollution-Derived Pathogens	Pathogens Introduced or Amplified Post-Harvest
Aeromonas hydrophila group	*Campylobacter* spp.	*Bacillus cereus*
Edwardsiella tarda		*Clostridium* spp.
Klebsiella spp.	**Enteric viruses**	
Mycobacterium spp.	*Listeria* spp.	**Enteric viruses**
	Salmonella spp.	*Listeria* spp.
Parasites	*Shigella* spp.	*Salmonella* spp.
Plesiomonas shigelloides	*Staphylococcus* spp.	*Shigella* spp.
Pseudomonas spp.		*Staphylococcus* spp.
Streptococcus iniae		*Vibrio parahemolyticus*
Vibrio spp.		

Table 2. Comparative Pathogenicity of Seafood-Borne Pathogens and Communicability to Humans

	Shellfish		Finfish		Human	
	Harbor	Disease	Harbor	Disease	Communicable	Disease
Aeromonas hydrophila group	+	–	+	+	+/s/f	+
Aeromonas salmonicida	–	–	+	+	–	–
Bacillus cereus	+	–	+	–	+/s/f	+
Campylobacter spp.	+	–	–	–	+/s	+
Citrobacter freundii	–	–	+	+	–	+
Clostridium spp.	+	–	+	–	+/s/f	+
Edwardsiella ictaluri	–	–	+	+	–	–
Edwardsiella tarda	–	–	+	+	+	+
Enteric viruses	+	–	–	–	+/s	+
Escherichia coli	+	–	+	–	b	+
Klebsiella spp.	+	–	+	–	b	+
Lactobacillus spp.	+	–	+	+	b	+
Listeria spp.	+	–	+	–	b	+
Mycobacterium spp.	a	–	+	+	+/f	+
Parasites	+	+/–	+	+/–	+/f	+
Plesiomonas shigelloides	+	–	+	+	+/s/f	+
Pseudomonas spp.	+	–	+	+	b	+
Salmonella spp.	+	–	+	–	+/s/f	+
Shigella spp.	+	–	+	–	+/s	+
Staphylococcus spp.	+	–	+	–	+/s/f	+
Streptococcus iniae	?	–	+	+	+/f	+
Vibrio spp.	+	+	+	+	+/s/f	+
Yersinia enterocolytica	+	–	–	–	+/s	+
Yersinia ruckeri	–	–	+	+	b	c

[a] Rare isolations of *Mycobacterium* spp. from shellfish have been recorded.
[b] Communicability from aquatic organisms to humans is not well established.
[c] Rare reports of *Yersinia ruckeri* disease in humans.
+/s: transmission occurs primarily via shellfish; +/f: transmission occurs primarily via finfish.
+/s/f: transmission occurs via either shellfish or finfish.

to their tendency to inoculate humans via dermal defects versus ingestion (of microorganisms or toxins), and whether associated disease conditions typically remain localized (i.e., confined to either the skin or the alimentary tract) or spread systemically (Table 3).

The following sections highlight various clinically significant human pathogens found in seafood products. Each segment incorporates a brief description of the agent, its natural distribution, typical vectors, comparative pathogenicity, and pathogenic mechanisms, followed by an account of attributed diseases in shellfish, finfish, and humans. Available methods to minimize human exposure are suggested. Concerning specific pathogens, it should be noted that incidence reports of seafood-related human morbidity usually contain an intrinsic bias toward exposures that are associated with profound enteric disease (Table 4). There are two basic explanations for this bias. First, wound infections and milder forms of gastrointestinal disease are comparatively underreported to state and national health agencies. Second, it is difficult to establish a firm relationship between these types of diseases and suspected pathogens. For instance, culture results in cases of human cellulitis are often non-diagnostic (Hook et al., 1986).

Several types of seafood- or aquaculture-associated agents will not be covered or emphasized in this chapter. These include: disease agents that historically are occupational hazards of aquaculturists and fish handlers rather than seafood consumers (e.g., *Erysipelothrix rhusiopathiae* (Douglas, 1995; Gorby and Peacock, 1988; Wood, 1975)

Table 3. Seafood-Borne Human Pathogens by Inoculation Route and Extent of Disease

	Dermal		GI	
	Localized	Systemic	Localized	Systemic
Aeromonas hydrophila group	+	+	+	+
Bacillus cereus	–	–	+	–
Campylobacter spp.	–	–	+	–
Clostridium botulinum	–	–	–	+
Clostridium perfringens	–	–	+	–
Edwardsiella tarda	+	+	+	+
Hepatitis A	–	–	+	+
Listeria spp.	–	–	+	+
Mycobacterium spp.	+	+[a]	–	–
Norwalk and related viruses	–	–	+	–
Parasites	–	–	+	+
Plesiomonas shigelloides	+	+	+	+
Salmonella spp.	–	–	+	+
Shigella spp.	–	–	+	+
Staphylococcus spp.	–	–	+	–
Streptococcus iniae	+	+	–	–
Vibrio spp.	+	+	+	+

[a] Systemic infection predominantly in immuno-suppressed individuals.

and *Leptospira* spp. (Douglas, 1995; Gill et al., 1985; Minette, 1983; Anderson et al., 1982; Robertson et al., 1981); agents uncommonly conveyed through seafoods (e.g., *Mycobacterium* spp.); and agents for which natural transmission from shellfish or finfish to humans has been rarely if ever documented (e.g., *Citrobacter* spp., *Klebsiella* spp., *Pasteurella* spp., *Proteus* spp., *Pseudomonas* spp., and *Yersinia* spp). Additionally, diseases caused by preformed toxins present at the time of harvest (e.g., dinoflagellate-generated biotoxins within shellfish; ciguatera and tetrodotoxin in finfish), histamine-forming bacteria (scombroid fish poisoning), and non-biological toxins (e.g., methylmercury and

Table 4. CDC-Reported Seafood-Borne Disease Outbreaks by Bacterial, Viral, and Unknown Etiologies, 1988–1992

	Shellfish	Finfish	Total Seafood	Percent[a]
Bacillus cereus	0	1	1	0.6%
Clostridium botulinum	0	11	11	6.1%
Salmonella spp.	4	4	8	4.4%
Shigella spp.	2	1	3	1.7%
Staphylococcus aureus	0	1	1	0.6%
Vibrio cholera	1	0	1	0.6%
Vibrio parahemolyticus	4	0	4	2.2%
Vibrio vulnificus	1	0	1	0.6%
Hepatitis A	1	0	1	0.6%
Etiology unknown	16	12	28	15.6%
Total	29	30	59	32.8%
Percent	49%	51%		

[a] Percentages based upon all reported disease derived from shellfish and finfish consumption, including chemical and parasitic etiologies.

polychlorinated biphenyls) will not be elaborated upon here, as comprehensive commentary on these subjects is provided elsewhere in this book and in several recent reviews (Mines et al., 1997; Eto, 1997; Grant, 1997; Fitzgerald et al., 1996; Underman and Leedom, 1993; Fang et al., 1991; Eastaugh and Shepherd, 1989). As a final note regarding this chapter, the term "seafood" will incorporate all forms of freshwater, brackish, and saltwater mollusks, crustaceans, and fish that are made commercially available for human consumption.

DISEASE AGENTS

Pathogens Ubiquitous to Aquatic Animals and Aquatic Environments

Bacteria

Infectious organisms transmitted by finfish to humans are basically restricted to bacteria and various forms of helminth and protozoan parasites. Conversely, bacteria take a backseat to human enteric viruses when it comes to shellfish-borne disease, at least in incidence if not severity (Rippey, 1994).

It is well known that the bacterial flora of aquatic animals tends to mirror the microbial population of their surroundings (Ghittino, 1972). Reflecting the extensive variety of bacteria that are indigenous to aquatic environments, pathogens in this collection are extremely diverse. Representatives are Gram-positive or Gram-negative, rods or cocci, hydrophilic or zoophilic, enteric or non-enteric, freshwater and/or marine, and are either tolerated by, or are harmful to, invertebrates and fish. Concerning humans, most of these bacteria are considered to be opportunists rather than primary pathogens. Accordingly, human exposure to these agents generates only sporadic individual cases and occasional outbreaks (the latter most often associated with consumption of undercooked or raw seafood). Among seafood-borne bacteria, a common theme is their ability to invade humans through puncture wounds and lesser skin breaks. For pathogens such as *Streptococcus iniae*, whose primary transmission route is transdermal rather than oral, increased human risk is associated with the offering to consumers of live fish versus processed products.

Aeromonas spp.

Within the family Vibrionaceae, members of the genus *Aeromonas* are currently partitioned into 13 hybridization groups that more or less correspond to various *Aeromonas* species. These Gram-negative, predominately motile, facultative anaerobic rods are commonly isolated from surface freshwaters, estuarine seawater, tap water (both raw and chlorinated), fish tanks, soil, freshwater and marine inhabitants, and various animal and vegetable foods of both aquatic and terrestrial origin (Hanninen et al., 1997; Tsai and Chen, 1996; Hassan et al., 1994; Holmberg, 1988; Sanyal et al., 1987; Karam et al., 1983). Given their cosmopolitan distribution, it is not surprising that *Aeromonas* spp. are recognized as normal microbial flora of both aquatic and terrestrial creatures (Kaper et al., 1981; Trust and Sparrow, 1974), although they are generally not considered to be normal inhabitants of the human alimentary tract (Davis et al., 1978). Seasonal occurrence of disease related to aeromonads has been correlated with accelerated multiplication of these bacteria in warm water (Holmberg, 1988; Williams and LaRock, 1985). *Aeromonas* spp. have been reported to elaborate a host of pathogenic factors, including hemolysins, cytotoxins, proteases, elastases, enterotoxins, heat-labile toxins, agglutinating substances, serum resistance factor, dermonecrotic factor, and adhesive factors (Tsai and Chen, 1996; Hassan et al., 1994; Santos et al., 1988; Karam et al., 1983). As testimony to the hardiness of these bacteria, aeromonads have been known to survive five weeks of freezing (Flynn and Knepp, 1987). They have also been shown to proliferate in a 36 percent carbon dioxide (CO_2) environment (Kaper et al., 1981), at 4 to 5°C (Palumbo, 1988), and in vacuum-packed food (Hudson et al., 1994). It is suspected, however, that post-processing contamination may be responsible for the isolation of aeromonads from heat-treated, aseptically packaged, or deep-frozen seafood products (Hanninen et al., 1997).

As a group, the motile aeromonads are well established as opportunistic pathogens of poikilotherms, especially in freshwater (Tsai and Chen, 1996; Santos et al., 1988). Conversely, the obligate fish pathogen *A. salmonicida* is not considered pathogenic for humans (Nemetz and Shotts, 1993). Whether isolated from the gastrointestinal tract of healthy fish (Santos et al., 1988), or from diseased fish tissues, *Aeromonas* species commonly reflect the distribution of bacteria within the surrounding water environment (Hanninen et al., 1997). In fish, motile aeromonad infections are known to manifest in peracute, acute, chronic, or subclinically latent forms. Lesions in fish associated with "motile aeromonas septicemia" are typical of fish bacteremias in general, and include external and internal hemorrhage, dermal ulcers, exophthalmia, ascites, and multifocal visceral necrosis. Additionally, motile aeromonads may combine with *Epistylis* spp. protozoa and other pathogens to produce the dermal and muscular excavations characteristic of "red sore."

Human exposure to pathogenic *Aeromonas* spp. has occurred through wound infections, and via ingestion of contaminated water or foods (Hanninen et al., 1997; Holmberg, 1988; Flynn and Knepp, 1987; Karam et al., 1983). Bivalve shellfish, such as raw or undercooked oysters, appear to be especially important vehicles of aeromonad-associated human disease, both as food items, and from puncture wounds (Hanninen et al., 1997; Tsai and Chen, 1996; Abeyta et al., 1986). There is some suggestion that *Aeromonas* strains isolated from foods of aquatic animal origin tend to be more toxigenic than strains recovered from poultry or red meats (Fricker and Thompsett, 1989). Alternatively, epidemiological evidence linking seafood-associated aeromonads to human disease may be circumstantial in some cases, as indicated by one study in which *Aeromonas* spp. hybridization groups commonly associated with human diarrhea were found to be uncommon among aeromonads obtained from samples of fish, fish eggs, and frozen shrimp (Hanninen et al., 1997). In addition to factors such as the particular species and strain of aeromonad implicated, the extent and severity of human illness secondary to *Aeromonas* spp. infection appears to depend heavily upon both the route of inoculation and the immunologic status of the host. There is evidence of increased risk to immune-compromised individuals, such as those with underlying liver disease or malignant neoplasms, although fulminate infections in immunocompetent hosts are also known to occur (Karam et al., 1983). The *Aeromonas* species that are most commonly implicated in human infections include *Aeromonas hydrophila*, *A. sobria*, and *A. caviae* (Palumbo et al., 1989; Holmberg, 1988). Clinical consequences of nonfatal and fatal aeromoniasis in humans include diarrhea, gastroenteritis, wound infections, endocarditis, meningitis, cellulitis, myositis, osteomyelitis, corneal ulcer, and septicemia (Tsai and Chen, 1996; Hassan et al., 1994; Flynn and Knepp, 1987; Karam et al., 1983). It is suggested that human aeromonad infections may be prevented by washing seafood prior to consumption (Flynn and Knepp, 1987).

Edwardsiella tarda

This small, motile, Gram-negative, facultative anaerobic, rod-shaped member of the family Enterobacteriaceae may be found worldwide in freshwater environments and within the alimentary tracts of at least four vertebrate classes (Humphrey et al., 1986). The major natural reservoirs of *E. tarda* are reported to be reptiles and freshwater fishes (Wyatt et al., 1979).

Although considered part of the normal flora of the fish intestinal tract (Vandepitte et al., 1983), E. tarda is responsible for serious parenteral disease in a wide range of fish species (Humphrey et al., 1986). Signs of septicemic edwardsiellosis in fish include lethargy, abnormal and erratic swimming, ascites, cutaneous ulcers, deep muscle abscesses, and focal suppurative, granulomatous, or necrotic lesions within major viscera. *E. tarda* additionally causes "emphysematous putrefactive disease of catfish," an aesthetically detrimental condition of profound economic importance to the catfish aquaculture industry. Ornamental tropical fishes are also susceptible to *E. tarda* infection and human disease has been associated with latently-infected aquarium specimens (Vandepitte et al., 1983).

Edwardsiella tarda is the only member of its genus that is a recognized human pathogen (Wilson et al., 1989). Human *E. tarda* infections are uncommon, especially outside of tropical or subtropical regions (Vandepitte et al., 1983). Predisposed persons include neonates, infants, and individuals with preexisting illnesses (Wilson et al., 1989; Vandepitte et al., 1983). Typical of bacteria that are endemic to aquatic environments, *E. tarda* infections can be caused by penetrating wounds, or by ingestion of contaminated water or infected fish

(Koshi and Lalitha, 1976). Subsequent to either route of inoculation, human edwardsiellosis may remain localized (e.g., deep soft-tissue abscesses or gastroenteritis) or become generalized (e.g., septicemia with meningitis) (Wilson et al., 1989; Humphrey et al., 1986; Farmer et al., 1985). Comparable to other types of enteric bacteria, the relationship of *E. tarda* infection to puncture wounds is far more easily established than its role as a gastrointestinal pathogen (Vandepitte et al., 1983).

Plesiomonas shigelloides

Plesiomonas shigelloides (formerly *Aeromonas shigelloides*) is a motile, facultative anaerobic, Gram-negative bacillus that is currently classified within the family Vibrionaceae. This classification status may soon change, however, as recent RNA sequence analysis studies have demonstrated a closer kinship to *Proteus* spp. within the Enterobacteriaceae (MacDonell and Colwell, 1985). Widely distributed in water and soil, *P. shigelloides* can be isolated from the alimentary tracts of animals and humans (Gopal and Burns, 1991; Brenden et al., 1988), although it is not considered to be a commensal organism in people (Arai et al., 1980). *P. shigelloides* is primarily an inhabitant of fresh and brackish waters within tropical, subtropical, and temperate regions, including coastal waters of the Gulf of Mexico (Brenden et al., 1988; Holmberg, 1988; Arai et al., 1980). Replication of *P. shigelloides* is facilitated by temperatures above 8°C, a factor that explains the seasonal pattern (summer months) of human outbreaks attributed to this organism (Brenden et al., 1988).

Although probably capable of causing septicemic disease in fish, *P. shigelloides* is not a common fish pathogen, much less a primary one. Within warmer climates, *P. shigelloides* has been known to colonize shellfish and fish (Arai et al., 1980). *P. shigelloides* has been isolated from the intestines, feces, gills, and external surfaces of fishes [Atlanta Centers for Disease Control (CDC), 1989], and from tropical fish aquarium water (Sanyal et al., 1987).

Human exposure to *P. shigelloides* occurs via water or insufficiently cooked seafoods. Alimentary disease is thought to be the primary outcome of infection (Hassan et al., 1994; Gopal and Burns, 1991; Brenden et al., 1988). Three reported gastrointestinal forms include secretory, invasive (*Shigella*-like), and cholera-like (Brenden et al., 1988). Despite a reasonable amount of accumulated data on plesiomonad-associated diarrhea, the enteropathogenicity of this organism remains controversial (Brenden et al., 1988). Compared to members of the *Vibrionaceae* such as *Vibrio* spp. and other *Aeromonas* spp., *P. shigelloides* is generally regarded as a less important gastrointestinal pathogen (Brenden et al., 1988; Holmberg, 1988). There is evidence that *P. shigelloides*-related disease is underrecognized. Such evidence includes a recent increase in numbers of reported cases, and the difficulty experienced by some commercial laboratories in isolating and identifying these bacteria (Brenden et al., 1988). Another problem in establishing a role for this organism in enteric illness is an inability to adequately characterize universally-accepted diarrheogenic mechanisms (Brenden et al., 1988). Virulence factors such as invasiveness appear to be strain-dependent, and the debate over enterotoxin production is unsettled (Brenden et al., 1988; Holmberg, 1988). Presently, a suitable animal model for *P. shigelloides*-induced disease does not exist, although it is noteworthy that human volunteers fed *P. shigelloides* did not develop diarrhea (Herrington et al., 1987). Extraintestinal inflammatory conditions attributed to *P. shigelloides* invasion and dissemination include meningitis, sepsis, cellulitis (due to puncture wounds), arthritis, endophthalmitis, pleural effusion, pseudoappendicitis, and cholecystitis (Gopal and Burns, 1991; Holmberg, 1988). Systemic infection may occur secondary to hematogenous spread from sites of cutaneous or oral inoculation (Ellner and McCarthy, 1973). *Plesiomonas*-induced disease has been linked to underlying immuno-compromising conditions. Additionally, it has been demonstrated that certain lifestyle choices can increase the susceptibility of immunocompetent individuals to infection. Specifically identified risk factors include raw oyster consumption, crabbing, travel (especially to Mexico), and self-medication with antacids. Occupations at elevated risk include fish handlers, aquaculturists, veterinarians, zoo keepers, and water sports performers (Brenden et al., 1988). In addition to oysters, seafood items implicated as vehicles for plesiomonad infection include salt mackerel (Hori et al., 1966) and cuttlefish salad (Miller and Koburger, 1985). It has been stated that low cooking temperatures should be lethal for *P. shigelloides* (Hackney et al., 1992).

Streptococcus iniae

This facultative anaerobic, Gram-positive, non-motile, non-sporulating coccus was first reported as a cause of abscesses in freshwater dolphins (*Inia geoffrensis*) in 1976 (Pier and Madin, 1976). A more commonly encountered milieu for *Streptococcus iniae* appears to be fish skin. Although the observation of recurrent *S. iniae* infections in certain aquaculture situations implies the presence of an extra-piscine reservoir, recovery of this bacterium from the aquatic environment is not uniformly successful (Perera et al., 1997). *S. iniae* appears to be tolerant of a wide range of ambient conditions (Perera et al., 1997). In one study, environmental components such as increased water temperature, increased salinity, and pH above neutral seemed to enhance infectivity in experimentally inoculated hybrid tilapia (Perera et al., 1997).

Unfortunately for fish, *S. iniae* is not necessarily limited to surface colonization. In commercially-reared tilapia (*Oreochromis* spp.), which appear to be common hosts for this organism, either dermal colonization or invasive disease may occur (Weinstein et al., 1997; Eldar et al., 1994). Experimental infection of tilapia has been accomplished via gavage or directly through the water (Perera et al., 1997). Ensuing internal spread of disease in food fishes such as tilapia, rainbow trout, and coho salmon typically manifests as meningoencephalitis (Perera et al., 1994). Other septicemic-type lesions such as ascites, multifocal cutaneous petechial hemorrhage, exophthalmia, and corneal opacity were noted in experimentally-infected tilapia (Perera et al., 1997), and were observed to a lesser extent in hybrid striped bass (Perera et al., 1997). Mortality rates of 30 to 50 percent have been recorded for *S. iniae*-infected fishponds (Eldar et al., 1994).

S. iniae is not the only fish-associated *Streptococcus* sp. that is potentially transferable to humans, despite the publicity associated with recent events. *S. iniae* became newsworthy in early 1996 when the cleaning of cultured whole tilapia by consumers was linked to three cases of human cellulitis, plus a single case of septicemic arthritis, meningitis, and endocarditis (Weinstein et al., 1996). Subsequently, retrospective investigation identified five additional consumer victims whose *S. iniae* infections were all caused by puncture wounds or lacerations acquired during whole fish preparation (Weinstein et al., 1997). In each of these instances, the fish had been purchased live, and then killed, gutted, and placed in cold storage for up to two days (Weinstein et al., 1997). It has been suggested that this type of handling practice may increase the potential for trauma-associated *S. iniae* infection (Weinstein et al., 1997). Prior to these incidents, *S. iniae* had not been known to cause disease in people. Subsequently, discussion evolved as to whether this bacterium represented a new or unrecognized human pathogen (Weinstein et al., 1996). Factors cited as supportive of the latter claim are the low frequency of routine cultures performed in human cellulitis cases, and the failure of some diagnostic laboratories to correctly identify *S. iniae*. Conversely, the recent emergence of *S. iniae* as a commercially-relevant fish pathogen has been presented as an argument for its neoteric origin (Weinstein et al., 1997).

Vibrio spp.

Vibrio is a rather large genus of motile, non-sporeforming, facultative anaerobic, Gram-negative, straight or curved rods within the family *Vibrionaceae*. Although *Vibrio* spp. may be isolated from both fresh and estuarine water (Hlady and Klontz, 1996; Hassan et al., 1994), the majority of these halophilic (salt-loving) species are ubiquitous in marine environments where their density is dependent upon temperature and salinity (Rippey, 1994; Hackney et al., 1992). Similar to motile aeromonads, *Vibrio* spp. often demonstrate a tendency to proliferate in the warmer summer months (Eastaugh and Shepherd, 1989; Holmberg, 1988). In coastal areas, *Vibrio* spp. may be commonly found on the external surfaces and within the gastrointestinal tract of finfish, and within the flesh of shellfish. Unlike some other seafood-borne bacterial human pathogens, shellfish accumulation of *Vibrio* spp. is not correlated with levels of human sewage pollution (Dalton, 1997; Hackney et al., 1992).

Numerous *Vibrio* species (e.g., *V. salmonicida*, *V. anguillarum*, *V. damsela*, *V. alginolyticus*, *V. vulnificus*, *V. ordalii*, et al.) are important pathogens for various marine fishes, where primary or secondary invasion of fish tissues can result in ulcerative cutaneous disease and/or septicemia. *Vibrio* spp. isolated from fish have been shown to produce an assortment of virulence factors including hemolysins, cytotoxins, and heat labile toxins (Hassan et al., 1994). In general, bacterial disease of bivalve mollusks does not reach the level of diversity observed in fish. Two *Vibrio* species, *V. anguillarum* and *V. alginolyticus*, are the agents of "bacillary necrosis" in cultured oyster larvae (Tubiash et al., 1965). In this sporadically devastating affliction, culpable bacteria have been demonstrated to produce a cytolytic toxin (Brown and Losee, 1978).

Vibrio spp. infections in humans are most often acquired through the ingestion of contaminated food or water, or by the direct inoculation of either traumatically-induced or preexisting skin defects (Hlady and Klontz, 1996). Shellfish are common vehicles of *Vibrio*-induced human illness. Of infectious diseases transmitted by mollusks, only enteric human viruses have a higher disease incidence (Hackney et al., 1992). In the United States, the number of outbreaks and cases of human vibriosis linked to raw oyster consumption vastly exceed infections attributed to other bivalves (Hlady and Klontz, 1996; Rippey, 1994). Gastroenteritis, wound infections, and septicemia are the three most common clinical presentations of *Vibrio* spp. infections in humans (Morris and Black, 1986). Other reported *Vibrio*-induced ailments include conjunctivitis, otitis media, respiratory tract infections, cerebrospinal fluid infections, and hepatic abscesses (Hassan et al., 1994).

The manner and degree to which a *Vibrio* bacterium manifests disease in a human individual is dependent upon variables such as the contaminating medium, the route of infection, and the immune status of the host. Also important is the particular species and strain of *Vibrio* isolated. Of an array of seafood-harbored *Vibrio* spp. that are potential human pathogens, the four species or groups that are usually incriminated in the United States are *V. cholera*-01 [currently the *El tor* biotype in the Western Hemisphere (Hackney et al., 1992)], *V. cholera* non-01, *V. parahaemolyticus*, and *V. vulnificus* (Hlady and Klontz, 1996; Fang et al., 1991). Any of these species (and other vibrios) can cause mild to severe enteric or parenteric disease. For *V. cholera* serotype 01, the severity of the associated diarrhea is based upon strain-dependent generation of the classic cholera toxin. Acting via the stimulation of excess cyclic AMP production within the human enterocyte luminal cell membrane, this toxin (and analogous substances uncommonly encountered in other *Vibrio* spp.) is responsible for the voluminous "rice water" diarrhea that can lead to severe dehydration and circulatory collapse (Hackney et al., 1992; Eastaugh and Shepherd, 1989). Conversely, the domestically-prevalent non-toxigenic strains of *V. cholera* 01 tend to provoke milder diarrhea and wound infections (Murphree and Tamplin, 1995; Hackney et al., 1992).

Considered to be less catastrophic than classic cholera, enteric disease attributed to the strains grouped as *V. cholera* non-01 can be more invasive, and consequently, bloody stools (dysentery) are sometimes encountered (Eastaugh and Shepherd, 1989). A high correlation exists between the occurrence of non-01 gastroenteritis and the ingestion of raw or undercooked shellfish originating from the Gulf of Mexico (Levine et al., 1993; Morris et al., 1981).

V. parahaemolyticus was previously thought to be a significant source of crustacean shellfish-associated gastroenteritis and food poisoning in Japan, but rarely in the United States (Hackney et al., 1992; Eastaugh and Shepherd, 1989). Antithetically, a recent report indicates that *V. parahaemolyticus* is the species most likely to cause human disease in Florida (Hlady and Klontz, 1996), and epidemiologic data indicates that it may now be the most common cause of mollusk-derived vibriosis in the United States (Rippey, 1994; Hackney et al., 1992). Contrary to the last statement are surveys which indicate that very few seafood isolates of *V. parahaemolyticus* possess the appropriate virulence factors necessary for gastroenteritis induction in humans (Blake, 1980; Fujino et al., 1974).

Currently, *V. vulnificus* is perhaps the most dangerous *Vibrio* species in the United States (Holmberg, 1988). Like other *Vibrio* spp., *V. vulnificus* infection is associated with the consumption of raw shellfish or the occurrence of penetrating wounds (Vartian and Septimus, 1990; Eastaugh and Shepherd, 1989; Holmberg, 1988). Dissimilar to other members of its genus, however, *V. vulnificus* is more liable to cause chronic deep-seated soft-tissue infections and septicemia than gastroenteritis, with an accompanying mortality rate as high as 50-90 percent in untreated patients, and 25 percent in medically-managed individuals (Eastaugh and Shepherd, 1989; Klontz et al., 1988). Well-defined factors predisposing to *V. vulnificus* infection and disease include the ingestion or shelling of Gulf Coast shellfish, summer seasonality, foreign travel (e.g., Mexico), antacid consumption, underlying liver disease (e.g., hepatic cirrhosis), diabetes, alcoholism, immunosuppressive disorders, and hemochromatosis (the last reflecting a possible preference of *Vibrio* spp. for transferrin-rich environments) (Hlady and Klontz, 1996; Rippey, 1994; Fang et al., 1991; Vartian and Septimus, 1990; Holmberg, 1988).

It would seem that vibriosis should be generally preventable. Risk factors for human vibriosis have been outlined, monitoring and decontamination procedures for shellfish-associated bacterial pathogens are currently in effect, and *Vibrio* spp. are readily inactivated by cooking (Hackney et al., 1992; Fang et al., 1991). There are several explanations for the continued emergence of human cases each year. First, although point-of-sale warnings regarding raw seafood consumption are mandatory in a number of states (Amaro and Biosca, 1996; Hlady and Klontz, 1996), posted notices have had limited detectable effect upon consumers (Amaro and Biosca, 1996). Second, current monitoring techniques and decontamination practices (such as depuration and relaying) are based upon indicator (coliform) bacterial counts, and therefore do not adequately reflect or alter shellfish *Vibrio* spp. concentrations (Murphree and Tamplin, 1995; Hackney et al., 1992; Fang et al., 1991). In actuality, *Vibrio* spp. have been shown to multiply up to tenfold during two days of depuration (Murphree and Tamplin, 1995). Lastly, parenteral vaccination against human vibriosis is presently considered impractical and ineffective (1980).

Parasites

A variety of nematode, cestode, trematode, and protozoan parasites may be present in fish and shellfish; however, only a few are actually infective and harmful to humans. Parasite infection generally occurs through the ingestion of infected fish, especially undercooked or raw specimens (Ahmed, 1991). In most seafood-vectored parasitic disease, humans are terminal hosts, accidentally interrupting normal maturation events in parasite development. Factors governing the actual incidence of such infections include specific features of the parasite's life cycle, the endemic geographic location of the parasite, and representative lifestyles of the human population at risk.

Nematodes

Worldwide, the most important zoonotic nematode disease acquired from fish is anisakidosis, caused by larvae of *Anisakis simplex*, and less commonly by *Pseudoterranova decipiens*. The larval stages of these parasites may be present in a wide variety of marine hosts, but are most commonly found in herring, mackerel, sardines, and walleye pollock (Ogawa, 1996; Bouree et al., 1995; Ahmed, 1991). Human anisakidosis occurs through the consumption of improperly cooked or raw fish, which have been responsible for over 3000 cases annually in Japan (Fountaine, 1985). Human infections are associated with acute abdominal distress accompanied by a severe inflammatory response, and symptoms include epigastric or diffuse abdominal pain, nausea, vomiting and chronic diarrhea (Bouree et al., 1995). Other reported agents in fish-borne human nematode infections include members of the genera *Contraceacum* (Ahmed, 1991), *Eustrongyloides* (Wittner et al., 1989), and *Capillaria* (Cross and Basaca-Sevilla, 1991).

A human nematode infection acquired from eating raw snails (or less commonly, undercooked crabs, prawns, or fish) is *Angiostrongylus cantonensis*. Historically, humans infected with *A. cantonensis* larvae have developed eosinophilic meningoencephalitis (Rosen et al., 1967). From sporadic cases to full-blown epidemics, human *A. cantonensis* infections have been reported from the South Pacific, Southeast Asia, Taiwan (Alicata, 1988; Kliks et al., 1982), and most recently, Puerto Rico (Anderson et al., 1986).

Cestodes

Only two species of fish-borne cestodes, *Diphyllobothrium latum* and *D. pacificum*, are commonly associated with human illness (Von Bondsdorff, 1977). Both are acquired through the ingestion of the larval plerocercoid stage in raw freshwater finfish, most notably salmon. Infections have been reported worldwide and include the Baltic Sea countries, China, Europe, Africa, Japan, Australia, the Great Lakes region of the United States, Alaska, and Canada (Ahmed, 1991). Human diphyllobothriasis may cause symptoms of abdominal discomfort, fatigue, diarrhea, constipation and megaloblastic anemia, though many infected individuals apparently remain asymptomatic (Ruttenber et al., 1984). Other cestodes that are reported to infect humans following fish consumption include members of the genera *Digramma brauni* and *Ligula intestinalis* (Jensen and Greenlees, 1988).

Trematodes

At least four species of digenetic trematode occasionally produce disease in humans (Harinasuta et al., 1993; Ahmed, 1991; Brier, 1990). These include *Clonorchis sinensis* (the Chinese liver fluke), *Opisthorchis* spp., *Paragonimus westermani* (the Oriental lung fluke) and *Heterophyes* sp. (an intestinal fluke). All are transmitted to humans by the consumption of the metacercarial larval stage in raw or improperly prepared fish (freshwater, brackish, or marine), crabs, or crayfish. All of these parasites are endemic to Asia and some occur sporadically in the Mediterranean, the lower Nile valley and western India (Ahmed, 1991). The majority of human trematode infections are asymptomatic, but some individuals may display clinical signs of abdominal discomfort, diarrhea, fever, myalgia, and hepatitis (Harinasuta et al., 1993).

Additional reports of trematode-induced disease involve a number of species of *Echinostoma*, and *Nanophyetus salmincola*. *Echinostoma* spp. trematode infections are usually acquired by eating raw or undercooked snails, tadpoles, or frogs containing the encysted metacercariae (Tangtrongchitr and Monzon, 1991; Waikagul, 1991). *Nanophyetus salmincola* is well-recognized as a canine pathogen, whereas it is only recently that humans were discovered to be potential hosts. The range of clinical complaints associated with *N. salmincola* infection includes increased frequency of bowel movements, abdominal discomfort, diarrhea, nausea and vomiting, weight loss, fatigue, and significant eosinophilia (Eastburn et al., 1987). *N. salmincola* is usually ingested in raw or undercooked fish containing the encysted metacercarial stage; however, there is a single case report of a patient infection caused by hand-to-mouth contamination. This episode occurred during the handling of naturally infected salmon (Harrell and Deardorff, 1990).

Protozoans

The literature includes rare reports of human protozoan infections derived from fish or shellfish. In one example, improperly prepared fish may have been the source of human giardiasis (Osterholm et al., 1981), whereas bivalves have been incriminated in human infections of *Cryptosporidia parvum* (Jensen and Greenlees, 1988).

Pollution-Derived Pathogens

Bacteria

As previously indicated, the majority of seafood-borne infections in the United States are thought to be of viral origin, and this statement is especially valid for pathogens associated with environmental contamination. Rationalizations for the comparatively lower incidence of pollution-derived bacterial disease include the efficacy of chlorine against enteric bacteria as used in wastewater treatment facilities (Rippey, 1994), and the relative success of coliform counts and mollusk decontamination procedures to detect and eliminate harmful concentrations of bacterial pathogens. Bacterial genera that have been linked to human sewage and/or production animal waste effluents include *Aeromonas, Campylobacter, Escherichia, Plesiomonas, Salmonella, Shigella* (Rippey, 1994), *Listeria*, and *Staphylococcus*. One potential pathway for *Salmonella* spp. contamination is via wastewater runoff due to the application of raw chicken manure as pond fertilizer (Buras, 1993). As yet, there is no conclusive evidence connecting animal wastes to human illnesses derived from shellfish (Stelma and McCabe, 1992). Not surprisingly, serologic evidence of exposure to human bacterial pathogens has been documented in fish proximate to heavily populated coastal areas (Janssen and Meyers, 1968). In other instances, it should be remembered that isolation of enteric bacteria from raw or processed seafood may very likely reflect product handling or storage rather than habitat pollution.

Campylobacter spp.

Campylobacter spp. are motile, pleomorphic, non-sporulating, microaerophilic, Gram-negative, curved-to-spiral rods that are grouped with genera such as *Alteromonas* and *Helicobacter*. Unlike *Vibrionaceae* bacteria, *Campylobacter* spp. are incapable of carbohydrate fermentation. Pathogenic species for humans include *C. jejuni*, *C. coli*, and *C. laridis*. Members of the *Campylobacter* genus vary in pathogenicity, and they often tend to be zoophilic. Typical habitats in both animals and humans include oral and reproductive tract mucus membranes and the intestinal tract (Holt et al., 1994). Waterborne *Campylobacter* spp. infections can occur, and these may be related to fecal contamination (Hackney et al., 1992). Fish and shellfish are not considered to be important vehicles for human campylobacteriosis (Eastaugh and Shepherd, 1989; Griffin et al., 1983), but low numbers of shellfish-related outbreaks have occurred in recent years (Hackney et al., 1992).

Salmonella spp.

Grouped within the Family Enterobacteriacea, *Salmonella* spp. are motile, facultative anaerobic, Gram-negative, straight rods that inhabit the alimentary systems of a vast array of homeothermic and poikilothermic vertebrates and invertebrates. This entourage includes virtually all domestic animals and pets, ornamental aquarium fish, mollusks, crustaceans, and even insects (Sanyal et al., 1987; Sasaki and Minette, 1985; Minette, 1984). Although not considered pathogenic for fish or shellfish (Sanyal et al., 1987; Minette, 1986) *Salmonella* spp. have been shown to persist in teleosts for up to 38 weeks (Heuschmann-Brunner, 1974). There is evidence that stress-induced shedding of *Salmonella* spp. may occur in latent carrier fish and other aquatic animals (Sanyal et al., 1987; Sasaki and Minette, 1985). Both freshwater and seawater can support *Salmonella* spp. growth, and certain members of this genus may survive for considerable periods of time in brackish or marine environments (Sasaki and Minette, 1985; Minette, 1984). *Salmonella* spp. have been isolated from marine fish, fish meal, oysters, and shrimp (Minette, 1986; Sasaki and Minette, 1985). Recovery of *Salmonella* spp. from natural waters is generally associated with human or animal pollution. It has been observed that fecal coliform counts and levels of *Escherichia coli* are satisfactory indicators for *Salmonella* spp. contamination, and for shellfish safety with respect to salmonellosis (Hackney et al., 1992; Andrews et al., 1976). Conversely, postharvest introduction and amplification of *Salmonella* spp. may also occur, as a proportion of food animals captured in comparatively pristine aquatic environments yield numerous salmonellae only following commercial slaughter and processing (Minette, 1984).

Salmonellosis is the most frequently reported zoonotic disease in the United States (Sasaki and Minette, 1985), and numbers of known cases continue to rise yearly (Fang et al., 1991). As an illustration of this point, *Salmonella* spp. accounted for 4.4 percent of all causes of seafood-borne disease outbreaks reported to the Centers for Disease Control and Prevention between 1988 and 1992 (Table 4) (CDC, 1996). Of the bacterial and viral agents reported in this survey, *Salmonella* spp. were second only to *Clostridium botulinum* in etiologic incidence. Symptoms of human *Salmonella* spp. infection generally develop 12 to 14 hours after food ingestion, producing disease that usually lasts two to three days (Hackney et al., 1992). Manifestations include nausea, diarrhea, vomiting, abdominal pain, headaches, and chills (Hackney et al., 1992). Human enteric salmonellosis is usually acquired through ingestion of bacteria-laden food products (Minette, 1984), although contamination via direct contact with animal wastes is an additional concern for persons in animal-handling occupations (Sasaki and Minette, 1985). Raw foods of animal origin are the major vectors for human infections, and in the United States, the outstanding candidates in this category are beef, turkey, homemade ice cream, pork, and chicken (Cox and Bailey, 1987; Jay, 1986). In comparison to these food items, salmonellosis secondary to seafood consumption is much less prevalent (CDC, 1996; Hackney et al., 1992; Eastaugh and Shepherd, 1989). Prior to 1950, the most common illness attributable to the ingestion of raw shellfish was typhoid fever (*S. typhi*). Human typhoid cases have not been reported in the United States during the past 40 years (Rippey, 1994); however, other *Salmonella* spp. pathogens continue to be isolated from shellfish (Hackney et al., 1992). Thus shellfish remain a persistent, albeit diminished, threat for human salmonellosis in the United States (Frasier et al., 1988).

Shigella spp.

Representing another genus in the Enterobactericeae, *Shigella* spp. are non-motile, non-sporulating, facultative anaerobic, Gram-negative, straight rods that can be found in the alimentary tracts of humans and non-human primates. Analogous to *Salmonella* spp. contamination, the introduction of *Shigella* spp. to shellfish may occur before or after harvest. Raw shellfish appear to be the predominant vehicle for *Shigella* spp. infections that are due to seafood consumption. Compared to salmonellosis, however, the incidence of shellfish-derived shigellosis is typically low (CDC, 1996; Hackney et al., 1992).

Clinical shigellosis is basically limited to humans and non-human primates. *Shigella* spp. possess the ability to both adhere to, and invade, the host intestinal epithelium. Therefore, symptoms in humans often include fever and dysentery, in addition to the more mundane enteric manifestations of vomiting, diarrhea, abdominal cramps, and tenesmus (Hackney et al., 1992).

Viruses

Raw molluscan shellfish are one of the most hazardous food items on the planet based upon the reality that 1/2000 meals results in human disease [Murphree and Tamplin, 1995; National Academy of Science (NAS), 1991]. Accumulated data indicates a dramatic rise in cases and outbreaks of shellfish-associated infectious disease within the past two decades (Rippey, 1994). The propensity for these filter feeders to transmit human pathogens is linked to several characteristics of shellfish animals and the industry they support. Such characteristics include the ability of bivalve mollusks to accumulate and concentrate microbes, bivalve settlement of coastal waters contaminated by the products of human habitation, illicit harvesting of shellfish from regulated sites, and inadequate seafood handling and processing procedures (Murphree and Tamplin, 1995; Rippey, 1994; Fang et al., 1991). It has been documented that enteric virus transmission can occur despite proper hygiene and the application of recommended handling procedures. Such situations provide evidence that the most significant aspect of shellfish-borne disease is the tendency for humans to eat uncooked or undercooked oysters and hard clams (Chalmers and McMillan, 1995; Fang et al., 1991).

Currently in the United States, human enteric viruses account for the majority of infections attributable to shellfish (Hackney et al., 1992). Paradoxically, published incidence reports of seafood-borne human outbreaks often do not defend this statement. Acknowledged underreporting of shellfish-associated viral gastroenteritis can be ascribed to the mildness of many cases, and to the lack of widely available assays used to detect enteric viruses in stool samples (Rippey, 1994). In one surveillance report, the etiology was undetermined in 16 percent of all seafood-associated outbreaks (Table 4) (CDC, 1996). It is plausible that enteric viral infections comprised a significant portion of these unknowns.

Marine bivalve mollusks may concentrate human viruses up to 100-fold within their alimentary tracts and digestive glands (Fang et al., 1991). Once accumulated, such viruses are incapable of multiplication within invertebrates (Hackney et al., 1992). Shellfish-transmitted cases of viral gastroenteritis tend to occur in the late spring and late fall, a seasonal pattern that has been attributed to maximum bivalve bioaccumulation and greater seafood consumption (Rippey, 1994). Among the more than 100 enteric viruses found in human feces, several pathogens alleged to be communicable through shellfish include the causes of Hepatitis A and E, Norwalk and Norwalk-like agents, the Snow Mountain agent, and small round viruses (Hackney et al., 1992). Conversely, polioviruses isolated from shellfish are considered to be benign vaccine strains (Hackney et al., 1992). The most prominent perpetrators in this assemblage, Hepatitis A virus and the Norwalk virus group, are discussed individually below.

Hepatitis A

One of the most serious shellfish-associated pathogens, the Hepatitis A agent is a 24- to 27-nanometer (nm) virus in the Family Picornaviridae. The route of transmission is usually fecal to oral through contaminated food, water, and interpersonal contact (Hackney et al., 1992). Not surprisingly, this enteric pathogen is endemic in underdeveloped countries and in regions where poor sanitation is the rule. The mishandling of food by infected persons is the most common cause of Hepatitis A outbreaks (Cliver, 1988). Nevertheless, within the past four decades, several large outbreaks of Hepatitis A in the United States have been tied to raw oysters and clams procured from contaminated areas, including illegally harvested bivalves from restricted sites (Fang et al., 1991).

Subsequent to ingestion, Hepatitis A virus proliferates in the intestinal epithelium, followed by dissemination to the liver via the portal blood supply (Hackney et al., 1992). Clinical manifestations of Hepatitis A are usually mild, but can be severe, and include fatigue, anorexia, fever, myalgia, nausea, abdominal discomfort, jaundice, hepatomegaly, and alterations of biomedical parameters consistent with hepatitis (Hackney et al., 1992; Fang et al., 1991). The prolonged incubation period of up to six or eight weeks between ingestion and onset of symptoms can obfuscate epidemiological investigations (Rippey, 1994; Hackney et al., 1992; Eastaugh and Shepherd, 1989). On the positive side, Hepatitis A virus can be inactivated by boiling in water for greater than 60 seconds (Dupont, 1986; Hughes et al., 1969).

Norwalk and Norwalk-like Agents

This group of unclassified 25- to 30-nm calicivirus-like agents may be the most common cause of shellfish-vectored acute gastroenteritis. Various names of Norwalk-related viruses include Montgomery County, Snow Mountain, Hawaii, Ditchling, Southampton, Taunton, Sapporo, and Otofuke (Gouvea et al., 1994). Human infection occurs through ingestion of contaminated water and shellfish, through ingestion of other raw or lightly-cooked foods tainted through the preparation process, or through interpersonal contact (Gouvea et al., 1994; Hackney et al., 1992). Norwalk virus can persist in chlorine levels that inactivate most enteric bacteria (Keswick et al., 1985), and, like Hepatitis A virus (Richards, 1988), this pathogen appears to be resistant to shellfish depuration procedures (Gouvea et al., 1994).

The onset time and duration of illness are each approximately 24 to 48 hours. Abdominal cramps, vomiting, and/or diarrhea are typical manifestations of Norwalk virus infection (Gouvea et al., 1994; Rippey, 1994; Fang et al., 1991). Symptom-based diagnosis is common, as Norwalk virus has not yet been cultivated, and tests are not widely available (Gouvea et al., 1994; Hackney et al., 1992). Infection can be prevented by steaming shellfish for at least four to six minutes (Fang et al., 1991; Eastaugh and Shepherd, 1989).

Pathogens Introduced or Amplified during Product Handling and Processing

For some of the pathogens isolated from shellfish and finfish products, transmission to humans may be virtually eliminated by precautions of proper hygiene, sanitation, and refrigeration following harvest. Members of this subset include *Bacillus cereus*, *Clostridium* spp., *Listeria* spp., and *Staphylococcus* spp. The seafood-vectored incidence of other zoonoses, such as salmonellosis, shigellosis, and enteric viral infections, would also be diminished by appropriate handling practices. Essentially, there are two mechanisms for the accretion of hazardous numbers of bacteria in harvested seafood items: 1) inoculation of the product by contaminated humans, equipment, or holding facilities (e.g., *Listeria* spp., *Staphylococcus* spp., *Salmonella* spp., *Shigella* spp., and enteric viruses), and 2) improper or inadequate storage conditions that favor the proliferation of bacterial pathogens (e.g., *B. cereus* and *Clostridium* spp.).

Bacteria

Bacillus cereus

In addition to its role as a mastitis agent and abortifacient in ruminants, spores of this aerobic or facultative anaerobic, Gram-positive rod have been known to germinate in a variety of foods including meats, desserts, sauces, soups, and fried rice (Carter et al., 1995). *Bacillus cereus* food poisoning usually takes one of two forms: emetic or diarrheal. The predominance of one form over another depends upon the presence of heat-stabile (emetic) or heat-labile (diarrheal) protein metabolites (Hackney et al., 1992). Reportedly, the *B. cereus* emetic syndrome resembles *Staphylococcus aureus* intoxication, whereas the diarrheal form, which is more commonly associated with shellfish consumption, is similar to *Clostridium perfringens* food poisoning (Hackney et al., 1992).

Clostridium spp.

Clostridium species are anaerobic or microaerophilic, spore-forming, generally motile, Gram-positive large rods. The two clostridial bacteria indicted as seafood-borne disease agents are *C. perfringens* and *C. botulinum*. *C. perfringens* can be found in most samples of soil and raw meat (Shandera et al., 1983), as well as the intestines of animals (Hackney et al., 1992). Proteinaceous materials provide optimal substrate for this bacillus (Hackney et al., 1992). *C. perfringens* spores can survive cooking, and it is not unusual for bacterial proliferation to occur in foods that are kept inadequately warm post-preparation (Bishai and Sears, 1993; Hackney et al., 1992). The most common strain of *C. perfringens* associated with food poisoning is type A (Hackney et al., 1992). Unlike *Staphylococcus aureus* food poisoning, which involves the consumption of preformed toxins, vegetative cells of *C. perfringens* are ingested and sporulate in the small intestine, releasing enterotoxin (Shandera et al., 1983). *C. perfringens* gastroenteritis tends to be mild, and low numbers of cases associated with seafood consumption are reported each year (Hackney et al., 1992). The disease usually presents acutely following ingestion of tainted food, and generally lasts less than 24 hours (Mines et al., 1997). Clinical symptoms include abdominal discomfort and watery stools (Mines et al., 1997; Hackney et al., 1992).

Spores of the obligate anaerobe *Clostridium botulinum* are typically found in soil, in marine sediment, and within fish intestines, but they are not considered to be a normal part of human gastrointestinal flora (CDC, 1991). Accordingly, toxin production usually occurs within the food source rather than the alimentary tract of victims (an exception is infant botulism in humans). Of the seven or eight different neurotoxins known to be produced by *C. botulinum*, the three most commonly implicated in human disease are A, B, and E (Mines et al., 1997). Botulism associated with fish products is most often caused by type E toxin (CDC, 1992). Conditions that tend to favor toxin production include low ambient oxygen concentration, pH > 4.6, lack of competing bacteria, warmth, and increased water in the medium (CDC, 1995). Food contaminated by *C. botulinum* may or may not show obvious signs of adulteration (Mines et al., 1997). In the United States, most cases of botulism are associated with home canning of vegetables, fruits, and condiments (CDC, 1991). Similarly, *C. botulinum* type E toxicosis associated with fish products is almost exclusively a consequence of suboptimal processing and storage procedures. Botulism toxin causes a potentially fatal peripheral autonomic neuropathy via blockade of presynaptic acetylcholine release. Following an incubation period of 12 to 36 hours, affected individuals manifest gastrointestinal symptoms that are either attended or succeeded by progressive cranial nerve abnormalities and symmetric descending paralysis (Weber et al., 1993; Eastaugh and Shepherd, 1989). Unlike *C. botulinum* spores, which can survive boiling (but not pressure-cooking), the heat-labile botulism toxin can be destroyed by heating to 100°C for 10 minutes (Mines et al., 1997).

Listeria spp.

These non-sporeforming, motile, facultative anaerobic, Gram-positive rods are ubiquitous in the environment, possibly as soil saphrophytes. In freshwater or marine settings, *Listeria* spp. typically concentrate in areas where human fecal pollution or high organic matter levels are prevalent (Loncarevic et al., 1996; Ben Embarek, 1994). Of the bacteria that comprise this genus, the two species most commonly found in finfish and shellfish are *Listeria monocytogenes* and *L. innocua*. *L. monocytogenes* tends to populate temperate fish and a variety of foods destined for human consumption, whereas *L. innocua* is more frequently isolated from shellfish and tropical fish (Jeyasekaran et al., 1996). Notwithstanding the catholic distribution of *Listeria* spp. in nature, direct human contamination is probably the most likely source of *Listeria* spp. in consumed fish (Ben Embarek, 1994). Previous work has determined that *L. monocytogenes* bacteria found in raw fish are not necessarily identical to strains isolated from processed products (Boerlin et al., 1997; Rorvik et al., 1995). Epidemiologic investigation suggests that reservoirs for *L. monocytogenes* contamination may include sites such as fishing vessels, boat landings, processing plants, and smokehouses (Boerlin et al., 1997; Jeyasekaran et al., 1996; Rorvik et al., 1995).

L. monocytogenes has no specific nutrient requirements, and can grow aerobically or anaerobically in temperatures as low as 1°C and in sodium chloride concentrations as high as 10 percent (Loncarevic et al., 1996). Viable and propagating *L. monocytogenes* bacilli have been isolated from a wide spectrum of both uncooked and processed food products, such as cheeses, milk products, meats, vegetables, and seafoods (Farber and Losos, 1988). Foods characteristically sold as "ready-to-eat," especially raw and lightly-cooked items such as chicken products, present the greatest risk to consumers (Loncarevic et al., 1996; Wilson, 1995; Ben Embarek, 1994). Unfortunately, even extensive processing does not entirely prevent food-borne listeriosis. Despite studies demonstrating the bactericidal efficacy of pasteurization on *L. monocytogenes*, these bacteria are regularly identified in seafoods that have been cooked and pasteurized (Ben Embarek, 1994). Foods other than fish and shellfish are more likely to be associated with human listeriosis cases (Boerlin et al., 1997; Loncarevic et al., 1996; Ben Embarek, 1994); however, a specific seafood that has received recent attention as a potential *L. monocytogenes* vehicle is cold-smoked salmon. Unlike hot-smoking procedures, the low temperatures and short exposure times of cold-smoking methods are not lethal for *L. monocytogenes* (Ben Embarek, 1994; Rorvik et al., 1991; Farber and Losos, 1988). Although the ability of *L. monocytogenes* to survive cold-smoking is cited as a major reason for its consistent isolation (up to 10 percent) from salmon prepared in this manner, its comparable prevalence in some hot-smoked salmon supports the theory that post-processing contamination may also be important (Ben Embarek, 1994). Two other relevant factors are the ability of *L. monocytogenes* to grow to high levels in refrigerated temperatures (e.g., 10°C) (Loncarevic et al., 1996), and the potential for bacterial growth during the extended shelf-life of the product (Loncarevic et al., 1996; Ben Embarek, 1994). Recommendations to limit *L. monocytogenes* proliferation in smoked salmon include a reduction in shelf-time, and storage of the product at 4°C or below (Loncarevic et al., 1996).

Listeria spp. are not commonly isolated from shellfish (Ben Embarek, 1994). Of shellfish, *L. monocytogenes* is usually only found in live shrimp (Motes, 1991), and *L. innocua* may be found in bivalves (Ben Embarek, 1994). Studies of finfish may suggest an increased prevalence of infection in freshwater species, as *L. monocytogenes* was isolated from 4/4 freshwater channel catfish but 0/10 live salmon (Ben Embarek, 1994; Leung and Huang, 1992). *Listeria* spp. appear to be of minimal threat to shellfish and finfish themselves. Experimentally, *L. monocytogenes* demonstrated an inability to multiply in live fish, and the 50 percent lethal dose (LD_{50}) of *L. monocytogenes* in zebrafish was much higher than LD_{50} levels recorded for mice (Menudier et al., 1996).

Human listeriosis is a rare, albeit severe, disease, with an overall mortality of 30 percent (Farber and Losos, 1988). The primary route of *L. monocytogenes* infection in humans is food-borne (Farber and Losos, 1988), and the major manifestations are meningitis, septicemia, and spontaneous abortion (Farber and Losos, 1988). There is evidence of a role for host resistance factors in the acquisition of listeriosis (Palumbo et al., 1989), and predisposed individuals include pregnant women and immunocompromised persons (Farber and Losos, 1988).

Staphylococcus spp.

These discrete, non-motile, non-sporeforming, facultative anaerobic, Gram-positive cocci primarily inhabit the mucous membranes and skin of warm-blooded vertebrates, although *Staphylococcus* spp. may also be isolated from the environment. Staphylococcosis is rare in shellfish and finfish, where it typically manifests as a non-specific septicemic disorder.

Seafood contamination by *S. aureus* or other *Staphylococcus* spp. is almost universally associated with product handling (Wentz et al., 1985). In this regard, human nasopharyngeal secretions provide a convenient source for seafood inoculation (Holmberg and Blake, 1984). Human *Staphylococcus* spp. food poisoning is caused by consumption of a preformed heat-stable compound, usually within high-protein foods (Mines et al., 1997). In addition to a source of contamination, enterotoxin production within the product requires a prolonged period of inadequate refrigeration (Mines et al., 1997). *Staphylococcus* spp. toxins are not absorbed by the host, and do not elicit an immune response (Bishai and Sears, 1993). Disease onset is approximately two to six hours, and the acute nausea, abdominal pain, vomiting, and diarrhea typically last less than 24 hours (Tranter, 1990; Holmberg and Blake, 1984). Generally speaking, *Staphylococcus*-related intoxication is truly dangerous only to geriatric or infirm individuals (Holmberg and Blake, 1984).

PREVENTIVE MEASURES

According to a CDC survey based upon data accumulated during a recent five-year period, the four most important contributing factors to food-borne disease outbreaks were improper holding temperatures (33 percent), inadequate cooking (26 percent), poor personal hygiene (18 percent), and contaminated equipment (12 percent) (CDC, 1996). In only 7 percent of outbreaks was the acquisition of food from an unsafe source incriminated (CDC, 1996). As applied to the seafood industry, these figures suggest that processors, retailers, and consumers must share the primary responsibility for the production and preparation of wholesome shellfish and finfish products. For the processors and retailers, measures to curtail hazardous levels of infectious organisms in seafoods include: the handling of raw and processed seafoods by persons trained in proper hygienic procedures in a sanitary workplace; the determination and maintenance of optimal storage and refrigeration conditions for various seafoods; and limitations on shelf-life for some ready-to-eat seafood products. Additionally, these interest groups must make all efforts to ensure that seafoods are not obtained from contaminated sources, that suitable surveillance methods for ubiquitous and pollution-derived pathogens are established and adopted, and that decontamination procedures for raw shellfish become routine and efficacious. Strict adherence to federally-mandated HACCP principles and regulations is obviously a step in the appropriate direction. Whereas attention is often focused upon the ability of agronomists and food manufacturers to present wholesome products to the public, the role of the consumer in controlling her or his own destiny cannot be ignored. It is imperative that the seafood industry take the lead in educating consumers in the utilization of appropriate seafood handling, storage, and cooking practices, and in the identification of human demographic groups and specific aquatic products that are associated with a higher risk of seafood-borne disease.

REFERENCES

Abeyta, C., C. A. Kaysner, M. M. Wekell, J. J. Sullivan, and G. N. Stelma. 1986. Recovery of *Aeromonas hydrophila* from oysters implicated in an outbreak of foodborne illness. *J. Food Protection* **49**: 643-646, 650.

Ahmed, F.E. (ed.) 1991. *Seafood Safety*. Food and Nutrition Board, Institute of Medicine, National Academy of Sciences.

Alicata, J.E. 1988. *Angiostrongylus cantonensis* (eosinophilic meningitis): Historical events in its recognition as a new parasitic disease of man. *J. Wash. Acad. Sci.* **78**: 38-46.

Amaro, C. and E. G. Biosca. 1996. *Vibrio vulnificus* biotype 2, pathogenic for eels, is also an opportunistic pathogen for humans. *Appl. Environ. Microbiol.* **62**: 1454-1457.

Anderson, E., D. J. Gruber, K. Sorensen, J. Beddard, and L. R. Ash. 1986. First report of *Angiostrongylus cantonensis* in Puerto Rico. *Am. J. Trop. Med. Hyg.* **35**: 319-322.

Andrews, W.H., C. D. Diggs, J. J. Miescier, C. R. Wilson, W. N. Adams, S. A. Furfari, and J. F. Musselman. 1976. Validity of members of the total coliform and fecal coliform groups for indicating the presence of *Salmonella* in the quahog, *Mercenaria mercenaria*. *J. Milk and Food Technol.* **39**: 322-324.

Arai, T., N. Ikejima, T. Itoh, S. Sakai, T. Shimada, and R. Sakazaki. 1980. A survey of *Plesiomonas shigelloides* from aquatic environments, domestic animals, pets and humans. *J. Hyg. (London)* **84**: 203-211.

Ben Embarek, P.K. 1994. Presence, detection and growth of *Listeria monocytogenes* in seafoods: a review. *Int. J. Food. Microbiol.* **23**: 17-34.

Bishai, W.R. and C. L. Sears. 1993. Food poisoning syndromes. *Gastroenterol. Clin. North Am.* **22**: 579.

Blake, P. 1980. Diseases of humans (other than cholera) caused by *Vibrios*. *Ann. Rev. Microbiol.* **34**: 341-352.

Boerlin, P., F. Boerlin-Petzold, E. Bannerman, J. Bille, and T. Jemmi. 1997. Typing *Listeria monocytogenes* isolates from fish products and human listeriosis cases. *Appl. Environ. Microbiol.* **63**: 1338-1343.

Bouree, P., A. Paugam, and J. C. Petithory. 1995. Anisakidosis: report of 25 cases and review of the literature. *Comp. Immunol. Microbiol. Infect. Dis.* **18**: 75-84.

Brenden, R., M. Miller, and J. Janda. 1988. Clinical disease spectrum and pathogenic factors associated with *Plesiomonas shigelloides* infections in humans. *Rev. Infect. Dis.* **10**: 303-316.

Brier, J.W. 1990. Comment: Emerging problems in seafood-borne parasitic zoonoses. Fifteenth Annual Confer-

Table 5. Relative Human Health Hazard from Pathogens in Shellfish and Finfish Food Products

Higher Risk	Moderate Risk	Lower Risk	Minimal Risk
Clostridium spp. **Enteric viruses** *Salmonella* spp. *Vibrio* spp.	*Aeromonas hydrophila* group *Bacillus cereus* *Campylobacter* spp. *Plesiomonas shigelloides* *Shigella* spp. *Staphylococcus* spp.	*Edwardsiella tarda* *Listeria* spp. *Streptococcus iniae*	*Citrobacter freundii* *Klebsiella* spp. *Lactobacillus* spp. *Mycobacterium* spp. **Parasites** *Yersinia* spp.

ence: Tropical and Subtropical Fisheries Technological Conference of the Americas. Second Joint Meeting with Atlantic Fisheries Technology Conference, Orlando, FL.

Brown, C. and E. Losee. 1978. Observations on natural and induced epizootics of vibriosis in *Crassotrea virginica* larvae. *J. Invert. Pathol.* **31**: 41-47.

Buras, N. 1993. Microbial safety of produce from wastewater-fed aquaculture. Environment and aquaculture in developing countries. *Proceedings of the 31st International Center for Living Aquatic Resources Management Conference.* Manila, Philippines.

Carter, G.R., M. M. Chengappa, A. W. Roberts, G. W. Claus, and Y. Rikihisa. 1995. *Essentials of Veterinary Microbiology*, 5th ed. Williams and Wilkins, Baltimore, MD.

Centers for Disease Control. 1992. Outbreak of type E botulism associated with uneviscerated, salt-cured fish product: New Jersey 1992. *MMWR.* **41**: 521.

Centers for Disease Control. 1995. Type B botulism associated with roasted eggplant in oil—Italy, 1993. *MMWR* **44**: 33.

Centers for Disease Control. 1991. Fish botulism: Hawaii, 1990. *MMWR* **40**: 412.

Centers for Disease Control. 1989. Aquarium-associated *Plesiomonas shigelloides* infection: Missouri. *MMWR* **38**: 617-619.

Chalmers, J.W. and J. H. McMillan. 1995. An outbreak of viral gastroenteritis associated with adequately prepared oysters. *Epidemiol. Infect.* **115**: 163-167.

Cliver, D.O. 1988. Virus transmission via foods. A scientific status summary by the Institute of Food Technologists' expert panel on food safety and nutrition. *Food Tech.* **42**: 241-247.

Cox, N. and J. Bailey. 1987. Pathogens associated with processed poultry. In *Microbiology of Poultry Meat Products*, F. Cunningham and N.A. Cox (eds.), Academic Press, Orlando, FL.

Cross, J.H. and V. Basaca-Sevilla. 1991. *Capillariasis philippinensis*: a fish-borne parasitic zoonosis. *Southeast Asian J .Trop. Med. Public Health* **22** Suppl:153-157.

Dalton, C. 1997. Commentary. An outbreak of Norwalk virus gastroenteritis following consumption of oysters. *Commun. Dis. Intell.* **21**: 321-322.

Davis, W.A., J. G. Kane, and V. F. Garagusi. 1978. Human *Aeromonas* infections: A review of the literature and a case report of endocarditis. *Medicine* **57**: 267-277.

Douglas, J.D. 1995. Salmon farming: occupational health in a new rural industry. *Occup. Med. (Oxford)* **45**: 89-92.

DuPont, H. 1986. Consumption of raw shellfish: is the risk now acceptable? *N. Engl. J. Med.* **314**: 707-708.

Eastaugh, J. and S. Shepherd. 1989. Infectious and toxic syndromes for fish and shellfish consumption: A review. *Arch. Intern. Med.* **149**: 1735.

Eastburn, R.L., T. R. Fritsche, and C. A. Terhune Jr. 1987. Human intestinal infection with *Nanophyetus salmincola* from salmonid fishes. *Am. J. Trop. Med. Hyg.* **36**: 586-591.

Eldar, A., Y. Bejerano, and H. Bercovier. 1994. *Streptococcus shiloi* and *Streptococcus difficile*: two new streptococcal species causing a meningoencephalitis, in fish. *Curr. Microbiol.* **28**: 139-143.

Ellner, P.D. and L. R. McCarthy. 1973. *Aeromonas shigelloides* bacteremia: a case report. *Am. J. Clin. Pathol.* **59**: 216-218.

Eto, K. 1997. Pathology of Minimata disease. *Toxicologic Pathology* **25**: 614-623.

Fang, G., V. Araujo, and R. L. Guerrant. 1991. Enteric infections associated with exposure to animals or animal products. *Infect. Dis. Clin. North. Am.* **5**: 681-701.

Farber, J.M. and J. Z Losos. 1988. *Listeria monocytogenes*: a foodborne pathogen. *Cmaj.* **138**: 413-418.

Farmer, J.J., B. R. Davis, F. C. Hickman-Brenner et al. 1985. Biochemical identification of new species and biogroups of *Enterobacteriaceae* isolated from clinical specimens. *J. Clin. Microbiol.* **21**: 46-76.

Fitzgerald, E.F., K. A. Brix, D. A. Deres, S. A. Hwang, B. Bush, G. Lambert, and A. Tarbell. 1996. Polychlorinated biphenyl (PCB) and dichlorodiphenyl dichloroethylene (DDE) exposure among Native American men from contaminated Great Lakes fish and wildlife. *Toxicol. Ind. Health* **12**: 361-368.

Flynn, T. and I. Knepp. 1987. Seafood shucking as an etiology for *Aeromonas hydrophila* infection. *Arch. Intern. Med.* **147**: 1816-1817.

Fountaine, R.E. 1985. Anisakiasis from the American perspective. *J. Am. Med. Assoc.* **253**: 1024-1025.

Frasier, M.B., J. A. Koburger, and C. I. Wei. 1988. Incidence of *Salmonellae* in clams, oysters, crabs and mullet. *J. Food Protection* **51**: 110-112.

Fricker, C.R. and S. Thompsett. 1989. *Aeromonas* spp. in foods: a significant cause of food poisoning? *Int. J. Food Microbiol.* **9**: 17-23.

Fujino, T., G. Sakaguchi, R. Sakazaki, and Y. Takeda. 1974. International symposium on *Vibrio parahemolyticus*. Saikon Pub. Co., Tokyo, Japan.

Ghittino, P. 1972. Aquaculture and associated diseases of fish of public health importance. *J. Am. Vet. Med. Assoc.* **161**: 1476-1485.

Gill, O.N., J. D. Coghlan, and I. M. Calder. 1985. The risk of leptospirosis in United Kingdom fish farm workers. Results from a 1981 serological survey. *J. Hyg. (London)* **94**: 81-86.

Gopal, V. and F. E. Burns. 1991. Cellulitis and compartment syndrome due to *Plesiomonas shigelloides*: a case report. *Mil. Med.* **156**: 43.

Gorby, G. and J. Peacock. 1988. *Erysipelothrix rhusiopathiae* endocarditis: Microbiologic, epidemiologic, and clinical features of an occupational disease. *Rev. Infect. Dis.* **10**: 317-325.

Gouvea, V., N. Santos, M.D.C.Timenetsky, and M. K. Estes. 1994. Identification of Norwalk virus in artificially-seeded shellfish and selected foods. *J. Virol. Methods* **48**: 177-187.

Grant, I.C. 1997. Ichthyosarcotoxism: poisoning by edible fish. *J. Accid. Emerg. Med.* **14**: 246-251.

Griffin, M.R., E. Dalley, M. S. Fitzpatrick, and S. H. Austin. 1983. *Campylobacter* gastroenteritis associated with raw clams. *J. Medical Soc. New Jersey* **80**: 607-609.

Hackney, C.R., M. B. Kilgen, and H. Kator. 1992. Public health aspects of transferring mollusks. *J. Shellfish Res.* **11**: 521-533.

Hanninen, M.L., P. Oivanen, and V. Hirvela-Koski. 1997. *Aeromonas* species in fish, fish-eggs, shrimp and freshwater. *Int. J. Food Microbiol.* **34**: 17-26.

Harinasuta, T., S. Pungpak, and J. S. Keystone. 1993. Trematode infections. Opisthorchiasis, clonorchiasis, fascioliasis, and paragonimiasis [published erratum appears in *Infect. Dis. Clin. North Am.* 1994 Mar 8(1), following table of contents]. *Infect. Dis. Clin. North Am.* **7**: 699-716.

Harrell, L.W. and T. L. Deardorff. 1990. Human nanophyetiasis: transmission by handling naturally infected coho salmon (*Oncorhynchus kisutch*). *J. Infect. Dis.* **161**: 146-148.

Hassan, M.M., K. M. Rahman, and A. Nahar. 1994. Studies on the bacterial flora of fish which are potential pathogens for humans. Isolation of various potential human pathogenic organisms from different parts of fish and their significance in initiating human diseases. *Bangladesh Med. Res. Counc. Bull.* **20**: 43-51.

Hassan, M.M., K. M. Rahman, and S. Tzipori. 1994. Studies on the bacterial flora of fish which are potential pathogens for humans. Virulence factors of potential human pathogen isolated. *Bangladesh Med. Res. Counc. Bull.* **20**: 86-98.

Herrington, D.A., S. Tzipori, R. M. Robins-Browne, B. D. Tall, and M. M. Levine. 1987. *In vitro* and *in vivo* pathogenicity of *Plesiomonas shigelloides*. *Infect. Immun.* **55**: 979-985.

Heuschmann-Brunner, G. 1974. Experimentelle untersuchungen uber moglichkeiten und verlauf einer infektion mit *Salmonella enteriditis* und *Salmonella typhimurium* bei subwasserfischen. *Zentralblatt fur Bakteriologie, Mikrobiologie und Hygiene. I. Abteile Originale B* **158**: 412-431.

Hlady, W.G. and K. C. Klontz. 1996. The epidemiology of *Vibrio* infections in Florida, 1981-1993. *J. Infect. Dis.* **173**: 1176-1183.

Holmberg, S.D. 1988. *Vibrios* and *Aeromonas*. *Infect. Dis. Clin. North Am.* **2**: 655-676.

Holmberg, S.D. and P. A. Blake. 1984. Staphylococcal food poisoning in the United States: New facts and old misconceptions. *JAMA* **251**: 487.

Holt, J.G., N. R. Krieg, P.H.A. Sneath, J. T. Staley, and S. T. Williams. 1994. *Bergey's Manual of Determinative Bacteriology*, Ninth ed. Williams & Wilkins, Baltimore, MD.

Hook, E.W. III, T. M. Hootonl, C. A. Horton, P. G. Ramsey, and M. Turck. 1986. Microbiologic evaluation of cutaneous cellulitis in adults. *Arch. Intern. Med.* **146**: 295-297.

Hori, M., K. Hayashi, K. Maeshima, M. Kigawa, T. Miyasato, Y. Yoneda, Y. and Hagihara. 1966. Food poisoning caused by *Aeromonas shigelloides*, with an antigen common to *Shigella dysenteriae* 7. *J.Japanese Assoc. Inf. Dis.* **39**: 433-441.

Hudson, J.A., S. J. Mott, and N. Penney. 1994. Growth of *Listeria monocytogenes*, *Aeromonas hydrophila*, and *Yersinia enterocolitica* on vacuum and saturated carbon dioxide controlled atmosphere-packaged sliced roast beef. *J. Food Prot.* **57**: 204-208.

Hughes, H., M. Merson,. and E. Gangarosa. 1969. The safety of eating shellfish. *JAMA* **208**: 649-655.

Humphrey, J.D., C. Lancaster, N. Gudkovs, and W. McDonald. 1986. Exotic bacterial pathogens *Edwardsiella tarda* and *Edwardsiella ictaluri* from imported ornamental fish *Betta splendens* and *Puntius conchonius*, respectively: isolation and quarantine significance. *Aust. Vet. J.* **63**: 369-371.

Janssen, W.A. and C. D. Meyers. 1968. Fish: serologic evidence of infection with human pathogens. *Science* **159**: 547-548.

Jay, J.M. (ed.) 1986. *Modern Food Microbiology*. Van Nostrand Reinhold Co., New York.

Jensen, G.L. and K. J. Greenlees. 1988. Public health issues in aquaculture. *Rev. Sci Tech. Off. Int. Epiz.* **16**: 641-651.

Jeyasekaran, G., I. Karunasagar, and I. Karunasagar. 1996. Incidence of *Listeria* spp. in tropical fish. *Int. J. Food Microbiol.* **31**: 333-340.

Kaper, J.B., H. Lockman, R. R. Colwell, and S. W. Joseph. 1981. *Aeromonas hydrophila*. Ecology and toxigenicity of isolates from an estuary. *J. Appl. Bacteriol.* **50**: 359-377.

Karam, G., A. Ackley, and W. Dismukes. 1983. Posttraumatic *Aeromonas hydrophila* osteomyelitis. *Arch. Intern. Med.* 143: 2073-2074.

Keswick, B.H., T. K. Satterwhite, P. C. Johnson, H. L. DuPont, S. L. Secor, J. A. Bitsura, G. W. Gary, and J. C. Hoff. 1985. Inactivation of Norwalk virus in drinking water by chlorine. *Appl. Environ. Microbiol.* **50**: 261-264.

Kliks, M.M., K. Kronenke, and J. M. Hardman. 1982. Eosinophilic radiculomyeloencephalitis: an angiostrongyliasis outbreak in American Samoa related to the ingestion of *Achantina fulica* snails. *Am. J. Trop. Med. Hyg.* **31**: 1114-1122.

Klontz, K., S. Lieb, M. Schreiber, H. Janowski, L. Baldy, and R. Gunn. 1988. Syndromes of *Vibrio vulnificus* infections: Clinical and epidemiologic features in Florida cases, 1981-1987. *Ann. Intern. Med.* **109**: 318.

Koshi, G. and M. K. Lalitha. 1976. *Edwardsiella tarda* in a variety of human infections. *Indian J. Med. Res.* **64**: 1753-1759.

Leung, C. K. and Y. W. Huang. 1992. Bacterial pathogens and indicators in catfish and pond environments. *J. Food Prot.* **55**: 425-427.

Levine, W.C., P. M. Griffin, and G. C. W. Group. 1993. *Vibrio* infections on the Gulf Coast: results of first year of regional surveillance. *J. Infect. Dis.* **167**: 479-483.

Loncarevic, S., W. Tham. and M. L. Danielsson-Tham. 1996. Prevalence of *Listeria monocytogenes* and other *Listeria* spp. in smoked and 'gravad' fish. *Acta Vet. Scand.* **37**: 13-18.

MacDonell, M.T. and R. R. Colwell. 1985. Phylogeny of the *Vibrionaceae*, and recommendation for two new genera, *Listonella* and *Shewanella*. *System Appl. Microbiol.* **6**: 171-182.

Menudier, A., F. P. Rougier, and C. Bosgiraud. 1996. Comparative virulence between different strains of *Listeria* in zebrafish (*Brachydanio rerio*) and mice. *Pathol. Biol. (Paris)* **44**: 783-9.

Miller, M.A. and J. A. Koburger. 1985. *Plesiomonas shigelloides*: an opportunistic food and waterborne pathogen. *J. Food Prot.* **48**: 449-457.

Mines, D., S. Stahmer, and S. M. Shepherd. 1997. Poisonings: food, fish, shellfish. *Emerg. Med. Clin. North Am.* **15**: 157-177.

Minette, H. 1983. Leptospirosis in poikilothermic vertebrates: A review. *Int. J. Zoon.* **10**: 111-121.

Minette, H.P. 1984. Epidemiologic aspects of salmonellosis in reptiles, amphibians, mollusks and crustaceans — a review. *Int. J. Zoonoses* **11**: 95-104.

Minette, H.P. 1986. Salmonellosis in the marine environment: A review and commentary. *Int. J . Zoonoses* **13**: 71-75.

Morris, J.G. and R. E. Black. 1986. Cholera and other vibrioses in the United States. *N. Engl. J. Med.* **312**: 343-350.

Morris, J.G.J., R. Wilson, B. R. Davis, et al. 1981. Non-0 group 1 *Vibrio cholerae* gastroenteritis in the United States. Clinical, epidemiologic and laboratory characteristics of sporadic cases. *Ann. Intern. Med.* **94**: 656-658.

Motes, M.L. Jr. 1991. Incidence of *Listeria* in shrimp, oysters, and estuarine waters. *J. Food Prot.* **54**: 170-173.

Murphree, R.L. and M. L. Tamplin. 1995. Uptake and retention of *Vibrio* cholerae O1 in the Eastern oyster, *Crassostrea virginica*. *Appl. Environ. Microbiol.* **61**: 3656-3660.

National Academy of Sciences. 1991. *Seafood Safety*, ed. National Academy Press, Washington, DC.

Nemetz, T.G. and E. M. Shotts. 1993. Zoonotic Diseases In *Fish Medicine*, M.K. Stoskopf (ed.). W.B. Saunders Company, Philadelphia, PA.

Ogawa, K. 1996. Marine parasitology with special reference to Japanese fisheries and mariculture. *Vet. Parasitol.* **64**: 95-105.

Osterholm, M.T., J. C. Forfang, T. L. Ristinen, A. G. Dean, J. W. Washburn, J. R. Godes, R. A. Rude, and J. G. McCullough. 1981. An outbreak of foodborne giardiasis. *N. Engl. J. Med.* **304**: 24-28.

Palumbo, S.A. 1988. The growth of *Aeromonas hydrophila* K144 in ground pork at 5°C. *Int. J. Food Microbiol.* **7**: 41-48.

Palumbo, S.A., M.M. Bencivengo, F. Del Corral, A. C. Williams, and R. L. Buchanan. 1989. Characterization of the *Aeromonas hydrophila* group isolated from retail foods of animal origin. *J. Clin. Microbiol.* **27**: 854-859.

Perera, R.P., S. K. Johnson, M. D. Collins, and D. H. Lewis. 1994. *Streptococcus iniae* associated with mortality of *Tilapia nilotica* x *T. aurea* hybrids. *J. Aquat. Animal Health.* **6**: 335-340.

Perera, R.P., S. K. Johnson, and D.H. Lewis. 1997. Epizootiological aspects of *Streptococcus iniae* affecting tilapia in Texas. *Aquaculture* **152**: 25-33.

Pier, G.B. and S. H. Madin. 1976. *Streptococcus iniae* sp. nov., a beta-hemolytic streptococcus isolated from an Amazon freshwater dolphin, *Inia geoffrensis*. *Int. J. Sys. Bacteriol.* **26**: 545-553.

Richards, G.P. 1988. Microbial purification of shellfish: A review of depuration and relaying. *J. Food Prot.* **51**: 218-251.

Rippey, S.R. 1994. Infectious diseases associated with molluscan shellfish consumption. *Clin. Microbiol. Rev.* **7**: 419-425.

Robertson, M.H., I. R. Clarke, J. D. Cohlan, and O. N. Gill. 1981. Leptospirosis in trout farmers. *Lancet* **2**: 626-627.

Rorvik, L.M., D. A. Caugant, and M. Yndestad. 1995. Contamination pattern of *Listeria monocytogenes* and other *Listeria* spp. in a salmon slaughterhouse and smoked salmon processing plant. *Int. J. Food Microbiol.* **25**: 19-27.

Rorvik, L.M., M. Yndestad, and E. Skjerve. 1991. Growth of *Listeria monocytogenes* in vacuum-packed, smoked salmon, during storage at 4°C. *Int. J. Food Microbiol.* **14**: 111-118.

Rosen, L., G. Loison, J. Laigret, and G. D. Wallace. 1967. Studies on eosinophilic meningitis. 3. Epidemiologic and clinical observations on Pacific islands and the possible etiologic role of *Angiostrongylus cantonensis*. *Am. J. Epidemiol.* **85**: 17-44.

Ruttenber, A.J., B. G. Weniger, F. Sorvillo, R. A. Murray, and S. L. Ford. 1984. Diphyllobothriasis associated with salmon consumption in Pacific Coast states. *Am. J. Trop. Med. Hyg.* **33**: 455-459.

Santos, Y., A. E. Toranzo, J. L. Barja, T. P. Nieto, and T. G. Villa. 1988. Virulence properties and enterotoxin production of *Aeromonas* strains isolated from fish. *Infect. Immun.* **56**: 3285-3293.

Sanyal, D., S. H. Burge, and P. G. Hutchings. 1987. Enteric pathogens in tropical aquaria, in *Epidemiol. Infect.* **99**: 635-640.

Sasaki, D.M. and H. P. Minette. 1985. Isolation of *Salmonella* in a public marine aquarium. *J. Am. Vet. Med. Assoc.* **187**: 1221-1222.

Shandera, W.X., C. O. Tacket, and P. A. Blake. 1983. Food poisoning due to *Clostridium perfringens* in the United States. *J. Infect. Dis.* **147**: 167.

Stelma, G.N. Jr. and L. J. McCabe. 1992. Nonpoint pollution from animal sources and shellfish sanitation. *J. Food Prot.* **55**: 649-656.

Tangtrongchitr, A. and R. B. Monzon. 1991. Eating habits associated with *Echinostoma malayanum* infections in the Philippines. *Southeast Asian J. Trop. Med. Public Health* **22** Suppl: 212-216.

Tranter, H.S. 1990. Foodborne staphylococcal illness. *Lancet* **336**: 1044.

Trust, T.J. and R. A. H. Sparrow. 1974. The bacterial flora in the alimentary tract of freshwater salmonid fishes. *Can. J. Microbiol.* **20**: 1219-1228.

Tsai, G.J. and T. H. Chen. 1996. Incidence and toxigenicity of *Aeromonas hydrophila* in seafood. *Int. J. Food Microbiol.* **31**: 121-131.

Tubiash, H.S., P. E. Chanley, and E. Leifson. 1965. Bacillary necrosis, a disease of larval and juvenile bivalve mollusks. *J. Bacteriol.* **90**: 1036-1044.

Underman, A.E. and J. M. Leedom. 1993. Fish and shellfish poisoning. *Current Clin. Top. Infect. Dis.* **13**: 203-225.

Vandepitte, J., P. Lemmens, and L. Swert. 1983. Human edwardsiellosis traced to ornamental fish. *J. Clin. Microbiol.* **17**: 165-167.

Vartian, C. and E. Septimus. 1990. Osteomyelitis caused by *Vibrio vulnificus*. *J. Infect. Dis.* **161**: 363.

Von Bondsdorff, B. 1977. *Diphyllobothriasis in Ma.* Academic Press, New York.

Waikagul, J. (1991. Intestinal fluke infections in Southeast Asia. *Southeast Asian J. Trop. Med. Public Health.* **22** Suppl: 158-162.

Weber, J.T., R. G. Hibbs Jr., A. Darwish, B. Mishu, A. L. Corwin, M. Rakha, C. L. Hatheway, S. el Sharkawy, S. A. el-Rahim, M. F. al-Hamd, et al. 1993. A massive outbreak of type E botulism associated with traditional salted fish in Cairo. *J. Infect. Dis.* **167**: 451-454.

Weinstein, M., D. Low, A. McGeer, B. Willey, D., Rose, M. Coulter, P. Wyper, A. Borczyk, M. Lovgren, and R. Facklam. 1996. Invasive infection due to *Streptococcus iniae*: a new or previously unrecognized disease: Ontario, 1995-1996. *Can. Commun. Dis. Rep.* **22**: 129-131; discussion 131-132.

Weinstein, M.R., M. Litt, D. A. Kertesz, P. Wyper, D. Rose, M. Coulter, A. McGeer, R. Facklam, C. Ostach, B. M. Willey, A. Borczyk, and D. E. Low. 1997. Invasive infections due to a fish pathogen, *Streptococcus iniae*. *S. iniae* Study Group. *N. Engl. J. Med.* **337**: 589-94.

Wentz, B.A., A. P. Duran, A. Scuartzentruber, A. H. Schwab, F. D. McClure, D. Archer, and R. B. Read. 1985. Microbiological quality of crabmeat during processing. *J. Food Prot.* **48**: 44-49.

WHO. 1980. World Health Organization Scientific Working Group: Cholera and other vibrio-associated diarrheas. *Bull. WHO* **58**: 353-374.

Williams, L.A. and P. A. LaRock. 1985. Temporal occurrence of *Vibrio* species and *Aeromonas hydrophila* in estuarine sediments. *Appl. Environ. Microbiol.* **50:** 1490-1495.

Wilson, I.G. 1995. Occurrence of *Listeria* species in ready-to-eat foods. *Epidemiol. Infect.* **115**: 519-526.

Wilson, J., R. Waterer, J. Wofford, and S. Chapman. 1989. Serious infections with *Edwardsiella tarda*: A case report and review of the literature. *Arch. Intern. Med.* **149**: 208-210.

Wittner, M., J. W. Turner, G. Jacquette, L. R. Ash, M. P. Salgo, and H. B. Tanowitz. 1989. Eustrongylidiasis: a parasitic infection acquired by eating sushi. *N. Engl. J. Med.* **320**: 1124-1126.

Wood, R. 1975. *Erysipelothrix* infection, in *Diseases Transmitted from Animals to Man*. W. Hubbert, W. McCullough, and P. Schnurrenberger (eds.). pp. 271-281. Charles C. Thomas, Springfield, IL.

Wyatt, L.E., R. Nickelson II, and C. Vanderzant. 1979. *Edwardsiella tarda* in freshwater catfish and their environment. *Appl. Environ. Microbiol.* **38**: 710-714.

Finfish Toxins

Peter D. R. Moeller

INTRODUCTION

Ciguatera, scombroid poisoning, and puffer fish toxicity are discussed in this section. All of these illnesses are manifested by a rapid onset of neurological complications. Despite their similarities, the illnesses can easily be differentiated by the type of fish consumed, since most of these toxins are species-specific. Ciguatoxin is sporadically found in coral reef fish. Scombrotoxin can accumulate in dark-fleshed fish when they are improperly stored or handled. And tetrodotoxin, the toxin which causes puffer fish toxicity, is found only in that species of fish.

CIGUATERA

Ciguatera is the most frequently reported nonbacterial illness associated with the consumption of seafood (Bean et al., 1990; Morris, 1980). It is caused by eating tropical and subtropical coral reef fish that have accumulated toxins. The fish most often implicated in cases of ciguatera are: barracuda, grouper (Figure 1), snapper, surgeonfish, amberjack, and parrot fish (Bryan, 1988, 1987; Tosteson et al., 1988; Kantha, 1987; Miyahara et al., 1987; Craig, 1980; Lawrence et al., 1980). Toxicity in fish is sporadic. Not all fish of the same variety and caught in the same area are necessarily toxic (Bryan, 1987; Hokama et al., 1983). Ciguatoxic fish are found worldwide between 35°N and 34°S latitude (Craig, 1980). There are only a few areas of the United States in which ciguatoxic fish are native: Florida, Hawaii, Puerto Rico, and the Virgin Islands. Most of the cases of ciguatera in the United States are reported from these areas; however, sporadic cases have also been reported in non-tropical areas, and are associated with a history of travel to the tropics, or commercial transport of tropical species.

The mechanism by which fish become toxic is similar to that of the shellfish toxins discussed in another chapter. The fish themselves are not inherently toxic, but become poisonous by eating naturally-occurring toxic dinoflagellates. Both herbivorous and carnivorous fish can become toxic. Herbivorous fish become poisonous by eating the toxic algae itself. Carnivorous fish accumulate the toxins by consuming other toxic fish. Large fish are generally more toxic than small fish of the same species because they consume greater amounts of food, and therefore sequester greater amounts of toxin (Craig, 1980). Muscle tissue, the part of the fish most frequently consumed, is the least toxic tissue, while the liver, intestines, testes, and ovaries are usually the most toxic (Bryan, 1987). The toxins are slowly purged from the fish when the population of toxic microorganisms decreases and the fish resumes feeding on non-toxic food.

The dinoflagellate species most often associated with ciguatera in fish is *Gambierdiscus toxicus* (Campbell et al., 1987; Adachi and Fukuyo, 1979;

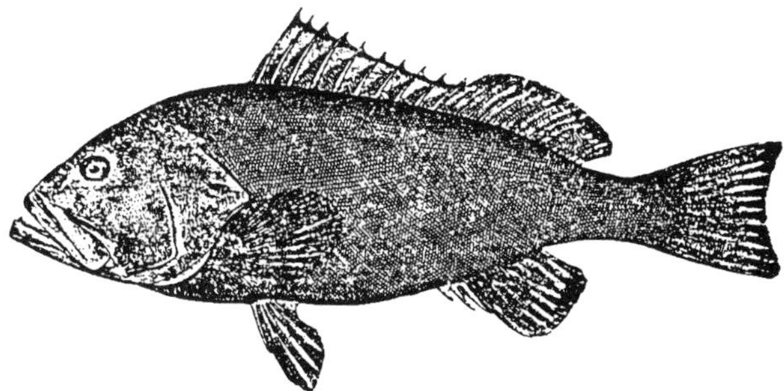

Figure 1a. Epinephelus morio, *red grouper (from Jordan and Eigenmann, 1890, plate 61).*

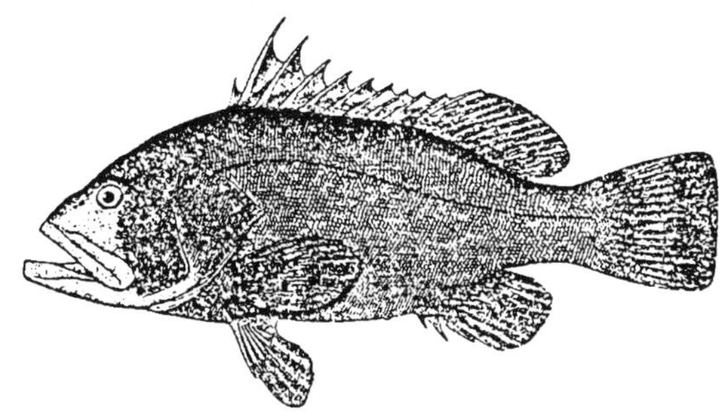

Figure 1b. Epinephelus nigritus, *warsaw grouper (from Stevenson, 1893, plate 19).*

Yasumoto et al., 1977). Other algal species that are suspected of contributing to ciguatera include: *Prorocentrum mexicanum, P. concavum, P. lima, P. emarginatum, Coolia monotis, Ostreopis lenticularis, O. ovata,* and *O. siamensis* (Juranovic and Park, 1991; Carlson and Tindall, 1985).

The conditions that lead to toxic dinoflagellate blooms are not well understood. One theory is that the toxic algae reproduce rapidly after major disturbances, such as storms, construction, and dredging (Craig, 1980). It is not known if the toxic algal populations fluctuate with regularity. Seasonal trends in benthic microalgal density have been reported in the Virgin Islands (Carlson and Tindall, 1985), Puerto Rico (Ballantine et al., 1988), and Australia (Gillespie et al., 1985, as cited in Scheuer, 1989), but have not been observed in Hawaii (Morris, 1980). The parameters that have been reported to cause seasonal changes in density have been different in each case. In the Virgin Islands, the algae are said to exhibit a bimodal pattern of abundance: population maxima occurs in conjunction with the peak periods of rainfall: April to May, and August to October (Carlson and Tindall, 1985). In Puerto Rico, peak populations of *G. toxicus* and *O. lenticularis* occur during the late summer and fall, and do not appear to be correlated to rainfall (Ballantine et al., 1988). And finally, in Southern Queensland, Australia, *G. toxicus* populations are reported to peak in September, coinciding with low water temperature (Gillespie et al., 1985, as cited in Scheuer, 1989). Clearly, more research is necessary before conclusions can be drawn about the conditions that cause blooms and seasonal trends in toxic microalgal populations.

At least two toxins are involved in causing ciguatera. Ciguatoxin, the principal toxin, is lipid soluble (Scheuer et al., 1967). It is mostly found in fish viscera and musculature, and can be extracted in minute quantities from *G. toxicus* cells (Jacyno and Miller, 1991). Studies indicate that ingestion of as little as 0.1 μg (11 Mouse Units) of ciguatoxin could cause illness in an adult human (Yasumoto, 1985). A secondary toxin, maitotoxin, is water soluble and approximately three times more toxic than ciguatoxin (Yasumoto, 1985). This toxin is found in large quantities in cell cultures of *G. toxicus* and in small quantities in fish viscera (Tindall et al., 1984), but its role in ciguatera poisoning remains unsubstantiated (Lewis and Holmes, 1993). Some references also include a third toxin, scaritoxin, which appears to be a metabolite of ciguatoxin (Kantha, 1987; Nukina et al., 1984). It is not known whether the conversion from ciguatoxin to scaritoxin occurs within the fish, or if it is induced by purification techniques in the laboratory (Nukina et al., 1984).

Ciguatera victims can exhibit gastrointestinal, neurological, and cardiovascular symptoms. The onset of symptoms is highly variable among individuals, even when the same fish is consumed (Calvert, 1991). Gastrointestinal symptoms, usually the first to be manifested, occur within hours of consuming the toxic fish. In some cases, the incubation period can be as short as one hour (Horwitz, 1977). The most common gastrointestinal symptoms include: diarrhea, abdominal pain, nausea, and vomiting (Calvert, 1991; Bryan, 1987; Horwitz, 1977). Victims usually recover from gastroenteritis within 24 hours (Calvert, 1991). Secondary cardiovascular symptoms, such as decreased pulse and low blood pressure, usually appear subsequently (Calvert, 1991). Cardiovascular complications, the least frequently observed symptoms, usually subside within 24 to 48 hours, but can persist as long as one week (Calvert, 1991). Neurological symptoms, usually the last to be manifested, include: oral and extremity cold-to-hot sensory reversal, itching (especially during activities that increase skin temperature and blood flow), sensitivity of the skin, vertigo, headache, weakness, lack of muscle coordination, muscular pain, dental pain, and joint pain (Calvert, 1991; Kantha, 1987; Lawrence et al., 1980). Although uncommon, severe cases of ciguatera can result in shock, seizures, coma, respiratory depression, and death (Calvert, 1991). Neurological symptoms usually resolve within three to four weeks (Calvert, 1991), but may recur intermittently with gradually diminishing severity for months or even years (Horwitz, 1977). Consumption of alcohol has been reported to exacerbate and/or cause the reemergence of neurological symptoms (Cameron and Capra, 1991).

Persons affected once should avoid eating potentially toxic fish for several months, since a second episode might be more severe. Repeated exposure may cause extreme sensitivity to the toxin, resulting in the onset of symptoms even when fish containing only trace amounts of toxin are consumed (Lawrence et al., 1980).

The early stages of ciguatera are generally treated by non-specific symptomatic and supportive therapy. Treatment with intravenous mannitol within 24 to 48 hours after exposure has been reported to be effective in relieving acute and chronic symptoms (Blythe et al., 1994; Bagnis et al., 1992; Stewart, 1991; Williams, 1990; Williams and Palafox, 1990; Pearn et al., 1989; Palafox et al., 1988). Controlled, double-blind clinical trials are needed to validate this treatment; however, the relative safety of mannitol and the observed dramatic recovery would seem justified in patients whose illness is moderate or severe, particularly within 24 hours of the onset of symptoms (Ruff and Lewis, 1994). Although the mechanism of action of mannitol is unclear, an osmotic effect reducing Schwann cell edema, thereby ameliorating neurological dysfunction (Pearn et al., 1989), is the most likely mode of action (Ruff and Lewis, 1994).

In the United States between 1983 and 1987, 87 outbreaks (332 cases) of ciguatera were reported to the Centers for Disease Control (CDC), making it the most frequently reported, non-bacterial, seafood-borne illness (Bean et al., 1990). According to the CDC, ciguatera is responsible for approximately 37 percent of the seafood-borne outbreaks in the United States (Bean et al., 1990). It is difficult to estimate the incidence of ciguatera worldwide since many of the areas affected by the illness are Third World countries where disease statistics are generally not available. It has been estimated that there may be as many as 50,000 cases of ciguatera worldwide per year (Gervais, 1985).

Mortality rates are also 7 to 20 percent (Craig, 1980), while another reference set the number considerably lower, at 0.1 percent (Bagnis et al., 1979, as cited in Calvert, 1991). No deaths from ciguatera have occurred in the United States (Morris, 1980).

Unsuspecting travelers to the tropics are prone to ciguatera. It has been reported that approximately 80 percent of the cases are due to weekend fishermen who are unfamiliar with the types of fish commonly shown to be ciguatoxic (Hokama, 1988). A similar statistic cited by the CDC is that 80 percent of the cases of ciguatera occur in the home (Bean et al., 1990). These statistics clearly demonstrate the need to educate persons traveling to the tropics about ciguatera and, more important, about the steps that should be taken to prevent the illness.

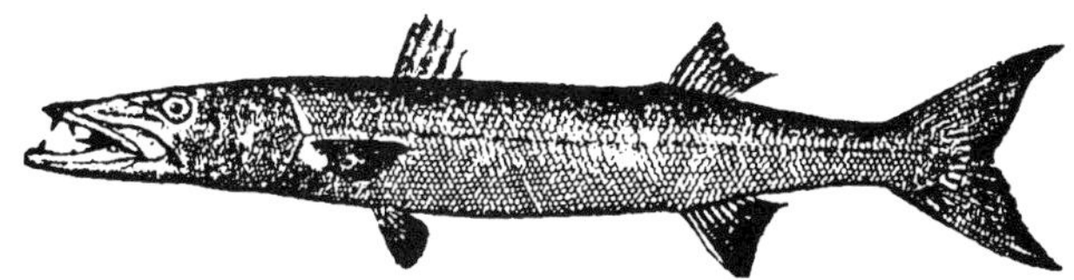

Figure 2. Sphyraena barracuda, *great barracuda (from Evermann and Marsh, 1902, figure 26).*

Prevention of Ciguatera

Ciguatoxic fish do not smell, taste, or look different from non-toxic fish (Craig, 1980), and they cannot be made safe to eat by cooking, freezing, drying, or smoking (Tosteson et al., 1988). To avoid becoming ill, consumers should refrain from eating the types of fish that are most often toxic. This conservative approach is endorsed by public health regulations that ban the sale of frequently toxic fish. For example, the Miami City Code prohibits the sale of barracuda (Figure 2) (Concon, 1988; Lawrence et al., 1980).

If one does choose to eat a potentially toxic species, there are a number of ways to decrease the chances of ingesting toxins, although these suggestions should not be considered foolproof methods of preventing illness. Avoid eating large fish since they tend to be more toxic than small fish of the same variety (Craig, 1980; Lawrence et al., 1980). Do not eat the viscera and roe of reef fishes, especially during the reproductive season (Craig, 1980). And finally, eat a small piece of fish and wait several hours in order to determine if any signs of poisoning occur before consuming the whole fish.

Bioassays utilizing a variety of animals including mice (Banner et al., 1960), mongooses (Banner et al., 1961), cats (Bagnis, 1973), and mosquitoes (Chungue et al., 1984) were the first methods used to detect ciguatoxic fish. However, none of the bioassays is practical, sensitive, or specific enough to be used in ciguatoxin monitoring (Hokama et al., 1987a; Hokama et al., 1987b). A group of scientists in Hawaii is currently working on an immunological method for detecting ciguatoxic fish to replace the bioassays. The first immunological method developed was a radioimmunoassay (RIA) using sheep anti-ciguatoxin serum labeled with iodine-125 (Hokama et al., 1987). Although the RIA was reported to be sensitive and relatively specific to ciguatoxin, some problems were associated with the procedure (Hokama et al., 1983). The methodology required expensive equipment, involved handling radioisotopes, was slow (approximately four hours), and could not be used in the field (Scheuer, 1989). Furthermore, the cost of running the RIA prohibited its use for fish smaller than nine kilograms (Hokama et al., 1983). The next methodology developed by the Hawaiian group was an enzyme immunoassay (EIA) using sheep anti-ciguatoxin serum coupled with horseradish peroxidase (Hokama et al., 1983). This technique was reported to be faster and simpler than the RIA, but still needed to be run in a laboratory with expensive equipment (Scheuer, 1989). The final immunologically-based procedure still in development is the stick enzyme immunoassay (S-EIA) or, simply, the stick test (Hokama et al., 1989, 1987a, 1987b, 1985). In this procedure, bamboo sticks pretreated to aid absorption of the toxin are stuck into the fish flesh for one second. The sticks are then fixed with methyl alcohol and immersed in a solution of sheep anti-ciguatoxin coupled with horseradish peroxidase. A positive result will change the bamboo stick to a dark-blue or purple color within 10 minutes. Although visual evaluation of a color change is subjective, this procedure can be used in the field, gives rapid results, and is relatively inexpensive.

SCOMBROID POISONING

Scombroid poisoning (also referred to as scombroid toxicity, histamine poisoning, or histamine toxicity) is an illness caused by eating certain types of fish that have been improperly handled or stored. The name scombroid poisoning comes from the fish families Scombridae and Scomberesocidae (Figure 3), which often cause illness. Some of the Scombridae fish that frequently are toxic include: tuna, mackerel, bonito, and skipjack. However, toxicity is not strictly limited to these fish families. The illness is frequently caused by non-scombroid fish. For example, between 1977 and 1984, 68 percent of the outbreaks of scombroid poisoning were caused by mahi mahi and bluefish, both non-scombroid fish (Bryan, 1988).

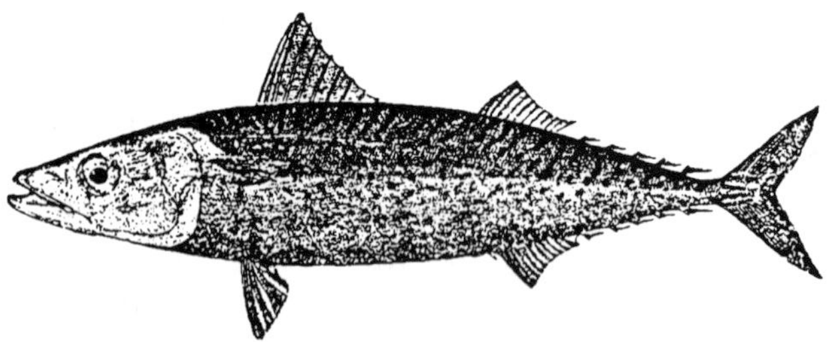

Figure 3. Scomber japonicus *(*Scomber colias*),* chub mackerel *(from Dresslar and Fesler, 1889, plate 2.)*

The toxin, or toxins, that cause scombroid poisoning are not completely known. It is generally true that a high level of histamine in fish is associated with toxicity. However, fish that have high levels of histamine do not always cause illness, and illness is sometimes caused by fish with low levels of histamine. These facts suggest that the toxin is not histamine alone, but histamine in combination with some other substances. It is believed that the toxin is actually composed of histamine, putrescine, and cadaverine.

In order for histamine to be produced, the following three conditions must be met: first, free histidine, the harmless precursor of histamine, must be present in the fish. Dark-fleshed fish, such as those in the Scombridae family, have an abundance of free histidine in their muscles, thus accounting for the high incidence of illness caused by these fish. Second, certain naturally-occurring bacteria must also be present in the fish. Only bacteria that produce histidine decarboxylase, the enzyme that causes the conversion of histidine to histamine, are significant in causing toxicity. *Proteus morganii* is probably the most significant of the histamine-forming bacteria (Chen et al., 1988; Taylor and Sumner, 1986; Eitenmiller et al., 1980; Arnold and Brown, 1978; Omura et al., 1978). *Hafnia alvei* and *Klebsiella pneumoniae* have also been isolated from toxic fish (Chen et al., 1988; Omura et al., 1978), although they tend to convert histidine to histamine at a slower rate than does *P. morganii* (Arnold and Brown, 1978). Other species of bacteria — such as *Escherichia coli, Clostridium perfringens,* and *Enterobacter aerogenes* — are capable of converting histidine to histamine, but are insignificant in causing toxicity since they are unlikely to be found in spoiling fish (Eitenmiller et al., 1980; Arnold and Brown, 1978).

The final criterion that must be met for histamine production is a high enough temperature for the enzymatic reaction to take place. The optimal and minimal temperatures required for histamine production are dependent on the species of bacteria. In general, the lowest temperatures that allow production of histamine range from 7° to 20°C (45° to 68°F) (Chen et al., 1988), and the optimal temperature for the conversion is from 20° to 30°C (68° to 86°F) (Kantha, 1987). Production of histamine can be fairly rapid. In one outbreak, threshold toxin levels were reached in mackerel after only three to four hours of defrosting at room temperature (about 30°C) (Kow-Tong and Malison, 1987).

Histamine and toxins related to histamine are commonly found in a variety of foods such as fish, cheese, sauerkraut, wine, yeast extracts, and putrid, aged, or fermented meats (Eitenmiller et al., 1980). Considering that these foods are common components of many peoples' diets, the human body presumably has a mechanism for breaking the toxins down. The mechanism for metabolizing histamine is by reaction with the enzymes diamine oxidase (DAO) and histamine N-methyltransferase (HMT). Since humans can metabolize histamine, it generally does not represent a hazard unless large amounts are ingested, or if the body's mechanism for breaking down histamine is not functioning properly (Taylor and Sumner, 1986).

Putrescine and cadaverine, the other components of the scombroid toxin, are formed from the decarboxylation of ornithine and lysine respectively. They are not particularly toxic on their own, but act by inhibiting the metabolism of histamine by DAO and HMT (Taylor and Sumner, 1986). DAO metabolizes cadaverine and putrescine at approximately three times the rate of histamine, thereby retarding the breakdown of histamine. Cadaverine can also act as an inhibitor of histamine metabolism by HMT. Theoretically, the presence of putrescine or cadaverine decreases the dose of histamine required to cause illness. In this way, putrescine and cadaverine are said to "potentiate" the toxicity of histamine.

Symptoms of scombroid poisoning usually begin within 10 minutes to four hours after consuming toxic fish. It is relatively common for victims to note that the fish had a "peppery," "metallic," "sharp," or "Cajun" taste. One of the first symptoms of scombroid poisoning is often flushing of the face and neck, accompanied by a feeling of overheating. This symptom is caused by the dilation of small blood vessels in the facial region. Facial flushing can be followed by a severe headache, dizziness, heart palpitations, rapid and weak pulse, numbness or burning of the mouth and throat, difficulty swallowing, and itching. Gastrointestinal symptoms such as nausea, vomiting, abdominal cramps, and diarrhea are experienced by approximately 25 percent of the victims of scombroid poisoning (MMWR, 1989; Kantha, 1987; Bryan, 1987; Murray et al., 1982; Eitenmiller et al., 1980; Arnold and Brown, 1978).

Most cases of scombroid poisoning are mild and subside in less than six hours with or without medical treatment. For those individuals who require medical care, administration of antihistamines has resulted in immediate improvement of the patient's condition (Kow-Tong and Malison, 1987; Taylor and Sumner, 1986). Even in severe cases, complete recovery is expected within 24 hours (Eitenmiller et al., 1980).

The dose of histamine required to cause scombroid poisoning in humans is not known. Two severe cases of scombroid poisoning occurred in New Mexico in 1987 from mahi mahi that had a histamine level of only 20 mg/100 g (MMWR, 1988). On the other hand, volunteers who were given 100 to 180 mg pure histamine orally showed only mild symptoms of poisoning (Motil and Scrimshaw, 1979). In part, this variability may be due to increased susceptibility in individuals with allergies, asthma, or peptic ulcers (Blackwell et al., 1969). However, a more compelling explanation for the discrepancy would be the presence of potentiators such as putrescine and cadaverine found in the fresh fish (but not in the pure histamine) that lower the dose of histamine required to cause illness (Taylor and Sumner, 1986; Murray et al., 1982; Arnold and Brown, 1978).

Scombroid toxicity is one of the most common illnesses associated with seafood. Between 1983 and 1987, the CDC reported 83 outbreaks involving 306 cases of scombroid toxicity (Bean et al., 1990). This statistic represents 35 percent of the seafood-borne outbreaks reported in the United States. The incidence of scombroid poisoning in the United States is lower than in those parts of the world where there is inadequate refrigeration and a greater dependency on fish in the diet (Eitenmiller et al., 1980).

Prevention of Scombroid Poisoning

Cooking, freezing, and smoking are ineffective in removing scombrotoxins from fish flesh. The best way to avoid scombroid poisoning is to prevent production of the toxin. Fresh fish should be stored in the coldest part of the refrigerator, preferably below 3°C (37°F) (Chen et al., 1988). Recreational fishermen should gut, bleed, and ice or refrigerate potentially toxic fish as quickly as possible. Low temperatures must be maintained until this fish is ready to be cooked. Frozen fish should be defrosted in the refrigerator or under cold running water, not at room temperature. Good hygienic practices should also be used to prevent introduction of histamine-producing bacteria (and other spoilage bacteria) to the fish.

The United States Food and Drug Administration (FDA) has established an action level for histamine in albacore, skipjack (Figure 4), and yellowfin tuna. Fish that exceed 50 mg histamine/100 grams fish cannot be sold. Furthermore, any fish which is found to exceed 5 mg histamine/100 grams may be more thoroughly investigated by the FDA before it is allowed to be sold, since a low level of histamine in fish indicates that some decomposition has occurred (FDA, 1982).

The method most commonly used to detect histamine, and the method approved by the Association of Official Analytical Chemists (AOAC), is a fluorometric assay (Taylor and Sumner, 1986; Arnold and Brown, 1978). The several different fluorometric procedures are based on spectrofluorometric measurement after reaction with ophthalaldehyde (Taylor et al., 1978). The methods differ in their procedure for cleaning and preparing the sample for testing (Taylor and Sumner, 1986). Other less commonly used histamine detection techniques include: an enzymatic assay, chromatography (thin layer, paper, gas-liquid, or high-pressure liquid chromatography), and guinea pig ileum bioassay (Taylor and Sumner, 1986).

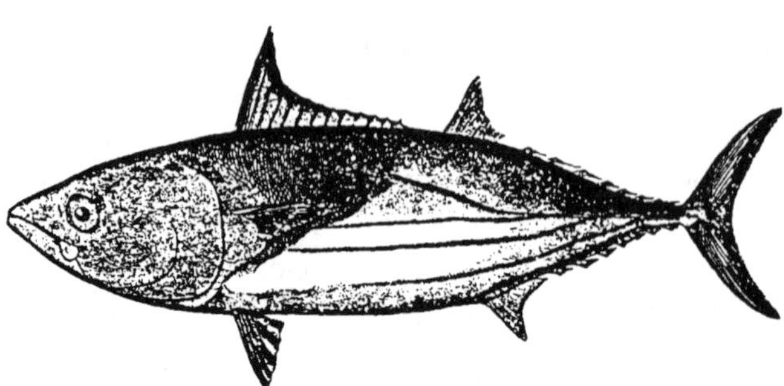

Figure 4. Katsuwonus pelamis *[*Gymnosarda pelamis*], skipjack tuna (from Dresslar and Fesler, 1889, plate 4).*

PUFFER FISH TOXICITY

The puffer fish, named for its ability to inflate itself when disturbed, is a poisonous fish found in the warm regions of the Pacific, Atlantic, and Indian oceans (Kantha, 1987). Puffer fish, also called blowfish, fugu, or sea squab (Figure 5), contain tetrodotoxin, a potent toxin that acts on the central and peripheral nervous systems of humans. Fugu is considered a delicacy in Japan despite the fact that consumption of an improperly prepared puffer can be fatal within minutes. The quantity and location of the toxin in the body is species-specific. Some of the more commonly toxic organs include: the entrails, liver, roe, ovaries, skin, and muscle (Liener, 1974). It is now believed that bacteria associated with puffer fish actually produce the tetrodotoxin, which is then accumu-

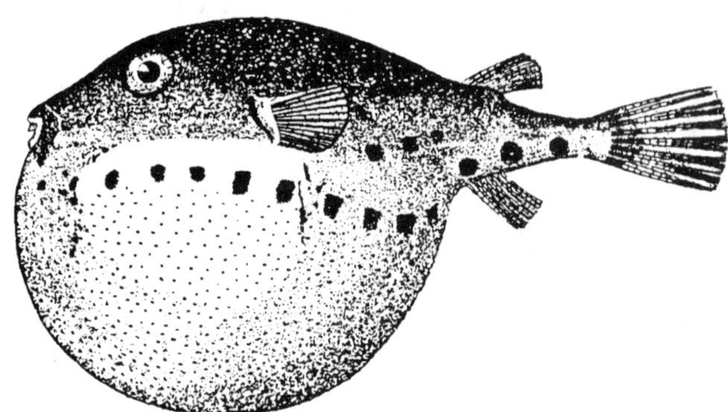

Figure 5a. Sphoeroides spengleri *(*Spheroides spengleri*), bandtail puffer, inflated (Evermann and Marsh, 1902, figure 79).*

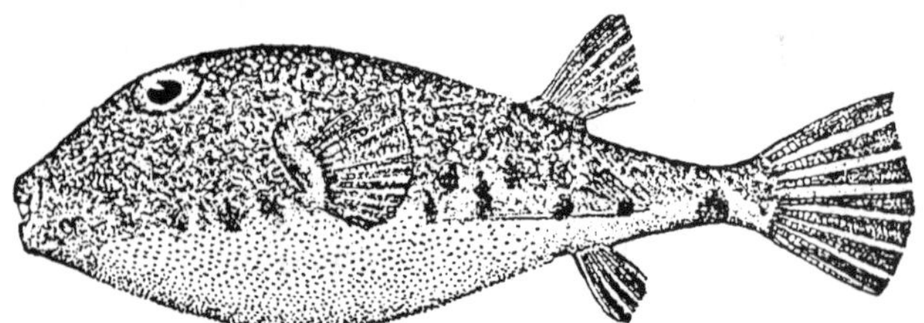

Figure 5b. Sphoeroides spengleri *(*Spheroides spengleri*), bandtail puffer, deflated (from Evermann and Bean, 1898, plate 57).*

lated to high concentrations in the fish organs (Rausch de Traubenberg and Lassus, 1991; Tamplin, 1990).

Symptoms of poisoning usually begin with 10 minutes of consuming puffer fish. The individual first experiences numbness and tingling of the lips, tongue, and inner surfaces of the mouth. This numbness can be followed by weakness, paralysis of limb and chest muscles, decreased blood pressure, and rapid and weak pulse. Death can occur within 30 minutes (Kantha, 1987; Horwitz, 1977). The symptoms of puffer fish poisoning should be treated directly, with specific attention to maintaining adequate respiration, circulation, and renal functions.

All fugu sold in Japan is carefully tested for tetrodotoxin before it reaches the consumer. Japanese chefs are required to undergo extensive training before they are allowed to handle fugu. According to Japanese reports, no cases of poisoning have been attributed to puffers prepared by certified chefs; problems only occur when inexperienced individuals try to prepare fugu. However, FDA sources report that there is an average of 19 fatalities from puffer fish poisoning in Japan each year.

The domestic species of puffer, sometimes called sea squab, is much less poisonous than the Japanese species. However, even this species should be regarded with care. From 1972 to 1974, the CDC reported one outbreak of puffer fish poisoning involving two cases that occurred in Florida (Horwitz, 1977).

In 1989, the FDA lifted an import ban on tiger fish. This species of puffer is less toxic than other Asian puffers, and the toxin is restricted to the liver. Before importation to the United States, the FDA requires that all tiger fish be cleaned, toxic tissues be removed, and fish be laboratory certified to be tetrodotoxin-free.

REFERENCES

Adachi, R. and Y. Fukuyo. 1979. The thecal structure of a marine toxic dinoflagellate *Gambierdiscus toxicus* gen. et. sp. nov. collected in a ciguatera endemic area. *Bulletin of the Japanese Society of Science and Fisheries.* **45**:67-71.

Arnold, S.H. and W.D. Brown. 1978. Histamine (?) toxicity from fish products. In C.O. Chichester, E.M. Mark, and G.F. Stewart (eds.), pp. 113-147. *Advances in Food Research.* Academic Press Inc. New York.

Bagnis, R. 1968. Clinical aspects of ciguatera (fish poisoning) in French Polynesia. *Hawaii Medical Journal.* **28**:25.

Bagnis, R. 1973. La ciguatera aux iles Marquises: aspects cliniques et epidemiologiques. *Bulletin of the World Health Organization.* **49**:67.

Bagnis, R.A., T. Kuberski, and S. Laugier. 1979. Clinical observations of 3,009 cases of ciguatera (fish poisoning) in the South Pacific. *American Journal of Medicine and Hygiene.* **28**:1067.

Bagnis, R., A. Spiegel, J.P. Boutin, C. Burducoa, L. Nguyen, J.L. Cartel, P. Capdevielle, P. Imbert, D. Prigent, C. Gras, and J. Roux. 1992. Evaluation de L'efficacite du mannitol dans le traitement de le ciguatera on Polynesie Francaise. *Medicine Tropicale.* **52**:67-73.

Ballantine, D.L., T.R. Tosteson, A.T. Bardales. 1988. Population dynamics and toxicity of natural populations of benthic dinoflagellates in southwestern Puerto Rico. *Journal of Experimental Marine Biology and Ecology.* **119**:201-212.

Banner, A.H., P.J. Scheuer, S. Sasaki, P. Helfrich and C.B. Alender. 1960. Observations of ciguatera-type toxin in fish. Annals of the New York Academy of Science. **90**:770-787.

Banner, A.H., S. Sasaki, P. Helfrich and C.B. Alender. 1961. Bioassay of ciguatera toxin. Nature. **189**:229-230.

Bean, N.H., P.M. Griffin, J.S. Goulding, and C. B. Ivey. 1990. Foodborne disease outbreaks, 5-year summary, 1983-1987. Morbidity and Mortality Weekly Review, CDC Surveillance Summaries. Volume 39, Number SS-1.

Blackwell, B., L.A. Mabbitt and E. Marley. 1969. Histamine and tyramine content of yeast products. *Journal of Food Science.* **34**:47-51.

Blythe, D.G., L.E. Fleming, D.R. Ayyar, D. deSylva, D. Baden, and K. Schrank. 1994. 08 01: Mannitol therapy for acute and chronic ciguatera fish poisoning. *Memoirs of the Queensland Museum.* **34** (3): 465-470.

Bryan, F.L. 1988. Risks associated with vehicles of foodborne pathogens and toxins. *Journal of Food Protection.* **51**: 498-508.

Bryan, F.L. 1987. Seafood-transmitted infections and intoxications in recent years. In D.E. Kramer and J. Liston (eds.), pp. 319-337. *Seafood Quality Determination.* Elsevier Science Publishing. Amsterdam.

Calvert, G.M. 1991. The recognition and management of ciguatera fish poisoning. In D.M. Miller (ed.), pp. 1-11. *Ciguatera Seafood Toxins.* CRC Press Inc. Boca Raton, FL.

Campbell, B., L.K. Nakagawa, M.N. Kobayashi, and Y. Hokama. 1987. *Gambierdiscus toxicus* in gut content of the surgeonfish *Ctenochaetus strigosus* (herbivore) and its relationship to toxicity. *Toxicon.* **25**: 1125-1127.

Cameron, J. and M.F. Capra. 1991. Neurological studies on the effects of ciguatoxin on mammalian nerve. In D.M. Miller (ed.), pp. 21-32. *Ciguatera Seafood Toxins.* CRC Press Inc. Boca Raton, FL.

Carlson, R.D. and D.R. Tindall. Distribution and periodicity of toxic dinoflagellates in the Virgin Islands. In D.M. Anderson, A.W. White, and D.G. Baden (eds.), pp. 171-176. *Toxic Dinoflagellates, Proceedings of the*

Third International Conference. Elsevier/North Holland, New York.

Chen, C-M., M.R. Marshall, J.A. Koberger, W.S. Otwell, and C.I. Wei. 1988. Determination of minimal temperatures for histamine production by five bacteria. *Proceedings of the 12th Annual Conference of the Tropical and Subtropical Fisheries Technology Society of the Americas.* Florida Sea Grant College Program.

Chungue, E., R. Bagnis, and F. Parc. 1984. The use of mosquitoes (*Aedes aegypt*) to detect ciguatoxin in surgeon fishes (*Ctenochaetus striatus*). *Toxicon.* **22**: 161-164.

Craig, C.P. 1980. It's always the big ones that should get away. *Journal of the American Medical Association.* **244**: 272-273.

Deichmann, W.B., W.E. MacDonald, D.A. Cubit, C.E. Wunsch, J.E. Bartels, and F.R. Merritt. 1977. Pain in jawbones and teeth in ciguatera intoxications. *Florida Scientist.* **40**:227-238.

Eitenmiller, R.R., J.H. Orr, and W.W. Wallis. 1984. Histamine formation in fish: microbiological and biochemical conditions. In R.E. Martin (ed.), pp. 39-50. *Chemistry and Biochemistry of Marine Food Products.* A.V.I. Publishing Company. Westport, CT.

Farn, G. and G.G. Sims. 1986. Chemical indices of decomposition in tuna. In D.E. Krammer and J. Liston (eds.), pp. 175-183. *Seafood Quality Determination.* Elsevier, Amsterdam.

FDA. 1982. Defect action levels for histamine in tuna; availability of guide. Federal Register. **47**:40478.

Food Protection Report. **5**(4):3-4. FDA lifts ban on importing puffer fish. 1989.

Gervais, A.J. 1985. Management of fisheries and public health problems associated with toxic dinoflagellates. In D.M. Anderson, A.W. White and D.G. Baden (eds.), pp. 530-533. *Toxic Dinoflagellates, Proceedings of the Third International Conferences.* Elsevier/North Holland. New York.

Gillespie, N., R. Lewis, J. Bourke, and M. Holmes. 1985. The significance of the absence of ciguatoxin in a wild population of *G. toxicus.* In *Proceedings of the Fifth International Coral Reef Congress, Tahiti.* Volume **4**, pp. 437-441. Antenne Museum-Ephe, Moorea, French Polynesia.

Hokama, Y., A.H. Banner, and D.B. Boylan. 1977. A radioimmunoassay for the detection of ciguatoxin. *Toxicon.* **15**: 317-325.

Hokama, Y., M.A. Abad, and L.H. Kimuar. 1983. A rapid enzyme-immunoassay for the detection of ciguatoxin in contaminated fish tissues. *Toxicon.* **21**: 817-824.

Hokama, Y., A.M. Osugi, S.A.A. Honda, and M.K. Matsuo. 1985. Monoclonal antibodies in the detection of ciguatoxin and other toxic polyethers in fish tissues by a rapid poke stick test. *Proceedings of the Fifth International Coral Reef Congress.* Tahiti. **4**: 449-455.

Hokama, Y., L.K. Shirai, L.M. Iwamoto, M.N. Kobayashi, C.S. Goto, and L.N. Nakagawa. 1987a. Assessment of a rapid enzyme immunoassay stick test for the detection of ciguatoxin and related polyether toxins in fish tissues. *Biology Bulletin.* **172**: 144-153.

Hokama, Y., S. Honda, M. Kobayashi, L. Nakagawa, J. Kurihara, and J. Miyahara. 1987b. Monoclonal antibodies in the detection of ciguatoxin (CTX) and related polyethers in contaminated fish tissues. In P. Gopalakrishnakone and C.K. Tan (eds.), pp. 385-393. *Progress in Venom and Toxin Research.* National University of Singapore.

Hokama, Y., S.A.A. Honda, M.N. Kobayashi, L.K. Nakagawa, A.A. Asahina, and J.T. Miyahara. 1989. Monoclonal antibody (MAB) in detection of ciguatoxin (CTX) and related polyethers by the stick-enzyme immunoassay (S-EIA) in fish tissues associated with ciguatera poisoning. In *IUPAC.* Tokyo, Japan. Elsevier/Amsterdam.

Horwitz, M.A. 1977. Specific diagnosis of foodborne disease. *Gastroenterology.* **73**: 375-381.

Jacyno, M. and D.M. Miller. 1991. Extraction methods for marine toxins. In D.M. Miller (ed.), pp. 53-72. *Ciguatera Seafood Toxins.* CRC Press Inc. Boca Raton, FL.

Jurnanovic, L.R. and D.L. Park, 1991. Foodborne toxins of marine origin: ciguatera. *Rev. Environ. Contam. Toxicol.* **117:** 51-94.

Kantha, S.S. 1987. Ichthyotoxins and their implications to human health. *Asian Medical Journal.* **30**: 458-470.

Kimua, L.H., Y. Hokama, M.A. Abad, M. Oyama, and J.T. Miyahara. 1982. Comparison of three different assays for the assessment of ciguatoxin in fish tissues: radioimmunoassay, mouse bioassay and in vitro guinea pig atrium assay. *Toxicon.* **20**: 907-912.

Kow-Ton, C. and M.D. Malison. 1987. Outbreak of scombroid fish poisoning, Taiwan American. *Journal of Public Health.* 77: 1335-1336.

Lawrence, D.N., M.B. Enriquez, R.M. Lumish, and A. Maceo. 1980. Ciguatera fish poisoning in Miami. *Journal of the American Medical Association.* **244**: 254-258.

Lewis, R.J. and M.J. Holmes, 1993. Origin and transfer of toxins involved in ciguatera. *Comp. Biochem. Physiol.* **106C**: 615-628.

Liener, I.E. 1974. *Toxic Constituents of Animal Foodstuff.* Pp. 73-110. New York Academic Press. New York, NY.

Lobel, P.S., D.M. Anderson, and M. Durand-Clement. 1988. Assessment of ciguatera dinoflagellate populations: sample variability and algal substrate selection. *Biology Bulletin.* **175**:94-101.

Miyahara, J.T., C.K. Kamibayashi, L.K. Shirai, and Y. Hokama. 1987. The similarity of toxins in *Ctenochaetus strigosus* and cultured *Gambierdiscus toxicus.* In P. Gopalakrishnakone and C.K. Tan (eds.), pp. 363-371. *Progress in Venom and Toxin Research.* National University of Singapore.

Morris, J.G. 1980. Ciguatera fish poisoning. *Journal of the American Medical Association.* **244**: 272-274.

Motil, K.J. and N.S. Scrimshaw. 1979. The role of exogenous histamine in scombroid poisoning. *Toxicology Letter.* **3**: 219-222.

MMWR. 1989. Scombroid-poisoning — Illinois, South Carolina. *Morbidity and Mortality Weekly Review.* **38**: 140-147.

MMWR. 1988. Scombroid-poisoning. *Morbidity and Mortality Weekly Review.* **37**(29).

Murray, C.K., G. Hobbs, and R.J. Gilbert. 1982. Scombrotoxin and scombrotoxin-like poisoning from canned fish. *Journal of Hygiene.* **88**: 215-220. Cambridge, MA.

Nukina, M., L.M. Koyanagi, and P.J. Scheuer. 1984. Two interchangeable forms of ciguatoxin. *Toxicon.* **22**:169.

Olson, D.A., D.W. Nellis, and R.S. Wood. 1984. Ciguatera in the eastern Caribbean. *Marine Fisheries Review.* **46**: 13.

Omura, Y., R.J. Price, and H.S. Olcott. 1978. Histamine-forming bacteria isolated from spoiled skipjack tuna and jack mackerel. *Journal of Food Service.* **43**: 1779-1781.

Palafox, N.A., Jain, L.G., Pinano, A.Z., Gulick, T.M., Williams, R.K., and Schatz, I.J. 1988. Successful treatment of ciguatera fish poisoning with intravenous manitol. *Journal of the American Medical Association.* **259**: 2740-2742.

Pearn, J.H., R.J. Lewis, T. Ruff, M. Tait, J. Quinn, W. Murtha, G. King, A. Mallett, and N.C. Gillespie. 1989. Ciguatera and mannitol: experience with a new treatment regimen. *Medical Journal of Australia.* **151**:77-80.

Rausch de Traubenberg, C. and P. Lassus, 1991. Dinoflagellate toxicity: are marine bacteria involved? Evidence from the literature. *Mar. Microbial Food Webs.* **5**:205-226.

Rice, S.L., R.R. Eitenmiller, and P.E. Koehler. 1976. Biologically active amines in food: a review. *Journal of Milk and Food Technology.* **39**:353-358.

Ruff, T.A. and R.J. Lewis. 1994 08 01: Clinical aspects of ciguatera: an overview. *Memoirs of the Queensland Museum.* **34**: 609-619.

Scheuer, P.J., W. Takahashi, J. Tsutsumi, and T. Yoshida. 1967. Ciguatoxin: Isolation and chemical nature. *Science.* **21**: 1267.

Stewart, M.P.M. 1991. Ciguatera fish poisoning treatment with intravenous mannitol. *Tropical Doctor.* **21**: 54-55.

Tamplin, M.L. 1990. A bacterial source of tetrodotoxins and saxitoxins. In Hall, S. and G. Strichartz (eds.), pp. 78-85. *Marine Toxins: Origin, Structure, and Molecular Pharmacology, ACS Symp. Ser. 418.* Amer. Chem. Soc. Washington, DC.

Taylor, F.J.R. 1979. A description of the benthic dinoflagellates associated with ciguatoxin, including observations on Hawaiian material. In Taylor, D.L. and H.H. Seliger (eds.). *Toxic Dinoflagellate Blooms.* Elsevier/North Holland.

Taylor, S.L. 1986. Histamine food poisoning: toxicology and clinical aspects. *Critical Reviews of Environmental Control, Critical Review of Toxicology.* **17**: 91-128.

Taylor, S.L. and S.S. Sumner. 1986. Determination of histamine, putrescine and cadaverine. In D.E. Kramer, and J. Liston (eds.), pp. 235-242. *Seafood Quality Determination.* Elsevier, New York.

Taylor, S.L., M. Leatherwood, and E.R. Lieber. 1978. A survey of histamine levels in sausages. *Journal of Food Protection.* **41**: 634-637.

Tindall, D.R., R.W. Dickey, R.D. Carlson and G. Morey-Gaines. 1984. Ciguatoxigenic dinoflagellates in the Caribbean Sea. In E.P. Ragelis (ed.). *Seafood Toxins.* American Chemical Society. Washington, DC.

Todd, E.C.D. 1985. Ciguatera poisoning in Canada. In Anderson, D.M., A.W. White, and D.G. Baden (eds.), pp. 505-510. *Toxic Dinoflagellates, Proceedings of the Third International Conference.* Elsevier/North Holland, New York.

Torda, T.A., E. Sinclair, and D.B. Ulyatt. 1973. Puffer fish (tetrodotoxin) poisoning, clinical record and suggested management. *Medical Journal of Australia.* **60**: 599-602.

Tosteson, T.R., D.L. Ballantine, and H.D. Durst. 1988. Seasonal frequency of ciguatoxic barracuda in Southwest Puerto Rico. *Toxicon.* **26**: 795-801.

Tosteson, T.R., D.L. Ballantine, C.G. Tosteson, V. Hensley, A.T. Bardales. 1989. Associated bacterial flora, growth, and toxicity of cultured benthic dinoflagellates *Ostreopsis lenticularis* and *Gambierdiscus toxicus. Applied and Environmental Microbiology.* **55**: 137-141.

Williams, R.K., Palafox, N.A. 1990. Treatment of pediatric cigratera fish poisoning. *American Journal of Diseases of Children.* **144**: 747-748.

Williamson, J. 1990. Ciguatera and mannitol: a successful treatment (letter). *Medical Journal of Australia.* **153**: 306-307.

Yasumoto, T. 1985. Recent progress in the chemistry of dinoflagellate toxins. In Anderson, D.M., A.W. White, and D.G. Baden (eds.), pp. 259-270. *Toxic Dinoflagellates, Proceedings of the Third International Conference.* Elsevier/North Holland, New York.

Yasumoto, T., I. Nakajima, R.A. Bagnis, and R. Adachi. 1977. Finding of a dinoflagellate as a likely culprit of ciguatera. *Bulletin of the Japanese Society of Science and Fisheries.* **43**: 1021-1026.

Shellfish Toxins

Peter D. R. Moeller

SHELLFISH TOXINS

Filter-feeding bivalve mollusks — such as clams, oysters, and mussels — can become toxic to humans during what are popularly called "red tides." A red tide is caused by the blooming or dramatic population increase of certain phytoplankton species. The types of phytoplankton most frequently implicated in red tides are dinoflagellates and diatoms. When the algal density gets high enough, the water can actually appear red (as well as brown, green, or yellow) — thus the name "red tide." A number of toxins can be produced by red tide phytoplankton, although it should be noted that not all red tides are toxic, and not all toxic blooms make the water red. A common misconception is that if the water is not red, then the shellfish are safe to eat. However, water color is not a good indicator of shellfish toxicity, since it is common for mollusks to become toxic even at phytoplankton concentrations below that causing discoloration of the water (Anderson and Morel, 1979). The muscle tissue of lobsters, crabs, shrimp, and finfish are safe to eat during red tides because they do not accumulate the shellfish toxins.

Scallops, although filter-feeding bivalves, are a special case. In the past it was generally accepted that the scallop adductor muscle remained free of toxins while the other body parts, which are not normally consumed, became highly toxic. Since the adductor muscle is the only part of the scallop that is traditionally sold in the United States, it was assumed that scallops were safe to eat during red tides. However, recent evidence has shown that under extremely toxic conditions scallop adductor muscles may not remain completely free of toxins. Another potential problem is the recent pressure to harvest whole or roe-on scallops. Whole scallops can be highly toxic and the levels of toxicity are unpredictable, making the management of harvesting whole scallops a difficult task. Although scallops are sold whole in other parts of the world, it would appear from the available evidence that this is not a wise practice.

Four human illnesses are associated with shellfish and toxic algal blooms: paralytic (PSP), neurotoxic (NSP), amnesic (ASP), and diarrhetic (DSP) shellfish poisonings.

DETECTION OF SHELLFISH TOXINS

Consumers are protected from shellfish toxins by the regulations imposed by the National Shellfish Sanitation Program (NSSP). The NSSP Manual of Operations states that "In locations known to be affected by marine biotoxins, the state shellfish control agency shall establish a marine biotoxin control plan which defines administrative procedures, laboratory support, surveillance strategies, and patrol procedures needed to provide public health control" (NSSP, 1989). The NSSP specifically addresses the control of PSP and NSP, and, if the need arises, can encompass the control of ASP and DSP as well. Under the guidelines of the NSSP, shellfish are periodically tested for biotoxins. Areas that are found to exceed quarantine levels are closed to harvesting. The shellfish beds are reopened to harvesting after toxin levels remain below quarantine levels for a period of time specified by the state.

The standard method for detecting the presence of biotoxins in shellfish, and the method approved by the NSSP for monitoring PSP and NSP toxins, is the mouse bioassay. This technique involves intraperitoneal injection of a 20-gram mouse with crude shellfish extracts. The bioassay is based on the relationship of dose to time of death of the mouse (Delaney, 1984). Advantages of the mouse assay are its speed and simplicity. Historical use for monitoring of PSP toxins has demonstrated that it provides adequate protection for consumers of shellfish. However, for some toxin classes, the mouse bioassay lacks adequate sensitivity. In addition, ethical issues relating to the use of animal testing have contributed to the desire to develop alternative, reliable assay methods. Alternative approaches under development include cytotoxicity assays, instrumental methods, immunoassays, and receptor assays. At the time of this writing, no alternative methods have yet undergone rigorous collaborative testing; thus, the mouse bioassay currently remains the standard method of detection for quarantine levels of shellfish toxins set by the NSSP.

AMNESIC SHELLFISH POISONING

In the late fall of 1987, 107 Canadians became ill after consuming cultured blue mussels harvested off Prince Edward Island (Todd, 1990). The illness was called amnesic shellfish poisoning (ASP) since some of the severely ill patients experienced persistent short-term memory loss. The mussels that caused the outbreak were found to contain domoic acid, a toxin not previously found in shellfish. Mussel toxicity was caused by ingestion of large quantities of the diatom *Pseudonitzschia* (formerly *Nitzschia) pungens* multiseries (Wright, 1989), an alga which, before the outbreak of 1987, was generally considered innocuous (Figure 1). *Pseudonitzschia* is a common coastal diatom of the Atlantic, Pacific, and Indian oceans and ranges between 62°N and 41°S latitude (Bird and Wright, 1989). It has a broad thermal tolerance and can thrive in the low salinities of estuaries (Bird and Wright, 1989).

Although *Pseudonitzschia* spp. are commonly found in coastal waters worldwide, its presence does not necessarily mean that cases of ASP will occur. Not all forms of *P. pungens* are capable of producing domoic acid. So far, the only type found to produce domoic acid is *P. pungens* multiseries, the organism implicated in the 1987 outbreak (Bird and Wright, 1989). Appreciable quantities of domoic acid are associated only with blooms of the diatom. This characteristic is supported by observations of natural *Pseudonitzschia* populations (Bird and Wright, 1989) as well as non-axenic laboratory cultures (toxins were produced only after the phytoplankton reached the stationary phase) (Bates et al., 1991; Subba Rao et al., 1990). The concentration of *Pseudonitzschia* during the 1987 bloom reached approximately 10^7 cells/l and resulted in a toxin level of 300-900 mouse units/gram (μg/g) mussel tissue (Wright, 1989).

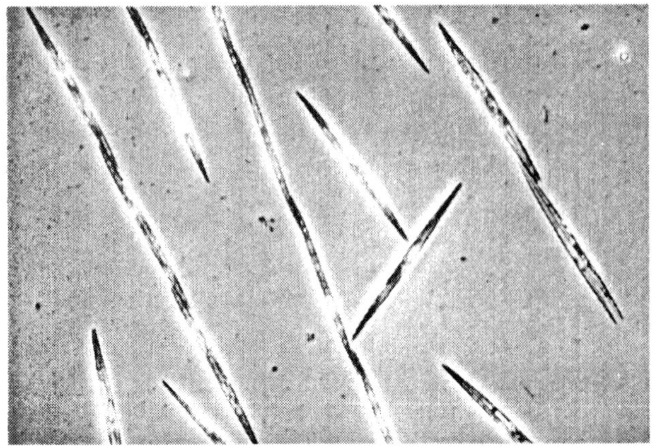

Figure 1. Pseudonitzschia multiseries *(400x).*

Following the 1987 Canadian ASP outbreak, an extensive domoic acid-related event occurred on the West Coast of the United States in 1991 with a mass mortality of pelicans in Monterey Bay, California (Work et al., 1992). In this case, the domoic acid source was identified as another species of diatom, *Pseudonitzschia australis* (Garrison et al., 1992). Later that same year, the recreational and commercial razor clam industries in Oregon and Washington were shut down due to contamination with domoic acid (Wood et al., 1994; Horner et al., 1993). Both states experienced closures of razor clam harvesting again in 1992 (Wood et al., 1994). While *Pseudonitzschia* spp. were suspected to be the causative organisms in the Oregon and Washington incidents, the source of the domoic acid has yet to be positively confirmed.

Blue mussels and razor clams are the primary shellfish implicated in cases of ASP to date; however, it is possible that other bivalves are capable of causing ASP. Canadian researchers have found low levels of domoic acid in soft-shell clams (*Mya arenari*) and in the digestive gland of sea scallops (Gilgan et al., 1990). The scallop adductor muscles were free of toxin and, on rare occasions, small amounts of toxin were found in the gonadal tissue. Domoic acid has also been found in the New England area, with levels as high as 568 to 595 μg/g tissue having been found in the digestive gland of some sea scallops (Shumway, 1990).

Canadian field studies conducted in 1988 showed that there is a lag time of approximately nine days between the peak domoic acid levels in the phytoplankton and the peak in mussels. It has been suggested that the lag time could be a useful monitoring tool. By testing the phytoplankton for an increase in domoic acid, authorities would have an early warning for the rise of toxin levels in shellfish. Mussel depuration rates during the same study did not exhibit a lag period. Domoic acid levels in mussels followed the decline in phytoplankton, a process which took two months to complete (Smith et al., 1990).

ASP causes both gastrointestinal and neurological disorders. In the early stages of ASP, the individual usually experiences gastrointestinal symptoms. The onset of gastroenteritis is usually within 24 hours of consuming toxic shellfish and can include: vomiting, abdominal cramps, diarrhea, anorexia, and nausea. Severe cases of ASP also cause neurological symptoms, usually within 48 hours of consumption. These symptoms may include: headache, dizziness, facial grimace or chewing motion, seizures, disorientation, short-term memory loss, excessive bronchial secretions, difficulty breathing, and coma (Shumway, 1990; Todd, 1990; Grey, 1988).

Epidemiological data from the 1987 outbreak showed that individuals with mild cases involving only gastroenteritis had ingested <75 mg of domoic acid. More severely ill individuals who experienced neurological as well as gastrointestinal symptoms had consumed 115 to 290 mg of the toxin (Todd, 1990).

Most of the victims of the 1987 outbreak recovered completely. However, 12 individuals experienced persistent disorientation and short-term memory loss, and three elderly persons died. Autopsies of the individuals who died showed brain lesions (Shumway, 1990; Smith et al., 1990; Todd, 1990; Wright, 1990; Grey, 1988).

High performance liquid chromatography (HPLC) is the primary method used to detect domoic acid in shellfish. The minimum detection level of domoic acid by HPLC is 0.2 µg/g, well below the quarantine level of 20 µg/g set by the Canadian government (Gilgan et al., 1990).

Since the outbreak in 1987, the Division of Fisheries and Oceans in Canada has been monitoring shellfish harvesting areas for domoic acid. The effort involves both testing water samples for the presence of *Pseudonitzschia pungens*, and testing the shellfish themselves for domoic acid. When the domoic acid level reaches 20 µg/g (20 ppm), the areas are closed to harvesting (Stephen, 1990). The State of Maine and the United States Food and Drug Administration (FDA) have begun testing mussels, soft-shell clams, and scallop digestive glands for domoic acid, and they are prepared to institute closures if the need arises.

PARALYTIC SHELLFISH POISONING

Red tides that cause paralytic shellfish poisoning (PSP) have occurred throughout the world and appear to be spreading (Shumway, 1990; Anderson, 1989). Many species of dinoflagellates cause paralytic shellfish poisoning. The species that commonly bloom in the New England area are *Alexandrium tamarense* (Figure 2) and *A. fundyense* (also known as *Gonyaulax tamarensis*, *G. excavata*, *G. tamarensis* var. *excavata*, *Gessnerium tamarensis*, and *Protogonyaulax tamarensis*) (Cembella et al., 1988). In New England, blooms most often occur in May, June, September, and October (Shumway and Cucci, 1987). The principal toxic species in the United States Pacific waters is *A. catenella* (Price and Kizer, 1990), and, to a lesser degree, *A. tamarense* (Delaney, 1984). *Pyrodinium bahamense* var. *compressa* (Canahui et al., 1989; White, 1988) and some species of *Gymnodinium* (Anderson, 1989; LaBarbera et al., 1989) have also been implicated in PSP toxicity in other parts of the world.

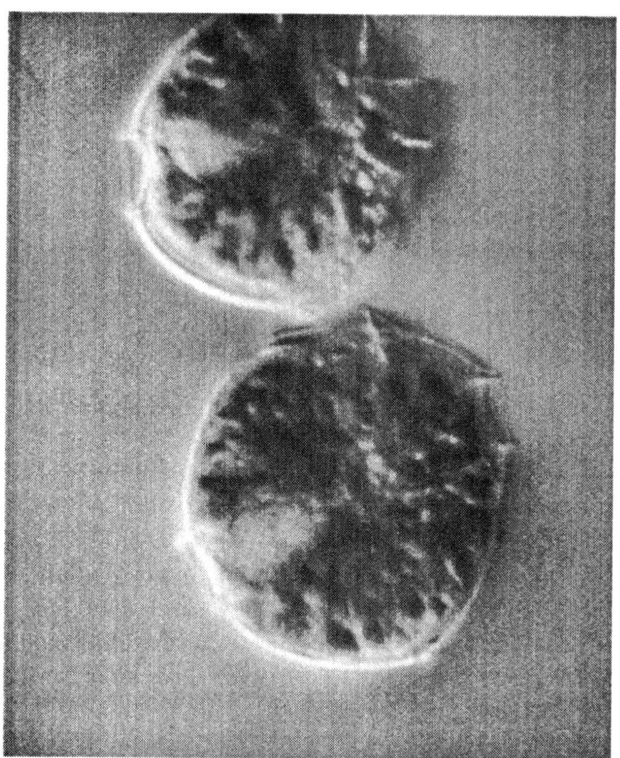

Figure 2. Alexandrium tamarense *(1250x).*

Alexandrium spp., as well as other PSP-causing dinoflagellates, are capable of producing non-motile cysts, a dormant stage that can withstand harsh conditions (Anderson and Wall, 1978). These cysts settle to the sediment and go through a mandatory dormancy period (Anderson, 1980). Germination occurs when conditions improve, and is dependent on appropriate temperature, oxygen content, and light (Anderson et al., 1987; Anderson, 1980). Areas that experience one red tide are often likely to experience recurrent blooms due to the germination of resuspended cysts (Anderson, 1989).

Cysts in the ballast water of ships may play a role in the dispersal of red tide dinoflagellates. Most phytoplankton cannot tolerate the cold and dark conditions of a ballast tank, or the shock of being released into water of different salinity, temperature, etc. However, cysts can tolerate these harsh conditions and can germinate after they are released from the ballast tank. This statement is supported by studies in Australia which showed that *A. tamarense* cysts remained viable after being held in ballast sediment for six months (Hallegraeff et al., 1990).

All filter-feeding bivalve mollusks accumulate and eliminate paralytic shellfish toxins. Mussels become highly toxic within a few days of the onset of a red tide, but also lose their toxin load rapidly (Shumway, 1990). They can become extremely toxic without apparent alert, and are often used as sentinel species in monitoring programs. For example, mussel (*Mytilus edulis* and *Modiolus modiolus*) toxin levels in Maine rose from the detection level to more than 8000 µg/100g in two days (Shumway et al., 1988). Calculations based on laboratory feeding experiments suggest that during blooms of highly toxic dinoflagellates the level of toxins in mussels can exceed quarantine levels (183 µg saxitoxin/100 g tissue) in less than one hour (Bricelj et al., 1990).

Soft-shell clams (*Mya arenaria*) and oysters (*Crassostrea virginica* and *Ostrea edulis*) require more time to accumulate toxins and are generally less toxic than are mussels. *Mya* become toxic approximately one week after mussels and also require longer to cleanse themselves of toxins (Shumway and Cucci, 1987). Field experiments in West Boothbay Harbor, Maine, showed that oysters exposed to *A. tamarense* became only slightly toxic approximately two weeks after toxin appeared in mussels. Toxin was depurated from oysters in approximately two to four weeks (Shumway, 1990).

Hard clams do not become as toxic as other mollusks. Laboratory studies have shown that *Mercenaria mercenaria* exposed to *A. tamarense* exhibit a pronounced valve closure (Shumway, 1990). However, Lee and Bricelj (1989) demonstrated that *M. mercenaria* will readily ingest low toxicity strains of *A. fundyense*, although only when a good food source is also present (Bricelj et al., 1990). These researchers suggested that the rapid rise in *A. fundyense* cell densities and its dominance during red tides may prohibit hard clams from accumulating significant levels of PSP toxins in nature.

Surf clams (*Spisula solidissima*) and butter clams (*Saxidomus giganteus*) exhibit rapid uptake and extremely slow elimination of toxins. Both of these species of clams have been reported to retain toxins for a period of years (Beitler and Liston, 1990; Shumway et al., 1988; Shumway and Cucci, 1987). In some areas of the Pacific Northwest and Alaska, butter clams are considered permanently toxic (Price and Kizer, 1990).

Scallops can be extremely toxic even during periods when blooms are not evident (Shumway et al., 1988). Most of the toxins are stored in the digestive gland and gonads, parts of the scallop that are not normally consumed in the United States. However, it appears that under highly toxic conditions, scallop adductor muscles can accumulate PSP toxins. During a survey of highly toxic offshore areas, a few scallops were found to have toxin concentrations above quarantine levels in the adductor muscle (183 µg saxitoxin/100 g tissue; J. Hurst and S. Shumway, unpublished data — June 28, 1990, station #6).

In the past, it was believed that toxic dinoflagellates did not harm or affect shellfish. However, recent evidence has shown that in the presence of *A. tamarense*, mollusks exhibit species-specific responses that include alterations in: shell valve activity, oxygen consumption, heart rate, filtration rate, and byssus production (Gainey and Shumway, 1988; Shumway and Cucci, 1987; Cucci et al., 1985; Shumway et al., 1985).

Humans who consume toxic shellfish usually become ill within 15 minutes to three hours (Hughes, 1979). The individual initially experiences a numbness, burning, or tingling sensation of the lips, tongue, face, and fingertips. These symptoms lead to general muscular uncoordination of arms, legs, and neck. Other symptoms may include: weakness, dizziness, malaise, prostration, headache, salivation, rapid pulse, thirst, dysphagia, perspiration, impairment of vision or temporary blindness, ataxia with a "floating" sensation, nausea, vomiting, diarrhea, and convulsions. Severe cases of PSP can result in respiratory paralysis, and — although rarely — PSP can be fatal. The symptoms of PSP must be treated directly since there is no known antidote (Price and Kizer, 1990; Concon, 1988; Bryan, 1987; Hughes, 1979).

The dose of paralytic shellfish toxins required to cause illness is not known. Human susceptibility to paralytic shellfish toxins is variable, and can depend on the individual's health, weight, age, etc. Price and Kizer have estimated that "ingestion of 200 to 500 µg is likely to cause at least mild symptoms; ingestion of 500 to 2,000 µg is likely to cause moderate to severe symptoms; and consumption of more than 2,000 µg is likely to produce serious, and possibly lethal, consequences."

Paralytic shellfish poisoning can be a serious illness, especially in Third World countries where monitoring is nonexistent and medical care may be poor. For example, a bloom of *Pyrodinium bahamense* var. *compressum* in the Philippines in June of 1987 was responsible for 691 cases of PSP and 14 deaths (Corrales and Gomex, 1990). Cases of PSP are rare in the United States, since harvesting of shellfish is regulated by an intense monitoring system. Six outbreaks of PSP were reported to the Centers for Disease Control from 1979 to 1984 (Bryan, 1987).

In addition to causing PSP in humans, blooms of *Alexandrium* spp. have caused fish kills that have been linked to the presence of PSP toxins in zooplankton (White, 1984). Laboratory experiments have shown that *Alexandrium* spp. is toxic to larval stages of fish as well (White et al., 1989). All evidence indicates that the toxins are confined to the viscera and that the flesh is consistently nontoxic (Anderson and White, 1989). Therefore, the only human health concerns would be those related to consumption of whole fish or particular internal organs, such as the liver, neither of which is a common way of eating fish in the United States.

PSP can be caused by a combination of saxitoxin and its 20 known chemical derivatives (Oshima et al., 1993). The toxin extent and composition can vary with dinoflagellate strains (Maranda et al., 1985), as well as with stage of growth (Anderson, 1990; Cembella et al., 1990; Boczar et al., 1988). Maranda et al. (1985) also demonstrated that *A. tamarense* toxicity in the Northeast follows a decreasing trend from north to south. At least 12 toxins can be produced by dinoflagellates. These include the carbamate toxins (saxitoxin, neosaxitoxin, and gonyautoxins 1-4), and the sulfocarbamoyl toxins (B1, B2, and C1-4). In addition, six decarbamoyl toxins, which are derivatives of the carbamate toxins, may be present in shellfish due to enzymatic action on the corresponding toxins (dc-saxitoxin) (Sullivan, 1988; Sullivan and Wekell, 1987).

The mouse bioassay is the standard method for detecting PSP toxins. A brief description of the methodology has been given earlier in this chapter. Purified saxitoxin is used as a standard so that mouse units can be converted to micrograms toxin/100 grams shellfish tissue (Delaney, 1984). The NSSP has set a quarantine level of 80 µg/100 grams shellfish tissue (NSSP, 1989). Some specific problems in detecting PSP toxins with the mouse bioassay are listed below:

1. The mouse bioassay is significantly limited in its minimum detection level of approximately 40 µg/100 g tissue, which is close to the maximum allowed level of 80 µg/100 g meat (Yang et al., 1987).
2. The Schantz "salt effect" may cause an underestimate of the toxin concentration by as much as a factor of three, especially at low toxin levels (Sullivan, 1988, citing Schantz et al., 1958).
3. The accuracy of the mouse bioassay can be +/- 20 percent (Sullivan and Wekell, 1987).

HPLC can also be used to detect PSP toxins (Oshima et al., 1988; Sullivan and Wekell, 1987; Sullivan et al., 1985). This methodology determines individual toxin levels and is more sensitive and accurate than the mouse bioassay (Hall et al., 1985). The disadvantages of using HPLC are the high cost of the HPLC unit itself, and the difficulties of obtaining a sufficient supply of the PSP toxin standard (Stabell and Cembella, 1990).

A number of immunologically-based methods, such as the radioimmunoassay and the enzyme-linked immunoassay (ELISA), have also been developed for detecting PSP toxins in shellfish. These immunological methods are based on competitive binding of the labeled antibodies to either free antigen (from the shellfish sample being analyzed) or antigen fixed to the surface of the test container. After an incubation period, the solution phase is washed away, leaving behind the fraction of antibody which was bound to the fixed antigen. The quantity of antibodies remaining is then determined. Detection of small quantities of antibody in the final process signifies that the shellfish sample has a high level of toxins (Vanderlaan et al., 1988). The radioimmunoassay has been found to be very sensitive to saxitoxin, but does not detect the other PSP toxins well (Yang et al., 1987; Hurst et al., 1985). The enzyme-linked immunoassay is reported to be at least four orders of magnitude more sensitive than HPLC when saxitoxin is the major toxin. Aside from saxitoxin, the ELISA method can detect neosaxitoxin and gonyautoxins-2 and -3 to a limited degree. Other PSP toxins are not detected by this method (Cembella et al., 1990).

More recently, radioreceptor assays have shown promise for use as monitoring tools. Receptor assays are analogous to immunoassays in that radiolabeled toxin standard competes with toxin in the sample for binding to their pharmacological target, the sodium channel. Unlike immunoassays, receptor assays recognize all cases of PSP toxins, in direct proportion to their toxic potency (Van Dolah et al., 1994).

DIARRHETIC SHELLFISH POISONING

The first reported cases of diarrhetic shellfish poisoning (DSP) occurred in Japan in 1976 (Yasumoto, 1978). By and large, DSP is a problem in Japan and Europe, but incidents have been reported throughout the world (Alvito et al., 1990; Edler and Hageltorn, 1990; Haamer et al., 1990; Hagel, 1990; Marcaillou-Le Baut and Masselin, 1990; Kat, 1983).

Although the United States has not had any confirmed cases of DSP, it should be considered a potential hazard for consumers of shellfish harvested from United States coastal waters. A confirmed outbreak of DSP has occurred as close as Nova Scotia, Canada (Quilliam et al., 1990). Within the United States, several species of *Dinophysis* (Figure 3) have been identified from water samples (from Narragansett Bay, Rhode Island), although no toxic shellfish have been found (Maranda and Shimizu, 1987). There is also circumstantial evidence that cases of DSP have occurred in the Mid-Atlantic region of the United States (Freudenthal and Jijina, 1988; 1985). These probable cases are based on symptoms, time of onset, negative results from conventional testing, and correlation with seasonal and spatial distribution of *Dinophysis* from monitoring data in the harvesting areas.

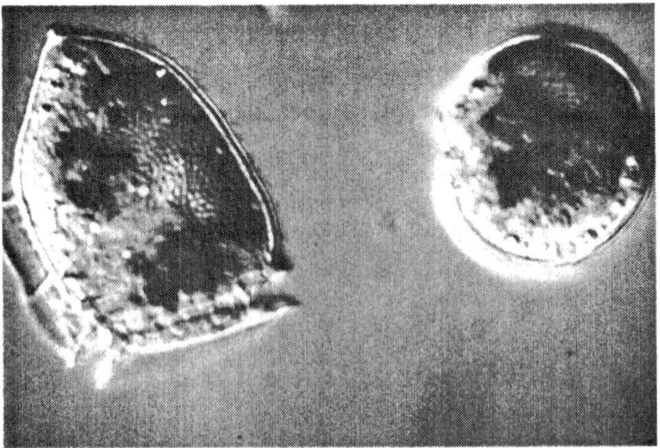

Figure 3. Left: Dinophysis norvegica; *right:* D. alexandrium *(1000x).*

A number of species of *Dinophysis* have been implicated in cases of DSP. *D. fortii* is the primary dinoflagellate responsible for DSP in Japan (Stamman et al., 1987; Yasumoto et al., 1980), while *D. acuminata* is the most important cause of DSP in Europe (Kat, 1983). Other species of *Dinophysis* that may cause DSP include: *D. acuta* (Elder and Hageltorn, 1990; Yasumoto, 1985); *D. norvegica* (Kat, 1985; Freudenthal and Jijina, 1985); *D. mitra*, *D. rotundata*, and *D. tripos* (Yasumoto, 1990). *Prorocentrum lima* (Figure 4) has been found to produce okadaic acid and its derivatives; however, its significance in causing DSP has not yet been determined (Yasumoto, 1990).

As the name implies, the symptoms of diarrhetic shellfish poisoning are gastrointestinal in nature. Symptoms usually begin within 30 minutes to a few hours after consuming contaminated

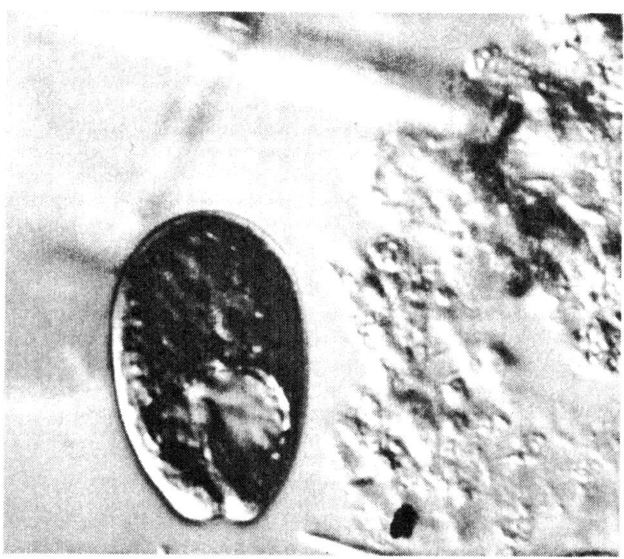

Figure 4. Prorocentrum lima *(1000x).*

shellfish and include: diarrhea, nausea, vomiting, chills, and moderate to severe abdominal pain and cramps. DSP is a self-limiting illness. Total recovery is expected within three days, with or without medical assistance. No known fatalities have occurred (Freudenthal and Jijina, 1988; Stamman et al., 1987; Freudenthal and Jijina, 1985; Yasumoto, 1985).

The toxins responsible for causing DSP differ throughout the world, as well as among shellfish species. The principal toxin in European mussels has been okadaic acid (Yasumoto, 1985). Derivatives of okadaic acid, dinophysostoxin-1 and -3, have been the primary toxins isolated from Japanese shellfish. Epidemiological data indicate that as little as 12 mouse units (equivalent to 48 μg okadaic acid) (Yasumoto, 1985) of dinophysostoxin-1 or okadaic acid is sufficient to cause illness in humans (Stamman et al., 1987; Yasumoto et al., 1980). Other DSP toxins that have been isolated, mostly from the digestive gland of Japanese scallops, include the pectinotoxins and the yessotoxins (Yasumoto, 1990).

Diarrhetic shellfish toxins are inhibitors of protein phosphatases, which are intimately involved in many physiological functions, including ion and water balance. Diarrhea associated with these toxins may be caused by the inhibition of protein phosphatases in the cells of the intestinal wall, resulting in ion and water imbalance (Cohen and Tsukitani, 1989).

The mouse bioassay can be used to detect diarrhetic shellfish toxins, although it is not a common method. False positives are not unusual when the bioassay is used to detect toxins in scallops (Lee et al., 1987; Hamano et al., 1985). The Netherlands currently uses the rat bioassay to detect DSP. In this method, white rats are fed the hepatopancreas of suspect shellfish and then observed. Consistency of the feces and refusal to consume the shellfish are used to test for the presence of DSP toxins (Hagel, 1990).

An HPLC method has also been developed for detecting DSP toxins (Lee et al., 1987). Shellfish extracts are esterified with 9-anthryldiazomethane (ADAM) and analyzed by HPLC. This method is most effective when okadaic acid and dinophysistoxin-1 are the principal toxins. The problems associated with this test are the difficulty in obtaining ADAM, the instability of ADAM, (Stabell and Cembella, 1990), and the methodology's limited ability to detect the other DSP toxins.

An ELISA test kit has been developed by Ube Industries, Limited, of Japan. Toxin extraction, sample preparation, and the actual assay can be completed in less than an hour (Uda et al., 1989). The ELISA kits can detect okadaic acid and dinophysistoxin-1 only, and the detection limit for okadaic acid is reported to be 10 ng/ml (Usagawa et al., 1989). An evaluation of the ELISA test kit by the Massachusetts Department of Public Health (MDPH) indicated that the kits may be useful as a screening tool, but are not precise enough to be used as a quantitative measure of DSP toxins in shellfish (Nassif, 1994).

Assays based on the pharmacological action of DSP toxins, the inhibition of protein phosphatases, have been employed, and show promise for accurately assessing DSP toxin activity in shellfish samples (Holmes, 1991).

NEUROTOXIC SHELLFISH POISONING

Blooms of *Gymnodinium breve* (formerly *Ptychodiscus brevis*) are usually associated with massive fish kills and human eye and respiratory irritations from exposure to sea spray containing lysed *G. breve* cells. This unarmored, or "naked," dinoflagellate is also the causative agent for neurotoxic shellfish poisoning (NSP). Neurotoxic shellfish poisoning is rare, and until recently was confined to the Gulf Coast of Florida. In 1987 a red tide occurred in Onslow Bay, North Carolina. The North Carolina bloom is believed to have been caused by the transportation of *G. breve* cells out of the Gulf of Mexico by the Loop Current, and then north by the Gulf Stream (Tester et al., 1990; Pietrafesa et al., 1987).

NSP resembles a mild case of ciguatera. Symptoms begin within 30 minutes to three hours of consuming contaminated shellfish (Hughes, 1979) and usually include: tingling of the face that spreads to other parts of the body, cold-to-hot sensory reversal, bradycardia, dilation of the pupils, and a feeling of inebriation. Less commonly, victims may experience prolonged diarrhea, nausea, and poor coordination (Concon, 1988; Hughes, 1979). Complete recovery is expected within 48 hours (Hughes, 1979). No deaths have been reported from NSP in the United States (Delaney, 1984).

Cases of NSP are rare. Between 1971 and 1977, the CDC reported only two outbreaks of NSP involving five individuals (Hughes, 1979). The Florida Department of Natural Resources has been monitoring its waters since the early 1960s. Shellfish beds are closed to harvesting when *G. breve* cell counts exceed 5000 cells/l. Mouse bioassays are also run to determine toxin concentrations (Gervais, 1985; Hunt and Tufts, 1979), and are used to determine re-opening of shellfish beds at 20 mouse units (mg)/100g shellfish.The unexpected bloom off the coast of North Carolina in 1987 resulted in 48 cases of NSP before officials were able to institute controls (Fowler and Tester, 1989). It is unlikely that this situation will occur again in North Carolina since officials now routinely test their shellfish for brevetoxins (Tester and Fowler, 1990).

G. breve requires high salinity (>24 ppt) for growth (Tester and Fowler, 1990). Blooms are generally initiated 18 to 74 km offshore and are transported inshore (Steidinger and Haddad, 1981). Due to the requirement for high salinity, *G. breve* is rarely found in the lower salinity estuaries where most shellfishing occurs. Laboratory cultures of *G. breve* produce at least eight toxins, collectively called brevetoxins (Schulman et al., 1990). Toxin content tends to fluctuate with age and stage of the culture (Roszell et al., 1990).

The mouse bioassay is the standard method for determining the level of toxins in shellfish. Shellfish toxicity is expressed in mg rather than in micrograms toxin, since NSP standards are not available for the conversion calculations. For a six-hour continual observation period, the assay's limit of detection is 20 mg/100 grams shellfish (Delaney, 1984). The NSSP Manual of Operation has set the brevetoxin quarantine level at any detectable amount of toxin (NSSP, 1989).

Two immunoassay techniques that have shown some promise for detecting neurotoxic shellfish toxins include a radioimmunoassay (Baden et al., 1985) and an enzyme immunoassay (Trainer and Baden, 1990). In addition, receptor binding competition assays for the NSP toxins show promise for routine testing of shellfish and algae samples (Van Dolah et al., 1994).

PREVENTION OF SHELLFISH POISONINGS

Toxins in shellfish cannot be destroyed by normal cooking, freezing, or smoking. The best prevention of shellfish poisoning is elimination of toxic shellfish before they reach the consumer. The NSSP Marine Biotoxin Monitoring Program effectively eliminates toxic shellfish from commercial distribution. Ensuring that recreational harvesters of shellfish are not exposed to marine biotoxins is more difficult. Some recreational harvesters are unaware of the threat of toxins or do not know how to get updated closure information. A common misconception, especially among recreational harvesters, is that if the water is not red, then the shellfish are safe to eat. Water color is not a good indicator of shellfish toxicity, since it is common for filter-feeding mollusks to become toxic even at phytoplankton concentrations below that necessary to discolor the water (Anderson and Morel, 1979).

REFERENCES

Alvito, P., I. Sousa, S. Franca, and M.A. De M. Sampaya. 1990. Diarrhetic shellfish toxins in bivalve mollusks along the coast of Portugal. In *Toxic Marine Phytoplankton*. E. Graneli, B. Sundstrom, L. Edler, and D.M. Anderson (eds.), pp. 443-448. Elsevier, New York.

Anderson, D.M. 1990. Toxin variability in *Alexandrium* species. In *Toxic Marine Phytoplankton*. E. Graneli, B. Sundstrom, L. Edler, and D.M. Anderson (eds.), pp. 44-51. Elsevier, New York.

Anderson, D.M. 1989. Toxic algal blooms and red tides: a global perspective. In *Red Tides: Biology, Environmental Science and Technology.* T. Okaichi, D.M. Anderson, and T. Nemoto (eds.), pp. 11-16. Elsevier, New York.

Anderson, D.M. 1980. Effects of temperature conditioning on development and germination of *Gonyaulax tamarensis* (Dinophyceae) hypnozygotes. *Journal of Phycology.* **16**:166-172.

Anderson, D.M. and F.M.M. Morel. 1979. Toxic dinoflagellate blooms in the Cape Cod region of Massachusetts. In *Toxic Dinoflagellate Blooms.* D.L. Taylor and H.H. Seliger (eds.), pp. 145-150. Elsevier/North Holland, New York.

Anderson, D.M. and D. Wall. 1978. Potential importance of benthic cysts of *Gonyaulax tamarensis* and *G. excavata* in initiating toxic dinoflagellate blooms. *Journal of Phycology.* **14**:224-234.

Anderson, D.M. and A.W. White. 1989. Toxic dinoflagellates and marine mammal mortalities: proceedings of an expert consultation held at the Woods Hole Oceanographic Institute. May 8-9, 1989. Coastal Resources Center, Woods Hole, MA. *Technical Report* CRC-89-6.

Anderson, D.M., C.D. Taylor, and E.V. Armbrust. 1987. The effects of darkness and anaerobiosis on dinoflagellate cyst germination. *Limnology and Oceanography.* **32**:340-351.

Baden, D.G., T.J. Mende, and L.E. Brand. 1985. Cross-reactivity in immunoassays directed against toxins isolated from *Ptychodiscus brevis.* In *Toxic Dinoflagellates, Proceedings of the Third International Conference.* D.M. Anderson, A.W. White, and D.G. Baden (eds.), pp. 363-368.Elsevier/North Holland, New York.

Bates, S.S., A.S.W. deFreitas, J.E. Milley, R. Pocklington, M.A. Quilliam, J.C. Smith, and J. Worms. 1991. Controls on domoic acid production by the diatom *Nitzschia pungens* f. multiseries in culture: nutrients and irradiance. *Can. J. Fish. Aquat. Sci.* **48**:1136-1144.

Beitler, M.K. and J. Liston. 1990. Uptake and tissue distribution of PSP toxins in butter clams. In *Toxic Marine Phytoplankton.* E. Graneli, B. Sundstrom, L. Edler, and D.M. Anderson (eds.), pp. 257-262. Elsevier, New York.

Bird, C.J. and J.L.C. Wright. 1989. The shellfish toxin domoic acid. *World Aquaculture.* **20**(1):40-41.

Boczar, B.A., M.K. Beitler, J. Liston, J.J. Sullivan, and R.A. Cattolico, 1988. Paralytic shellfish toxins in *Protogonyaulax tamarensis* and *Protogonyaulax catenella* in axenic culture. *Plant Physiology.* **88**:1285-1290.

Bricelj, V.M., J.H. Lee, A.D. Cembella, and D.M. Anderson. 1990. Uptake of *Alexandrium fundyense* by *Mytilus edulis* and *Mercenaria mercenaria* under controlled conditions. In *Toxic Marine Phytoplankton.* E. Graneli, B. Sundstrom, L. Edler, and D.M. Anderson (eds.), pp. 269-274. *Toxic Marine Phytoplankton.* Elsevier, New York.

Bryan, F.L. 1987. Seafood-transmitted infections and intoxications in recent years. In *Seafood Quality Determination.* D.E. Kramer and J. Liston (eds.), pp. 319-337. Elsevier Science Publishing. The Netherlands.

Canahui, E., F.R. Loessener, L. Miller, E.W. King, and S. Hall. 1989. *Pyrodinium bahamense* and PSP along the Pacific Coast of Central America (Abstract). *Journal of Shellfish Research.* **8**:440.

Cembella, A., C. Destombe, and J. Turgeon. 1990a. Toxin composition of alternative life history stages of *Alexandrium*, as determined by high performance liquid chromatography. In *Toxic Marine Phytoplankton.* E. Graneli, B. Sundstrom, L. Edler, and D.M. Anderson (eds.), pp. 333-338. Elsevier, New York.

Cohen, P. and Tsukitani, Y. 1989. Okadaic acid: a new probe for the study of cellular regulation. *TIBS* **15:** 98:102.

Cembella, A., Y. Parent, D. Jones, and G. Lamoureus. 1990b. Specificity and cross-reactivity of an absorption-inhibition enzyme-linked immunoassay for the detection of paralytic shellfish toxins. In Graneli, E., B. Sundstrom, L. Edler, and D.M. Anderson (eds.), pp. 339-344. *Toxic Marine Phytoplankton.* Elsevier, New York.

Cembella, A.D., J. Turgeon, J. C. Therriault, and P. Beland. 1988. Spatial distribution of *Protogonyaulax tamarensis* resting cysts in nearshore sediments along the north coast of the lower St. Lawrence estuary. *Journal of Shellfish Research.* **7**:597-610.

Concon, J.M. 1988. In *Food Toxicology.* J.M. Concon (ed.), Marcel Dekker Inc. New York.

Corrales, R.A. and E.D. Gomez. 1990. Red tide outbreaks and their management in the Philippines. In *Toxic Marine Phytoplankton.* E. Graneli, B. Sundstrom, L. Edler, and D.M. Anderson (eds.), pp. 453-458. Elsevier, New York.

Cucci, T.L., S.E. Shumway, R.C. Newell, and C.M. Yentsch. 1985. A preliminary study of the effects of *Gonyaulax tamarensis* on feeding in bivalve mollusks. In *Toxic Dinoflagellates.* D.M. Anderson, A.W. White and D.G. Baden (eds.), pp. 395-400. Elsevier, New York.

Delaney, J.E. 1984. Bioassay procedures for shellfish toxins. In A.E. Greenberg and D.A. Hunt (eds), pp. 64-80. Laboratory procedures for the examination of seawater and shellfish. The American Public Health Association. Washington, DC.

Duerden, C. 1989. Domoic acid probably annual event, says DFO. *Canadian Aquaculture.* March/April.

Edler, L. and M. Hageltorn. 1990. Identification of the causative organism of a DSP-outbreak on the Swedish West Coast. In *Toxic Marine Phytoplankton.* E. Graneli, B. Sundstrom, L. Edler, and D.M. Anderson (eds.), pp. 345-349. Elsevier, New York.

Fowler, P.K. and P.A. Tester. 1989. Impacts of the 1987-88 North Carolina red tide (Abstract). *Journal of Shellfish Research.* **8**:440.

Freudenthal, A.R. and J.L. Jijina. 1988. Potential hazards of *Dinophysis* to consumers and shellfisheries. *Journal of Shellfish Research.* **7**:695-701.

Freudenthal, A.R. and J.L. Jijina. 1985. Shellfish poisoning episodes involving or coincidental with dinoflagellates. In *Toxic Dinoflagellates, Proceedings of the Third International Conference.* D.M. Anderson, A.W. White, and D.G. Baden (eds.), pp. 461-466. Elsevier/North Holland, New York.

Gainey, L.F. Jr. and S.E. Shumway. 1988a. A compendium of the responses of bivalve mollusks to toxic dinoflagellates. *Journal of Shellfish Research.* **7**:623-628.

Gainey, L.F. Jr. and S.E. Shumway. 1988b. Physiological effects of *Protogonyaulax tamarensis* on cardiac activity in bivalve mollusks. *Comparative Biochemistry and Physiology.* **91**:159-164.

Garrison, D.L., S.M. Conrad, P.P. Eilers, and E.M. Waldron. Confirmation of domoic acid production by *Pseudonitzschia australis* (Bacillariophyceae) cultures. *J. Phycol.* **28**: 604-607.

Gervais, A.J. 1985. Management and public health problems associated with toxic dinoflagellates. In *Toxic Dinoflagellates, Proceedings of the Third International Conference.* D.M. Anderson, A.W. White, and D.G. Baden (eds.), pp. 530-533. Elsevier/North Holland, New York.

Gilgan, M.W., B.G. Burns, and G.J. Landry. 1990. Distribution and magnitude of domoic acid contamination in Atlantic Canada during 1988. In *Toxic Marine Phytoplankton.* E. Graneli, B. Sundstrom, L. Edler, and D.M. Anderson (eds.), pp. 469-473. Elsevier, New York.

Grey, C. 1987. *Canadian Medical Association Journal.* **138**: 350-351.

Haamer, J., P.O. Andersson, O. Lindahl, S. Lange, X.P. Li, and L. Edebo. 1990. Geographic and seasonal variation of okadaic acid content in farmed mussel, *Mytilus edulis* (Linnaeus), 1758, along the Swedish West Coast. *Journal of Shellfish Research.* **9**: 103-108.

Hagel, P. 1989. Monitoring program for DSP in Dutch shellfish growing areas (Abstract). *Journal of Shellfish Research.* **8**:441.

Hall, S., Y. Shimizu, J.J. Sullivan, B. Underal, R. Bagnis, S.R. Davio, M.R. Ross, and D.G. Baden. 1985. Toxin analysis and assay methods. In *Toxic Dinoflagellates, Proceedings of the Third International Conference.* D.M. Anderson, A.W. White, and D.G. Baden (eds.), pp. 545-548. Elsevier/North Holland, New York.

Hallegraeff, G.M., C.J. Bolch, J. Bryan, and B. Koerbiin. 1990. Microalgal spores in a ship's ballast water: a danger o aquaculture. In *Toxic Marine Phytoplankton.* E. Graneli, B. Sundstrom, L. Edler, and D.M. Anderson (eds.), pp. 475-480. Elsevier, New York.

Hamano, Y., Y. Kinoshita, and T. Yasumoto. 1985. Suckling mice assay for diarrhetic shellfish toxins. In *Toxic Dinoflagellates, Proceedings of the Third International Conference.* D.M. Anderson, A.W. White and D.G. Baden (eds.), pp. 382-387. Elsevier/North Holland, New York.

Holmes, D.F.B. 1991. Liquid chromatography-linked protein phosphatase bioassay; a highly sensitive marine bioscreen for okadaic acid and related diarrhetic shellfish toxins. *Toxicon.* **29**: 469-477.

Horner, R.A., M.B. Kusske, B.P. Moynihan, R.N. Skinner, and J.C. Wekell. 1993 Retention of domoic acid by Pacific razor clams, *Silqua patula* (Dixon, 1993): preliminary study. *J. Shellfish Res.* **12**: 451-456.

Hughes, J.M. 1979. Epidemiology of shellfish poisoning in the United States, 1971-1977. In *Toxic Dinoflagellate Blooms.* D.L. Taylor, and H.H. Seliger (eds.), pp. 23-28. Elsevier/North Holland, New York.

Hunt, D., N. Tufts, and J. Hughes. 1979. Monitoring programs and epidemiology. In *Toxic Dinoflagellate Blooms.* D.L. Taylor, and H.H. Seliger (eds.), pp. 489-492. Elsevier/North Holland, New York.

Hurst, J.W. Jr. 1979. Shellfish monitoring in Maine. In *Toxic Dinoflagellate Blooms.* D.L. Taylor and H.H. Seliger (eds.), pp. 231-233. Elsevier/North Holland, New York.

Hurst, J.W.R. Selvin, J.J. Sullivan, C.M. Yentsch, and R.R. L. Guillard. 1985. Intercomparison of various assay methods for the detection of shellfish toxins. In *Toxic Dinoflagellates, Proceedings of the Third International Conference.* D.M. Anderson, A.W. White, and D.G. Baden (eds.), pp. 427-342. Elsevier/North Holland, New York.

Kat, M. 1983. *Dinophysis acuminata* blooms in the Dutch coastal area related to diarrhetic mussel poisoning in the Dutch Waddensea. *Sarsia.* **68**:81-84.

Kat, M. 1985. *Dinophysis acuminata* blooms, the distinct cause of Dutch mussel poisoning. In *Toxic Dinoflagellates, Proceedings of the Third International Conference.* D.M. Anderson, A.W. White, and D.G. Baden (eds.), pp. 73-77. Elsevier/North Holland, New York.

LaBaraera, A., G. Estrella, L. Miller, E.W. King, and S. Hall. 1989. *Alexandrium* sp., *Gymnodium catenatum* and PSP in Venezuela (Abstract). *Journal of Shellfish Research.* **8**: 442.

LeDoux, M., M. Fremy, E. Nezam, and E. Erard. 1989. Recent occurrence of paralytic shellfish poisoning (PSP) toxins from North Western Coasts of France (Abstract). *Journal of Shellfish Research.* **8**: 486.

Lee, J. and V.M. Bricelj. 1989. Uptake and depuration of PSP toxins from the red tide dinoflagellate *Alexandrium fundyense* by *Mercenaria mercenaria* (Abstract). *Journal of Shellfish Research.* **8**: 442.

Lee, J.S., T. Yanagi, R. Kenma, and T. Yasumoto. 1987. Fluorometric determination of diarrhetic shellfish toxins by high-performance liquid chromatography. *Agricultural Biology and Chemistry.* **51**: 877-881.

Lutz, R.A. and L.S. Incze. 1979. Impact of toxic dinoflagellate blooms on the North American shellfish industry. In *Toxic Dinoflagellate Blooms.* D.L. Taylor, and H.H. Seliger (eds.), pp. 476-483. Elsevier/North Holland, New York.

Maranda, L. and Y. Shimizu. 1987. Diarrhetic shellfish poisoning in Narragansett Bay. *Estuaries.* **10**: 298-302.

Maranda, L., D.M. Anderson, and Y. Shimizu. 1985. Comparison of toxicity between populations of *Gonyaulax tamarensis* of eastern North American waters. *Estuaries, Coastal and Shelf Science.* **21**:401-410.

Marcaillou-Le Baut, C. and P. Masselin. 1990. Recent data on diarrhetic shellfish poisoning in France. In *Toxic Marine Phytoplankton.* E. Graneli, B. Sundstrom, L. Edler, and D.M. Anderson (eds.), pp. 487-492. Elsevier, New York.

Nassif, J. Massachusetts Department of Public Health. Personal communication. 1994.

National Shellfish Sanitation Program Manual of Operations. 1989 Revision. United States Department of

Health and Human Services. Food and Drug Administration. Washington, DC.

Oshima, Y., T. Hayakawa, M. Hashimoto, Y. Kotaki, and T. Yasumoto. 1982. Classification of *Protogonyaulax tamarensis* from northern Japan into three strains by toxin composition. *Bulletin of the Japanese Society of Science and Fisheries.* **48**:851-854.

Oshima, Y., K. Sugino, and T. Yasumoto. 1988. Latest advances in HPLC analysis of paralytic shellfish toxins. In *Mycotoxins and Phycotoxins '88.'* S. Natori, K. Hashimoto, and Y. Ueno (eds.), pp. 319-326.

Oshima, Y., S.I. Blackburn, and G.M. Hallegaeff. 1993. Comparative study on paralytic shellfish toxin profiles of the dinoflagellate *Gymnodinium catenatum* from three different countries. *Mar. Biol.* **116**: 471-476.

Pietrafesa, L.J., G.S. Janowitz, D.S. Brown, F. Askari, C. Gabriel, and L.A. Salzillo. 1988. The invasion of the red tide in North Carolina coastal waters. UNC Sea Grant College Program. Paper #88-1. Raleigh, NC.

Price, D.W. and K.W. Kizer. 1990. California's paralytic shellfish poisoning prevention program, 1927-89. California Department of Health Services, pp. 1-36. Sacramento, CA.

Quilliam, M.A., M.W. Gilgan, S. Pleasance, A.S.W. deFreitas, D. Douglas, L. Fritz, T. Hu, J.C. Marr, C. Smith, and J.L.C. Wright. 1990. Confirmation of an incident of diarrhetic shellfish poisoning in Eastern Canada (Abstract). October, 1990. *Workshop on Harmful Marine Algae.*

Ross, M.R., A. Siger, B.C. Abbott, and A. Hancock. 1985. The house fly: an acceptable subject for paralytic shellfish toxin bioassay. In *Toxic Dinoflagellates, Proceedings of the Third International Conference.* D.M. Anderson, A.W. White, and D.G. Baden (eds.), pp. 433-438. Elsevier/North Holland, New York.

Roszell, L.E., L.S. Schulman, and D.G. Baden. 1990. Toxin profiles are dependent on growth stages in cultured *Ptychodiscus brevis.* In *Toxic Marine Phytoplankton.* E. Graneli, B. Sundstrom, L. Edler, and D.M. Anderson (eds.), pp. 403-406. Elsevier, New York.

Schantz, E.J., E.F. McFarren, M.L. Schafer, and K.H. Lewis. 1958. Purified poison for bioassay standardization. *Journal of the Association of Official Analytical Chemists.* **41**:160-168.

Schulman, L.S., L.E. Roszell, T.J. Mende, R.W. King, and D.G. Baden. 1990. A new polyether toxin from Florida's red tide dinoflagellate *Ptychodiscus brevis.* In *Toxic Marine Phytoplankton.* E. Graneli, B. Sundstrom, L. Edler, and D.M. Anderson (eds.), pp. 407-412. Elsevier, New York.

Shumway, S.E. 1990. A review of the effects of algal blooms on shellfish and aquaculture. *World Aquaculture.* **21**: 65-104.

Shumway, S.E. and T.L. Cucci. 1987. The effects of the toxic dinoflagellate *Protogonyaulax tamarensis* on the feeding and behavior of bivalve mollusks. *Aquatic Toxicology.* **10**: 9-27.

Shumway, S.E., J. Barter, and S. Sherman-Caswell. 1990. Auditing the impact of toxic algal blooms on oysters. *Environmental Auditor.* **2**: 41-56.

Shumway, S.E., S. Sherman-Caswell, and J.W. Hurst. 1988. Paralytic shellfish poisoning in Maine: monitoring a monster. *Journal of Shellfish Research.* **7**: 643-652.

Shumway, S.E., F.C. Pierce, and K. Knowlton. 1987. The effect of *Protogonyaulax tamarensis* on byssus production in *Mytilus edulis* (L.), *Modiolus modiolus* (Linnaeus, 1758) and *Geukensia demissa* (Dillwyn). *Comparative Biochemistry and Physiology.* **87**: 1021-1023.

Shumway, S.E., T.L. Cucci, L. Gainey, and C.M. Yentsch. 1985. A preliminary study of the behavioral and physiological effects of *Gonyaulax tamarensis* on bivalve mollusks. In *Toxic Dinoflagellates, Proceedings of the Third International Conference.* D.M. Anderson, A.W. White, and D.G. Baden (eds.), pp. 389-394. Elsevier, New York.

Smith, J.C., R. Cormier, J. Worms, C.J. Bird, M.A. Quilliam, R. Pocklington, R. Angus, and L. Hanic. 1990. Toxic blooms of the domoic acid containing diatom *Nitzschia pungens* in the Cardigan River, Prince Edward Island, in 1988. In *Toxic Marine Phytoplankton.* E. Graneli, B. Sundstrom, L. Edler, D.M. Anderson (eds.), pp. 227-232. Elsevier, New York.

Stabell, O.B. and A.D. Cembella. 1990. Standardizing extraction and analysis techniques for marine phytoplankton toxins. In *Toxic Marine Phytoplankton.* E. Graneli, B. Sundstrom, L. Edler, D.M. Anderson (eds.), pp. 518-521. Elsevier, New York.

Stamman, E., D.A. Segar, and P.G. Davis. 1987. A preliminary epidemiological assessment of the potential for diarrhetic shellfish poisoning in the Northeast United States. *NOAA Technical Memorandum NOS OMA.* 34.

Stephen, K. 1990. Domoic acid follow up. *Aquaculture Association of Canada Bulletin.* **990**: 35-36.

Steidinger, K.A. and K. Haddad. 1981. Biological and hydrographic aspects of red tides. *Bio Science.* **31**:814-819.

Subba Rao, D.V., A.S.W. deFreitas, M.A. Quillian, R. Pocklington, and S.S. Bates. 1990. Rates of production of domoic acid, a neurotoxic amino acid in the pennate marine diatom *Nitzschia pungens.* In *Toxic Marine Phytoplankton.* E. Graneli, B. Sundstrom, L. Edler, and D.M. Anderson (eds.), pp. 413-417. Elsevier, New York.

Sullivan, J.J., J. Jonas-Davis, and L.L. Kentala. 1985. The determination of PSP toxins by HPLC and autoanalyzer. In *Toxic Dinoflagellates, Proceedings Third International Conference.* D.M. Andersor White, and D.G. Baden (eds.), pp. 275-280 New York.

Sullivan, J.J. and M.M. Wekell. 1987. The applic. .on of high performance liquid chromatography in a paralytic shellfish poisoning monitoring program. In *Seafood Quality Determination.* D.E. Kramer and J. Liston (eds.), pp. 357-371. Elsevier, New York.

Tester, P.A. and P.K. Fowler. 1990. Brevetoxin contamination of *Mercenaria mercenaria* and *Crassostrea virginica*: a management issue. In *Toxic Marine Phytoplankton.* E. Graneli, B. Sundstrom, L. Edler, and D.M. Anderson (eds.), pp. 499-503. Elsevier, New York.

Tester, P.A., P.K. Fowler, and J.T. Turner. 1989. Gulf stream transport of the toxic red tide dinoflagellate *Ptychodiscus brevis* from Florida to North Carolina. In *Novel phytoplankton blooms: causes and impacts of recurrent brown tides and other unusual blooms.* E.M. Cosper, V.M. Bricelj, and E.J. Carpenter (eds.), pp. 349-358. Springer-Verlag, Berlin.

Todd, E.C.D. 1990. Amnesic shellfish poisoning — a new seafood toxin syndrome. In *Toxic Marine Phytoplankton.* E. Graneli, B. Sundstrom, L. Edler, and D.M. Anderson (eds.), pp. 504-508. Elsevier, New York.

Trainer, V.L. and D.G. Baden. 1990. Enzyme immunoassay of brevetoxins. In *Toxic Marine Phytoplankton.* E. Graneli, B. Sundstrom, L. Edler, and D.M. Anderson (eds.), pp. 430-435. Elsevier, New York.

Uda, T., Y. Itoh, M. Nishimura, T. Usagawa, M. Murata, and T. Yasumoto. 1989. Enzyme immunoassay using monoclonal antibody specific for diarrhetic shellfish poisons. In *Mycotoxins and Phycotoxins '88.'* S. Natori, K. Hashimoto, and Y. Ueno (eds.), pp. 335-342. Elsevier, Amsterdam.

Usagawa, T., M. Nishimura, Y. Itoh, T. Uda, and T. Yasumoto. 1989. Preparation of monoclonal antibodies against okadaic acid prepared from the sponge *Halichondria okadai. Toxicon.* **27**: 1323-1330.

Van Dolah, F.M., E.L. Finley, B.L. Haynes, G.J. Doucette, P.D. Moeller, and J.S. Ramsdell. 1994. Development of rapid and sensitive high throughout pharmacological assays for marine phycotoxins. *Natural Toxins.* **2**: 189-196.

Vanderlaan, M., B.E. Watkins, and L. Stanker. 1988. Environmental monitoring by immunoassay. *Environmental Science and Technology.* **22**: 247-254.

White, A.W. 1984. Paralytic shellfish toxins and finfish. In *Seafood Toxins.* E.P. Ragelis (ed.), pp. 171-180, Number 262, American Chemical Society Symposium Series.

White, A.W. 1988. Blooms of toxic algae worldwide: their effects on fish farming and shellfish resources. Proceedings of the International Conference on Impact of Toxic Algae on Mariculture. *Aqua-Nor '87 International Fish Farming Exhibition.* Trondeim, Norway.

White, A.W., O. Fukuhara, and M. Anraku. 1989. Mortality of fish larvae from eating toxic dinoflagellates or zooplankton containing dinoflagellate toxins. In *Red Tides: biology, environmental science, and toxicology.* T. Okaichi, D.M. Anderson and T. Nemoto (eds.), pp. 395-398. Elsevier Science Publishing. New York.

Wood, A.M., L.P. Shapiro, and S.S. Bates (eds.). ORESU-WO4-001. 1994. Domoic acid: final report of the workshop. Oregon Inst. Mar. Biol. OSU Tech. Rep.

Work, T.M., A.M. Beale, L. Fritz, M.A. Quilliam, M. Silver, K. Buck, and J.L.C. Wright, 1992. Domoic acid intoxication of brown pelicans and cormorants in Santa Cruz, California. In *Toxic Phytoplankton Blooms in the Sea.* J.T. Smayda, and Y. Shimizu. (eds.), pp. 643-649. Elsevier Sci. Publ., B.V Amsterdam.

Wright, J.L.C. 1989. Domoic acid, a new shellfish toxin: the Canadian experience (Abstract). *Journal of Shellfisheries Research.* **8**: 444.

Yang, G.C., S.J. Imagire, P. Yasaei, E.P. Ragelis, D.L. Park, S.W. Page, R.E. Carlson, and P.E. Guire. 1987. Radioimmunoassay of paralytic shellfish toxins in clams and mussels. *Bulletin of Environmental Contamination and Toxicology.* **39**:264-271.

Yasumoto, T. 1990. Marine microorganisms, pp. 3-8. *Toxic Marine Phytoplankton.* Elsevier, New York.

Yasumoto, T., Y. Oshima, W. Sugawara, Y. Fukoyo, H. Oguri, T. Igarash, and N. Fujita. 1980. Identification of *Dinophysis fortii* as the causative organism in diarrhetic shellfish poisoning. *Bulletin of the Japanese Society of Science and Fisheries.* **46**: 1405-1411.

Yasumoto, T., A. Inoue, T. Ochi, K. Fujimoto, Y. Oshima, Y. Fukuyo, R. Adachi, and R. Bagnis. 1980. *Bulletin of the Japanese Society of Science and Fisheries.* **46**: 1397.

Yasumoto, T., Y. Oshima, and M. Yamaguchi. 1978. Occurrence of a new type of shellfish poisoning in the Tohoku District. *Bulletin of the Japanese Society of Science and Fisheries.* **44**: 1249-1255.

Health-Related Bacteria in Seafoods

Paul G. Comar

While illness reports linked to most prepared seafoods are rare, consumption of certain fish and raw shellfish can pose some health concerns. Seafood can become contaminated with pathogenic bacteria from three sources: (1) fecal pollution of the aquatic environment, (2) naturally occurring pathogens in the aquatic system, and (3) poor hygiene during harvesting, distribution, and food preparation (NAS, 1991). Viruses, parasites, and marine biotoxins are other environmental contaminants in some fishery products. These non-bacterial agents and their effects on human health are addressed in other parts of this section.

This chapter on bacterial pathogens is divided into two parts. The first concerns the vibrios and related bacteria that naturally inhabit coastal waters. The second, entitled "Other Bacteria," includes brief descriptions of bacteria responsible for foodborne disease in many foods, including seafoods. Illnesses from these bacteria are mainly caused by poor food handling and preparation practices. Methods to limit exposure and reduce the risk of infection and disease are presented.

VIBRIO BACTERIA

Vibrios are a group of naturally occurring bacteria that live primarily in salty coastal waters. Consumption of vibrio bacteria or contact through open wounds may cause several illnesses in humans. Fish and shellfish, especially filter-feeding bivalves, can transmit vibrio bacteria to humans. The *Vibrio* species of most concern to seafood consumers are *V. vulnificus*, *V. cholerae* (O1 and non-O1), *V. parahaemolyticus*, and *V. mimicus*.

VIBRIO VULNIFICUS

Vibrio vulnificus is a common part of the bacterial flora of nearshore environments. It has been isolated from the Gulf of Mexico (Kelly, 1982), the Atlantic (O'Neill et al., 1992; Oliver et al., 1983), and the Pacific oceans (Kaysner, 1989). This distribution in United States waters has been confirmed by various investigators, including Tamplin (1994), who coordinated a national study of the bacterium in shellfish, seawater, and sediments. While isolations have been made from all coasts, *V. vulnificus* densities are highest in warmer waters with low to moderate salinities. The seasonal occurrence of this bacterium is well recognized (O'Neill et al., 1992; Tamplin, 1990; Oliver et al., 1983; Tamplin et al., 1982; Kelly, 1982). High salinities reduce the numbers of *V. vulnificus* (Kaspar and Tamplin, 1993). Environmental conditions favoring growth include water temperatures of 20° to 30°C, salinities of 5 to 20 ppt, and neutral-to-alkaline pH (Tamplin, 1994). Highest densities are usually found when water temperature exceeds 25°C for several months.

V. vulnificus is not detected regularly from waters less than 15°C; however, there is evidence that it "over-winters" in sediment and shellfish. Oysters from the cold winter waters of the Gulf of Mexico have been analyzed and found to be free of *V. vulnificus*. Yet, when placed into tanks at 25°C, oysters from the same collection exhibited high levels of the bacteria within 24 hours (Kaspar and Tamplin, 1993). Possible explanations include resuscitation of large numbers that are viable but non-culturable at low temperatures and rapid replication of fewer numbers that survive cold periods.

V. vulnificus illnesses are rare and are caused almost entirely by wound infections or consumption of raw bivalve shellfish. While the illness incidence is low, its severity and associated high mortality make it of great concern to public health authorities and the affected shellfish industry. The primary food vehicle in the United States is raw oysters from the Gulf of Mexico [Centers for Disease Control (CDC), 1997]. These illnesses appear as single cases, not as outbreaks, in certain medically compromised individuals at higher risk for the disease. Infections are manifested in three ways: gastroenteritis, wound infection, and primary septicemia.

Gastroenteritis

Gastroenteritis, the least reported syndrome, is manifested by vomiting, diarrhea, and abdominal cramping. These symptoms alone may lead to hospitalization but not death.

Wound Infection

Wound infections may occur when an open wound comes in contact with the organism, sometimes as a cut from stepping on a shell. The infections begin as red, swollen, painful areas. Blisters may develop, and the disease may progress rapidly to severe tissue necrosis. Fifty percent of such infections require debridement or amputation. If the infection spreads to the bloodstream, death can occur. Mortality from wound infections is approximately 25 percent, but is 50 percent for those in the medically compromised high-risk group (Whitman, 1994).

Primary Septicemia

The most severe clinical syndrome, primary septicemia (blood poisoning), results in the mortality of 50 percent of those who become infected (CDC, 1997). Symptoms of septicemia usually begin within 16 to 24 hours of consuming contaminated seafood (Bryan, 1987; Oliver, 1981; Blake et al., 1980b). The symptoms include fever, chills, hypotension, malaise, prostration, myalgia, and occasionally abdominal pain, nausea, vomiting, and diarrhea (Bryan, 1987; Morris and Black, 1985; Tacket et al., 1984; Bachman et al., 1983; Oliver, 1981; Blake et al., 1980b). *V. vulnificus* enters the blood through the intestinal wall and causes damage to blood vessels. This damage in turn causes fluid and proteins to leave the blood and collect in adjacent tissues, forming bulbous skin lesions (Oliver, 1981). More than 70 percent of patients with septicemia have these characteristic skin lesions, usually on their lower extremities (Morris and Black, 1985; Bachman et al., 1983; Oliver, 1981).

Healthy individuals are at very low risk of becoming infected by *V. vulnificus* through raw shellfish consumption. However, certain health factors put some individuals at increased risk for the infection and its progression to septicemia. Up to 75 percent of the cases resulting in primary septicemia occur in those with liver disease or heavy alcohol intake. Nearly all the remaining cases are in (1) individuals with an iron storage disease (hemochromatosis), (2) those with a disease that affects the immune system (such as diabetes or cancer), or (3) those who take immunosuppressive drugs (Whitman, 1994; Tacket et al., 1984; Oliver, 1981; Blake et al., 1980b). Individuals with alcohol-related liver disease and hemochromatosis commonly have higher than normal iron levels in their blood (Morris and Black, 1985; Tacket et al., 1984; Oliver, 1981). Unusually high levels of iron are a growth promoter for *V. vulnificus*, increasing a person's susceptibility to septicemia.

The particular disease and each individual's stage of that disease are key factors in determining the susceptibility of a person to infection by *V. vulnificus*. In addition to host susceptibility, other determinants include the variable virulence of strains (there are hundreds) and the dose of the organism through consumption. Because of these different factors, it is currently impractical to project or estimate an infectious dose. The rate of *V. vulnificus* infection in the Gulf states has been estimated at 0.5 to 0.9 per one million persons (Whitman, 1994; Desenclos et al., 1991). However, a behavioral risk factor survey of nearly 1,500 adults in Florida in 1988 indicated that the risk to raw oyster eaters with liver disease is much higher: 70 per one million population (Desenclos et al., 1991). Even within this high-risk population, the rate of infection is not extreme, very likely because of the variable factors previously noted.

The peak incidence of human illness in the United States occurs from May to October, coinciding with the rise in *V. vulnificus* densities in the marine environment (Cook and Ruple, 1989; Tacket et al., 1984; Bachman et al., 1983; Blake, 1983; Blake et al., 1980b). January and February are the only months where no such illnesses have been reported. For those reports linked to shellfish consumption, over 70 percent have been traced back to the source state. Oysters from the Gulf of Mexico are almost exclusively the vehicle for disease transmission, though on occasion clams from the Atlantic Coast of Florida have been implicated (CDC, 1997).

In 1988 the CDC began a voluntary illness-reporting system in states where the majority of cases had been reported, including all the Gulf states and California. From 1989 through 1996 there have been 149 illnesses resulting in 75 deaths confirmed and attributed to the consumption of raw shellfish (CDC, 1997). Over the five-year period from 1992 through 1996, the CDC data indicate 121 cases and 66 deaths. The average number of reported illnesses and deaths was higher during 1992-1996, due to either an actual increase in illnesses or better awareness and reporting.

At present, neither regulatory nor educational outreach steps have been attempted to reduce wound infections by limiting physical contact with estuarine water or organisms, but measures to reduce risk from raw shellfish consumption have been initiated. In the United States safety in the consumption of raw molluscan shellfish is maintained through the testing of water for the presence of fecal bacteria, followed by harvest restrictions when levels of these bacteria exceed certain standards. Nationally this testing is done through a cooperative state-federal program, the National Shellfish Sanitation Program (NSSP), with the Food and Drug Administration (FDA) having federal oversight responsibilities. The presence and density of *V. vulnificus* is not related to fecal pollution of shellfish growing waters; thus current harvesting regulations do not prevent these illnesses. However, federal and state health authorities along with the shellfish industry through the Interstate Shellfish Sanitation Conference (ISSC) are cooperatively developing and implementing several strategies to reduce the risk of this illness from shellfish consumption. Their focus is on education and risk communication to shellfish consumers and those in the medically compromised groups, improved post-harvest shellfish handling practices, shellfish processing to mitigate levels present in shellfish prior to consumption, and research on *V. vulnificus* and factors affecting its virulence. It is hoped that these multiple approaches will combine to reduce health concerns about raw shellfish consumption.

The FDA has distributed health advisories to the medical community and to high-risk disease associations, implemented a seafood hotline, and developed media releases published in newspapers for the general public. The ISSC, FDA, National Oceanic and Atmospheric Administration (NOAA), Sea Grant universities, state public health agencies, and others have cooperated in developing educational brochures and messages. A few states have begun educating shellfish consumers and high-risk populations through targeted programs and messages at restaurants and other points-of-sale (Welch, 1994). The ISSC, through a grant from NOAA in 1997, is developing and producing educational kits for states and coordinating the distribution of health advisories to improve the effectiveness of this educational effort.

V. vulnificus multiplies in shellfish exposed to elevated temperatures after harvest. Bacterial densities peaked after 12 hours post-harvest at mean ambient air temperatures of 24° to 28°C in the Gulf of Mexico during the summer months (Cook, 1994). The level of *V. vulnificus* in oysters at 12 hours was approximately 200 times that at harvest. Refrigeration below 13°C stops the growth of *V. vulnificus*. In 1995 and 1996, the Gulf states and the shellfish industry instituted voluntary and then mandatory practices to reduce the maximum time allowed between harvest and refrigeration. Analytical studies are testing the effectiveness of these procedures over previous handling methods in reducing *V. vulnificus* densities. Though the infective dose is not known and likely varies with the host and the percentage of the more virulent strains of the bacterium present, these steps seem prudent.

Raw oysters are the market form that currently demands the highest value. However, various investigations and marketing attempts are underway to process oysters to reduce the hazard to high-risk consumers. In-shell bivalves treated by heating, then rapid chilling, flash freezing, and hydrostatic pressure are new market forms developed since 1996. These new processes are designed to greatly reduce the level of *V. vulnificus* while maintaining many edibility attributes of live raw shellfish.

Vibrio cholerae

Vibrio cholerae are divided into two serotypes: Ogawa and Inaba, and two biotypes: Classical and El Tor (Morris and Black, 1985). The type of *V. cholerae* found most frequently in the United States, serotype Inaba and biotype El Tor (*V. cholerae* non-O1), is far less pathogenic than the type (*V. cholera* O1) that commonly causes epidemic cholera in parts of Asia (Morris and Black, 1985).

V. cholerae can be found in brackish water, estuaries, and salt marshes within the temperate zone (Hood and Ness, 1982; Colwell et al., 1981). Highest densities of *V. cholerae* are usually found during the summer months (Blake, 1983). Unlike other *Vibrio* bacteria, *V. cholerae* and *V. mimicus* do not require salt for growth (Blake, 1983; Blake et al., 1980a). *V. cholerae* produces chitinase (an enzyme that breaks down chitin) and exhibits seasonal fluctuations similar to copepods, suggesting that this bacterium may exist in association with copepods (Shandera et al., 1983a).

Cholera caused by *V. cholerae* O1 was first recognized in the United States in 1832. The disease was believed to have been eradicated since there were no reported cases of cholera from 1911 to 1973. However, in 1973 a case of cholera of unknown origin was reported, the first case in the United States in more than 60 years (Shandera et al., 1983a). Although raw oysters are the seafood most often associated with illness, cholera has also been attributed to crabs, shrimp, and finfish [Morbidity and Mortality Weekly Report (MMWR), 1989; Bryan, 1987; Klontz et al., 1987; MMWR, 1986; Blake, 1983; Shandera et al., 1983a]. Eleven cases occurred in Louisiana in 1978 caused by undercooked crabs (Blake et al., 1980a). From 1973 to 1992, the CDC reported 65 cases of cholera from food consumption, with 26 of these cases from one outbreak linked to reef fish in Guam and a number of others traced to seafood imported to the United States (MMWR, 1996; Adams et al., 1988). Sporadic cases of cholera have been associated with seafood harvested from the Gulf of Mexico, leading researchers to speculate that *V. cholerae* is indigenous to the Gulf (Morris and Black, 1985; Shandera et al., 1983a; Colwell et al., 1981; Blake et al., 1980a). *V. cholerae* has also been recovered from Chesapeake Bay water samples (Colwell et al., 1981).

Symptoms of cholera can begin within six hours to five days of contact with *V. cholerae* O1. The disease is characterized by a rapid onset of profuse watery diarrhea and dehydration. Other symptoms can include anorexia, abdominal cramps, vomiting and, in rare cases, low-grade fever. In severe cases of cholera, diarrhea can become gray and mucoid, a condition often described as "rice water." This symptom can result in rapid dehydration and death if not treated promptly (Brown, 1991). However, illness severe enough to require hospitalization is rare in *V. cholerae* non-O1 El Tor infections, the strain most common in the United States (Klontz et al., 1987; Morris and Black, 1985; Shandera et al., 1983a; Horwitz, 1977). People with decreased stomach acidity are more susceptible to cholera than healthy individuals, presumably because large numbers of *V. cholerae* survive transit through the stomach (Bryan, 1987). Cholera is treated by replacement of fluids and electrolytes administered orally or intravenously. Antibiotics shorten the duration of the illness.

In 1991 and 1992, a more toxigenic strain of *V. cholerae* O1 was detected at levels of 10^1 to 10^7 per gram of oysters in several shellfish from Mobile Bay, Alabama (Motes et al., 1994). This strain closely resembled one originally isolated in Peru in 1991 that was responsible for a cholera epidemic and thousands of deaths in South and Central America through the early 1990s (Motes et al., 1994). Alabama's shellfish harvesting grounds were closed for several months, and no illnesses attributed to shellfish in the area were reported. The source of the bacteria was traced to ship ballast water from vessels in port at Mobile. Samples of ballast water taken from five ships, with last ports registered as South American, were positive for the toxigenic strain. Some of these ships hold 500,000 gallons of ballast water, providing a significant source of contamination (McCarthy and Khambaty, 1994). New international policies on ballast water exchange are being considered to address both this and the accidental transport of other undesirable marine animal and plant species.

Non-O1 *Vibrio cholerae*

Vibrio cholerae non-O1 has also been referred to in some early publications as "non-agglutinating vibrio" and "non-cholerae vibrio" (Morris et al., 1981). It is biochemically similar to the epidemic strain of *V. cholerae* but does not agglutinate in *V. cholerae* O-group 1 antiserum (Morris et al., 1981; Blake et al., 1980b). Some strains of non-O1 produce a heat-labile toxin similar to cholerae toxin; however, most strains isolated from sick persons in the United States lack this cholera-like activity (Morris et al., 1981). Illness can occur from accidental consumption of both toxigenic and nontoxigenic strains of *V. cholerae*, although the toxigenic variety produces more severe illness. As with other vibrios, *V. cholerae* non-O1 is commonly found in estuaries, bays, and brackish waters — both sewage-contaminated ones and those apparently free of pollution (Bryan, 1987; Morris and Black, 1985; Hood and Ness, 1982; Blake et al., 1980b; Kaper et al., 1979). Densities generally increase during the warmer months (DePaola et al., 1983; Blake et al., 1980b).

Illness from *V. cholerae* non-O1 is usually associated with consumption of raw oysters; however, shrimp and blue crabs have also been found to cause illness (Bryan, 1987; Blake, 1983; Davis and Sizemore, 1982; Morris et al., 1981; Blake et al., 1980b). *V. cholerae* non-O1 primarily causes mild to moderate gastroenteritis, with symptoms beginning within 48 hours after consuming contaminated shellfish and persisting for two to 12 days. The most common symptoms include diarrhea, dehydration, and abdominal cramps. Fever, chills, nausea, vomiting, and bloody or mucoid diarrhea can occur, but are more variable (Morris and Black, 1985; Morris et al., 1981; Blake et al., 1980b). If necessary, gastroenteritis can be treated with oral or intravenous rehydration (Blake et al., 1980b).

Unlike *V. cholerae* O1, non-O1 *V. cholerae* can also cause septicemia. It has been isolated from a number of human sources other than feces, including blood, infected wounds, ears, gallbladder, sputum, appendix, peritoneal fluid, and cerebrospinal fluid. The source of these extraintestinal infections is unknown, although some affected persons have reported a history of recent exposure to seawater. Cases of septicemia usually involve alcoholics or individuals with pre-existing immunocompromising diseases (Morris and Black, 1985; Blake et al., 1980b).

Illness from non-O1 *V. cholerae* occurs worldwide (Blake et al., 1980b). From 1973 to 1992, there were two seafood-associated outbreaks (one shrimp and one oyster) of non-O1 *V. cholerae* illness in the United States (MMWR, 1996; Bean and Griffin, 1990; Bryan, 1987). Gastroenteritis from this bacterium is infrequent compared to the high frequency with which the organism is isolated from the marine environment (Morris and Black, 1985). Scientists have speculated that illness may be infrequent for one or more of the following reasons: all environmental isolates may not be capable of causing disease, a very high inoculation may be required for infection, or infection may seldom result in illness severe enough to require medical attention (Morris and Black, 1985).

VIBRIO PARAHAEMOLYTICUS

Gastroenteritis and wound infections from *Vibrio parahaemolyticus* have been reported throughout the world (Blake et al., 1980b). *V. parahaemolyticus* was first recognized as a pathogen in the early 1950s in Japan (Morris and Black, 1985; Blake et al., 1980b). It is a naturally-occurring bacterium of estuaries and other coastal areas throughout most of the world (Morris and Black, 1985; Kaneko and Colwell, 1975). Like *V. cholerae*, *V. parahaemolyticus* produces chitinase and is more often found attached to copepods than free-living in the marine environment (Kaneko and Colwell, 1975). It is presumed that *V. parahaemolyticus* bacteria derive essential nutrients from their association with copepods (Kaneko and Colwell, 1975). In most areas, bacterial densities in water and shellfish increase during warmer months (Paille et al., 1987; Eyles and Davey, 1984; Davis and Sizemore, 1982; Hackney et al., 1980). As a result, most outbreaks of *V. parahaemolyticus* illness in the United States occur during the summer when environmental densities peak (Watkins and Cabelli, 1985; Blake, 1983).

The onset of gastroenteritis usually occurs within 12 to 48 hours of consuming contaminated seafood. There are two distinct gastrointestinal syndromes caused by *V. parahaemolyticus*. The more common of the two syndromes in the United States is characterized by nausea, vomiting, abdominal cramps, and prominent watery diarrhea. Headache, fever, and chills may also be present, although they are less common. The second syndrome, which resembles dysentery, occurs mostly in India and Bangladesh and is rarely found in the United States. Symptoms of dysentery include fever, abdominal pain, and bloody or mucoid stools. Although severe illness and death can occur, gastroenteritis in the United States is usually self-limited, lasting an average of three days (Bryan, 1987; Morris and Black, 1985; Blake et al., 1980b; Horwitz, 1977).

In Japan, where it is the leading cause of seafood-borne illness, infection is frequently caused by consumption of raw finfish (Bryan, 1987; Blake et al., 1980b). In the United States, gastroenteritis is usually associated with consumption of contaminated crabs, oysters, shrimp, and lobster (Bryan, 1987; Blake, 1983; Thompson et al., 1976a), sometimes through recontamination and outgrowth of the bacterium after cooking (Cook and Ruple, 1988; Son and Fleet, 1980). The first confirmed outbreak of *V. parahaemolyticus* gastroenteritis in the United States occurred in 1971 (Thompson et al., 1976a). From 1973 to 1992, the CDC reported 22 outbreaks of seafood-associated gastroenteritis caused by *V. parahaemolyticus* (MMWR, 1996; Bean and Griffin, 1990; Bryan, 1987). The actual incidence of such illnesses is unknown since, like many foodborne illnesses, this disease is self limiting and those affected may not seek treatment. Also, states vary considerably in their *V. parahaemolyticus* investigation and reporting practices. However, four outbreaks in 1997 and 1998 resulting in more than 700 cases have been traced to consumption of raw oysters and some additional seafood products from Texas, New York, and the Pacific Northwest.

Not all strains of *V. parahaemolyticus* are enteric pathogens. In fact, nearly all isolates obtained from shellfish, sediments, and waters are non-pathogenic. Therefore merely determining the total concentration of this species in seafood is inadequate for assessing the risk of illnesses. A thermostable direct hemolysin (tdh) is produced by most pathogenic strains of *V. parahaemolyticus*. It provides a reasonably reliable means of differentiating pathogenic from non-pathogenic isolates.

The risk of illness from *V. parahaemolyticus* can be significantly reduced by storing and handling seafood below the temperature that allows the bacteria to replicate (Cook and Ruple, 1988; Bryan, 1987; Boutin et al., 1985; Blake et al., 1980b). Rapidly chilling and storing seafood at 4°C, or on ice, will ensure that replication of *V. parahaemolyticus* (and other vibrios) does not occur (Cook and Ruple, 1989; Boutin et al., 1985; Oliver, 1981).

Vibrio mimicus

Vibrio mimicus was originally incorrectly identified as *V. cholerae*. However, it differs from *V. cholerae* in its inability to ferment sucrose (Morris and Black, 1985). Unlike the other *Vibrio* bacteria, *V. mimicus* (and *V. cholerae*) does not require salt for growth (Blake, 1983; Shandera et al., 1983b). *V. mimicus* is likely a part of the normal marine flora of the Atlantic and Gulf coasts.

V. mimicus can cause both gastroenteritis and ear infections in humans. Gastrointestinal illness has been associated with consumption of raw oysters and boiled crayfish (Bryan, 1987; Morris and Black, 1985; Blake, 1983; Shandera et al., 1983b). Ear infections are associated with exposure to seawater. Symptoms of gastroenteritis normally begin within 24 hours of consuming the bacteria. Diarrhea, nausea, vomiting, and abdominal cramps are the most common symptoms. Some infected individuals have also experienced fever, headache, and bloody diarrhea. Between 1977 and 1981, 21 cases (19 gastroenteritis and two ear infections) were reported to the CDC (Shandera et al., 1983b).

OTHER VIBRIONACEAE

Other *Vibrio* species, including *V. hollisae*, *V. fluvialis*, and *V. furnissii*, inhabit coastal waters and exhibit environmental growth and distribution characteristics similar to the previously mentioned vibrios. These species have been confirmed as vehicles in sporadic cases of disease. Two other species in the Family Vibrionaceae, which are found in freshwater and estuarine environments, are probable agents of sporadic cases of gastrointestinal disease in the United States. *Aeromonas hydrophila* and *Plesiomonas shigelloides* have been found in the stools of some patients with diarrhea, but clear causative data through feeding studies have not been established, possibly due to different pathogenicities of various strains (NAS, 1991). *A. hydrophila* infection can result through water contact with open wounds or by ingestion. Gastrointestinal symptoms can persist for several days, with septicemia evidenced in some immunocompromised individuals. Most infections from *P. shigelloides* occur during summer months and are predominantly waterborne. Symptoms are usually milder than those of *A. hydrophila* and typically last one to two days, but severe gastrointestinal illness and septicemia is possible in children and those with other complicating medical conditions (FDA, 1992).

OTHER BACTERIA

The bacteria described in this section are principally "handling bacteria." All foods are subject to bacterial contamination by poor hygienic practices. Unless otherwise noted, illnesses caused by these bacteria are gastrointestinal in nature and include symptoms such as nausea, vomiting, abdominal cramps, and diarrhea. As with other protein foods, nearly all bacterial illnesses associated with seafood can be prevented by thorough cooking. Most bacteria are killed when the internal temperature of the fish reaches 66°C (Harbell, 1988). After seafood is properly cooked, care should be taken to avoid cross-contamination of cooked and raw seafood (Bryan, 1987; Blake et al., 1980b). Cooked food should be eaten shortly after preparation or held above 60°C to prevent replication of bacteria. Individuals who are medically compromised due to a variety of illnesses should refrain from eating raw fish and shellfish.

Campylobacter jejuni

Campylobacter jejuni is a disease-causing agent principally associated with illness from poultry, milk, and water (West, 1989; Griffin et al., 1983). The first outbreak in the United States was reported to the CDC in 1978; from 1978 to 1992, a total of 80 foodborne outbreaks of the disease were confirmed (MMWR, 1996; Bean and Griffin, 1990). It is suspected, through recent surveys, that *C. jejuni* is the leading cause of foodborne illness in the United States, but the gastrointestinal symptoms lasting two to five days go largely unreported (FDA, 1992). Contaminated raw clams were implicated in one outbreak of gastroenteritis (Griffin et al., 1983). In one study, cooked hand-picked blue crab meat was analyzed for a variety of pathogens, and 36 of 240 samples were found positive for the organism, all at very low levels (Reinhard et al., 1996). Epidemiological data indicate *C. jejuni* presents little risk to seafood consumers.

Clostridium botulinum

There have been 291 foodborne outbreaks of botulism from 1973 to 1992 in a wide array of foods (MMWR, 1996; Bean and Griffin, 1990). However, botulism is still considered a rare disease in the United States since many of these outbreaks are single cases. Through 1992 this was the only foodborne bacterial infection for which single cases were collated and reported by the CDC as outbreaks (Bean and Griffin, 1990). Although cases are infrequent, mortality in untreated patients can reach 70 percent (NAS, 1991). Spores of *C. botulinum* can be introduced into seafood from sediment and seawater, but are not considered a health hazard at this point. However, if contaminated seafood is inadequately cooked and then stored under anaerobic conditions (such as cans), the spores can germinate and a lethal toxin can be produced (West, 1989; Bryan, 1987). Similarly, when spores survive under vacuum or in modified-atmosphere packaged seafood, toxin can be elaborated when the product is held for prolonged periods above 4°C. Insufficient heating and poor storage conditions lead to toxin production and illnesses in other foods as well. The symptoms of botulism include double vision, headache, vertigo, dizziness, loss of reflexes to light, ataxia, dysphagia, fatigue, and dry mouth (Bryan, 1987). Death can result from respiratory failure and airway obstruction. Botulism can be prevented by providing adequate heat to kill spores before the food is canned. Home-canned foods are more often the cause of botulism than are commercially prepared foods (Horwitz, 1977). The risk of acquiring botulism from seafood is high for Native Alaskans who frequently eat fermented seafoods, but low or nonexistent for other ethnic groups (Bryan, 1988).

Listeria monocytogenes

In the early 1900s *Listeria monocytogenes* was recognized as a bacterium that caused illness in farm animals. More recently, it has been identified as the causative agent of listeriosis in humans. *L. monocytogenes* is ubiquitous in nature and has been isolated from soil, vegetation, marine sediments, and water throughout the world (Peters, 1989). The bacterium has also been isolated from dairy products (MMWR, 1989), vegetables (Lechowich, 1988), beef (Peters, 1989), and poultry (Gellin and Broome, 1989). Seafoods that have tested positive for *L. monocytogenes* include raw fish, cooked crabmeat, raw and cooked shrimp, raw lobster, langostinos, scallops, squid, surimi, and smoked fish [National Fisheries Institute (NFI), 1989; Weagant et al., 1988; *Food Chemical News*, 1987]. Though *L. monocytogenes* has been isolated from seafood, listeriosis has never been directly associated with the consumption of fish or shellfish in the United States. Vegetables and dairy products are the predominant foods that have been conclusively implicated in outbreaks of listeriosis. It is not understood why there is a relatively high incidence of *L. monocytogenes* in food but a low incidence of illness. It is possible that some strains of the bacteria are avirulent or are less virulent than others.

Most healthy individuals are either unaffected by *L. monocytogenes* or experience only mild flu-like symptoms (Peters, 1989). Victims of listeriosis are usually pregnant women, newborn infants, or immunocompromised adults (Lennon et al., 1984). Conditions that increase susceptibility to listeriosis in adults include cancer, diabetes, liver disease, achlorhydria, AIDS, having received a transplant, and taking immunosuppressive drugs (Gellin and Broome, 1989; Mascola et al., 1988; Ho et al., 1986; Stamm et al., 1982; Louria et al., 1967). The manifestations of listeriosis are distinct among the three most commonly affected groups. Listeriosis in pregnant women occurs most frequently during the third trimester. Pregnant women may experience a flu-like illness with fever, headache, myalgia, diarrhea, abdominal cramps, and lower back pain. Approximately 30 percent of women are asymptomatic. Premature labor or abortion can result within three to seven days, after which the woman's illness is usually self-limited. Some infected newborns are asymptomatic, while others may exhibit meningitis, respiratory distress, septicemia, and diarrhea. Listeriosis in immunocompromised adults can result in meningitis, fever, and septicemia. As many as one quarter of these patients also have non-specific symptoms, such as fatigue, malaise, nausea, cramps, vomiting, and diarrhea. The mortality rate is highest among newborns and immunocompromised adults, ranging from 20 to 40 percent (Gellin and Broome, 1989; Peters, 1989; Lennon et al., 1984).

This bacterium is rapidly killed during normal cooking procedures. The bacterium has a growth range from 0° to 45°C, with optimal growth occurring from 30° to 37°C (Lechowich, 1988). Given this bacterium's ability to multiply at low temperatures, it is possible for a small number of organisms to grow to an infectious dose, even when food is properly stored in the refrigerator. Since *L. monocytogenes* grows slowly at refrigeration temperatures, the greatest threat of listeriosis is from inadequately processed ready-to-eat products that require no further cooking in the home.

Salmonella and *Shigella* spp.

The primary habitats for Salmonellae are the intestinal tracts of birds, reptiles, farm animals, and humans (Jay, 1978), while Shigellae are host-specific to humans and other primates (NAS, 1991). Because these organisms can be excreted in large numbers by infected carriers, they are commonly found in the environment. Bacteria of the genera *Salmonella* and *Shigella* can be introduced to seafood by sewage in the marine environment and by poor handling practices after harvest (Bryan, 1987). Molluscan shellfish that can concentrate these bacteria from contaminated water have been the source of occasional disease outbreaks. Cases of salmonellosis and shigellosis are often unreported; however, severe cases can result in death, especially in the very young, the very old, and the infirm. Illnesses attributed to seafoods and all other foods can be significantly reduced by sanitary food handling, thorough cooking, and adequate refrigeration.

Salmonellosis is one of the most common forms of foodborne illness reported in the United States, with an estimated two to four million cases annually, although most of these cases go unreported. The disease has been linked to consumption of contaminated meat, poultry, eggs, pork, seafood, dairy products, vegetables, and prepared salads (egg, tuna, or potato salads) (Bryan, 1988; Horwitz, 1977). *Salmonella* is frequently isolated from imported frozen shrimp and froglegs that have been harvested from contaminated waters or processed and packed under unsanitary conditions. Gastroenteritis is the most common clinical manifestation of infection by Salmonellae; however, more severe symptoms arise if the infection leads to septicemia or enteric fever. *Salmonella typhi* is the bacterium responsible for typhoid fever. The duration of gastrointestinal symptoms from salmonellosis is typically two to three days after an incubation period of up to 48 hours.

An estimated 300,000 cases of shigellosis (bacillary dysentery) occur annually in the United States. Though the number of cases attributable to food is unknown, the low infectious dose of 10 to 100 bacteria make food a likely vehicle for illnesses (FDA, 1992). Seafood is rarely implicated in transmitting the infection.

STAPHYLOCOCCUS AUREUS

The primary sources of *S. aureus* are the skin and nose of humans (Bryan, 1988), indicating that this bacterium is usually introduced to food by human handling (West, 1989; Bryan, 1988). *S. aureus* can replicate and produce toxins when contaminated foods are stored at room temperature or in large containers in the refrigerator (West, 1989; Bryan, 1988). Because the illness is caused by a toxin rather than an infection by the organism, the onset of illness may occur in less than an hour and its duration may last for just hours rather than days. Foods that have caused *S. aureus* gastroenteritis include meat, pork, poultry, cream-filled pastries, seafood, and prepared salads (Bryan, 1988; Horwitz, 1977).

REFERENCES

Adams, L.B., M.C. Hertic, and R.J. Siebeling. 1988. Detection of *V. cholerae* with monoclonal antibodies specific for serovar 01 lipopolysaccharide. *Journal of Clinical Microbiology.* **26**(9):1801-1809.

Bachman, B., W.P. Boyd, S. Lieb, and G.E. Rodrick. 1983. Marine non-cholera infections in Florida. *Southern Medical Journal.* **76**(3):296-299.

Bean, N.H. and P.M. Griffin. 1990. Foodborne disease outbreaks in the United States, 1973-1987: pathogens, vehicles and trends. *Journal of Food Protection.* **53**(9):804-817.

Blake, P.A. 1983. Vibrios on the half-shell: what the walrus and the carpenter didn't know. *Annals of Internal Medicine.* **99**(4):558-559.

Blake, P.A., D.T. Allegra, J.D. Snyder, T.J. Barrett, L. McFarland, C.T. Caraway, J.C. Feely, J.P. Craig. J.V. Lee, N.D. Puhr, and R.A. Feldman. 1980a. Cholera — a possible endemic focus in the United States. *New England Journal of Medicine.* **302**:305-309.

Blake, P.A., R.E. Weaver, and D.C. Hollis. 1980b. Diseases of humans (other than cholera) caused by Vibrios. *Annual Review of Microbiology.* **34**:341-367.

Boutin, B.K., A.L. Reyes, J.T. Peeler, and R.M. Twedt. 1985. Effect of temperature and suspending vehicle on survival of *Vibrio parahaemolyticus* and *Vibrio vulnificus*. *Journal of Food Protection.* **48**(10):875-878.

Brown, P. 1991. Latin America struggles as cholera spreads. *New Scientist.* **130**(1767):12.

Bryan, F.L. 1987. Seafood-transmitted infections and intoxications in recent years. In *Seafood Quality Determinations.* D.E. Kramer and J. Liston (eds.). Elsevier Science Publishing. Pp. 319-337.

Bryan, F.L. 1988. Risks associated with vehicles of foodborne pathogens and toxins. *Journal of Food Protection.* **51**(6):498-508.

CDC. 1997. Shellfish-related *Vibrio vulnificus* cases (unpublished database report). Centers for Disease Control and Prevention. Atlanta, GA.

Colwell, R.R., R.J. Seidler, J. Kaper, S.W. Joseph, S. Garges, H. Lockman, D. Maneval, H. Bradford, N. Robens, E. Remmers, I. Huq and A. Huq. 1981. Occurrence of *Vibrio cholera* serotype O1 in Maryland and Louisiana estuaries. *Applied and Environmental Microbiology.* **41**(2):555-558.

Cook, D.W. 1994. Effect of time and temperature on multiplication of *Vibrio vulnificus* in post-harvest Gulf Coast shellstock oysters. *Applied and Environmental Microbiology.* **50**(9):3483-3484.

Cook, D.W. and A.D. Ruple. 1988. Microflora modification in temperature-abused shellstock oysters (Final Report). Mississippi-Alabama Sea Grant Consortium. Ocean Springs, MS. Report Number MASGP-87-047.

Cook, D.W. and A.D. Ruple. 1989. *Vibrio vulnificus* in post-harvest shellstock and processed Gulf Coast oysters. *Journal of Shellfish Research.* **8**(2):449.

Davis, J.W. and R. K. Sizemore. 1982. Incidence of *Vibrio* species associated with blue crabs (*Callinectes sapidus*) collected from Galveston Bay, Texas. *Applied and Environmental Microbiology.* **43**(5):1092-1097.

DePaola, A., M.W. Presnell, M.L. Motes, R.M. McPhearson, R.M. Twedt, R.E. Becker, and S. Zywno. 1983. Non-O1 *Vibrio cholerae* in shellfish, sediment and waters of the U.S. Gulf Coast. *Journal of Food Protection.* **46**(9):802-806.

Desenclos, J.A., K.C. Klontz, L.E. Wolfe, and S. Hoercherl. 1991. The risk of *Vibrio* illness in the Florida raw oyster eating population, 1981-1988. *American Journal of Epidemiology.* **134**:290-297.

Eyles, M.J. and G.R. Davey. 1984. Microbiology of commercial depuration of the Sydney rock oyster, *Crassostrea commercialis*. *Journal of Food Protection.* **47**(9):703-706.

FDA. 1992. Foodborne Pathogenic Microorganisms and Natural Toxins. Food and Drug Administration, Center for Food Safety and Applied Nutrition. Washington, DC.

Food Chemical News. 1987. FDA checking imported, domestic shrimp, crabmeat for *Listeria*. August 17, pp. 7-8.

Gellin, B.G. and C.V. Broome. 1989. Listeriosis. *Journal of the American Medical Association.* **261**(9):1313-1320.

Griffin, M.R., E. Dalley, M. Fitzpatrick, and S.H. Austin. 1983. *Campylobacter* gastroenteritis associated with raw clams. *Journal of the Medical Society of New Jersey.* **80**:607-609.

Hackney, C.R., B. Ray, and M.L. Speck. 1980. Incidence of *Vibrio parahaemolyticus* in the microbiological quality of seafood in North Carolina. *Journal of Food Protection.* **43**(10):769-773.

Harbell, S. 1988. Controlling seafood spoilage. Washington Sea Grant, Seafood Processing Series. Seattle, WA.

Ho, J.H., K.N. Shands and G. Friedland. 1986. An outbreak of type 4b *Listeria monocytogenes* infection involving patients from eight Boston hospitals. *Archives of Internal Medicine.* **146**:520-524.

Hood, M.A. and G.E. Ness. 1982. Survival of *Vibrio cholerae* and *Escherichia coli* in estuarine waters and sediment. *Applied and Environmental Microbiology.* **43**(3):578-584.

Horwitz, M.A. 1977. Specific diagnosis of foodborne disease. *Gastroenterology*. **73**(2):375-381.

Jay, J.M. 1978. Modern Food Microbiology. Second Edition. D. Van Nostrand. New York, NY.

Kaneko, T. and R.R. Colwell. 1975. Adsorption of *Vibrio parahaemolyticus* onto chitin and copepods. *Applied Microbiology*. **29**(2):269-274.

Kaper, S.H., H. Lockman, R.R. Colwell, and S.W. Joseph. 1979. Ecology, seriology and enterotoxin production of *Vibrio cholerae* in Chesapeake Bay. *Applied and Environmental Microbiology*. **37**:91-102.

Kaspar, C.W. and M.L. Tamplin. 1993. The effects of temperature and salinity on the survival of *Vibrio vulnificus* in seawater and shellfish. *Applied and Environmental Microbiology*. **59**:2425-2429.

Kaysner, C.A. 1989. *Vibrio* species of the U.S. West Coast (Abstract). *Journal of Shellfish Research*. **8**(2):449.

Kelly, M.T. 1982. Effect of temperature and salinity on *Vibrio* (Beneckea) *vulnificus* occurrence in the Gulf Coast environment. *Applied and Environmental Microbiology*. **44**(4):820-824.

Klontz, K.C., R.V. Tauxe, W.L. Cook, W.H. Riley, and I.K. Wachsmuth. 1987. Cholera after the consumption of raw oysters: a case report. *Annals of Internal Medicine*. **107**(6):846-848.

Lechowich, R.V. 1988. Microbiological challenges of refrigerated foods. *Food Technology*. December:84-94.

Lennon, D., B. Lewis, C. Mantell, D. Becroft, B. Dove, K. Farmer, S. Tonkin, N. Yeates, R. Stamp, and K. Mickleson. 1984. Epidemic perinatal listeriosis. *Pediatric Infectious Disease*. **3**(1):30-34.

Louria, D.B., T.H. Hensic, and D. Armstrong. 1967. Listeriosis complicating malignant disease: a new association. *Annals of Internal Medicine*. **67**:261-281.

Mascola, L., L. Lieb, and J. Chiu. 1988. Listeriosis: an uncommon opportunistic infection in patients with acquired immunodeficiency syndrome, a report of five cases and review of the literature. *American Journal of Medicine*. **84**:162-164.

McCarthy, S.A. and F.M. Khambaty. 1994. International dissemination of epidemic *Vibrio cholerae* by cargo ship ballast and other nonpotable water. *Applied and Environmental Microbiology*. **60**:2597-2601.

MMWR. 1986. *Morbidity and Mortality Weekly Report*. **35**:606-607.

MMWR. 1989. *Morbidity and Mortality Weekly Report*. **38**(2):18-19.

MMWR. 1996. Surveillance for foodborne disease outbreaks — United States, 1988-1992. *Morbidity and Mortality Weekly Report*. **45**(5):1-71.

Morris, J.G. and R.E. Black. 1985. Cholera and other Vibrios in the United States. *New England Journal of Medicine*. **312**(6):343-350.

Morris, J.G., R. Wilson, B.R. Davis, I.K. Wachsmuth, C.F. Riddle, H.G. Wathen, R.A. Pollard and P.A. Blake. 1981. Non-O group 1 *Vibrio cholerae* gastroenteritis in the United States. *Annals of Internal Medicine*. **94**(5):656-658.

Motes, M., A DePaola, S. Zywno-Van Ginkel, and M. McPhearson. 1994. Occurrence of toxigenic *Vibrio cholerae* O1 in oysters in Mobile Bay, Alabama: An ecological investigation. *Journal of Food Protection*. **57**(11):975-980.

NAS, 1991. Seafood Safety. F.E. Ahmed, editor. National Academy of Sciences, Committee on Evaluation of the Safety of Fishery Products. National Academy Press, Washington, DC.

NFI. 1989. Some considerations for control of *Listeria*. National Fisheries Institute. Washington, D.C.

Oliver, J.D. 1981. The pathogenicity and ecology of *Vibrio vulnificus*. *Marine Technology Society Journal*. **15**(2):45-52.

Oliver, J.D., R.A. Warner, and D.R. Cleland. 1983. Distribution of *Vibrio vulnificus* and other lactose-fermenting Vibrios in the marine environment. *Applied and Environmental Microbiology*. **45**:985-998.

O'Neill, K.R., S.H. Jones, and D.J. Grimes. 1992. Seasonal incidence of *Vibrio vulnificus* in the Great Bay estuary of New Hampshire and Maine. *Applied and Environmental Microbiology*. **58**:3257-3262.

Paille, D., C. Hackney, L. Reily, M. Cole, and M. Kilgen. 1987. Seasonal variation in the fecal coliform population of Louisiana oysters and its relationship to microbiological quality. *Journal of Food Protection*. **50**(7):545-549.

Peters, J.B. 1989. *Listeria monocytogenes*: a bacterium of increasing concern. Washington Sea Grant, Seafood Processing Series. Seattle, WA.

Reinhard, R.G., T.J. McAdam, G.J. Flick, R.E. Croonenberghs, R.F. Wittman, A.A. Diallo, and C. Fernandes. 1996. Analysis of *Campylobacter jejuni*, *Campylobacter coli*, *Salmonella*, *Klebsiella pneumoniae*, and *Escherichia coli* O157:H7 in fresh hand-picked blue crab (*Callinectes sapidus*) meat. *Journal of Food Protection*. **59**(8):803-807.

Shandera, W.X., B. Hafkin, D.L. Martin, J.P. Taylor, D.L. Maserang, J.G. Wells, M. Kelly, K. Ghandi, J.B. Kaper, J.V. Lee, and P.A. Blake. 1983a. Persistence of cholera in the United States. *American Journal of Tropical Medicine and Hygiene*. **32**(4):812-817.

Shandera, W.X., J.M. Jeffrey, M. Johnston, B.R. Davis, and P.A. Blake. 1983b. Disease from infection with *Vibrio mimicus*, a newly recognized *Vibrio* species. *Annals of Internal Medicine*. **99**(2):169-171.

Son, N.J. and G.H. Fleet. 1980. Behavior of pathogenic bacteria in the oyster *Crassostrea commercialis* during depuration, relaying and storage. *Applied and Environmental Microbiology*. **40**(6):994-1002.

Stamm, A.M., W.E. Dismukes, and B.P. Simmons. 1982. Listeriosis in renal transplant recipients: report of an outbreak and review of 102 cases. *Review of Infectious Disease*. **4**:665-682.

Tacket, C.O., F. Brenner, and P.A. Blake. 1984. Clinical features and an epidemiological study of *Vibrio vulnificus* infections. *Journal of Infectious Diseases*. **149**(4):558-561.

Tamplin, M.L. 1990. The ecology of *Vibrio vulnificus* in *Crassostrea virginica*. *Journal of Shellfish Research*. **9**:254.

Tamplin, M.L. 1994. The ecology of *Vibrio vulnificus*. In *Proceedings of the 1994 Vibrio vulnificus Workshop*. Food and Drug Administration, Office of Seafood. Washington, DC, pp. 75-85.

Tamplin, M., G.E. Rodrick, H.L. Blake, and T. Cuba. 1982. Isolation and characterization of *Vibrio vulnificus* from two Florida estuaries. *Applied and Environmental Microbiology.* **44**:1466-1470.

Thompson, C.A. Jr., C. Vanderzant, and S.M. Ray. 1976a. Relationship of *Vibrio parahaemolyticus* in oysters, water and sediment, and bacteriological and environmental indices. *Journal of Food Science.* **41**:118-122.

Thompson, C.A. Jr., C. Vanderzant and S.M. Ray. 1976b. Serological and hemolytic characteristics of *Vibrio parahaemolyticus* from marine sources. *Journal of Food Protection.* **41**:204-205.

Watkins, W.D. and V.J. Cabelli. 1985. Effect of fecal pollution on *Vibrio parahaemolyticus* densities in an estuarine environment. *Applied and Environmental Microbiology.* **49**(5):1307-1313.

Weagant, S.D., P.N. Sado, K.G. Colburn, J.D. Torkelson, F.A. Stanley, M.H. Krane, S.C. Shields, and C.F. Thayer. 1988. The incidence of *Listeria* species in frozen seafood products. *Journal of Food Protection.* **51**(8):655-657.

Welch, R. 1994. Consumer education and health advisory programs. In *Proceedings of the 1994 Vibrio vulnificus Workshop.* Food and Drug Administration, Office of Seafood. Washington, DC, pp.119-125.

West, P.A. 1989. Human pathogens and public health indicator organisms in shellfish. In *Methods for the microbiological examination of fish and shellfish.* B. Austin, and D. A. Austin (eds.). Ellis Horwood Limited, Chichester, England.

Whitman, C. 1994. Epidemiology. In *Proceedings of the 1994 Vibrio vulnificus Workshop.* Food and Drug Administration, Office of Seafood. Washington, DC, pp.13-23.

Viral Diseases Associated with Seafood

Cheryl M. Woodley

INTRODUCTION

Human enteric viruses are the leading cause of molluscan shellfish-associated viral disease. The enteric viruses are animal viruses that are principally transmitted by the fecal-oral route and replicate in the intestines. Though currently more than 120 enteric viruses are known from human feces (Murphy and Kingsbury, 1990) and ultimately find their way into sewage, two groups of these viruses, hepatitis A (HAV) and the Norwalk-family of viruses (also referred to as small round-structured viruses, SRSV), pose the greatest health risk to seafood consumers (for review see Metcalf et al., 1995; Jaykus et al., 1994; Goyal, 1984). Although HAV and Norwalk-like viruses are the predominant causes of shellfish-associated disease, caliciviruses, rotaviruses, astroviruses and hepatitis E virus (HEV) have also been associated with water- and/or food-transmitted diseases [for review see Metcalf et al., 1995; Jaykus et al., 1994; Morbidity and Mortality Weekly Report (MMWR), 1990; Richards, 1987].

Seafood-associated viral illnesses were first recognized in 1955, with a documented outbreak of oyster-associated hepatitis (Lindberg-Braman, 1956). Since then, epidemiological evidence from numerous outbreaks of shellfish-related illness has shown that shellfish can and do serve as effective vehicles for the transmission of viral hepatitis and gastroenteritis (for review see Jaykus et al., 1994). Although studies have shown that other types of fish and Crustacea harbor human pathogenic bacteria and viruses, only shellfish have been implicated in the transmission of enteric viral illness (Gerba, 1988).

Exposure of human populations to enteric viruses can occur via several routes: shellfish grown in contaminated marine waters, food crops grown in land irrigated with wastewater or fertilized with sludge, recreational waters, contaminated drinking water, and contamination of food by infected food handlers (Metcalf et al., 1995; Smith and Fratamico, 1995; Jaykus et al., 1994). Human sewage is the most important source of these pathogens in the marine environment (Metcalf et al., 1995; Jaykus et al., 1994; Berg and Metcalf, 1978). Sewage provides the vehicle by which viruses may contaminate surface water, ground water, wastewater-irrigated or sludge-amended soils, and marine and estuarine waters. These viruses enter the environment by various routes, including direct discharge of treated and untreated sewage into marine waters, via sewage-contaminated rivers and streams, through ground waters from solid waste landfills, through the ocean disposal of sewage sludge, and from boat wastes (Metcalf et al., 1995; Gerba, 1988; Metcalf et al., 1988; Goyal, 1984).

Raw sewage commonly carries virus concentrations of $5x10^3$ to $2.8x10^4$ plaque-forming units (PFU)/liter, but can be reduced to 50 PFU/liter after treatment (Metcalf et al., 1995; Melnick, 1984). Secondary treatment of sewage — i.e., treatment that does not include disinfection with chlorine — results in inactivation of only 90 percent of viruses (Berger, 1982). Even sewage which has been chlorinated may not be completely free of viral pathogens because viruses can form aggregates with suspended solids that are then protected from chlorination (Goyal, 1984; Liu et al., 1971). While malfunctioning septic systems are obvious sources of viruses to the marine environment, it has also been found that properly functioning septic systems can contribute to viral pollution in some areas through ground waters. Septic systems in areas with large grain, sandy soils may effectively trap pathogenic and fecal indicator bacteria, but may allow viruses, which are considerably smaller than bacteria, to pass through (Heufelder, 1989), while soils with high clay content readily adsorb viruses and promote their removal from ground waters (Metcalf et al., 1995).

Marine sediments can also play a major role in viral distribution throughout the marine environment (Goyal, 1984; LaBelle et al., 1980). Solids-associated viruses, once discharged into the aquatic environment, move through the water column into bottom sediments where they can be concentrated 10^1 to 10^4 times more than in surrounding waters. As a reservoir of viruses, sediments are often resuspended by turbulence due to natural (storms, currents, tides, waves, seasonal turnover, increased river discharge, changes in water quality, and normal activities of aquatic organisms) and anthropogenic (dredging, boating, swimming, fishing, skin- and scuba diving) factors (Metcalf et al., 1995; Goyal, 1984; LaBelle et al., 1982, 1980; Goyal et al., 1979). Once resuspended, the viruses become available for bioaccumulation in shellfish.

Human enteric viruses are not known to replicate in seawater or sediment, so their presence indicates that they can survive for long periods in the marine environment (LaBelle et al., 1980). Generally, viruses survive better at low temperatures, in low salinity, and in association with solids (Gerba and Goyal, 1978). Field and laboratory studies have demonstrated that enteric viruses can survive from a few days to more than 130 days in marine water (Gerba and Goyal, 1978) and up to 17 months in marine sediment (Goyal et al., 1984), with generally higher concentrations in sediment than in the overlying water (Goyal et al., 1984; LaBelle et al., 1980). Enteric viruses have also been shown to survive during sludge treatments for 30 days of digestion at 50°C, with HAV surviving even longer due to their higher thermoresistance (Metcalf et al., 1995).

UPTAKE AND ELIMINATION OF VIRUSES BY SHELLFISH

Many commercially valuable species of shellfish such as oysters (*Crassostrea virginica, C. gigas, C. glomerata, Ostrea edulis, O. lurida*), clams (*Mercenaria mercenaria, Mya arenaria, Tapes japonica*), mussels (*Mytilus edulis, M. galloprovincialis*), cockles (*Cerastoderma edule*), and Hawaiian bivalves of the families Pinnidae and Isognomonidae have been documented to rapidly accumulate viruses when exposed to contaminated water (Jaykus et al., 1994; Noble, 1990). Molluscan bivalves are filter feeders and, as such, tend to concentrate, or bioaccumulate, viruses from the water in which they grow. Bivalves sieve out suspended food particles from a current of water passing through the shell cavity. This feeding mechanism has been reported to allow concentration of viruses from surrounding waters by a factor that varies from less than one- to over 1000-fold higher than in the overlying water (Jaykus et al., 1994; Gerba, 1988; Gerba and Goyal, 1978). Maximum accumulation of viruses takes place within a few hours, and the digestive tract harbors most of the accumulated viruses (Atmar et al., 1995; Metcalf et al., 1979). Virus levels are maintained as long as sufficient quantities of viruses are present in the water (Gerba and Goyal, 1978). Uptake is greatest when the viruses are associated with solids (Landry et al., 1983; Metcalf et al., 1979) and is higher from resuspended sediment than from undisturbed sediment (Landry et al., 1983). Although introduction within the growing water is probably the major cause of viral contamination in shellfish (DuPont, 1986), contamination of shellfish by poor personal hygiene of those handling the shellfish should not be overlooked (Cliver, 1994).

Depuration and relaying are two "self-purification" methods used to eliminate microorganisms from shellfish (Richards, 1988). Both rely on the ability of the mollusks to eliminate contaminating microbes through normal feeding, digestion, and excretion. Relaying involves transferring shellfish from contaminated waters to clean waters where they filter-feed for a determined period of time. Depuration is a controlled purification accomplished by placing shellfish in tanks containing purified flowing seawater and allowing the shellfish to "purge" themselves for two to three days, after which samples must meet federal fecal coliform standards. It is often assumed that depurated shellfish are bacteriologically and virologically safe; however, this is not always true (Richards, 1990, 1988). Long-term laboratory experiments using *Crassostrea gigas* (Scotti et al., 1983) and *Mercenaria mercenaria* (Canzonier, 1971) have shown that bacterial depuration rates do not accurately predict viral contamination levels. Further, studies with *C. gigas* have shown that viruses can be present in oyster tissue even after 64 hours of depuration (Hay and Scotti, 1986). Today, it is generally accepted that the coliform standard established in the United States under the National Shellfish Sanitation Program (NSSP) for molluscan shellfish harvesting does not correlate with the reduction of viruses in shellfish (Ellender et al., 1989; Goyal et al., 1984, 1979; LaBelle et al., 1980; Gerba et al., 1979; Gerba and Goyal, 1978). Outbreaks of viral disease occurring from shellfish harvested from approved waters (Jaykus et al., 1994; Richards, 1988) highlight the fact that bacterial elimination is not a reliable index of viral elimination during depuration or relaying. To date there are no reliable indicators of virus elimination from shellfish or their growing waters. Gamma irradiation of live shellfish has been investigated as an alternative means to improve the margin of safety for the shellfish consumer against virally-transmitted disease (Mallett et al., 1991), but has not been adopted by the industry nor approved by United States regulatory agencies.

CHARACTERISTICS OF SHELLFISH-ASSOCIATED VIRUSES

The enteric viruses implicated in shellfish-associated outbreaks are represented in the families Picornaviridae, Caliciviridae, and Astroviridae. These viruses each share a common mode of transmission — the fecal-oral route — and replicate primarily in the intestinal tract (Murphy and Kingsbury, 1990). They are classified into several groups (genera) on the basis of morphologic, physicochemical, antigenic, and genetic differences. Although there is a wide variety of clinical symptoms associated with enteric viruses, those symptoms accompanying shellfish-associated viral infections generally are limited to either hepatitis, hepatomegaly, fever, malaise, severe vomiting, nausea, diarrhea, and/or abdominal pain/cramps, depending on the particular virus involved in the illness (Mast and Krawczynski, 1996; Middleton, 1996; Gerba, 1988).

Hepatitis A

Infectious viral hepatitis type A (HAV) is one of the most serious of the viral illnesses associated with contaminated shellfish (Richards, 1985). The first shellfish-borne hepatitis was reported in the 1950s (Lindberg-Braman, 1956; Roos, 1956). Since these early reports, a variety of shellfish have been implicated in cases of infectious hepatitis, including hard clams (both raw and steamed), oysters, mussels, soft-shell clams, and cockles (Figure 1) (Bryan, 1987; Gerba and Goyal, 1978; Mackowiak et al., 1976, 1975; Portnoy et al., 1975; Feingold, 1973). One of the largest shellfish-associated outbreaks of HAV occurred in Shanghai, China, in 1988, where 300,000 cases (4 percent of the total population) were reported in a two-month period. This epidemic occurred approximately one month after the introduction of thousands of tons of presumably contaminated clams. Subsequent laboratory tests of clams taken from the market and from the harvest sites isolated HAV (Xu et al., 1992).

A systemic infection, viral hepatitis A is characterized by gastrointestinal symptoms and injury to the liver (Hollinger and Ticehurst, 1990; Bryan, 1987). Infectious hepatitis has an incubation period of approximately four weeks, with a range of two to six weeks. The initial symptoms are usually weakness, fever, anorexia, nausea, malaise, lassitude, and abdominal pain. As the illness progresses, the individual may develop jaundice and may excrete bile in the urine, causing a dark color. The illness may last for a few weeks to several months. The illness can range from very mild (young children are often asymptomatic) to severe, requiring hospitalization. The fatality rate is low (<0.1 percent), and deaths primarily occur among the elderly and individuals with underlying diseases (North Carolina Division of Health Services, 1989; Bryan, 1987; Feingold, 1973).

The disease agent HAV is a non-enveloped 27- to 32-nanometer (nm) particle with a positive single-stranded RNA genome of approximately 7,480 base pairs and belongs to the hepatovirus group in the Picornaviridae (Hollinger and Ticehurst, 1990; Coulepis et al., 1982). Detection of HAV in clinical or environmental samples has been hampered due to (1) the lack of a reliable cell culture assay for the wild-type virus, (2) low quantities of virus shed in feces, and (3) the frequency of toxins and inhibitors in environmental samples. Recently, some of the obstacles have been overcome. Immunoassays are now available, allowing clinical diagnosis to become routine. These assays, however, lack sensitivity in the detection of HAV in shellfish or environmental samples, due to low levels of the virus. Alternatively, the nucleic acid-based reverse transcription/ polymerase chain reaction (RT-PCR) assays and assays which use a combined immunoassay and PCR (antigen capture PCR, AC-PCR) (Deng et al., 1994) have increased assay specificity and sensitivity of virus detection, allowing detection of virus in shellfish and water samples associated causally with outbreaks of disease (for review see Metcalf et al., 1995; Jaykus et al., 1994).

Currently, immune globulin prophylaxis is used against HAV, but is impractical for repeated exposure. HAV vaccine development of inactivated, attenuated, and recombinant vaccines is underway. Two HAV vaccines are now available in the United States; both are inactivated vaccines. Although a recombinant HAV vaccine has not yet been successful, molecular approaches are being used to evaluate vaccine candidates and to pursue an effective recombinant vaccine (Hollinger and Ticehurst, 1990).

Norwalk Virus and Other SRSVs

Norwalk virus belongs in the group of viruses known as small round-structured viruses (SRSV), or the Norwalk-like family of agents (LeBaron et al., 1990; Kapikian and Chanock, 1990). The Norwalk agent was first recognized as a pathogen in 1968, during an outbreak of gastroenteritis in Norwalk, Ohio — hence the name Norwalk virus (Adler and Zicki, 1969). Molluscan shellfish including clams (both raw and steamed), oysters, and cockles have since been associated with Norwalk or other related SRSV gastroenteritis (Kirkland et al., 1996; Bean et al., 1990; Bryan, 1987; Porter and Parkin, 1987; Morse et al., 1986; Richards, 1985; Gunn et al., 1982; Eyles et al., 1981; Grohmann et al., 1981).

Gastroenteritis due to the Norwalk agent is not unique to ingestion of contaminated shellfish. The virus can be transmitted by drinking water, food, aerosols, person-to-person contact (LeBaron et al., 1990), and unintentional ingestion of contaminated water during swimming (Levine and Craun, 1990). The Norwalk agents play a major role in waterborne gastroenteritis in the United States. Combined data from the Centers for Disease Control (CDC) and the Environmental Protection Agency (EPA) for the period 1986 to 1988 indicate that Norwalk virus was responsible for the second largest number of cases of waterborne illness (Levine and Craun, 1990). One outbreak in 1987, which was linked to commercial ice made from contaminated well water, caused more than 5,000 cases of gastroenteritis (Levine and Craun, 1980). In 1993 an estimated six million oysters from Louisiana were bathed in Norwalk after ships with ill crews dumped waste overboard. This outbreak affected 20,000 to 30,000 oyster consumers. Poor hygienic practices of sick food handlers have also been identified as an important cause of Norwalk gastroenteritis (LeBaron et al., 1990).

Norwalk virus primarily causes a gastrointestinal illness accompanied by symptoms such as nausea, vomiting, non-bloody diarrhea, and abdominal cramps. Less frequently, individuals have complained of chills, low-grade fever, headache, muscular pain, and anorexia. Symptoms usually begin within 24 to 48 hours of contact with the virus. Gastroenteritis caused by Norwalk virus is a self-limiting illness that usually persists less than 48 hours, but can last as long as one week. No prolonged illness or long-term symptoms have been observed in individuals participating in volunteer studies. The main risks of viral gastroenteritis are dehydration and electrolyte imbalance, which can be controlled by increased intake of fluids (Middleton, 1996; Kapikian and Chanock, 1990; LeBaron et al., 1990; Bryan, 1987; Porter and Parkin, 1987; Morse et al., 1986; Gerba et al., 1985; Gunn et al., 1982; Eyles et al., 1981; Grohmann et al., 1981). Immunity following Norwalk infection is short-lived.

The Norwalk group of viruses are icosahedral viruses that range in size from 23 to 34 nm and have a positive-strand RNA genome, approximately 7.6 kilobases in length (Dingle et al., 1995; Jiang et al., 1990). Genetic analyses have recently divided this group of viruses into at least three genogroups. The first genogroup is represented by Norwalk and South Hampton virus; the second genogroup includes Snow Mountain, Hawaii, Bristol, and Toronto agents; and the third genogroup includes the Sapporo strain of human caliciviruses (LeGuyader et al., 1996; Dingle et al., 1995; Metcalf et al., 1995).

Characterization of this group of viruses has been challenging in part because these agents cannot be propagated in cell culture or in experimental animal models; thus, initial studies were limited to human volunteers. Norwalk was originally detected using Norwalk antigen and acute and convalescent sera from volunteers in radioimmunoassays and enzyme-linked immunosorbent assays (ELISAs). Subsequent cloning and sequencing of the Norwalk genome as well as other human caliciviruses have resulted in the classification of Norwalk, the prototype strain for this group of viruses, as a calicivirus (Jiang et al., 1993) and in the development of nucleic acid-based assays to detect viral nucleic acid (for review see Metcalf et al., 1995). Expression of recombinant viral proteins has resulted in an antibody to the virus and new diagnostic assays. Unfortunately, these recombinant antigen ELISA-based assays are type-specific and unable to detect related agents. This problem has been circumvented by using RT-PCR methods. This approach can be adapted to detect a broad range of caliciviruses for general screening purposes or to detect specific virus strains, depending on assay design. Currently, RT-PCR is the method of choice for detection of viruses in shellfish. The sensitivity of this method has been calculated between one and 10 virus particles in stools and between nine and 90 in shellfish, based on estimates from electron microscopy. Interlaboratory testing of one RT-PCR method resulted in a test specificity of 100 percent and test sensitivity of 79 percent, similar to many clinical assays in use today (Atmar et al., 1996).

Hepatitis E

Hepatitis E virus (HEV) was first detected in serologic tests in which HAV and hepatitis B virus (HBV) were excluded — thus, the original designation, non-A non-B hepatitis. HEV is found worldwide and is endemic in many developing countries, with outbreaks reported in Asia, Africa, the Middle East, Mexico, and Central America; however, only a few cases have been documented in Western Europe and the United States (Mast and Krawczynski, 1996; Walter, 1994). Although contaminated drinking water appears to be the leading cause of outbreaks, transmission of HEV by consumption of contaminated shellfish has been documented (for review see Jaykus et al., 1994).

HEV resembles HAV clinically, but has a higher mortality rate and is less infectious, with some evidence that humans become immunized once they have contracted the virus. The modal incubation period is 40 days, with the highest rates of jaundice in young to middle-aged adults. Infection in pregnant women results in an unexplained high mortality approaching 20 percent, and transmission from mother to fetus has been reported in women who recovered after infection during the third trimester (Mast and Krawczynski, 1996).

The HEV genome is single-stranded, positive-sense RNA, approximately 7.5 kilobases in length, and tentatively classified in the Caliciviridae (Mast and Krawczynski, 1996). Three strains (Burmese, Pakistani, and Mexican) have been isolated and their genomes cloned and sequenced. Nucleic acid analyses indicate that they contain highly conserved regions that are being used to develop diagnostic RT-PCR tests for detection of HEV strains across their geographic range (Metcalf et al., 1995). The cloning and expression of recombinant viral proteins has also provided a means for more sensitive immunologic assays for detection of HEV antibody in patients with acute hepatitis during outbreaks.

Astrovirus

Astroviruses are star-shaped, 28 nm RNA viruses. They are second only to rotaviruses as causes of childhood diarrhea, and have been implicated in outbreaks in England due to contaminated water and oysters (Glass et al., 1996; Jaykus et al., 1994). In Japan, outbreaks due to eating raw oysters have led to the isolation of a cytopathic small round virus, the Aichi strain, that most closely resembles an astrovirus (Yamashita et al., 1993). The incubation period is two to four days and the illness duration is usually one to four days. The symptoms include vomiting, diarrhea, fever, and abdominal pain. At least five serotypes have been identified that share a common group antigen. Cloning and sequencing of astroviruses have led to more sensitive and specific RT-PCR assays. The assays have focused on the 3' terminal sequences where serotype specific tests have been possible; however, additional target sequences with broader serotype reactivity would likely be more advantageous for detection of all serotypes that might be present in contaminated shellfish samples (Metcalf et al., 1995).

Poliovirus

Poliovirus is probably the most frequently recovered virus from United States shellfish growing waters because of the common practice of immunizing American children against polio (Larkin and Hunt, 1982). This fact should not cause alarm since viruses of vaccinal origin do not present a health hazard. The polio vaccine consists of live attenuated viruses that replicate in the intestine but produce few or no clinical symptoms. Children who have been immunized excrete viruses (from 10^3 to 10^6 viruses/gram feces) for several days after the vaccine is administered (Larkin and Hunt, 1982). Examination of 20 percent of the polioviruses isolated during a survey of the Texas Gulf showed that all were of vaccinal origin and therefore did not represent a health risk (Goyal et al., 1979). Due to the availability of cell culture-adapted strains and nucleic acid sequence data, polio has, however, been used successfully as a model system for the development of many RT-PCR formatted assays for the detection of other shellfish-associated human viral pathogens (Lees et al., 1994).

Other Viruses

In addition to the above-mentioned viruses associated with shellfish-related outbreaks, rotaviruses are the leading cause of infant and childhood viral enteritis worldwide. Members of the Reoviridae, rotaviruses are non-enveloped, 70 nm wheel-like viruses with a double-stranded RNA genome that consists of 11 different segments. There are at least five serotypes, but numerous strains have been detected genetically (Middleton, 1996). Several methods have been developed for detection of rotavirus in environmental samples; however, rotavirus has been detected in food and water associated with outbreaks of illness only a few times and has not been involved in any shellfish-transmitted illness (Metcalf et al., 1995).

In addition to viral hepatitis and gastroenteritis, two studies have implicated Creutzfeldt-Jakob disease with raw shellfish consumption, but a linkage has not been conclusive (Gerba, 1988). There are also conflicting reports that mollusks may transmit hepatitis B (Jaykus et al., 1994). This remains speculative.

PREVENTION OF INFECTIOUS HEPATITIS AND VIRAL GASTROENTERITIS

Epidemiological studies from numerous reports of shellfish-associated disease outbreaks have shown a definite risk associated with the ingestion of raw shellfish harvested from sewage-polluted waters (Mallett et al., 1991). As previously discussed, the standards set by the NSSP may not completely protect consumers of raw or steamed shellfish from viral illness. Illegal harvesting of mollusks from closed beds can also play a role in causing illness. Those eating raw mollusks should be aware of the risks involved, and should buy their shellfish from a dealer with high standards to ensure that they were legally harvested. Recreational harvesters should be certain that the area is open before harvesting.

Thorough cooking is the only way to ensure that molluscan shellfish are free of viral pathogens. Consumers should be warned that some of the traditional methods of cooking mollusks are not sufficient to kill viruses. Cases of infectious hepatitis and viral gastroenteritis have been linked to undercooked shellfish. Steaming mollusks only until the shells open is not sufficient time to kill viruses. Experiments have shown that soft-shell clams will open after only 60 seconds exposure to steam (Koff and Sear, 1967). The oysters that appeared "done" after at least 12 minutes of steaming had not reached adequate temperatures to inactivate Norwalk-like viruses (Kirkland et al., 1996). To be sure that all viruses are killed, shellfish should be cooked to an internal temperature of 100°C (212°F) (Giusti and Gaeta, 1981; Koff and Sear, 1967). Freezing is not an effective method of inactivating hepatitis A (Gerba et al., 1985) or Norwalk virus (LeBaron et al., 1990).

It should be mentioned that carnivorous shellfish, such as crabs and lobsters, can accumulate viruses by contact with contaminated seawater and sediment, and/or by consuming contaminated bivalves (Hejkal and Gerba, 1981). The risk of viral illness from consumption of crabs and lobsters is far less significant than it is from molluscan shellfish for several reasons. Unlike mollusks, crabs and lobsters do not concentrate viruses (Hejkal and Gerba, 1981). The highest concentrations of viruses are found in the hemolymph and digestive tract, portions of the crab which are not usually consumed (Goyal et al., 1984; Hejkal and Gerba, 1981), and unlike molluscan shellfish, crabs and lobsters are not consumed raw. Laboratory cooking experiments have shown that boiling blue crabs for eight minutes will inactivate 99.9 percent of tested viruses (poliovirus, simian rotavirus, and echovirus were the viruses tested) (Hejkal and Gerba, 1981). Although the risk of viral illness from eating crabs and lobsters is slight, crabs and lobsters should be thoroughly cooked and handled properly after cooking to avoid exposure.

REFERENCES

Adler, J. and R. Zicki. 1969. Winter vomiting disease. *J. Infect. Dis.* **119**:668.

Atmar R.L., F.H. Neill, J. L. Romalde, F. Le Guyader, C.M. Woodley, T.G. Metcalf, and M.K. Estes. 1995. Detection of Norwalk virus and hepatitis A virus in shellfish tissues using the polymerase chain reaction. *Appl. Environ. Microbiol.* **61**:3014-3018.

Atmar, R.L., F.H. Neill, C.M. Woodley, R. Manger, S. Fout, W. Burkhardt, E.R. McGovern, F. Le Guyader, T.G. Metcalf, and M.K. Estes. 1996. A collaborative evaluation of a method for the detection of Norwalk virus in shellfish tissues using the polymerase chain reaction. *Appl. Environ. Microbiol.* **62**:254-258.

Bean, N.H., P.M. Griffin, J.S. Goulding, and C.B. Ivey. 1990. Foodborne disease outbreaks, 5-year summary, 1983-1987. *Morbidity and Mortality Weekly Rep.* **39**:SS-1.

Berg, G. and T. G. Metcalf. 1978. Indicators of viruses in water. In *Indicators of Viruses in Water and Food*. G. Berg (ed.), pp. 267-96. Ann Arbor Science Publishers. Ann Arbor, MI.

Berger, B.B. 1982. Water and wastewater quality control and the public health. *Annu. Rev. of Public Health.* 3:359.

Bryan, F.L. 1987. Seafood-transmitted infections and intoxications in recent years. In *Seafood Quality Determination*, D.E. Kramer and J. Liston (eds.), pp. 319-337. Amsterdam: Elsevier Science Publishers.

Canzonier, W.J. 1971. Accumulation and elimination of coliphage S-13 by the hard clam *Mercenaria mercenaria*. *Appl. Microbiol.* **21**:1024-1031.

Cliver, D.O. 1994. Epidemiology of viral foodborne disease. *J. Food Prot.* **57**:263-266.

Coulepis, A.G., S.A. Locarnini, E.G. Westaway, G.A. Tannock, and I.D. Gust. 1982. Biophysical and biochemical characterization of hepatitis A virus. *Intervirol.* **18**:107-127.

Deng, M.Y., S.P. Day, and D.O. Cliver. 1994. Detection of hepatitis A virus in environmental samples by antigen-capture PCR. *Appl. Environ. Microbiol.* **60**:1927-1933.

Dingle, K.E., P.R. Lambden, E.O. Caul, and I.N. Clarke. 1995. Human enteric *Caliciviridae*: the complete genome sequence and expression of virus-like particles from a genetic group II small round-structured virus. *J. Gen. Virol.* **76**:2349-2355.

DuPont, H.L. 1986. Consumption of raw shellfish — is the risk now unacceptable? *New England J. Med.* **314**:707-708.

Ellender, R.D., F.G. Howell, and W. Isphording. 1989. Role of suspended solids in the survival and transport of enteric viruses in the estuarine environment. Mississippi-Alabama Sea Grant Consortium. Ocean Springs, MS. Report Number MASGP-88-050. Pp. 1-28.

Eyles, M.J., G.R. Davey, and E.J. Huntley. 1981. Demonstration of viral contamination of oysters responsible for an outbreak of viral gastroenteritis. *J. Food Protect.* **44**:294-296.

Feingold, A.O. 1973. Hepatitis from eating steamed clams. *J. Am. Med. Assoc.* **225**:526-527.

Gerba, C.P. 1988. Viral disease transmission by seafoods. *Food Technol.* **42**:99-103.

Gerba, C.P. and S.M. Goyal. 1978. Detection and occurrence of enteric viruses in shellfish: a review. *J. Food Protect.* **41**:743-754.

Gerba, C.P., S.M. Goyal, R.L. LaBelle, I. Cech, and G.F. Bodgan. 1979. Failure of indicator bacteria to reflect the occurrence of enteroviruses in marine waters. *Am. J. Pub. Health.* **69**:1116-1119.

Gerba, C.P., J.B. Rose, and S.N. Singh. 1985. Waterborne gastroenteritis and viral hepatitis. *Crit. Rev. Environ. Control.* **15**:213-236.

Giusti, G. and G.B. Gaeta. 1981. Doctors in the kitchen: experiments with cooking bivalve mollusks. *N. Engl. J. Med.* **304**:1371-1372.

Glass, R.I., J. Noel, D. Mitchell, J.E. Herrmann, N.R. Blacklow, L.K.Pickering, P. Dennehy, G. Ruiz-Palacios, M.L. de Guerrero, and S.S. Monroe. 1996. The changing epidemiology of astrovirus-associated gastroenteritis: a review. *Arch. Virol. Suppl.* **12**:287-300.

Goyal, S.M. 1984. Viral pollution of the marine environment. *Crit. Rev. Environ. Control.* **14**:1-32.

Goyal, S.M., W.N. Adams, M.L. O'Malley, and D.W. Lear. 1984. Human pathogenic viruses at sewage sludge disposal sites in the Middle Atlantic region. *Appl. Environ. Microbiol.* **48**:758-763.

Goyal, S.M., C.P. Gerba, and J.L. Melnick. 1979. Human enteroviruses in oysters and their overlying waters. *Appl. Environ. Microbiol.* **37**:572-581.

Grohmann, G.S., A.M. Murphy, P.J. Christopher, E. Auty, and H.B. Greenberg. 1981. Norwalk virus gastroenteritis in volunteers consuming depurated oysters. *Aust. J. Exp. Biol. Med. Sci.* **59**:219-228.

Gunn, R.A., H.T. Janowski, S. Lieb, E.C. Prather, and H.B. Greenberg. 1982. Norwalk virus gastroenteritis following raw oyster consumption. *Am. J. Epidemiol.* **115**:348-351.

Hay, B. and P. Scotti. 1986. Evidence for intracellular absorption of virus by the Pacific oyster, *Crassostrea gigas*. *N. Zealand J. Mar. Freshwater Res.* **20**:665-659.

Hejkal, T.W. and C.P. Gerba. 1981. Uptake and survival of enteric viruses in the blue crab, *Callinectes sapidus*. *Appl. Environ. Microbiol.* **41**:207-211.

Heufelder, G. 1989. Pollution sources in Buttermilk Bay: keeping it all in perspective. In *Shellfish Closures in Massachusetts: Status and Options*, A.W. White and L.A. Campbell (eds.), pp. 7-13. Woods Hole Oceanographic Institution Sea Grant Program, Woods Hole, MA. Report Number WHOI-89-35.

Hollinger, F.B. and J. Ticehurst. 1990. Hepatitis A Virus. In *Fields' Virology*, Vol. 1, B.N. Fields, D.M. Knipe, et al. (eds.), pp. 631-667. Raven Press Ltd., New York.

Jaykus, L-A., M.T. Hemard, and M.D. Sobsey. 1994. Human enteric pathogenic viruses. In *Environmental Indicators and Shellfish Safety*, C.R. Hackney and M.D. Pierson (eds.), pp. 92-153. New York: Chapman & Hall.

Jiang, X., D.Y. Graham, K. Wang, and M.K. Estes. 1990. Norwalk virus genome cloning and characterization. *Science*. **250**:1580-1583.

Jiang, X., M. Wang, K. Wang, and M.K. Estes. 1993. Sequence and genomic organization of Norwalk virus. *Virology*. **195**:51-61.

Kapikian, A.Z. and R. M. Chanock. 1990. Norwalk Group of Viruses. In *Fields' Virology*, Vol. 1, B.N. Fields, D.M. Knipe et al. (eds.), pp. 671-693. Raven Press Ltd., New York.

Kirkland, K.B., R. A. Merriwether, J. K. Leiss, and W. R. MacKenzie. 1996. Steaming oysters does not prevent Norwalk-like gastroenteritis. *Pub. Health Rep.* **111**:527-530.

Koff, R.S. and H.S. Sear. 1967. Internal temperature of steamed clams. *N. Engl. J. Med.* **276**:737-739.

LaBelle, R.L. and C.P. Gerba. 1982. Investigations into the protective effect of estuarine sediment on virus survival. *Water Res.* **16**:469-478.

LaBelle, R.L., C.P. Gerba, S.M. Goyal, J.L. Melnick, I. Cech, and G.F. Bogdan. 1980. Relationships between environmental factors, bacterial indicators, and the occurrence of enteric viruses in estuarine sediments. *Appl. Environ. Microbiol.* **39**:588-596.

Landry, E.F., J.M. Vaughn, T.J. Vicale, and R. Mann. 1983. Accumulation of sediment associated viruses in shellfish. *Appl. Environ. Microbiol.* **45**:238-247.

Larkin, E.P. and D.A. Hunt. 1982. Bivalve mollusks: control of microbial contaminants. *Bioscience*. **32**:193-197.

LeBaron, C.W., N.P. Furutan, J.F. Lew, J.A. Allen, V. Gouvea, C. Moe, and S.S. Monroe. 1990. Viral agents of gastroenteritis: public health importance and outbreak management. *Morbid. Mortal. Wkly. Rep.* **39**: 1-23.

Lees, D.N., K. Henshilwood, and W.J. Dore. Development of a method for detection of enteroviruses in shellfish by PCR with poliovirus as a model. *Appl. Environ. Microbiol.* **60**:2999-3005.

LeGuyader, F., M.K. Estes, M.E. Hardy, F.H. Neill, J. Green, D.W.G. Brown, and R.L. Atmar. 1996. Evaluation of a degenerate primer for the PCR detection of human caliciviruses. *Arch. Virol.* **141**:2225-2235.

Levine, W.C. and G.F. Craun. 1990. Waterborne disease outbreaks, 1986-1988. *Morbid. Mortal. Wkly. Rep.* **39**:1-13.

Lindberg-Braman, A.M. 1956. Clinical observations on the so-called oyster hepatitis. *Am. J. Pub. Health.* **53**:1003-1011.

Liston, J. 1989. Current issues in food safety — especially seafoods. *J. Am. Dietetic Assoc.* **89**:911-913.

Liu, O.C., H.R. Seraichekas, E.W. Akin, D.A. Brashear, E.L. Katz, and W.J. Hill Jr. 1971. Relative resistance of twenty human enteric viruses to free chlorine in Potomac water. *Proc. 13th Water Qual. Conf.* University of Illinois Press. Urbana/Champaign.

Mackowiak, P.A., C.T. Caraway, and B.L. Portnoy. 1976. Oyster-associated hepatitis: lessons from the Louisiana experience. *Am. J. Epidemiol.* **103**:181-191.

Mallett, J.C., L.E. Beghian, T.G. Metcalf, and J.D. Kaylor. 1991. Potential of irradiation technology for improved shellfish sanitation. *J. Food Safety*. **11**:231-245.

Mast, E.E. and K. Krawczynski. 1996. Hepatitis E: An overview. *Annu. Rev. Med.* **47**:257-266.

Melnick, J.L. (ed.). 1984. Monographs in Virology. Vol. 15, *Enteric Viruses in Water*, p. 235. Basel: Karger.

Metcalf, T.G., X. Jiang, M.K. Estes, and J.L. Melnick. 1988. Nucleic acid probes and molecular hybridization for detection of viruses in environmental samples. *Progr. Med. Virol.* **35**:186-214.

Metcalf, T.G., J.L. Melnick, and M.K. Estes. 1995. Environmental virology: From detection of water by isolation to identification by molecular biology — A trip of over 50 years. *Annu. Rev. Microbiol.* **49**:461-87.

Metcalf, T.G., B. Mullin, D. Eckerson, E. Moulton, and E.P. Larkin. 1979. Bioaccumulation and depuration of enteroviruses by the soft-shelled clam, *Mya arenaria*. *Appl. Environ. Microbiol.* **38**:275-282.

Middleton, P.J. 1996. Viruses that multiply in the gut and cause endemic and epidemic gastroenteritis. *Clin. Diag. Virol.* **6**:93-101.

Morbidity and Mortality Weekly Rep 1990. Viral agents of gastroenteritis, public health importance and outbreak management. **39**RR-5:1-24.

Morse, D.L., J.J. Guzewich, J.P. Hanrahan, R. Stricof, M. Shayegani, R. Deibel, J.C. Grabau, N.A. Nowak, J.E. Herrmann, G. Cukor, and N.R. Blacklow. 1986. Widespread outbreaks of clam- and oyster-associated gastroenteritis: role of Norwalk virus. *N. Engl. J. Med.* **314**:678-681.

Murphy, F.A. and D. W. Kingsbury. 1990. Virus Taxonomy. In *Fields' Virology*, Vol. 1, B.N. Fields, D.M. Knipe et al. (eds.), p. 14. Raven Press Ltd., New York.

Noble, R.C. 1990. Death on the half-shell: The health hazards of eating shellfish. *Perspec. Biol. Med.* **33**:313-322.

North Carolina Division of Health Services. 1989. Epidemiology of hepatitis A in North Carolina in 1988. Epi Notes, Epidemiology Section. Raleigh. Report No. 89-1, pp. 1-9.

Porter, J. and W. Parkin. 1987. Outbreaks of clam-associated gastroenteritis in New Jersey: 1983-1984. *New Jersey Med.* **84**:649-651.

Portnoy, B.L., P.A. Mackowiak, C.T. Caraway, J.A. Walker, T.W. McKinley, and C.A. Klein. 1975. Oyster-associated hepatitis failure of shellfish certification programs to prevent outbreaks. *J. Am. Med. Assoc.* **233**:1065-1068.

Richards, G.P. 1985. Outbreaks of shellfish-associated enteric virus illness in the United States: Requisite for development of viral guidelines. *J. Food Prot.* **48**:815-823.

Richards, G.P. 1987. Shellfish-associated enteric virus illness in the United States, 1934-1984. *Estuaries*. **10**:84-85.

Richards, G.P. 1988. Microbial purification of shellfish: A review of depuration and relaying. *J. Food Prot.* **51**:218-251.

Richards, G.P. 1990. Shellfish Depuration. In *Microbiology of Marine Food Products*, D.R. Ward and C.R. Hackney (eds.), pp. 395-428. Van Nostrand Reinhold, New York.

Roos, R. 1956. Hepatitis epidemic conveyed by oysters. *Svenska Lakartidningen.* **53**:989-1003.

Scotti, P.D., G.C. Fletcher, D.H. Buisson, and S. Fredericksen. 1983. Virus depuration of the Pacific oyster (*Crassostrea gigas*) in New Zealand. *New Zealand J. Sci.* **26**:9-13.

Smith, J.L. and P.M. Fratamico. 1995. Factors involved in the emergence and persistence of food-borne diseases. *J. Food Protect.* **58**:696-708.

Walter, E. 1994. Hepatitis E—epidemiology and clinical aspects. *Schweiz. Rundsch. Med. Prax.* **83:**1008-1010.

Xu, Z.Y., Z-H. Li, J-X. Wang, Z-P Xiao, and D-X Dong. 1992. Ecology and prevention of a shellfish-associated hepatitis A epidemic in Shanghai, China. *Vaccine.* **10S**:S67-S68.

Yamashita, T., K. Sakae, Y. Ishihara, S. Isomura, and E. Utagawa. 1993. Prevalence of newly isolated, cytopathic small round virus (Aichi strain) in Japan. *J. Clin. Microbiol.* **31**:2938-2943.

Health-Related Parasites in Seafoods

Patricia A. Fair

INTRODUCTION

A parasite by definition is metabolically dependent on another organism, the host, which to some degree is injured by the parasite. Parasitic diseases transmitted from wild or domestic animals are called zoonoses. Marine zoonotic infections in humans can result from consumption and physical contact with contaminated seafood. Human parasitic infections associated with seafoods occur primarily by eating raw or undercooked fish. Generally, most are harmless to humans. The presence of parasites in fishery products is unappealing, and some can cause infections. Parasites caused only 4 percent of all food-borne disease outbreaks reported to the Centers for Disease Control (CDC) during 1983 through 1987 (Bean et al., 1990). Parasitic infections from seafood are much less common than are bacterial or viral infections, as indicated in a study by the National Academy of Science (NAS) (Ahmed, 1991). Only 39 cases from two outbreaks of fish-associated parasitic illnesses were reported to the CDC between 1978-1987. The NAS study found that parasitic infections in the United States are rare. Consumption of raw fish is relatively common in Japan, the Netherlands, and parts of South America. In Japan, where seafood is a dietary staple and often consumed raw, the incidence of anisakiasis has generally been estimated to exceed 1,000 symptomatic cases per year (Oshima and Kliks, 1986). There were specific estimates of 3,141 cases in 1984 (Oshima, 1987), and 11,232 cases in 1988 (Asaishi et al., 1989). The prevalence estimates for ascariasis worldwide is one billion with 1,550 deaths from intestinal obstruction (Markell et al., 1992). Parasitic infections commonly occur in certain ethnic groups that consume raw or partially cooked seafood. Some of the fish preparations that are more frequently associated with illness include: sushi, sashimi, ceviche, sunomano, marinated fish, lomi lomi, gravlox, poisson crux, cold smoked fish, and Dutch green herring. Raw fish dishes have only recently become popular in the United States. A second, less obvious, cause for human parasitic infection is undercooked fish. Although the incidence of infection in the United States is rising, it is still considerably lower than in other parts of the world. Parasitic infections in the United States are principally confined to coastal regions, especially Hawaii, Alaska, and California. With improved transportation and technology, fresh seafoods are becoming widely available from all areas; thus, the potential for infections from parasites also increases.

While many parasites have been reported in fish, only a few species are capable of infecting humans. The most important helminths acquired from fish by humans are the anisakid nematodes, cestodes of the genus *Diphyllobothrium*, and digenetic trematodes of the families Heterophyidae, Opisthorchiidae, and Nanophyetidae (Adams et al., 1997). Helminth (worm) parasites can cause aesthetic problems in seafood, or medical problems when transmitted to man from seafood. Roundworms and tapeworms pose the greatest health concerns. In the United States, human parasitic infections from marine fish are most often caused by nematodes (roundworms). The two most common roundworms in fish are *Anisakis simplex* and *Pseudoterranova decipiens*. *Eustrongylides* is a third genus of nematodes that has infrequently caused human illness. Freshwater fish can become infected with numerous species of cestodes (tapeworms) and trematodes (flatworms). Marine fish are generally free of these two kinds of parasites, but there are exceptions. Anadromous fish, especially salmonids, which travel between fresh- and saltwater for breeding purposes, can become infected with some freshwater parasites. Two examples of predominantly freshwater fish parasites that can also infect salmonids are cestodes in the genus *Diphyllobothrium* and the trematode *Nanophyetus salmincola*. Both freshwater and marine parasites may be patho-

Table 1. Major Parasites Transmitted to Humans from Seafood

Parasite	Vector	Infective Form
NEMATODES		
Anisakis simplex	salmon, tuna, herring, mackerel, squid, anchovies	third-stage larval
Pseudoterranova decipiens	cod, pollock, haddock	third-stage larval
Eustrongylides spp.	killifish, estuarine	fourth-stage larval
Angiostrongylus cantonensis	freshwater fish	third-stage larval
Gnathostoma spinigerum	freshwater fish	third-stage larval
CESTODES		
Diphyllobothrium latum	salmon, pike, perch, turbot	pleroceroid in fish
Diphyllobothrium pacificum	marine fishes	pleroceroid in fish
TREMATODES		
Nanophyetus salmincola	salmonids	metacercaria
Opisthorchis sinensis	freshwater fish	metacercaria
Heterophyes heterophyes	mullet, tilapia, mosquito fish	metacercaria
Paragonimus westermani	freshwater crabs, crayfish	metacercaria
ACANTHOCEPHALANS		
Corynosoma strumosum	salmon	acanthetta

genic to fish but incapable of harming humans. While parasites that affect seafood quality will be covered, the focus of this section will be on major parasites that have human health implications (Table 1).

Most fish parasites are not adapted to living in human hosts, and usually die without the individual being aware of their presence. For many parasites humans are not the definitive hosts. An intermediate host is ordinarily one that harbors the asexual or immature stages, whereas the parasite undergoes sexual reproduction in the definitive or final host. Several reviews are available on parasite transmission from seafood (Sakanari et al., 1995; Deardorff and Overstreet, 1991; Deardorff et al., 1987; Olson, 1987; Higashi 1985; Williams and Jones, 1976), as well as general and medical parasitology reviews (Markell et al., 1992; Bogitsh and Cheng, 1990; Cliver, 1990; Zaman and Keong, 1990; Noble et al., 1989; Roderick and Cheng, 1989; Cheng, 1986; Garcia and Bruckner, 1977). Risks associated with exposure depend upon geographic location of harvest. Endemic areas for specific helminth zoonoses in the United States include the Great Lakes and Florida for *Diphyllobothrium latum*; the Pacific Northwest and marine areas for *Diphyllobothrium pacificum*; the Pacific Islands, New Orleans, and Puerto Rico for *Angiostrongylus cantonensis*; the Atlantic and Pacific coasts for *Anisakis simplex*; the Pacific Coast for *Contraceacum osculatum*; the Gulf of Mexico for *Contraceacum* sp.; the Northern Atlantic and Pacific Ocean for *Pseudoterranova decipiens*; and the Pacific islands for *Paragonimus westermani* (Ahmed, 1991; Bryan, 1986; Healy and Juranek, 1979; Chitwood, 1970).

PREVENTION OF PARASITIC INFECTIONS

There are no infallible methods of detecting and removing parasites from fish. Although unsightly, a dead parasite cannot cause illness. Therefore, the best way to prevent illness is to ensure that parasites are killed before eating. Cooking fish to an internal temperature of 60°C (140°F) will kill nematodes, cestodes, and trematodes [United States Food and Drug Administration (FDA), 1987]. If the fish will be eaten raw, it is advisable to freeze it first. Commercial blast freezing at -35°C (31°F) for 15 hours has been found to inactivate all *Anisakis* and *Pseudoterranova* larvae (Deardorff and Throm, 1988). For freezing fresh fish at home, the temperature and time required to kill parasites depends on the species of fish, the depth of penetration, and the physiological condition of the larvae. The FDA recommends freezing fish at -23°C (10°F) for at least 168 hours (FDA, 1987) to ensure that all parasites are killed. Jackson et al. (1990) concurred with the FDA advisory as the most practical safeguard. It is estimated that only 10 percent of all food-borne illness in the United States is attributed to seafood, and the implementation of Hazard Analysis and Critical Control Points (HACCP) should help to further increase the safety of seafoods (Durborow, 1999). However, the wide publicity of food-borne disease outbreaks, emerging pathogenic species, and an increased number of immuno-compromised individuals require new strategies and technologies to protect consumer health. Research since the mid-'50s has shown the beneficial effects of irradiation on maintaining quality and freshness of foods. Irradiation processing has been widely used for seafood products in Asia and Europe; this technology is still being considered by the USFDA (Andrews et al., 1998). However, most of the information available on irradiation of seafoods has been related to microorganisms, and there is little information on this technique to inactivate parasites found in marine and freshwater fish and shellfish. The use of irradiation was not effective in inactivating anisakids in herring without the use of high doses, which produced an unacceptable product (Van Mameren et al., 1969). An electromagnetic method for detection of parasites using changes in current flow may have potential for automated industrial applications (Choudhury and Bublitz, 1994). Improved diagnostic technology (radiologic imaging, immunologic diagnoses, and molecular biology) are providing new options for the clinical management of parasitic infections (Chantz and Kramer, 1995). Overcoming problems with faster methods, such as polymerase chain reaction (PCR), will increase the detection and identification of parasites and food-borne pathogens (Hill, 1996). It has recently become popular to cook fish very lightly. This trend should be discouraged, especially when using species of fish prone to parasitic infection. It is recommended that consumers cook seafood sufficiently to destroy parasites before consuming it. Individuals choosing to consume raw seafood should be informed about the potential risks, especially among immuno-compromised individuals with such diseases as liver dysfunction and diabetes.

NEMATODES

The anisakid nematodes in the genera *Anisakis, Phoconema, Terranova*, and *Hysterothylacium* are significant parasites since they are widely distributed geographically as well as within numerous fish hosts. They are potentially a concern for public health as well as for aesthetic reasons. The adult forms and their definitive hosts include: *Anisakis* (porpoise), *Pseudoterranova* (seals), *Terranova* (whales and elasmobranchs), and *Hysterothylacium* (fishes). The larvae occur in invertebrates and fishes, many of commercial value. *Anisakis* species and *Pseudoterranova decipiens* infect consumers of raw seafood. Anisakiasis was first found in the Netherlands with green herring and in Japan with sashimi consumption.

ANISAKIS SIMPLEX AND *PSEUDOTERRANOVA DECIPIENS*

Description

Anisakiasis is a parasitic disease caused by ingestion of *Anisakis* larva present in raw or undercooked fish. Anisakiasis is common in the Netherlands and Japan, and cases are increasing in the United States and Europe. In the United States all the cases reported are from the third stage juvenile of two species, *Anisakis simplex* and *Pseudoterranova decipiens*. While 50 cases have been reported in the United States (Schantz, 1989), this number should be considered an underestimate since it is not regarded as a reportable disease by the CDC and it is easily misdiagnosed. In the United States, disease transmission can occur from Pacific salmon (*Oncorhynchus* spp.), rockfishes (*Sebastes* spp.), yellowfin tuna (*Thunnus albacores*), and skipjack tuna (*Euthynnus pelamis*), and from meals served both at home and in some restaurants (Deardorff et al., 1987). The majority of *Anisakis* infections are acquired from dishes prepared at home. Restaurants have rarely been incriminated (Schantz, 1989). The first case of anisakiasis in the United States was reported in 1958 (McKerrow et al., 1988). Both *Anisakis simplex* and *Pseudoterranova decipiens* are members of the family Anisakidae. These two species are morphologically similar and cause infections that are difficult to distinguish. Their life cycles are similar, although the final and intermediate hosts differ with each nematode. Because of these similarities, they are often confused, making the documentation and reporting somewhat ambiguous. To avoid redundancy and to present the family Anisakidae as a whole, *Anisakis* and *Pseudoterranova* are described together with their subtle differences noted.

The common name of *Anisakis simplex* is herringworm, after one of the more frequently infected fish. The larval stage of *Anisakis* can be found in the viscera and musculature of a variety of fish; larvae are usually 18 to 36 mm long, 0.24 to 0.69 mm wide, and whitish in color (Pinkus et al., 1975). It is this stage that can present a health risk to humans. Within infected fish, herringworm larvae *Pseudoterranova decipiens* (= *Phocanema*) (Olson, 1983) is more commonly called codworm or sealworm, after its most common intermediate and final hosts. The third larval stage infects a variety of fish and can cause human illness if accidentally consumed. Larval codworms are 5-58 mm long, 0.3 to 1.2 mm wide and yellowish, brownish, or reddish in color (Hafsteinsson and Rizvi, 1987).

Life Cycle

The life cycle of the herringworm is outlined in Figure 1. Adult herringworms are found in the digestive tract of various marine mammals, including whales, dolphins, and porpoises. Eggs from a mature female worm are released in the feces and develop to second-stage larvae within the water column. The second stage emerges from the egg to a free-swimming juvenile (coracidium) that is consumed by euphausiids, mainly krill. Molting is the third stage (procercoid). At this point, the cycle can branch. The infected euphausiids may be consumed by whales, where they continue development into adults, or the larvae may be transferred from the euphausiids to an intermediate host, either fish or squid. If consumed by an acceptable definitive host, the worm migrates to the hemocoel or mesentery where it develops into a third-stage juvenile (plerocercoid) infective to marine mammals or humans. Within the intermediate hosts, herringworms migrate to the viscera and musculature and remain in the third larval stage. The larval worms (1 centimeter long) reside in the posterior body cavity of fish, in a coiled and encapsulated state (McGladdery, 1986; Smith and Wooten, 1975). Herringworm larvae can be sequentially passed from prey fish to predator fish. Fish and squid, infected with the third-stage larvae, are consumed by piscivorous marine mammals, the final hosts. In mammals, the larvae attach to the stomach, develop into fourth-stage larvae, and then to an adult stage. Humans are considered accidental hosts since transfer of parasites to humans cannot result in a complete cycle (Oshima, 1987). Some species of other nematode groups use fish as paratenic hosts and may infect

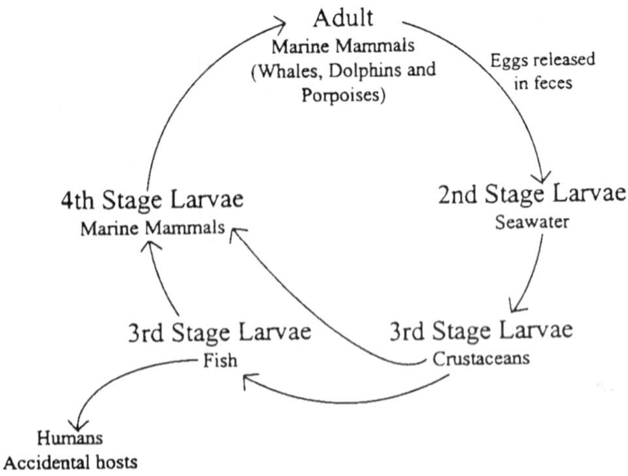

Figure 1. Life cycle of Anisakis simplex.

humans. Humans may then be definitive hosts; or paratenic (transport) hosts, in which no development occurs; or accidental hosts, in which the worms do not survive.

Anisakis juveniles are cream-colored and thus difficult to detect. Fishes are more infective in areas where marine mammals are present. Surveys for helminth distributions show the highest abundance along the Pacific Coast, a fact that coincides with the fact that 86 percent of *Anisakis* human infections are in the western United States (Deardorff and Overstreet, 1991). The increasing populations of marine mammals may increase the number of infected fish available to humans. Brattery (1989) found more than 10,000 adult roundworms in the stomach of a grey seal. Since the enactment of the Marine Mammal Protection Act of 1972, the population of marine mammals has increased (Bonnell et al., 1983, as cited in McKerrow et al., 1988), as have the number of cases of anisakiasis. Between 1980 and 1988 there was a 70 percent increase in the number of reported cases (McKerrow et al., 1988). However, it is not clear whether (1) this reflects an increase in the number of human infections or (2) the number is higher due to increased physician awareness, improved diagnosis, and reporting of the illness. *Anisakis* is more prevalent in the Pacific Ocean than in the Atlantic, partly because the Pacific has a larger population of whales, one of the main final hosts of *Anisakis* (Myers, 1979). Generally, herringworm infections are more prevalent and more intense in larger and older fish

(McGladdery, 1986) and in pelagic, deep water fish (Templeman et al., 1957). The types of fish most commonly infected with *Anisakis* larvae include herring, Pacific salmon, mackerel, Pacific rockfish, Atlantic and Pacific cod, walleye pollock, whiting, bonito, and squid.

The area of worm burden within the fish tends to differ from species to species. Fish that feed on euphausiids, such as herring and mackerel, tend to have larvae in their body cavity. Piscivorous fish, such as whiting and cod, have larvae distributed in both the body cavity and the musculature (Smith, 1984). Storing ungutted fish on ice after capture may result in postmortem migration of *Anisakis* larvae from the viscera to the flesh of some species (Myers, 1979). To reduce the health hazard, it is recommended that all fishes be eviscerated immediately after capture, except salmon where worms are more abundant in the musculature (Deardorff and Overstreet, 1991). An extensive epidemiological study (Van Thiel et al., 1960) demonstrated no cases of aniakiasis prior to 1955 in the Netherlands when herring fishermen were gutting and curing their catch at sea. The occurrence of *Anisakis* increased dramatically when herring began being gutted ashore in 1955, and then a significant reduction occurred when legislation required freezing to kill the worms.

The life cycle of *Pseudoterranova decipiens* is similar to that of *Anisakis*, with these differences: the species that act as intermediate and final hosts, and the inability of the larvae to be directly transferred from crustaceans to the final hosts. The final hosts of codworm include seals (especially grey seals), sea lions, and walruses. The eggs are passed in the feces and develop to second-stage larvae in seawater. The larvae are ingested by crustaceans and molt to third-stage larvae. Fish can become infected by consuming crustaceans, and by serially passing the larvae from one fish to another. When infected fish are eaten by marine mammals, the larvae attach to the stomach, develop to fourth-stage larvae, and finally to adults. Again, humans are considered accidental hosts since transfer of parasites to humans cannot result in a complete life cycle (Hafsteinsson and Rizvi, 1987; Oshima, 1987; Margolis, 1977). *P. decipiens* are more prevalent in the Atlantic Ocean than in the Pacific, since the Atlantic has a large population of grey seals (Myers, 1979). Like *Anisakis*, the incidence of *P. decipiens* infection in fish generally increases with the length, weight, and age of the fish host (Margolis, 1977; Scott and Martin, 1957; Templeman et al., 1957). Bottom- and near-bottom-dwelling fish of the continental shelf are more likely to be infected with *P. decipiens* than are fish that dwell in the open ocean (Margolis, 1977; Templeman et al., 1957). The infection rate in fish can be highly variable from one geographic area to another (Templeman et al., 1957). Infection rates of cod have been found to vary from less than one percent in some offshore areas of Atlantic Canada, to as high as 90 percent in the southern Gulf of St. Lawrence (Margolis, 1977). Consumption of salmon and Pacific rockfish on the United States West Coast have caused *P. decipiens* infections (Kliks, 1983; Myers, 1979; Margolis, 1977). The species of fish most often infected with sealworm larvae include cod, walleye, pollock, halibut, greenling, squid, and some species of flatfish. A study comparing infection rates of *A. simplex* found that wild salmon had 87 percent infection rates while pen-reared salmon lacked any infection, indicating that farmed salmon may increase the safety margin among consumers eating raw salmon (Deardorff and Kent, 1989).

Symptoms/Treatment

Cases of human infection are rare relative to the quantities of herringworms accidentally consumed (Pinkus et al., 1975). In the unusual event that a larva is consumed alive and survives the chewing process, it can invade the mucosa and cause a condition called anisakiasis. Due to the non-specific symptoms of anisakiasis, it is often difficult to diagnose. The symptoms of anisakiasis often mimic those of acute appendicitis, gastric ulcers, gastrointestinal cancer, and food poisoning. Consumption of raw or inadequately prepared seafood infected with juveniles results primarily in penetration of the gastrointestinal tract of the host, but larvae may also penetrate the pancreas and other tissues (Yokogawa and Yoshimura, 1967). Penetration in the stomach, called gastric anisakiasis, usually causes symptoms within one to 12 hours of eating infected fish (Deardorff et al., 1986a). The acute stage includes sudden, severe epigastric pain, sometimes accompanied by nausea, vomiting, and occasionally fever. The onset of symptoms varies depending on the location of larval penetration. The onset of intestinal anisakiasis usually occurs within one to seven days of eating infected fish (Wittner et al., 1989). Surgical removal is necessary in the rare event that a larva penetrates the mucosa. Patients are expected to recover completely. Antihelminthic drugs are not useful in treating cases of anisakiasis (Schantz, 1989). Gastroscopy can reveal the lesion and the larvae attached to the mucosa; X-ray examination may show single or multiple ulcers. Definite diagnosis can be made on morphological characteristics of the whole worm (Sugimachi et al., 1985), although diagnosis can be difficult when worms migrate to extragastric sites (Oshima, 1987). Serological tests are now available in some areas such as Japan. Sonographic examination should be applied in patients with acute abdominal pain who have had a recent history of ingesting raw fish (Shirahama et al., 1992). Treatment for acute infection is removal of larvae through a gastroscope. Chronic symptoms include vague epigastric distress, occult blood in gastric juice or stool, and a high peripheral eosinophil count. The wide variety of clinical signs makes diagnosis difficult and an endoscopic identification is needed for reliable diagnosis and subsequent removal (Deardorff et al. 1986a). Since the nematodes do not develop to full maturity and produce eggs in humans, stool examinations are not useful for diagnosis (Schantz, 1989; Deardorff et al., 1986a).

It is generally accepted that *P. decipiens* infections are less severe and invasive than are *Anisakis* infections (McKerrow et al., 1988; Olson, 1987). Specimens are often coughed up (Deardorff et al., 1987; Olson et al., 1983), manually removed from the mouth, or vomited by the individual within 48 hours of consumption. Often a tingling sensation in the back of the throat is reported before the parasite is expelled (Oshima, 1987). Although most cases are asymptomatic, codworms are capable of causing gastric anisakiasis. The symptoms may include: epigastric pain, nausea, vomiting, and abdominal discomfort (Margolis, 1977). Animal experiments support the idea that *P. decipiens* is capable of penetrating the gastric mucosa, although penetration is usually superficial and involves only the head end of the worm (Young and Lowe, 1969). There has been only one case of human *P. decipiens* in the United States in which tissue penetration has occurred, and twelve cases in which a larva was either coughed up or manually removed from the mouth (Hafsteinsson and Rizvi, 1987). Codworms are red and grow to a large size, so they are noticeable, thereby discouraging consumer purchase (Deardorff and Overstreet, 1991). Many studies have been performed investigating the bioactivity of excretory products from juvenile *Anisakis* (reviewed by Bier et al., 1987). Petithory et al. (1990) suggest that *A. simplex* may be a co-factor of gastric cancer based on epidemiological, experimental, and histopathological data.

Detection of *Anisakis* and *P. decipiens* Larvae

Processing plants often use "candling" to detect nematode larvae in fillets. Candling is done by passing the fillets over a brightly back-lit table. Parasites show up as dark spots against the lighter fish flesh. Not only is this technique costly and labor intensive, it is not completely effective in detecting larvae. Candling can only be used to detect parasites in thin fillets and is useless for whole fish or steaks. *Anisakis* larvae are especially difficult to detect since they are light in color and tend to blend in with the fish flesh. Even *P. decipiens*, which tends to be larger and darker in color, is not easily detected by candling. Studies have shown that candling cannot detect *P. decipiens* embedded deeper than 6 millimeters in the fish musculature (Hafsteinsson and Rizvi, 1987). Candling was only 50 percent effective in detecting codworms (Honans and MacFarlane, 1957). Some research has been done towards developing a more reliable method of detecting and removing nematodes from fish. An acoustic detection technique had shown promise; however, this technique has not been commercially viable (Margolis, 1977). Studies by Adams et al. (1999) suggested that the parasite *Anisakis simplex* was inactivated by a thermal microwave treatment at 150°–170°F depending upon sample size. Nematode infection is reduced when fish are gutted soon after capture; salting may also reduce the incidence. Smoking or light salting does not affect larvae, but freezing at -20°C for 72 hours or heating at 55°C for 10 seconds will kill adult parasites (Khalil, 1969).

EUSTRONGYLIDES SPP.

Description

Human infections by nematodes in the genus *Eustrongylides* are extremely rare, since these parasites are not found in species of fish commonly eaten by humans (Schantz, 1989). It is a parasite of brackish and freshwater bait fish, especially minnows (Gunby, 1982). In some areas of the East Coast, *Eustrongylides* infections can be intense in the minnow population. Two studies in the Baltimore area have shown that 33 percent (Shirazian et al., 1984) to 48 percent (Gunby, 1982) of the minnows can be infected with the larva. Dicotophymatids can negatively impact fisheries since infected fish can have reddish juveniles in the musculature, in lesions, or encapsulated in the body cavity. Infections occur in both freshwater and marine fishes. There have been only five reported cases of *Eustrongylides* infection in humans. In 1982, three Baltimore area fishermen became ill after consuming live bait fish (Shirazian et al., 1984; Gunby, 1982), a practice that is not recommended. Two more cases were reported in 1989, one involving a New Jersey fisherman who ingested live bait fish (Eberhard, 1989), and the other a New York man who ate homemade sushi of an uncertain fish variety (Wittner et al., 1989).

Life Cycle

Typically, the final hosts of *Eustrongylides* are fish-eating birds, frequently great blue herons and egrets (Figure 2). Eggs are passed in the feces and are ingested by aquatic oligochaetes, the first intermediate host. The larvae are then transferred to a variety of fish, especially minnows, which serve as the second intermediate host. In the advanced larval stage, *Eustrongylides* is pinkish-red in color and can be as large as 40 mm long and one mm wide (Wittner et al., 1989). At this point, the life cycle may take one of two routes. Infected fish may be eaten by birds, thereby transferring the parasite directly to the final host. Or, the larvae may first be transferred to a paratenic, or transport host, before being transferred to the final host. The most common paratenic host of *Eustrongylides* larvae are reptiles, amphibians, and fish (Cooper et al., 1978). The green-backed heron (*Butorides striatus*) and mergansus (*Mergus* spp.) in southeastern ponds are vectors; screening should prevent bird use and reduce infections.

Symptoms/Treatment

In humans, *Eustrongylides* larvae are capable of perforating the intestinal wall and entering the abdominal cavity. This causes severe abdominal pain and tenderness. Infected individuals usually require surgery to remove the parasite (Schantz, 1989; Wittner et al., 1989; Gunby, 1982), although one patient has been successfully treated with drugs (Shirazian et al., 1984).

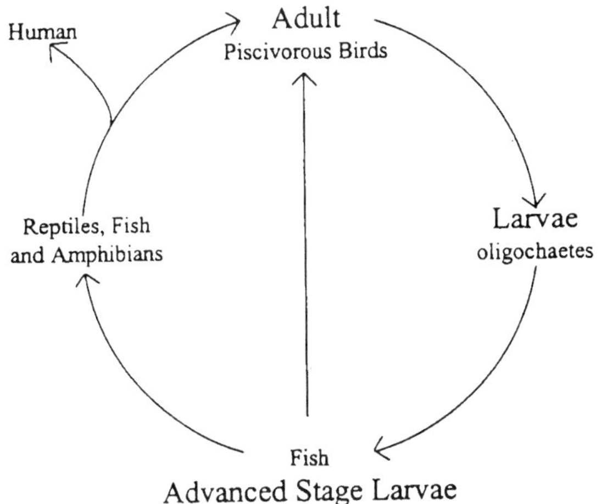

Figure 2. Life cycle of Eustrongylides.

ANGIOSTRONGYLUS CANTONENSIS

Description

Angiostrongylus cantonensis, a lungworm of rats, can cause eosinophilic meningitis and eye disease in humans. The nematode has become established in Hawaii (Beaver et. al., 1984; Alicata, 1962) as well as in Cuba, Puerto Rico, and New Orleans (Campbell and Little, 1988). Other species (*A. costaricensis*, for example) occur in Texas in a natural host, the cotton rat, *Sigmodon hispidus*. With the penetration of this host-parasite into tropical and sub-tropical regions of Africa, India, the Caribbean, and the tropical Gulf Coast of the United States, it is considered a significant public health risk (Kliks and Palumbo, 1990).

Life Cycle

Adult *Angiostrongylus cantonensis* inhabit pulmonary arteries of several genera of rodents — mainly *Rattus* and *Bandicota*. A wide variety of mollusks feed on rodent feces. The parasite develops to the third infective stage in rats, humans, and other mammals. In addition to molluscan intermediate hosts, crustacea and planaria serve as paratenic hosts. In unnatural hosts, such as man, development is usually incomplete. Rats near food supplies at shrimp and fish ponds may infect terrestrial gastropods that can be ingested by the cultured species and can then become infective. Infections transmitted by marine or estuarine fishes have not been confirmed, but some shrimp and possibly fishes serve as paratenic or carrier hosts. *Angiostrongylus cantonensis* is acquired by eating raw snails, which serve as the intermediate host or the transport host. In Tahiti, human infections were traced to ingestion of raw prawns (*Macrobrachium*) (Alicata and Brown, 1962).

Symptoms/Treatment

Patients develop severe headaches, and may exhibit convulsions, arm or leg weakness, paresthesia, vomiting, facial paralysis, neck stiffness, and fever (Beaver et al., 1984). The disease can be fatal to humans and is transmitted by various paratenic hosts. Heavy infections can cause transient meningitis or a more severe disease involving brain, spinal cord, and nerve roots with a characteristic eosinophilia of the peripheral blood or central spinal fluid. Epidemics and sporadic infections occur most commonly in the South Pacific, Southeast Asia, and Taiwan (Beaver et al., 1984; Alicata, 1962). The parasite has also appeared in the Caribbean and in rat and snail populations in Puerto Rico (Anderson et al., 1986). There are no specific treatment (Kliks and Palumbo, 1992).

OTHER NEMATODES

GNATHOSTOMA SPINIGERUM

Some other nematodes are transmitted via seafood. Raw or undercooked freshwater fish in Asia may serve as vehicles of *Gnathostoma spinigerum* when waters are contaminated with feces of carnivores such as dogs and cats. The third-stage larvae can cause gnathostomiasis, or "creeping eruption." The adult worms, ranging from 30 to 50mm in length, embed in the stomach of domestic and wild carnivores such as cats, tigers, and dogs. Eggs pass into the water with feces and the first-stage, free-swimming larvae are released. Larvae are consumed by copepods that become hosts to second-stage larva. When these copepods are eaten by an appropriate vertebrate, the infective third stage develops. Fish, amphibians, birds, and reptiles serve as intermediate hosts. The infected intermediate host is eaten by a wild carnivore to complete the life cycle. *Gnathostoma spinigerum* is acquired by eating raw or undercooked freshwater fish. Clinically it shows larval migrans, a granulomatous lesion, or stationary abscess. Tissue and organ damage can occur to both the brain and the eye. The larvae, which can reach 10mm, can live for many years in humans (Bunnag and Harinasuta, 1989). Southeast Asia, India, and Israel have reported infections (Healy and Juranek, 1979). Due to the Japanese custom of eating sliced raw fish (sashimi), human gnathostomiasis is one of the important zoonoses in Japan (Sato et al., 1992).

CESTODES

The only two cestodes associated with parasitic infections from seafood are *Diphyllobothrium latum* and *D. pacificum*. Transmission occurs through ingestion of plerocercoids from raw freshwater fish. While cases have been reported among Jewish cooks who sample "gefilte fish," there are few data to estimate United States infection rates. *D. latum* is indigenous to the former Soviet Union, the Baltic Sea countries, central and southeastern Europe, Lake N'gami, Africa, northern Manchuria, Japan, New South Wales, Australia, Canada, southern Chile, and Argentina. In the United States, infections have been reported in northern Minnesota and Alaska.

DIPHYLLOBOTHRIUM LATUM AND *DIPHYLLOBOTHRIUM PACIFICUM*

Description

The broad fish tapeworm of humans is distributed worldwide, occurring in regions where pickled or insufficiently cooked freshwater fish is common. Cestodes, or tapeworms, in the genus *Diphyllobothrium* are mainly parasites of freshwater fish; however, anadromous fish, especially Pacific salmonids, can also carry the parasite (Schantz, 1989; Olson, 1987). The larval stage, the stage found in fish, can range from a few millimeters to several centimeters in length (Olson, 1986). De Carneri and Vita (1973) estimated fewer than 100,000 cases of diphyllobothriasis in North America. Common vectors for *D. latum* are freshwater fish (pike, burbot, perch), as well as salmonid fishes. Although the incidence of diphyllobothriasis in the United States is not reported to CDC, an indirect measure can be based on the drug niclosamide used to treat the disease, which has increased (Deardorff and Kent, 1989).

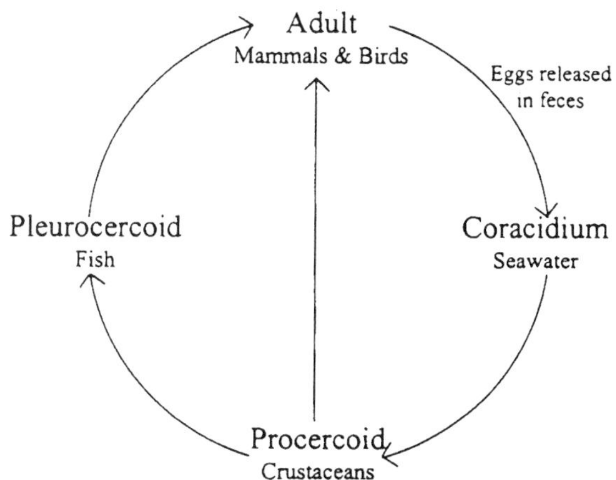

Figure 3. Life cycle of Diphyllobothrium.

Life Cycle

The life cycle involves three hosts: copepod, fish, and mammal. Adult *Diphyllobothrium* live in the intestines of a variety of fish-eating mammals and birds (Figure 3). The tapeworm eggs pass out with the feces and develop to coracidia, or ciliated larvae, in the water. In the egg, a hexacanth embryo (oncosphere) develops which then hatches into a free-swimming ciliated coracidium. The larvae are consumed by crustaceans, usually copepods, and develop to the procercoid stage. By consuming infected crustaceans, fish become the second intermediate hosts and the procercoid migrates onto the viscera or muscle fibers within the fish, where the parasites develop to the plerocercoid stage. Birds and mammals, including humans, become the final hosts by consuming fish containing the plerocercoid (Olson, 1987; Barnes, 1980). When the contaminated fish is eaten, the plerocercoid develops into a mature adult tapeworm up to 10 meters in length. A single worm can discharge up to one million ova per day. The tapeworm that forms in the human intestines may live as long as 25 years and attain infection intensities of 200 worms per person (Sakanari et al., 1995). Sparaganosis is the infection of humans by plerocercoid larvae of diphyllobothrid tapeworms. The infection can be acquired by drinking water containing copepods infected with the larval stage of the parasite. The larva penetrates the intestinal wall and enters muscles or subcutaneous tissues and grows into a sparganum larva. Infections in humans have also occurred when snakes or tadpoles infected with plerocercoids are consumed or used as poultices on open wounds.

Symptoms/Treatment

Most cases of human infection, called diphyllobothriasis, are asymptomatic (Olson, 1987; Schmidt and Roberts, 1985). However, some individuals can experience abdominal pain, diarrhea, constipation, and occasionally anemia (Schantz, 1989). Treatment of diphyllobothriasis by administration of antihelminthic drugs is recommended (Schantz, 1989). The broad fish tapeworm, *Diphyllobothrium latum*, in humans causes an anemia similar to pernicious anemia because of the parasite's affinity for vitamin B12. This tapeworm can absorb 10 to 50 times the amount of B12 as other tapeworms. Since B12 is important in blood formation, infection by *D. latum* often results in anemia (von Bonsdoroff, 1977). With marginal diets, parasitic infections can seriously undermine health. Diagnosis of diphyllobothriasis is performed by finding large characteristic eggs in the feces. Niclosamide (Niclocide) is an effective treatment (Bogitsh and Cheng, 1990). The drug's mode of action is the inhibition of oxidative phosphorylation in the mitochondria, resulting in death of the worm (Markell et al., 1992). Praziquantel is also highly effective. Since humans are the primary hosts of *D. latum*, proper disposal of human feces is needed to control the infection, along with thoroughly cooking fish.

TREMATODES

Digeneans from the phylum Platyhelminthes are commonly known as flatworms, flukes, or trematodes. Although rare in the United States, several trematodes can be transmitted via seafood and thus cause significant health problems in humans. Flukes are more common outside North America, mainly in China and Asia, resulting from a of lack of sanitation and the consumption of raw fish. Only digenetic trematodes have produced disease in humans, and data indicate no substantive health problem in the United States (Ahmed, 1991). A few of these include the human liver fluke *(Opisthorchis sinensis)*, transmitted via freshwater fish such as carp culture operations where human wastes are used; *Nanophyetus salmincola*, the salmon-poisoning fluke; and *Paragonimus westermani*, the Oriental lung fluke, predominantly in China, Japan, Korea, and the Philippines, and transmitted by eating freshwater crabs or crawfish either raw or pickled. *Paragonimus westermani* form metacercarial cysts in the gills, muscles, and hepatopancreas of the crab. Most trematode species that infect people in the United States are in the family Heterophyidae that have a metacercaria in fishes.

General Life Cycle

Most digeneans use a mollusk as the first intermediate host to undergo asexual reproduction, but only a few species of trematodes of the family Heterophyidae can encyst or encapsulate in fish and infect humans. A snail serves as the first intermediate host, releasing cercariae which penetrate fish and then encyst. A proper avian or mammalian host eats the prey with a developed encysted metacercaria, and the worm matures to 1 to 2 mm in the vertebrate's intestine. Most heterophyids can infect a variety of hosts, and their eggs may filter through the intestinal wall and lodge in the cardiac tissue (Africa, 1937) and brain (Deschiens et al., 1958). Mullet has been reported as the primary marine fish of public health significance with regard to heterophyiasis (Beaver et al., 1984) in Hawaii, as well as in the southeast United States (Welberry and Pacetti, 1954). The metacercariae of *Phagicola nana* encyst in the muscle tissue of bass and sunfishes and have been considered as a potential health hazard in the coastal eastern and southern United States (Font et al., 1984). Since mullet have been observed to contain 6,000 metacercariae per gram of flesh (Paperna and Overstreet, 1981) and some can survive minimal heating, even a small number may cause disease in humans. Many genera of the Heterophyidae are known, and Cheng (1983) summarized the species that were transmissible to humans by eating raw or poorly cooked fish. Heterophyiasis may involve more than parasitism of the intestinal tract and may be carried in the blood to the heart, becoming encapsulated in cardiac muscle. Cardiac arrest and death to the host can occur. Among people of Filipino descent in Hawaii, the "mystery death" is suspected to be due to this condition, with 14.6 percent of cardiac deaths attributed to heterophyid myocarditis (Kean and Breslau, 1964). An intestinal fluke, *Heterophyes*, is transmitted by eating fresh- or brackish-water fish as raw, salted, or dried product. The area of prevalence for this organism is in the lower Nile valley near the Mediterranean Coast, as well as Asia and western India. While members of the genus *Heterophyes* do not seem to occur naturally in the United States, transmission of heterophyes-like intestinal flukes from sushi have been reported (Adams et al., 1986). Moderate infections occur with symptoms of mild irritation, colicky pain, and mucous diarrhea. *Metagonimus yokogawai* may be present in freshwater trout in Asia, the Balkans, and northern Siberia.

Symptoms/Treatment

When the definitive host, which can be a human, ingests the flesh of infected crustaceans, the parasite encysts in the intestine. It then bores through into the coelom, into the diaphragm and lungs where it becomes encapsulated. Adults of this species do not enter the lungs but lodge in the spleen, liver, urinary system, eyes, muscles, brain, and other areas. Within the cysts host cells and eggs are infiltrated in a reddish fluid. Infected lungs appear peppery and individuals have coughing, difficulty in breathing, slight fever, and blood-stained sputum. Discovery of eggs in sputum is a reliable diagnosis. Yokogawa (1965) provides a critical review of paragonimiasis.

Nanophyetus salmincola

Description

Nanophyetus salmincola is important because it is the vector of the salmon-poisoning disease causing rickettsia, fatal to dogs (Phillips, 1955). The metacercariae are pathogenic to fish. It is called the salmon-poisoning fluke since it serves as the vector for the rickettsial organism, *Neorickettsia helminthoeca*, which is pathogenic to canine hosts that ingest raw salmon harboring this parasite. About 90 percent of dogs infected with salmon-poisoning disease die (Millemann and Knapp, 1970). Interestingly, the cause of salmon poisoning is not *Nanophyetus* itself, but the rickettsia carried by the trematode. Illness in humans has been associated only with the trematode, not the rickettsial organism (Milleman and Knapp, 1970). Humans can be infected, although nanophyetiasis is not serious. *Nanophyetus salmincola* is extensively distributed in Oregon, and 100 percent of the snail hosts during June through October harbor mature cercariae (Gebhardt et al., 1966).

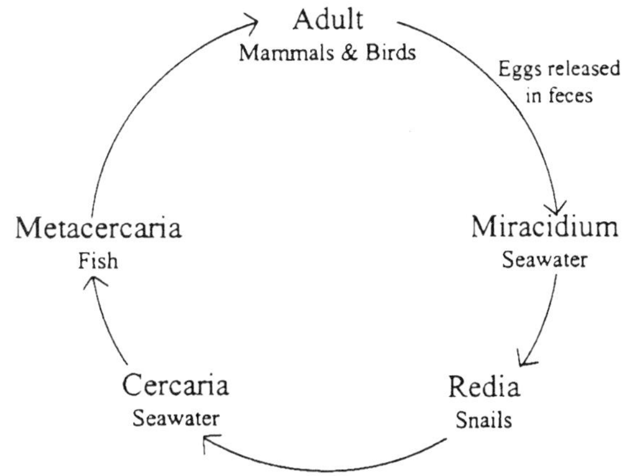

Figure 4. Life cycle of Nanophyetus salmincola.

Life Cycle

The life cycle of *Nanophyetus salmincola* is shown in Figure 4. Raccoons, skunks, and minks are the primary definitive hosts of *Nanophyetus*, but a variety of other piscivorous animals and birds can also act as final hosts. Eggs are passed in feces and develop to miracidia in the water column. Eggs passed from the host require 185-200 days to develop into free-swimming miracidia. The molluscan host is the stream snail, *Oxytreme silicula*, which gives rise to mother rediae to daughter rediae and cercariae. The cercariae penetrate and encyst under the skin and kidneys of many fish, primarily salmon and trout. Fish can show exophalmia; damaged gills, eyes, and other organs; and impaired swimming ability. When infected fish are eaten by other animals, the metacercariae encyst and develop into adults in the small intestine. The miracidia penetrate the flesh of a specific species of snail, *Juga plicifera*, and develop to rediae. Snails shed cercariae, which are capable of penetrating fish skin on contact. Within the fish flesh and internal organs, the parasite develops to the metacercarial stage. Infected fish are consumed by piscivorous mammals and birds, where the parasite attaches to the walls of the small intestine and matures to an adult. When the metacercariae are consumed by fish-eating mammals, including humans, they become infected, but only canids are affected by the rickettsia *Neorickettsia helminthoeca*.

Symptoms/Treatment

Most canines infected with this parasite will die unless treated. *Nanophyetus* can cause disease in humans, although usually it isn't fatal. Symptoms include: diarrhea, abdominal discomfort, nausea, vomiting, weight loss, peripheral blood eosinophilia, and fatigue (Eastburn et al., 1987). Treatment was most effective with bithional (Markell et al., 1992). The reason for increased numbers of infections of *N. salmincola* in the United States is the consumption of raw, incompletely cooked and smoked salmon and steelhead trout containing metacercariae. The consumption of eggs from these fishes can also cause infection. Since the early 1930s, the fluke *Nanophyetus salmincola* has been known to infect the people of Siberia (Skrjabin and Podjapolskaja, 1931, as cited in Eastburn et al., 1987). More recently, this parasite has been associated with illness in parts of the western United States (Eastburn et al., 1987). Cases in the United States have been associated with eating Pacific salmon and steelhead trout that were raw, cold smoked, or partially cooked. Human infections are manifested by increased frequency of bowel movements or diarrhea, abdominal discomfort, nausea, vomiting, weight loss, fatigue, and diminished appetite. Asymptomatic cases have been reported. Some patients have been successfully treated with antihelminthic drugs. Patients receiving no medical treatment have recuperated slowly over a period of one to four months (Eastburn et al., 1987). Preventive measures consist of proper sewage disposal of human feces, treatment of infected persons, health education, and thorough cooking of fish.

Opisthorchis (*Clonorchis*) *sinensis*

Description

Opisthorchis sinensis is a parasite problem in developing countries where human feces are used to fertilize freshwater fish ponds, and where fish products are not properly preserved (pickled, salted, cooked, etc.). It is estimated that 20 million people are infected in mainland China, 4.5 million in Korea, and 7 million in Thailand, with a 90 percent infection rate in some villages (Bunnag and Harinasuta, 1989). Only the cyst stage of the Chinese liver fluke is transmitted by eating raw or undercooked freshwater fish. The endemic area for this parasite is from Japan to Vietnam. Testing for intestinal parasites in the U.S. population during 1987 revealed a parasite incidence rate of 20 percent for all parasites, *Clonorchis* and *Opisthorchis* species contributed 0.6 percent of that total (Kappus et al., 1991).

Life Cycle

Adult worms of *Opisthorchis sinensis* live in the liver of definitive hosts and release their eggs through the bile ducts into the host's intestine. The adult worm may live to 30 years (Bunnag and Hariasuta, 1989) and produce 4,000 eggs per day (Schmidt and Roberts, 1985). Eggs are transmitted via the feces to freshwater ponds, consumed by aquatic snails, and undergo several generations of asexual reproduction until larvae are released (Schmidt and Roberts, 1985). The larvae attach to the scales or tissues of a fish and encyst. An infected fish may have more than 1,000 encysted larvae. Humans or other appropriate hosts are infected when these encysted larvae are consumed. The larval flukes are released in the intestine and migrate to the liver via the bile ducts. There they develop into adult worms. Definitive hosts are humans, cats, tigers, or other fish-eating mammals.

Symptoms/Treatment

Severe symptoms can result from the consumption of 500 to 1,000 worms; 21,000 worms have been reported from an individual upon autopsy (Noble et al., 1989). Antihelminthic drugs are effective therapy.

ACANTHOCEPHALANS

Corynosoma strumosum

Description

Acanthocephalans are spiny-headed worms with larvae requiring arthropods as first or only intermediate hosts. Adult acanthocephalans infect the intestines of pinnipeds, such as the sea lion. The eggs are released with feces into seawater and consumed by crustaceans. These crustaceans can be eaten by fishes, resulting in larvae development, or by a pinniped — in which case the parasite will develop into an adult.

Life Cycle

Some species use crustaceans as the first or only intermediate host, while some use piscine paratenic hosts. In addition, some cycles require a host to ensure transmission to a proper warm- or cold-blooded vertebrate definitive host. Larvae in fish are found in the internal organs rather than in the musculature; since few individuals consume the internal organs, the infection rate is low. Consumption of infected salmon by Alaskan Eskimos resulted in infection by *C. strumosum* and *Acanthocephalus rauschi* (Deardorff and Overstreet, 1991; Schmidt and Roberts, 1985). *Bulbosoma* sp., a thorny-headed worm, also infects humans and can cause serious intestinal problems. It has been found in American waters (Deardorff and Overstreet, 1991). Schmidt (1971) suggested a high rate of accidental parasitism from *Corynosoma strumosum* among Eskimos. The juveniles of several species of *Corynosoma* occur in salmon and can probably temporarily infect humans, although the transmission is insignificant due to the scarcity of juveniles and their location in viscera rather than the flesh (Margolis, 1977). Schmidt and Roberts (1985) concluded that acanthocephalans were more of a threat to wild and captive animals and less of a human health concern. There are other potential acanthocephalan infections in Americans, since fishes serve as paratenic hosts of species of *Bulbosoma*.

CONCLUSION

Other potential parasitic agents may also be transmitted to humans through fishes. As cuisines are expanded to new locales and fish products from different locations are available, new zoonotic infections can be expected. As United States seafood imports increase, the range of places where parasites are potentially harmful increases, and the potential public health problems also expand. There is concern that a parasite population may become established in the United States by introduction of one or more stages of that parasite within animals or humans that comprise its life cycle.

Human infection can also result from direct contact with uncooked, infected fish. Fishermen and seafood handlers may be exposed to parasitic infections either through direct skin contact (Harrel and Deardorff, 1990; Deardorff et al., 1986b) or accidental ingestion. Fishes in culture operations are prone to disease conditions through crowding. Many different parasites can thrive, some of which may pose public health risks. However, infections can be eliminated more readily than in the wild, with elimination of an essential host and effective treatment. For example, birds that feed on cultured fish may defecate into the water and serve as a source of parasites. Netting ponds can discourage the taking of stock and prevent parasite transmission; *Anisakis simplex*, *Diphyllobothrium*, and *N. salmincola* were found only in wild salmon when compared to pen-reared salmon (Deardorff and Kent, 1989). Distribution of information on parasitic diseases, and continued improvement in effective diagnosis and treatment, will be key factors in controlling and minimizing parasitic infections from seafood. Public awareness of the possible health risks that can result from consumption of raw and lightly cooked seafood is of critical importance. With the introduction of new species as food items and new methods of preparation, there is a need for constant monitoring of potential health risks.

REFERENCES

Adams, A.M., K.S. Miller, M.M. Wekell, and R.M. Dong. 1999. Survival of *Anisakis simplex* in microwave-processed arrowtooth flounder (*Atheresthes stomias*). *Journal of Food Protection*. 403-409.

Adams, A.M., K.D. Murrell, and J.H. Cross. 1997. Parasites of fish and risks to public health. *Review Scientific Technology*. **16**(2): 652-60.

Adams, K.O., D.L. Jungkind, E.J Bergquist, and C.W. Wirts. 1986. Intestinal fluke infections as a result of eating sushi. *American Journal of Clinical Pathology*. **86**:688-689.

Africa, C.M., W. de Leon, and E.Y. Garcia. 1937. Heterophyidiasis, V: ova in the spinal cord of man. *Philippine Journal of Sciences*. **62**:393-403.

Ahmed, F.E. 1991. *Seafood Safety*. Committee on Evaluation of the Safety of Fishery Products. National Academy Press. Washington, DC.

Alicata, J.E 1962. *Angiostrongylus cantonensis* (Nematoda: Mestastrongylidae) as a causative agent of eosinophilic meningoencephalitis of man in Hawaii and Tahiti. *Canadian Journal of Zoology*. **40**:5-8.

Alicata, J.E. and R.W Brown. 1962. Observations on the method of human infection with *Angiostrongylus cantonensis* in Tahiti. *Canadian Journal of Zoology*. **40**:755-760.

Alonso, A., A. Moreno-Ancillo, A. Daschner, and M.C. Lopez-Serrano. 1999. Dietary assessment in five cases of allergic reactions due to gastroallergic anisakiasis. *Allergy* **54**(5): 517-520.

Anderson, E., D.J. Gubler, K. Sorensen, J. Beddard, and L.R. Ash. 1986. First report of *Angiostrongylus cantonensis* in Puerto Rico. *Tropical Medicine and Hygiene*. **35** (2):319-322.

Andrews, L.S., M. Ahmedna, R.M. Grodner, J.A. Luzzo, P. S. Murano, E.A. Murano, R.M. Rao, S. Shane, and P.W. Wilson. 1998. Food preservation using ionizing radiation. *Reviews in Environmental Contamination and Toxicology* **154**:1-53.

Asaisha, K., C. Nishino, and H. Hyasha. 1989. Geographical distribution and epidemiology. In H. Ishikura and M. Namiki (eds.), *Gastric Anisakiasis in Japan*. Springer-Verlag, pp. 31-36. Berlin.

Bean, N.H., P.M. Griffin, J.S. Goulding, and C.B. Ivey. 1990. Food borne disease outbreaks, 5-year summary, 1983-1987. *Morbidity and Mortality Weekly Report CDC Surveillance Summary*. **39**(1): 15-57.

Barnes, R.D. 1980. *Invertebrate Zoology*. Fourth edition. W.B. Saunders Company. Philadelphia, PA.

Beaver, P.C., R.C. Jung, and E.W. Cupp. 1984. *Clinical Parasitology*, ninth edition. Lea and Febiger, Philadelphia, PA.

Bier, J.W. 1976. Experimental anisakiasis: cultivation and temperature tolerance determinations. *Journal of Milk and Food Technology*. **39**:132-137.

Bier, J.W., T. L. Deardorff, G.J. Jackson, and R. B. Raybourne. 1987. Human anisakiasis. In Zbigniew S. and Pawlowski (eds.), *Baillier's Clinical Tropical Medicine and Communicable Diseases*, Vol 2. Harcourt Brace Jovanovich, London. Pp. 723-733.

Bogitsh, B.J. and Cheng, T.C. 1990. *Human Parasitology*. Saunders College Publ., NY.

Bonnel, M.L., M. D. Pierson, and G. D. Farrens. 1983. Pinnipeds and sea otters of central and northern California, 1983: status, abundance and distribution. Part 1, Vol III. Investigators' Report, University of California-Santa Cruz, prepared for OCS Region, Minerals Management Service, Department of Interior. Washington, DC.

Brattery, J. 1989. The life cycle of sealworm, *Pseudoterranova decipiens*, in the North Atlantic. Abstract presented at Nematode Problems in North Atlantic Fish. A Workshop in Kiel, 3-4 April, ed. H. Moller, International Council for Exploration of the Sea. C.M. 1989/F6 Mariculture Committee.

Bryan, F.L. 1986. Seafood-transmitted infections and intoxications in recent years. In D.E. Dramer and J. Liston (eds.). *Seafood Quality Determination. Proceedings of an International Symposium Coordinated by the University of Alaska, November 10-14*, Elsevier Science Publishers, Amsterdam, The Netherlands.

Bunnag, D. and K. T.Harinasuta. 1989. Liver, lung and intestinal trematode disease. In Goldsmith and D. Heyneman (eds.), *Tropical Medicine and Parasitology*. San Mateo: Appleton and Lange, pp. 464-472.

Cali, A. and Owen, R.L. 1988. Microsporidiosis. In A. Balows, W.J., Jr., Hausler, E.H. Lennette, P. Halonen, F.A. Murphy, M. Ohashi, and A. Turano (eds.). *Laboratory Diagnosis of Infectious Diseases Principles and Practice, Vol. I: Bacterial, Mycotic, and Parasitic Diseases*, and Vol. II: *Viral, Rickettsial, and Chlamydial Diseases.*, pp. 929-950. Springer-Verlag, NY.

Campbell, B.G. and Little, M.D. 1988. The finding of *Angiostrongylus cantonensis* in rats in New Orleans. *American Journal of Tropical Medicine and Hygiene.* **38**:568-573.

Cheng, T.C. 1976. The natural history of anisakiasis in animals. *Journal of Milk and Food Technology.* **39**(1):32.

Cheng, T.C. 1983. Human parasites transmissible by seafood — and related problems. In. C.O. Chichester and H. Graham (eds.), *Microbial Safety of Fishery Products,* pp. 163-189. Academic Press, NY.

Cheng, T. 1986. *General Parasitology.* Academic Press Inc., NY.

Chitwood, M. 1970. Nematodes of medical significance found in market fish. *American Journal of Tropical Medicine and Hygiene.* **19**:599-602.

Cliver, D.O. 1990. Parasites. In *Foodborne Diseases,* D.O. Cliver (ed.). Pp. 293-305. Academic Press. San Diego, CA.

Choudbury, G.S.and C.G. Bublitz. 1994. Electromagnetic method for detection of parasites in fish. *Journal of Aquatic Food Production and Technology.* **3** (1):49-64.

Cooper, C.L., J.L. Crites, and D.J. Sprinkle-Fastzkie. 1978. Population biology and behavior of larval *Eustrongylides tubifex* (Nematoda: Dioctophymatida) in poikilothermous hosts. *Journal of Parasitology.* **64**:102-107.

Deardorff, T.L., J. Altman, C. M. Nolan. 1987. Human anisakiasis: two case reports from the State of Washington. *Proceedings of Helminthology Society.* **54**:274-275.

Deardorff, T.L., T. Fukumura, and R. B. Raybourne. 1986a. Invasive anisakiasis: a case report from Hawaii. *Gastroenterology.* **90**:1047-1050.

Deardorff, T.L., S.G. Kayes, and T. Fukumura. 1991. Human anisakiasis transmitted by marine food products. *Hawaii Medical Journal,* **50**(1): 9-16.

Deardorff, T.L., and M.L. Kent. 1989. Prevalence of larval *Anisakis simplex* in pen-reared and wild-caught salmon (Salmonidae) from Puget Sound, Washington. *Journal of Wildlife Diseases.* **25**:416-419.

Deardorff, T.L. and R.M Overstreet. 1991. Seafood-Transmitted Zoonoses in the United States: the Fishes, the Dishes, the Worms. In *Microbiology of Marine Food Products,* Ward, D.B. and C. Hackney (eds.), pp. 211-265. Van Nostrand Reinhold, NY.

Deardorff, T.L., R. M. Overstreet., M. Okihiro, and R. Tam. 1986b. Piscine adult nematode invading an open lesion in a human hand. *American Journal of Tropical Medicine and Hygiene.* **35**:827-830.

Deardorff, T.L. and R. Throm. 1988. Commercial fast-freezing of third-stage *Anisakis simplex* larvae encapsulated in salmon and rockfish. *Journal of Parasitology.* **74** (4):600-603.

de Carneri, J. and G. Vita 1973. Drugs used in cestode diseases. In Caviers, R. (ed.), *Chemotherapy of Helminthiasis, Vol. 1, Section 64. International Encyclopedia of Pharmacology and Therapeutics Series,* pp. 145-160. Pergamon Press. Elmsford, NY.

Deschiens, R., H. Collomb, and J. Demarchi. 1958. Distomastose celebrale a *Heterophyes.* In *Abstracts of the Sixth International Congress of Tropical Medicine and Malaria.*

Durborow, R.M. 1999. Health and safety concerns in fisheries and aquaculture. *Occupational Medicine* **14**(2): 373-406.

Eastburn, R.L., T.R. Fritsche, and C. A. Terhune Jr. 1987. Human infection with *Nanophyetus salmincola* from salmonid fishes. *American Journal of Tropical Medicine and Hygiene.* **36** (3):586-591.

Eberhard, M.L. 1989. Intestinal perforation caused by *Eustrongylides* in New Jersey. *American Journal of Tropical Medicine and Hygiene.*

FDA. 1987. Food preparation — raw, marinated or partially cooked fishery products. In "Retail Food Protection Program Information Manual," part 6, chapter 1, number 2-403. Center for Food Safety and Applied Nutrition, Retail Food Protection Branch.

Font, W.F., R. M. Overstreet, and R. W. Heard. 1984. Taxonomy and biology of *Phagicola nana* (Digenea: Heterophyidae). *Transactions of the American Microscopy Society.* **103**:408-422.

Garcia, L.S., and D.A. Bruckner. 1997. *Diagnostic Medical Parasitology,* ASM Press, pp. 1-937.

Gebhardt, G.A., R.E Millemann, S.E. Knapp, and P. A. Nyberg. 1966. Salmon poisoning disease. II. Second intermediate host susceptibility studies. *Journal of Parasitology.* **52**:54-59.

Gunby, P. 1982. One worm in the minnow equals too many in the gut. *Journal of the American Medical Association.* **248** (2):163.

Hafsteinsson, H. and S. S. H. Rizvi. 1987. A review of the sealworm problem: biology, implications and solutions. *Journal of Food Protection.* **50** (1):70-84.

Harrell, L.W. and T. L. Deardorff. 1990. Human nanophyetiasis: treatment by handling naturally infected coho salmon (*Oncorhynchus kisutch*). *Journal of Infectious Diseases.* **161**:146-148.

Hauck, A.K. 1977. Occurrence and survival of the larval nematode *Anisakis* sp. in the flesh of frozen, brined and smoked Pacific herring, *Clupea harengus pallasi. Journal of Parasitology.* **63** (3):515-519.

Healy, G.R. and G. Juranek. 1979. In *Food-Borne Infections and Intoxications,* H. Riemann and F. L. Bryan (eds.). 2nd ed. Academic Press, NY.

Higashi, G.I. 1985. Foodborne parasites transmitted to man from fish and other aquatic foods. *Food Technology.* **39**: 69-74.

Hill, W.E. 1996. The polymerase chain reaction: applications for the detection of food borne pathogens. *Critical Reviews in Food Science Nutrition,* **36**(1-2): 123-173.

Honans, R.E.S., and A. S. MacFarlene. 1957. Preliminary report on the occurrence of codworms in flounders of the maritimes and on the efficiency of present candling methods in fish plants. Pamphlet H. Inspection Branch, Department of Fisheries, Maritimes Area, Halifax, Canada.

Jackson, G.J., J. W. Bier, and T. L. Schwarz. 1990. More on making sushi safe (letter). *New England Journal of Medicine.* **322** (14):1011.

Kappus, K.K., D.D. Jaranek, and J.M. Roberts. 1991. Results of testing for intestinal parasites by state diagnostic laboratories, United States, 1987. *Morbidity and Mortality Weekly Reports CDC Surveillance Summary*, 40(4): 25-45.

Khalil, L.F. 1969. Larval nematodes in the herring (*Clupea harengus*) from British coastal waters and adjacent territories. *Journal of Marine Biological Association United Kingdom*. **49**:641-659.

Kean, B.H., and R. C. Breshlau. 1964. Cardiac heterophydiasis. In *Parasites of the Human Heart*, pp. 95-103. Grune and Stratton, NY.

Kliks, M.M. 1983. Anisakiasis in the western United States: four new case reports from California. *American Journal of Tropical Medicine and Hygiene*. **32**:536-532.

Kliks, M.M. and N. E. Palumbo. 1992. Eosinophilic meningitis beyond the Pacific Basin: the global dispersal of a peridomestic zoonosis caused by *Angiostrongylus cantonensis*, the nematode lungworm of rats. *Social Science and Medicine*. **34** (2):199-212.

Lehrer, S.B., W.E. Horner, and G. Reese. 1996. Why are some proteins allergenic? Implications for biotechnology. *Critical Reviews in Food Science Nutrition*, **36**(6): 553-564.

Leung, P.S., Y.C. Chen, D.L. Mykles, W.K. Chow, C.P. Li, and K.H. Chu. 1998. Molecular identification of the lobster muscle protein tropomosin as a seafood allergen. *Molecular Marine Biotechnology*, 7(1): 12-20.

Margolis, L. 1977. Public health aspects of "codworm" infection: a review. *Journal of the Fisheries Research Board of Canada*. **34**:887-898.

Markell, E.K, M. Voge, and D. T. John. 1992. *Medical Parasitology*. 7th Ed., pp. 5-21, 226-293. W.B. Saunders Co. Philadelphia, PA.

McGladdery, S.E. 1986. *Anisakis simplex* (Nematoda: Anisakidae) infection of the musculature and body cavity of Atlantic herring (*Clupea harengus*). *Canadian Journal of Fisheries and Aquatic Science*. **43**:1312-1317.

McKerrow, J.H., J. Sakananari, and T. L. Deardorff. 1988. Anisakiasis: revenge of the sushi parasite. *New England Journal of Medicine*. **319** (18):1228-1229.

Milleman, R.E. and S. E. Knapp. 1970. Spread of *Nanophyetus salmincola* and "salmon poisoning" disease. *Advances in Parasitology*. **8**:1-41.

Moreno-Ancillo, A., M.T. Caballero, R. Cabanas, J. Conteras, J.A. Martin-Barroso, P. Barranco, M.C. Lopez-Serrano. 1997. *Annals of Allergy, Asthma & Immunology*, 247-250.

Montoro, A., M.J. Perteguer, T. Chivato, R. Laguna, and Cuellar. 1997. Recidivous acute urticaria caused by *Anisakis simplex*. *Allergy (Copenhagen)*, 985-991.

Morikawa, A., M. Kato, K. Tokuyama, T. Kuroume, M. Minoshima, and S. Iwata. 1990. Anaphylaxis to grand keyhole limpet (abalone-like shellfish) and abalone. *Annals of Allergy*. **65**:415-417.

Myers, B.J. 1979. Anisdakine nematodes in fresh commercial fish from waters along the Washington, Oregon and California coasts. *Journal of Food Protection*. **42**:380-384.

Noble, E.R., G. A. Noble, G. A. Schad, and A. J. MacInnes. 1989. *Parasitology: The Biology of Animal Parasites*, 6th Ed., Lea and Febiger, Philadelphia, PA.

Olson, R.E. 1987. Marine fish parasites of public health importance. In *Seafood Quality Determination*. D.E. Kramer and J. Liston (eds.). Elsevier, The Netherlands, pp.339-355.

Olson, A.C., Jr., M. D. Lewis, and M. L. Hauser. 1983. Proper identification of anisakine worms. *American Society of Medical Technology*. **49**:111-114.

Oshima T. 1987. Anisakiasis — Is the sushi bar guilty? *Parasitology Today*. **3** (2):44-48.

Oshima T. and Kliks, M. 1986. Effects of marine parasites on human health. *International Journal of Parasitology*. **17**:415-421.

Paperna, I. and R. M. Overstreet. 1981. Parasites and diseases of mullets (Mugilidae). In: O.H. Oren (ed.), *Aquaculture of Grey Mullets*. International Biological Programme 26, pp. 411-493. Cambridge University Press, England.

Petithory, J.C., B. Paugam, P. Buyet-Rousset, and A. Paugam. 1990. *Anisakis simplex*, a co-factor of gastric cancer? *Lancet*. **336** (8721):1002.

Phillips, C.B. 1955. There's always something new under the "parasitological sun" (the unique story of helminth-borne salmon-poisoning disease). *Journal of Parasitology*. **41**:125-148.

Pinkus, G.S., C. Coolidge, and M. D. Little. 1975. Intestinal anisakiasis: first case report from North America. *American Journal of Medicine*. **59**:114-120.

Sakanari, J.A., M. Moser, and T. L. Deardorff. 1995. Fish Parasites and Human Health. Report No. T-CSGCP-034, pp. 1-27. California Sea Grant College, La Jolla, CA.

Sata, H., K. Harvo and K. Hanada. 1992. Five confirmed human cases of *Gnathostomiasis nipponica* recently found in northern Japan. *Journal of Parasitology*. **78** (6):1006-1010.

Schantz, P.M. 1989. The dangers of eating raw fish. *New England Journal of Medicine*. **320** (7):1143-1145.

Schantz, P.M., and M.H.J. Kramer. 1995. Larval cestode infections: cysticercosis and echinococcosis. *Current Opinion in Infectious Diseases*. **8**:342-350.

Schmidt, G.D. 1971. Acanthocephalan infections of man, with two new records. *Journal of Parasitology*. **57**:582-584.

Schmidt, G.D. and L. S. Roberts. 1985. In *Foundations of Parasitology*, 3rd Ed., C.V. Mosby College Publishing. St. Louis, MO.

Scott, D.M. and W. R. Martin. 1957. Variations in the incidence of larval nematodes in Atlantic cod fillets along the Southern Canadian mainland. *Journal of the Fisheries Research Board of Canada*. **14**:975-996.

Shirahama, M., K. Takafumi, I. Hiromi, V. Satoshi, O. Yoshiro, and S. Yuichiro. 1992. Intestinal anisakiasis: U.S. in Diagnosis. *Radiology*. **185**:789-793.

Shirazian, D., E. L. Schiller, C. A. Glaser, and S. L.Vanderfecht. 1984. Pathology of larval *Eustrongylides* in the rabbit. *Journal of Parasitology*. **70** (5):803-806.

Skrjabin, K.J. and W. P. Podjapolskaja. 1931. *Nanophyetus schikhobalowi*, n. sp., *ein neuer Trematode aus Darm des Menschen. Zlb. Bakt.* I. Orig. **119**:294-297.

Smith, J.W. 1984. The abundance of *Anisakis simplex* L3 in the body cavity and flesh of marine teleosts. *International Journal of Parasitology*. **14** (5):491-495.

Smith, J.W. and R. Wootten. 1975. Experimental studies on migration of *Anisakis* sp. Larvae (Nematode: Ascaridida) into the flesh of herring, *Clupea harengus* L. *International Journal of Parasitolology*. **5**:133-136.

Sugimachi, K., K. Inokuchi, T. Ooiwa, T. Fujino, and Y. Ishii. 1985. Acute gastric anisakiasis. *Journal of American Medical Association*. **253**:1012-1013.

Templeman, W., H. J. Squires, and A. M. Fleming. 1957. Nematodes in the fillets of cod and other fishes in Newfoundland and neighboring areas. *Journal of the Fisheries Research Board of Canada*. **14**:831-897.

Van Mameren, J., K. J. Van Spreekens, H. Houwing, and D. A. A. Mossell. 1969. Effect of irradiation on the keeping quality of packaged cod flesh. In *Freezing and Irradiation of Fish*. Fishing News Books, London.

Van Thiel, P., F. C. Kuipers, and R. T. Roskam. 1960. A nematode parasitic to herring, causing acute abdominal syndromes in man. *Tropical Geogr. Medicine*. **2**:97-111.

von Bonsdoroff, B. 1977. *Diphyllobothriasis in Man*. Academic Press, NY.

Welberry, A.E. and W. Pacetti. 1954. Intestinal fluke infestation in a native negro child. *Bulletin of the Dade County Medical Association*. Pp. 34-35.

Williams, H.H. and A. Jones. 1976. Marine helminths and human health. In CIH Miscellaneous Publications No. 3. Farnham Royal, Slough, U.K. Commonwealth Agricultural Bureaux, pp.1-47.

Wittner, M., J.W. Turner, G. Jacquette, L. R. Ash, M. P. Salgo, and H. B. Tanowitz. 1989. Eustrongylidiasis — a parasitic infection acquired by eating sushi. *New England Journal of Medicine*. **320** (17):1124-1126.

Yokogawa, M. 1965. Paragonimus and paragonimiasis. *Advances in Parasitology*. **3**:99-158.

Yokogawa, M. and H. Yoshimura. 1967. Clinicopathologic studies on larval anisakiasis in Japan. *American Journal of Tropical Medicine and Hygiene*. **16**:723-728.

Young, P.C. and D. Lowe. 1969. Larval nematodes from fish of the subfamily Anisakinae and gastrointestinal lesions in mammals. *Journal of Comparative Pathology*. **79**:301-313.

Zaman, V. and L. A. Keong. 1990. *Handbook of Medical Parasitology*. 2nd Edition, pp. 108-215. K.C. Ang. Publ. Pte. Ltd., Singapore.

Seafood Allergies and Intolerances

Patricia A. Fair

INTRODUCTION

There are two types of adverse food reactions: allergy and intolerance. Often, the term allergy causes confusion. A food allergy is defined by the American Academy of Allergy and Immunology (1984) as an "immunologic reaction resulting from the ingestion of a food or food additive." The term "immunologic" indicates that antibodies are produced in response to an allergen. "Food intolerance" refers to an abnormal physiologic response to an ingested food or food additive that is not proved to be immunologic in nature, including pharmacologic, metabolic, or toxic responses to food or food additives. There are many references on allergies and intolerance that include information relating to seafood (David, 1993; Husby et al., 1993; Metcalfe, 1991; Perkin, 1990; Brostoff and Challacombe, 1987; Breneman, 1984).

In general, food intolerances and allergies have become an increasing problem. Americans are experiencing increased exposure to possible food allergies as well as a decreasing tolerance to foods because of the tremendous increase in food varieties and food additives. The actual incidence of food allergy is unknown; however, it is estimated that 0.1 to 0.7 percent of the population have immunoglobulin E (IgE)-mediated food allergies (Farrell, 1988). In general, known food allergens have molecular weights between 10 and 70 kDa, stimulate the immune response by inducing the production of allergen-specific IgE, and are stable molecules that are resistant to processing, cooking, and digestion (Lehrer et al., 1996). Food reactions that are allergic in nature account for less than 20 percent of all reactions, and only one-half to one-third of these are proven in controlled, double-blind studies (Anderson, 1986). Food allergies and intolerance are more frequent in infants and children than adults. With most foods there is a tendency toward remission in later years of life (Sampson and Scanlon, 1989), with the exception of fish, which may increase in severity with age (Dannaeus and Inganas, 1981). Fish hypersensitivity is common in countries with high fish consumption (Sampson and Metcalfe, 1991); for example, an estimated one per 1000 Norwegians have fish allergies (Aas, 1987). Many factors — such as age, sex, heredity, stress, health, season, and nutrition — contribute to production of allergic disease and symptoms.

DIAGNOSIS

Diagnosis of adverse reactions to foods is a difficult, complicated process, beginning with a thorough evaluation of past history, followed by diagnostic tests such as skin testing, elimination diets, food challenges, and special tests for IgE (Metcalfe, 1984). True allergies are uncommon and can be misdiagnosed, and several means of diagnosis may be needed. Goldman (1963) established criteria still used as the standard for the diagnosis of food allergy/intolerance. It basically consists of eliminating the offending food, followed by a challenge, and then elimination again. There are no simple laboratory methods for diagnosing food allergy. Many methods are available, including skin tests, cytotoxic tests, food diaries, intracolonic topical test, radioallergosorbent tests for measuring IgE-associated antibody against a specific antigen — a method first found useful in confirming cod allergy (Aas and Lundkvist, 1973) — and enzyme-linked immunoabsorbent assay (ELISA) — a color method for detecting specific IgG4 and IgE, used to measure monoclonal antibodies reactive with soluble antigens. Lab procedures coupled with elimination diet techniques and food challenges can achieve a high degree of accuracy.

Most hypersensitivity reactions to foods tend to be IgE-mediated; positive hypersensitivity skin tests are useful predictive tests. Double-blind, placebo-controlled food challenges were performed on individuals with adverse reaction to shrimp (Daul et al., 1988). Food allergic reactions from shrimp have demonstrated type I, IgE-mediated immediate hypersensitivity reactions (Daul et al., 1988). Immunologic tests have limited value, and diet avoidance followed by specific food challenge confirms that specific foods cause clinical symptoms. Avoidance of specified items identified can be utilized to manage the problem.

SYMPTOMS

The most common clinical symptoms of food allergy involve the gastrointestinal tract (nausea, vomiting, diarrhea) and skin [urticaria (hives), eczema or atopic dermatitis, angioedema]; less observed are respiratory symptoms (asthma and rhinitis). Other symptoms include laryngeal edema, anaphylactic shock, hypotension, and headache. Urticaria and angioedema are acute early symptoms; gastrointestinal symptoms such as vomiting and diarrhea follow; eczema and bronchitis/asthma may also occur. In addition to these symptoms, food allergy has also been implicated in a variety of chronic disorders such as rheumatoid arthritis, irritable bowel syndrome, epilepsy, migraine, and fatigue-tension syndrome (Taylor, 1985).

Reactions between food proteins and antibodies (the specialized proteins produced in response to the antigen or food protein) can be serious. However, they are usually less severe. Food antigens are derived from animal and vegetable sources and are usually proteins or glycoproteins. Allergic reactions to food involve the production of IgE, which binds to certain cells called mast cells. Consumption of food or antigens causes the antigen to bind to IgE, triggering chemical reactions in the mast cells leading to clinical symptoms (Metcalfe, 1984; Atkins, 1983).

Immune mechanisms occurring in food allergy are of the type I or immediate hypersensitivity reaction, characterized by short onset times. Subsequent exposure to the allergen results in release of allergic mediators, including histamine. Some nonimmune-mediated intolerance reactions are termed anaphylactoid reactions since they look similar to food anaphylaxis. This reaction involves non-immune release or formation of chemical mediators as seen in histamine poisoning from ingestion of scombroid fish. Contamination of fish by decarboxylating bacteria and inadequate refrigeration are requisites for the release of large quantities of histamine, causing a chemical reaction. Anaphylaxis can develop within minutes to a few hours with both mild and more severe symptoms. Mild symptoms include: generalized skin itching (erythmia), urticaria, and also angioedema (swelling of face, hands, feet, or genitalia). Serious systemic reactions include: fever-like symptoms (sneezing, runny nose, swelling), asthma-like symptoms (wheezing, coughing, respiratory distress), and possible vascular shock, vomiting, diarrhea, and cardiovascular collapse and death (Settipane and Settipane, 1991). Cod is one of the most frequently causative seafood agents, but reactions have included haddock, sprat, halibut, plaice, mackerel, trout, salmon, bonito, tuna, sardines, bluefish, and mahi mahi (Taylor, 1989; Aas, 1966b). Symptoms usually appear within minutes or up to three hours after ingestion and include headache, flushing, and throbbing pain in the neck (Taylor, 1989; Finn, 1987).

Treatment of food allergies demands avoidance of foods containing the offending allergen. With seafood allergies, individuals are often advised to avoid all seafood species even though cross-reactions have not been reported among finfish, crustaceans, and mollusks (Taylor et al., 1986). However, significant cross-reactivity occurs within Crustacea, and individuals with sensitivities to shrimp, crab, or lobster should avoid all crustaceans (Waring et al., 1985). Pharmacologic agents, such as antihistamines, are useful in treating symptoms after exposure, but not in preventing the reaction.

SEAFOOD-RELATED ALLERGIES AND INTOLERANCES

Allergy to fish is not common worldwide, but most often occurs in fish-processing and fish-eating communities (Aas, 1984). Sensitive individuals may react similarly to foods of different groups or several foods within the same group. Sensitivity to shrimp may also be shown to other crustaceans, while others are only sensitive to shrimp. Allergenic foods may also contain more than one allergen. Sensitive persons may exhibit sensitivity to any or all allergens in a particular food. Cod, for example, has a major protein named allergen M but also contains a minor allergen which can also cause sensitivity. Codfish antigen studies have served as a model for allergic reactions, and the major allergens have been isolated and purified from cod (Essayed and Aas, 1971, 1969; Aas, 1966) and shrimp (Hoffman and Day, 1981). Fish allergens have yielded significant knowledge on the interaction between allergen and IgE antibodies. *Anisakis simplex* can cause allergic reactions in sensitized patients, even after the fish has been cooked, and some are related to acute parasitism such as gastroallergic anisakiasis. The antigens of the live parasite probably cause the allergic symptoms in gastroallergic anisakiasis, and patients with this disease can tolerate deep-frozen seafood in which the parasites are dead (Alonso et al., 1999). A cause of recidivous urticaria in 25 Spanish patients who had eaten fish or seafood revealed that *A. simplex* was the main cause in patients who eat fish and are not sensitized to it (Monero-Ancillo et al., 1997; Montoro et al., 1997). Allergen M in codfish was the first food allergen to be extensively characterized (Essayed et al., 1974; Aas, 1966b), and the major allergenic determinant has been identified and synthesized (Essayed et al., 1974). Antigenic differences have been found among various species of fish (Aas, 1987), although several species have antigens closely related to allergen M.

Allergens in fish are highly resistant to destruction from heat, acid, and enzymes. Therefore, processing and cooking does not impart less allergic properties. Lehrer et al. (1990) found that shrimp allergens were released into the water during boiling. Breast-fed infants can develop allergic symptoms after the mother has consumed fish, as shown in the codfish allergen (Aas, 1966). While most allergies result from eating the food, inhalation and contact can also produce some symptoms. Inhalation from steam-cooked fish or dust from dried fish has been shown to cause allergic asthma (Aas, 1984; Gaddie et al., 1980), and contact dermatitis has been observed in fish processors (Beck and Nissen, 1983).

Shellfish, a popular cuisine item in many areas, are frequently the cause of adverse reactions in sensitized individuals either from ingestion or occupational exposure. Some food antigens may be more potent than others. For example, shrimp sensitivity requires only 1 to 2 grams while most foods require 20 g (FDA, 1986). Patients with allergic reactions to crustaceans have common and possibly cross-reactive IgE-reactive epitopes in lobster and shrimp (Leung et al., 1998). Anaphylactic reactions to snail, mussel, oyster (Lehrer et al., 1987), and grand keyhole limpet (Carrillo et al. 1991; Morikawa et al., 1990) have been reported. Seafood factory workers exposed to aerosols from prawns have also developed respiratory symptoms and elevated IgE and IgG antibodies (McSharry, 1994).

Seafood intolerance is rare and most likely occurs in individuals with certain risk categories predisposed to other health complications. Much more prevalent are seafood allergies, which are difficult to diagnose and are distinguished as immunological reactions rather than an inability to digest. Both intolerance and allergies to seafood can be due to food additives, causing confusion with diagnosis. Additional studies are needed to characterize seafood allergies.

A National Academy of Sciences study on seafood safety (Ahmed, 1991) indicated that regulatory response was based on the proper labeling to distinguish seafood species and type, ingredients in formulated seafood (such as surimi), and processing (such as sulfites in shrimp to reduce melanosis). Treatment of shrimp with sulfite must be labeled if residual concentration exceeds 10 parts per million (ppm). The continued practice of sulfite use for melanosis in shrimp and the potential for causing allergic reactions is questionable (Lecos, 1985). Substitution of fish species could also present adverse effects in individuals subject to reactions.

Increased awareness of adverse reactions to seafood is needed at all levels. Seafood processors need to be aware of the potential for allergic reactions among seafood handlers and should employ reliable diagnosis of workers claiming allergies (Ahmed, 1991). Food products containing seafood require accurate labels describing ALL ingredients. Use of fish proteins may enrich the major allergenic fraction and pose a risk for sensitized individuals (Aas, 1987). Statements such as "may contain" one or more ingredients is a warning to sensitive consumers to avoid such products. The ingredients listed should also contain the presence of other potentially allergenic substances such as sulfites. Explicit food labels are needed to alert sensitive individuals to the presence of potentially hazardous substances. Finally, consumers need to be educated in terms of potential hazards, especially if they are sensitized or have experienced adverse reactions to seafood.

REFERENCES

Aas, K. 1966. Studies of hypersensitivity of fish. Allergological and serological differentiation between various species of fish. *International Archives Allergy and Applied Immunology*. **30**:257-267.

Aas, K. 1966b. Studies of hypersensitivity to fish. Characterization of different allergen extracts from fish with respect to content of protein antigens and allergenic activity. *International Archives of Allergy*. **30**:1-14.

Aas, K. 1984. Antigens in food. *Nutrition Review*. **42**:85-91.

Aas, K. 1987. Fish allergy and the codfish allergen model. In *Food Allergy and Intolerance*. J. Brostoff and S.J. Challacombe (eds.), Balliere-Tindall, pp. 356-366. London.

Aas, K. and S. Essayed. 1969. Characterization of a major allergen (cod). Effect of enzymic hydrolysis on the allergenic activity. *Journal of Allergy*. **44**:333.

Aas, K. and U. Lundkvist, 1973. Radioallergosorbent test with purified allergen from codfish. *Clinical Allergy*. **3**:255-261.

Ahmed, F.E. 1991. *Seafood Safety*. Committee on Evaluation of the Safety of Fishery Products. National Academy Press. Washington, DC.

Alonso, A., A. Moreno-Ancillo, A. Daschner, and M.C. Lopez-Serrano. 1999. Dietary assessment in five cases of allergic reactions due to gastroallergic anisakiasis. *Allergy* 54(5): 517-520.

American Academy of Allergy and Immunology. 1984. Adverse Reactions to Foods. Committee on Adverse Reactions to Foods. National Institutes of Health, NIH Publication No. 84-2442.

Anderson, J.S. 1986. The establishment of common language concerning adverse reactions to foods and food additives. *Journal of Allergy and Clinical Immunology*. **78**(1):140-144.

Atkins, F.M. 1983. The basis of immediate hypersensitivity reactions to foods. *Nutrition Review*. **41**.

Beck, H.I. and B.K. Nissen. 1983. Contact urticaria to commercial fish in atopic persons. *Acta Dermatovenereology* **63**:257.

Breneman, J.C. 1984. *Basics of Food Allergy*. 2nd Edition. Charles C. Thomas Publisher, Springfield, IL.

Carrillo, T., F.R. DeCastro, M. Cuevas, J. Caminero, and P. Cabrera. 1991. Allergy to limpet. *Allergy*. **46**:515-519.

Dannaeus, A. and M. Inganas. 1981. A follow-up study of children with food allergy. Clinical course in relation to serum IgE- and IgG- antibody levels to milk, egg and fish. *Clinical Allergy*. **11**:533-539.

Daul, C.B., J.E. Morgan, J. Hughes, and S.B Lehrer. 1988. Provocation-challenge studies in shrimp-sensitive individuals. *Journal of Allergy and Clinical Immunology*. **81**(6): 1180-1186.

David, T.J. 1993. Food and Food Intolerance in Childhood. Blackwell Scientific Publications, London.

Essayed, S.M., H.V. Bahr-Lindstrom, and H. Bennich. 1974. The primary structure of fragment TM2 of allergen M from cod. *Scandinavian Journal of Immunology*. **3**:3313-3320.

Essayed, S. and K. Aas. 1971. Characterization of a major allergen (cod). Observations on effect of denaturation on the allergenic activity. *Journal of Allergy*. **47**:17.

Farrell, M. K. 1988. Food Allergy. In *Manual of Allergy and Immunology*, 2nd ed. G. Lawlor and T. Fishers (eds.), Little, Brown. Boston.

FDA Ad Hoc Committee on Hypersensitivity to Food Constituents. 1986. Report. U.S. Food and Drug Administration. Washington, DC.

Finn, R. Pharmacologic actions in foods. 1987. In *Food Allergy and Intolerance*. J. Brostoff, S. J. Challacombe (eds.), Bailliere-Tindall/WB Saunders, pp. 425-430. London.

Gaddie, J., J. D.W. Anderson, W.A. Sellers, S. Saperstein, W.T. Kniker, and S.R. Halpern. 1963. Milk allergy, I: Oral challenge with milk and isolated milk proteins in allergic children. *Pediatrics*. **32**:425-443.

Hoffman, D.R. and E.D. Day. 1981. The major heat stable allergen of shrimp. *Annuals of Allergy*. **47**:17.

Husby, S., S. Halken, and A. Host. 1993. Food Allergy. In *Nutrition and Immunology. Human Nutrition. A Comprehensive Treatise*. D.M. Klurfeld (ed.). Plenum Press, pp. 25-41. NY.

Lecos, C. 1985. Reacting to sulfites. *FDA Consumer*. **19**:22-25.

Lehrer, S.B. and M.L. McCants. 1987. Reactivity of IgE antibodies with crustacea and oyster allergens: Evidence for common antigenic structures. *Journal of Allergy and Clinical Immunology*. **80**(2):133-139.

Lehrer, S.B., D. Ibanez, M.L. McCants, C.B, Daul, and J.E. Morgan. 1990. Characterization of water-soluble shrimp allergens released during boiling. *Journal of Allergy and Clinical Immunology*. **85**(6):1005-1013.

Leung, P.S., Y.C. Chen, D.L. Mykles, W.K. Chow, C.P. Li, and K.H. Chu. 1998. Molecular identification of the lobster muscle protein tropomosin as a seafood allergen. *Molecular Marine Biotechnology*, **7**(1): 12-20.

McSharry, C., K. Anderson, I.C. McKay, M.J. Colloff, C. Feyerabend, and R.B. Wilson. 1994. The IgE and IgG antibody responses to aerosols of *Nephrops norvegicus* (prawn) antigens: the association with clinical hypersensitivity and with cigarette smoking. *Clinical Experimental Immunology*. **97**:499-504.

Metcalfe, D.D. 1984. Diagnostic procedures for immunologically-mediated food sensitivity. *Nutrition Reviews*. **42**(3):92-97.

Metcalfe, D.D. 1991. Review, in *Current Opinions in Immunology* **6**:881-886.

Monero-Ancillo, A., M.T. Caballero, R. Cabanas, J. Conteras, J.A. Martin-Barroso, P. Barranco, M.C. Lopez-Serrano. 1997. *Annals of Allergy, Asthma & Immunology*, 247-250.

Montoro, A., M.J. Perteguer, T. Chivato, R. Laguna, and Cuellar. 1997. Recidivous acute urticaria caused by *Anisakis simplex*. *Allergy (Copenhagen)*, 985-991.

Morikawa, A., M. Kato, K. Tokuyama, T. Kuroume, M. Minoshima, and S. Iwata. 1990. Anaphylaxis to grand keyhole limpet (abalone-like shellfish) and abalone. *Annals of Allergy.* **65**:415-417.

Perkin, J.E. 1990. *Food Allergies and Adverse Reactions.* Aspen Publications, Gaithersburg, MD.

Sampson, H.A. and S.M. Scanlon. 1989. Natural history of food hypersensitivity in children with atopic dermatitis. *Journal of Pediatrics.* **115**:23-27.

Sampson, H.A. and D.D. Metcalfe. 1991. Immediate reactions to foods. In D.D. Metcalfe, H.A. Sampson, and R.A Simon (eds.), *Food Allergy and Adverse Reactions to Foods and Food Additives.* Blackwell-Scientific Publishing, Boston.

Settipane, R.A. and G.A. Settipane. 1991. Anaphylaxis and Food Allergy. In *Food Allergy and Adverse Reactions to Foods and Food Additives.* D.D. Metcalfe, H.A. Sampson, and R.A Simon (eds.), Blackwell-Scientific Publishing, Boston.

Taylor, S.L. 1985. Food allergies. *Food Technology.* **39**:98-105.

Taylor, S.L., R.K. Bush, and W.W. Busse. 1986. Avoidance diets — How selective should we be? *New England Reg. Allergy Proceedings.* **7**:527-532.

Taylor, S.L, J.E. Stratton, and J.A. Nordlee. 1989. Histamine poisoning (scombroid fish poisoning): An allergy-like intoxication. *Journal of Toxicology and Clinical Toxicology.* **27**(4-5):225-240.

Waring, N.P., C. B. Daul, R. D. deShazo, M. L. McCants, and S. B. Lehrer. 1985. Hypersensitivity reactions to ingested crustacea: clinical evaluation and diagnostic studies in shrimp-sensitive individuals. *Journal of Allergy and Clinical Immunology.* **76**:440-445.

Section VII. Consumer Acceptance

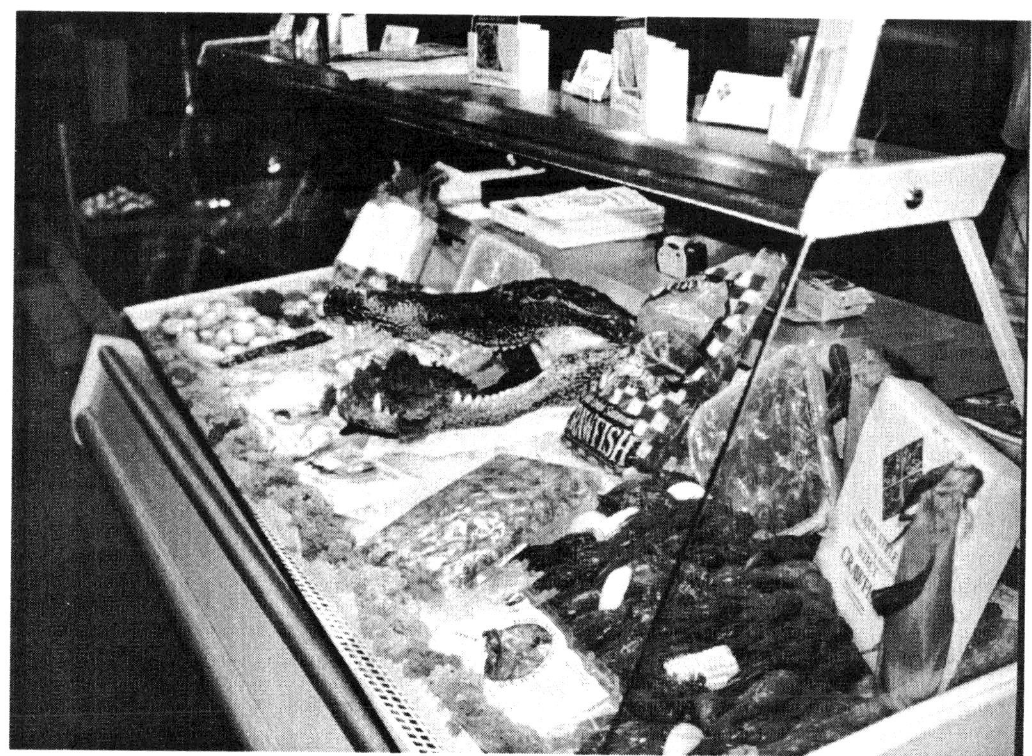

Marketing Fish and Shellfish

Charles W. Coale Jr.

INTRODUCTION

There are significant differences between the marine and freshwater fisheries in customer acceptance, merchandising, product selection, promotion, product form, species availability, and cost. Accordingly, marketing methods for the marine and freshwater categories differ. Freshwater and marine species have shared market visibility since the days of Captain John Smith, and now the arrival of aquaculture methods and biotechnology offer additional promise for product supply to the market. Problems and opportunities will be discussed and components identified that make up a successful marketing plan.

Profitability from marketing edible fish and shellfish requires a well-planned and well-executed marketing program. Fishery statistics are a vital component in understanding the market. To that end, this chapter will cover not only factors that impact profitability, but also various marketing strategies. During 1998, Americans consumed about 14.9 pounds of fish and shellfish on a per capita basis [United States Department of Commerce (USDC), 1999]. In recent years, the United States news media have devoted considerable attention to broadcasting information about the declining stocks of natural fish species in the marine environment (USDC, 1993). Foreign trade has contributed imports and domestic aquaculture marketing has provided commercial fish and shellfish to fill these product deficits.

How big is the seafood (edible supplies) market? During 1997, American retail consumer expenditures for fishery products amounted to $46.5 billion, of which $31.3 billion resulted from the food service industry and about $14.8 billion from the retail food industry (USDC, 1998). The catch from recreational fishing and consumption also adds to the market statistics.

Foreign trade (importation) supplies a portion of these consumer products from continents including North and South America, Europe, Asia, Africa, Australia, and the countries of Oceania. For 1998, the estimated volume of imported products amounted to some 3.6 billion pounds, with a value of about $8.2 billion (USDC, 1999). Shrimp was the largest volume of product imported, representing about 700 million pounds valued at $3.1 billion (USDC, 1999).

The United States domestic catch for 1998 amounted to about 9.2 billion pounds, with a value of $3.1 billion (USDC, 1999). For aquaculture, the principal domestic products include catfish, trout, tilapia, and salmon. For 1997, catfish production totaled 524 million pounds, with an industry value of $372 million [United States Department of Agriculture (USDA), 1998]. The largest volumes of product came from the catfish, trout, and salmon industries. For 1998, marine food product exports amounted to about 1.7 billion pounds, with a value of about $2.3 billion (USDC, 1999). Annually, watermen harvest a significant tonnage of ocean and tributary fish for industrial and recreational purposes. These harvest pressures reduce the biomass for a commercial harvest.

The food sector for edible fish and shellfish is dynamic, but competes in the muscle (protein) food group of consumer purchases. Consumption of substitute muscle foods on a per capita basis provides a challenge to fish and shellfish marketers. Given the 1997 per capita consumption data, domestic consumers ate 190.3 pounds of red meat, poultry, and seafood. Of this quantity, consumption included 64.8 pounds of poultry, 63.8 pounds of beef, 45.6 pounds of pork, 0.8 pounds of lamb, and 0.9 pounds of veal (Putnam and Allshouse, 1999). Consumption of fish and shellfish has remained fairly stable during the past decade. Consumption data for 1997 show that fish and shellfish products make up about 7.6 percent (14.9 lbs.) of annual muscle food intake. Other inedible fish and aqua-plant products have significant value in the market, but will not be discussed in this chapter.

FLOW OF FISH AND SHELLFISH IN MARKETS

Who supplies the fish and shellfish market? Fish and shellfish move from domestic and worldwide producers to domestic and international consumers (Figure 1). The conceptual design of Figure 1 was based on data cited in the introduction. Many different products from aquaculture and marine sources make up the international edible food market. The flow of aquaculture and marine products has many entry and exit points in the marketing channel. The diagram in Figure 1 shows alternative marketing institutions and channels. Figure 1 consists of two or more vertical channels and horizontal entry and exit points for product flow. As products flow through a marketing channel, they generate costs, returns, and profit. Additional handlers in a market channel mean more costs and more pressure on profit margins.

There are many markets and options for buyers and sellers in a given fish and shellfish market. Successful marketing strategies focus on three ideas: (1) niche markets, (2) product development and resource utilization, and (3) channel management alternatives. Such strategies for selecting a species, market form, and channel depend on product characteristics and its market demand.

MARKET SITUATION FOR FISH AND SHELLFISH

Market Organization

Fish and shellfish products flow from producers to consumers (Figure 1). The market is international in scope. The production sector in the fish and shellfish industry organizes itself by commodity groups, such as watermen specializing in harvesting crabs, oysters, clams, finfish in pound nets, etc. For many years watermen had a traditional practice of harvesting marine resources on a seasonal basis. Similarly, an aquaculture producer usually produces one species in a pond, raceway, or recirculating system. A large number of supply firms have developed specialty processes and markets, and stock retail-oriented markets.

The market for fish and shellfish is highly competitive, resulting from several factors. A large number of buyers and sellers buy in the market, with freedom of entry into the industry; in most cases the market is served with *undifferentiated* products. Price is difficult to influence in a market outlet because of the undifferentiated product base. The producers' and marketers' success in the marketplace is based on controlling costs.

Those enterprising persons wishing to enter or expand the fish and shellfish market should have a well-developed plan. A written plan should answer certain fundamental questions that should guide a proprietor into a successful venture: What is the consumer demand for species? What product and services should be produced? What is the product supply? What products are available and will compete? What marketing strategies and channels should be used? A case study illustrating sunshine bass marketing (see later in chapter) will pinpoint these questions.

Market Niches

Physical and economic variables characterize a niche market that describes the consumer behavior in a market. The equations of niche market variables include price and quantity of goods on the demand side, and cost(s) of inputs for production, marketing, and distribution. Cost budgets for each economic segment contribute to the total costs of a market. For a market to be profitable, revenues generated by price and market volume must cover total costs. Price levels in the niche market gauge the perception of value of a product and its likely substitutes. Producers will furnish quantities of goods that they feel will provide them a profit.

Marketers create niches for all fish and shellfish. Fish and shellfish flow directly to consumers or through vendors to consumers (Figure 1, number 1). For instance, a sports fisherman may catch five large bluefish and sell them to his neighbors. He is using a direct marketing channel for his fish. In Figure 1, the chart shows a recreational fisherman (lower right-hand side) as a recreational supplier and his neighbors as consumers. In a more complicated arrangement, a commercial seafood marketer may buy a boatload of fish from a marine harvester. The marketer is also a fish processor and distributor. He may ship one-half of the boatload to a retail supermarket chain and one-half to a food service firm. In a recreational case, the niche market consists of one transaction.

In another channel, a commercial marketer's niche market consists of up to five transactions and many physical movements in a market channel (see Figure 1, number 2). Entrepreneurs create

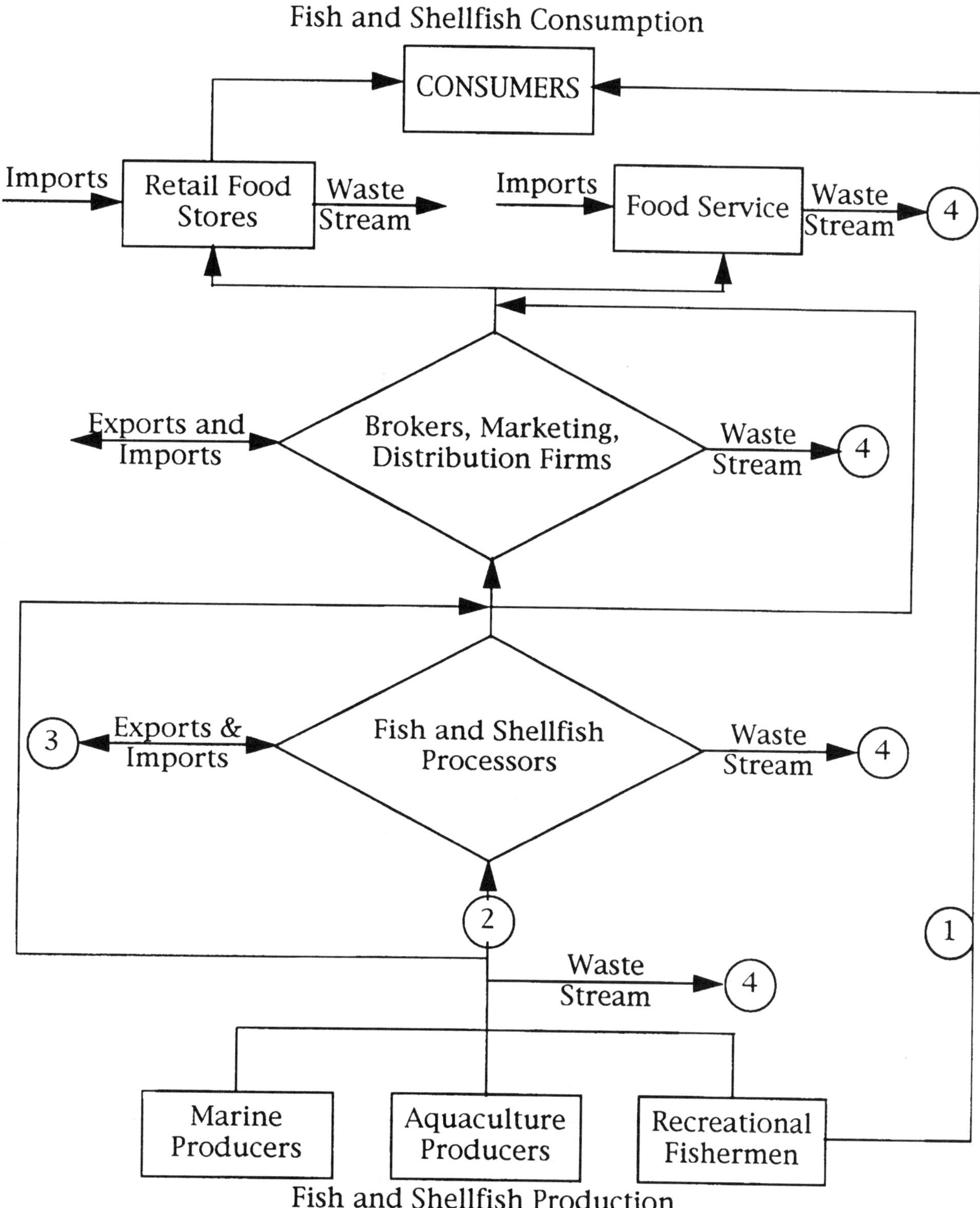

Figure 1. Market flows for fish and shellfish, 1992.

a market niche at different sector levels by supplying the market. In the market niche, economic transactions, communications, and customer behavior occur. For fish and shellfish, note the market channel niches for local, regional, national, and international products (see Figure 1, number 3, for an international focus).

Marketing Strategy

The marketing strategy focuses on a planning process for achieving a successful marketing campaign based on a profitable outcome. As portrayed in Table 1, the marketing manager has at least four concerns in developing a marketing strategy. He must keep the price and cost in mind when budgeting the marketing program. Price will fall into a range based on factors described earlier.

Three significant cost factors include the market form, the market channel, and the geographical location of the market. Budgeting methods bring a marketing strategy into focus for an enterprise. For an example, a producer might develop a market for a product in a fresh form, marketed directly to consumers, in a local geographical market area (Table 1). The producer must estimate a price level to cover the costs involved in product development, production, preparation, packaging, promotion for the market channel, and distribution. With market channel distribution, the product price level may be set in a medium price-range. The consumers' willingness to accept the price offered will depend on other factors of demand, such as the price of substitute products, and consumers' incomes, tastes, and preferences. The budget is a good tool for defining all related expenses. If the cost budget shows that $1,000.00 is required to produce 500 pounds of product, then the "going-out" price for the producer must equal $2.00 per pound just to break even. The budget tool should account for all costs, if the exercise is to have any meaning.

Market Channel

A market channel guides the flow of products from producer to consumer; information flows from consumer to producer (Figure 1). The importance of Figure l is to remind the market participants that many different routes exist in the seafood market from consumer to producer. The product, its characteristics, and price provide the stimulus for the continued flow of products. The marketing channel provides consumer-producer access for different fish species and product forms. A retail marketing outlet may provide customers access to about 30 to 40 different fish and shellfish species. Specifically, customers who patronize a retail outlet may prefer fillet products, while customers planning a crab feast may prefer unpicked cooked crabs. The retail market outlet and supporting channel illustrates that a given marketing channel is a market niche. The market niche indicates similar characteristics for product price and revenue, and shows the costs of production, marketing, and distribution.

Market Promotion

Market promotion communicates the value and benefits of the products and services incorporated into the product mix. Marketers promote fish by either branded or generic methods. For the fish and shellfish industry, promotion has tended toward the generic in its support of the marine industry. The market channel will determine the type of promotion based on the customer base and other factors.

Table 1. Selected Marketing Strategy Variables for Fish and Shellfish

Price	Market form	Market channel	Place
Low	Fresh	Direct	Local
Medium	Frozen	Food Service	Regional
High	Processed	Retail/Wholesale	National
	Prepared	Military	International

(Modified from information provided by Paul Lynn and Dr. James Daniels, Mrs. Paul's Kitchens, Campbell's Soup Inc., 1990.)

Market Reconnaissance

One of the best ways to execute a successful marketing program is to understand customer wants and information from the market, and provide desired products in the marketplace. How can all the economic variables be described and quantified? Market reconnaissance takes many forms. Marketers collect market intelligence data by various methods, including observing market outlet pricing, product characteristics, marketing channel attributes, and studying marketing and economic statistics.

Once collected, what does the economic data mean? A spreadsheet is a convenient way to display data and track market opportunities. A market development spreadsheet — called a marketing reconnaissance spreadsheet — structures and analyzes data.

This marketing tool describes the product form, unit price, quantity, and value relationships in a marketing channel (Coale, Spittle, and McDowell, 1990). Given a price estimate for sunshine bass, the spreadsheet simulates a market niche by calculating prices, costs, and product values for the marketing channel, including product distribution, processing, and assembly (Virginia Agriculture Experiment Station Bulletin 93-3, 1993). Once calculated, the spreadsheet illustrates a residual value to the producer (Table 2). In the example, the retail marketing channel shows the impact of each sector: production, physical distribution, processing, and retailing. The data in Table 2 show the impact of product shrinkage in the marketing channel. The physical distribution connects the producing and processing sectors, and the processing and retailing sectors. The retail market manager sets the product price based on the quantity and price of competing products.

Table 2. Marketing Reconnaissance Spreadsheet*
Commodity = Whole Gutted Sunshine Bass

8	A Product Flow Section	B Product Form	C Unit Price	D Quantity (pounds)	E Value (dollars)
9	RETAILING				
10	Price of Retail	Gutted Fish	$5.99	11,333	$67,887
11	Price from Processing	Gutted Fish	3.73	11,333	42,259
16	PROCESSING				
17	Price of Processed Product	Gutted Fish	3.73	11,333	42,259
18	Price after distribution	Gutted Fish	3.16	12,222	38,167
19	Shrinkage %	.07		889	
21	Variable Costs		.33		
23	Cost/Pound (Processing)	—	.33	—	—
25	PHYSICAL DISTRIBUTION				
26	Cost to Processing	Fish	3.12	12,222	38,167
27	Price at the Farm	Fish	3.06	12,500	37,375
28	Shrinkage %	.02		278	
30	Physical Distribution				
31	Cost/Pound (Distribution)	—	.06	—	—
33	PRODUCTION				
34	Residual Price at Production	Fish	2.99	12,500	$37,375

Based on Retail Gross Margin of 37.75 percent.
*Calculated values based on assumptions: 20-foot refrigerated truck (fully loaded, 12,500 pounds), on an average route of 300 miles, fuel at $1.20/gallon, and processing at $0.30/pound.
Note: Commodity is market-driven and price is set at retail. The producer is the residual price claimant.

Scouting the Market

The market reconnaissance spreadsheet for whole-gutted fish illustrates the application of the marketing data in a case study analysis, with data coming from a market research study (Table 2) (Coale, 1993). The food retailer establishes a price for the fish marketed based on the marketing conditions, and/or prices of competing fish, shellfish, and other muscle foods. Although a single producer cannot influence the market, he can estimate the expected values in the channel, and select the best marketing alternatives. The researchers traced the flow of product from retail unit back to production; because of the structure in a food-marketing economy, the producer accepts the price and quantity demanded from the market.

PRODUCT DEVELOPMENT AND RESOURCE UTILIZATION

Product development and resource utilization focus on getting the most value from the resources of edible fish or shellfish produced. The key to profitability in product marketing is cost recovery from both muscle tissue and waste products. The composition of fish or shellfish includes edible muscle tissue, inedible byproducts, and — from aquaculture — two waste streams, solid and liquid. The price, species of fish, market form, and market channel determine the value of the edible muscle tissue and the byproducts; these variables determine whether a sustainable industry exists.

Examine the Market

Entrepreneurs considering a fishery or aquaculture venture should develop a strategy and prepare a market reconnaissance spreadsheet. There are many marketing development combinations that involve fish species, market forms, and market channels. Two examples illustrate the point. One: a producer markets tilapia directly to customers. The customer buys the tilapia as whole fish and disposes of the inedible byproduct. While the producer does not incur a cost of disposal, he also loses any potential value of the byproducts. Two: a producer markets directly to the local consumer, but changes the market form of the fish into fillets before the sale. The producer sells the edible muscle tissue and retains the value of the byproduct. Depending on the byproduct options, the producer may incur revenue from the byproduct or may incur a cost of disposal. Here, the important issue is to define a value of the waste streams. Depending on market form and market channel selection, the value of the whole fish at the producer level will vary based on the marketing niche pursued.

The business owner must compare alternatives and develop a program, with objective seafood standards and merchandising variables factored into the market channel, and learn to maximize the qualities in alternative products. For example, *The Seafood Handbook* offers insights into merchandising tilapia, which has gained popularity in the seafood market (Straus, 1991). The many marketing opportunities for tilapia include: live, fresh, frozen, smoked, or combined with further processing for a value-added product. One variable in the market niche is market form. Fillet form adds to the value. For those in the fillet market, yield is very important. A fillet yield for tilapia (skin-on) amounts to about 39 percent of the whole fish weight. If the value of whole tilapia is $1.00 per pound, its fillet (skin-on) is worth $2.56 per pound. Reduced yields result if the marketer sells the same tilapia fillet with the skin off. At a 35 percent yield, the skinless fillet is worth $2.86. The difference in value between skin-off and skin-on fillets is $0.30 per pound. Aquacultured fish with high production costs need a marketing strategy to maximize the fish value to justify a higher price. The market reconnaissance spreadsheet helps with comparisons among species, product or market forms, and distribution channels.

Product Form

Customers have different needs for specific product forms of fish and shellfish. Marketable fish begin as a food commodity (live fish). Marketers change the form to meet customer needs. In a market channel, vendors may add value several times before a consumer buys it.

The edible muscle tissue generates revenues, while inedible carcass parts (scales, fins, etc.) generate cost at harvest. Preliminary estimates of the yield of the edible muscle tissue for hybrid striped bass (one-pound fish) were 24 to 26 percent of carcass weight (Coale, 1993). Currently, fish byproducts described have a considerably lower economic value, if any. The waste products described in the recirculating system produce liquid and solid waste from the production process. Any economic analysis must value joint products in the total plan. Research to account for all resources in a product will determine if the production is feasible.

Product Shrinkage

Product shrinkage has an impact on market channel profit (Table 2). Product shrinkage results from two sources: further processing and after processing. During the marketing process, further processing modifies the edible fish form from a whole round fish to a product form desired for each fish. Recent studies show that a whole-gutted fish will retain about 90 percent of its original weight, while fish fillets from the same fish species will retain only 25 to 30 percent of its whole weight. The handling practices and sanitary conditions in a market channel influence the shelf life.

A procedure for evaluating a marketable product is as follows: The spreadsheet entries calculated the value of waste tissue (shrinkage in head, guts, etc.) as the product moved through a marketing channel. Shrinkage associated with product alteration changes the product value as the product moves from handler to handler in the channel. As a result of the shrinkage calculations, estimates of the product value show the estimated value of the fish to the production unit.

Product Utilization Economics

The economics of product utilization plays an important role in industry survival. (Discussions of the waste stream profile and analysis originated with Dr. Paul Hoepner and the author during the development of a recirculating aquaculture system research proposal.) In particular, fishery economics deals with resource utilization questions throughout an aquaculture system. Fishery production systems generate four principal products. They are: (1) live fish, which include edible muscle tissue, (2) fish byproducts, (3) liquid waste, and (4) solid waste (see Figure 1, number 4, for waste stream flows). On face value, edible muscle tissue generates revenues, while fish byproducts (liquid and solid wastes) generate costs. Managers must give careful consideration to finding value in these fish byproducts. Yield data of aquacultured fish has shown a high cost associated with marketing only the edible muscle tissue and discarding related waste products.

Whole Fish and Shellfish

Aquaculture systems produce live fish or shellfish. In this process, the system generates liquid and solid waste products as the fish grow. Live fish and shellfish have value in the market based upon their utilization in a variety of market forms. Ethnic customers demand tilapia species in a live market form. Rockfish, a cross used in hybrid striped bass, are in demand in a filleted form. Consumers demand shellfish on the half-shell and shucked. These market forms and their corresponding market value target the economic value of the species produced and utilization of the product in the market. In aquaculture, the fish life cycle generates liquid and solid waste products, occurring on a continual basis. The more dense the biomass, the greater the waste streams.

Fish Byproducts

Gutted fish and filleted products will have residual byproducts (non-edible). To measure, the byproduct tissue is weighed and classified. Byproducts of fish will remain in batches corresponding to the gutted-and-filleted samples. The byproduct composition will be determined for gutted-and-filleted hybrid striped bass products.

Liquid Waste

The composition of liquid waste from the production cycle can be classified and measured. The liquid waste management will occur in the system and its impact will be accounted for in a separate arrangement.

Solid Waste

The composition of the solid waste generated from the production cycle will also be classified and measured. The solid waste is accounted for in the management of the system. As the biomass grows, the solid waste problem intensifies.

Data from whole fish divided into edible muscle tissue and fish byproduct define the stream of product resources. Among those resources, a value can be attached to each component of the resource stream.

CHANNEL MANAGEMENT ALTERNATIVES

Marketing Alternatives

A marketer of fish or shellfish has a number of choices in marketing his products. Those choices depend on resource base, reputation, unique processes, or products. Figure 1 shows many fish marketing opportunities for marketers who plan ahead. *In choosing a merchandising strategy, a seafood marketer needs information, so that he can target a market niche.* For example, the scenario includes high-price product, prepared and ready-to-eat, sold directly to customers in a local market (Table 1). Establishing a budget for such a niche market will show the level of price needed to cover the cost services provided in the market form, the channel, and the geographical distribution pattern. Additional refinements are made to the specific market.

If a fish or shellfish producer focuses his attention on the market channel, he has four options in Table 1. By selecting a direct market channel, he probably will market in a local geographical area to local customers. He has several scenarios available for local customers depending on his location (Lacey et al., 1990).

Marketing Scenarios

Fish and shellfish marketers have a variety of direct marketing outlets. Aquaculture gives the producer harvest control of the resource. Live aquacultured fish and shellfish provide for flexibility in timing the harvest. By preplanning, a marketer has several options available. *As shown in Table 1, developing a strategy enables the marketer to target customers and markets.* Cultivating a local market provides an understanding of customer needs. In a case study developed, direct marketing may have as many as five different scenarios. The composition of the marketing program depends on customer demands and the products and services available. Each scenario is defined and includes expected prices and price ranges. Setting prices comes from experience in the local area, prices of substitutes, and an effective marketing plan.

Scenarios Supporting Direct Marketing

- Scenario 1. Marketing live fish to customers by "fee-fishing" privileges at pondside
- Scenario 2. Marketing fresh fish to customers at pondside
- Scenario 3. Marketing live fish by fee fishing at pondside, and fresh fish at pondside
- Scenario 4. Marketing fresh fish to civic organizations
- Scenario 5. Marketing fresh fish to civic organizations, and prepared fish to catered events

The five scenarios detailed for direct marketers provide a steppingstone sequence for moving from a single enterprise to a very diversified marketing plan. A direct fish or shellfish marketer might begin with fee-fishing sales at pondside, while coordinating production and marketing decisions. When Scenario 1 reaches market capacity, a second scenario may be added to boost revenue and utilize the resources of the business. Customers may not wish to "catch" their fish but may want to buy dressed fresh fish at the pondside. By combining scenarios 1 and 2 into 3, the direct marketer has a combination of fee-fishing and sales at pondside. On a different tack, a marketer could supply fish to civic organizations for benefit dinners. The civic organization could procure the products directly and sponsor dinners prepared by organizations. Adding a twist to organizational sales, a marketer could offer a catered fish dinner program to organizations.

Scenarios Supporting Commercial Marketing

- Scenario 6. Marketing whole fish to restaurant chains
- Scenario 7. Marketing processed fish to restaurant chains
- Scenario 8. Marketing whole fish to commercial processors

Marketing fish or shellfish in the commercial channel provides a different scope of operation. The commercial market usually demands larger volumes of product. The scale of operation is much more demanding, and moving into a commercial market presents more risk.

A feasible solution of any of the eight scenarios depends on the market niche created. Developing a budget and projecting the costs and returns will direct the fish or shellfish marketer in the direction to make a profit. The scenario and scenario combination budgets overlaying the niche market will provide information for the proper direction for market development.

SUMMARY

Profitable marketing of fish and shellfish is a constant challenge to the seafood industry. Although the domestic population has increased slightly over the past ten years, the per capita consumption of seafood and seafood products has remained constant at about 15.0 pounds per year. During the decade of the 1990s, domestic landings increased slightly from 1990 to1995 but had declined by the end of the decade to about the same level as in 1990. During 1997, about 122 million metric tons were produced worldwide using both aquaculture and commercial production methods. The import and export markets remain dynamic and active. During 1997, the top seven fishery commodity groups worldwide had an import value of $56 billion and an export value of about $51 billion.

Aquaculture production accounted for about 20 percent of the total harvest of seafood. Scientists continue to refine aquaculture production methods and expect aquaculture to play an important future role in the supply of fish and shellfish products. During 1997, two of the top five valued species were shrimp and salmon. These two species have potential for being cultured by aquaculture methods. Other fish (trout and catfish) and shellfish products will be developed based on the market demand and technology available, to create a verycompetitive industry.

Seafood industry managers have many options in developing market niches for fish and shellfish. By setting up a marketing framework as illustrated in this chapter, marketing managers can assess the market and evaluate their profit options. Competition will continue to become more intense from the global marketing of fish and shellfish. By comparing alternative marketing strategies and by estimating revenues and costs of predetermined actions, marketing managers can make profitable decisions. The case study method illustrated in this chapter shows the value of using a budgeting process along with other spreadsheet tools for analysis. In general, seafood marketing operations are much more complicated than those illustrated in the case study. But the ideas and marketing methods illustrated provide an overview of the fish and shellfish marketing process, illustrate marketing tools for evaluating marketing channel costs, and identify some of the options that will most benefit producers. The seafood market is dynamic. Good market research and statistics, analysis, and program implementation are the keys to survival.

REFERENCES

Based on research data contained in *Marketing Aquaculture Products: A Retail Market Case Study for Sunshine Bass.* Bulletin 903-3. Virginia Agricultural Experiment Station, Blacksburg, VA.

Carroll Trosclair & Associates Inc. *Water Farming Journal.* **9:** 7 May-June 1994. Metairie, LA.

Coale, Charles W. Jr. et al. 1993. *Marketing Aquaculture Products: A Retail Market Case Study for Sunshine Bass.* Bulletin 93-3. Virginia Agricultural Experiment Station, Blacksburg, VA.

Coale, Charles W. Jr., Gerald D. Spittle, and George R. McDowell. Design of original market reconnaissance model during 1990. Department of Agricultural and Applied Economics, Virginia Tech, Blacksburg, VA.

Putnam, Judith Jones and Jane E. Allshouse. 1999. *Food Consumption, Prices, and Expenditures, 1970-97.* Food and Rural Economics Division, Economic Research Service, U.S. Department of Agriculture. Statistical Bulletin No. 965.

Hoepner, Paul and Charles W. Coale Jr. Discussions of the waste stream profile and analysis during the development of a recirculating aquaculture system research proposal.

Lacey, Patricia F., H. Russell Perkinson, and Charles W. Coale Jr. June 1990. *Financial Planning and Analysis for Aquaculture Enterprises.* Virginia Cooperative Extension and Virginia Sea Grant Program, SP90-11. Department of Agricultural Economics, Virginia Tech, Blacksburg, VA.

Livestock and Poultry Situation and Outlook. 1993.

NBC News. *Today Show.* Friday, August 5, 1994.

Showalter, O. Franklin. June 1994. *Aquaculture Potentials for the East Coast.* AGEC 4974, Independent Study. Charles W. Coale Jr. (instructor).

Straus, Karen (ed.). *The Seafood Handbook: Seafood Standards.* Seafood Business Magazine: Rockland, ME. 1991.

United States Department of Agriculture. *Aquaculture: Situation and Outlook Report.* Economic Research Service. AQS-12. March 1994.

United States Department of Commerce. 1999. Fisheries of the United States 1998. USDC/NOAA/NMFS Publication. Current Fishery Statistics No. 9800.

United States Department of Commerce. 1993; 1992. *Fisheries of the United States.* National Marine Fisheries Service. Current Fishery Statistics, No. 9200.

Cookery

Laurie M. Dean

INTRODUCTION

"I would eat more fish, but I really don't know how to prepare it."

"I couldn't find Yucatan Scamp like the recipe called for, so we had hamburgers again instead."

"I don't know how to tell if fish or shellfish is fresh."

"Is buying shrimp with the heads on really a better buy than buying headless shrimp?"

"When eating boiled crawfish, are you really supposed to suck the heads?"

These are just a few of the concerns that consumers have about the purchase, preparation, and consumption of the vast array of seafood available on the market today. The purpose of this chapter is to explain the basics of seafood handling and preparation and to give the consumer some easy and tasty recipe suggestions for use with the species covered. Because literally hundreds of comprehensive volumes have been written about seafood cookery, it is the intent of the author only to pique the interest of the potential seafood cook who, it is hoped, will research the myriad of seafood cookbooks available in local libraries and bookstores around the country. It is not the intent of the author to make this chapter anything other than a text covering the absolute basics of buying and cooking the more commonly available species of finfish and shellfish.

The chapter is divided into: (A) buying guides and cookery methods appropriate for finfish; (B) buying guides and cookery methods appropriate for a variety of shellfish; (C) serving suggestions.

FINFISH

Much of the fish sold in the markets today is cleaned and processed into dressed fish, fillets, and steaks, or even further processed into frozen or canned products. Many fish markets do, however, carry fresh whole fish. Fresh whole or dressed fish (scales, entrails, and usually head, tail, and fins removed) have the following characteristics: (1) firm flesh, (2) fresh, mild odor, (3) bright, clear eyes that protrude from the head, (4) red gills that are free of slime, and (5) shiny, unfaded skin color. Fresh fillets and steaks should have flesh that ap-

Fishery products prepared under approved, sanitary conditions.

Seafood products that are uniform in size, are free of blemishes and defects, are in excellent condition, and possess good flavor and all the characteristics of the species.

Figure 1. Seafood Quality standards.

pears fresh-cut, with no browning or signs of drying out, and should have a fresh, mild odor (Figure 2). If encased in a wrap, the wrap should be moisture- and vapor-proof, with no air space between the wrap and the fish.

To help you decide how much fresh fish to buy per serving, use the following as a guide:

whole fish: ¾ pound per serving (Figure 3)
dressed or pan-dressed : ½ pound per serving
fillets, steaks, or other fresh portions such as chunks (Figure 4): 1/3 pound per serving

Frozen fish should be solidly frozen when purchased. There should be no dry or discolored spots on the flesh and the flesh should have little or no odor. The packaging of the fish should be moisture-proof (to prevent drying out), and vapor-proof (to prevent rancidity). If it appears that the fish has been previously thawed and then refrozen, avoid buying it, because the quality (especially the texture) will be inferior.

Finally, when buying fish, remember that often one species or form of finfish can be substituted for another in a given recipe. You should, however, keep in mind that white-meated fish are usually milder in flavor than fish that have a higher proportion of dark muscle meat. (The dark fish muscle contains a higher proportion of enzymes and fat than the lighter fish muscle, and this helps to account for the stronger flavor.) Safe rules of thumb to follow are that white-meated

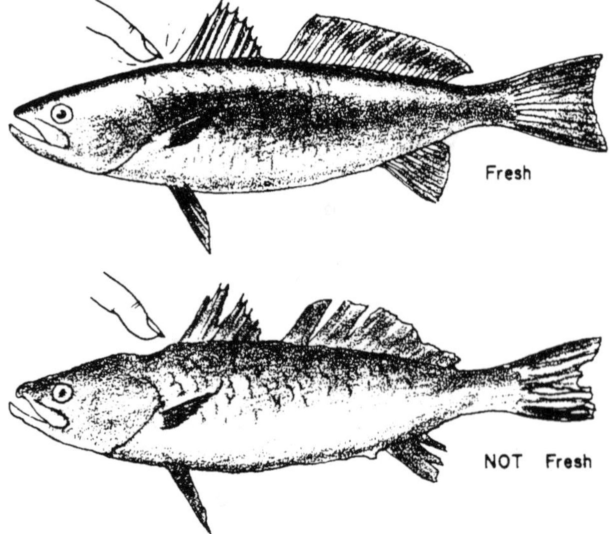

Figure 2. Check points for choosing a fresh fish: 1–Luster; 2–Eyes; 3–Gills, color and odor; 4–Odor; 5–Flesh; 6–Cavity; 7–Vent.

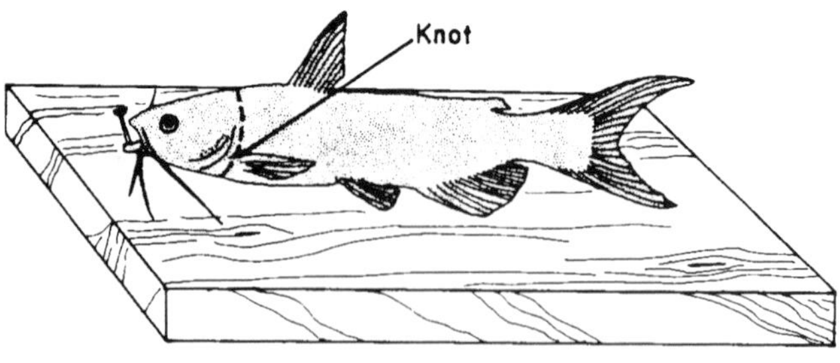

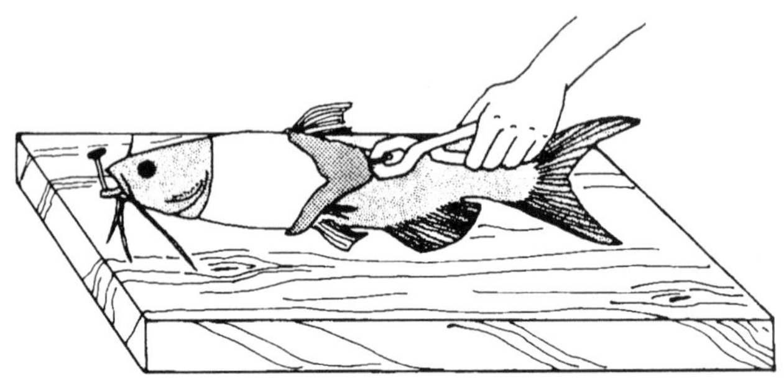

Figure 3. Skinning a catfish.

fish can be easily substituted for other white-meated fish, and dark-meated fish substituted for other dark-meated fish in most recipes. Additionally, dark-meated fish tend to be better when prepared by a method using dry heat (baking, broiling, grilling), as opposed to moist heat cooking (soups, stews, poaching). Some common examples of white-meated or lean fish include haddock, pollock, cod, flounder, halibut, sole, snapper, croaker, spot, drum, sea bass, sea trout, tilefish, monkfish, and mahi-mahi. Common examples of dark-meated or fatty fish are swordfish, salmon, eel, mackerel, tuna, bluefish, mullet, pompano, sturgeon, shad, herring, and smelt.

STOCKS, SOUPS, STEWS, AND CHOWDERS

Seafood soups go by many names, but all make hearty main dishes or appetizer portions that are high in nutritive value and appetite appeal. As an added bonus, they also stretch the seafood dollar.

Seafood soups range in variety from the simplest and most economical of fish stocks to a more complicated bouillabaisse or gumbo. All of these, however, when properly prepared, are gourmet fare.

The most basic of all seafood soups is fish stock, made by cooking fish trimmings with selected vegetables, spices, and usually wine or vin-

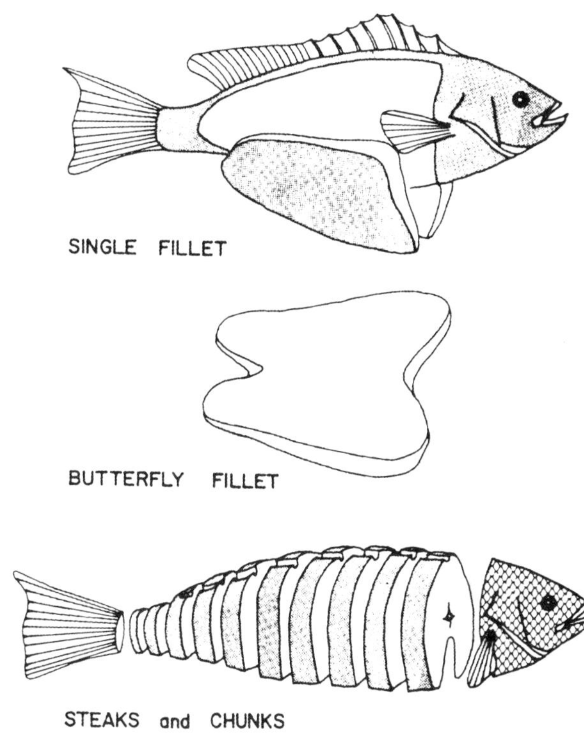

Figure 4. Types of cuts.

egar. After straining, this clear, thin liquid is used as a "base" for many soups and sauces.

Chowders are usually a mixture of finfish or shellfish, vegetables, and spices, with either a tomato or a cream base. Chowder recipes often call for fish stock as a basic ingredient.

A bisque is a rich cream-based soup made with shellfish and spices.

Creole bouillabaisse and gumbo are hearty seafood stews made of a combination of seafoods and vegetables, usually made with a roux of oil and flour. The roux for gumbo is generally dark chocolate-brown, whereas the roux for bouillabaisse is light brown.

Unlike meat stews which require long cooking times to tenderize the meat, seafood soups and stews require a relatively short time for preparation. The broth and vegetables are cooked for 30 minutes to an hour to blend flavors. The seafood is added at the end of the cooking period.

When selecting finfish for a soup, take into account the following guidelines:

1. Use a firm-textured fish that will hold its shape.
2. Unless the recipe specifically calls for them, avoid fish high in fat, because they tend to have a stronger flavor and fall apart easily.
3. Remove the skin and dark muscle underneath for a milder flavor.
4. Remove all bones before adding fish to the soup. A pair of tweezers works wonders for those "hard-to-get" bones.
5. Add the seafood toward the end of the cooking period.

To make a basic rich stock, save the trimmings from the fish. The following recipe is a good one.

Fumet or Fish Stock

3 to 4 tablespoons margarine
½ cup white wine or 2 tablespoons white wine vinegar
½ cup shallots or onions
2 cups cold water
¼ cup finely chopped carrot
bouquet garni (parsley, thyme, bay leaf tied in cheesecloth)
½ cup finely chopped celery
½ to 1 pound washed fish bones, heads, tails, and skins
5 to 6 peppercorns
3 to 4 cloves

Melt margarine in stock pot or large saucepan. Sauté onions, carrots, and celery until onions are translucent but not brown (about 4 to 5 minutes). Add the remaining ingredients, except peppercorns; bring to a simmer and cook uncovered for about 20 to 30 minutes. Add peppercorns for the last few minutes of cooking. Cooking the stock longer will result in a bitter taste developing from the bones.

Strain the stock through wet cheesecloth and use for sauces and soups. This stock will keep for several days covered in the refrigerator or for a month or so when frozen. *(Source: Dean)*

Hearty Fish Gumbo

1 pound skinless fish fillets, fresh or frozen
¼ cup margarine or butter, melted
1 cup chopped onions
1 cup thinly sliced celery
¾ cup chopped green pepper
1 tablespoon finely chopped parsley
1 clove garlic, peeled and minced
1 tablespoon all-purpose flour

1½ teaspoons chili powder
1½ teaspoons salt
1 teaspoon paprika
1/8 teaspoon cayenne pepper
1 can (1 pound) tomato wedges, or whole tomatoes, undrained
1 can (8 ounces) tomato sauce
½ cup water
1 package (10 ounces) frozen whole okra, thawed
2 cups cooked rice
chopped parsley (garnish)

Thaw fish if frozen. Cut fish into 1½ inch pieces. In a 5-quart Dutch oven, sauté onions, celery, green pepper, parsley, and garlic in margarine until tender but not brown, stirring occasionally. Combine flour, chili powder, salt, paprika, and cayenne pepper. Stir into vegetables. Add tomato wedges, tomato sauce, and water; simmer 4 to 6 minutes. Add fish and okra; reduce heat; cover and simmer for 10 to 15 minutes or until fish flakes easily when tested with a fork and okra is done. Serve over cooked rice. Garnish with parsley. Makes 6 servings. *(Source: Dean)*

POACHING

Poaching is cooking fish in just enough liquid to cover it. Poached fish can be served hot with a sauce or garnish, or can be flaked and then used in salads or casseroles. Although there are beautiful fish poachers on the market, they tend to be expensive and are not absolutely necessary for poaching. Fish can be poached in almost any shallow, wide pan (such as a large frying pan) that has a lid. Suitable fish for poaching are firm-textured and on the lean side, although salmon is a notable exception. High-fat fish, such as herring and mackerel, and soft-fleshed fish are also not good choices.

Poaching liquids are a matter of taste. You can use anything from water with salt added; to a mixture of water, aromatic vegetables, and wine or wine vinegar (called a court bouillon); to a mixture of water and milk with selected spices added. You also can use a fish or shellfish stock such as the one described in the previous section. Any number of court bouillon recipes can be found in fish cookbooks.

No matter what liquid you choose for poaching, do not overcook the fish! The standard time for cooking fish by most methods is 10 minutes for each inch of thickness, with the fish being measured in the thickest part. Timing begins when the liquid reaches a simmer. Also be careful that the poaching medium remains at the simmering point and does not boil. Check the liquid frequently by lifting the lid. Larger, thicker fish should be started in cold liquid to prevent the outer areas from drastically overcooking before the interior of the fish is done. Test for doneness by flaking with a fork.

If you have chosen to poach a whole fish, be sure to remove the skin immediately after you remove the fish from the cooking liquid. Small-to-medium-sized fish may be wrapped in cheesecloth to help them hold their shape. The remaining court bouillon may be used as a base for a sauce to serve with your fish.

Poached Fillet of Fish

Make a *Court Bouillon:*
1 cup water
1 cup dry white wine
a few celery leaves
1 small onion, sliced
4 to 5 peppercorns
1 bay leaf
2 sprigs parsley
1 slice lemon

Combine all of above ingredients in a heavy skillet. Simmer for 10 minutes.

Use 1 pound fresh or frozen lean fish fillets or steaks. (Frozen fish fillets are very good cooked in this manner.) Allow to thaw, if necessary, then place 2 or 3 fillets in the simmering broth. Cook only until the fish becomes opaque. Do not overcook. Remove to warm serving platter and repeat until all fillets have been poached.

Strain and measure sauce. If liquid is more than ¾ cup, return to skillet and boil rapidly to reduce that amount. Use this to make sauce.
(Source: Dean)

White Wine Sauce

¾ cup fish broth
¼ cup dry white wine
1 cup cream
4 tablespoons flour
4 tablespoons butter
salt, white pepper

Melt butter in heavy saucepan, add flour, and cook until bubbling. Remove from heat and add liquids all at once. Return to heat and cook over high heat, stirring constantly until sauce is thickened. Correct seasonings and pour over fish which has been thoroughly drained. *(Source: Dean)*

STEAMING

Any fish suitable for poaching is also suitable for steaming. The difference between steaming and poaching is that the fish is cooked above, and not in, the liquid. A variety of equipment can be used for steaming: traditional steamers are convenient and easy to use, but depending on the size of the fish, you can also use Chinese bamboo steamers (use a plate to hold the fish), inexpensive collapsible metal steam baskets, or even a heatproof platter set on empty cans in a large boiler. Take care that the vessel chosen has a tight-fitting lid to maintain the steam and, if using a Chinese steamer or a platter set in a boiler, be sure that there is sufficient space around the edge of the plate or platter to allow steam to circulate.

The liquid used for steaming can be plain water, salted water, water with herbs or spices added, or water with wine added. As the water boils, the steam with its added flavorings will add to the delicate flavor of the fish. If you are steaming fish on a plate or platter, herbs or other seasonings can be added to a sauce applied directly to the fish, so that the fish steams in a seasoned liquid. Whatever the liquid medium you choose, you must first bring the liquid in the covered pot to a boil. When the pot is full of steam, add your fish.

As with other forms of fish cookery, steam fish for 10 minutes for each inch of thickness, with the fish being measured in the thickest part. The fish is done when it flakes easily when tested with a fork. Steamed fish can be served as is, with a sauce or other garnish, or flaked and used in salads or casseroles.

Steamed Fish

1½ pounds fish fillets, steaks, or pan-dressed fish, fresh or frozen
1 quart boiling water
1½ teaspoons salt

Thaw frozen fish. Place fish in a well-greased steamer insert pan. Sprinkle fish with salt. Cover and cook over boiling water for 5 to 10 minutes or until fish flakes easily when tested with a fork. Cool. Remove skin and bones. Makes 2 cups cooked fish.

[Source: United States Government Printing Office (GPO), "Let's Cook Fish!"]

Fish Salad Supreme

2 pounds poached fish
2 tablespoons finely chopped onion
2 tablespoons finely chopped green pepper
½ cup salad oil
¼ cup wine vinegar
1/8 teaspoon garlic powder
¼ teaspoon oregano leaves
2 teaspoons parsley flakes
salt and pepper, to taste
¼ cup sour cream

Cut fish into very small pieces. Put into a bowl and add all ingredients except sour cream; toss gently. Refrigerate until thoroughly chilled. Just before serving, stir in sour cream. Makes 4 cups salad. *[Source: Maryland Seafood Marketing Authority (SMA)]*

Variations of steaming are cooking fish in paper ("en papillote") and cooking fish enclosed within a dough. For these variations, the fish may be partially cooked beforehand or may even be canned. The fish is not steamed over boiling water, but actually continues to cook itself in the steam formed within the covering.

Fillet of Fish in Parchment

Preheat oven to 350°F.

4 to 6 fresh or partially thawed lean fish fillets
3 tablespoons olive oil
1 onion, finely chopped
2 teaspoons chopped parsley
salt and pepper, to taste
1 to 2 tablespoons butter for greasing paper
1½ cup butter for sauce
1½ cups finely chopped cooked mushrooms
juice of 1 lemon
3 egg yolks, well-beaten
parchment paper

Place fillets in baking dish and coat with olive oil. Sprinkle with onion, parsley, salt, and pepper. Cook fish in 350°F oven until it is opaque on the bottom and translucent on the top. Reduce heat to 325°F. Add wine, cover, and bake for 5 to 10 minutes longer.

Meanwhile, cut 4 to 6 squares of the parchment paper and butter them on one side. (If kitchen parchment paper is unavailable, foil may be used.) Scrape the onion off each fillet and lay one fillet on each piece of parchment. Close the paper around the fish, leaving one flap open. Into the pan in which the fillets were cooked put butter (for sauce), mushrooms, lemon juice, and egg yolks. Stir well to blend and then simmer slowly, stirring occasionally, until the sauce is thick. Put 1 to 2 tablespoons of sauce on each fillet and then close the last flap on each envelope. Heat for 10 to 15 minutes in a 350°F oven. Serve in the envelope. *(Source: Dean)*

Fish en Croute

Preheat oven to 425°F.

1 recipe of a short-crust pastry dough -- or for a richer crust, use a brioche dough
2 lean fish fillets (each no more than 1/2 inch thick)
Stuffing of your choice: crab stuffing, duxelles (below), or a savory vegetable-and-herb stuffing are good choices.
1 egg white with 1 to 2 tablespoons cold water added

Roll out dough into 2 rectangles each about ¼ inch thick. Rectangles should be larger than the fillets by about 2 inches all around. Place first rectangle on a greased baking dish, top with one fillet, spread about ½ inch of topping on the fillet, and place the second fillet on top, sandwich-style. Measure height of 2 fillets with stuffing included. Lay the second pastry rectangle over the fish, seal the dough all around, brush with the egg-white wash, and bake for 15 minutes at 425°F. Reduce the heat to 350°F and cook for 10 minutes for each inch of thickness. *(Source: Dean)*

Duxelles

½ pound mushrooms, finely chopped
3 shallots, finely chopped
2 tablespoons butter
salt and pepper, to taste

Place mushrooms in a cheesecloth square, draw up the sides, and squeeze as much juice as possible out of them. Lightly sauté shallots in butter, then add mushrooms and cook slowly until the moisture has evaporated and the mixture is dark. *(Source: Dean)*

BAKING

Baking is one of the simplest methods of cooking both fat and lean fish. It involves placing fish, covered or uncovered, in a preheated moderate (usually 350°F) oven for a time equivalent to 10 minutes per inch of thickness, measured at the thickest part. Any form of fish —whole, dressed, fillets, or steaks — can be cooked this way. For best flavor, and when possible, leave the head on the fish when baking. Lean fish should be basted with melted fat or oil to prevent drying.

Baked Stuffed Fish

Preheat oven to 350°F.

1, 3- or 4-pound whole fish, scaled, gilled and gutted
½ lemon cut into rounds
½ onion cut into thin rounds
1 recipe crab stuffing or other stuffing of your choice
melted butter *(Source: Dean)*

Crab Stuffing

½ pound crab meat (preferably claw meat)
¼ cup Italian-seasoned bread crumbs
about 2 tablespoons mayonnaise
¼ teaspoon salt
pinch of freshly-ground black pepper
½ teaspoon Worcestershire sauce
½ teaspoon dry mustard

Scrub the interior of the fish to remove any traces of blood. Remove the cartilage from the crab meat and mix with the remainder of the ingredients. Lightly sprinkle the interior of the fish with salt and pepper. Loosely spoon the crab meat into the fish cavity (if necessary, truss the belly flaps with thin cotton string). Make diagonal ½ inch-deep slashes on the top side of the fish. Slice 1 lemon and 1 medium onion in thin rounds. Place ½ of a lemon slice and ½ of an onion slice in each slash. Brush with melted butter and bake at 350°F for 10 minutes per inch of thickness of fish measured at the thickest part. *(Source: Dean)*

Lighthouse Fish Roll-ups

1½ pounds skinless thin fish fillets
½ pound Maryland crab meat
½ cup dry bread crumbs
2 (10½ ounce) cans Newburg sauce (or 2cans condensed cream of mushroom soup)

salt, for sprinkling
lemon-and-pepper seasoning, for sprinkling
paprika, for sprinkling
¼ cup (½ stick) margarine or butter

Wash and dry fish thoroughly. Cut into about 3-inch strips. Remove cartilage from crab meat.

In a small bowl, mix crab meat, bread crumbs, and ½ cup Newburg sauce (or soup). Divide stuffing among fish by placing lengthwise along each fillet. Roll up firmly and place, cut-side down, on foil-lined baking pan. Sprinkle fish rolls with salt, lemon-and-pepper seasoning, and paprika. Put margarine or butter over top of each fish. Bake at 350°F until fish flakes easily when tested with a fork, about 20 to 30 minutes. Heat remaining Newburg sauce (or soup) and serve with fish. Makes 6 servings. *(Source: Maryland SMA)*

A variation of baking is planking. To plank a fish, place the seasoned fish on a preheated oiled plank or well-greased "bake-and-serve" platter (about 18x13 inches). Brush the fish with fat and bake at 350°F until fish flakes easily when tested in the thickest part. Planked fish is usually served surrounded by vegetables. Around parts of the Chesapeake Bay in Virginia, annual "Shad-Plankings" are timed to coincide with yearly spawning migrations, and are virtually synonymous with outdoor political gatherings and fundraisers.

Planked Pan-Dressed Fish

3 pounds pan-dressed fish, fresh or frozen
2 tablespoons melted fat or oil
2 tablespoons lemon juice
1½ teaspoons salt
½ teaspoons paprika
dash of pepper
seasoned hot mashed potatoes
seasoned hot cooked vegetables (asparagus, broccoli, carrots, cauliflower, onions, peas, or tomatoes)

Thaw frozen fish. Clean, wash, and dry fish. Place fish in a single layer on a preheated, oiled plank or well-greased bake-and-serve platter, 18x13 inches. Combine remaining ingredients and mix well. Brush fish with sauce. Bake in a moderate oven, 350°F, for 25 to 30 minutes or until the fish flake easily when tested with a fork. Remove from oven and arrange the hot mashed potatoes and two or more hot vegetables around the fish. Makes 6 servings.

(Source: U.S. GPO, "Let's Cook Fish!")

BROILING

Broiling is another type of dry heat cooking method; but, in broiling, the heat is intense and direct. Appropriate fish for broiling are lean or fatty fish that are about one inch thick, whether processed in dressed, fillet, or steak form. Thinner fish tend to dry out during broiling. The broiler pan should be greased and the fish basted with fat both before placing them under the broiler and at least once during cooking. Position the fish 3-4 inches from the broiler and cook until the fish flakes easily when tested with a fork. Unless the fish is quite thick, it should not be necessary to turn the fish during cooking.

Spicy Broiled Fish

2 pounds fish fillets or steaks
¼ cup (½ stick) margarine or butter
½ teaspoon prepared mustard
½ teaspoon salt
¼ teaspoon lemon-and pepper seasoning
¼ teaspoon seafood seasoning
1/8 teaspoon tarragon
1/8 teaspoon rosemary
½ cup dry white wine (or water)

Wash and dry fish thoroughly. Place in single layer, skin-side down, on foil-lined shallow baking pan. Melt margarine or butter in a small pan. Add rest of ingredients and cook over low heat until seasonings are blended and mixture is warm.

Pour ½ of sauce over fish. Broil, about 4 inches from source of heat, for 5 minutes. Pour remaining sauce over fish. Broil until fish flakes easily when tested with a fork, about 5 minutes more. Be careful not to overcook. Remove fish to serving platter and pour pan juices over fish. Makes 6 servings.

(Source: Maryland SMA)

GRILLING

One of the easiest and most enjoyable methods of cooking fish is to grill it. Whether you use a charcoal, gas, or electric grill, the results of grilling fish are wonderful -- if you keep in mind a few tips. First, be sure that the cuts are an inch or more in thickness whether you use dressed fish, fillets, or steaks. Thinner cuts will dry out more easily. Second, allow about ¾ pound whole fish, ½ pound of pan-dressed fish, or 1/3 pound of fillets or steaks per person. Frozen fish may be used, but should be thawed completely in the refrigerator (18 to 20 hours per pound) before grilling.

There are long-handled, hinged wire grills on the market that are perfect for grilling fish. Some of these are somewhat oval-shaped and are designed for cooking whole fish, while others are square and can be used for other cuts. Both should be greased well before cooking. If you do not have one of these, then grease the rack of your grill thoroughly. You will find that leaving the scales on the fish (if using dressed, pan-dressed, or fillets) will make removing the fish from the grill much easier.

Once you have taken care of the grill, the next thing you need is a good basting sauce which contains some type of fat. This can be a simple sauce of butter and herbs, commercial Italian dressing, or a barbecue sauce such as the one that is contained in the recipe given below. Baste the fish frequently to prevent drying. After basting, place the fish about four inches from the coals and cook until the fish flakes easily when tested with a fork (about 10 to 20 minutes). Make sure you don't overcook the fish.

Barbecued Fillets or Steaks

2 pounds fish fillets or steaks, fresh or frozen
¼ cup chopped onion
2 tablespoons chopped green pepper
1 clove garlic, finely chopped
2 tablespoons melted fat or oil
1 can (8 ounce) tomato sauce
2 tablespoons lemon juice
1 tablespoon Worcestershire sauce
1 tablespoon sugar
2 tablespoons salt
¼ teaspoon pepper

Thaw frozen fish. Cook onion, green pepper, and garlic in fat until tender. Add remaining ingredients and simmer 5 minutes, stirring occasionally. Cool. Cut fish into 6 portions. Place fish in a single layer in a shallow baking dish. Pour sauce over fish and let stand for 30 minutes, turning once. Remove fish, reserving sauce for basting. Place fish in well-greased, hinged wire grill. Cook about 4 inches from moderately hot coals for 5 to 8 minutes. Baste with sauce. Turn and cook for 5 to 8 minutes longer or until fish flakes easily when tested with a fork. Makes 6 servings.

(Source: Dean)

SMOKING

Smoking (for flavor only) is a simple technique that requires a minimum of effort and equipment, and fish smoked in this manner can be used in various recipes from appetizers to salads and casseroles. It is not a method of preserving fish, however. Items needed for smoking are a hooded or covered grill (either gas, electric, or charcoal); briquettes (if a charcoal grill is to be used); 1 pound of hickory or other hardwood chips; water; salt; oil; and fish. The best smoked fish is produced from "fat" fish such as bluefish, mullet, mackerel, herring, and shad, to name a few; however, other species can be used.

To smoke fish for flavor:

1. Soak the chips in 2 quarts of water until the fire is ready (or at least as long as the fish marinates).
2. Marinate the fish in a brine of 1 cup of salt dissolved in 1 gallon of water for the length of time shown in Table 1.
3. Start the fire using fewer briquettes than for an average broiling fire. Adjust the tempera-

Table 1. Timetable for Smoking Fish

Size and Shape	Marinate	Cook (°F)	Time
Fillets or Steaks (1/2-inch thick)	30 min.	150-175	60 min.
		200	30 min.
		250	20 min.
Fillets or Steaks (3/4-inch thick)	45 min.	150-175	90 min.
		200	30-45 min.
		250	30 min.
Fillets or Steaks (1 1/2-inch thick)	60 min.	150-175	120 min.
		200	75 min.
		250	45-50 min.

ture on gas or electric grills according to Table 1. When the coals have burned to a red color, spread evenly over the bottom of the grill.

4. Cover the charcoal with 1/3 of the wet chips, which not only produce the smoke, but also lower the temperature.
5. Grease the grill generously and keep oil handy for basting.
6. Drain and dry the fish and place it on the grill skin-side down.
7. Baste the fish at the start and as needed during cooking to prevent the fish from drying out.
8. Cover the grill with the hood.
9. Smoke the fish until it is golden in color and flakes when tested with a fork in the thickest part.
10. Add the remainder of the chips as needed to produce smoke.
11. Fish are done when they turn to a golden brown and flake easily when tested with a fork.

Smoked Bluefish Spread

1½ pounds smoked bluefish or other fish fillets
2 teaspoons minced onion
2 teaspoons finely chopped celery
1 clove minced garlic
2 tablespoons finely chopped sweet pickles
1¼ cups mayonnaise
1 tablespoon mustard
dash Worcestershire sauce
2 tablespoons chopped parsley

Remove skin and bones from fish; flake fish well. Mix all ingredients together and chill at least one hour. Makes approximately 3½ cups.

(Source: Townsend)

DRYING IN A DEHYDRATOR

Many different electric food dehydrators are on the market today, and it is possible to produce fish jerky in them. If you have one of these dehydrators, you should follow the manufacturer's instructions carefully for cleaning, processing (usually into ¼- to 3/8-inch strips), brining, seasoning with a dry cure, refrigeration, and finally, drying (usually 12 to 14 hours at 140°F to 160°F) until dry.

FRYING

Frying is a method of cooking food in fat. For frying, choose a fat that may be heated to a high temperature without danger of smoking. A smoking fat begins to decompose and will give the food an unpleasant flavor. Vegetable oils and fats are preferable to animal fats. Frozen fish must be thawed before frying. Separate the pieces and cut to uniform size.

The temperature of the fat is extremely important. Heat that is too high will brown the outside of the fish before the centers are cooked.Heat that is too low will give a pale, greasy, and fat-soaked product. The most satisfactory frying temperature for fish is 350°F to 374°F.

After frying, drain the fish immediately on absorbent paper to remove excess fat. Keep the fish warm in a low oven until all pieces are cooked, then serve immediately.

Deep-Fat Frying

Deep-fat frying is a quick and excellent way to cook tender foods and pre-cooked foods. Use enough fat to float the fish but do not fill the fryer more than half full. You must allow room for the fish and for the bubbling fat.

The fish may be dipped in a liquid and coated with a breading, or dipped in batter. The coating will keep the fish moist during frying and will give them a delicious crispness.

Place only one layer of fish in the fry basket at a time and allow enough room so that the pieces do not touch. This prevents the temperature of the fat from dropping suddenly and assures thorough cooking and even browning. When the fat has heated to the proper temperature, lower the basket into the fryer slowly to prevent excess bubbling. If the fat is at the right temperature when the fish are added, a crust forms almost immediately, holding in the juices and preventing the fat from soaking in. Fry until the fish are golden brown and flake easily, usually about 3 to 5 minutes.

Deep-Fat Fried Fillets or Steaks

2 pounds fish fillets or steaks, fresh or frozen
¼ cup milk
1 egg, beaten
1 teaspoon salt

dash of pepper
1½ cups dry bread, cereal, or cracker crumbs
fat for frying

Thaw frozen fish. Cut fish into 6 portions. Combine milk, egg, salt, and pepper. Dip fish in milk and roll in crumbs. Place in a single layer in fry basket. Fry in deep fat, 350°F, for 3 to 5 minutes or until fish are brown and flake easily when tested with a fork. Drain on absorbent paper. Makes 6 servings.

Note: A commercial breading may be used. Follow the directions on the package.

(Source: U.S. GPO, "Let's Cook Fish!")

Pan Frying

Of all the ways of cooking fish, pan-frying or cooking in a small amount of fat in a frying pan is probably the most frequently used -- and most frequently abused -- method. It is an excellent way of cooking pan-dressed fish, fillets, and steaks.

Generally, the procedure is to heat about 1/8-inch of fat in a heavy frying pan to about 350°F. Place one layer of breaded fish in the hot fat, taking care not to overload the pan and thus cool the fat. Fry until brown on one side; then turn and brown on the other side. Cooking time will vary with the thickness of the fish, generally 8 to 10 minutes.

Quick-and-Easy Maryland Fried Fish

½ pound pan-dressed fish or 1/3 pound fish fillets, per person
salt, for sprinkling
lemon-and-pepper seasoning, for sprinkling
1 to 2 cups dry pancake mix
fat or oil, for frying

Wash and dry fish. Dip fish into clean, cool water; sprinkle lightly with salt and lemon-and-pepper seasoning; then coat lightly with pancake mix. Fry in deep fat at 350°F for 4 to 5 minutes, or fry in 1½ inches hot fat in fry pan, four to five minutes on each side. Fish is done when browned on both sides and flakes easily when tested with a fork. Be careful not to overcook.

Remove fish from pan and drain on paper towel. Serve with cocktail or tartar sauce.

(Source: Maryland SMA)

Oven-Frying

Oven-frying, as the name implies, involves baking fish in the oven with the resulting product resembling fried fish. In order for the process to work, a very high oven temperature (500°F) should be used, the fish should be cut in serving-sized portions, and the fish should be well-coated with seasoned crumbs. Each portion is then placed on a shallow, well-greased pan, drizzled with melted fat, and baked until done.

Oven-Fried Fillets or Steaks

2 pounds fish fillets or steaks, fresh or frozen
½ cup milk
1 teaspoon salt
1½ cups cereal crumbs or toasted dry bread crumbs
¼ cup melted fat or oil

Thaw frozen fish. Cut fish into 6 portions. Combine milk and salt. Dip fish in milk and roll in crumbs. Place fish in a single layer, skin side down, on a well-greased baking pan, 15 x 10 x 1 inches. Pour fat over fish. Bake in an extremely hot oven, 500°F, for 10 to 15 minutes or until fish are golden brown and flake easily when tested with a fork. Makes 6 servings.

(Source: U.S. GPO, "Let's Cook Fish!")

Stir-Frying

Fish can be stir-fried in a traditional wok over a flame, an electric wok, or even a heavy cast-iron skillet. To stir-fry, choose a fish with firm flesh to withstand the rapid stirring characteristic of the method. From the fillet stage, cut the fish into bite-sized pieces or chunks. Marinate the fish in a mixture of dry sherry, cornstarch, salt, and perhaps soy sauce and sugar if desired for 10 to 15 minutes; then drain. Choose an oil that has a high smoke point such as peanut oil, and then heat the oil until a drop of water "dances" on the top. Add flavor-producing spices such as garlic and fresh sliced ginger root to flavor the oil; cook a minute or so and then remove them from the oil. Add the fish and cook and stir until opaque; remove fish from the pan or push up onto the side of the wok. Add vegetables in the order of cook time required (longest to shortest) and stir-fry as above, pushing the cooked foods up on the sides of the wok (or removing them from the pan) before adding the next item. When stir-frying is finished, add the fish and vegetables back to the pan and then add a combination of fish or chicken stock and cornstarch. Cook until cornstarch is clarified.

Fish with Broccoli

½ pound firm textured fish (thawed if previously frozen)
3 tablespoons soy sauce
1 teaspoon sugar
1 tablespoon cornstarch
1 tablespoon dry sherry
½ teaspoon salt
4 thin slices each of ginger and garlic
½ cup onions, cut vertically
2 cups small broccoli florets
2 tablespoons cornstarch
1½ cups fish stock (chicken or vegetable may be substituted)
About 4 tablespoons cooking oil

Cut fish into strips and marinate in the soy sauce, sugar, cornstarch, sherry, and salt for 10 to15 minutes. Add cornstarch to the stock. Place 2 tablespoons oil in the pan and heat. Add garlic and ginger. Cook on high for a few seconds to flavor the oil and then remove them. Add the onion and stir-fry about 30 seconds; remove from the pan. Add the broccoli and stir-fry; remove from the pan. If needed, add more oil to the pan, heat, and add the drained fish. Cook until opaque. Add vegetables back to the pan and pour in the stock/cornstarch mixture. Cook until the cornstarch clarifies. Serve with rice. *(Source: Dean)*

MICROWAVE COOKING

Seafoods can be cooked rapidly with delicious results in your microwave oven. As a matter of fact, seafood is an ideal product for microwave cooking. Although different brands of microwave ovens may vary in power and features, the following are a few general tips for cooking seafood:

- Cook seafood in a covered container to hold in moisture and further reduce cooking times. Microwave-safe plastic wrap makes a good covering, but be sure to pierce it or turn back one edge before removing it to allow excess steam to escape.
- When cooking more than one piece of seafood, spread the pieces out to allow space between each and try to put the thickest portion to the outside. Do not stack or layer pieces in the pan.
- Remember that microwave cooking involves residual cooking, and, therefore, most recipes allow for a "standing" period to complete the cooking process. The seafood normally should remain covered during this time.
- Seafood cooks so rapidly and is so delicate that, when preparing a whole meal, it is probably best to cook it last to avoid reheating it.
- The shells of shellfish are suitable cooking vessels for microwaving.
- When baking a whole fish, the head and tails can be shielded with a very small piece of aluminum foil during cooking to prevent excess drying.
- Before cooking, fish may be brushed with butter or a dilute solution of a gravy-browning agent to enhance the color of the cooked product. Fish is done when the flesh turns opaque and flakes easily when tested with a fork in the thickest part. Shellfish is generally done when the flesh turns opaque and, if cooked in the shell, the shell turns red.

SHELLFISH

CRABS

There are many species of crab that are edible, but the most common are the blue crab, the stone crab, and the king and snow crabs. Of all the crabs, however, the blue crab is the most important commercial species. Blue crabs are caught and marketed in both hard-shelled and soft-shelled stages. If you buy fresh hard- or soft-shelled crabs, be sure that they are alive at the time of purchase. If you prefer male crabs known as "jimmies" over female crabs, sometimes called "sooks," you can determine gender by turning the crab over and inspecting the "apron." Male crabs have a slender, pointed apron and mature females have a beehive-shaped apron. In most crabbing states, crabs should be a minimum of 5 inches from point to point across the back.

If you do not want to cook your own crabs and pick the meat, you can purchase fresh-cooked, pasteurized, or canned crabmeat. There is a difference among these forms. Fresh-cooked meat is simply packed into the container and sealed with a lid; it should be used within three to five days. Pasteurized crabmeat has been packed, hermetically sealed in cans, and heat-processed to prolong shelf life. Pasteurized meat will last for up to six months in the refrigerator; however, once opened, it should be used within three to five days. It can be substituted easily for fresh. Both fresh and pasteurized crabmeat should have a fresh odor. Shelf-stable canned crabmeat can also be substituted for fresh, but the texture is somewhat different and salt has been added to the contents, so keep this in mind when substituting.

Laurie's Cocktail Quiche

Pastry:
1 3-ounce package cream cheese
½ cup butter
1 cup flour

Mix cream cheese and butter; add flour and mix well. Form into 1-inch ball and press into miniature muffin tins (bottom and sides).

Note: Conventional pastry may be used, and the mixture may be put into a 9-inch pie pan.

Filling:

½ cup mayonnaise
2 eggs, beaten
1 2/3 cups crabmeat
1/3 cup sliced onion
2 tablespoons flour
½ cup milk
8 ounces Swiss cheese, sliced or grated

Combine mayonnaise, flour, beaten eggs, and milk; mix until blended. Stir in crabmeat, cheese, and onion. Pour or spoon into individual muffin cups. Bake at 350°F until set. *(Source: Dean)*

Stone Crab

What you will find in the store or market is only the claws from the stone crab. Because crabs have the ability to regenerate appendages, when the stone crab is caught, one or both of the powerful black-tipped claws is removed and the crab is returned to the water. The claws are immediately cooked and sold either refrigerated, or most often, frozen. Freshness is determined on the basis of mild odor. If purchased freshly-cooked and frozen, the shelf-life of stone crab claws is about six months. To prepare stone crab claws, first thaw them in the refrigerator for 12 to 18 hours (thawing under cold running water or at room temperature reduces their quality).

For warm claws, place thawed claws in a steamer basket over — not in — boiling water and steam just until heated through. For cold or warm claws, crack the shell with a hammer or nutcracker; remove the shell and moveable pincer, leaving the meat attached to the remaining pincer. Serve with drawn butter or warm lemon butter or the mustard sauce below:

Mustard Sauce

½ cup sour cream
1½ tablespoons prepared mustard
2 teaspoons melted butter
2 teaspoons parsley flakes
1/8 teaspoon salt

Combine all ingredients. Heat on low just until warm, stirring occasionally. Do not boil.

(Source: Florida Extension Service)

King Crab and Snow Crab

The king crab and its slightly smaller cousin the snow crab are mostly caught commercially in the cold, deep waters of the Pacific Northwest and Alaska. The king crab is so named for its large size, and can measure several feet across, from leg-tip to leg-tip. Like the stone crab, king and snow crabs are processed immediately when caught and are generally available in the frozen-cooked form. The prime meat of these crabs is located in the claws, legs, and shoulders, and can be served cold or warm with drawn butter and herbs, or can be used in quiches, in casseroles, or as a stuffing for fish. When purchasing, beware of signs of freezer burn or other damage (for example, cracked claws, which may allow dehydration). The legs should have a fresh, mild odor.

To warm the crab legs, you can either steam them (see "stone crab") if you have a large enough steamer, or broil or bake them in the oven with a little garlic and herb butter.

Crab with Garlic and Herb Butter

2-3 pounds king or snow crab
1 cup butter, melted
2 large cloves garlic, finely minced
2 tablespoons fresh dill weed
1 tablespoon fresh chopped parsley

2 teaspoons fresh chopped basil
1 teaspoon Worcestershire sauce
salt and pepper to taste

Mix butter, garlic, herbs, and spices together. Add crabmeat and heat in 350°F oven until warm. Serve with butter sauce. *(Source: Dean)*

SHRIMP

There are several hundred species of shrimp in the waters of the world. They range in size from less than one inch up to 12 inches in length (prawns). Colors (when alive) are pale pink, pale brown, pale gray, or white. When cooked, the shells turn pinkish red, and the meat becomes white with reddish tinges. Shrimp can be purchased raw or cooked, peeled or unpeeled, and fresh or frozen. They are usually sold according to the size based on the number of headless shrimp per pound (i.e., shrimp with the designation "26-30s" will have that many headless shrimp in one pound).

Fresh shrimp are firm in texture, have little or no black spotting in the shell segments, and have a mild odor. Frozen shrimp should be solidly frozen, have little or no odor, no brown spots, little or no black spotting, and no sign of freezer burn. If purchasing raw shrimp, remember that if you buy shrimp with the heads on, approximately 50 percent of the weight of your purchase is contained in the head and shell. Fifteen percent of the weight of raw headless shrimp is in the shell. Cooking will further reduce the yield by about 35 percent. Roughly speaking then, about two pounds of raw headless, unpeeled shrimp, properly cooked, will yield one pound of cooked, peeled, and de-veined shrimp. Therefore, you can make these purchases more cost effective by saving the shrimp heads and shells and substituting them for fish heads and bones to make a court bouillon (see Finfish: Soups, Stews, Sauces, and Chowders).

Fresh shrimp should be cooked within two days of purchase or, if that is not possible, should be frozen on the day of purchase. Frozen shrimp should be thawed in the refrigerator for 18 to 24 hours or under cold running water — never at room temperature.

Shrimp can be cooked in a variety of ways: simmering, boiling, steaming, broiling, grilling, frying, etc. Larger shrimp lend themselves to singular presentations such as grilling, frying, or broiling, while smaller shrimp are good for use in soups, stews, casseroles, quiches, sauces, etc.

Boiled Shrimp

2 to 3 pounds raw, headless, unpeeled shrimp -- fresh or frozen
1 bag commercial "crab & shrimp boil" (liquid may be substituted if desired)
1 lemon, sliced in rounds
2 tablespoons salt

Thaw frozen shrimp. Add crab boil, lemons, and salt to several quarts of water. Bring to a boil. Add shrimp all at once. Cook just until shrimp meat begins to separate from the shell. Pour into a colander and rinse shrimp under cold running water for 1 to 2 minutes. *(Source: Dean)*

"To de-vein or not to de-vein, that is the question," to paraphrase Shakespeare. The "vein" in question, also known as the "sand vein" because it tends to be gritty, runs down the center of the back of the shrimp. In general, the larger the shrimp, the more necessary it becomes to remove it. For small shrimp, it is a matter of choice, adding quite a bit of labor time to preparation.

Shrimp and Tomato Quiche

¾ cup mayonnaise
1 egg
2 egg yolks
¼ pound grated Jack cheese
¾ pound grated Swiss cheese
4 fresh tomatoes, chopped
1½ pounds cooked shrimp
1 small onion, diced
1 large green pepper, diced
6 tablespoons butter or margarine
salt and pepper to taste
1 tablespoon basil
½ cup parsley, chopped
2 tablespoons chives, chopped
3 large garlic cloves, minced
2 unbaked 9" pie shells
bread crumbs
melted butter or margarine

Beat together mayonnaise, egg, and yolks. Add cheeses, tomatoes, and shrimp. Sauté onion and pepper in butter. Add spices, parsley, chives, and garlic to sautéed vegetables. Add all to mayonnaise mixture and mix. Pour into pie shells.

Sprinkle with crumbs and drizzle with butter. Bake at 450°F for 10 minutes. Reduce heat to 350°F and bake 40 to 45 minutes or longer until set.

(Source: Dean)

Harriet's Fried Shrimp

2 pounds medium to large headless shrimp, fresh or frozen
1 envelope commercial pancake mix
1 package Pepperidge Farm Cornbread stuffing mix
salt and pepper
oil for frying

Thaw frozen shrimp. Peel shrimp, leaving the tail on. Cut down the top curve of the shrimp almost, but not quite all the way, through. Open out the "butterflied" shrimp and remove the sand vein under running water. Pat shrimp dry and season with salt and pepper. Mix the pancake batter according to package directions (it should be thick enough to coat the shrimp). Place the stuffing mix in a blender or food processor and process until fine. Dip the shrimp into the pancake batter and then into the crumbs. Fry at 350°F until done.

(Source: Dean)

Rock Shrimp

If you've visited your local seafood market lately, you may have seen a product that resembled a miniature lobster tail. This delectable seafood morsel is the tail section of the rock shrimp, which derives its name from the rigid exoskeleton or shell encasing its body.

The rock shrimp tail resembles the lobster tail in appearance and texture. Its flavor could be placed somewhere between shrimp and lobster.

Because rock shrimp are highly perishable, most are marketed in the raw, frozen state as either whole or split tails. Like other shrimp, rock shrimp are sold by the size or number per pound.

When buying rock shrimp, check the color and odor of the flesh. Good quality rock shrimp have transparent or clear, white flesh with no discoloration and a mild odor.

The yield for properly cleaned and cooked rock shrimp is about one-half the weight of the raw tail. One pound of cooked tails (two pounds raw) is enough to serve four to six people.

If they are available, the split tails are the easiest to prepare. Whole tails need to be cleaned, but the cleaning process is not difficult. Because they are extremely perishable, thaw and clean the rock shrimp under cold running water, and cook them immediately. To prepare rock shrimp for boiling, frying, or baking, hold the tail section in one hand with the swimmerets (legs) down towards the palm of the hand. Using kitchen shears, insert one blade of the scissors into the sand vein opening and cut through the shell along the outer curve to the end of the tail. Pull the sides of the shell apart and remove the meat. Wash thoroughly to remove the sand vein (this step is particularly important in rock shrimp).

If you want to broil the rock shrimp in the shell, use the following procedure: Place the rock shrimp tail on a cutting board with the swimmerets up. With a sharp knife, make a cut between the swimmerets through the meat to the hard shell. Spread the shell until it lies flat; wash thoroughly in cold water to remove the sand vein.

Rock shrimp cook more quickly than other shrimp, and should be closely attended in order to avoid overcooking. To boil, place rock shrimp in boiling, salted water and simmer for 30 to 45 seconds. Drain and immediately rinse in cold water to stop further cooking.

To broil, lay rock shrimp flat on a baking pan with meat exposed. Brush with melted margarine and sprinkle with garlic salt and paprika. Broil four inches from the source of heat for about two minutes or until the tails turn upward.

To fry, bread rock shrimp with your favorite breading and fry just long enough to lightly brown — about 30 seconds.

Curried Rock Shrimp on Rice

1 pound rock shrimp
2 tablespoons margarine or butter
2 tablespoons all-purpose flour
¼ teaspoon salt
½ teaspoon powdered curry
dash of pepper
1 cup milk

Peel and de-vein rock shrimp and set it aside. Melt the butter or margarine in a saucepan over low heat. Blend in flour, salt, pepper, and powdered curry. Add milk at once. Cook quickly, stirring constantly until mixture thickens and bubbles. Add shrimp and cook approximately 3 minutes or until shrimp meat turns white. Stir frequently. Serve over cooked rice. *(Source: Dean)*

LOBSTER

There are two species of "lobster" that are commonly found in seafood markets and grocery stores. The first is the Northern or true lobster. The primary production of this type in the United States is centered in Maine; therefore, it is often referred to as the Maine lobster. These lobsters have two claws: a large one for crushing and a smaller one for cutting. Most of the meat of these lobsters is found in the claws and the tail.

The second species is the spiny or rock lobster which is actually not a lobster at all, but rather a sea crawfish. The spiny lobster does not have the heavy claws of the Northern lobster. It is covered with spines on both body and legs, and it has long slender antennae. It is mainly caught off the coast of Florida, with a similar species caught off the coast of California. The meat of the spiny lobster is concentrated in the tail and it is this species that constitutes the bulk of frozen lobster tails sold in markets and restaurants.

Lobsters can be purchased live, whole cooked in the shell, frozen raw, cooked and frozen (this market form is less desirable because of the deterioration caused from toughening of the meat and loss of flavor in storage), fresh cooked meat, frozen cooked meat, and canned cooked meat.

When buying whole raw lobster, be sure that the animal is alive, shows movement in the legs when touched, and that the tail curls under when picked up. Cooked whole lobster should be bright red and should have a fresh odor. Frozen lobster and lobster tails should be hard frozen and have no odor.

Lobster in the shell is sold by weight as follows: "chickens" average one pound; "quarters" average 1¼ pounds; "halves" average 1½ pounds; "large" average 1¾ to 3 pounds; and "jumbos" weigh over 3 pounds. Lobster tails usually run from four ounces to one pound. A one-pound lobster will yield approximately 2/3 cup of meat.

To cook whole lobster, it is best to simmer it just below the boiling point. Boiling tends to toughen the meat and destroy some of the delicate flavor. The following recipe is a good basic one for "boiled" lobster.

Boiled Lobster

2 live lobsters (1 to 1¼ pounds each)
3 quarts boiling water
3 tablespoons salt
clarified butter

Thaw lobster if frozen. In a 6-quart saucepan, bring water and salt to a boil. Plunge live lobster head first (or place thawed lobster) into boiling salted water. Cover and return to boiling point. Reduce heat and SIMMER for 12 to 15 minutes. Larger lobsters may require slightly more cooking time. Drain. Rinse with cold water for 1 to 2 minutes. Serve with clarified butter.

Whole lobster can also be baked in the oven; however, it should be frequently basted or completely covered with stuffing during baking. The following is a simple stuffed lobster recipe.
(Source: Dean)

Baked Stuffed Lobster

2 live or frozen lobsters (1 to 1¼ pounds each)
1½ cup soft bread crumbs
½ cup grated cheddar cheese
2 tablespoons melted butter
1 tablespoon grated onion
paprika

Preheat oven to 400°F. Thaw lobster under cold running water if frozen. If live, place lobster in the freezer for 20 to 30 minutes to make the following step easier. Cut the lobster in half lengthwise and remove the stomach and intestinal vein. Rinse and clean the body cavity. Combine crumbs, cheese, butter, and onion and place this in the body cavity and over the surface of the tail meat. Sprinkle with paprika and bake in a hot (400°F) oven for 15 to 20 minutes or until lightly browned.
(Source: Florida Extension Service)

OYSTERS

With the coming of fall comes also the season for plump, delectable oysters. Although oysters are available year-round, it is in the cooler months that they reach the zenith of their culinary goodness. (The belief that oysters should not be eaten in months that do not contain the letter "r" is a myth.)

Table 2. Guidelines for Purchasing Oysters

Name	Size	# in 1 Gallon	Use
Counts	Extra large	under 160	Any variation of use on the half-shell
Extra	Large	160-210	Same as counts or selects
Selects	Medium	211-300	Some use in half-shell variations; fried, poached, etc.
Standards	Small	301-500	Soups, stews, chowders, fritters, casseroles
Standards	Very small	over 500	Same as small standards

Oysters are available in several forms: alive in the shell, fresh-shucked, frozen in various forms, and canned. When buying oysters in the shell, be sure that the shells are tightly closed, or that they will close when tapped lightly. Shells that are open and will not close indicate that the oyster is dead.

If you choose to purchase already-shucked oysters, look for the following signs of quality: The oyster meats should be plump and generally creamy to beige to gray in color, although there are color variations which are harmless. Occasionally, you may find oysters that appear green in color. This green is probably chlorophyll from green plants the oyster had been eating before being caught. Sometimes, too, especially in late fall and early winter, oysters or their liquor may appear red. Once again, this coloration is due to plant pigment from the oyster's food.

A pink color in oysters is more than likely due to the pigment from the eggs of the pea crab, a tiny animal that lives in the gills of the oyster and feeds on the same food that the oyster filters for itself. Pea crabs are a gourmet delicacy, so instead of throwing them away, cook them along with your oysters.

The liquor surrounding the oysters should be clear or slightly opalescent, free from shell particles, and should have a mild, clean odor.

Oysters are sold by size with the price increasing from small to large. Because of the variation in price, be sure to purchase oysters according to their intended use (Table 2).

Live shell oysters will remain alive for 7 to 10 days if stored in the refrigerator at 35°F to 40°F (*no* ice on them). Freshly shucked oysters will maintain quality for about a week, provided they are properly handled and their container is packed in ice in the refrigerator. Frozen oysters should be thawed in the refrigerator or under cold running water. As with other seafood, once thawed, oysters should not be re-frozen and should be used immediately.

Oysters can be cooked by a variety of methods: broiled, roasted, simmered, baked, fried, poached, etc.

Once you have made your purchase and you are ready to prepare the oysters, remember the cardinal rule of seafood cooking: Don't Overcook! Oysters are done when the gills curl. Overcooking tends to make the oyster tough and rubbery in texture.

Try the following oyster recipes the next time your palate yearns for a rich, hearty dish.

Old-Fashioned Oyster Chowder

1 pint oysters, fresh or frozen, undrained
8 strips bacon, diced
2 tablespoons margarine
2 cups cooked potatoes, coarsely chopped
1 medium onion, chopped, or 1/2 cup sliced green onion
1 medium carrot, coarsely shredded
½ cup diced celery
½ cup water
2 cups milk
2 cups half & half
1 can (12 ounce) whole kernel corn, drained
1½ teaspoons salt
1/8 teaspoon white pepper
2 dashes liquid hot pepper sauce
chopped parsley

Thaw oysters if frozen. Fry bacon over moderate heat in large Dutch oven until crisp. Remove bacon; set aside. Remove all but 2 tablespoons of bacon drippings from the pan; add margarine. Sauté potatoes in drippings until lightly browned.

Add carrot, celery, and water; cover and simmer about 5 minutes or until vegetables are tender. Add milk, half-&-half, corn, salt, pepper, and hot pepper sauce. Simmer. Add oysters, oyster liquid and bacon. Heat just until edges of oysters curl. Add more hot pepper sauce if desired. Ladle into soup bowls; sprinkle parsley over the top. Makes about 10 cups, 6 to 8 servings. *(Source: Dean)*

Oysters Rockefeller Variation

1 pint oysters, selects or counts, fresh or frozen
¼ cup margarine or butter
¼ cup chopped celery
¼ cup chopped green onion
2 tablespoons chopped parsley
1 (10-ounce) package frozen chopped spinach
1 tablespoon anisette
¼ teaspoon salt
24 pastry shells
¼ cup dry bread crumbs
1 tablespoon melted butter or margarine

Thaw oysters if frozen. Cook oysters in their natural liquor until edges curl. Drain. In a small saucepan melt ¼ cup margarine. Add celery, green onion, and parsley. Cover and cook 5 minutes or until tender. Combine cooked vegetables with spinach in blender container. Add anisette and salt. Chop vegetables in blender until almost pureed, stopping once or twice to push vegetables into knife blades. (Vegetables may be run through a food mill.) Make pastry shells approximately 1½ inches in diameter. Bake at 450°F for 5 to 8 minutes or until lightly browned. Place one oyster in each pastry shell; top with spinach mixture. Bake in hot oven at 450°F for 5 to 8 minutes. Serve immediately. Makes 24 hors d'oeuvres.

(Source: Dean)

CLAMS

Clams are available alive in the shell, fresh-shucked, frozen, or canned. Both hard and soft fresh clams are available on the market. Buying hard clams is a process of understanding sizes in relation to the intended use, and knowing the signs of quality for both shell and shucked clams.

Hard clams are generally sold under market names that correspond to their relative size. The largest (and usually the cheapest) are the "chowders." As their name indicates, these clams are used in chowders, soups, and stews, or are the main ingredient in fritters, stuffed clams, minced clams, and other recipes. Their toughness makes it necessary to use tenderizing techniques such as chopping, grinding, or mincing.

The next size of hard clam is called a "cherrystone." There are about 300 to 325 cherrystones in a bushel, as opposed to about 125 chowders. Cherrystones are more tender and are used in baking and, to some degree, in the raw half-shell trade.

"Littlenecks" are the smallest, most tender, and most expensive of the hard clams. They are eaten on the half-shell and steamed. When you buy shell clams, be sure that they are alive. Gaping shells that do not close when handled mean that the clams are dead and therefore not usable.

To shuck hard clams, wash them thoroughly and hold the clam in the palm of one hand with the hinged end against the palm. Insert a clam knife or other slender knife between the halves of the shell and cut around the clam, twisting the knife slightly to pry open the shell. Cut the two muscles free from the two halves of the shell. If the clams are to be served on the half-shell, remove one-half of the shell. If they are to be used in recipes requiring removal from the shell, clean the meat thoroughly, and strain the liquor through cheesecloth or a fine strainer or sieve.

Shucked clams are generally sold by the pint or quart. Look for plump meat, clear liquid, and a minimum of shell particles. The quantity of clams to buy depends on how they are to be served. Generally, for six servings you will need three dozen shell clams or one quart of shucked meats.

Soft-shell clams have elongated shells that are thin and brittle. They cannot close tightly because their necks extend beyond the shell. When buying fresh soft clams, look for signs that the clam is alive by touching the neck (it should "twitch" when touched).

Clams can be steamed, stewed, poached, fried, broiled, baked, etc. The following are good, basic recipes for clams.

Steamed Clams

4 dozen small soft-shell clams
salt and pepper
seafood seasoning (optional)
melted butter

Put washed clams in a pan. Sprinkle with salt, pepper, and seafood seasoning. Add water; cover and bring to a boil. Reduce heat and steam 10 to 15 minutes or until shells open wide. Drain, reserving liquid. Strain liquid. Serve clams hot with clam liquid and melted butter in separate bowls. Serves 4 as an appetizer. *(Source: Dean)*

Clams Oregano

2 dozen clams (cherrystone)
2 tablespoons olive oil
1 medium onion
1 teaspoon basil
2 tablespoons lemon juice
1 cup fresh bread crumbs
2 tablespoons butter
1 clove garlic
1 tablespoon oregano
½ cup chopped parsley

Sauté onion and garlic in olive oil and butter. Combine remaining ingredients. Spread on clams; bake 10 minutes in 350°F oven or freeze for later use. *(Source: Dean)*

Clam Corn Chowder

1 quart whole clams, minced, or 2 (7 ounce) cans minced clams
1 cup clam liquor and water
3 slices bacon, chopped
1 cup chopped onion
2 cups diced raw potato
1½ cups drained whole-kernel corn
3 cups milk
2 tablespoons flour
1 tablespoon butter or margarine
1 teaspoon celery salt
1 teaspoon salt
dash of pepper
½ cup coarse cracker crumbs (optional)

Drain clams; chop if using whole clams. Pour liquor into measuring cups and add water as needed to fill cup to one-cup level. Fry bacon until crisp; add onion and cook until tender. Add potatoes, clam liquor, and water. Cover; simmer gently until potatoes are tender. Add corn and milk. Blend flour and butter or margarine and stir into soup. Cook slowly until mixture thickens slightly, stirring constantly. Add seasonings and clams; simmer 5 minutes. Top with cracker crumbs. Serve hot. Makes 6 servings.

(Source: Dean)

SCALLOPS

The two main varieties of scallops available on the market today are the bay scallop, which is about ½ inch in diameter, and the sea scallop, which may be as large as two inches in diameter. Because scallops perish so quickly after being removed from the water, they are usually shucked onboard the scallop boat. For that reason, the term "scallop" here is used to refer to the "eye" or the adductor muscle. This is the muscle that opens and closes the shell.

When purchasing sea scallops, look for whitish meat, very little liquid, and a fresh, even slightly sweetish, odor. Bay scallops should have meat that may be creamy white, light tan or pinkish; should be practically free of liquid; and should have a fresh odor.

Fresh scallops should be stored on ice in a refrigerator at a temperature between 35°F and 40°F, but they are best used the day of purchase. Raw frozen scallops should be held at 0°F or below and will last three to four months. Frozen scallops should be thawed under cold running water or in the refrigerator.

Scallops can be poached, broiled, charcoal grilled (on kabobs), baked, fried, etc. As with other seafood, avoid overcooking. Bay scallops can be substituted for sea scallops, and vice versa.

The following recipes are crowd-pleasers.

Coquilles St. Jacques

1 cup dry white wine (not cooking wine)
¾ cup water
1 pound bay scallops
½ pound mushrooms, finely sliced
1 bay leaf
1 tablespoon lemon juice
1 tablespoon chopped parsley
3 peeled, whole shallots
4 tablespoons butter
4 tablespoons flour
1¾ cups liquid reserved from simmering scallops
2 egg yolks
4 tablespoons cream
fresh bread crumbs
2 tablespoons butter (for dotting)
salt and pepper

Pre-heat oven to 450°F. Simmer scallops and mushrooms in wine, 3/4 cup water, lemon juice, bay leaf, parsley, and shallots for 5 minutes. Remove and strain, reserving liquid.

Make a medium white sauce with 4 tablespoons butter, flour, and 1¾ cups liquid reserved from simmering. Stir 2 egg yolks into 4 tablespoons cream and then add to sauce. Season to taste. Place scallops into ovenproof shells or ramekins and spoon sauce over. Add mushrooms around the edges. Sprinkle with bread crumbs, dot with butter, and heat at 450°F until lightly browned. Serve at once. Serves 4. *(Source: Dean)*

Laurie's Easy Scallop-and-Shrimp Kabobs

1 pound sea scallops
1 pound large shrimp, headed, peeled, and cleaned
several pieces of bacon
dry white wine
salt and pepper

Marinate scallops and shrimp in enough white wine to cover them for about an hour in the refrigerator. Begin skewering in the following fashion: skewer one end of a slice of bacon, the "head end" of a shrimp, then a scallop, then the "tail end" of the shrimp, and then the bacon again; continue skewering in this fashion. Season with salt and pepper and grill over hot coals until shrimp and scallops are opaque. *(Source: Dean)*

MUSSELS

Mussels, like other bivalves, should be alive when purchased and cooked; that is, the shells should be closed tightly and not broken. Before cooking, the shells should be scrubbed with a stiff brush and the "beard" or hair-like strands emerging from the shell should be pulled off. Mussels generally are steamed in a covered pot over boiling water, saltwater, or, for extra flavor, over a court bouillon (see Finfish: Soups, Stews, etc.). Mussels are done when the shells open and should not be overcooked. Serve them with a bit of the steaming liquid.

For variety, if you have larger mussels, you can open them, stuff them with the stuffing of your choice, re-close them by tying a string around them, and simmer in a court bouillon.

SQUID

Calamari is the Italian word for squid, long considered a delicacy in many countries of the world. Although it is not as popular in the United States, it is finding its way onto the menus of a variety of restaurants across the country.

Squid can be cooked by a number of methods, but it tends to be very tough if improperly cooked. Very short cooking times (2 to 3 minutes) as in deep frying or longer cooking times (30 minutes or longer), as in stewing, in baking, in sauces, etc., seem to work best for squid. Marinating also helps soften the connective tissue of squid.

Squid is generally sold whole fresh or frozen, or canned with or without its ink, in brine, in oil, or in tomato sauce.

Fresh and thawed frozen squid should smell clean and fresh. The skin should be creamy in color, with very tiny red flecks. As the squid begins to spoil, pigments are released into the flesh, causing a deep reddish-purple coloration.

To clean squid:

1. Thaw, if frozen.
2. Hold the tube-like body (called the mantle) in one hand and pull and twist off the head with the other hand.
3. There is a clear, chitinous "pen" (the last vestige of a shell) that runs down the "back" of the squid. Pull this out, making sure to get it all.
4. Grasp one of the wing-like fins located at the narrow end of the mantle and slowly pull to remove the speckled skin.
5. Clean out the mantle, wash it thoroughly and drain.
6. Remove the arms and tentacles by cutting just on the tentacle side of the eyes. Remove the hard beak in the center of the arms and tentacles.
7. You can now either cut the mantle into rings, cut strips or pieces from the mantle after slitting it down the center, or leave the mantle as is for stuffing.

The following recipes are both tasty and economical.

Calamari Vinaigrette

2 pounds whole squid, fresh or frozen
2 cups boiling water
1 teaspoon salt
1 lemon
¼ cup olive oil
2 tablespoons vinegar
1½ teaspoons lemon juice
½ teaspoon salt
¼ teaspoon freshly ground black pepper
salad greens
chopped parsley

Thaw squid if frozen. Clean squid in manner described above, and cut the mantle into rings. Combine boiling water, salt, and squid rings and tentacles, simmer for 3 minutes. Remove the "zest" from the lemon (the zest is the outer layer of the skin; the inner white membrane is very bitter, so try to avoid it) and add the zest to the simmering squid. Continue simmering for no more than 5 minutes longer. Drain and rinse squid in cold water. Combine olive oil, vinegar, lemon juice, salt, and pepper. Pour this vinaigrette over the squid cover and marinate until chilled. Drain squid and serve on greens. Garnish with chopped parsley. Serves 6.

(Source: U.S. GPO, "A Seafood Heritage")

Italian Stuffed Squid

6 large squid (about 1 pound)
¼ cup Italian-style bread crumbs
2 tablespoons minced parsley
2 teaspoons freshly grated Parmesan cheese
2 teaspoons minced garlic
1 egg, beaten
¼ cup vegetable oil
salt and pepper
2 garlic cloves, sliced
1 can (about 15 ounces) whole peeled tomatoes, chopped
1 teaspoon basil
½ teaspoon each ground oregano and rosemary
¼ cup finely chopped green pepper
¼ cup dry white wine

Clean squid. Chop tender parts of the tentacles and mix them with the bread crumbs, parsley, cheese, 1½ teaspoons of minced garlic, egg, and 1 tablespoon salad oil. Blend well; add salt and pepper to taste.

Spoon stuffing loosely into each squid and close with a toothpick. Do not overstuff, because the squid will shrink during cooking.

Add the remaining 3 tablespoons oil to a skillet large enough to hold the squid in a single layer. Heat the oil and cook the sliced garlic until golden brown, then discard the garlic. Arrange the squid in oil and brown all sides lightly.

Add tomatoes, green pepper, and remaining minced garlic, basil, oregano, rosemary, wine, and salt and pepper to taste. Cover tightly and cook 25 to 30 minutes. Remove toothpick from squid and serve whole or sliced, alone or with pasta. Serves 3 to 4. *(Source: Virginia Extension Service)*

CONCH

Conch is generally marketed in the fresh, fresh-frozen, or canned state. The edible portion of the conch is the muscular foot, which is very tough and, therefore, must be tenderized by pounding, grinding, stewing, or marinating. The skin, ranging from orange in places to almost black, must also be removed (this is done prior to tenderizing). The simplest way to peel the skin is to start at the wide end and peel downward. A paring knife will most certainly come in handy to help in the peeling process.

Beer-Battered Conch

3 pounds conch meat, fresh or frozen
1 tablespoon sugar
1¼ cups self-rising flour
1 teaspoon dry mustard
¼ teaspoon cayenne pepper
1 egg, beaten
1 cup beer
vegetable oil for deep frying

Thaw conch if frozen. Peel skin off and discard. Tenderize conch by pounding with a meat mallet or edge of a plate. Mix flour, sugar, mustard, and cayenne. Combine egg and beer carefully and then slowly stir into flour mixture. Dip conch into batter and fry 1 or 2 at a time in hot oil (450°F) for 3 to 5 minutes or until golden brown. Serve with a slice of lemon or lime, and cocktail or tartar sauce. *(Source: Dean)*

Conch Chowder

24 conch
½ pound butter
4 cups water
6 large carrots, diced
2 green peppers, diced
8 potatoes, diced
3 onions, diced
4 cups canned tomatoes
2 bay leaves
½ cup flour
¼ cup salad oil
salt and pepper to taste

Peel the skin off the conch and then grind the cleaned meat. Sauté the ground conch in the butter until tender. Mix water, vegetables, and bay leaves in a large kettle and simmer until vegetables are tender. Add conch meat and butter to the kettle and continue to simmer. Brown flour in the oil, making a roux. Stir roux into the chowder and simmer until thickened. Season to taste. Serves 6.

(Source: Dean)

ABALONE

The abalone is a univalve mollusk characterized by an oval, ear-shaped shell and a large "foot" which constitutes the edible portion. Fresh abalone, seldom marketed except in California where it is harvested commercially, may be eaten raw, sautéed as steak, or cut into chunks and eaten in chowder.

Like conch, abalone must be tenderized. To tenderize, pound slices with a meat mallet (smooth side) until the meat is limp and velvety (like a limp pancake). The abalone may then be breaded and pan-fried, or battered and deep-fried.

Fried Abalone

3 pounds abalone
beer batter (see "conch")
oil for frying

Slice and tenderize abalone and cut into thin strips. Heat oil to 370°F. Dip strips of abalone into batter and fry until delicately browned. Season with salt and pepper. *(Source: Dean)*

CRAWFISH

Generally, when we think of crawfish, we think of Louisiana. Crawfish are found in abundance in the low-lying areas around the Mississippi Delta, particularly in areas of rice fields and swamps. Most are commercially harvested in the spring season and are featured in numerous local fairs and festivals, one of the most popular being in New Iberia, Louisiana. Crawfish have moved from being a regional dish to one that has found a national following with the recent surge in popularity in "Cajun Cooking" and New Orleans cuisine. Crawfish are marketed fresh in some areas of the country, or cooked and then individually quick frozen (IQF) in other areas.

Crawfish is wonderful when used as the main ingredient in bisque, etouffee, jambalaya, salad, or fried or boiled and eaten with drawn butter. It is obviously a matter of individual taste, but when serving boiled crawfish, they are boiled whole and served head-on; true aficionados of this delicacy do, in fact, suck the heads to savor the unique piquant taste. As noted in the next recipe, however, the "heads" could be more accurately described as the body shell to which the head is attached.

The following recipe calls for raw crawfish; if you buy the IQF crawfish, you would simply leave off the steps involving purging and boiling.

Crawfish Bisque

8 pounds live crawfish
1½ cups salt

Wash live crawfish in cold water. In large container dissolve salt in about 3 gallons water and soak crawfish for 15 minutes to purge. In 10-quart pot, bring to boil 6 quarts of water. With tongs, drop in crawfish and boil 5 minutes. Remove crawfish; cool. Shell crawfish as follows: Break off tail; snap in half lengthwise; lift out meat in one piece; discard tail shell. Snap off large claws (if desired, break claws with nutcracker and remove bits of meat) and smaller legs; discard. Cut off top of head just behind the eyes; discard. Scoop body shell clean, carefully remove and reserve yellow fat or "butter"; discard intestinal matter. Clean and wash thoroughly 48 body shells, which in Louisiana are referred to as crawfish "heads." Finely chop all the tail meat or put it through finest blade of food grinder. There should be about 3 cups ground crawfish tail meat, 48 "heads" for stuffing, and the reserved crawfish fat.

Bisque

1 cup ground crawfish tail meat
reserved crawfish fat
¼ cup bacon fat
¼ cup margarine or butter
½ cup all-purpose flour
2 cups finely chopped onion
1 cup finely chopped celery
½ cup finely chopped green pepper
1 clove garlic, minced
4 cups hot water
2 cans (15 ounces each) tomato sauce with tomato bits
¼ cup chopped parsley
2 tablespoons lemon juice
2 bay leaves
1 teaspoon dried thyme leaves, crushed
1 teaspoon salt
¾ teaspoon cayenne pepper
8 whole allspice
48 stuffed "heads"(recipe below)
3 cups cooked rice

In a 4- to 5-quart Dutch oven, melt bacon fat and margarine. Blend in flour. Cook, stirring constantly, over medium low heat until brown in color, about 15 to 20 minutes. Add onion, celery, green pepper, and garlic. Cover and cook 5 minutes or until tender. Gradually stir in water. Add tomato sauce, parsley, lemon juice, bay leaves, thyme, salt, cayenne, and allspice. Stir in crawfish meat and fat. Cover, bring to a boil and simmer for 1 hour. To serve, ladle into individual soup plates over boiled rice and drop in 5 or 6 stuffed "heads." Makes 6 to 8 entree servings.

Stuffed Heads

½ cup margarine or butter
1 cup finely chopped onion
½ cup finely chopped celery
1 clove garlic, minced
¼ cup chopped parsley
1 teaspoon salt
¼ teaspoon cayenne pepper
2 cups ground crawfish tail meat
2 cups soft bread crumbs
48 empty shells of crawfish "heads"
½ cup all-purpose flour
fat for deep frying

In a 10-inch fry pan, melt margarine. Add onion, celery, and garlic. Cover and cook 5 minutes or until tender. Stir in parsley, salt, cayenne, and crawfish meat. Combine with bread crumbs. Stuff mixture into empty shells of crawfish heads. Roll in flour. Place in single layer in fry basket. Fry in deep fat, 350°F, for 3 minutes or until lightly browned. Drain on absorbent paper. Keep warm until ready to serve.

(Source: U.S. GPO, "A Seafood Heritage")

SEAWEED

As a food, seaweed is not widely accepted in Western countries, but is used extensively in Asia, especially in Japan. It is used in a variety of dishes, from soups and salads to vegetable dishes. Generally it is purchased dry either in sheets or tangles. The most popular forms are kelp, *nori, wakame*, and *hijiki*.

Kelp (*kombu*) soup stock is the Japanese equivalent of beef or chicken soup stock. To make the stock, the kelp is dropped into hot water, boiled for a few minutes, and then taken out.

Wakame is sold dried, but restored to the fresh green state by briefly soaking in warm water. It is used in soups and salads.

Hijiki is a small black seaweed which must be soaked for a few hours until soft. It can be fried in oil with sliced carrots or onion, with soy sauce and sugar added at the end of frying.

Nori is pressed, sun-dried algae sold in the form of paper-thin sheets. It is purplish-black in color. Nori is used in the following "Westernized" version of Sushi.

Sushi Rolls

(Seaweed stuffed with vinegared rice)

Vinegared Rice

¼ cup rice vinegar
3 tablespoons sugar
2 teaspoons salt
1 tablespoon sherry (or rice wine if available)
2 cups raw rice (washed until the water runs clear)

Mix vinegar, sugar, salt, and sherry together. Wash and cook the rice with an equal amount of water. When the rice is done and still hot, place on a shallow platter, pouring the vinegar mixture over the rice and mixing with a few, swift strokes at the same time. Allow the rice to cool.

Sushi

6 cups vinegared rice (recipe above)
6 dried mushrooms
9 shrimp
2 eggs
6 sheets nori cut in half lengthwise
18 sprigs coriander or spinach leaves

Soak mushrooms in 1 cup warm water until soft (about 20 minutes). Squeeze lightly; remove stem; slice in ½ inch pieces; and cook in 2 tablespoons soy sauce, 1 tablespoon sugar, and ½ cup of the soaking liquid. Parboil coriander or spinach for 30 seconds or so. Drain and cool. Skewer the raw cleaned shrimp lengthwise with a toothpick to prevent curling and cook in boiling water just until done. Drain and cool. Beat eggs one at a time and fry in approximately 2 teaspoons oil to make crepes. Cool and slice in ½-inch strips.

To assemble: Place a nori sheet flat. Spread a fistful of rice on, leaving 1½ to 2 inches on the far side. Lay egg slices on, then coriander or spinach, then mushrooms, then whole shrimp. Roll up tightly, pushing hard to eliminate air spaces (a bamboo mat will help). Refrigerate for 10 minutes. Slice in rounds and serve at room temperature with soy sauce. *(Source: Dean)*

SERVING FINFISH AND SHELLFISH

Sauces

There are numerous sauces that make tasty accompaniments to finfish and shellfish. A simple sauce of melted butter and chopped fresh parsley accompanied with a lemon wedge is all you need to enhance the delicate flavor of poached fish. Other butter-based, egg-based, or tomato-based sauces lend themselves to a variety of fish and shellfish. Choose one of the sauces below or create your own.

Newburg Sauce

½ cup butter
¼ cup all-purpose flour
½ teaspoon salt
1/8 teaspoon cayenne pepper
3 cups half-&-half
6 egg yolks, beaten
1/3 cup sherry

In a saucepan, melt butter. Stir in flour, salt, and cayenne pepper. Add half-&-half gradually and cook until thick and smooth, stirring constantly. Stir a little of the hot sauce into the egg yolks; add to remaining sauce, stirring constantly. Remove from heat and slowly stir in sherry. *(Source: Dean)*

Blender Hollandaise Sauce

3 egg yolks
2 tablespoons lemon juice
dash cayenne pepper
½ cup butter

Place egg yolks, lemon juice, and cayenne pepper in blender container. Cover; quickly turn blender on and off. Heat butter until melted and almost boiling. Turn blender on high speed; slowly pour in butter, blending until thick and fluffy, about 30 seconds. Heat over warm, not hot, water until ready to serve. Makes 1 cup. *(Source: Dean)*

Remoulade Sauce

¼ cup tarragon vinegar
2 tablespoons prepared brown mustard
1 tablespoon catsup
1½ teaspoons paprika
¼ teaspoon salt
¼ teaspoon cayenne pepper
½ cup salad oil
¼ cup chopped celery
¼ cup chopped green onion
1 tablespoon chopped parsley

In a small bowl combine vinegar, mustard, catsup, paprika, salt, and cayenne. Slowly add salad oil, beating constantly. Stir in celery, green onion, and parsley. Allow to stand 3 to 4 hours to blend flavors. Makes 1¼ cups. *(Source: Dean)*

Cucumber Butter

1 medium cucumber, peeled and seeded
½ cup margarine or butter
2 tablespoons lemon juice
4 drops liquid hot pepper sauce
2 teaspoons dill weed
1 teaspoon grated onion
¼ teaspoon salt

Cut up cucumber and blend in a blender until smooth. In a mixing bowl, whip margarine until light and fluffy. Gradually beat in cucumber (about ½ cup) and continue beating until combined and smooth. Blend in lemon juice, dill weed, onion, salt, and liquid hot pepper sauce. Serve at room temperature with fried fish. Makes 1¼ cups. *(Source: Dean)*

Spicy Garden Tomato Sauce

2 tablespoons margarine or butter
¼ cup minced onion
¼ cup chopped green pepper
1 clove garlic, crushed
1 can (1 pound) tomatoes, undrained, chopped
2 tablespoons tomato paste
2 tablespoons lemon juice
½ teaspoon liquid hot pepper sauce
½ teaspoon leaf thyme
½ teaspoon salt
¼ teaspoon basil leaves

In a small saucepan, cook onion, green pepper, and garlic in margarine until tender. Stir in chopped tomatoes, tomato paste, lemon juice, liquid hot pepper sauce, thyme, salt, and basil. Heat to simmering. Simmer 5 minutes to blend flavors. Serve hot or cold with fried fish. Makes 2½ cups.

(Source: Dean)

Horseradish Sauce

½ cup prepared horseradish
1 tablespoon all-purpose four
¼ teaspoon paprika
½ teaspoon salt
1 cup half-&-half

In a small saucepan, combine horseradish, flour, paprika, and salt. Stir in half-&-half. Cook until thickened, stirring constantly. Makes approximately 1½ cups. *(Source: Dean)*

Cool Blender Sauce

1 egg
1 teaspoon salt
1 teaspoon sugar
1 teaspoon mustard
2 drops hot pepper sauce
dash of pepper
1 teaspoon instant minced onion
3 tablespoons lemon juice
¾ cup mayonnaise
¾ cup salad oil
1/3 cup chopped parsley
1 clove minced garlic
1 tablespoon horseradish

Put first 9 ingredients into blender jar. Cover; blend a few seconds; uncover; add oil slowly, keeping motor running. Blend until thick and smooth. Add parsley, garlic, and horseradish, and blend until you get a smooth sauce. Serve with smoked fish. Makes approximately 2¼ cups.

(Source: Dean)

HERBS

The wonders of fresh (and dried) herbs should not be underestimated when cooking fish or shellfish. Grilling or broiling fish or shellfish using such herbs as basil, oregano, parsley, rosemary, dill, fennel, tarragon, or sage either sprinkled on or packed into the body cavity yields a tasty treat. In addition, if you have a place to grow your own herbs, you can even save the dried stalks (of basil and fennel especially), soak them in water, and put them on the coals when grilling fish.

SOURCES OF RECIPES

Dean, Laurie. Private recipe collection.
Florida Extension Department. *Florida Seafood: Basics and Beyond*. Pamphlet. Tallahassee, FL.
Maryland Seafood Marketing Authority. *Maryland Seafood Cookbook*. Annapolis, MD.
Townsend, Patsy E. (compiler). *Foods from the Sea*. Sea Grant Program, Virginia Tech, Blacksburg, VA.
United States Government Printing Office. *Let's Cook Fish!* Washington, DC.
United States Government Printing Office. *A Seafood Heritage: From the Plains to the Pacific*. Washington, DC.
Virginia Extension Service. *Calamari: One of the Sea's Delicacies!* Pamphlet. Hampton, VA.

Factors Relating to Finfish Flavor, Odor, and Quality Changes

John Spinelli

John Spinelli (1925-1996)

John Spinelli was an unusually gifted and perceptive research chemist. After graduating in chemistry from the University of Washington in 1949, he spent 14 years as an analytical and research chemist with Food Chemical and Research Foods in Seattle, Washington. For the next 26 years, he worked as a research chemist and then as division director with the National Marine Fisheries Service. Most of his professional life was dedicated to the study of the proteins of fish and fishery products. Much of what he did related to practical applications, but he always proceeded from a well-researched and scientific basis. His unusual combination of the theoretical and practical approaches yielded a continual flow of results valuable to the fish-food industry and to many others, much of which was published in the United States and overseas in more than 60 research papers. He also contributed review articles to several technical books and proceedings of technical conferences, and was in frequent demand as a speaker worldwide.

Spinelli was an Emeritus member and past president (1956) of the Puget Sound Section of the Institute of Food Technology and a member of the American Chemical Society.

His knowledge and interests were wide-ranging. Upon his retirement, he became interested in the concept of chaos — about which he wrote a science fiction novel. He will be greatly missed by his family, his many friends, and his associates across the world.

— Roland Finch

INTRODUCTION

Finfish offer a wide diversity of proteinaceous food. Unlike their landroving counterparts that have been efficiently crossbred and domesticated to less than 100 major species, the number of finfish species consumed by humans easily exceeds several hundred. These seafoods are captured primarily in the wild from oceans, lakes, and rivers that circle the globe.

In general, fish prior to processing have a mild flavor and are a very perishable product. Fish are captured by a wide variety of means ranging from primitive to highly-developed technological methods; the factors that govern their flavor and quality from capture to consumer are both diverse and complex. Added to these inherent factors are other influences such as the location of the catch (water temperature and food supply), and environmental conditions – both natural and those induced by human intervention.

GENERAL SENSORY CHANGES IN FRESH STORED FISH

If we assume that the edibility of fish is related to four main attributes – odor, flavor, texture, and appearance – then fish could not qualify as a quality product if any one of these criteria were not met. Even freshly caught fish would fail if the flesh were bruised or parasitized, or if odors or flavor deviated because of environmental factors. Catch methods also can cause an imbalance in the physiological condition of fish and render it an inferior product. For example, chalkiness in halibut (Patashnik et al., 1964) is related to the extent of struggle in long-line fishing, during which the halibut may struggle for more than a day before being landed. "Burnt" tuna (Cramer et al., 1977) was related to physiological stress and post mortem handling.

Basically, it is recognized that fish, after landing, are altered from their original condition because of three changes that occur in sequence (Figure 1). Assuming the fish are ice-stored, the first quality changes (Phase I) are quite rapid and can generally be recognized as a loss of flavor. These changes, termed autolytic, alter some flavor compounds and/or enhancers associated with fresh fish. Any weekend fisherman, for example, will attest to the superior flavor of freshly caught fish as compared to those stored on ice for two or three days.

In the early part of Phase II (four to eight days of storage), fish lose their characteristically mild flavor and gradually deteriorate in flavor, odor, and texture until they are "stale." In the final stages of Phase II (nine to 12 days), microbial attack is superimposed on the autolytic changes, and the flavor and odor is characterized as "fishy" and ammoniacal. As the fish enter Phase III, organoleptic changes become clearly evident. The eyes become distorted and shrunken, and the red color of the gills deteriorates to dark grey.

PHASE I: AUTOLYTIC CHANGES

The initial organoleptic quality of fish is lost quickly and is somewhat difficult for the average consumer to experience; in most instances the fish enter Phase II even before they are purchased. Myriad volatile compounds have been associated with the organoleptic properties of fresh fish (Jo-

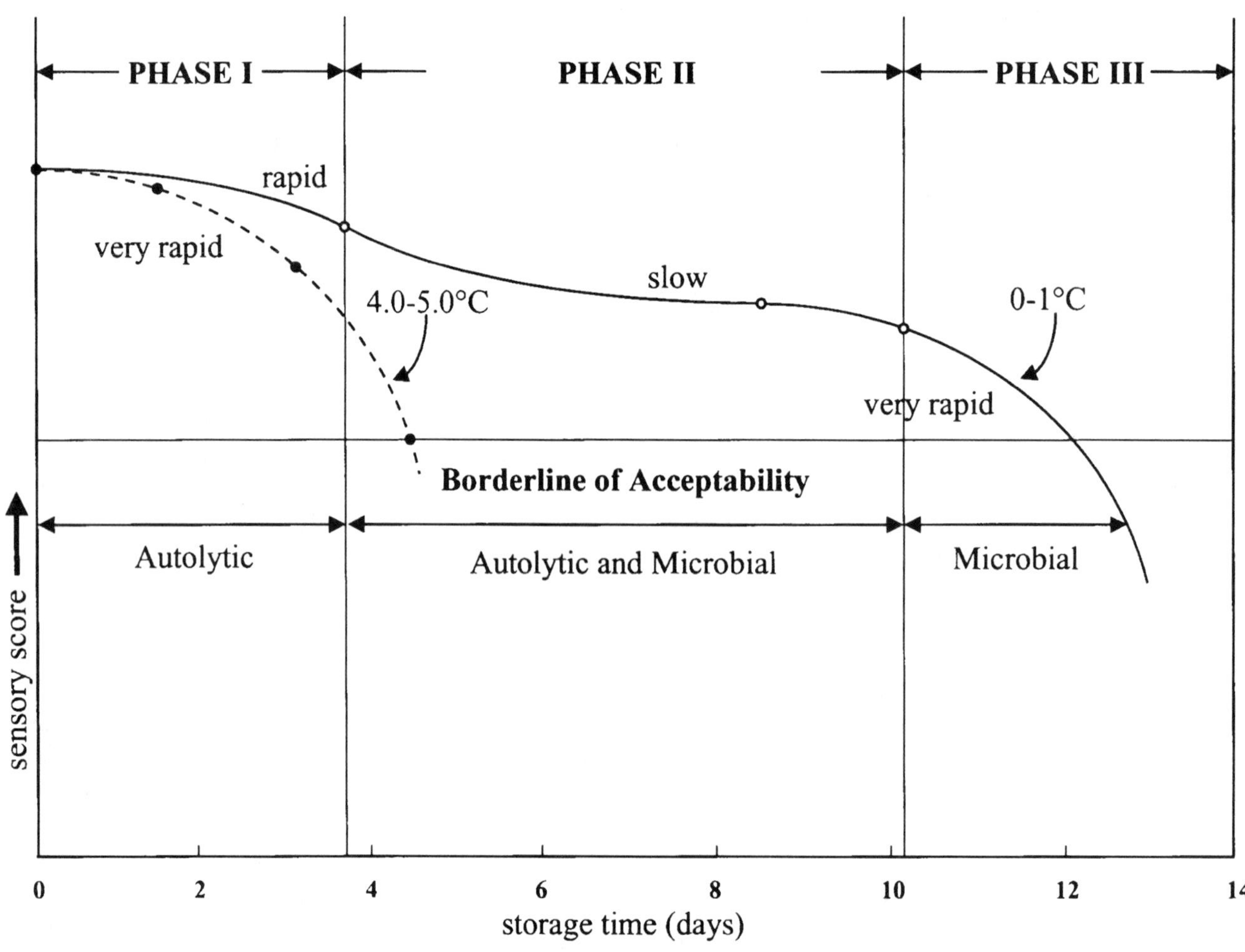

Figure 1. Sensory changes that generally occur in fish stored at 0–1°C and 4.0–5.0°C.

sephson et al., 1987; Diemer, 1965; Mangen, 1959; Mendelsohn and Steinberg, 1962). The identity of these individual organic compounds may be categorized into three major types: alcohols, carbonyls (straight-chained, branched, cyclic, and mixed), and similarly-structured esters.

The organoleptic properties of some of the compounds vary among investigators and different species. For example, Josephson and his coworkers related the volatiles of fresh whitefish (a freshwater species) to at least 13 readily recognized odors. Such terms as "butterlike" were related to 2,3,-butandione; "mushroom" to 1-octen-3-01; "geranium leaves" to 1-5-octadien-3-one; and "cardboardlike" to 2-nonenal. Other investigators (Jellinek and Stansby, 1971; Shewan et al., 1953) have used such descriptive terms as "seaweedy and sweet-honeylike" to convey the odor characteristics of fresh fish.

The inherent dangers and confusion occasioned when each investigator uses his own descriptive terms as a means of ensuring uniformity among panel members had been recognized shortly before the work of Shewan et al. (1953). Cairncross and Sjostrom (1950) noted the need and suggested a flavor-profiling method for standardized terminology that described odor and flavors in food products. Flavor-profiling methods were further suggested by Sawyer and Cardello (1988), Martin et al. (1983), Cardello et al. (1982), King et al. (1980), and Caul (1957). A compilation of this work resulted in the addition of 12 different flavor notes and taints superimposed on the four basic flavor units used to describe the organoleptic characteristics of foods: sweet, sour, salty, and bitter (Table 1).

Because the autolytic changes in post mortem fish are extremely rapid, it is difficult to design an experiment measuring the changes in volatiles in fresh fish. Josephson et al. (1987), for example, reported that the headspace gases of whitefish stored in sealed cans contained significant quantities of high-molecular weight alcohols during the first three days of storage, then dropped dramatically throughout the remainder of the storage period. Short-chain alcohols, however, were hardly in evidence until after four days of storage and did not show significant changes until the fish were well into the spoilage phase. These observations may indeed provide valuable information, but in reality fresh fish are not provided to the consumer in this manner.

Table 1. Twelve Attributes Identified by Consumer and Trained Panelists for Characterizing Finfish Flavors

Flavor Attributes
Overall flavor intensity
Delicate or fresh fish
Heavy or gamey fish
Fish (old fish)
Seaweed
Fish oil
Buttery
Nutty
Musty
Ammonia
Metallic
Shellfish

In his chapter "Fish Flavors," Jones (1967) stated that his collaborators detected metallic flavors when fish were cooked pre-rigor [adenosine triphosphate (ATP) present], but found the metallic note absent when the same fish were cooked post-rigor (ATP absent). He also reported that in some species, specifically plaice, there is an improvement in flavor when fish is cooked shortly after post-rigor.

It is perhaps the absence of odors that consumers can more readily identify as fish freshness. In this respect the flavor influences on fish have been more firmly identified with autolytically-derived, non-volatile compounds, microbial metabolites, and their interactions among themselves and the continually changing substrate in which they reside.

PHASE II: AUTOLYTIC AND MICROBIAL EFFECTS

Autolytic Effects

Since the early 1940s it has been suspected that the catabolism of muscle ATP (Figure 2) played a role in the flavor of fish. It was first reported by Kuninaka (1964) that as early as 1913 Japanese investigators had identified inosine monophosphate (IMP) as a flavor ingredient in dried bonito. It was fortunate these investigators chose to study bonito, because phosphomonoesterases are almost inactive in this species and prodigious amounts (approximately 8.0 micromoles per gram) of IMP

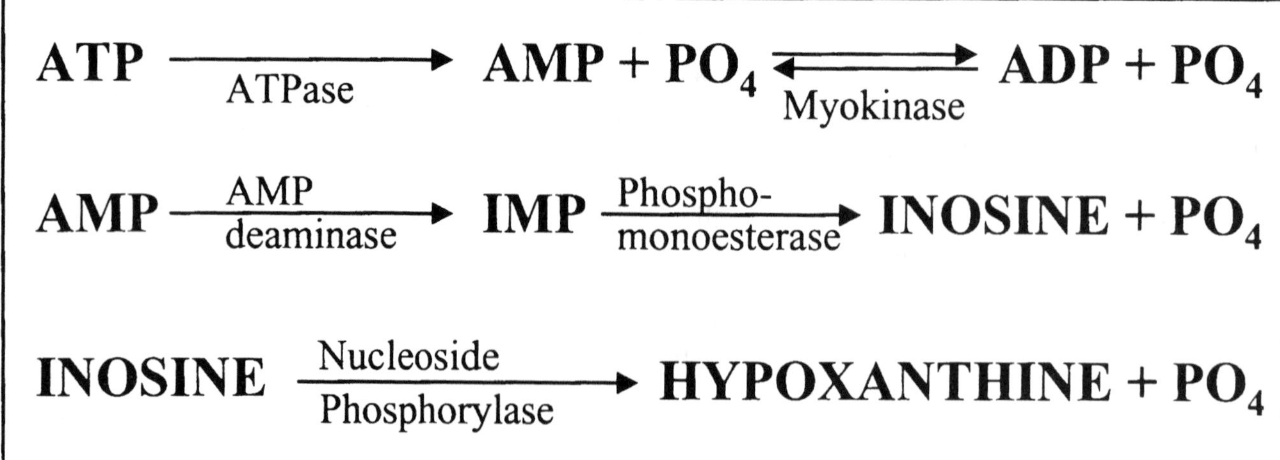

Figure 2. Catabolism of ATP to hypoxanthine and ribose via IMP and inosine.

remain in the fish after 24 days of storage on ice (Endo and Takeda, 1965).

Research in this area was largely ignored until Tarr (1949) reported on the acid-soluble phosphorous compounds in fish muscle. Though his work was primarily concerned with the involvement of bound and unbound sugars in browning reactions, it did prompt other investigators to focus on dephosphorylation reactions and their relation to flavor.

Subsequently, Tarr (1955) extended his investigations on dephosphorylation reactions to show that the rate of IMP degradation and the formation of inosine and free ribose varied significantly among species and set the stage for a worldwide study on the rates of these reactions among species. Other than their effect on flavor, there was interest in measuring the amounts of IMP and its derivatives inosine and hypoxanthine (Hx) as indices of quality. One of the first attempts in this area was the work reported by Shewan and Jones (1957). They measured the amount of free ribose in fish stored at ice temperatures and correlated it with storage time. In a classic paper by Kassemsarn et al. (1963), the pathways of dephosphorylation reaction were clearly elucidated. Today the rate of IMP degradation in almost every commercially important fish has been determined.

Whether nucleotides and their derivatives could be correlated with flavor was still somewhat controversial during the early and mid-1960s. Jones (1965) reported that the flavor of cod when stored under relatively sterile conditions degrades to a flavorless or woolly taste. Fraser et al. (1963) found that in red fish IMP was barely measurable after three days of iced storage, and suggested that IMP was not related to the flavor of this species. Most of the doubt regarding the flavor-contributing factors of IMP was removed when Kuninaka et al. (1964) defined the structural conditions necessary for the two nucleotides IMP and guanosine monophosphate (GMP) to become flavor enhancers (Figure 3). It should be noted that the two-position (designated by X) of the purine moiety must contain either an H, NH_2, or OH group. The substitution of NH_2 on the two-position converts IMP to GMP. Both IMP and GMP appear in today's commercial markets as flavor enhancers.

Further evidence of the flavor-enhancing properties of IMP was supplied by Spinelli and Miyauchi (1968). These investigators irradiated English sole and Pacific ocean perch fillets at 0.25 megarads, sealed them in pouches and stored them at 33° to 35°F for 14 to 19 days. At the end of these storage periods, IMP levels were undetectable in both species. The fillets were divided into two groups and treated thus: 258 milligrams of salt per 100 grams of fillet were added to the control group, and a similar amount of salt containing one µmole of IMP (disodium 2 H 0 equivalent) per gram of fillet was added to the second group. Immediately after treatment the fillets were deep fat-fried and presented to a consumer panel of about 70 individuals. The results (Figure 4)

I

X = H for IMP; NH_2 for GHP

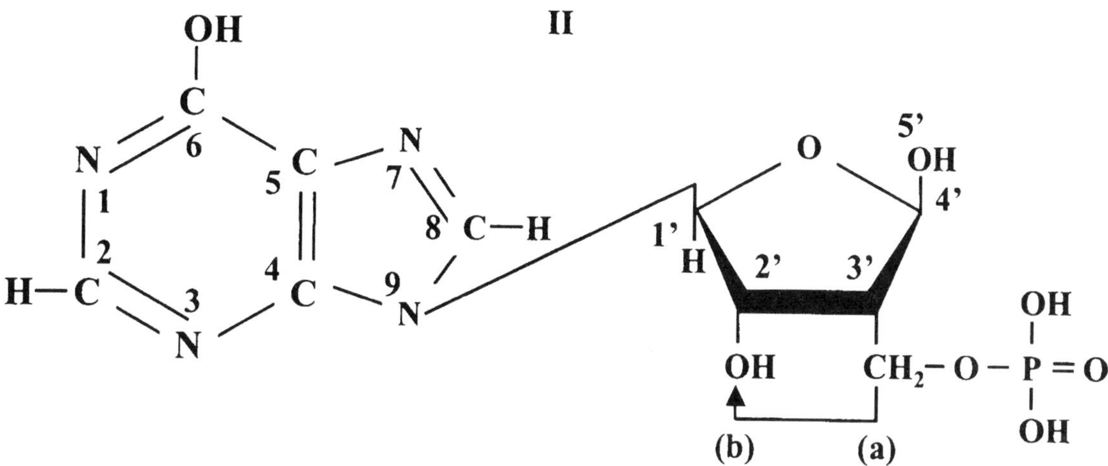

Figure 3. Structural requirements for 5′ flavor-contributing nucleotides (I). Non-flavoring structure of 2′ nd 3′ (II).

show that the fillets treated with IMP were preferred over the controls of Pacific ocean perch and petrale sole, but there was no statistical difference in preference with English sole fillets.

Other data given in this report showed that the threshold level of IMP in Pacific Ocean perch and petrale sole was 0.7 μmoles per gram, and that in bland fish such as English sole or Pacific hake, IMP has less flavor impact than on more flavorful fish.

Further evidence that IMP is a flavor enhancer in fish was provided by Groninger and Spinelli (1968). These investigators inhibited the dephosphorylation of 5 IMP by treating fish fillets with EDTA (ethylenediamine tetraacetic acid), and found that panelists preferred the treated fillets.

These experiments support the observations of Kuninaka (1964) and perhaps reconcile the differences in the results obtained with English sole and those of Fraser et al. (1963), who found little or no flavor in red fish sole and red fish. Both have extremely rapid rates of IMP degradation.

Controversial aspects also surround the impact of IMP derivatives on the flavor of fish. The primary derivatives of IMP are inosine, Hx, free ribose, and phosphate. Kazeniac (1963) believed inosine to be a bitter component in chicken, and Jones (1963) reported it (depending on pH and concentration) to be tasteless in chill-stored cod. Hashimoto (1964) stated that the Japanese investigators Udo and Sato (1962) found it tasteless.

Cann et al. (1962) attributed the bitterness formed in chill-stored cod to Hx. Spinelli (1965) found Hx to be bitter in aqueous concentrations approaching 0.0005 percent, but not in irradiated petrale sole fillets with maximum levels of Hx (5.5 micrograms/grams).

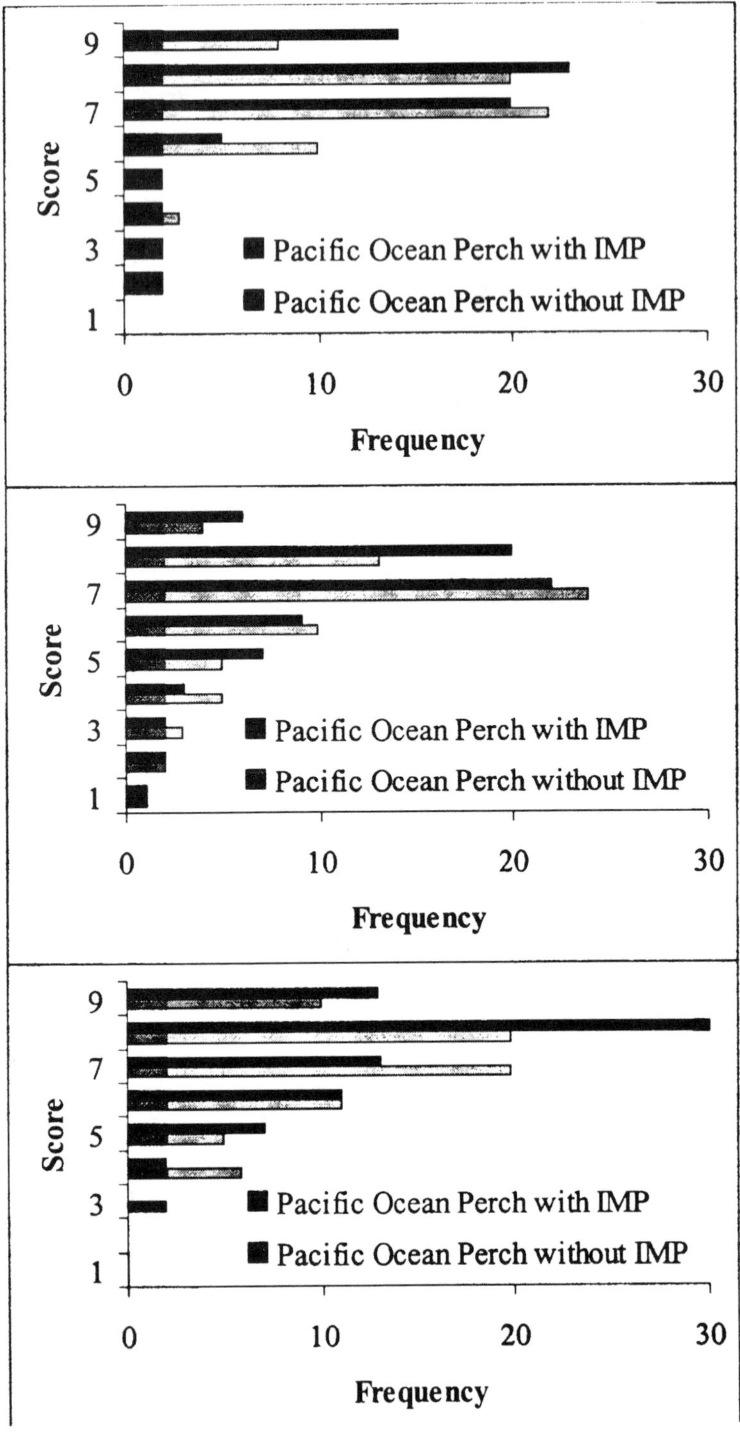

Figure 4. Comparison of scoring distribution of preference panel (9-point hedonic scale) on three species of 14- to 19-day-old irradiated fish with 1μM/g of added IMP and without IMP.

In more concentrated solutions, flavor notes of salty and sour were often detected by panelists.

Spinelli and Miyauchi (1968) reported that when fish such as petrale sole, English sole, and Pacific ocean perch were irradiated, trained panelists rarely identified bitterness as a flavor change until bacterial counts exceeded 10^6 per gram of fish. Jones (1967), in comments on these findings, related them to pH changes induced by microbial growth in chill-stored cod (not irradiated). Although no further definitive work could be found in this area, Jones' explanation seemed reasonable, but it should also be recognized that differences in the same-but-altered substrate or different substrates may also contribute to the flavor impact of Hx.

Josephson et al. (1987) reported that sweet notes appear in white fish as it approaches spoilage. Similarly, the amino acid taurine (which is prevalent in both fish and meats) has been associated with the "cooked" flavor of meat (Hornstein, 1967), "astringent" in chicken (Kazeniac, 1961), and "bitter" or "tasteless" in marine products (Konosu and Hashimoto, 1964; Udo and Sato, 1962).

Indeed, the relation between substrate and flavor-producing substances becomes readily apparent when one notes the many fish analogue products that have succeeded in the marketplace during the past 15 years. As early as 1984, Russo reported that on a worldwide basis fish analogues and minced blocks made from Alaskan pollock and Pacific hake — which were once considered scrap fish — exceed the dollar value of wild salmon. The production of the fish paste surimi, the starting material from which fish analogues are made, is described by Okada and Tamoto (1986), and Lanier and Park (in another chapter of this book).

IMP Catabolism and Its Relation to Quality

This chapter would not be complete without a brief discussion of the relation between nucleotides and fish quality. The Japanese investigators Ehira and Uchiyama (1986) present a comprehensive report on research leading to what they refer to as the "K" value, or an index for estimating the freshness of fish. The K value is described by the mathematical equation:

$$\frac{\text{Hx} + \text{inosine} \times 100}{\text{inosine} + \text{Hx} + \text{IMP}}$$

Examining the equation shows that as Hx and inosine increase, IMP must decrease; otherwise, K would remain near zero, indicating high freshness. This anomaly is rare, and Ehira and Uchiyama provide a brief explanation of this observation by dividing fish into groups, each showing K values ranging from 0 to 100 percent. They contend that K values are superior to the simple Hx values proposed by Jones et al. (1964). The K value, which includes measurements of all three compounds, provides a clearer picture of freshness, but the measurement of either IMP or Hx may well suffice for a simple quality control method.

Wekell and Barnett (1992) report K values are not used in the United States because the determination uses expensive equipment and is not automated. The first automated freshness meter (KV-101), however, was developed by the Oriental Electric Company Ltd. in 1982 and distributed by the Oriental Yeast Company in both the United States and Canada. The tester was expensive and relied on older technology than is available today.

Paradoxically, even a laboratory of modest means can quickly determine total nucleotides using the method described by Jones and Murray (1964). In this method the neutralized perchloric acid extract of fish tissue is removed by passing an aliquot through a small column containing Dowex 1 (formate form). Absorbency readings made at 250 millimicrons before and after the extracts had passed through the column are related to the nucleotide content of the fish and are referred to as "apparent IMP." Actually, apparent IMP values can be obtained by simply adding 0.5 grams of the resin (Cl-form) to 25 millileters of the neutralized extract, and stirring the mixture for five minutes, before and after making the absorbency readings.

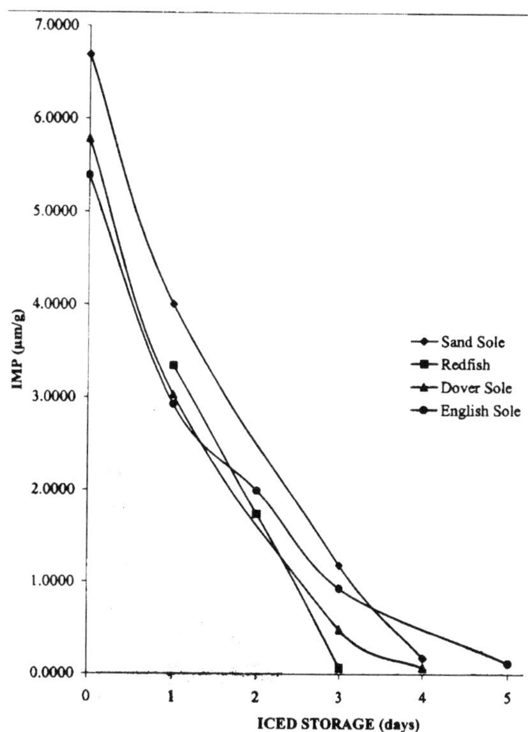

Figure 5. Four different species of fish in which dephosphorylation of IMP is very rapid.

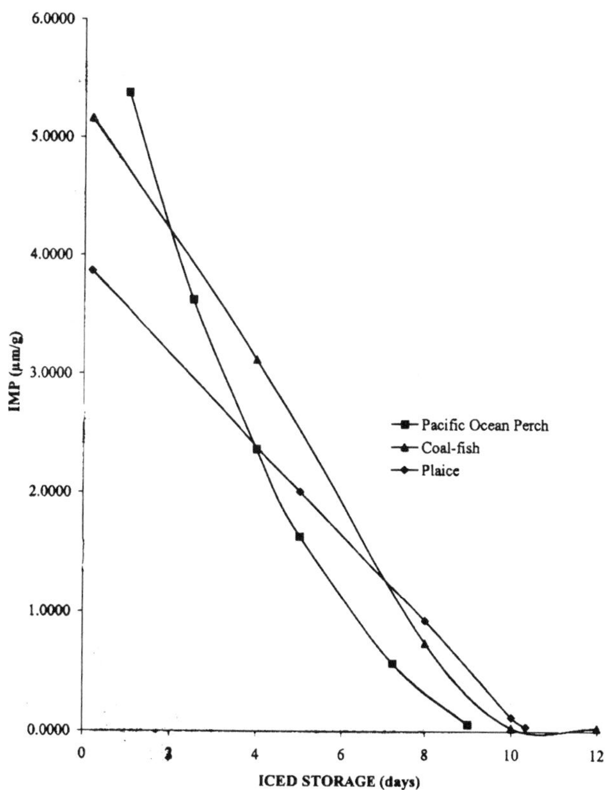

Figure 6. Three different species of fish with intermediate rates of IMP dephosphorylation.

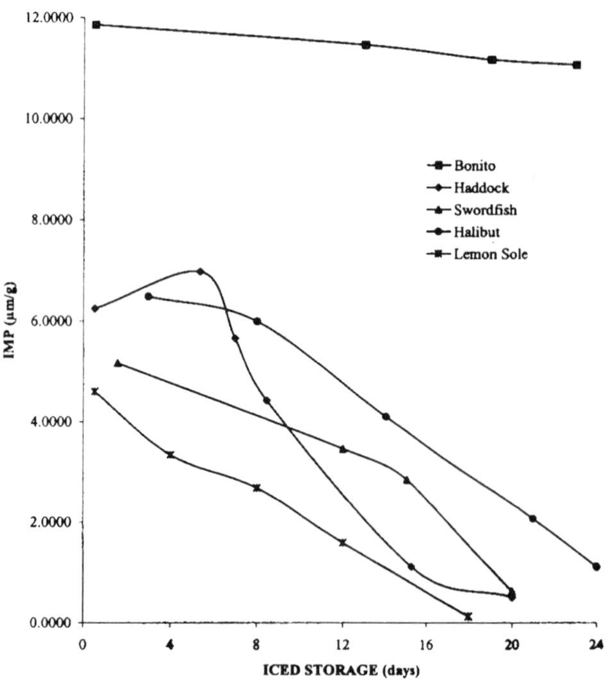

Figure 7. Five different species of fish with slow rates of IMP.

Of course, it is important to know the rate of degradation of IMP. Examples of how these rates vary in different fish are shown in Figures 5, 6, and 7. As an objective measure of freshness, the above tests are deemed more accurate than other electronic testers that are subject to many problems: excess electrolytes, freezing, and pressures exerted on the fish after capture (Barnett, 1994; Schneriger, 1994).

MICROBE-INDUCED FLAVOR CHANGES

As the fish approach eight days of storage (Phase II, Figure 1), odor and flavor changes become apparent. These changes are induced by microbial attack and are superimposed on those resulting from autolysis. The most conspicuous organoleptic changes are formed by metabolic end products produced by the prevalent microorganisms. The nature of the microorganisms is related to: the geographical areas of capture, sanitary conditions aboard the vessel and processing plants, and in some instances handling practices in retail outlets. As a group, these volatiles are predominantly classified as nitrogenous and sulfurous. The nitrogenous fractions associated with ammoniacal odors include ammonia (NH_3) and the three amines: tri-, di-, and mono-methyl amines. There is little doubt that ammonia and these amines are the first olfactory changes that become apparent even to the least discriminating consumer. Trimethylamine (TMA) has been suspected of reacting with partially oxidized lipids (peroxides) to form "fishy" flavors and odors (Stansby, 1962).

TMA is derived from microbial attack on the endogenous compound TMAoxide (Groninger, 1959; Watson, 1939) found in most saltwater finfish and crustacea, but rarely in freshwater species (Reay and Shewan, 1949). Other investigators (Dyer, 1952; Beatty, 1939; Smith, 1936) attribute this phenomenon to the rapid elimination of TMAoxide. Bilinski (1964) reported that there is no evidence of any metabolic pathway leading to the formation of TMA in fish. Goldstein et al. (1967) and Yamada (1967) apparently agree with Bilinski, arguing that TMAoxide is a detoxified form of TMA. Fugiwara-Arasaki and Mino (1972) related the presence of both TMA and its oxide to food sources such as algae when they tested and found their presence in 20 different species of green and red algae.

Regenstein et al. (1982) reported that he found varying contents of TMAoxide in six different fish species; he included data on their relative abundance in the respective organs of the fish. He also reported that the odor threshold of ammonia was 200 times that of TMA. Hebard et al. (1982) provide one of the most extensive reports on the occurrence and significance of amines in fish and crustacea, and is recommended reading for those interested in this area of research. Other higher molecular weight-nitrogenous compounds such as histamine and putrescine have been found in significant amounts in tunas (Farn and Sims, 1987; Hui and Taylor, 1983). Histamine is a highly physiologically-active compound and is formed via the decarboxylation of the amino acid histidine. Several species of bacteria are able to decarboxylate histidine, the most predominant being *Proteus morganii* and *Klebsiella pneumoniae* (Taylor and Speckhard, 1983; Omura et al., 1978).

Although there is little doubt that these compounds contribute to organoleptic changes in the later phases of spoilage and are being considered as important spoilage indicators for tunas, their contribution to early changes is doubtful. In this respect it should be noted that Day (1967) suggested that bleu cheese, which characteristically contains more amino acids and peptides than do fish, does not impart cheese flavor as such, but appears to function by providing a brothy background on which flavor-producing substances reside. The implication here is that the combined effect between the microbial metabolites and altered substrate provides the total flavor characteristics of the cheese. Similarly, changes in the fish substrate such as texture and autolytic and atmospheric alterations of tissue lipids, at first subtly, and then distinctly, alter the organoleptic properties of the fish. These effects are enhanced by time, temperature, storage, and processing conditions. Depending on these conditions, incipient spoilage generally becomes recognizable when total bacterial counts exceed 10^6/grams of fish. It should be noted, however, that fish processed under unsanitary conditions can exceed these bacterial counts before organoleptic changes become evident, and a strong distinction should be made between food spoilage and food poisoning organisms (Dack, 1956).

Contingent on the storage conditions, pre- and post-processing, it is the composition of the microflora contaminating the fish that alerts the consumer to incipient spoilage (Lerke et al., 1965; Shewan et al., 1960). It has been firmly established that specific genera of bacteria are more effective than others in altering the odor of fresh fish. Strangely enough, the majority of microorganisms contaminating fish captured in temperate waters are neither proteolytic or TMA-producing. In the latter portion of Phase II and the beginning of Phase III, *Pseudomonas* are primarily responsible for the production of the highly odoriferous nitrogenous and sulfhydryl compounds such as NH_3, TMA, hydrogen sulfide, and the mercaptans (Pelroy and Ekland, 1966; Shewan, 1960). In irradiation studies Pelroy et al. (1964) found the initial microflora of petrale sole fillets was composed of: 10 percent *Pseudomonas*, 19 percent *Achromobacter*, 22 percent *Micrococcus*, 31 percent *Coryneforms*, 12 percent lactic acid bacteria, and 6 percent *Flavobacteria*. After seven days storage, incipient spoilage was detected and the microflora were composed of 94 percent *Pseudomonas*, 5 percent lactic acid bacteria and 1 percent *Flavobacterium*. Similar shifts in bacterial populations with other species have been reported (Hobbs, 1982; Kazanas and Emerson, 1968; Liston, 1960). There was no disagreement among the aforementioned investigators that *Pseudomonas* bacteria were the most biochemically active organisms and that the least active were lactic acid bacteria. At 33°F *Micrococcus* and *Bacillus* contribute little, if any, activity.

The highly odoriferous sulfur-based compounds occur in the final phases of storage where the odor is often described as vile and putrid (Mendelson and Steinberg, 1962; Shewan, 1960). Similarly, carbonyls such as aldehydes and ketones, along with phenols, esters, ethers, etc., cannot be distinctively identified as organoleptic because of the overwhelming odor of the nitrogenous- and sulfur-based compounds.

The preceding discussion was limited because it is unusual for one to consume fresh fish that has reached this condition. These studies are extremely useful, however, in developing spoilage indices for processed fish where these compounds are not as readily detectable. Ethanol, which forms in the beginning of the spoilage phase in white fish and salmonids (Hollingsworth et al., 1987; Josephson et. al., 1987), serves as a spoilage index because its contribution to flavor has not been established. Similarly, carbonyls, such as aldehydes and ketones, along with phenols, esters, ethers, etc., cannot be distinctly identified because of the overwhelming odors related to *pseudomonas* growth. These studies are quite useful, however, because they form a basis for more accurately developing quality and condition indices. In this context we have come a long way since Farber and Cederquist (1958) believed that the quality of fish could simply be determined by measuring the total volatile-reducing substances in both fresh and processed fish.

Even with the introduction of today's advanced instrumentation, which can readily identify highly complex individual chemical compounds, there is no assurance of their organoleptic impact when fish is tested by trained panelists. For example, Kunosu and Yamaguchi (1982) published an excellent report on the individual flavor and odor compounds in fish and shellfish. They provide the reader with numerous compounds in these marine species by classifying them into distinct groups, and in many instances their chemical structures. Their contribution to flavor, however, is scant because they omit the effect of lipids and instead concentrate their studies on the water-soluble components – such as amino acids, peptides, dephosphorylated compounds, etc. – and assess their relation to flavor. Their conclusions are that it is the water-soluble constituents that are primarily related to flavor. However, it would seem that these investigators face the same dilemma offered by Hall (1968), who sensibly put forth his description of flavor as the sum of those flavor and odor characteristics of any material taken in by the mouth, their impingement on the tactile receptors (temperature, pungency, condiments, etc.), and the perceptive integration of these factors developed by the brain. Hall's remarks were certainly not completely new to Day (1967) and Horstein (1967), who both referred to the fact that when flavor and odor substances are present in different mixtures, quantities, and substrates, a mere change in cooking temperature or pH will change the basic state of flavor. Added to these basics are the 12 subtle flavor notes or taints shown in Table 1.

Actually, the study of autolytic, biochemical, and microbial changes as they relate to flavor has been fairly well established and controlled, even if not completely understood. Research in this area, however, is beginning to turn away from the objective organoleptic characterization criteria of compounds and their relation to flavor and is turning towards flavor profiling — a method that was first grounded on the scientific disciplines outlined by Sjostrom, 1950 (also Sawyer et al., 1988; Caul, 1957). The reason for this shift in research objectives is three-fold: the erosion of the abundance of well-recognized species, the concomitant emergence of new species and fish analogues entering the marketplace, and the relative consumer preferences and economics related to these changes (Kinnucan et al., 1993; Edwards, 1992).

FLAVOR CHANGES IN PROCESSED FISH

For the same reasons we preserve fish today, early mankind preserved fish and other foodstuffs for the purpose of providing a reservoir of food in times of scarcity. Lacking today's technological skills, early methods began with simple expedients such as sun-drying and short-term chill storage. Advances were made by such techniques as smoking, salting, pickling, and uncontrolled fermentation. Modern techniques were ushered in during the mid-nineteenth century when the French confectioner and inventor Nicolas Appert demonstrated the thermal sterilization of vegetables. Appert's preserved peas in wine bottles have led to today's canning industry.

Effect of Canning

Thermal processing produces dramatic organoleptic changes in fish, the most notable being a change in texture. Again, these changes are related to species, their proximate composition and condition at the time of processing, and preparation of the fish prior to final processing. These include skinning, boning, and precooking. The addition of salt, condiments, and other flavor-altering ingredients such as tomato paste, mustard, protein hydrolysates, condensed phosphates, etc., contribute to an almost-new food form when the product reaches the marketplace.

Even though autolytic, microbial, and enzymatic activity are eliminated by the long and high temperatures used to ensure a sterile product, chemical changes have been shown to occur in canned fishery products. Stansby found that TMA accumulates in canned salmon (1962). Pre-cooking was found to increase the DMA and TMA content of sardines (Hughes, 1959). In canned tuna, scorch (an orange-yellow color) produces a burnt flavor. It most often develops where tuna muscle touches the lid of the can. Other unsightly colors also develop in tuna, such as pink pigments (Brown et al., 1958; Brown and Tappel, 1958). Greening in tuna is quite common and develops during thermal processing. The chemistry of its formation is related to a reaction between TMA and cysteine; a method for predicting its appearance was reported by Nogaoka et al. (1971). Landgraf (1976) expressed the conventional wisdom – "the ultimate product form will be of no better quality than the raw material from which it comes"; however, the ultimate product can certainly be of poorer quality than the raw material. Pre-cooking can result in a significant loss of flavor and odor-producing substances contained in such products as sardines and tuna (Slabyj, 1978; Finch and Courtney, 1976). Greening, for example, occurs in perfectly good fish; other visually distracting products such as curd form in canned salmon.

Curd is a form of coagulated soluble proteins that is most obvious on the surface and often streaks throughout the product. Curd itself does not produce an "off-flavor," but its presence leads to the perception of an inferior product. Successful attempts to prevent its formation have been reported by Yamamoto and Mackey (1981), who added papain to pink salmon tins prior to sealing, and Wekell and Teeny (1988), who dipped pink salmon steaks for 30 to 120 seconds in 5 percent sodium tripolyphosphate containing 2 percent salt. It can be seen, therefore, that changes induced by thermal processing could disguise the quality of marginal fish to that displayed by higher quality fish. The need for objective tests is obvious.

Effect of Freezing

Though freezing proteinaceous products represents the best method for maintaining a fresh-like quality, it is axiomatic that it will not improve that quality of the fish. It has been well established that if good freezing procedures are employed, but poor or fluctuating temperatures are encountered during distribution and storage, detrimental quality changes will definitely occur. The salient factors related to the changes were presented by Pedraja (1972). These include:

1. Condition of the fish at time of capture — e.g., the state of nutrition, environmental factors, season, and age of the fish.
2. Methods of capture, handling, and processing prior to freezing — i.e., bleeding.
3. The state of rigor prior to processing. He emphasizes that fish should pass through rigor prior to freezing, but tempers his remarks by stating that shore-based processing plants should allow fish to pass through rigor at lower temperatures; with on-board vessel processing higher temperatures followed by fast-freezing is recommended.
4. Frozen fish and fishery products should be held as close to constant temperature as possible (preferably below -10°F). He points to fishsticks becoming rancid within three months if they are stored at fluctuating temperatures between -20° and +10°F.

Chemical Reactions

Today, most consumers are familiar with the term "fresh-frozen." Contrary to the quality information distributors may wish to convey, fastidious or even infrequent consumers are aware that high-quality fresh fish differ from high-quality frozen fish. For example, fresh halibut has a texture somewhat similar to poached egg, while frozen halibut, held in storage for even a short time, begins to exhibit a texture that is fibrous.

The flavor of frozen fish held in storage for more than six months is affected by two major changes: oxidative and texture alteration. These changes, to a large degree, can be controlled by procedures outlined by Pedraja and by pre-treatment of the fish with antioxidants (Anderson and Danielson, 1961; Tarr, 1947), before or after freezing. Practically all frozen fish is glazed with aqueous solutions that may or may not contain additives. Some processors may add from 0.5 to 2.0 percent corn syrup solids to their glazing solutions to prevent cracking and exposing the fish surface to atmospheric oxygen (Nelson, 1994). Tarr (1949) suggested that antioxidants added to glazes improved their effectiveness by preventing lipid degradation, which leads to rancid flavors. Whether antioxidants are as useful for glazing whole fish as they are for fillets is open to some question. Tarr (1947) tested ascorbic acid and ethyl and propyl gallate in minced salmon and herring flesh. After 176 days in storage (-20°C), salmon treated with 0.05 percent ascorbic acid had peroxide values of 0.2, and those treated with ethyl and propyl gallate had peroxide values of 2.0 and 1.5 respectively. After 150 days storage, herring treated with ascorbic acid reached values of 2.6, while those treated with ethyl and propyl gallate were 0.09 and 2.9. Both untreated salmon and herring controls had similar peroxide values of 5.5 and 5.6.

Research on treating fresh fish with antioxidants (ethoxyguin and tocopherols) has never produced promising results. Stansby and Olcott (1976) believed that the difficulty resided in the inability of the antioxidants to penetrate the fish tissue and inhibit the catalyzing effect of heme substances (Tappel, 1962; Watts, 1954). Attempts to overcome this problem were outlined by Spinelli and Miller (1983), who developed a type of ham-stitching device in which the tip of the needle is blocked and

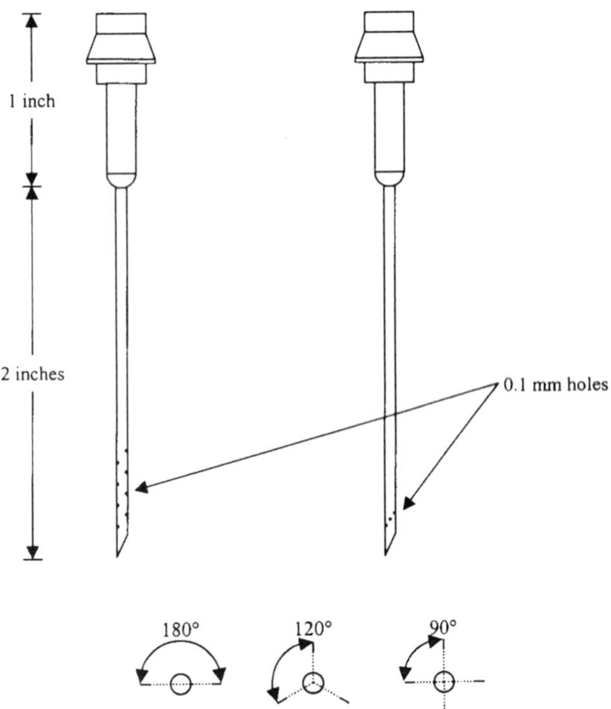

Figure 8. Experimental needles used in evaluating a high-pressure injection machine.

the antioxidant solution injected at high pressure from the small holes drilled along the barrel of the needle (Figure 8). A laboratory prototype using a block of 12 needles to study the distribution of aqueous additives showed good results, but a commercial unit constructed by the Intermediate Technology Manufacturing Company (1986) did not prove adequate for industry requirements.

Other attempts to solve flavor problems arising from microbial growth and oxidative rancidity turned to storing fresh fish in modified atmospheres — nitrogen and carbon dioxide (CO_2) — and/or storing fresh fish in refrigerated seawater saturated with CO_2 (Gill and Tan, 1980; Nelson and Barnett, 1971). Storing fresh fish and fillets in modified atmospheres has produced moderate to good results (Parkin and Brown, 1982), while storing fish in refrigerated seawater saturated with CO_2 has not met with success. Fish stored in refrigerated seawater is still used extensively, particularly in the salmon industry.

Texture Alteration

Frozen fish, particularly the gadoids, show distinct textural changes during storage that can range from the loss of flakiness to outright toughness. These changes are more prominent in fish stored at about -20°C. The reactions causing these changes have been attributed to the reaction of fatty acids and/or formaldehyde (FA) with the structural (myofibrillar) proteins. The two reactions should be considered distinct, with most investigators favoring the protein-FA reaction. Another plausible explanation for texture alteration is attributed to the concentration of salts during freezing; this concentration follows cellular breakdown by the denaturing effect of the salts on structural proteins (Love et al., 1965). Tanaka (1965) and Liljemark (1965) examined fish tissue by electron microscopy with the aim of describing textural alteration by changes in the biochemical structure of the protein. Aitken and Connell (1977) followed this work with the hypothesis that freezing changes the interfilament distances among the myofibrillar proteins, which in turn results in lower water-holding capacity and toughness. None of these investigations yielded positive results, and Connell was forced to conclude that texture alteration results from an increase in the hydrophobicity of the proteins.

In other studies conducted primarily at the National Marine Fisheries Laboratory, Gloucester, Massachusetts, the reaction between free fatty acids and structural proteins was extensively examined (Anderson et al., 1965; Anderson and King, 1963; Anderson and Steinberg, 1963; King et al., 1962). This research at Gloucester was an expansion of the work started by Dyer and Dingle (1961), who speculated that toughness in frozen fish resulted from free fatty acids formed in fresh fish before freezing reacted with structural proteins. Because most of the work at Gloucester was done with model systems, the study was not considered conclusive. However, Dyer's and Dingle's research did verify that the protein in fish such as gadoids was less resistant to protein alteration than fatter species such as plaice, ocean perch, and salmon.

These observations motivated other investigators to study textural alteration by placing particular emphasis on the reaction between FA and proteins (Castell et al., 1973; Tokunaga, 1965; Amano and Yamada, 1964). These investigators noted considerable amounts of FA and DMA in frozen cod, and Babbitt et al. (1972) suggested that DMA should be used as a quality index of frozen fish rather than TMA. Prior to 1979 most investigators believed that DMA was produced from TMAO by the specific enzyme TMOase. Actually, most of this research was done on fish organs such as the liver, pyloric seca, kidney, and spleen, and no evidence has been presented that the specific enzyme TMOase has been isolated from fish tissue.

Spinelli and Koury (1981, 1979) expressed considerable doubt whether TMAOase was responsible for TMAO degradation in gadoid tissue. They demonstrated that two catabolites of cysteine – i.e., cysteine sulfinic acid and hypotaurine – were responsible for TMAO degradation and that the two typically follow the reaction described by Vaisey (1956): cysteine + Fe^{+2} —> DMA + FA.

Effect of Curing

Cured fish, as described by Stansby (1976), are fish products in which preservation is achieved by reducing the moisture content and adding chemical preservatives, or both. He adds that, early in the twentieth century and even in the United States, curing was the principal method of preserving fish. Salt (NaCl) or a combination of salt and vinegar (brines), with or without chemical additives or condiments, were the preservation methods of choice. Generally, and particularly with fatty species, the fish are kept below the surface of the brine during the curing phase. This forms a natural barrier to prevent atmospheric oxidation of the lipids. Though mild curing and smoking still command a significant portion of the seafood industry, pickling has not enhanced its position in the marketplace. Of the 17 countries that are attempting to efficiently upgrade their fishery resources, only one, Korea, included pickling as a method of preservation (Da Casta, 1974). The flavor characteristics of pickled fish are "built in" to the product and are tailored to fit the desires of various ethnic groups. These products, depending on the salt, vinegar, and assorted condiments used, possess a moderate-to-strong flavor distinctly related to those of marine origin: herring, anchovies, sardines, and shrimp. Flavor alterations of concern relate to the quality and preparation of the fish prior to processing and the development of rancidity.

Effect of Drying and Smoking

It is commonly known that the removal of water retards the growth of microorganisms and limits enzymatic activity. Because fish oils contain a much larger percent of unsaturated fatty acids than do animal and vegetable oils, they are more susceptible to oxidative and heme (substance) catalyzed reactions. The ease of non-enzymatic free-radical oxidation is generally proportional to the number of ethylene bonds (-C=C-C-C=C)n, and their oxidation products yield a bewildering array of individual compounds (Dratz and Deese, 1986; Olcott, 1962). These reactive products cause a number of organoleptic changes ranging from acridness to rancidity, and in interactions of lipid-protein-amine types, the term "lingering aftertaste" is linked to the description of the product.

Hardy and McGill (1979) provide a brief description of the historical and modern aspects of drying, salting, and smoking and suggest that smoking was a "spin-off" of early drying methods. Modern kilns were introduced in the 1930s, and controls to improve their smoking parameters were added during the 1950s. Though liquid smoke flavor (a condensate obtained from wood smoke) may be added to semi-cured fish, it is suggested that only a small increase in shelf life is gained from the antibacterial and antioxidant effects of the additive; the main aim of modern smoking processes is flavoring, not preservation. Smoked products do, however, contain a wide variety of forms whose moisture contents range from high to extremely low. They are provided to the consumer in wrapped (permeable and impermeable), frozen, and canned forms and are often subject to temperature restrictions related to processing conditions. Salts (NaCl or combinations of NaCl and KCl, where KCl is used as a sodium diluent) are used most effectively as antimicrobials (Pelroy et al., 1985). In this context the preservation of dried, salted, and smoked fish has a relation to the amount of bound water (water activity) remaining in the products (Labuza et al., 1972; Martinez and Labuza, 1968). In contrast to frozen fish, where free water exists in substantial quantities even at temperatures below -27.5°C, hard-smoked and dried fishery products can contain low (about 2 percent) amounts of free water (Sussman and Chin, 1965). This is particularly true with products containing high quantities of salt, various disaccharides, and polyhydric alcohols in which water is immobilized by hydrogen-bonding and not free to act as a vehicle for the transport of enzymes leading to organoleptic changes. Freeze-dried products, however, do not enjoy wide consumer acceptance because of the considerable loss of volatile flavors and textural alterations due to their poor rehydration properties (Koury and Spinelli, 1975).

SUMMARY AND CONCLUSIONS

In 1994, Americans consumed 15 pounds of fish per capita, or four more pounds than they did in 1980. Because the price of fish has not decreased, it is only logical to assume that at least three factors are responsible for this per capita increase: the nutritional benefits of fish consumption, the improvement of product quality, and the introduction of new species and prepared fishery products entering the marketplace.

Fisheries have advanced on a broad front, utilizing the technological data gathered and developed by government and industry inquiry. The industry has now arrived at a point where the array of new products and species entering the market requires quality control systems to be as rigid as for any other segment of the food industry. Consider the fact that in the late 1960s and early 1970s, pollock and hake were considered scrap fish, Pacific rock fishes a "poor man's" food, and most squid relegated to bait stature. Aquaculture in the marine environment in some quarters was still considered a "pipe dream," as were fish analogues.

The days have passed where fish were primarily critiqued by subjective judgments that indicated fish were either "good or not so good." Not all of the chemical reactions and interactions are known, but it is known how to control the fundamentals to allow industry to reproduce quality products on a reasonably reliable basis. Though product quality in fresh and frozen fish can be assessed before they enter the marketplace, there is still room for improvement in their storage and distribution, and in conditions in the marketplace. Unfortunately, not enough action has been taken to address the detrimental effects of low and fluctuating temperatures. Bearing in mind that most fish are still captured in the wild, we are still at the mercy of nature for the quantity and quality of available fish. Training panelists and inspectors to use standardized descriptive terms has not proceeded with the same zeal as when it was first initiated by the National Marine Fisheries Service (NMFS) in the mid-1980s (Wekell and Barnett, 1992). Constant attention must be given to maintaining product quality and uniformity, since the industry is dependent on the harvest cycles of a resource that must be hunted — the exception being aquaculture. In the final analysis, the desires and concerns of the ultimate judge — the consumer — must be met.

REFERENCES

Aitken, A. and J.J. Connell. 1977. The effects of frozen storage on muscle filament spacing in fish. *Bulletin International Institution of Refrigeration*. Kor Isruhe, Fed. Rep. of Germany. **1**: 187-191.

Amano, K. and K. Yamada. 1964. A biological formation of formaldehyde in the muscle tissue of gadoid fish. *Bulletin of Japanese Society Science Fisheries*. **30**:430-435.

Anderson, M.L., M. Steinberg, and F.J. King. 1965. Some physical effects of freezing fish muscle and their relation to protein-fatty acid interaction. *The Technology of Fish Utilization*. Rudolph Kreuzer (ed.), pp. 105-111. Fishing News (Books) Ltd. London.

Anderson, M.L., F.J. King, and M. Steinberg. 1963. Effect of linolenic, linoleic and oleic acids on measuring protein extractability from code skeletal muscle with the solubility test. *J. Food Sci.* **28**:286-288.

Anderson, K. and C.E. Danielson. 1961. Storage changes in frozen fish. A comparison of subjective and objective tests. *Food Technol.* **15**:(2)55-57.

Babbitt, J.K., D.L. Crawford, and D.K. Law. 1972. Decomposition of trimethylamine oxide and changes in protein extractability during frozen storage of minced and intact hare muscle. *J. Agric. Food Chem.* **20**:1052-1054.

Barnett, H. 1994. Personal communication. National Marine Fisheries Service. Seattle, WA.

Beatty, S.A. 1939. Studies on fish spoilage, III. The trimethyloxide content of the muscles on Nova Scotia fish. *Journal of Fisheries Research Board of Canada*. **4**:229-232.

Bilinski, E. 1964. Biosynthesis of trimethylammonium compounds in aquatic animals, IV. Precursors of trimethylamine oxide and betaines in marine teleosts. *J. Fisheries Research Board of Canada*. **21**:765-770.

Brown, W.P., A.L. Tappel, and H.S. Olcott. 1958. Pigment off-color in cooked tuna meat. *Food Research*. **23**:262-265.

Brown, W.D. and A.L. Tappel. 1958. Oxidative changes in hematin pigments of meats and fish. *Walterstein Laboratory Communications*. **21**:299-308.

Cairncross, S.E. and L. Sjostrom. 1950. Flavor profiles — new approach to flavor problems. *Food Technology*. **4**:308-311.

Cardello, A.V., F.M. Sawyer, O. Maller, and L. Higman. 1982. Sensory evaluation of texture and appearance of 17 species of North Atlantic fish. *J. Food Sci.* **47**:1818-1823.

Castell, C.H. and B. Smith. 1973. Measurement of formaldehyde in fish muscle using tea extraction and nash reagent. *Journal Fisheries Research Board of Canada.* **30**:91-108.

Castell, C.H., B. Smith, and W.J. Dyer. 1973. Effects of formaldehyde on salt extractable proteins of gadoid muscle. *Journal Fisheries Research Board of Canada.* **30**:1052-1054.

Caul, J.F. 1957. The profile method of flavor analysis. *Advances in Food Research.* **7**:1-40.

Cramer, J.I., R.S. Shomura, and S.H.H. Yuan. 1977. The problem of burnt tuna in the Hawaiian fishery. *Indo-Pacific Fisheries Commission.* **18**:213-223.

Dack, G.M. 1956. Evaluation of microbiological standards for foods. *Food Technology.* **10**:507-509.

Day, E.A. 1967. Cheese flavor. *Symposium on Foods: The Chemistry and Physiology of Flavors.* H.W. Schultz, E.A. Day, and L.M. Libbey (eds.), pp. 331-361. AVI Publishing Co. Inc. Westport, CT.

Diemar, W. 1965. Gas chromatography in the analysis of volatile odor and flavor compounds in fish flesh. *Technology of Fish Utilization.* Rudolph Kreuzer (ed.). Fishing News (Books) Ltd. London.

Dratz, E.A. and A.J. Deese. 1986. The role of dossosahexaenoic acid (22:6W3) in biological membranes: examples from photoreceptors and model membrane bilayers. In *Health Effects of Polyunsaturated Fatty Acids in Seafoods.* A.P. Simpopoulos, R.E. Kifer, and R.E. Martin (eds.), pp. 319-351.

Dyer, W.J. and J.R. Dingle. 1961. Fish proteins with special reference to freezing. In *Fish as Food I.* George Borgstrom (ed.), pp. 275-327. Academic Press. New York.

Dyer, W.J. 1952. Amines in Fish Muscle VI. Trimethylamine oxide content of fish and marine invertebrates. *Journal of Fisheries Research Board of Canada.* **8**:314-324.

Edwards, S.F. 1992. Evidence of structural and preference change for seafood. *Marine Resources Economics.* **7**:141-152.

Endo, K. and M. Takeda. 1965. Studies on the extracts of muscle tissue of tuna. *Tuna Fishing Majuro Gyogya.* **36**:38-45.

Farns G., and G.F. Sims. 1987. Chemical indices of decomposition in tuna. In *Seafood Quality Determination.* D.E. Kramer and J. Liston (eds.), pp. 175-183. Elsevier Publishing Co. Inc. New York.

Farber, L. and A. Cederquist. 1958. The determination of volatile reducing substances for the chemical assessment of the freshness of fish and fish products. *Food Technology.* **12**:677-680.

Finch, R. and G. Courtney. 1976. *The Tuna Industry: Industrial Fishery Technology.* M.E. Stansby (ed.), pp. 91-109. Robert E. Krieger Publishing Co. Huntington, NY.

Fraser, D.I., S.G. Simpson, and W. J. Dyer 1963. Very rapid accumulation of hypoxanthine in the muscle of redfish stored in ice. *J. Food Science.* **25**:817-821.

Fugiwara-Arasaki, T. and N. Mino. 1972. The distribution of trimethylamine and trimethylamine oxide in marine algae. *Proceedings 7th International Seaweed Symposium.* Sapporo, Japan, pp. 506-510.

Gill, C.O. and K.H. Tan. 1980. Effect of carbon dioxide on growth of spoilage bacteria. *Applied Environ. Microbiol.* **39**:317-319.

Goldstein, L., S.C. Hartman and R.P. Forster. 1967. On the origin of trimethylamine oxide in spring dogfish. *Compendium Biochemical Physiology.* **21**:719-722.

Groninger, H.S. and J. Spinelli. 1968. EDTA inhibition of inosine monophosphate dephosphorylation in refrigerated fishery products. *Agric. and Food Chem.* **16**:97-99.

Groninger, H.S. 1959. The occurrence and significance of trimethamine oxide in marine animals. U.S. Fish and Wildlife Service. Special Science Report: Fisheries. **333**.

Hall, R.L. 1968. Food flavors: benefits and problems. *Food Technology.* **22**(11):1388-1392.

Hardy, R. and A.S. McGill. 1979. Smoking of foods: methods and some toxicological aspects. In *Processing Biochemistry.* Wheatland Journals Ltd. Watford, England. 2. Tory Memoir No. 598.

Hebard, C.E., G.J. Flick, and R.E. Martin. 1982. *Occurrence and Significance of Trimethylamine Oxide and Its Derivatives in Fish and Shellfish.* R.E. Martin, G.J. Flick, C.E. Hebard and D.R. Ward (eds.), pp. 1149-304. AVI Publishing Co. Westport, CT.

Hobbs, G. 1982. Changes in fish after catching. In *Fish Handling and Processing.* A. Aiken, I.M. Machie and J.H. Merritt (eds.), pp. 20-27. Edinburgh.

Hollingsworth, T.A. Jr., H.R. Throm, and M.M. Wekell. 1987. Determination of ethanol in canned salmon. In *Seafood Quality Determination.* D.E. Kramer and J. Liston (eds.), pp. 153-160. Science Publishing Co., Elsevier, NY.

Hornstein, I. 1967. Flavor of red meats. In *Symposium on Foods: The Chemistry and Physiology of Flavors.* H.W. Shultz, E. A. Day, and L. M. Libbey (eds.), pp. 228-250. AVI Publishing Co. Inc. Westport, CT.

Hui, J.Y. and S.L. Taylor. 1983. High pressure liquid chromatographic determination of putrefactive amines in foods. *J. Assoc. of Official Analytical Chemists.* **66**:853-857.

Intermediate Technology Manufacturing Co. 1986. Pamphlet. Bellevue, WA.

Jellinek, G. and M. E. Stansby. 1971. Masking undesirable flavors in fish oils. *Fisheries Bulletin.* **69**:215-222.

Jones, N.R. 1967. Fish flavors. In *Symposium on Foods: Chemistry and Physiology of Flavors.* H.W. Shultz, E.A. Day, and L.M. Libbey (eds.), pp. 267-295. AVI Publishing Co. Inc. Westport, CT.

Jones, N.R. and J. Murray. 1964. Rapid measures of nucleotide dephosphorylation in iced fish muscle: their value as indices of freshness and of inosine 5'inosine monophosphate concentration. *J. Science Food Agric.* **15**:684-690.

Jones, N.R. 1963. Interconversion of flavorous nucleotide catabolites in chilled and frozen fish. *Proceedings XI.*

International Congress of Refrigeration. Munich. **11**:917-922.

Josephson, D.B., R.C. Lindsay, and G. Olafsdottir. 1987. Measurement of volatile aroma constituents as a means for following sensory deterioration of fresh and fishery products. In *Seafood Quality Determination. Proceedings of the International Symposium on Seafood Quality Determination.* D. E. Kramer and J. Liston (eds.), pp. 27-47. Elsevier, New York.

Kassemsarn, B.O., B.S. Peroz, J. Murray, and N.R. Jones. 1963. Nucleotide degradation in the muscle of iced haddock (*Gadus aeglefinus*), lemon sole (*Pleuronectes microcephalus*) and (*Pleuronectes platessa*). *Journal Food Sci.* **28**:28-37.

Kazanas, N. and J.A. Emerson. 1968. Effect of g irradiation on the microflora of freshwater fish III. Spoilage patterns and extension of refrigerated storage life of yellow perch fillets irradiated to 0.1 and 0.2 megarad. *Applied Microbiol.* **16**:242-247.

Kazeniac, S.J. 1961. Chicken flavor. In *Proceedings of Flavor Chemistry Symposium.*, pp. 137-159. Campbell Soup Co. Camden, NJ.

King, F.J., J. Kapsalis, A.V. Cardello, and J. R. Brooker. 1980. Consumer and instrumental edibility measures for grouping fish species. In *Advances in Fish Science and Technology.* J.J. Connell (ed.). Fishing News (Books) Ltd. London.

King, F.J., M. L. Anderson, and M. A. Steinberg. 1962. Reaction of cod actomyosin with linoleic and linolenic acids. *J. Food Sci.* **27**:63-66.

Kinnucan, H.W., R.G. Nelson, and J. Hiariey. 1993. *Marine Resources Economics.* **8**:273-291.

Knamen, D.E. and J. Liston. 1987. Determination of fish freshness using K value and comments on some other biochemical changes in relation to freshness. In *Seafood Quality Determination.* Ehira, S. H. Uchiyama (eds.), pp. 185-207. Elsevier. New York.

Konosu, S. and K. Yamaguchi. 1982. The flavor components in fish and shellfish. In *Chemistry and Biochemistry of Marine Food Products.* R.E. Martin, G.J. Flick, C. Hebard and D.R. Ward (eds.), pp. 367-404. AVI Publishing Co. Westport, CT.

Konosu, S., M. Ozay, and Y. Hashimoto. 1964. Free amino acids in the muscle of a few species of fish. *Bulletin Japanese Society Fisheries.* **30**:930-934.

Koury, B.J. and J. Spinelli. 1975. Effect of moisture, carbohydrate, and atmosphere on the functional stability of fish protein isolates. *J. Food Sci.* **40**:58-61.

Kuninaka, A., K. Masajiro, and K. Sakaguchi. 1964. History and development of flavor nucleotides. *Food Technol.* **18**(3):287-293.

Labuza, T.P., L. McNally, D. Gallagher, J. Hawkes, and F. Hurtado. 1972. Stability of intermediate moisture foods 1. Lipid oxidation. *J. Food Sci.* **37**:154-159.

Landgraf, R. Jr. 1976. Canned fishery products. In *Industrial Fishery Technology.* M.E. Stansby (ed.), pp. 302-314. Robert E. Krieger Publishing Co. Huntington, NY.

Lerke, P., R. Adams, and L. Farber. 1965. Bacteriology of spoilage of fish muscle III. Characterization of spoilers. *Applied Microbiol.* **13**:625-630.

Liljemark, A. 1965. Influence of freezing and cold storage on the submicroscopical structure of fish muscle. In *Freezing and Irradiation of Fish.* Rudolph Kreuzer (ed.), pp. 140-146.

Liston, J. 1960. The bacterial flora of fish caught in the Pacific. *Journal of Applied Microbiol.* **23**:469-470.

Love, R.M., M.M. Aref, M.K. Elerian, J.I.M. Ironside, E.M. Mackey, and M.G. Varela. 1965. Protein denaturation in frozen fish X: changes in cod muscle in the unfrozen state, with some further observations on the principles underlying the cell fragility method. *J. Food Sci. and Food Agric.* **16**:259-267.

Mangen, G.F. Jr. 1959. Discarbonyl compounds as components of fish odor. *Com. Fisheries Rev.* **21**(7):21-23.

Martin, R.E., W.H. Doyle, and J.R. Brooker. 1983. Toward an improved seafood nomenclature system. *Marine Fish Review.* **45**(7-9):1-20.

Martinez, F. and T.P. Labuza. 1968. Rate of deterioration of freeze-dried salmon as a function of relative humidity. *J. Food Sci.* **33**:241-247.

Mendelson, J.M. and M.A. Steinberg. 1962. Development of volatile carbonyls in haddock (*Melanogrammus aeglefinus*) flesh during storage at 2°C. *Food Technol.* **16**:113-115.

Nelson, R.W. 1994. Personal communication. Seattle, WA.

Nelson, R.W. and H.J. Barnett. 1971. Fish preservation in refrigerated seawater modified with carbon dioxide. In *Progress in Refrigeration Science and Technology, Proceedings XIII in International Congress of Refrigeration.* **3**:57-64.

Nogaoka, C. M., M. Yumagata, and K. Horimoto. 1971. A method for predicting "greening" of tuna before cooking. In *Fish Inspection and Quality Control.* R. Kreuzer (ed.), pp. 97-99. Fishing News (Books) Ltd., London.

Okada, M. and Tamoto. 1986. Introduction to surimi manufacturing technology. Overseas Fishery Corporation Foundation, pp. 1-64. Tokyo.

Olcott, H.S. 1962. Marine products. In *Symposium on Foods: Lipids and Their Oxidation.* H.W. Schultz, E.A. Day, and R.O. Sinnhuber (eds.), pp. 173-189. The AVI Publishing Co. Inc. Westport, CT.

Omura, Y., R.J. Price, and H.S. Olcott. 1978. Histamine-forming bacteria isolated from spoiled skipjack tuna and jack mackeral. *J. of Food Sci.* **43**:1779-1781.

Oriental Electric Co. Ltd. 1986. Freshness Meter KV 101. Tokyo Pamphlet.

Parkin, K.L. and W.D. Brown. 1982. Preservation of seafood with modified atmospheres. In *Chemistry and Biochemistry of Marine Food Products.* R.E. Martin, G.J. Flick, C.E. Hebard, and D. R. Ward (eds.), pp. 453-465. AVI Publishing Co. Westport, CT.

Patashink, M. and H.S. Groniger. 1964. Observations on the milky condition in some Pacific Coast fisheries. *J. Fish. Res. Board of Canada.* **21**:335-346.

Pedraja, R.R. 1972. Quality aspects of refrigerated and frozen fishery products. 1972. In American Society of Heating, Refrigerating and Air Conditioning Engineers Inc. Nov. Reprint, unnumbered.

Pelroy, G.A., A. Sherer, M. E. Peterson, R. Paranjye, and M.W. Eklund. 1985. Inhibition of *Clostridium botuli-*

num, type E toxic formation by potassium chloride and sodium chloride in hot-processed (smoked) whitefish (*Coregous chipeaformis*). *J. Food Protection.* **48**:971-975.

Pelroy, G.A. and M.W. Eklund. 1966. Changes in the microflora of vacuum-packaged, irradiated petrale sole (*Eopsetta jordani*) fillets stored at O.5°C. *Applied Microbiol.* **14**:921-927.

Reay, G.A. and J.M. Shewan. 1949. *The Spoilage of Fish and Its Preservation by Chilling.* E.M. Mark and G.F. Stewart (eds.), pp. 343-398. Academic Press. New York.

Regenstein, J.M., M.A. Schlosser, A. Sampson, and M. Fey. 1982. Chemical changes of trimethylamine oxide during fresh and frozen storage of fish. In *Chemistry and Biochemistry of Marine Food Products.* R.E. Martin, G.J. Flick, C.E. Hebard, and D.R. Ward (eds.), pp. 137-148. AVI Publishing Co. Westport, CT.

Russo, J.R. 1984. The surimi revolution. *Prepared Foods.* October 3, 1984, p.9.

Schneriger, D. 1994. Personal communication. Seattle, WA.

Shewan, J.M., G. Hobbs, and W. Hodgkiss. 1960. A determinative scheme for the identification of certain genera of gram-negative bacteria, with special reference to pseudomonadaceae. *J. of Applied Microbiol.* **23**:379-390.

Shewan J.M. 1962. The bacteriology of fresh and spoiling fish and some related chemical changes. In *Recent Advances in Food Science (Vol. I) Commodities.* J. Hawthorne and J.M. Leitch (eds.), pp. 167-193. London.

Shewan, J.M. and N.R. Jones. 1957. Chemical changes occurring in cod muscle during chill storage and their possible use as objective indices of quality. *J. Sci. and Food Agric.* **8**:491-498.

Shewan, J.M., R. G. Macintosh, C.G. Tucker, A.S.C. Ehrenberg. 1953. The development of a numerical scoring system for the sensory assessment of the spoilage of wet whitefish stored in ice. *J. Sci. Food Agric.* **4**:283-298.

Slabyj, B.M. 1978. Effect of processing holding on the quality of canned sardines. *J. of Food Science.* **46**:656-660.

Smith, H.W. 1936. The retention and physiological role of urea in the elasmobranchii. *Biological Revues.* **22**:49-82.

Spinelli, J. and R. Miller. 1983. Development of a high pressure injection device. Unpublished. Presented at Pacific Fisheries Technologists. March. Union, WA..

Spinelli, J. and B. J. Koury. 1981. Some new observations on the pathways of formation of dimethylamine in fish muscle and liver. *J. Agric. and Food Chem.* **29**:327-331.

Spinelli, J. and B. J. Koury. 1979. Nonenzymic formation of dimethylamine in dried fishery products. *J. Agric. and Food Chem.* **27**:1104-1108.

Spinelli, J. and D. Miyauchi. 1968. 5. The effect of 5'inosine monophosphate on the flavor of irradiated fish fillets. *Food Technol.* **22**(6):123-125.

Spinelli, J. 1965. Effect of hypoxanthine on the flavor of fresh and stored low-dose-irradiated petrale sole fillets. *J. Food Sci.* **30**:1063-1067.

Stansby, M.E. 1976. Cured fishery products. In *Industrial Fishery Technology.* M.E. Stansby (ed.), pp. 323-335. Robert E. Krieger Publishing Co. Huntington, NY.

Stansby, M.E. and H.S. Olcott. 1976. Composition of fish. In *Industrial Fishery Technology.* M.E. Stansby (ed.), pp. 339-349. Robert E. Krieger Publishing Co. Huntington, NY.

Stansby, M.E. 1962. Speculations on fishy odors and flavors. *Food Technology.* **16**:32.

Sussman, N.V. and L. Chin. 1965. Liquid water in frozen tissue: Study by nuclear magnetic resonance. *Science.* **151**:324-325.

Tanaka, T. 1965. Electron-microscopic studies on toughness in frozen fish. In *The Technology of Fish Utilization.* Rudolph Kreuzer (ed.). pp. 121-125. Fishing News (Books) Ltd., London.

Tokunaga, T. 1965. Studies on the development of dimethylamine and formaldehyde in Alaskan pollack muscle during frozen storage II factors affecting formation of dimethylamine and formaldehyde. *Bull. Hokkaido Reg. Fish. Res. Lab.* **30**:90-107.

Tarr, H.L.A. 1955. Fish muscle riboside hydrolases. *Biochemical Journal.* **59**:386-397.

Tarr, H.L.A. 1949. The acid-soluble phosphorus compounds of fish skeletal muscle. *J. Fisheries Research Board of Canada.* **7**:608-612.

Tarr, H.L.A. 1947. Control of rancidity in fish flesh. I. Chemical antioxidants. *Journal Fisheries Research Board of Canada.* **7**:137-154.

Taylor, S.L. and M.W. Speckhard. 1983. Isolation of histamine-producing bacteria from frozen tuna. *Marine Fisheries Revue.* **45**(4-6):35-39.

True, R.H. and B. M. Slabyj. 1978. Effect of preprocessing holding on the quality of canned Maine sardines. *J. Food Sci.* **43**:1172-1176.

Vaisey, E.B. 1956. The non-enzymic production of trimethylamine oxide to dimethylamine and formaldehyde. *Canadian Journal of Biochemistry and Physiology.* **34**:1085-1090.

Watson, D.W. 1939. Studies on fish spoilage V. The role of trimethylamine oxide in the respiration of achromobacter. *Journal Fisheries Research Board of Canada.* **4**:267-280.

Watts, B.M. 1954. Oxidative rancidity and discoloration in meat. *Advances in Food Research.* **5**:1-52. Academic Press Inc. New York.

Wekell, J.C. and H.J. Barnett. 1992. Seafood flavor and quality. In *Encyclopedia of Food Science and Scientific Technology.* W.H. Hui (ed.), pp. 2300-2323. John Wiley and Son. New York.

Wekell, J.C. and F.M. Teeny. 1988. Canned salmon curd reduced by use of phosphates. *J. Food Sci.* **53**:1009-1013.

Yamamoto, M., and J. Mackey. 1981. An enzymatic method for reducing curd formation in canned salmon. *J. Food Sci.* **46**:656-659.

Yamada, K. 1967. Occurrence and origin of trimethylamine oxide in fishes and marine invertebrates. *Bulletin Japanese Society and Science Fisheries.* **33**:591-603.

VIII. Future Outlook

World Aquaculture Potential

William G. Gordon

Aquaculture -- the farming of finfish, shell fish, crustaceans, and seaweed, and the aquatic ranching of anadromous fish -- is a rapidly growing industry in many parts of the world. Aquaculture has been practiced since the earliest records of human history; yet, unlike agriculture, it has not flourished in the United States.

Despite its antiquity, aquaculture throughout most of the world contributes little as a source of food compared to agriculture. World aquaculture production reached 26 million metric tons (mmT) in 1996, an increase of about 40 percent over the previous two decades [Food and Agriculture Organization of the United States (FAO), 1996]. This represents almost a doubling each decade; however, part of the increase may reflect better reporting, as the number of countries reporting aquaculture statistics to the FAO has increased during the same period.

Table 1 provides an overview for world aquaculture production for 1987 by type of seafood and region. Marine and freshwater production are combined in the statistics available on world production. Until recently world production has been dominated by pond culture of freshwater finfish, usually carp species (common, Chinese, Indian), grown throughout Asian countries and some European nations. More recently other freshwater species have contributed to freshwater aquaculture. Such species as tilapia (0.6 mmT), rainbow trout (0.4 mmT), channel catfish (0.2 mmT), and eel (0.1 mmT) now are commonly raised in many countries. The carps account for about 9 mmT. Today crawfish culture of *Macrobrachium* (0.3 mmT) brings the total freshwater production to about 11 mmT, or 45 percent of the world's total aquaculture production. (See other chapters for further discussion of individual species.)

Table 1. World Aquaculture Production in 1987. Figures are in metric tons.

Region	Finfish	Crustaceans	Mollusks	Seaweed	Other	Total	% of World Total
Africa, North & Northeast	51,397	2	286	—	50	51,685	0.39
Africa, South & Sahara	10,461	77	229	—	—	10,817	0.08
North America	226,672	44,480	138,841	—	—	449,993	3.41
Central America	9,485	6,564	50,719	—	—	66,768	0.50
South America	21,674	79,759	9,178	—	—	113,067	0.86
Caribbean	17,725	800	1,503	210	31	20,269	0.15
Europe	399,037	3,285	645,271	—	—	1,047,593	7.93
Former USSR	288,970	—	159	3,459	—	292,588	2.22
Near East	23,816	18	—	—	—	23,834	0.18
Oceania	2,730	174	27,186	1,710	140	31,939	0.24
East Asia	4,421,638	326,436	1,720,092	3,064,916	27,461	9,561,353	72.39
West Asia	1,279,836	113,311	84,843	60,000	20	1,538,010	11.64
Total	6,793,441	574,906	2,672,394	3,139,473	27,702	13,207,916	
% of World Total	51.4	4.4	20.2	23.8	0.2	100	

Source: Food and Agriculture Organization, 1989.

Marine aquaculture production until recently has lagged behind freshwater aquaculture. For many years some Asian countries have captured juvenile milkfish from the wild and grown them out in shallow estuarine ponds (about 0.4 mmT). The growing techniques are relatively crude, as the fish are fed primarily natural food.

By far the most sophisticated marine finfish culture is the growing of salmonids in protected coastal waters. Salmonids (trout and salmon) have been successfully spawned and reared as juveniles in hatcheries for many years. At smolt size they are released into rivers and streams, where they migrate to saltwater to grow as adults. Beginning in the early 1970s, Pacific salmon were held in pens for growth to market size. Later the Norwegians perfected net-cage culture of Atlantic salmon in their coastal fjord systems. The Norwegians carefully control diet and growing conditions, producing 5 to 10 tons of salmon in a 12 x 12 meter cage during an 18-month growing period. Cage culture of salmonids has spread to the United Kingdom, Denmark, France, Spain, Canada, Chile, Australia, New Zealand, and the United States. The practice of cage culture is spreading to other countries and other species of marine finfish. As scientists unlock the crucial knowledge of spawning, nutrition, and environmental requirements, significant production will occur in a wide variety of marine species. Currently, species under investigation for intensive culture include cod, halibut, dolphin, and flounder.

The trend in developed countries will become more intensive, with breeding, rearing, and harvesting within carefully controlled growing facilities. This will be necessary to assure a high-quality product, as well as to control environmental pollution from the rearing facilities. Growers will use high densities and formulated food to obtain the fastest growing rates independent of season. Already markets are responding to this consistent high quality source of seafood. Buyers in Europe, Japan, and the United States often receive daily shipments of product from the producing countries. To the chagrin of many wild-capture harvesters, finfish aquaculture ventures now control the market on quality and price.

The culture of crustaceans, particularly the marine shrimps (genus *Penaeus*) is one of the fastest growing and economically successful aquaculture sectors. Technology for spawning of fertile female *Penaeus* and rearing of the larva in captivity was developed in Japan and spread throughout Southeast Asia. Knowledge was gained quickly about growing postlarvae to a marketable adult in shallow estuarine ponds. Shrimp farming spread quickly throughout coastal Asia and found its way to the Western Hemisphere, particularly Ecuador's coasts, and soon thereafter into virtually all tropical maritime countries of Central and South America. Worldwide the production of marine shrimps from aquaculture has increased from about two percent in 1980 to about 35 percent in 1996 (FAO, 1996). About 80 percent of this aquaculture production is concentrated in Southeast Asia, with China the leading nation (Table 2). In the Western Hemisphere, Ecuador is the leading producer due to a much greater hatchery capability and the economic incentive to produce a high-value commodity for the world market.

Freshwater aquaculture of the giant freshwater prawn *Macrobrachium* also contributes to crustacean aquaculture. Although relatively easy to culture, the species has not drawn intense interest, due to the competitive advantage of marine shrimp. They are grown successfully in Malaysia, and several other Asian nations. Overall production is about 0.02 mmT.

The continued growth of shrimp aquaculture will match or exceed demand. Wild capture fisheries have reached their potential and/or are declining due to habitat deterioration and overfishing. Although the future growth may not match that of the late 1980s, predictions by those knowledgeable about shrimp aquaculture indicate significant growth within China and other Asian countries, as well as South and Central America.

Molluscan aquaculture, particularly oysters, has flourished in the world for many years. Cultivation of seed in hatcheries is a well-developed technology initiated in the United States during the 1920s and now practiced around the world. Although hatcheries are used to produce seed oysters, the most common practice is to collect wild seed in oyster-growing areas during the three-dimensional growing modes, suspending the attached oysters from the surface to bottom on ropes hanging from floats or rafts. In this fashion the oysters are less subject to predation and

Table 2. World Commercial Catch of Fish, Crustaceans, and Mollusks, by Countries, 1992–96 (Does not include marine mammals and aquatic plants.)

Country	1992[1]	1993[1]	1994[1]	1995[1]	1996
	- - - - - - - Thousand Metric Tons Live Weight - - - - - -				
China	16,579	19,708	23,834	28,418	31,937
Peru	7,508	9,008	12,005	8,943	9,522
Chile	6,502	6,035	7,839	7,591	6,911
Japan	8,502	8,081	7,398	6,787	6,793
United States[2]	5,604	5,940	5,926	5,638	5,394
India	4,233	4,546	4,738	4,906	5,260
Russian Federation	5,611	4,461	3,781	4,374	4,729
Indonesia	3,439	3,685	3,917	4,145	4,402
Thailand	3,246	3,385	3,522	3,756	3,648
Norway	2,561	2,588	2,551	2,803	2,963
South Korea	2,696	2,649	2,701	2,688	2,772
Philippines	2,272	2,226	2,233	2,221	2,133
Iceland	1,577	1,718	1,560	1,616	2,064
North Korea[3]	1,780	1,782	1,802	1,850	1,800
Denmark	1,997	1,657	1,916	2,044	1,723
Mexico	1,248	1,201	1,267	1,359	1,499
Spain[3]	1,255	1,220	1,310	1,370	1,289
Bangladesh	967	1,047	1,091	1,173	1,264
Malaysia	1,105	1,155	1,182	1,245	1,240
Argentina	705	932	950	1,149	1,239
Taiwan	1,314	1,411	1,249	1,288	1,230
Viet Nam[3]	1,080	1,100	1,150	1,100	1,000
United Kingdom	870	929	964	1,004	978
Canada	1,337	1,183	1,076	933	971
Burma	800	837	924	832	873
Brazil[3]	790	780	820	805	850
France	821	830	854	882	828
Ecuador	361	328	345	612	794
Morocco	548	623	752	846	640
Italy	558	564	576	614	560
Pakistan	553	622	552	542	555
Turkey	454	559	603	652	555
New Zealand	503	471	492	613	493
Venezuela	334	397	441	505	490
Ghana	427	376	336	354	477
Netherlands	487	533	529	522	463
Ukraine	526	371	311	414	450
South Africa	696	565	525	578	440
Senegal	370	382	350	359	436
Iran	334	318	332	368	382
All others	9,178	9,003	8,754	9,379	8,963
Total	101,728	105,206	113,458	117,278	121,010

[1] Revised

[2] Includes the weight of clam, oyster, scallop, and other mollusk shells. This weight is not included in U.S. landings statistics shown elsewhere.

[3] Data estimated by FAO.

Source: Food and Agriculture Organization of the United Nations (FAO). Fisheries of the United States. 1997 (Current Fishing Statistics). September 1998, U.S. Dept. of Commerce No. 9700.

food is more readily available than if the animals were on the bottom. In addition, many more oysters can be grown per surface area. Raft culture of bivalves is practiced in most of the leading producing countries: Spain, France, Japan, Taiwan, and North and South Korea. In most of these countries, the growing waters are protected from sources of human pollution.

Of the bivalves cultured around the world, the Pacific cupped oyster *Crassostrea gigas* is the leading species. It is grown most abundantly in Japan, but has been introduced successfully to most oyster-producing countries of the world.

Several other bivalves are now being successfully cultured. Various clams, scallops, and several species of mussel are now grown in various locations around the world. Scallops, which command a high market value and grow quickly, may soon pass all other bivalves as the favored culture species. Scallop farms are now in production in Ireland, Peru, Chile, Canada, Japan, Korea, China, and the United States.

The culture of bivalves is, and undoubtedly will continue to be, the most successful form of marine aquaculture. This will occur for a number or reasons: wild stocks are fully harvested, many wild growing areas are closed to harvesting due to pollution, some wild stocks are depleted, market demand is high, and there is an excellent economic return.

Different species of seaweed are also cultured in many parts of the world for use as staple foods or supplements to foods or drugs. In Japan, the red alga *Porphyra* is grown and marketed as more is used as food. The kelp *Laminaria* grown in Japan and China is used as a food supplement since it contains significant amounts of iodine. In other countries, seaweeds are grown and used for cattle feeds. In addition to their direct use as human and animal foods, seaweeds are grown and harvested for use in drugs and food ingredients. Seaweeds contain polysaccharide agar, algenic acid, or carrageenan. When extracted from the plants, these products have widespread use in the food, drug, and cosmetic industries.

Dwindling supplies and depleted stocks of wild populations of seaweed, fish, and shellfish have led to greater culture of desirable species in many countries. As demand for products continues to expand, the culture of such plants and fish will likewise expand and spread to other countries. Developing countries in particular will see that aquaculture offers great trade and creates employment for their populations. Foreign buyers eager to obtain high quality seafood products frequently provide capital and scientific and technological expertise to initiate aquaculture ventures. The governments of the countries frequently adopt policies and practices which expedite developments. Examples of this may be found in Peru, Chile, Argentina, China, Mexico, and other Asian and South American countries. One important feature in many of these countries — important for the future of aquaculture — is the commitment to preserve high water quality in coastal environments. Another important aspect is that many countries have established stringent quality assurance programs to guide the growing, harvesting, processing, and marketing of cultured products. As a result, seafood buyers throughout the world frequently select aquaculture products over wild-harvest products. The future for aquaculture is most promising.

FISHERIES

Total world commercial landings were 121 mmT in 1996. Of this total at least 26 mmT, or about 21 percent, of the commercial harvest was derived from aquaculture. China was the leading nation with 26 percent of the total catch; Peru, second with 8 percent; Chile, third with 5.7 percent; Japan, fourth with 5.6 percent; United States, fifth with 4.5 percent; and India, sixth with 4.4 percent [National Marine Fisheries Service (NMFS), 1997]. The 1996 landings were the highest in the last decade. The statistics do not include marine mammals or aquatic plants. Table 2 presents the United States and world fish catches for the years 1992-1996.

Table 3 provides information of the world catch by marine area and inland waters for the years 1992-1996. During this period the catch by ocean area increased by approximately 12 percent, with only minor downward adjustments. These changes are believed to be a result of changes in abundance of coastal stocks or to reflect changes taking place in world fishing fleets (breakup of the former USSR), and various governments' policies regarding access to fishing grounds. Many coastal nations having extended their economic zones seaward to 200 nautical miles during the 1970s have now

Table 3. World Commercial Catch of Fish, Crustaceans, and Mollusks, by Major Fishing Areas, 1992–1996. (Does not include marine mammals and aquatic plants.)

Area	1992	1993	1994	1995	1996
	- - - - - - - - - - 1000 metric tons (live weight) - - - - - - - - - -				
Marine Areas					
Pacific Ocean	54,378	56,393	63,057	63,201	64,903
Atlantic Ocean	24,343	23,690	23,648	24,827	24,706
Indian Ocean	7,363	7,869	7,737	8,010	8,242
Total	86,084	87,952	94,442	96,038	97,851
Inland Waters					
North America	584	579	573	540	561
South America	357	375	403	425	402
Europe	504	497	509	527	513
Former USSR	667	544	460	416	416
Asia	11,668	13,372	15,245	17,335	19,326
Africa	1,839	1,864	1,805	1,974	1,919
Oceania	25	23	21	23	22
Total	15,644	17,254	19,016	21,240	23,159
Grand Total	101,728	105,206	113,458	117,278	121,010

Source: FAO, 1996.

phased out virtually all foreign fisheries to accommodate their domestic fishing interests.

Most world experts in fisheries believe that the potential yield of fish from the sea is about 100-125 mmT (Martin, 1992). Many feel that yields from wild-caught fisheries have peaked. Most fishery experts concede that virtually all established major world fisheries and most of the recently discovered and exploited resources are already fished at or beyond long-term sustainable yield and have concluded that, with the exception of unconventional resources (e.g., Antarctic krill, lantern fish), no major unexploited resource of economic or human food value remains in the sea. The world's human population has more than doubled since 1950 (from 2.5 to 5.5 billion), while fish production has grown more than six times (from 21 to 120 mmT). As the human population reaches six billion, an annual seafood production of 138 mmT would be required. Most capture fisheries, believed to have already reached natural limits, cannot meet this demand level. Aquaculture (which has significantly grown over the past decade, now producing 26 mmT) could meet this anticipated demand and effectively supplement the yield from wild-capture fisheries.

Globally, the marine catch accounts for 16 percent of the animal protein consumption and is the most important source of protein in developing countries. In Asia, one billion people rely on fish as their primary source, as do many people in island nations and the coastal nations of Africa (FAO, 1993). Seafood accounts for about 80 million tons of protein per year, followed by pork at 70 million tons, and 52 million tons of beef. Seafood will remain an important component of the world's protein supply.

The world catch of fish and shellfish by species groups is given in Table 4 for the years 1992-1996. Of these major groups, the pelagic species (herring, sardines, and anchovies) constitute nearly 20 percent of the total commercial harvest. This total is marketed fresh, frozen, canned, or cured as human food, or reduced to meal and oil for aquaculture and animal feedstock and other industrial purposes.

Seafood products are important commodities of world trade. The FAO world trade lists 176 countries or areas engaged as exporters or importers of seafood products. Trade in seafood products in recent years has reached about $109 billion in value. For many countries exports of seafood may constitute the highest valued portion of their total exports (FAO, 1996). Fisheries are also large employers of workers and in some coastal areas the sole employer; thus, it is important socially

Table 4. Fisheries: World Commercial Catch of Fish, Crustaceans, and Mollusks, by Species Groups, 1992–1996 (Does not include marine mammals and aquatic plants.)

Species Group	1992[1]	1993[1]	1994[1]	1995[1]	1996
	---------- 1000 metric tons (live weight) ----------				
Carps, barbels, cyprinids	7,081	8,185	9,530	10,970	12,340
Cods, hakes, haddocks	10,435	9,915	9,641	10,606	10,712
Flatfish	1,170	1,105	992	925	940
Herrings, sardines, anchovies	21,196	21,895	25,836	21,975	22,323
Jacks, mullets, sauries	10,541	10,129	10,061	11,050	11,329
Mackerel, snoeks, cutlassfishes	3,452	4,008	4,514	4,688	5,137
Redfish, basses, congers	6,060	5,819	6,445	7,031	6,827
River eels	209	203	206	202	229
Salmons, trouts, smelts	1,467	1,700	1,803	2,089	2,102
Shads	694	656	634	685	723
Sharks, rays, chimaeras	728	741	752	757	759
Sturgeons, paddlefish	14	9	8	7	6
Tilapias	1,064	1,084	1,129	1,251	1,319
Tunas, bonitos, billfishes	4,512	4,556	4,623	4,708	4,587
Other fishes	16,409	17,351	17,490	18,277	19,215
Crabs	1,062	1,068	1,259	1,267	1,339
Krill	305	89	84	119	101
Lobsters	214	211	222	228	223
Shrimp	2,970	2,924	3,128	3,217	3,385
Other crustaceans	909	1,159	1,299	1,515	1,681
Abalones, winkles, conchs	85	94	100	99	107
Clams, cockles, arkshells	2,125	2,473	2,635	2,729	2,703
Mussels	1,394	1,368	1,321	1,416	1,382
Oysters	1,711	2,032	2,821	3,248	3,224
Scallops	1,056	1,459	1,634	1,653	1,739
Squids, cuttlefishes, octopus	2,737	2,715	2,773	2,861	3,038
Other mollusks	1,511	1,833	2,012	3,075	2,813
Sea urchins, other echinoderms	102	105	117	127	118
Miscellaneous	515	320	389	503	609
Total	101,728	105,206	113,458	117,278	121,010

[1] Revised

Source: Food and Agriculture Organization of the United Nations (FAO), 1996.

and politically to maintain fisheries at high levels. Recently the FAO has classified nearly every part of the world's oceans as fully exploited. In addition, many areas of the ocean are classified as being depleted of certain species, while others are over-exploited and headed for serious depletion. Furthermore, the FAO estimates that about 60 percent of the fish types tracked by the organization are considered as fully exploited, over-exploited, or depleted.

In the United States, the situation is equally serious. The long-term potential harvest involving the United States exceeds 9 percent of the total potential. Although the United States has owned and been responsible for these resources since 1977, of the known 163 stock groups, 48 percent are fully utilized, 40 percent are over-exploited, and 17 percent are under-utilized. Of these 17 percent, very few have significant economic value.

The roots of the current crises reach back to the mid-1970s. When nations of the world began to extend national jurisdictions seaward, most coastal nations claimed sovereignty to 200 nautical miles beyond their shores as the sole way to protect their coastal stocks from over-exploitation by foreign fleets. These actions were later ratified by the 1982 Law of the Sea. Having gained exclusive autonomy over about 90 percent of the world's fishery resources, nations began to expand their fishing fleets. From 1970 to 1989, the size of the global fleet about doubled in numbers; modern technology increased efficiency four- to sevenfold. The undeniable fact is that national fishing fleets have grown too big for existing stocks. The FAO conservatively estimates that globally annual expenditures on fishing amount to $124 billion, in order to catch about $100 billion worth of fish (FAO, 1993). Presumably, national and regional governments make up the $24 billion difference with low interest loans and direct subsidies for new vessels and operations. These government actions keep more people in fisheries than the oceans can support. For the fisheries of the future to prosper, decision-makers need to confront the proverbial "too many fishers chasing too few fish."

Open access to fishing grounds contributes to the great increases in size of the fishing fleets. Without limitations on access, investments in new vessels will continue to be made well after the sustainable catch has been reached. Once investments are made, fishermen will only leave the fisheries for new grounds where they can use the equipment; otherwise, they remain in the overfished grounds until forced out of business. Government assistance exacerbates this problem, creating a situation that can lead to stock collapse or prolong the period of recovery. Since most fishing grounds are already fully exploited, there literally is no place to go.

Governments must stop carrying their fishing industries as a budgetary burden. As in the management of grazing, logging, or mining on public lands where fees are essential to limiting exploitation and compensating government for use of commonly held resources, extraction fees should be collected for use of the fishing grounds. In Australia, for example, fees for use of the fisheries range from 11 to 60 percent of the gross value of the landed catch (FAO, 1993).

The first step in stock recovery is to halt overfishing by re-establishing the basic tenets of fishery management, many of which are still applicable and (if applied rigorously by national governments) could provide workable means of restoring fisheries to sustainable levels. Worldwide, the economic benefits of improved management would be on the order of billions of dollars. If stocks were allowed to recover according to FAO (1993) estimates, the total annual catch could increase by as much as 20 million tons, worth about $16 billion at today's prices. In addition, governments could save some $24 billion per year by eliminating financial assistance activities and perhaps earn another $25 billion per year in landing fees. Although this does not take into account the adjustment cost to societies which will be spent to re-direct former fishermen into other occupations, it does convey the magnitude of present day economic mismanagement of the global fisheries, which in turn continues to wreak ecological damage to the ocean.

Although "sustainable yield" is an elusive objective because natural fluctuations in fish stocks are part of the system and difficult to account for, the world's supply of seafood could be enhanced by approaching the problems of depletion with such as the final goal. Globally more of the human population will depend upon governments to do so.

REFERENCES

Angel, Martin V. 1992. Managing bioadversity in the ocean. In *Diversity of Oceanic Life*, Melvin N. A. Petersen (ed.). Center for Strategic International Studies.

FAO. 1996. *Yearbook of Fishery Statistics*. Vol. 78, Food and Agriculture Organization, Rome.

FAO. 1993. *Marine Fisheries and the Law of the Sea: A Decade of Change*. FAO Circular No. 853, Rome.

NOAA. *Fisheries of the United States*. 1997. Current Fishery Statistics No. 9700. United States Department of Commerce.

Aquaculture

Richard T. Lovell

INTRODUCTION

Aquaculture is the fastest-growing food-producing industry in the world today. It will continue to grow as an industry and to provide a higher percentage of the world's needs for fishery products. Reasons for this are increasing demand for fish worldwide, especially in the more developed countries; diminishing supplies and increasing costs of sea-caught fish; greater consistency in supply and quality of cultured fish; utilization of resources unsuitable for other types of food production; and attractive investment opportunities in aquaculture.

Fish farming has demonstrated extremely rapid growth during the last two decades. Channel catfish farming in the United States is an example of the rapidly growing aquaculture industry. It has grown from almost obscurity in 1970 to an annual production of over 180,000 tons in 1993 (USDA, 1994). Farming of penaeid (marine) shrimp – primarily in Central and South America and Asia – is the fastest growing aquaculture enterprise worldwide, supplying approximately 25 percent of the world's consumption of shrimp. Ocean pen culture of salmon is a thriving industry in Norway, Chile, and areas of Western Europe, where it provides 90 percent of the salmon consumed in that area and is a valuable export commodity. High-value marine species, such as sea bream and turbot, are being cultured on a large commercial scale in Europe and Japan. Among the many marine and freshwater species emerging in commercial importance in the United States and around the world are redfish; hybrid, striped bass; and tilapia.

Aquaculture is more than a science in its infancy; it is now recognized as a viable and profitable enterprise worldwide. It will continue to grow and supply an increasingly larger percentage of fishery products. The purpose of this chapter is to introduce the concept of aquaculture for food production and to describe some of the products that are available from domestic or imported sources for consumers in the United States.

EVOLUTION OF AQUACULTURE

Fish farming is believed to have been practiced in China as early as 2000 B.C.; a classical account of the culture of common carp was written by Fan Lei in 475 B.C. (Villas, 1953). The Romans built fish ponds during the first century A.D., and during the Middle Ages fishpond-building by religious orders was widespread throughout Europe (Lovell et al., 1978). Carp farming in Eastern European countries was popular in the twelfth and thirteenth centuries. In Southeast Asia, fishponds were believed to have evolved naturally along with salt-making in coastal areas; the salt beds were utilized to grow milkfish during the rainy season (Schuster, 1952).

Early in the twentieth century, several forms of fish culture were fairly well-established, such as milkfish farming in Southeast Asia, carp polyculture in China, carp monoculture in Europe, tilapia culture in tropical Africa, culture of indigenous finfish and crustaceans in estuarine impoundments in Asian and Southeast Asian coastal areas, and hatchery-rearing of salmonids in North America and Western Europe. With the exception of salmonid culture, these forms of aquaculture were generally extensive; where the nutrient inputs into the system were restricted or limited to fertilizers and crude sources of foods, the yields were low.

Aquaculture made its greatest advancement during the latter part of the twentieth century. Technology began to develop from a scientific base. Research programs for aquaculture were established in specialized areas, such as genetics, nutrition, diseases, water management, and engineering. New species were examined for their aquaculture potential by trying to "close the life cycle" – that is, to reproduce the fish and grow them to market size in a controlled environment.

With the present base of research and technology, yields and risks for a number of aquaculture enterprises are now predictable, which makes them attractive investment opportunities.

BENEFITS OF AQUACULTURE

The supply, price, and quality of sea-caught fish fluctuate considerably because the ocean is an unmanaged resource whose yield is difficult to predict. But when fish are cultured, as is corn in a field, the supply can be controlled more effectively. High quality can be maintained because the production and harvest conditions are controlled and farmed fish usually go into the processing plant alive.

Resources that are unsuitable or unused for other food production purposes may be adaptable to fish farming. On the Mississippi River floodplain, where catfish farming in the United States is the most concentrated, ponds are usually sited on land too poorly drained for most crops. Crawfish have been successfully farmed in Louisiana swampland simply by enclosing such land with a levee and flooding the area. Shrimp farms are built on salt flats and other coastal lands that are unsuitable for other uses. Salmon and other marine species are grown in net pens which are suspended in offshore ocean water.

Fish convert feeds into body tissue more efficiently than do farm animals. Cultured catfish gain approximately 0.75 gram of weight per gram of diet, whereas chickens, the most feed-efficient warm-blooded food animal, gain about 0.48 gram of weight per gram of diet (National Research Council, 1983) (Table 1). The reason for the superior food conversion efficiency of fish is that they are fed diets with higher percentages of protein, which is economical because of the lower dietary energy requirement of fish, not because fish convert protein more efficiently. Thus, the primary advantage of fish over land animals is the lower energy cost of protein gain rather than superior food conversion. Energy requirement per gram of protein gain is 21 kilocalories for channel catfish versus 43 for the broiler chicken (Table 1).

LEVELS OF FISH FARMING

Generally, fish with the highest market value are cultured most intensively, using more expensive facilities and food; species with lower market value are produced under less modified culture conditions with greater dependence on natural foods, and, consequently, with lower yields per unit of culture space. For convenience, the level or intensity of fish farming may be equated with a food source, as presented in the following.

Production of Fish Exclusively from Natural Aquatic Foods

Several important species of food fish are capable of obtaining their food from plankton. Some of them, such as the silver carp, *Hypophthalmichthys molitrix*, accept artificial (supplemental) food reluctantly. Others, such as tilapia, have the ability to feed on plankton but also feed on bottom materials (detritus) and readily accept artificial feeds. The common carp, *Cyprinus carpio*, which is cultured in many areas of the world, is an efficient bottom feeder. Some fish have herbivorous appetites and consume large quantities of higher aquatic plants. In this group are the white amur, *Ctenopharyngodon idella*, and some tilapia species. All of these fish have been cultured without artificial feeds, principally in areas outside of the

Table 1. Comparison of Efficiency of Utilization of Feed, Dietary Protein, and Metabolizable Energy (ME) by Fish, Chicken, and Cattle[1]

Animal	Feed composition			Efficiency		
	Protein (percent)	ME (kcal/g)	ME protein ratio (kcal/g)	Weight gain per g food consumed (g)	Protein gain per g protein consumed (g)	ME required per g wt. gain (kcal ME)
Channel catfish	32	2.7	8.5	0.75	0.36	21
Broiler chicken	18	2.8	16.0	0.48	0.33	43
Beef cattle	11	2.6	24.0	0.13	0.15	167

[1] Adapted from National Research Council (1983).

United States, in areas where supplemental feeds are expensive or unavailable. Usually, yields are low but production costs are less where supplemental feeding is not practiced.

Supplementing Natural Foods with Artificial Feed

This level of fish farming involves taking full advantage of natural productivity in the pond and using artificial feeds as a supplement to increase yield further. Usually, the increased yield of fish resulting from the additional feeding is profitable. For example, the yield of common carp in fertilized ponds is 390 kilograms/hectare [2.471 acres (ha)] (Yashouv, 1959); the addition of grain or grain byproducts increases yields to 1530 kg/ha; and where high-quality supplemental fish feeds are added, yields of 3300 kg/ha are obtained (Sarig, 1974). Tilapia (*Oreochromis* spp.) and shrimp (*Penaeus* sp.) are efficient users of natural pond foods, whereas channel catfish (*Ictalurus punctatus*) and salmonoids (*Salmo* and *Oncorhynchus* spp.) are relatively inefficient.

Intensive Farming of Fish Using Artificial Feeds

At this level of production, the primary concern is maximum yield per unit of space and effort, with feed cost being secondary. This type of farming requires a high level of management. Examples are rainbow trout (*Oncorhynchus mykiss*) cultured in spring-fed raceways, channel catfish (*Ictalurus punctatus*) produced in intensively-stocked ponds, and tilapia cultured in recirculating raceways. Production costs are high because water quality must be maintained by oxygenation or continuous replacement of water, and nutritionally complete diets must be provided.

CULTURED SPECIES OF FISH

Desired characteristics in cultured fish are good market value, reproducible in captivity, rapid growth, acceptance of prepared feeds, resistance to diseases, and tolerance to variable conditions of dissolved oxygen and temperature. The major cultured species in the world generally have most of these characteristics. Table 2 represents recent estimates of total harvest yields for the major cultured marine and freshwater species.

Table 2. Metric Tons of Cultured Fish and Crustaceans Harvested in 1990[1]

Species	Area	Production
Shrimp	World	633,000
Salmon	World	250,000
Channel catfish	United States	165,000
Freshwater crawfish	United States	30,000
Rainbow trout	United States	27,000

[1] Sources: shrimp - Rosenberry (1990); salmon - Fish Farm Int. (1991); channel catfish and rainbow trout - USDA (1991); crawfish - Huner and Romaire (1990).

Channel Catfish

Culture of channel catfish is the largest aquacultural industry in the United States. Years ago freshwater catfish was held in low esteem by many consumers in the United States; however, energetic market development has increased the demand for this fish in all areas. It presently is fourth in popularity among United States consumers of fishery products, behind tuna, cod, and shrimp. Catfish flesh is mostly white muscle, is free of intramuscular bones, and has a delicate, non-fishy flavor.

Catfish farming (Figure 1a) is almost exclusively in large earthen ponds. It began in the southeastern United States in ponds used for sport fishing. Because channel catfish have long been popular in that area, they were grown and processed for retail food markets. Eastern ponds were stocked with 2,500 to 5,000 fingerlings per hectare in early spring; the fish were fed pelleted, concen-

Figure 1a. Catfish farming.

Figure 1b. Example of "pond" aquaculture.

trated feeds and were harvested the following fall. Early yields were 1,000 to 2,000 kg/ha. By increasing stocking densities, improving nutrition, compensating for water quality problems, and using multiple harvesting techniques (harvesting the large fish several times per season and simultaneously restocking with small fish, without draining the pond), yields have been increased to 4,000 to 15,000 kg/ha per year. On the Mississippi River flood plain, where large ponds can be built economically, catfish farming has become a major enterprise. Many farms are several hundred hectares in size, with individual ponds of five to 10 hectares (Figure 1b).

The channel catfish has many desirable traits for intensive culture. It can be spawned in captivity and can be cultured in ponds or in densely-stocked cages or raceways. It grows rapidly; a 10-gram fingerling reaches a harvest size of 0.5 kilogram in six months if the water temperature remains above 23°C. It accepts a variety of supplemental feeds and is relatively disease-free when environmental stresses are minimized. It tolerates daily and seasonal variations in pond water quality and temperature (from near freezing to 34°C).

The first problem in the culture of catfish in ponds is insufficient dissolved oxygen. Phytoplankton cannot produce enough oxygen during prolonged periods of diminished sunlight to maintain a large catfish population. Fish farmers can prepare for oxygen depletion by predicting oxygen consumption rates in their ponds and using emergency aeration equipment. Many farmers have permanent electrical aerators in their ponds that can be used daily (usually from midnight until dawn) during heavy feeding. The second yield-limiting factor is nitrogenous waste products, namely nitrites. Farmers can add salt (NaCl) to the ponds, thereby minimizing nitrite toxicity; the chloride ion impedes the absorption of nitrite through the gills.

For maximum growth, the farmer is interested in a high rate of food consumption by the fish; however, because uneaten feed cannot be recovered, overfeeding represents not only an economic waste, but it creates a greater oxygen demand on the culture system. A useful management tool in feeding catfish is the use of extruded or floating-type feeds. Extruded fish feeds have advantages in that they allow the feeder to observe the fish feeding. By being able to see the fish eat, the feeder can feed the fish closer to their maximum rate of consumption without overfeeding, and disease and water quality problems can be detected more easily. Extruded feeds are also more water stable than pelleted feeds. However, their benefits must be evaluated against additional manufacturing cost (10 to 20 percent) when compared with compression pelleted (sinking) feeds. More energy is required to make extruded feeds and equipment cost is considerably higher.

On large commercial farms, feed is delivered from the plant in bulk. The feed is dispensed by a vehicle driving along all sides of the pond. Optimum time of day for feeding catfish in ponds is influenced by dissolved oxygen and water temperature. Because of low dissolved oxygen (DO) in the pond water late at night and early in the morning, catfish should not be fed until well after sunrise when the DO has risen to a level for active feeding by the fish. Also, feeding should not be done at night because the fish's increased oxygen requirement, which occurs in four to eight hours subsequent to ingesting food, should not coincide with the decreased DO in the pond.

Daily feed allowance is usually based upon visual observation of the feeding activity of the fish in the pond. The maximum daily feed allowance is dependent upon the amount of feed that the pond can "metabolize" per day, and thus influences the amount of fish that can be produced. Dr. Homer Swingle, in his pioneer research on pond culture of catfish at Auburn University, found that approximately 32 kilograms of feed per hectare could be fed per day without serious risk of oxygen depletion. Later, with better understanding of pond dissolved oxygen dynamics and pond aeration equipment, catfish farmers have been able to increase this feeding rate up to 100 to 200 kg/ha.

A popular practice for catfish farmers is to keep a population of mixed sizes of fish in the pond at all times and periodically "harvest off" the larger fish. A readily available supply of fingerlings to replace the harvested fish is necessary. Channel catfish will grow from a 20-gram seedling to a 0.5-kilogram harvestable fish during a six-month growing season or to a 1.5-kg fish if allowed to grow for a second season. The most desirable market size is 0.8 to 1.0 kg live weight. Catfish are harvested by seining and transported to processing plants alive.

Live harvested fish are immediately slaughtered or held live in tanks with flowing water until slaughtered. The processing flow scheme is as follows: heading, eviscerating, skinning, chilling, sizing, filleting, icing or freezing, and packaging. Most products are bulk-packed (5 to 10 kg). Approximately 40 percent of the processed products are ice-packed, and 60 percent are frozen. Frozen fillets are injected with a polyphosphate solution as a preservative and glazed.

Major finished products are fillets (0.1 to 0.3 kg) and whole sectional cuts from large fish, nuggets (belly wall), and a small quantity of further-processed products. Size-graded catfish fillets account for 47 percent of the marketed product; whole dressed fish, size-graded, account for 31 percent; and other products, such as nuggets, steaks, and value-added products, account for 22 percent.

Catfish farming has grown at the rate of 20 percent per year over the past five years and has become a major food-producing industry in the United States. Interestingly, since its inception as a commercial food fish, it has not commanded high market prices compared to other fish. This may have been a blessing in that technology has been developed with emphasis on efficiency and low production cost. The current cost of production is estimated at \$1.30 to \$1.40/kg, which is remarkably lower than that of other intensively cultured aquaculture species.

OTHER POND-CULTURED SPECIES

Marine shrimp farming is the largest and fastest-growing aquaculture enterprise in the world. This is because of its high market value, availability of culture technology, and favorable growing environments in tropical areas around the world. Shrimp farmers now produce 25 percent of the shrimp placed on the world market; in 1980, shrimp farming supplied only ten percent of the world's shrimp needs (Rosenberry, 1990). Shrimp culture systems, which are almost exclusively in ponds, are generally characterized as extensive, semi-intensive, and intensive. Extensive farming, which is often in impounded estuarine areas, means low production density (one to three shrimp/meter^2 and yields around 300 to 400 kg/ha) with no supplemental feeding or water pumping. Semi-intensive culture is a higher level of production and involves feeding and pumping water into constructed ponds five to 20 hectares in size. Yields are usually 1,000 to 5,000 kg/ha in two crops/year. Intensive culture involves smaller rearing areas (ponds or raceways), more water exchange, aeration, and waste removal. The technology for intensive culture is more sophisticated and labor and energy requirements are higher; however, yields of 10,000 kg/ha/year are obtainable.

Tilapias are tropical fish. They have been cultured in tropical areas under primitive conditions for many years. They feed on plankton and pond-bottom organisms, grow rapidly, and respond well to organic fertilization. They also respond to supplemental feeds. At present tilapias are not important culture fish in the United States because they are sensitive to low temperatures. However, various intensive culture systems involving recirculating water or water from geothermal sources have been developed. The light, mild-flavored, low-fat flesh appeals to consumers in the United States. There is a significant import of tilapia cultured in the tropical Americas.

Most of the crawfish (*Procambarus* spp.) grown in the United States are produced and consumed in Louisiana. These crustaceans are raised in large shallow impoundments that can be drained in the summer for growth of vegetation and then flooded in the fall. Crawfish are often produced as a rotation crop in rice fields. In 1989 there were 60,000 hectares in production in the southern United States (Huner and Romaire, 1990), with a yield of 30,000 tons. They eat decaying plant materials, organisms associated with the decomposition, and growing aquatic plants. Annual yields average 600 kg/ha but can reach 2,000 kg/ha with supplemental feeding (Huner and Romaire, 1990).

Carp species are the most extensively cultured fish in the world. They grow well under a variety of cultural conditions, use natural foods efficiently, and respond well to supplemental feeding. However, consumers in the industrialized countries object to the taste and boniness, so carp is not considered a dining delicacy.

Species Cultured in Net Pens and Raceways

Because rainbow trout have been produced in hatcheries in the United States for release for sport fishing for many years, much is known about their husbandry. Strains have been bred to grow well under culture conditions. They are usually cultured in cold (10° to 15°C), flowing water. Because of the absence of natural food, the trout farmer must take care to ensure that the diet is nutritionally complete. Trout grow to a marketable size of 0.3 to 0.5 kilograms in five to seven months and are popular with consumers. The big drawback is the need for large quantities of flowing water at the proper temperature (Figure 2).

Salmon are also cultured primarily in net pens on offshore rafts in cool coastal seawaters in both the Northern and Southern hemispheres. Norway and Chile are the world leaders, but other areas in northern Europe, the Pacific coast of Canada, and

Figure 2. Trout farm aquaculture in Virginia's mountains.

the southern shores of Australia and South America support a growing industry. Optimum market size for salmon is three to five kilograms, which requires two to three years to produce. There is interest in net pen culture of salmon in the United States, but development is limited because of social and political roadblocks to the use of coastal waters for fish culture. The United States imports a significant amount of cultured salmon from Europe, Canada, and Chile.

Recently, marine species of finfish with high consumer appeal, such as sea bream, sea bass, and turbot, are being cultured intensively in various parts of the world, mainly western Europe. This has come about only because the life cycle has been closed; i.e., the fish have spawned from cultured broodstock and the hatchlings grown under controlled conditions. Limiting factors in culture of marine finfishes are difficulties in seedstock production, expensive feed costs (some species require moist feeds), and expensive pumping of seawater. Generally, these marine species command a high market price.

NUTRITIONAL VALUE

The percentage of edible, lean tissue in fish is higher than that in beef, pork, or poultry because fish contains less bone, adipose tissue, and connective tissue. Table 3 compares the yield of lean edible tissue in the dressed carcass of different food animals. Using channel catfish as an example, 81 percent of the dressed carcass (less head, skin, and viscera) is edible muscle tissue; the remainder is bone with no waste, connective tissue, or trim fat.

The caloric value of cultured fish is related to its fat content and varies with species, size, diet, and season. For example, fillets from channel catfish ranging in live-weight from 0.4 to 2.0 kilograms may vary in fat content from 6 to 12 percent (Li, 1991). Fillets from channel catfish harvested in the fall (end of feeding period) contained 24 percent more fat than those harvested in spring (fasted over winter) (Nettleton et al., 1990). Fillets from channel catfish fed a high energy diet (6 percent supplemental fat) contained 40 percent more fat than those fed a traditional catfish feed (Mohammed, 1989). Shrimp contain less than 1 percent fat in the tail muscle because fat is stored in the hepatopancreas, which is in the head. Tilapia, which are indigenous to the tropics, contain only 1 to 1.5 percent fat in the flesh, which is much less than freshwater fish from temperate regions, such as catfish and carp. Cultured salmon at a marketable size of 5 kilograms will contain 10 percent or more of muscle fat. Thus, the caloric value of fish flesh will vary from less than 100 to over 200 kilocalories/100 grams.

Protein percentage in fish flesh ranges from 14 to 18 percent, varying inversely with fat content. Essential amino acid content of fish muscle has a favorable profile for the human diet. It is relatively similar to that of muscle protein of other animal species; however, Lovell and Ammerman

Table 3. Dressing Percentage and Carcass Characteristics of Various Food Animals[1]

Source of flesh	Dressing percentage[2]	Characteristics of dressed carcass		
		Refuse[3] (percent)	Lean (percent)	Fat (percent)
Channel catfish	60	14	81	7
Beef	61	15	60	25
Pork	72	21	54	26
Chicken	72	30	65	9

[1] Sources: Channel catfish, Lovell (1979); beef, Browning et al. (1988) and United States Department of Agriculture (1986); pork, Prince et al. (1987) and United States Department of Agriculture (1983); poultry, E. T. Moran Jr., Poultry Science Department, Auburn University (personal communication, 1989).

[2] The marketable percentage of the animal after slaughter.

[3] In fish, bones only; in beef and pork, bones, trim fat, and tendons; in poultry, bones only.

(1974) reported that the sulphur amino acid (methionine and cystine) content of channel catfish was lower than that of poultry, which coincides with a lower dietary requirement for these amino acids by channel catfish (2.3 percent of the dietary protein) as compared to poultry (3.3 percent of the protein) (National Research Council, 1991, 1987).

Fatty acid composition of fish varies with diet. Lipids from ocean-caught fish contain 15 to 27 percent omega-3 (n-3) poly-unsaturated fatty acids (PUFA) (Bimbo, 1987). Cultured fish that are fed diets containing a high concentration of marine fish oil, such as salmonids, would also have relatively high amounts of n-3 PUFA in their lipids. Conversely, lipids of grain-fed freshwater fish like the channel catfish contain only about 2.5 percent n-3 PUFA (Nettleton et al., 1990). This difference is because n-3 PUFA comes primarily from marine algae and moves through the food chain to the fish, whereas grain-based catfish feeds are very limited in n-3 PUFA. Nettleton et al. (1990) reported that the average percentages of saturated, monounsaturated, and polyunsaturated fatty acids in farm-raised channel catfish were 24, 56, and 18, respectively, which compares with 32, 31, and 31 in a marine fish oil (menhaden) (Bimbo, 1987). Most of the polyunsaturated fatty acids in farm-raised catfish is 18:2 n-6, while in marine fish oil it is PUFA n-3 (Nettleton et al., 1990).

The n-3 PUFA composition of fish flesh can be enhanced by feeding with marine fish oil. Channel catfish fed feeds supplemented with 0, 2, 4, and 6 percent menhaden oil showed a marked increase in n-3 PUFA in the fillets as the amount of menhaden oil in the diet increased (Mohammed, 1989). Adding 6 percent menhaden oil to catfish feed increased n-3 PUFA in the lipids to 67 percent of that of ocean salmon but greatly increased the total fat content. Sensory tests revealed that channel catfish fed 4 percent or more of menhaden oil had a "fishy" flavor which was considered undesirable for the normally mild-flavored cultured catfish.

Cultured fish are useful sources of several B vitamins, phosphorus, and trace minerals. Table 4 shows the vitamin and mineral content for farm-raised catfish. Fish flesh is low in calcium, vitamin C, and vitamins A and E. Table 5 shows proximate and lipid composition of typical farm-raised channel catfish fillets.

QUALITY

Aquaculture provides an opportunity for a high level of quality control over the processed fish. Time intervals between harvest and slaughter, and between death of the animal and preservation of the processed product are relatively short. For example, rainbow trout and channel catfish enter the processing line alive and are slaughtered, chilled, and frozen within minutes. Farm-raised shrimp are removed alive from the ponds and ice-packed on the pond bank to immediately reduce temperature. At the pond-site the shrimp are beheaded, washed in chlorinated wa-

Table 4. Vitamin and Mineral Contents of Raw Fillets of Cultured Channel Catfish Fillets[1]

Vitamin	Content (mg/100g)	Mineral	Content (mg/100g)
Niacin	2.30	Potassium	315.000
Pantothenic acid	0.57	Sodium	33.000
Thiamin	0.34	Magnesium	23.000
Vitamin B-6	0.19	Calcium	7.100
Riboflavin	0.07	Phosphorus	1.840
Ascorbic acid	< 1	Zinc	0.570
Vitamin A	< 100 IU	Iron	0.340
Vitamin E	< 1 IU	Selenium	0.013
		Copper	< 0.060
		Manganese	< 0.830
		Chromium	< 1

[1] Adapted from Nettleton et al., 1990. Fish size was 0.45 to 0.68 kilograms. Values represent means of four collections made at different times during the year.

Table 5. Proximate and Lipid Composition of Cultured Channel Catfish Fillets[1]

Item	G/100g raw fillet
Calories	128.00
Protein	15.60
Water	76.40
Ash	1.00
Total lipids	6.90
Saturated fatty acids	1.51
Polyunsaturated fatty acids	1.22
Total n-3 fatty acids	0.16
Highly unsaturated n-3 fatty acids	0.10
Cholesterol	33.40

[1] Adapted from Nettleton et al., 1990. Fish size was 0.45 to 0.68 kilograms. Values represent means of four collections made at different times during the year.

ter, and iced again for transfer to a freezing plant. Time from harvest to freezing is less than 24 hours and the temperature of the product is continuously low.

Cultured fish have not been recognized as major sources of microorganisms that cause foodborne illness, although, as in the case of other food animals, environmental contamination could occur. Byrd (1973) surveyed processed catfish from plants in the southern United States and found relatively low surface total plate counts of 10^2 to $10^3/cm^2$.

Processing of aquaculture products in the United States is presently not under continual government inspection as is poultry and red meat processing. Officials in the Food and Drug Administration (FDA) state that federal inspection of seafoods and aquaculture products using the new Hazard Analysis Critical Control Point (HACCP) program will be mandatory within a short time. The processors are presently subject to regulations of state health agencies for plant design and operation, and to the FDA for product safety. In many cases the processors use voluntary inspection or quality assurance procedures, provided by the National Marine Fisheries Service (NMFS). Industry associations do provide quality assurance guidelines. For example, much of the catfish processed in the United States is processed under procedures defined and monitored by the Catfish Institute (a state marketing association) and carry a label indicating such.

Aquaculture products have an advantage over many wild-caught fishery products in that they are produced and harvested under controlled conditions. This fact alone does not ensure quality, but does provide an opportunity to monitor the quality and control of the environment from which the fish were harvested prior to final processing. Aquaculture products, as all fishery products, are highly perishable and require the same high level of temperature control to preserve quality and ensure safety.

ENVIRONMENT-RELATED OFF-FLAVOR IN CULTURED FISH

A serious quality problem in some aquaculture systems is off-flavors which can be absorbed from the culture environment by the fish in freshwater or low salinity (less than 3 parts per thousand) seawater. Most notable is the musty, muddy flavor that is usually caused by geosmin [trans-1,10-dimethyl-trans-(9)-decanol] or MIB (2-methylisoborneol) (Martin et al., 1987; Lovell, 1983). These compounds are synthesized by species of the blue-green algae and actinomycetes. Fish absorb these compounds through the food chain and directly from the water through the gills and skin. Sensory thresholds of geosmin and MIB in fish flesh are 8.4 and 0.8 parts per billion, respectively (Chan, 1989). Lelana (1988) found that sensorily detectable amounts of geosmin could concentrate in the flesh of channel catfish with 15 minutes exposure to 2.5 parts per billion of geosmin in the surrounding water (this concentration is often found in water when off-flavor occurs). Although the musty flavor usually disappears within several weeks, it is a serious inconvenience to the industry. There is no satisfactory control for production of the odorous compounds in the culture system; however, off-flavor can be purged from the fish relatively quickly by holding the fish in clean water for three to seven days, depending upon temperature and flavor intensity (Lovell, 1983). There are no objective tests for off-flavor, so processors use trained personnel to taste the fish prior to processing. Environment-related off-flavor has been found worldwide with many species and under a variety of conditions, including warmwater, coldwater, and marine species. Lovell and Broce (1985) reported intense musty flavor and high geosmin concentration in cultured shrimp which occurred during a decrease in salinity in shrimp ponds to less than three parts per thousand, thereby allowing geosmin-producing algae to grow.

Other than fish oils, commonly-used feed ingredients have negligible effect on fish flavor. Johnson and Dupree (1990) evaluated a variety of commercial feed ingredients for their effect on flavor of channel catfish. They added each ingredient singly to a purified diet and found that none had an important effect on the flavor of the fish.

FLESH COLOR

Color of flesh is an important marketing attribute for many fish. Salmon must have a deep pink color. In wild salmon this pigment is provided by the carotenoid astaxanthin that comes through the food chain from micro-crustaceans in the marine environment. However, in culture systems a source of astaxanthin or canthaxanthin (a synthetic) must be included in the diet.

Interestingly, consumers object to any amount of pigment in light-fleshed fish, such as channel catfish. Dietary sources of xanthophylls (zeaxanthin and lutein), commonly found in yellow corn and corn gluten meal, cause objectionable concentrations of yellow color in catfish muscle. Lee (1987) found that the maximum concentration of xanthophylls in catfish feed that would not cause sensorial detectable yellow color was 11 milligrams/kilogram.

MARKETABLE AQUACULTURE PRODUCTS

Farm-raised shrimp are marketed in similar forms as ocean-caught shrimp. They are usually shipped from the production site, headless and frozen, to a further processing plant where they are prepared into a variety of marketable products. Channel catfish are processed in plants near the farms, primarily into fillets and whole dressed fish (less head, skin, viscera). About 40 percent of the processed products are ice-packed and 60 percent are frozen. Most of the products are bulk-packed. There is relatively little further processing of catfish products at present. Most other farm-raised finfish are marketed in fresh or frozen forms without further processing. As aquaculture foods increase in quantity, they undoubtedly will be marketed more as value-added products.

REFERENCES

Bimbo, A. P. 1987. Marine oils. *J. Amer. Oil Chem. Soc.* **64** (5)9.

Browning, M.A., D.L. Huffman, W.R. Egbert, and W.R. Jones. 1988. Composition of beef, pp.166-168. *Proc. Recip. Meat Conf.* Chicago, IL.

Byrd, L .A. 1973. The microbiology of pond-raised catfish. Master's Thesis. Auburn University, AL.

Chan, Carmen M. 1989. Taste threshold and consumer response to geosmin and 2-methyliso-borneol in farm-raised channel catfish. Master's Thesis. Auburn University, AL.

Huner, J.V. and R.P. Romaire. 1990. Crawfish culture in the southeastern USA. *World Aquaculture.* **21**(4) 58-65.

Johnson, Peter and H.K. Dupree. 1990. Least-cost ingredients work well in catfish feeds. *Feedstuffs.* **62**(54):15-17.

Lee, P. 1987. Carotenoids in channel catfish. Ph.D. Dissertation. Auburn University, AL.

Lelana, Ewan. 1988. Geosmin and off-flavor in channel catfish. Ph.D. Dissertation. Auburn University, AL.

Lovell, R.T. 1979. Fish culture in the United States. *Science.* **206**:1368-1372.

Lovell, R.T. 1983. Off-flavor in pond-cultured channel catfish. *Wat. Sci. Tech.* **15**: 67-73.

Lovell, R.T. 1989a. Nutrition and feeding of fish. Van Nostrand Reinhold Inc. New York.

Lovell, R.T. 1989b. Content of omega-3 fatty acids can be increased in farm-raised catfish. Highlights of Agr. Res. *Ala. Agric. Exp. Stat.* **35**:16.

Lovell, R.T. and Ammerman, G. R. 1974. Processing farm-raised catfish. *So. Coop. Ser. Bull.* 193.

Lovell, R.T. and D. Broce. 1985. Cause of musty flavor in pond-cultured shrimp. *Aquaculture.* **50**:169-174.

Lovell, R.T., E.W. Shell, and R.O. Smitherman. 1978. Progress and prospects in fish farming, pp. 262-290. Academic Press Inc. New York.

Li, Menghe. 1991. Response of channel catfish to various concentrations of dietary protein under satiate and restricted feeding regimes. Ph.D. Dissertation. Auburn University, AL.

Martin, J.F., C.P. McCoym, W. Greenleaf, and L. Bennett. 1987. Analysis of 2-methyliso-borneol in water, mud and channel catfish from commercial ponds in Mississippi. *Con. J. Fish. Aquat. Sci.* **44**:909.

Mohammed, Tahya. 1989. Effect of feeding menhaden oil on fatty acid composition and sensory qualities in channel catfish. Master's Thesis. Auburn University, AL.

National Research Council. 1987. Nutrient requirements of poultry. National Academy of Sciences. Washington, DC.

National Research Council. 1983. Nutrient requirements of warmwater fish and shellfish. National Academy of Sciences. Washington, DC.

National Research Council. 1991. Nutrient Requirements of fish. National Academy of Sciences. Washington, DC.

Nettleton, J. A., W.H.Lori Jr., V. Klatt, W.M.N. Ratnayake, and R.G. Ackman. 1990. Nutrients and chemical residues in one-to-two pound Mississippi farm-raised catfish. *J. Food Sci.* **55**: 954-958.

Prince, T. J., D. L. Huffman, P. M. Brown and I. R. Gillespie. 1987. Effects of ractopamine on growth and carcass composition of finishing pigs. *J. Am. Sci. Suppl.* 1, **65**.

Rosenberry, Robert. 1990. World shrimp farming 1990. *Aqua. Digest, Sp. Rep.*

Salmon production. 1991. *Fish Farm. Int.* **18** (1): 21.

Schuster, W. H. 1952. Milkfish farming in southeast Asia. *Proceedings Indo-Pacific Fish Council.* Southeast Asian Fish. Dev. Cent., Sp. Public.

USDA. 1983. Composition of foods: pork products. *Agric. Handbook No. 8-10.*

USDA. 1986. Composition of foods: pork products. *Agric. Handbook No. 8-13.*

USDA. 1990. National Aquaculture Statistical Service Sp. Cr. United States Department of Agriculture. Washington, DC.

Villas, D.K. 1953. Fish Farming in the Philippines. Boakman Bo. Manila, Philippines.

Wyatt, L.E., R. Nickelson, and C. Vanderzart. 1979. Occurrence and control of *Salmonella* in freshwater catfish. *J. Food Sci.* **44**:1067-1069.

Fish and Seafood Import and Export: Imports and Importing

Richard E. Gutting Jr.

INTRODUCTION

The United States buys about half its fish and seafood from other countries and is the largest seafood market in the world. While United States tariffs are low, especially for products which are not canned or otherwise prepared, many different import restrictions exist and strict health and product labeling standards apply. This chapter describes these restrictions and requirements.

THE ENTRY PROCESS

When seafood reaches the United States, the consignee must file entry documents for the shipment with the United States Customs Service. Imported seafood is not legally entered into the United States until the shipment has arrived at the port of entry, delivery of the seafood has been authorized by the United States Customs Service, and any estimated duties have been paid. The importer must arrange for any examination and the release of the shipment by customs officials.

Seafood may be entered for consumption, entered for warehousing at the port of arrival, or transported in-bond to another port of entry and entered there under the same conditions as at the port of arrival. Arrangements for transporting the shipment to another port of entry in-bond may be made by the consignee, by a customhouse broker, or by any other person having a sufficient interest in the shipment for that purpose.

Entry documents must be prepared carefully because strict penalties apply. Section 592 of the Tariff Act (16 U.S.C. 1592) generally provides that any person who by fraud, by gross negligence, or by negligence, enters, introduces, or attempts to introduce merchandise into the commerce of the United States by means of any material and false written or oral statement or document, or by any omission which is material, will be subject to a monetary penalty. The person's merchandise may be seized, in certain circumstances, to insure payment of the penalty, and forfeited if the penalty is not paid. Criminal penalties also apply for intentional fraud.

For those companies not using a customs broker, a bond must be posted for each entry over $1,250. The amount of the bond is equal to the value of the shipment plus any duty owed. An informal entry (less than $1,250) requires no bond.

Fish Landed by United States Fishing Vessels

Fish caught by United States-flag fishing vessels and landed in the United States by those vessels are considered to be domestic production, whether the fish were caught in United States waters, on the high seas, or in foreign waters. No Customs Service clearance is needed and there are no duties owed on this production.

Fish caught by United States-flag vessels in international waters and landed in a foreign port for transshipment to the United States must be entered into the United States as "imports." These shipments, however, are considered to be products from an "American fishery" and are eligible for free entry [Harmonized Tariff Schedule of the United States (HTSUS) Heading 9815]. The term "American fishery" is defined as a "fishing enterprise conducted under the American flag by vessels of the United States on the high seas or in foreign waters in which such vessels have the right, by treaty or otherwise, to take fish or other marine products and may include a shore station operated in conjunction with such vessels by the owner or master thereof" (United States Note 1 of Subchapter XV of the HTSUS).

Foreign Trade Zones

A seafood shipment which is placed in a foreign or "free" trade zone is not considered to be "entered" into the United States. These zones are secured areas legally outside the United States customs territory. Their purpose is to attract and promote international trade and commerce. Foreign trade zones are operated as public utilities by states, political subdivisions, or corporations charged for that purpose and usually are located in or near a customs port of entry, at industrial parks or terminal warehouse facilities. A Foreign-Trade Zone Board, created by the Foreign-Trade Zones Act of 1934, reviews and approves applications to establish, operate, and maintain foreign-trade zones.

Foreign exporters may forward their goods to a Foreign Trade Zone to be held for an unlimited period without being subject to customs entry, payment of duty or tax, or bond. Seafood lawfully brought into a zone may be stored, sold, exhibited, broken up, repacked, assembled, distributed, sorted, graded, cleaned, mixed with foreign or domestic products, or otherwise manipulated or processed. The Foreign-Trade Zone Board may determine, however, that a particular processing operation is not in the public interest.

If a product is entered for consumption from a Foreign Trade Zone, duties and taxes will be assessed on the entered product, according to the condition of the foreign product at the time it entered into the zone, whether it has been placed in the status of privileged foreign product prior to manipulation or processing, or on the basis of its condition at the time of entry for consumption, if the foreign product was placed under nonprivileged status at the time of entry into the zone.

Production of seafood products in zones by the combined use of domestic and foreign materials makes unnecessary either the sending of the domestic materials abroad for processing, or the duty-paid or bonded importation of raw materials. Duties on the foreign product involved in such processing are payable only on the actual quantity of such foreign goods incorporated in the final processed product transferred from a zone for entry into the commerce of the United States. If there is any unrecoverable waste resulting from processing, allowances are made for it, thereby eliminating payment of duty except on the products which are actually entered. If there is any recoverable waste, it is dutiable only in its condition as such and in the quantity entered.

FDA Clearance

Imported seafood is regulated by the Food and Drug Administration (FDA) and is subject to FDA inspection at the time of entry. Shipments found not to comply with FDA requirements are subject to detention and must be brought into compliance, destroyed, or re-exported. At the discretion of the FDA, an importer may be permitted to bring a nonconforming shipment into compliance if it is possible to do so. Any sorting, reprocessing, or labeling must be supervised by the FDA at the expense of the importer.

The FDA will not examine pre-entry samples and it does not approve labels or packaging. In an effort to prevent delays or rejections at the border, it does, however, offer informal comments as work loads permit.

When a shipment of seafood enters the United States, it is only then subject to FDA sampling and examination. Although shipments are free to proceed in most cases to the consignee or customer, the shipment must remain intact until samples (if drawn) are examined. Normally, eight to ten working days are needed for this process to take place. If the FDA happens to be at the port of entry, the shipment could be released immediately. However, depending on the laboratory work required, this process can take upward of three to four weeks. If the FDA does not wish to examine the shipment, a "may proceed notice" will be issued.

CUSTOMS DUTIES

The classification and valuation of imports are the two most important steps in determining the extent to which United States Customs duties are owed. Classifications and valuations must be provided by seafood importers when an entry is filed with the United States Customs Service.

Classifying the Shipment

Classifications under the Harmonized Tariff Schedule of the United States (HTSUS) must be furnished even though this information is not pertinent to dutiable status. (An annotated, looseleaf edition of the HTSUS may be purchased from the United States Government Printing Office, Washington, DC 20402.) Classification is initially the responsibility of the importer, customs broker or other person preparing entry papers.

The HTSUS is divided into various sections and chapters dealing separately with products in broad product categories. For example, these categories separately place fish and shellfish in Chapter 3, and preparations of fish and shellfish in Chapter 16.

Seafood products are classifiable within these two chapters: (1) under items or descriptions which name them; (2) under provisions of general description; (3) under provisions which identify them by component material; or (4) under provisions which describe the product in accordance with its actual or principal use. When two or more provisions seem to cover the same product, the prevailing provision is determined in accordance with the legal notes and General Rules of Interpretation for the HTSUS. Also applicable are tariff classification principles contained in administrative precedents or in the case law of the United States Court of International Trade (formerly the United States Customs Court) or the United States Court of Appeals for the Federal Circuit (formerly the United States Court of Customs and Patent Appeals).

United States Customs District offices will make classification rulings on specific products when requested by importers. These rulings are binding as of the date of the ruling and are to be accepted at all ports of entry unless revoked by the Office of Regulations and Rulings.

To obtain such a ruling, a request must be made in writing and must concern prospective customs transactions: that is, transactions which are not already pending before a customs office by reason of arrival, entry, or otherwise. Questions arising in connection with ongoing customs transactions, whether they are current or completed, may not be the subject of ruling requests.

If a ruling request does not provide sufficient information, a letter will be sent to the requester within three calendar days detailing the additional information needed. Otherwise, the United States Customs Service will have 30 days to issue a ruling.

Assessing the Duty Rate

When products are dutiable, *ad valorem*, specific or compound rates may be assessed. An *ad valorem* rate, which is the type of rate most often applied to seafood, is a percentage of the value of the product, such as 5 percent *ad valorem*. A specific duty rate is a specified amount per unit of weight or other quantity, such as $0.059 per pound. A compound duty rate is a combination of both an *ad valorem*-rate and a specific rate, such as $0.07 per pound plus 10 percent *ad valorem*.

Tariff-rate quotas provide for the entry of a specified quantity of a product at a reduced rate of duty during a given period. There is no limitation on the amount of the product which may be entered during the quota period, but quantities entered in excess of the quota for the period are subject to higher duty rates.

Rates of duty for imported seafood vary depending upon the country of origin. Seafood imported from most nations is dutiable under the Most-Favored-Nation (MFN) rates (reported in the "General" column under Column 1 of the HTSUS). Seafood from countries to which the MFN rates have not been extended is dutiable at the full or "statutory" rates (reported in Column 2 of the HTSUS). The United States grants MFN status to most of its trading partners and only a few countries are required to pay the higher general tariffs.

Free rates are provided for many seafood products under the MFN rates. Duty-free status, however, is also available under various exemptions (reflected in the "Special" Column of the HTSUS). One of the more frequently applied exemptions from duty is provided by the Generalized System of Preferences (GSP) program. GSP rates are identified either by "A" or "A*" in the "Special" Column under Column 1 of the HTSUS. Seafood classifiable under a Subheading in this manner may qualify for duty free entry if imported into the United States directly from a designated country and territory. Seafood from one or more of these countries, however, may be excluded from the exemption if there is an "A*" in the Special Column.

GSP-eligible seafood qualifies for duty-free entry when it is from a beneficiary developing country and meets other requirements: Eligibility for GSP benefits is based on three factors: (1) the general level of development in the beneficiary country; (2) the product's competitiveness in the United States market (not to exceed a certain dollar level or 50 percent of total United States imports of that product); and (3) the country's practices in the areas of market access for goods and services, export policies, trade-related investment, protection of intellectual property rights and internationally recognized worker rights. The GSP program is a discretionary, unilateral program designed to assist economic development in developing countries. Several other developed countries maintain similar, independent GSP programs. The United States currently grants duty-free treatment under the GSP to approximately 3,000 products. However, the GSP system is subject to both annual and periodic general reviews, at which time products may lose their duty-free status, new products may be granted this status, or countries may gain or lose full GSP privileges.

The Caribbean Basin Initiative (CBI) is a program providing for the duty-free entry of products from designated beneficiary countries or territories. This program was enacted in the Caribbean Basin Economic Recovery Act, and became effective on January 1, 1984.

The following countries and territories have been designated as CBI beneficiary countries:

Antigua and Barbuda	Honduras
Aruba	Jamaica
Bahamas	Montserrat
Barbados	Netherlands Antilles
Belize	Panama
Costa Rica	Saint Kitts and Nevis
Dominica	Saint Lucia
Dominican Republic	Saint Vincent and the Grenadines
El Salvador	Trinidad and Tobago
Guatemala	Virgin Islands, British
Haiti	

This list changes from time to time. Therefore, it is necessary to consult General Note (3) (c) (vi) in the latest edition of the HTSUS which will contain the most up-to-date information. CBI duty-free items are identified by either an "E" or "E*" in the "Special" Column under Column 1 of the HTSUS. Seafood classifiable under a subheading designated in this manner many qualify for duty-free entry if imported into the United States directly from any of the designated countries and territories. Seafood from one or more of these countries, however, may be excluded from the exception if there is an "E*" in the "Special" Column.

The Compact of Free Association (FAS) provides for the duty-free entry of merchandise from designated freely associated states of the United States. This FAS program became effective on October 18, 1989, and has no termination date. The present beneficiary countries are the Marshall Islands and Federated States Micronesia.

Customs Brokers

Clearing goods through United States Customs can be difficult, time-consuming, and frustrating. Improper or incomplete documentation is often the cause of delays or rejections. United States Customs officials may not act as agents or forwarders for importers. They may, however, give advice and assistance. Only the owner of goods, the United States purchaser (or his authorized regular employees), or a licensed customs broker may enter goods into the United States.

Customs brokers specialize in import documentation, regulations, and procedures. They are United States citizens, residents or private United States firms which are licensed to act as agents for importers in transacting their customs business. A customs broker should not be confused with a "broker," "agent," or "manufacturer's representative" or other persons retained by the trading community for promoting its marketing activities.

Customs brokers provide a range of services related to the entry and clearance of goods into the United States. They can, for example, assist with country-of-origin markings, the labeling of products, and matters that concern seafood safety. They are also up-to-date on changes in duty rates, import quotas, and anti-dumping or countervailing duty measures.

United States customs brokers also provide advice regarding the proper tariff classification of goods, applicable tariff rates, and value for duty. They assist in making claims or filing appeals, and in obtaining binding tariff classification or other rulings for products.

FREE TRADE AGREEMENTS

A free trade agreement is an arrangement between two or more countries in which each removes tariff and other restrictions on trade between those countries. The General Agreement on Tariffs and Trade (GATT) permits free trade agreements as a deviation from the general principle that nations should not discriminate among nations if the free trade agreement meets certain criteria. Most significantly: (1) the free trade area must eliminate duties and other restrictive measures on "substantially all" trade between the parties; and (2) duties and other regulations of commerce maintained by the parties may not be higher or more restrictive to the trade of third countries than they were prior to the agreement.

The GATT is the principal international treaty that delineates the rules the United States and most other nations must follow in international trade.

The United States has also entered into three different free trade agreements with Israel, Canada, and Mexico.

Israel

The United States-Israel Free Trade Area agreement (FTA) provides for free or reduced rates of duty for merchandise from Israel to stimulate trade between the two countries. The program was authorized by the United States in the Trade and Tariff Act of 1984 and implemented by the United States-Israel Free Trade Area Implementation Act of 1985 and Presidential Proclamation 5365 of August 30, 1985. The FTA became effective September 1, 1985, and has no termination date.

The FTA governs most tariff items listed in the HTSUS. These items are identified by "I" in the "Special" Column under Column 1 of the HTSUS. As of January 1, 1995, all currently eligible reduced rate importations from Israel were accorded duty-free treatment.

Canada

The United States-Canada Free Trade Agreement (CFTA) entered into force in January 1989. It provides a broad arrangement governing international trade and investment between Canada and the United States.

The centerpiece of the CFTA is the mutual phase-out of United States and Canadian tariffs on goods, including all fish and seafood products. Products were placed on one of three lists to become duty free on January 1 of either 1989, 1993, or 1998. The CFTA also provided for accelerated tariff reductions, if both sides agree. Negotiations to accelerate tariff reductions are held each year.

The CFTA includes new rules of origin in order to prevent Third World countries from gaining duty-free access to either Canada or the United States by transshipping goods. Under these rules, imported fish must be processed in ways that are commercially and physically significant in order to be treated as a product of Canada or the United States for purposes of the CFTA.

A Canada-United States Trade Commission was established under the CFTA to supervise the implementation of the agreement and to resolve disputes. If consultations are unsuccessful, the Commission refers issues to expert panels or binding arbitration.

Both countries continue to apply their own national antidumping and countervailing duty laws. However, appeals from final rulings are made to independent bi-national panels, rather than the courts.

North America

Under the North American Free Trade Agreement, or NAFTA, Mexican and United States tariffs on all fish and seafood products either were eliminated immediately, or will be phased out in equal annual amounts within five years, 10 years, or 15 years. This phaseout of tariffs, however, may be accelerated through mutual agreement. Tariffs on fish and seafood products shipped between the United States and Canada continue to phase out under the Canada-United States Free Trade Agreement.

NAFTA requires that a product originate in Mexico, Canada, or the United States in order to obtain the tariff reductions and other benefits of NAFTA. For seafood harvested by vessels, the country of origin is determined by the flag of the harvesting vessel. Otherwise, the territory in which seafood is harvested determines its country of origin. When a product is processed, its country of origin changes to the nation where the processing occurs, if the processing operation causes the product to be reclassified in certain specific ways under the HTSUS. NAFTA trade benefits do not extend to products shipped outside the territory of NAFTA nations for further processing, other than "unloading, reloading, or any other operation necessary to preserve it in good condition or to transport" it to a NAFTA nation.

To obtain NAFTA trade benefits for a shipment, exporters must complete and sign a Certificate of Origin and importers must make a written declaration of the country of origin based upon this certificate. Severe penalties may be imposed for false statements.

Pre-NAFTA tariffs may be reinstated temporarily on a product imported from Mexico, if United States trade officials find that imports have increased because of NAFTA and the United States industry is harmed. Similar safeguards against "import surges" from Canada under the Canada-United States Free Trade Agreement are continued under NAFTA. United States laws against the unfair dumping and subsidization of Mexican products are continued under NAFTA, except that disputes over final antidumping and countervailing duty rulings are referred to a binational review process similar to that under the Canada-United States Free Trade Agreement.

Customs user fees were eliminated between Mexico and the United States by June 30, 1999. Similar fees between the United States and Canada were eliminated January 1, 1994. Mexico may continue to levy export fees on basic foodstuffs, including canned sardines and tuna, to assure adequate domestic supplies. Temporary entry for business purposes is made easier, so that managers and experts may be transferred to affiliated businesses.

Private investors are accorded most-favored-nation, national and nondiscriminatory treatment, and the right to arbitrate disputes. Restrictions on the equity ownership of seafood firms are prohibited. Government screening of acquisitions under $150 million is phased out over 10 years.

Two separate NAFTA side agreements were also signed to promote the effective enforcement of domestic labor and environmental laws. Each agreement establishes a three-nation Commission headed up by a Council to investigate evidence of non-enforcement of certain domestic laws and to promote cooperation.

The labor agreement governs those domestic laws concerning occupational safety, child labor, minimum wages, and the resolution of labor disputes. The environmental agreement governs those laws concerning pollution, environmental protection, and the protection of animals such as the Marine Mammal Protection Act and the Endangered Species Act, but does not include fishery management laws such as the Magnuson Act.

If a NAFTA nation is exhibiting a persistent pattern of non-enforcement of either the labor or environment agreement, the appropriate commission may refer the matter to a dispute settlement panel. Based on a panel's recommendation, the commission, by a two-to-one vote, may impose monetary fines against an offending NAFTA nation to be used by the commission to resolve the problem. If fines are not paid, tariff benefits under NAFTA may be suspended to recover the amount owed.

MARKING AND LABELING REQUIREMENTS

Several federal agencies regulate how imported fish and seafood products must be labeled and marked. Failure to comply with these requirements can lead to product seizures and severe penalties.

Food Labeling

The requirements regarding the labeling of seafood which are enforced by the Food and Drug Administration under the Fair Packaging and Labeling Act (15 U.S.C. 1452-1461), the Federal Food, Drug, and Cosmetic Act (21 U.S.C. 301-392) and the Nutrition Labeling and Education Act of 1990 (Public Law 101-535) are described in another chapter.

Lacey Act Marking

Fish and seafood products imported into the United States must comply with the marking requirements of Section 7(a)(2) of the Lacey Act, which are enforced by the National Marine Fisheries Service [16 U.S.C. 3376(a)(2)]. The applicable rules require containers to be marked on the outside with the names and addresses of the shipper and consignee and with an accurate list of its contents by species and number of each species [50 C.F.R. 246.2(a)]. These requirements are primarily intended as a regulatory tool to assist in identifying shipments of fish. Congress limited the penalty for "simple, unintentional clerical violations of regulations" to a maximum $250 penalty. Nevertheless, if a package is falsely marked in order to evade government scrutiny such mislabeling will not be treated as a mere marking offense.

Fish and seafood products in retail packages with labeling in compliance with the requirements of the Federal Food, Drug, and Cosmetic Act are exempt from this Lacey Act marking requirement [50 C.F.R. 246.2(b)(2)]. It also is acceptable if a shipping container is marked with the word "fish" or "seafood" and is physically accompanied with an invoice or other document showing the required information (50 C.F.R. 246.2(a).

Country-of-Origin Marking

The United States Customs Service requires that imported fish and seafood products be marked with their appropriate country of origin. This marking requirement, which applies to all products, is separate and apart from any marking or labeling requirement of other federal laws.

Section 304 of the Tariff Act (19 U.S.C. 1304) requires that imported articles be marked so that the "ultimate purchaser" in the United Sates will know its country of origin. Seafood and other natural products are exempt from individual marking, but the containers in which they reach the ultimate purchaser must be marked [19 U.S.C. 1304(a)(3)(j) and 19 C.F.R. 134.33].

The rules provide that the country of origin of a product is "the country of manufacture, production, or growth"[19 C.F.R. 134.1(b)]. No mention is made about the harvesting of fish. Nevertheless, the courts have ruled that the country of origin of fish caught on the high seas is determined by the flag of the catching vessel.

The marking of the country of origin on a seafood package must be "legible, indelible, and permanent" (19 C.F.R. 134.410). Markings must include the English name of the country of origin, unless another marking is specifically authorized by the Commissioner of Customs. The degree of permanence should be at least sufficient to ensure that in any reasonably foreseeable circumstance the marking will remain on the container until it reaches the ultimate purchaser, unless it is deliberately removed. The marking must survive normal distribution and handling. The ultimate purchaser in the United States must be able to find the marking easily and read it without strain.

More than one country-of-origin label may be required on a seafood package if the words "United States," "American," the letters "U.S.A.," or any variation of such words or letters, or the name of any city or locality in the United States, or the name of any foreign country or locality other than the country or locality in which the product was manufactured or produced, appear on the package. In this instance, the name of the country of origin preceded by "made in," "product of," or other words of similar meaning, must appear, legibly and permanently, in close proximity to such words, and in at least a comparable size. Care should be taken to ensure that package labels required by the FDA, which include listing the address of the manufacturer, packager, or distributor, comply with this country-of-origin marking rule.

A 10 percent *ad valorem* penalty is assessed for improperly marked shipments. Persons intentionally removing a country-of-origin label to conceal required information face a $5,000 fine and up to one year in prison.

Repackaging

The Customs Service requires that when imported fish or shellfish are "repackaged" in the United States, the new containers must be marked with the country of origin (19 C.F.R.134.5).

To minimize the practice of not disclosing country-of-origin information on the new containers, Customs requires importers of fish and shellfish and other articles incapable of being marked to certify that: (a) if the importer repacks the article, he shall do so in accordance with the marking requirements; or (b) if the article is sold or transferred, the importer shall notify the subsequent purchaser or repacker, in writing, at the time of sale or transfer, that any repacking must conform to these requirements.

The certification may appear as a typed or stamped statement on an appropriate entry document or commercial invoice, or on a preprinted attachment to such entry or invoice; or it may be submitted in blanket form to cover all importations of a particular product for a given period (e.g., calendar year). If the blanket procedure is used, a certification must be filed at each port where the article is entered.

If the article is sold or transferred to a subsequent purchaser or repacker, the importer must give the following notice to the purchaser or repacker:

> **Notice to Subsequent Purchaser or Packer**
>
> These articles are imported. The requirements of 19 U.S.C.1304 and 19 C.F.R. part 134 provide

> that the articles or their containers must be marked in a conspicuous place as legibly, indelibly, and permanently as the nature of the article or container will permit, in such a manner as to indicate to an ultimate purchaser in the United States the English name or the country of origin of the article.

Processing

It is a well-settled principle of customs law that when processing in the United States "substantially transforms" an imported product, the processed product is exempt from country-of-origin marking requirements. Country-of-origin labeling is not required on these processed products, because in those situations where a substantial transformation occurs, the United States processing company is considered to be the "ultimate purchaser" of the imported product for purposes of section 304 of the Tariff Act.

The basic test for "substantial transformation" is whether the processing operation in the United States creates "a new and different article of commerce with a new name, character, or use." This test has been followed by the courts for many years and has been applied in hundreds of cases. Whether a particular processing operation meets this test is decided on a case-by-case basis, depending upon an analysis of several different factors, including whether the name of the product changed, the product's tariff code classification changed, significant value was added or costs were incurred, and whether a wholesale product was converted to a retail product.

The Customs Service has ruled that the procedure of thawing, sorting, peeling, deveining, icing, and packaging of imported green headless frozen shrimp does *not* "substantially transform" the imported shrimp under the Tariff Act and that therefore the processed product must be marked with the country-of-origin of the imported shrimp. The Customs Service explains:

> Both before and after the shelling and deveining, the product is still essentially the same, i.e., raw shrimp. While these processing operations change the physical appearance of the shrimp to a certain degree, and render the product ready for eating, in our opinion, these changes are minor ones which do not fundamentally change the character of the shrimp. The quality, size, etc., of the end product is attributable to the imported product and not the domestic processing.

In another ruling the Custom Service has found that cooking does not substantially transform shrimp. Persons may obtain a written ruling from the Customs Service to clarify the applicability of the substantial transformation rule to particular situations.

Products of the United States

There are no federal requirements that any seafood product made in the United States be labeled as a "Product of the United States" or "Made in the United States." The use of this type of marking, however, must comply with guidelines established by the Federal Trade Commission (FTC) and the FDA.

Under FTC rules any product marked with "Made in U.S.A." may not include foreign-made components. A product processed in the United States, however, might properly be marked "Made in U.S.A." if the marking is accompanied by appropriate qualifying words (e.g., "of X-country components" or "of X-country materials"), provided this additional disclosure is made as conspicuously as the claim "Processed in the U.S.A." and is in close proximity thereto.

The Federal Food, Drug, and Cosmetic Act provides that food is misbranded if its labeling is false or misleading (21 U.S.C. 343). FDA officials have explained that labeling seafood products as a "Product of the U.S.A." violates this misbranding prohibition if a country-of-origin label would be required for the product.

Dolphin-Safe Labels

The labeling of certain retail tuna products as "dolphin-safe" is subject to detailed rules enforced by the National Marine Fisheries Service under the Dolphin Protection Consumer Information Act (16 U.S.C. 1385). These rules require that tuna products which are labeled "dolphin-safe" be accompanied by documentation proving that the products were caught in a manner not harmful to dolphin.

The rules define the term "dolphin-safe" to mean tuna that is not caught either by large-scale driftnets on the high seas, or by purse seine vessels greater than 400 short tons carrying capacity in the eastern tropical Pacific Ocean, unless the tuna shipment is accompanied by an observer statement and a captain's statement that the tuna was not harvested by intentionally deploying the net on or around dolphins (50 C.F.R. Part 247). Each subsequent buyer of such products must endorse the certificate accompanying the shipment. Congress is considering legislation to amend these requirements.

Protected-Species Labels

The Convention on International Trade in Endangered Species of Wild Flora and Fauna (CITES) is an international agreement designed to control trade in animal and plant species that are, or may become, threatened with extinction. The CITES came into force in 1975 and today more than 100 nations are parties.

Different levels of trade restrictions apply to products made from species listed in the three appendices to the CITES. "Endangered" species are listed in Appendix I and "threatened" species are listed in Appendix II. Nations wanting special protection for species within their territory may have those species listed in Appendix III even though they are not endangered or threatened. Decisions to add, delete, or transfer species from one CITES Appendix to another are made at the biennial meeting of the member nations. These changes must be formally proposed by a member nation and require the approval of two-thirds of the nations present and voting. A CITES member nation may enter a "reservation" about an individual listing and thus exempt itself from CITES requirements for that species. See the later discussion of the other restrictions which apply to imports of species listed in the CITES appendices.

Those fish and seafood products which are made in whole or in part from a species listed in Appendices II or III of the CITES must be shipped in containers marked on the outside with the names and addresses of the sender and receiver, and a statement of the species (and the numbers of each species) in the container (50 C.F.R. Part 133 Subpart B).

Trademarks and Trade Names

Seafood products bearing counterfeit trademarks, or marks which copy or simulate a registered trademark are prohibited from entry into the United States, provided a copy of the United States trademark is filed with the Customs Service and recorded (19 C.F.R. Part 133, Subpart B). It also is unlawful to import or sell articles bearing trademarks owned by a United States citizen without permission of the owner (15 U.S.C. 1124; 19 U.S.C. 1526).

A trademark is a word, symbol, design, or slogan (or combination of these) which identifies or distinguishes a product and which can (with renewal) last indefinitely. The rights to a trademark can result either from usage or filing an application with the United States Commissioner of Patents and Trademarks.

RESTRICTED PRODUCTS

Various United States laws restrict the importation of fish and seafood products into the United States.

Contraband

The Lacey Act prohibits any person to "import, export, transport, sell, receive, acquire, or purchase any fish . . . taken or possessed in violation of any law, treaty, or regulation of the United States or in violation of any Indian tribal law," and "to import, export, transport, sell, receive, acquire, or purchase in interstate or foreign commerce" any fish that are "taken, possessed, transported, or sold in violation of any law or regulation of any State or in violation of any foreign law" (16 U.S.C. 3401).

Penalties for violating this prohibition are severe. Civil penalties up to $10,000 may be assessed for offenses if a person "in the exercise of due care should know that the fish . . . were taken, possessed, transported or sold" in violation of an underlying law, treaty, or regulation. The Act also provides for felony violations, carrying penalties up to $20,000 and five years in jail, and for misdemeanor violations, carrying fines up to $10,000 and one year in jail. Finally, the Act makes any fish or seafood imported, exported, transported, sold, received, acquired, or purchased contrary to the act (other than marking prohibitions) subject to forfeiture, notwithstanding any civil criminal culpability requirements, and the government may proceed *in rem* against an item without seeking criminal or civil penalties. Administrative forfeiture of items valued at $10,000 or less is permitted.

Violations of some foreign laws will not trigger a Lacey Act offense. In a controversial case, a wildlife importer was indicted under the Lacey Act and the court found that the Act covered only foreign laws "designed and intended for the protection of wildlife in those countries." Viewing the foreign law as a revenue law, the court said that it did not trigger the applicability of the Lacey Act. Congress, however, disapproved of the court's interpretation, labeling the quoted language as "too restrictive." Instead, Congress laid down this rule:

> The Act's reference to 'any law, treaty or regulation' is not intended to include laws, treaties or regulations that are plainly and solely revenue laws with no specific reference to wildlife.

Congress meant the Lacey Act to cover laws that are revenue-producing or that concern safety if they also relate to wildlife protection. Courts have not accepted the argument that the Lacey Act is unconstitutional because it takes into account the laws of foreign nations.

Endangered Species

The Endangered Species Act (ESA) prohibits the importation of seafood produced from those endangered species which are listed in Appendix I of the CITES. About 675 species are presently listed, including most whales, dolphins, and seals; many sea turtles and crocodilians; several species of fish, including the short-nose sturgeon, Baltic sturgeon, and totoaba; and numerous mussels and clams (50 C.F.R. 23.23).

Products from threatened species of fish or shellfish which are listed in Appendix II of the CITES may be imported into the United States, but only under a permit issued by the United States Fish and Wildlife Service (FWS). CITES responsibilities in the United States have been delegated to three offices in the FWS. The Office of Management Authority is responsible for policymaking and the biennial meetings of CITES members. The Office of Scientific Authority reviews proposals to change the CITES Appendices. The Wildlife Permit Office is responsible for issuing needed permits. Appendix II of the CITES contains more than 27,000 species, including the Asian and Indian bullfrog (*Rana hexadactyla* and *Rana tigerina*), the Atlantic sturgeon, and the American alligator.

FWS regulations require that each import of a threatened species product must have a valid foreign "export permit" issued by "the managing authority of the country of reexport." These documents must be obtained prior to importation (50 C.F.R. Part 23). Permits or certificates must be displayed, but are not collected at United States ports of entry. Imports of protected species from a nation that is not a party to the CITES must be accompanied with documents from that nation that contain all of the information normally required.

Imports of threatened species also must enter the United States through a designated Custom port unless an "Exception to Designation Port" permit is obtained (New York, NY; Miami, FL; New Orleans, LA; Dallas/Ft. Worth, TX; Los Angeles, CA; San Francisco, CA; Seattle, WA; and Honolulu, HI). Importers may apply for a permit to allow for importation at a nondesignated port or ports during a specified period of time, not to exceed two years. Permits may be granted either to "minimize deterioration or loss" or "to alleviate undue economic hardship." Finally, a "Declaration for Importation or Exportation of Fish and Wildlife" (Form 3-177) must be filed at the Customs port of entry. This form is available from the Customs Service or the FWS.

The ESA also prohibits imports of species which are listed as "endangered" or "threatened" under the ESA even if they are not listed on a CITES appendix [16 U.S.C. 1538 (1)(A)]. The ESA defines an "endangered" species as one which is in danger of extinction throughout all or a significant portion of its range. "Threatened" species are defined as any species that is likely to become endangered throughout all or a significant portion of its range.

The lists of species in the CITES appendices which have been accepted by the United States (50 C.F.R. 23.23) are distinct from the lists of endangered and threatened species under the ESA (50 C.F.R. 115). For example, the leatherneck, loggerhead and green sea turtles are listed under the ESA, but not under the CITES. Thus a species might not be listed under the CITES, but would still be protected by the ESA.

Penalties for violating the ESA are severe. Civil penalties range from a maximum of $500 for unknowing violations, to a maximum of $25,000 where there is "knowing" unlawful conduct [(16

U.S.C. 1540(a)]. Civil penalties are assessed administratively, with an opportunity for a hearing and later judicial review (50 C.F.R. Part 11). Criminal penalties can be imposed for knowing violations, with a maximum of one year imprisonment and a $50,000 fine. Strict liability also can be imposed on any person engaged in the business of importing fish for violations, and a civil penalty of up to $12,000 may be imposed [16 U.S.C. 1540(a)(1)].

Marine Mammal Products

The importation of marine mammal products is prohibited under Section 102 of the Marine Mammal Protection Act (16 U.S.C. 1372).

Live Seafood

The importation of live animals is subject to certain prohibitions, restrictions, permit and quarantine requirements of several agencies. Importations of wildlife, parts, or their products must be entered at certain designated Customs ports of entry unless an exception is granted by the United States Fish and Wildlife Service (50 C.F.R. Parts 17 and 23).

The importation of live turtles and turtle eggs is subject to the requirements of the United States Public Health Service and the Veterinary Services of the Animal and Plant Health Inspection Service (42 C.F.R. 71.52). The importation of turtles with a carapace length of less than four inches and psittacine birds is subject to the requirements of the Food and Drug Administration (FDA) (21 C.F.R.1240.62).

There are also United States regulations which require that fish of the salmonidae family be certified free from viral hemorrhagic septicemia and whirling disease before they are imported into the United States. The provisions of these regulations require that all live or dead fish or eggs of salmonids are prohibited entry into the United States for any purpose, unless such importations are by direct shipment accompanied by a certification that the importation is free of the protozoan *myxosoma cerebralis* (whirling disease) and the virus-causing viral hemorrhagic septicemia (also referred to as Egtved disease, named after the city in Denmark where it was first identified).

The certification must be signed in the country-of-origin by a designated official acceptable to the Secretary of the Interior as being qualified in fish pathology, or in the United States by a qualified fish pathologist designated for this purpose by the Secretary of the Interior. This restriction, however, does not apply to salmonid fish or eggs that have been processed by canning, pickling, smoking, or otherwise prepared in a manner whereby the spores of *myxosoma cerebralis* and the virus-causing viral hemorrhagic septicemia have been killed. Fish so prepared are not required to be accompanied by a disease certificate.

Fish Landed by Foreign Fishing Vessels

The Nicholson Act (46 U.S.C. 451) prohibits foreign fishing vessels from landing their catch in United States ports. The United States Customs Service has interpreted "land" to include transshipment and "port" to include any place within the United States Customs Territory (i.e., the territorial sea). The effect is to prohibit transshipments within the territorial waters of the United States as well as offloadings in port.

Fishery Management Restrictions

Section 307 of the Magnuson Act [16 U.S.C. 1857(1)(G)] prohibits the importation of any fish taken or retained contrary to rules promulgated under the Act. Several of these rules ban or restrict the importation of fish to aid in the enforcement of requirements for fish caught subject to United States jurisdiction. For example, rules prohibit the importation of undersized New England ground fish [50 C.F.R. 651.7(6)]; the commercial possession of any Atlantic billfish (50 C.F.R. 644.7), and the possession of undersized Atlantic scallops [50 C.F.R. 649.20(b)]. In addition, the so-called Mitchell Amendment to the Magnuson Act banned the importation of eggbearing whole live American lobsters, or those smaller than the minimum possession size in effect under the American lobster plan [50 C.F.R. 649.20(b)].

All 50 states regulate the harvesting and possession of fish products within their geographical jurisdiction. In some instances states prohibit the possession of fish products caught outside the state.

Driftnet-Caught Fish

Controversies over the use of large-scale driftnets on the high seas prompted the United Nations to ban their use (United Nations General

Assembly Resolution 44/225). An amendment to Section 101 of the Marine Mammal Protection Act in 1990 also banned the importation of any fish caught with high seas driftnets in the South Pacific beginning July 1, 1991, and from anywhere on the high seas after July 1, 1992.

Products of Convict or Forced Labor

Fish and seafood products which are produced or processed by means of the use of convict labor, forced labor, or indentured labor under penal sanctions are prohibited from importation, provided a finding has been published (17 U.S.C. 1307).

Foreign Assets Control

The Office of Foreign Assets Control in the Department of Treasury administers laws which prohibit the importation of goods from several nations deemed to be the "enemy" of the United States. Importers should consult the applicable regulations for current restrictions (31 C.F.R. Chapter V). Specific licenses are required to bring restricted merchandise into the United States, but they are rarely granted.

SEAFOOD EMBARGOES

Several different laws authorize the President, at his discretion, to embargo fish and seafood imports for the purpose of encouraging foreign nations to comply with various domestic and international conservation programs.

The Pelly Amendment

The first, and most important, of these laws is the so-called Pelly Amendment to the Fishermen's Protective Act (22 U.S.C. 1978). Under his law the Secretary of Commerce is directed to determine and certify whether:

> . . . nationals of a foreign country, directly or indirectly, are conducting fishing operations in a manner or under circumstances which diminish the effectiveness of an international fishery conservation program.

The term "international fishery conservation program" means "any ban, restriction, regulation, or other measure in effect pursuant to a multilateral agreement which is in force with respect to the United States, the purpose of which is to conserve or protect the living resources of the sea" [22 U.S.C. 1789(h)(3)].

The phrase "diminish the effectiveness" is not defined. The Supreme Court, however, has ruled that certification is not required of every nation whose fishing operations exceed international harvest quotas. Instead, the Pelly Amendment calls for a judgment to be made on a case-by-case basis [*Japan Whaling Assoc. v. American Cetacean Society*, 478 United States 221 (1986)].

After certification, the Pelly Amendment provides that ". . . the President may . . . prohibit the bringing or the importation into the United States of any product . . . from the offending country for such duration as the President determines appropriate and to the extent that such prohibition is sanctioned by the General Agreement on Tariffs and Trade." (The reference in the Pelly amendment to actions to which are "sanctioned" by the General Agreement on Tariffs and Trade (GATT) is discussed later in this section.)

The Pelly Amendment was broadened in 1978 to give the President discretionary authority to ban imports of fish and wild animals taken within a foreign nation or packed, possessed or prepared within a country, if the nation was certified as "diminishing the effectiveness of any international program for endangered or threatened species" [22 U.S.C. 1978 (a)(2)]. The potential sanctions were broadened from "fish and wildlife products" to "any products" in 1992.

The President is required to notify Congress of any action taken pursuant to a certification. If the President does not impose the sanctions authorized, he must give reasons to Congress for his failure to do so [22 U.S.C. 1978 (b)].

Eight nations have been certified under the Pelly Amendment, but trade sanctions have been imposed only once. Nevertheless, the threat of sanctions has been used repeatedly in negotiating concessions.

Penalties for violations of the Pelly Amendment include fines up to $10,000 for the first offense and $20,000 for subsequent violations. Any illegal shipments are subject to forfeiture [22 U.S.C. 1978 (e)]. In 1984, however, Congress increased the possible fine for all crimes, both misdemeanors and felonies, in the Omnibus Crime Control Act, raising it to $250,000 for individuals and $500,000 for persons other than individuals.

Tuna Agreements

Two laws authorize discretionary embargoes to encourage foreign compliance with international management measures concerning tuna. Section 6(c) of the Inter-American Tropical Tuna Act authorizes embargoes of fish regulated by the Inter-American Tropical Tuna Commission from nations determined to be fishing in a ". . . manner or in such circumstances as would tend to diminish the effectiveness of the conservation recommendation of the Commission . . ." (16 U.S.C. 955). Section 6(c) of the Atlantic Tunas Convention Act contains a similar provision with respect to the International Convention on the Conservation of Atlantic Tunas (16 U.S.C. 971).

The phrase "tend to" used in these tuna laws is not found in the Pelly Amendment. Also, unlike the Pelly Amendment, neither of these tuna laws is limited to embargoes "sanctioned" by the GATT. These differences suggest that import embargoes would be easier to impose than under the Pelly Amendment. No embargoes, however, have been imposed under these laws.

Fish from Driftnet Nations

Section 206(f) of the Magnuson Act [16 U.S.C. 1826 (f)] was amended in 1990 to direct the Secretary of Commerce to certify "nations that conduct, or authorize their nationals to conduct, large-scale driftnet fishing beyond the exclusive economic zone of any nation in a manner that diminishes the effectiveness of or is inconsistent with any international agreement governing large scale driftnet fishing to which the United States is a party or otherwise subscribes" (PL 101-267).

Such a certification is deemed to be a certification under the Pelly Amendment, thus authorizing the President to embargo any fish imported from a driftnet nation.

Protecting Dolphins

United States import bans also have been aimed at encouraging foreign nations to comply with fishing standards developed by the United States to protect dolphins. These trade embargoes were enacted after similar fishing restrictions were imposed on United States vessels in an effort to "level the playing field" with foreign competitors.

Yellowfin tuna in the eastern tropical Pacific Ocean swim under schools of dolphin. To catch these tuna, fishing vessels surround dolphin (and the tuna swimming beneath them) with purse seine nets. Since its enactment in 1972, the Marine Mammal Protection Act (MMPA) has required United States vessels to take special measures to reduce the incidental mortality of these dolphin.

Amendments to the MMPA in 1984 and 1988 imposed four different embargoes:

1. A mandatory "primary" embargo of yellowfin tuna against harvesting nations violating United States dolphin protection standards;
2. A mandatory "intermediary" embargo of yellowfin tuna against nations receiving yellowfin tuna from nations subject to a "primary" embargo;
3. A discretionary embargo against all fish products from nations subject to a "secondary" embargo; and
4. A mandatory embargo of any fish product caught with fishing technology harmful to mammals.

The "primary" yellowfin tuna embargo provisions require the President to embargo imports of yellowfin tuna caught with purse seines in the eastern tropical Pacific unless the Secretary of Commerce determines that the country involved has a dolphin protection program: (1) "comparable" to that of the United States, and (2) resulting in an "average kill per set" no more than 1.25 times that of the United States fleet. The standards also include separate maximum mortality limits for vessels of a nation for two specific stocks of dolphins, determined as a percentage of total mammal mortality by those nation's vessels for the year.

This "primary" embargo prohibits imports directly from a harvesting nation found not to be in compliance with United States standards. The applicable rules require that annual findings be made regarding the fishing operations of foreign nations based upon the annual submission of evidence from the nations concerned. The rules governing these primary embargoes have changed several times in the past few years in response to a series of court rulings and legislative amendments. (50 C.F.R. Part 216).

A "secondary" or "intermediary" embargo provision requires the President to prohibit yellowfin tuna imports from being shipped through "intermediary nations." Section 101 (a)(2)(C) of the

MMPA provides that for purposes of enforcing these "intermediary" embargoes, the Secretary of Commerce "shall require the Government of any intermediary nation from which yellowfin tuna or tuna products will be exported to the United States to certify and provide reasonable proof that it has acted to prohibit the importation of such tuna and tuna products from any nation from which direct export to the United States . . . is banned . . . within sixty days." Unless the intermediary nation's ban is effective within 60 days of the United States import ban, and the United States receives proof within ninety days of the date of the United States ban, the tuna imports from the intermediary nation are prohibited.

Under Section 101(a)(2)(D) of the MMPA, six months after a primary or intermediary embargo is declared, an automatic certification occurs under the Pelly Amendment. Thus, either a primary or secondary embargo may be expanded by the President to any fish product under the Pelly Amendment six months after the embargo is declared.

In recent years, the United States has prohibited the importation of yellowfin tuna (and products thereof) harvested by vessels from several Central and South American nations. These primary embargoes have resulted in additional secondary embargoes.

Seafood importers of tuna must file a "Fisheries Certificate of Origin" (NOAA Form 370) when the shipment enters the United States. In addition, a responsible governmental official of the exporting nation must certify that the tuna in the shipment were not caught with driftnets. This certification may be by a separate letter, or through signature of the Fisheries Certificate of Origin.

Protecting Other Marine Mammals

While most of the activity under the MMPA has focused on purse seine fishing for tuna on dolphin and large-scale driftnets, Section 101(a)(2) of the MMPA [16 U.S.C. 1371 (a)(2)] also provides that:

> The Secretary of the Treasury shall ban the importation of commercial fish or products from fish which has been caught with commercial fishing technology which results in the incidental kill or incidental serious injury of ocean mammals in excess of United States standards. For purposes of applying the preceding sentence, the Secretary—
>
> (A) shall insist on reasonable proof from the government of any nation from which fish or fish products will be exported to the United States of the effects on ocean mammals of the commercial fishing technology in use for such fish or fish products exported from such nation to the United States . . . .

In addition, Section 102(c)(3) of the MMPA prohibits the import of:

> Any fish, whether fresh, frozen, or otherwise prepared, if such fish was caught in a manner which the Secretary has proscribed for persons subject to the jurisdiction of the United States, whether or not any marine mammals were in fact taken incident to the catching of the fish.

These import bans could be a major factor in the future, depending upon the domestic restrictions imposed upon United States fishing vessels.

Sea Turtles

Another law requiring import embargoes is intended to encourage nations to comply with United States sea turtle protection requirements for shrimp trawlers.

Shrimp trawlers in the Gulf of Mexico unintentionally capture several species of sea turtles protected under the Endangered Species Act. When regulations were published requiring the use of turtle excluder devices in United States shrimp nets, Congress enacted a law which provides that shrimp harvested with technology that may adversely affect certain sea turtles may not be imported into the United States, unless the Secretary of State certifies by May 1 each year that the nation concerned has turtle protection requirements "comparable" to those imposed on United States shrimpers (Section 609 of Public Law 101-162).

The Secretary of State has determined that the import restrictions of this section do not apply to shrimp produced from aquaculture operations and has established guidelines for compliance by foreign nations (50 Federal Register 1051). Shrimp imports from several nations have been embargoed under these provisions.

Protecting United States Fishing Vessels

Under the Magnuson Fishery Conservation and Management Act (MFCMA), imports of fish products are embargoed from those countries with which the United States cannot conclude an international fishery agreement allowing United States fishing vessels equitable access to fisheries over which that country asserts exclusive fishery management authority, as recognized by the United States.

United States imports of tuna products from Mexico were embargoed under these provisions in 1980 (45 Federal Register 137) because of the seizure of a United States tuna-fishing vessel by the Mexican government. The vessel was fishing in waters claimed by the Mexican government as part of its territorial fishery zone, but not recognized by the United States as such with respect to jurisdiction over tuna resources.

International Legal Restraints

Some of the import embargoes and restrictions described in this section have been challenged in the World Trade Organization (WTO) under the General Agreement on Tariffs and Trade (GATT).

A WTO member nation may challenge the legality of an import ban or restriction of another WTO member nation under special dispute resolution procedures. These procedures call for a legal hearing before a dispute resolution panel which recommends a solution to the WTO for its approval. If a member nation refuses to comply with a WTO decision against its trade restriction, the injured nation may retaliate.

One of the fundamental rules of the GATT is Article X(1), which prohibits nations from imposing quantitative restrictions (quotas) or quantitative prohibitions (embargoes) on imports. No prohibitions or restrictions other than duties, taxes or other charges, whether made effective through quotas, import or export licenses or other measures, shall be instituted or maintained by any contracting party on the importation of any product of the territory by any other contracting party or on the exportation or sale of export of any product destined for the territory of any other contracting party. Seafood embargoes are clearly inconsistent with the free flow of goods which is one of the basic objectives of the GATT.

There are numerous exceptions to Article XI permitted by the GATT. The two exceptions where trade embargoes might be justified for environmental purposes are Articles XX(b) and (g):

> Subject to the requirement that such measures are not applied in a manner which would constitute a means of arbitrary or unjustifiable discrimination between countries where the same conditions prevail, or a disguised restriction on international trade, nothing in this Agreement shall be construed to prevent the adoption of enforcement by any contracting party of measures:
>
> (b) necessary to protect human, animal or plant life or health; . . .
>
> (g) relating to the conservation of exhaustible natural resources if such measures are made effective in conjunction with restrictions on domestic production or consumption . . . .

Article XX(b) allows governments to impose trade measures that are "necessary to protect human, animal or plant life or health." Article XX(g) permits governments to impose trade measures, such as import embargoes, where the measures "relate" to the conservation of exhaustible natural resources if such measures are made effective in conjunction with restrictions on domestic production or consumption." Both article XX(b) and (g) are qualified by the requirements that any trade measures taken must not constitute arbitrary or unjustified discrimination and must not be a disguised restriction on trade [See generally: J. Klabbers, Jurisprudence in International Trade Law; Article XX of GATT, 26 No. 2 J. World Trade 63-94 (1992)].

Vessel Seizures by Canada

In 1980, Canada won a GATT protest over a United States embargo against its tuna exports which had been prompted by the Canadian seizure of a United States albacore vessel fishing in Canadian waters. Such an embargo was authorized under Section 205 of the Magnuson Act, but was declared contrary to GATT because the United States embargo was too broad; i.e., it applied to all tuna and not just the "like product" (albacore) which was under dispute.

The First Tuna-Dolphin Ruling

In August of 1991, another GATT panel ruled in the first tuna-dolphin case that a United States embargo against tuna imported from Mexico was not an appropriate restriction under GATT Article III, because it was aimed against the fishing operations producing the tuna, and not tuna as a product. The panel said that the embargo could not be justified as a measure needed to protect either animal or human health under Article XX(b), or to protect natural resources under Article XX (g), because GATT Article XX defenses do not apply to measures against activities taking place outside the geographical jurisdiction of a nation. The panel's reasoning was that one nation should not be allowed to unilaterally determine environmental policies from which other GATT nations could not deviate losing rights under GATT.

The panel went on to say that even if GATT Article XX(b) and (g) could be applied extrajurisdictionally, they were not justified as being "necessary to protect . . . animal . . . life or health" under GATT Article XX (b), because the embargo was based on an "unpredictable" United States requirement that the kill-per-set of Mexican vessels meet the kill per set rate of United States vessels. Likewise, the embargo was not justified under Article XX (g) as a measure "primarily aimed at the conservation of dolphins," because it was based on "unpredictable conditions."

Since the primary embargo against Mexico was GATT-inconsistent, the GATT panel said that any United States tuna embargo against intermediary nations was also GATT-inconsistent. The panel also noted that the President's discretion to embargo fishery products under the Pelly Amendment did not, by itself, violate the GATT.

The GATT panel in the first tuna-dolphin case said that the MMPA requirement that importers of tuna products labeled "Dolphin Safe" have documentation to prove their shipment was in compliance with "Dolphin Safe" requirements did not violate the GATT, because these requirements were internal measures which did not discriminate between countries.

Canadian Export Restrictions

Article 1201 of the United States-Canada Free Trade Agreement (FTA), negotiated in 1986-1987, incorporates the basic provisions against import bans and restrictions of the GATT. In 1989, a FTA legal panel ruled that a Canadian restriction on the export of unprocessed herring and salmon was an embargo under GATT Article XI, which was not excused as a conservation measure under GATT Article XX(g). The panel noted that a measure must meet four conditions to qualify for an Article XX(g) exemption: (1) It must relate to an exhaustible renewable resource; (2) Domestic production or consumption of the product must be limited; (3) The measure must not create arbitrary or unjustifiable discrimination between nations; and (4) It must be primarily aimed at conservation.

United States Lobster Restrictions

In 1990, Canada unsuccessfully challenged a United States prohibition against the possession of small lobsters in the United States set forth in Section 205 of the Magnuson Act. Under the FTA dispute resolution procedure, Canada claimed this United States lobster restriction was contrary to FTA Article 407, which incorporates the prohibition against embargoes in GATT Article XI. In its ruling, the FTA panel said the restriction was an import ban under GATT Article XI, but was a requirement "affecting the internal sale . . . or use of products" allowed under GATT Article III.

The Second Tuna-Dolphin Ruling

On May 20, 1994, a second GATT panel ruled that the secondary United States embargoes of tuna under the Marine Mammal Protection Act violated the GATT. These embargoes banned the import of tuna from countries whose dolphin protection measures were inconsistent with United States requirements.

The panel ruled that the United States could embargo imports to protect natural resources outside its borders, but not in cases where the embargo user intended to force other countries to change their policies. The panel also noted that embargoes must be based on a product's characteristics, and not on the policies of the country of origin of the product.

Unfair Trade

Several United States trade laws are designed to counter the effect of foreign imports that benefit from unfair trade practices such as subsidies, or those which cause serious injury or threat to a

United States industry. On several occasions, various shrimp, salmon, ground fish, and tuna industry groups have petitioned federal officials in an attempt to raise tariff rates on imports, so far with limited success.

Antidumping Actions

The antidumping provisions of the Trade Act are designed to prevent unfair price discrimination and below-cost sales that cause material injury to United States industries (19 U.S.C. 160 & 1673). The Department of Commerce (DOC) and the International Trade Commission (ITC) jointly administer this law. Under the program, offsetting duties are imposed when the DOC determines a class or kind of imported product is being sold at "less-than-fair-value" and the ITC determines that dumped imports are causing or threatening material injury to a United States industry.

An investigation typically starts when a United States industry, trade association, or union files a petition. If the DOC decides the petition contains sufficient support for its allegations, it will initiate an antidumping investigation. After a case is initiated by the DOC, the ITC makes a preliminary decision on whether there is a reasonable indication that imports are causing or threaten material injury, or are materially retarding an industry's establishment.

After the ITC makes its preliminary decision, the DOC sends out questionnaires to foreign producers, inquiring into whether the product is being sold unfairly. Based on the responses, the DOC issues a preliminary ruling. If it determines that unfair sales are being made, liquidation of customs entries is suspended and companies are required to post a bond equal to the estimated average antidumping margin. After questionnaire responses are verified on site and interested party comments are evaluated, the DOC issues a final determination. If the final DOC determination is affirmative, the case shifts back to the ITC for a final decision on material injury.

Dumping investigations are conducted under strict statutory deadlines:

Day 1: Petition filed with the DOC and the ITC.
Day 20: DOC decision to initiate.
Day 45: ITC preliminary injury determination.
Day 160: DOC preliminary determination of dumping margin.
Day 235: DOC final determination.
Day 286: ITC final injury determination.
Day 287: DOC issues antidumping duty order if both the DOC and ITC have issued affirmative final determinations.

Countervailing Duty Actions

Section 701 of the Tariff Act offsets the benefits of foreign government subsidies (17 U.S.C. 1671). The procedures followed are very similar to antidumping procedures. The DOC normally initiates cases based upon a petition submitted by an interested party. The ITC is responsible for investigating material injury.

For a case to proceed, the ITC must first make a preliminary finding of "reasonable indication" of material injury or threat of material injury, or material retardation of an industry's establishment. The DOC then issues preliminary and final determinations on subsidization. If the decision is affirmative, the ITC makes a final injury determination.

In 1983, the United States tuna industry obtained a ruling against the Philippine government, which imposed a very small additional tariff on tuna imports to offset the subsidization of products. There also was a small countervailing duty (5.82 percent) on certain species of fresh groundfish imported from Canada, which was revoked on January 1, 1991 (56 Federal Register 28530).

In 1978, the Fishermen's Marketing Association filed a petition concerning fish and fish products from Canada, including whole cod, fresh, chilled, or frozen; salted, pickled, smoked, or kippered cod, cusk, haddock, hake, or pollock; cod and flatfish (except turbot) frozen in blocks of 10 pounds or more each; and flatfish fillets, fresh, chilled, or frozen (except halibut or turbot) (*Certain Fish From Canada*, Investigation No. 303-TA-3, USITC Publication 9119, September 1978). The ITC determined by a vote of five to zero that an industry in the United States was not being injured, was not likely to be injured, and was not prevented from being established, by reason of the importation of certain duty-free fish from Canada. Although the ITC determined total United States imports from Canada of ground fish and ground fish products increased as a percent of apparent United States consumption, the percentage of apparent United States consumption accounted for by United States domestic producers also increased; therefore, the ITC found the impact of the imports from Canada was primarily on imports from other sources, which declined as a percentage of apparent domestic consumption. Increased landings and fishing prices for most categories of ground fish and ground fish products indicated to the ITC that the financial situation of the United States producers was improving. The ITC held that the bounties and grants found by the Treasury had been virtually eliminated and the remaining bounties and grants were not likely to have an injurious impact on the United States industry.

In 1979, the ITC investigated the importation from Canada of whole cusk, haddock, hake and pollock, whether fresh, chilled, or frozen; fish blocks made of Atlantic ocean perch, haddock, whiting, and other fish blocks except those made of cod, flatfish, or pollock; live lobsters; and scallops (*Certain Fish and Certain Shellfish from Canada*, Investigation No. 303-TA-9, USITC Publication 966, April 1979). It determined by a vote of five to zero that the industry in the United States was not being injured, was not likely to be injured, and was not prevented from being established, by reason of the importation. The ITC found that United States landings of all four species of whole fish increased, but the ratio of imports from Canada to apparent United States consumption fell. Despite the fact that the United States dollar had appreciated in relation to the Canadian dollar, the ITC had no information that the prices for whole fish from Canada sold in the United States differed from those for United States whole fish. As for fish blocks, the Commission determined that virtually all fish landed in the United States are sold fresh and fresh fish commands a higher price per pound than frozen fish. The ITC also found that United States landings of sea scallops and the number of scallop dredges over five tons had increased, leading to increased employment opportunities for the scallop fishermen, whereas Canada's share of the United States scallop market had remained fairly steady. Also, the ITC held that the fishery management plans initiated in conjunction with the establishment of the 200-mile limit would provide a comprehensive program for expanding United States production of the subject ground fish and shellfish.

In August 1979, an amended petition was filed by the Fishermen's Marketing Association of Washington, asking the ITC to determine whether fresh, chilled, or frozen cod, cusk, haddock, hake, pollock, whiting, Atlantic ocean perch, Pacific rockfish (including Pacific ocean perch), flounder, turbot, and all other flatfish except halibut were being imported into the United States in such increased quantities as to be a substantial cause of serious injury, or the threat thereof, to the domestic industry, producing articles like or directly competitive with the imported articles (*Certain Fish...*, Investigation No. TA-201-41, USITC Publication 1028, January 1980). In 1980, the ITC determined by a vote of three to zero that the above-mentioned ground fish were not being imported into the United States in such increased quantities as to be a substantial cause of serious injury, or threat of serious injury, to the domestic industry producing the like or directly competitive products. The ITC determined that fresh or chilled ground fish products were not likely or directly competitive with frozen ground fish products. The ITC determined imports had increased within the meaning of section 201(b)(1); however, they also determined the bulk of the subject imports were frozen, but nearly all the production of the domestic industry was marketed fresh. The only issue that remained was whether increased imports of whole fish and fresh fillets were a substantial cause of serious injury or threat thereof to the United States industry. The ITC determined that imports might well have been a problem to the United States industry, but that an overly rapid expansion of the fishing fleet on the West Coast was the primary problem of West Coast United States fishermen, and that the conservation quotas on the East Coast were the primary problem for East Coast fishermen.

Injury from Increased Imports

Sections 201 to 204 of the Trade Act of 1974 concern investigations by the ITC as to whether a product is being imported into the United States in such increased quantities as to be a substantial cause of serious injury, or the threat thereof, to a domestic industry (17 U.S.C. 2251-54). If the ITC makes an affirmative determination, it recommends to the President the action that will facilitate positive adjustment by the industry to import competition. After considering the ITC recommendation, the President may take action in the form recommended by the ITC or may take certain other action. Sections 201 and 204 also provide for the filing of industry adjustment plans and commitments in connection with an investigation, for provisional relief in the case of perishable agricultural articles or critical circumstances, and for Commission monitoring of action and reports on the effectiveness of the action taken.

Other Unfair Practices

Section 301 of the Trade Act concerns investigations by the United States Trade Representative (USTR) into allegations that foreign countries are denying benefits to the United States under trade agreements or are otherwise engaged in unjustifiable, unreasonable, or discriminatory acts that burden or restrict commerce of the United States.

The USTR may initiate investigations upon petition by an interested person, or upon its own initiative. If petitioned, the USTR decides within 45 days whether or not to initiate the investigation. If the USTR decides to initiate the investigation, the petitioner and other interested persons are afforded an opportunity to present their views — including a public hearing if requested.

In the investigation, the USTR determines whether United States rights under trade agreements are being denied or whether acts or policies of the foreign country are unjustifiable, unreasonable, or discriminatory and burden or restrict United States commerce. If the USTR finds such circumstances, the USTR then considers what actions are appropriate to enforce the rights of the United States under the agreements or to eliminate the acts or policies. With certain exceptions, the USTR is required to take action, subject to any direction of the President, if the USTR finds that United States rights under trade agreements are being denied, or that acts or policies of the foreign country are "unjustifiable," as defined by statute, and burden or restrict United States commerce. If the USTR finds an act or policy to be "unreasonable" or "discriminatory," as defined by statute, and to burden or restrict United States commerce, the USTR has discretion over whether to take action.

The USTR has authority to (1) suspend trade agreement concessions, (2) impose duties or other import restrictions, (3) impose fees or restrictions on services, (4) enter into agreements with the subject country to eliminate the offending practice or to provide compensatory benefits for the United States, and (5) restrict service sector authorizations. The actions may be taken against all countries or solely against the subject country. Most actions may be taken against any goods or economic sectors, without regard to whether the goods or economic sectors were the subject of the investigation. The USTR is to give preference to duties over types of import restrictions.

In 1986 a group of Pacific Northwest seafood processors filed a section 301 petition complaining of Canadian prohibition of unprocessed salmon and herring exports. In 1987 the United States requested establishment of a GATT dispute settlement panel, and in late November 1987 the GATT panel ruled in favor of the United States. The panel report was adopted in February 1988.

Instead of removing the export restrictions, in April 1989 Canada implemented regulations requiring all Pacific roe herring and salmon caught in Canadian waters to be landed in British Columbia before export. In May 1989, the United States challenged Canada's landing requirements before a CFTA Chapter 18 dispute settlement panel.

In October 1989, the panel found that Canadian landing requirements violated CFTA Article 407, which prohibits GATT-inconsistent export restrictions. Taking into account the panel findings, the United States and Canada negotiated an interim solution to the dispute in February 1990. As a result, Canada agreed to permit up to 25 percent of Canadian-caught salmon and herring to be exported to the United States without landing in Canada. In June 1990, the USTR terminated its section 301 investigation and announced it would monitor Canadian implementation of the Interim Agreement.

Section 337 of the Tariff Act (19 U.S.C. 1337) declares unlawful:

> (a) The importation, sale for importation, or sale after importation by the owner, importer, consignee, or by the agent of any such person, that infringes a valid and enforceable United States patent, or a registered trademark, copyright, or mask work, for which an indus-

> try exists or is in the process of being established in the United States, and
>
> (b) Unfair methods of competition and unfair acts in importation or sale, by the owner, importer, consignee, or by the agent of any such person, if such methods or acts have the threat or effect of destroying or substantially injuring an industry in the United States, of preventing the establishment of such an industry, or of restraining or monopolizing trade and commerce in the United States.

The ITC is authorized, upon the filing of a complaint or on its own initiative, to investigate alleged violations of Section 337 and to determine whether such violations exist. In lieu of a determination, the ITC may terminate an investigation, in whole or in part, on the basis of a settlement agreement or consent order. In appropriate circumstances, the ITC may issue limited or general exclusion orders, which direct that certain goods be denied entry into the United States, and/or may issue cease and desist orders, which enjoin a person from further violation of Section 337.

If the ITC has reason to believe that a complaint or investigation under Section 337 is based solely on alleged facts falling under the antidumping or countervailing duty laws, the ITC must decline to institute or must terminate, as the case may be, its investigation under Section 337. If the complaint or investigation before the ITC is based in part on Section 337 and in part on the antidumping countervailing duty laws, the ITC may institute or continue an investigation.

Under Section 332 of the Tariff Act (19 U.S.C. 1332), the ITC conducts investigations into trade and tariff matters upon request of the President, the Committee on Ways and Means of the House of Representatives, the Committee on Finance of the Senate, either branch of the Congress, or upon the ITC's own initiative. The ITC has broad authority to investigate matters pertaining to the customs laws of the United States, foreign competition with domestic industry, and international trade relations.

Unlike other investigations by the ITC, there is no established procedure for the initiation of investigations under Section 332 by public petition. However, in the course of an investigation under Section 332, the ITC generally seeks written submission and makes its reports under Section 332 available to the public, with the exception of confidential business information or reports (or portions of reports) classified as confidential under national security criteria.

Fish and Seafood Import and Export:
Exports and Exporting

Roy E. Martin

INTRODUCTION

The United States is presently the world's second largest exporter of fishery products, but this was not always the case. United States exports increased dramatically after 1976. Before the extension of our national fishery jurisdiction, the annual value of our exports ranged from about $100 to $300 million. In 1977, the 200-mile limit came into effect, giving American fishermen priority access to the fisheries of what is now called the Exclusive Economic Zone. What the foreign buyer once could harvest directly, he now had to buy from a United States operator if the United States fishing capacity existed.

At the same time, many other nations extended their fishery management jurisdictions to 200 miles, displacing the distant water fleets of various consuming or marketing countries. The resulting shifts in terms of resource access produced changes in the volume and patterns of international trade in fishery products.

United States exports for 1997 of edible fishery products were 2.0 billion pounds valued at $2.7 billion — a decrease of 93.2 million pounds and $319.2 million when compared to 1996 [National Marine Fisheries Service (NMFS), 1997]. See Tables 1–9 and Figures 1 and 2.

Other problems for any exporter are interpreting and meeting the special needs of overseas markets. Foreign government rules may require particular labeling, restrict the use of certain additives, and otherwise necessitate product changes before exporting. But a more difficult and subtle source of problems is the foreign market itself. Different cultural preferences for the way a product looks or tastes are probably the easiest to identify and adapt to, if one is willing to make the changes necessary. But subtle problems may be

Table 1. Fishery products exports, 1988–97[1]

	Edible			Non-edible	Total
Year	Thousand Pounds	Metric Tons	Thousand Dollars		
1988	1,085,935	492,577	2,213,326	125,061	2,338,387
1989	1,405,977	637,747	2,355,603	2,582,538	4,938,141
1990	1,947,292	883,286	2,881,262	3,084,677	5,965,939
1991	2,058,594	933,772	3,155,771	3,386,037	6,541,808
1992	2,087,606	946,932	3,465,667	3,653,965	7,119,632
1993	1,986,027	900,856	3,076,813	3,847,911	6,924,724
1994	1,978,507	897,445	3,126,120	4,254,741	7,380,861
1995	2,047,181	928,595	3,262,242	5,005,878	8,268,120
1996	2,112,055	958,022	3,032,282	5,621,169	8,653,451
1997	2,018,889	915,762	2,713,082	6,640,533	9,353,615

[1] Figures reflect both domestic and foreign (re-exports).

Note: The increase in the nonedible value beginning in 1989 is due to re-examination of commodities that are considered to be based on fishery products including fish, shellfish, aquatic plants and animals and any products thereof, including processed and manufactured products.

Source: United States Department of Commerce, Bureau of the Census.

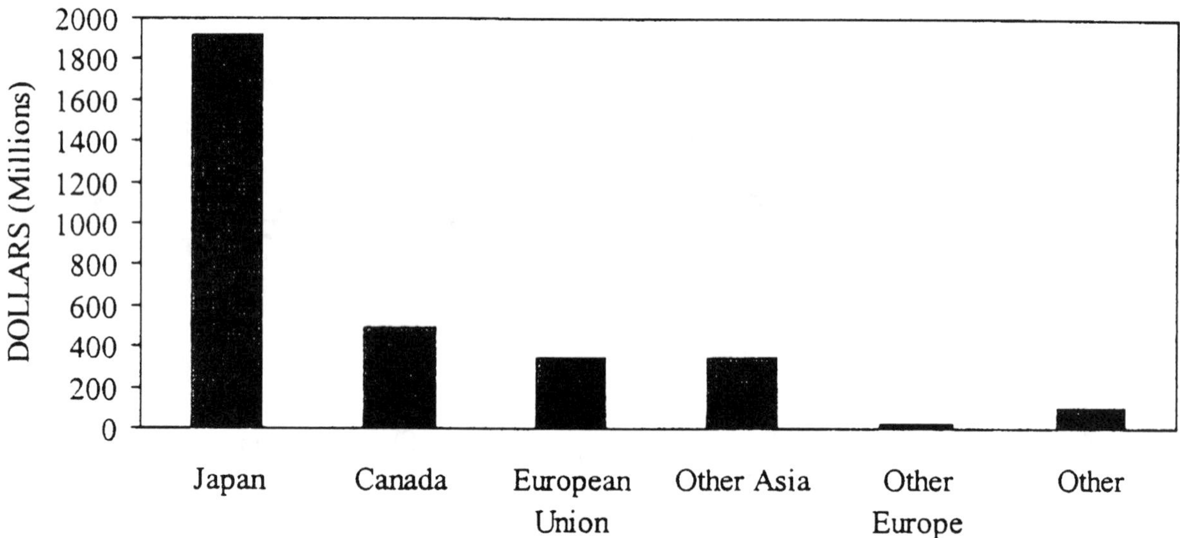

Figure 1. U.S. exports of edible fishery products, 1997, to major markets. (Source: U.S. Department of Commerce, Bureau of the Census.)

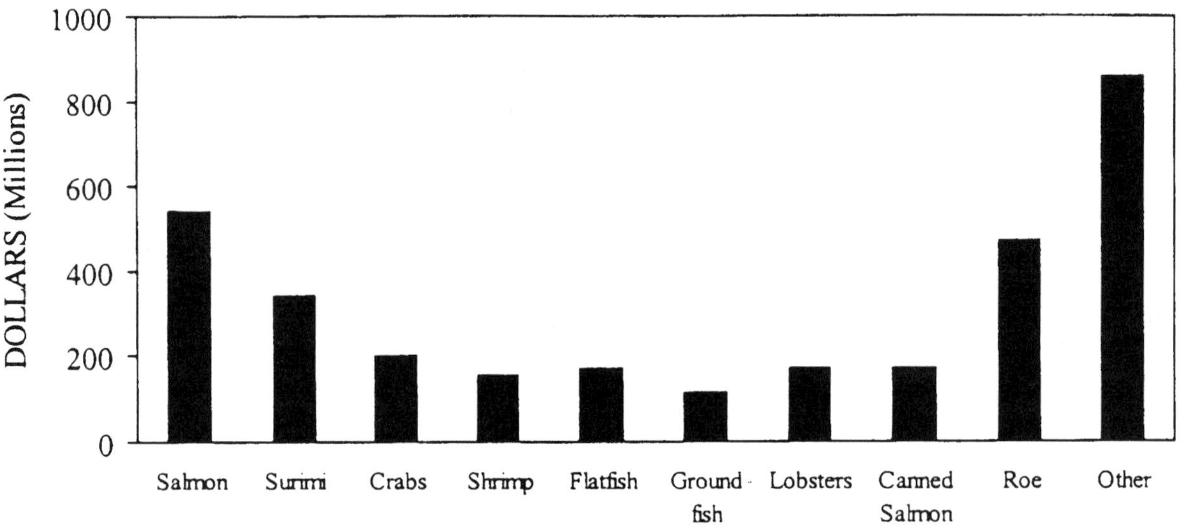

Figure 2. U.S. exports of edible fishery products, 1997, by major group. (Source: U.S. Department of Commerce, Bureau of the Census.)

discovered after shipment of the product overseas. Different distribution systems may require different packaging, such as smaller units to fit on the narrower handcarts used in the more-constricted Japanese warehouse aisles. Cultural anomalies, such as the Japanese superstition about the character "4" or the western taboo on "13," can oblige one to pack in different quantities for different markets. And, of course, the term "quality" has some very subjective elements that may make squid frozen in bulk completely unacceptable in Spain, despite its apparent high quality in the eyes of the United States.

Just after World War II, when the American economy was strong and growing, our need for export markets led to a national policy of free trade. The most visible and quantifiable trade barrier was the tariff. The United States negotiated the General Agreement on Tariffs and Trade (GATT) in 1947, an umbrella treaty setting forth an agreed basis on which the free world would conduct its international trade. A key element of

Table 2. Fresh and Frozen Salmon Exports, Whole or Eviscerated, by Country of Destination 1996 and 1997[1]						
	1996			1997		
Country	Thousand Pounds	Metric Tons	Thousand Dollars	Thousand Pounds	Metric Tons	Thousand Dollars
Japan	176,441	80,033	344,756	113,477	51,473	208,038
Canada	55,565	25,204	73,794	35,289	16,007	52,136
France	11,180	5,071	11,068	11,058	5,016	12,750
Belgium	4,559	2,068	5,288	3,662	1,661	4,383
Denmark	5,337	2,421	5,973	3,366	1,527	3,566
Spain	2,293	1,040	1,866	3,186	1,445	3,231
Taiwan	2,901	1,316	5,333	1,382	627	3,230
Germany	2,057	933	2,337	2,899	1,315	2,849
China	2,156	978	3,182	2,533	1,149	2,513
Other	14,484	6,570	16,215	13,089	5,937	14,825
Total	276,973	125,634	469,812	189,942	86,157	307,521

[1]Figures reflect both domestic and foreign (re-exports).
Source: United States Department of Commerce, Bureau of the Census.

Table 3. Canned Salmon Exports, by Country of Destination, 1996 and 1997[1]						
	1996			1997		
Country	Thousand Pounds	Metric Tons	Thousand Dollars	Thousand Pounds	Metric Tons	Thousand Dollars
United Kingdom	40,809	18,511	73,252	35,075	15,910	67,228
Canada	29,297	13,289	42,845	20,763	9,418	29,277
Australia	12,225	5,545	18,103	11,565	5,246	18,415
Netherlands	6,329	2,871	10,148	6,479	2,939	10,192
New Zealand	672	305	881	1,664	755	2,140
Belgium	1,270	576	1,826	1,612	731	2,108
France	611	277	1,085	831	377	1,139
Japan	522	237	761	589	267	910
Ireland	650	295	966	388	176	668
Other	3,144	1,426	4,226	2,654	1,204	3,360
Total	95,530	43,332	154,093	81,621	37,023	135,437

[1]Figures reflect both domestic and foreign (re-exports).
Source: United States Department of Commerce, Bureau of the Census.

the GATT was the commitment not to raise tariffs above their present levels except under certain exceptions. Every so often, the GATT-member countries meet in a multilateral trade negotiation (MTN) and agree on further tariff reductions. This has led to a situation in which, in many cases, tariffs are not serious obstacles to trade. They still have major effects on relative competitive advantage, but they do not function as absolute barriers with the frequency that they used to. Trade barriers have received considerable attention in the past few years. This was a natural outcome of the increase in seafood exporting activity as American businesses began to do more business overseas.

Two engines will drive future trade: the first is termed the agreement on sanitary and phytosanitary measures (SPS), and the second, the agreement on technical barriers to trade (TBT). The GATT precipitated the World Trade Organization (WTO), which will be the arbitrator of trade disagreements in the future.

Table 4. Fresh and Frozen Shrimp Exports by Country of Destination, 1996 and 1997[1]

Country	1996 Thousand Pounds	1996 Metric Tons	1996 Thousand Dollars	1997 Thousand Pounds	1997 Metric Tons	1997 Thousand Dollars
Canada	19,015	8,625	74,828	17,381	7,884	71,724
Mexico	7,211	3,271	24,123	8,023	3,639	24,772
Japan	2,284	1,036	11,286	1,620	735	9,366
China	1,975	896	5,071	1,495	678	4,119
Honduras	904	410	3,869	935	424	3,558
Brazil	340	154	752	988	448	3,261
Thailand	1,687	765	6,086	809	367	2,958
Norway	820	372	3,567	628	285	2,684
South Korea	500	227	2,027	267	121	1,258
Other	3,902	1,770	13,455	2,668	1,210	9,742
Total	38,638	17,526	145,054	34,813	15,791	133,442

[1]Figures reflect both domestic and foreign (re-export).
Source: United States Department of Commerce, Bureau of the Census.

SPECIAL REQUIREMENTS

The European Union (EU) is a group of countries that have agreed to harmonize their commodity requirements to facilitate commerce among member states. On March 15, 1993, the EU issued Decision 93/185 prescribing certain measures concerning the certification of fishery products from non-member (third) countries. Beginning July 1, 1993, the EU required that each shipment of fishery products (the EU definition of "fishery products" includes molluscan shellfish) from a third country must be accompanied by an EU Health Certificate. The EU Health Certificate must state that the shipment was produced in an establishment covered under a regulatory oversight program equivalent to that in place in the EU, and signed by a competent authority.

The EU has designated the United States Food and Drug Administration (FDA) as a competent authority to sign EU Health Certificates. The FDA Export Health Certificate Program has been developed to meet the current EU requirements. There are no fees associated with the program. [Note: The National Marine Fisheries Service has also been designated as a competent authority to sign certificates (Food and Drug Administration, 1995).]

FDA EXPORT HEALTH CERTIFICATE PROGRAM

The FDA has designed the Export Health Certificate Program to meet the current requirements for EU Health Certificates; it includes: (1) the maintenance of an EU Export Health Certificate List that will include all establishments accepted into the program and will be provided to the EU quarterly; (2) the assignment of a unique seven-digit central file number to each establishment accepted into the program that must be included on the EU Health Certificate to identify the establishment that processed the product; and (3) the signature of EU Health Certificates for products processed by establishments in the program.

Establishments that substantially transform domestic and imported products are eligible for the FDA's EU Export Health Certificate List. Substantial transformation means the product is eviscerated, portioned, breaded, canned, stuffed, value-added, glazed, or otherwise significantly processed. Repacking the product from one container to another, warehousing, and transporting are not considered substantial transformation. No inspection of the establishment or product by the FDA will be necessary upon receipt of an EU Health Certificate for signature. Signature of the health certificate verifies the establishment is covered under an oversight program and is in compliance with the FDA laws and regulations. If an establishment wants to participate in the program, it shall submit a written request to the FDA that addresses the following six points:

1. The official name and address of the establishment site that produces the products intended

Table 5. Canned Shrimp Exports by Country of Destination, 1996 and 1997[1]

	1996			1997		
Country	Thousand Pounds	Metric Tons	Thousand Dollars	Thousand Pounds	Metric Tons	Thousand Dollars
Canada	963	437	4,739	423	192	2,462
Belgium	132	60	818	247	112	1,460
Thailand	437	198	2,281	229	104	1,271
Indonesia	57	26	348	163	74	577
Ecuador	207	94	1,315	75	34	403
China	71	32	197	49	22	218
Taiwan	–	–	–	24	11	206
Greece	538	244	427	220	100	175
South Korea	84	38	520	13	6	74
Other	176	80	936	26	12	183
Total	2,665	1,209	11,581	1,470	667	7,029

[1]Figures reflect both domestic and foreign (re-export).
Source: United States Department of Commerce, Bureau of the Census.

for export to the EU. For companies with more than one site that want to participate in the program, a separate written request is required for each location. This information is necessary to determine whether the firm is currently in the FDA's inventory of establishments subject to inspection and has been assigned a central file number. The central file number is the unique identification number that will be used for the program.

2. A list of the products and packaging types processed at the establishment site and intended for export to the EU — for example, canned salmon, frozen halibut, fresh king crab, raw oysters. This information is necessary to determine whether the products processed by the establishment are covered under the FDA's laws and regulations.
3. The name, mailing address, telephone number, and FAX number of a designated contact person for the establishment. This individual can be at the establishment's site or any other company location, and should be prepared to provide additional information if questions arise during the processing of the written request and EU Health Certificates.
4. The name of any state and other Federal agencies that inspect the establishment and analyze its products. The written request shall include a statement that the FDA will be provided access to these inspection and analytical records upon request. Inspection and product history records from other agencies may be necessary to supplement the FDA's information about the establishment during the review of the written request.
5. Assurance that the individuals representing the establishment and submitting EU Health Certificates to the FDA for signature realize that they are subject to the provisions of Title 18, United States Code of Federal Regulations. Title 18 states that it is a criminal offense to willfully make false statements to a United States official in the performance of their duties, or to alter or counterfeit official documents. The written request shall include a statement acknowledging this fact. This statement is necessary to protect the integrity of the program and to ensure that potential participants realize they are subject to criminal prosecution if they submit false information or fraudulent documents to the FDA.
6. A description of the establishment's programs to ensure that its products meet existing FDA laws and regulations, including their quality control and sanitation programs. The written request must state that the establishment operates under a Hazard Analysis Critical Control Point (HACCP) or equivalent program (as of December 18, 1997). Further, the written request shall include a statement that the FDA may have access to the establish-ment's quality control,

Table 6. Fish and Marine Animal Oil Exports, by Country of Destination, 1996 and 1997[1]

	1996			1997		
Country	Thousand Pounds	Metric Tons	Thousand Dollars	Thousand Pounds	Metric Tons	Thousand Dollars
Netherlands	99,416	45,095	16,248	116,249	52,730	26,420
Canada	55,904	25,358	13,013	46,806	21,231	13,244
Mexico	14,888	6,753	3,229	22,716	10,304	5,131
South Korea	8,719	3,955	1,905	10,452	4,741	2,688
China	1,003	455	631	5,099	2,313	2,068
Hong Kong	51	23	91	3,018	1,369	1,619
Japan	3,673	1,666	657	6,762	3,067	1,251
Spain	2,864	1,299	395	2,643	1,199	772
United Kingdom	–	–	–	185	84	108
Other	776	352	254	1,325	601	415
Total	187,294	84,956	36,423	215,255	97,639	53,716

[1]Figures reflect both domestic and foreign (re-export).
Source: United States Department of Commerce, Bureau of the Census.

sanitation, production, and HACCP records upon request. This information is needed to provide the FDA with some confidence that the establishment and its products comply with the laws and regulations enforced by the FDA, over-and-above that provided by its own inspection and sampling programs.

The FDA will review the information provided in the written request for participation in the program along with the establishment's recent inspection and product history records, including any information received from state and other Federal agencies. The establishment will be accepted and placed on the FDA's EU Export Health Certificate List if the records show the establishment and its products comply with the laws and regulations of the FDA, and with its state and Federal agency counterparts. The establishment's request will be denied if it does not meet the program criteria, or if the review of the inspection and product history finds the establishment is subject to regulatory action. An establishment may also be excluded from the program if the FDA's records show an ongoing ineffective recall, an unresolved consumer complaint involving injury or illness, the submission of false information, or ongoing legal action as the result of state and other Federal agency inspection and sampling activities.

If the establishment is accepted into the program, it will be notified in writing and provided with a unique seven-digit central file number. The establishment will then be placed on the FDA's EU Export Health Certificate List. If acceptance into the program is denied, the establishment will be notified in writing of the reason(s) and provided additional instructions.

The establishment will remain on the list indefinitely unless there is a change in its status that disqualifies it from the program. If subsequently disqualified, written notification with the reasons(s) for removal will be sent to the establishment.

EU Health Certificates in the nine official languages, along with instructions, will be made available to the establishments entered onto the EU Export Health Certificate List. For each shipment, a completed EU Health Certificate must be submitted to the FDA's seafood coordinator for review and processing. The establishment's status will be updated during the review of the EU Health Certificate and, if the establishment is still qualified, the health certificate will be assigned a unique shipment number, signed, and returned to the establishment.

If the EU Health Certificate is not accepted, a written notification with the reason(s) will be sent to the establishment.

Table 7. Fresh and Frozen Crab Exports, by Country of Destination, 1996 and 1997[1]						
	1996			1997		
Country	Thousand Pounds	Metric Tons	Thousand Dollars	Thousand Pounds	Metric Tons	Thousand Dollars
Japan	35,044	15,896	137,529	28,955	13,134	82,230
Canada	8,702	3,947	15,535	10,571	4,795	17,712
Thailand	216	98	951	897	407	2,658
China	1,962	890	5,849	1,349	612	2,228
Taiwan	117	53	167	421	191	722
Hong Kong	51	23	284	163	74	418
Mexico	79	36	244	123	56	401
United Kingdom	104	47	528	55	25	332
Belgium	7	3	65	49	22	313
Other	461	209	1,251	549	249	1,647
Total	46,742	21,202	162,403	43,133	19,565	108,661

[1]Figures reflect both domestic and foreign (re-export).
Source: United States Department of Commerce, Bureau of the Census.

Brokers, dealers, and other companies are not eligible for the EU Export Health Certificate List because they do not have a processing site that can be inspected. However, they can submit EU Health Certificates for products processed by other establishments on the EU Export Health Certificate List, provided they obtain the unique seven-digit central file number from the product source establishment and include it on the health certificate. EU Health Certificates and instructions are available to these companies upon request.

The EU Health Certificate can be submitted to the FDA by the United States Postal Service, express package service, or courier. For the return of the signed EU Health Certificate, the establishment shall provide a self-addressed, postage-paid return envelope, and an express package airbill form under their account number, or they may arrange for pickup by courier service.

The FDA advises that any United States fish or fishery product processor exporting to the EU consider using two FDA publications to prepare to meet the EU requirements. They are: the FDA HACCP regulaton of December 18, 1997, "Rule to Establish Procedures for the Safe Processing and Importing of Fish and Fishery Products," and the January 1998 "Fish and Fishery Products Hazards and Controls Guide." The Guide includes an explanation of the internationally recognized principles of HACCP and how they can be applied by a processor.

REQUIREMENTS FOR SUCCESSFUL EXPORT MARKETING

Development of export markets requires that particular attention be given to commodity and market characteristics. Those characteristics that contribute to successful export marketing depend on the type of market and commodity being exported. This discussion is only a brief overview of some of the more important characteristics that can foster successful exporting of seafood products. The characteristics discussed can be categorized into the four groups of (1) commodity characteristics, (2) market information, (3) business marketing practices, and (4) institutional support.

Commodity Characteristics

Because fresh seafood products are highly perishable, they require special care and facilities in export trade. Developing markets for highly-processed products tends to reduce potential problems relating to spoilage losses. Dependable means of rapid transportation with suitable preservation capabilities lessens the possibility of product loss. Product insurance further lessens the exporter's risk with perishable products.

Table 8. Fresh and Frozen Crabmeat Exports, by Country of Destination, 1996 and 1997[1]

Country	1996 Thousand Pounds	1996 Metric Tons	1996 Thousand Dollars	1997 Thousand Pounds	1997 Metric Tons	1997 Thousand Dollars
Japan	853	387	2,636	1,497	679	3,813
Russian Federation	547	248	499	875	397	1,206
China	97	44	130	183	83	598
Hong Kong	236	107	245	159	72	459
Canada	126	57	397	150	68	443
Philippines	—	—	—	51	23	200
Switzerland	2	1	3	15	7	168
Mexico	26	12	86	86	39	149
South Korea	198	90	331	15	7	102
Other	1,021	463	1,767	368	167	680
Total	3,106	1,409	6,094	3,399	1,542	7,818

[1]Figures reflect both domestic and foreign (re-export).
Source: United States Department of Commerce, Bureau of the Census.

A second characteristic of seafood products is the limited control producers have over annual supply and the highly seasonal nature of production. Storage is one means of reducing the impact of seasonal variations. This in turn requires inventory capital and firm market commitments to reduce risk and make inventory capital available at lower rates. The uncontrollable variation in annual supply often prevents export trade because either large quantity requests by buyers cannot be filled or economically efficient-size shipments are not possible. Polling arrangements among exporters may resolve the problem of limited supply.

Market Information

Certain types of market information are necessary for successful export trade activities. Species or product identification is necessary to communicate market prices and coordinate export supply with demand. Foreign consumers often desire products exporters have available but do not recognize the product because of differences in nomenclature. Seafood exporters should also be aware of foreign consumers' concept of quality. An established set of grades and standards will facilitate export trade, providing they reflect foreign consumer tastes and preferences. Supplies and demands are more easily communicated with these items of market information. Foreign consumer tastes and preferences may be determined through reviews of trade statistics, market surveys, trade missions, and trade shows (Pochaska and Cato, 1981).

Business Marketing Practices

In establishing export markets, some of the major aspects of business marketing that are associated with high success levels are market development, transportation, buyer-seller relations, packaging, and financing. Success in these areas often depends on the size of the firm and the support facilities provided by the market system, including industry associations and governmental bodies.

Large firms or groups of smaller export firms often engage in market promotion and development activities in order to identify and contact potential foreign buyers. These activities, in addition to discovering existing markets, often develop new markets and/or increase prices received. Associated with these market development activities should be the establishment of committed buyer-seller arrangements. Some of the more notable trade problems stem from delays in receiving payments for export shipments and a lack of firm commitments for export shipments. Without these assurances, exporters take considerable risks and incur substantial interest and storage costs waiting for final trade transactions.

Table 9. Fish Meal Exports, by Country of Destination, 1996 and 1997[1]

Country	1996 Thousand Pounds	1996 Metric Tons	1996 Thousand Dollars	1997 Thousand Pounds	1997 Metric Tons	1997 Thousand Dollars
China	24,486	11,107	7,653	73,318	33,257	13,995
Taiwan	49,456	22,433	13,882	43,651	19,800	10,897
Canada	31,274	14,186	7,224	39,169	17,767	10,463
Japan	20,719	9,398	5,683	22,123	10,035	6,108
Hong Kong	6,261	2,840	1,688	10,373	4,705	2,421
Philippines	19,868	9,012	4,567	7,474	3,390	2,386
Thailand	1,164	528	2,651	494	224	2,190
Mexico	2,981	1,352	755	1,887	856	2,093
Saudi Arabia	—	—	—	5,615	2,547	1,721
Other	30,203	13,700	8,828	12,185	5,527	4,923
Total	186,412	84,556	52,931	216,289	98,108	57,197

[1]Figures reflect both domestic and foreign (re-export).
Source: United States Department of Commerce, Bureau of the Census.

In order to fully take advantage of all trade possibilities, adequate transportation systems are a necessity. Ideally, transportation systems should provide access to all or most market areas with freight rates at levels that do not impair the nation's competitive advantage in foreign trade. Assembly of adequate volumes for shipment and establishment of large-volume markets tend to lower freight rates. Vertical integration in the production and processing of seafoods facilitates assembly of larger volumes. Especially important are transportation systems that protect highly perishable seafood products.

Another marketing practice important in export trade is packaging and labeling. United States exporters must be aware that the metric system is used in most seafood-importing countries. Packaging used in export trade usually must be more durable to withstand more lengthy and difficult trips. At the same time, consumer or buyer appeal must be considered in choosing packaging materials.

A final factor to be considered is capital requirements and financing. The cost of international trade is often considerable. Efficient business practices tend to make financing more readily available. Governmental assurances are often necessary due to added risks in international trade.

Institutional Support

Institutional support facilitates international trade. Institutional support comes in the form of government programs, foundation activities, and industry associations.

Group action is often needed to reduce tariffs, quotas, and other trade barriers. Government action, when necessary, requires industry support or pressures. The same is true where domestic product standards enforced by government agencies are stronger than those required in importing countries or by competing exporting nations. Higher or stiffer standards tend to put domestic exporters at a comparative disadvantage.

In order to compete with many seafood-exporting countries, United States exporters need more favorable financing arrangements. Many foreign competitors are wholly or partly financed by their government.

Market development programs sponsored by government agencies or industry groups are often necessary because of the small size of many seafood exporters. The scale of operation for many exporters is simply too small for them to incur the tremendous expense of trade missions or other such activities. Institutionally sponsored market-development programs reduce the competitive advantage that competing exporters have when their governments provide these activities.

EXPORT MARKETING STRATEGY

The target marketing strategy for determining potential export markets for domestic seafood and in making buyer contacts in those areas has consisted of three components. These have been market personnel training, trade shows, and trade missions.

Personnel Training

Most marketing personnel in the universities and the various state market development agencies have been trained and are specialists in domestic seafood marketing. Very few had knowledge of the financial requirements, legal aspects, the "customs," and logistical (among other) problems associated with export market development. This knowledge is necessary for successful export-market development by a marketing specialist.

Trade Shows

A trade show usually focuses on seafood, or various food commodities, and represents an atmosphere in which prospective buyers and sellers become acquainted. Product samples are normally presented to trade show audiences from a display booth manned by marketing personnel. enabling an on-site discussion of the attributes of the product. Each participant is normally shown and/or presented potential supplier lists and information about the product such as seasonality, price ranges, and other desired or requested product attributes. Usually, trade shows are regularly-scheduled events, and represent a fairly low-cost method for a large number of seafood suppliers from a region to contact a larger number of potential product-buyers.

Trade Missions

Trade missions, in contrast to trade shows, are usually accomplished by a delegation of industry and government representatives who visit a particular trade area or country with a particular mission or objective. This objective may consist of informing the country about the potential for seafood exports from the United States and of discovering the needs of that particular country regarding seafood products that the United States can produce. Contacts are made with all industry segments in the country including shippers, marketing agents, consumers, etc. Contacts are also made with government representatives to discuss and inform the government of the intent of the trade mission, and to fully understand the trading customs of the country or area.

INSPECTION SERVICES

Voluntary Federal inspection services are available to help United States exporters gain ready acceptance of their products on arrival at their ports of destination. Inspection and certification documents can be used to promote the marketability of the products in other countries (NMFS, 1985).

Inspection Services for Exporters

The National Marine Fisheries Service helps United States exporters maintain and verify the quality of their products through voluntary Federal inspection services. These are basically the same as the fee-for-service programs for domestic processors.

The NMFS sets grade standards that indicate the quality level of fishery products (Grade A, B, or C). Grading is not required for exported products. However, quality standards provide important technical information which foreign buyers can use in determining the acceptability of products.

United States exporters may arrange for contract inspection at their plants or for lot inspection at any point before shipment. Products that are processed under contract inspection and controlled conditions can qualify to use the Federal inspection mark on their labels.

The exporter may set his own specifications for inspection and certification. Inspections will be made and documents issued which verify that the product was processed in compliance with the specifications set by the exporter or by the buyer in the foreign country.

Fishery products can also be inspected and certified to be in compliance with "Recommended General Principles of Good Hygiene." These principles are a general requirement for environmental hygiene and sanitation in the processing of food. These and other international standards can serve as useful specifications for the United States exporter and his customer in foreign countries.

The NMFS offers additional assistance to exporters. It will review product labels upon request and will furnish technical information regarding definitions and standards. Finally, the NMFS can arrange to certify that export products have met requirements of the country of destination, thereby simplifying procedures after arrival there.

Benefits for Exporters

Fishery product inspection and a document of certification provide an unbiased statement of what a shipment contains, its condition, or the conditions under which it was processed. The document normally eases the acceptance of the product at the buying country. Prompt acceptance minimizes product delay at the point of entry in the foreign country. Routine acceptance avoids most of the specific examinations that a foreign country may elect to carry out; thus, most of the non-tariff barriers to the importation of the product can be eliminated.

Inspection services for export products also help processors upgrade the quality control in their plants. Better quality control aids the consistent production of safe and uniformly high-quality products and, in turn, improves the marketability and acceptability of the product both abroad and at home.

The following may be considered to be the 10 most common mistakes and pitfalls to be avoided by new seafood exporters:

1. Lack of knowledge/understanding of foreign buyer specifications for seafood products.
2. Insufficient knowledge of various countries' import regulations, requirements, and practices pertaining to seafood.
3. Trying to fill on a year-round basis an export order for a product that has too much variation in supply.
4. Failure to obtain qualified counseling and to develop a master international marketing plan before starting an export business.
5. Insufficient commitment by senior company management to overcome the initial difficulties and financial requirements.
6. Insufficient care in selecting overseas distributors.
7. Improperly completed documentation and improper use of documents.
8. Chasing orders for a seafood product from around the world instead of establishing a basis for profitable operations and orderly growth.
9. Neglecting export business, orders, and clients when the United States domestic market for seafood booms.
10. Unwillingness to modify the seafood products to meet foreign regulations or cultural preferences of other countries (Carroll, 1983).

EXPORT GLOSSARY

Some Common Terms Used in International Trade

Acceptance: This term has several related meanings:

1. A time draft (or bill of exchange) which the drawee (the Payer) has accepted and is unconditionally obligated to pay at maturity. The draft must be presented first for acceptance — the drawee becomes the "acceptor" — then for payment. The word "accepted" and the date and place of payment must be written on the face of the draft.

2. The drawee's act in receiving a draft and thus entering into the obligation to pay its value at maturity.

(Broadly speaking) Any agreement to purchase goods under specified terms.

Ad Valorem: "According to value." See *Duty*.

Advisory Capacity: A term indicating that a shipper's agent or representative is not empowered to make definitive decisions or adjustments without approval of the group or individual represented. Compare *Without Reserve*.

Affreightment (Contract of): An agreement between a steamship line (or similar carrier) and an importer or exporter in which cargo space is reserved on a vessel for a specified time and at a specified price. The importer/exporter is obligated to make payment whether or not the shipment is made.

After Date: A phrase indicating that the date of maturity of a draft or other negotiable instrument is fixed by the date on which it drawn. The date of maturity does not, therefore, depend on acceptance by the drawee. Compare *After Sight, At Sight*.

After Sight: A phrase indicating that payment on a draft or other negotiable instrument is due a specified number of days after presentation of the draft to the drawee or payee. Compare *After Date, At Sight*.

Agent: See *Foreign Sales Agent*.

Air Waybill: A bill of lading that covers both domestic and international flights transporting goods to a specified destination. Technically, it is a non-negotiable instrument of air transport that serves as a receipt for the shipper, indicating that the carrier has accepted the goods listed therein and obligates itself to carry the consignment to the airport of destination according to specified conditions. Compare *Inland Bill of Lading, Ocean Bill of Lading, Through Bill of Lading.*

Alongside: A phrase referring to the side of a ship. Goods to be delivered "alongside" are to be placed on the dock or lighter within reach of the transport ship's tackle so that they can be loaded aboard the ship.

Antidiversion Clause: See *Destination Control Statement.*

Arbitrage: The process of buying foreign exchange, stocks, bonds, and other commodities in one market and immediately selling them in another market at higher prices.

At Sight: A phrase indicating that payment on a draft or other negotiable instrument is due upon presentation or demand. Compare *After Sight, After Date.*

Barratry: Negligence or fraud on the part of a ship's officers or crew resulting in injury or loss to the ship's owners.

Barter: Trade in which merchandise is exchanged directly for other merchandise without use of money. Barter is an important means of trade with countries using currency that is not readily convertible.

Bill of Exchange: See *Draft.*

Bill of Lading: A document that establishes the terms of a contract between a shipper and a transportation company under which freight is to be moved between specified points for a specified charge. Usually prepared by the shipper on forms issued by the carrier, it serves as a document of title, a contract of carriage, and a receipt for goods. Also see *Air Waybill, Inland Bill of Lading, Ocean Bill of Lading, Through Bill of Lading.*

Bonded Warehouse: A warehouse authorized by customs authorities for storage of goods on which payment of duties is deferred until the goods are removed.

Booking: An arrangement with a steamship company for the acceptance and carriage of freight.

Brussels Tariff Nomenclature (BTN): See *Nomenclature of the Customs Cooperation Council.*

Buying Agent: See *Purchasing Agent.*

Carnet: A customs document permitting the holder to carry or send merchandise temporarily into certain foreign countries (for display, demonstration, or similar purposes) without paying duties or posting bonds (not needed for seafood in most countries unless samples are brought back to the United States).

Cash Against Documents (C.A.D.): Payment for goods in which a commission house or other intermediary transfers title documents to the buyer upon payment in cash.

Cash In Advance (C.I.A.): Payment for goods in which the buyer pays when ordering and in which the transaction is binding on both parties.

Certificate of Inspection: A document certifying that merchandise (such as perishable goods) was in good condition immediately prior to its shipment.

Certificate of Manufacture: A statement (often Notarized) in which a producer of goods certifies that the manufacturing has been completed and the goods are now at the disposal of the buyer.

Certificate of Origin: A document, required by certain foreign countries for tariff purposes, certifying as to the country of origin of specified goods.

C. & F. "Cost and Freight": A pricing term indicating that these costs are included in the quoted price.

Chamber of Commerce: An association of businesspeople organized to promote local business interests.

Charter Party: A written contract, usually on a special form, between the owner of a vessel and a "charterer" who rents use of the vessel or a part of its freight use. The contract generally includes the freight rates and the ports involved in the transportation.

C. & I. "Cost and Insurance": A pricing term indicating that these costs are included in the quoted price.

C.I.F. "Cost, Insurance, Freight": A pricing term indicating that these costs are included in the quoted price.

C.I.F. & C. "Cost, Insurance, Freight, and Commission": A pricing term indicating that these costs are included in the quoted price.

C.I.F. & E. "Cost, Insurance, Freight and (Currency) Exchange": A pricing term indicating that these costs are included in the quoted price.

Clean Bill of Lading: A receipt for goods issued by a carrier with an indication that the goods were received in "apparent good order and condition," without damages or other irregularities. Compare *Foul Bill of Lading.*

Clean Draft: A draft to which no documents have been attached.

Collection Papers: All documents (invoices, bills of lading, etc.) submitted to a buyer for the purpose of receiving payment for a shipment.

Commercial Invoice: An itemized list of goods shipped, usually included among an exporter's Collection Papers.

Commission Agent: See *Purchasing Agent.*

Common Carrier: An individual, partnership, or corporation that transports persons or goods for compensation.

Confirmed Letter of Credit: A letter of credit, issued by a foreign bank, whose validity has been confirmed by an American bank. An exporter whose payment terms are a confirmed letter of credit is assured of payment even if the foreign buyer or the foreign bank defaults. See *Letter of Credit.*

Consignment: Delivery of merchandise from an exporter (the consignor) to an agent (the consignee) under agreement that the agent sell the merchandise for the account of the exporter. The consignor retains title to the goods until the consignee has sold them. The consignee sells the goods for commission and remits the net proceeds to the consignor.

Consular Declaration: A formal statement, made to the consul of a foreign country, describing goods to be shipped.

Consular Invoice: A document, required by some foreign countries, describing a shipment of goods and showing information such as the consignor, consignee, and value of the shipment. Certified by a consular official of the foreign country, it is used by the country's customs officials to verify the value, quantity, and nature of the shipment.

Countervailing Duty: An extra duty imposed by the Secretary of the Treasury to offset export grants, bounties, or subsidies paid to foreign suppliers in certain countries by the governments of those countries as an incentive to exports.

Credit Risk Insurance: Insurance designed to cover risks of nonpayment for delivered goods. Compare *Marine Insurance.*

Customs: The authorities designated to collect duties levied by a country on imports and exports. The term also applies to the procedures involved in such collection.

Customhouse Broker: An individual or firm licensed to enter and clear goods through Customs.

Date Draft: A draft that matures a specified number of days after the date it is issued, without regard to the date of *Acceptance* (definition 2). Compare *Sight Draft, Time Draft.*

Demurrage: Excess time taken for loading or unloading a vessel. Demurrage refers only to situations in which the charterer or shipper, rather than the vessel's operator, is at fault.

Destination Control Statement: Any of various statements that the United States Government requires to be displayed on export shipments and that specify the destinations for which export of the shipment has been authorized.

Devaluation: The official lowering of the value of one country's currency in terms of one or more foreign currencies. Thus, if the United States dollar is devalued in relation to the French franc, one dollar will "buy" fewer francs than before.

DISC: Domestic International Sales Corporation.

Dispatch: An amount paid by a vessel's operator to a charterer if loading or unloading is completed in less time than stipulated in the charter party.

Distributor: A foreign agent who sells directly for a supplier and maintains an inventory of the supplier's products.

Dock Receipt: A receipt issued by an ocean carrier to acknowledge receipt of a shipment at the carrier's dock or warehouse facilities. Also see Warehouse Receipt.

Documents Against Acceptance (D/A): Instructions given by a shipper to a bank indicating that documents transferring title to goods should be delivered to the buyer (or drawee) only upon the buyer's acceptance of the attached draft.

Documents Against Payment (D/P): Instructions given by a shipper to a bank indicating that documents transferring title to goods should be delivered to the buyer (or drawee) only upon the buyer's payment of the attached draft.

Draft (or Bill of Exchange): An unconditional order in writing from one person (the drawer) to another (the drawee), directing the drawee to pay a specified amount to a named payee at a fixed or determinable future date.

Drawback: A refund of duties paid on imported goods that is provided at the time of their reexportation.

Drawee: The individual or firm on whom a draft is drawn and who owes the indicated amount. Compare *Drawer.* Also see *Draft.*

Drawer: The individual or firm that issues or signs a draft and thus stands to receive payment of the indicated amount from the drawee. Compare *Drawee.* Also see *Draft.*

Dumping: Importing merchandise into a country (e.g., the United States) at low prices that are detrimental to local producers of the same kind of merchandise.

Duty: A tax imposed on imports by the customs authority of country. Duties are generally based on the value of the goods (*ad valorem* duties), some other factor such as weight or quantity (specific duties), or a combination of value and other factors (compound duties).

EMC: See *Export Management Company.*

Eurodollars: United States dollars placed on deposit in banks outside the United States (primarily in Europe).

Ex ("from"): When used in pricing terms such as "Ex Plant" or "Ex Dock," it signifies that the price quoted applies only at the point of origin (in the two examples, at the seller's plant or a dock at the import point). In practice, this kind of quotation indicates that the seller agrees to place the goods at the disposal of the buyer at the specified place within a fixed period of time.

Exchange Rate: The price of one currency in terms of another — i.e., the number of units of one currency that may be exchanged for one unit of another currency.

Export: To send or transport goods out of a country for sale in another country. In international sales, the exporter is usually the seller or the seller's agent. Compare *Import.*

Export Broker: An individual or firm that brings together buyers and sellers for a fee but does not necessarily take part in actual sales transactions.

Export License: A government document that permits the "licensee" to engage in the export of designated goods to certain destinations.

Export Management Company: A private firm that serves as the export department for several processors, soliciting and transacting export business on behalf of its clients in return for a commission, salary, or retainer plus commission.

Export Merchant: A company that buys products directly from processors, then packages and marks the merchandise for resale under its own name.

Export Trading Company: A firm that purchases foreign goods for resale in its own local market.

F.A.S. "Free Alongside": A pricing term indicating that the quoted price includes the cost of delivering the goods alongside a designated vessel.

F.I. "Free In": A pricing term indicating that the charterer of a vessel is responsible for the cost of loading goods into the vessel.

F.I.O. "Free In and Out": a pricing term indicating that the charterer of a vessel is responsible for the cost of loading and unloading goods from the vessel.

F.O. "Free Out": A pricing term indicating that the charterer of a vessel is responsible for the cost of unloading goods from the vessel.

F.O.B. "Free On Board": A pricing term indicating that the quoted price includes the cost of loading the goods into transport vessels at the specified place.

Force Majeure: The title of standard clause in marine contracts exempting the parties for nonfulfillment of their obligations as a result of conditions beyond their control, such as earthquakes, floods, or war.

Foreign Exchange: The currency or credit instruments of a foreign country. Also, transactions involving purchase and/or sale of currencies.

Foreign Freight Forwarder: See *Freight Forwarder.*

Foreign Sales Agent: An individual or firm that serves as the foreign representative of a domestic supplier and seeks sales abroad for the supplier.

Foreign Trade Zone: See *Free Trade Zone.*

Foul Bill of Lading: A receipt for goods issued by a carrier with an indication that the goods were damaged when received. Compare *Clean Bill of Lading.*

F.P.A. "Free of Particular Average": The title of a clause used in marine insurance, indicating that partial loss or damage to a foreign shipment is not covered. (Note: Loss resulting from certain conditions, such as the sinking or burning of the ship, may be specifically exempted from the effect of the clause.). Compare *W.P.A.*

Free Port: An area such as a port city into which product may legally be moved without payment of duties.

Free Trade Zone: A port designated by the government of a country for duty-free entry of any non-prohibited goods. Product may be stored, displayed, used for processing, etc., within the zone and reexported without duties being paid. Duties are imposed on the product (or items processed from the product) only when the goods pass from the zone into an area of the country subject to the Customs Authority.

Freight Forwarder: An independent business that handles export shipments for compensation. Your freight forwarder is among the best sources of information and assistance on United States export regulations and documentation, shipping methods, and foreign import regulations.

F.S.C.: Foreign Sales Corporation.

GATT "General Agreement on Tariffs and Trade": A multilateral treaty whose purpose is to help reduce barriers between the signatory countries and to promote trade through tariff concessions.

General Export License: Any of various export licenses covering export commodities for which Validated Export Licenses are not required. No formal application or written authorization is needed to ship exports under a general Export License. Seafood exporters fall into this licensing category, designated as G-DEST.

Gross Weight: The full weight of a shipment, including goods and packaging. Compare *Tare Weight.*

Import: To bring foreign goods into a country. In international sales, the importer is usually the buyer or an intermediary who accepts and transmits goods to the buyer. Compare *Export.*

Import License: A document required and issued by some national governments authorizing the importation of goods into their individual countries.

Inherent Vice: An insurance term referring to any defect or other characteristic of a product that could result in damage to the product without external cause (for example, instability in a chemical that could cause it to explode spontaneously). Insurance policies may specifically exclude losses caused by inherent vice.

Inland Bill of Lading: A bill of lading used in transporting goods overland to the exporter's international carrier. Although a through bill of lading can sometimes be used, it is usually necessary to prepare both an inland bill of lading and an ocean bill of lading for export shipments. Compare *Air Waybill, Ocean Bill of Lading, Through Bill of Lading.*

International Freight Forwarder: See *Freight Forwarder.*

Irrevocable Letter of Credit: A letter of credit in which the specified payment is guaranteed by the bank if all terms and conditions are met by the drawee. Compare *Revocable Letter of Credit.*

Joint Venture: A business undertaking in which more than one firm shares ownership and control, or when United States fishing vessels sell their catch to a foreign processor vessel at sea (which may or may not also involve product landed shoreside).

Letter of Credit (L/C): A document, issued by a bank per instructions by a buyer of goods, authorizing the seller to draw a specified sum of money under specified terms, usually the receipt by the bank of certain documents within a given time.

Licensing: A business arrangement in which the processor of a product (or a firm with proprietary rights over certain technology, trademarks, etc.) grants permission to some other group or individual to process that product (or make use of that proprietary material) in return for specified royalties or other payment.

Lighter: An open or covered barge towed by a tugboat and used mainly in harbors and inland waterways (sometimes used in exporting fish meal in the South).

Marine Insurance: Broadly, insurance covering loss or damage of goods at sea. Marine insurance will typically compensate the owner of merchandise for losses sustained from fire, shipwreck, piracy, and various other causes, but will excludes losses that can be legally recovered from the carrier. Compare *Credit Risk Insurance.*

Marking (or Marks): Letters, numbers, and other symbols placed on cargo packages to facilitate identification.

Nomenclature of the Customs Cooperation Council: The customs tariff used by many countries worldwide, including most European nations but not the United States. It is also known as the Brussels Tariff Nomenclature. Compare *Standard Industrial Classification, Standard International Trade Classification.*

Ocean Bill of Lading: A bill of lading (B/L) indicating that the exporter consigns a shipment to an international carrier for transportation to a specified foreign market. Unlike an Inland B/L, the ocean B/L also serves as a collection document. If it is a "Straight B/L," the foreign buyer can obtain the shipment from the carrier by simply showing proof of identity. If a "Negotiable B/L" is used, the buyer must first pay for the goods, post a bond, or meet other conditions agreeable to the seller. Compare *Air Waybill, Inland Bill of Lading, Through Bill of Lading.*

Open Account: A trade agreement in which goods are shipped to a foreign buyer without guarantee of payment. The obvious risk this method poses to the supplier makes it essential that the buyer's integrity be unquestionable.

Open Insurance Policy: A marine insurance policy that applies to all shipments made by an exporter over a period of time rather than to one shipment only.

Packing List: A list showing the number and kinds of items being shipped, as well as other information needed for transportation purposes.

Parcel Post Receipt: The postal authorities' signed acknowledgment of delivery to them of a shipment made by parcel post.

Proforma Invoice: An invoice provided by a supplier prior to the shipment of merchandise, informing the buyer of the kinds and quantities of goods to be sent, their value, and important specifications (weight, size, etc.).

Purchasing Agent: An agent who purchases goods in his/her own country on behalf of foreign importers such as government agencies and large private concerns.

Quota: The quantity of goods of a specific kind that a country will permit to be imported without restriction or imposition of additional duties. (In many countries, seafood import quotas are based on how well their domestic fleet performs.)

Quotation: An offer to sell goods at a stated price and under specified conditions.

Representative: See *Foreign Sales Agent.*

Revocable Letter of Credit: A letter of credit which can be canceled or altered by drawee (buyer) after it has been issued by the drawee's bank. Compare *Irrevocable Letter of Credit.*

S.A. (*Societe Anonyme*): French expression meaning a corporation (a similar abbreviation is used for Spanish corporations).

Shipper's Export Declaration: A form required for all shipments by the United States Treasury Department and prepared by a shipper (or his freight forwarder), indicating the value, weight, destination, and other basic information about an export shipment.

Ship's Manifest: An instrument in writing, signed by the captain of a ship, that lists the individual shipments constituting a ship's cargo.

SIC: See *Standard Industrial Classification.*

Sight Draft: A draft that is payable upon presentation to the drawee. Compare *Date Draft, Time Draft.*

SITC: See *Standard International Trade Classification.*

Sport Exchange: The purpose or sale of foreign exchange for immediate delivery.

Standard Industrial Classification (SIC): A standard numerical code system used by the United States Government to classify products and services. Compare *Nomenclature of the Customs Cooperation Council, Standard International Trade Classification.*

Standard International Trade Classification (SITC). A standard numerical code system developed by the United Nations to classify commodities used in international trade. Compare *Nomenclature of the Customs Cooperation Council, Standard Industrial Classification.*

State-Controlled Trading Company: In a country with a state trading monopoly, a trading entity empowered by the country's government to conduct export business.

Steamship Conference: A group of steamship

operators that operate under mutually agreed-upon freight rates.

Tare Weight: The weight of a container and/or packing materials without the weight of the goods it contains. Compare *Gross Weight.*

Through Bill of Lading: A single bill of lading covering both the domestic and international carriage of an export shipment. An air waybill, for instance, is essentially a through bill of lading used for air shipments. Ocean shipments, on the other hand, usually require two separate documents — an inland bill of lading for domestic carriage and an ocean bill of lading for international carriage. Through bills of lading, therefore, cannot be used. Compare *Air Waybill, Inland Bill of Lading, Ocean Bill of Lading.*

Time Draft: A draft that matures either a certain number of days after acceptance or a certain number of days after the date of the draft. Compare *Date Draft, Sight Draft.*

Tramp Steamer: A ship not operating on regular routes or schedules.

Trust Receipt: Release of product by a bank to a buyer in which the bank retains title to the product. The buyer, who obtains the goods for processing or sales purposes, is obligated to maintain the goods (or the proceeds from their sale) distinct from the remainder of his/her assets and to hold them ready for possession by the bank.

Validated Export License: A document issued by the United States Government authorizing the export of commodities for which written export authorization is required by law. Not needed for exporting seafood. Compare *General Export License.*

W.A. "With Average": A marine insurance term meaning that a shipment is protected from partial damage whenever the damage exceeds 3 percent (or some other percentage).

Warehouse Receipt: A receipt issued by a warehouse listing goods received for storage.

Wharfage: A charge assessed by a pier or dock owner for handling incoming or outgoing cargo.

Without Reserve: A term indicating that a shipper's agent or representative is empowered to make definitive decisions and adjustments abroad without approval of the group or individual represented. Compare *Advisory Capacity.*

REFERENCES

Carroll, W. F. 1983. A basic guide to exporting seafood. Mid-Atlantic Fisheries Development Foundation. Annapolis, MD.

Food and Drug Administration. 1995. Industry advisory: European union requirements. United States Department of Health and Human Services. Washington, DC.

Food and Drug Administration. 1998. Fish and Fishery Products Hazards and Controls Guide. 2nd ed. United States Department of Health and Human Services. Washington, DC.

National Marine Fisheries Service. 1997. Current Fishery Statistics (No. 9700). United States Department of Commerce. Silver Spring, MD.

National Marine Fisheries Service. 1985. United States export of fishery products. In *Food Fish Facts* (No. 55). United States Department of Commerce. Silver Spring, MD.

Pochaska, F. J. and J. C. Cato. 1981. Developing export markets for Gulf of Mexico and South Atlantic Seafood Products. In *Proceedings — Sixth Annual Tropical and Subtropical Fisheries Technological Conferences of the Americas*, Texas A&M University. College Station.

The Future of the Industry

Richard E. Gutting Jr.

The future of the United States fish and seafood industry depends upon the complex interaction of many different forces and events. Attempting to forecast exactly what will happen is impossible. Nevertheless, it is possible to identify some of the trends which may impact United States seafood companies in the future.

FACTORS INFLUENCING DEMAND

Consumption of seafood in the United States has been increasing due not only to a growing population, but also to a rise in per capita consumption. Recent trends are summarized in Figure 1.

The United States population, although slowing in growth, is projected to increase for several more decades. Strong demographic trends in the United States also suggest that demand for seafood will continue in the future. The United States population is growing older. As we enter the next century, the fastest-growing segments of our society are those aged 45 to 54. This age group has high levels of disposable income and is more likely to eat out or buy higher-priced foods such as seafood.

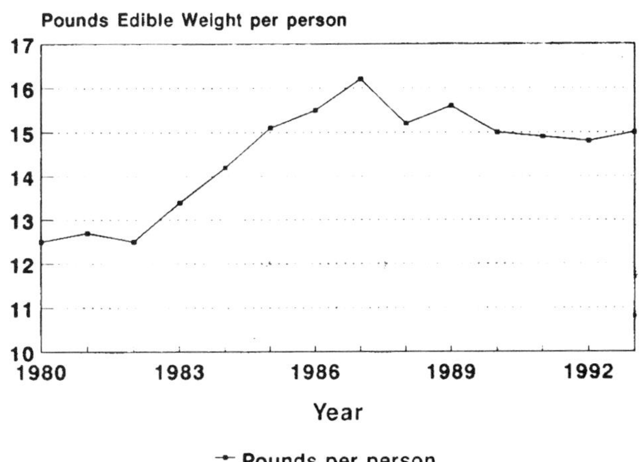

Figure 1. U.S. per capita consumption of edible meat (in pounds).

As Americans age, they are becoming increasingly aware of the relation between diet and good health. Books on nutrition and government dietary guidelines cite fish and seafood as low in fat, easily digestible, and a good source of many important minerals and vitamins. Scientific studies suggest that there are health advantages to eating fish for its special omega-3 fatty acid content. New labeling requirements also should raise public awareness about nutrition. Many restaurants have introduced special menus featuring meals low in fat, cholesterol, and sodium. These health messages should expand markets for seafood at the expense of red meat.

Consumers, on the other hand, continue to feel a "lack of competence and comfort" in preparing and storing fish. This attitude partially explains the high consumption of seafood at restaurants. There is a trend, however, for consumers to rely upon others to prepare the food they eat at home. Most adult men and women now work outside the home. This growth in two-income couples has increased disposable income, but has left more Americans with less time to spend it. Studies suggest that consumers are spending less time in the kitchen. As a result, value-added products, ready-to-eat items, and microwave entrees have become more popular. Supermarket deli departments also are evolving into food service operations, competing with fast-food and takeout restaurants. The seafood industry can only benefit from these lifestyle changes.

Seafood prices, however, have been rising, and there are limits to what people will spend when alternatives are readily available at lower cost. Rising prices have been offset somewhat by higher standards of living and growth in the real disposable income of consumers. Whether these trends will continue, however, is uncertain.

Another factor that has helped offset rising prices is the extensive positive publicity seafood received in the 1980s as a source of good nutrition. In the late 1980s, however, the public media turned its attention to seafood contaminants, ocean pollution, and the question of adequacy of government inspection programs. A series of media-inflamed "seafood scandals" ensued. Similar media "scares" were directed at other foods as well.

Food safety will continue to be one of the key consumer issues in the future. Media attention has sensitized consumers to food contamination and the effect of chemicals and pollution upon public health. The tendency of fish to absorb and concentrate some pollutants, the natural perishability of fish, and the harvest of fish from uncontrolled ocean environments make it a likely target for further media stories.

Consumer apprehension about the safety of seafood should diminish, however, as the United States and other nations adopt more stringent inspection programs based upon Hazard Analysis Critical Control Point (HACCP) principles. Industry use of quality-assurance systems, such as certification under the International Accepted Standard (ISO 9000 Series) and Total Quality Management, also are expanding. As a result, consumer confidence in seafood should improve.

Consumers also are becoming more concerned about protecting the environment. Ocean fisheries are renewable and enjoy many advantages over other foods from an environmental perspective. In the past decade, however, animal protection groups have initiated several boycotts against seafood products in an effort to pressure commercial fishermen into modifying their harvesting operations. More recently, sensational media coverage of overfishing and wasteful fishing practices have been eroding the goodwill of consumers that fishermen have long enjoyed.

Although environmental concerns have come in and out of fashion among consumers, the long-term growth and interest in environmental and animal protection issues have been dramatic. While these issues are unlikely to become the dominant factor in consumer decisions, controversies and boycotts are likely to become more popular and effective in the future.

The marketing of seafood also continues to suffer from the fragmentation of the industry and the commodity nature of many products. Traditionally, the company that caught and packed the most fish made the most money. Today, many seafood firms continue to be production-oriented, certainly relative to those firms producing packaged goods. One result is that seafood is not heavily advertised at the consumer level, except by restaurants and supermarkets.

Marketing efforts in the future should increase, however, as more large consumer-product companies enter the seafood business and seek to establish loyal consumer bases for their brands. More firms are beginning to realize that the better approach to selling a product is "what does the consumer want?"

FACTORS INFLUENCING SUPPLY

Imports supply approximately half of the fish and seafood consumed in the United States. The remainder is provided by domestic fishermen and aquaculture producers. Recent trends in the United States supply are summarized in Figures 2 and 3.

IMPORTS

The composition of world trade in fish and seafood products has changed notably the past two decades as a result of extended fisheries jurisdiction, the rapid expansion of aquaculture, and the introduction of new products. Export volume and the number of countries exporting products have increased significantly, particularly among developing nations. World imports, on the other hand, remain concentrated on the three largest markets of the United States, Japan, and Europe which together account for more than 70 percent of sales.

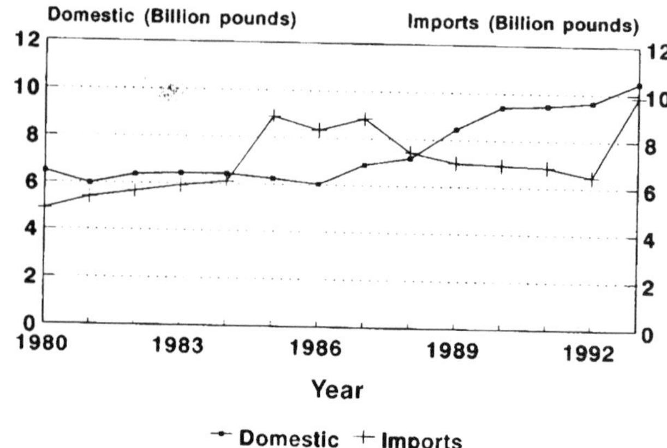

Figure 2. Edible and industrial U.S. fishery products supply, 1980–1983.

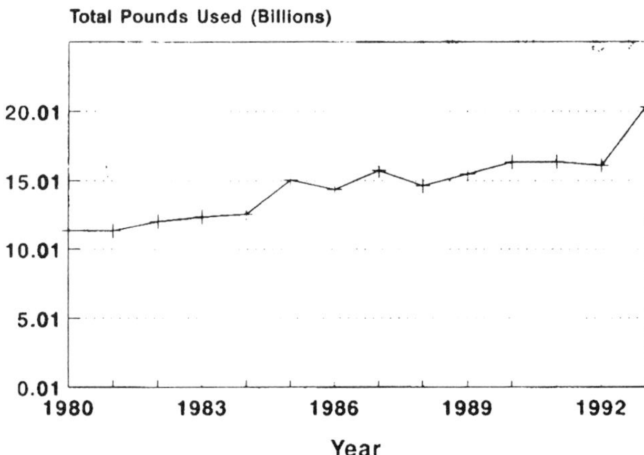

Figure 3. Edible and non-edible round weight, total U.S. supply.

At the present time seafood imports from more than 100 countries play a critical role in supplying the United States marketplace. Whether this flow of imports will continue to grow in the future depends upon many factors, including the changing status of global fisheries, future growth in aquaculture production, the relative strength of the United States economy and currency, and global trade policies.

Future Global Fisheries Production

The significant buildup in fishing during the last half-century is having a tremendous impact on ocean fishery resources. During the 1950s, reported landings from ocean fisheries increased at an average rate of 6.8 percent per year. In the 1960s this growth rate increased to 7.4 percent, but then declined to 1.7 percent in the 1970s and 3.6 percent in the 1980s. In the first three years of the 1990s landings declined at an annual rate of 1.5 percent (Garcia and Newton, 1994).

Unlike agriculture and livestock commodities, whose production can be expanded or reduced to meet changes in consumer demand, wild fish stocks have finite biological limits and vary in size each year due to natural biological cycles and environmental conditions. The proliferation of fishing vessels and advances in fishing technology throughout the world have boosted fishing power significantly and put more fish stocks at risk.

Fishermen have managed to keep the total global catch climbing by abandoning fished-out stocks and pursuing new species. These typically have been lower-value species that were previously undesirable and unwanted because they were deemed too small, bony, unappetizing, or otherwise not good for eating.

In recent years, worldwide landings from ocean fisheries have tended to become asymptotic around 90 million metric tons. While this may not represent the absolute upper limit for world fisheries, it does suggest that future landings may not be appreciably greater unless present fishery practices change.

In 1971, a study sponsored by the Food and Agriculture Organization (FAO) of the United Nations estimated that the marine environment unaided could sustain yields of about 100 million tons of fish per year, 10 million tons more than the present catch. Although this projection is inherently fraught with uncertainty, the recent leveling off of the world catch and the state of the major fishing grounds suggests that it could be optimistic. The FAO now estimates that 17 of the world's major fishing areas have either reached or exceeded their natural limits, and that nine are in serious decline. FAO scientists now say that the world catch is unlikely to reach and maintain 100 million tons unless stocks are better managed (FAO Fisheries Circular No. 853, 1993).

The new regime of the sea negotiated in the 1970s provided coastal nations with the authority to conserve stocks and to manage harvests within 200 miles of the coastline. This transfer of authority, however, did not always lead to successful conservation, and fishing fleet over-capacity remains a serious problem in many areas.

Most harvesting occurs within 200 miles of the coastline and thus is subject to the management measures of coastal nations. Whether these nations will institute the harvest restraints needed to rebuild depleted stocks in the future is unclear. Overfishing, however, has become a global concern and many nations are cutting back on fishing quotas and stepping up enforcement and surveillance efforts.

Threats to fishery habitats around the world also have become of greater concern. The relative importance of habitat loss to fish production is not well-documented. Certainly, there are many examples that show that serious alterations of nearshore habitat have caused local fisheries to decline. But the overall picture is less clear.

Most fish stocks have at least one stage during their life cycle when they must utilize a specific area to the exclusion of other habitats. Many stocks, particularly those using inshore and estuarine areas, may need a series of these critical habitats at different stages of their development. (Critical habitat, used in this chapter, refers to locations such as marshes, mangrove swamps, seagrass beds, kelp forests, and estuaries which are indispensable to a species during at least a portion of its life history.) Conversely, as fish mature, their habitat requirements often become less specific and they are able to exploit a wider range of habitats. Most important, fish stocks have at least one period in their life cycle during which the absence of critical habitat will result in lowered year-class survival, and ultimately, lowered annual production. Not surprisingly, inshore and estuarine habitats, such as marshes and seagrass beds, are more vulnerable to human influences than others, such as the oceanic water column or the outer continental shelf, which are much less easily disturbed.

Human population trends around the world suggest that future development will occur in the more vulnerable coastal regions where estuarine-dependency is highest. Although there is little concrete scientific data demonstrating direct cause/effect relationships, the kind of perturbations resulting from this development, such as dredging and filling, runoff and waste disposal, are not conducive to the production of seafood.

Overfishing and habitat degradation do not mean that seafood supplies cannot be increased from ocean fisheries in the future. Reduced harvest quotas and better management could increase production in the long term and there still are fisheries that are not fully developed. The FAO estimates that better management of global fisheries might boost production approximately 20 million metric tons (FAO, 1992).

Elimination of discards and more efficient processing with less spoilage and waste could also boost seafood production significantly in the future. (The terms "catch" and "landing" are frequently used interchangeably. But a substantial part of the fish caught by fishermen is not landed. It is discarded at sea for one reason or another, usually because of market considerations and sometimes because of "conservation" laws or regulations. Few, if any, of these discards are reported in the annual statistics published by government agencies.) In 1983, for example, the FAO estimated at least 6.72 million metric tons of fish and shellfish were being discarded by fishing vessels. More recently, global discards have been estimated to be approximately 27 million metric tons, or more than 25 percent of the fish caught. Growing concerns over the bycatch of non-target species, and the need to increase production, are prompting the industry to reduce bycatch and more fully utilize its catch.

There also is, as in all foods, a considerable amount of waste and spoilage after fish is landed — how much is impossible to estimate. Poor handling, storage, and processing of seafood products, however, is believed to result in enormous losses, particularly in developing countries. For example, Bahar Munip (1993) estimates losses as high as 26 percent in the Association of SouthEast Asian Nations (ASEAN) member nations. The National Research Council (1985) estimated that one-fourth of the world's food supply was lost through microbial activity. Significant gains are possible if ways can be found to avoid these losses.

Future Global Aquaculture Production

Aquaculture, or fish farming, also should boost the global supply of seafood in the future. The fresh- and saltwater rearing of fish and shellfish in captivity is included under the term aquaculture, as well as the open seas ranching of such species as salmon, where juvenile fish are released into the ocean. Aquaculture activities are classified as either extensive or intensive. Extensive cultures are those with low-density populations where nutritional and environmental requirements are provided, largely with supplement, through the natural productivity of the water systems — catfish, for example. Intensive cultures are those systems with densities that require significant supplemental food/environmental manipulation — such as salmon hatcheries.

Aquaculture production has more than doubled in the past two decades and now accounts for more than 10 percent of the global seafood supply. Future expansion could increase harvests without the constraints on wild harvests.

World production of farm-raised shrimp, for example, has increased rapidly. Production has declined in the past few years due to a poorly defined disease and to pollution problems in several of the leading shrimp-farming nations. Although shrimp production continues to grow in several nations, it is becoming apparent that indiscriminate growth leads to problems. Future growth is likely to be less dramatic than in the past decade.

Aquaculture of salmon also has profoundly changed the marketing and trade of salmon as well as salmon fisheries management. Pen-raised salmon has grown rapidly from being virtually nonexistent in the late 1970s to comprising over 30 percent of the global harvests. It has become a year-round staple in United States fish markets and is still growing. Salmon enhancement or ranching also has become the dominant source of "wild" salmon in many areas, such as Japanese chum runs, and the Columbia River pink runs in Prince William Sound, Alaska.

Future growth also is expected in the production of other aquaculture species. Tilapia, for example, may be the next to break through to a mass market. This growth, however, will be constrained by competition for clean water, appropriate nearshore sites, and the cost of feed. As fish farms have expanded, becoming an industry rather than a novelty, they have sparked opposition from groups concerned with coastal pollution, the purity of wild stocks, and other ecological issues. Local fishermen, for example, sometimes object to the waste generated by farmed fish or to pens that block access to fishing grounds. Landowners and tourist organizations have objected to the installation of farms on scenic stretches of coast that are important for tourism. Many fishermen fear that pen-raised fish will escape and kill off wild fish. Ecologists are concerned about the use of toxic chemicals or antibiotics that are introduced into the ocean. Environmentalists are worried about marine mammals becoming attracted to the pens as a source of food and then becoming entangled in the netting. Even bird lovers are upset when they hear stories of herons, egrets, or osprey being killed to keep them from feeding at grow-out ponds and pens.

These fears have generated strong opposition to fish farms in several nations. While the potential for environmental damage is often exaggerated, future growth in aquaculture will depend upon advances in technology, medicine, and fish farming techniques that address these concerns.

Fish feeds may also be a significant restraint in the future. Feed, the highest single cost in most aquaculture operations, can consist of such varied products as alfalfa meal, animal flesh, corn gluten, fish meal, cottonseed meal, soybeans, wheat, and brewer's yeast. Feed development entails many unanswered questions involving fish dietary and nutritional requirements, effects of temperature and water quality on feeds, the rate of water exchange, the effect of uneaten feeds on the health of fish, and feed particle size and density.

Future International Trade Restraints

The seafood marketplace is global, and buyers in the United States must compete for product with buyers in Japan, Europe, and other major markets. This competition is growing more intense each year. Global demand for fish and seafood is growing as populations and standards of living in major market nations expand; the medium population projection of the United Nations shows world population growing from 5.4 billion today and leveling off at 11.5 billion in 2150 (United Nations, 1992). Rising incomes in the Pacific Rim and Southeast Asia regions, in particular, will translate into greater demand. Emerging markets in these areas in the future are likely to compete much more intensely for supplies with the traditional major seafood markets in Japan, the United States, and Europe.

Changes in currency exchange rates also will continue to significantly impact the movement of seafood. Since the United States dollar was allowed to float against other currencies in 1971, it has fallen in relative value more often than it has risen. The growing strengthening of the Japanese yen in the 1990s, in particular, has had a major impact on seafood markets. A strong yen and a weak dollar have tended to divest seafood away from the United States market. As a result, imports have been more expensive to buy and more domestic production has been exported.

Over the past two decades the major importing nations have reduced their tariffs and use of import quotas through a series of multilateral negotiations under the auspices of the General Agreement on Tariffs and Trade (GATT), with the rate of reduction somewhat higher for unprocessed products than for processed products. These have spurred an increase in international trade.

Future changes in trade policies are difficult to project. It is reasonable to assume, however, that as the world depends more and more upon international trade, tariffs and other protectionist measures will continue to abate. With most nations dropping political barriers to trade, and trying to improve the standard of living for growing numbers of people, a substantial increase in the overseas demand for fish and seafood should result.

The ability of United States companies to secure supplies overseas in the future also depends upon United States trade policy. In the past two decades, United States officials have begun to use the threat of United States import embargoes to coerce foreign nations into complying with United States fishery requirements. Indeed, the number of laws authorizing the use of trade embargoes to protect ocean resources is increasing. Statutes that use the threat of trade sanctions to enforce international agreements include the Pelly Amendment; Packwood Magnuson Amendment; Driftnet Impact Monitoring, Assessment and Control Act of 1987; Driftnet Act Amendments of 1990; 1990 Driftnet Amendments to the Marine Mammal Protection Act; International Dolphin Conservation Act of 1992; Section 205 of the Magnuson Fishery Conservation and Management Act of 1976; Section 801 of the Fishery Conservation Amendments of 1990 (Anadromous Fish Certification Amendments); Atlantic Tunas Convention Act of 1975; Tuna Conventions Act of 1950; Endangered Species Act of 1973; and Lacey Act Amendments of 1981.

This trend is likely to continue. As domestic fishermen become more tightly restricted, the demands to "level the playing field" with foreign fishermen will grow. United States officials are likely to be constrained by a lack of funds to encourage compliance by other nations and will probably turn to the use of import bans to force foreign compliance with tighter fishing restrictions.

Growing food safety concerns also will impact imports. Present food-safety inspections of seafood imports focus on spoilage and labeling. Public concerns over seafood safety and contamination, however, are prompting a reexamination of issues regarding the long-term exposure to chemicals and the way seafood is processed and inspected overseas before it is exported to the United States. Stricter tolerances are possible and future inspections of overseas suppliers could reduce the available supply of imports in the future.

Future Domestic Production

Although total United States production has grown during the past two decades, it now seems to be leveling off. Most stocks off Alaska are fully utilized, while many stocks in the Atlantic are overutilized. In the Southeast and Pacific there are many stocks whose status is unknown. Only a few stocks, such as Atlantic mackerel, are underdeveloped.

The long-term potential yield of United States ocean fisheries is estimated to be approximately 9.2 million metric tons [National Oceanic and Atmospheric Association (NOAA), 1993]. Recent production levels amount to approximately 6.3 million metric tons, or about 68 percent of this potential. Government fishery biologists insist, however, that present fishery practices must change if this potential is to be realized.

The regulatory climate for commercial fishing in the United States has evolved from one of exploratory fishing and expansion in the 1960s, to the "Americanization" of fisheries in the 1970s and 1980s, to a new focus on the problems of overfishing (conservation), overcapacity (economic viability) and bycatch (waste). Fishery managers are beginning to cut back on harvest quotas to conserve stocks. As is evident in recent media reports, the general public has become aware that United States fisheries are depressed and that solutions must be found.

This effort to rebuild depressed stocks likely will keep harvests at lower levels for the next few years. Officials predict, however, that rebuilding will boost seafood production in the long term. A much larger yield also could be obtained from those stocks which are underutilized, but market conditions have kept harvests low. These conditions, however, are likely to change in favor of development as global demand continues to increase.

The way in which fish are allocated to fishermen also is changing. Fish traditionally have been "common property" and available to anyone who wanted to fish. Government officials say this "open access" caused overcapacity (too many vessels fishing), inefficiency, and the depletion of fish stocks (Sissenwine and Rosenberg, 1993).

Historically, officials have tried to prevent this "tragedy of the commons" by imposing restrictions on fishing days, areas, and types of gear. This approach, however, is under considerable criticism because it has created inefficiencies and has failed to halt an over-abundance of fishing power relative to available catch. Many officials now believe that access to fisheries must be closed.

United States fisheries, however, have been remarkably resistant to the economic arguments in favor of limited entry schemes. Not so many years ago, limited entry licensing was presented as a simple solution to overcapacity and excessive harvests. The real-life experience with limited entry regimes, however, has been sobering. Now the use of the individual fishing quotas (IFQs) is being advocated by academic economists and officials as a way of preventing overfishing and overcapitalization.

In contrast to "common property" systems where fishermen do not own the resources until they catch it, in IFQ systems fishermen own shares in the right to harvest. In some instances these shares are transferable and can be bought, sold, leased, or inherited just as other property. Unlike a true private property scheme, however, the government retains the right to determine an overall quota and other harvest restrictions.

Canada has allocated individual quotas in several fisheries, though generally not (or not yet) on a fully transferable basis. In New Zealand individual quotas are established in several fisheries and are transferable. The concept also is being pursued in Australia and Iceland. The use of quota systems in the United States, so far, has been limited (IFQ systems have been adopted in the United States for the surf clam, wreckfish, and Alaskan halibut and sablefish fisheries). As a result, a general assessment of their effectiveness is not yet available.

From a theoretical perspective, IFQ systems have been viewed as an ideal scheme, leading to the generation of maximum net economic returns. Attempts to establish IFQ systems have been controversial, however, largely because of the difficulty in allocating shares among fishermen and the resulting windfall profits.

Bycatch and waste also have become important concerns. A variety of techniques have been attempted by managers, scientists, and engineers to reduce bycatch and discard levels. These include efforts at improving gear selectively, the development of gear taking advantage of different species behavior, and various time and area restrictions. Emerging ideas include incentive programs and individual vessel quotas that place responsibility for reducing discards and bycatch on individual fishermen. This strategy, however, relies upon observer programs to audit progress; which may not be feasible in many fisheries.

For many years, animal protection groups have been interested in reducing the bycatch of marine mammals, birds, and sea turtles in fishing gear. Struggles to reduce or eliminate mortalities of dolphin in tuna nets, sea turtles in shrimp trawls, and porpoise, birds, and sea turtles in driftnets have focused the public's attention on bycatch problems and the potential impact of fishing on ocean animals and the marine ecosystem. Fishery managers increasingly are acknowledging the demands of preservationists and animal rights activists. As a result, more restrictions on fishing are expected in the future.

The recovery of many formerly-depleted mammal populations, such as harbor seals and California sea lions, has prompted increased concern that growing populations of mammals are now depleting important fish stocks (Olelsiuk, 1993). United States fishery laws do not allow the harvesting of marine mammals to increase the food available to humans. Instead, the taking of mammals by fishermen is allowed only for unintentional and incidental takings that occur as a necessary part of commercial fishing.

Competition between commercial and recreational fishermen, although certainly not new, also will continue to intensify as a result of increased demand for seafood and recreational opportunities. In recent years, federal officials have begun to designate species as "game fish." Redfish, coho and chinook salmon, and Atlantic billfish, for example, have been set aside for the exclusive use of recreational fishermen. Allocation of other shared species, such as sharks, may also occur.

Similar trends are evident at the state level. California, Florida, and Texas, three of the most populous states, have instituted net fishing bans in recent years. In Texas, for example, trout and redfish have been taken off the commercial market and are now classified as game fish. These events suggest that recreational fishermen are becoming more successful in pressing their interests. In terms of numbers — 15 million marine recreational fishermen, versus 300,000 commercial fishermen — their potential influence is significant, and while their numbers are not growing rapidly, they seem to be gaining political strength. The result is likely to be fewer fish for the seafood industry in the future.

Aquaculture in the United States has grown at approximately 20 percent annually through the last decade and currently provides more than 10 percent of the domestic supply. [Freshwater aquaculture accounts for approximately three quarters of total United States production. Coastal aquaculture (clams, mussels, salmon, shrimp) in the United States remains a relatively small industry.] Despite advances in technology, marine aquaculture ventures in the United States are expected to grow slowly in the future. The regulatory milieu for aquaculture operations along the coastline is complex and getting more so. Political considerations — problems with permits, effluent guidelines, and so on — will continue to depress domestic growth. Freshwater aquaculture, on the other hand, does not suffer from the same constraints and should continue to expand, particularly in raising catfish and such newer species as tilapia.

Future Technological Developments

Hundreds of seafood research projects are underway in the United States directed toward discovering new and better ways to grow, harvest, process, store, and market fish and seafood. Emphasis in the future will be on achieving greater resource efficiency and productivity and on developing more consumer-oriented products.

Fishing gear development in the past has concentrated on increasing the amount of fish caught in a given time. Greater effort in the future will be directed toward methods to reduce bycatch and more fully use the catch. Changes in on-board handling practices also will be sought to improve the value of landed products.

While it is clearly not profitable to keep every fish that is caught, the industry is expected to devote increased efforts to find viable alternatives to discarding fish. Also, while the prevailing practice of icing fish will not be replaced as the primary means of on-board preservation, small on-board freezing and refrigeration systems will grow in use and icing practices will be refined.

Further advances in communication, electronics, and computing will enhance vessel efficiency and performance. Use of satellite data to navigate and locate fish concentrations will increase. New electronics will greatly enhance engine performance and fuel efficiency. Sophisticated hydroacoustical fish detection aids such as image intensifiers and various specialized sensors will continue to evolve. More vessels will use on-board computers to process data from various sensors to guide their fishing gear.

Advances also are expected to continue in aquaculture. Aquaculture engineering will offer a greater variety of culture systems, waste treatment or removal, and water and feed delivery. Reformulation of aquaculture feeds will allow nutritional needs to be met, while decreasing nutrient loadings. Development of improved disease diagnostic tools and vaccines will improve yields and decrease risks, while decreasing possible impacts on wild stocks. Additional therapeutics will increase yield and decrease the reliance on particular antibiotics, decreasing concerns about antibiotic resistance. Development of domesticated aquaculture stocks will improve yield, decrease reliance on collecting seed stock from wild sources, and decrease the attractiveness of culturing exotic species. Development and commercialization of improved methods for producing reproductively sterile aquaculture stocks will minimize genetic impacts of cultured stocks on wild stocks.

Better material handling systems — more efficient and faster — will better preserve quality during processing. Advances in automation will allow processors to handle more types and sizes of fish and recover more byproducts through mincing and better recovery. The advent of robotics systems for the processing of different sizes of fish on the same processing line may eliminate the need for large amounts of hand labor and for large numbers of expensive, single-purpose machinery.

Environmental concerns will foster the development of enzyme-based systems to convert food byproducts into useful products. These systems not only will lessen the load on the waste stream, they will yield value-added products. Prototypes of these technologies have demonstrated their feasibility.

Future research into the nutritional aspects of fish and seafood will lead to a better understanding of the connection between various food components and health and disease. This new knowledge will be used to create seafoods that are nutritionally designed to promote health, as well as meet the special dietary needs of pregnant women, older people, and those with special needs. Efforts will be directed toward helping people better control allergies and genetic and other diseases.

The future implementation of HACCP-based inspection systems will foster future research on rapid pathogen detection and identification methods. Promising work already underway suggests that substantial breakthroughs will greatly increase the quality and safety of seafood products. In the future, sophisticated biosensors will detect spoilage and measure minute levels of contaminants.

Advancing information technology will allow marketers to become more sophisticated in targeting market niches and segments. The amount of available data will grow exponentially, and those firms that can access and use it effectively will have the edge in finding customers.

Future Industry Developments

Seafood firms in the future will face a variety of competing demands: keeping costs low and quality high, staying competitive in a global marketplace, and meeting consumer preferences for more healthy and environmentally benign products.

Companies will take steps to minimize the risks inherent in fishing activities. At the harvesting level, ownership of the medium and large vessels will continue to evolve away from the traditional single-boat owner. Financial risk will be spread over more owners per vessel. More multiple vessel ventures will appear, though the harvesting sector will remain highly unconcentrated.

The industry will attempt to protect its margins by improving efficiency and by making structural changes to reduce the high variability of the business. Capital and technology will be substituted for labor to increase productivity. The inherent risk of the industry will be reduced by diversification. Harvesting efficiency will be improved marginally by regulations that better match the size of a fishery with the number of harvesting units (by restricted access).

Processing operations, as opposed to smaller fresh/frozen operations, will continue to become more concentrated. More processors will seek multifishery capabilities in order to minimize seasonal fluctuations and guarantee steadier supplies to customers. Again, higher capital requirements, brought about by the need of more efficient operations, will force more economies of scale.

Environmental rules and laws will continue expanding at all levels of government, requiring increased public and private expenditures. The Environmental Protection Agency (EPA) estimates that industry costs will double in the next ten years as a result. Among the challenges facing the seafood industry are new restrictions on the use of refrigerants such as chlorofluorocarbons, cleaners such as chlorine, the discharge of wastewaters, the use of packaging materials, and the disposal of solid waste.

United States seafood firms, along with business in general, also will face labor shortages and consequently higher costs as the numbers of new workers entering the workforce declines. Firms will continue to be asked to pay for the costs of solving various social problems, further driving up the cost of labor.

New value-added forms will be developed as firms make an effort to diversify their product mix, reduce waste, and respond to consumer demand for convenience. Yields also will be improved by both faster handling and better preservation. Smoothing the cycles of fish delivery to processors (via resource control) is not likely, although some small advances may reduce at-the-dock spoilage before unloading. More companies will try to increase their processing capacity through automation. Processors will continue to stabilize their supply of raw material by importing partially processed fish in quantity. There also will be a tendency to merge with large and diversified food-processing companies. Processors and distributors of fish products will also seek to reduce their risks by diversifying operations and sources of supply.

The trend toward vertical integration is likely to continue, again in an effort to lower the costs associated with business variability. More linkages will take place between fishermen and processors, with the latter providing funds for the fishermen with whom they contract.

Traditional market channels have distributed fish commodities through long and complex sales channels. More recently, movements of value-added products have grown and distribution channels have become shorter and more specialized. Trade shows and advances in communication have made it easier for buyers and producers to find each other. Retailers and restaurant chains are working much more closely with producers and processors. Larger customers are insisting on consistent quality.

The emergence of the jumbo jet and refrigerated vans has changed seafood markets around the world. Here in the United States fish from New Zealand or South America is as likely to be found on supermarket counters as fish from New England and Alaska. The infrastructure that supports the distribution of seafood is capital intensive and changes tend to be evolutionary rather than revolutionary. Further advances, however, should continue to open new markets and sources of supply. More value-added products than bulk commodities will be traded internationally.

CONCLUSION

Consumer expenditures for seafood in the United States should remain strong in the future, due to a growing population, a growing awareness of the healthy benefits of eating fish, and consumer demand for convenience and variety. More United States seafood companies will compete on quality and with brand products.

There will be no let-up of media and public attention to food contaminants, pollution, and threats to the environment, and the industry must be prepared to endure the scrutiny of a better informed consumer. Consumers will continue to be influenced by various environmental and food-safety media "campaigns" and will pay more attention to labels, freshness, and ingredients. They also will want more information about their seafood.

An increasing body of data indicates that worldwide, harvests are leveling off, perhaps as a precursor to decline. Mounting evidence of overfishing and habitat decline raises serious questions about whether ocean harvests can be boosted further on a sustainable basis. Catch has increased in some areas and declined in others, but the historical trend of increased catches may soon be over. Trends in United States fisheries are a bit murky, but the overall pattern seems to reflect those emerging at the global level.

Past increases in production have been supported by remarkable technological progress in vessel design, fishing gear, on-board positioning and detection equipment, and on-board fish preservation and processing. To expand harvests in the future, it will be necessary to rebuild depleted stocks, increase use of the catch, and expand aquaculture production. As ocean fisheries become more valuable and competition from overseas buyers strengthens, technological advances will accelerate.

REFERENCES

Garcia, S.M., and C. Newton. 1994. Current situation, trends and prospects in world capture fisheries. Paper presented at the Conference on Fisheries Management. Seattle, Washington. June, 1994.

FAO. Marine fisheries and the law of the sea: a decade of change. Fisheries Circular No. 853. Rome, 1993.

FAO. World fisheries situation. May, 1992.

Munip, Bahar. 1993. Postharvest handling of agricultural produce in ASEAN.

National Research Council. An evaluation of the role of microbiological press (1985).

NOAA. Our Living Oceans: Report on the status of United States living marine resources, 1993.

Olelsiuk, Peter F. 1993. Annual prey consumption by harbor seals (*Phoca vitalina*) in the State of Georgia, British Columbia. *Fishery Bulletin*. **91**:491-515.

Sissenwine and Rosenberg. Marine fisheries at a critical juncture. *Fisheries*. **18**: 10 October, 1993.

United Nations. Department of economic and social affairs. 1992. *Long-Range World Population Projections*. New York.

Resource List of Web Sites

Thomas E. Rippen

Site Name / Info	Institutions and Organizations	Address

Government

ACCSP - The Atlantic Coastal Cooperative Statistics Program
.... http://www.safmc.nmfs.gov/ACCSPHM/accsp.html
Agricultural Systems Home Page .... http://www.reeusda.gov/paa
Aquaculture Information Center (AIC) .... http://www.nal.usda.gov/afsic/
Biological Resource Division (BRD) .... http://www.nbs.gov/
Center for Disease Control .... http://www.cdc.gov/
Cooperative State Research, Education and Extension Service .... http://www.reeusda.gov/

Environmental Protection Agency (EPA) .... http://www.epa.gov/
EPA Office of Science & Technology Beach Closures .... http://www.epa.gov/ost/beaches
EPA-OST Fish Consumption Advisory Site .... http://www.epa.gov/ost/fish/
EPA Office of Wetlands, Oceans and Watersheds .... http://www.epa.gov/owow/
EPA-OWOW Pfiesteria Piscicida Information .... http://www.epa.gov/owow/estuaries/pfiesteria/

ERS Situation and Outlook Reports .... http://usda.mannlib.cornell.edu/mor_start.html
FDA Home Page .... http://www.fda.gov/fdahomepage.html
FDA Center for Food Safety and Applied Nutrition .... http://vm.cfsan.fda.gov/list.html
FDA Center for Veterinary Medicine .... http://www.cvm.fda.gov/
Fisheries Statistics and Economics Division http://www.st.nmfs.gov/st1/index.html

Florida Caribbean Science Center .... http://www.nfrcg.gov/
Government Printing Office (GPO) .... http://www.access.gpo.gov/su_docs/
HACCP Alliance Training Program Schedule via SeafoodNIC
.... http://www-seafood.ucdavis.edu/haccp/training/schedule.htm
HACCP Training Programs and Resources Database
.... http://www.nal.usda.gov/fnic/foodborne/haccp/index.shtml
Joint Subcommittee on Aquaculture .... http://ag.ansc.purdue.edu/aquanic/jsa
Leetown Research and Development Laboratory .... http://www.lsc.usgs.gov/rdl/lsc-rdl.htm
National Agricultural Library .... http://www.nalusda.gov/
National Biological Information Infrastructure .... http://www.nbii.gov/
National Marine Fisheries Service .... http://www.nmfs.gov/

National Geospatial Data Clearinghouse (USGS) .... http://nsdi.usgs.gov/nsdi/
National Oceanic & Atmospheric Administration (NOAA) .... http://www.noaa.gov/
National Oceanic & Atmospheric Administration (NOAA) Central Library http://www.lib.noaa.gov/
NOAA Environmental Information Services .... http://www.esdim.noaa.gov/
NOAA Research — Environmental Science Division .... http://www.oar.noaa.gov/

National Operational Hydrologic Remote Sensing Center .... http://www.nohrsc.nws.gov/
National Research Institute .... http://www.reeusda.gov/nri/
National Science Foundation (NSF) .... http://www.nsf.gov/
National Sea Grant Depository .... http://nsgd.gso.uri.edu/
National Sea Grant Program .... http://www.mdsg.umd.edu/NSGO/

National Wetlands Inventory .... http://www.nwi.fws.gov/
Northeast Fisheries Science Center .... http://www.wh.whoi.edu/
Northwest Fisheries Center .... http://listeria.nwfsc.noaa.gov/
Oak Ridge National Laboratory .... http://www.esd.ornl.gov/
Patuxent Wildlife Research Center Monitoring Program .... http://www.im.nbs.gov/
Seafood Network Information Center .... http://www-seafood.ucdavis.edu/
Sea Grant Non-indigenous Species Site .... http://www.ansc.purdue.edu/sgnis/
Sea Grant - Northeast Regional Home Page .... http://seagrant.gso.uri.edu/region/
US Army Corps of Engineers .... http://www.wes.army.mil/
United States Department of Agriculture (USDA) .... http://www.usda.gov/
USDA CRIS Search http://cristel.nal.usda.gov:8080/
USDA's Pfiesteria Page .... http://www.nal.usda.gov/wqic/pfiest.html
USDA/FDA Food-borne Illness Education Information Center .... http://www.nal.usda.gov/fnic/foodborne/foodborn.htm
US FDA/Center for Food Safety & Applied Nutrition web site on HACCP .... http://vm.cfsan.fda.gov/~lrd/haccpsub.html
US Fish and Wildlife Service .... http://www.fws.gov/
U.S. Geological Survey (USGS) .... http://www.usgs.gov/
U.S. Geological Survey - Marine Geology .... http://marine.usgs.gov/
Water Quality Information Center .... http://www.nal.usda.gov/wqic/

State and Local

Aquaculture in Hawaii .... http://www.aloha.com/~aquacult/
Arizona Aquaculture .... http:// ag.arizona.edu/azaqua/
California Aquaculture .... http://aqua.ucdavis.edu/
Chesapeake Bay Program .... http://www.chesapeake.org/
Chesapeake Research Consortium .... http://www.chesapeake.org/crc/crc.html
Chesapeake Information Management System .... http://www.chesapeakebay.net
Columbia Basin Fish and Wildlife Authority .... http://www.cbfwf.org/
Fishes of the Sacramento - San Joaquin Estuary and Adjacent Waters http://elib.cs.berkeley.edu/kopec/tr9/html/home.html
Fish.NET .... http://www.newsdata.com/enernet/fishnet
Fish Passage Center .... http://www.fpc.org/
Florida Game and Fish Commission .... http://fcn.state.fl.us/gfc/gfchome.html
Illinois-Indiana Sea Grant Program .... http://flag.ansc.purdue.edu/il-in-sg/home.htm
Lobster Institute .... http://www.lobster.um.maine.edu/lobster/
Maryland Department of Natural Resources Fish Health .... http://www.dnr.state.md.us/fishhealth.html
Natural Energy Laboratory of Hawaii .... http://bigisland.com/nelha/
New England Aquarium .... http://www.neaq.org/
Northwest Power Planning Council .... http://www.nwppc.org/
Northeastern Regional Aquaculture Center .... http://www.umassd.edu/specialprograms/nrac/
Pacific Regional Aquaculture Information Service for Educators .... http://lama.kcc.hawaii.edu/praise/
Pacific States Marine Fisheries Commission .... http://www.psmfc.org/
Pond Dynamics / Aquaculture Collaborative Research Support Program .... http://www.orst.edu/Dept/crsp/homepage.html
Rural Development Center .... http://www.umes.umd.edu/dept/rudept.html
Scripps Institution of Oceanography .... http://sio.ucsd.edu/
SkipJack Net .... http://skipjack.net/

Virginia Institute of Marine Science .... http://www.vims.edu/
Wisconsin Sea Grant Program .... http://www.seagrant.wisc.edu/

Foreign and International

ACZISC .... http://is.dal.ca/~mbutler/aczisc
ALCOM .... http://www.fao.org/waicent/faoinfo/fishery/alcom/broch.htm
ANGFA - Australia New Guinea Fishes Association
.... http://www.ozemail.com.au/~fisher/angfa.htm
Aquatic Conservation Network .... http://www.acn.ca/
ASEAN Fisheries Post-Harvest Technology Information Network
.... http://www.asean.fishnet.gov.sg/
Canadian Aquaculture Institute .... http://www.upei.ca/~cai/
Canadian Department of Fisheries and Oceans .... http://www.ncr.dfo.ca/
Cooperative Research Center for Freshwater Ecology
.... http://lake.canberra.edu.au/crcfe/crchome.html
Department of Fish Culture and Fisheries .... http://www.zod.wau.nl/~www-venv/
Environment Canada .... http://www.ns.doe.ca/
Fish Health .... http://www.upei.ca/~fishhlth/fish.htm
Food and Agriculture Organization (FAO) .... http://www.fao.org/
Great Lakes Fishery Commission .... http://www.glfc.org/
HACCP Introduction for Fish Processors http://www.asean.fishnet.gov.sg/p2O.html
Institute of Aquaculture .... http://www.stir.ac.uk/aqua/
JASON .... http://www.seagrant.wisc.edu/Education/Jason/
Laboratory of Aquaculture & Artemia Reference Center
.... http://allserv.rug.ac.be/~booghe/index.html
Lobster Health Research Centre .... http://www.upei.ca/-lobster/
North Atlantic Fisheries College .... http://www.nafc.ac.uk/
SBC & Associates .... http://www.fishace.demon.co.uk/
University of Texas, Department of Aquaculture
.... http://info.utas.edu.au/docs/aquaculture/Home-Page.html

Academic or Scientific

American Fisheries Society .... http://www.fisheries.org/
American Society of Agricultural Engineers .... http://asae.org/
American Society of Limnology and Oceanography (ASLO)
.... http://aslo.org/
BIOSIS .... http://www.biosis.org/
Auburn University Aquaculture and Fisheries Home Page
.... http://www.ag.auburn.edu/dept/faa/faa1.html
Biosystems Analysis Group .... http://biosys.bre.orst.edu/aquacult/aquacult.htm
Colorado State University, College of Natural Resources .... http://www.cnr.colostate.edu/
Columbia River Basin Research .... http://www.cqs.washington.edu/index.html
Compendium of Fish and Fishery Product Processes, Hazards and Controls
.... http://www-seafood.ucdavis.edu/haccp/compendium/compend.htm
Cornell University Biodiversity and Biological Collection .... http://www.bio.cornell.edu/
Cornell University Center for the Environment .... http://www.cfe.cornell.edu/
Dissolved Gas Abatement Study
.... http://www.nww.usace.army.mil/html/offices/en/db/hy/disgas.htm
EXtension TOXicology NETwork .... http://ace.ace.orst.edu/info/extoxnet/
FishBase .... http://www.cgiar.org/ICLARM/fishbase/index.htm
International Association of Milk, Food and Environmental Sanitarians Inc.
.... http://www.iamfes.org/
Marine Biological Laboratory .... http://www.mbl.edu/

Marine Science Institute .... http://wwwutmsi.zo.utexas.edu/
MOTE Marine Laboratory .... http://www.marinelab.sarasota.fl.us/
National Center for Ecological Analysis and Synthesis .... http://www.nceas.ucsb.edu/
North Temperate Lates Long Term Ecological Research .... http://limnosun.limnology.wisc.edu/
Oceanography on the Net .... http://scilib.ucsd.edu/sio/
Performance Standards .... http://www.nbiap.vt.edu/perfstands/psmain.html
Provasoli-Guillard National Center for Culture of Marine Phytoplankton
.... http://ccmp.bigelow.org/GI/GI_OO.html
SIUC Fisheries Research and Aquaculture Demonstration Center
.... http://131.230.57.1/FishWeb/CoopFish.htm
Systematic Botany and Mycology Fungal Databases .... http://nt.ars-grin.gov/fungaldatabases/
University of Idaho Aquaculture .... http://www.uidaho.edu/ag/aquaculture/
Unsteady Dissolved Gas Transport Modeling http://terrassa.pnl.gov:2080/DGAS/
World Wide Web Virtual Library: Biosciences .... http://golgi.harvard.edu/htbin/biopages
Woods Hole Oceanographic Institution .... http://www.whoi.edu/

Trade

American Tilapia Association .... http://ag.arizona.edu/azaqua/ata.html
Indiana Aquaculture Association .... http://ag.ansc.purdue.edu/aquanic/iaa/
National Fisheries Institute .... http://www.nfi.org/

Special Interest

Algal Culture Center .... http://ccmp.bigelow.org/
American Nishikigoi Promotion Society .... http://www.aa.net/~koi/anpa.html
American Zoo and Aquarium Association .... http://www.aza.org/
Breeder's Registry .... http://www.breeders-registry.gen.ca.us/
Cichlid Homepage .... http://trans4.neep.wisc.edu/~gracy/fish/
EnviroLink Network .... http://www.envirolink.org/
Fish INformation Service: FINS .... http://www.actwin.com/fish/
Photo Gallery .... http://www.scubacentral.com/photogallery/photogal.shtml
Zebra Mussel Information Resources .... http://www.nfrcg.gov/zebra.mussel/

Regulations and Statistics

ACCSP - The Atlantic Coastal Cooperative Statistics Program
.... http://www.safmc.nmfs.gov/ACCSPHM/accsp.html
FDA/Center for Veterinary Medicine Page .... http://www.fda.gov/cvm/
FDA HACCP Page .... http://vm.cfsan.fda.gov/~lrd/haccpsub.html
FDA/CFSAN: Food Labeling Page .... http://vm.cfsan.fda.gov/label.html
EPA Laws and Regulations Page .... http://www.epa.gov/epahome/rules.html
HACCP Material Available in Spanish
.... http://www.nal.usda.gov/fnic/cgi-bin/haccpres.pl?spanish
National Agricultural Statistics Service .... http://www.usda.gov/nass/

NMFS Fisheries Statistics and Economics Page http://www.st.nmfs.gov/st1/index.html

Magazines

Aquaculture Magazine On-line .... http://www.ioa.com/home/aquamag/
Aquaculture On-line .... http://www.aquaculture-online.org/
Deep-sea Newsgroup Archive .... http://www.bio.net/hypermail/DEEPSEA/
The Journal of Extension .... http://www.joe.org/
Morbidity and Mortality Weekly Report .... http://www.cdc.gov/epo/mmwr/mmwr.html
Northern Aquaculture Magazine .... http://www.naqua.com/
NW Fishletter.................................... http://www.newsdata.com/enernet/fishletter/index.html
Progressive Farm On-line .... http://pathfinder.com/PF/

Other

AquaNIC Home .... http://ag.ansc.purdue.edu/aquanic/
Useful Conversion Programs .... http://ag.ansc.purdue.edu/aquanic/images/tools/conversi.htm
Aquatic Network .... http://www.aquanet.com/
Bibliography of Genetic Variation in Natural Populations
.... http://www.lib.umt.edu/guide/allendorf.htm
Bonneville Power Administration.................................. http://www.efw.bpa.gov/Environment/
EE Link .... http://www.nceet.snre.umich.edu/
The Electronic Zoo Fish Links Section .... http://netvet.wustl.edu/fish.htm
Fishing News Books .... http://www.blacksci.co.uk/fnb/
FishJobs... http://www.fishjobs.com/
Nanoworld .... http://www.uq.oz.au/nanoworld/
National Library for the Environment .... http://www.cnie.org/nle/
National Snow and Ice Data Center .... http://www-nsidc.colorado.edu/
Ocean Voice International .... http://www.ovi.ca/
Seafood Datasearch .... http://www.seafood.com/
SUNY Morrisville College Library Links - Aquaculture
.... http://www.snymor.edu/pages/library/course/aquaculture.htmlx
Tragedy of the Coastal Commons http://www.kenyon.edu/projects/envs6l/
Water Resources Consulting Services .... http://www.waterengr.com/

Index

Numerals in italics indicate illustrations.

B

D

E

F

G

H

I

L

M

N

O

P

Q

R

S

T

U

V

W

X

Y

Z